▶ CONVERSION FACTORS AND CONSTANTS

LENGTH
1 in. = 2.54 cm
1 ft = 12 in. = 0.3048 m
1 yd = 3 ft = 0.9144 m
1 mi = 5280 ft = 1.609 km
1 cm = 0.3937 in.
1 m = 100 cm = 3.281 ft
1 km = 1000 m = 0.6214 mi
1 lightyear = 9.461×10^{12} km

AREA
1 in.^{2} = 6.452 cm^2
1 ft^2 = 144 in.^{2} = 0.09290 m^2
1 m^2 = 10^4 cm^2 = 10.76 ft^2

VOLUME
1 ft^3 = 1728 in.^{3} = 2.832×10^4 cm^3
1 m^3 = 10^6 cm = 6.102×10^4 in.^{3}
1 liter = 1000 cm^3 = 0.03531 ft^3 = 0.2644 gal.
   1 gal = 3.786 liter = 231 in.^{3}

WEIGHT AND MASS
1 kg = 2.205 lb
1 kg = 1000 g = 0.06852 slug
1 slug = 14.59 kg

FORCE
1 lb = 4.448 N
1 N = 0.2248 lb
1 dyne = 10^{-5} N = 2.248×10^{-6} lb

VELOCITY
1 ft/sec = 0.6818 mi/hr = 0.3048 m/sec
1 mi/min = 60 mi/hr = 88 ft/sec
1 mi/hr = 1.467 ft/sec = 1.609 km/hr
1 m/sec = 3.281 ft/sec

ACCELERATION
1 ft/sec^2 = 0.3048 m/sec^2
1 m/sec^2 = 3.281 ft/sec^2

TIME
1 day = 24 hr = 8.640×10^4 sec
1 year = 365.25 day = 3.1558×10^7 sec

PRESSURE, ENERGY, AND POWER
1 atm = 14.70 lb/in.^{2} = 1.013×10^5 N/m^2
1 joule = 10^7 erg = 0.7376 ft-lb
1 hp = 550 ft-lb/sec = 746 joules/sec

PHYSICAL CONSTANTS
Acceleration from the earth's gravity (g): 32 ft/sec^2 = 9.8 m/sec^2
Speed of light (c): 1.86×10^5 mi/sec = 3.00×10^8 m/sec
Universal gravitational constant (G): 6.67×10^{-11} N-m^2/kg^2

MATHEMATICAL CONSTANTS
π = 3.14159265358979323846 . . . e = 2.71828182845904523536 . . .

▶ ALGEBRA

QUADRATIC FORMULA
$$ax^2 + bx + c = 0 \quad \text{if } x = \frac{-b + \sqrt{b^2 - 4ac}}{2a} \quad \text{or} \quad x = \frac{-b - \sqrt{b^2 - 4ac}}{2a}$$

BINOMIAL THEOREM
$$(a + b)^n = a^n + na^{n-1}b + \frac{n(n-1)}{2}a^{n-2}b^2 + \cdots + nab^{n-1} + b^n$$

EXPONENTIAL AND LOGARITHM RULES

$a^p = e^{p \ln a}$ $\log_a p = \dfrac{\ln p}{\ln a}$

$a^{p+q} = a^p a^q$ $\log_a(pq) = \log_a p + \log_a q$

$a^{-p} = \dfrac{1}{a^p}$ $\log_a p^{-1} = -\log_a p$

$(a^p)^q = a^{pq}$ $\log_a(p^q) = q \log_a p$

Calculus

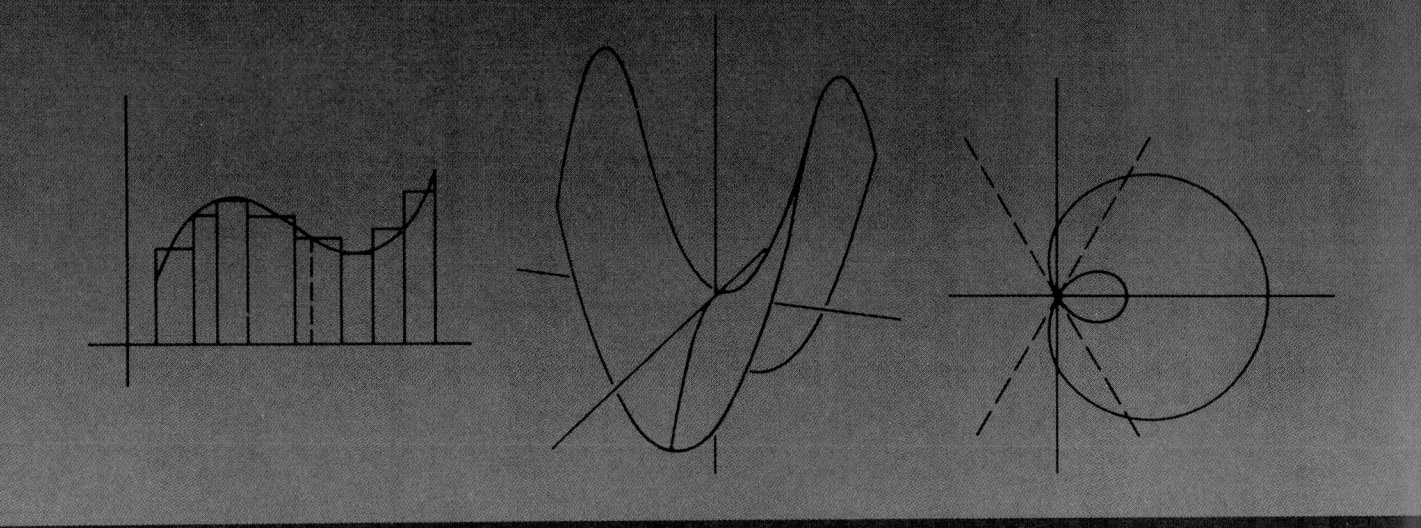

THE RANDOM HOUSE/BIRKHÄUSER MATHEMATICS SERIES

J. Douglas Faires
Youngstown State University

Barbara T. Faires
Westminster College

Calculus
SECOND EDITION

RANDOM HOUSE

New York

About the cover: The karst hills of the Kweilin region of China form natural hyperbolic curves. Their appearance in this cover design expresses the function of calculus as a gateway to analysis of the natural world.

cover photo credit: LEO DE WYS

Second Edition
987654321
Copyright © 1988 by Random House, Inc.
All rights reserved under International and Pan-American Copyright Conventions. No part of this book may be reproduced in any form or by any means, electronic or mechanical, including photocopying, without permission in writing from the publisher. All inquiries should be addressed to Random House, Inc., 201 East 50th Street, New York, N.Y. 10022. Published in the United States by Random House, Inc., and simultaneously in Canada by Random House of Canada, Limited, Toronto.

Library of Congress Cataloging-in-Publication Data

Faires, J. Douglas.
    Calculus.

    (The Random House/Birkhäuser mathematics series)
    First ed. published under title: Calculus and analytic geometry.
    Includes index.
    1. Calculus. 2. Geometry, Analytic. I. Faires, Barbara Trader. II. Faires, J. Douglas. Calculus and analytic geometry. III. Title. IV. Series.
QA303.F294   1987   515   87-20468
ISBN 0-394-36624-7

Preface

The study of calculus has been a core subject area of science and engineering for many years and is now included as an important topic in most other quantitative disciplines. Calculus has a wide application due to its ability to study objects in a continuous state of change. To consider the change of one quantity with respect to another we use the notion of a function. In effect, then, calculus is concerned with the behavior and application of functions; so it is with the definition of a function that we begin the book. The standard precalculus review material has been incorporated within the discussion of functions so that it is available but not obtrusive.

New topics are presented in small segments with frequent reference to previously studied topics. This can be seen as early as Section 1.1. Problems that will be studied as physical applications of the derivative and integral in Chapters 3 and 5 are used in the introduction to functions. These applications can be difficult topics to master since two relatively new concepts are used simultaneously: A physical problem must be given a mathematical description, and the mathematical problem must be solved using recently discovered calculus techniques. Much of the difficulty in these applications can be avoided by previewing the problem when functions are first introduced.

Organization

Concept organization is the same for the three major topics of calculus: the limit, the derivative, and the integral. In each case the topic is presented using the basic outline:

1. Geometric motivation.
2. Intuitive examples discussed analytically and graphically.
3. A precise analytical definition.
4. Examples that connect the intuitive notion and the definition.
5. Applications of the concept.

This presentation gives the reader the opportunity to view a new concept intuitively and to see how the definition evolves before the concept is applied.

Flexibility

The text has been written to serve a wide variety of curricula. Sections that might be omitted or discussed in a different order have been organized so that subsequent material can be presented easily. For example, infinite limits and limits at infinity are considered in Chapter 1 where their definitions can be easily compared with those of other limits. Vertical and horizontal asymptotes are also discussed at this time. These topics could, however, be postponed until the other graphing techniques are considered at the end of Chapter 2.

L'Hôpital's Rule is presented in Chapter 3, since it is an application of one of the most important results of calculus, the Mean Value Theorem. It could be postponed until it is applied to improper integrals or infinite series.

The natural logarithm function has been presented early in the text so that all the trigonometric functions can be integrated before considering the applications of the integral. This topic can, however, be easily postponed until Chapter 6 when the natural exponential function is discussed.

Conic sections are presented in Chapter 10 since they are used extensively as cross sections of three-dimensional surfaces, but this material could be presented any time after the second section of Chapter 1.

Graphing

New concepts are more easily understood when given a familiar geometric representation, so material on graphing is presented early and extensively throughout the book. All the graphs in both the text and the answer section have been redrawn for this edition using state-of-the-art computer graphic techniques. This permits a combination of accuracy and artistic representation that was unavailable just a few years ago. In addition, the three-dimensional surfaces that are most difficult to visualize have been given a full-color representation so the various portions can be readily distinguished.

Examples and Exercises

The first third of the problems at the end of each section follow the pattern of the examples in the section. Students can usually solve these problems with minimal help from an instructor. The remaining exercises increase gradually in difficulty to those that require a thorough knowledge of the concepts discussed in the section and often some ingenuity. The later problems are designed to present applications, extend concepts considered in the section, and challenge the clever. Included among these are applications from virtually every discipline that uses calculus techniques as well as problems from the William Lowell Putnam examination. This examination is given each year to more than 2000 undergraduate students in the United States and Canada. The Putnam problems are generally quite difficult and are included primarily to illustrate the power of the topic being studied.

The review exercises at the end of each chapter emphasize basic concepts rather than presenting extensions or generalizations of the chapter material. They have been extensively rewritten and expanded for this edition. A student who can work all the review exercises for a chapter can feel confident about knowing the basic material in that chapter.

Chapter Introductions and Summaries

The chapter introductions provide an overview of the material of the chapter. They relate the material about to be presented to topics that have been previously discussed and describe some of the applications to be considered in the chapter.

The summary at the end of the chapter lists the important results, concepts, and formulas in the chapter. Page references are given in the summary so that unfamiliar topics can be found easily when reviewing material.

Other Features of the Book

New techniques are viewed as refinement procedures whenever possible. For example, one specific function is graphed three times in Chapter 2: initially, when finding the maximum and minimum values of a function; next, when information about the increasing and decreasing nature of functions is considered; finally, when concavity is used for fine-tuning.

Applications of the integral are discussed immediately after the integral is introduced to show the importance of this concept and to provide practice with the basic integration techniques. Integral applications are also included in the exercise sets of later chapters as reinforcement when new functions and integration techniques are introduced.

The presentation of the calculus for vector-valued functions and functions of several variables parallels the treatment in the single variable case. The first chapter on multivariable functions considers some common functions and their graphs, similar to the treatment of single-variable functions in the first part of Chapter 1. Each topic is motivated geometrically before an analytical definition is given. Now, however, the motivation includes specific reference to the corresponding notion for functions of a single variable. The results that hold in the new situation are presented in a form similar to the single-variable result. This method of presentation is followed throughout the later chapters of the book, including Chapter 16, which gives a treatment of line and surface integrals consistent with the integral discussion in the other chapters.

Computer applications are presented in the book at places where the computer can be an effective tool. The Bisection Method is introduced in Chapter 1 as an application of the Intermediate Value Theorem.

Newton's Method is discussed early in the book, with a brief explanation of its accuracy characteristics. The Trapezoidal and Simpson's Rules for numerical integration are presented in Chapter 7, and the extension of Simpson's Rule to multiple integrals is discussed in Chapter 15. Chapter 17 includes an introduction to numerical methods for approximating the solution to initial-value problems in differential equations. BASIC programs for these techniques, as well as for Simpson's method modified for double and triple definite integrals, are listed in Appendix C.

Accuracy

The review and development process of this book has been intensive, as is evident from the list of contributors given in the acknowledgments. It is a fundamental Random House policy to produce books as free from error as possible and a number of steps have been used to ensure this. Every exercise was reworked in the final draft by each of the authors as well as by David Ruppel and Phillip Schmidt of the University of Akron. Each routine example and exercise was checked using a symbolic manipulation program, and all those that are numerically oriented were solved using both a computer and a calculator.

In addition, at the final stage of production each example and exercise was checked by a team of accuracy reviewers:

Carlton Evans, *New Mexico State University*

Darel Hardy, *Colorado State University*

Terry Herdman, *Virginia Polytechnic Institute and State University*

Gail Kaplan, *United States Naval Academy*

Roger Nelson, *Ball State University*

James Osterburg, *University of Cincinnati*

Karen Zak, *United States Naval Academy*

While we are confident that this process removed nearly all the inaccuracies and points of ambiguity, we are not so naive as to believe that every error has been uncovered. We would be most grateful to know of any error you find so that it can be corrected in subsequent printings.

Acknowledgments

It is a pleasure to express our appreciation to the many people who aided in the preparation of this edition. Particularly beneficial was the reviewing process, for which we thank the following:

Arnold Adelberg, *Grinnell College*
Calvin D. Ahlbrandt, *University of Missouri — Columbia*
Daniel D. Anderson, *University of Iowa*
Nancy Angle, *Colorado School of Mines*
Fred Brauer, *University of Wisconsin — Madison*
Sydney Bulman-Fleming, *Wilfrid Laurier University*
John Buoni, *Youngstown State University*
Robert Carson, *University of Akron*
Duane Deal, *Ball State University*
Dennis R. Dunninger, *Michigan State University*
Bruce Edwards, *University of Florida*
Joe Elich, *Utah State University*
Carlton Evans, *New Mexico State University*
August Garver, *University of Missouri — Rolla*
Nathaniel Grossman, *University of California — Los Angeles*
Steven Hanzely, *Youngstown State University*
Darel Hardy, *Colorado State University*
Melvin Hausner, *New York University — Courant Institute*
Terry Herdman, *Virginia Polytechnic Institute and State University*
John Hosack, *Colby College*
Gail Kaplan, *United States Naval Academy*
William Kirby, *Bowling Green State University*
John Klippert, *James Madison University*
Elton Lacey, *Texas A & M University*
Robert Lohman, *Kent State University*
Robert McFadden, *Northern Illinois University*

Daniel Messerschmidt, *Lynchburg College*
Burnett C. Meyer, *University of Colorado*
Carl David Minda, *University of Cincinnati*
Roger Nelson, *Ball State University*
Carol O'Dell, *Ohio Northern University*
James Osterburg, *University of Cincinnati*
Horatio Porta, *University of Illinois at Urbana Champaign*
Neal Raber, *University of Akron*
Jerry Reed, *Mississippi State University*
Mac Rugheimer, *Montana State University*
John T. Scheick, *Ohio State University*
Charles W. Schelin, *University of Wisconsin — La Crosse*
Phillip H. Schmidt, *University of Akron*
Thomas Schwartzbauer, *Ohio State University*
Richard J. Shores, *Lynchburg College*
David R. Stone, *Georgia Southern College*
Olaf Stackelberg, *Kent State University*
Willy Taylor, *Texas Southern University*
Russell Thompson, *Utah State University*
Jerry Uhl, *University of Illinois at Urbana Champaign*
Beverly West, *Cornell University*
Kenneth Yanosko, *Humboldt State University*
Karen Zak, *United States Naval Academy*
Joseph Zund, *New Mexico State University*

We are especially grateful to Fred Rickey of Bowling Green State University and Joe Zund of New Mexico State University for their valuable contributions to the historical notes and to Phil Schmidt and Dave Ruppel of the University of Akron for checking the answers to all the exercises. We would also like to thank Dick Burden of Youngstown State University for generating the BASIC computer programs to accompany the numerical methods in the book.

The art in the text was produced by Brian and George Morris at Scientific Illustrators and the graphs in the full-color section were provided by James T. Hoffman and David Hoffman of the University of Massachusetts at Amherst. We are most grateful for the fine illustrations that they produced.

We were fortunate to have John Martindale as editor of this project. His extensive expertise in mathematical publishing was evident throughout and was particularly helpful in providing meaningful reviews of the manuscript. In addition we would like to express our appreciation to the developmental editor on this project, Alexa Barnes, for her many helpful remarks. We also thank a fine group of student assistants at Youngstown State University for helping to prepare the manuscript for production: JoAnn DeSantis, Jeff Kubina, Diane Pappada, Brigitte Ramos, and Joan Walsh.

Finally, we would like to express our appreciation for the temporary home provided for us by the University of Sussex during the 1985–1986 academic year and to the students, faculty, and administrators at Westminster College and Youngstown State University for their cooperation and encouragement.

Contents

Introduction to Calculus   xix

1

Functions, Limits, and Continuity   1

- **1.1** The Definition of a Function   2
- **1.2** Linear and Quadratic Functions   14
- **1.3** Graphs of Some Common Functions   24
- **1.4** Combining Functions   32
- **1.5** The Limit of a Function: The Intuitive Notion   41
- **1.6** The Limit of a Function: The Definition   47
- **1.7** Limits and Continuity   53
- **1.8** One-sided Limits and Continuity on an Interval   61
- **1.9** Limits at Infinity and Horizontal Asymptotes   69
- **1.10** Infinite Limits and Vertical Asymptotes   75

   Important Terms and Results   83
   Review Exercises   83

2

The Derivative   87

- **2.1** The Slope of a Curve   88
- **2.2** The Derivative of a Function   95
- **2.3** Formulas for Differentiation   104
- **2.4** Differentiation of Trigonometric Functions   115
- **2.5** The Derivative of a Composite Function: The Chain Rule   122
- **2.6** Implicit Differentiation   130
- **2.7** Maxima and Minima of Functions   138
- **2.8** The Mean Value Theorem   146
- **2.9** Increasing and Decreasing Functions: The First Derivative Test   153
- **2.10** Higher Derivatives: Concavity and the Second Derivative Test   160

2.11 Comprehensive Graphing 168

    Important Terms and Results 178
    Review Exercises 178

3

Applications of the Derivative   181

3.1 Rectilinear Motion 182
3.2 Applications Involving Maxima and Minima 191
3.3 Related Rates 204
3.4 Differentials 214
3.5 Indeterminate Forms: l'Hôpital's Rule 220
3.6 Newton's Method 231
3.7 Applications to Business and Economics 237

    Important Terms and Results 244
    Review Exercises 245

4

The Integral   249

4.1 Area 250
4.2 The Definite Integral 258
4.3 The Fundamental Theorem of Calculus 270
4.4 The Indefinite Integral 278
4.5 Integration by Substitution 286
4.6 The Natural Logarithm Function 294
4.7 Improper Integrals 306
4.8 The Discovery of Calculus 313

    Important Terms and Results 316
    Review Exercises 317

5

Applications of the Definite Integral   319

5.1 Areas of Regions in the Plane 320
5.2 Volumes of Solids with Known Cross Sections: Disks 328
5.3 Volumes of Solids of Revolution: Shells 339
5.4 Arc Length and Surfaces of Revolution 347
5.5 Moments and the Center of Mass 357
5.6 Work 367
5.7 Fluid Pressure 374

    Important Terms and Results 379
    Review Exercises 380

6

The Calculus of Inverse Functions   383

6.1 Inverse Functions 384
6.2 Inverse Trigonometric Functions 394

- 6.3 The Natural Exponential Function   406
- 6.4 General Exponential and Logarithm Functions   413
- 6.5 Additional Indeterminate Forms   420
- 6.6 Exponential Growth and Decay   424
- 6.7 Separable Differential Equations   431
- 6.8 Hyperbolic Functions   442
- 6.9 Inverse Hyperbolic Functions   447

Important Terms and Results   452
Review Exercises   453

7

Techniques of Integration   455

- 7.1 Integration by Parts   457
- 7.2 Integrals of Products of Trigonometric Functions   466
- 7.3 Trigonometric Substitution   475
- 7.4 Partial Fractions   481
- 7.5 Integrals Involving Quadratic Polynomials   491
- 7.6 Numerical Integration   495

Important Terms and Results   505
Review Exercises   506

8

Sequences and Series   509

- 8.1 Infinite Sequences   510
- 8.2 Infinite Series   522
- 8.3 Infinite Series with Positive Terms   531
- 8.4 Alternating Series   541
- 8.5 Absolute Convergence   545
- 8.6 Power Series   554
- 8.7 Differentiation and Integration of Power Series   561
- 8.8 Taylor Polynomials and Taylor Series   567
- 8.9 Applications of Taylor Polynomials and Taylor Series   577

Important Terms and Results   585
Review Exercises   585

9

Polar Coordinates and Parametric Equations   587

- 9.1 The Polar Coordinate System   588
- 9.2 Graphing in Polar Coordinates   592
- 9.3 Areas of Regions Using Polar Coordinates   599
- 9.4 Parametric Equations   603
- 9.5 Tangent Lines to Curves   610

9.6 Lengths of Curves   615

Important Terms and Results   620
Review Exercises   621

10

Conic Sections 623

10.1 Parabolas   625
10.2 Ellipses   634
10.3 Hyperbolas   640
10.4 Rotation of Axes   646
10.5 Polar Equations of Conic Sections   653

Important Terms and Results   661
Review Exercises   661

11

Vectors 663

11.1 The Rectangular Coordinate System in Space   664
11.2 Vectors in Space   668
11.3 The Dot Product of Vectors   676
11.4 The Cross Product of Vectors   683
11.5 Planes   690
11.6 Lines in Space   695

Important Terms and Results   702
Review Exercises   703

12

Vector-Valued Functions 705

12.1 The Definition of a Vector-Valued Function   706
12.2 The Calculus of Vector-Valued Functions   710
12.3 Unit Tangent and Unit Normal Vectors   716
12.4 Velocity and Acceleration of Objects in Space   722
12.5 Curvature   729
12.6 Newton's and Kepler's Laws of Motion   736

Important Terms and Results   745
Review Exercises   745

13

Multivariable Functions 747

13.1 Functions of Several Variables   748
13.2 Functions of Two Variables: Level Curves   752
13.3 Functions of Three Variables: Level Surfaces   759
13.4 Quadric Surfaces   763

13.5 Cylindrical and Spherical Coordinates in Space 771

    Important Terms and Results 778
    Review Exercises 779

14

The Differential Calculus of Multivariable Functions   781

14.1 Limits and Continuity 782
14.2 Partial Derivatives 790
14.3 Differentiability of Multivariable Functions 797
14.4 The Chain Rule 805
14.5 Directional Derivatives and Gradients 812
14.6 Tangent Planes and Normals 821
14.7 Extrema of Multivariable Functions 827
14.8 Lagrange Multipliers 840

    Important Terms and Results 848
    Review Exercises 848

15

Integral Calculus of Multivariable Functions   851

15.1 Double Integrals 852
15.2 Iterated Integrals 857
15.3 Double Integrals in Polar Coordinates 865
15.4 Center of Mass and Moments of Inertia 873
15.5 Surface Area 878
15.6 Triple Integrals 883
15.7 Triple Integrals in Cylindrical and Spherical Coordinates 894
15.8 Applications of Triple Integrals 901

    Important Terms and Results 908
    Review Exercises 908

16

Line and Surface Integrals   911

16.1 Vector Fields 912
16.2 Line Integrals 917
16.3 Physical Applications of Line Integrals 926
16.4 Line Integrals Independent of Path 932
16.5 Green's Theorem 943
16.6 Surface Integrals 954
16.7 The Divergence Theorem 962
16.8 Stokes's Theorem 967

16.9 Applications of the Divergence and Stokes's Theorems 975

Important Terms and Results 982
Review Exercises 983

17

Ordinary Differential Equations 985

17.1 Introduction 986
17.2 Homogeneous Differential Equations 990
17.3 Exact Differential Equations 994
17.4 Linear First-Order Differential Equations 1000
17.5 Second-Order Linear Equations: Homogeneous Type 1007
17.6 Second-Order Linear Equations: Nonhomogeneous Type 1014
17.7 Numerical Methods for First-Order Initial-Value Problems 1023

Important Terms and Results 1028
Review Exercises 1028

Appendix A

Review Material A1

A.1 The Real Line A2
A.2 The Coordinate Plane A6
A.3 Trigonometric Functions A9
A.4 Mathematical Induction A18

Appendix B

Additional Calculus Theorems and Proofs A22

Appendix C

Basic Computer Programs A39

Appendix D

Tables A48

Answers to Odd-Numbered Exercises S1
Credits C1
Index I1

Introduction to Calculus

Precalculus courses supply you with the mathematical tools of arithmetic, algebra, geometry, and trigonometry. In these courses you encounter applications to certain everyday situations. Included are problems such as finding the average speed of a car and determining the area contained within a region in the shape of a triangle, rectangle, or other polygon. To describe the basic distinction between calculus and precalculus topics, let us consider speed and area problems from both viewpoints.

Consider first the speed of a car. If we know the distance traveled by the car over a specific time interval we can calculate the average speed over the time interval. For example, if the car has traveled 150 miles between 8:00 AM and 11:00 AM, the *average* speed is

$$\frac{150 \text{ miles}}{3 \text{ hours}} = 50 \text{ miles per hour.}$$

This calculation involves elementary arithmetic to tell what has happened over a time interval, but it does not give information about the speed of the car at any specific instant within the time interval. If we knew the car had averaged 50 mph over a smaller interval of time we could assume with greater confidence that the actual, or instantaneous, speed of the car at a particular instance was close to 50 mph, but we would never be able to find the exact instantaneous speed of the car by this process.

The speedometer of the car tells this instantaneous speed at the precise instant the speedometer is read. Alternatively, the instantaneous speed can be computed by determining the average speed over increasingly small time intervals and observing what happens as the length of the time intervals approaches zero. If the distance the car has traveled is known as a function of time, differential calculus can be applied to give this instantaneous speed.

Differential calculus is also used to determine the answers to questions such as: What amount of sales of a product will produce the maximum profit for a company? How can a formula for the volume of a geometric object be used to determine quickly the amount of paint needed to coat the object? Problems of this type are considered in

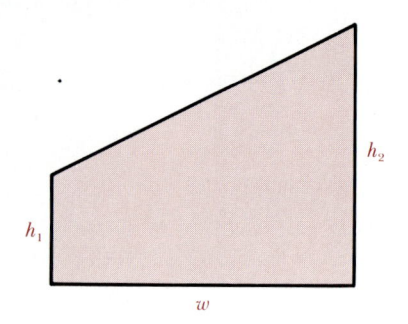

Figure 1

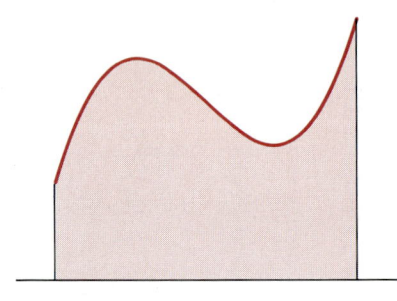

Figure 2

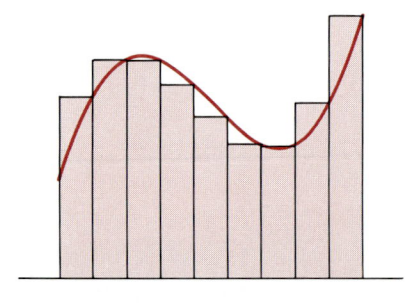

Figure 3

Chapter 3 as applications of one of the fundamental notions of calculus — the derivative.

Area problems also have a precalculus–calculus distinction. The area of a polygon can be obtained from a geometric formula, a precalculus tool. For example, the area of the trapezoid shown in Figure 1 is the product of the width of the base and the average of the height of the sides:

$$\text{Area} = \tfrac{1}{2}(h_1 + h_2) \cdot w.$$

The exact area of the region shown in Figure 2 cannot be found by algebraic means, but it can be approximated by summing the area of rectangles such as those shown in Figure 3. The area of the region in Figure 2 is the limit of sums of this type. Again, the methods of calculus are required to make this notion and application precise. In this instance we use the other fundamental concept of calculus — the integral.

The integral also enables us to answer questions about the length, volume, and surface area of geometric figures. For example, what is the surface area of a doughnut? How long is the path of a baseball as it travels from the bat to the player's glove? What is the volume of ice cream in an ice cream cone? Applications of this type are discussed in Chapters 4 and 5.

The common feature of problems that makes calculus applicable is the need to consider concepts that are in a state of change. Differential calculus is used to describe the instantaneous rate of change of one quantity with respect to another. This is the tool needed to describe the instantaneous speed of the car, since this is the instantaneous rate of change of the distance with respect to time. Integral calculus is used to describe the area of the region shown in Figure 2, a region whose upper boundary is a continually changing curve.

These are examples of applications of calculus to single variable problems. Later in the text you will discover the power of calculus when applied to problems in the world in which we live, three-dimensional space. You will find, however, that the calculus in this setting is a relatively simple extension of the basic concepts you studied in the first few chapters of the book.

It took over 2500 years for calculus to progress from the early notions of the subject to the stage we study today. For most of this period the concepts of differential and integral calculus were considered distinct. It was not until the latter part of the seventeenth century that mathematicians, led by Sir Isaac Newton in England and Gottfried Wilhelm von Leibniz in Europe, discovered the connection between these fundamental ideas. In the past three hundred years calculus has been put on a firm mathematical foundation and refined to the point that it now follows logically from a few basic notions and principles. It is of exceptional importance for applications in many areas that were unknown at the time of its development, applications in every area of quantitative study.

Calculus

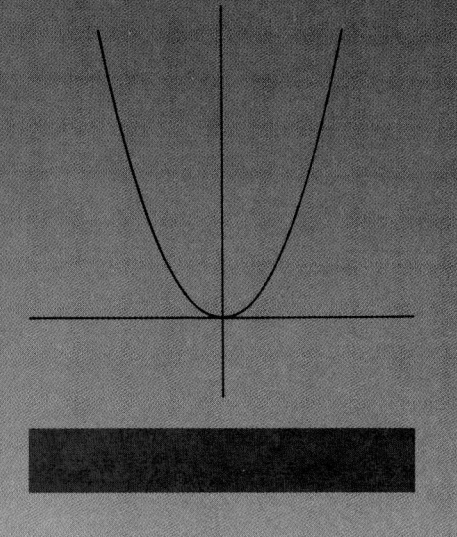

1

Functions, Limits, and Continuity

CHAPTER 1 FUNCTIONS, LIMITS, AND CONTINUITY

Calculus is concerned with the properties, applications, and behavior of functions. The first part of this chapter considers some of the common functions used throughout the text, functions that will be familiar from precalculus courses. The emphasis here is on graphing techniques and applications that are encountered frequently in the study of calculus.

The latter portion of the chapter considers the limit of a function, a concept that distinguishes calculus from algebra, geometry, and other precalculus topics. Calculus was developed to study such topics as the velocity of a moving particle, the tangent to a curve, and the area of a region bounded by a curve. At the center of each of these problems lies the idea of a limit.

1.1
The Definition of a Function

The notion of a function is one of the fundamental tools of mathematics. A function is a means of associating each element in a given set with a specific element of another set in such a way that every element in the first set corresponds to precisely one element in the second. Figure 1.1 illustrates such a correspondence, where X represents the first, or given, set and Y represents the second set.

(1.1)
DEFINITION
A **function** from a set X into a set Y is a rule of correspondence that assigns each element in X to precisely one element in Y.

If f is used to denote a function from a set X into a set Y, then $f(x)$ denotes the unique element in Y that corresponds to the element x in X. The set X is called the **domain** of f. The **range** of f is the set of elements in Y that are associated with some element in X. To indicate that f is a function with domain X and range in Y, we write $f: X \to Y$ and say "f maps X into Y."

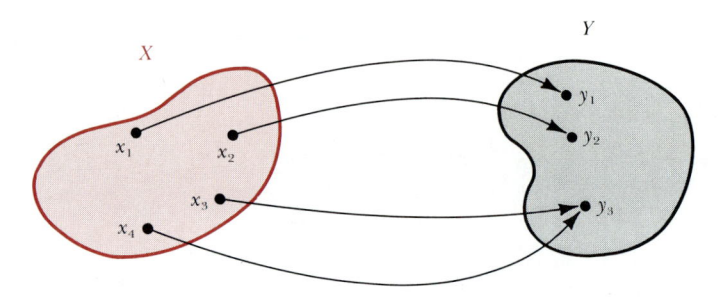

Figure 1.1
Function from X to Y.

If x is in the domain of a function f, we say that "f is defined at x" or that "$f(x)$ exists." The terminology "f is defined on X'" means that $f(x)$ exists for every element x in X', that is, X' is a subset of the domain of f.

Although each element in the domain of a function corresponds to a unique element in the range, more than one element in the domain can be associated with the same element of the range. In Figure 1.1, for example, the distinct elements x_3 and x_4 in X are associated with the same element y_3 in Y.

The study of calculus is concerned primarily with functions whose domain and range are both contained in the set of real numbers, denoted by $\mathbb{R}$. Functions of this type are usually described by simply giving their rule of correspondence. For example, $f(x) = x^3 + 1$ describes a function f with domain and range both contained in $\mathbb{R}$. Unless otherwise specified, the domain of the function is assumed to be the largest subset of $\mathbb{R}$ for which the correspondence produces a real number. The range is the set of real numbers associated with some number in the domain.

Example 1

Find the domain and range of the function f whose rule of correspondence is $f(x) = \sqrt{x} + 1$.

Solution

The square root of x, $\sqrt{x}$, is a real number if and only if $x \geq 0$. Consequently, the domain of f is the set of all nonnegative real numbers.

Since the symbol $\sqrt{}$ indicates the principal, or nonnegative, square root, $f(x) = \sqrt{x} + 1 \geq 1$, for all $x \geq 0$. The range of f is the set of real numbers greater than or equal to 1. ▶

There is no special significance to the variable x; in fact, it might be better to describe the function in Example 1 by writing $f(\underline{}) = \sqrt{\underline{}} + 1$. This form indicates more clearly that whatever value is used to fill the space on the left side of the equation must also be used on the right side. For example, 9 is in the domain of f and $f(\underline{9}) = \sqrt{\underline{9}} + 1 = 3 + 1 = 4$.

It is important to be comfortable with function notation since in Chapter 2 we consider in detail quotients of the form

$$\frac{f(x+h) - f(x)}{h},$$

where $h \neq 0$.

Example 2

Consider the function f described by $f(x) = x^2 + x$. Simplify

$$\frac{f(x+h) - f(x)}{h},$$

where $h \neq 0$.

Solution
Since
$$f(x+h) = (x+h)^2 + (x+h),$$
$$\frac{f(x+h) - f(x)}{h} = \frac{(x+h)^2 + (x+h) - (x^2+x)}{h}$$
$$= \frac{x^2 + 2xh + h^2 + x + h - x^2 - x}{h}$$
$$= \frac{h(2x+h+1)}{h} = 2x + h + 1,$$

provided that $h \neq 0$. ▶

The domain and range of functions from $\mathbb{R}$ into $\mathbb{R}$ are often described using interval notation. For real numbers $a < b$, the **open interval** (a, b) is defined as

$$(a, b) = \{x | a < x < b\},$$

and read: "the set of real numbers x such that x is greater than a and less than b."

When the endpoints of the interval are included, the interval is called a **closed interval**, denoted by

$$[a, b] = \{x | a \leq x \leq b\}.$$

Half-open intervals contain one endpoint but not the other. Thus,

$$(a, b] = \{x | a < x \leq b\} \quad \text{and} \quad [a, b) = \{x | a \leq x < b\}$$

are half-open intervals.

The **interior** of an interval consists of all the numbers in the interval that are not endpoints. The intervals (a, b), $[a, b]$, $(a, b]$, and $[a, b)$ all have the same interior, that is, the open interval (a, b).

The infinity symbol, ∞, is used to indicate that an interval is unbounded. The intervals $[a, \infty)$ and (a, ∞) are not bounded above, whereas the intervals $(-\infty, a]$ and $(-\infty, a)$ are not bounded below. The interval $(-\infty, \infty)$ represents the set $\mathbb{R}$ of all real numbers.

In general, a square bracket indicates that the number next to it is in the set, and a parenthesis indicates that the number next to it does not belong to the set. One obvious implication of interval notation is that the symbols ∞ and $-\infty$ are never next to a square bracket since they do not represent real numbers. Table 1.1 summarizes the interval notation.

HISTORICAL NOTE

This functional notation was first introduced in 1734 by **Leonhard Euler** (1707–1783). Euler wrote the most complete calculus books of his day, including in the text a number of his original results. The volumes on differential calculus and differential equations were published in 1755, but the three extensive volumes on integral calculus did not appear until the period 1768–1770.

Table 1.1

INTERVAL NOTATION	SET NOTATION	GRAPHIC REPRESENTATION
(a, b)	$\{x \mid a < x < b\}$	open at a, open at b
$[a, b]$	$\{x \mid a \leq x \leq b\}$	closed at a, closed at b
$[a, b)$	$\{x \mid a \leq x < b\}$	closed at a, open at b
$(a, b]$	$\{x \mid a < x \leq b\}$	open at a, closed at b
$[a, \infty)$	$\{x \mid a \leq x < \infty\}$	closed at a
(a, ∞)	$\{x \mid a < x < \infty\}$	open at a
$(-\infty, b]$	$\{x \mid -\infty < x \leq b\}$	closed at b
$(-\infty, b)$	$\{x \mid -\infty < x < b\}$	open at b
$(-\infty, \infty)$	$\{x \mid -\infty < x < \infty\}$	all reals, through 0

Example 3

Given $f(x) = \dfrac{1}{x - 1}$, find each of the following.

a) $f(0)$ and $f(-1)$
b) A value of x for which $f(x) = 3$
c) The domain of f

Solution

a) $f(0) = \dfrac{1}{0 - 1} = \dfrac{1}{-1} = -1$ and $f(-1) = \dfrac{1}{-1 - 1} = \dfrac{1}{-2} = -\dfrac{1}{2}$

b) If
$$\frac{1}{x - 1} = 3,$$

then
$$\frac{1}{3} = x - 1 \quad \text{and} \quad x = 1 + \frac{1}{3} = \frac{4}{3}.$$

Thus,
$$f\left(\frac{4}{3}\right) = \frac{1}{\frac{4}{3} - 1} = \frac{1}{\frac{4-3}{3}} = 3.$$

c) The quotient $1/(x-1)$ is undefined when $x = 1$. This implies that the domain of f is the set of all real numbers except $x = 1$, and is given in interval notation by $(-\infty, 1) \cup (1, \infty)$. ▶

The symbol $\cup$ in Example 3(c) is used to denote the union of the two intervals $(-\infty, 1)$ and $(1, \infty)$. In general, the *union* $A \cup B$ of sets A and B is the collection of all elements in A *or* in B. The *intersection* $A \cap B$ of A and B is the collection of all elements in *both* A and B.

Example 4

Find the domain of the function given by
$$g(x) = \sqrt{x^2 - 4x + 3}.$$

Solution

The domain of g is the set of all real numbers such that $x^2 - 4x + 3 \geq 0$. Since
$$x^2 - 4x + 3 = (x - 3)(x - 1),$$

x is in the domain of g if the factors $(x - 3)$ and $(x - 1)$ are both nonnegative or the factors are both nonpositive.

The top line in Figure 1.2 indicates that the factor $(x - 3)$ is nonnegative when $x \geq 3$ and is nonpositive when $x \leq 3$. The second line gives similar information about the factor $(x - 1)$. The third line shows that the product $(x - 3)(x - 1)$ is nonnegative if either $x \leq 1$ or $x \geq 3$. Thus the domain of g is $(-\infty, 1] \cup [3, \infty)$. ▶

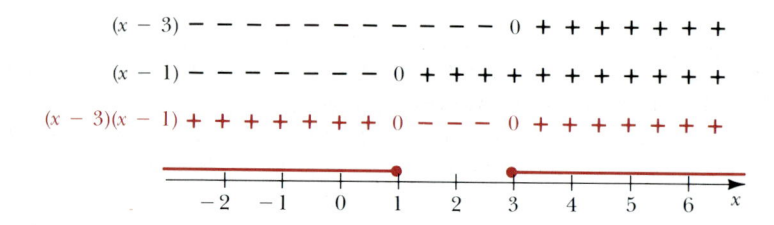

Figure 1.2
$\sqrt{x^2 - 4x + 3}$ is defined if x is in $(-\infty, 1] \cup [3, \infty)$.

Example 5

Find $f(-2)$, $f(0)$, and $f(5)$ for the function defined by

$$f(x) = \begin{cases} -x, & \text{if } x < 0, \\ 3, & \text{if } x \geq 0. \end{cases}$$

Solution

Since $-2 < 0$ and $f(x) = -x$ when $x < 0$, it follows that $f(-2) = -(-2) = 2$. When $x \geq 0$, $f(x) = 3$, so $f(0) = 3$ and $f(5) = 3$.

$$f(x) = \begin{cases} -x, & \text{if } x < 0, \\ 3, & \text{if } x \geq 0 \end{cases}$$

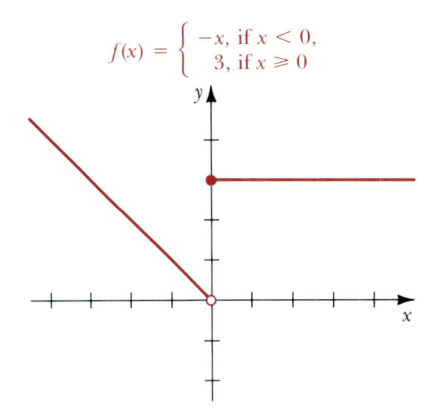

Figure 1.3

The set $\mathbb{R} \times \mathbb{R}$ (also denoted by $\mathbb{R}^2$) consists of all ordered pairs (x, y) for x and y in $\mathbb{R}$. A subset of $\mathbb{R} \times \mathbb{R}$ is associated with each function whose domain and range are contained in $\mathbb{R}$. This subset consists of all pairs of the form $(x, f(x))$, where x is in the domain of f. The collection of points in the coordinate plane corresponding to this subset is called the **graph** of f.

To *sketch the graph* of f means to distinguish, in the coordinate plane, those points in the graph of f from the remainder of the points in the plane. For example, the graph of the function described in Example 5 is sketched in Figure 1.3. The small open circle at $(0, 0)$ indicates that this point does *not* lie on the sketch of the graph, and the solid circle at $(0, 3)$ means that this point *is* on the sketch of the graph.

The *graph of an equation* in the variables x and y consists of the points in the plane whose coordinates satisfy the equation. The graph of the function f, then, is the graph of the equation $y = f(x)$. The terminology is generally simplified by saying "graph the equation" instead of "sketch the graph of the equation," and the sketch is referred to as the graph of the equation.

When a function f is described by an equation $y = f(x)$, the variable x is called the **independent variable** and y is called the **dependent variable.** The independent variable x represents values in the domain of f and the dependent variable y represents values in the range of f, so each value of x must correspond to precisely one value of y. We say that *the equation describes y as a function of x.*

Example 6

Determine whether each of the following equations describes y as a function of x.

a) $y = x^2$ **b)** $y^2 = x$ **c)** $x^2 + y^2 = 1$

Solution

a) The equation $y = x^2$ describes y as a function of x because each value of x corresponds to precisely one value of y, namely, the square of x. (See Figure 1.4a on the following page.)

b) The equation $y^2 = x$ does not describe y as a function of x since all nonzero values of x correspond to two distinct values of y: $y = \sqrt{x}$ and $y = -\sqrt{x}$. For example, $x = 4$ corresponds to both $y = 2$ and $y = -2$.

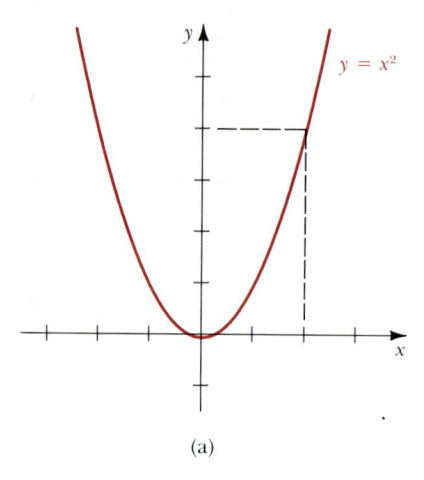

Figure 1.4

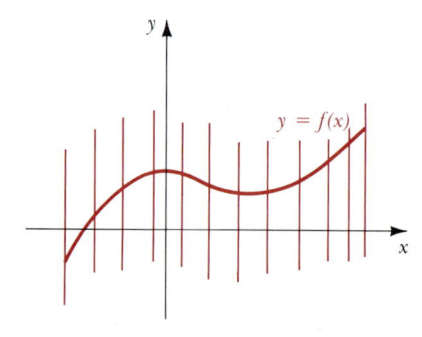

Figure 1.5
Each vertical line intersects the graph of a function at most once.

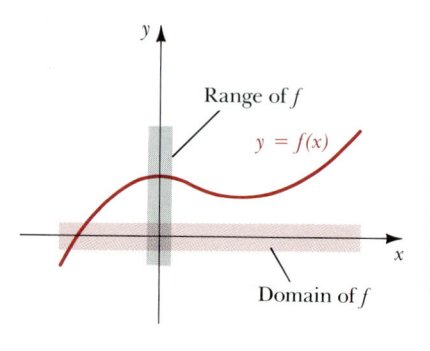

Figure 1.6

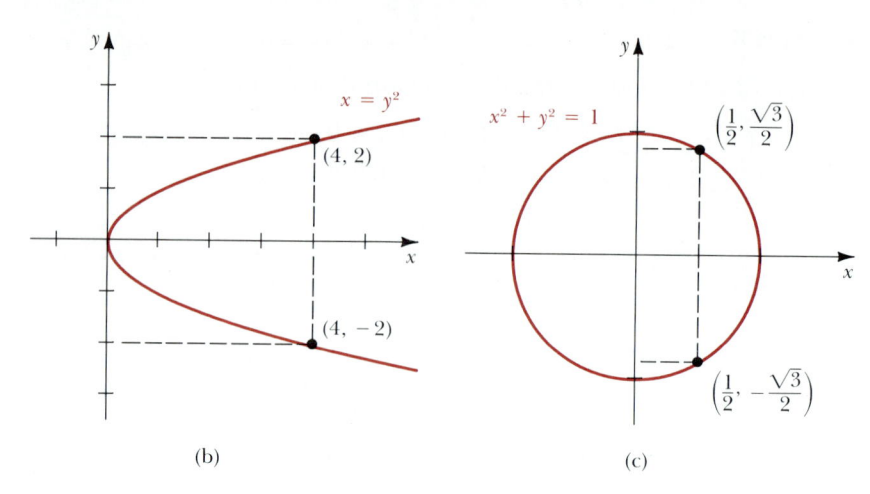

Figure 1.4(b) shows that both (4, 2) and (4, −2) are on the graph of the equation $y^2 = x$.

c) The equation $x^2 + y^2 = 1$ does not describe y as a function of x since, for example, $x = 0$ corresponds to both $y = 1$ and $y = -1$. (See Figure 1.4c.) ▶

The graphs of the equations in Example 6 illustrate a test to distinguish equations that represent functions from those that do not represent functions (see Figure 1.5).

(1.2)
An equation describes y as a function of x precisely when every vertical line intersects its graph in at most one point. In addition, when the graph is that of a function, the following are true (see Figure 1.6):

(1.3)
The domain of the function is described by those values on the horizontal axis (the axis of the independent variable) through which a vertical line intersects the graph.

(1.4)
The range of the function is described by those values on the vertical axis (the axis of the dependent variable) through which a horizontal line intersects the graph.

The graphs of the equations in Example 6 also illustrate a feature known as *symmetry* with respect to a coordinate axis. A graph has this property when the portion of the graph on one side of the axis is the mirror image of the portion on the other side. This axis symmetry is illustrated in Figure 1.7 and is defined as follows.

(1.5)
The graph of an equation is **symmetric with respect to the y-axis** provided that whenever (x, y) is on the graph, $(-x, y)$ is also on the graph.

Figure 1.7
(a) An even function: y-axis symmetry;
(b) not a function: x-axis symmetry.

(1.6)
The graph of an equation is **symmetric with respect to the x-axis** provided that whenever (x, y) is on the graph, $(x, -y)$ is also on the graph.

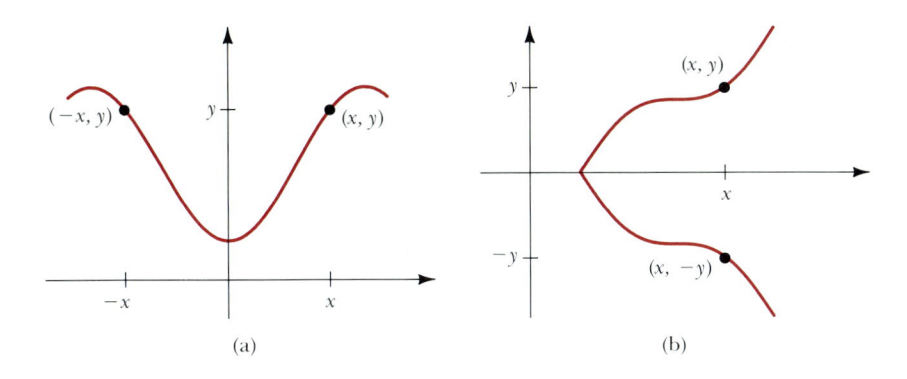

In Example 6, the graph of $y = x^2$ is symmetric with respect to the y-axis and the graph of $x = y^2$ is symmetric with respect to the x-axis. The graph of the circle with equation $x^2 + y^2 = 1$ is symmetric with respect to both the x-axis and the y-axis.

Symmetry is also defined with respect to the origin. A graph having this symmetry is shown in Figure 1.8. The circle with equation $x^2 + y^2 = 1$, shown in Figure 1.4(c), also has this property.

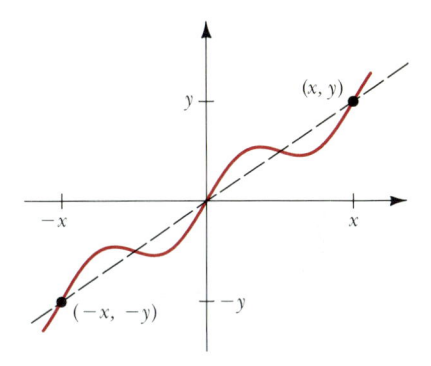

Figure 1.8
An odd function: origin symmetry.

(1.7)
The graph of an equation is **symmetric with respect to the origin** provided that whenever (x, y) is on the graph, $(-x, -y)$ is also on the graph.

The graph of a *function* cannot be symmetric with respect to the x-axis unless all values of the function are zero, since (x, y) and $(x, -y)$ can both be on the graph only when $y = 0$.

The following definition characterizes those functions whose graphs are symmetric with respect to the y-axis or with respect to the origin.

(1.8)
DEFINITION

i) A function f is said to be **even** if whenever x is in the domain of f, $-x$ is also in the domain of f and $f(-x) = f(x)$.
ii) A function f is said to be **odd** if whenever x is in the domain of f, $-x$ is also in the domain and $f(-x) = -f(x)$.

Definition 1.8 implies that the graph of an even function is symmetric with respect to the y-axis, for if $(x, f(x))$ is on the graph, then $(-x, f(x))$ is on the graph as well. Similarly, the graph of an odd function is symmetric with respect to the origin since if $(x, f(x))$ is on the graph, then $(-x, -f(x))$ is also on the graph.

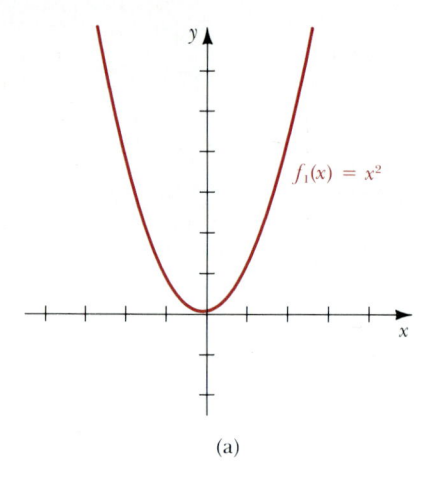

Figure 1.9
(a) y-axis symmetry; (b) origin symmetry; (c) no axis or origin symmetry.

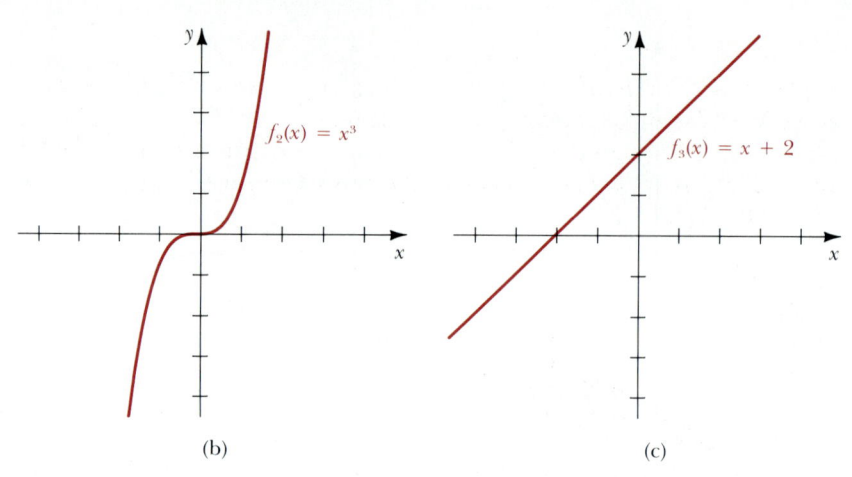

Example 7

Determine whether each of the following functions is even, odd, or neither even nor odd.

a) $f_1(x) = x^2$ **b)** $f_2(x) = x^3$ **c)** $f_3(x) = x + 2$

Solution

The graphs of these functions are shown in Figure 1.9.

a) For $f_1(x) = x^2$,
$$f_1(-x) = (-x)^2 = x^2 = f_1(x),$$
so f_1 is an even function.

b) For $f_2(x) = x^3$,
$$f_2(-x) = (-x)^3 = -x^3 = -f_2(x),$$
so f_2 is an odd function.

c) For $f_3(x) = x + 2$,
$$f_3(-x) = -x + 2.$$
Since this expression differs from both
$$f_3(x) = x + 2 \quad \text{and} \quad -f_3(x) = -x - 2,$$
the function f_3 is neither even nor odd.

Note that the graph of f_1 is symmetric with respect to the y-axis and the graph of f_2 is symmetric with respect to the origin, but the graph of f_3 has no axis or origin symmetry. ▶

Our final example illustrates the use of function notation to express real-life problems in a mathematical form. Generally, the most difficult part of a problem is translating the written description of the problem to a form using mathematical equations. The "Know–Find" outline used in Example 8 can be helpful in making this translation.

Example 8

Two ships sail from the same port. The first ship leaves port at 1:00 A.M. and travels eastward at a rate of 15 knots (nautical miles per hour). The second ship leaves port at 2:00 A.M. and travels northward at a rate of 10 knots. Find the distance d between the ships as a function of the time after 2:00 A.M.

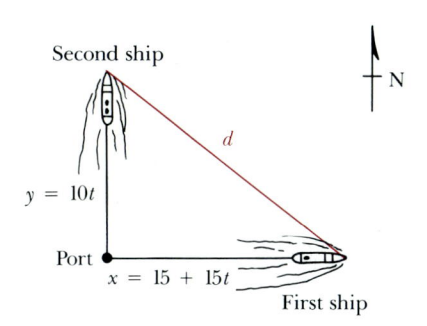

Figure 1.10

Solution

We must first introduce a time variable into the problem. Let t denote the time in hours after 2:00 A.M. Since the second ship travels at a rate of 10 nautical miles per hour, it is $10t$ nautical miles from port at time t.

The first ship is traveling at a rate of 15 nautical miles per hour and has traveled for $(1+t)$ hours at time t. Thus its distance from port is $15(1+t) = 15 + 15t$ nautical miles at time t. Figure 1.10 shows the position of the ships at a given time t.

The following is a concise description of the problem.

KNOW	FIND
1) The distance of the first ship from port: $x = 15 + 15t$, $t \geq 0$.	a) The distance d separating the ships as a function of t.
2) The distance of the second ship from port: $y = 10t$, $t \geq 0$.	

Since the paths of the ships are perpendicular, we use the Pythagorean theorem to determine the answer. The ships are

$$d(t) = \sqrt{x^2 + y^2} = \sqrt{(15 + 15t)^2 + (10t)^2} = \sqrt{325t^2 + 450t + 225}$$

nautical miles apart t hours after 2:00 A.M. ▶

Note that the domain of the function described in Example 8 has been restricted by the physical conditions of the problem to the interval $[0, \infty)$. In fact, a reasonable approximation to the actual distance between the ships is given by $d(t)$ only for small values of t. This is due to variations in the actual speed of the ships, deviation from the assumed course, the curvature of the earth, and many other factors. Keep in mind that when mathematical expressions are used to solve physical problems, the answer to the mathematical problem can vary from the answer to the physical problem due to such neglected technicalities.

▶ EXERCISE SET 1.1

1. If $f(x) = 4x^2 + 5$, find each of the following.
   a) $f(2)$
   b) $f(\sqrt{3})$
   c) $f(2 + \sqrt{3})$
   d) $f(2) + f(\sqrt{3})$

2. If $f(x) = 3x - 5x^2$, find each of the following.
   a) $f(7)$
   b) $f(\sqrt{2})$
   c) $f(7 + \sqrt{2})$
   d) $f(7) + f(\sqrt{2})$

In Exercises 3–10, a function is described. Find the domain of the function.

3. $f(x) = x^2$

4. $f(x) = \dfrac{1}{x - 3}$

5. $f(x) = \dfrac{x^2 - 1}{x^4 - x^2}$

6. $f(x) = \dfrac{1}{\sqrt{x - 3}}$

7. $f(x) = \sqrt{x(x-2)}$
8. $f(x) = \dfrac{1}{\sqrt{3x - x^2}}$
9. $f(x) = \sqrt{x^2 - x - 6}$
10. $f(x) = \dfrac{1}{\sqrt{x^2 - x - 6}}$

In Exercises 11–16, determine whether the equation describes y as a function of x.

11. $3x + 2y = 5$
12. $y = x^2 + 3$
13. $x = y^2 + 3$
14. $x^2 + y^2 = 16$
15. $x^2 + 4x + y^2 = 12$
16. $x^2 - y^3 = 4$

In Exercises 17–20, the correspondence rule of a function f is given. Describe the function given by $f(-x)$, $-f(x)$, $f(1/x)$, $1/f(x)$, $f(\sqrt{x})$, and $\sqrt{f(x)}$.

17. $f(x) = x^2 + 2$
18. $f(x) = x^2 + 2x + 3$
19. $f(x) = 1/x$
20. $f(x) = \sqrt{x}$

In Exercises 21–26, a function is described and a number in the range of the function is given. Find a number in the domain of the function corresponding to this value in the range.

21. $f(x) = x^2 - 1$; 0
22. $f(x) = x^2 - 1$; 2
23. $f(x) = x^2 - 1$; $\tfrac{1}{2}$
24. $h(x) = \sqrt{x}$; $\tfrac{3}{4}$
25. $g(x) = x^2 + 2x + 2$; 1
26. $f(x) = \dfrac{1}{x^2 + 2x + 2}$; $\tfrac{1}{2}$

Find $f(x+h)$ and $\dfrac{f(x+h) - f(x)}{h}$, if $h \neq 0$, for the functions described in Exercises 27–34.

27. $f(x) = 2x - 4$
28. $f(x) = 7x + 5$
29. $f(x) = \tfrac{3}{2}x - \tfrac{1}{4}$
30. $f(x) = -\tfrac{3}{2}x + 5$
31. $f(x) = x^2$
32. $f(x) = x^2 - 1$
33. $f(x) = 2x^2 - x$
34. $f(x) = 4x^2 + 3x + 1$

35. Determine whether each of the following functions is odd, even, or neither odd nor even.
    a) $f(x) = x^2 + 1$
    b) $f(x) = x^3 - 1$
    c) $f(x) = x^3 + 3x$
    d) $f(x) = \sqrt{x}$
    e) $f(x) = x^2 + x$
    f) $f(x) = x^2 + x^3$
    g) $f(x) = x^4 - x^2$
    h) $f(x) = 1/x$

36. Determine which of the functions described in Exercise 35 have graphs that are (i) symmetric with respect to the y-axis and (ii) symmetric with respect to the origin.

37. Show that a function cannot be both odd and even unless it is zero at every number in its domain.

38. A rectangular plot of ground containing 432 ft² is to be fenced off within a large lot. Express the perimeter of the plot as a function of the width. What is the domain of this function?

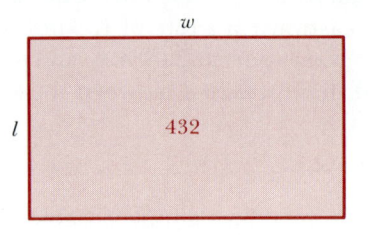

39. A rectangular plot of ground containing 432 ft² is to be fenced off within a large lot, and a fence is to be constructed down the middle. Express the amount of fence required as a function of the length of the dividing fence. What is the domain of this function?

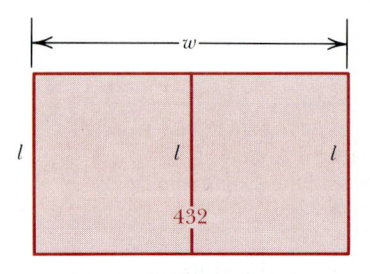

40. Two ships sail from the same port. The first ship leaves at noon and travels eastward at 10 knots. The second ship leaves at 3:00 P.M. and travels southward at 15 knots. Find the distance d between the ships as a function of the time after 3:00 P.M.

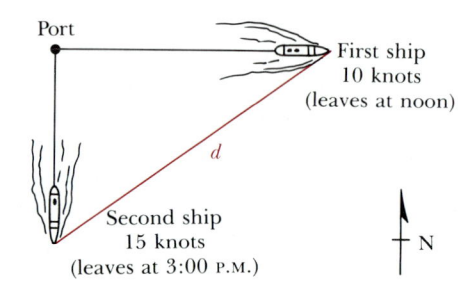

41. Two tankers are traveling in the midst of the Atlantic Ocean (see the figure). The first tanker is 100 nautical miles due north of the second at 1:00 GMT (Greenwich Mean Time) and is traveling east at a rate of 20 knots. The second tanker is traveling north at 15 knots. Express the distance d between the two tankers as a function of the number of hours after 1:00 GMT.

42. A rectangle is to be placed in a circle of radius 5 in. with its corners on the boundary of the circle. Express the area of the rectangle as a function of the width of one side, and determine the domain of this function.

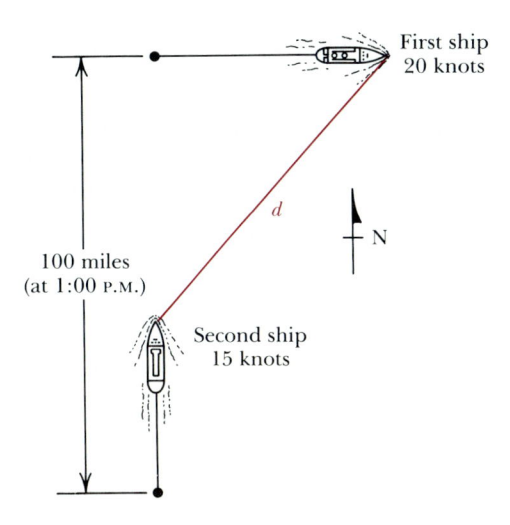

Figure for Exercise 41.

43. A rectangular beam is to be cut from a circular log of radius 5 in. in such a way that the corners of the beam all lie on the circumference of the circle. Define a function that describes the area of a cross section of the end of a beam of this type in terms of the width of the beam. What is the domain of this function?

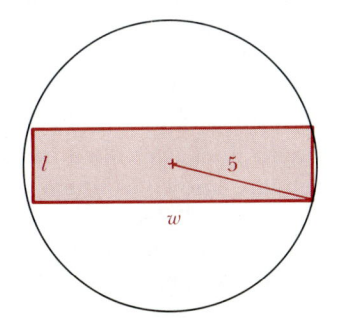

44. A standard can contains a volume of 900 cm³. The can is in the shape of a right circular cylinder with a top and a bottom. Express the surface area of the can as a function of the radius of the bottom.

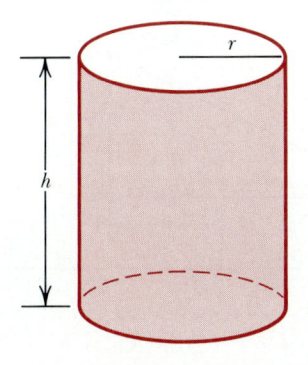

45. An open rectangular box is to be made from a sheet of metal 8 cm wide and 11 cm long by cutting a square from each corner, bending up the sides, and welding the edges. Express the volume of the box as a function of the length of a side of the square cut from each corner, and determine the domain of this function.

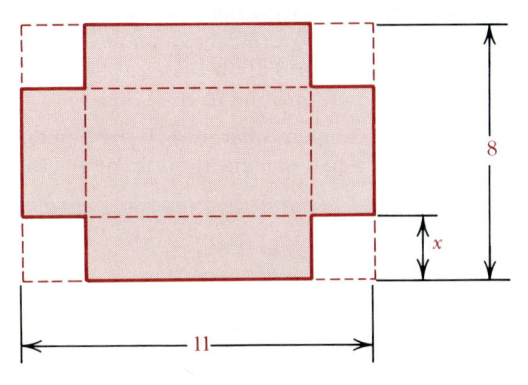

46. A one-mile-long race track is to be built with two straight sides and with semicircles at the ends. Express the area enclosed by the track as a function of the diameter of the semicircles.

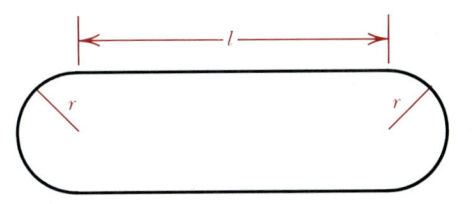

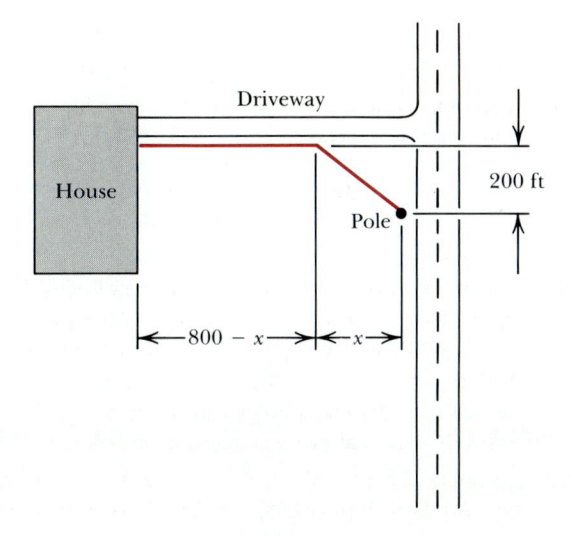

47. A house is built with a straight driveway 800 ft long, as shown in the figure. A utility pole on a line perpendicular to the driveway and 200 ft from the end of the driveway is the closest point from which electricity can be furnished. The utility company will furnish power with underground cable at $2 per foot and with overhead lines at no charge. However, for overhead lines the company requires that a strip 30 ft wide be cleared. The owner of the house estimates that to clear a strip this wide will cost $3 for each foot of overhead wire used. How much will it cost to run the lines if they are run overhead to a point on the driveway x feet from the end of the driveway ($0 \leq x \leq 800$) and then run underground to the house?

48. A charter bus company charges $10 per person for a round trip to a ball game with a discount given for group fares. A group purchasing more than 10 tickets receives a reduction per ticket of 25¢ times the number of tickets purchased in excess of 10. If 11 people go to the game, each of the 11 tickets cost $9.75; if 16 people go, each ticket costs $8.50; and so on. Express the amount of money the company receives as a function of the number of tickets it sells to a group.

49. Reconsider the problem described in Exercise 48 with the added restriction that the price per person is never less than $6.25. Express the amount of money that the company now receives as a function of the number of tickets sold to the group.

1.2

Linear and Quadratic Functions

Two of the most elementary classes of functions are linear functions, whose graphs are straight lines, and quadratic functions, whose graphs are parabolas. We will use these familiar functions frequently to illustrate the various concepts of calculus. In this brief review we present their properties in a manner consistent with their use in calculus.

Linear Functions

Functions whose graphs are straight lines are called **linear functions** and the equations describing them are known as **linear equations.**

Consider a straight line l that is not parallel to the y-axis, and suppose $P(x_1, y_1)$ and $Q(x_2, y_2)$ are two distinct points that lie on l, as illustrated in Figure 1.11. The **slope** m of the line l is defined by

(1.9) $$m = \frac{y_2 - y_1}{x_2 - x_1},$$

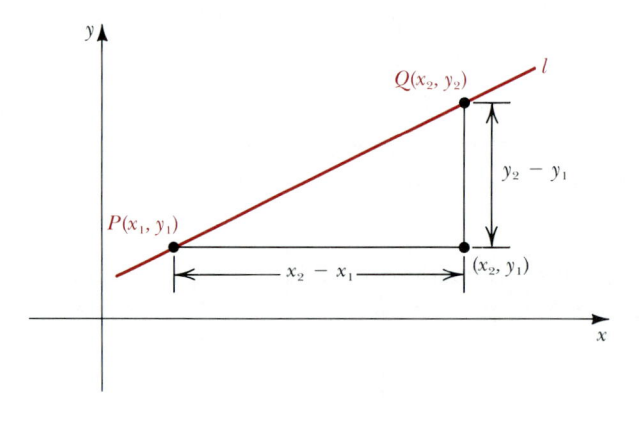

Figure 1.11
The slope of l is $\frac{y_2 - y_1}{x_2 - x_1}$.

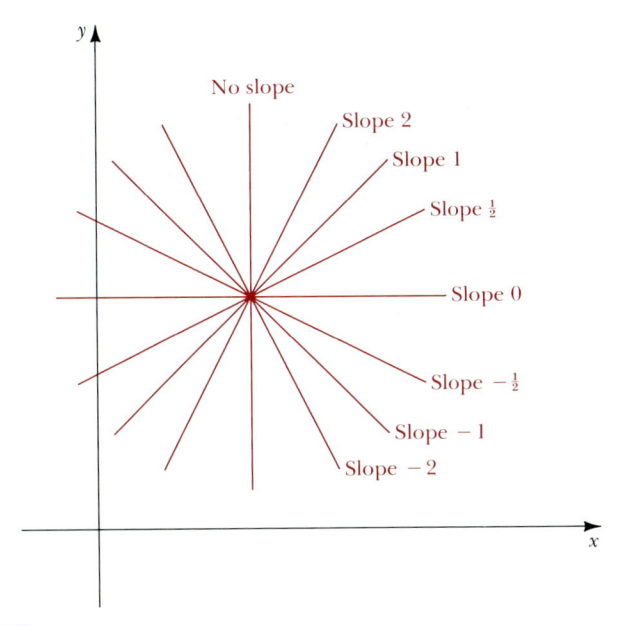

Figure 1.12

a number that is independent of the choice of the points P and Q on the line.

A vertical line, one that is parallel to the y-axis, has the property that all points on the line have the same x-coordinate. For lines of this type the slope is undefined. (Note that the calculation in (1.9) would require division by zero.)

Figure 1.12 shows various lines passing through a common point and lists their slopes. The slope determines the direction of a line, in the sense that:

(1.10)

i) A line with a positive slope is directed upward (looking from left to right). The values of y increase as the values of x increase, and the linear function is said to be increasing.

ii) A line with a negative slope is directed downward (looking from left to right). The values of y decrease as the values of x increase, and the linear function is said to be decreasing.

iii) On a line with a large positive slope, the values of y increase more rapidly than on one with a smaller positive slope. On a line with a negative slope that is large in magnitude, the values of y decrease more rapidly than on one with a negative slope of a smaller magnitude.

To determine an equation of the line l that passes through the point $P(x_1, y_1)$ and has slope m, let $Q(x, y)$ be any other point on l. Then

$$\frac{y - y_1}{x - x_1} = m,$$

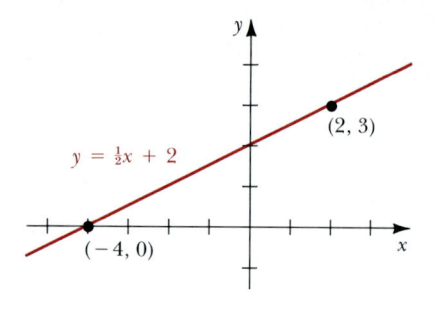

Figure 1.13

so the line has equation

(1.11) $$y - y_1 = m(x - x_1).$$

This equation is known as a **point–slope** form of the equation of a line. Equation (1.11) can be rewritten as

$$y = mx + y_1 - mx_1$$

or as

(1.12) $$y = mx + b,$$

where $b = y_1 - mx_1$. This equation is called the **slope–intercept** form of the equation of a line. Since $(0, b)$ is the point where the line crosses the y-axis, the number b is called the **y-intercept** of the line.

Example 1

The line passing through the points $(2, 3)$ and $(-4, 0)$ is shown in Figure 1.13.

a) Find a point–slope form of the equation of this line.
b) Find the slope–intercept form of the equation of this line.

Solution

The slope of this line is

$$m = \frac{3 - 0}{2 - (-4)} = \frac{3}{6} = \frac{1}{2}.$$

a) Using Equation (1.11) and the point $(2, 3)$, we find that the point–slope form is

$$y - 3 = \frac{1}{2}(x - 2).$$

If we use Equation (1.11) and the point $(-4, 0)$, the point–slope form is

$$y - 0 = \frac{1}{2}(x - (-4)).$$

b) When changed to the slope–intercept form (1.12), both equations reduce to

$$y = \frac{1}{2}x + 2.$$ ▶

Note in Example 1 that although there are different point–slope equations for the line, they reduce to the same, or *unique*, slope–intercept form.

Example 2

Sketch the graph of the linear function described by $f(x) = 2 - \frac{2}{3}x$.

Solution
The equation
$$y = 2 - \frac{2}{3}x$$
describes the line with slope $-\frac{2}{3}$ and y-intercept 2, so the point (0, 2) lies on the line. To find another point, set $y = 0$. The x-value of this point is called the **x-intercept** since this is where the line crosses the x-axis. When $y = 0$,
$$0 = 2 - \frac{2}{3}x, \quad \text{so} \quad \frac{2}{3}x = 2 \quad \text{and} \quad x = 3.$$

The straight line through (0, 2) and (3, 0) is the graph of the function f, as shown in Figure 1.14. ▶

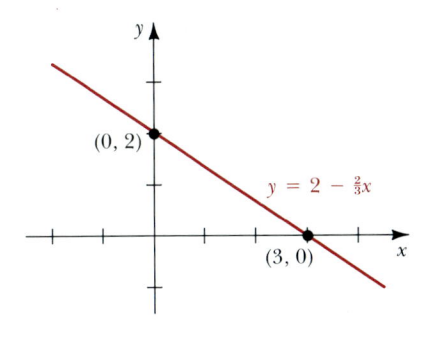

Figure 1.14

When two lines l_1 and l_2 intersect, an angle with radian measure in $[0, \pi)$ is generated by rotating l_1 counterclockwise about the point of intersection until it coincides with l_2. This angle is called the *angle from l_1 to l_2*; it is labeled θ in Figure 1.15(a). The angle between two parallel lines is defined to be zero.

The angle α, where $0 \leq \alpha < \pi$, from the x-axis to the line l is known as the *angle of inclination* of l (see Figure 1.15b).

The definition of the tangent function (see, in particular, Equation A.23 of Appendix A.3) implies that if a line has slope m and angle of inclination α, then $m = \tan \alpha$. Figure 1.15(c) illustrates this situation when $0 < \alpha < \pi/2$.

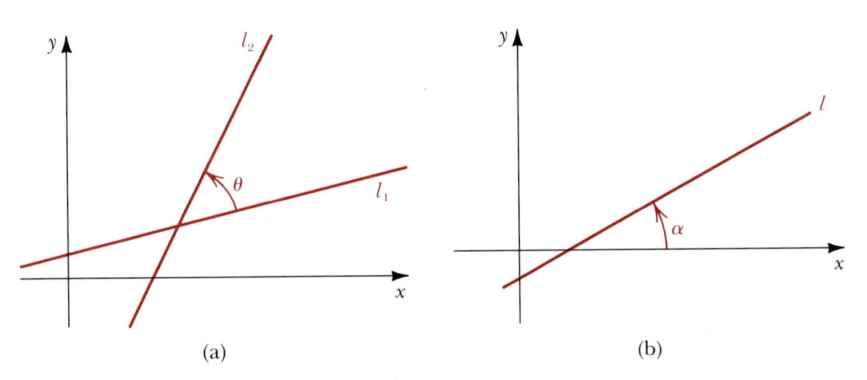

Figure 1.15
(a) The angle between the lines l_1 and l_2; (b) the angle α of inclination of l; (c) the tangent of α is the slope of l.

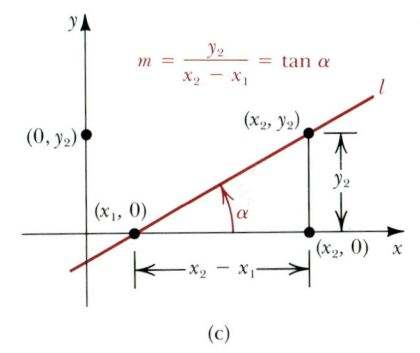

(a)

(b)

(c)

The slope of a line is used to determine when lines are parallel and when they are perpendicular (see Figure 1.16 on the following page). The following theorem describes this connection. (Its proof is in Appendix B.)

(1.13)
THEOREM

a) Two nonvertical lines are parallel if and only if they have the same slope.
b) Two nonvertical lines are perpendicular if and only if the product of their slopes is -1.

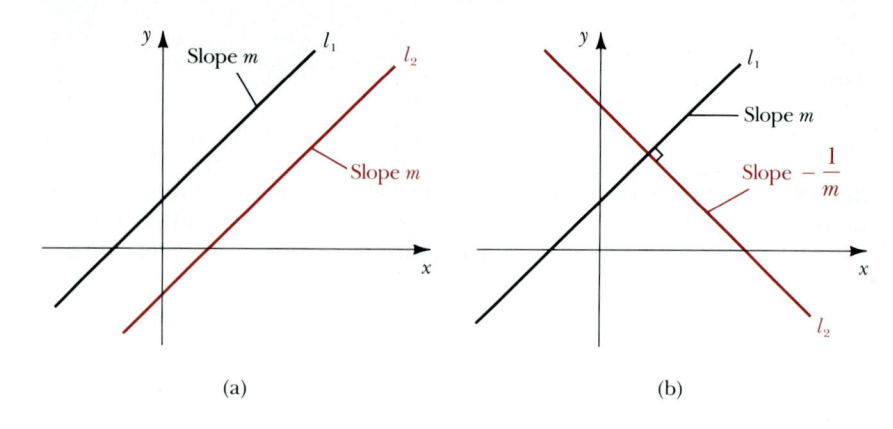

Figure 1.16
(a) Parallel lines; (b) perpendicular lines.

Example 3

Find an equation of the line passing through (3, 1) and:

a) parallel to the line with equation $y = 2x - 1$;
b) perpendicular to the line with equation $y = 2x - 1$.

Solution

a) The slope of the line with equation $y = 2x - 1$ is 2 so, by Theorem 1.13(a), the slope of any line parallel to this line is also 2. Since the required line passes through (3, 1), as shown in Figure 1.17(a), it has equation

$$y - 1 = 2(x - 3), \quad \text{or} \quad y = 2x - 5.$$

b) Let m be the slope of the line perpendicular to the line with equation $y = 2x - 1$. It follows from Theorem 1.13(b) that $2 \cdot m = -1$. Thus $m = -\frac{1}{2}$ and the line passing through (3, 1) perpendicular to $y = 2x - 1$ has equation

$$y - 1 = -\tfrac{1}{2}(x - 3), \quad \text{or} \quad x + 2y - 5 = 0.$$

This line is shown in Figure 1.17(b). ▶

The equation of the line in Example 3(b) is expressed in the form

(1.14) $$Ax + By + C = 0,$$

where A, B, and C are constants. Any linear equation can be written in this form. When $B \neq 0$, the equation describes a line with slope $-A/B$. When $B = 0$ and $A \neq 0$, the equation describes a vertical line.

Quadratic Functions

Quadratic functions are described by **quadratic equations** of the form

(1.15) $$y = ax^2 + bx + c, \quad \text{where } a \neq 0.$$

Graphs of quadratic functions are called *parabolas*. The basic parabola is the graph of $y = x^2$, the quadratic equation generated when $a = 1$, $b = 0$, and $c = 0$. The graph of this equation is shown in Figure 1.18.

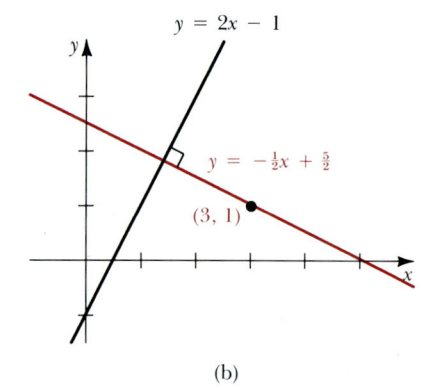

Figure 1.17
(a) Parallel lines; (b) perpendicular lines.

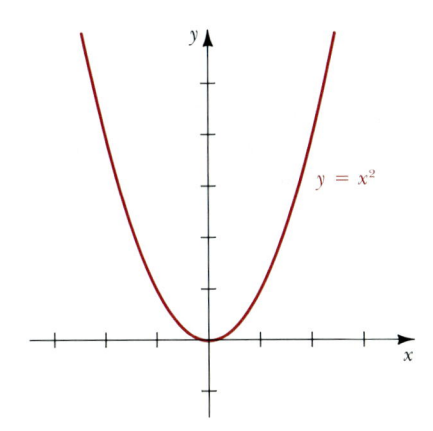

1.2 LINEAR AND QUADRATIC FUNCTIONS

Figure 1.18
The graph is a parabola.

Figure 1.19
Horizontal translation.

The graph of $f(x) = x^2$ can be used to determine the graphs of other quadratic functions, as illustrated in the following examples.

Example 4

Sketch the graph of the function given by $g(x) = (x - 1)^2$.

Solution

The correspondence described by g is similar to that of $f(x) = x^2$, but one unit is subtracted before the squaring operation is performed. Consequently, $g(1) = (1 - 1)^2 = 0^2 = f(0)$, $g(2) = 1^2 = f(1)$, $g(3) = f(2)$, and so on. For this reason, the graph of g is the same as the graph of f except that it is moved one unit to the right, as shown in Figure 1.19. ▶

The shifting technique described in Example 4 can be applied in general, as illustrated in Figure 1.20. Suppose that the graph of $y = f(x)$

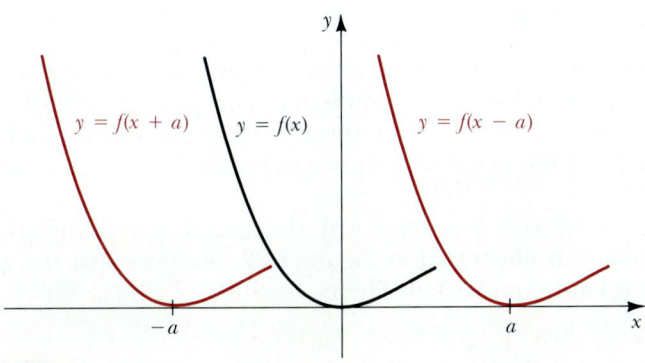

Figure 1.20
Horizontal translation.

is known and $a > 0$. Then:

(1.16)

i) The graph of $y = f(x - a)$ is the graph of f shifted a units to the right.
ii) The graph of $y = f(x + a)$ is the graph of f shifted a units to the left.

Any quadratic equation of the form
$$ax^2 + bx + c$$
can be written as the sum of a constant and a squared term involving the variable x. This technique is called *completing the square*. First write
$$ax^2 + bx + c = a\left(x^2 + \frac{b}{a}x\right) + c.$$

Now add $(b/2a)^2$ inside the parentheses and $-a(b/2a)^2 = -b^2/4a$ outside to compensate. Then
$$ax^2 + bx + c = a\left[x^2 + \frac{b}{a}x + \left(\frac{b}{2a}\right)^2\right] + c - \frac{b^2}{4a}$$
and

(1.17) $$ax^2 + bx + c = a\left(x + \frac{b}{2a}\right)^2 + \frac{4ac - b^2}{4a}.$$

Example 5

Sketch the graph of $h(x) = x^2 - 2x + 3$.

Solution
To complete the square in this quadratic equation, add and subtract the term
$$\left(\frac{-2}{2}\right)^2 = 1$$
to obtain
$$h(x) = x^2 - 2x + 3 = (x^2 - 2x + 1) + 3 - 1 = (x - 1)^2 + 2.$$

When $h(x)$ is written in this form, we see that it is the same as $g(x) + 2$, for $g(x)$ given in Example 4. The graph of h is the graph of g shifted 2 units upward. Consequently, the graph of h shown in Figure 1.21 has the same shape as the graph of g shown in Figure 1.19. ▶

The general rule associated with the graphing technique described in Example 5 is illustrated in Figure 1.22. Suppose that the graph of $y = g(x)$ is known and $b > 0$. Then:

(1.18)

i) The graph of $y = g(x) + b$ is the graph of g shifted b units upward.
ii) The graph of $y = g(x) - b$ is the graph of g shifted b units downward.

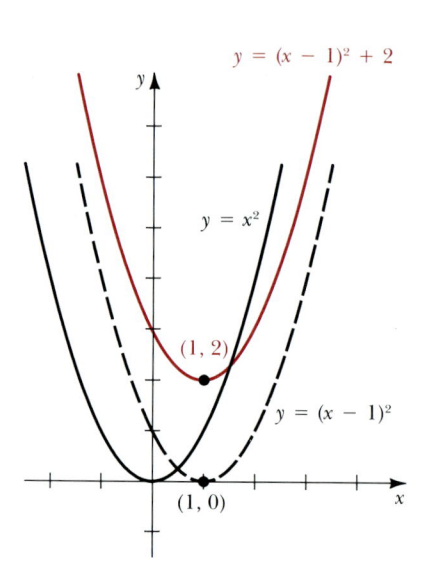

Figure 1.21
A horizontal and vertical translation of a parabola.

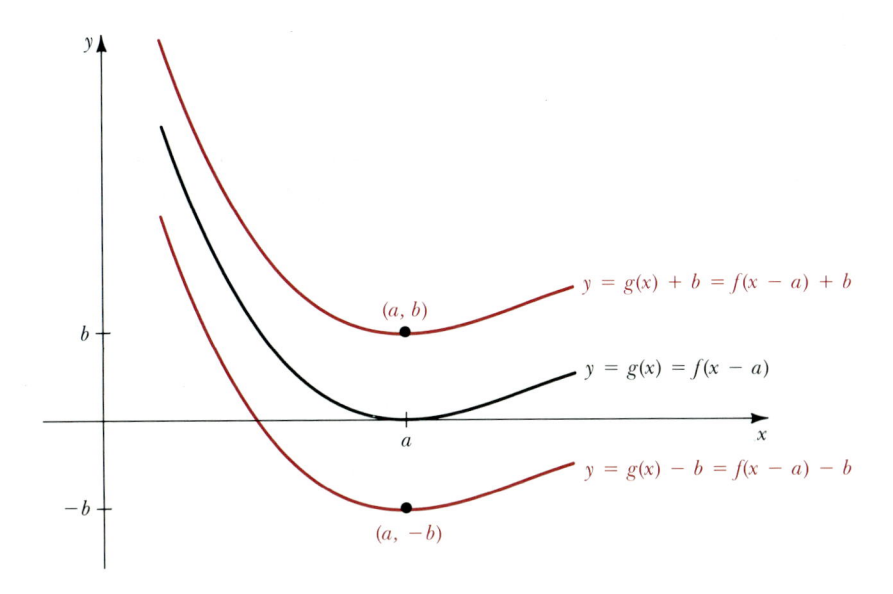

Figure 1.22
Horizontal and vertical translation.

The graph of a quadratic function with equation $y = ax^2$, for an arbitrary constant $a \neq 0$, has a shape similar to the graph of $y = x^2$ (see Figure 1.23). The sign of a determines whether the graph opens upward ($a > 0$) or opens downward ($a < 0$). In addition, the greater the magnitude of a, the narrower the opening of the graph.

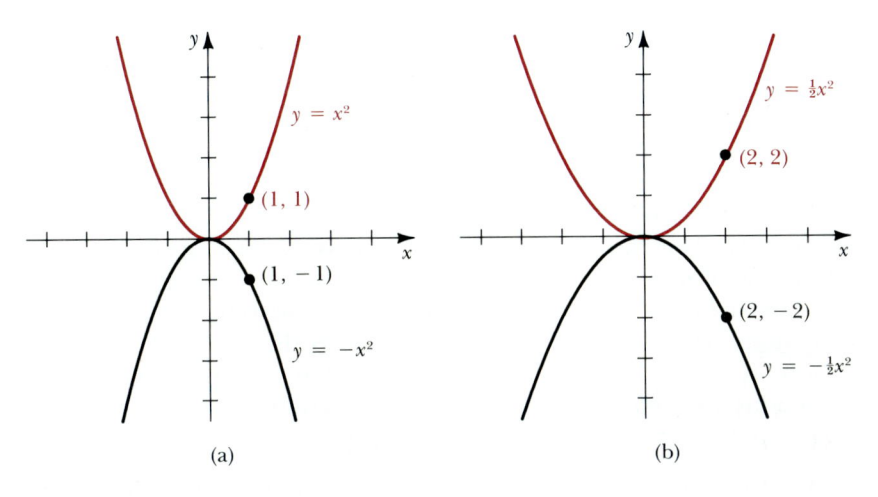

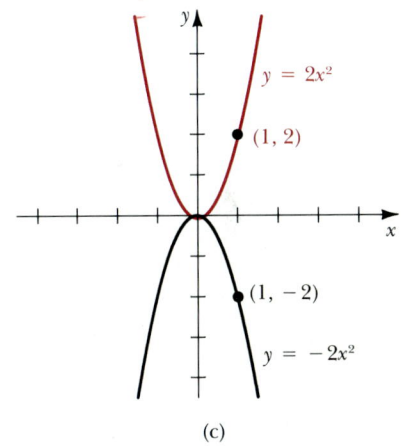

Figure 1.23

Example 6

Sketch the graph of $f(x) = 2x^2 + 12x + 17$.

Solution

To use the shifting techniques described in (1.16) and (1.18), complete the square on the right side of the equation. First factor the coefficient of x^2 from the first two terms of the expression. Then

$$f(x) = 2(x^2 + 6x) + 17 = 2(x^2 + 6x + 9) - 2 \cdot 9 + 17$$
$$= 2(x + 3)^2 - 1.$$

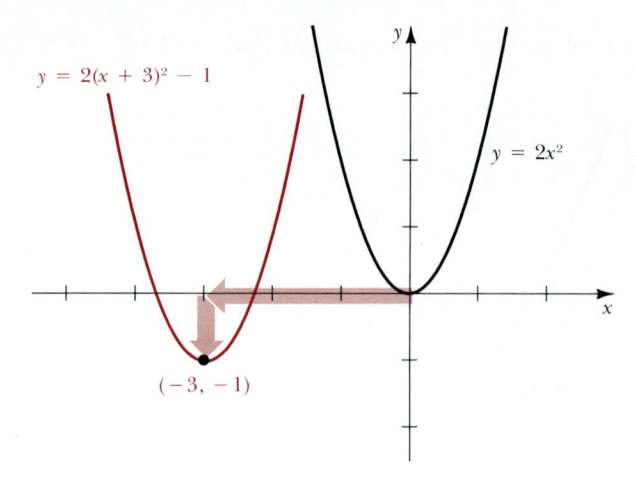

Figure 1.24
A horizontal and vertical translation of $y = 2x^2$.

Using the technique in (1.16), we find that the graph of $y = 2(x + 3)^2$ is the graph of $y = 2x^2$ shifted 3 units to the left. Then applying (1.18), we see that the graph of $y = 2(x + 3)^2 - 1$ is the graph of $y = 2(x + 3)^2$ shifted 1 unit downward.

Consequently, the graph of $y = 2x^2 + 12x + 17 = 2(x + 3)^2 - 1$ has the same shape as the graph of $y = 2x^2$ but it is shifted 3 units to the left and 1 unit downward, as shown in Figure 1.24. ▸

The functions considered in this section are examples of polynomial functions. A **polynomial** is an expression of the form

(1.19) $\qquad a_n x^n + a_{n-1} x^{n-1} + \cdots + a_2 x^2 + a_1 x + a_0,$

where n is a nonnegative integer and $a_0, a_1, a_2, \ldots, a_n$ are constants. If $a_n \neq 0$, the polynomial is said to have **degree** n. For example, $x^4 + 3x^2 + 1$ is a polynomial of degree four, and $7x^{39} - \pi x^2$ is a polynomial of degree 39. The expression $2x^4 - 3x^{-1}$ is not a polynomial, however, since one of its terms has a power of -1, a negative integer.

Linear functions that are not constant are first-degree polynomial functions because they can be expressed in the form $y = a_1 x + a_0$, where a_1 is the slope of the line and a_0 is the y-intercept. Quadratic functions are second-degree polynomial functions because they can be expressed as $y = a_2 x^2 + a_1 x + a_0$.

▸ **EXERCISE SET 1.2**

1. Plot the pair of points given in each of the following, and sketch the straight line determined by these points. Find the slope of the line.

   a) $(0, 0), (1, 2)$ b) $(-1, 2), (3, -2)$
   c) $(1, -3), (5, 1)$ d) $(-3, -1), (1, 2)$
   e) $(-1, 3), (-5, -1)$ f) $(3, 1), (-1, -2)$

2. Find an equation of the line determined by the pair of points in each part of Exercise 1.

3. Find equations of the lines that pass through the point $(3, 2)$ and have slope:

   a) 1 b) -1 c) 0
   d) 3 e) 10 f) π

4. Sketch the graphs of the lines given in Exercise 3 on the same coordinate axes.

5. Sketch the graph of the line associated with each of the following linear equations. Tell which pairs of lines are parallel.
   a) $y = 2$
   b) $y = x + 1$
   c) $y = -x + 1$
   d) $y = x + 4$
   e) $y = 2x - 5$
   f) $y = -4$
   g) $x + y = 0$
   h) $4x - 2y - 4 = 0$
   i) $-x + y + 4 = 0$
   j) $2x - 2y + 1 = 0$

6. Sketch the graph of the line associated with each of the following linear equations. Tell which pairs of lines are perpendicular.
   a) $y = -1$
   b) $x = -2$
   c) $y = 2x + 3$
   d) $y = 3x + 5$
   e) $y = \frac{1}{3}x - \frac{1}{3}$
   f) $y = -x - 3$
   g) $y = -2x - 5$
   h) $y = -3x - 7$
   i) $x + 3y + 5 = 0$
   j) $x + 2y + 1 = 0$

In Exercises 7–14, the equation of a line is given together with a point not on the line. Find the slope–intercept form of the equation of the line that passes through the given point and is (a) parallel to the given line and (b) perpendicular to the given line.

7. $y = 2x + 1$; $(0, 0)$
8. $y = 4x + 3$; $(0, 0)$
9. $y = 3x - 2$; $(1, 2)$
10. $y = \frac{1}{2}x + 2$; $(1, 1)$
11. $y = -2x + 3$; $(-1, 2)$
12. $y = -7x - 5$; $(-1, -3)$
13. $x + y + 1 = 0$; $(0, 0)$
14. $2x - 3y - 4 = 0$; $(1, -1)$

In Exercises 15–40, a quadratic equation is given. Sketch the graph of the equation and find the range of the function described by the equation.

15. $y = x^2 + 1$
16. $y = x^2 - 2$
17. $y = (x - 3)^2$
18. $y = (x + 3)^2$
19. $y = x^2 - 4x + 4$
20. $y = x^2 - 4x$
21. $y = x^2 - 4x + 3$
22. $y = x^2 - 4x + 5$
23. $y = x^2 + 4x$
24. $y = x^2 + 4x + 5$
25. $y = -x^2$
26. $y = -x^2 + 1$
27. $y = -x^2 - 1$
28. $y = -(x - 1)^2$
29. $y = 2x^2$
30. $y = 2x^2 - 6$
31. $y = 2(x - 3)^2$
32. $y = 3x^2 + 6x$
33. $y = 2x^2 + 8x + 6$
34. $y = 2x^2 + 5x + 2$
35. $y = \frac{1}{2}x^2$
36. $y = \frac{1}{2}x^2 + 2$
37. $y = \frac{1}{2}x^2 - 2x + 2$
38. $y = \frac{1}{2}x^2 - 2x + 1$
39. $y = \frac{1}{2}x^2 - 3x + 1$
40. $y = \frac{1}{3}x^2 - 2x + 1$

41. Determine which of the graphs sketched in Exercises 15–40 are symmetric with respect to the y-axis?

42. Sketch the graph of f if
$$f(x) = \begin{cases} x + 2, & \text{if } x < -1, \\ x^2, & \text{if } x \geq -1. \end{cases}$$

43. Sketch the graph of g if
$$g(x) = \begin{cases} 0, & \text{if } x < 0, \\ x^2, & \text{if } 0 \leq x \leq 2, \\ -x + 6, & \text{if } x > 2. \end{cases}$$

44. Use the graph of the function f shown in the figure to sketch the graph of each of the following.
   a) $y = f(x + 2)$
   b) $y = f(x + 2) + 1$
   c) $y = f(x - 1)$
   d) $y = f(x - 1) + 1$
   e) $y = f(x + 2) - 1$
   f) $y = f(x - 1) - 2$

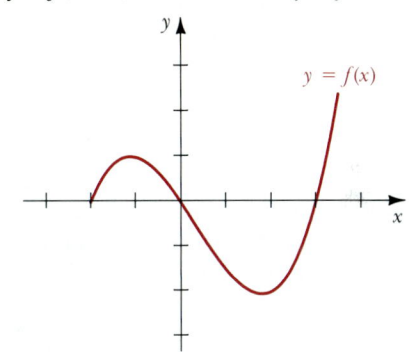

45. Show that the line with x-intercept $a \neq 0$ and y-intercept $b \neq 0$ has equation
$$\frac{x}{a} + \frac{y}{b} = 1.$$

46. Show that the quadratic function with x-intercepts $a \neq 0$ and $-a$ and y-intercept $b \neq 0$ has equation
$$\frac{x^2}{a^2} + \frac{y}{b} = 1.$$

47. The function defined by $s(t) = 784 - 16t^2$ describes the height, in feet, of a rock t seconds after it has been dropped from the top of a certain 60-story building, as shown in the figure on the following page.
   a) Sketch the graph of s.
   b) Determine physically reasonable definitions for the domain and range of s.
   c) How long does it take the rock to reach the bottom of the building?

48. The function defined by $v(t) = -32t$ describes the velocity in feet per second of the rock t seconds after it has been dropped from the building described in Exercise 47.
   a) Sketch the graph of v.
   b) Determine physically reasonable definitions for the domain and range of v.

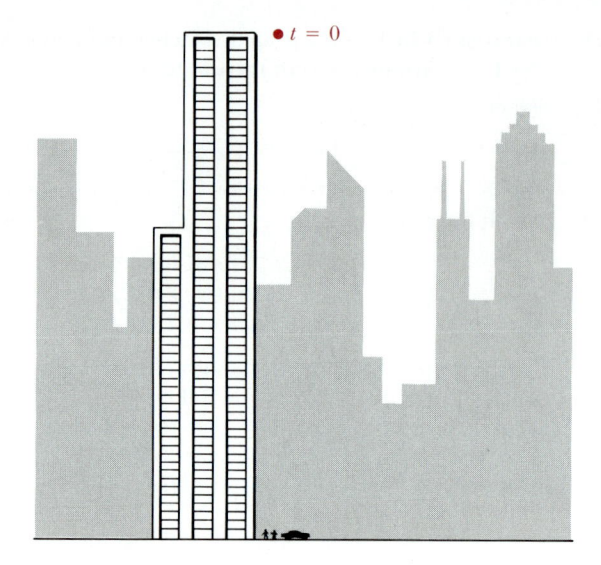

Figure for Exercise 47.

c) What is the velocity of the rock when it reaches the bottom of the building?

49. Determine the linear function that relates the temperature in degrees Celsius to the temperature in degrees Fahrenheit, and use this function to determine the Fahrenheit temperature corresponding to 28° C. (*Hint:* At sea level, water freezes at 32° F (0° C) and boils at 212° F (100° C).)

50. A sweet potato farmer in North Carolina finds that 200 bushels of yams can be produced at an average cost of $3 per bushel and 1000 bushels can be produced at an average cost of $2 per bushel. Assuming that the function describing the average cost is linear, find the average cost function and sketch its graph.

51. A hotel normally charges $40 for a room. However, special group rates are advertised: If the group requires more than 5 rooms, the price for each room is decreased by one dollar times the number of rooms exceeding five, with the restriction that the minimum price per room is $20.
    a) Express the charge per room as a function of the number of rooms used by a group, and sketch the graph of this function.
    b) Express the revenue received from a group as a function of the number of rooms used by the group, and sketch the graph of this function.

52. The average weight W (in grams) of a fish in a particular pond depends on the total number n of fish in the pond according to the model
$$W(n) = 500 - 0.5\, n.$$
    a) Sketch the graph of the function W.
    b) Express the total fish weight production (in grams) as a function of the number of fish in the pond. Sketch the graph of this function.
    c) What do you think happens when $n \geq 1000$?

53. When a solid rod is heated, its length increases by a certain amount depending on its coefficient of linear expansion. This coefficient, α, is assumed to be a constant, depending only on the material of the rod. The amount of increase in length is the product of the length, the change in temperature, and α. Suppose a steel rod has a length of 2 m at 0° C and that the coefficient of linear expansion for this material is $\alpha = 11 \times 10^{-6}$.
    a) Find a function that describes the length of the rod in terms of its temperature above 0° C.
    b) Determine its length when the temperature is 1000° C.

54. Use the technique of completing the square to show that when $a \neq 0$, the equation $0 = ax^2 + bx + c$ has solutions
$$x = \frac{-b \pm \sqrt{b^2 - 4ac}}{2a}.$$
This is known as the *quadratic formula*.

1.3
Graphs of Some Common Functions

We now consider the graphs of some functions that are used frequently in the study of calculus. Each function will be used in Chapters 1 and 2 to emphasize at least one important concept of calculus.

1.3 GRAPHS OF SOME COMMON FUNCTIONS

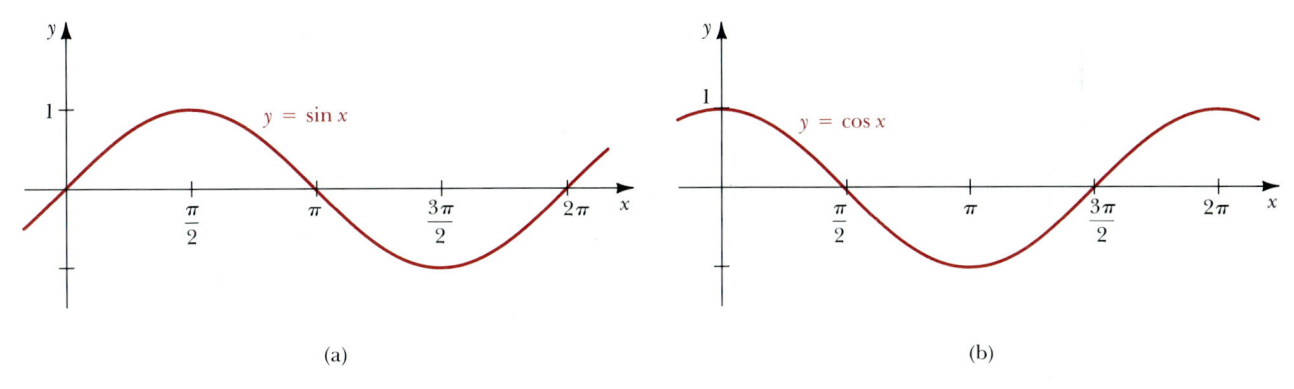

(a) $y = \sin x$

(b) $y = \cos x$

Figure 1.25

The Basic Trigonometric Functions

Appendix A.3 contains a comprehensive review of trigonometry, which begins with the definitions of the **trigonometric functions.** This appendix also considers the basic properties and identities used in the study of calculus. Included in the review are the graphs of the sine and cosine functions, which are shown in Figure 1.25.

First note that the graph of the sine function is symmetric with respect to the origin and that the graph of the cosine function is symmetric with respect to the y-axis. This is a result of the trigonometric identities

(1.20) $\qquad \sin(-x) = -\sin x \quad \text{and} \quad \cos(-x) = \cos x,$

which imply that the sine function is odd and that the cosine function is even.

Note too that shifting the graph of the sine function $\pi/2$ units to the left produces the graph of the cosine function. This is a consequence of applying the shifting technique shown in (1.16) and the trigonometric identity

(1.21) $\qquad \cos x = \sin(x + \pi/2).$

Similarly, the graph of the sine function is produced when the graph of the cosine function is shifted $\pi/2$ units to the right because of the identity

(1.22) $\qquad \sin x = \cos(x - \pi/2).$

Example 1 concerns the graph of the tangent function, shown in Figure 1.26. This trigonometric function is defined by

(1.23) $\qquad \tan x = \dfrac{\sin x}{\cos x}, \quad \text{provided } \cos x \neq 0.$

The tangent function is an odd function, so its graph is symmetric with respect to the origin.

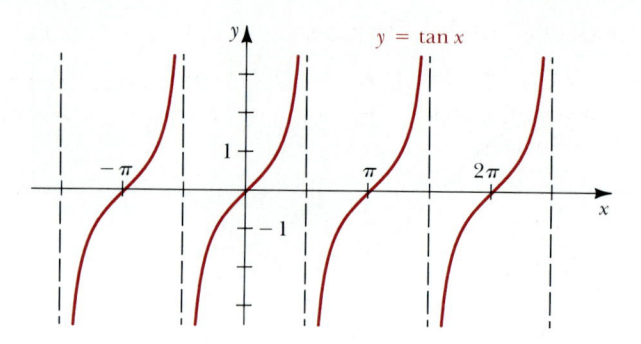

Figure 1.26

Example 1

Use the graph of the tangent function shown in Figure 1.26 to sketch the graph of

$$f(x) = \tan(x - \pi/4) + 1.$$

Solution

The graph of $y = \tan(x - \pi/4)$, shown in black in Figure 1.27, is the graph of $y = \tan x$ shifted $\pi/4$ units to the right.

The graph of $y = f(x) = \tan(x - \pi/4) + 1$ is the graph of $y = \tan(x - \pi/4)$ shifted one unit upward, and is shown in color in Figure 1.27.

Note that the graph of $y = f(x)$ appears to pass through the origin. We can easily verify that this is indeed the case:

$$f(0) = \tan(0 - \pi/4) + 1 = -1 + 1 = 0. \blacktriangleright$$

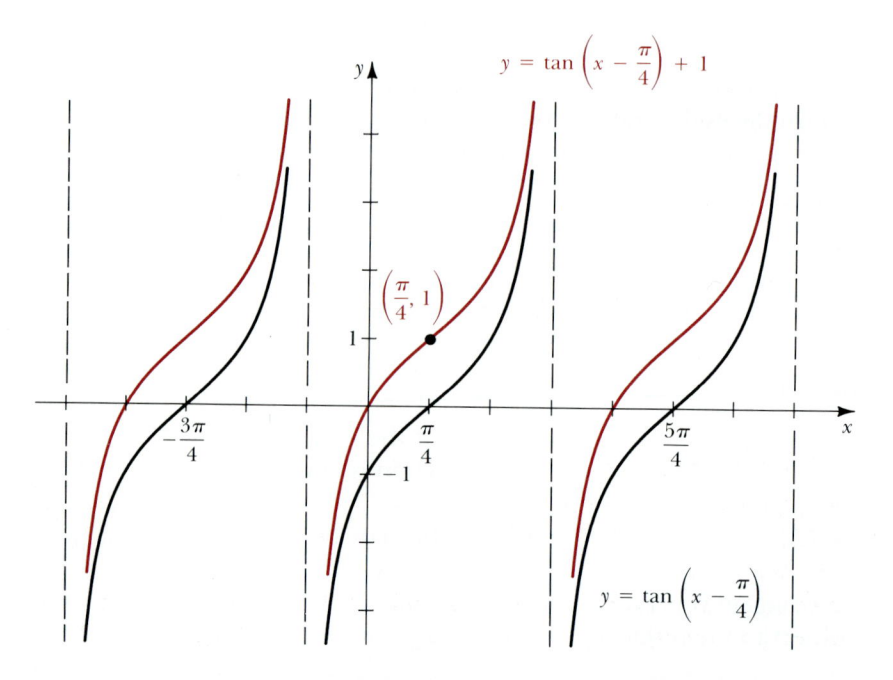

Figure 1.27
A translation of the tangent function.

The Absolute Value Function

The **absolute value function**, $f(x) = |x|$, associates with each real number x its magnitude, that is, the distance from x to the origin. The function is defined by

(1.24) $$|x| = \begin{cases} x, & \text{if } x \geq 0, \\ -x, & \text{if } x < 0, \end{cases}$$

and its graph is shown in Figure 1.28. When $x \geq 0$, the graph coincides with the line $y = x$; when $x < 0$, the graph coincides with the line $y = -x$. Note that the graph is not "smooth" at $(0, 0)$, but turns at this point to form an abrupt angle. This feature of the graph is of particular interest when the derivative is introduced in Chapter 2.

Since $|x| = |-x|$ for any real number x, the absolute value function is an even function, and its graph is symmetric with respect to the y-axis.

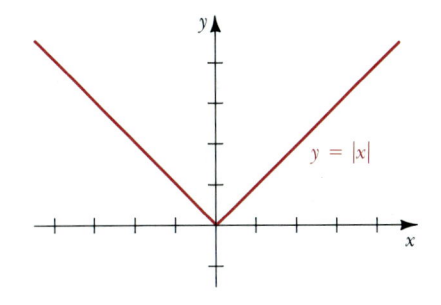

Figure 1.28
The absolute value function.

Example 2

Use the graph of $y = |x|$ to sketch the graph of $y = |x - 1| - 2$.

Solution
The procedure outlined in (1.16) in Section 1.2 implies that the graph of $y = |x - 1|$ is the graph of $y = |x|$ shifted one unit to the right. Principle (1.18) implies that the graph of the equation $y = |x - 1| - 2$ is the graph of $y = |x - 1|$ shifted two units downward. (See Figure 1.29.) ▶

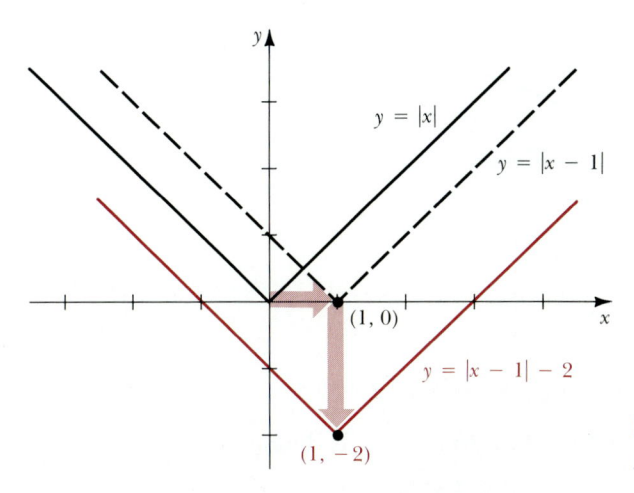

Figure 1.29
A translation of $y = |x|$.

Example 3

Sketch the graph of $y = |2x - 1|$.

Solution
Since $|2x - 1| = 2|x - \frac{1}{2}|$, the graph of $y = |2x - 1|$ is that of $y = 2|x|$ shifted $\frac{1}{2}$ unit to the right. These graphs are shown in Figure 1.30. ▶

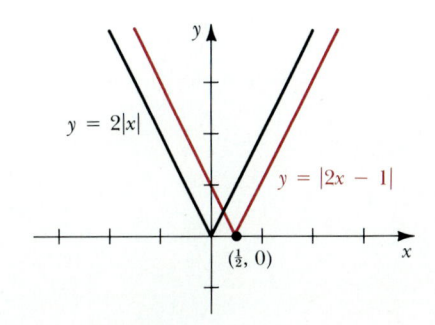

Figure 1.30

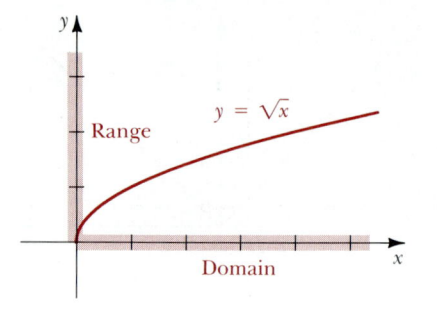

Figure 1.31
The square root function.

The Square Root Function

The graph of the **square root function,** $f(x) = \sqrt{x}$, is shown in Figure 1.31. Both the domain and the range of the square root function are the set of all nonnegative real numbers, the interval $[0, \infty)$. Note that the graph of the square root function appears to approach the y-axis vertically as x approaches 0. After the concepts of limits and the derivative have been introduced, we can show that this is indeed the case.

Example 4

Sketch the graph of $f(x) = \sqrt{x-2} - 1$ and determine the domain and range of f.

Solution

The graph of $y = \sqrt{x-2}$, shown in black in Figure 1.32, is the graph of $y = \sqrt{x}$ shifted two units to the right.

The graph of $y = \sqrt{x-2} - 1$ is the graph of $y = \sqrt{x-2}$ shifted one unit downward. The domain of f is $[2, \infty)$ and the range of f is $[-1, \infty)$, as we can see from the graph of f, shown in color in Figure 1.32. ▶

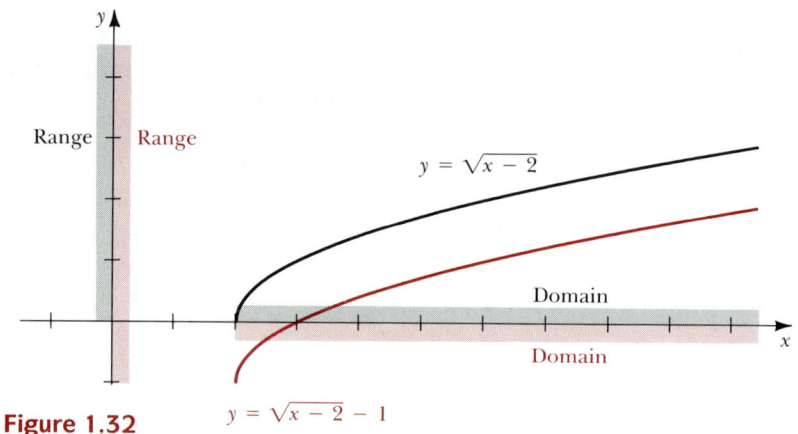

Figure 1.32
A translation of $y = \sqrt{x}$.

The Greatest Integer Function

The **greatest integer function,** $f(x) = [\![x]\!]$, is defined by

(1.25) $$[\![x]\!] = m,$$

where m is the unique integer satisfying $m \leq x < m + 1$. In essence, the greatest integer function "rounds down" a real number to the next lowest integer. For example,

$$[\![1.2]\!] = 1, \quad [\![-1.5]\!] = -2, \quad [\![2]\!] = 2, \quad \text{and} \quad [\![.3]\!] = 0.$$

This function assumes the constant integer value m on each interval $[m, m+1)$, which implies that the graph of f consists of horizontal line segments beginning at each of the integers. For example, if x is in the interval $[-2, -1)$, then $-2 \leq x < -1$ and $f(x) = -2$; if $0 \leq x < 1$, $f(x) = 0$; and so on. The graph of the greatest integer function is shown in Figure 1.33.

Note that the domain of the greatest integer function is the set $\mathbb{R}$ of all real numbers, but its range contains only integers.

WHEN	$-2 \leq x < -1$	$-1 \leq x < 0$	$0 \leq x < 1$	$1 \leq x < 2$	$2 \leq x < 3$
$[\![x]\!]$ IS	-2	-1	0	1	2

Since the greatest integer function "jumps" at each integer, it is useful for describing certain real-world situations, as illustrated in the following example.

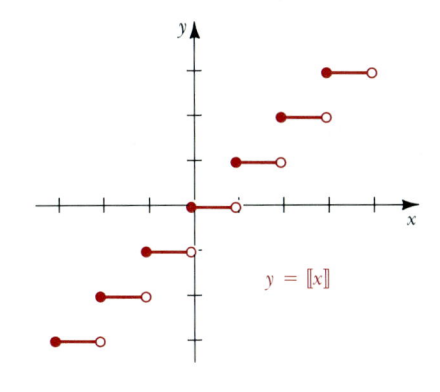

Figure 1.33
The greatest integer function.

Example 5

A cab driver charges two dollars for the first mile or fraction thereof and one dollar for each additional mile or fraction thereof. Describe graphically the relationship between the distance traveled and the cost, and find a function that relates the cost to the distance traveled.

Solution

If no charge is assessed until the trip begins, the domain of the function contains only positive numbers. The costs for some distances are listed in Figure 1.34(a).

The charge "jumps" from $2 to $3 as the meter registers 1 mile and remains at $3 until the meter registers 2 miles. This is shown graphically in Figure 1.34(b). We can use the greatest integer function to express the cost c, in dollars, as a function of the distance x, in miles:

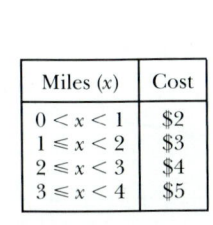

$$c(x) = 2 + [\![x]\!] \text{ dollars}, \quad \text{for } x > 0. \quad \blacktriangleright$$

Miles (x)	Cost
$0 < x < 1$	$2
$1 \leq x < 2$	$3
$2 \leq x < 3$	$4
$3 \leq x < 4$	$5

(a)

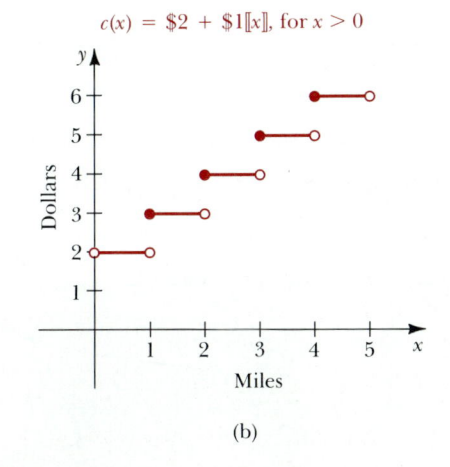

(b)

Figure 1.34

▶ EXERCISE SET 1.3

In Exercises 1–12, use the graphs of the sine and cosine functions to sketch the graph of each of the functions.

1. $f(x) = 2 + \sin x$
2. $f(x) = -3 + \cos x$
3. $f(x) = \cos(x + \pi/2)$
4. $f(x) = \sin(x + \pi)$
5. $f(x) = 2\sin x$
6. $f(x) = -3\cos x$
7. $h(x) = 3 + 2\sin x$
8. $h(x) = 2 - 3\cos x$
9. $g(x) = \cos 2x$
10. $g(x) = \cos 3x$
11. $f(x) = 4\sin^2 x$
12. $f(x) = 3\cos^2 x$

[Hint: $\sin^2 x = (1 - \cos 2x)/2$.]

In Exercises 13–22, use the graph of $y = |x|$ to sketch the graph of each of the functions.

13. $f(x) = |x - 3|$
14. $f(x) = |x + 1|$
15. $f(x) = |x - 3| + 3$
16. $f(x) = |x + 1| - 1$
17. $f(x) = |x - 2| + 2$
18. $f(x) = |x + 2| - 2$
19. $f(x) = |2x|$
20. $f(x) = |2x + 4|$
21. $f(x) = |2x + 5|$
22. $f(x) = |3x - 7|$

In Exercises 23–30:

a) Use the graph of $y = \sqrt{x}$ to sketch the graph of each of the functions.
b) Find the domain and range of each of the functions.

23. $g(x) = \sqrt{x} + 2$
24. $g(x) = \sqrt{x + 2}$
25. $g(x) = \sqrt{x + 2} - 2$
26. $g(x) = \sqrt{x - 2} + 2$
27. $g(x) = -\sqrt{x + 2}$
28. $g(x) = -\sqrt{x} - 2$
29. $g(x) = 2 - \sqrt{x + 2}$
30. $g(x) = 2 - \sqrt{x - 2}$

31. Use the graph of $f(x) = x^3$, shown in the figure at left below, to sketch the graph of g if g is defined by each of the following.
   a) $g(x) = x^3 + 1$
   b) $g(x) = (x + 1)^3$
   c) $g(x) = x^3 - 1$
   d) $g(x) = (x - 1)^3$
   e) $g(x) = 2x^3$
   f) $g(x) = -2x^3$

32. Use the graph of $f(x) = \sqrt[3]{x}$, shown in the figure at right below, to sketch the graph of g if g is defined by each of the following.
   a) $g(x) = \sqrt[3]{x} + 1$
   b) $g(x) = \sqrt[3]{x + 1}$
   c) $g(x) = \sqrt[3]{x} - 1$
   d) $g(x) = \sqrt[3]{x - 1}$
   e) $g(x) = 2\sqrt[3]{x}$
   f) $g(x) = -2\sqrt[3]{x}$

33. Sketch the graph of each of the following functions on the same set of coordinate axes.
   a) $f_1(x) = \sin x$
   b) $f_2(x) = \sin(x - \pi/4)$
   c) $f_3(x) = -\sin(x - \pi/4)$
   d) $f_4(x) = 1 - \sin(x - \pi/4)$

34. Sketch the graph of each of the following functions on the same set of coordinate axes.
   a) $f_1(x) = \cos x$
   b) $f_2(x) = -\cos x$
   c) $f_3(x) = -\cos(x + \pi/4)$
   d) $f_4(x) = 2 - \cos(x + \pi/4)$

In Exercises 35–40, use the graph of $y = [\![x]\!]$ to sketch the graph of each of the functions.

35. $h(x) = [\![x - 2]\!]$
36. $h(x) = [\![x + 2]\!]$
37. $h(x) = [\![x]\!] - 2$
38. $h(x) = [\![x]\!] + 2$
39. $h(x) = [\![x + 2]\!] - 2$
40. $h(x) = -[\![x]\!]$

Figure for Exercise 31.

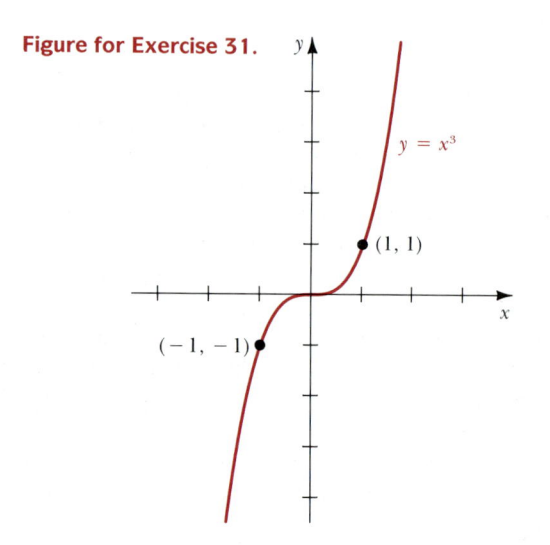

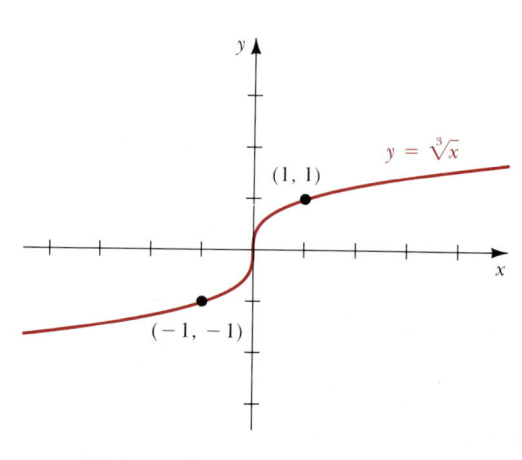

Figure for Exercise 32.

41. The distance

$$d((x_1, y_1), (x_2, y_2)) = \sqrt{(x_1 - x_2)^2 + (y_1 - y_2)^2}$$

between two points in a plane is derived in Appendix A.2. Use this formula to find the distance between each of the following pairs of points.
- **a)** (0, 0), (1, 2)
- **b)** (1, 1), (3, 8)
- **c)** (1, 1), (1, 2)
- **d)** (−1, 4), (0, −5)

42. A circle C with center (h, k) and radius r is the set of all points that are a distance r from the point (h, k) (see the figure). Show that (x, y) is on C precisely when

$$(x - h)^2 + (y - k)^2 = r^2.$$

This is called the *standard form* of the equation of the circle C.

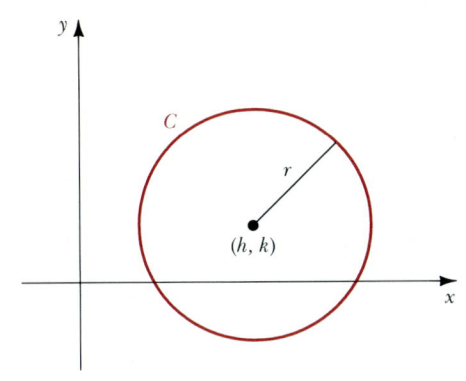

In Exercises 43–48, find the standard form of the equation of each of the circles with center and radius as given, and sketch the graph of each.

43. Center (0, 0); radius 1
44. Center (2, 3); radius 1
45. Center (−2, 3); radius 2
46. Center (−2, −3); radius 3
47. Center (0, 2); radius 3
48. Center (1, −4); radius 4

In Exercises 49–58, the equation of a circle is given. (a) Find the center and radius of each circle, and (b) sketch its graph. [*Hint:* If necessary, complete the square in both variables to express the equation in standard form.]

49. $x^2 + y^2 = 9$
50. $x^2 + y^2 = 16$
51. $x^2 + (y - 1)^2 = 1$
52. $(x - 1)^2 + y^2 = 1$
53. $(x - 2)^2 + (y - 1)^2 = 9$
54. $(x + 2)^2 + (y - 1)^2 = 9$
55. $x^2 - 4x + y^2 - 2y = 4$
56. $x^2 + y^2 + 4x + 2y = 4$
57. $x^2 + y^2 + 3x - 4y = 1$
58. $2x^2 + 2y^2 + 3x - 4y = 1$

59. Suppose that (x_1, y_1) and (x_2, y_2) are distinct points on the line l. Show that $(\frac{1}{2}(x_1 + x_2), \frac{1}{2}(y_1 + y_2))$ is the midpoint of the line segment from (x_1, y_1) to (x_2, y_2).

60. Find the points (x, y) in the plane that are 5 units from (0, 0) and 3 units from the x-axis.

61. Find the points (x, y) in the plane that are 5 units from (1, 2) and 5 units from the y-axis.

62. Find the perpendicular distance from the point (1, 3) to the line $y = (x/2) - 1$ by completing the following.
- **a)** Find an equation of the line through the point (1, 3) and perpendicular to the line $y = (x/2) - 1$.
- **b)** Find the point of intersection of the line found in part (a) and the line $y = (x/2) - 1$.
- **c)** Find the distance between the point (1, 3) and the point found in part (b).

63. Find an equation of the circle that passes through (3, 1), (3, 5), and (5, 3).

64. Find an equation of the circle that passes through (1, 3), (−1, 2), and (3, 1).

65. Find the general form of an equation for a circle that passes through (0, 0).

66. Express the distance from the point (0, 1) to a point on the parabola $y = x^2$ as a function of the x-coordinate of the point on the curve.

67. Express the distance from the point (0, 1) to a point on the circle $x^2 + y^2 = 1$ as a function of the y-coordinate of the point on the curve.

68. What geometric figure is described by the equation

$$d((x, y), (1, 1)) = d((x, y), (3, 3))?$$

69. A camera televising the return of the opening kickoff of a football game is located 5 yd from the east edge of the field and in line with the goal line. The player with the football runs down the east edge (just in bounds) for a touchdown. Express the tangent of the angle θ shown in the figure on the following page in terms of the distance x of the player from the goal line.

70. A 3-ft-high picture is placed on a wall with its base 3 ft above an observer's eye level. Express the tangent of the angle θ shown in the figure on the following page in terms of the distance of the observer from the wall.

71. A cab driver charges $5.00 for the first two miles or fraction thereof and $1.00 for each additional mile or fraction thereof. Describe a function that relates cost to the distance traveled, and sketch the graph of this function.

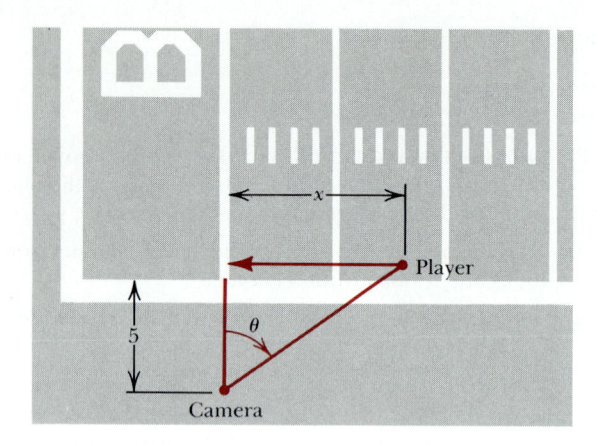

Figure for Exercise 69.

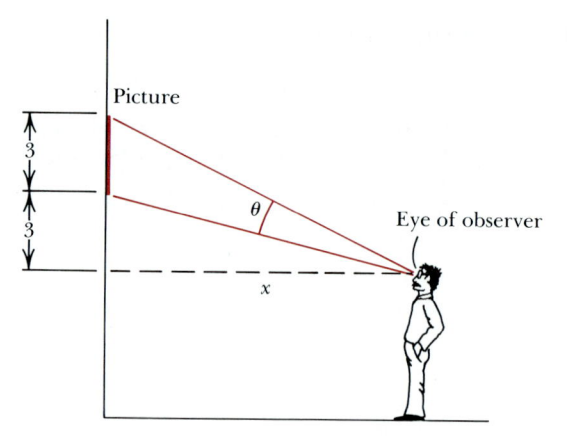

Figure for Exercise 70.

72. In Boston, cab drivers charge $1.10 for the first $\frac{2}{7}$ mile or fraction thereof and 20 cents for each additional $\frac{1}{7}$ mile or fraction thereof.
    a) Describe a function that relates cost to the distance traveled.
    b) Suppose the driver tells you that it costs about $10.00 to travel from your hotel to Logan International Airport and you know that the distance is 6 miles. How much tip has been included in the driver's approximation?

73. The cost of first-class postage is 22 cents for the first ounce or fraction thereof and 17 cents for each additional ounce or fraction thereof. Describe this function in terms of the greatest integer function, and sketch its graph.

74. The Ohio Turnpike is 241 mi in length and has service plazas located 75 and 160 mi from Eastgate, the entrance to the Turnpike at the Pennsylvania line. Express the distance of a car from the nearest service plaza as a function of the car's distance from Eastgate, and sketch the graph of this function.

75. Whooping cranes spend the winter at feeding grounds in the Aransas Bay in Texas and the summer at feeding grounds 2500 mi away in Wood Buffalo National Park, Northwest Territories, Canada. When migrating north in the spring, they stop to rest at feeding grounds in Valentine, Nebraska, a distance 1200 mi from Aransas Bay. Find a function that describes the minimal distance of the cranes from one of these feeding grounds as a function of their distance from Aransas Bay, assuming the cranes fly in a straight line from Aransas Bay to Wood Buffalo Park.

Putnam exercise

76. If f and g are real-valued functions of one real variable, show that there exist numbers x and y such that $0 \leq x \leq 1$, $0 \leq y \leq 1$, and $|xy - f(x) - g(y)| \geq \frac{1}{4}$. (This exercise was problem 4, part I of the twentieth William Lowell Putnam examination given on November 21, 1959. The examination and its solution are in the January 1961 issue of the *American Mathematical Monthly*, pp. 27–33.)

1.4

Combining Functions

The profit for a business is the difference between the revenue, or the amount of money taken in, and the cost of operating the business. Let x represent the number of items sold by a business, and suppose that R and C are revenue and cost functions, respectively, depending on x.

Then the profit function P for this business is the difference:
$$P(x) = R(x) - C(x).$$

This is an example of a function that is defined by combining functions using an arithmetic operation, subtraction. In this section we describe both arithmetic and nonarithmetic methods of combining functions.

(1.26)
DEFINITION
If f and g are functions, then the functions $f + g$, $f - g$, $f \cdot g$, and f/g are defined by the following:

$$(f + g)(x) = f(x) + g(x),$$
$$(f - g)(x) = f(x) - g(x),$$
$$(f \cdot g)(x) = f(x) \cdot g(x), \quad \text{and}$$
$$\left(\frac{f}{g}\right)(x) = \frac{f(x)}{g(x)}.$$

In order for any of these operations to be defined at a number x, both $f(x)$ and $g(x)$ must be defined. The domains of $f + g$, $f - g$, and $f \cdot g$ consist of those real numbers that are common to both the domain of f and the domain of g.

The domain of the quotient f/g consists of those real numbers x that are in both the domain of f and the domain of g, and that also satisfy $g(x) \neq 0$.

Example 1

Find $f + g$, $f - g$, $f \cdot g$, f/g, and their domains if $f(x) = 1/(x^2 - 1)$ and $g(x) = x/\sqrt{x + 2}$.

Solution
We have

$$(f + g)(x) = f(x) + g(x) = \frac{1}{x^2 - 1} + \frac{x}{\sqrt{x + 2}},$$

$$(f - g)(x) = f(x) - g(x) = \frac{1}{x^2 - 1} - \frac{x}{\sqrt{x + 2}},$$

$$(f \cdot g)(x) = f(x) \cdot g(x) = \frac{1}{x^2 - 1} \cdot \frac{x}{\sqrt{x + 2}} = \frac{x}{(x^2 - 1)\sqrt{x + 2}}, \quad \text{and}$$

$$\left(\frac{f}{g}\right)(x) = \frac{f(x)}{g(x)} = \frac{\frac{1}{x^2 - 1}}{\frac{x}{\sqrt{x + 2}}} = \frac{\sqrt{x + 2}}{x(x^2 - 1)}.$$

The domain of f is the set of all real numbers except ± 1, and the domain of g is the interval $(-2, \infty)$. Thus the domain of $f + g$, $f - g$, and $f \cdot g$ is the set of real numbers x that satisfy $x > -2$, but $x \neq \pm 1$.

In order for x to be in the domain of the quotient f/g, it must also be true that $g(x) \neq 0$, that is, $x \neq 0$. Therefore, the domain of f/g is the set of real numbers x that satisfy $x > -2$, but $x \neq \pm 1$ and $x \neq 0$.

The simplified form on the right side of the equation describing the function f/g seems to imply that $x = -2$ is in the domain of f/g. This is not the case, however, since $x = -2$ is not in the domain of g. ▶

Example 2

Sketch the graph of $f(x) = \sin x + \cos x$.

Solution

A point (x, y) is on the graph of f precisely when

$$y = y_1 + y_2, \quad \text{where} \quad y_1 = \sin x \quad \text{and} \quad y_2 = \cos x.$$

So, the y-coordinates on the graph of f are obtained by adding the corresponding y-coordinates on the graphs of the sine and cosine functions, as shown in Figure 1.35.

The graph of f has the appearance of a sine or cosine curve that has been magnified and shifted. That this is indeed the case follows from the identity

$$f(x) = \sin x + \cos x = \sqrt{2} \cos (x - \pi/4),$$

which you are asked to verify in Exercise 66(a). ▶

Special cases of the product and quotient of functions are also useful as graphing techniques. In particular, if f is defined by $f(x) = c$, then $(f \cdot g)(x) = f(x) \cdot g(x) = c \cdot g(x)$.

Suppose the graph of g is known. To graph the equation $y = c \cdot g(x)$, we use the same technique used in Section 1.2, where we obtained the graph of $y = ax^2$ from the graph of $y = x^2$. The graph of a function g and its modifications for various values of c are shown in Figure 1.36.

Figure 1.37 illustrates the typical effect that is produced on a graph when translation is combined with multiplication by a constant.

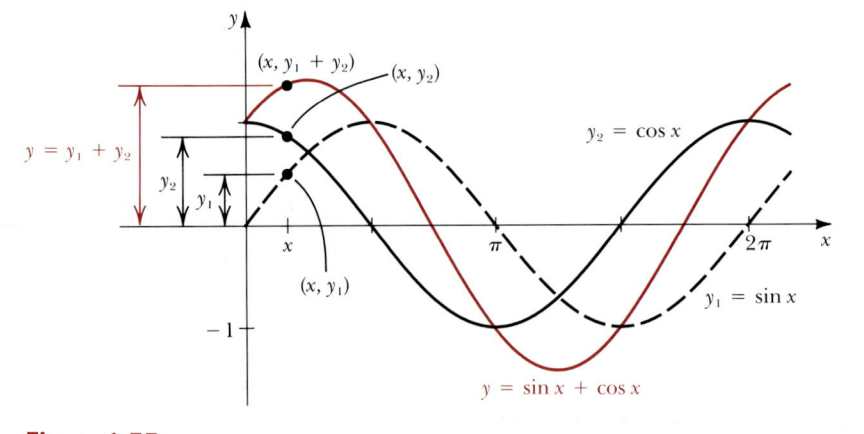

Figure 1.35
$\sin x + \cos x = \sqrt{2} \cos (x - \pi/4)$.

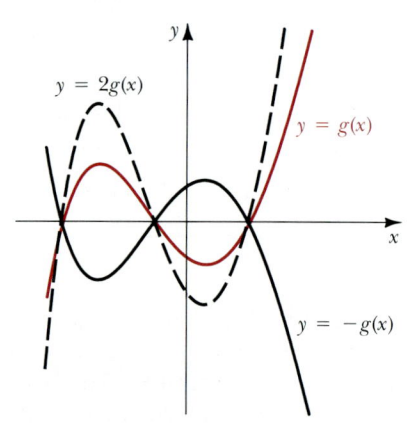

Figure 1.36

1.4 COMBINING FUNCTIONS   35

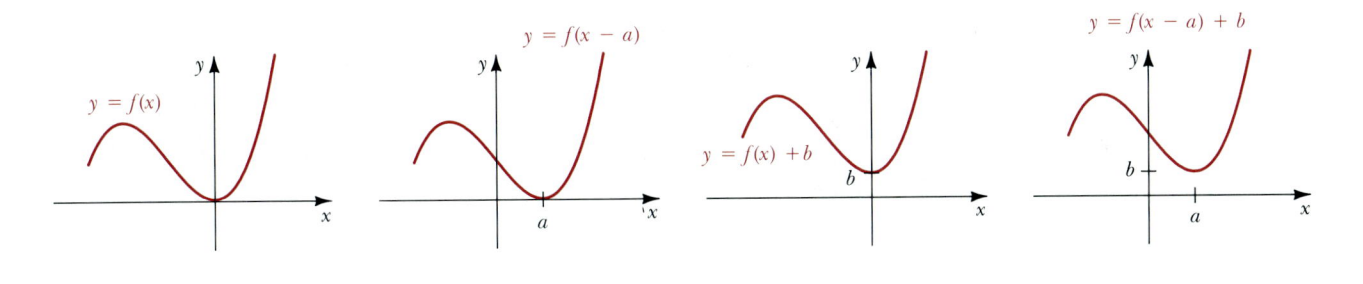

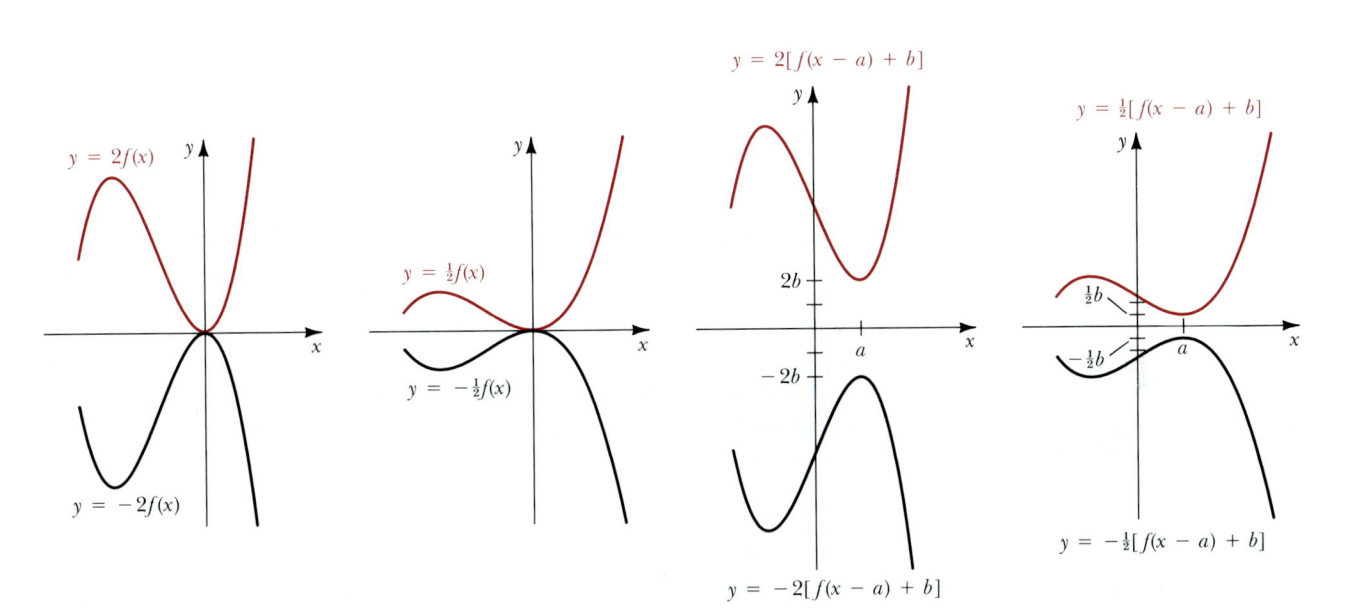

Figure 1.37

Translation and magnification of graphs.

The *reciprocal* of a function g, defined by $(1/g)(x) = 1/g(x)$, is a special case of the quotient of two functions that can be used in graphing. This is illustrated in the next example and elaborated on in Exercise 45.

Example 3

Use the graph of $g(x) = x$ to determine the graph of $h(x) = (1/g)(x) = 1/x$.

Solution

The graph of g is the straight line through $(0, 0)$ with slope 1. Consequently, the function h is undefined at $x = 0$ and its values increase in magnitude as x approaches zero. In addition, $h(x) = 1/x$ approaches zero as x increases in magnitude. Noting that $h(x)$ and $g(x)$ always have the same sign and that both graphs pass through the points $(1, 1)$ and $(-1, -1)$ leads to the graph shown in Figure 1.38. ▶

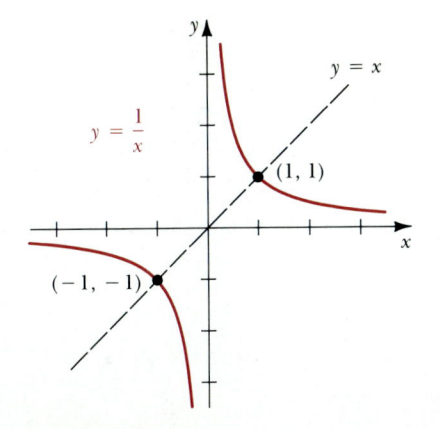

Figure 1.38

The reciprocal of $y = x$.

(a) (b) (c)

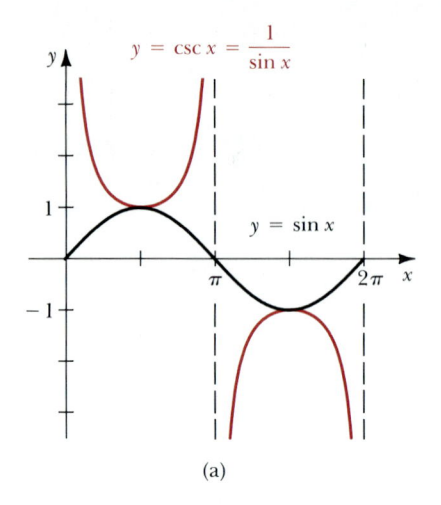

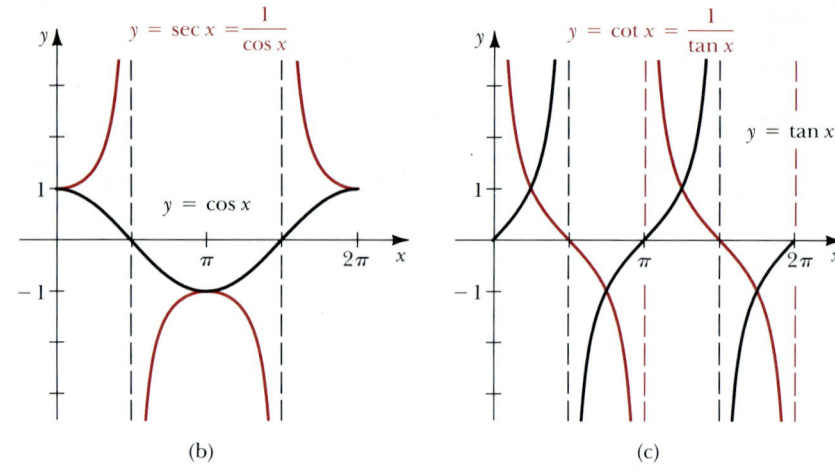

Figure 1.39
The reciprocals of (a) the sine, (b) the cosine, and (c) the tangent functions.

An important application of the reciprocal graphing technique is to produce the graphs of the secant, cosecant, and cotangent functions from the graphs of the cosine, sine, and tangent functions, respectively. The relationship between the graphs of these functions is shown in Figure 1.39.

Another method of combining functions is called the **composition** of functions, a type of building process that makes complicated functions from more elementary ones.

(1.27) DEFINITION

Suppose that g is a function mapping a set X into a set Y and that f is a function mapping the set Y into a set Z. The **composition** of f and g, written $f \circ g$, is a function mapping the set X into the set Z, defined for each x in X by

$$(f \circ g)(x) = f(g(x)).$$

The composition of a pair of functions f and g is illustrated in Figure 1.40.

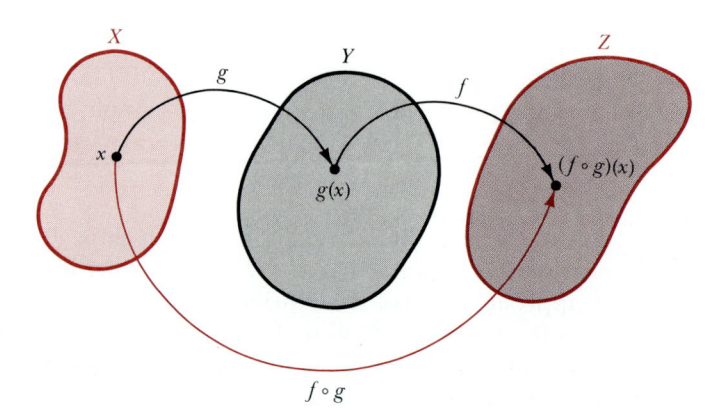

Figure 1.40
Composition of functions.

Example 4

Find $(f \circ g)(2)$ and $(g \circ f)(2)$ if $f(x) = x - 1$ and $g(x) = x^2 + x - 2$.

Solution

As shown in Figure 1.41,
$$(f \circ g)(2) = f(g(2)) = f(4) = 3$$
and
$$(g \circ f)(2) = g(f(2)) = g(1) = 0.$$

Note that in Example 4,
$$(f \circ g)(2) \neq (g \circ f)(2).$$
It is generally true for functions f and g that
$$(f \circ g)(x) \neq (g \circ f)(x).$$

The following example shows that for functions whose domain and range consist of real numbers, the domain of the composition $f \circ g$ consists of those values x for which $f(g(x))$ is defined, that is, the values of x in the domain of g for which $g(x)$ is in the domain of f.

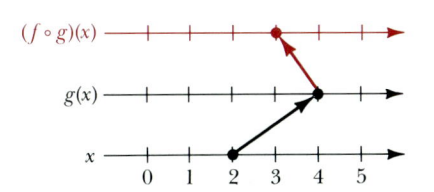

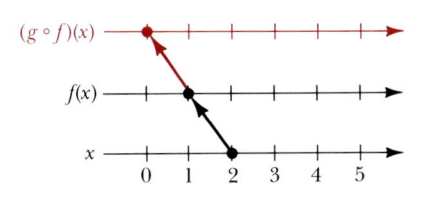

Figure 1.41
$(f \circ g)(x) \neq (g \circ f)(x).$

Example 5

Let $f(x) = \sqrt{x - 1}$ and $g(x) = 1/x^2$.

a) Find $(f \circ g)(x)$ and the domain of $f \circ g$.
b) Find $(g \circ f)(x)$ and the domain of $g \circ f$.

Solution

a) We have
$$(f \circ g)(x) = f(g(x)) = f\left(\frac{1}{x^2}\right) = \sqrt{\frac{1}{x^2} - 1}.$$

The domain of g is the set of nonzero real numbers, and the domain of f is the interval $[1, \infty)$. Consequently, the domain of $f \circ g$ is the set of nonzero real numbers x such that $g(x) = 1/x^2 \geq 1$, that is, $x \neq 0$ and $x^2 \leq 1$. In interval notation, the domain of $f \circ g$ is $[-1, 0) \cup (0, 1]$.

b) We find that
$$(g \circ f)(x) = g(f(x)) = g(\sqrt{x - 1}) = \frac{1}{(\sqrt{x - 1})^2} = \frac{1}{x - 1}.$$

The domain of $g \circ f$ is the set of real numbers x such that $x \geq 1$ and $f(x) = \sqrt{x - 1} \neq 0$, that is, the interval $(1, \infty)$.

Example 6

Write the function described by $h(x) = \sin(x^2 - 4x)$ as the composition of two functions.

Solution

One way to decompose h is to observe the chain:

$$x \longrightarrow (x^2 - 4x) \longrightarrow \sin(x^2 - 4x).$$

Let $f(x) = \sin x$ and $g(x) = x^2 - 4x$. Then

$$(f \circ g)(x) = f(g(x)) = f(x^2 - 4x) = \sin(x^2 - 4x)$$

and

$$h = f \circ g.$$

This representation of h is not unique; there are many different compositions that give $h(x)$. For example, the composition $\hat{f} \circ \hat{g}$, where $\hat{f}(x) = \sin(x - 4)$ and $\hat{g}(x) = (x - 2)^2$, also gives h, since

$$\sin[(x-2)^2 - 4] = \sin[(x^2 - 4x + 4) - 4] = \sin(x^2 - 4x). \quad \blacktriangleright$$

The procedure shown in Example 6 decomposes a complicated function into a sequence of more elementary or familiar functions, a technique that is extremely valuable in calculus. It is often useful to illustrate a composition by a chain, as we did in Example 6.

Composition of functions will cause no difficulty if you remember that function notation such as $f(x) = x^2 - 1$ means simply that for any element x in the domain, the value $f(x)$ is obtained by squaring x and subtracting 1. In particular, keep in mind that there is nothing special about the variable x used in the description of f. The notation $f(\_\_) = (\_\_)^2 - 1$ is helpful when considering the composition since it emphasizes that $f(g(x)) = (g(x))^2 - 1$.

The final example in this section illustrates the use of the composition of functions to describe a problem involving variation in time. Such applications are discussed in detail in Section 3.3.

Example 7

A boat is leaking oil into a lake. Suppose the shape of the oil spill is approximately circular and at any time t minutes after the leak has begun, the radius of the circle is $r(t) = \sqrt{t} + \sqrt[3]{t}$. Find the area of the spill at any time t after the leak has begun.

Solution

The area of a circle of radius r is $A(r) = \pi r^2$. The area of the spill depends on the radius, and the radius depends on time. Thus A is a function of t given by

$$(A \circ r)(t) = A(r(t)) = A(\sqrt{t} + \sqrt[3]{t}) = \pi(\sqrt{t} + \sqrt[3]{t})^2. \quad \blacktriangleright$$

▶ EXERCISE SET 1.4

In Exercises 1–8, a pair of functions f and g is described. Determine the equation describing the following functions and the domain in each case: (a) $f + g$; (b) $f - g$; (c) $f \cdot g$; (d) (f/g); (e) $f \circ g$; and (f) $g \circ f$.

1. $f(x) = x$, $g(x) = x + 1$
2. $f(x) = x^2$, $g(x) = x + 1$
3. $f(x) = 1/x$, $g(x) = \sqrt{x-1}$
4. $f(x) = x^2 + 1$, $g(x) = 1/\sqrt{x}$
5. $f(x) = x^2 + 1$, $g(x) = x - 2$
6. $f(x) = |x|$, $g(x) = 2x - 1$
7. $f(x) = x^2 + x$, $g(x) = \sin x$
8. $f(x) = \cos x$, $g(x) = \tan x$

In Exercises 9–36, use the graphs of the functions discussed in Section 1.3 to sketch the graph of the function described.

9. $f(x) = \sin x - \cos x$
10. $f(x) = \cos x - \sin x$
11. $f(x) = 2\sin x - \cos x$
12. $f(x) = \cos x - 2\sin x$
13. $f(x) = 2\sin x - 3\cos x$
14. $f(x) = 3\cos x - 2\sin x$
15. $f(x) = -|x|$
16. $f(x) = 2|x|$
17. $f(x) = |x| + |x+1|$
18. $f(x) = 2|x| + |x+1|$
19. $f(x) = |x| + 2|x+1|$
20. $f(x) = 2|x| + 2|x+1|$
21. $f(x) = |x| - |x+1|$
22. $f(x) = |x+1| - |x|$
23. $f(x) = 2\sqrt{x}$
24. $f(x) = 2 - 3\sqrt{x}$
25. $f(x) = \sqrt{x} + \sqrt{x-1}$
26. $f(x) = 2\sqrt{x} - \sqrt{x-1}$
27. $f(x) = \sqrt{x} + |x+1|$
28. $f(x) = |x| - \sqrt{x-1}$
29. $f(x) = |\sin x|$
30. $f(x) = |\cos x|$
31. $f(x) = \sin |x|$
32. $f(x) = \cos |x|$
33. $f(x) = -[\![x]\!]$
34. $f(x) = 2[\![x]\!]$
35. $f(x) = x + [\![x]\!]$
36. $f(x) = |x| - [\![x]\!]$

In Exercises 37–42, express the function given as the composition of two functions, one of which is a polynomial.

37. $f(x) = \sqrt{x^2 + 3x}$
38. $f(x) = (x^3 - 5x + 1)^5$
39. $f(x) = \sin(x-3)^2$
40. $f(x) = \sin^2(x-3)$
41. $f(x) = \cos^2 x + 3\cos x + 4$
42. $f(x) = \sin^2 x + \cos^3 x + 5\cos x + 3$
43. Express $f(x) = (\sqrt{x^2 + 1} + 2)^4$ as a composition of three functions, two of which are polynomials.
44. Functions f and g are defined by $f(x) = x^2 - 9$ and $g(x) = x + 3$. Sketch the graph of f/g. Sketch the graph of h if $h(x) = x - 3$. How do the two graphs differ?
45. Answer the following concerning a function g and its reciprocal $1/g$.

a) $\dfrac{1}{g(x)}$ is undefined when $g(x) = $ _____.

b) $g(x)$ and $\dfrac{1}{g(x)}$ have the same value when $g(x) = $ _____ or when $g(x) = $ _____.

c) Do $g(x)$ and $\dfrac{1}{g(x)}$ always have the same sign?

d) When the magnitude of $g(x)$ is small, what is the magnitude of $\dfrac{1}{g(x)}$?

e) When the magnitude of $g(x)$ is large, what is the magnitude of $\dfrac{1}{g(x)}$?

46. Use the results of Exercise 45 and the graph of g to sketch the graph of $h(x) = 1/g(x)$ when:
   a) $g(x) = 2x - 1$;
   b) $g(x) = |x|$;
   c) $g(x) = x^2 - 1$;
   d) $g(x) = x^2 - 4x + 3$.

In Exercises 47–50, sketch the graphs of the functions in the order given, and observe the difference in the graph that each successive complication introduces.

47. a) $f_1(x) = x - 1$ b) $f_2(x) = (x-1)^2$
   c) $f_3(x) = x^2 - 2x$ d) $f_4(x) = |x^2 - 2x|$
   e) $f_5(x) = \dfrac{1}{|x^2 - 2x|}$ f) $f_6(x) = \dfrac{-1}{|x^2 - 2x|}$

48. a) $f_1(x) = x - 2$ b) $f_2(x) = (x-2)^2$
   c) $f_3(x) = x^2 - 4x + 2$ d) $f_4(x) = |x^2 - 4x + 2|$
   e) $f_5(x) = \dfrac{1}{|x^2 - 4x + 2|}$ f) $f_6(x) = \dfrac{2}{|x^2 - 4x + 2|}$

49. a) $f_1(x) = x - \pi$
   b) $f_2(x) = 2(x - \pi)$
   c) $f_3(x) = \sin 2(x - \pi)$
   d) $f_4(x) = -\sin 2(x - \pi)$
   e) $f_5(x) = 1 - \sin 2(x - \pi)$
   f) $f_6(x) = |1 - \sin 2(x - \pi)|$

50. a) $f_1(x) = x + \pi/4$
   b) $f_2(x) = 3(x + \pi/4)$
   c) $f_3(x) = \cos 3(x + \pi/4)$
   d) $f_4(x) = -2\cos 3(x + \pi/4)$
   e) $f_5(x) = 1 - 2\cos 3(x + \pi/4)$
   f) $f_6(x) = |1 - 2\cos 3(x + \pi/4)|$

51–54. For the functions described in Exercises 47–50, (a) find a function g_1 with the property that $g_1 \circ f_1 = f_2$, a function g_2 with the property that $g_2 \circ f_2 = f_3$, and so on. (b) Show that if the functions g_1, g_2, g_3, g_4, and g_5 are chosen in this manner, then we can write $f_6 = g_5 \circ g_4 \circ g_3 \circ g_2 \circ g_1 \circ f_1$.

55. Show that for any function f, $g(x) = (f(x) + f(-x))/2$ describes an even function.

56. Show that for any function f, $h(x) = (f(x) - f(-x))/2$ describes an odd function.

57. Use Exercises 55 and 56 to show that any function f can be written as the sum of an odd and an even function, provided that when x is in the domain of f, $-x$ is also in the domain of f.

58. Use the results in Exercises 55, 56, and 57 to write each of the following functions as the sum of an odd and an even function.
    a) $f(x) = x^2 + x$
    b) $f(x) = 1/x + 1$
    c) $f(x) = |x| + 1$
    d) $f(x) = |x + 1|$

59. Show that the composition of two odd functions is an odd function.

60. Show that the composition of an odd and an even function, in either order, is an even function.

61. A metal sphere is heated so that t sec after the heat has been applied, the radius $r(t)$ is given by $r(t) = 3 + 0.01t$ cm. Express the volume of the sphere as a function of t.

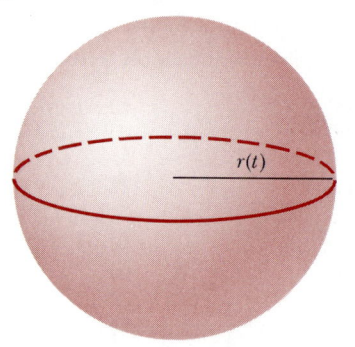

62. Sand is poured onto a conical pile whose radius and height are always equal, although both increase with time. The height of the pile t sec after the pouring begins is given by $h(t) = 10 + 0.25t$ ft. Express the volume of the pile as a function of t.

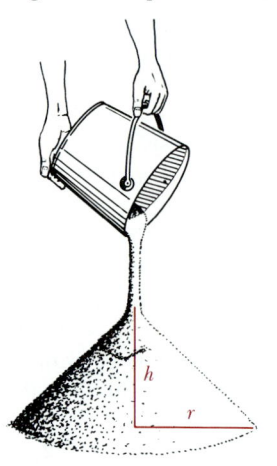

63. A spherical balloon is inflated so that its radius at the end of t sec is $r(t) = 3\sqrt{t} + 5$ cm, $0 \le t \le 4$. Express the volume and surface area as a function of time. What are the units of these quantities?

64. In Exercise 45 of Exercise Set 1.2, the height of a rock above the ground t sec after it has been dropped from the top of a building was stated to be $s(t) = 784 - 16t^2$. The domain of the function s is the closed interval whose left endpoint is zero and whose right endpoint is the time it takes the rock to reach the ground. Define a function $\tilde{s}$ that describes the height of the rock above the ground, assuming that the rock is dropped at $t = 2$ instead of at $t = 0$. What are the domain and range of $\tilde{s}$?

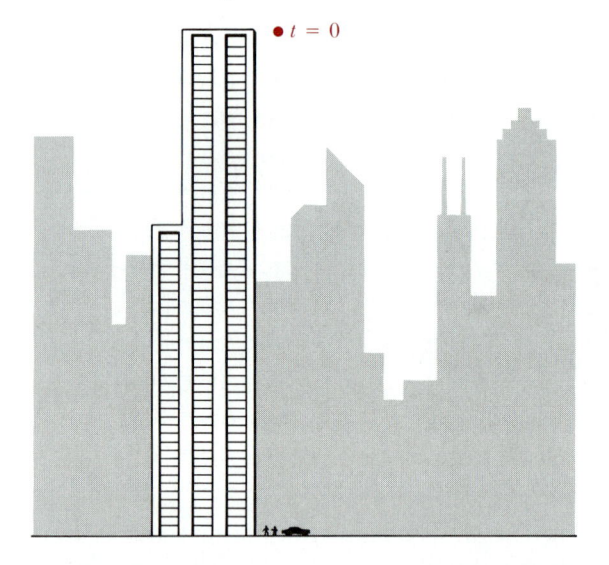

65. Newton's law of gravitational attraction states that the attraction between an object of mass m_1 and an object of mass m_2 is directly proportional to the product of the masses m_1 and m_2 of the objects and inversely proportional to the square of the distance r between the centers of mass of the objects. Write a functional relationship expressing this force in terms of the distance r, assuming that the masses remain constant. What restrictions must be put on the domain of this function if it is to describe the physical situation? Sketch the graph of the function.

66. Use a trigonometric identity to show each of the following.
    a) $\sqrt{2} \cos(x - \pi/4) = \sin x + \cos x$
    b) $A \cos(x - \delta) = a \sin x + b \cos x$, where $A = \sqrt{a^2 + b^2}$, $\sin \delta = a/\sqrt{a^2 + b^2}$, and $\cos \delta = b/\sqrt{a^2 + b^2}$, provided $A \ne 0$

67. If P_n and Q_n are both polynomials of degree n, what can you say about the degree of each of the following polynomials?
   a) $P_n + Q_n$ **b)** $P_n - Q_n$
   c) $P_n \cdot Q_n$ **d)** $3P_n^2 - 2Q_n^3$
   e) R, where $R(x) = xP_n(x) - Q_n(x)$

Putnam exercises

68. Determine all polynomials P such that $P(x^2 + 1) = (P(x))^2 + 1$ and $P(0) = 0$. (This exercise was Problem A–2 of the thirty-second William Lowell Putnam examination given on December 4, 1971. The examination and its solution are in the February 1973 issue of the *American Mathematical Monthly*, pp. 172–179.)

69. Let F be a real-valued function defined for all real x except for $x = 0$ and $x = 1$ and satisfying the functional equation $F(x) + F\{(x - 1)/x\} = 1 + x$. Find all functions F satisfying these conditions. (This exercise was Problem B–2 of the thirty-second William Lowell Putnam examination given on December 4, 1971. The examination and its solution are in the February 1973 issue of the *American Mathematical Monthly*, pp. 172–179.)

1.5 The Limit of a Function: The Intuitive Notion

The limit is the concept that distinguishes calculus from algebra, geometry, and other precalculus topics. In a real sense, then, our study of calculus begins at this point.

Although much of calculus was known in the seventeenth and early eighteenth centuries, it was not until the latter part of the nineteenth century that a precise definition of the limit (such as the one given in Section 1.6) was developed. In this section we consider the intuitive notion of the limit of a function at a number. Although this notion is not sufficient to prove rigorously the results of calculus, it was enough to give mathematicians and scientists of the seventeenth and eighteenth centuries the insight to develop most of the power of the subject.

The limit describes what happens to the values $f(x)$ of the function as x *approaches* the number a, as opposed to $f(a)$, which gives the value of the function when x is *equal* to a. We say that $f(x)$ has the limit L as x approaches the number a provided that $f(x)$ becomes and remains close to L as x becomes close, but not equal, to a. This is expressed by writing

$$\lim_{x \to a} f(x) = L.$$

When such a number L exists, we say that L is the limit of $f(x)$ as x approaches a, or simply that L is the limit of f at a.

Example 1

Find $\lim_{x \to 2} f(x)$ for the function defined by $f(x) = x^2$.

Solution
The questions are as follows.

Table 1.2

x	$f(x) = x^2$	x	$f(x) = x^2$
1.9	3.61	2.1	4.41
1.99	3.9601	2.01	4.0401
1.999	3.996001	2.001	4.004001
1.9999	3.99960001	2.0001	4.00040001
1.99999	3.9999600001	2.00001	4.0000400001

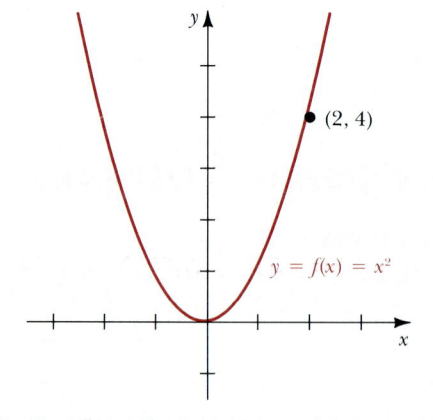

Figure 1.42
$\lim_{x \to 2} f(x) = 4$ and $f(2) = 4$.

i) Is there a number L with the property that x^2 becomes and remains close to L as x approaches 2?
ii) If such a number L exists, what is it?

To answer these questions, consider the values of f at some numbers close to 2 (see Table 1.2 and Figure 1.42). It seems reasonable to conclude from the table that

$$\lim_{x \to 2} f(x) = \lim_{x \to 2} x^2 = 4.$$ ▶

In Example 1, the limit of $f(x)$ as x approaches a is the same as the value $f(a)$. In other situations, however, what happens to the values of the function as x approaches a might be quite different from what happens to the function at a. The examples in this section have been chosen to illustrate some of the possibilities that can occur, and each is reconsidered as more information about the limit is introduced.

Example 2 shows that a function can have a limit at a number that is not in the domain of the function.

Example 2

Find $\lim_{x \to 2} g(x)$ for the function g defined by

$$g(x) = \frac{x^3 - 2x^2}{x - 2}.$$

Solution
Since

$$x^3 - 2x^2 = x^2(x - 2),$$

$g(x)$ can be expressed as

$$g(x) = x^2, \quad \text{provided } x \neq 2.$$

Note that $g(x)$ is undefined when $x = 2$ (see Figure 1.43). However, in determining the limit at 2 we are concerned with only the values of $g(x)$ when x is close to 2, but $x \neq 2$. Thus the limit in this example is the same as the limit in Example 1:

$$\lim_{x \to 2} g(x) = \lim_{x \to 2} x^2 = 4.$$ ▶

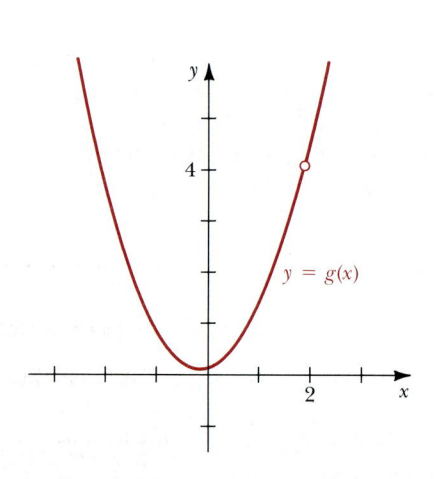

Figure 1.43
$\lim_{x \to 2} g(x) = 4$ and $g(2)$ does not exist.

The function in Example 3 illustrates a situation in which both the limit of a function at a number and the value of the function at that number exist, but are unequal.

Example 3

Determine $\lim_{x \to 2} h(x)$ for the function h defined by

$$h(x) = \begin{cases} x^2, & \text{if } x \neq 2, \\ 1, & \text{if } x = 2. \end{cases}$$

Solution

The graph of h is shown in Figure 1.44. When $x \neq 2$, the values of $h(x)$ are those of x^2. Since the number 2 is explicitly excluded when considering the limit,

$$\lim_{x \to 2} h(x) = \lim_{x \to 2} x^2 = 4,$$

even though $h(2) = 1$. ▶

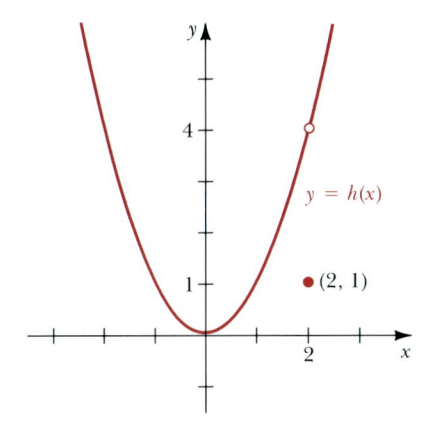

Figure 1.44

$\lim_{x \to 2} h(x) = 4$ and $h(2) = 1$.

In the next example, the limit of the function fails to exist at certain numbers in the domain of the function.

Example 4

Find $\lim_{x \to 2} [\![x]\!]$.

Solution

In Section 1.3 we saw that the graph of the greatest integer function consists of "steps" at the integers, as shown in Figure 1.45. By examining this graph, we see that for x close to 2, $[\![x]\!]$ can be either 1 or 2, depending on whether $x \leq 2$ or $x > 2$. No single number L exists with all values of $[\![x]\!]$ close to L when x is close to 2. So

$$\lim_{x \to 2} [\![x]\!] \text{ does not exist,}$$

even though $[\![2]\!] = 2$ does exist.

The same reasoning can be used to show that the greatest integer function does not have a limit at any integer. ▶

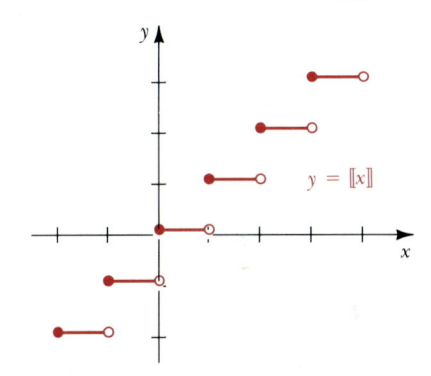

Figure 1.45

$\lim_{x \to 2} i(x)$ does not exist and $i(2) = 2$.

Examples 2, 3, and 4 illustrate why we use the words "x becomes close, but not equal, to a" when describing the limit. Remember,

when finding $\lim_{x \to a} f(x)$, we are interested only in the behavior of the function *near* a, not *at* a.

Example 5 illustrates how a calculator or computer can be used to gain intuition about the limit of a function. It should be emphasized, however, that this technique only indicates (but does not prove) what the limit is *likely* to be. Because calculators and computers use only a finite number of values to represent all the real numbers, examples can be constructed for which this process gives misleading information.

Example 5

Find $\lim\limits_{x \to 0} \dfrac{\sin x}{x}$.

Solution

This problem is different from the preceding examples because although both the numerator and the denominator of the quotient approach 0, they have no common factor. Table 1.3 lists some values of sin x for x close to 0. (If you are verifying these values using a calculator, be sure to use radian mode rather than degrees for the values of x. Radian measure is the standard mode for expressing trigonometric functions in the quantitative sciences.)

The entries in the table indicate that as x approaches 0, the values of $(\sin x)/x$ approach 1. This suggests that

$$\lim_{x \to 0} \frac{\sin x}{x} = 1. \blacktriangleright$$

Table 1.3

x	$\sin x$	x	$\sin x$
0.1	0.099833417	-0.1	-0.099833417
0.01	0.009999833	-0.01	-0.009999833
0.001	0.001000000	-0.001	-0.001000000
0.0001	0.000100000	-0.0001	-0.000100000

Our final example gives a geometric application of the limit. This application is used in Chapter 2 to introduce one of the basic concepts of calculus: the derivative.

Example 6

Consider the function defined by $f(x) = x^2$. If $h \neq 0$, a unique line joins the distinct points $(1, f(1)) = (1, 1)$ and $(1 + h, f(1 + h)) = (1 + h, (1 + h)^2)$. This line is called a *secant line* to the graph of f at $(1, 1)$. A number of secant lines are shown in Figure 1.46.

a) Find the slope of a typical secant line joining $(1, 1)$ and $(1 + h, (1 + h)^2)$.

b) Find the limit of the slopes of the secant lines as h approaches 0.

Solution

a) The slope of the secant line joining $(1, 1)$ and $(1 + h, (1 + h)^2)$, if $h \neq 0$, is

$$\frac{f(1 + h) - f(1)}{(1 + h) - 1} = \frac{(1 + h)^2 - 1}{h} = \frac{1 + 2h + h^2 - 1}{h}$$

$$= \frac{2h + h^2}{h} = 2 + h.$$

1.5 THE LIMIT OF A FUNCTION: THE INTUITIVE NOTION

Figure 1.46
Secant lines at (1, 1).

Figure 1.47
Tangent line at (1, 1).

b) The slope of the secant line is $2 + h$ whenever $h \neq 0$, so the limit of the slopes of the secant lines as h approaches 0 is

$$\lim_{h \to 0} \frac{f(1+h) - f(1)}{(1+h) - 1} = \lim_{h \to 0} (2 + h) = 2.$$

▶

Figure 1.47 shows secant lines to the graph at (1, 1) together with the line through (1, 1) whose slope is the limit of the slopes of the secant lines. This line is called the *tangent line* to the graph at (1, 1).

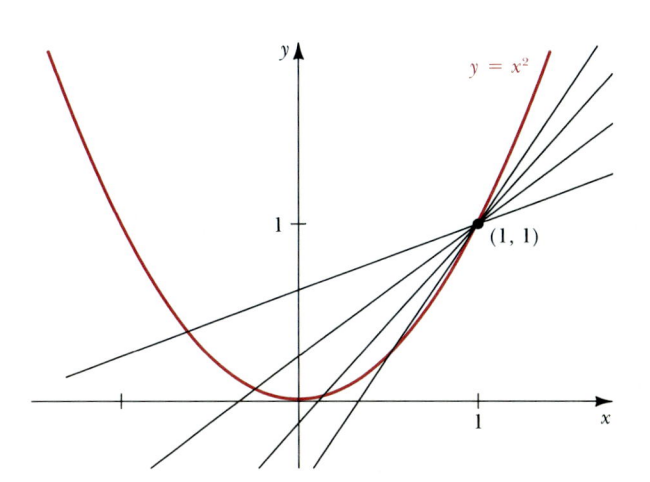

Tangent lines are fundamental to the study of calculus. They are used to introduce the derivative at the beginning of Chapter 2.

▶ EXERCISE SET 1.5

In Exercises 1 and 2, use the graph of the function f to determine the limit of f at $a = 1$, if it exists.

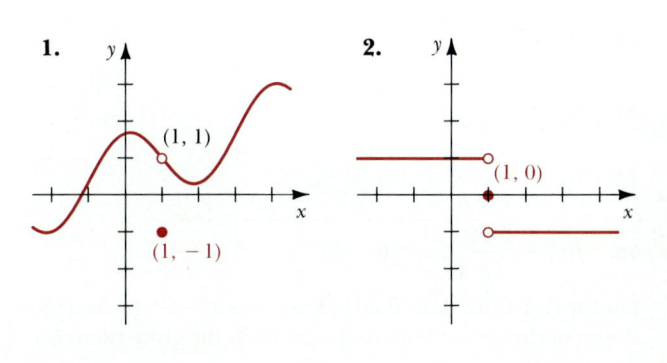

1.

2.

In Exercises 3–40, a function f is described and a number a is given. Determine the limit of f at a, if it exists.

3. $f(x) = 5x - 7$, $a = 2$

4. $f(x) = 3 - 2x^2 + 5x$, $a = 4$

5. $f(x) = \dfrac{1}{x^2}$, $a = \dfrac{1}{2}$

6. $f(x) = \dfrac{x^2 + 1}{x^3 + 2}$, $a = -1$

7. $f(x) = \dfrac{x^2 - 9}{x - 3}$, $a = 3$

8. $f(x) = \dfrac{x^2 - 9}{x + 3}$, $a = -3$

9. $f(x) = \dfrac{x^2 - 16}{x + 4}, a = -4$

10. $f(x) = \dfrac{x^2 - 16}{x - 4}, a = 4$

11. $f(x) = x + \dfrac{x^2 - 4}{x - 2}, a = 2$

12. $f(x) = x - \dfrac{x^2 - 4}{x - 2}, a = 2$

13. $f(x) = \dfrac{x^2 - 5x + 6}{x - 2}, a = 2$

14. $f(x) = \dfrac{2x^2 - 5x + 3}{2x - 3}, a = \dfrac{3}{2}$

15. $f(x) = \dfrac{2x^2 - 3x - 2}{6x^2 + 5x + 1}, a = -0.5$

16. $f(x) = \dfrac{3x^2 - 17x + 20}{6x^2 - 7x - 5}, a = \dfrac{5}{3}$

17. $f(x) = \dfrac{x^3 - 1}{x - 1}, a = 1$

18. $f(x) = \dfrac{x^3 - 27}{x - 3}, a = 3$

19. $f(x) = \dfrac{x^3 - 3x^2 - 4x + 12}{x - 3}, a = 3$

20. $f(x) = \dfrac{x^3 - 3x^2 - 4x + 12}{x - 2}, a = 2$

21. $f(x) = \dfrac{x^3 - 3x^2 - 4x + 12}{x^2 - 5x + 6}, a = 3$

22. $f(x) = \dfrac{x^3 - 3x^2 - 4x + 12}{x^2 - 5x + 6}, a = 2$

23. $f(x) = \dfrac{|x|}{x}, a = 1$

24. $f(x) = \dfrac{|x|}{x}, a = -\sqrt{3}$

25. $f(x) = \dfrac{|x|}{x}, a = 0$

26. $f(x) = \dfrac{1}{x - 1}, a = 1$

27. $f(x) = [\![x]\!], a = 4$

28. $f(x) = [\![x]\!], a = \dfrac{3}{2}$

29. $f(x) = \dfrac{x - 4}{x - 2}, a = 2$

30. $f(x) = \dfrac{x - 4}{x - 2}, a = 4$

31. $f(x) = \sin x, a = \dfrac{\pi}{2}$

32. $f(x) = \cos x, a = 3\pi$

33. $f(x) = \sec x, a = 0$

34. $f(x) = \tan x, a = \dfrac{\pi}{4}$

35. $f(x) = \dfrac{\tan x}{\sec x}, a = 0$

36. $f(x) = \dfrac{\tan x}{\sec x}, a = \dfrac{\pi}{2}$

37. $f(x) = \tan x \cos x, a = \dfrac{\pi}{2}$

38. $f(x) = \cot x \sin x, a = 0$

39. $f(x) = \begin{cases} x, & \text{if } x \text{ is not an integer,} \\ 2, & \text{if } x \text{ is an integer,} \end{cases} a = 3$

40. $f(x) = \begin{cases} \sin x, & \text{if } 0 \le x < 2\pi, \\ \cos x, & \text{if } 2\pi \le x < 4\pi, \end{cases} a = 2\pi$

41. Let
$$f(x) = \begin{cases} 2x, & \text{if } x < 2, \\ 0, & \text{if } x = 2, \\ x^2, & \text{if } x > 2. \end{cases}$$
Find each of the following.
a) $f(2)$
b) $\lim_{x \to 2} f(x)$
c) $\lim_{x \to 1} f(x)$
d) $\lim_{x \to 3} f(x)$

42. Let
$$f(x) = \begin{cases} \sin x, & \text{if } 2n\pi \le x < 2(n+1)\pi \text{ for } n \\ & \text{an even integer,} \\ \cos x, & \text{if } 2n\pi \le x < 2(n+1)\pi \text{ for } n \\ & \text{an odd integer.} \end{cases}$$
a) Sketch the graph of f.
b) Find $f(\pi)$, $f(2\pi)$, and $f(3\pi)$.
c) Find $\lim_{x \to 2\pi} f(x)$, if it exists.
d) Find $\lim_{x \to 3\pi} f(x)$, if it exists.

In Exercises 43–46, a function f is described and a number a is given. Use a calculator to find the values of the function for x near a, and determine the limit of f at a, if it exists.

43. $f(x) = \dfrac{\cos x - 1}{x}, a = 0$

44. $f(x) = \dfrac{\tan x}{x}, a = 0$

45. $f(x) = \dfrac{\cos x}{x - \pi/2}, a = \dfrac{\pi}{2}$

46. $f(x) = \dfrac{\sin 2x}{x}, a = 0$

For the functions described in Exercises 47–54, (a) find the slopes of the secant lines to the graph at the given point for arbitrary values of $h \neq 0$, and (b) find the limit of the slopes of the secant lines as h approaches 0.

47. $f(x) = x^2$, (2, 4)
48. $f(x) = x^2$, (0, 0)
49. $f(x) = x^2$, (−3, 9)
50. $f(x) = x^2$, (a, a^2)
51. $f(x) = x^3$, (1, 1)
52. $f(x) = x^3$, (a, a^3)
53. $f(x) = 1/x$, (1, 1)
54. $f(x) = 1/x$, $(a, 1/a)$

55. Air is pumped into a spherical balloon at the rate of 30 in^3/sec. Precisely 75 min after the pumping begins, the balloon bursts.
   a) What is the limiting amount of air that was pumped into the balloon?
   b) What was the radius at that instant?

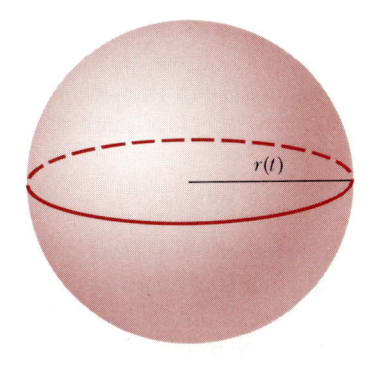

56. Suppose that $f(x) = 1000/(x + 1)$ gives the reading of a thermometer in degrees Fahrenheit when the thermometer is x inches from a flame and that the thermometer will immediately burst if it touches the flame. Find $\lim_{x \to 0} (1000/(x + 1))$, which gives the limiting temperature reading as the thermometer approaches the flame.

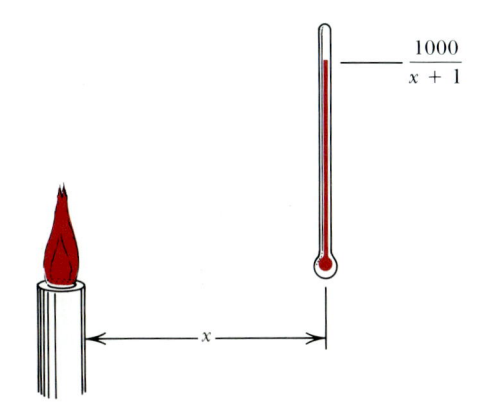

57. In Example 5 we found that it is probable that

$$\lim_{x \to 0} \frac{\sin x}{x} = 1.$$

A warning was given in that problem to use radian mode rather than degrees to evaluate the sine function. Suppose that this warning is ignored. Find the probable limit and explain the discrepancy.

1.6

The Limit of a Function: The Definition

In Section 1.5 we found properties that the limit should possess and discussed situations in which the limit should and should not exist. We now introduce a definition that preserves these properties but can be used analytically, without referring to such vague notions as "close to" and "approaching."

In defining the limit of a function we are dealing with a relative concept. We want to have $\lim_{x \to a} f(x) = L$ precisely when $f(x)$ becomes "close to" L as x becomes "close to" a. But the closeness of $f(x)$ to L is, in general, relative to how close x is to a. The following definition, illustrated in Figure 1.48, uses tolerances about both the number a and the number L to relate these two notions of closeness.

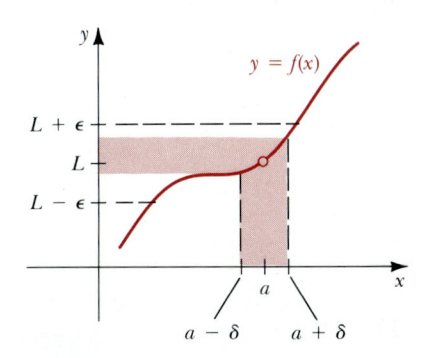

Figure 1.48
$\lim_{x \to a} f(x) = L.$

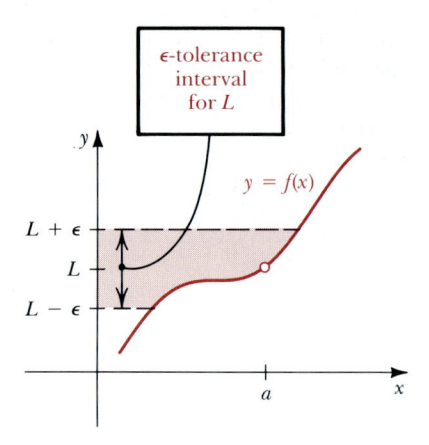

Figure 1.49
An ε-tolerance interval for L.

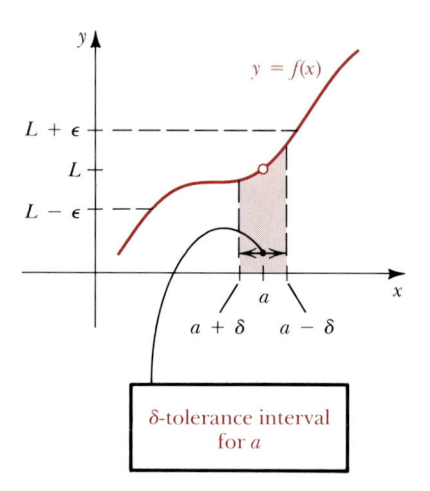

Figure 1.50
A δ-tolerance interval for a.

(1.28)
DEFINITION
Suppose f is defined on an open interval containing a, except possibly at a itself. The **limit of $f(x)$ as x approaches a is L,** provided that for every number $\varepsilon > 0$, a number $\delta > 0$ exists with the property that

$$|f(x) - L| < \varepsilon \quad \text{whenever} \quad 0 < |x - a| < \delta.$$

When such a number L exists, we write $\lim_{x \to a} f(x) = L$.

Observe how Definition 1.28 evolves from our intuitive notion of the limit. Consider the phrase "$f(x)$ is close to L." We can express "close" by choosing an appropriate tolerance and stating that the value of $f(x)$ differs from L by less than this tolerance. The tolerance about L is usually denoted by ε (epsilon), and represents an arbitrary, but generally small, positive real number. Figure 1.49 shows an ε tolerance interval about L. The phrase "$f(x)$ is close to L" can then be replaced by the phrase "the distance between $f(x)$ and L is less than ε" and concisely written as

(1.29)
$$|f(x) - L| < \varepsilon.$$

The next step is to express the statement "x is close, but not equal, to a" by an analytical expression and provide a link with the statement "$f(x)$ is close to L." We again use a tolerance, denoted by δ (delta), representing a positive real number and illustrated in Figure 1.50. The phrase "x is close, but not equal, to a" can be replaced by the phrase "the distance between x and a is less than δ and greater than 0," or by writing

(1.30)
$$0 < |x - a| < \delta.$$

The link between statements (1.29) and (1.30) is provided by requiring that for *every* tolerance $\varepsilon > 0$, a tolerance $\delta > 0$ exists with the property that "the distance from $f(x)$ to L is less than ε" for every number x provided that "the distance from x to a is less than δ, and greater than 0." Using statements (1.29) and (1.30) to express this requirement leads to Definition 1.28.

We will see numerous situations in which the limit of a function fails to exist. When the limit *does* exist, it is unique. The proof of this is given in Appendix B.

(1.31)
THEOREM
If $\lim_{x \to a} f(x) = L$ and $\lim_{x \to a} f(x) = M$, then $L = M$.

Examples 1 and 2 demonstrate the definition of the limit when applied to linear functions.

Example 1

The graph of $f(x) = 2x - 1$ is given in Figure 1.51. Show that $\lim_{x \to 2} f(x) = 3$.

1.6 THE LIMIT OF A FUNCTION: THE DEFINITION

Solution
Given the tolerance $\varepsilon > 0$, we must show that a number $\delta > 0$ exists with

$$|(2x - 1) - 3| < \varepsilon$$

for all x satisfying $0 < |x - 2| < \delta$.

We can see from Figure 1.51 that such a δ can be found. In fact, given a horizontal strip of width ε on each side of the line $y = 3$, we can find a vertical strip of width δ on each side of the line $x = 2$, with the property that the graph of f lies entirely within the rectangle formed by the intersection of the two strips.

To give an analytic proof that $\lim_{x \to 2} f(x) = 3$, suppose that x satisfies the inequality $0 < |x - 2| < \delta$. Then

$$|f(x) - 3| = |(2x - 1) - 3| = 2|x - 2| < 2\delta.$$

For any given number $\varepsilon > 0$, choose $\delta > 0$ to satisfy $2\delta \leq \varepsilon$; for example, choose

$$\delta = \varepsilon/2.$$

If δ satisfies this condition and $0 < |x - 2| < \delta$, then

$$|f(x) - 3| = |(2x - 1) - 3| = 2|x - 2| < 2\delta = 2(\varepsilon/2) = \varepsilon.$$

This completes the proof. ▶

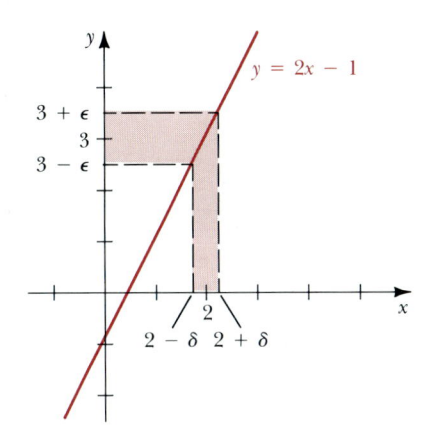

Figure 1.51
$\lim_{x \to 2} (2x - 1) = 3.$

The value of $\delta = \varepsilon/2$ in Example 1 could have been anticipated because the graph of f is a line with slope 2. This means that any change in the dependent variable x produces twice this change in the values of the function.

Example 2

The graph of $f(x) = -3x + 3$ is given in Figure 1.52. Show that $\lim_{x \to -1} f(x) = 6.$

Solution
Given a number $\varepsilon > 0$, we must show that a number $\varepsilon > 0$ exists with

$$|(-3x + 3) - 6| < \varepsilon$$

for all x satisfying $0 < |x + 1| < \delta$. If $0 < |x + 1| < \delta$, then

$$|f(x) - 6| = |(-3x + 3) - 6| = |-3(x + 1)| = 3|x + 1| < 3\delta.$$

Consequently, for any given $\varepsilon > 0$, choose a positive number δ to satisfy $3\delta \leq \varepsilon$; for example, choose

$$\delta = \varepsilon/3.$$

If δ satisfies this condition and $0 < |x + 1| < \delta$, then

$$|f(x) - 6| = |(-3x + 3) - 6| = 3|x + 1| < 3(\varepsilon/3) = \varepsilon.$$

This shows that

$$\lim_{x \to -1} f(x) = 6.$$

▶

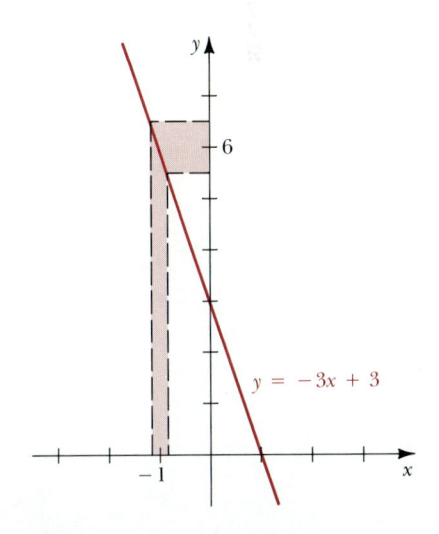

Figure 1.52
$\lim_{x \to -1} (-3x + 3) = 6.$

50   CHAPTER 1   FUNCTIONS, LIMITS, AND CONTINUITY

Note that in each of these examples three steps were required for the proof that $\lim_{x \to a} f(x) = L$:

i) the simplification of $|f(x) - L|$ to involve $|x - a|$;
ii) the choice of an appropriate value for δ; and
iii) the verification that the choice of δ suffices.

The following example illustrates how Definition 1.28 can be used to show that certain limits do not exist.

Example 3

Show that the function defined by $f(x) = \dfrac{|x|}{x}$ does not have a limit at 0.

Solution

This function can be expressed alternatively as

$$f(x) = \begin{cases} 1, & \text{if } x > 0, \\ -1, & \text{if } x < 0. \end{cases}$$

The function is undefined at $x = 0$.

Let L be an arbitrary real number. We will show that L cannot be the limit of the function at 0. This is done by finding a number ε, specifically $\varepsilon = 0.5$, with the property that there is no number $\delta > 0$ satisfying

$$|f(x) - L| < 0.5 \quad \text{whenever} \quad 0 < |x - 0| < \delta.$$

Consider the region between the horizontal lines $y = L + 0.5$ and $y = L - 0.5$. If $L \geq 0$, this region cannot contain any portion of the graph for which $x < 0$ (see Figure 1.53a). Thus the limit, if it did exist, could not be nonnegative.

On the other hand, if $L < 0$, the region between $y = L + 0.5$ and $y = L - 0.5$ cannot contain any portion of the graph for which $x > 0$ (see Figure 1.53b). This rules out the possibility of the limit being negative. Since the limit L can be neither negative nor nonnegative, L does not exist. ▶

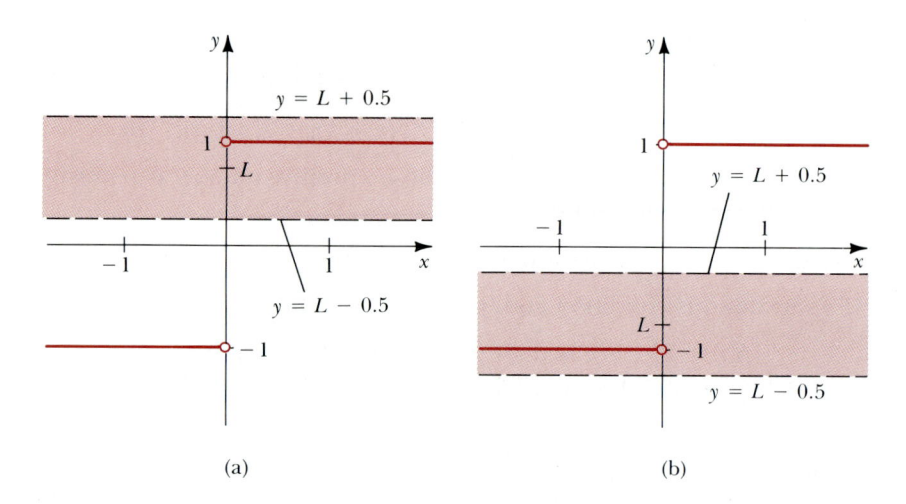

Figure 1.53

$\lim_{x \to 0} |x|/x$ does not exist.

(a)   (b)

Note that the number δ in Definition 1.28 depends on both a and ε, as well as on the function f, and that δ is not unique. In fact, if a specific number δ is found, *any* number $\tilde{\delta} > 0$ that is smaller than δ also suffices, for if x satisfies the condition $0 < |x - a| < \tilde{\delta}$, and $\tilde{\delta} \leq \delta$, then x also satisfies $0 < |x - a| < \delta$.

In Examples 1 and 2, the functions are linear and the application of the limit definition is straightforward. For more complex functions the demonstration that the limit exists is usually more difficult, as illustrated in the next example.

Example 4

Consider the function given by $f(x) = x^2$. Show that $\lim\limits_{x \to 2} f(x) = 4$.

Solution

For a given number $\varepsilon > 0$, we must show that a number $\delta > 0$ exists with the property that

$$|x^2 - 4| < \varepsilon \quad \text{whenever} \quad 0 < |x - 2| < \delta.$$

Proceeding as in the previous examples, we know that if

$$0 < |x - 2| < \delta,$$

then

$$|x^2 - 4| = |(x - 2)(x + 2)| = |x - 2||x + 2| < \delta|x + 2|.$$

Solving the equation $\delta|x + 2| = \varepsilon$ for δ produces

$$\delta = \frac{\varepsilon}{|x + 2|},$$

which is not a constant.

To obtain a constant that will suffice, recall the freedom we have in choosing δ. The only stipulation on δ is that if

$$0 < |x - 2| < \delta, \quad \text{then} \quad |x^2 - 4| < \varepsilon.$$

In our present situation, we first choose δ small enough so that if $|x - 2| < \delta$, then $|x + 2|$ is bounded by a known value. For example, suppose we require that $\delta \leq 1$. If x satisfies $|x - 2| < \delta$, then x also satisfies $|x - 2| < 1$. Thus,

$$-1 < x - 2 < 1.$$

The size of $x + 2$ can be established by adding 4 throughout the inequality

$$3 < x + 2 < 5.$$

Thus, choosing $\delta \leq 1$ and requiring x to satisfy $|x - 2| < \delta$ forces

$$|x + 2| < 5.$$

We must now satisfy $|f(x) - 4| < \varepsilon$. To do this note that if $|x - 2| < \delta$

and $\delta \leq 1$, then
$$|f(x) - 4| = |x^2 - 4| = |x + 2||x - 2| < 5 \cdot |x - 2| < 5\delta.$$

This implies that we should also choose δ to satisfy $\delta \leq \varepsilon/5$.

To satisfy both $\delta \leq 1$ and $\delta \leq \varepsilon/5$, we choose $\delta > 0$ to be the minimum of $\varepsilon/5$ and 1. This ensures that if
$$0 < |x - 2| < \delta,$$
then
$$|f(x) - 4| = |x^2 - 4| = |x + 2||x - 2| < 5\delta \leq 5(\varepsilon/5) = \varepsilon.$$

This completes the proof. ▶

The proof of the limit result in Example 4 is complicated even though the function is relatively simple. It should be clear from this example that before we can work efficiently with the limits of functions, machinery must be developed for determining and demonstrating limits without continually applying the definition. This topic is considered in Section 1.7.

▶ EXERCISE SET 1.6

In Exercises 1–12, a function f is described and values of a and L are given with $\lim_{x \to a} f(x) = L$. For the given value of ε, find a number $\delta > 0$ with the property that $|f(x) - L| < \varepsilon$ whenever $0 < |x - a| < \delta$.

1. $f(x) = 2x + 3$, $a = 2$, $L = 7$, $\varepsilon = 1$
2. $f(x) = 2x + 3$, $a = 2$, $L = 7$, $\varepsilon = 0.5$
3. $f(x) = x - 2$, $a = 2$, $L = 0$, $\varepsilon = 0.005$
4. $f(x) = x + 1$, $a = 1$, $L = 2$, $\varepsilon = 10^{-4}$
5. $f(x) = 2x + 1$, $a = 1$, $L = 3$, $\varepsilon = 0.1$
6. $f(x) = 3x + 1$, $a = 1$, $L = 4$, $\varepsilon = 0.1$
7. $f(x) = 3 - 2x$, $a = 2$, $L = -1$, $\varepsilon = 0.1$
8. $f(x) = 3 - 2x$, $a = 2$, $L = -1$, $\varepsilon = 10^{-10}$
9. $f(x) = \dfrac{x^2 - 4}{x + 2}$, $a = 2$, $L = 0$, $\varepsilon = 0.01$
10. $f(x) = \dfrac{x^2 - 4}{x - 2}$, $a = 2$, $L = 4$, $\varepsilon = 0.01$
11. $f(x) = x^2$, $a = 3$, $L = 9$, $\varepsilon = 0.1$
12. $f(x) = x^2 + 1$, $a = -1$, $L = 2$, $\varepsilon = 0.1$

In Exercises 13–24, use the definition of the limit to prove each assertion.

13. $\lim_{x \to 3} (2x + 5) = 11$
14. $\lim_{x \to 2} (2x + 3) = 7$
15. $\lim_{x \to -1} (3x + 4) = 1$
16. $\lim_{x \to -2} (-3x + 4) = 10$
17. $\lim_{x \to -2} (-2x + 5) = 9$
18. $\lim_{x \to 3} \dfrac{x^2 - 4}{x - 2} = 5$
19. $\lim_{x \to 2} |x - 2| = 0$
20. $\lim_{x \to 2} |x + 3| = 5$
21. $\lim_{x \to 0} x^2 = 0$
22. $\lim_{x \to -1} (x^2 + 1) = 2$
23. $\lim_{x \to 2} 1/x = 1/2$
24. $\lim_{x \to 4} \sqrt{x} = 2$

25. Show that $\lim_{x \to a} f(x) = 0$ if and only if $\lim_{x \to a} |f(x)| = 0$.

26. Suppose that $\lim_{x \to a} f(x) = L$. Must $\lim_{x \to a} |f(x)| = |L|$? If so, prove it; if not, find a counterexample to the statement. (A *counterexample* is an example for which the statement is false.)

27. Suppose that $\lim_{x \to a} |f(x)| = |L|$. Must $\lim_{x \to a} f(x) = L$? If so, prove it; if not, find a counterexample to the statement.

28. Show that $\lim_{x \to a} f(x) = L$ if and only if $\lim_{x \to a} (f(x) - L) = 0$.

29. Suppose that numbers $\delta > 0$ and M exist with $f(x) \leq M$ for all values of x in $(a - \delta, a + \delta)$. Show that $\lim_{x \to a} f(x) \leq M$, if this limit exists.

30. Consider the function described by
$$f(x) = \begin{cases} 1, & \text{if } x = \pm 1/n,\ n = 1, 2, 3, 4, \ldots, \\ 0, & \text{if } x \neq \pm 1/n. \end{cases}$$

Does $\lim_{x \to 0} f(x)$ exist?

31. Show that if $\lim_{x\to a} f(x) = 0$ and $\varepsilon > 0$ is given, then an interval containing a exists with $|f(x)| < \varepsilon$ for all $x \neq a$ in the interval.

32. Show that if $\lim_{x\to a} f(x) > 0$, then an interval containing a exists with $f(x) > 0$ for all $x \neq a$ in the interval.

33. Show that if $\lim_{x\to a} f(x) < 0$, then an interval containing a exists with $f(x) < 0$ for all $x \neq a$ in the interval.

1.7 Limits and Continuity

The common usage of *continuous* signifies behavior without break or interruption. The term *continuous* is used in calculus to describe functions whose graphs have this type of behavior. Consider the functions used in Section 1.5 to introduce the notion of limit. Their graphs are shown in Figure 1.54.

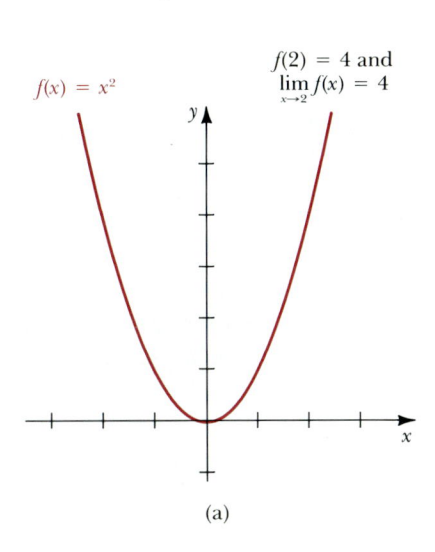

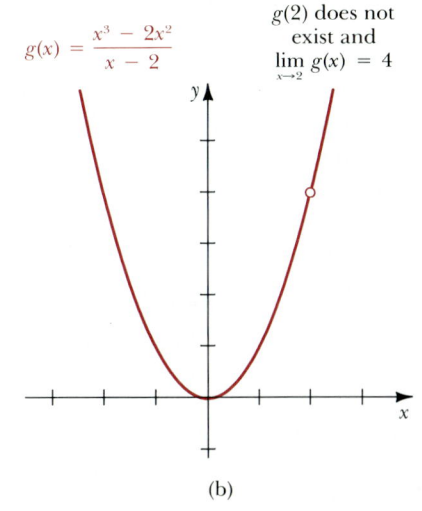

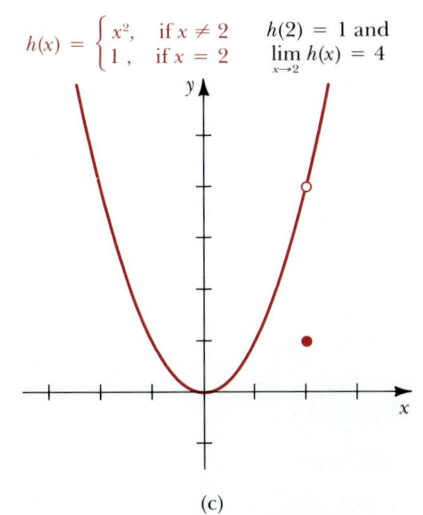

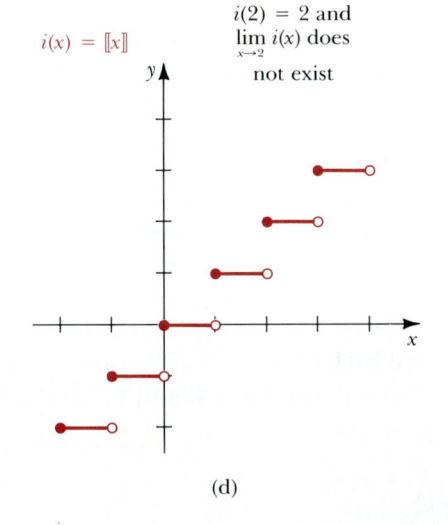

$f(x) = x^2$; $f(2) = 4$ and $\lim_{x\to 2} f(x) = 4$

$g(x) = \dfrac{x^3 - 2x^2}{x - 2}$; $g(2)$ does not exist and $\lim_{x\to 2} g(x) = 4$

(a)   (b)

$h(x) = \begin{cases} x^2, & \text{if } x \neq 2 \\ 1, & \text{if } x = 2 \end{cases}$; $h(2) = 1$ and $\lim_{x\to 2} h(x) = 4$

$i(x) = [\![x]\!]$; $i(2) = 2$ and $\lim_{x\to 2} i(x)$ does not exist

(c)   (d)

Figure 1.54

Of these four functions, $f(x) = x^2$ is the only one for which the value of the function at $x = 2$ is equal to its limit there. Because of this, only the graph of f is continuous in the ordinary sense at $x = 2$; the graphs of g, h, and i all have some type of break or interruption at $x = 2$. The following definition ensures that a function f is *continuous* at a number a in its domain precisely when the graph of f is continuous in the ordinary sense at the point $(a, f(a))$.

(1.32)
DEFINITION
A function f is **continuous** at a if:

i) $f(a)$ exists, that is, a is in the domain of f;
ii) $\lim_{x \to a} f(x)$ exists; and
iii) $\lim_{x \to a} f(x) = f(a)$.

If a function f is not continuous at a, then f is said to be **discontinuous** at a. Thus f is discontinuous at a if any of the three criteria in Definition 1.32 is not met.

If it is known that a is in the domain of a function f, then to show that f is continuous at a requires demonstrating that the limit of f at a exists and that

$$\lim_{x \to a} f(x) = f(a).$$

A slight modification of this equation will be used frequently in Chapter 2. For this modification, a variable h is defined by $h = x - a$. Then $x = h + a$, and h approaches 0 if and only if x approaches a. Thus:

(1.33)
If a is in the domain of f, then f is continuous at a precisely when the limit of f at a exists and

$$\lim_{h \to 0} f(a + h) = f(a).$$

As indicated at the end of Section 1.6, the ε, δ process involved with limits is too tedious to use in most situations. We now introduce some results whose proofs depend on this ε, δ process. Once these results have been proved, however, the process can be avoided for finding most of the limits we will need.

The first result concerns the limits and continuity of linear functions. Note that the proof of this result mimics the solutions to Examples 1 and 2 in Section 1.6.

(1.34)
THEOREM
For any constants m and b and any real number a,

$$\lim_{x \to a} (mx + b) = ma + b.$$

Proof. Suppose $\varepsilon > 0$ is given. To show that $\lim_{x \to a} (mx + b) = ma + b$, we must show that there is a number δ with the property that

$$|(mx + b) - (ma + b)| < \varepsilon$$

whenever

$$0 < |x - a| < \delta.$$

To find δ we consider two cases:

i) If $m = 0$, then

$$|(mx + b) - (ma + b)| = 0 < \varepsilon,$$

regardless of the value of x. In this case, any positive number δ will suffice.

ii) If $m \neq 0$, then

$$|(mx + b) - (ma + b)| = |m| \cdot |x - a|.$$

Choose δ to satisfy $0 < \delta \leq \varepsilon/|m|$. If

$$0 < |x - a| < \delta \leq \varepsilon/|m|,$$

then

$$|(mx + b) - (ma + b)| = |m| \cdot |x - a| < |m| \cdot \delta \leq |m| \cdot \varepsilon/|m| = \varepsilon. \quad \triangleright$$

It follows from Theorem 1.34 that if f is any linear function and a is a real number, then

$$\lim_{x \to a} f(x) = f(a).$$

Therefore:

(1.35)
COROLLARY
A linear function is continuous at each real number.

A special case of Corollary 1.35 occurs when $m = 0$:

(1.36)
Every constant function is continuous at each real number.

The following theorem is a powerful result for determining limits of complicated functions. It provides a constructive process for splitting a complicated problem into manageable parts. A proof of this theorem is given in Appendix B.

(1.37)
THEOREM
If f and g are functions with $\lim_{x \to a} f(x) = L$ and $\lim_{x \to a} g(x) = M$, then:

a) $\lim_{x \to a} [f(x) + g(x)] = L + M;$

b) $\lim_{x \to a} [f(x) - g(x)] = L - M;$

c) $\lim_{x \to a} [f(x) \cdot g(x)] = L \cdot M;$

d) $\lim_{x \to a} cf(x) = cL,$ for any constant $c;$

e) $\lim_{x \to a} \dfrac{1}{f(x)} = \dfrac{1}{L},$ provided $L \neq 0;$

f) $\lim_{x \to a} \dfrac{g(x)}{f(x)} = \dfrac{M}{L},$ provided $L \neq 0;$

g) $\lim_{x \to a} \sqrt[n]{f(x)} = \sqrt[n]{L}$ if n is an odd positive integer or if n is an even positive integer and $L > 0.$

Example 1

Use Theorems 1.34 and 1.37 to show that $\lim_{x \to 2} x^2 = 4.$

Solution

By Theorem 1.34, $\lim_{x \to 2} x = 2.$ Using this result and applying Theorem 1.37(c) with $f(x) = x$ and $g(x) = x$ gives us

$$\lim_{x \to 2} x^2 = \lim_{x \to 2} (x \cdot x) = \lim_{x \to 2} x \cdot \lim_{x \to 2} x = 2 \cdot 2 = 4.$$ ▶

The ease with which we found the limit in Example 1, once Theorems 1.34 and 1.37 are known, contrasts sharply with the ε, δ proof given in Example 4 of Section 1.6.

If f and g are functions that are continuous at a, then

$$L = \lim_{x \to a} f(x) = f(a) \quad \text{and} \quad M = \lim_{x \to a} g(x) = g(a).$$

This implies the following corollary to Theorem 1.37.

(1.38) COROLLARY

If f and g are functions that are continuous at a, then:

a) $f + g$ is continuous at $a;$
b) $f - g$ is continuous at $a;$
c) $f \cdot g$ is continuous at $a;$
d) cf is continuous at a, for any constant $c;$
e) $1/f$ is continuous at a, provided $f(a) \neq 0;$
f) g/f is continuous at a, provided $f(a) \neq 0;$
g) $f^{1/n}$, defined by $f^{1/n}(x) = \sqrt[n]{f(x)}$, is continuous at a if n is an odd positive integer or if n is an even positive integer and $f(a) > 0.$

Example 2

Show that $f(x) = 7x^2 - 7x + 1$ is continuous at each real number.

Solution
The function can be expressed as
$$f(x) = 7(g(x))^2 + h(x),$$
where
$$g(x) = x \quad \text{and} \quad h(x) = -7x + 1.$$
The functions g and h are linear, hence continuous at each real number by Corollary 1.35. It follows from Corollary 1.38 that the product g^2 is continuous, as are $7g^2$ and the sum $7g^2 + h$. Thus f is continuous at each real number. ▶

Every polynomial can be decomposed into sums and products of linear functions in the manner outlined in Example 2. This implies the following result.

(1.39)
THEOREM
A polynomial function is continuous at each real number.

A **rational function** is the quotient of two polynomials, a function of the form

(1.40) $$f(x) = \frac{a_n x^n + a_{n-1} x^{n-1} + \cdots + a_1 x + a_0}{b_m x^m + b_{m-1} x^{m-1} + \cdots + b_1 x + b_0},$$

where $a_0, a_1, \ldots, a_n$ and $b_0, b_1, \ldots, b_m$ are constants. Since polynomials are continuous at each real number, Corollary 1.38(f) implies that a rational function is continuous at each number for which the denominator is not zero. Thus:

(1.41)
A rational function is continuous at each real number in its domain.

The following example shows how rational functions can have limits at points of discontinuity.

Example 3
Consider the function defined by
$$f(x) = \frac{x^2 - 4}{x^2 - 3x + 2}.$$

a) Show that f is discontinuous at 1 and at 2.
b) Show that $\lim_{x \to 2} f(x)$ exists.

Solution

a) The denominator of $f(x)$ factors into
$$x^2 - 3x + 2 = (x-1)(x-2),$$
so the denominator is zero at 1 and at 2. Therefore, f is discontinuous at $x = 1$ and at $x = 2$.

b) The numerator of $f(x)$ factors into
$$x^2 - 4 = (x+2)(x-2).$$
Thus when $x \neq 2$,
$$f(x) = \frac{x^2 - 4}{x^2 - 3x + 2} = \frac{(x+2)(x-2)}{(x-1)(x-2)} = \frac{x+2}{x-1}.$$
Since $x = 2$ is not considered when calculating the limit at 2,
$$\lim_{x \to 2} f(x) = \lim_{x \to 2} \frac{x+2}{x-1} = \frac{2+2}{2-1} = 4. \quad \blacktriangleright$$

An obvious question remaining from Example 3 concerns the limit at the other point of discontinuity, $x = 1$. Note that this discontinuity remains even in the factored and simplified form of $f(x)$. In fact, as x approaches 1 the numerator of $(x+2)/(x-1)$ approaches 3 while the denominator approaches 0. We will consider problems of this type in Section 1.10.

Theorem 1.37 provides limit results for all the arithmetic operations involving functions, and these results seem quite natural. We now consider the limit of a composition of functions. For composite functions you might *conjecture* (a conjecture is an educated guess) that if $\lim_{x \to a} g(x) = b$ and $\lim_{x \to b} f(x) = L$, then $\lim_{x \to a} f(g(x)) = L$. In this case your intuition would be incorrect (see Exercise 58). If f is continuous at b, however, the result is valid. Appendix B contains a proof of this result.

(1.42)
THEOREM
If $\lim_{x \to a} g(x) = b$ and f is continuous at b, then
$$\lim_{x \to a} f(g(x)) = f(\lim_{x \to a} g(x)) = f(b).$$

When the function g in Theorem 1.42 is continuous at a, the limit of g at a exists and
$$\lim_{x \to a} g(x) = g(a).$$
This provides the following important special case of Theorem 1.42.

(1.43)
THEOREM
If g is continuous at a and f is continuous at $b = g(a)$, then $f \circ g$ is continuous at a.

The final result that we consider in this section is called the *Squeeze Theorem*. The name is derived from the fact that it gives results about a function that is squeezed between two other functions (see Figure 1.55). The proof of this theorem follows directly from the definition of the limit.

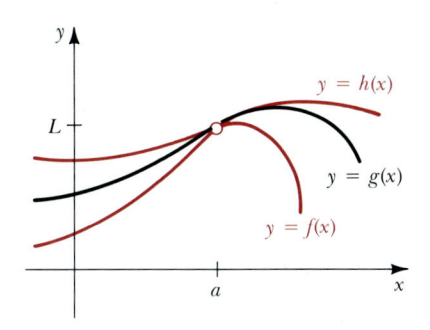

(1.44)
THEOREM: SQUEEZE THEOREM
Suppose that f, g, and h are functions with

$$f(x) \leq g(x) \leq h(x)$$

for each $x \neq a$ in an open interval containing a. If $\lim_{x \to a} f(x) = L$ and $\lim_{x \to a} h(x) = L$, then $\lim_{x \to a} g(x) = L$.

Figure 1.55
$\lim_{x \to a} g(x) = L$.

Example 4

Determine

$$\lim_{x \to 0} \frac{x^2 \cos x}{x^2 + 1}.$$

Solution
Since $-1 \leq \cos x \leq 1$ for all x,

$$\frac{-x^2}{x^2 + 1} \leq \frac{x^2 \cos x}{x^2 + 1} \leq \frac{x^2}{x^2 + 1}.$$

However,

$$\lim_{x \to 0} \frac{-x^2}{x^2 + 1} = 0 \quad \text{and} \quad \lim_{x \to 0} \frac{x^2}{x^2 + 1} = 0,$$

so the Squeeze Theorem implies that

$$\lim_{x \to 0} \frac{x^2 \cos x}{x^2 + 1} = 0. \quad \blacktriangleright$$

▶ EXERCISE SET 1.7

In Exercises 1–6, the graph of a function is given. From the graph, determine the points at which the function is discontinuous, and give the criterion of Definition 1.32 that is not met.

1.

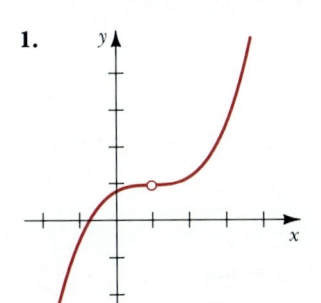

2.

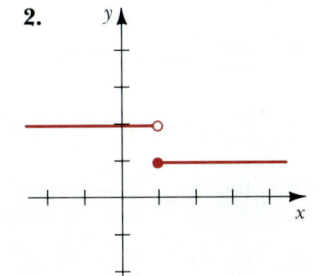

3.

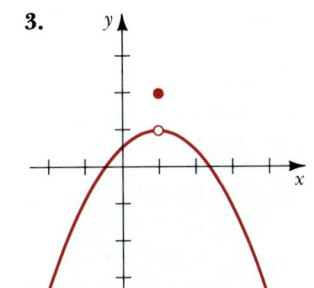

4.

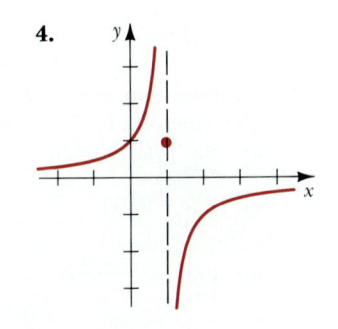

5.

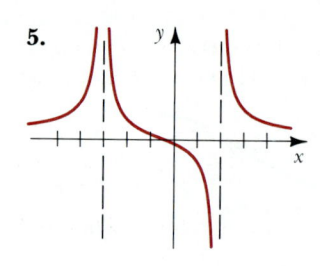

6.

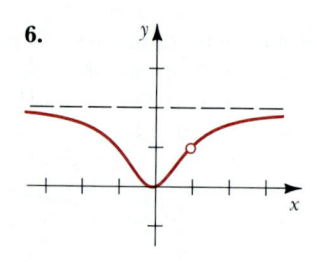

In Exercises 7–21, sketch the graph of the function described, and determine any numbers at which the function is discontinuous. For each discontinuity, give the criterion of Definition 1.32 that is not met.

7. $f(x) = \dfrac{x^2 - 4}{x - 2}$

8. $f(x) = \begin{cases} \dfrac{x^2 - 4}{x - 2}, & \text{if } x \neq 2 \\ 1, & \text{if } x = 2 \end{cases}$

9. $f(x) = \dfrac{|x|}{x}$

10. $f(x) = \dfrac{|x + 4|}{x + 4}$

11. $f(x) = \begin{cases} \dfrac{x^3 - 4x}{x^2 - 4}, & \text{if } x \neq 2, -2, \\ 2, & \text{if } x = 2, \\ 3, & \text{if } x = -2 \end{cases}$

12. $f(x) = \begin{cases} \dfrac{x^3 - 3x^2 - 4x + 12}{x^2 - 5x + 6}, & \text{if } x \neq 2, 3, \\ 0, & \text{if } x = 2, \\ 5, & \text{if } x = 3 \end{cases}$

13. $f(x) = [\![x - 1]\!]$
14. $f(x) = [\![2x]\!]$
15. $f(x) = \tan x$
16. $f(x) = \cot x$
17. $f(x) = \sec x$
18. $f(x) = \csc x$

19. $f(x) = \begin{cases} x, & \text{if } x \text{ is not an integer,} \\ 0, & \text{if } x \text{ is an integer} \end{cases}$

20. $f(x) = \begin{cases} 1, & \text{if } x \text{ is not an integer,} \\ (-1)^x, & \text{if } x \text{ is an integer} \end{cases}$

21. $f(x) = \begin{cases} x, & \text{if } x \text{ is not an integer,} \\ (-1)^x, & \text{if } x \text{ is an integer} \end{cases}$

22. Suppose that $\lim_{x \to 3} f(x) = 7$ and $\lim_{x \to 3} g(x) = -3$. Use Theorem 1.37 to find each of the following limits.
   a) $\lim_{x \to 3} [f(x) + g(x)]$
   b) $\lim_{x \to 3} f(x)g(x)$
   c) $\lim_{x \to 3} -8f(x)$
   d) $\lim_{x \to 3} [f(x)]^2$
   e) $\lim_{x \to 3} \sqrt{f(x) - g(x)}$
   f) $\lim_{x \to 3} \dfrac{f(x)}{-7g(x)}$

In Exercises 23–32, use the limit theorems in this section to determine the limit at the specified number. Determine whether the function is continuous at the number.

23. $\lim_{x \to -3} (x^4 - 5x^3 + 7)$

24. $\lim_{x \to 4} (x^3 - 3x^2 + 5)$

25. $\lim_{x \to 4} (x^2 + 3)(x - 2)$

26. $\lim_{x \to -1} (3x + 2x^2 + 1)(4x^2 - 3)$

27. $\lim_{x \to 3} \sqrt{x^3 + 22}$

28. $\lim_{x \to -1} (x^3 - 4x)^{1/4}$

29. $\lim_{x \to -2} \dfrac{1}{\sqrt{x^2 + 12} - \sqrt[3]{x^2 + 4}}$

30. $\lim_{x \to 0} \dfrac{1}{\sqrt{x^2 + 10} - \sqrt{5 - 3x^2}}$

31. $\lim_{x \to 6} \dfrac{3x - 7}{x^2 + 2}$

32. $\lim_{x \to -3} \dfrac{x^2 - 6x}{x^3 + 4x + 4}$

In Exercises 33–38, functions f and g are described. Find $f \circ g$ and $g \circ f$, and determine where these functions are continuous.

33. $f(x) = \dfrac{1}{x}$, $g(x) = x - 2$

34. $f(x) = x^3$, $g(x) = \sqrt{x}$

35. $f(x) = \dfrac{x - 1}{x + 1}$, $g(x) = \sqrt[4]{x}$

36. $f(x) = \dfrac{x - 1}{x + 1}$, $g(x) = \sqrt[3]{x}$

37. $f(x) = x^2 - 5x + 6$, $g(x) = \sqrt{x}$

38. $f(x) = x^3 - 3x^2 - 4x + 12$, $g(x) = \sqrt{x}$

In Exercises 39–50, determine the limits or tell why they do not exist. (These exercises generally require an algebraic simplification before the limit can be determined.)

39. $\lim_{x \to -1} \dfrac{x^2 - 3x - 4}{5x + 5}$

40. $\lim_{x \to 2} \dfrac{x^3 - 8}{x - 2}$

41. $\lim_{x \to -1} \dfrac{x^2 - 3x - 4}{3x + 5}$

42. $\lim_{x \to 2} \dfrac{x^2 - 5x + 6}{x^3 - 2x^2 + 4x - 8}$

43. $\lim_{x \to -5/2} \dfrac{3x^2 - 2x + 1}{2x + 5}$

44. $\lim_{x \to -1} \dfrac{3x + 5}{x^2 - 3x - 4}$

45. $\lim_{x \to 4} \dfrac{2x + 5}{3x^2 - 2x + 1}$

46. $\lim_{x \to -5/2} \dfrac{2x + 5}{3x^2 - 2x + 1}$

47. $\lim_{x \to 0} \dfrac{(2 + x)^2 - 4}{x}$

48. $\lim_{x \to 0} \dfrac{(2 + x)^3 - 8}{x}$

49. $\lim_{x \to 2} \dfrac{(1/x) - (1/2)}{x - 2}$

50. $\lim_{x \to 0} \dfrac{1/(2 + x) - (1/2)}{x}$

51. A function is said to be *bounded* if a number M exists with $|f(x)| \leq M$ for all x in the domain of f. Suppose a bounded function f is defined on an open interval about zero. Use the Squeeze Theorem to show that $\lim_{x \to 0} x \cdot f(x) = 0$.

52. Determine $\lim_{x \to 0} x \sin(1/x)$.

53. Find the constant c that will make

$$f(x) = \begin{cases} \dfrac{x^2 - 1}{x - 1}, & \text{if } x \neq 1, \\ c, & \text{if } x = 1 \end{cases}$$

continuous at $a = 1$.

54. Find the constant c that will make

$$f(x) = \begin{cases} \dfrac{x^3 - 8}{x - 2}, & \text{if } x \neq 2, \\ c, & \text{if } x = 2 \end{cases}$$

continuous at $a = 2$.

55. What must be true about a rational function f of the form

$$f(x) = \dfrac{ax^2 + bx + c}{rx^2 + sx + t}$$

if it is known that f has a limit at x_0 and $rx_0^2 + sx_0 + t = 0$?

56. When a function f is discontinuous at a number a but $\lim_{x \to a} f(x)$ exists, the discontinuity is called a *removable discontinuity*. A discontinuity that is not removable is called an *essential discontinuity*. Classify the discontinuities in Exercises 1–21 as either removable or essential.

57. Use Theorem 1.42 to show that $\lim_{x \to a} |f(x)| = |L|$ whenever $\lim_{x \to a} f(x) = L$.

58. Let functions f and g be defined by

$$g(x) = 2 \quad \text{and} \quad f(x) = \begin{cases} x^2, & \text{if } x \neq 2, \\ 0, & \text{if } x = 2. \end{cases}$$

Show that $\lim_{x \to 1} g(x) = 2$ and $\lim_{x \to 1} f(g(x)) \neq \lim_{x \to 2} f(x)$. Why does this not contradict Theorem 1.42?

In Exercises 59–62, find functions f and g that satisfy the specified conditions.

59. The function $f + g$ is continuous at 1, but neither f nor g is continuous at 1.

60. The function $f \cdot g$ is continuous at 1, but neither f nor g is continuous at 1.

61. Both of the functions $f + g$ and $f \cdot g$ are continuous at 1, but neither f nor g is continuous at 1.

62. The function f/g is continuous at 1, but neither f nor g is continuous at 1.

63. An ε, δ definition of continuity is as follows:

A function f is continuous at the number a provided that $f(a)$ exists and that whenever a number $\varepsilon > 0$ is given, a number $\delta > 0$ can be found with the property that $|f(x) - f(a)| < \varepsilon$ whenever $|x - a| < \delta$.

Use this definition to prove that the function f defined by $f(x) = 2x - 3$ is continuous at each real number.

1.8
One-Sided Limits and Continuity on an Interval

In order for the limit of a function f to exist at a, Definition 1.28 requires that f be defined on an open interval containing a, except possibly at a itself. This condition is too restrictive to be useful in all circumstances.

Consider, for example, the graph of $f(x) = \sqrt{x} + 1$ shown in Figure 1.56. This function has no negative real numbers in its domain so the

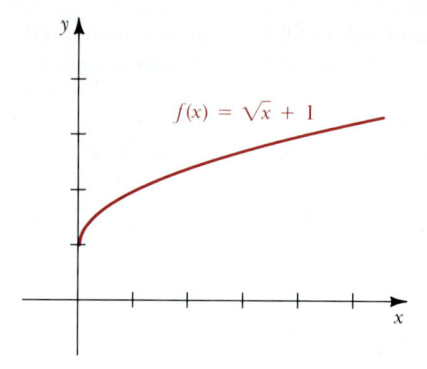

Figure 1.56

limit of f at 0 does not exist. For $x > 0$, however, $f(x) = \sqrt{x} + 1$ approaches 1 as x approaches 0. If the values of x could be restricted to the positive side of 0, a limit would exist. The following definition of **one-sided limits** relaxes part of the requirements for a limit to enable us to analyze some situations when the limit does not exist.

(1.45)
DEFINITION
The **limit from the right of the function f at a is L,** written

$$\lim_{x \to a^+} f(x) = L,$$

provided that for every number $\varepsilon > 0$, a number $\delta > 0$ exists with the property that

$$|f(x) - L| < \varepsilon \quad \text{whenever} \quad 0 < x - a < \delta;$$

that is, when $a < x < a + \delta$ (see Figure 1.57).

The **limit from the left of the function f at a is L,** written

$$\lim_{x \to a^-} f(x) = L,$$

provided that for every number $\varepsilon > 0$ a number $\delta > 0$ exists with the property that

$$|f(x) - L| < \varepsilon \quad \text{whenever} \quad -\delta < x - a < 0;$$

that is, when $a - \delta < x < a$ (see Figure 1.58).

Definition 1.45 implies that

$$\lim_{x \to 0^+} (\sqrt{x} + 1) = 1,$$

but

$$\lim_{x \to 0^-} (\sqrt{x} + 1) \text{ does not exist.}$$

The next theorem follows immediately from Definition 1.45 and the definition of $\lim_{x \to a} f(x)$.

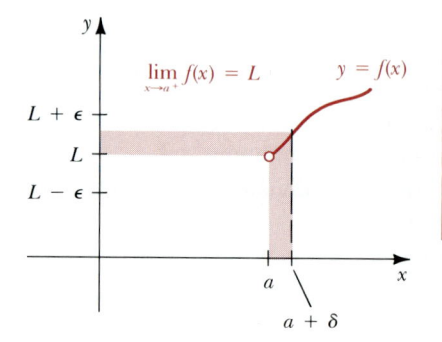

Figure 1.57
$\lim_{x \to a^+} f(x) = L.$

(1.46)
THEOREM
$\lim_{x \to a} f(x)$ exists and is L if and only if the following conditions hold:

i) $\lim_{x \to a^+} f(x)$ exists;

ii) $\lim_{x \to a^-} f(x)$ exists; and

iii) $\lim_{x \to a^+} f(x) = \lim_{x \to a^-} f(x) = L.$

Example 1

For the function h defined by $h(x) = [\![x]\!]$, find $\lim_{x \to 2^+} h(x)$, $\lim_{x \to 2^-} h(x)$, and $\lim_{x \to 2} h(x)$, if they exist.

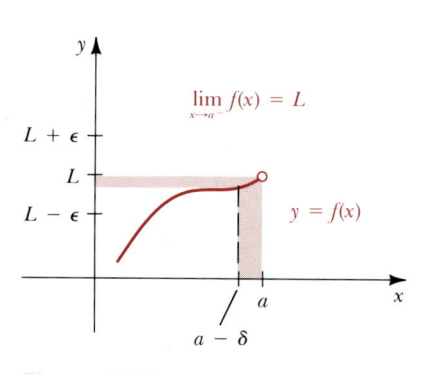

Figure 1.58
$\lim_{x \to a^-} f(x) = L.$

Solution
For x in the interval $[2, 3)$, $[\![x]\!] = 2$, so
$$\lim_{x \to 2^+} h(x) = 2.$$
For x in the interval $[1, 2)$, $[\![x]\!] = 1$, so
$$\lim_{x \to 2^-} h(x) = 1.$$
Since $\lim_{x \to 2^+} h(x) \neq \lim_{x \to 2^-} h(x)$ (see Figure 1.59), Theorem 1.46 implies that $\lim_{x \to 2} h(x)$ does not exist. ▶

Since the definition of continuity is based directly on that of the limit, corresponding definitions can be made for continuity from the right and from the left.

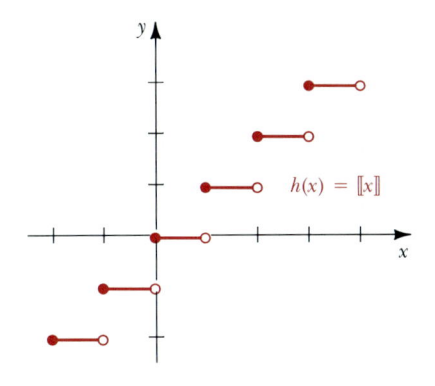

Figure 1.59
$\lim_{x \to 2^-} [\![x]\!] = 1$, $\lim_{x \to 2^+} [\![x]\!] = 2$.

(1.47)
DEFINITION
A **function f is continuous from the right at the number a** provided

i) $f(a)$ exists;
ii) $\lim_{x \to a^+} f(x)$ exists; and
iii) $\lim_{x \to a^+} f(x) = f(a)$.

Likewise, a **function f is continuous from the left at the number a** provided that

i) $f(a)$ exists;
ii) $\lim_{x \to a^-} f(x)$ exists; and
iii) $\lim_{x \to a^-} f(x) = f(a)$.

As might be expected, this definition leads to a result similar to Theorem 1.46.

(1.48)
THEOREM
A function f is continuous at a if and only if both of the following statements hold:

i) f is continuous from the right at a, and
ii) f is continuous from the left at a.

Theorems 1.37 and 1.42 in Section 1.7 gave limit results concerning arithmetic and composition operations of functions. These results, as well as the Squeeze Theorem, are equally valid for one-sided limits. Similarly, the results in Section 1.7 concerning continuity hold as well for continuity from the left and right. We will use these results even though they have not been stated explicitly.

The concepts of continuity from the right and left can be used to define continuity of a function on an interval.

(1.49)
DEFINITION
i) A function *f* **is said to be continuous on the open interval** (a, b) if *f* is continuous at every number in (a, b).
ii) A function *f* **is said to be continuous on the closed interval** $[a, b]$ if *f* is continuous on (a, b), is continuous from the right at a, and is continuous from the left at b.

Definitions for continuity on half-open intervals are made in a similar manner, as illustrated in the following example.

Example 2

Determine the largest interval on which $f(x) = \sqrt{x} + 1$ is continuous.

Solution
We considered this function at the beginning of the section; its graph is shown in Figure 1.60. Since

$$\lim_{x \to 0^+} f(x) = 1 = f(0),$$

f is continuous from the right at 0. The square root function is continuous at any positive real number by Corollary 1.38(g), so *f* is continuous on the interval $[0, M)$ for any positive real number M. Hence the largest interval on which *f* is continuous is $[0, \infty)$, the domain of *f*. ▶

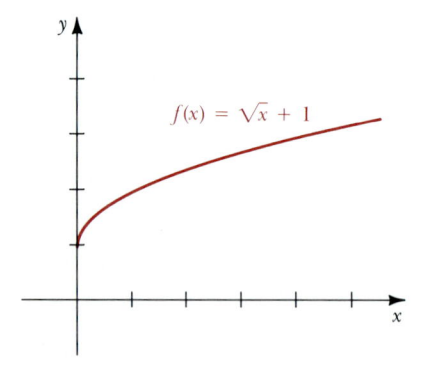

Figure 1.60
$\lim_{x \to 0^+} \sqrt{x} + 1 = 1.$

We conclude this section with a discussion of the Intermediate Value Theorem and one of its applications. Geometrically, the *Intermediate Value Theorem* states that if *f* is continuous on the interval $[a, b]$ and k is a number between $f(a)$ and $f(b)$, then the graph of *f* intersects the line $y = k$ at least once, as shown in Figure 1.61.

(1.50)
THEOREM: THE INTERMEDIATE VALUE THEOREM
Suppose *f* is a function that is continuous on the interval $[a, b]$ and k is any number between $f(a)$ and $f(b)$. Then there is at least one number c between a and b with $f(c) = k$.

Figure 1.61 shows that more than one intersection can occur. The Intermediate Value Theorem guarantees that there is at least one.

The proof of this theorem uses the Completeness Property of the real numbers, which is considered in Chapter 8.

Example 3

Use the Intermediate Value Theorem to show that for $f(x) = x^3 - 2x - 5$, there is a number c in $(2, 3)$ with $f(c) = 0$.

Solution
Since *f* is a polynominal function, *f* is continuous on $[2, 3]$. Also, $f(2) = -1$ and $f(3) = 16$, so $f(2) < 0 < f(3)$. The Intermediate Value Theorem ensures that there is a number c in $(2, 3)$ with $f(c) = 0$. ▶

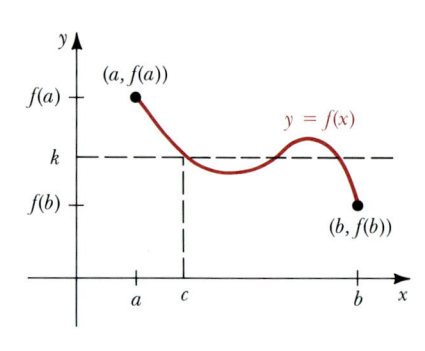

Figure 1.61
$f(c) = k.$

1.8 ONE-SIDED LIMITS AND CONTINUITY ON AN INTERVAL

The Intermediate Value Theorem is the basis for the **Bisection Method,** an elementary technique to approximate the solution to an equation of the form $f(x) = 0$.

Suppose that f is continuous on $[a, b]$ and that $f(a)$ and $f(b)$ are nonzero and have different signs. In this case, zero lies between $f(a)$ and $f(b)$, and the Intermediate Value Theorem ensures that a solution to $f(x) = 0$ exists in (a, b) (see Figure 1.62). The Bisection Method can be used to obtain an approximate solution c in (a, b) to any specified accuracy, provided the calculations are done with sufficient accuracy.

To begin the Bisection Method, let $a_0 = a$ and $b_0 = b$ and choose c_0 as the midpoint of the interval $[a, b]$:

$$c_0 = \frac{a + b}{2}.$$

If $f(c_0) = 0$, then c_0 is a solution. Suppose $f(c_0) \neq 0$. If the signs of $f(a)$ and $f(c_0)$ differ, as shown in Figure 1.63, let

$$a_1 = a_0 \quad \text{and} \quad b_1 = c_0.$$

If the signs of $f(a)$ and $f(c_0)$ agree, let

$$a_1 = c_0 \quad \text{and} \quad b_1 = b_0.$$

In either case, the interval $[a_1, b_1]$ has half the length of the interval $[a_0, b_0]$ and contains a number c satisfying $f(c) = 0$.

We reapply the process to construct a sequence of intervals $[a_0, b_0]$, $[a_1, b_1], [a_2, b_2], \ldots, [a_n, b_n]$, with

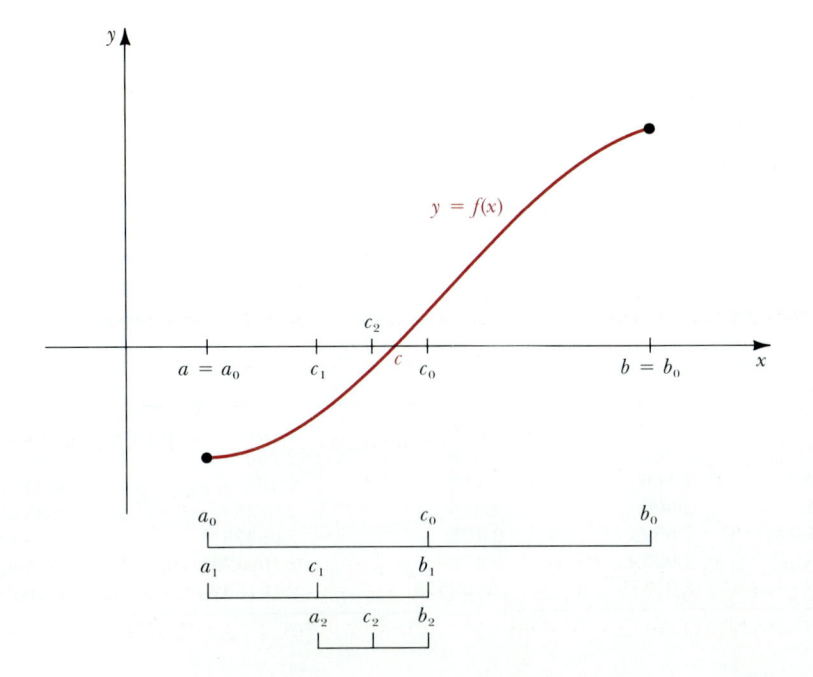

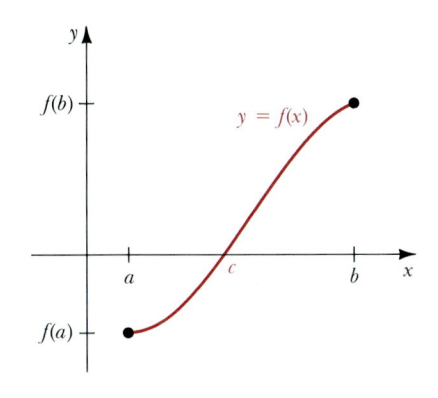

$$a_0 \leq a_1 \leq a_2 \leq \cdots \leq a_n \leq c \leq b_n \leq \cdots \leq b_2 \leq b_1 \leq b_0.$$

Figure 1.62
Object: Find c with $f(c) = 0$.

Figure 1.63
Successive applications of the Bisection Method.

Since each application reduces the length of the interval by half, the midpoint c_n of the nth interval approaches c as n becomes large. In fact, it can be shown (see Exercise 23 in Appendix A.4) that for each integer n,

$$|c_n - c| < \frac{b-a}{2^n}.$$

Appendix C contains a BASIC computer program for the Bisection Method.

Example 4

Find an approximation to a solution of the equation

$$f(x) = x^3 - 2x - 5 = 0$$

that is accurate to within $0.01 = 10^{-2}$.

Solution

In Example 3 it was shown that a solution to this equation exists in the interval $(2, 3)$, so we will apply the Bisection Method with $a_0 = 2$ and $b_0 = 3$. Then

$$c_0 = \tfrac{1}{2}(2 + 3) = 2.5$$

and

$$f(c_0) = f(2.5) = (2.5)^3 - 2(2.5) - 5 = 5.625.$$

Since $f(a_0) < 0$ and $f(c_0) > 0$,

$$a_1 = a_0 = 2 \quad \text{and} \quad b_1 = c_0 = 2.5.$$

Table 1.4 lists the results obtained for the first six applications. Since the width of the interval $[a_6, b_6]$ is

$$2.109375 - 2.09375 = 0.015625$$

and c_6 is the midpoint of this interval, c_6 is an approximation accurate to within $0.015625/2 = 0.0078125 < 0.01$. Thus the approximation 2.10 is sufficiently accurate for this problem.

From the column of values of $f(c_i)$, we might suspect that c_4 is the most accurate entry. In Section 3.6 we will see that this is indeed the case. ▶

Table 1.4

i	a_i	b_i	c_i	$f(c_i)$
0	2.000	3.000	2.500	5.6250
1	2.000	2.500	2.250	1.8906
2	2.000	2.250	2.125	0.3457
3	2.000	2.125	2.0625	−0.3513
4	2.0625	2.125	2.09375	−0.0089
5	2.09375	2.125	2.109375	0.1668
6	2.09375	2.109375	2.1015625	−0.7969

EXERCISE SET 1.8

1. The graph of a function f is shown in the figure.
   a) Find $\lim_{x \to 2^+} f(x)$ and $\lim_{x \to 2^-} f(x)$.
   b) Does $\lim_{x \to 2} f(x)$ exist? Why?

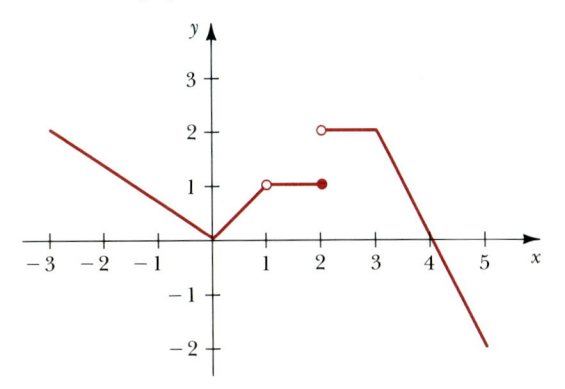

2. Use the graph of the function shown in Exercise 1 to determine each of the following, if they exist.
   a) $\lim_{x \to 1^+} f(x)$ b) $\lim_{x \to 1^-} f(x)$ c) $\lim_{x \to 1} f(x)$

In Exercises 3–34, evaluate the limit, if it exists.

3. $\lim_{x \to 0^-} (2 - \sqrt{x})$

4. $\lim_{x \to 0^+} (2 - \sqrt{x})$

5. $\lim_{x \to 2^+} \sqrt{x^2 - 4}$

6. $\lim_{x \to -2^+} \sqrt{x^2 - 4}$

7. $\lim_{x \to -3^-} \sqrt{x^2 - 9}$

8. $\lim_{x \to -3^+} \sqrt{x^2 - 9}$

9. $\lim_{x \to 3^+} \sqrt{x^2 - 4}$

10. $\lim_{x \to 3^-} \sqrt{x^2 - 4}$

11. $\lim_{x \to 0^+} \frac{|x|}{x}$

12. $\lim_{x \to 0^-} \frac{|x|}{x}$

13. $\lim_{x \to 0} \frac{|x|}{x}$

14. $\lim_{x \to 0} \frac{\sqrt{x^2}}{|x|}$

15. $\lim_{x \to 0^+} \left(\frac{1}{x} - \frac{1}{|x|} \right)$

16. $\lim_{x \to 0^-} \left(\frac{1}{x} + \frac{1}{|x|} \right)$

17. $\lim_{x \to 2^+} \frac{x-2}{\sqrt{x^2 - 4}}$

18. $\lim_{x \to 2^-} \frac{x-2}{\sqrt{4 - x^2}}$

19. $\lim_{x \to 2^+} \left(\frac{1}{x-2} - \frac{4}{x^2-4} \right)$

20. $\lim_{x \to 2^-} \sqrt{\frac{1}{x-2} - \frac{4}{x^2-4}}$

21. $\lim_{x \to 0^+} \frac{\sqrt{x^2 + 4} - 2}{x}$

    $\left[\text{Hint: Multiply by } \frac{\sqrt{x^2+4}+2}{\sqrt{x^2+4}+2}. \right]$

22. $\lim_{x \to 3^+} \frac{\sqrt{x^2 + 16} - 5}{x - 3}$

23. $\lim_{x \to 4} \frac{(x-4)^2}{\sqrt{x+12} - 4}$

24. $\lim_{x \to 3} \frac{\sqrt{6+x} - 3}{x - 3}$

25. $\lim_{x \to 4^-} \frac{(1/\sqrt{x}) - (1/2)}{x - 4}$

26. $\lim_{x \to 2^+} \frac{1/(\sqrt{x+7}) - (1/3)}{x - 2}$

27. $\lim_{x \to 3^+} (x - [\![x]\!])$

28. $\lim_{x \to 3^-} (x - [\![x]\!])$

29. $\lim_{x \to 3} (x - [\![x]\!])$

30. $\lim_{x \to 3/2} (x - [\![x]\!])$

31. $\lim_{x \to 0^+} \frac{\sqrt{1+x} - \sqrt{1-x}}{x}$

32. $\lim_{x \to 0^-} \frac{\sqrt{1+x} - \sqrt{1-x}}{x}$

33. $\lim_{x \to 1} f(x)$, if $f(x) = \begin{cases} x^2, & \text{when } x < 1, \\ -x, & \text{when } x \geq 1 \end{cases}$

34. $\lim_{x \to -1} f(x)$, if $f(x) = \begin{cases} x^2, & \text{when } x < -1, \\ 0, & \text{when } x \geq -1 \end{cases}$

In Exercises 35–38, the graph of a function is given. Use the graph to find the intervals on which the function is continuous.

35.

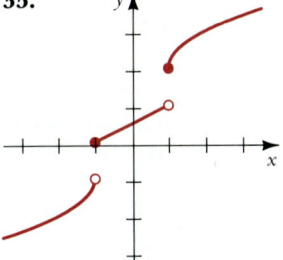

36.

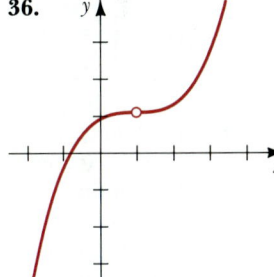

37.

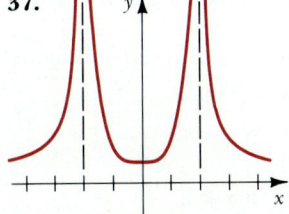

38.

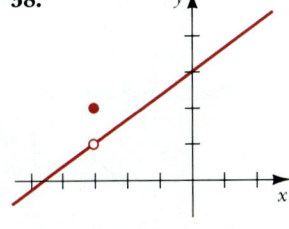

In Exercises 39–49, determine the intervals on which the function described is continuous.

39. $f(x) = x^3 + 2$
40. $f(x) = 2x^2 - 1$
41. $f(x) = \dfrac{1}{x}$
42. $f(x) = \sqrt{x}$
43. $f(x) = \sqrt{x - 3}$
44. $g(x) = \dfrac{1}{\sqrt{x - 3}}$
45. $f(x) = |x|$
46. $g(x) = \dfrac{1}{|x|}$
47. $h(x) = \dfrac{|x|}{x}$
48. $g(x) = \begin{cases} 1, & \text{if } x \text{ is not an integer,} \\ 0, & \text{if } x \text{ is an integer} \end{cases}$
49. $f(x) = \begin{cases} 1, & \text{if } x \text{ is not an integer,} \\ (-1)^x, & \text{if } x \text{ is an integer} \end{cases}$

50. The graphs of the trigonometric functions are shown in Sections 1.3 and 1.4. Use the graphs to determine the intervals on which each of the trigonometric functions is continuous.

51. Suppose
$$f(x) = \begin{cases} x^2 + 1, & \text{if } x < -1, \\ 2x - 3, & \text{if } -1 \leq x \leq 2, \\ x + 2, & \text{if } x > 2. \end{cases}$$
   a) Sketch the graph of f.
   b) Find $\lim_{x \to 2^+} f(x)$ and $\lim_{x \to 2^-} f(x)$, and determine whether f is continuous from the right or left at 2. Is f continuous at 2?
   c) Find $\lim_{x \to -1^+} f(x)$ and $\lim_{x \to -1^-} f(x)$, and determine whether f is continuous from the right or left at -1. Is f continuous at -1?
   d) Determine the intervals on which f is continuous.

52. Suppose
$$f(x) = \begin{cases} -\tfrac{1}{3}x - 1, & \text{if } x \leq 0, \\ x^2 - 1, & \text{if } 0 < x < 1 \text{ or } 1 < x < 2, \\ \tfrac{1}{2}x + 2, & \text{if } x \geq 2. \end{cases}$$
   a) Sketch the graph of f.
   b) Find $\lim_{x \to 0^+} f(x)$ and $\lim_{x \to 0^-} f(x)$, and determine if f is continuous from the right or left at 0. Is f continuous at 0?
   c) Find $\lim_{x \to 1^+} f(x)$ and $\lim_{x \to 1^-} f(x)$, and determine if f is continuous from the right or left at 1. Is f continuous at 1?
   d) Find $\lim_{x \to 2^+} f(x)$ and $\lim_{x \to 2^-} f(x)$, and determine if f is continuous from the right or left at 2. Is f continuous at 2?
   e) Determine the intervals on which f is continuous.

53. Find constants a and b that will make
$$g(x) = \begin{cases} x^2, & \text{if } x \leq 1, \\ ax + b, & \text{if } 1 < x < 2, \\ x^3, & \text{if } x \geq 2 \end{cases}$$
continuous for all x.

54. a) Use the figure to convince yourself that if $0 < t < \pi/2$, then $0 < \sin t < t$.
   b) Use the Squeeze Theorem for one-sided limits to show that $\lim_{t \to 0^+} \sin t = 0$.
   c) Show that $\lim_{t \to 0} \sin t = 0$.
   d) Use the fact that if $-\pi/2 < t < \pi/2$, then $\cos t = \sqrt{1 - \sin^2 t}$ to show that $\lim_{t \to 0} \cos t = 1$.

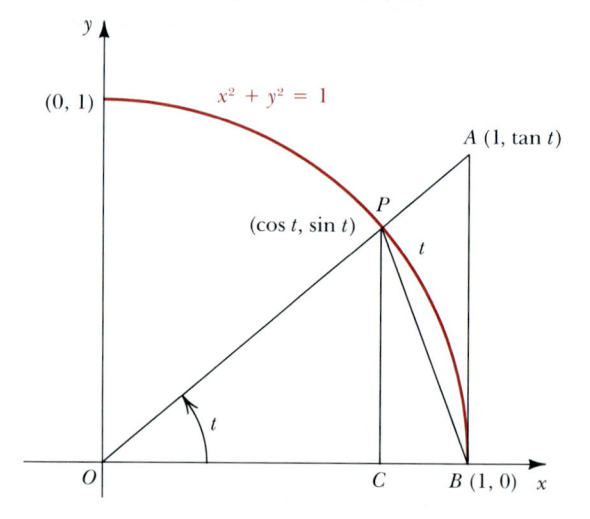

In Exercises 55–58, a function f, an interval $[a, b]$, and a number k between $f(a)$ and $f(b)$ are given. Verify that the conclusion of the Intermediate Value Theorem holds by finding a number c in (a, b) with $f(c) = k$.

55. $f(x) = 7x - 4$, $[1, 3]$, $k = 15$
56. $f(x) = x^2 - 1$, $[-1, 2]$, $k = 2$
57. $f(x) = x^3 + 1$, $[-2, 3]$, $k = 9$
58. $f(x) = x^2 + 5x + 3$, $[-3, 1]$, $k = 1$

In Exercises 59–64, (a) use the Intermediate Value Theorem to show that $f(x) = 0$ has a solution on the given interval, and (b) use the Bisection Method on the interval to determine an approximate solution that is accurate to within 10^{-1}.

59. $f(x) = x^3 - 2x^2 - 5$, $[2, 3]$
60. $f(x) = x^3 - x - 1$, $[1, 2]$
61. $f(x) = x^3 - 9x^2 + 12$, $[-2, 0]$
62. $f(x) = x^3 - 9x^2 + 12$, $[8, 9]$
63. $f(x) = \cos x - x$, $[0, \pi/2]$
64. $f(x) = x - 0.2 \sin x - 0.8$, $[0, \pi/2]$

65. Use the Bisection Method to determine an approximate x-intercept of the graph of $f(x) = 2x^3 - 9x^2 + 12x - 2$ accurate to within 10^{-1}.

66. Use the Bisection Method to determine approximate x-intercepts (there are two) of the graph of $f(x) = x^4 + 2x^2 - 1$ accurate to within 10^{-1}.

67. Find a function f with the property that $\lim_{x\to 1^+} f(x)$, $\lim_{x\to 1^-} f(x)$, and $\lim_{x\to 1} |f(x)|$ all exist, but $\lim_{x\to 1} f(x)$ does not exist.

68. If a function f has an essential discontinuity (see Exercise 55 of Section 1.7) at the number a and both $\lim_{x\to a^+} f(x)$ and $\lim_{x\to a^-} f(x)$ exist, then f is said to have a *jump discontinuity* at a. Which, if any, of the discontinuities in Exercises 7–21 of Section 1.7 are jump discontinuities?

69. Find a function f that is continuous on the interval $(0, 1)$ and has the property that for any positive number M, numbers x_1 and x_2 exist in $(0, 1)$ with $f(x_1) < -M$ and $f(x_2) > M$.

70. Suppose that g is a continuous function on $[a, b]$ and that $a \leq g(x) \leq b$ for all x in $[a, b]$. Show that there is a number p in $[a, b]$ such that $g(p) = p$. (p is called a *fixed point* of g.) [*Hint*: Define $h(x) = g(x) - x$ and apply the Intermediate Value Theorem to h.]

71. Prove Theorem 1.46.

72. Prove Theorem 1.48.

73. A function is said to be *Lipschitz continuous* on an interval $[a, b]$ if a positive constant M exists with $|f(x_1) - f(x_2)| < M|x_1 - x_2|$ for all x_1 and x_2 in $[a, b]$. Show that a function that is Lipschitz continuous on $[a, b]$ is also continuous on $[a, b]$.

74. Suppose that $f(x) \leq 0$ for all x in (a, b) and that $\lim_{x\to a^+} f(x)$ exists.
    a) Show that $\lim_{x\to a^+} f(x) \leq 0$.
    b) Give an example to show that it is possible to have $\lim_{x\to a^+} f(x) = 0$ even if $f(x) < 0$ for all x in (a, b).

75. Suppose that $f(x) \geq 0$ for all x in (a, b) and that $\lim_{x\to a^+} f(x)$ exists.
    a) Show that $\lim_{x\to a^+} f(x) \geq 0$.
    b) Give an example to show that it is possible to have $\lim_{x\to a^+} f(x) = 0$ even if $f(x) > 0$ for all x in (a, b).

76. Suppose f, g, and h are functions with the property that $f(x) \leq g(x) \leq h(x)$ for each $x \neq a$ in the open interval $(a, a + \delta)$, for some constant $\delta > 0$. If
$$\lim_{x\to a^+} f(x) = L \quad \text{and} \quad \lim_{x\to a^+} h(x) = L,$$
prove that $\lim_{x\to a^+} g(x) = L$. (This is the Squeeze Theorem for limits from the right.)

77. Suppose f, g, and h are functions with the property that $f(x) \leq g(x) \leq h(x)$ for each $x \neq a$ in the open interval $(a - \delta, a)$, for some constant $\delta > 0$. If
$$\lim_{x\to a^-} f(x) = L \quad \text{and} \quad \lim_{x\to a^-} h(x) = L,$$
prove that $\lim_{x\to a^-} g(x) = L$. (This is the Squeeze Theorem for limits from the left.)

1.9

Limits at Infinity and Horizontal Asymptotes

A metal bar with an initial temperature of 200° F is placed in a freezing unit kept at the constant temperature of 0° F. At the end of each subsequent hour the temperature of the bar is reduced to half its previous value (see Figure 1.64). As time passes the temperature of the metal decreases, but does it ever reach 0° F?

Regardless of your answer to this question, it is clear that the temperature of the bar approaches 0° F as t increases. In other words, the *limiting* temperature of the bar is zero as time increases without bound. The notation

$$\lim_{x\to\infty} f(x) = L$$

is used to express that the values of f approach the real number L as x increases without bound. Although the symbol ∞ does not represent a real number, $\lim_{x\to\infty} f(x) = L$ is often read "$f(x)$ approaches L as x approaches ∞" or "the limit of f at ∞ is L."

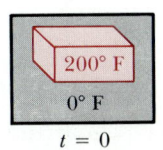

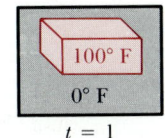

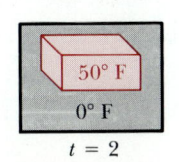

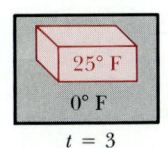

Figure 1.64

In a similar manner, $\lim_{x \to -\infty} f(x) = L$ is used to express that the values of f approach L as x decreases without bound, and is read "the limit of f at $-\infty$ is L."

In the metal bar problem, the temperature of the bar (initially 200° F) was 100° F at the end of 1 hr, 50° F at the end of 2 hr, and so on. Thus

$$T(t) = \frac{200}{2^t},$$

describes the bar's temperature, in degrees Fahrenheit, t hours after being placed in the freezing unit. The limiting temperature of the bar is given by

$$\lim_{t \to \infty} T(t).$$

As t increases, 2^t increases without bound, so $1/2^t$ approaches 0. This suggests that

$$\lim_{t \to \infty} T(t) = \lim_{t \to \infty} \frac{200}{2^t} = 0.$$

Note, however, that no matter how large t becomes, $T(t) = 200/2^t > 0$; so $T(t)$ approaches 0 as $t \to \infty$, but is not equal to 0 for any value of t.

To describe precisely the limit at infinity, we need to express "increases without bound" in an analytical manner. This is done in the following definition.

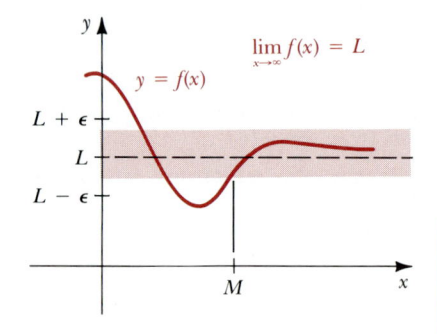

Figure 1.65
$\lim_{x \to \infty} f(x) = L.$

(1.51)
DEFINITION

Suppose f is a function defined on the interval (a, ∞) for some real number a. The **limit of $f(x)$ as x approaches infinity is the number L,** written $\lim_{x \to \infty} f(x) = L$, provided that for every number $\varepsilon > 0$, a number M exists with the property that

$$|f(x) - L| < \varepsilon \quad \text{whenever} \quad x > M.$$

A function with the limit L as x approaches ∞ is shown in Figure 1.65. Note that the graph of f stays between the horizontal lines $y = L - \varepsilon$ and $y = L + \varepsilon$ whenever $x > M$.

The limit of a function as x approaches $-\infty$ is defined in a similar manner.

(1.52)
DEFINITION

Suppose f is a function defined on the interval $(-\infty, a)$ for some real number a. The **limit of $f(x)$ as x approaches negative infinity is the number L,** written $\lim_{x \to -\infty} f(x) = L$, provided that for every number $\varepsilon > 0$, a number M exists with the property that

$$|f(x) - L| < \varepsilon \quad \text{whenever} \quad x < M.$$

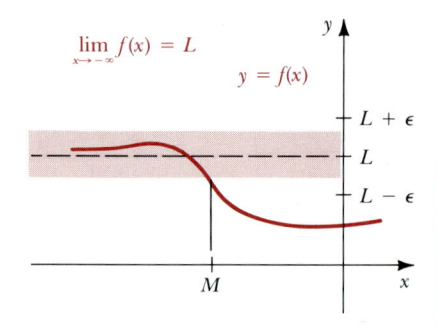

Figure 1.66
$\lim_{x \to -\infty} f(x) = L.$

A function with the limit L as x approaches $-\infty$ is shown in Figure 1.66.

In Example 3 of Section 1.4 we saw that the graph of $h(x) = 1/x$ appears to approach 0 as x increases in magnitude (see Figure 1.67). This implies that $\lim_{x \to \infty} 1/x = 0$, which we can now show precisely:

For any number $\varepsilon > 0$, let $M = 1/\varepsilon$. When $x > M$,
$$|1/x - 0| = |1/x| < 1/M = \varepsilon.$$

This proves that $\lim_{x \to \infty} 1/x = 0$.

In a similar manner it can be shown that $\lim_{x \to -\infty} 1/x = 0$. In fact, for any positive rational number r,

(1.53) $\quad \lim_{x \to \infty} 1/x^r = 0 \quad$ and when x^r is defined, $\quad \lim_{x \to -\infty} 1/x^r = 0.$

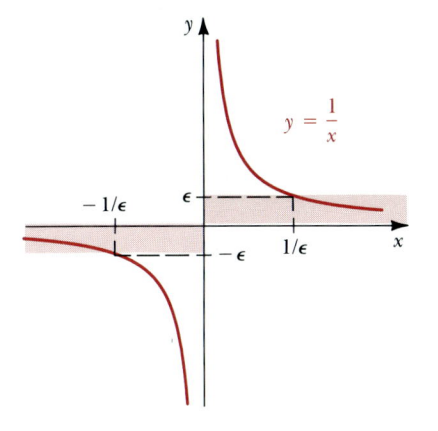

Figure 1.67
$\lim_{x \to \infty} 1/x = 0$ and $\lim_{x \to -\infty} 1/x = 0$.

These results are useful for determining limits of other functions.

Limit theorems analogous to those in Section 1.7 concerning the limits of sums, products, quotients, and roots of functions are also true for limits at ∞ and $-\infty$. In particular, all the arithmetic limit results given in Theorem 1.37 as well as the Squeeze Theorem remain valid if we assume that a is replaced by the symbol ∞ or $-\infty$.

Example 1

Determine
$$\lim_{x \to \infty} \frac{3x^2 + 7x + 2}{x^2 - 2x - 8}.$$

Solution

Since we are interested only in large values of x, we can assume that $x \neq 0$ and divide the numerator and denominator by x^2, the highest power in the denominator. Most of the terms will then be in a form to which (1.53) can be applied. Thus

$$\lim_{x \to \infty} \frac{3x^2 + 7x + 2}{x^2 - 2x - 8} = \lim_{x \to \infty} \frac{3 + \dfrac{7}{x} + \dfrac{2}{x^2}}{1 - \dfrac{2}{x} - \dfrac{8}{x^2}}$$

$$= \frac{\lim\limits_{x \to \infty} 3 + 7 \lim\limits_{x \to \infty} \dfrac{1}{x} + 2 \lim\limits_{x \to \infty} \dfrac{1}{x^2}}{\lim\limits_{x \to \infty} 1 - 2 \lim\limits_{x \to \infty} \dfrac{1}{x} - 8 \lim\limits_{x \to \infty} \dfrac{1}{x^2}} = \frac{3 + 0 + 0}{1 - 0 - 0} = 3. \quad \blacktriangleright$$

Example 2

Determine
$$\lim_{x \to \infty} \frac{\sqrt{x^2 + 3}}{\sqrt[3]{x^4 - 4x^3 + 5}}.$$

Solution

In Example 1 we solved the problem by dividing the polynomials in the numerator and denominator by the highest power of x in the denominator. We can apply a similar technique here even though the numerator and denominator are not polynomials.

The denominator involves a cube root, $\sqrt[3]{x^4 - 4x^3 + 5}$, and the highest power under the radical is x^4. Thus the highest power in this denominator is

$$\sqrt[3]{x^4} = x^{4/3}.$$

Dividing both the numerator and denominator by $x^{4/3}$ gives

$$\lim_{x \to \infty} \frac{\sqrt{x^2 + 3}}{\sqrt[3]{x^4 - 4x^3 + 5}} = \lim_{x \to \infty} \frac{\sqrt{\dfrac{x^2+3}{x^{8/3}}}}{\sqrt[3]{\dfrac{x^4 - 4x^3 + 5}{x^4}}} = \lim_{x \to \infty} \frac{\sqrt{\dfrac{1}{x^{2/3}} + \dfrac{3}{x^{8/3}}}}{\sqrt[3]{1 - \dfrac{4}{x} + \dfrac{5}{x^4}}},$$

so

$$\lim_{x \to \infty} \frac{\sqrt{x^2 + 3}}{\sqrt[3]{x^4 - 4x^3 + 5}} = \frac{\sqrt{\lim\limits_{x \to \infty} \dfrac{1}{x^{2/3}} + \lim\limits_{x \to \infty} \dfrac{3}{x^{8/3}}}}{\sqrt[3]{1 - \lim\limits_{x \to \infty} \dfrac{4}{x} + \lim\limits_{x \to \infty} \dfrac{5}{x^4}}} = \frac{\sqrt{0+0}}{\sqrt[3]{1 - 0 + 0}} = 0. \; \blacktriangleright$$

Example 3

Determine

$$\lim_{x \to -\infty} \frac{x+7}{\sqrt{x^2 - 3}}.$$

Solution

The denominator is again a radical, $\sqrt{x^2 - 3}$, and the highest power under the radical is x^2. In this case we divide the numerator and denominator by $\sqrt{x^2}$. Care is required in doing this since we are considering the limit at $-\infty$. As x approaches $-\infty$, x is negative and

$$\sqrt{x^2} = |x| = -x.$$

With this in mind, we divide the numerator and denominator by $\sqrt{x^2}$ to obtain

$$\lim_{x \to -\infty} \frac{x+7}{\sqrt{x^2 - 3}} = \lim_{x \to -\infty} \frac{\dfrac{x+7}{\sqrt{x^2}}}{\dfrac{\sqrt{x^2 - 3}}{\sqrt{x^2}}} = \lim_{x \to -\infty} \frac{\dfrac{x+7}{-x}}{\sqrt{\dfrac{x^2 - 3}{x^2}}}$$

Thus

$$\lim_{x \to -\infty} \frac{x+7}{\sqrt{x^2 - 3}} = \lim_{x \to -\infty} \frac{-1 - \dfrac{7}{x}}{\sqrt{1 - \dfrac{3}{x^2}}} = -1. \; \blacktriangleright$$

Example 4

The monthly sales s (in dollars) of a new product depend on the time t (in months) after the product has been introduced according to the rule:

$$s(t) = \frac{2000(1+t)}{2+t}.$$

a) What is $\lim_{t \to \infty} s(t)$?
b) What information concerning the sales of the new product does this limiting value provide?

Solution
a) Dividing both the numerator and denominator by t and proceeding as in Example 1, we have

$$\lim_{t \to \infty} s(t) = \lim_{t \to \infty} \frac{2000(1+t)}{2+t} = \lim_{t \to \infty} \frac{\frac{2000}{t} + 2000}{\frac{2}{t} + 1} = 2000.$$

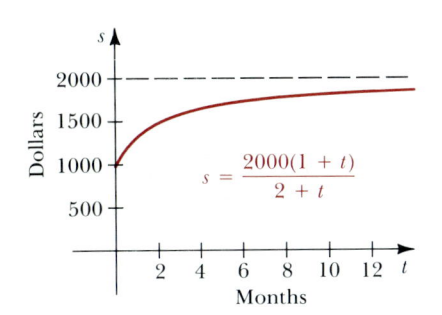

Figure 1.68
$\lim_{t \to \infty} s(t) = 2000.$

The graph of the function s for $t > 0$ is shown in Figure 1.68.

b) The limiting value of $2000 in sales per month is approximately the monthly sales expected once the product has been accepted by the public. If this model is accurate, the monthly sales for the product should approach but not exceed $2000. ▸

In the preceding example the horizontal line, $s = 2000$, illustrates the limiting value of sales, and the distance between this line and the curve represents the amount that the sales differ from this value at any particular time. This line is called a *horizontal asymptote* of the graph.

(1.54) DEFINITION

A line $y = b$ is a **horizontal asymptote** of the graph of the function f if either $\lim_{x \to \infty} f(x) = b$ or $\lim_{x \to -\infty} f(x) = b$.

Example 5

Find the horizontal asymptotes of the graph of

$$f(x) = \frac{x^2}{x^2 + 1}.$$

Solution
To find the limit at ∞, divide the numerator and denominator by the highest power in the denominator, x^2. Then

$$\lim_{x \to \infty} f(x) = \lim_{x \to \infty} \frac{x^2}{x^2 + 1} = \lim_{x \to \infty} \frac{1}{1 + \frac{1}{x^2}} = 1.$$

This implies that the line $y = 1$ is a horizontal asymptote. We can compute the limit at $-\infty$ in a similar manner, or we can use the fact that f is an

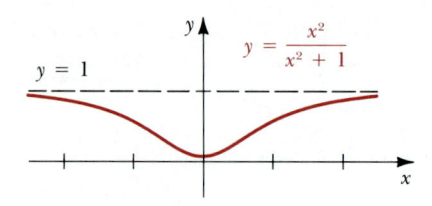

Figure 1.69

$$\lim_{x \to \infty} \frac{x^2}{x^2 + 1} = 1 \text{ and } \lim_{x \to -\infty} \frac{x^2}{x^2 + 1} = 1.$$

even function since $f(-x) = f(x)$. Thus the graph is symmetric with respect to the y-axis and

$$\lim_{x \to -\infty} f(x) = \lim_{x \to \infty} f(-x) = \lim_{x \to \infty} f(x) = 1.$$

By either method we find that $y = 1$ is the only horizontal asymptote.

The graph showing the horizontal asymptote is sketched in Figure 1.69. In Chapter 2 we will see how the exact shape of the graph is determined. ▶

Example 6

Find the horizontal asymptotes of the graph of

$$f(x) = \frac{x}{\sqrt{x^2 + 1}}.$$

Solution

Let us first consider $\lim_{x \to \infty} f(x)$. Suppose we divide the numerator and the denominator by $\sqrt{x^2}$. Since $\sqrt{x^2} = x$ when x is positive,

$$\lim_{x \to \infty} f(x) = \lim_{x \to \infty} \frac{x/\sqrt{x^2}}{\sqrt{x^2 + 1}/\sqrt{x^2}} = \lim_{x \to \infty} \frac{x/x}{\sqrt{(x^2 + 1)/x^2}} = \lim_{x \to \infty} \frac{1}{\sqrt{1 + 1/x^2}} = 1.$$

In this case f is an odd function since $f(-x) = -f(x)$, and the graph is symmetric with respect to the origin. Therefore,

$$\lim_{x \to -\infty} f(x) = \lim_{x \to \infty} f(-x) = \lim_{x \to \infty} -f(x) = -1.$$

The graph showing horizontal asymptotes at $y = 1$ and $y = -1$ is given in Figure 1.70. The exact shape can be determined by methods considered in Chapter 2. ▶

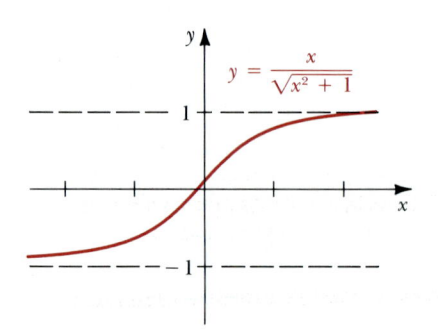

Figure 1.70

$$\lim_{x \to \infty} \frac{x}{\sqrt{x^2 + 1}} = 1 \text{ and}$$
$$\lim_{x \to -\infty} \frac{x}{\sqrt{x^2 + 1}} = -1.$$

▶ EXERCISE SET 1.9

In Exercises 1–38, find the limit, if it exists.

1. $\lim_{x \to -\infty} \dfrac{2x - 3}{3x + 5}$

2. $\lim_{x \to -\infty} \dfrac{5x - 7}{x + 300}$

3. $\lim_{x \to \infty} \dfrac{1}{x - 2}$

4. $\lim_{x \to -\infty} \dfrac{1}{x - 2}$

5. $\lim_{x \to -\infty} \dfrac{x^2}{x^3 + 1}$

6. $\lim_{x \to \infty} \dfrac{x + 2}{x^2}$

7. $\lim_{x \to \infty} \dfrac{x^2 + 3}{x^2 + 1}$

8. $\lim_{x \to -\infty} \dfrac{x^3 - 2}{x^3 + 4}$

9. $\lim_{x \to \infty} \dfrac{3x^2 + 2x + 1}{4x^2 - 3}$

10. $\lim_{x \to \infty} \dfrac{142x^3 + 1}{0.1x^4}$

11. $\lim_{x \to -\infty} \dfrac{50 - x^3}{-x^3}$

12. $\lim_{x \to \infty} \dfrac{50 - x^3}{-x^3}$

13. $\lim_{x \to \infty} \dfrac{\sin x}{x}$

$\left[\text{Hint: } \left|\dfrac{\sin x}{x}\right| \leq \dfrac{1}{|x|}.\right]$

14. $\lim_{x \to \infty} \dfrac{1 + \sin x}{x}$

15. $\lim_{x \to \infty} \dfrac{\cos x}{x + 2}$

16. $\lim_{x \to \infty} \dfrac{x^2 \cos x}{x^3 + 2}$

17. $\lim_{x \to -\infty} \dfrac{x^2 + \sin x \cos x}{x^2 + 2x - 1}$

18. $\lim_{x \to \infty} \dfrac{1 + \sqrt[4]{x}}{1 - \sqrt[4]{x}}$

19. $\lim_{x \to -\infty} \dfrac{1 + \sqrt[3]{x}}{1 - \sqrt[3]{x}}$

20. $\lim_{x \to \infty} \dfrac{4 + \sqrt[3]{8x}}{4 - \sqrt[3]{x}}$

21. $\lim_{x \to \infty} \dfrac{\sqrt{x^2 - 4}}{5x + 2}$

22. $\lim_{x \to -\infty} \dfrac{\sqrt{x^2 - 4}}{2x + 5}$

23. $\lim_{x \to \infty} \dfrac{\sqrt{4x^2 + 3x - 9}}{x - 2}$

24. $\lim_{x \to \infty} \dfrac{3x^2 - 4}{\sqrt{x^5 - 4x^2}}$

25. $\lim_{x \to -\infty} \dfrac{3x^2 - 4}{\sqrt{x^5 - 4x^2}}$

26. $\lim_{x \to -\infty} \dfrac{|1 - x|}{\sqrt{x^2 - 3x + 4}}$

27. $\lim_{x \to \infty} \dfrac{(2x + 5)(x^2 - 2x + 4)}{(9x^2 + 3)(2x - 7)}$

28. $\lim_{x \to -\infty} \dfrac{(x^3 + 3x^2 + 2)(5x - 6)}{(x^2 + 2x + 1)(4 - x^2)}$

29. $\lim_{x \to -\infty} \dfrac{(4 - x^2)\sqrt{x^2 + 2}}{3x^3 - 9x + \pi}$

30. $\lim_{x \to \infty} \dfrac{(3x^2 - 5)\sqrt{x^2 + x + 1}}{(4x - 3)\sqrt{9x^4 + 3x^2 + 1}}$

31. $\lim_{x \to \infty} (\sqrt{x^2 + 1} - x)$

   $\left[\text{Hint: Multiply by } \dfrac{\sqrt{x^2 + 1} + x}{\sqrt{x^2 + 1} + x}.\right]$

32. $\lim_{x \to \infty} (\sqrt{4x^2 + 1} - 2x)$

33. $\lim_{x \to \infty} (\sqrt{x^2 + x} - x)$

34. $\lim_{x \to \infty} (\sqrt{x^2 + 2x + 2} - x)$

35. $\lim_{x \to \infty} \dfrac{\sqrt{x^2 + \sqrt{x^2 + 1}}}{x}$

36. $\lim_{x \to -\infty} \dfrac{\sqrt{x^2 + \sqrt{x^2 + 1}}}{\sqrt{x^2 + 1}}$

37. $\lim_{x \to \infty} \dfrac{x}{[\![x]\!]}$

38. $\lim_{x \to -\infty} \dfrac{[\![x]\!]}{x}$

In Exercises 39–42, find the horizontal asymptotes of the graph of the function defined and sketch the graph.

39. $f(x) = \dfrac{1}{x^2 + 1}$

40. $f(x) = \dfrac{x}{x^2 + 1}$

41. $f(x) = \dfrac{x}{\sqrt{x^2 + 1}}$

42. $f(x) = \dfrac{\sin x}{x^2 + 1}$

43. We can see that $\lim_{x \to \infty} 1/x^3 = 0$. How large must x be to ensure that $1/x^3$ is less than each of the following?
    a) .001
    b) 10^{-6}

44. We can see that $\lim_{x \to \infty} 1/(x + 1) = 0$. How large must x be to ensure that $1/(x + 1)$ is less than each of the following?
    a) .01
    b) .001

45. Use Definition 1.51 to prove that $\lim_{x \to \infty} 1/x^2 = 0$.

46. If r is a positive rational number and c is any real number, prove that $\lim_{x \to \infty} c/x^r = 0$.

47. Suppose f is an even function. Show that $\lim_{x \to \infty} f(x) = \lim_{x \to -\infty} f(x)$, provided these limits exist.

48. Suppose f is an odd function. Show that $\lim_{x \to \infty} f(x) = -\lim_{x \to -\infty} f(x)$, provided these limits exist.

49. The Dull Calculator Company estimates that the cost of producing x calculators is

    $$C(x) = 2000 + 10x + 100/x \text{ dollars.}$$

    Find the limit of the average cost per calculator as x approaches ∞. What does this limit indicate?

50. The intensity of illumination at a point is proportional to the product of the strength of the light source and the inverse of the square of the distance from the source. If a light source has strength s, what happens to the illumination as the distance from the source increases?

51. The relationship between the gross photosynthetic rate P of a leaf and the light intensity I is given by

    $$P(I) = \left(a + \dfrac{b}{I}\right)^{-1},$$

    where a and b are positive constants that depend on the species and levels of other external factors. Find the maximum attainable photosynthetic rate. Sketch the graph of P when $a = b = 1$.

52. A test is known to give reliable results with probability

    $$R(t) = \dfrac{tr}{1 + (t - 1)r},$$

    where $t > 0$ is the number of hours the test is run and r is a constant satisfying $0 < r < 1$.
    a) Find $\lim_{t \to \infty} R(t)$ and interpret the result.
    b) Find $\lim_{t \to 0^+} R(t)$ and interpret the result.

1.10

Infinite Limits and Vertical Asymptotes

We now consider the case of a function f whose values increase without bound as x approaches some number a. The limit of f at a does not exist in this situation (it is not a real number). Rather, it fails to exist in a

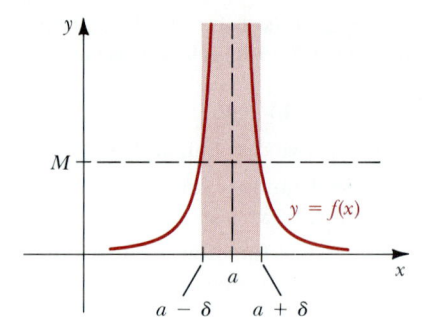

Figure 1.71
$\lim_{x \to a} f(x) = \infty$.

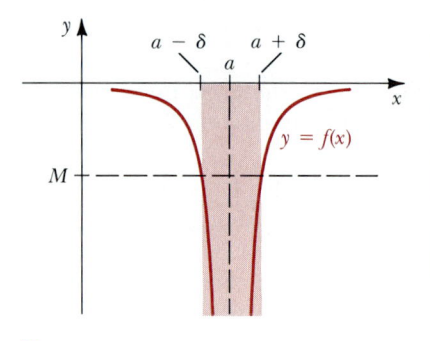

Figure 1.72
$\lim_{x \to a} f(x) = -\infty$.

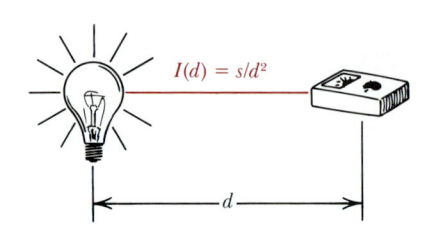

Figure 1.73

special way. Even so, it is common to indicate that the values of f increase without bound as x approaches a by writing

$$\lim_{x \to a} f(x) = \infty$$

and stating that f has an *infinite limit* at a.

To define an infinite limit, we need a precise way to express that as x approaches a, the values of f become and remain greater than any preassigned bound. Such a description is given in the following definition and illustrated in Figure 1.71.

(1.55)
DEFINITION
Let f be a function defined on an open interval containing a, except possibly at a itself. The **limit of $f(x)$ as x approaches a is infinite,** written $\lim_{x \to a} f(x) = \infty$, provided that for every number M, a number $\delta > 0$ exists with the property that

$$f(x) > M \quad \text{whenever} \quad 0 < |x - a| < \delta.$$

Negatively infinite limits are defined in a similar manner and illustrated in Figure 1.72.

(1.56)
DEFINITION
Let f be a function that is defined on an open interval containing a, except possibly at a itself. The **limit of $f(x)$ as x approaches a is negatively infinite,** written $\lim_{x \to a} f(x) = -\infty$, provided that for every M, a number $\delta > 0$ exists with the property that

$$f(x) < M \quad \text{whenever} \quad 0 < |x - a| < \delta.$$

One-sided infinite limits can also be defined and are often the form in which infinite limits are used. For example, we write $\lim_{x \to a^+} f(x) = \infty$, provided that for every number M, a number $\delta > 0$ exists with the property that

$$f(x) > M \quad \text{whenever} \quad 0 < x - a < \delta,$$

that is, whenever $a < x < a + \delta$. Other definitions are considered in the exercises.

As an application of a one-sided infinite limit, consider the question of determining the intensity of the illumination of a light near the source of the light. If a particular light source has strength s, then the illumination $I(d)$ of the light as a function of its distance $d > 0$ from the source is given by

$$I(d) = \frac{s}{d^2}$$

(see Figure 1.73). As the distance from the source decreases, the illumination increases without bound (when $d = 0.01$, $I = 10{,}000s$; when $d = 0.0001$, $I = 100{,}000{,}000s$; and so on), so

$$\lim_{d \to 0^+} I(d) = \infty.$$

1.10 INFINITE LIMITS AND VERTICAL ASYMPTOTES

It is important to realize that in writing $\lim_{d \to 0^+} I(d) = \infty$ we are not stating that the illumination is "infinite." The notation means only that the formula $I(d) = s/d^2$ implies that there is no bound to the amount of illumination from the light source as the distance from the light source becomes small.

Example 1

Find:

a) $\lim_{x \to 1^+} \dfrac{1}{x - 1}$;

b) $\lim_{x \to 1^-} \dfrac{1}{x - 1}$.

c) Sketch the graph of $f(x) = \dfrac{1}{x - 1}$.

Solution

a) As x approaches 1 and $x > 1$, $x - 1$ approaches 0 and $x - 1 > 0$. Thus, as x approaches 1 from the right, $1/(x - 1)$ increases without bound and

$$\lim_{x \to 1^+} \frac{1}{x - 1} = \infty.$$

b) As x approaches 1 and $x < 1$, $x - 1$ approaches 0 and $x - 1 < 0$, so $1/(x - 1)$ is negative and increasing in magnitude. That is,

$$\lim_{x \to 1^-} \frac{1}{x - 1} = -\infty.$$

c) The function f is not defined at 1, the y-intercept of the graph is -1, and the graph does not intersect the x-axis. In addition,

$$\lim_{x \to \infty} \frac{1}{x - 1} = 0 \quad \text{and} \quad \lim_{x \to -\infty} \frac{1}{x - 1} = 0.$$

Plotting the points $(2, 1)$ and $(0, -1)$ and using the preceding analysis gives the graph shown in Figure 1.74. Note that the graph of this function is simply the graph of $y = 1/x$ moved one unit to the right. ▶

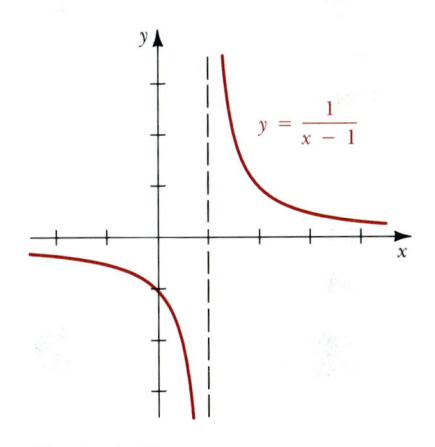

Figure 1.74

$\lim_{x \to 1^+} \dfrac{1}{x - 1} = \infty$ and $\lim_{x \to 1^-} \dfrac{1}{x - 1} = -\infty$.

A number of theorems can be established concerning the arithmetic of infinite limits. A typical example is the following.

(1.57) THEOREM

If a is any real number and f and g are functions with $\lim_{x \to a} f(x) = \infty$ and $\lim_{x \to a} g(x) = L$ where L is a real number, then:

i) $\lim_{x \to a} (f(x) \pm g(x)) = \infty$;

ii) $\lim_{x \to a} (f(x) \cdot g(x)) = \begin{cases} \infty, & \text{if } L > 0, \\ -\infty, & \text{if } L < 0; \end{cases}$

iii) $\lim_{x \to a} \dfrac{f(x)}{g(x)} = \begin{cases} \infty, & \text{if } L > 0, \\ -\infty, & \text{if } L < 0; \end{cases}$

iv) $\lim_{x \to a} \dfrac{g(x)}{f(x)} = 0.$

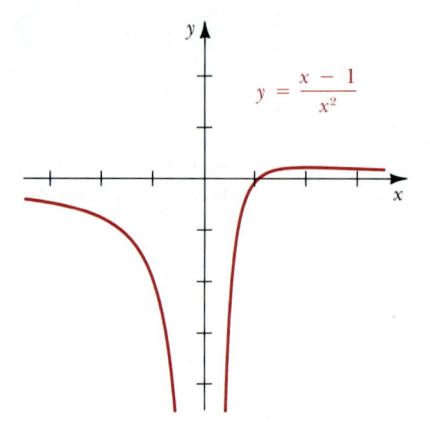

Figure 1.75

This theorem also holds if in each case a is replaced by a^+, a^-, ∞, or $-\infty$. A similar result holds when $\lim_{x \to a} f(x) = -\infty$.

Example 2

Find $\lim_{x \to 0} h(x)$ for the function h defined by $h(x) = (x-1)/x^2$.

Solution

Observe that $h(x) = f(x) \cdot g(x)$ where $f(x) = 1/x^2$ and $g(x) = x - 1$. Since $\lim_{x \to 0} f(x) = \infty$ and $\lim_{x \to 0} g(x) = -1$, Theorem 1.57(ii) implies that $\lim_{x \to 0} h(x) = -\infty$.

Another way to see that this limit is $-\infty$ is: The numerator $x - 1$ is approximately -1 when x is close to 0, but the denominator x^2 is a positive number close to 0. Consequently, the quotient $(x-1)/x^2$ becomes negatively infinite as x approaches 0. ▶

The graph of $h(x) = (x-1)/x^2$ approaches the line $x = 0$ as x approaches 0, as shown in Figure 1.75. The line $x = 0$ is called a *vertical asymptote* of the graph of h.

(1.58) DEFINITION

The line $x = a$ is a **vertical asymptote** of the graph of the function f if one or more of the following are satisfied:

$$\lim_{x \to a^+} f(x) = \infty, \quad \lim_{x \to a^+} f(x) = -\infty, \quad \lim_{x \to a^-} f(x) = \infty, \quad \text{or} \quad \lim_{x \to a^-} f(x) = -\infty.$$

Example 3

Find any horizontal and vertical asymptotes and sketch the graph of the function defined by

$$f(x) = \frac{x^2 + 3x + 2}{x^2 - 2x - 8}.$$

Solution

To find the horizontal asymptotes first determine $\lim_{x \to \infty} f(x)$. Divide the numerator and the denominator by the highest power in the denominator, x^2:

$$\lim_{x \to \infty} f(x) = \lim_{x \to \infty} \frac{x^2 + 3x + 2}{x^2 - 2x - 8} = \lim_{x \to \infty} \frac{1 + \dfrac{3}{x} + \dfrac{2}{x^2}}{1 - \dfrac{2}{x} - \dfrac{8}{x^2}} = 1.$$

Using the same procedure shows that $\lim_{x \to -\infty} f(x) = 1$, so the only horizontal asymptote is $y = 1$. This line is shown in Figure 1.76.
Since

$$f(x) = \frac{x^2 + 3x + 2}{x^2 - 2x - 8} = \frac{(x+2)(x+1)}{(x+2)(x-4)},$$

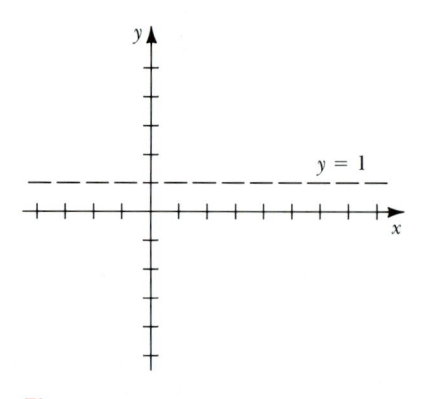

Figure 1.76

1.10 INFINITE LIMITS AND VERTICAL ASYMPTOTES

f is undefined at $x = -2$ and $x = 4$. The lines $x = -2$ and $x = 4$ are thus the only candidates for vertical asymptotes. To determine if they are indeed vertical asymptotes, we consider the one-sided limits at $x = -2$ and $x = 4$:

$$\lim_{x \to -2^+} f(x) = \lim_{x \to -2^+} \frac{(x+2)(x+1)}{(x+2)(x-4)}$$

$$= \lim_{x \to -2^+} \frac{x+1}{x-4} = \frac{-1}{-6} = \frac{1}{6}.$$

Similarly, $\lim_{x \to -2^-} f(x) = \frac{1}{6}$, so a vertical asymptote does *not* occur at $x = -2$. In fact, there is a "hole" in the graph at $(-2, \frac{1}{6})$, as shown in Figure 1.77. However,

$$\lim_{x \to 4^+} f(x) = \lim_{x \to 4^+} \frac{(x+2)(x+1)}{(x+2)(x-4)} = \lim_{x \to 4^+} \frac{x+1}{x-4}$$

and since

$$\lim_{x \to 4^+} (x+1) = 5 \quad \text{and} \quad \lim_{x \to 4^+} \frac{1}{x-4} = \infty,$$

it follows that

$$\lim_{x \to 4^+} \frac{x+1}{x-4} = \infty.$$

This implies that $x = 4$ is a vertical asymptote of the graph, illustrated by the dashed vertical line in Figure 1.77. Also,

$$\lim_{x \to 4^-} f(x) = \lim_{x \to 4^-} \frac{x+1}{x-4} = -\infty.$$

When $x \neq -2$, $f(x)$ is $(x+1)/(x-4)$, which describes a continuous function on the intervals $(-\infty, 4)$ and $(4, \infty)$ and has intercepts at $(0, -\frac{1}{4})$ and $(-1, 0)$. The graph is shown in Figure 1.78. To indicate that -2 is not in the domain of f, a small open circle has been drawn on the graph at $(-2, \frac{1}{6})$.

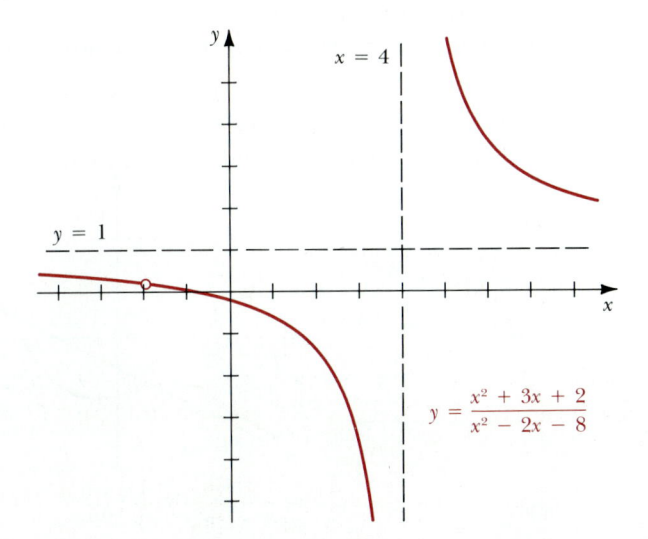

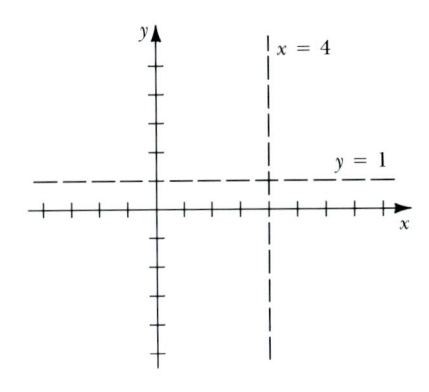

Figure 1.77

Figure 1.78

The graph of $y = \dfrac{x^2 + 3x + 2}{x^2 - 2x - 8}$.

We have again made assumptions about the graph that we cannot presently justify: The graph has no oscillatory behavior and moves smoothly on the intervals $(-\infty, -2)$, $(-2, 4)$, and $(4, \infty)$. The validity of these assumptions can be established after we study Section 2.10. ▶

It is frequently necessary to use infinite limits and limits at infinity in the same expression. The definitions required to describe these concepts are a combination of the infinite limit definitions of this section and those given in Section 1.9.

For example, if a function is defined on some unbounded interval (a, ∞), the values of f are said to become positively infinite as x increases without bound, written

$$\lim_{x \to \infty} f(x) = \infty,$$

provided that for any number M, a number $\hat{M}$ exists with the property that $f(x) > M$ whenever $x > \hat{M}$.

The definitions of $\lim_{x \to \infty} f(x) = -\infty$, $\lim_{x \to -\infty} f(x) = \infty$, and $\lim_{x \to -\infty} f(x) = -\infty$ are made in a similar manner.

Frequently functions that become infinite as x increases without bound do so in a manner that can be approximated by a known curve. For example, the graph of the function described by

$$f(x) = x + \frac{1}{x^2 + 1}$$

is shown in Figure 1.79. As x approaches either ∞ or $-\infty$, the term

$$\frac{1}{x^2 + 1}$$

approaches 0, so the graph of f approaches the line with equation $y = x$. This line is called an *oblique* or *slant asymptote* to the graph of

$$f(x) = x + \frac{1}{x^2 + 1}.$$

In general, a line with equation $y = mx + b$, where $m \neq 0$, is called a **slant asymptote** of the graph of $y = f(x)$ (see Figure 1.80) if either

$$\lim_{x \to \infty} [f(x) - (mx + b)] = 0 \quad \text{or} \quad \lim_{x \to -\infty} [f(x) - (mx + b)] = 0.$$

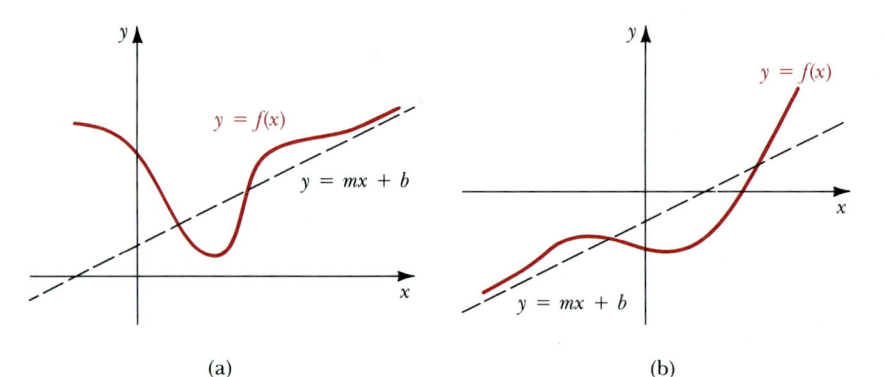

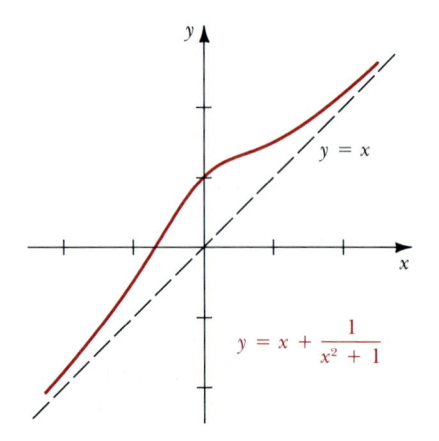

Figure 1.79
Slant asymptote: $y = x$.

Figure 1.80
Slant asymptote: $y = mx + b$.

(a)

(b)

▶ EXERCISE SET 1.10

In Exercises 1–34, find the limit or tell why it does not exist.

1. $\lim_{x \to 2^+} \dfrac{1}{x-2}$
2. $\lim_{x \to 2^-} \dfrac{1}{x-2}$
3. $\lim_{x \to 2} \dfrac{1}{x-2}$
4. $\lim_{x \to 2} \dfrac{1}{(x-2)^2}$
5. $\lim_{x \to 2^+} \dfrac{x+2}{x-2}$
6. $\lim_{x \to 2^-} \dfrac{x+2}{x-2}$
7. $\lim_{x \to 1^+} \dfrac{x+1}{x^2-1}$
8. $\lim_{x \to 1} \dfrac{x+1}{x^2-1}$
9. $\lim_{x \to 1} \dfrac{x^2-1}{x-1}$
10. $\lim_{x \to -1} \dfrac{x^2-1}{x+1}$
11. $\lim_{x \to -3} \dfrac{x^2+2x-8}{x^2+x-6}$
12. $\lim_{x \to 2} \dfrac{x^2+2x-8}{x^2+x-6}$
13. $\lim_{x \to 5^+} \dfrac{3x^2+1}{(x-2)(x-5)}$
14. $\lim_{x \to 5^-} \dfrac{3x^2+1}{(x-2)(x-5)}$
15. $\lim_{x \to 2^+} \dfrac{3x^2+1}{(x-2)(x-5)}$
16. $\lim_{x \to 2^-} \dfrac{3x^2+1}{(x-2)(x-5)}$
17. $\lim_{x \to \pi/2^-} \tan x$
18. $\lim_{x \to \pi/2^+} \tan x$
19. $\lim_{x \to 0^+} \csc x$
20. $\lim_{x \to 0^+} \sec x$
21. $\lim_{x \to 0^-} \dfrac{|x|}{x^2}$
22. $\lim_{x \to 2} \dfrac{x}{|x-2|}$
23. $\lim_{x \to 1^+} \sqrt{\dfrac{x}{x-1}}$
24. $\lim_{x \to 1^+} \dfrac{x}{x^2-1}$
25. $\lim_{x \to \infty} \dfrac{x^3+1}{x^2}$
26. $\lim_{x \to \infty} \dfrac{x-3x^2+2}{x}$
27. $\lim_{x \to -\infty} \dfrac{x^3+1}{x^2}$
28. $\lim_{x \to -\infty} \dfrac{(x-3)(x+1)}{x}$
29. $\lim_{x \to \infty} (x - \sin x)$
30. $\lim_{x \to -\infty} (x - 3 \sin x)$
31. $\lim_{x \to \infty} x \sin x$
32. $\lim_{x \to \infty} \sin x$
33. $\lim_{x \to \infty} f(x)$, where $f(x) = \begin{cases} x, & \text{if } x \text{ is not an integer}, \\ 1, & \text{if } x \text{ is an integer} \end{cases}$
34. $\lim_{x \to \infty} f(x)$, where $f(x) = \begin{cases} 5/x, & \text{if } x \text{ is not an integer}, \\ 0, & \text{if } x \text{ is an integer} \end{cases}$

In Exercises 35–56, determine the equations of any vertical and horizontal asymptotes to the graphs of the functions and equations described. Sketch each of these graphs.

35. $f(x) = \tan x$
36. $f(x) = \cot x$
37. $f(x) = \sec x$
38. $f(x) = \csc x$
39. $f(x) = \dfrac{x-1}{x+1}$
40. $f(x) = \dfrac{2x-3}{1-x}$
41. $f(x) = \dfrac{1}{x^2-4x+3}$
42. $f(x) = \dfrac{1}{x^2-4x+4}$
43. $f(x) = \dfrac{1}{x^2-4x+5}$
44. $f(x) = \dfrac{1}{\sqrt{x^2-4}}$
45. $f(x) = \dfrac{x}{\sqrt{x^2-4}}$
46. $f(x) = \dfrac{x}{\sqrt{4-x^2}}$
47. $f(x) = \dfrac{1}{\sqrt{1-x^2}}$
48. $f(x) = \dfrac{x^2-5x+6}{x^2-9}$
49. $f(x) = \dfrac{x^2-1}{x^2-2x+1}$
50. $f(x) = \dfrac{1+x^2}{x^2}$
51. $f(x) = \dfrac{4x^2}{9-x^2}$
52. $f(x) = \dfrac{x^2-4}{9-x^2}$
53. $x^2 y^2 - x^2 - y^2 = 0$
54. $xy^2 + 3x - 6 = 0$
55. $yx^2 + 4y - x = 0$
56. $xy - y^2 - 1 = 0$

Exercises 57–62 describe conditions that the graph of a function is to satisfy. In each case, sketch such a graph.

57. A vertical asymptote at $x = 0$ and a horizontal asymptote at $y = -1$
58. Vertical asymptotes at $x = 1$ and $x = -1$ and a horizontal asymptote at $y = 1$
59. Vertical asymptotes at $x = 1$ and $x = -1$ and no horizontal asymptotes
60. Horizontal asymptotes at $y = 1$ and $y = -1$ and a vertical asymptote at $x = 1$
61. Horizontal asymptotes at $y = 1$ and $y = -1$ and no vertical asymptotes
62. Horizontal asymptotes at $y = 1$ and $y = -1$ and vertical asymptotes at $x = 1$ and $x = -1$
63. We can see that $\lim_{x \to 0^+} 1/x = \infty$. How small must x be to ensure that $1/x$ is greater than each of the following?
    a) 100
    b) 10,000
64. We can see that $\lim_{x \to 0} 1/x^4 = \infty$. How small in absolute value must x be to ensure that $1/x^4$ is greater than each of the following?
    a) 100
    b) 10,000
65. a) Use the fact that $|\sin x| \leq 1$ to prove that $\lim_{x \to \infty} (x - \sin x) = \infty$.

b) How large must x be to ensure that $x - \sin x$ is greater than 100?

66. a) Prove that $\lim_{x \to \infty} (-x + 5 \cos x) = -\infty$.
    b) How large must x be to ensure that $-x + 5 \cos x$ is less than -1000?

67. Suppose f is a rational function of the form
$$f(x) = \frac{a_m x^m + a_{m-1} x^{m-1} + \cdots + a_1 x + a_0}{b_n x^n + b_{n-1} x^{n-1} + \cdots + b_1 x + b_0},$$
where $a_m \neq 0$ and $b_n \neq 0$. What can be said about $\lim_{x \to \infty} f(x)$ and $\lim_{x \to -\infty} f(x)$ if:
a) $m \geq n$? b) $m < n$? c) $m \geq n$?

68. Suppose P is a polynomial. What possibilities exist for $\lim_{x \to \infty} P(x)$ and $\lim_{x \to -\infty} P(x)$, and when do these possibilities occur?

In Exercises 69–76, construct functions f, g, and h with $\lim_{x \to \infty} f(x) = 0$, $\lim_{x \to \infty} g(x) = \infty$, and $\lim_{x \to \infty} h(x) = -\infty$ that satisfy the given condition.

69. $\lim_{x \to \infty} f(x) g(x) = 1$
70. $\lim_{x \to \infty} f(x) g(x) = \infty$
71. $\lim_{x \to \infty} f(x) g(x) = 0$
72. $\lim_{x \to \infty} (g(x) + h(x)) = 1$
73. $\lim_{x \to \infty} (g(x) + h(x)) = 0$
74. $\lim_{x \to \infty} (g(x) + h(x)) = -1$
75. $\lim_{x \to \infty} \frac{g(x)}{h(x)} = -\infty$
76. $\lim_{x \to \infty} \frac{g(x)}{h(x)} = 0$

77. Give a definition for $\lim_{x \to a^+} f(x) = -\infty$, and use this definition to show that $\lim_{x \to 0^+} -1/x = -\infty$.

78. Give a definition for $\lim_{x \to a^-} f(x) = \infty$, and use this definition to show that $\lim_{x \to 0^-} 1/|x| = \infty$.

79. Determine any slant asymptotes to the graph of each of the following functions.
a) $f(x) = x - \dfrac{1}{x}$
b) $f(x) = \dfrac{1}{x} - 2x$
c) $f(x) = \dfrac{2 - x^2 + x}{x}$
d) $f(x) = \dfrac{2x^2 + 3x + 1}{x}$

80. Show that a slant asymptote to the graph of any function of the form
$$f(x) = \frac{ax^2 + bx + c}{mx}, \quad a \neq 0,$$
is
$$y = \frac{a}{m} x + \frac{b}{m}.$$

81. Show that the graph of
$$f(x) = \frac{x^2 + |x| + 1}{x}$$
has a different slant asymptote at ∞ than it does at $-\infty$.

82. In Example 2 we found that $\lim_{x \to 0} (x - 1)/x^2 = -\infty$ by showing that $\lim_{x \to 0} (x - 1) = -1$ while $\lim_{x \to 0} 1/x^2 = \infty$. Suppose instead that we had rewritten $(x - 1)/x^2 = (1/x) - (1/x^2)$ and considered $\lim_{x \to 0} 1/x$ and $\lim_{x \to 0} 1/x^2$. Why can we not use these results to show that $\lim_{x \to 0} (x - 1)/x^2 = -\infty$?

83. Living tissue is excited by an electric current only if the current reaches or exceeds a certain threshold. A function describing the dependence of the threshold on the duration t of current flow is given by Weiss' Law:
$$f(t) = \frac{a}{t} + b,$$
where a and b are positive constants. Describe the behavior of the threshold when t approaches 0 and when t approaches ∞.

84. Van der Waals equation of state for one mole of a gas is
$$\left(P + \frac{a}{V^2}\right)(V - b) = RT,$$
where P is the pressure, V is the volume, and T is the temperature of the gas, and R, a, and b are positive constants. What does this law predict will happen to the temperature of the gas as V approaches 0 and as V approaches ∞, assuming that the pressure remains constant?

85. The rate at which an enzyme-catalyzed reaction proceeds depends on the concentration of the enzyme and substrate. This rate is described by the equation
$$v(x) = \frac{Vx}{x + K},$$
where $v(x)$ is the velocity of the reaction, x is the substrate concentration, V is the maximum value of the velocity, and K is a constant, called the *Michaelis constant*, that depends on the reaction. This equation is known variously as the *Michaelis–Menton* and the *Briggs–Haldane* equation. Sketch the graph of v and label any asymptotes. Find the value of x corresponding to $v(x) = V/2$.

86. The special theory of relativity states that the mass of an object relative to a system depends on its velocity relative to that system. If we let m_0 denote the mass of an object when it is at rest and $m(v)$ denote the mass of the object when it has velocity v, then
$$m(v) = m_0 \left(1 - \frac{v^2}{c^2}\right)^{-1/2},$$
where c is the speed of light, a constant. Sketch the graph of the function m. Describe what must occur physically as an object approaches the speed of light by considering $\lim_{v \to c^-} m(v)$.

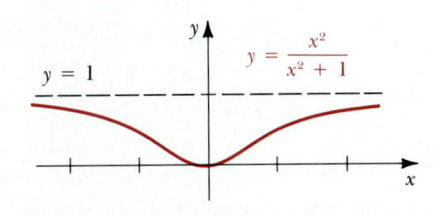

▶ IMPORTANT TERMS AND RESULTS

CONCEPT	PAGE	CONCEPT	PAGE
Function	2	Continuous	53
Symmetry	8	Rational function	57
Linear function	14	Squeeze Theorem	59
Quadratic function	18	One-sided limit	62
Polynomial	22	Intermediate Value Theorem	64
Trigonometric functions	25	Bisection Method	65
Absolute value function	27	Limit at infinity	70
Square root function	28	Horizontal asymptote	73
Greatest integer function	28	Infinite limit	76
Composition	36	Vertical asymptote	78
Limit of f at a	41	Slant asymptote	80

▶ REVIEW EXERCISES

For the functions in Exercises 1–4, sketch the graph of f and find each of the following.

a) $f(\sqrt{2})$
b) $f(\sqrt{2} + 1)$
c) $f(\sqrt{2}) + f(1)$
d) $f(\sqrt{x})$
e) $\sqrt{f(x)}$
f) $\dfrac{f(x + h) - f(x)}{h}$

1. $f(x) = 2x - 3$
2. $f(x) = x^2 + 1$
3. $f(x) = x^2 - 7x + 10$
4. $f(x) = x^2 + 2x + 4$

In Exercises 5–14, find an equation of the line. Sketch the graph of each line.

5. The line through the point $(0, 2)$ with slope -1.
6. The line through the point $(1, -2)$ and parallel to the line with equation $y - x - 1 = 0$.
7. The line through the point $(1, -2)$ and perpendicular to the line with equation $y - x - 1 = 0$.
8. The line through the points $(-2, 0)$ and $(1, 3)$.
9. A horizontal line through the point $(1, -3)$.
10. A vertical line through the point $(1, -3)$.
11. A line with x-intercept 2 and slope -3.
12. A line with y-intercept -3 and slope -2.
13. A line with x-intercept 1 and y-intercept 2.
14. A line perpendicular to the line joining $(-1, 2)$ and $(2, 0)$ and passing through $(\tfrac{3}{2}, 1)$.

In Exercises 15–22, sketch the graph of the function.

15. $f(x) = x^2 - 2x - 1$
16. $f(x) = x^2 - 2x + 4$
17. $f(x) = x^2 - 2x - 4$
18. $f(x) = x^2 - 2x$
19. $f(x) = -x^2 + 6x$
20. $f(x) = 2x^2 + 4x + 3$
21. $f(x) = |x^2 - 2x|$
22. $f(x) = \dfrac{1}{x^2 - 2x}$

The graph of $f(x) = x^2/(x^2 + 1)$ is shown in the figure. Use this graph to sketch the graph of each of the functions described in Exercises 23–28.

23. $f(x) = \dfrac{x^2}{x^2 + 1} + 1$
24. $h(x) = \dfrac{x^2}{x^2 + 1} - 1$
25. $k(x) = \dfrac{-x^2}{x^2 + 1}$
26. $l(x) = \dfrac{3x^2}{x^2 + 1}$
27. $u(x) = \dfrac{x^2 + 1}{x^2}$
28. $v(x) = -\dfrac{x^2 + 1}{x^2}$

In Exercises 29–34, use the graphs of the trigonometric functions to sketch the graph of the function.

29. $f(x) = -2 \sin x$ **30.** $g(x) = -2 + \sin x$
31. $h(x) = 1 + \csc x$ **32.** $f(x) = -1 + \csc x$
33. $f(x) = \tan(x + \pi/2)$ **34.** $h(x) = \sec 2x$

For the functions described in Exercises 35–38, find (a) $f \circ g$, (b) $g \circ f$, and (c) the domain of each.

35. $f(x) = \sqrt{x^2 - 4}$, $g(x) = \dfrac{1}{x}$

36. $f(x) = |x|$, $g(x) = x^2 - 2x$

37. $f(x) = \dfrac{1}{x^2 - 1}$, $g(x) = x + 1$

38. $f(x) = \dfrac{x}{x+1}$, $g(x) = \sqrt{x}$

39. Express the function described by $f(x) = (x^2 - 7x + 1)^3$ as a composition of two functions.

40. Express the function described by $f(x) = \sin \sqrt{x^3 + 1}$ as a composition of three functions.

In Exercises 41–68, evaluate the limit.

41. $\lim\limits_{x \to 5} \dfrac{x^2 - 25}{x^2 - 6x + 5}$ **42.** $\lim\limits_{x \to 5^-} \dfrac{1}{x - 5}$

43. $\lim\limits_{x \to 6^+} \sqrt{x - 6}$ **44.** $\lim\limits_{x \to \infty} \dfrac{1}{x^2 + 2}$

45. $\lim\limits_{x \to -\infty} \dfrac{1}{x^2 - 1}$ **46.** $\lim\limits_{x \to 1^-} \dfrac{|x - 1|}{x - 1}$

47. $\lim\limits_{x \to 1^+} \dfrac{|x - 1|}{x - 1}$ **48.** $\lim\limits_{x \to 1} \dfrac{|x - 1|}{x - 1}$

49. $\lim\limits_{x \to 1^+} \dfrac{x}{x - 1}$ **50.** $\lim\limits_{x \to \infty} \dfrac{x}{x - 1}$

51. $\lim\limits_{x \to \infty} \dfrac{x^5 - 3x^3 + x - 1}{x^2 + 3}$ **52.** $\lim\limits_{x \to -\infty} \dfrac{x^5 + 2x^2 - 7x}{x^3 - 1}$

53. $\lim\limits_{x \to \infty} \dfrac{\sin x}{x^2 + 1}$ **54.** $\lim\limits_{x \to \infty} (-x + \sin x)$

55. $\lim\limits_{x \to 2} \dfrac{\sqrt{2 + x} - 2}{x - 2}$ **56.** $\lim\limits_{x \to 3} \dfrac{\sqrt{10 - \sqrt{x^2 + 1}}}{x - 3}$

57. $\lim\limits_{x \to \infty} \dfrac{x^3 - 17x}{4x^3 + 1}$ **58.** $\lim\limits_{x \to -\infty} \dfrac{x^3 - 17x}{4x^3 + 1}$

59. $\lim\limits_{x \to \infty} \dfrac{2 - x}{x^2}$ **60.** $\lim\limits_{x \to -\infty} \dfrac{x^3 - 7}{x^2 + 2x}$

61. $\lim\limits_{x \to 0} \dfrac{1}{|x|}$ **62.** $\lim\limits_{x \to 1} \dfrac{\sqrt{x} - 1}{x - 1}$

63. $\lim\limits_{x \to 2^-} \dfrac{1}{x^2 - 4}$ **64.** $\lim\limits_{x \to 2^+} \dfrac{1}{x^2 - 4}$

65. $\lim\limits_{h \to 0} \dfrac{\dfrac{1}{\sqrt{x+h}} - \dfrac{1}{\sqrt{x}}}{h}$ **66.** $\lim\limits_{h \to 0} \dfrac{(x+h)^2 - x^2}{h}$

67. $\lim\limits_{x \to \infty} (\sqrt{x^2 + x^{3/2}} - x)$ **68.** $\lim\limits_{x \to \infty} (\sqrt{x^2 + \sqrt{x}} - x)$

In Exercises 69–76, determine any numbers at which the function is discontinuous, and give the reason for the discontinuity.

69. $f(x) = \dfrac{1}{x^2 - 4}$

70. $f(x) = \dfrac{|x - 1|}{x - 1}$

71. $f(x) = \begin{cases} x^2, & \text{if } x \le 0, \\ 2x, & \text{if } x > 0 \end{cases}$

72. $f(x) = \begin{cases} x + 1, & \text{if } x < 0, \\ x^3, & \text{if } x \ge 0 \end{cases}$

73. $f(x) = \dfrac{x^2 - 5x + 6}{x^2 - 7x + 12}$

74. $f(x) = \dfrac{x^2 - 1}{(x^2 + 4)(x - 1)}$

75. $f(x) = \tan x$

76. $f(x) = \csc x$

In Exercises 77–82, determine the intervals on which the function described is continuous.

77. $f(x) = \dfrac{1}{4x - 3}$ **78.** $g(x) = \sqrt{2x - 1}$

79. $h(x) = \sqrt{x^2 - 1}$ **80.** $f(x) = \dfrac{1}{1 - x^2}$

81. $f(x) = \sec x$ **82.** $f(x) = \dfrac{|x - 1|}{x - 1}$

In Exercises 83–90, (a) find any horizontal, vertical, and slant asymptotes of the graph of the function and (b) sketch the graph.

83. $f(x) = \dfrac{1}{(x - 1)^2}$ **84.** $f(x) = \dfrac{1}{x^2 + 2x}$

85. $f(x) = \dfrac{x^2 - 2x + 1}{x^2 - 2x}$ **86.** $f(x) = \dfrac{x^3 + x^2}{x^2 - 4}$

87. $f(x) = \dfrac{3x^2}{x^2 + 3}$ **88.** $f(x) = \sqrt{\dfrac{x^3 + x^2}{x^2 + 1}}$

89. $f(x) = \dfrac{x - 2}{x^2 - 2x}$ **90.** $f(x) = \dfrac{x}{|x - 4|}$

91. Use the Bisection Method to find an approximation, accurate to within 10^{-1}, to a value of x in $[-1, 2]$ that satisfies $2x^3 - x - 2 = 0$.

92. Use the Bisection Method to find an approximation, accurate to within 10^{-1}, to a value of x in $[0, 1]$ that satisfies $x^2 - 2 \cos \pi x = 0$.

93. Use the definition of the limit to show that $\lim_{x \to 1} (3x + 5) = 8$.

94. Use the definition of the limit to show that $\lim_{x \to 3} 1/(x - 1) = \frac{1}{2}$.

95. If
$$f(x) = \begin{cases} x + 2, & \text{if } x < -2, \\ 2, & \text{if } -2 \leq x \leq 2, \\ 2 - x, & \text{if } x > 2, \end{cases}$$
find each of the following, if they exist.

a) $\lim_{x \to -2^-} f(x)$ **b)** $\lim_{x \to -2^+} f(x)$

c) $\lim_{x \to -2} f(x)$ **d)** $\lim_{x \to 2^-} f(x)$

e) $\lim_{x \to 2^+} f(x)$ **f)** $\lim_{x \to 2} f(x)$

96. Find a constant a that will make the function f continuous at $x = 1$ if
$$f(x) = \begin{cases} \dfrac{x^3 - 3x^2 + 2}{x^2 - 1}, & \text{for } x \neq 1, \\ a, & \text{for } x = 1. \end{cases}$$

97. We can see that $\lim_{x \to \infty} 1/x^2 = 0$. How large must x be to ensure that $1/x^2$ is less than each of the following?

a) 0.01 **b)** 0.001

98. We can see that $\lim_{x \to 0} 1/x^2 = \infty$. How small in absolute value must x be in order that $1/x^2 > 10{,}000$?

99. Suppose that f and g are functions and that $\lim_{x \to a} g(x) = \infty$ and $\lim_{x \to a} f(x) = 2$. Find each of the following.

a) $\lim_{x \to a} (f(x) + g(x))$ **b)** $\lim_{x \to a} (f(x) - g(x))$

c) $\lim_{x \to a} f(x)g(x)$ **d)** $\lim_{x \to a} \dfrac{f(x)}{g(x)}$

100. If $\lim_{x \to a} (f(x) + g(x))$ exists, does this imply that both $\lim_{x \to a} f(x)$ and $\lim_{x \to a} g(x)$ exist?

101. If $\lim_{x \to a} f(x)g(x)$ exists, does this imply that both $\lim_{x \to a} f(x)$ and $\lim_{x \to a} g(x)$ exist?

102. Give an example of two functions f and g and a number a with the property that $\lim_{x \to a} f(x)/g(x)$ exists but neither $\lim_{x \to a} f(x)$ nor $\lim_{x \to a} g(x)$ exists.

2
The Derivative

The derivative is a fundamental concept of calculus that permits us to consider instantaneous rates of change in functional values. Applications of this concept occur in virtually every area of scientific study. The derivative is used by physicists to study the motion of particles; by biologists to study the growth rate of organisms; and by engineers to study a multitude of subjects, including heat flow, circuit theory, and the effects of chemical reactions. Economists analyze marginal cost and revenue using the derivative, and psychologists use it to study the response to stimuli.

Indeed, any subject that depends on methods of approximation or statistics frequently makes use of the derivative. Although the applications come from diverse areas, they are linked by the common need to measure the rate of change in a certain quantity relative to the change in another quantity.

2.1

The Slope of a Curve

The slope of a line is a constant that describes the direction of the line. In Section 1.2 we saw that if a line with equation $y = mx + b$ has a positive slope m, the values of y increase as x increases. If the line has a negative slope, the values of y decrease as x increases. The magnitude of m describes the rate at which the values of y are increasing or decreasing, as shown in Figure 2.1.

The increasing and decreasing behavior of an arbitrary curve is more

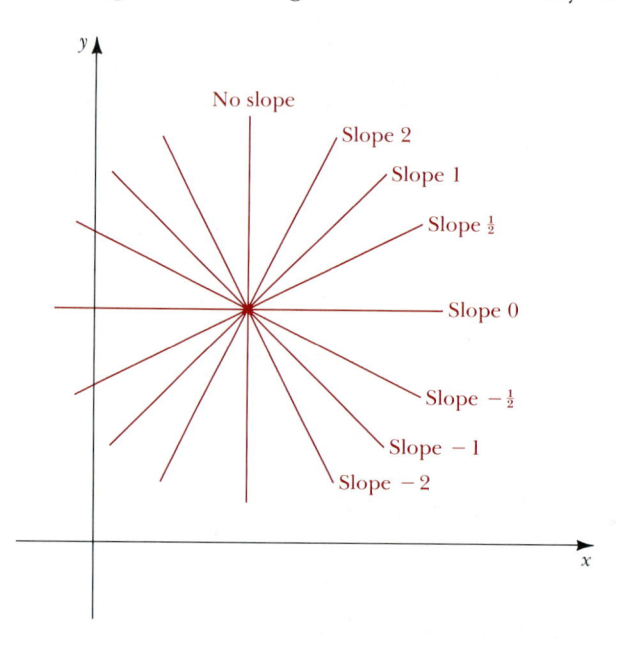

Figure 2.1

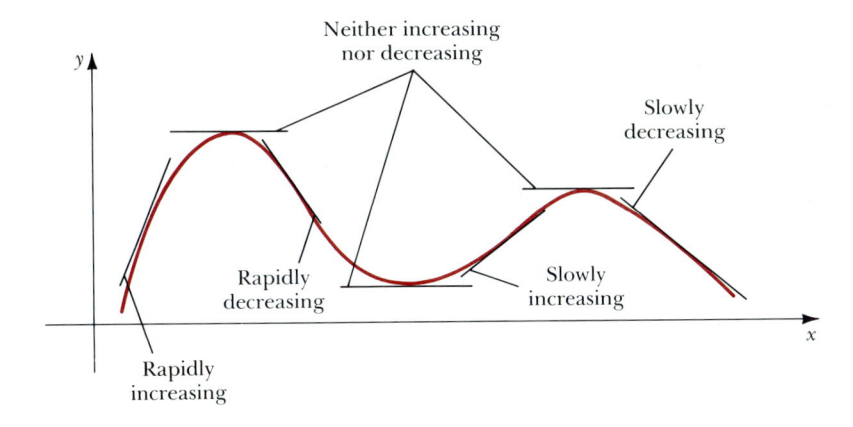

Figure 2.2

difficult to describe because this behavior can vary with the points on the curve (see Figure 2.2). To describe the slope, or direction, of the graph of an arbitrary function at a particular point, we use the notion of a *tangent line* to the graph of the function at the point.

In an intuitive sense, the tangent line to the graph of a function f at a point $(a, f(a))$ is the line that:

a) passes through the point $(a, f(a))$, and
b) has a slope that indicates at what rate the graph of f is increasing or decreasing at $(a, f(a))$

(see Figure 2.3).

The first condition is well defined and easy to fulfill: A nonvertical line with slope m passing through $(a, f(a))$ has an equation of the form

$$y - f(a) = m(x - a).$$

To see how the second condition is satisfied, let us return to a problem discussed in Example 6 of Section 1.5.

Consider the function described by $f(x) = x^2$. The graph of f near the point $(1, 1)$ is crudely approximated by the *secant lines* shown in Figure 2.4 on the following page. These lines are obtained by choosing small numbers $h \neq 0$ and constructing lines that pass through $(1, 1)$ and $(1 + h, (1 + h)^2)$. It appears from the figure that the smaller the magnitude of h, the better the slope of the secant line approximates the slope of the curve at $(1, 1)$. In fact, if the slopes of the secant lines approach a finite limit as h approaches zero, then we say that this limit is the slope of the tangent line to the graph at $(1, 1)$, as shown in Figure 2.5. The slope of the secant line joining $(1, 1)$ and $(1 + h, (1 + h)^2)$ is

$$\frac{(1 + h)^2 - 1}{(1 + h) - 1} = \frac{(1 + h)^2 - 1}{h},$$

so the slope of the tangent line at $(1, 1)$ is

$$\lim_{h \to 0} \frac{(1 + h)^2 - 1}{h} = \lim_{h \to 0} \frac{1 + 2h + h^2 - 1}{h}$$
$$= \lim_{h \to 0} (2 + h) = 2.$$

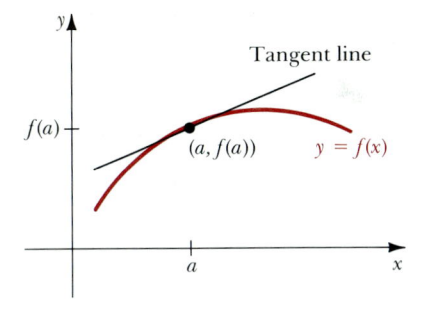

Figure 2.3
Tangent line at $(a, f(a))$.

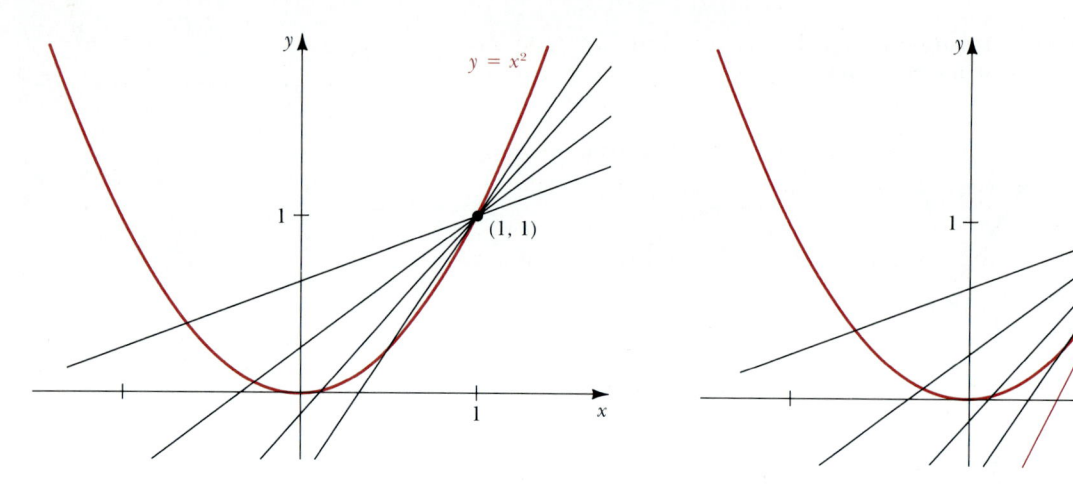

Figure 2.4

Figure 2.5

Since the tangent line passes through (1, 1), it has equation

$$y - 1 = 2(x - 1).$$

This procedure is used to define the tangent line to the graph of an arbitrary function. It is described precisely in the following definition.

(2.1) DEFINITION

The **tangent line** to the graph of a function f at the point $(x, f(x))$ is the line passing through $(x, f(x))$ with slope

$$\lim_{h \to 0} \frac{f(x + h) - f(x)}{h},$$

provided this limit exists.

The slope of the tangent line to the graph of f at $(x, f(x))$ is also called the slope of the curve described by $y = f(x)$ at $(x, f(x))$.

Example 1

Find (a) the slope and (b) an equation of the tangent line to the graph of $f(x) = x^2 + x$ at the point $(-2, 2)$.

Solution

a) By Definition 2.1, the slope of the tangent line at $(-2, 2)$ is

$$\lim_{h \to 0} \frac{f(-2 + h) - f(-2)}{h} = \lim_{h \to 0} \frac{[(-2 + h)^2 + (-2 + h)] - [(-2)^2 - 2]}{h}$$

$$= \lim_{h \to 0} \frac{4 - 4h + h^2 - 2 + h - 4 + 2}{h}$$

$$= \lim_{h \to 0} \frac{-3h + h^2}{h} = \lim_{h \to 0} (-3 + h) = -3.$$

b) Since $(-2, 2)$ is on the tangent line, the tangent line has equation
$y - 2 = -3(x + 2)$ or $y = -3x - 4$ (see Figure 2.6). ▶

Example 2

a) Find the slope of the tangent line to the graph of $f(x) = x^2$ at an arbitrary point (x, x^2).
b) Use the result in part (a) to find an equation of the line tangent to the graph at $(-1, 1)$.

Solution
a) This problem is no more difficult than if a specific point on the graph were given. The slope of the tangent line at (x, x^2) is

$$\lim_{h \to 0} \frac{f(x+h) - f(x)}{h} = \lim_{h \to 0} \frac{(x+h)^2 - x^2}{h}$$
$$= \lim_{h \to 0} \frac{x^2 + 2xh + h^2 - x^2}{h}$$
$$= \lim_{h \to 0} (2x + h) = 2x.$$

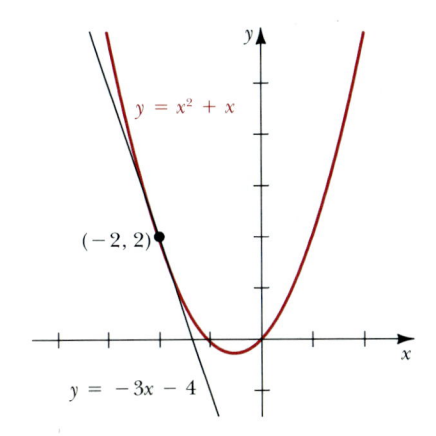

Figure 2.6
Tangent line at $(-2, 2)$.

b) When $x = -1$, the slope of the tangent line is $2(-1) = -2$. The tangent line at $(-1, 1)$ therefore has equation

$$y - 1 = -2[x - (-1)] \quad \text{or} \quad y = -2x - 1. \quad ▶$$

Note in Figure 2.7 that at the point $(0, 0)$, the tangent line to the graph is horizontal. In addition, the tangent line has a negative slope when $x < 0$ and a positive slope when $x > 0$. Later in this chapter we will see that by carefully examining the signs of the slopes of the tangent lines to the graph of a function, we can determine a great deal about the behavior of the function.

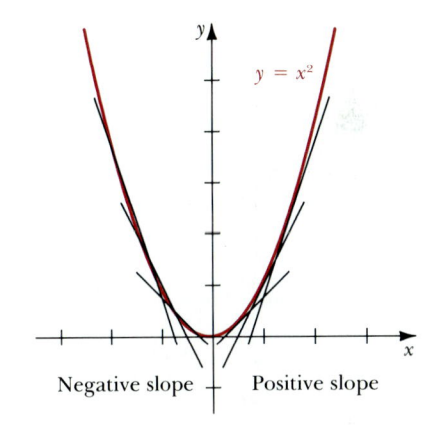

Figure 2.7

Example 3
Find the slope of the tangent line to the graph of the function described by $f(x) = 1/x$ at $(x, f(x))$, $x \neq 0$.

Solution
The slope is given by

$$\lim_{h \to 0} \frac{f(x+h) - f(x)}{h} = \lim_{h \to 0} \frac{\frac{1}{x+h} - \frac{1}{x}}{h}$$
$$= \lim_{h \to 0} \frac{\frac{x - (x+h)}{(x+h)x}}{h}$$
$$= \lim_{h \to 0} \frac{-h}{h(x+h)x} = \lim_{h \to 0} \frac{-1}{(x+h)x} = \frac{-1}{x^2}.$$

Consequently, the slope of the tangent line to the graph of $f(x) = 1/x$ is always negative, as shown in Figure 2.8. ▶

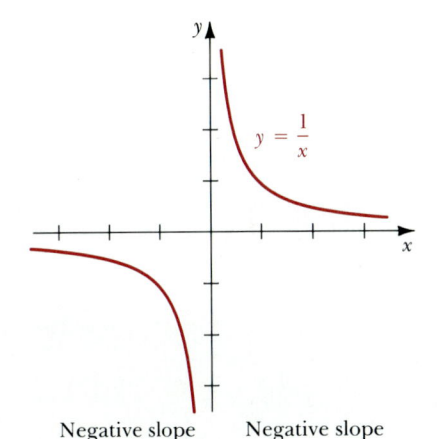

Figure 2.8

The **normal line** to the graph of a function at a point is the line perpendicular to the tangent line at the point. From Theorem 1.13, the slopes m_1 and m_2 of nonvertical lines that are perpendicular satisfy the equation $m_1 = -1/m_2$.

Example 4

Find an equation of the normal line to the graph of the function described by $f(x) = x^2 + x$ at $(-2, 2)$.

Solution

The tangent line at $(-2, 2)$ was shown in Example 1 to have slope -3. The slope of the normal line to the graph of f at $(-2, 2)$ is consequently $-1/(-3) = 1/3$. The point $(-2, 2)$ lies on the normal line, so the line has equation

$$y - 2 = \tfrac{1}{3}(x + 2) \quad \text{or} \quad y = \tfrac{1}{3}x + \tfrac{8}{3}$$

(see Figure 2.9). ▶

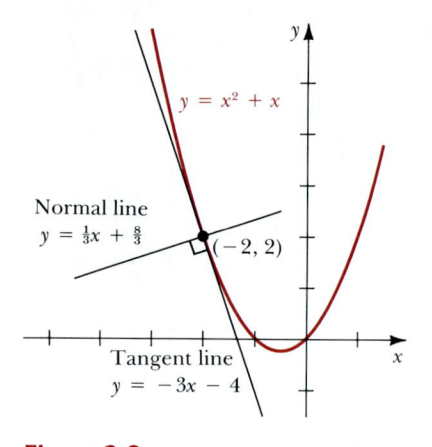

Figure 2.9
Tangent and normal lines at $(-2, 2)$.
Normal line $y = \tfrac{1}{3}x + \tfrac{8}{3}$
Tangent line $y = -3x - 4$
$y = x^2 + x$

Definition 2.1 states that the slope of the tangent line to the graph of a function f at the point $(x, f(x))$ is

$$\lim_{h \to 0} \frac{f(x + h) - f(x)}{h},$$

provided this limit exists. This definition does not allow for the case of vertical tangent lines, since the slope is undefined for vertical lines. The concept of a vertical tangent line to the graph of a function is described in the following definition and illustrated in Figure 2.10.

(2.2) DEFINITION

Suppose f is continuous at a. The line $x = a$ is a **vertical tangent** to the graph of f at the point $(a, f(a))$ if

$$\lim_{h \to 0} \left| \frac{f(a + h) - f(a)}{h} \right| = \infty.$$

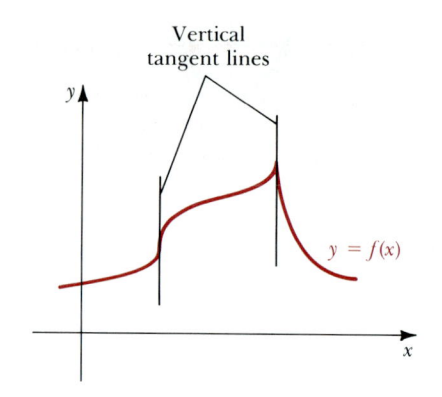

Figure 2.10
Vertical tangent lines.

Example 5

Show that the line $x = 1$ is a vertical tangent to the graph of $f(x) = (x - 1)^{2/3} + 2$ at the point $(1, 2)$.

Solution

The function f is continuous on the entire real line. Since

$$\frac{f(1 + h) - f(1)}{h} = \frac{\{[(1 + h) - 1]^{2/3} + 2\} - [(1 - 1)^{2/3} + 2]\}}{h}$$

$$= \frac{(h^{2/3} + 2) - (0^{2/3} + 2)}{h} = \frac{h^{2/3}}{h} = h^{-1/3},$$

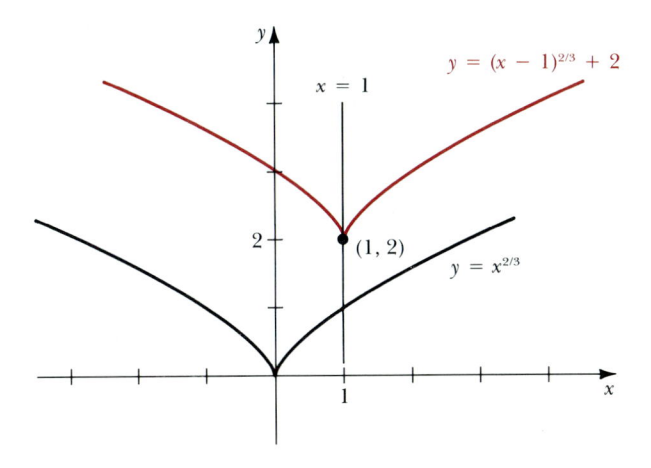

Figure 2.11
Vertical tangent line at $x = 1$.

we have

$$\lim_{h \to 0} \left| \frac{f(1+h) - f(1)}{h} \right| = \lim_{h \to 0} |h^{-1/3}| = \infty.$$

By Definition 2.2, the line $x = 1$ is a vertical tangent to the graph. The graph of $y = (x - 1)^{2/3} + 2$ and the vertical tangent line are shown in Figure 2.11. Note that the graph of $y = (x - 1)^{2/3} + 2$ is the graph of $y = x^{2/3}$ shifted one unit to the right and two units upward.

▶ EXERCISE SET 2.1

In Exercises 1–10, a function f is described and a point on its graph is given.

a) Sketch the graph of the function.
b) Find the slope of the tangent line to the graph at the given point.
c) Find an equation of the tangent line to the graph at the point.
d) Find an equation of the normal line to the graph at the point.

1. $f(x) = 2x^2 + 1$, $(2, 9)$
2. $f(x) = 2x^2 - 1$, $(2, 7)$
3. $f(x) = x - 3x^2$, $(1, -2)$
4. $f(x) = 3 - 4x^2$, $(2, -13)$
5. $f(x) = x^2 + 2x + 3$, $(-1, 2)$
6. $f(x) = 2x^2 + 4x + 1$, $(2, 17)$
7. $f(x) = x^3$, $(1, 1)$
8. $f(x) = x^3$, $(2, 8)$
9. $f(x) = 1/x$, $(2, \frac{1}{2})$
10. $f(x) = 1/x$, $(-2, -\frac{1}{2})$

In Exercises 11–16, (a) find the slope of the tangent line to the graph of the function f at an arbitrary point $(x, f(x))$, and (b) determine any points on the graph of f at which the tangent line is horizontal.

11. $f(x) = x^2 + 2x + 2$
12. $f(x) = 3 - 4x - x^2$
13. $f(x) = 2x^2 - 3x + 4$
14. $f(x) = x^3 - 3x$
15. $f(x) = x^3 + 3x^2 + 3x + 2$
16. $f(x) = x + 1/x$

17–22. For each of the functions in Exercises 11–16, determine when the slope of the line tangent to the graph at $(x, f(x))$ is (a) positive, and (b) negative.

23. Show that the graph of each of the functions described below has a vertical tangent line at the indicated point. Give an equation of the tangent line.

    a) $f(x) = x^{1/3}$, $(0, 0)$ b) $f(x) = (x - 1)^{1/5}$, $(1, 0)$
    c) $f(x) = \begin{cases} \sqrt{x}, & \text{if } x \geq 0, \\ \sqrt{-x}, & \text{if } x < 0, \end{cases}$ $(0, 0)$
    d) $f(x) = x^2 - 3x^{2/3}$, $(0, 0)$

24. The graph of a function f and numbers a_i (for $i = 0, 1, 2, 3, 4, 5$) on the x-axis are given in the figure. Tell whether the slope of the tangent line at each point $(a_i, f(a_i))$ is zero, positive, or negative, or if the tangent line is vertical.

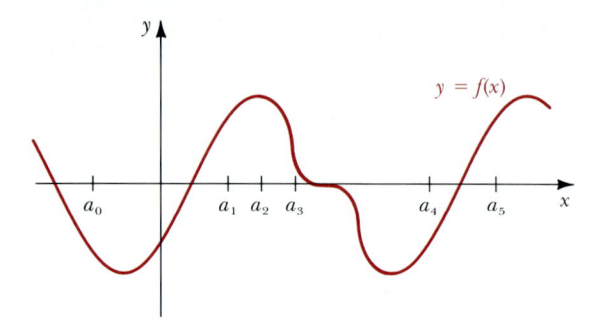

25. Find a point on the graph of $f(x) = x^2 + 2x + 2$ at which the tangent line has slope 1.

26. Find a point on the graph of $f(x) = 2x^2 - 3x + 4$ at which the tangent line has slope -1.

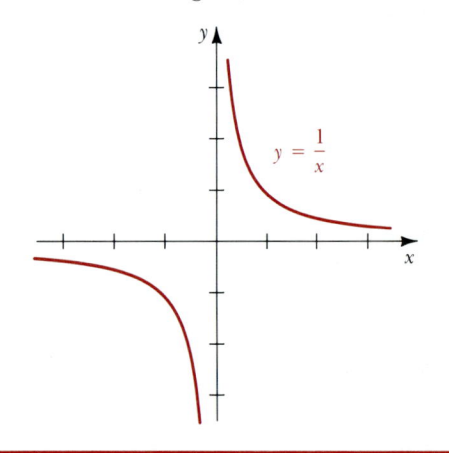

27. Find a point on the graph of $f(x) = 2x^2 + 3x$ at which the tangent line is parallel to the line with equation $x + y = 4$.

28. Find a point on the graph of $f(x) = 3x^2 + 2x - 1$ at which the tangent line is parallel to the line with equation $y - 4x = 3$.

29. Find a point on the graph of $f(x) = 2x^2 + 3x$ at which the normal line is parallel to the line with equation $x - y = 4$.

30. Find a point on the graph of $f(x) = 3x^2 + 2x - 1$ at which the normal line is parallel to the line with equation $y - x = 2$.

31. Find two points on the graph of $f(x) = 1/x$ at which the tangent line is parallel to the line with equation $y = 3 - 4x$ (see the figure).

32. Find a point on the graph of $f(x) = 1/x^2$ at which the tangent line is parallel to the line with equation $y = 3 - 4x$ (see the figure).

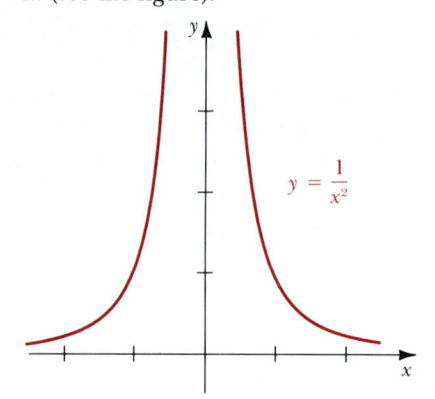

33. Find a number a with the property that the line tangent to the graph of $f(x) = x^2$ at $(a, f(a))$ passes through $(-1, 0)$.

34. Show that no number a exists with the property that the line tangent to the graph of $f(x) = x^2$ at the point $(a, f(a))$ passes through the point $(0, 1)$.

35. On a walk over a gentle hill, we find that the altitude, in feet, t minutes after the walk has begun is given by $A(t) = 300 + 40t - 2t^2$, where $0 \leq t \leq 20$.
    a) Find the slope of the tangent line to the graph of A at an arbitrary point $(t, A(t))$.
    b) Use the result in part (a) to decide for which values of t the walk is uphill.
    c) Use the result in part (a) to decide for which values of t the walk is downhill.
    d) Determine the altitude at the summit.

36. Sketch the graph of f, and determine the slope of the tangent line at an arbitrary point $(a, f(a))$ if
$$f(x) = \begin{cases} x^2, & \text{if } x \geq 0, \\ x^3, & \text{if } x < 0. \end{cases}$$

37. Consider the function f described by
$$f(x) = \begin{cases} x, & \text{if } x = 1/n \text{ for an integer } n, \\ -x, & \text{if } x \neq 1/n. \end{cases}$$
Show that the graph of f does not have a tangent line at the point $(0, 0)$.

38. Consider the function g described by
$$g(x) = \begin{cases} x^2, & \text{if } x = 1/n \text{ for an integer } n, \\ -x^2, & \text{if } x \neq 1/n. \end{cases}$$
Show that the graph of g has a tangent line at the point $(0, 0)$.

2.2
The Derivative of a Function

In Section 2.1 we saw that the slope of a tangent line to the graph of a function is the limiting value of a quotient. This quotient describes the rate of change in the values of the function relative to the change in the independent variable.

The need to determine the rate of change of a certain quantity relative to the change in another quantity occurs in many situations. The physicist studying the velocity of a particle is concerned with the change in distance relative to the change in time. The biologist studying the growth rate of a colony of bacteria is interested in the change in the number of colony members relative to the change in time. The economist studying a marginal revenue problem is studying the change in the amount of money received from the sale of items relative to the change in the demand for the items. Mathematically, these problems have the same format:

1) Find a function that associates one quantity with another.
2) Determine the rate of change of the function values with respect to the change in the independent variable.

In this section we begin the study of a concept that describes this rate of change. It is one of the most fundamental concepts in calculus: the *derivative*.

(2.3)
DEFINITION
The **derivative** of a function f is a function f', defined by

$$f'(x) = \lim_{h \to 0} \frac{f(x+h) - f(x)}{h}.$$

The domain of f' consists of all x in the domain of f for which this limit exists. If x is in the domain of f', then f is said to be **differentiable** at x.

In Section 2.1 we found that the slope of the tangent line to the graph of a function f at a point $(x, f(x))$ is

$$\lim_{h \to 0} \frac{f(x+h) - f(x)}{h},$$

provided this limit exists. Comparing this limit to the limit in Definition 2.3, we see that:

(2.4)
If f is differentiable at x, then $f'(x)$ is the slope of the tangent line to the graph of the function f at $(x, f(x))$.

Reviewing the results of Examples 1, 2, and 3 in Section 2.1 in light of

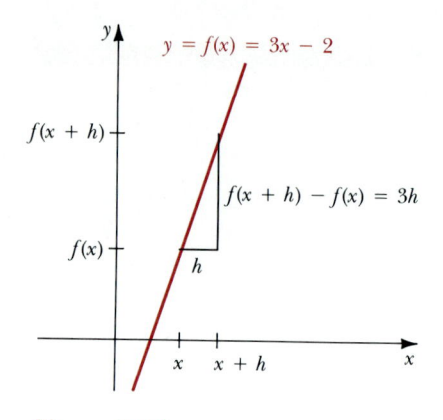

Figure 2.12
For all x, $f'(x) = 3$.

the definition of $f'(x)$, we find that:

If $f(x) = x^2 + x$, then $f'(-2) = -3$.
If $f(x) = x^2$, then $f'(x) = 2x$.
If $f(x) = 1/x$, then $f'(x) = -1/x^2$, if $x \neq 0$.

Example 1

Find $f'(x)$ if $f(x) = 3x - 2$.

Solution
Before applying Definition 2.3, consider the graph of f shown in Figure 2.12. For any value of x, the derivative should be 3, the slope of the line. Definition 2.3 shows that this is indeed the case:

$$f'(x) = \lim_{h \to 0} \frac{f(x+h) - f(x)}{h} = \lim_{h \to 0} \frac{[3(x+h) - 2] - (3x - 2)}{h},$$

so

$$f'(x) = \lim_{h \to 0} \frac{3x + 3h - 2 - 3x + 2}{h} = \lim_{h \to 0} \frac{3h}{h} = 3.$$ ▶

Example 2

a) Find $f'(x)$ if $f(x) = 1/x^2$ and $x \neq 0$.
b) Find an equation of the tangent line to the graph of f at $(2, \frac{1}{4})$.

Solution
a) We have

$$f'(x) = \lim_{h \to 0} \frac{f(x+h) - f(x)}{h}$$

$$= \lim_{h \to 0} \frac{\frac{1}{(x+h)^2} - \frac{1}{x^2}}{h}$$

$$= \lim_{h \to 0} \frac{x^2 - (x+h)^2}{h(x+h)^2 x^2}$$

$$= \lim_{h \to 0} \frac{x^2 - x^2 - 2xh - h^2}{h(x+h)^2 x^2}$$

$$= \lim_{h \to 0} \frac{-2x - h}{(x+h)^2 x^2} = \frac{-2x}{x^2 \cdot x^2} = \frac{-2}{x^3}.$$

b) The slope of the tangent line at $(2, \frac{1}{4})$ is

$$f'(2) = \frac{-2}{(2)^3} = \frac{-1}{4},$$

so the tangent line (see Figure 2.13) has equation

$$y - \frac{1}{4} = -\frac{1}{4}(x - 2) \text{ or } y = -\frac{1}{4}x + \frac{3}{4}$$ ▶

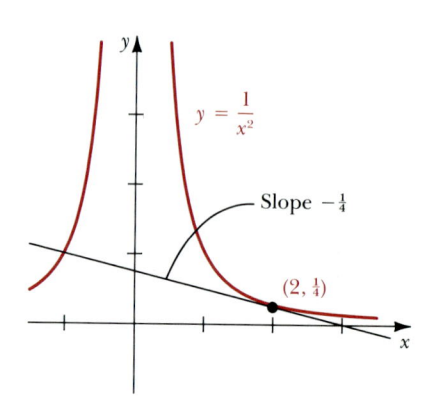

Figure 2.13
The tangent line at $(2, \frac{1}{4})$ has equation $y - \frac{1}{4} = -\frac{1}{4}(x - 2)$.

The quotient

(2.5) $$\frac{f(x+h)-f(x)}{h}$$

is called a **difference quotient.** The difference quotient describes the **average rate of change** of the values of the function with respect to the change in the independent variable, *from x to x + h*.

The derivative of f at x is the limit of the difference quotient as h approaches zero. Thus the derivative $f'(x)$ represents the **instantaneous rate of change** of the function values with respect to the change in the independent variable *at x*. The distinction between average rate of change and instantaneous rate of change is illustrated in the next example.

Example 3

A stone dropped into a still pond causes circular ripples to radiate outward. The area of the disturbed water t seconds after the stone has hit is $A(t) = 4\pi t^2$ m².

a) Find the average rate of change of the area of the disturbed water from $t = 1$ to $t = 1.5$ sec.
b) Find the instantaneous rate of change of the area of the disturbed water at $t = 1$ sec.

Solution

a) The average rate of change from $t = 1$ to $t = 1.5$ sec is

$$\frac{A(1.5)-A(1)}{0.5} = \frac{4\pi(1.5)^2 - 4\pi(1)^2}{0.5} = 8\pi(2.25-1) = 10\pi \ \frac{m^2}{sec}.$$

b) The instantaneous rate of change at $t = 1$ sec is

$$A'(1) = \lim_{h\to 0} \frac{A(1+h)-A(1)}{h} = \lim_{h\to 0} \frac{4\pi(1+h)^2 - 4\pi(1)^2}{h},$$

so

$$A'(1) = \lim_{h\to 0} \frac{4\pi(1+2h+h^2-1)}{h} = \lim_{h\to 0} 4\pi(2+h) = 8\pi.$$

For any $h \ne 0$, the units of $[A(1+h) - A(1)]/h$ are m²/sec, so the units of $A'(1)$ are also m²/sec. Thus the instantaneous rate of change of the area at $t = 1$ sec is 8π m²/sec. ▶

In some cases, an equivalent limit definition for the derivative is more convenient. Suppose that x is fixed and the variable z is defined by $z = x + h$. As h approaches zero, z approaches x (see Figure 2.14). Since $h = z - x$, we can rewrite

$$f'(x) = \lim_{h\to 0} \frac{f(x+h)-f(x)}{h}$$

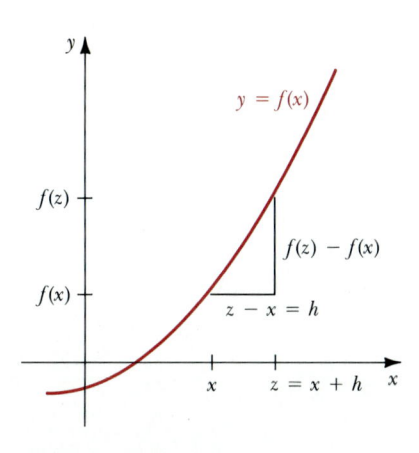

Figure 2.14

$$f'(x) = \lim_{z\to x} \frac{f(z)-f(x)}{z-x}.$$

using the variable z as

(2.6) $$f'(x) = \lim_{z \to x} \frac{f(z) - f(x)}{z - x}.$$

It is important to be able to recognize and use each form of the definition of the derivative. The following example illustrates the use of the two forms to solve the same problem.

Example 4

Find $f'(x)$ if $f(x) = x^3$ using (a) Definition 2.3 and (b) Equation (2.6).

Solution
a) By Definition 2.3,

$$f'(x) = \lim_{h \to 0} \frac{f(x+h) - f(x)}{h}$$
$$= \lim_{h \to 0} \frac{(x+h)^3 - x^3}{h}$$
$$= \lim_{h \to 0} \frac{x^3 + 3x^2h + 3xh^2 + h^3 - x^3}{h}$$
$$= \lim_{h \to 0} \frac{h(3x^2 + 3xh + h^2)}{h}$$
$$= \lim_{h \to 0} (3x^2 + 3xh + h^2) = 3x^2.$$

b) By the alternative definition given in (2.6),

$$f'(x) = \lim_{z \to x} \frac{f(z) - f(x)}{z - x} = \lim_{z \to x} \frac{z^3 - x^3}{z - x}.$$

But $z^3 - x^3$ can be factored as

$$z^3 - x^3 = (z - x)(z^2 + zx + x^2),$$

so

$$f'(x) = \lim_{z \to x} \frac{(z - x)(z^2 + zx + x^2)}{z - x}$$
$$= \lim_{z \to x} (z^2 + zx + x^2) = 3x^2.$$

This is, of course, the same result we obtained in part (a) using the original definition. ▶

The notation for the derivative can also assume different forms, depending on both the application and historical practice. Common alternative notations for $f'(x)$ are

$$\frac{d}{dx} f(x), \quad D_x f(x),$$

and, when the function f is expressed using the independent variable t to represent time, $\dot{f}(t)$. When f is defined by $y = f(x)$, it is also common to use

$$\frac{dy}{dx} \quad \text{and} \quad y'$$

to denote the derivative.

There is another notation that is historically associated with the derivative. It is common to let Δx (read "delta x") denote a change in the variable x and, if $y = f(x)$, to let Δy denote the corresponding change in the function values. Consequently, if $\Delta x = h$, then $\Delta y = f(x + h) - f(x)$, and we can write $f'(x)$ as

$$(2.7) \qquad f'(x) = \lim_{\Delta x \to 0} \frac{f(x + \Delta x) - f(x)}{\Delta x} = \lim_{\Delta x \to 0} \frac{\Delta y}{\Delta x}.$$

Equation (2.7) is the basis for the dy/dx, or *Leibniz*, notation for the derivative.

Example 5

Differentiate the function defined by $f(x) = \sqrt{x}$.

Solution

We know that

$$D_x f(x) = \lim_{h \to 0} \frac{f(x + h) - f(x)}{h} = \lim_{h \to 0} \frac{\sqrt{x + h} - \sqrt{x}}{h}.$$

To simplify this expression, we multiply both numerator and denominator by $(\sqrt{x + h} + \sqrt{x})$, which, in effect, "rationalizes" the numerator:

$$D_x f(x) = \lim_{h \to 0} \left(\frac{\sqrt{x + h} - \sqrt{x}}{h} \right) \left(\frac{\sqrt{x + h} + \sqrt{x}}{\sqrt{x + h} + \sqrt{x}} \right)$$

$$= \lim_{h \to 0} \frac{x + h - x}{h(\sqrt{x + h} + \sqrt{x})} = \lim_{h \to 0} \frac{h}{h(\sqrt{x + h} + \sqrt{x})}$$

$$= \lim_{h \to 0} \frac{1}{\sqrt{x + h} + \sqrt{x}} = \frac{1}{2\sqrt{x}}.$$

Note that $f(0)$ is defined, but $f'(0)$ is not. ▶

HISTORICAL NOTE

The notation $f'(x)$ first appeared in 1770 and was popularized in *Mécanique Analytique*, a treatise applying calculus techniques to mechanics, published in 1788 by **Joseph-Louis Lagrange** (1736–1813). This book has the interesting feature of containing no diagrams or drawings, just text and formulas. Lagrange was honored by Napoleon, who proclaimed him "the great pyramid of the mathematical sciences." Nevertheless Lagrange was a modest person who when pressed for his opinion, usually began his response with the phrase "I do not know. . . ."

The dy/dx notation for the derivative is called Leibniz notation because it has its basis with one of the mathematicians credited with founding calculus, Gottfried Leibniz. Isaac Newton, the other founder of calculus, used a notation on which $\dot{f}(t)$ is based.

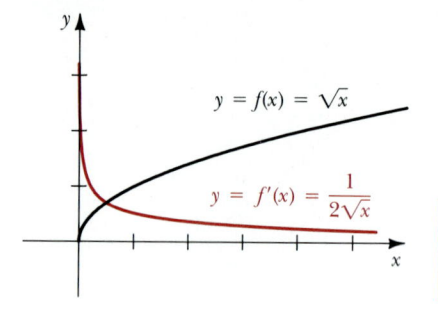

Figure 2.15
The domain of f is $[0, \infty)$, and the domain of f' is $(0, \infty)$.

The domain of f' must be contained in the domain of f. The function considered in Example 5 shows, however, that the domain of a function and the domain of its derivative need not be equal (see Figure 2.15).

An important connection existing between the concepts of differentiability and continuity is established in the following theorem.

(2.8) THEOREM

If f is differentiable at a, then f is continuous at a.

Proof. The definition of $f'(a)$ implies that $f(a)$ exists. To complete the proof that f is continuous at a, we must show that $\lim_{x \to a} f(x)$ exists and is equal to $f(a)$. In Section 1.7 we found that this is equivalent to showing that

$$\lim_{h \to 0} f(a + h) = f(a).$$

Since $f'(a)$ exists,

$$\lim_{h \to 0} [f(a + h) - f(a)] = \lim_{h \to 0} \left[\frac{f(a + h) - f(a)}{h} \cdot h \right]$$

$$= \lim_{h \to 0} \frac{f(a + h) - f(a)}{h} \cdot \lim_{h \to 0} h$$

$$= f'(a) \cdot 0 = 0.$$

Thus

$$0 = \lim_{h \to 0} [f(a + h) - f(a)] = \lim_{h \to 0} f(a + h) - \lim_{h \to 0} f(a).$$

But $f(a)$ is constant, so $\lim_{h \to 0} f(a) = f(a)$ and

$$\lim_{h \to 0} f(a + h) = \lim_{h \to 0} f(a) = f(a),$$

so the continuity at a is established. ▷

The following logical equivalent of Theorem 2.8 is also used:

If f is not continuous at a, then f is not differentiable at a.

This is the contrapositive of the statement in Theorem 2.8. In general, the *contrapositive* of the statement "p implies q" is the statement "not q implies not p." The original statement is true if and only if its contrapositive is true.

The converse of Theorem 2.8 does not hold, that is, there are functions continuous at numbers at which they are not differentiable. For example,

the absolute value function, $f(x) = |x|$, is continuous at zero, but is not differentiable there.

To see that $f'(0)$ does not exist, consider the difference quotient:

$$\frac{f(0 + h) - f(0)}{h} = \frac{|h| - |0|}{h} = \frac{|h|}{h}.$$

Since
$$\lim_{h \to 0^+} \frac{|h|}{h} = 1 \quad \text{and} \quad \lim_{h \to 0^-} \frac{|h|}{h} = -1,$$
the limit of the difference quotient does not exist as h approaches zero, so the derivative at zero does not exist. In fact,

(2.9) $$D_x |x| = \begin{cases} 1, & \text{if } x > 0, \\ \text{undefined}, & \text{if } x = 0, \\ -1, & \text{if } x < 0. \end{cases}$$

Figure 2.16
The absolute value function has no derivative at zero.

The graph of the absolute value function, shown in Figure 2.16, illustrates one of the common geometric features of the graph of a function at a point where the derivative fails to exist. The graph changes direction abruptly at (0, 0) to form a "corner" there.

The function whose graph is shown in Figure 2.17 is not differentiable at x_1, x_2, x_3, or x_4.

i) The derivative cannot exist at x_1 since the function is discontinuous at that point.
ii) A corner occurs at $(x_2, f(x_2))$, which implies that the one-sided limits of the difference quotient differ at x_2 (as they do for the absolute value function at zero), so the derivative does not exist.
iii) The derivative fails to exist at both x_3 and x_4 because the limit of the difference quotient becomes infinite, as indicated by the vertical tangent lines at $(x_3, f(x_3))$ and $(x_4, f(x_4))$.

Listed in Table 2.1 are the derivative results we have seen thus far. We will be extending this knowledge greatly in the next few sections.

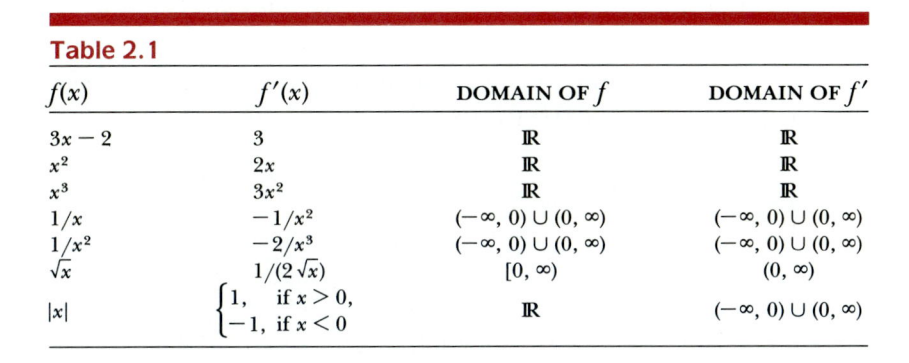

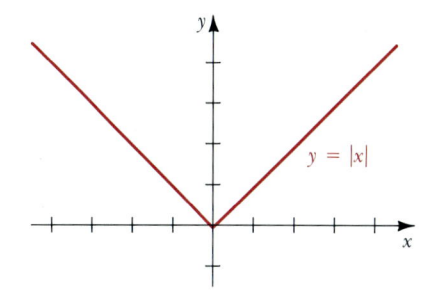

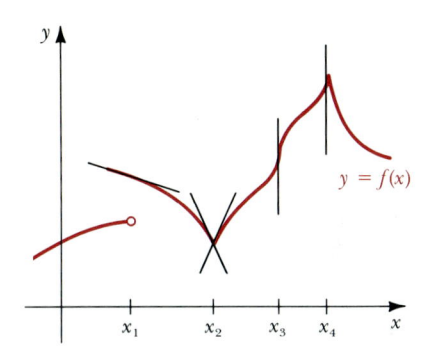

Figure 2.17
This function has no derivative at x_1, x_2, x_3, or x_4.

Table 2.1

$f(x)$	$f'(x)$	DOMAIN OF f	DOMAIN OF f'
$3x - 2$	3	$\mathbb{R}$	$\mathbb{R}$
x^2	$2x$	$\mathbb{R}$	$\mathbb{R}$
x^3	$3x^2$	$\mathbb{R}$	$\mathbb{R}$
$1/x$	$-1/x^2$	$(-\infty, 0) \cup (0, \infty)$	$(-\infty, 0) \cup (0, \infty)$
$1/x^2$	$-2/x^3$	$(-\infty, 0) \cup (0, \infty)$	$(-\infty, 0) \cup (0, \infty)$
$\sqrt{x}$	$1/(2\sqrt{x})$	$[0, \infty)$	$(0, \infty)$
$|x|$	$\begin{cases} 1, & \text{if } x > 0, \\ -1, & \text{if } x < 0 \end{cases}$	$\mathbb{R}$	$(-\infty, 0) \cup (0, \infty)$

▶ EXERCISE SET 2.2

In Exercises 1–16, use the formula $f'(x) = \lim_{h \to 0} \dfrac{f(x+h) - f(x)}{h}$ to find $f'(x)$.

1. $f(x) = 3x$, $x = 1$
2. $f(x) = 5x + 2$, $x = 1$
3. $f(x) = x^2 + x$, $x = 3$
4. $f(x) = x^2 - x$, $x = 3$

5. $f(x) = 2x^2 + 4$, $x = 1$
6. $f(x) = -2x^2 + 3x$, $x = -1$
7. $f(x) = x^2 + 2x + 1$, $x = -1$
8. $f(x) = 4x^2 - 2x + 3$, $x = 0.003$
9. $f(x) = x^3$, $x = -4$
10. $f(x) = -x^3$, $x = 4$
11. $f(x) = \dfrac{1}{\sqrt{x}}$, $x = 4$
12. $f(x) = \dfrac{1}{\sqrt{x}}$, $x = 9$
13. $f(x) = \dfrac{1}{\sqrt{x+2}}$, $x = 2$
14. $f(x) = \dfrac{1}{\sqrt{x+1}}$, $x = 8$
15. $f(x) = x^2 + \dfrac{1}{x}$, $x = 2$
16. $f(x) = x + \sqrt{x}$, $x = 4$

In Exercises 17–22, use the formula

$$f'(x) = \lim_{z \to x} \frac{f(z) - f(x)}{z - x}$$

to find $f'(x)$. Compare this method with the corresponding exercises in 1 through 16 and decide which method you prefer in each case.

17. $f(x) = x^2 + x$, $x = 3$
18. $f(x) = x^2 - x$, $x = 3$
19. $f(x) = x^2 + 2x + 1$, $x = -1$
20. $f(x) = -x^3$, $x = 4$
21. $f(x) = \dfrac{1}{\sqrt{x+2}}$, $x = 2$
22. $f(x) = \dfrac{1}{\sqrt{x+1}}$, $x = 8$

In Exercises 23–34, a function f is described. Find f' and its domain, and sketch the graphs of f and f'.

23. $f(x) = 4x - 2$
24. $f(x) = 3x + 2$
25. $f(x) = x^2 + 2x + 1$
26. $f(x) = x^2 + 2x - 1$
27. $f(x) = x^3$
28. $f(x) = x^3 + 1$
29. $f(x) = (x + 1)^3$
30. $f(x) = |x - 1|$
31. $f(x) = 1/x$
32. $f(x) = 1/x^2$
33. $f(x) = x^{1/3}$
34. $f(x) = x^{2/3}$

35. Find the instantaneous rate of change of the function described by $f(x) = x^2 + x$ at $x = 1$.

36. Find the instantaneous rate of change of the function described by $f(x) = x^3 + x$ at $x = 1$.

37. The area of a circle with radius r is given by $A(r) = \pi r^2$. Find the instantaneous rate of change of the area with respect to the radius.

38. The volume of a cube with side s is given by $V(s) = s^3$. Find the instantaneous rate of change of the volume with respect to s.

39. A simple model for population growth uses the statement that the rate of change of the population with respect to time is proportional to the population. Express this statement as an equation involving a derivative.

40. The rate of change in the amount of a radioactive material is known to be proportional to the amount of material present. Express this statement as an equation involving a derivative.

The right- and left-hand derivatives of a function are defined by using the definitions for right- and left-hand limits (see the figure). The **right-hand derivative**, f'_+, of the function f at x is defined by

$$f'_+(x) = \lim_{h \to 0^+} \frac{f(x+h) - f(x)}{h},$$

provided this limit exists. The **left-hand derivative**, f'_-, of the function f at x is defined by

$$f'_-(x) = \lim_{h \to 0^-} \frac{f(x+h) - f(x)}{h},$$

provided this limit exists. Use these definitions to find $f'_+(x)$ and $f'_-(x)$ in Exercises 41–46.

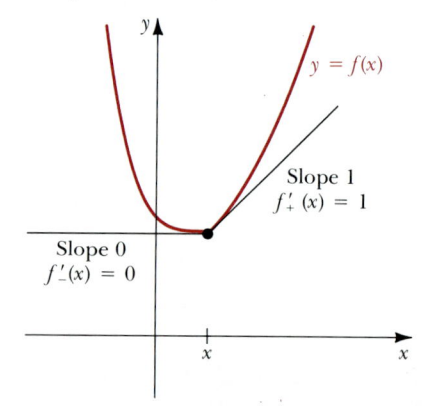

41. $f(x) = x^2 + x$, $x = 3$
42. $f(x) = |x|$, $x = 1$
43. $f(x) = |x|$, $x = 0$
44. $f(x) = x^{1/2}$, $x = 0$
45. $f(x) = x^{3/2}$, $x = 0$
46. $f(x) = x^{5/6}$, $x = 0$

47–52. A function is said to be **differentiable on an open interval** (a, b) if f is differentiable at each number in (a, b). Also, f is said to be **differentiable on a closed interval** $[a, b]$ provided that f is differentiable on (a, b) and that $f'_+(a)$ and $f'_-(b)$ exist. Similar definitions hold for differentiability on intervals of the form $(a, b]$ and $[a, b)$. Determine the largest intervals on which each of the functions in Exercises 41–46 are differentiable.

53. Prove that $f'(x)$ exists if and only if $f'_+(x)$ and $f'_-(x)$ exist and $f'_+(x) = f'_-(x)$.

54. The volume of a yeast starter for making sourdough bread is approximated by

$$V = \begin{cases} 1, & \text{if } t \leq 2, \\ 1 + \sqrt{t - 2}, & \text{if } 2 < t < 6, \\ 3, & \text{if } t \geq 6, \end{cases}$$

where t is measured in days and V in cups.
  a) What is the instantaneous rate of change in the volume of the starter with respect to time when $t = 3$?
  b) Find $V'_+(2)$ and $V'_-(2)$, if they exist.
  c) Find $V'_+(6)$ and $V'_-(6)$, if they exist.
  d) Is V differentiable at $t = 2$?
  e) Is V differentiable at $t = 6$?

55. Suppose both $f'_+(x)$ and $f'_-(x)$ exist. Must f be continuous at x?

56. Suppose $f'(x)$ exists. Show that

$$f'(x) = \lim_{h \to 0} \frac{f(x + h) - f(x - h)}{2h}.$$

57. Consider the quadratic function described by $f(x) = ax^2 + bx + c$ (see the figure). Show that the slope of the line joining the points $(x - h, f(x - h))$ and $(x + h, f(x + h))$ is equal to $f'(x)$.

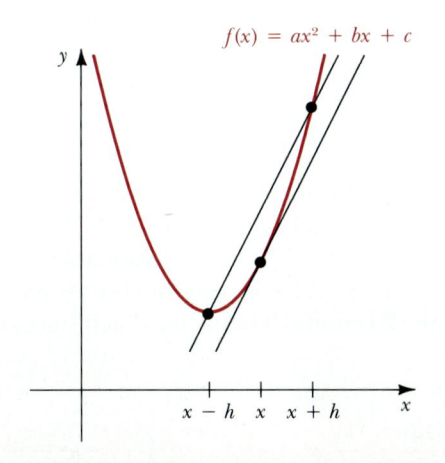

58. Suppose that f is an odd function and f' is defined at every real number. Show that f' is an even function.

59. Suppose that f is an even function and f' is defined at every real number. Show that f' is an odd function and $f'(0) = 0$.

60. Find the y-intercept and x-intercept of the tangent line to the graph of a function f at a point $(a, f(a))$ (see the figure).

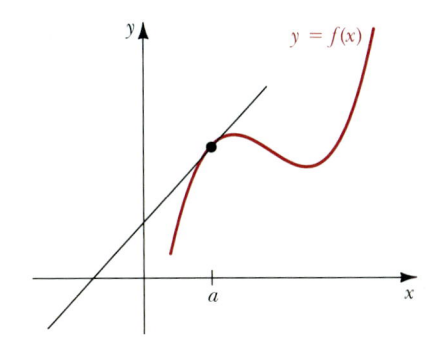

61. Find the y-intercept and x-intercept of the normal line to the graph of a function f at a point $(a, f(a))$ (see the figure).

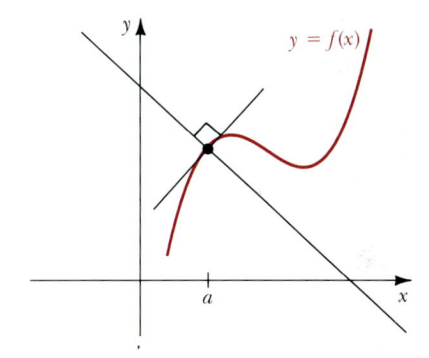

62. Suppose that f is a function with $f(0) \neq 0$ and with the property that for all real numbers x_1 and x_2, $f(x_1 + x_2) = f(x_1) f(x_2)$. Show that $f(0) = 1$. Use this to show that if $f'(0)$ exists, then $f'(x)$ exists for all x in $\mathbb{R}$ and $f'(x) = f(x) \cdot f'(0)$.

63. Suppose that f is a function with the property that for all positive real numbers x_1 and x_2, $f(x_1 x_2) = f(x_1) + f(x_2)$. Show that $f(1) = 0$ and use this to show that if $f'(1)$ exists, then $f'(x) = f'(1)/x$ for any positive real number x.

64. Suppose f is a function that satisfies (i) $f(0) = 0$ and (ii) $f'(x)$ is defined for all x in $(-1, 1)$. Show that $|f|$ is differentiable at 0 if and only if $f'(0) = 0$.

2.3
Formulas for Differentiation

Before we can use the derivative effectively, we need a way to determine easily the derivatives of a large class of functions. This section is the first of four devoted to deriving rules for differentiation. The verification of these rules uses the definition of the derivative, but once the rules are verified, we can find the derivatives of many common functions without evaluating a limit.

The first result to consider is the case of a linear function, one whose graph is a straight line (see Figure 2.18). For any real number x and any $h \neq 0$, the difference quotient

$$\frac{f(x+h) - f(x)}{h}$$

is the slope of the line. Consequently, the derivative of a linear function is a constant, the slope of the line.

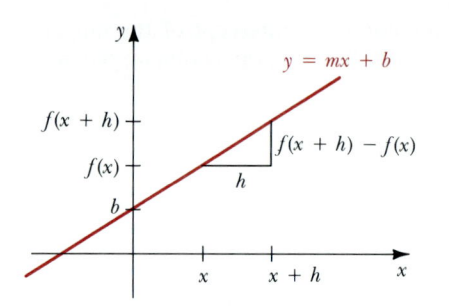

Figure 2.18
For any constants m and b,
$D_x(mx + b) = m$.

(2.10)
THEOREM
If f is a function described by $f(x) = mx + b$, then $f'(x) = m$ for any real number x.

Proof

$$f'(x) = \lim_{h \to 0} \frac{f(x+h) - f(x)}{h}$$
$$= \lim_{h \to 0} \frac{[m(x+h) + b] - (mx + b)}{h} = \lim_{h \to 0} \frac{mh}{h} = m \quad \triangleright$$

Theorem 2.10 holds for any pair of constants m and b. In particular, if $m = c$ and $b = 0$, then

(2.11)
$$D_x(cx) = c;$$

and if $m = 0$ and $b = c$, then

(2.12)
$$D_x c = 0.$$

The following theorem generalizes results from two previous examples. From Example 2 of Section 2.1 we have $D_x x^2 = 2x$, and from Example 4 of Section 2.2 we have $D_x x^3 = 3x^2$. A reasonable conjecture from these results is that for any positive integer n, $D_x x^n = nx^{n-1}$. This result is established using the **Binomial Theorem,** which states that for any positive integer n:

(2.13)
$$(x+h)^n = x^n + nx^{n-1}h + \frac{n(n-1)}{2}x^{n-2}h^2 + \cdots + nx\,h^{n-1} + h^n.$$

(2.14)
THEOREM: THE POWER RULE
If $f(x) = x^n$, where n is a positive integer, then $f'(x) = nx^{n-1}$.

Proof

$$f'(x) = \lim_{h \to 0} \frac{f(x+h) - f(x)}{h}$$

$$= \lim_{h \to 0} \frac{(x+h)^n - x^n}{h}$$

$$= \lim_{h \to 0} \frac{[x^n + nx^{n-1}h + \frac{n(n-1)}{2}x^{n-2}h^2 + \cdots + nx\, h^{n-1} + h^n] - x^n}{h}$$

$$= \lim_{h \to 0} \frac{h\left[nx^{n-1} + \frac{n(n-1)}{2}x^{n-2}h + \cdots + nx\, h^{n-2} + h^{n-1}\right]}{h}$$

$$= \lim_{h \to 0} \left[nx^{n-1} + \frac{n(n-1)}{2}x^{n-2}h + \cdots + nx\, h^{n-2} + h^{n-1}\right]$$

All terms except the first include a factor of h and approach 0 as h approaches 0. Thus, $f'(x) = nx^{n-1}$. ▷

Example 1

If $f(x) = x^9$, find $f'(-1)$.

Solution
By the Power Rule,
$$f'(x) = 9x^{9-1} = 9x^8,$$
so
$$f'(-1) = 9(-1)^8 = 9.$$ ▶

The first arithmetic derivative result concerns the derivative of the sum and difference of two differentiable functions.

(2.15)
THEOREM: THE SUM RULE
If f and g are differentiable at x, then
$$D_x(f+g)(x) = D_x f(x) + D_x g(x),$$
and
$$D_x(f-g)(x) = D_x f(x) - D_x g(x).$$

Proof

$$D_x(f+g)(x) = \lim_{h \to 0} \frac{(f+g)(x+h) - (f+g)(x)}{h}$$

$$= \lim_{h \to 0} \frac{f(x+h) + g(x+h) - f(x) - g(x)}{h}$$

$$= \lim_{h \to 0} \left(\frac{f(x+h) - f(x)}{h} + \frac{g(x+h) - g(x)}{h} \right)$$

$$= \lim_{h \to 0} \frac{f(x+h) - f(x)}{h} + \lim_{h \to 0} \frac{g(x+h) - g(x)}{h}$$

$$= D_x f(x) + D_x g(x)$$

The proof that $D_x(f-g)(x) = D_x f(x) - D_x g(x)$ can be handled in the same manner (see Exercise 53). ▷

Example 2

Find $D_x(x^7 + x^5 - x^3 + 2x)$.

Solution

From repeated applications of the Sum and Power Rules, we find that

$$D_x(x^7 + x^5 - x^3 + 2x) = D_x x^7 + D_x x^5 - D_x x^3 + D_x(2x)$$
$$= 7x^6 + 5x^4 - 3x^2 + 2. \quad \blacktriangleright$$

In Chapter 1 we found that if c is a constant and $\lim_{x \to a} f(x) = L$ exists, then $\lim_{x \to a} cf(x)$ also exists and is equal to cL. This result forms the basis for the proof of the following theorem.

(2.16) THEOREM

If f is differentiable at x, then for any constant c,

$$D_x(cf)(x) = cD_x f(x).$$

Proof

$$D_x(cf)(x) = \lim_{h \to 0} \frac{cf(x+h) - cf(x)}{h}$$

$$= \lim_{h \to 0} \frac{c[f(x+h) - f(x)]}{h}$$

$$= c \lim_{h \to 0} \frac{f(x+h) - f(x)}{h}$$

$$= cD_x f(x) \quad \triangleright$$

Example 3

Find $D_x f(x)$ if $f(x) = 3x^2 + 2x - 1$.

Solution

By the Sum Rule,
$$D_x f(x) = D_x (3x^2 + 2x - 1) = D_x (3x^2) + D_x (2x - 1).$$

Theorem 2.16 implies that $D_x (3x^2) = 3 D_x x^2$, so the Power Rule and Theorem 2.10 give
$$D_x f(x) = 3(2x) + 2 = 6x + 2. \qquad \blacktriangleright$$

The rules of differentiation used in Example 3 applied to an arbitrary polynomial show the following.

(2.17)
If
$$P(x) = a_n x^n + a_{n-1} x^{n-1} + \cdots + a_1 x + a_0,$$
then
$$P'(x) = n a_n x^{n-1} + (n-1) a_{n-1} x^{n-2} + \cdots + 2 a_2 x + a_1.$$

We have established that when two functions are differentiable the derivative of the sum or difference is obtained by taking the sum or difference of their derivatives. A natural question to ask is whether the same type of result is true for products and quotients. To see that it is *not*, let $f(x) = 1$ and $g(x) = x^2$. Then $D_x (fg)(x) = D_x (x^2) = 2x$ and $[D_x f(x)] \cdot [D_x g(x)] = 0 \cdot 2x = 0$. So, in this example, if $x \neq 0$, then
$$D_x (fg)(x) \neq D_x f(x) \cdot D_x g(x).$$

A rule for multiplication does exist and is easily applied. It is called the *Product Rule*.

(2.18)
THEOREM: THE PRODUCT RULE
If f and g are differentiable at x, then
$$D_x (fg)(x) = f(x) D_x g(x) + g(x) D_x f(x).$$

Proof. We have
$$D_x (fg)(x) = \lim_{h \to 0} \frac{f(x+h) g(x+h) - f(x) g(x)}{h}.$$

To change this into a form involving the quotients associated with $D_x f(x)$ and $D_x g(x)$,
$$\frac{f(x+h) - f(x)}{h} \quad \text{and} \quad \frac{g(x+h) - g(x)}{h},$$

we add and subtract the quantity $f(x+h)\,g(x)$ in the numerator:

$$D_x\,(fg)(x) = \lim_{h \to 0} \frac{f(x+h)\,g(x+h) - f(x+h)\,g(x) + f(x+h)\,g(x) - f(x)g(x)}{h}$$

$$= \lim_{h \to 0} \left[f(x+h)\,\frac{g(x+h) - g(x)}{h} + g(x)\,\frac{f(x+h) - f(x)}{h} \right]$$

$$= \lim_{h \to 0} f(x+h) \lim_{h \to 0} \left[\frac{g(x+h) - g(x)}{h} \right] + \lim_{h \to 0} g(x) \lim_{h \to 0} \left[\frac{f(x+h) - f(x)}{h} \right]$$

$$= \lim_{h \to 0} f(x+h)\,D_x\,g(x) + \lim_{h \to 0} g(x)\,D_x\,f(x).$$

Since f has a derivative at x, Theorem 2.8 implies that f is continuous at x; so $\lim_{h \to 0} f(x+h) = f(x)$. Moreover, $g(x)$ is independent of h, so $\lim_{h \to 0} g(x) = g(x)$. Therefore,

$$D_x\,(f \cdot g)(x) = D_x\,[f(x) \cdot g(x)] = f(x)\,D_x\,g(x) + g(x)\,D_x\,f(x). \quad \triangleright$$

Example 4

Calculate $df(x)/dx$ if $f(x) = (x^2 + x - 1)(2x + 4)$.

Solution
We have

$$\frac{df(x)}{dx} = \frac{d}{dx}\,[(x^2 + x - 1)(2x + 4)]$$

$$= (x^2 + x - 1)\,\frac{d}{dx}\,(2x + 4) + (2x + 4)\,\frac{d}{dx}\,(x^2 + x - 1)$$

$$= (x^2 + x - 1)(2) + (2x + 4)\left[\frac{d}{dx}\,(x^2) + \frac{d}{dx}\,(x - 1)\right]$$

$$= 2x^2 + 2x - 2 + (2x + 4)(2x + 1)$$

$$= 2x^2 + 2x - 2 + 4x^2 + 10x + 4$$

$$= 6x^2 + 12x + 2.$$

We could also obtain this result by performing the multiplication $(x^2 + x - 1)(2x + 4) = 2x^3 + 6x^2 + 2x - 4$ and then taking the derivative of each term in succession. ▶

The formula for the derivative of the product of three or more differentiable functions can be derived by applying the product rule as many times as necessary. For example, suppose f, g, and h are all differentiable at x. Then

$$D_x\,(fgh)(x) = D_x\,[(fg)(x) \cdot h(x)] = (fg)(x)\,D_x\,h(x) + h(x)\,D_x\,(fg)(x).$$

But

$$D_x\,(fg)(x) = f(x)\,D_x\,g(x) + g(x)\,D_x\,f(x),$$

so

(2.19)
$$D_x(fgh)(x) = f(x)\,g(x)\,D_x\,h(x) + f(x)\,h(x)\,D_x\,g(x) + g(x)\,h(x)\,D_x\,f(x).$$

Example 5

Find $f'(x)$ if $f(x) = (x^2 + 1)(x^3 + 2x - 1)(4x - 2)$.

Solution
By (2.19),

$$\begin{aligned}
f'(x) &= (x^2 + 1)(x^3 + 2x - 1)\,D_x\,(4x - 2) \\
&\quad + (x^2 + 1)(4x - 2)\,D_x\,(x^3 + 2x - 1) \\
&\quad + (x^3 + 2x - 1)(4x - 2)\,D_x\,(x^2 + 1) \\
&= (x^2 + 1)(x^3 + 2x - 1)(4) + (x^2 + 1)(4x - 2)(3x^2 + 2) \\
&\quad + (x^3 + 2x - 1)(4x - 2)(2x).
\end{aligned}$$
▶

The final arithmetic derivative rule for a pair of functions is the *Quotient Rule*, which can be established easily once the following result has been proved.

(2.20)
THEOREM: THE RECIPROCAL RULE
If g is differentiable at x and $g(x) \neq 0$, then

$$D_x\left(\frac{1}{g}\right)(x) = -\frac{D_x\,g(x)}{[g(x)]^2}.$$

Proof

$$\begin{aligned}
D_x\left(\frac{1}{g}\right)(x) &= \lim_{h \to 0} \frac{\left(\frac{1}{g}\right)(x+h) - \left(\frac{1}{g}\right)(x)}{h} \\
&= \lim_{h \to 0} \frac{\frac{1}{g(x+h)} - \frac{1}{g(x)}}{h} = \lim_{h \to 0} \frac{g(x) - g(x+h)}{g(x+h)g(x)\,h} \\
&= \lim_{h \to 0} \left(\frac{g(x+h) - g(x)}{h}\right)\left(\frac{-1}{g(x+h)g(x)}\right) \\
&= \lim_{h \to 0} \frac{g(x+h) - g(x)}{h} \lim_{h \to 0} \frac{-1}{g(x+h)g(x)} \\
&= D_x\,g(x)\,\frac{-1}{\lim_{h \to 0} g(x+h) \cdot g(x)}.
\end{aligned}$$

Since $D_x\,g(x)$ exists, g is continuous at a and $\lim_{h \to 0} g(x+h) = g(x)$. With this,

$$D_x\left(\frac{1}{g}\right)(x) = D_x\,g(x)\,\frac{-1}{g(x)g(x)} = -\frac{D_x\,g(x)}{[g(x)]^2}.$$
▷

The Product and Reciprocal Rules can now be combined to give the Quotient Rule.

(2.21)
THEOREM: THE QUOTIENT RULE
If f and g are differentiable at x and $g(x) \neq 0$, then
$$D_x\left(\frac{f}{g}\right)(x) = \frac{g(x)\, D_x f(x) - f(x)\, D_x g(x)}{[g(x)]^2}.$$

Proof. We first use the Product Rule and then the Reciprocal Rule:
$$D_x\left(\frac{f}{g}\right)(x) = D_x\left(f \cdot \frac{1}{g}\right)(x)$$
$$= f(x)\, D_x\left(\frac{1}{g}\right)(x) + \left(\frac{1}{g}\right)(x)\, D_x f(x)$$
$$= f(x)\left(\frac{-D_x g(x)}{[g(x)]^2}\right) + \left(\frac{1}{g(x)}\right) D_x f(x)$$
$$= \frac{-f(x)\, D_x g(x) + g(x)\, D_x f(x)}{[g(x)]^2}$$
$$= \frac{g(x)\, D_x f(x) - f(x)\, D_x g(x)}{[g(x)]^2}. \quad \triangleright$$

Example 6

Find $D_x f(x)$ if
$$f(x) = \frac{x^4 + 2x - 1}{3x^2 + 5}.$$

Solution
$$D_x f(x) = D_x\left(\frac{x^4 + 2x - 1}{3x^2 + 5}\right)$$
$$= \frac{(3x^2 + 5)\, D_x\,(x^4 + 2x - 1) - (x^4 + 2x - 1)\, D_x\,(3x^2 + 5)}{(3x^2 + 5)^2}$$
$$= \frac{(3x^2 + 5)(4x^3 + 2) - (x^4 + 2x - 1)(6x)}{(3x^2 + 5)^2}$$
$$= \frac{6x^5 + 20x^3 - 6x^2 + 6x + 10}{(3x^2 + 5)^2} \quad \blacktriangleright$$

Example 7

Use the Quotient and Product Rules to find $f'(t)$ if
$$f(t) = \frac{(t + 1)(t + 2)}{t - 1}.$$

Solution

$$f'(t) = \frac{(t-1) \, D_t \, [(t+1)(t+2)] - (t+1)(t+2) \, D_t \, (t-1)}{(t-1)^2}$$

$$= \frac{(t-1)[(t+1) \, D_t \, (t+2) + (t+2) \, D_t \, (t+1)] - (t+1)(t+2)(1)}{(t-1)^2}$$

$$= \frac{(t-1)[(t+1) + (t+2)] - (t+1)(t+2)}{(t-1)^2}$$

$$= \frac{(t-1)(2t+3) - (t+1)(t+2)}{(t-1)^2}$$

$$= \frac{2t^2 + t - 3 - t^2 - 3t - 2}{(t-1)^2} = \frac{t^2 - 2t - 5}{t^2 - 2t + 1} \quad \blacktriangleright$$

Example 8

Find the derivative of

$$h(x) = \frac{1}{x^2}.$$

Solution

We can apply the Quotient Rule with $f(x) = 1$ and $g(x) = x^2$:

$$h'(x) = \frac{x^2 \, D_x \, (1) - 1 \cdot D_x \, x^2}{(x^2)^2} = \frac{0 - 2x}{x^4} = \frac{-2}{x^3};$$

or we can use the Reciprocal Rule with $g(x) = x^2$:

$$h'(x) = \frac{-g'(x)}{[g(x)]^2} = \frac{-2x}{(x^2)^2} = \frac{-2}{x^3}. \quad \blacktriangleright$$

Example 8 implies that $D_x \, (x^{-2}) = -2x^{-3}$, so the Power Rule holds for $n = -2$. In fact the rule holds for all negative as well as positive integers.

(2.22)
COROLLARY

If $f(x) = x^n$, where n is a negative integer, then $f'(x) = nx^{n-1}$.

Proof. If n is a negative integer, then $-n$ is a positive integer. Since $x^n = 1/x^{-n}$, the Power and Reciprocal Rules can be applied to x^{-n}:

$$f'(x) = D_x \, (x^n) = D_x \left(\frac{1}{x^{-n}}\right) = \frac{-D_x \, (x^{-n})}{(x^{-n})^2}$$

$$= \frac{-(-nx^{-n-1})}{x^{-2n}} = nx^{-n-1} \, x^{2n}$$

$$= nx^{n-1}. \quad \triangleright$$

Example 9

Find
$$D_x\left(\frac{2}{x^3} + \frac{1}{x^5}\right).$$

Solution

$$\begin{aligned}
D_x\left(\frac{2}{x^3} + \frac{1}{x^5}\right) &= 2\,D_x\,(x^{-3}) + D_x\,(x^{-5}) \\
&= 2(-3x^{-4}) + (-5x^{-6}) \\
&= -6x^{-4} - 5x^{-6}
\end{aligned}$$ ▶

Example 10

Find $D_x\,(x+1)^{-2}$.

Solution

Rewrite $(x+1)^{-2}$ as
$$(x+1)^{-2} = \frac{1}{(x+1)^2} = \frac{1}{x^2 + 2x + 1}.$$

Then
$$\begin{aligned}
D_x\,(x+1)^{-2} &= D_x\left(\frac{1}{x^2 + 2x + 1}\right) \\
&= \frac{-D_x\,(x^2 + 2x + 1)}{(x^2 + 2x + 1)^2} \\
&= \frac{-(2x+2)}{(x^2 + 2x + 1)^2} = \frac{-2(x+1)}{(x+1)^4} = \frac{-2}{(x+1)^3}.
\end{aligned}$$ ▶

Note the similarity between the result in Example 10 and the corresponding result in Example 8 when the Power Rule is applied to x^{-2}. In Section 2.5 we will examine this connection in detail.

Table 2.2 is a summary of the rules discussed in this section. You will be using these rules extensively throughout the remainder of your study of calculus. It will save a great deal of time later if you commit them to memory now.

Table 2.2

1. $D_x\,(mx + b) = m$
2. $D_x\,x^n = nx^{n-1}$, for any nonzero integer n
3. $D_x\,(f \pm g)(x) = D_x\,f(x) \pm D_x\,g(x)$
4. $D_x\,(fg)(x) = f(x) \cdot D_x\,g(x) + g(x) \cdot D_x\,f(x)$
5. $D_x\,(1/g)(x) = \dfrac{-D_x\,g(x)}{[g(x)]^2}$, provided $g(x) \neq 0$
6. $D_x\,(f/g)(x) = \dfrac{g(x) \cdot D_x\,f(x) - f(x) \cdot D_x\,g(x)}{[g(x)]^2}$, provided $g(x) \neq 0$

EXERCISE SET 2.3

In Exercises 1–38, use the results of this section to find the derivative of the function described.

1. $f(x) = x^2 - 2x + 7$
2. $f(x) = 2x^2 - 3x + 4$
3. $f(x) = \frac{1}{2}x^2 + 4x + 9$
4. $f(x) = \frac{1}{4}x^4 + \frac{1}{3}x^3 + \frac{1}{2}x^2 + x + 1$
5. $f(x) = 17x^{31} + 14x^{29} + x^4$
6. $f(x) = 21x^{16} + 14x^{12} + 11x^9$
7. $f(t) = \pi t^2 + (3 + 4\sqrt{2})t + 9$
8. $f(s) = (\pi^2 + 1)s^2 + \sqrt{3}(s + 1)$
9. $h(x) = (x^2 + 4)(x^2 - 4)$
10. $f(x) = (2x + 5)(3x - 7)$
11. $r(u) = (u^2 + u + 2)(u^4 + u^3)$
12. $r(t) = (7t^3 - t^2 + 5)(15t + 4)$
13. $f(x) = \frac{x^2}{3} + \frac{3}{x^2}$
14. $f(x) = \frac{x^4}{4} + \frac{4}{x^4}$
15. $h(t) = (t^4 - 3t^2 + 5)(3t^4 - t^{-2})$
16. $g(z) = (z^3 - 3z^{-2} + z)(5 + z^{-6})$
17. $f(z) = \frac{z}{z + 1}$
18. $r(z) = \frac{z + 1}{z}$
19. $H(x) = \frac{3x + 4}{5x^2 + 1}$
20. $g(w) = \frac{w^2 + 3w - 2}{w^2 + 3w + 2}$
21. $H(t) = \frac{t^2 - t - 6}{t^2 + 3t - 4}$
22. $f(x) = \frac{x^3 - x^2 + 2x - 2}{3x + 4}$
23. $F(s) = \left(\frac{s^2 + 1}{s^2 + 2}\right)(s^2 + 3)$
24. $g(t) = \left(\frac{t^2 + 1}{t^2 + 2}\right)\left(\frac{t^2 + 3}{t^2 + 4}\right)$
25. $g(x) = \frac{x(x^2 - 1)}{(x^2 + 1)}$
26. $h(t) = \frac{t^2 + 1}{(t^2 + 2)(t^2 + 3)}$
27. $f(x) = (x + 1)(x + 2)(x + 3)$
28. $f(x) = (x^2 - 1)(x^2 - 2)(x^2 - 3)$
29. $f(x) = x(x^2 + 2x + 1)(x - 5)$
30. $g(x) = x(x^2 + 1)(x^3 + 2)$
31. $f(x) = \frac{7x^3 - 4x^2 + 3x + 5}{7}$
32. $g(x) = \frac{7}{7x^3 - 4x^2 + 3x + 5}$
33. $g(x) = (2x)^{-4}$
34. $f(x) = (3x)^3$
35. $h(x) = (3x + 5)^2$
36. $f(x) = (7x + 1)^{-2}$
37. $g(t) = (2t + 1)^3$
38. $f(x) = \left[(x^2 + 1)\left(\frac{x^2 + 3}{x + 2}\right)\right]^{-1}$

39. Find the point on the graph of $y = x^2 + 3x - 4$ at which the slope of the tangent line is zero.

40. Find the point on the graph of $y = x^2 - x$ at which the tangent line is horizontal.

41. Find an equation of the tangent line to the graph of $y = (x^2 - 1)/(x^2 + 1)$ at $(2, \frac{3}{5})$.

42. Find an equation of the normal line to the graph of $y = (x^2 - 1)/(x^2 + 1)$ at $(2, \frac{3}{5})$.

43. Water is pumped into a reservoir so that the volume V (in gallons) t min after the pumping has begun is given by $V = 50t^2$.
    a) Find the average rate of change of V with respect to t from $t = 1$ to $t = 2$ min.
    b) Find the instantaneous rate of change of V with respect to t when $t = 1$ min.

44. The size (in square centimeters) of a bacteria population residing on a nutrient agar at time t (in hours) is given by $P(t) = 1000 + 100t - 10t^2$. Find the instantaneous rate of growth (a) when $t = 2$ hr, (b) when $t = 5$ hr, and (c) when $t = 7$ hr.

45. The size of an insect population at time t (measured in days) is given by $P(t) = 10000 - 9000/(t + 1)$.
    a) What is the initial population $P(0)$?
    b) Find the instantaneous rate of growth at any time $t > 0$, assuming that P is a continuous function of time.

46. The heat capacity of oxygen, measured in cal/deg-mole, depends on temperature according to the formula

$$C_p = 8.27 + 2.6 \times 10^{-4} T - 1.87 \times 10^{-5} T^2,$$

where T is measured in °C, and $25 < T < 650$. Find the instantaneous rate of change of the heat capacity with respect to temperature when $T = 500°$ C.

47. For temperatures close to absolute zero ($-273°$ C) the atomic heat of aluminum is accurately described by Debye's equation

$$H = 464.6 \left(\frac{T}{375}\right)^3 \frac{\text{cal}}{\text{deg-mole}},$$

where T is the temperature in degrees Kelvin (°K = °C + 273). Find the instantaneous rate of change in the heat with respect to T when the temperature is $-250°$ C.

48. A liquid flowing through a cylindrical tube has varying velocity depending on its distance r from the center of the tube:

$$v(r) = k(R^2 - r^2),$$

where R is the radius of the tube and k is a constant that depends on the length of the tube and the velocity of the fluid at the ends (see the figure). Find the rate of change of $v(r)$ with respect to the distance from the center of the tube (a) at the center of the tube, (b) halfway between the center of the tube and the wall of the tube, and (c) at the wall of the tube.

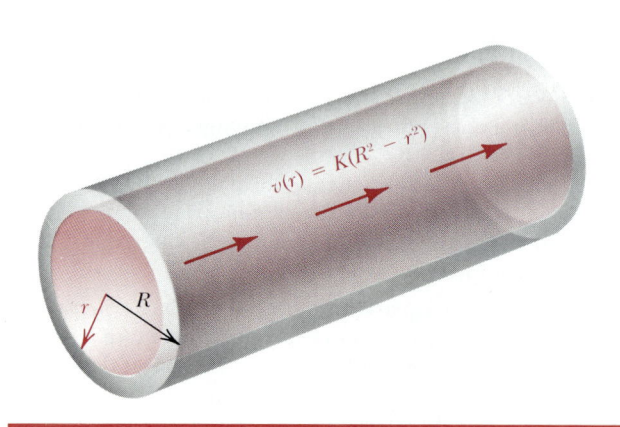

49. In Example 5 of Section 2.2 we showed that $D_x \sqrt{x} = 1/(2\sqrt{x})$, which is equivalent to stating that $D_x x^{1/2} = (1/2)x^{-1/2}$. Use this fact and the Product or Reciprocal Rule to find (a) $D_x x^{3/2}$, (b) $D_x x^{5/2}$, and (c) $D_x x^{-1/2}$.

50. Using the results of Exercise 49, find a formula for the derivative of the function described by $f(x) = x^{n/2}$, where n is a positive odd integer. Derive this formula by writing $n = 2k + 1$ for some positive integer k and expressing $x^{n/2}$ as $x^k \cdot x^{1/2}$.

51. Use the formula you deduced in Exercise 50 to find a formula for $D_x x^{-n/2}$ that is valid for any positive integer n.

52. Suppose f, g, and h are differentiable at x.
   a) Show that $D_x (fg/h)(x)$ is

   $$\frac{f'(x) g(x) h(x) + f(x)g'(x) h(x) - f(x) g(x) h'(x)}{[h(x)]^2},$$

   provided $h(x) \neq 0$.
   b) Show that $D_x (f/gh)(x)$ is

   $$\frac{f'(x) g(x) h(x) - f(x) g'(x) h(x) - f(x) g(x) h'(x)}{[g(x) h(x)]^2},$$

   provided $g(x) \neq 0$ and $h(x) \neq 0$.

53. Prove that if f and g are differentiable at x, then $D_x (f - g)(x) = D_x f(x) - D_x g(x)$.

54. a) Suppose $f'(x)$ exists. Show that for any positive integer n

   $$\lim_{h \to 0} \frac{[f(x + h)]^n - [f(x)]^n}{h} = n[f(x)]^{n-1} f'(x).$$

   [*Hint:* First factor the numerator.]
   b) Use the result in part (a) to find $D_x (2x + 1)^4$.

55. What is faulty about the following reasoning? When $x \neq 0$,

$$x^2 = x + x + \cdots + x \quad \text{(written } x \text{ times)}$$

so

$$D_x x^2 = D_x (x + \cdots + x).$$

But then

$$2x = 1 + 1 + \cdots + 1,$$

which implies that $2x = x$, and $2 = 1$.

2.4
Differentiation of Trigonometric Functions

Trigonometric applications discussed in precalculus courses are usually geometric in nature; for example, applications in navigation and surveying. With the introduction of calculus the trigonometric functions can be given a larger role. In subsequent courses you will see that these functions are fundamental to the study of heat flow, vibrations, electrical circuits, and, in fact, any topic associated with periodic oscillations.

The first step in making full use of trigonometric functions is to determine the derivatives of the two basic functions: the sine and the cosine. We can then find the derivatives of the remaining trigonometric functions easily since they are reciprocals or quotients of this pair.

The calculation of the derivative of the sine function involves finding

$$D_x \sin x = \lim_{h \to 0} \frac{\sin(x+h) - \sin x}{h}.$$

From a basic trigonometric identity, we have

$$\sin(x + h) = \sin x \cos h + \sin h \cos x,$$

so

$$D_x \sin x = \lim_{h \to 0} \frac{\sin x \cos h + \sin h \cos x - \sin x}{h}$$

$$= \lim_{h \to 0} \left(\frac{\sin x (\cos h - 1)}{h} + \cos x \frac{\sin h}{h} \right)$$

$$= \sin x \lim_{h \to 0} \frac{\cos h - 1}{h} + \cos x \lim_{h \to 0} \frac{\sin h}{h}.$$

Determining the derivative of this basic trigonometric function hinges on finding values for

$$\lim_{h \to 0} \frac{\sin h}{h} \quad \text{and} \quad \lim_{h \to 0} \frac{\cos h - 1}{h}.$$

These limits are found using the Squeeze Theorem.

(2.23)
THEOREM

$$\lim_{t \to 0} \frac{\sin t}{t} = 1$$

Proof. First we show that $\lim_{t \to 0^+} (\sin t)/t = 1$. Since we are interested in the limit as t approaches zero from the right, we can restrict t to the interval $(0, \pi/2)$.

The demonstration of the result is geometric (see Figure 2.19a). It requires comparing the areas in the triangles *OBP* and *OBA* to the area

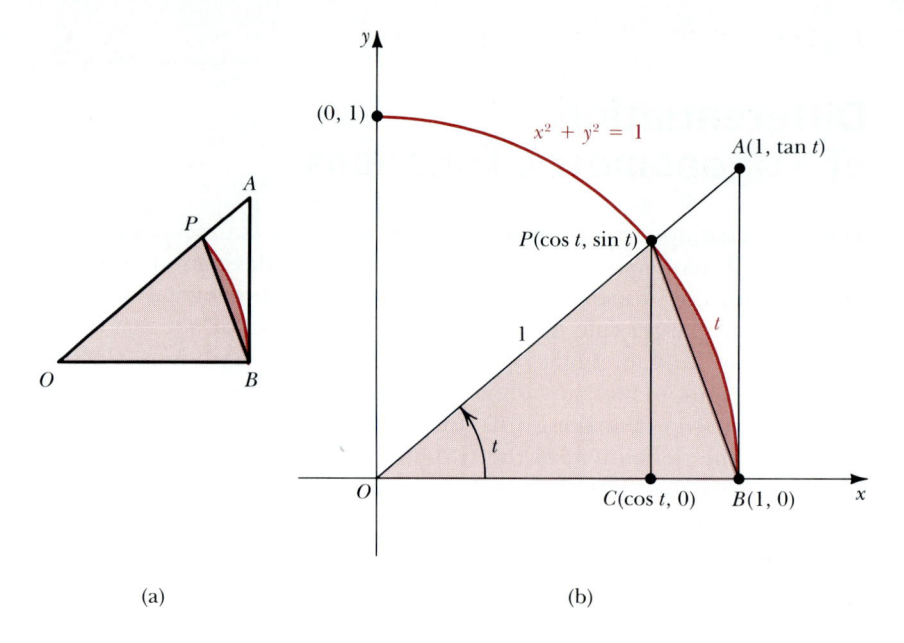

Figure 2.19
$\frac{1}{2}\sin t < t < \frac{1}{2}\tan t$.

(a)

(b)

in the sector OBP. This sector is bounded by a portion of the unit circle and by the rays of the angle with radian measure t. The coordinates of P are given by the definition of the sine and cosine functions. Before proceeding further, you should satisfy yourself that the coordinates of A, B, C, and P are as given in Figure 2.19(b).

We know that the area of a triangle is one half the product of the base and the altitude. Thus

$$\text{Area of triangle } OBP = \frac{1}{2} d(O, B) \, d(C, P) = \frac{1}{2} \cdot 1 \cdot \sin t = \frac{1}{2} \sin t$$

and

$$\text{Area of triangle } OBA = \frac{1}{2} d(O, B) \, d(B, A) = \frac{1}{2} \cdot 1 \cdot \tan t = \frac{1}{2} \frac{\sin t}{\cos t}.$$

To determine the area of the sector OBP, we use the fact that the area inside the unit circle is π and that the area of the sector OBP constitutes the portion $t/2\pi$ of the area inside the circle:

$$\text{Area of sector } OBP = \frac{t}{2\pi} \cdot \pi = \frac{1}{2} t.$$

From Figure 2.19(a) we see that

Area of triangle OBP < Area of sector OBP < Area of triangle OBA,

so

$$\frac{1}{2} \sin t < \frac{1}{2} t < \frac{1}{2} \frac{\sin t}{\cos t}.$$

Since $0 < t < \pi/2$, $\sin t > 0$ and

(2.24) $$0 < \sin t < t < \frac{\sin t}{\cos t}.$$

Multiplying inequality (2.24) by $1/\sin t$ produces

$$1 < \frac{t}{\sin t} < \frac{1}{\cos t}$$

and taking reciprocals gives

$$1 > \frac{\sin t}{t} > \cos t.$$

However, $0 < t < \pi/2$, so

$$\cos t = \sqrt{1 - \sin^2 t}.$$

In addition, inequality (2.24) implies that

$$0 < \sin^2 t < t^2, \quad \text{so} \quad \sqrt{1 - \sin^2 t} > \sqrt{1 - t^2}.$$

Consequently,

(2.25) $$1 > \frac{\sin t}{t} > \cos t = \sqrt{1 - \sin^2 t} > \sqrt{1 - t^2}.$$

But

$$\lim_{t \to 0^+} 1 = 1 \quad \text{and} \quad \lim_{t \to 0^+} \sqrt{1 - t^2} = 1,$$

so the Squeeze Theorem for limits from the right implies that

$$\lim_{t \to 0^+} \frac{\sin t}{t} = 1.$$

To show that the limit from the left agrees with this result, recall that the sine function is odd, so $\sin(-t) = -\sin t$. Thus,

$$\lim_{t \to 0^-} \frac{\sin t}{t} = \lim_{t \to 0^+} \frac{\sin(-t)}{-t} = \lim_{t \to 0^+} \frac{-\sin t}{-t} = \lim_{t \to 0^+} \frac{\sin t}{t} = 1.$$

Both right and left limits at 0 are 1, so

$$\lim_{t \to 0} \frac{\sin t}{t} = 1. \qquad \triangleright$$

One consequence of Theorem 2.23 is that the sine and cosine functions are continuous at zero:

$$\lim_{t \to 0} \sin t = \lim_{t \to 0} \left[\frac{\sin t}{t} \cdot t \right]$$

$$= \lim_{t \to 0} \frac{\sin t}{t} \cdot \lim_{t \to 0} t = 1 \cdot 0 = 0$$

$$= \sin 0.$$

Then since $\cos t = \sqrt{1 - \sin^2 t}$ for t in $(-\pi/2, \pi/2)$,

$$\lim_{t \to 0} \cos t = \lim_{t \to 0} \sqrt{1 - \sin^2 t} = \sqrt{1 - (\lim_{t \to 0} \sin t)^2} = \sqrt{1 - 0} = 1 = \cos 0.$$

The other limit needed to determine the derivative of the sine function is found in the following corollary to Theorem 2.23.

(2.26) COROLLARY

$$\lim_{t \to 0} \frac{\cos t - 1}{t} = 0$$

Proof

$$\lim_{t \to 0} \frac{\cos t - 1}{t} = \lim_{t \to 0} \left(\frac{\cos t - 1}{t} \cdot \frac{\cos t + 1}{\cos t + 1} \right)$$

$$= \lim_{t \to 0} \frac{\cos^2 t - 1}{t(\cos t + 1)} = \lim_{t \to 0} \frac{-\sin^2 t}{t(\cos t + 1)}$$

$$= \lim_{t \to 0} \frac{\sin t}{t} \cdot \frac{-\sin t}{\cos t + 1}$$

$$= \lim_{t \to 0} \frac{\sin t}{t} \cdot \lim_{t \to 0} \frac{-\sin t}{\cos t + 1}$$

Since $\lim_{t \to 0} (\sin t)/t = 1$, $\lim_{t \to 0} (-\sin t) = 0$, and $\lim_{t \to 0} \cos t = 1$,

$$\lim_{t \to 0} \frac{\cos t - 1}{t} = 1 \cdot \frac{0}{2} = 0. \qquad \triangleright$$

The basic result concerning the **derivatives of trigonometric functions** can now be derived.

(2.27) THEOREM

For any real number x,

$$D_x \sin x = \cos x \quad \text{and} \quad D_x \cos x = -\sin x.$$

Proof. At the beginning of the section we saw that

$$D_x \sin x = \lim_{h \to 0} \frac{\sin(x + h) - \sin x}{h}$$

$$= \lim_{h \to 0} \frac{\sin x \cos h + \sin h \cos x - \sin x}{h}$$

$$= \sin x \lim_{h \to 0} \frac{\cos h - 1}{h} + \cos x \lim_{h \to 0} \frac{\sin h}{h}.$$

Using Theorem 2.23 and Corollary 2.26, we have

$$D_x \sin x = \sin x \cdot 0 + \cos x \cdot 1 = \cos x.$$

In addition, since $\cos(x + h) = \cos x \cos h - \sin x \sin h$,

$$D_x \cos x = \lim_{h \to 0} \frac{\cos(x+h) - \cos x}{h}$$

$$= \lim_{h \to 0} \frac{\cos x \cos h - \sin x \sin h - \cos x}{h}$$

$$= \lim_{h \to 0} \left[\cos x \left(\frac{\cos h - 1}{h} \right) - \sin x \left(\frac{\sin h}{h} \right) \right]$$

$$= \cos x \lim_{h \to 0} \frac{\cos h - 1}{h} - \sin x \lim_{h \to 0} \frac{\sin h}{h}$$

Theorem 2.23 and Corollary 2.26 imply that

$$D_x \cos x = \cos x \cdot 0 - \sin x \cdot 1 = -\sin x. \quad \triangleright$$

Example 1

Find $D_x (x \sin x - \cos x)$.

Solution

$$D_x (x \sin x - \cos x) = (x D_x \sin x + \sin x D_x x) - D_x \cos x$$

$$= x \cos x + \sin x - (-\sin x)$$

$$= x \cos x + 2 \sin x \quad \blacktriangleright$$

Example 2

Find

$$D_x \left(\frac{x^2 + 2}{1 + \sin x} \right).$$

Solution

$$D_x \left(\frac{x^2 + 2}{1 + \sin x} \right) = \frac{(1 + \sin x) D_x (x^2 + 2) - (x^2 + 2) D_x (1 + \sin x)}{(1 + \sin x)^2}$$

$$= \frac{(1 + \sin x) 2x - (x^2 + 2) \cos x}{(1 + \sin x)^2} \quad \blacktriangleright$$

We can obtain the derivatives of the remaining trigonometric functions by using Theorem 2.27 and the Reciprocal and Quotient Rules.

(2.28)
COROLLARY

For each real number x for which the functions are defined:

a) $D_x \tan x = \sec^2 x$;
b) $D_x \sec x = \sec x \tan x$;
c) $D_x \cot x = -\csc^2 x$; and
d) $D_x \csc x = -\csc x \cot x$.

Proof

a) $$D_x \tan x = D_x\left(\frac{\sin x}{\cos x}\right) = \frac{\cos x \, D_x \sin x - \sin x \, D_x \cos x}{(\cos x)^2}$$

So

$$D_x \tan x = \frac{(\cos x)^2 + (\sin x)^2}{(\cos x)^2} = \frac{1}{\cos^2 x} = \sec^2 x.$$

b) $$D_x \sec x = D_x\left(\frac{1}{\cos x}\right) = \frac{-D_x \cos x}{(\cos x)^2} = \frac{-(-\sin x)}{(\cos x)^2}$$

So

$$D_x \sec x = \frac{1}{\cos x} \cdot \frac{\sin x}{\cos x} = \sec x \tan x$$

Parts (c) and (d) can be verified in a similar manner (see Exercises 41 and 42). ▷

Example 3

Find

$$D_t\left(\frac{\tan t}{1 + 2 \sec t}\right).$$

Solution

$$D_t\left(\frac{\tan t}{1 + 2 \sec t}\right) = \frac{(1 + 2 \sec t) D_t \tan t - \tan t \, D_t (1 + 2 \sec t)}{(1 + 2 \sec t)^2}$$

$$= \frac{(1 + 2 \sec t)(\sec^2 t) - \tan t \, (2 \sec t \tan t)}{(1 + 2 \sec t)^2}.$$ ▶

Example 4

Find the slope of the tangent line to the graph of $f(x) = x^2 \tan x + \csc x$ at $(\pi/4, f(\pi/4))$.

Solution
We have

$$f'(x) = D_x (x^2 \tan x) + D_x (\csc x)$$
$$= x^2 D_x \tan x + \tan x \, D_x x^2 + (-\csc x \cot x)$$
$$= x^2 \sec^2 x + 2x \tan x - \csc x \cot x.$$

The slope of the tangent line is given by $f'(\pi/4)$:

$$f'\left(\frac{\pi}{4}\right) = \left(\frac{\pi}{4}\right)^2 \sec^2 \frac{\pi}{4} + 2\left(\frac{\pi}{4}\right)\left(\tan \frac{\pi}{4}\right) - \csc \frac{\pi}{4} \cot \frac{\pi}{4}.$$

2.4 DIFFERENTIATION OF TRIGONOMETRIC FUNCTIONS

Using the triangle in Figure 2.20, we have

$$f'\left(\frac{\pi}{4}\right) = \frac{\pi^2}{16}(\sqrt{2})^2 + \frac{\pi}{2}(1) - \sqrt{2}(1) = \frac{\pi^2}{8} + \frac{\pi}{2} - \sqrt{2}.$$

Table 2.3 lists the derivatives of the trigonometric functions for ease of reference.

Table 2.3

$D_x \sin x = \cos x$	$D_x \cos x = -\sin x$
$D_x \tan x = \sec^2 x$	$D_x \cot x = -\csc^2 x$
$D_x \sec x = \sec x \tan x$	$D_x \csc x = -\csc x \cot x$

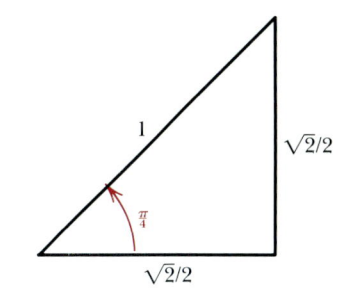

Figure 2.20

▶ EXERCISE SET 2.4

In Exercises 1–30, use the results of this section to find the derivative of the function.

1. $f(x) = \sin x + \tan x$
2. $f(x) = \cos x - \cot x$
3. $f(x) = x^3 - \cos x$
4. $f(x) = 2x^2 + \tan x$
5. $g(t) = t^3 \cos t$
6. $h(z) = 2z^4 \sin z$
7. $h(x) = \sin x \cos x$
8. $f(x) = \sec x \csc x$
9. $r(t) = \tan t \cot t$
10. $h(x) = \sec x \cos x$
11. $f(x) = \sin^2 x$
12. $f(x) = \tan^2 x$
13. $f(x) = \dfrac{2x}{\sin x}$
14. $g(r) = \dfrac{2r^2 + 1}{\cos r}$
15. $f(x) = 2x \csc x$
16. $g(r) = (2r^2 + 1) \sec r$
17. $f(x) = x \tan x + x^2 \cot x$
18. $f(x) = (x^2 + 1) \sin x + 2x \sec x$
19. $f(\theta) = \dfrac{\theta^2}{1 + \cos \theta}$
20. $g(\theta) = \dfrac{2\theta^2}{1 + \tan \theta}$
21. $h(\theta) = \dfrac{\sec \theta + 1}{\tan \theta}$
22. $h(\theta) = \dfrac{\theta + \sin \theta}{\cos \theta}$
23. $f(t) = \dfrac{\sin t \tan t}{\sin t + \cos t}$
24. $y(t) = \dfrac{\sin t + 1}{\cos t \cot t}$
25. $g(x) = (\csc x + \sec x) \sin x \cos x$
26. $h(x) = (x^2 - 1) \sin x \cos x$
27. $g(x) = x \tan^2 x$
28. $g(x) = (2x^2 - 1) \sin^2 x$
29. $r(t) = \dfrac{\csc t - \cot t}{t + 2}$
30. $s(t) = \dfrac{\csc t - \sec t}{1 - \cot t}$

In Exercises 31–40, use Theorem 2.23 and Corollary 2.26 to find the limit.

31. $\lim\limits_{t \to 0} \dfrac{\sin t}{2t}$
32. $\lim\limits_{t \to 0} \dfrac{\sin 2t}{2t}$
33. $\lim\limits_{t \to 0} \dfrac{\sin^2 t}{t}$
34. $\lim\limits_{t \to 0} \dfrac{\sin^2 t}{t^2}$
35. $\lim\limits_{t \to 0} \dfrac{\tan t}{t}$
36. $\lim\limits_{t \to 0} \dfrac{\cos^2 t - 1}{t}$
37. $\lim\limits_{t \to 0} \dfrac{t}{\sin t}$
38. $\lim\limits_{t \to 0} \dfrac{t}{1 - \cos t}$
39. $\lim\limits_{t \to 0} \left(\dfrac{t^2 + 1}{t}\right) \sin t$
40. $\lim\limits_{t \to 0} \dfrac{1 - \cos t}{\sin t}$

41. Show that $D_x \cot x = -\csc^2 x$:
    a) by using Theorem 2.27 and the Quotient Rule;
    b) by using Corollary 2.28(a) and the Reciprocal Rule.

42. Show that $D_x \csc x = -\csc x \cot x$.

43. Find an equation of the tangent line to the graph of $y = \sin x$ when $x = \pi/3$.

44. Find an equation of the normal line to the graph of $y = \sin x$ when $x = \pi/3$.

45. Given that $f(x) = \sin x$ for x in the interval $[0, 2\pi]$:
    a) Find the x-coordinate of all points at which the slope of the tangent line to the graph of f is 0.
    b) Find all values of x for which the slope of the tangent line at $(x, f(x))$ is positive.
    c) Sketch the graph of f.

46. Find the point on the graph of $y = \sin x + \cos x$, $0 \le x \le \pi$, at which the tangent line is horizontal.

47. Show that the slope of the tangent line to the graph of $y = 2x + \sin x$ is never less than 1.

48. Show that for no value of x is the tangent line to $y = \sin x$ perpendicular to the tangent line to $y = \cos x$.

49. The figure shows a triangle in which x, y, and θ are changing, but the third side remains fixed. Express x in terms of θ, and determine the instantaneous rate of change of x with respect to θ when $\theta = \pi/4$.

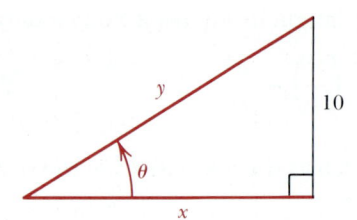

50. Use the figure for Exercise 49 to express y in terms of θ, and determine the instantaneous rate of change of y with respect to θ when $\theta = \pi/4$.

51. Consider the function described by $f(x) = \sin(1/x)$.
    a) If n is an even positive integer, what is
    $$f\left(\frac{1}{(n-\tfrac{1}{2})\pi}\right)?$$
    b) If n is an odd positive integer, what is
    $$f\left(\frac{1}{(n-\tfrac{1}{2})\pi}\right)?$$
    c) Find $\lim_{x \to 0^+} f(x)$ if this limit exists.

2.5

The Derivative of a Composite Function: The Chain Rule

In Example 10 of Section 2.3 we found that

$$D_x (x+1)^{-2} = -2(x+1)^{-3}$$

and mentioned the similarity between this result and the Power Rule:

$$D_x x^n = nx^{n-1}.$$

Consider now

$$\begin{aligned} D_x (2x^2 + 3)^2 &= D_x [(2x^2 + 3)(2x^2 + 3)] \\ &= (2x^2 + 3)(4x) + (2x^2 + 3)(4x) \\ &= 8x(2x^2 + 3). \end{aligned}$$

This is perhaps not quite the result that was expected. However, rewriting this as

$$D_x (2x^2 + 3)^2 = 2(2x^2 + 3)^1 (4x),$$

2.5 THE DERIVATIVE OF A COMPOSITE FUNCTION: THE CHAIN RULE

we see that it follows somewhat the pattern of $D_x x^n = nx^{n-1}$. The difference is that it also contains the factor $4x$, which is the derivative of $2x^2 + 3$.

The pattern of the preceding illustrations leads us to expect that

$$D_x (\sin x)^2 = 2(\sin x)^1 D_x \sin x = 2 \sin x \cos x.$$

Using the Product Rule, we can verify that this is true:

$$\begin{aligned} D_x (\sin x)^2 &= D_x [(\sin x)(\sin x)] \\ &= (\sin x)(\cos x) + (\cos x)(\sin x) \\ &= 2 \sin x \cos x. \end{aligned}$$

Each of these examples consists of the composition of basic functions whose derivatives have been studied in the first four sections of this chapter. In this section we introduce the *Chain Rule* for differentiating the composition of functions. This rule and the arithmetic rules presented in Section 2.3 provide the tools that enable us to differentiate a wide range of functions.

(2.29)
THEOREM: THE CHAIN RULE
If g is differentiable at x and f is differentiable at $g(x)$, then $f \circ g$ is differentiable at x and

$$(f \circ g)'(x) = f'(g(x)) \cdot g'(x).$$

To see why we expect the Chain Rule to be true, consider

$$(f \circ g)'(x) = \lim_{h \to 0} \frac{(f \circ g)(x + h) - (f \circ g)(x)}{h}$$

$$= \lim_{h \to 0} \left[\frac{f(g(x + h)) - f(g(x))}{h} \cdot \frac{g(x + h) - g(x)}{g(x + h) - g(x)} \right],$$

provided $g(x + h) - g(x) \neq 0$. Thus,

$$(f \circ g)'(x) = \lim_{h \to 0} \left[\frac{f(g(x + h)) - f(g(x))}{g(x + h) - g(x)} \cdot \frac{g(x + h) - g(x)}{h} \right]$$

$$= \lim_{h \to 0} \frac{f(g(x + h)) - f(g(x))}{g(x + h) - g(x)} \cdot \lim_{h \to 0} \frac{g(x + h) - g(x)}{h}.$$

Since g is differentiable at x, Theorem 2.8 implies that g is continuous at x and that

$$\lim_{h \to 0} g(x + h) = g(x).$$

If k is defined by

$$k = g(x + h) - g(x),$$

then

$$g(x + h) = g(x) + k,$$

and k approaches zero as h approaches zero. Thus we can rewrite $(f \circ g)'(x)$ as

$$\begin{aligned}(f \circ g)'(x) &= \lim_{h \to 0} \frac{f(g(x+h)) - f(g(x))}{g(x+h) - g(x)} \cdot \lim_{h \to 0} \frac{g(x+h) - g(x)}{h} \\ &= \lim_{k \to 0} \frac{f(g(x) + k) - f(g(x))}{k} \cdot \lim_{h \to 0} \frac{g(x+h) - g(x)}{h} \\ &= f'(g(x)) \cdot g'(x).\end{aligned}$$

The difficulty with constructing a proof of the Chain Rule on the basis of this motivation is that we cannot proceed beyond the first step unless we know that $g(x+h) - g(x) \neq 0$ whenever $h \neq 0$. This is not true in general. For example, if g is a constant function, then $g(x+h) - g(x) = 0$ for *all* values of x and h. Therefore, the proof must proceed on a different course. (A proof is given in Appendix B; it has been deferred to the appendix because it does not aid in the intuitive understanding of the Chain Rule.)

Example 1

Use the Chain Rule to find $D_x (x+1)^{-2}$.

Solution

Let $g(x) = x + 1$ and $f(x) = x^{-2}$. Then

$$\begin{aligned}D_x (x+1)^{-2} = (f \circ g)'(x) &= f'(g(x)) \cdot g'(x) \\ &= -2(x+1)^{-3} \cdot 1 \\ &= -2(x+1)^{-3}.\end{aligned}$$ ▶

An easy way to remember the Chain Rule is to use Leibniz notation for the derivative. Suppose $u = g(x)$ and $y = (f \circ g)(x) = f(g(x)) = f(u)$. Then,

$$\frac{dy}{dx} = (f \circ g)'(x), \quad \frac{dy}{du} = f'(u) = f'(g(x)), \quad \text{and} \quad \frac{du}{dx} = g'(x).$$

Consequently, the Chain Rule can be expressed in the form

(2.30) $$\frac{dy}{dx} = \frac{dy}{du} \cdot \frac{du}{dx},$$

a form that makes the rule seem quite natural.

The following examples use the form of the Chain Rule given in Equation (2.30).

Example 2

Find $D_x (2x^2 + 3)^2$ using the form of the Chain Rule

$$\frac{dy}{dx} = \frac{dy}{du} \cdot \frac{du}{dx}.$$

Solution

The composition can be described by the following chain:
$$x \longrightarrow 2x^2 + 3 \longrightarrow (2x^2 + 3)^2.$$

Let $u = 2x^2 + 3$, the first part of this composition, and $y = (2x^2 + 3)^2$. Then $y = u^2$, so
$$\frac{dy}{du} = 2u = 2(2x^2 + 3) \quad \text{and} \quad \frac{du}{dx} = 4x.$$

The Chain Rule implies that
$$D_x (2x^2 + 3)^2 = \frac{dy}{dx} = \frac{dy}{du} \cdot \frac{du}{dx} = 2(2x^2 + 3)(4x) = 8x(2x^2 + 3). \quad \blacktriangleright$$

Example 3

Differentiate
$$h(x) = \frac{1}{(x^2 + x + 1)^{12}}.$$

Solution

The following chain describes the composition:
$$x \longrightarrow x^2 + x + 1 \longrightarrow (x^2 + x + 1)^{-12}.$$

Let $u = x^2 + x + 1$, the first part of the composition, and $y = h(x)$. Then $y = u^{-12}$, so
$$\frac{dy}{du} = -12u^{-13} = -12(x^2 + x + 1)^{-13} \quad \text{and} \quad \frac{du}{dx} = 2x + 1.$$

The Chain Rule implies that
$$h'(x) = D_x (x^2 + x + 1)^{-12} = -12(x^2 + x + 1)^{-13}(2x + 1). \quad \blacktriangleright$$

Example 4

Determine $D_x \sin^5 x$.

Solution

This composition can be described as:
$$x \longrightarrow \sin x \longrightarrow \sin^5 x.$$

Let $u = \sin x$ and $y = \sin^5 x$. Then $y = u^5$, so
$$\frac{dy}{du} = 5u^4 = 5 \sin^4 x \quad \text{and} \quad \frac{du}{dx} = \cos x.$$

The Chain Rule implies that
$$D_x \sin^5 x = \frac{dy}{du} \frac{du}{dx} = 5 \sin^4 x \cos x. \quad \blacktriangleright$$

126   CHAPTER 2 THE DERIVATIVE

Note that in all of these examples, we have used the special case of the Chain Rule: If $u = f(x)$, then

(2.31) $$D_x u^n = nu^{n-1} \cdot D_x u$$

for any nonzero integer n. The following examples illustrate the Chain Rule when other types of composition are involved.

Example 5

Find $D_x \sin x^5$.

Solution

Note the change of order in this composition,

$$x \longrightarrow x^5 \longrightarrow \sin x^5,$$

from the chain described in Example 4. Here we let $u = x^5$, the first part of the composition, and $y = \sin x^5$. Then $y = \sin u$, so

$$\frac{dy}{du} = \cos u \quad \text{and} \quad \frac{du}{dx} = 5x^4.$$

The Chain Rule gives

$$D_x \sin x^5 = \frac{dy}{du}\frac{du}{dx} = (\cos x^5)(5x^4) = 5x^4 \cos x^5.$$ ▶

Example 6

Find $D_x \tan (3x^2 + 1)$.

Solution

The following chain describes the composition:

$$x \longrightarrow 3x^2 + 1 \longrightarrow \tan (3x^2 + 1).$$

Let $u = 3x^2 + 1$, the first part of the composition, and $y = \tan (3x^2 + 1) = \tan u$. Then

$$\frac{dy}{du} = \sec^2 u = \sec^2 (3x^2 + 1) \quad \text{and} \quad \frac{du}{dx} = 6x,$$

so

$$D_x \tan (3x^2 + 1) = \frac{dy}{du}\frac{du}{dx} = [\sec^2(3x^2 + 1)]6x = 6x \sec^2 (3x^2 + 1).$$ ▶

Example 7

Find $D_x [\cos (x^2 + 1)]^3$.

Solution

This function involves two compositions that can be expressed as

$$x \longrightarrow x^2 + 1 \longrightarrow \cos (x^2 + 1) \longrightarrow [\cos (x^2 + 1)]^3,$$

so the Chain Rule must be applied twice. It is first applied to the final composition, the one on the right of the composition chain:

$$D_x [\cos(x^2+1)]^3 = 3[\cos(x^2+1)]^2 D_x \cos(x^2+1).$$

Then since

$$D_x \cos(x^2+1) = -\sin(x^2+1) D_x (x^2+1) = -\sin(x^2+1)(2x),$$

we have

$$D_x [\cos(x^2+1)]^3 = 3[\cos(x^2+1)]^2 \{[-\sin(x^2+1)](2x)\}$$
$$= -6x \sin(x^2+1)[\cos(x^2+1)]^2. \quad \blacktriangleright$$

In the next example we use the Chain Rule in conjunction with the Quotient Rule. Problems of this type can be complicated, but they are not hard if you proceed in logical steps, as outlined in the solution, and have a good working knowledge of the basic rules of differentiation.

Example 8

Determine $f'(x)$ if

$$f(x) = \left(\frac{\sin(2x+1)}{\cos(3x^2-4)} \right)^4.$$

Solution

We can "construct" $f(x)$ as follows:

$$x \to 2x+1 \to \sin(2x+1) \searrow \frac{\sin(2x+1)}{\cos(3x^2-4)} \to \left(\frac{\sin(2x+1)}{\cos(3x^2-4)} \right)^4.$$
$$x \to 3x^2-4 \to \cos(3x^2-4) \nearrow$$

As in Example 7, the last operation used to construct $f(x)$ is the first operation to which the derivative rules are applied. Since the last operation is a composition, we use the Chain Rule:

$$f'(x) = 4 \left(\frac{\sin(2x+1)}{\cos(3x^2-4)} \right)^3 D_x \left(\frac{\sin(2x+1)}{\cos(3x^2-4)} \right).$$

Next we apply the Quotient Rule and then the Chain Rule to obtain $D_x [\sin(2x+1)/\cos(3x^2-4)]$:

$$D_x \left(\frac{\sin(2x+1)}{\cos(3x^2-4)} \right) = \frac{\cos(3x^2-4) D_x \sin(2x+1) - \sin(2x+1) D_x \cos(3x^2-4)}{[\cos(3x^2-4)]^2}$$

$$= \frac{\cos(3x^2-4)[\cos(2x+1) D_x (2x+1)] - \sin(2x+1)[-\sin(3x^2-4) D_x (3x^2-4)]}{[\cos(3x^2-4)]^2}$$

$$= \frac{\cos(3x^2-4) \cos(2x+1)(2) + \sin(2x+1) \sin(3x^2-4)(6x)}{[\cos(3x^2-4)]^2}.$$

Thus

$$f'(x) = 4 \left(\frac{\sin(2x+1)}{\cos(3x^2-4)} \right)^3 \left[\frac{2 \cos(3x^2-4) \cos(2x+1) + 6x \sin(2x+1) \sin(3x^2-4)}{[\cos(3x^2-4)]^2} \right]. \quad \blacktriangleright$$

Table 2.4

If $u = g(x)$ is a differentiable function of x, then:

$$D_x u^n = nu^{n-1} D_x u, \text{ for any integer } n \neq 0.$$

$D_x \sin u = \cos u \, D_x u$	$D_x \cos u = -\sin u \, D_x u$
$D_x \tan u = \sec^2 u \, D_x u$	$D_x \cot u = -\csc^2 u \, D_x u$
$D_x \sec u = \sec u \tan u \, D_x u$	$D_x \csc u = -\csc u \cot u \, D_x u$

The use of the Chain Rule for differentiation is so common that the derivative rules for functions are often given in a form that incorporates the Chain Rule. Listed in Table 2.4 are derivatives written in this generalized manner.

The final example in this section shows how the Chain Rule can be used in applications involving the instantaneous rate of change of a quantity with respect to time. We will consider problems of this type extensively in Section 3.3.

Example 9

The area of a circle with radius r ft is given by $A = \pi r^2$ ft². Suppose the radius of the circle depends on time t (in seconds) in accordance with the formula $r = t^3 + 1$ (see Figure 2.21).

a) Find dA/dt, the instantaneous rate of change of the area with respect to time.
b) Find dA/dt when $t = 2$ sec.

Solution

a) The Chain Rule is applied to find

$$\frac{dA}{dt} = \frac{dA}{dr} \cdot \frac{dr}{dt} = \frac{d(\pi r^2)}{dr} \cdot \frac{d(t^3 + 1)}{dt} = (2\pi r)(3t^2).$$

Expressing dA/dt in terms of t, we have

$$\frac{dA}{dt} = 2\pi(t^3 + 1)(3t^2) = 6\pi t^2(t^3 + 1).$$

Since A is given in square feet and t is in seconds, dA/dt is described in ft²/sec.

b) If we use the notation

$$\left.\frac{dA}{dt}\right|_{t=2}$$

to represent dA/dt (when $t = 2$), then

$$\left.\frac{dA}{dt}\right|_{t=2} = 6\pi 2^2(2^3 + 1) = 216\pi \, \frac{\text{ft}^2}{\text{sec}}.$$

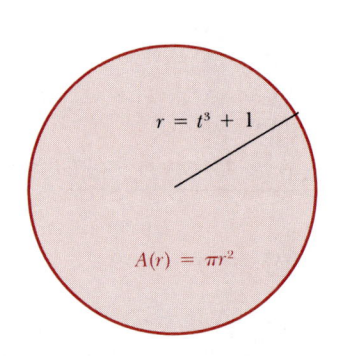

Figure 2.21

▶ EXERCISE SET 2.5

In Exercises 1–36, find the derivative of the function described.

1. $f(x) = (x + 1)^4$
2. $f(x) = (x^2 + 1)^4$
3. $f(x) = (x^2 - 3x + 4)^7$
4. $g(x) = (x^3 - x^2 + 1)^3$
5. $r(s) = (6s + 4)^{-2}$
6. $f(x) = (6x)^{-2} + 4^{-2}$
7. $f(x) = \sin \pi x$
8. $r(w) = 3 \tan 2w$
9. $f(x) = \cos(x^2 + 1)$
10. $g(x) = \cos^3 x$
11. $f(x) = \dfrac{1}{(6x^2 - 2x)^2}$
12. $h(w) = \dfrac{5}{(3w^2 + 1)^4}$
13. $h(x) = \dfrac{(3x^2 + 1)^4}{5}$
14. $f(x) = (x^2 + 3x)^{-2}$
15. $f(x) = (3x^2 + 13)^{-7}$
16. $f(t) = (3t^2 + 13t)^{-7}$
17. $g(s) = \left(\dfrac{s-1}{s+1}\right)^3$
18. $f(x) = \left(\dfrac{8x+3}{1+9x}\right)^4$
19. $f(x) = \dfrac{(x-1)^2}{(x+1)^3}$
20. $h(x) = (4x - 5)(x^2 + 2x)^2$
21. $f(x) = (\cos x - 1)^3$
22. $f(x) = \cos(x^3 - 1)$
23. $g(x) = (\cos x)^3 - 1$
24. $f(x) = \cos x^3 - 1$
25. $f(x) = \tan\left(\dfrac{x+1}{x-1}\right)$
26. $f(x) = \cos\left(\dfrac{x}{\sin x}\right)$
27. $h(s) = s^2 \cos(s^2 + 1)$
28. $g(x) = 2x^3 \sec(x^2 - 1)$
29. $f(x) = (x + x^{-1})^{-1}$
30. $k(x) = (x + x^{-1})^4$
31. $h(x) = (\tan 2x)^3 \sin(1 - x^2)$
32. $f(x) = [\sin(2x + 1)]^3 \cos(x^2 + 1)$
33. $f(x) = (\sin^3 x + 1)^2$
34. $f(x) = (\sec^2 x + 2)^3$
35. $r(t) = \left(\dfrac{\cos(2t - 1)}{\cot(t^2 + 1)}\right)^3$
36. $s(t) = \left(\dfrac{\sec(2t^2 - 1) + 2}{\csc(t^3 + 1) - 1}\right)^4$

37. Find the derivative of each of the following functions.
    a) $f_1(x) = (4x - 5)^3$
    b) $f_2(x) = (4x - 5)^3 + 2x + 1$
    c) $f_3(x) = [(4x - 5)^3 + 2x + 1]^2$
    d) $f_4(x) = \{[(4x - 5)^3 + 2x + 1]^2 - 7x\}^4$

38. Find the derivative of each of the following functions.
    a) $f_1(x) = (x^2 - 1)^2$
    b) $f_2(x) = (x^2 - 1)^2 + 4x^3 - 3$
    c) $f_3(x) = [(x^2 - 1)^2 + 4x^3 - 3]^3$
    d) $f_4(x) = \{[(x^2 - 1)^2 + 4x^3 - 3]^3 - 3x^4\}^2$

39. Find the derivative of each of the following functions.
    a) $f_1(x) = x^2 - x$
    b) $f_2(x) = \sin(x^2 - x)$
    c) $f_3(x) = [x^3 + \sin(x^2 - x)]^2$
    d) $f_4(x) = \cos[x^3 + \sin(x^2 - x)]^2$

40. Find the derivative of each of the following functions.
    a) $f_1(x) = x^2 - x$
    b) $f_2(x) = (x^2 - x)^2$
    c) $f_3(x) = \sin[x^3 + (x^2 - x)^2]$
    d) $f_4(x) = \{\sin[x^3 + (x^2 - x)^2]\}^2$

41. Find the derivative of each of the following functions.
    a) $f_1(x) = \sin x$
    b) $f_2(x) = \sin(\cos x)$
    c) $f_3(x) = \sin[\cos(x^2 + \tan x)]$
    d) $f_4(x) = \sin\{\cos[\sin^2 x + \tan(\sin x)]\}$

42. Find an equation of the tangent line to the graph of $f(x) = [(x - 1)/(x + 2)]^3$ at $(2, 1/64)$.

43. Find an equation of the normal line to the graph of $f(x) = [(x - 1)/(x + 2)]^3$ at $(2, 1/64)$.

44. Find an equation of a tangent line to the graph of $f(x) = [(x + 1)/(x - 1)]^2$ and parallel to the line with equation $y = 4x - 1$.

45. Find an equation of a tangent line to the graph of $f(x) = [(x + 1)/(x - 1)]^2$ and perpendicular to the line with equation $x = 1$.

46. The volume V of a sphere with radius r is $V = 4\pi r^3/3$. Suppose the radius of the sphere depends on time t in accordance with the formula $r(t) = t^2 + 3$. Find $D_t V$, the derivative of the volume of the sphere with respect to time.

47. The volume V of a cone with radius r and altitude h is $V = \pi r^2 h/3$. Suppose the radius and altitude of the cone are always equal and depend on time t according to the formula $r = h = 3t$. Find dV/dt, the derivative of the volume of the cone with respect to time.

48. Suppose the radius of a circle is increasing at the rate of 5 cm/sec, that is, $dr/dt = 5$ cm/sec.
 a) Find the instantaneous rate of change of its area with respect to time.
 b) How fast is the area increasing when the radius is 30 cm?

49. Suppose a metal sphere is heated so that its radius increases at the rate of 2 mm/sec.
 a) Find the instantaneous rate of change of its volume with respect to time.
 b) How fast is the volume increasing when the radius is 6 cm?

50. Sand is poured onto a conical pile whose radius and height are always equal, although both increase with time. The height of the pile is increasing at the rate of 3 in./sec. Find the rate at which the volume is increasing.

51. A spherical ice ball is melting so that its radius decreases at the rate of 0.1 cm/hr. Find the rate at which the volume is decreasing when the diameter is 10 cm.

52. The monthly sales s (in dollars) of a new product depend on the time t (in months) after the product has been introduced according to the rule:

$$s(t) = \frac{2000(1 + t)}{2 + t}.$$

The monthly profit is given by $P(s) = 0.25s$. Find the rate of change of the profit with respect to the time t, for any time t after the new product has been introduced.

53. The volume of a solid varies with temperature. For pure silver, the volume of a 1-g bar is

$$V = 0.095[1 + 5.83 \times 10^{-5}(T - 25.0)] \text{ cm}^3$$

at T °C. Suppose the temperature of a bar of silver depends on time t (in hours) according to the formula

$$T = 7.5 + 3t + 0.04t^2 + \frac{17.5}{t + 1} \text{ °C}.$$

Find the rate of change of the volume with respect to time when $t = 3$ hr.

The derivative of the absolute value function is

$$D_x |x| = \frac{|x|}{x}, \quad \text{if } x \neq 0.$$

In Exercises 54–57, use the Chain Rule to find $f'(x)$ for the function described.

54. $f(x) = |2x|$
55. $f(x) = |x + 3|$
56. $f(x) = |x^2 - 1|$
57. $f(x) = |1 - x^2|$

58. Suppose f and g are differentiable functions with the property that $(f \circ g)(x) = x$. Use the Chain Rule to show that $f'(g(x)) = 1/g'(x)$.

59. Use the result of Exercise 58 to find $g'(x)$ when $g(x) = \sqrt[3]{x}$, assuming that g is differentiable.

2.6

Implicit Differentiation

The relationship between two variables is generally described by an equation. In the preceding sections of this chapter we considered the derivative of y with respect to x, dy/dx, when the equation relating y to x is expressed in the explicit form

(2.32) $$y = f(x).$$

In this section we will show how to find dy/dx, assuming it exists, when the relationship between x and y is given by an equation that is not in the form of (2.32).

2.6 IMPLICIT DIFFERENTIATION

An equation involving x and y is said to define y **explicitly** as a function of x if the equation is written in the form (2.32) for some specific function f. The equation is said to define y **implicitly** as a function of x whenever a function f exists and $y = f(x)$ satisfies the equation. For example, the equation

$$y = f(x) = \sqrt{4 - x^2}$$

defines y explicitly as a function of x. The equation

$$x^2 + y^2 = 4,$$

however, defines y only *implicitly* as a function of x. In this case, y has a variety of functional relationships that satisfy the equation, including, as shown in Figure 2.22,

$$y = f(x) = \sqrt{4 - x^2}, \qquad y = g(x) = -\sqrt{4 - x^2},$$

and even

$$y = h(x) = \begin{cases} -\sqrt{4 - x^2}, & \text{if } -2 \leq x < 0, \\ \sqrt{4 - x^2}, & \text{if } 0 \leq x \leq 2. \end{cases}$$

Figure 2.22

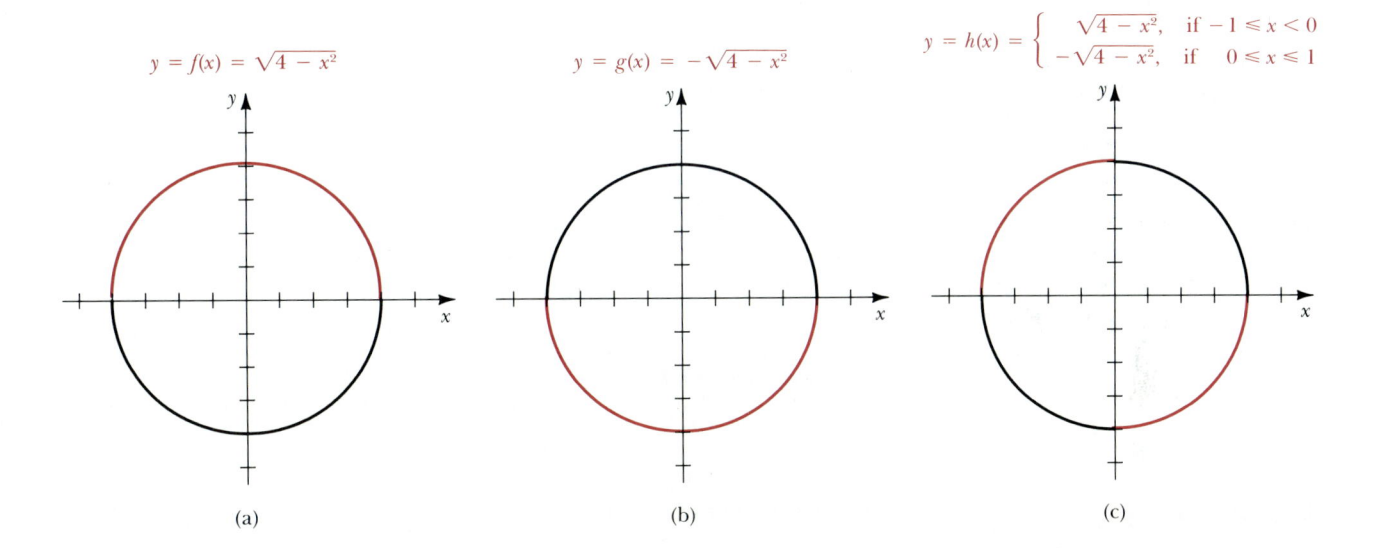

$y = f(x) = \sqrt{4 - x^2}$

$y = g(x) = -\sqrt{4 - x^2}$

$y = h(x) = \begin{cases} \sqrt{4 - x^2}, & \text{if } -1 \leq x < 0 \\ -\sqrt{4 - x^2}, & \text{if } 0 \leq x \leq 1 \end{cases}$

(a) (b) (c)

Note that the functions f and g are differentiable at $x = 0$, but the function h is not even continuous at $x = 0$. We will not consider continuity and differentiability of functions defined implicitly; these questions are best left to a course in advanced calculus. We will assume that functions defined implicitly are differentiable, if correctly restricted, and consider how these derivatives are determined.

Before demonstrating the technique called implicit differentiation, let us consider the following problem.

Find the slope of the tangent line to the circle described by $x^2 + y^2 = 4$ at the point $(1, -\sqrt{3})$.

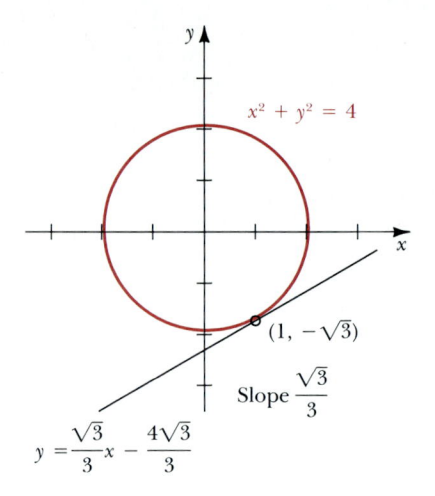

Figure 2.23

The tangent line to $x^2 + y^2 = 2$ at $(1, -\sqrt{3})$.

The equation $x^2 + y^2 = 4$ does not describe y explicitly as a function of x. However, when $y \leq 0$, the function

$$y = f(x) = -\sqrt{4 - x^2}$$

is described and $(1, -\sqrt{3})$ lies on the graph of f. In Example 5 of Section 2.2 we found that

$$D_x \sqrt{x} = \frac{1}{2\sqrt{x}}.$$

Consequently, we can use the Chain Rule to show that

$$y' = f'(x) = \frac{-1}{2\sqrt{4-x^2}} D_x (4 - x^2) = \frac{-1(-2x)}{2\sqrt{4-x^2}} = \frac{x}{\sqrt{4-x^2}}.$$

When $x = 1$, $y' = 1/\sqrt{3} = \sqrt{3}/3$, which is the slope of the tangent line to the circle at $(1, -\sqrt{3})$, as shown in Figure 2.23.

The rather lengthy solution to the problem above involved replacing the equation $x^2 + y^2 = 4$, which describes y *implicitly* as a function of x, with the equation $y = -\sqrt{4-x^2}$, which describes y *explicitly* as a function of x. Let us now solve the problem without first finding an explicit representation of y as a function of x. The Chain Rule implies that if y is a differentiable function of x, then

$$D_x y^n = ny^{n-1} D_x y.$$

Using this result and differentiating both sides of $x^2 + y^2 = 4$ with respect to x, we find

$$D_x (x^2 + y^2) = D_x (4)$$
$$D_x (x^2) + D_x (y^2) = 0$$
$$2x + 2y D_x y = 0.$$

We now solve for $D_x y$,

$$D_x y = \frac{-2x}{2y} = -\frac{x}{y},$$

and evaluate at $(1, -\sqrt{3})$:

$$D_x y \text{ (at } (1, -\sqrt{3})) = \frac{-1}{-\sqrt{3}} = \frac{1}{\sqrt{3}} = \frac{\sqrt{3}}{3}.$$

This procedure is called **implicit differentiation.** When applying this technique, we assume that the given equation defines y as a differentiable function of x, if correctly restricted, even though we may not be able to solve algebraically for y in terms of x.

Example 1

Use implicit differentiation to find dy/dx if $xy^2 + 6y + x = 0$.

Solution

When the equation $xy^2 + 6y + x = 0$ implicitly defines y as a differentiable function of x,

$$D_x(xy^2 + 6y + x) = D_x(0),$$

so

$$x\, D_x(y^2) + y^2\, D_x\, x + 6\, D_x\, y + D_x\, x = 0.$$

Since $D_x\, y^2 = 2y\, D_x\, y$ and $D_x\, x = 1$,

$$x(2y\, D_x\, y) + y^2 + 6\, D_x\, y + 1 = 0$$
$$y^2 + (2xy + 6)\, D_x\, y + 1 = 0.$$

Now solve for $D_x\, y$:

$$\frac{dy}{dx} = D_x\, y = \frac{-1 - y^2}{2xy + 6}.$$

▶

Example 2

The graph of the equation $3(x^2 + y^2)^2 = 25(x^2 - y^2)$ is the **lemniscate** shown in Figure 2.24. Find an equation of the tangent line to this curve at $(2, 1)$.

Solution

Assuming that the given equation defines y as a differentiable function of x, we take the derivative of both sides with respect to x:

$$D_x(3(x^2 + y^2)^2) = D_x(25(x^2 - y^2))$$
$$3 \cdot 2(x^2 + y^2)\, D_x(x^2 + y^2) = 25(D_x(x^2 - y^2))$$
$$6(x^2 + y^2)(2x + 2y\, D_x\, y) = 25(2x - 2y\, D_x\, y).$$

Dividing both sides by 2 and solving for $D_x\, y$ gives

$$6x(x^2 + y^2) + 6y(x^2 + y^2)\, D_x\, y = 25x - 25y\, D_x\, y.$$

so

$$(6y(x^2 + y^2) + 25y)\, D_x\, y = 25x - 6x(x^2 + y^2),$$

and

$$D_x\, y = \frac{25x - 6x(x^2 + y^2)}{25y + 6y(x^2 + y^2)}.$$

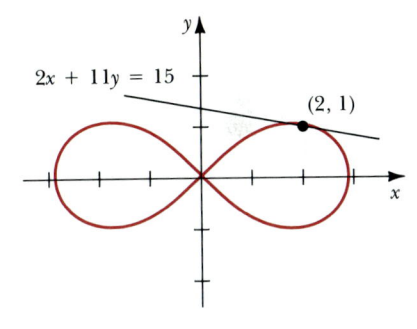

Figure 2.24

The tangent line to the lemniscate at $(2, 1)$.

The slope of the tangent line at $(2, 1)$ is

$$D_x\, y \text{ (at } (2, 1)) = \frac{(25)(2) - 6(2)(4 + 1)}{25 + 6(4 + 1)} = \frac{50 - 60}{55} = -\frac{2}{11}.$$

The tangent line to the lemniscate at $(2, 1)$ has equation

$$y - 1 = -\frac{2}{11}(x - 2) \quad \text{or} \quad 2x + 11y = 15.$$

▶

Example 3

Find $D_x y$ if $\cos xy + \sin x = 1$.

Solution

When the equation $\cos xy + \sin x = 1$ defines y as a differentiable function of x,
$$D_x (\cos xy + \sin x) = D_x (1),$$
so
$$-\sin xy \, D_x (xy) + \cos x = 0.$$
The Product Rule implies that
$$D_x (xy) = x D_x y + y D_x x = x D_x y + y,$$
so
$$-\sin xy \, (x D_x y + y) + \cos x = 0$$
and
$$-x \sin xy \, D_x y = y \sin xy - \cos x.$$
Thus,
$$D_x y = \frac{y \sin xy - \cos x}{-x \sin xy} = \frac{\cos x - y \sin xy}{x \sin xy}. \qquad \blacktriangleright$$

Implicit differentiation can also be used to extend the Power Rule.

(2.33)
THEOREM
If $f(x) = x^r$, where r is any nonzero rational number, then $f'(x) = rx^{r-1}$, whenever the derivative exists.

Proof. The result has been shown to be true when r is a nonzero integer.

If r is a nonzero rational number, then r can be written as $r = p/q$, where p and q are nonzero integers. Since
$$y = x^r = x^{p/q},$$
taking y to the qth power implies that
$$y^q = x^p.$$
Both y and x now have integral exponents, and implicit differentiation gives
$$D_x y^q = D_x x^p, \quad \text{so} \quad qy^{q-1} D_x y = px^{p-1}$$
and
$$D_x y = \frac{px^{p-1}}{qy^{q-1}}.$$

Since $y = x^{p/q}$,

$$D_x y = \frac{p}{q} \frac{x^{p-1}}{(x^{p/q})^{q-1}} = \frac{p}{q} \frac{x^{p-1}}{x^{p-p/q}} = \frac{p}{q} x^{p-1} x^{-p+p/q} = \frac{p}{q} x^{p/q-1} = rx^{r-1}. \quad \blacktriangleright$$

An alternative proof of Theorem 2.33, given in Appendix B, shows in addition that x^r is differentiable provided x^{r-1} exists. In Chapter 6 it is shown that the Power Rule also holds for irrational exponents.

Example 4

Find $f'(x)$ for $f(x) = \sqrt{x} + \dfrac{1}{\sqrt{x}}$.

Solution
We have

$$f(x) = \sqrt{x} + \frac{1}{\sqrt{x}} = x^{1/2} + x^{-1/2},$$

so

$$f'(x) = \frac{1}{2} x^{-1/2} - \frac{1}{2} x^{-3/2} = \frac{1}{2\sqrt{x}} - \frac{1}{2(\sqrt{x})^3}. \quad \blacktriangleright$$

The Power Rule is used so frequently in conjunction with the Chain Rule that we list the following result for ease of reference.

(2.34) COROLLARY
If f is a differentiable function and r is a nonzero rational number, then $D_x [f(x)]^r$ exists and

$$D_x [f(x)]^r = r[f(x)]^{r-1} D_x f(x),$$

provided $[f(x)]^{r-1}$ is defined.

Example 5
Find $D_x (\sqrt[3]{x^2 + 1})^2$.

Solution

$$\begin{aligned}
D_x (\sqrt[3]{x^2 + 1})^2 &= D_x (x^2 + 1)^{2/3} \\
&= \frac{2}{3} (x^2 + 1)^{-1/3} D_x (x^2 + 1) \\
&= \frac{2}{3} (x^2 + 1)^{-1/3} (2x) \\
&= \frac{4x}{3\sqrt[3]{x^2 + 1}}. \quad \blacktriangleright
\end{aligned}$$

▶ EXERCISE SET 2.6

In Exercises 1–16, use implicit differentiation to find $D_x y$.

1. $x^3 + y^3 = 0$
2. $x^4 - y^4 = 0$
3. $xy = 1$
4. $x^2 = y^3$
5. $\sin y = x$
6. $\tan y = x$
7. $x^2 - xy + y^2 = 0$
8. $x^2 + xy + y^2 = 0$
9. $2x^2 - x^2y^2 + y^{-3} = 4$
10. $4x^3 - 5x^2y^2 + 4y^3 = 3y$
11. $\sin x + \cos y = 1$
12. $\sin xy = \cos x$
13. $y^2 - \sin x = 0$
14. $\tan x + xy + y^2 = 0$
15. $y = \tan xy$
16. $y = \sec xy$

In Exercises 17–48, find $D_x y$.

17. $y = x^{1/3}$
18. $y = (x + 1)^{1/3}$
19. $y = 3x^{4/5}$
20. $y = (3x)^{4/5}$
21. $y = (x + 1)^{2/3} + x$
22. $y = ((x + 1)^{2/3} + x)^{1/3}$
23. $y = \cos \sqrt{x}$
24. $y = \cos \sqrt{x^2 + 1}$
25. $y = \sqrt[4]{x^2 + 2x + 2}$
26. $y = \sqrt[3]{\dfrac{3x - 1}{2x + 5}}$
27. $y = \sqrt[3]{(3x^2 + 4x)^2}$
28. $y = \sqrt[3]{x} + \dfrac{1}{\sqrt[3]{x}}$
29. $y = \sec \sqrt{x^2 + 1}$
30. $y = \sqrt{\sec(x^2 + 1)}$
31. $y = \sqrt{x^2 + \tan x}$
32. $y = \sqrt[3]{\csc x + \cot x}$
33. $y = x^{1/2} + x^{3/4} + x^{-6/5}$
34. $y = [x^2 + (x^2 + 4)^{1/2}]^{1/2}$
35. $y = (x^2 + 2)^{1/3}(x^2 - 2)^{1/4}$
36. $y = (x^2 - 2)^{1/3}(x^2 - 2)^{1/4}$
37. $y = 6x^2 + \dfrac{3}{x} - \dfrac{2}{3\sqrt[3]{2x^2}}$
38. $y = x\sqrt{4x^2 + 1} + \dfrac{5x^{5/3}}{4x - 2}$
39. $y = \sqrt{x^2 + \sqrt{x^2 + \sqrt{x + 1}}}$
40. $y = \dfrac{(x - 1)^{1/2}(x + 1)^{1/2}}{(x - 2)^{1/2}(x + 2)^{1/2}}$
41. $y = \sqrt[4]{\dfrac{1 + \sin x^2}{\cos(x^2 + 1)}}$
42. $y = \sqrt[4]{\dfrac{\sin(x^2 + 1)}{1 + \cos x^2}}$
43. $x^{1/3} + y^{2/3} = 5$
44. $\sqrt{xy} = x + y$
45. $y^2 = \dfrac{x}{xy + 1}$
46. $x^2y + xy^2 = 6(x^2 + y^2)$
47. $\sin \dfrac{x}{y} = x$
48. $\sin \dfrac{x}{y} = y$

49. Find an equation of the tangent line to the graph of $f(x) = 3x^{1/3}$ at $(1, 3)$.
50. Find an equation of the normal line to the graph of $f(x) = 3x^{1/3}$ at $(1, 3)$.
51. Find the slope of the tangent line to the graph of $x^2y - 3x + y = 4$ at $(1, \tfrac{7}{2})$.
52. Find the slope of the normal line to the graph of $x^2y^2 - 2xy + x = 1$ at $(1, 2)$.
53. Find an equation of the tangent line to the graph of $x^3 + y^3 = 2$ at $(1, 1)$.
54. Find an equation of the normal line to the graph of $x^3 + y^3 = 2$ at $(1, 1)$.
55. Show that if $\sin y = x$ and $-\pi/2 < y < \pi/2$, then
$$\frac{dy}{dx} = \frac{1}{\sqrt{1 - x^2}}.$$
56. Show that if $\tan y = x$ and $-\pi/2 < y < \pi/2$, then
$$\frac{dy}{dx} = \frac{1}{1 + x^2}.$$

57. The graph of the equation $x^2 + y^2 = r^2$ for a positive constant r is a circle of radius r with center at $(0, 0)$. Find the slope of the tangent line to the circle at an arbitrary point (x_0, y_0) on the circle.

58. Use the result of Exercise 57 to prove that the tangent line to a circle at any point on the circle is perpendicular to the line joining the point and the center of the circle.

59. The graph of $x^2/a^2 + y^2/b^2 = 1$ is called an *ellipse* (see the figure). Show that an equation for the tangent line at an arbitrary point (x_0, y_0) on the ellipse is
$$\frac{xx_0}{a^2} + \frac{yy_0}{b^2} = 1.$$

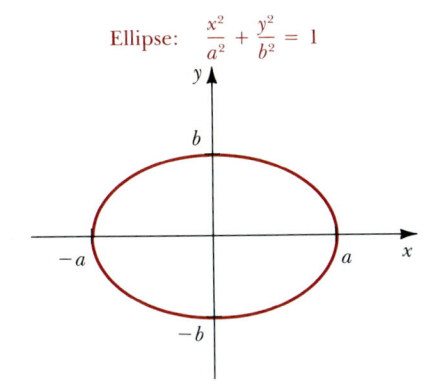

Ellipse: $\dfrac{x^2}{a^2} + \dfrac{y^2}{b^2} = 1$

60. The graph of $x^2/a^2 - y^2/b^2 = 1$ is called a *hyperbola* (see the figure). Show that an equation for the tangent line at an arbitrary point (x_0, y_0) on the hyperbola is
$$\frac{xx_0}{a^2} - \frac{yy_0}{b^2} = 1.$$

Hyperbola: $\dfrac{x^2}{a^2} - \dfrac{y^2}{b^2} = 1$

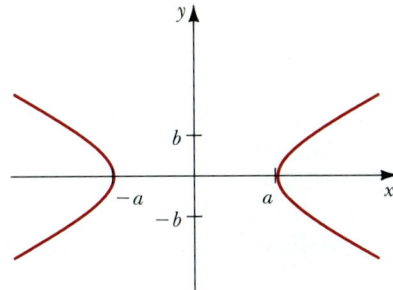

61. a) Sketch the graph of the equation $x^4 + y^4 = 1$.
    b) Find all points on the graph that have the property that the tangent line to the graph is perpendicular to the line joining the point and $(0, 0)$.

62. The graph of $x^{2/3} + y^{2/3} = 1$ is called an *astroid* (see the figure). Find an equation of the line tangent to the astroid at $(-1/8, 3\sqrt{3}/8)$.

Astroid: $x^{2/3} + y^{2/3} = 1$

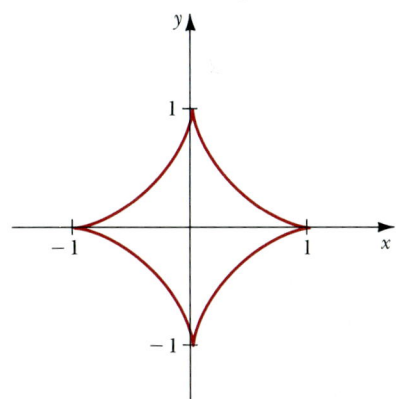

63. The graph of $x^3 + y^3 = 9xy$ is called a *folium of Descartes* (see the figure). Find an equation of the line tangent to the folium of Descartes at $(4, 2)$.

Folium of Descartes: $x^3 + y^3 = 9xy$

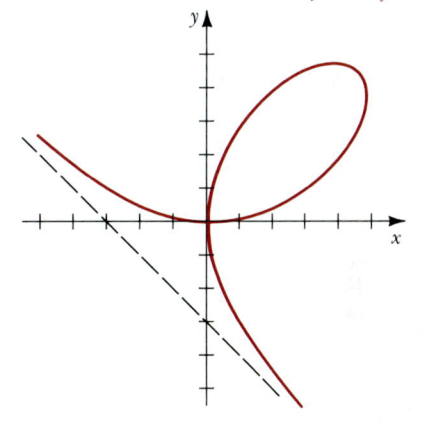

64. Suppose that for each positive rational number r, a function f_r is defined by
$$f_r(x) = \begin{cases} 0, & \text{if } x \leq 0, \\ x^r, & \text{if } x > 0. \end{cases}$$
For which values of r is f_r differentiable at every real number?

65. Recall that $|x| = \sqrt{x^2}$ for any real number x. Use this representation to find $f'(x)$ if:
    a) $f(x) = |x|$,
    b) $f(x) = |x^2 - 1|$,
    c) $f(x) = \sqrt{|x + 3|}$,
    d) $f(x) = |\sin x|$.

66. The production cost x and profit $p(x)$ of a firm are related by the equation $3x + 0.1x\, p(x) + (p(x))^2 = 130$, $0 \leq x \leq 40$, where x and $p(x)$ are measured in hundreds of dollars per week. Find the instantaneous rate of change of the profit with respect to the production cost when the production cost is \$3000 per week and the profit is \$500 per week.

67. Van der Waal's equation of state for a gas is

$$\left(P + \frac{a}{V^2}\right)(V - b) = RT,$$

written for one mole of the gas, where R, a, and b are constants depending on the gas and T, V, and P represent, respectively, temperature, volume, and pressure. Suppose that the volume and temperature vary while the pressure is held constant. Use implicit differentiation to find the rate of change of the volume with respect to the temperature.

2.7

Maxima and Minima of Functions

One of the great advantages of calculus is the ability to use the derivative to help find maximum and minimum values of functions. This feature has made calculus applicable in many areas of science, engineering, and business. In business, for example, it is used to decide what price a firm should set for its product in order to produce the maximum profit. In engineering, it is used to determine the maximum stress on a proposed structure before the structure is designed to ensure a proper margin of safety. Once this maximum stress is known, it may also be used to determine the minimum cost necessary to meet the specifications.

(2.35)
DEFINITION
A function f defined on an interval I is said to have an **absolute maximum** at c in I if $f(x) \leq f(c)$ for all x in I. Similarly, f is said to have an **absolute minimum** at c in I if $f(x) \geq f(c)$ for all x in I.

If f has either an absolute maximum or an absolute minimum at c, then f is said to have an **absolute extremum** at c.

Before attacking the problem of determining extreme values for functions we consider a result called the *Extreme Value Theorem,* which ensures, under suitable conditions, that such extreme values exist. Although the statement of the theorem is quite elementary, the proof involves theory that is generally deferred to a course in advanced calculus.

(2.36)
THEOREM: THE EXTREME VALUE THEOREM
If f is continuous on a closed interval $[a, b]$, then there exist numbers c_1 and c_2 in $[a, b]$ with the property that for all x in $[a, b]$,

$$f(c_1) \leq f(x) \leq f(c_2).$$

In Theorem 2.36, $f(c_1)$ is the absolute minimum value of $f(x)$ on $[a, b]$ and $f(c_2)$ is the absolute maximum value on this interval (see Figure 2.25).

Although the Extreme Value Theorem ensures that absolute maximum and minimum values exist for functions continuous on closed intervals, it does not provide a constructive method for determining these values. Methods of calculus are used to determine these values for many functions. The following definition is needed to describe these methods.

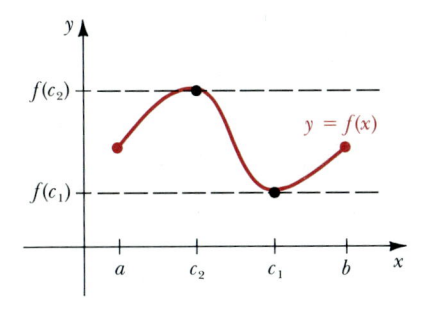

Figure 2.25
Extreme values of f on $[a, b]$.

(2.37)
DEFINITION
A function f is said to have a **relative, or local, maximum** at c if there exists an open interval I, containing c, with the property that $f(x) \leq f(c)$ for all x in I. Similarly, f is said to have a **relative, or local, minimum** at c if there exists an open interval I, containing c, with the property that $f(x) \geq f(c)$ for all x in I.

If f has either a relative maximum or a relative minimum at c, then f is said to have a **relative extremum** at c.

The function f whose graph is shown in Figure 2.26 has relative minima at c_1 and c_3 and relative maxima at c_2 and c_4. The graph suggests that the tangent lines at c_1, c_2, and c_3 have slope 0 and that there is no unique tangent line to the graph at c_4; that is, $f'(c_1) = f'(c_2) = f'(c_3) = 0$ and $f'(c_4)$ does not exist.

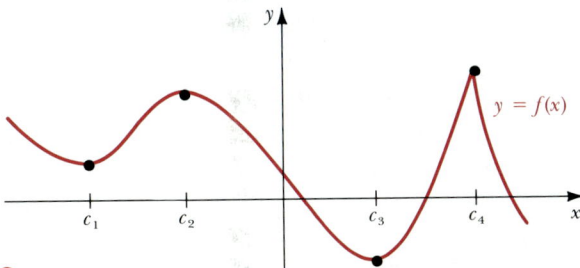

Figure 2.26
$f'(c_1) = f'(c_2) = f'(c_3) = 0$; $f'(c_4)$ does not exist.

The following theorem gives an important connection between the derivative of a function and the location of the numbers at which a function has a relative maximum or relative minimum. This result provides the basis for many applications involving the derivative.

(2.38)
THEOREM
Suppose f is a function that has a relative minimum or relative maximum at c. If $f'(c)$ exists, then $f'(c) = 0$.

Proof. We will show the result for the case when f has a relative maximum at c. The case when f has a relative minimum at c is handled similarly.

Since a relative maximum occurs at c, there is an open interval I containing c with the property that

$$f(x) \leq f(c) \quad \text{for all } x \text{ in } I.$$

If $f'(c)$ exists, both one-sided limits of the difference quotient exist and are equal to $f'(c)$; that is,

$$f'(c) = \lim_{h \to 0^+} \frac{f(c+h) - f(c)}{h} \quad \text{and} \quad f'(c) = \lim_{h \to 0^-} \frac{f(c+h) - f(c)}{h}.$$

We restrict the values of h so that $c + h$ is in I. Then

$$f(c+h) \leq f(c) \quad \text{and} \quad f(c+h) - f(c) \leq 0.$$

When $h > 0$,

$$\frac{f(c+h) - f(c)}{h} \leq 0.$$

Therefore,

$$\lim_{h \to 0^+} \frac{f(c+h) - f(c)}{h} \leq 0.$$

However, when $h < 0$,

$$\frac{f(c+h) - f(c)}{h} \geq 0,$$

so

$$\lim_{h \to 0^-} \frac{f(c+h) - f(c)}{h} \geq 0.$$

This implies that

$$0 \leq \lim_{h \to 0^-} \frac{f(c+h) - f(c)}{h} = f'(c) = \lim_{h \to 0^+} \frac{f(c+h) - f(c)}{h} \leq 0$$

and that $f'(c) = 0$. ▷

It follows from Theorem 2.38 that:

(2.39)
A function can have a relative maximum or a relative minimum only at those numbers in its domain at which the derivative is undefined or is zero. These numbers are called *critical numbers* or, more commonly, **critical points.** Consequently,

> the only numbers at which a function can have a relative maximum or a relative minimum are the critical points.

Example 1

Find all relative maxima and minima of the function given by $f(x) = 2x^3 - 9x^2 + 12x - 2$. Sketch the graph of f.

Solution

The derivative given by

$$f'(x) = 6x^2 - 18x + 12$$

is defined for all values of x, so relative maxima or minima of f can occur only when $f'(x) = 0$. Since

$$f'(x) = 6x^2 - 18x + 12 = 6(x^2 - 3x + 2) = 6(x-1)(x-2),$$

$f'(x) = 0$ at $x = 1$ and $x = 2$. The only critical points are $x = 1$ and $x = 2$.

Evaluating the function at the critical points, we have $f(1) = 3$ and $f(2) = 2$. Also, since $f(0) = -2$, the graph intersects the y-axis at $(0, -2)$.

Using the knowledge that $\lim_{x \to \infty} f(x) = \infty$ and $\lim_{x \to -\infty} f(x) = -\infty$, we find the graph of f shown in Figure 2.27. The function f has a relative maximum at $x = 1$, with $f(1) = 3$, and a relative minimum at $x = 2$, with $f(2) = 2$. ▶

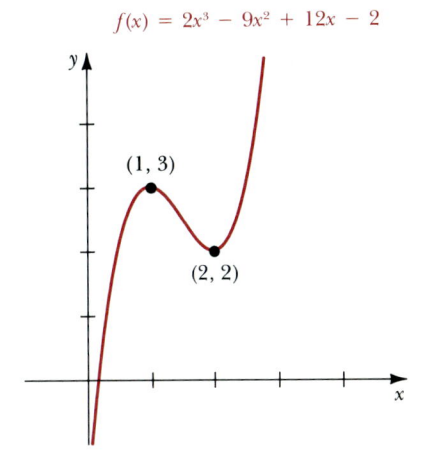

Figure 2.27
Critical points: $x = 1$, $x = 2$.

Example 2

Determine all relative maxima and minima of the function f, where $f(x) = x^2 - 3x^{2/3}$. Sketch the graph of f.

Solution

This function has a derivative given by

$$f'(x) = 2x - 2x^{-1/3},$$

which is undefined at $x = 0$. So $x = 0$ is a critical point.

To determine any remaining critical points, we set $f'(x) = 0$ and solve for x:

$$0 = 2x - 2x^{-1/3} = 2x^{-1/3}(x^{4/3} - 1),$$

so

$$x^{4/3} = 1.$$

Solutions to this equation are $x = \pm 1$, so f has critical points $x = -1$ and $x = 1$ in addition to the critical point $x = 0$.

Evaluating f at the critical points, we see that $f(0) = 0$, $f(-1) = -2$, and $f(1) = -2$.

The graph is symmetric with respect to the y-axis and intersects the x-axis only when

$$x^2 - 3x^{2/3} = x^{2/3}(x^{4/3} - 3) = 0,$$

that is, when $x = 0$ and $x = \pm 3^{3/4} \approx \pm 2.3$.

Since $\lim_{x \to \infty} f(x) = \infty$, the graph is as shown in Figure 2.28. The shape of the graph at $(0, 0)$ reflects the fact that $f'(x)$ does not exist at $x = 0$. The function f has relative (and absolute) minima at $x = \pm 1$, with $f(1) = f(-1) = -2$, and a relative maximum at $x = 0$, with $f(0) = 0$. ▶

It is *not* true that a relative maximum or minimum occurs at every critical point. Consider, for example, $f(x) = x^3$. Since $D_x x^3 = 3x^2$, the

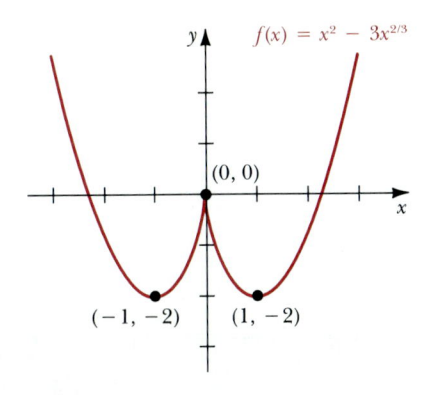

Figure 2.28
Critical ponts: $x = -1$, $x = 0$, $x = 1$.

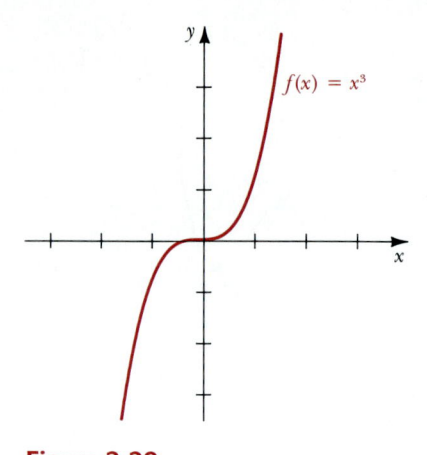

Figure 2.29
No extrema at the critical point $x = 0$.

only critical point is $x = 0$. However,

$$\text{if} \quad x < 0, \quad \text{then} \quad f(x) < f(0) = 0,$$

but

$$\text{if} \quad x > 0, \quad \text{then} \quad f(x) > f(0) = 0.$$

Thus f has neither a relative maximum nor a relative minimum at $x = 0$, as we can see from the graph in Figure 2.29.

Examples 3 and 4 illustrate a method for determining the absolute maximum and absolute minimum values of a function that is continuous on a closed interval:

(2.40) EXTREMA OF A FUNCTION ON A CLOSED INTERVAL

i) Find the critical points of the function.
ii) Find the values of the function at the critical points.
iii) Find the values of the function at the endpoints.
iv) The absolute maximum value of the function is the largest of those found in (ii) and (iii).
v) The absolute minimum value of the function is the smallest of those found in (ii) and (iii).

Example 3

Determine the absolute maximum and absolute minimum values of $f(x) = 2x^3 - 9x^2 + 12x - 2$ on the interval $[0, 4]$.

Solution

i) In the solution to Example 1 we found that the critical points of f are $x = 1$ and $x = 2$.
ii) The values of the function at the critical points are

$$f(1) = 3 \quad \text{and} \quad f(2) = 2.$$

iii) The values of the function at the endpoints are

$$f(0) = -2 \quad \text{and} \quad f(4) = 30.$$

iv) The absolute maximum value of $f(x)$ on $[0, 4]$ is $f(4) = 30$.
v) The absolute minimum value of $f(x)$ on $[0, 4]$ is $f(0) = -2$.

The graph of f shown in Figure 2.30 confirms these results. ▶

Example 4

Determine the absolute maximum and absolute minimum values of $f(x) = x^2 - 3x^{2/3}$ on $[0, 8]$.

Solution
In Example 2 we found that f has a relative minimum at $x = \pm 1$ and $f(1) = f(-1) = -2$. Also, f has a relative maximum at 0 and $f(0) = 0$.

Figure 2.30

Considering the values of f at the endpoints we see that $f(0) = 0$ and $f(8) = 52$. So f has an absolute minimum of -2 at the critical point $x = 1$ and an absolute maximum of 52 at the endpoint $x = 8$. ▶

Absolute maximum and minimum values need not exist for functions defined on intervals that are not closed. For example, the function considered in Example 4, $f(x) = x^2 - 3x^{2/3}$, assumes no absolute maximum on the interval $(0, 8)$ since $\lim_{x \to 8^-} f(x) = 52$, but $f(x) < 52$ for all x in $(0, 8)$. It does have an absolute minimum at $x = 1$, however.

The maximization and minimization of quantities, particularly absolute maximum or minimum values, has application in many areas. The following example illustrates a simple business application. Applications involving extrema are considered in detail in Section 3.2.

Example 5

The profit realized from selling x tons of sand per day, where $4 \leq x \leq 40$, is given in dollars per day by the function

$$P(x) = 44x - x^2 - 160.$$

What number of tons of sand sold per day produces the maximum profit?

Solution
The mathematical interpretation of this question is: What number in the interval $[4, 40]$ produces an absolute maximum for the function P?

First find the critical points of P:

$$P'(x) = 44 - 2x = 2(22 - x).$$

Thus the only critical point occurs at $x = 22$. Since $P(22) = 324$, $P(4) = 0$, and $P(40) = 0$, P has an absolute maximum at $x = 22$. The maximum profit that can be realized is $P(22) = 324$ dollars per day.

The domain of the function in this example was restricted to the interval $[4, 40]$. Note that the profit $P(x)$ is positive only for x in this interval (see Figure 2.31). ▶

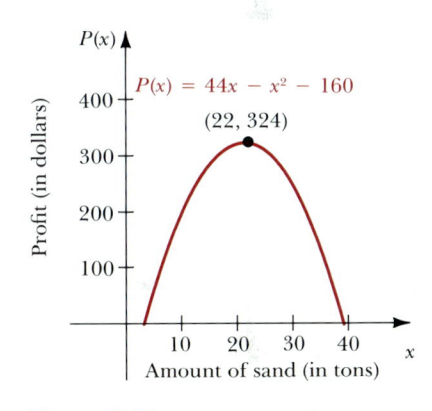

Figure 2.31

▶ EXERCISE SET 2.7

In Exercises 1–32, find the critical points of the function.

1. $f(x) = x^2 - 4x + 4$
2. $f(x) = 9 - 3x^2$
3. $f(x) = x^3 - 6x^2 + 9x + 1$
4. $f(x) = x^3 - 7x^2 + 14x - 8$
5. $f(x) = x^3 + 3x^2 - 24x + 7$
6. $f(t) = t^3 + t^2 + t + 1$
7. $h(s) = s^3 + s^2 - s - 1$
8. $f(z) = z^3 + z^2 + z - 1$
9. $r(w) = 3w^5 - 25w^3 + 60w + 7$
10. $g(t) = 3t^4 - 8t^3 - 30t^2 + 72t + 13$
11. $f(x) = \dfrac{1}{x}$

12. $f(x) = \dfrac{1}{x+1}$

13. $f(x) = \dfrac{x}{x-1}$

14. $g(x) = \dfrac{x+1}{x-1}$

15. $h(t) = \dfrac{t}{t^2+4}$

16. $f(t) = \dfrac{t}{t^2-4}$

17. $g(s) = \dfrac{s-2}{s^2-4}$

18. $r(t) = \dfrac{t+4}{t^2-16}$

19. $h(s) = \dfrac{s^2-3s+2}{s^2+s-2}$

20. $f(t) = \dfrac{2t^2-t-3}{4t^2-4t-3}$

21. $g(x) = \dfrac{2x^2-x-3}{x^2+x-2}$

22. $f(x) = \dfrac{x^2-3x+2}{4x^2-4x-3}$

23. $f(x) = \sqrt{x+2}$

24. $h(x) = \sqrt{x^2-4}$

25. $g(t) = \sqrt{t^2-3t+2}$

26. $h(s) = \sqrt{6s^2+s-12}$

27. $g(x) = x - \sin x$

28. $f(t) = \tan t - t$

29. $r(s) = \sin s - \cos s$

30. $f(x) = \tan x - \sec x$

31. $h(x) = \sin^2 x + \cos x$

32. $f(x) = \csc x + \cot x$

In Exercises 33–54, (a) find the critical points of the function, and use this information to sketch the graph of each function. (b) Where do relative and absolute extrema occur in each case?

33. $f(x) = x^2 - 6x + 4$
34. $f(x) = x^2 - 6x + 3$
35. $f(x) = 4 + 6x - x^2$
36. $h(x) = x^3 - 3x^2 + 3$
37. $v(t) = t^4 - 6t^2 + 8t$
38. $g(s) = s^3 - 12s + 4$
39. $f(x) = \sin(x + \pi/2)$
40. $f(x) = \cos(x - \pi/2)$
41. $f(x) = x + \dfrac{1}{x}$
42. $f(t) = \dfrac{t}{t-4}$
43. $f(x) = x^2 + \dfrac{1}{x^2}$
44. $g(x) = \dfrac{x+4}{x-4}$
45. $g(x) = \dfrac{x+4}{x^2-4}$
46. $h(x) = \dfrac{x^2+4}{x^2-4}$
47. $f(s) = s - \cos s$
48. $h(x) = 1 + \sin x$
49. $g(x) = \sin 4x$
50. $g(t) = \sin t + \cos t$
51. $f(x) = \sqrt{4 - x^2}$
52. $f(x) = \dfrac{1}{\sqrt{4-x^2}}$
53. $f(x) = |x^3 - x^2|$
54. $f(x) = |\sin x|$

In Exercises 55–70, the domain of the function is restricted to the interval given. Sketch the graph of the function on this interval, and determine the absolute maximum and absolute minimum values of the function, if these values exist.

55. $f(x) = x^2 - 4x + 3$, $[0, 4]$
56. $f(x) = x^3 - 3x$, $[-2, 2]$
57. $f(x) = \dfrac{1}{x}$, $[-4, -2]$
58. $f(x) = \dfrac{1}{x}$, $[-1, 1]$, $x \neq 0$
59. $f(x) = \dfrac{2}{x-3}$, $[4, 5]$
60. $f(x) = \dfrac{x}{(x-2)^2}$, $[-3, 1]$
61. $f(x) = 2x^3 - 3x^2 - 12x + 15$, $[0, 3]$
62. $f(x) = 2x^3 - 3x^2 - 12x + 15$, $[-2, 4]$
63. $f(x) = \sin x$, $[-\pi, \pi]$
64. $f(x) = \sin 2x$, $[-\pi, \pi]$
65. $f(x) = \cos x$, $[-1, 1]$
66. $f(x) = \sin x$, $[-\pi/4, \pi/4]$
67. $f(x) = |x^3 - 4x|$, $[-2, 2]$
68. $f(x) = |x^4 - x|$, $[-2, 2)$
69. $f(x) = \begin{cases} 2x + 7, & \text{if } -2 \leq x < -1, \\ 4 - x, & \text{if } -1 \leq x \leq 5 \end{cases}$
70. $f(x) = \begin{cases} x^2 + 2x + 1, & \text{if } -3 \leq x \leq 0, \\ 2x - x^2 - 3, & \text{if } 0 < x < 2, \\ 3 - x, & \text{if } 2 \leq x \leq 3 \end{cases}$

71. Use the result of Exercise 41 to find the minimum value possible for the sum of a positive number and its reciprocal.

72. What is the maximum slope of a tangent line to $f(x) = x^3 - 3x^2 + 4x - 5$ for x in $[-1, 2]$?

73. The cost of producing $1000x$ yd^2 of carpet per week is given by $C(x) = x^2 - 10x + 30$, where $C(x)$ is given in hundreds of dollars. Because of factory size, at most 15,000 yd^2 can be produced in one week. Sketch the

graph of this function. Find the absolute maximum and absolute minimum weekly costs.

74. The size (in square centimeters) of a bacteria population residing on a nutrient agar at time t (in hours) is given by $P(t) = 1000 + 100t - 10t^2$. Give the domain of P, and find the absolute maximum and absolute minimum size of the bacteria population.

75. The solubility of sucrose in water is temperature-dependent according to the formula

$$S = 64.542 + 7.982 \times 10^{-2}T + 1.9658 \times 10^{-3}T^2 - 9.691 \times 10^{-6}T^3,$$

where T is measured in degrees Celsius and S is given in grams of sucrose per 100 grams of solution. Find the lowest temperature that produces a maximum solubility.

76. Another equation that describes the solubility of sucrose in water is

$$S = \begin{cases} 64.53 + 9.37 \times 10^{-2}T + 1.2 \times 10^{-3}T^2, \\ \qquad \text{if } T \leq 38.04, \\ 61.15 + 2.25 \times 10^{-1}T + 8.4 \times 10^{-5}T^2, \\ \qquad \text{if } T > 38.04. \end{cases}$$

Find the lowest temperature that produces a maximum solubility.

77. The power P of a steam turbine depends on the peripheral speed of the wheel surrounding the turbine blades. If S_1 is the speed of the steam entering the turbine and S_2 is the peripheral speed of the wheel, then

$$P = kS_2(S_1 - S_2)$$

for some constant k. Suppose the turbine blades can be tilted to vary S_2 while S_1 remains constant. What value of S_2 produces the maximum power?

78. The velocity of a particular class of chemical reaction obeys the law $v = a(b + x)(c - x)$, where a, b, and c are positive constants and x is the amount of substrate that has decomposed. Find the value of x at which the velocity is a maximum.

79. The solubility of cuprous chloride in solutions containing excess chloride is

$$S = \frac{1.8 \times 10^{-7}}{[\text{Cl}^-]} + 1.0 \times 10^{-5} + 7.7 \times 10^{-2}\,[\text{Cl}^-],$$

where $[\text{Cl}^-]$ is the concentration of free chloride moles per liter. Find the concentration that minimizes S.

80. Prove Theorem 2.38 for the case when f has a local minimum at c.

81. Use the fact that a polynomial has no more distinct roots than its degree to show that a polynomial of degree n has at most $n - 1$ critical points.

82. Suppose f is defined by

$$f(x) = \frac{ax + b}{cx + d}$$

for some constants a, b, c, and d and that f has a critical point. What can be said about the constants a, b, c, and d?

83. The Extreme Value Theorem states that any function continuous on a closed interval $[a, b]$ assumes both its maximum and minimum values on that interval. This result need not be true unless the interval is closed.
   a) Find a function that is continuous on the interval $(0, 1]$ and does not assume a maximum on this interval.
   b) Find a function that is continuous on $(0, 1]$ and does not assume a minimum on this interval.
   c) Find a function that is continuous on $(0, 1)$ and assumes neither a maximum nor a minimum on this interval.
   d) Find a function that is continuous on $(0, 1]$ and assumes neither a maximum nor a minimum on this interval. (One possibility: Modify the function $f(x) = \sin(1/x)$ and use the fact that the sine function is continuous.)

Putnam exercises

84. Let $f(x) = a_1 \sin x + a_2 \sin 2x + \cdots + a_n \sin nx$, where $a_1, a_2, \ldots, a_n$ are real numbers and where n is a positive integer. Given that $|f(x)| \leq |\sin x|$ for all real x, prove that

$$|a_1 + 2a_2 + \cdots + na_n| \leq 1.$$

(This exercise was problem A–1 of the twenty-eighth William Lowell Putnam examination given on December 2, 1967. The examination and its solution are in the September 1968 issue of the *American Mathematical Monthly*, pp. 734–739.)

85. Consider a polynomial $f(x)$ with real coefficients having the property $f(g(x)) = g(f(x))$ for every polynomial $g(x)$ with real coefficients. Determine and prove the nature of $f(x)$.

(This exercise was problem 5, part I of the twenty-first William Lowell Putnam examination given on December 3, 1960. The examination and its solution are in the September 1961 issue of the *American Mathematical Monthly*, pp. 632–637.)

2.8
The Mean Value Theorem

The main topic of this section is the Mean Value Theorem. This theoretical result ensures that the graph of a differentiable function defined on a closed interval assumes its mean, or average, slope at some point within the interval. The Mean Value Theorem permits us to relate the derivative of a function to the behavior of the function without directly involving a limit. Since virtually every application of calculus has a connection with the Mean Value Theorem, it is perhaps the single most important theorem in the study of calculus.

The proof of the Mean Value Theorem is facilitated by first considering the following lemma. (A **lemma** is a result that is useful in its own right, but whose primary purpose is to prove a major theorem.)

(2.41)
LEMMA: ROLLE'S THEOREM
Suppose f is continuous on the interval $[a, b]$ and differentiable on (a, b). If $f(a) = f(b)$, then there exists at least one number c, $a < c < b$, with $f'(c) = 0$.

Proof. There are two possibilities for the function f.

i) Suppose f is a constant function on $[a, b]$ (see Figure 2.32). In this case, $f'(x) = 0$ for all x in (a, b) and c can be chosen to be any number between a and b.

ii) Suppose f is not constant on $[a, b]$ (see Figure 2.33). In this case, there must exist an x in (a, b) with either $f(x) > f(a)$ or $f(x) < f(a)$. For the sake of argument, suppose $f(x) > f(a)$. The Extreme Value Theorem guarantees that there is a number c in $[a, b]$ at which f assumes its maximum value. Since $f(c) \geq f(x) > f(a) = f(b)$, c cannot be either a or b. But an absolute maximum occurs at c, and an absolute maximum can occur only at an endpoint of the interval or at a critical point of f. Therefore, c must be a critical point of f. Since f' exists at each number in (a, b), $f'(c) = 0$. ▷

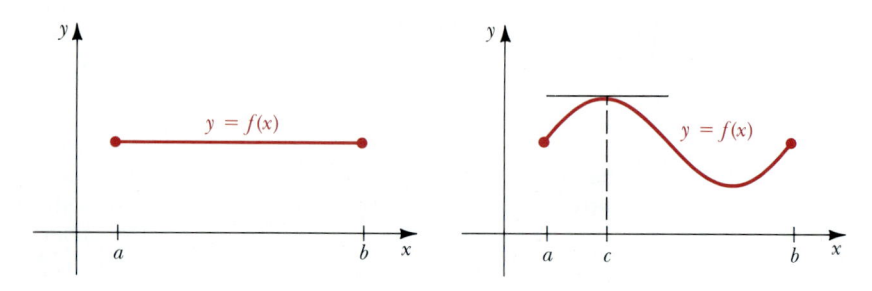

Figure 2.32

Figure 2.33

Rolle's Theorem states that if f is continuous on $[a, b]$ and differentiable on (a, b), and $f(a) = f(b)$, then at some point $(c, f(c))$, with $a < c < b$, the tangent line to the graph of f is horizontal. That is, the tangent line is parallel to the line joining the points $(a, f(a))$ and $(b, f(b))$. The Mean Value Theorem makes a similar geometric statement when $f(a)$ and $f(b)$ are not necessarily equal.

(2.42)
THEOREM: THE MEAN VALUE THEOREM
If a function f is continuous on the interval $[a, b]$ and differentiable on (a, b), then there exists at least one number c, with $a < c < b$, and

$$f'(c) = \frac{f(b) - f(a)}{b - a}.$$

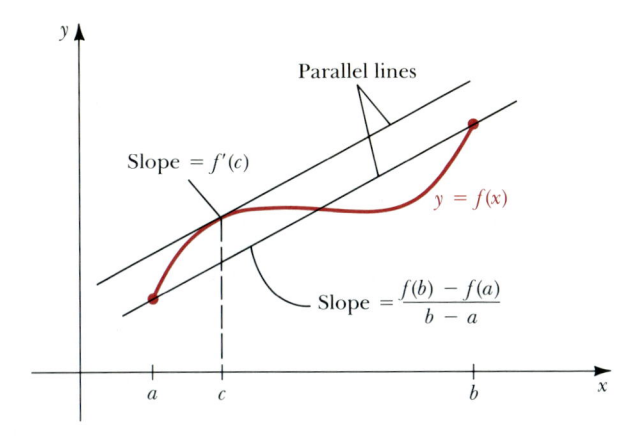

Figure 2.34
$f'(c) = \frac{f(b) - f(a)}{b - a}.$

Before proving the Mean Value Theorem, consider the graph shown in Figure 2.34 of a function continuous on $[a, b]$ and differentiable on (a, b). Since

$$\frac{f(b) - f(a)}{b - a}$$

is the slope of the line joining the points $(a, f(a))$ and $(b, f(b))$, and $f'(c)$ is the slope of the tangent line to the graph of f at $(c, f(c))$, the Mean Value Theorem asserts that a number c exists between a and b at which these two slopes are equal. That is, the tangent line to the graph of f at $(c, f(c))$ is parallel to the line joining the points $(a, f(a))$ and $(b, f(b))$. (In our sketch, there are actually two choices for c.)

HISTORICAL NOTE

Michel Rolle (1652–1719) was a respected member of the Académie des Sciences in France who initially criticized the "new" calculus proposed by Isaac Newton and Gottfried Leibniz. The theorem that bears his name appeared in a little-known treatise on geometry and algebra entitled *Méthode pour résoudre les égalités* published in 1691. It is ironic that one of the basic results in the theory of calculus was proved by a person who was vigorously opposed to the calculus methods of his contemporaries. In later life, Rolle acknowledged that the calculus techniques were of value and basically sound.

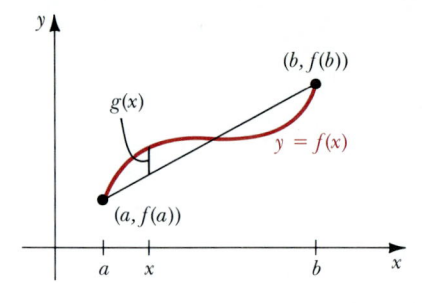

Figure 2.35

$$g(x) = f(x) - \left[\frac{f(b) - f(a)}{b - a} (x - a) + f(a) \right].$$

Proof. We begin by defining a function g that satisfies the hypotheses of Rolle's Theorem on $[a, b]$. Consider first the line joining the points $(a, f(a))$ and $(b, f(b))$. Since this line has slope $(f(b) - f(a))/(b - a)$, its equation can be expressed as

$$y = \frac{f(b) - f(a)}{b - a} (x - a) + f(a).$$

For any x in $[a, b]$, let $g(x)$ be the difference between $f(x)$ and the y-coordinate of the point on this line whose first coordinate is x (see Figure 2.35). Thus g is defined by

$$g(x) = f(x) - \left[\frac{f(b) - f(a)}{b - a} (x - a) + f(a) \right]$$

for x in $[a, b]$. We can easily verify that $g(a) = 0$ and $g(b) = 0$, so $g(a) = g(b)$.

Since f is continuous on $[a, b]$, g is also continuous on $[a, b]$. Moreover,

$$g'(x) = f'(x) - \frac{f(b) - f(a)}{b - a},$$

so g' exists whenever f' exists. This implies that g is differentiable on (a, b). Therefore, g satisfies the hypothesis of Rolle's Theorem and a number c, with $a < c < b$, exists with $g'(c) = 0$. That is, for some c in (a, b),

$$0 = g'(c) = f'(c) - \frac{f(b) - f(a)}{b - a}.$$

Consequently,

(2.43) $$f'(c) = \frac{f(b) - f(a)}{b - a}. \qquad \triangleright$$

The Mean Value Theorem is similar to the Intermediate Value and Extreme Value Theorems in the sense that it is an *existence* theorem. The Mean Value Theorem states that a number c satisfying Equation (2.43) *exists*, but the theorem does not explain how to determine c. Sometimes a value for c can be found algebraically, as shown in the following example, but this is not always the case.

Example 1

Given $f(x) = x^{2/3}$, find a value of c in $(0, 8)$ that is ensured by the conclusion of the Mean Value Theorem.

Solution

The graph of f is shown in Figure 2.36. We first show that the hypotheses of the Mean Value Theorem are satisfied on the interval $[0, 8]$.

The function given by $f(x) = x^{2/3}$ is continuous on $[0, 8]$. Its derivative, $f'(x) = \frac{2}{3} x^{-1/3}$, exists except when $x = 0$, so f is differentiable on

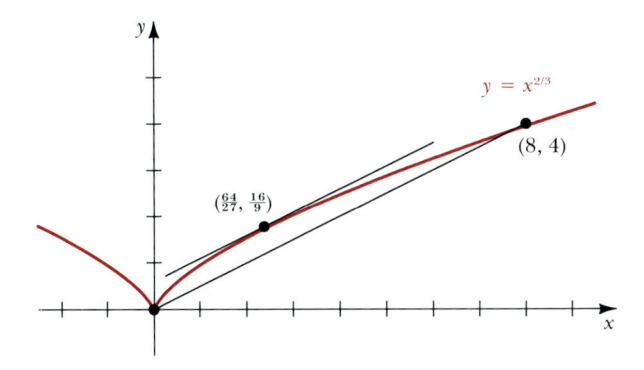

Figure 2.36
$f'\left(\dfrac{64}{27}\right) = \dfrac{f(8) - f(0)}{8 - 0}.$

$(0, 8)$. We must find c, $0 < c < 8$, satisfying

$$\frac{2}{3} c^{-1/3} = f'(c) = \frac{f(8) - f(0)}{8 - 0} = \frac{4 - 0}{8} = \frac{1}{2}.$$

Although the Mean Value Theorem does not give a constructive method of finding c, we can obtain the value of c in this example by solving the equation

$$\frac{2}{3} c^{-1/3} = \frac{1}{2}.$$

Since $c^{-1/3} = \dfrac{3}{4}$,

$$c = \left(\frac{3}{4}\right)^{-3} = \left(\frac{4}{3}\right)^{3} = \frac{64}{27}.$$

▶

The conclusion reached in the Mean Value Theorem does *not* hold for $f(x) = x^{2/3}$ on the interval $[-1, 1]$ (see Figure 2.37). Since $f(-1) = f(1)$, the quotient

$$\frac{f(b) - f(a)}{b - a} = \frac{f(1) - f(-1)}{1 + 1} = 0,$$

but the derivative $f'(x) = \frac{2}{3} x^{-1/3}$ is never zero. Note, however, that the hypotheses of the Mean Value Theorem do not hold for f on the interval $[-1, 1]$ since f is not differentiable at $x = 0$.

In Section 2.3 we found that the derivative of a constant function is always zero. The following application of the Mean Value Theorem implies that the converse of this result is also true. The statement of this theorem uses the notion of the *interior* of an interval. Recall from Chapter 1 that the interior of an interval consists of all numbers in the interval that are not endpoints.

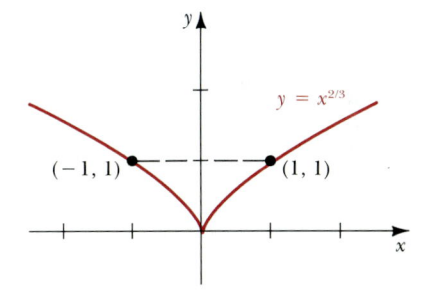

Figure 2.37

(2.44)
THEOREM
Suppose f is continuous on an interval I and differentiable on the interior of I. If $f'(x) = 0$ at each number x in the interior of I, then f is a constant function on I.

Proof. Let x_1 and x_2 denote arbitrary distinct numbers in the interval I and suppose $x_1 < x_2$. Since $[x_1, x_2]$ is contained within I, f is continuous on the interval $[x_1, x_2]$ and differentiable on (x_1, x_2). The hypotheses of the Mean Value Theorem hold for f on $[x_1, x_2]$, so a number c in (x_1, x_2) exists with

$$f'(c) = \frac{f(x_2) - f(x_1)}{x_2 - x_1}.$$

By hypothesis, $f'(c) = 0$, which implies that $f(x_1) = f(x_2)$. Thus f assigns the same value to arbitrary points in the interval I and f is constant on I. ▷

Theorem 2.44 is the basic result needed to study the inverse notion associated with the derivative:

> Given a function F, determine, if possible, a function whose derivative is F.

This concept, called **antidifferentiation,** is used in the discussion of motion problems in Section 3.1 and is studied extensively when we consider integral calculus beginning in Chapter 4.

(2.45)
DEFINITION
Any function F is said to be an **antiderivative** of the function f if $F' = f$.

Suppose F_1 and f are defined by $F_1(x) = x^3 + \sin 2x$ and $f(x) = 3x^2 + 2\cos 2x$. Then

$$D_x F_1(x) = 3x^2 + 2\cos 2x = f(x),$$

so F_1 is an antiderivative of f. Note, however, that F_2 and F_3 defined by

$$F_2(x) = x^3 + \sin 2x - 4 \quad \text{and} \quad F_3(x) = x^3 + \sin 2x + 6$$

have the same derivative as F_1, so they are also antiderivatives of f. The antiderivative of a function, then, is clearly *not* unique. The following result shows how antiderivatives of the same function are related.

(2.46)
COROLLARY
$F_1'(x) = F_2'(x)$ for every number x in an interval I if and only if a constant C exists so that for all x in I,

$$F_1(x) = F_2(x) + C.$$

Proof. If a constant C does exist with $F_1(x) = F_2(x) + C$, then

$$F_1'(x) = F_2'(x) + 0 = F_2'(x).$$

Conversely, if $F_1'(x) = F_2'(x)$ for all x in an interval I, then an application of Theorem 2.44 to the function $F_1 - F_2$ shows that a constant C exists with $F_1(x) - F_2(x) = C$, that is,

$$F_1(x) = F_2(x) + C. \qquad \triangleright$$

Corollary 2.46 implies that:

(2.47)
Two functions are antiderivatives of the same function on an interval precisely when they differ by a constant.

Example 2

Find a function F with the property that $F'(x) = 2x$ and $F(0) = 4$.

Solution
Since $D_x x^2 = 2x$, the required function F must have the property that
$$F(x) = x^2 + C, \quad \text{for some constant } C.$$
However,
$$4 = F(0) = 0^2 + C = C, \quad \text{so } C = 4$$
and
$$F(x) = x^2 + 4. \quad \blacktriangleright$$

Each of the rules for derivatives that we have seen in this chapter has a counterpart for antiderivatives. For example, we know that if $n \neq 0$ is a rational number, then
$$D_x x^n = nx^{n-1}.$$
Equivalently, when $n \neq -1$,
$$D_x \frac{x^{n+1}}{n+1} = x^n,$$
so any function of the form
$$F(x) = \frac{x^{n+1}}{n+1} + C$$
is an antiderivative of $f(x) = x^n$, provided $n \neq -1$.

In a similar manner, we find that for any constant C,

$-\cos x + C$ describes an antiderivative of $\sin x$

and

$\sin x + C$ describes an antiderivative of $\cos x$.

Example 3

Determine the form of the antiderivatives of
$$f(x) = 12x^2 + 5 \sin x.$$

Solution
Since $D_x x^3 = 3x^2$ and $D_x \cos x = -\sin x$,
$$D_x (4x^3 - 5 \cos x) = 12x^2 + 5 \sin x.$$
A function F is an antiderivative of f precisely when
$$F(x) = 4x^3 - 5 \cos x + C, \quad \text{for some constant } C. \quad \blacktriangleright$$

▶ EXERCISE SET 2.8

In Exercises 1–10, determine whether the hypotheses of Rolle's Theorem are satisfied for the given function and interval. If the hypotheses are satisfied, find a number c in the interval with $f'(c) = 0$.

1. $f(x) = x^2 - 3x + 2$, $[1, 2]$
2. $f(x) = x^2 + 9$, $[-3, 3]$
3. $f(x) = x^3 - 2x^2 - x + 2$, $[-1, 2]$
4. $f(x) = x^3 - 3x$, $[0, 3]$
5. $f(x) = \dfrac{x^2 - 1}{x}$, $[-1, 1]$
6. $f(x) = \dfrac{x^2 - 1}{x^2 + 1}$, $[-1, 1]$
7. $f(x) = \sin x$, $[0, \pi]$
8. $f(x) = \tan x$, $[0, \pi]$
9. $f(x) = \sin x - \cos x$, $[\pi/4, 5\pi/4]$
10. $f(x) = 1 - \cos x$, $[0, 2\pi]$

In Exercises 11–22, determine whether the hypotheses of the Mean Value Theorem are satisfied for the given function and interval. If the hypotheses are satisfied, find a number c in the interval that is ensured by the conclusion of the Mean Value Theorem.

11. $f(x) = x^2 + 4x - 5$, $[-1, 2]$
12. $f(x) = 2x^2 + 3x - 5$, $[1, 4]$
13. $f(x) = x + \dfrac{1}{x}$, $[1, 2]$
14. $f(x) = x + \dfrac{1}{x}$, $[-2, -1]$
15. $f(x) = x^{4/5}$, $[0, 1]$
16. $f(x) = x^{4/5}$, $[-1, 1]$
17. $f(x) = \sqrt{1 - x^2}$, $[-1, 1]$
18. $f(x) = \sqrt{1 - x^2}$, $[-1, 0]$
19. $f(x) = \dfrac{x^2 + 3x}{x - 1}$, $[-1, 0]$
20. $f(x) = \dfrac{x^2 + 3x}{x - 1}$, $[0, 2]$
21. $f(x) = \sin x$, $[0, \pi/2]$
22. $f(x) = \cos x$, $[0, \pi]$

In Exercises 23–42, find a function f that satisfies the given conditions.

23. $f(0) = 3$, $f'(x) = 0$ for all x
24. $f(0) = -1$, $f'(x) = 2$ for all x
25. $f(0) = 5$, $f'(x) = 2x$ for all x
26. $f(1) = 3$, $f'(x) = -2$ for all x
27. $f(0) = 2$, $f'(x) = \sin x$ for all x
28. $f(\pi/2) = 3$, $f'(x) = 2\cos x$ for all x
29. $f(0) = -2$, $f'(x) = 2x - 1$ for all x
30. $f(1) = 1$, $f'(x) = 4x^3$ for all x
31. $f(-2) = 4$, $f'(x) = 3x^2 - 5$ for all x
32. $f(0) = -3$, $f'(x) = x^2 + x$ for all x
33. $f(0) = -1$, $f'(x) = 3\sin 3x$ for all x
34. $f(0) = -1$, $f'(x) = \sin 3x$ for all x
35. $f(1) = 0$, $f'(x) = 1/x^2$ for all $x > 0$
36. $f(1) = 1$, $f'(x) = x^{-3}$ for all $x > 0$
37. $f(0) = 2$, $f'(0) = 3$, $f''(x) = 2$ for all x
38. $f(1) = 1$, $f'(1) = 2$, $f''(x) = 6x$ for all x
39. $f(0) = 2$, $f'(0) = -2$, $f''(x) = 2x + 3$ for all x
40. $f(1) = 2$, $f'(1) = -1$, $f''(x) = x^2 + x$ for all x
41. $f(0) = 0$, $f'(0) = 0$, $f''(x) = \sin x - \cos x$ for all x
42. $f(\pi/2) = 1$, $f'(\pi/2) = -2$, $f''(x) = \sin x - \cos x$ for all x

43. Explain why the hypotheses of the Mean Value Theorem are not satisfied for the given function and interval in each of the following.
    a) $f(x) = |x|$, $[-1, 1]$
    b) $f(x) = \tan x$, $[0, \pi]$
    c) $f(x) = [\![x]\!]$, $[0, 2]$
    d) $f(x) = \sec x$, $[0, \pi]$

44. Use Rolle's Theorem and the Intermediate Value Theorem to show that the graph of f intersects the x-axis exactly once in the given interval for each of the following.
    a) $f(x) = x^3 - 3x - 1$, $[1, 2]$
    b) $f(x) = \cos x - x$, $[0, \pi/4]$
    c) $f(x) = 2x^3 - 3x^2 - 12x$, $[3, 4]$
    d) $f(x) = x^7 - 4x^3 + 1$, $[0, 1]$

45. Use Rolle's Theorem to show that the graph of $f(x) = x^3 + 2x + k$ crosses the x-axis exactly once, regardless of the value of the constant k.

46. Suppose that f is a function whose second derivative is defined at each number in the interval $[a, b]$ and for which three distinct numbers x_1, x_2, and x_3 exist in $[a, b]$ with $f(x_1) = f(x_2) = f(x_3) = 0$. Use Rolle's

Theorem to show that $f''(c) = 0$ for some value of c in (a, b).

47. Give an example of a function that is continuous and differentiable on the interval $(0, 1)$, but for which the conclusion of the Mean Value Theorem does not hold for the interval $[0, 1]$.

48. Give an example of a function that is continuous on $[0, 1]$ and differentiable except at $x = \frac{1}{2}$, but for which the conclusion of the Mean Value Theorem does not hold.

49. Suppose f is a quadratic function described by $f(x) = a_2 x^2 + a_1 x + a_0$, for arbitrary constants a_0, a_1, and a_2. Show that the hypotheses of the Mean Value Theorem hold for any interval $[a, b]$ and that the number c ensured by the theorem is unique and is equal to $(a + b)/2$.

50. Use the Mean Value Theorem to show that if $f'(x) \neq 0$ for all values of x, then $a \neq b$ implies that $f(a) \neq f(b)$.

51. Use the Mean Value Theorem to show that $|\sin a - \sin b| \leq |a - b|$, and deduce from this result that $|\sin a + \sin b| \leq |a + b|$.

52. In Exercise 70 of Section 1.8, we found that if g is continuous and $a \leq g(x) \leq b$ whenever $a \leq x \leq b$, then a fixed point p of g exists in $[a, b]$, that is, $a \leq p \leq b$ and $g(p) = p$. Suppose, in addition, that $|g'(x)| < 1$ on $[a, b]$. Show that there is exactly one fixed point in $[a, b]$.

53. Suppose $f'(x) > 0$ on (a, b) and $f'(x) < 0$ on (b, c). Show that if x_1 is in (a, b) there is at most one number x_2 in (b, c) with $f(x_1) = f(x_2)$.

54. Suppose f and g are continuous on $[a, b]$ and differentiable on (a, b). Show that if $f(a) = g(a)$ and $f(b) = g(b)$, then there is a number c in (a, b) with $f'(c) = g'(c)$.

55. Suppose f' exists on $[a, b]$ and is continuous at a and at b. Show that if $f'(a)f'(b) < 0$, then there is a number c in (a, b) with $f'(c) = 0$.

Putnam exercise

56. If $a_0, a_1, \ldots, a_n$ are real numbers satisfying

$$\frac{a_0}{1} + \frac{a_1}{2} + \cdots + \frac{a_n}{n+1} = 0,$$

show that the equation $a_0 + a_1 x + a_2 x^2 + \cdots + a_n x^n = 0$ has at least one real root.

(This exercise was problem 1, part I of the eighteenth William Lowell Putnam examination given on February 8, 1958. The examination and its solution are in the January 1961 issue of the *American Mathematical Monthly*, pp. 18–22.)

2.9

Increasing and Decreasing Functions: The First Derivative Test

We began the chapter with an intuitive discussion of the increasing and decreasing nature of the graph of a function. This concept is related to the slope of the tangent line to the graph and this, in turn, is related to the derivative. The first definition in this section describes what we mean when we say that a *function* is increasing or is decreasing on an interval. The Mean Value Theorem is used to determine when this occurs.

(2.48)
DEFINITION

i) A function f is **increasing** on an interval I in its domain if

$$x_1 < x_2 \quad \text{implies} \quad f(x_1) < f(x_2)$$

whenever x_1 and x_2 are in I.

ii) A function f is **decreasing** on an interval I in its domain if
$$x_1 < x_2 \quad \text{implies} \quad f(x_1) > f(x_2)$$
whenever x_1 and x_2 are in I.

From the graph shown in Figure 2.38, it appears that on intervals where the slopes of the tangent lines are positive, the function is increasing; on those where the slopes of the tangent lines are negative, the function is decreasing. This is established in the following theorem.

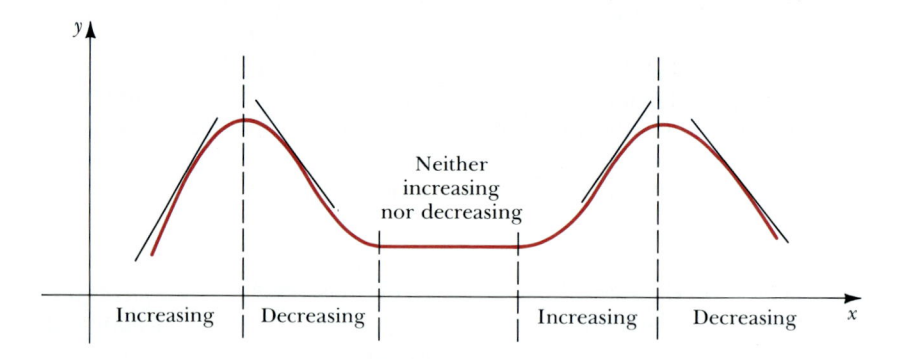

Figure 2.38

(2.49) THEOREM

Suppose f is a function that is continuous on an interval I and differentiable on the interior of I.

i) If $f'(x) > 0$ on the interior of I, then f is increasing on I.
ii) If $f'(x) < 0$ on the interior of I, then f is decreasing on I.

Proof

i) Suppose that x_1 and x_2 are arbitrary numbers in I and that $x_1 < x_2$. By hypothesis, f is continuous on $[x_1, x_2]$ and differentiable on (x_1, x_2). The Mean Value Theorem implies that there exists a number c in (x_1, x_2) with
$$f'(c) = \frac{f(x_2) - f(x_1)}{x_2 - x_1}.$$
Since $x_2 - x_1 > 0$ and $f'(c) > 0$,

(2.50) $$f(x_2) - f(x_1) > 0$$

and

(2.51) $$f(x_1) < f(x_2).$$

Thus f is increasing on I.

ii) The proof of case (ii) is the same as that for (i) except $f'(c) < 0$, so the inequalities (2.50) and (2.51) are reversed. ▷

Example 1

If $f(x) = 2x^3 - 9x^2 + 12x - 2$, find the intervals on which f is increasing and the intervals on which f is decreasing.

Solution

In Example 1 of Section 2.7 we found critical points of this function to be $x = 1$ and $x = 2$ and sketched its graph. Since f' is continuous, the critical points are the only places at which $f'(x) = 6(x - 1)(x - 2)$ can change sign. If we consider the intervals $(-\infty, 1)$, $(1, 2)$, and $(2, \infty)$ and use Theorem 2.49, we can determine when f is increasing and when f is decreasing.

The sign chart below shows the sign of each of the factors of $f'(x)$ on these intervals. Once the sign of the individual factors has been given we can determine the sign of $f'(x)$ and, as a consequence, the increasing and decreasing behavior of the graph.

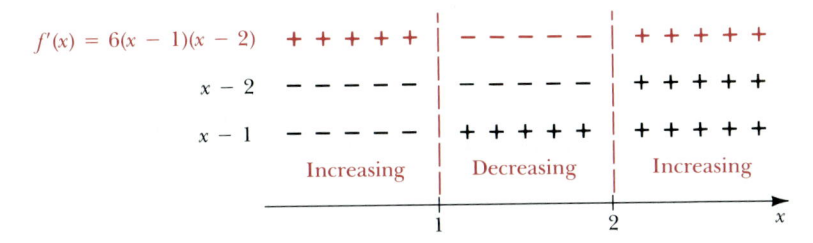

$f(x) = 2x^3 - 9x^2 + 12x - 2$

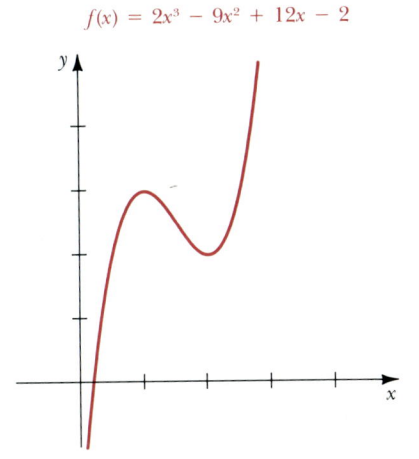

Figure 2.39

$f'(x) = 6(x - 1)(x - 2)$	$+ + + + +$	$- - - - -$	$+ + + + +$
$x - 2$	$- - - - -$	$- - - - -$	$+ + + + +$
$x - 1$	$- - - - -$	$+ + + + +$	$+ + + + +$
	Increasing	Decreasing	Increasing
	1	2	x

Since $f'(x) > 0$ on the intervals $(-\infty, 1)$ and $(2, \infty)$, f is increasing on $(-\infty, 1]$ and on $[2, \infty)$. However, $f'(x) < 0$ on the interval $(1, 2)$, so f is decreasing on $[1, 2]$. The conclusions drawn from the chart agree with our previous knowledge of the graph (see Figure 2.39). ▶

Example 2

(a) For $f(x) = x + 1/x$, find the intervals on which f is increasing and the intervals on which f is decreasing. (b) Sketch the graph of f.

Solution

(a) Note that $f(x)$ is undefined at $x = 0$. The derivative of f at x, when $x \neq 0$, is

$$f'(x) = 1 - \frac{1}{x^2} = \frac{x^2 - 1}{x^2},$$

which is 0 if $x = 1$ or $x = -1$. We partition the real line into intervals determined by these critical points and the point of discontinuity, $x = 0$. These are the only values at which f' can change sign.

The sign chart that follows shows the sign of the numerator $x^2 - 1$, the denominator x^2, and the resulting sign of $f'(x)$. The chart implies that f is increasing on the intervals $(-\infty, -1]$ and $[1, \infty)$, but is decreasing on $[-1, 0)$ and $(0, 1]$. (Since x^2 is never negative, the inclusion of the sign of this factor makes no contribution. In the future such factors will not be included in the chart.)

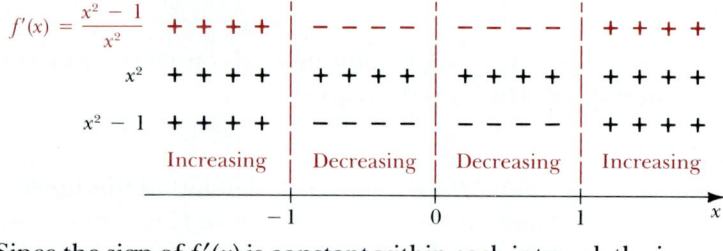

$$f'(x) = \frac{x^2-1}{x^2} \quad \substack{+\,+\,+\,+ \\ \\ \\ \text{Increasing}} \Bigg| \substack{-\,-\,-\,- \\ \\ \\ \text{Decreasing}} \Bigg| \substack{-\,-\,-\,- \\ \\ \\ \text{Decreasing}} \Bigg| \substack{+\,+\,+\,+ \\ \\ \\ \text{Increasing}}$$

$$x^2 \quad +\,+\,+\,+\ \big|\ +\,+\,+\,+\ \big|\ +\,+\,+\,+\ \big|\ +\,+\,+\,+$$

$$x^2-1 \quad +\,+\,+\,+\ \big|\ -\,-\,-\,-\ \big|\ -\,-\,-\,-\ \big|\ +\,+\,+\,+$$

Since the sign of $f'(x)$ is constant within each interval, the increasing and decreasing behavior of f can also be determined by examining the value of $f'(x)$ at one arbitrary point in each of the intervals $(-\infty, -1)$, $(-1, 0)$, $(0, 1)$, and $(1, \infty)$. For example, when:

$x = -2$, $f'(x) = \frac{3}{4}$, so $f'(x) > 0$ for every x in $(-\infty, 1)$;
$x = -\frac{1}{2}$, $f'(x) = -3$, so $f'(x) < 0$ for every x in $(-1, 0)$;
$x = \frac{1}{2}$, $f'(x) = -3$, so $f'(x) < 0$ for every x in $(0, 1)$;
$x = 3$, $f'(x) = \frac{8}{9}$, so $f'(x) > 0$ for every x in $(1, \infty)$.

(b) To sketch the graph, we plot $(-1, f(-1)) = (-1, -2)$ and $(1, f(1)) = (1, 2)$ and use the increasing and decreasing information. In addition, note that the graph of f is symmetric with respect to the origin and that a vertical asymptote occurs at $x = 0$, since

$$\lim_{x \to 0^+}\left(x + \frac{1}{x}\right) = \infty \quad \text{and} \quad \lim_{x \to 0^-}\left(x + \frac{1}{x}\right) = -\infty.$$

The graph of f must appear as shown in Figure 2.40; it has the line $y = x$ as a slant asymptote. ▶

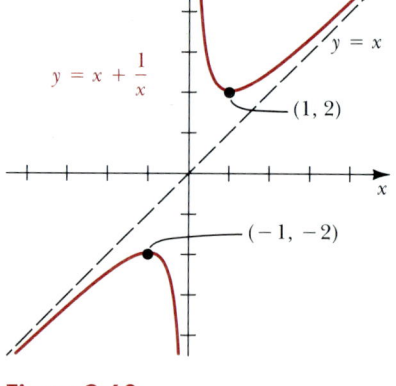

Figure 2.40

In Section 2.7 we found that relative extrema can occur only at critical points. We also saw that there are critical points at which relative extrema do not exist. The object of the following theorem is to separate the critical points of a function into three categories:

i) those at which a relative maximum occurs,
ii) those at which a relative minimum occurs, and
iii) those at which neither a relative maximum nor a relative minimum occurs.

The results of this theorem can be anticipated by considering Figure 2.41.

(2.52)
THEOREM: THE FIRST DERIVATIVE TEST
Suppose f is continuous on an open interval (a, b) containing c.

i) If $f'(x) > 0$ on (a, c) and $f'(x) < 0$ on (c, b), then f has a relative maximum at $x = c$.
ii) If $f'(x) < 0$ on (a, c) and $f'(x) > 0$ on (c, b), then f has a relative minimum at $x = c$.
iii) If $f'(x) > 0$ on both (a, c) and (c, b), or if $f'(x) < 0$ on both (a, c) and (c, b), then f has neither a relative maximum nor a relative minimum at $x = c$.

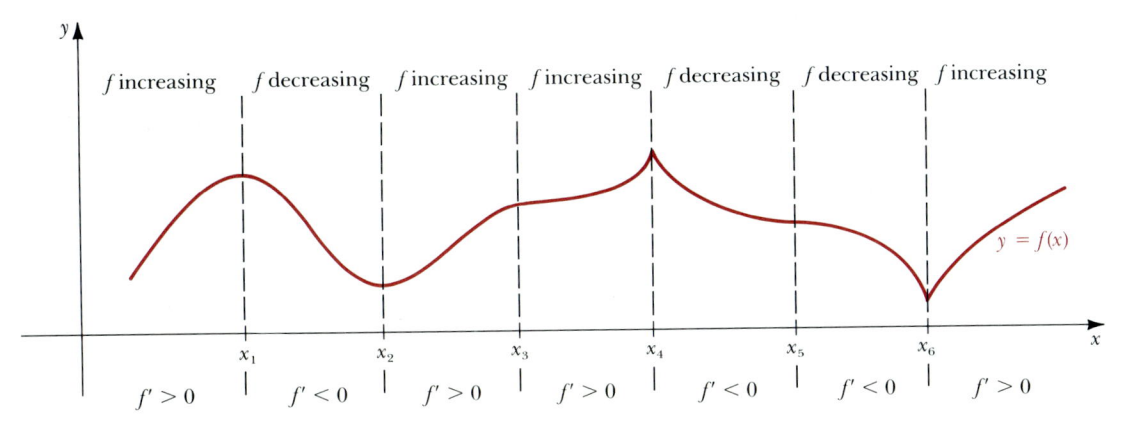

Figure 2.41

Proof. (i) By Theorem 2.49, f is increasing on the interval $(a, c]$ and decreasing on $[c, b)$. Thus for all x in (a, b), $f(x) \leq f(c)$. Therefore, f has a relative maximum at c.

The proofs of parts (ii) and (iii) are similar and are considered in Exercise 56. ▷

Example 3

Use the First Derivative Test to find relative maxima and minima of f if $f(x) = x^2 - 3x^{2/3}$.

Solution

We found in Section 2.7 (Example 2) that

$$f'(x) = 2x - 2x^{-1/3} = 2x^{-1/3}(x^{4/3} - 1).$$

The critical points of f are $x = 0$ (since $f'(0)$ is undefined), $x = 1$, and $x = -1$ (since $f'(1) = 0$ and $f'(-1) = 0$). These are the only values of x at which f' can change sign. The sign chart indicates that

$$f'(x) < 0 \text{ on } (-\infty, -1) \text{ and on } (0, 1)$$

and

$$f'(x) > 0 \text{ on } (-1, 0) \text{ and on } (1, \infty).$$

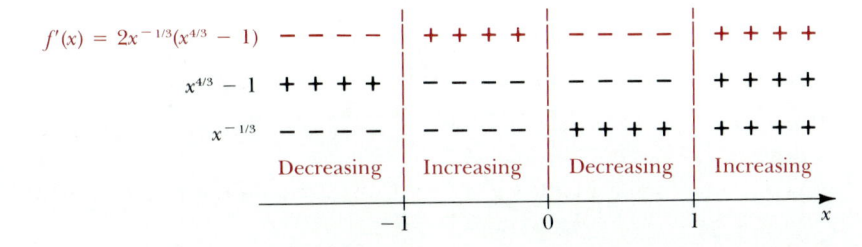

The First Derivative Test implies that f has a relative minimum at $x = -1$. A relative minimum also occurs at $x = 1$.

$f(x) = x^2 - 3x^{2/3}$

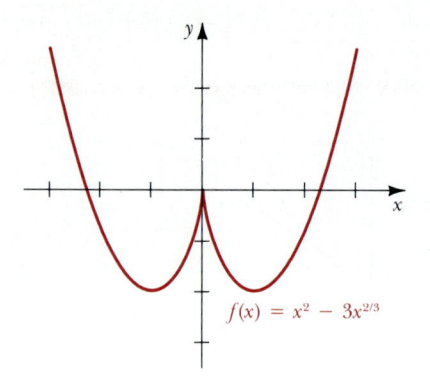

Figure 2.42

Since
$$f'(x) > 0 \text{ on } (-1, 0) \quad \text{and} \quad f'(x) < 0 \text{ on } (0, 1),$$

a relative maximum occurs at $x = 0$. This agrees with the conclusions reached in Example 2 of Section 2.7 (see Figure 2.42). ▶

Example 4

Use the First Derivative Test to sketch the graph of $f(x) = x(4 - x^2)^{1/2}$.

Solution
The domain of f is the interval $[-2, 2]$, since $4 - x^2$ must be nonnegative. By the Product Rule,

$$f'(x) = (4 - x^2)^{1/2} + x(\tfrac{1}{2})(4 - x^2)^{-1/2}(-2x)$$
$$= (4 - x^2)^{1/2} - x^2(4 - x^2)^{-1/2}$$
$$= (4 - x^2)^{-1/2}[(4 - x^2) - x^2]$$
$$= 2(4 - x^2)^{-1/2}(2 - x^2).$$

Critical points are $x = 2$ and $x = -2$, where $f'(x)$ is undefined, and $x = \sqrt{2}$ and $x = -\sqrt{2}$, where $f'(x) = 0$.

The sign chart and the First Derivative Test imply that a relative minimum occurs at $x = -\sqrt{2}$ and a relative maximum at $x = \sqrt{2}$.

$f'(x) = 2(2 - x^2)(4 - x^2)^{-1/2}$	– – –	+ + + + + +	– – –
$(4 - x^2)^{-1/2}$	+ + +	+ + + + + +	+ + +
$2 - x^2$	– – –	+ + + + + +	– – –
	Decreasing	Increasing	Decreasing
	$-2 \quad -\sqrt{2}$		$\sqrt{2} \quad 2$

The graph is symmetric with respect to the origin. It approaches the x-axis vertically at $x = 2$ and at $x = -2$ because

$$\lim_{x \to 2^-} f'(x) = \lim_{x \to 2^-} 2(4 - x^2)^{-1/2}(2 - x^2) = -\infty$$

and $\lim_{x \to -2^+} f'(x) = -\infty$. The graph is shown in Figure 2.43. ▶

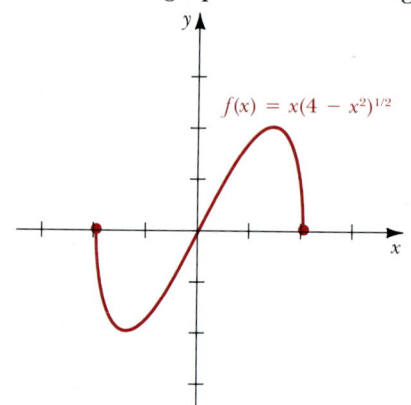

$f(x) = x(4 - x^2)^{1/2}$

Figure 2.43

EXERCISE SET 2.9

In Exercises 1–6, the graph of a function is given.

a) Use the graph to determine the intervals on which the function is increasing and those on which it is decreasing.
b) Where do relative and absolute extrema occur? What is the derivative at these points.
c) Find the absolute extrema when they exist.

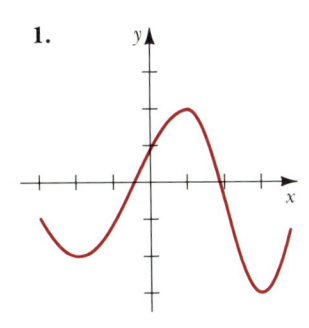

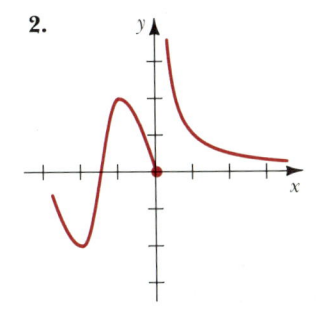

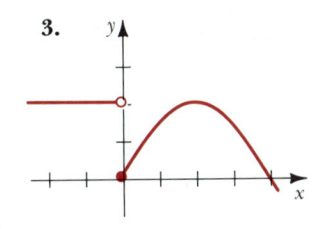

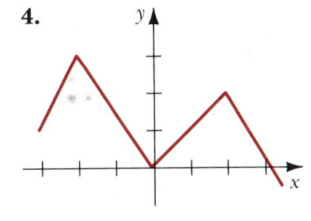

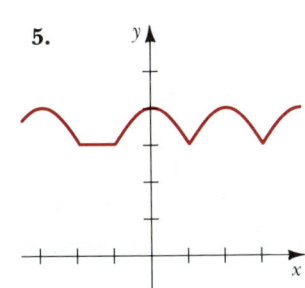

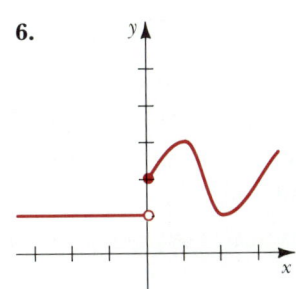

1.
2.
3.
4.
5.
6.

In Exercises 7–48, (a) determine the intervals on which the function is increasing and those on which the function is decreasing; (b) find the relative extrema, and sketch the graph of the function.

7. $f(x) = 4x - 3$
8. $f(x) = -2x + 3$
9. $f(x) = x^2 - 4x + 4$
10. $f(x) = 4 + 4x - x^2$
11. $f(x) = 2x^2 - 9x + 3$
12. $f(x) = x^3 - 6x^2$
13. $f(x) = x^3 - 3x$
14. $f(x) = x^3 - x^2$
15. $f(x) = x^4 - 2x^2$
16. $f(x) = x^4 + 2x^2$
17. $f(x) = 2x + \dfrac{1}{2x}$
18. $f(x) = 2x - \dfrac{1}{2x}$
19. $f(x) = x^3(1 - x)$
20. $f(x) = x^2(1 - x)^2$
21. $f(x) = \sin x$
22. $f(x) = \cos x$
23. $f(x) = \tan x$
24. $f(x) = \sec x$
25. $f(x) = x^2 + \dfrac{1}{x^2}$
26. $f(x) = x + \dfrac{1}{\sqrt{x}}$
27. $f(x) = \dfrac{x + 1}{x - 1}$
28. $f(x) = \dfrac{2x - 3}{x - 2}$
29. $f(x) = \dfrac{x^2 + 1}{x^2 - 1}$
30. $f(x) = \dfrac{x^2 - 1}{x^2 + 1}$
31. $f(x) = x\sqrt{9 - x^2}$
32. $f(x) = x\sqrt{x^2 - 9}$
33. $f(x) = (x - 2)^2(x + 1)^2$
34. $f(x) = (3x - 5)^2(4 - x)^3$
35. $f(x) = (x - 1)^2(x - 2)^3(x - 3)^4$
36. $f(x) = (x^2 - 1)^2(x^2 + 1)^2$
37. $f(x) = |x^2 - 2|$
38. $f(x) = x|x^2 - 2|$
39. $f(x) = x^{3/2} - 3x^{1/2}$
40. $f(x) = x^{3/2} + 3x^{1/2}$
41. $f(x) = x^{7/6} - x^{5/3}$
42. $f(x) = x^{7/3} + x^{4/3} - 3x^{1/3}$
43. $f(x) = \sqrt[4]{x^4 - 2x^2}$
44. $f(x) = \sqrt[3]{x^3 - 3x}$

45. $f(x) = \sin x - x/2$
46. $f(x) = x - \sin x$
47. $f(x) = \sin x - \cos x$
48. $f(x) = \sin^2 x$
49. The cost of producing x gallons, $500 \leq x \leq 4000$, of maple syrup is $C(x) = 15x - 0.002x^2$ dollars. Find the values of x for which the cost is increasing and those for which the cost is decreasing. What is the maximum cost? What is the minimum cost?
50. The amount of sales of a new product t months after the product has been introduced is given by $s(t) = 2000(1 + t)/(10 + t)$. Show that the amount of sales is increasing. Is there a maximum amount of sales?
51. A ball is thrown upward from the earth (see the figure). Its distance above the ground is given by $s(t) = 88t - 16t^2 + 3$ (in feet). This equation is valid from time (in seconds) $t = 0$ until the ball returns to the ground.
    a) Find the values of t for which the distance is increasing.
    b) What is the maximum distance of the ball above the ground; that is, how high does the ball go?

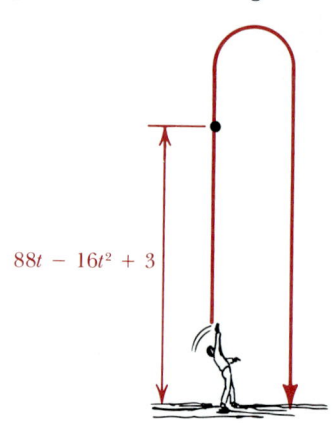

52. A baseball is hit (see the figure) so that its distance above the ground is given by $s(t) = 96t - 16t^2 + 2$ (in feet). This equation is valid from time (in seconds) $t = 0$ until the ball returns to the ground.
    a) Find the values of t for which the distance above the earth is increasing.
    b) Find the values of t for which this distance is decreasing.
    c) What is the maximum height of the baseball above the ground?

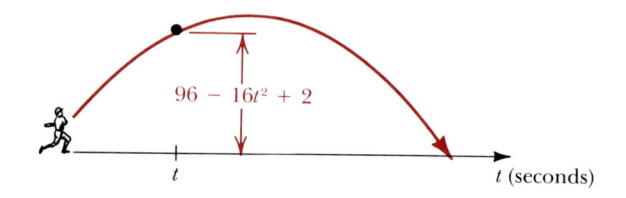

53. Suppose f and g are functions defined on $[a, b]$ and both $f'(x) > 0$ and $g'(x) > 0$ for every x in $[a, b]$. Must the function $h = f \cdot g$ be increasing on $[a, b]$?

54. Suppose f and g are increasing functions on their respective domains and $f \circ g$ exists for all x in $[a, b]$. Show that $f \circ g$ is increasing on $[a, b]$.

55. Prove Theorem 2.49 (ii): If $f'(x) < 0$ on the interior of an interval I and f is continuous on I, then f is decreasing on I.

56. a) Prove part (ii) of the First Derivative Test.
    b) Prove part (iii) of the First Derivative Test.

In Exercises 57–60, find values of a, b, and c that ensure the function f is increasing on the interval $(-\infty, -1]$, decreasing on the interval $[-1, 1]$, and increasing on the interval $[1, \infty)$, or show that no such constants exist.

57. $f(x) = ax^2 + bx + c$
58. $f(x) = ax^3 + bx + c$
59. $f(x) = ax^3 + bx^2 + cx$
60. $f(x) = ax^3 + bx^2 + c$

2.10

Higher Derivatives: Concavity and the Second Derivative Test

The derivative has enabled us to determine when a function is increasing and when it is decreasing and, as a consequence, to find its maximum and minimum values. To obtain additional geometric information about the graph of a function, however, we need **higher derivatives.**

2.10 HIGHER DERIVATIVES

A differentiable function f defines a new function f'. If f' is also differentiable, its derivative $(f')'$ can be defined. Similarly, if $(f')'$ is differentiable, we have $((f')')'$, and so on. This process can be continued so long as the defined functions are differentiable.

To express this concept more concisely, we generally write $(f')'$, called the **second derivative** of f, as

$$f'', \qquad D_x^2 f, \quad \text{or} \quad \frac{d^2}{dx^2} f.$$

Similarly,

$$f''', \qquad D_x^3 f, \quad \text{or} \quad \frac{d^3}{dx^3} f$$

represents the **third derivative** of f, $((f')')'$, and so on.

The notation $f^{(n)}$ is also used to represent the nth derivative of f, particularly when n is greater than 3. For example, the 10th derivative of f is written

$$f^{(10)}, \qquad D_x^{10} f, \quad \text{or} \quad \frac{d^{10}}{dx^{10}} f.$$

When an equation involving the variables x and y defines y as a function of x, the notations $y^{(n)}$, $D_x^{(n)} y$, and $d^n y/dx^n$ are used to represent the nth derivative.

After the following examples, which will clarify the concept of higher derivatives, we will see some interesting graphing applications of the second derivative of a function.

Example 1

If $f(x) = x^3 + 3x^2 - 2x + \sin x$, find the first four derivatives of f.

Solution

$$f'(x) = 3x^2 + 6x - 2 + \cos x,$$
$$f''(x) = 6x + 6 - \sin x,$$
$$f'''(x) = 6 - \cos x, \text{ and}$$
$$f^{(4)}(x) = \sin x.$$ ▶

Example 2 demonstrates finding higher derivatives for functions expressed implicitly.

Example 2

Find $D_x y$, $D_x^2 y$, and $D_x^3 y$ if $x^2 + y^2 = 1$.

Solution

Using implicit differentiation, we have

$$2x + 2y \, D_x y = 0$$

so

$$D_x y = \frac{-x}{y}.$$

162 CHAPTER 2 THE DERIVATIVE

Thus,
$$D_x^2 y = D_x\left(\frac{-x}{y}\right) = -\frac{y D_x x - x D_x y}{y^2} = -\frac{y - x D_x y}{y^2}.$$

Replacing $D_x y$ by its value $-x/y$ gives
$$D_x^2 y = -\frac{y - x\left(\frac{-x}{y}\right)}{y^2} = -\frac{y^2 + x^2}{y^3} = -\frac{1}{y^3},$$

since $x^2 + y^2 = 1$.

Similarly,
$$D_x^3 y = D_x\left(-\frac{1}{y^3}\right) = \frac{3}{y^4} D_x y = \frac{3}{y^4}\left(-\frac{x}{y}\right) = \frac{-3x}{y^5}. \quad \blacktriangleright$$

The First Derivative Test was introduced in Section 2.9 as a method of determining which critical points of a function give relative maxima and relative minima. Another test uses the *second derivative* of the function for this purpose. This result is more restricted in its application than the First Derivative Test, but it is easy to apply if the second derivative of the function is known. Figure 2.44 illustrates the result.

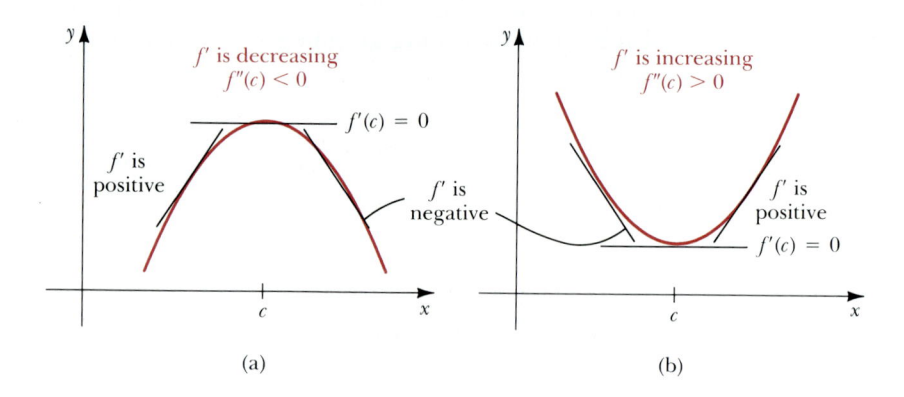

Figure 2.44
(a) Relative maximum at $x = c$;
(b) relative minimum at $x = c$.

(2.53)
THEOREM: THE SECOND DERIVATIVE TEST

Suppose that f is differentiable on an open interval containing c and that $f'(c) = 0$.

i) If $f''(c) < 0$, then f has a relative maximum at c.
ii) If $f''(c) > 0$, then f has a relative minimum at c.

Proof. (i) If
$$f''(c) = \lim_{h \to 0} \frac{f'(c + h) - f'(c)}{h} < 0,$$

then an interval (a, b) about c exists with
$$\frac{f'(c + h) - f'(c)}{h} < 0,$$

whenever $c + h$ is in (a, b) and $h \neq 0$. Since $f'(c) = 0$, we have

$$\frac{f'(c + h)}{h} < 0,$$

whenever $c + h$ is in (a, b) and $h \neq 0$. This inequality implies that:

$$\text{if } h < 0, f'(c + h) > 0, \quad \text{so } f \text{ is increasing on } (a, c]$$

and

$$\text{if } h > 0, f'(c + h) < 0, \quad \text{so } f \text{ is decreasing on } [c, b).$$

By the First Derivative Test, a relative maximum occurs at c.

Part (ii) is proved in a similar manner. The proof is considered in Exercise 78. ▷

Example 3

Use the Second Derivative Test to find the relative maxima and minima of $f(x) = x^4 - 4x^3 - 2x^2 + 12x + 1$.

Solution
Since

$$f'(x) = 4x^3 - 12x^2 - 4x + 12 = 4(x^3 - 3x^2 - x + 3),$$

f has no points of discontinuity and the only critical points occur when $f'(x) = 0$. To find the critical points we will factor $f'(x)$. First note that $f'(1) = 0$, so $(x - 1)$ is a factor of $f'(x)$. In fact,

$$f'(x) = 4(x^3 - 3x^2 - x + 3)$$
$$= 4(x - 1)(x^2 - 2x - 3)$$
$$= 4(x - 1)(x - 3)(x + 1),$$

which implies that the critical points are $x = 1$, $x = 3$, and $x = -1$.

The second derivative is

$$f''(x) = 12x^2 - 24x - 4,$$

so

$f''(1) = -16 < 0$ and f has a relative maximum at $x = 1$;
$f''(3) = 32 > 0$ and f has a relative minimum at $x = 3$; and
$f''(-1) = 32 > 0$ and f has a relative minimum at $x = -1$.

The graph of f is shown in Figure 2.45. ▶

Nothing can be inferred from the Second Derivative Test when both $f'(c) = 0$ and $f''(c) = 0$. This is evident from considering the functions whose graphs are shown in Figure 2.46. The first derivative of each of these functions is 0 at $x = 0$, and the second derivative is also 0 at $x = 0$. However, although f_1 and f_2 have neither a relative maximum nor a relative minimum at $x = 0$, f_3 has a relative minimum at $x = 0$ and f_4 has a relative maximum at $x = 0$.

The second derivative of a function can also be used to determine when the graph has an upward curving shape, such as that shown in

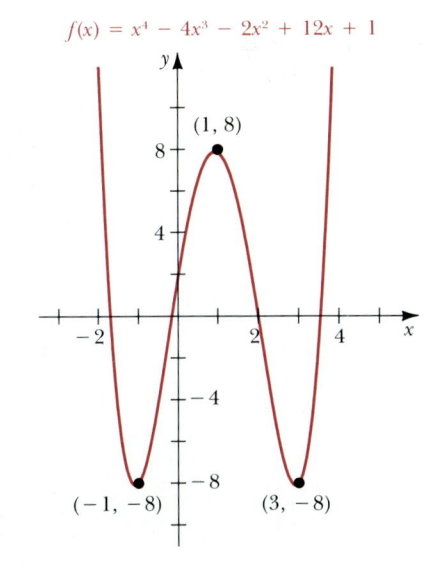

Figure 2.45

$f_1(x) = x^3$ $f_2(x) = -x^3$ $f_3(x) = x^4$ $f_4(x) = -x^4$

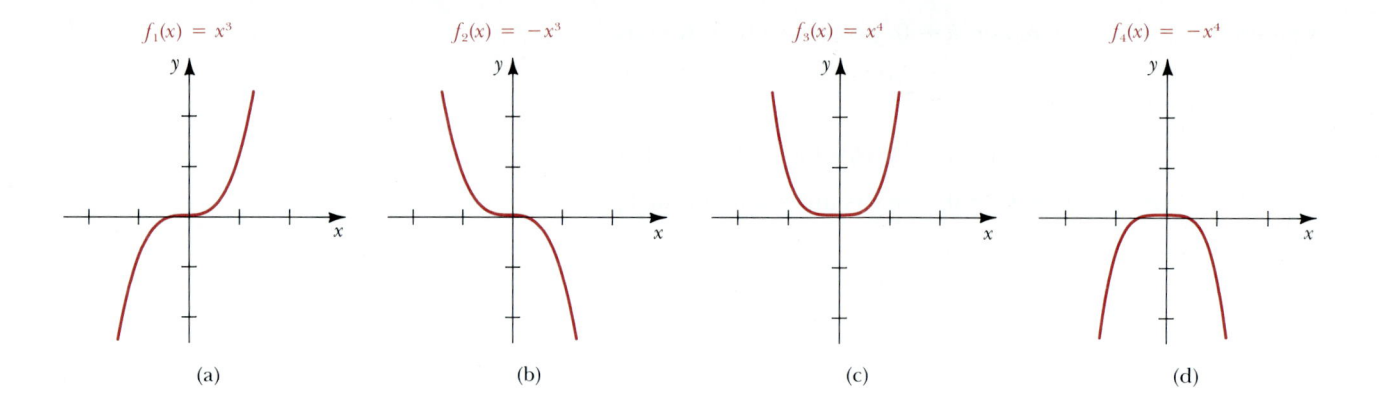

(a)   (b)   (c)   (d)

Figure 2.46

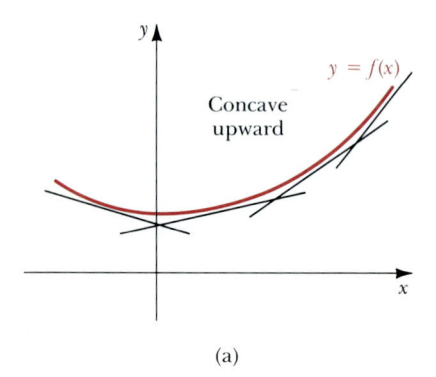

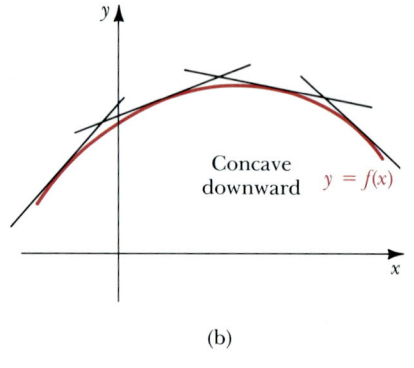

(a) Concave upward, $y = f(x)$

(b) Concave downward, $y = f(x)$

Figure 2.47

Figure 2.47(a) and when it has a downward curving shape, such as that shown in Figure 2.47(b). The graph of the function in Figure 2.47(a) is *concave upward.* It has the property that every tangent line to the curve lies below the curve. The graph of the function in Figure 2.47(b) is *concave downward:* Every tangent line to the curve lies above the curve.

Note that the slopes of the tangent lines to the concave-upward curve in Figure 2.47(a) increase as x increases; this implies that the derivative f' is an increasing function. On the other hand, the slopes of the tangent lines to the concave-downward curve in Figure 2.47(b) decrease as x increases, so in this case f' is a decreasing function. This observation permits us to define *concavity* of the graph of a function in terms of the behavior of the derivative of the function.

(2.54) DEFINITION

Suppose f is a function that is continuous at c.

i) The graph of f is said to be **concave upward** at $(c, f(c))$ if there exists an open interval I containing c with the property that f' is increasing on I.

ii) The graph of f is said to be **concave downward** at $(c, f(c))$ if there exists an open interval I containing c with the property that f' is decreasing on I.

The graph of f is said to be concave upward on the interval I if f is concave upward at $(x, f(x))$ for every x in I. Similarly, the graph of f is said to be concave downward on the interval I if f is concave downward at $(x, f(x))$ for every x in I.

The following theorem concerning the concavity of a function follows by applying Theorem 2.49 to the function f'.

(2.55) THEOREM

Let I be an open interval.

i) If $f''(x) > 0$ on I, then the graph of f is concave upward on I.

ii) If $f''(x) < 0$ on I, then the graph of f is concave downward on I.

Proof. (i) If $f''(x) > 0$ on the open interval I, then f' is an increasing function on I. By Definition 2.54, f is concave upward on I.

Part (ii) is shown in a similar manner. ▷

Example 4

For $f(x) = 2x^3 - 9x^2 + 12x - 2$, find intervals on which the graph is concave upward and intervals on which the graph is concave downward.

Solution

The graph of this function was previously considered in Example 1 of Section 2.7 and Example 1 in Section 2.9. Here we are adding the refinement of concavity:

$$f'(x) = 6x^2 - 18x + 12$$

and

$$f''(x) = 12x - 18 = 6(x - \tfrac{3}{2}).$$

Since

$$f''(x) < 0 \quad \text{when} \quad x < \tfrac{3}{2},$$

the graph is concave downward on $(-\infty, \tfrac{3}{2})$; and since

$$f''(x) > 0 \quad \text{when} \quad x > \tfrac{3}{2},$$

the graph is concave upward on $(\tfrac{3}{2}, \infty)$. ▶

The points at which the concavity of the graph changes from concave upward to concave downward (and conversely) are naturally of interest. These are *points of inflection* for the graph, as shown in Figure 2.48.

(2.56) DEFINITION

A point $(c, f(c))$ is called a **point of inflection** for the graph of f if an interval (a, b) containing c exists with either of the following properties:

i) f is concave upward on (a, c) and concave downward on (c, b); or
ii) f is concave downward on (a, c) and concave upward on (c, b).

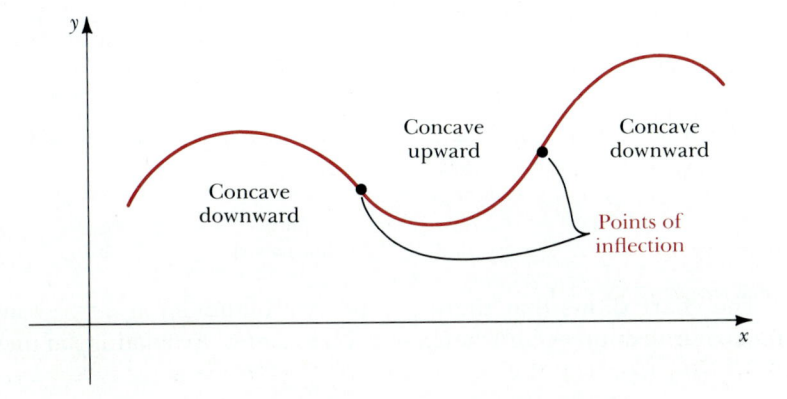

Figure 2.48

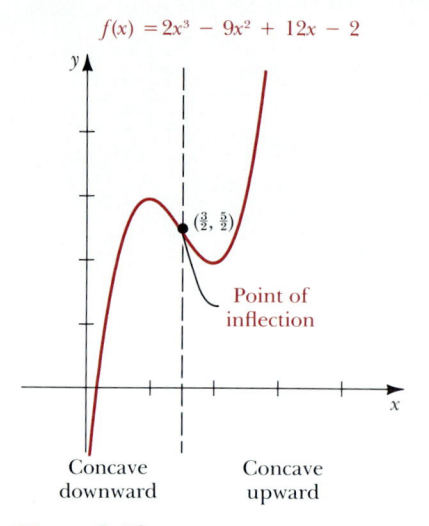

Figure 2.49: $f(x) = 2x^3 - 9x^2 + 12x - 2$, with point of inflection at $(\frac{3}{2}, \frac{5}{2})$, concave downward then concave upward.

The graph of $f(x) = 2x^3 - 9x^2 + 12x - 2$ considered in Example 4 is concave downward on $(-\infty, \frac{3}{2})$ and concave upward on $(\frac{3}{2}, \infty)$. Thus, $(\frac{3}{2}, f(\frac{3}{2})) = (\frac{3}{2}, \frac{5}{2})$ is a point of inflection for the graph (see Figure 2.49).

If f'' is continuous, it can change sign at a number c only if $f''(c)$ is 0. This observation used in conjunction with Theorem 2.55 tells us where to look for points of inflection.

(2.57) COROLLARY
Suppose that $(c, f(c))$ is a point of inflection for the graph of f and that for some constants a and b, f'' is continuous on both (a, c) and (c, b). If $f''(c)$ exists, then $f''(c) = 0$.

The converse of Corollary 2.57 does *not* hold, which we can see by examining the graph of $f(x) = x^4$ shown in Figure 2.50. The second derivative is 0 at $x = 0$. However, no point of inflection occurs at $x = 0$, since the concavity of the graph does not change.

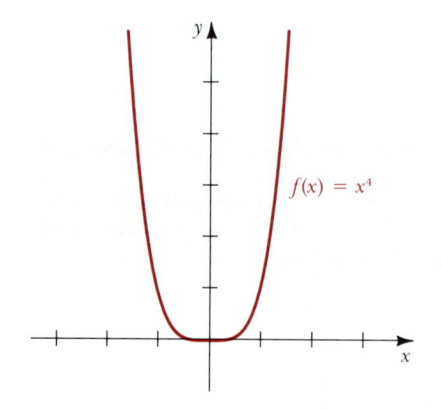

Figure 2.50: $f(x) = x^4$.

Example 5
Use the critical points and points of inflection to sketch the graph of $f(x) = x^4 + 2x^3 - 1$.

Solution
Since
$$f'(x) = 4x^3 + 6x^2 = 2x^2(2x + 3),$$
critical points are $x = 0$ and $x = -\frac{3}{2}$. Also,
$$f''(x) = 12x^2 + 12x = 12x(x + 1).$$

Thus, $f''(x) = 0$ at $x = 0$ and at $x = -1$, so possible points of inflection are $(0, f(0))$ and $(-1, f(-1))$. Partitioning the real line using the critical points and possible points of inflection gives the information in the chart.

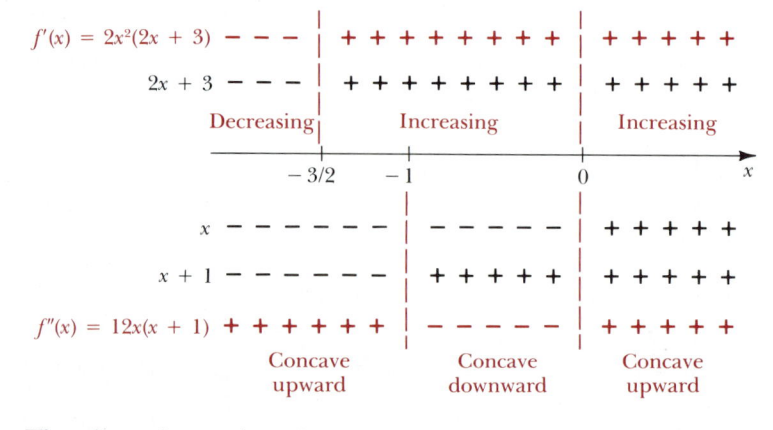

	$-3/2$	-1	0	
$f'(x) = 2x^2(2x+3)$	$---$	$+++++++$	$+++++$	
$2x + 3$	$---$	$+++++++$	$+++++$	
	Decreasing	Increasing	Increasing	
x	$------$	$-----$	$+++++$	
$x + 1$	$------$	$+++++$	$+++++$	
$f''(x) = 12x(x+1)$	$++++++$	$-----$	$+++++$	
	Concave upward	Concave downward	Concave upward	

The chart shows that there is a relative minimum at $x = -\frac{3}{2}$ and points of inflection occur when $x = -1$ and $x = 0$. Evaluating f at these

$f(x) = x^4 + 2x^3 - 1$

$(0, -1)$

$(-1, -2)$

$(-\frac{3}{2}, -\frac{43}{16})$

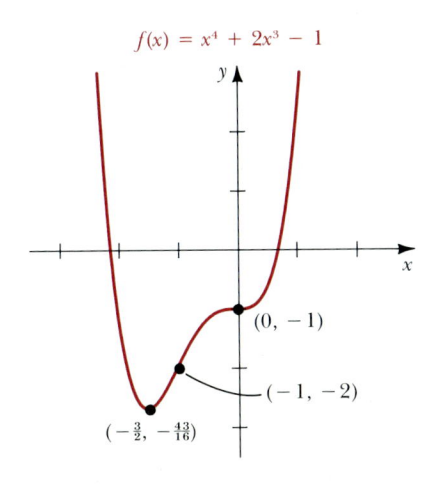

Figure 2.51

numbers, we find that the points $(-\frac{3}{2}, -\frac{43}{16})$, $(-1, -2)$, and $(0, -1)$ lie on the graph of f. Since $f(-3) > 0$ and $f(-2) < 0$, the graph has an x-intercept in the interval $(-3, -2)$. It also has an x-intercept in $(0, 1)$. Using this information with the observation that $\lim_{x \to -\infty} f(x) = \infty$ and $\lim_{x \to \infty} f(x) = \infty$ gives the graph shown in Figure 2.51. ▶

▶ **EXERCISE SET 2.10**

In Exercises 1–10, find $D_x y$ and $D_x^2 y$.

1. $y = 3x^2 + 4x$
2. $y = 3x^3 - 2x^2 + x - 1$
3. $y = x^4 + 1 + x^{-4}$
4. $y = x^3/3 - x^{-1}$
5. $y = 3\sqrt{x} + 2x^2$
6. $y = 7x^5 - 2\sqrt[3]{x}$
7. $y = \sin 2x$
8. $y = \sin x \cos x$
9. $y = \sin(x^2 + 1)$
10. $y = \sin^3 x$

In Exercises 11–14, a function f is described. Find f' and f''.

11. $f(x) = \sqrt{x} + \sqrt{x^3}$
12. $f(x) = \dfrac{1}{\sqrt{x}} + \dfrac{1}{\sqrt{x^3}}$
13. $f(x) = \tan 2x$
14. $f(x) = \cos \pi x$

15.–18. Find $f^{(3)}$ and $f^{(4)}$ for the functions described in Exercises 11–14.

In Exercises 19–26, use implicit differentiation to find $D_x^2 y$.

19. $x^2 + y^2 = 4$
20. $x^{1/2} + y^{1/2} = 4$
21. $x^2 + 2xy + y^2 = 16$
22. $x^2 + xy + y^2 = 4$
23. $x = \sin y$
24. $\cos xy = 1$
25. $x^2 + xy = y$
26. $x^2 y = 1 + xy$

In Exercises 27–38, (a) use the Second Derivative Test, when applicable, to determine relative extrema, and (b) sketch the graphs of the function.

27. $f(x) = x^2 - 4x + 2$
28. $f(x) = -x^2 + x$
29. $f(x) = x^3 + 3x$
30. $f(x) = 2 - 4x + x^3$
31. $f(x) = x^3 + 6x^2 + 2$
32. $f(x) = x^4 + 6x^2$
33. $f(x) = \sin x$
34. $f(x) = \cos x$
35. $f(x) = x + \dfrac{1}{x}$
36. $f(x) = 2x^2 + \dfrac{1}{2x^2}$
37. $f(x) = x^{3/2} - 3x^{1/2}$
38. $f(x) = 2x^{5/3} - 5x^{2/3}$

In Exercises 39–56:
a) determine intervals on which the graph of the given function is concave upward and intervals on which the graph is concave downward;
b) determine any points of inflection of the function; and
c) sketch the graph of the function.

39. $f(x) = x^4 - 6x^2$
40. $f(x) = x^5 - 10x^3$
41. $f(x) = \dfrac{x}{1 - x}$
42. $f(x) = \dfrac{x}{1 - x^2}$
43. $f(x) = \tan x$
44. $f(x) = \cot x$

45. $f(x) = \dfrac{x+1}{x-1}$
46. $f(x) = \dfrac{x^2+1}{x^2-1}$
47. $f(x) = x + \sin x$
48. $f(x) = \cos x - x$
49. $f(x) = x\sqrt{x^2 - 4}$
50. $f(x) = \dfrac{x}{\sqrt{x^2-4}}$
51. $f(x) = (x^2 - 9)^2$
52. $f(x) = (x^2 - 16)^2$
53. $f(x) = (x-2)^2(x+1)^2$
54. $f(x) = (x-2)^2(x+3)^3$
55. $f(x) = \sin 2x + 8 \sin x$
56. $f(x) = \cos 2x + 4 \cos x$

In Exercises 57–64, find a general formula expressing $f^{(n)}(x)$ for any positive integer n for the function described.

57. $f(x) = \dfrac{1}{x}$
58. $f(x) = \dfrac{1}{x-1}$
59. $f(x) = \dfrac{1}{x+1}$
60. $f(x) = \dfrac{1}{(x-1)^2}$
61. $f(x) = x^{-1/2}$
62. $f(x) = x^{1/3}$
63. $f(x) = \sin x$
64. $f(x) = \cos x$

65. Find h'' in terms of the derivatives of f and g if $h = f \circ g$.
66. Find a function f with the property that the domain of f is not equal to the domain of f' and the domain of f' is not equal to the domain of f''.
67. Suppose $P(x) = a_2 x^2 + a_1 x + a_0$, where $a_0, a_1,$ and a_2 are constants. Find $P(0), P'(0),$ and $P''(0)$.
68. Use the result in Exercise 67 to find a polynomial P of degree two or less such that $P(0) = 1, P'(0) = 4,$ and $P''(0) = 3$. Is this polynomial unique?
69. Suppose that P is a polynomial of degree n and that
$$P(x) = a_n x^n + a_{n-1} x^{n-1} + \cdots + a_1 x + a_0.$$
Find a relationship between the coefficients $a_0, a_1, \ldots, a_n$ of P and the evaluation of P and its derivatives at $x = 0$.

70. Show that the graph of a cubic polynomial must have exactly one point of inflection.
71. Show that the graph of a quartic (fourth-degree) polynomial can have two points of inflection or no points of inflection, but it cannot have just one point of inflection.
72. Suppose the function described by $f(x) = a_3 x^3 + a_2 x^2 + a_1 x + a_0$ for arbitrary constants $a_0, a_1, a_2,$ and a_3 has critical points at x_1 and x_2. Show that the graph of f has a point of inflection at $(x_0, f(x_0))$ where $x_0 = (x_1 + x_2)/2$.
73. Suppose that f'' exists on an interval $[a, b]$ and that c_1 and c_2 are critical points of f lying in (a, b). Show that if f' is not constant on (a, b), then f has a point of inflection between c_1 and c_2.
74. Suppose that f''' is defined on an open interval containing c and that $f''(c) = 0$. Show that if $f'''(c) \neq 0$, then $(c, f(c))$ is a point of inflection for the graph of f.
75. Find a cubic polynomial that passes through the point $(0, 0)$ and has a relative maximum at $x = 1$ and whose graph has a point of inflection at $x = 0$.
76. Suppose f is a cubic polynomial that has both a horizontal tangent and point of inflection at $(1, 1)$. Find the most general form of the equation of f.
77. A function f has the property that $f''(x) > 0$ if $x < 0$ and $f''(x) > 0$ if $x > 0$. In addition, $f(-1) = 0$, $f(0) = 2$, and $f(1) = 1$. What must be true about f' at $x = 0$?
78. Prove part (ii) of the Second Derivative Test.

Putnam exercise

79. On the domain $0 \leq \theta \leq 2\pi$: Prove that $(\sin \theta)^2 \cdot \sin(2\theta)$ takes its maximum at $\pi/3$ and $4\pi/3$ (and hence its minimum at $2\pi/3$ and $5\pi/3$).

(This exercise was problem B–6, part (a) of the thirty-fourth William Lowell Putnam examination given on December 1, 1973. The examination and its solution are in the December 1974 issue of the *American Mathematical Monthly*, pp. 1089–1095.)

2.11

Comprehensive Graphing

Techniques have been introduced throughout the first two chapters to aid in sketching the graphs of functions. In this section we summarize these results and show how they can be applied in a systematic manner.

In Chapter 1 we found that:

INTERCEPTS
The graph of f has a y-intercept at b if $f(0) = b$.
The graph of f has an x-intercept at a if $f(a) = 0$.

(See Figure 2.52.)

The concepts of symmetry and translation were also discussed in Chapter 1 (see Figures 2.53 and 2.54).

Figure 2.52
The intercept of the x-axis is at $(a, 0)$; the intercept of the y-axis is at $(0, b)$.

SYMMETRY
The graph of f is symmetric with respect to the y-axis if $f(-x) = f(x)$ (see Figure 2.53a).

The graph of f is symmetric with respect to the origin if $f(-x) = -f(x)$ (see Figure 2.53b).

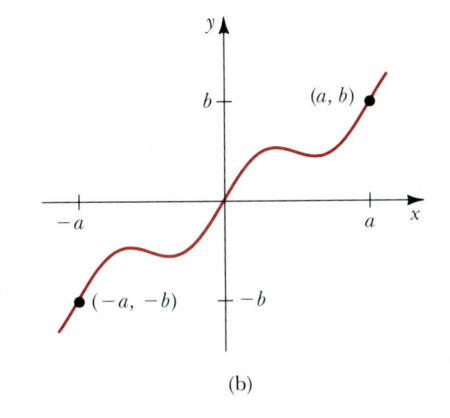

Figure 2.53
(a) y-axis symmetry; (b) origin symmetry.

TRANSLATION
— the graph of $y = f(x)$
— the graph of $y = f(x) + b$

— the graph of $y = f(x - a)$
— the graph of $y = f(x - a) + b$

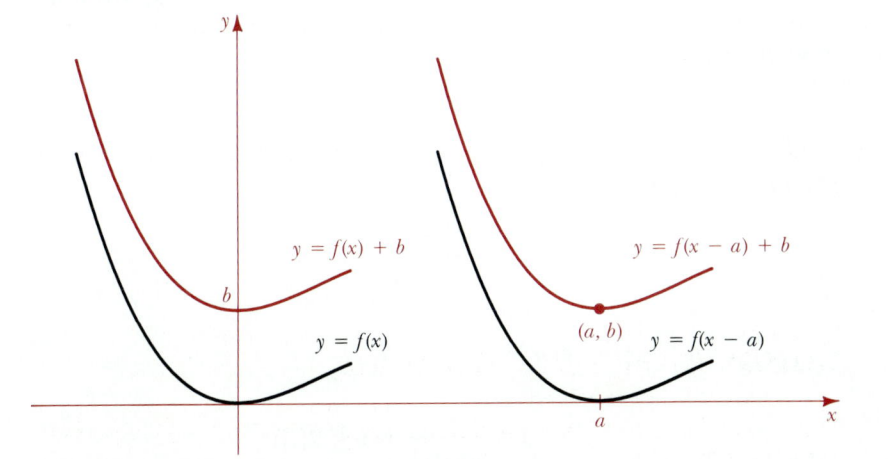

Figure 2.54

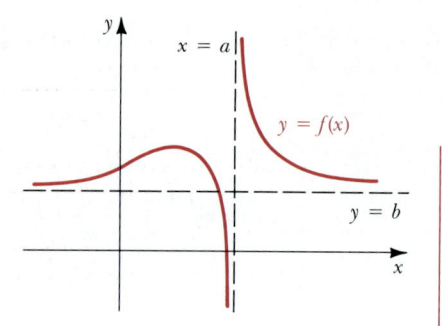

Figure 2.55
Horizontal asymptote $y = b$ and vertical asymptote $x = a$.

In the latter sections of Chapter 1 the concepts of infinite limits and limits at infinity were used to define vertical and horizontal asymptotes (see Figure 2.55):

ASYMPTOTES

The line $x = a$ is a vertical asymptote of the graph of f if

$$\lim_{x \to a^+} f(x) = \pm\infty \quad \text{or} \quad \lim_{x \to a^-} f(x) = \pm\infty.$$

The line $y = b$ is a horizontal asymptote of the graph of f if

$$\lim_{x \to \infty} f(x) = b \quad \text{or} \quad \lim_{x \to -\infty} f(x) = b.$$

The development of the derivative has provided a number of results that are important for graphing:

If f is continuous on an interval I, then f is:

i) increasing on I if $f'(x) > 0$ on the interior of I;
ii) decreasing on I if $f'(x) < 0$ on the interior of I.

If the interval I is open, then the graph of f on I is:

i) concave upward if $f''(x) > 0$ on I;
ii) concave downward if $f''(x) < 0$ on I.

These results provide answers to the question of when relative extrema and points of inflection occur.

i) If $f'(c)$ exists, a relative extremum can occur at c only if $f'(c) = 0$.
ii) If $f''(c)$ exists, a point of inflection can occur at $(c, f(c))$ only if $f''(c) = 0$.

Generally, not all these results can be used on a given problem, but they provide an impressive collection from which to choose appropriate techniques. The examples in this section are designed to help develop guidelines for making reasonable choices.

Example 1

Sketch the graph of

$$f(x) = \frac{x^2}{x^2 + 1} + 1.$$

Solution

    i) Since

$$1 \leq \frac{x^2}{x^2 + 1} + 1 < 2,$$

the values of f are in the interval $[1, 2)$ and there is no x-axis intercept. The y-axis intercept occurs at $(0, 1)$, since $f(0) = 1$.

ii) The graph is symmetric with respect to the y-axis because

$$f(-x) = \frac{(-x)^2}{(-x)^2 + 1} + 1 = \frac{x^2}{x^2 + 1} + 1 = f(x).$$

iii) The line $y = 2$ is a horizontal asymptote to the graph of f since

$$\lim_{x \to \infty} \frac{x^2}{x^2 + 1} + 1 = 2.$$

There are no vertical asymptotes to the graph because f is continuous at each real number.

iv) Since

$$f'(x) = \frac{(x^2 + 1)2x - x^2(2x)}{(x^2 + 1)^2} = \frac{2x}{(x^2 + 1)^2},$$

$f'(x) = 0$ at $x = 0$, the only critical point.

v) To determine the concavity of the graph, we need to find f'':

$$f''(x) = \frac{2(x^2 + 1)^2 - 2x[2(x^2 + 1)(2x)]}{(x^2 + 1)^4}$$

$$= \frac{2(x^2 + 1)(x^2 + 1 - 4x^2)}{(x^2 + 1)^4}$$

$$= \frac{2(1 - 3x^2)}{(x^2 + 1)^3}.$$

Since $(x^2 + 1)^3 > 0$ for all values of x, the only possible points of inflection occur when $1 - 3x^2 = 0$, that is, when

$$x = -\sqrt{3}/3 \quad \text{or} \quad x = \sqrt{3}/3.$$

We can use the critical points and the possible points of inflection to partition the domain of f as shown in the chart.

This analysis implies that the graph of f is as shown in Figure 2.56. Note that the graph of f is the graph of $y = x^2/(x^2 + 1)$ translated one unit upward.

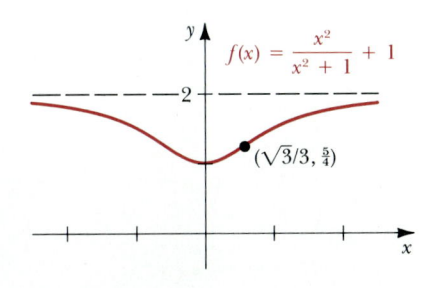

▶ **Figure 2.56**

Example 2

Sketch the graph of
$$f(x) = \frac{\sqrt{x^2 + 1}}{x}.$$

Solution

i) This graph has no x-intercepts since $f(x)$ is never 0. In addition, the graph has no y-intercept, since $x = 0$ is not in the domain of f.

ii) The graph is symmetric with respect to the origin because
$$f(-x) = \frac{\sqrt{(-x)^2 + 1}}{(-x)}$$
$$= -\frac{\sqrt{x^2 + 1}}{x} = -f(x).$$

iii) If $x > 0$, $x = \sqrt{x^2}$ and $f(x)$ can be rewritten as
$$f(x) = \frac{\sqrt{x^2 + 1}}{\sqrt{x^2}}$$
$$= \sqrt{1 + \frac{1}{x^2}}.$$

With this representation we see that
$$\lim_{x \to \infty} f(x) = 1, \quad \text{so } y = 1 \text{ is a horizontal asymptote}$$

and
$$\lim_{x \to 0^+} f(x) = \infty, \quad \text{so } x = 0 \text{ is a vertical asymptote.}$$

The graph is symmetric with respect to the origin, so these results also imply that
$$\lim_{x \to -\infty} f(x) = \lim_{x \to \infty} f(-x) = \lim_{x \to \infty} -f(x) = -1$$

(so $y = -1$ is also a horizontal asymptote) and that
$$\lim_{x \to 0^-} f(x) = \lim_{x \to 0^+} f(-x) = \lim_{x \to 0^+} -f(x) = -\infty.$$

iv) Since
$$f'(x) = \frac{x \cdot (1/2)(x^2 + 1)^{-1/2} \cdot 2x - \sqrt{x^2 + 1} \cdot (1)}{x^2}$$
$$= \frac{x^2 - (x^2 + 1)}{x^2 \sqrt{x^2 + 1}}$$
$$= -\frac{1}{x^2 \sqrt{x^2 + 1}},$$

$f'(x) < 0$ for all $x \neq 0$. Therefore, f is decreasing on both $(-\infty, 0)$ and $(0, \infty)$.

v) The Reciprocal Rule implies that

$$f''(x) = -\frac{1}{[x^2\sqrt{x^2+1}]^2} \cdot \{-D_x[x^2\sqrt{x^2+1}]\}$$

$$= \frac{1}{[x^2\sqrt{x^2+1}]^2} \cdot [2x\sqrt{x^2+1} + x^2(1/2)(x^2+1)^{-1/2}(2x)]$$

$$= \frac{1}{x^4(x^2+1)} \cdot \left[\frac{2x(x^2+1) + x^3}{\sqrt{x^2+1}}\right]$$

$$= \frac{(3x^2+2)}{x^3(x^2+1)^{3/2}}.$$

Since $(3x^2+2)/(x^2+1)^{3/2}$ is always positive, the sign of $f''(x)$ is the same as the sign of $1/x^3$.

Consequently,

when $x > 0$, $f''(x) > 0$ and the graph is concave upward, but

when $x < 0$, $f''(x) < 0$ and the graph is concave downward.

The graph is shown in Figure 2.57. ▶

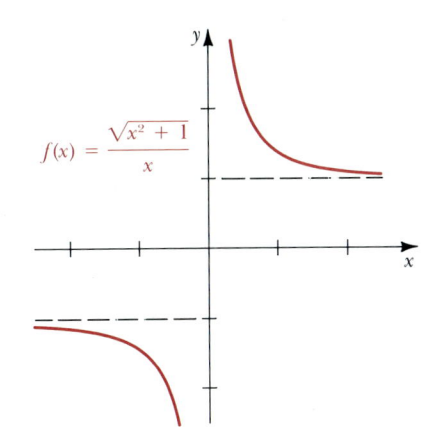

Figure 2.57

Example 3

Sketch the graph of

$$f(x) = \frac{x^2+x-2}{x^2}.$$

Solution

i) This graph has x-intercepts when

$$0 = x^2 + x - 2$$
$$= (x-1)(x+2),$$

that is, at $x = 1$ and at $x = -2$. Since $x = 0$ is not in the domain of f, there are no y-intercepts.

ii) There is no symmetry since

$$f(-x) = \frac{x^2 - x - 2}{x^2}$$

is neither $f(x)$ nor $-f(x)$.

iii) A horizontal asymptote occurs at $y = 1$ because both

$$\lim_{x \to \infty} \frac{x^2+x-2}{x^2} = 1 \quad \text{and} \quad \lim_{x \to -\infty} \frac{x^2+x-2}{x^2} = 1.$$

A vertical asymptote occurs at $x = 0$ with

$$\lim_{x \to 0^+} \frac{x^2+x-2}{x^2} = -\infty \quad \text{and} \quad \lim_{x \to 0^-} \frac{x^2+x-2}{x^2} = -\infty.$$

iv) Writing $f(x)$ as

$$f(x) = 1 + \frac{1}{x} - \frac{2}{x^2},$$

we have

$$f'(x) = -\frac{1}{x^2} + \frac{4}{x^3} = \frac{4-x}{x^3}$$

and the only critical point is $x = 4$.

v) Since

$$f''(x) = \frac{2}{x^3} - \frac{12}{x^4} = \frac{2(x-6)}{x^4},$$

the only possible point of inflection can occur when $x = 6$.

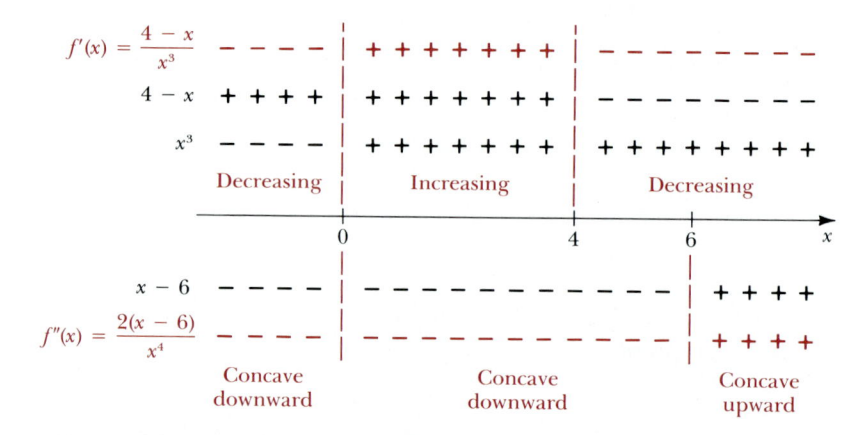

$f'(x) = \frac{4-x}{x^3}$	$----$	$++++++$	$----------$
$4-x$	$++++$	$++++++$	$----------$
x^3	$----$	$++++++$	$++++++++$
	Decreasing	Increasing	Decreasing
	0	4	6
$x-6$	$----$	$----------$	$++++$
$f''(x) = \frac{2(x-6)}{x^4}$	$----$	$----------$	$++++$
	Concave downward	Concave downward	Concave upward

Partitioning the real line using the point of discontinuity, the critical points, and the possible point of inflection gives the information in the chart. This implies that the graph of f is as shown in Figure 2.58; the insert exaggerates the local behavior at $(4, \frac{9}{8})$ and $(6, \frac{10}{9})$ to show the relative maximum and point of inflection. ▶

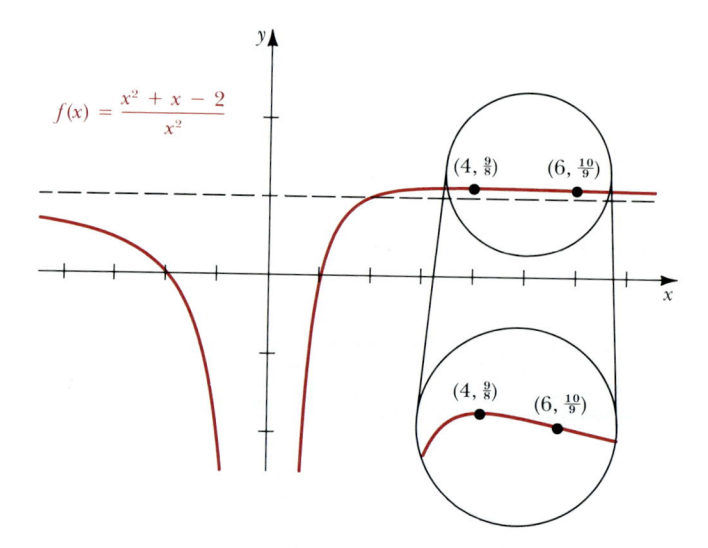

Figure 2.58

Example 4

Sketch the graph of $f(x) = x + \sin x$.

Solution

i) Since $f(0) = 0$, the graph intersects both axes at $(0, 0)$. It is difficult to determine whether there are other x-intercepts since this requires solving for x in the equation

$$0 = x + \sin x.$$

We defer this question to a later point in the analysis.

ii) The graph is symmetric with respect to the origin since

$$f(-x) = (-x) + \sin(-x)$$
$$= -x - \sin x = -f(x).$$

iii) There are no vertical asymptotes to the graph because f is continuous at each real number. Since $|\sin x| \leq 1$ for all x,

$$\lim_{x \to \infty} (x + \sin x) = \infty,$$

while

$$\lim_{x \to -\infty} (x + \sin x) = -\infty.$$

Thus there are also no horizontal asymptotes to the graph.

iv) Since

$$f'(x) = 1 + \cos x,$$

the only critical points occur when $\cos x = -1$, that is, when

$$x = (2k + 1)\pi, \quad \text{for some integer } k.$$

Note, too, that $f'(x) \geq 0$ for all values of x, so the graph is never decreasing. One result of this is that the graph can have no x-intercepts except at $x = 0$.

v) Also,

$$f''(x) = -\sin x,$$

so possible points of inflection occur only when

$$x = k\pi, \quad \text{for some integer } k.$$

Because the sine function, and hence $f''(x)$, changes sign at each multiple of π, the concavity of the graph changes at

$$(k\pi, f(k\pi)), \quad \text{for each integer } k.$$

The graph of f is shown in Figure 2.59. It oscillates about the line $y = x$. ▶

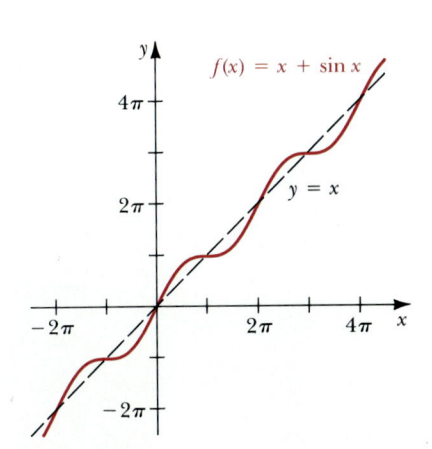

Figure 2.59

Example 5

Sketch the graph of $x^2y^2 - y^2 + x^2 = 0$.

Solution

This equation does not describe y as a function of x since

$$y^2(x^2 - 1) = -x^2 \quad \text{implies} \quad y^2 = \frac{x^2}{1 - x^2}$$

and

$$y = \pm \frac{\sqrt{x^2}}{\sqrt{1 - x^2}}.$$

However, the graph is symmetric with respect to the x-axis and the y-axis because if (x, y) is on the graph, then $(-x, y)$ and $(x, -y)$ are also on the graph. This implies that the complete graph of the equation can be deduced easily from the graph in any one quadrant.

In the first quadrant, $x \geq 0$ and $y \geq 0$ and y is described by the function

$$y = f(x) = \frac{x}{\sqrt{1 - x^2}}.$$

We will first sketch the graph of this function when $x \geq 0$.

Note that x must be less than 1 in order for $f(x)$ to be defined. Thus the domain of f is $[0, 1)$. A vertical asymptote for the graph occurs at $x = 1$ because

$$\lim_{x \to 1^-} \frac{x}{\sqrt{1 - x^2}} = \infty.$$

Since

$$f'(x) = \frac{1 \cdot \sqrt{1 - x^2} - x[(1/2)(1 - x^2)^{-1/2}(-2x)]}{(\sqrt{1 - x^2})^2}$$

$$= \frac{(1 - x^2)^{-1/2}[(1 - x^2) + x^2]}{(1 - x^2)}$$

$$= (1 - x^2)^{-3/2},$$

there are no critical points. Also, $f'(x) > 0$ for all x in $[0, 1)$, so f is always increasing. In addition,

$$f''(x) = -\frac{3}{2}(1 - x^2)^{-5/2}(-2x)$$

$$= 3x(1 - x^2)^{-5/2},$$

and the graph of f is concave upward on $[0, 1)$.

This information implies that the graph of the equation in the first quadrant is as shown in Figure 2.60(a). The x-axis and y-axis symmetry gives the complete graph shown in Figure 2.60(b). ▶

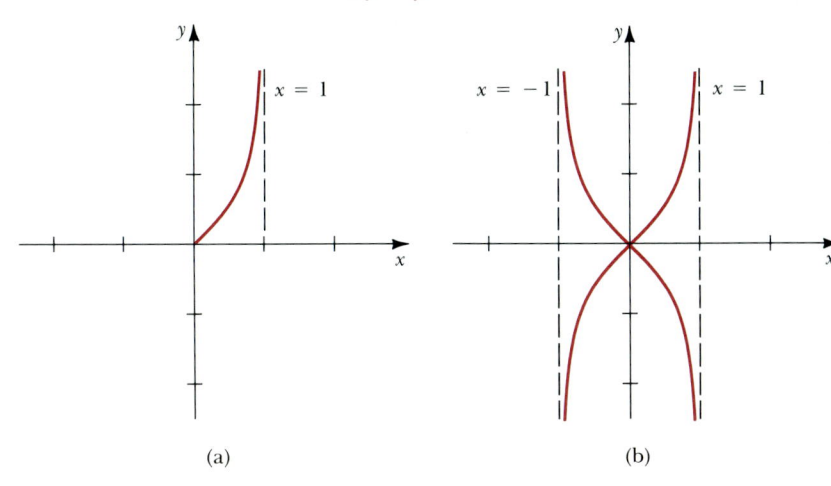

$$x^2y^2 - y^2 + x^2 = 0$$

(a) (b) **Figure 2.60**

▶ EXERCISE SET 2.11

In Exercises 1–32, sketch the graph of the function as completely as possible. When appropriate, describe symmetry, intercepts, asymptotes, relative extrema, points of inflection, and concavity.

1. $f(x) = x^3 - 6x^2$
2. $f(x) = x^3 - 3x$
3. $f(x) = x^4 - 2x^2$
4. $f(x) = x^4 + 2x^2$
5. $f(x) = x(x - 2)^2$
6. $f(x) = x^2(x - 2)$
7. $f(x) = x^2 + \dfrac{1}{x^2}$
8. $f(x) = x + \dfrac{1}{\sqrt{x}}$
9. $f(x) = x\sqrt{x - 1}$
10. $f(x) = x\sqrt{x + 1}$
11. $f(x) = x + \cos x$
12. $f(x) = 2x - \sin x$
13. $f(x) = (x - 2)^2(x + 1)^2$
14. $f(x) = x^2(x - 2)^2$
15. $f(x) = x^{2/3} + x^{5/3}$
16. $f(x) = x^{7/6} - x^{5/3}$
17. $f(x) = \dfrac{x + 2}{x - 3}$
18. $f(x) = \dfrac{x - 3}{x + 2}$
19. $f(x) = \dfrac{1}{x^2 - 1}$
20. $f(x) = \dfrac{1}{x^2 - 5x + 6}$
21. $f(x) = \dfrac{x^2 - 4}{x^2 - 5x + 6}$
22. $f(x) = \dfrac{x^2 - 5x + 6}{x^2 - 1}$
23. $f(x) = \dfrac{3 - 2x - 5x^2}{x^2}$
24. $f(x) = \dfrac{x^2 - 4}{x^2 - 2x + 1}$
25. $f(x) = \dfrac{x^2 - 1}{x^3}$
26. $f(x) = \dfrac{x^3 - 1}{x^2}$
27. $f(x) = \dfrac{\sqrt{x^2 + 1}}{x}$
28. $f(x) = \dfrac{x^2 + 1}{\sqrt{x^2 - 1}}$
29. $f(x) = |x^2 - 2|$
30. $f(x) = x|x^2 - 2|$
31. $f(x) = \cos^2 x + 2$
32. $f(x) = \sin^2 x - 3$

In Exercises 33–40, sketch the graph of the equation as completely as possible. When appropriate, describe symmetry, intercepts, asymptotes, relative extrema, points of inflection, and concavity.

33. $x^2y^2 - x^2 - y^2 = 1$
34. $xy^2 + 3x = 6$
35. $x^2 - y^2 = x^2y^2$
36. $yx^2 + 4y = x$
37. $xy - y^2 = 1$
38. $x^2y + 4x + y = 0$
39. $x^2 - xy + 1 = 0$
40. $xy - 2y^2 = 2$

▶ IMPORTANT TERMS AND RESULTS

CONCEPT	PAGE	CONCEPT	PAGE
Tangent line	90	Absolute extrema	138
Vertical tangent	92	Relative extrema	139
Derivative	95	Critical point	140
Power Rule	105	Rolle's Theorem	146
Sum Rule	105	Mean Value Theorem	147
Product Rule	107	Antiderivative	150
Reciprocal Rule	109	Increasing, decreasing	153
Quotient Rule	110	First Derivative Test	156
Derivatives of trigonometric		Higher derivatives	160
functions	118	Second Derivative Test	162
Chain Rule	123	Concavity	164
Implicit differentiation	132	Point of inflection	165

DERIVATIVE FORMULAS

1. $D_x(mx + b) = m$
2. $D_x(cf)(x) = c\, D_x f(x)$
3. $D_x(f \pm g)(x) = D_x f(x) \pm D_x g(x)$
4. $D_x(fg)(x) = f(x)\, D_x g(x) + g(x)\, D_x f(x)$
5. $D_x\left(\dfrac{f}{g}\right)(x) = \dfrac{g(x)\, D_x f(x) - f(x)\, D_x g(x)}{[g(x)]^2}$
6. $D_x f(g(x)) = f'(g(x)) \cdot g'(x)$

If $u = g(x)$ is a differentiable function of x, then:

7. $D_x u^r = r u^{r-1} \cdot D_x u$, for any nonzero rational number r.
8. $D_x \sin u = \cos u\, D_x u$
9. $D_x \cos u = -\sin u\, D_x u$
10. $D_x \tan u = \sec^2 u\, D_x u$
11. $D_x \cot u = -\csc^2 u\, D_x u$
12. $D_x \sec u = \sec u \tan u\, D_x u$
13. $D_x \csc u = -\csc u \cot u\, D_x u$

▶ REVIEW EXERCISES

In Exercises 1–36, find the derivative of the function.

1. $f(x) = 3x^4 - 2x^3 + x - 5$
2. $f(x) = 13x^{14} - 16x^4 + 2x - 1$
3. $f(x) = 3\sqrt{x} + \dfrac{1}{3\sqrt{x}}$
4. $f(x) = \sqrt{3x} + \dfrac{1}{\sqrt{3x}}$
5. $g(t) = \dfrac{16t^2 - 32}{t}$
6. $r(z) = \dfrac{3z^2 + 2z - 5}{z^2}$
7. $f(x) = (x^3 - 7x + 5)\sin x$
8. $g(x) = \dfrac{\sin x}{x + 1}$
9. $h(w) = (w^2 + 2w + 1)^2$
10. $f(w) = (w^3 - w + 1)^4$
11. $f(x) = (x^2 - 7)^3(x^4 + 1)$
12. $h(t) = (t^2 + t)^3(t^3 - 1)^2$
13. $g(x) = \dfrac{x + 2}{\sqrt{x^2 - 4}}$
14. $h(x) = \dfrac{\sqrt{x^2 - 4}}{x + 2}$

15. $g(x) = [(x^2 +)^3 - 7x]^5$
16. $f(x) = [(x^2 + 1)^3 - (x + 1)^2]^5$
17. $g(s) = [(2s + 1)^3 + s^{1/3}]^4$
18. $f(x) = [(2x + 1)^{1/3} + x^3]^4$
19. $h(u) = u^3 - \dfrac{1}{u^2} + \dfrac{1}{u} - 2$
20. $r(s) = \sqrt{s} + \dfrac{1}{2\sqrt{s}} - \dfrac{3}{s\sqrt{s}}$
21. $g(x) = \left(\dfrac{x^3 - 8}{x^2 + 4}\right)^5$
22. $h(x) = \left(\dfrac{x^4 - 1}{x^2 + 2x + 1}\right)^{-3}$
23. $f(x) = \left(\dfrac{x}{x^3 + 1}\right)^{-4}$
24. $f(x) = \left(\dfrac{x^3 - 3x + 1}{x^2 - 2}\right)^{-5}$
25. $f(x) = \sin(2x + 3)$
26. $g(x) = \sin(2x + 3)^3$
27. $h(x) = [\sin(2x + 1)]^2$
28. $r(x) = [\sin(2x + 1)^3]^2$
29. $f(x) = \tan(x^2 + 1)^3$
30. $g(x) = [\tan(x^2 + 1)^3]^{1/2}$
31. $h(t) = \sqrt{\sec(t^2 + 1)}$
32. $f(t) = \sec\sqrt{t^2 + 1}$
33. $g(x) = (x^2 + 1)(2x - 1)(x + 3)$
34. $g(w) = 2w^5(w^3 + 1)^2(w + 2)^4$
35. $g(x) = x \tan x \sec(x + 1)$
36. $h(x) = (x^2 + 1)\cot(x - 1)\sin x^2$

37–46. Find the second and third derivatives of the functions described in Exercises 1–10.

In Exercises 47–58, find dy/dx.

47. $xy + x^2y = 1$
48. $xy^2 - x^2y = 7$
49. $2xy - xy^2 + x = 0$
50. $x^2y + xy^2 + y^2 = 2$
51. $(x^2 + y^2)^2 = xy$
52. $(x^2 + y^2)^{1/2} = xy$
53. $\sqrt{xy} + \dfrac{1}{\sqrt{xy}} = 3$
54. $\sqrt{xy} + \dfrac{1}{\sqrt{xy}} = 3$
55. $y \sin x + xy = 0$
56. $y \sin x + x \sin y = 0$
57. $\tan xy + y \tan x = x$
58. $\tan xy + x \tan y = x$

In Exercises, 59–62, find (a) dy/dx and (b) d^2y/dx^2.

59. $y^3 + x^3 = 3xy$
60. $x^2 + 3xy + y^2 = 1$
61. $\sin y + xy = 0$
62. $x \tan y + y = 1$
63. Find the first five derivatives of f if:
    a) $f(x) = x^4 - 7x^2$;
    b) $f(x) = (x + 1)^{1/2}$;
    c) $f(x) = \sin x$;
    d) $f(x) = x \sin x$.
64. Find the first five derivatives of f if:
    a) $f(x) = x^5 - 3x^3 + 7x^2 + x - 1$;
    b) $f(x) = (2x + 3)^{1/2}$;
    c) $f(x) = \cos x$;
    d) $f(x) = x^2 \cos x$.

For the functions described in Exercises 65–76, find the critical points, intervals on which the function is decreasing and increasing, relative extrema, intervals on which the function is concave upward or downward, and points of inflection. Find any horizontal or vertical asymptotes to the graph, and sketch the graph of the function.

65. $f(x) = 2x^3 - 3x^2 - 12x + 13$
66. $f(x) = x^3 - x$
67. $f(x) = x^{3/2} - 3x^{1/2}$
68. $f(x) = \dfrac{x^2}{8} - \dfrac{1}{x}$
69. $f(x) = 2\sqrt{x} - x$
70. $f(x) = 3x^4 - 4x^3$
71. $f(x) = 3 \sin 2x$
72. $f(x) = x + \sin x$
73. $f(x) = \dfrac{1}{x^2 - 2x}$
74. $f(x) = \dfrac{2x^2 - 1}{x^2 - 1}$
75. $f(x) = \dfrac{x + 1}{\sqrt{x - 1}}$
76. $f(x) = \dfrac{x - 1}{\sqrt{x + 1}}$

In Exercises 77–82, use the definition of the derivative to verify the statement.

77. $D_x(x^2 - x) = 2x - 1$
78. $D_x(x^2 + 3x + 3) = 2x + 3$
79. $D_x\left(\dfrac{2}{x} + x - 1\right) = -\dfrac{2}{x^2} + 1$
80. $D_x\left(x^3 + \dfrac{1}{x}\right) = 3x^2 - \dfrac{1}{x^2}$

81. $D_x(\sqrt{x} + 3x) = \dfrac{1}{2\sqrt{x}} + 3$

82. $D_x \sqrt{x+2} = \dfrac{1}{2\sqrt{x+2}}$

83. Find an equation of the tangent line to the graph of $f(x) = x(4 - x^2)^{1/2}$ at $(0, 0)$.

84. Find an equation of the tangent line to the graph of $f(x) = \sqrt[3]{4 - x}$ that is parallel to the line with equation $x + 12y - 13 = 0$.

85. Find an equation of the normal line to the graph of $f(x) = x(4 - x^2)^{1/2}$ at $(0, 0)$.

86. Find an equation of the normal line to the graph of $f(x) = \sqrt[3]{4 - x}$ that is parallel to the line with equation $12x - y - 13 = 0$.

87. Find an equation of the tangent line to the graph of $f(x) = x^3 - 3x + 2$ that passes through $(0, 0)$.

88. Find equations of the tangent lines to the graph of $f(x) = x^2 - 2x$ that pass through $(3, 2)$.

89. Give an example of a function that is continuous at $x = 1$, but not differentiable there.

90. Given an example of a function that is continuous at every real number and differentiable at every number except at $x = 0$ and $x = 1$.

91. Find the absolute maximum and minimum values of $f(x) = 2 - |1 - x|$ on the interval $[0, 2]$.

92. Find absolute extrema of $f(x) = (x - 2)^{1/3}(2x - 2)^{2/3}$ on the interval $[0, 2]$.

93. Show that if $k > 0$, then the function described by $f(x) = x^3 + 3kx - 5$ has no relative extrema.

94. Sketch the graph of a function that has a relative maximum at $(0, 2)$, a relative minimum at $(2, 0)$, and a point of inflection at $(1, 1)$.

95. Sketch the graph of a function that is increasing on the intervals $(-\infty, -1)$, $(3, 6)$, and $(6, \infty)$, and satisfies $\lim_{x \to 6^+} f(x) = -\infty$, $\lim_{x \to 6^-} f(x) = \infty$.

96. Let f be such that $f(0) = 1$ and $f'(x) = 2$ for all x. What is $f(x)$?

97. Let f be such that $f(0) = 1$ and $f'(x) = 3x^2$ for all x. What is $f(x)$?

98. Show that the graph of a quadratic polynomial cannot have a point of inflection.

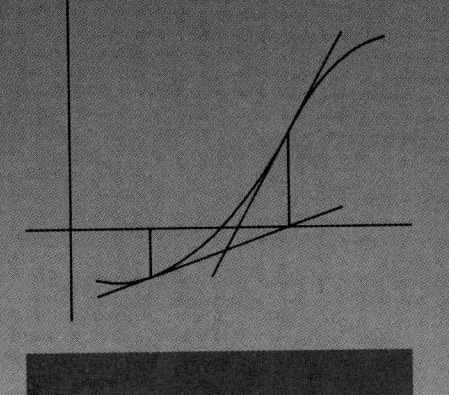

3
Applications of the Derivative

In Chapter 2 the derivative was introduced, with an emphasis on its use in sketching the graphs of functions. This chapter considers a few of the other diverse applications of the derivative. The common thread in these applications is the need to measure the instantaneous rate of change of one quantity with respect to another.

The first three sections of this chapter are concerned primarily with physical applications of the derivative: these are applications that relate to problems in science, engineering, and business. Section 3.4 discusses the differential, a concept that has mathematical and physical applications.

Sections 3.5 and 3.6 present other mathematical applications of the derivative. The first of these sections introduces l'Hôpital's Rule for evaluating limits. Section 3.6 concerns Newton's method for approximating the roots of equations. This technique serves as an introduction to numerical analysis, which is an area of computational mathematics.

The final section of the chapter shows how the derivative is used to solve problems in business and economics. You will undoubtedly see these applications again if you study quantitative subjects in these fields.

3.1

Rectilinear Motion

An interest in methods to study the nature of continuous motion was one of the principal reasons for the development of calculus. In this section we use the derivative to describe the motion of a particle that moves in a straight line. This is called **rectilinear motion.**

Suppose a rock is dropped from a height of 1600 ft and that the height, in feet, of the rock above the ground t sec after it has been dropped is given by

$$s(t) = 1600 - 16t^2, \quad \text{where } 0 \leq t \leq 10$$

(see Figure 3.1). How fast is the rock moving 5 sec after it has been dropped?

The *average* rate at which an object travels is found by dividing the distance traveled by the time required to travel that distance. **Average velocity** describes this rate as well as the direction in which the object is moving. The **average speed** of an object is the magnitude, or absolute value, of its average velocity.

For the falling rock in our example, the distance traveled is the difference between the height of the rock at two different times. Because the rock is falling to earth, the distance from the ground is decreasing with time, and the average velocity is always negative. For example, the average velocity of the rock from $t = 5$ sec to $t = 6$ sec is

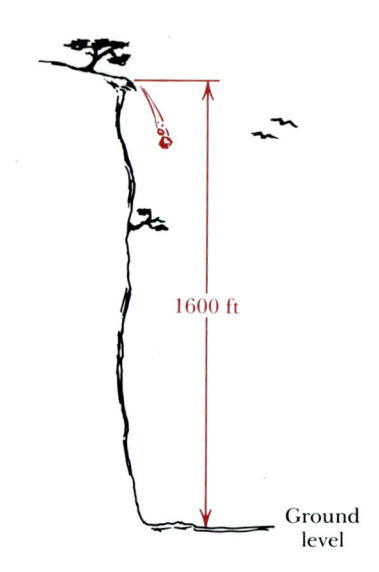

Figure 3.1

$$\text{Average velocity} \atop (t = 5 \text{ to } t = 6) = \frac{s(6) - s(5) \text{ ft}}{(6-5) \text{ sec}}$$

$$= \frac{(1600 - 16(6)^2) - (1600 - 16(5)^2) \text{ ft}}{1 \text{ sec}}$$

$$= (1024 - 1200)\frac{\text{ft}}{\text{sec}} = -176 \frac{\text{ft}}{\text{sec}}.$$

This calculation gives the *average* velocity of the rock for the 1-sec time interval from $t = 5$ sec to $t = 6$ sec. The question "How fast is the rock moving 5 sec after it has been dropped?" is asking for the *instantaneous* velocity at $t = 5$ sec. This velocity, denoted $v(t)$, is the instantaneous rate of change of $s(t)$ with respect to the change in t.

In Chapter 2 we found that the *derivative* of a function describes the instantaneous rate of change in the values of a function with respect to the change in its independent variable. Thus the velocity of the rock, in feet per second, at $t = 5$ sec is

$$v(5) = s'(5) = \lim_{h \to 0} \frac{s(5+h) - s(5)}{h}.$$

Since $s(t) = 1600 - 16t^2$,

$$v(t) = s'(t) = -32t,$$

so

$$v(5) = -160 \text{ ft/sec}.$$

The negative value for $v(t)$ implies that $s(t)$ is decreasing with time, which is, of course, the case since the rock is falling.

The rate at which the velocity of an object changes with respect to time is also of interest in physical problems. The derivative of the velocity gives this instantaneous rate of change, which is called the *acceleration* and denoted $a(t)$. For the falling rock, $v(t) = -32t$ ft/sec, so

$$a(t) = v'(t) = s''(t) = -32 \frac{\text{ft/sec}}{\text{sec}} = -32 \frac{\text{ft}}{\text{sec}^2}.$$

This connection between the derivative and the *motion equations* of objects is one of the fundamental discoveries in the history of science and is the beginning of the study of physics as we know it. It permits us to make the following definition.

(3.1)
DEFINITION
Suppose $s(t)$ describes the rectilinear (straight-line) motion of an object.

The **velocity** of the object at time t, $v(t)$, is defined by $v(t) = s'(t)$, provided $s'(t)$ exists.

The **speed** of the object is defined to be the magnitude of the velocity, $|v(t)|$.

The **acceleration** at time t, $a(t)$, is defined by $a(t) = v'(t) = s''(t)$, provided $s''(t)$ exists.

In problems concerning rectilinear motion a positive direction must be given for the motion. This direction is generally chosen to be upward if the motion is along a vertical line. If the motion is along a horizontal line, the positive direction is usually assumed to be to the right.

Example 1

Suppose a particle moves along a straight line so that its position at time $t \geq 0$ is given by

$$s(t) = t^3 - 12t^2 + 36t,$$

where t is measured in minutes (min) and $s(t)$ in centimeters (cm). Find the velocity and acceleration of the particle and describe its motion.

Solution

The velocity in cm/min at any time t is

$$v(t) = s'(t) = 3t^2 - 24t + 36,$$

and the acceleration in cm/min^2 is

$$a(t) = v'(t) = s''(t) = 6t - 24.$$

We can use $v(t)$ to determine the direction of the motion. Since

$$v(t) = 3(t^2 - 8t + 12) = 3(t - 6)(t - 2),$$

$v(2)$ and $v(6)$ are both zero. The particle is instantaneously stopped at $t = 2$ min and at $t = 6$ min.

When t is less than 2 or greater than 6, $v(t) > 0$, so $s(t)$ is increasing and the motion is to the right. When t is between 2 and 6, $v(t) < 0$ and the motion is to the left.

Calculating $s(t)$ for $t = 0$, $t = 2$, $t = 6$, and $t = 8$, we can sketch the pattern of motion of the particle, as shown in Figure 3.2. Although the actual path of the particle lies along the line s, we show the motion above this line so that overlapping paths can be distinguished.

Since $a(t) = 6t - 24$ is negative when $0 \leq t < 4$, the velocity of the particle is decreasing during that time. The velocity is increasing when $t > 4$. At $t = 4$, the velocity is not changing since $v'(4) = a(4) = 0$. ▶

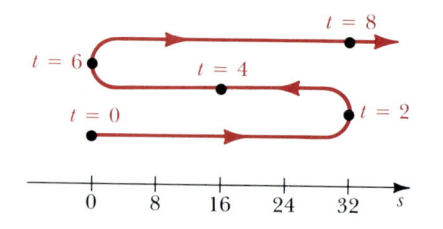

Figure 3.2

Example 2

A ball thrown upward from the earth with an initial velocity of 88 ft/sec is $s(t) = 88t - 16t^2$ ft above the earth t sec after it has been thrown. This equation is valid from the time the ball is thrown until the time it returns to earth. Find:

a) the velocity and acceleration of the ball at any time t;
b) how many seconds it takes the ball to reach its highest point;
c) how high the ball goes;
d) how many seconds it takes the ball to reach the ground; and
e) the velocity of the ball when it hits the ground.

Solution

a) $v(t) = s'(t) = 88 - 32t \dfrac{\text{ft}}{\text{sec}}$ and $a(t) = s''(t) = -32 \dfrac{\text{ft}}{\text{sec}^2}$.

b) At the ball's highest altitude $s(t)$ is a maximum. A maximum can occur only at an endpoint of the domain of s or at a critical point. At the endpoints of the domain of s, the ball is on the ground, so the ball's highest altitude must occur at a critical point. Thus the highest altitude occurs at the time when

$$0 = s'(t) = v(t) = 88 - 32t, \quad \text{that is, when} \quad t = \frac{88}{32} = \frac{11}{4} \text{ sec.}$$

Since $v(\frac{11}{4}) = 0$, the ball is instantaneously stopped when it reaches its maximum height.

c) The highest altitude is found by evaluating $s(t)$ at $t = \frac{11}{4}$ sec:

$$s\left(\frac{11}{4}\right) = 88\left(\frac{11}{4}\right) - 16\left(\frac{11}{4}\right)^2 = 121 \text{ ft.}$$

d) The ball is on the ground precisely when $s(t) = 0$. Solving

$$0 = s(t) = 88t - 16t^2 = 8t(11 - 2t)$$

for t, we have $t = 0$, the time when the ball was thrown, and $t = \frac{11}{2}$, the time when it returns to earth.

e) The velocity when the ball hits the ground is

$$v\left(\frac{11}{2}\right) = 88 - 32\left(\frac{11}{2}\right) = -88 \dfrac{\text{ft}}{\text{sec}},$$

the negative of the velocity at which the ball was thrown.

A sketch of the height of the ball as a function of time is shown in Figure 3.3.

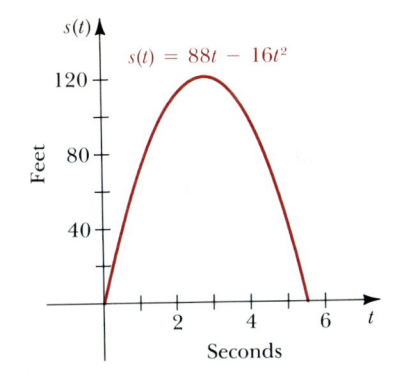

▶ **Figure 3.3**

Note in Example 2 that the acceleration of the ball is constant, the same value as that obtained for the acceleration of the falling rock at the beginning of the section. This constant, -32 ft/sec^2, or -9.8 m/sec^2, is approximately the acceleration due to the gravitational force of the earth on an object at the surface of the earth. By making the assumption that this constant acceleration holds for objects that remain close to the surface of the earth, we can determine equations that describe the velocity, $v(t)$, of the object and its distance, $s(t)$, above the surface.

Suppose $a(t) = -32$. Since $D_t(-32t) = -32$, both

$$D_t v(t) = a(t) = -32 \quad \text{and} \quad D_t(-32t) = -32.$$

In Section 2.8 we saw that two functions have the same derivative on an interval precisely when they differ by a constant. Thus a constant C must exist with

$$v(t) = -32t + C.$$

To determine C, let $t = 0$. Then

$$v(0) = -32 \cdot 0 + C,$$

so
$$C = v(0) \quad \text{and} \quad v(t) = -32t + v(0).$$
Similarly, both
$$D_t\, s(t) = v(t) = -32t + v(0) \quad \text{and} \quad D_t\,[-16t^2 + v(0)t] = -32t + v(0).$$
Thus a constant K exists with
$$s(t) = -16t^2 + v(0)t + K.$$
Evaluating at $t = 0$ gives
$$s(0) = -16(0)^2 + v(0) \cdot 0 + K,$$
so
$$K = s(0) \quad \text{and} \quad s(t) = -16t^2 + v(0)t + s(0).$$
In summary, the assumption that the acceleration is
$$a(t) = -32 \text{ ft/sec}^2$$
implies that

(3.2) $\quad v(t) = -32t + v(0) \text{ ft/sec} \quad \text{and} \quad s(t) = -16t^2 + v(0)t + s(0) \text{ ft}.$

If the acceleration is given in the equivalent metric units, then
$$a(t) = -9.8 \text{ m/sec}^2$$
implies that

(3.3)
$$v(t) = -9.8t + v(0) \text{ m/sec} \quad \text{and} \quad s(t) = -4.9t^2 + v(0)t + s(0) \text{ m}.$$

Example 3

Derive the motion and velocity equations for the problem described in Example 2.

Solution

In Example 2 the ball was thrown upward from the earth with an initial velocity of 88 ft/sec. This implies that
$$v(0) = 88 \text{ ft/sec} \quad \text{and} \quad s(0) = 0.$$
Therefore,
$$v(t) = -32t + 88 \text{ ft/sec}$$
and
$$s(t) = -16t^2 + 88t + 0 = 88t - 16t^2 \text{ ft}. \quad \blacktriangleright$$

Example 4

A child stands at the top of a cliff and throws a rock straight down into a pond 120 m below. The rock hits the pond 3 sec later. At what velocity is the rock thrown?

Solution

The distance is given in meters and

$$a(t) = -9.8 \text{ m/sec}^2$$

implies that

$$v(t) = -9.8t + v(0) \text{ m/sec} \quad \text{and} \quad s(t) = -4.9t^2 + v(0)t + s(0) \text{ m}.$$

In this problem $v(0)$ is not given, but two values of $s(t)$ are known:

$$s(0) = 120 \quad \text{and} \quad s(3) = 0.$$

Substituting $s(0) = 120$ into the equation for $s(t)$ gives

$$s(t) = -4.9t^2 + v(0)t + 120.$$

When this equation is evaluated at $t = 3$, we have

$$0 = s(3) = -4.9(3)^2 + v(0) \cdot 3 + 120.$$

This is an equation in which $v(0)$ is the only unknown. Solving for $v(0)$, we have

$$v(0) = \frac{(4.9)(9) - 120}{3} = -25.3 \text{ m/sec}.$$

The initial velocity is -25.3 m/sec, so the rock was thrown downward with an initial speed of 25.3 m/sec. ▶

The final example shows that general rectilinear-motion problems are solved in the same manner as free-fall problems.

Example 5

The screen of a television set is a cathode-ray tube 28.0 cm long. An electric field at the back of the tube accelerates an electron from rest at 5.00×10^{16} cm/sec². This field continues for 2.00 cm, after which the electron travels at constant velocity to the front of the tube (see Figure 3.4).

a) How long does it take the electron to pass through the electric field?
b) How long does it take the electron to reach the screen?

Solution

a) Let $s(t)$ be the distance from the back of the tube t sec after the electron has been emitted. Then $s(0) = 0$ and since the electron starts from rest, $v(0) = 0$. For the first 2.00 cm, $a(t) = 5.00 \times 10^{16}$, so

$$v(t) = 5.00 \times 10^{16} t + v(0) = 5.00 \times 10^{16} t$$

and

$$s(t) = 5.00 \times 10^{16} (t^2/2) + s(0) = 2.50 \times 10^{16} t^2.$$

The distance to the end of the field is 2.00 cm, so the electron will reach this point when

$$2.00 = 2.50 \times 10^{16} t^2,$$

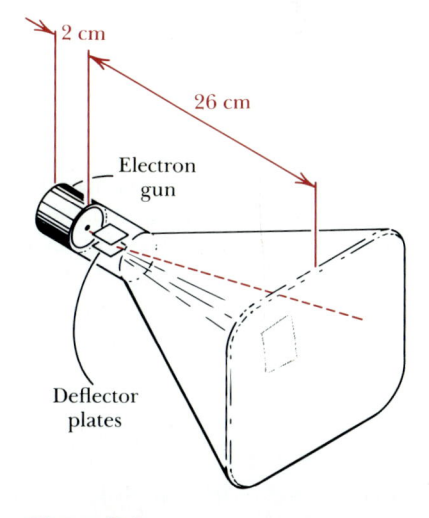

Figure 3.4
A cathode-ray tube.

that is, when
$$t = \left(\frac{2.00}{2.50 \times 10^{16}}\right)^{1/2} = 8.94 \times 10^{-9} \text{ sec.}$$

b) When the electron passes through the field, its velocity is
$$v(8.94 \times 10^{-9}) = (5.00 \times 10^{16})(8.94 \times 10^{-9}) = 4.47 \times 10^8 \text{ cm/sec,}$$
so to travel the remaining 26.0 cm to the front of the screen takes
$$\frac{26.0}{4.47 \times 10^8} = 5.82 \times 10^{-8} \text{ sec.}$$
The total time for the electron to reach the screen is
$$8.94 \times 10^{-9} + 5.82 \times 10^{-8} = 6.71 \times 10^{-8} \text{ sec.} \quad \blacktriangleright$$

▶ **EXERCISE SET 3.1**

In Exercises 1–8, $s(t)$ describes the distance of an object from a specified point at the end of time t. Find the instantaneous velocity and acceleration of the object at the time given.

1. $s(t) = 32t^2 + 3$, $t = 4$
2. $s(t) = 16t^2 + 300$, $t = 4$
3. $s(t) = t + \sin t$, $t = \pi/2$
4. $s(t) = \sqrt{t^2 + 1}$, $t = 6$
5. $s(t) = \sqrt[3]{t^2 + 2t + 1}$, $t = 0$
6. $s(t) = \dfrac{t+1}{\sqrt{t^2+1}}$, $t = 1$
7. $s(t) = \dfrac{\sin t}{t+1}$, $t = \pi$
8. $s(t) = \sin t + t \cos t$, $t = \dfrac{\pi}{4}$

In Exercises 9–16, $s(t)$ describes the distance of an object from a specified point at the end of time $t \geq 0$. Sketch a figure of the motion of the object similar to the figure given in Example 1. Determine if the object is ever momentarily stopped.

9. $s(t) = t^2 - 2t + 2$
10. $s(t) = t^2 - t$
11. $s(t) = t^3 - 3t^2 + 4$
12. $s(t) = t^3 + 2$
13. $s(t) = t^4 - 4t^3 + 4t^2 + 1$
14. $s(t) = 2t^3 - 9t^2 + 12t + 12$
15. $s(t) = \dfrac{t}{t+1}$
16. $s(t) = \sin t$, $0 \leq t \leq 2\pi$

17. A ball thrown upward at 24 ft/sec from a platform 16 ft above the ground is $s(t) = 16 + 24t - 16t^2$ ft above the ground t sec after it has been thrown. Find:
    a) the velocity and acceleration of the ball at any time t;
    b) how many seconds it takes the ball to reach its highest point;
    c) how high the ball will go;
    d) how many seconds it takes the ball to reach the ground; and
    e) the velocity of the ball when it hits the ground.

18. A ball thrown upward from the roof of a building 245 m high with an initial velocity of 24.5 m/sec is $s(t) = 245 + 24.5t - 4.9t^2$ m above the ground t sec after it has been thrown. Determine:
    a) the velocity and acceleration of the ball at any time t;
    b) how many seconds it takes the ball to reach its highest point;
    c) how high the ball will go;
    d) how many seconds it takes the ball to reach the ground; and
    e) the velocity of the ball when it hits the ground.

19. a) What is the minimal initial velocity necessary to throw a ball 100 ft high?
    b) How long does it take the ball to reach 100 ft when thrown at this minimal initial velocity?

20. Suppose a ball is thrown upward with twice the initial velocity determined in Exercise 19.
    a) How long does it take the ball to reach its maximum height?
    b) What will be the maximum height of the ball?

21. A student with a water balloon is at the top of a 42-ft building. A mathematics professor who is 6 ft tall is walking below. How long before the professor arrives should the student release the missile, if the object is to score a direct hit on the top of the professor's head? (Do *not* experimentally verify your answer!)

22. The Sears Tower in Chicago, with a height of 1450 ft, is the tallest building in the world (see the figure).
    a) How long does it take an object dropped from the top of this building to reach the ground?
    b) What is the velocity of the object when it reaches the ground?

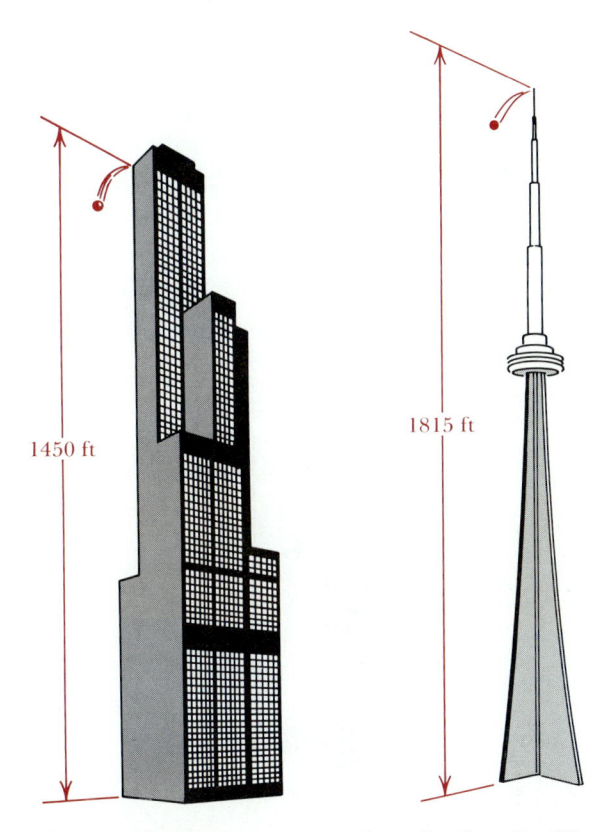

Figure for Exercise 22.   Figure for Exercise 23.

23. The world's tallest freestanding structure is Toronto's CN Tower, which rises 1815 ft above the ground (see the figure).
    a) How much longer does it take an object to reach the ground from the top of the CN Tower than from the top of the Sears Tower described in Exercise 22?
    b) At what velocity does an object hit the ground when it is dropped from the top of the CN Tower?

24. The child described in Example 4 throws another rock in a vertical direction so that it lands in the pond. This rock takes 4 sec to reach the water. At what velocity is the rock thrown?

25. An unintelligent archer fires an arrow directly upward with a velocity of 20 m/sec. How much time does the archer have to find cover before being skewered?

26. A cat jumps vertically upward to reach the top of a ledge that is 5 ft above the original position of the cat. What is the minimal initial velocity the cat must have in order to reach the ledge?

27. A high-jumper can clear a bar set at a height of 7 ft.
    a) What initial velocity must the jumper have in order to clear this bar?
    b) What is the minimal time that the jumper will be in the air?

28. The acceleration due to gravity on the moon is approximately one-sixth that on earth. Suppose the high-jumper in Exercise 27 can jump with the same initial velocity on the moon as on earth.
    a) How high would the jump be on the moon?
    b) How long would the jumper be above the surface of the moon?

29. Suppose the high-jumper described in the preceding exercises made a jump with the same initial velocity from the surface of a planet with approximately the same mass as the sun. Such a planet would have an acceleration due to gravity of approximately twenty-four times that of earth.
    a) How high would the jump be on this planet?
    b) How long would the jumper be above the surface of the planet?

30. A railroad company must install a signal to instruct a railway engineer whether to stop at an upcoming station. What is the least distance from the station that the signal can be placed if it is to stop a train traveling at 60 mph, given that the train's brakes will slow the train at 2 ft/sec^2 (that is, the acceleration is -2 ft/sec^2)?

31. Suppose the train in Exercise 30 travels at a maximum of 30 mph. By what amount does this decrease the distance required for the placement of the signal?

32. A Lufthansa–Airport–Express train runs from Düsseldorf to the Frankfort Airport, a distance of

93.6 km. What is the minimal time possible for this run if the acceleration and deceleration rates for the train are 0.9 m/sec² and the maximum speed of the train is 200 km/hr?

33. A track star figures that he starts with zero initial velocity and accelerates for the first 16 yd at a constant rate. He then runs the remainder of a 100-yd dash with a constant velocity. He also knows he can run the 100 yd in 10 sec.
   a) What is his acceleration for the first 16 yd?
   b) How long does it take to run the first 16 yd?

34. The track star described in Exercise 33 has set a goal to run a mile at the same pace that he runs the 100-yd dash; that is, he will run the first 100 yd of the mile in 10 sec and will cover the remaining 1660 yd with the same constant velocity that he ran the last 84 yd of the dash.
   a) If he succeeds, what will be his time for the mile?
   b) How likely is he to succeed?

35. An open construction elevator is ascending on the outside of a building at the rate of 8 ft/sec (see the figure). A ball is dropped from the floor of the elevator when the elevator's floor is 80 ft above the ground. How long does it take the ball to reach the ground?

36. An open construction elevator is ascending on the outside of a building at the rate of 8 ft/sec. When the bottom of the elevator is 80 ft above the ground, a ball is dropped from 160 ft above the floor of the elevator to the floor of the elevator. How long does it take the ball to reach the floor of the elevator?

37. A building consists of 140 stories, each story being the same height. An object is dropped from the roof of the building and is observed to take 2 sec to pass from the 105th story to the 70th story. What is the height of the building?

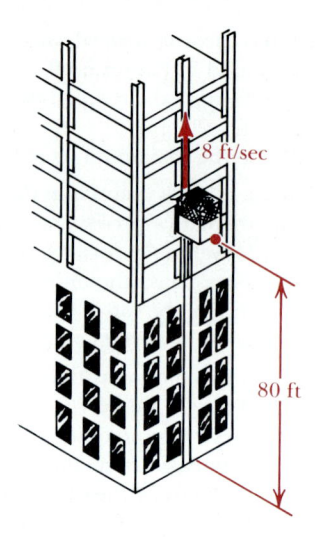

Figure for Exercise 35.

38. A child shoots a ball bearing in a slingshot directly up in the air (see the figure). A man on the third floor of a building sees the ball bearing ascend past his 4-ft window in 0.25 sec. How long from the time he first sees the projectile is it safe for him to put his head out the window to yell at the child?

Putnam exercises

39. A particle moving on a straight line starts from rest and attains a velocity v_0 after traversing a distance s_0. If the motion is such that the acceleration was never increasing, find the maximum time for the traverse.

(This exercise was problem B–2 of the thirty-third William Lowell Putnam examination given on December 2, 1972. The examination and its solution are in the November 1973 issue of the *American Mathematical Monthly,* pp. 1019–1028.)

40. A circle stands in a plane perpendicular to the ground, and a point *A* lies in this plane exterior to the circle and higher than its bottom. A particle starting from rest at *A* slides without friction down an inclined straight line until it reaches the circle. Which straight line allows descent in the shortest time? (The starting point *A* and the circle are fixed; the stopping point *B* is allowed to vary over the circle.)

   (This exercise was Problem A–2 of the thirty-fifth William Lowell Putnam examination given on December 7, 1974. The examination and its solution are in the November 1975 issue of the *American Mathematical Monthly,* pp. 907–912.)

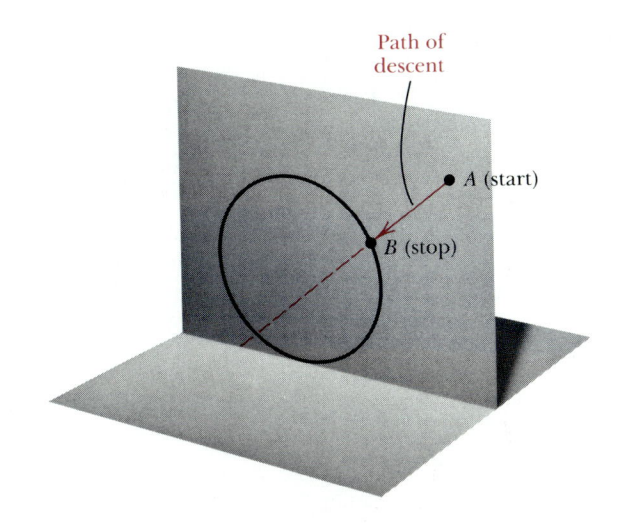

3.2
Applications Involving Maxima and Minima

One of the most common applications of the derivative involves problems that call for maximizing or minimizing a function. Businesses are concerned with maximizing profit and minimizing loss; engineers need to design systems to produce maximum performance; even the ordinary motorist is concerned with driving so that the distance traveled or the travel time is minimized.

If variables are introduced to represent certain quantities, each of these problems becomes a particular case of the general mathematical problem of finding an absolute maximum or an absolute minimum. The most difficult part of the problem is translating the ordinary language used to describe the problem into a form that can be handled mathematically. The "Know–Find" format introduced in Example 7 of Section 1.1 is used to make the translation easier.

Guidelines for solving applied extrema problems will be given after considering a few examples.

Example 1

Farmer MacDonald has 300 ft of chicken wire with which to construct a rectangular pen to hold a flock of chickens. A 400-ft–long chicken coop will be used to form one side of the pen, so the wire is needed only for the remaining three sides. How can the pen be constructed so that the birds have the maximum space in which to roam?

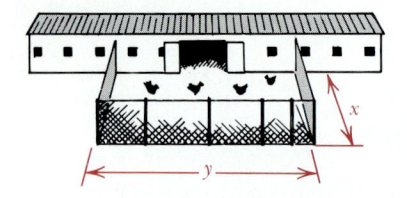

Figure 3.5

Solution

We first sketch a picture of the situation confronting MacDonald and introduce variables to label the unknown quantities. In this way we can write precisely what is known and determine the mathematical problem that must be solved. It also gives us time to reflect on the problem and consider any logical flaws it might have.

In Figure 3.5 we have chosen to use y to denote the length of the fence that is parallel to the chicken coop and x to denote the length of the sides perpendicular to the coop. The choice of variables is arbitrary.

The quantity we wish to maximize is the area, A, inside the pen, where
$$A = x \cdot y.$$

This expression for A involves both the variables x and y. To use the methods of Chapter 2 we first express A as a function of just one of these variables.

The length of the wire provides the necessary relationship between x and y. There is 300 ft of wire, all of which must be used if the area is to be maximized, so
$$2x + y = 300, \quad \text{which gives} \quad y = 300 - 2x.$$
Thus,
$$A = x \cdot y = x(300 - 2x) = 300x - 2x^2.$$

The physical dimensions of the pen in Figure 3.5 imply that $0 \le x \le 150$. The essential elements of the problem are expressed concisely below:

KNOW	FIND
1) $2x + y = 300$,	a) the value of x that maximizes $A(x)$,
2) the area of the pen is $A = xy$.	b) the value of y corresponding to the value of x found in (a).
From (1) and (2):	
3) $A(x) = x(300 - 2x)$, $0 \le x \le 150$	

We have now reduced the problem to finding an absolute maximum for the function defined by
$$A(x) = 300x - 2x^2$$
on the interval [0, 150]. The Extreme Value Theorem ensures that the absolute maximum is assumed on [0, 150]. This absolute maximum can occur only at a critical point or at one of the endpoints of the interval. Since
$$A'(x) = 300 - 4x,$$
the only critical point of A occurs when $A(x) = 0$, that is, at
$$x = \frac{300}{4} = 75.$$

At the endpoints of the domain of A, we have $A(0) = 0$ and $A(150) = 0$.

At the critical point,
$$A(75) = 300(75) - 2(75)^2 = 11{,}250.$$

So $x = 75$ ft gives the maximum area. The corresponding value for y is
$$y = 300 - 2(75) = 150 \text{ ft.}$$

Farmer MacDonald can maximize the area inside the pen by making the sides perpendicular to the chicken coop 75 ft long and the side parallel to the coop 150 ft long. ▶

Example 2

Find two positive real numbers whose product is 16 and whose sum is a minimum.

Solution

Let x represent one of the positive real numbers and y represent the other. The problem is to determine x and y so that
$$x \cdot y = 16 \quad \text{and} \quad S = x + y \text{ is a minimum.}$$

Solving the equation $x \cdot y = 16$ for y in terms of x and substituting into $S = x + y$ gives $y = 16/x$ and
$$S(x) = x + 16/x, \quad \text{where } x > 0.$$

The essential elements of the problem are listed below.

KNOW	FIND
1) $y = \dfrac{16}{x}$,	a) the value of x that minimizes $S(x)$,
2) $S = x + y$.	b) the value of y corresponding to the value of x found in (a).
From (1) and (2):	
3) $S(x) = x + \dfrac{16}{x}, x > 0.$	

This problem is slightly more complicated than Example 1 because the domain of the function S, $(0, \infty)$, is not a closed interval. Since the Extreme Value Theorem holds only for closed intervals, it is not immediately clear that the function S has an absolute minimum on $(0, \infty)$. However,
$$S'(x) = 1 - 16/x^2,$$

and the only positive solution to $S'(x) = 0$ is the critical point $x = 4$. If
$$0 < x < 4, \quad S'(x) < 0 \quad \text{and} \quad S \text{ is decreasing,}$$

whereas if
$$4 < x < \infty, \quad S'(x) > 0 \quad \text{and} \quad S \text{ is increasing.}$$

We can see from the graph of S shown in Figure 3.6 that an absolute minimum on $(0, \infty)$ occurs at $x = 4$.

The positive real numbers whose product is 16 and whose sum is a minimum are $x = 4$ and $y = 16/4 = 4$. ▶

After working through these two examples we can list some common steps in the solution to applied extrema problems.

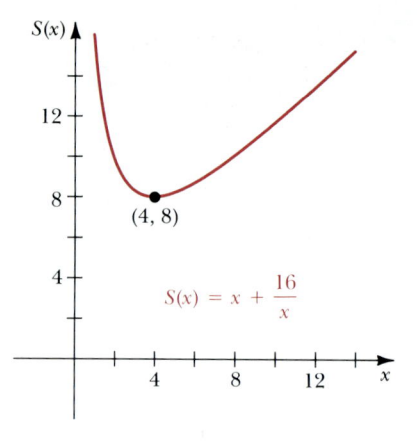

Figure 3.6
Absolute minimum at $x = 4$.

(3.4)
GUIDELINES FOR SOLVING APPLIED EXTREMA PROBLEMS

i) If possible, draw a picture to illustrate the problem and label the pertinent parts.

ii) Write down any relationships between the variables and the constants in the problem.

iii) Express the quantity to be maximized or minimized as a function of just *one* of the variables, making sure that the domain of the function is physically reasonable.

iv) Find the critical points of the function.

v) If the domain of the function is a closed interval, find the values of the function at the critical points and at the endpoints of the domain. The absolute maximum is the largest of these values, and the absolute minimum is the smallest.

vi) If the domain of the function is not a closed interval, the problem may not have a solution. If a solution does exist, it must occur at one of the critical points or at an endpoint that is in the domain. Use the first and second derivatives of the function to help analyze the behavior of its graph and determine any absolute extrema.

Example 3

A rectangle is to be inscribed in a semicircle with radius 2 in. Find the dimensions of the rectangle that encloses the maximum area.

Solution

The problem is illustrated in Figure 3.7, where l denotes the length of the rectangle and w denotes the width. The area A of the rectangle is $A = lw$. To express A as a function of a single variable, we apply the Pythagorean Theorem to the triangle ABC:

$$(l/2)^2 + w^2 = 4.$$

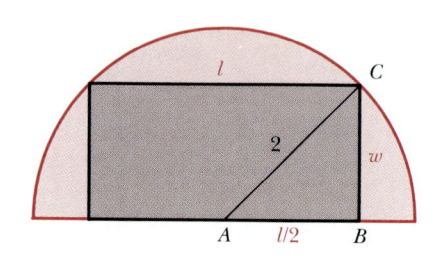

Figure 3.7

The essential elements of the problem are summarized below.

KNOW	FIND
1) area of the rectangle is $A = lw$,	a) the value of w that maximizes $A(w)$,
2) $(l/2)^2 + w^2 = 4$ so $l = 2\sqrt{4 - w^2}$.	b) the value of l corresponding to the value of w found in (a).

From (1) and (2):

3) $A(w) = 2w\sqrt{4 - w^2}, 0 \leq w \leq 2$.

3.2 APPLICATIONS INVOLVING MAXIMA AND MINIMA

To maximize A, we first find any critical points of A:

$$A'(w) = 2\sqrt{4-w^2} + 2w(\tfrac{1}{2})(4-w^2)^{-1/2}(-2w)$$
$$= 2(4-w^2)^{-1/2}[(4-w^2) - w^2]$$
$$= \frac{4(2-w^2)}{\sqrt{4-w^2}}.$$

Note that $A'(w)$ is undefined at $w = \pm 2$ and $A'(w) = 0$ at $w = \pm\sqrt{2}$. However, the domain of A is $[0, 2]$, so the only critical points are $w = 2$ and $w = \sqrt{2}$.

The absolute maximum for $A(w)$ can occur only at a critical point or at an endpoint. Since

$$A(0) = 0, \quad A(2) = 0, \quad \text{and} \quad A(\sqrt{2}) = 2\sqrt{2}\sqrt{4-2} = 4,$$

the maximum occurs at $w = \sqrt{2}$ in. By (2) in the table, the corresponding value for l is

$$l = 2\sqrt{4-2} = 2\sqrt{2} \text{ in.} \qquad \blacktriangleright$$

Example 4

Find the ratio of the radius to the height of a right circular cylinder that has minimal surface area, assuming that the volume of the cylinder is fixed and that the cylinder has a top and a bottom.

Solution

The cylinder shown in Figure 3.8 has volume $V = \pi r^2 h$. The circumference of the circular top is $2\pi r$, so the side of the cylinder has surface area $2\pi rh$. The surface areas of the circular top and bottom are both πr^2. The total surface area of the cylinder is the sum of the areas of the top, bottom, and sides: $S = 2\pi rh + 2\pi r^2$. Thus we have the following.

KNOW	FIND
1) $V = \pi r^2 h$, (V is a constant),	a) the value of r that minimizes $S(r)$,
2) $S = 2\pi rh + 2\pi r^2$.	b) the value of h corresponding to the value of r found in (a),
From (1) and (2):	
3) $S(r) = 2\pi r\left(\dfrac{V}{\pi r^2}\right) + 2\pi r^2$	c) the ratio of the values of r and h found in (a) and (b).
$= \dfrac{2V}{r} + 2\pi r^2,\ 0 < r < \infty.$	

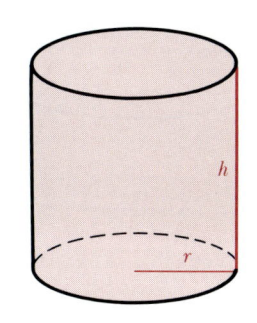

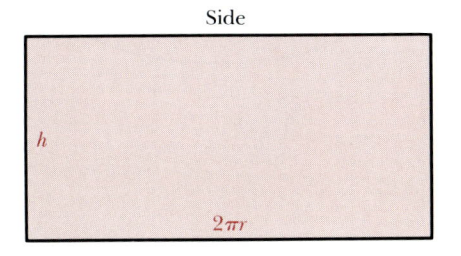

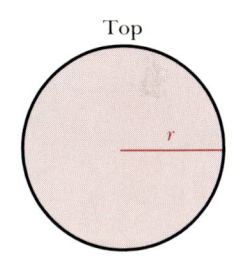

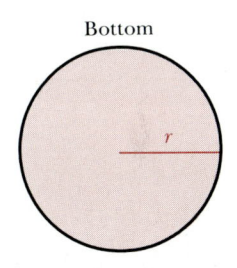

Figure 3.8

Calculating

$$S'(r) = \frac{-2V}{r^2} + 4\pi r,$$

we see that $S'(r)$ exists for all r in the domain of S. If $S'(r) = 0$, then

$$\frac{-2V}{r^2} + 4\pi r = 0, \quad \text{which implies that} \quad 4\pi r^3 = 2V.$$

The only critical point is

$$r = \sqrt[3]{\frac{V}{2\pi}} = \left(\frac{V}{2\pi}\right)^{1/3}.$$

Also,

$$\text{if} \quad 0 < r < \left(\frac{V}{2\pi}\right)^{1/3}, \quad S'(r) < 0 \quad \text{and } S' \text{ is decreasing;}$$

and

$$\text{if} \quad \left(\frac{V}{2\pi}\right)^{1/3} < r < \infty, \quad S'(r) > 0 \quad \text{and } S \text{ is increasing.}$$

This implies that S has an absolute minimum at $r = (V/(2\pi))^{1/3}$. The value of h corresponding to this value of r is

$$h = \frac{V}{\pi r^2} = \frac{V}{\pi \left(\sqrt[3]{V/(2\pi)}\right)^2} = \frac{V}{\pi \left(V/(2\pi)\right)^{2/3}} = \sqrt[3]{\frac{4V}{\pi}},$$

and the ratio of the radius to the height (see Figure 3.9) is

$$\frac{r}{h} = \frac{\sqrt[3]{V/(2\pi)}}{\sqrt[3]{4V/\pi}} = \sqrt[3]{\frac{V}{2\pi} \cdot \frac{\pi}{4V}} = \sqrt[3]{\frac{1}{8}} = \frac{1}{2}. \quad \blacktriangleright$$

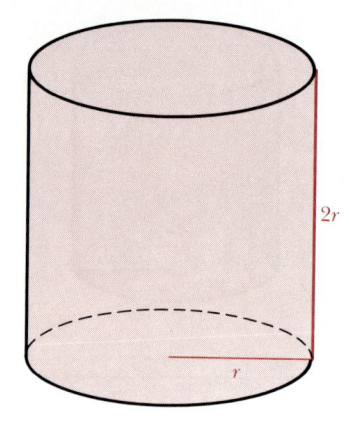

Figure 3.9

Example 5

A distributor has determined that 200 gal of Old Horse Light beer can be sold to Leo's Tavern if the price is $2 per gallon. For each cent per gallon that the price is lowered, 10 more qt of beer are sold. At what price should beer be sold in order to maximize the revenue received by the distributor?

Solution

Let x in [0, 200] denote the number of cents per gallon by which the price is lowered. Then the price is $(2.00 - 0.01x)$ dollars/gallon, and the amount of Old Horse sold is

$$200 \text{ gal} + 10x \text{ qt} = (200 + 2.5x) \text{ gal}.$$

The revenue earned is the product of the selling price and the amount sold at that price. Summarizing this information, we have the following:

KNOW	FIND
1) price is $(2.00 - 0.01x)$ dollars/gallon,	a) the value of x that maximizes $R(x)$,
2) amount sold is $(200 + 2.5x)$ gallons,	b) the price corresponding to the value of x found in (a),
3) revenue earned is $R(x) = (2.00 - 0.01x)(200 + 2.5x)$ dollars, $0 \leq x \leq 200$.	

3.2 APPLICATIONS INVOLVING MAXIMA AND MINIMA

To maximize R, we first find the critical points. Since

$$R'(x) = -0.01(200 + 2.5x) + 2.5(2.00 - 0.01x)$$
$$= -2 - 0.025x + 5 - 0.025x$$
$$= 3 - 0.05x,$$

the only critical point occurs when

$$0 = 3 - 0.05x, \quad \text{that is,} \quad x = 60.$$

The price corresponding to $x = 60$ is

$$2.00 - (0.01)(60) = \$1.40,$$

and the revenue produced at that price is

$$R(60) = (2.00 - 0.60)[200 + 2.5(60)] = \$490.$$

Checking for endpoint extrema we find that the revenue at $x = 0$ is $400 and is $0 when $x = 200$ gal. Therefore, the maximum revenue occurs when the price is $1.40 per gallon. ▶

Example 6

The problem in Example 5 would be more realistic if instead of maximizing the revenue received, we could determine the price that would maximize the distributor's profit. Suppose the distributor must pay $0.90 per gallon to the brewery. What price would maximize the profit, assuming that the other conditions in Example 5 are the same?

Solution
Let x in [0, 200] denote the number of cents per gallon by which the price is lowered. The profit per gallon is the price per gallon charged by the distributor minus the cost per gallon to the distributor: $[(2.00 - 0.01x) - 0.90]$ dollars/gallon.

KNOW	FIND
1) price is $(2.00 - 0.01x)$ dollars/gallon,	a) the value of x that maximizes $P(x)$,
2) profit earned is $[(2.00 - 0.01x) - 0.90]$ dollars/gallon,	b) the price corresponding to the value of x found in (a).
3) amount sold is $(200 + 2.5x)$ gallons,	
4) total profit earned is $P(x) = (1.10 - 0.01x)(200 + 2.5x)$ dollars, $0 \leq x \leq 200$.	

Since

$$P'(x) = -0.01(200 + 2.5x) + 2.5(1.10 - 0.01x)$$
$$= -2 - 0.025x + 2.75 - 0.025x$$
$$= 0.75 - 0.05x,$$

a critical point of P occurs when

$$0 = 0.75 - 0.05x, \quad \text{that is, when} \quad x = 15.$$

$P(x) = (1.10 - 0.01x)(200 + 2.5x)$

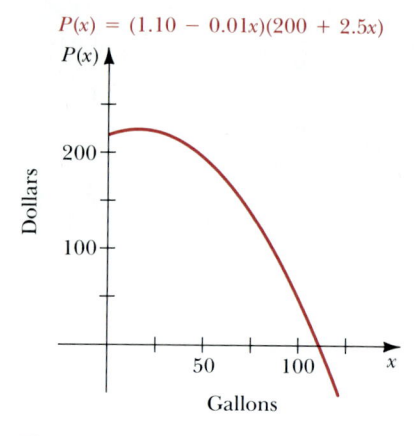

Figure 3.10
Absolute maximum at $x = 15$.

The price corresponding to $x = 15$ is

$$[2.00 - 0.01(15)] = \$1.85 \text{ per gallon},$$

and the corresponding profit earned is

$$P(15) = (1.10 - 0.15)(200 + 37.50) = \$225.63.$$

The domain of P is $[0, 200]$. Note, however, that when $x > 110$, the profit is negative so there is a loss. At $x = 0$ the profit is $220, which is less than the profit at the critical point $x = 15$. Consequently, the selling price that maximizes the profit for the distributor is $1.85 per gallon. The graph of the profit function is shown in Figure 3.10. ▶

Suppose the distributor in Example 6 had reduced the price by $0.60, the amount shown in Example 5 that would produce the maximum *revenue*. The profit realized would be

$$P(60) = (1.10 - 0.60)(200 + 150) = \$175.$$

This represents a reduction of more than 22% from the maximum profit obtainable, which illustrates the elementary fact of commerce that producing maximum revenue is often not equivalent to producing maximum profit.

Example 7

The mathematically inclined Farmer MacDonald needs to build a trough to slop hogs. The trough is to be made from three rough-sawn oak 1×10s (1 in. by 10 in.), each 8 ft long.

a) At what angle should the sides of the trough be sloped in order to enclose the maximum amount of swill?

b) How much will this trough hold?

Solution

Let x, h, and θ be as shown in Figure 3.11, and let A denote the cross-sectional area of the trough, measured in ft². Since the length of the trough is fixed at 8 ft, the volume of the trough is $V = 8A$ ft³. We will maximize the volume if we maximize the cross-sectional area A. It can be seen from Figure 3.12 that this area is

$$A = 10h + 2(\tfrac{1}{2}xh) = (10h + xh).$$

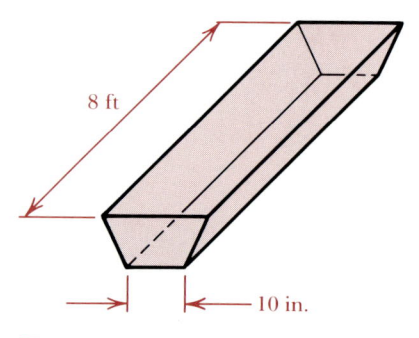

Figure 3.11

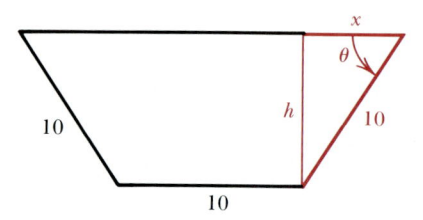

Figure 3.12

3.2 APPLICATIONS INVOLVING MAXIMA AND MINIMA

We will solve the problem by first writing x and h in terms of the angle θ shown in Figure 3.12. The problem can also be solved using the relationship between x and h given by the Pythagorean Theorem (see Exercise 33).

KNOW	FIND
1) $A = 10h + xh;$	a) the value of θ that maximizes $A(\theta)$;
2) $h = 10 \sin \theta;$	b) the maximum value of A;
3) $x = 10 \cos \theta.$	c) the maximum volume V.

From (1), (2), and (3),

4) $A(\theta) = 100 \sin \theta + 100 \sin \theta \cos \theta,$
   $0 \le \theta \le \pi/2,$
5) $V(\theta) = 8A(\theta), 0 \le \theta \le \pi/2.$

Since

$$A'(\theta) = 100 \cos \theta + 100(\cos^2 \theta - \sin^2 \theta)$$
$$= 100 \cos \theta + 100 \cos^2 \theta - 100(1 - \cos^2 \theta)$$
$$= 200 \cos^2 \theta + 100 \cos \theta - 100,$$

critical points of A occur when

$$0 = 200 \cos^2 \theta + 100 \cos \theta - 100$$
$$= (2 \cos^2 \theta + \cos \theta - 1)(100)$$
$$= [(2 \cos \theta - 1)(\cos \theta + 1)](100).$$

Thus,

$$\cos \theta = \frac{1}{2}, \quad \text{so } \theta = \frac{\pi}{3} \text{ or } \theta = \frac{5\pi}{3}$$

or

$$\cos \theta = -1, \quad \text{so } \theta = \pi.$$

Since π and $5\pi/3$ are not in the domain of A, the only critical point is $\pi/3$ and

$$A\left(\frac{\pi}{3}\right) = 100 \sin \frac{\pi}{3} + 100 \sin \frac{\pi}{3} \cos \frac{\pi}{3}$$
$$= 100 \left(\frac{\sqrt{3}}{2}\right) + 100 \left(\frac{\sqrt{3}}{2}\right)\left(\frac{1}{2}\right) = 75 \sqrt{3} \approx 130.$$

The values of A at the endponts $\theta = 0$ and $\theta = \pi/2$ are

$$A(0) = 0 \quad \text{and} \quad A(\pi/2) = 100,$$

so the absolute maximum occurs at $\theta = \pi/3$.

The maximum cross-sectional area of the trough is

$$A\left(\frac{\pi}{3}\right) = 75 \sqrt{3} \text{ in}^2 = \frac{75 \sqrt{3}}{144} \text{ ft}^2,$$

so the maximum volume of the trough is

$$V\left(\frac{\pi}{3}\right) = 8A\left(\frac{\pi}{3}\right) = 8 \cdot \frac{75\sqrt{3}}{144} \approx 7.22 \text{ ft}^3.$$

Since 1 ft³ is approximately 7.48 gal, the maximum volume is approximately 54 gal. (This is a quantity sufficient to feed 71 adult hogs if the trough is filled to capacity once a day.) ▶

▶ EXERCISE SET 3.2

1. Find the positive number with the property that the sum of the number and its reciprocal is a minimum.

2. Find the positive number with the property that the sum of the square of the number and the square of its reciprocal is a minimum.

3. A rectangular plot of ground containing 432 ft² is fenced off in a large lot. Find the dimensions of the plot that requires the least amount of fence.

4. A rectangular plot of ground containing 432 ft² is fenced off in a large lot, and a fence is constructed down the middle of the lot to separate it into equal parts. Find the dimensions of the plot that requires the minimal amount of fencing.

5. Suppose the fence used to enclose the plot of ground described in Exercise 4 costs $10 per foot and the fence used to divide the plot into parts costs $5 per foot. Find the dimensions of the plot that requires the least expense for fencing.

6. A rectangular dog run is to contain 864 ft².
   a) If the dog's owner must pay for the fencing, what should be the dimensions of the run to minimize cost?
   b) Suppose a neighbor has agreed to let the owner use an already constructed fence for one side of the run. What should the dimensions of the run be in this situation if the owner's cost is to be a minimum?

7. A rectangular box with no top is to contain 2250 in³. Find the dimensions to minimize the amount of material used to construct the box if the length of the base is three times the width.

8. Suppose the box described in Exercise 7 is constructed with a top. What dimensions would minimize the amount of material required?

9. The United States Postal Service has recently decreed that no rectangular-shaped parcel can be mailed if the total of its length and girth (perimeter of a cross section) exceeds 108 in. (see the figure). Find the maximum volume in a rectangular parcel that can be mailed if a cross section of the parcel is a square.

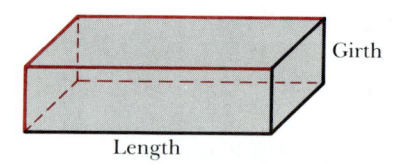

10. The speed of a point on the rim of a flywheel t sec after the flywheel has started to turn is given by the formula $v = 36t^2 - t^3$ ft/sec.
    a) Find its greatest speed.
    b) How long does it run before it reaches this speed?

11. The turning effect of a ship's rudder is found to be $T = k \cos \theta (\sin \theta)^2$, where k is a positive constant and θ is the angle that the direction of the rudder makes with the keel line of the ship ($0 \leq \theta \leq \pi/2$). For what value of θ is the rudder most effective?

12. A 1-mi race track is to be built with two straight sides and semicircles at the ends (see the figure). What is the maximum amount of area needed to construct the track?

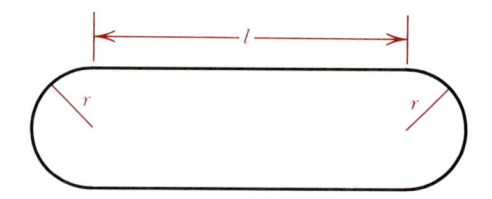

13. A standard can contains a volume of 900 cm³. The can is in the shape of a right circular cylinder with a top and bottom. Find the dimensions of the can that minimize the amount of material needed for construction.

14. In constructing a can in the shape of a right circular cylinder, no waste is produced when the side of the can is cut, but the top and bottom are each stamped from a square sheet and the remainder is wasted (see

the figure). Find the relative dimensions of the can that uses the least amount of material with this construction method.

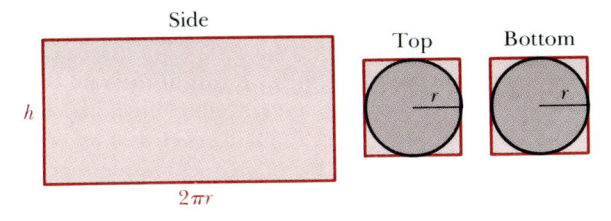

15. An open rectangular box is to be made from a piece of cardboard 8 in. wide and 8 in. long by cutting a square from each corner and bending up the sides. Find the dimensions of the box with the largest volume.

16. A rectangular plot that will contain a vineyard of one acre in area (43,560 ft²) is to be laid out. The vineyard must have a boundary of 8 ft on all sides in order for equipment to pass and an 8-ft pathway down the middle. What is the minimal acreage required for this plot?

17. A charter bus company charges $10 per person for a round trip to a ball game with a discount given for group fares. A group purchasing more than 10 tickets at one time receives a reduction per ticket of $0.25 times the number of tickets purchased in excess of 10. Determine the maximum revenue that can be received by the bus company.

18. A hotel with 25 rooms normally charges $40 for a room; however, special group rates are advertised: If the group requires more than 5 rooms, the price for each room is decreased by $1 times the number of rooms exceeding 5. Find the maximum revenue that the hotel can receive from a group.

19. A rectangle is placed inside a circle of radius r with its corners on the boundary of the circle (see the figure). Of all such rectangles, find the dimensions of the one that encloses the maximum area.

20. A rectangle is placed inside a circle of radius r with its corners on the boundary of the circle. What dimensions should be given to the rectangle to maximize the sum of its perimeter and the length of its two diagonals?

21. An isosceles triangle is placed inside a circle of radius r with its vertices on the boundary of the circle (see the figure). How should this be accomplished if the area of the triangle is to be maximized?

22. "The isosceles triangle with two fixed equal sides and maximum area is not an equilateral triangle." Show that this statement is true by finding the length of the base of an isosceles triangle that maximizes the area over all such triangles.

23. The area of the print on a book page is 42.5 in². The margins are 1 in. on the sides and bottom and 1/2 in. at the top. What should be the dimensions of a page of this book if the only object is to use the minimal amount of paper?

24. A field is fenced off in the form of a rectangle containing 10,000 ft². In addition to the fencing required for the perimeter of the field, an isosceles triangle is fenced off in one corner by running a fence from the midpoint on the shortest side to the adjacent side enclosing the corner of the rectangle (see the figure). Find the dimensions of the field that minimizes the amount of fencing required.

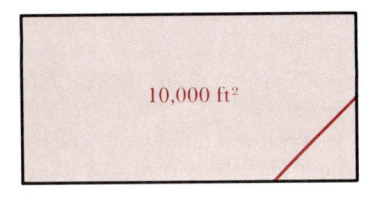

25. A wire 1 ft long is cut in two pieces: One piece is used to construct a square, the other to construct an equilateral triangle. Where should the cut be made in order to minimize the sum of the areas of the figures?

26. A wire 1 ft long is cut in two pieces: One piece is used to construct a square, the other to construct a circle. Where should the cut be made in order to minimize the sum of the areas of the figures?

27. Find the volume of the largest right circular cylinder that can be placed inside a sphere of radius 1.

28. Find the volume of the largest right circular cone that can be placed inside a sphere of radius 1.

29. The strength of a rectangular beam is directly proportional to the product of the width of the beam and the square of its depth. Find the dimensions of the strongest beam that can be cut from a log with radius r.

30. The stiffness of a rectangular beam is directly proportional to the product of the width of the beam and the cube of its depth. Find the dimensions of the stiffest beam that can be cut from a log with radius r.

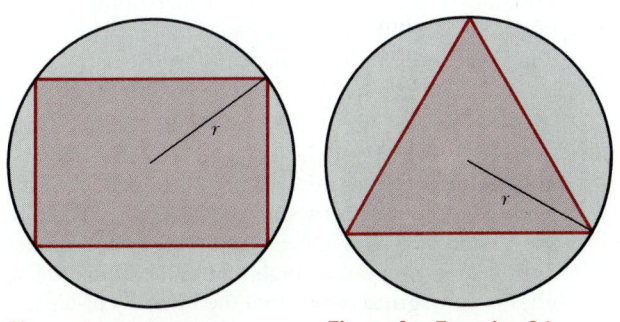

Figure for Exercise 19.   **Figure for Exercise 21.**

31. Suppose (x_0, y_0) is a point that does not lie on the circle $x^2 + y^2 = 1$. Show that the shortest distance from (x_0, y_0) to the circle is along a line that passes through the center of the circle.

32. Show that the shortest distance from the point (x_0, y_0) to the line with equation $Ax + By + C = 0$ is
$$\frac{|Ax_0 + By_0 + C|}{\sqrt{A^2 + B^2}}.$$

33. The maximum volume in the problem in Example 7 can be found without using trigonometry by expressing the area A totally as a function of x. Show that this method of solution gives the same result as that found in Example 7.

34. A house is built with a straight driveway 800 ft long, as shown in the figure. A utility pole on a line perpendicular to the driveway and 200 ft from the end of the driveway is the closest point from which electricity can be furnished. The utility company will furnish power with underground cable at $2 per foot and with overhead lines at no charge. However, for overhead lines the company requires that a strip of 30 ft wide be cleared. The owner of the house estimates that to clear a strip this wide will cost $3 for each foot of overhead wire used. At what point on the driveway should the switch from overhead to underground be made in order to minimize the cost?

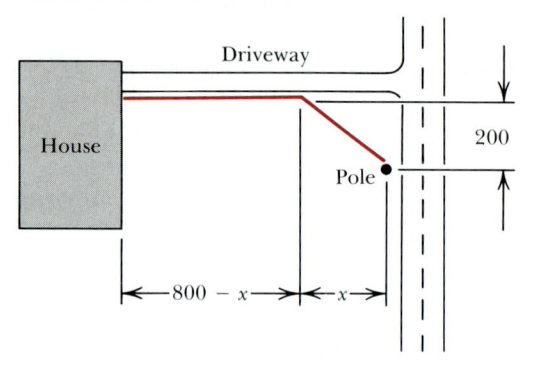

35. A crew of painters is assigned to paint a second-floor wall on the outside of a building along a busy sidewalk. They must leave a corridor, for unsuperstitious pedestrians, between the wall and their ladders. The corridor is 6 ft wide and 8 ft high. What is the minimal length of ladder they can use to reach the wall, and how far from the base of the wall should it be placed?

36. A bully armed with a knowledge of calculus is planning an attack on his next victim. The attack must be made on a sidewalk between two lights that are 200 ft apart, one of which is twice as bright as the other. Before dropping out of high school the bully took a physics course and recalls that the intensity of illumination from a light varies inversely as the square of the distance from the light. Where will he attack if he always attacks at the darkest point between the lights?

37. Two tankers are traveling in the midst of the Atlantic Ocean. The first tanker is 100 nautical miles due north of the second at 1:00 P.M. GMT (Greenwich Mean Time) and traveling due east at the rate of 20 knots. The second tanker is traveling due north at 15 knots. At what time are the tankers closest together, and what is the minimal distance separating them?

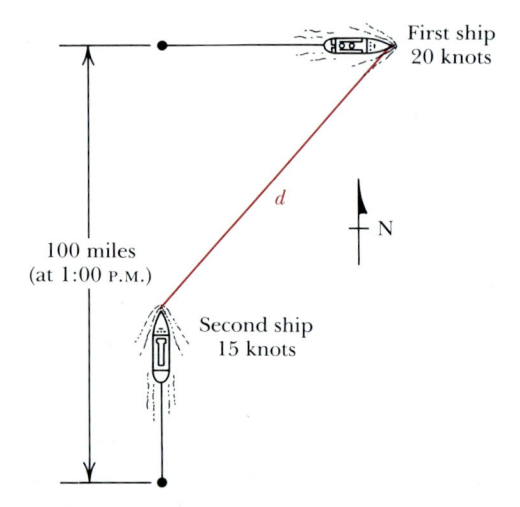

38. A warehouse is to be built beside a long straight highway running north and south (see the figure at the top of page 203). This warehouse will house equipment produced in two factories and sent there by air for storage. The northern factory lies 80 mi east of the highway; the other lies on the highway. The point on the highway that is closest to the northern factory is 100 mi north of the second factory. Where should the

Figure for Exercise 35.

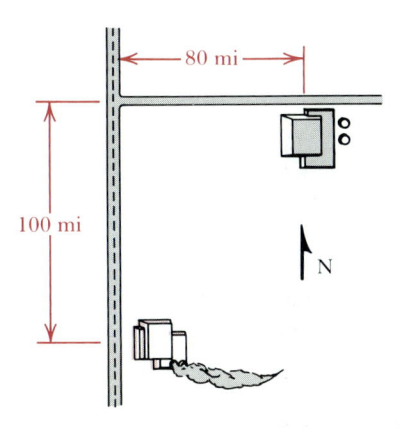

warehouse be located if it is to minimize the sum of the air distances from the warehouse to the two factories?

39. Consider the problem described in Exercise 38 with the following modification: The southern factory lies 80 mi west of the highway instead of beside the highway. Where should the warehouse now be located along the highway if the sum of the air distances from the warehouse to the two factories is to be minimized?

40. The Boondockia Outfitting Corporation has decided to produce and market a small backpacking tent. Their first task is to determine the dimensions of the tent requiring the minimal amount of material that satisfies the following conditions (see the figure): (a) the tent must have two sides, two ends, and a bottom; (b) the volume of the tent must be at least 100 ft³; (c) the tent must be 8 ft long; (d) cross sections of the tent parallel to each end must be congruent isosceles triangles. Find the dimensions of the tent using the minimal amount of material that satisfies these requirements.

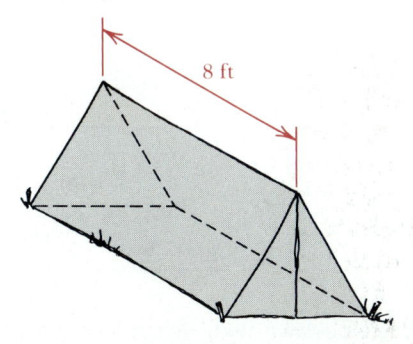

41. Boondockia, Inc., has decided that the tent described in Exercise 40 using the minimal amount of material is not marketable because it does not have enough space. They have decided instead to design the tent so that it is 6 ft wide at the bottom and has congruent cross sections in the form of rectangles topped by isosceles triangles. The minimal height of the rectangle is to be 1 ft and the total height to the center of the tent at least 30 in. Assuming all other specifications are the same as those in Exercise 40, how should the tent be designed in order to minimize the amount of material required?

42. The law of reflection proposed by Euclid in *Catoptrica* states that when light strikes a flat reflecting surface, the angle of incidence θ_i is the same as the angle of reflection θ_r (see the figure). Prove this law by applying Fermat's principle that the reflected light travels from point P to point Q along the path that requires the shortest time. [*Hint:* Show that $\theta_i(x) = \theta_r(x)$ when $d_1(x) + d_2(x)$ is a minimum.]

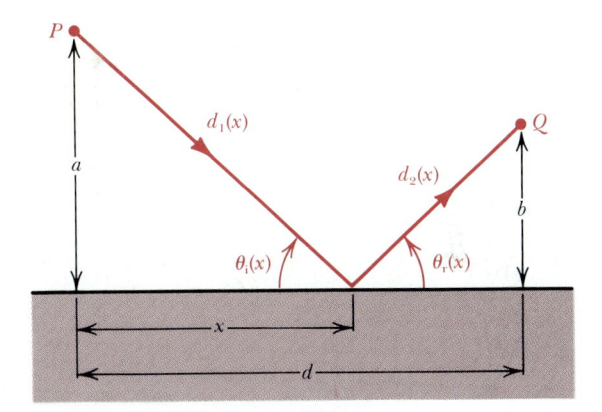

43. Snell's Law states that when light travels between two different mediums with speeds c_1 and c_2, respectively, the angle of incidence θ_i and the angle of refraction θ_r are related by $\sin \theta_i / c_i = \sin \theta_r / c_r$ (see the figure). Use Fermat's principle, referred to in Exercise 42, to prove Snell's Law.

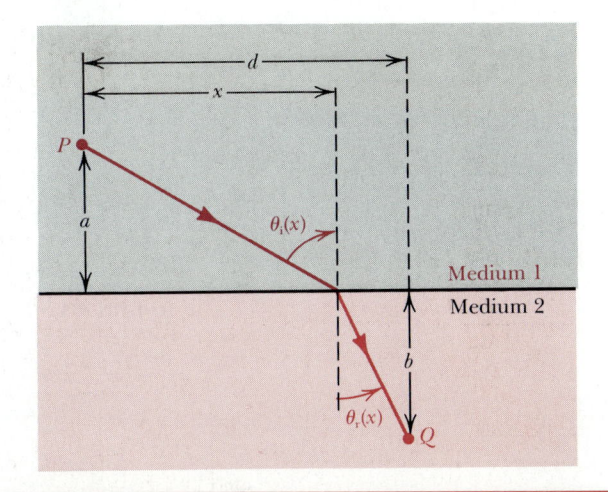

3.3
Related Rates

Two applications of the derivative we have seen are:

i) find the slope of the tangent line to the graph of a function, and
ii) determine the velocity of an object that travels in a straight line.

The feature common to these applications is that the derivative gives the instantaneous rate of change of one quantity with respect to another. The slope of the tangent line to the graph of a function measures the instantaneous rate of change in the function values with respect to the change in the independent variable. Velocity measures the instantaneous rate of change in the distance traveled with respect to time.

In a large group of physical problems, functions are involved in which both the independent variable and the value of the function are time-related. Suppose y is described by a function of x and t is a time variable on which both x and y depend. Problems of **related rates** involve the relationship between the instantaneous rate of change of x with respect to time, dx/dt, and the instantaneous rate of change of y with respect to time, dy/dt.

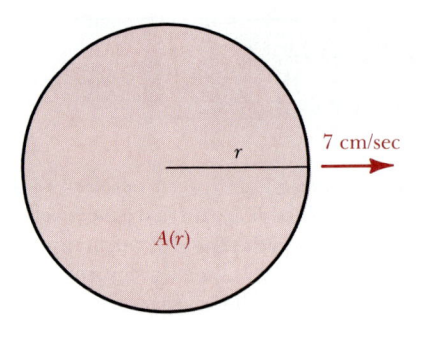

Figure 3.13

Example 1

Suppose the radius of a circle is increasing at 7 cm/sec. How fast is the area increasing when the radius is 20 cm?

Solution

If r denotes the radius of the circle, then the area is $A = \pi r^2$ (see Figure 3.13). Since r depends on the time variable t, the area A also depends on t. The rate at which the radius is changing with respect to time is dr/dt, and the rate at which the area is changing is dA/dt. The important information in the problem is summarized in the following table.

KNOW		FIND
1) $A = \pi r^2$;	a)	$\dfrac{dA}{dt}$;
2) $\dfrac{dr}{dt} = 7 \dfrac{\text{cm}}{\text{sec}}$.	b)	$\dfrac{dA}{dt}$ when $r = 20$ cm.

Since $A = \pi r^2$ and dr/dt are both known, we can use the Chain Rule to find dA/dt:

$$\frac{dA}{dt} = \frac{dA}{dr} \cdot \frac{dr}{dt} = (2\pi r)(7) \frac{\text{cm}^2}{\text{sec}}$$

$$= 14\pi r \frac{\text{cm}^2}{\text{sec}}.$$

Thus

$$\left.\frac{dA}{dt}\right|_{r=20} = 14\pi(20) = 280\pi \, \frac{\text{cm}^2}{\text{sec}}.$$ ▶

It is important not to begin by substituting specific data into a related-rate problem. In Example 1, we do not use the fact that $r = 20$ cm until we have determined the appropriate derivative dA/dt in a general form. If we had substituted $r = 20$ cm into the problem initially, it would have produced $A = \pi(20)^2 = 400\pi$ and $d(400\pi)/dt = 0$.

The following can be of help in solving problems of related rates.

(3.5)
GUIDELINES FOR SOLVING RELATED RATES PROBLEMS

i) If possible, draw a picture to illustrate the problem and label the pertinent parts.
ii) Write down any relationships between the variables and the constants in the problem.
iii) Determine an equation that when differentiated with respect to time contains, as its only unknown, the rate you are to find.
iv) Differentiate the equation, and solve for the unknown rate. Use the Chain Rule and implicit differentiation, if appropriate.
v) Introduce any specific data into the problem.

Note the similarities and differences between the methods of solution for related-rates problems and the extrema problems discussed in Section 3.2. In each case we draw, if possible, a picture to illustrate the problem and use variables to label the unknowns. Then we write down any relationships between the variables and constants in the problem. In the extrema problems we eliminate variables to obtain a function that can be maximized or minimized. In the related-rates problems, we differentiate an equation with respect to time to determine how the variables are related. Then the unknown rate is found and the specific data entered.

Example 2

A painter is painting a house using a ladder 15 ft long. A dog runs by the ladder dragging a leash that catches the bottom of the ladder and drags it directly away from the house at a rate of 22 ft/sec. Assuming the ladder continues to be pulled away from the wall at this speed, how fast is the top of the ladder moving down the wall when the top is 5 ft from the ground?

Solution
The problem is illustrated in Figure 3.14, where y denotes the distance from the top of the ladder to the ground and x denotes the distance

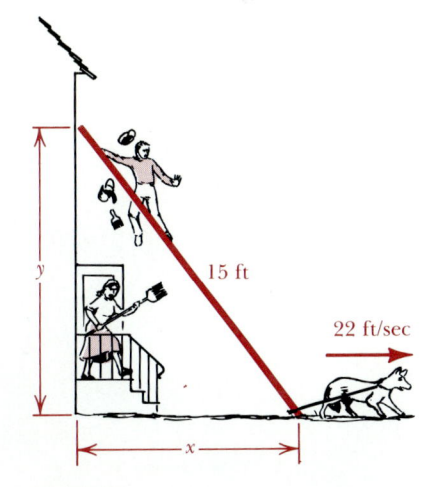

Figure 3.14

from the bottom of the ladder to the house. It follows from the Pythagorean Theorem that $x^2 + y^2 = (15)^2 = 225$. The information in the problem can be expressed as follows:

KNOW	FIND
1) $x^2 + y^2 = 225$;	a) $\dfrac{dy}{dt}$;
2) $\dfrac{dx}{dt} = 22 \dfrac{\text{ft}}{\text{sec}}$;	b) $\dfrac{dy}{dt}$ when $y = 5$ ft.
3) when $y = 5$ ft, $x = \sqrt{225 - 25} = \sqrt{200}$ ft.	

From (1) in the table, we have $y = \sqrt{225 - x^2}$. The Chain Rule implies that

$$\frac{dy}{dt} = \frac{dy}{dx}\frac{dx}{dt} = \frac{1}{2}(225 - x^2)^{-1/2}(-2x)\frac{dx}{dt}$$

$$= \frac{-x}{\sqrt{225 - x^2}}\frac{dx}{dt},$$

so

$$\frac{dy}{dt} = \frac{-x}{y} \cdot 22 \frac{\text{ft}}{\text{sec}}.$$

The velocity of the top of the ladder when it is 5 ft above the ground is therefore

$$\left.\frac{dy}{dt}\right|_{y=5} = \frac{-\sqrt{200}}{5}(22) \approx -62.23 \frac{\text{ft}}{\text{sec}}.$$

The speed of the ladder when $y = 5$ ft is 62.23 ft/sec. The negative value for the velocity indicates, correctly, that the distance between the top of the ladder and the ground is decreasing with time. ▶

Implicit differentiation can also be used to determine dy/dt in Example 2. From (1) in the table, we have

$$\frac{d}{dt}(x^2 + y^2) = \frac{d}{dt}(225),$$

so

$$2x\frac{dx}{dt} + 2y\frac{dy}{dt} = 0$$

and

$$\frac{dy}{dt} = \frac{-x}{y}\frac{dx}{dt}.$$

Implicit differentiation is a useful tool to employ in many problems involving related rates. It is required in the next example.

3.3 RELATED RATES

Example 3

For the situation described in Example 2, find how fast the angle between the ladder and the ground is changing when the top of the ladder is 5 ft from the ground.

Solution

Let θ denote the angle between the ladder and ground, as shown in Figure 3.15. The problem is to find $d\theta/dt$ given dx/dt, so we need an equation that relates θ and x.

KNOW	FIND
1) $\cos \theta = \dfrac{x}{15}$;	a) $\dfrac{d\theta}{dt}$;
2) $\dfrac{dx}{dt} = 22 \dfrac{\text{ft}}{\text{sec}}$.	b) $\dfrac{d\theta}{dt}$ when $y = 5$ ft.

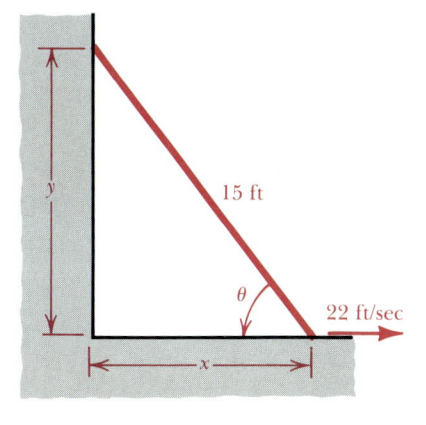

Figure 3.15

Taking the derivative of both sides of the equation in (1) with respect to t, we have

$$\frac{d \cos \theta}{dt} = \frac{d\left(\dfrac{x}{15}\right)}{dt}, \quad \text{so} \quad -\sin \theta \frac{d\theta}{dt} = \frac{1}{15} \cdot \frac{dx}{dt}.$$

Consequently,

$$\frac{d\theta}{dt} = -\frac{1}{15 \sin \theta} \frac{dx}{dt}$$

$$= \frac{-22}{15 \sin \theta}.$$

When $y = 5$ ft, $\sin \theta = \frac{1}{3}$, so

$$\frac{d\theta}{dt} = \frac{-22}{15(\frac{1}{3})} = -\frac{22}{5}.$$

Since θ is given in radians, the units for $d\theta/dt$ are radians/sec. ▶

Example 4

A revolving beacon located 3 mi from a straight shoreline makes 2 revolutions per minute. Find the speed of the spot of light along the shore when it is 2 mi from the point on the shore nearest the light (see Figure 3.16).

Solution

Since the beacon makes 2 revolutions per minute and 1 revolution is 2π radians, $d\theta/dt = 4\pi$ radians/min. The rate at which the light is moving along the shore is given by dx/dt, the instantaneous rate of change of the

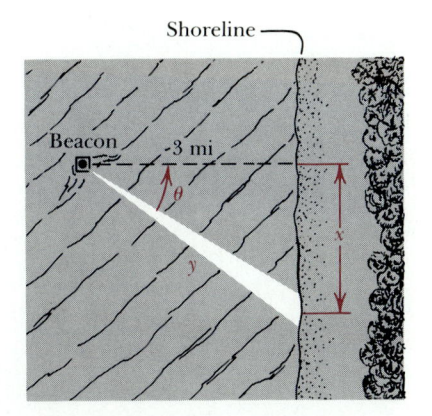

Figure 3.16

distance x with respect to time t.

KNOW	FIND
1) $\dfrac{d\theta}{dt} = 4\pi;$	a) $\dfrac{dx}{dt};$
2) $\tan \theta = \dfrac{x}{3}.$	b) $\dfrac{dx}{dt}$, when $x = 2$.

We first take the derivative of both sides of the equation in (2):

$$\sec^2 \theta \, \frac{d\theta}{dt} = \frac{1}{3} \frac{dx}{dt}.$$

We then solve for dx/dt:

$$\frac{dx}{dt} = 3 \sec^2 \theta \cdot \frac{d\theta}{dt} = (3 \sec^2 \theta) \, 4\pi.$$

To find dx/dt when $x = 2$, we first find $\sec^2 \theta$ when $x = 2$:

$$\tan \theta = \frac{x}{3} = \frac{2}{3} \quad \text{and} \quad \tan^2 \theta + 1 = \sec^2 \theta,$$

so

$$\sec^2 \theta = \left(\frac{2}{3}\right)^2 + 1 = \frac{13}{9}.$$

Consequently,

$$\frac{dx}{dt} = \left(3 \, \frac{13}{9}\right) 4\pi = \frac{52}{3} \pi \approx 54.4 \, \frac{\text{mi}}{\text{min}}. \quad \blacktriangleright$$

Note that although the beacon described in Example 4 is rotating quite slowly, the light beam is moving very rapidly. This is an example of a phenomenon called *optical leverage,* a concept commonly used in measuring instruments.

Example 5

Two ships sail from the same port. The first ship leaves port at 1:00 A.M. and travels eastward at a rate of 15 knots (nautical miles per hour). The second ship leaves port at 2:00 A.M. and travels northward at a rate of 10 knots. Determine the rate at which the ships are separating at 3:00 A.M.

Solution

Figure 3.17 illustrates the relevant information in the problem. Note that both x and y are changing with respect to time, with

$$\frac{dx}{dt} = 15 \quad \text{and} \quad \frac{dy}{dt} = 10.$$

From the triangle in Figure 3.17, we have

$$z^2 = x^2 + y^2.$$

The facts of the problem are summarized in the following table.

KNOW	FIND
1) $z^2 = x^2 + y^2$;	a) $\dfrac{dz}{dt}$;
2) $\dfrac{dx}{dt} = 15$;	b) $\dfrac{dz}{dt}$ at 3:00 A.M.
3) $\dfrac{dy}{dt} = 10$;	
4) At 3:00 A.M., $x = 30$ and $y = 10$.	

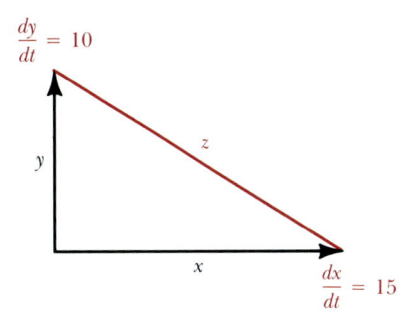

From (1) in the table, we have

$$2z\frac{dz}{dt} = 2x\frac{dx}{dt} + 2y\frac{dy}{dt} \quad \text{and} \quad \frac{dz}{dt} = \frac{x}{z}\frac{dx}{dt} + \frac{y}{z}\frac{dy}{dt}.$$

At 3:00 A.M.,

$$z = \sqrt{(30)^2 + (10)^2} = \sqrt{1000} = 10\sqrt{10} \text{ nautical miles.}$$

The separation rate at 3:00 A.M. is thus

$$\frac{dz}{dt} = \frac{30}{10\sqrt{10}}(15) + \frac{10}{10\sqrt{10}}(10) = 55\frac{\sqrt{10}}{10} \approx 17.4 \text{ knots.} \quad \blacktriangleright$$

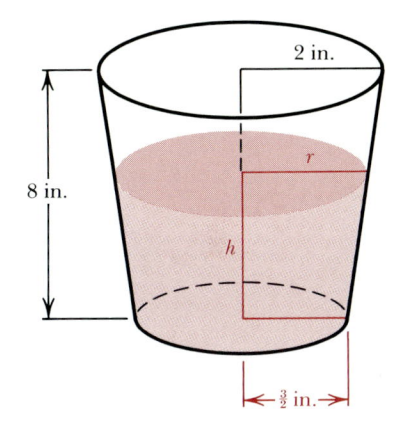

The problem in Example 5 can also be solved by first finding the separation distance between the ships explicitly as a function of time. Exercise 15 asks that you solve the problem in this way, which uses a distance formula found in Example 7 of Section 1.1.

Example 6

A straw is used to drink water from a straight-sided cup that has a diameter of 3 in. at the bottom and 4 in. at the top and a height of 8 in. The liquid is being consumed at the rate of 4 in³/sec. How fast is the level of the water dropping when there is a 3-in. depth of water left in the cup?

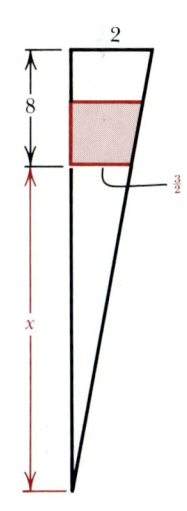

Solution

The shape of the cup is called the **frustum** of a right circular cone. The volume of such a solid is $V = \pi h\,(r_1^2 + r_1 r_2 + r_2^2)/3$, where r_1 and r_2 are the radii of the two ends of the frustum and h is the height. (This formula is derived in Exercise 73 of Section 5.2.) In this problem r_2 is always $\frac{3}{2}$, the radius of the bottom of the cup, but the radius $r_1 = r$ of the top of the frustum depends on the depth of the water, h, and hence on time. We can obtain the relationship between r and h by considering Figure 3.18 and extending this into a triangle, as shown in Figure 3.19.

Figure 3.17

Figure 3.18

Figure 3.19

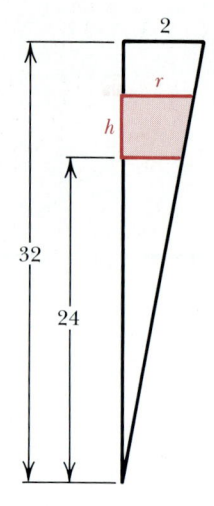

Figure 3.20

We first use the similar triangles in Figure 3.19 to find x:

$$\frac{2}{8+x} = \frac{\frac{3}{2}}{x}, \quad 2x = 12 + \frac{3x}{2}, \quad \text{and} \quad x = 24.$$

Now we use the similar triangles in Figure 3.20 to determine the relationship of r to h:

$$\frac{r}{24+h} = \frac{2}{32} \quad \text{so} \quad r = \frac{24+h}{16} = \frac{3}{2} + \frac{h}{16}.$$

The information used in the problem is as follows:

KNOW	FIND
1) $V = \frac{1}{3}\pi h\left[r^2 + r\left(\frac{3}{2}\right) + \left(\frac{3}{2}\right)^2\right]$;	a) $\frac{dh}{dt}$;
2) $r = \frac{3}{2} + \frac{h}{16}$;	b) $\frac{dh}{dt}$, when $h = 3$.
3) $\frac{dV}{dt} = -4 \, \frac{\text{in}^3}{\text{sec}}$.	

Combining (1) and (2) from the table gives

$$V = \frac{1}{3}\pi h\left[\left(\frac{3}{2} + \frac{h}{16}\right)^2 + \left(\frac{3}{2} + \frac{h}{16}\right)\left(\frac{3}{2}\right) + \left(\frac{3}{2}\right)^2\right]$$

$$= \frac{1}{3}\pi h\left(\frac{9}{4} + \frac{3}{16}h + \frac{h^2}{256} + \frac{9}{4} + \frac{3}{32}h + \frac{9}{4}\right)$$

$$= \frac{1}{3}\pi\left(\frac{h^3}{256} + \frac{9}{32}h^2 + \frac{27}{4}h\right).$$

Thus,

$$\frac{dV}{dt} = \frac{1}{3}\pi\left(\frac{3h^2}{256} + \frac{9}{16}h + \frac{27}{4}\right)\frac{dh}{dt}$$

and

$$\frac{dh}{dt} = \frac{\frac{dV}{dt}}{\frac{1}{3}\pi\left(\frac{3h^2}{256} + \frac{9}{16}h + \frac{27}{4}\right)} = \frac{-4}{\frac{1}{3}\pi\left(\frac{3h^2}{256} + \frac{9}{16}h + \frac{27}{4}\right)}.$$

When $h = 3$,

$$\frac{dh}{dt} = \frac{-4}{\frac{1}{3}\pi\left(\frac{27}{256} + \frac{27}{16} + \frac{27}{4}\right)} = \frac{-1024}{729\,\pi} \approx -0.45 \, \frac{\text{in.}}{\text{sec}}.$$

The depth of the water is decreasing at the rate of 0.45 in./sec when the depth is 3 in. ▶

EXERCISE SET 3.3

1. A metal sphere is heated so that its radius increases at the rate of 1 mm/sec. How fast is its volume changing when its radius is 30 mm?

2. The radius of a circle is increasing at the rate of 3 cm/sec. At what rate is the area of the circle increasing when the radius is 20 cm?

3. The sides of a square are increasing at the rate of 2 in./min. At what rate is the area of the square increasing when the sides are 4 in.?

4. The length of a rectangle is three times its width and the length is increasing at the rate of 9 in./sec. How fast is the area of the rectangle changing?

5. The edges of a cube are increasing at the rate of 2 in./min. At what rate is the volume of the cube increasing when the edges are 4 in.?

6. Reconsider the cube described in Exercise 5. At what rate is the total surface area of the cube increasing when the edges are 4 in.?

7. The sides of an equilateral triangle are increasing at the rate of 1 cm/sec. At what rate is the area of the triangle increasing when the sides are 4 cm?

8. The two equal sides of an isosceles triangle are increasing at the rate of 1 cm/sec while the third side is held at 4 cm. At what rate is the area of the triangle increasing when the sides are all 4 cm?

9. A stone is dropped into a pool of still water from a height of 150 ft. Circular ripples radiate at the rate of 3 in./sec from the spot where the stone hits the water.
   a) What is the area of the disturbed water 4 sec after the stone hits?
   b) How fast is the area changing at this time?

10. A certain yeast culture grows in a circular colony. As it grows the surface area it covers is directly proportional to its population and contains 10^5 members when the area is 1 cm². How fast is the population increasing when the radius of the circle is 12 cm if the radius of the circle is increasing at the rate of 3 cm/hr?

11. Gas is pumped into a spherical balloon at the rate of 1 ft³/min. How fast is the diameter of the balloon increasing when the balloon contains 36 ft³ of gas?

12. Flour sifted onto waxed paper forms a conical pile whose radius and height are always equal, although both increase with time (see the figure). The volume of flour on the waxed paper is increasing at the rate of 7.26 in³/sec. How fast is the height of the flour increasing when the volume is 29 in³?

13. Find the rate of change of the area of a circle with respect to its radius. Compare this with the circumference of the circle.

14. Find the rate of change of the volume of a sphere with respect to its radius. Compare this with the surface area of a sphere.

15. Work the problem in Example 5 by first expressing the distance z between the ships explicitly as a function of t.

16. A 15-ft ladder is leaning against a wall, and its base is pushed toward the wall at the rate of 2 ft/sec. How fast is the top of the ladder moving up the wall when the top is 9 ft from the ground?

17. An 8-ft 2 × 4 is leaning against a 10-ft wall. The lower end of the 2 × 4 is pulled away from the wall at the rate of 1 ft/sec. How fast is the top of the 2 × 4 moving toward the ground (a) when it is 5 ft from the ground and (b) when it is 4 ft from the ground?

18. An 8-ft 2 × 4 is leaning against a 5-ft wall with the remainder of the 2 × 4 hanging over the wall. The lower end of the 2 × 4 is pulled away from the wall at a rate of 1 ft/sec. How fast is the top of the 2 × 4 moving toward the ground when this end is 5 ft from the ground, that is, when the upper end just reaches the wall?

19. A rectangular swimming pool 50 ft long and 30 ft wide is being filled with water to a depth of 8 ft at the rate of 3 ft³/min.
   a) How long does it take to fill the pool?
   b) At what rate is the depth of water in the pool increasing when the pool is half full of water?

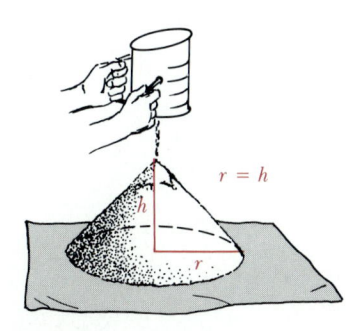

Figure for Exercise 12.

20. A rectangular swimming pool 50 ft long and 30 ft wide has a depth of 8 ft for the first 20 ft of its length and a depth of 3 ft on the last 20 ft of its length, and tapers linearly for the 10 ft in the middle of its length (see the figure). The pool is being filled with water at the rate of 3 ft³/min.
   a) How long does it take to fill the pool?
   b) At what rate is the depth of water in the pool increasing when the pool is half full of water?

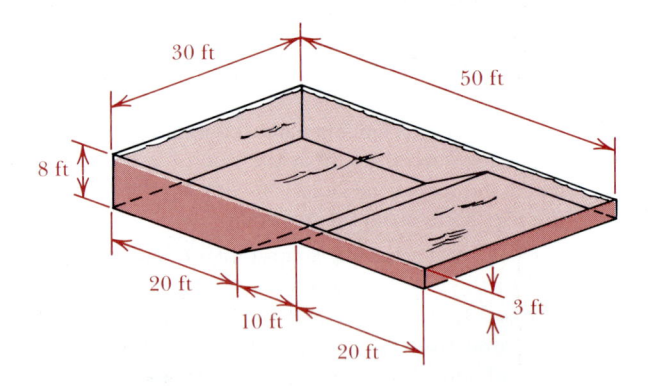

21. A woman on a dock is using a rope to pull in a canoe. The rope is pulled at the rate of 2 ft/sec, 3 ft above the point level with the connection of the rope to the canoe (see the figure). How fast is the canoe approaching the dock when the length of rope from her hands to the canoe is 10 ft?

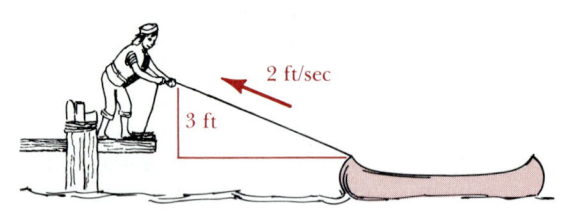

22. A kite is flying at an altitude of 80 m and is carried horizontally at this altitude by the wind at the rate of 2 m/sec (see the figure). At what rate is the string released to maintain this flight when 100 m of string have been released?

23. A kite is carried horizontally at the rate of 1.5 m/sec and is rising at 2.0 m/sec. How fast is the string released to maintain this flight when 100 m of string have been released and the kite is at an altitude of 80 m?

24. A single-engine airplane passes over a beacon and heads east at the rate of 100 mph. Two hours later a jet passes over the beacon at the same altitude traveling north at 400 mph. Assuming that the planes stay on these courses, how fast are they separating 1 hr after the jet has passed over the beacon?

25. Two ships meet at a point in the ocean with one of the ships traveling south at the rate of 15 mph and the other traveling west at 20 mph. At what rate are the ships separating 2 hr after they meet?

26. Two cars approach an intersection at right angles. One car is traveling at the rate of 50 mph and the other is traveling 40 mph. How fast are the cars approaching each other when the first car is 30 ft from the intersection and the second is 40 ft from the intersection?

27. A fisherman sitting on the end of a pier with his pole 3 m above the water snags what he assumes to be a large fish and reels in his line at the steady rate of 1 m/sec. He does not realize that the object is actually an old log lying just below the surface until the log is 5 m from the end of the rod. How fast is the log approaching the pier at this time?

28. A camera televising the return of the opening kickoff of a football game is located 5 yd from the east edge of the field and in line with the goal line. The player with the football runs down the east edge (just in bounds) for a touchdown. When he is 10 yd from the goal line, the camera is turning at a rate of 0.5 radian/sec. How fast is the player running?

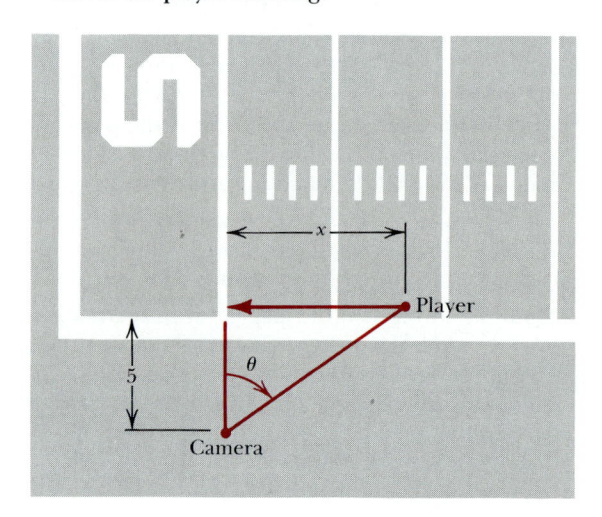

29. A revolving beacon located 1 mi from a straight shoreline turns at 1 revolution per minute. Find the speed of the spot of light along the shore when it is 2 mi away from the point on the shore nearest the light.

30. A metal cylinder contracts as it cools, the height of the cylinder decreasing at 4.5×10^{-4} cm/sec and the radius decreasing at 3.75×10^{-5} cm/sec. At what rate is the volume of the cylinder decreasing when its height is 200 cm and its radius is 10 cm?

31. A woman 5 ft 6 in. tall walks at the rate of 6 ft/sec toward a street light that is 16 ft above the ground.
   a) At what rate is the tip of her shadow moving?
   b) At what rate is the length of her shadow changing when she is 10 ft from the base of the light?

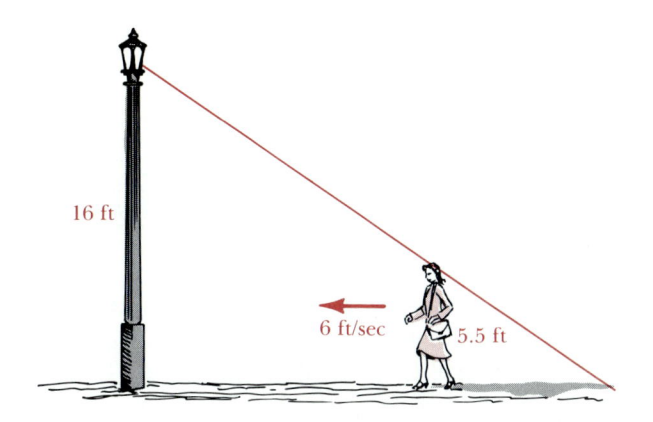

32. Suppose Farmer MacDonald constructs a hog trough to maximize capacity in the manner described in Example 7 of Section 3.2. If the hogs continuously consume the slop at the rate of 1.2 ft³ per hour, how fast is the height of the slop decreasing when the height of the slop in the trough is 6 in.?

33. A horse trough 10 ft long has a cross section in the shape of an inverted equilateral triangle with an altitude of 2 ft. The trough leaks water through a crack in the bottom at the rate of 1 ft³/hour.
   a) At what rate is the height of the water in the trough decreasing when the depth of the water is 1 ft?
   b) At what rate is the height of the water in the trough decreasing when the trough contains 10 ft³ of water?

34. A picture with height 3 ft is placed on a wall with its base 3 ft above an observer's eye level. The observer approaches the wall at the rate of 1 ft/sec. How fast is the angle θ, shown in the figure, changing when the observer is 10 ft from the wall?

35. An object that weighs w_0 lb on the surface of the earth weighs approximately

$$w(r) = w_0 \left(\frac{3960}{3960 + r} \right)^2 \text{ lb}$$

when lifted a distance of r mi from the earth's surface. Find the rate at which the weight of an object weighing 2000 lb on the earth's surface is changing when it is 100 mi above the earth's surface and is being lifted at the rate of 10 mi/sec.

36. The owner of a dog kennel reads in *Dog's Life* that the surface area of a dog is approximately related to its weight by the equation

$$s = 0.1 w^{2/3},$$

where the weight w of the dog is measured in kilograms and the surface area s of the dog is measured in square meters. The amount of flea powder the owner must purchase is directly proportional to the surface area of the dogs. If the average pup in the kennel gains weight at the approximate rate of 0.8 kg/wk, at what rate is the purchase of powder increasing when there are 23 dogs, the average dog weighs 20 kg, and a can of powder covers 3 m²?

37. Oil is leaking from an ocean tanker at the rate of 5000 L/sec. The leakage results in a circular oil slick with a depth of 5 cm. (*Note:* 1 liter = 1000 cm³.)
   a) How fast is the radius of the oil slick increasing when the radius is 300 m?
   b) How fast is the radius of the oil slick increasing 4 hr after the leakage has begun?

38. In actual practice an oil slick like the one described in Exercise 37 does not have a constant depth; the depth of the slick decreases as the oil moves from the point of spillage and depends primarily on the turbulence of the water and the viscosity of the oil. Suppose the

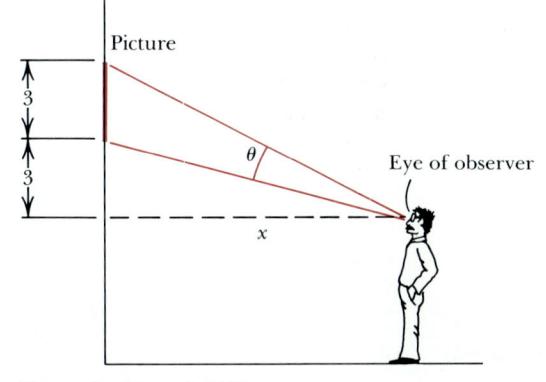

Figure for Exercise 34.

depth of the oil varies linearly from a maximum of 5 cm at the point of leakage to a minimum of 0.5 cm at the outside edge of the slick, and that oil is leaking from the tanker at the rate of 5000 L/sec. How fast is the radius of the slick increasing 4 hr after the leakage has begun?

39. A Cessna Citation III flying at the rate of 530 mph passes over Logan, Utah, at 7:00 A.M. Mountain Standard Time heading due south to Phoenix. Fifteen minutes later an F-15 Eagle fighter aircraft passes over Logan traveling at the rate of 1535 mph in a course 24° west of due south heading home to Nellis Air Force Base in Nevada.
   a) Are the planes approaching each other or separating at 7:30 A.M.?
   b) At what rate? [*Hint:* Use the law of cosines.]

3.4

Differentials

The slope of the tangent line to the graph of a differentiable function f at a point $(x, f(x))$ is $f'(x)$ (see Figure 3.21a). The slopes of the secant lines joining the points $(x, f(x))$ and $(x + h, f(x + h))$ on the graph,

$$\frac{f(x + h) - f(x)}{h},$$

approximate the derivative $f'(x)$ for small values of h (see Figure 3.21b).

In this section we consider applications that arise when this process is reversed. Instead of using the change in the function values to approximate the derivative, we use the derivative to approximate the change in function values. In our discussion, the Leibniz notation Δx

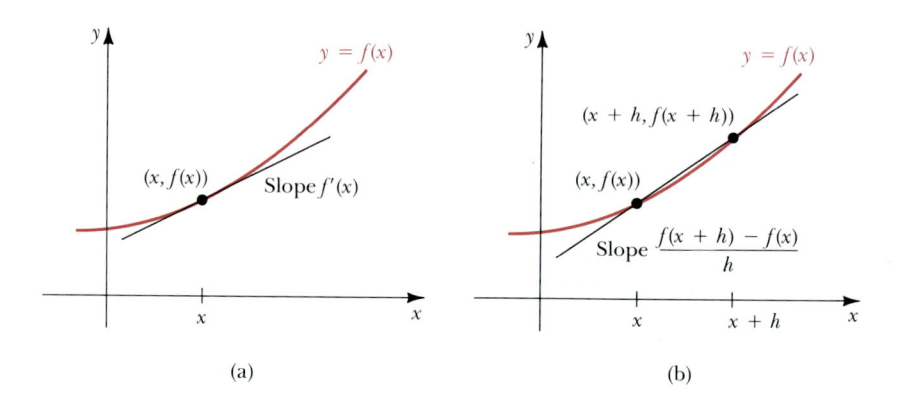

Figure 3.21
$f'(x) \approx \dfrac{f(x + h) - f(x)}{h}.$

(delta x) is used, instead of h, to denote the change in the independent variable x. This is done in part for notational convenience and in part for tradition.

Consider a function defined by the equation $y = f(x)$. If the independent variable is changed from x to $x + \Delta x$, the value Δx is called an **increment of x**. The corresponding change in the dependent variable (see Figure 3.22),

(3.6) $$\Delta y = f(x + \Delta x) - f(x),$$

is called an **increment of y**.

If f is differentiable at x, then

$$f'(x) = \lim_{\Delta x \to 0} \frac{f(x + \Delta x) - f(x)}{\Delta x} = \lim_{\Delta x \to 0} \frac{\Delta y}{\Delta x}$$

exists. Thus, for small values of Δx, Δy can be approximated by $f'(x)\,\Delta x$, that is,

(3.7) $$\Delta y \approx f'(x)\,\Delta x.$$

For a fixed value of x, Δy and $f'(x)\,\Delta x$ are both functions whose independent variable is the increment Δx. The function of Δx defined by $f'(x)\,\Delta x$ is denoted dy and called the **differential of y** with respect to the fixed value x and the increment Δx:

(3.8) $$dy = f'(x)\,\Delta x.$$

The increment Δx is called the **differential of x** and denoted dx. Thus we can write $dy = f'(x)\,dx$ consistent with the Leibniz notation for the derivative $dy/dx = f'(x)$.

The relationship between Δy and dy is shown in Figure 3.23:

i) The increment Δy gives the exact change in the function values corresponding to the change from x to $x + \Delta x$ (see Figure 3.23a).
ii) The differential dy gives the corresponding change along the tangent line to the graph at the point $(x, f(x))$ (see Figure 3.23b).

When the differential dy is used to approximate the increment Δy, the tangent line at $(x, f(x))$ is being used to approximate the curve.

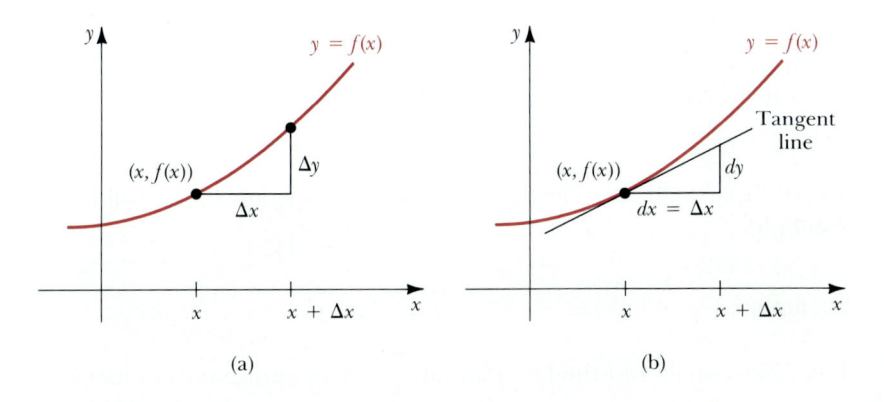

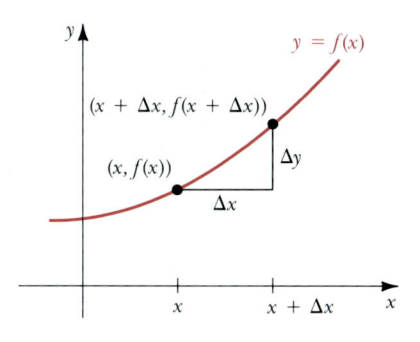

Figure 3.22

Figure 3.23

$$f'(x) = \frac{dy}{dx} \approx \frac{\Delta y}{\Delta x}.$$

Example 1

For the function described by $y = f(x) = 3x^2 - 4x + 5$:

a) Find dy, Δy, and $dy - \Delta y$.
b) Evaluate dy, Δy, and $dy - \Delta y$ when $x = 3$ and $\Delta x = 0.01$.

Solution

a) Since $f'(x) = 6x - 4$,
$$dy = f'(x)\,\Delta x = (6x - 4)\,\Delta x.$$

On the other hand,
$$\begin{aligned}\Delta y &= f(x + \Delta x) - f(x) \\ &= 3(x + \Delta x)^2 - 4(x + \Delta x) + 5 - (3x^2 - 4x + 5) \\ &= 3x^2 + 6x\,\Delta x + 3(\Delta x)^2 - 4x - 4\,\Delta x + 5 - 3x^2 + 4x - 5 \\ &= (6x - 4)\,\Delta x + 3(\Delta x)^2.\end{aligned}$$

Thus,
$$dy - \Delta y = -3(\Delta x)^2.$$

b) When $x = 3$ and $\Delta x = 0.01$,
$$dy = (18 - 4)(0.01) = 0.14,$$
$$\Delta y = (18 - 4)(0.01) + 3(0.01)^2 = 0.1403,$$

and
$$dy - \Delta y = 0.14 - 0.1403 = -0.0003. \quad\blacktriangleright$$

Note in Example 1 that the difference between dy and Δy is much smaller than the corresponding value of $\Delta x = 0.01$. In general, as Δx becomes small, the difference $dy - \Delta y$ becomes even smaller. This is true because

$$\lim_{\Delta x \to 0} \frac{dy - \Delta y}{\Delta x} = \lim_{\Delta x \to 0} \frac{f'(x)\,\Delta x - [f(x + \Delta x) - f(x)]}{\Delta x}$$
$$= \lim_{\Delta x \to 0} \left[f'(x) - \frac{f(x + \Delta x) - f(x)}{\Delta x} \right]$$
$$= f'(x) - f'(x) = 0.$$

Since the differential dy can be used to approximate the increment $\Delta y = f(x + \Delta x) - f(x)$, we also have the approximation

(3.9) $\quad f(x + \Delta x) = f(x) + \Delta y \approx f(x) + dy = f(x) + f'(x)\,\Delta x,$

when Δx is small. We illustrate the use of this result in the following example.

Example 2

Use differentials and the fact that $\tan \dfrac{\pi}{4} = 1$ to approximate $\tan 48°$.

Solution

Let $f(x) = \tan x$. Then
$$dy = f'(x)\,\Delta x = \sec^2 x\,\Delta x.$$

The next step is to convert the degree measure to radians:
$$48° = 48\left(\frac{\pi}{180}\right) = \frac{4\pi}{15} \text{ radians}.$$

Let
$$x = \frac{\pi}{4} \quad \text{and} \quad x + \Delta x = \frac{4\pi}{15}.$$

Then
$$\Delta x = \frac{4\pi}{15} - \frac{\pi}{4} = \frac{\pi}{60}$$

and
$$dy = \left(\sec^2 \frac{\pi}{4}\right)\left(\frac{\pi}{60}\right) = (\sqrt{2})^2\left(\frac{\pi}{60}\right) = \frac{\pi}{30}.$$

By Equation (3.9),
$$\tan 48° = \tan \frac{4\pi}{15} \approx \tan \frac{\pi}{4} + dy = 1 + \frac{\pi}{30} \approx 1.105. \quad \blacktriangleright$$

One use of differentials is to determine approximate bounds for the values of a function on the basis of a bound for its independent variable.

Example 3

The diameter of the sphere in Figure 3.24 is measured as 16 cm, with a maximal measurement error of ± 0.01 cm. Use differentials to estimate the maximum error in computing the volume and surface area of this sphere.

Solution

If r denotes the radius and D the diameter of the sphere, then the volume is
$$V = \frac{4}{3}\pi r^3 = \frac{1}{6}\pi D^3$$

and the surface area is
$$S = 4\pi r^2 = \pi D^2.$$

Therefore,
$$\frac{dV}{dD} = \frac{1}{2}\pi D^2, \quad \text{so} \quad dV = \frac{1}{2}\pi D^2\,dD$$

and
$$\frac{dS}{dD} = 2\pi D, \quad \text{so} \quad dS = 2\pi D\,dD.$$

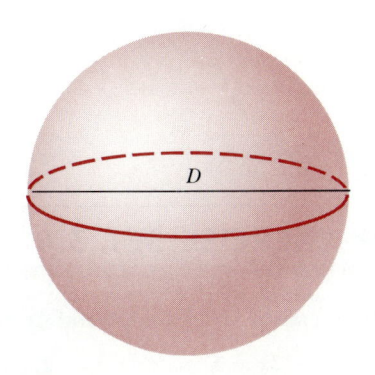

Figure 3.24

With $D = 16$ cm and $|dD| \leq 0.01$ cm, the maximum error for the volume is approximately

$$|dV| = \left|\frac{1}{2}\pi D^2\, dD\right| = \frac{1}{2}\pi D^2 |dD| \leq \frac{1}{2}\pi(16)^2(0.01) \approx 1.28\,\pi \text{ cm}^3.$$

An approximation for the maximum error for the surface area is

$$|dS| = |2\pi D\, dD| = 2\pi D|dD| \leq 2\pi(16)(0.01) = 0.32 \text{ cm}^2. \quad \blacktriangleright$$

The *relative error* in an approximation is defined to be the magnitude of the quotient of the approximation error and the true value:

$$\text{relative error} = \left|\frac{\text{error}}{\text{true value}}\right|.$$

In Example 3 we saw how the differential dy can be used to approximate the maximum error Δy in a calculation. In such a situation, we can use the quotient dy/y to approximate the relative error:

$$\text{relative error} = \left|\frac{\Delta y}{y}\right| \approx \left|\frac{dy}{y}\right|.$$

Example 4

Use differentials to approximate the maximum relative errors for the volume and surface area calculations performed in Example 3.

Solution

The volume and surface area of a perfect sphere of diameter 16 cm are

$$V = \frac{1}{6}\pi(16)^3 = \frac{2048}{3}\pi \text{ cm}^3 \quad \text{and} \quad S = \pi(16)^2 = 256\pi \text{ cm}^2.$$

The maximum relative errors in the calculations for the volume and surface area are approximately

$$\left|\frac{dV}{V}\right| = \frac{1.28\pi}{\left(\frac{2048}{3}\pi\right)} = 0.001875 = 0.1875\%$$

and

$$\left|\frac{dS}{S}\right| = \frac{0.32\pi}{256\pi} = 0.00125 = 0.125\%. \quad \blacktriangleright$$

▶ EXERCISE SET 3.4

In Exercises 1–12, find dy and Δy in terms of x and Δx.

1. $y = 3x - 5$
2. $y = 3 - 2x$
3. $y = x^2 - 2x + 3$
4. $y = 1 - x + 2x^2$

5. $y = x^3 - x$
6. $y = x^3 - 3x^2 + 3x - 1$
7. $y = \sqrt{x + 1}$
8. $y = \dfrac{1}{x - 1}$
9. $y = x + \dfrac{1}{x}$
10. $y = \dfrac{1}{\sqrt{x}}$
11. $y = \tan x$
12. $y = x + \sin x$

In Exercises 13–16, find dy and Δy if $y = f(x)$ and (a) $x = 2$, $\Delta x = 0.1$, and (b) $x = -1$, $\Delta x = -0.2$.

13. $f(x) = x^2 - 4x + 1$
14. $f(x) = 3 - 2x^2$
15. $f(x) = x^3 - 2x$
16. $f(x) = x^3 - 2x^2$

In Exercises 17–22, (a) find dy, (b) find Δy, and (c) approximate $f(x + \Delta x)$, assuming $y = f(x)$.

17. $f(x) = \sin x$, $x = \dfrac{\pi}{2}$, $\Delta x = \dfrac{\pi}{4}$
18. $f(x) = \cos x$, $x = \dfrac{\pi}{2}$, $\Delta x = \dfrac{\pi}{4}$
19. $f(x) = \dfrac{1}{x}$, $x = 1$, $\Delta x = -0.02$
20. $f(x) = \dfrac{x}{x + 1}$, $x = 0$, $\Delta x = 0.1$
21. $f(x) = \sqrt{x}$, $x = 9$, $\Delta x = 0.03$
22. $f(x) = \dfrac{1}{\sqrt[3]{x + 1}}$, $x = 7$, $\Delta x = -0.1$

In Exercises 23–28, use differentials and an easily calculated value (as in Example 2) to approximate the value required.

23. $\sin 31°$
24. $\cos 58°$
25. $\sqrt{9.03}$
26. $\sqrt{50}$
27. $\sqrt[3]{65}$
28. $\sqrt[3]{0.0011}$

29. Use differentials to approximate the maximum possible error produced when calculating the area of a circle, if it is known that the radius of the circle is 2 ± 0.001 ft.

30. Use differentials to approximate the maximum possible error produced when calculating the circumference of a circle, if it is known that the radius of the circle is 2 ± 0.001 ft.

31. Use differentials to approximate the maximum possible error produced when calculating (a) the volume and (b) the surface area of a sphere, if the radius of the sphere is 2 ± 0.001 ft. (c) Approximate the maximum relative errors in these calculations.

32. Use differentials to approximate the maximum possible error produced when calculating (a) the volume and (b) the surface area of a cube, if the length of an edge is 1 ± 0.0005 ft. (c) Approximate the maximum relative error in these calculations.

33. A hole $\frac{1}{4}$ in. in diameter has been drilled through a wooden 4×4. The hole is enlarged to $\frac{5}{16}$ in. in diameter. Use differentials to approximate how much additional wood is removed. (The dimensions of a wooden 4×4 are 3.5 in. by 3.5 in.)

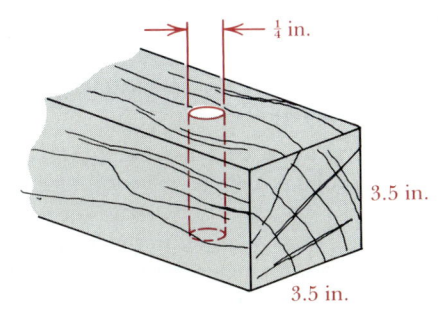

34. An open cylindrical pipe for pumping fuel oil has an outside diameter of 3 in. and a length of 8 ft. The outside of the pipe is painted with a coating $\frac{1}{16}$ in. thick. Use differentials to determine the number of gallons of paint needed to paint 100 pipes of this type. (One U.S. gallon is equivalent to 231 in^3.)

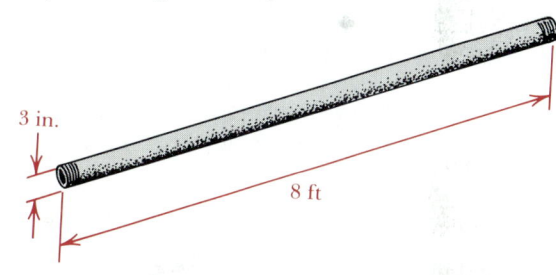

35. In Exercise 29 of Exercise Set 3.2 the dimensions were found for the strongest rectangular beam that can be cut from a circular log of radius r. Use this result and differentials to approximate the percentage of increase in the strength of the strongest rectangular beam that can be cut from a circular log if the radius of the log increases from r to $r + \Delta r$.

36. In Exercise 30 of Exercise Set 3.2 the dimensions were found for the stiffest rectangular beam that can be cut from a circular log of radius r. Use this result and differentials to approximate the percentage increase in the stiffness of the stiffest rectangular beam that can be cut from a circular log if the radius of the log is increased from r to $r + \Delta r$.

37. A metal spherical ball with a volume of 36 cm^3 is given a chrome plating 0.2 cm thick. Use differentials to approximate the volume of the chromium required to plate the ball.

38. A string is placed around the earth at the equator. (Assume that the earth's surface at the equator is a circle with radius 3960 mi.) We cut the string and raise it above the surface of the earth exactly 1 ft at each point (so that small dogs can pass beneath, perhaps). By some magical means the string stays in this position. When the string is raised 1 ft the ends of the string no longer meet. How much string is needed to complete the circle 1 ft above the earth?
   a) Use differentials to solve this problem.
   b) Solve the problem directly by determining the circumference of each circle and subtracting to find the difference.

39. The spherical metal balls used in a pinball machine must have a volume of 65.5 cm^3, with a deviation of no more than 0.1 cm^3, in order to have the correct mass for activating certain targets in the machine. Use differentials to approximate the interval in which the diameters of the balls must lie if they are to meet this specification.

40. Suppose, in addition, that the balls described in Exercise 39 must have a surface area of 78.5 cm^2, with a deviation of no more than 0.1 cm^2. Use differentials to approximate the interval in which the diameters of the balls must lie if they are to meet both specifications.

41. Temperature on weather forecasts is generally given to the nearest degree. Suppose you are told that the present temperature is 20° C, which means that it is somewhere between 19.5° C and 20.5° C. Use differentials to approximate the maximum error in degrees Fahrenheit that can be produced if the 20° C figure is converted to 68° F.

42. Two thermometers are placed in a room. Both read the temperature in degrees Celsius with an error of no more than ±0.51° C. The thermometers are read simultaneously to the nearest degree Celsius and then converted to a Fahrenheit scale. Use differentials to approximate the maximum number of degrees Fahrenheit by which the thermometers can disagree.

43. A 5-ft stick is placed vertically in the ground 100 ft from the base of a tree. A line joining the top of the tree and the top of the stick touches the ground at a point 10 ft from the base of the stick.
   a) How high is the tree?

   Suppose an error of no more than 1 in. is made in determining the spot where the line joining the top of the tree and the top of the stick hits the ground; the other measurements are exact.
   b) Use differentials to approximate the maximum error produced in calculating the height of the tree.

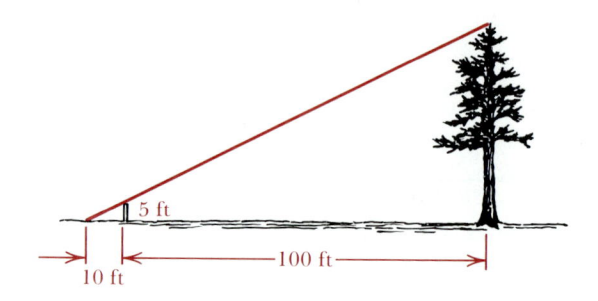

44. Reconsider the situation in Exercise 43 but assume instead that an error of no more than 1 in. is made in measuring the distance from the base of the tree to the base of the stick; the other measurements are exact. Use differentials to approximate the maximum error produced in calculating the height of the tree.

3.5

Indeterminate Forms: l'Hôpital's Rule

Early in the study of the limit we established that

$$\lim_{x \to a} \frac{f(x)}{g(x)} = \frac{\lim_{x \to a} f(x)}{\lim_{x \to a} g(x)},$$

provided that $\lim_{x \to a} f(x)$ and $\lim_{x \to a} g(x)$ both exist and that $\lim_{x \to a} g(x) \neq 0$. However, this result was not sufficient to answer many questions concerning the limit of a quotient of two functions. For example, the introductory section on limits, Section 1.5, considered two limits of this type:

$$\lim_{x \to 2} \frac{x^3 - 2x^2}{x - 2} \quad \text{and} \quad \lim_{x \to 0} \frac{\sin x}{x},$$

where in each case both the numerator and the denominator have the limit zero.

These are examples of limits that are said to have the *indeterminate form* $0/0$. The term "indeterminate form" is used to describe limits of combinations of functions to which the arithmetic limit rules do not apply directly.

The first indeterminate forms we consider are those of the type $0/0$. Later in the section we will discuss indeterminate forms of the types ∞/∞, $0 \cdot \infty$, and $\infty - \infty$. Additional limits of indeterminate forms are presented in Section 6.5.

(3.10)
DEFINITION
The quotient $f(x)/g(x)$ is said to have the **indeterminate form $0/0$** at a if both

$$\lim_{x \to a} f(x) = 0 \quad \text{and} \quad \lim_{x \to a} g(x) = 0,$$

where a represents a real number, ∞, or $-\infty$.

By this definition,

$$\frac{x^3 - 2x^2}{x - 2} \quad \text{has the indeterminate form } 0/0 \text{ at } 2,$$

and

$$\frac{\sin x}{x} \quad \text{has the indeterminate form } 0/0 \text{ at } 0.$$

In Section 1.5, we used factoring to show that

$$\lim_{x \to 2} \frac{x^3 - 2x^2}{x - 2} = \lim_{x \to 2} \frac{x^2(x - 2)}{x - 2} = \lim_{x \to 2} x^2 = 4.$$

The first time we encountered an indeterminate form that an algebraic manipulation would not resolve was in evaluating

$$\lim_{x \to 0} \frac{\sin x}{x} = 1.$$

In this case, we used a geometric technique to determine the limit.

To introduce a general method for evaluating indeterminate forms, we first consider a special case of the form $0 \neq 0$. Suppose that $f(x)/g(x)$ assumes this form (so $\lim_{x \to a} f(x) = 0$ and $\lim_{x \to a} g(x) = 0$), that both f and g are differentiable at a, and that $g'(a) \neq 0$. Differentiability at a implies continuity at a, so $f(a) = 0$ and $g(a) = 0$. Thus,

$$f(x) = f(x) - f(a) \quad \text{and} \quad g(x) = g(x) - g(a).$$

Since we are considering only values of x satisfying $x \neq a$, the ratio of $f(x)$ to $g(x)$ can be written

$$\frac{f(x)}{g(x)} = \frac{f(x) - f(a)}{g(x) - g(a)} = \frac{\dfrac{f(x) - f(a)}{x - a}}{\dfrac{g(x) - g(a)}{x - a}}, \quad \text{provided } g(x) \neq g(a).$$

This implies that if $g(x) \neq g(a)$ near a, then

$$\lim_{x \to a} \frac{f(x)}{g(x)} = \lim_{x \to a} \frac{f(x) - f(a)}{g(x) - g(a)} = \frac{\displaystyle\lim_{x \to a} \frac{f(x) - f(a)}{x - a}}{\displaystyle\lim_{x \to a} \frac{g(x) - g(a)}{x - a}} = \frac{f'(a)}{g'(a)}.$$

In the problem $\lim_{x \to 2} (x^3 - 2x^2)/(x - 2)$,

$$f'(x) = D_x (x^3 - 2x^2) = 3x^2 - 4x \quad \text{and} \quad g'(x) = D_x (x - 2) = 1,$$

so

$$\lim_{x \to 2} \frac{x^3 - 2x^2}{x - 2} = \frac{f'(2)}{g'(2)} = \frac{12 - 8}{1} = 4.$$

For the problem $\lim_{x \to 0} (\sin x)/x$,

$$D_x \sin x = \cos x \quad \text{and} \quad D_x x = 1,$$

so

$$\lim_{x \to 0} \frac{\sin x}{x} = \frac{\cos 0}{1} = 1.$$

(There is a logical difficulty associated with this result. It is discussed in Exercise 48.)

Although this is the basic procedure for evaluating indeterminate forms, the requirement that both f and g be differentiable at a is an overly restrictive condition. For example, the condition could never be satisfied when considering the value of an indeterminate form at infinity.

The following extension of the Mean Value Theorem is used to establish a more general result.

(3.11)
THEOREM: THE CAUCHY MEAN VALUE THEOREM
Suppose f and g are continuous on $[a, b]$ and differentiable on (a, b). If g' is nonzero on (a, b), then a number c exists, $a < c < b$, with

$$\frac{f(b) - f(a)}{g(b) - g(a)} = \frac{f'(c)}{g'(c)}.$$

Proof. The proof of this theorem is similar to that of the Mean Value Theorem: A new function h is defined on $[a, b]$ that satisfies the hypotheses of Rolle's Theorem.

Let
$$h(x) = [f(b) - f(a)][g(x) - g(a)] - [g(b) - g(a)][f(x) - f(a)].$$
Then h is continuous on $[a, b]$ and differentiable on (a, b). In addition,
$$h(a) = [f(b) - f(a)][g(a) - g(a)] - [g(b) - g(a)][f(a) - f(a)] = 0$$
and
$$h(b) = [f(b) - f(a)][g(b) - g(a)] - [g(b) - g(a)][f(b) - f(a)] = 0,$$
so Rolle's Theorem implies that a number c, $a < c < b$, exists with
$$0 = h'(c) = [f(b) - f(a)]g'(c) - [g(b) - g(a)]f'(c).$$
Since g' is never zero on (a, b), Rolle's Theorem implies that $g(b) - g(a)$ is nonzero. Also, $g'(c)$ is nonzero, so we can divide by $g'(c)[g(b) - g(a)]$, which implies that
$$0 = \frac{f(b) - f(a)}{g(b) - g(a)} - \frac{f'(c)}{g'(c)}.$$
Thus,
$$\frac{f(b) - f(a)}{g(b) - g(a)} = \frac{f'(c)}{g'(c)}. \qquad \triangleright$$

Note that the Mean Value Theorem is the special case of the Cauchy Mean Value Theorem that results when $g(x) = x$ for each x in $[a, b]$.

The basic theorem involving indeterminate forms can now be proved. This theorem and its various modifications are collectively known as **l'Hôpital's Rule.**

(3.12)
THEOREM
Suppose f and g are differentiable and g' is nonzero on an open interval I containing a, except possibly at a. If $f(x)/g(x)$ has the indeterminate form $0/0$ at a and
$$\lim_{x \to a} \frac{f'(x)}{g'(x)} = L, \quad \text{then} \quad \lim_{x \to a} \frac{f(x)}{g(x)} = L.$$

HISTORICAL NOTE

Guillaume François Antoine l'Hôpital, Marquis de St. Mesme (1661–1704), introduced this method in the first calculus textbook, *Analyse des infiniments petits,* published in Paris in 1696. This book had a wide circulation and brought the differential notation into general usage in France as well as making it known throughout Europe. The rule was evidently a discovery of Johann Bernoulli, who sent all his mathematical discoveries to l'Hôpital in exchange for a regular salary. Controversy developed regarding whether this arrangement was clandestine, with l'Hôpital attempting to receive credit for Bernoulli's work.

In his preface, l'Hôpital takes special note of the contribution of all the members of the Bernoulli family, particularly Johann. It was only after l'Hôpital's death that Bernoulli accused him of plagiarism, an accusation that at the time was generally dismissed but now seems to be well founded. In any case, the book was a major contribution to the spread of the knowledge of calculus throughout Europe, particularly in France.

Proof. There is no loss of generality in assuming that f and g are continuous at $x = a$ since the limits consider only the values of $f(x)$ and $g(x)$ for x near a, not at $x = a$. Therefore, we assume that $f(a) = 0$ and $g(a) = 0$.

Suppose that x is in I and $x \neq a$. If $x < a$, we apply the Cauchy Mean Value Theorem to f and g on the interval $[x, a]$. If $x > a$, we apply this result on $[a, x]$. In either case, the Cauchy Mean Value Theorem implies that there is a number c between x and a with

$$\frac{f'(c)}{g'(c)} = \frac{f(x) - f(a)}{g(x) - g(a)} = \frac{f(x)}{g(x)}.$$

Since c is between x and a,

$$\lim_{x \to a} \frac{f(x)}{g(x)} = \lim_{x \to a} \frac{f'(c)}{g'(c)} = \lim_{c \to a} \frac{f'(c)}{g'(c)} = L.$$ ▷

The verification that f and g are differentiable and that $g'(a) \neq 0$ has not been included in the following examples. It is usually straightforward. Note, however, that the verification of the indeterminate form is *always* included.

Example 1

Evaluate $\lim\limits_{x \to -1} \dfrac{x^6 - x^4}{x^5 + 2x^2 - 1}$.

Solution
Since

$$\lim_{x \to -1} (x^6 - x^4) = 0 = \lim_{x \to -1} (x^5 + 2x^2 - 1),$$

l'Hôpital's Rule can be applied:

$$\lim_{x \to -1} \frac{x^6 - x^4}{x^5 + 2x^2 - 1} = \lim_{x \to -1} \frac{D_x (x^6 - x^4)}{D_x (x^5 + 2x^2 - 1)}$$

$$= \lim_{x \to -1} \frac{6x^5 - 4x^3}{5x^4 + 4x} = \frac{-6 + 4}{5 - 4} = -2.$$ ▶

Example 2

Evaluate $\lim\limits_{x \to 9} \dfrac{\sqrt{x} - 3}{x - 9}$.

Solution
Since

$$\lim_{x \to 9} (\sqrt{x} - 3) = 0 = \lim_{x \to 9} (x - 9),$$

l'Hôpital's Rule can be applied:

$$\lim_{x \to 9} \frac{\sqrt{x} - 3}{x - 9} = \lim_{x \to 9} \frac{D_x (\sqrt{x} - 3)}{D_x (x - 9)} = \lim_{x \to 9} \frac{\frac{1}{2\sqrt{x}}}{1} = \frac{1}{6}.$$

Try to find two other methods of determining this limit. ▶

Example 3

Evaluate $\lim\limits_{x \to \pi} \dfrac{1 + \cos x}{\sin x}$.

Solution
We have

$$\lim_{x \to \pi} (1 + \cos x) = 0 = \lim_{x \to \pi} \sin x,$$

so

$$\lim_{x \to \pi} \frac{1 + \cos x}{\sin x} = \lim_{x \to \pi} \frac{D_x (1 + \cos x)}{D_x (\sin x)} = \lim_{x \to \pi} \frac{-\sin x}{\cos x} = 0.$$ ▶

Example 4

Evaluate $\lim\limits_{x \to 0} \dfrac{x \cos x - \sin x}{x - \sin x}$.

Solution
We have

$$\lim_{x \to 0} (x \cos x - \sin x) = 0 = \lim_{x \to 0} (x - \sin x)$$

so

$$\lim_{x \to 0} \frac{x \cos x - \sin x}{x - \sin x} = \lim_{x \to 0} \frac{\cos x - x \sin x - \cos x}{1 - \cos x}$$
$$= \lim_{x \to 0} \frac{-x \sin x}{1 - \cos x}.$$

But

$$\lim_{x \to 0} -x \sin x = 0 = \lim_{x \to 0} (1 - \cos x)$$

so we apply l'Hôpital's Rule again:

$$\lim_{x \to 0} \frac{x \cos x - \sin x}{x - \sin x} = \lim_{x \to 0} \frac{-x \sin x}{1 - \cos x}$$
$$= \lim_{x \to 0} \frac{-\sin x - x \cos x}{\sin x}.$$

Once again we have the form 0/0; this time l'Hôpital's Rule resolves the problem:

$$\lim_{x \to 0} \frac{x \cos x - \sin x}{x - \sin x} = \lim_{x \to 0} \frac{-\sin x - x \cos x}{\sin x}$$

$$= \lim_{x \to 0} \frac{-\cos x - \cos x + x \sin x}{\cos x}$$

$$= \frac{-2}{1} = -2. \quad \blacktriangleright$$

When both $\lim_{x \to a} f(x)$ and $\lim_{x \to a} g(x)$ are infinite (either ∞ or $-\infty$), the quotient $f(x)/g(x)$ is said to have the indeterminate form ∞/∞ at a. L'Hôpital's Rule for the case of the indeterminate form ∞/∞ at a also holds, but the proof is more complicated.

(3.13) THEOREM

Suppose f and g are differentiable and g' is nonzero on an open interval containing a, except possibly at a. If $f(x)/g(x)$ has the indeterminate form ∞/∞ at a and

$$\lim_{x \to a} \frac{f'(x)}{g'(x)} = L, \quad \text{then} \quad \lim_{x \to a} \frac{f(x)}{g(x)} = L.$$

L'Hôpital's Rule also holds for one-sided limits and for the case when either or both of a and L are infinite. We will assume these results.

Example 5

Use l'Hôpital's Rule to determine $\lim_{x \to \infty} \dfrac{3x^3 + 2x - 1}{2x^3 + x^2 - 5}$.

Solution

This limit has the indeterminate form ∞/∞, so

$$\lim_{x \to \infty} \frac{3x^3 + 2x - 1}{2x^3 + x^2 - 5} = \lim_{x \to \infty} \frac{9x^2 + 2}{6x^2 + 2x} = \lim_{x \to \infty} \frac{18x}{12x + 2} = \lim_{x \to \infty} \frac{18}{12} = \frac{3}{2}. \quad \blacktriangleright$$

We can also evaluate the limit in Example 5 by first dividing the numerator and the denominator by x^3, the highest power in the denominator, and then applying the limit results of Section 1.9.

Example 6

Evaluate $\lim_{x \to \frac{\pi}{2}^-} \dfrac{2x + \sec x}{3 + \tan x}$.

Solution

This limit has the indeterminate form ∞/∞, since

$$\lim_{x \to \frac{\pi}{2}^-} (2x + \sec x) = \infty \quad \text{and} \quad \lim_{x \to \frac{\pi}{2}^-} (3 + \tan x) = \infty.$$

Consequently,
$$\lim_{x \to \frac{\pi}{2}^-} \frac{2x + \sec x}{3 + \tan x} = \lim_{x \to \frac{\pi}{2}^-} \frac{2 + \sec x \tan x}{\sec^2 x}.$$

L'Hôpital's Rule can be applied to the new quotient since
$$\lim_{x \to \frac{\pi}{2}^-} (2 + \sec x \tan x) = \infty \quad \text{and} \quad \lim_{x \to \frac{\pi}{2}^-} \sec^2 x = \infty,$$
but it again leads to an indeterminate form ∞/∞. However, by writing $\tan x$ and $\sec x$ in terms of sines and cosines, we can simplify the expression:
$$\lim_{x \to \frac{\pi}{2}^-} \frac{2x + \sec x}{3 + \tan x} = \lim_{x \to \frac{\pi}{2}^-} \frac{2 + \sec x \tan x}{\sec^2 x} = \lim_{x \to \frac{\pi}{2}^-} [2 \cos^2 x + \sin x] = 1.$$
▶

A simplification of the type shown in Example 6 should be used whenever possible. This makes the differentiation easier even when the problem can be solved by repeated application of l'Hôpital's Rule.

Example 7

Evaluate $\lim_{x \to 0} \dfrac{x}{1 + \sin x}$.

Solution
This is *not* a problem requiring l'Hôpital's Rule since
$$\lim_{x \to 0} \frac{x}{1 + \sin x} = \frac{\lim_{x \to 0} x}{\lim_{x \to 0} (1 + \sin x)} = \frac{0}{1} = 0.$$

In fact, if l'Hôpital's Rule were mistakenly applied to this limit problem, an incorrect answer would result:
$$\lim_{x \to 0} \frac{D_x x}{D_x (1 + \sin x)} = \lim_{x \to 0} \frac{1}{\cos x} = 1.$$
▶

Beware of the pitfalls that arise from the careless use of l'Hôpital's Rule. The technique can be applied *only* to indeterminate forms of the type $0/0$ or ∞/∞.

Other types of limits lead to indeterminate forms. If $\lim_{x \to a} f(x) = 0$ and $\lim_{x \to a} g(x) = \infty$, then $f(x) \cdot g(x)$ has the indeterminate form $0 \cdot \infty$ at a. Quite often this limit can be resolved by expressing $f(x) g(x)$ as
$$f(x) g(x) = \frac{f(x)}{\left(\dfrac{1}{g(x)}\right)} \quad \text{or} \quad f(x) g(x) = \frac{g(x)}{\left(\dfrac{1}{f(x)}\right)}$$
and applying l'Hôpital's Rule to the functions f and $1/g$ or to g and $1/f$.

Example 8

Evaluate $\lim\limits_{x\to\infty} x \sin \dfrac{1}{x}$.

Solution

This limit has indeterminate form $0 \cdot \infty$, since

$$\lim_{x\to\infty} x = \infty \quad \text{and} \quad \lim_{x\to\infty} \sin \frac{1}{x} = 0.$$

However,

$$\lim_{x\to\infty} x \sin \frac{1}{x} = \lim_{x\to\infty} \frac{\sin \dfrac{1}{x}}{\dfrac{1}{x}},$$

which is of the form $0/0$. Thus,

$$\lim_{x\to\infty} x \sin \frac{1}{x} = \lim_{x\to\infty} \frac{\sin \dfrac{1}{x}}{\dfrac{1}{x}} = \lim_{x\to\infty} \frac{\left(-\dfrac{1}{x^2}\right)\cos \dfrac{1}{x}}{-\dfrac{1}{x^2}}$$

$$= \lim_{x\to\infty} \cos \frac{1}{x} = 1. \qquad \blacktriangleright$$

The problem in Example 8 can be solved alternatively by using the change of variable $z = 1/x$. As x becomes large, z approaches zero from the right. Consequently,

$$\lim_{x\to\infty} x \sin \frac{1}{x} = \lim_{z\to 0^+} \frac{1}{z} \sin z = \lim_{z\to 0^+} \frac{\sin z}{z} = 1.$$

An indeterminate form of the type $\infty - \infty$ occurs when we are evaluating $\lim_{x\to a} (f(x) - g(x))$, where both $\lim_{x\to a} f(x) = \infty$ and $\lim_{x\to a} g(x) = \infty$. The following example shows how to manipulate problems of this type into a form appropriate for applying l'Hôpital's Rule.

Example 9

Evaluate $\lim\limits_{x\to 0^+} (\csc x - \cot x)$.

Solution
Since

$$\lim_{x\to 0^+} \csc x = \infty \quad \text{and} \quad \lim_{x\to 0^+} \cot x = \infty$$

this limit has the indeterminate form $\infty - \infty$. However,

$$\lim_{x\to 0^+} (\csc x - \cot x) = \lim_{x\to 0^+} \left(\frac{1}{\sin x} - \frac{\cos x}{\sin x}\right) = \lim_{x\to 0^+} \frac{1 - \cos x}{\sin x}.$$

This final limit has the indeterminate form 0/0, so l'Hôpital's Rule can now be applied:

$$\lim_{x \to 0^+} (\csc x - \cot x) = \lim_{x \to 0^+} \frac{1 - \cos x}{\sin x} = \lim_{x \to 0^+} \frac{D_x (1 - \cos x)}{D_x \sin x}$$

$$= \lim_{x \to 0^+} \frac{\sin x}{\cos x} = 0.$$

▶

Other indeterminate forms of the type 1^∞, 0^0, and ∞^0 are considered in Chapter 6 after general exponential functions have been introduced.

We end this section with another caution against attempting to apply l'Hôpital's Rule when it is inappropriate. In such a case it rarely leads to a correct conclusion (especially when prearranged by a devious calculus instructor!).

▶ **EXERCISE SET 3.5**

In Exercises 1–40, determine the limit, if it exists.

1. $\lim_{x \to 1} \dfrac{x^3 - 2x^2 + 2x - 1}{x^2 - 1}$

2. $\lim_{x \to -2} \dfrac{x^3 + 2x^2 - x - 2}{3x^2 + 10x + 8}$

3. $\lim_{x \to 3} \dfrac{x^4 - 6x^2 - 5x - 12}{3x^2 - 7x - 6}$

4. $\lim_{x \to 1} \dfrac{x^4 - x^3 + 3x^2 + x - 4}{x^3 + x^2 - 7x + 5}$

5. $\lim_{x \to -2} \dfrac{2 + 9x - 2x^2 - 3x^3}{2x^4 + 7x^3 + 10x^2 + 7x - 2}$

6. $\lim_{x \to -4} \dfrac{x^3 + 8x^2 + 11x - 20}{2x^3 + 5x^2 - 8x + 16}$

7. $\lim_{x \to 0} \dfrac{1 - \cos x}{2x}$

8. $\lim_{x \to 0} \dfrac{2x}{\tan x}$

9. $\lim_{x \to 0} \dfrac{\sin 2x}{3x}$

10. $\lim_{x \to 0} \dfrac{\cos x + x - 1}{3x}$

11. $\lim_{x \to -2} \dfrac{\sqrt{x + 6} - 2}{x^2 - 4}$

12. $\lim_{x \to 2} \dfrac{\sqrt{x + 7} - 3}{x^2 - 4}$

13. $\lim_{x \to -2} \dfrac{\sqrt{6 + x} - \sqrt{2 - x}}{x + 2}$

14. $\lim_{x \to 1} \dfrac{2 - \sqrt{x + 3}}{1 - x}$

15. $\lim_{x \to 2} \dfrac{\sqrt{4x + 1} - \sqrt[3]{13x + 1}}{x + 2}$

16. $\lim_{x \to 4} \dfrac{2 - \sqrt[3]{x + 4}}{4 + \sqrt{12 + x}}$

17. $\lim_{x \to \infty} \dfrac{x^2 + 5}{3x^2 + 1}$

18. $\lim_{x \to \infty} \dfrac{3x^2 - 7}{2x^2 + x + 3}$

19. $\lim_{x \to \infty} \dfrac{3x^2 + 4x + 5}{2x^2 + 3x - 2}$

20. $\lim_{x \to -\infty} \dfrac{2x^3 + 7x - 5}{7x^3 + 2x - 5}$

21. $\lim_{x \to -\infty} \dfrac{5 - 2x^3 - 3x^4}{3 + 2x + 7x^4}$

22. $\lim_{x \to \infty} \dfrac{3x + 4x^2 - 5x^4}{2 + x + 4x^2 - 3x^4}$

23. $\lim_{x \to \frac{\pi}{2}} \dfrac{1 - \sin x}{\cos x}$

24. $\lim_{x \to 0} \dfrac{1 - \sin x}{\cos x}$

25. $\lim_{x \to 0} \dfrac{\csc x}{\cot x}$

26. $\lim_{x \to \frac{\pi}{2}^-} \dfrac{\sec x}{\tan x}$

27. $\lim_{x \to 0} \dfrac{\sin x - x}{x^3}$

28. $\lim_{x \to 0} \dfrac{\sin x^2}{\sin^2 x}$

29. $\lim_{x \to 0} \dfrac{4x^2 \cos 3x}{\sin^2 3x}$

30. $\lim_{x \to 0} \dfrac{1 - \cos x^2}{x^3}$

31. $\lim_{x \to 0^+} \left(\dfrac{1}{x} - \dfrac{1}{\sin x} \right)$

32. $\lim_{x \to 0} \left(\dfrac{1}{1 - \cos x} - \dfrac{1}{x^2} \right)$

33. $\lim_{x \to \infty} \left(\dfrac{1}{x^2 + 2} - \dfrac{1}{x} \right)$

34. $\lim_{x \to \frac{\pi}{2}^+} \left(\dfrac{1}{\sec x} - \dfrac{1}{x - \pi/2} \right)$

35. $\lim_{x \to -\infty} \left(\dfrac{x}{x^2 + 1} - \dfrac{x}{x^2 - 1} \right)$

36. $\lim_{x \to \frac{\pi}{2}} (\tan x - \sec x)$

37. $\lim_{x \to \infty} x^2 \sin \dfrac{1}{x}$

38. $\lim_{x \to 0^-} x^2 \cot x$

39. $\lim_{x \to \frac{\pi}{2}} \left(x - \dfrac{\pi}{2} \right) \sec x$

40. $\lim_{x \to 0^+} 4x^2 \cot 3x$

41. The first published example of l'Hôpital's Rule was given in *Analyse des infiniments petits*. In Example I, l'Hôpital found
$$\lim_{x \to a} \dfrac{\sqrt{2a^3x - x^4} - a\sqrt[3]{a^2x}}{a - \sqrt[4]{ax^3}}.$$
Determine this limit.

In Exercises 42–45, find all values of c that satisfy the conclusions of the Cauchy Mean Value Theorem if the hypotheses of the theorem hold for the given function and interval.

42. $f(x) = x$, $g(x) = x^2$, $[1, 2]$

43. $f(x) = x - 2$, $g(x) = x^3$, $[-1, 0]$

44. $f(x) = x^2$, $g(x) = x^3$, $[-1, 1]$

45. $f(x) = x^2 - 3x$, $g(x) = x^2 - 1$, $[0, 2]$

46. What must a be if
$$g(x) = \begin{cases} \dfrac{x^3}{\tan x - x}, & \text{if } x \neq 0, \\ a, & \text{if } x = 0 \end{cases}$$
is continuous at zero?

47. Determine
$$\lim_{x \to \infty} \dfrac{\sin x - x}{\sin x + x}.$$
Does l'Hôpital's Rule resolve this problem?

48. L'Hôpital's Rule was used to show that $\lim_{x \to 0} (\sin x)/x = 1$. In Section 2.4, this limit was used to show that the derivative of the sine function is the cosine function. Explain why the use of l'Hôpital's Rule involves circular reasoning in this instance.

49. Show that the limit in Exercise 28 can be found without applying l'Hôpital's Rule by using the fact that $\lim_{x \to 0} (\sin x)/x = 1$ and observing that
$$\dfrac{\sin x^2}{(\sin x)^2} = \dfrac{\sin x^2}{x^2} \left(\dfrac{x}{\sin x} \right)^2.$$

50. Find a pair of differentiable functions f and g and a real number a with the property that $\lim_{x \to a} f(x)/g(x)$ exists and is finite, but $\lim_{x \to a} f'(x)/g'(x)$ does *not* exist.

51. Find a pair of differentiable functions f and g with the property that $\lim_{x \to \infty} f(x)/g(x)$ exists and is finite, but $\lim_{x \to \infty} f'(x)/g'(x)$ does *not* exist.

52. Show that all the hypotheses of the Cauchy Mean Value Theorem are necessary, that is:
   a) Find a pair of functions that are not continuous on an interval $[a, b]$ for which the conclusion does not hold.
   b) Find a pair of functions that are not differentiable on (a, b), but are continuous on $[a, b]$, for which the conclusion does not hold.
   c) Find a pair of functions that are continuous on $[a, b]$ and differentiable on (a, b) for which the conclusion does not hold.

53. Suppose $f'(a)$ and $\lim_{x \to a} f'(x)$ exist. Show that f' must be continuous at a.

54. Suppose f'' exists at a. Show that
$$f''(a) = \lim_{h \to 0} \dfrac{f(a + h) + f(a - h) - 2f(a)}{h^2}.$$

55. The proof of the Cauchy Mean Value Theorem states that since g' is nonzero on (a, b), $g(b) - g(a) \neq 0$. Show that this statement is true.

3.6
Newton's Method

Suppose that f is a continuous function on the interval $[a, b]$ and that $f(a)$ and $f(b)$ are nonzero and have differing signs. It was shown in Section 1.8 that the Bisection Method can be used to find an approximation in (a, b) to a solution, or root, of the equation $f(x) = 0$ to any specified accuracy. For practical purposes, however, the Bisection Method often takes too much computation time to be competitive with other techniques.

In this section we consider a powerful root-finding method called the *Newton–Raphson Method,* or simply **Newton's Method.** This technique requires that the function f be differentiable on an interval containing the root and uses the derivative to improve the approximations.

Suppose that the graph of a function f crosses the x-axis precisely once in the interval $[a, b]$ and that the derivative of f exists and does not change sign in $[a, b]$. The graph of such a function might have a form similar to the one shown in Figure 3.25. Newton's Method is described geometrically as follows.

To approximate the point c where the graph of f crosses the x-axis, that is, the root of $f(x) = 0$:

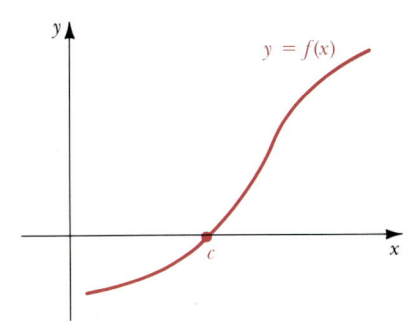

Figure 3.25
Object: Find c with $f(c) = 0$.

(3.14)
NEWTON'S METHOD

i) Assume an initial approximation x_0 to c.
ii) Construct the tangent line to the graph of f at $(x_0, f(x_0))$ and call the x-intercept of this line x_1.
iii) Draw the tangent line to the graph of f at $(x_1, f(x_1))$ and call the x-intercept of this line x_2.
iv) Continue the process to generate the successive approximations x_3, x_4, x_5, and so on.

Figure 3.26 shows the first few approximations of a typical application.

To determine analytically the value of x_1 given the initial approximation x_0, first note that the tangent line to the graph of f at $(x_0, f(x_0))$ has equation

$$y - f(x_0) = f'(x_0)(x - x_0).$$

Since $(x_1, 0)$ is a point on this tangent line,

$$0 - f(x_0) = f'(x_0)(x_1 - x_0).$$

If $f'(x_0) \neq 0$, we can solve this equation for x_1:

$$x_1 = x_0 - \frac{f(x_0)}{f'(x_0)}.$$

Once x_1 has been determined, x_2 is computed in a similar manner:

$$x_2 = x_1 - \frac{f(x_1)}{f'(x_1)}, \quad \text{provided } f'(x_1) \neq 0.$$

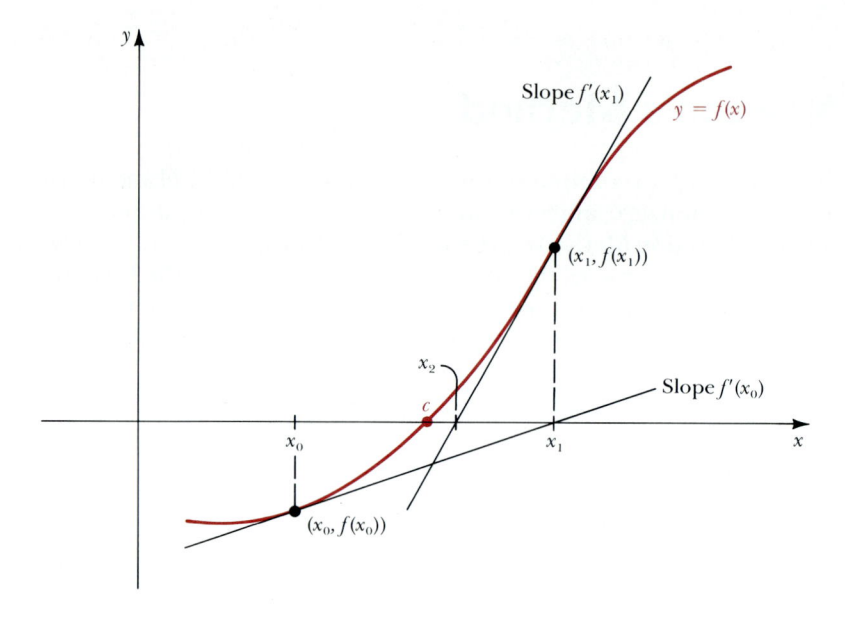

Figure 3.26

Newton's Method: $x_{n+1} = x_n - \dfrac{f(x_n)}{f'(x_n)}$.

In general, for each positive integer n,

(3.15) $$x_{n+1} = x_n - \frac{f(x_n)}{f'(x_n)}, \quad \text{provided } f'(x_n) \neq 0.$$

A technique, such as Newton's Method, that applies the same process repeatedly is called an **iterative method.** The values x_0, x_1, x_2, . . . obtained from the application of this process are called **iterates.** Iterative processes are important approximation techniques because they adapt easily to computer implementation. Listed in Appendix C is a BASIC program for Newton's Method. If you examine this program you will see that very few steps are required for its implementation.

Before considering additional theoretical aspects of Newton's Method, let us apply the technique to the problem used to illustrate the Bisection Method in Example 4 of Section 1.8.

Example 1

Use Newton's Method to approximate a root of the equation

$$f(x) = x^3 - 2x - 5 = 0.$$

Solution

In Example 4 of Section 1.8 we saw that $f(2) = -1$ and $f(3) = 16$, so the Intermediate Value Theorem implies that a root lies in the interval $(2, 3)$. Moreover,

$$f'(x) = 3x^2 - 2 > 0 \quad \text{for all } x \text{ in } [2, 3],$$

so f is increasing on $[2, 3]$ and there is only one root in $(2, 3)$ (see Figure 3.27).

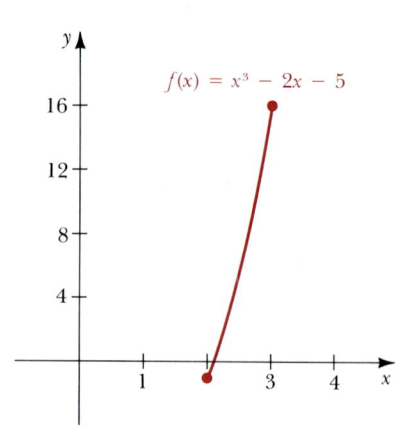

Figure 3.27

Since the magnitude of $f(2)$ is much less than the magnitude of $f(3)$, it seems likely that the root is close to 2. If we choose $x_0 = 2$, then

$$x_1 = x_0 - \frac{f(x_0)}{f'(x_0)} = x_0 - \frac{x_0^3 - 2x_0 - 5}{3x_0^2 - 2}$$

so

$$x_1 = 2 - \frac{2^3 - 2(2) - 5}{3(2)^2 - 2} = 2 - \frac{-1}{10} = 2.1.$$

Reapplying Newton's Method gives the new approximation:

$$x_2 = x_1 - \frac{f(x_1)}{f'(x_1)} = 2.1 - \frac{(2.1)^3 - 2(2.1) - 5}{3(2.1)^2 - 2} = 2.0945683.$$

Applying the method once more leads to the approximation

$$x_3 = x_2 - \frac{f(x_2)}{f'(x_2)}$$
$$= 2.0945683 - \frac{(2.0945683)^3 - 2(2.0945683) - 5}{3(2.0945683)^2 - 2}$$
$$= 2.0945516.$$

Assuming that we are using a device that gives accurate results to seven decimal places, subsequent applications continue to give the value 2.0945516. ▶

Since all approximations beyond x_2 in Example 1 produce the same result, one is led to suspect that the approximation $x_3 = 2.0945516$ is accurate to the places listed. This is indeed the case, as we can show by a result proved in Section 8.9. This result implies that, in most applications, once an approximation to the root is sufficiently accurate, Newton's Method tends to essentially double the number of correct decimal places with each succeeding application. For our purposes, we will assume that if consecutive iterations of Newton's Method agree to n decimal places, then the latter iteration is accurate to at *least* n decimal places.

By contrast, six iterations of the Bisection Method applied to the function and interval given in Example 1 produced an approximation that was accurate to only two decimal places.

In 1685, **John Wallis** (1616–1703) used Example 1 in his book, *De algebra tractatus* (1685), to illustrate the method Isaac Newton had devised for approximating solutions of equations. A special case of Newton's Method for approximating square roots had been known from Babylonian times, and was also known to apply to problems involving polynomials. In his *De analysi per aequationes numero terminorum infinitas* and *Methods of fluxions*, which were published in 1711 and 1736, respectively, Newton worked out only one example, $y^3 - 2y - 5 = 0$.

The first systematic account of the method was provided by **Joseph Raphson** (1648–1715) in 1690 in *Analysis aequationum universalis*. He indicated the use of such expressions as our Equation (3.15).

HISTORICAL NOTE

Example 2

Find an approximation to a root of the equation

$$f(x) = \cos x - x = 0$$

that is accurate to six decimal places.

Solution

Since $f(0) = 1 > 0$ and $f(\pi) = -1 - \pi < 0$, a root lies in the interval $(0, \pi)$. If we let x_0 be the midpoint of this interval, $x_0 = \pi/2$, then the approximations generated by Newton's Method are

$$x_1 = 0.785398, \quad x_2 = 0.739536, \quad x_3 = 0.739085, \quad x_4 = 0.739085.$$

Since x_3 and x_4 agree to the decimal places listed, we assume that $x_4 = 0.739085$ is accurate to six decimal places. ▶

In Example 1, the initial approximation x_0 was chosen as the endpoint at which the magnitude of the function was smallest. In Example 2, the choice for x_0 was the midpoint of the interval. Both of these choices seem reasonable since they produced accurate results very quickly. If we had drawn a sketch of the graph of $f(x) = \cos x - x$, such as the one shown in Figure 3.28, we might have chosen x_0 closer to the actual root, but the improvement would probably not have been dramatic.

In some problems Newton's Method converges for any choice of the initial approximation. The following example demonstrates rather clearly that this is not always true.

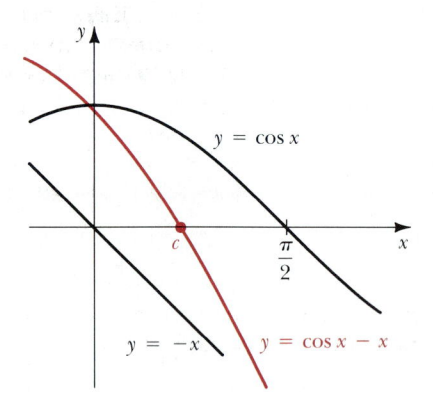

Figure 3.28

Example 3

The graph of $f(x) = 2x^3 - 9x^2 + 12x - 2$ has precisely one x-intercept. Approximate this intercept to an accuracy of three decimal places.

Solution

If we evaluate the function f at $x = 0$ and $x = 2$, we find that $f(0) = -2$ and $f(2) = 2$. If we use only this observation, a reasonable choice for x_0 is the midpoint of the interval $[0, 2]$, that is, $x_0 = 1$.

Before attempting to apply Newton's Method with this choice of x_0, let us recall that the graph of this function was progressively refined using the techniques in Chapter 2. Figure 3.29 shows the graph as found in Example 4 of Section 2.10. Note that a critical point occurs at $x = 1$ with $f'(1) = 0$. The tangent line to the graph of f at $(1, 3)$ is horizontal so it has no x-intercept. Thus the choice $x_0 = 1$ will not permit us to compute any further iterates.

The graph shows that a reasonable initial approximation is $x_0 = 0$. With this choice of x_0, Newton's Method gives

$$x_1 = 0.167, \quad x_2 = 0.193, \quad \text{and} \quad x_3 = 0.193.$$

The intercept is approximately 0.193. ▶

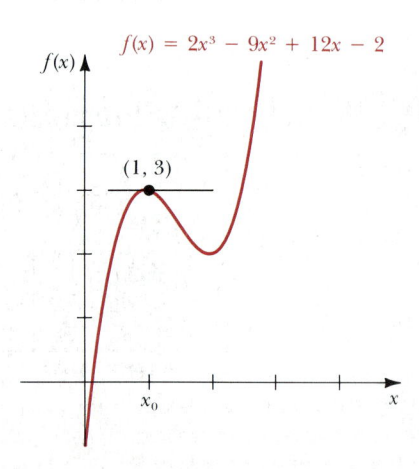

Figure 3.29
When $x_0 = 1$, $f'(x_0) = 0$.

Even a poor choice for x_0 often leads eventually to an accurate solution. For example, the choices $x_0 = 3$ and $x_0 = 1.5$ in Example 3 will produce sufficiently accurate solutions, although they require 9 and 18 iterations, respectively, to do so.

As these examples show, Newton's Method often gives very accurate approximations in just a few iterations. Since this method uses knowledge about both the function and its derivative, it generally obtains a sufficiently accurate result more rapidly than does the Bisection Method, which considers only function values. This speed, however, is obtained at a price. The Bisection Method has easily verifiable conditions under which one can be sure that a sufficiently accurate result will be found. No easily verifiable conditions are available for Newton's Method. The best advice for use of Newton's Method is to try the procedure for a reasonable initial approximation. If there are going to be difficulties applying the method with this initial approximation, it is generally obvious within a few iterations.

One disadvantage of Newton's Method concerns the need to evaluate the derivative of the function at each step in the process. In Exercise 26, we describe a method called the **Secant Method,** which has most of the advantages of Newton's Method but overcomes this weakness.

▶ EXERCISE SET 3.6

For these exercises, assume that if two consecutive approximations agree to a certain number of decimal places, the latter approximation is accurate to at least that number of places.

In Exercises 1–10, approximate to four decimal places the value of x lying in the given interval for which $f(x) = 0$.

1. $f(x) = x^3 - 2x^2 - 5$, $[2, 3]$
2. $f(x) = x^3 - x - 1$, $[1, 2]$
3. $f(x) = x^3 - 9x^2 + 12$, $[-2, 0]$
4. $f(x) = x^3 - 9x^2 + 12$, $[0, 2]$
5. $f(x) = x^3 - 9x^2 + 12$, $[8, 9]$
6. $f(x) = x^3 + 3x^2 - 1$, $[-4, -1]$
7. $f(x) = \cos x - x$, $[0, \pi/2]$
8. $f(x) = x - 0.2 \sin x - 0.8$, $[0, \pi/2]$
9. $f(x) = x^4 - 2x^3 - 5x^2 + 12x - 5$, $[0, 1]$
10. $f(x) = 2x^4 - 8x^3 + 8x^2 - 1$, $[1, 2]$
11. Consider the equation $0 = x^m - a$, where a is a positive constant. The only positive solution to this equation is $\sqrt[m]{a}$, the positive mth root of a. Show that Newton's Method applied to $f(x) = x^m - a$ produces the approximations

$$x_{n+1} = \frac{(m-1)x_n^m + a}{mx_n^{m-1}}.$$

12. Use the result of Exercise 11 to approximate each of the following to within five decimal places.
    a) $\sqrt{2}$, $x_0 = 1.5$
    b) $\sqrt{5}$, $x_0 = 2$
    c) $\sqrt[3]{25}$, $x_0 = 3$
    d) $\sqrt[5]{1000}$, $x_0 = 4$

13. Consider the equation $0 = 1/x - a$, where a is a nonzero constant. The only solution to this equation is $1/a$, the reciprocal of a. Show that Newton's Method applied to $f(x) = 1/x - a$ produces the approximations

$$x_{n+1} = 2x_n - ax_n^2.$$

14. Use the result of Exercise 13 to approximate each of the following to within five decimal places.
    a) $1/3$, $x_0 = 0.3$
    b) $1/7$, $x_0 = 0.1$
    c) $1/9$, $x_0 = 0.1$
    d) $-2/3$, $x_0 = -0.6$

15. Use Newton's Method to approximate to within three decimal places the x-intercepts of $f(x) = x^4 - 4x^3 - 2x^2 + 12x + 1$, whose graph is given in Example 3 of Section 2.10.

16. Use Newton's Method to approximate to within three decimal places both x-intercepts of $f(x) = x^4 + 2x^3 - 1$, whose graph is given in Example 5 of Section 2.10.

17. Find an approximation, accurate to within six decimal places, to the nonzero intercept of the graphs of $y = x^2$ and $y = \sin x$.

18. Find approximations, accurate to within six decimal places, to the intercepts of the graphs of $y = x^2$ and $y = \cos x$.

19. Approximate to within four decimal places the smallest positive intercept of the graphs of $y = x$ and $y = \tan x$.

20. Repeat Exercise 19 finding instead the fourth smallest positive intercept.

21. Newton's Method applied to $f(x) = x + \sqrt{x} - 1 = 0$ can be expressed as
$$x_{n+1} = x_n - \frac{f(x_n)}{f'(x_n)} = x_n - \frac{2x_n\sqrt{x_n} + 2x_n - 2\sqrt{x_n}}{2\sqrt{x_n} + 1}.$$
Apply Newton's Method to this function on the interval $[0, 1]$ with (a) $x_0 = 1$ and (b) $x_0 = 0$. Explain any apparent difficulties.

22. Use Newton's Method to find an approximation, accurate to three decimal places, to the point on the graph of $y = x^2$ that is closest to $(1, 0)$.

23. Use Newton's Method to find an approximation, accurate to three decimal places, to the point on the graph of $y = 1/x$ that is closest to the point $(1, 0)$.

24. The sum of two numbers is 20. If each number is added to its square root, the product of the two sums equals 155.55. Determine the two numbers to within four decimal places.

25. Exercise 14 of Exercise Set 3.2 describes the process involved in constructing a can with a fixed volume in the shape of a right circular cylinder. The actual construction of such a can requires that the circular top and bottom be cut with a radius greater than the radius of the can so that the excess material can be used to form a seal with the side of the can. Assume that circles of diameter $2r + 0.5$ cm are required to form the top and bottom of a can with radius r (see the figure). Approximate, to three decimal places, the minimum amount of material needed to construct such a can containing 1500 cm³.

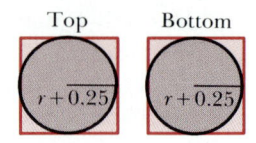

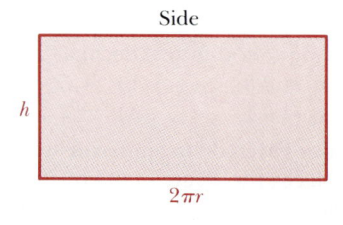

26. The *Secant Method* approximates the solution of an equation of the form $f(x) = 0$ without using the derivative of f. The method proceeds by assuming two initial approximations x_0 and x_1 to the solution of the equation and computing a new approximation x_2 by determining where the line joining $(x_0, f(x_0))$ and $(x_1, f(x_1))$ crosses the x-axis. The points $(x_1, f(x_1))$ and $(x_2, f(x_2))$ are then used to construct another line that crosses the x-axis at another approximation x_3, and so on (see the figure).

   a) Show that for $n \geq 2$,
   $$x_n = x_{n-1} - \frac{f(x_{n-1})(x_{n-1} - x_{n-2})}{f(x_{n-1}) - f(x_{n-2})}.$$

   b) Show that the Secant Method is a modification of Newton's Method that is produced by replacing $f'(x_{n-1})$ with an approximation to that value.

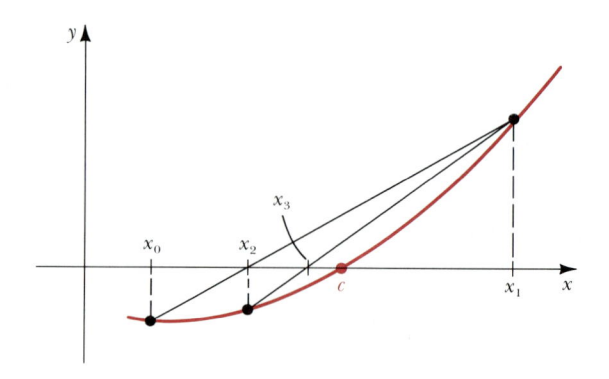

27–36. Use the Secant Method to find the approximations requested in Exercises 1–10.

3.7
Applications to Business and Economics

Determining the rate at which certain quantities change is a subject of primary interest to financial planners. Company executives need to know how a price change relates to the change in the demand, how a demand change relates to the change in revenue and profit, and so on. In economics, the adjective "marginal" is used to express the instantaneous rate of change of a quantity. The reason for using the term marginal is seen more easily if you consider it in light of its synonym, "extra." The marginal cost, for example, is the rate at which the cost changes relative to the change in the number of units produced. As such, it provides an estimate of the cost of producing one extra, or marginal, unit of the item.

Before considering the application of the derivative to economic problems, we need the definitions of some common economic functions.

> **Cost,** or **total cost, function** (denoted C or, by economists, TC) describes the cost $C(x)$ of producing x units of an item.
>
> **Average cost function** (denoted c or AC) describes the cost $c(x)$ of producing a single unit of an item if x units are produced: $c(x) = C(x)/x$.
>
> **Demand function** (denoted p) describes the price $p(x)$ per unit of an item when x units are demanded, or sold. (This function is also called the **price function.**)
>
> **Revenue,** or **total revenue, function** (denoted R or TR) describes the revenue $R(x)$ produced when x units of an item are sold: $R(x) = xp(x)$.
>
> **Profit,** or **total profit, function** (denoted P or TP) describes the profit $P(x)$ realized from the sale of x units of an item: $P(x) = R(x) - C(x)$.

To see how the marginal concepts are developed, consider the case of marginal cost. If $C(x)$ is the cost of producing x units of an item, then the cost of producing one additional item is $\Delta C = C(x+1) - C(x)$. In this case, $\Delta x = dx = 1$. For large values of x it is reasonable to approximate ΔC by the differential, dC; thus,

$$C'(x) = \frac{dC}{dx} \approx \frac{C(x+1) - C(x)}{1} = C(x+1) - C(x).$$

The marginal cost at x is consequently defined to be $C'(x)$. In fact, associated with each differentiable economic function is a marginal function.

> **Marginal cost function** (denoted C' or, by economists, MC) describes the rate of change of the cost relative to the change in the number of units produced.

Marginal average cost function (denoted c' or MAC) describes the rate of change of the cost of a single unit of an item relative to the change in the number of units produced.

Marginal demand function (denoted p' or Mp) describes the rate of change of the demand for an item relative to the change in the number of units produced.

Marginal revenue function (denoted R' or MR) describes the rate of change of the revenue produced relative to the number of units sold.

Marginal profit function (denoted P' or MP) describes the rate of change of the profit produced relative to the number of units sold.

Example 1

Maple Leaf Moccasins, Ltd., can produce between 100 and 1000 pairs of moccasins per day. The cost of producing x pairs of moccasins, $100 \leq x \leq 1000$, is given by

$$C(x) = 295 + 3.28x + 0.003x^2 \text{ dollars.}$$

The demand function for the company is

$$p(x) = 7.47 + \frac{321}{x} \text{ dollars per pair.}$$

a) Describe the marginal cost, marginal average cost, marginal demand, marginal revenue, and marginal profit functions.
b) Determine the number of pairs that should be produced if the average cost per unit is to be a minimum.
c) Determine the number of pairs that should be produced in order to maximize the profit.

Solution

a) The average cost is given by

$$c(x) = \frac{C(x)}{x} = \frac{295}{x} + 3.28 + 0.003x,$$

the revenue by

$$R(x) = xp(x) = 7.47x + 321,$$

and the profit by

$$P(x) = R(x) - C(x) = 4.19x + 26 - 0.003x^2.$$

The marginal functions are as follows:

Marginal cost: $C'(x) = 3.28 + 0.006x$;
Marginal average cost: $c'(x) = -295/x^2 + 0.003$;
Marginal demand: $p'(x) = -321/x^2$;
Marginal revenue: $R'(x) = 7.47$;
Marginal profit: $P'(x) = 4.19 - 0.006x$.

b) The minimum average cost can occur only at the endpoints of the domain of c (when $x = 100$ or $x = 1000$) or when $c'(x) = 0$. If $c'(x) = 0$, then

$$\frac{295}{x^2} = 0.003, \quad \text{so} \quad x = \sqrt{\frac{295}{0.003}} \approx 313.6.$$

The integers closest to the critical point are 313 and 314. Since $c(313) = c(314) = 5.16$, $c(100) = 6.53$, and $c(1000) = 6.58$, either 313 or 314 pairs should be produced in order to obtain the minimum average cost.

c) The maximum profit can occur only when $x = 100$, $x = 1000$, or $P'(x) = 0$. If $P'(x) = 0$, then

$$0.006x = 4.19, \quad \text{so} \quad x = \frac{4.19}{0.006} \approx 698.3.$$

At the integers closest to the critical point, $P(698) = P(699) = 1489$. Since $P(100) = 415$ and $P(1000) = 1216$, the production of either 698 or 699 pairs produces the maximum profit. ▶

To see the significance of the marginal functions described in Example 1, suppose that $x = 200$. Then the 201st pair of moccasins will approximately:

i) have a production cost of $C'(200) = \$4.48$;
ii) lower the average production cost per pair by $c'(200) = -\$0.004375$;
iii) decrease the selling price per pair by $p'(200) = -\$0.008025$;
iv) produce a revenue of $R'(200) = \$7.47$; and
v) generate a profit of $P'(200) = \$2.99$.

Business executives frequently use charts and graphs to determine trends and as a basis for business decisions. Shown in Figure 3.30 are the graphs of the average cost and marginal cost functions from Example 1. Note that in Example 1 the minimum average cost of production occurs at the intersection of the graphs for the average cost and marginal cost. To see that this is true in general, note that since

$$c(x) = \frac{C(x)}{x}, \quad \text{we have} \quad c'(x) = \frac{xC'(x) - C(x)}{x^2}.$$

Thus $c'(x)$ is zero precisely when $xC'(x) - C(x) = 0$, that is, when

$$C'(x) = \frac{C(x)}{x} = c(x).$$

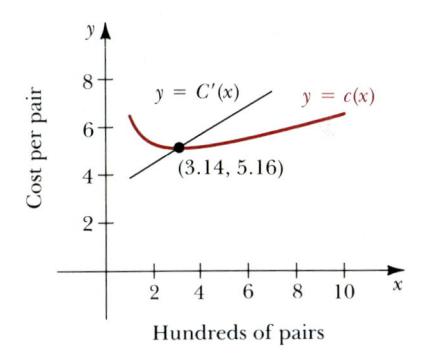

Figure 3.30

Consequently, a relative minimum for $c(x)$ can occur only when the marginal cost and average cost functions coincide.

In a practical business situation, it may be unrealistic to determine explicit cost functions. However, as production increases it is relatively easy to determine values of $c(x)$ and estimates for $C'(x)$ on the basis of the approximation

$$C'(x) \approx C(x+1) - C(x).$$

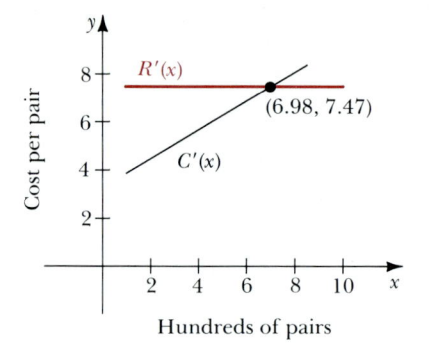

Figure 3.31

In general, when $c(x)$ is decreasing, $C'(x)$ is less than $c(x)$ and is increasing. When $c(x)$ and $C'(x)$ agree, production should increase no further if production levels are based on minimum average cost of production.

In a similar manner, the number of units that should be produced and sold in order to maximize profit can be estimated since

$$P'(x) = R'(x) - C'(x).$$

If $P'(x) = 0$, then $R'(x) = C'(x)$, that is, the profit function P can have a relative maximum only if the marginal revenue is equal to the marginal cost. Both these quantities can be estimated from sales and production data by

$$R'(x) \approx R(x+1) - R(x) \quad \text{and} \quad C'(x) \approx C(x+1) - C(x).$$

The graphs of the marginal cost and marginal revenue functions from Example 1 intersect when $x = 698$ pairs, which is the production that yields a maximum profit (see Figure 3.31).

The following example illustrates how a relatively minor change in a firm's cost function with no accompanying change in the demand function can influence the profitability of the company.

Example 2

Suppose that the government levies an additional tax of 6% on the cost of producing the moccasins in Example 1 and that the demand function for the firm remains

$$p(x) = 7.47 + \frac{321}{x} \text{ dollars per pair.}$$

a) Determine the number of pairs that should be produced if the average cost per unit is to be a minimum.

b) Determine the number of pairs that should be produced in order to maximize the profit.

Solution

a) The cost function after the tax has been added is 6% more than the cost function given in Example 1. Thus,

$$C(x) = 1.06(295 + 3.28x + 0.003x^2)$$

and

$$c(x) = 1.06\left(\frac{295}{x} + 3.28 + 0.003x\right).$$

Consequently,

$$c'(x) = 1.06\left(\frac{-295}{x^2} + 0.003\right),$$

and the only critical point is again $x = \sqrt{295/0.003} \approx 313.6$. Since $c(313) = c(314) = 5.47$, $c(100) = 6.92$, and $c(1000) = 6.97$, the minimum average cost, although increased by 6%, still occurs when either 313 or 314 pairs are produced.

b) The profit function is now

$$P(x) = xp(x) - C(x)$$
$$= 7.47x + 321 - 1.06(295 + 3.28x + 0.003x^2)$$
$$= 3.9932x + 8.30 - 0.00318x^2.$$

Thus,

$$P'(x) = 3.9932 - 0.00636x,$$

and the only critical point is

$$x = \frac{3.9932}{0.00636} \approx 627.86.$$

Because $P(627) = P(628) \approx 1262$, $P(100) \approx 376$, and $P(1000) \approx 822$, the maximum profit occurs when either 627 or 628 pairs are produced.

Looking at the maximum profit of $1489 found in Example 1, we see that the 6% increase in the cost of production decreased the maximum profit by $227, or over 15%. The 6% increase in cost also caused a decrease in production of over 10%, if the maximum profit is to be realized.

This decrease in profit and production could easily have such an adverse effect on the economy that the government might receive less total revenue than it did before the additional tax was assessed. ▶

It is generally assumed in a free market economy that an increase in the price of an object is accompanied by a decrease in its demand, whereas a decrease in price leads to an increase in demand. If the unit price is increased, more revenue per unit will result; but if the price increase leads to a significant decrease in demand, the total revenue can decrease.

To make sound pricing decisions, one needs a quantitative method to measure the public's responsiveness to a price change. Such a measure is given by the ratio of the percentage of change in demand to the percentage of change in price. The ratio when a units are sold is

(3.16)
$$-\frac{\left(\dfrac{x-a}{a}\right)}{\left(\dfrac{p(x)-p(a)}{p(a)}\right)} = -\frac{p(a)}{a} \cdot \frac{1}{\left(\dfrac{p(x)-p(a)}{x-a}\right)}$$

The minus sign is added to make the ratio positive: Since p is a decreasing function of x, the quotient $[p(x) - p(a)]/(x - a)$ is negative.

When x is close to a, $(p(x) - p(a))/(x - a)$ is approximately $p'(a)$, so the ratio in Equation (3.16) is approximately $-p(a)/ap'(a)$.

The **price elasticity** of demand is described by a function E, where

(3.17)
$$E(x) = \frac{-p(x)}{xp'(x)}.$$

This function measures the change in demand relative to a change in the selling price when x units are sold.

Elasticity is used to measure the responsiveness of total revenue to price changes. Since $R(x) = xp(x)$,

$$R'(x) = p(x) + xp'(x) = p(x)\left(1 + \frac{xp'(x)}{p(x)}\right) = p(x)\left(1 - \frac{1}{E(x)}\right).$$

i) When $E(x) > 1$, $1/E(x) < 1$ and $R'(x) > 0$, so R is increasing. In this case, the demand function is said to be **elastic,** indicating that an increase in x produces an increase in total revenue. A decrease in price results in an increase in x, so if the demand function is elastic, a decrease in price will increase the total revenue. The larger the value of $E(x)$, the more dramatic the change.

ii) When $E(x) < 1$, $1/E(x) > 1$ and $R'(x) < 0$, so R is decreasing. In this case, the demand function is said to be **inelastic,** indicating that a decrease in price, and consequently an increase in x, are accompanied by a decrease in total revenue.

iii) When $E(x) = 1$, the demand function is said to have **unit elasticity,** that is, changes in price do not influence total revenue.

Example 3

Suppose that a demand function for food processors is

$$p(x) = 200 - 0.02x \text{ dollars per unit.}$$

a) Will a rise in price increase or decrease the revenue received if 1000 food processors are currently sold?

b) At what level of sales will it not make a difference whether the price is increased or decreased?

Solution

a) We have

$$E(x) = \frac{-p(x)}{xp'(x)} = \frac{-(200 - 0.02x)}{x(-0.02)} = \frac{200 - 0.02x}{0.02x};$$

so,

$$E(1000) = \frac{200 - 20}{20} = 9.$$

Since $E(1000) > 1$, there will be a decrease in revenue if prices rise.

b) To determine the level of sales at which a price change will not influence the total revenue, we determine the number of units x for which $E(x) = 1$:

$$E(x) = \frac{200 - 0.02x}{0.02x} = 1 \quad \text{if } 0.04x = 200, \text{ that is, if } x = 5000.$$

If fewer than 5000 units are currently sold, an increase in price produces a decrease in total revenue. An increase in revenue is produced if more than 5000 units are currently sold and prices rise. There is unit elasticity at $x = 5000$. ▶

It is important to realize that elasticity measures the relative change in total revenue with respect to the change in price. Elasticity does not involve the cost function, so it does not measure the change in profit with respect to the change in price. Only when profit can be assumed to be a fixed proportion of the revenue can elasticity be used to determine pricing strategies for maximizing profit. (This was not the case, for example, in the problem considered in Examples 5 and 6 of Section 3.2.)

▶ **EXERCISE SET 3.7**

For each of the cost functions given in Exercises 1–4 find (a) the average cost function, (b) the marginal cost function, (c) the marginal average cost function, (d) the average cost per unit if 1000 units of an item are produced, and (e) the cost of the 1001st unit produced.

1. $C(x) = 1500 + 2x - 0.0003x^2$
2. $C(x) = 875 + 2.6x + 0.01x^2$
3. $C(x) = 15500 + 77x - 0.00001x^3$
4. $C(x) = 38 + 4.75x - 0.01\sqrt{x}$

In Exercises 5–8, $R(x)$ denotes the revenue realized from the sale of x units of an item. Find the marginal revenue function.

5. $R(x) = 3.49 - 0.0002x$
6. $R(x) = 43.52 - 0.01x + 0.00002x^2$
7. $x^2 R(x) - 500x + 300R(x) - 1900 = 0$
8. $x^2 R(x) + 10R(x) + 0.01\,[R(x)]^2 - 25 = 0$

Trasho Manufacturing, Inc., figures that the total cost of producing x plastic wastebaskets per week is

$$C(x) = 1700 + 0.76x + 0.0001x^2 \text{ dollars}$$

and that the total revenue from their sale is

$$R(x) = 2.85x - 0.00008x^2 \text{ dollars.}$$

9. Determine the marginal cost, marginal revenue, and marginal profit functions for the Trasho Corporation.
10. Determine the number of wastebaskets that Trasho should produce if the average cost per basket is to be minimized.
11. a) How many wastebaskets per week should be produced if Trasho wishes to maximize profit?
    b) How much profit is made if this number is produced?
12. Sketch the graphs of the marginal cost and marginal revenue functions, and show that they intersect when the maximum profit is produced.

The employees of Trasho Manufacturing are given a raise that adds $0.09x$ to the weekly cost function. This raise is passed on to the distributor, which changes the revenue function to

$$R(x) = 2.94x - 0.00008x^2 \text{ dollars.}$$

13. How does this change modify the maximum profit the company can realize and the number of wastebaskets that should be produced in order to maximize the profit?
14. Suppose that the total wages paid to all Trasho employees is proportional to the maximum profit that can be realized. Have the employees collectively benefited from the wage increase?

Suppose that instead of adding $0.09x$ to the weekly Trasho cost function, the management determines that the amount of the raise should be incorporated into the fixed cost. The new cost function in this case is

$$C(x) = 2400 + 0.76x + 0.0001x^2 \text{ dollars.}$$

The revenue function remains

$$R(x) = 2.85x - 0.00008x^2 \text{ dollars.}$$

15. What is the maximum profit that can now be realized and how many wastebaskets should be produced? Compare this result with those in Exercises 11 and 13.

16. Suppose that the total wages paid to all Trasho employees is proportional to the maximum profit that can be realized. Should the employees prefer this cost-function approach to the one described before Exercise 13?

17. The cost of producing x genuine birchbark canoes is
$$C(x) = 300x + 500 \text{ dollars.}$$
Each canoe is sold for $1000. Find the revenue function, profit function, and marginal profit function.

18. The Dull Calculator Company estimates that the cost of producing x calculators is
$$C(x) = 2000 + 10x + \frac{10{,}000}{x} \text{ dollars.}$$
   a) Find the marginal cost function.
   b) Find the average cost function.
   c) Find the marginal cost when $x = 1000$.
   d) Find the approximate cost of producing the 1001st calculator.
   e) Find the number of calculators that should be produced in order to minimize the average cost.
   f) Sketch the graphs of the marginal cost and average cost functions.

19. Let $p(x) = (600 - x^2)/25$ be the selling price when x units of an item are sold.
   a) Find the price elasticity of demand.
   b) Find the value of x that gives unit elasticity.
   c) Will a rise in price increase or decrease revenue if 10 units are sold? if 20 units are sold?

20. Consider a demand function
$$p(x) = \frac{36}{x - 36} + 12,$$
where $p(x)$ is the selling price when x units of an item are sold.
   a) Find the price elasticity of demand.
   b) Will a rise in price increase or decrease revenue if 40 units are in demand?

21. The demand function for a certain item is
$$p(x) = 80 - 0.01x \text{ dollars per unit.}$$
Find the number of units at which a price change will not influence total revenue.

▶ IMPORTANT TERMS AND RESULTS

CONCEPT	PAGE
Velocity	183
Speed	183
Acceleration	183
Motion equations	186
Applied extrema	191
Related rates	204
Differentials	215
Indeterminate forms	221
l'Hôpital's Rule	223
Newton's Method	231
Marginal functions	237
Price elasticity	241

REVIEW EXERCISES

Consider the functions described in Exercises 1–4.

  a) Find the slope of the tangent line to the graph of f at any point $(x, f(x))$.
  b) If $f(x)$ describes the distance of an object from a fixed point at the end of time x, find the instantaneous velocity of the object at time x.
  c) If $f(x)$ gives the profit realized from the sale of x units of an item, find the marginal profit function.
  d) If $y = f(x)$, find dy.

1. $f(x) = x^3 - 2x^2 + 1$
2. $f(x) = x + \sqrt{x}$
3. $f(x) = \sin x + x$
4. $f(x) = 2\sqrt{x} + \sec x$

In Exercises 5–8, (a) find dy, (b) find Δy, and (c) approximate $f(x + \Delta x)$, assuming $y = f(x)$.

5. $y = x^2 + 2x$, $x = 3$, $\Delta x = 0.3$
6. $y = x^3 - 4$, $x = -1$, $\Delta x = 0.2$
7. $y = \tan x$, $x = \pi/4$, $\Delta x = 0.01$
8. $y = \cot x$, $x = \pi/4$, $\Delta x = -0.01$

In Exercises 9–18, use l'Hôpital's Rule, if applicable, to determine the limits.

9. $\lim\limits_{x \to 1} \dfrac{x^3 - 3x^2 + 3x - 1}{x^2 - 1}$

10. $\lim\limits_{x \to -1} \dfrac{x^3 - 3x^2 + 3x - 1}{x^2 - 1}$

11. $\lim\limits_{x \to 3} \dfrac{\sqrt{x+1} - 2}{x^2 - 9}$

12. $\lim\limits_{x \to 2} \dfrac{3 - \sqrt{x+7}}{x^2 - 4x + 4}$

13. $\lim\limits_{x \to 0^+} \dfrac{\sin x}{\sqrt{x}}$

14. $\lim\limits_{x \to 0} \dfrac{\sin x - x}{x^3}$

15. $\lim\limits_{x \to \frac{\pi}{2}} \dfrac{\cot x}{x - \dfrac{\pi}{2}}$

16. $\lim\limits_{x \to 0} \dfrac{x \cos x}{\sin x}$

17. $\lim\limits_{x \to 0^+} \left(\dfrac{1}{x} - \dfrac{1}{\tan x} \right)$

18. $\lim\limits_{x \to 0^-} \left(\dfrac{1}{\sin x} - \dfrac{1}{\tan x} \right)$

In Exercises 19–24, find an approximation, accurate to 10^{-3}, to a value of x in the given interval for which $f(x) = 0$ by using (a) Newton's Method and (b) the Secant Method.

19. $f(x) = x^3 - 2x - 1$, $[1, 2]$
20. $f(x) = 2x^3 - x^2 - 2$, $[1, 2]$
21. $f(x) = x^4 - x^3 - 1$, $[0, 2]$
22. $f(x) = x^4 - 3x^2 - 3$, $[-3, -1]$
23. $f(x) = x^2 - \cos x$, $[0, \pi/2]$
24. $f(x) = x^3 - 0.8 - 0.2 \sin x$, $[0, \pi/2]$

25. The distance of an object from a specified point at the end of time t is given by $s(t) = 4t - t^2$.
  a) Find the average velocity from $t = 0$ to $t = 2$.
  b) Find the instantaneous velocity when $t = 2$.
  c) Find the acceleration when $t = 2$.

26. The height of an object thrown from the top of a building at time $t = 0$ is given by $s(t) = 200 - 20t - 16t^2$.
  a) Find the average velocity from $t = 0$ to $t = 2$.
  b) Find the instantaneous velocity when $t = 1$ and when $t = 2$.
  c) Find the acceleration when $t = 1$ and when $t = 2$.

27. A rock is dropped from the top of a 10-story, 120-ft–high building. What is the velocity of the rock when it hits the ground?

28. A rock is thrown directly upward at 100 ft/sec from the top of the building described in Exercise 27. What is the velocity of this rock when it hits the ground?

29. A ball is thrown upward from a point 3 ft above the ground with an initial velocity of 88 ft/sec. How long will the ball remain in the air?

30. What is the minimum initial velocity needed to throw a ball into the air if the ball must remain in the air for 4 sec before returning to earth?

31. A child on the roof of a building throws a ball directly downward to another child on the ground. The second child immediately throws the ball back up to the first. If the building is 36 ft high and the children cannot throw the ball faster than 48 ft/sec, what is the minimum time it takes for the ball to return to the roof?

32. A car traveling at 90 mph decelerates to a stop in 6 sec. Assuming that the deceleration rate is constant, describe the velocity and distance functions for the motion.

33. The winner of the 1987 NHRA Winternationals was the perennial champion Don Garlits. His top fuel rag-

dragster had a time of 5.29 sec for the quarter mile from a standing start and a final speed of 270 mph. Assume that the acceleration for the run was constant. Determine (a) the acceleration and (b) the initial velocity.

34. On the planet Tauranious a wrench dropped 200 ft from the top of a space platform takes 8 sec to reach the surface of the planet. Can a worker on the surface throw the wrench back up to the platform, assuming a maximum throwing speed of 50 ft/sec?

35. Determine the point on the line with equation $2x + y = 3$ that is closest to the point (2, 1).

36. Find the shortest distance from the point (0, 5) to the parabola $4y = x^2$.

37. Find the positive number with the property that the sum of the number and twice the square of its reciprocal is a minimum.

38. A can in the shape of a right circular cylinder with a bottom but no top contains a volume of 900 cm³. Find the dimensions of the can that minimize the amount of material needed for construction.

39. Find the dimensions of the largest rectangle that can be placed inside a right triangle with sides of 5, 12, and 13 in., if one vertex of the rectangle is at the right angle and another is on the hypotenuse.

40. A rectangle is placed inside an equilateral triangle with its base on the base of the triangle. Show that the maximum area in the rectangle is one half the area of the triangle.

41. A closed rectangular box with a square base is constructed with a top costing twice as much as the sides and bottom. The box is to contain 96 in³. What should be the dimensions of the box in order to minimize the building cost?

42. An open water tank in the shape of an inverted right circular cone is designed to hold 576 ft³. Find the dimensions of the cone that has the least lateral surface area.

43. Find the dimensions of the right circular cylinder of greatest volume that can be placed inside a right circular cone with a radius of 6 in. and a height of 12 in.

44. Find the dimensions of the right circular cone of least volume that encloses a right circular cylinder with a radius of 6 in. and a height of 12 in.

45. A *Norman window* is a window that consists of a rectangle surmounted by a semicircle. (These windows were commonly constructed in English architecture for about 100 years after the Norman conquest of Britain, circa 1066–1190). How can the area of the window be maximized if the perimeter of the window is to be 25 ft?

46. Sand is poured onto a conical pile whose radius and height are always equal, although both increase with time. The height of the pile is increasing at the rate of 3 in./sec. How fast is the volume of the pile increasing when the height of the pile is 10 ft?

47. Two trees 21 ft apart are to be reinforced by a nylon rope connected to the trunks and tied at ground level to a stake in the ground between the trees. If the rope is tied 15 ft above the ground on one tree and 20 ft above the ground on the other, find the location of the stake that will minimize the amount of rope used.

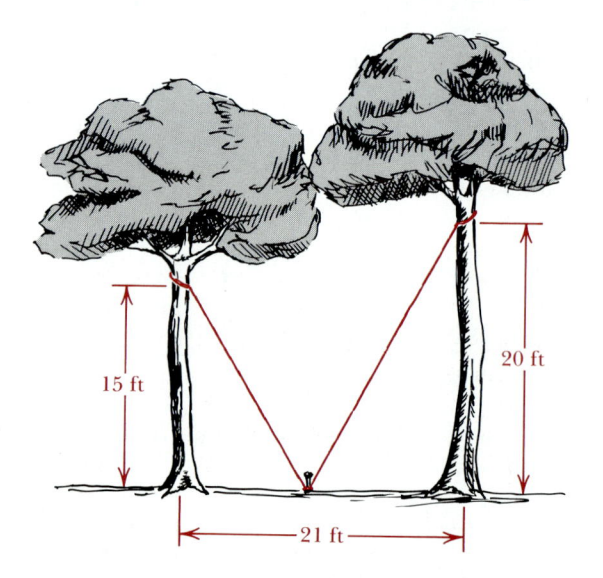

Figure for Example 45.

48. A ranger in a rubber raft 2 mi from a straight sandy beach must travel to a point 6 mi from the point on the beach closest to the raft. The ranger can paddle the raft at 4 mph and can run on the sandy beach at 6 mph. How close to the destination point should the raft land in order to minimize the time required to reach this point?

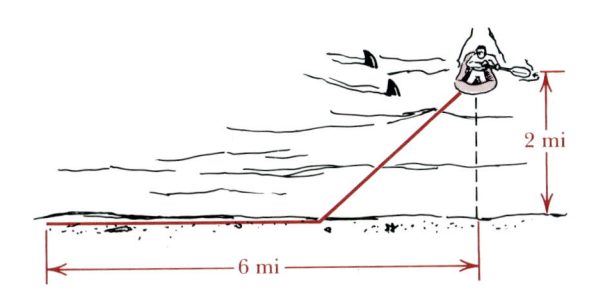

49. An angler has a fish at the end of a line. The line is reeled in at the rate of 2 ft/sec from a bridge 30 ft above the water. At what rate is the fish moving through the water when the length of the line is 50 ft?

50. A car traveling 60 mph northward on a straight road crosses a railroad track perpendicular to the road. A train going 80 mph directly eastward crosses the road 15 min later. At what rate are the car and train separating 30 min after the train has crossed the road?

51. Water flows at the rate of 2 ft^3/min into a tank in the shape of an inverted right circular cone of altitude 6 ft and radius 2 ft. At what rate is the surface of the water rising when the tank is half full?

52. A baseball diamond is a 90-ft square. A player hits a ball along the thirdbase line at 100 ft/sec and runs to first base at 25 ft/sec.
  a) At what rate is the distance between the ball and first base changing when the ball is halfway to third base?
  b) At what rate is the distance between the ball and the player changing when the ball is halfway to third base?

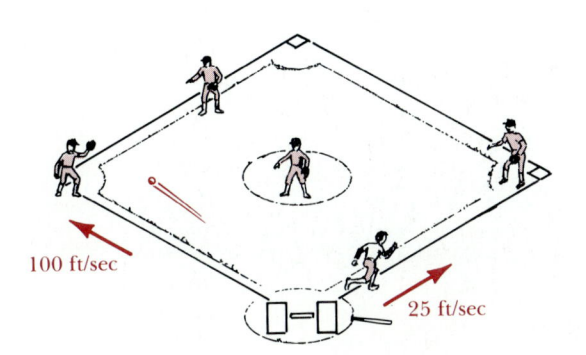

53. When a circular metal plate is heated, its radius increases at the rate of 1 mm/sec. At what rate is the area of the plate increasing when the radius is 4 cm?

54. A man 6 ft tall walks toward a mercury vapor light on the top of a 24-ft pole.
  a) How fast is the tip of his shadow moving toward the pole when he is 36 ft from the pole and walking toward it at the rate of 4 ft/sec?
  b) How fast is the length of his shadow decreasing at that time?

55. The edges of an equilateral triangle are increasing at the rate of 1 mm/sec. How fast is the area of the triangle increasing when the area is 14 mm^2?

56. A circle sits inside the triangle considered in Exercise 55 and increases so that points on its surface always touch the sides of the triangle. How fast is the perimeter of the circle increasing when the area of the triangle is 14 mm^2?

57. Water is pumped into a bucket at the rate of 3 L/min. The bucket is in the shape of a frustum of a cone with bottom diameter 20 cm, top diameter 30 cm, and height 40 cm. How fast is the level of water rising when the height of the water in the bucket is 10 cm?

58. A spherical water tank has many coats of old paint and a radius, with the paint, of exactly 3 ft. The paint is ground off and the tank repainted with a net decrease of 0.2 in. in the radius. Use differentials to approximate the net volume of paint removed.

59. A spherical balloon is inflated with air. Use differentials to approximate the change in the volume and the surface area of the balloon as the radius increases from 5 to 5.5 in.

60. A USSR "ALFA" class attack submarine is traveling due south at 37 knots. At 20:00 GMT (Greenwich Mean Time) the submarine is at a point 200 nautical miles northwest of the Pearl Harbor Naval shipyard. At this time the US Navy aircraft carrier CARL VINSON leaves the yard traveling 28 knots due west toward Guam.
  a) At what time are the ships closest to each other?
  b) What is the minimum distance between the ships?
  c) At what rate are the ships approaching each other at 21:00 GMT?

toward Vladivostok on a path 35° north of that of the airliner.
- **a)** At what time are the planes closest to each other?
- **b)** How fast are the planes separating at 11:45, when the Tu-144 begins its descent for the Alma-Ata airport?

62. A potter estimates that the cost of making x mugs is

$$C(x) = 500 + 3x + \frac{10}{x} \text{ dollars.}$$

Find:
- **a)** the marginal cost function;
- **b)** the average cost function;
- **c)** the approximate cost of producing the 101st mug.

63. Let $p(x) = (400 - x)/15$ be the selling price when x units are sold.
- **a)** Find the price elasticity of demand.
- **b)** Find the value of x that gives unit elasticity.
- **c)** Will a rise in price increase or decrease revenue if 150 units are sold? if 250 units are sold?

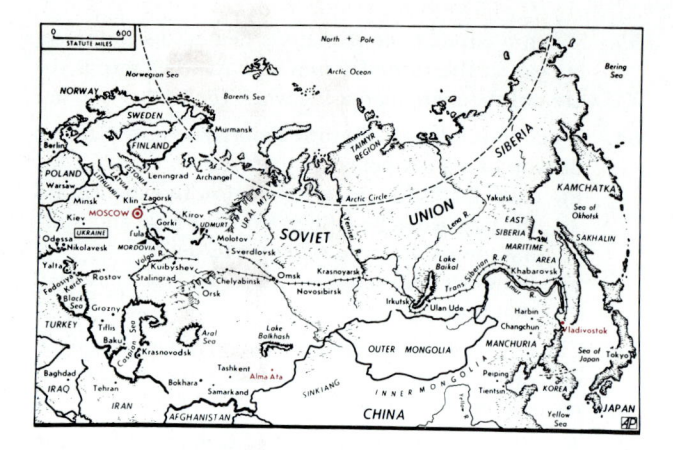

61. At 10:15 A.M. Moscow time, a Tu-144 supersonic aircraft carrying 134 passengers heads from Moscow's Domodedovo airport toward Alma-Ata, the capital of Kazakhstan, traveling at 1314 mph. Ten minutes later a MiG-25 flies over Moscow at 1730 mph heading

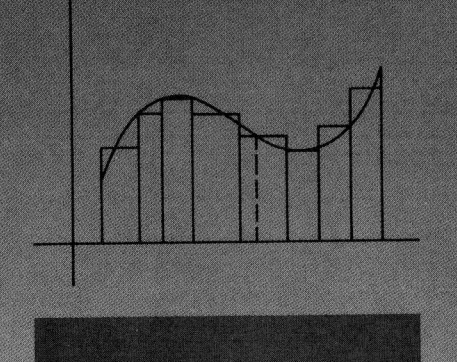

4

The Integral

The two fundamental concepts of calculus are the derivative and the integral. The derivative was introduced in Chapter 2 as a means of determining whether the graph of a function is increasing or decreasing and at what rate. There are many other applications of the derivative, however, in virtually every field of quantitative study. Some of these were considered in Chapter 3.

Chapter 4 discusses the second basic concept of calculus, which is the integral. The integral is introduced to solve the problem of determining the area of a region bounded by the *x*-axis and the graph of a continuous and nonnegative function defined on a closed interval. In Chapter 5 we will see that this is only one of the many applications of the integral.

Although the principles underlying both the derivative and the integral have been known since antiquity, a link between these seemingly distinct subjects was not discovered until the latter part of the seventeenth century. The Fundamental Theorem of Calculus, presented in Section 4.3, details this connection and is considered one of the most important results in the history of science. The events leading to its discovery are detailed in Section 4.8.

4.1
Area

There are a limited number of geometric figures for which the area can be determined easily. We can calculate the area of a rectangle, trapezoid, triangle, and such other geometric figures as a pentagon or octagon. In general, however, there is no algebraic method for calculating the area A of a region such as the one shown in Figure 4.1, where f is a continuous and nonnegative function defined on the closed interval $[a, b]$. In this section we show that an approximation and limiting process leads to a definition for the area A. In Section 4.2 this notion is extended to define the definite integral of a function.

A crude approximation A_1 to the area of the region shown in Figure 4.1 is the area of the rectangle whose base is the interval $[a, b]$ and whose height is the value of the function f at b, as shown in Figure 4.2.

$$A_1 = f(b)(b - a) = f(x_1)(b - a).$$

To find a better approximation to A, we divide the interval $[a, b]$ into two equal subintervals, as shown in Figure 4.3. The midpoint of $[a, b]$ is

$$x_1 = a + \frac{b - a}{2},$$

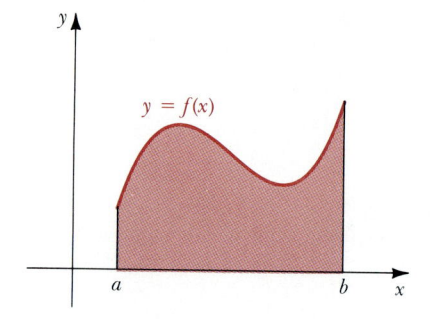

Figure 4.1

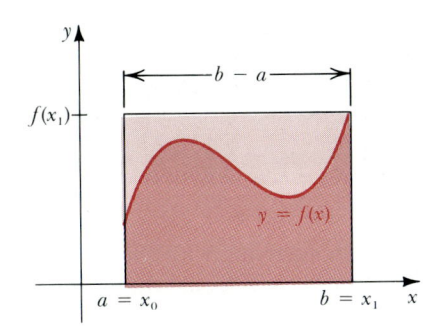

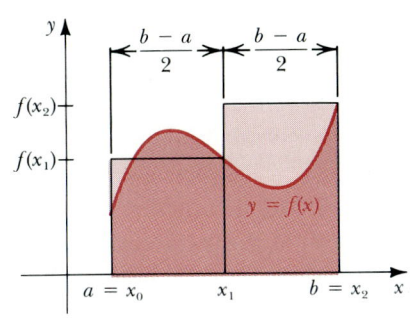

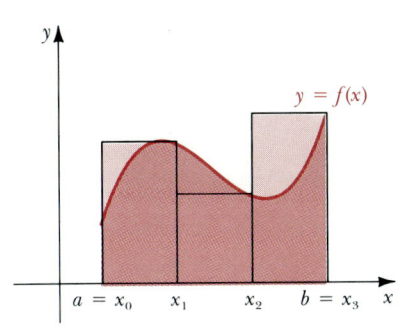

Figure 4.2

Figure 4.3

Figure 4.4

so the sum A_2 of the areas of the rectangles in Figure 4.3 is

$$A_2 = f(x_1)\left(\frac{b-a}{2}\right) + f(x_2)\left(\frac{b-a}{2}\right)$$
$$= [f(x_1) + f(x_2)]\left(\frac{b-a}{2}\right).$$

Dividing the interval $[a, b]$ into three equal subintervals, as shown in Figure 4.4, and letting

$$x_1 = a + (b-a)/3 \quad \text{and} \quad x_2 = a + 2(b-a)/3$$

gives the sum of the areas of the three rectangles in this figure:

$$A_3 = [f(x_1) + f(x_2) + f(x_3)]\left(\frac{b-a}{3}\right).$$

At the nth stage, $[a, b]$ is divided into n equal subintervals (see Figure 4.5) and an approximation to A is

(4.1) $\quad A_n = [f(x_1) + f(x_2) + \cdots + f(x_n)]\left(\dfrac{b-a}{n}\right),$

where

$$x_i = a + \frac{i(b-a)}{n}, \quad \text{for each } i = 1, 2, \ldots, n.$$

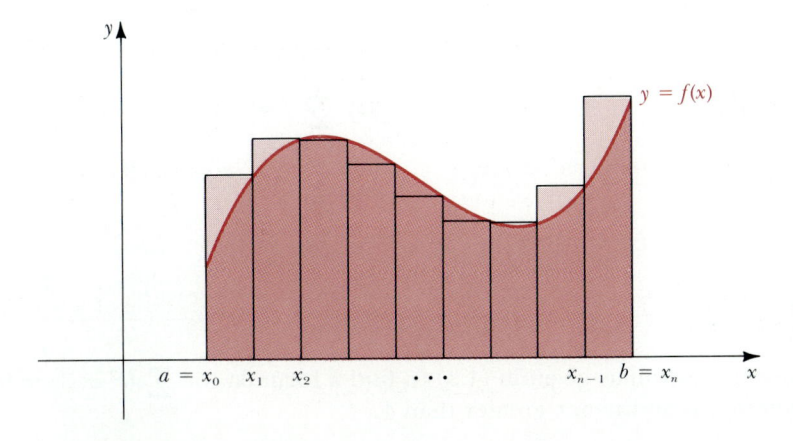

Figure 4.5

When n is large, the notation involved with this summation is cumbersome. To condense the notation, the **summation symbol** Σ (the Greek capital letter sigma) is used to represent sums of consecutive terms.

(4.2)
DEFINITION
The sum of the consecutive quantities a_i with i starting with the index p and ending with the index n, $p < n$, is denoted by $\Sigma_{i=p}^{n} a_i$, that is,

$$\sum_{i=p}^{n} a_i = a_p + a_{p+1} + a_{p+2} + \cdots + a_{n-1} + a_n.$$

Example 1

Find the value of each of the following.

a) $\displaystyle\sum_{i=1}^{4} (i^2 + 3i + 5)$ b) $\displaystyle\sum_{i=2}^{5} \sqrt{3i + 2}$

Solution

a) $\displaystyle\sum_{i=1}^{4} (i^2 + 3i + 5) = (1^2 + 3 \cdot 1 + 5) + (2^2 + 3 \cdot 2 + 5)$
$\phantom{\displaystyle\sum_{i=1}^{4} (i^2 + 3i + 5) =} + (3^2 + 3 \cdot 3 + 5) + (4^2 + 3 \cdot 4 + 5)$
$\phantom{\displaystyle\sum_{i=1}^{4} (i^2 + 3i + 5) } = 9 + 15 + 23 + 33 = 80$

b) $\displaystyle\sum_{i=2}^{5} \sqrt{3i + 2} = \sqrt{3 \cdot 2 + 2} + \sqrt{3 \cdot 3 + 2} + \sqrt{3 \cdot 4 + 2} + \sqrt{3 \cdot 5 + 2}$
$\phantom{\displaystyle\sum_{i=2}^{5} \sqrt{3i + 2} } = \sqrt{8} + \sqrt{11} + \sqrt{14} + \sqrt{17} \approx 14.01$ ▶

The following summation results will be used to solve problems involving area. The results in the first three parts are easy to see. The results in parts (iv), (v), and (vi) are considered in the exercises.

(4.3)

i) $\displaystyle\sum_{i=p}^{n} (a_i + b_i) = \sum_{i=p}^{n} a_i + \sum_{i=p}^{n} b_i$ ii) $\displaystyle\sum_{i=p}^{n} ca_i = c \sum_{i=p}^{n} a_i,$
 for any constant c

iii) $\displaystyle\sum_{i=1}^{n} 1 = n$ iv) $\displaystyle\sum_{i=1}^{n} i = \frac{n(n + 1)}{2}$

v) $\displaystyle\sum_{i=1}^{n} i^2 = \frac{n(n + 1)(2n + 1)}{6}$ vi) $\displaystyle\sum_{i=1}^{n} i^3 = \frac{n^2(n + 1)^2}{4}$

Example 2

Use the formulas given in (4.3) to find a formula for $\displaystyle\sum_{i=1}^{n} (i^2 + 3i + 5)$, where n is an integer greater than 1.

Solution
We have

$$\sum_{i=1}^{n}(i^2 + 3i + 5) = \sum_{i=1}^{n} i^2 + 3\sum_{i=1}^{n} i + 5\sum_{i=1}^{n} 1$$

$$= \frac{n(n+1)(2n+1)}{6} + 3\frac{n(n+1)}{2} + 5n$$

$$= \frac{n(n+1)(2n+1) + 9n(n+1) + 30n}{6}$$

$$= \frac{2n^3 + 3n^2 + n + 9n^2 + 9n + 30n}{6}$$

$$= \frac{2n^3 + 12n^2 + 40n}{6} = \frac{n^3 + 6n^2 + 20n}{3}.$$

Note that when $n = 4$, this formula gives

$$\sum_{i=1}^{n}(i^2 + 3i + 5) = \frac{64 + 6 \cdot 16 + 20 \cdot 4}{3} = \frac{240}{3} = 80,$$

the same result as shown in part (a) of Example 1. ▶

In Equation (4.1) we found that A_n, the sum of the areas of the n rectangles shown in Figure 4.5, is

$$A_n = [f(x_1) + f(x_2) + \cdots + f(x_n)]\left(\frac{b-a}{n}\right),$$

where

$$x_i = a + \frac{i(b-a)}{n}, \quad \text{for each } i = 1, 2, \ldots, n.$$

The summation notation introduced in Definition 4.2 permits us to write A_n as

$$A_n = \sum_{i=1}^{n} f(x_i)\left(\frac{b-a}{n}\right).$$

The collection of numbers $\{A_n\}$ is said to be a *sequence*. Sequences are studied in detail in Chapter 8. For the purpose of this discussion, it suffices to know that the sequence $\{A_n\}$ converges to a number A if A_n approaches A as n increases without bound. This is written

$$\lim_{n \to \infty} A_n = A.$$

As the number of rectangles increases, the approximation is expected to improve. Thus it is reasonable to expect that the area is

(4.4) $\quad A = \lim_{n \to \infty} A_n = \lim_{n \to \infty} \sum_{i=1}^{n} f\left(a + \frac{i(b-a)}{n}\right)\left(\frac{b-a}{n}\right),$

assuming that this limit exists.

In Section 4.2 we will examine this summation process in some detail

CHAPTER 4 THE INTEGRAL

and explain why this particular choice of rectangles produces satisfactory results. For the purposes of our discussion here, let us see what Formula (4.4) gives when applied to a few examples.

Example 3

Use Formula (4.4) to find the area of the region bounded by the graph of $f(x) = x^2$, the x-axis, and the lines $x = 1$ and $x = 3$.

Solution

Figure 4.6 shows the region whose area we wish to find. If the interval $[1, 3]$ is divided into n equal subintervals, the width of each subinterval is

$$\frac{b-a}{n} = \frac{3-1}{n} = \frac{2}{n}.$$

The right endpoint of the ith subinterval occurs at

$$x_i = a + \left(\frac{b-a}{n}\right)i = 1 + \frac{2i}{n},$$

so the ith approximating rectangle has height

$$f(x_i) = f\left(1 + \frac{2i}{n}\right) = \left(1 + \frac{2i}{n}\right)^2$$

and width $2/n$. The area of this ith rectangle is

$$\left(1 + \frac{2i}{n}\right)^2 \left(\frac{2}{n}\right).$$

Consequently,

$$A_n = \sum_{i=1}^{n} f\left(1 + \frac{2i}{n}\right)\left(\frac{2}{n}\right) = \sum_{i=1}^{n} \left(1 + \frac{2i}{n}\right)^2 \left(\frac{2}{n}\right)$$

$$= \sum_{i=1}^{n} \left(1 + \frac{4i}{n} + \frac{4i^2}{n^2}\right)\left(\frac{2}{n}\right) = \sum_{i=1}^{n} \left(\frac{2}{n} + \frac{8i}{n^2} + \frac{8i^2}{n^3}\right)$$

$$= \sum_{i=1}^{n} \frac{2}{n} + \sum_{i=1}^{n} \frac{8}{n^2} i + \sum_{i=1}^{n} \frac{8}{n^3} i^2$$

$$= \frac{2}{n} \sum_{i=1}^{n} 1 + \frac{8}{n^2} \sum_{i=1}^{n} i + \frac{8}{n^3} \sum_{i=1}^{n} i^2.$$

The summation formulas in (4.3) imply that

$$A_n = \frac{2}{n}(n) + \frac{8}{n^2} \cdot \frac{n(n+1)}{2} + \frac{8}{n^3} \cdot \frac{n(n+1)(2n+1)}{6}$$

$$= 2 + \frac{4}{n^2}(n^2 + n) + \frac{4}{3n^3}(2n^3 + 3n^2 + n)$$

$$= \frac{26}{3} + \frac{8}{n} + \frac{4}{3n^2}.$$

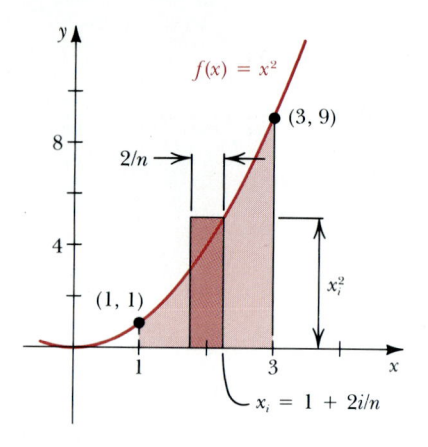

Figure 4.6

The area of the region bounded by the x-axis, the lines $x = 1$ and $x = 3$, and the graph of $f(x) = x^2$ is therefore

$$A = \lim_{n \to \infty} A_n = \lim_{n \to \infty} \left(\frac{26}{3} + \frac{8}{n} + \frac{4}{3n^2} \right) = \frac{26}{3}.$$

▶

Example 4

Use Formula (4.4) to find the area of the region bounded by the graph of $f(x) = \frac{1}{2}(x^3 + 2)$, the x-axis, and the lines $x = -1$ and $x = 2$.

Solution

The region is shown in Figure 4.7. If we divide the interval $[-1, 2]$ into n equal subintervals, the common width of the subintervals is

$$\frac{2 - (-1)}{n} = \frac{3}{n}.$$

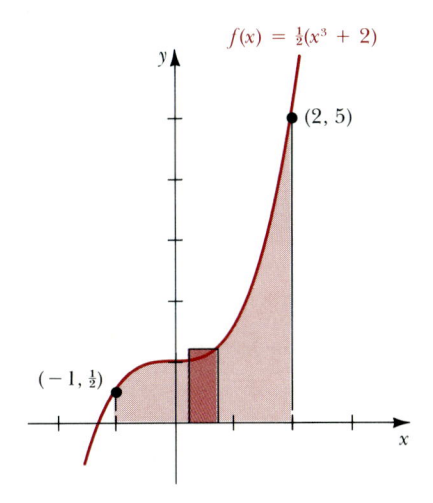

Figure 4.7

The right endpoint of the ith subinterval is $-1 + (3/n)i$, so

$$A_n = \sum_{i=1}^{n} f\left(-1 + \frac{3}{n}i\right)\left(\frac{3}{n}\right)$$

$$= \sum_{i=1}^{n} \frac{1}{2}\left[\left(-1 + \frac{3}{n}i\right)^3 + 2\right]\left(\frac{3}{n}\right)$$

$$= \frac{3}{2n} \sum_{i=1}^{n} \left[(-1)^3 + 3(-1)^2\left(\frac{3i}{n}\right) + 3(-1)\left(\frac{3i}{n}\right)^2 + \left(\frac{3i}{n}\right)^3 + 2\right]$$

$$= \frac{3}{2n} \sum_{i=1}^{n} \left[-1 + \frac{9i}{n} - \frac{27i^2}{n^2} + \frac{27i^3}{n^3} + 2\right]$$

$$= \frac{3}{2n} \left[\sum_{i=1}^{n} 1 + \frac{9}{n}\sum_{i=1}^{n} i - \frac{27}{n^2}\sum_{i=1}^{n} i^2 + \frac{27}{n^3}\sum_{i=1}^{n} i^3\right]$$

$$= \frac{3}{2n} \left[n + \frac{9}{n}\left(\frac{n(n+1)}{2}\right) - \frac{27}{n^2}\left(\frac{n(n+1)(2n+1)}{6}\right) + \frac{27}{n^3}\left(\frac{n^2(n+1)^2}{4}\right)\right]$$

$$= \frac{3}{2} + \frac{27}{4}\left(\frac{n+1}{n}\right) - \frac{27}{4}\left(\frac{(n+1)(2n+1)}{n^2}\right) + \frac{81}{8}\left(\frac{(n+1)^2}{n^2}\right)$$

$$= \frac{3}{2} + \frac{27}{4}\left(1 + \frac{1}{n}\right) - \frac{27}{4}\left(2 + \frac{3}{n} + \frac{1}{n^2}\right) + \frac{81}{8}\left(1 + \frac{2}{n} + \frac{1}{n^2}\right).$$

The area of the region is thus

$$A = \lim_{n \to \infty} A_n = \frac{3}{2} + \frac{27}{4}(1 + 0) - \frac{27}{4}(2 + 0 + 0) + \frac{81}{8}(1 + 0 + 0)$$

$$= \frac{39}{8}.$$

▶

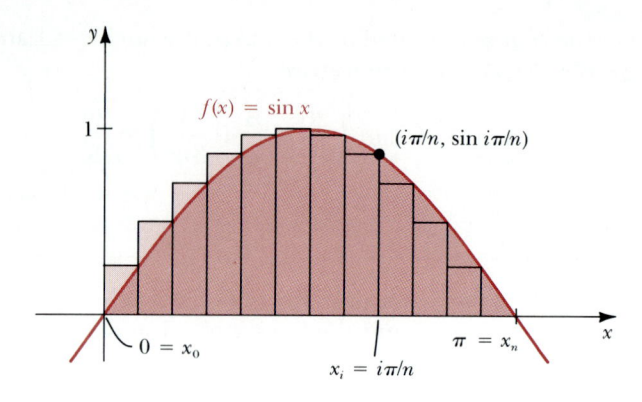

Figure 4.8

Example 5

Use Formula (4.4) to express the area of the region bounded by the graph of $f(x) = \sin x$ and the x-axis for x in the interval $[0, \pi]$.

Solution

We divide the interval $[0, \pi]$ into n equal subintervals, as shown in Figure 4.8. The width of each subinterval is $(\pi - 0)/n = \pi/n$, and the right endpoint of the ith subinterval is

$$x_i = 0 + \left(\frac{\pi}{n}\right)i = \frac{\pi i}{n}.$$

By (4.4), the area of the region is

$$A = \lim_{n \to \infty} A_n = \lim_{n \to \infty} \sum_{i=1}^{n} \left(\sin \frac{\pi i}{n}\right)\frac{\pi}{n} = \lim_{n \to \infty} \frac{\pi}{n} \sum_{i=1}^{n} \sin \frac{\pi i}{n}.$$

We presently have no convenient way to evaluate this limit. Table 4.1 lists values obtained from a programmable calculator. This table suggests that the limit is 2. This is indeed the case, as we will see after we have studied the Fundamental Theorem of Calculus in Section 4.3. ▶

Example 5 demonstrates that to solve the problem of finding the area of regions in the plane, we must either expand our knowledge of summation formulas or find a way to circumvent the limit process. We choose the latter course, beginning in Section 4.2.

Table 4.1

n	A_n
1	0
5	1.933766
10	1.983524
20	1.995886
50	1.999342
100	1.999836
1000	1.999998

▶ EXERCISE SET 4.1

In Exercises 1–12, find the value of the sum.

1. $\displaystyle\sum_{i=1}^{4}(i+2)$

2. $\left(\displaystyle\sum_{i=1}^{4}i\right) + 2$

3. $\displaystyle\sum_{i=1}^{7}(i^2 - 5)$

4. $\displaystyle\sum_{i=2}^{5}(3 - 2i^2)$

5. $\displaystyle\sum_{j=3}^{7}(j^2 + 2j + 2)$

6. $\left(\displaystyle\sum_{j=3}^{7}(j^2 + j)\right) + 2$

7. $\displaystyle\sum_{i=1}^{6}(-1)^i\, i$

8. $\displaystyle\sum_{j=3}^{7}(-1)^j\, j^2$

9. $\sum_{i=1}^{7} (i+1)^2$ 10. $\sum_{i=1}^{7} (i^2 + 2i)$

11. $\sum_{j=2}^{4} (ij + kj^2)$ 12. $\sum_{i=2}^{4} (ij + kj^2)$

In Exercises 13–18, use Formulas (4.3) to evaluate the sum.

13. $\sum_{i=1}^{11} (i^2 + 2i)$ 14. $\sum_{i=1}^{15} (3i^3 - i^2)$

15. $\sum_{i=1}^{n} (i^2 + i)$ 16. $\sum_{i=1}^{n} i(i-1)$

17. $\sum_{i=10}^{20} (i^3 - i^2)$ 18. $\sum_{i=5}^{15} (i^3 + i^2 + 1)$

In Exercises 19–24, express the given sum using summation notation.

19. $4 + 9 + 16 + \cdots + 64$
20. $1 + 3 + 5 + 7 + \cdots + 15$
21. $2 + 5 + 8 + \cdots + 23$
22. $2 - 4 + 6 - 8 + \cdots + 1002$
23. $1 - \frac{1}{2} + \frac{1}{3} - \frac{1}{4} + \cdots + \frac{1}{11}$
24. $\frac{1}{2} + \frac{2}{3} + \frac{3}{4} + \cdots + \frac{9}{10}$

In Exercises 25–36, find the area of the region bounded by the x-axis, the graph of f, and the given lines.

25. $f(x) = 2x + 3$, $x = -1$, $x = 1$
26. $f(x) = x + 1$, $x = 0$, $x = 1$
27. $f(x) = x^2$, $x = 0$, $x = 1$
28. $f(x) = x^2$, $x = 2$, $x = 4$
29. $f(x) = x^2 + x$, $x = 2$, $x = 4$
30. $f(x) = 2x^2$, $x = -1$, $x = 1$
31. $f(x) = (x+1)^2$, $x = -1$, $x = 1$
32. $f(x) = (x+1)^2$, $x = -3$, $x = -1$
33. $f(x) = 4 - x^2$, $x = -2$, $x = 0$
34. $f(x) = x^2 + 3x + 2$, $x = 0$, $x = 2$
35. $f(x) = x^3 + 1$, $x = 0$, $x = 2$
36. $f(x) = 3x^3 - 1$, $x = 1$, $x = 2$

37. Find the area of the region bounded by the lines $x = 1$ and $x = 3$, the x-axis, and the graph of f, where f is described below. Compare the answers in (c) and (d) to those in (a) and (b).
    a) $f(x) = 3$
    b) $f(x) = x$
    c) $f(x) = 3 + x$
    d) $f(x) = 3x$

38. The brothers of Gamma Gamma Gamma fraternity are planning an empty beer can stacking contest in which the cans must be stacked in a triangular fashion: the top row will have one can, the second row two cans, and so on.
    a) How many cans must they empty if they want to provide enough to reach 15 rows?
    b) Find a formula that will tell them how many cans they need to reach n rows, for any integer n.
    c) Find the total volume of the stacked cans if there are 15 rows and each can is 2 in. in diameter and 6 in. tall.

39. A grocer stacks oranges in pyramid form with 150 on the bottom level in 15 rows of 10 each. How many oranges are in this pyramid if they are stacked until only one row is on top?

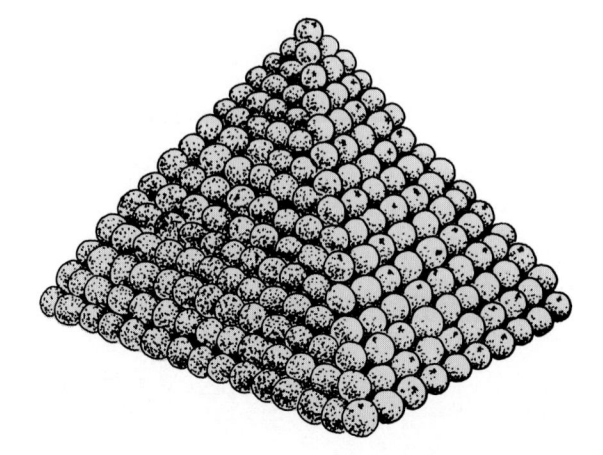

40. The following sums have the *telescoping* property that each term cancels a part of the next term. Compute

each sum by writing out the sums and canceling where possible.

a) $\sum_{k=5}^{15} \left(\frac{1}{k} - \frac{1}{k+1} \right)$

b) $\sum_{i=1}^{4} (2^i - 2^{i+1})$

c) $\sum_{i=10}^{100} (10^{i+1} - 10^i)$

d) $\sum_{j=1}^{10} (\sqrt{2j+1} - \sqrt{2j-1})$

41. Show that for any numbers $a_0, a_i, \ldots, a_n$,
$$\sum_{k=1}^{n} (a_k - a_{k-1}) = a_n - a_0.$$

42. Derive the formula for $\sum_{i=1}^{n} i$ in (4.3) using the fact that the sum can be written as both

i) $\sum_{i=1}^{n} i = 1 + 2 + \cdots + (n-1) + n$, and

ii) $\sum_{i=1}^{n} i = n + (n-1) + \cdots + 2 + 1$.

Then add equations (i) and (ii).

43. Derive a formula for $\sum_{i=1}^{n} r^i$ using the fact that

i) $\sum_{i=1}^{n} r^i = r + r^2 + r^3 + \cdots + r^n$, and

ii) $r \sum_{i=1}^{n} r^i = r^2 + r^3 + \cdots + r^{n+1}$.

Then subtract equation (ii) from equation (i).

44. Derive the formula for $\sum_{i=1}^{n} i^2$ in (4.3) by verifying each of the following.

a) $i^3 - (i-1)^3 = 3i^2 - 3i + 1$

b) $\sum_{i=1}^{n} i^2 = \frac{1}{3} \left\{ \sum_{i=1}^{n} [i^3 - (i-1)^3] + 3 \sum_{i=1}^{n} i - \sum_{i=1}^{n} 1 \right\}$

c) The results in Exercises 41 and 42 imply that
$$\sum_{i=1}^{n} i^2 = \frac{n(n+1)(2n+1)}{6}.$$

45. Derive the formula for $\sum_{i=1}^{n} i^3$ in (4.3) by verifying each of the following.

a) $i^4 - (i-1)^4 = 4i^3 - 6i^2 + 4i - 1$

b) $\sum_{i=1}^{n} i^3 = \frac{1}{4} \left\{ \sum_{i=1}^{n} [i^4 - (i-1)^4] + 6 \sum_{i=1}^{n} i^2 - 4 \sum_{i=1}^{n} i + \sum_{i=1}^{n} 1 \right\}$

c) The results in Exercises 41, 42, and 44 imply that
$$\sum_{i=1}^{n} i^3 = \frac{n^2(n+1)^2}{4}.$$

46. In Example 5 we found that the area of the region bounded by the graph of $y = \sin x$ and the x-axis on the interval $[0, \pi]$ required $\lim_{n \to \infty} A_n$, where
$$A_n = \frac{\pi}{n} \sum_{i=1}^{n} \sin \frac{i\pi}{n}.$$

Evaluate this limit by showing each of the following.

a) For each $i = 1, 2, \ldots, n$,
$$\sin \frac{i\pi}{n} \sin \frac{\pi}{2n} = \frac{1}{2} \left[\cos \frac{(2i-1)\pi}{2n} - \cos \frac{(2i+1)\pi}{2n} \right].$$

b) $\sum_{i=1}^{n} \sin \frac{i\pi}{n} \sin \frac{\pi}{2n} = \frac{1}{2} \left[\cos \frac{\pi}{2n} - \cos \frac{(2n+1)\pi}{2n} \right]$

c) $A_n = \frac{(\pi/2n)}{\sin (\pi/2n)} \left[\cos \frac{\pi}{2n} - \cos \frac{(2n+1)\pi}{2n} \right]$

d) $\lim_{h \to \infty} A_n = 2$

47. Suppose $a_1, a_2, \ldots, a_n$ are constants and that f is a function defined by $f(x) = \sum_{i=1}^{n} (x - a_i)^2$. Show that f has an absolute minimum when $x = (\sum_{i=1}^{n} a_i)/n$, the average of the numbers $a_1, a_2, \ldots, a_n$.

4.2

The Definite Integral

The repetitive summation process used in Section 4.1 occurs frequently in applications. For example, the volume of a solid is approximated by summing the volumes of discs that resemble slices of the solid (see

4.2 THE DEFINITE INTEGRAL

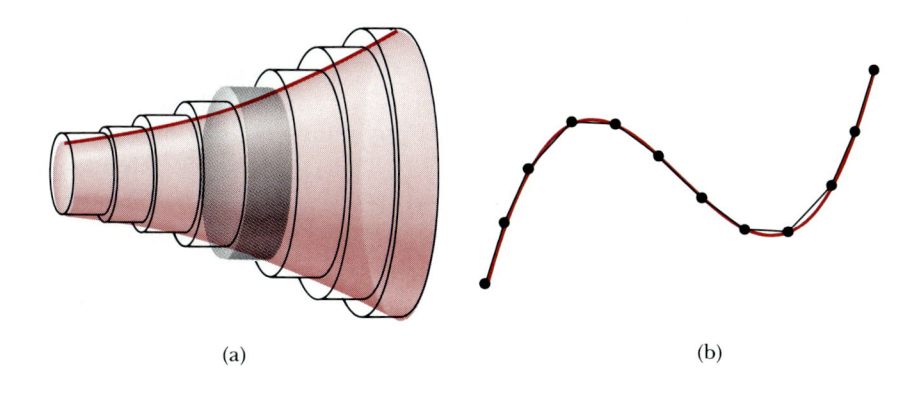

(a)   (b)

Figure 4.9

Figure 4.9a). The length of a curve is approximated by summing the lengths of line segments joining points on the curve (see Figure 4.9b). The mass of a rod with varying density is approximated by summing the mass of small segments of the rod on which the density is assumed to be constant. Each of these applications, as well as many others, involve the same mathematical concept: finding the limiting value of approximating sums. Because of the wide variety of applications, we must study this concept in a more general setting than the one chosen for the discussion of area in Section 4.1. For this purpose we need the *definite integral*, the subject of this section.

Let f be defined on a closed interval $[a, b]$. To define the definite integral of f on $[a, b]$, we introduce some new terminology and notation.

(4.5)

i) A finite set of numbers $\mathcal{P} = \{x_0, x_1, \ldots, x_n\}$ is called a **partition** of $[a, b]$ provided $a = x_0 < x_1 < \cdots < x_n = b$.

ii) For each $i = 1, 2, \ldots, n$, define $\Delta x_i = x_i - x_{i-1}$ (see Figure 4.10).

iii) The **norm** of the partition $\mathcal{P}$, denoted $\|\mathcal{P}\|$, is the largest of Δx_1, $\Delta x_2, \ldots, \Delta x_n$, that is,

$$\|\mathcal{P}\| = \text{maximum } \{\Delta x_1, \Delta x_2, \ldots, \Delta x_n\}.$$

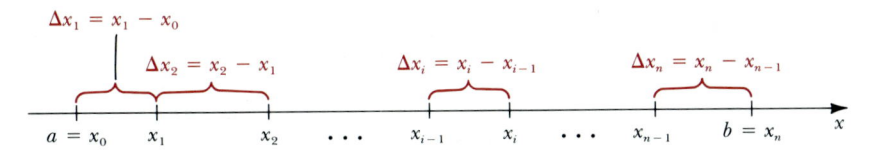

Figure 4.10
The partition of $[a, b]$.

For example, $\mathcal{P} = \{0, \frac{1}{4}, \frac{3}{4}, 1\}$ is a partition of the interval $[0, 1]$ with $\Delta x_1 = \frac{1}{4}$, $\Delta x_2 = \frac{1}{2}$, and $\Delta x_3 = \frac{1}{4}$ (see Figure 4.11). So the norm of $\mathcal{P}$ is

$$\|\mathcal{P}\| = \text{maximum } \{\tfrac{1}{4}, \tfrac{1}{2}, \tfrac{1}{4}\} = \tfrac{1}{2}.$$

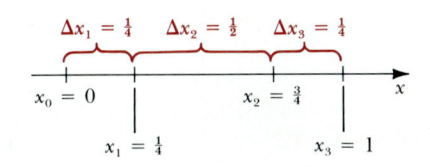

Figure 4.11

(4.6)
DEFINITION

Let f be defined on $[a, b]$, and let $\mathcal{P} = \{x_0, x_1, \ldots, x_n\}$ be a partition of $[a, b]$. Suppose $z_1, z_2, \ldots, z_n$ are any numbers in $[a, b]$ with the property that $x_{i-1} \leq z_i \leq x_i$, for each $i = 1, 2, \ldots, n$. The **Riemann sum** of f with respect to the partition $\mathcal{P}$ and $z_1, z_2, \ldots, z_n$ is defined by

$$S(f, \mathcal{P}, \{z_i\}) = \sum_{i=1}^{n} f(z_i)\,\Delta x_i.$$

Example 1

Suppose $f(x) = x^3$ and $\mathcal{P}$ is the partition $\{-1, 0, 0.4, 1, 1.25, 2\}$ of $[-1, 2]$. Find (a) the norm of $\mathcal{P}$ and (b) the Riemann sum associated with f and $\mathcal{P}$, if $z_1 = -0.6$, $z_2 = 0.2$, $z_3 = 0.5$, $z_4 = 1$, and $z_5 = 1.7$.

Solution

The graph of $f(x) = x^3$ and a geometric illustration of the Riemann sum are shown in Figure 4.12.

a) For the partition $\mathcal{P}$, $\Delta x_1 = 1$, $\Delta x_2 = 0.4$, $\Delta x_3 = 0.6$, $\Delta x_4 = 0.25$, and $\Delta x_5 = 0.75$, so

$$\|\mathcal{P}\| = \text{maximum } \{1, 0.4, 0.6, 0.25, 0.75\} = 1.$$

b) The Riemann sum is

$$\begin{aligned}\sum_{i=1}^{5} f(z_i)\,\Delta x_i &= (-0.6)^3(1) + (0.2)^3(0.4) \\ &\quad + (0.5)^3(0.6) + (1)^3(0.25) + (1.7)^3(0.75) \\ &= -0.216 + 0.0032 + 0.075 + 0.25 + 3.68475 \\ &= 3.79695.\end{aligned}$$

A geometric interpretation of a Riemann sum for a continuous and nonnegative function f on the closed interval $[a, b]$ is the sum of the areas of the rectangles shown in Figure 4.13. Note that for such a

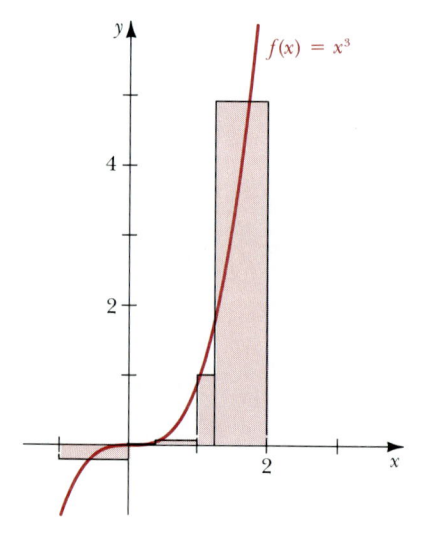

Figure 4.12

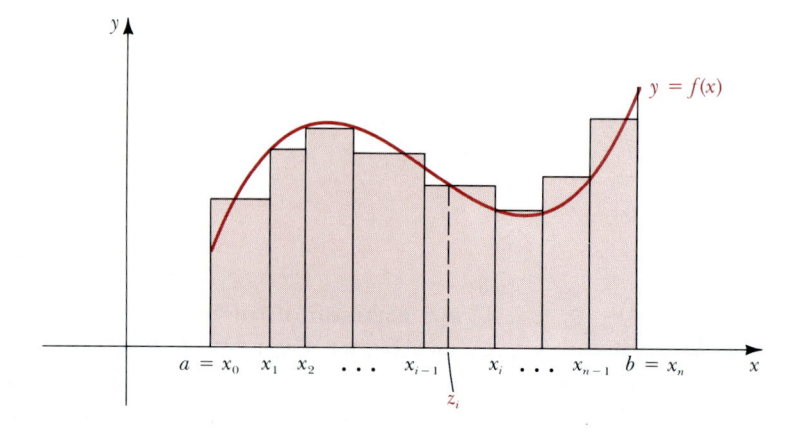

Figure 4.13

A general Riemann sum for a nonnegative continuous function.

4.2 THE DEFINITE INTEGRAL

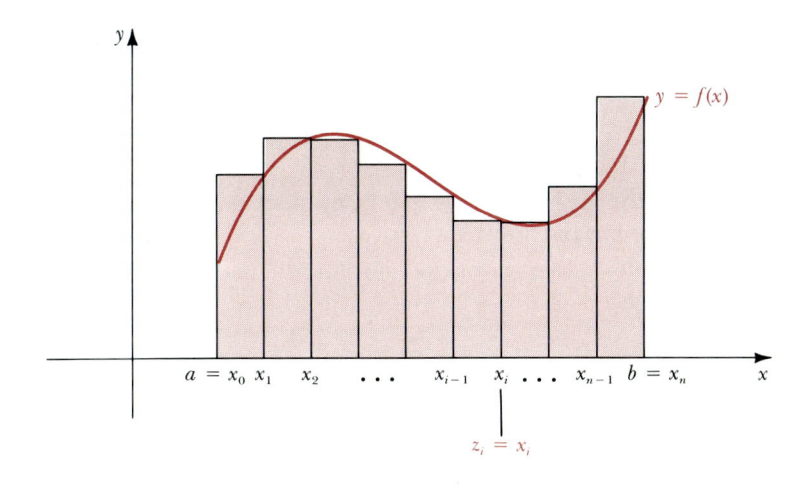

Figure 4.14
The Riemann sum used in Section 4.1.

function the concept of the Riemann sum includes the type of area-approximating sums A_n that were considered in Section 4.1. These occur when the partition is chosen so that

$$x_i = a + \frac{i(b-a)}{n}, \quad \text{for each } i = 1, 2, \ldots, n$$

and when $z_i = x_i$ for each i, as shown in Figure 4.14.

However, the definition of a general Riemann sum of a function f differs from our discussion of area in a number of ways. For a Riemann sum:

i) the function need not be continuous on $[a, b]$;
ii) the function can assume negative values;
iii) the definition of a partition allows the interval $[a, b]$ to be divided into unequal parts; and
iv) the value z_i at which the function is to be evaluated can be chosen arbitrarily within the ith interval.

(4.7)
DEFINITION
If f is defined on $[a, b]$ and L is a number with the property that

$$L = \lim_{\|\mathcal{P}\| \to 0} \sum_{i=1}^{n} f(z_i)\, \Delta x_i,$$

where $\sum_{i=1}^{n} f(z_i)\, \Delta x_i$ is a Riemann sum with respect to the partition $\mathcal{P}$, then L is called the **definite integral** of f on $[a, b]$ and is denoted

$$\int_a^b f(x)\, dx.$$

When $\int_a^b f(x)\, dx$ exists, f is said to be **integrable** on $[a, b]$.

The limit in the definition of the definite integral is more complicated than the limit discussed in Chapter 1. In this case, the limit must be independent of both the choice of the partition $\mathcal{P}$ and the manner in which the numbers $z_1, z_2, \ldots, z_n$ are chosen from the partition. To be precise we define

$$\int_a^b f(x)\, dx = \lim_{\|\mathcal{P}\| \to 0} \sum_{i=1}^n f(z_i)\, \Delta x_i$$

provided that, for any number $\varepsilon > 0$, a number $\delta > 0$ exists with the property that for *any* partition $\mathcal{P} = \{x_0, x_1, \ldots, x_n\}$, with $\|\mathcal{P}\| < \delta$, and *any* set of numbers $z_1, z_2, \ldots, z_n$, chosen with $x_{i-1} \leq z_i \leq x_i$ for each $i = 1, 2, \ldots, n$,

$$\left| \int_a^b f(x)\, dx - \sum_{i=1}^n f(z_i)\, \Delta x_i \right| < \varepsilon.$$

Example 2

Evaluate the definite integral $\int_0^3 -2\, dx$.

Solution

Since $f(z_i) = -2$ for all choices of z_i (see Figure 4.15), then

$$\sum_{i=1}^n f(z_i)\, \Delta x_i = \sum_{i=1}^n -2\, \Delta x_i = -2 \sum_{i=1}^n \Delta x_i$$
$$= -2\, [(x_1 - x_0) + (x_2 - x_1) + (x_3 - x_2) + \cdots + (x_n - x_{n-1})]$$
$$= -2\, (x_n - x_0) = -2\, (3 - 0) = -6.$$

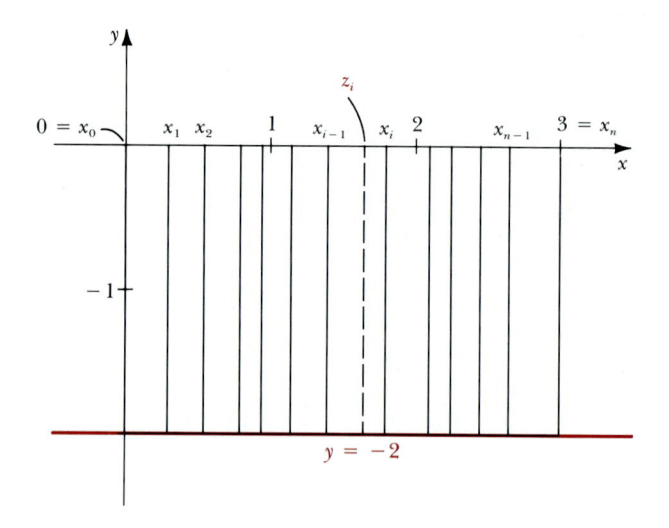

Figure 4.15
$\int_0^3 -2\, dx = -6.$

HISTORICAL NOTE

Georg Friedrich Bernhard Riemann (1826–1866) made many of the important discoveries classifying functions that have integrals. He constructed a function having an infinite number of points of discontinuity in an interval on which its integral exists, and determined precise conditions for bounded functions to be integrable. He also did fundamental work in geometry and complex function theory. He is regarded as one of the most brilliant and profound mathematicians of the nineteenth century.

In this case, all associated Riemann sums have the same value, independent of the partition and the choice of the numbers z_i. Consequently,

$$\int_0^3 -2\, dx = \lim_{\|\mathcal{P}\|\to 0} \sum_{i=1}^n f(z_i)\, \Delta x_i = -6. \quad\blacktriangleright$$

We can apply the method illustrated in Example 2 to any constant function to give the first general result concerning the definite integral.

(4.8) THEOREM
If f is a constant function described by $f(x) = k$ on the interval $[a, b]$, then $\int_a^b f(x)\, dx = k(b - a)$, that is,

$$\int_a^b k\, dx = k(b - a).$$

The definite integral of a function on an interval is a number that depends solely on the function f, called the **integrand**, and the interval $[a, b]$. The numbers a and b are called the **limits of integration:** a is the lower limit, b the upper. The symbol $\int$ used in representing the definite integral is derived from an elongated S and indicates the connection between the definite integral and the Riemann sum.

The definition of the definite integral may seem to impose rather restrictive conditions for a function to be integrable on an interval. The following theorem, given here without proof, shows that the class of integrable functions is quite large.

(4.9) THEOREM
If f is continuous on $[a, b]$, then f is integrable on $[a, b]$.

The scope of Theorem 4.9 is quite broad, since most of the functions we have considered are continuous on every closed interval in their domain. The converse of the theorem is not true, however. There are many functions that are integrable on intervals containing points of discontinuity. The greatest integer function, for example, is discontinuous at every integer and yet is integrable on every closed interval (see Exercise 41). The following result, also given without proof, shows, however, that not all functions are integrable.

(4.10) THEOREM
If f is integrable on $[a, b]$, then f is bounded on $[a, b]$, that is, a constant M exists with $|f(x)| \leq M$ for all x in $[a, b]$.

Theorem 4.10 implies, for example, that the function defined by

$$f(x) = \begin{cases} 1/x, & \text{when } x \neq 0, \\ c, & \text{when } x = 0 \end{cases}$$

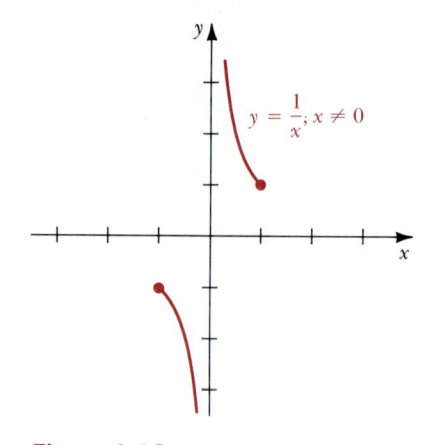

Figure 4.16

A function that is not integrable on $[-1, 1]$.

cannot be integrated on the interval $[-1, 1]$, no matter what value c is assigned to $f(0)$, since f is unbounded on $[-1, 1]$ (see Figure 4.16).

The definition of the definite integral ensures that when a function f is integrable on an interval $[a, b]$, the limit

$$\lim_{\|\mathscr{P}\| \to 0} \sum_{i=1}^{n} f(z_i) \, \Delta x_i$$

is independent of the way in which the Riemann sums are chosen. In particular, the integral is the limit of the Riemann sums defined by choosing equal subintervals for the partition and, in each subinterval, selecting z_i as the right endpoint. In this case, the norm of the partition is

$$\|\mathscr{P}\| = \frac{b-a}{n},$$

and the specification $\|\mathscr{P}\| \to 0$ can be replaced by the specification that $n \to \infty$. Consequently, when f is integrable on $[a, b]$,

$$\int_a^b f(x) \, dx = \lim_{n \to \infty} \sum_{i=1}^{n} f\left(a + \frac{i(b-a)}{n}\right)\left(\frac{b-a}{n}\right).$$

This limit was considered earlier in the discussion on area in Section 4.1; it motivates the following definition.

(4.11) DEFINITION

If f is continuous and nonnegative on $[a, b]$, then the **area** A bounded by the graph of f, the x-axis, and the lines $x = a$ and $x = b$ is

$$A = \int_a^b f(x) \, dx.$$

Example 3

Find the area bounded by the graph of $f(x) = x^3$, the x-axis, and the lines $x = 0$ and $x = 2$ (see Figure 4.17).

Solution

Since f is continuous and nonnegative on $[0, 2]$, Definition 4.11 implies that the area is $\int_0^2 x^3 \, dx$.

We choose $\Delta x_i = (2 - 0)/n = 2/n$ for each i and $z_i = 0 + (2/n)i = 2i/n$. Applying the summation results discussed in Section 4.1 gives

$$\int_0^2 x^3 \, dx = \lim_{\|\mathscr{P}\| \to 0} \sum_{i=1}^{n} f(z_i) \, \Delta x_i = \lim_{n \to \infty} \sum_{i=1}^{n} \left(\frac{2i}{n}\right)^3 \left(\frac{2}{n}\right)$$

so

$$\int_0^2 x^3 \, dx = \lim_{n \to \infty} \frac{16}{n^4} \sum_{i=1}^{n} i^3 = \lim_{n \to \infty} \frac{16}{n^4} \left(\frac{n^2(n+1)^2}{4}\right) = 4.$$ ▶

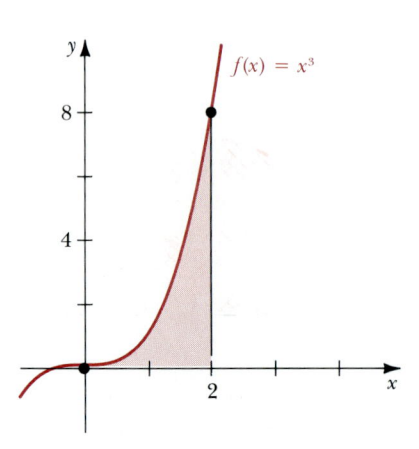

Figure 4.17

Example 4

Find $\int_{-1}^{3/2} 2x \, dx$.

Solution
The function described by $f(x) = 2x$ is continuous on $[-1, \frac{3}{2}]$ so the integral exists. Suppose the interval $[-1, \frac{3}{2}]$ is partitioned into n equal subintervals. Then

$$\Delta x_i = \frac{(\frac{3}{2}) - (-1)}{n} = \frac{5}{2n}.$$

The right endpoint of the ith subinterval is $z_i = -1 + (5/2n) \, i$. Thus,

$$\int_{-1}^{3/2} 2x \, dx = \lim_{n \to \infty} \sum_{i=1}^{n} 2\left[-1 + \frac{5}{2n} i\right]\left(\frac{5}{2n}\right)$$

$$= \lim_{n \to \infty} \frac{5}{n}\left[-\sum_{i=1}^{n} 1 + \frac{5}{2n} \sum_{i=1}^{n} i\right],$$

so

$$\int_{-1}^{3/2} 2x \, dx = \lim_{n \to \infty} \frac{5}{n}\left[-n + \frac{5}{2n} \cdot \frac{n(n+1)}{2}\right]$$

$$= \lim_{n \to \infty} \left[-5 + \frac{25}{4} \cdot \frac{n+1}{n}\right] = -5 + \frac{25}{4} = \frac{5}{4}.$$

The function f is negative on a portion of the interval $[-1, \frac{3}{2}]$, as shown in Figure 4.18. Thus, we cannot interpret this result as the area of a region, as we did in Example 3. ▶

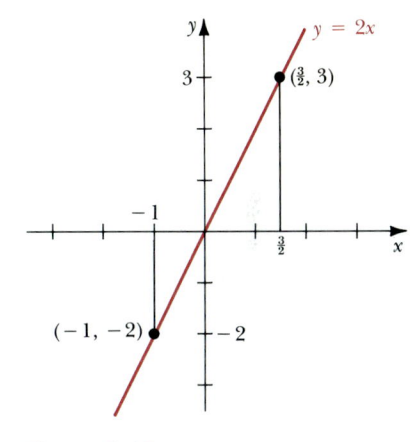

Figure 4.18

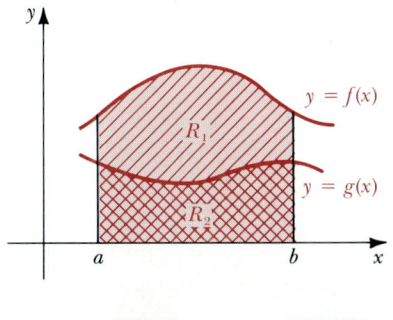

The examples in Sections 4.1 and 4.2 show that the summation and limit procedure is long and often complicated even for rather elementary functions. In Section 4.3, we are introduced to results that eliminate the need to consider Riemann sums for most integral problems. To obtain these important results, we first need some general properties of definite integrals. The next property follows easily from the definition of the definite integral.

(4.12)
THEOREM
If f and g are integrable on $[a, b]$ and $f(x) \geq g(x)$ for each x in $[a, b]$, then

$$\int_{a}^{b} f(x) \, dx \geq \int_{a}^{b} g(x) \, dx.$$

The special case when f and g are continuous and nonnegative has the area interpretation shown in Figure 4.19. In this figure

$$\int_{a}^{b} f(x) \, dx = \text{area of } R_1 \geq \text{area of } R_2 = \int_{a}^{b} g(x) \, dx.$$

Figure 4.19

An important special case of Theorem 4.12 results when g is the constant zero function. In this case, Theorem 4.12 implies that

$$\int_a^b g(x)\,dx = \int_a^b 0\,dx = 0 \cdot (b-a) = 0,$$

so we now have the following.

(4.13)
COROLLARY
If f is integrable and $f(x) \geq 0$ on $[a, b]$, then

$$\int_a^b f(x)\,dx \geq 0.$$

Corollary 4.13 is also a special case of the following important result that can be used to bound the definite integral of a function. Figure 4.20 is an illustration of the result when the function is continuous and nonnegative.

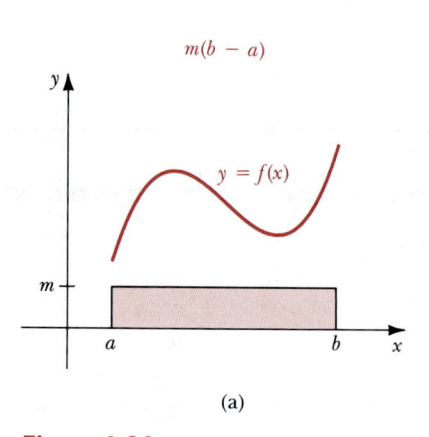

Figure 4.20
$m(b-a) \leq \int_a^b f(x)\,dx \leq M(b-a).$

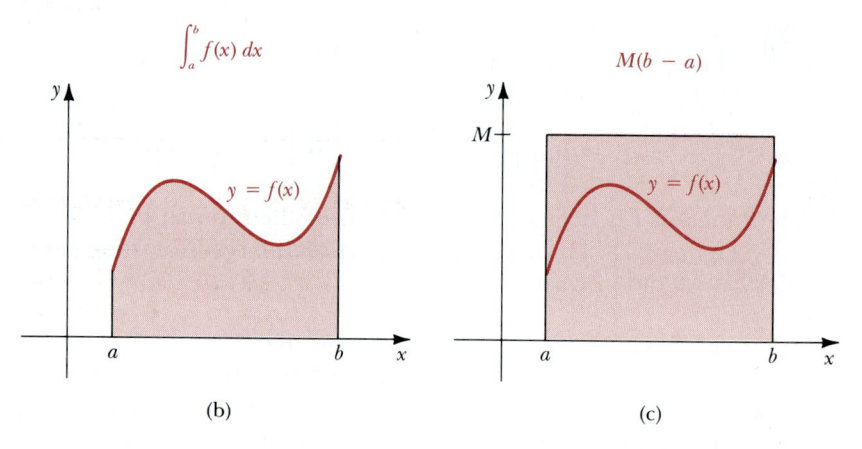

(4.14)
COROLLARY
If f is integrable on $[a, b]$ and constants m and M exist with $m \leq f(x) \leq M$ for all x in $[a, b]$, then

$$m(b-a) \leq \int_a^b f(x)\,dx \leq M(b-a).$$

Example 5

Find upper and lower bounds for

$$\int_0^2 (2x^3 - 9x^2 + 12x - 2)\,dx.$$

Solution

To apply Corollary 4.14 to this problem we need bounds for the values of $f(x) = 2x^3 - 9x^2 + 12x - 2$ for x in the interval $[0, 2]$. Since f is continuous, the Extreme Value Theorem ensures that f assumes absolute maximum and absolute minimum values on $[0, 2]$. In addition, the absolute maximum and minimum values must occur at either a critical point or an endpoint.

In Example 1 of Section 2.7 we found that this function has critical points $x = 1$ and $x = 2$ (see Figure 4.21). Since $f(0) = -2$, $f(1) = 3$, and $f(2) = 2$,

$$-2 \leq f(x) \leq 3 \quad \text{for all } x \text{ in } [0, 2],$$

and

$$-4 = -2(2-0) \leq \int_0^2 (2x^3 - 9x^2 + 12x - 2)\, dx \leq 3(2-0) = 6,$$

Thus -4 is a lower bound for the integral and 6 is an upper bound. ▶

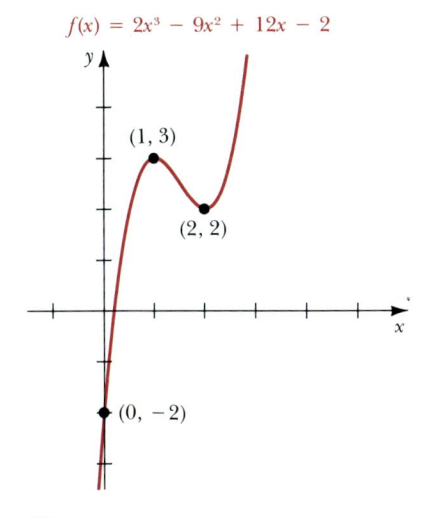

Figure 4.21

$-4 \leq \int_0^2 f(x)\, dx \leq 6.$

Before presenting the final result in this section, we need to extend the notion of the definite integral.

The definition of the definite integral $\int_a^b f(x)\, dx$ requires that $a < b$. We would like to remove this restriction and define this integral when $b \leq a$. When a and b are the same, the area interpretation of the integral requires that the integral be zero, so we make the definition that whenever $f(a)$ exists,

(4.15)
$$\int_a^a f(x)\, dx = 0.$$

When $b < a$ and $\int_a^b f(x)\, dx$ exists, we define

(4.16)
$$\int_a^b f(x)\, dx = -\int_b^a f(x)\, dx.$$

(4.17)
THEOREM

If f is integrable on an interval containing a, b, and c, then

$$\int_a^b f(x)\, dx = \int_a^c f(x)\, dx + \int_c^b f(x)\, dx,$$

regardless of the order of a, b, and c in the interval.

A proof of Theorem 4.17 is included in Appendix B. Figure 4.22 shows a geometric interpretation of this result when f is continuous and nonnegative and $a < c < b$.

Area of $R_1 = \int_a^c f(x)\, dx$
Area of $R_2 = \int_c^b f(x)\, dx$

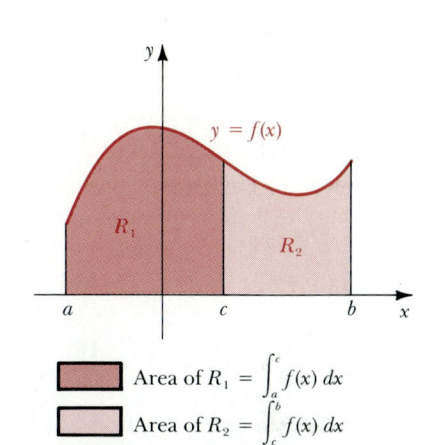

Figure 4.22

(Area of R_1) + (Area of R_2) = $\int_a^b f(x)\, dx.$

EXERCISE SET 4.2

In Exercises 1–12, find the value of the Riemann sum associated with the function, partition, and points described.

1. $f(x) = 2x + 3$, $\mathcal{P} = \{0, 1, 2, 3\}$, $z_1 = 0$, $z_2 = 1$, $z_3 = 2$
2. $f(x) = 2x + 3$, $\mathcal{P} = \{0, 1, 2, 3\}$, $z_1 = 1$, $z_2 = \frac{3}{2}$, $z_3 = 2$
3. $f(x) = x^2$, $\mathcal{P} = \{0, \frac{1}{4}, \frac{1}{2}, \frac{3}{4}, 1\}$, $z_1 = 0$, $z_2 = \frac{1}{4}$, $z_3 = \frac{1}{2}$, $z_4 = \frac{3}{4}$
4. $f(x) = x^2$, $\mathcal{P} = \{0, \frac{1}{4}, \frac{1}{2}, \frac{3}{4}, 1\}$, $z_1 = \frac{1}{4}$, $z_2 = \frac{1}{2}$, $z_3 = \frac{3}{4}$, $z_4 = 1$
5. $f(x) = x^2$, $\mathcal{P} = \{0, \frac{1}{4}, \frac{1}{2}, \frac{3}{4}, 1\}$, $z_1 = 0$, $z_2 = \frac{1}{2}$, $z_3 = \frac{1}{2}$, $z_4 = 1$
6. $f(x) = x^2$, $\mathcal{P} = \{0, \frac{1}{4}, \frac{1}{2}, \frac{3}{4}, 1\}$, $z_1 = \frac{1}{8}$, $z_2 = \frac{3}{8}$, $z_3 = \frac{5}{8}$, $z_4 = \frac{7}{8}$
7. $f(x) = x^2$, $\mathcal{P} = \{0, \frac{1}{3}, \frac{1}{2}, \frac{2}{3}, 1\}$, $z_1 = 0$, $z_2 = \frac{1}{2}$, $z_3 = \frac{1}{2}$, $z_4 = 1$
8. $f(x) = x^2$, $\mathcal{P} = \{0, \frac{1}{3}, \frac{1}{2}, \frac{7}{9}, 1\}$, $z_1 = \frac{1}{8}$, $z_2 = \frac{3}{8}$, $z_3 = \frac{5}{8}$, $z_4 = \frac{7}{8}$
9. $f(x) = x^2 - x$, $\mathcal{P} = \{-1, -0.5, 0.5, 1\}$, $z_1 = -1$, $z_2 = 0$, $z_3 = 0.75$
10. $f(x) = \frac{1}{x}$, $\mathcal{P} = \{-2, -1, 0, 1, 2\}$, $z_1 = -1.5$, $z_2 = -0.5$, $z_3 = 0.5$, $z_4 = 1.5$
11. $f(x) = \sin x$, $\mathcal{P} = \left\{0, \frac{\pi}{4}, \frac{2\pi}{3}, \pi\right\}$, $z_1 = \frac{\pi}{4}$, $z_2 = \frac{\pi}{3}$, $z_3 = \frac{3\pi}{4}$
12. $f(x) = \tan x$, $\mathcal{P} = \left\{0, \frac{\pi}{4}, \frac{2\pi}{3}, \pi\right\}$, $z_1 = \frac{\pi}{6}$, $z_2 = \frac{\pi}{4}$, $z_3 = \frac{5\pi}{6}$

In Exercises 13–22, find the value of the definite integral by taking the limit of the appropriate Riemann sums.

13. $\int_0^2 3\, dx$
14. $\int_1^4 -2\, dx$
15. $\int_0^1 (2x + 1)\, dx$
16. $\int_{-1}^1 (2x + 3)\, dx$
17. $\int_0^1 x^2\, dx$
18. $\int_0^3 (x^2 + x)\, dx$
19. $\int_0^1 x^3\, dx$
20. $\int_{-1}^1 (x^2 + x)\, dx$
21. $\int_{-1}^1 x^3\, dx$
22. $\int_{-2}^3 -x^2\, dx$

In Exercises 23–26, use the application of the definite integral as an area to evaluate the definite integral. [Hint: Sketch the graph of the integrand.]

23. $\int_0^2 |x|\, dx$
24. $\int_{-2}^2 |x|\, dx$
25. $\int_{-1}^1 \sqrt{1 - x^2}\, dx$
26. $\int_0^2 \sqrt{4 - x^2}\, dx$

In Exercises 27–34, determine upper and lower bounds for the integrals by finding the extrema of the integrand and using Corollary 4.14.

27. $\int_0^3 (x^2 - 2x + 2)\, dx$
28. $\int_{-2}^1 (4x^2 + 12x + 5)\, dx$
29. $\int_{-1}^2 (2x^3 - 3x^2 + 4)\, dx$
30. $\int_{-1}^2 (x^4 - 8x^2 + 3)\, dx$
31. $\int_0^\pi (\sin x + \cos x)\, dx$
32. $\int_{-\pi/4}^{\pi/4} (\sec x - \tan x)\, dx$
33. $\int_1^3 \frac{2x}{x^2 + 5x + 4}\, dx$
34. $\int_0^{\pi/3} (\tan x - \frac{4}{3}x)\, dx$

35. Without evaluating the integrals, show that
$$\int_0^1 x\, dx \geq \int_0^1 x^2\, dx$$
and that
$$\int_1^2 x\, dx \leq \int_1^2 x^2\, dx.$$
Interpret this geometrically.

36. Use Theorem 4.12 to show that
$$\left|\int_a^b f(x)\, dx\right| \leq \int_a^b |f(x)|\, dx$$
for every continuous function f and any interval $[a, b]$.

37. Construct a simple example showing that, in general,
$$\left|\int_a^b f(x)\,dx\right| \neq \int_a^b |f(x)|\,dx.$$

38. Construct a simple example showing that, in general,
$$\int_a^b f(x)g(x)\,dx \neq \left(\int_a^b f(x)\,dx\right)\left(\int_a^b g(x)\,dx\right).$$

39. Find the area bounded by the graph of $f(x) = x^3$, the x-axis, and the lines $x = 1$ and $x = 3$ using a procedure similar to that outlined in Section 4.1. Take A_n to be the sum of the areas of the regions described below.
    a) Rectangles whose heights are the values of f at the left endpoint of each subinterval.
    b) Rectangles whose heights are the values of f at the right endpoint of each subinterval.
    c) Rectangles whose heights are the values of f at the midpoint of each subinterval.

40. Theorem 4.9 states that a function continuous on a closed interval has a definite integral on that interval. Show that the converse of this theorem is not true by showing that the function described by
$$f(x) = \begin{cases} 1, & \text{if } x \geq 0, \\ -1, & \text{if } x < 0 \end{cases}$$
has a definite integral on the interval $[-1, 1]$, but is not continuous at $x = 0$.

41. Show that the discontinuous function described by $f(x) = [\![x]\!]$ has a definite integral on $[\frac{1}{2}, \frac{3}{2}]$.

42. Suppose that for x in $[1, 5]$, $f(x) = 3$ and g is defined by
$$g(x) = \int_1^x f(t)\,dt.$$
    a) Describe g analytically.
    b) Sketch the graph of g.
    c) Determine $g'(x)$.
    d) Sketch the graph of g'.

43. The graph of a function f, defined by
$$f(x) = \begin{cases} 0, & \text{if } -2 \leq x < -1, \\ x + 1, & \text{if } -1 \leq x < 0, \\ 1 - x, & \text{if } 0 \leq x < 1, \\ 0, & \text{if } 1 \leq x \leq 2 \end{cases}$$
is shown in the figure. For x in $[-2, 2]$, a function g is defined by
$$g(x) = \int_{-2}^x f(t)\,dt.$$

a) Describe g analytically.
b) Sketch the graph of g.
c) Determine $g'(x)$.
d) Sketch the graph of g'.

44. Suppose f is a continuous function and a and b are constants with $0 < a \leq f(x) \leq b$ when $0 \leq x \leq 1$. Show that
$$2a \leq \int_0^1 f(x)\,dx + ab\int_0^1 \frac{dx}{f(x)} \leq 2b.$$

Putnam exercises

45. Find all continuous positive functions $f(x)$, for $0 \leq x \leq 1$, such that
$$\int_0^1 f(x)\,dx = 1, \quad \int_0^1 f(x)x\,dx = \alpha, \quad \int_0^1 f(x)x^2\,dx = \alpha^2,$$
where α is a given real number. (This exercise was problem 2, part I of the twenty-fifth William Lowell Putnam examination given on December 5, 1964. The examination and its solution are in the September 1965 issue of the *American Mathematical Monthly*, pp. 734–739.)

46. Prove that if a function f is continuous on the closed interval $[0, \pi]$ and if
$$\int_0^\pi f(\theta)\cos\theta\,d\theta = \int_0^\pi f(\theta)\sin\theta\,d\theta = 0,$$
then there exist points α and β such that
$$0 < \alpha < \beta < \pi \quad \text{and} \quad f(\alpha) = f(\beta) = 0.$$

(This exercise was problem 5, part I of the twenty-fourth William Lowell Putnam examination given on December 7, 1963. The examination and its solution are in the June–July 1964 issue of the *American Mathematical Monthly*, p. 636.)

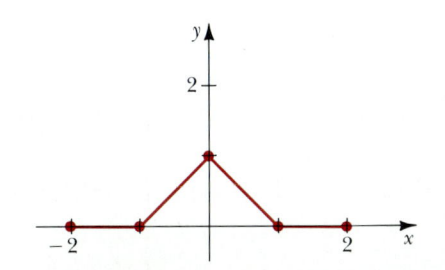

Figure for Exercise 43.

4.3

The Fundamental Theorem of Calculus

This section has a rather imposing title, but one that is completely justified. It will be shown that the seemingly distinct concepts of the derivative and the definite integral are in fact intimately related. This connection can be used to calculate the definite integral of many functions without referring to Riemann sums.

The Fundamental Theorem of Calculus climaxes a series of results concerning the definite integral. The first of these is the Mean Value Theorem for Integrals.

(4.18)

THEOREM: THE MEAN VALUE THEOREM FOR INTEGRALS
If f is continuous on $[a, b]$, then there is at least one number z in (a, b) with

$$f(z) = \frac{1}{b-a} \int_a^b f(x)\, dx.$$

Proof. Because f is continuous on $[a, b]$, the Extreme Value Theorem implies that numbers c_1 and c_2 exist in $[a, b]$ at which f assumes its minimum and maximum values; that is, $f(c_1) \leq f(x) \leq f(c_2)$ for each value of x in $[a, b]$. However, $f(c_1)$ and $f(c_2)$ are constant, so Corollary 4.14 in Section 4.2 implies that

$$f(c_1)(b-a) \leq \int_a^b f(x)\, dx \leq f(c_2)(b-a).$$

Thus,

$$f(c_1) \leq \frac{1}{b-a} \int_a^b f(x)\, dx \leq f(c_2).$$

The number

$$\frac{1}{b-a} \int_a^b f(x)\, dx$$

lies between two values of the function f. Since f is continuous on $[a, b]$, the Intermediate Value Theorem ensures that there is at least one number z between c_1 and c_2 with

$$f(z) = \frac{1}{b-a} \int_a^b f(x)\, dx.$$

Both c_1 and c_2 belong to $[a, b]$, so z must be in (a, b). ▷

The number

(4.19)
$$\frac{1}{b-a}\int_a^b f(x)\,dx$$

is called the **average value** of the function f on the interval $[a, b]$.

In the special case when the function f is continuous and nonnegative, the definite integral $\int_a^b f(x)\,dx$ gives the area of the region bounded by the graph of f, the x-axis, and the lines $x = a$ and $x = b$. The Mean Value Theorem for Integrals implies that there is a number z in (a, b) such that this area is equal to the area of the rectangle with base of width $b - a$ and height $f(z)$ (see Figure 4.23). That is,

$$\int_a^b f(x)\,dx = f(z)(b-a).$$

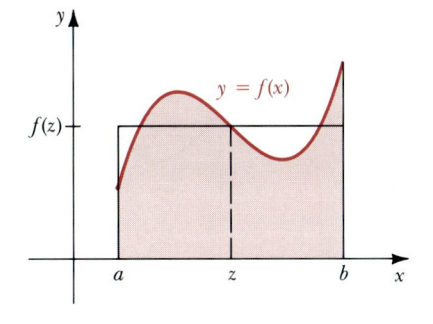

Figure 4.23

$f(z) = \dfrac{1}{b-a}\int_a^b f(x)\,dx.$

Example 1

Consider $f(x) = x^3$ on the interval $[0, 2]$. Use the result in Example 3 of Section 4.2 to:

a) determine the average value of f on $[0, 2]$; and
b) find a number z guaranteed by the Mean Value Theorem for Integrals.

Solution

a) In Example 3 of Section 4.2 we found that

$$\int_0^2 x^3\,dx = 4,$$

so the average value of f on $[0, 2]$ (see Figure 4.24) is

$$\frac{1}{2-0}\int_0^2 x^3\,dx = \frac{4}{2} = 2.$$

b) The Mean Value Theorem for Integrals states that there is a number z in $(0, 2)$ for which $f(z)$ is equal to the average value of f on $[0, 2]$, that is,

$$z^3 = 2.$$

Although the Mean Value Theorem for Integrals does not provide a means for finding z, it is clear that for this example,

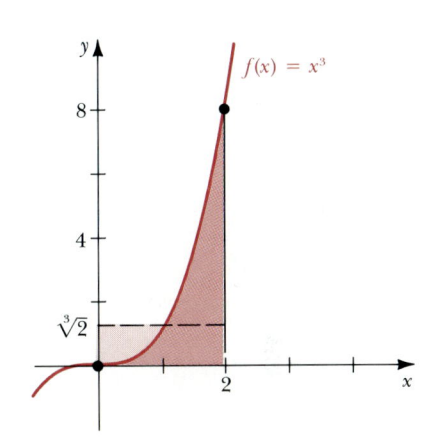

$$z = \sqrt[3]{2}. \qquad \blacktriangleright$$

Figure 4.24

$(\sqrt[3]{2})^3 = \dfrac{1}{2}\int_0^2 x^3\,dx.$

The other preliminary result to the Fundamental Theorem of Calculus needs a few introductory remarks. Suppose f is continuous on $[a, b]$. Then f is continuous, and hence integrable, on the subinterval $[a, x]$ for each x in $[a, b]$. Thus the expression

$$G(x) = \int_a^x f(t)\,dt$$

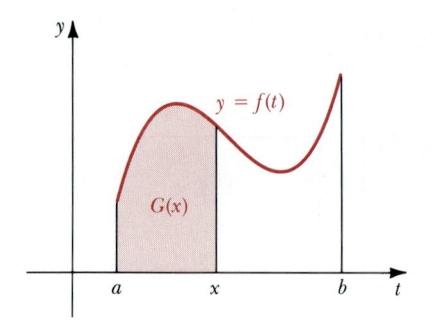

Figure 4.25

$$G(x) = \int_a^x f(t)\, dt.$$

defines a function G whose domain contains the interval $[a, b]$. The variable t has been used in the integrand to eliminate possible confusion with the variable in the upper limit of integration, x, which is the independent variable defining G (see Figure 4.25).

The most important result in calculus concerns the derivative of the function G. This result shows that the two seemingly distinct concepts of calculus, the derivative and the integral, are in fact inseparably linked.

(4.20)
LEMMA: THE FUNDAMENTAL LEMMA OF CALCULUS
If f is continuous on $[a, b]$ and G is defined by

$$G(x) = \int_a^x f(t)\, dt$$

for each x in $[a, b]$, then G is differentiable on $[a, b]$ and

$$G'(x) = f(x)$$

for each x in (a, b).

Proof. Let x be an arbitrary number in (a, b). By the definition of the derivative of G at x,

$$G'(x) = \lim_{h \to 0} \frac{G(x+h) - G(x)}{h} = \lim_{h \to 0} \frac{1}{h}\left[\int_a^{x+h} f(t)\, dt - \int_a^x f(t)\, dt\right].$$

Theorem 4.17 implies that

$$\int_a^{x+h} f(t)\, dt = \int_a^x f(t)\, dt + \int_x^{x+h} f(t)\, dt,$$

so

$$G'(x) = \lim_{h \to 0} \frac{1}{h}\left[\int_a^x f(t)\, dt + \int_x^{x+h} f(t)\, dt - \int_a^x f(t)\, dt\right]$$

$$= \lim_{h \to 0} \frac{1}{h} \int_x^{x+h} f(t)\, dt.$$

If h is restricted so that $x + h$ is also in the interval $[a, b]$, the Mean Value Theorem for Integrals implies that there is a number z between x and $x + h$ with

$$f(z) = \frac{1}{h}\int_x^{x+h} f(t)\, dt.$$

Thus,

$$G'(x) = \lim_{h \to 0} \frac{1}{h}\int_x^{x+h} f(t)\, dt = \lim_{h \to 0} f(z).$$

But z is between x and $x + h$ and f is continuous, so, as shown in Figure 4.26, $\lim_{h \to 0} f(z) = f(x)$. This implies that

$$G'(x) = f(x).$$

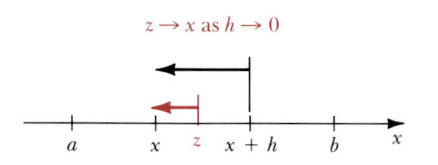

Figure 4.26

Since x is an arbitrary number in (a, b), f is differentiable on (a, b), and for each x in (a, b), $G'(x) = f(x)$.

To extend the result to the closed interval $[a, b]$, we must show that

$$\lim_{h \to 0^+} \frac{G(a + h) - G(a)}{h} \quad \text{and} \quad \lim_{h \to 0^-} \frac{G(b + h) - G(b)}{h}$$

exist and are equal, respectively, to $f(a)$ and $f(b)$. (See Exercises 47–52 in Section 2.2.) These results are considered in Exercise 68. ▷

Example 2

Find

a) $D_x \int_1^x \sqrt{t^2 + 1} \, dt$
b) $D_x \int_1^{x^2} \sqrt{t^2 + 1} \, dt$

Solution

a) Since $f(t) = \sqrt{t^2 + 1}$ describes a continuous function, the Fundamental Lemma of Calculus tells us that

$$D_x \int_1^x \sqrt{t^2 + 1} \, dt = \sqrt{x^2 + 1}.$$

b) The Chain Rule is required for this differentiation:

$$x \longrightarrow x^2 \longrightarrow \int_1^{x^2} \sqrt{t^2 + 1} \, dt.$$

If

$$y = \int_1^{x^2} \sqrt{t^2 + 1} \, dt \quad \text{and} \quad u = x^2, \quad \text{then } y = \int_1^u \sqrt{t^2 + 1} \, dt.$$

Thus,

$$\frac{dy}{du} = \sqrt{u^2 + 1}, \quad \text{and} \quad \frac{du}{dx} = 2x.$$

The Chain Rule implies that

$$D_x y = D_x \int_1^{x^2} \sqrt{t^2 + 1} \, dt = \frac{dy}{du} \frac{du}{dx} = \sqrt{u^2 + 1} \, (2x) = 2x \sqrt{x^4 + 1}. \quad \blacktriangleright$$

The concept of the *antiderivative* of a function was first introduced in Section 2.8:

A function F is an **antiderivative** of a function f if

$$F' = f.$$

As an application of the Mean Value Theorem we found that:

$F_1'(x) = F_2'(x)$ for every number x in an interval if and only if a constant C exists with $F_1(x) = F_2(x) + C$.

Consequently:

Two antiderivatives of the same function on an interval differ by a constant.

The Fundamental Lemma of Calculus states that every continuous function f has an antiderivative G of the form

$$G(x) = \int_a^x f(t)\, dt,$$

where a is an arbitrarily chosen real number in the domain of f. Thus *every* antiderivative F of the continuous function f can be expressed in the form

$$F(x) = G(x) + C = \int_a^x f(t)\, dt + C,$$

for some constant C. The proof of the Fundamental Theorem of Calculus relies on this fact.

(4.21)
THEOREM: THE FUNDAMENTAL THEOREM OF CALCULUS
If f is continuous on $[a, b]$ and F is any antiderivative of f, then

$$\int_a^b f(t)\, dt = F(b) - F(a).$$

Proof. Since F is an antiderivative of the continuous function f,

$$F(x) = \int_a^x f(t)\, dt + C,$$

for some constant C. Evaluating this expression when $x = a$, we find that

$$F(a) = \int_a^a f(t)\, dt + C = 0 + C = C.$$

Hence, $C = F(a)$ and

$$F(x) = \int_a^x f(t)\, dt + F(a).$$

Evaluating F when $x = b$ shows that

$$F(b) = \int_a^b f(t)\, dt + F(a)$$

so

$$\int_a^b f(t)\, dt = F(b) - F(a). \quad \triangleright$$

Note that if $F(x) + C$ is used in place of $F(x)$ in the Fundamental Theorem, the same result is obtained:

$$\int_a^b f(x)\, dx = [F(b) + C] - [F(a) + C] = F(b) - F(a).$$

This verifies the statement in the theorem that *any* antiderivative of f can be used in the evaluation of $\int_a^b f(x)\, dx$.

The Fundamental Theorem gives the important connection between the definite integral and the derivative. We can now find the value of many definite integrals without referring to Riemann sums.

Example 3

Find $\int_{-2}^{1} 2x \, dx$.

Solution
To apply the Fundamental Theorem, first observe that $D_x x^2 = 2x$. So $F(x) = x^2$ defines an antiderivative that can be used in the evaluation of the definite integral:

$$\int_{-2}^{1} 2x \, dx = F(1) - F(-2) = 1^2 - (-2)^2 = -3.$$

▶

Example 4

Use the Fundamental Theorem of Calculus to find

$$\int_{1}^{3} (2x^2 - 5) \, dx.$$

Solution
To apply the Fundamental Theorem, we need an antiderivative of $f(x) = 2x^2 - 5$. Since $D_x x^3 = 3x^2$ and $D_x x = 1$,

$$D_x \left(\frac{2}{3} x^3 - 5x \right) = \frac{2}{3} (3x^2) - 5 \cdot 1 = 2x^2 - 5 = f(x).$$

Thus F defined by $F(x) = \frac{2}{3}x^3 - 5x$ is an antiderivative of f and

$$\int_{1}^{3} (2x^2 - 5) \, dx = F(3) - F(1)$$

$$= \left[\frac{2}{3} (3)^3 - 5(3) \right] - \left[\frac{2}{3} (1)^3 - 5(1) \right]$$

$$= \frac{54}{3} - 15 - \frac{2}{3} + 5 = \frac{22}{3}.$$

▶

If F is an antiderivative of a continuous function f, a notation commonly used to summarize the Fundamental Theorem is

(4.22)
$$\int_{a}^{b} f(x) \, dx = F(x) \Big]_{a}^{b} = F(b) - F(a).$$

For example,

$$\int_{-2}^{1} 2x \, dx = x^2 \Big]_{-2}^{1} = 1^2 - (-2)^2 = -3.$$

To conclude this section, we return to the area problem considered in Example 5 of Section 4.1.

Example 5

Let $f(x) = \sin x$. Find A, the area of the region bounded by the graph of f and the x-axis, for x in the interval $[0, \pi]$.

Solution

Since $y = \sin x \geq 0$ for x in the interval $[0, \pi]$, the area shown in Figure 4.27 is given by the definite integral

$$A = \int_0^\pi \sin x \, dx.$$

An antiderivative of $f(x) = \sin x$ is given by $F(x) = -\cos x$, so

$$A = \int_0^\pi \sin x \, dx = -\cos x \Big]_0^\pi = -\cos \pi - (-\cos 0) = 1 + 1 = 2.$$

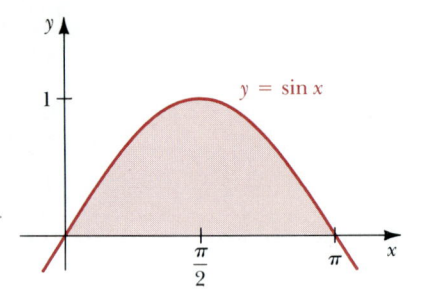

Figure 4.27
The area is $\int_0^\pi \sin x \, dx$.

The answer in Example 5 was conjectured from the table of values given with this problem in Section 4.1. At that point, however, our only means of determining this result was a summation process. With the help of the Fundamental Theorem of Calculus, a formidable problem (see Exercise 46 in Section 4.1) has been reduced to an easy exercise.

▶ EXERCISE SET 4.3

In Exercises 1–28, use the Fundamental Theorem of Calculus to evaluate the definite integrals.

1. $\int_2^5 2x \, dx$
2. $\int_{-2}^1 2x \, dx$
3. $\int_{-1}^1 3x^2 \, dx$
4. $\int_1^3 4x^3 \, dx$
5. $\int_0^4 -3 \, dx$
6. $\int_{-1}^2 5 \, dx$
7. $\int_1^3 (3x^2 - 2x) \, dx$
8. $\int_4^5 (3x^2 + 2) \, dx$
9. $\int_0^{\pi/2} \cos t \, dt$
10. $\int_{-\pi/2}^{\pi/2} \cos w \, dw$
11. $\int_0^{\pi/2} \sin x \, dx$
12. $\int_{-\pi/2}^0 \sin x \, dx$
13. $\int_{-\pi/4}^{\pi/4} \sec^2 x \, dx$
14. $\int_0^{\pi/4} \sec x \tan x \, dx$
15. $\int_0^3 y^2 \, dy$
16. $\int_{-2}^3 (z^2 + z) \, dz$
17. $\int_0^2 (6x^2 - 5) \, dx$
18. $\int_{-1}^2 (2x^2 - 5) \, dx$
19. $\int_1^2 t^{-3} \, dt$
20. $\int_{-1}^1 \sqrt[3]{x} \, dx$
21. $\int_1^{16} z^{3/4} \, dz$
22. $\int_{-1}^{-8} t^{2/3} \, dt$
23. $\int_{-\pi/4}^{-\pi/2} \cos y \, dy$
24. $\int_{-\pi/4}^{-\pi/2} \sin y \, dy$
25. $\int_1^2 (x^3 - x^{-2}) \, dx$
26. $\int_{-2}^{-3} (z^4 + z^{-4}) \, dz$
27. $\int_{\pi/2}^{3\pi/2} (\sin x + \cos x) \, dx$
28. $\int_{-\pi/4}^{-\pi/3} (\sec^2 x - \csc^2 x) \, dx$

In Exercises 29–38, use the Fundamental Theorem of Calculus to find the area of the region bounded by the x-axis, the graph of f, and the given lines. Make a sketch to show the region in each case.

29. $f(x) = 3$, $x = 0$, $x = 4$
30. $f(x) = 2x$, $x = 0$, $x = 4$
31. $f(x) = 3x^2 + 1$, $x = -1$, $x = 1$

32. $f(x) = 3x^2 + 1$, $x = 0$, $x = 1$
33. $f(x) = 3x^2 + 2x$, $x = 0$, $x = 3$
34. $f(x) = 4x^3$, $x = 0$, $x = 2$
35. $f(x) = \cos x$, $x = -\pi/2$, $x = \pi/2$
36. $f(x) = \cos x$, $x = 0$, $x = \pi/2$
37. $f(x) = \sin x$, $x = 0$, $x = \pi/2$
38. $f(x) = \sin x$, $x = 3\pi/4$, $x = \pi$

In Exercises 39–44, find the average value of the function on the given interval.

39. $f(x) = 2x + 1$, $[0, 3]$
40. $f(x) = 2x + 3$, $[-1, 1]$
41. $f(x) = 3x^2$, $[-1, 1]$
42. $f(x) = x^2$, $[1, 4]$
43. $f(x) = \sin x$, $[0, \pi]$
44. $f(x) = \cos x$, $[0, \pi]$

In Exercises 45–52, use the Fundamental Lemma of Calculus to find $D_x F(x)$ for the function F.

45. $F(x) = \int_0^x t\sqrt{t^2 + 9} \, dt$
46. $F(x) = \int_0^x \sqrt{t^2 + 9} \, dt$
47. $F(x) = \int_0^3 t\sqrt{t^2 + 9} \, dt$
48. $F(x) = \int_1^4 \sqrt{t^2 + 9} \, dt$
49. $F(x) = x \int_1^x \sqrt{t^2 + 9} \, dt$
50. $F(x) = x + \int_0^x \sqrt{t^2 + 9} \, dt$
51. $F(x) = \int_x^1 t\sqrt{t^2 + 9} \, dt$
52. $F(x) = \int_x^1 \sqrt{t^2 + 9} \, dt$

53. Suppose that f is continuous on $[a, b]$ and g is a differentiable function whose range is contained in $[a, b]$. Use the Chain Rule and the Fundamental Lemma of Calculus to show that

$$D_x \left(\int_a^{g(x)} f(t) \, dt \right) = f(g(x)) \cdot g'(x).$$

In Exercises 54–57, determine the derivatives of the function.

54. $F(x) = \int_2^{x^2} t^2 \sqrt{t^2 + 9} \, dt$
55. $F(x) = x^2 \int_2^{x^2} \sqrt{t^2 + 9} \, dt$
56. $F(x) = \int_1^{\sqrt{x}} t \cos t^3 \, dt$
57. $F(x) = \int_x^{\sqrt{x}} t \cos t^3 \, dt$

58. Show that the graph of

$$F(x) = \int_0^{\tan x} \frac{1}{1 + t^2} \, dt$$

is increasing on $(-\pi/2, \pi/2)$.

59. Sketch the graph of the function described by $f(x) = x$. Find the average value of this function on the interval $[0, 10]$. Does this agree with your intuitive idea of the average value of a function on an interval?

60. The specific heat of oxygen in cal/deg-mole depends on temperature according to the formula

$$C_p = 8.27 + 2.6 \times 10^{-4} T - 1.87 \times 10^{-5} T^2,$$

where T is measured in $°$ C. Find the average specific heat for T in the interval $[25°$ C, $650°$ C$]$.

61. Use the Mean Value Theorem for Integrals to show that if f is continuous and $\int_a^b f(x) \, dx = 0$, then $f(c) = 0$ for some number c in (a, b).

62. Suppose f is a continuous function and $g(x) = x \int_a^x f(t) \, dt$. Show that

$$g'(x) = \frac{g(x)}{x} + x f(x), \quad \text{provided } x \neq 0.$$

63. Use l'Hôpital's Rule and the Fundamental Lemma of Calculus to find

$$\lim_{x \to 0} \frac{1}{x^2} \int_0^x t\sqrt{3 + t^2} \, dt.$$

64. Use l'Hôpital's Rule and the Fundamental Lemma of Calculus to find

$$\lim_{x \to \infty} \frac{1}{x^3 + 1} \int_1^x (t^2 + t) \, dt.$$

65. Suppose f' is continuous on $[0, 2]$. What is

$$\lim_{x \to 1} \frac{1}{x - 1} \int_1^x f'(t) \, dt?$$

66. Show that if f is continuous and positive on $[a, b]$, then the function G defined in Lemma 4.20 is an increasing function on $[a, b]$. What conditions on f would imply that G is decreasing on $[a, b]$?

67. Suppose f is a function that is differentiable on $[a, b]$ with $f'(x) > 0$ for all x in $[a, b]$. Show that the graph of the function G defined in Lemma 4.20 is concave upward on $[a, b]$.

68. Complete the proof of the Fundamental Lemma of Calculus by showing that

$$\lim_{h \to 0^+} \frac{G(a+h) - G(a)}{h} = f(a)$$

and

$$\lim_{h \to 0^-} \frac{G(b+h) - G(b)}{h} = f(b).$$

69. Show that if f' is continuous on $[a, b]$, then

$$\int_a^b f(x) f'(x) \, dx = \frac{(f(b))^2 - (f(a))^2}{2}.$$

70. Suppose f' and g' are continuous on $[-1, 1]$. What is

$$\lim_{x \to 0} \frac{1}{x} \left[\int_0^x f(t) \, g'(t) \, dt + \int_0^x f'(t) \, g(t) \, dt \right]?$$

71. Determine $F''(\pi/4)$, if

$$F(x) = \int_x^1 f(t) \, dt \quad \text{and} \quad f(t) = \int_1^{2t} \frac{\sin u}{u} \, du.$$

72. Determine $F''(\pi)$, if

$$F(x) = \int_1^{\sin x} f(t) \, dt \quad \text{and} \quad f(t) = \int_2^{t^3} \sqrt{1 + u^3} \, du.$$

4.4
The Indefinite Integral

The Fundamental Theorem of Calculus cannot be used to any great extent until we have an efficient procedure for constructing antiderivatives of functions. Fortunately, the machinery is already available for this purpose; it needs only the concept of the indefinite integral to describe it completely.

(4.23) DEFINITION
The set of all antiderivatives of the continuous function f is called the **indefinite integral** of f.

If F is an antiderivative of f, then the indefinite integral of f is written

(4.24) $$\int f(x) \, dx = F(x) + C,$$

where C denotes an arbitrary constant called a **constant of integration.**

There is some imprecision in the notation (4.24) since the indefinite integral of f is a set of functions. The notation should be

$$\int f(x) \, dx = \{F + C \mid F \text{ is an antiderivative of } f \text{ and } C \text{ is a constant function}\},$$

but this is too cumbersome for common usage.

The Fundamental Theorem of Calculus states that *any* antiderivative of a continuous function can be used to evaluate a definite integral. With the notation presented in (4.24), this implies that

$$\int_a^b f(x) \, dx = \left[F(x) + C \right]_a^b = \left[\int f(x) \, dx \right]_a^b.$$

Example 1

Find (a) $\int 3(x+1)^2 \, dx$; (b) $\int_0^2 3(x+1)^2 \, dx$.

Solution

a) Since $D_x (x+1)^3 = 3(x+1)^2$,

$$\int 3(x+1)^2 \, dx = (x+1)^3 + C.$$

b) $\int_0^2 3(x+1)^2 \, dx = [(x+1)^3 + C]_0^2$
$$= [(2+1)^3 + C] - [(0+1)^3 + C]$$
$$= 27 - 1 = 26. \qquad \blacktriangleright$$

Note that the constant C of integration does not make a contribution to the final result shown in Example 1(b). This agrees with the Fundamental Theorem of Calculus, which states that *any* antiderivative can be used to evaluate a definite integral. It is customary, though not necessary, to choose the constant to be zero when evaluating definite integrals.

Many of the formulas involving derivatives can be expressed using the concept of the indefinite integral and the fact that if f is continuous,

$$D_x F(x) = f(x) \quad \text{if and only if} \quad \int f(x) \, dx = F(x) + C.$$

For example, for any rational number $r \neq -1$,

$$\int x^r \, dx = \frac{x^{r+1}}{r+1} + C \quad \text{because} \quad D_x \left(\frac{x^{r+1}}{r+1} + C \right) = x^r.$$

These results are summarized in Table 4.2 on the following page.

Note that the Fundamental Theorem of Calculus can be applied only to a function whose antiderivative is known. For example, we can evaluate $\int \sec^2 x \, dx$ using property (6) in Table 4.2 and $\int \sec x \tan x \, dx$ using property (8), but until we determine a function whose derivative is sec x we cannot evaluate $\int \sec x \, dx$.

There is a significant difference between finding the derivative of a function and the antiderivative of a function. Finding the derivative of a function is generally straightforward, even though at times it becomes quite complicated. For example, if $f(x) = (3x^2 + 4)^6 (2x^2 + 1)^{3/2}$, we can find $f'(x)$ by applying the elementary arithmetic and composition rules. Determining an antiderivative of a function such as f is a much more difficult problem. Now we are searching for a function F such that $F'(x) = (3x^2 + 4)^6 (2x^2 + 1)^{3/2}$, and we have no elementary rules to apply.

Keep in mind that the functions used for examples in the next few chapters have been chosen specifically because their integrals can be obtained readily. More difficult integration problems will be considered later, particularly in Chapter 7.

Table 4.2

DERIVATIVE RESULTS	CORRESPONDING INTEGRAL RESULTS
1. $D_x (f \pm g)(x) = D_x f(x) \pm D_x g(x)$	$\int (f \pm g)(x)\, dx = \int f(x)\, dx \pm \int g(x)\, dx$
2. $D_x (kf)(x) = k\, D_x f(x)$ for any constant k	$\int k f(x)\, dx = k \int f(x)\, dx$
3. $D_x x^{r+1} = (r+1)x^r$, for any rational number $r \neq -1$	$\int x^r\, dx = \dfrac{x^{r+1}}{r+1} + C$
4. $D_x \sin x = \cos x$	$\int \cos x\, dx = \sin x + C$
5. $D_x \cos x = -\sin x$	$\int \sin x\, dx = -\cos x + C$
6. $D_x \tan x = \sec^2 x$	$\int \sec^2 x\, dx = \tan x + C$
7. $D_x \cot x = -\csc^2 x$	$\int \csc^2 x\, dx = -\cot x + C$
8. $D_x \sec x = \sec x \tan x$	$\int \sec x \tan x\, dx = \sec x + C$
9. $D_x \csc x = -\csc x \cot x$	$\int \csc x \cot x\, dx = -\csc x + C$

Example 2

Find $\int (4x^3 - 5\sqrt[3]{x^2})\, dx$.

Solution
Integral properties (1) and (2) in Table 4.2 can be used to rewrite the integral as

$$\int (4x^3 - 5\sqrt[3]{x^2})\, dx = 4 \int x^3\, dx - 5 \int x^{2/3}\, dx,$$

so (3) in Table 4.2 implies that

$$\int (4x^3 - 5\sqrt[3]{x^2})\, dx = 4 \left(\frac{x^4}{4} + C_1 \right) - 5 \left(\frac{x^{5/3}}{\frac{5}{3}} + C_2 \right)$$
$$= x^4 - 3x^{5/3} + 4C_1 - 5C_2.$$

Since C_1 and C_2 are both arbitrary constants, the linear combination $4C_1 - 5C_2$ also assumes arbitrary values and can be replaced by a single arbitrary constant denoted simply by C. Thus,

$$\int (4x^3 - 5\sqrt[3]{x^2})\, dx = x^4 - 3x^{5/3} + C.$$

To check this solution, we find

$$D_x (x^4 - 3x^{5/3}) = 4x^3 - 5x^{2/3} = 4x^3 - 5\sqrt[3]{x^2}.$$

▶

In the future we will not keep track of the arbitrary constants from each part of a linear combination of integrals, but will instead combine them initially into one arbitrary constant. This constant is generally denoted C.

Example 3

Find $\int \tan^2 x \, dx$.

Solution

The indefinite integral of $\tan^2 x$ is not listed in Table 4.2. However, $\tan^2 x = \sec^2 x - 1$, so

$$\int \tan^2 x \, dx = \int (\sec^2 x - 1) \, dx$$

$$= \int \sec^2 x \, dx - \int 1 \, dx = \tan x - x + C. \quad \blacktriangleright$$

Example 4

Find $\int_1^4 \left(t\sqrt{t} + \frac{3}{t^2} \right) dt$.

Solution

We first find the indefinite integral using result (3) of Table 4.2:

$$\int \left(t\sqrt{t} + \frac{3}{t^2} \right) dt = \int (t^{3/2} + 3t^{-2}) \, dt = \int t^{3/2} \, dt + \int 3t^{-2} \, dt$$

so

$$\int \left(t\sqrt{t} + \frac{3}{t^2} \right) dt = \frac{t^{5/2}}{\frac{5}{2}} + 3 \frac{t^{-1}}{-1} + C = \frac{2}{5} t^{5/2} - \frac{3}{t} + C.$$

This indefinite integral solution can be verified as follows:

$$D_t \left[\frac{2}{5} t^{5/2} - 3t^{-1} + C \right] = t^{3/2} + 3t^{-2}.$$

The definite integral is

$$\int_1^4 \left(t\sqrt{t} + \frac{3}{t^2} \right) dt = \frac{2}{5} t^{5/2} - \frac{3}{t} \Big]_1^4$$

$$= \left[\frac{2}{5} (4)^{5/2} - \frac{3}{4} \right] - \left(\frac{2}{5} - 3 \right)$$

$$= \frac{64}{5} - \frac{3}{4} - \frac{2}{5} + 3 = \frac{293}{20}. \quad \blacktriangleright$$

Note in Example 4 that the constant of integration is omitted in the evaluation of the definite integral because it does not contribute to the

Example 5

Find an equation of the curve that passes through (0, 1) and whose tangent line at any pont (x, y) has slope $x^2 - 2$.

Solution

The slope of the tangent line to a curve $y = f(x)$ at a point (x, y) is given by $f'(x)$. Thus the problem is to find a function f that satisfies

$$f'(x) = x^2 - 2 \quad \text{and} \quad f(0) = 1.$$

Therefore,

$$f(x) = \int (x^2 - 2) \, dx = \frac{x^3}{3} - 2x + C,$$

for some constant C. The curve passes through (0, 1), so $f(0) = 1$ and

$$1 = f(0) = \frac{0}{3} - 2 \cdot 0 + C \quad \text{implies that} \quad C = 1.$$

The curve is described by

$$y = f(x) = \frac{x^3}{3} - 2x + 1.$$

▶

The concluding example in this section is similar to the problems considered in Section 3.1. Here the notation associated with the indefinite integral is used in the solution.

Example 6

A ball is thrown upward with an initial velocity of 80 ft/sec from the top of a building that is 96 ft high (see Figure 4.28). Find:

a) $s(t)$, the distance from the ball to the ground at time t;
b) the time it takes the ball to reach the ground; and
c) the maximum height of the ball.

Solution

a) We assume that the acceleration of the ball due to the gravitational force of the earth is constant, $a(t) = -32$ ft/sec². The velocity $v(t)$ satisfies $v'(t) = a(t)$, and

$$\int a(t) \, dt = \int -32 \, dt = -32t + C, \quad \text{so} \quad v(t) = -32t + C,$$

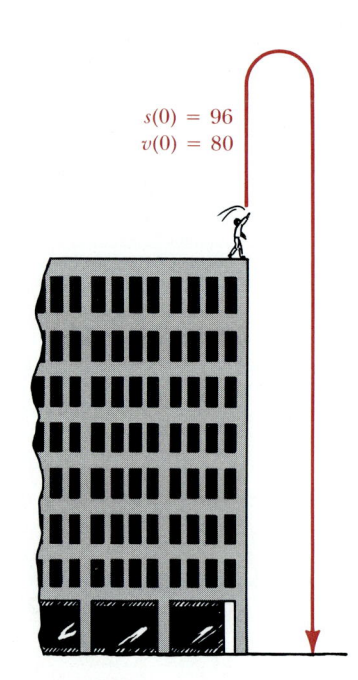

$s(0) = 96$
$v(0) = 80$

Figure 4.28

for some constant C. But $v(0) = 80$, so $C = 80$ and
$$v(t) = -32t + 80 \text{ ft/sec.}$$

Since $s'(t) = v(t)$ and
$$\int v(t) \, dt = \int (-32t + 80) \, dt = -16t^2 + 80t + K,$$

the distance above the ground at any time t after the ball has been thrown is
$$s(t) = -16t^2 + 80t + K,$$

for some constant K. At $t = 0$, $s(t) = 96$, so $K = 96$. Thus,
$$s(t) = -16 t^2 + 80t + 96 \text{ ft.}$$

b) When the ball hits the ground, $s(t) = 0$. The only positive solution to
$$0 = -16t^2 + 80t + 96 = -16(t^2 - 5t - 6) = -16(t - 6)(t + 1)$$

is $t = 6$, so the ball hits the ground in 6 sec.

c) At its maximum height, $v(t) = s'(t)$ is zero. This occurs when
$$0 = -32t + 80, \quad \text{so} \quad t = \frac{80}{32} = \frac{5}{2} \text{ sec.}$$

The maximum height of the ball is
$$s\left(\frac{5}{2}\right) = -16\left(\frac{5}{2}\right)^2 + 80\left(\frac{5}{2}\right) + 96 = 196 \text{ ft.} \quad \blacktriangleright$$

When finding indefinite integrals, *always* check your result by taking the derivative of the answer. If this derivative agrees with the original integrand, the answer is correct; if it does not, an error has been made. This is a much better practice than checking the answers at the back of the book. In addition, it often eliminates tedious algebraic simplification, since two correct answers can assume forms that appear at first to be different. For example, in Example 1 we found that

$$\int 3(x + 1)^2 \, dx = (x + 1)^3 + C, \quad \text{since} \quad D_x (x + 1)^3 = 3(x + 1)^2.$$

However, this indefinite integral can also be evaluated by first expanding $(x + 1)^2$ to give

$$\int 3(x + 1)^2 \, dx = \int 3(x^2 + 2x + 1) \, dx$$
$$= \int 3x^2 \, dx + \int 6x \, dx + \int 3 \, dx$$
$$= x^3 + 3x^2 + 3x + C.$$

The two seemingly different antiderivatives $(x + 1)^3$ and $x^3 + 3x^2 + 3x$ are both valid because they differ only by a constant:
$$(x + 1)^3 = (x^3 + 3x^2 + 3x) + 1.$$

▶ EXERCISE SET 4.4

In Exercises 1–30, (a) find the indefinite integral and (b) check the answer by differentiating your result.

1. $\int 4x^3\, dx$

2. $\int 3x^2\, dx$

3. $\int x^3\, dx$

4. $\int y^5\, dy$

5. $\int (2y^3 + 3y^2)\, dy$

6. $\int (5x^4 - 4x^5)\, dx$

7. $\int (3x^2 + 4x - 2)\, dx$

8. $\int (4x^3 + 7x^2 - 6x + 1)\, dx$

9. $\int (x^{2/3} + 3x^{1/3} + 4x^2)\, dx$

10. $\int (z^{3/4} - 2z^{1/2} + z^3)\, dz$

11. $\int (\cos t + 1)\, dt$

12. $\int (\sin x - x)\, dx$

13. $\int \cot^2 u\, du$

14. $\int \dfrac{1}{\cos^2 x}\, dx$

15. $\int 3t(t^3 + 1)\, dt$

16. $\int y\sqrt{y}\, dy$

17. $\int (\sqrt{x} + \cos x)\, dx$

18. $\int (x^{3/2} - 2\sin x)\, dx$

19. $\int (1 + \sec^2 x)\, dx$

20. $\int (x + \tan^2 x)\, dx$

21. $\int 3(3x + 1)^2\, dx$

22. $\int 3t(t^3 + 4)\, dt$

23. $\int (\sqrt[5]{t} - \sqrt[5]{t^2})\, dt$

24. $\int \left(x^3\sqrt{x} + \dfrac{1}{x^{-3}}\right)\, dx$

25. $\int \dfrac{y^2 - 3y}{y}\, dy$

26. $\int \dfrac{3x^3 - 2x^2 + x}{x}\, dx$

27. $\int \dfrac{x^3 - 2x^2 + 4}{x^5}\, dx$

28. $\int \dfrac{z^{1/2} + z^3 - z^4}{z}\, dz$

29. $\int (x^3 - x^5)^2\, dx$

30. $\int x^3(x + 1)(3x + 2)\, dx$

In Exercises 31–48, find the definite integral.

31. $\int_2^4 (4x^2 + 3x + 1)\, dx$

32. $\int_0^1 (4x^5 + 5x^3)\, dx$

33. $\int_0^\pi \sin x\, dx$

34. $\int_0^\pi \cos x\, dx$

35. $\int_1^4 \left(\sqrt{x} + \dfrac{1}{\sqrt{x}}\right)\, dx$

36. $\int_8^{27} \left(\sqrt[3]{z} - \dfrac{1}{\sqrt[3]{z}}\right)\, dz$

37. $\int_2^4 x(x + 5)\, dx$

38. $\int_{-1}^0 (x + 3)(x + 4)\, dx$

39. $\int_1^2 \left(z + \dfrac{1}{z}\right)^2\, dz$

40. $\int_{-2}^{-4} \left(t^2 - \frac{1}{t^2}\right)^2 dt$

41. $\int_{-1}^{1} 2(3x+4)^3 dx$

42. $\int_{-1}^{1} (x^3 - x^2)^2 dx$

43. $\int_{1}^{2} \frac{\sqrt{x}+1}{\sqrt{x}} dx$

44. $\int_{1}^{4} \frac{t^{1/4} - t^2}{t} dt$

45. $\int_{0}^{\pi/4} \sec^2 x \, dx$

46. $\int_{\pi/4}^{\pi/2} \csc^2 t \, dt$

47. $\int_{0}^{\pi/4} \sec t \, (\sec t + \tan t) \, dt$

48. $\int_{\pi/4}^{\pi/2} \csc z \, (\cot z - \csc z) \, dz$

In Exercises 49–54, determine the area of the region bounded by the x-axis and the graph of the given function on the specified interval. Draw a sketch of this region.

49. $f(x) = \sqrt{x}$, $[1, 2]$
50. $f(x) = x^2 + 1/x^2$, $[1, 2]$
51. $f(x) = 1 + 2 \sin x$, $[0, \pi/2]$
52. $f(x) = 2 \sin x + \cos x$, $[0, \pi/2]$
53. $f(x) = x^2 - 5x + 4$, $[-1, 1]$
54. $f(x) = x + \sin x$, $[0, \pi]$

In Exercises 55–62, the derivative of a function and the value of the function at one point are given. Determine the function.

55. $f'(x) = 3$, $f(0) = 2$
56. $f'(x) = 2x$, $f(0) = 1$
57. $f'(x) = x$, $f(1) = 0$
58. $f'(x) = \cos x$, $f(0) = 0$
59. $f'(x) = x^2 - x$, $f(0) = 0$
60. $f'(x) = x^4$, $f(1) = 1$
61. $f'(x) = \sin x - x$, $f(0) = -1$
62. $f'(x) = x + 1/x^2$, $f(1) = 2$

63. Find an equation of the curve that passes through $(0, 0)$ and whose tangent line at (x, y) has slope $x + 2$.

64. Find an equation of the curve that passes through $(1, 0)$ and whose tangent line at (x, y) has slope $1 - 3x^2$.

65. Find an equation of the curve passing through $(0, 2)$ and $(1, 3)$ with the property that $y'' = 6x - 2$.

66. Find an equation of the curve passing through $(0, 1)$ and $(1, 0)$ with the property that $y'' = 12x^2 - 4$.

67. By observing production and sales data, the Trasho Company has found that its marginal profit from selling wastebaskets is given by

$$P'(x) = 2.8 - 0.006x^2$$

when x units are sold. It is known that Trasho loses $1000 when nothing is sold. Find the profit function of this company.

68. The marginal cost of a certain company is given by $C'(x) = 5.25 + 0.02x$ when x units are sold. Find the cost function of this company if the cost is $100 when no units are sold.

69. An object is moving in a straight line and has acceleration $a(t) = 3t^2$ cm/sec^2. Find its velocity at the end of 2 sec if it has an initial velocity of 4 cm/sec.

70. Suppose the velocity of a ball t sec after it is thrown into the air is given by $v(t) = 80 - 32t$ and that when the ball is thrown, its distance above the ground is 64 ft. Find its distance $s(t)$ above the ground at any time t after it is thrown.

71. A rock is thrown vertically into the air with an initial velocity of 60 mph. (a) How high will the rock go? (b) What is the velocity of the rock after 3 sec?

72. The CN tower in Toronto, Ontario, is the tallest free-standing structure in the world. The top of this tower is 1815 ft above the ground, and the concrete structure supporting the tower is 1464 ft high. Suppose an object is dropped from the top of the tower (see the figure). How much later should another object be dropped from the top of the concrete supporting structure if the two objects are to reach the ground simultaneously?

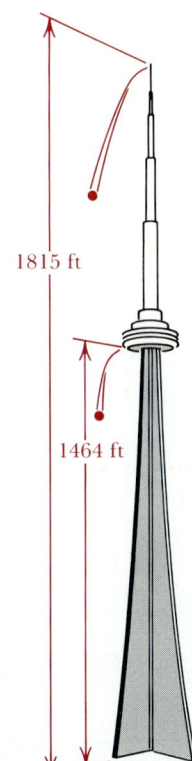

Figure for Exercise 72.

73. An open construction elevator is rising at the rate of 3 ft/sec. A hammer falls from the elevator when it is 100 ft above the ground. How long does it take the hammer to reach the ground?

74. Drops of water fall down a mine shaft at a uniform rate of 1 drop/sec. A mine elevator moving up the shaft at the rate of 10 m/sec is struck by a drop of water when it is 100 m below ground level (see the figure). When and where will the next drop of water strike the elevator?

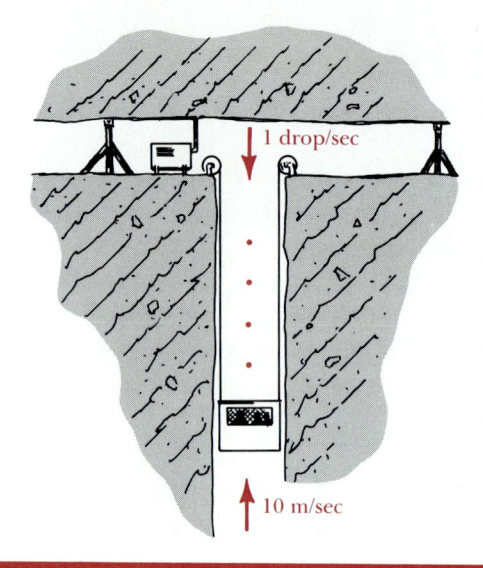

4.5

Integration by Substitution

The Chain Rule plays an important role in determining the derivatives of functions. In this section we show that it is equally valuable as the basis for a technique to evaluate certain common integrals.

Suppose F is an antiderivative of the continuous function f and that g is a differentiable function whose range is contained in the domain of F. The Chain Rule states that

$$D_x\,(F(g(x))) = F'(g(x))\,g'(x) = f(g(x))\,g'(x).$$

The corresponding result for indefinite integrals is

(4.25) $$\int f(g(x))\,g'(x)\,dx = F(g(x)) + C.$$

The reason for using the differential notation in the integral can now be appreciated. If $u = g(x)$, then the differential du is $du = g'(x)\,dx$. With this notation, Equation (4.25) becomes

(4.26) $$\int f(g(x))\,g'(x)\,dx = \int f(u)\,du = F(u) + C = F(g(x)) + C.$$

The process of choosing u in a given problem is called **integration by substitution.** To make an appropriate choice for u requires a sound knowledge of the basic integration rules listed in Table 4.2 together with a few general rules that are described in the following examples.

Example 1

Find $\int (3x + 4)^4 \, 3 \, dx$.

Solution

First note that the integrand involves the composition

$$x \longrightarrow \overbrace{3x + 4}^{u} \longrightarrow (3x + 4)^4.$$

Let

$u = 3x + 4$, the first part of the composition. Then $du = 3 \, dx$.

Making this substitution produces

$$\int (3x + 4)^4 \, 3 \, dx = \int u^4 \, du = \frac{u^5}{5} + C = \frac{(3x + 4)^5}{5} + C.$$

The final answer is expressed in terms of x, the variable involved in the original integral. ▶

Example 2

Find $\int 2x \cos (x^2 + 2) \, dx$.

Solution
Let

$$u = x^2 + 2. \quad \text{Then} \quad du = 2x \, dx.$$

Thus,

$$\int 2x \cos (x^2 + 2) \, dx = \int \cos u \, du = \sin u + C = \sin (x^2 + 2) + C.$$

Note that again the substitution u is made for the first part of a composition:

$$x \longrightarrow \overbrace{x^2 + 2}^{u} \longrightarrow \cos (x^2 + 2).$$

The solution can be verified by differentiating:

$$D_x \, [\sin (x^2 + 2) + C] = \cos (x^2 + 2)(2x) = 2x \cos (x^2 + 2). \quad ▶$$

Example 3

Find (a) $\int (4x^2 - 3)^9 \, x \, dx$; (b) $\int_{0.5}^{1} (4x^2 - 3)^9 \, x \, dx$.

Solution
a) Let u be the first part of the composition:

$$x \longrightarrow \overbrace{4x^2 - 3}^{u} \longrightarrow (4x^2 - 3)^9.$$

Then

$$u = 4x^2 - 3 \quad \text{and} \quad du = 8x \, dx.$$

Since the exact form of du does not appear in the integral, the substitution is made as follows:

$$du = 8x\, dx \quad \text{implies that} \quad x\, dx = \frac{1}{8} du,$$

so

$$\int (4x^2 - 3)^9 x\, dx = \int u^9 \frac{1}{8} du = \frac{1}{8} \int u^9\, du = \frac{1}{8}\left(\frac{u^{10}}{10}\right) + C$$

and

$$\int (4x^2 - 3)^9 x\, dx = \frac{(4x^2 - 3)^{10}}{80} + C.$$

b) From part (a) we have

$$\int_{0.5}^{1} (4x^2 - 3)^9 x\, dx = \frac{(4x^2 - 3)^{10}}{80}\bigg]_{0.5}^{1}$$

$$= \frac{(4 \cdot 1^2 - 3)^{10}}{80} - \frac{(4(0.5)^2 - 3)^{10}}{80}$$

$$= \frac{1}{80} - \frac{1024}{80} = -\frac{1023}{80}. \quad \blacktriangleright$$

The definite integral $\int_{0.5}^{1}(4x^2 - 3)^9 x\, dx$ in Example 3(b) was calculated by first finding the corresponding indefinite integral $\int (4x^2 - 3)^9 x\, dx$ and then evaluating at the limits of integration $x = 1$ and $x = 0.5$. Another method for evaluating a definite integral of this type involves substituting for the limits of integration when a substitution for the variable is made. This changes the problem into a definite integral involving the substitution variable. To see how and why this can be done, recall from Equation (4.26) that if F is an antiderivative of f, then

$$\int_a^b f(g(x))\, g'(x)\, dx = F(g(x))\bigg]_a^b = F(g(b)) - F(g(a)).$$

However, the Fundamental Theorem of Calculus also implies that for $u = g(x)$,

$$\int_{g(a)}^{g(b)} f(u)\, du = F(g(b)) - F(g(a)).$$

Thus, if $u = g(x)$,

(4.27)
$$\int_a^b f(g(x))\, g'(x)\, dx = \int_{g(a)}^{g(b)} f(u)\, du.$$

When a definite integral is evaluated in this manner, it is good practice to add the appropriate variables to the limits of integration to avoid confusion. For example, Equation (4.27) is written

$$\int_{x=a}^{x=b} f(g(x))\, g'(x)\, dx = \int_{u=g(a)}^{u=g(b)} f(u)\, du.$$

Example 4

Evaluate the integral given in Example 3(b),

$$\int_{0.5}^{1} (4x^2 - 3)^9 \, x \, dx,$$

by changing the limits of integration when the variable substitution is made.

Solution

The substitution is made as in Example 3:

$$u = 4x^2 - 3, \quad \text{so} \quad du = 8x \, dx \quad \text{and} \quad x \, dx = \frac{1}{8} du.$$

When

$$x = 0.5, \quad u = 4(0.5)^2 - 3 = -2,$$

and when

$$x = 1, \quad u = 4(1)^2 - 3 = 1.$$

Thus,

$$\int_{x=0.5}^{x=1} (4x^2 - 3)^9 \, x \, dx = \frac{1}{8} \int_{u=-2}^{u=1} u^9 \, du = \frac{u^{10}}{80} \Big]_{u=-2}^{u=1}$$

$$= \frac{1^{10}}{80} - \frac{(-2)^{10}}{80} = -\frac{1023}{80}.$$

This, of course, agrees with the result in Example 3. ▶

The following problems are somewhat more complicated than those in the preceding examples. Note, however, that the method of solution is the same:

A substitution is made for the first part of a composition.

Example 5

Find $\int \dfrac{x+1}{\sqrt{x^2 + 2x}} \, dx.$

Solution

The term $x^2 + 2x$ is the first part of the composition described by the chain

$$x \longrightarrow \overbrace{x^2 + 2x}^{u} \longrightarrow (x^2 + 2x)^{-1/2}.$$

If

$$u = x^2 + 2x, \quad \text{then} \quad du = (2x + 2) \, dx = 2(x + 1) \, dx$$

and
$$\int \frac{x+1}{\sqrt{x^2+2x}}\, dx = \int \frac{1}{\sqrt{u}} \frac{1}{2}\, du = \frac{1}{2} \int u^{-1/2}\, du = \frac{1}{2} \frac{u^{1/2}}{\frac{1}{2}} + C.$$
Thus,
$$\int \frac{x+1}{\sqrt{x^2+2x}}\, dx = \sqrt{x^2+2x} + C.$$ ▶

Example 6

Find $\int \tan^3 x \sec^2 x \, dx.$

Solution

There are two compositions in this problem, one involving the term $\tan^3 x$, the other involving $\sec^2 x$. Consider the resulting expression if we make the substitution $u = \tan x$. In this case, $du = \sec^2 x\, dx$ and
$$\int \tan^3 x \sec^2 x\, dx = \int u^3\, du = \frac{u^4}{4} + C = \frac{\tan^4 x}{4} + C.$$ ▶

Example 6 can also be solved by making the substitution $u = \sec x$, but the trigonometric identity $\tan^2 x + 1 = \sec^2 x$ must be used. It is instructive, however, to carry out this substitution, hence its inclusion as Exercise 69.

There are also two compositions in the integrand of the following example. In this integral, however, care must be taken; only one choice for the substitution will resolve the problem.

Example 7

Find $\int (x+1)^2 \cos(x^3 + 3x^2 + 3x)\, dx.$

Solution

Two compositions are present in the integrand, one involving $(x+1)^2$,
$$x \longrightarrow x + 1 \longrightarrow (x+1)^2,$$
and one involving $\cos(x^3 + 3x^2 + 3x)$,
$$x \longrightarrow x^3 + 3x^2 + 3x \longrightarrow \cos(x^3 + 3x^2 + 3x).$$
If we use the first composition and make the substitution
$$u = x + 1, \quad \text{then} \quad du = dx$$
and it does not simplify the integrand.

On the other hand, if
$$u = x^3 + 3x^2 + 3x, \quad \text{then} \quad du = (3x^2 + 6x + 3)\, dx$$
and
$$\frac{1}{3} du = (x^2 + 2x + 1)\, dx = (x+1)^2\, dx.$$

So
$$\int (x+1)^2 \cos(x^3 + 3x^2)dx = \frac{1}{3}\int \cos u \, du = \frac{1}{3}\sin u + C$$
and
$$\int (x+1)^2 \cos(x^3 + 3x^2 + 3x) \, dx = \frac{1}{3}\sin(x^3 + 3x^2 + 3x) + C. \quad \blacktriangleright$$

Example 8

Find $\int_0^1 x\sqrt{x+1} \, dx$.

Solution

The only composition in the integrand is given by
$$x \longrightarrow x+1 \longrightarrow \sqrt{x+1}.$$

If we substitute for the first part of this composition, then
$$u = x+1 \quad \text{and} \quad du = dx.$$

To express the integrand completely in terms of the new variable u, we must also solve for x in terms of u: $x = u - 1$. Thus,
$$\int x\sqrt{x+1} \, dx = \int (u-1)\sqrt{u} \, du = \int (u^{3/2} - u^{1/2}) \, du.$$

Changing the limits from x to u gives
$$\int_{x=0}^{x=1} x\sqrt{x+1} \, dx = \int_{u=1}^{u=2} (u-1)\sqrt{u} \, du = \int_1^2 (u^{3/2} - u^{1/2}) \, du$$
$$= \frac{2}{5}u^{5/2} - \frac{2}{3}u^{3/2}\Big]_1^2.$$

Thus,
$$\int_0^1 x\sqrt{x+1} \, dx = \left[\frac{2}{5}(2)^{5/2} - \frac{2}{3}(2)^{3/2}\right] - \left(\frac{2}{5} - \frac{2}{3}\right)$$
$$= \frac{8}{5}\sqrt{2} - \frac{4}{3}\sqrt{2} + \frac{4}{15} = \frac{4}{15}(\sqrt{2} + 1). \quad \blacktriangleright$$

▶ EXERCISE SET 4.5

1. Use integration by substitution with $u = x^2 + 5$ and $du = 2x \, dx$ to evaluate each of the following integrals.

    a) $\int 2x(x^2 + 5)^5 \, dx$
    b) $\int \frac{x}{(x^2+5)^3} \, dx$
    c) $\int 3x \sin(x^2 + 5) \, dx$
    d) $\int \frac{x}{\cos^2(x^2+5)} \, dx$

2. Use integration by substitution with $u = x^3 + 3x$ and $du = (3x^2 + 3)\,dx$ to evaluate each of the following integrals.

a) $\displaystyle\int \frac{3x^2 + 3}{(x^3 + 3x)^4}\,dx$

b) $\displaystyle\int (x^2 + 1) \sin(x^3 + 3x)\,dx$

c) $\displaystyle\int (2x^2 + 2)\sqrt{x^3 + 3x}\,dx$

d) $\displaystyle\int \frac{(x^2 + 1)}{\sin^2(x^3 + 3x)}\,dx$

In Exercises 3–68, evaluate the integral.

3. $\displaystyle\int 2(2x - 3)^2\,dx$

4. $\displaystyle\int 5(5x + 4)^5\,dx$

5. $\displaystyle\int 3t^2(t^3 + 4)\,dt$

6. $\displaystyle\int 3t^2(t^3 + 4)^5\,dt$

7. $\displaystyle\int 3t^2(t^3 + 4)^{1/2}\,dt$

8. $\displaystyle\int 3t^2(t^3 + 4)^{-1/2}\,dt$

9. $\displaystyle\int \sqrt{4x - 5}\,dx$

10. $\displaystyle\int (x + 1)\sqrt{x^2 + 2x}\,dx$

11. $\displaystyle\int \frac{dx}{(x - 1)^4}$

12. $\displaystyle\int \frac{dx}{(x + \pi)^6}$

13. $\displaystyle\int \sin 2x\,dx$

14. $\displaystyle\int_0^\pi \cos 3x\,dx$

15. $\displaystyle\int_0^{\,\prime} \sin \pi x\,dx$

16. $\displaystyle\int \cos(2x + 1)\,dx$

17. $\displaystyle\int \frac{dx}{(2x - 1)^3}$

18. $\displaystyle\int \frac{dx}{(3x - 1)^{1/3}}$

19. $\displaystyle\int \frac{2x}{(x^2 + 7)^4}\,dx$

20. $\displaystyle\int \frac{x\,dx}{(x^2 + 3)^3}$

21. $\displaystyle\int_0^2 (t + 2)\sqrt{t^2 + 4t + 1}\,dt$

22. $\displaystyle\int (x + 1)\sin(x^2 + 2x + 3)\,dx$

23. $\displaystyle\int_{-1}^1 x^2 \sqrt{x^3 + 1}\,dx$

24. $\displaystyle\int_0^1 5x^3(x^4 + 1)^4\,dx$

25. $\displaystyle\int_3^4 \frac{2}{\sqrt{3t - 7}}\,dt$

26. $\displaystyle\int \frac{x^2}{(x^3 + 1)^4}\,dx$

27. $\displaystyle\int \frac{3x^2 + 1}{\sqrt[3]{x^3 + x}}\,dx$

28. $\displaystyle\int \frac{2x^3 + x}{\sqrt{x^4 + x^2}}\,dx$

29. $\displaystyle\int \sin 3x \cos^3 3x\,dx$

30. $\displaystyle\int_0^\pi \cos x \sqrt{\sin x}\,dx$

31. $\displaystyle\int_0^{\pi/4} \tan x \sec^2 x\,dx$

32. $\displaystyle\int_{\pi/12}^{\pi/9} \cot 3x \csc 3x\,dx$

33. $\displaystyle\int \sin^{10} x \cos x\,dx$

34. $\displaystyle\int \frac{\sin x}{\cos^5 x}\,dx$

35. $\displaystyle\int (x^3 + x^2)^4(3x^2 + 2x)\,dx$

36. $\displaystyle\int \sqrt{x^3 + x^2}\,(3x^2 + 2x)\,dx$

37. $\displaystyle\int_{-1}^1 \frac{x + 1}{(x^2 + 2x + 2)^3}\,dx$

38. $\displaystyle\int \frac{x}{\cos^2 x^2}\,dx$

39. $\displaystyle\int \frac{(\sqrt{x} - 1)^2}{\sqrt{x}}\,dx$

40. $\displaystyle\int \frac{(\sqrt{x} - 1)^{1/2}}{\sqrt{x}}\,dx$

41. $\displaystyle\int \left(1+\frac{1}{t}\right)^3 \frac{1}{t^2}\, dt$

42. $\displaystyle\int_1^4 \frac{1}{\sqrt{x}\,(\sqrt{x}+1)^2}\, dx$

43. $\displaystyle\int (x+\sin x \cos x)\, dx$

44. $\displaystyle\int (\sqrt{x} - 2\csc^2 x)\, dx$

45. $\displaystyle\int (z+1)\sqrt{z-1}\, dz$

46. $\displaystyle\int_1^2 (z-1)\sqrt{z+1}\, dz$

47. $\displaystyle\int_0^1 \frac{x}{\sqrt{x+1}}\, dx$

48. $\displaystyle\int x\sqrt{2x+1}\, dx$

49. $\displaystyle\int_1^2 \left(1+\frac{1}{t^2}\right)^3 \frac{1}{t^3}\, dt$

50. $\displaystyle\int_4^9 \left(2+\frac{1}{\sqrt{z}}\right)^4 \frac{1}{z\sqrt{z}}\, dz$

51. $\displaystyle\int \frac{\sin x}{(2+3\cos x)^2}\, dx$

52. $\displaystyle\int \frac{\sec^2 x}{(1+\tan x)^3}\, dx$

53. $\displaystyle\int 3x^3\sqrt{x^2+1}\, dx$

54. $\displaystyle\int x^5 \sqrt{2x^2+1}\, dx$

55. $\displaystyle\int_1^2 \frac{\sqrt{\sqrt{x}+1}}{\sqrt{x}}\, dx$

56. $\displaystyle\int_0^4 \sqrt{x}\sqrt{x\sqrt{x}+1}\, dx$

57. $\displaystyle\int \frac{\sec^2 \sqrt{s}}{\sqrt{s}}\, ds$

58. $\displaystyle\int_{\pi^2/16}^{\pi^2/4} \frac{\cos\sqrt{x}}{\sqrt{x}\sin^3\sqrt{x}}\, dx$

59. $\displaystyle\int (x^2+1)\sqrt{x-2}\, dx$

60. $\displaystyle\int x^3\sqrt{x^2-1}\, dx$

61. $\displaystyle\int \frac{1}{x^2+2x+1}\, dx$

62. $\displaystyle\int \frac{1}{x^3-3x^2+3x-1}\, dx$

63. $\displaystyle\int \frac{1}{x^2+6x+9}\, dx$

64. $\displaystyle\int_0^2 \frac{1}{x^2+2x+1}\, dx$

65. $\displaystyle\int \frac{x^2+2x}{x^2+2x+1}\, dx$

[*Hint:* First divide the denominator into the numerator.]

66. $\displaystyle\int \frac{x^3-3x^2+3x}{x^3-3x^2+3x-1}\, dx$

67. $\displaystyle\int \frac{4x^2+4x+5}{4x^2+4x+1}\, dx$

68. $\displaystyle\int \frac{27x^3+54x^2+36x}{27x^3+54x^2+36x+8}\, dx$

69. Find $\int \tan^3 x \sec^2 x\, dx$ by making the substitution $u = \sec x$. What is the difference between this solution and the one given in Example 6?

70. Find the area of the region bounded by the graph of $f(x) = \sin x \cos x$, the x-axis, and the lines $x = 0$ and $x = \pi/2$.

In Exercises 71–74, the derivative of a function and its value at one point are given. Find the function.

71. $f'(x) = \sin x \cos x$, $f(0) = 0$

72. $f'(x) = x\cos(x^2+\pi)$, $f(0) = 1$

73. $f'(x) = (x+2)\sqrt{x^2+4x}$, $f(0) = 0$

74. $f'(x) = \dfrac{x}{\sqrt{x^2+1}}$, $f(0) = 1$

In Exercises 75–82, determine one of the integrals (a) or (b) that can be evaluated and evaluate it.

75. a) $\displaystyle\int (x+1)\cos(x+1)\, dx$

 b) $\displaystyle\int \cos(x+1)\, dx$

76. a) $\displaystyle\int t\sin t^2\, dt$

 b) $\displaystyle\int \sin t^2\, dt$

77. a) $\displaystyle\int x^{10}\sin x^9\, dx$

 b) $\displaystyle\int x^9 \sin x^{10}\, dx$

78. a) $\displaystyle\int \sqrt{z}\,(\sqrt{z}+1)^{13}\, dz$

 b) $\displaystyle\int (\sqrt{z}+1)^{13}\, dz$

79. a) $\displaystyle\int \sqrt{t}\, \sec^2 \sqrt{t}\, dt$

    b) $\displaystyle\int (1/\sqrt{t}) \sec^2 \sqrt{t}\, dt$

80. a) $\displaystyle\int x^{2/3} \sqrt{2 + 3x^{1/3}}\, dx$

    b) $\displaystyle\int x^{-2/3} \sqrt{2 + 3x^{1/3}}\, dx$

81. a) $\displaystyle\int \frac{\sin x}{\cos^2 x}\, dx$

    b) $\displaystyle\int \frac{\sin^2 x}{\cos x}\, dx$

82. a) $\displaystyle\int 2y^3 \sqrt{y^2 + 1}\, dy$

    b) $\displaystyle\int 2y^2 \sqrt{y^2 + 1}\, dy$

83. Show that if f is an even integrable function defined on an interval $[-a, a]$, then

$$\int_{-a}^{a} f(x)\, dx = 2 \int_{0}^{a} f(x)\, dx$$

(see the figure).

84. Show that if f is an odd integrable function defined on an interval $[-a, a]$, then

$$\int_{-a}^{a} f(x)\, dx = 0$$

(see the figure).

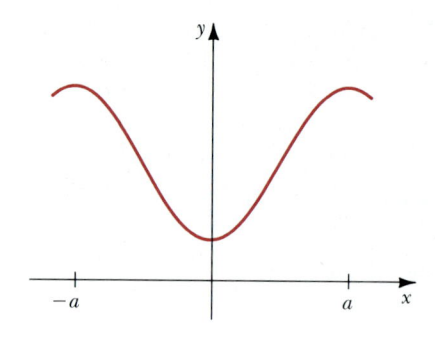

Even function
$$\int_{-a}^{a} f(x)\, dx = 2 \int_{0}^{a} f(x)\, dx$$

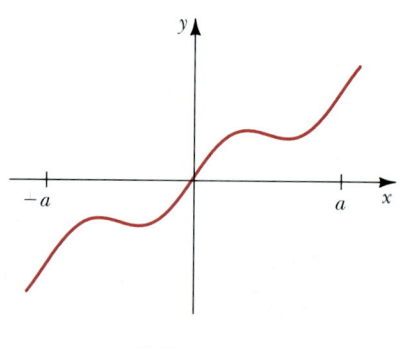

Odd function
$$\int_{-a}^{a} f(x)\, dx = 0$$

4.6

The Natural Logarithm Function

If you examine Table 4.2 back at the beginning of Section 4.4, you will see some gaps in the column of integral results. Although we can integrate the sine and cosine functions, we cannot integrate the tangent, secant, cotangent, and cosecant functions. In addition, the derivative rule

$$D_x\, x^r = rx^{r-1}, \quad \text{provided } r \neq 0,$$

leads to the integration formula

$$\int x^r\, dx = \frac{x^{r+1}}{r+1} + C, \quad \text{provided } r \neq -1.$$

4.6 THE NATURAL LOGARITHM FUNCTION

But this formula cannot be extended to include the case when $r = -1$, since division by zero would then occur.

In this section we introduce a function associated with the integral of x^{-1}. This function also provides the means for integrating the tangent, secant, cotangent, and cosecant functions.

Since $f(x) = x^{-1}$ is continuous on any interval that does not contain zero, f has antiderivatives on the interval $(0, \infty)$. One particular antiderivative of f on this interval is called the *natural logarithm* function. It will become clear why this function is called a logarithm after some of its properties are given.

(4.28) DEFINITION

The **natural logarithm** function ln is defined for all real numbers $x > 0$ by

$$\ln x = \int_1^x \frac{1}{t}\, dt.$$

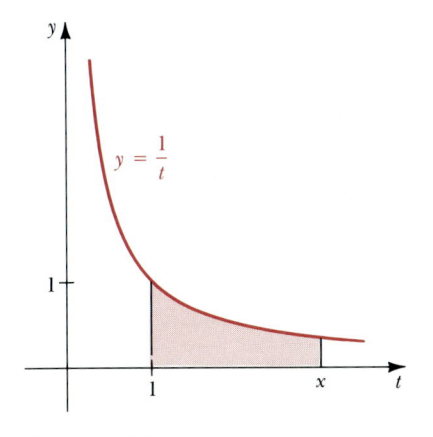

Figure 4.29
If $x > 1$, ln x is this area.

The variable t in the integrand in this definition is used to avoid possible confusion between the variable of integration and the independent variable x used to describe the natural logarithm function.

Note that the domain of the natural logarithm function is $(0, \infty)$; ln x is not defined if $x \leq 0$.

When $x > 1$, ln x is the positive number that describes the area of the region shown in Figure 4.29. When $0 < x < 1$, the fact that

$$\ln x = \int_1^x \frac{1}{t}\, dt = -\int_x^1 \frac{1}{t}\, dt$$

implies that ln x is the negative number that describes the negative of the area of the region shown in Figure 4.30.

Some properties of the natural logarithm function follow easily from the definition: ln 1 is a definite integral whose upper and lower limits of integration are the same, so

(4.29) $\ln 1 = 0.$

Since

$$D_x \ln x = D_x \int_1^x \frac{1}{t}\, dt,$$

the Fundamental Lemma of Calculus implies that

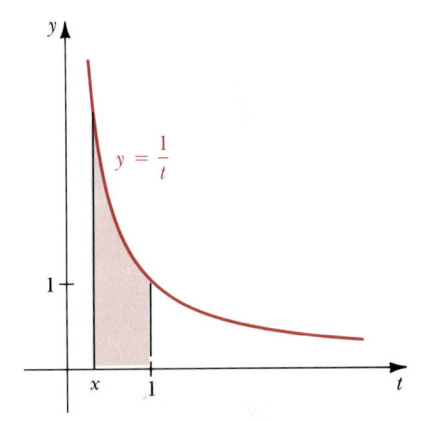

(4.30) $D_x \ln x = \dfrac{1}{x}.$

Figure 4.30
If $0 < x < 1$, $-\ln x$ is this area.

HISTORICAL NOTE

Nicolaus Mercator (1620–1687) first used this description of the natural logarithm in 1668 in his book *Logarithmotechnia*. Mercator's given name was Nicolaus Kaufmann; he was not related to the famous Flemish cartographer Gerhardus Mercator.

If $x < 0$, then $-x > 0$ and $\ln(-x)$ is defined. The Chain Rule implies that

$$D_x \ln(-x) = \frac{1}{-x} D_x(-x)$$

$$= \frac{1}{-x}(-1) = \frac{1}{x}.$$

Thus, if $x \neq 0$,

(4.31)
$$D_x \ln|x| = \frac{1}{x}.$$

Hence,

(4.32)
$$\int \frac{1}{x} dx = \ln|x| + C,$$

provided $x \neq 0$.

Example 1

Determine $\int \frac{x}{x^2 + 3} dx$.

Solution

Let $u = x^2 + 3$. Then $du = 2x\, dx$ and $\frac{1}{2} du = x\, dx$. So

$$\int \frac{x}{x^2 + 3} dx = \frac{1}{2} \int \frac{1}{u} du$$

$$= \frac{1}{2} \ln|u| + C$$

$$= \frac{1}{2} \ln(x^2 + 3) + C.$$

Note that the absolute value can be omitted in the final result since $x^2 + 3 > 0$ for all values of x. ▶

Applying the Chain Rule to Equation (4.31) leads to the general differentiation rule for the natural logarithm:

(4.33)
If $u = g(x)$ is a differentiable function of x and $g(x)$ is never zero, then

$$D_x \ln|u| = \frac{1}{u} D_x u.$$

Example 2

Find $D_x \ln \frac{\sqrt{x^2 + 3x}}{x}$.

Solution
By Equation (4.33),

$$D_x \ln \frac{\sqrt{x^2+3x}}{x} = \frac{1}{\left(\frac{\sqrt{x^2+3x}}{x}\right)} D_x \frac{\sqrt{x^2+3x}}{x}$$

$$= \frac{x}{\sqrt{x^2+3x}} \cdot \frac{x D_x \sqrt{x^2+3x} - \sqrt{x^2+3x}\, D_x x}{x^2}$$

$$= \frac{x}{\sqrt{x^2+3x}} \cdot \frac{x[\frac{1}{2}(x^2+3x)^{-1/2}(2x+3)] - \sqrt{x^2+3x}}{x^2}$$

$$= \frac{x}{\sqrt{x^2+3x}} \cdot \frac{\frac{x(2x+3) - 2(\sqrt{x^2+3x})^2}{2\sqrt{x^2+3x}}}{x^2}$$

$$= \frac{2x^2+3x-2x^2-6x}{2x(x^2+3x)} = \frac{-3}{2(x^2+3x)}.$$

▶

The properties of the natural logarithm function given in the following theorem provide the justification for calling this function a logarithm. These properties are true for all logarithm functions and should be familiar to you if you have studied common logarithms.

(4.34)
THEOREM
If a and b are positive real numbers and r is a rational number, then:

a) $\ln ab = \ln a + \ln b$; b) $\ln \frac{a}{b} = \ln a - \ln b$; c) $\ln a^r = r \ln a$.

Proof

a) We have

$$\ln ab = \int_1^{ab} \frac{1}{t}\, dt = \int_1^a \frac{1}{t}\, dt + \int_a^{ab} \frac{1}{t}\, dt = \ln a + \int_a^{ab} \frac{1}{t}\, dt.$$

It remains to show that $\int_a^{ab}(1/t)\, dt = \ln b$. To do so we first let $u = t/a$. Then $du = (1/a)\, dt$ and $dt = a\, du$. So

$$\int_{t=a}^{t=ab} \frac{1}{t}\, dt = \int_{u=1}^{u=b} \frac{a\, du}{ua} = \int_1^b \frac{du}{u} = \ln b.$$

Thus, $\ln ab = \ln a + \ln b$.

b) This property follows from part (a) by noting that since $a = (a/b) \cdot b$,

$$\ln a = \ln \frac{a}{b} \cdot b = \ln \frac{a}{b} + \ln b.$$

Therefore,

$$\ln \frac{a}{b} = \ln a - \ln b.$$

c) This property can be proved in a manner similar to the proof of part (a), but it is instructive to see the use of an alternative method. It follows from Equation (4.33) that for all $x > 0$ and all rational numbers r,

$$D_x (\ln x^r) = \frac{1}{x^r} D_x x^r = \frac{1}{x^r} (rx^{r-1}) = \frac{r}{x} = D_x (r \ln x).$$

Consequently, $\ln x^r$ and $r \ln x$ are antiderivatives of the same function. This implies that there is a constant C with

$$\ln x^r = r \ln x + C.$$

But when $x = 1$,

$$\ln 1^r = \ln 1 = 0 \quad \text{and} \quad r \ln 1 = r \cdot 0 = 0,$$

so C must be zero and $\ln x^r = r \ln x$. Replacing x by a in this equation gives the rule in the form listed in part (c) of the theorem. ▷

The properties in Theorem 4.34 are used to simplify problems involving the natural logarithm function.

Example 3

Find $D_x \ln \dfrac{\sqrt{x^2 + 3x}}{x}$ by first using Theorem 4.34 to simplify the expression.

Solution
By Theorem 4.34,

$$\ln \frac{\sqrt{x^2 + 3x}}{x} = \ln (x^2 + 3x)^{1/2} - \ln x = \frac{1}{2} \ln (x^2 + 3x) - \ln x.$$

Thus,

$$D_x \ln \frac{\sqrt{x^2 + 3x}}{x} = \frac{1}{2} D_x \ln (x^2 + 3x) - D_x \ln x$$

$$= \frac{1}{2} \frac{D_x (x^2 + 3x)}{x^2 + 3x} - \frac{1}{x}$$

$$= \frac{2x + 3}{2(x^2 + 3x)} - \frac{1}{x}.$$

This can be simplified to

$$D_x \ln \frac{\sqrt{x^2 + 3x}}{x} = \frac{x(2x + 3) - 2(x^2 + 3x)}{2x(x^2 + 3x)} = \frac{-3}{2(x^2 + 3x)}.$$

Note how much simpler the steps in this example are compared with those in Example 2, where this derivative was found without applying Theorem 4.34. ▶

The natural logarithm function can be used to simplify the differentiation of a function that involves combinations of exponents, quotients,

4.6 THE NATURAL LOGARITHM FUNCTION

and products. This procedure, known as **logarithmic differentiation,** is illustrated in the following example.

Example 4

Find $D_x y$ if $y = \dfrac{x\sqrt[3]{x+4}}{\sqrt{x^2-1}}$.

Solution

We first take the natural logarithm of both sides of the equation and simplify the expression using the properties of the natural logarithms:

$$\ln y = \ln \frac{x\sqrt[3]{x+4}}{\sqrt{x^2-1}}$$

$$= \ln (x\sqrt[3]{x+4}) - \ln \sqrt{x^2-1}$$

$$= \ln x + \ln \sqrt[3]{x+4} - \ln \sqrt{x^2-1}$$

$$= \ln x + \frac{1}{3}\ln(x+4) - \frac{1}{2}\ln(x^2-1).$$

Next we differentiate both sides with respect to x to obtain

$$\frac{1}{y}D_x y = \frac{1}{x} + \frac{1}{3(x+4)} - \frac{2x}{2(x^2-1)}$$

and then solve for $D_x y$:

$$D_x y = y\left(\frac{1}{x} + \frac{1}{3(x+4)} - \frac{2x}{2(x^2-1)}\right)$$

$$= \frac{x\sqrt[3]{x+4}}{\sqrt{x^2-1}}\left(\frac{1}{x} + \frac{1}{3(x+4)} - \frac{x}{x^2-1}\right). \quad \blacktriangleright$$

In Example 4, if $y \leq 0$, then $\ln y$ is not defined. However, if $y < 0$, we can use $\ln |y|$ to obtain the same result.

Logarithmic differentiation can be useful in finding $f'(x)$ whenever the logarithm of $f(x)$ exists and can be simplified by the arithmetic rules in Theorem 4.34. The steps involved are listed below.

(4.35) LOGARITHMIC DIFFERENTIATION

i) Let $y = f(x)$.
ii) Consider $\ln y = \ln f(x)$.
iii) Apply the results in Theorem 4.34 to simplify $\ln f(x)$.
iv) Use implicit differentiation to find $\dfrac{1}{y} D_x y = D_x \ln f(x)$.
v) Then

$$f'(x) = D_x y = y\, D_x \ln f(x) = f(x)\, D_x \ln f(x).$$

Integration formula (4.32) can be used in conjunction with the arithmetic logarithm rules to find integration formulas for the tangent, secant, cotangent, and cosecant functions.

To integrate the tangent function

$$\int \tan x \, dx = \int \frac{\sin x}{\cos x} \, dx,$$

we use the substitution $u = \cos x$. Then $du = -\sin x \, dx$ and

$$\int \tan x \, dx = -\int \frac{du}{u} = -\ln |u| + C = -\ln |\cos x| + C.$$

Since

$$-\ln |\cos x| = \ln |\cos x|^{-1} = \ln |(\cos x)^{-1}| = \ln |\sec x|,$$

this formula can be rewritten as

(4.36) $$\int \tan x \, dx = -\ln |\cos x| + C = \ln |\sec x| + C.$$

In a similar manner, the substitution $u = \sin x$ applied to the cotangent function gives

$$\int \cot x \, dx = \int \frac{\cos x}{\sin x} \, dx = \int \frac{du}{u} = \ln |u| + C,$$

so

(4.37) $$\int \cot x \, dx = \ln |\sin x| + C = -\ln |\csc x| + C.$$

Example 5

Find $\int \tan (x + 2) \, dx$.

Solution
If we let $u = x + 2$, then $du = dx$. By (4.36),

$$\int \tan (x + 2) \, dx = \int \tan u \, du = \ln |\sec u| + C$$
$$= \ln |\sec (x + 2)| + C. \quad \blacktriangleright$$

A rather unusual substitution is used to integrate the secant function. First multiply the integrand by $(\sec x + \tan x)/(\sec x + \tan x)$.

$$\int \sec x \, dx = \int \sec x \left(\frac{\sec x + \tan x}{\sec x + \tan x} \right) dx = \int \frac{\sec^2 x + \sec x \tan x}{\sec x + \tan x} \, dx.$$

Now let $u = \sec x + \tan x$. Then $du = (\sec x \tan x + \sec^2 x) \, dx$ and

$$\int \sec x \, dx = \int \frac{1}{u} \, du$$
$$= \ln |u| + C$$

Thus,

(4.38) $$\int \sec x \, dx = \ln |\sec x + \tan x| + C.$$

A similar substitution is used to integrate the cosecant function. In this case, the integrand is multiplied by $(\csc x + \cot x)/(\csc x + \cot x)$. The substitution $u = \csc x + \cot x$ produces the formula

$$\int \csc x \, dx = -\ln |\csc x + \cot x| + C.$$

Using trigonometric identities, it can be shown that

$$(\csc x + \cot x)^{-1} = \csc x - \cot x,$$

so

(4.39) $$\int \csc x \, dx = -\ln |\csc x + \cot x| + C = \ln |\csc x - \cot x| + C.$$

You should complete the details of this substitution to verify that (4.39) is correct.

Example 6

Determine $\int x \csc (3x^2 + 4) \, dx$.

Solution

If we let $u = 3x^2 + 4$, then $du = 6x \, dx$ and $x \, dx = \frac{1}{6} du$. Thus,

$$\int x \csc (3x^2 + 4) \, dx = \int \csc u \, \frac{1}{6} \, du$$

$$= \frac{1}{6} \ln |\csc u - \cot u| + C$$

and

$$\int x \csc (3x^2 + 4) \, dx = \frac{1}{6} \ln |\csc (3x^2 + 4) - \cot (3x^2 + 4)| + C. \quad \blacktriangleright$$

We complete this section with several additional observations about the natural logarithm function that help to sketch its graph. First note that:

(4.40)
The natural logarithm function is increasing and its graph is concave downward.

This is a consequence of the fact that for $x > 0$

$$D_x \ln x = \frac{1}{x} > 0 \quad \text{and} \quad D_x^2 \ln x = -\frac{1}{x^2} < 0.$$

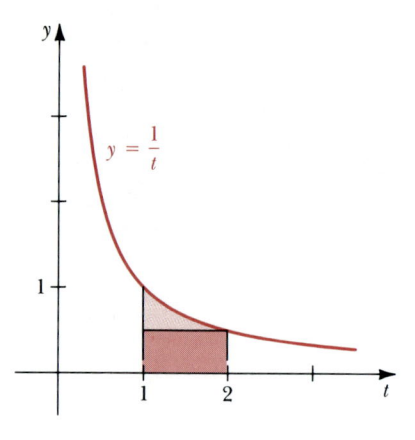

Figure 4.31
$\frac{1}{2} \le \ln 2$.

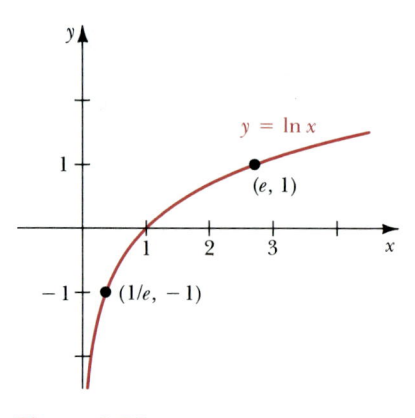

Figure 4.32
The natural logarithm function.

Next observe that

(4.41) $$\lim_{x \to \infty} \ln x = \infty.$$

To prove (4.41), first consider the portion of the graph of $f(t) = 1/t$ shown in Figure 4.31. Since $1/t \ge 1/2$ on $[1, 2]$,

(4.42) $$\ln 2 = \int_1^2 \frac{1}{t}\, dt \ge \int_1^2 \frac{1}{2}\, dt = \frac{1}{2}(2 - 1) = \frac{1}{2}.$$

Now suppose that $x = 2^n$ for some positive integer n. Then

$$\ln x = \ln 2^n = n \ln 2 \ge n/2.$$

Consequently, $\ln x$ can be made larger than any specified bound. Since the natural logarithm function is increasing on $(0, \infty)$, this implies that

$$\lim_{x \to \infty} \ln x = \infty.$$

Our final observation is that

(4.43) $$\lim_{x \to 0^+} \ln x = -\infty.$$

Since $x > 0$ approaches zero precisely when its reciprocal, $z = x^{-1}$, approaches infinity,

$$\lim_{x \to 0^+} \ln x = \lim_{z \to \infty} \ln z^{-1} = \lim_{z \to \infty} (-\ln z) = -\infty.$$

These observations, together with the fact that $\ln 1 = 0$, imply that the shape of the graph of the natural logarithm function is as shown in Figure 4.32.

We have shown that the natural logarithm function is differentiable, hence continuous on its domain $(0, \infty)$. Since

$$\ln 1 = 0 \quad \text{and, by (4.42),} \quad \ln 4 = \ln 2^2 = 2 \ln 2 \ge 1,$$

the Intermediate Value Theorem ensures the existence of a number, denoted e, whose natural logarithm is 1. This number is one of the most important mathematical constants. It is an irrational number whose value is approximately

(4.44) $$e \approx 2.718281828459045.$$

Example 7

In (4.41) we found that $\lim_{x \to \infty} \ln x = \infty$. Show, however, that the natural logarithm approaches infinity slower than any positive rational power of

HISTORICAL NOTE

The symbol e was first used in 1728 by **Leonhard Euler**, who also employed the Σ notation (both for finite and infinite summations) in 1755. In 1737 he introduced the symbol π to represent $3.141592653589793 \ldots$; it became a common notation after he used it in *Introductio in analysin infinitorum* in 1748. However, the first use of the symbol π was by William Jones in 1706 in *Synopsis palmariorum matheseas*.

x in the sense that

(4.45)
$$\lim_{x \to \infty} \frac{\ln x}{x^r} = 0$$

for all rational numbers $r > 0$.

Solution
Since

$$\lim_{x \to \infty} \ln x = \infty \quad \text{and} \quad \lim_{x \to \infty} x^r = \infty,$$

we can apply l'Hôpital's Rule. Thus,

$$\lim_{x \to \infty} \frac{\ln x}{x^r} = \lim_{x \to \infty} \frac{D_x (\ln x)}{D_x x^r} = \lim_{x \to \infty} \frac{(1/x)}{rx^{r-1}} = \lim_{x \to \infty} \frac{1}{rx^r} = 0.$$

▶

▶ EXERCISE SET 4.6

In Exercises 1–30, find the derivative of the function.

1. $f(x) = \ln(x + 2)$
2. $f(x) = \ln 2x$
3. $f(x) = \ln x^2$
4. $f(x) = \ln(x^2 + 3)$
5. $f(x) = \ln(x^2 + 3)^2$
6. $f(x) = \ln(x^2 + 2x)$
7. $f(x) = \ln x^{1/4}$
8. $f(x) = (\ln x)^{1/4}$
9. $f(x) = \ln(\ln x)$
10. $f(x) = \ln(1/\ln x)$
11. $f(x) = \ln(\ln x^{1/4})$
12. $f(x) = \ln(\ln x)^{1/4}$
13. $f(x) = \ln(x + \sqrt{x^2 - 1})$
14. $f(x) = \ln(x - \sqrt{x^2 - 1})$
15. $f(x) = \ln\left(\frac{1 + x}{1 - x}\right)$
16. $f(x) = \ln\left(\frac{x^3 + 4x + 5}{x^2 + 4x + 1}\right)$
17. $f(x) = \left(\ln \frac{1+x}{1-x}\right)^{1/2}$
18. $f(x) = \ln \sqrt[4]{\frac{x^3 + 4x + 5}{x^2 + 4x + 1}}$
19. $f(x) = \ln(\sin x)$
20. $f(x) = \ln(\csc x)$
21. $f(x) = x \ln(x^2 + 1)$
22. $f(x) = \frac{\ln(x^2 + 1)}{x}$
23. $f(x) = \frac{x^2 + 1}{\ln x}$
24. $f(x) = x \ln x - x$
25. $f(x) = \ln|2x + 1|$
26. $f(x) = \ln|9x^2 - 4|$
27. $f(x) = (\ln \sqrt{x^2 + 2})^3$
28. $f(x) = (\ln \sqrt{x} + 1)^{1/2}$
29. $f(x) = \cos(\ln x)$
30. $f(x) = \tan(\ln x)$

In Exercises 31–36, find $D_x y$.

31. $\ln xy = 1$
32. $\ln(x + y) = x$
33. $\ln(x + y) = y$
34. $\ln \frac{x}{y} + \ln \frac{y^2}{x^2} = 5$
35. $\ln\left(\frac{x}{y}\right) + x^2 + y^2 = 3$
36. $\ln(x + y) = \ln(x - y) + 1$

In Exercises 37–54, evaluate the integral.

37. $\int \frac{dx}{x - 3}$

38. $\int \dfrac{dx}{2x+5}$

39. $\int_{-3}^{-1} \dfrac{dx}{x-1}$

40. $\int_{-5}^{-1} \dfrac{dx}{2x+1}$

41. $\int_0^{1/4} \tan \pi x \, dx$

42. $\int_0^{\pi^2/16} 3x \sec x^2 \, dx$

43. $\int (\csc 2x + \cot 3x) \, dx$

44. $\int_1^2 \left(\dfrac{1}{x+1} + \cot \dfrac{\pi x}{6} \right) dx$

45. $\int \dfrac{x+1}{x^2+2x-3} \, dx$

46. $\int \dfrac{2x}{(x^2+4)^2} \, dx$

47. $\int_2^3 \dfrac{\ln x}{x} \, dx$

48. $\int \dfrac{(\ln x)^3}{x} \, dx$

49. $\int \dfrac{(\ln x^2)^3}{x} \, dx$

50. $\int_1^4 \dfrac{\cos (\ln x)}{x} \, dx$

51. $\int_0^3 \dfrac{dx}{\sqrt{x+1}}$

52. $\int \dfrac{dx}{\sqrt{x}(\sqrt{x}+1)}$

53. $\int \dfrac{\cos x}{1+\sin x} \, dx$

54. $\int \dfrac{\sec^2 x}{2+\tan x} \, dx$

In Exercises 55–60, use logarithmic differentiation to find $D_x y$.

55. $y = (x^2+8)^3 (3x+7)^5$

56. $y = \dfrac{(x^2+7x)^5 \sqrt{x^3+1}}{(x^3-2x+1)^{1/5}}$

57. $y = \sqrt{x \sqrt{(x+1) \sqrt{x}}}$

58. $y = \dfrac{(x^3-5)^{2/3} (x^2+2)^{1/3}}{\sqrt{x}}$

59. $y = \sqrt{x^2+1} \, (x^3-1)^5 (2x^2+1)^7$

60. $y = \sqrt{x^2 \sqrt{(3x-7) \sqrt{x^3+1}}}$

In Exercises 61–70, evaluate the limit.

61. $\lim\limits_{x \to \infty} \dfrac{\ln \sqrt{x}}{x}$

62. $\lim\limits_{x \to \infty} \dfrac{\ln x^5}{\sqrt{x}}$

63. $\lim\limits_{x \to \infty} \dfrac{\ln \sqrt{x}}{x^5}$

64. $\lim\limits_{x \to \infty} \dfrac{\sqrt{\ln x^5}}{x^5}$

65. $\lim\limits_{x \to 0^+} x \ln x$

66. $\lim\limits_{x \to 0^+} \sqrt{x} \ln x$

67. $\lim\limits_{x \to 0^+} x^2 \ln \sqrt{x}$

68. $\lim\limits_{x \to 0^+} \sqrt{x} \ln x^2$

69. $\lim\limits_{x \to \infty} (x - \ln x)$

70. $\lim\limits_{x \to 0^+} (\ln x + 1/x)$

71. Show that for every positive rational number r, $\lim_{x \to 0^+} x^r \ln x = 0$.

In Exercises 72–81, sketch the graph of the function.

72. $f(x) = \ln (x+1)$
73. $f(x) = \ln (x-1)$
74. $f(x) = 1 + \ln x$
75. $f(x) = \ln (-x)$
76. $f(x) = -\ln x$
77. $f(x) = \ln |x|$
78. $f(x) = \ln \dfrac{x+1}{x}$
79. $f(x) = x \ln x$
80. $f(x) = x^2 \ln x$
81. $f(x) = \sqrt{x} \ln x$

82. Find the area of the region bounded by the graph of $f(x) = 2/(x-1)$, the x-axis, and the lines $x = 2$ and $x = 3$.

83. Find an equation of the tangent line to the graph of $y = x + 3 \ln x$ at $(1, 1)$.

84. Verify that $\int \ln x \, dx = x \ln x - x + C$, and use this to find the following.

 a) $\int_1^3 \ln x \, dx$

 b) $\int (3 \ln x - 2x) \, dx$

 c) $\int_2^4 (\ln x - 2x) \, dx$

 d) $\int 2x \ln x^2 \, dx$

 e) The area of the region bounded by the graph of $f(x) = \ln x$, the x-axis, and the line $x = 3$.

 f) An equation of the curve passing through $(1, 1)$ whose tangent line at (x, y) has slope $\ln x$.

85. Use

$$\dfrac{x}{1-x^2} = \dfrac{1}{2} \left[\dfrac{1}{1-x} - \dfrac{1}{1+x} \right]$$

to find
$$\int \frac{x}{1-x^2}\, dx.$$
Compare this result with that obtained by making the substitution $u = 1 - x^2$.

86. Use
$$\frac{1}{x^2 - 5x + 6} = \frac{1}{x-3} - \frac{1}{x-2}$$
to find
$$\int \frac{1}{x^2 - 5x + 6}\, dx.$$

87. a) Use
$$\frac{1}{1-x^2} = \frac{1}{2}\left[\frac{1}{1-x} + \frac{1}{1+x}\right]$$
to find
$$\int \frac{1}{1-x^2}\, dx.$$
b) Verify the identity $\sec x = \cos x/(1 - \sin^2 x)$.
c) Use the result in part (a) and the identity in part (b) to evaluate $\int \sec x\, dx$.
d) Derive an identity for $\csc x$ in a form similar to the identity in part (b), and use this to evaluate $\int \csc x\, dx$.

88. If r is a nonzero rational number, use the substitution $u = t^{1/r}$ to show that $\ln a^r = r \ln a$, in a manner similar to that used in proving property (a) of Theorem 4.34.

89. a) Show that if $0 < a < b$, then
$$\frac{1}{b} < \frac{\ln b - \ln a}{b - a} < \frac{1}{a}.$$
b) By choosing a and b appropriately, show that for all $x > 0$ and all positive integers n,
$$\frac{1}{(1 + x/n)} < \frac{\ln (1 + x/n)^n}{x} < 1.$$
c) Show that for all $x > 0$,
$$\lim_{n \to \infty} \ln (1 + x/n)^n = x.$$

90. Use Newton's Method to find an approximation, accurate to within three decimal places, to a number that satisfies $\ln x = x/10$.

91. Suppose that g is continuous on $[0, 1]$, $f(x) > 0$, and $f'(x) = f(x) g(x)$ for x in $[0, 1]$. Show that
$$\int_0^1 g(x)\, dx = \ln \frac{f(1)}{f(0)}.$$

92. The vapor pressure P of water measured in millimeters of mercury depends on the temperature of the water according to the equation
$$\ln P = 21.020 - \frac{5.319 \times 10^3}{T}, \qquad T > 1000°\ \text{C}.$$
Find an expression for dP/dT.

93. The molar absorption coefficient in spectroscopy, α, is defined by
$$\alpha(x) = \frac{-0.4343}{cx} \ln\left(\frac{I(x)}{I_0}\right),$$
where c is the molar concentration of the substance in solution, x is the thickness of the solution layer that absorbs the light, I_0 is the initial intensity of the light beam, and $I(x)$ is the intensity of the light after it has passed through the location. Find an expression for the rate at which $I(x)$ changes with respect to x.

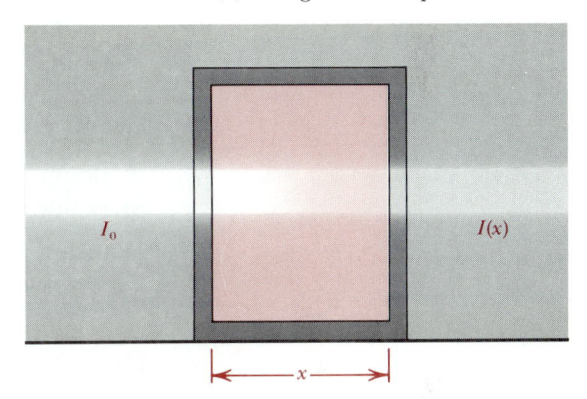

94. When a gas is expanded or compressed from a volume V_1 to a volume V_2, the work W done is $W = \int_{V_1}^{V_2} P(V)\, dV$, where P is the pressure exerted by the gas as a function of the volume V. Suppose a gas satisfies the equation of state $P(V)V = RT + \alpha P(V)$, where α is a function of the temperature T and R is a constant. Show that if the gas expands isothermally (without changing T) from V_1 to V_2, then
$$W = RT \ln\left|\frac{P(V_1)}{P(V_2)}\right|.$$

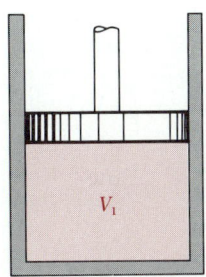

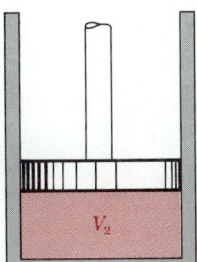

95. A company must make a decision whether to purchase a new machine for shredding scrap metal. It is estimated that the machine will have a useful life of 5 years and will be worth $500 at the end of that time. The machine costs $4500 new and the savings in dollars from the machine during its useful life are given by $s(t) = 3500/(t + 2)$, where t is time in years. Will the machine pay for itself?

96. Show that for all $x > 0$, $x > \sqrt{1 + x} \ln (1 + x)$.

4.7

Improper Integrals

The definition of a definite integral was given in Section 4.2 for a function f defined on a closed interval $[a, b]$ by considering the limit of Riemann sums. The function f need not be continuous for this integral to exist, but, by Theorem 4.10, f must be bounded. Later in that section the definition of the definite integral was extended to consider $\int_a^b f(x)\, dx$ when $a = b$ and when $a > b$.

In this section the concept of the definite integral is extended to include:

  i) definite integrals with infinite intervals of integration; and
  ii) definite integrals of unbounded functions.

These integrals are called *improper integrals* to distinguish them from the standard Riemann integral defined in Section 4.2.

Definite Integrals with Infinite Intervals of Integration

The functions described by

$$f(x) = 1/x \quad \text{and} \quad g(x) = 1/x^2$$

have similar properties when restricted to the interval $[1, \infty)$. Both:

  i) have the value 1 when $x = 1$;
  ii) are positive and decreasing for $x > 1$; and
  iii) approach zero as x approaches infinity.

This can be seen in Figure 4.33.

The definite integrals of f and g on an interval $[1, M]$, where $M > 1$, are

$$\int_1^M f(x)\, dx = \int_1^M \frac{1}{x}\, dx = \ln M$$

and

$$\int_1^M g(x)\, dx = \int_1^M \frac{1}{x^2}\, dx = -\frac{1}{x}\Big]_1^M = 1 - \frac{1}{M}.$$

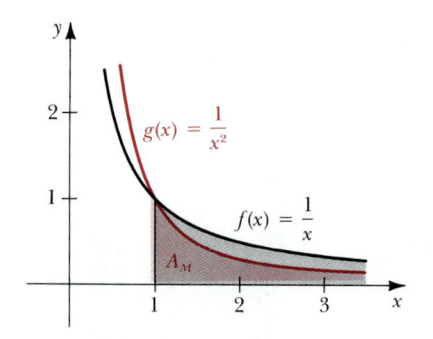

Figure 4.33

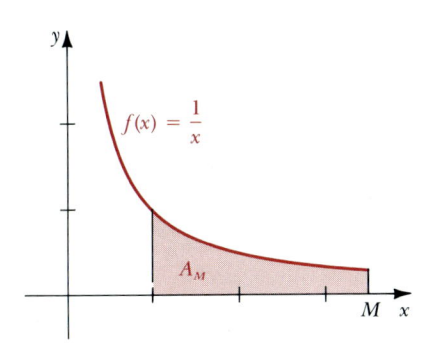

Figure 4.34
$\lim_{M \to \infty} A_M = \infty.$

Figure 4.35
$\lim_{M \to \infty} A_M = 1.$

We now see a significant difference between $f(x) = 1/x$ and $g(x) = 1/x^2$:

$$\lim_{M \to \infty} \int_1^M \frac{1}{x} \, dx = \lim_{M \to \infty} \ln M = \infty,$$

as shown in Figure 4.34, whereas

$$\lim_{M \to \infty} \int_1^M \frac{1}{x^2} \, dx = \lim_{M \to \infty} \left(1 - \frac{1}{M}\right) = 1,$$

as shown in Figure 4.35. Thus the area of the region bounded by the x-axis, the graph of $f(x) = 1/x$, and the lines $x = 1$ and $x = M$ becomes unbounded as M approaches infinity. On the other hand, the area of the region bounded by the x-axis, the graph of $g(x) = 1/x^2$, and the line $x = 1$ could logically be defined to be one.

(4.46)
DEFINITION
If f is continuous on $[a, \infty)$, then

$$\int_a^\infty f(x) \, dx = \lim_{M \to \infty} \int_a^M f(x) \, dx.$$

Similarly, if f is continuous on $(-\infty, a]$, then

$$\int_{-\infty}^a f(x) \, dx = \lim_{M \to -\infty} \int_M^a f(x) \, dx.$$

These types of integrals are called **improper integrals.** An improper integral is said to **converge** if the limit is finite and is said to **diverge** otherwise.

By the discussion preceding Definition 4.46, we see that

$$\int_1^\infty \frac{1}{x^2} = 1 \quad \text{converges,} \quad \text{but} \quad \int_1^\infty \frac{1}{x} \, dx \quad \text{diverges.}$$

Example 1

Determine whether the improper integral $\int_2^\infty \frac{1}{\sqrt{x}}\,dx$ converges or diverges.

Solution
We have
$$\int_2^\infty \frac{1}{\sqrt{x}}\,dx = \lim_{M\to\infty}\int_2^M x^{-1/2}\,dx = \lim_{M\to\infty} 2x^{1/2}\Big]_2^M$$
$$= \lim_{M\to\infty}(2\sqrt{M} - 2\sqrt{2}) = \infty.$$

The improper integral diverges. ▶

Example 2

Determine whether the improper integral $\int_0^\infty \frac{x}{(1+x^2)^2}\,dx$ converges or diverges.

Solution
We have
$$\int_0^\infty \frac{x}{(1+x^2)^2}\,dx = \lim_{M\to\infty}\int_0^M \frac{x}{(1+x^2)^2}\,dx.$$

To evaluate this integral we let $u = 1 + x^2$, so $du = 2x\,dx$. Then, $\frac{1}{2}du = x\,dx$ and
$$\int \frac{x}{(1+x^2)^2}\,dx = \int \frac{1}{u^2}\frac{1}{2}\,du = -\frac{1}{2u} + C = -\frac{1}{2(1+x^2)} + C.$$

The integral converges since
$$\int_0^\infty \frac{x}{(1+x^2)^2}\,dx = \lim_{M\to\infty} -\frac{1}{2(1+x^2)}\Big]_0^M$$
$$= \lim_{M\to\infty}\left(-\frac{1}{2(1+M^2)} + \frac{1}{2}\right) = \frac{1}{2}.$$ ▶

The improper integral over the set of all real numbers, the interval $(-\infty, \infty)$, can also be defined.

(4.47) DEFINITION

If f is continuous on $(-\infty, \infty)$ and a is a real number, then
$$\int_{-\infty}^\infty f(x)\,dx = \int_{-\infty}^a f(x)\,dx + \int_a^\infty f(x)\,dx.$$

The improper integral $\int_{-\infty}^\infty f(x)\,dx$ is said to **converge** if *both*
$$\int_{-\infty}^a f(x)\,dx \quad \text{and} \quad \int_a^\infty f(x)\,dx$$
converge, and is said to **diverge** otherwise.

4.7 IMPROPER INTEGRALS

Although a specific real number a is introduced in the definition of $\int_{-\infty}^{\infty} f(x)\, dx$, the convergence and value of this improper integral are independent of the number a. Exercise 46 concerns a demonstration of this fact.

Example 3

Determine whether the improper integral

$$\int_{-\infty}^{\infty} \frac{x}{(1+x^2)^2}\, dx$$

converges or diverges.

Solution

This integral can be rewritten as

$$\int_{-\infty}^{\infty} \frac{x}{(1+x^2)^2}\, dx = \int_{-\infty}^{0} \frac{x}{(1+x^2)^2}\, dx + \int_{0}^{\infty} \frac{x}{(1+x^2)^2}\, dx.$$

The second integral in this sum was found in Example 2 to have the value $\tfrac{1}{2}$. It was also found that

$$\int \frac{x}{(1+x^2)^2}\, dx = \frac{-1}{2(1+x^2)} + C.$$

Thus the first integral is

$$\int_{-\infty}^{0} \frac{x}{(1+x^2)^2}\, dx = \lim_{M \to -\infty} \int_{M}^{0} \frac{x}{(1+x^2)^2}\, dx = \lim_{M \to -\infty} \frac{-1}{2(1+x^2)} \Big]_{M}^{0}$$

$$= \lim_{M \to -\infty} \left(-\frac{1}{2} + \frac{1}{2(1+M^2)}\right) = -\frac{1}{2}.$$

Consequently,

$$\int_{-\infty}^{\infty} \frac{x}{(1+x^2)^2}\, dx = -\frac{1}{2} + \frac{1}{2} = 0.$$

The graph of $f(x) = x/(1+x^2)^2$ is shown in Figure 4.36.

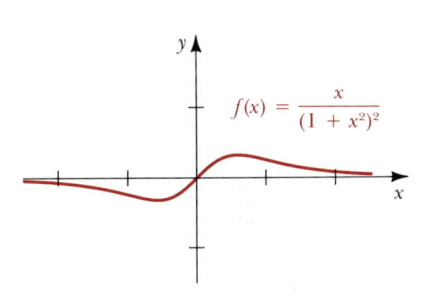

▶ **Figure 4.36**

Definite Integrals of Unbounded Functions

Consider the definite integral

$$\int_{0}^{2} \frac{1}{(x-1)^2}\, dx.$$

If the Fundamental Theorem of Calculus is applied blindly to this integral, then

$$\int_{0}^{2} \frac{1}{(x-1)^2}\, dx = -\frac{1}{(x-1)} \Big]_{0}^{2}$$

$$= -\frac{1}{2-1} + \frac{1}{0-1} = -2.$$

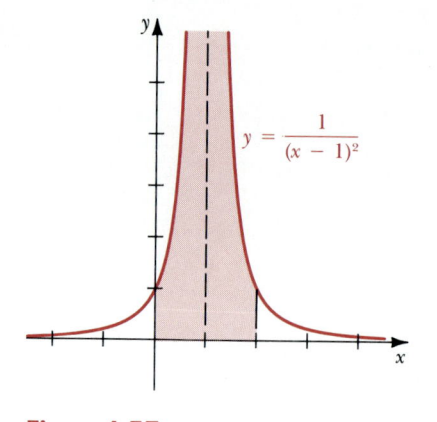

Figure 4.37
$$\lim_{x \to 1} \frac{1}{(x-1)^2} = \infty.$$

This is clearly false, since, as shown in Figure 4.37, the integrand, $1/(x-1)^2$, is never negative. The error occurs because the Fundamental Theorem of Calculus is used where it does not apply.

Theorem 4.10 stated that if the definite integral of a function exists on an interval, then the function is bounded on that interval. However,

$$\lim_{x \to 1} \frac{1}{(x-1)^2} = \infty,$$

so the function defined by $f(x) = 1/(x-1)^2$ is unbounded on the interval $[0, 2]$.

This brings us to the second type of improper integral: the extension of the concept of the definite integral to an interval on which the integrand is unbounded.

(4.48)
DEFINITION
If f is continuous on $(a, b]$ and has an infinite discontinuity at a, that is, $\lim_{x \to a^+} f(x) = \pm\infty$, then

$$\int_a^b f(x)\, dx = \lim_{M \to a^+} \int_M^b f(x)\, dx.$$

Similarly, if f is continuous on $[a, b)$ and has an infinite discontinuity at b, then

$$\int_a^b f(x)\, dx = \lim_{M \to b^-} \int_a^M f(x)\, dx.$$

An integral involving an infinite discontinuity is also called an **improper integral**; it is said to **converge** if the limit is finite and is said to **diverge** otherwise.

In addition, if f is continuous on $[a, c)$ and on $(c, b]$ but has an infinite discontinuity at c, then

(4.49) $$\int_a^b f(x)\, dx = \int_a^c f(x)\, dx + \int_c^b f(x)\, dx.$$

This improper integral is said to **converge** if *both*

$$\int_a^c f(x)\, dx \quad \text{and} \quad \int_c^b f(x)\, dx$$

converge, and is said to **diverge** otherwise.

Example 4

Determine whether each of the following improper integrals converges or diverges.

a) $\displaystyle\int_0^1 \frac{dx}{(x-1)^2}$ b) $\displaystyle\int_0^2 \frac{dx}{(x-1)^2}$

Solution

a) An infinite discontinuity occurs at $x = 1$ so we must examine

$$\lim_{M \to 1^-} \int_0^M \frac{dx}{(x-1)^2} = \lim_{M \to 1^-} \left[-\frac{1}{x-1} \right]_0^M = \lim_{M \to 1^-} \left[-\frac{1}{M-1} - 1 \right] = \infty.$$

This improper integral diverges.

b) The improper integral

$$\int_0^2 \frac{dx}{(x-1)^2}$$

also diverges, since this integral cannot converge unless *both*

$$\int_0^1 \frac{dx}{(x-1)^2} \quad \text{and} \quad \int_1^2 \frac{dx}{(x-1)^2}$$

converge. ▶

Example 5

Determine whether the improper integral $\int_0^4 \frac{1}{\sqrt{x}} \, dx$ converges or diverges.

Solution

The function $f(x) = 1/\sqrt{x}$ has an infinite discontinuity at $x = 0$:

$$\lim_{M \to 0^+} \int_M^4 \frac{dx}{\sqrt{x}} = \lim_{M \to 0^+} 2\sqrt{x} \Big]_M^4 = \lim_{M \to 0^+} (2\sqrt{4} - 2\sqrt{M}) = 4.$$

Thus the improper integral converges and

$$\int_0^4 \frac{dx}{\sqrt{x}} = 4.$$

Figure 4.38 shows the graph of this function. ▶

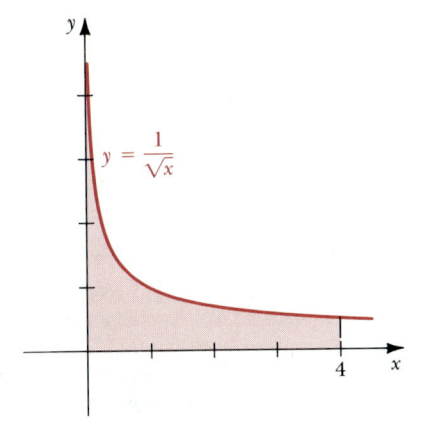

Figure 4.38

$\int_0^4 \frac{1}{\sqrt{x}} \, dx = 4.$

Example 6

Determine whether the improper integral $\int_{-1}^8 x^{-2/3} \, dx$ converges or diverges.

Solution

Since $f(x) = x^{-2/3}$ has an infinite discontinuity at zero, the integral must be considered as

$$\int_{-1}^8 x^{-2/3} \, dx = \int_{-1}^0 x^{-2/3} \, dx + \int_0^8 x^{-2/3} \, dx.$$

Evaluating each improper integral, we have

$$\int_{-1}^0 x^{-2/3} \, dx = \lim_{M \to 0^-} \int_{-1}^M x^{-2/3} \, dx$$

$$= \lim_{M \to 0^-} 3x^{1/3} \Big]_{-1}^M = \lim_{M \to 0^-} [3M^{1/3} - 3(-1)] = 3$$

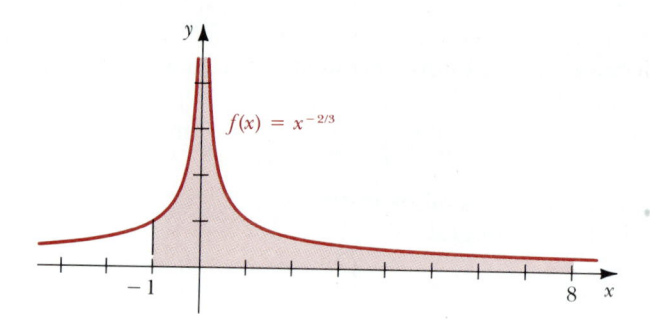

Figure 4.39
$$\int_{-1}^{1} x^{-2/3}\, dx = 6.$$

and

$$\int_{0}^{8} x^{-2/3}\, dx = \lim_{M \to 0^+} \int_{M}^{8} x^{-2/3}\, dx = \lim_{M \to 0^+} 3x^{1/3}\Big]_{M}^{8}$$
$$= \lim_{M \to 0^+} (3 \cdot 8^{1/3} - 3M^{1/3}) = 6.$$

Thus,

$$\int_{-1}^{8} x^{-2/3}\, dx = 3 + 6 = 9.$$

The graph of f is shown in Figure 4.39. ▶

▶ EXERCISE SET 4.7

In Exercises 1–34, determine whether the improper integrals converge or diverge, and evaluate those that converge.

1. $\displaystyle\int_{1}^{\infty} \frac{1}{x^{3/4}}\, dx$

2. $\displaystyle\int_{-\infty}^{-1} \frac{dx}{x^3}$

3. $\displaystyle\int_{1}^{\infty} \frac{x^2}{(x^3+1)^2}\, dx$

4. $\displaystyle\int_{100}^{\infty} \frac{1}{x+5}\, dx$

5. $\displaystyle\int_{3}^{\infty} \frac{1}{x(\ln x)^2}\, dx$

6. $\displaystyle\int_{0}^{\infty} \frac{x}{1+x^2}\, dx$

7. $\displaystyle\int_{1}^{\infty} \frac{x}{(x^2+1)^3}\, dx$

8. $\displaystyle\int_{1}^{\infty} \frac{1}{x\sqrt{x}}\, dx$

9. $\displaystyle\int_{1}^{\infty} \frac{1}{\sqrt{x}(\sqrt{x}+1)}\, dx$

10. $\displaystyle\int_{-\infty}^{0} \frac{x}{\sqrt{x^2+1}}\, dx$

11. $\displaystyle\int_{1}^{\infty} \left(\frac{1}{x^3} + \frac{1}{x^2}\right) dx$

12. $\displaystyle\int_{1}^{\infty} \left(\frac{1}{x^3} + \frac{1}{x^4}\right) dx$

13. $\displaystyle\int_{0}^{\infty} \cos x\, dx$

14. $\displaystyle\int_{5}^{\infty} x \sin x^2\, dx$

15. $\displaystyle\int_{-1}^{0} \frac{1}{x^2}\, dx$

16. $\displaystyle\int_{0}^{1} \frac{dx}{(x-1)^{2/3}}$

17. $\displaystyle\int_{1}^{2} \frac{1}{(x-1)^2}\, dx$

18. $\displaystyle\int_{0}^{1} \frac{1}{\sqrt{1-x}}\, dx$

19. $\displaystyle\int_{0}^{\pi/2} \sec^2 x\, dx$

20. $\displaystyle\int_{0}^{\pi/2} \tan^2 x\, dx$

21. $\displaystyle\int_{0}^{\pi/2} \sec x\, dx$

22. $\displaystyle\int_{0}^{\pi/2} \tan x\, dx$

23. $\displaystyle\int_{0}^{2} \frac{1}{(x-1)^2}\, dx$

24. $\displaystyle\int_{-1}^{1} \frac{1}{x^3}\, dx$

25. $\displaystyle\int_{0}^{1} \frac{1}{\sqrt[3]{3x-1}}\, dx$

26. $\displaystyle\int_{-1}^{0} \frac{dx}{\sqrt[3]{2x+1}}$

27. $\displaystyle\int_{-\infty}^{\infty} \frac{x}{(1+x^2)^3}\, dx$

28. $\displaystyle\int_{-\infty}^{\infty} \frac{x^2}{1+x^3}\, dx$

29. $\displaystyle\int_{0}^{2} \left(\frac{1}{x} + \frac{1}{x^2}\right) dx$

30. $\displaystyle\int_{1}^{\infty} \left(\frac{1}{x} + \frac{1}{x^2}\right) dx$

31. $\displaystyle\int_{0}^{\infty} \frac{1}{\sqrt[3]{x-1}}\, dx$

32. $\displaystyle\int_{-1}^{\infty} \frac{1}{\sqrt[3]{x}}\, dx$

33. $\displaystyle\int_{-\infty}^{0} \frac{1}{(x+2)^2}\, dx$

34. $\displaystyle\int_{-\infty}^{0} \frac{x}{(x^2-4)^3}\, dx$

In Exercises 35–38, the boundaries of a region are given. Determine whether the region has finite area and if so, calculate the area.

35. $y = \dfrac{1}{x^2}$, x-axis, $x = 0$, $x = 1$

36. $y = \tan x$, x-axis, $x = 0$, $x = \dfrac{\pi}{2}$

37. $y = \dfrac{1}{x^3}$, x-axis, $x = 1$, $x \geq 1$

38. $y = \dfrac{1}{\sqrt{x+1}}$, x-axis, $x = 3$, $x \geq 3$

39. Show that for a rational number r, the integral $\int_1^\infty x^{-r}\, dx$ converges if $r > 1$ and diverges if $r \leq 1$.

40. Show that for a rational number r, the integral $\int_0^1 x^{-r}\, dx$ converges if $0 \leq r < 1$ and diverges if $r \geq 1$.

41. Find all positive rational numbers r for which the integral $\int_{-\infty}^{-1} x^{-r}\, dx$ converges.

42. Find all positive rational numbers r for which the integral $\int_{-1}^{0} x^{-r}\, dx$ converges.

43. a) Show that $\lim_{M \to \infty} \int_{-M}^{M} x\, dx$ exists.
    b) Show that $\int_{-\infty}^{\infty} x\, dx$ diverges.
    Parts (a) and (b) show the reason for defining $\int_{-\infty}^{\infty} f(x)\, dx$ as in (4.47) rather than as $\lim_{M \to \infty} \int_{-M}^{M} f(x)\, dx$.

44. Show that if $\int_1^\infty f(x)\, dx$ converges and c is a constant, then $\int_1^\infty cf(x)\, dx$ converges.

45. Suppose that f and g are continuous on $[1, \infty)$ and $0 \leq g(x) \leq f(x)$ for $x \geq 1$. Show that if $\int_1^\infty g(x)\, dx$ diverges, then $\int_1^\infty f(x)\, dx$ diverges.

46. Suppose $\int_{-\infty}^{a} f(x)\, dx = I_1$ and $\int_a^\infty f(x)\, dx = I_2$. Use Theorem 4.17 in Section 4.2 to show that, for any other real number b,

$$\int_{-\infty}^{b} f(x)\, dx + \int_b^\infty f(x)\, dx = I_1 + I_2.$$

4.8 The Discovery of Calculus

Gottfried Wilhelm von Leibniz (1646–1716) and Isaac Newton (1642–1727) are generally credited with independently discovering the calculus during the last third of the seventeenth century. But which part of calculus did they discover: the differential calculus, the integral calculus, or the important link between these two concepts, the Fundamental Theorem of Calculus? The answer to this question is one that appears all too frequently on multiple-choice examinations: none of the above. To place this statement in perspective, let us consider some important mathematical results that led to the unified theory of calculus.

The underlying concept of integral calculus was used by Greek mathematicians at least as early as the time of Eudoxus of Cnidus (408–355 B.C.) and Archimedes (circa 287–212 B.C.). Archimedes determined the area of a circle by computing the area of inscribed and circumscribed polygons of increasing numbers of sides, as shown in Figure 4.40. In this and similar applications, he was using the concept behind Riemann sums. Archimedes also devised ways of determining the tangent lines to certain curves, including parabolas and the curve that bears his name: the spiral of Archimedes, shown in Figure 4.41. In some sense, then, Archimedes could be considered the founder of calculus. He did not, however, have a notion of a unified theory that could be applied to more than a few specific cases, nor did he recognize a connection between the differential and integral concepts of calculus.

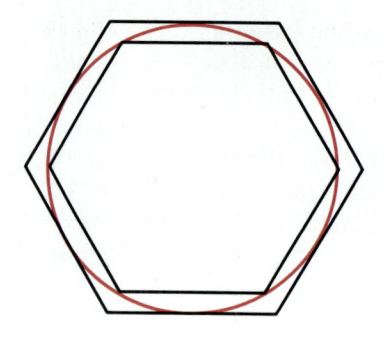

Figure 4.40

Figure 4.41

Little progress was made toward the discovery of the unified theory of calculus until the beginning of the seventeenth century, when the techniques of Eudoxus and Archimedes were used by Bonaventura Cavalieri (1598–1647) and Galileo Galilei (1564–1642). Cavalieri was able to compute the areas and volumes of a number of common figures. His techniques were extended in Italy by Evangelista Torricelli (1608–1647), in France by Pierre de Fermat (1601–1665), Christiaan Huygens (1629–1695), and Gilles Personier de Roberval (1602–1675), and in England by John Wallis (1616–1703) and Isaac Barrow (1630–1677).

The problem of constructing tangents to curves was also revived during this period, first by Fermat and his contemporary and rival René Descartes (1596–1650) and later by Roberval and Barrow. With such active research, it was only a matter of time until the unified theory was discovered.

The first published statement concerning the Fundamental Theorem of Calculus appears in *Lectiones geometricae,* a treatise published by Barrow in 1670. The theorem, however, is believed to have been recognized intuitively by Galileo 50 years earlier in connection with his study of motion.

This brings us to the time of Newton, a young student at Cambridge in the 1660s, and to Leibniz, who was born in Leipzig and was self-trained in mathematics. These two men systematically unified and codified the known results of calculus, giving, in essence, algorithmic procedures for the use of these results. Each gave a proof of the Fundamental Theorem of Calculus and each clearly demonstrated the importance of this new theory.

Newton developed most of his calculus, called the "method of fluxions and fluents," during the period from 1664 to 1669 and compiled his results in the tract *De analysi per equationes numero terminorum infinitas* in 1669. Although this manuscript was circulated and studied by a number of his English contemporaries, it did not appear in print until 1711, over 40 years later. In fact, Newton probably used his calculus to develop many important discoveries regarding gravitation and the motion of objects, but his treatise on this subject, *Philosophiae naturalis principia mathematica* (1687), contains only classical geometric demonstrations.

It is difficult to determine precisely when Leibniz first became interested in the calculus, but it was probably shortly before he traveled to France and England in 1673 as a political envoy. While visiting the London home of John Collins (1625–1683), he saw Newton's 1669 tract. He probably did not have a sufficient mathematical background to follow Newton's arguments at this time but he was nevertheless excited by the results, particularly those dealing with series. After studying Descartes' fundamental work *La géométrie,* he communicated with Newton regarding the discoveries Newton had made. The two exchanged several letters during 1676–1677, by which time Leibniz had developed his own theory of calculus. The letters generally describe the extent of their work, but often omit crucial details necessary for the methods of discovery.

Leibniz understandably expected that Newton would soon publish a treatise on calculus. When it became obvious that this work was not forthcoming, Leibniz began in 1682 to publish his own discoveries in a series of papers in the *Acta eruditorum,* a journal published in Berlin with a wide circulation in Europe. In 1684 Leibniz published the first work on differential calculus and in 1686 the first on integral calculus. His articles are often vague and sketchy and were never collected in a definitive treatise.

Because of Leibniz's prior publication, his calculus became the version known to the mathematical public of the time, particularly the European scientific community. We use his differential notation, dy/dx, for differentiation and his elongated S symbol, $\int$, to represent integration. He called his integral calculus *calculus summatorius;* the term *integral* was introduced by Jakob Bernoulli in 1690. Newton's notation was generally more cumbersome, although his symbol $\dot{y}$ to denote the derivative of y is still commonly used to indicate differentiation with respect to time.

Many reasons have been suggested for Newton's failure to capitalize on his discovery of calculus: his reticence, his preoccupation with other research, and his lack of interest in publishing. Certainly he had a complex personality and a sensitivity to criticism. Nevertheless, spurred on by their friends and colleagues, Newton and Leibniz were locked in a bitter controversy for nearly 20 years over who deserved the credit for discovering the differential calculus. This is one of the saddest chapters in the history of mathematics: Leibniz complained that Newton's attitude was the malicious interpretation of a man who was looking for a quarrel, while Newton said that second inventors count for nothing! Of course, there was more than enough honor to go around, and the effect of ther quarrel has been only to tarnish the images of both these mathematical giants.

It should be kept in mind that neither Newton nor Leibniz established their results with anything resembling modern mathematical rigor. We have seen that the limit concept is basic to the study of both the differential and the integral calculus. Although this concept is intuitively clear, its definition is quite sophisticated. It was not until 1870 that Heinrich Eduard Heine (1821–1881) published the definition for the limit of a function that we use today. Heine's work was strongly influenced by that of Karl Weierstrass (1815–1897), who was one of the leaders in the movement to place function theory on a firm and rigorous basis. With Heine's definition, the results of calculus can be proved as rigorously as the geometric demonstrations that were accepted by the mathematicians of the seventeenth century.

▶ IMPORTANT TERMS AND RESULTS

CONCEPT	PAGE	CONCEPT	PAGE
Summation	252	Fundamental Lemma of Calculus	272
Partition	259	Fundamental Theorem of Calculus	274
Riemann sum	260	Indefinite integral	278
Definite integral	261	Integration by substitution	286
Area	264	Natural logarithm	295
Mean Value Theorem for Integrals	270	Logarithmic differentiation	299
Average value	271	Improper integrals	306

DERIVATIVE FORMULA

If $u = g(x)$ is a positive differentiable function of x, then

$$D_x \ln u = \frac{1}{u} D_x u.$$

INTEGRAL FORMULAS

1. $\int_a^b f(x)\, dx = \int_a^c f(x)\, dx + \int_c^b f(x)\, dx$

2. If $f(x) \geq g(x)$ on $[a, b]$, then $\int_a^b f(x)\, dx \geq \int_a^b g(x)\, dx$.

3. $\int_a^b f(x)\, dx = -\int_b^a f(x)\, dx$

4. $\int_a^a f(x)\, dx = 0$

5. $\int (f \pm g)(x)\, dx = \int f(x)\, dx \pm \int g(x)\, dx$

6. $\int k f(x)\, dx = k \int f(x)\, dx$

7. $\int x^r\, dx = \frac{x^{r+1}}{r+1} + C,$ for any rational $r \neq -1$

8. $\int x^{-1}\, dx = \ln |x| + C$

9. $\int \sin x\, dx = -\cos x + C$

10. $\int \cos x\, dx = \sin x + C$

11. $\int \sec^2 x\, dx = \tan x + C$

12. $\int \csc^2 x\, dx = -\cot x + C$

13. $\int \sec x \tan x\, dx = \sec x + C$

14. $\int \csc x \cot x\, dx = -\csc x + C$

15. $\int \tan x\, dx = \ln |\sec x| + C$

16. $\int \cot x\, dx = \ln |\sin x| + C$

17. $\int \sec x\, dx = \ln |\sec x + \tan x| + C$

18. $\int \csc x\, dx = \ln |\csc x - \cot x| + C$

▶ REVIEW EXERCISES

In Exercises 1–8, find the value of the sum.

1. $\sum_{i=1}^{4} 2^i$

2. $\sum_{i=1}^{5} \frac{1}{2}\left(1 + \frac{i}{2}\right)^2$

3. $\sum_{i=3}^{6} (2i + 1)$

4. $\sum_{j=1}^{3} (j^2 + 2j - 1)$

5. $\sum_{i=1}^{n} (3i - 5)$

6. $\sum_{i=1}^{n} (3i^2 + 4i - 5)$

7. $\sum_{i=1}^{n} (2i - 1)$

8. $\sum_{i=1}^{n} (3i^2 - 3i + 1)$

9. Use the definition of the definite integral to show that
$$\int_{1}^{3} (x^2 - 2x) \, dx = \frac{2}{3}.$$

10. Use the definition of the definite integral to find the value of
$$\int_{2}^{5} (3x^2 - 1) \, dx.$$

11. Express each of the following limits as a definite integral over the interval [0, 2], and find its value.

    a) $\lim_{n \to \infty} \sum_{i=1}^{n} 3(x_i)^2 \frac{2}{n}$
    b) $\lim_{n \to \infty} \sum_{i=1}^{n} (2x_i + 1) \frac{2}{n}$

12. Determine a function for which each of the following is a Riemann sum.

    a) $\sum_{i=1}^{n} (x_i^3 + x_i - 1) \Delta x_i$
    b) $\sum_{i=1}^{n} \sqrt{1 + x_i^2} \, \Delta x_i$

In Exercises 13–50, evaluate the integral.

13. $\int x^3 \, dx$

14. $\int (2x - 15) \, dx$

15. $\int_{-3}^{5} (2x^3 + 5) \, dx$

16. $\int \left(\sqrt{x} + \frac{1}{x^2}\right) dx$

17. $\int 4(4x + 7)^3 \, dx$

18. $\int (4x + 7)^3 \, dx$

19. $\int \left(x\sqrt{x} + \frac{1}{x}\right) dx$

20. $\int (x^3 - 2)^2 \, dx$

21. $\int \sqrt[3]{8x^8} \, dx$

22. $\int \frac{x^2 + 2x}{x^3 + 3x^2 + 1} \, dx$

23. $\int \left(\frac{1}{x^4} + \frac{1}{\sqrt[4]{x}}\right) dx$

24. $\int_{-2}^{2} 3w\sqrt{4 - w^2} \, dw$

25. $\int_{0}^{\pi} (\cos t - t) \, dt$

26. $\int x \cos(x^2 - 1) \, dx$

27. $\int \frac{x}{\sqrt{4 - x^2}} \, dx$

28. $\int_{2}^{3} \frac{x + 1}{x^2 + 2x + 3} \, dx$

29. $\int w^3 \sqrt{w^2 + 1} \, dw$

30. $\int x^2 \sqrt{x - 1} \, dx$

31. $\int \frac{1}{\sqrt{x}(1 - \sqrt{x})^2} \, dx$

32. $\int_{2}^{5} \frac{1}{x} \, dx$

33. $\int \frac{1}{x - 2} \, dx$

34. $\int \frac{x}{x^2 + 2} \, dx$

35. $\int \frac{x - 3}{x^2 - 6x + 5} \, dx$

36. $\int \frac{1}{x^2 + 2x + 1} \, dx$

37. $\int_{0}^{2} 2u^2 \sqrt{u^3 + 1} \, du$

38. $\int x^5 \sqrt{x^2 + 4} \, dx$

39. $\int_{1}^{2} \frac{s}{(1 + 2s)^3} \, ds$

40. $\displaystyle\int \frac{t^3 - 7t^2 + 1}{t^2}\, dt$

41. $\displaystyle\int \frac{x^2}{x^2 - 2x + 2}\, dx$

42. $\displaystyle\int \frac{x^2 - 2x}{x^2 - 5x + 6}\, dx$

43. $\displaystyle\int \cos 2x\, dx$

44. $\displaystyle\int_0^1 \sin \pi x\, dx$

45. $\displaystyle\int \tan 4x\, dx$

46. $\displaystyle\int \frac{\sin x}{1 - \cos x}\, dx$

47. $\displaystyle\int \sin^2 x \cos x\, dx$

48. $\displaystyle\int \frac{\sin \sqrt{x}}{\sqrt{x}}\, dx$

49. $\displaystyle\int \frac{\cos (\ln x)}{x}\, dx$

50. $\displaystyle\int (\sin^2 x + \sin x) \cos x\, dx$

In Exercises 51–64, determine whether the integral converges or diverges, and evaluate those that converge.

51. $\displaystyle\int_1^\infty \frac{3}{x^2}\, dx$

52. $\displaystyle\int_0^2 \frac{1}{\sqrt{x}}\, dx$

53. $\displaystyle\int_0^\infty \frac{1}{(x+1)^2}\, dx$

54. $\displaystyle\int_0^\infty \frac{1}{10x}\, dx$

55. $\displaystyle\int_2^\infty \frac{1}{x(\ln x)^2}\, dx$

56. $\displaystyle\int_{-\infty}^\infty \frac{1}{x^2 - 2x + 1}\, dx$

57. $\displaystyle\int_3^\infty \frac{1}{\sqrt{3x}}\, dx$

58. $\displaystyle\int_0^9 \frac{1}{(x-1)^{2/3}}\, dx$

59. $\displaystyle\int_{-3}^3 \frac{x\, dx}{\sqrt{9 - x^2}}$

60. $\displaystyle\int_0^1 \frac{1}{\sqrt{1 - x}}\, dx$

61. $\displaystyle\int_{-1}^8 x^{-2/3}\, dx$

62. $\displaystyle\int_0^\infty \frac{x}{\sqrt{x^2 + 1}}\, dx$

63. $\displaystyle\int_{-\infty}^\infty \cos x\, dx$

64. $\displaystyle\int_{-\infty}^\infty \frac{1}{1 - x}\, dx$

In Exercises 65–72, find the derivative of the function.

65. $f(x) = (\ln x)^3$

66. $f(x) = x \ln x - x$

67. $f(x) = \ln (x^3 - 7x + 1)$

68. $g(t) = \dfrac{\ln (\sqrt{t} + 1)}{\sqrt{t}}$

69. $h(x) = [\ln (x^{1/3} + 1)]^{1/3}$

70. $h(x) = (\ln x^4)(x^3 - 2x)$

71. $f(x) = x \displaystyle\int_1^x \ln t\, dt$

72. $f(x) = \displaystyle\int_1^x \frac{1}{t^2 + 1}\, dt$

73. Find an equation of the curve that passes through $(4, 0)$ and whose tangent line at (x, y) has slope $2x - 1$.

74. Find an equation of the curve that has a horizontal tangent line at $(0, 1)$ and for which $y'' = 3x^2 - 1$.

75. Find the area of the region bounded by the curve $y = (x - 2)^{2/3}$, the x-axis, and the lines $x = 1$ and $x = 10$. Make a sketch of the region.

76. Find the area of the region bounded by the curve $y = 9 - x^2$ and the x-axis. Make a sketch of the region.

77. Find the area of the region bounded by the graph of $f(x) = x^3 - 3x^2 - x + 3$, the lines $x = -1$ and $x = 1$, and the x-axis. Make a sketch of the region.

78. Use l'Hôpital's Rule to evaluate each of the following limits.

a) $\displaystyle\lim_{x \to \infty} \frac{\ln (x^3 + 1)}{x - 1}$

b) $\displaystyle\lim_{x \to 0^+} x \ln (x^3 + 1)$

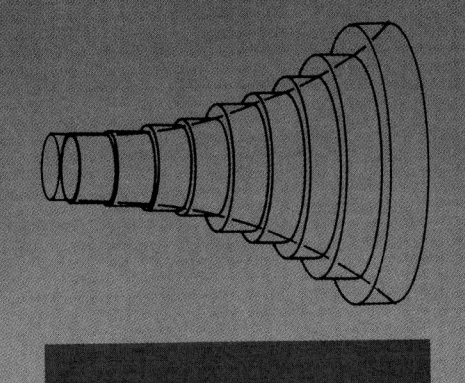

5
Applications of the Definite Integral

CHAPTER 5 APPLICATIONS OF THE DEFINITE INTEGRAL

The definite integral was introduced in Chapter 4 with the problem of determining the area of a region bounded by the *x*-axis and the graph of a continuous nonnegative function defined on a closed interval. In this chapter we consider a few of the other applications of the definite integral. The common thread in these applications is that each involves the determination of the limit of Riemann sums, which leads to the definite integral.

The first application considers finding the area of more general regions of the plane than those discussed in Chapter 4. The next three sections show that we can apply the definite integral to find the volume and surface area of some solids in space as well as the length of certain curves in the plane.

The later sections of the chapter discuss some of the many applications of the integral to problems in engineering and physics. These include determining the work done by a varying force, finding the moments and center of mass of a solid, and finding the force produced by fluid pressure.

5.1

Areas of Regions in the Plane

In Section 4.1 we discussed the problem of determining the area A of the region bounded by the graph of a continuous nonnegative function f, the *x*-axis, and the lines $x = a$ and $x = b$ (see Figure 5.1a). The area A is approximated by sums of the form

$$A_n = \sum_{i=1}^{n} f(x_i)\, \Delta x_i,$$

where $x_i = a + i(b - a)/n$ and $\Delta x_i = x_i - x_{i-1}$. Geometrically, A_n is the sum of the areas of the rectangles shown in Figure 5.1(b).

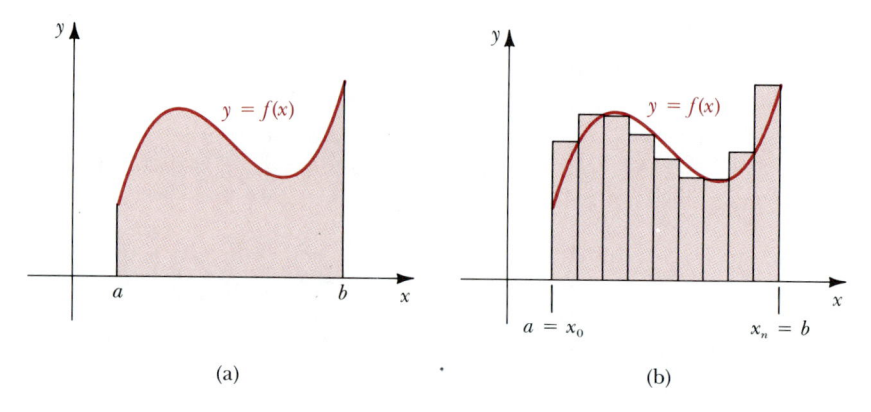

Figure 5.1

The area is $\int_a^b f(x)\, dx$.

5.1 AREAS OF REGIONS IN THE PLANE

In Section 4.2 we found that for each n, A_n is a Riemann sum for f on $[a, b]$. The area A was defined as the limit of these Riemann sums, which is the definite integral of f on $[a, b]$:

(5.1) $$A = \int_a^b f(x)\, dx.$$

We now show how the integral can be used to find the area of more general regions.

Example 1

Find the area of the region bounded by the x-axis, the line $x = 2$, and the graph of $f(x) = x^3 - 2x^2 + 3$.

Solution

The region whose area is to be found is shown in Figure 5.2. To construct this sketch, we use the methods of Chapter 2 to find that the critical points are $x = 0$ and $x = 1$ and to determine the increasing and decreasing behavior of the graph. When $x = -1$, $f(x) = 0$, so the limits of integration for finding the area are $x = -1$ and $x = 2$.

Since f is nonnegative on the interval $[-1, 2]$, the area is

$$\int_{-1}^{2} f(x)\, dx = \int_{-1}^{2} (x^3 - 2x^2 + 3)\, dx = \left(\frac{x^4}{4} - \frac{2x^3}{3} + 3x\right)\Big]_{-1}^{2}$$

$$= \left[\frac{16}{4} - \frac{16}{3} + 6\right] - \left[\frac{1}{4} + \frac{2}{3} - 3\right] = \frac{27}{4}.$$ ▶

Suppose f is a function that is continuous and nonpositive on an interval $[a, b]$. The area of the region bounded by $x = a$, $x = b$, the x-axis, and the graph of f is found by considering the continuous nonnegative function $-f$ on $[a, b]$ (see Figure 5.3).

The area of the region below the x-axis bounded by the graph of f is the same as the area above the x-axis bounded by the graph of $-f$. Thus the area of this region is

$$\int_a^b -f(x)\, dx = -\int_a^b f(x)\, dx.$$

The following example shows how this result is used to find the area of a region bounded by the graph of a continuous function that changes sign on an interval:

i) Determine the subintervals on which $f(x) \geq 0$ and those on which $f(x) \leq 0$.
ii) Integrate f over those on which $f(x) \geq 0$.
iii) Integrate $-f$ over those on which $f(x) \leq 0$.
iv) Sum the values of the integrals in (ii) and (iii).

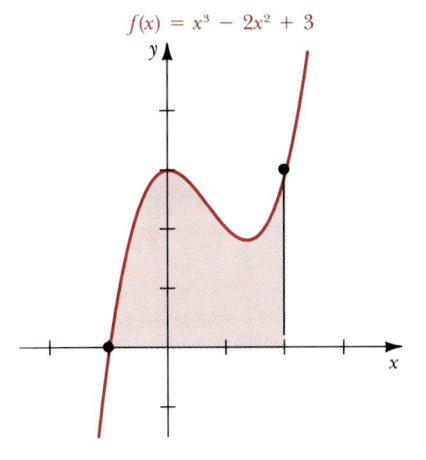

Figure 5.2

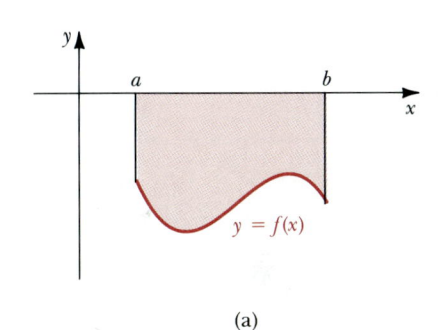

Figure 5.3
The area is $-\int_a^b f(x)\, dx$.

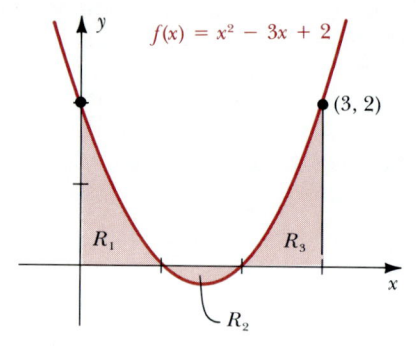

Figure 5.4

Example 2

Find the area of the region bounded on [0, 3] by the x-axis and the graph of $f(x) = x^2 - 3x + 2$.

Solution

Since $x^2 - 3x + 2 = (x - 2)(x - 1)$,

$$f(x) \geq 0 \quad \text{when } 0 \leq x \leq 1 \text{ and when } 2 \leq x \leq 3$$

and

$$f(x) \leq 0 \quad \text{when } 1 \leq x \leq 2.$$

The graph of f is shown in Figure 5.4.

To find the area, we divide the region into three parts, R_1, R_2, and R_3. The area A is the sum of the areas of these three regions. Since $f(x) \leq 0$ on [1, 2], the area of R_2 is found by integrating $-f(x) = -(x^2 - 3x + 2)$ on [1, 2]:

$$A = \int_0^1 (x^2 - 3x + 2)\, dx + \int_1^2 [-(x^2 - 3x + 2)]\, dx$$
$$+ \int_2^3 (x^2 - 3x + 2)\, dx$$
$$= \left(\frac{x^3}{3} - \frac{3x^2}{2} + 2x\right)\Big]_0^1 - \left(\frac{x^3}{3} - \frac{3x^2}{2} + 2x\right)\Big]_1^2 + \left(\frac{x^3}{3} - \frac{3x^2}{2} + 2x\right)\Big]_2^3$$
$$= \left[\left(\frac{1}{3} - \frac{3}{2} + 2\right) - 0\right] - \left[\left(\frac{8}{3} - 6 + 4\right) - \left(\frac{1}{3} - \frac{3}{2} + 2\right)\right]$$
$$+ \left[\left(9 - \frac{27}{2} + 6\right) - \left(\frac{8}{3} - 6 + 4\right)\right]$$
$$= \frac{11}{6}.$$

▶

The regions described in Examples 1 and 2 are bounded between the x-axis and the graph of a continuous function f. We next consider the area of a region that is bounded between the graphs of two continuous functions f and g on an interval $[a, b]$, where $f(x) \geq g(x)$. Such a region is shown in Figure 5.5(a).

Suppose f and g are continuous on $[a, b]$ and $f(x) \geq g(x)$ for each x in $[a, b]$. Consider a partition

$$\mathcal{P} = \{x_0, x_1, x_2, \ldots, x_n\}$$

of $[a, b]$ and for each i, let $z_i = x_i$. Construct approximating rectangles for the region between the graphs of f and g on the subinterval $[x_{i-1}, x_i]$ for each $i = 1, 2, \ldots, n$ (see Figure 5.5b). The width of the ith rectangle is Δx_i and the height is $f(x_i) - g(x_i)$, so the area of the ith rectangle is

$$[f(x_i) - g(x_i)]\, \Delta x.$$

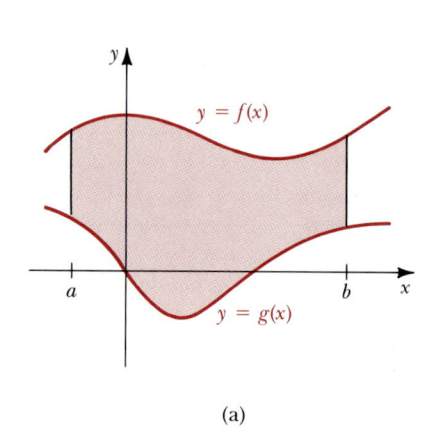

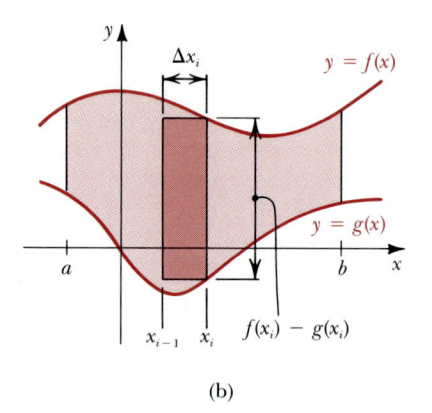

Figure 5.5
The area is $\int_a^b [f(x) - g(x)]\, dx$.

The total area of the approximating rectangles is

$$\sum_{i=1}^{n} [f(x_i) - g(x_i)] \, \Delta x_i.$$

This is a Riemann sum for the function $f - g$ on the interval $[a, b]$. The area of the region is the limit of these Riemann sums, which is the definite integral of $f - g$ on the interval $[a, b]$.

(5.2)
DEFINITION
The **area** of the region bounded by the graphs of $y = f(x)$ and $y = g(x)$ on the interval $[a, b]$, where $f(x) \geq g(x)$, is

$$A = \int_a^b [f(x) - g(x)] \, dx.$$

Note that the result in (5.2) is valid whenever $f(x) \geq g(x)$ for all x in $[a, b]$, regardless of the sign of $f(x)$ and $g(x)$. Exercise 56 outlines an approach to the development of (5.2) that emphasizes this fact. Also note that if $g(x) = 0$ for all x in $[a, b]$, Definition (5.2) states that the area of the region bounded by the x-axis and the graph of the nonnegative function f on the interval $[a, b]$ is simply the area described in (5.1).

Example 3

Find the area of the region bounded by the graphs of $f(x) = \sqrt{x}$ and $g(x) = -x$ on the interval $[1, 4]$.

Solution

First we draw the graphs of f and g on the same set of coordinate axes and observe that $f(x) > g(x)$ for all x in $[1, 4]$, as shown in Figure 5.6. The area of the shaded region is

$$A = \int_1^4 [f(x) - g(x)] \, dx = \int_1^4 (\sqrt{x} - (-x)) \, dx = \int_1^4 (\sqrt{x} + x) \, dx,$$

so

$$A = \left(\frac{2}{3} x^{3/2} + \frac{x^2}{2} \right) \Big]_1^4 = \left[\left(\frac{16}{3} + 8 \right) - \left(\frac{2}{3} + \frac{1}{2} \right) \right] = \frac{73}{6}.$$

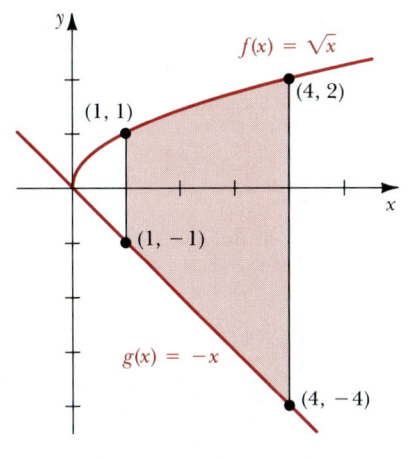

▶ **Figure 5.6**

The interval of integration was specified for the problem in Example 3. More often, area problems ask us simply to find the area of the region bounded between two graphs. In this case, the interval of integration is determined by the intersection of the graphs of the functions.

Example 4

Determine the area of the region bounded by the graphs of $f(x) = \sqrt{x}$ and $g(x) = x^2$.

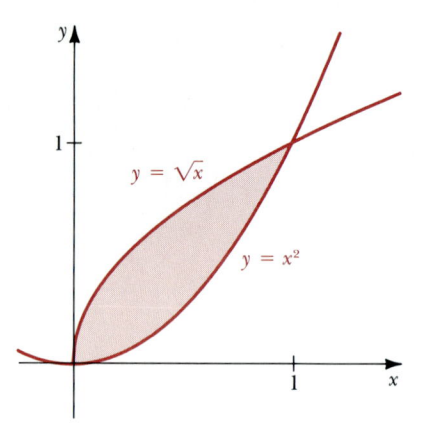

Figure 5.7

Solution
The graphs of f and g intersect when $x = 1$ and when $x = 0$, as shown in Figure 5.7. Since $f(x) = \sqrt{x} \geq x^2 = g(x)$ on the interval $[0,1]$, the area is

$$A = \int_0^1 (\sqrt{x} - x^2) \, dx = \left[\frac{2}{3} x^{3/2} - \frac{1}{3} x^3 \right]_0^1.$$

Thus,

$$A = \left[\frac{2}{3} - \frac{1}{3} \right] - [0 - 0] = \frac{1}{3}.$$

The following example shows how Definition (5.2) is used to find the area bounded between the graphs of two continuous functions f and g when $f(x) \geq g(x)$ for some values of x but $g(x) \geq f(x)$ for others.

Example 5

Find the total area of all the regions bounded by the graphs of $f(x) = x^3 - 2x^2$ and $g(x) = x^2 - 2x$.

Solution
No interval boundaries have been given in the statement of this problem. As in Example 4, the intervals of integration are determined by the points of intersection of the two graphs. To find these points, we set $f(x) = g(x)$ and solve for x:

$$x^3 - 2x^2 = x^2 - 2x$$

implies that

$$0 = x^3 - 3x^2 + 2x = x(x^2 - 3x + 2) = x(x - 1)(x - 2).$$

The graphs intersect when $x = 0$, $x = 1$, and $x = 2$.

The graphing techniques of Chapter 2 can be used to show that the graphs of f and g are as illustrated in Figure 5.8. The regions bounded between these graphs occur when x is in the interval $[0, 2]$.

On $[0, 1]$, $f(x) \geq g(x)$ and the area of R_1 is

$$A_1 = \int_0^1 [(x^3 - 2x^2) - (x^2 - 2x)] \, dx = \int_0^1 (x^3 - 3x^2 + 2x) \, dx,$$

so

$$A_1 = \frac{1}{4} x^4 - x^3 + x^2 \Big]_0^1 = \frac{1}{4} - 1 + 1 = \frac{1}{4}.$$

On $[1, 2]$, $f(x) \leq g(x)$ and the area of R_2 is

$$A_2 = \int_1^2 [(x^2 - 2x) - (x^3 - 2x^2)] \, dx = \int_1^2 (3x^2 - x^3 - 2x) \, dx,$$

so

$$A_2 = x^3 - \frac{1}{4} x^4 - x^2 \Big]_1^2 = (8 - 4 - 4) - \left(1 - \frac{1}{4} - 1 \right) = \frac{1}{4}.$$

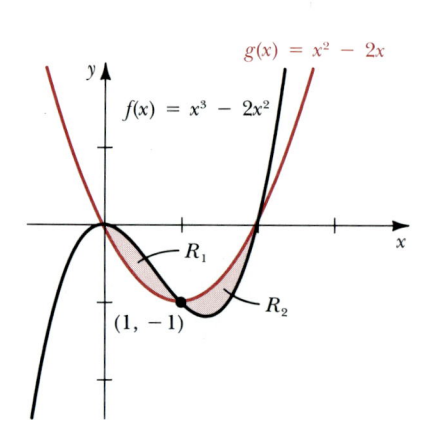

Figure 5.8

The total area is

$$A = A_1 + A_2 = \frac{1}{4} + \frac{1}{4} = \frac{1}{2}.$$

Example 6 involves a region that is bounded by distinct curves on different intervals.

Example 6

Find the area of the region bounded by the graphs of $y^2 = x + 1$ and $x + y = 1$.

Solution

The equation $y^2 = x + 1$ does not describe y as a function of x since $y = \pm\sqrt{x + 1}$. However, the dark-shaded region R_1 to the left of the y-axis in Figure 5.9 is bounded by the graphs of $y = \sqrt{x + 1}$ and $y = -\sqrt{x + 1}$, so the area of R_1 is

$$A_1 = \int_{-1}^{0} (\sqrt{x+1} - (-\sqrt{x+1}))\, dx$$

$$= 2 \int_{-1}^{0} (x+1)^{1/2}\, dx$$

$$= 2\left(\frac{2}{3}(x+1)^{3/2}\right)\Big]_{-1}^{0} = \frac{4}{3}.$$

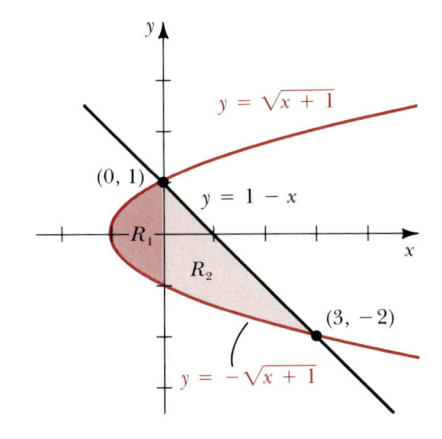

Figure 5.9

The light-shaded region R_2 to the right of the y-axis in Figure 5.9 is bounded above by the graph of $y = 1 - x$ and below by the graph of $y = -\sqrt{x + 1}$, so the area of R_2 is

$$A_2 = \int_{0}^{3} [(1-x) - (-\sqrt{x+1})]\, dx = \int_{0}^{3} (1 - x + (x+1)^{1/2})\, dx$$

$$= \left(x - \frac{x^2}{2} + \frac{2}{3}(x+1)^{3/2}\right)\Big]_{0}^{3}$$

$$= \left[3 - \frac{9}{2} + \frac{2}{3}(8) - \frac{2}{3}\right] = \frac{19}{6}.$$

The sum of the areas of R_1 and R_2 gives the total area:

$$A = A_1 + A_2 = \frac{4}{3} + \frac{19}{6} = \frac{9}{2}.$$

The problem in Example 6 can also be solved using the following modification of Definition (5.2):

The area of the region bounded by the graphs of $x = h(y)$ and $x = k(y)$, where $h(y) \geq k(y)$ for all y in $[c, d]$, is

(5.3) $$A = \int_{c}^{d} [h(y) - k(y)]\, dy.$$

Example 7

Use (5.3) to find the area of the region bounded by the graphs of $y^2 = x + 1$ and $x + y = 1$.

Solution

On the linear boundary, $x + y = 1$, so $x = 1 - y$. On the parabolic boundary, $y^2 = x + 1$, so $x = y^2 - 1$. The intersection points $(3, -2)$ and $(0, 1)$ imply that in order for (x, y) to be in the region R, shown in Figure 5.10, the y-coordinate must be in the interval $[-2, 1]$. For y in this interval,

$$y^2 - 1 \le 1 - y,$$

so the area of the region is

$$A = \int_{-2}^{1} [(1 - y) - (y^2 - 1)] \, dy$$

$$= \int_{-2}^{1} (2 - y - y^2) \, dy = \left(2y - \frac{y^2}{2} - \frac{y^3}{3} \right) \Big]_{-2}^{1}.$$

Thus,

$$A = \left[\left(2 - \frac{1}{2} - \frac{1}{3} \right) - \left(-4 - 2 + \frac{8}{3} \right) \right] = \frac{9}{2},$$

which is, of course, the same result as that obtained in Example 6.

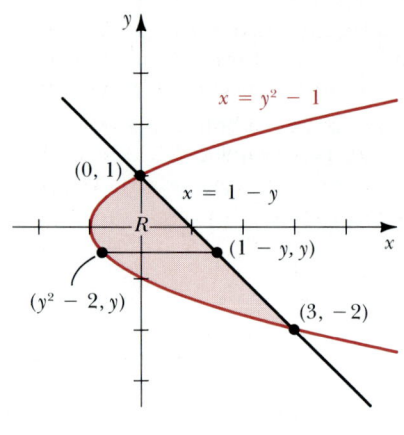

Figure 5.10

▶ EXERCISE SET 5.1

In Exercises 1–46: a) sketch the graphs of the equations, and b) find the area of the region bounded by these graphs.

1. $y = x^2 - 2x$, $x = 2$, $x = 4$, $y = 0$
2. $y = 6x - x^2 - 5$, $x = 2$, $x = 4$, $y = 0$
3. $y = x^2 - 2x$, $y = 0$
4. $y = 6x - x^2 - 5$, $x = 0$, $x = 1$, $y = 0$
5. $y = 1/x$, $x = 1$, $x = 5$, $y = 0$
6. $y = 1/x$, $x = 0.1$, $x = 1$, $y = 0$
7. $y = 1/x$, $x = -1$, $x = -5$, $y = 0$
8. $y = 1/x$, $x = -0.1$, $x = -1$, $y = 0$
9. $y = x^2$, $y = \sqrt{x}$, $x = 1$, $x = 2$
10. $y = x^2$, $y = \sqrt{x}$, $x = 2$
11. $y = x^3 - x$, $y = 0$
12. $y = x - x^5$, $y = 0$
13. $y = \tan x$, $x = -\pi/4$, $x = \pi/4$, $y = 0$
14. $y = |\sin x|$, $x = 0$, $x = 2\pi$, $y = 0$
15. $y = x^3$, $y = x^2$
16. $y = x^3$, $y = \sqrt{x}$
17. $y = 1 - x^2$, $y = -3$
18. $y = x^2 - 3$, $y = 4$
19. $y = x^2 - 1$, $y = 1 - x^2$
20. $y = x^2 - 2$, $y = x$
21. $y = x^{3/2}$, $y = x$
22. $y = -x^3$, $y = x$, $y = 1$
23. $y = \cos x$, $y = \sin x$, $x = -\pi/2$, $x = \pi/2$
24. $y = \sec^2 x$, $y = \cos x$, $x = -\pi/4$, $x = \pi/4$
25. $y = x$, $y = 1 - x$, $y = -x/2$
26. $y = x + 1$, $y = (2x + 1)/3$, $y = (2x + 8)/5$
27. $y = 2x - x^2$, $y = 2 - x$
28. $y = x^2$, $y = 4x - x^2$
29. $y = \sin 2x$, $y = \sin x$, $x = 0$, $x = \pi$
30. $y = \sin 2x$, $y = \sin x$, $x = 0$, $x = 2\pi$

31. $y = x/2$, $y = \sqrt[4]{x}$
32. $y = x/2$, $y = \sqrt[3]{x}$
33. $y = \sin x$, $y = \tan x$, $x = \pi/4$
34. $y = \cos x$, $y = \cot x$, $x = \pi/4$
35. $y = x^3 - x$, $y = 2x^3 - 2x$
36. $y = x^3 - 3x^2 - x + 3$, $y = x^2 - 1$
37. $y = x^3$, $y = (16x + 8x^2 - 3x^3)/5$
38. $y = x^2 + 2x$, $y = 2x\sqrt{x+2}$
39. $y = 1 + \sin x$, $y = \cos^2 x$, $x = 0$, $x = 3\pi/2$
    [*Hint:* $2\cos^2 x = 1 + \cos 2x$.]
40. $y = 2\tan x$, $y = \sec^2 x$, $x = \pi/6$, $x = \pi/3$
41. $y = |x|$, $y = x + 2$, $y = 2 - x$
42. $y = |x|$, $y = 2 - x^2$
43. $x = y^2$, $x + y = 2$
44. $x = y^2$, $y = x - 6$
45. $y^2 = x + 2$, $y = -x$
46. $y^2 = x + 1$, $y^2 = 1 - x$
47. Use integration to find the area of the triangle with vertices (0, 0), (1, 1), and (1, 0).
48. Use integration to find the area of the triangle with vertices (1, 1), (3, 2), and (2, 4).
49. A line through (0, 0) intersects the curve $y = x^2$ at a point (a, a^2). The area of the region bounded above by the line and below by the curve is 27. Find a.
50. Consider the region bounded by $y = x^2 - 1$ and $y = 1 - x^2$.
    a) Show that the line $y = x$ divides the region into two parts with equal area.
    b) Show that any line through the origin divides the region into two parts with equal area.
51. Suppose $A(a)$ denotes the area of the region bounded by $y = x^2 - a^2$ and $y = a^2 - x^2$ (see the figure). Find the rate of change of this area with respect to the change in a. 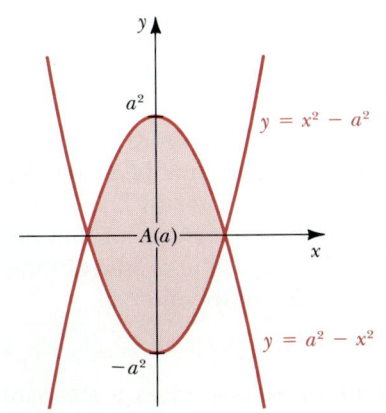
52. Suppose $A(a)$ denotes the area of the region bounded by $y = x^2 - a$ and $y = a - x^2$ for $a > 0$. Find the rate of change of this area with respect to the change in a.
53. Farmer MacDonald's pasture is bounded on the south and the east by a pair of perpendicular dirt roads that meet at the southeast corner of the pasture. The pasture runs 390 ft along the south-bounding road and 1430 ft along the east-bounding road. The remaining boundary is a parabola with vertex at the north point of the east-bounding road and symmetric with respect to the line made by this road. Each cow requires $\frac{1}{2}$ acre of pasture (an acre is 43,560 ft²). How many cows can MacDonald pasture in this field? 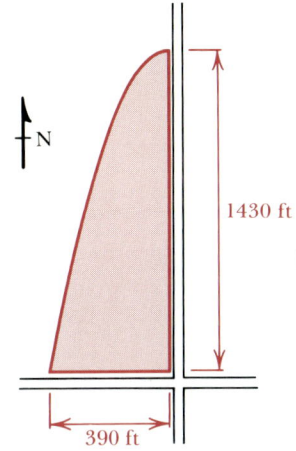
54. Find the area of the region described in Example 3 by the following procedure.
    i) Find the area of the region R_1 bounded by the graph of f, the x-axis, and the lines $x = 1$ and $x = 4$.
    ii) Find the area of the region R_2 bounded by the graph of g, the x-axis, and the lines $x = 1$ and $x = 4$.
    iii) Add the area of R_1 and the area of R_2. Compare your answer to that of Example 3.
55. The irrational number π is defined as the area of the unit circle.
    a) Show that $\pi = 4\int_0^1 \sqrt{1 - x^2}\,dx$.
    b) Use the result in part (a) to show that the area of the circle with radius r and equation $x^2 + y^2 = r^2$ is πr^2.
    c) Use the result in either part (a) or part (b) to show that the area of the ellipse with equation $b^2 x^2 + a^2 y^2 = a^2 b^2$ is πab.
56. Tell why each of the following statements is true. The statements give an alternative way of deriving Definition (5.2).
    a) If g is continuous on $[a, b]$, then a number m exists with $m \leq g(x)$ for all x in $[a, b]$.

b) The area bounded by the x-axis, the lines $x = a$ and $x = b$, and the graph of $y = g(x) - m$ is $\int_a^b g(x)\,dx - m(b - a)$.

c) If f is continuous and $f(x) \geq g(x)$ for all x in $[a, b]$, then the area bounded by the x-axis, the lines $x = a$ and $x = b$, and the graph of $y = f(x) - m$ is $\int_a^b f(x)\,dx - m(b - a)$.

d) Parts (b) and (c) imply that the area bounded by the lines $x = a$ and $x = b$ and by the graphs of f and g is $\int_a^b [f(x) - g(x)]\,dx$.

5.2

Volumes of Solids with Known Cross Sections: Disks

The rectangular parallelepiped shown in Figure 5.11(a) and the right circular cylinder shown in Figure 5.11(b) are examples of solids that have a constant cross-sectional area with respect to some axis. From geometry we know that the volume of such a solid is the product of the cross-sectional area and the length of the solid along the axis.

The cone shown in Figure 5.11(c) does not have a constant cross-sectional area, so this simple formula will not suffice to determine its volume. In this section we show how the definite integral is used to determine the volume of solids, such as the cone, that have varying cross-sectional area.

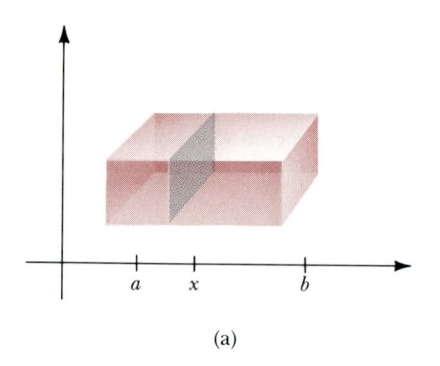

Figure 5.11

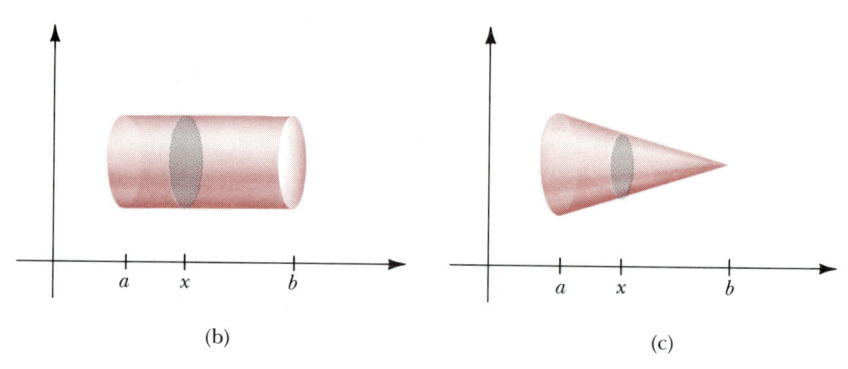

Suppose S is a solid lying between the planes that pass through $x = a$ and $x = b$ and are perpendicular to the x-axis. In addition, suppose A is a continuous function defined on $[a, b]$ that describes the cross-sectional area of S perpendicular to the x-axis (see Figure 5.12).

Let $\mathcal{P} = \{x_0, x_1, \ldots, x_n\}$ be a partition of $[a, b]$ and for each $i = 1, 2, \ldots, n$, choose z_i arbitrarily in the interval $[x_{i-1}, x_i]$. Figure 5.13 shows that an approximation ΔV_i to the volume of that portion of the solid intersecting the interval $[x_{i-1}, x_i]$ is the product of the area of the cross section $A(z_i)$ and the width Δx_i of the interval:

(5.4) $$\Delta V_i = A(z_i)\,\Delta x_i.$$

Summing n such elements of volume gives a Riemann sum that

5.2 VOLUMES OF SOLIDS WITH KNOWN CROSS SECTIONS: DISKS

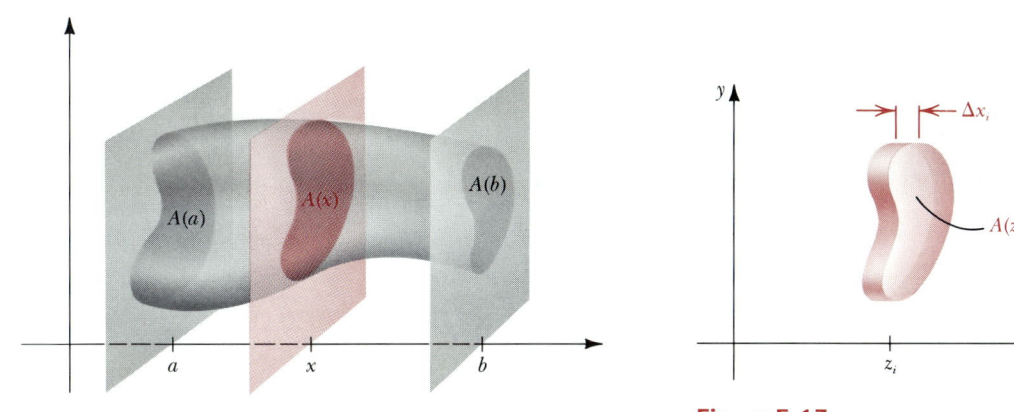

Figure 5.12
The area of the cross section is $A(x)$.

Figure 5.13

approximates the volume of the solid S,

$$(5.5) \qquad \sum_{i=1}^{n} A(z_i)\, \Delta x_i,$$

an approximation that should improve as the norm of the partition approaches zero.

As the norm of the partition approaches zero, these Riemann sums approach the definite integral of the function A on the interval $[a, b]$. This leads to the following definition for the volume of S.

(5.6) DEFINITION

Suppose S is a solid lying between the planes that pass through $x = a$ and $x = b$ and are perpendicular to the x-axis. Let A be a continuous function defined on $[a, b]$ that describes the cross-sectional area of S perpendicular to the x-axis. Then the **volume** of S is

$$V = \int_a^b A(x)\, dx.$$

Note that when the cross-sectional area of a solid is constant, $A(x) = A$ for all x in $[a, b]$, Definition (5.6) implies that the volume is the product of the cross-sectional area A and the length, $(b - a)$, of the solid. This is consistent with the results illustrated in Figure 5.11(a) and (b).

Example 1 shows how to use Definition (5.6) to determine the volume of a cone.

Example 1

Suppose S is a cone lying between the planes that pass through $x = 0$ and $x = 6$ and that every cross section of S perpendicular to the x-axis is a circle with radius $x/2$ (see Figure 5.14). Find the volume of S.

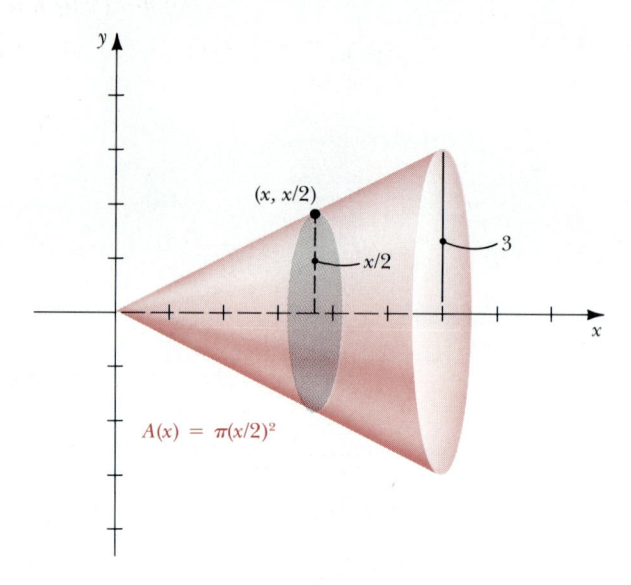

Figure 5.14
$A(x) = \pi(x/2)^2$.

Solution

For each x in $[0, 6]$, the cross section of S is a circle with radius $x/2$, so the cross-sectional area is

$$A(x) = \pi \left(\frac{x}{2}\right)^2 = \frac{\pi x^2}{4}.$$

Definition (5.6) implies that the volume of the cone is

$$V = \int_0^6 \frac{\pi}{4} x^2 \, dx = \frac{\pi}{4} \left[\frac{x^3}{3}\right]_0^6 = \frac{216}{12} \pi = 18\pi.$$ ▶

The formula for the volume of a cone with base radius r and height h is shown in Exercise 57 to be $V = \frac{1}{3}\pi r^2 h$. The result in Example 1 is the special case that occurs when $r = 3$ and $h = 6$.

Example 2

The base of a solid is bounded in the xy-plane by the graphs of $x = y^2$ and $x = 4$. Each cross section perpendicular to the x-axis is an isosceles triangle with altitude 1. Find the volume of the solid.

Solution

The base of the solid is shown in Figure 5.15(a), and some typical cross sections are shown in Figure 5.15(b). Figure 5.15(b) is redrawn in Figure 5.15(c) using a three-dimensional perspective.

Since $A(x)$ is the area of a triangle with height 1 and base $2\sqrt{x}$,

$$A(x) = \frac{1}{2} \text{ (base)(height)} = \frac{1}{2} (2\sqrt{x}) \cdot 1 = \sqrt{x}.$$

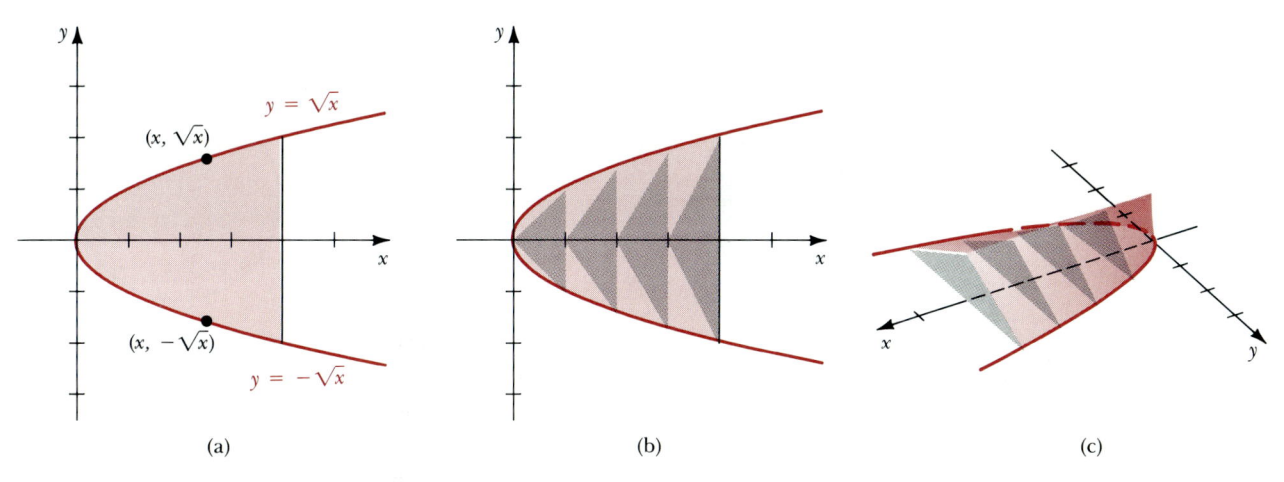

Figure 5.15

By Definition (5.6), the volume of the solid is

$$V = \int_0^4 \sqrt{x}\, dx = \frac{2}{3} x^{3/2} \Big]_0^4 = \frac{2}{3}(8-0) = \frac{16}{3}.$$

Definition (5.6) implies that if two solids are bounded between a pair of parallel planes and if the cross sections of the solids cut by any plane parallel to and between the bounding planes are equal, then the volumes of the solids are equal. This result is known as **Cavalieri's Principle** since it was first expressed by Bonaventura Cavalieri in 1635. It is illustrated in Figure 5.16 where all the solids have the same volume.

Volumes of Revolution: Disks

A solid of revolution is generated when a region in the plane is revolved about a line that does not intersect the interior of the region. For example, the cone in Example 1 is a solid of revolution produced by revolving about the x-axis the region bounded on the interval $[0, 6]$ by the x-axis and the line $y = x/2$. We can use (5.6) to find a general formula for the volume of a solid of revolution.

Suppose R is a region in the plane with boundaries $x = a$, $x = b$, the x-axis, and the graph of the continuous nonnegative function f, as shown in Figure 5.17(a). Revolving R about the x-axis generates the solid shown in Figure 5.17(b). For each x in $[a, b]$, the cross section of this solid is the disk shown in Figure 5.17(c) with area

$$A(x) = \pi [f(x)]^2.$$

When applied to this situation, Definition (5.6) gives the following result.

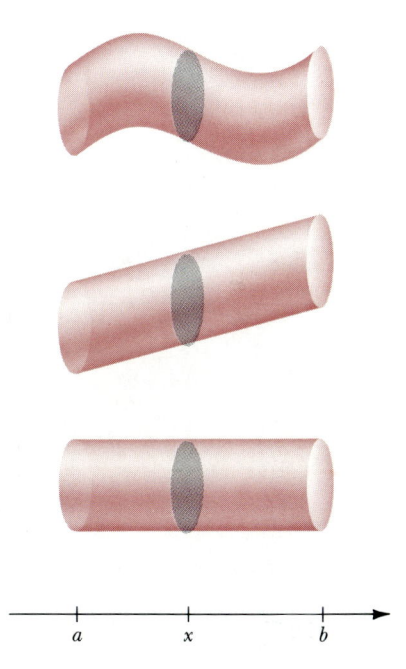

Figure 5.16
The volumes of the solids are equal.

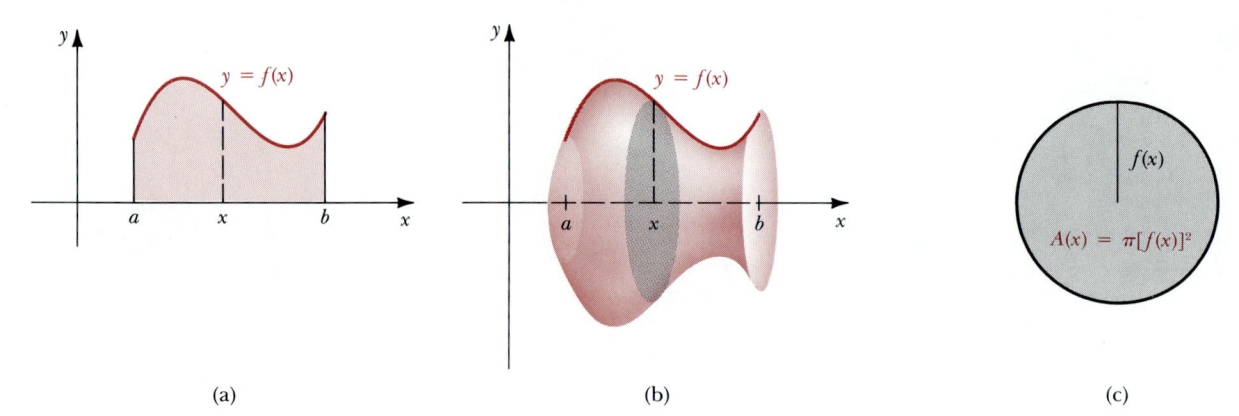

Figure 5.17
The area in (c) is $\pi[f(x)]^2$, so the volume is $\pi \int_a^b [f(x)]^2 \, dx$.

(5.7) THEOREM
If f is continuous and nonnegative on the interval $[a, b]$, then the volume of the solid generated by revolving about the x-axis the region bounded on $[a, b]$ by the x-axis and the graph of f is
$$V = \pi \int_a^b [f(x)]^2 \, dx.$$

Example 3

Find the volume of the solid generated by revolving about the x-axis the region bounded on $[1, 4]$ by the x-axis and the graph of $f(x) = \sqrt{x}$.

Solution
The solid generated by revolving this region about the x-axis is shown in Figure 5.18. The volume of this solid is
$$V = \pi \int_1^4 (\sqrt{x})^2 \, dx = \pi \int_1^4 x \, dx = \frac{\pi}{2} x^2 \Big]_1^4 = \frac{15}{2} \pi.$$
▶

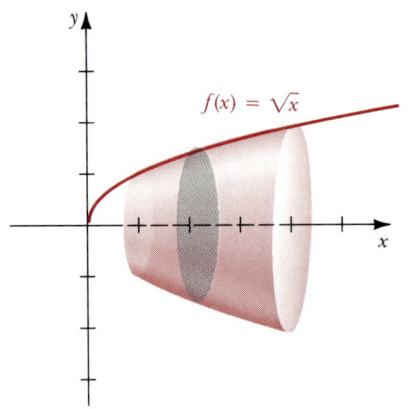

Figure 5.18

The volume generated by revolving the area of a region bounded on an interval by the graphs of two functions is obtained in a similar manner. Suppose that f and g are continuous with $f(x) \geq g(x) \geq 0$ for each x in $[a, b]$ and that R is the region bounded on $[a, b]$ by the graphs of f and g (see Figure 5.19a). If R is revolved about the x-axis, the solid generated is of the type shown in Figure 5.19(b).

For each x in $[a, b]$, the cross section of the solid is a washer such as the one shown in Figure 5.19(c). The area of this washer is the difference between the area of the circle with radius $f(x)$ and the area of the circle with radius $g(x)$:
$$A(x) = \pi[f(x)]^2 - \pi[g(x)]^2.$$
Thus we have the following.

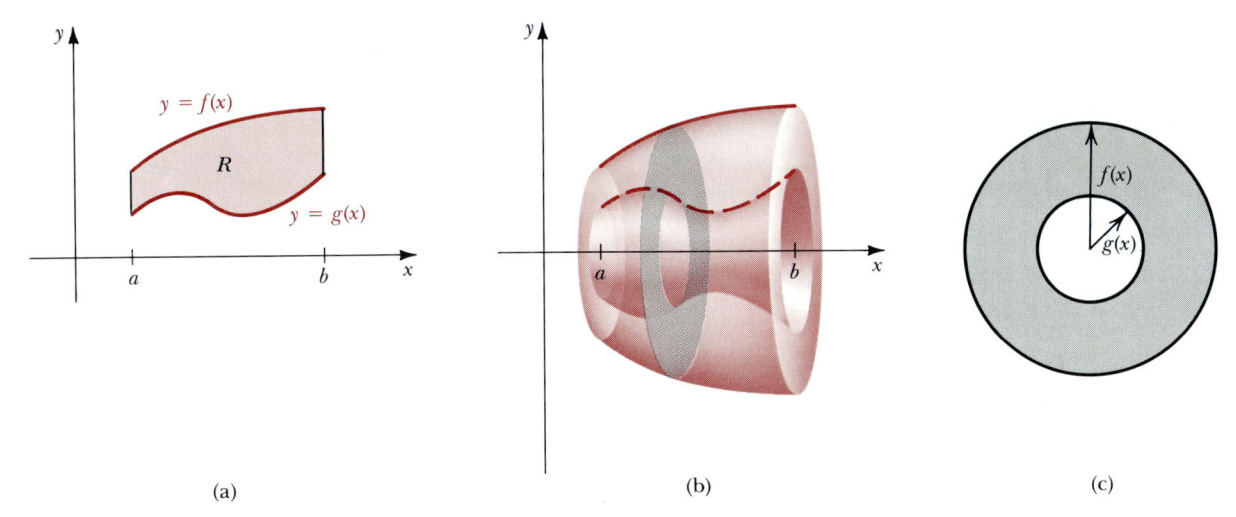

5.2 VOLUMES OF SOLIDS WITH KNOWN CROSS SECTIONS: DISKS

(a)   (b)   (c)

Figure 5.19
The area in (c) is $\pi\{[f(x)]^2 - [g(x)]^2\}$, so the volume is $\pi \int_a^b \{[f(x)]^2 - [g(x)]^2\} \, dx$.

(5.8)
THEOREM

If f and g are continuous and $f(x) \geq g(x) \geq 0$ on $[a, b]$, then the volume of the solid generated by revolving about the x-axis the region bounded on $[a, b]$ by the graphs of f and g is

$$V = \pi \int_a^b \{[f(x)]^2 - [g(x)]^2\} \, dx.$$

Example 4

Find the volume of the solid generated by revolving about the x-axis the region in the first quadrant bounded by the graphs of $y = x + 1$ and $y = x^3 + 1$.

Solution

The interval boundaries for the region have not been stated in the problem. As in Section 5.1, these boundaries are determined by the intersections of the graphs. The points of intersection for the graphs occur when

$$x + 1 = x^3 + 1,$$

that is, when

$$0 = x^3 - x = x(x^2 - 1)$$
$$= x(x - 1)(x + 1).$$

This implies that $x = -1$, $x = 0$, or $x = 1$.

Since the region is in the first quadrant, x must be in the interval $[0, 1]$, as shown in Figure 5.20(a). The solid of revolution is shown in

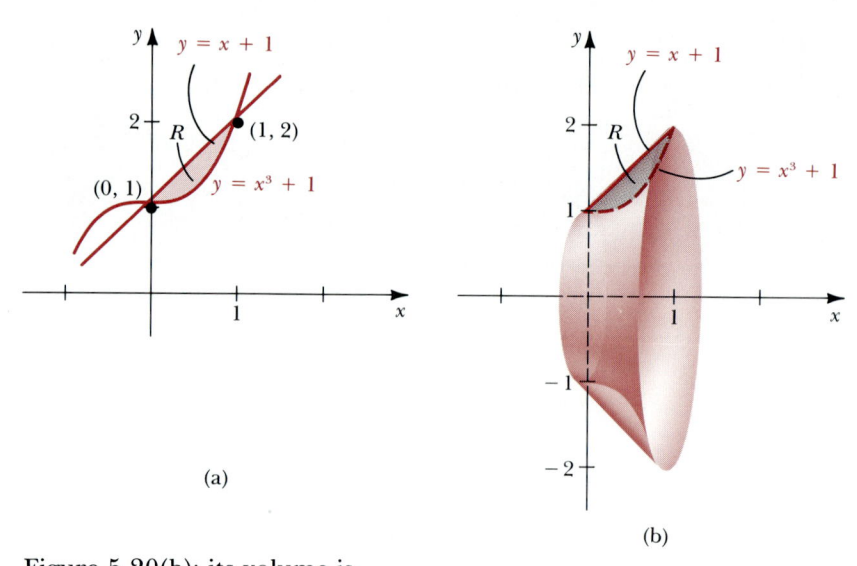

Figure 5.20

(a)

(b)

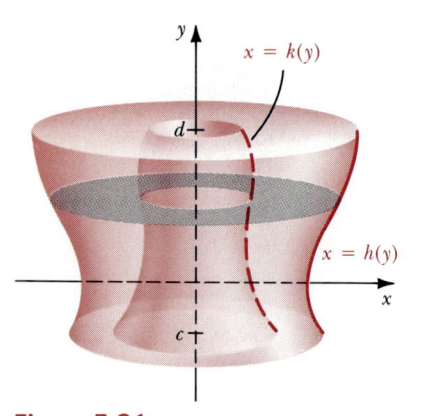

Figure 5.21

The volume is $\pi \int_c^d \{[h(y)]^2 - [k(y)]^2\}\, dy$.

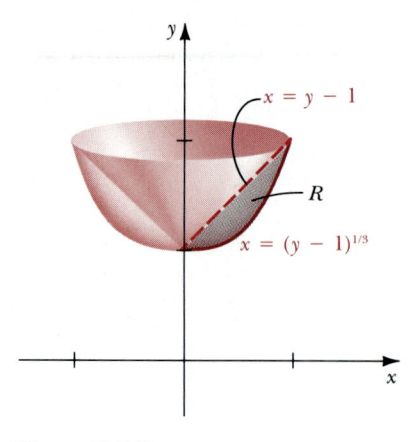

Figure 5.22

Figure 5.20(b); its volume is

$$V = \pi \int_0^1 [(x+1)^2 - (x^3+1)^2]\, dx$$

$$= \pi \int_0^1 (x+1)^2\, dx - \pi \int_0^1 (x^6 + 2x^3 + 1)\, dx$$

$$= \pi \left. \frac{(x+1)^3}{3} \right]_0^1 - \pi \left[\frac{x^7}{7} + \frac{2x^4}{4} + x \right]_0^1$$

$$= \pi \left[\frac{8}{3} - \frac{1}{3} \right] - \pi \left[\frac{1}{7} + \frac{1}{2} + 1 \right] = \frac{29}{42}\pi.$$

▶

The volume of a solid generated by revolving a region about the y-axis can be determined by the method of disks, provided the boundaries of the region can be described with x written as a function of y. A region of this type is illustrated in Figure 5.21.

If h and k are continuous and $h(y) \geq k(y) \geq 0$ on $[c, d]$, then the volume of the solid generated by revolving about the y-axis the region bounded on $[c, d]$ by the graphs of h and k is

(5.9) $$V = \pi \int_c^d \{[h(y)]^2 - [k(y)]^2\}\, dy.$$

Example 5

Find the volume of the solid generated when the region described in Example 4 is revolved about the y-axis.

Solution

Since the revolution is about the y-axis (see Figure 5.22), the first step is to express the boundaries by writing x as a function of y.

The equation
$$y = x^3 + 1 \quad \text{implies that} \quad x = (y-1)^{1/3},$$
and the equation
$$y = x + 1 \quad \text{implies that} \quad x = y - 1.$$
In Example 4 we found that the graphs intersect at $(0, 1)$ and $(1, 2)$. For y in $[1, 2]$,
$$y - 1 \le (y - 1)^{1/3}.$$
Thus,
$$\begin{aligned} V &= \pi \int_1^2 \{[(y-1)^{1/3}]^2 - (y-1)^2\}\, dy \\ &= \pi \int_1^2 [(y-1)^{2/3} - (y-1)^2]\, dy \\ &= \pi \left[\frac{3}{5}(y-1)^{5/3} - \frac{(y-1)^3}{3} \right]_1^2 \\ &= \pi \left(\frac{3}{5} - \frac{1}{3} \right) = \frac{4}{15}\pi. \end{aligned}$$

▶

The final example in this section considers a problem in which a plane region is revolved about a line parallel to a coordinate axis.

Example 6

Find the volume generated when the region bounded on $[-1, 0]$ by the x-axis and the graph of $y = x^2 + 1$ is revolved about the line $y = 2$.

Solution

The region to be revolved is shown in Figure 5.23(a), and the solid of revolution in Figure 5.23(b). The typical cross-sectional segment highlighted in Figure 5.23(b) is reproduced in Figure 5.23(c).

The outside radius of the segment being revolved is always 2. The inside radius $r(x)$ depends on the value of x in $[-1, 0]$:
$$r(x) = 2 - (x^2 + 1) = 1 - x^2.$$
The total volume generated is
$$\begin{aligned} V &= \pi \int_{-1}^0 [2^2 - (1 - x^2)^2]\, dx = \pi \int_{-1}^0 (3 + 2x^2 - x^4)\, dx \\ &= \pi \left(3x + \frac{2x^3}{3} - \frac{x^5}{5} \right)\bigg]_{-1}^0 = \frac{52}{15}\pi. \end{aligned}$$

▶

The problem in Example 6 can also be solved by translating the region to be revolved two units downward and revolving this new region about the x-axis. The solution by this method is considered in Exercise 70.

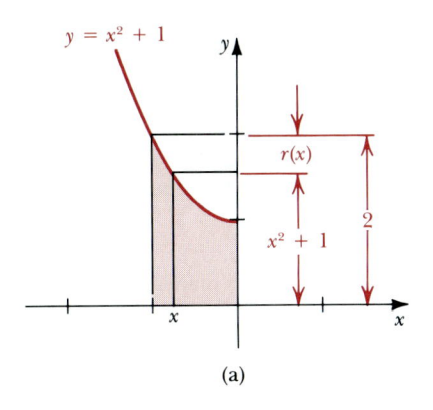

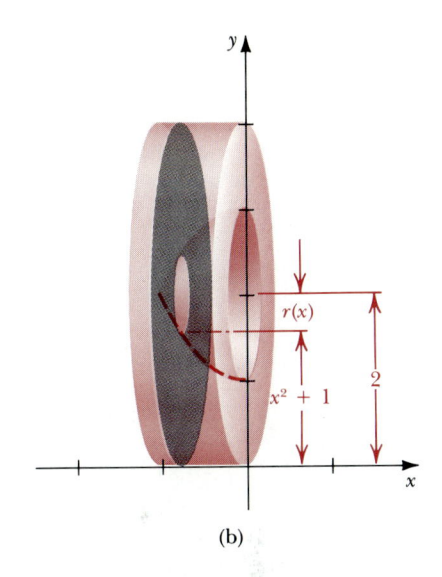

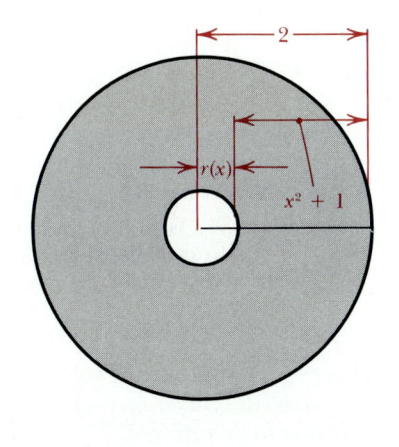

Figure 5.23

EXERCISE SET 5.2

1. Suppose S is a pyramid lying between the planes that pass through $x = 0$ and $x = 6$ and that every cross section of S perpendicular to the x-axis is a square with width $6 - x$, $0 \leq x \leq 6$. Find the volume of S.

2. Suppose the frustum S of a right circular cone lies between the planes that pass through $x = 0$ and $x = 5$. Every cross section of S perpendicular to the x-axis is a circle with radius $(x + 1)/2$, $0 \leq x \leq 5$. Find the volume of S.

3. A sphere S of radius 2 lies between the planes that pass through $x = -2$ and $x = 2$. Every cross section of S perpendicular to the x-axis is a circle of radius $\sqrt{4 - x^2}$, $-2 \leq x \leq 2$. Show that the volume of S is $32\pi/3$.

4. The base of a solid S is bounded in the xy-plane by the isosceles triangle shown in the accompanying figure. Every cross section perpendicular to the x-axis is a semicircle. Find the volume of S.

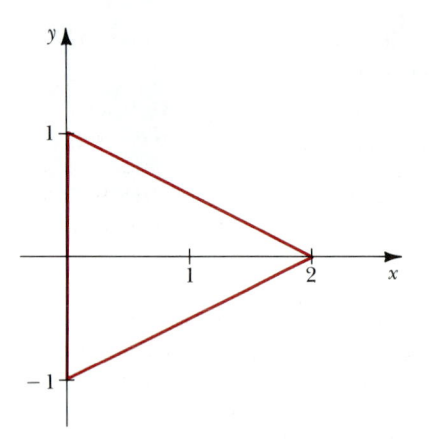

5. Suppose a solid D has the same base as that described for the solid in Exercise 4. Every cross section perpendicular to the x-axis is a rectangle with height 2. Find the volume of D.

6. The base of a solid is bounded in the xy-plane by the graph of $x = 9 - y^2$ and the y-axis. Every cross section perpendicular to the x-axis is a semicircle. Find the volume of the solid.

In Exercises 7–38, find the volume generated if the region bounded by the curves is revolved about the x-axis. Sketch the region to be revolved.

7. $y = x^2, y = 0, x = 3$
8. $y = x^3, y = 0, x = 2$
9. $y = x^3, y = 0, x = 1, x = 2$
10. $y = x^2, y = 0, x = 1, x = 3$
11. $y = x^2 + 2x + 1, x = 1, y = 0$
12. $y = 4x - x^2, y = 0$
13. $y = 1/x, y = 0, x = 1, x = 2$
14. $y = 1/\sqrt{x}, x = 1, x = 5, y = 0$
15. $y = x^2, y = x^3$
16. $y = x^2, y = \sqrt{x}$
17. $y = x^3, y = 1, x = 2$
18. $y = x^2, y = 8 - x^2$
19. $y = 2x^2, y = 3 - x^2$
20. $y = x^2, y = 1, x = 3$
21. $y = x^2 - x, y = 0$
22. $y = x^3 - x, y = 0$
23. $y = \sec x, x = -\pi/4, x = \pi/4, y = 0$
24. $y = \cot x, x = \pi/4, x = 3\pi/4, y = 0$
25. $y = \sqrt{\sin x}, x = 0, x = \pi, y = 0$
26. $y = \sqrt{\cos x}, x = -\pi/2, x = \pi/2, y = 0$
27. $y = \sin x, x = 0, x = \pi, y = 0$
    $\left[\text{Hint: } \sin^2 x = \dfrac{1 - \cos 2x}{2}.\right]$
28. $y = \cos x, x = 0, x = \pi, y = 0$
29. $y = \sin x, x = -\pi, x = 0, y = 0$
30. $y = \sin x, x = 0, x = 2\pi, y = 0$
31. $y = \sqrt{\tan x}, x = 0, x = \pi/4, y = 0$
32. $y = \sqrt{\csc x}, x = \pi/4, x = \pi/2, y = 0$
33. $y = \dfrac{\sqrt{x}}{\sqrt[4]{x^2 + 1}}, x = 0, x = 1, y = 0$
34. $y = \dfrac{\sqrt{x}}{\sqrt[3]{x^2 + 1}}, x = 0, x = 1, y = 0$
35. $y = 1/x, y = x, x = 2, y = 0$
36. $y = 1/\sqrt{x}, y = x, y = 0, x = 5$
37. $y = x + 1, y = 3x - 5, y = 1$
38. $y = x + 1, y = 3x - 5, y = x/2 + 1$

In Exercises 39–48, find the volume generated if the region bounded by the curves is revolved about the y-axis.

39. $y = x^3, x = 0, y = 1$
40. $y = \sqrt{x}, x = 0, y = 2$
41. $y = x^2, y = x$

42. $y = x^2$, $y = x^3$
43. $y = 1/x$, $y = 1$, $y = 2$, $x = 0$
44. $y = 2x - 1$, $y = -2x + 7$, $y = 1$
45. $y = x^2 - 2x + 1$, $y = x + 1$
46. $y = x^2 - x$, $y = 0$
47. $y = 2x - 1$, $y = -2x + 7$, $y = x$
48. $y = x^3$, $y = x$
49. Find the volume of the solid generated if the region in the first quadrant bounded by the graph of $y = \sqrt{x}$ and the lines $x = 0$ and $y = 1$ is revolved about the line:
    a) $x = 1$;   b) $y = 1$;
    c) $x = -1$; d) $y = 2$.
50. Find the volume of the solid generated if the region bounded by the graph of $x = \sqrt{y}$ and the lines $x = 0$ and $y = 1$ is revolved about the line:
    a) $x = 1$;   b) $y = 1$;
    c) $x = -1$; d) $y = 2$.
51. The base of a solid is bounded by the circle in the xy-plane with equation $x^2 + y^2 = r^2$, and each cross section perpendicular to the x-axis is a square. Find the volume of the solid.
52. The base of a solid is bounded by the circle in the xy-plane with equation $x^2 + y^2 = r^2$, and each cross section perpendicular to the x-axis is an equilateral triangle. Find the volume of the solid.
53. A solid has its base in the xy-plane bounded by the graphs of $y = x^2$ and $y = 1$. Find the volume of the solid if each cross section perpendicular to the x-axis is a square.
54. A solid has its base in the xy-plane bounded by the graphs of $y = x^2$ and $y = 1$. Find the volume of the solid if each cross section perpendicular to the y-axis is a square.
55. Revolve about the x-axis the region bounded by the x-axis and the upper portion of the circle with equation $x^2 + y^2 = r^2$ to generate a sphere of radius r. Show that the volume of this sphere is $4\pi r^3/3$.
56. Revolve about the x-axis the region bounded by the x-axis and the lines $x = 0$, $x = h$, and $y = r$ to generate a right circular cylinder of radius r and height h. Show that the volume of this cylinder is $\pi r^2 h$.
57. Revolve about the x-axis the region bounded by the x-axis, the line $x = h$, and the line segment joining the point $(0, 0)$ and the point (h, r) to generate a cone of radius r and height h. Show that the volume of this cone is $\pi r^2 h/3$.
58. A pyramid has a base in the shape of a square of length r and has an altitude h. Find the volume of the pyramid.

59. A water tower has the shape of a sphere with radius 50 ft. The height of the water in the tank is 75 ft. How many gallons of water are in the tank? (One U.S. gallon is equivalent to 231 cubic inches.)
60. A loaf of French bread has a circular cross section whose area in square inches is given by
$$A(x) = \pi(256 - x^2)/64$$
for x in the interval $[-16, 16]$. Draw a sketch of the loaf and find its volume.
61. A toothpick is a cylinder of diameter 2 mm and length 50 mm tipped on the ends with cones of altitude 7 mm. If 70% of the wood used to make a toothpick is wasted in construction, how much wood is required to make a box of 750 toothpicks?
62. The center of a universal joint on an automobile driveshaft is in the form of two right circular cylinders of the same radii, $r = 0.5$ in., intersecting at right angles. What is the volume contained in this intersection?
63. Suppose the circular cylinders that make up the universal joint described in Exercise 62 are both 2.5 in. in length. How much metal is required to make this universal joint?
64. A bull's horn is described by drawing the curve $y = x^2/8$ for x in $[0, 8]$ and then drawing circles perpendicular to the x-axis whose centers are along this curve and where the radius of the circle with center $(x, x^2/8)$ is $r = 1 - x^2/64$ in. Find the volume in the bull's horn.
65. A swimming pool is 30 ft wide and 50 ft long with straight sides (see the figure). The depth of the water at one end is 8 ft, and the pool continues at this depth for 20 ft. The depth of the pool then decreases linearly for the next 10 ft to a depth of 3 ft and continues at this depth for the final 20 ft of the pool. Find the volume of water in the pool.

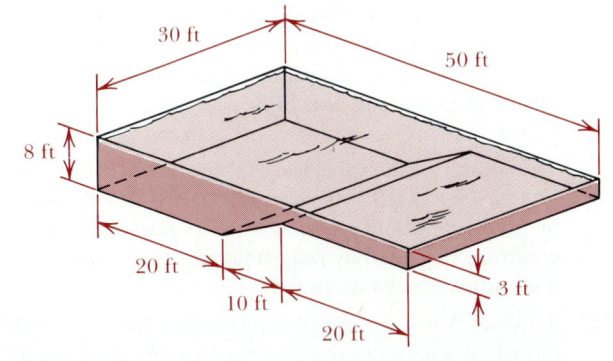

66. The recommended procedure to follow when felling a tree is first to make a notch on the fall side of the tree,

that is, the side on which the tree will fall. This should be done by making a horizontal cut one third of the way through the tree. Then a cut is made one third as far above the horizontal cut as the tree is thick, meeting the horizontal cut at its end (see the figure). Set up the integral that describes the volume of wood removed from the tree by this notch, if the tree is 18 in. in diameter. Do not try to evaluate this integral.

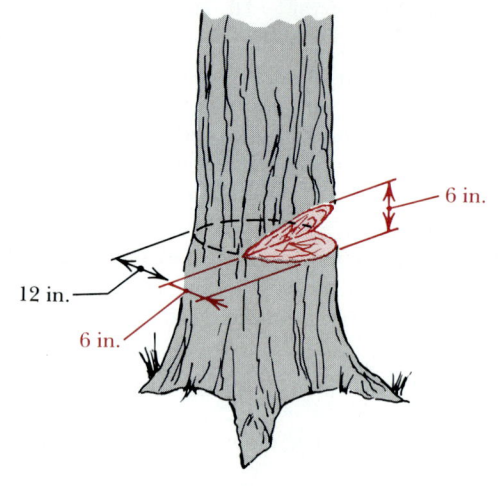

67. A hole of radius 3 is bored through the center of a sphere with radius 4. How much volume has been removed?

68. A hole of radius r is bored through the center of a sphere of radius R, removing half the volume of the sphere. What is r?

69. Solve Exercise 68 using Cavalieri's principle as follows: Suppose a plane perpendicular to the axis of the cylindrical hole cuts both the original sphere and the sphere with radius $[R^2 - r^2]^{1/2}$ a distance h from the center of each sphere. Show that the cross-sectional area cut from each solid is equal, and use the volume of the smaller sphere to deduce the result.

70. Determine the volume described in Example 6 by first translating the curve two units downward and then revolving about the x-axis the region bounded by the resulting curve and the lines $y = -2$, $x = -1$, and $x = 0$.

71. The volume of the spherical cap shown in the figure is $V = \pi h^2(3r - h)/3$, where $h \leq r$. Derive this formula by revolving about the x-axis the region in the first quadrant bounded by the graph of $y = \sqrt{r^2 - x^2}$ and the lines $x = r - h$ and $y = 0$.

72. A pyramid is a geometric solid with a flat base and linearly sloping sides that meet in an apex, as shown in the figure. The base of the pyramid is usually in the form of a triangle, square, or other polygon, but that need not be the case. Suppose a pyramid has base area B and height h. For each x in $[0, h]$, the area of the cross section x units from the apex has the form $A(x) = Kx^2$, for some constant K.
a) Find the constant K.
b) Show that the pyramid has volume $V = Bh/3$.

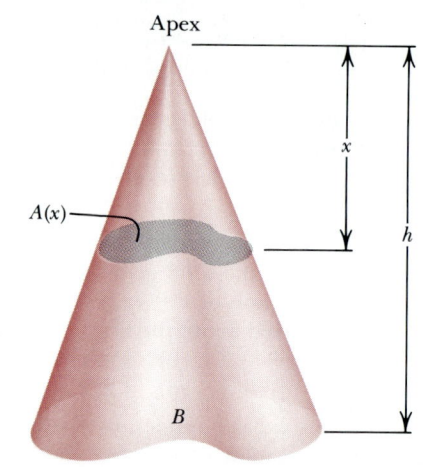

73. A frustum of a right cone is shown in the figure. Use the method of disks to show that the volume of this solid is

$$V = \frac{\pi h}{3}(r_1^2 + r_1 r_2 + r_2^2).$$

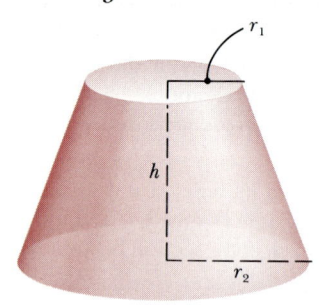

Figure for Exercise 73.

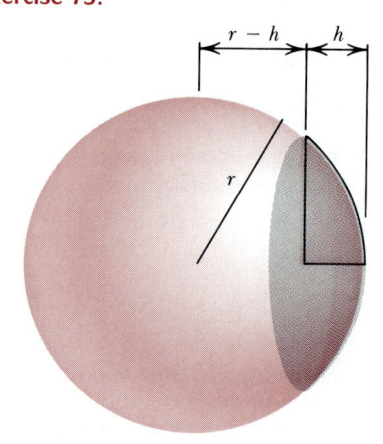

Figure for Exercise 71.

74. A sphere has a hole bored vertically through its center. The height of the solid that remains is 6 in. What is the volume of this remaining solid? [*Note:* You do not need to know either the radius of the sphere or the radius of the hole to solve this problem.]
75. For each $M > 1$, consider the region R_M bounded by the graph of $y = 1/x$, the x-axis, and the lines $x = 1$ and $x = M$.
    a) Determine the area A_M of the region R_M.
    b) Determine the volume V_M generated when the region R_M is revolved about the x-axis.
    c) Show that the improper integral describing the limiting area as M approaches infinity diverges.
    d) Show that the improper integral describing the limiting volume as M approaches infinity has the value π.
    e) Resolve the following paradox: The area of the region R bounded by the x-axis, the graph of $f(x) = 1/x$, and the line $x = 1$ is unbounded so no finite amount of paint could cover R. If R is revolved about the x-axis, however, a solid of revolution is generated with a finite volume. Filling this volume with paint *would* cover R.

5.3

Volumes of Solids of Revolution: Shells

In Example 5 of Section 5.2 we found the volume of a solid generated by revolving a region in the plane about the y-axis. The method used there requires that the boundaries of the region first be expressed with x as a function of y. In this section we introduce another method for finding the volume of a solid of revolution, one that enables us to find the volume of a solid whose axis of rotation is the y-axis and whose boundaries are expressed with y as a function of x.

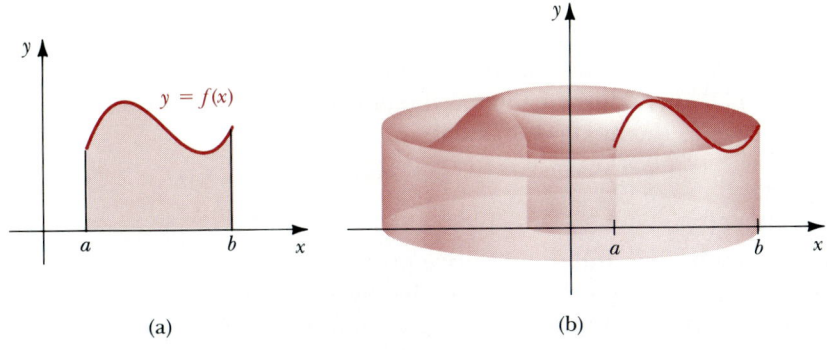

Figure 5.24

The volume is $2\pi \int_a^b xf(x)\,dx$.

Suppose R is a region bounded by the x-axis and the graph of the continuous nonnegative function f on the interval $[a, b]$, where $a \geq 0$ (see Figure 5.24a). A solid of revolution, shown in Figure 5.24(b), is generated when this region is revolved about the y-axis. To find the volume of this solid, consider a partition $\mathcal{P} = \{x_0, x_1, x_2, \ldots, x_n\}$ of $[a, b]$ and for each $i = 1, 2, \ldots, n$, let z_i be a number in the interval $[x_{i-1}, x_i]$. The cylindrical shell in Figure 5.25(b) is generated when the rectangle in Figure 5.25(a) is revolved about the y-axis.

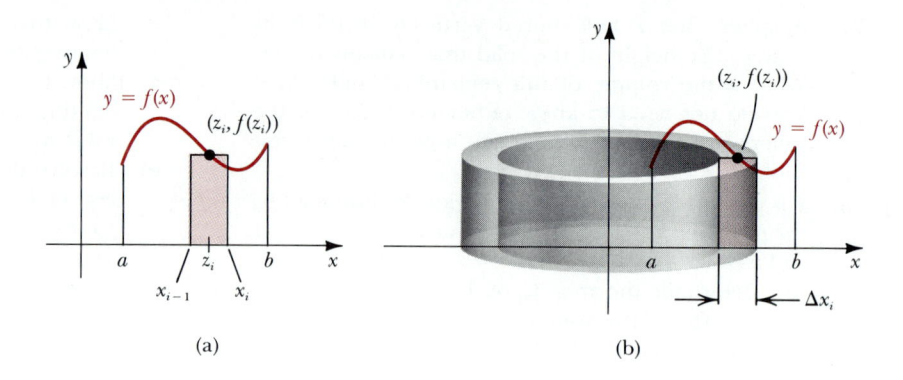

Figure 5.25

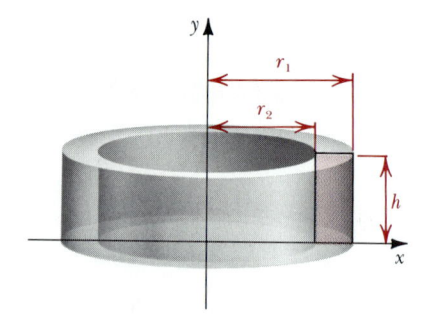

(a) (b)

Figure 5.26
The volume is $\pi r_1^2 h - \pi r_2^2 h$.

The volume of a typical cylindrical shell, such as the one shown in Figure 5.26, with outer radius r_1, inner radius r_2, and height h is

(5.10) $\quad V = \pi r_1^2 h - \pi r_2^2 h = \pi(r_1^2 - r_2^2)h = \pi(r_1 + r_2)(r_1 - r)h.$

Since the shell in Figure 5.25(b) has outer radius x_i, inner radius x_{i-1}, and height $f(z_i)$, it follows from (5.10) that the volume ΔV_i of this shell is

$$\Delta V_i = \pi(x_i + x_{i-1})(x_i - x_{i-1})f(z_i) = \pi(x_i + x_{i-1})f(z_i)\,\Delta x_i.$$

When constructing a Riemann sum, we can arbitrarily choose the number z_i in the interval $[x_{i-1}, x_i]$. If z_i is chosen as the midpoint of this interval, $z_i = \frac{1}{2}(x_i + x_{i-1})$, then

$$\Delta V_i = 2\pi[\tfrac{1}{2}(x_i + x_{i-1})]\,f(z_i)\,\Delta x_i = 2\pi z_i f(z_i)\,\Delta x_i.$$

Summing the volumes ΔV_i of the shells for each $i = 1, 2, \ldots, n$ gives the total approximating volume

$$\sum_{i=1}^{n} 2\pi z_i f(z_i)\,\Delta x_i.$$

This is a Riemann sum for the function described by $2\pi x f(x)$ on the interval $[a, b]$. Taking the limit as the norm of the partition approaches zero leads to the method of shells.

(5.11) DEFINITION

If f is continuous and nonnegative on the interval $[a, b]$, where $a \geq 0$, then the volume of the solid generated by revolving about the y-axis the region bounded on $[a, b]$ by the x-axis and the graph of f is

$$V = 2\pi \int_a^b x f(x)\,dx.$$

This definition is consistent with Definition (5.6) in Section 5.2 in the sense that when both methods can be applied to a problem, the same result is obtained regardless of the method used. This will be shown specifically in Example 2 for the solid generated in Example 5 of Section 5.2.

Example 1

Find the volume of the solid generated by revolving about the y-axis the region bounded on $[1, 2]$ by the x-axis and the graph of $f(x) = x^2$.

Solution

The region to be revolved is shown in Figure 5.27(a) and the resulting solid in Figure 5.27(b). It follows from Definition (5.11) that

$$V = 2\pi \int_1^2 x\, f(x)\, dx = 2\pi \int_1^2 x(x^2)\, dx = 2\pi \int_1^2 x^3\, dx.$$

So

$$V = 2\pi \left(\frac{x^4}{4}\right)\Big]_1^2 = 2\pi \left[\frac{16}{4} - \frac{1}{4}\right] = \frac{15}{2}\pi.$$

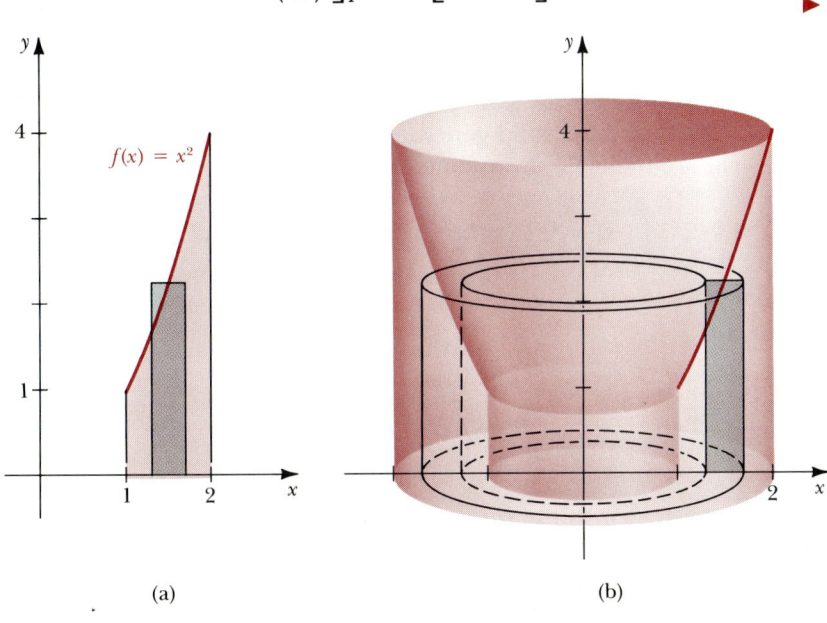

Figure 5.27

The volume generated by revolving about the y-axis the region bounded between the graphs of two continuous functions is obtained in a similar manner (see Figure 5.28).

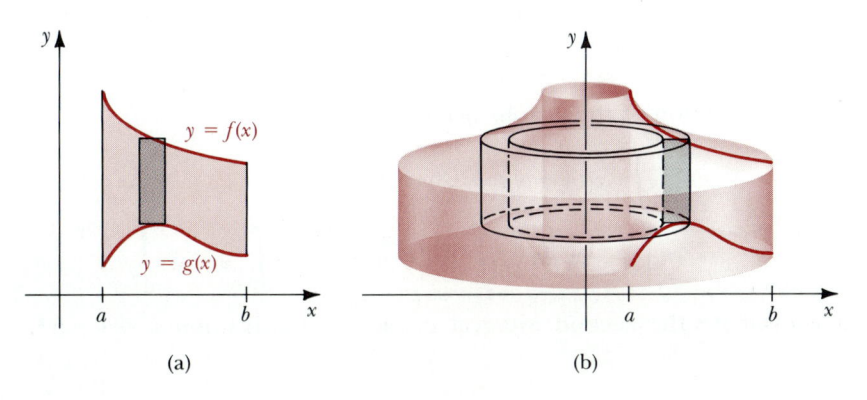

Figure 5.28

The volume is $2\pi \int_a^b x[f(x) - g(x)]\, dx$.

(5.12)
THEOREM

Suppose f and g are continuous and $f(x) \geq g(x)$ for all x in $[a, b]$, where $a \geq 0$. The volume of the solid generated by revolving about the y-axis the region bounded on $[a, b]$ by the graphs of f and g is

$$V = 2\pi \int_a^b x[f(x) - g(x)]\, dx.$$

Example 2

Use Theorem (5.12) to find the volume of the solid generated by revolving about the y-axis the region bounded by the graphs of $y = x + 1$ and $y = x^3 + 1$.

Solution

This solid was considered first in Example 5 of Section 5.2, where we found the volume to be $\frac{4}{15}\pi$. The region to be revolved is shown in Figure 5.29(a) and the solid generated in Figure 5.29(b).

By Theorem (5.12), the volume of the solid is

$$V = 2\pi \int_0^1 x[(x+1) - (x^3+1)]\, dx = 2\pi \int_0^1 (x^2 - x^4)\, dx,$$

so

$$V = 2\pi \left(\frac{x^3}{3} - \frac{x^5}{5}\right)\Big]_0^1$$

$$= 2\pi \left(\frac{1}{3} - \frac{1}{5}\right) = \frac{4}{15}\pi.$$

This is consistent with the result in Example 5 of Section 5.2. ▶

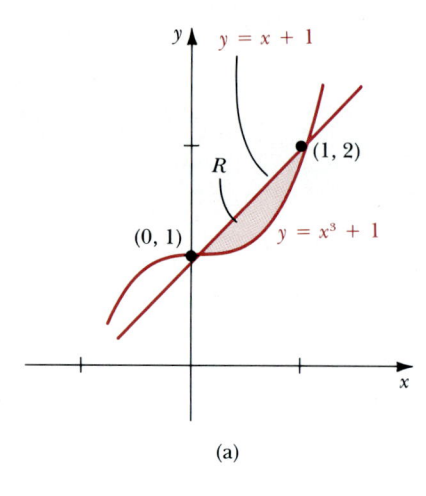

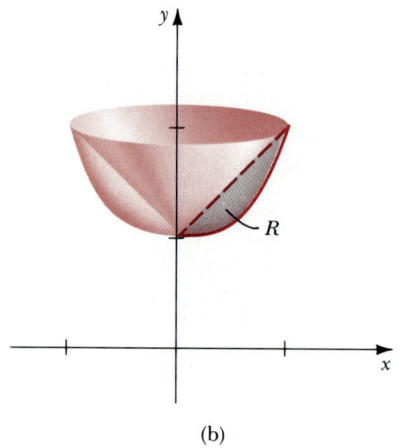

Figure 5.29

Example 3

An ice cream cone has the shape approximated by revolving about the y-axis the region in the first quandrant bounded by the y-axis and the graphs of $y = 4x$ and $y = 4 + \sqrt{1 - x^2}$ (see Figure 5.30). Find the amount of ice cream in the cone.

Solution

The intersection of the graphs occurs at $(1, 4)$, so

$$V = 2\pi \int_0^1 x[(4 + \sqrt{1-x^2}) - 4x]\, dx$$

$$= 2\pi \left[\int_0^1 4x\, dx + \int_0^1 x\sqrt{1-x^2}\, dx - \int_0^1 4x^2\, dx\right].$$

We evaluate the second integral using the substitution $u = 1 - x^2$,

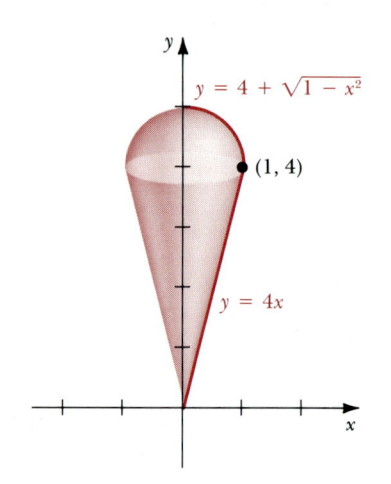

Figure 5.30

$du = -2x \, dx$:

$$\int x\sqrt{1-x^2} \, dx = -\frac{1}{2} \int u^{1/2} \, du = -\frac{1}{2} \frac{u^{3/2}}{(3/2)} + C$$

$$= -\frac{1}{3}(1-x^2)^{3/2} + C.$$

Thus,

$$V = 2\pi \left[2x^2 - \frac{1}{3}(1-x^2)^{3/2} - \frac{4}{3}x^3 \right]_0^1$$

$$= 2\pi \left[\left(2 - 0 - \frac{4}{3}\right) - \left(0 - \frac{1}{3} - 0\right) \right] = 2\pi \text{ in}^3.$$

There are 231 in³ in a U.S. gallon, so one gallon of ice cream will make $231/2\pi \approx 36$ cones of the type described in Example 3.

The last example in this section shows how the method of shells is used to find the volume of a solid generated when a region is revolved about the *x*-axis. The example uses the following modification of Theorem (5.12), which is illustrated in Figure 5.31.

Suppose h and k are continuous and $h(y) \geq k(y)$ for all y in $[c, d]$, where $c \geq 0$. The volume of the solid generated by revolving about the *x*-axis the region bounded on $[c, d]$ by the graphs of $x = h(y)$ and $x = k(y)$ is

(5.13) $$V = 2\pi \int_c^d y[h(y) - k(y)] \, dy.$$

Example 4

Find the volume of the solid generated by revolving about the *x*-axis the region bounded by the graphs of

$$x = 2y - y^2 \quad \text{and} \quad x = y^2 - y.$$

Solution

To use the method of disks to solve this problem would require writing y as a function of x on the boundaries of each of the distinct regions R_1, R_2, and R_3 shown in Figure 5.32(a). Since the boundaries of these regions are different, we would need to evaluate three integrals.

We can solve the problem more easily by considering x as a function of y and using the method of shells. The curves intersect when

$$y^2 - y = 2y - y^2,$$

that is, when

$$0 = 2y^2 - 3y = y(2y - 3).$$

Thus $y = 0$ or $y = \frac{3}{2}$ (see Figure 5.32a).

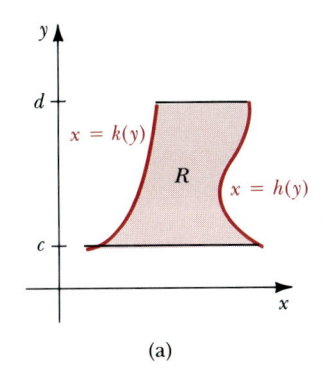

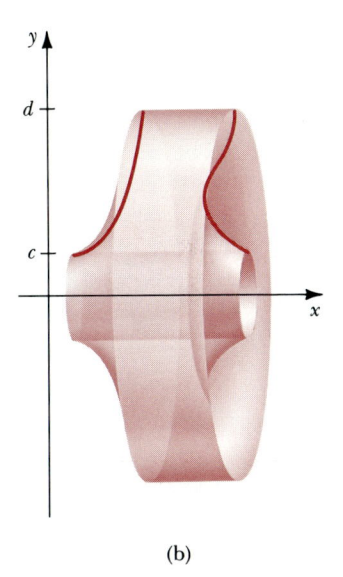

Figure 5.31
The volume is $2\pi \int_c^d y[h(y) - k(y)] \, dy.$

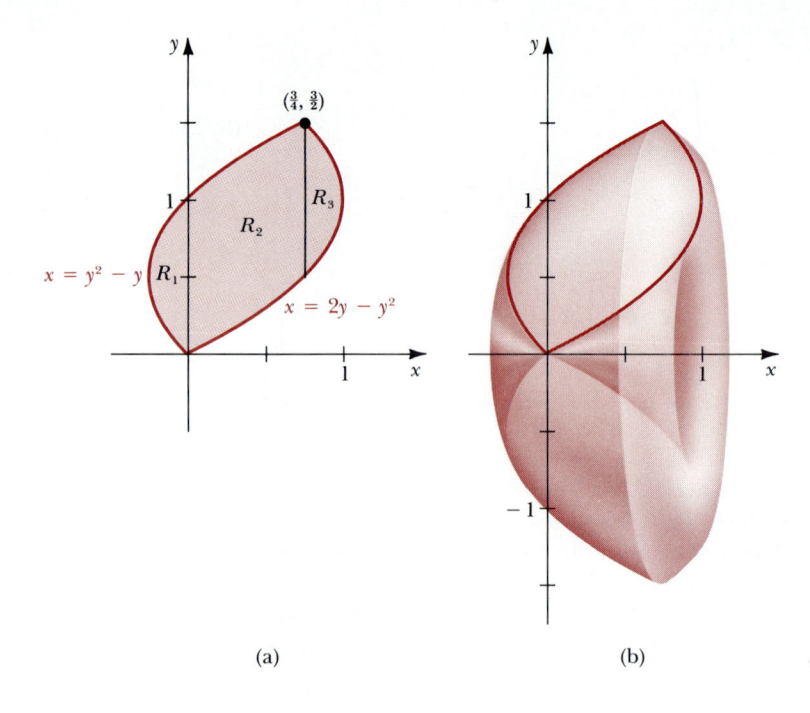

Figure 5.32

(a) (b)

For y in the interval $[0, \frac{3}{2}]$,
$$y^2 - y \leq 2y - y^2.$$
By (5.13), the volume V of the solid in Figure 5.32(b) is

$$V = 2\pi \int_0^{3/2} y[(2y - y^2) - (y^2 - y)] \, dy$$

$$= 2\pi \int_0^{3/2} y(3y - 2y^2) \, dy$$

$$= 2\pi \int_0^{3/2} (3y^2 - 2y^3) \, dy$$

$$= 2\pi \left(y^3 - \frac{y^4}{2} \right) \Big]_0^{3/2}$$

$$= 2\pi \left[\frac{27}{8} - \frac{81}{32} \right]$$

$$= \frac{27\pi}{16}.$$
▶

Figure 5.33 gives a summary to help decide which method, disk or shell, is likely to be most appropriate for finding a particular volume of revolution. The appropriate method usually depends on the ease with which the boundaries of the revolved region can be represented. This summary will be particularly helpful when you are working Exercises 31–34, where the solid of revolution is described, but it is not specified which method should be used to find the volume.

5.3 VOLUMES OF SOLIDS OF REVOLUTION: SHELLS

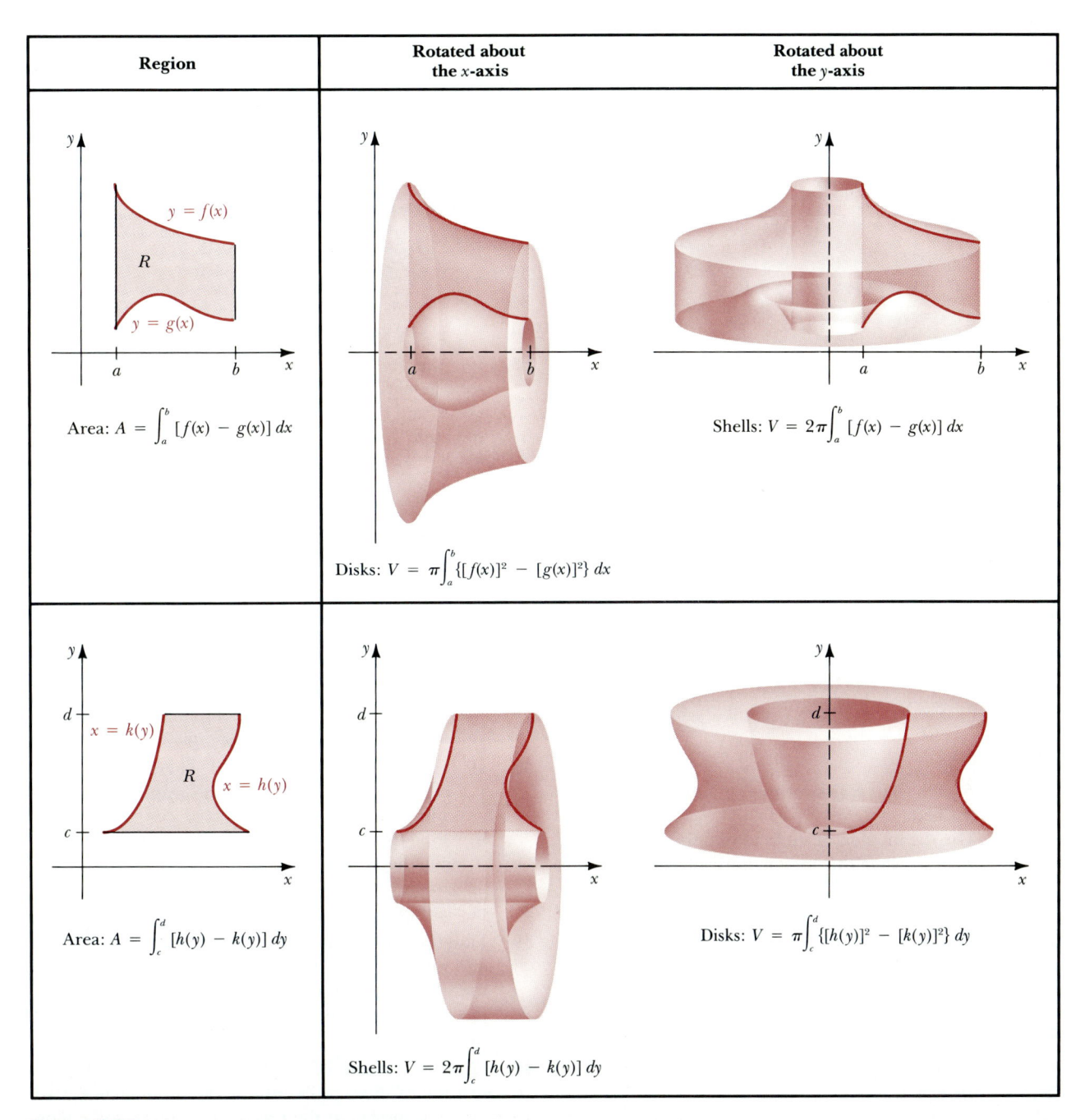

Region	Rotated about the x-axis	Rotated about the y-axis
Area: $A = \int_a^b [f(x) - g(x)]\, dx$	Disks: $V = \pi \int_a^b \{[f(x)]^2 - [g(x)]^2\}\, dx$	Shells: $V = 2\pi \int_a^b [f(x) - g(x)]\, dx$
Area: $A = \int_c^d [h(y) - k(y)]\, dy$	Shells: $V = 2\pi \int_c^d [h(y) - k(y)]\, dy$	Disks: $V = \pi \int_c^d \{[h(y)]^2 - [k(y)]^2\}\, dy$

Figure 5.33

► EXERCISE SET 5.3

In Exercises 1–18:
- **a)** Sketch the region whose boundaries are described.
- **b)** Use the method of shells to find the volume generated when the region is revolved about the y-axis.

1. $y = x^2$, $x = 1$, $y = 0$
2. $y = x$, $x = 5$, $y = 0$
3. $y = x^3$, $x = 1$, $x = 2$, $y = 0$
4. $y = \sqrt{x^2 + 1}$, $x = 1$, $x = 3$, $y = 0$
5. $y = x^2$, $y = \sqrt[3]{x}$
6. $y = x^3$, $y = x$
7. $y = \cos x^2$, $x = 0$, $y = 0$, $x = \sqrt{\pi/2}$
8. $y = x - x^2$, $y = 0$
9. $y = x - x^3$, $y = 0$
10. $y = x^2$, $x = 1$, $x = 3$, $y = \frac{1}{2}x + \frac{1}{2}$
11. $y = \frac{1}{2}x + 1$, $x + y = 4$, $y = 1$
12. $y = \frac{1}{2}x + 1$, $x + y = 4$, $x + 4y = 4$
13. $y = x + 2/x$, $x = 1$, $y = \frac{11}{6}(x - 1)$
14. $y = x + 2/x$, $x = 1$, $y = \frac{1}{6}(x + 23)$
15. $y = \sqrt{x + 1}$, $3y = 2x$, $x = 0$
16. $y = 2\sqrt{4 - x}$, $3y = 2x$, $x = 0$
17. $y = 3/\sqrt{x^2 + 1}$, $x = 1$, $x = 2$, $y = 0$
18. $y = 3/\sqrt{x + 1}$, $x = 1$, $x = 2$, $y = 0$

In Exercises 19–30:
- **a)** Sketch the region whose boundaries are described.
- **b)** Use the method of shells to find the volume generated when the region is revolved about the x-axis. (This requires that x be expressed in terms of y.)

19. $x = \sqrt{y}$, $x = 0$, $y = 4$
20. $y = x^{2/3}$, $x = 0$, $y = 4$
21. $y = 1/x$, $y = 1$, $y = 2$, $x = 0$
22. $x = \sqrt{4 - y}$, $x = 0$, $y = 0$
23. $y = x^2$, $y = x$
24. $y = \sqrt[3]{x}$, $y = x^2$
25. $y = x^3$, $y = x^2$
26. $y = 2/x$, $x + y = 3$
27. $x = y - y^2$, $x = y^2 - y$
28. $x = 2y - y^2$, $x + y = 0$
29. $x = 1 - \sqrt{1 - y}$, $x = 0$, $x + y = 2$
30. $y = x + 1$, $2x + y = 4$, $x = 0$

In Exercises 31–34, use either the method of disks or the method of shells to find the volume of the solid generated if the region bounded by the curves described is revolved about (a) the x-axis; (b) the y-axis.

31. $y = x^2 - 4x + 5$, $y = 5 - x^2 + 4x$
32. $y = 2 - \sqrt{x}$, $y = 1 + \sqrt{x + 1}$, $x = 4$
33. $y = 1 + \sqrt{x}$, $x = 2y - 2$
34. $x = y^3 - y + 2$, $x = 2$

35. Use the method of shells to find the volume of the solid generated if the region bounded by the graph of $y = x^3$ and the lines $x = 0$ and $y = 1$ is revolved about the line:
   - **a)** $x = 1$;
   - **b)** $y = 1$;
   - **c)** $x = -1$;
   - **d)** $y = 2$.

36. Use the method of shells to find the volume of the solid generated if the region bounded by the graph of $y = \sqrt{x}$ and the lines $x = 0$ and $y = 1$ is revolved about the line:
   - **a)** $x = 1$;
   - **b)** $y = 1$;
   - **c)** $x = -1$;
   - **d)** $y = 2$.

37. Find the volume of the solid generated by revolving the region bounded by the graphs of $x = y^2 - 2y$ and $x = y - y^2$ about the x-axis. Compare this region and volume with those given in Example 4.

38.
   - **a)** Find the volume of the solid generated by revolving the region bounded by the graphs of $f(x) = x^2$ and $g(x) = \sqrt{x}$ about the y-axis.
   - **b)** Find the volume of the solid generated by revolving the region described in part (a) about the x-axis.
   - **c)** Compare the results of parts (a) and (b).

39. Find the volume of a right circular cylinder of height h and radius r using the method of shells to determine the volume generated by revolving about the y-axis the region bounded by the lines $x = r$ and $y = h$ and the x- and y-axes.

40. Find the volume of a sphere of radius r using the method of shells to determine the volume generated by revolving about the y-axis a portion of the region bounded by the circle $x^2 + y^2 = r^2$.

41. Find the volume of a cone of height h and radius r using the method of shells to determine the volume generated by revolving about the y-axis the region

bounded by the x- and y-axes and the line segment joining the points $(r, 0)$ and $(0, h)$.

42. A cylindrical hole is bored through the center of a sphere of radius 4 cm (see the figure). Find the volume removed from the sphere if the height of the hole is 4 cm.

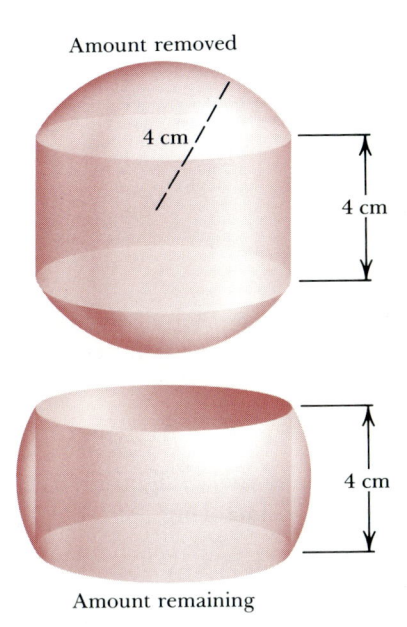

Amount removed

4 cm

4 cm

4 cm

Amount remaining

43. When a liquid flows through a cylindrical tube, friction at the walls of the tube tends to slow its motion. This results in liquid flowing faster near the center of the tube than near the walls. In fact, the liquid can be considered as flowing in circular layers, or laminae, which have the constant velocity in each layer given by *Poiseuille's Law*:

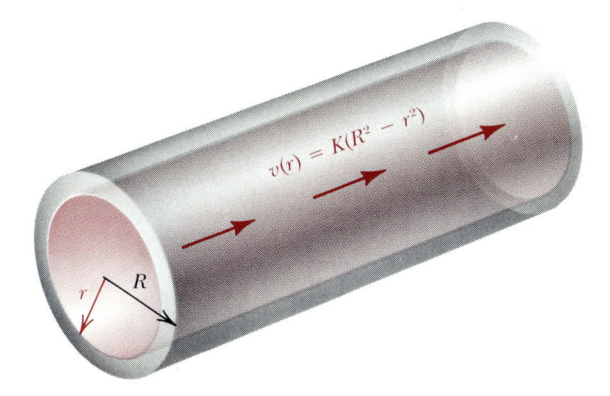

$$v(r) = K(R^2 - r^2),$$

where R is the radius of the tube, r is the distance from the center of the tube, and K is a constant. Partition the radius of the tube and use this partition to construct a Riemann sum that will lead to the formula $F = K\pi R^4/2$, called *Poiseuille's Equation*, that describes the volume of flow through the tube per unit time.

44. In Exercise 71 of Section 5.2, the volume of a spherical cap is derived by revolving about the x-axis the region bounded by the graph of $y = \sqrt{r^2 - x^2}$ and the lines $x = r - h$ and $y = 0$. Derive this formula using the method of shells.

5.4

Arc Length and Surfaces of Revolution

The concept of the length of a curve is intuitive. Actually determining the length of a specific curve, however, is often difficult. In this section we see that the length of certain curves can be approximated by Riemann sums in a manner similar to the area and volume applications studied in the first three sections of this chapter.

Suppose C denotes the graph of a differentiable function f from $(a, f(a))$ to $(b, f(b))$ (see Figure 5.34a). Let $\mathcal{P} = \{x_0, x_1, \ldots, x_n\}$ be a partition of the interval $[a, b]$. The curve C is approximated by the collection of straight-line segments joining $(x_{i-1}, f(x_{i-1}))$ to $(x_i, f(x_i))$ for each $i = 1, 2, \ldots, n$, as shown in Figure 5.34(b). The sum of the lengths of these line segments approximates the length of C.

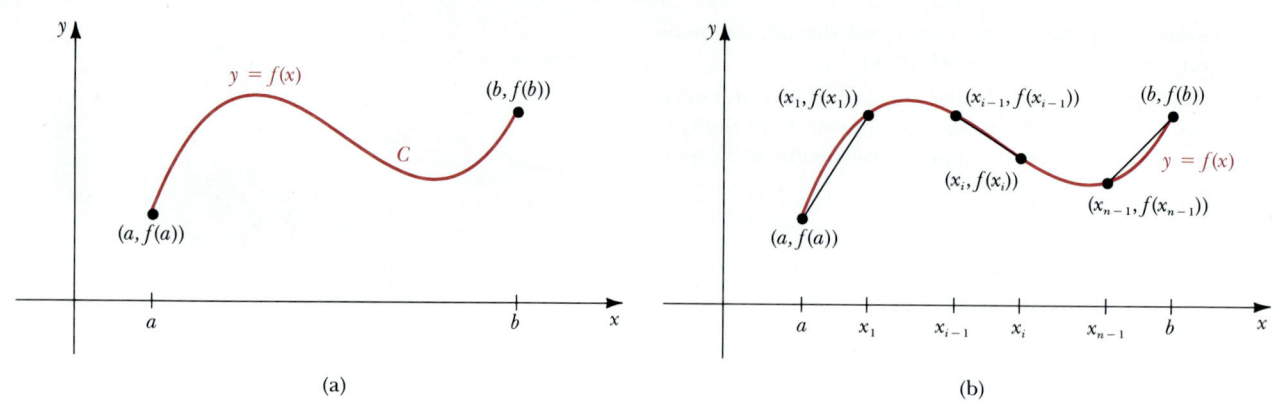

Figure 5.34

(a)

(b)

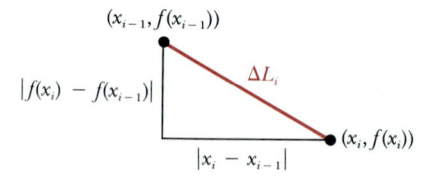

Figure 5.35

The length ΔL_i of the segment from $(x_{i-1}, f(x_{i-1}))$ to $(x_i, f(x_i))$ (see Figure 5.35) is

$$\Delta L_i = \sqrt{(x_i - x_{i-1})^2 + (f(x_i) - f(x_{i-1}))^2}.$$

To express the sum of the lengths of the straight-line segments as a Riemann sum, we first rewrite ΔL_i by factoring

$$\sqrt{(x_i - x_{i-1})^2} = |x_i - x_{i-1}|$$
$$= x_i - x_{i-1}$$
$$= \Delta x_i$$

from the square root. Then

(5.14) $$\Delta L_i = \sqrt{1 + \left[\frac{f(x_i) - f(x_{i-1})}{x_i - x_{i-1}}\right]^2} \Delta x_i.$$

Since f is differentiable on the interval $[a, b]$, the Mean Value Theorem guarantees the existence of a number z_i with $x_{i-1} < z_i < x_i$ and

$$f'(z_i) = \frac{f(x_i) - f(x_{i-1})}{x_i - x_{i-1}}.$$

Thus Equation (5.14) can be rewritten as

$$\Delta L_i = \sqrt{1 + [f'(z_i)]^2}\, \Delta x_i.$$

Summing the ΔL_i gives a Riemann sum,

$$\sum_{i=1}^{n} \Delta L_i = \sum_{i=1}^{n} \sqrt{1 + [f'(z_i)]^2}\, \Delta x_i,$$

for the function described by $\sqrt{1 + [f'(x)]^2}$. This Riemann sum approximates the length of the curve C, and the approximations should improve as the norm of the partition approaches zero. Since the limit of the Riemann sums is the definite integral of the function given by $\sqrt{1 + [f'(x)]^2}$, this leads to the following definition (see Figure 5.36).

(5.15)
DEFINITION
If f is continuous on $[a, b]$ and differentiable on (a, b), the **arc length** of the graph of f from $(a, f(a))$ to $(b, f(b))$ is

$$L_{[a, b]}(f) = \int_a^b \sqrt{1 + [f'(x)]^2} \, dx,$$

provided this integral exists.

When this integral exists, f is said to be **rectifiable** on the interval $[a, b]$.

A condition sufficient to guarantee that a function f is rectifiable on an interval is that f' is continuous on the interval. Such a function is said to be **smooth** on that interval. When f is smooth on $[a, b]$, $\sqrt{1 + [f'(x)]^2}$ describes a continuous function, so it follows from Theorem (4.9) that the integral in Definition (5.15) exists.

If f is not differentiable at the endpoints of the interval $[a, b]$, then an improper integral is used to determine the arc length on $[a, b]$.

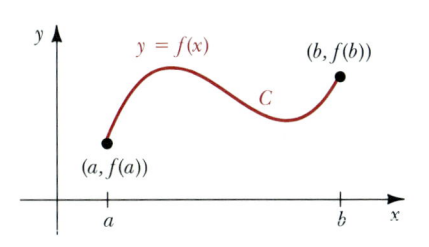

Figure 5.36

The length is $\int_a^b \sqrt{1 + [f'(x)]^2} \, dx.$

Example 1

Find the arc length of the graph of $f(x) = x^{3/2}$ from $(1, 1)$ to $(9, 27)$.

Solution

Since $f'(x) = \frac{3}{2} x^{1/2}$ is continuous on $[0, \infty)$, f is smooth on $[1, 9]$ and hence rectifiable. Thus,

$$L_{[1, 9]}(f) = \int_1^9 \sqrt{1 + \left[\frac{3}{2} x^{1/2}\right]^2} \, dx = \int_1^9 \sqrt{1 + \frac{9}{4} x} \, dx.$$

To determine $\int \sqrt{1 + \frac{9}{4} x} \, dx$, let $u = 1 + \frac{9}{4} x$. Then $du = \frac{9}{4} dx$ and

$$\int \sqrt{1 + \frac{9}{4} x} \, dx = \int \sqrt{u} \, \frac{4}{9} \, du = \frac{4}{9} \cdot \frac{2}{3} u^{3/2} + C = \frac{8}{27}\left(1 + \frac{9}{4} x\right)^{3/2} + C.$$

Thus,

$$L_{[1, 9]}(f) = \frac{8}{27}\left(1 + \frac{9}{4} x\right)^{3/2} \Big]_1^9 = \frac{8}{27}\left[\left(1 + \frac{81}{4}\right)^{3/2} - \left(1 + \frac{9}{4}\right)^{3/2}\right] \approx 27.3$$

This result appears reasonable from the graph shown in Figure 5.37. ▶

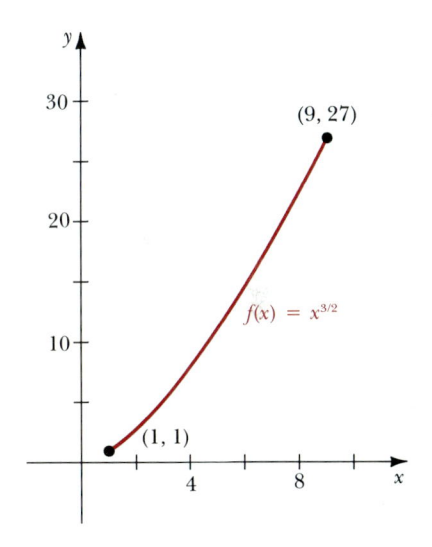

Figure 5.37

The function in Example 1, $f(x) = x^{3/2}$, might seem a complicated function to use for a first example, but this was the first algebraic curve shown to be rectifiable. It was called a semi-cubical parabola in the seventeenth century and shown to be rectifiable by William Neil (1637–1670) in 1657, Heinrich van Heureat (1633–1660) in 1659 and Pierre de Fermat in 1660. All of the proofs were geometric rather than analytic, but are in essence equivalent to our current procedure. The importance of this result is that it showed that the rectification problem could be reduced to a quadrature or area problem. Showing that one class of problems could be reduced to another was an important step toward the fundamental theorem of calculus.

HISTORICAL NOTE

Example 2

Show that the function described by $f(x) = x^2$ is rectifiable on the interval $[0, 1]$.

Solution

Since $f'(x) = 2x$ describes a continuous function on the interval $[0, 1]$, f is rectifiable on $[0, 1]$. The obvious question to ask in return is: What is the arc length of the graph of $f(x) = x^2$ from $(0, 0)$ to $(1, 1)$?

The arc length is given by

$$L_{[0, 1]}(f) = \int_0^1 \sqrt{1 + [2x]^2}\, dx$$

$$= \int_0^1 \sqrt{1 + 4x^2}\, dx.$$

We have not yet seen a technique for evaluating this integral and will postpone its evaluation until Section 7.3. ▶

The preceding example uncovers one of the difficulties in determining the arc length of a curve. Even for relatively elementary functions, an integral having an integrand of the form

$$\sqrt{1 + [f'(x)]^2}$$

can be difficult or even impossible to evaluate exactly. The functions in the exercise set have been chosen so that the definite integral can be evaluated using at most a substitution technique of the type discussed in Chapter 4. Occasionally, however, the evaluation requires some algebraic simplification before the integration can be done. In the following example, we can easily evaluate the integral once the square under the radical in the integrand is observed. It would be a very difficult problem, indeed, without this simplification.

In Chapter 7 we introduce additional techniques for evaluating definite integrals, and present some methods for approximating the definite integral of those we cannot evaluate exactly.

Example 3

Determine the arc length of the graph of $f(x) = \dfrac{x^3}{6} + \dfrac{1}{2x}$ for x in the interval $[1, 2]$.

Solution

Since $f'(x) = x^2/2 - 1/2x^2$,

$$L_{[1, 2]}(f) = \int_1^2 \left[1 + \left(\frac{x^2}{2} - \frac{1}{2x^2}\right)^2\right]^{1/2} dx$$

$$= \int_1^2 \left[1 + \left(\frac{x^4}{4} - \frac{1}{2} + \frac{1}{4x^4}\right)\right]^{1/2} dx$$

$$= \int_1^2 \left[\frac{x^4}{4} + \frac{1}{2} + \frac{1}{4x^4}\right]^{1/2} dx.$$

The expression in the integrand can be factored as a perfect square:

$$\frac{x^4}{4} + \frac{1}{2} + \frac{1}{4x^4} = \left(\frac{x^2}{2} + \frac{1}{2x^2}\right)^2.$$

Therefore,

$$L_{[1,\,2]}(f) = \int_1^2 \left(\frac{x^2}{2} + \frac{1}{2x^2}\right) dx = \left[\frac{x^3}{6} - \frac{1}{2x}\right]_1^2$$

$$= \left(\frac{8}{6} - \frac{1}{4}\right) - \left(\frac{1}{6} - \frac{1}{2}\right) = \frac{17}{12}. \quad \blacktriangleright$$

The topic of arc length is reconsidered in Chapters 9 and 16, where more general results are obtained.

Area of Surfaces of Revolution

A surface is generated when a curve is revolved about a line. For example, the surface of a cone is generated if the hypotenuse of a right triangle is revolved about one of the legs forming the right angle, as shown in Figure 5.38. In the remainder of this section we consider how to determine the area of a surface of revolution. The procedure is similar to that of determining the arc length of a curve, with a few added complications.

Suppose that f is nonnegative and continuous on $[a, b]$, that f is differentiable on (a, b), and that the graph of f between $(a, f(a))$ and $(b, f(b))$ is revolved about the x-axis. Let $\mathscr{P} = \{x_0, x_1, \ldots, x_n\}$ be a partition of $[a, b]$, and consider the surface generated when a typical line segment joining $(x_{i-1}, f(x_{i-1}))$ to $(x_i, f(x_i))$ is revolved about the x-axis (see Figure 5.39). The solid bounded by this surface is called the *frustum* of a cone. We need to determine its surface area.

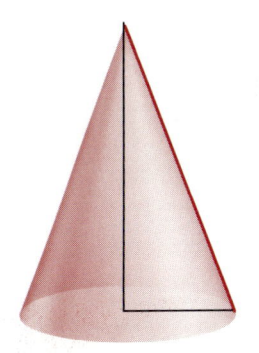

Figure 5.38

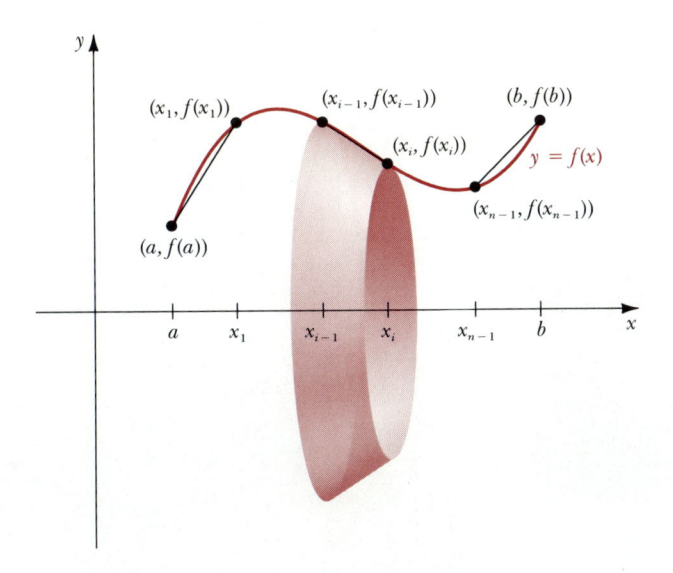

Figure 5.39

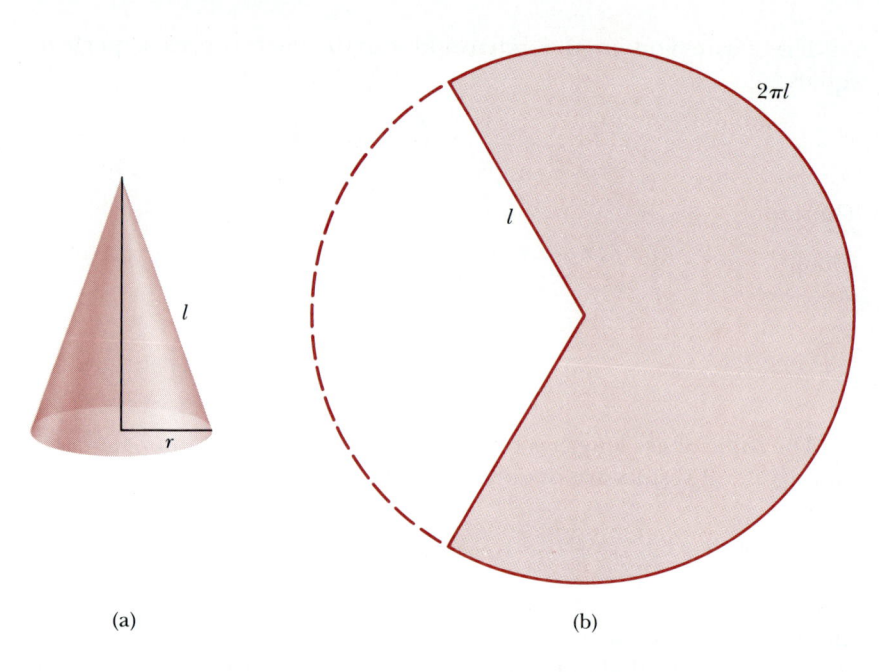

Figure 5.40

(a)   (b)

Consider the problem of determining the surface area of the cone with radius r and slant height l shown in Figure 5.40(a). If we cut this cone down its side and spread out the surface, we find that it forms a sector of a circle with radius l and circumference $2\pi l$, as shown in Figure 5.40(b). The entire circle has area πl^2.

The circumference of the base of the cone is $2\pi r$, so the area of the sector and, hence, the surface area of the cone is

(5.16)
$$\text{SA} = \left(\frac{2\pi r}{2\pi l}\right) \pi l^2 = \pi l r.$$

The surface area of the frustum of the cone in Figure 5.41(a) is the

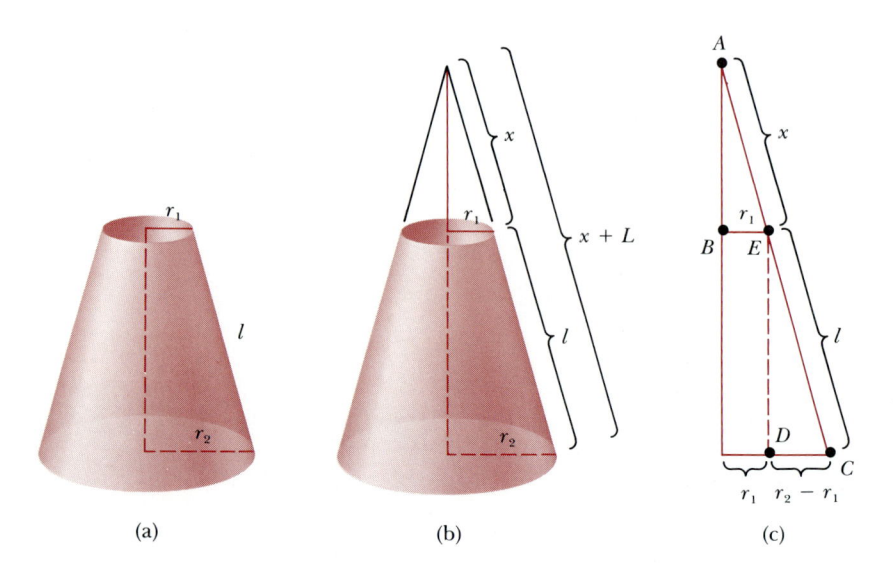

Figure 5.41   (a)   (b)   (c)

difference between the surface areas of the two cones in Figure 5.41(b):
$$SA = \pi(x+l)r_2 - \pi x r_1.$$

To find this surface area in terms of the radii r_1 and r_2 we must first find the slant height x of the top cone. The similar triangles ABC and CDE shown in Figure 5.41(c) imply that

$$\frac{x}{r_1} = \frac{l}{r_2 - r_1} \quad \text{so} \quad x = \left(\frac{r_1}{r_2 - r_1}\right) l$$

and

$$x + l = \left(\frac{r_1}{r_2 - r_1}\right) l + l = \left(\frac{r_1 + r_2 - r_1}{r_2 - r_1}\right) l = \left(\frac{r_2}{r_2 - r_1}\right) l.$$

The surface area of the frustum of the cone shown in Figure 5.40(a) is thus

$$SA = \pi(x+l)r_2 - \pi x r_1 = \pi \left(\frac{r_2}{r_2 - r_1}\right) l r_2 - \pi \left(\frac{r_1}{r_2 - r_1}\right) l r_1 = \pi \left(\frac{r_2^2 - r_1^2}{r_2 - r_1}\right) l,$$

that is,

(5.17) $$SA = \pi(r_2 + r_1)l.$$

The frustum of the cone shown in Figure 5.39 has radii $f(x_{i-1})$ and $f(x_i)$ and slant height

$$l = \sqrt{(x_i - x_{i-1})^2 + (f(x_{i-1}) - f(x_i))^2},$$

so Formula (5.17) implies that the surface area generated by revolving about the x-axis the line segment joining $(x_{i-1}, f(x_{i-1}))$ to $(x_i, f(x_i))$ is

(5.18) $$\Delta SA_i = \pi[f(x_{i-1}) + f(x_i)] \sqrt{(x_i - x_{i-1})^2 + [f(x_i) - f(x_{i-1})]^2}.$$

Following the procedure used to develop the arc length formula, we rewrite Equation (5.18) as

$$\Delta SA_i = \pi[f(x_{i-1}) + f(x_i)] \sqrt{1 + \left[\frac{f(x_i) - f(x_{i-1})}{x_i - x_{i-1}}\right]^2} \Delta x_i,$$

and the Mean Value Theorem implies that there is a number z_i in the interval (x_{i-1}, x_i) with

(5.19) $$\Delta SA_i = \pi[f(x_{i-1}) + f(x_i)] \sqrt{1 + [f'(z_i)]^2} \, \Delta x_i.$$

Note that (5.19) can be rewritten as

(5.20) $$\Delta SA_i = 2\pi \left[\frac{f(x_{i-1}) + f(x_i)}{2}\right] \sqrt{1 + [f'(z_i)]^2} \, \Delta x_i,$$

where $[f(x_{i-1}) + f(x_i)]/2$ is simply the average of the numbers $f(x_{i-1})$ and $f(x_i)$. Since f is a continuous function, the Intermediate Value Theorem implies that there is a number w_i in the interval $[x_{i-1}, x_i]$ with $f(w_i) = [f(x_{i-1}) + f(x_i)]/2$. Equation (5.20) can now be rewritten as

$$\Delta SA_i = 2\pi f(w_i) \sqrt{1 + f'(z_i)^2} \, \Delta x_i.$$

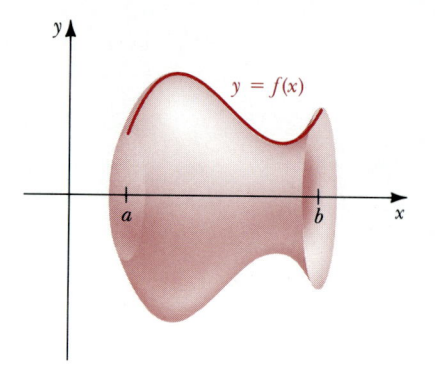

Figure 5.42
The surface area is
$$2\pi \int_a^b f(x)\sqrt{1+[f'(x)]^2}\,dx.$$

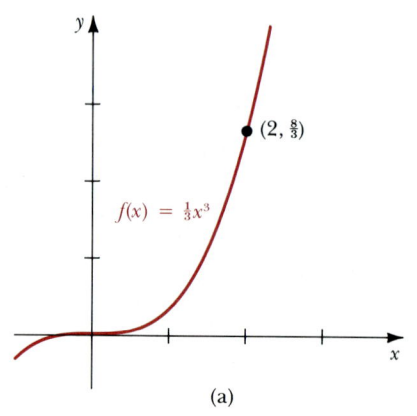

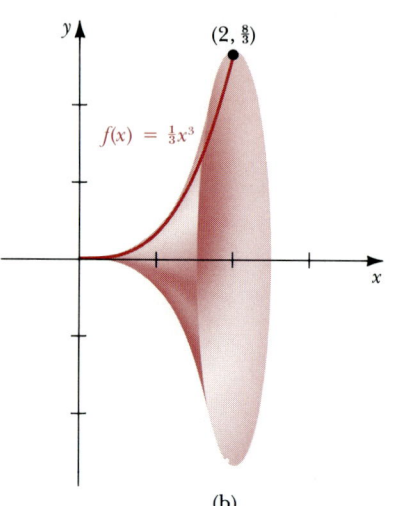

Figure 5.43

Summing the surface area on each of the frustums gives an approximation to the total area of the surface generated by rotating the graph of f between $(a, f(a))$ and $(b, f(b))$ about the x-axis:

$$\text{SA} \approx \sum_{i=1}^n \Delta\text{SA}_i = \sum_{i=1}^n 2\pi f(w_i)\sqrt{1+[f'(z_i)]^2}\,\Delta x_i.$$

This sum is not a Riemann sum, however, since it involves two values w_i and z_i in each interval $[x_{i-1}, x_i]$. Fortunately, even if $w_i \neq z_i$,

$$\lim_{\|\mathcal{P}\|\to 0} \sum_{i=1}^n 2\pi f(w_i)\sqrt{1+[f'(z_i)]^2}\,\Delta x_i = \int_a^b 2\pi f(x)\sqrt{1+[f'(x)]^2}\,dx.$$

This is a result of a theorem stated by Gilbert Ames Bliss (1876–1951) in 1914. (The more general Duhamel's Theorem is used in Chapter 16.)

It leads to the following definition for the area of a surface of revolution (see Figure 5.42).

(5.21) DEFINITION

Suppose f is nonnegative and continuous on $[a, b]$ and differentiable on (a, b). The **surface area** generated by revolving the graph of f between $(a, f(a))$ and $(b, f(b))$ about the x-axis is

$$\text{SA}_{[a,b]}(f) = 2\pi \int_a^b f(x)\sqrt{1+[f'(x)]^2}\,dx,$$

provided this integral exists.

As in the case of the arc length, an improper integral is used to determine the surface area when f is not differentiable at the endpoints of $[a, b]$.

Example 4

Find the area of the surface generated by revolving about the x-axis the graph of $f(x) = x^3/3$ for x in $[0, 2]$.

Solution

The curve is shown in Figure 5.43(a) and the surface generated in Figure 5.43(b).

Since f is differentiable on $[0, 2]$ and $f'(x) = x^2$, Definition (5.21) gives

$$\text{SA}_{[0,2]}(f) = 2\pi \int_0^2 \frac{x^3}{3}\sqrt{1+[x^2]^2}\,dx$$

$$= \frac{2\pi}{3}\int_0^2 x^3\sqrt{1+x^4}\,dx.$$

If $u = 1 + x^4$, then $du = 4x^3 \, dx$ and

$$\text{SA}_{[0,2]}(f) = \frac{2\pi}{3} \int_{x=0}^{x=2} x^3 \sqrt{1+x^4} \, dx$$

$$= \frac{2\pi}{3} \int_{u=1}^{u=17} \frac{1}{4} u^{1/2} \, du$$

$$= \frac{\pi}{6} \left(\frac{2}{3} u^{3/2} \right) \Big]_1^{17}$$

$$= \frac{\pi}{9} [(17)^{3/2} - 1] \approx 24.1.$$

▶

Example 5

Show that the surface area of a sphere of radius r is $4\pi r^2$.

Solution

The sphere of radius r with center at the origin is generated when the graph of $f(x) = \sqrt{r^2 - x^2}$ for x in $[-r, r]$ is revolved about the x-axis (see Figure 5.44). Note, however, that f is not differentiable at the endpoints of the interval, so the surface area is determined using an improper integral. Because of the y-axis symmetry of the graph, this surface area is twice that generated by revolving the portion of the curve $f(x) = \sqrt{r^2 - x^2}$ for x in the interval $[0, r]$.

$$\text{SA}_{[-r,r]} = 2 \left\{ 2\pi \int_0^r \sqrt{r^2 - x^2} \left[1 + \left(\frac{-2x}{2\sqrt{r^2 - x^2}} \right)^2 \right]^{1/2} dx \right\}$$

$$= 4\pi \lim_{M \to r^-} \int_0^M \sqrt{r^2 - x^2} \left[\frac{r^2 - x^2 + x^2}{r^2 - x^2} \right]^{1/2} dx$$

$$= 4\pi \lim_{M \to r^-} \int_0^M r \, dx$$

$$= 4\pi \lim_{M \to r^-} r(M - 0) = 4\pi r^2.$$

▶

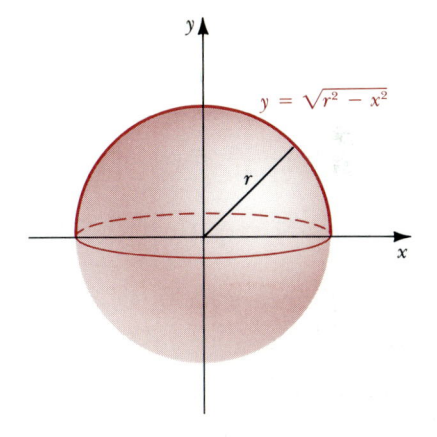

Figure 5.44

Calculating surface area is as complicated as calculating arc length. The integral is difficult to evaluate except in special cases, like the examples above. Even for $f(x) = x^2$, the determination of the surface area obtained by revolving the graph of f about the x-axis involves the integral,

$$2\pi \int x^2 \sqrt{1 + 4x^2} \, dx,$$

which we cannot evaluate at this time. For more complicated functions the difficulties generally increase.

▶ EXERCISE SET 5.4

In Exercises 1–22, (a) sketch the curve and (b) find the length of the curve.

1. $y = 3x + 4$, $[2, 4]$
2. $y = -2x + 1$, $[0, 2]$
3. $y = x^{3/2}$, $[3, 15]$
4. $y = (x + 1)^{3/2}$, $[0, 1]$
5. $y = 3x^{3/2}$, $[0, 1]$
6. $y = (x - 4)^{3/2}$, $[4, 7]$
7. $y = \left(x^2 - \frac{2}{3}\right)^{3/2}$, $[1, 2]$
8. $y = \frac{x^3}{6} + \frac{1}{2x}$, $[-3, -1]$
9. $y = \frac{x^4 + 12}{12x}$, $[1, 3]$
10. $y = \frac{x^6 + 32}{32x^2}$, $[2, 3]$
11. $y = x^{3/2} - \frac{\sqrt{x}}{3}$, $[1, 4]$
12. $y = \frac{\sqrt{x}}{3}(1 - 3x)$, $[1, 4]$
13. $y = x^{2/3}$, $[1, 8]$
14. $x = 3y^{3/2} - 1$, $[2, 23]$
15. $y = \ln \sec x$, $x = 0$, $x = \frac{\pi}{4}$
16. $y = \ln \sin x$, $x = \frac{\pi}{4}$, $x = \frac{\pi}{2}$
17. $(y + 1)^3 = (x - 1)^2$, $[1, 2]$
18. $(y + 1)^2 = (x - 1)^3$, $[1, 2]$
19. $x^{2/3} + y^{2/3} = 1$, $[\frac{1}{2}, 1]$, $y \geq 0$
20. $x^{2/3} + y^{2/3} = 1$, $[-1, -\frac{1}{2}]$, $y \geq 0$
21. $x^{2/3} + y^{2/3} = 1$, $[0, 1]$, $y \geq 0$
22. $x^{2/3} + y^{2/3} = 1$, $[-1, 1]$, $y \geq 0$

In Exercises 23–30, (a) sketch the graph of the curve and (b) find the surface area generated when the curve is revolved about the x-axis.

23. $y = x^3$, $[1, 3]$
24. $y = x^{1/2}$, $[2, 6]$
25. $y = x + 1$, $[0, 2]$
26. $y = 2x + 3$, $[1, 3]$
27. $y = \sqrt{x + 1}$, $[1, 5]$
28. $y = \sqrt{3x - 1}$, $[3, 12]$
29. $y = \frac{x^3}{6} + \frac{1}{2x}$, $[1, 2]$
30. $y = \sqrt{4 - x^2}$, $[0, 1]$

31. Find a formula for the surface area of a cylinder of height h and radius r by revolving about the x-axis the line segment joining $(0, r)$ and (h, r).

32. Find a formula for the surface area of a cone by revolving about the x-axis the line segment joining $(0, 0)$ and (h, r).

33. Find the coordinates of the point on the graph of $f(x) = x^{3/2}$ that is midway on this graph from the points $(0, 0)$ and $(1, 1)$.

34. Find the perimeter of the region bounded by the curves $y = x^{3/2}$, $y = x + 4$, and $x = 0$.

35. A local scout troop wants to make a canvas tepee. The tepee will be conical with a base of diameter 12 ft and a vertex 10 ft above the ground (see the figure). They want to leave the tepee open at the top from 1 ft below the vertex in order for supporting poles to extrude. How much canvas should they purchase?

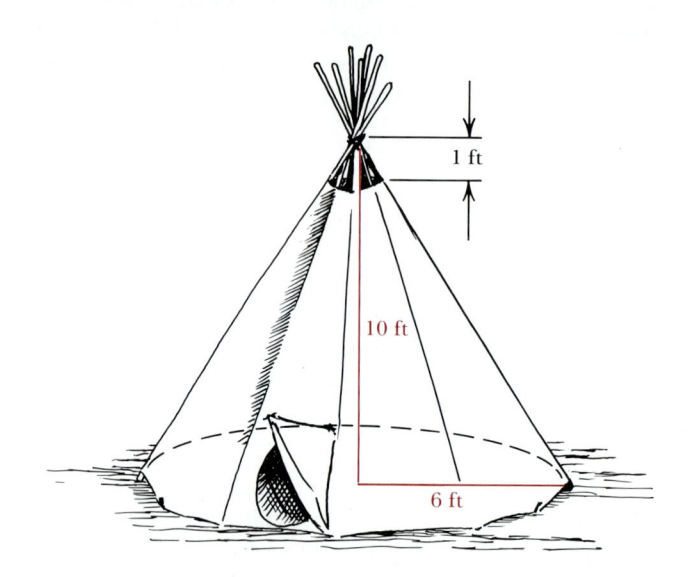

36. A ball hit from the ground travels along a curve described by $y = 88t - 16t^2$ for $0 \leq t \leq \frac{11}{2}$. Set up the integral that gives the actual distance the ball travels.

37. A potter is designing a pot to have the shape of the surface obtained by revolving about the x-axis the curve shown in the accompanying figure, with the

bottom of the pot to be the end with radius 1. Sketch the pot and find its surface area.

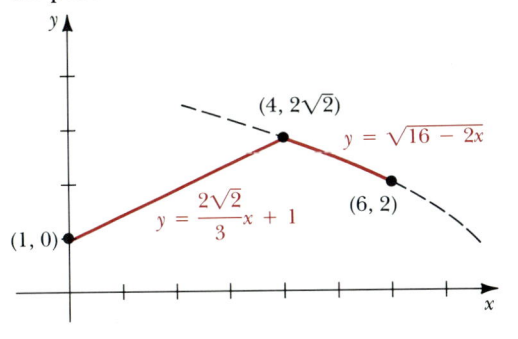

38. The arc length of a curve from $(0, 1)$ to $(1, 0)$ is
$$\int_0^1 \sqrt{\frac{x^2}{1-x^2} + 1}\, dx = \int_0^1 \frac{1}{\sqrt{1-x^2}}\, dx.$$
Find a possible equation for the curve. Is the curve uniquely defined by these conditions?

39. Show that
$$\int_0^1 \sqrt{1-x^2}\, dx = \frac{1}{2} \int_0^1 \frac{1}{\sqrt{1-x^2}}\, dx$$
without evaluating either integral. [*Hint:* Consider Exercise 38.]

5.5
Moments and the Center of Mass

Suppose that n objects with masses $m_1, m_2, \ldots, m_n$ lie on a coordinate line at the points $x_1, x_2, \ldots, x_n$. There is precisely one point along the line at which the system is balanced, in the sense that the system does not rotate if it is supported only at this point (see Figure 5.45). This balance point $\bar{x}$ is called the **center of mass** of the system.

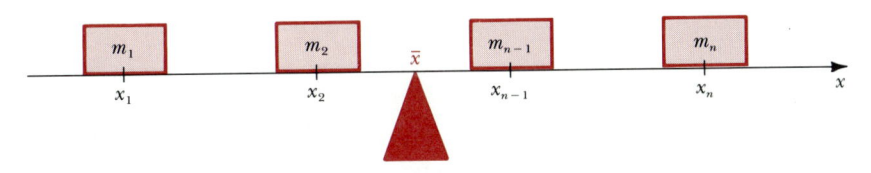

Figure 5.45

To determine the center of mass, we first define the moment M_p of the system about p, for any point p along a coordinate line, by
$$M_p = \sum_{i=1}^{n} m_i(x_i - p).$$

Physically, M_p describes the tendency of the system to rotate when a pivot is placed at the point p. If $M_p > 0$, the rotation is clockwise (see Figure 5.46a); if $M_p < 0$, the rotation is counterclockwise (see Figure 5.46b).

Since the center of mass $\bar{x}$ is located at the point at which no rotation occurs, $M_{\bar{x}} = 0$. Thus,
$$0 = M_{\bar{x}} = \sum_{i=1}^{n} m_i(x_i - \bar{x}) = \sum_{i=1}^{n} m_i x_i - \bar{x} \sum_{i=1}^{n} m_i$$

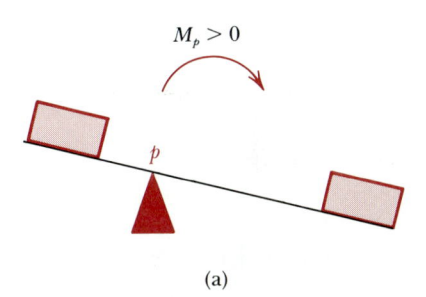

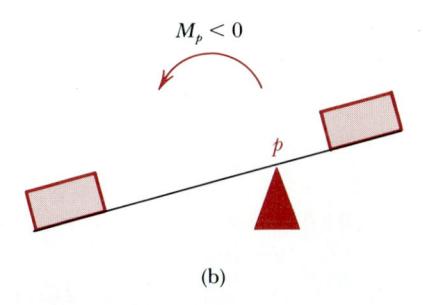

Figure 5.46

and

(5.22)
$$\bar{x} = \frac{\sum_{i=1}^{n} m_i x_i}{\sum_{i=1}^{n} m_i}.$$

Equation (5.22) implies that the center of mass of the system is the moment of the system about the origin, M_0, divided by the total mass of the system, $M = \sum_{i=1}^{n} m_i$:

$$\bar{x} = \frac{M_0}{M}.$$

Example 1

Two blocks of equal mass m are placed on opposite ends of a 12-ft seesaw. What is the moment of the system about the pivot if the distance of the pivot from the left end of the seesaw is (a) 2 ft? (b) 6 ft? (c) 10 ft?

Rotation clockwise

$M_p = -2m + 10m$
$= 8m > 0$

(a)

No rotation

$M_p = -6m + 6m$
$= 0$

(b)

Rotation counterclockwise

$M_p = -10m + 2m$
$= -8m < 0$

(c)

Figure 5.47

Solution

The solutions to (a), (b), and (c) are illustrated in Figure 5.47. The center of mass is located 6 ft from the left end, since the moment at this point is zero. ▶

The concepts of moment and center of mass are extended easily to a collection of objects in the plane. Suppose n objects with masses m_1, $m_2, \ldots, m_n$ lie in a coordinate plane at the points with coordinates $(x_1, y_1), (x_2, y_2), \ldots, (x_n, y_n)$, as shown in Figure 5.48. Moments of this system are defined about both coordinate axes.

The **moment about the y-axis**, denoted M_y, describes the tendency of the system to rotate about the y-axis. It is defined by

(5.23)
$$M_y = \sum_{i=1}^{n} m_i x_i.$$

The **moment about the x-axis**, M_x, describes the tendency of the system to rotate about the x-axis, and is defined by

(5.24)
$$M_x = \sum_{i=1}^{n} m_i y_i.$$

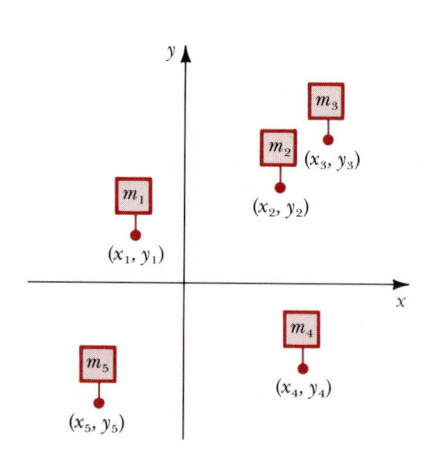

Figure 5.48

5.5 MOMENTS AND THE CENTER OF MASS

If $M = \sum_{i=1}^{n} m_i$ is the total mass of the system, then the center of mass of the system is the point with coordinates $(\bar{x}, \bar{y})$, where

(5.25) $$\bar{x} = \frac{M_y}{M} \quad \text{and} \quad \bar{y} = \frac{M_x}{M}.$$

Example 2

Particles with masses 2, 3, 5, and 6 units are located at the points $(1, -1)$, $(2, 1)$, $(-1, 1)$, and $(-2, -2)$, respectively (see Figure 5.49). Find the center of mass of this system.

Solution

The total mass of the system is $M = 2 + 3 + 5 + 6 = 16$, and the moments about the x- and y-axes are

$$M_x = 2(-1) + 3(1) + 5(1) + 6(-2) = -6$$

and

$$M_y = 2(1) + 3(2) + 5(-1) + 6(-2) = -9.$$

The center of mass is $(\bar{x}, \bar{y})$, where

$$\bar{x} = \frac{M_y}{M} = -\frac{9}{16} \quad \text{and} \quad \bar{y} = \frac{M_x}{M} = -\frac{6}{16} = -\frac{3}{8}.$$

▶

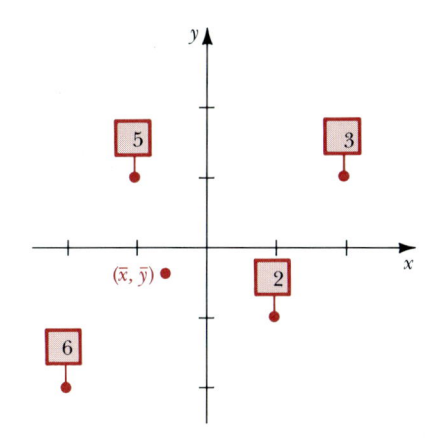

Figure 5.49

The definite integral can be applied to extend the concepts of moment and center of mass to regions in the plane. A region in the plane is considered a thin sheet of material called a **lamina**. A lamina with constant, or uniform, area density is called a **homogeneous lamina**, and its center of mass is called the **centroid** of the lamina.

Suppose a homogeneous lamina with constant density σ lies in the plane and is bounded on $[a, b]$ by the x-axis and the graph of the continuous nonnegative function f (see Figure 5.50a). Let $\mathscr{P} = \{x_0, x_1, \ldots, x_n\}$ be a partition of $[a, b]$ and for each $i = 1, 2, \ldots, n$, choose z_i as the midpoint of the interval, $z_i = \frac{1}{2}(x_{i-1} + x_i)$. The centroid of the rectangle shown in Figure 5.50(b) with base the interval $[x_{i-1}, x_i]$ and height $f(z_i)$ is the geometric center of the rectangle:

$$(z_i, \tfrac{1}{2}f(z_i)).$$

The area of the ith rectangle is $f(z_i)\,\Delta x_i$, so the mass of this rectangle is $\sigma f(z_i)\,\Delta x_i$. Since the mass of the rectangle can be concentrated at its centroid without changing the moments, the moment about the x-axis of this rectangle is

$$(\sigma f(z_i)\,\Delta x_i)\left(\frac{1}{2}f(z_i)\right) = \frac{\sigma}{2}[f(z_i)]^2\,\Delta x_i.$$

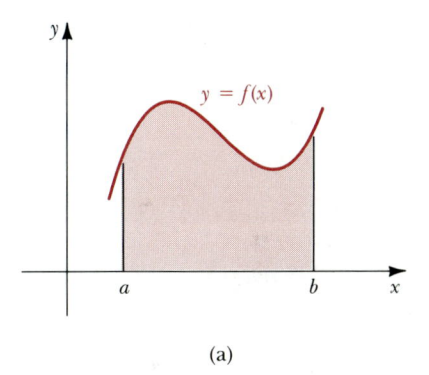

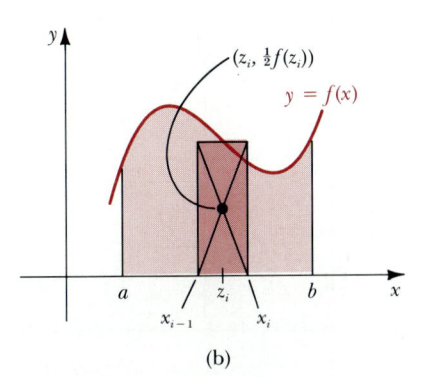

Figure 5.50

The moment about the x-axis of the system consisting of n rectangles is

$$\sum_{i=1}^{n} \frac{\sigma}{2} [f(z_i)]^2 \, \Delta x_i.$$

This is a Riemann sum for the function described by $(\sigma/2)[f(x)]^2$ on the interval $[a, b]$.

Similarly, the moment of the system of n rectangles about the y-axis is

$$\sum_{i=1}^{n} [\sigma f(z_i) \, \Delta x_i] z_i = \sum_{i=1}^{n} \sigma z_i f(z_i) \, \Delta x_i,$$

a Riemann sum for the function described by $\sigma x f(x)$ on the interval $[a, b]$.

Since the total mass of this homogeneous lamina is the product of its density σ and its area $\int_a^b f(x) \, dx$,

$$M = \sigma \int_a^b f(x) \, dx$$

we have the following definition.

(5.26)
DEFINITION
Suppose R is a homogeneous lamina with density σ, bounded on $[a, b]$ by the x-axis and the graph of a continuous nonnegative function f. The **moments of R about the x-axis, M_x, and about the y-axis, M_y,** are

$$M_x = \frac{\sigma}{2} \int_a^b [f(x)]^2 \, dx \quad \text{and} \quad M_y = \sigma \int_a^b x f(x) \, dx.$$

The **centroid** (or *center of mass*) of R is $(\bar{x}, \bar{y})$, where

$$\bar{x} = \frac{M_y}{M} = \frac{\int_a^b x f(x) \, dx}{\int_a^b f(x) \, dx} \quad \text{and} \quad \bar{y} = \frac{M_x}{M} = \frac{\int_a^b [f(x)]^2 \, dx}{2 \int_a^b f(x) \, dx}.$$

Note that the centroid of the lamina is independent of the constant σ. It is common practice to assume that $\sigma = 1$ when the centroid of the lamina is calculated and the moments themselves are not of importance.

Example 3

Find the centroid of the homogeneous lamina with constant density $\sigma = 1$ bounded on $[0, 2]$ by the x-axis and the graph of $f(x) = x^2$ (see Figure 5.51).

Solution
By Definition (5.26), the moment about the x-axis is

$$M_x = \frac{1}{2} \int_0^2 (x^2)^2 \, dx = \frac{x^5}{10} \bigg]_0^2 = \frac{16}{5},$$

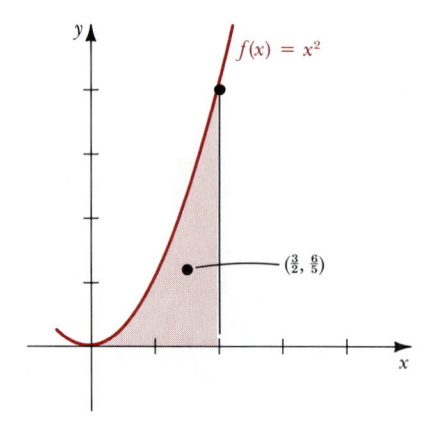

Figure 5.51

the moment about the y-axis is

$$M_y = \int_0^2 x(x^2)\, dx = \frac{x^4}{4}\Big]_0^2 = 4,$$

and the mass of the lamina is the area of the region

$$M = \int_0^2 x^2\, dx = \frac{x^3}{3}\Big]_0^2 = \frac{8}{3}.$$

The centroid of the lamina is $(\bar{x}, \bar{y})$, where

$$\bar{x} = \frac{M_y}{M} = \frac{4}{8/3} = \frac{3}{2} \quad \text{and} \quad \bar{y} = \frac{M_x}{M} = \frac{16/5}{8/3} = \frac{6}{5}.$$

The centroid of a lamina bounded between the graphs of two continuous functions is determined in a similar manner. Suppose a homogeneous lamina with density σ is bounded on $[a, b]$ by the graphs of the continuous functions f and g, where $f(x) \geq g(x)$ on $[a, b]$ (see Figure 5.52a). If $[a, b]$ is partitioned as before and z_i is again chosen to be the midpoint of the interval $[x_{i-1}, x_i]$, rectangles of the form shown in Figure 5.52(b) are produced.

(5.27)
$$M = \int_a^b \sigma[f(x) - g(x)]\, dx,$$

and the moments about the x-axis and y-axis are

(5.28)
$$M_x = \int_a^b \sigma[f(x) - g(x)]\left[\frac{f(x) + g(x)}{2}\right] dx = \int_a^b \frac{\sigma}{2}\{[f(x)]^2 - [g(x)]^2\}\, dx$$

and

(5.29)
$$M_y = \int_a^b \sigma x[f(x) - g(x)]\, dx.$$

So the centroid is $(\bar{x}, \bar{y})$, where

(5.30)
$$\bar{x} = \frac{\int_a^b x[f(x) - g(x)]\, dx}{\int_a^b [f(x) - g(x)]\, dx} \quad \text{and} \quad \bar{y} = \frac{\int_a^b \{[f(x)]^2 - [g(x)]^2\}\, dx}{2\int_a^b [f(x) - g(x)]\, dx}.$$

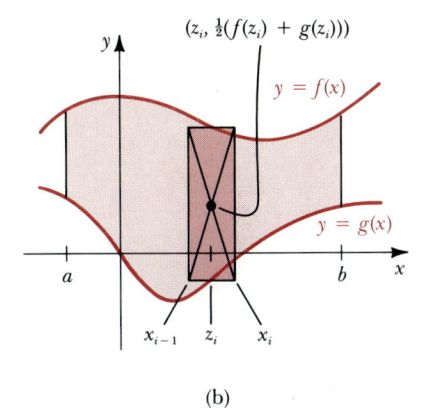

Figure 5.52

Example 4

Find the centroid of the homogeneous lamina bounded by the graphs of $f(x) = \sqrt{x}$ and $g(x) = x/2$ (see Figure 5.53).

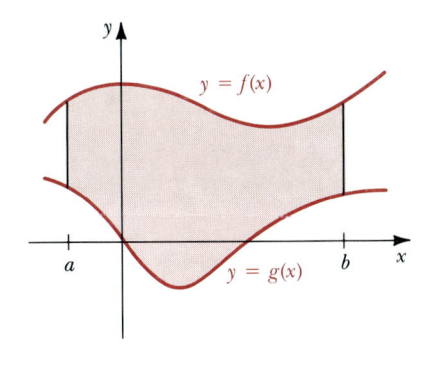

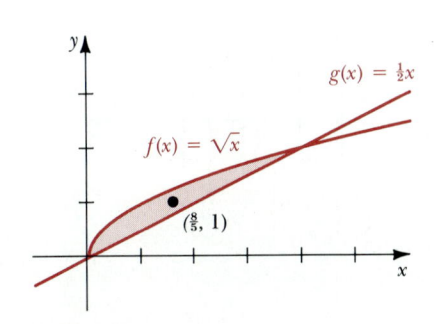

Figure 5.53

Solution
The curves intersect when $x = 0$ and $x = 4$, and for x in $[0, 4]$, $\sqrt{x} \geq x/2$. If $\sigma = 1$, the mass of the lamina is

$$M = \int_0^4 \left[\sqrt{x} - \frac{x}{2} \right] dx = \left[\frac{2}{3} x^{3/2} - \frac{x^2}{4} \right]_0^4 = \frac{16}{3} - 4 = \frac{4}{3}.$$

The moment about the x-axis is

$$M_x = \frac{1}{2} \int_0^4 \left[(\sqrt{x})^2 - \left(\frac{x}{2}\right)^2 \right] dx = \frac{1}{2} \left[\frac{x^2}{2} - \frac{x^3}{12} \right]_0^4 = \frac{4}{3},$$

and the moment about the y-axis is

$$M_y = \int_0^4 x \left[\sqrt{x} - \frac{x}{2} \right] dx = \left[\frac{2}{5} x^{5/2} - \frac{x^3}{6} \right]_0^4 = \frac{32}{15}.$$

The centroid of the lamina is $(\bar{x}, \bar{y})$, where

$$\bar{x} = \frac{M_y}{M} = \frac{(32/15)}{(4/3)} = \frac{8}{5} \quad \text{and} \quad \bar{y} = \frac{M_x}{M} = \frac{(4/3)}{(4/3)} = 1.$$ ▶

Theorems of Pappus

Centroids can be used to determine the volume and surface area of a solid of revolution. Consider first the volume. If $(\bar{x}, \bar{y})$ is the centroid of the region R shown in Figure 5.54(a), then by (5.30),

$$\bar{x} = \frac{\int_a^b x\, [f(x) - g(x)]\, dx}{\int_a^b [f(x) - g(x)]\, dx} \quad \text{and} \quad \bar{y} = \frac{\int_a^b \{[f(x)]^2 - [g(x)]^2\}\, dx}{2 \int_a^b [f(x) - g(x)]\, dx}.$$

The method of disks implies that the volume V_x of the solid generated by

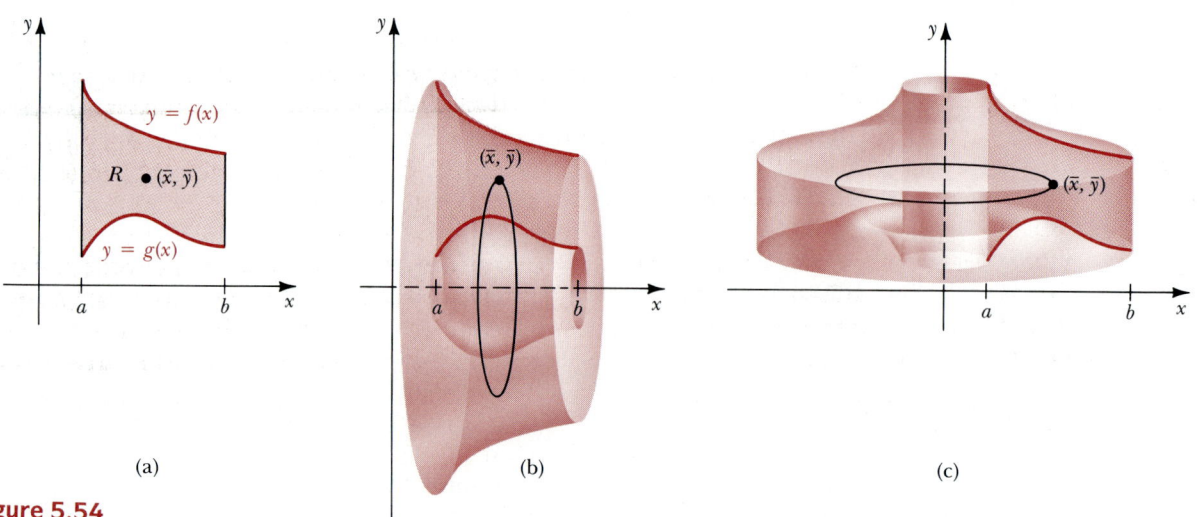

(a)   (b)   (c)

Figure 5.54

revolving R about the x-axis (see Theorem (5.8) and Figure 5.54b) is

$$V_x = \pi \int_a^b \{[f(x)]^2 - [g(x)]^2\} \, dx = 2\pi \bar{y} \int_a^b [f(x) - g(x)] \, dx.$$

Also, the method of shells implies that the volume V_y of the solid generated by revolving R about the y-axis (see Theorem 5.12 and Figure 5.54c) is

$$V_y = 2\pi \int_a^b x \, [f(x) - g(x)] \, dx = 2\pi \bar{x} \int_a^b [f(x) - g(x)] \, dx.$$

The area of R is $\int_a^b [f(x) - g(x)] \, dx$, so in each case, the volume of the solid generated by revolving the region about the axis is the product of the area of the region and the distance traveled by the centroid during the revolution. These results are special cases of the following theorem.

(5.31)
THEOREM: THE FIRST THEOREM OF PAPPUS
The volume of the solid generated by revolving a plane region about a line that does not intersect the region is the product of the area of the region and the distance traveled by the centroid during the revolution.

Example 5

Determine the volume generated if the region bounded by the graphs of $f(x) = \sqrt{x}$ and $g(x) = \dfrac{x}{2}$ is revolved about $x = -1$.

Solution
In Example 4 we found that the centroid of this region is $(\tfrac{8}{5}, 1)$ (see Figure 5.55). The distance traveled by the centroid as the region is revolved about $x = -1$ is the circumference of a circle with radius $\tfrac{8}{5} + 1 = \tfrac{13}{5}$, that is, $2\pi(\tfrac{13}{5}) = \tfrac{26}{5}\pi$.

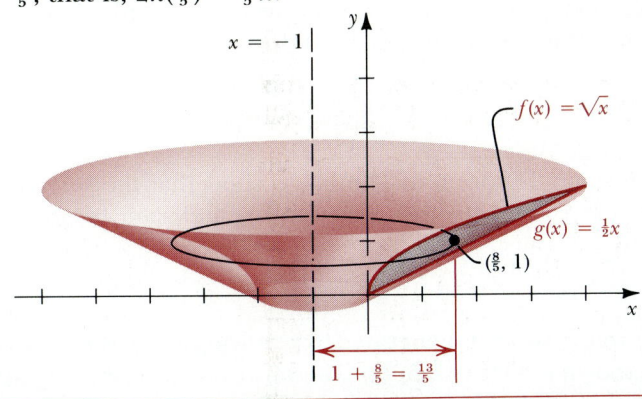

Figure 5.55

HISTORICAL NOTE

Pappus of Alexandria (300 A.D.) compiled a set of 8 volumes that contain the bulk of the Greek mathematics known in his time. The first and second theorems of Pappus are included in this collection and appear to be his own contribution. He also did extensive work on conic sections and wrote valuable commentaries on Euclid's *Data* and *Elements* and on Ptolemy's *Almagest*.

The area of the region was found to be $\frac{4}{3}$ in Example 4, so the volume generated by revolving the region about the line $x = -1$ is

$$V = \frac{26}{5}\pi \cdot \frac{4}{3} = \frac{104}{15}\pi.$$ ▶

The centroid of a curve described by the graph of a continuous function f on an interval $[a, b]$ is defined by assuming that the curve is the shape of a rod with uniform density. By following steps similar to those used for a homogeneous lamina, it can be shown (see Exercise 34) that the centroid of the curve is $(\bar{x}, \bar{y})$, where

(5.32)

$$\bar{x} = \frac{\int_a^b x\sqrt{1+[f'(x)]^2}\,dx}{\int_a^b \sqrt{1+[f'(x)]^2}\,dx} \quad \text{and} \quad \bar{y} = \frac{\int_a^b f(x)\sqrt{1+[f'(x)]^2}\,dx}{\int_a^b \sqrt{1+[f'(x)]^2}\,dx}.$$

In Section 5.4 we saw that if the graph of f is revolved about the x-axis, the area of the surface generated is

$$\text{SA} = \int_a^b 2\pi f(x)\sqrt{1+[f'(x)]^2}\,dx = 2\pi\bar{y}\int_a^b \sqrt{1+[f'(x)]^2}\,dx.$$

Since the length of the curve is $\int_a^b \sqrt{1+[f'(x)]^2}\,dx$, the surface area is the product of the length of the curve and the distance traveled by the centroid during the revolution. This a special case of the Second Theorem of Pappus.

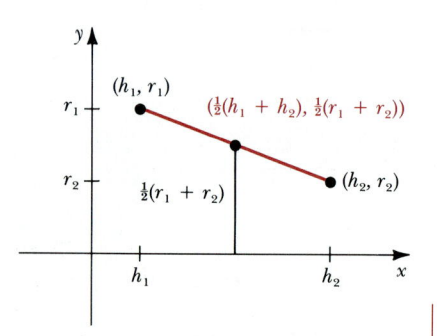

(5.33)

THEOREM: THE SECOND THEOREM OF PAPPUS

The surface area generated by revolving a plane curve about a line that does not intersect the curve is the product of the length of the curve and the distance traveled by the centroid during the revolution.

Example 6

Use the Second Theorem of Pappus to show that the surface area of the frustum of a cone with radii r_1 and r_2 and slant height l is SA $= \pi(r_1 + r_2)l$.

Solution

Consider this frustum as generated by revolving about the x-axis the line segment shown in Figure 5.56(a). The centroid of the line segment is its midpoint: $(\frac{1}{2}(h_1 + h_2), \frac{1}{2}(r_1 + r_2))$. As the line segment is revolved about the x-axis, the centroid travels a distance

$$2\pi[\tfrac{1}{2}(r_1 + r_2)] = \pi(r_1 + r_2).$$

Since the length of the line segment is l, the Second Theorem of Pappus

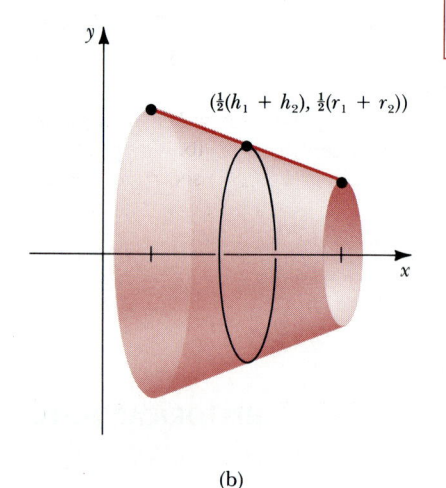

Figure 5.56

▶ EXERCISE SET 5.5

Objects are placed along a line with the coordinates and relative masses given in Exercises 1–4. Find (a) the moment of the system with respect to the origin and (b) the center of mass of the system.

1. $x_1 = 1$, $m_1 = 1$; $x_2 = 3$, $m_2 = 2$; $x_3 = 5$, $m_3 = 3$
2. $x_1 = -2$, $m_1 = 1$; $x_2 = 1$, $m_2 = 1$; $x_3 = 4$, $m_3 = 1$; $x_4 = 6$, $m_4 = 1$
3. $x_1 = -1$, $m_1 = 3$; $x_2 = 0$, $m_2 = 1$; $x_3 = 2$, $m_3 = 4$; $x_4 = 4$, $m_4 = 3$
4. $x_1 = -5$, $m_1 = 3$; $x_2 = -1$, $m_2 = 4$; $x_3 = 0$, $m_3 = 5$; $x_4 = 3$, $m_4 = 3$; $x_5 = 4$, $m_5 = 2$

Objects are placed in the plane with the coordinates and relative masses given in Exercises 5–8. Find (a) the moments of the system with respect to the axes and (b) the center of mass of the system.

5. $P_1(1, 0)$, $m_1 = 1$; $P_2(2, 3)$, $m_2 = 3$; $P_3(-1, 1)$, $m_3 = 2$
6. $P_1(-2, 1)$, $m_1 = 1$; $P_2(0, 0)$, $m_2 = 1$; $P_3(1, 3)$, $m_3 = 1$; $P_4(5, 7)$, $m_4 = 1$
7. $P_1(3, 2)$, $m_1 = 3$; $P_2(2, -1)$, $m_2 = 2$; $P_3(5, -2)$, $m_3 = 4$; $P_4(-2, 0)$, $m_4 = 1$
8. $P_1(-4, -3)$, $m_1 = 4$; $P_2(-2, 2)$, $m_2 = 6$; $P_3(-5, 0)$, $m_3 = 2$; $P_4(4, 4)$, $m_4 = 3$; $P_5(6, 2)$, $m_5 = 1$

In Exercises 9–20, (a) sketch the region and (b) find the centroid of the homogeneous lamina described by the region.

9. $y = 4 - x^2$, $y = 0$
10. $y = x^2 + 1$, $x = -1$, $x = 1$, $y = 0$
11. $y = x^2$, $y = 1$
12. $y = x^2$, $y = x$
13. $y = x^3$, $y = x$, $x \geq 0$
14. $y = 1 - x$, $x = 0$, $y = 0$
15. $y = x^2$, $y = \sqrt{x}$
16. $y = x^3$, $y = x^2$
17. $y = 1/x^2$, $x = 1$, $x = 4$, $y = 0$
18. $y = 1/x$, $x = 1$, $x = 4$, $y = 0$
19. $y = x$, $y = 4 - x$, $y = 0$
20. $y = \dfrac{x}{2}$, $y = 1 - x$, $y = 0$

21. Find the center of mass of the system shown in the figure, if the top rectangle has a uniform density of 2 kg/m² and the bottom rectangle has a uniform density of 1 kg/m².

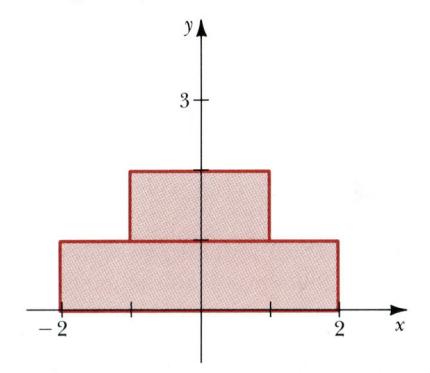

22. A square lamina with sides of length 1 ft and uniform density 1 lb/ft² sits on the center of the top of a rectangular lamina with height 3 ft, width 2 ft, and a uniform density of 2 lb/ft². Find the center of mass of this system.

23. A 2-lb mass is located at a point with coordinates $(-1, 1)$ and a 5-lb mass is located at $(2, 4)$. Where should a 1-lb mass be located if the system is to be balanced at the origin?

24. Two children are playing on a seesaw 16 ft long. One child weighs 60 lb, the other 67 lb. Suppose the lighter child sits at one end of the seesaw. How far from the opposite end must the other child sit if the seesaw is to be balanced in the middle?

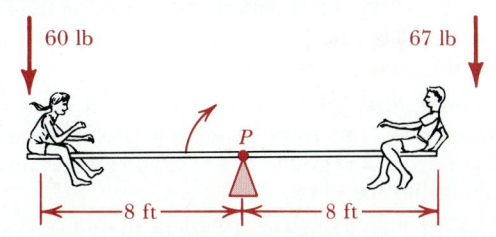

25. Suppose in Exercise 24 that instead of moving the children on the seesaw to balance the weight, we can

add a weight to the seesaw at a point 3 ft from one end. How much weight would be required to balance the seesaw if the children sit at opposite ends?

26. Suppose the pivotal point of the seesaw described in Exercise 24 can be moved. Where should it be placed so that the children can sit at opposite ends of the seesaw and be balanced?

27. Suppose a rod has the dimension of length only and is of length l. In this case, the mass at each point is determined by a function that describes its linear density, that is, its mass per unit length. Let σ be the continuous function with domain $[0, l]$ that describes the density at a point on the rod as a function of its distance from a fixed end of the rod. The total mass of the rod is

$$M = \int_0^l \sigma(x)\, dx$$

and the moment of the rod about p for p in $[0, l]$ is

$$M_p = \int_0^l (x - p)\, \sigma(x)\, dx.$$

A straight rod has the density given by the function in each of the following. Find the moment of the rod with respect to the origin and its center of mass.
a) $\sigma(x) = x$, $[0, 2]$ b) $\sigma(x) = x^2$, $[0, 2]$
c) $\sigma(x) = x$, $[1, 3]$ d) $\sigma(x) = x - x^2$, $[0, 1]$

28. A solid pole 16 ft long has uniform density and tapers linearly from a diameter of 6 in. to a diameter of 4 in. How far from the larger end of the pole is the center of mass?

29. Suppose that instead of being solid, the pole described in Exercise 28 is a hollow pipe made of half-inch-thick stock. Would this change the position of the center of mass?

30. Use the First Theorem of Pappus to find the volume of the solid generated by revolving the region bounded by the graphs of $y = x^2$ and $y = x$ about (a) the x-axis; (b) the y-axis.

31. Use the First Theorem of Pappus to determine the volume of the solid generated by revolving the region bounded by the graphs of $y = x$ and $y = \sqrt{x}$ about (a) the x-axis; (b) the y-axis.

32. Use the First Theorem of Pappus to determine the volume of the solid generated when the region bounded by the graphs of $y = x^2$ and $y = x$ is revolved about the line $y = 2$.

33. Use the First Theorem of Pappus to find the volume of the solid generated by revolving a circle of radius r centered at $(R, 0)$ about the y-axis, where $r < R$. This solid is called a *torus*, and is the shape of a donut.

34. Suppose a rod of uniform density has the shape given by the graph of a continuously differentiable function f on the interval $[a, b]$. Use a derivation similar to that used to find the centroid of a homogeneous lamina to deduce that the center of mass of the rod is $(\bar{x}, \bar{y})$, as given in (5.32).

35. Use the Second Theorem of Pappus to find the surface area of the surface generated by revolving about the x-axis the graph of $y = 2x + 1$ for x in $[0, 2]$.

36. Use the Second Theorem of Pappus to find a formula for the surface area of a cone in terms of its radius r and its altitude h.

37. The surface area of a sphere of radius r is $4\pi r^2$. Use this fact and the Second Theorem of Pappus to find the center of mass of a rod of uniform mass that is bent in the shape of a semicircle of radius 1.

38. Show that the centroid of a circular homogeneous lamina is located at the center of the circle.

39. Show that the centroid of a homogeneous lamina in the shape of a rectangle is the same point as the center of mass of a system of four uniform masses placed at the vertices of the rectangle.

40. Show that the centroid of a homogeneous lamina in the shape of a triangle is the same point as the center of mass of a system of three uniform masses placed at the vertices of the triangle.

41. A sandwich is constructed of two identical pieces of bread and a piece of provolone cheese. The cheese extends beyond the boundary of the bread by varying amounts. Show that the sandwich can be divided by one cut in such a manner that both the cheese and the bread are split into equal amounts. Is it still possible to divide evenly the various ingredients if we add a slice of ham to the sandwich?

42. In statistics, moments are used to describe the shape of the distribution of a random variable. If the distribution of a random variable is described by a continuous function f, then the moment about the origin of the distribution is called the *mean* of the distribution, and is given by

$$\mu = \int_{-\infty}^{\infty} x f(x)\, dx.$$

Find the mean of the distribution given by

$$f(x) = \begin{cases} \frac{1}{2}x, & \text{for } 0 < x < 2, \\ 0, & \text{elsewhere.} \end{cases}$$

43. The *second moment about the mean* of the distribution of a random variable described by a continuous function f is called the *variance*. This second moment is defined by

$$\sigma^2 = \int_{-\infty}^{\infty} (x - \mu)^2 f(x)\, dx.$$

This moment measures the spread or dispersion of the distribution. The square root of the variance, σ, is called the *standard deviation* of the distribution. Find the variance and standard deviation for the distribution in Exercise 42.

Putnam exercise

44. Let V be the region in the Cartesian plane consisting of all points (x, y) satisfying the simultaneous conditions

$$|x| \leq y \leq |x| + 3 \quad \text{and} \quad y \leq 4.$$

Find the centroid $(\bar{x}, \bar{y})$ of V. (This exercise was Problem A–1 of the forty-third William Lowell Putnam examination given on December 4, 1982. The examination and its solution are in the October 1983 issue of the *American Mathematical Monthly*, pp. 546–553.)

5.6

Work

The concept of work is used to describe the transfer of energy. In the British engineering system, the system in everyday use in the United States, distance is commonly measured in feet (ft) and force is given in pounds (lb), so the units of work are foot-pounds (ft-lb). In the metric system, used universally for scientific work and in common use in most countries, distance is measured in meters (m), and force is given in newtons (N) (the force required to give a mass of 1 kilogram (kg) an acceleration of 1 meter per second squared (m/sec²) on a frictionless surface). In this case, the units of work are newton-meters (N-m), also called *joules*. A conversion table for working between the two systems is given inside the front cover of the book.

When a constant force F acts through a distance x, the work done by the force is defined as

$$W = F \cdot x.$$

Suppose we compute the work required to lift an object weighing 10 lb from the ground to a point 3 ft above the ground. The weight of the object describes the force exercised by the gravitational force of the earth; an equivalent force must be applied to the object in order to lift it from the ground. Consequently, the work done in lifting the object is the product of the force, $F = 10$ lb, and the distance moved, 3 ft, or

$$W = (10 \text{ lb})(3 \text{ ft}) = 30 \text{ ft-lb}.$$

Transfer of energy is also required to stretch or compress a spring. The farther the spring is stretched from its natural length, the more difficult it is to stretch the spring further. This indicates that the force required to stretch a spring is variable. In fact, from physical experiments it can be shown that, within the elastic limit of the spring, the force required to stretch a spring x units beyond its natural length is

$$F(x) = kx,$$

for some constant k, called the *spring constant*. This formula is known as **Hooke's Law** and is illustrated in Figure 5.57.

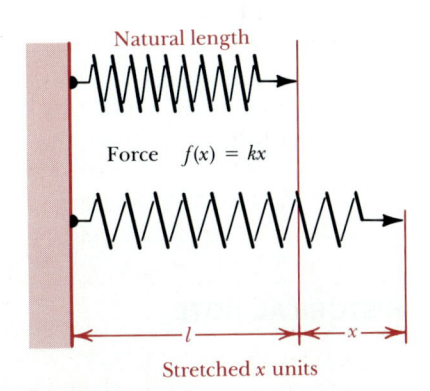

Figure 5.57

The work done by a variable force is defined as a definite integral by using a procedure similar to that used in the other applications in this chapter. The interval over which the force acts is partitioned into subintervals, and the force is assumed to be constant on each subinterval. This leads to a Riemann sum approximating the work, which in turn leads to a definite integral.

Suppose a force given by a function F acts on an object as it moves along the straight-line segment described by the interval $[a, b]$. Let $\mathcal{P} = \{x_0, x_1, \ldots, x_n\}$ be a partition of $[a, b]$ and for each $i = 1, 2, \ldots, n$, choose a number z_i in $[x_{i-1}, x_i]$. The work done by the force on the subinterval $[x_{i-1}, x_i]$ is approximately

$$\Delta W_i = F(z_i)\,\Delta x_i,$$

so the total work done by the force F in moving the object from a to b is approximated by the Riemann sum

$$\sum_{i=1}^{n} F(z_i)\,\Delta x_i.$$

The limit of these Riemann sums as the norm of the partition approaches zero is the definite integral in the following definition. Although our motivation of this definition through the use of Riemann sums requires $a \leq b$, no such stipulation is made in the definition.

(5.34)
DEFINITION

Suppose F is a function describing the force acting on an object as the object moves linearly from point a to point b. Then the **work** done by the force is

$$W = \int_a^b F(x)\,dx,$$

provided this integral exists.

Example 1

Use Definition (5.34) to determine the work required to (a) lift a 10-lb weight 3 ft above the ground, and (b) lower a 10-lb weight to the ground from 3 ft above the ground.

Solution
a) Since $F = 10$ lb,

$$W = \int_0^3 10\,dx = 10x \Big]_0^3 = 30 \text{ ft-lb},$$

which agrees with the result at the beginning of the section.

HISTORICAL NOTE

Robert Hooke (1635–1703) was a contemporary of Isaac Newton. His work was primarily in experimental physics where he did fundamental research in the applications of balance springs and pendulums. He anticipated Newton's law of gravitation, but was unable to establish the relationship mathematically.

b) In this case,

$$W = \int_3^0 10 \, dx = 10x \Big]_3^0 = -30 \text{ ft-lb}.$$

The negative result indicates that the force is acting in the direction of the motion of the object, that is, the force helps rather than hinders the motion. ▶

Example 2

A force of 6 N is required to stretch a spring a distance of 0.5 m from its natural length. Determine the work done when the spring is stretched this distance.

Solution
By Hooke's Law, a constant k exists with $F(x) = kx$ (see Figure 5.58.) Since

$$6 = F(0.5) = k \cdot (0.5), \quad k = \frac{6}{0.5} = 12 \quad \text{and} \quad F(x) = 12x.$$

Definition (5.34) implies that the work done is

$$W = \int_0^{0.5} 12x \, dx = 6x^2 \Big]_0^{0.5} = \frac{3}{2} \text{ N-m}.$$

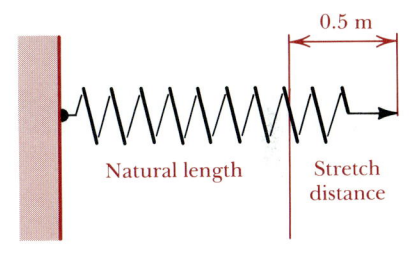

▶ **Figure 5.58**

Example 3

A 5-lb hammer is lying at the bottom of an empty well that is 16 ft deep. A rope weighing 1 oz/ft is used to lift the hammer to the surface. How much work is done?

Solution
When the hammer is at the point labeled x in Figure 5.59, it is hanging from $(16 - x)$ ft of rope. The weight of the rope when the hammer is at this point is consequently

$$(16 - x) \text{ ft } (1 \text{ oz/ft}) = (16 - x) \text{ oz} = \frac{16 - x}{16} \text{ lb}.$$

The combined weight of the hammer and rope is therefore

$$F(x) = 5 \text{ lb} + \frac{16 - x}{16} \text{ lb} = \frac{6 - x}{16} \text{ lb}.$$

The work required to lift the hammer from the bottom of the well is

$$W = \int_0^{16} \left(6 - \frac{x}{16}\right) dx = 6x - \frac{x^2}{32} \Big]_0^{16} = 6(16) - \frac{(16)^2}{32} = 88 \text{ ft-lb}.$$

▶

In the study of motion problems in Chapter 3, we made the simplifying assumption that acceleration of an object due to the gravita-

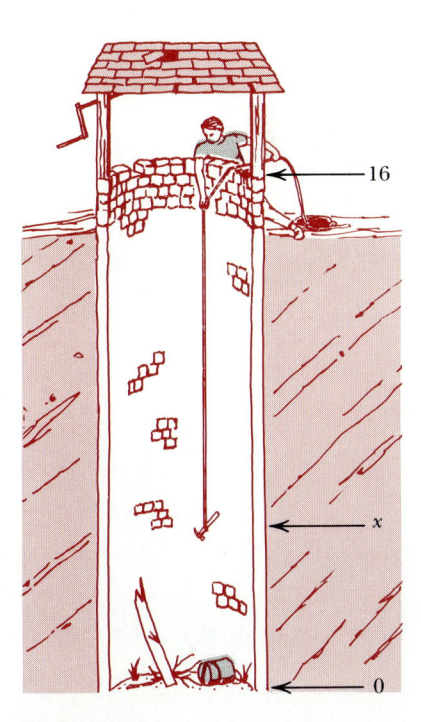

▶ **Figure 5.59**

tional force of the earth is constant, that is, $a(t) = -32$ ft/sec². This is a reasonable assumption only when the object is close to the surface of the earth. In more general situations, we must allow for the gravitational force between two objects to vary as the distance between the objects changes.

Newton's Law of Gravitation states that the gravitational force between two objects with masses m_1 and m_2 is

(5.35)
$$F(x) = \frac{Gm_1m_2}{x^2},$$

where x is the distance between the centers of mass of the objects and G is a constant called the *universal gravitational constant*. The following example shows how we can use this result to determine the work required to lift a satellite into orbit.

Example 4

A NASA space shuttle orbiter (see Figure 5.60) complete with maximum payload has a weight on earth of 215,000 lb. How much work is required to lift the orbiter from the earth to its orbit 170 mi above the surface of the earth?

Figure 5.60
Space shuttle orbiter Atlantis being launched from the Kennedy Space Center in Florida.

Solution

Rather than determine G, m_1, and m_2 in (5.35) individually, we find them together as a single constant. Since the surface of the earth is approximately 3960 mi from its center of mass and the gravitational force of the orbiter on the surface of the earth is its weight we have

$$215,000 = \frac{Gm_1m_2}{(3960)^2}$$

and $Gm_1m_2 = (215,000)(3960)^2$ lb–mi². Thus, the gravitational force on the satellite when it is x mi from the center of the earth is

$$F(x) = \frac{(215,000)(3960)^2}{x^2} \text{ lb.}$$

The work required to lift the shuttle into orbit is the integral of F from the surface of the earth, 3960 mi from the earth's center, to the height of the orbit, $3960 + 170 = 4130$ mi from the earth's center:

$$W = \int_{3960}^{4130} \frac{(215,000)(3960)^2}{x^2} \, dx$$

$$= \left[-\frac{(215,000)(3960)^2}{x} \right]_{3960}^{4130}$$

$$= 3.505 \times 10^7 \text{ mi-lb} = 1.850 \times 10^{11} \text{ ft-lb.} \quad \blacktriangleright$$

The result in Example 4 does not include the work required to lift the first-stage booster rockets with their solid fuel or the liquid fuel used by the engines of the orbiter during launch. Adding these considerations would make the problem more difficult, since the relationship between fuel usage and distance is not easily determined.

5.6 WORK

The preceding examples show that it is easy to determine the work done by a force when the representation of the force is known. Many problems, however, involve forces whose representation is not immediately recognizable. The remaining examples in this section are of this type. In each case, we approximate the work over a small interval and use this approximation to derive an appropriate definite integral.

Example 5

A reservoir full of water has the shape of a hemisphere with radius 10 ft. Determine the work required to pump this water to the top of the reservoir.

Solution

The hemisphere is shown in Figure 5.61(a), and a vertical cross section is sketched in Figure 5.61(b). Our plan is to approximate the work required to pump a thin layer of water to the top and then sum the approximations. This leads to the definite integral given in Definition (5.34).

Let $\mathscr{P} = \{x_0, x_1, \ldots, x_n\}$ be a partition of $[0, 10]$ and for each $i = 1, 2, \ldots, n$, choose z_i arbitrarily in $[x_{i-1}, x_i]$. Since the weight-density (force per unit volume) of water is approximately 62.4 lb/ft³ and each cross section is a circle with radius $\sqrt{100 - z_i^2}$, the force ΔF_i on the ith layer of water is

$$\Delta F_i = (62.4 \text{ lb/ft}^3) \cdot (\text{volume of water in the } i\text{th layer})$$
$$= (62.4 \text{ lb/ft}^3) \cdot [\pi(\sqrt{100 - z_i^2})^2 \, \Delta x_i \text{ ft}^3]$$
$$= 62.4\pi(100 - z_i^2) \, \Delta x_i \text{ lb}.$$

The work required to pump the ith layer of water to the top is approximately

$$\Delta W_i = \Delta F_i \text{ (distance)} = [(62.4)\pi(100 - z_i^2) \, \Delta x_i] z_i \text{ ft-lb},$$

so the work required to pump all the water to the top is approximated by

$$W \approx \sum_{i=1}^{n} \Delta W_i = \sum_{i=1}^{n} 62.4\pi z_i (100 - z_i^2) \, \Delta x_i \text{ ft-lb}.$$

The total work is the limit of these Riemann sums, which leads to a definite integral:

$$W = \int_0^{10} 62.4\pi x(100 - x^2) \, dx = 62.4\pi \left[50x^2 - \frac{1}{4}x^4 \right]_0^{10}$$
$$= 62.4\pi(5000 - 2500)$$
$$= 156,000\pi \text{ ft-lb}. \quad \blacktriangleright$$

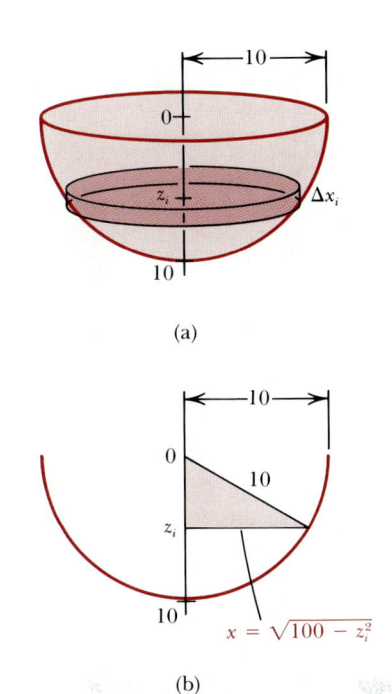

Figure 5.61

Example 6

How much work is required to pump the water in the reservoir described in Example 5 to a point 5 ft above the reservoir?

Solution

The only difference between this example and the preceding one is the distance that the water must be pumped. With the same notation used in Example 5, we see from Figure 5.62 that this distance is $(x + 5)$ and

$$W = \int_0^{10} 62.4\pi(x + 5)(100 - x^2)\, dx$$

$$= 62.4\pi \int_0^{10} (100x + 500 - x^3 - 5x^2)\, dx$$

$$= 62.4\pi \left[50x^2 + 500x - \frac{1}{4}x^4 - \frac{5}{3}x^3 \right]_0^{10}$$

$$= 62.4\pi \left(5000 + 5000 - 2500 - \frac{5000}{3} \right)$$

$$= 364{,}000\pi \text{ ft-lb.}$$

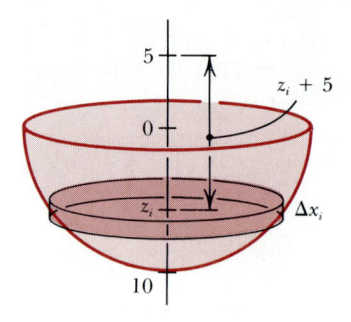

Figure 5.62

▶ EXERCISE SET 5.6

1. Determine the work required to lift a 50-lb bag of dog food 3 ft above the ground.

2. How much work is done in lifting a 110-lb barbell from the ground to a point 14 in. above your head?

3. A diesel locomotive on a freight train exerts the constant force of 8 tons on a train while moving the train at 40 mph. How much work is done by the locomotive if the train travels at this speed for 1 mi?

4. A speedboat exerts a force of 1200 lb when moving through the water at 63 mph. How much work is done if the boat travels at this speed for 5 mi?

5. A force of 6 N stretches a spring 40 cm. How much work is done if the spring is stretched 80 cm?

6. A force of 3 lb stretches a spring 4 in. How much work is done when the spring is stretched 1 ft?

7. A force of 5 N stretches a spring 20 cm.
   a) How far will a force of 10 N stretch the spring?
   b) How much work is done when the force of 5 N is applied?
   c) How much work is done when the force of 10 N is applied?

8. A spring is 6 in. long and is stretched 1 in. by a 5-lb weight.
   a) What is the work done in stretching this spring from 6 to 8 in.?
   b) What is the work done in stretching this spring from 8 to 10 in.?

9. A spring is 10 in. long, and a force of 5 lb is required to compress it to a length of 6 in. How much work is required to stretch it from its natural length of 10 in. to a length of 20 in.?

10. A spring has a natural length of 20 cm, and a force of 4 N is required to compress it to a length of 10 cm. How much work is required to stretch the spring from 20 cm to 40 cm?

11. A 3-gal bucket full of water is at the bottom of a 100-ft well. How much work is required to lift this bucket to the top of the well, assuming the bucket weighs 3 lb when empty and the weight of the rope is insignificant? (1 gal ≈ 0.134 ft³)

12. A 50-ft rope weighing 2 oz/ft is hanging from the top of a 100-ft bridge. How much work is required to pull this rope to the top of the bridge?

13. A 3-gal bucket full of water is at the bottom of a 100-ft well. How much work is required to lift this bucket to the top of the well, assuming the bucket weighs 3 lb when empty and the rope weighs 4 oz/ft?

14. A 50-ft rope weighing 2 oz/ft is hanging from the top of a 25-ft bridge with 25 ft of the rope coiled at the bottom. How much work is required to pull this rope to the top of the bridge?

15. The cable of a crane used to move heavy equipment weighs 1.5 lb/ft, and the hook at the end of the cable weighs 35 lb. How much work is done in lifting a 500-lb object from the ground to the top of the crane, which is 6 ft above the ground?

16. The anchor on the *U.S.S. Constitution* (*Old Ironsides*) was forged in the 1790s and weighs 5300 lb. The ship is presently moored at Boston Naval Yard in Charlestown, Mass., under the care of the U.S. Navy. The depth in the harbor is approximately 40 ft (depending on the tide).

a) How much work is required to raise the anchor from the bottom of the harbor to a point 16 ft above the water, assuming that the weight of the chain is negligible?

b) If the chain actually weighs 50 lb/ft, how much work does this add to the task?

17. The Tiros is a meteorological satellite system that has been in operation for over two decades. Determine the work required to lift a 1620-lb TIROS-N satellite to its orbit 517 mi above the earth.

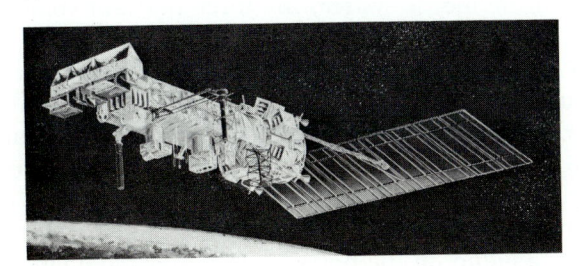

TIROS-N meteorological satellite in earth orbit.

18. An Intelstat 5A communication satellite weighing 4300 lb is launched into a geosynchronous orbit 22,144 mi above the earth. How much work is required to lift the satellite from the earth to its orbit? (In a geosynchronous orbit, the satellite will remain above the same location on the earth at all times.)

19. A swimming pool that is 30 ft by 50 ft and 8 ft deep is filled with water.
    a) How much work is required to pump all the water from the pool?
    b) How much work is required to pump all the water to a point 3 ft above its surface?

20. Suppose the water in the swimming pool described in Exercise 19 is only 7 ft deep. How much work is required to pump all the water from the pool?

21. A cylindrical tank of radius 2 ft and height 6 ft is full of oil weighing 55 lb/ft^3.
    a) How much work is necessary to pump the oil to the top of the tank?
    b) Suppose the tank is placed on its side and the oil is pumped to the top of the tank. Set up the integral that gives the work required in this case.

22. Suppose a one-horsepower motor is used to pump the oil from the tank described in Exercise 21(a). (One horsepower equals 550 ft-lb/sec.) How long will it take to pump the oil into a tank truck 8 ft above the original level of the oil?

23. In Exercise 65 of Section 5.2, we described a swimming pool with dimensions 30 ft by 50 ft that is 8 ft deep for 20 ft, then slopes linearly upward for 10 ft, and is 3 ft deep for the remaining 20 ft. How much work is required to pump all the water from the pool to a point 3 ft above its original surface?

24. A 10,000-gal tank full of water has the shape of a cone with its vertex at the bottom and its height and top diameter equal (see the figure). The water is pumped into a cylindrical tank with 10,000-gal capacity and equal diameter and height. The tops of the two tanks are in the same plane.
    a) How much work is done to pump the water from the inverted cone into the cylinder?
    b) If the procedure is reversed, with the cylindrical tank full and the inverted cone empty, how much work is necessary to pump the water?

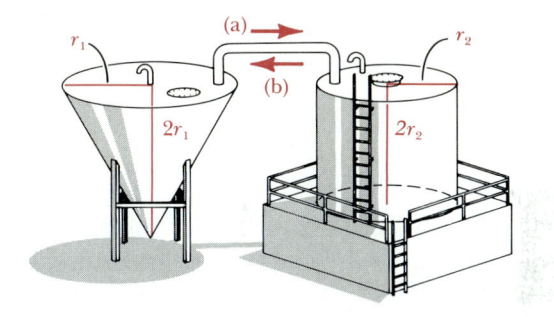

25. A service station has a 2000-gal cylindrical tank whose top is buried 2 ft below the ground, with the axis of the cylinder parallel to the ground. Once a week this station is serviced by the supplier. In one week, 800 gal of gasoline are sold, so the service station has the choice of filling the tank every other week or topping it off each week. Which method of purchase requires the least work for the service station pumps?

26. Show that when a spring is stretched from a distance x_1 beyond its natural length to a distance x_2 beyond its natural length, the work required is the product of the distance $(x_2 - x_1)$ and the average of the forces $F(x_1)$ and $F(x_2)$, that is,
$$W = \frac{F(x_1) + F(x_2)}{2}(x_2 - x_1).$$

27. A weight w stretches a spring S_1 a distance d beyond its natural length. This same weight stretches a spring S_2 a distance $2d$ beyond its natural length. How are the spring constants of S_1 and S_2 related?

28. A spring S with length l and spring constant k is cut into two springs of length $l/2$. What are the spring constants of the new springs?

29. Suppose the spring S described in Exercise 28 is cut instead into two springs of lengths $l/3$ and $2l/3$. What are the spring constants of these springs?

5.7

Fluid Pressure

Pressure is defined as force per unit area. The pressure exerted by a liquid on an object immersed in the liquid depends on the volume density (force per unit volume) of the liquid and the depth of the liquid. In the British engineering system, the force per unit volume is generally described using lb/ft^3 and the depth is given in feet, so the unit of pressure is lb/ft^2. In the metric system, the force per unit volume is usually given in Newtons per cubic meter (N/m^3) and the depth in meters, so the unit of pressure is N/m^2. Other units of measure, such as *atmospheres* (atm) and *inches of mercury* (in-Hg) are also used to measure pressure, but we will not consider these. Pressures given in these units can be converted to either lb/ft^2 or N/m^2 by referring to the table inside the front cover.

A somewhat surprising, but easily verified, physical fact is that the pressure exerted by a liquid on the bottom of an open container depends only on the height of the liquid in the container, not on the volume of liquid in the container. A simple experiment to demonstrate this fact consists of placing water in a glass apparatus of the type shown in Figure 5.63. Since the water level in each of the containers above the reservoir is the same, the pressure of the water on the bottom of the containers must be equal, even though the containers vary in size and shape.

The physical principle governing this situation is called **Pascal's Law.** This law states that each point in a stationary liquid is subject to forces of the same magnitude in every direction (or the point would move in the direction dictated by the greatest force; see Figure 5.64). This implies that the forces on each point at the same depth are equal; thus all points at the same depth are subject to the same pressure.

The pressure p exerted by a liquid on any horizontal surface in the liquid is the product of the weight per unit volume w of the liquid and the height h of the liquid above the surface; that is,

$$p = wh.$$

Since pressure is force per unit area, the total force F exerted by a liquid on a horizontally submerged plate of area A is

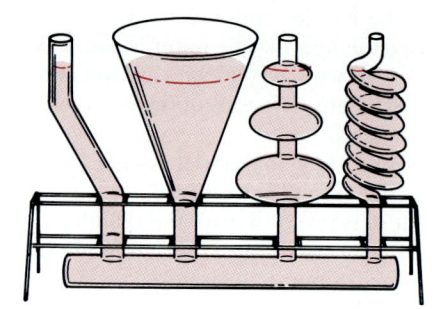

(5.36) $$F = pA = whA.$$

Figure 5.63
There is equal pressure at the bottom of each vessel.

Figure 5.64
There is equal pressure in each direction.

HISTORICAL NOTE

Blaise Pascal (1623–1662) began his mathematical research career at the age of 16, obtaining an important result concerning the geometry of conic sections. Together with Pierre de Fermat, he is considered a founder of not only analytic geometry but also the theory of probability. He initiated his study of probability at the request of Chevalier de Méré, who needed a method for computing the payoff for an interrupted dice game.

Pascal had a religious experience on November 23, 1654, that turned his interest from science and mathematics to philosophy and religion. He returned to mathematics only briefly in 1658–1659, when he did fundamental work on the study of cycloids and in the theory of hydrostatics.

Example 1

A 55-gal cylindrical barrel half full of water is sitting on an end. The barrel is 3 ft high with a radius of 0.88 ft. Find the total force exerted by the water on the bottom of the barrel (see Figure 5.65).

Solution

The force on the bottom of the barrel depends on the pressure exerted by the water and the area of an end of the barrel. The area is

$$A = \pi r^2 = \pi(0.88)^2 \approx 2.43 \text{ ft}^2.$$

Since water weighs approximately 62.4 lb/ft³ and the depth of the water in the half-full barrel is 1.5 ft, the total force exerted by the water on the bottom of the barrel is

$$F = \left(62.4 \frac{\text{lb}}{\text{ft}^3}\right)(1.5 \text{ ft})(2.43 \text{ ft}^2) \approx 227 \text{ lb}.$$

▶

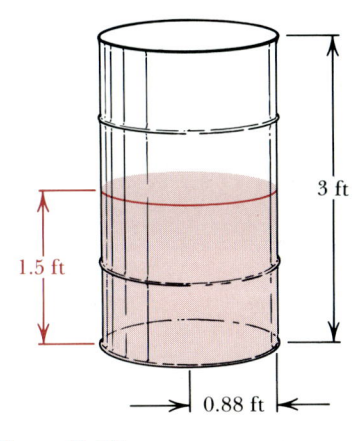

Figure 5.65

When a plate is submerged other than horizontally, the situation is more complicated because the depth of the liquid in contact with the plate is not constant. We can handle this case in a manner similar to that used in the previous applications of the definite integral:

i) Partition the plate into narrow horizontal strips.
ii) Assume that the force is constant on each strip.
iii) Calculate the force on each strip.
iv) Sum each of the individual forces to approximate the total force.
v) Introduce the definite integral as the limit of the approximations.

Suppose a plate is submerged vertically from a depth $x = a$ to a depth $x = b$ in a liquid with weight-density w and that f is a function describing the width of the plate in terms of its depth. To find the total force exerted by the liquid on the plate, first let $\mathscr{P} = \{x_0, x_1, \ldots, x_n\}$ be a partition of $[a, b]$ and for each $i = 1, 2, \ldots, n$, choose a number z_i in $[x_{i-1}, x_i]$. Then determine the force exerted on a typical rectangle with area $f(z_i) \Delta x_i$ at a depth z_i (see Figure 5.66):

$$\Delta F_i = (\text{weight-density})(\text{depth})(\text{area}) = w z_i f(z_i) \Delta x_i.$$

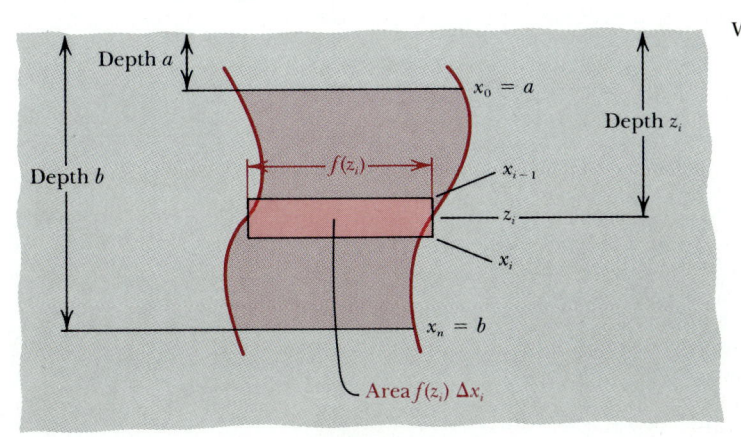

Figure 5.66

The total force exerted on the plate can therefore be approximated by the Riemann sum

$$F \approx \sum_{i=1}^{n} \Delta F_i = \sum_{i=1}^{n} wz_i\, f(z_i)\, \Delta x_i,$$

and the total force on the plate is defined as the integral of the function described by $wx\, f(x)$.

(5.37) DEFINITION

Suppose a plate is submerged vertically in a liquid whose weight-density is w. If f is a function with domain $[a, b]$ that describes the width of the plate as a function of its depth below the surface of the liquid, then the total force F exerted on the plate by the **fluid pressure** is

$$F = \int_a^b wx\, f(x)\, dx,$$

provided this integral exists.

Example 2

Suppose the 55-gal barrel described in Example 1 is half full of water and lying on its side. What is the total force exerted by the water on an end of the barrel?

Solution

From Example 1 we know that the radius of the barrel is 0.88 ft. Figure 5.67 shows that the width of the end of the barrel x ft below the surface of the water is $f(x) = 2\sqrt{(0.88)^2 - x^2}$ ft.

Since water weighs 62.4 lb/ft^3, the total force exerted on an end of the barrel is

$$F = \int_0^{0.88} 2(62.4)\, x\, \sqrt{(0.88)^2 - x^2}\, dx$$

$$= \frac{-124.8}{3} \left[(0.88)^2 - x^2\right]^{3/2} \Big]_0^{0.88} \approx 28.3 \text{ lb.}$$

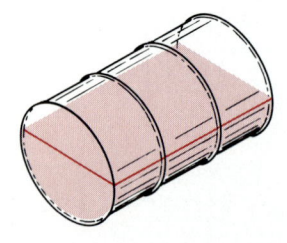

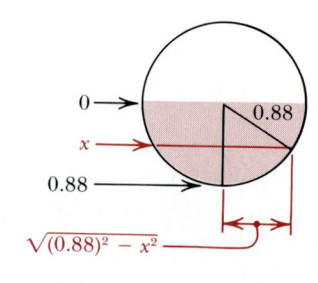

Figure 5.67

(a)   (b)

In the next example, the region on which the force is exerted is described using a coordinate system with the depth given in terms of y. This is often a convenient way to set up fluid pressure problems.

Example 3

A vertical gate on a dam has the shape of a parabola with height 8 ft and width 8 ft at its top. Find the force of the water on the gate when the water is level with the top of the gate.

Solution

The region representing the gate is shown in Figure 5.68. A rectangular coordinate system has been placed so that the depth of the water is measured along the y-axis and the origin is at the bottom of the gate.

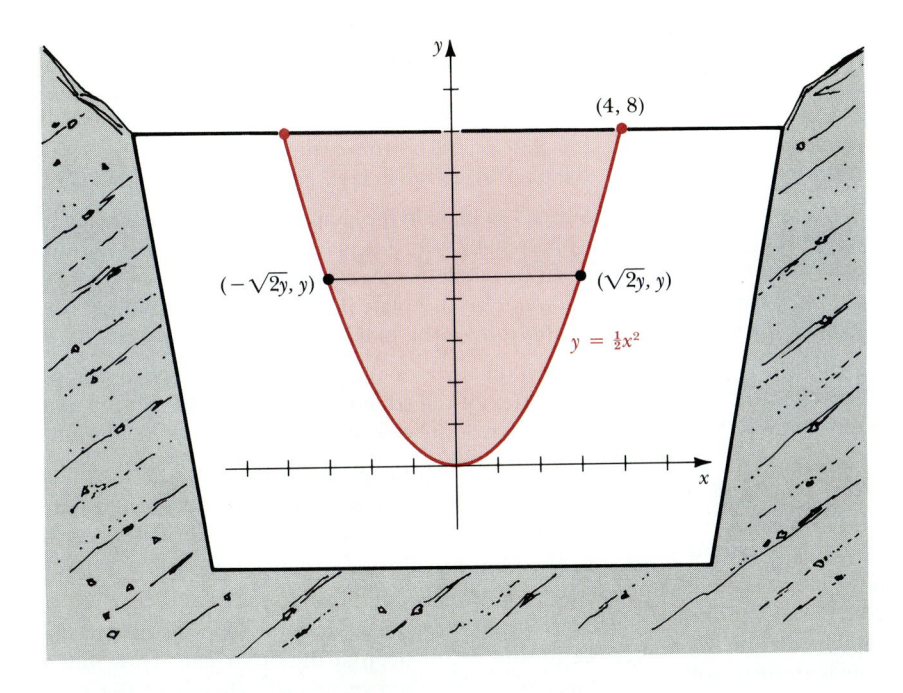

Figure 5.68

The boundary of the region is a parabola, a curve described by an equation of the form $y = ax^2$ for some constant a. Since $y = 8$ when $x = 4$, we have

$$a = \frac{y}{x^2} = \frac{8}{16}$$
$$= \frac{1}{2},$$

and the lateral boundary of the gate is described by $y = \frac{1}{2} x^2$, or $x = \pm\sqrt{2y}$.

For y in $[0, 8]$, the depth of the water is $(8 - y)$ ft and the width of the gate is $2\sqrt{2y}$ ft. Since the water weighs 62.4 lb/ft^3, the total force on the gate is

$$\int_0^8 62.4(8-y)2\sqrt{2y}\,dy = 124.8\sqrt{2}\int_0^8 (8y^{1/2} - y^{3/2})\,dy$$

$$= 124.8\sqrt{2}\left[\frac{16}{3}y^{3/2} - \frac{2}{5}y^{5/2}\right]_0^8$$

$$= 249.6\sqrt{2}\left(\frac{8}{3}(8)^{3/2} - \frac{1}{5}(8)^{5/2}\right)$$

$$= 249.6\sqrt{2}\cdot 64\sqrt{8}\left(\frac{1}{3} - \frac{1}{5}\right) \approx 8520 \text{ lb.}$$

▶ **EXERCISE SET 5.7**

1. A plate with area 6 ft² is lying flat on the bottom of a tank of water 8 ft deep. What is the pressure on the plate if the plate is in the shape of (a) a rectangle 2 ft wide and 3 ft long? (b) a circle? (c) an equilateral triangle? What is the total force exerted on each of the plates by the water?

2. A metal plate in the shape of a rectangle 2 ft high and 3 ft wide is placed vertically in a tank of water 8 ft deep, with the width of the rectangle at the bottom of the tank. What is the total force exerted by the water on the plate?

3. What would be the force on the plate described in Exercise 2 if it is rotated 90°, so that the 2-ft side is on the bottom of the tank?

4. Suppose the plate described in Exercise 2 is placed in the tank with its 3-ft width lying on the surface of the water. What is the force on the plate in this situation?

5. A vertical gate on a dam is in the shape of a rectangle 10 ft wide and 5 ft deep with its width parallel to the surface of the water. Find the force of the water on the gate when the level of water in the dam is at the top of the gate.

6. Reconsider the dam gate discussed in Exercise 5. How much force is exerted by the water on this gate when the top of the gate is 5 ft below the surface of the water?

7. Reconsider the dam gate discussed in Exercise 5. How much force is exerted by the water on this gate when the water is 1 ft below the top of the gate?

8. A vertical gate on a dam is in the shape of the region described in feet by $y = x^2$, $y = 9$. The top of the dam is described by $y = 9$. Find the force of the water on the gate when the level of water is 3 ft below the top of the dam.

9. A metal plate in the shape of the region bounded by the graphs of $y = 4 - x^2$ and $y = 0$ is submerged vertically in a tank of water 6 ft deep with the flat edge of the plate on the bottom of the tank. The dimensions of the plate are given in feet. Find the force on the plate.

10. Suppose a plate in the shape of an equilateral triangle with area $4\sqrt{3}/3$ ft² is placed vertically at the bottom of an 8-ft tank of water. What is the force exerted by the water on this plate if one side of the triangle is on the bottom of the tank?

11. What is the force on the plate described in Exercise 10 if a vertex of the triangle is touching the bottom and the side opposite this vertex is parallel to the bottom?

12. A circular plate with an area of 6 ft² is placed vertically on the bottom of a tank of water 8 ft deep. Set up the integral that describes the force on this plate.

13. In Exercise 65 of Section 5.2, we discussed a swimming pool with dimensions 30 ft by 50 ft that is 8 ft deep for 20 ft on one end, slopes linearly upward for 10 ft, and is 3 ft deep at the other end. How much force does the water exert on each side of the pool when the pool is filled with water?

14. How much force is exerted on each side of the pool described in Exercise 13 when the pool is half-full of water?

15. How much force is acting on the bottom of the swimming pool described in Exercise 13 when the pool is filled with water?

16. How much force is acting on the bottom of the pool described in Exercise 13 when the pool is half-full of water?

17. A pig trough 8 ft long has a cross section in the shape of an isosceles trapezoid with a lower base of 1 ft, an

upper base of 2 ft, and an altitude of 1 ft (see the figure). The trough is filled with swill of density 64 lb/ft³.
  a) What is the total force on an end of the trough?
  b) What is the total force on the bottom of the trough?

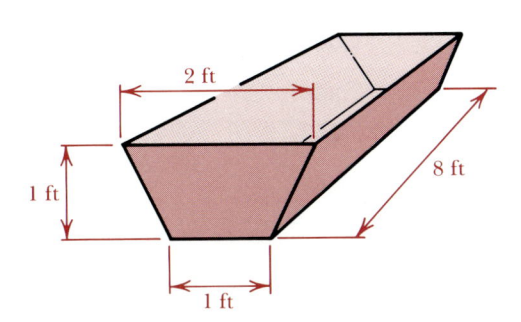

18. A glass in the shape of a right circular cylinder with height 6 in. and radius 2 in. is filled with water (see the figure). What is the total force acting on the side of the glass?

19. Suppose the filled glass described in Exercise 18 is placed in a basin of water and that the water in the basin just reaches the top of the glass. What is the total force acting on the sides of the glass?

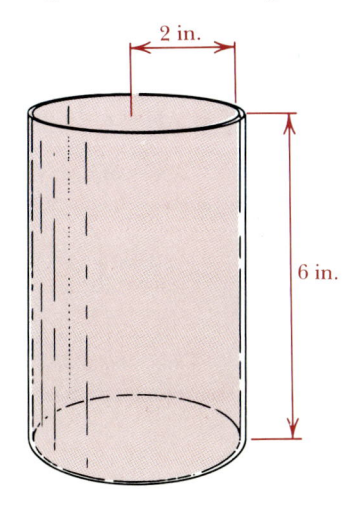

Figure for Exercises 18 and 19.

▶ IMPORTANT TERMS AND RESULTS

APPLICATION	BASIC FORMULA	PAGE
Area	$A = \int_a^b [f(x) - g(x)]\, dx$	323
Volume		
Cross sections	$V = \int_a^b A(x)\, dx$	329
Disks	$V = \pi \int_a^b [(f(x))^2 - (g(x))^2]\, dx$	333
Shells	$V = 2\pi \int_a^b x[f(x) - g(x)]\, dx$	342
Arc length	$L_{[a,b]}(f) = \int_a^b \sqrt{1 + [f'(x)]^2}\, dx$	349
Surface area	$SA_{[a,b]}(f) = 2\pi \int_a^b f(x)\sqrt{1 + [f'(x)]^2}\, dx$	354
Moments		
About the x-axis	$M_x = \dfrac{\sigma}{2} \int_a^b [f(x)]^2\, dx$	360
About the y-axis	$M_y = \sigma \int_a^b x f(x)\, dx$	360

▶ IMPORTANT TERMS AND RESULTS (cont.)

APPLICATION	BASIC FORMULA	PAGE
Moments (cont.)		
Mass	$M = \sigma \int_a^b f(x)\, dx$	360
Centroid	$\bar{x} = M_y/M,\ \bar{y} = M_x/M$	360
Theorems of Pappus		362
Work	$W = \int_a^b F(x)\, dx$	368
Fluid pressure	$F = \int_a^b wx\, f(x)\, dx$	376

▶ REVIEW EXERCISES

In Exercises 1–12, (a) sketch the region and (b) find the area of the region.

1. $y = \sqrt{x},\ y = x^3$
2. $y = \sqrt{x-1},\ y = \sqrt{2-x},\ y = 0$
3. $y = 4 - x^2,\ y = x^2 - 4$
4. $y = x - x^3 + 1,\ y = 1$
5. $y^2 = 4 - x,\ y = x + 2$
6. $y = 4x - x^2,\ y = x$
7. $y = x^2 - 1,\ y = 7 - x^2$
8. $y = \sin \pi x,\ y = \cos \pi x,\ x = 0,\ x = 1$
9. $y = 2x^3 - 9x^2 + 12x - 2,\ y = -2,\ x = 3$
10. $y = 2x^3 - 3x^2 - 12x + 13,\ x = -1,\ x = 2,\ y = 0$
11. $y = x,\ y = x + \sin x,\ x = 0,\ x = \pi$
12. $y = x,\ y = x + \cos x,\ x = 0,\ x = \pi$

In Exercises 13–20:
   a) Sketch the region.
   b) Use the method of disks to find the volume generated when the region is revolved about the indicated line.

13. $y = 6/x,\ x = 2,\ x = 4,\ y = 0$; about the x-axis
14. $y = 6/x,\ x = 2,\ x = 4,\ y = 0$; about the y-axis
15. $y = x^2 + 2x + 2,\ x = 0,\ x = 2,\ y = 0$; about the x-axis
16. $y = \csc x,\ x = \pi/4,\ x = \pi/2$; about the x-axis
17. $y^2 = 4x,\ x = 4$; about the x-axis
18. $y^2 = 4x,\ x = 4$; about the y-axis
19. $y^2 = 4x,\ x = 4$; about $x = 4$
20. $y^2 = 4x,\ x = 4$; about $x = 6$

In Exercises 21–28:
   a) Sketch the region.
   b) Use the method of shells to find the volume generated when the region is revolved about the indicated line.

21. $y = 6/x,\ x = 2,\ x = 6,\ y = 0$; about the y-axis
22. $y = 6/x,\ y = 3/2,\ x = 2$; about the x-axis
23. $y = x^2 + 2x + 2,\ x = 0,\ x = 2,\ y = 0$; about the y-axis
24. $y = \csc x^2,\ x = \sqrt{\pi/4},\ x = \sqrt{\pi/2},\ y = 0$; about the y-axis
25. $y = 2\sqrt{x},\ x = 4,\ y = 0$; about the y-axis
26. $y = 2\sqrt{x},\ x = 4,\ y = 0$; about the x-axis
27. $y = 2\sqrt{x},\ x = 4,\ y = 0$; about $x = 4$
28. $y = 2\sqrt{x},\ x = 4,\ y = 0$; about $y = 4$
29. The base of a solid is bounded by the circle in the xy-plane with equation $x^2 + y^2 = 25$ and each cross section perpendicular to the x-axis is an isosceles triangle with altitude 6. Set up the integral that describes the volume of this solid.
30. The base of a solid is bounded by the circle in the

xy-plane with equation $x^2 + y^2 = 25$ and each cross section perpendicular to the y-axis is an equilateral triangle. Set up the integral that describes the volume of this solid.

31. A solid has its base in the xy-plane bounded by the graphs of $x = y^2$ and $x = 4$. Find the volume of the solid if each cross section perpendicular to the x-axis is a square.

32. A solid has its base in the xy-plane bounded by the graphs of $x = y^2$ and $x = 4$. Find the volume of the solid if each cross section perpendicular to the y-axis is a square.

33. Find the length of the curve described by $y = 7x + 2$ from $x = -1$ to $x = 3$.

34. Find the length of the curve described by $y = 2x^{3/2} + 2$ from $x = 4$ to $x = 9$.

35. Find the length of the curve described by $y = \ln(\cos x)$ from $x = -\pi/4$ to $x = \pi/4$.

36. Find the length of the curve described by $y = x^3/6 + 1/(2x)$ from $x = 1$ to $x = 3$.

37. Find the area of the surface generated by revolving about the x-axis the curve described by $y = \sqrt{2x - 1}$ for $\frac{1}{2} \leq x \leq 1$.

38. Find the area of the surface generated by revolving about the x-axis the curve described by $y = (x^{3/2}/3) - x^{1/2}$ for $1 \leq x \leq 4$.

39. A 3-lb mass is located at a point with coordinates (1, 0), a 5-lb mass is located at (2, 3), and a 4-lb mass is located at (4, 5). Where is the center of mass of this system?

40. Is it possible to add an additional mass to the system described in Exercise 39 at a point on the line $x = 5$ so that the center of mass of the new system is (4, 2)? If so, how much mass should be added and where?

41. Find the centroid of the homogeneous lamina bounded by the graphs of $y = x^2 - 1$ and $y = 0$.

42. Find the centroid of the homogeneous lamina bounded by the graphs of $y = x^2 - 2x + 1$ and $y = 4$.

43. Find the centroid of the homogeneous lamina bounded by the graphs of $y = \sqrt{x}$ and $y = x$.

44. Find the centroid of the homogeneous lamina bounded by the graph of $y^2 = 4x$ and the lines $x = 0$ and $y = 4$.

45. Use the First Theorem of Pappus to find the volume of the solid generated when the region bounded by the graphs of $y = x^2 - 1$ and $y = 0$ is revolved about the x-axis. [*Hint:* Use your solution from Exercise 41.]

46. Use the First Theorem of Pappus to find the volume of the solid generated when the triangle with vertices at (1, 1), (−1, 1), and (0, 2) is revolved about the x-axis.

47. How much work is done stretching a spring 4 in. if it takes 250 lb of force to stretch it that amount?

48. A force of 5 lb stretches a spring 1 ft.
    a) How much work is done stretching the spring this distance?
    b) How much work is done if the spring is stretched from 1 ft to 2 ft beyond its natural length?

49. How much work is required to lift a USSR Soyuz satellite weighing 14,900 lb to an orbit 122 mi above the earth?

50. A ship's anchor weighs 2000 lb, and the anchor chain weighs 30 lb/ft. How much work is required to pull this anchor 50 ft vertically up through the hawsehole?

51. A conical tank with the vertex at the bottom has radius 5 ft and altitude 20 ft. If the water in the tank is 12 ft deep, find the work required to pump the water (a) over the top; (b) to a point 10 ft above the top.

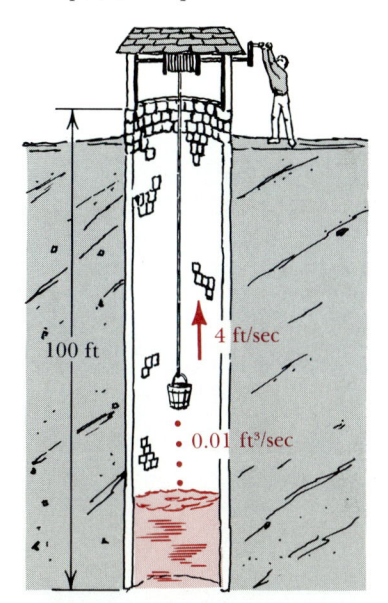

52. A bucket full of water is pulled to the top of a 100-ft well at the rate of 4 ft/sec (see the figure). When the bucket left the bottom of the well it contained 4 ft³ of water, but there is a hole in the bucket that causes the water to leak out at the rate of 0.01 ft³/sec. Assuming that the weight of the bucket and the rope is negligible, how much work is required to lift the bucket to the top of the well? (Assume that water weighs 62.4 lb/ft³.)

53. Rework Exercise 52 with the additional assumption that the empty bucket weighs 10 lb and that the rope weighs 0.25 lb/ft.

54. A metal plate in the shape of a square with sides 2 ft is placed vertically at the bottom of a tank containing 10 ft of water. What is the total force exerted by the water on the plate?

55. A floodgate has a parabolic shape described by $y = x^2$. If the gate is 4 ft deep and 4 ft wide at the top, find the force of the water on the gate when the level of water is at the top of the gate.

56. A right circular cylinder with base area A and height h is filled with fluid of density σ. Show that the work done in pumping all the fluid to the top of the cylinder is given by $W = \sigma h^2 A/2$.

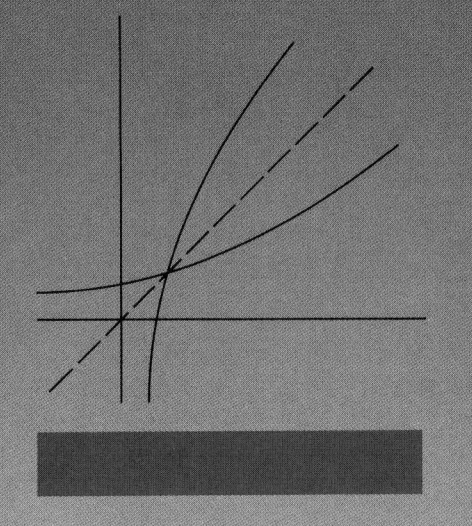

6

The Calculus of Inverse Functions

CHAPTER 6 THE CALCULUS OF INVERSE FUNCTIONS

The term "inverse" is introduced in precalculus courses to describe the relationship of logarithms to exponentials and to study the inverses to the trigonometric functions. The first section of this chapter reviews the notion of an inverse function and shows how calculus is used to add significantly to the knowledge of inverse functions. These results are applied in Section 6.2 to a class of familiar functions: the inverse trigonometric functions.

Section 6.3 begins a study of the natural exponential function (the inverse to the natural logarithm function). This function leads to definitions of general exponential and logarithm functions and, later in the chapter, to the hyperbolic functions.

Sections 6.6 and 6.7 consider some powerful applications of the natural exponential function. The range of these applications shows why this is perhaps the most important of all functions for calculus applications.

6.1

Inverse Functions

Since the term *inverse* is used to convey the notion of reversal, an inverse function is one that reverses the correspondence of some other function. If a function f with domain X and range Y has an inverse, then this inverse function has domain Y and range X. In addition, if g represents the inverse function for f and x is in the domain of f with $y = f(x)$, then g has the property that

$$g(y) = g(f(x)) = x$$

(see Figure 6.1).

This relationship breaks down, however, if some element y in Y is the image of two different elements x_1 and x_2 in X. In this case, both $f(x_1) = y$ and $f(x_2) = y$, and the reversal does not produce a function. To have an inverse, a function must be of a special type, called a *one-to-one* function.

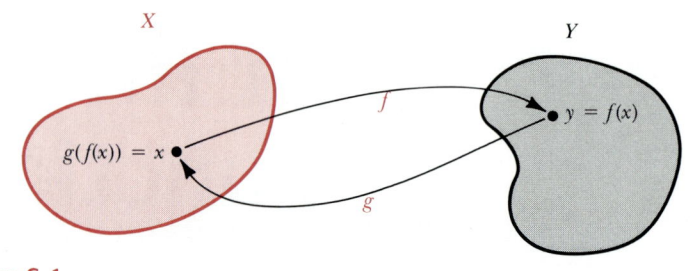

Figure 6.1

If g is the inverse of f, then $g(f(x)) = x$.

(6.1)
DEFINITION
A function f is said to be a **one-to-one** function provided that for every x_1 and x_2 in the domain of f,

$$\text{if } f(x_1) = f(x_2), \text{ then } x_1 = x_2.$$

Any function f has the property that if $f(x_1) \neq f(x_2)$, then $x_1 \neq x_2$. Therefore, Definition (6.1) implies that:

If f is a one-to-one function and x_1 and x_2 are in the domain of f,

(6.2) $f(x_1) = f(x_2)$ holds if and only if $x_1 = x_2$.

Note that $f(x) = x^3$ describes a one-to-one function, since every number in the range corresponds to a single number in the domain, its unique cube root. However, the function described by $f(x) = x^2$ is not a one-to-one function because, for example, $f(2) = (2)^2 = 4$ and $f(-2) = (-2)^2 = 4$, and certainly $2 \neq -2$. In fact, every nonzero number y in the range of $f(x) = x^2$ corresponds to two distinct numbers in the domain, $x_1 = \sqrt{y}$ and $x_2 = -\sqrt{y}$.

Geometrically, f is a one-to-one function precisely when every horizontal line intersects its graph at most once. Figure 6.2 shows that the functions described by $f(x) = x^3$ and $f(x) = 1/x$ are one-to-one functions. Figure 6.3 illustrates that the functions described by $f(x) = x^2$ and $f(x) = \sin x$ are *not* one-to-one functions.

(6.3)
DEFINITION
Suppose f is a one-to-one function with domain X and range Y. A function g with domain Y and range X is called the **inverse function** for f if

$$g(f(x)) = x \quad \text{for every } x \text{ in } X.$$

The fact that f is a one-to-one function ensures that:

i) a unique inverse function exists;
ii) the domain of the inverse function is the range of f; and
iii) the inverse function is itself a one-to-one function.

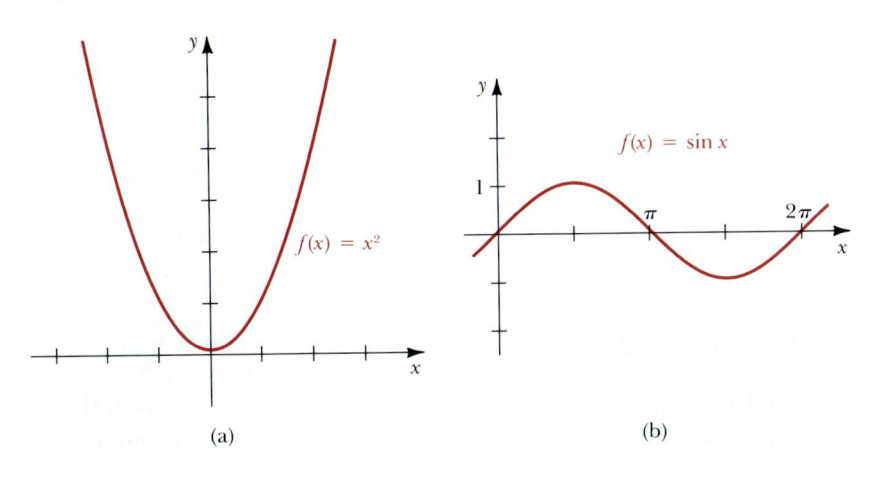

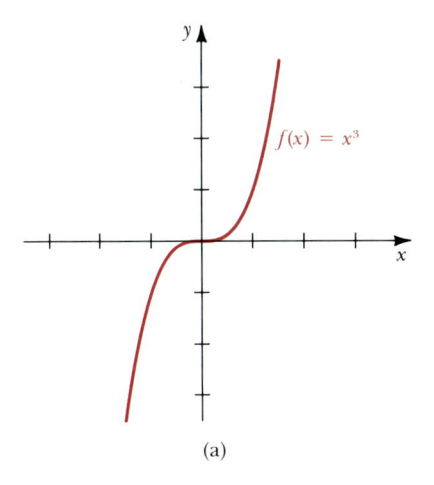

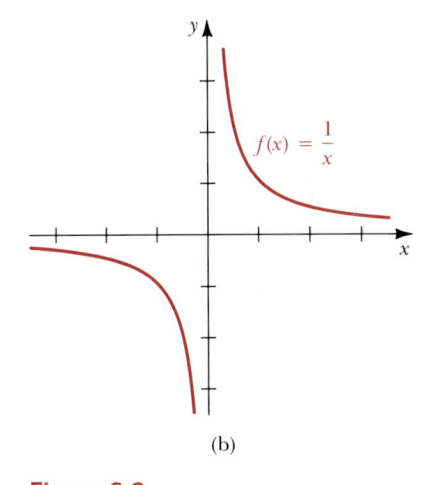

Figure 6.2
These are one-to-one functions.

Figure 6.3
These are *not* one-to-one functions.

The inverse function for f is often denoted f^{-1}, so the equation in Defintion (6.3) is

$$f^{-1}(f(x)) = x \quad \text{for every } x \text{ in } X.$$

Since f^{-1} is a one-to-one function, it also has an inverse, which is $(f^{-1})^{-1} = f$. Thus,

$$f(f^{-1}(y)) = y \quad \text{for every } y \text{ in } Y.$$

The relationship between the variables expressing the function f and its inverse f^{-1} is also given by

(6.4) $\qquad f(x) = y \quad \text{if and only if} \quad f^{-1}(y) = x$

for every x in the domain of f (see Figure 6.4). This relationship is useful in describing inverse functions explicitly.

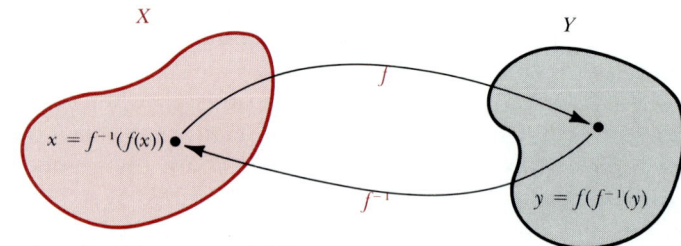

Figure 6.4

The functions f and f^{-1} are inverse to each other.

Example 1

Suppose $f(x) = 2x - 3$. Show that f is a one-to-one function, and describe f^{-1}.

Solution

If $f(x_1) = f(x_2)$, then

$$2x_1 - 3 = 2x_2 - 3, \quad \text{so} \quad 2x_1 = 2x_2 \quad \text{and} \quad x_1 = x_2.$$

Thus f is a one-to-one function and f^{-1} exists.

Let $y = f(x) = 2x - 3$. Since $y = 2x - 3$ implies that $x = \frac{1}{2}(y + 3)$, we see that f^{-1} is described by $f^{-1}(y) = \frac{1}{2}(y + 3)$.

The variable used to describe the function f^{-1} is irrelevant, so the function f^{-1} can also be described using the variable x:

$$f^{-1}(x) = \tfrac{1}{2}(x + 3).$$

The graphs of $y = f(x)$ and $y = f^{-1}(x)$ are shown in Figure 6.5. ▶

The following example shows how the domain of a function that is not a one-to-one function can be restricted to produce a one-to-one function.

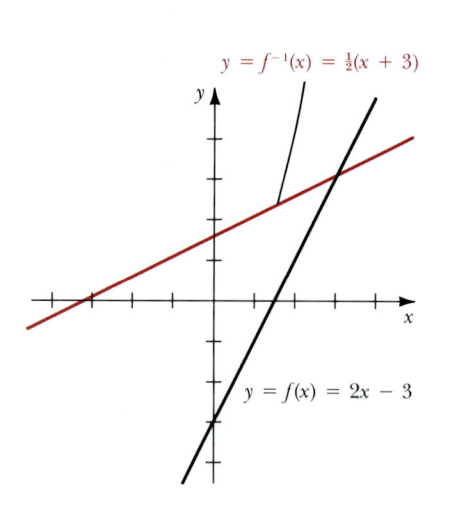

Figure 6.5

Example 2

Show that the function described by

$$f(x) = x^2,$$

with domain $[0, \infty)$, is a one-to-one function, and describe f^{-1}.

Solution
By restricting the domain of $f(x) = x^2$ to $[0, \infty)$, we have the relation

$$y = x^2 \quad \text{and} \quad x \geq 0 \quad \text{if and only if} \quad x = \sqrt{y}.$$

The range of f is $[0, \infty)$, and each number y in $[0, \infty)$ corresponds to precisely one number in the domain. Thus f is a one-to-one function and has an inverse. Moreover, $f^{-1}(y) = \sqrt{y}$, or equivalently,

$$f^{-1}(x) = \sqrt{x}.$$

Figure 6.6 shows the graphs of $y = f(x)$ and $y = f^{-1}(x)$. ▶

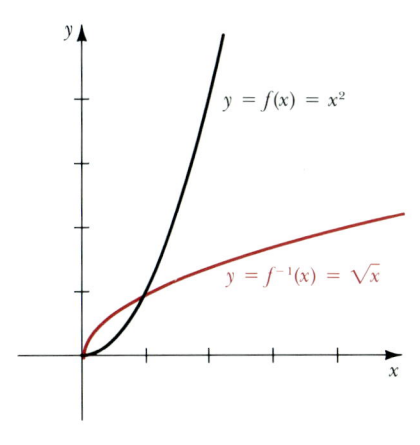

Figure 6.6

Figures 6.5 and 6.6 illustrate an important relationship between the graph of a one-to-one-function and the graph of its inverse.

(6.5)
The graph of $y = f^{-1}(x)$ is the reflection of the graph of $y = f(x)$ about the line $y = x$.

This geometric feature follows from (6.4). A point (a, b) is on the graph of f precisely when $b = f(a)$. Since this is equivalent to $a = f^{-1}(b)$, the point (a, b) is on the graph of f precisely when (b, a) is on the graph of f^{-1} (see Figure 6.7).

The following theorem is the first in a series of results about the inverse of an always increasing or always decreasing function. Reading the statement ignoring the parentheses gives the theorem for increasing functions. The corresponding theorem for decreasing functions is read using the remarks in parentheses, while deleting the references to increasing.

(6.6)
THEOREM
If f is a function that is increasing (decreasing) on its domain, then f^{-1} exists and is also increasing (decreasing).

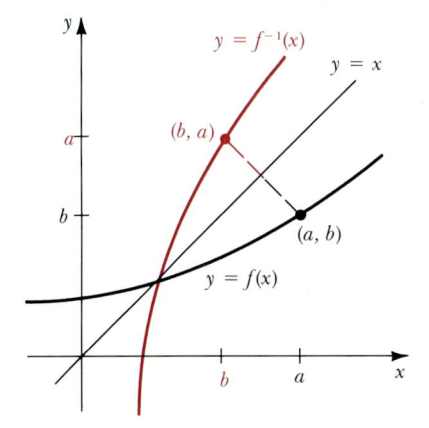

Figure 6.7
The graph of $y = f^{-1}(x)$ is the reflection about the line $y = x$ of the graph of $y = f(x)$.

Proof. Suppose f is increasing. To show that f has an inverse, we must show that f is a one-to-one function.

Let x_1 and x_2 be in the domain of f with $x_1 \neq x_2$. If $x_1 < x_2$, then, since f is increasing, $f(x_1) < f(x_2)$. On the other hand, if $x_2 < x_1$, then $f(x_2) < f(x_1)$. In either case, $x_1 \neq x_2$ implies that $f(x_1) \neq f(x_2)$. Thus, $f(x_1) = f(x_2)$ implies $x_1 = x_2$, so f is a one-to-one function and has an inverse f^{-1}.

To show that f^{-1} is increasing, let y_1 and y_2 be in the domain of f^{-1} with

$$y_1 < y_2.$$

The domain of f^{-1} is the range of f, so numbers x_1 and x_2 exist with

$$f(x_1) = y_1 < y_2 = f(x_2).$$

Since f is an increasing function, the inequality $f(x_1) < f(x_2)$ holds precisely when $x_1 < x_2$. Using (6.4) to express this inequality in terms of f^{-1} gives

$$x_1 = f^{-1}(y_1) < f^{-1}(y_2) = x_2.$$

Thus $y_1 < y_2$ implies that $f^{-1}(y_1) < f^{-1}(y_2)$ and f^{-1} is increasing.

The proof for decreasing functions follows the same line with the reversal of some of the inequalities. ▷

If the function f in Theorem (6.6) is known to be continuous, then the inverse function is also continuous. The proof of this result, given in Appendix B, is not difficult, but requires considerable manipulation between the function and its inverse.

(6.7)
THEOREM

If f is continuous and increasing (decreasing) on its domain, then f^{-1} is also continuous and increasing (decreasing).

Figure 6.7 illustrates the result in Theorem (6.7) for increasing functions, and Figure 6.8 illustrates the result for decreasing functions.

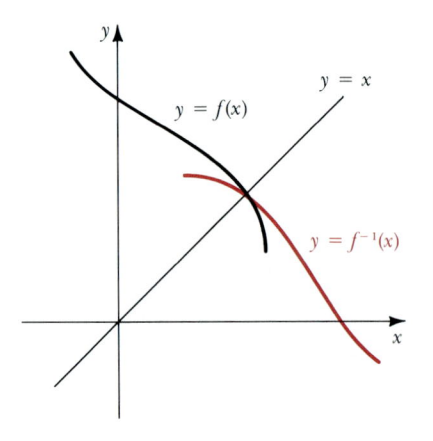

Figure 6.8

Example 3

Verify that the conclusions of Theorem (6.7) hold for $f(x) = x^2 - 2$ with domain $[1, 8]$.

Solution

Since $f'(x) = 2x > 0$ for all x in $[1, 8]$, f is increasing on $[1, 8]$. Theorem (6.6) implies that f^{-1} exists and is increasing.

Since f is increasing on $[1, 8]$, the range of f is contained in the interval $[f(1), f(8)] = [-1, 62]$. But f is continuous on $[1, 8]$, so the Intermediate Value Theorem implies that for every number k in $[-1, 62]$, there is a number c in $[1, 8]$ with $f(c) = k$. Thus k is in the domain of f^{-1}, and the domain of f^{-1} is precisely $[f(1), f(8)] = [-1, 62]$.

To determine f^{-1}, let $y = x^2 - 2$ and solve for x in terms of y:

$$y = x^2 - 2 \quad \text{implies that} \quad x = \pm\sqrt{y + 2}.$$

The domain of f is $[1, 8]$, so the positive sign must be chosen and $f^{-1}(y) = \sqrt{y + 2}$, or equivalently,

$$f^{-1}(x) = \sqrt{x + 2}.$$

Since

$$(f^{-1})'(x) = \tfrac{1}{2}(x + 2)^{-1/2}$$

exists for all x in $[-1, 62]$, f^{-1} is continuous on $[-1, 62]$. ▶

In Example 3, the continuous function f with domain $[1, 8]$ has an inverse with domain $[f(1), f(8)]$. It is true in general that:

(6.8)
If f is continuous and increasing (decreasing) on $[a, b]$, then f^{-1} is continuous and increasing on $[f(a), f(b)]$ (decreasing on $[f(b), f(a)]$).

The final result concerning increasing and decreasing functions and their inverses adds the hypothesis of differentiability to the function to produce the conclusion of differentiability of the inverse function. This theorem will be called the *Inverse Function Theorem*.

(6.9)
THEOREM: THE INVERSE FUNCTION THEOREM
Suppose f is a differentiable and increasing (decreasing) function whose domain is an open interval I. If $f'(x) \neq 0$ for all x in I, then:

i) f^{-1} is differentiable and increasing (decreasing), and
ii) if $y = f(x)$, then

$$D_y f^{-1}(y) = \frac{1}{D_x f(x)}.$$

Proof. We will show that the result holds when f is increasing. The proof is similar if f is decreasing.

Theorem (6.7) ensures that f^{-1} exists, is increasing, and is continuous. It remains to show that $(f^{-1})'(y)$ exists and is as specified. To eliminate confusion between changes in the variables x and y, we use the Leibniz notation to determine the derivative of f^{-1}. Thus, if the limit exists,

$$D_y f^{-1}(y) = \lim_{\Delta y \to 0} \frac{f^{-1}(y + \Delta y) - f^{-1}(y)}{\Delta y}.$$

Let

$$\Delta x = f^{-1}(y + \Delta y) - f^{-1}(y).$$

Then

$$f^{-1}(y + \Delta y) = f^{-1}(y) + \Delta x$$
$$= x + \Delta x,$$

so

$$y + \Delta y = f(x + \Delta x)$$

and

$$\Delta y = f(x + \Delta x) - y$$
$$= f(x + \Delta x) - f(x).$$

Since f is increasing, Δx is nonzero precisely when Δy is nonzero. In addition, f^{-1} is continuous, so Δx approaches zero as Δy approaches zero and

$$D_y f^{-1}(y) = \lim_{\Delta y \to 0} \frac{f^{-1}(y + \Delta y) - f^{-1}(y)}{\Delta y}$$

$$= \lim_{\Delta x \to 0} \frac{(x + \Delta x) - x}{f(x + \Delta x) - f(x)}$$

$$= \lim_{\Delta x \to 0} \frac{1}{\frac{f(x + \Delta x) - f(x)}{\Delta x}} = \frac{1}{D_x f(x)},$$

provided $D_x f(x) \neq 0$. This result is illustrated in Figure 6.9(a) when f is increasing and in Figure 6.9(b) when f is decreasing. ▷

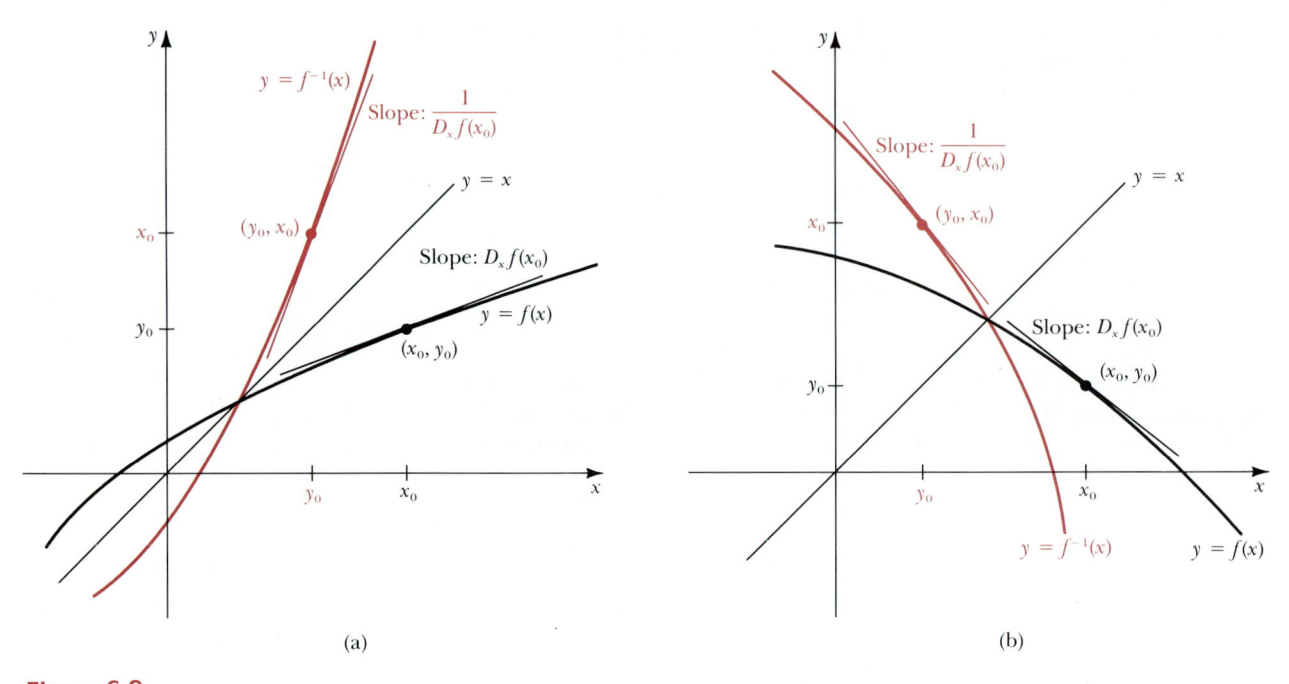

Figure 6.9

Example 4

Find $(f^{-1})'(y)$ for $f(x) = x^3 + 1$:

a) directly, without using the derivative relation in the Inverse Function Theorem;
b) using the derivative relation in the Inverse Function Theorem.

Solution

a) Let $y = x^3 + 1$. Then $x^3 = y - 1$ and

$$x = f^{-1}(y) = (y - 1)^{1/3}.$$

Thus,
$$(f^{-1})'(y) = \frac{1}{3}(y-1)^{-2/3},$$
provided $y \neq 1$.

b) The derivative relation in the Inverse Function Theorem gives
$$(f^{-1})'(y) = D_y f^{-1}(y) = \frac{1}{D_x f(x)} = \frac{1}{3x^2}.$$

To express this in terms of the variable y used to describe f^{-1}, we again use the fact that $y = x^3 + 1$ implies that $x = (y-1)^{1/3}$. Thus, if $y \neq 1$,
$$(f^{-1})'(y) = \frac{1}{3x^2} = \frac{1}{3 \cdot [(y-1)^{1/3}]^2} = \frac{1}{3}(y-1)^{-2/3}. \quad \blacktriangleright$$

The derivative relation in the Inverse Function Theorem is used repeatedly in this chapter. In the following example this relation *must* be used since an explicit representation of the inverse function cannot be found.

Example 5

Determine an equation of the tangent line to the graph of f^{-1} at $(4, 1)$ if $f(x) = x^5 + 3x^3 + x - 1$.

Solution

First note that $f(1) = 4$, so $f^{-1}(4) = 1$ and the point $(4, 1)$ is on the graph of f^{-1} (see Figure 6.10). We cannot find an explicit representation for f^{-1} since this would require solving for x in terms of y in the equation $y = x^5 + 3x^3 + x - 1$. However,
$$f'(x) = 5x^4 + 9x^2 + 1 > 0$$
for all x, so f is always increasing. Hence f^{-1} exists and is differentiable. By the Inverse Function Theorem,
$$(D_y f^{-1})(y) = \frac{1}{D_x f(x)} = \frac{1}{5x^4 + 9x^2 + 1}.$$

The slope of the tangent line to the graph of f^{-1} at $(4, 1)$ is
$$(D_y f^{-1})(4) = \frac{1}{D_x f(1)} = \frac{1}{15}.$$

The line with slope $\frac{1}{15}$ passing through $(4, 1)$ has equation
$$y - 1 = \frac{1}{15}(x - 4). \quad \blacktriangleright$$

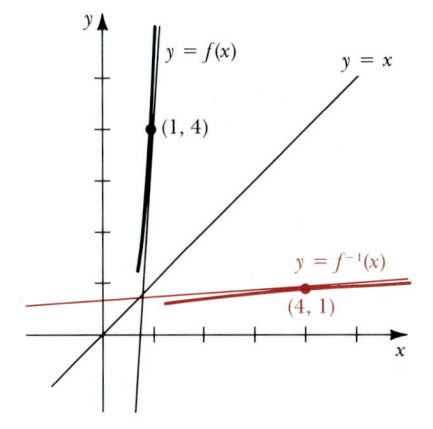

Figure 6.10

Recall that if f is a one-to-one function, then $(f^{-1})^{-1} = f$. The Implicit Function Theorem, together with this fact, implies that a one-to-one function has a nonzero derivative precisely when its inverse function has a nonzero derivative. Thus the roles of f and its inverse f^{-1} can be reversed in the Inverse Function Theorem. In particular, the equation in part (ii) of the theorem can also be expressed as

(6.10) $$D_x f^{-1}(x) = \frac{1}{D_y f(y)},$$

which is the form used most often in the remainder of the chapter.

▶ EXERCISE SET 6.1

In Exercises 1–10, determine whether the function is a one-to-one function.

1. $f(x) = 3x + 4$
2. $f(x) = 1 - 2x$
3. $f(x) = x^3$
4. $f(x) = 2x^3 + 1$
5. $f(x) = x^2 + 4$
6. $f(x) = x^2 - 1$
7. $f(x) = \sqrt{x + 1}$
8. $f(x) = \sqrt{x} + 1$
9. $f(x) = \dfrac{1}{x - 1}$
10. $f(x) = \dfrac{1}{x^2 - 1}$

11–20. For the one-to-one functions in Exercises 1–10:
  a) Give the domain and range of f.
  b) Find f^{-1} and give its domain and range.
  c) Sketch the graph of $y = f(x)$ and $y = f^{-1}(x)$ on the same set of axes.

In Exercises 21–28, a function is described that is not a one-to-one function, so no inverse function exists. Determine a new function that is a one-to-one function by modifying the domain of the given function. Sketch the graph of the new function and its inverse on the same set of axes.

21. $f(x) = x^2 - 4$
22. $f(x) = 2x^2 + 1$
23. $f(x) = |x + 1|$
24. $f(x) = \cos x$
25. $f(x) = \sin x$
26. $f(x) = \tan x$
27. $f(x) = 1/(x^2 + 1)$
28. $f(x) = 1/(x^2 - 1)$

In Exercises 29–36, a differentiable one-to-one function is described on a particular interval. Find $(f^{-1})'(a)$ directly and (b) using the derivative relation in the Inverse Function Theorem.

29. $f(x) = 3x - 1$, $(-\infty, \infty)$
30. $f(x) = 2x + 5$, $(-\infty, \infty)$
31. $f(x) = x^3$, $[1, 2]$
32. $f(x) = x^2 + 3$, $[1, 9]$
33. $f(x) = 1/x$, $(0, \infty)$
34. $f(x) = 1/x$, $(-\infty, 0)$
35. $f(x) = \dfrac{x + 1}{x}$, $[1, 2]$
36. $f(x) = \dfrac{x - 4}{x + 1}$, $[0, 1]$

In Exercises 37–40, use the Inverse Function Theorem to find the slope of the tangent line to the graph of f^{-1} at the point $(a, f^{-1}(a))$.

37. $f(x) = x^3 + x + 1$, $(3, 1)$
38. $f(x) = 2x^5 + 3x^2$, $(5, 1)$
39. $f(x) = \sqrt{x + 4} + x^3$, $(2, 0)$
40. $f(x) = \dfrac{x^3 + 8}{x^5 - 1}$, $(0, -2)$

In Exercises 41–46, show that the function f described has the property that $f = f^{-1}$.

41. $f(x) = x$
42. $f(x) = -x$
43. $f(x) = 1/x$
44. $f(x) = -1/x$
45. $f(x) = \sqrt{1 - x^2}$, $0 \le x \le 1$
46. $f(x) = -\sqrt{4 - x^2}$, $-2 \le x \le 0$
47. By examining the graphs of the functions described in Exercises 41–46, make a conjecture concerning the symmetry of the graph of a function f with the property that $f = f^{-1}$.
48. Suppose f is a linear function described by $f(x) = mx + b$, where $m \ne 0$. Show that f^{-1} exists and is also a linear function.
49. Suppose f is a linear function described by $f(x) = mx + b$ and that $f = f^{-1}$. What can be said about m and b?
50. Describe a function T that converts temperature readings given in Fahrenheit into readings in Celsius. (a) Determine T^{-1}. (b) What application does T^{-1} have? (c) What do T' and $(T^{-1})'$ describe?

Each of the functions in Exercises 51–54 has an inverse. Sketch the graph of the function and its inverse, and define $f^{-1}(x)$.

51. $f(x) = \begin{cases} -x + 2, & \text{if } x < 1, \\ -\dfrac{1}{2}x + \dfrac{3}{2}, & \text{if } x \geq 1 \end{cases}$

52. $f(x) = \begin{cases} -x + 2, & \text{if } x < 1, \\ \dfrac{1}{x}, & \text{if } x \geq 1 \end{cases}$

53. $f(x) = \begin{cases} x, & \text{if } x < 0, \\ x^2, & \text{if } 0 \leq x \leq 1, \\ \sqrt{x}, & \text{if } x > 1 \end{cases}$

54. $f(x) = \begin{cases} \dfrac{1}{x^2}, & \text{if } x < 0, \\ 0, & \text{if } x = 0, \\ -\dfrac{1}{x^2}, & \text{if } x > 0 \end{cases}$

55. Suppose f is a function defined by $f(x) = x^k$ for some rational number k. What values are possible for k if f has an inverse and $f(x) = f^{-1}(x)$ for all x in the domain of f?

56. Suppose f and g are one-to-one functions with domain and range $\mathbb{R}$. Give examples to show that (a) $f + g$, (b) $f - g$, (c) $f \cdot g$, and (d) f/g need not be one-to-one functions.

57. Suppose f and g are one-to-one functions with domain and range $\mathbb{R}$. Show that (a) $f \circ g$ is also a one-to-one function with domain and range $\mathbb{R}$ and (b) $(f \circ g)^{-1} = g^{-1} \circ f^{-1}$.

58. Suppose f is a differentiable one-to-one function and that the tangent line to the graph of f at $(c, f(c))$ has equation $y = mx + b$, where $m \neq 0$. Show that the tangent line to the graph of f^{-1} at $(f(c), c)$ has equation $y = (1/m)x - (1/m)b$.

59. Suppose the tangent lines to the graph of f at (a, b) and to the graph of f^{-1} at (b, a) intersect. Show that the intersection point must lie on the line $y = x$.

60. Suppose f is a differentiable one-to-one function and that f^{-1} is known to be differentiable. Apply the Chain Rule to the equation $x = f^{-1}(f(x))$ to obtain the derivative relation in the Inverse Function Theorem:

$$D_y f^{-1}(y) = \frac{1}{D_x f(x)}.$$

61. Show that the following formula holds, provided the necessary differentiability conditions on f and f^{-1} are satisfied and that $D_x f(x) \neq 0$:

$$D_y^2 f^{-1}(y) = \frac{-D_x^2 f(x)}{[D_x f(x)]^3}.$$

62. Use the result in Exercise 61 to state conclusions about the concavity of the graph of $y = f^{-1}(x)$ based on the concavity of the graph of $y = f(x)$.

63. a) Show that $f(x) = x^3 + x + 1$ describes a one-to-one function.
   b) Use Newton's Method to determine, within 10^{-3}, $f^{-1}(2)$.
   c) Use the value found in part (b) to approximate $(f^{-1})'(2)$.

64. Show that an even function cannot be a one-to-one function unless its domain contains only the number zero.

65. Given an example of a continuous one-to-one odd function with domain $(-\infty, \infty)$.

66. Give an example of a continuous non–one-to-one odd function with domain $(-\infty, \infty)$.

67. Find a function f with the property that whenever g is defined by $g(x) = f(x) + c$ for any constant c, then $g = g^{-1}$.

68. Find a function f with the property that whenever g is defined by $g(x) = cf(x)$ for any nonzero constant c, then $g = g^{-1}$.

69. Suppose f is a one-to-one function and c is a constant. (a) Show that g defined by $g(x) = f(x + c)$ is a one-to-one function, and (b) determine the relationship between f^{-1} and g^{-1}.

70. Suppose f is a one-to-one function on $[a, b]$ and $y = f(x)$.
   a) Show that for all x in $[a, b]$,
   $$D_x\left[\int_a^x f(t)\, dt + \int_{f(a)}^{f(x)} f^{-1}(t)\, dt\right] = f(x) + x f'(x).$$
   b) Use part (a) to show that for all x in $[a, b]$,
   $$\int_{f(a)}^{f(x)} f^{-1}(t)\, dt = x f(x) - a f(a) - \int_a^x f(t)\, dt.$$
   c) Use part (b) to derive the indefinite integral formula
   $$\int f^{-1}(y)\, dy = y f^{-1}(y) - \int f(x)\, dx.$$

6.2

Inverse Trigonometric Functions

The sine function is not a one-to-one function. Its graph, shown in black in Figure 6.11, illustrates that *every* number in the range of the sine function corresponds to more than one value in the domain.

Suppose, however, that we restrict the domain to an interval on which the sine function is a one-to-one function and takes on all the values in $[-1, 1]$, the range of the sine function. The interval $[-\pi/2, \pi/2]$ satisfies these conditions, as we can see from the graph shown in color in Figure 6.11. The sine function restricted to $[-\pi/2, \pi/2]$ is an increasing, hence, one-to-one function. This restricted sine function has an inverse, the *inverse sine* or *arcsine* function.

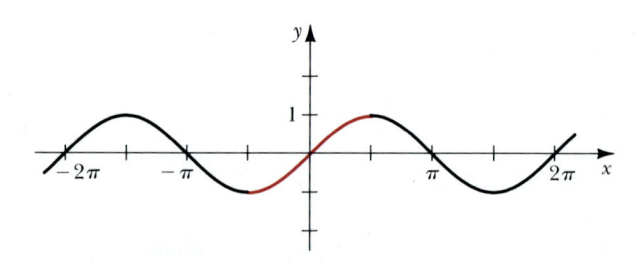

Figure 6.11
The domain of $y = \sin x$ is restricted to $[-\pi/2, \pi/2]$.

The arcsine function is denoted arcsin or $\sin^{-1}$. We will use the notation arcsin since the $\sin^{-1}$ notation can be confused with the reciprocal of the sine function. The arcsin notation also agrees with the treatment in most higher mathematics and science texts. It has the further advantage of being closely related to the notation used in modern computer languages, where the arcsine function is generally denoted by some abbreviation of arcsin, such as ASIN or ASN.

(6.11)
DEFINITION
The **arcsine** function, denoted **arcsin**, is the function with domain $[-1, 1]$ and range $[-\pi/2, \pi/2]$ defined by

$$\arcsin (\sin x) = x, \quad \text{for every } x \text{ in } [-\pi/2, \pi/2].$$

Using (6.5) in Section 6.1 we can easily sketch the graph of the arcsine function. The graph of $y = \arcsin x$ is the reflection about the line $y = x$ of the graph of the restricted sine function (see Figure 6.12).

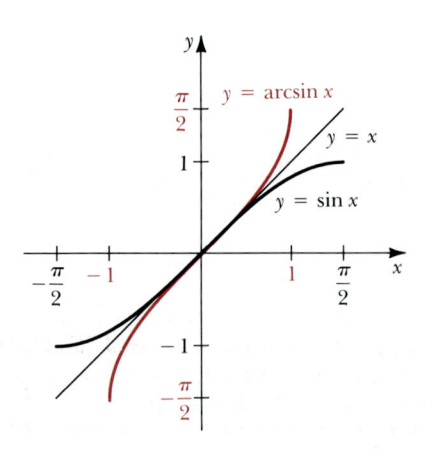

Figure 6.12

Example 1

Find (a) arcsin 1; (b) arcsin (-1); (c) arcsin $\frac{1}{2}$.

Solution

a) $\sin \dfrac{\pi}{2} = 1$ and $\dfrac{\pi}{2}$ is in $\left[-\dfrac{\pi}{2}, \dfrac{\pi}{2}\right]$, so arcsin $1 = \dfrac{\pi}{2}$.

6.2 INVERSE TRIGONOMETRIC FUNCTIONS

b) $\sin -\dfrac{\pi}{2} = -1$ and $-\dfrac{\pi}{2}$ is in $\left[-\dfrac{\pi}{2}, \dfrac{\pi}{2}\right]$, so $\arcsin(-1) = -\dfrac{\pi}{2}$.

c) $\sin \dfrac{\pi}{6} = \dfrac{1}{2}$ and $\dfrac{\pi}{6}$ is in $\left[-\dfrac{\pi}{2}, \dfrac{\pi}{2}\right]$, so $\arcsin \dfrac{1}{2} = \dfrac{\pi}{6}$. ▶

The sine function is continuous and increasing on $[-\pi/2, \pi/2]$ and has a derivative, the cosine function, which is nonzero on $(-\pi/2, \pi/2)$. The Inverse Function Theorem ensures that the arcsine function is increasing and differentiable on the interval $(-1, 1)$.

To determine the derivative of the arcsine function, we use (6.4) to relate the arcsine function to its inverse, the restricted sine function:

(6.12) $y = \arcsin x$ if and only if $x = \sin y$ and $-\pi/2 \leq y \leq \pi/2$.

Applying (6.10) in this case gives

$$D_x \arcsin x = \dfrac{1}{D_y \sin y} = \dfrac{1}{\cos y}, \qquad -\dfrac{\pi}{2} < y < \dfrac{\pi}{2}.$$

However,

$$\cos^2 y + \sin^2 y = 1$$

so

$$\cos y = \pm \sqrt{1 - \sin^2 y} = \pm \sqrt{1 - x^2}.$$

Since y is in $(-\pi/2, \pi/2)$ and in this interval $\cos y > 0$, we have

(6.13) $$D_x \arcsin x = \dfrac{1}{\sqrt{1 - x^2}},$$

provided $|x| < 1$.

Example 2

Determine $D_x \arcsin(x^2 - 1)$.

Solution

Applying (6.13) and the Chain Rule gives

$$D_x \arcsin(x^2 - 1) = \dfrac{1}{\sqrt{1 - (x^2 - 1)^2}} D_x(x^2 - 1) = \dfrac{2x}{\sqrt{2x^2 - x^4}}. \qquad ▶$$

The indefinite integral formula associated with (6.13) is

(6.14) $$\int \dfrac{dx}{\sqrt{1 - x^2}} = \arcsin x + C.$$

Example 3

Evaluate $\displaystyle\int \dfrac{dx}{\sqrt{4 - x^2}}$.

Solution
We have

$$\int \frac{dx}{\sqrt{4-x^2}} = \int \frac{dx}{\sqrt{4\left(1-\frac{x^2}{4}\right)}} = \frac{1}{2}\int \frac{dx}{\sqrt{1-\frac{x^2}{4}}}.$$

Let $u = x/2$. Then $du = dx/2$, so $2\,du = dx$ and

$$\int \frac{dx}{\sqrt{4-x^2}} = \frac{1}{2}\int \frac{2\,du}{\sqrt{1-u^2}}$$

$$= \int \frac{du}{\sqrt{1-u^2}} = \arcsin u + C$$

$$= \arcsin\left(\frac{x}{2}\right) + C. \qquad \blacktriangleright$$

By making the type of variable substitution illustrated in Example 3, we have a general integration formula that holds whenever $a > 0$:

(6.15)
$$\int \frac{dx}{\sqrt{a^2 - x^2}} = \arcsin\left(\frac{x}{a}\right) + C.$$

General integration formulas will not always be stated. They are, however, the type of formulas given in integral tables such as those appearing in the summary table at the end of the chapter and in Appendix D.

The inverses of the other trigonometric functions are defined by making domain restrictions similar to the one made for the sine function. The domain of the cosine function is restricted to the interval $[0, \pi]$ (see Figure 6.13). On this interval the cosine function is a one-to-one, continuous, and decreasing function, and has the same range as the unrestricted cosine function, the interval $[-1, 1]$.

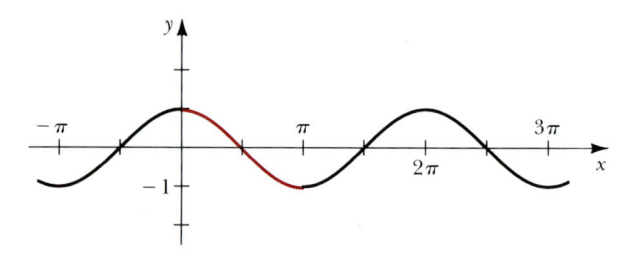

Figure 6.13
The domain of $y = \cos x$ is restricted to $[0, \pi]$.

(6.16)
DEFINITION
The **arccosine** function, denoted **arccos,** is the function with domain $[-1, 1]$ and range $[0, \pi]$ defined by

$$\arccos(\cos x) = x, \quad \text{for } x \text{ in } [0, \pi].$$

The graphs of $y = \arccos x$ and the restricted cosine function are shown in Figure 6.14.

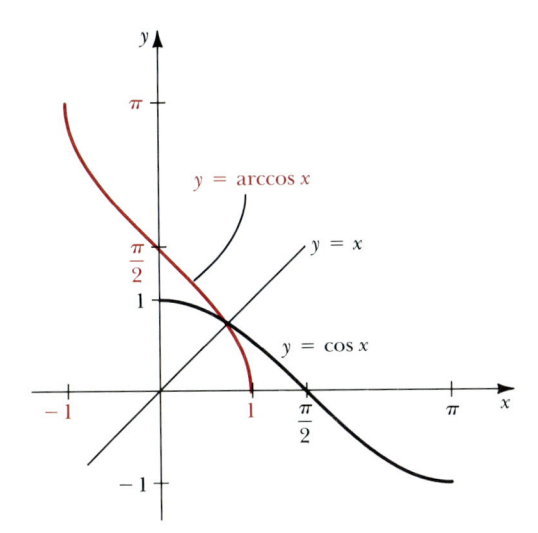

Figure 6.14

Example 4

Find $\sin[\arccos(-\tfrac{1}{2})]$.

Solution

Let $y = \arccos(-\tfrac{1}{2})$. Since $\cos y = -\tfrac{1}{2}$ and $0 \le y \le \pi$, we must have $y = 2\pi/3$. Figure 6.15 shows that

$$\sin\left[\arccos\left(-\frac{1}{2}\right)\right] = \sin\frac{2\pi}{3} = \frac{\sqrt{3}}{2}.$$

Example 5

Find $\cos[\arccos \tfrac{5}{13} + \arcsin \tfrac{3}{5}]$.

Solution

The trigonometric identity

$$\cos(a+b) = \cos a \cos b - \sin a \sin b$$

with $a = \arccos \tfrac{5}{13}$ and $b = \arcsin \tfrac{3}{5}$ implies that

$$\cos[\arccos \tfrac{5}{13} + \arcsin \tfrac{3}{5}] = \tfrac{5}{13}\cos[\arcsin \tfrac{3}{5}] - \tfrac{3}{5}\sin[\arccos \tfrac{5}{13}].$$

Since $b = \arcsin \tfrac{3}{5}$, $\sin b = \tfrac{3}{5}$ and $-\pi/2 \le b \le \pi/2$. Thus,

$$\cos b = \sqrt{1 - \sin^2 b} = \sqrt{1 - \tfrac{9}{25}} = \sqrt{\tfrac{16}{25}} = \tfrac{4}{5}.$$

This relationship can also be seen in the triangle shown in Figure 6.16. The relative lengths of the hypotenuse and vertical side are obtained from the fact that $\sin b = \tfrac{3}{5}$. The Pythagorean Theorem implies that the horizontal side has length 4, so

$$\cos[\arcsin \tfrac{3}{5}] = \cos b = \tfrac{4}{5}.$$

Figure 6.15

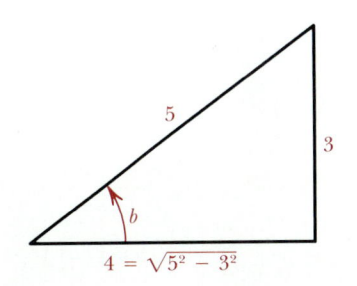

Figure 6.16

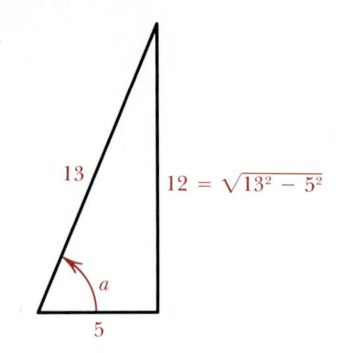

In a similar manner, the triangle in Figure 6.17 shows that

$$\sin [\arccos \tfrac{5}{13}] = \sin a = \tfrac{12}{13}.$$

So

$$\cos [\arccos \tfrac{5}{13} + \arcsin \tfrac{3}{5}] = \tfrac{5}{13} \cdot \tfrac{4}{5} - \tfrac{3}{5} \cdot \tfrac{12}{13} = -\tfrac{16}{65}. \quad \blacktriangleright$$

We can apply the Inverse Function Theorem to the arccosine function to show that

$$D_x \arccos x = -\frac{1}{\sqrt{1-x^2}}.$$

Since

$$D_x \arcsin x = \frac{1}{\sqrt{1-x^2}},$$

We have

$$D_x \arcsin x = D_x (-\arccos x),$$

and a constant C exists with

$$\arcsin x = -\arccos x + C.$$

Evaluating this expression at $x = 0$ gives $C = \pi/2$, that is,

(6.17) $\arcsin x + \arccos x = \pi/2,$ for every x in $[-1, 1]$.

A major application of the inverse trigonometric functions in calculus involves the evaluation of integrals. Since the arccosine and arcsine functions are used to evaluate integrals of the same form, we can dispense with the arccosine function for the purpose of integration. The other inverse trigonometric functions that are used for evaluating integrals are the *arctangent* and *arcsecant* functions.

(6.18)
DEFINITION
The **arctangent** function, denoted **arctan,** is the function with domain $(-\infty, \infty)$ and range $(-\pi/2, \pi/2)$ defined by

$$\arctan (\tan x) = x, \quad \text{for every } x \text{ in } (-\pi/2, \pi/2).$$

The graphs of the arctangent and restricted tangent functions are shown in Figure 6.18.

To find the derivative of the arctangent function, we first use (6.4) to relate the arctangent to its inverse, the restricted tangent:

(6.19) $y = \arctan x$ if and only if $x = \tan y$ and $-\pi/2 < y < \pi/2$.

We then apply (6.10):

$$D_x \arctan x = \frac{1}{D_y \tan y} = \frac{1}{\sec^2 y} \quad \text{for } -\frac{\pi}{2} < y < \frac{\pi}{2}.$$

6.2 INVERSE TRIGONOMETRIC FUNCTIONS

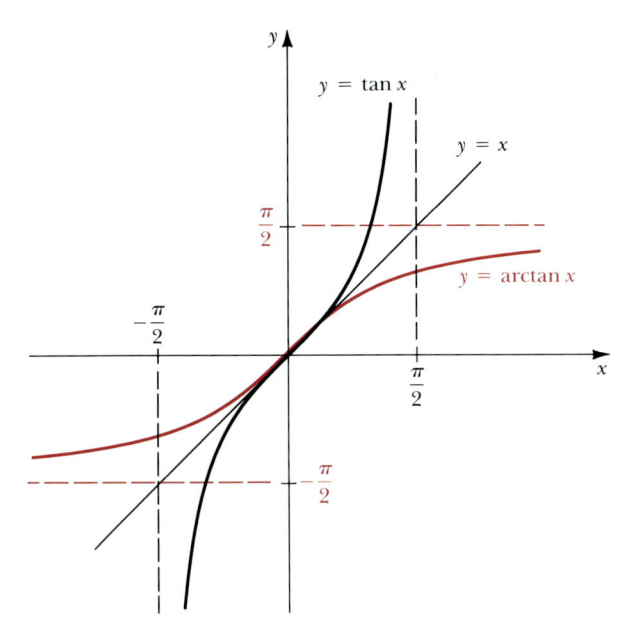

Figure 6.18

However, $\sec^2 y = 1 + \tan^2 y = 1 + x^2$, so

(6.20) $$D_x \arctan x = \frac{1}{1 + x^2}.$$

The associated integral formula is

$$\int \frac{1}{1 + x^2} \, dx = \arctan x + C.$$

The following example illustrates how a variable substitution leads to the general integration formula

(6.21) $$\int \frac{1}{a^2 + x^2} \, dx = \frac{1}{a} \arctan \left(\frac{x}{a}\right) + C, \quad a \neq 0.$$

Example 6

Find the area of the region bounded by the graph of $f(x) = \dfrac{1}{x^2 + 4}$, the x-axis, the y-axis, and the line $x = 2$.

Solution

The region described is shown in Figure 6.19. Its area is

$$\int_0^2 \frac{dx}{x^2 + 4} = \int_0^2 \frac{dx}{4\left(\dfrac{x^2}{4} + 1\right)}$$

$$= \frac{1}{4} \int_0^2 \frac{dx}{\left(\dfrac{x}{2}\right)^2 + 1}.$$

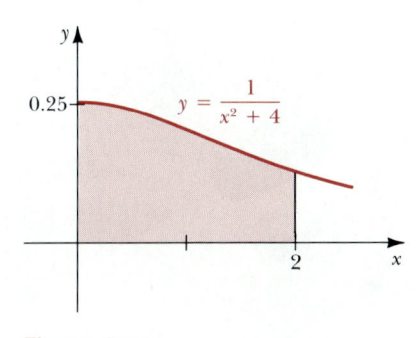

Figure 6.19

Let $u = x/2$. Then $du = \frac{1}{2} dx$ and

$$\int \frac{1}{x^2 + 4} dx = \frac{1}{2} \int \frac{1}{u^2 + 1} du = \frac{1}{2} \arctan u + C = \frac{1}{2} \arctan \frac{x}{2} + C,$$

which agrees with formula (6.21) when $a = 2$. The area is

$$\int_0^2 \frac{1}{x^2 + 4} dx = \frac{1}{2} \arctan \frac{x}{2} \Big]_0^2$$

$$= \frac{1}{2} [\arctan 1 - \arctan 0] = \frac{1}{2} \left[\frac{\pi}{4} - 0\right] = \frac{\pi}{8}. \quad \blacktriangleright$$

Example 7

Determine $\int \frac{dx}{x^2 + 2x + 2}$.

Solution

Completing the square in the denominator leads to

$$\int \frac{dx}{x^2 + 2x + 2} = \int \frac{dx}{(x^2 + 2x + 1) + 1} = \int \frac{dx}{(x + 1)^2 + 1}.$$

If $u = x + 1$, then $du = dx$ and

$$\int \frac{dx}{x^2 + 2x + 2} = \int \frac{dx}{(x + 1)^2 + 1} = \int \frac{du}{u^2 + 1} = \arctan u + C.$$

Thus,

$$\int \frac{dx}{x^2 + 2x + 2} = \arctan (x + 1) + C. \quad \blacktriangleright$$

Example 8

A night watchman wants the greatest possible view of the second-story windows of a building as he makes his rounds. The windows are 4 ft high and begin 11 ft from the ground. Assume the watchman's eyes are precisely 6 ft from the ground. How far from the building should the watchman walk in order to maximize the angle of view, that is, the angle θ shown in Figure 6.20?

Solution

From the triangles in Figure 6.20, we see that

$$\tan \theta_1 = \frac{9}{x} \quad \text{and} \quad \tan \theta_2 = \frac{5}{x}.$$

Thus,

$$\tan \theta = \tan (\theta_1 - \theta_2) = \frac{\tan \theta_1 - \tan \theta_2}{1 + \tan \theta_1 \tan \theta_2} = \frac{\dfrac{9}{x} - \dfrac{5}{x}}{1 + \dfrac{9}{x} \cdot \dfrac{5}{x}} = \frac{4x}{x^2 + 45}.$$

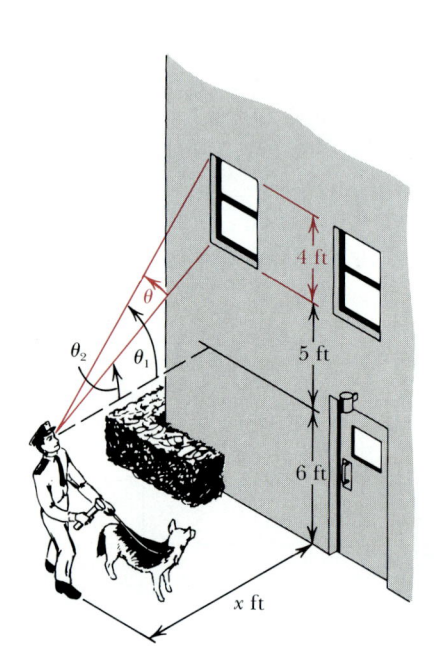

Figure 6.20

Consequently,

$$\theta(x) = \arctan\left(\frac{4x}{x^2 + 45}\right), \quad \text{where} \quad 0 < x < \infty.$$

Since

$$\theta'(x) = \frac{1}{1 + \left(\frac{4x}{x^2 + 45}\right)^2} D_x\left(\frac{4x}{x^2 + 45}\right)$$

$$= \frac{(x^2 + 45)^2}{(x^2 + 45)^2 + 16x^2} \cdot \frac{4(x^2 + 45) - 4x(2x)}{(x^2 + 45)^2}$$

$$= \frac{4(45 - x^2)}{(x^2 + 45)^2 + 16x^2},$$

the only critical point is $x = \sqrt{45}$. When $0 < x < \sqrt{45}$, $\theta'(x) > 0$; and when $x > \sqrt{45}$, $\theta'(x) < 0$. A relative and absolute maximum occurs at $x = \sqrt{45}$, so the optimal distance from the wall for the watchman to walk is

$$x = \sqrt{45} \approx 6.71 \text{ ft.} \quad \blacktriangleright$$

(6.22)
DEFINITION

The **arcsecant** function, denoted **arcsec,** is the function with domain $(-\infty, -1] \cup [1, \infty)$ and range $[0, \pi/2) \cup (\pi/2, \pi]$ defined by

arcsec (sec x) = x,   for every $x \neq \pi/2$ in $[0, \pi]$.

The graphs of the arcsecant and restricted secant functions are shown in Figure 6.21.

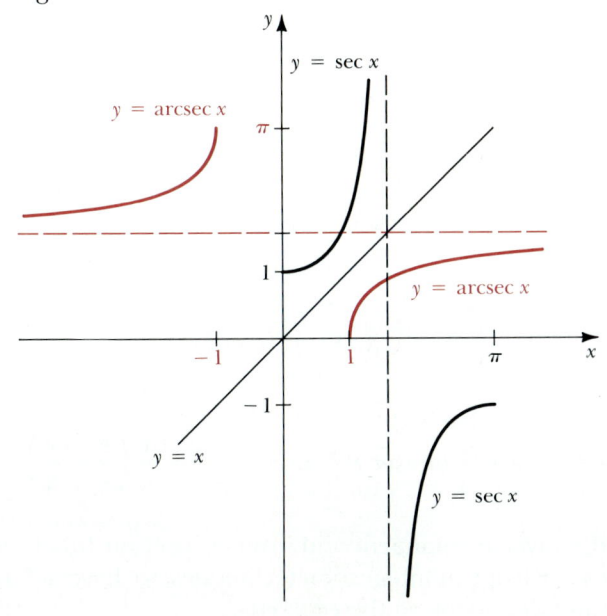

Figure 6.21

To find the derivative of the arcsecant, let $y = \operatorname{arcsec} x$. Then $x = \sec y$, where y is in $[0, \pi]$, but $y \neq \pi/2$. This implies that

$$\cos y = \frac{1}{\sec y} = \frac{1}{x}, \quad \text{and} \quad y = \arccos\left(\frac{1}{x}\right).$$

Thus,

(6.23) $$\operatorname{arcsec} x = \arccos\left(\frac{1}{x}\right), \text{ if } |x| \geq 1.$$

The derivative of the arcsecant follows easily from this equation:

$$D_x \operatorname{arcsec} x = D_x \arccos\left(\frac{1}{x}\right) = -\frac{1}{\sqrt{1 - \frac{1}{x^2}}}\left(-\frac{1}{x^2}\right),$$

so

$$D_x \operatorname{arcsec} x = \frac{1}{\frac{x^2}{\sqrt{x^2}}\sqrt{x^2 - 1}} = \frac{1}{\sqrt{x^2}\sqrt{x^2 - 1}}.$$

Since $|x| = \sqrt{x^2}$, the derivative of the arcsecant is

(6.24) $$D_x \operatorname{arcsec} x = \frac{1}{|x|\sqrt{x^2 - 1}}, \quad \text{if } |x| > 1.$$

The integral formula associated with the inverse secant function is generally expressed in the form

(6.25) $$\int \frac{dx}{x\sqrt{x^2 - a^2}} = \frac{1}{a} \operatorname{arcsec} \frac{|x|}{a} + C, \quad \text{provided } |x| > a > 0.$$

This formula can be verified by differentiation.

Example 9

Find $\int_{3\sqrt{2}}^{6} \frac{dx}{x\sqrt{x^2 - 9}}$.

Solution

We have
$$\int \frac{dx}{x\sqrt{x^2 - 9}} = \frac{1}{3} \operatorname{arcsec} \frac{|x|}{3} + C,$$

so

$$\int_{3\sqrt{2}}^{6} \frac{dx}{x\sqrt{x^2 - 9}} = \frac{1}{3}(\operatorname{arcsec} 2 - \operatorname{arcsec} \sqrt{2}) = \frac{1}{3}\left(\frac{\pi}{3} - \frac{\pi}{4}\right) = \frac{\pi}{36}. \quad \blacktriangleright$$

Since the inverse cotangent and inverse cosecant functions do not add to our knowledge of integration techniques, we have deferred their definition and discussion to the exercises.

EXERCISE SET 6.2

In Exercise 1–18, find the exact value of the expression given.

1. $\arccos\left(\dfrac{\sqrt{2}}{2}\right)$
2. $\arcsin(1)$
3. $\arcsin\left(\dfrac{\sqrt{3}}{2}\right)$
4. $\arccos\left(-\dfrac{\sqrt{2}}{2}\right)$
5. $\arctan(1)$
6. $\operatorname{arcsec}\left(\dfrac{2\sqrt{3}}{3}\right)$
7. $\operatorname{arcsec}(-1)$
8. $\arctan(\sqrt{3})$
9. $\sin\left[\arccos\left(\dfrac{5}{13}\right)\right]$
10. $\cos\left[\arcsin\left(\dfrac{1}{3}\right)\right]$
11. $\tan[\arccos(1)]$
12. $\sec\left[\arcsin\left(-\dfrac{\sqrt{2}}{2}\right)\right]$
13. $\arcsin\left[\sin\left(\dfrac{\pi}{4}\right)\right]$
14. $\arccos\left[\cos\left(\dfrac{\pi}{2}\right)\right]$
15. $\cos[\operatorname{arcsec} 2 + \arctan 2]$
16. $\tan\left[\arctan 2 + \arctan\left(\dfrac{1}{4}\right)\right]$
17. $\sin\left[\arccos\left(\dfrac{1}{2}\right) + \arctan\left(\dfrac{3}{4}\right)\right]$
18. $\csc\left[\arcsin\left(\dfrac{\sqrt{3}}{2}\right) + \arctan\sqrt{3}\right]$

19. a) Explain why $\arccos\left[\cos\left(\dfrac{4\pi}{3}\right)\right] \neq \dfrac{4\pi}{3}$.
    b) Find $\arccos\left[\cos\left(\dfrac{4\pi}{3}\right)\right]$.

20. a) Explain why $\arcsin\left[\sin\left(\dfrac{3\pi}{4}\right)\right] \neq \dfrac{3\pi}{4}$.
    b) Find $\arcsin\left[\sin\left(\dfrac{3\pi}{4}\right)\right]$.

21. If $y = \arcsin(1)$, find each of the following.
    a) $\sin y$ b) $\cos y$
    c) $\sec y$ d) $\cot y$

22. If $y = \arccos\left(\dfrac{1}{2}\right)$, find each of the following.
    a) $\sin y$ b) $\sec y$
    c) $\tan y$ d) $\csc y$

23. If $y = \arctan(-1)$, find each of the following.
    a) $\sin y$ b) $\cos y$
    c) $\cot y$ d) $\csc y$

24. If $y = \operatorname{arcsec}(\sqrt{2})$, find each of the following.
    a) $\cos y$ b) $\csc y$
    c) $\cot y$ d) $\tan y$

In Exercises 25–36, find the derivative of the function.

25. $f(x) = \arcsin(x + 1)$
26. $f(x) = \arccos(x + 1)$
27. $f(x) = \arctan(x^2 - x + 3)$
28. $f(x) = \dfrac{\arcsin(3x + 5)}{x + 2}$
29. $f(x) = \arcsin(x \arccos x)$
30. $f(x) = \dfrac{\arcsin x}{\arccos x}$
31. $f(x) = \sin x \arcsin x$
32. $f(x) = \sin(\arcsin x)$
33. $f(x) = \arcsin\left[\dfrac{(x^3 + 7)}{(x^2 - x)}\right]$
34. $f(x) = \sqrt{\arctan \sqrt{x}}$
35. $f(x) = \operatorname{arcsec}(\arctan \sqrt{x} + 1)$
36. $f(x) = \arctan(\operatorname{arcsec} \sqrt{x} + 1)$

In Exercises 37–54, evaluate the integral.

37. $\displaystyle\int \dfrac{dx}{x^2 + 36}$
38. $\displaystyle\int \dfrac{dx}{\sqrt{1 - 9x^2}}$
39. $\displaystyle\int_0^1 \dfrac{dx}{\sqrt{4 - x^2}}$
40. $\displaystyle\int \dfrac{dx}{x\sqrt{x^2 - 16}}$
41. $\displaystyle\int \dfrac{dx}{4x^2 + 9}$
42. $\displaystyle\int \dfrac{dx}{(x + 3)^2 + 1}$
43. $\displaystyle\int \dfrac{dx}{x\sqrt{16x^2 - 1}}$
44. $\displaystyle\int_3^4 \dfrac{dx}{\sqrt{x^4 - x^2}}$
45. $\displaystyle\int \dfrac{dx}{x\sqrt{x^4 - 1}}$
46. $\displaystyle\int \dfrac{2x\,dx}{\sqrt{1 - x^4}}$
47. $\displaystyle\int_{1/4}^{1/2} \dfrac{dx}{\sqrt{x - x^2}}$
48. $\displaystyle\int_2^{\infty} \dfrac{dx}{x\sqrt{x^2 - 1}}$
49. $\displaystyle\int_0^1 \dfrac{x}{x^2 + 1}\,dx$
50. $\displaystyle\int \dfrac{2x + 3}{x^2 + 4}\,dx$
51. $\displaystyle\int \dfrac{5 - x}{\sqrt{25 - x^2}}\,dx$
52. $\displaystyle\int \dfrac{2x + 1}{x^2 + x + 3}\,dx$
53. $\displaystyle\int \dfrac{dx}{x^2 + 4x + 8}$
54. $\displaystyle\int \dfrac{x\,dx}{x^2 + 4x + 8}$

Show that the inverse trigonometric functions satisfy the identities listed in Exercises 55–60.

55. $\arcsin(-x) = -\arcsin x$
56. $\arctan(-x) = -\arctan x$
57. $\arccos(-x) = \pi - \arccos x$
58. $\cos(\arcsin x) = \sqrt{1-x^2}$
59. $\sin(\arccos x) = \cos(\arcsin x)$
60. $\sec(\arctan x) = \sqrt{x^2+1}$

In Exercises 61–68, sketch the graph of the function.

61. $y = 2\arcsin x$
62. $y = \arcsin(2x)$
63. $y = \dfrac{\pi}{2} + \arcsin x$
64. $y = \arccos x - \dfrac{\pi}{2}$
65. $y = -\arctan x$
66. $y = \dfrac{\pi}{2} + \arctan x$
67. $y = \arctan \dfrac{1}{x}$
68. $y = \arctan x + \arctan \dfrac{1}{x}$

In Exercises 69–72, use l'Hôpital's Rule to find the limit.

69. $\lim\limits_{x \to 0} \dfrac{\arctan x}{x}$
70. $\lim\limits_{x \to 0} \dfrac{\arcsin x}{(\pi/2) - \arccos x}$
71. $\lim\limits_{x \to 0^+} \csc x \arcsin x$
72. $\lim\limits_{x \to \infty} \left(\dfrac{\pi}{2} - \arctan x\right) \cot\left(\dfrac{1}{x}\right)$

73. a) Find $D_x \left[\arctan \dfrac{1}{x}\right]$.
   b) Use the result of part (a) and $D_x(\arctan x)$ to show that if $x \neq 0$, then
   $$\arctan x + \arctan \dfrac{1}{x} = C$$
   for some constant C.
   c) Determine the value of C.

74. Verify formula (6.25) by differentiating the right side of the equation.

75. The **arccotangent** function, denoted **arccot**, is defined by $\operatorname{arccot}(\cot x) = x$, for all x satisfying $0 < x < \pi$. Show that:

   a) $\operatorname{arccot} x = \arctan(1/x)$
   b) $D_x \operatorname{arccot} x = -\dfrac{1}{x^2+1}$;
   c) $\displaystyle\int \dfrac{1}{x^2+a^2}\, dx = -\dfrac{1}{a} \operatorname{arccot} \dfrac{x}{a} + C$, if $a \neq 0$.

76. The **arccosecant** function, denoted **arccsc**, is defined by $\operatorname{arccsc}(\csc x) = x$, for all x satisfying $0 < |x| \leq \pi/2$. Show that:

   a) $\operatorname{arccsc} x = \arcsin(1/x)$, if $|x| \geq 1$
   b) $D_x \operatorname{arccsc} x = \dfrac{-1}{|x|\sqrt{x^2-1}}$;
   c) $\displaystyle\int \dfrac{1}{x\sqrt{x^2-a^2}}\, dx = -\dfrac{1}{a} \operatorname{arccsc} \dfrac{|x|}{a} + C$, if $|x| > a > 0$.

77. Show that for all $|x| \geq 1$, $\operatorname{arcsec} x + \operatorname{arccsc} x = \pi/2$.

78. Find the area of the region bounded by the y-axis, the x-axis, and the graph of $f(x) = 1/(x^2+1)$.

79. In high school geometry one is told that the circumference of a circle of radius r is $2\pi r$. Verify the truth of this statement by finding the length of the curve that is the graph of $x^2 + y^2 = r^2$.

80. Prove that if $|ab| \neq 1$, then:

   a) $\arctan a + \arctan b = \arctan\left[\dfrac{a+b}{1-ab}\right]$;
   b) $\arctan a - \arctan b = \arctan\left[\dfrac{a-b}{1+ab}\right]$.

81. Use the result in Exercise 80 to show that:

   a) $\arctan\left(\dfrac{1}{2}\right) + \arctan\left(\dfrac{1}{3}\right) = \dfrac{\pi}{4}$;
   b) $4\arctan\left(\dfrac{1}{5}\right) - \arctan\left(\dfrac{1}{239}\right) = \dfrac{\pi}{4}$.

82. Most scientific calculators provide keys for evaluating the arcsine, arccosine, and arctangent functions. Show that the values of only the arctangent function need be given by showing that if $0 < x < 1$:

   a) $\arccos x = \arctan(\sqrt{1-x^2}/x)$;
   b) $\arcsin x = \pi/2 - \arctan(\sqrt{1-x^2}/x)$.

83. Derive expressions for the arcsecant, arccotangent, and arccosecant functions similar to those given for the arcsine and arccosine in Exercise 82.

84. Exercise 70 in Section 6.1 shows that if f is a differentiable one-to-one function and $y = f(x)$, then
$$\int f^{-1}(y)\, dy = y f^{-1}(y) - \int f(x)\, dx.$$
Use this result to derive integral formulas for (a) the arcsine; (b) the arctangent; (c) the arcsecant.

85. A camera on a tripod is filming the launch of a rocket at a distance of 6 mi from the launch pad. Five seconds after launch, the rocket is 2 mi directly over the pad and rising at the rate of 3 mi/sec (see the figure). At what rate must the camera rotate in order to keep the rocket in view?

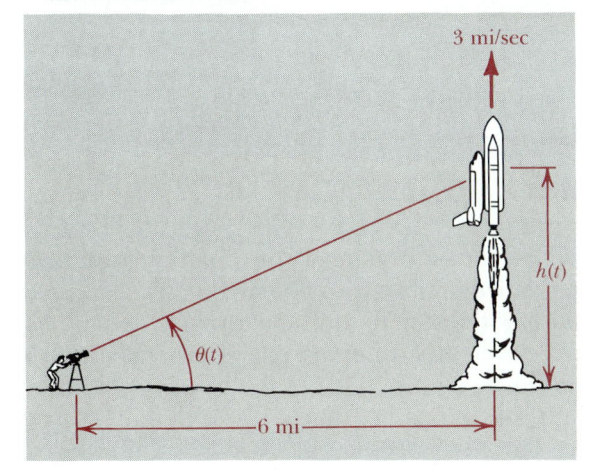

86. A Coast Guard cutter with a revolving searchlight is 100 m from a straight shoreline. The beam of light is moving at the rate of 8000π m/min when the beam passes a point on the shore that is 200 m from the point on the shore closest to the cutter (see the figure). How fast is the searchlight rotating?

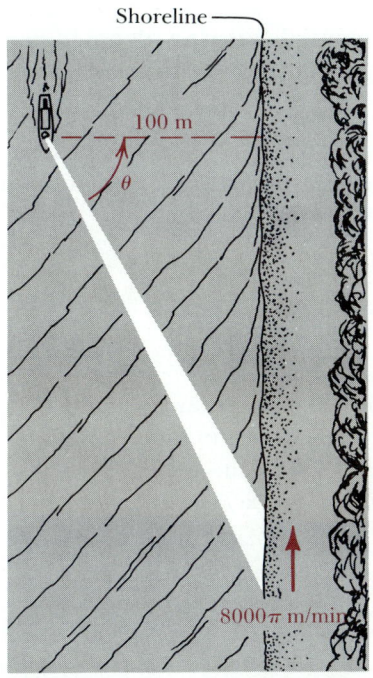

87. A museum plans to hang a rare masterpiece for public viewing. The painting is 5 ft high, and the viewers pass the picture 6 ft from the wall on which the painting will be mounted (see the figure). How high should the painting be mounted to afford the average 5-ft, 6-in. eyelevel viewer the best view? (The best view occurs when the angle formed by the viewer's eyes and the top and bottom of the picture is a maximum.)

88. Suppose the painting discussed in Exercise 87 is mounted with its base 8 ft above the floor. How far should the average viewer stand from the painting to ensure the best view if there is no restriction on the viewer's distance from the painting?

89. Farmer MacDonald plans to use three pieces of straight fencing each 10 ft long to make a pen for pigs along the side of a 120-ft barn (see the figure). The barn is to be one side of the pen; another side is to be parallel to the barn. What angle should the other sides make with the side of the barn in order to maximize the area enclosed?

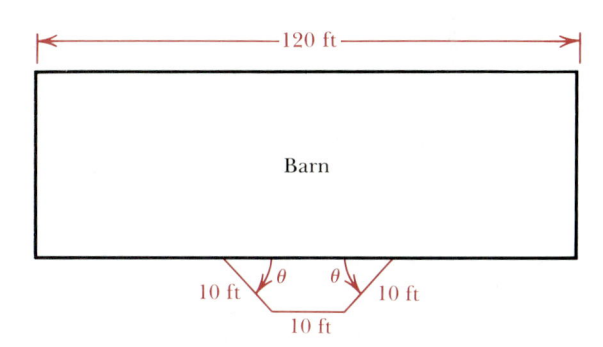

90. Show that if $0 \leq x \leq 1$, (a) $\sin(\cos x) < \cos(\sin x)$; and (b) $\cos(\arcsin x) < \arcsin(\cos x)$.

91. Show that

$$\int_0^{\pi/2} \sin x \, dx + \int_0^1 \arcsin x \, dx = \frac{\pi}{2}$$

without evaluating the integrals.

Putnam exercise

92. Prove

$$\frac{22}{7} - \pi = \int_0^1 \frac{x^4(1-x)^4}{1+x^2} \, dx.$$

(This exercise was problem A–1 of the twenty-ninth William Lowell Putnam examination given on December 7, 1968. The examination and its solution are in the October 1969 issue of the *American Mathematical Monthly*, pp. 911–915.)

6.3
The Natural Exponential Function

In Section 4.6 we saw that the natural logarithm function is increasing and has a nonzero derivative on its entire domain, $(0, \infty)$. The Inverse Function Theorem implies that the natural logarithm function has an inverse that is also increasing and differentiable. This inverse is called the *natural exponential function*.

(6.26)
DEFINITION
The **natural exponential function,** denoted **exp,** is the function with domain $(-\infty, \infty)$ and range $(0, \infty)$ defined by

$$\exp(\ln x) = x, \quad \text{for every } x \text{ in } (0, \infty).$$

Since the inverse of the natural exponential function is the natural logarithm, we also have

(6.27) $\quad \ln(\exp x) = x, \quad$ for every real number x.

Consequently,

(6.28) $\quad y = \exp x \quad$ if and only if $\quad x = \ln y$, where $y > 0$.

The graph of $y = \exp x$ is found by reflecting the graph of $y = \ln x$ about the line $y = x$, as shown in Figure 6.22.

We saw in Section 4.6 that $\ln 1 = 0$, so $\exp 0 = 1$. We also found that there exists a number $e \approx 2.7182818$, with $\ln e = 1$. Thus $\exp 1 = e$. These relations,

(6.29) $\quad \exp 0 = 1 \quad$ and $\quad \exp 1 = e$,

give the points $(0, 1)$ and $(1, e)$ on the graph of the natural exponential function in Figure 6.22. The graph also shows that

(6.30) $\quad \lim_{x \to \infty} \exp x = \infty \quad$ and $\quad \lim_{x \to -\infty} \exp x = 0,$

which is a consequence of $\lim_{x \to \infty} \ln x = \infty$ and $\lim_{x \to 0^+} \ln x = -\infty$.

Before continuing the study of the natural exponential function, we introduce some simplifying notation. This notation requires defining a^x for all real numbers x and $a > 0$. To see the form this definition should assume, let us consider the arithmetic properties of the natural loga-

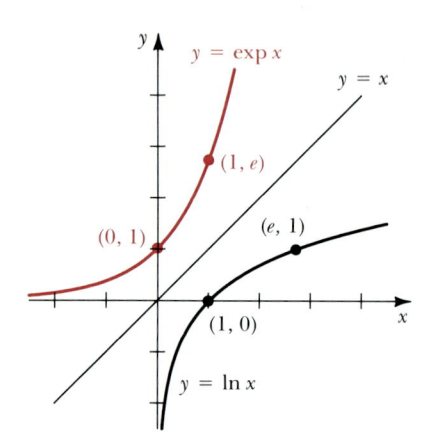

Figure 6.22
The graph of $y = \exp x$ is the reflection about the line $y = x$ of the graph of $y = \ln x$.

rithm function given in Theorem (4.34) of Section 4.6:

(4.34)
THEOREM
If a and b are positive real numbers, then:

i) $\ln ab = \ln a + \ln b$;
ii) $\ln (a/b) = \ln a - \ln b$;
iii) $\ln a^r = r \ln a$, for any rational number r.

It follows from Property (iii) that for any rational number r and any real number $a > 0$,

$$\exp (\ln a^r) = \exp (r \ln a).$$

But the natural exponential and logarithm functions are inverses of each other, so $\exp (\ln a^r) = a^r$ and

$$a^r = \exp (r \ln a).$$

The following definition of a^x extends this result.

(6.31)
DEFINITION
For any real numbers x and $a > 0$, the exponential with base a and power x, denoted a^x, is defined by

$$a^x = \exp (x \ln a).$$

One immediate consequence of this definition is that it extends the result in part (iii) of Theorem 4.34 to

(6.32) $\qquad \ln a^x = \ln (\exp x \ln a) = x \ln a,$

for all real numbers x and $a > 0$. The reason for introducing a^x at this time is the situation when a is the real number e. In this case,

(6.33) $\qquad e^x = \exp (x \ln e) = \exp (x \cdot 1) = \exp x.$

The notation e^x is the most common way of describing the natural exponential function and the one we generally use in the remainder of the text. Keep the original notation in mind, however, since most computer programming languages use EXP to denote this function.

The e^x notation enables us to express the defining relation for the natural exponential function given in Definition (6.26) as

(6.34) $\qquad e^{\ln x} = x, \quad \text{whenever } x > 0.$

In addition, Equation (6.27) becomes

(6.35) $\qquad \ln e^x = x, \quad \text{for every real number } x,$

and Equation (6.28) can be written as

(6.36) $\qquad y = e^x \quad \text{if and only if} \quad x = \ln y, \text{ where } y > 0.$

Arithmetic properties of the natural exponential function can be established using Theorem (4.34) with its extension, given in (6.32), and the inverse relation with the natural logarithm function.

> **(6.37)**
> **THEOREM**
> If x_1 and x_2 are real numbers, then:
>
> **i)** $e^{(x_1+x_2)} = e^{x_1} \cdot e^{x_2}$; **ii)** $e^{(x_1-x_2)} = e^{x_1}/e^{x_2}$; **iii)** $e^{(x_1 x_2)} = (e^{x_1})^{x_2}$.

Proof. First let
$$M = e^{x_1} \quad \text{and} \quad N = e^{x_2}.$$
The inverse relationship with the natural logarithm function implies that
$$x_1 = \ln M \quad \text{and} \quad x_2 = \ln N.$$

i) Using part (i) of Theorem (4.34) gives
$$x_1 + x_2 = \ln M + \ln N = \ln MN,$$
and invoking the inverse relationship again implies that
$$e^{(x_1+x_2)} = MN = e^{x_1} \cdot e^{x_2}.$$

ii) Using part (ii) of Theorem (4.34) gives
$$x_1 - x_2 = \ln M - \ln N = \ln M/N,$$
so
$$e^{(x_1-x_2)} = \frac{M}{N} = \frac{e^{x_1}}{e^{x_2}}.$$

iii) Also, by the extension (6.32) of part (iii) of Theorem (4.34),
$$x_1 x_2 = (\ln M) x_2 = x_2 \ln M = \ln M^{x_2},$$
so
$$e^{(x_1 x_2)} = M^{x_2} = (e^{x_1})^{x_2}. \qquad \triangleright$$

The natural exponential function is differentiable since its inverse, the natural logarithm function, has a derivative that is never zero. To determine the derivative of the natural exponential function, we again apply the derivative relation given in Equation (6.10):
$$D_x e^x = \frac{1}{D_y \ln y}$$
$$= \frac{1}{(1/y)} = y = e^x.$$

Thus,

(6.38) $$D_x e^x = e^x.$$

Applying the Chain Rule to (6.38) gives the following.

(6.39)
THEOREM
If $u = g(x)$ is a differentiable function of x, then
$$D_x e^u = e^u D_x u.$$

Equation (6.38) also tells us that the natural exponential function is its own *antiderivative*, so

(6.40) $$\int e^x \, dx = e^x + C.$$

The importance of the natural exponential function for calculus applications is due to these results. Some of these applications are examined in detail in Sections 6.6 and 6.7.

Example 1

Find $D_x e^{x^2}$.

Solution
$$D_x e^{x^2} = e^{x^2} D_x x^2 = 2xe^{x^2} \quad \blacktriangleright$$

Example 2

Find $D_x e^{x \sin x}$.

Solution
$$D_x e^{x \sin x} = e^{x \sin x} \cdot D_x (x \sin x) = e^{x \sin x} (\sin x + x \cos x) \quad \blacktriangleright$$

Example 3

Find $D_x \sin(e^{x^2})$.

Solution
$$D_x \sin(e^{x^2}) = \cos(e^{x^2}) \cdot D_x e^{x^2} = (\cos(e^{x^2})) \cdot (e^{x^2} \cdot 2x) = 2x \, e^{x^2} \cos(e^{x^2}) \quad \blacktriangleright$$

Example 4

Sketch the graph of $f(x) = xe^{-x}$.

Solution
Since $e^{-x} = 1/e^x > 0$ for all x, $f(x)$ is positive when $x > 0$ and negative when $x < 0$. The only axis intercept occurs at the origin. Since $\lim_{x \to -\infty} e^x = 0$, we have $\lim_{x \to -\infty} e^{-x} = \lim_{x \to -\infty} (1/e^x) = \infty$ and
$$\lim_{x \to -\infty} xe^{-x} = -\infty.$$

However, $xe^{-x} = x/e^x$ has the indeterminate form ∞/∞ as x approaches infinity. In order to find the limit, we use l'Hôpital's Rule:

$$\lim_{x \to \infty} \frac{x}{e^x} = \lim_{x \to \infty} \frac{D_x \, x}{D_x \, e^x} = \lim_{x \to \infty} \frac{1}{e^x} = 0.$$

So the x-axis, $y = 0$, is a horizontal asymptote to the graph.
Since

$$f'(x) = 1 \cdot e^{-x} + x \cdot e^{-x}(-1) = (1-x)e^{-x},$$

we have

$$f'(x) > 0 \quad \text{if } x < 1 \quad \text{and} \quad f'(x) < 0 \quad \text{if } x > 1.$$

Therefore, the graph of f is increasing on $(-\infty, 1]$ and decreasing on $[1, \infty)$. At the critical point $x = 1$, $f(1) = e^{-1} \approx 0.37$.
In addition

$$f''(x) = (-1) \cdot e^{-x} + (1-x) \cdot e^{-x}(-1) = (x-2)e^{-x}.$$

This implies that:

if $x < 2$, $f''(x) < 0$ and the graph is concave downward, and
if $x > 2$, $f''(x) > 0$ and the graph is concave upward.

Thus a point of inflection occurs at $(2, f(2)) = (2, 2e^{-2}) \approx (2, 0.27)$.
The graph of f is shown in Figure 6.23. ▶

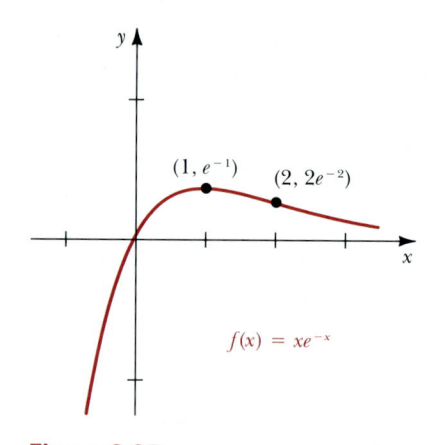

Figure 6.23

In Example 4 we used l'Hôpital's Rule to show that $\lim_{x \to \infty} x/e^x = 0$. By applying this technique repeatedly, we can show that for any positive integer n,

(6.41) $$\lim_{x \to \infty} \frac{x^n}{e^x} = 0.$$

This implies that as x becomes large,

e^x grows faster than any power of x.

The final examples in this section illustrate the integral property of the natural exponential function.

Example 5

Evaluate $\int e^{4x-1} \, dx$.

Solution
Let $u = 4x - 1$. Then $du = 4 \, dx$ and

$$\int e^{4x-1} \, dx = \int e^u \frac{1}{4} \, du = \frac{1}{4} e^u + C = \frac{1}{4} e^{4x-1} + C.$$ ▶

Example 6

Evaluate $\int_0^\pi e^{3\sin x} \cos x \, dx$.

Solution
Let $u = 3 \sin x$. Then $du = 3 \cos x \, dx$ and

$$\int e^{3\sin x} \cos x \, dx = \int e^u \frac{1}{3} du = \frac{1}{3} e^u + C = \frac{1}{3} e^{3\sin x} + C.$$

Thus,

$$\int_0^\pi e^{3\sin x} \cos x \, dx = \frac{1}{3}[e^{3\sin \pi} - e^{3\sin 0}] = \frac{1}{3}[e^0 - e^0] = 0. \quad \blacktriangleright$$

▶ EXERCISE SET 6.3

In Exercises 1–22, find $D_x y$.

1. $y = e^{3x}$
2. $y = e^{x^2-x}$
3. $y = e^{\sqrt{x}}$
4. $y = \sqrt{e^x}$
5. $y = x/(e^x + e^{-x})$
6. $y = (e^x + e^{-x})^2$
7. $y = e^x \ln x$
8. $y = e^{2 \ln x}$
9. $y = e^x \sin x + e^x \cos x$
10. $y = (e^{x^2+1})^{1/2}$
11. $y = e^{\arcsin x}$
12. $y = e^{\arctan x}$
13. $y = e^{\sin x}$
14. $y = \arcsin e^{-x^2}$
15. $y = e^{x^3} \arcsin x$
16. $y = \arctan(e^x + 1)$
17. $y = \ln(e^x)^2$
18. $y = \ln(2e^x)$
19. $ye^x + xe^y = 1$
20. $x^2 e^y + e^x \sin y = 2$
21. $e^{xy} = \ln x$
22. $e^{xy} = e^x e^{-y} + e^y e^{-x}$

In Exercises 23–36, evaluate the integral.

23. $\int e^{3x-2} \, dx$
24. $\int xe^{x^2} \, dx$
25. $\int_0^1 \frac{e^x + e^{-x}}{2} \, dx$
26. $\int \frac{e^x + 1}{e^{2x}} \, dx$
27. $\int \frac{e^x}{e^x + 1} \, dx$
28. $\int \frac{e^x - e^{-x}}{e^x + e^{-x}} \, dx$
29. $\int e^{\sin x} \cos x \, dx$
30. $\int_0^1 xe^{2-x^2} \, dx$
31. $\int e^{2x} e^{5x} \, dx$
32. $\int \frac{e^x}{e^{2x} + 1} \, dx$
33. $\int \frac{e^x}{\sqrt{1 - e^{2x}}} \, dx$
34. $\int \frac{dx}{\sqrt{e^{2x} - 1}}$
35. $\int \frac{xe^{\arctan x^2}}{1 + x^4} \, dx$
36. $\int \frac{1}{1 + e^{-x}} \, dx$

In Exercises 37–48, sketch the graph of the function.

37. $f(x) = e^{-x}$
38. $f(x) = e^{2x}$
39. $f(x) = e^x + e^{-x}$
40. $f(x) = e^x - e^{-x}$
41. $f(x) = xe^x$
42. $f(x) = e^x/x$
43. $f(x) = x^{-1} \ln x$
44. $f(x) = x \ln x$
45. $f(x) = x + \ln x$
46. $f(x) = x - \ln x$
47. $f(x) = x^2 e^x$
48. $f(x) = x^2 e^{-x}$

In Exercises 49–54:
a) Sketch the graphs of each pair of functions f and g.
b) Determine when $f = g^{-1}$.

49. $f(x) = \ln x^2$, $g(x) = \exp\left(\dfrac{x}{2}\right)$
50. $f(x) = \ln \dfrac{x}{2}$, $g(x) = \exp(2x)$
51. $f(x) = \ln |x|$, $g(x) = \exp |x|$
52. $f(x) = -\ln x$, $g(x) = \exp(-x)$
53. $f(x) = 1 + \ln x$, $g(x) = \exp(x - 1)$
54. $f(x) = 2 \ln x$, $g(x) = \dfrac{1}{2} \exp x$

In Exercises 55–62, use l'Hôpital's Rule to evaluate the limit.

55. $\lim\limits_{x \to 0} \dfrac{e^x - 1}{x}$
56. $\lim\limits_{x \to \infty} \dfrac{e^x}{x^2}$
57. $\lim\limits_{x \to \infty} \dfrac{x^2}{e^x}$
58. $\lim\limits_{x \to \infty} \dfrac{\ln x}{x}$
59. $\lim\limits_{x \to \infty} \dfrac{x^3}{e^{x^2}}$
60. $\lim\limits_{x \to 0} \dfrac{e^x - 1 - x}{\sin^2 x}$
61. $\lim\limits_{x \to -\infty} (e^{-x} + x)$
62. $\lim\limits_{x \to 1^+} \left(\dfrac{1}{e^{x-1} - 1} - \dfrac{1}{\ln x}\right)$

63. Find an equation of the tangent line to the graph of $f(x) = e^x$ at $(1, e)$.

64. Find an equation of the tangent line to the graph of $f(x) = e^{2x}$ that passes through $(2, 0)$.

65. Find the area of the region in the plane bounded by the x-axis, the y-axis, the line $x = 1$, and the graph of $y = e^x$.

66. Find the area of the region bounded by the graphs of $x = 0$, $y = e^x$, and $y = e^{2x} - 2$.

67. Find the area of the region bounded by the x-axis, the y-axis, the line $x = a$, and the graph of $f(x) = e^x/(e^{2x} + 1)$.

68. Show that the area of the region bounded by the x-axis, the lines $x = -a$ and $x = a$, and the graph of $f(x) = e^{|x|}/(e^{2|x|} + 1)$ is twice the area of the region found in Exercise 67.

69. Find the volume of the solid generated when the region bounded by the graph of $f(x) = e^{-x^2}$, the lines $x = 0$ and $x = 1$, and the x-axis is revolved about the y-axis. Attempt, for a short while, to find the area of this region.

70. Use the derivative of the function described by $f(x) = e^x - \ln x$ to demonstrate that the graphs of $y = \ln x$ and $y = e^x$ cannot intersect.

71. Does the graph of $f(x) = e^x(1 - x) + x^2$ cross the x-axis between $x = 0$ and $x = 1$?

72. The Laplace transform of a continuous function f, described in the variable t, is a new function, described in the variable s, defined by

$$L\{f\}(s) = F(s) = \int_0^\infty e^{-st} f(t)\, dt,$$

provided this improper integral exists. Show that if the Laplace transforms of f and g exist and if a and b are constants, then

$$L\{af + bg\} = aL\{f\} + bL\{g\}.$$

73. Use the definition of the Laplace transform given in Exercise 72 to show that for any constant a, $L(e^{at}) = 1/(s - a)$.

74. The first study of how the time rate of a chemical reaction changes with respect to temperature was made by Svante August Arrhenius (1859–1927) in 1889. It is generally assumed that the time rate k has the form

$$k = ze^{-Q/RT},$$

where T is the temperature, Q is the activation energy of the chemical reaction, R is the gas constant of the chemical substance, and z is a constant. Show that k satisfies the equation

$$\frac{d \ln k}{dT} = \frac{Q}{RT^2}.$$

75. Only a small proportion of seedlings planted on a piece of bare ground reach maturity. In an experiment with foxglove (*Digitalis purpurea*) plants, the number y surviving at time t (measured in months from the emergence of the seedlings) was found to be

$$y = 100e^{-0.2310t}.$$

a) What was the original number of foxglove seedlings?

b) What was the half-life of this group of seedlings? (That is, the time when half the original number of seedlings have died.)

c) If it takes 15 months for the seedlings to flower, how many seedlings are likely to survive to flower?

76. The number n of fruit flies (*Drosophila melanogaster*) in a colony after t days of breeding is

$$n(t) = \frac{230}{1 + 6.9310e^{-0.1702t}}.$$

a) What is the maximum number of flies in the colony?

b) How many flies were originally in the colony?

c) When would the number of flies be twice the original number?

77. The concentration $C(t)$ of a drug in the bloodstream of a patient is described by

$$C(t) = 0.03te^{-0.01t} \text{ mg/cm}^3,$$

where t is measured in minutes after an injection. Use Newton's Method to determine, to the nearest second, when the concentration is 0.001 mg/cm^3.

78. A drug is infused into the circulation system of a cat at time $t = 0$. The concentration $C(t)$ of the drug for $t > 0$ is given by

$$C(t) = c_1 e^{-k_1 t} + c_2 e^{-k_2 t},$$

where k_1, k_2, c_1, and c_2 are positive constants. Suppose the volume V of the circulating fluid is constant. Determine the rate at which the amount of the drug is changing with respect to time and the limit of this rate as t approaches infinity.

79. Sketch the graph of $f(x) = e^{1/x}$, and find the largest intervals on which this function is differentiable. (Differentiability on an interval is discussed in Exercises 47–52 of Section 2.2.)

80. Suppose the quadratic equation $ar^2 + br + c = 0$ has

two distinct real roots r_1 and r_2. Show that for any constants c_1 and c_2, $y = c_1 e^{r_1 x} + c_2 e^{r_2 x}$ satisfies $ay'' + by' + cy = 0$.

81. Functions of the form $f(x) = e^{ax} \cos bx$ arise frequently in the study of harmonic motion. (Some simple problems involving this motion are considered in Chapter 17.) Show that if $y = e^{ax} \cos bx$, then $y'' - 2ay' + (a^2 + b^2)y = 0$.

82. Use the graphs of $y = e^{-x}$ and $y = \cos x$, together with the fact that $|\cos x| \leq 1$, to produce a sketch of the graph of $y = e^{-x} \cos x$.

83. Use the relative maximum of the function $f(x) = xe^{-x}$ to show that for all positive real numbers x, $\ln x \leq x - 1$.

84. The *geometric mean* (GM) of the positive real numbers $a_1, a_2, \ldots, a_n$ is defined as $GM = (a_1 \cdot a_2 \cdot \cdots \cdot a_n)^{1/n}$, and the *arithmetic mean* (AM) is defined as $AM = (\sum_{i=1}^{n} a_i)/n$. Apply the inequality found in Exercise 83 with $x = a_i/AM$ for each $i = 1, 2, \ldots, n$, and show that the geometric mean never exceeds the arithmetic mean.

Putnam exercises

85. Let C be the class of all real-valued continuously differentiable functions f on the interval $0 \leq x \leq 1$ with $f(0) = 0$ and $f(1) = 1$. Determine the largest real number u such that

$$u \leq \int_0^1 |f'(x) - f(x)| \, dx$$

for all f in C. (This exercise was problem A–6 of the forty-first William Lowell Putnam examination given on December 6, 1980. The examination and its solution are in the October 1981 issue of the *American Mathematical Monthly*, pp. 605–612.)

86. Evaluate $\int_0^\infty t^{-1/2} e^{-1985(t+t^{-1})} \, dt$. You may assume that $\int_{-\infty}^\infty e^{-x^2} \, dx = \sqrt{\pi}$. (This exercise was problem B–5 of the forty-sixth William Lowell Putnam examination given on December 7, 1985. The examination and its solution are in the October 1986 issue of the *American Mathematical Monthly*, pp. 620–626.)

6.4
General Exponential and Logarithm Functions

The exponential with base $a > 0$ and power x was defined in Section 6.3 as

$$a^x = \exp(x \ln a)$$

to provide a convenient way to represent the natural exponential function:

$$\exp x = e^x.$$

When $a \neq 1$, $f(x) = a^x$ describes a function with the same domain and range as those of the natural exponential function.

(6.42)
DEFINITION
For a fixed positive real number $a \neq 1$, the function with domain $(-\infty, \infty)$ and range $(0, \infty)$, defined by

$$a^x = \exp(x \ln a) = e^{x \ln a},$$

is called the **exponential function with base a**.

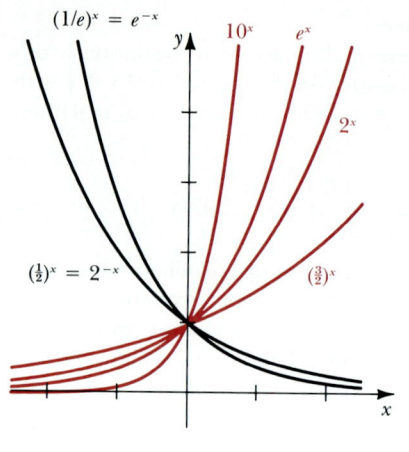

Figure 6.24

In this section we consider the properties of general exponential functions and those of their inverses, the general logarithm functions.

Figure 6.24 shows the graphs of some general exponential functions. When $a > 1$, $\ln a > 0$ and the graph of a^x has a form similar to the graph of e^x. When $0 < a < 1$, $\ln a < 0$, so the graph of a^x assumes a form similar to the graph of $e^{-x} = 1/e^x$. The exponential function is not considered when $a = 1$, since in this case a^x describes a constant function $f(x) = 1^x \equiv 1$.

The following theorem lists common rules for exponents that are discussed in algebra courses. They are now extended to include all real numbers as exponents.

(6.43)
THEOREM

If x_1, x_2, a, and b are real numbers with $a > 0$ and $b > 0$, then:

i) $a^{x_1+x_2} = a^{x_1} a^{x_2}$;   ii) $a^{x_1-x_2} = \dfrac{a^{x_1}}{a^{x_2}}$;

iii) $a^{x_1 x_2} = (a^{x_1})^{x_2}$;   iv) $(ab)^{x_1} = a^{x_1} b^{x_1}$.

Proof. Since $e^{x_1+x_2} = e^{x_1} \cdot e^{x_2}$ for every pair of real numbers, we have

$$a^{x_1+x_2} = e^{[(x_1+x_2)\ln a]} = e^{(x_1 \ln a + x_2 \ln a)} = e^{(x_1 \ln a)} e^{(x_2 \ln a)} = a^{x_1} a^{x_2}.$$

The proofs of the other parts are similar and are considered in Exercise 53.   ▷

The derivative and integral results for the natural exponential function lead to similar information about the general exponential functions.

For any positive real number $a \neq 0$,

$$D_x a^x = D_x e^{x \ln a} = e^{x \ln a} D_x (x \ln a) = e^{x \ln a} \ln a,$$

so

(6.44)
$$D_x a^x = a^x \ln a.$$

Since $\ln a$ is a constant, this also implies that

$$D_x \frac{a^x}{\ln a} = \frac{1}{\ln a} D_x a^x$$

$$= \frac{1}{\ln a} (a^x \ln a) = a^x.$$

So

(6.45)
$$\int a^x \, dx = \frac{a^x}{\ln a} + C.$$

Example 1

Find each of the following.

a) $D_x [\sec^2 x \cdot 2^{\tan x}]$

b) $\displaystyle\int \sec^2 x \cdot 2^{\tan x} \, dx.$

Solution

a) $D_x [\sec^2 x \cdot 2^{\tan x}] = \sec^2 x \, D_x \, 2^{\tan x} + 2^{\tan x} \, D_x \sec^2 x$

$\qquad = \sec^2 x \, [2^{\tan x} \ln 2 \, D_x \tan x]$
$\qquad \quad + 2^{\tan x} \, [2 \sec x \, D_x \sec x]$

$\qquad = \sec^2 x \, [2^{\tan x} (\ln 2 \, \sec^2 x)]$
$\qquad \quad + 2^{\tan x} \, [2 \sec x \, (\sec x \tan x)]$

$\qquad = \sec^2 x \cdot 2^{\tan x} \, [(\ln 2) \sec^2 x + 2 \tan x]$

b) Let $u = \tan x$. Then $du = \sec^2 x \, dx$ and

$$\int \sec^2 x \cdot 2^{\tan x} \, dx = \int 2^u \, du = \frac{2^u}{\ln 2} + C = \frac{2^{\tan x}}{\ln 2} + C. \quad \blacktriangleright$$

In Chapter 2, a series of results showed that the Power Rule, $D_x x^r = r x^{r-1}$, holds whenever r is a rational number and $x > 0$. Since $x^r = \exp(r \ln x) = e^{r \ln x}$ is now defined for any real number r, provided $x > 0$, the Power Rule can be extended further.

(6.46)
THEOREM: THE GENERAL POWER RULE
For any real number r,
$$D_x x^r = r x^{r-1}, \quad \text{for every } x > 0.$$

Proof. Since $x^r = e^{r \ln x}$,

$$D_x x^r = D_x e^{r \ln x} = e^{r \ln x} [D_x (r \ln x)] = x^r \left[r \left(\frac{1}{x} \right) \right].$$

Thus,
$$D_x x^r = r x^{r-1}. \quad \triangleright$$

Example 2

Find $D_x \, x^{\sqrt{2}}$.

Solution
$$D_x \, x^{\sqrt{2}} = \sqrt{2} \, x^{\sqrt{2} - 1} \quad \blacktriangleright$$

Example 3

Find $D_x (\sin x)^\pi$.

Solution
Using the Chain Rule, we obtain
$$D_x (\sin x)^\pi = \pi (\sin x)^{(\pi - 1)} D_x \sin x = \pi (\sin x)^{(\pi - 1)} \cos x. \quad \blacktriangleright$$

Definition (6.42) also enables us to define functions of the form $[f(x)]^{g(x)}$, provided $f(x) > 0$, as

(6.47) $\qquad\qquad [f(x)]^{g(x)} = e^{g(x) \ln f(x)}.$

Example 4

Find $D_x\, x^{\cos x}$, when $x > 0$.

Solution

The function defined by $x^{\cos x}$ is neither a power function nor an exponential function, since the power and base both vary with x. However, by (6.47),

$$x^{\cos x} = e^{\cos x \ln x},$$

so

$$D_x\, x^{\cos x} = D_x\, e^{\cos x \ln x} = e^{\cos x \ln x}\, [D_x\, (\cos x \ln x)].$$

Thus,

$$D_x\, x^{\cos x} = x^{\cos x} \left[(-\sin x) \ln x + (\cos x) \left(\frac{1}{x}\right) \right].$$

General Logarithm Functions

The exponential function a^x is an increasing function if $a > 1$, and is a decreasing function if $0 < a < 1$ (see Figure 6.24 on page 414). In either case the exponential function has domain $(-\infty, \infty)$ and range $(0, \infty)$ and is differentiable for all x. By the Inverse Function Theorem the general exponential function has a differentiable inverse. This inverse is called the *logarithm function with base a*.

(6.48)
DEFINITION

The **logarithm function** with base the positive real number $a \neq 1$, denoted $\log_a$, is the function with domain $(0, \infty)$ and range $(-\infty, \infty)$ defined by

$$\log_a (a^x) = x, \quad \text{for every real number } x.$$

For $a > 1$, the graph of the logarithm function with base a assumes a form such as that shown in Figure 6.25.

Since the exponential function with base a is the inverse of the logarithm function with base a, we also have

(6.49) $\qquad a^{\log_a x} = x, \quad \text{for every } x > 0,$

and

(6.50) $\quad \log_a x = y, \quad \text{where } x > 0 \quad \text{if and only if} \quad a^y = x.$

One immediate consequence of Definition (6.48) is that

(6.51) $\qquad \log_e x = \ln x, \quad \text{for every } x > 0.$

Thus the natural logarithm function is simply the logarithm function with base e.

The special case of the general logarithm that occurs when $a = 10$ is

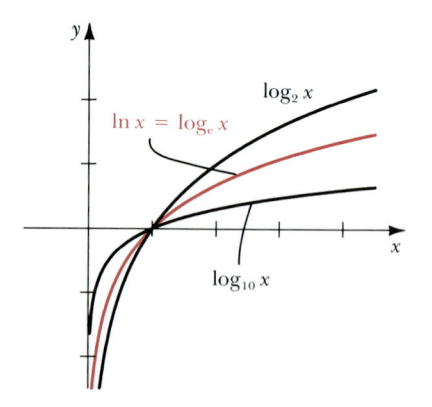

Figure 6.25

often called the **common logarithm** and is often denoted with the base deleted, that is, $\log x \equiv \log_{10} x$.

An interesting relationship exists between logarithms with different bases. To establish this identity, suppose a and b are distinct positive numbers and that for a fixed, but arbitrary, positive real number x,

$$M = \log_a x \quad \text{and} \quad N = \log_b x.$$

It follows from (6.50) that $a^M = x$ and $b^N = x$, so

$$a^M = b^N.$$

Thus,

$$M \ln a = \ln a^M = \ln b^N = N \ln b \quad \text{and} \quad M = \frac{\ln b}{\ln a} N.$$

Recalling the definition of M and N yields the identity

(6.52) $$\log_a x = \frac{\ln b}{\ln a} \log_b x.$$

Equation (6.52) is particularly useful when $b = e$. In this case, $\ln b = \ln e = 1$ and $\log_b x = \log_e x = \ln x$, so (6.52) reduces to

(6.53) $$\log_a x = \frac{\ln x}{\ln a}.$$

This relationship implies that every general logarithm function is a multiple of the natural logarithm function and that for any positive $a \neq 1$ and every $x > 0$,

(6.54) $$D_x \log_a x = \frac{1}{x \ln a}.$$

Example 5

Find $D_x \log_{10}(x^2 + 4x + 4)$.

Solution

$$\begin{aligned} D_x \log_{10}(x^2 + 4x + 4) &= D_x \left(\frac{\ln(x^2 + 4x + 4)}{\ln 10} \right) \\ &= \frac{D_x(x^2 + 4x + 4)}{(x^2 + 4x + 4)} \cdot \frac{1}{\ln 10} \\ &= \frac{2x + 4}{(x + 2)^2 \ln 10} = \frac{2}{(x + 2) \ln 10} \quad \blacktriangleright \end{aligned}$$

The arithmetic properties for the general logarithm functions are listed in the following theorem. Because of the relation in (6.54), this theorem follows from Theorem (4.34) in Section 4.6, assuming the extension of part (iii) given in (6.32).

(6.55)
THEOREM

If x_1, x_2, and a are positive real numbers and $a \neq 1$, then:

i) $\log_a (x_1 x_2) = \log_a x_1 + \log_a x_2$; **ii)** $\log_a (x_1/x_2) = \log_a x_1 - \log_a x_2$; **iii)** $\log_a (x_1^{x_2}) = x_2 \log_a x_1$.

We have seen an interesting transition in our study of the logarithmic and exponential functions. In Section 4.6, the natural logarithm function was defined in terms of a definite integral. This led, in Section 6.3, to the natural exponential function, the inverse of the natural logarithm function. Once the natural logarithm and exponential functions were defined, they were used to define the general exponential functions. Each of the general exponential functions has an inverse, which is a general logarithm function.

Figure 6.26 lists the steps in the development of this study and a few of the important results that were obtained. The symbol $\leftrightarrow$ in this diagram is used in place of the phrase "if and only if."

Natural Logarithm Function

$\ln x = \int_1^x \frac{1}{t}\, dt, x > 0$

$D_x \ln x = \frac{1}{x}$

$\int \frac{1}{x}\, dx = \ln |x| + C$

Inverse Function Theorem

Natural Exponential Function

$\ln x = y \leftrightarrow \exp y = x$

$D_x \exp x = \exp x$

$\int \exp x\, dx = \exp x + C$

General Exponential Function

$a^x = \exp (x \ln a)$

$D_x a^x = a^x \ln a$

$\int a^x\, dx = \frac{a^x}{\ln a} + C$

$e^x \equiv \exp x$

Inverse Function Theorem

General Logarithm Function

$a^x = y \leftrightarrow \log_a y = x$

$D_x \log_a x = \frac{1}{x \ln a}$

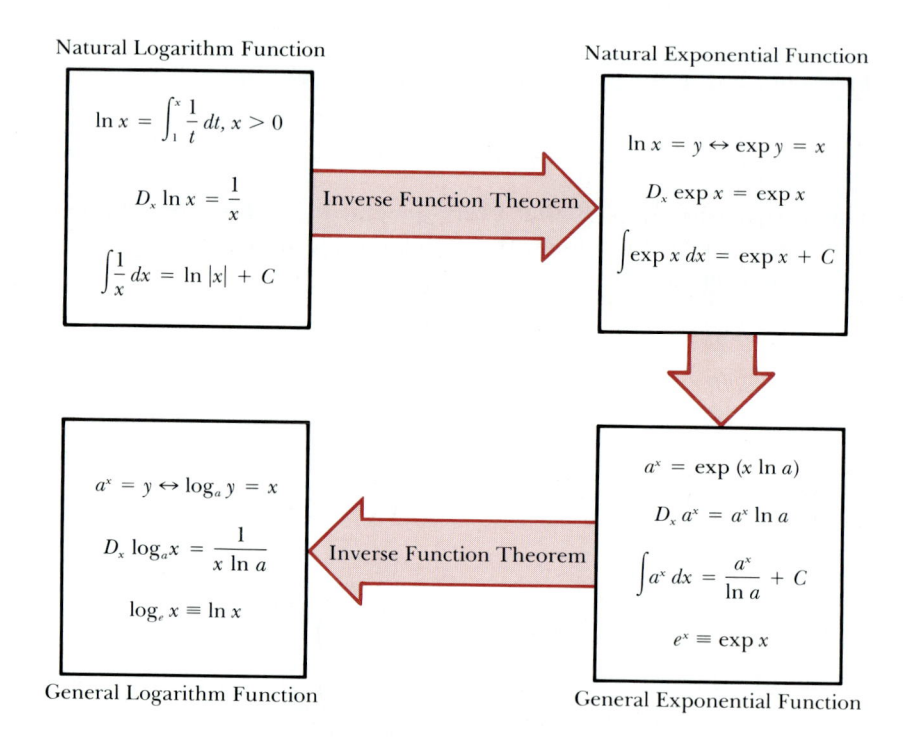

$\log_e x \equiv \ln x$

Figure 6.26

▶ EXERCISE SET 6.4

In Exercises 1–32, find $D_x y$.

1. $y = 2^x$
2. $y = 2^{x^2}$
3. $y = 3^x + 2^{x^2}$
4. $y = 3^x \cdot 2^{x^2}$
5. $y = (\sin x)^e$
6. $y = (\sin \pi x)^\pi$
7. $y = (3 + \pi)^x$
8. $y = x^{3+\pi}$
9. $y = x^{(\pi + \sqrt{2})^2}$
10. $y = (\pi + \sqrt{2})^{x^2}$
11. $y = \log_{10} (x^2 + 1)$
12. $y = \log_3 (x^2 + 2x + 3)$

13. $y = \log_2\left(\dfrac{x^2+1}{x^2+2}\right)$ 14. $y = \log_{10}(\log_{10} x + 3)$

15. $y = \log_{10}(\ln x)$ 16. $y = \ln(\log_{10} x)$

17. $y = \log_2\sqrt{x^2+3x-1}$ 18. $y = \log_2(x^{-1} + 2x^{-3})$

19. $y = \log_4 \dfrac{\sqrt{x^2+2x}}{\sqrt[3]{x^4-5}}$

20. $y = \log_{10}(\sqrt{x^2+3x}\,\sqrt[3]{3x-5})$

21. $y = (x+1)^x$ 22. $y = x^{x+1}$

23. $y = x^{\pi x}$ 24. $y = (\pi x)^x$

25. $y = (\cos x)^x$ 26. $y = (\sin x)^x$

27. $y = x^{\sin x}$ 28. $y = (\sin x)^{\sin x}$

29. $y \log_{10} x = x \log_{10} y$ 30. $\log_{10}(xy) = \log_9(xy)$

31. $x^y + y^x = 1$ 32. $x^y + (\sin y)^x = \pi$

In Exercises 33–38, evaluate the integral.

33. $\displaystyle\int_0^1 3^{2x}\, dx$ 34. $\displaystyle\int 2^{3x}\, dx$

35. $\displaystyle\int x\, 3^{x^2}\, dx$ 36. $\displaystyle\int_1^2 2^x 3^x\, dx$

37. $\displaystyle\int x^e\, dx$ 38. $\displaystyle\int (x \ln 2)^{-1}\, dx$

In Exercises 39–48, sketch the graph of the function.

39. $f(x) = 2^x$ 40. $f(x) = 3^x$

41. $f(x) = 2^x + 3^x$ 42. $f(x) = 3^x - 2^x$

43. $f(x) = \log_2 x$ 44. $f(x) = \log_3 x$

45. $f(x) = \log_3 x - \log_2 x$ 46. $f(x) = x \log_{10} x$

47. $f(x) = \log_2(\log_2 x)$ 48. $f(x) = \dfrac{\log_2 x}{x}$

In Exercises 49–52, use l'Hôpital's Rule to evaluate the limit.

49. $\displaystyle\lim_{x \to 0^+} \dfrac{2^x - x - 1}{x^2}$ 50. $\displaystyle\lim_{x \to \infty} \dfrac{x}{3^x}$

51. $\displaystyle\lim_{x \to \infty} \dfrac{\log_2 x}{x}$ 52. $\displaystyle\lim_{x \to 0^+} x \log_2 x$

53. Prove parts (ii) and (iii) of Theorem (6.43): If x_1, x_2, and a are real numbers with $a > 0$, then:

   ii) $a^{x_1-x_2} = \dfrac{a^{x_1}}{a^{x_2}}$;   iii) $a^{x_1 x_2} = (a^{x_1})^{x_2}$.

54. Suppose $f(x) = x^x$ and $g(x) = f(2x)$. Find $g'(x)$.
55. Suppose $f'(x) = x^x$ and $g(x) = f(2x)$. Find $g'(x)$.
56. Find $D_x \log_x e$.
57. Show that the graphs of $y = \log_a x$ and $y = \log_b x$, for $a \neq b$, intersect only at $(1, 0)$.
58. Suppose $a > 1$ is a constant. Sketch the graphs of $y = \log_a x$ and $y = \log_{1/a} x$ on the same set of axes.

59. Show that if f and g are differentiable and $f(x) > 0$, then
$$D_x f(x)^{g(x)} = [f(x)^{g(x)} \ln f(x)]\, g'(x) + [g(x) f(x)^{g(x)-1}]\, f'(x).$$

60. Interpret the result in Exercise 59:
   a) when f is a nonzero constant function;
   b) when g is a nonzero constant function.

61. Show that a solution to the equation
$$x + \log_a x = a$$
is $x = a^a/c$, where $c^c = a^{a^a}$.

62. The *gamma function* is defined by
$$\Gamma(x) = \int_0^\infty t^{x-1} e^{-t}\, dt \quad \text{for } x > 0.$$

   a) Determine $\Gamma(1)$.
   b) It is shown in Exercise 41 of Section 15.3 that $\int_0^\infty e^{-x^2}\, dx = \sqrt{\pi}/2$. Use this result to show that $\Gamma(\tfrac{1}{2}) = \sqrt{\pi}$.

63. Show that $f(x) = (1 + 1/x)^x$ is an increasing function when x is positive.

Putnam exercises

64. The graph of the equation $x^y = y^x$ in the first quadrant (that is, the region where $x > 0$ and $y > 0$) consists of a straight line and a curve. Find the coordinates of the intersection point of the line and the curve. (This exercise was problem 1, part I of the twenty-second William Lowell Putnam examination given on December 2, 1961. The examination and its solution are in the October 1962 issue of the *American Mathematical Monthly*, pp. 762–767.)

65. Evaluate $\displaystyle\int_0^{\pi/2} \dfrac{dx}{1 + (\tan x)^{\sqrt{2}}}.$

(This exercise was Problem A–3 of the forty-first William Lowell Putnam examination given on December 6, 1980. The examination and its solution are in the October 1981 issue of the *American Mathematical Monthly*, pp. 605–612.)

66. Find, with proof, all real-valued functions $y = g(x)$ defined and *continuous* on $[0, \infty)$, positive on $(0, \infty)$, such that for all $x > 0$ the y-coordinate of the centroid of the region
$$R_x = \{(s, t)\,|\, 0 \le s \le x,\ 0 \le t \le g(s)\}$$
is the same as the average of g on $[0, x]$. (This exercise was Problem B–4 of the forty-fifth William Lowell Putnam examination given on December 1, 1984. The examination and its solution are in the October 1985 issue of the *American Mathematical Monthly*, pp. 560–567.)

6.5

Additional Indeterminate Forms

L'Hôpital's Rule was introduced in Section 3.6 to evaluate limits with indeterminate forms of the type $0/0$ and ∞/∞. These forms appear frequently in problems involving the logarithm, exponential, and inverse trigonometric functions. For example, we used the rule in Section 6.3 to show that the graph of $f(x) = xe^{-x}$ has the horizontal asymptote $y = 0$. In this problem, the limit of $f(x)$ as x approaches infinity is of the form $0 \cdot \infty$, which can be transformed into an indeterminate form of the type ∞/∞. Although this application of l'Hôpital's Rule involves the natural exponential function, it follows the pattern described in Section 3.6.

In this section we first review the process of evaluating an indeterminate form of the type $0 \cdot \infty$, and then show that l'Hôpital's Rule can be applied to indeterminate forms of the type 0^0, 1^∞, and ∞^0 as well.

Example 1

Determine $\lim\limits_{x \to 0^+} x \ln x$.

Solution
Since
$$\lim_{x \to 0^+} x = 0 \quad \text{and} \quad \lim_{x \to 0^+} \ln x = -\infty,$$

the limit in this problem has the indeterminate form $0 \cdot \infty$ as x approaches zero from the right. To use l'Hôpital's Rule we need to convert the problem into one of the form $0/0$ or ∞/∞. The derivative of $1/x$ is easily found and is not complicated, so we write

$$\lim_{x \to 0^+} x \ln x = \lim_{x \to 0^+} \frac{\ln x}{(1/x)}$$

and apply l'Hôpital's Rule to this ∞/∞ indeterminate form:

$$\lim_{x \to 0^+} x \ln x = \lim_{x \to 0^+} \frac{\ln x}{(1/x)} = \lim_{x \to 0^+} \frac{D_x \ln x}{D_x (1/x)}$$
$$= \lim_{x \to 0^+} \frac{(1/x)}{(-1/x^2)} = -\lim_{x \to 0^+} x = 0. \quad \blacktriangleright$$

Example 2

Determine $\lim\limits_{x \to \infty} x(2^{1/x} - 1)$.

Solution
Since
$$\lim_{x \to \infty} x = \infty \quad \text{and} \quad \lim_{x \to \infty} (2^{1/x} - 1) = 0,$$

the limit has the indeterminate form $0 \cdot \infty$ as x approaches infinity. To use l'Hôpital's Rule, we change x to $1/(1/x)$ and rewrite the problem in the indeterminate form $0/0$:

$$\lim_{x \to \infty} x(2^{1/x} - 1) = \lim_{x \to \infty} \frac{2^{1/x} - 1}{(1/x)}.$$

Since

$$D_x (2^{1/x} - 1) = 2^{1/x} \ln 2 \, D_x (1/x),$$

we have

$$\lim_{x \to \infty} x(2^{1/x} - 1) = \lim_{x \to \infty} \frac{D_x (2^{1/x} - 1)}{D_x (1/x)} = \lim_{x \to \infty} 2^{1/x} \ln 2 = \ln 2. \quad \blacktriangleright$$

Functions written in the form $[f(x)]^{g(x)}$ can have indeterminate forms of the type 0^0, ∞^0, or 1^∞. For example, the function defined by $h(x) = x^x$ has the indeterminate form 0^0 as x approaches zero from the right, since both the base and the exponent approach zero. For a positive number x, $0^x = 0$ and $x^0 = 1$, so it is not clear what will happen to x^x as x approaches zero.

To evaluate a limit with an indeterminate form of the type ∞^0, 0^0, or 1^∞, we use the relation given in (6.47):

$$[f(x)]^{g(x)} = e^{g(x) \ln f(x)}.$$

Since the natural exponential function is continuous,

(6.56) $$\lim_{x \to a} [f(x)]^{g(x)} = e^{\lim_{x \to a} g(x) \ln f(x)},$$

and we can easily find $\lim_{x \to a} [f(x)]^{g(x)}$ once we know $\lim_{x \to a} g(x) \ln f(x)$. The following examples show how this technique is used to evaluate indeterminate forms of the type 0^0, ∞^0, and 1^∞.

Example 3

Find $\lim\limits_{x \to 0^+} x^x$.

Solution

This problem has the indeterminate form 0^0 as x approaches zero from the right. However,

$$x^x = e^{x \ln x},$$

and $x \ln x$ has the indeterminate form $0 \cdot \infty$ at zero. In Example 1 we found that $\lim_{x \to 0^+} x \ln x = 0$, so

$$\lim_{x \to 0^+} x^x = \lim_{x \to 0^+} e^{x \ln x} = e^{\lim_{x \to 0^+} x \ln x} = e^0 = 1. \quad \blacktriangleright$$

Example 4

Find $\lim\limits_{x \to \infty} x^{1/x}$.

Solution

Since
$$\lim_{x \to \infty} x = \infty \quad \text{and} \quad \lim_{x \to \infty} 1/x = 0,$$

this problem has the indeterminate form ∞^0 at infinity. It follows from (6.56) that
$$\lim_{x \to \infty} x^{1/x} = \lim_{x \to \infty} e^{\frac{1}{x} \ln x} = e^{\lim_{x \to \infty} \left(\frac{1}{x} \ln x\right)}.$$

Since $(1/x) \ln x$ has an indeterminate form ∞/∞ at infinity, we apply l'Hôpital's Rule:
$$\lim_{x \to \infty} \frac{\ln x}{x} = \lim_{x \to \infty} \frac{D_x \ln x}{D_x x} = \lim_{x \to \infty} \frac{1/x}{1} = 0.$$

Consequently,
$$\lim_{x \to \infty} x^{1/x} = e^{\lim_{x \to \infty} \left(\frac{1}{x} \ln x\right)} = e^0 = 1. \quad \blacktriangleright$$

Example 5

Find $\lim_{x \to 0^+} (e^x - 1)^x$.

Solution

Since
$$\lim_{x \to 0^+} (e^x - 1) = 0 \quad \text{and} \quad \lim_{x \to 0^+} x = 0,$$

this problem has the indeterminate form 0^0 as x approaches zero from the right. We first use (6.56) to write
$$\lim_{x \to 0^+} (e^x - 1)^x = e^{\lim_{x \to 0^+} [x \ln (e^x - 1)]}.$$

Note that $x \ln (e^x - 1)$ has the indeterminate form $0 \cdot \infty$ and can be rewritten as
$$\frac{\ln (e^x - 1)}{1/x}.$$

We now apply l'Hôpital's Rule to this ∞/∞ indeterminate form:
$$\lim_{x \to 0^+} x \ln (e^x - 1) = \lim_{x \to 0^+} \frac{\ln (e^x - 1)}{1/x} = \lim_{x \to 0^+} \frac{D_x \ln (e^x - 1)}{D_x (1/x)},$$

so
$$\lim_{x \to 0^+} x \ln (e^x - 1) = \lim_{x \to 0^+} \frac{\left(\frac{1}{e^x - 1}\right) e^x}{(-1/x^2)} = \lim_{x \to 0^+} \frac{-e^x x^2}{e^x - 1}.$$

This has the indeterminate form $0/0$, so we apply l'Hôpital's Rule again:
$$\lim_{x \to 0^+} x \ln (e^x - 1) = \lim_{x \to 0^+} \frac{D_x (-e^x x^2)}{D_x (e^x - 1)} = \lim_{x \to 0^+} \frac{-e^x x^2 - 2xe^x}{e^x} = 0.$$

Thus,
$$\lim_{x\to 0^+} (e^x - 1)^x = e^{\lim_{x\to 0^+} [x\ln(e^x-1)]} = e^0 = 1.$$

▶

Example 6

Show that $\lim_{x\to\infty} (1 + 1/x)^x = e$.

Solution
Since
$$\lim_{x\to\infty} (1 + 1/x) = 1 \quad \text{and} \quad \lim_{x\to\infty} x = \infty,$$

$\lim_{x\to\infty} (1 + 1/x)^x$ has the indeterminate form 1^∞ at infinity. We rewrite the limit using (6.56):

$$\lim_{x\to\infty} \left(1 + \frac{1}{x}\right)^x = e^{\lim_{x\to\infty} x \ln(1+(1/x))}$$

Since $x \ln [1 + (1/x)]$ has the indeterminate form $\infty \cdot 0$,

$$\frac{\ln\left(1 + \dfrac{1}{x}\right)}{(1/x)}$$

has the indeterminate form $0/0$. Applying l'Hôpital's Rule to this expression gives

$$\lim_{x\to\infty} x \ln\left(1 + \frac{1}{x}\right) = \lim_{x\to\infty} \frac{\ln\left(1 + \dfrac{1}{x}\right)}{(1/x)} = \lim_{x\to\infty} \frac{D_x \ln\left(1 + \dfrac{1}{x}\right)}{D_x (1/x)},$$

so

$$\lim_{x\to\infty} x \ln\left(1 + \frac{1}{x}\right) = \lim_{x\to\infty} \frac{\left(1 / \left(1 + \dfrac{1}{x}\right)\right)\left(-\dfrac{1}{x^2}\right)}{(-1/x^2)} = \lim_{x\to\infty} \frac{x}{x+1} = 1.$$

Hence,

$$\lim_{x\to\infty} \left(1 + \frac{1}{x}\right)^x = e^{\lim_{x\to\infty} x \ln(1+(1/x))} = e^1 = e.$$

▶

Since

(6.57)
$$e = \lim_{x\to\infty} \left(1 + \frac{1}{x}\right)^x,$$

approximations to e can be found by evaluating $[1 + (1/x)]^x$ for large values of x. Table 6.1 lists some of these approximations. Note that x must be quite large before the approximation produces accurate results. (The correct value of e to the places listed is 2.7182818.)

Table 6.1

x	$\left(1 + \dfrac{1}{x}\right)^x$
100	2.7048138
1,000	2.7169239
10,000	2.7181459
100,000	2.7182682
1,000,000	2.7182805

EXERCISE SET 6.5

In Exercises 1–20, evaluate the limit.

1. $\lim\limits_{x \to 0^+} x^{x^2}$
2. $\lim\limits_{x \to 0^+} (1 + x^2)^{1/x^2}$
3. $\lim\limits_{x \to 0^+} x^{\sin x}$
4. $\lim\limits_{x \to 0^+} (\sin x)^x$
5. $\lim\limits_{x \to 0^+} (x + e^x)^{1/x}$
6. $\lim\limits_{x \to \infty} (x + e^x)^{1/x}$
7. $\lim\limits_{x \to 0^+} (1 + x)^{\ln x}$
8. $\lim\limits_{x \to \infty} \left(\sin \frac{1}{x}\right)^x$
9. $\lim\limits_{x \to 0^+} (x + \sin x)^x$
10. $\lim\limits_{x \to \infty} (x + \sin x)^{-x}$
11. $\lim\limits_{x \to 2^+} (x^2 - 4)^{(x-2)}$
12. $\lim\limits_{x \to 2^+} (x - 2)^{(x^2-4)}$
13. $\lim\limits_{x \to 1^+} (x - 1)^{\ln x}$
14. $\lim\limits_{x \to 1^+} (\ln x)^{(x-1)}$
15. $\lim\limits_{x \to \frac{\pi}{2}^-} (1 + \tan x)^{\cos x}$
16. $\lim\limits_{x \to \frac{\pi}{2}^-} (1 + \cos x)^{\tan x}$
17. $\lim\limits_{x \to \infty} (e^{-x})^{e^x}$
18. $\lim\limits_{x \to \infty} (e^x)^{e^{-x}}$
19. $\lim\limits_{x \to 0^+} (x^x)^x$
20. $\lim\limits_{x \to 0^+} x^{(x^x)}$

In Exercises 21–28, use $\lim\limits_{x \to \infty} [1 + (1/x)]^x = e$ to determine the limit.

21. $\lim\limits_{x \to 0^+} (1 + x)^{1/x}$
22. $\lim\limits_{x \to \infty} \left(1 + \frac{1}{x}\right)^{2x}$
23. $\lim\limits_{x \to \infty} \left(1 + \frac{1}{2x}\right)^x$
24. $\lim\limits_{x \to \infty} \left(1 + \frac{2}{x}\right)^x$
25. $\lim\limits_{x \to -\infty} \left(1 - \frac{1}{x}\right)^x$
26. $\lim\limits_{x \to -\infty} \left(1 - \frac{2}{x}\right)^{-3x}$
27. Sketch the graph of $f(x) = x^x$.
28. Sketch the graph of $f(x) = x^{-x}$.
29. If the interest on an investment $A(0)$ is compounded m times a year and the interest rate is i, the total amount of investment at the end of t years is

$$A(t) = A(0)\left(1 + \frac{i}{m}\right)^{mt}.$$

Find $A(t)$ if the compounding is continuous (that is, if $m \to \infty$).

30. a) Show that for every positive integer n, $\lim\limits_{x \to \infty} (x^n/e^x) = 0$.
    b) Show that $\lim\limits_{x \to \infty} x^x/e^x = \infty$.

31. Show that for any real numbers a and $b > 0$,
    a) $\lim\limits_{x \to \infty} \left(1 + \frac{a}{x}\right)^x = e^a$;
    b) $\lim\limits_{x \to \infty} x(b^{1/x} - 1) = \ln b$;
    c) $\lim\limits_{x \to \infty} \frac{b^{1/x} - 1}{a^{1/x} - 1} = \log_a b$;
    d) $\lim\limits_{x \to \infty} (1 - a + ab^{1/x})^x = b^a$.

32. Exercise 31(a) implies that e^a is approximated by $(1 + a/2^n)^{2^n}$ for positive integers n. Approximate e using this formula with $n = 2, 4, 8,$ and 16.

33. Exercise 31(b) implies that $\ln b$ is approximated by $2^n(b^{(1/2^n)} - 1)$ for positive integers n. Approximate $\ln 2$ using this formula with $n = 2, 4, 8,$ and 16.

34. Determine

$$\lim\limits_{x \to \infty} (x^{x/(x-1)} - x).$$

Putnam exercise

35. Evaluate

$$\lim\limits_{t \to 0^+} \left\{\int_0^1 [bx + a(1 - x)]^t \, dx\right\}^{1/t},$$

where $0 < a < b$ are constants. (This exercise was Problem B–2 of the fortieth William Lowell Putnam examination given on December 1, 1979. The examination and its solution are in the October 1980 issue of the *American Mathematical Monthly*, p. 639.)

6.6

Exponential Growth and Decay

The natural exponential function has application in problems involving population growth, continuous rates of interest, radioactive decay, the change in the dilution or contamination of a liquid, and many other

6.6 EXPONENTIAL GROWTH AND DECAY

subjects. In this section we examine a few of these applications and see that in each case the mathematical problem to be solved is the same:

Find a function y in the variable x satisfying

(6.58) $$y'(x) = ky(x), \quad \text{where } k \text{ is a constant.}$$

To find $y(x)$ we first rewrite (6.58) as

$$\frac{y'(x)}{y(x)} = k$$

and then integrate with respect to x. Since

$$D_x \ln |y(x)| = \frac{y'(x)}{y(x)},$$

we have

$$\ln |y(x)| = \int \frac{y'(x)}{y(x)} \, dx = \int k \, dx = kx + C,$$

for some constant C. Thus,

$$|y(x)| = e^{kx+C} = e^C e^{kx},$$

which implies that

$$y(x) = \pm e^C e^{kx}.$$

But

$$y(0) = \pm e^C e^0 = \pm e^C,$$

so the function satisfying Equation (6.58) is given by

(6.59) $$y(x) = y(0) e^{kx}.$$

Radioactive Decay

The first application we consider involves the decay of a radioactive substance. The rate of decay changes with time and is proportional to the amount of the substance present at that time. If $A(t)$ denotes the amount at time t, then the situation is described by

$$\frac{dA(t)}{dt} = kA(t), \quad \text{where } A(t) > 0.$$

Example 1

Suppose 100 mg of radioactive carbon-14, denoted $_6^{14}C$, is present at time $t = 0$, and only 50 mg remains at $t = 5730$ yr.

a) Find an expression for the amount $A(t)$ present at any time $t \geq 0$.
b) Determine the amount present when $t = 2000$ yr.

Solution

The problem is summarized as shown in the table.

KNOW	FIND
1) $\dfrac{dA(t)}{dt} = kA(t),$	a) $A(t),$
2) $A(0) = 100,$	b) $A(2000).$
3) $A(5730) = 50.$	

Since
$$\frac{dA(t)}{dt} = kA(t),$$
it follows from (6.59) that
$$A(t) = A(0)e^{kt} = 100e^{kt}.$$
To find the constant k, we use (3) from the table:
$$50 = A(5730) = 100e^{5730k}, \quad \text{so} \quad \frac{1}{2} = e^{5730k}.$$
Thus,
$$5730k = \ln\left(\frac{1}{2}\right) = \ln 2^{-1} = -\ln 2 \quad \text{and} \quad k = \frac{-\ln 2}{5730}.$$
The answer to part (a) is
$$A(t) = 100e^{-t(\ln 2)/5730},$$
and the answer to part (b) is
$$A(2000) = 100e^{-2000(\ln 2)/5730} \approx 78.51 \text{ mg}. \quad \blacktriangleright$$

The length of time it takes a radioactive substance to decay to one-half of its original amount is called the **half-life** of the substance. The half-life of the carbon-14 described in Example 1 is 5730 yr. The arithmetic properties of the exponential and logarithm functions provide an alternative expression for $A(t)$ that emphasizes more clearly the half-life:
$$A(t) = 100e^{-t(\ln 2)/5730} = 100e^{\ln 2^{(-t/5730)}} = 100(2)^{-t/5730},$$
so
$$A(t) = 100\,(1/2)^{t/5730}.$$

Archaeologists have used the decay property of ^{14}C to establish the age of relics and artifacts since the Nobel laureate Willard Libby (1908–1980) introduced the technique in the early 1950s. Living tissue contains three carbon isotopes. Approximately 98.89% of this carbon consists of the stable isotope $^{12}_{6}$C and most of the remainder is the stable isotope $^{13}_{6}$C, but a small amount, about one part in one trillion, is the radioactive isotope $^{14}_{6}$C. This isotope decays over time to produce nitrogen, $^{14}_{7}$N.

Radioactive carbon dating makes the assumption that the percentage of $^{14}_{6}$C and $^{12}_{6}$C in all *living* things has remained constant throughout history. The date at which an ancient organism ceased to live can be estimated by comparing the current $^{14}_{6}$C to $^{12}_{6}$C proportion to the original proportions of these isotopes.

In recent years, the assumption that the $^{14}_{6}$C to $^{12}_{6}$C proportion has remained constant throughout history has been shown to be in error. Nuclear testing tends to increase the $^{14}_{6}$C proportion, while smokestack emissions tend to decrease the $^{14}_{6}$C proportion. In addition, it has been shown that before 1000 B.C., the proportion of $^{14}_{6}C$ was less than the present proportion, and before 6000 B.C. the level was only about 8% of the present proportion. This is thought to be due to the bombardment of the earth's upper atmosphere by cosmic rays.

Continuous Rates of Interest

Lending institutions pay interest on savings accounts at a specified rate compounded over a certain period of time. The most common compounding periods are quarterly (4 periods per year), monthly (12 periods per year), and daily (365 periods per year).

Suppose that an amount $A(0)$ is deposited in a savings account paying an annual interest rate i compounded over a period h, $0 < h < 1$, and that $A(t)$ represents the amount in the account at time $t \geq 0$. There are $1/h$ such time periods in one year, so the interest rate in each time period is ih. The interest paid on the account for the time period from t to $t + h$ is $ihA(t)$. This implies that

$$A(t + h) = A(t) + ihA(t),$$

so

$$\frac{A(t + h) - A(t)}{h} = iA(t).$$

If we take the limit of this quotient as h approaches zero, we obtain

(6.60) $\quad A'(t) = iA(t)$, which has the solution $A(t) = A(0)e^{it}$.

An account that grows in this manner is said to have interest *compounded continuously*. Continuous compounding gives the maximum effective interest rate that can occur from compounding. For example, the maximum effective interest that an account earning a stated annual interest rate of 6% can earn is

$$e^{0.06(1)} - 1 = 1.0618 - 1 = 0.0618 = 6.18\%,$$

no matter how frequently the interest is compounded.

In the following example the growth in an account when the interest is compounded daily is approximated quite accurately by assuming instead that the interest is compounded continuously.

Example 2

Suppose $1000 is deposited in a savings account that pays $7\frac{1}{4}\%$ per annum compounded daily. By assuming the interest is compounded continuously, estimate:

a) the amount in the account at the end of 4 yr;
b) the time when the account will contain $1500.

Solution

Let $A(t)$ denote the amount of money in the account at the end of t yr.

Then the problem is stated concisely as follows:

KNOW	FIND
1) $\dfrac{dA(t)}{dt} = 0.0725 A(t),$	a) $A(4)$,
2) $A(0) = \$1000.$	b) t when $A(t) = \$1500.$

a) It follows from (6.60) that
$$A(t) = A(0)e^{0.0725t} = 1000 e^{0.0725t},$$
so
$$A(4) = 1000 e^{(0.0725)(4)} = 1000 e^{0.29} = \$1336.43.$$

The actual amount that accumulates if the interest is compounded daily is
$$1000\left(1 + \frac{0.0725}{365}\right)^{4(365)} = \$1336.39.$$

b) When $A(t) = 1500$,
$$1500 = 1000 e^{0.0725t} \quad \text{and} \quad 1.5 = e^{0.0725t}.$$
So
$$\ln 1.5 = 0.0725 t \quad \text{and} \quad t = \frac{\ln 1.5}{0.0725} \approx 5.6.$$

The account will reach $1500 in approximately 5.6 yr. ▶

Population Growth

In the two preceding examples, the same type of equation
$$\frac{dA(t)}{dt} = kA(t)$$
described each situation. The difference between these two problems is the sign of k. In the decay problem, $k < 0$. In the deposit problem, which is a growth situation, $k > 0$.

The growth (or attrition) rate of an unrestricted population is often assumed to be proportional to the number of individuals in the population. Although the actual population is given in integer values, it is easier to solve the mathematical problem associated with population change if we allow nonintegral values for the population. In this case we have the same type of problem we discussed earlier.

Example 3

Suppose 10,000 people presently live in a certain community. Census figures show that the influx rate (the rate at which people enter) in this community is 55 persons per 1000 and that the attrition rate (the rate at which people leave) is 14 persons per 1000.

a) The sewage treatment facilities are to be expanded to ensure that they will handle the needs of the community for the next 15 yr. How many people should the facility support?
b) How many should the facility support if it is designed to handle the needs for the next 100 yr?

KNOW	FIND
1) $\dfrac{dP(t)}{dt} = \dfrac{41}{1000} P(t),$	a) $P(t),$
2) $P(0) = 10,000.$	b) $P(15),$
	c) $P(100).$

Solution
The net rate of increase in the population each year is $55 - 14 = 41$ persons per 1000. If $P(t)$ represents the population in t years, the problem can be summarized as shown in the table.

Since $P'(t) = 0.041 P(t)$, it follows from (6.59) that

$$P(t) = P(0)e^{0.041t} = 10,000 e^{0.041t}.$$

Thus,

a) $$P(15) = 10,000 e^{0.041(15)} = 18,496$$

and

b) $$P(100) = 10,000 e^{0.041(100)} = 603,403. \quad \blacktriangleright$$

A city planner might accept the population prediction for 15 yr in the preceding example, but would hardly be expected to accept the 100-yr figure. Although both figures were calculated by the same mathematical formula, this formula is probably accurate only for relatively small values of t. When t becomes large, the assumption of unrestricted growth is no longer valid, and the equation $dP(t)/dt = kP(t)$ is not an accurate model of the population-growth problem.

Change in Dilution of a Liquid

Example 4

A tank contains 500 gal of brine in which 150 lb of salt has been dissolved (see Figure 6.27). Fresh water runs into the tank at the rate of 5 gal/min, and the well-stirred mixture is drained from the tank at the same rate. Find:

a) the amount of salt in the tank at any time t;
b) the amount of salt in the tank at the end of 1 hr;
c) the amount of time required to reduce the amount of salt in the tank to 10 lb.

Solution
Let $A(t)$ denote the number of pounds of salt in the tank t min after the fresh water is introduced to the tank. Then $A'(t)$, the rate of change in the amount of salt with respect to time, is the difference between the rate at which salt enters the tank, 0 lb/min, and the rate at which salt

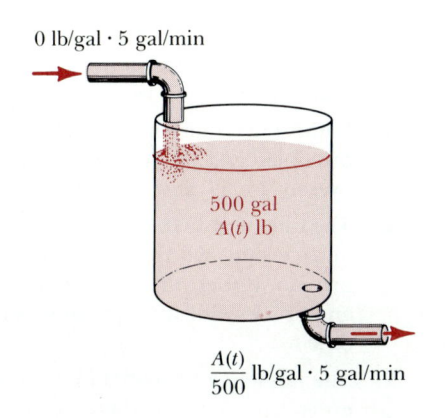

Figure 6.27

leaves the tank. Thus,

$$A'(t) = \text{Rate in} - \text{Rate out} = 0 \, \frac{\text{lb}}{\text{gal}} \cdot 5 \, \frac{\text{gal}}{\text{min}} - \frac{A(t)}{500} \, \frac{\text{lb}}{\text{gal}} \cdot 5 \, \frac{\text{gal}}{\text{min}}$$
$$= -\frac{A(t)}{100} \, \frac{\text{lb}}{\text{min}}.$$

KNOW	FIND
1) $\dfrac{dA(t)}{dt} = -\dfrac{A(t)}{100}$,	a) $A(t)$,
2) $A(0) = 150$.	b) $A(60)$,
	c) t, with $A(t) = 10$.

The problem is expressed concisely as shown in the table.

a) Since $A'(t) = -0.01 A(t)$, it follows from (6.59) that
$$A(t) = A(0)e^{-0.01t} = 150e^{-0.01t}.$$

b) Thus,
$$A(60) = 150e^{-0.6} \approx 82.3 \text{ lb}.$$

c) To find t with $A(t) = 10$ requires solving for t in the equation
$$10 = 150e^{-0.01t}.$$

Thus,
$$e^{-0.01t} = \frac{10}{150} \quad \text{and} \quad -0.01t = \ln\frac{10}{150} = -\ln 15.$$

So
$$t = 100 \ln 15 \approx 270.8 \text{ min.} \qquad \blacktriangleright$$

▶ **EXERCISE SET 6.6**

1. The radioactive isotope thorium $^{234}_{90}$Th has a half-life of approximately 590 hr. If there is 50 mg present at time $t = 0$, find:
   a) an expression for the amount of thorium present at any time t;
   b) how much thorium will remain at the end of 100 hr.

2. Suppose 100 mg of the radioactive isotope thorium-234 described in Exercise 1 is present initially. How much thorium will remain at the end of one week?

3. Radioactive radiothorium $^{228}_{90}$Th has a half-life of approximately 1.90 yr. If 100 mg of radiothorium is present today, how much was present one year ago?

4. The rate of growth of the number of bacteria present in a certain culture is proportional to the number present. If there are 500 at time $t = 0$ and 1000 at the end of 5 hr, find an expression for the number at any time t.

5. The number of bacteria in a certain culture doubles every 4 hr. Suppose there are 1000 at time $t = 0$.
   a) Find an expression for the number at any time t.
   b) How many bacteria will be present after 7 hr?

6. Under ideal conditions, a cell of the bacteria *Escherichia coli* divides in approximately 22 mins. Suppose the initial amount $A(0)$ of bacteria is large enough to assume that the bacteria multiply continuously.
   a) Find an expression for $A(t)$, where t is measured in minutes.
   b) Determine the rate at which the bacteria is increasing at the end of 1 hr.

7. One bank advertises that it pays an interest rate of $5\frac{1}{4}\%$ compounded continuously, while a second bank advertises that it pays $5\frac{1}{2}\%$ compounded semiannually. Which rate is most advantageous to the investor?

8. An amount of money doubles at the end of 9 yr when invested at a certain rate of interest compounded continuously. What is that rate of interest?

9. A census of the United States population is taken every 10 yr. The table at the top of the next page gives, in thousands of people, the population from 1930 to 1980. Assume the population changes by an amount proportional to the amount of population present. Predict the population in the year 2000 on the basis of

YEAR	POPULATION (IN THOUSANDS)
1930	123,203
1940	131,669
1950	150,697
1960	179,323
1970	203,212
1980	226,505

the population in the years (a) 1930 and 1940; (b) 1970 and 1980. Compare your results and make conclusions.

10. Consider the population table given in Exercise 9. Predict the population in 1980 on the basis of the population in the years (a) 1930 and 1940; (b) 1960 and 1970. Do these results lead you to different conclusions about your prediction in Exercise 9?

11. The rate at which sugar dissolves in a liquid kept at a constant temperature is proportional to the amount that remains to be dissolved. Suppose a home brewer adds 10 lb of sugar to a vat of water and that 5 lb dissolves in the first 5 min. How long does it take for the next 4 lb to dissolve?

12. A 220-L oak barrel that previously held red wine has been drained and refilled with water. The residue of wine remaining in the barrel mixes with the water to form a solution that is 1 part wine to 400 parts water. Fresh water is then run into the barrel at a rate of 9 L/min, and the uniform solution runs out the bung hole at the same rate. How long should the water continue to run if the final mixture is to contain no more than 0.1 L of wine?

13. Five hundred gallons of pesticide is accidentally spilled into a lake with a volume of 8×10^7 gal and uniformly mixes with the water. A river flows into the lake bringing 10,000 gal of fresh water per minute, and the uniform mixture flows out of the lake at the same rate. How long will it take to reduce the pesticide in the lake to a safe level of 1 part per billion?

14. Rework Exercise 13 for a lake containing 10 times the volume, but with an inflow/outflow rate of only 1000 gal/min.

15. In a chemical reaction, one chemical X is converted into another chemical at a rate proportional to the amount of X present at any given time. Suppose that 50 g of X is present initially and 1 hr later 12 g is present.
    a) What percent of X has been converted at the end of 30 min?
    b) When will 90% of X have been converted into the other chemical?

16. A towel hung on a line loses moisture at a rate directly proportional to its moisture content. If the towel loses half of its moisture content in 2 hr, how long will it take to be 95% dry?

17. Tests show that the size of a fire in an enclosed room grows exponentially according to Equation (6.59) from the beginning of the fire until flashover, the point at which fire spontaneously consumes all remaining combustible material in the room. In a test conducted by Professor Howard Emmons at Harvard University, a fire was started at time zero and flashover occurred in 425 sec. Assume that the size of the fire at $t = 0$ is given the unit value 1 and that the flashover has a size 3000 times greater. Determine the size of the fire at 1-min intervals from the time it begins.

18. Bankers often approximate the time it takes to double the amount of an investment made at a fixed interest rate by dividing the percent of annual interest into 70. For example, $10,000 invested at 8.75% per year will become $20,000 in approximately $70/8.75 = 8$ yr. Show that this formula has mathematical validity.

19. A sample of wood is found to contain carbon with one part $^{14}_{6}C$ to 1.35×10^{12} parts $^{12}_{6}C$. Assuming that the ratio at the time the tree died was one part $^{14}_{6}C$ to 10^{12} parts $^{12}_{6}C$ and that the half-life of $^{14}_{6}C$ is 5730 yr, estimate the age of the sample.

6.7

Separable Differential Equations

In Section 6.6 we considered applications of the natural exponential function to problems that involve an equation of the form

$$y'(x) = ky(x), \quad \text{for some constant } k.$$

This equation is an example of a **first-order ordinary differential equation,** an equation that involves a function and its first derivative.

Differential equations are used frequently in applications to express how the change in the amount of a quantity, a derivative, is related to the amount of the quantity. The differential equations considered in this section are closely related to the exponential growth and decay problems. Other common differential equations are discussed in Chapter 17.

A differential equation is called **separable** if it has the form

(6.61) $$y' = \frac{p(x)}{q(y)},$$

where p and q are continuous functions and q is nonzero. The term *separable* indicates that the equation can be written with all terms involving the dependent variable y on one side of the equation and all those involving the independent variable x on the other side:

$$q(y)y' = p(x).$$

The equation $y' = ky$ considered in Section 6.6 is separable whenever $y \neq 0$, with $p(x) = k$ and $q(y) = 1/y$. The differential equations

$$y' = \frac{\cos(x^2 + 1)}{3 + \sin y} \quad \text{and} \quad y' = (x^3 + 3)(y^4 + y^2 + 1)$$

are also separable, but the equations

$$y' = x + y \quad \text{and} \quad y' = \sin(xy)$$

are not.

The separable equation (6.61) can be rewritten as

$$q(y)\frac{dy}{dx} = p(x), \quad \text{so} \quad 0 = q(y)\frac{dy}{dx} - p(x).$$

Integrating both sides of the equation with respect to x implies that for an arbitrary constant C,

$$C = \int q(y)\frac{dy}{dx}\, dx - \int p(x)\, dx$$

and

$$C = \int q(y)\, dy - \int p(x)\, dx.$$

This produces a **general solution** of the separable differential equation, that is, a solution containing an arbitrary constant.

Example 1

Find a general solution of the separable equation

$$y' = \frac{2x}{3y^2}.$$

Solution

Since $\quad 3y^2 \dfrac{dy}{dx} = 2x, \quad$ we have $\quad 0 = 3y^2 \dfrac{dy}{dx} - 2x.$

Integrating gives

$$C = \int \left(3y^2 \frac{dy}{dx}\right) dx - \int 2x \, dx$$
$$= \int 3y^2 \, dy - \int 2x \, dx = y^3 - x^2,$$

so

$$y = \sqrt[3]{x^2 + C}. \qquad \blacktriangleright$$

Example 2

Find a general solution of the differential equation

$$y' = 3x^2 y.$$

Solution
This equation is separable only if $y \neq 0$. In this case,

$$\frac{1}{y}\frac{dy}{dx} = 3x^2, \quad \text{so} \quad 0 = \frac{1}{y}\frac{dy}{dx} - 3x^2.$$

Thus,

$$C = \int \frac{1}{y} dy - \int 3x^2 \, dx = \ln|y| - x^3,$$

so

$$\ln|y| = x^3 + C$$

and

$$|y| = e^{x^3 + C} = e^C e^{x^3}.$$

Since e^C can assume any positive value, the solution can be written as

$$y(x) = Ce^{x^3}$$

for any constant $C \neq 0$. Note that $y(x)$ is always positive or always negative, depending on the sign of C.

We can easily verify that $y(x) \equiv 0$ is also a solution to the differential equation $y' = 3x^2 y$, so the general solution does not need the stipulation that $C \neq 0$. The equation is not separable when $y = 0$, however, so the solution corresponding to $C = 0$ must be obtained by an alternative method. $\blacktriangleright$

Example 3

Find a solution of the differential equation $y' = 1/x^2 y^2$ that satisfies the condition $y(1) = 3$ and is valid when $x > 0$.

Solution
Since

$$\frac{dy}{dx} = \frac{1}{x^2 y^2}, \quad \text{we have} \quad 0 = y^2 \frac{dy}{dx} - \frac{1}{x^2}$$

and
$$C = \int y^2\, dy - \int \frac{1}{x^2}\, dx = \frac{y^3}{3} + \frac{1}{x}.$$

Since $y = 3$ when $x = 1$,
$$C = \frac{27}{3} + \frac{1}{1} = 10.$$

Consequently,
$$10 = \frac{y^3}{3} + \frac{1}{x}, \quad \text{which implies} \quad y^3 = 30 - \frac{3}{x},$$

and, for $x > 0$,
$$y = \left(30 - \frac{3}{x}\right)^{1/3}. \quad \blacktriangleright$$

Newton's Law of Cooling states that the rate of change of the temperature of an object is proportional to the difference between the temperature of the object and the temperature of its surrounding medium. Let $T(t)$ denote the temperature of the object at time t, and let L denote the temperature of the surrounding medium. Then Newton's Law of Cooling can be expressed as

$$\frac{dT(t)}{dt} = k(T(t) - L).$$

If $T = T(t) \neq L$, then
$$\frac{1}{T - L}\frac{dT}{dt} = k,$$

which is a separable equation that can be solved by writing
$$C = \int \frac{dT}{T - L} - \int k\, dt = \ln|T - L| - kt.$$

But
$$\ln|T - L| = C + kt$$

implies that
$$|T - L| = e^{C + kt} = e^C e^{kt}.$$

Replacing the positive constant e^C by the arbitrary nonzero constant C gives

$$T - L = Ce^{kt}, \quad \text{that is,} \quad T(t) = Ce^{kt} + L.$$

Evaluating $T(t)$ when $t = 0$ produces
$$T(0) = Ce^0 + L = C + L, \quad \text{so} \quad C = T(0) - L$$

and
$$T(t) = (T(0) - L)e^{kt} + L$$

describes the temperature of the object at any time t.

Example 4

An outdoor thermometer reading $-3°$ C is brought into a $20°$ C room. One minute later the thermometer reads $5°$ C. How long will it take to reach $19.5°$ C?

Solution

Let $T(t)$ represent the reading on the thermometer t min after it is brought into the room (see Figure 6.28). The problem is expressed concisely below.

KNOW	FIND
1) $T(t) = (T(0) - L)e^{kt} + L$,	a) k,
2) $T(0) = -3°$ C,	b) t so that $T(t) = 19.5°$ C.
3) $L = 20°$ C,	
4) $T(1) = 5°$ C.	

It follows from (1), (2), and (3) in the table that

$$T(t) = -23e^{kt} + 20.$$

To find k, we use this result with (4) to obtain

$$5 = T(1) = -23e^k + 20.$$

Solving for k gives

$$e^k = 15/23 \quad \text{so} \quad k = \ln(15/23).$$

Thus,

$$T(t) = -23e^{t\ln(15/23)} + 20.$$

The temperature reaches $19.5°$ C when

$$19.5 = -23e^{t\ln(15/23)} + 20.$$

This implies that

$$e^{t\ln(15/23)} = 0.5/23, \quad \text{so} \quad t\ln(15/23) = \ln(0.5/23)$$

and

$$t = \frac{\ln(0.5/23)}{\ln(15/23)} \approx 8.96 \text{ min.} \quad \blacktriangleright$$

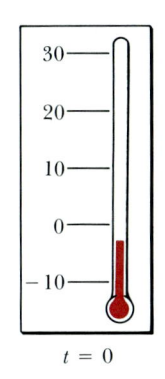

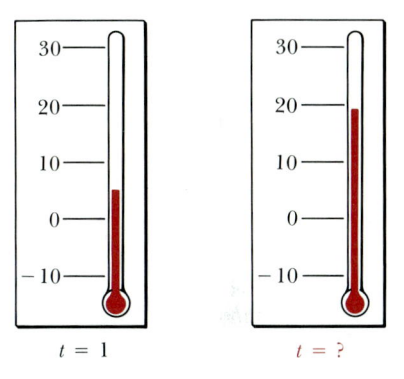

Figure 6.28

In Section 6.6 we considered the growth of an unrestricted population, a population $P(t)$ described by the equation

$$\frac{dP(t)}{dt} = kP(t),$$

where k is the growth rate. In Example 3 of Section 6.6 we found that this model of population growth is likely to give an unreasonable prediction unless the time period is quite small.

A modification of this differential equation produces better approximations in many situations. Suppose certain community considerations make it reasonable to assume that the maximum population that can be supported is a constant L. The differential equation

(6.62) $\quad \dfrac{dP(t)}{dt} = k(L - P(t))P(t), \quad \text{where } k > 0 \text{ and } 0 < P(t) < L,$

expresses the more reasonable situation that:

i) the rate of growth $k(L - P(t))$ depends directly on the population,
ii) as $P(t)$ approaches the maximum population L, the rate at which the population changes goes to zero.

The differential equation (6.62) is called the **logistic equation** for population growth. To solve this separable equation, we first rewrite it as

$$0 = \dfrac{1}{(L-P)P} \dfrac{dP}{dt} - k.$$

Integrating with respect to t gives

$$C = \int \dfrac{1}{(L-P)P} \, dP - kt.$$

It is easily verified that

$$\dfrac{1}{(L-P)P} = \dfrac{1}{L}\left[\dfrac{1}{P} + \dfrac{1}{L-P}\right],$$

so

$$C = \dfrac{1}{L} \int \left[\dfrac{1}{P} + \dfrac{1}{L-P}\right] dP - kt$$
$$= \dfrac{1}{L}\left[\ln P - \ln(L-P)\right] - kt = \dfrac{1}{L} \ln \dfrac{P}{L-P} - kt$$

and

$$C = \dfrac{1}{L} \ln \dfrac{P(t)}{L - P(t)} - kt.$$

When $t = 0$,

$$C = \dfrac{1}{L} \ln \dfrac{P(0)}{L - P(0)},$$

so

$$\dfrac{1}{L} \ln \dfrac{P(0)}{L - P(0)} = \dfrac{1}{L} \ln \dfrac{P(t)}{L - P(t)} - kt.$$

Thus,

$$kLt = \ln \dfrac{P(t)}{L - P(t)} - \ln \dfrac{P(0)}{L - P(0)} = \ln \left[\dfrac{P(t)}{L - P(t)} \dfrac{L - P(0)}{P(0)}\right]$$

and

$$\dfrac{P(t)(L - P(0))}{P(0)(L - P(t))} = e^{kLt}.$$

Solving this equation for $P(t)$ gives

$$P(t) = \frac{P(0)Le^{kLt}}{(L-P(0))+P(0)e^{kLt}} = \frac{P(0)L}{P(0)+(L-P(0))e^{-kLt}}.$$

The graph of the solution to the logistic equation is shown in Figure 6.29. Initially, the graph is concave upward, and $P(t)$ appears to grow exponentially. Then the graph becomes concave downward, and $P(t)$ approaches the limiting population. Other features of the graph of the logistic equation are considered in Exercise 31. The logistic equation can also be applied when $P(t) > L$. Exercise 32 discusses one example.

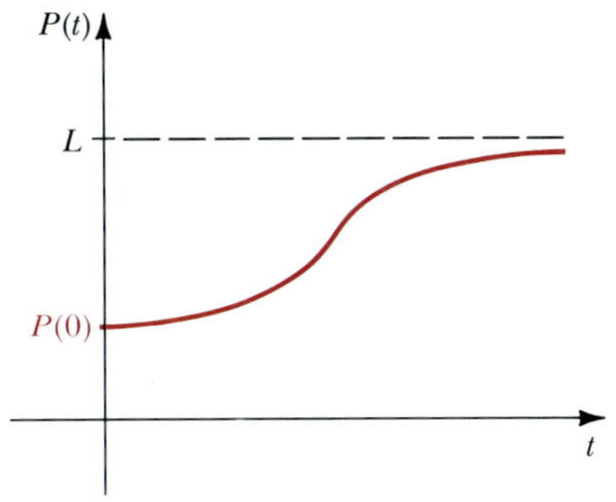

Figure 6.29
Logistic growth.

Example 5

Consider the population problem given in Example 3 of Section 6.6, with the restriction that the maximum supportable population is 100,000 persons. In that example the initial population was 10,000 and the yearly increase rate was 0.041. Use the logistic equation with initial growth rate 0.041 to predict the population in (a) 15 yr and (b) 100 yr.

Solution
The constant k in the logistic equation is chosen so that the term $k(L-P(t))$ is the initial growth rate, that is, the rate when $t = 0$. Thus,

$$0.041 = k(L - P(0)) = k(100{,}000 - 10{,}000),$$

so

$$k = \frac{0.041}{90{,}000}.$$

Since $L = 100{,}000$, the logistic equation predicts that the population in t years will be

$$P(t) = \frac{P(0)L}{P(0)+(L-P(0))e^{-0.41t/9}}$$

$$= \frac{(10{,}000)(100{,}000)}{(10{,}000)+(90{,}000)e^{-0.41t/9}}.$$

Evaluating when $t = 15$ and $t = 100$ gives

a) $P(15) = 18{,}036$ and b) $P(100) = 91{,}359$.

The predicted population in 15 yr is quite close to our previous prediction of 18,496 persons. A significant difference occurs in the 100-yr figure: The prediction in Example 3 of Section 6.6 was 603,403. ▶

Example 6

An object is propelled from the surface of the earth with an initial velocity v_0. What is the minimal value of v_0 that ensures the object does not return to earth?

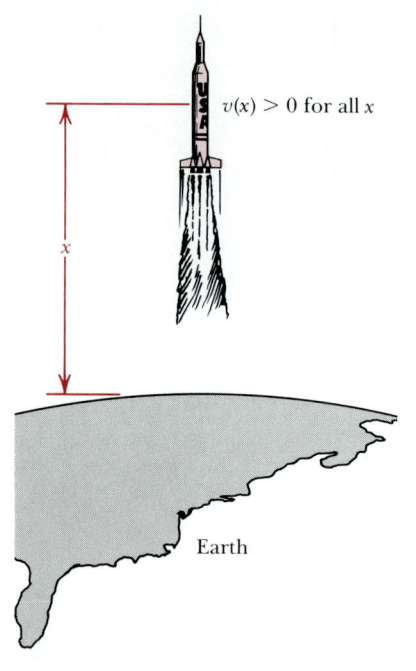

Solution
In Section 5.6, we used Newton's Law of Gravitation to determine the work required to put the Space Shuttle into orbit. This law states that the force of attraction between two bodies is inversely proportional to the square of the distance between their centers of mass. When an object with mass m is a distance x above the surface of the earth, as shown in Figure 6.30, this force of attraction can be written in the form

$$F(x) = \frac{mm_e G}{(x+R)^2},$$

where m_e is the mass of the earth and G is the universal gravitational constant. On the surface of the earth, that is, when $x = 0$, the force of attraction is the product of the mass m of the object and the gravitational constant $g = -32$ ft/sec². Thus,

$$mg = F(0) = \frac{mm_e G}{R^2}, \quad \text{which implies that} \quad m_e G = gR^2$$

and

$$F(x) = \frac{mgR^2}{(x+R)^2} \quad \text{when } x \geq 0.$$

Newton's Second Law of Motion states that the product of the mass of an object and its acceleration is equal to the sum of all the forces acting on the object. If we assume that the only force acting on the object is the gravitational force of the earth and use the fact that the acceleration is given by d^2x/dt^2, then

$$m\frac{d^2x}{dt^2} = \frac{mgR^2}{(x+R)^2},$$

so

(6.63)
$$\frac{d^2x}{dt^2} = \frac{gR^2}{(x+R)^2}.$$

This is a second-order differential equation (it involves a second derivative), one that we cannot solve at present. To change (6.63) into a separable first-order equation, we use the fact that the velocity v is dx/dt, so $dv/dt = d^2x/dt^2$. The Chain Rule implies that

$$\frac{d^2x}{dt^2} = \frac{dv}{dt} = \frac{dv}{dx}\frac{dx}{dt} = \frac{dv}{dx}v.$$

So (6.63) can be written as

$$v\frac{dv}{dx} = \frac{gR^2}{(x+R)^2},$$

a separable equation in x and v. Consequently,

$$C = \int v\, dv - \int \frac{gR^2}{(x+R)^2}\, dx = \frac{v^2}{2} + \frac{gR^2}{(x+R)}$$

and, since $v = v_0$ when $x = 0$,

$$C = \frac{v_0^2}{2} + gR. \quad \text{Thus,} \quad \frac{v_0^2}{2} + gR = \frac{v^2}{2} + \frac{gR^2}{(x+R)}$$

and

(6.64) $$\frac{(v(x))^2}{2} = \frac{v_0^2}{2} + gR - \frac{gR^2}{(x+R)}.$$

In order for the object to reverse direction and return to earth, $v(x)$ must be zero for some $x \geq 0$. The object will *not* return to earth if the velocity $v(x)$ is positive for all values of $x \geq 0$.

To ensure that $v(x) > 0$ for all $x \geq 0$, we choose v_0 so that the right side of Equation (6.64) is always positive. Since x and R are positive and g is negative,

$$\frac{gR^2}{x+R} < 0, \quad \text{for all } x \geq 0.$$

But

$$\lim_{x \to \infty} \frac{gR^2}{x+R} = 0, \quad \text{so} \quad \frac{v_0^2}{2} + gR - \frac{gR^2}{x+R}$$

will be positive for all values of x precisely when

$$\frac{v_0^2}{2} + gR > 0.$$

This implies that

$$v_0 > \sqrt{-2gR} = \left[-2\left(-32\frac{\text{ft}}{\text{sec}^2}\right)\left(\frac{1 \text{ mi}}{5280 \text{ ft}}\right)(3960 \text{ mi})\right]^{1/2} \approx 6.93\frac{\text{mi}}{\text{sec}},$$

a rather large initial velocity. ▶

▶ EXERCISE SET 6.7

In Exercises 1–12, find the general solution of the differential equation.

1. $y' = 7y$
2. $y' = y^2$
3. $y' = y^2 + 1$
4. $y' = x(y^2 + 1)$
5. $y' = \dfrac{\sin x + x^2}{\cos y - 1}$
6. $y' = e^{x+3y}$
7. $y' + y^2 \cos x = 0$
8. $xy' - \sqrt{1-y^2} = 0$
9. $e^x y' + e^y = 0$
10. $\tan x + (\sec y)^2 y' = 0$
11. $y + e^x y' = 0$
12. $(\cos y \csc x)y' - e^{\sin y} \sin x = 0$

In Exercises 13–18, find the solution of the differential equation that satisfies the given condition.

13. $y' + y^2 \cos x = 0$, $y(0) = 2$
14. $xy' - \sqrt{1-y^2} = 0$, $y(1) = -1$
15. $y' = x(y+1)$, $y(0) = 1$
16. $y' = \dfrac{\sin x}{y}$, $y(0) = \pi$
17. $y' = e^y \cos x$, $y(\pi) = 0$
18. $y(1-x^2)y' + x(y^2 + 1) = 0$, $y(\sqrt{2}) = 2$
19. An oven thermometer reads 65° F. It is placed in an

oven set at 375° F. At the end of 2 min the thermometer reads 300° F, and at the end of 3 min it reads 345° F. Assuming the thermometer is accurate, is the oven temperature actually 375° F? If not, what is the actual oven temperature?

20. A cup of tea is made by pouring boiling water (212° F) into a cup. Three minutes later the tea is brewed, but at 180° F is too hot to drink. Assuming that the air temperature is a constant 68° F, how long will it take for the tea to become a drinkable temperature of 160° F or less?

21. A thermometer reading 0° C is placed in a room where the temperature is 20° C. In 3 min the thermometer reads 12° C. How long will it take for this thermometer to read 19° C?

22. An automobile cooling system has a capacity of 17 qt. Initially, the system contains a mixture of 2 gal of antifreeze and 9 qt of water. Water runs into the system at the rate of 1 gal/min, and the homogeneous mixture runs out a petcock at the same rate. How much antifreeze remains in the system after 5 min? [Hint: See Example 4 in Section 6.6.]

23. A tank contains 300 gal of salt-free water. A brine containing 0.5 lb of salt per gallon of water runs into the tank at the rate of 2 gal/min, and the well-stirred mixture runs out at the rate of 2 gal/min. What is the concentration of salt in the tank at the end of 10 min?

24. Rework Exercise 23 with the following modifications.
   a) The brine runs into and out of the tank at the rate of 3 gal/min.
   b) The brine runs into and out of the tank at the rate of 1 gal/min.

25. The dispersion of information among a population can be modeled by the differential equation

$$\frac{dP}{dt} = k(1 - P),$$

where P denotes the proportion of the population aware of the information at time t and k is a positive constant. Suppose 10% of the population learns of a tax rebate on the 7:00 A.M. news and 50% is aware of the rebate by noon. What percentage will know about the rebate before the 6:00 P.M. news?

26. An indoor/outdoor thermometer has two sensing units, one placed outside, the other inside. The outdoor thermometer reads 5° F and the indoor thermometer reads 65° F when the thermometer is placed in a wine cellar held at 45° F. One minute later the outdoor thermometer reads 25° F and the indoor thermometer reads 55° F.

   a) Which of the two thermometers will first be within 5° of the correct temperature?
   b) Can you conclude anything significant about the thermometers from this problem?

27. A thermometer reading 0° C is brought into a 20° C room. In 3 min the thermometer reads 12° C. The thermometer is then placed in a freezer for 3 min where the temperature is 0° C and immediately brought back into the room where the temperature is 20° C. What is the reading on the thermometer 3 min after it has been brought back into the room?

28. Suppose the cycle described in Exercise 27 is repeated: for 3 min the thermometer is placed in the freezer and then for 3 min it is placed in the room. What is the temperature at the end of this time?

29. A body is found floating face down in Lake Gotchaheny (a lake with a constant temperature of 62° F). When the body was taken from the water at 11:50 A.M. its temperature was 66° F. The temperature of the body when first found at 11:00 A.M. was 67° F. When did the victim meet his demise? (Assume that the victim was a normal 98.6° F before going to the watery grave.)

30. It is estimated that the limiting population the United States can support is 500,000,000 people. Assume that the population of the United States satisfies the logistic equation with $L = 5 \times 10^8$. Predict the population in the year 2000 on the basis of the population in the years (a) 1930 and 1940; (b) 1970 and 1980. Compare these predictions with the corresponding predictions in Exercise 9 of Section 6.6.

YEAR	POPULATION (IN THOUSANDS)
1930	123,203
1940	131,669
1950	150,697
1960	179,323
1970	203,212
1980	226,505

31. Suppose a population satisfies the logistic equation

$$\frac{dP(t)}{dt} = k(L - P(t))P(t), \quad \text{where } 0 < P(0) < L.$$

   a) Show that the graph of P must be as shown in Figure 6.29.
   b) Find the values of t and $P(t)$ that correspond to a point of inflection.

32. Consider the logistic equation (6.62) under the assumption that $P(0) > L$.

a) Determine $\lim_{t \to \infty} P(t)$.
b) Sketch the graph of P.

33. Another mathematical model for describing the restricted growth of a population uses the *Gompertz function*, given by
$$f(t) = ae^{-be^{-kt}}, \quad t \geq 0,$$
for positive constants a, b, and k.
   a) Sketch the graph of f noting, in particular, any points of inflection.
   b) What are the initial and limiting populations in this model?

34. Show that the Gompertz function described in Exercise 33 satisfies the differential equation
$$f'(t) = kf(t) \ln \frac{a}{f(t)}.$$
Interpret this differential equation as a rate of change, and decide what the constant a represents.

35. Use the Gompertz function to predict the population of the United States in the year 2000 on the basis of the assumptions and data described in Exercise 30.

36. A model for the dispersion of information was considered in Exercise 25. Another model uses the differential equation
$$\frac{dP}{dt} = kP(1 - P),$$
where P denotes the proportion of the population aware of the information at time t and k is a positive constant. Suppose 10% of the population learns of a tax rebate on the 7:00 A.M. news and 50% is aware of the rebate by noon. What percentage will know about the rebate before the 6:00 P.M. news?

37. The production of the enzyme trypsin from its inactive precursor trypsinogen is an autocatalytic reaction (the production of trypsin from trypsinogen does not begin unless trypsin is present, and the production rate increases with the increased production of trypsin). The production rate is described by a separable differential equation of the form
$$\frac{dA(t)}{dt} = k(A_1 + A(t))(A_2 - A(t)),$$
where $A(t)$ denotes the amount of trypsin present at time t, A_1 and A_2 are the amounts of trypsinogen and trypsin, respectively, initially present, and k is a positive constant.
   a) Find the general solution to this differential equation.

b) When is the reaction rate a maximum?

38. Early work in quantitative psychology by Ernst Heinrich Weber (1795–1878) and Gustav Fechner (1801–1887) suggested that the reaction R of an organism to a stimulus S is inversely proportional to the intensity of the stimulus. The differential equation
$$\frac{dR}{dS} = \frac{k}{S}$$
expressing this relationship is known as the *Weber–Fechner Law*. The threshold level of a stimulus S_0 is defined to be the minimal level at which a reaction is obtained. Find the solution to the differential equation describing the Weber–Fechner Law.

39. *Fick's First Law* states that the diffusion of a substance across a cellular membrane is directly proportional to the difference in the fixed concentration C outside the cell and the varying concentration $c(t)$ inside the cell. In addition, the diffusion is directly proportional to the surface area of the membrane and inversely proportional to the volume of the cell.
   a) Derive the differential equation described by Fick's Law.
   b) Find its general solution.

40. The rate at which an enzyme reaction converts a substrate into a product is described by the *Michaelis–Menten equation*
$$\frac{dS(t)}{dt} = -\frac{A\,S(t)}{B + S(t)},$$
where $S(t)$ is the amount of substrate at time t and A and B are constants depending on the maximum velocity of the reaction and the experimental conditions. Find the general solution to the Michaelis–Menten equation.

41. A culture of Heriff Dansk Ølgær yeast is introduced into a mixture of malt, hops, sugar, and water. The yeast converts the sugar to alcohol, producing a liquid commonly known as beer. This good beer yeast will produce a 9% alcohol beverage; at this point, the alcohol retards the growth of the yeast and there is no further activity. The logistic differential equation is an excellent model for yeast and alcohol production in beer. Suppose that there is initially 10 g of yeast introduced, that three days later this amount has increased to 320 g, and that the limiting amount of yeast produced is 500 g. To ensure a certin amount of foam, or head, the beer is to be bottled when 98.5% of the fermentation is complete. When should the beer be bottled?

6.8
Hyperbolic Functions

Certain arithmetic combinations of the natural exponential function occur sufficiently often that special functions are defined to discuss them more easily. These are known as the **hyperbolic functions.**

The most common hyperbolic functions are the **hyperbolic sine,** denoted sinh, and the **hyperbolic cosine,** denoted cosh, defined by

$$\text{(6.65)} \qquad \sinh x = \frac{e^x - e^{-x}}{2} \quad \text{and} \quad \cosh x = \frac{e^x + e^{-x}}{2}.$$

Solving equations (6.65) simultaneously for e^x and e^{-x} gives

$$\text{(6.66)} \qquad e^x = \cosh x + \sinh x \quad \text{and} \quad e^{-x} = \cosh x - \sinh x.$$

Consequently,

$$\cosh^2 x - \sinh^2 x = (\cosh x + \sinh x)(\cosh x - \sinh x) = (e^x)(e^{-x}) = e^0$$

and

$$\text{(6.67)} \qquad \cosh^2 x - \sinh^2 x = 1.$$

The graphs of the hyperbolic sine and cosine functions can be obtained by referring to the graphs of $y = e^x/2$, $y = e^{-x}/2$, and $y = -e^{-x}/2$. The graph of $y = \sinh x$ is shown in Figure 6.31(a) as the sum of the graphs of $y = e^x/2$ and $y = -e^{-x}/2$. The graph of $y = \cosh x$ is shown in Figure 6.31(b) as the sum of the graphs of $y = e^x/2$ and $y = e^{-x}/2$.

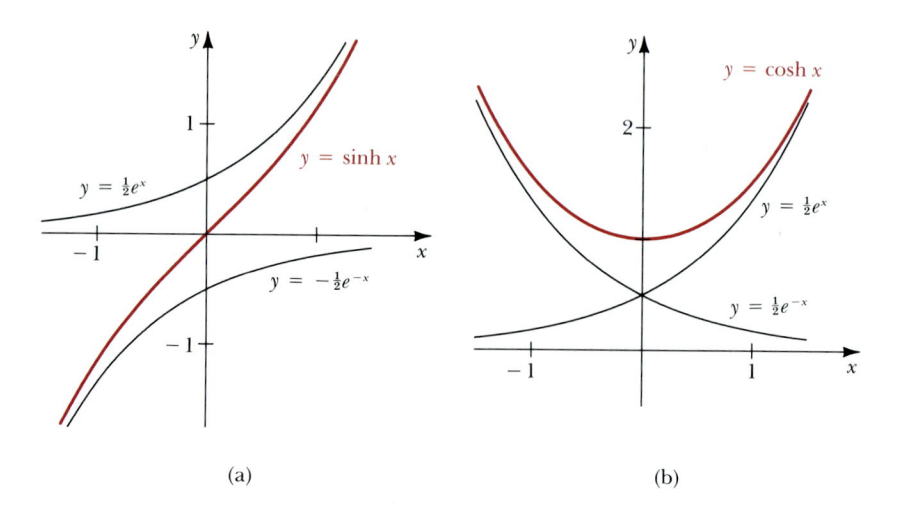

Figure 6.31

(a)   (b)

HISTORICAL NOTE

The concept of hyperbolic functions was suggested in 1727 by **Friedrich Christoph Mayer** (1697–1729), and was formally introduced in 1757 by **Vincenzo Riccati** (1707–1775). The current notation for these functions is due to **Johann Heinrich Lambert** (1728–1777), who used this notation in a paper published in 1771.

The names suggest that the hyperbolic functions have properties in common with the trigonometric functions. Some of these properties are listed below.

i) The trigonometric sine and cosine functions can be derived by mapping the real line onto the unit circle $x^2 + y^2 = 1$ (see Figure 6.32). The hyperbolic sine and cosine functions can be derived by mapping the real line onto the portion of the hyperbola $x^2 - y^2 = 1$ corresponding to $x \geq 1$ (see Figure 6.33).

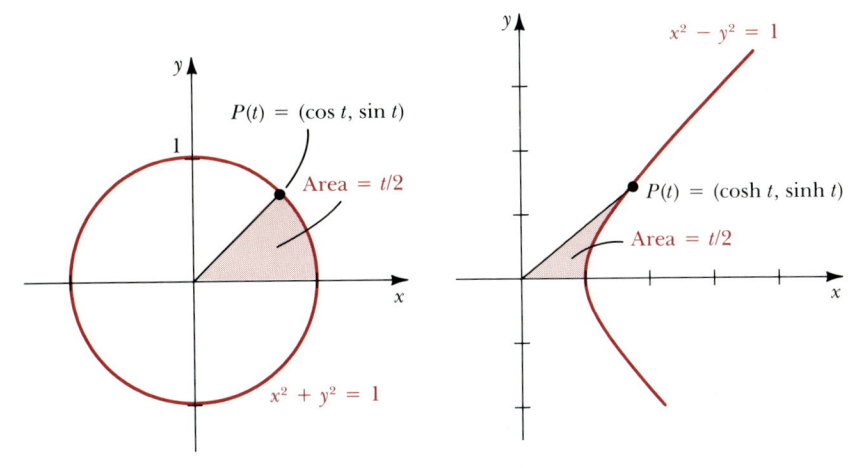

Figure 6.32

Figure 6.33

ii) The area of the region within the unit circle $x^2 + y^2 = 1$ bounded by the rays through the origin intersecting the circle at $P(0)$ and at $P(t)$ is $A = \pi(t/2\pi) = t/2$. If you solve Exercise 49, you will find that the region bounded by the hyperbola $x^2 - y^2 = 1$ and the rays through the origin intersecting the hyperbola at $P(0)$ and at $P(t)$ also has area $t/2$.

iii) The identity $\cosh^2 x - \sinh^2 x = 1$ given in (6.67) is similar to the trigonometric identity $\cos^2 x + \sin^2 x = 1$.

Another relationship of the hyperbolic sine and cosine functions that is reminiscent of the trigonometric functions is the property of their derivatives and integrals:

$$D_x \sinh x = D_x \left(\frac{e^x - e^{-x}}{2} \right) = \frac{e^x + e^{-x}}{2} = \cosh x$$

and

$$D_x \cosh x = D_x \left(\frac{e^x + e^{-x}}{2} \right) = \frac{e^x - e^{-x}}{2} = \sinh x,$$

so

$$\int \cosh x \, dx = \sinh x + C \quad \text{and} \quad \int \sinh x \, dx = \cosh x + C.$$

Example 1

Evaluate $\int \cosh^3 x \sinh x \, dx$.

Solution

Let $u = \cosh x$. Then $du = \sinh x \, dx$ and

$$\int \cosh^3 x \sinh x \, dx = \int u^3 \, du = \frac{u^4}{4} + C = \frac{\cosh^4 x}{4} + C.$$ ▶

The other hyperbolic functions are defined in a manner similar to that used to define their trigonometric counterparts:

$$\tanh x = \frac{\sinh x}{\cosh x} = \frac{e^x - e^{-x}}{e^x + e^{-x}};$$

$$\coth x = \frac{\cosh x}{\sinh x} = \frac{e^x + e^{-x}}{e^x - e^{-x}}, \text{ provided } x \neq 0;$$

$$\operatorname{sech} x = \frac{1}{\cosh x} = \frac{2}{e^x + e^{-x}};$$

$$\operatorname{csch} x = \frac{1}{\sinh x} = \frac{2}{e^x - e^{-x}}, \text{ provided } x \neq 0.$$

The graphs of these functions are shown in Figure 6.34.

The derivatives and corresponding indefinite integral formulas for these functions follow. It is interesting to note the similarities, as well as the differences, between these results and the corresponding results for the trigonometric functions.

$$D_x \tanh x = \operatorname{sech}^2 x, \quad \int \operatorname{sech}^2 x \, dx = \tanh x + C$$

$$D_x \coth x = -\operatorname{csch}^2 x, \quad \int \operatorname{csch}^2 x \, dx = -\coth x + C$$

$$D_x \operatorname{sech} x = -\operatorname{sech} x \tanh x, \quad \int \operatorname{sech} x \tanh x \, dx = -\operatorname{sech} x + C$$

$$D_x \operatorname{csch} x = -\operatorname{csch} x \coth x, \quad \int \operatorname{csch} x \coth x \, dx = -\operatorname{csch} x + C$$

Figure 6.34

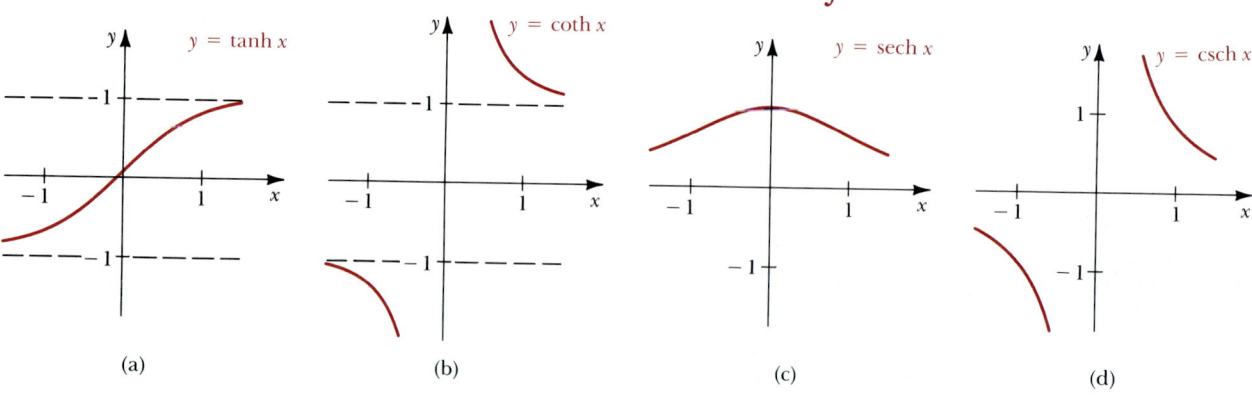

(a) $y = \tanh x$

(b) $y = \coth x$

(c) $y = \operatorname{sech} x$

(d) $y = \operatorname{csch} x$

There are identities relating the other hyperbolic functions analogous to the trigonometric identities. Two of these are:

(6.68) $\text{sech}^2 x + \tanh^2 x = 1$ and $\coth^2 x - \text{csch}^2 x = 1$.

Example 2

Find $D_x (\tanh x^3)(\text{sech } x^3)$.

Solution

Applying the Product and Chain Rules, we have

$$D_x (\tanh x^3)(\text{sech } x^3) = (\text{sech}^2 x^3)(3x^2)(\text{sech } x^3)$$
$$- (\tanh x^3)(\text{sech } x^3)(\tanh x^3) \, 3x^2$$
$$= 3x^2 \, \text{sech } x^3 [\text{sech}^2 x^3 - \tanh^2 x^3]. \quad \blacktriangleright$$

Example 3

Evaluate $\int \tanh x \, dx$.

Solution

We have
$$\int \tanh x \, dx = \int \frac{\sinh x}{\cosh x} dx.$$

With $u = \cosh x$, $du = \sinh x \, dx$, so

$$\int \tanh x \, dx = \int \frac{du}{u} = \ln|u| + C$$
$$= \ln \cosh x + C. \quad \blacktriangleright$$

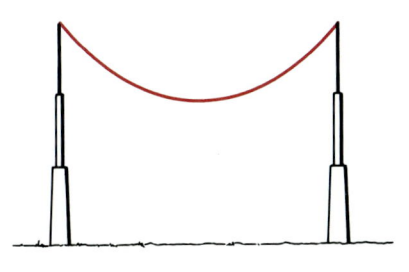

Figure 6.35
A catenary.

The hyperbolic cosine function has an important application in the representation of a curve commonly occurring in physical problems. Suppose a flexible cable of uniform density w is supported at its ends and hangs under its own weight as shown in Figure 6.35. Place an xy-coordinate system so that the x-axis lies parallel to the ground level and the y-axis is midway between the supports. The position of the cable is described by the graph of the equation

$$y = \frac{T_0}{w} \cosh \frac{wx}{T_0},$$

where T_0 is the horizontal tension on the cable at its lowest point, a distance T_0/w above the x-axis (see Figure 6.36). A curve of this type is called a **catenary**. (The name is derived from the Latin word *catena*, meaning chain.) The shape assumed by a power line between two poles, a stationary jump rope, or an unweighted necklace is a catenary. Exercises 51 and 52 consider some practical applications of catenary design.

$y = (T_0/w) \cosh (wx/T_0)$

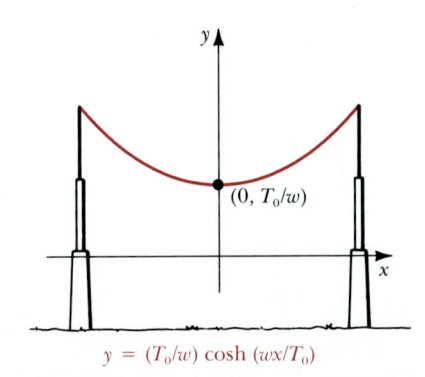

Figure 6.36
The catenary described by $y = (T_0/w) \cosh (wx/T_0)$.

EXERCISE SET 6.8

In Exercises 1–16, differentiate the function.

1. $f(x) = \sinh x^2$
2. $f(x) = \cosh(3x^2 + 2)$
3. $f(x) = \sinh\sqrt{x^2 + 2x}$
4. $f(x) = \tanh e^x$
5. $f(x) = \cosh(\ln x)$
6. $f(x) = \sinh x^3 - \cosh x^3$
7. $f(x) = \text{sech}(e^{x^3+3x})$
8. $f(x) = \ln(\coth\sqrt{x})$
9. $f(x) = (\tanh\sqrt{x^2 - x})^3$
10. $f(x) = \sin x^2 \sinh x^2$
11. $f(x) = e^{\sinh x^2}$
12. $f(x) = [\ln(\cosh x)]^2$
13. $f(x) = \dfrac{\cosh x}{x + \sinh x}$
14. $f(x) = \dfrac{e^x + e^{-x}}{\cosh x}$
15. $f(x) = \dfrac{\tanh x}{1 + x^2}$
16. $f(x) = \tanh\left(\dfrac{x}{1 + x^2}\right)$

In Exercises 17–26, evaluate the integral.

17. $\displaystyle\int \dfrac{\cosh\sqrt{x}}{\sqrt{x}}\, dx$
18. $\displaystyle\int_0^1 e^x \sinh(e^x)\, dx$
19. $\displaystyle\int (\text{sech } 2x)^2\, dx$
20. $\displaystyle\int (\tanh 2x)^2\, dx$
21. $\displaystyle\int \sinh x \cosh x\, dx$
22. $\displaystyle\int x \tanh x^2\, dx$
23. $\displaystyle\int \dfrac{\cosh x}{(\sinh x)^2}\, dx$
24. $\displaystyle\int \dfrac{x \cosh x^2}{(\sinh x^2)^2}\, dx$
25. $\displaystyle\int (\text{sech } x)^2\, e^{\tanh x}\, dx$
26. $\displaystyle\int (\text{sech } x)^2 \sqrt{\tanh x}\, dx$

In Exercises 27–32, evaluate the limit.

27. $\displaystyle\lim_{x\to 0} \dfrac{\sinh x}{\sin x}$
28. $\displaystyle\lim_{x\to 0} \dfrac{\cosh x - 1}{x^2}$
29. $\displaystyle\lim_{x\to 0} \dfrac{\ln \cosh x}{x}$
30. $\displaystyle\lim_{x\to\infty} x\, \text{sech } x$
31. $\displaystyle\lim_{x\to\infty} x \ln(\tanh x)$
32. $\displaystyle\lim_{x\to 0^+} (\coth x)(\ln x)$

In Exercises 33–44, verify the identity.

33. $\cosh x + \sinh x = e^x$
34. $\cosh x - \sinh x = e^{-x}$
35. $\text{sech}^2 x + \tanh^2 x = 1$
36. $\coth^2 x - \text{csch}^2 x = 1$
37. $\sinh(a \pm b) = \sinh a \cosh b \pm \cosh a \sinh b$
38. $\cosh(a \pm b) = \cosh a \cosh b \pm \sinh a \sinh b$
39. $\tanh(a \pm b) = \dfrac{\tanh a \pm \tanh b}{1 \pm \tanh a \tanh b}$
40. $\cosh\dfrac{x}{2} = \sqrt{\dfrac{\cosh x + 1}{2}}$
41. $\sinh\dfrac{x}{2} = \begin{cases} \sqrt{\dfrac{\cosh x - 1}{2}}, & \text{if } x \geq 0, \\ -\sqrt{\dfrac{\cosh x - 1}{2}}, & \text{if } x < 0 \end{cases}$
42. $\cosh(\ln x) + \sinh(\ln x) = x$
43. $(\cosh x + \sinh x)^n = \cosh nx + \sinh nx$
44. $(\cosh x - \sinh x)^n = \cosh nx - \sinh nx$

45. Find the volume generated by revolving the region bounded by the graph of $f(x) = \cosh x$, the x-axis, and the lines $x = 0$ and $x = a$ about the x-axis.

46. Find the volume generated by revolving the region bounded by the graph of $f(x) = \sinh x$, the x-axis, and the lines $x = 0$ and $x = a$ about the x-axis.

47. Show that the length of the curve $y = \cosh x$ from $(0, 1)$ to $(a, \cosh a)$ is $\sinh a$.

48. Show that the surface area generated by revolving the graph of $f(x) = \cosh x$ from $(0, 1)$ to $(a, \cosh a)$ about the x-axis is $(\pi/2)(2a + \sinh 2a)$.

49. Show that the area of the region bounded by the hyperbola $x^2 - y^2 = 1$ and the rays through the origin intersecting the hyperbola at $P(0) = (\cosh 0, \sinh 0)$ and $P(t) = (\cosh t, \sinh t)$ is $t/2$.

50. Show that a catenary is concave upward at each point.

51. A cable hangs between two 40-ft power poles whose bases are on the same level. The sag in the middle of the cable is 5 ft and the towers are 100 ft apart.
    a) Find the length of the cable in terms of the quotient T_0/w, where T_0/w is as shown in Figure 6.36.
    b) Use Newton's Method to approximate T_0/w and find the length of the cable accurate to within 0.1 ft.

52. The Gateway Arch in St. Louis, Missouri, has the shape of an inverted catenary with both height and width at ground level 630 ft. The arch sits on the west bank of the Mississippi river and contains elevators that take visitors through the hollow arches to an observation post at the top of the arch.
   a) Use Newton's Method to find an approximate equation for the arch. [*Hint:* Place an *xy*-coordinate system with the origin T_0/w units above the top of the arch and the positive *y*-direction directed downward.]
   b) Find the length of this arch.

The Gateway Arch in St. Louis

6.9
Inverse Hyperbolic Functions

The graph of the hyperbolic sine function shown in Figure 6.37 indicates that this is a one-to-one and increasing function. This is easily verified by observing that

$$D_x \sinh x = \cosh x > 0$$

for all values of *x*.

It follows from the Inverse Function Theorem that sinh has a differentiable inverse function that is also increasing. We call this function the **inverse hyperbolic sinh function,** denoted **arcsinh:**

(6.69) $y = \text{arcsinh } x$ if and only if $x = \sinh y$.

The Inverse Function Theorem also implies that

$$D_x \text{arcsinh } x = \frac{1}{D_y \sinh y} = \frac{1}{\cosh y}.$$

Since
$$\cosh^2 y - \sinh^2 y = 1, \quad \text{we have} \quad \cosh^2 y = 1 + \sinh^2 y$$

and

$$\cosh y = \pm \sqrt{1 + \sinh^2 y}.$$

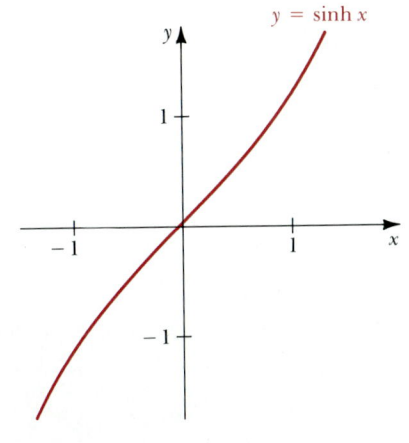

Figure 6.37

But the hyperbolic cosine function is always positive, so
$$\cosh y = \sqrt{1 + \sinh^2 y} = \sqrt{1 + x^2}$$
and

(6.70) $$D_x \operatorname{arcsinh} x = \frac{1}{\sqrt{1 + x^2}}.$$

The corresponding indefinite integral formula is
$$\int \frac{dx}{\sqrt{1 + x^2}} = \operatorname{arcsinh} x + C.$$

Note the similarity between this result and the integration formula involving the arcsine:
$$\int \frac{dx}{\sqrt{1 - x^2}} = \arcsin x + C.$$

Example 1

Find $D_x \operatorname{arcsinh}(x^2 - 1)$.

Solution
Using Equation (6.70) and the Chain Rule, we have
$$D_x \operatorname{arcsinh}(x^2 - 1) = \frac{1}{\sqrt{1 + (x^2 - 1)^2}} \cdot D_x(x^2 - 1) = \frac{2x}{\sqrt{x^4 - 2x^2 + 2}}. \blacktriangleright$$

Example 2

Show that $\operatorname{arcsinh} x = \ln(x + \sqrt{x^2 + 1})$.

Solution
Let $y = \operatorname{arcsinh} x$. Then
$$x = \sinh y = \frac{e^y - e^{-y}}{2} \quad \text{and} \quad e^y - 2x - e^{-y} = 0.$$

To solve for y in terms of x, we first multiply this equation by e^y. This gives a quadratic equation in the variable e^y,
$$(e^y)^2 - 2xe^y - 1 = 0,$$
to which the quadratic formula can be applied:
$$e^y = \frac{2x \pm \sqrt{(-2x)^2 - 4(1)(-1)}}{2} = x \pm \sqrt{x^2 + 1}.$$

The left side of this equation is always positive, which can occur only if the positive sign is chosen on the right side. Thus,
$$e^y = x + \sqrt{x^2 + 1}$$
and
$$\operatorname{arcsinh} x = y = \ln(x + \sqrt{x^2 + 1}). \blacktriangleright$$

6.9 INVERSE HYPERBOLIC FUNCTIONS

The graphs of the hyperbolic functions, tanh, coth, and csch, indicate that these are also one-to-one functions and consequently have inverse functions. If the domains of the hyperbolic cosine and secant functions are restricted to $[0, \infty)$, they also have inverse functions. The graphs of the inverse hyperbolic functions are shown in Figure 6.38.

A procedure similar to that used to find the derivative of the arcsinh can be used to find the derivatives of the other inverse hyperbolic functions. The derivative formulas for these functions are given below.

$$D_x \operatorname{arccosh} x = \frac{1}{\sqrt{x^2-1}}, \quad \text{for } x > 1$$

$$D_x \operatorname{arctanh} x = \frac{1}{1-x^2}, \quad \text{for } |x| < 1$$

$$D_x \operatorname{arccoth} x = \frac{1}{1-x^2}, \quad \text{for } |x| > 1$$

$$D_x \operatorname{arccsch} x = \frac{-1}{|x|\sqrt{1+x^2}}, \quad \text{for } x \neq 0$$

$$D_x \operatorname{arcsech} x = \frac{-1}{x\sqrt{1-x^2}}, \quad \text{for } 0 < x < 1$$

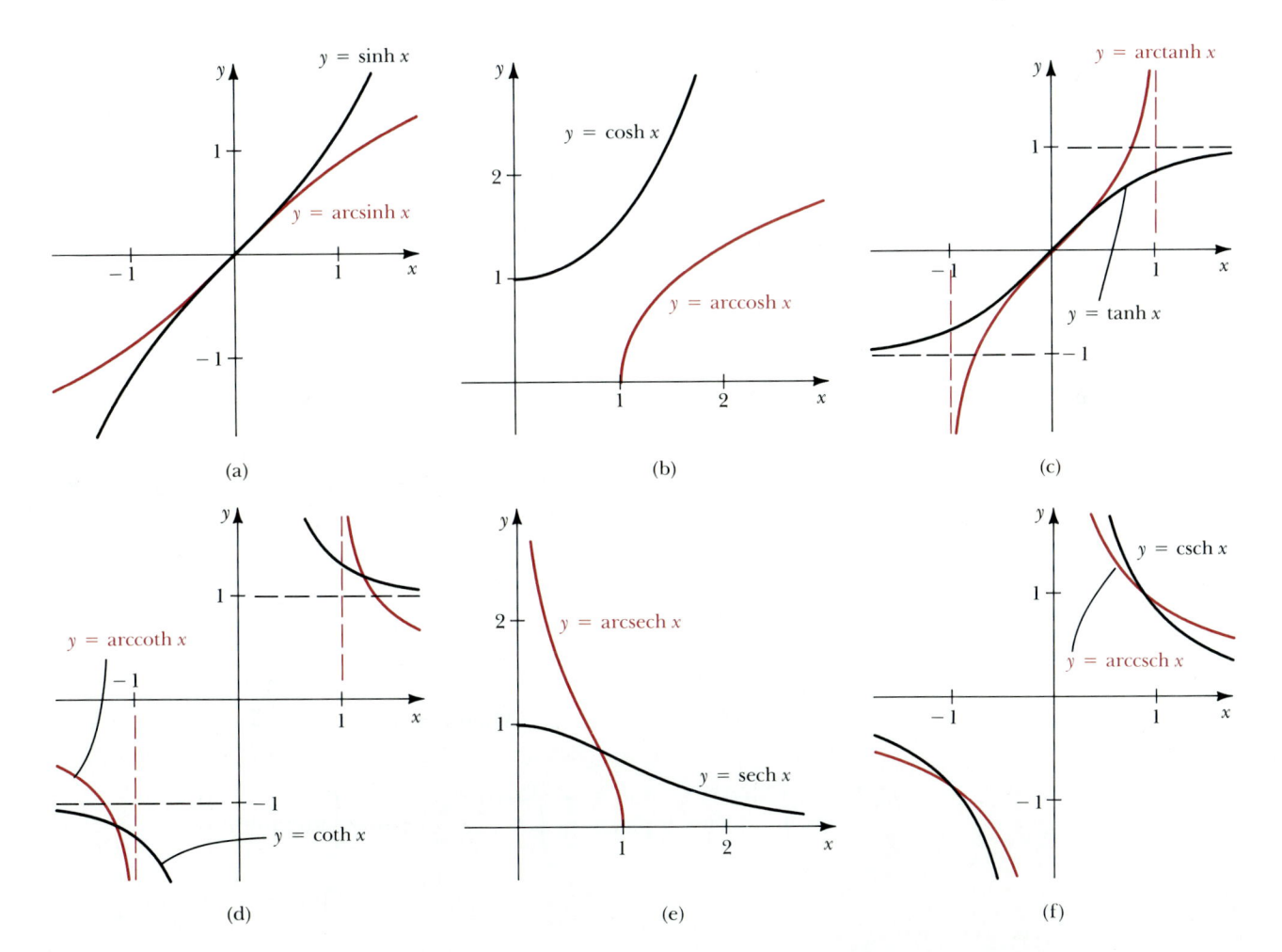

Figure 6.38

The inverse hyperbolic functions.

(a) $y = \sinh x$, $y = \operatorname{arcsinh} x$

(b) $y = \cosh x$, $y = \operatorname{arccosh} x$

(c) $y = \operatorname{arctanh} x$, $y = \tanh x$

(d) $y = \operatorname{arccoth} x$, $y = \coth x$

(e) $y = \operatorname{arcsech} x$, $y = \operatorname{sech} x$

(f) $y = \operatorname{csch} x$, $y = \operatorname{arccsch} x$

In Example 2 we found that the inverse hyperbolic sine function has the logarithmic representation arcsinh $x = \ln(x + \sqrt{x^2 + 1})$. The other inverse hyperbolic functions also have logarithmic representations, which can be verified in a manner similar to the demonstration in Example 2. The following integral formulas associated with the inverse hyperbolic functions also list the equivalent logarithmic representations. In Chapter 7 we will discuss a general technique for evaluating these integrals.

$$\int \frac{dx}{\sqrt{x^2 + 1}} = \text{arcsinh } x + C = \ln(x + \sqrt{x^2 + 1}) + C$$

$$\int \frac{1}{\sqrt{x^2 - 1}} dx = \text{arccosh } x + C = \ln(x + \sqrt{x^2 - 1}) + C, \quad \text{for } x \geq 1$$

$$\int \frac{1}{1 - x^2} dx = \begin{cases} \text{arctanh } x + C, & \text{for } |x| < 1 \\ \text{arccoth } x + C, & \text{for } |x| > 1 \end{cases} = \frac{1}{2} \ln \left| \frac{x + 1}{x - 1} \right| + C$$

$$\int \frac{1}{x\sqrt{1 + x^2}} dx = -\text{arccsch } |x| + C = \ln \frac{|x|}{1 + \sqrt{1 + x^2}} + C, \quad \text{for } |x| > 0$$

$$\int \frac{1}{x\sqrt{1 - x^2}} dx = -\text{arcsech } x + C = \ln \frac{x}{1 + \sqrt{1 - x^2}} + C, \quad \text{for } 0 < x \leq 1$$

Example 3

Evaluate $\displaystyle\int \frac{e^x}{\sqrt{1 + e^{2x}}} dx$.

Solution
Let $u = e^x$. Then $du = e^x dx$ and

$$\int \frac{e^x}{\sqrt{1 + e^{2x}}} dx = \int \frac{du}{\sqrt{1 + u^2}} = \text{arcsinh } u + C = \text{arcsinh } e^x + C.$$

Alternatively, the logarithmic representation gives

$$\int \frac{e^x}{\sqrt{1 + e^{2x}}} dx = \ln(e^x + \sqrt{e^{2x} + 1}) + C.$$ ▶

Example 4

Evaluate $\displaystyle\int_0^{1/2} \frac{x}{1 - x^4} dx$.

Solution
Let $u = x^2$; then $du = 2x\, dx$. When $x = 0$, $u = 0$, and when $x = \frac{1}{2}$, $u = \frac{1}{4}$, so

$$\int_{x=0}^{x=1/2} \frac{x}{1 - x^4} dx = \frac{1}{2} \int_{u=0}^{u=1/4} \frac{du}{1 - u^2} = \frac{1}{2} \left[\text{arctanh } u \right]_0^{1/4}$$

$$= \frac{1}{2} \left[\text{arctanh } \frac{1}{4} - \text{arctanh } 0 \right] = \frac{1}{2} \text{arctanh } \frac{1}{4}.$$

To evaluate $\operatorname{arctanh} \frac{1}{4}$, we can use the logarithmic equivalence

$$\operatorname{arctanh} x = \frac{1}{2} \ln \frac{1+x}{1-x} \quad \text{when } |x| < 1.$$

Thus,

$$\operatorname{arctanh} \frac{1}{4} = \frac{1}{2} \ln \frac{(5/4)}{(3/4)} = \frac{1}{2} \ln \frac{5}{3} \approx 0.2554$$

and

$$\int_0^{1/2} \frac{x}{1-x^4} \, dx = \frac{1}{2} \operatorname{arctanh} \frac{1}{4} \approx 0.1277.$$

▶

▶ EXERCISE SET 6.9

In Exercises 1–10, differentiate the function.

1. $f(x) = \operatorname{arcsinh} x^3$
2. $f(x) = \operatorname{arctanh}(3x^2 + 1)$
3. $f(x) = \operatorname{arccosh} \sqrt{x}$
4. $f(x) = \sqrt{\operatorname{arccosh} x}$
5. $f(x) = \sinh x^2 \operatorname{arcsinh} x^2$
6. $f(x) = \sinh(\operatorname{arcsinh} x^2)$
7. $f(x) = \operatorname{arctanh}(\tan x)$
8. $f(x) = \tan(\operatorname{arctanh} x)$
9. $f(x) = \dfrac{\operatorname{arcsinh} x}{\operatorname{arccosh} x}$
10. $f(x) = (\operatorname{arccosh} x)^2 - (\operatorname{arcsinh} x)^2$

In Exercises 11–20, evaluate the integral.

11. $\displaystyle\int_0^4 \frac{dx}{\sqrt{x^2+9}}$
12. $\displaystyle\int_0^1 \frac{dx}{\sqrt{9x^2+4}}$
13. $\displaystyle\int \frac{x \, dx}{\sqrt{x^4+9}}$
14. $\displaystyle\int \frac{\cos x \, dx}{\sqrt{1+\sin^2 x}}$
15. $\displaystyle\int \frac{dx}{\sqrt{9x^2-4}}$
16. $\displaystyle\int \frac{dx}{x\sqrt{4-9x^2}}$
17. $\displaystyle\int \frac{e^x \, dx}{1-e^{2x}}$
18. $\displaystyle\int \frac{dx}{x\sqrt{x^2+9}}$
19. $\displaystyle\int \frac{dx}{4x-x^2}$
20. $\displaystyle\int \frac{dx}{\sqrt{x^2+4x+5}}$

In Exercises 21–28, verify the identity.

21. $\operatorname{arccosh} x = \ln(x + \sqrt{x^2-1})$, for $x \geq 1$
22. $\operatorname{arctanh} x = \dfrac{1}{2} \ln \dfrac{1+x}{1-x}$, for $|x| < 1$
23. $\operatorname{arccsch} x = \operatorname{arcsinh}(1/x)$, for $x > 0$
24. $\operatorname{arcsech} x = \operatorname{arccosh}(1/x)$, for $0 < x < 1$
25. $\operatorname{arccoth} x = \operatorname{arctanh}(1/x)$, for $|x| > 1$
26. $\operatorname{arccsch} x = \ln \dfrac{1 + \sqrt{x^2+1}}{x}$, for $x > 0$
27. $\operatorname{arcsech} x = \ln \dfrac{1 + \sqrt{1-x^2}}{x}$, for $0 < x < 1$
28. $\operatorname{arccoth} x = \dfrac{1}{2} \ln \dfrac{x+1}{x-1}$, for $|x| > 1$

In Exercises 29–34, evaluate the limit.

29. $\displaystyle\lim_{x \to 0^-} \frac{\operatorname{arcsinh} x}{x}$
30. $\displaystyle\lim_{x \to 1^+} \frac{\operatorname{arccosh} x}{x-1}$
31. $\displaystyle\lim_{x \to 0^+} x \operatorname{arcsech} x$
32. $\displaystyle\lim_{x \to \infty} x \operatorname{arccsch} x$
33. $\displaystyle\lim_{x \to \infty} (\operatorname{arcsinh} x - \ln x)$
34. $\displaystyle\lim_{x \to 1^-} [2 \operatorname{arctanh} x + \ln(1-x)]$
35. Show that $\operatorname{arctanh} \frac{3}{5} = \operatorname{arcsinh} \frac{3}{4} = \operatorname{arccosh} \frac{5}{4}$.

▶ IMPORTANT TERMS AND RESULTS

CONCEPT	PAGE
One-to-one function	385
Inverse function	385
Inverse Function Theorem	389
Inverse trigonometric functions	394
Natural exponential function	406
General exponential function	413
General Power Rule	415
General logarithm function	416
Indeterminate forms 0^0, 1^∞, ∞^0	420
Exponential growth and decay	424
Separable differential equations	431
Hyperbolic functions	442
Catenary	445
Inverse hyperbolic functions	447

MAJOR DERIVATIVE FORMULAS

1. $D_x f^{-1}(x) = \dfrac{1}{D_y f(y)}$ — 392

2. $D_x \arcsin x = \dfrac{1}{\sqrt{1-x^2}}$ — 395

3. $D_x \arctan x = \dfrac{1}{1+x^2}$ — 399

4. $D_x \operatorname{arcsec} x = \dfrac{1}{|x|\sqrt{x^2-1}}$ — 402

5. $D_x e^x = e^x$ — 408
6. $D_x a^x = a^x \ln a$ — 414
7. $D_x x^r = rx^{r-1}$, for $x > 0$ and any real number r — 415
8. $D_x \log_a x = \dfrac{1}{x \ln a}$ — 417
9. $D_x \sinh x = \cosh x$ — 443
10. $D_x \cosh x = \sinh x$ — 443

MAJOR INTEGRAL FORMULAS

1. $\int \dfrac{1}{\sqrt{a^2-x^2}}\, dx = \arcsin \dfrac{x}{a} + C$ — 396

2. $\int \dfrac{1}{a^2+x^2}\, dx = \dfrac{1}{a} \arctan \dfrac{x}{a} + C$ — 399

3. $\int \dfrac{1}{x\sqrt{x^2-a^2}}\, dx = \dfrac{1}{a} \operatorname{arcsec} \dfrac{|x|}{a} + C$ — 402

4. $\int e^x\, dx = e^x + C$ — 409

5. $\int a^x\, dx = \dfrac{a^x}{\ln a} + C,\ a \neq 1$ — 414

REVIEW EXERCISES

In Exercises 1–8:
  a) Show that the function is one to one.
  b) Find f^{-1} and give its domain and range.
  c) Sketch the graphs of $y = f(x)$ and $y = f^{-1}(x)$ on the same set of axes.
  d) Use the derivative relation in the Inverse Function Theorem to determine the derivative of f^{-1}.

1. $f(x) = 2x + 3$
2. $f(x) = 1 - 2x$
3. $f(x) = x^3 - 1$
4. $f(x) = 2x^3 + 1$
5. $f(x) = x^2 + 1$, $x \geq 0$
6. $f(x) = 1 - 2x^2$, $x \geq 0$
7. $f(x) = \dfrac{1}{x+1}$
8. $f(x) = \dfrac{1}{x^2+1}$, $x > 0$

In Exercises 9–34, find dy/dx.

9. $y = \arcsin(x/3)$
10. $y = \arctan(2x - 1)$
11. $y = e^{x^3}$
12. $y = e^x \arccos x$
13. $y = 2^{x^3}$
14. $y = (e^{1/x})^2$
15. $y = e^{3 \ln x}$
16. $y = 3^{x^2} 2^x$
17. $y = xe^{\tan x}$
18. $y = \log_{10} 4x$
19. $y = \log_{10} \dfrac{x+1}{x^2+1}$
20. $y = \arccos(e^x + 2^x)$
21. $y = \ln e^{x^3}$
22. $y = \dfrac{1}{10^{2x}}$
23. $y = x - \ln(1 + e^x)$
24. $y = x^{\sqrt{2}}$
25. $y = x^e$
26. $y = \arctan(e^{1/x})$
27. $y = \dfrac{1 + e^{3x}}{\cosh x}$
28. $y = \cosh(x^2 - 1)$
29. $y = e^x \sinh(2x - 1)$
30. $y = (\tanh e^x)^3$
31. $y = x^{\cos x}$
32. $y = (\cos x)^x$
33. $\ln(x + y) = \arctan(x/y)$
34. $\cosh(x + y) = \sinh(xy)$

In Exercises 35–48, evaluate the integral.

35. $\displaystyle\int \dfrac{dx}{x^2 + 9}$
36. $\displaystyle\int_0^7 \dfrac{dx}{x^2 + 49}$
37. $\displaystyle\int_1^2 x^2 e^{x^3} \, dx$
38. $\displaystyle\int \dfrac{xe^{x^2}}{e^{x^2} + 1} \, dx$
39. $\displaystyle\int \dfrac{dx}{\sqrt{9 - x^2}}$
40. $\displaystyle\int \dfrac{dx}{x\sqrt{x^2 - 4}}$
41. $\displaystyle\int \dfrac{e^x \, dx}{\sqrt{1 - e^{2x}}}$
42. $\displaystyle\int \dfrac{dx}{x\sqrt{4x^2 - 1}}$
43. $\displaystyle\int \dfrac{dx}{x^2 - 2x + 2}$
44. $\displaystyle\int_{-2}^{-1} \dfrac{dx}{x^2 + 4x + 5}$
45. $\displaystyle\int_{-1}^1 2^x \, dx$
46. $\displaystyle\int \dfrac{e^x}{e^{2x} + 1} \, dx$
47. $\displaystyle\int \dfrac{\sinh(\ln x)}{x} \, dx$
48. $\displaystyle\int e^x \cosh e^x \, dx$

49. Find the exact value of the following expressions.
   a) $\arcsin\left(-\dfrac{1}{2}\right)$ b) $\arctan(-1)$ c) $\arccos \dfrac{\sqrt{3}}{2}$

50. Find the exact value of the following expressions.
   a) $\sin\left(\arccos \dfrac{\sqrt{2}}{2}\right)$ b) $\operatorname{arcsec} \sqrt{2}$ c) $\arctan \dfrac{\sqrt{3}}{3}$

In Exercises 51–60, find the limit.

51. $\displaystyle\lim_{x \to 0^+} x^{\ln x}$
52. $\displaystyle\lim_{x \to 0^+} (2x)^x$
53. $\displaystyle\lim_{x \to 0^+} x^{\tan x}$
54. $\displaystyle\lim_{x \to -\infty} x \tan \dfrac{1}{x}$
55. $\displaystyle\lim_{x \to \infty} (1 + x)^{1/x}$
56. $\displaystyle\lim_{x \to \infty} \left(\dfrac{x+2}{x}\right)^x$
57. $\displaystyle\lim_{x \to \infty} \left(\tan \dfrac{1}{x}\right)^x$
58. $\displaystyle\lim_{x \to \infty} (e^x)^{e^{-x}}$
59. $\displaystyle\lim_{x \to 0} \dfrac{\arcsin x}{x}$
60. $\displaystyle\lim_{x \to \infty} x \arcsin\left(\dfrac{1}{x}\right)$

In Exercises 61–66, sketch the graph of the function.

61. $f(x) = \arcsin(x - 1)$
62. $f(x) = x + \arcsin x$
63. $f(x) = 1 + 2^x$
64. $f(x) = e^{x-1}$
65. $f(x) = \sinh 2x$
66. $f(x) = \sinh(x - 1)$

In Exercises 67–72, find the general solution of the differential equation.

67. $y' = x^2 + 2x$
68. $y' = \dfrac{y}{2}$
69. $y' = e^{x+y}$
70. $y' = y \tan x$
71. $y' = \dfrac{1 + y^2}{1 + x^2}$
72. $y' = \dfrac{y}{x} \ln x$

In Exercises 73–76, find the solution to the differential equation that satisfies the given condition.

73. $y' = 2xy,\ y(0) = 5$
74. $y' = y(x + 1),\ y(0) = 1$
75. $2x(y + 1) - y\,y' = 0,\ y(1) = 0$
76. $y' = x^2 y,\ y(0) = 3$

77. Find an equation of the tangent line to the graph of $y = f^{-1}(x)$ at $(5, 1)$ if $f(x) = x^3 - x^2 + 4x + 1$.

78. Find an equation of the normal line to the graph of $y = f^{-1}(x)$ at $(3, 2)$ if $f(x) = x^3 - 2x^2 + 3$.

79. A bacteria culture is known to grow at a rate proportional to the amount present. After 1 hr, 1000 bacteria are present and after 4 hr, 3000 are present. Find (a) an expression for the number of bacteria at any time t and (b) the number of bacteria originally in the culture.

80. The radioactive isotope uranium $^{235}_{92}\text{U}$ has a half-life of 8.8×10^8 years. If 1 g was initially present, find the amount decayed by the end of 1000 yr.

81. A tank originally contains 100 gal of pure water. Starting at time $t = 0$, brine containing 3 lb of dissolved salt per gallon runs into the tank at the rate of 4 gal/min. The mixture is kept uniform by stirring, and the well-stirred mixture runs out of the tank at the same rate.
  a) Find the amount of salt remaining in the tank at the end of 30 min.
  b) Find the amount of time required to raise the amount of salt in the tank to 10 lb.

82. How long does it take for an amount of money to double if it is deposited at 6% compounded continuously?

83. A thermometer reading 70° F is taken outside, where the temperature is 0° F. After 3 min outside, the thermometer reading is 15° F. Express the reading on the thermometer as a function of time.

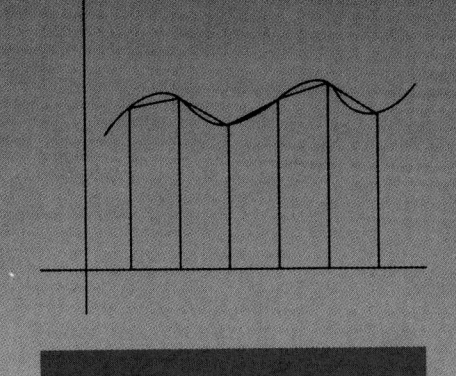

7
Techniques of Integration

In Chapters 4, 5, and 6 integral problems were solved using integration by substitution involving one or more of the basic integral formulas shown in Table 7.1:

Table 7.1

$$\int x^n \, dx = \frac{x^{n+1}}{n+1} + C, \text{ if } n \neq -1;$$

$$\int \csc^2 x \, dx = -\cot x + C;$$

$$\int \frac{1}{x} \, dx = \ln |x| + C;$$

$$\int \sec x \tan x \, dx = \sec x + C;$$

$$\int \sin x \, dx = -\cos x + C;$$

$$\int \csc x \cot x \, dx = -\csc x + C;$$

$$\int \cos x \, dx = \sin x + C;$$

$$\int e^x \, dx = e^x + C;$$

$$\int \tan x \, dx = \ln |\sec x| + C;$$

$$\int a^x \, dx = \frac{a^x}{\ln a} + C;$$

$$\int \cot x \, dx = \ln |\sin x| + C;$$

$$\int \frac{1}{\sqrt{1-x^2}} \, dx = \arcsin x + C;$$

$$\int \sec x \, dx = \ln |\sec x + \tan x| + C;$$

$$\int \frac{1}{x^2+1} \, dx = \arctan x + C;$$

$$\int \csc x \, dx = \ln |\csc x - \cot x| + C;$$

$$\int \frac{1}{x\sqrt{x^2-1}} \, dx = \text{arcsec } |x| + C.$$

$$\int \sec^2 x \, dx = \tan x + C;$$

In Section 5.4, however, we found that to determine the arc length of $y = x^2$ requires the evaluation of

$$\int \sqrt{1 + 4x^2} \, dx,$$

an integral we could not find using an elementary substitution. This is not an isolated example; there are countless other commonly occurring integrals whose values cannot be determined easily using integration by substitution.

This chapter examines some of the more frequently used techniques for evaluating integrals. Other methods are considered in the exercises. The final section in the chapter discusses two numerical methods for approximating the definite integral of a function. The widespread use of powerful computing machinery has made methods of numerical approximation increasingly important.

7.1

Integration by Parts

When the integration formulas were developed in Chapter 4, the formula corresponding to the product rule for differentiation was not given. This rule does, however, provide a useful technique of integration.

Since the product rule for differentiation is

$$D_x [f(x)\, g(x)] = f(x)\, g'(x) + g(x)\, f'(x),$$

the corresponding indefinite integral equation is

$$f(x)\, g(x) = \int D_x f(x)\, g(x)\, dx$$

$$= \int f(x)\, g'(x)\, dx + \int g(x)\, f'(x)\, dx.$$

For integration purposes this integral formula is expressed as

(7.1) $$\int f(x)\, g'(x)\, dx = f(x)\, g(x) - \int g(x)\, f'(x)\, dx$$

and called the **integration-by-parts** formula. The name indicates that the application involves splitting the integrand into two parts: $f(x)$ and $g'(x)$.

Example 1

Determine $\int x \cos x\, dx$.

Solution
We split the integrand into two parts. Let

$$f(x) = x \quad \text{and} \quad g'(x) = \cos x.$$

Then

$$f'(x) = 1 \quad \text{and} \quad g(x) = \sin x + C.$$

The integration-by-parts formula implies that

$$\int x \cos x\, dx = x(\sin x + C) - \int (\sin x + C) \cdot 1 \cdot dx$$

$$= x \sin x + Cx + \cos x - Cx + K$$

$$= x \sin x + \cos x + K. \qquad \blacktriangleright$$

Before considering more applications of the integration-by-parts technique, we introduce notation to simplify the procedure. Let

$$u = f(x) \quad \text{and} \quad v = g(x).$$

Then
$$du = f'(x)\,dx \quad \text{and} \quad dv = g'(x)\,dx,$$
so the integration-by-parts technique is given by

(7.2) $$\int u\,dv = uv - \int v\,du.$$

Example 2

Determine $\int x \ln x \, dx$.

Solution

Let $\quad u = \ln x \quad \text{and} \quad dv = x\,dx.$

Then $\quad du = \dfrac{1}{x}\,dx \quad \text{and} \quad v = \dfrac{x^2}{2} + C.$

By (7.2),
$$\int x \ln x \, dx = \ln x \left(\frac{x^2}{2} + C\right) - \int \left(\frac{x^2}{2} + C\right)\frac{1}{x}\,dx$$
$$= \frac{x^2}{2}\ln x + C \ln x - \int \frac{x}{2}\,dx - \int \frac{C}{x}\,dx$$
$$= \frac{x^2}{2}\ln x + C \ln x - \frac{x^2}{4} - C \ln x + K$$
$$= \frac{x^2}{2}\ln x - \frac{x^2}{4} + K. \quad\blacktriangleright$$

Notice in Examples 1 and 2 that the constant of integration C that arises in determining $v = g(x)$ from $dv = g'(x)\,dx$ makes no contribution to the final result. This is always the case when using integration by parts. If v is any solution to $dv = g'(x)\,dx$, then all solutions are of the form $v + C$, for some constant C. But

$$u(v+C) - \int (v+C)\,du = uv + uC - \int v\,du - C\int du$$
$$= uv + uC - \int v\,du - Cu$$
$$= uv - \int v\,du.$$

Consequently, the constant C can be chosen to be zero, as we will do in the remainder of this section.

Making the correct choice for u and dv depends on experience and a few general rules. The basic rule is:

You must be able to determine the antiderivative of dv after making your choice.

Other guidelines will be discussed as we proceed through the section.

The following example shows the use of integration by parts in an integrand where there is no obvious product.

Example 3

Determine $\int \ln x \, dx$.

Solution

Let
$$u = \ln x \quad \text{and} \quad dv = dx.$$
Then
$$du = \frac{1}{x} dx \quad \text{and} \quad v = x.$$
Thus,
$$\int \ln x \, dx = x \ln x - \int x \frac{1}{x} dx$$
$$= x \ln x - x + C. \quad \blacktriangleright$$

Since ln is one of our basic functions, the result in Example 3 is highlighted here for easy reference:

(7.3)
$$\int \ln x \, dx = x \ln x - x + C.$$

Example 4

Evaluate $\int_0^1 xe^x \, dx$.

Solution

We first determine the indefinite integral $\int x e^x \, dx$. Let
$$u = x \quad \text{and} \quad dv = e^x \, dx.$$
Then
$$du = dx \quad \text{and} \quad v = e^x.$$
Thus,
$$\int xe^x \, dx = xe^x - \int e^x \, dx = xe^x - e^x + C$$
and
$$\int_0^1 xe^x \, dx = xe^x - e^x \Big]_0^1 = (e - e) - (0 - 1) = 1. \quad \blacktriangleright$$

To illustrate the importance of making the right choice for u and dv, suppose that in Example 4 we had chosen

$$u = e^x \quad \text{and} \quad dv = x\,dx.$$

Then

$$du = e^x\,dx \quad \text{and} \quad v = \frac{x^2}{2},$$

so

$$\int xe^x\,dx = \frac{x^2}{2}e^x - \int \frac{x^2}{2}e^x\,dx.$$

Although this result is correct, it is not of use, since the integral that remains to be evaluated is more complicated than the original integral.

Example 5

Evaluate $\int_0^\pi x^2 \sin x\,dx$.

Solution

Let

$$u = x^2 \quad \text{and} \quad dv = \sin x\,dx.$$

Then

$$du = 2x\,dx \quad \text{and} \quad v = -\cos x.$$

Thus,

$$\int x^2 \sin x\,dx = -x^2 \cos x + 2\int x \cos x\,dx.$$

To evaluate $\int x \cos x\,dx$, we use integration by parts again. Let

$$u = x \quad \text{and} \quad dv = \cos x\,dx.$$

Then

$$du = dx \quad \text{and} \quad v = \sin x.$$

So

$$\int x \cos x\,dx = x \sin x - \int \sin x\,dx = x \sin x + \cos x + C.$$

Thus,

$$\int x^2 \sin x\,dx = -x^2 \cos x + 2(x \sin x + \cos x) + C.$$

The definite integral is

$$\int_0^\pi x^2 \sin x\,dx = (-x^2 \cos x + 2x \sin x + 2\cos x)\Big]_0^\pi$$

$$= [(-\pi^2(-1) - 2) - (2)] = \pi^2 - 4. \quad \blacktriangleright$$

Some problems solved using integration by parts tend to "cycle" with repeated application of the procedure. This is illustrated in the following example.

Example 6

Evaluate $\int e^x \cos x \, dx$.

Solution

Let
$$u = e^x \quad \text{and} \quad dv = \cos x \, dx.$$
Then
$$du = e^x \, dx \quad \text{and} \quad v = \sin x.$$
So
$$\int e^x \cos x \, dx = e^x \sin x - \int e^x \sin x \, dx.$$

We apply the technique again to determine $\int e^x \sin x \, dx$. Let
$$u = e^x \quad \text{and} \quad dv = \sin x \, dx.$$
Then
$$du = e^x \, dx \quad \text{and} \quad v = -\cos x.$$
So
$$\int e^x \sin x \, dx = -e^x \cos x + \int e^x \cos x \, dx.$$

The original integral is
$$\int e^x \cos x \, dx = e^x \sin x - \int e^x \sin x \, dx$$
$$= e^x \sin x + e^x \cos x - \int e^x \cos x \, dx$$

or, since the term $\int e^x \cos x \, dx$ appears on both sides of the equation,

(7.4)
$$2 \int e^x \cos x \, dx = e^x (\sin x + \cos x) + C.$$

Note that when the indefinite integral is moved from the right side of the equation, an arbitrary constant C is added. Since Equation (7.4) gives the value of twice the integral required, the solution to the problem is

$$\int e^x \cos x \, dx = \frac{e^x}{2} (\sin x + \cos x) + C.$$

▶

The following example illustrates some pitfalls to be avoided when using integration by parts. In our previous examples the choices of u and dv have been relatively straightforward. This is not the case in Example 7.

Example 7

Evaluate $\int x^3 e^{x^2} dx$.

Solution
Following the lead of Example 4: Let

$$u = x^3 \quad \text{and} \quad dv = e^{x^2} dx.$$

Then

$$du = 3x^2 dx \quad \text{and} \quad v = \int e^{x^2} dx.$$

Pitfall #1: $\int e^{x^2} dx$ cannot be evaluated.

Try again: Let

$$u = e^{x^2} \quad \text{and} \quad dv = x^3 dx.$$

Then

$$du = 2xe^{x^2} \quad \text{and} \quad v = \frac{x^4}{4}.$$

So

$$\int x^3 e^{x^2} dx = \frac{x^4}{4} e^{x^2} - \frac{1}{2} \int x^5 e^{x^2} dx.$$

Pitfall #2: A more complicated integral results.

Although there are a number of alternative choices for u and dv, all lead to the same type of difficulties, except: Let

$$u = x^2 \quad \text{and} \quad dv = x e^{x^2} dx.$$

Then

$$du = 2x\, dx \quad \text{and} \quad v = \int x e^{x^2} = \frac{1}{2} e^{x^2}.$$

So

$$\int x^3 e^{x^2} dx = \frac{x^2}{2} e^{x^2} - \int x e^{x^2} dx = \frac{x^2}{2} e^{x^2} - \frac{1}{2} e^{x^2} + C. \quad \blacktriangleright$$

The preceding example shows how successful integration-by-parts selections are generally made. The selection $dv = xe^{x^2} dx$ is the most complicated portion of the product $x^3 e^{x^2} dx$ that can be integrated readily. Reviewing the other examples in this section you can see that this rule prevails there as well.

When using integration by parts the priority of selection is for dv rather than u, and dv is generally chosen to be the most complicated portion of the product that can be integrated readily.

The following example shows the integration-by-parts technique applied to an improper integral.

Example 8

Determine whether the improper integral

$$\int_0^e \frac{\ln x}{x^2}\, dx$$

converges or diverges, and evaluate it if it converges.

Solution

First we determine the indefinite integral $\int \frac{\ln x}{x^2}\, dx$.

Let

$$u = \ln x \quad \text{and} \quad dv = \frac{1}{x^2}\, dx.$$

Then

$$du = \frac{1}{x}\, dx \quad \text{and} \quad v = -\frac{1}{x}.$$

So

$$\int \frac{\ln x}{x^2}\, dx = -\frac{\ln x}{x} + \int \frac{1}{x^2}\, dx = -\frac{\ln x}{x} - \frac{1}{x} + C.$$

Since the integrand has an infinite discontinuity at zero,

$$\int_0^e \frac{\ln x}{x^2}\, dx = \lim_{M \to 0^+} \int_M^e \frac{\ln x}{x^2}\, dx = \lim_{M \to 0^+} \left[-\frac{\ln x}{x} - \frac{1}{x} \right]_M^e$$

$$= \lim_{M \to 0^+} \left[\left(-\frac{1}{e} - \frac{1}{e} \right) - \left(-\frac{\ln M}{M} - \frac{1}{M} \right) \right]$$

$$= -\frac{2}{e} + \lim_{M \to 0^+} \frac{\ln M + 1}{M} = -\infty.$$

The integral diverges. ▶

Example 9

Use integration by parts to derive the formula

$$\int x^n e^x\, dx = x^n e^x - n \int x^{n-1} e^x\, dx.$$

Solution

Let
$$u = x^n \quad \text{and} \quad dv = e^x \, dx.$$
Then
$$du = nx^{n-1} \quad \text{and} \quad v = e^x.$$
So
$$\int x^n e^x \, dx = x^n e^x - n \int x^{n-1} e^x \, dx. \quad \blacktriangleright$$

The formula in Example 9 is called a **reduction formula** because it reduces the power of the variable in the integrand. You will find a number of these reduction formulas listed in the integral tables in Appendix D. They can all be derived using integration by parts.

▶ EXERCISE SET 7.1

In Exercises 1–48, evaluate the integral.

1. $\int xe^{-x} \, dx$
2. $\int_0^{\pi/2} x \sin x \, dx$
3. $\int_0^{\pi} x \sin 2x \, dx$
4. $\int_{-1}^{0} 3xe^{2x} \, dx$
5. $\int_1^e x \ln x \, dx$
6. $\int_1^e \sqrt{x} \ln x \, dx$
7. $\int \arcsin x \, dx$
8. $\int \arctan x \, dx$
9. $\int \arccos x \, dx$
10. $\int \text{arccot}\, x \, dx$
11. $\int x^2 e^x \, dx$
12. $\int (x^2 + 1)e^{2x} \, dx$
13. $\int_{\pi/2}^{\pi} x^2 \cos x \, dx$
14. $\int_0^1 4x^2 e^{-3x} \, dx$
15. $\int \ln x^2 \, dx$
16. $\int (\ln x)^2 \, dx$
17. $\int x^3 \ln x^2 \, dx$
18. $\int x^3 (\ln x)^2 \, dx$
19. $\int_0^4 x\sqrt{2x+1} \, dx$
20. $\int x(x-1)^5 \, dx$
21. $\int x \sec^2 x \, dx$
22. $\int \sin x \ln \cos x \, dx$
23. $\int x^3 (1-x^2)^{-1/2} \, dx$
24. $\int x^3 \sin x^2 \, dx$
25. $\int e^x \sin x \, dx$
26. $\int e^{2x} \cos x \, dx$
27. $\int e^{3x} \cos 2x \, dx$
28. $\int e^{2x} \cos 3x \, dx$
29. $\int \sec^3 x \, dx$
30. $\int_0^{\pi/2} \sin x \cos 2x \, dx$
31. $\int_0^{\pi/6} \sin 2x \sin 3x \, dx$
32. $\int x \tan x \sec x \, dx$
33. $\int \cos \sqrt{x} \, dx$
34. $\int \sin \sqrt{2x} \, dx$
35. $\int e^{\sqrt{x}} \, dx$
36. $\int_0^4 e^{\sqrt{2x+1}} \, dx$
37. $\int_0^1 x \arctan x \, dx$
38. $\int_{\sqrt{2}}^{2} x \, \text{arcsec}\, x \, dx$
39. $\int x \, 3^x \, dx$
40. $\int x \, 2^{-x} \, dx$
41. $\int x^2 \, 2^x \, dx$
42. $\int x^3 \, 3^x \, dx$
43. $\int \sin \ln x \, dx$
44. $\int \cos \ln x \, dx$
45. $\int \ln(x + \sqrt{x^2+1}) \, dx$
46. $\int (1+x)e^x \ln x \, dx$
47. $\int \sin^2 x \, dx$
48. $\int \cos^2 x \, dx$

In Exercises 49–54, (a) sketch the region described by the graphs of the equations and (b) find the area of the region.

49. $y = \ln x, y = 0, x = e^2$
50. $y = x \sin x, y = 0, x = 0, x = \pi$
51. $y = \arctan x, x = -1, x = 1, y = 0$
52. $y = \ln x, y = \dfrac{x-1}{e-1}$
53. $y = \arcsin x, y = \arccos x, y = 0$
54. $y = \arctan x, y = \text{arccot } x, y = 0, x = \sqrt{3}$

In Exercises 55–58:
a) Sketch the region described by the graphs of the equations.
b) Use the method of disks to find the volume generated when the region is revolved about the x-axis.

55. $y = \sin x, y = 0, x = \pi/4$
56. $y = \ln x, y = 0, x = e$
57. $y = \ln x, y = \dfrac{x-1}{e-1}$
58. $y = 2^x, y = 3^x, x = 1$

59–62:
Use the method of shells to find the volumes generated when the regions described in Exercises 55–58 are revolved about the y-axis.

In Exercises 63–68, derive the reduction formula.

63. $\displaystyle\int x^n e^{-x}\, dx = -x^n e^{-x} + n \int x^{n-1} e^{-x}\, dx$

64. $\displaystyle\int x^n \ln x\, dx = \dfrac{1}{n+1} x^{n+1} \ln x - \dfrac{1}{(n+1)^2} x^{n+1} + C$

65. $\displaystyle\int (\ln x)^n\, dx = x(\ln x)^n - n \int (\ln x)^{n-1}\, dx$

66. $\displaystyle\int \dfrac{1}{x(\ln x)^n}\, dx = \dfrac{-1}{(n-1)(\ln x)^{n-1}} + C, n \ne 1$

67. $\displaystyle\int x^n \sin x\, dx = -x^n \cos x + n \int x^{n-1} \cos x\, dx$

68. $\displaystyle\int x^n \cos x\, dx = x^n \sin x - n \int x^{n-1} \sin x\, dx$

69. Find the centroid of the homogeneous lamina bounded by the graphs of $y = e^x$, $x = 1$, and $y = 1 - x$.

70. Determine the fallacy in the following argument.
a) Integration by parts with $u = 1/x$ and $dv = dx$ implies that
$$\int \dfrac{1}{x}\, dx = \left(\dfrac{1}{x}\right) \cdot x - \int \left(\dfrac{-1}{x^2}\right) \cdot x\, dx = 1 + \int \dfrac{1}{x}\, dx.$$
b) Part (a) implies that
$$\int \dfrac{1}{x}\, dx - \int \dfrac{1}{x}\, dx = 1.$$
c) Part (b) implies that $0 = 1$.

71. A distribution in statistics known as the *gamma distribution* has probability density given by
$$f(x) = \begin{cases} kx^{\alpha-1} e^{-x/\beta}, & \text{for } x > 0, \\ 0, & \text{for } x \le 0, \end{cases}$$
where $\alpha > 0$ and $\beta > 0$ and k must be such that $\int_{-\infty}^{\infty} f(x)\, dx = 1$. For $\alpha = \beta = 2$, find k.

72. An *exponential distribution* has density function described by
$$f(x) = \begin{cases} e^{-x}, & \text{for } x > 0, \\ 0, & \text{for } x \le 0. \end{cases}$$
Find the mean μ and standard deviation σ of this distribution. (See Exercises 42 and 43 in Section 5.5 for the definitions of μ and σ.)

73. The *gamma function* was defined in Exercise 62 of Section 6.4 by
$$\Gamma(x) = \int_0^{\infty} t^{x-1} e^{-t}\, dt, \quad \text{for } x > 0.$$
In that exercise it was shown that $\Gamma(1) = 1$ and $\Gamma(\tfrac{1}{2}) = \sqrt{\pi}$.
a) Show that $\Gamma(x) = (x - 1)\Gamma(x - 1)$ for each $x > 1$.
b) Show that $\Gamma(n + 1) = n(n - 1) \cdots \cdots 2 \cdot 1 \equiv n!$ for each positive integer n.

74. Verify the reduction formula
$$\int_0^{\infty} x^n e^{-x^2}\, dx = \dfrac{n-1}{2} \int_0^{\infty} x^{n-2} e^{-x^2}\, dx.$$

75. Use the reduction formula in Exercise 74 together with the results in Exercise 73 and in Exercise 62 of Section 6.4 to show that
$$\int_0^{\infty} x^n e^{-x^2}\, dx = \dfrac{1}{2}\Gamma\left(\dfrac{n+1}{2}\right), \text{ for any integer } n \ge 0.$$

76. The Laplace transform of a function f was defined in Exercise 72 of Section 6.3 by $L(f) = F(s) = \int_0^{\infty} e^{-st} f(t)\, dt$. Suppose f' is continuous on $[0, \infty)$ and $|f(t)| \le Ce^{kt}$ for all $t > 0$, where C and k are positive constants. (a) Use integration by parts to show that $L(f') = sL(f) - f(0)$, whenever $s \ge k$. Assume that a is constant. Determine (b) $L(\sin at)$ and (c) $L(\cos at)$.

7.2

Integrals of Products of Trigonometric Functions

Integrals involving products and powers of trigonometric functions occur frequently in their own right as well as in the evaluation of other integrals. The first type we consider involves the products of sine and cosine functions.

The technique required to evaluate the integral

$$\int \sin^n x \cos^m x \, dx$$

for positive integers m and n depends on whether the integers are both even or at least one of the integers is odd.

(7.5)
If the power n of the sin x term is odd:

i) Make the substitution $u = \cos x$ and $du = -\sin x \, dx$.
ii) Express the integrand as $\sin^{n-1} x \cos^m x \, (\sin x \, dx)$.
iii) Use the identity $\sin^2 x = 1 - \cos^2 x$ to change $\sin^{n-1} x$ into powers of $u = \cos x$.

Example 1

Evaluate $\int \sin^3 x \cos^2 x \, dx$.

Solution
Since the power of sin x is odd, let $u = \cos x$. Then $du = -\sin x \, dx$ and

$$\int \sin^3 x \cos^2 x \, dx = \int \sin^2 x \sin x \cos^2 x \, dx$$

$$= \int (1 - \cos^2 x) \cos^2 x \sin x \, dx$$

$$= \int (\cos^2 x - \cos^4 x) \sin x \, dx$$

$$= \int (u^2 - u^4)(-du)$$

$$= -\frac{u^3}{3} + \frac{u^5}{5} + C.$$

So,

$$\int \sin^3 x \cos^2 x \, dx = -\frac{\cos^3 x}{3} + \frac{\cos^5 x}{5} + C. \quad \blacktriangleright$$

(7.6)
If the power m of the $\cos x$ term is odd:

> **i)** Make the substitution $u = \sin x$ and $du = \cos x \, dx$.
> **ii)** Express the integrand as $\sin^n x \cos^{m-1} x \, (\cos x \, dx)$.
> **iii)** Use the identity $\cos^2 x = 1 - \sin^2 x$ to change $\cos^{m-1} x$ into powers of $u = \sin x$.

Example 2

Evaluate $\int \sin^2 x \cos^5 x \, dx$.

Solution
Since the power of $\cos x$ is odd, let $u = \sin x$. Then $du = \cos x \, dx$ and

$$\int \sin^2 x \cos^5 x \, dx = \int \sin^2 x \cos^4 x \cos x \, dx$$

$$= \int \sin^2 x \, (1 - \sin^2 x)^2 \cos x \, dx$$

$$= \int u^2 (1 - u^2)^2 \, du$$

$$= \int u^2 (1 - 2u^2 + u^4) \, du$$

$$= \int (u^2 - 2u^4 + u^6) \, du$$

$$= \frac{u^3}{3} - \frac{2u^5}{5} + \frac{u^7}{7} + C.$$

So,

$$\int \sin^2 x \cos^5 x \, dx = \frac{\sin^3 x}{3} - \frac{2 \sin^5 x}{5} + \frac{\sin^7 x}{7} + C. \quad \blacktriangleright$$

(7.7)
If both powers n and m are even, use the identities

$$\sin^2 x = \frac{1}{2}(1 - \cos 2x) \quad \text{and} \quad \cos^2 x = \frac{1}{2}(1 + \cos 2x)$$

to reduce the powers of the sine and cosine in the integrand.

If both powers n and m are odd, choose (7.5) if $n < m$ and choose (7.6) if $m < n$. If $n = m$, the choice is immaterial.

Example 3

Evaluate $\int \sin^2 x \cos^2 x \, dx$.

Solution

Since both powers are even, we use (7.7) and the identities

$$\sin^2 x = \frac{1}{2}(1 - \cos 2x) \quad \text{and} \quad \cos^2 x = \frac{1}{2}(1 + \cos 2x).$$

We then have

$$\int \sin^2 x \cos^2 x \, dx = \frac{1}{4} \int (1 - \cos 2x)(1 + \cos 2x) \, dx$$

$$= \frac{1}{4} \int (1 - \cos^2 2x) \, dx.$$

However,

$$\cos^2 2x = \frac{1}{2}(1 + \cos 4x),$$

so

$$\int \sin^2 x \cos^2 x \, dx = \frac{1}{4} \int \left[1 - \frac{1}{2}(1 + \cos 4x)\right] dx$$

$$= \frac{1}{4} \int \left(\frac{1}{2} - \frac{1}{2}\cos 4x\right) dx$$

$$= \frac{1}{4}\left(\frac{x}{2} - \frac{\sin 4x}{8}\right) + C$$

$$= \frac{1}{8} x - \frac{1}{32} \sin 4x + C. \quad \blacktriangleright$$

Example 4

Evaluate $\int \dfrac{\cos^3 x}{\sqrt{\sin x}} \, dx.$

Solution

Although this integral contains a nonintegral power of sin x, the technique is the same as in the previous examples. First note that

$$\int \frac{\cos^3 x}{\sqrt{\sin x}} \, dx = \int \frac{\cos^2 x}{\sqrt{\sin x}} \cos x \, dx = \int \frac{1 - \sin^2 x}{\sqrt{\sin x}} \cos x \, dx.$$

Since the power of the cosine is odd, we use the method described in (7.6). Let $u = \sin x$; then $du = \cos x \, dx$ and

$$\int \frac{\cos^3 x}{\sqrt{\sin x}} \, dx = \int \frac{1 - u^2}{\sqrt{u}} \, du$$

$$= \int (u^{-1/2} - u^{3/2}) \, du$$

$$= 2u^{1/2} - \frac{2}{5} u^{5/2} + C.$$

So
$$\int \frac{\cos^3 x}{\sqrt{\sin x}}\, dx = 2\,(\sin x)^{1/2} - \frac{2}{5}\,(\sin x)^{5/2} + C.$$
▶

The technique involved in the integration of
$$\int \tan^n x \sec^m x\, dx$$
is similar to that used for the powers of sines and cosines. Three approaches are again needed, depending on whether n and m are odd or even.

(7.8)
If the power m of the sec x term is even:

i) Make the substitution $u = \tan x$ and $du = \sec^2 x\, dx$.
ii) Express the integrand as $\tan^n x \sec^{m-2} x\, (\sec^2 x\, dx)$.
iii) Use the identity $\sec^2 x = 1 + \tan^2 x$ to change $\sec^{m-2} x$ into powers of $u = \tan x$.

(7.9)
If both powers n and m are odd:

i) Make the substitution $u = \sec x$ and $du = \sec x \tan x\, dx$.
ii) Express the integrand as $\tan^{n-1} x \sec^{m-1} x\, (\sec x \tan x\, dx)$.
iii) Use the identity $\tan^2 x = \sec^2 x - 1$ to change $\tan^{n-1} x$ into powers of $u = \sec x$.

(7.10)
If the power n of the tan x term is even and the power m of the sec x term is odd, use integration by parts with
$$dv = \sec x \tan x\, dx$$
to reduce the powers of the tangent and secant in the integrand.

Example 5

Evaluate $\int \tan x \sec^4 x\, dx$.

Solution
Since the power of sec x is even, use (7.8) and let $u = \tan x$. Then $du = \sec^2 x\, dx$ and
$$\int \tan x \sec^4 x\, dx = \int u \sec^2 x\, du.$$
Since $\sec^2 x = 1 + \tan^2 x = 1 + u^2$,
$$\int \tan x \sec^4 x\, dx = \int u(1 + u^2)\, du = \frac{u^2}{2} + \frac{u^4}{4} + C.$$

So,
$$\int \tan x \sec^4 x \, dx = \frac{\tan^2 x}{2} + \frac{\tan^4 x}{4} + C.$$ ▶

Example 6

Evaluate $\int \tan^5 x \sec x \, dx$.

Solution
Since both powers are odd, we use (7.9) and let $u = \sec x$. Then $du = \sec x \tan x \, dx$ and

$$\int \tan^5 x \sec x \, dx = \int \tan^4 x \tan x \sec x \, dx$$
$$= \int [\sec^2 x - 1]^2 \tan x \sec x \, dx$$
$$= \int [u^2 - 1]^2 \, du$$
$$= \int (u^4 - 2u^2 + 1) \, du = \frac{u^5}{5} - \frac{2u^3}{3} + u + C.$$

So,
$$\int \tan^5 x \sec x \, dx = \frac{\sec^5 x}{5} - \frac{2 \sec^3 x}{3} + \sec x + C.$$ ▶

Example 7

Evaluate $\int \tan^2 x \sec x \, dx$.

Solution
Since the power of $\tan x$ is even and the power of $\sec x$ is odd, use (7.10) and integration by parts. Let

$$u = \tan x \quad \text{and} \quad dv = \tan x \sec x \, dx.$$

Then
$$du = \sec^2 x \, dx \quad \text{and} \quad v = \sec x,$$

so
$$\int \tan^2 x \sec x \, dx = \sec x \tan x - \int \sec^3 x \, dx.$$

But
$$\sec^3 x = \sec^2 x \sec x = (1 + \tan^2 x) \sec x = \sec x + \tan^2 x \sec x.$$

Thus,

$$\int \tan^2 x \sec x \, dx = \sec x \tan x - \int \sec x \, dx - \int \tan^2 x \sec x \, dx$$

and

$$2 \int \tan^2 x \sec x \, dx = \sec x \tan x - \int \sec x \, dx.$$

Hence,

$$\int \tan^2 x \sec x \, dx = \frac{1}{2} \left(\sec x \tan x - \int \sec x \, dx \right)$$

$$= \frac{1}{2} (\sec x \tan x - \ln |\sec x + \tan x|) + C. \quad \blacktriangleright$$

The following example illustrates how to evaluate integrals of the type

$$\int \cot^n x \csc^m x \, dx.$$

Note the similarity between this procedure and the one used to evaluate

$$\int \tan^n x \sec^m x \, dx.$$

Example 8

Evaluate $\int \csc^4 x \cot^2 x \, dx$.

Solution
Since $D_x \cot x = -\csc^2 x$ and $\csc^2 x = \cot^2 x + 1$, we make the substitution $u = \cot x$. Then $du = -\csc^2 x \, dx$ and

$$\int \csc^4 x \cot^2 x \, dx = \int \csc^2 x \cot^2 x \, (\csc^2 x \, dx)$$

$$= - \int (\cot^2 x + 1) \cot^2 x \, (\csc^2 x \, dx)$$

$$= - \int (u^2 + 1) u^2 \, du$$

$$= - \int (u^4 + u^2) \, du$$

$$= - \left(\frac{u^5}{5} + \frac{u^3}{3} \right) + C.$$

So,

$$\int \csc^4 x \cot^2 x \, dx = - \frac{\cot^5 x}{5} - \frac{\cot^3 x}{3} + C. \quad \blacktriangleright$$

Some common applications require the evaluation of integrals of the form

$$\int \sin ax \cos bx \, dx, \quad \int \sin ax \sin bx \, dx, \quad \text{and} \quad \int \cos ax \cos bx \, dx.$$

When $a = b$, these integrals are evaluated using the method described for products of sines and cosines. This method is also appropriate when $a = -b$, since the sine is an odd function and the cosine is an even function.

Let us consider

$$\int \sin ax \cos bx \, dx \quad \text{when} \quad a \neq \pm b.$$

We first replace the product $\sin ax \cos bx$ in the integrand using the trigonometric identities:

$$\sin (a + b)x = \sin ax \cos bx + \sin bx \cos ax$$

and

$$\sin (a - b)x = \sin ax \cos bx - \sin bx \cos ax.$$

Adding these identities gives

$$\sin (a + b)x + \sin (a - b)x = 2 \sin ax \cos bx,$$

so

$$\int \sin ax \cos bx \, dx = \int \frac{1}{2} [\sin (a + b)x + \sin (a - b)x] \, dx$$

and

(7.11) $$\int \sin ax \cos bx \, dx = -\frac{1}{2} \left[\frac{\cos (a + b)x}{a + b} + \frac{\cos (a - b)x}{a - b} \right] + C.$$

In a similar manner, we use the trigonometric identities

$$\cos (a + b)x = \cos ax \cos bx - \sin ax \sin bx$$

and

$$\cos (a - b)x = \cos ax \cos bx + \sin ax \sin bx$$

to show that when $a \neq \pm b$,

(7.12) $$\int \sin ax \sin bx \, dx = \frac{1}{2} \left[\frac{\sin (a - b)x}{a - b} - \frac{\sin (a + b)x}{a + b} \right] + C$$

and

(7.13) $$\int \cos ax \cos bx \, dx = \frac{1}{2} \left[\frac{\sin (a - b)x}{a - b} + \frac{\sin (a + b)x}{a + b} \right] + C.$$

Example 9

Evaluate $\int \sin 3x \cos 7x \, dx$.

Solution

$$\int \sin 3x \cos 7x \, dx = \int \frac{1}{2} [\sin 10x + \sin(-4x)] \, dx$$

$$= \int \frac{1}{2} [\sin 10x - \sin 4x] \, dx$$

$$= \frac{1}{20}(-\cos 10x) - \frac{1}{8}(-\cos 4x) + C$$

$$= \frac{1}{8} \cos 4x - \frac{1}{20} \cos 10x + C \qquad \blacktriangleright$$

▶ **EXERCISE SET 7.2**

In Exercises 1–46, evaluate the integral.

1. $\int \sin^2 x \cos x \, dx$

2. $\int \cos^3 x \, dx$

3. $\int_0^{\pi/2} \sin^2 2x \cos^3 2x \, dx$

4. $\int \sin^5(x+1) \, dx$

5. $\int \sin 3x \cos^4 3x \, dx$

6. $\int_{\pi/4}^{\pi/2} \sin^2 x \cos^2 x \, dx$

7. $\int_0^{\pi/6} \sin^2 3x \cos^4 3x \, dx$

8. $\int_{\pi/4}^{\pi/2} \sin^4 2x \cos^3 2x \, dx$

9. $\int_0^{\pi} \cos^4 5x \, dx$

10. $\int \sin^4 2x \, dx$

11. $\int \sqrt{\sin x} \cos x \, dx$

12. $\int \sqrt{\sin x} \cos^3 x \, dx$

13. $\int_{-\pi/2}^{\pi/2} (\sin x + \cos x)^3 \, dx$

14. $\int_{-\pi/2}^{\pi/2} (\sin x + \cos x)^2 \, dx$

15. $\int \sec^4 x \tan^3 x \, dx$

16. $\int \tan^3 x \sec^2 x \, dx$

17. $\int_{\pi/4}^{\pi/2} \cot^2 x \, dx$

18. $\int_0^{\pi/2} \left(\tan \frac{x}{2}\right)^2 dx$

19. $\int \cot x \csc^4 x \, dx$

20. $\int \csc^2 5x \cot^2 5x \, dx$

21. $\int \cot^5 x \csc x \, dx$

22. $\int \cot x \csc^3 x \, dx$

23. $\int \cot^3 2x \, dx$

24. $\int x(\cot x^2 \csc x^2)^2 \, dx$

25. $\int (\tan x + \cos x)^2 \, dx$

26. $\int (\tan 2x + \cot 2x)^2 \, dx$

27. $\int \cos^3 x \ln(\sin x) \, dx$

28. $\int \sec^4 x \ln(\tan x) \, dx$

29. $\displaystyle\int e^{\sin x} \cos^3 x \, dx$

30. $\displaystyle\int e^{\tan x} \sec^4 x \, dx$

31. $\displaystyle\int \cos^2 x \tan x \, dx$

32. $\displaystyle\int \cos^2 x \tan^5 x \, dx$

33. $\displaystyle\int \frac{\tan x}{\sec x} \, dx$

34. $\displaystyle\int \frac{\tan x}{\sec^3 x} \, dx$

35. $\displaystyle\int \frac{\sec x}{\tan x} \, dx$

36. $\displaystyle\int \frac{\sec^2 x}{\tan^3 x} \, dx$

37. $\displaystyle\int \tan^3 x \, dx$

38. $\displaystyle\int \sin^2 x \cot^3 x \, dx$

39. $\displaystyle\int \frac{\cot x}{\csc^4 x} \, dx$

40. $\displaystyle\int \csc^3 x \, dx$

41. $\displaystyle\int \sin x \cos 2x \, dx$

42. $\displaystyle\int \sin 2x \cos x \, dx$

43. $\displaystyle\int \sin 2x \cos 3x \, dx$

44. $\displaystyle\int \sin 2x \sin 4x \, dx$

45. $\displaystyle\int \cos 3x \cos 5x \, dx$

46. $\displaystyle\int \sin 7x \cos x \, dx$

In Exercises 47–50, (a) sketch the region bounded by the graphs of the equations and (b) find the area of the region.

47. $y = \cos^2 x$, $x = 0$, $x = \pi/4$, $y = 0$
48. $y = \sin^2 x$, $y = \cos^2 x$, $x = 0$, $x = \pi/4$
49. $y = \sin^2 x$, $y = \sin^2 2x$, $x = \pi/4$, $x = \pi/2$
50. $y = \sin^3 x$, $y = \sin^3 2x$, $x = \pi/4$, $x = \pi/2$

In Exercises 51–54:
a) Sketch the region bounded by the graphs of the equations.
b) Use the method of disks to find the volume generated when the region is revolved about the x-axis.

51. $y = \sin x$, $y = 0$, $x = 0$, $x = \pi$
52. $y = \sin x$, $y = \cos x$, $x = 0$, $x = \pi/2$
53. $y = \tan x$, $y = 0$, $x = \pi/4$
54. $y = \tan x$, $y = \sec x$, $x = 0$, $x = \pi/4$

The reduction formulas in Exercises 55–60 are listed in the integration tables in Appendix D. Use the methods of this section and integration by parts to derive these formulas.

55. $\displaystyle\int (\sin x)^n \, dx = -\frac{1}{n} (\sin x)^{n-1} \cos x$
$\displaystyle\qquad + \frac{n-1}{n} \int (\sin x)^{n-2} \, dx$

56. $\displaystyle\int (\cos x)^n \, dx = \frac{1}{n} (\cos x)^{n-1} \sin x$
$\displaystyle\qquad + \frac{n-1}{n} \int (\cos x)^{n-2} \, dx$

57. $\displaystyle\int (\tan x)^n \, dx = \frac{1}{n-1} (\tan x)^{n-1} - \int (\tan x)^{n-2} \, dx$

58. $\displaystyle\int (\cot x)^n \, dx = \frac{-1}{n-1} (\cot x)^{n-1} - \int (\cot x)^{n-2} \, dx$

59. $\displaystyle\int (\sec x)^n \, dx = \frac{1}{n-1} (\sec x)^{n-2} \tan x$
$\displaystyle\qquad + \frac{n-2}{n-1} \int (\sec x)^{n-2} \, dx$

60. $\displaystyle\int (\csc x)^n \, dx = \frac{-1}{n-1} (\csc x)^{n-2} \cot x$
$\displaystyle\qquad + \frac{n-2}{n-1} \int (\csc x)^{n-2} \, dx$

61. Find the centroid of the homogenous lamina bounded by the graphs of $y = \sin x$, $y = \cos x$, $x = 0$, and $x = \pi/4$.

62. A particle is traveling with velocity in ft/sec of $v(t) = (\cos (\pi t/3))^2$. Find the distance traveled by the particle from $t = 0$ to $t = 1$.

63. An application of the identity $\sin 2x = 2 \sin x \cos x$ can be used to simplify the integration of $\int \sin^n x \cos^n x \, dx$ when n is even. Use this technique on the integral in Example 3.

64. A set of functions $\{f_1, f_2, \ldots, f_n\}$ is *orthogonal* on an interval $[a, b]$ if $\int_a^b f_i(x) f_j(x) \, dx = 0$ whenever $i \ne j$. Consider the set of functions defined by $\{1, \cos x, \ldots, \cos nx, \sin x, \ldots, \sin nx\}$. Show that for each positive integer n, this set is orthogonal on $[-\pi, \pi]$.

65. A set of functions $\{f_1, f_2, \ldots, f_n\}$ is *orthonormal* on $[a, b]$ if, in addition to being orthogonal on $[a, b]$, $\int_a^b [f_i(x)]^2 \, dx = 1$ for each i. Show that, for a positive integer n, the set of functions defined by $\{1/\sqrt{2\pi}, (1/\sqrt{\pi}) \cos x, \ldots, (1/\sqrt{\pi}) \cos nx, (1/\sqrt{\pi}) \sin x, \ldots, (1/\sqrt{\pi}) \sin nx\}$ is orthonormal on $[-\pi, \pi]$.

7.3
Trigonometric Substitution

Integrals containing an expression of the form
$$\sqrt{a^2-x^2}, \quad \sqrt{a^2+x^2}, \quad \text{or} \quad \sqrt{x^2-a^2},$$
for a constant a, can often be simplified by making an appropriate trigonometric substitution to eliminate the radical. The substitutions are based on the identity
$$a^2 = a^2(\sin^2\theta + \cos^2\theta)$$
or its equivalent form involving tangents and secants,
$$a^2(1 + \tan^2\theta) = a^2\sec^2\theta.$$

When an integral involves an expression of the form
$$\sqrt{a^2-x^2}, \quad a > 0,$$
we make the substitution
$$x = a\sin\theta, \quad \text{where } -\frac{\pi}{2} \le \theta \le \frac{\pi}{2}.$$

This changes the radical to
$$\sqrt{a^2-x^2} = \sqrt{a^2 - a^2\sin^2\theta} = \sqrt{a^2\cos^2\theta}.$$
Since $-\pi/2 \le \theta \le \pi/2$, $\cos\theta \ge 0$. So $\sqrt{a^2\cos^2\theta} = a\cos\theta$ and
$$\sqrt{a^2-x^2} = a\cos\theta, \quad \text{where } -\frac{\pi}{2} \le \theta \le \frac{\pi}{2}.$$

This substitution is interpreted geometrically in Figure 7.1. In this figure, the value of x is positive when $0 < \theta \le \pi/2$, but is negative when $-\pi/2 \le \theta < 0$.

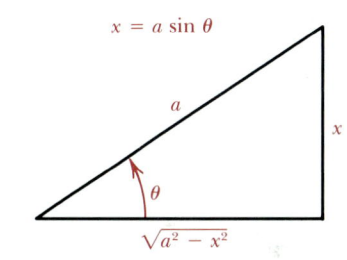

Figure 7.1

Example 1

Evaluate $\int \sqrt{4-x^2}\, dx$.

Solution
Let $x = 2\sin\theta$. Then $dx = 2\cos\theta\, d\theta$ and
$$\int \sqrt{4-x^2}\, dx = \int \sqrt{4 - 4\sin^2\theta}\, 2\cos\theta\, d\theta$$
$$= \int (2\cos\theta)\, 2\cos\theta\, d\theta$$
$$= 4\int \cos^2\theta\, d\theta.$$

The trigonometric identity

$$\cos^2 \theta = \frac{1}{2}(1 + \cos 2\theta)$$

reduces the integral to

$$\int \sqrt{4 - x^2}\, dx = 4 \int \frac{1}{2}(1 + \cos 2\theta)\, d\theta$$
$$= 2\theta + \sin 2\theta + C$$
$$= 2\theta + 2 \sin \theta \cos \theta + C.$$

The solution will be complete when we rewrite this result in terms of the variable x. Since

$$x = 2 \sin \theta, \quad \text{we have} \quad \theta = \arcsin\left(\frac{x}{2}\right).$$

The triangle in Figure 7.2 can be used to find the value of $\cos \theta$:

$$\cos \theta = \frac{\sqrt{4 - x^2}}{2}.$$

Thus,

$$\int \sqrt{4 - x^2}\, dx = 2 \arcsin\left(\frac{x}{2}\right) + \frac{x}{2}\sqrt{4 - x^2} + C. \quad \blacktriangleright$$

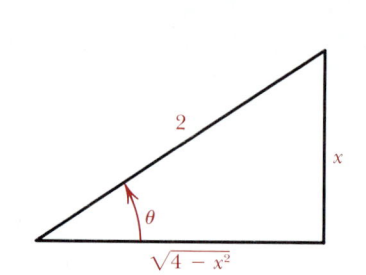

Figure 7.2

Example 2

Evaluate $\int \frac{\sqrt{9 - x^2}}{x}\, dx$.

Solution
Let $x = 3 \sin \theta$. Then $dx = 3 \cos \theta\, d\theta$ and

$$\int \frac{\sqrt{9 - x^2}}{x}\, dx = \int \frac{\sqrt{9 - 9 \sin^2 \theta}}{3 \sin \theta} 3 \cos \theta\, d\theta$$
$$= \int \frac{3 \cos \theta}{\sin \theta} \cos \theta\, d\theta$$
$$= 3 \int \frac{\cos^2 \theta}{\sin \theta}\, d\theta = 3 \int \frac{1 - \sin^2 \theta}{\sin \theta}\, d\theta$$
$$= 3 \int \csc \theta\, d\theta - 3 \int \sin \theta\, d\theta$$
$$= 3 \ln|\csc \theta - \cot \theta| + 3 \cos \theta + C.$$

Figure 7.3 shows that

$$\csc \theta = \frac{3}{x}, \quad \cot \theta = \frac{\sqrt{9 - x^2}}{x}, \quad \text{and} \quad \cos \theta = \frac{\sqrt{9 - x^2}}{3}.$$

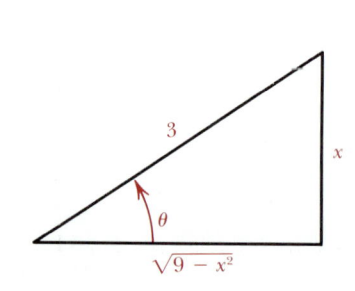

Figure 7.3

Thus,

$$\int \frac{\sqrt{9-x^2}}{x} dx = 3 \ln |\csc \theta - \cot \theta| + 3 \cos \theta + C$$

$$= 3 \ln \left| \frac{3}{x} - \frac{\sqrt{9-x^2}}{x} \right| + \sqrt{9-x^2} + C.$$ ▶

When an integral involves an expression of the form

$$\sqrt{a^2 + x^2}, \quad a > 0,$$

we make the substitution

$$x = a \tan \theta, \quad \text{where } -\frac{\pi}{2} < \theta < \frac{\pi}{2}.$$

This simplifies the radical to

$$\sqrt{a^2 + x^2} = \sqrt{a^2 + a^2 \tan^2 \theta} = \sqrt{a^2 \sec^2 \theta}.$$

Since $\sec \theta > 0$ for θ satisfying $-\pi/2 < \theta < \pi/2$,

$$\sqrt{a^2 + x^2} = a \sec \theta.$$

A geometric interpretation of this substitution is shown in Figure 7.4. In this figure, we see that the value of x is positive when $0 < \theta < \pi/2$, but is negative when $-\pi/2 < \theta < 0$.

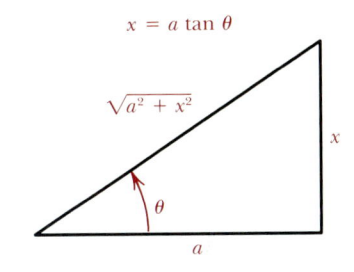

Figure 7.4

Example 3

Evaluate $\int \frac{1}{\sqrt{9+x^2}} dx$.

Solution
Let $x = 3 \tan \theta$. Then $dx = 3 \sec^2 \theta \, d\theta$ and

$$\int \frac{1}{\sqrt{9+x^2}} dx = \int \frac{3 \sec^2 \theta}{3 \sec \theta} d\theta = \int \sec \theta \, d\theta = \ln |\sec \theta + \tan \theta| + C.$$

Figure 7.5 shows that

$$\sec \theta = \frac{\sqrt{9+x^2}}{3} \quad \text{and} \quad \tan \theta = \frac{x}{3},$$

so

$$\int \frac{1}{\sqrt{9+x^2}} dx = \ln \left| \frac{1}{3} \sqrt{9+x^2} + \frac{x}{3} \right| + C = \ln |\sqrt{9+x^2} + x| - \ln 3 + C.$$

Since C is an arbitrary constant and $-\ln 3$ is also a constant, it is standard practice to replace the sum $-\ln 3 + C$ by the arbitrary constant symbol C. In addition, $\sqrt{9+x^2} + x$ is always positive so

$$\int \frac{1}{\sqrt{9+x^2}} dx = \ln (\sqrt{9+x^2} + x) + C.$$ ▶

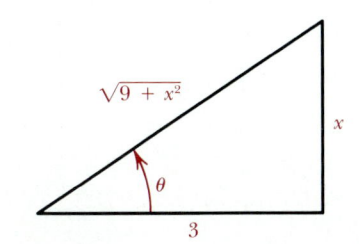

Figure 7.5

When the integral involves an expression of the form
$$\sqrt{x^2 - a^2}, \quad a > 0,$$
we make the substitution
$$x = a \sec \theta, \quad \text{where } 0 \leq \theta \leq \pi, \quad \theta \neq \frac{\pi}{2}.$$

This simplifies the radical to
$$\sqrt{x^2 - a^2} = \sqrt{a^2 \sec^2 \theta - a^2} = \sqrt{a^2(\sec^2 \theta - 1)} = \sqrt{a^2 \tan^2 \theta},$$
so
$$\sqrt{x^2 - a^2} = \begin{cases} a \tan \theta, & \text{for } 0 \leq \theta < \dfrac{\pi}{2}, \\ -a \tan \theta, & \text{for } \dfrac{\pi}{2} < \theta \leq \pi. \end{cases}$$

The difference in sign occurs because $\tan \theta \geq 0$ for $0 \leq \theta < \pi/2$ and $\tan \theta \leq 0$ for $\pi/2 < \theta \leq \pi$.

Example 4

Evaluate $\displaystyle\int_2^4 \frac{\sqrt{x^2 - 4}}{x} \, dx$.

Solution

Let $x = 2 \sec \theta$. Then $dx = 2 \sec \theta \tan \theta \, d\theta$. When
$$x = 2, \quad 1 = \sec \theta, \quad \text{so} \quad \theta = \text{arcsec } 1 = 0;$$
and when
$$x = 4, \quad 2 = \sec \theta, \quad \text{so} \quad \theta = \text{arcsec } 2 = \pi/3.$$
Thus
$$\int_{x=2}^{x=4} \frac{\sqrt{x^2 - 4}}{x} \, dx = \int_{\theta=0}^{\theta=\pi/3} \frac{\sqrt{4 \sec^2 \theta - 4}}{2 \sec \theta} \, 2 \sec \theta \tan \theta \, d\theta$$
$$= 2 \int_0^{\pi/3} \sqrt{\tan^2 \theta} \tan \theta \, d\theta.$$

But $\tan \theta$ is positive for θ in the interval $[0, \pi/3]$, so $\sqrt{\tan^2 \theta} = \tan \theta$ and
$$\int_{x=2}^{x=4} \frac{\sqrt{x^2 - 4}}{x} \, dx = 2 \int_{\theta=0}^{\theta=\pi/3} \tan^2 \theta \, d\theta$$
$$= 2 \int_0^{\pi/3} (\sec^2 \theta - 1) \, d\theta$$
$$= 2(\tan \theta - \theta) \Big]_0^{\pi/3}$$
$$= 2[(\tan \pi/3 - \pi/3) - (\tan 0 - 0)]$$
$$= 2\sqrt{3} - 2\pi/3.$$

▶

Integration by trigonometric substitutions provides an easy method for determining the general integration formulas associated with the inverse trigonometric functions. For example, to evaluate

$$\int \frac{1}{a^2 + x^2} \, dx,$$

let $x = a \tan \theta$ and $dx = a \sec^2 \theta \, d\theta$. Then

$$\int \frac{1}{a^2 + x^2} \, dx = \int \frac{a \sec^2 \theta \, d\theta}{a^2 + a^2 \tan^2 \theta}$$

$$= \int \frac{a \sec^2 \theta \, d\theta}{a^2 \sec^2 \theta} = \int \frac{1}{a} \, d\theta = \frac{1}{a} \theta + C.$$

Thus,

$$\int \frac{1}{a^2 + x^2} \, dx = \frac{1}{a} \arctan \frac{x}{a} + C.$$

The examples in this section demonstrate the basic procedure used to evaluate integrals requiring trigonometric substitution. The restrictions on θ in the trigonometric substitutions are made to ensure that x/a is in the domain of the appropriate inverse trigonometric function. Keeping these restrictions in mind, we have listed the substitutions, for $a > 0$, in Table 7.2.

Table 7.2

TERM IN THE ORIGINAL INTEGRAND	TRIGONOMETRIC SUBSTITUTION	SIMPLIFIED TERM IN THE INTEGRAND
$\sqrt{a^2 - x^2}$	$x = a \sin \theta$	$\sqrt{a^2 - a^2 \sin^2 \theta} = \sqrt{a^2 \cos^2 \theta} = a \cos \theta$
$\sqrt{a^2 + x^2}$	$x = a \tan \theta$	$\sqrt{a^2 + a^2 \tan^2 \theta} = \sqrt{a^2 \sec^2 \theta} = a \sec \theta$
$\sqrt{x^2 - a^2}$	$x = a \sec \theta$	$\sqrt{a^2 \sec^2 \theta - a^2} = \sqrt{a^2 \tan^2 \theta} = \pm a \tan \theta$

▶ EXERCISE SET 7.3

In Exercises 1–28, evaluate the integral.

1. $\int \sqrt{1 - x^2} \, dx$
2. $\int \sqrt{100 - x^2} \, dx$
9. $\int_0^1 \frac{x^2}{x^2 + 1} \, dx$
10. $\int \frac{x^2}{\sqrt{1 + x^2}} \, dx$

3. $\int_0^4 \sqrt{16 - x^2} \, dx$
4. $\int_{-3}^3 \sqrt{9 - x^2} \, dx$
11. $\int_1^2 x \sqrt{x^2 - 1} \, dx$
12. $\int_2^3 2x \sqrt{x^2 - 4} \, dx$

5. $\int \frac{3}{x^2 + 9} \, dx$
6. $\int \frac{3}{9x^2 + 1} \, dx$
13. $\int_1^2 \frac{\sqrt{x^2 - 1}}{x} \, dx$
14. $\int_{-2}^{-1} \frac{x}{\sqrt{x^2 - 1}} \, dx$

7. $\int \frac{1}{\sqrt{16 + x^2}} \, dx$
8. $\int_0^5 \frac{1}{\sqrt{25 + x^2}} \, dx$
15. $\int_{-1}^1 \sqrt{16 - 9x^2} \, dx$
16. $\int_{-9/2}^{9/2} \sqrt{81 - 4x^2} \, dx$

17. $\int \sqrt{4-(x+1)^2}\, dx$
18. $\int \dfrac{1}{1+(2x-3)^2}\, dx$
19. $\int_0^\infty \dfrac{1}{(x+1)^2+1}\, dx$
20. $\int_0^1 \dfrac{1}{\sqrt{1-x^2}}\, dx$
21. $\int \dfrac{x^3}{\sqrt{4x^2-9}}\, dx$
22. $\int \dfrac{x^3}{\sqrt{4-9x^2}}\, dx$
23. $\int \dfrac{1}{\sqrt{x}(x+1)}\, dx$
24. $\int \dfrac{\sqrt{1-x}}{\sqrt{x}}\, dx$

[Hint: Let $\sqrt{x} = \tan\theta$.]

25. $\int e^x \sqrt{1-e^{2x}}\, dx$
26. $\int_0^{\ln\sqrt{3}} \dfrac{e^x}{e^{2x}+1}\, dx$
27. $\int \dfrac{e^{x/2}}{e^x+1}\, dx$
28. $\int \dfrac{1}{\sqrt{e^{2x}+1}}\, dx$

29. Find the area of the region bounded by the graph of $y = 1/(x^2+1)$, the x-axis, and the lines $x = -2$ and $x = 2$.

30. Show that the area enclosed by the circle $x^2 + y^2 = r^2$ is πr^2.

31. Find the area of the region enclosed by the ellipse $x^2/16 + y^2/9 = 1$.

32. Show that the area of the region enclosed by the ellipse $x^2/a^2 + y^2/b^2 = 1$ is πab.

33. Find the area of the region bounded on the right by the ellipse $x^2/16 + y^2/9 = 1$ and on the left by the line $x = 1$.

In Exercises 34–37, find the arc length of the graph of the function.

34. $f(x) = x^2/2$, $[0, 2]$
35. $f(x) = \sqrt{4-x^2}$, $[-2, 2]$
36. $f(x) = \tfrac{1}{2}(x^2 - 2x)$, $[-1, 1]$
37. $f(x) = \ln x$, $[1, e]$

38. Use the result of Exercise 37 to find the arc length of the graph of $f(x) = e^x$ from $x = 0$ to $x = 1$.

39. Show that the circumference of the circle $x^2 + y^2 = r^2$ is $2\pi r$.

40. Find the volume of the solid obtained by revolving about the y-axis the region bounded by the graph of $y = x/\sqrt{1+x^2}$, the line $x = 3$, and the x-axis.

41. The velocity of an object moving along a straight line is given by $v(t) = \sqrt{25 - t^2}$ cm/sec. Find the distance traveled by the object during the time interval $[0, 5]$.

42. Determine the area of the surface generated by revolving about the x-axis the graph of $f(x) = \sin x$ for x in $[0, \pi/2]$.

43. In Example 2 of Section 5.4 we found that the arc length of the graph of $f(x) = x^2$ on $[0, 1]$ is $\int_0^1 \sqrt{1+4x^2}\, dx$. At that time we could not evaluate this integral. Evaluate it now.

44. Exercise 66 of Section 5.2 describes a wedge cut out of a tree (see the figure). That exercise asks that the integral describing the volume of the wedge be set up, but not evaluated. Evaluate this integral now.

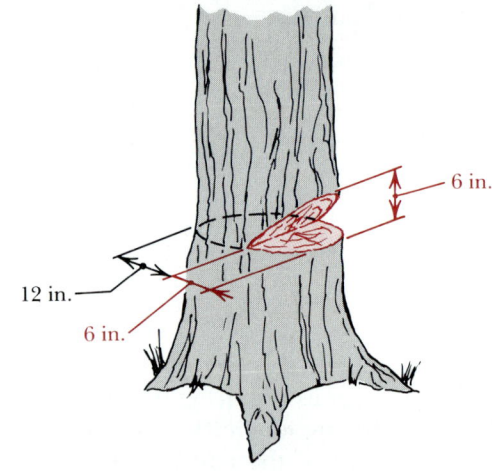

45. A *torus* is a solid generated by revolving a circle about a line that does not intersect the circle (see the figure). Find the volume of the torus generated by revolving about the y-axis the circle with equation $(x-2)^2 + y^2 = 1$.

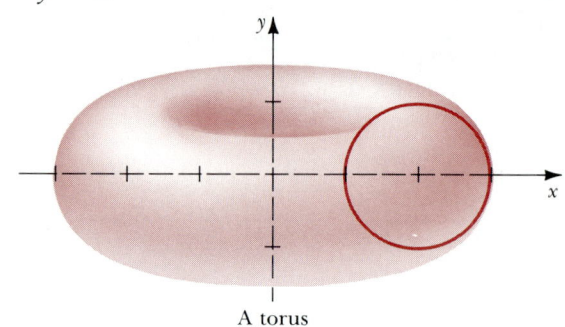

A torus

46. A baker has been making, and selling for 20¢ each, plain bagels in the shape of a torus. The bagel can be described by revolving a circle of diameter 1 in. about

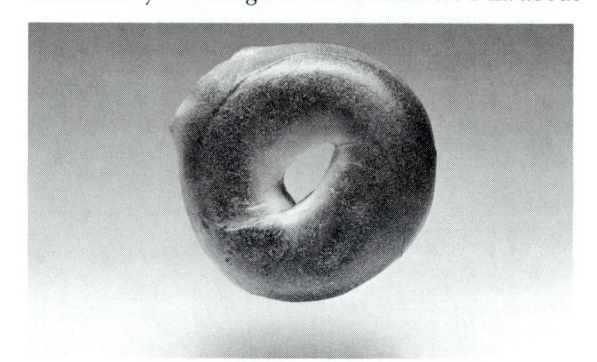

a line 1 in. from the center of the circle. This baker decides to make a giant bagel in the shape of a torus generated by revolving a circle of radius 1 in. about a line 1.25 in. from the center of the circle. What should be the selling price of the giant bagels?

47. Find the area of the region bounded by the hyperbola $x^2/16 - y^2/9 = 1$ and the line $x = 8$.

48. In Exercise 36 of Section 5.4 you were asked to set up the integral to find the actual distance traveled by a ball hit from the ground (see the figure). The ball travels along the curve described by $y = 88t - 16t^2$, $0 \leq t \leq \frac{11}{2}$. Evaluate the integral, that is, find the length of the curve.

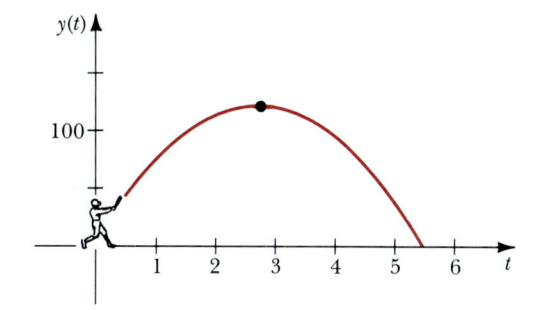

Figure for Exercise 48.

7.4
Partial Fractions

Consider

$$\int \frac{2}{1-x^2} \, dx.$$

By making the assumption that $|x| < 1$, we can use a trigonometric substitution to evaluate this integral. Let

$$x = \sin \theta \quad \text{and} \quad dx = \cos \theta \, d\theta.$$

Then

$$\int \frac{2}{1-x^2} \, dx = \int \frac{2}{1-\sin^2 \theta} \cos \theta \, d\theta = 2 \int \frac{\cos \theta}{\cos^2 \theta} \, d\theta = 2 \int \frac{1}{\cos \theta} \, d\theta$$

and

$$\int \frac{2}{1-x^2} \, dx = 2 \int \sec \theta \, d\theta = 2 \ln |\sec \theta + \tan \theta| + C.$$

The triangle in Figure 7.6 shows that

$$\sec \theta = \frac{1}{\sqrt{1-x^2}} \quad \text{and} \quad \tan \theta = \frac{x}{\sqrt{1-x^2}},$$

so

$$\int \frac{2}{1-x^2} \, dx = 2 \ln \left| \frac{1}{\sqrt{1-x^2}} + \frac{x}{\sqrt{1-x^2}} \right| + C = 2 \ln \left| \frac{1+x}{\sqrt{1-x^2}} \right| + C.$$

We can rewrite this solution using the fact that

$$\frac{1+x}{\sqrt{1-x^2}} = \sqrt{\frac{(1+x)^2}{(1+x)(1-x)}} = \sqrt{\frac{1+x}{1-x}} = \left(\frac{1+x}{1-x}\right)^{1/2}.$$

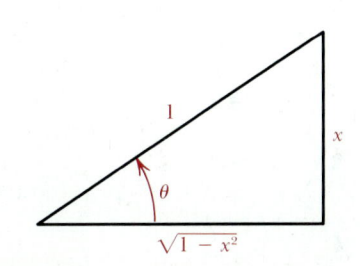

Figure 7.6

and the arithmetic properties of the natural logarithm:

$$\int \frac{2}{1-x^2} dx = 2 \cdot \frac{1}{2} \ln \left| \frac{1+x}{1-x} \right| + C = \ln|1+x| - \ln|1-x| + C.$$

Although the trigonometric substitution $x = \sin \theta$ required that $|x| < 1$, if we check this result by differentiation we see that

$$D_x [\ln|1+x| - \ln|1-x|] = \frac{1}{1+x} - \frac{1}{1-x}(-1)$$

$$= \frac{(1-x)+(1+x)}{1-x^2} = \frac{2}{1-x^2}$$

holds whenever $x \ne \pm 1$.

If we are observant, this check does more than verify the correctness of the result; it shows that if we had initially written the integrand as

$$\frac{2}{1-x^2} = \frac{1}{1+x} + \frac{1}{1-x},$$

we could have performed the integration easily:

$$\int \frac{2}{1-x^2} dx = \int \frac{1}{1+x} dx + \int \frac{1}{1-x} dx = \ln|1+x| - \ln|1-x| + C.$$

This observation leads us to the integration technique considered in this section: the method of **integration by partial fractions.**

Let us first review briefly rational functions. The quotient of two polynomial functions is called a **rational function.** For example, if

$$Q(x) = a_n x^n + \cdots + a_1 x + a_0 \quad \text{and} \quad P(x) = b_m x^m + \cdots + b_1 x + b_0$$

are polynomials, then a rational function is defined by

$$(7.14) \quad R(x) = \frac{Q(x)}{P(x)} = \frac{a_n x^n + a_{n-1} x^{n-1} + \cdots + a_1 x + a_0}{b_m x^m + b_{m-1} x^{m-1} + \cdots + b_1 x + b_0}.$$

A rational function with the degree of its numerator less than the degree of its denominator is called a **proper rational function.** The rational function described in (7.14) is proper precisely when $m > n$.

Polynomial division can be used to write any rational function as the sum of a polynomial function and a proper rational function. In addition, any *proper* rational function can be expressed as a sum of fractions of the form

$$\text{(i)} \ \frac{A}{(x+a)^n}, \quad \text{(ii)} \ \frac{Bx}{(x^2+bx+c)^n}, \quad \text{(iii)} \ \frac{C}{(x^2+bx+c)^n},$$

where A, B, C, a, b, and c represent constants and the quadratic $x^2 + bx + c$ is **irreducible,** that is, it has no real linear factors. A fraction of the form (i), (ii), or (iii) is called a **partial fraction.**

The method of partial fractions reduces the expression for an arbitrary rational function to the sum of a polynomial and partial fractions. The polynomial and the partial fractions are then integrated by techniques considered earlier.

The general rules for determining the correct partial-fraction decomposition of a *proper* rational function described by $R(x) = Q(x)/P(x)$ are as follows.

1) Factor $P(x)$ into a product of linear and irreducible quadratic factors.
2) Each linear factor of $P(x)$ of the form $(x + a)^n$ contributes a sum of the form

$$\frac{A_1}{x+a} + \frac{A_2}{(x+a)^2} + \cdots + \frac{A_n}{(x+a)^n}$$

to the decomposition, where $A_1, A_2, \ldots, A_n$ are constants.

3) Each irreducible quadratic factor of $P(x)$ of the form $(x^2 + bx + c)^n$ contributes a sum of the form

$$\frac{B_1 x + C_1}{x^2 + bx + c} + \frac{B_2 x + C_2}{(x^2 + bx + c)^2} + \cdots + \frac{B_n x + C_n}{(x^2 + bx + c)^n}$$

to the decomposition, where $B_1, B_2, \ldots, B_n$ and $C_1, C_2, \ldots, C_n$ are constants.

4) The total partial-fraction decomposition of $R(x)$ is the sum of all partial fractions contributed from (2) and (3).

The procedures used to determine the specific values of the appropriate constants are illustrated in the following examples.

Example 1

Evaluate $\int \dfrac{x}{x^2 + 5x + 6}\, dx$.

Solution

The rational function is proper and the denominator factors as $x^2 + 5x + 6 = (x + 3)(x + 2)$, so by Rules (2) and (4), the partial-fraction decomposition has the form

$$\frac{x}{(x+3)(x+2)} = \frac{A_1}{x+3} + \frac{A_2}{x+2}$$

for some constants A_1 and A_2. Note that these are the most general partial fractions that contribute to a common denominator of the form $(x + 3)(x + 2)$.

To determine A_1 and A_2, we multiply by the original denominator $(x + 3)(x + 2)$ to obtain

$$x = A_1(x + 2) + A_2(x + 3).$$

We then evaluate this equation at the zeros of the original denominator:

$x = -3$: $\quad -3 = A_1(-3 + 2) + A_2(-3 + 3) = -A_1$, so $A_1 = 3$;

$x = -2$: $\quad -2 = A_1(-2 + 2) + A_2(-2 + 3) = A_2$, so $A_2 = -2$.

This implies that

$$\frac{x}{(x+3)(x+2)} = \frac{3}{x+3} - \frac{2}{x+2}$$

and

$$\int \frac{x}{x^2+5x+6}\,dx = \int \left[\frac{3}{x+3} - \frac{2}{x+2}\right] dx$$

$$= 3\ln|x+3| - 2\ln|x+2| + C$$
$$= \ln|(x+3)^3| - \ln|(x+2)^2| + C$$
$$= \ln\frac{|(x+3)^3|}{(x+2)^2} + C. \quad \blacktriangleright$$

The integrand in Example 1 had two distinct linear factors. Example 2 illustrates the method of partial fractions when the linear factors are not distinct.

Example 2

Evaluate $\int \dfrac{x^2+3x+3}{x(x+2)^2}\,dx$.

Solution
Rules (2) and (4) imply that the form of the partial-fraction decomposition for this proper rational function is

$$(7.15) \qquad \frac{x^2+3x+3}{x(x+2)^2} = \frac{A_1}{x} + \frac{A_2}{x+2} + \frac{A_3}{(x+2)^2}.$$

The terms in this decomposition involve the most general partial fractions that can contribute to a common denominator of the form $x(x+2)^2$. In this example, as in Example 1, the number of constants to be determined is the same as the degree of the denominator of the original rational function. This is always the case regardless of the form of the denominator.

Multiplying both sides of Equation (7.15) by $x(x+2)^2$ gives

$$(7.16) \qquad x^2 + 3x + 3 = A_1(x+2)^2 + A_2(x+2)x + A_3 x.$$

Evaluating (7.16) at the zeros of $x(x+2)^2$ gives

$x = 0$: $\quad 3 = A_1(4) + A_2(0) + A_3(0), \quad$ so $A_1 = 3/4$;
$x = -2$: $\quad 1 = (3/4)(0) + A_2(0) + A_3(-2), \quad$ so $A_3 = -1/2$.

There is no value of x that can be used to reduce the coefficients of both A_1 and A_3 to zero. To find A_2, we evaluate Equation (7.16) at any number other than 0 and -2, for example,

$x = 1$: $\quad 7 = 3(3)^2/4 + A_2(3)(1) - (1/2)(1), \quad$ so $A_2 = 1/4$.

Hence,

$$\frac{x^2+3x+3}{x(x+2)^2} = \frac{3}{4x} + \frac{1}{4(x+2)} - \frac{1}{2(x+2)^2}$$

and

$$\int \frac{x^2 + 3x + 3}{x(x+2)^2}\, dx = \frac{3}{4}\int \frac{dx}{x} + \frac{1}{4}\int \frac{dx}{x+2} - \frac{1}{2}\int \frac{dx}{(x+2)^2}$$

$$= \frac{3}{4}\ln|x| + \frac{1}{4}\ln|x+2| + \frac{1}{2}\left(\frac{1}{x+2}\right) + C. \quad \blacktriangleright$$

An alternative method for determining the appropriate partial-fraction decomposition is to determine the constants that give equality for each power of x. For example, in Equation (7.16), we have

$$x^2 + 3x + 3 = A_1(x+2)^2 + A_2(x+2)x + A_3 x,$$

so

$$x^2 + 3x + 3 = (A_1 + A_2)x^2 + (4A_1 + 2A_2 + A_3)x + 4A_1.$$

If the coefficients of each power of x are to agree, then for

$$x^2: \quad 1 = A_1 + A_2,$$
$$x: \quad 3 = 4A_1 + 2A_2 + A_3,$$

and constants: $\quad 3 = 4A_1.$

Thus,

$$A_1 = 3/4;\ A_2 = 1 - A_1 = 1 - 3/4 = 1/4;$$

and

$$A_3 = 3 - 2A_2 - 4A_1 = 3 - 1/2 - 3 = -1/2.$$

This alternative technique for evaluating the constants can be used in any problem involving partial fractions.

The following example demonstrates the partial-fraction procedure when the rational function is not proper.

Example 3

Evaluate $\int R(x)\, dx$, where

$$R(x) = \int \frac{2x^4 - x^3 - 5x^2 - 3x + 10}{x^3 - x^2 - 4x + 4}.$$

Solution

We first use polynomial division to express the integrand as the sum of a polynomial and a proper rational function:

$$
\begin{array}{r}
2x + 1 \\
x^3 - x^2 - 4x + 4 \overline{\smash{\big)}\, 2x^4 - x^3 - 5x^2 - 3x + 10}\\
\underline{2x^4 - 2x^3 - 8x^2 + 8x }\\
x^3 + 3x^2 - 11x + 10\\
\underline{x^3 - x^2 - 4x + 4}\\
4x^2 - 7x + 6
\end{array}
$$

and
$$R(x) = 2x + 1 + \frac{4x^2 - 7x + 6}{x^3 - x^2 - 4x + 4}.$$

So
$$\int R(x)\,dx = \int (2x+1)\,dx + \int \frac{4x^2 - 7x + 6}{x^3 - x^2 - 4x + 4}\,dx$$

or

(7.17) $$\int R(x)\,dx = x^2 + x + \int \frac{4x^2 - 7x + 6}{x^3 - x^2 - 4x + 4}\,dx.$$

Completely factoring the denominator, $P(x) = x^3 - x^2 - 4x + 4$ relies on the fact that:

(7.18)
If $P(x)$ is a polynomial, then $x - a$ is a factor of $P(x)$ if and only if $P(a) = 0$.

Since $P(1) = 0$, $x - 1$ is a factor of $P(x)$. Dividing $P(x)$ by $x - 1$ leads to the complete factorization of the denominator:
$$P(x) = (x-1)(x^2 - 4)$$
$$= (x-1)(x-2)(x+2).$$

The proper rational function in (7.17) remaining to be integrated can be expressed as
$$\frac{4x^2 - 7x + 6}{x^3 - x^2 - 4x + 4} = \frac{4x^2 - 7x + 6}{(x-1)(x-2)(x+2)}$$
$$= \frac{A_1}{x-1} + \frac{A_2}{x-2} + \frac{A_3}{x+2}.$$

To determine the constants A_1, A_2, and A_3, we first multiply both sides of the equation by the original denominator $P(x) = (x-1)(x-2)(x+2)$:
$$4x^2 - 7x + 6 = A_1(x-2)(x+2) + A_2(x-1)(x+2) + A_3(x-1)(x-2)$$

and then find the values of the constants that satisfy this polynomial equation. In particular, for this equation to be satisfied at the zeros of P:

$x = 1$: $\quad 3 = A_1(-1)(3) + A_2(0)(3) + A_3(0)(-1), \quad$ so $A_1 = -1$;
$x = 2$: $\quad 8 = -1(0)(4) + A_2(1)(4) + A_3(1)(0), \quad$ so $A_2 = 2$;
$x = -2$: $\quad 36 = -1(-4)(0) + 2(-3)(0) + A_3(-3)(-4), \quad$ so $A_3 = 3$.

This implies that
$$\frac{4x^2 - 7x + 6}{x^3 - x^2 - 4x + 4} = \frac{-1}{x-1} + \frac{2}{x-2} + \frac{3}{x+2}.$$

and

$$\int \frac{4x^2 - 7x + 6}{x^3 - x^2 - 4x + 4} dx = -\int \frac{dx}{x-1} + 2\int \frac{dx}{x-2} + 3\int \frac{dx}{x+2}$$

$$= -\ln|x-1| + 2\ln|x-2| + 3\ln|x+2| + C$$

$$= -\ln|x-1| + \ln(x-2)^2 + \ln|x+2|^3 + C$$

$$= \ln\left|\frac{(x-2)^2(x+2)^3}{x-1}\right| + C.$$

Finally,

$$\int R(x)\,dx = \int \frac{2x^4 - x^3 - 5x^2 - 3x + 10}{x^3 - x^2 - 4x + 4} dx$$

$$= \int \left[2x + 1 + \frac{4x^2 - 7x + 6}{x^3 - x^2 - 4x + 4}\right] dx$$

$$= x^2 + x + \ln\left|\frac{(x-2)^2(x+2)^3}{x-1}\right| + C.$$ ▶

We consider next the method of partial fractions applied to a problem with an irreducible quadratic in the denominator.

Example 4

Evaluate $\int \frac{1}{(x-1)(x^2+1)}\,dx$.

Solution

Since $x^2 + 1$ has no real factors, Rule (3) implies that this term contributes a term of the form $(Bx + C)/(x^2 + 1)$ to the decomposition. Thus, the total decomposition is of the form

$$\frac{1}{(x-1)(x^2+1)} = \frac{A}{x-1} + \frac{Bx + C}{x^2 + 1}.$$

Note that, as in the previous examples, the number of constants to be found agrees with the degree of the original denominator. To determine the constants A, B, and C, we multiply by $(x-1)(x^2+1)$:

$$1 = A(x^2 + 1) + (Bx + C)(x - 1).$$

If we evaluate this expression at $x = 1$, the zero of $(x - 1)$, then:

$$x = 1: \quad 1 = A(2), \quad \text{so } A = 1/2.$$

To determine B and C, we evaluate the equation at another pair of numbers and use the known value, $A = 1/2$. For example,

$$x = 0: \quad 1 = (1/2)(1) + C(-1), \quad \text{so } C = -1/2;$$
$$x = -1: \quad 1 = (1/2)(2) + [B(-1) + (-1/2)](-2), \quad \text{so } B = -1/2.$$

Thus,
$$\frac{1}{(x-1)(x^2+1)} = \frac{1/2}{x-1} - \frac{(1/2)x + 1/2}{x^2+1}$$
and
$$\int \frac{1}{(x-1)(x^2+1)} dx = \frac{1}{2} \int \left(\frac{1}{x-1} - \frac{x+1}{x^2+1} \right) dx$$
$$= \frac{1}{2} \left[\ln|x-1| - \int \frac{x}{x^2+1} dx - \int \frac{1}{x^2+1} dx \right]$$
$$= \frac{1}{2} \left[\ln|x-1| - \frac{1}{2} \ln(x^2+1) - \arctan x \right] + C. \quad \blacktriangleright$$

The final example in this section illustrates the method of partial fractions when the denominator contains an irreducible quadratic factor of the form $(x^2 + bx + c)^n$, where $n > 1$.

Example 5

Evaluate $\int \dfrac{1}{x(1+x^2)^2} dx$.

Solution

The rules of decomposition imply that the partial-fraction decomposition is of the form

(7.19) $$\frac{1}{x(1+x^2)^2} = \frac{A}{x} + \frac{B_1 x + C_1}{1+x^2} + \frac{B_2 x + C_2}{(1+x^2)^2}$$

Note again that the number of constants, five, agrees with the degree of the original denominator. To find the constants, we write (7.19) as
$$1 = A(1+x^2)^2 + B_1 x^2(1+x^2) + C_1 x(1+x^2) + B_2 x^2 + C_2 x.$$
When $x = 0$, $1 = A$, so
$$1 = (1+x^2)^2 + B_1 x^2(1+x^2) + C_1 x(1+x^2) + B_2 x^2 + C_2 x.$$

We could evaluate this equation at four arbitrary nonzero values of x and solve the resulting systems of equations to find the remaining constants. Instead we will find the constants by the method of equating the coefficients for each power of x, as we did in the alternative method described following Example 2.

We expand the right side of the last equation and combine similar terms:
$$1 = 1 + 2x^2 + x^4 + B_1 x^2 + B_1 x^4 + C_1 x + C_1 x^3 + B_2 x^2 + C_2 x$$
or
$$1 = (1 + B_1)x^4 + C_1 x^3 + (2 + B_1 + B_2)x^2 + (C_1 + C_2)x + 1.$$

Since the constant terms already agree, we equate the coefficients for each power of x:

$$x^4: 0 = 1 + B_1, \quad \text{which implies that } B_1 = -1;$$
$$x^3: 0 = C_1, \quad \text{which implies that } C_1 = 0;$$
$$x^2: 0 = 2 + B_1 + B_2, \quad \text{which implies that } B_2 = -1;$$
and $\quad x: 0 = C_1 + C_2, \quad \text{which implies that } C_2 = 0.$

The partial-fraction decomposition is

$$\frac{1}{x(1+x^2)^2} = \frac{1}{x} - \frac{x}{1+x^2} - \frac{x}{(1+x^2)^2}$$

and

$$\int \frac{dx}{x(1+x^2)^2} = \int \frac{dx}{x} - \int \frac{x}{1+x^2}\,dx - \int \frac{x}{(1+x^2)^2}\,dx$$

$$= \ln|x| - \frac{1}{2}\ln(1+x^2) + \frac{1}{2(1+x^2)} + C.$$

▶ **EXERCISE SET 7.4**

In Exercises 1–32, evaluate the integral.

1. $\displaystyle\int \frac{1}{x^2-1}\,dx$

2. $\displaystyle\int \frac{1}{x^2-5x+6}\,dx$

3. $\displaystyle\int_3^4 \frac{1}{x^2-3x+2}\,dx$

4. $\displaystyle\int_0^1 \frac{x}{x^2-2x-3}\,dx$

5. $\displaystyle\int \frac{x}{x^2-1}\,dx$

6. $\displaystyle\int \frac{x^2}{x^2-1}\,dx$

7. $\displaystyle\int_{-1}^1 \frac{x^2+1}{4-x^2}\,dx$

8. $\displaystyle\int \frac{x^3+2}{x^2+7x+12}\,dx$

9. $\displaystyle\int \frac{x^3+x^2+x}{x^2+2x+1}\,dx$

10. $\displaystyle\int \frac{x^2}{(x-1)^2}\,dx$

11. $\displaystyle\int \frac{x^2+1}{x^3+x^2-4x-4}\,dx$

12. $\displaystyle\int \frac{1}{x^3+2x^2-x-2}\,dx$

13. $\displaystyle\int \frac{2x^2-9x+12}{x^3-4x^2+4x}\,dx$

14. $\displaystyle\int \frac{5x^2-19x+18}{x^3-5x^2+3x+9}\,dx$

15. $\displaystyle\int_{-1/2}^{1/2} \frac{3}{x^4-5x^2+4}\,dx$

16. $\displaystyle\int \frac{1}{x^4-3x^3-7x^2+27x-18}\,dx$

17. $\displaystyle\int \frac{1}{x^3+x^2+x+1}\,dx$

18. $\displaystyle\int \frac{1}{x^3-x^2+x-1}\,dx$

19. $\displaystyle\int \frac{x^2}{(x^2+1)^2}\,dx$

20. $\displaystyle\int_0^3 \frac{2x^5-5x}{(x^2+2)^2}\,dx$

21. $\displaystyle\int \frac{2x-3}{(2x^2+5)(x^2+3)}\,dx$

22. $\displaystyle\int \frac{8x^3+3x^2+4x+12}{(x^2+4)(2x^2+1)}\,dx$

23. $\displaystyle\int \frac{dx}{1+\sqrt{x}}$

24. $\displaystyle\int \frac{\sqrt{x}-1}{\sqrt{x}+1}\,dx$

25. $\displaystyle\int \frac{e^x\,dx}{e^{2x}-4}$

26. $\displaystyle\int \frac{e^{2x}\,dx}{e^{4x}-5e^{2x}+6}$

27. $\displaystyle\int \frac{\cos x}{\sin x\,(\sin x - 1)}\,dx$

28. $\displaystyle\int \frac{\sec^2 x}{\tan x\,(\tan x + 1)}\,dx$

29. $\displaystyle\int \frac{\cos x}{\sin^2 x + \sin x}\,dx$

30. $\displaystyle\int \frac{\csc^2 x}{\cot^2 x + 4\cot x}\,dx$

31. $\displaystyle\int \frac{dx}{1+\sqrt{x+1}}$

32. $\displaystyle\int \frac{dx}{x\sqrt{x+2}}$

33. a) Sketch the graph of the function described by
$$f(x) = \frac{1}{x^2 - x - 6}$$
for x in $[0, 2]$.
b) Find the area of the region bounded by the graph of f, the lines $x = 0$ and $x = 2$, and the x-axis.

34. a) Sketch the graph of the function described by $f(x) = 1/(x^2 + x)$ for x in $[1, \infty)$.
b) Find, if possible, the area of the region bounded by the graph of f and the x-axis on the interval $[1, \infty)$.

35. Find the volume of the solid obtained if the region described in Exercise 33 is revolved about the x-axis.

36. Find, if possible, the volume of the solid obtained if the region described in Exercise 34 is revolved about the x-axis.

37. Find the volume of the solid obtained if the region described in Exercise 33 is revolved about the y-axis.

38. Find, if possible, the volume of the solid obtained if the region described in Exercise 34 is revolved about the y-axis.

39. Show that if $u = \tan x/2$, $-\pi < x < \pi$, then
$$\sin x = \frac{2u}{1+u^2}, \quad \cos x = \frac{1-u^2}{1+u^2}, \quad \text{and} \quad dx = \frac{2\,du}{1+u^2}.$$

Use the substitution and relations in Exercise 39 to determine the integrals in Exercises 40–47.

40. $\displaystyle\int_0^{\pi/2} \frac{dx}{1+\sin x}$

41. $\displaystyle\int_0^{\pi/2} \frac{dx}{1+\cos x}$

42. $\displaystyle\int \frac{dx}{\cos x - \sin x + 1}$

43. $\displaystyle\int \frac{dx}{\cos x + \sin x + 1}$

44. $\displaystyle\int \frac{dx}{\sin x + \cos x}$

45. $\displaystyle\int \frac{dx}{\sin x + \tan x}$

46. $\displaystyle\int \frac{dx}{3\cos x - 4\sin x}$

47. $\displaystyle\int_0^{\pi/2} \frac{dx}{(3+2\cos x)^2}$

Partial-fraction methods appear frequently in various areas of engineering because of their connection with Laplace transforms. The Laplace transform changes a differential equation into an algebraic equation whose solution can be found by ordinary algebraic manipulation. Reversing the transform procedure produces the solution to the differential equation. If L denotes the Laplace transform operation, then $L(\sin at) = a/(s^2 + a^2)$, $L(\cos at) = s/(s^2 + a^2)$, and $L(e^{at}) = 1/(s-a)$, where a is a constant. (This was discussed in Exercise 76 of Section 7.1 and Exercise 72 of Section 6.3.) Use these results together with the arithmetic result in Exercise 73 of Section 6.3 to find the functions whose Laplace transforms are given in Exercises 48–51.

48. $\displaystyle\frac{s^2 + s - 4}{s^3 - s^2 + s - 1}$

49. $\displaystyle\frac{2s^2 + 2s + 2}{s^3 - 3s^2 - 4s + 12}$

50. $\displaystyle\frac{s^3 - 5s - 26}{(s^2 + s - 2)(s^2 + 4)}$

51. $\displaystyle\frac{6s^2 - 10s + 24}{s^4 + 8s^2 - 9}$

7.5

Integrals Involving Quadratic Polynomials

In this section we consider integrals whose integrands have quadratic terms of the form $ax^2 + bx + c$, for some nonzero constants a, b, and c. By completing the square, we can change the term $ax^2 + bx + c$ into one of the forms $u^2 \pm a^2$ or $a^2 - u^2$, forms for which integration techniques have already been considered.

Example 1

Evaluate $\int \dfrac{dx}{\sqrt{x^2 + 2x + 5}}$.

Solution

We first complete the square in $x^2 + 2x + 5$:

$$x^2 + 2x + 5 = (x^2 + 2x + 1) + 4 = (x + 1)^2 + 4.$$

Thus,

$$\int \frac{dx}{\sqrt{x^2 + 2x + 5}} = \int \frac{dx}{\sqrt{(x+1)^2 + 4}}.$$

Since $(x + 1)^2 + 4$ is of the form $u^2 + a^2$, the substitution $x + 1 = 2 \tan \theta$ is used. Then $dx = 2 \sec^2 \theta \, d\theta$ and

$$\int \frac{dx}{\sqrt{x^2 + 2x + 5}} = \int \frac{dx}{\sqrt{(x+1)^2 + 4}}$$

$$= \int \frac{2 \sec^2 \theta \, d\theta}{\sqrt{4 (\tan \theta)^2 + 4}} = \int \frac{2 \sec^2 \theta \, d\theta}{2 \sec \theta}$$

$$= \int \sec \theta \, d\theta = \ln |\sec \theta + \tan \theta| + C.$$

The triangle in Figure 7.7 implies that

$$\sec \theta = \frac{\sqrt{x^2 + 2x + 5}}{2},$$

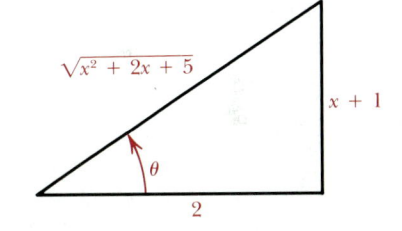

Figure 7.7

so

$$\int \frac{dx}{\sqrt{x^2 + 2x + 5}} = \ln \left| \frac{\sqrt{x^2 + 2x + 5}}{2} + \frac{x + 1}{2} \right| + C$$

$$= \ln |\sqrt{x^2 + 2x + 5} + x + 1| - \ln 2 + C$$

$$= \ln (\sqrt{x^2 + 2x + 5} + x + 1) + C,$$

since $\sqrt{x^2 + 2x + 5} + x + 1$ is always positive and $\ln 2$ is a constant. ▶

Example 2

Determine $\int_1^2 \sqrt{3 + 2t - t^2} \, dt$.

Solution

We first complete the square under the radical in the integrand:

$$\int_1^2 \sqrt{3 + 2t - t^2} \, dt = \int_1^2 \sqrt{4 - (t^2 - 2t + 1)} \, dt = \int_1^2 \sqrt{4 - (t - 1)^2} \, dt.$$

In this case, since $4 - (t - 1)^2$ is of the form $a^2 - u^2$, we let $u = t - 1$. Then $du = dt$. When $t = 1$, $u = 0$, and when $t = 2$, $u = 1$, so

$$\int_{t=1}^{t=2} \sqrt{3 + 2t - t^2} \, dt = \int_{u=0}^{u=1} \sqrt{4 - u^2} \, du.$$

Now let $u = 2 \sin \theta$. Then $du = 2 \cos \theta \, d\theta$. When $u = 0$, $\sin \theta = 0$, so $\theta = 0$; and when $u = 1$, $\sin \theta = 1/2$, so $\theta = \pi/6$. Thus,

$$\int_{t=1}^{t=2} \sqrt{3 + 2t - t^2} \, dt = \int_{\theta=0}^{\theta=\pi/6} \sqrt{4 - 4 \sin^2 \theta} \, 2 \cos \theta \, d\theta$$

$$= 4 \int_{\theta=0}^{\theta=\pi/6} \cos^2 \theta \, d\theta$$

$$= 4 \int_{\theta=0}^{\theta=\pi/6} \frac{1 + \cos 2\theta}{2} \, d\theta$$

$$= 2 \left[\theta + \frac{1}{2} \sin 2\theta \right]_{\theta=0}^{\theta=\pi/6}$$

$$= 2 \left[\left(\frac{\pi}{6} + \frac{1}{2} \cdot \frac{\sqrt{3}}{2} \right) - (0 + 0) \right]$$

$$= \frac{\pi}{3} + \frac{\sqrt{3}}{2}. \qquad \blacktriangleright$$

Example 3

Evaluate $\int \frac{4x^2 - x + 5}{(x - 5)(x^2 + 4x + 5)} \, dx$.

Solution

Since $x^2 + 4x + 5$ is irreducible, the partial-fraction form of the integrand is

$$\frac{4x^2 - x + 5}{(x - 5)(x^2 + 4x + 5)} = \frac{A}{x - 5} + \frac{Bx + C}{x^2 + 4x + 5}.$$

As in previous partial-fraction examples, the number of constants to be determined is the same as the degree of the original denominator. To find the constants, we multiply by $(x - 5)(x^2 + 4x + 5)$ to obtain

$$4x^2 - x + 5 = A(x^2 + 4x + 5) + Bx(x - 5) + C(x - 5).$$

If $\quad x = 5:\quad 100 = 50A,\qquad$ so $A = 2$;
$\quad\quad\quad x = 0:\quad 5 = 10 - 5C,\qquad$ so $C = 1$.

To find B let x be any number other than 0 and 5, for example, let $x = 1$.

$$x = 1:\quad 8 = 20 - 4B - 4,\quad \text{so } B = 2.$$

The partial-fraction form of the integrand is therefore

$$\frac{4x^2 - x + 5}{(x - 5)(x^2 + 4x + 5)} = \frac{2}{x - 5} + \frac{2x + 1}{x^2 + 4x + 5}$$

and

$$\int \frac{4x^2 - x + 5}{(x - 5)(x^2 + 4x + 5)}\, dx = \int \left(\frac{2}{x - 5} + \frac{2x + 1}{x^2 + 4x + 5}\right) dx$$

$$= 2 \ln |x - 5| + \int \frac{2x + 1}{x^2 + 4x + 5}\, dx.$$

To evaluate

$$\int \frac{2x + 1}{x^2 + 4x + 5}\, dx,$$

first let $u = x^2 + 4x + 5$. Then $du = (2x + 4)\, dx$ and

$$\int \frac{2x + 1}{x^2 + 4x + 5}\, dx = \int \frac{2x + 4 - 3}{x^2 + 4x + 5}\, dx$$

$$= \int \frac{2x + 4}{x^2 + 4x + 5}\, dx - \int \frac{3}{x^2 + 4x + 5}\, dx$$

$$= \int \frac{1}{u}\, du - \int \frac{3}{x^2 + 4x + 5}\, dx$$

$$= \ln |u| - \int \frac{3}{x^2 + 4x + 5}\, dx$$

$$= \ln |x^2 + 4x + 5| - 3 \int \frac{dx}{x^2 + 4x + 5}.$$

To evaluate the integral

$$\int \frac{dx}{x^2 + 4x + 5},$$

we complete the square in the denominator as we did in Examples 1 and 2:

$$x^2 + 4x + 5 = (x^2 + 4x + 4) + 1$$
$$= (x + 2)^2 + 1.$$

Let $u = x + 2$. Then $du = dx$ and

$$\int \frac{dx}{x^2 + 4x + 5} = \int \frac{du}{u^2 + 1} = \arctan u + C = \arctan (x + 2) + C.$$

Thus,

$$\int \frac{4x^2 - x + 5}{(x-5)(x^2+4x+5)} dx = 2\ln|x-5| + \ln|x^2+4x+5| - 3\arctan(x+2) + C.$$

▶ EXERCISE SET 7.5

In Exercises 1–20, evaluate the integral.

1. $\displaystyle\int \frac{1}{x^2 + 2x + 10} dx$

2. $\displaystyle\int \frac{1}{x^2 + 4x + 5} dx$

3. $\displaystyle\int \frac{1}{(x^2 + 2x + 10)^2} dx$

4. $\displaystyle\int \frac{1}{(x^2 + 4x + 5)^3} dx$

5. $\displaystyle\int \sqrt{2x - x^2}\, dx$

6. $\displaystyle\int_{-1}^{0} \sqrt{x^2 - 2x}\, dx$

7. $\displaystyle\int \sqrt{3 - x^2 - 2x}\, dx$

8. $\displaystyle\int \sqrt{5 - x^2 - 2x}\, dx$

9. $\displaystyle\int \frac{1}{\sqrt{18 + 6x + x^2}} dx$

10. $\displaystyle\int \frac{1}{\sqrt{16 + 6x - x^2}} dx$

11. $\displaystyle\int \frac{x}{x^2 + 2x + 2} dx$

12. $\displaystyle\int_{0}^{2} \frac{x}{x^2 + 6x + 10} dx$

13. $\displaystyle\int_{3}^{4} \frac{3x^2 - 2x - 1}{x^3 - x^2 - x - 2} dx$

14. $\displaystyle\int \frac{x^2 + 2x + 2}{x^3 + 3x^2 + 3x + 2} dx$

15. $\displaystyle\int \frac{x - 3}{(x^2 + 1)(x^2 + x + 2)} dx$

16. $\displaystyle\int \frac{1}{x^3 - x^2 - x - 2} dx$

17. $\displaystyle\int_{-2}^{-1} \frac{x + 2}{(x^2 + x + 1)^2} dx$

18. $\displaystyle\int \frac{x^3}{(x^2 - 3x + 9)^2} dx$

19. $\displaystyle\int \frac{1}{x^5 + x^3 - x^2 - 1} dx$

20. $\displaystyle\int \frac{x}{x^5 - x^4 - 3x^2 - x - 2} dx$

21. Use the substitution $z = \sqrt[6]{x}$ to evaluate

$$\int \frac{\sqrt{x}}{1 + \sqrt[3]{x}} dx.$$

22. Use the substitution $z = \sqrt[12]{x}$ to evaluate

$$\int \frac{dx}{\sqrt[3]{x} + \sqrt[4]{x}}.$$

23. Make a substitution based on the idea in Exercise 21 to evaluate

$$\int \frac{e^{x/2}}{1 + e^{x/3}} dx.$$

24. Make a substitution based on the idea in Exercise 22 to evaluate

$$\int \frac{dx}{e^{x/3} + e^{x/4}}.$$

25. Find the area of the region bounded by the graph of $f(x) = 1/(x^2 + 2x + 2)$ and the x-axis on the interval $[-2, 0]$.

26. Find the volume generated by revolving about the y-axis the region bounded by the graph of $f(x) = 1/(x^2 - 6x + 10)$ and the x-axis on the interval $[1, 5]$.

In Exercises 48–51 in Section 7.4, the partial-fraction technique was used to find functions whose Laplace transforms were specified. A result known as the *First Shifting Theorem* for Laplace transforms states that if $L\{f(t)\} = F(s)$ and a is a constant, then $L\{e^{at} f(t)\} = F(s-a)$. For example,

$$L\{\cos 2t\} = \frac{s}{(s^2+4)}, \quad \text{so} \quad L\{e^{3t} \cos 2t\} = \frac{(s-3)}{[(s-3)^2+4]}.$$

Use this result to find the functions whose Laplace transforms are given in Exercises 27–30.

27. $\dfrac{s-3}{s^2-6s+10}$

28. $\dfrac{s+2}{s^2-6s+10}$

29. $\dfrac{s-4}{(s^2-2s+2)(s-2)}$

30. $\dfrac{6s^2+14}{(s^2-4s+5)(s^2+9)}$

7.6 Numerical Integration

In this chapter we have examined a number of methods for evaluating indefinite integrals. When an integral has a form other than those we have seen, the usual procedure is to consult a list of *integral tables*. A list of the most common integrals is in Appendix D. Handbooks of mathematical tables, such as the *CRC Standard Mathematical Tables*, contain more comprehensive lists. This particular work contains over 700 definite and indefinite integrals.

In this section we consider methods for finding *approximate* values of definite integrals. These methods can be applied to a variety of problems since they do not require knowing an antiderivative of the integrand. You might expect approximation methods to be a last resort, used only on problems for which no integration technique can be applied. However, these methods are often applied to definite integrals of elementary functions, especially in a problem requiring a computer solution. To halt a computer program in order to evaluate an integral is generally inconvenient and inefficient. In addition, the error in the approximation can often be controlled so that it does not affect significantly the accuracy of the result.

Numerical integration uses information about the integrand to approximate integrals at specific values. In practical applications, this information may be obtained experimentally, in which case an explicit representation of the integrand is unlikely to be known.

The methods discussed in this section involve determining a polynomial that agrees with the integrand at certain specified values, and then integrating the polynomial.

Suppose f is integrable on the interval $[x_{i-1}, x_i]$. The equation of the line passing through the points $(x_{i-1}, f(x_{i-1}))$ and $(x_i, f(x_i))$ can be written in the form $y = L_1(x)$, where $L_1(x)$ is the linear polynomial defined by

(7.20) $$L_1(x) = \frac{x - x_i}{x_{i-1} - x_i} f(x_{i-1}) + \frac{x - x_{i-1}}{x_i - x_{i-1}} f(x_i).$$

This is easily verified since

$$L_1(x_{i-1}) = 1 \cdot f(x_{i-1}) + 0 \cdot f(x_i) = f(x_{i-1})$$

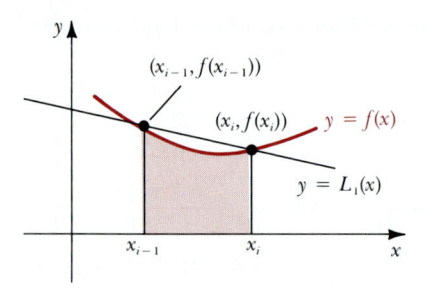

Figure 7.8

and
$$L_1(x_i) = 0 \cdot f(x_{i-1}) + 1 \cdot f(x_i) = f(x_i).$$

The graph of $y = L_1(x)$ is the line shown in Figure 7.8. Integrating $L_1(x)$ over $[x_{i-1}, x_i]$ gives the following approximation to the integral of f over this interval, where h is defined as $h = x_i - x_{i-1}$:

$$\int_{x_{i-1}}^{x_i} f(x)\, dx \approx \int_{x_{i-1}}^{x_i} L_1(x)\, dx$$

$$= \frac{1}{h} \int_{x_{i-1}}^{x_i} [-(x - x_i) f(x_{i-1}) + (x - x_{i-1}) f(x_i)]\, dx$$

$$= \frac{1}{h} \left[-\frac{(x - x_i)^2}{2} f(x_{i-1}) + \frac{(x - x_{i-1})^2}{2} f(x_i) \right]_{x_{i-1}}^{x_i}$$

$$= \frac{1}{h} \left[\frac{(x_{i-1} - x_i)^2}{2} f(x_{i-1}) + \frac{(x_i - x_{i-1})^2}{2} f(x_i) \right]$$

$$= \frac{h}{2} [f(x_{i-1}) + f(x_i)].$$

If f is continuous and nonnegative on the interval $[x_{i-1}, x_i]$, this approximation is the area of the trapezoid shown in Figure 7.8 which approximates the area bounded by the graph of f and the x-axis on $[x_{i-1}, x_i]$.

Suppose now that f is integrable on an interval $[a, b]$ and that $\mathscr{P} = \{x_0, x_1, x_2, \ldots, x_n\}$ is an equally spaced partition of $[a, b]$ with $h = x_i - x_{i-1}$ for each $i = 1, 2, \ldots, n$. The integral of f on $[a, b]$ is approximated by summing the trapezoidal approximations on each subinterval $[x_{i-1}, x_i]$ (see Figure 7.9):

$$\int_a^b f(x)\, dx = \int_{x_0}^{x_1} f(x)\, dx + \int_{x_1}^{x_2} f(x)\, dx + \cdots + \int_{x_{n-1}}^{x_n} f(x)\, dx$$

$$\approx \frac{h}{2} [f(x_0) + f(x_1)] + \frac{h}{2} [f(x_1) + f(x_2)] + \cdots$$

$$+ \frac{h}{2} [f(x_{n-1}) + f(x_n)].$$

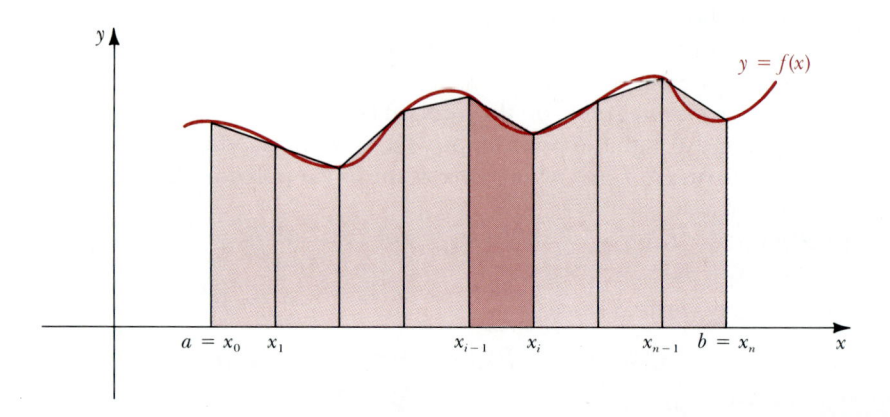

Figure 7.9
The Trapezoidal Rule.

Combining the approximations gives

(7.21)
THE TRAPEZOIDAL RULE

$$\int_a^b f(x)\, dx \approx \frac{h}{2} [f(x_0) + 2f(x_1) + 2f(x_2) + \cdots + 2f(x_{n-1}) + f(x_n)]$$

Example 1

Approximate $\int_{-1}^{1} e^x\, dx$ using the partition with $x_i = -1 + i/4$, for each $i = 0, 1, \ldots, 8$ and:

a) the Trapezoidal Rule;
b) the Riemann sum with $z_i = x_i$, for each i;
c) the Riemann sum with $z_i = x_{i-1}$, for each i.

Solution

a) Since $h = 1/4$, the Trapezoidal Rule (see Figure 7.10) gives

$$\int_{-1}^{1} e^x\, dx \approx \frac{\frac{1}{4}}{2} [e^{-1} + 2e^{-0.75} + 2e^{-0.5} + 2e^{-0.25} + 2e^0$$
$$+ 2e^{0.25} + 2e^{0.5} + 2e^{0.75} + e^1]$$
$$= 2.36263.$$

b) The Riemann sum with $z_i = x_i$ for each i (see Figure 7.11) gives

$$\int_{-1}^{1} e^x\, dx \approx \frac{1}{4}\left[e^{-0.75} + e^{-0.5} + e^{-0.25} + e^0 + e^{0.25} \right.$$
$$\left. + e^{0.5} + e^{0.75} + e^1 \right]$$
$$= 2.65643.$$

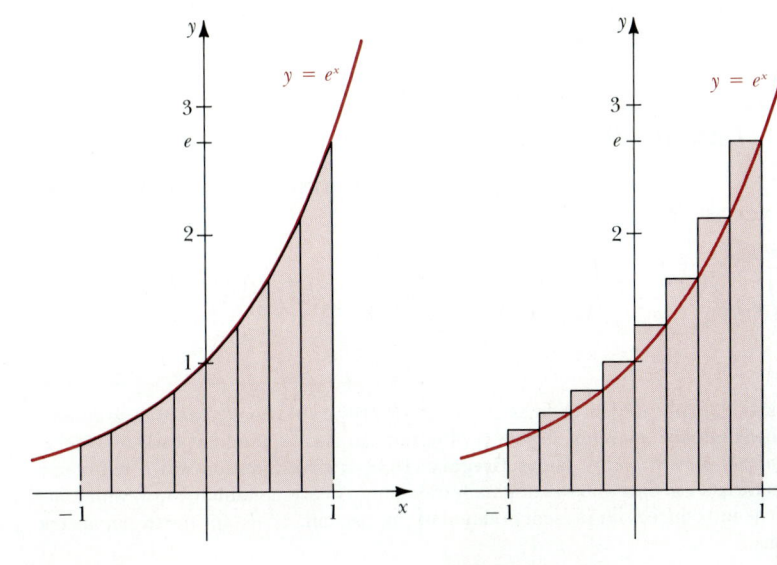

Figure 7.10

Figure 7.11

c) The Riemann sum with $z_i = x_{i-1}$ for each i (see Figure 7.12) gives

$$\int_{-1}^{1} e^x \, dx \approx \frac{1}{4}\left[e^{-1} + e^{-0.75} + e^{-0.5} + e^{-0.25} \right.$$
$$\left. + e^0 + e^{0.25} + e^{0.5} + e^{0.75} \right]$$
$$= 2.06883.$$

Since, to the accuracy listed,

$$\int_{-1}^{1} e^x \, dx = e^x \Big]_{-1}^{1} = e - e^{-1} = 2.35040,$$

the errors in the approximations are (a) 0.01223, (b) 0.30603, and (c) -0.28157. ▶

In Example 1, the Trapezoidal Rule approximation is clearly superior to either of the approximations from the Riemann sums. The next method we discuss uses quadratic polynomials to approximate the function and generally leads to even better results. This method is called **Simpson's Rule**.

Suppose f is integrable on $[a, b]$ and $\mathcal{P} = \{x_0, x_1, \ldots, x_n\}$ is an equally spaced partition of $[a, b]$ for some *even* integer n. The polynomial defined by

(7.22)

$$L_2(x) = \frac{(x - x_i)(x - x_{i+1})}{(x_{i-1} - x_i)(x_{i-1} - x_{i+1})} f(x_{i-1}) + \frac{(x - x_{i-1})(x - x_{i+1})}{(x_i - x_{i-1})(x_i - x_{i+1})} f(x_i)$$
$$+ \frac{(x - x_{i-1})(x - x_i)}{(x_{i+1} - x_{i-1})(x_{i+1} - x_i)} f(x_{i+1})$$

agrees with f at x_{i-1}, x_i, and x_{i+1}, since

$$L_2(x_{i-1}) = 1 \cdot f(x_{i-1}) + 0 \cdot f(x_i) + 0 \cdot f(x_{i+1}) = f(x_{i-1}),$$
$$L_2(x_i) = 0 \cdot f(x_{i-1}) + 1 \cdot f(x_i) + 0 \cdot f(x_{i+1}) = f(x_i),$$

and

$$L_2(x_{i+1}) = 0 \cdot f(x_{i-1}) + 0 \cdot f(x_i) + 1 \cdot f(x_{i+1}) = f(x_{i+1}).$$

Letting $h = x_{i+1} - x_i = x_i - x_{i-1}$ and integrating over $[x_{i-1}, x_{i+1}]$ gives (see Exercise 20):

$$\int_{x_{i-1}}^{x_{i+1}} f(x) \, dx \approx \int_{x_{i-1}}^{x_{i+1}} L_2(x) \, dx = \frac{h}{3}\left[f(x_{i-1}) + 4f(x_i) + f(x_{i+1}) \right].$$

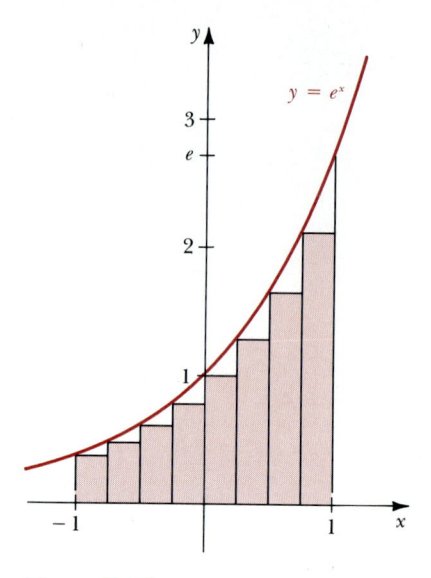

Figure 7.12

HISTORICAL NOTE

Formula (7.23) appeared in **Thomas Simpson's** (1710–1761) book *Mathematical dissertations on physical and analytical subjects* (1743), but can also be found as early as 1668 in *Exercitationes geometricae* by **James Gregory** (1638–1675). Simpson was a self-taught mathematician, a genius who was originally trained as a weaver. Simultaneously with Euler in 1748 he introduced our present notation of sin, cos, tan, . . . , for the trigonometric functions.

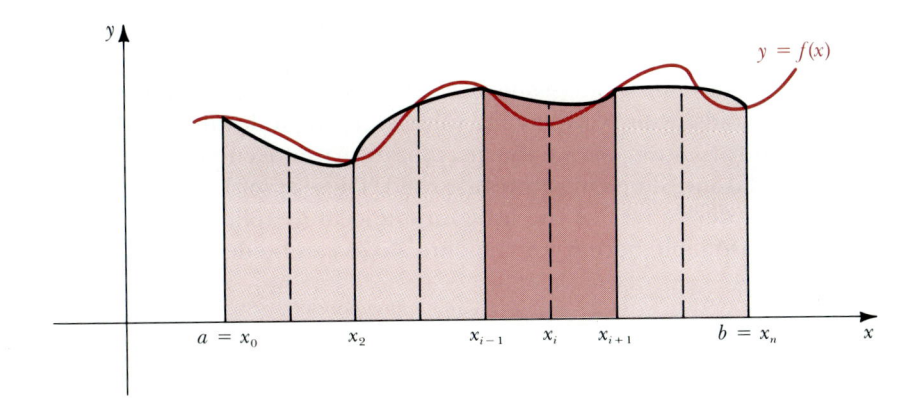

Figure 7.13
Simpson's Rule.

An approximation to the integral of f is found by summing the approximations on each subinterval (see Figure 7.13):

$$\int_a^b f(x)\,dx = \int_{x_0}^{x_2} f(x)\,dx + \int_{x_2}^{x_4} f(x)\,dx + \cdots + \int_{x_{n-2}}^{x_n} f(x)\,dx$$

$$\approx \frac{h}{3}[f(x_0) + 4f(x_1) + f(x_2)] + \frac{h}{3}[f(x_2) + 4f(x_3) + f(x_4)]$$

$$+ \cdots + \frac{h}{3}[f(x_{n-2}) + 4f(x_{n-1}) + f(x_n)].$$

Combining the approximations gives the following.

(7.23) SIMPSON'S RULE

$$\int_a^b f(x)\,dx \approx \frac{h}{3}[f(x_0) + 4f(x_1) + 2f(x_2) + 4f(x_3) + \cdots + 2f(x_{n-2})$$

$$+ 4f(x_{n-1}) + f(x_n)]$$

Example 2

Use Simpson's Rule with $n = 8$ to approximate $\int_{-1}^{1} e^x\,dx$. Compare this result to the Trapezoidal-Rule approximation found in Example 1.

Solution
Simpson's Rule with $n = 8$ gives

$$\int_{-1}^{1} e^x\,dx \approx \frac{(1/4)}{3}[e^{-1} + 4e^{-0.75} + 2e^{-0.5} + 4e^{-0.25}$$

$$+ 2e^0 + 4e^{0.25} + 2e^{0.5} + 4e^{0.75} + e^1]$$

$$= 2.35045.$$

The error in this approximation is 0.00005 compared to an error of 0.01223 using the Trapezoidal Rule. Simpson's Rule is significantly superior for this problem. ▶

Appendix C contains BASIC programs for implementing the Trapezoidal and Simpson's Rules.

In most instances, Simpson's Rule gives better approximations than the Trapezoidal Rule for the same value of n. The following results give specific information about the maximum error occurring when the Trapezoidal and Simpson's Rules are used. These error formulas rely on the higher derivatives of the integrand and are explicitly useful only when a reasonable bound for the derivatives can be determined.

(7.24)
THEOREM: THE TRAPEZOIDAL RULE
If f is defined on $[a, b]$ and $|f''(x)| \leq M$ for all x in $[a, b]$, then

$$\left| \int_a^b f(x)\, dx - \frac{h}{2} [f(x_0) + 2f(x_1) + \cdots + 2f(x_{n-1}) + f(x_n)] \right| \leq \frac{(b-a)M}{12} h^2,$$

where $\{x_0, x_1, \ldots, x_n\}$ is an evenly spaced partition of $[a, b]$ and $h = x_i - x_{i-1}$ for each $i = 1, 2, \ldots, n$.

(7.25)
THEOREM: SIMPSON'S RULE
If f is defined on $[a, b]$ and $|f^{(4)}(x)| \leq M$ for all x in $[a, b]$, then

$$\left| \int_a^b f(x)\, dx - \frac{h}{3} [f(x_0) + 4f(x_1) + 2f(x_2) + \cdots + f(x_n)] \right| \leq \frac{(b-a)M}{180} h^4,$$

provided n is even, where $\{x_0, x_1, \ldots, x_n\}$ is an evenly spaced partition of $[a, b]$ and $h = x_i - x_{i-1}$ for each $i = 1, 2, \ldots, n$.

The expressions on the right side of the inequalities in Theorems (7.24) and (7.25) are called **error bounds.** When h is small and the bounds on the second and fourth derivatives of f are nearly the same, the term h^4 in Simpson's Rule produces a much smaller error bound than the term h^2 in the Trapezoidal Rule.

Example 3

Compare the error bounds for the Trapezoidal Rule and Simpson's Rule when applied to approximate $\int_{-1}^{1} e^x\, dx$ with $n = 8$.

Solution
Since $f(x) = e^x$, $f''(x) = f^{(4)}x = e^x$, and both derivatives are bounded on $[-1, 1]$ by $M = e^1 = e$. The bound for the Trapezoidal Rule with $n = 8$ is therefore

$$\frac{2}{12} e \left(\frac{1}{4}\right)^2 \approx 0.02832,$$

whereas the bound for Simpson's Rule with $n = 8$ is

$$\frac{2}{180} e \left(\frac{1}{4}\right)^4 \approx 0.00012.$$

Note that the actual errors in approximation, 0.01223 for the Trapezoidal Rule and 0.00005 for Simpson's Rule, are well within the error bounds. ▶

Example 4

Find an approximate value for the arc length of the curve described by $f(x) = \sin x$ from $(0, 0)$ to $(\pi/4, \sqrt{2}/2)$.

Solution

The arc length of the curve is

$$L = \int_0^{\pi/4} \sqrt{1 + [D_x f(x)]^2}\, dx = \int_0^{\pi/4} \sqrt{1 + \cos^2 x}\, dx.$$

This integral belongs to the class of *elliptic integrals,* integrals that cannot be evaluated exactly by any integration technique. Using the Trapezoidal and Simpson's Rules, we can generate approximations to this integral, as shown in Table 7.3.

Table 7.3

n	TRAPEZOIDAL RULE	SIMPSON'S RULE
2	1.0527995	1.0582938
4	1.0567808	1.0581079
6	1.0575120	1.0580979
8	1.0577674	1.0580963
10	1.0578856	1.0580958

Even without determining the error bounds, we would suspect from the agreement of the Simpson's-Rule approximations that the approximation 1.0581 is accurate to the number of decimal places listed. ▶

The study of statistics is concerned with the way in which randomly chosen values are distributed. To describe a continuous numerical distribution, statisticians use a **probability density function:** a nonnegative function with the property that the area bounded by the curve and the x-axis is 1. Integrating the probability density function over an interval $[a, b]$ gives the probability (a number in $[0, 1]$) that a randomly chosen value will fall between a and b.

An important family of probability density functions has the form

(7.26) $$f(x) = \frac{1}{\sigma \sqrt{2\pi}}\, e^{-\frac{1}{2}\left(\frac{x-\mu}{\sigma}\right)^2},$$

where μ and σ are constants and $\sigma > 0$. The probability distribution described by this type of function is called a **normal distribution;** its graph is a bell-shaped curve of the type shown in Figure 7.14.

Normal distributions are associated with problems involving the distribution of errors in experimental data, tolerances of standard machined parts, scores on standardized examinations, and characteristic features of large populations, features such as the height, weight, or

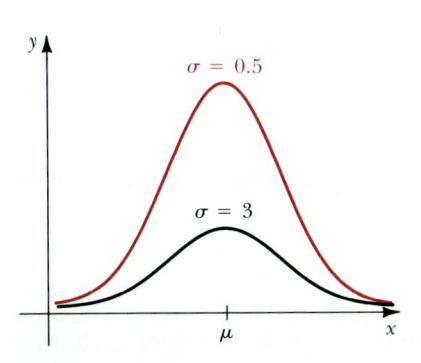

Figure 7.14

Normal distributions.

voting preferences of individuals. The constant μ in the probability density function describes the **mean,** or average, value of the distribution. The constant σ, called the **standard deviation,** describes the tendency of the values to cluster about the mean. A large value of σ indicates that the distribution is widely spread, whereas a small value of σ implies that the values are tightly clustered (see Figure 7.14).

Integrals of normal probability density functions are of interest because of their applications to probability, but these functions have no elementary antiderivative, so the Fundamental Theorem of Calculus cannot be applied. To find the probability that a randomly chosen value from a normal distribution lies in an interval $[a, b]$, we must determine numerically the value of

$$\frac{1}{\sigma\sqrt{2\pi}} \int_a^b e^{-\frac{1}{2}\left(\frac{x-\mu}{\sigma}\right)^2} dx.$$

Actually this probability can be determined by consulting extensive tables associated with the probability density function of the **standard normal** distribution, the normal distribution with $\mu = 0$ and $\sigma = 1$. These tabulated values, however, were determined by numerical approximations.

Example 5

The scores on a standard college mathematical aptitude examination are normally distributed with mean $\mu = 500$ and standard deviation $\sigma = 100$. Find the probability that a randomly chosen examination score lies between 300 and 700.

Solution

The probability that a score lies in the interval $[300, 700]$ is

$$P = \frac{1}{100\sqrt{2\pi}} \int_{300}^{700} e^{-\frac{1}{2}\left(\frac{x-500}{100}\right)^2} dx.$$

Using the variable substitution $z = (x - 500)/100$ and $dz = dx/100$ transforms the problem into

$$P = \frac{1}{100\sqrt{2\pi}} \int_{x=300}^{x=700} e^{-\frac{1}{2}\left(\frac{x-500}{100}\right)^2} dx = \frac{1}{\sqrt{2\pi}} \int_{z=-2}^{z=2} e^{-\frac{z^2}{2}} dz.$$

Since e^{-z^2} is symmetric with respect to the vertical axis,

$$P = \frac{1}{\sqrt{2\pi}} \int_{-2}^{2} e^{-\frac{z^2}{2}} dz = \frac{2}{\sqrt{2\pi}} \int_{0}^{2} e^{-\frac{z^2}{2}} dz.$$

Using Simpson's Rule with various choices of n produces the approximations P_n in Table 7.4. From this table we conclude that the probability that a randomly chosen score lies between 300 and 700 is approximately 0.954, or over 95% of the time. ▶

Table 7.4

n	P_n
2	0.94721
4	0.95440
6	0.95448
8	0.95449

The final example of numerical integration concerns the evaluation of an integral when the integrand is not explicitly known.

Example 6

A surveyor has found the width of a pond at equally spaced points along one side, as shown in Figure 7.15. Use Simpson's Rule to approximate the area of the surface of the pond, assuming all distances in the figure are given in feet.

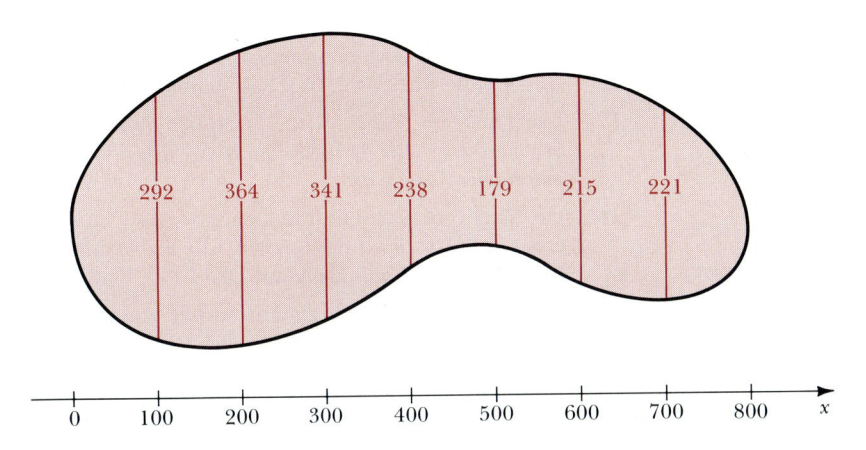

Figure 7.15

Solution

Let $w(x)$ denote the width of the pond at each point on the coordinate line. Then the area A of the surface of the pond is $A = \int_0^{800} w(x)\, dx$. Simpson's Rule implies that

$$A \approx \left(\frac{100}{3}\right)[w(0) + 4w(100) + 2w(200) + 4w(300) + 2w(400)$$
$$+ 4w(500) + 2w(600) + 4w(700) + w(800)]$$
$$= 192{,}200 \text{ ft}^2.$$

One U.S. acre is $43{,}560 \text{ ft}^2$, so the area of the pond is approximately 4.41 acres. ▶

HISTORICAL NOTE

Most of the integration techniques in this chapter were systematically worked out by the Bernoullis and Euler. The use of substitutions was pioneered by Johann Bernoulli in his lectures of 1691–1692 and were used extensively by Euler in his *Institutiones calculi integralis* (1768–1794). The idea underlying integration by parts appeared in 1715 in Brook Taylor's *Methodus incrementorum directa et inversa*; however, the name "integration by parts" appeared first in 1797–1798 in the *Traité du calcul différentiel et du calcul intégral* of **Sylvestre François Lacroix** (1765–1843).

EXERCISE SET 7.6

1. Use (a) the Trapezoidal Rule and (b) Simpson's Rule to approximate $\int_1^3 \ln x \, dx$, using 4 subintervals. (c) Compare the approximations to the actual value.

2. Approximate $\int_0^{0.1} x^{1/3} \, dx$ using (a) the Trapezoidal Rule and (b) Simpson's Rule with 2 subintervals. (c) Compare the approximations to the actual value.

Approximate the integrals in Exercises 3–8 using (a) the Trapezoidal Rule and (b) Simpson's Rule, with the indicated number n of subintervals.

3. $\int_0^1 e^{x^2} \, dx, \; n = 4$

4. $\int_{-1}^0 \sqrt{1 + x^3} \, dx, \; n = 4$

5. $\int_{-1}^1 \sin \pi x^2 \, dx, \; n = 4$

6. $\int_0^{\pi/2} \sqrt{1 + \cos x} \, dx, \; n = 4$

7. $\int_1^2 \frac{\sin x}{x} \, dx, \; n = 8$

8. $\int_1^5 \frac{\ln x}{x} \, dx, \; n = 8$

9. Use the Trapezoidal Rule with $n = 4$ to find an approximation to the length of the cosine curve from $x = 0$ to $x = \pi/2$.

10. Use the Trapezoidal Rule with 8 subintervals to approximate the area of the region bounded by the standard normal curve,

    $$y = \frac{1}{\sqrt{2\pi}} e^{-x^2/2},$$

    and the x-axis on the interval $[-1, 1]$.

11. Repeat Exercise 9 using Simpson's Rule.

12. Repeat Exercise 10 using Simpson's Rule.

13. Find an approximation to the area of the region bounded by the normal curve

    $$y = \frac{1}{\sigma\sqrt{2\pi}} e^{-(x/\sigma)^2/2}$$

    and the x-axis on the interval $[-\sigma, \sigma]$ using the Trapezoidal Rule with $n = 8$.

14. Repeat Exercise 13 using Simpson's Rule with $n = 8$.

15. To obtain a more accurate approximation to the area of the surface of the pond in Example 6, the width of the pond was found at the additional points given below.

x	50	150	250	350	450	550	650	750
$w(x)$	181	342	368	289	193	186	232	188

Use Simpson's Rule with the data in Example 6 and the additional data to approximate the area of the surface of the pond.

16. Use Theorem (7.24) to find an integer n such that the error in the approximation of $\int_1^3 \ln x \, dx$ by the Trapezoidal Rule with $h = 2/n$ is less than 10^{-3}.

17. Use Theorem (7.24) to find an upper bound for the error in using the Trapezoidal Rule with $h = \frac{1}{2}$ to approximate $\int_1^3 \ln x \, dx$.

18. Use Theorem (7.25) to find an integer n such that the error in the approximation of $\int_1^3 \ln x \, dx$ by Simpson's Rule with $h = 2/n$ is less than 10^{-3}.

19. Use Theorem (7.25) to find an upper bound for the error in using Simpson's Rule with $h = \frac{1}{2}$ to approximate $\int_1^3 \ln x \, dx$.

20. Show that

    $$\int_{x_{i-1}}^{x_{i+1}} L_2(x) \, dx = \frac{h}{3} \left[f(x_{i-1}) + 4f(x_i) + f(x_{i+1}) \right].$$

    [*Hint*: Use the substitution $x = x_{i-1} + t \cdot h$.]

21. A *superegg* is described by Martin Gardner in his "Mathematical Games" column of the *Scientific American*, September 1965, as the solid of revolution obtained by revolving the *super ellipse*

    $$\frac{|x|^{2.5}}{a^{2.5}} + \frac{|y|^{2.5}}{b^{2.5}} = 1$$

    about the x-axis (see the figure). Use Simpson's Rule with $n = 4$ to approximate the volume of the superegg when $a = 3$ and $b = 2$.

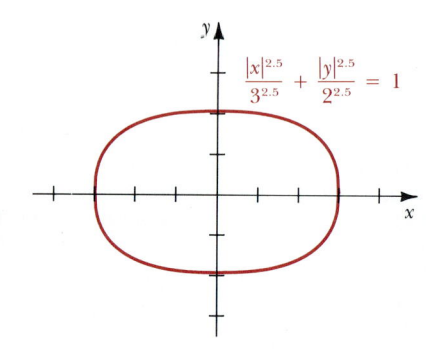

22. In an electrical circuit containing an impressed voltage ε, a capacitor with capacitance C farads, and a resistance of R ohms, the following relationship holds:

$$\varepsilon = Ri + \frac{1}{C}\int_0^t i\, dt,$$

where i is the current, in amperes, in the circuit at time t, in seconds (see the figure). Suppose that in a particular circuit $R = 0.1$ ohm, $C = 1$ microfarad (10^{-6} farads), and at time t, $i = \sqrt{1 + (\sin \pi t/4)^2}$ amperes. Use Simpson's Rule with $n = 4$ to find the voltage after 2 sec.

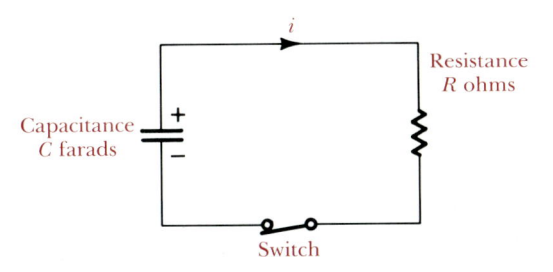

23. Suppose the grade-point averages of all college students in the United States are normally distributed with mean 2.4 and standard deviation 0.8. Find, to within 10^{-3}, the probability that a randomly chosen student has a grade-point average (a) between 1.6 and 3.2; (b) between 2.4 and 3.2; (c) between 2.4 and 3.0; (d) greater than 3.6.

24. a) Show that the average of the Riemann-sum approximations found in parts (b) and (c) of Example 1 is the same as the Trapezoidal Rule approximation found in part (a) of that example.
    b) Show that it is always true that if the Riemann sums are chosen as in parts (b) and (c) of Example 1, then their average is the same as the Trapezoidal-Rule approximation.

25. Show that the Trapezoidal Rule gives the exact result on any interval when applied to a polynomial of degree 1 or less.

26. Show that Simpson's Rule gives the exact result on any interval when applied to a polynomial of degree 3 or less.

The polynomials L_1 and L_2 defined in this section are special cases of the *Lagrange interpolatory polynomials*. In general, if f is a function defined at the $n + 1$ numbers $x_0, x_1, \ldots, x_n$, then the nth Lagrange polynomial for f at $x_0, x_1, \ldots, x_n$ is defined by $L_n(x) =$

$$\sum_{i=0}^{n} \frac{(x - x_0)\cdots(x - x_{i-1})(x - x_{i+1})\cdots(x - x_n)}{(x_i - x_0)\cdots(x_i - x_{i-1})(x_i - x_{i+1})\cdots(x_i - x_n)} f(x_i).$$

The following exercises concern the polynomials L_n.

27. Show that the polynomial $L_n(x)$ has the property that $L_n(x_i) = f(x_i)$ for each $i = 0, 1, \ldots, n$.

28. The Fundamental Theorem of Algebra implies that a polynomial of degree n has at most n distinct zeros. Use this fact to show that if P is a polynomial of degree at most n whose graph passes through $(x_0, f(x_0))$, $(x_1, f(x_1))$, ..., $(x_n, f(x_n))$, then $P \equiv L_n$.

▶ IMPORTANT TERMS AND RESULTS

INTEGRATION METHOD	WHEN APPROPRIATE	PAGE
Parts	$\int u\, dv = uv - \int v\, du$	457
Trigonometric products	$\int \sin^n x \cos^m x\, dx$	466
	$\int \tan^n x \sec^m x\, dx$	469
	$\int \cot^n x \csc^m x\, dx$	471

▶ IMPORTANT TERMS AND RESULTS (cont.)

INTEGRATION METHOD	WHEN APPROPRIATE	PAGE
Trigonometric substitutions	$\sqrt{a^2 - x^2}$. Let $x = a \sin \theta$.	479
	$\sqrt{a^2 + x^2}$. Let $x = a \tan \theta$.	479
	$\sqrt{x^2 - a^2}$. Let $x = a \sec \theta$.	479
Partial fractions	$\int \dfrac{Q(x)}{P(x)} dx$, where P and Q are polynomials	482
Quadratic polynomials	$\int f(x)\, dx$, where f is a function involving a quadratic, $ax^2 + bx + c$.	491

APPROXIMATION METHOD		PAGE
Trapezoidal Rule	$\int_a^b f(x)\, dx \approx \dfrac{h}{2} [f(x_0) + 2f(x_1) + \cdots + 2f(x_{n-1}) + f(x_n)]$	497
Simpson's Rule	$\int_a^b f(x)\, dx \approx \dfrac{h}{3} [f(x_0) + 4f(x_1) + 2f(x_2) + \cdots + 2f(x_{n-2}) + 4f(x_{n-1}) + f(x_n)]$	499

▶ REVIEW EXERCISES

In Exercises 1–102, evaluate the integral.

1. $\int (x - 1) \sin x\, dx$
2. $\int 2x \sqrt{x + 1}\, dx$
3. $\int xe^{2x}\, dx$
4. $\int x\, 2^x\, dx$
5. $\int \dfrac{x}{\sqrt{1 + x^2}}\, dx$
6. $\int \dfrac{\sqrt{x^2 - 9}}{x}\, dx$
7. $\int \dfrac{dx}{x^2 - 4}$
8. $\int \dfrac{x^3 + 2x + 1}{x^2 + 1}\, dx$
9. $\int (\cot x + \tan x)^2\, dx$
10. $\int \dfrac{1}{x\sqrt{x^2 + 4}}\, dx$
11. $\int \cos^5 x \sqrt{\sin x}\, dx$
12. $\int \dfrac{x^2}{(1 - x^2)^{3/2}}\, dx$
13. $\int \dfrac{x + 2}{x^3 + 2x^2 - 3x}\, dx$
14. $\int (x^2 + 1)\, 3^x\, dx$
15. $\int \sin^3 2x \cos^2 2x\, dx$
16. $\int x^2 \ln x\, dx$
17. $\int \dfrac{x^3}{\sqrt{x^2 - 1}}\, dx$
18. $\int \sqrt{x^2 + 4}\, dx$
19. $\int x(x^2 + 1)^5\, dx$
20. $\int \arctan 4x\, dx$
21. $\int \sec^3 x \tan^3 x\, dx$
22. $\int (\sin x)^e \cos^3 x\, dx$
23. $\int \dfrac{x^3 + 2x + 1}{x^3 + x}\, dx$
24. $\int (x^3 + 2x)\, e^{x^2}\, dx$
25. $\int \dfrac{\sqrt{x^2 + 1}}{x^3}\, dx$
26. $\int \operatorname{arccot} 3x\, dx$
27. $\int \dfrac{1}{x\sqrt{x^2 - 1}}\, dx$
28. $\int \dfrac{x}{x^2 + 5x + 4}\, dx$
29. $\int \dfrac{x^3}{\sqrt{1 - x^2}}\, dx$
30. $\int \dfrac{1}{x^2 - 8x + 25}\, dx$
31. $\int x^3 \sin 2x\, dx$
32. $\int \dfrac{1}{x^2 + 4x + 5}\, dx$
33. $\int \dfrac{2x}{x^2 + 6x - 7}\, dx$
34. $\int \dfrac{x - 1}{x^2 + x}\, dx$
35. $\int \ln \sqrt{2x - 1}\, dx$
36. $\int \dfrac{e^x}{1 + e^x}\, dx$
37. $\int x^5 e^{-x^2}\, dx$
38. $\int (\sin 4x)^3 (\cos 4x)^\pi\, dx$
39. $\int \tan^5 x \sec^4 x\, dx$
40. $\int \dfrac{dx}{x^2 - 4x}$

41. $\displaystyle\int \frac{e^x}{1+e^{2x}}\,dx$

42. $\displaystyle\int \frac{e^{\tan x}}{1-\sin^2 x}\,dx$

43. $\displaystyle\int \frac{dx}{x^3-4x^2}$

44. $\displaystyle\int \frac{x}{x^2+6x+10}\,dx$

45. $\displaystyle\int x^3(16-x^2)^{3/2}\,dx$

46. $\displaystyle\int \frac{dx}{x^3-4x}$

47. $\displaystyle\int \tan^5 x\,dx$

48. $\displaystyle\int \ln(x^2+1)\,dx$

49. $\displaystyle\int \frac{\ln x^2}{x}\,dx$

50. $\displaystyle\int (\cos x)\ln\sin x\,dx$

51. $\displaystyle\int (x^2+x+2)\ln x\,dx$

52. $\displaystyle\int (\sin x)^3 \ln\cos x\,dx$

53. $\displaystyle\int \frac{\sin(\ln x)\cos(\ln x)\,dx}{x}$

54. $\displaystyle\int \frac{x^2-1}{x^3+x^2-4x-4}\,dx$

55. $\displaystyle\int \sqrt{9x^2-16}\,dx$

56. $\displaystyle\int \sqrt{16-9x^2}\,dx$

57. $\displaystyle\int \frac{x+1}{x^3+2x^2+2x}\,dx$

58. $\displaystyle\int \tan 2x\,\sec^4 2x\,dx$

59. $\displaystyle\int (\tan x+\sec x)^2\,dx$

60. $\displaystyle\int e^x \sin 3x\,dx$

61. $\displaystyle\int e^{3x}\sin 2x\,dx$

62. $\displaystyle\int \arctan\sqrt{x}\,dx$

63. $\displaystyle\int \frac{1}{\sqrt{7-6x-x^2}}\,dx$

64. $\displaystyle\int \frac{x^3-4x}{(x^2+1)^2}\,dx$

65. $\displaystyle\int \sin(\ln 2x)\,dx$

66. $\displaystyle\int \frac{x^3+x^2-4x-5}{x^2-4}\,dx$

67. $\displaystyle\int x\arctan\sqrt{x}\,dx$

68. $\displaystyle\int (\sin x+\csc x)^2\,dx$

69. $\displaystyle\int \frac{3x^2}{\sqrt{4x^2-1}}\,dx$

70. $\displaystyle\int \frac{x^3+x^2}{x^2+2x+2}\,dx$

71. $\displaystyle\int \frac{x^3+6x^2+3x+6}{x^3+2x^2}\,dx$

72. $\displaystyle\int \frac{x\,dx}{\sqrt{x-x^2}}$

73. $\displaystyle\int \frac{x^3+5x^2+2x-4}{x^4-1}\,dx$

74. $\displaystyle\int e^x(\cot e^x)^2\,dx$

75. $\displaystyle\int x^2\arctan x\,dx$

76. $\displaystyle\int \frac{x^3+x^2-4x-5}{x^2-3x+2}\,dx$

77. $\displaystyle\int \tan^3 2x\,dx$

78. $\displaystyle\int \frac{x^2}{\sqrt{1+4x^2}}\,dx$

79. $\displaystyle\int \frac{x^5}{\sqrt{1-2x^3}}\,dx$

80. $\displaystyle\int \frac{x^4+3x^3-4x+1}{x^3+4x^2+4x}\,dx$

81. $\displaystyle\int \cos 2x\sin 3x\,dx$

82. $\displaystyle\int \sin 4x\cos 5x\,dx$

83. $\displaystyle\int \sin 3x\cos x\,dx$

84. $\displaystyle\int \operatorname{arcsinh} x\,dx$

85. $\int \ln\left(\dfrac{\sqrt{x+4}}{x+1}\right) dx$

86. $\int \dfrac{\sin 2x}{1+\sin^2 x} dx$

87. $\int \dfrac{\cosh \ln x}{x} dx$

88. $\int \ln(x - \sqrt{x^2+1})\, dx$

89. $\int \dfrac{1}{(x+1)(x^2+x+1)} dx$

90. $\int \dfrac{dx}{x - \sqrt{x}}$

91. $\int \dfrac{1}{2 - \sqrt{3x}} dx$

92. $\int \dfrac{\sqrt{x}}{1+\sqrt{x}} dx$

93. $\int \dfrac{\sqrt{x}}{1+\sqrt[4]{x}} dx$

94. $\int \dfrac{\sqrt[3]{x}}{1+\sqrt[4]{x}} dx$

(*Hint*: See Exercise 22 of Section 7.5.)

95. $\int \dfrac{e^{x/3}}{1+e^{x/2}} dx$

(*Hint*: See Exercise 23 of Section 7.5.)

96. $\int \dfrac{1}{e^{x/3}+e^{x/2}} dx$

97. $\int \dfrac{1}{\sin x \cos x} dx$

98. $\int \dfrac{1}{\sin^2 x \cos^2 x} dx$

99. $\int \dfrac{2}{\cos x + \cot x} dx$

100. $\int \dfrac{1+\sin x}{1+\cos x} dx$

(*Hint*: See Exercise 39 in Section 7.4.)

101. $\int \dfrac{4}{\cos x + 3} dx$

102. $\int \dfrac{3}{2+\sin x} dx$

103. Find the area of the region bounded by the x-axis, the line $x = e$, and the graph of $y = \ln x$.

104. Find the area bounded by the graphs of $y = e^{x-1}$, $y = (x + \ln x)/x$, and $x = e$.

105. Find the volume of the solid generated by revolving about the x-axis the region bounded by the x-axis, the lines $x = -2$ and $x = 2$, and the graph of $y = 4/(x^2 + 4)$.

106. Find the volume of the solid generated by revolving about the y-axis the region described in Exercise 105.

107. Find the volume of the solid generated by revolving about the y-axis the region bounded by the x-axis, the lines $x = 1$ and $x = 3$, and the graph of $y = e^x$.

108. Find the volume of the solid generated by revolving about the x-axis the region described in Exercise 107.

109. Find the centroid of the homogeneous lamina bounded by the x-axis, the line $x = e$, and the graph of the function described by $f(x) = \ln x$.

110. Find the centroid of the homogeneous lamina bounded by the x-axis and the graph of $y = \sin x$ for x in $[0, \pi]$.

In Exercises 111–114, approximate the integral using (a) the Trapezoidal Rule and (b) Simpson's Rule, with the indicated number n of subintervals.

111. $\int_0^1 \sqrt{1-x^3}\, dx,\ n = 4$

112. $\int_0^{\pi/3} \sqrt{\tan x}\, dx,\ n = 2$

113. $\int_1^3 \dfrac{\sin^2 x}{x} dx,\ n = 8$

114. $\int_{-\pi/2}^{\pi/2} \sqrt{1+\sin x}\, dx,\ n = 4$

115. Use the Trapezoidal Rule with $n = 8$ to find an approximation to the arc length of the graph of the function described by $f(x) = x^3$ from $x = 0$ to $x = 2$.

116. Use Simpson's Rule with $n = 8$ to find an approximation to the length of the sine curve from $x = -\pi/2$ to $x = \pi/2$.

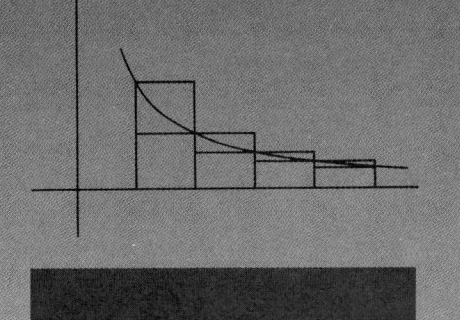

8
Sequences and Series

A sequence is an ordered collection of numbers. In Chapter 4 we used this notion to find approximations A_n to the area A bounded on a closed interval by the x-axis and the graph of a continuous nonnegative function (see Figure 8.1). The sequence of approximations A_1, A_2, A_3, . . . approaches A as n becomes large.

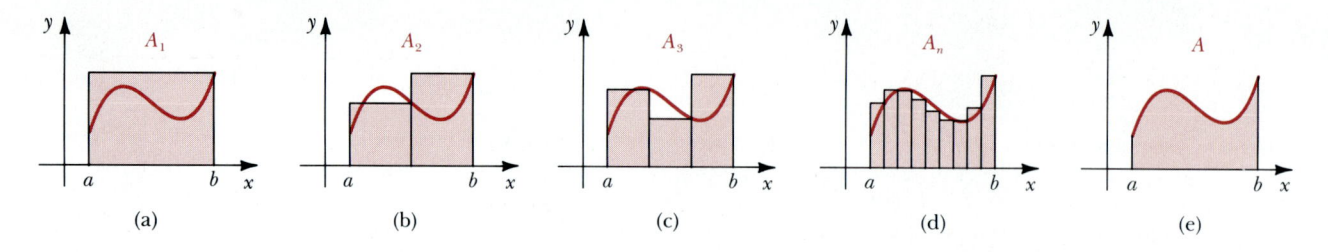

Figure 8.1
$\lim_{n \to \infty} A_n = A$.

In Section 8.1 we consider properties of sequences. We then see how sequences are used to define series, which are infinite sums of numbers. Series of polynomials are introduced in the later sections of the chapter. These provide a link between such seemingly distinct classes of functions as the trigonometric, exponential, and polynomial functions. They give us a new tool, one used by the early developers of the calculus, to determine many derivative and integral results.

The topics developed in this chapter form a basis for future study in such subjects as differential equations, numerical analysis, and complex analysis.

8.1

Infinite Sequences

In the preceding chapters we considered real-valued functions whose domains are intervals. A sequence is a real-valued function whose domain consists of a set of integers.

> **(8.1)**
> **DEFINITION**
> A function whose domain consists of an infinite set of consecutive nonnegative integers is called an **infinite sequence,** or simply a **sequence.**

The number corresponding to a particular integer n is called the **nth term** of the sequence. This number is denoted by a subscripted index such as a_n, instead of using functional notation such as $f(n)$. The sequence itself is denoted $\{a_n\}$. The index of a sequence normally begins with either $n = 1$ or $n = 0$.

Example 1

List the first four terms of the sequence $\{a_n\}$, $n \geq 1$, where

$$a_n = \frac{1}{2^n}.$$

Solution

The first four terms of this sequence are

$$a_1 = \frac{1}{2}, \quad a_2 = \frac{1}{4}, \quad a_3 = \frac{1}{8}, \quad \text{and} \quad a_4 = \frac{1}{16}.$$

Figure 8.2 gives two geometric representations of the first four terms of this sequence. In Figure 8.2(a), the graph of the sequence is given on a set of coordinate axes. In Figure 8.2(b), the points corresponding to the terms are plotted and labeled on a horizontal axis. ▶

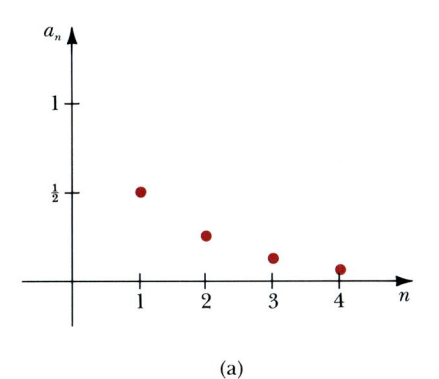

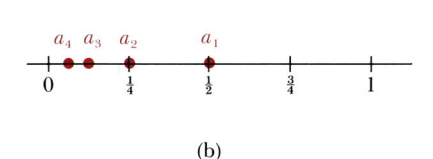

Figure 8.2

$a_n = \dfrac{1}{2^n}.$

Example 2

List the first four terms of the sequence $\{b_n\}$, $n \geq 1$, where

$$b_n = 1 - \frac{1}{2^n}.$$

Solution

The first four terms of this sequence are

$$b_1 = \frac{1}{2}, \quad b_2 = \frac{3}{4}, \quad b_3 = \frac{7}{8}, \quad \text{and} \quad b_4 = \frac{15}{16}.$$

Figure 8.3 shows the positions of these points in the interval $[0, 1]$. ▶

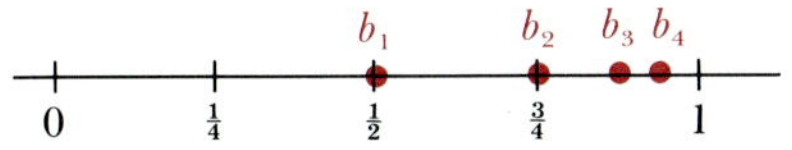

Note the connection between the sequences defined in Examples 1 and 2:

$$a_1 = \frac{1}{2} = b_1, \quad a_1 + a_2 = \frac{3}{4} = b_2, \quad a_1 + a_2 + a_3 = \frac{7}{8} = b_3,$$

and, in general,

$$a_1 + a_2 + \cdots + a_n = 1 + \frac{1}{2} + \frac{1}{4} + \cdots + \frac{1}{2^n} = 1 - \frac{1}{2^n} = b_n.$$

This connection is examined further in Section 8.2.

Figure 8.3

$b_n = 1 - \dfrac{1}{2^n}.$

Example 3

Find the first six terms of the sequence

$$\left\{(-1)^{n+1} \frac{2n}{n^2 + 1}\right\}, \quad n \geq 1.$$

Solution

The terms of this sequence alternate in sign because of the factor $(-1)^{n+1}$. The first six terms are illustrated in Figure 8.4:

$$a_1 = 1, \quad a_2 = \frac{-4}{5}, \quad a_3 = \frac{3}{5}, \quad a_4 = \frac{-8}{17}, \quad a_5 = \frac{5}{13}, \quad a_6 = \frac{-12}{37}. \quad \blacktriangleright$$

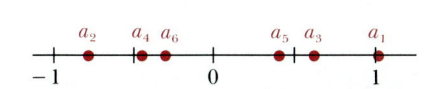

Figure 8.4

$a_n = (-1)^{n+1} \dfrac{2n}{n^2 + 1}.$

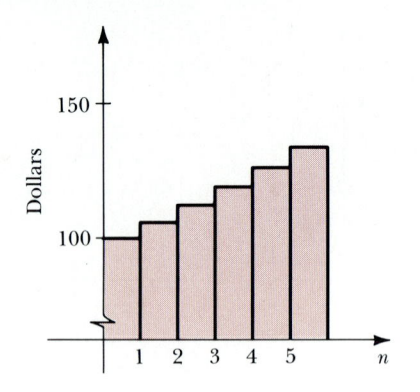

Figure 8.5
$a_n = 100(1 + 0.06)^n$.

Example 4

The nth term in the sequence $\{100(1 + 0.06)^n\}$, $n \geq 0$, is the amount in a savings account at the end of n years if $100 is deposited at 6% interest per year compounded annually. How much is in the account at the end of 5 years?

Solution

Since the fifth term gives the amount in the account at the end of 5 years, this amount is

$$100(1 + 0.06)^5 = \$133.82.$$

The amount of accumulated savings is shown on the bar graph in Figure 8.5. The bar graph is a common way to represent financial data. ▶

Another way to describe the sequence in Example 4 is to specify that $A_0 = 100$ and that the amount A_n in the account at the end of n years is related to the amount A_{n-1} in the account at the end of $n - 1$ years by the equation

$$A_n = A_{n-1} + 0.06 A_{n-1} = 1.06 A_{n-1}.$$

This is a recursive definition for the sequence. In general, a sequence $\{a_n\}$ is defined **recursively** by giving the first terms of the sequence and a formula that defines a_n using preceding terms.

Example 5

Determine the first five terms of the sequence $\{a_n\}$ if

$$a_1 = 1, \quad a_2 = 1, \quad \text{and} \quad a_n = \frac{a_{n-1}}{n-1} - \frac{a_{n-2}}{n-2}, \quad \text{for } n \geq 3.$$

Solution

The first two terms are given. The next three are

$$a_3 = \frac{a_2}{2} - \frac{a_1}{1} = \frac{1}{2} - \frac{1}{1} = -\frac{1}{2},$$

$$a_4 = \frac{a_3}{3} - \frac{a_2}{2} = \frac{(-\frac{1}{2})}{3} - \frac{1}{2} = -\frac{2}{3},$$

and

$$a_5 = \frac{a_4}{4} - \frac{a_3}{3} = \frac{(-\frac{2}{3})}{4} - \frac{(-\frac{1}{2})}{3} = 0.$$

▶

The terms of the sequence $\{1/2^n\}$ in Example 1 approach 0 as n increases without bound, though no term has the value 0. In Example 2, the terms of the sequence $\{1 - 1/2^n\}$ approach 1 as n increases; and in Example 3, the terms $(-1)^{n+1}[2n/(n^2 + 1)]$ approach 0. Situations of this type are described using the limit concept. Since the statement "as n increases without bound" can be replaced by "as n approaches infinity," a new definition is not required. We simply modify our original definition of the limit of a function at infinity:

> The real number L is the limit of the function f as x approaches ∞, written $\lim_{x \to \infty} f(x) = L$, provided that for every number $\varepsilon > 0$, a number N exists with the property that
>
> $$|f(x) - L| < \varepsilon \quad \text{whenever } x > N.$$

Definition (8.2) follows from this statement if we replace x with n and $f(x)$ with a_n. The definition is illustrated in Figure 8.6.

Figure 8.6
$\lim_{n \to \infty} a_n = L.$

(8.2) DEFINITION

The real number L is the limit of the sequence $\{a_n\}$, written

$$\lim_{n \to \infty} a_n = L,$$

provided that for every number $\varepsilon > 0$, a number N exists with the property that

$$|a_n - L| < \varepsilon \quad \text{whenever } n > N.$$

The definitions needed to describe infinite limits of sequences also follow from the corresponding definitions for functions. For example, we say that

$$\lim_{n \to \infty} a_n = \infty,$$

provided that for any number M, a positive integer N exists with the property that

$$a_n > M \quad \text{whenever } n > N.$$

The definition for $\lim_{n \to \infty} a_n = -\infty$ is similar and is considered in Exercise 51.

A sequence that has a finite limit is called a **convergent sequence.** If $\lim_{n \to \infty} a_n = \pm \infty$ or if $\lim_{n \to \infty} a_n$ does not exist, then the sequence $\{a_n\}$ is called a **divergent sequence.**

Example 6

Prove that $\lim_{n \to \infty} \dfrac{1}{n} = 0$.

Solution

Given $\varepsilon > 0$, we must show that a number M exists with the property that

$$\left| \frac{1}{n} - 0 \right| = \frac{1}{n} < \varepsilon \quad \text{whenever } n > M.$$

For any given $\varepsilon > 0$, we choose $M > 0$ to satisfy $1/M \le \varepsilon$, that is, $M \ge 1/\varepsilon$. If $n > M$, then

$$\left| \frac{1}{n} - 0 \right| = \frac{1}{n} < \frac{1}{M} \le \varepsilon.$$

This completes the proof that $\lim_{n \to \infty} 1/n = 0$. ▶

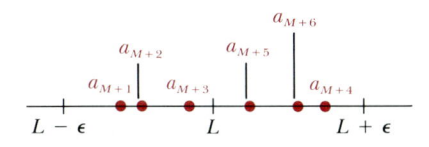

In a manner similar to that used in Example 6, it can be shown that

$$\lim_{n \to \infty} \frac{1}{n^p} = 0, \text{ for any } p > 0.$$

Example 7

Determine whether each of the following sequences is convergent or divergent, and give the limit of the sequence if it converges.

(a) $\left\{\left(\frac{1}{5}\right)^n\right\}$ (b) $\left\{\left(\frac{3}{2}\right)^n\right\}$ (c) $\{(-1)^n\}$

Solution

a) This sequence can be written as $\{1/5^n\}$. As $n \to \infty$, $5^n \to \infty$, so $1/5^n \to 0$. Consequently,

$$\lim_{n \to \infty} \left(\frac{1}{5}\right)^n = 0.$$

b) Since $(\frac{3}{2}) > 1$, $\lim_{n \to \infty} (\frac{3}{2})^n = \infty$, so the sequence $\{(\frac{3}{2})^n\}$ is divergent.

c) The terms of this sequence, $-1, 1, -1, 1, -1, 1, \ldots$, do not approach a unique number. Thus, $\lim_{n \to \infty} (-1)^n$ does not exist, and $\{(-1)^n\}$ is a divergent sequence. ▶

The sequences in Example 7 are of the type r^n, where r is a constant. In (a), $|r| < 1$ and the sequence converges. In (b), $|r| > 1$ and the sequence diverges. In fact, it is true in general that

(8.3)
$$\lim_{n \to \infty} r^n = 0 \quad \text{if } |r| < 1 \qquad \text{and} \qquad \{r^n\} \text{ diverges} \quad \text{if } |r| > 1.$$

When $r = -1$, the divergent sequence in Example 7(c) results. When $r = 1$, the series is the convergent constant sequence all of whose terms are 1.

In general, a constant sequence is a sequence $\{a_n\}$ for which $a_n = c$, for some constant c. For such a sequence, $\lim_{n \to \infty} a_n = c$.

The limit theorems discussed in Chapter 1 are equally valid for sequences and eliminate the need to use the definition in most limit problems.

(8.4)
THEOREM
If $\{a_n\}$ and $\{b_n\}$ are convergent sequences with $\lim_{n \to \infty} a_n = L$ and $\lim_{n \to \infty} b_n = M$, then:

a) $\lim_{n \to \infty} (a_n \pm b_n) = L \pm M;$ b) $\lim_{n \to \infty} (a_n b_n) = LM;$

c) $\lim_{n \to \infty} \left(\frac{a_n}{b_n}\right) = \frac{L}{M}$, provided $M \neq 0$ and $b_n \neq 0$ for all n;

d) $\lim_{n \to \infty} c a_n = cL$, for any constant c;

e) $\lim_{n\to\infty} \sqrt[m]{a_n} = \sqrt[m]{L}$, if m is an odd positive integer, or if m is an even positive integer and $a_n \geq 0$ for all n.

Example 8

Show that $\lim_{n\to\infty} (1 - 1/2^n) = 1$.

Solution

The constant sequence whose terms are 1 has the limit 1, and the sequence whose terms are $1/2^n = (1/2)^n$ has the limit 0. So part (a) of Theorem (8.4) implies that

$$\lim_{n\to\infty} \left(1 - \frac{1}{2^n}\right) = \lim_{n\to\infty} 1 - \lim_{n\to\infty} \left(\frac{1}{2}\right)^n = 1 - 0 = 1. \quad \blacktriangleright$$

Example 9

Determine $\lim_{n\to\infty} \dfrac{n^3 + n^2 + 1}{4n^3 + 2n + 3}$.

Solution

This limit is found using Theorem (8.4) and the fact that if $p > 0$, then $\lim_{n\to\infty} 1/n^p = 0$. We first divide the numerator and denominator by n^3, the highest power in the denominator, and then take the limit of each term:

$$\lim_{n\to\infty} \frac{n^3 + n^2 + 1}{4n^3 + 2n + 3} = \lim_{n\to\infty} \frac{1 + \dfrac{1}{n} + \dfrac{1}{n^3}}{4 + \dfrac{2}{n^2} + \dfrac{3}{n^3}}$$

$$= \frac{\lim_{n\to\infty} 1 + \lim_{n\to\infty} \dfrac{1}{n} + \lim_{n\to\infty} \dfrac{1}{n^3}}{\lim_{n\to\infty} 4 + 2\lim_{n\to\infty} \dfrac{1}{n^2} + 3\lim_{n\to\infty} \dfrac{1}{n^3}} = \frac{1}{4}. \quad \blacktriangleright$$

The procedure used in Example 9 should seem familiar; it was used in Chapter 1 to find limits of functions at infinity.

The following theorem is analogous to the Squeeze Theorem studied in Section 1.7

(8.5) THEOREM

Suppose $\{a_n\}$, $\{b_n\}$, and $\{c_n\}$ are sequences with $\lim_{n\to\infty} a_n = L$ and $\lim_{n\to\infty} b_n = L$. If for some $M > 0$, $a_n \leq c_n \leq b_n$ when $n > M$, then $\lim_{n\to\infty} c_n = L$.

Example 10

Find $\lim_{n\to\infty} \left(\dfrac{\sin n}{n}\right)$.

Solution
For all n,
$$-1 \leq \sin n \leq 1, \quad \text{so for } n \neq 0, \quad -\frac{1}{n} \leq \frac{\sin n}{n} \leq \frac{1}{n}.$$

However,
$$\lim_{n \to \infty} \frac{1}{n} = 0 \quad \text{and} \quad \lim_{n \to \infty}\left(-\frac{1}{n}\right) = -\lim_{n \to \infty} \frac{1}{n} = 0,$$

so Theorem (8.5) implies that $\lim_{n \to \infty} \frac{\sin n}{n} = 0$. ▶

An interesting special case of Theorem (8.5) occurs when $a_n = c_n$ for $n > M$. In this case, the theorem implies that:

(8.6)
Convergent sequences that differ by only a finite number of terms have the same limit.

The close relationship between the limit of a function at infinity and the limit of a sequence produces the next result.

(8.7)
THEOREM
Suppose f is a function with $\lim_{x \to \infty} f(x) = L$. If $f(n) = a_n$ for each positive integer n, then $\lim_{n \to \infty} a_n = L$.

Proof. Let $\varepsilon > 0$ be given. Since $\lim_{x \to \infty} f(x) = L$, there is a number M such that $|f(x) - L| < \varepsilon$ whenever $x > M$. Thus, if $n > M$,
$$|f(n) - L| = |a_n - L| < \varepsilon.$$

Consequently, $\lim_{n \to \infty} a_n = L$. ▷

Although the proof of Theorem (8.7) is given only for the case when the limit is a real number, the result is also true when L is replaced by ∞ or $-\infty$.

The converse of Theorem (8.7) is not true. Consider, for example, $f(x) = \sin \pi x$. The graph of f continually oscillates between 1 and -1, as shown in Figure 8.7, so $\lim_{x \to \infty} f(x)$ does not exist. For each n, however, $a_n = f(n) = \sin \pi n = 0$, so $\lim_{n \to \infty} a_n = 0$.

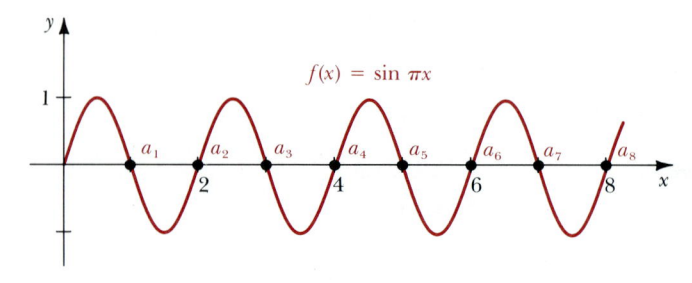

Figure 8.7
$\lim_{n \to \infty} a_n = 0$, but $\lim_{x \to \infty} f(x)$ does not exist.

Example 11

Find $\lim_{n \to \infty} \dfrac{\ln n}{n}$.

Solution
Since $\lim_{x \to \infty} \ln x = \infty$ and $\lim_{x \to \infty} x = \infty$, we can apply l'Hôpital's Rule to find $\lim_{x \to \infty} (\ln x / x)$.

$$\lim_{x \to \infty} \frac{\ln x}{x} = \lim_{x \to \infty} \frac{D_x \ln x}{D_x x} = \lim_{x \to \infty} \frac{1/x}{1} = 0.$$

Theorem (8.7) implies that

$$\lim_{n \to \infty} \frac{\ln n}{n} = 0. \qquad \blacktriangleright$$

Bounded Sequences

(8.8)
DEFINITION
A sequence $\{a_n\}$ is said to be **bounded above** if a number M exists with $a_n \leq M$ for each n. In this case, M is called an **upper bound** for $\{a_n\}$. Similarly, $\{a_n\}$ is **bounded below** if a number m exists with $a_n \geq m$ for each n. In this case, m is called a **lower bound** for $\{a_n\}$. A sequence that is bounded both above and below is called a **bounded sequence** (see Figure 8.8).

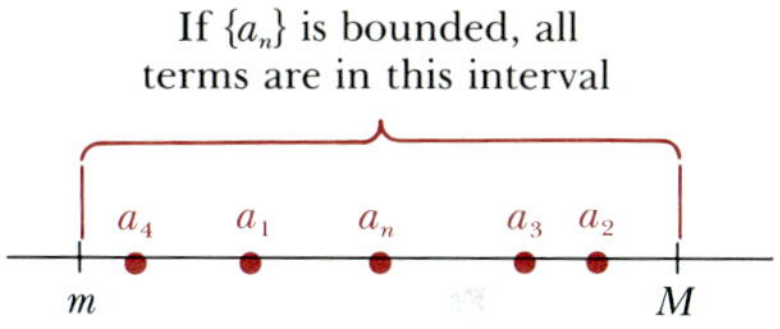

Figure 8.8
If $\{a_n\}$ is bounded, all its terms are within a closed interval.

Example 12

Give upper and lower bounds for the sequence

$$\left\{ \frac{n}{n+1} \right\}, \qquad n \geq 1.$$

Solution
An upper bound for this sequence with terms $\tfrac{1}{2}, \tfrac{2}{3}, \tfrac{3}{4}, \tfrac{4}{5}, \ldots$ is any number greater than or equal to 1, and a lower bound for this sequence is any number less than or equal to $\tfrac{1}{2}$. $\blacktriangleright$

(8.9)
THEOREM
If $\{a_n\}$ is a convergent sequence, then $\{a_n\}$ is bounded.

Proof. Suppose L is the limit of this sequence. Then for any $\varepsilon > 0$, and in particular for $\varepsilon = 1$, an integer M exists with the property that

$$|a_n - L| < \varepsilon = 1 \quad \text{whenever } n > M.$$

Consequently, for $n > M$, all terms of the sequence satisfy the inequality

$$-1 < a_n - L < 1, \quad \text{so} \quad L - 1 < a_n < L + 1$$

and these terms are bounded. There is only a finite number of terms of the sequence, the terms $a_1, a_2, \ldots, a_M$, not covered by this inequality. If we let

$$M_1 = \text{maximum } \{L + 1, a_1, a_2, \ldots, a_M\}$$

and

$$M_2 = \text{minimum } \{L - 1, a_1, a_2, \ldots, a_M\},$$

then for any n, $M_2 \leq a_n \leq M_1$, so the sequence is bounded (see Figure 8.9). ▷

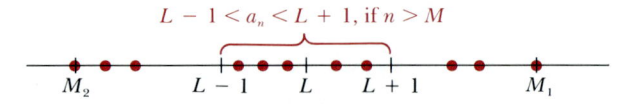

Figure 8.9

The converse of Theorem (8.9) is not true. For example, the sequence $\{(-1)^n\}$ is bounded above by 1 and below by -1, but this sequence is *not* convergent. To ensure that a bounded sequence is convergent, an additional condition is needed.

Similar to the definition for increasing and decreasing functions, we have:

(8.10)
A sequence $\{a_n\}$ is

i) **increasing** if $a_n < a_{n+1}$ for each n;
ii) **decreasing** if $a_n > a_{n+1}$ for each n.

A sequence is said to be **monotonic** if it is either an increasing or a decreasing sequence. The following statement concerning bounded monotonic sequences is a fundamental property of the set of real numbers. It is called the **Completeness Property** of the real numbers.

(8.11)
Every bounded monotonic sequence of real numbers converges.

The following example gives an application of the Completeness Property. The sequence in this example uses *factorial notation*.

If n is a positive integer, **n factorial,** written $n!$, is defined as

(8.12)
$$n! = 1 \cdot 2 \cdot 3 \cdot \cdots \cdot n.$$

In addition, we define $0! = 1$. This definition implies that

(8.13)
$$(n + 1)! = (n + 1) \cdot n!$$

Example 13

Use the Completeness Property to show that, if $a_n = 2^n/n!$, $n \geq 1$, then $\{a_n\}$ converges.

Solution

The $(n+1)$st term of the sequence can be written recursively:

$$a_{n+1} = \frac{2^{n+1}}{(n+1)!} = \frac{2}{(n+1)} \cdot \frac{2^n}{n!} = \frac{2}{n+1} \cdot a_n.$$

When $n \geq 2$, $2/(n+1) < 1$, so the sequence is decreasing after the second term. The sequence $\{a_n\}$ is also bounded, since for all n,

$$0 < a_n \leq a_1 = 2.$$

The Completeness Property implies that the sequence converges. It does not, however, give the value of the limit. ▶

▶ EXERCISE SET 8.1

The nth term of a sequence is given in Exercises 1–44.
- **a)** List the first three terms of the sequence. (Assume that $n \geq 1$ unless otherwise specified.)
- **b)** Determine if the sequence is convergent.
- **c)** If the sequence is convergent, give its limit. (It might help to review the infinite limit problems in Section 1.10 and the exercises on l'Hôpital's Rule in Sections 3.5 and 6.5.)
- **d)** Determine if the sequence is bounded.

1. $a_n = \dfrac{2}{n+1}$
2. $a_n = \dfrac{n}{n^2+3}$
3. $a_n = (-1)^n \dfrac{1}{n}$
4. $a_n = (-1)^n n,\ n \geq 0$
5. $a_n = \dfrac{n}{n+4},\ n \geq 0$
6. $a_n = (-1)^n \dfrac{n}{n+4},\ n \geq 0$
7. $a_n = \dfrac{n^2+2n+3}{n^2-4n+3},\ n \geq 4$
8. $a_n = \dfrac{n^2}{n^2-4},\ n \geq 3$
9. $a_n = \dfrac{n^3+2n^2-1}{n^2+n}$
10. $a_n = \dfrac{n^3+2n^2-1}{n^4+n}$
11. $a_n = \dfrac{\pi}{n}$
12. $a_n = \ln n$
13. $a_n = \dfrac{e^n}{n}$
14. $a_n = \dfrac{n}{e^n}$
15. $a_n = \dfrac{e^n}{n^p},\ p > 0$
16. $a_n = \dfrac{\sin n}{e^n}$
17. $a_n = \dfrac{\ln n}{n^2} - 1$
18. $a_n = \dfrac{\ln n}{e^n}$
19. $a_n = \sin \dfrac{\pi}{n}$
20. $a_n = \cos \dfrac{\pi}{n}$
21. $a_n = \dfrac{e^n}{n^2+3n-4},\ n \geq 2$
22. $a_n = \dfrac{n^3-1}{n-2},\ n \geq 3$
23. $a_1 = 1,\ a_n = \dfrac{a_{n-1}}{2},$ if $n \geq 2$
24. $a_1 = 1,\ a_n = \dfrac{2}{a_{n-1}},$ if $n \geq 2$
25. $a_n = (-1)^n \dfrac{n^2+3}{\sqrt[4]{n^9+3n^3+4}}$
26. $a_n = \dfrac{n^3+3}{\sqrt[4]{n^9+3n^3+4}}$
27. $a_n = \dfrac{1}{n} - \dfrac{1}{n+1}$
28. $a_n = \dfrac{1}{n+1} - \dfrac{1}{n+2}$
29. $a_n = \sqrt{n^2+1} - n$
30. $a_n = \sqrt{n+1} - \sqrt{n}$
31. $a_n = \dfrac{1}{\sqrt{n^2+1}-n}$
32. $a_n = \dfrac{1}{\sqrt{n^2+n}-n}$
33. $a_n = \dfrac{1}{n} \sin \dfrac{\pi}{n}$
34. $a_n = n \sin \dfrac{\pi}{n}$
35. $a_n = \dfrac{1}{n} \cos n\pi$
36. $a_n = n \cos n\pi$
37. $a_n = \dfrac{n}{n^2+1} \cos n\pi$
38. $a_n = \dfrac{n}{n^2+1} \sin \dfrac{n\pi}{2}$
39. $a_n = \dfrac{e^n}{1+e^n},\ n \geq 0$
40. $a_n = \dfrac{2^n}{1+2^n},\ n \geq 0$
41. $a_n = \left(1 + \dfrac{1}{n}\right)^n$
42. $a_n = \left(1 + \dfrac{1}{n}\right)^{2n}$
43. $a_1 = 1,\ a_n = \dfrac{a_{n-1}}{n^2},$ if $n \geq 2$
44. $a_1 = 1,\ a_2 = \dfrac{1}{2},\ a_n = \dfrac{a_{n-1}+a_{n-2}}{n},\ n \geq 3$

45. Consider the sequence whose first terms are 1, 0, 1, 0, 0, 1, 0, 0, 0, 1, 0, 0, 0, 0, 1, 0, 0, 0, 0, 0, 1, 0. . . . Suppose the terms continue in this manner. Does the sequence converge?

46. In *Liber abaci*, published in 1202 by Leonardo Fibonacci (circa 1180–1250), also known as Leonardo of Pisa, one finds the following problem: "How many pairs of rabbits are produced from one pair in a year if every month each pair produces a new pair that from the second month itself becomes productive?" If F_n denotes the number of pairs of rabbits at the end of the nth month, then $F_1 = 1$, $F_2 = 1$, $F_3 = 2$ and, in general, $F_n = F_{n-1} + F_{n-2}$ for $n \geq 3$. Find the number of pairs of rabbits at the end of the first year.

47. The population of a city was 250,000 at the end of 1987. It has been estimated that in each year following 1987, 50,000 people will enter the city, and 15% of the people living in the city at the beginning of the year will leave the city during the year. Let x_n denote the population of the city at the end of n years after 1987. Find a formula for x_n in terms of n.

48. Inscribe a regular polygon with n sides within a unit circle (see the figure).
   a) Show that the perimeter of the polygon is $p_n = 2n \sin \pi/n$.
   b) Find $\lim_{n \to \infty} p_n$. [*Hint:* Bisect the angle $2\pi/n$ and find a pair of congruent triangles, one of which clearly has a side with length equal to one half the length of a side of the polygon.]

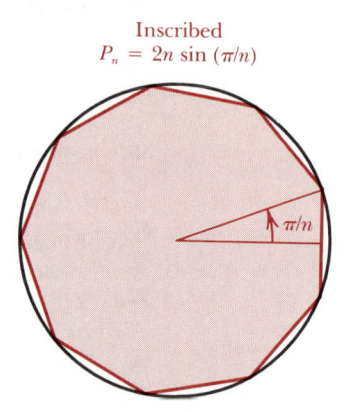

Inscribed
$P_n = 2n \sin (\pi/n)$

49. Circumscribe a regular polygon with n sides outside a unit circle (see the figure).
   a) Show that the perimeter of the polygon is $p_n = 2n \tan \pi/n$.
   b) Find $\lim_{n \to \infty} p_n$.

50. An equation of the form $x_n = ax_{n-1} + b_n$, $n \geq 1$ where a and x_0 are known constants and $\{b_n\}$ is a known sequence, is called a *linear first-order difference equation*. Verify that

$$x_n = x_0 a^n + \sum_{m=1}^{n} a^{n-m} b_m$$

satisfies this equation.

51. Give a definition for $\lim_{n \to \infty} a_n = -\infty$ that has the property that $\lim_{n \to \infty} a_n = -\infty$ precisely when $\lim_{n \to \infty} (-a_n) = \infty$.

52. Use Definition (8.2) to prove that if $\lim_{n \to \infty} a_n = L$, then $\lim_{n \to \infty} |a_n| = |L|$.

53. Find a sequence $\{a_n\}$ with the property that $\lim_{n \to \infty} |a_n|$ exists, but $\lim_{n \to \infty} a_n$ does not exist.

54. Show that $\lim_{n \to \infty} r^n = 0$ if $|r| < 1$, and that $\{r^n\}$ diverges if $|r| > 1$.

55. Use Definition (8.2) to prove that if $\lim_{n \to \infty} a_n = L$ and $\lim_{n \to \infty} b_n = M$, then:
   a) $\lim_{n \to \infty} (a_n + b_n) = L + M$;
   b) $\lim_{n \to \infty} ca_n = cL$, for any constant c.

56. Find sequences $\{a_n\}$ and $\{b_n\}$ with the property that $\{a_n + b_n\}$ is a convergent sequence, but both $\{a_n\}$ and $\{b_n\}$ are divergent sequences.

57. Show that if $\{a_n + b_n\}$ and $\{a_n - b_n\}$ both converge, then $\{a_n\}$ and $\{b_n\}$ converge.

58. Show that if $\{a_n\}$ is a convergent sequence, then for any number $\varepsilon > 0$, there exists a positive integer M such that $|a_n - a_m| < \varepsilon$ whenever $n, m > M$. (The condition described in this problem is known as the *Cauchy criterion*. It is proved in advanced calculus texts that a sequence is convergent if and only if it satisfies the Cauchy criterion.)

59. Suppose the sequence $\{a_n\}$, $n \geq 1$, converges to L. Use Definition (8.2) to show that for any integer p, the sequence $\{a_{n+p}\}$ also converges to L.

60. Consider the sequence $\{a_n\}$ defined recursively by

Circumscribed
$P_n = 2n \tan (\pi/n)$

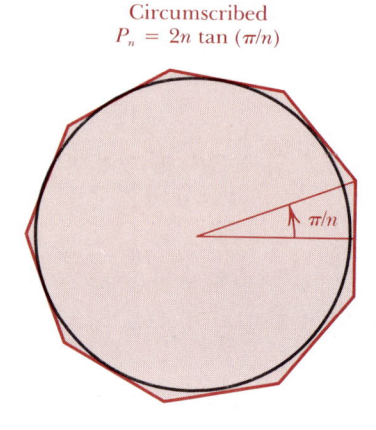

Figure for Exercise 49

$a_1 = \sqrt{2}$, $a_n = \sqrt{2a_{n-1}}$, for each $n \geq 2$. It can be shown by mathematical induction that $a_n < a_{n+1} < 2$ for each integer n. Show that the sequence converges, and use the result in Exercise 59 to determine the limit.

61. In Exercise 52 of Section 2.8 it was shown that: If a function g satisfies the following,
   i) g is differentiable on $[a, b]$,
   ii) a number L exists with $|g'(x)| \leq L < 1$ for all x in $[a, b]$, and
   iii) $a \leq g(x) \leq b$, for all x in $[a, b]$,
   then g has a unique fixed point p in $[a, b]$ (that is, $g(p) = p$). Show that if x_0 is chosen arbitrarily in $[a, b]$ and $x_n = g(x_{n-1})$ for each $n \geq 1$, then $\lim_{n \to \infty} x_n = p$.

62. a) Use Exercise 61 to show that the sequence defined by
$$x_n = \frac{1}{2} x_{n-1} + \frac{1}{x_{n-1}}, \quad \text{for } n \geq 1$$
converges to $\sqrt{2}$ whenever $x_0 > \sqrt{2}$.
   b) Use the fact that $0 < (x_0 - \sqrt{2})^2$ whenever $x_0 \neq \sqrt{2}$ to show that if $0 < x_0 < \sqrt{2}$, then $x_1 > \sqrt{2}$.
   c) Use the results of parts (a) and (b) to show that the sequence in (a) converges to $\sqrt{2}$ whenever $x_0 > 0$.

63. a) Show that if A is any positive number, then the sequence defined by
$$x_n = \frac{1}{2} x_{n-1} + \frac{A}{2x_{n-1}}, \quad \text{for } n \geq 1$$
converges to $\sqrt{A}$ whenever $x_0 > 0$.
   b) What happens if $x_0 < 0$?

64. A *prime number* is a positive integer that is evenly divided only by 1 and itself. Euclid demonstrated around 300 B.C. that there is an infinite sequence of prime numbers by assuming that the number of primes $p_1, p_2, \ldots, p_n$ is finite and constructing $p = p_1 \cdot p_2 \cdot \ldots \cdot p_n + 1$. Show that p is not divisible by any of $p_1, p_2, \ldots, p_n$.

65. The sequence $\{F_n\}$ described in Exercise 46 is called a *Fibonacci sequence*. Its terms occur naturally in many botanical species, particularly those with petals or scales arranged in the form of a logarithmic spiral. (Sunflower seed arrangement, scales on pine cones and pineapples, bud formation on alders and birches are examples of this phenomenon. The study of the arrangement of scales, leaves, and such is known as *phyllotaxiology*.) Consider the sequence $\{x_n\}$, where $x_n = F_{n+1}/F_n$. Assuming that $\lim_{n \to \infty} x_n = x$ exists, show that $x = (1 + \sqrt{5})/2$. This number is called the *golden ratio* and, among its many applications, gives the average monthly increase rate in the rabbit population described in Exercise 46. [*Hint:* Divide $F_{n+1} = F_n + F_{n-1}$ by F_n to obtain a relationship between x_n and x_{n-1}. Then show that $x = 1 + 1/x$ and solve for x.]

Putnam exercises

66. For each $x > e^e$, define a sequence $S_x = u_0, u_1, u_2, \ldots$ recursively as follows: $u_0 = e$, while for $n \geq 0$, u_{n+1} is the logarithm of x to the base u_n. Prove that S_x converges to a number $g(x)$ and that the function g defined in this way is continuous for $x > e^e$. (This exercise was problem B–5 of the forty-third William Lowell Putnam examination given on December 3, 1983. The examination and its solution are in the October 1984 issue of the *American Mathematical Monthly*, pp. 487–495.)

67. Let $\|u\|$ denote the distance from the real number u to the nearest integer (for example, $\|2.8\| = 0.2 = \|3.2\|$). For positive integers n, let
$$a_n = \frac{1}{n} \int_1^n \left\|\frac{n}{x}\right\| dx.$$
Determine $\lim_{n \to \infty} a_n$. You may assume the identity
$$\frac{2}{1} \cdot \frac{2}{3} \cdot \frac{4}{3} \cdot \frac{4}{5} \cdot \frac{6}{5} \cdot \frac{6}{7} \cdot \frac{8}{7} \cdot \frac{8}{9} \cdots = \frac{\pi}{2}.$$
(This exercise was problem B–5 of the forty-fourth William Lowell Putnam examination given on December 3, 1983. The examination and its solution are in the October 1984 issue of the *American Mathematical Monthly*, pp. 487–495.)

68. Let n be a positive integer, and define
$$f(n) = 1! + 2! + \cdots + n!.$$
Find polynomials $P(x)$ and $Q(x)$ such that
$$f(n + 2) = P(n)f(n + 1) + Q(n)f(n),$$
for all $n \geq 1$. (This exercise was problem B–1 of the forty-fifth William Lowell Putnam examination given on December 1, 1984. The examination and its solution are in the October 1985 issue of the *American Mathematical Monthly*, pp. 560–567.)

69. Let d be a real number. For each integer $m \geq 0$, define a sequence $\{a_m(j)\}$, $j = 0, 1, 2, \ldots$ by the condition
$$a_m(0) = d/2^m \quad \text{and} \quad a_m(j + 1) = (a_m(j))^2 + 2a_m(j),$$
$j \geq 0$. Evaluate $\lim_{n \to \infty} a_n(n)$. (This exercise was problem A–3 of the forty-sixth William Lowell Putnam examination given on December 7, 1985. The examination and its solution are in the October 1986 issue of the *American Mathematical Monthly*, pp. 620–626.)

8.2

Infinite Series

In Section 8.1 we found that

$$\frac{1}{2} + \frac{1}{4} + \frac{1}{8} + \cdots + \frac{1}{2^n} = 1 - \frac{1}{2^n}$$

and that

$$\lim_{n \to \infty} \left(1 - \frac{1}{2^n}\right) = 1.$$

In a sense, then, the infinite sum

$$\frac{1}{2} + \frac{1}{4} + \frac{1}{8} + \frac{1}{16} + \cdots + \frac{1}{2^n} + \cdots$$

has the value 1. The usual summation process, however, requires the number of terms summed to be finite. To consider infinite sums we need the following definition.

(8.14) DEFINITION
Let $\{a_n\}$ be a sequence. An expression of the form

$$\sum_{n=1}^{\infty} a_n = a_1 + a_2 + a_3 + \cdots + a_n + \cdots$$

is called an **infinite series,** or simply a **series,** and a_n is the **nth term** of the series.

Associated with each infinite series $\sum_{n=1}^{\infty} a_n$ is a sequence $\{S_n\}$ defined by

$$S_1 = a_1, \quad S_2 = a_1 + a_2, \quad S_3 = a_1 + a_2 + a_3,$$

and, in general,

$$S_n = a_1 + a_2 + a_3 + \cdots + a_n.$$

The sequence $\{S_n\}$ is called the **sequence of partial sums** of the series $\sum_{n=1}^{\infty} a_n$. For each $n \geq 2$,

(8.15) $$S_n = S_{n-1} + a_n.$$

The sequence of partial sums $\{S_n\}$ associated with the infinite series considered at the beginning of this section,

$$\sum_{n=1}^{\infty} \frac{1}{2^n} = \frac{1}{2} + \frac{1}{4} + \frac{1}{8} + \cdots,$$

is

$$S_1 = \frac{1}{2}, \quad S_2 = \frac{1}{2} + \frac{1}{4} = \frac{3}{4}, \quad S_3 = \frac{1}{2} + \frac{1}{4} + \frac{1}{8} = \frac{7}{8},$$

and, in general,

$$S_n = \frac{1}{2} + \frac{1}{4} + \frac{1}{8} + \frac{1}{16} + \cdots + \frac{1}{2^n} = 1 - \frac{1}{2^n},$$

for each integer $n \geq 1$.

The following definition describes the sense in which

$$\sum_{n=1}^{\infty} \frac{1}{2^n} = 1.$$

(8.16)
DEFINITION

Let $\sum_{n=1}^{\infty} a_n$ be an infinite series with sequence of partial sums $\{S_n\}$.

i) If $\{S_n\}$ converges to a real number L, then the series $\sum_{n=1}^{\infty} a_n$ is said to be a **convergent series.** The number L is called the **sum** or **limit** of the series. This is written $\sum_{n=1}^{\infty} a_n = L$.

ii) If $\{S_n\}$ diverges, then the series $\sum_{n=1}^{\infty} a_n$ is said to be a **divergent series.**

Example 1

Find the sequence of partial sums of the series $\sum_{n=1}^{\infty} \frac{1}{n(n+1)}$ and determine if this series converges.

Solution

For each n,

$$S_n = \frac{1}{1 \cdot 2} + \frac{1}{2 \cdot 3} + \frac{1}{3 \cdot 4} + \cdots + \frac{1}{n(n+1)}.$$

It is easy to verify, and to derive by the procedure used for integration by partial fractions, that

$$\frac{1}{n(n+1)} = \frac{1}{n} - \frac{1}{n+1}.$$

So

$$S_n = \left(1 - \frac{1}{2}\right) + \left(\frac{1}{2} - \frac{1}{3}\right) + \cdots + \left(\frac{1}{n-1} - \frac{1}{n}\right) + \left(\frac{1}{n} - \frac{1}{n+1}\right)$$

$$= 1 - \frac{1}{n+1}$$

and

$$\lim_{n \to \infty} S_n = \lim_{n \to \infty} \left(1 - \frac{1}{n+1}\right) = 1.$$

Consequently, the series converges and

$$\sum_{n=1}^{\infty} \frac{1}{n(n+1)} = 1.$$

▶

When all but the first and last terms of the partial sums cancel, as they do in Example 1, the series is said to be **telescoping.**

Example 2

Does $\sum_{n=1}^{\infty} (-1)^n$ converge?

Solution
The sequence of partial sums $\{S_n\}$ is given by
$$S_1 = -1, \quad S_2 = -1 + 1 = 0, \quad S_3 = -1 + 1 - 1 = -1,$$
and, in general,
$$S_n = \begin{cases} -1, & \text{if } n \text{ is odd,} \\ 0, & \text{if } n \text{ is even.} \end{cases}$$

The sequence $\{S_n\}$ does not converge, so $\sum_{n=1}^{\infty} (-1)^n$ is divergent. ▶

(8.17)
THEOREM

If the infinite series $\sum_{n=1}^{\infty} a_n$ converges, then $\lim_{n \to \infty} a_n = 0$.

Proof. If $\sum_{n=1}^{\infty} a_n$ converges, then a number L exists with $L = \lim_{n \to \infty} S_n$. Since S_n is the sum of the first n terms and S_{n-1} is the sum of the first $n-1$ terms, $S_n = S_{n-1} + a_n$ and $S_n - S_{n-1} = a_n$. But,
$$\lim_{n \to \infty} S_n = L, \quad \text{so} \quad \lim_{n \to \infty} S_{n-1} = L$$
and
$$\lim_{n \to \infty} a_n = \lim_{n \to \infty} (S_n - S_{n-1}) = \lim_{n \to \infty} S_n - \lim_{n \to \infty} S_{n-1} = 0. \quad \triangleright$$

The following logical equivalent of Theorem (8.17) is the *contrapositive* of that statement.

(8.18)

If $\{a_n\}$ is a sequence that does not have the limit zero, then $\sum_{n=1}^{\infty} a_n$ diverges.

Example 3

Does $\sum_{n=1}^{\infty} [1 - (\frac{1}{2})^n]$ converge or diverge?

Solution
Since $\lim_{n \to \infty} [1 - (\frac{1}{2})^n] = 1$, it follows from (8.18) that this infinite series diverges. ▶

The converse of Theorem (8.17) is not true:

$$\lim_{n \to \infty} a_n = 0 \quad \text{does not imply that} \quad \sum_{n=1}^{\infty} a_n \text{ converges.}$$

For example:

(8.19)
The series $\sum_{n=1}^{\infty} \dfrac{1}{n}$ diverges even though $\lim_{n \to \infty} \dfrac{1}{n} = 0$.

This series is called the **harmonic series.** The name derives from the fact that strings of the same tension produce harmonic musical tones if their lengths are proportional to the terms of this series.

To see that the harmonic series diverges, we will show that the sequence of partial sums of this series is unbounded. This implies that the sequence of partial sums diverges, which, by Definition (8.16), means that the series diverges.

Consider the partial sums of the form S_{2^n}:

$$S_2 = 1 + \frac{1}{2} = \frac{3}{2},$$

$$S_4 = S_2 + \frac{1}{3} + \frac{1}{4} > S_2 + \frac{1}{4} + \frac{1}{4} = \frac{3}{2} + \frac{1}{2} = \frac{4}{2},$$

$$S_8 = S_4 + \frac{1}{5} + \frac{1}{6} + \frac{1}{7} + \frac{1}{8} > S_4 + \frac{1}{8} + \frac{1}{8} + \frac{1}{8} + \frac{1}{8} > \frac{4}{2} + \frac{1}{2} = \frac{5}{2},$$

and, in general,

$$S_{2^n} > \frac{n+2}{2}.$$

Since these partial sums are unbounded, the series diverges. This important result is established using another method in Section 8.3. Other demonstrations are given in the exercises.

The following theorems are used often in the study of series. Since the convergence of a series is equivalent to the convergence of its sequence of partial sums, the proofs of these theorems follow from the corresponding results for convergence of sequences.

(8.20)
THEOREM
i) If $\sum_{n=1}^{\infty} a_n = L$ and $\sum_{n=1}^{\infty} b_n = M$ converge, then

$$\sum_{n=1}^{\infty} (a_n + b_n) = L + M \quad \text{and} \quad \sum_{n=1}^{\infty} (a_n - b_n) = L - M.$$

ii) $\sum_{n=1}^{\infty} a_n$ converges if and only if $\sum_{n=1}^{\infty} ca_n$ converges for any real number $c \neq 0$. In this case,

$$\sum_{n=1}^{\infty} ca_n = c \sum_{n=1}^{\infty} a_n.$$

Definition (8.14) defines series only for sequences of terms whose indices begin with $n = 1$. This specification is made initially to keep the notation associated with series to a minimum. The definition is extended easily to define the series associated with any sequence:

If M is any nonnegative integer and $\{a_n\}$ is a sequence defined for $n \geq M$, then

$$\sum_{n=M}^{\infty} a_n = a_M + a_{M+1} + \cdots + a_{M+n} + \cdots.$$

This series converges precisely when the sequence of its partial sums $\{S_n\}$ converges, where

$$S_1 = a_M, \qquad S_2 = a_M + a_{M+1}, \qquad S_3 = a_M + a_{M+1} + a_{M+2}$$

and, in general,

$$S_n = a_M + a_{M+1} + a_{M+2} + \cdots + a_{M+n-1}.$$

The following theorem uses this extension to state that the fact of convergence or divergence of a series does not depend on the value of the first M terms of the series, no matter how large M might be.

(8.21) THEOREM

For any nonnegative integer, the series $\sum_{n=1}^{\infty} a_n$ converges if and only if the series $\sum_{n=M}^{\infty} a_n$ converges, and in this case,

$$\sum_{n=1}^{\infty} a_n = \sum_{n=1}^{M-1} a_n + \sum_{n=M}^{\infty} a_n.$$

For example, we have seen that $\sum_{n=1}^{\infty} 1/2^n = 1$, so

$$\sum_{n=4}^{\infty} \frac{1}{2^n} = 1 - \left(\frac{1}{2} + \frac{1}{4} + \frac{1}{8}\right) = \frac{1}{8}.$$

(8.22) COROLLARY

If an integer M exists with $a_n = b_n$ for all $n \geq M$, then $\sum_{n=1}^{\infty} a_n$ and $\sum_{n=1}^{\infty} b_n$ both converge or both diverge.

We conclude this section by considering a class of series for which the question of convergence is easily determined. These series are the *geometric* series, the terms of which are generated by multiplying the previous term by a fixed constant called the *ratio* of the series. One example of a geometric series is the series

$$\frac{1}{2} + \frac{1}{4} + \frac{1}{8} + \cdots + \frac{1}{2^n},$$

where the ratio is $1/2$. In general, a **geometric series** has the form

(8.23) $$\sum_{n=0}^{\infty} ar^n = a + ar + ar^2 + ar^3 + \cdots + ar^n + \cdots,$$

where $a \neq 0$ is called the *first term* and r is the *ratio*.

(8.24)
THEOREM
A geometric series $\sum_{n=0}^{\infty} ar^n$ is convergent if and only if $|r| < 1$. When $|r| < 1$,
$$\sum_{n=0}^{\infty} ar^n = a + ar + ar^2 + \cdots = \frac{a}{1-r}.$$

Proof. If $|r| \geq 1$, $\lim_{n \to \infty} ar^n \neq 0$ so by (8.18) the series diverges.

Suppose $|r| < 1$. Then the nth partial sum S_n of the series can be written as
$$S_n = a + ar + ar^2 + \cdots + ar^{n-2} + ar^{n-1},$$
so
$$rS_n = ar + ar^2 + ar^3 + \cdots + ar^{n-1} + ar^n.$$
The terms of S_n and rS_n are the same except for the first term in S_n, a, and the last term in rS_n, ar^n. Subtracting rS_n from S_n gives
$$S_n - rS_n = a - ar^n$$
and
$$S_n = \frac{a - ar^n}{1 - r} = a\left[\frac{1 - r^n}{1 - r}\right].$$
Since $|r| < 1$, $\lim_{n \to \infty} r^n = 0$. Thus,
$$\sum_{n=0}^{\infty} ar^n = \lim_{n \to \infty} S_n = \lim_{n \to \infty}\left[a \cdot \frac{1 - r^n}{1 - r}\right] = \frac{a}{1 - r},$$
and the series converges. ▷

Example 4

Determine whether the series $\sum_{n=0}^{\infty} 3\left(\frac{4^n}{5^{n+1}}\right)$ converges or diverges.

Solution
We can rewrite this series as
$$\sum_{n=0}^{\infty} 3\left(\frac{4^n}{5^{n+1}}\right) = \sum_{n=0}^{\infty} \frac{3}{5}\left(\frac{4^n}{5^n}\right) = \sum_{n=0}^{\infty} \frac{3}{5}\left(\frac{4}{5}\right)^n,$$
which is a geometric series with first term $a = \frac{3}{5}$ and ratio $r = \frac{4}{5}$. By Theorem (8.24), this series converges; in fact,
$$\sum_{n=0}^{\infty} 3\left(\frac{4^n}{5^{n+1}}\right) = \frac{\frac{3}{5}}{1 - \frac{4}{5}} = 3. \qquad \blacktriangleright$$

The term "convergence" was introduced by James Gregory in 1668. Nicolaus Bernoulli used the term "divergence" in 1713.

HISTORICAL NOTE

Example 5

Determine whether the series $\sum_{n=0}^{\infty} \frac{3^n + 2^n}{4^n}$ converges or diverges.

Solution
The geometric series

$$\sum_{n=0}^{\infty} \frac{3^n}{4^n} = \sum_{n=0}^{\infty} \left(\frac{3}{4}\right)^n \quad \text{and} \quad \sum_{n=0}^{\infty} \frac{2^n}{4^n} = \sum_{n=0}^{\infty} \left(\frac{2}{4}\right)^n$$

both converge and have sums

$$\frac{1}{1-\frac{3}{4}} = 4 \quad \text{and} \quad \frac{1}{1-\frac{2}{4}} = 2,$$

respectively. Consequently, the series converges and

$$\sum_{n=0}^{\infty} \frac{3^n + 2^n}{4^n} = \sum_{n=0}^{\infty} \frac{3^n}{4^n} + \sum_{n=0}^{\infty} \frac{2^n}{4^n} = 4 + 2 = 6.$$

EXERCISE SET 8.2

In Exercises 1–14, determine whether the series converges or diverges.

1. $\sum_{n=1}^{\infty} \frac{n}{n+1}$
2. $\sum_{n=2}^{\infty} \frac{2n^2 + 1}{n^2 - 1}$
3. $\sum_{n=1}^{\infty} \left(\frac{2}{3}\right)^n$
4. $\sum_{n=1}^{\infty} \left(\frac{3}{2}\right)^n$
5. $\sum_{n=1}^{\infty} \frac{e^n}{n}$
6. $\sum_{n=1}^{\infty} \sin n\pi$
7. $\sum_{n=1}^{\infty} \frac{1}{(n+5)(n+6)}$
8. $\sum_{n=1}^{\infty} \frac{1}{n^2 + 6n + 8}$
9. $\sum_{n=1}^{\infty} \frac{n^2}{(n+5)(n+6)}$
10. $\sum_{n=1}^{\infty} \frac{n^2}{n^2 + 6n + 8}$
11. $\sum_{n=1}^{\infty} \left(\frac{1}{2^n} - \frac{1}{2^{n+1}}\right)$
12. $\sum_{n=1}^{\infty} \left(\frac{1}{3^n} - \frac{1}{3^{n+2}}\right)$
13. $\sum_{n=1}^{\infty} n \sin \frac{1}{n}$
14. $\sum_{n=1}^{\infty} n \tan \frac{1}{n}$

In Exercises 15–28, find the sum of the infinite series.

15. $\sum_{n=1}^{\infty} \left(\frac{1}{3}\right)^n$
16. $\sum_{n=2}^{\infty} \left(\frac{3}{4}\right)^n$
17. $\sum_{n=3}^{\infty} \left(\frac{2}{5}\right)^{n-1}$
18. $\sum_{n=1}^{\infty} \left(-\frac{1}{2}\right)^n$
19. $\sum_{n=0}^{\infty} 3^n \cdot 5^{-n}$
20. $\sum_{n=1}^{\infty} 4 \left(\frac{3}{\pi}\right)^n$
21. $\sum_{n=1}^{\infty} \frac{1}{(n+1)(n+2)}$
22. $\sum_{n=1}^{\infty} \frac{1}{(n+2)(n+3)}$
23. $\sum_{n=2}^{\infty} \frac{1}{n^2 - 1}$
24. $\sum_{n=3}^{\infty} \frac{1}{n^2 - 4}$
25. $\sum_{n=1}^{\infty} \left(\frac{1}{2^n} - \frac{1}{3^n}\right)$
26. $\sum_{n=1}^{\infty} (-1)^n \, 2^{n+1} \, 3^{-n+2}$
27. $\sum_{n=2}^{\infty} \left[\left(\frac{2}{3}\right)^n + \frac{1}{n^2 - 1}\right]$
28. $\sum_{n=1}^{\infty} \frac{4^n - 3^n}{12^n}$

29. Show that the series

$$\sum_{n=1}^{\infty} \ln\left(\frac{n}{n+1}\right)$$

diverges.

30. Show that for any positive real number a, the series $\sum_{n=1}^{\infty} a^{1/n}$ diverges.

31. Theorem (8.20) considers sums and constant multiples of convergent series. Show that similar results are not true for products and quotients, even for finite sums, by calculating

$$\sum_{n=1}^{4} 2^n, \quad \sum_{n=1}^{4} 3^n, \quad \sum_{n=1}^{4} 6^n, \quad \text{and} \quad \sum_{n=1}^{4} \left(\frac{2}{3}\right)^n.$$

32. An *arithmetic sequence* is a sequence of the form $a_n = a + nd$, $n \geq 0$, where a is the first term of the sequence and d is called the common difference. Show that the series $\sum_{n=1}^{\infty} a_n$, where $\{a_n\}$ is an arithmetic sequence, diverges except in the case when $a = 0$ and $d = 0$.

33. Show that the infinite repeating decimal 0.19191919 ... can be expressed in infinite series form as
$$0.19191919\ldots = \sum_{n=1}^{\infty} 19\left(\frac{1}{100}\right)^n.$$
Use this expression to determine the rational number whose decimal expansion is 0.191919. ...

34. Find a pair of integers whose quotient is the rational number with the infinite repeating decimal expansion 0.345734573457. ...

35. Give an example of infinite series $\sum_{n=1}^{\infty} a_n$ and $\sum_{n=1}^{\infty} b_n$ where both diverge, yet $\sum_{n=1}^{\infty} (a_n + b_n)$ converges.

36. Prove that if $\sum_{n=1}^{\infty} a_n = L$ and c is a constant, then $\sum_{n=1}^{\infty} ca_n = cL$.

37. Suppose the infinite series $\sum_{n=1}^{\infty} a_n$, where $a_n \neq 0$ for each n, converges. What can be said about the infinite series $\sum_{n=1}^{\infty} 1/a_n$?

38. The Rhind papyrus, written in Egypt about 1650 B.C., is a text by the scribe Ahmes that contains 85 problems thought to be transcripted from an earlier Egyptian work. Problem 79 in this work is a forerunner to the old English rhyme: As I was going to St. Ives, I met a man with seven wives; each wife had seven sacks; each sack had seven cats; each cat had seven kits. Kits, cats, sacks, and wives, how many are going to St. Ives? Solve this problem.

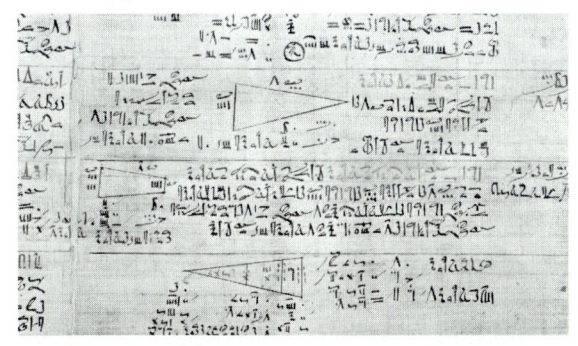

A portion of the Rhind papyrus transcribed by Ahmes about 1650 B.C. The papyrus is on display in the British Museum in London.

39. The government claims to be able to stimulate the economy substantially by giving each taxpayer a $50 tax rebate. They reason that 90% of this amount will be spent, that 90% of the amount spent will again be spent, and so on. If this is true, how much total expenditure will result from this $50 rebate?

40. Suppose the government levies a 20% tax on all money spent. (a) How does this tax influence the total expenditure from the tax rebate described in Exercise 39? (b) How much of its rebate will the government eventually recover with this tax?

41. A toy company introduces a new ball called a "Whizball." When dropped from a height h onto a hard surface, it rebounds approximately $0.8h$. Suppose this ball is dropped from 20 ft above a hard surface (see the figure).
    a) How far does it travel before it hits the surface four times?
    b) Assuming that it bounces continually when dropped from this height, what is the total distance the ball travels?

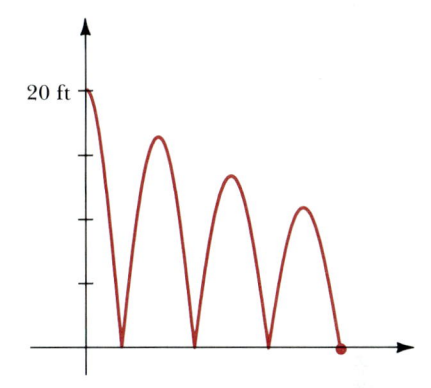

42. How long does it take the ball described in Exercise 41 to complete bouncing when dropped from the height of 20 ft?

43. In practice the ball described in Exercise 41 bounces only as long as the force of rebound exceeds the force of friction between the ball and the surface on which it bounces. A ball dropped from 20 ft actually bounces 19 times before coming to rest on a concrete pavement.
    a) How far does the ball travel during this time?
    b) How long does it take the ball to travel this distance?

44. Suppose the Whizball described in Exercise 41 is dropped from a height of 40 ft rather than 20 ft and bounces continually.
    a) How would this change the distance the ball travels?
    b) How would this change the time required to complete the bouncing?

45. An old Greek paradox due to Zeno concerns Achilles and the tortoise. Achilles can run ten times as fast as the tortoise, who has been given a 100-m head start. By the time Achilles reaches the starting position of the tortoise, the tortoise has moved ahead 10 m. By the time Achilles travels this 10-m distance, the tortoise had moved ahead another meter. Since this process continues indefinitely, Achilles can never quite catch the tortoise. Show the fallacy in this logic by finding precisely how far Achilles must run in order to catch the tortoise.

46. A famous story about the outstanding mathematician John von Neumann (1903–1957) concerns the following problem: Two bicyclists start 20 mi apart and head toward each other, each going 10 mph. At the same time, a fly traveling 15 mph leaves the front wheel of one bicycle, flies to the front wheel of the other bicycle, turns around and flies back to the wheel of the first bicycle, and so on, continuing in this manner until trapped between the two wheels. What total distance did the fly fly? There is a quick way to solve this problem. However, von Neumann allegedly solved the problem instantly by summing an infinite series. Solve this problem using both methods. (An interesting article on John von Neumann, written by P. R. Halmos, appears in the April 1973 issue of the *American Mathematical Monthly*.)

47. A white square is divided into 25 equal squares. One of these squares is blackened. Each of the remaining smaller white squares is then divided into 25 equal squares and one of each is blackened (see the figure). If this process is continued indefinitely, how many sizes of black squares will be necessary in order to blacken a total of at least 80% of the original white square?

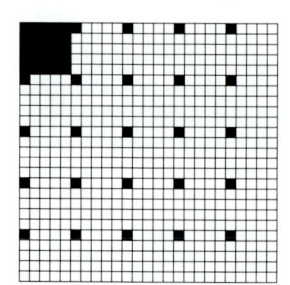

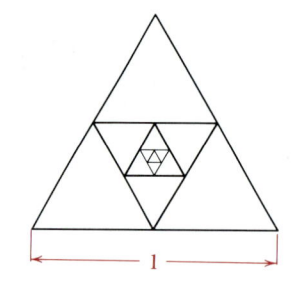

Figures for Exercises 47 (left) and 48 (right).

48. An equilateral triangle has sides of unit length. Another equilateral triangle is constructed within the first by placing its vertices at the midpoints of the sides of the first triangle. A third equilateral triangle is constructed within the second in the same manner, and so on (see the figure).
    a) What is the sum of the perimeters of the triangles?
    b) What is the sum of the areas of the triangles?

49. Two students, Pat and Mike, flip a coin to see who buys lunch. The first to flip a head must buy. Mike suggests that Pat flip first. Pat protests saying that one half of the time he will lose on the first flip, but relents when Mike explains to him that he will have this same likelihood of buying if Pat's flip is a tail. On average, who gets the better of this deal and by how much?

50. Show that $\sum_{n=1}^{\infty} 1/n^2 < 2$ by grouping the terms of the series in increasing powers of two, that is, consider the first term of the series, then the next two terms, the next four terms, and so on. Show that the mth grouping is less than 2^{1-m}.

51. Complete the steps in the following proof that the harmonic series $\sum_{n=1}^{\infty} 1/n$ diverges. Suppose $\sum_{n=1}^{\infty} 1/n = S$. Find $\sum_{n=1}^{\infty} 1/2n$ and show that

$$\sum_{n=1}^{\infty} \frac{1}{2n} < \sum_{n=1}^{\infty} \frac{1}{2n-1}.$$

Use this to construct the contradiction that $S > \frac{1}{2}S + \frac{1}{2}S$.

Putnam exercises

52. $S_1 = \ln a$, and $S_n = \sum_{i=1}^{n-1} \ln(a - S_i)$, $n > 1$. Show that

$$\lim_{n \to \infty} S_n = a - 1.$$

(This exercise was problem 6, part I of the seventeenth William Lowell Putnam examination given on March 2, 1957. The examination and its solution are in the November 1957 issue of the *American Mathematical Monthly*, pp. 649–654.)

53. If $f(x)$ is a real-valued function defined for $0 < x < 1$, then the formula $f(x) = o(x)$ is an abbreviation for the statement that

$$\frac{f(x)}{x} \to 0 \quad \text{as } x \to 0.$$

Keeping this in mind, prove the following: If

$$\lim_{x \to 0} f(x) = 0 \quad \text{and} \quad f(x) - f\left(\frac{x}{2}\right) = o(x),$$

then $f(x) = o(x)$. (This exercise was problem 5, part I of the fourteenth William Lowell Putnam examination given on March 6, 1954. The examination and its solution are in the October 1954 issue of the *American Mathematical Monthly*, pp. 542–549.)

54. Express

$$\sum_{k=1}^{\infty} \frac{6^k}{(3^{k+1} - 2^{k+1})(3^k - 2^k)}$$

as a rational number. (This exercise was problem A–2 of the forty-fifth William Lowell Putnam examination given on December 1, 1984. The examination and its solution are in the October 1985 issue of the *American Mathematical Monthly*, pp. 560–567.)

8.3
Infinite Series with Positive Terms

The definition of convergence of an infinite series is generally difficult to apply to a specific problem. In Section 8.2 we saw that if $\lim_{n \to \infty} a_n \neq 0$, then the series $\sum_{n=1}^{\infty} a_n$ diverges, and that a geometric series $\sum_{n=0}^{\infty} ar^n$ diverges if $|r| \geq 1$ and converges if $|r| < 1$. In this and succeeding sections, we develop additional results to determine convergence without directly applying the definition.

We first consider the class of series with all positive terms. A series of this type has the property that its sequence of partial sums is increasing. The Completeness Property (8.11) implies that convergence of the series depends entirely on whether the sequence of partial sums is bounded.

(8.25)
THEOREM
Suppose $\sum_{n=1}^{\infty} a_n$ is a series with positive terms, and $\{S_n\}$ is the associated sequence of partial sums. The series $\sum_{n=1}^{\infty} a_n$ converges if and only if the sequence $\{S_n\}$ is bounded.

Proof. By definition, the series $\sum_{n=1}^{\infty} a_n$ is convergent, and $\sum_{n=1}^{\infty} a_n = L$ if and only if $\lim_{n \to \infty} S_n = L$. If $\sum_{n=1}^{\infty} a_n$ converges, then the sequence $\{S_n\}$ of partial sums converges, and Theorem (8.9) implies that the sequence $\{S_n\}$ is bounded.

Conversely, suppose the sequence $\{S_n\}$ is bounded. Since $a_n > 0$ for each n, $\{S_n\}$ is also increasing. The Completeness Property (8.11) implies that the bounded increasing sequence $\{S_n\}$ converges. Thus, $\sum_{n=1}^{\infty} a_n$ converges. ▷

Example 1

Show that the series $\sum_{n=1}^{\infty} \frac{1}{(n+1)!}$ is convergent.

Solution

This series has only positive terms so its sequence of partial sums, $\{S_n\}$, is increasing. To show that $\{S_n\}$ converges, we need to show that $\{S_n\}$ is bounded. For all n,

$$\frac{1}{(n+1)!} = \frac{1}{1 \cdot 2 \cdot 3 \cdot \cdots \cdot n(n+1)} \leq \frac{1}{1 \cdot \underbrace{2 \cdot 2 \cdot \cdots \cdot 2}_{n \text{ times}}} = \frac{1}{2^n}.$$

Consequently, for each m,

$$S_m = \frac{1}{2!} + \frac{1}{3!} + \cdots + \frac{1}{(m+1)!}$$
$$< \frac{1}{2} + \frac{1}{2^2} + \cdots + \frac{1}{2^m} = \sum_{n=1}^{m} \frac{1}{2^n} < \sum_{n=1}^{\infty} \frac{1}{2^n}.$$

The series $\sum_{n=1}^{\infty} 1/2^n$ is a geometric series with ratio $r = \frac{1}{2}$ that converges to 1. This implies that $0 < S_m \le 1$ for all m. Thus, $\{S_m\}$ is bounded and Theorem (8.25) implies that $\sum_{n=1}^{\infty} 1/(n+1)!$ is convergent.

In fact,
$$0 < \sum_{n=1}^{\infty} \frac{1}{(n+1)!} \le 1.$$

Theorem (8.25) does not, however, give the precise value. ▶

In Example 1 we implicitly applied the following theorem when comparing the series $\sum_{n=1}^{\infty} 1/(n+1)!$ to the series $\sum_{n=1}^{\infty} 1/2^n$.

(8.26)
THEOREM: THE COMPARISON TEST
Suppose $\sum_{n=1}^{\infty} a_n$ and $\sum_{n=1}^{\infty} b_n$ are series with positive terms and M is a positive integer.

i) If $\sum_{n=1}^{\infty} b_n$ converges and $a_n \le b_n$ for all $n > M$, then $\sum_{n=1}^{\infty} a_n$ converges.

ii) If $\sum_{n=1}^{\infty} b_n$ diverges and $a_n \ge b_n$ for all $n > M$, then $\sum_{n=1}^{\infty} a_n$ diverges.

In essence, the Comparison Test states that for positive-term series:

i) If the terms of $\{a_n\}$ do not exceed those of $\{b_n\}$ and the terms of $\{b_n\}$ are small enough to ensure convergence, then the terms of $\{a_n\}$ are also small enough to ensure convergence; and

ii) If the terms of $\{a_n\}$ exceed those of $\{b_n\}$ and the partial sums of $\{b_n\}$ are unbounded, then the partial sums of $\{a_n\}$ are unbounded as well.

The proof of the Comparison Test follows directly from this logic.

The following is a useful, but slightly more complicated, version of the Comparison Test.

(8.27)
THEOREM: THE LIMIT COMPARISON TEST
Suppose $\sum_{n=1}^{\infty} a_n$ and $\sum_{n=1}^{\infty} b_n$ are series with positive terms and that

$$\lim_{n \to \infty} \frac{a_n}{b_n} = L,$$

where L represents either a nonnegative real number or infinity.

i) If $\sum_{n=1}^{\infty} b_n$ converges and $0 \le L < \infty$, then $\sum_{n=1}^{\infty} a_n$ converges.

ii) If $\sum_{n=1}^{\infty} b_n$ diverges and $0 < L \le \infty$, then $\sum_{n=1}^{\infty} a_n$ diverges.

Proof. Suppose first that $0 < L < \infty$. If we choose $\varepsilon = L/2 > 0$, an integer M exists with the property that

$$\left| \frac{a_n}{b_n} - L \right| < \frac{L}{2},$$

whenever $n > M$. Thus, when $n > M$,

$$-\frac{L}{2} < \frac{a_n}{b_n} - L < \frac{L}{2}, \quad \text{so} \quad \frac{L}{2} < \frac{a_n}{b_n} < \frac{3L}{2}$$

8.3 INFINITE SERIES WITH POSITIVE TERMS

and
$$\frac{L}{2} b_n < a_n < \frac{3L}{2} b_n.$$

This inequality together with the Comparison Test implies that $\sum_{n=1}^{\infty} a_n$ converges if $\sum_{n=1}^{\infty} (3L/2) b_n$ converges, and $\sum_{n=1}^{\infty} a_n$ diverges if $\sum_{n=1}^{\infty} (L/2) b_n$ diverges. We know from Theorem (8.20) that $\sum_{n=1}^{\infty} (3L/2) b_n$ converges precisely when $\sum_{n=1}^{\infty} b_n$ converges and $\sum_{n=1}^{\infty} (L/2) b_n$ diverges precisely when $\sum_{n=1}^{\infty} b_n$ diverges. Consequently, $\sum_{n=1}^{\infty} a_n$ converges precisely when $\sum_{n=1}^{\infty} b_n$ converges.

If $\sum_{n=1}^{\infty} b_n$ converges and $L = \lim_{n \to \infty} (a_n/b_n) = 0$, then an integer M exists when $a_n/b_n < 1$ when $n > M$. Consequently, when $n > M$, $a_n < b_n$. The Comparison Test implies that $\sum_{n=1}^{\infty} a_n$ converges.

Similarly, if $\sum_{n=1}^{\infty} b_n$ diverges and $L = \lim_{n \to \infty} (a_n/b_n) = \infty$, an integer M exists with $a_n/b_n > 1$ and $a_n > b_n$ when $n > M$. By the Comparison Test, $\sum_{n=1}^{\infty} a_n$ diverges. ▷

Example 2

Determine whether the series $\sum_{n=1}^{\infty} \frac{3}{2^n + 1}$ converges or diverges.

Solution

The series $\sum_{n=1}^{\infty} 1/2^n$ converges, so the series $\sum_{n=1}^{\infty} 3/2^n$ converges as well. Since
$$\frac{3}{2^n + 1} < \frac{3}{2^n},$$

The Comparison Test implies that
$$\sum_{n=1}^{\infty} \frac{3}{2^n + 1}$$

converges. ▶

Example 3

Determine whether the series $\sum_{n=1}^{\infty} \frac{3}{2^n - 1}$ converges or diverges.

Solution

Noting the similarity between this series and the series considered in Example 2, we should expect this series to be convergent. However, $1/(2^n - 1)$ is greater than the term $1/2^n$ of the series to which we would naturally compare, so we cannot use the Comparison Test. Instead we use the Limit Comparison Test. Since $\sum_{n=1}^{\infty} 1/2^n$ converges and

$$\lim_{n \to \infty} \frac{\frac{3}{2^n - 1}}{\frac{1}{2^n}} = \lim_{n \to \infty} \frac{3 \cdot 2^n}{2^n - 1} = 3,$$

the Limit Comparison Test implies that $\sum_{n=1}^{\infty} \dfrac{3}{2^n - 1}$ converges. ▶

Before we can apply the comparison tests effectively, we need a larger collection of convergent and divergent series with which to compare. An important application of the next theorem is to determine convergence of a type of series that can be used easily for comparison.

The following theorem is called the *Integral Test* since it uses convergence information about a particular improper integral to obtain convergence information about a series. It might be helpful to review the beginning of Section 4.7 at this time to ensure that the concept of an improper integral is clear. In particular, recall that if f is continuous on the interval $[a, \infty)$, then the improper integral

$$\int_a^\infty f(x)\, dx$$

is said to *converge* if

$$\lim_{M \to \infty} \int_a^M f(x)\, dx$$

is finite and is said to *diverge* otherwise.

(8.28)
THEOREM: THE INTEGRAL TEST
If f is continuous and decreasing on $[1, \infty)$ and $a_n = f(n)$ for each term in the positive-term series $\sum_{n=1}^{\infty} a_n$, then

$$\sum_{n=1}^{\infty} a_n \text{ converges if and only if } \int_1^\infty f(x)\, dx \text{ converges.}$$

Proof. Since f is decreasing on each interval $[i, i+1]$, the inequalities

$$0 < a_{i+1} = f(i+1) \leq f(x) \leq f(i) = a_i$$

hold for each integer i and each x in $[i, i+1]$.

Consequently, for each integer i,

$$a_{i+1} = \int_i^{i+1} a_{i+1}\, dx \leq \int_i^{i+1} f(x)\, dx \leq \int_i^{i+1} a_i\, dx = a_i.$$

These inequalities can be expressed geometrically by considering the rectangles R_{i+1} and R_i with base $[i, i+1]$ and heights a_{i+1} and a_i shown in Figure 8.10:

area of R_{i+1} ≤ area below the graph f on $[i, i+1]$ ≤ area of R_i.

Adding the terms in these inequalities gives

$$(8.29) \quad 0 < \sum_{i=1}^{n} a_{i+1} \leq \sum_{i=1}^{n} \int_i^{i+1} f(x)\, dx = \int_1^{n+1} f(x)\, dx \leq \sum_{i=1}^{n} a_i,$$

as illustrated in Figure 8.11. The inequality

$$\int_1^{n+1} f(x)\, dx \leq \sum_{i=1}^{n} a_i$$

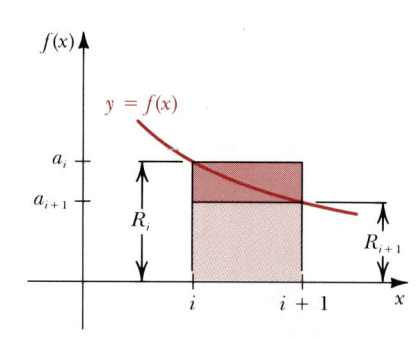

Figure 8.10

$a_{i+1} \leq \int_i^{i+1} f(x)\, dx \leq a_i.$

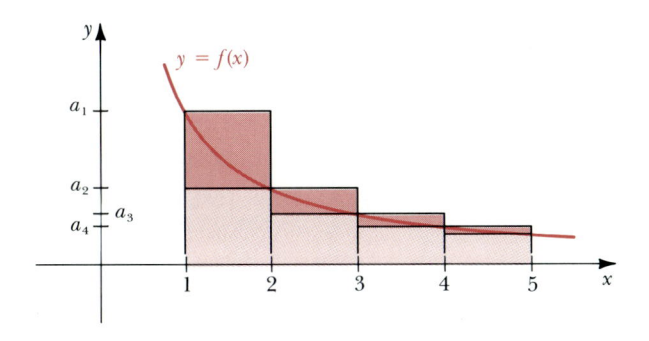

Figure 8.11
$$\sum_{i=1}^{n} a_{i+1} \leq \int_{1}^{n+1} f(x)\, dx \leq \sum_{i=1}^{n} a_i.$$

implies that if $\int_{1}^{\infty} f(x)\, dx = \lim_{n \to \infty} \int_{1}^{n+1} f(x)\, dx = \infty$, then $\sum_{i=1}^{\infty} a_i$ is infinite as well. So if $\int_{1}^{\infty} f(x)\, dx$ diverges, then $\sum_{n=1}^{\infty} a_n$ diverges. Thus,

if $\sum_{n=1}^{\infty} a_n$ converges, then $\int_{1}^{\infty} f(x)\, dx$ also converges.

Suppose now that $\int_{1}^{\infty} f(x)\, dx$ converges. Then the sequence defined by $b_n = \int_{1}^{n+1} f(x)\, dx$ is bounded. By (8.29),

$$0 < \sum_{i=1}^{n} a_{i+1} \leq \int_{1}^{n+1} f(x)\, dx,$$

so the sequence of partial sums $S_n = \sum_{i=1}^{n} a_{i+1}$ is also bounded. Theorem (8.25) implies that the series $\sum_{i=1}^{\infty} a_{i+1}$ converges. But

$$\sum_{n=1}^{\infty} a_n = a_1 + \sum_{i=1}^{\infty} a_{i+1},$$

so $\sum_{n=1}^{\infty} a_n$ also converges. ▷

Example 4

Use the Integral Test to show that the harmonic series $\sum_{n=1}^{\infty} \frac{1}{n}$ diverges.

Solution
The function defined by $f(x) = 1/x$ is continuous and decreasing on the interval $[1, \infty)$, and $a_n = 1/n = f(n)$. Since

$$\int_{1}^{\infty} \frac{1}{x}\, dx = \lim_{M \to \infty} \int_{1}^{M} \frac{1}{x}\, dx = \lim_{M \to \infty} \ln |x| \Big]_{1}^{M} = \lim_{M \to \infty} (\ln M - \ln 1) = \infty,$$

the Integral Test implies that $\sum_{n=1}^{\infty} 1/n$ diverges. ▶

Example 5

Determine whether the series $\sum_{n=1}^{\infty} \frac{n}{e^n}$ converges or diverges.

Solution
The function defined by $f(x) = xe^{-x}$ is differentiable, hence continuous

on the interval $[1, \infty)$. Moreover,
$$D_x f(x) = e^{-x} - xe^{-x} = (1-x)e^{-x} < 0$$
on $(1, \infty)$, so f is decreasing on $[1, \infty)$ and the Integral Test can be applied. To evaluate $\int_1^\infty xe^{-x}\,dx$, we use integration by parts. Let
$$u = x \quad \text{and} \quad dv = e^{-x}\,dx.$$
Then
$$du = dx \quad \text{and} \quad v = -e^{-x},$$
so
$$\int xe^{-x}\,dx = -xe^{-x} + \int e^{-x}\,dx = -xe^{-x} - e^{-x} + C$$
and
$$\begin{aligned}\int_1^\infty xe^{-x}\,dx &= \lim_{M\to\infty} \int_1^M xe^{-x}\,dx \\ &= \lim_{M\to\infty} \left[-xe^{-x} - e^{-x}\right]_1^M \\ &= \lim_{M\to\infty} (-Me^{-M} - e^{-M}) - (-e^{-1} - e^{-1}) \\ &= 2e^{-1} - \lim_{M\to\infty} \frac{M+1}{e^M}.\end{aligned}$$

By l'Hôpital's Rule,
$$\lim_{M\to\infty} \frac{M+1}{e^M} = \lim_{M\to\infty} \frac{1}{e^M} = 0,$$
so
$$\int_1^\infty xe^{-x}\,dx = 2e^{-1} < \infty.$$

The Integral Test implies that
$$\sum_{n=1}^\infty ne^{-n} = \sum_{n=1}^\infty \frac{n}{e^n}$$
converges. ▶

An important application of the Integral Test is to determine the behavior of series of the form
$$\sum_{n=1}^\infty \frac{1}{n^p},$$
where p is a real number. These **p-series** are used frequently for comparison.

When $p > 0$, the function defined by
$$f(x) = \frac{1}{x^p}$$

is continuous and decreasing on $[1, \infty)$. The Integral Test implies that the convergence of a p-series depends on whether f has a finite integral on $[1, \infty)$. We saw in Example 4 that if $p = 1$, then the harmonic series $\sum_{n=1}^{\infty} 1/n$ diverges.

When $p \neq 1$,

$$\int_1^\infty \frac{1}{x^p} dx = \lim_{M \to \infty} \int_1^M \frac{1}{x^p} dx$$

$$= \lim_{M \to \infty} \left(\frac{1}{-p+1}\right) \frac{1}{x^{p-1}} \Big]_1^M = \lim_{M \to \infty} \frac{1}{1-p} \left[\frac{1}{M^{p-1}} - 1\right].$$

This limit is finite when $p > 1$ (in fact, it is $1/(p-1)$), but is infinite when $p < 1$. The results of the Integral Test applied to the p-series are listed in Theorem (8.30). The extension of the result to the case $p \leq 0$ is obtained by observing that in this case the terms of the series do not converge to zero, so the series must diverge.

(8.30) THEOREM

The p-series $\sum_{n=1}^{\infty} \frac{1}{n^p}$ converges if $p > 1$ and diverges if $p \leq 1$.

Often the terms in a positive-term series behave in a manner similar to a specific p-series as n becomes large. The convergence or divergence of the original series can then be determined by comparison with that p-series.

Example 6

Determine whether $\sum_{n=1}^{\infty} \frac{n}{(n+1)^{5/2}}$ converges or diverges.

Solution

For large values of n, $n + 1$ is approximately the same as n. Thus, as n becomes large,

$$\frac{n}{(n+1)^{5/2}} \approx \frac{n}{n^{5/2}} = \frac{1}{n^{3/2}}.$$

Since $p = \frac{3}{2} > 1$, the series $\sum_{n=1}^{\infty} 1/n^{3/2}$ converges. The Limit Comparison Test gives

$$\lim_{n \to \infty} \frac{\frac{n}{(n+1)^{5/2}}}{\frac{1}{n^{3/2}}} = \lim_{n \to \infty} \frac{n}{(n+1)^{5/2}} \cdot n^{3/2} = \lim_{n \to \infty} \left(\frac{n}{n+1}\right)^{5/2} = 1,$$

so the series $\sum_{n=1}^{\infty} \frac{n}{(n+1)^{5/2}}$ converges. ▶

Example 7

Determine whether $\sum_{n=1}^{\infty} \dfrac{5n}{2\sqrt[3]{n^5 + n}}$ converges or diverges.

Solution

For large values of n, n^5 is much larger than n, and, in a relative sense, $n^5 + n$ is approximately n^5. Thus,

$$\frac{5n}{2\sqrt[3]{n^5 + n}} \approx \frac{5n}{2\sqrt[3]{n^5}} = \frac{5}{2n^{2/3}},$$

so we compare this series with the divergent p-series $\sum_{n=1}^{\infty} 1/n^{2/3}$. Applying the Limit Comparison Test gives

$$\lim_{n \to \infty} \frac{\dfrac{5n}{2\sqrt[3]{n^5 + n}}}{\dfrac{1}{n^{2/3}}} = \lim_{n \to \infty} \frac{5n}{2\sqrt[3]{n^5 + n}} \cdot \sqrt[3]{n^2}$$

$$= \lim_{n \to \infty} \frac{5}{2} \frac{\sqrt[3]{n^5}}{\sqrt[3]{n^5 + n}} = \frac{5}{2},$$

so the series $\sum_{n=1}^{\infty} \dfrac{5n}{2\sqrt[3]{n^5 + n}}$ diverges. ▶

▶ EXERCISE SET 8.3

In Exercises 1–20, use either the Comparison Test or the Limit Comparison Test to determine whether the series converges or diverges.

1. $\sum_{n=1}^{\infty} \dfrac{1}{2n + 1}$
2. $\sum_{n=1}^{\infty} \dfrac{2}{3n - 1}$
3. $\sum_{n=1}^{\infty} \dfrac{1}{n^2 + 2n + 2}$
4. $\sum_{n=1}^{\infty} \dfrac{1}{n^2 - 2n + 2}$
5. $\sum_{n=5}^{\infty} \dfrac{1}{\sqrt{n - 2}}$
6. $\sum_{n=1}^{\infty} \dfrac{n^3 + 3n^2 + 3}{n^5 + 2n - 5}$
7. $\sum_{n=1}^{\infty} \dfrac{1}{n(n + 4)}$
8. $\sum_{n=1}^{\infty} \dfrac{1}{n + \sqrt{n}}$
9. $\sum_{n=1}^{\infty} \dfrac{1}{\sqrt{n^3 + 4}}$
10. $\sum_{n=1}^{\infty} \dfrac{n}{\sqrt{n^2 + 4}}$
11. $\sum_{n=1}^{\infty} \dfrac{\cos n\pi + 2}{n^2}$
12. $\sum_{n=1}^{\infty} \dfrac{\sin n\pi}{n^2}$
13. $\sum_{n=1}^{\infty} \dfrac{1}{n2^n}$
14. $\sum_{n=1}^{\infty} \dfrac{1}{\sqrt{n}3^n}$
15. $\sum_{n=1}^{\infty} \dfrac{n!}{(n + 1)!}$
16. $\sum_{n=1}^{\infty} \dfrac{n!}{(n + 2)!}$
17. $\sum_{n=1}^{\infty} \dfrac{n!}{(2n)!}$
18. $\sum_{n=1}^{\infty} \dfrac{(n!)^2}{(2n)!}$
19. $\sum_{n=1}^{\infty} \sin \dfrac{\pi}{n^2}$
20. $\sum_{n=1}^{\infty} \dfrac{1}{n} \tan \dfrac{\pi}{n}$

In Exercises 21–30, use the Integral Test to determine whether the series converges or diverges.

21. $\sum_{n=1}^{\infty} \dfrac{1}{n + 1}$
22. $\sum_{n=1}^{\infty} \dfrac{1}{(2n - 1)^2}$
23. $\sum_{n=1}^{\infty} \dfrac{\ln n}{n}$
24. $\sum_{n=2}^{\infty} \dfrac{1}{n\sqrt{\ln n}}$
25. $\sum_{n=2}^{\infty} \dfrac{1}{n \ln n}$
26. $\sum_{n=2}^{\infty} \dfrac{1}{n (\ln n)^2}$
27. $\sum_{n=1}^{\infty} \dfrac{n}{e^{n^2}}$
28. $\sum_{n=1}^{\infty} \dfrac{1}{1 + n^2}$
29. $\sum_{n=1}^{\infty} \dfrac{n}{(1 + n^2)^2}$
30. $\sum_{n=3}^{\infty} \dfrac{1}{n \ln n \ln (\ln n)}$

In Exercises 31–40, determine whether the series converges or diverges.

31. $\sum_{n=0}^{\infty} \dfrac{2^n + 1}{3^n + 2}$

32. $\sum_{n=1}^{\infty} \dfrac{n}{n^2 + 1}$

33. $\sum_{n=1}^{\infty} \left[\left(\dfrac{2}{3}\right)^n + \dfrac{1}{n} \right]$

34. $\sum_{n=1}^{\infty} \dfrac{2n + 1}{n^2 + n}$

35. $\sum_{n=0}^{\infty} \dfrac{n + 2}{2^n + 1}$

36. $\sum_{n=1}^{\infty} \dfrac{\ln n}{n^{3/2}}$

37. $\sum_{n=0}^{\infty} \dfrac{2^n + n^2}{3^n + n}$

38. $\sum_{n=0}^{\infty} \dfrac{\arctan n}{n^2 + 1}$

39. $\sum_{n=2}^{\infty} (\ln n)^{-n}$

40. $\sum_{n=1}^{\infty} \left(\dfrac{n}{n+1} \right)^n$

41. Show that for all $x > 0$, $\ln x < x$, and conclude from this that $\sum_{n=2}^{\infty} (1/\ln n)$ diverges.

42. Determine all values of p for which the series
$$\sum_{n=2}^{\infty} \dfrac{1}{n (\ln n)^p}$$
converges.

43. Show that
$$\sum_{n=3}^{\infty} \dfrac{1}{n \ln n (\ln (\ln n))^p}$$
converges if and only if $p > 1$.

44. Determine whether
$$\sum_{n=1}^{\infty} \dfrac{1}{n^{1+1/n}}$$
converges or diverges.

45. A biologist examines a circular plate for a certain type of bacteria by drawing concentric circles of radius n, for positive integers n. The number of bacteria between the $(n - 1)$st circle and the nth circle is inversely proportional to the area of the nth circle. (The constant of proportionality is independent of n.) Show that the number of bacteria on the plate is finite without assuming that the plate has finite radius.

46. Suppose the biologist in Exercise 45 finds that the number of bacteria between the $(n - 1)$st circle and the nth circle is inversely proportional to the radius of the nth circle, rather than its area. Can it still be deduced that the number of bacteria on the plate is finite, without assuming that the plate has finite radius?

47. a) Show that for any positive real number r,
$$\lim_{n \to \infty} \dfrac{\ln n}{n^r} = 0.$$

b) Show that for any positive real number r, the relationship
$$\dfrac{1}{n^{1+r}} < \dfrac{1}{n \ln n} < \dfrac{1}{n}$$
holds for sufficiently large n.

c) Conclude from (a) and (b) that the convergence or divergence of the series
$$\sum_{n=2}^{\infty} \dfrac{1}{n \ln n}$$
cannot be determined by comparison to any of the p-series.

48. a) Show that if $\sum_{n=1}^{\infty} a_n$ is a convergent series of positive terms, then $\sum_{n=1}^{\infty} a_n^2$ is a convergent series.

b) If $\sum_{n=1}^{\infty} a_n$ is a divergent series of positive terms, must $\sum_{n=1}^{\infty} a_n^2$ also diverge?

49. Use the identity $2a_n b_n = (a_n + b_n)^2 - a_n^2 - b_n^2$ and the result in Exercise 48(a) to show that if $\sum_{n=1}^{\infty} a_n$ and $\sum_{n=1}^{\infty} b_n$ are both convergent series with positive terms, then $\sum_{n=1}^{\infty} a_n b_n$ converges.

50. Use the Limit Comparison Test and the fact that the terms of a convergent series are bounded to show that if $\sum_{n=1}^{\infty} a_n$ and $\sum_{n=1}^{\infty} b_n$ are convergent series with positive terms, then $\sum_{n=1}^{\infty} a_n b_n$ converges.

51. Suppose $\sum_{n=1}^{\infty} a_n$ is a convergent series of positive terms and that $\{b_n\}$ is a sequence with $\lim_{n \to \infty} b_n = b > 0$. Show that the series $\sum_{n=1}^{\infty} a_n b_n$ converges.

52. a) Show that $\sum_{n=1}^{\infty} \tan (1/n)$ diverges.
b) Show that $\sum_{n=1}^{\infty} (\tan (1/n))^2$ converges.
c) For which values of k does $\sum_{n=1}^{\infty} (\tan (1/n))^k$ converge?

53. Show that
$$\sum_{n=1}^{\infty} \arctan \left[\dfrac{1}{n^2 + n + 1} \right] = \dfrac{\pi}{4}.$$

[*Hint:* Show first that $\arctan x - \arctan y = \arctan [(x - y)/(1 + xy)]$; then cleverly choose x and y.]

54. Evaluate
$$\sum_{k=0}^{\infty} \dfrac{k^2 + 3k + 1}{(k + 2)!}.$$

55. Use the inequalities in (8.29) to show that if f is continuous and decreasing and $f(i) = a_i > 0$ for each i, then for $n \geq 2$,
$$\int_1^{n+1} f(x)\, dx \leq \sum_{i=1}^{n} a_i \leq \int_1^{n} f(x)\, dx + a_1.$$

56. Suppose f is continuous and decreasing and $f(i) = a_i > 0$ for each i, where $\sum_{i=1}^{\infty} a_i = S$ is convergent with the sequence of partial sums $\{S_n\}$. Show that for all $n \geq 1$.

$$\int_{n+1}^{\infty} f(x)\, dx \leq S - S_n \leq \int_n^{\infty} f(x)\, dx.$$

57. Use the result of Exercise 56 to determine the number of terms required to approximate the sum of each of the following convergent series to within 10^{-3}.

a) $\sum_{n=1}^{\infty} \frac{1}{n^2}$ b) $\sum_{n=1}^{\infty} \frac{1}{n^3}$ c) $\sum_{n=1}^{\infty} \frac{1}{e^n}$

58. In Exercise 25 you found that $\sum_{n=2}^{\infty} (1/n \ln n)$ diverges. The number of subatomic particles in the known universe is estimated to be about 10^{125}. Use the result in Exercise 55 to find an upper bound for

$$\sum_{n=2}^{10^{125}} \frac{1}{n \ln n}.$$

59. In many applications, it is important to have an approximation for $n!$ when n is large. A crude approximation is obtained by noting that $\ln n! = \sum_{i=1}^{n} \ln i$ and using an inequality obtained in a similar manner to that in Exercise 55 to deduce that

$$e \left(\frac{n}{e} \right)^n < n! < e \left(\frac{n+1}{e} \right)^{n+1}.$$

Verify this inequality.

60. A better approximation to $n!$ is known as *Stirling's formula*:

$$n! \approx \sqrt{2\pi n}(n/e)^n.$$

Show that for $n \geq 2$ this approximation is within the upper and lower bounds for $n!$ given in Exercise 59.

61. It seems intuitively reasonable that if one begins stacking cards at the edge of a table, at no time can any card in the stack totally extend beyond the edge of the table. Not true. In fact, the cards can be arranged to extend any finite distance beyond the edge of the table. Show that this is true by showing that:

a) the first card can be placed to extend $\frac{1}{2}$ its length beyond the table;

b) the first two cards can be placed to extend a total of $\frac{3}{4}$ a card length beyond the table.

c) the first n cards can be placed to extend a total of $\sum_{i=1}^{n} 1/2i$ card lengths beyond the table

(see the figure). [*Hint:* Assume that the ith card from the top of the stack extends $1/2i$ times its length beyond the $(i+1)$st in the stack. Sum the total amount of card length extending beyond the table and compare to the amount lying above the table.]

Putnam exercises

62. Suppose $u_0, u_1, u_2 \ldots$ is a sequence of real numbers such that

$$u_n = \sum_{k=1}^{\infty} u_{n+k}^2, \quad \text{for } n = 0, 1, 2, \ldots.$$

Prove that if $\sum_{n=0}^{\infty} u_n$ converges, then $u_k = 0$ for all k. (This exercise was problem 6, part I of the fourteenth William Lowell Putnam examination given on March 6, 1954. The examination and its solution are in the October 1954 issue of the *American Mathematical Monthly*, pp. 542–549.)

63. Let $\{a_n\}$ be a sequence of real numbers satisfying the inequalities

$$0 \leq a_k \leq 100 a_n \quad \text{for } n \leq k \leq 2n \text{ and } n = 1, 2, \ldots,$$

and such that the series $\sum_{n=0}^{\infty} a_n$ converges. Prove that $\lim_{n \to \infty} n a_n = 0$.

(This exercise was problem 5, part II of the twenty-fourth William Lowell Putnam examination given on December 7, 1963. The examination and its solution are in the June–July 1964 issue of the *American Mathematical Monthly*, pp. 636–641.)

64. For positive real x, let

$$B_n(x) = 1^x + 2^x + 3^x + \cdots + n^x.$$

Prove or disprove the convergence of

$$\sum_{n=2}^{\infty} \frac{B_n(\log_n 2)}{(n \log_2 n)^2}.$$

(This exercise was problem A–2 of the forty-third William Lowell Putnam examination given on December 4, 1982. The examination and its solution are in the October 1983 issue of the *American Mathematical Monthly*, pp. 546–553.)

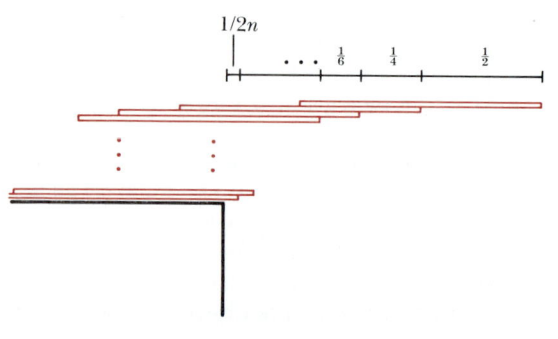

Figure for Exercise 61.

8.4 Alternating Series

An arbitrary series can contain both positive and negative terms without any systematic order. In applications, however, two special types of series frequently occur: series with all positive terms and series with alternating positive and negative terms.

In Section 8.3 we discussed series of positive terms. In this section we consider the other special type, series whose terms alternate in sign. These series are called **alternating series** and are commonly expressed as either

$$\sum_{n=1}^{\infty} (-1)^{n+1} a_n = a_1 - a_2 + a_3 - a_4 + \cdots$$

or

$$\sum_{n=1}^{\infty} (-1)^n a_n = -a_1 + a_2 - a_3 + a_4 - \cdots,$$

where the a_n terms are assumed to be nonnegative. For example,

$$\sum_{n=1}^{\infty} (-1)^n \frac{1}{n} \quad \text{and} \quad \sum_{n=1}^{\infty} (-1)^{n+1} \frac{1}{\ln(n+1)}$$

are alternating series. A convergence test for this class of series is given in the following theorem.

(8.31)
THEOREM: THE ALTERNATING SERIES TEST
If $a_n > a_{n+1} > 0$, for each $n \geq 1$, and if $\lim_{n \to \infty} a_n = 0$, then the alternating series

$$\sum_{n=1}^{\infty} (-1)^{n+1} a_n \quad \text{and} \quad \sum_{n=1}^{\infty} (-1)^n a_n$$

both converge.

Proof. We will show that under the conditions given, the alternating series $\sum_{n=1}^{\infty} (-1)^{n+1} a_n$ converges. The fact that the alternating series

---

HISTORICAL NOTE

The Alternating Series Test was discovered by Leibniz and described by him in letters in 1705 and 1714.

As an example of the confusion that accompanied the early nonrigorous work on infinite series, we note that Leibniz, as well as both Jakob and Johann Bernoulli, had no doubt that

$$\tfrac{1}{2} = 1 - 1 + 1 - 1 + \cdots,$$

which we know to be false. Guido Grandi (who we will meet in the Historical Note of Section 9.5) took this argument one step further and concluded that

$$\tfrac{1}{2} = 0 + 0 + 0 + \cdots.$$

He concluded from this that God could create the world out of nothing.

$\sum_{n=1}^{\infty}(-1)^n a_n$ also converges follows from Theorem (8.20 ii) and the fact that

$$\sum_{n=1}^{\infty}(-1)^{n+1}a_n = -\sum_{n=1}^{\infty}(-1)^n a_n.$$

Consider the partial sums for the series $\sum_{n=1}^{\infty}(-1)^{n+1}a_n$. For the even-indexed partial sums, a typical term S_{2n} can be written either as

$$S_{2n} = (a_1 - a_2) + (a_3 - a_4) + \cdots + (a_{2n-1} - a_{2n})$$

or as

$$S_{2n} = a_1 - (a_2 - a_3) - (a_4 - a_5) - \cdots - (a_{2n-2} - a_{2n-1}) - a_{2n}.$$

Since all the expressions within parentheses are positive, the first equation implies that the sequence of even partial sums is increasing. The second equation implies that the sequence is bounded above by a_1. Consequently, by the Completeness Property, the limit of the sequence of even partial sums exists. We denote this limit by S:

$$S = \lim_{n \to \infty} S_{2n}.$$

If we now consider the odd-indexed partial sums, we have

$$S_{2n+1} = S_{2n} + a_{2n+1}.$$

But

$$\lim_{n \to \infty} a_{2n+1} = 0, \quad \text{so} \quad \lim_{n \to \infty} S_{2n+1} = \lim_{n \to \infty} S_{2n} = S$$

and the series $\sum_{n=1}^{\infty}(-1)^{n+1}a_n$ converges.

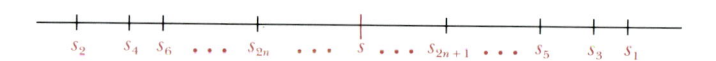

Figure 8.12
$\lim_{n \to \infty} S_n = S.$

Figure 8.12 shows the ordering of the partial sums $\{S_n\}$; the even partial sums form an increasing sequence, and the odd partial sums form a decreasing sequence. ▷

Example 1

Is the series $\sum_{n=1}^{\infty}(-1)^n \dfrac{1}{\sqrt{n}}$ convergent or divergent?

Solution
For all $n \geq 1$,

$$0 < a_{n+1} = \frac{1}{\sqrt{n+1}} < \frac{1}{\sqrt{n}} = a_n.$$

In addition, $\lim_{n \to \infty} a_n = 0$. It follows from Theorem (8.31) that the series converges. ▶

By examining the alternating behavior of the series $\sum_{n=1}^{\infty} (-1)^{n+1} a_n$ more closely, we can determine a bound for the difference between the nth partial sum and the sum of the series. The bound for this difference is also a bound for the error that occurs if S_n is used to approximate the actual sum $S = \sum_{n=1}^{\infty} (-1)^{n+1} a_n$.

(8.32)
COROLLARY
If $\sum_{n=1}^{\infty} (-1)^{n+1} a_n$ is a convergent alternating series with $0 < a_{n+1} < a_n$ for each integer n and S is its limit, then

$$|S_m - S| < a_{m+1}$$

for each integer m.

Proof. First note that for each m,

$$S - S_m = \sum_{n=1}^{\infty} (-1)^n a_n - \sum_{n=1}^{m} (-1)^n a_n = \sum_{n=m+1}^{\infty} (-1)^n a_n,$$

so

$$S - S_m = (-1)^{m+1}[a_{m+1} - a_{m+2} + a_{m+3} - a_{m+4} + \cdots].$$

For each integer n, $0 < a_{n+1} < a_n$, so $0 < a_n - a_{n+1}$ and

$$a_{m+1} - a_{m+2} + a_{m+3} - a_{m+4} + \cdots = (a_{m+1} - a_{m+2}) + (a_{m+3} - a_{m+4}) + \cdots > 0.$$

Also,

$$a_{m+1} - a_{m+2} + a_{m+3} - a_{m+4} + \cdots = a_{m+1} - (a_{m+2} - a_{m+3}) - (a_{m+4} - a_{m+5}) - \cdots < a_{m+1},$$

so

$$|S - S_m| = a_{m+1} - (a_{m+2} - a_{m+3}) - (a_{m+4} - a_{m+5}) - \cdots < a_{m+1}. \triangleright$$

Example 2

a) Show that

$$\sum_{n=1}^{\infty} (-1)^{n+1} \frac{1}{n} = 1 - \frac{1}{2} + \frac{1}{3} - \frac{1}{4} + \cdots$$

converges.

b) Find m so that S_m approximates the limit S to within 10^{-1}.

Solution
a) Since

$$\lim_{n \to \infty} \frac{1}{n} = 0 \quad \text{and} \quad 0 < \frac{1}{n+1} < \frac{1}{n},$$

the series

$$\sum_{n=1}^{\infty} (-1)^{n+1} \frac{1}{n}$$

converges.

b) To determine the limit of the infinite series to within 10^{-1}, we need to find the value of a finite sum S_m with the property that $|S - S_m| < 10^{-1}$. It follows from Corollary (8.32) that for each integer m,

$$|S - S_m| = \left| \sum_{n=1}^{m} (-1)^{n+1} \frac{1}{n} - S \right| < \frac{1}{m+1}.$$

To find m so that $|S - S_m| < 10^{-1}$, it suffices to find m with

$$\frac{1}{m+1} = 10^{-1} = \frac{1}{10}.$$

So $m = 9$.

An approximation to S that is accurate to within 10^{-1} is

$$S_9 = 1 - \frac{1}{2} + \frac{1}{3} - \frac{1}{4} + \frac{1}{5} - \frac{1}{6} + \frac{1}{7} - \frac{1}{8} + \frac{1}{9} \approx 0.75.$$

In Section 8.7 we will find that the limit of this infinite series is $\ln 2 \approx 0.6931$. ▶

Example 3

How many terms of the convergent series

$$\sum_{n=1}^{\infty} (-1)^{n+1} \frac{1}{n!}$$

are required to guarantee that $|S - S_m| < 10^{-3}$?

Solution

We need to find m so that

$$\left| S - \sum_{n=1}^{m} (-1)^{n+1} \frac{1}{n!} \right| < \frac{1}{(m+1)!} \leq 10^{-3}.$$

In this case, we cannot solve for m algebraically. However, the values in Table 8.1 indicate that $m = 6$ is the first integer with the property that $1/(m+1)! \leq 10^{-3}$. Hence $|S - S_6| < 10^{-3}$.

Since $|S - S_6| < 1/(7!) < 0.0002$,

$$0.6317 = S_6 - 0.0002 < S < S_6 + 0.0002 = 0.6321.$$

Thus the limit S correct to three decimal places is 0.632. ▶

Table 8.1

m	$\dfrac{1}{(m+1)!}$	S_m
1	0.5000	1.0000
2	0.1667	0.5000
3	0.0417	0.6667
4	0.0083	0.6250
5	0.0014	0.6333
6	0.0002	0.6319

▶ EXERCISE SET 8.4

In Exercises 1–36, determine whether the series converges or diverges.

1. $\displaystyle\sum_{n=1}^{\infty} (-1)^{n+1} \frac{1}{n+1}$
2. $\displaystyle\sum_{n=1}^{\infty} \frac{(-1)^n}{3n+2}$
3. $\displaystyle\sum_{n=5}^{\infty} (-1)^{n+1} \frac{1}{\sqrt{n}-2}$
4. $\displaystyle\sum_{n=1}^{\infty} \frac{(-1)^n}{n\sqrt{n}}$

5. $\sum_{n=1}^{\infty} (-1)^{n-1} \dfrac{n}{n^2+1}$

6. $\sum_{n=1}^{\infty} \dfrac{(-1)^{n+1}}{n^2}$

7. $\sum_{n=0}^{\infty} (-1)^n \dfrac{n+2}{4n+5}$

8. $\sum_{n=1}^{\infty} \dfrac{(-1)^n}{(n+1)\sqrt{n}}$

9. $\sum_{n=1}^{\infty} \dfrac{(-1)^{n-1} n^2}{n^2+1}$

10. $\sum_{n=1}^{\infty} (-1)^{n+1} \dfrac{n^{3/2}}{n+4}$

11. $\sum_{n=1}^{\infty} (-1)^n \dfrac{\ln n}{n}$

12. $\sum_{n=2}^{\infty} \dfrac{(-1)^n n}{\ln n}$

13. $\sum_{n=2}^{\infty} (-1)^n \dfrac{1}{\ln n}$

14. $\sum_{n=2}^{\infty} \dfrac{(-1)^n}{\ln(\ln n)}$

15. $\sum_{n=1}^{\infty} (-1)^{n-1} \dfrac{n}{e^n}$

16. $\sum_{n=1}^{\infty} (-1)^n \dfrac{e^n}{n}$

17. $\sum_{n=1}^{\infty} (-1)^n \dfrac{e^n}{n!}$

18. $\sum_{n=1}^{\infty} (-1)^{n-1} \dfrac{e^n}{n^e}$

19. $\sum_{n=1}^{\infty} \dfrac{(-1)^{2n+1}}{n}$

20. $\sum_{n=1}^{\infty} (-1)^{2n} \dfrac{\sqrt{n+2}}{n^2+3n+1}$

21. $\sum_{n=1}^{\infty} \dfrac{(-1)^n (1000)^n}{n!}$

22. $\sum_{n=1}^{\infty} \dfrac{(-1)^{n+1} n!}{(1000)^n}$

23. $\sum_{n=0}^{\infty} (-1)^n \dfrac{n^2}{\pi^n}$

24. $\sum_{n=1}^{\infty} \dfrac{(-1)^n e^n}{n^4}$

25. $\sum_{n=1}^{\infty} (-1)^{n-1} \dfrac{\pi^n}{ne^n}$

26. $\sum_{n=1}^{\infty} (-1)^{n-1} \dfrac{e^n}{n\pi^n}$

27. $\sum_{n=1}^{\infty} \dfrac{n(-2)^{2n}}{5^n}$

28. $\sum_{n=1}^{\infty} \dfrac{n(-4)^{3n}}{5^n}$

29. $\sum_{n=1}^{\infty} (-1)^n \dfrac{n^3}{n^3+e^n}$

30. $\sum_{n=1}^{\infty} \dfrac{(-1)^n 2e^n}{n^3+e^n}$

31. $\sum_{n=1}^{\infty} (-1)^n \dfrac{\cos n\pi}{n}$

32. $\sum_{n=1}^{\infty} \dfrac{(-1)^n \sin \dfrac{n\pi}{2}}{(n^3+n)^{1/2}}$

33. $\sum_{n=1}^{\infty} \dfrac{1}{\sqrt{n}} \sin \dfrac{n\pi}{2}$

34. $\sum_{n=1}^{\infty} \dfrac{1}{n^2} \cos \dfrac{n\pi}{2}$

35. $\sum_{n=1}^{\infty} \dfrac{(-1)^n (n!)^2}{(2n)!}$

36. $\sum_{n=1}^{\infty} \dfrac{(-1)^n (3n)!}{(n!)^3}$

In Exercises 37–42, determine the number of terms of the series that are required to produce an approximation to the infinite sum that is accurate to within 10^{-4}.

37. $\sum_{n=1}^{\infty} \dfrac{(-1)^n}{n}$

38. $\sum_{n=1}^{\infty} \dfrac{(-1)^{n+1}}{n^2}$

39. $\sum_{n=2}^{\infty} \dfrac{(-1)^n}{\ln n}$

40. $\sum_{n=0}^{\infty} \dfrac{(-1)^n}{5^n+3}$

41. $\sum_{n=1}^{\infty} \dfrac{(-1)^{n-1}}{n!}$

42. $\sum_{n=1}^{\infty} \dfrac{(-1)^{n+1}}{n^n}$

In Exercises 43–48, determine an approximation to the sum of the series to within the accuracy specified.

43. $\sum_{n=1}^{\infty} \dfrac{(-1)^{n+1}}{n^2}$; 10^{-1}

44. $\sum_{n=1}^{\infty} \dfrac{(-1)^n}{n^3}$; 10^{-2}

45. $\sum_{n=0}^{\infty} \dfrac{(-1)^n}{4^n+1}$; 10^{-4}

46. $\sum_{n=0}^{\infty} \dfrac{(-1)^n}{5^n+3}$; 10^{-4}

47. $\sum_{n=1}^{\infty} \dfrac{(-1)^{n-1}}{n!}$; 10^{-5}

48. $\sum_{n=1}^{\infty} \dfrac{(-1)^{n+1}}{n^n}$; 10^{-5}

49. In Exercise 48 of Section 8.3 it was stated that if $\sum_{n=1}^{\infty} a_n$ is a convergent series of positive terms, then $\sum_{n=1}^{\infty} a_n^2$ is also convergent. Show that this need not be true if the terms of the series alternate in sign.

8.5

Absolute Convergence

In the preceding sections we found convergence tests for two special types of series: positive-term series and alternating series. This section is concerned with the convergence of general series and the relationship between a general series $\sum_{n=1}^{\infty} a_n$ and the positive-term series of its absolute values

$$\sum_{n=1}^{\infty} |a_n| = |a_1| + |a_2| + |a_3| + \cdots .$$

(8.33) DEFINITION

A series $\sum_{n=1}^{\infty} a_n$ is said to be **absolutely convergent** if the series of the absolute values of the terms,

$$\sum_{n=1}^{\infty} |a_n| = |a_1| + |a_2| + |a_3| + \cdots$$

converges.

Example 1

Show that the series $\sum_{n=1}^{\infty} (-1)^n \frac{1}{n^2}$ is absolutely convergent.

Solution
Since

$$\sum_{n=1}^{\infty} \left|(-1)^n \frac{1}{n^2}\right| = \sum_{n=1}^{\infty} \frac{1}{n^2}$$

is a convergent p-series,

$$\sum_{n=1}^{\infty} (-1)^n \frac{1}{n^2}$$

is absolutely convergent. ▶

The following theorem provides an important reason for considering absolute convergence.

(8.34) THEOREM

If a series is absolutely convergent, then it is convergent.

Proof. Suppose $\sum_{n=1}^{\infty} |a_n|$ is a convergent series. Then the multiple of this series $\sum_{n=1}^{\infty} 2|a_n|$ is also convergent. Moreover, for each integer n,

$$-|a_n| \le a_n \le |a_n|, \qquad \text{so} \qquad 0 \le a_n + |a_n| \le 2|a_n|.$$

By the Comparison Test, the series

$$\sum_{n=1}^{\infty} (a_n + |a_n|)$$

is convergent. So $\sum_{n=1}^{\infty} a_n$ can be expressed as the difference of two convergent series

$$\sum_{n=1}^{\infty} a_n = \sum_{n=1}^{\infty} [(a_n + |a_n|) + (-|a_n|)] = \sum_{n=1}^{\infty} (a_n + |a_n|) - \sum_{n=1}^{\infty} |a_n|.$$

It follows from Theorem (8.20) that $\sum_{n=1}^{\infty} a_n$ is convergent. ▷

Example 2

Is $\sum_{n=1}^{\infty} \frac{\cos n}{n^{3/2} + 1}$ a convergent series?

Solution
This is neither a positive-term series nor an alternating series, so the tests in Sections 8.3 and 8.4 do not apply. However,

$$\left| \frac{\cos n}{n^{3/2} + 1} \right| \leq \frac{1}{n^{3/2} + 1} < \frac{1}{n^{3/2}}$$

and $\sum_{n=1}^{\infty} (1/n^{3/2})$ is a convergent p-series.

It follows from the Comparison Test that

$$\sum_{n=1}^{\infty} \left| \frac{\cos n}{n^{3/2} + 1} \right|$$

converges. Therefore,

$$\sum_{n=1}^{\infty} \frac{\cos n}{n^{3/2} + 1}$$

is absolutely convergent, hence convergent. ▶

In Example 2 of Section 8.4 we saw that the series

$$\sum_{n=1}^{\infty} (-1)^{n+1} \frac{1}{n}$$

converges. This series is not absolutely convergent, however, since

$$\sum_{n=1}^{\infty} \left| (-1)^{n+1} \frac{1}{n} \right| = \sum_{n=1}^{\infty} \frac{1}{n},$$

the divergent harmonic series. The series $\sum_{n=1}^{\infty} (-1)^{n+1} \frac{1}{n}$ is an example of a *conditionally convergent series*.

(8.35)
DEFINITION
A convergent series $\sum_{n=1}^{\infty} a_n$ that is not absolutely convergent is said to be **conditionally convergent**.

Note that a series with positive terms is absolutely convergent precisely when it is convergent. It is *never* conditionally convergent.

Example 3

Determine whether the series

$$\sum_{n=1}^{\infty} (-1)^{n+1} \frac{1}{\sqrt[3]{n}}$$

is divergent, conditionally convergent, or absolutely convergent.

Solution
Since

$$\sum_{n=1}^{\infty} \left| (-1)^{n+1} \frac{1}{\sqrt[3]{n}} \right| = \sum_{n=1}^{\infty} \frac{1}{\sqrt[3]{n}}$$

is a divergent p-series with $p = \frac{1}{3} < 1$, the series does not converge absolutely. However,

$$0 < a_{n+1} = \frac{1}{\sqrt[3]{n+1}} < \frac{1}{\sqrt[3]{n}} = a_n, \quad \text{for each } n \geq 1,$$

and $\lim_{n \to \infty} 1/\sqrt[3]{n} = 0$, so the Alternating Series Test implies that

$$\sum_{n=1}^{\infty} (-1)^{n+1} \frac{1}{\sqrt[3]{n}}$$

converges. Therefore, this series is conditionally convergent. ▶

We now introduce two tests for absolute convergence that follow from the Comparison Test for positive-term series. The first of these is the *Ratio Test*.

(8.36)
THEOREM: THE RATIO TEST
Suppose $\sum_{n=1}^{\infty} a_n$ is a series of nonzero terms and

$$\lim_{n \to \infty} \left| \frac{a_{n+1}}{a_n} \right| = L,$$

where L represents either a nonnegative real number or infinity.

i) If $0 \leq L < 1$, the series is absolutely convergent.
ii) If $L > 1$, the series is divergent.
iii) If $L = 1$, no conclusion about convergence can be made. The series might be absolutely convergent, conditionally convergent, or divergent.

Proof
i) For $0 \leq L < 1$, let

$$r = \frac{L+1}{2} \quad \text{and} \quad \varepsilon = 1 - r.$$

Then $0 < r < 1$ and $\varepsilon > 0$. Since $\lim_{n \to \infty} |a_{n+1}/a_n| = L$, an integer M exists with

$$\left| \left| \frac{a_{n+1}}{a_n} \right| - L \right| < \varepsilon$$

whenever $n \geq M$. Consequently,

$$\left| \frac{a_{n+1}}{a_n} \right| < L + \varepsilon = L + (1-r) = L + 1 - \left(\frac{L+1}{2} \right) = \frac{L+1}{2} = r,$$

so

$$|a_{n+1}| < |a_n| r$$

whenever $n \geq M$. In particular,

$$|a_{M+1}| < |a_M| r,$$
$$|a_{M+2}| < |a_{M+1}| r < |a_M| r^2,$$
$$|a_{M+3}| < |a_{M+2}| r < |a_M| r^3,$$

and, in general, $|a_{M+n}| < |a_M| r^n$. The series $\sum_{n=1}^{\infty} |a_M| r^n$ is geometric with ratio r, where $0 < r < 1$, and hence converges. The Comparison Test implies that $\sum_{n=1}^{\infty} |a_{M+n}|$ also converges. Since $\sum_{n=1}^{\infty} |a_n|$ differs from $\sum_{n=1}^{\infty} |a_{M+n}|$ in only the first M terms, $\sum_{n=1}^{\infty} |a_n|$ converges.

ii) This statement is shown in a similar manner by choosing r to be a number between 1 and L. If $\varepsilon = L - r$, then an integer M exists with

$$\left| \left| \frac{a_{n+1}}{a_n} \right| - L \right| < \varepsilon$$

whenever $n \geq M$. So

$$\left| \frac{a_{n+1}}{a_n} \right| > L - \varepsilon = L - (L - r) = r > 1,$$

for $n \geq M$. Thus for all $n \geq M$, $|a_{n+1}| > |a_n| > 0$. Consequently, $\lim_{n \to \infty} |a_n| \neq 0$ and $\lim_{n \to \infty} a_n \neq 0$, which implies that the series is divergent.

Series that demonstrate the validity of statement (iii) in the Ratio Test are given after Example 5. ▷

Example 4

Determine whether the series

$$\sum_{n=1}^{\infty} (-1)^n \frac{n}{2^n}$$

is divergent, conditionally convergent, or absolutely convergent.

Solution
Applying the Ratio Test to the series, we have

$$\lim_{n \to \infty} \left| \frac{a_{n+1}}{a_n} \right| = \lim_{n \to \infty} \left| \frac{(-1)^{n+1} \frac{n+1}{2^{n+1}}}{(-1)^n \frac{n}{2^n}} \right| = \lim_{n \to \infty} \frac{n+1}{2^{n+1}} \cdot \frac{2^n}{n} = \lim_{n \to \infty} \frac{n+1}{n} \cdot \frac{1}{2} = \frac{1}{2},$$

so the series is absolutely convergent. ▶

Example 5

Use the Ratio Test to discuss the behavior of the series

$$\sum_{n=1}^{\infty} \frac{n!}{n^n}.$$

Solution
The ratio of the terms of the series is

$$\left| \frac{a_{n+1}}{a_n} \right| = \frac{\frac{(n+1)!}{(n+1)^{n+1}}}{\frac{n!}{n^n}} = \frac{(n+1)!}{(n+1)^{n+1}} \cdot \frac{n^n}{n!} = (n+1) \frac{n^n}{(n+1)^{n+1}} = \frac{n^n}{(n+1)^n} = \frac{1}{\left(1 + \frac{1}{n}\right)^n}.$$

In Section 6.5 we found that $\lim_{x \to \infty} (1 + 1/x)^x$ is an indeterminate form of type 1^∞, and we used l'Hôpital's Rule to show that the value of this limit is e. Thus, $\lim_{n \to \infty} (1 + 1/n)^n = e$ and

$$\lim_{n \to \infty} \left| \frac{a_{n+1}}{a_n} \right| = \lim_{n \to \infty} \frac{1}{\left(1 + \frac{1}{n}\right)^n} = \frac{1}{e} < 1.$$

The Ratio Test implies that the series is absolutely convergent. ▶

To see the validity of statement (iii) in Theorem (8.36), consider applying the Ratio Test to the alternating series

$$\sum_{n=1}^{\infty} \frac{(-1)^{n+1}}{n^p}.$$

For any real number p,

$$\lim_{n \to \infty} \left| \frac{a_{n+1}}{a_n} \right| = \lim_{n \to \infty} \frac{\frac{1}{(n+1)^p}}{\frac{1}{n^p}} = \lim_{n \to \infty} \left(\frac{n}{n+1}\right)^p = 1.$$

However:

i) when $p = 2$, the series $\sum_{n=1}^{\infty} \frac{(-1)^n}{n^2}$ converges absolutely;

ii) when $p = 1$, the series $\sum_{n=1}^{\infty} \frac{(-1)^n}{n}$ converges conditionally; and

iii) when $p = -1$, the series $\sum_{n=1}^{\infty} \frac{(-1)^n}{n^{-1}} = \sum_{n=1}^{\infty} n(-1)^n$ diverges.

This demonstrates the fact that:

> No information about the convergence of a series is obtained from the Ratio Test when
>
> $$\lim_{n \to \infty} \left| \frac{a_{n+1}}{a_n} \right| = 1.$$

In this case some other method must be used.

The following test for convergence is known as the *Root Test*. It is most useful for series $\sum_{n=1}^{\infty} a_n$ when a_n has the form $(b_n)^n$.

(8.37)
THEOREM: THE ROOT TEST

Suppose $\sum_{n=1}^{\infty} a_n$ is a series and $\lim_{n \to \infty} |a_n|^{1/n} = L$, where L represents either a nonnegative real number or infinity.

i) If $0 \le L < 1$, the series is absolutely convergent.
ii) If $L > 1$, the series is divergent.
iii) If $L = 1$, no conclusion about convergence can be made. The series might be absolutely convergent, conditionally convergent, or divergent.

The proof of the Root Test is similar to the proof of the Ratio Test. For example, to show part (i) we assume that $L < 1$ and choose $r = (L + 1)/2$. Then $L < r < 1$. Since

$$\lim_{n \to \infty} |a_n|^{1/n} = L \quad \text{and} \quad L < r,$$

an integer M exists with the property that

$$|a_n|^{1/n} < r, \quad \text{whenever } n \geq M.$$

Thus,

$$|a_n| < r^n, \quad \text{whenever } n \geq M.$$

Since $r < 1$, the geometric series $\sum_{n=1}^{\infty} r^n$ converges and, by the Comparison Test, the series $\sum_{n=1}^{\infty} |a_n|$ also converges.

Example 6

Does the series

$$\sum_{n=1}^{\infty} \left(-\frac{1}{n}\right)^n$$

converge or diverge?

Solution

We apply the Root Test:

$$\lim_{n \to \infty} \left|\left(-\frac{1}{n}\right)^n\right|^{1/n} = \lim_{n \to \infty} \frac{1}{n} = 0 < 1.$$

The series converges absolutely, hence converges. ▶

Elementary properties of algebra ensure that when the terms of a finite sum are rearranged, the value of the sum remains the same. For example,

$$6 = 11 + 4 - 2 - 7 = -2 + 11 + 4 - 7 = -7 + 4 + 11 - 2.$$

The following theorem, given without proof, states that for convergent

---

HISTORICAL NOTE

The first known appearance of infinite series occurs in the work of Archimedes. Infinite series were freely used in the late seventeenth century by Newton, Leibniz, and others. However, little or no attention was given to the general question of establishing rigorous tests for their convergence or divergence.

The rigorous foundations of both the theories of infinite sequences and series were established by **Augustin Louis Cauchy** (1789–1857) in his *Cours d'analyse de l'école royale polytechnique, Part I Analyse algébrique* in 1821 and *Exercices des mathématiques* (four volumes, 1826–1830). In these volumes he proved the Comparison Test (8.26) and the Integral Test (8.28). The concept of absolute convergence as well as the Ratio Test (8.36), is due to him.

The story is told that the distinguished mathematical astronomer and physicist Pierre Simon Laplace (1749–1827) was present when Cauchy read his first paper on infinite series. Laplace was so impressed by Cauchy's work that he immediately set about to check that all the series he had used in his multivolume *Mécanique céleste* were convergent. Laplace was fortunate: All his series converged!

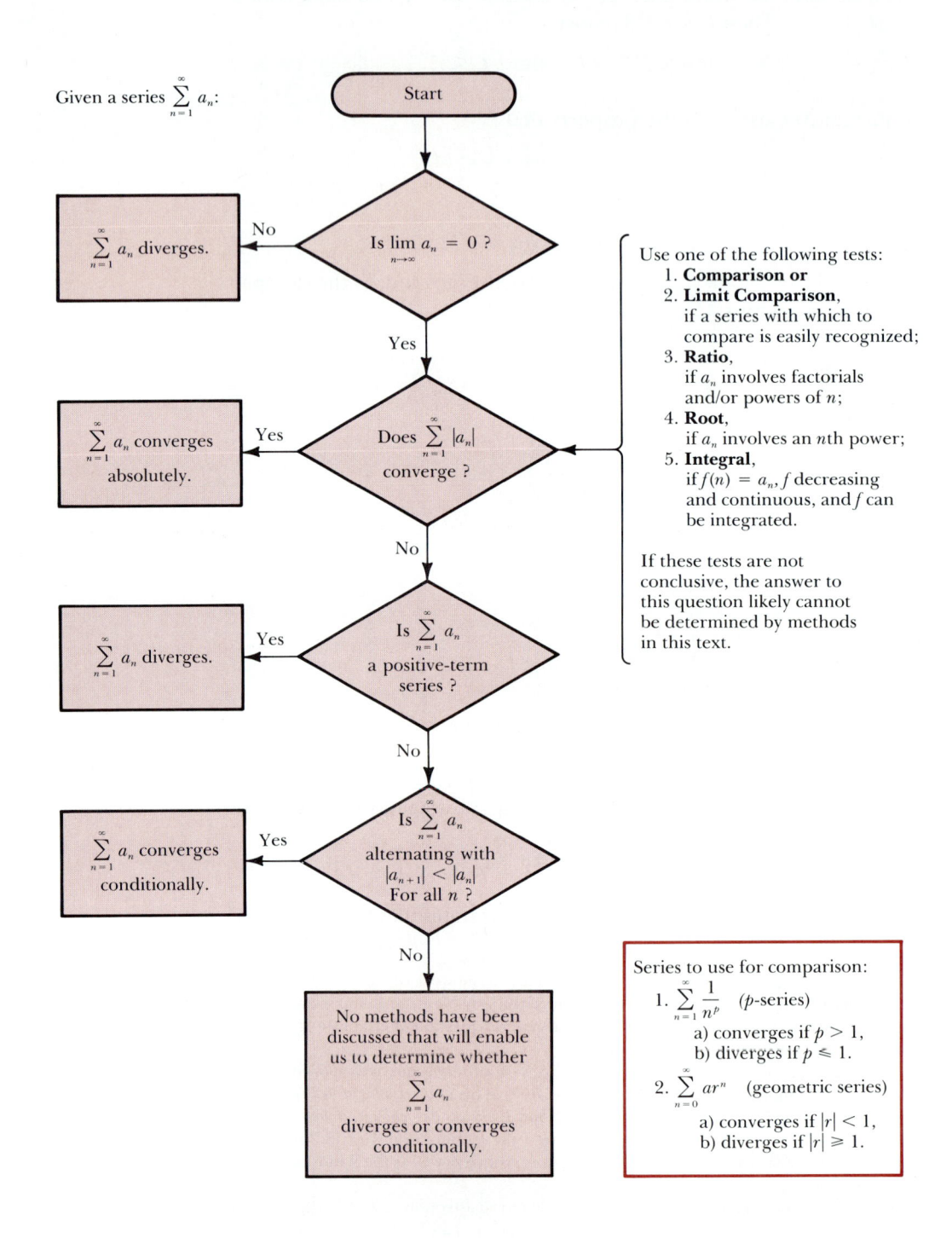

CHAPTER 8 SEQUENCES AND SERIES

Given a series $\sum_{n=1}^{\infty} a_n$:

Start

Is $\lim_{n\to\infty} a_n = 0$?
- **No** → $\sum_{n=1}^{\infty} a_n$ diverges.
- **Yes** ↓

Does $\sum_{n=1}^{\infty} |a_n|$ converge?
- **Yes** → $\sum_{n=1}^{\infty} a_n$ converges absolutely.
- **No** ↓

Use one of the following tests:
1. **Comparison** or
2. **Limit Comparison**, if a series with which to compare is easily recognized;
3. **Ratio**, if a_n involves factorials and/or powers of n;
4. **Root**, if a_n involves an nth power;
5. **Integral**, if $f(n) = a_n$, f decreasing and continuous, and f can be integrated.

If these tests are not conclusive, the answer to this question likely cannot be determined by methods in this text.

Is $\sum_{n=1}^{\infty} a_n$ a positive-term series?
- **Yes** → $\sum_{n=1}^{\infty} a_n$ diverges.
- **No** ↓

Is $\sum_{n=1}^{\infty} a_n$ alternating with $|a_{n+1}| < |a_n|$ For all n?
- **Yes** → $\sum_{n=1}^{\infty} a_n$ converges conditionally.
- **No** ↓

No methods have been discussed that will enable us to determine whether $\sum_{n=1}^{\infty} a_n$ diverges or converges conditionally.

Series to use for comparison:
1. $\sum_{n=1}^{\infty} \dfrac{1}{n^p}$ (*p*-series)
   a) converges if $p > 1$,
   b) diverges if $p \leq 1$.
2. $\sum_{n=0}^{\infty} ar^n$ (geometric series)
   a) converges if $|r| < 1$,
   b) diverges if $|r| \geq 1$.

Figure 8.13

infinite series, equality of sums under the rearrangement of terms holds precisely when the series is absolutely convergent. An interesting example concerning part (ii) of the theorem is given in Exercise 51.

(8.38)
THEOREM
Suppose $\sum_{n=0}^{\infty} a_n$ is a convergent series.

i) If the series is absolutely convergent, then any rearrangement of the terms produces a series that is absolutely convergent and converges to the same value.
ii) If the series is conditionally convergent and L represents either an arbitrary real number, ∞, or $-\infty$, then there is a rearrangement of the terms to produce a series that converges to L.

We close this section with a review of the tests for convergence discussed in Sections 8.2–8.5. This review is presented in the flowchart in Figure 8.13, which can be used as a guide to determine the convergence property of a series. The chart correctly indicates that some questions regarding series convergence have been left unanswered.

▶ EXERCISE SET 8.5

In Exercises 1–42, determine whether the series is divergent, conditionally convergent, or absolutely convergent.

1. $\sum_{n=1}^{\infty} (-1)^{n+1} \dfrac{1}{n+1}$
2. $\sum_{n=1}^{\infty} \dfrac{(-1)^n}{3n+2}$
3. $\sum_{n=5}^{\infty} (-1)^{n+1} \dfrac{1}{\sqrt{n-2}}$
4. $\sum_{n=1}^{\infty} \dfrac{(-1)^n}{n\sqrt{n}}$
5. $\sum_{n=1}^{\infty} (-1)^{n-1} \dfrac{n}{n^2+1}$
6. $\sum_{n=1}^{\infty} \dfrac{(-1)^{n+1}}{n^2}$
7. $\sum_{n=0}^{\infty} (-1)^n \dfrac{n+2}{4n+5}$
8. $\sum_{n=1}^{\infty} \dfrac{(-1)^n}{(n+1)\sqrt{n}}$
9. $\sum_{n=1}^{\infty} \dfrac{(-1)^{n-1} n^2}{n^2+1}$
10. $\sum_{n=1}^{\infty} (-1)^{n+1} \dfrac{n^{3/2}}{n+4}$
11. $\sum_{n=1}^{\infty} (-1)^n \dfrac{\ln n}{n}$
12. $\sum_{n=2}^{\infty} \dfrac{(-1)^n}{\ln n}$
13. $\sum_{n=2}^{\infty} (-1)^n \dfrac{1}{\ln n}$
14. $\sum_{n=2}^{\infty} \dfrac{(-1)^n}{\ln (\ln n)}$
15. $\sum_{n=1}^{\infty} (-1)^{n-1} \dfrac{n}{e^n}$
16. $\sum_{n=1}^{\infty} (-1)^n \dfrac{e^n}{n}$
17. $\sum_{n=1}^{\infty} (-1)^n \dfrac{e^n}{n!}$
18. $\sum_{n=1}^{\infty} (-1)^{n-1} \dfrac{e^n}{n^e}$
19. $\sum_{n=1}^{\infty} \dfrac{(-1)^{2n+1}}{n}$
20. $\sum_{n=1}^{\infty} (-1)^{2n} \dfrac{\sqrt{n+2}}{n^2+3n}$
21. $\sum_{n=1}^{\infty} \dfrac{(-1)^n (1000)^n}{n!}$
22. $\sum_{n=1}^{\infty} \dfrac{(-1)^{n+1} n!}{(1000)^n}$
23. $\sum_{n=0}^{\infty} (-1)^n \dfrac{n^2}{\pi^n}$
24. $\sum_{n=1}^{\infty} \dfrac{(-1)^n e^n}{n^4}$
25. $\sum_{n=1}^{\infty} (-1)^{n-1} \dfrac{\pi^n}{ne^n}$
26. $\sum_{n=1}^{\infty} (-1)^{n-1} \dfrac{e^n}{n\pi^n}$
27. $\sum_{n=1}^{\infty} \dfrac{n(-2)^{2n}}{5^n}$
28. $\sum_{n=1}^{\infty} \dfrac{n(-4)^{3n}}{5^n}$
29. $\sum_{n=1}^{\infty} (-1)^n \dfrac{n^3}{n^3+e^n}$
30. $\sum_{n=1}^{\infty} \dfrac{(-1)^n 2e^n}{n^3+e^n}$
31. $\sum_{n=1}^{\infty} (-1)^n \dfrac{\cos n\pi}{n}$
32. $\sum_{n=1}^{\infty} \dfrac{(-1)^n \sin (n\pi/2)}{(n^3+n)^{1/2}}$
33. $\sum_{n=1}^{\infty} \dfrac{\sin (n\pi/2)}{\sqrt{n}}$
34. $\sum_{n=1}^{\infty} \dfrac{\cos (n\pi/4)}{n^2}$
35. $\sum_{n=1}^{\infty} \dfrac{(-1)^n (n!)^2}{(2n)!}$
36. $\sum_{n=1}^{\infty} \dfrac{(-1)^n (n!)^3}{(3n)!}$
37. $\sum_{n=1}^{\infty} \dfrac{(-1)^n n^n}{(2n)!}$
38. $\sum_{n=1}^{\infty} (-1)^n \ln \left(\dfrac{n}{n+1} \right)$
39. $\sum_{n=1}^{\infty} \dfrac{n!}{1 \cdot 3 \cdot 5 \cdot 7 \cdots (2n-1)}$
40. $\sum_{n=1}^{\infty} \dfrac{1 \cdot 3 \cdot 5 \cdot 7 \cdots (2n-1)}{2 \cdot 4 \cdot 6 \cdot 8 \cdots (2n)}$

41. $\sum_{n=1}^{\infty} \left(\dfrac{-n}{n^2+1}\right)^n$

42. $\sum_{n=1}^{\infty} \left(\dfrac{-2n}{n+1}\right)^n$

In Exercises 43–46, use the Root Test to test the convergence of the series.

43. $\sum_{n=2}^{\infty} \dfrac{1}{(\ln n)^n}$

44. $\sum_{n=1}^{\infty} \dfrac{1}{(\ln 2)^n}$

45. $\sum_{n=1}^{\infty} \left(\dfrac{n-1}{3n+1}\right)^n$

46. $\sum_{n=1}^{\infty} \dfrac{2n}{n^2}$

47.
a) Show that for every real number p, $\sum_{n=1}^{\infty} n^p r^n$ converges if $|r| < 1$ and diverges if $|r| > 1$.
b) What happens to the series in part (a) if $|r| = 1$?

48. Show that
$$\sum_{n=1}^{\infty} (-1)^n \left[\left(1+\dfrac{1}{n}\right)^{n+1} - \left(1+\dfrac{1}{n}\right)^n\right]$$
converges conditionally.

49. Show that if $\sum_{n=1}^{\infty} a_n$ is a series that is absolutely convergent, then $\sum_{n=1}^{\infty} a_n^2$ is also absolutely convergent.

50. Suppose the series $\sum_{n=0}^{\infty} a_n$ is conditionally convergent. What can be said about the series
$$\sum_{n=0}^{\infty} \dfrac{|a_n| + a_n}{2} \quad \text{and} \quad \sum_{n=0}^{\infty} \dfrac{|a_n| - a_n}{2} ?$$

51. Let S denote the sum of the conditionally convergent alternating series
$$S = \sum_{n=1}^{\infty} \dfrac{(-1)^{n+1}}{n}.$$
By adding the nth term of
$$\dfrac{1}{2} \sum_{n=1}^{\infty} \dfrac{(-1)^{n+1}}{n} = \sum_{n=1}^{\infty} \dfrac{(-1)^{n+1}}{2n}$$
to the $(2n)$th term of
$$\sum_{n=1}^{\infty} \dfrac{(-1)^{n+1}}{n},$$
show that a rearrangement of the alternating harmonic series results whose value is $3S/2$.

52. Find the flaw(s) in the following logic:
$$0 < 1 - \tfrac{1}{2} + \tfrac{1}{3} - \tfrac{1}{4} + \tfrac{1}{5} - \tfrac{1}{6} + \tfrac{1}{7} - \tfrac{1}{8} + \cdots$$
$$= (1 + \tfrac{1}{3} + \tfrac{1}{5} + \cdots) - (\tfrac{1}{2} + \tfrac{1}{4} + \tfrac{1}{6} + \tfrac{1}{8} + \cdots)$$
$$= [(1 + \tfrac{1}{3} + \tfrac{1}{5} + \tfrac{1}{7} + \cdots)$$
$$+ (\tfrac{1}{2} + \tfrac{1}{4} + \tfrac{1}{6} + \tfrac{1}{8} + \cdots)]$$
$$- 2(\tfrac{1}{2} + \tfrac{1}{4} + \tfrac{1}{6} + \tfrac{1}{8} + \cdots)$$
$$= (1 + \tfrac{1}{2} + \tfrac{1}{3} + \tfrac{1}{4} + \cdots)$$
$$- (1 + \tfrac{1}{2} + \tfrac{1}{3} + \tfrac{1}{4} + \cdots) = 0.$$
So $0 < 0$.

8.6

Power Series

A polynomial $P(x)$ has the property that
$$P(x) = a_0 + a_1 x + \cdots + a_n x^n$$
for some integer n and collection of constants $a_0, a_1, \ldots a_n$. We have seen that all polynomials are continuous, differentiable, and integrable and that derivatives and indefinite integrals of polynomials are also polynomials. In this section we consider an extension of polynomials that results from considering a series with an infinite number of terms of the form $a_n x^n$. This produces an infinite series of constants and powers of x called a **power series** in the variable x. A power series, then, is an expression of the form

(8.39) $$\sum_{n=0}^{\infty} a_n x^n = a_0 + a_1 x + a_2 x^2 + \cdots + a_n x^n + \cdots,$$

where $a_0, a_1, a_2, \ldots$ represent constants and x represents a variable.

Examples of power series are

a) $\sum_{n=0}^{\infty} x^n = 1 + x + x^2 + \cdots$ and

b) $\sum_{n=0}^{\infty} \dfrac{(-1)^n x^n}{n+1} = 1 - \dfrac{1}{2}x + \dfrac{1}{3}x^2 - \dfrac{1}{4}x^3 + \cdots .$

To apply the techniques of calculus to a function that is expressed in power-series form,

(8.40) $$f(x) = \sum_{n=0}^{\infty} a_n x^n,$$

we first consider the domain of f, which consists of those values of x for which the series converges. Since only the constant term a_0 remains when a power series

$$a_0 + a_1 x + a_2 x^2 + \cdots = \sum_{n=0}^{\infty} a_n x^n$$

is evaluated at zero, the power series in (8.40) always converges at zero.

Example 1

Determine the values of x for which each of the following power series converges.

a) $\sum_{n=0}^{\infty} x^n$
b) $\sum_{n=0}^{\infty} \dfrac{(-1)^n x^n}{n+1}$

Solution

a) The power series $\sum_{n=0}^{\infty} x^n$ is simply the geometric series with ratio x. We know from Section 8.2 that this series converges when $|x| < 1$ and diverges when $|x| \geq 1$. In fact, when $|x| < 1$,

$$\sum_{n=0}^{\infty} x^n = \dfrac{1}{1-x}.$$

b) The series converges when $x = 0$. To determine the nonzero values of x for which this series converges, we apply the Ratio Test. For a fixed value of $x \neq 0$,

$$\lim_{n \to \infty} \left| \dfrac{\dfrac{(-1)^{n+1} x^{n+1}}{n+2}}{\dfrac{(-1)^n x^n}{n+1}} \right| = \lim_{n \to \infty} \dfrac{|x^{n+1}|}{n+2} \dfrac{n+1}{|x^n|} = \lim_{n \to \infty} \dfrac{n+1}{n+2} |x| = |x|.$$

The Ratio Test implies that the series converges if $|x| < 1$ and diverges if $|x| > 1$.

The Ratio Test gives no information about convergence when the ratio is 1, so the question of convergence at $x = 1$ and at $x = -1$ remains. However, at $x = 1$, the power series is

$$\sum_{n=0}^{\infty} \dfrac{(-1)^n}{n+1},$$

an alternating series with

$$0 < a_{n+1} = \frac{1}{n+2} < \frac{1}{n+1} = a_n$$

and $\lim_{n\to\infty} [1/(n+1)] = 0$. Hence it converges.

At $x = -1$, the power series is

$$\sum_{n=0}^{\infty} \frac{(-1)^n(-1)^n}{n+1} = \sum_{n=0}^{\infty} \frac{1}{n+1} = 1 + \frac{1}{2} + \frac{1}{3} + \cdots,$$

which is the divergent harmonic series. Consequently, the power series

$$\sum_{n=0}^{\infty} \frac{(-1)^n x^n}{n+1}$$

converges for $-1 < x \leq 1$ and diverges otherwise. Note that this implies that the function defined by

$$f(x) = \sum_{n=0}^{\infty} \frac{(-1)^n x^n}{n+1}$$

has $(-1, 1]$ as its domain. ▶

In order for the continuity, differentiability, or integrability of a function to be considered, the domain of the function must consist of an interval or collection of intervals. We found in Example 1 that the functions defined by the power series

$$\sum_{n=0}^{\infty} x^n \quad \text{and} \quad \sum_{n=0}^{\infty} \frac{(-1)^n x^n}{n+1}$$

do have intervals for their domains. The following lemma implies that this is true in general of functions defined by power series.

(8.41)
LEMMA

Suppose $\sum_{n=0}^{\infty} a_n x^n$ is a power series.

i) If the series converges at c, then the series converges absolutely at x whenever $|x| < |c|$.

ii) If the series diverges at b, then the series diverges at x whenever $|x| > |b|$.

Proof

i) Since $\sum_{n=0}^{\infty} a_n c^n$ converges, $\lim_{n\to\infty} a_n c^n = 0$. A convergent sequence is bounded, so a number M exists with $|a_n c^n| \leq M$ for all integers $n \geq 0$. Suppose x is arbitrarily chosen with $|x| < |c|$. Then $|x/c| < 1$ and the geometric series with first term M and ratio $|x/c|$,

$$\sum_{n=0}^{\infty} M \left|\frac{x}{c}\right|^n,$$

converges. However,

$$|a_n x^n| = |a_n c^n| \cdot \left|\frac{x^n}{c^n}\right| \leq M \left|\frac{x}{c}\right|^n,$$

so the Comparison Test implies that

$$\sum_{n=0}^{\infty} |a_n x^n|$$

converges and that the series is absolutely convergent when $|x| < |c|$.

ii) This part of the lemma can be proved directly, but it is easier to apply part (i). Suppose the series diverges at b but converges for some x with $|x| > |b|$. This implies, by part (i), that $\sum_{n=0}^{\infty} a_n b^n$ converges absolutely, which contradicts the assumption of divergence at b. Thus if the series diverges at b and $|x| > |b|$ the series diverges at x. ▷

Lemma (8.41) implies that if a power series converges for some real numbers but diverges for others, then a number R exists with the property that the series converges absolutely if $|x| < R$ and diverges if $|x| > R$. This permits us to partition power series into three classes, as outlined in the following theorem.

(8.42)
THEOREM

If $\sum_{n=0}^{\infty} a_n x^n$ is a power series, then precisely one of the following statements holds:

i) $\sum_{n=0}^{\infty} a_n x^n$ converges only at $x = 0$.
ii) $\sum_{n=0}^{\infty} a_n x^n$ converges for all real numbers x.
iii) There is a number $R > 0$ such that
   a) $\sum_{n=0}^{\infty} a_n x^n$ converges absolutely if $|x| < R$;
   b) $\sum_{n=0}^{\infty} a_n x^n$ diverges if $|x| > R$.

The number R in part (iii) is called the **radius of convergence** of the series. A series of type (ii) is said to have an infinite radius of convergence. A series of type (i) has radius of convergence zero.

The interval on which the series converges is called the **interval of convergence**. The radius of convergence and, consequently, the interior of the interval of convergence can usually be found by applying the Ratio Test. The behavior at the endpoints of this interval is determined by applying one of the tests for convergence of a series of constants, as was done in Example 1(b).

Example 2

Find the radius of convergence for the power series

$$\sum_{n=0}^{\infty} n! x^n = 1 + x + 2x^2 + 6x^3 + \cdots.$$

Solution

For any fixed nonzero value of x, we apply the Ratio Test:

$$\lim_{n \to \infty} \left| \frac{(n+1)! \, x^{n+1}}{n! \, x^n} \right| = \lim_{n \to \infty} (n+1) |x| = \infty.$$

Consequently, for all $x \neq 0$, this series diverges. The radius of convergence is therefore zero, and the series converges only at $x = 0$. ▶

Example 3

Find the domain of the function defined by the power series

$$f(x) = \sum_{n=0}^{\infty} \frac{3^n}{n^3 + 1} x^n = 1 + \frac{3}{2} x + x^2 + \frac{27}{28} x^3 + \cdots.$$

Solution

The domain of f is the interval of convergence for the power series. To determine this, we apply the Ratio Test:

$$\lim_{n \to \infty} \left| \frac{\frac{3^{n+1}}{(n+1)^3 + 1} x^{n+1}}{\frac{3^n}{n^3 + 1} x^n} \right| = \lim_{n \to \infty} \frac{n^3 + 1}{(n+1)^3 + 1} \, 3|x| = 3|x|.$$

Consequently, the series converges if $3|x| < 1$, that is, if $|x| < \frac{1}{3}$. The series diverges if $|x| > \frac{1}{3}$. This implies that the series converges if $-\frac{1}{3} < x < \frac{1}{3}$, and diverges if $x > \frac{1}{3}$ or if $x < -\frac{1}{3}$.

The Ratio Test gives no convergence information at $x = \frac{1}{3}$ and at $x = -\frac{1}{3}$ since the limit of the ratio in these cases is 1. To determine whether the series converges at these values, we examine them individually and apply other tests.

At $x = \frac{1}{3}$,

$$\sum_{n=0}^{\infty} \frac{3^n \left(\frac{1}{3}\right)^n}{n^3 + 1} = \sum_{n=0}^{\infty} \frac{1}{n^3 + 1},$$

which converges by comparison with the p-series $\sum_{n=1}^{\infty} 1/n^3$.

At $x = -\frac{1}{3}$,

$$\sum_{n=0}^{\infty} \frac{3^n (-\frac{1}{3})^n}{n^3 + 1} = \sum_{n=0}^{\infty} \frac{(-1)^n}{n^3 + 1}$$

converges absolutely, hence converges.

The interval of convergence of the series, and consequently the domain of f, is $[-1/3, 1/3]$. ▶

A **power series in $x - a$**, where a is a constant, has the form

(8.43) $$\sum_{n=0}^{\infty} a_n (x - a)^n.$$

This power series is said to be *centered at a*; the power series we have considered thus far in the section have been centered at zero.

Example 4

Find the interval of convergence of the power series

$$\sum_{n=1}^{\infty} (-1)^{n+1} \frac{(x-1)^n}{n} = (x-1) - \frac{1}{2}(x-1)^2 + \frac{1}{3}(x-1)^3 - \cdots.$$

Solution

We apply the Ratio Test to the series:

$$\lim_{n\to\infty} \left| \frac{(-1)^{n+2} \frac{(x-1)^{n+1}}{n+1}}{(-1)^{n+1} \frac{(x-1)^n}{n}} \right| = \lim_{n\to\infty} \frac{n}{n+1} |x-1| = |x-1|.$$

Thus the series converges if $|x-1| < 1$; that is, if x is in $(0, 2)$. The series diverges if $|x-1| > 1$, that is, if $x < 0$ or if $x > 2$. To determine the convergence at $x = 0$ and at $x = 2$, we examine these cases individually.

At $x = 0$, the series diverges since

$$\sum_{n=1}^{\infty} (-1)^{n+1} \frac{(-1)^n}{n} = \sum_{n=1}^{\infty} (-1)^{2n+1} \frac{1}{n} = -\sum_{n=1}^{\infty} \frac{1}{n}.$$

At $x = 2$, the series converges since

$$\sum_{n=1}^{\infty} \frac{(-1)^{n+1} (1)^n}{n} = \sum_{n=1}^{\infty} \frac{(-1)^{n+1}}{n}.$$

The interval of convergence is $(0, 2]$. ▶

Example 5

Find the domain of the function defined by

$$f(x) = \sum_{n=0}^{\infty} \frac{(2x-1)^n}{3^n}.$$

Solution

Since

$$\lim_{n\to\infty} \left| \frac{\frac{(2x-1)^{n+1}}{3^{n+1}}}{\frac{(2x-1)^n}{3^n}} \right| = \lim_{n\to\infty} \frac{|2x-1|}{3} = \frac{|2x-1|}{3},$$

the series converges if

$$\frac{|2x-1|}{3} < 1, \quad \text{that is, if} \quad |2x-1| < 3.$$

The inequality $|2x - 1| < 3$ is equivalent to $-3 < 2x - 1 < 3$, hence to $-2 < 2x < 4$. The series converges if $-1 < x < 2$ and diverges if $x < -1$ or $x > 2$.

At $x = -1$,
$$\sum_{n=0}^{\infty} \frac{(-3)^n}{3^n} = \sum_{n=0}^{\infty} (-1)^n$$
diverges since the terms of the series do not approach zero.

Similarly, at $x = 2$,
$$\sum_{n=0}^{\infty} \frac{3^n}{3^n} = \sum_{n=0}^{\infty} 1$$
diverges. Consequently, the domain of f is $(-1, 2)$. ▶

The result in Example 5 can also be found by first finding the domain of the function g, where g is defined by the power series centered at zero:
$$g(x) = \sum_{n=0}^{\infty} \frac{(2x)^n}{3^n} = \sum_{n=0}^{\infty} \left(\frac{2}{3}\right)^n x^n.$$

Since
$$f(x) = \sum_{n=0}^{\infty} \frac{(2x-1)^n}{3^n} = \sum_{n=0}^{\infty} \left(\frac{2}{3}\right)^n \left(x - \frac{1}{2}\right)^n = g\left(x - \frac{1}{2}\right),$$
the domain of f is the translation of the domain of g one half unit to the right. Applying the Ratio Test to $g(x)$ gives
$$\lim_{n \to \infty} \left| \frac{\left(\frac{2}{3}\right)^{n+1} x^{n+1}}{\left(\frac{2}{3}\right)^n x^n} \right| = \frac{2}{3}|x|.$$

This series converges if $|x| < \frac{3}{2}$ and diverges if $|x| > \frac{3}{2}$. At $x = -\frac{3}{2}$,
$$\sum_{n=1}^{\infty} \left(\frac{2}{3}\right)^n \left(-\frac{3}{2}\right)^n = \sum_{n=1}^{\infty} (-1)^n \quad \text{diverges}$$
and at $x = \frac{3}{2}$,
$$\sum_{n=1}^{\infty} \left(\frac{2}{3}\right)^n \left(\frac{3}{2}\right)^n = \sum_{n=1}^{\infty} 1 \quad \text{also diverges.}$$

Thus the domain of g is $(-\frac{3}{2}, \frac{3}{2})$, which implies that the domain of f is $(-\frac{3}{2} + \frac{1}{2}, \frac{3}{2} + \frac{1}{2}) = (-1, 2)$.

▶ EXERCISE SET 8.6

In Exercises 1–26, find (a) the radius of convergence and (b) the interval of convergence of the power series.

1. $\sum_{n=0}^{\infty} \frac{x^n}{n+1}$
2. $\sum_{n=1}^{\infty} \frac{x^n}{\sqrt{n}}$
3. $\sum_{n=1}^{\infty} \frac{(\ln n) x^n}{n^2}$
4. $\sum_{n=1}^{\infty} \frac{x^n}{n(n+1)}$

5. $\sum_{n=0}^{\infty} \dfrac{x^n}{n!}$

6. $\sum_{n=1}^{\infty} \dfrac{(1001)^n x^n}{n!}$

7. $\sum_{n=1}^{\infty} \dfrac{x^{n+1}}{n^2}$

8. $\sum_{n=0}^{\infty} \dfrac{x^{2n}}{n+1}$

9. $\sum_{n=0}^{\infty} \dfrac{(-1)^n x^n}{n!}$

10. $\sum_{n=0}^{\infty} \dfrac{(-1)^{n+1} x^{2n}}{(2n)!}$

11. $\sum_{n=0}^{\infty} \dfrac{(-1)^{n+1} x^{2n+1}}{(2n+1)!}$

12. $\sum_{n=2}^{\infty} \dfrac{x^{2n+1}}{\ln n}$

13. $\sum_{n=1}^{\infty} \dfrac{2^n x^n}{n}$

14. $\sum_{n=0}^{\infty} \dfrac{n^n x^n}{(2n)!}$

15. $\sum_{n=5}^{\infty} \dfrac{x^{n-4}}{\sqrt{n+4}}$

16. $\sum_{n=1}^{\infty} \dfrac{(3x)^n}{\arctan n}$

17. $\sum_{n=0}^{\infty} (x-2)^n$

18. $\sum_{n=0}^{\infty} \dfrac{(x-2)^n}{n+1}$

19. $\sum_{n=0}^{\infty} (n+2)^2 (x-2)^n$

20. $\sum_{n=0}^{\infty} n!(x-2)^n$

21. $\sum_{n=1}^{\infty} \dfrac{(x+3)^{2n}}{n}$

22. $\sum_{n=1}^{\infty} \dfrac{(\ln n)(x+1)^{3n}}{n(n+1)}$

23. $\sum_{n=1}^{\infty} \dfrac{(3x-1)^n}{n^2}$

24. $\sum_{n=0}^{\infty} \dfrac{n(2x-5)^n}{n+1}$

25. $\sum_{n=1}^{\infty} \dfrac{(2x+1)^n}{\sqrt{n}}$

26. $\sum_{n=0}^{\infty} \dfrac{(3x+4)^{2n}}{2^n}$

In Exercises 27–32, find the domain of the function.

27. $f(x) = \sum_{n=2}^{\infty} \dfrac{x^n}{n \ln n}$

28. $f(x) = \sum_{n=1}^{\infty} \dfrac{(-1)^n x^{2n}}{n\sqrt{n+1}}$

29. $f(x) = \sum_{n=0}^{\infty} \dfrac{(x-1)^n}{n^2+2}$

30. $f(x) = \sum_{n=1}^{\infty} \dfrac{(3x+1)^n}{\ln(n+1)}$

31. $f(x) = \sum_{n=0}^{\infty} \dfrac{(2x-3)^n}{n^2+1}$

32. $f(x) = \sum_{n=1}^{\infty} \dfrac{(x+7)^{2n} \ln n}{(n+1) \arctan n}$

In Exercises 33–40, find the values of x for which the series converges.

33. $f(x) = \sum_{n=0}^{\infty} x^n (x-2)^n$

34. $f(x) = \sum_{n=0}^{\infty} (x-1)^n (x+1)^n$

35. $\sum_{n=0}^{\infty} \dfrac{1}{x^n}$

36. $\sum_{n=0}^{\infty} \dfrac{n}{x^n}$

37. $\sum_{n=0}^{\infty} \dfrac{2^n}{x^n}$

38. $\sum_{n=0}^{\infty} \dfrac{3^n}{(x-1)^n}$

39. $\sum_{n=0}^{\infty} (\cos x)^n$

40. $\sum_{n=1}^{\infty} \dfrac{(\tan x)^n}{n}$

41. Find the interval of convergence of the series $\sum_{n=0}^{\infty} [2 + (-1)^n] x^n$.

42. Suppose the Ratio Test applied to $\sum_{n=0}^{\infty} a_n x_0^n$ gives
$$\lim_{n \to \infty} \left| \dfrac{a_{n+1}}{a_n} x_0 \right| < 1.$$
Show that $\sum_{n=0}^{\infty} n^k a_n x^n$ converges absolutely for any integer k and every value of x with $|x| < |x_0|$.

43. Suppose the power series $\sum_{n=0}^{\infty} a_n x^n$ and $\sum_{n=0}^{\infty} b_n x^n$ have radius of convergence R_1 and R_2, respectively. What can be said about the radius of convergence R of $\sum_{n=0}^{\infty} (a_n + b_n) x^n$?

44. Suppose the series $\sum_{n=0}^{\infty} a_n x^n$ has the property that $1/L < a_n < L$ for all n, where L is a positive constant.
   a) Show that the radius of convergence of this series is 1.
   b) What is the interval of convergence of such a series?

45. Find the interval of convergence of the power series
$$\sum_{n=1}^{\infty} \dfrac{n^n x^n}{n!}.$$
[Hint: When considering convergence at the endpoints, use Stirling's Formula, given in Exercise 60 of Section 8.3.]

8.7

Differentiation and Integration of Power Series

Power series are generalizations of polynomials. A polynomial

$$P(x) = a_0 + a_1 x + \cdots + a_m x^m = \sum_{n=0}^{m} a_n x^n$$

can be differentiated and integrated easily:

$$D_x P(x) = a_1 + 2a_2 x + \cdots + m a_m x^{m-1} = \sum_{n=1}^{m} n a_n x^{n-1}$$

and

$$\int_0^x P(t)\, dt = a_0 x + \frac{a_1 x^2}{2} + \cdots + \frac{a_m x^{m+1}}{m+1} = \sum_{n=0}^{m} \frac{a_n x^{n+1}}{n+1}.$$

The same type of differentiation and integration properties hold for functions defined by power series.

(8.44)
THEOREM

If f is a function defined by the power series

$$f(x) = \sum_{n=0}^{\infty} a_n x^n = a_0 + a_1 x + a_2 x^2 + a_3 x^3 + \cdots$$

with radius of convergence R, then:

i) f is differentiable at each x in $(-R, R)$, and

$$D_x f(x) = D_x \sum_{n=0}^{\infty} a_n x^n = \sum_{n=1}^{\infty} n a_n x^{n-1} = a_1 + 2a_2 x + 3a_3 x^2 + \cdots.$$

ii) For each x in $(-R, R)$, $\int_0^x f(t)\, dt$ exists and

$$\int_0^x f(t)\, dt = \int_0^x \left[\sum_{n=0}^{\infty} a_n t^n \right] dt = \sum_{n=0}^{\infty} \frac{a_n}{n+1} x^{n+1}$$

$$= a_0 x + \frac{a_1}{2} x^2 + \frac{a_2}{3} x^3 + \cdots.$$

The proof of this theorem, especially part (i), is rather long but not beyond the scope of this course. This proof is included in Appendix B, and is instructive for reviewing the definition of the derivative and the Fundamental Theorem of Calculus.

Example 1

Use Theorem (8.44) and the fact that

$$\frac{1}{1-x} = \sum_{n=0}^{\infty} x^n, \quad \text{for } |x| < 1$$

to find a power-series representation for $f(x) = 1/(1-x)^2$.

Solution

Finding a power-series representation for $f(x)$ means finding a series $\sum_{n=0}^{\infty} a_n x^n$ such that $f(x) = \sum_{n=0}^{\infty} a_n x^n$ for each x in the interval of convergence of the power series. Since

$$D_x \left(\frac{1}{1-x} \right) = \frac{1}{(1-x)^2},$$

Thus $f(x)$ has the power-series representation

$$f(x) = \frac{e^x - 1}{x} = \sum_{n=0}^{\infty} \frac{x^n}{(n+1)!},$$

provided $x \neq 0$. ▶

Example 5

Approximate $\int_0^{0.1} e^{-x^2} dx$ to within 10^{-8}.

Solution
The function given by $f(x) = e^{-x^2}$ is continuous, hence integrable. We cannot use the Fundamental Theorem of Calculus for the evaluation of its integral, since f has no elementary antiderivative. However, by modifying the power series in Example 3, we can express this function as

$$e^{-x^2} = \sum_{n=0}^{\infty} \frac{(-x^2)^n}{n!} = \sum_{n=0}^{\infty} \frac{(-1)^n x^{2n}}{n!}$$

and

$$\int_0^x e^{-t^2} dt = \sum_{n=0}^{\infty} \frac{(-1)^n x^{2n+1}}{n!(2n+1)}.$$

Thus

$$\int_0^{0.1} e^{-t^2} dt = \sum_{n=0}^{\infty} \frac{(-1)^n (0.1)^{2n+1}}{n!(2n+1)},$$

which is an alternating series. By Corollary (8.32) the mth partial sum approximates the true value with error not exceeding the magnitude of the $(m+1)$st term:

$$\frac{(0.1)^{2m+3}}{(m+1)!(2m+3)}.$$

Table 8.2 shows that the first integer for which this value does not exceed 10^{-8} is $m = 2$. This approximation is

$$\int_0^{0.1} e^{-x^2} dx \approx \sum_{n=0}^{2} \frac{(-1)^n (0.1)^{2n+1}}{n!(2n+1)}$$

$$= 0.1 - \frac{0.001}{3} + \frac{0.00001}{10} = 0.09966767.$$

The error in the approximation is at most the magnitude of the third term of this alternating series:

$$\frac{(0.1)^{2(2)+3}}{(2+1)!(2 \cdot 2 + 3)} = \frac{(0.1)^7}{42} \approx 2.4 \times 10^{-9}. \quad ▶$$

Table 8.2

m	$\dfrac{(0.1)^{2m+3}}{(m+1)!(2m+3)}$
0	3.3×10^{-4}
1	1.0×10^{-6}
2	2.4×10^{-9}
3	9.3×10^{-12}

EXERCISE SET 8.7

In Exercises 1–20, find a power-series representation for the function. Use the fact that $1/(1-x) = \sum_{n=0}^{\infty} x^n$ for $|x| < 1$, and $e^x = \sum_{n=0}^{\infty} x^n/n!$ for all x, and the arithmetic and calculus properties of power series.

1. $f(x) = \dfrac{1}{1+x}$

2. $f(x) = \dfrac{1}{1-2x}$

3. $f(x) = \dfrac{1}{2-x}$

4. $f(x) = \dfrac{1}{3-2x}$

5. $f(x) = \dfrac{1}{1+x^2}$

6. $f(x) = \arctan x$

7. $f(x) = \dfrac{x}{1+x^2}$

8. $f(x) = \ln(1+x^2)$

9. $f(x) = \dfrac{1}{1-x^2}$

10. $f(x) = \dfrac{1+x^2}{1-x^2}$

11. $f(x) = \dfrac{1}{(1+x)^2}$

12. $f(x) = \ln(1+x)$

13. $f(x) = \dfrac{e^x - 1}{x}$

14. $f(x) = \dfrac{x}{1-x}$

15. $f(x) = x \ln(1+x)$

16. $f(x) = \ln\left(\dfrac{1+x}{1+x^2}\right)$

17. $f(x) = \dfrac{2x}{1-3x^2}$

18. $f(x) = \displaystyle\int_0^x e^{t^2} \, dt$

19. $f(x) = \displaystyle\int_0^x t e^{t^3} \, dt$

20. $f(x) = \displaystyle\int_0^x [e^t + \ln(t+1)] \, dt$

21. Show that if $0 < x < 1$, then
$$\left| \dfrac{1}{1+x} - \sum_{i=0}^{n} (-1)^i x^i \right| < x^{n+1}.$$

22. **a)** Show that
$$\int_0^1 \dfrac{1}{1+t^2} \, dt = \sum_{i=0}^{\infty} (-1)^i \dfrac{1}{2i+1}.$$

b) Show that
$$\left| \dfrac{\pi}{4} - \sum_{i=0}^{n} (-1)^i \dfrac{1}{2i+1} \right| < \dfrac{1}{2n+3}.$$

In Exercises 23–26, use a power-series representation to approximate the definite integral to within 10^{-4}.

23. $\displaystyle\int_0^{0.2} e^{-x^2} \, dx$

24. $\displaystyle\int_0^{0.5} \dfrac{1}{1+x^3} \, dx$

25. $\displaystyle\int_0^{0.1} \ln(x+1) \, dx$

26. $\displaystyle\int_{-0.01}^{0} \dfrac{(e^{-x} + x - 1)}{x^2} \, dx$

27. Find a power-series representation and the interval of convergence for each of the following.

a) $f_1(x) = \dfrac{1}{3+4x}$ **b)** $f_2(x) = \dfrac{x}{3+4x}$

c) $f_3(x) = \dfrac{1}{(3+4x)^2}$ **d)** $f_4(x) = \ln(3+4x)$

28. Find a power-series representation and the interval of convergence for each of the following.

a) $f_1(x) = \dfrac{1}{2+x^2}$ **b)** $f_2(x) = \dfrac{x}{2+x^2}$

c) $f_3(x) = \dfrac{x}{(2+x^2)^2}$ **d)** $f_4(x) = \ln(2+x^2)$

29. Use the technique of partial fractions to decompose $1/(1-x^2)$ and obtain a power-series representation. Compare this series with the series found in Exercise 9.

30. Use the technique of partial fractions to decompose $(1+x^2)/(1-x^2)$ and obtain a power-series representation. Compare this series with the series found in Exercise 10.

31. Use integration by parts to find $\int \ln(1+x) \, dx$, and use this result to find a power-series representation for $x \ln(1+x)$. Compare this result with the result in Exercise 15.

32. **a)** Use the series for e^x and e^{-x} to obtain series for $\cosh x$ and $\sinh x$.

b) Use the series obtained in part (a) to show that $D_x \cosh x = \sinh x$ and $D_x \sinh x = \cosh x$.

33. Suppose f is a function with a power-series representation of the form $f(x) = \sum_{n=0}^{\infty} a_n x^n$. If $f(0) = 0$ and $f'(x) = f(x)$ for all x, determine the constants in the power-series representation.

34. Suppose f is a function with a power-series representation of the form $f(x) = \sum_{n=0}^{\infty} a_n x^n$. If $f(0) = 0$, $f'(0) = 1$ and $f''(x) = -f(x)$ for all x, determine the constants in the power-series representation.

35. Consider the functions f and g defined by
$$f(x) = \sum_{n=0}^{\infty} \dfrac{(-1)^{n+1} x^{2n+1}}{(2n+1)!} \quad \text{and} \quad g(x) = \sum_{n=0}^{\infty} \dfrac{(-1)^n x^{2n}}{(2n)!}.$$
Compute $f'(x)$ and $g'(x)$ (and be observant).

36. Consider the series given in Example 2 for $\ln(1-x)$. Let $x = \tfrac{1}{2}$, and show that $\ln 2$ can also be expressed as
$$\ln 2 = \sum_{n=0}^{\infty} \dfrac{1}{2^{n+1}(n+1)}.$$

37. Since $\tan \pi/4 = 1$, π can be found by evaluating

4 arctan 1.
a) Find a power series for arctan x.
b) Determine the number of terms of the series that need to be summed in order to ensure that $|4 \arctan 1 - \pi| < 10^{-3}$.
c) The single precision version of the scientific programming language FORTRAN requires the value of π to be within 10^{-7}. How many terms of this series must be summed in order to obtain this degree of accuracy?

38. Exercise 37 details a rather inefficient means of obtaining an approximation to π. The method can be improved substantially by observing that $\pi/4 = \arctan \frac{1}{2} + \arctan \frac{1}{3}$ (see Exercise 81a in Section 6.2) and evaluating the series for arctan at $\frac{1}{2}$ and at $\frac{1}{3}$. Determine the number of terms that must be summed in order to ensure an approximation to π to within 10^{-3}.

39. Another formula for computing π can be deduced from the identity $\pi/4 = 4 \arctan \frac{1}{5} - \arctan \frac{1}{239}$ (see Exercise 81b of Section 6.2). Determine the number of terms that must be summed in order to ensure an approximation to π to within 10^{-3}.

40. The mathematician and astronomer Friedrich Wilhelm Bessel (1784–1846) used the differential equation $x^2 y'' + xy' + (x^2 - m^2)y = 0$, where m is a positive integer, in his study of planetary motion. The solution to this equation is known as the *Bessel function* of order m, defined by the series

$$J_m(x) = \sum_{n=0}^{\infty} \frac{(-1)^n x^{2n+m}}{n!(m+n)! 2^{2n+m}}.$$

Show that J_m satisfies the differential equation. (The Bessel functions are also important in the study of heat flow in cylinders, vibrations in circular membranes, the vibration of chains, and many other mathematical applications.)

Putnam exercises

41. For $0 < x < 1$, express

$$\sum_{n=0}^{\infty} \frac{x^{2n}}{1 - x^{2n+1}}$$

as a rational function of x. (This exercise was problem A–4 of the thirty-eighth William Lowell Putnam examination given on December 3, 1977. The examination and its solution are in the March 1979 issue of the *American Mathematical Monthly*, pp. 170–175.)

42. Evaluate in closed form

$$\sum_{k=1}^{n} \binom{n}{k} k^2.$$

Note that

$$\binom{n}{k} = \frac{n(n-1) \cdots (n-k+1)}{1 \cdot 2 \cdots k}.$$

(This exercise was problem 5, part I of the twenty-third William Lowell Putnam examination given on December 1, 1962. The examination and its solution are in the September 1963 issue of the *American Mathematical Monthly*, pp. 713–717.)

8.8
Taylor Polynomials and Taylor Series

Example 3 in Section 8.7 demonstrated that the natural exponential function can be represented by a power series that converges for all real numbers. In this section we discuss power-series representations for other functions. Questions to consider are:

i) If a function has a power-series representation, what is it?
ii) Which functions have power-series representations?

First consider Question (i): Suppose a function f has a power-series representation of the form

$$f(x) = \sum_{n=0}^{\infty} a_n x^n = a_0 + a_1 x + a_2 x^2 + \cdots$$

with radius of convergence $R > 0$. We need to determine the constants $a_0, a_1, a_2, \ldots$.

When f is evaluated at zero, all terms of the series but the first vanish, so

$$a_0 = f(0).$$

Using Theorem (8.44) to differentiate f gives

$$f'(x) = \sum_{n=1}^{\infty} n a_n x^{n-1} = a_1 + 2a_2 x + 3a_3 x^2 + \cdots,$$

so

$$a_1 = f'(0).$$

This procedure can be continued by repeated differentiation:

$$f''(x) = \sum_{n=2}^{\infty} n(n-1) a_n x^{n-2},$$

so

$$f''(0) = 2 \cdot 1 a_2 \quad \text{and} \quad a_2 = \frac{f''(0)}{2};$$

and

$$f'''(x) = \sum_{n=3}^{\infty} n(n-1)(n-2) a_n x^{n-3},$$

so

$$f'''(0) = 3 \cdot 2 \cdot 1 a_3 \quad \text{and} \quad a_3 = \frac{f'''(0)}{6}.$$

In general,

$$a_n = \frac{f^{(n)}(0)}{n!}$$

for all nonnegative integers n, if we define $0! = 1$ and let $f^{(0)}(x) = f(x)$.

This answers Question (i).

(8.45)
THEOREM
If a function f has a power-series representation $\sum_{n=0}^{\infty} a_n x^n$ with radius of convergence $R > 0$, then this power series is unique and is given by

$$f(x) = \sum_{n=0}^{\infty} \frac{f^{(n)}(0)}{n!} x^n$$

for x in $(-R, R)$.

The power-series representation given in Theorem (8.45) is called the **Maclaurin series** for f. The generalization to a power series centered at an arbitrary value a is contained in the following corollary. This series is called the **Taylor series** for f at a. Thus the Maclaurin series for f is the Taylor series centered at $a = 0$.

(8.46)
COROLLARY
If f has a power-series representation centered at a with radius of convergence $R > 0$, then the power series at a is unique and is given by

$$f(x) = \sum_{n=0}^{\infty} \frac{f^{(n)}(a)}{n!}(x-a)^n$$

for x in $(a - R, a + R)$.

Although these results are important, they do not completely answer the questions about functions and their power-series representations. They tell us how to find a power-series representation when one is known to exist, but not *when* we expect to find such a representation. The following result is more important in this regard and helps to answer the second question posed at the beginning of this section.

(8.47)
THEOREM: TAYLOR'S THEOREM WITH REMAINDER
If a function f has $n + 1$ derivatives on an open interval I about a and x is in this interval, then a number ξ between x and a exists with

$$f(x) = P_n(x) + R_n(x),$$

where

$$P_n(x) = \sum_{i=0}^{n} \frac{f^{(i)}(a)}{i!}(x-a)^i$$

and

$$R_n(x) = \frac{f^{(n+1)}(\xi)}{(n+1)!}(x-a)^{n+1}.$$

Proof. Assume that $P_n(x)$ is defined as stated in the theorem and that $R_n(x)$ is defined by

$$R_n(x) = f(x) - P_n(x)$$

for an arbitrary, but fixed, value of x in I. We will show that $R_n(x)$ must have the representation given in the theorem.

If $x = a$, then $f(x) = P_n(x)$ and $R_n(x) = 0$, so the result is established.

HISTORICAL NOTE

Brook Taylor (1685–1731) discussed series of this type in *Methodus incrementorum directa et inversa*, published in 1715. Special cases of this result, perhaps even the result itself, had been known much earlier to Newton, Gregory, and others.

Colin Maclaurin (1698–1746) is best remembered for his strong defense of the calculus techniques of Newton, which came under attack in 1734 by a nonmathematician, Bishop George Berkeley. The result bearing Maclaurin's name appeared in his work *Treatise on Fluxions* published in 1742, but was known at least as early as Taylor's work in 1715.

Although Maclaurin did not discover the series that bears his name, he did devise, in about 1729, a means of finding solutions to systems of linear equations. The method is known as Cramer's rule, although Cramer did not publish it until 1750.

If $x \neq a$, the proof proceeds by applying Rolle's Theorem to the function g in the independent variable t, where

$$g(t) = f(x) - \left[f(t) + f'(t)(x-t) + f''(t)\frac{(x-t)^2}{2} \right.$$
$$\left. + \cdots + f^{(n)}(t)\frac{(x-t)^n}{n!} \right] - R_n(x)\frac{(x-t)^{n+1}}{(x-a)^{n+1}}.$$

First note that if $t = x$, then all the terms involving $(x - t)$ are zero, so

$$g(x) = f(x) - f(x) = 0.$$

In addition, if $t = a$, then

$$g(a) = f(x) - \left[f(a) + f'(a)(x-a) + f''(a)\frac{(x-a)^2}{2} \right.$$
$$\left. + \cdots + f^{(n)}(a)\frac{(x-a)^n}{n!} \right] - R_n(x)\frac{(x-a)^{n+1}}{(x-a)^{n+1}}$$
$$= f(x) - P_n(x) - R_n(x) = 0.$$

Since f has $n + 1$ derivatives on I, g is differentiable, hence continuous on the closed interval with endpoints a and x. It follows from Rolle's Theorem that a number ξ between x and a exists with $g'(\xi) = 0$. Keeping in mind that x is fixed, we find that

$$g'(t) = 0 - \left\{ f'(t) + [f''(t)(x-t) - f'(t)] + \left[f'''(t)\frac{(x-t)^2}{2} \right.\right.$$
$$\left. -f''(t)(x-t) \right] + \left[f^{(4)}(t)\frac{(x-t)^3}{3!} - f'''(t)\frac{(x-t)^2}{2} \right]$$
$$+ \cdots + \left[f^{(n+1)}(t)\frac{(x-t)^n}{n!} - f^{(n)}(t)\frac{(x-t)^{n-1}}{(n-1)!} \right] \right\} + R_n(x)\frac{(n+1)(x-t)^n}{(x-a)^{n+1}}$$
$$= \frac{-f^{(n+1)}(t)(x-t)^n}{n!} + \frac{R_n(x)(n+1)(x-t)^n}{(x-a)^{n+1}} = -(x-t)^n\left[\frac{f^{(n+1)}(t)}{n!} - R_n(x)\frac{(n+1)}{(x-a)^{n+1}} \right].$$

So

$$0 = g'(\xi) = -(x-\xi)^n \left[\frac{f^{(n+1)}(\xi)}{n!} - R_n(x)\frac{(n+1)}{(x-a)^{n+1}} \right],$$

which implies, since $x \neq \xi$, that

$$\frac{f^{(n+1)}(\xi)}{n!} = R_n(x)\frac{(n+1)}{(x-a)^{n+1}}.$$

Thus,

$$R_n(x) = \frac{f^{(n+1)}(\xi)}{(n+1)!}(x-a)^{n+1}.$$

▷

8.8 TAYLOR POLYNOMIALS AND TAYLOR SERIES

The polynomial P_n, of degree at most n, defined by

(8.48) $$P_n(x) = \sum_{i=0}^{n} \frac{f^{(i)}(a)}{i!}(x-a)^i$$

is called the **nth Taylor polynomial for f at a,** and

(8.49) $$R_n(x) = \frac{f^{(n+1)}(\xi)}{(n+1)!}(x-a)^{n+1}$$

is called the **remainder** for this polynomial. When $a = 0$, P_n is also called the **nth Maclaurin polynomial for f.**

Example 1

Find the second Maclaurin polynomial for $f(x) = \cos x$, and use the remainder to find a bound for the error introduced when this polynomial is used to approximate $\cos 0.01$.

Solution
Since $f(x) = \cos x$, $f'(x) = -\sin x$, $f''(x) = -\cos x$, and $f'''(x) = \sin x$, we have

$$P_2(x) = f(0) + f'(0) \cdot x + \frac{f''(0)}{2}x^2 = 1 + 0 \cdot x + \frac{(-1)}{2}x^2 = 1 - \frac{x^2}{2}$$

and

$$R_2(x) = \frac{f'''(\xi)}{6}x^3 = \frac{x^3}{6}\sin \xi$$

for some number ξ between x and zero. (Figure 8.14 shows the graph of P_2 and its relation to the graph of f.) Thus,

$$\cos 0.01 \approx P_2(0.01) = 1 - \frac{(0.01)^2}{2} = 0.99995.$$

Since $|\sin \xi| < 1$, the maximum error in this approximation is

$$|R_2(0.01)| = \left|\frac{(0.01)^3}{6}\sin \xi\right| \le \frac{(0.01)^3}{6} \approx 1.67 \times 10^{-7}.$$

Consequently, $\cos 0.01 \approx 0.999950$, where all six decimal places listed are correct. ▶

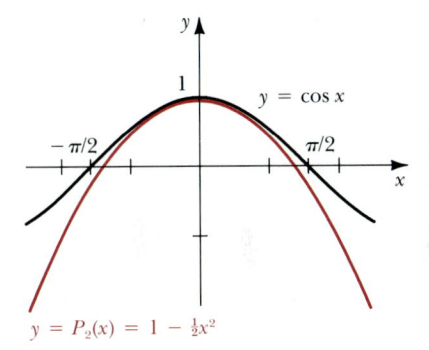

Figure 8.14
The second Maclaurin polynomial for $\cos x$.

Example 2

Repeat the calculations in Example 1 using instead the third Maclaurin polynomial for $f(x) = \cos x$.

Solution
The third Maclaurin polynomial is given by

$$P_3(x) = 1 - \frac{x^2}{2} + \frac{f'''(0)}{6}x^3 = 1 - \frac{x^2}{2} + \frac{0}{6}x^3 = 1 - \frac{x^2}{2},$$

the same expression as $P_2(x)$. Hence,
$$\cos 0.01 \approx P_3(0.01) = 0.99995.$$

Before dismissing this example as superfluous, however, consider the remainder term $R_3(x)$. Since $f^{(4)}(x) = \cos x$ and $|\cos \xi| \leq 1$,

$$|R_3(0.01)| = \left|\frac{(0.01)^4}{24} \cos \xi\right| \leq \frac{(0.01)^4}{24} \approx 4.17 \times 10^{-10}.$$

This shows that the approximation to $\cos 0.01$ is much more accurate than deduced in Example 1. In fact,

$$\cos 0.01 \approx 0.999950000$$

is correct to all nine decimal places listed. ▶

Example 3

Determine the first five Taylor polynomials for $f(x) = \ln x$ at $a = 1$.

Solution
Since
$$f(x) = \ln x, \qquad f(1) = 0;$$
$$f'(x) = 1/x, \qquad f'(1) = 1;$$
$$f''(x) = -1/x^2, \qquad f''(1) = -1;$$
$$f'''(x) = 2/x^3, \qquad f'''(1) = 2;$$

and
$$f^{(4)}(x) = -6/x^4, \qquad f^{(4)}(1) = -6.$$

The first five Taylor polynomials for f at $a = 1$ are

$P_0(x) = 0,$

$P_1(x) = 0 + 1 \cdot (x - 1) = x - 1,$

$P_2(x) = 0 + 1 \cdot (x - 1) + \dfrac{-1}{2!}(x - 1)^2 = (x - 1) - \dfrac{1}{2}(x - 1)^2,$

$P_3(x) = (x - 1) - \dfrac{1}{2}(x - 1)^2 + \dfrac{2}{3!}(x - 1)^3$

$ = (x - 1) - \dfrac{1}{2}(x - 1)^2 + \dfrac{1}{3}(x - 1)^3,$

and

$P_4(x) = (x - 1) - \dfrac{1}{2}(x - 1)^2 + \dfrac{1}{3}(x - 1)^3 + \dfrac{-6}{4!}(x - 1)^4$

$ = (x - 1) - \dfrac{1}{2}(x - 1)^2 + \dfrac{1}{3}(x - 1)^3 - \dfrac{1}{4}(x - 1)^4.$ ▶

The graphs of these polynomials, together with the graph of f, are shown in Figure 8.15. The higher-degree Taylor polynomials agree

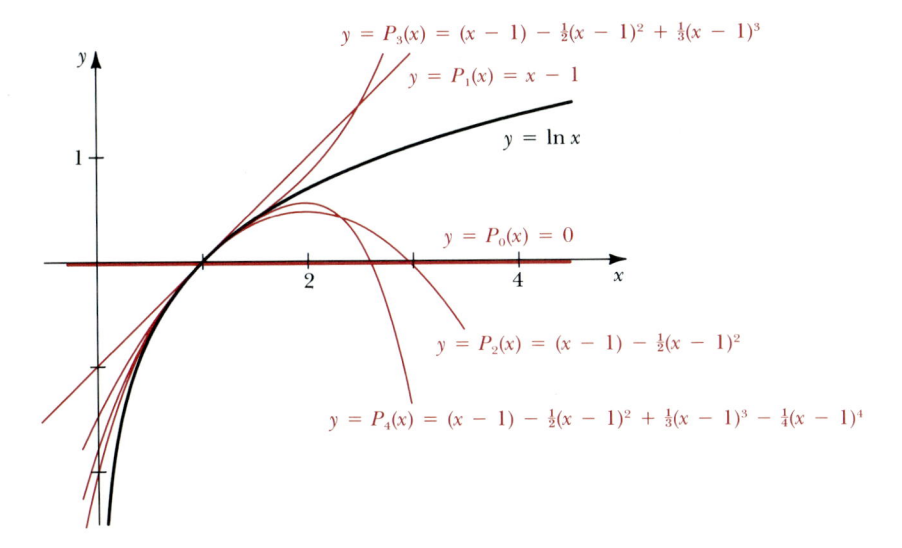

$$y = P_3(x) = (x - 1) - \tfrac{1}{2}(x - 1)^2 + \tfrac{1}{3}(x - 1)^3$$
$$y = P_1(x) = x - 1$$
$$y = \ln x$$
$$y = P_0(x) = 0$$
$$y = P_2(x) = (x - 1) - \tfrac{1}{2}(x - 1)^2$$
$$y = P_4(x) = (x - 1) - \tfrac{1}{2}(x - 1)^2 + \tfrac{1}{3}(x - 1)^3 - \tfrac{1}{4}(x - 1)^4$$

Figure 8.15

Taylor polynomials about $a = 1$ for $\ln x$.

quite closely with f on $(0, 2)$, but deviate dramatically outside this interval. This is due to the fact that the Taylor series for $\ln x$ at $a = 1$ has radius of convergence 1, as shown in Example 2 of Section 8.7.

The following corollary to Taylor's Theorem provides the answer to the second question posed at the beginning of the section:

ii) Which functions have power-series representations?

(8.50)
COROLLARY

If a function f has derivatives of all orders on an open interval about a and if

$$\lim_{n \to \infty} R_n(x) = 0$$

for all x in that interval, then f has the Taylor-series representation

$$f(x) = \sum_{n=0}^{\infty} \frac{f^{(n)}(a)}{n!} (x - a)^n$$

for all x in the interval.

Example 4

Find the Maclaurin series for the sine and cosine functions.

Solution
Letting $f(x) = \sin x$, we have

$$\begin{aligned}
f(x) &= \sin x, & f(0) &= 0; \\
f'(x) &= \cos x, & f'(0) &= 1; \\
f''(x) &= -\sin x, & f''(0) &= 0; \\
f'''(x) &= -\cos x, & f'''(0) &= -1; \\
f^{(4)}(x) &= \sin x, & f^{(4)}(0) &= 0.
\end{aligned}$$

In general, $f^{(n)}(0)$ is zero when n is even and is alternately 1 and -1 when n is odd. The Maclaurin series is consequently

$$x - \frac{x^3}{3!} + \frac{x^5}{5!} - \frac{x^7}{7!} + \cdots = \sum_{n=0}^{\infty} (-1)^n \frac{x^{2n+1}}{(2n+1)!},$$

a series that converges for all real numbers. (You can use the Ratio Test to verify this.) To show that this series represents $\sin x$ for all values of x, we must show that

$$\lim_{n \to \infty} R_n(x) = 0$$

for all values of x.

Taylor's Theorem implies that for a fixed but arbitrary value of x, there is a number ξ between x and zero with

$$R_n(x) = \frac{f^{(n+1)}(\xi)}{(n+1)!} x^{(n+1)}.$$

Since $|f^{(n+1)}(\xi)|$ is either $|\cos \xi|$ or $|\sin \xi|$, $|f^{(n+1)}(\xi)| \leq 1$ for every n and

$$|R_n(x)| \leq \frac{|x|^{n+1}}{(n+1)!}.$$

It was shown in Example 3 of Section 8.7 that

$$\sum_{n=0}^{\infty} \frac{x^{n+1}}{(n+1)!}$$

converges for all x. This implies that

$$\lim_{n \to \infty} \frac{x^{n+1}}{(n+1)!} = 0, \quad \text{so} \quad \lim_{n \to \infty} |R_n(x)| \leq \lim_{n \to \infty} \frac{|x|^{n+1}}{(n+1)!} = 0.$$

Thus, for all real numbers x,

$$\sin x = x - \frac{x^3}{3!} + \frac{x^5}{5!} - \frac{x^7}{7!} + \cdots = \sum_{n=1}^{\infty} (-1)^n \frac{x^{2n+1}}{(2n+1)!}.$$

Although the Maclaurin series for $\cos x$ can be derived in the same manner as $\sin x$, it is easier to apply Theorem (8.44) once the series representation for $\sin x$ is known. For all real numbers x,

$$\cos x = D_x \sin x = D_x \left(x - \frac{x^3}{3!} + \frac{x^5}{5!} - \frac{x^7}{7!} + \cdots \right)$$

$$= 1 - \frac{x^2}{2!} + \frac{x^4}{4!} - \frac{x^6}{6!} + \cdots = \sum_{n=0}^{\infty} (-1)^n \frac{x^{2n}}{(2n)!}. \blacktriangleright$$

The power-series representations of certain functions will be referred to in this and later mathematics courses. For ease of reference, we list the most frequently used series:

(8.51)

$$e^x = \sum_{n=0}^{\infty} \frac{x^n}{n!}, \quad \text{for all } x \text{ in } \mathbb{R};$$

$$\sin x = \sum_{n=0}^{\infty} (-1)^n \frac{x^{2n+1}}{(2n+1)!}, \quad \text{for all } x \text{ in } \mathbb{R};$$

$$\cos x = \sum_{n=0}^{\infty} (-1)^n \frac{x^{2n}}{(2n)!}, \quad \text{for all } x \text{ in } \mathbb{R};$$

$$\frac{1}{1-x} = \sum_{n=0}^{\infty} x^n, \quad \text{for } |x| < 1.$$

▶ EXERCISE SET 8.8

In Exercises 1–6, use Taylor's Theorem to find the first four terms of the Maclaurin series for the functions described.

1. $f(x) = e^{2x}$
2. $f(x) = e^{-x}$
3. $f(x) = \ln(x+1)$
4. $f(x) = \cos 2x$
5. $f(x) = \arctan x$
6. $f(x) = x^3 + 3x + 1$

7–12. List the first four Maclaurin polynomials for the functions given in Exercises 1–6.

In Exercises 13–22, (a) find the fourth Taylor polynomial for the function f about the number a and (b) find the remainder for the Taylor polynomial.

13. $f(x) = e^x$, $a = 1$
14. $f(x) = e^x$, $a = -2$
15. $f(x) = \sin x$, $a = \pi/2$
16. $f(x) = \cos x$, $a = \pi/2$
17. $f(x) = \tan x$, $a = \pi/4$
18. $f(x) = 1/x$, $a = 1$
19. $f(x) = \sqrt{x}$, $a = 4$
20. $f(x) = \ln x$, $a = 1$
21. $f(x) = x^2 + 2 + e^x$, $a = 1$
22. $f(x) = x^2 + 2x + 3$, $a = 1$

In Exercises 23–32, determine the Maclaurin series for the function described. Use summation notation to express the result.

23. $f(x) = e^{-x}$
24. $f(x) = e^{3x}$
25. $f(x) = \sin 3x$
26. $f(x) = x \sin 3x$
27. $f(x) = \dfrac{1}{x+1}$
28. $f(x) = \dfrac{x}{x+1}$
29. $f(x) = \dfrac{\sin x}{x}$
30. $f(x) = \dfrac{1}{1-x^3}$
31. $f(x) = 2^x$
32. $f(x) = x3^x$

In Exercises 33–40, determine the Taylor series about a for the function described. Use summation notation to express the result.

33. $f(x) = 1/x$; $a = 2$
34. $f(x) = 1/x$; $a = -3$
35. $f(x) = \sin x$; $a = \pi/2$
36. $f(x) = \cos x$; $a = \pi/2$
37. $f(x) = e^x$; $a = 1$
38. $f(x) = e^{2x}$; $a = -1$
39. $f(x) = \ln x$; $a = 1$
40. $f(x) = \ln 2x$; $a = \dfrac{1}{2}$

41. (a) Use the third Maclaurin polynomial to approximate $\sin 0.01$, and (b) find the maximum error for this approximation.

42. (a) Use the third Taylor polynomial about $a = 1$ to approximate $\ln 1.1$, and (b) find the maximum error for this approximation.

43. Use a Taylor polynomial for the function $f(x) = \sqrt{x}$ about $a = 4$ to find an approximation to $\sqrt{4.1}$ that is accurate to within 10^{-8}.

44. Use a Taylor polynomial for the function $f(x) = \ln x$ about $a = e$ to find an approximation to $\ln 3$ that is accurate to within 10^{-4}.

45. Use a Taylor polynomial to approximate $\sin 5°$ to an accuracy of 10^{-4}.

46. Use a Taylor polynomial to approximate $\cos 42°$ to an accuracy of 10^{-4}.

47. Show that if $x > 0$, then:

a) $\left| \sin x - \sum_{i=0}^{n} (-1)^i \dfrac{x^{2i+1}}{(2i+1)!} \right| < \dfrac{x^{2n+3}}{(2n+3)!};$

The expressions for $\sin x$ and $\cos x$ in (8.51) were known to Newton and included in *De analysi per equationes numero terminorum infinitas*, which was written in 1669, but not published until 1711. Leibniz was also aware of these results.

HISTORICAL NOTE

b) $\left|\cos x - \sum_{i=0}^{n} (-1)^i \frac{x^{2i}}{(2i)!}\right| < \frac{x^{2n+2}}{(2n+2)!}$.

48. a) Use the power-series representation of $\sin x$ to give a power-series representation of $f(x) = \sin x^2$.
    b) Give a power-series representation of
    $$\int_0^x \sin t^2 \, dt.$$
    c) Use the power series from part (b) to approximate
    $$\int_0^1 \sin t^2 \, dt$$
    to within 10^{-4}.

49. Suppose P is a polynomial of degree n. Show that the nth Maclaurin polynomial about $a = 0$ is precisely P.

50. Suppose P is a polynomial of degree n. Show that the nth Taylor polynomial about a is precisely P regardless of the value chosen for a.

51. An nth-degree polynomial $P_n(x) = a_0 + a_1 x + \cdots + a_n x^n$ in the variable x can be changed into a polynomial in the variable $x - a$ by expanding P_n in an nth Taylor polynomial about a. For example, if $P_n(x) = a_0 + a_1 x + a_2 x^2$, then $P_n'(x) = a_1 + 2a_2 x$ and $P_n''(x) = 2a_2$. So
    $$P_n(x) = (a_0 + a_1 a + a_2 a^2) + (a_1 + 2a_2 a)(x - a) + a_2(x - a)^2.$$
    Use this technique to change each of the following polynomials into a form involving the variable $x - a$.
    a) $P_2(x) = 2x^2 + 3x + 1$, $a = 1$
    b) $P_2(x) = 3x^2 - x + 4$, $a = -2$
    c) $P_3(x) = x^3 + x^2$, $a = -1$
    d) $P_3(x) = 3x^3 - 2x^2 + 1$, $a = 3$

52. The method for completing the square of a quadratic polynomial can be derived by using the result in Exercise 51. Suppose $f(x) = ax^2 + bx + c$. Expand f in a second Taylor polynomial about an arbitrary point d and then choose d so that the coefficient of $(x - d)$ is zero. Use this technique to complete the square of the quadratic polynomial $P(x) = 2x^2 + 3x + 1$.

53. Prove that the Maclaurin series of a function has only even powers if and only if the function is even. [*Hint:* Recall that the derivative of an even function is an odd function and that the derivative of an odd function is even.]

54. Prove that the Maclaurin series of a function has only odd powers if and only if the function is odd.

55. The nth Taylor polynomial for a function f at a is sometimes referred to as the polynomial of degree at most n that "best" approximates f near a.
    a) Explain why this description is accurate.
    b) Find the quadratic polynomial that best approximates a function f at $x = 1$, if the function has as its tangent line the line with equation $y = 4x - 1$ when $x = 1$ and has $f''(1) = 6$.

56. It is necessary to verify that $\lim_{n \to \infty} R_n(x) = 0$ to ensure that the Taylor series for a function actually represents the function. Consider the function
    $$f(x) = \begin{cases} e^{-1/x^2}, & \text{if } x \neq 0, \\ 0, & \text{if } x = 0. \end{cases}$$
    The Maclaurin series of this function is identically zero, so the series represents f only when $x = 0$. Show that this statement is true by constructing the Maclaurin series in the following manner.
    a) Show that $\lim_{x \to 0} e^{-1/x^2} x^{-n} = 0$ for each n.
    b) Use the definition of the derivative to find $f'(0)$.
    c) Use the definition of the derivative to find $f^{(n)}(0)$ for $n \geq 2$.
    d) Construct the Maclaurin series.
    e) Sketch the graph of f.

57. a) Show that for all $x > 0$,
    $$x - \frac{x^2}{2} < \ln(x + 1) < x - \frac{x^2}{2} + \frac{x^3}{3}.$$
    b) Use the result in part (a) to show that
    $$\lim_{n \to \infty} e^{-nx} \left(1 + \frac{x}{n}\right)^{n^2} = e^{-x^2/2}.$$

Putnam exercises

58. Assume that $|f(x)| \leq 1$ and $|f''(x)| \leq 1$ for all x on an interval of length at least 2. Show that $|f'(x)| \leq 2$ on the interval. (This exercise was problem 4, part I of the twenty-third William Lowell Putnam examination given on December 1, 1962. The examination and its solution are in the September 1963 issue of the *American Mathematical Monthly*, pp. 713–717.)

59. Let C be a real number, and let f be a function such that
    $$\lim_{x \to \infty} f(x) = C, \quad \lim_{x \to \infty} f'''(x) = 0.$$
    Prove that $\lim_{x \to \infty} f'(x) = 0$ and $\lim_{x \to \infty} f''(x) = 0$.

    (This exercise was problem 4, part II of the nineteenth William Lowell Putnam examination given on November 22, 1958. The examination and its solution are in the January 1961 issue of the *American Mathematical Monthly*, pp. 22–27.)

60. For which real numbers c is $(e^x + e^{-x})/2 \leq e^{cx^2}$ for all real x? (This exercise was problem B–1 of the forty-first William Lowell Putnam examination given on December 6, 1980. The examination and its solution are in the October 1981 issue of the *American Mathematical Monthly,* pp. 605–612.)

8.9

Applications of Taylor Polynomials and Taylor Series

In this section we briefly consider a few of the interesting results concerning Taylor polynomials and Taylor series.

The Binomial Series

In algebra courses we learn that integral powers of a sum can be expanded using the Binomial Theorem:

$$(a + b)^k = a^k + ka^{k-1}b + \frac{k(k-1)}{2} a^{k-2}b^2 + \cdots$$

$$+ \frac{k(k-1)\cdots(k-n+1)}{n!} a^{k-n}b^n + \cdots + kab^{k-1} + b^k$$

$$= \sum_{n=0}^{k} \frac{k!}{n!(k-n)!} a^{k-n}b^n,$$

where a and b are any real numbers and k is a positive integer.

We will consider this expansion when $a = 1$ and $b = x$, and relax the restriction on the power k to allow it to assume noninteger values. That is, we consider the expansion of $(1 + x)^k$ for real numbers x and k.

To obtain this expansion, we use the Maclaurin-series representation of the function f where $f(x) = (1 + x)^k$:

$$f(x) = (1 + x)^k, \qquad \text{so } f(0) = 1;$$
$$f'(x) = k(1 + x)^{k-1}, \qquad \text{so } f'(0) = k;$$
$$f''(x) = k(k-1)(1 + x)^{k-2}, \qquad \text{so } f''(0) = k(k-1);$$

and, in general, for any integer n

$$f^{(n)}(x) = k(k-1)\cdots(k-n+1)(1+x)^{k-n},$$

so

$$f^{(n)}(0) = k(k-1)\cdots(k-n+1).$$

Consequently, the Maclaurin series is

$$1 + \sum_{n=1}^{\infty} \frac{k(k-1)\cdots(k-n+1)}{n!} x^n.$$

The series has radius of convergence 1, since

$$\lim_{n\to\infty} \left| \frac{\frac{k(k-1)\cdots(k-n)}{(n+1)!}x^{n+1}}{\frac{k(k-1)\cdots(k-n+1)}{n!}x^n} \right| = \lim_{n\to\infty}\left|\frac{k-n}{n+1}x\right| = |x|.$$

This series,

(8.52) $\quad 1 + \sum_{n=1}^{\infty} \frac{k(k-1)\cdots(k-n+1)}{n!}x^n = (1+x)^k, \quad \text{for } |x| < 1$

is called the **binomial series.** It is shown in Appendix B that this series does represent $(1+x)^k$ when $|x| < 1$.

Example 1

Find the binomial series for $\sqrt{1+x}$, and use this series to approximate $\sqrt{1.1}$ to within 10^{-6}.

Solution
We have

$$(1+x)^{1/2} = 1 + \sum_{n=1}^{\infty} \frac{\frac{1}{2}\left(\frac{1}{2}-1\right)\cdots\left(\frac{1}{2}-n+1\right)}{n!} x^n$$

$$= 1 + \frac{1}{2}x + \frac{1}{2}\left(-\frac{1}{2}\right)\frac{x^2}{2} + \frac{1}{2}\left(-\frac{1}{2}\right)\left(-\frac{3}{2}\right)\frac{x^3}{6} + \cdots$$

$$= 1 + \frac{1}{2}x - \frac{1}{8}x^2 + \frac{3}{48}x^3 - \frac{15}{384}x^4 + \cdots.$$

When $x = 0.1$,

$$\sqrt{1.1} = (1 + 0.1)^{1/2}$$

$$= 1 + \frac{1}{2}(0.1) - \frac{1}{8}(0.1)^2 + \frac{3}{48}(0.1)^3 - \frac{15}{384}(0.1)^4 + \cdots.$$

After the first term, this is an alternating series whose terms decrease in magnitude and have the limit zero. The Alternating Series Test implies that the series converges, and the corollary to that test implies that the mth partial sum of the series differs from the sum of the series by less than the magnitude of a_{m+1}. Thus to find the number of terms needed to ensure an approximation that is accurate to within 10^{-6}, we need to determine m so that $|a_{m+1}| < 10^{-6}$.

Table 8.3 shows that this first occurs when $m = 4$, and

$$a_{m+1} = a_5 = \frac{\left(\frac{1}{2}\right)\left(-\frac{1}{2}\right)\left(-\frac{3}{2}\right)\left(-\frac{5}{2}\right)\left(-\frac{7}{2}\right)}{120}(0.1)^5 \approx 2.73 \times 10^{-7}.$$

Consequently, an approximation to $\sqrt{1.1}$ to within 10^{-6} is

$$\sqrt{1.1} \approx 1 + \frac{1}{2}(0.1) - \frac{1}{8}(0.1)^2 + \frac{3}{48}(0.1)^3 - \frac{15}{384}(0.1)^4 \approx 1.048809. \quad \blacktriangleright$$

Table 8.3

m	$\|a_m + 1\|$
0	5.00×10^{-2}
1	1.25×10^{-3}
2	6.25×10^{-5}
3	3.91×10^{-6}
4	2.73×10^{-7}

Example 2

Use Theorem (8.44) and the binomial series with $k = -\frac{1}{2}$ to derive a series for arcsin x that is valid when $|x| < 1$.

Solution
Since
$$\arcsin x = \int_0^x (1-t^2)^{-1/2}\, dt,$$

we first find the series representation for $(1-t^2)^{-1/2}$ and then integrate using Theorem (8.44). Since

$$(1+x)^{-1/2} = 1 - \frac{1}{2}x + \left(-\frac{1}{2}\right)\left(-\frac{3}{2}\right)\frac{x^2}{2!} + \left(-\frac{1}{2}\right)\left(-\frac{3}{2}\right)\left(-\frac{5}{2}\right)\frac{x^3}{3!} + \cdots$$
$$+ \left[\left(-\frac{1}{2}\right)\left(-\frac{3}{2}\right)\cdots\left(\frac{-(2n-1)}{2}\right)\right]\frac{x^n}{n!} + \cdots$$
$$= 1 - \frac{1}{2}x + \frac{1\cdot 3}{2^2}\frac{x^2}{2!} - \frac{1\cdot 3\cdot 5}{2^3}\frac{x^3}{3!} + \cdots$$
$$+ (-1)^n \frac{1\cdot 3\cdot 5 \cdots (2n+1)}{2^n}\frac{x^n}{n!} + \cdots,$$

replacing x by $-t^2$ gives

$$(1-t^2)^{-1/2} = 1 + \frac{1}{2}t^2 + \frac{1\cdot 3}{2^2}\frac{t^4}{2!} + \frac{1\cdot 3\cdot 5}{2^3}\frac{t^6}{3!} + \cdots$$
$$= 1 + \sum_{n=1}^{\infty} \frac{1\cdot 3\cdot 5 \cdots (2n-1)}{2^n}\frac{t^{2n}}{n!}.$$

Consequently,
$$\arcsin x = \int_0^x (1-t^2)^{-1/2}\, dt$$
$$= x + \sum_{n=1}^{\infty} \frac{1\cdot 3\cdot 5 \cdots (2n-1)}{2^n}\frac{x^{2n+1}}{(2n+1)n!}.$$

This series has radius of convergence $r = 1$ (see Exercise 9). ▶

Newton's Method

Newton's Method is used to find approximate solutions to an equation of the form $f(x) = 0$. This method was discussed in Section 3.6 as an application of the derivative. It begins by assuming an initial approximation x_0 to the solution of the equation. Subsequent approximations $x_1, x_2, \ldots$ are generated using the formula

(8.53)
$$x_{n+1} = x_n - \frac{f(x_n)}{f'(x_n)}$$

for $n > 0$, provided $f'(x_n)$ is defined and nonzero. The method often produces exceedingly accurate results. With the aid of Taylor polynomials we can show why this is true.

Suppose c denotes the solution to $f(x) = 0$ and f'' exists on an interval containing both c and the initial approximation x_0. Expanding f in its first Taylor polynomial at x_0 and evaluating at $x = c$ gives

$$0 = f(c) = f(x_0) + f'(x_0)(c - x_0) + \frac{f''(\xi)}{2}(c - x_0)^2,$$

where ξ lies between x_0 and c. Consequently, if $f'(x_0) \neq 0$,

$$c - x_0 + \frac{f(x_0)}{f'(x_0)} = \frac{-f''(\xi)}{2f'(x_0)}(c - x_0)^2.$$

Since

$$x_1 = x_0 - \frac{f(x_0)}{f'(x_0)},$$

this implies that

$$c - x_1 = -\frac{f''(\xi)}{2 f'(x_0)}(c - x_0)^2.$$

If a bound M is known for the second derivative of f on an interval about c and x_0 is within this interval, then

$$|c - x_1| \leq \frac{M}{|2 f'(x_0)|}|c - x_0|^2.$$

This inequality implies that Newton's Method has the tendency to approximately double the number of digits of accuracy with each successive approximation. For example, if the error in approximating c by x_0 is on the order of 10^{-n}, then the error in approximating c by x_1 is on the order of $C(10^{-n})^2 = C(10^{-2n})$, where $C = M/|2f'(x_0)|$.

Example 3

Find an approximation to the solution of the equation $x = 3^{-x}$ that is accurate to within 10^{-8}.

Solution

A solution to this equation corresponds to a solution to $f(x) = 0$, where

$$f(x) = x - 3^{-x}.$$

Since f is continuous and $f(0) = -1$ while $f(1) = \frac{2}{3}$, the Intermediate Value Theorem implies that a solution to the equation lies in the interval $(0, 1)$. As an initial approximation we will use the midpoint of this interval, $x_0 = 0.5$. Succeeding approximations are generated by applying the formula

$$x_{n+1} = x_n - \frac{f(x_n)}{f'(x_n)} = x_n - \frac{x_n - 3^{-x_n}}{1 + 3^{-x_n} \ln 3}.$$

8.9 APPLICATIONS OF TAYLOR POLYNOMIALS AND TAYLOR SERIES

Table 8.4

i	x_i	$\|x_i - x_{i-1}\|$
0	0.500000000	—
1	0.547329757	0.047329757
2	0.547808574	0.000478817
3	0.547808622	0.000000048

These approximations are listed in Table 8.4, together with differences between successive approximations.

The difference in successive approximations leads to the correct conclusion that x_3 is accurate to the places listed and that the solution to $x = 3^{-x}$ accurate to within 10^{-8} is 0.54780862. ▶

Euler's Formula

The Maclaurin-series representation for the sine, cosine, and natural exponential functions given in (8.51) can be used to extend the domain of these functions to the set of complex numbers. By a **complex number** we mean a number of the form

$$z = x + iy,$$

where x and y are real numbers and i is a symbol with the property that $(i)^2 = -1$. The number x is the **real part** of z, and the number y the **imaginary part** of z.

The plane provides a graphic representation of the complex numbers, as shown in Figure 8.16. The horizontal axis represents the real part of the complex number and the vertical axis the imaginary part. This representation is called an *Argand diagram*, named for Jean Robert Argand (1768–1822), who, in 1806, published a paper in which such a representation was used. Actually it was Caspar Wessel (1745–1818) who first used this graphic representation in a paper published by the Danish Academy of Sciences in 1798. The complex plane is also called the Gaussian plane, although Carl Friedrich Gauss did not publish his views until later.

The magnitude of a complex number $z = x + iy$ is given by $|z| = \sqrt{x^2 + y^2}$. With the notation in Figure 8.16 we see that

$$x = |z| \cos \theta \quad \text{and} \quad y = |z| \sin \theta,$$

so

$$z = x + iy = |z|(\cos \theta + i \sin \theta).$$

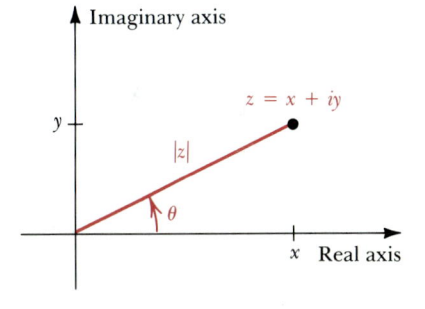

Figure 8.16

For complex numbers z, e^z is defined using the power-series representation

$$e^z = \sum_{n=0}^{\infty} \frac{z^n}{n!} = 1 + z + \frac{z^2}{2!} + \frac{z^3}{3!} + \frac{z^4}{4!} + \frac{z^5}{5!} + \cdots,$$

which can be shown to converge for any complex number z. In particular, for real numbers x, this implies that

$$e^{ix} = 1 + ix + \frac{(ix)^2}{2!} + \frac{(ix)^3}{3!} + \frac{(ix)^4}{4!} + \frac{(ix)^5}{5!} + \cdots.$$

Since $i^2 = -1$, $i^3 = -i$, $i^4 = 1$, $i^5 = i$, $i^6 = i^2 = -1$, and so on,

$$e^{ix} = 1 + ix - \frac{x^2}{2!} - \frac{ix^3}{3!} + \frac{x^4}{4!} + \frac{ix^5}{5!} + \cdots.$$

However, absolute convergence permits us to rewrite this series as

$$e^{ix} = \left(1 - \frac{x^2}{2!} + \frac{x^4}{4!} - \cdots\right) + i\left(x - \frac{x^3}{3!} + \frac{x^5}{5!} - \cdots\right).$$

Since

$$\cos x = 1 - \frac{x^2}{2!} + \frac{x^4}{4!} - \cdots$$

and

$$\sin x = x - \frac{x^3}{3!} + \frac{x^5}{5!} - \cdots,$$

we have the equation

(8.54) $$e^{ix} = \cos x + i \sin x.$$

This intriguing relation is known as **Euler's Formula.** It has a number of interesting applications. For example, suppose that the rules of exponents hold for complex as well as real numbers. Then for any real number x and any integer n,

$$(\cos x + i \sin x)^n = (e^{ix})^n = e^{i(nx)}$$
$$= \cos nx + i \sin nx,$$

so for every integer n,

(8.55) $$(\cos x + i \sin x)^n = \cos nx + i \sin nx.$$

This is known as **de Moivre's formula.**

Example 4

Use de Moivre's Formula to find all complex numbers that are cube roots of 8.

HISTORICAL NOTE

Abraham de Moivre (1667–1754) did his most lasting work in probability theory, extending the work of Pascal and Fermat. The formula that bears his name first appeared in a different form in 1707 and was included in *Miscellanea analytica de seriebus et quadratis* in 1730. It was explicitly stated and proved inductively by Euler in *Introductio in analysin infinitorum* of 1748, but with $\sqrt{-1}$ appearing in place of i. Euler first used i for $\sqrt{-1}$ in a memoir of 1777 that was not published until 1794.

Solution

If we express the real number 8 as

$$8 = 8(\cos 0 + i \sin 0)$$

and assume that, for some value of θ,

$$z = |z|(\cos \theta + i \sin \theta)$$

is a cube root of 8, then

$$8 = 8(\cos 0 + i \sin 0) = |z|^3(\cos \theta + i \sin \theta)^3 = z^3.$$

By de Moivre's Formula,

$$(\cos \theta + i \sin \theta)^3 = \cos 3\theta + i \sin 3\theta,$$

so we must have

$$8(\cos 0 + i \sin 0) = |z|^3(\cos 3\theta + i \sin 3\theta).$$

Consequently,

$$|z|^3 = 8, \quad \text{which implies that} \quad |z| = 2,$$

and

$$\cos 3\theta = \cos 0, \quad \text{and} \quad \sin 3\theta = \sin 0.$$

There is flexibility in the choice of θ since these equations are satisfied whenever $\theta = 2n\pi/3$ for some integer n.

When $n = 0$, $n = 1$, and $n = 2$, we have the following cube roots:

$$n = 0: z_0 = 2(\cos 0 + i \sin 0) = 2;$$

$$n = 1: z_1 = 2\left(\cos \frac{2\pi}{3} + i \sin \frac{2\pi}{3}\right)$$

$$= 2\left(-\frac{1}{2} + i \frac{\sqrt{3}}{2}\right) = -1 + i\sqrt{3};$$

and

$$n = 2: z_2 = 2\left(\cos \frac{4\pi}{3} + i \sin \frac{4\pi}{3}\right)$$

$$= 2\left(-\frac{1}{2} - i \frac{\sqrt{3}}{2}\right)$$

$$= -1 - i\sqrt{3}.$$

Other integral values of n produce repetitions of these results. The three complex cube roots of 8 are thus

$$2, \quad -1 + i\sqrt{3}, \quad \text{and} \quad -1 - i\sqrt{3}.$$

These roots are represented on the Argand diagram in Figure 8.17. They are equally spaced on the circle centered at the origin with radius $2 = 8^{1/3}$.

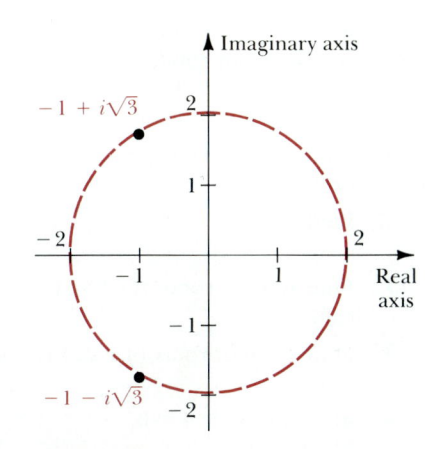

▶ **Figure 8.17**

► EXERCISE SET 8.9

In Exercises 1–6, find a power-series representation for each function, and give the radius of convergence of the series.

1. $f(x) = \sqrt[3]{1 + x}$
2. $f(x) = \sqrt{1 + x^2}$
3. $f(x) = \dfrac{1}{\sqrt{1 + x}}$
4. $f(x) = (1 + x)^{-3}$
5. $f(x) = (3x + 2)^{3/2}$
6. $f(x) = \sqrt{9 + 2x}$

In Exercises 7 and 8, approximate the definite integral to within 10^{-4}.

7. $\displaystyle\int_0^{1/2} \sqrt[3]{1 + x^2}\, dx$
8. $\displaystyle\int_0^{1/2} \dfrac{dx}{\sqrt[4]{1 + x^3}}$

9. Show that the series for arcsin x given in Example 2 has radius of convergence $R = 1$.

10. For the function described by $f(x) = x^3 - 2x - 1$, find an approximation, accurate to within 10^{-5}, to the x-intercept that is between 1 and 2.

11. Find an approximation in $[1, 2]$ to a solution of the equation $x^4 - 2x - 2 = 0$ that is accurate to within 10^{-5}.

12. Use Euler's Formula to determine each of the following.
    a) e^{13i}
    b) e^{2+13i}
    c) e^{-1+2i}

13. Write $e^{2+\pi i}$ in the form $a + bi$, where a and b are real numbers.

14. In each of the following, indicate on an Argand diagram the set of complex numbers $z = x + iy$ that satisfy the given condition.
    a) (Real part of z) > 0
    b) $|z| \le 1$
    c) z is a real number
    d) $|e^z| \ge 1$

15. Find all cube roots of 1 and represent them graphically.

16. Find all cube roots of -1 and represent them graphically.

17. Find all sixth roots of 64 and represent them graphically.

18. Find all sixth roots of i and represent them graphically.

19. a) Solve $x^4 + 4 = 0$.
    b) Use the solutions to factor $x^4 + 4$ into two real quadratic factors.

20. a) Solve $x^4 + 32 = 0$.
    b) Use the solutions to factor $x^4 + 32$ into two real quadratic factors.

21. a) Use Euler's Formula to show that
    $$\cos x = \dfrac{e^{ix} + e^{-ix}}{2} \quad \text{and} \quad \sin x = \dfrac{e^{ix} - e^{-ix}}{2i}.$$
    b) Use the results in part (a) to show that
    $$\cos ix = \cosh x \quad \text{and} \quad \sin ix = i \sinh x.$$

22. Show that if $n > 1$, $x > -1$, and $x \ne 0$, then:
    a) $(1 + x)^n > 1 + nx$;
    b) $(1 + x)^{1/n} < 1 + x/n$.

23. For which complex numbers $z = x + iy$ does the equality $|e^{-iz}| = 1$ hold?

24. Show that the square roots of i are $\pm(1 + i)/\sqrt{2}$.

25. The accumulated value of a savings account based on regular periodic payments can be determined from the *annunity-due equation*,
    $$A = \dfrac{P}{r}(1 + r)[(1 + r)^n - 1].$$
    In this equation A is the amount in the account, P is the amount regularly deposited, and r is the rate of interest per period for the n deposit periods.
    An engineer would like to have a savings account valued at \$75,000 upon retirement in 20 years, and can afford to put \$150 per month toward this goal. Use Newton's Method to answer the following question: What is the minimal interest rate at which this amount can be deposited, assuming that the interest is compounded monthly?

26. The amount of money required to pay off a mortgage over a fixed period of time is given by
    $$A = \dfrac{P}{r}[1 - (1 + r)^{-n}],$$
    known as an *ordinary-annuity equation*. In this equation A is the amount of the mortgage, P is the amount of each payment, and r is the interest rate per period for the n payment periods. Suppose a 30-year home mortgage in the amount of \$50,000 is needed and that the borrower can afford house payments of at most \$450 per month. Use Newton's Method to find the maximum interest rate that the borrower can afford to pay.

IMPORTANT TERMS AND RESULTS

CONCEPT	PAGE	CONCEPT	PAGE	CONCEPT	PAGE
Sequence	510	Limit Comparison Test	532	Radius of convergence	557
Convergent sequence	513	Integral Test	534	Interval of convergence	557
Divergent sequence	513	p-series	536	Power series in $x - a$	558
Bounded sequence	517	Alternating Series Test	541	Maclaurin series	568
Monotonic sequence	518	Absolute convergence	546	Taylor series	568
Completeness Property	518	Conditional convergence	547	Taylor polynomials	571
Series	522	Ratio Test	548	Binomial series	577
Convergent series	523	Root Test	550	Newton's Method	579
Harmonic series	525	Flowchart for testing		Euler's Formula	582
Geometric series	526	convergence	552	de Moivre's Formula	582
Comparison Test	532	Power series in x	554		

REVIEW EXERCISES

In Exercises 1–16, the nth term of a sequence is given.
  a) Determine if the sequence is bounded.
  b) Determine whether the sequence converges or diverges.
  c) If the sequence converges, find its limit.

1. $a_n = \dfrac{3n+2}{1-2n}$

2. $a_n = (-1)^n \dfrac{n\pi + 1}{n}$

3. $a_n = 3$

4. $a_n = \dfrac{\sin n\pi}{n}$

5. $a_n = \sqrt[n]{n+1}$

6. $a_n = \dfrac{\ln n}{n}$

7. $a_n = \sum_{i=1}^{n} \dfrac{1}{i}$

8. $a_n = \sum_{i=1}^{n} \left(\dfrac{1}{2}\right)^i$

9. $a_n = \dfrac{e^n}{n}$

10. $a_n = \dfrac{n}{e^n}$

11. $a_n = (-1)^n(\sqrt{n^2+1} - n)$

12. $a_n = \dfrac{n^2 - 2n + 5}{3n^2 - 8}$

13. $a_n = \dfrac{1}{\sqrt[n]{2}}$

14. $a_n = \left(\dfrac{1}{n}\right)^{1/n}$

15. $a_1 = 1$, $a_n = \dfrac{a_{n-1}}{3}$, if $n \geq 2$

16. $a_1 = 1$, $a_n = na_{n-1}$, if $n \geq 2$

17. Suppose $\{a_n\}$ is defined by $a_1 = 1$ and
$$a_n = \dfrac{n-1}{n} a_{n-1}, \quad \text{if } n \geq 2.$$
  a) Use the Completeness Property to show that $\{a_n\}$ is convergent.
  b) Find an explicit formula for a_n.
  c) Determine $\lim_{n \to \infty} a_n$.

18. Repeat Exercise 17 for the sequence $\{a_n\}$ defined by $a_1 = 1$ and
$$a_n = \dfrac{n^2 - 1}{n^2} a_{n-1}, \quad \text{if } n \geq 2.$$

In Exercises 19–38, determine whether the series is divergent, conditionally convergent, or absolutely convergent.

19. $\sum_{n=0}^{\infty} \dfrac{1}{1+\sqrt{n}}$

20. $\sum_{n=1}^{\infty} \dfrac{1}{n^2} \sin n$

21. $\sum_{n=0}^{\infty} \dfrac{1}{e^n}$

22. $\sum_{n=1}^{\infty} \dfrac{\ln n}{n^3}$

23. $\sum_{n=1}^{\infty} (-1)^{n+1} \dfrac{1+\sqrt{n}}{2^n}$

24. $\sum_{n=1}^{\infty} \dfrac{\arctan n}{n^2+1}$

25. $\sum_{n=0}^{\infty} \dfrac{(-1)^n}{2n+1}$

26. $\sum_{n=1}^{\infty} \dfrac{1}{\sqrt[n]{2}}$

27. $\sum_{n=1}^{\infty} \dfrac{n!}{n^n}$

28. $\sum_{n=1}^{\infty} (-1)^n \dfrac{n}{5^n}$

29. $\sum_{n=1}^{\infty} \dfrac{(-1)^n(1+4n)}{7n^2 - 1}$

30. $\sum_{n=1}^{\infty} (-1)^n \dfrac{2n}{n!}$

31. $\sum_{n=2}^{\infty} \dfrac{3}{n \ln n}$

32. $\sum_{n=1}^{\infty} (-1)^{n+1} \dfrac{2^n}{e^n}$

33. $\sum_{n=1}^{\infty} \dfrac{(-1)^n \ln n}{n+1}$

34. $\sum_{n=1}^{\infty} \dfrac{1}{n^n}$

35. $\sum_{n=1}^{\infty} \dfrac{(-1)^n n}{(2n+1)^2}$

36. $\sum_{n=2}^{\infty} \dfrac{1}{\ln n}$

37. $\sum_{n=1}^{\infty} (-1)^{n+1} \left(\dfrac{3}{4n}\right)$

38. $\sum_{n=1}^{\infty} \dfrac{n^2 - 4n + 4}{9n^2 + 3n - 2}$

In Exercises 39–52, find (a) the radius and (b) the interval of convergence of the power series.

39. $\sum_{n=1}^{\infty} \dfrac{x^n}{n^2}$

40. $\sum_{n=1}^{\infty} n^2 x^n$

41. $\sum_{n=1}^{\infty} \dfrac{x^n}{n^3 + n + 1}$

42. $\sum_{n=1}^{\infty} \dfrac{nx^n}{n^2 - 3}$

43. $\sum_{n=0}^{\infty} (x+2)^n$

44. $\sum_{n=1}^{\infty} \dfrac{x^n}{(2n+1)^2}$

45. $\sum_{n=0}^{\infty} \dfrac{n^2}{2^n} (x-2)^n$

46. $\sum_{n=1}^{\infty} \dfrac{\ln n}{n} x^n$

47. $\sum_{n=0}^{\infty} 3^n (x-3)^n$

48. $\sum_{n=0}^{\infty} \dfrac{2^n}{n!} (x-3)^n$

49. $\sum_{n=0}^{\infty} \left(\dfrac{x^2 - 1}{2}\right)^n$

50. $\sum_{n=1}^{\infty} \left(\dfrac{x}{n}\right)^n$

51. $\sum_{n=1}^{\infty} \dfrac{1 \cdot 3 \cdot 5 \cdots (2n-1)}{3 \cdot 6 \cdot 9 \cdots (3n)} x^n$

52. $\sum_{n=2}^{\infty} \dfrac{2 \cdot 5 \cdot 8 \cdots (3n-4)}{3 \cdot 6 \cdot 9 \cdots (3n-3)} x^{n-1}$

For the series in Exercises 53–55, find an approximation to the sum that is accurate to within 10^{-5}.

53. $\sum_{n=2}^{\infty} \dfrac{(-1)^{n-1}}{4^n(n-1)}$

54. $\sum_{n=1}^{\infty} \dfrac{(-1)^{n+1}}{(2n-1)^3}$

55. $\sum_{n=1}^{\infty} \dfrac{(-1)^{n-1}}{n 3^n}$

56. Find a power series centered at zero with radius of convergence $R = 1$ that:
   a) converges at 1 and -1;
   b) diverges at 1 and -1;
   c) converges at 1 and diverges at -1;
   d) diverges at 1 and converges at -1.

57. Find a power series centered at 2 with radius of convergence $R = 3$ that:
   a) converges at 5 and -1;
   b) diverges at 5 and -1;
   c) converges at 5 and diverges at -1;
   d) diverges at 5 and converges at -1.

58. Give an example of a bounded sequence that does not converge.

59. Give an example of a monotonic sequence that does not converge.

60. Show that the sequence $\{\ln n/(n+1)\}$ is a decreasing sequence for $n > 4$.

In Exercises 61–64, use the power-series representations (8.51) and the arithmetic and calculus properties of power series to find a power-series representation for the function.

61. $f(x) = \dfrac{\sin x}{x}$

62. $f(x) = \ln \dfrac{1}{1-x}$

63. $f(x) = \cosh x$

64. $f(x) = \sinh x$.

65. a) Find the fourth Maclaurin polynomial for $f(x) = x^6 - 7x^5 + 4x^2 - 2x - 3$.
    b) Find the remainder for the polynomial.

66. a) Find the fourth Taylor polynomial about 3 for $f(x) = 1/x^2$.
    b) Find the remainder for the polynomial.

67. a) Find the fourth Taylor polynomial about $\pi/4$ for $f(x) = \sec x$.
    b) Use the polynomial found in part (a) to approximate $\sec 0.8$.

68. a) Find the fourth Taylor polynomial about $\pi/4$ for $f(x) = \csc x$.
    b) Use the polynomial found in part (a) to approximate $\csc 0.8$.

69. Find the Maclaurin series for
$$f(x) = \dfrac{x-1}{x+1}.$$

70. Find the Maclaurin series for
$$\int_0^x \dfrac{e^t - 1}{t}\, dt.$$

71. Find the Taylor series for $f(x) = \ln 2x$ about 1.

72. Find a series representation for the definite integral $\int_0^1 e^{x^2}\, dx$.

73. Let $x_1 = 1$ and $x_{n+1} = \dfrac{x_n}{2} + \dfrac{3}{2x_n}$.

Find the fourth term of this sequence and compare it to the $\sqrt{3}$.

74. Find the values of x for which the series
$$\sum_{n=1}^{\infty} \left(\dfrac{\cos x}{n}\right)^n$$
converges.

75. Find the values of x for which the series
$$\sum_{n=1}^{\infty} \dfrac{1}{n}\left(\dfrac{x-1}{x}\right)^n$$
converges.

76. Find all fourth roots of 16.

77. Write $e^{(\pi/4)i - 2}$ in the form $a + bi$, where a and b are real numbers.

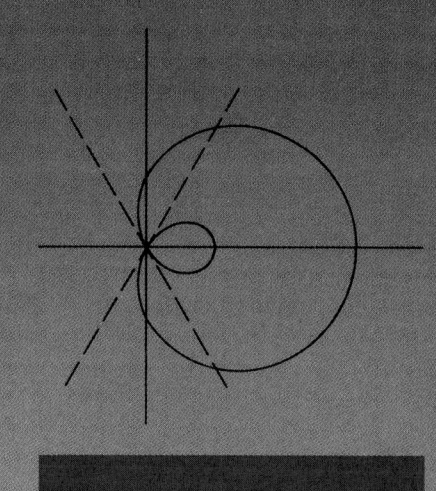

9
Polar Coordinates and Parametric Equations

In the rectangular, or Cartesian, coordinate system, a point in the plane is specified by an ordered pair (x, y) that describes the directed distances of the point from the x- and y- axes, as shown in Figure 9.1(a). A point can also be specified by the pair (r, θ), where θ is the angle shown in Figure 9.1(b) and r is the distance from the origin to the point. This method of specifying points is called the *polar coordinate system.*

In the first part of this chapter we consider curves that are conveniently represented using the polar coordinate system. The remainder of the chapter discusses the use of parametric equations for describing general curves in the plane. This description can be used to extend many of the applications of calculus.

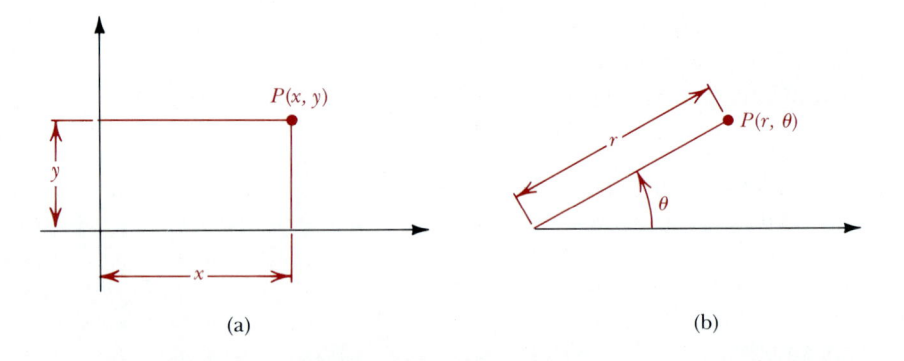

Figure 9.1
(a) The Cartesian coordinate system; (b) the polar coordinate system.

9.1
The Polar Coordinate System

The polar coordinate system is described using a fixed point and a fixed half-line, or **ray,** that originates at the point. The fixed point is called the **pole** of the polar coordinate system, and the **polar axis** is the scaled horizontal ray that originates at the pole and extends infinitely to the right, as shown in Figure 9.2(a).

To determine polar coordinates for a point P, we draw the line l through P and the pole. With the origin at the pole, we scale l with the same units as the polar axis. An arrow is used to denote the positive direction on l, as shown in Figure 9.2(b). Let θ be an angle with initial side the polar axis and terminal side l, and let r be the coordinate on l of the point P. Then (r, θ) is said to be a pair of **polar coordinates** for P.

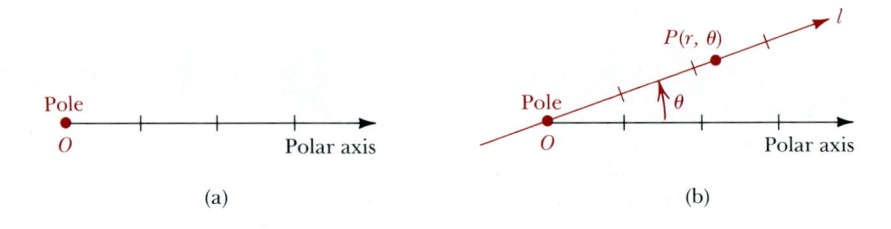

Figure 9.2

9.1 THE POLAR COORDINATE SYSTEM

(a)	(b)	(c)	(d)
(r, θ)	$(r, 2\pi + \theta)$	$(-r, -\pi + \theta)$	$(r, \theta - 2\pi)$

Figure 9.3

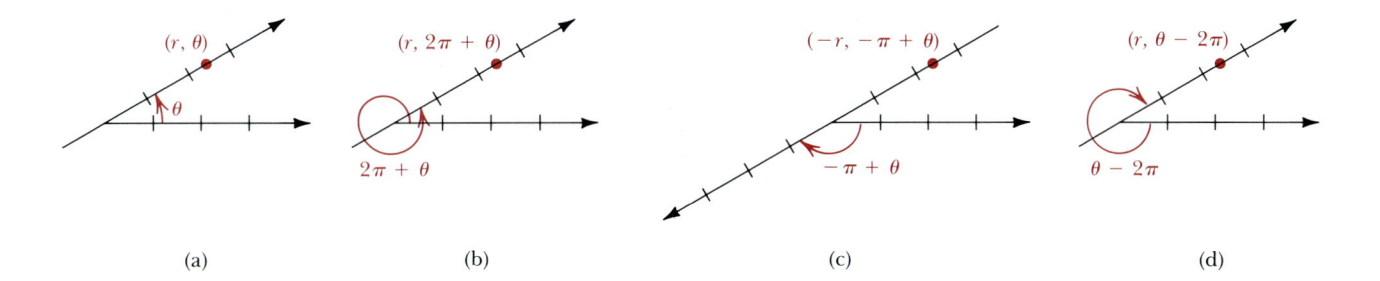

Points in the plane do not have a unique polar representation. A point with polar coordinates (r, θ) is also represented by $(r, \theta + 2\pi)$, $(-r, \theta + \pi)$, and $(r, \theta - 2\pi)$, as shown in Figure 9.3, as well as by an infinite number of other pairs of polar coordinates. In addition, the pole O is represented by the polar coordinates $(0, \theta)$, for any angle θ. This multiple polar representation of points does not generally cause difficulties, but it should be kept in mind.

To allow convenient conversion between the polar and Cartesian coordinate systems, we use a common scale for the systems and orient them so that the pole is at the origin and the polar axis coincides with the positive x-axis, as shown in Figure 9.4. The right triangle in Figure 9.4 shows that to change from polar coordinates to Cartesian coordinates, we can use

(9.1) $\qquad x = r \cos \theta \quad \text{and} \quad y = r \sin \theta.$

Although our figure demonstrates the situation only for P in the first quadrant and $r > 0$, the equations in (9.1) can be verified for any values of r and θ.

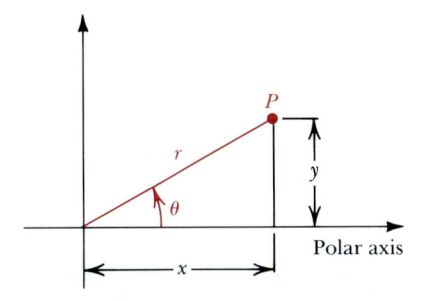

Figure 9.4

To change from Cartesian coordinates to polar coordinates, we modify the equations in (9.1):

$$x^2 + y^2 = r^2 (\cos \theta)^2 + r^2 (\sin \theta)^2 = r^2$$

and

$$\frac{y}{x} = \frac{r \sin \theta}{r \cos \theta} = \tan \theta, \quad \text{provided } x \neq 0.$$

So *one* set of polar coordinates describing the point with Cartesian coordinates (x, y) is given by

(9.2)
$$r = \pm \sqrt{x^2 + y^2}, \quad \text{and} \quad \theta = \arctan\left(\frac{y}{x}\right), \quad \text{provided } x \neq 0.$$

The sign for r is chosen to ensure that the point lies in the appropriate quadrant.

In the remainder of this chapter we will assume that both sets of coordinates are available when discussing a problem and change from one system to the other when convenient.

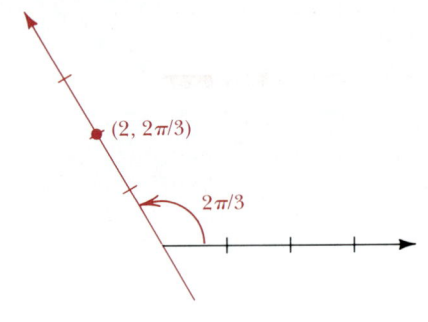

Figure 9.5

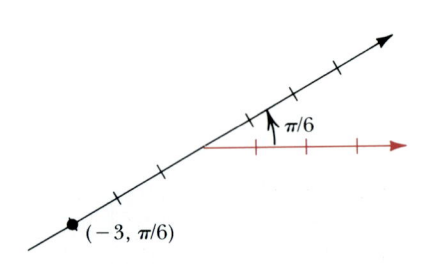

Figure 9.6

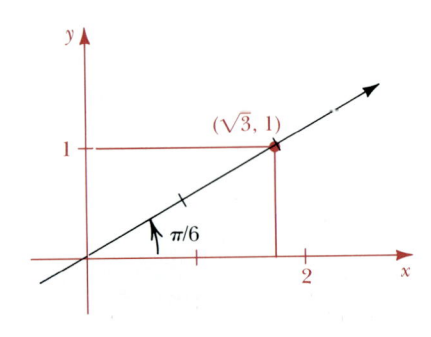

Figure 9.7

Example 1

Find the Cartesian coordinates of the points with polar coordinates (a) $(2, 2\pi/3)$ and (b) $(-3, \pi/6)$.

Solution

a) The point $(2, 2\pi/3)$ is shown in Figure 9.5. The equations in (9.1) imply that

$$x = 2 \cos \frac{2\pi}{3} = 2\left(-\frac{1}{2}\right) = -1$$

and

$$y = 2 \sin \frac{2\pi}{3} = 2\left(\frac{\sqrt{3}}{2}\right) = \sqrt{3};$$

so the Cartesian coordinates are $(-1, \sqrt{3})$.

b) The point with polar coordinates $(-3, \pi/6)$ is shown in Figure 9.6. It has Cartesian coordinates

$$x = -3 \cos \frac{\pi}{6} = -3 \frac{\sqrt{3}}{2}$$

and

$$y = -3 \sin \frac{\pi}{6} = -3\left(\frac{1}{2}\right) = -\frac{3}{2}.$$

▶

Example 2

Find polar coordinates for the point with Cartesian coordinates $(\sqrt{3}, 1)$.

Solution

Figure 9.7 indicates one possibility for r and θ:

$$r = \sqrt{(\sqrt{3})^2 + 1^2} = 2 \quad \text{and} \quad \theta = \arctan\left(\frac{1}{\sqrt{3}}\right) = \frac{\pi}{6},$$

so one pair of polar coordinates of this point is $(2, \pi/6)$.

Other pairs of polar coordinates for this point are $(2, 13\pi/6)$, $(2, -11\pi/6)$, $(-2, 7\pi/6)$, and $(-2, -5\pi/6)$. In fact, any pairs of points (r, θ) satisfying

$$r = 2 \quad \text{and} \quad \theta = \frac{\pi}{6} + 2n\pi, \quad n \text{ an integer}$$

or

$$r = -2 \quad \text{and} \quad \theta = \frac{7\pi}{6} + 2n\pi, \quad n \text{ an integer}$$

describes the point with Cartesian coordinates $(\sqrt{3}, 1)$.

▶

EXERCISE SET 9.1

1. Plot the points whose polar coordinates are each of the following.
   a) $(3, \pi/6)$
   b) $(1, 4\pi/3)$
   c) $(0, \pi/6)$
   d) $(2, -\pi/3)$

2. Plot the points whose polar coordinates are each of the following.
   a) $(-2, \pi/3)$
   b) $(2, 4\pi/3)$
   c) $(-2, -4\pi/3)$
   d) $(-2, -2\pi/3)$

3. Plot the points whose polar coordinates are each of the following.
   a) $(1, 375\pi/2)$
   b) $(1, 1)$
   c) $(0, 0)$
   d) $(\pi, 1)$

4. Find polar coordinates of each of the points whose Cartesian coordinates are given below.
   a) $(0, 4)$
   b) $(-2, 2)$
   c) $(\sqrt{3}, -1)$
   d) $(-2, -2\sqrt{3})$

5. Give four other sets of polar coordinates that describe the points with polar coordinates:
   a) $(2, 0)$
   b) $(-1, \pi/2)$
   c) $(1, 3\pi/4)$
   d) $(1, -\pi)$

6. Give four other sets of polar coordinates that describe the points with polar coordinates:
   a) $(2, 2\pi/3)$
   b) $(1, \pi)$
   c) $(0, 0)$
   d) $(1, 1)$

7. Find the Cartesian coordinates of each of the following points given in polar coordinates.
   a) $(2, \pi/4)$
   b) $(1, \pi/2)$
   c) $(-1, \pi/2)$
   d) $(2, 2\pi/3)$

8. Find the Cartesian coordinates of each of the following points given in polar coordinates.
   a) $(-1, \pi)$
   b) $(-3, 7\pi/6)$
   c) $(2, -7\pi/4)$
   d) $(-4, \pi)$

9. Show that the distance between points with polar coordinates (r_1, θ_1) and (r_2, θ_2) is
   $$d((r_1, \theta_1), (r_2, \theta_2)) = \sqrt{r_1^2 + r_2^2 - 2r_1 r_2 \cos(\theta_2 - \theta_1)}.$$

10. Determine a polar equation for the line passing through the points with polar coordinates (r_1, θ_1) and (r_2, θ_2).

11. Show that $x^2 - y^2 = 1$ has polar equation $r^2 = \sec 2\theta$.

12. A football field is 100 yd long and 160 ft wide (see the figure). Place a polar coordinate system with the pole at one corner of the field and the polar axis along the adjacent 100-yd side. What are the polar coordinates of the corner of the field diagonally opposite the pole?

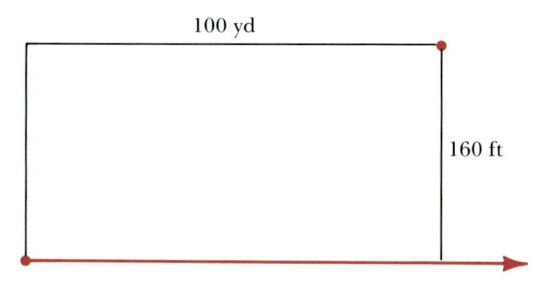

13. A surveyor sets a transit at one corner of a lot with four straight sides. An angle of 0° is set on a line through another corner 900 ft away. The two remaining corners have polar coordinates (630, 45°) and (200, 90°). Draw a sketch of the lot and determine its acreage (1 acre is 43,560 ft²).

14. A major league baseball infield is in the shape of a square with sides 90 ft. Suppose a polar coordinate system is placed with the pole at home plate and the polar axis parallel to the line from first to third base with positive direction on the first-base side. What are the polar coordinates of first, second, and third base?

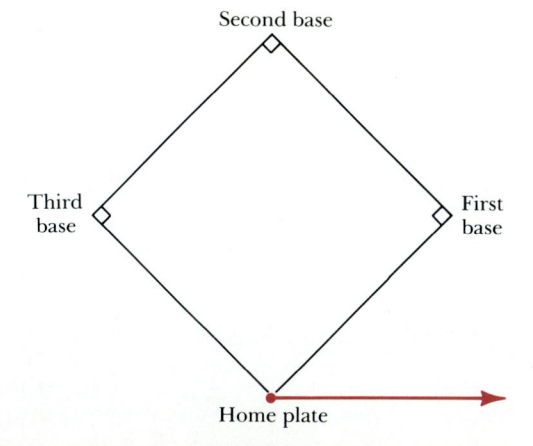

9.2
Graphing in Polar Coordinates

The circle shown in Figure 9.8 is described by the polar equation $r = 2$, since the points on the circle are precisely those whose distance from the origin is 2. This is a much simpler equation than its Cartesian counterpart, $x^2 + y^2 = 4$.

The polar equation
$$r = \theta, \qquad 0 \leq \theta \leq 2\pi$$

describes the spiraling curve shown in Figure 9.9. This curve is very difficult to express using Cartesian coordinates. For example, a Cartesian equation for the portion of the first loop of the graph lying in the first quadrant can be determined using the relations between polar and Cartesian coordinates discussed in the preceding section. The Cartesian equation for this portion of the curve is

$$\sqrt{x^2 + y^2} = \arctan\left(\frac{y}{x}\right), \qquad x > 0, y > 0,$$

a considerably more complicated representation.

Although there are exceptions, polar equations are generally appropriate for describing curves that have some systematic rotational behavior centered at the origin. Most other curves are represented more easily by Cartesian equations.

Many of the graphing techniques used for Cartesian equations can be applied to polar equations with only minor modifications. For example:

Figure 9.8

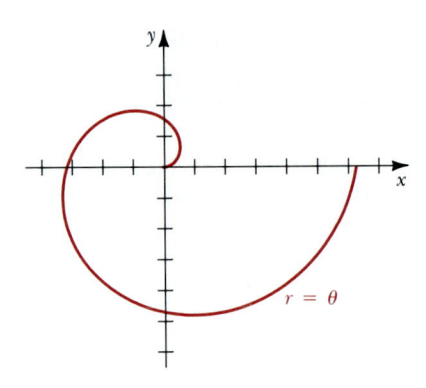

Figure 9.9

(9.3)
POLAR INTERCEPT TESTS

i) Intercepts of the x-, or polar, axis occur when θ is a multiple of π;
ii) Intercepts of the y-axis, the polar line $\theta = \pi/2$, occur when θ is an odd multiple of $\pi/2$.

Some common polar symmetry tests are given in (9.4) and illustrated in Figure 9.10. This is not a complete list since each point in the plane can be represented in an infinite number of ways. They are, however, the easiest tests to apply and are generally sufficient.

(9.4)
POLAR SYMMETRY TESTS

i) The graph is symmetric with respect to the x-, or polar, axis if an equivalent equation is obtained when (r, θ) is replaced by either $(r, -\theta)$ or $(-r, \pi - \theta)$ (Figure 9.10a).
ii) The graph is symmetric with respect to the y-axis, the line $\theta = \pi/2$, if an equivalent equation is obtained when (r, θ) is replaced by either $(r, \pi - \theta)$ or $(-r, -\theta)$ (Figure 9.10b).
iii) The graph is symmetric with respect to the pole if an equivalent

9.2 GRAPHING IN POLAR COORDINATES

Figure 9.10
(a) *x*-axis symmetry; (b) *y*-axis symmetry; (c) origin symmetry.

equation is obtained when (r, θ) is replaced by either $(r, \pi + \theta)$ or $(-r, \theta)$ (Figure 9.10c).

In addition, if $r = f(\theta)$ and f is differentiable we have the following.

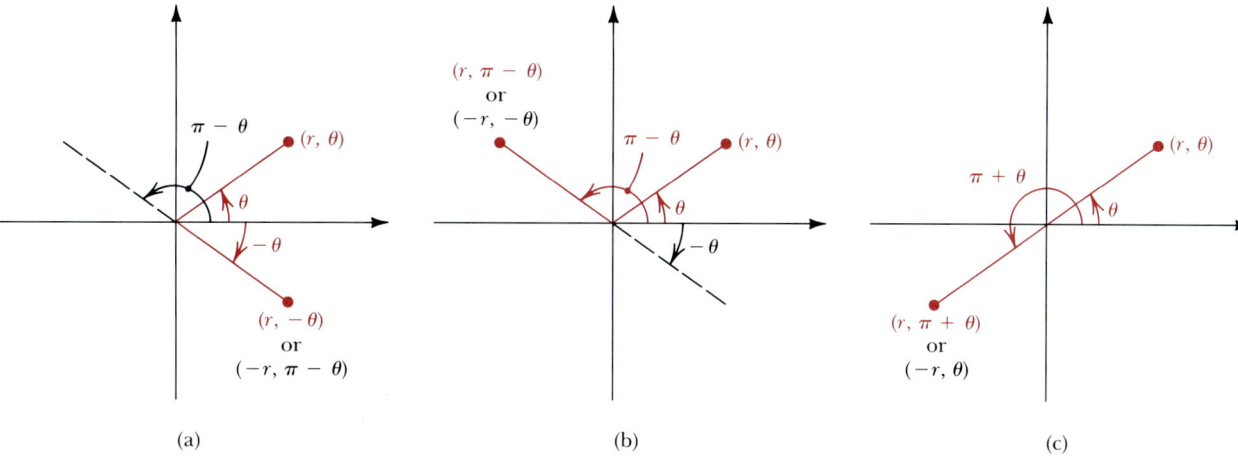

(9.5)

i) When $f'(\theta) > 0$, the values of r are increasing as θ moves counterclockwise.
ii) When $f'(\theta) < 0$, the values of r are decreasing as θ moves counterclockwise.
iii) The graph approaches the pole along the line θ if $f(\theta) = 0$.

Example 1

Sketch the graph of the polar equation

$$r = 2 \sin \theta$$

and find a corresponding Cartesian equation.

Solution

The sine function is periodic with period 2π, so we need consider only the graph on an interval of length 2π, such as $[-\pi, \pi]$. Since $\sin(-\theta) = -\sin\theta$, symmetry test (ii) implies that the graph is symmetric with respect to the *y*-axis. The complete graph can be determined from the graph when θ is in $[-\pi/2, \pi/2]$.

Table 9.1 lists representative values of θ and corresponding values of r. Since $r'(\theta) = 2\cos\theta > 0$ on $[-\pi/2, \pi/2]$, the values of r increase as θ increases, and the graph for θ in $[-\pi/2, \pi/2]$ is as shown in Figure 9.11. This is the complete graph, since it is symmetric with respect to the *y*-axis. The graph is a circle with radius 1 and center at the point with Cartesian coordinates $(0, 1)$.

To find a Cartesian equation, we first multiply both sides of the equation $r = 2\sin\theta$ by r to obtain $r^2 = 2r\sin\theta$. Using the relationships

Table 9.1

θ	r
$-\pi/2$	-2
$-\pi/3$	$-\sqrt{3}$
$-\pi/4$	$-\sqrt{2}$
$-\pi/6$	-1
0	0
$\pi/6$	1
$\pi/4$	$\sqrt{2}$
$\pi/3$	$\sqrt{3}$
$\pi/2$	2

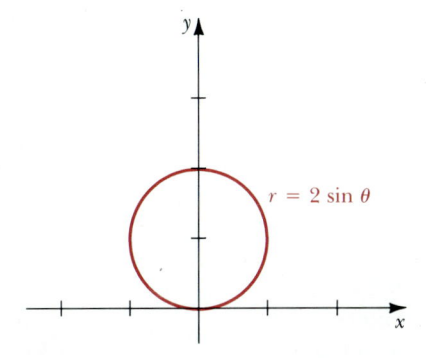

Figure 9.11

$r^2 = x^2 + y^2$ and $y = r \sin \theta$, we can write the equation in Cartesian coordinates:

$$x^2 + y^2 = 2y, \quad \text{so} \quad x^2 + y^2 - 2y = 0.$$

Completing the square we have

$$x^2 + y^2 - 2y + 1 = 1, \quad \text{so} \quad x^2 + (y-1)^2 = 1. \quad \blacktriangleright$$

Paper specifically designed for sketching graphs in polar coordinates is available in most college bookstores or can be constructed easily with a ruler and compass. Figure 9.12 is sketched on paper of this type. If you have difficulty sketching polar curves, it may help to use this special paper.

Example 2

Sketch the graph of the polar equation

$$r = 2(1 + \sin \theta).$$

Solution

As in Example 1, the periodicity of the sine function implies that the graph is determined by taking θ in the interval $[-\pi, \pi]$. This graph is also symmetric with respect to the y-axis, since

$$\sin(\pi - \theta) = \sin \pi \cos \theta - \sin \theta \cos \pi = \sin \theta$$

implies that an equivalent equation is obtained when (r, θ) is replaced by $(r, \pi - \theta)$.

Table 9.2 lists values of θ in the interval $[-\pi/2, \pi/2]$ and corresponding values of r. The values of r increase as θ increases, since $r'(\theta) = 2 \cos \theta > 0$ on $(-\pi/2, \pi/2)$.

The graph for θ in $[-\pi/2, \pi/2]$ is shown in Figure 9.12(a). Symmetry with respect to the y-axis gives the complete graph, shown in Figure

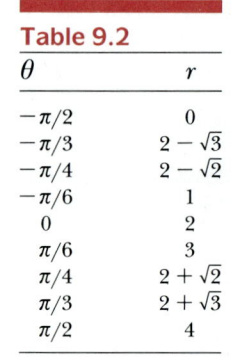

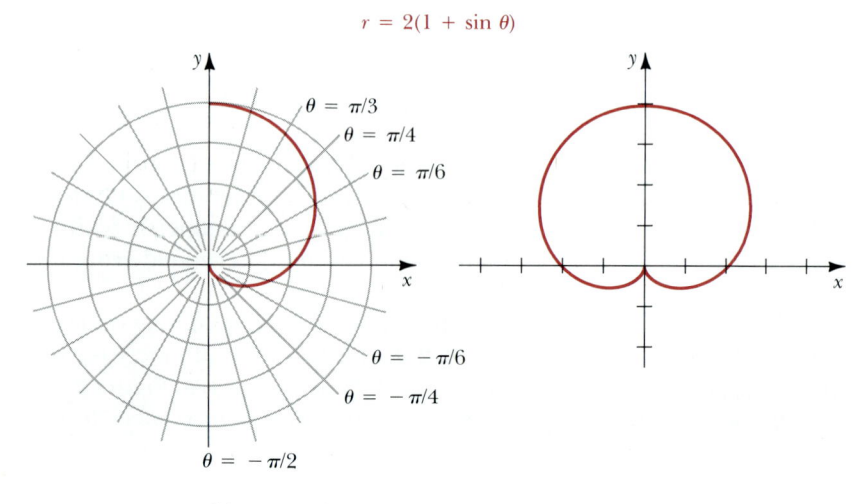

Table 9.2

θ	r
$-\pi/2$	0
$-\pi/3$	$2 - \sqrt{3}$
$-\pi/4$	$2 - \sqrt{2}$
$-\pi/6$	1
0	2
$\pi/6$	3
$\pi/4$	$2 + \sqrt{2}$
$\pi/3$	$2 + \sqrt{3}$
$\pi/2$	4

Figure 9.12
A cardioid.

(a)   (b)

9.12(b). The graph is called a **cardioid,** a name derived from the Greek word *kardia,* meaning heart.

An x-intercept occurs at the point with polar coordinates $(2, 0)$, and y-intercepts occur at $(4, \pi/2)$ and $(0, -\pi/2)$, the pole. The line with equation $\theta = -\pi/2$ is a ray along which the graph approaches the pole. ▶

Example 3

Sketch the graph of the polar equation $r = \cos 2\theta$.

Solution

The cosine function is even, so the graph is symmetric with respect to the x-, or polar, axis. The graph is also symmetric with respect to the y-axis since

$$\cos 2(\pi - \theta) = \cos (2\pi - 2\theta) = \cos (-2\theta) = \cos 2\theta.$$

Because of symmetry, it suffices to determine the behavior of the graph for θ in the interval $[0, \pi/2]$. Table 9.3 lists values of θ in this interval and corresponding values of r.

Since $r = 0$ when $\theta = \pi/4$, the graph approaches the pole along the line $\theta = \pi/4$. This line is illustrated in Figure 9.13(a), which also shows the graph of the equation for θ in the interval $[0, \pi/2]$.

Symmetry with respect to the x-axis enables us to extend the graph as shown in Figure 9.13(b), and symmetry with respect to the y-axis gives the completed graph, shown in Figure 9.13(c). The graph is called a **four-leafed rose.** ▶

Table 9.3

θ	r
0	1
$\pi/8$	$\sqrt{2}/2$
$\pi/6$	$1/2$
$\pi/4$	0
$\pi/3$	$-1/2$
$5\pi/12$	$-\sqrt{3}/2$
$\pi/2$	-1

$r = \cos 2\theta$

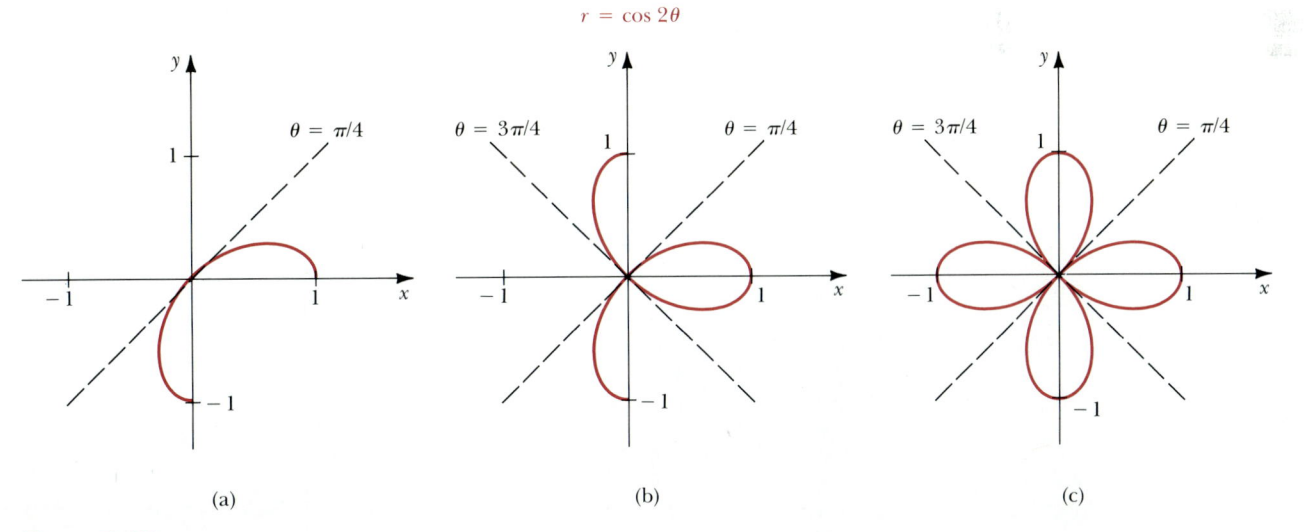

(a)   (b)   (c)

Figure 9.13
A four-leafed rose.

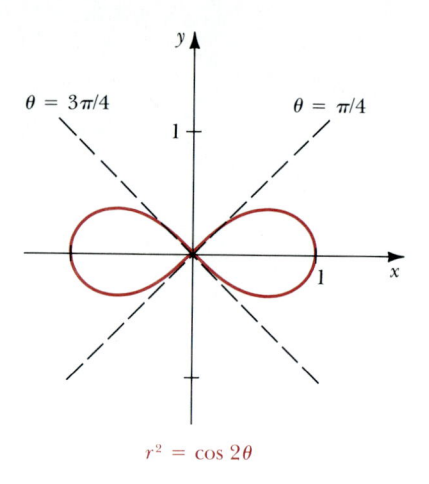

Figure 9.14
A lemniscate

Example 4

Sketch the graph of the polar equation
$$r^2 = \cos 2\theta.$$

Solution

The analysis of this equation is similar to that of the equation in Example 3, except that those values of θ in $[0, 2\pi]$ for which $\cos 2\theta < 0$ must be excluded; that is, θ must satisfy

$$-\frac{\pi}{2} < 2\theta < \frac{\pi}{2} \quad \text{or} \quad \frac{3\pi}{2} < 2\theta < \frac{5\pi}{2}.$$

Thus the graph is determined by values of θ satisfying either

$$-\frac{\pi}{4} \le \theta \le \frac{\pi}{4} \quad \text{or} \quad \frac{3\pi}{4} \le \theta \le \frac{5\pi}{4}.$$

The graph is called a **lemniscate** and is shown in Figure 9.14. Dotted lines have been used to indicate rays along which the graph approaches the pole. ▶

Example 5

Sketch the graph of the polar equation
$$r = 1 + 2\cos\theta.$$

Solution

The graph is symmetric with respect to the x-axis, so it suffices to determine the graph for $0 \le \theta \le \pi$. The only y-intercept in this interval occurs at $(1, \pi/2)$. The other intercepts are at the points with polar coordinates $(3, 0)$, $(-1, \pi)$, and $(0, 2\pi/3)$. The rays along which the graph approaches the pole are $\theta = 2\pi/3$ and $\theta = 4\pi/3$. The table of values for θ in $[0, \pi]$ (see Table 9.4) and symmetry with respect to the x-axis indicate that the graph is as shown in Figure 9.15. This figure is called a **limaçon**. ▶

Table 9.4

θ	r
0	3
$\pi/6$	$1 + \sqrt{3}$
$\pi/4$	$1 + \sqrt{2}$
$\pi/3$	2
$\pi/2$	1
$2\pi/3$	0
$3\pi/4$	$1 - \sqrt{2}$
$5\pi/6$	$1 - \sqrt{3}$
π	-1

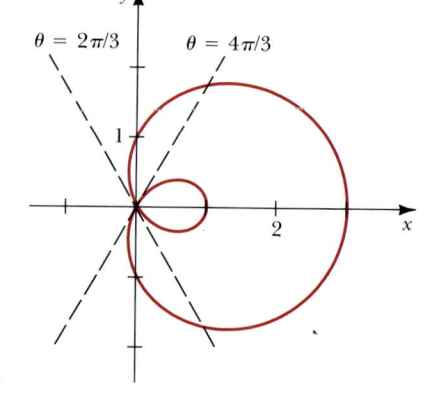

Figure 9.15
A limaçon.

9.2 GRAPHING IN POLAR COORDINATES

Table 9.5
Polar Graphing Summary

EQUATION	CURVE
$r = a$	Circle with center at $(0, 0)$ and radius $\|a\|$.
$\left.\begin{array}{l}r = a + b \cos \theta \\ r = a + b \sin \theta\end{array}\right\}$	Limaçon. If $\|a\| < \|b\|$, the limaçon has a loop. If $\|a\| = \|b\|$, the limaçon is heart-shaped and called a cardioid. (See Figure 9.16.)
$\left.\begin{array}{l}r^2 = \pm a^2 \cos 2\theta \\ r^2 = \pm a^2 \sin 2\theta\end{array}\right\}$	Lemniscate.
$\left.\begin{array}{l}r = a \sin n\theta \\ r = a \cos n\theta\end{array}\right\}$	n-leafed rose if n is odd; $2n$-leafed rose if n is even.
$r = f(\theta)$, where f is a monotonic function	Spiral.

The curves in Examples 1–5 are samples of certain general classes of polar curves, summarized in Table 9.5.

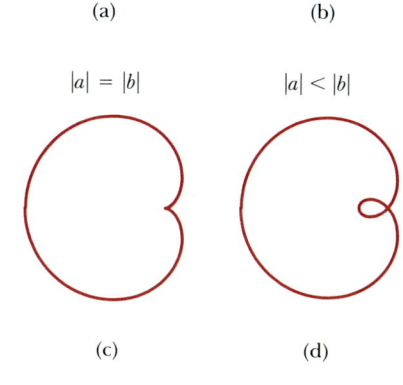

(a) $|a| \geq 2|b|$

(b) $a|b| > |a| > |b|$

(c) $|a| = |b|$

(d) $|a| < |b|$

Figure 9.16

Example 6

Sketch the graph of the polar equation $r = -\theta$.

Solution
The graph of this equation is a spiral similar to the graph shown in Figure 9.9. The graph is symmetric with respect to the y-axis since the equation is satisfied for (r, θ) precisely when it is satisfied for $(-r, -\theta)$. Thus the complete graph is determined from the graph when $\theta \geq 0$.

Since $dr/d\theta = -1$, r is a decreasing function of θ. As θ becomes large, r becomes large in magnitude, but negative. Plotting representative points and using the y-axis symmetry gives the graph shown in Figure 9.17. ▶

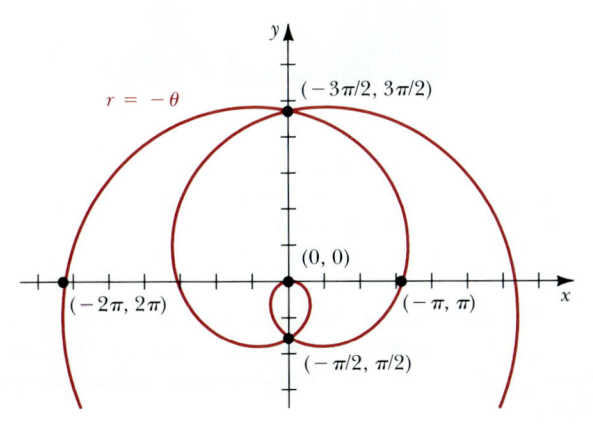

Figure 9.17
A spiral.

EXERCISE SET 9.2

In Exercises 1–34, sketch the graph of the polar equation.

1. $r = 3$
2. $r = 5\pi/4$
3. $\theta = 5\pi/4$
4. $\theta = 3$
5. $r = 3 \cos \theta$
6. $r = 6 \sin \theta$
7. $r = -3 \cos \theta$
8. $r = -6 \sin \theta$
9. $r = 1 + \cos \theta$
10. $r = 1 + \sin \theta$
11. $r = 2 + \sin \theta$
12. $r = 1 + 2 \sin \theta$
13. $r = 2 - \sin \theta$
14. $r = 1 - 2 \sin \theta$
15. $r = 3 \sin 3\theta$
16. $r = 4 \cos 3\theta$
17. $r = 4 \sin 4\theta$
18. $r = 3 \cos 4\theta$
19. $r = 2 \sin 2\theta$
20. $r^2 = 4 \sin 2\theta$
21. $r^2 = -9 \cos 2\theta$
22. $r^2 = 16 \cos 4\theta$
23. $r = 3 \sec \theta$
24. $r = -2 \csc \theta$
25. $r = \theta$
26. $r = -2^\theta$
27. $r = e^\theta$
28. $r = \ln \theta$
29. $r = 1/\theta$
30. $r = 2^{-\theta}$
31. $r = 2(\sin \theta + \cos \theta)$
32. $r = 2(\sin \theta - \cos \theta)$
33. $r = 2 \sin \theta + \cos \theta$
34. $r = 2 \sin \theta - \cos \theta$

Compare the graphs of the polar equations in each part of Exercises 35–43.

35. a) $r = 1 + \cos \theta$ b) $r = 1 - \cos \theta$
    c) $r = \cos \theta - 1$ d) $r = -\cos \theta - 1$
36. a) $r = 1 + \sin \theta$ b) $r = 1 - \sin \theta$
    c) $r = \sin \theta - 1$ d) $r = -\sin \theta - 1$
37. a) $r = \cos \theta$ b) $r = 1 + \cos \theta$
    c) $r = 2 + \cos \theta$ d) $r = 3 + \cos \theta$
38. a) $r = \sin \theta$ b) $r = \sin 2\theta$
    c) $r = \sin 3\theta$ d) $r = \sin 4\theta$
39. a) $r = \sin \theta$ b) $r = 1 + \sin \theta$
    c) $r = 2 + \sin \theta$ d) $r = 3 + \sin \theta$
40. a) $r = \sin \theta$ b) $r = \cos \theta$
    c) $r = \csc \theta$ d) $r = \sec \theta$
41. a) $r = 2^\theta$ b) $r = 2^{-\theta}$
    c) $r = -2^\theta$ d) $r = -2^{-\theta}$
42. a) $r = e^\theta$ b) $r = e^{-\theta}$
    c) $r = -e^\theta$ d) $r = -e^{-\theta}$
43. a) $r = \cos (\theta - \pi)$ b) $r = \cos (\theta - \pi/2)$
    c) $r = \sin (\theta - \pi)$ d) $r = \sin (\theta - \pi/2)$

44. Show that for constants a and b, the polar curve with equation $r = a \cos \theta + b \sin \theta$ is the circle through the origin with Cartesian equation
$$\left(x - \frac{a}{2}\right)^2 + \left(y - \frac{b}{2}\right)^2 = \frac{a^2 + b^2}{4}.$$

45. Suppose f is a differentiable function and $r = f(\theta)$. Find $dy/d\theta$ and $dx/d\theta$ in terms of $\theta, f(\theta),$ and $f'(\theta)$. Use these expressions and the Chain Rule to show that the graph of the equation has:
   a) a horizontal tangent at (r, θ) if $r = -f'(\theta) \tan \theta$;
   b) a vertical tangent at (r, θ) if $r = f'(\theta) \cot \theta$.

46. Use the result in Exercise 45 to find the horizontal and vertical tangents of each of the following.
   a) $r = \cos \theta, 0 \le \theta \le 2\pi$
   b) $r = 2 + 2 \sin \theta, 0 \le \theta \le 2\pi$
   c) $r = 1 - \cos \theta, 0 \le \theta \le 2\pi$
   d) $r = e^{-\theta}, -\pi \le \theta \le \pi$

47. The chambered nautilus *(Nautilus pompilus)* is a mollusk found in the Pacific and Indian oceans. The outside of its shell grows in the form of an exponential spiral. A typical equation of such a spiral is $r = 2e^{-0.2\theta}$. Sketch the graph of this spiral, and compare the resulting curve to the curve in the photograph.

A split view of the shell of a chambered nautilus *(Nautilus pompilus)*.

48. The lengthwise cross section of an apple can be reasonably approximated using a polar equation. What form would you expect the equation to assume if the polar axis is the stem of the apple and the pole is at the base of the stem?

49. A polar coordinate system on a major league baseball infield was described in Exercise 14 of Section 9.1. Use this coordinate system to describe the base paths of the field.

9.3

Areas of Regions Using Polar Coordinates

In this section we derive a result that allows us to use the polar-coordinate representation to find the area of regions in the plane. To find the area of a region whose boundaries are described using Cartesian coordinates, we summed the areas of approximating rectangles and took the limit as the width of the rectangles uniformly approached zero. To find the area of a region bounded by a polar equation, we divide the region into approximating sectors of circles, as shown in Figure 9.18, and sum the area of these sectors.

Suppose a curve is described by $r = f(\theta)$, for some continuous nonnegative function f defined on the interval $[\alpha, \beta]$, where $\alpha < \beta \leq \alpha + 2\pi$. Let R denote the region bounded by the graph of f and the lines $\theta = \alpha$ and $\theta = \beta$. (The restrictions that f be nonnegative and that $\alpha < \beta \leq \alpha + 2\pi$ are made to ensure that the region R is not overlapping.)

To find an approximation to the area of this region, we let

$$\mathcal{P} = \{\theta_0, \theta_1, \ldots, \theta_n\}$$

be a partition of the interval $[\alpha, \beta]$. We then choose ψ_i arbitrarily in the subinterval $[\theta_{i-1}, \theta_i]$ for each $i = 1, 2, \ldots, n$, and let $\Delta\theta_i = \theta_i - \theta_{i-1}$. The partition determines subregions R_i bounded by the rays $\theta = \theta_{i-1}$ and $\theta = \theta_i$ and by the graph of the function. An approximation to the area of R_i is the area of the sector of the circle with radius $f(\psi_i)$ and central angle $\Delta\theta_i$ (see Figure 9.19).

The area of a circle with radius $f(\psi_i)$ is $\pi[f(\psi_i)]^2$. The radian measure of the sector from θ_{i-1} to θ_i is $\Delta\theta_i$ and the radian measure of the entire circle is 2π, so the area of the sector is

$$\pi[f(\psi_i)]^2 \frac{\Delta\theta_i}{2\pi} = \frac{[f(\psi_i)]^2}{2} \Delta\theta_i.$$

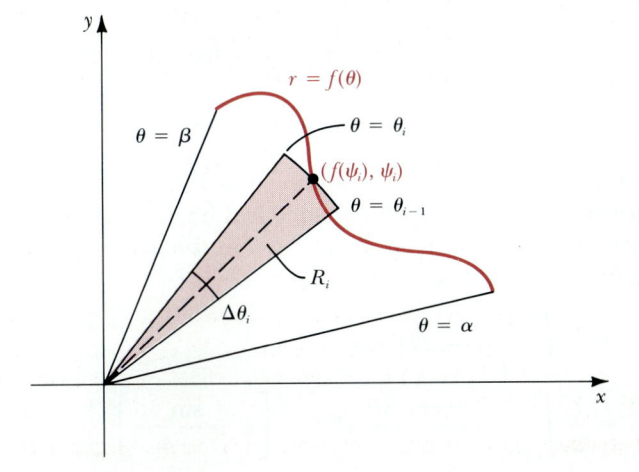

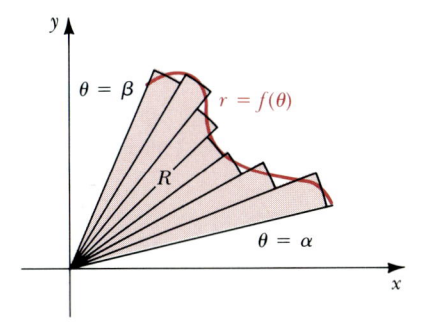

Figure 9.18

Figure 9.19

The area is $\dfrac{1}{2} \displaystyle\int_\alpha^\beta [f(\theta)]^2 \, d\theta$.

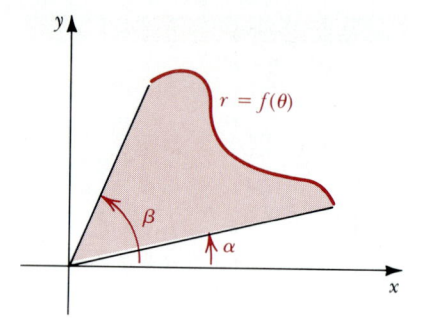

Figure 9.20

An approximation to the entire area of the region R is the sum of these approximations

$$\sum_{i=1}^{n} \frac{[f(\psi_i)]^2}{2} \Delta\theta_i,$$

a Riemann sum for the function described by $[f(\theta)]^2/2$ on the interval $[\alpha, \beta]$. As the norm of the partition approaches zero, this approximation leads to a definite integral describing the area shown in Figure 9.20:

(9.6) $$A = \frac{1}{2} \int_{\alpha}^{\beta} [f(\theta)]^2 \, d\theta.$$

Example 1

Find the area of the region bounded by the graph of the cardioid

$$r = 2(1 + \sin\theta).$$

Solution

The graph of this equation was discussed in Example 2 of Section 9.2 and is shown in Figure 9.21 together with a typical sector. Note that r is a continuous nonnegative function of θ. By (9.6), the area of the region is

$$A = \frac{1}{2} \int_0^{2\pi} [2(1 + \sin\theta)]^2 \, d\theta$$

$$= 2 \int_0^{2\pi} [1 + 2\sin\theta + (\sin\theta)^2] \, d\theta$$

$$= 2 \left[\theta - 2\cos\theta\right]_0^{2\pi} + 2 \int_0^{2\pi} \frac{1 - \cos 2\theta}{2} \, d\theta$$

$$= 4\pi + \left[\theta - \frac{\sin 2\theta}{2}\right]_0^{2\pi} = 6\pi.$$

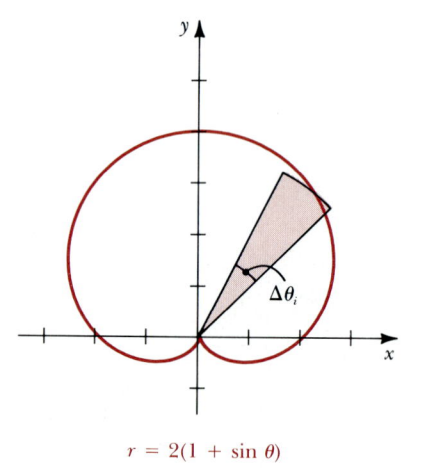

$r = 2(1 + \sin\theta)$

Figure 9.21

Example 2

Find the area of the region bounded by one leaf of the four-leafed rose $r = \cos 2\theta$.

Solution

The graph of this equation was discussed in Example 3 of Section 9.2 and is shown in Figure 9.22. The shaded leaf horizontal and to the right of the y-axis can be described by taking θ in the interval $[-\pi/4, \pi/4]$. The area of this region is

$$A = \frac{1}{2} \int_{-\pi/4}^{\pi/4} [\cos 2\theta]^2 \, d\theta$$

$$= \frac{1}{2} \int_{-\pi/4}^{\pi/4} \frac{1 + \cos 4\theta}{2} \, d\theta = \frac{1}{4} \left[\theta + \frac{\sin 4\theta}{4}\right]_{-\pi/4}^{\pi/4} = \frac{\pi}{8}.$$

$r = \cos 2\theta$

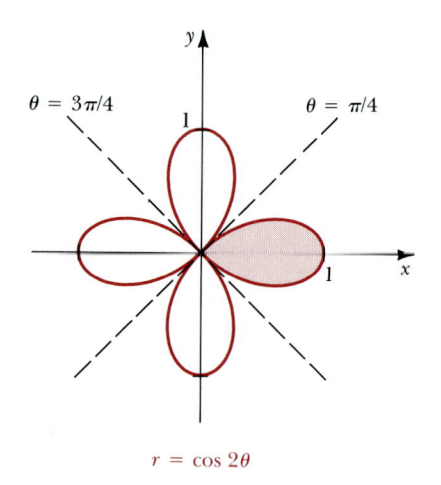

Figure 9.22

Because of the x-axis symmetry of the graph, the area is also given by

$$A = 2\left[\frac{1}{2}\int_0^{\pi/4}(\cos 2\theta)^2\,d\theta\right] = \int_0^{\pi/4}\cos^2 2\theta\,d\theta.$$ ▶

Example 3

Find the area of the region lying inside the graphs of both polar equations $r = 1$ and $r = 2\sin\theta$.

Solution

The first problem is to determine where the curves intersect. We see in Figure 9.23 that there are two points of intersection. Suppose (r, θ) describes a point of intersection, with $r = 1$ and $r = 2\sin\theta$. Solving these equations simultaneously gives

$$1 = 2\sin\theta, \quad \text{so} \quad \sin\theta = \frac{1}{2},$$

which is satisfied

$$\text{when} \quad \theta = \frac{\pi}{6} \quad \text{and when} \quad \theta = \frac{5\pi}{6}.$$

The region whose area is to be found is divided into three parts, R_1, R_2, and R_3, as shown in Figure 9.24. The region R_1 is bounded by the graph of $r = \sin\theta$ for θ in $[0, \pi/6]$. The region R_2 is bounded by the graph of $r = 1$ for θ in $[\pi/6, 5\pi/6]$, and R_3 is bounded by $r = \sin\theta$ for θ in $[5\pi/6, \pi]$. Thus the area is given by

$$A = \frac{1}{2}\int_0^{\pi/6}(2\sin\theta)^2\,d\theta + \frac{1}{2}\int_{\pi/6}^{5\pi/6}(1)^2\,d\theta + \frac{1}{2}\int_{5\pi/6}^{\pi}(2\sin\theta)^2\,d\theta.$$

However, the y-axis symmetry of the region implies that the area is also given by the simpler expression

$$A = 2\left[\frac{1}{2}\int_0^{\pi/6}(2\sin\theta)^2\,d\theta + \frac{1}{2}\int_{\pi/6}^{\pi/2}(1)^2\,d\theta\right]$$
$$= \int_0^{\pi/6}4\left(\frac{1-\cos 2\theta}{2}\right)d\theta + \int_{\pi/6}^{\pi/2}d\theta.$$

So

$$A = 2\left[\theta - \frac{\sin 2\theta}{2}\right]_0^{\pi/6} + \left(\frac{\pi}{2} - \frac{\pi}{6}\right) = \frac{\pi}{3} - \frac{\sqrt{3}}{2} + \frac{\pi}{3} = \frac{2\pi}{3} - \frac{\sqrt{3}}{2}.$$ ▶

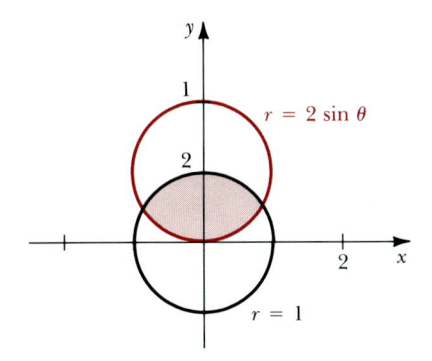

Figure 9.23

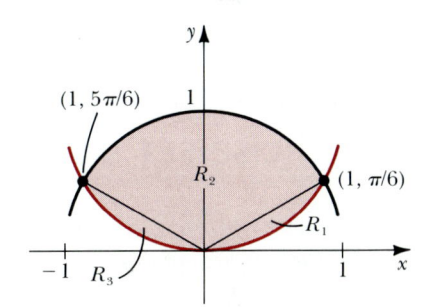

Figure 9.24

In Example 3 we found the points of intersection of two curves by solving the equations simultaneously. This method does not necessarily give *all* points of intersection of curves with polar equations. Because a point has an infinite number of sets of polar coordinates, it is possible for the intersection to occur at a point for which no single set of polar coordinates satisfies both equations. The following example shows this; other illustrations can be found in Exercises 31–36.

Example 4

Find the points of intersection of the cardioids $r = 1 - \cos \theta$ and $r = 1 + \sin \theta$.

Solution

The graphs of the cardioids are given in Figure 9.25, where three points of intersection are indicated. First we solve the equations simultaneously to obtain

$$1 - \cos \theta = 1 + \sin \theta \quad \text{so} \quad -\cos \theta = \sin \theta.$$

This equation is satisfied when $\theta = 3\pi/4$ and when $\theta = 7\pi/4$, so two points of intersection are $(1 + \sqrt{2}/2, 3\pi/4)$ and $(1 - \sqrt{2}/2, 7\pi/4)$. A third point of intersection is shown in Figure 9.25 to be at the origin. Polar coordinates of the origin that satisfy $r = 1 - \cos \theta$ are $(0, 0)$, because $1 - \cos 0 = 1 - 1 = 0$. However, $(0, 0)$ does not satisfy $r = 1 + \sin \theta$. The pole on the graph of $r = 1 + \sin \theta$ corresponds to the point $(0, 3\pi/2)$, since $1 + \sin (3\pi/2) = 1 + (-1) = 0$. ▶

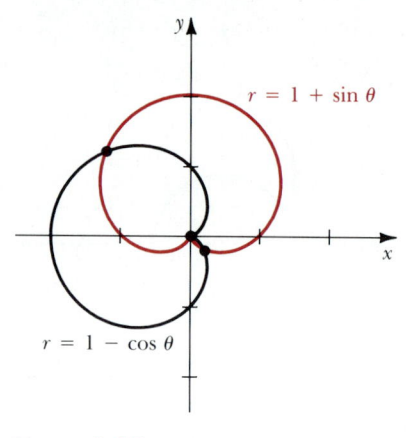

Figure 9.25

The preceding example shows that it is important to sketch the graphs of curves given in polar coordinates when finding points of intersection. It is quite possible for an intersection to occur at a point for which no single set of polar coordinates satisfies both equations.

▶ EXERCISE SET 9.3

In Exercises 1–16: (a) sketch the graph of the polar equation; (b) find the area of the region bounded by the graph.

1. $r = 2$
2. $r = \pi/2$
3. $r = 2 \sin \theta$
4. $r = 4 \cos \theta$
5. $r = -3 \cos \theta$
6. $r = -5 \sin \theta$
7. $r = 1 + \cos \theta$
8. $r = 2 - \sin \theta$
9. $r = 4 - 2 \sin \theta$
10. $r = 3 - 2 \cos \theta$
11. $r = \sin 2\theta$
12. $r = \sin 3\theta$
13. $r^2 = 4 \cos 2\theta$
14. $r^2 = 9 \sin 2\theta$
15. $r = \theta, -\pi/2 \leq \theta \leq \pi/2$
16. $r = \theta^2, -\pi < \theta < \pi$

In Exercises 17–24: a) Sketch the graphs of the pair of polar equations; b) Find the area of the region inside the graph of the first equation and outside the graph of the second equation.

17. $r = 2, r = 1$
18. $r = 2, r = 1 + \cos \theta$
19. $r = 1 + \cos \theta, r = 1$
20. $r = 1, r = 1 + \cos \theta$
21. $r = 2, r^2 = 4 \cos 2\theta$
22. $r = \sin \theta, r^2 = \sin 2\theta$
23. $r = 1 + \cos \theta, r = 1 + \sin \theta$
24. $r = 1 + \cos \theta, r = 1 - \cos \theta$

In Exercises 25–30:
  a) Sketch the graphs of the pair of polar equations.
  b) Find the area of the region inside the graphs of both equations.

25. $r = 1, r = 1 + \cos \theta$
26. $r = 1 + \cos \theta, r = 1 - \cos \theta$
27. $r = 1 + \cos \theta, r = 1 + \sin \theta$
28. $r = 2 + 2 \sin \theta, r = 3$
29. $r = \sin 2\theta, r = \cos 2\theta$
30. $r = 3 + 3 \cos \theta, r = -3 \cos \theta$

In Exercises 31–36: a) Show graphically that the pair of curves has intersections at which there is no single pair (r, θ) of polar coordinates that satisfies both equations; b) Find polar coordinates for the intersection points.

31. $r = \cos \theta, r = \sin \theta$
32. $r = \cos \theta + 1, r = -\cos \theta - 1$
33. $r = \cos \theta + 1, r = \cos \theta - 1$
34. $r = 1, r = 2 \sin 2\theta$
35. $r = \cos (\theta/2), r = \sin (\theta/2)$
36. $r = 2 \cos 2\theta, r = 2 \sin 2\theta$
37. a) Show that the graphs of $r = 1 + \cos \theta$ and $r^2 = 4 \cos \theta$ intersect at four distinct points.
    b) Determine the Cartesian coordinates of two of the points.
38. In Exercise 44 of Section 9.2, it was stated that the polar equation $r = a \cos \theta + b \sin \theta$ represents a circle through the origin with Cartesian equation
$$\left(x - \frac{a}{2}\right)^2 + \left(y - \frac{b}{2}\right)^2 = \frac{a^2 + b^2}{4}.$$
Use (9.6) to verify that the area of the circle is
$$\pi(a^2 + b^2)/4.$$
39. In Exercise 47 of Section 9.2, the equation $r = 2e^{-0.2\theta}$ is used to describe the outside of the shell of a chambered nautilus. Suppose θ_1 is a fixed positive real number, and A_n denotes the area enclosed by the spiral from $\theta = (n - 1)\theta_1$ to $\theta = n\theta_1$. Show that A_{n+1}/A_n is a constant depending only on θ_1.
40. a) Sketch the graph of the spiral of Archimedes given by the equation $r = a\theta$, where $a > 0$ is a constant and $\theta \geq 0$.
    b) Show that the area of the region lying outside the nth loop of the graph but inside the $(n + 1)$st loop of the graph is $8a^2\pi^3 n$.

9.4

Parametric Equations

In a number of earlier instances, we have expressed problems involving Cartesian-coordinate equations in terms of a third variable. For example, the definition of the basic trigonometric functions uses the Cartesian equation
$$x^2 + y^2 = 1$$
together with the specification that for every real number t,
$$x = \cos t \quad \text{and} \quad y = \sin t.$$

The related-rates problems in Section 3.3 are another example of this representation. They involve interrelated variables, each of which depends on time.

The following definition formally recognizes this type of expression.

(9.7)
DEFINITION
Suppose a **curve** C in the plane consists of the set of points of the form (x, y), where $x = f(t)$ and $y = g(t)$ for some pair of continuous functions f and g defined on an interval I. The equations $x = f(t), y = g(t)$ are said to give a **parametric representation** of C in the **parameter** t.

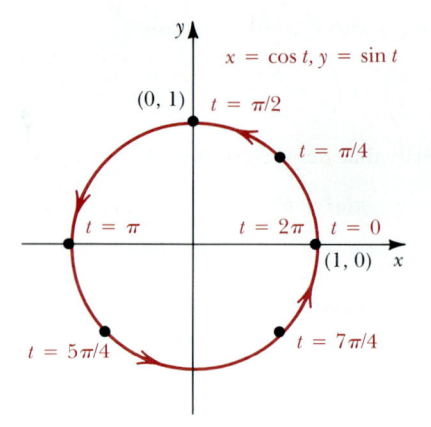

Figure 9.26

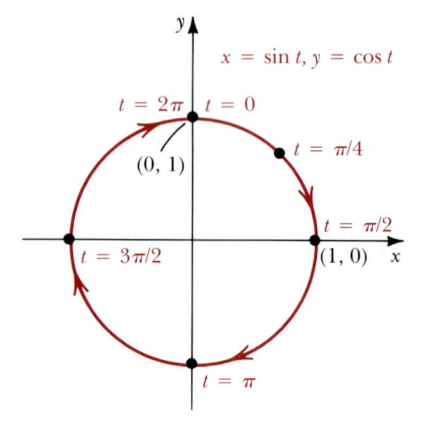

Figure 9.27

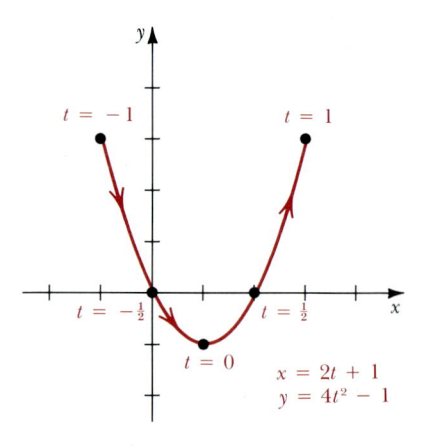

Figure 9.28

A parametric representation $x = f(t)$ and $y = g(t)$ for t in I is said to be a **smooth** representation for the curve C if f' and g' are continuous and never simultaneously zero on the interior of I. A curve having a smooth representation is called a **smooth curve**. Recall that the notion of smoothness was included in the discussion of arc length in Section 5.4.

The circle with center (0, 0) and radius 1 is a smooth curve; it is given, for example, by the smooth parametric representation:

$$x = \cos t \quad \text{and} \quad y = \sin t \quad \text{for } t \text{ in } [0, 2\pi].$$

The values of t shown on the graph in Figure 9.26 indicate that as t goes from 0 to 2π, points on the graph are traced counterclockwise beginning and ending at (1, 0). The arrowheads on the graph indicate the direction of increasing t.

Another set of parametric equations for the same circle is

$$x = \sin t \quad \text{and} \quad y = \cos t \quad \text{for } t \text{ in } [0, 2\pi].$$

With these equations, the points are traced clockwise beginning and ending at (0, 1), as shown in Figure 9.27.

Additional parametric representations for this circle are

$$x = \sin 2t \quad \text{and} \quad y = \cos 2t \quad \text{for } t \text{ in } [0, \pi]$$

and

$$x = \cos t^2 \quad \text{and} \quad y = \sin t^2 \quad \text{for } t \text{ in } [\sqrt{2\pi}, \sqrt{4\pi}].$$

In fact, the only conditions that must be satisfied to describe this curve are that

i) $x^2 + y^2 = 1$ and
ii) each point on the circle is represented by some value of the parameter.

Example 1

Describe and sketch the graph of the curve whose parametric equations are

$$x = 2t + 1, \quad y = 4t^2 - 1, \quad -1 \leq t \leq 1.$$

Solution

Eliminating the parameter t produces an equation in Cartesian coordinates,

$$t = \frac{x-1}{2}, \quad \text{so} \quad y = 4\left(\frac{x-1}{2}\right)^2 - 1 = (x-1)^2 - 1.$$

This is an equation of a parabola. As t varies from -1 to 1, $P(t)$ traces points on the curve from $(-1, 3)$ to $(3, 3)$. The curve is shown in Figure 9.28. ▶

Example 2

Find parametric equations for the line through (1, 2) with slope 1/2.

Solution

A Cartesian equation for this line is

$$y - 2 = \frac{1}{2}(x - 1).$$

We can find one set of parametric equations by letting $t = x - 1$. Then

$$y - 2 = \frac{1}{2}t, \quad \text{so} \quad y = \frac{1}{2}t + 2.$$

The line shown in Figure 9.29 is described by the parametric equations

$$x = t + 1, \quad y = \frac{1}{2}t + 2, \quad -\infty < t < \infty.$$

We can find another set of parametric equations that describe this line by letting

$$x = t, \quad y = \frac{1}{2}(t - 1) + 2, \quad -\infty < t < \infty.$$

A third set is found by letting

$$x = \ln t + 1, \quad y = \frac{1}{2}\ln t + 2 = \ln \sqrt{t} + 2, \quad 0 < t < \infty.$$

The only conditions that need be fulfilled are that

$$y - 2 = \frac{1}{2}(x - 1)$$

and that both x and y assume all real values. ▶

Note that the parametric equations

$$x = t^2, \quad y = \frac{1}{2}(t^2 + 3), \quad -\infty < t < \infty,$$

do not describe the entire line in Example 2. This set of equations describes only the portion of the line when $x \geq 0$ and traces this portion of the line twice for t in $(-\infty, \infty)$, as shown by the solid portion of the line in Figure 9.30.

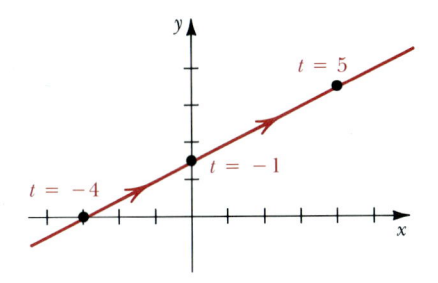

Figure 9.29

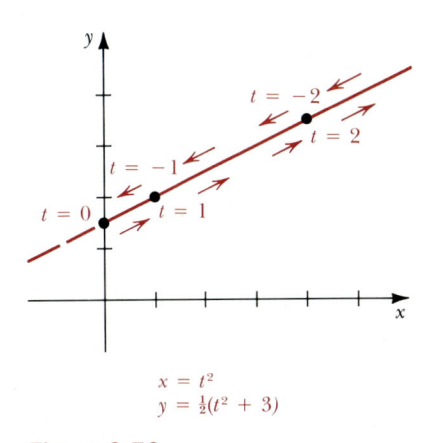

Figure 9.30
$x = t^2$
$y = \frac{1}{2}(t^2 + 3)$

Example 3

A circle of radius R rolls along a horizontal line. Find parametric equations of the curve traced by a fixed point P on the circumference of the circle. This curve is called a **cycloid.** (It is the curve described, for example, by the head of a nail embedded in a tire as the tire rolls along the ground.)

Solution
Suppose the horizontal line along which the circle rolls is the *x*-axis and that when $t = 0$ the point *P* is at the origin. Let *O* denote the center of the circle and *t* denote the radian measure of the angle through which the line *OP* has moved. Suppose $0 < t < \pi/2$ and that *P* has moved to the position shown in Figure 9.31.

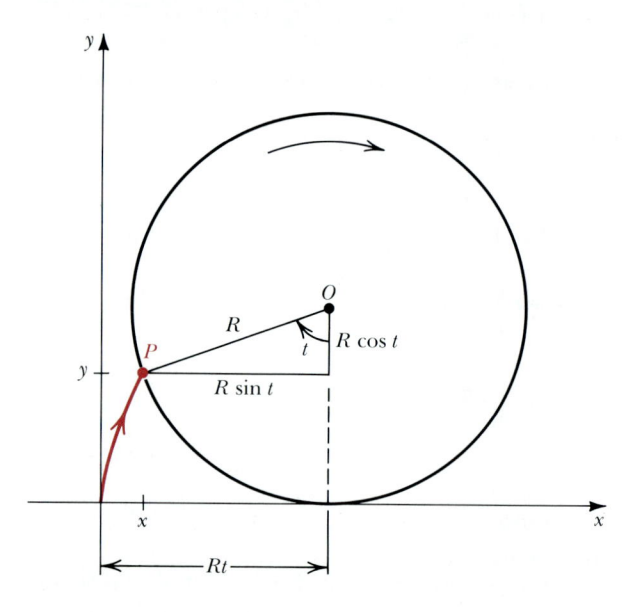

Figure 9.31
Generating a cycloid.

The coordinates (x, y) of *P* are

$$x = Rt - R \sin t \quad \text{and} \quad y = R - R \cos t.$$

In fact, these equations hold for any $t \geq 0$. The curve that this set of parametric equations describes is shown in Figure 9.32. ▶

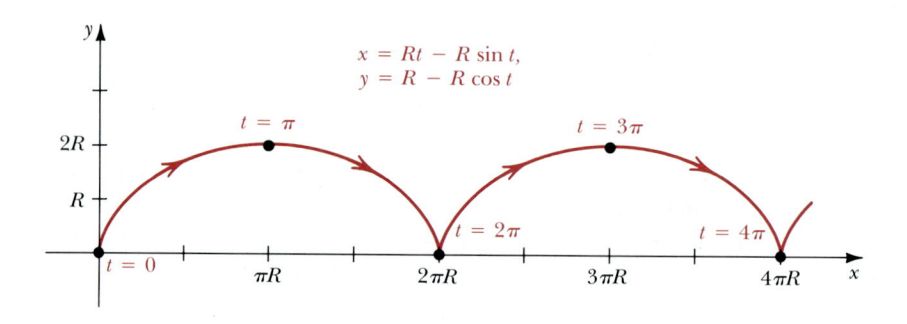

Figure 9.32
A cycloid.

The graph of the cycloid comes to a "point" when $t = 2\pi$, and in an intuitive sense it is not smooth there. Although the derivatives of the parametric equations

$$x'(t) = R - R \cos t \quad \text{and} \quad y'(t) = R \sin t$$

are both continuous functions, they are simultaneously zero when *t* is any integral multiple of 2π. This permits the graph to change direction abruptly at these points and demonstrates why the definition of smoothness requires that the derivatives not be simultaneously zero.

An interesting property of the cycloid is derived by considering the area bounded by the x-axis and one arc of the curve. One arc is described by restricting t to $[0, 2\pi]$. The area is given by

$$A = \int_{t=0}^{t=2\pi} y\, dx = \int_{t=0}^{t=2\pi} (R - R\cos t)\, dx.$$

Since $x = Rt - R\sin t$, we have $dx = (R - R\cos t)\, dt$ and

$$A = \int_0^{2\pi} (R - R\cos t)(R - R\cos t)\, dt$$

$$= R^2 \int_0^{2\pi} [1 - 2\cos t + (\cos t)^2]\, dt$$

$$= R^2 \left\{ \left[t - 2\sin t \right]_0^{2\pi} + \frac{1}{2} \int_0^{2\pi} (1 + \cos 2t)\, dt \right\}$$

$$= R^2 \left\{ 2\pi + \frac{1}{2} \left[t + \frac{1}{2} \sin 2t \right]_0^{2\pi} \right\} = 3\pi R^2.$$

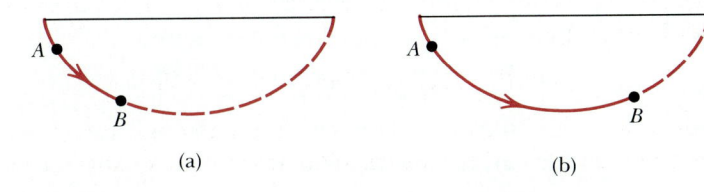

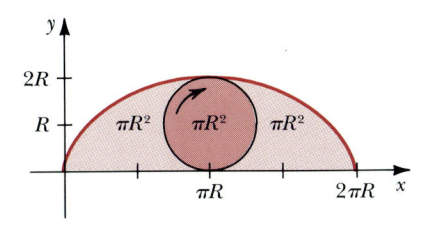

Figure 9.33

Figure 9.33 gives a geometric interpretation of this result. When the circle has completed half its revolution, the boundary of the circle divides the region bounded by the graph of the cycloid and the x-axis into three parts with equal area.

If the parametric equations for the cycloid are expressed with $R < 0$, the graph of the cycloid is as shown in Figure 9.34. This curve is the solution to two famous historical problems:

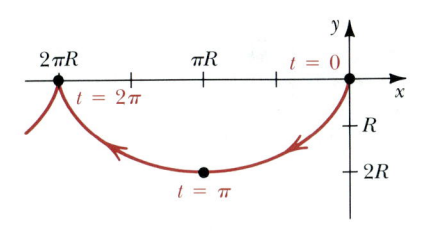

Figure 9.34

1. The brachistochrone or "shortest-time" problem

Find the curve that describes the shape a wire should assume in order to enable a frictionless bead to slide from a point A to a lower point B in the shortest time (see Figure 9.35a).

An interesting feature of the solution to this problem is that for some points B, the shortest time is accomplished by sliding to a point lower than B and then up to B, as shown in Figure 9.35(b).

Johann Bernoulli (1667–1748) posed this problem to the mathematical world in January 1697. Separate solutions to the problem were published in *Acta eruditorum* in May 1697 by Johann and his older brother Jakob (1654–1705). Leibniz and Newton also had solutions to the problem published in this journal. Newton received the problem in the late afternoon of January 29, 1697, and the next morning posted his

Figure 9.35
Brachistochrone problem.

solution to the Royal Society in London to be submitted anonymously. The elegance of the solution, however, left little doubt as to the author. Johann Bernoulli, upon seeing it, allegedly remarked "the lion is known by his paw."

2. The tautochrone or "equal-time" problem

Find a curve that describes the shape a wire should assume if the time it takes a frictionless bead to travel to the bottom of the curve is independent of the point on the curve at which the bead starts.

Three beads starting at the same time from the points A, B, and C on the cycloid shown in Figure 9.36 reach P simultaneously.

Christiaan Huygens published the solution to this problem in his book *Horologium oscillatorium* (1673). This was a study of mathematics and mechanics motivated by the search for a reliable clock. Using the tautochrone property of the cycloid, Huygens, in 1656/57, constructed a pendulum clock having a period independent of the amplitude of its oscillation.

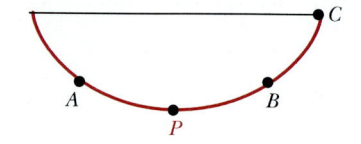

Figure 9.36
Tautochrone problem.

▶ **EXERCISE SET 9.4**

In Exercises 1–16, (a) sketch the graphs of the curves described by the parametric equations and (b) find a Cartesian equation describing each curve.

1. $x = 3t$, $y = t/2$
2. $x = 2t + 1$, $y = 3t - 2$
3. $x = \sqrt{t}$, $y = t + 1$
4. $x = t + 2$, $y = -3\sqrt{t}$
5. $x = \sin t$, $y = \cos^2 t$
6. $x = \sec t$, $y = \tan t$
7. $x = 3 \sin t$, $y = 4 \cos t$
8. $x = 3 \sec t$, $y = 4 \tan t$
9. $x = e^t$, $y = e^{-t}$
10. $x = \ln t$, $y = \ln \sqrt{t}$
11. $x = \sin t$, $y = \sec t$
12. $x = \sin t$, $y = \cot t$
13. $x = 4 + 2 \cos t$, $y = 6 + 2 \sin t$
14. $x = \sin t + 1$, $y = 2 \cos t - 1$
15. $x = 4t^3$, $y = 2t^2 + 1$
16. $x = e^{-t} - 2$, $y = e^{2t} + 3$

In Exercises 17–20, sketch the graphs of the parametric equations.

17. a) $x = t$, $y = t^2$
    b) $x = t^2$, $y = t$
    c) $x = t^2$, $y = t^4$
    d) $x = t^4$, $y = t^2$
18. a) $x = 2t + 1$, $y = 3t$
    b) $x = 3t + 1$, $y = 2t$
    c) $x = 2t$, $y = 3t + 1$
    d) $x = 3t$, $y = 2t + 1$
19. a) $x = \sin t + 1$, $y = \cos t$
    b) $x = \sin t$, $y = \cos t + 1$
    c) $x = \cos t + 1$, $y = \sin t$
    d) $x = \cos t$, $y = \sin t + 1$
20. a) $x = 1 - \sin t$, $y = 1 - \cos t$
    b) $x = \sin t - 1$, $y = \cos t - 1$
    c) $x = 1 - \cos t$, $y = 1 - \sin t$
    d) $x = \cos t - 1$, $y = \sin t - 1$

In Exercises 21–26, the graphs of the parametric equations in each part represent a portion of the same curve. Sketch the graphs of these equations, and label representative values of the parameter.

21. a) $x = \cos t$, $y = \sin t$; $0 \le t \le 2\pi$
    b) $x = \sin t$, $y = \cos t$; $0 \le t \le 2\pi$
    c) $x = t$, $y = \sqrt{1 - t^2}$; $-1 \le t \le 1$
    d) $x = -t$, $y = \sqrt{1 - t^2}$; $-1 \le t \le 1$
22. a) $x = t + 1$, $y = 4t + 5$

b) $x = t^2 - \frac{1}{4}, y = 4t^2$
c) $x = \ln t, y = 1 + \ln t^4; t > 0$
d) $x = \sin t, y = 1 + 4 \sin t; 0 \leq t \leq \pi$

23. a) $x = t, y = \ln t; t > 1$
    b) $x = e^t, y = t; t > 0$
    c) $x = t^2, y = 2 \ln t; t > 1$
    d) $x = 1/t, y = -\ln t; t > 1$

24. a) $x = t, y = 1/t; t > 0$
    b) $x = e^t, y = e^{-t}; t > 0$
    c) $x = \sin t, y = \csc t; 0 < t < \pi$
    d) $x = \tan t, y = \cot t; 0 < t < \pi/2$

25. a) $x = t, y = t^2$
    b) $x = \sqrt{t}, y = t$
    c) $x = \sin t, y = 1 - \cos^2 t$
    d) $x = -e^t, y = e^{2t}$

26. a) $x = t, y = \sqrt{t^2 - 1}; t > 1$
    b) $x = \sqrt{t^2 + 1}, y = t; t > 1$
    c) $x = \sec t, y = \tan t$
    d) $x = \cosh t, y = \sinh t$

27. Does the curve given parametrically by $x = (\sin t)/t$, $y = (\cos t)/t$ have a horizontal or vertical asymptote?

28. Consider the curves described by the following sets of parametric equations for t in $(-\infty, \infty)$.
    i) Show that each set of equations has a point at which the derivatives of the equations are simultaneously zero.
    ii) Sketch the graphs of the equations, and decide which curves are smooth in an intuitive sense.
    iii) Find a smooth parametric representation for the curves you think are smooth.
    a) $x = t^2, y = t^2$ b) $x = t^3, y = t^3$
    c) $x = t^2, y = t^3$ d) $x = t^3, y = t^2$
    e) $x = t^2, y = t^4$ f) $x = t^4, y = t^2$

29. The curve with the Cartesian equation $x^3 + y^3 = xy$ is called a *folium of Descartes*.
    a) Show that this curve has a parametric representation
    $$x = \frac{t^2}{1 + t^3}, \quad y = \frac{t}{1 + t^3}, \quad \text{for } t \neq -1.$$
    b) Sketch the graph of this curve, and use arrows to indicate increasing values of t.

30. A 20-ft ladder rests against a vertical wall. A cat starts to climb the ladder at the rate of 1 ft/sec, and the bottom of the ladder slides away from the wall at 1 ft/sec (see the figure). Find each of the following.
    a) The equation of the curve described by the cat
    b) The maximum distance of the cat from the floor
    c) The maximum distance of the cat from the wall
    d) The maximum distance of the cat from the corner
    e) The vertical velocity of the cat at the end of 16 sec

31. An *epicycloid* is a curve traced by a fixed point P on a circle of radius b as it rotates around a stationary circle of radius a (see the figure). Suppose that the fixed point P is at $A(a, 0)$ when $t = 0$, and that the circle rotates as shown in the figure.
    a) Show that the point C has coordinates
    $$((a + b) \cos t, (a + b) \sin t).$$
    b) Show that the arclength PB on the circle of radius b is at.
    c) Show that the angle BCP is $(a/b)t$.
    d) Show that the angle DCP is $\pi - ((a + b)/b)t$.
    e) Derive the parametric equations for the epicycloid:
    $$x = (a + b) \cos t - b \cos [((a + b)/b)t],$$
    $$y = (a + b) \sin t - b \sin [((a + b)/b)t].$$

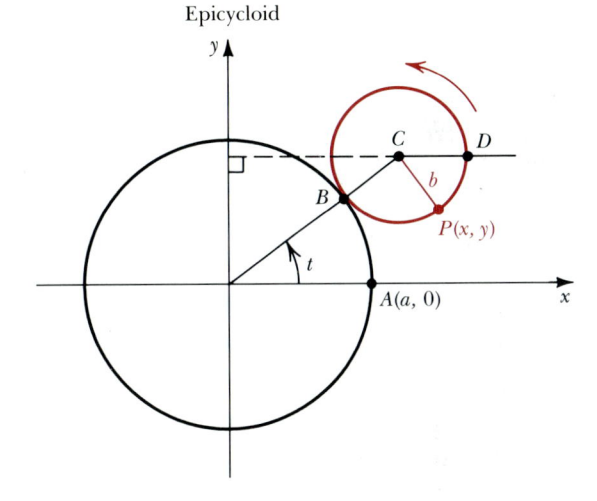

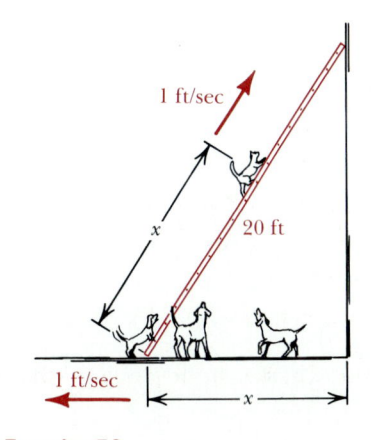

Figure for Exercise 30.

32. A *hypocycloid* is a curve traced by a fixed point P on a circle of radius b as it rotates inside a stationary circle of radius a, where $b < a$ (see the figure). Suppose that the fixed point P is at $A(a, 0)$ when $t = 0$, and that the circle rotates as shown in the figure. By completing steps similar to those in Exercise 31, derive the parametric equations for the hypocycloid:

$$x = (a - b) \cos t + b \cos [((a - b)/b)t],$$
$$y = (a - b) \sin t - b \sin [((a - b)/b)t].$$

33. Sketch the graphs of the epicycloid when (a) $b = a$ and (b) when $b = a/4$.

34. a) Sketch the graph of the hypocycloid when $b = a/4$.
   b) Show that in this case the hypocycloid also has equation $x^{2/3} + y^{2/3} = a^{2/3}$.

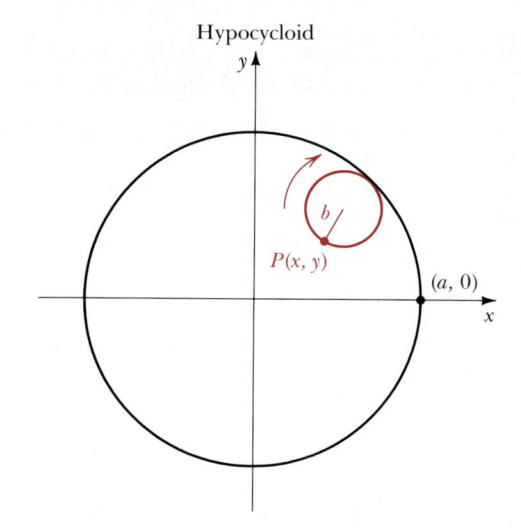

9.5

Tangent Lines to Curves

In this section we consider the problem of finding an equation of a tangent line to a smooth curve described parametrically. Suppose C is a curve described by

$$x = f(t), \quad y = g(t), \quad \text{for } t \text{ in } I,$$

where f and g are differentiable functions. If $f'(t) = dx/dt$ is never zero in I, then f has a continuous inverse and $y = g(t) = g(f^{-1}(x))$. So y is a function of x and, by the Chain Rule,

$$\frac{dy}{dt} = \frac{dy}{dx}\frac{dx}{dt}.$$

Thus,

(9.8) $$\frac{dy}{dx} = \frac{dy/dt}{dx/dt} = \frac{g'(t)}{f'(t)}.$$

Example 1

Find an equation of the tangent line to the curve described by $x = 3t + t^3$, $y = t^3 - 9t^2$ at the point when $t = 1$.

Solution

The slope of the tangent line at any point is given by the derivative:

$$\frac{dy}{dx} = \frac{dy/dt}{dx/dt} = \frac{3t^2 - 18t}{3 + 3t^2} = \frac{t^2 - 6t}{1 + t^2}.$$

When $t = 1$, the slope is $-\frac{5}{2}$ and the point on the curve is $(4, -8)$. Consequently, an equation of the tangent line is

$$y + 8 = -\tfrac{5}{2}(x - 4) \quad \text{or} \quad 5x + 2y - 4 = 0. \qquad \blacktriangleright$$

Example 2

Sketch the curve described by $x = t^2$, $y = t^3 - 3t$, $-\infty < t < \infty$.

Solution

We can use the first derivative $y'(x)$ to determine when the graph is increasing and when the graph is decreasing:

$$y'(x) = \frac{dy}{dx} = \frac{dy/dt}{dx/dt} = \frac{3t^2 - 3}{2t} = \frac{3(t - 1)(t + 1)}{2t}.$$

The derivative is zero at $t = 1$ and $t = -1$, and the curve has horizontal tangent lines at these points. Definition (2.2) implies that the graph has a vertical tangent line at $t = 0$ since

$$\lim_{t \to 0} |y'(x)| = \lim_{t \to 0} \left| \frac{3(t - 1)(t + 1)}{2t} \right| = \infty.$$

The second derivative $y''(x)$ will determine the concavity of the graph. Care must be taken when finding $y''(x)$ because the first derivative is expressed in terms of t. Since

$$y''(x) = \frac{d^2 y}{dx^2} = \frac{d(dy/dx)}{dx},$$

Equation (9.8) implies that

$$y''(x) = \frac{\dfrac{d(dy/dx)}{dt}}{\dfrac{dx}{dt}} = \frac{D_t \left(\dfrac{3t^2 - 3}{2t} \right)}{D_t t^2} = \frac{\dfrac{3}{2}(1 + t^{-2})}{2t} = \frac{3}{4} \left(\frac{t^2 + 1}{t^3} \right).$$

Thus $y''(x)$ is positive when $t > 0$, and the graph is concave upward. When $t < 0$, $y''(x)$ is negative, and the graph is concave downward. The results concerning these derivatives are summarized in the following chart.

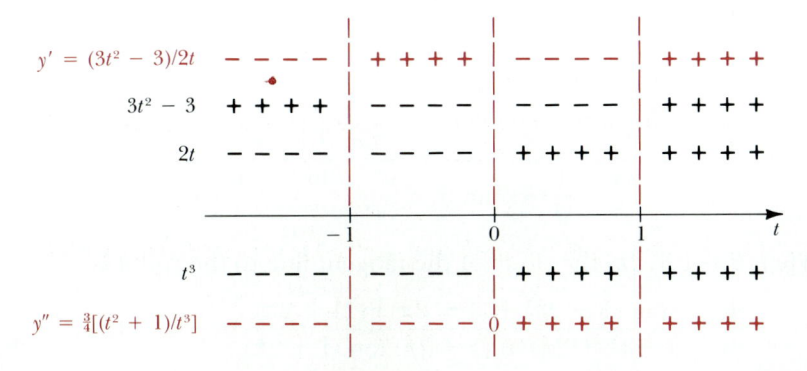

Setting $y = 0$ and solving for t implies that the curve intersects the x-axis when $t = 0$ and when $t = \pm\sqrt{3}$. The graph intersects the y-axis only when $t = 0$.

Plotting the points for $t = -2, -1, 0, 1, 2$ and using the information on the chart, we obtain the graph shown in Figure 9.37. ▶

Parametric equations are introduced in this chapter because of the natural relationship between this concept and functions defined using polar coordinates. If r is a continuous function on an interval I, then

$$x = r(\theta) \cos \theta \quad \text{and} \quad y = r(\theta) \sin \theta$$

are parametric equations with θ as the parameter. We implicitly used this fact in Section 9.2 when sketching the graphs of polar equations in a Cartesian coordinate system.

In this case, θ is the parameter and

$$\frac{dy}{dx} = \frac{dy/d\theta}{dx/d\theta} = \frac{D_\theta[r(\theta) \sin \theta]}{D_\theta[r(\theta) \cos \theta]} = \frac{r'(\theta) \sin \theta + r(\theta) \cos \theta}{r'(\theta) \cos \theta - r(\theta) \sin \theta},$$

or, dividing by $\cos \theta$, if $\cos \theta \neq 0$,

(9.9)
$$\frac{dy}{dx} = \frac{\dfrac{dr}{d\theta} \tan \theta + r}{\dfrac{dr}{d\theta} - r \tan \theta}.$$

Example 3

Find the slope of the tangent line to the spiral with polar equation

$$r = \ln \theta, \qquad \theta \geq 1,$$

at the end of one rotation about the origin.

Solution

The spiral is shown in Figure 9.38. One rotation is complete when $\theta = 1 + 2\pi$ and $r = \ln(1 + 2\pi)$. Since

$$\frac{dr}{d\theta} = \frac{1}{\theta},$$

we have

$$\frac{dy}{dx} = \frac{\dfrac{\tan \theta}{\theta} + r}{\dfrac{1}{\theta} - r \tan \theta} = \frac{\tan \theta + \theta \ln \theta}{1 - \theta(\ln \theta)(\tan \theta)}.$$

When $\theta = 1 + 2\pi$, the slope of the tangent line to the spiral is

$$\frac{dy}{dx} = \frac{\tan(1 + 2\pi) + (1 + 2\pi) \ln(1 + 2\pi)}{1 - (1 + 2\pi) \ln(1 + 2\pi) \tan(1 + 2\pi)} \approx -0.74429.$$ ▶

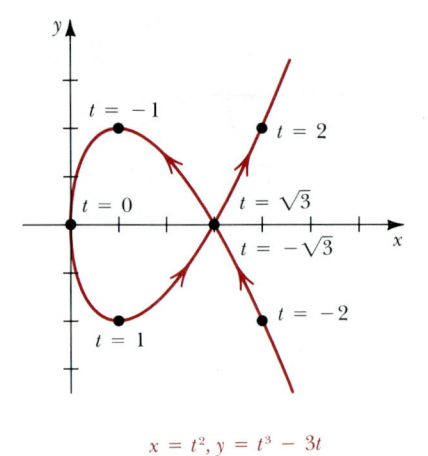

$x = t^2, y = t^3 - 3t$

Figure 9.37

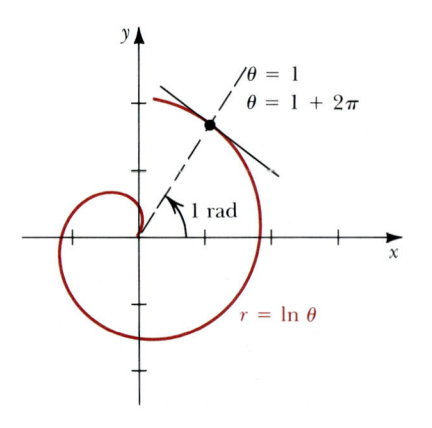

Figure 9.38

Example 4

Find a Cartesian equation of the tangent line to the cardioid $r = 1 - \cos \theta$ at the point with polar coordinates $(3/2, 2\pi/3)$.

Solution
Using (9.9), we see that the slope of the tangent line at any point (r, θ) is

$$\frac{dy}{dx} = \frac{\sin \theta \tan \theta + 1 - \cos \theta}{\sin \theta - (1 - \cos \theta) \tan \theta}.$$

Thus the slope at $(3/2, 2\pi/3)$ is

$$\left.\frac{dy}{dx}\right|_{\theta = \frac{2\pi}{3}} = \frac{\frac{\sqrt{3}}{2}(-\sqrt{3}) + 1 + \frac{1}{2}}{\frac{\sqrt{3}}{2} - \left(1 + \frac{1}{2}\right)(-\sqrt{3})} = 0,$$

and the tangent line is horizontal. The y-coordinate of the point with polar coordinates $(3/2, 2\pi/3)$ is

$$y = \frac{3}{2} \sin \frac{2\pi}{3} = \frac{3\sqrt{3}}{4}.$$

Consequently, the Cartesian equation of the tangent line shown in Figure 9.39 is

$$y = \frac{3\sqrt{3}}{4}.$$

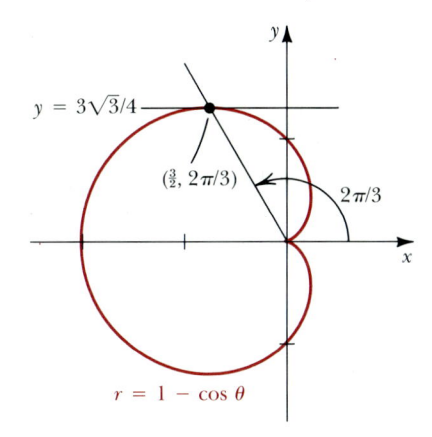

▶

Figure 9.39

▶ EXERCISE SET 9.5

In Exercises 1–16, find the slope, if it exists, of the tangent line to the curve at the specified value of t. Check this result with the sketch of the curve that was determined in the corresponding exercise in Exercise Set 9.4.

1. $x = 3t, y = t/2, t = 1$
2. $x = 2t + 1, y = 3t - 2, t = -2$
3. $x = \sqrt{t}, y = t + 1, t = 4$
4. $x = t + 2, y = -3\sqrt{t}, t = 9$
5. $x = \sin t, y = \cos^2 t, t = \pi/3$
6. $x = \sec t, y = \tan t, t = \pi/4$
7. $x = 3 \sin t, y = 4 \cos t, t = \pi$
8. $x = 3 \sec t, y = 4 \tan t, t = \pi/4$
9. $x = e^t, y = e^{-t}, t = 0$
10. $x = \ln t, y = \ln \sqrt{t}, t = e$
11. $x = \sin t, y = \sec t, t = \pi/4$
12. $x = \sin t, y = \cot t, t = \pi/2$
13. $x = 4 + 2 \cos t, y = 6 + 2 \sin t, t = \pi/2$
14. $x = \sin t + 1, y = 2 \cos t - 1, t = \pi$
15. $x = 4t^3, y = 2t^2 + 1, t = -1$
16. $x = e^{-t} - 2, y = e^{2t} + 3, t = 1$

In Exercises 17–30: a) Find the slope, if it exists, of the tangent line to the polar curve at the specified value of θ; b) Find a Cartesian equation of the tangent line.

17. $r = 1, \theta = \pi/4$
18. $r = 2, \theta = \pi/3$
19. $r = 2 \sin \theta, \theta = \pi/4$
20. $r = -3 \cos \theta, \theta = \pi/4$
21. $r = 1 + \cos \theta, \theta = \pi/2$
22. $r = 2 + \sin \theta, \theta = 0$
23. $r = 3 \sin 2\theta, \theta = \pi/4$
24. $r = 4 \cos 2\theta, \theta = \pi/4$
25. $r^2 = 4 \cos 2\theta, \theta = \pi/6$
26. $r^2 = 16 \sin 4\theta, \theta = \pi/16$
27. $r = 0, \theta = 5\pi/4$
28. $r = \ln \theta, \theta = e$
29. $r = 2^{-\theta}, \theta = -2$
30. $r = 3 \sec \theta, \theta = \pi/4$

In Exercises 31–40, find $D_x^2 y$.

31. $x = 3t, y = t/2$
32. $x = 2t + 1, y = 3t - 2$
33. $x = 3 \sin t, y = 4 \cos t$
34. $x = \sec t, y = \tan t$
35. $x = e^t, y = e^{-t}$
36. $x = 4t^3, y = 2t^2 + 1$
37. $r = \sin \theta$
38. $r = 2 + \cos \theta$
39. $r = 2^{-\theta}$
40. $r = \ln \theta$

In Exercises 41–46, a pair of intersecting polar curves is listed. Find equations of the tangent lines to the curves at the points of intersection.

41. $r = 1, r = 1 + \cos \theta$
42. $r = 1 + \cos \theta, r = 1 - \cos \theta$
43. $r = 1 + \cos \theta, r = 1 + \sin \theta; 0 \le \theta \le \pi/2$
44. $r = 3, r = 2 + 2 \sin \theta$
45. $r = \sin \theta, r = \cos \theta$
46. $r = \sin 2\theta, r = \cos 2\theta, 0 \le \theta \le \pi/2$
47. a) Sketch the graph of the curve described by
$$x = 1 - t^2, \quad y = t - t^3, \quad \text{for } -1 \le t \le 1.$$
   b) Find the area of the region bounded by the curve described in part (a).
48. a) Find the slope of the tangent line at each point of the *astroid* described by
$$x = \cos^3 t, \quad y = \sin^3 t, \quad \text{for } 0 \le t \le 2\pi.$$
   b) Sketch the graph of the astroid.
49. Find the values of θ that produce horizontal and vertical tangents to the graph of $r = e^\theta$.
50. The parametric equations
$$x = a \tan \theta, \quad y = a \cos^2 \theta$$
describe a curve known as the *witch of Agnesi*. Express y in terms of x and sketch this curve.

51. The *involute* of a circle is a curve generated by the end of a string kept taut while being unwound from a circle (see the figure). Show that this curve has parametric equations
$$x = a \cos t + at \sin t, \quad y = a \sin t - at \cos t.$$

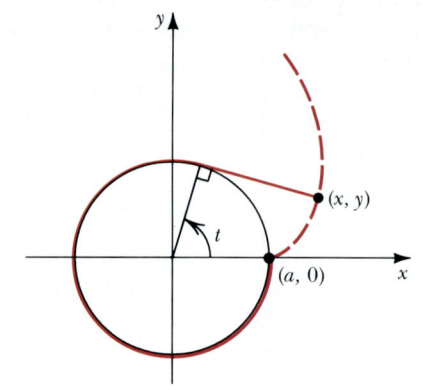

52. A goat is tied to a silo of radius a ft with enough rope to just reach the opposite side of the silo. Show that the grazing area available to the goat is $\frac{5}{6} \pi^3 a^2$. [*Hint:* Consider Exercise 51.]

53. To connect a circular segment of roadbed for a highway or railroad to a straight segment, engineers often use a cubic spiral of the form $r = a_3 \theta^3 + a_2 \theta^2 + a_1 \theta + a_0$. The constants are chosen so that both the segments and their tangents agree with the cubic spiral (see the figure). Find an equation of the cubic spiral shown in the figure.

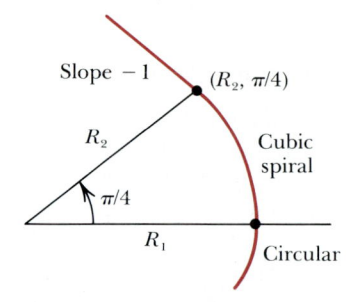

HISTORICAL NOTE

Maria Gaetana Agnesi (1718–1799) discussed the curve that is now called the witch of Agnesi in her calculus book *Instituzioni analitiche*, published in 1748. This was the first comprehensive calculus text published after the book of l'Hôpital, and was translated into English in 1801. The curve was called a *versiera*, which means "to turn," but was mistranslated into English as "witch," or "she-devil." However, the curve had been discovered earlier by Fermat and was studied by **Guido Grandi** (1671–1742) in his *Quadratura circuli et hyperbolae*, published in 1703 and 1710. Grandi investigated a number of flower-like curves in his *Flores geometrici* (1728).

9.6

Lengths of Curves

In this chapter we have seen a number of curves (cardioids, spirals, and cycloids, for example) that are not easily represented by a Cartesian equation, but can be represented using polar or parametric equations. In this section we derive a formula for the length of a curve described parametrically. An easily applied formula is then deduced for finding the length of a curve given in polar-equation form.

Suppose C is a smooth curve described by the parametric equations

$$x = f(t), \quad y = g(t), \quad \text{for } t \text{ in } [a, b].$$

If $\mathcal{P} = \{t_0, t_1, \ldots, t_n\}$ is a partition of $[a, b]$, then the straight-line segments joining the points $(f(t_{i-1}), g(t_{i-1}))$, and $(f(t_i), g(t_i))$ for each $i = 1, 2, \ldots, n$, approximate the curve C, as shown in Figure 9.40. The sum of their lengths approximates the length of C.

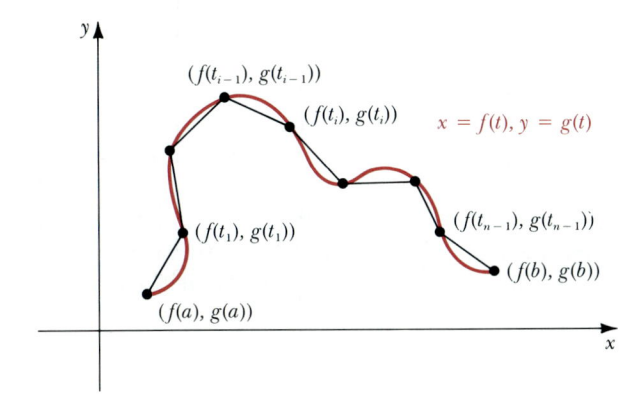

Figure 9.40

Since the length of the ith line segment is

$$\sqrt{[f(t_i) - f(t_{i-1})]^2 + [g(t_i) - g(t_{i-1})]^2},$$

the total length of the straight-line segments is

$$\sum_{i=1}^{n} \Delta L_i = \sum_{i=1}^{n} \sqrt{[f(t_i) - f(t_{i-1})]^2 + [g(t_i) - g(t_{i-1})]^2}$$

$$= \sum_{i=1}^{n} \left\{ \left[\frac{f(t_i) - f(t_{i-1})}{t_i - t_{i-1}} \right]^2 + \left[\frac{g(t_i) - g(t_{i-1})}{t_i - t_{i-1}} \right]^2 \right\}^{1/2} \Delta t_i.$$

The curve C is smooth on $[a, b]$ so the functions are both differentiable and satisfy the hypotheses of the Mean Value Theorem on $[a, b]$. This implies that for each $i = 1, 2, \ldots, n$, numbers w_i and z_i in $[t_{i-1}, t_i]$ exist with

$$\frac{f(t_i) - f(t_{i-1})}{t_i - t_{i-1}} = f'(w_i) \quad \text{and} \quad \frac{g(t_i) - g(t_{i-1})}{t_i - t_{i-1}} = g'(z_i).$$

Consequently, the length of the curve is approximated by

$$\sum_{i=1}^{n} \{[f'(w_i)]^2 + [g'(z_i)]^2\}^{1/2} \Delta t_i,$$

where w_i and z_i are in $[t_{i-1}, t_i]$ for each i. This is not a Riemann sum because it is possible and indeed likely that $w_i \ne z_i$. However, a result known as *Duhamel's principle for integrals* can be used to show that

$$\int_a^b \{[f'(t)]^2 + [g'(t)]^2\}^{1/2} \, dt = \lim_{\|\mathcal{P}\| \to 0} \sum_{i=1}^{n} \{[f'(w_i)]^2 + [g'(z_i)]^2\}^{1/2} \Delta t_i,$$

provided f' and g' are continuous on $[a, b]$. (The theorem of Bliss referred to in Section 5.4 is a special case of Duhamel's principle.)

(9.10)
DEFINITION

If C is a curve described by

$$x = f(t), \quad y = g(t), \quad a \le t \le b,$$

then the **arc length** of C is

$$L_C = \int_a^b \{[f'(t)]^2 + [g'(t)]^2\}^{1/2} \, dt,$$

provided the integral exists.

When the integral in Definition (9.10) exists, C is called a **rectifiable** curve (see Figure 9.41). Theorem (4.9) states that every function continuous on an interval is integrable on that interval. So, in particular, every *smooth* curve is rectifiable.

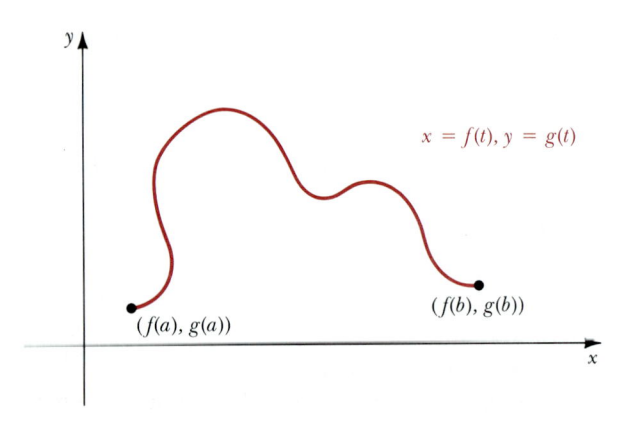

Figure 9.41
The length is $\int_a^b \{[f'(t)]^2 + [g'(t)]^2\}^{1/2} \, dt$.

Example 1

Find the length of one arch of the cycloid with parametric equations

$$x = t - \sin t, \quad y = 1 - \cos t.$$

Solution
The graph of the cycloid is shown in Figure 9.42. As t varies from 0 to 2π, the corresponding point $P(t)$ on the cycloid moves from the point $(0, 0)$ to the point $(2\pi, 0)$. Thus,

$$L = \int_0^{2\pi} \{[D_t(t - \sin t)]^2 + [D_t(1 - \cos t)]^2\}^{1/2} \, dt$$

$$= \int_0^{2\pi} [(1 - \cos t)^2 + (\sin t)^2]^{1/2} \, dt$$

$$= \int_0^{2\pi} [1 - 2\cos t + \cos^2 t + \sin^2 t]^{1/2} \, dt$$

$$= \int_0^{2\pi} [2 - 2\cos t]^{1/2} \, dt.$$

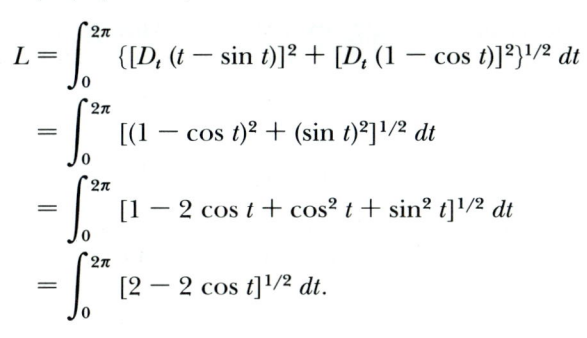

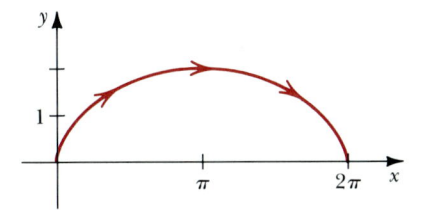

Figure 9.42

To evaluate this integral, first recall the identity $\sin^2 x = \frac{1}{2}[1 - \cos 2x]$. Thus,

$$\sin^2\left(\frac{t}{2}\right) = \frac{1 - \cos t}{2} \quad \text{and} \quad 2 - 2\cos t = 4\sin^2\left(\frac{t}{2}\right).$$

For t in $[0, 2\pi]$, $\sin(t/2) \geq 0$, so

$$L = \int_0^{2\pi} (2 - 2\cos t)^{1/2} \, dt$$

$$= \int_0^{2\pi} 2\sin\left(\frac{t}{2}\right) dt$$

$$= -4\cos\frac{t}{2}\Big]_0^{2\pi} = 8. \quad \blacktriangleright$$

If h is a function with a continuous derivative on $[a, b]$ and $y = h(x)$, then the parametric equations

$$x = t, \quad y = h(t),$$

describe a curve C whose length, by Definition (9.10), is

$$L_C = \int_a^b \{[1]^2 + [h'(t)]^2\}^{1/2} \, dt$$

$$= \int_a^b \sqrt{1 + [h'(x)]^2} \, dx.$$

This shows that when y is expressed as a function of x, Definition (9.10) reduces to the definition of the arc length of a curve given in Section 5.4.

When C is a curve described by a polar equation

$$r = f(\theta), \quad \alpha \leq \theta \leq \beta,$$

the parametric equations

$$x = f(\theta)\cos\theta, \quad y = f(\theta)\sin\theta, \quad \alpha \leq \theta \leq \beta$$

also describe C. In this case, Definition 9.10 implies that

$$L_C = \int_\alpha^\beta \{[D_\theta(f(\theta)\cos\theta)]^2 + [D_\theta(f(\theta)\sin\theta)]^2\}^{1/2}\,d\theta$$

$$= \int_\alpha^\beta \{[f'(\theta)\cos\theta - f(\theta)\sin\theta]^2 + [f'(\theta)\sin\theta + f(\theta)\cos\theta]^2\}^{1/2}\,d\theta$$

$$= \int_\alpha^\beta \{[f'(\theta)]^2(\cos\theta)^2 - 2f'(\theta)f(\theta)\cos\theta\sin\theta + [f(\theta)]^2(\sin\theta)^2$$
$$+ [f'(\theta)]^2(\sin\theta)^2 + 2f'(\theta)f(\theta)\sin\theta\cos\theta + [f(\theta)]^2(\cos\theta)^2\}^{1/2}\,d\theta$$

$$= \int_\alpha^\beta \{[f'(\theta)]^2(\cos^2\theta + \sin^2\theta) + [f(\theta)]^2(\sin^2\theta + \cos^2\theta)\}^{1/2}\,d\theta$$

$$= \int_\alpha^\beta \{[f'(\theta)]^2 + [f(\theta)]^2\}^{1/2}\,d\theta.$$

Since $r = f(\theta)$, the arc length of the curve, shown in Figure 9.43, is

(9.11) $$L_C = \int_\alpha^\beta [r^2 + (r')^2]^{1/2}\,d\theta.$$

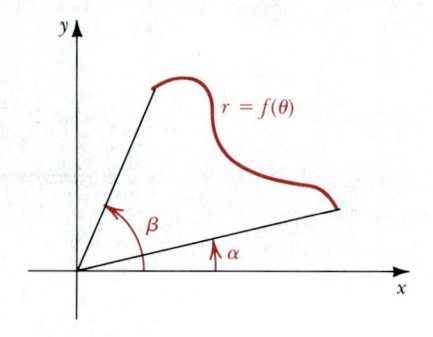

Figure 9.43

The length is $\int_\alpha^\beta [(r')^2 + r^2]^{1/2}\,d\theta$.

Example 2

Find the length of the cardioid $r = 2(1 + \cos\theta)$.

Solution

The cardioid, shown in Figure 9.44, is symmetric with respect to the x-axis. This implies that the total length of the graph is obtained by doubling the length for θ in $[0, \pi]$. So

$$L = 2\left[\int_0^\pi \{4(1 + \cos\theta)^2 + (-2\sin\theta)^2\}^{1/2}\,d\theta\right]$$

$$= 2\int_0^\pi \{4[1 + 2\cos\theta + (\cos\theta)^2] + 4(\sin\theta)^2\}^{1/2}\,d\theta$$

$$= 2\int_0^\pi \{4 + 8\cos\theta + 4(\cos^2\theta + \sin^2\theta)\}^{1/2}\,d\theta$$

$$= 2\int_0^\pi (8 + 8\cos\theta)^{1/2}\,d\theta$$

$$= 4\sqrt{2}\int_0^\pi (1 + \cos\theta)^{1/2}\,d\theta$$

$$= 8\int_0^\pi \cos\frac{\theta}{2}\,d\theta = 16\sin\frac{\theta}{2}\Big]_0^\pi = 16.\quad\blacktriangleright$$

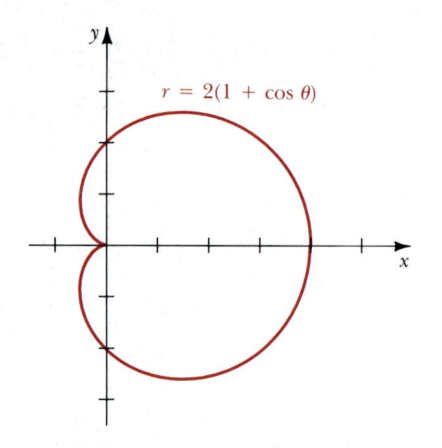

Figure 9.44

Example 3

Find the length of the spiral

$$r = e^{\theta/2\pi}, \qquad 0 \le \theta \le 2\pi.$$

Solution

The spiral is shown in Figure 9.45. We have

$$L = \int_0^{2\pi} \left[(e^{\theta/2\pi})^2 + \left(\frac{1}{2\pi} e^{\theta/2\pi}\right)^2 \right]^{1/2} d\theta$$

$$= \int_0^{2\pi} \left[e^{\theta/\pi} + \frac{1}{4\pi^2} e^{\theta/\pi} \right]^{1/2} d\theta$$

$$= \int_0^{2\pi} \left[e^{\theta/\pi} \left(1 + \frac{1}{4\pi^2}\right) \right]^{1/2} d\theta = \left(1 + \frac{1}{4\pi^2}\right)^{1/2} \int_0^{2\pi} e^{\theta/2\pi} d\theta$$

$$= \left[\left(\frac{4\pi^2 + 1}{4\pi^2}\right)^{1/2} (2\pi\, e^{\theta/2\pi})\right]_0^{2\pi}$$

$$= (4\pi^2 + 1)^{1/2} (e - 1) \approx 10.93216.$$

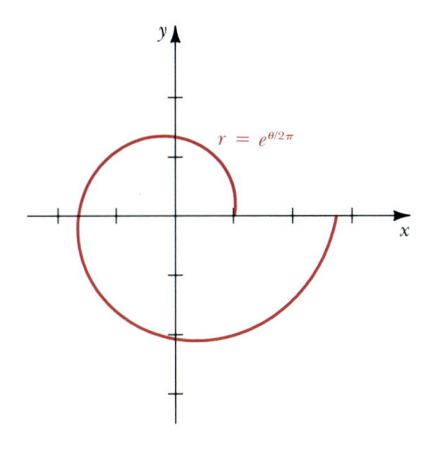

Figure 9.45

Although the integrals in these examples can be determined easily, there are many curves for which the exact value of the length is very difficult to determine. Recall from Section 5.4 that this is also true for curves described by Cartesian equations.

▶ EXERCISE SET 9.6

In Exercises 1–10, find the length of the curve described by the parametric equations.

1. $x = 3t - 1, y = 2t + 4, 1 \leq t \leq 3$
2. $x = 1 - t, y = 3t + 4, -1 \leq t \leq 5$
3. $x = t^3 + 1, y = 3t^2, 0 \leq t \leq 3$
4. $x = 3t^2 + 2, y = 4t^2 - 1, 0 \leq t \leq 2$
5. $x = t^2 + 1, y = 2t - 3, 0 \leq t \leq 1$
6. $x = \ln \sin 2t, y = 2t + 3, \pi/6 \leq t \leq \pi/3$
7. $x = t + 1, y = \ln \cos t, 0 \leq t \leq \pi/4$
8. $x = e^{-t} \cos t, y = e^{-t} \sin t, 0 \leq t \leq 2\pi$
9. $x = e^t \sin t, y = e^t \cos t, 0 \leq t \leq 2\pi$
10. $x = \cos t + t \sin t, y = t \cos t - \sin t, 0 \leq t \leq \pi$

In Exercises 11–16, find the length of the curve described by the polar equations.

11. $r = 3, 0 \leq \theta \leq \pi$
12. $r = 3\theta, 0 \leq \theta \leq \pi$
13. $r = e^{2\theta}, 0 \leq \theta \leq 1$
14. $r = 3^{-\theta}, -1 \leq \theta \leq 1$
15. $r = 1 - \cos \theta, 0 \leq \theta \leq 2\pi$
16. $r = 1 + \sin \theta, 0 \leq \theta \leq 2\pi$

17. Sketch the graphs of the parametric equations given below for $0 \leq t \leq \pi/2$, and find the length of each curve.
    a) $x = \cos t, y = \sin t$ b) $x = \cos^2 t, y = \sin^2 t$
    c) $x = \cos^3 t, y = \sin^3 t$

18. Sketch the graphs of the polar equations given below for $0 \leq \theta \leq \pi$, and find the length of each curve.
    a) $r = \cos \theta$ b) $r = \cos^2 \dfrac{\theta}{2}$ c) $r = \cos^3 \dfrac{\theta}{3}$

19. Show that the length of one arch of the cycloid
    $$x = R(t - \sin t), \quad y = R(1 - \cos t)$$
    is $8|R|$.

20. A piece of gravel is embedded in the tread of a bicycle tire 27 in. in diameter. How far does the piece of gravel travel during a 4-mi bicycle trip?

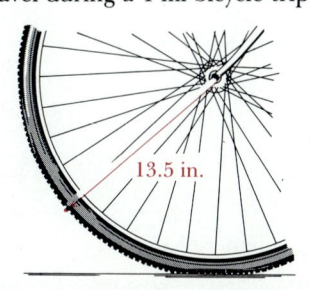

21. In Exercise 47 of Section 9.2, the equation $r = 2e^{-0.2\theta}$ was used to describe the outside of the shell of a chambered nautilus. Find the total length of the outside of the shell as a function of θ.

22. The study of light diffraction at a rectangular aperture involves the Fresnel integrals

$$C(t) = \int_0^t \cos \frac{\pi w^2}{2} \, dw$$

and

$$S(t) = \int_0^t \sin \frac{\pi w^2}{2} \, dw,$$

named in honor of the physicist Augustin Jean Fresnel (1788–1827). These integrals cannot be evaluated directly, but Marie Alfred Cornu (1841–1902) used the parametric representation $x = C(t)$, $y = S(t)$ to produce a geometric illustration of the *Fresnel integrals*. These curves, known as *Cornu spirals,* are shown in the accompanying figure.
   a) Find the length of the Cornu spiral for t in $[t_1, t_2]$.
   b) Find the slope of the tangent line to the spiral at t_1.

23. How much recording tape of one mil (0.001 in.) thickness is required to fill a reel with inside diameter 2 in. and outside diameter 7 in. [*Hint:* Make the assumption that on the first revolution the thickness varies linearly from zero to its actual thickness 0.001 in. and then remains 0.001 in. thick.]

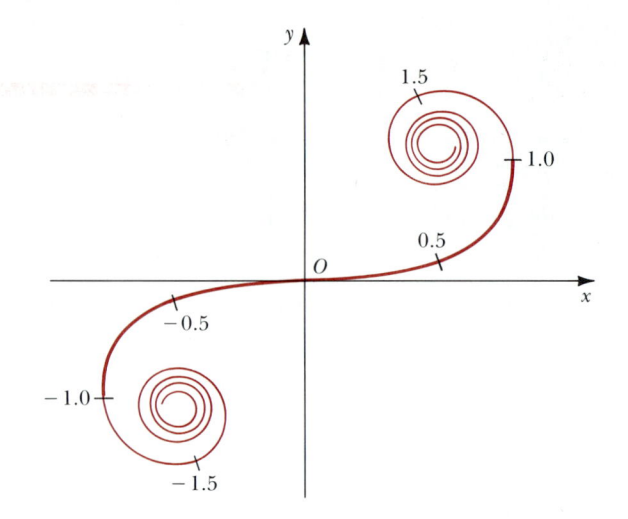

Figure for Exercise 22.

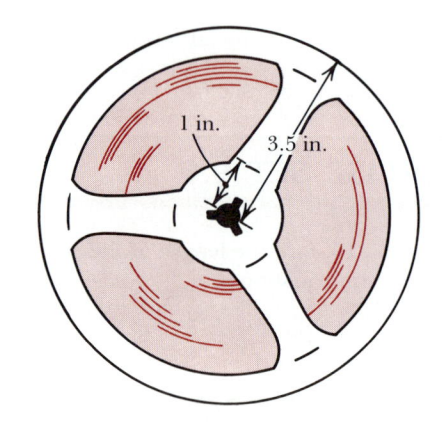

Figure for Exercise 23.

▶ IMPORTANT TERMS AND RESULTS

CONCEPT	PAGE	CONCEPT	PAGE
Polar coordinates	588	Spiral	597
Polar symmetry	592	Area using polar coordinates	600
Cardioid	595	Parametric equations	603
n-leafed rose	595	Smooth curve	604
Lemniscate	596	Cycloid	605
Limaçon	596	Tangent lines	610
Polar graphing summaries	597	Arc length	616

REVIEW EXERCISES

In Exercises 1–4, find the Cartesian coordinates for the point described by the polar coordinates.

1. $(3, \pi/6)$
2. $(1, -3\pi/4)$
3. $(0, -\pi)$
4. $(-2, -5\pi/4)$

In Exercises 5–8, find two different sets of polar coordinates for the point described by the Cartesian coordinates.

5. $(1, 1)$
6. $(-2, 0)$
7. $(0, -2)$
8. $(-4, -4)$

In Exercises 9–24, sketch the graph of the curve described by the polar equation.

9. $r = 4$
10. $r = -1$
11. $\theta = \pi/2$
12. $\theta = 0$
13. $r = 2\cos\theta$
14. $r = 3\sin\theta$
15. $r = 1 + 2\cos\theta$
16. $r = 3 + 3\sin\theta$
17. $r = 4\cos 2\theta$
18. $r = 2\cos 4\theta$
19. $r^2 = 4\cos 2\theta$
20. $r^2 = 2\cos 4\theta$
21. $r = -\theta$
22. $r = \theta^2$
23. $r = 1 + \ln\theta$
24. $r = 1 + e^\theta$

In Exercises 25–36, (a) sketch the graph of the curve and (b) find the slope of the tangent line to the curve at the indicated point.

25. $x = 2\cos t, y = 3\sin t; t = \pi/2$
26. $x = 2t + 1, y = 3t + 7; t = 1$
27. $x = e^t, y = 1 + e^{-t}; t = 0$
28. $x = \cos 2t, y = \sin t; t = \pi/6$
29. $x = \sin^2 t + 1, y = \cos^2 t; t = \pi/4$
30. $x = 3\sin^2 t + 2\cos^2 t, y = 2\sin^2 t + \cos^2 t; t = \pi/4$
31. $x = t^2, y = t; t = 2$
32. $x = -t^2, y = t + 1; t = -1$
33. $x = \ln t, y = \frac{1}{2}(t - 1/t); t = 1$
34. $x = \ln t, y = \frac{1}{2}(t + 1/t); t = e$
35. $r = 2\cos\theta, (\sqrt{2}, \pi/4)$
36. $r^2 = \cos\theta, (\sqrt{2}/2, \pi/3)$

In Exercises 37–46, (a) sketch the region and (b) determine the area of the region.

37. Inside $r = 3\cos\theta$
38. Inside $r = 5\sin\theta$
39. Inside $r = 3\cos 2\theta$
40. Inside $r = \sin 3\theta$
41. Inside $r = 1$ and $r = 1 + \sin\theta$
42. Inside $r = 2$ and $r = 2 + \sin\theta$
43. Inside $r = 1$ and outside $r = 1 + \sin\theta$
44. Inside $r = 2$ and outside $r = 2 + \sin\theta$
45. Inside $r = 1 + \sin\theta$ and outside $r = 1$
46. Inside $r = 2 + \sin\theta$ and outside $r = 2$

In Exercises 47–52, (a) sketch the graph of the curve described by the polar equation and (b) find the length of the curve.

47. $r = 2\theta, 0 \le \theta \le \pi/2$
48. $r = e^{-\theta}, 0 \le \theta \le 2\pi$
49. $r = 2 - 2\cos\theta$
50. $r = 3 - 3\sin\theta$
51. $r = 2\cos\theta - 2$
52. $r = -3\sin\theta - 3$

In Exercises 53–58, (a) sketch the graph of the curve described by the parametric equation and (b) find the length of the curve.

53. $x = 2t + 1, y = 3t - 2, 2 \le t \le 3$
54. $x = 1 - t, y = 2t + 5, 0 \le t \le 4$
55. $x = 4t^2, y = 3t^2 + 5, 0 \le t \le 2$
56. $x = 4t^2, y = 3t + 5, 0 \le t \le 2$
57. $x = \ln \sin t, y = t, \pi/6 \le t \le \pi/4$
58. $x = e^{2t}\cos t, y = e^{2t}\sin t, 0 \le t \le \pi$

In Exerises 59–62, (a) sketch the graph of the equation and (b) find the corresponding Cartesian equation.

59. $r = \sec\theta$
60. $r = \dfrac{1}{1 + \cos\theta}$
61. $r(2\cos\theta - \sin\theta) = 3$
62. $r = \dfrac{2}{1 - \sin\theta}$

63. Find a polar equation for the curve described by the parametric equations
$$x = \frac{\cos\theta}{\theta}, \quad y = \frac{\sin\theta}{\theta}.$$

64. Find the points on the cardioid $r = 2(1 - \cos\theta)$ at which the tangent line is (a) perpendicular to the polar axis and (b) parallel to the polar axis.

65. Give a set of parametric equations that describe the Cartesian equation $16x^2 + 9y^2 = 1$.

66. Give a set of parametric equations that describe the Cartesian equation $9x^2 + 16y^2 = 144$.

67. The position of a particle at time t is given by
$$x = 4t^2 - 1,\ y = 3t^2 + 4.$$
Find the distance the particle travels as t goes from 0 to 1.

68. The position of a particle at time t is given by $x = \sin t$, $y = \cos t$. Find the distance the particle travels as t goes from 0 to $\pi/2$.

69. a) Find $\lim_{\theta \to \pi^+/2} \tan \theta$, $\lim_{\theta \to \pi^-/2} \tan \theta$.
    b) If $r = \tan \theta$, find $dr/d\theta$.
    c) Sketch the graph of the polar equation $r = \tan \theta$.

10

Conic Sections

624   CHAPTER 10   CONIC SECTIONS

In this chapter we systematically examine the graphs associated with equations of the form

$$Ax^2 + Bxy + Cy^2 + Dx + Ey + F = 0,$$

where A, B, C, D, E, and F are constants.

We have seen several examples of equations of this type, for example, the parabola with equation $x^2 - y = 0$ and the circle with equation $x^2 + y^2 = 1$. The graphs of these equations are called **conic sections**, or simply **conics**, because they describe figures that result from intersecting a plane with a double-napped right circular cone. Figure 10.1 shows a double-napped cone with its component parts labeled.

In general, three distinct figures can be produced by such an intersection:

i) A *parabola*, if the plane intersects only one nappe of the cone and is parallel to a line lying entirely in the cone, called a *generator* of the cone (see Figure 10.2a);

ii) an *ellipse*, if the plane intersects only one nappe of the cone but is not parallel to a generator (see Figure 10.2b); and

iii) a *hyperbola*, if the plane intersects both nappes of the cone (see Figure 10.2c).

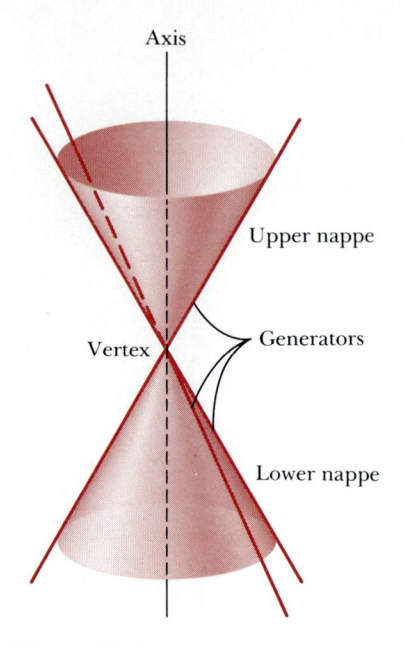

Figure 10.1

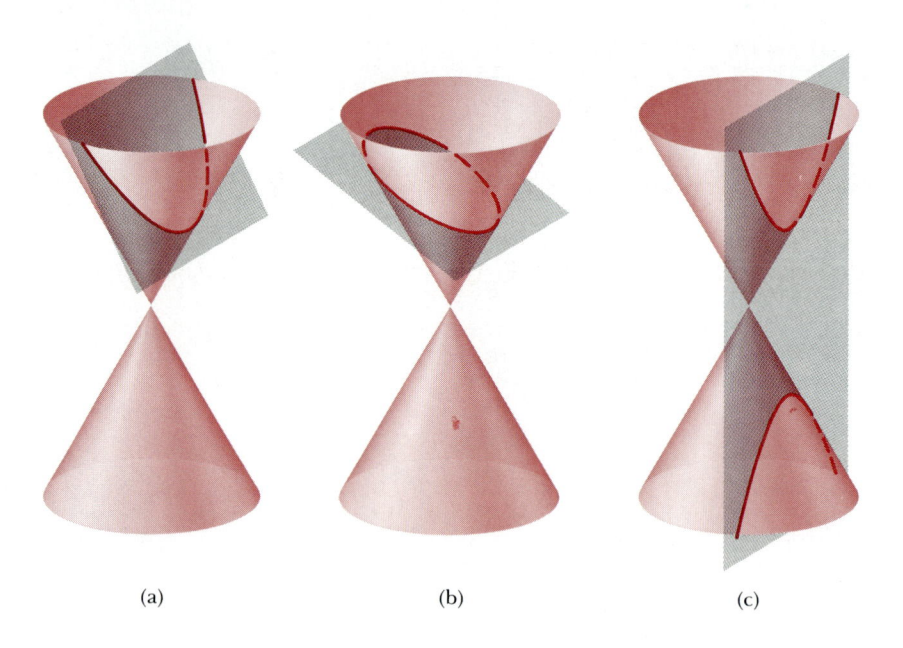

Figure 10.2
(a) A parabola; (b) an ellipse; (c) a hyperbola.

(a)   (b)   (c)

Figures such as a point, circle, line, or pair of intersecting lines can also be produced, as shown in Figure 10.3, but these are considered special, or *degenerate*, cases of the parabola, ellipse, and hyperbola.

We do not need this three-dimensional description of conic sections since we are interested in curves in the plane. Instead we will define the conic sections in terms of distances from fixed lines and points.

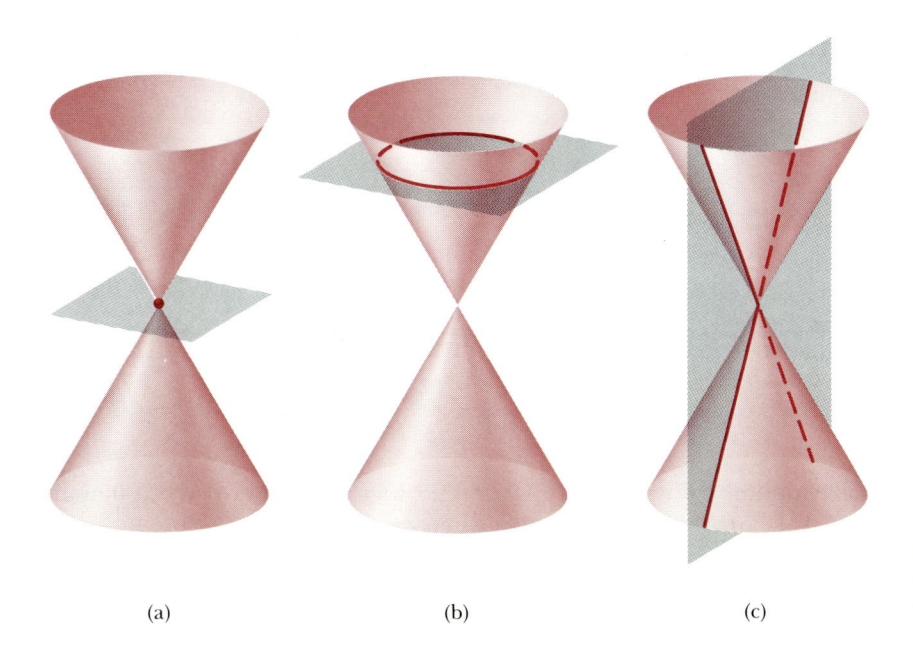

Figure 10.3
(a) A point; (b) a circle; (c) intersecting lines.

(a)   (b)   (c)

One reason for discussing conic sections at this time is to prepare for the graphing of solids and surfaces in space. Describing three-dimensional objects with a graph that, by necessity, is two-dimensional can be quite difficult. The usual practice is to determine the shape of the cross sections of the objects, and many common three-dimensional objects have cross sections that are conic sections.

10.1
Parabolas

(10.1)
DEFINITION
A **parabola** is the set of points in a plane that are equidistant from a given point, called the **focus** or **focal point,** and a given line, called the **directrix,** that does not contain the focal point.

The line through the focal point of the parabola perpendicular to the directrix is called the **axis** of the parabola. The axis intersects the parabola midway between the focal point and the directrix at the **vertex** of the parabola (see Figure 10.4).

Since we have discussed parabolas in the past without referring to this definition, we should establish that the definition describes the same figures that have been discussed previously. Suppose an xy-coordinate system is superimposed so that the vertex of a parabola is at the origin and the focal point lies on the y-axis at the point $(0, c)$. Then the directrix

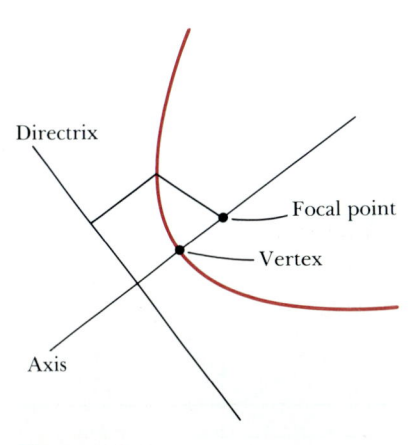

Figure 10.4
A parabola.

CHAPTER 10 CONIC SECTIONS

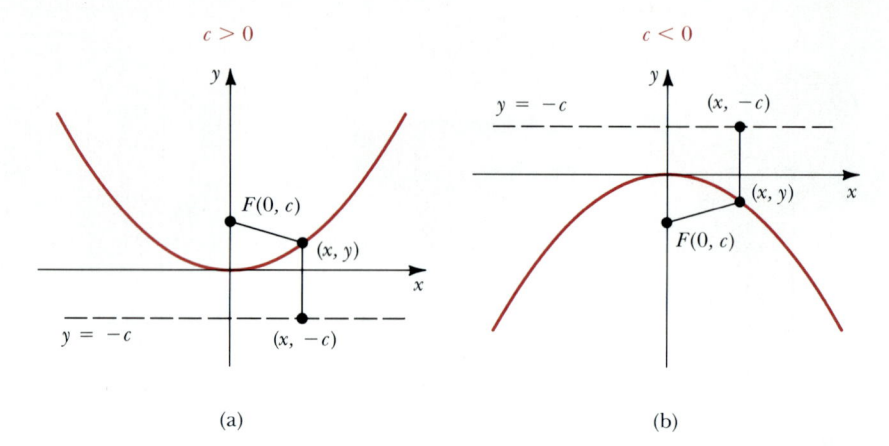

Figure 10.5

(a) $c > 0$
(b) $c < 0$

has equation $y = -c$. Figure 10.5(a) shows a parabola when $c > 0$, and Figure 10.5(b) shows a parabola when $c < 0$.

The distance from an arbitrary point (x, y) on the graph of the parabola to the focal point $(0, c)$ is the same as the distance from (x, y) to the directrix $y = -c$, that is,

$$d((x, y), (0, c)) = d((x, y), (x, -c)).$$

Thus,

$$\sqrt{(x-0)^2 + (y-c)^2} = \sqrt{(x-x)^2 + (y+c)^2},$$

so

$$x^2 + (y-c)^2 = (y+c)^2,$$

and

$$x^2 + y^2 - 2yc + c^2 = y^2 + 2yc + c^2.$$

This simplifies to

$$x^2 = 4yc.$$

Therefore, the parabola with focal point $(0, c)$ and directrix $y = -c$ has equation

(10.2) $$y = \frac{1}{4c} x^2,$$

which is a quadratic equation of the type considered in Section 1.3.

Example 1

Find the focal point and directrix of the parabola with equation $y = x^2$.

Solution

The focal point of this parabola lies at $(0, c)$, where

$$\frac{1}{4c} = 1, \quad \text{so} \quad c = \frac{1}{4}.$$

The equation of the directrix is consequently $y = -\frac{1}{4}$. The graph of the parabola is shown in Figure 10.6. ▶

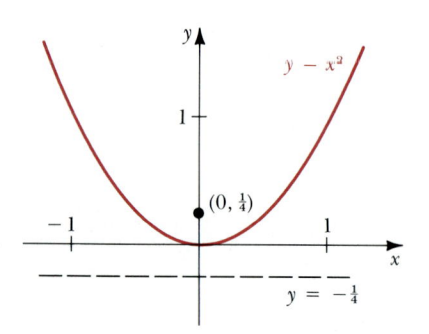

Figure 10.6

(a) $c > 0$

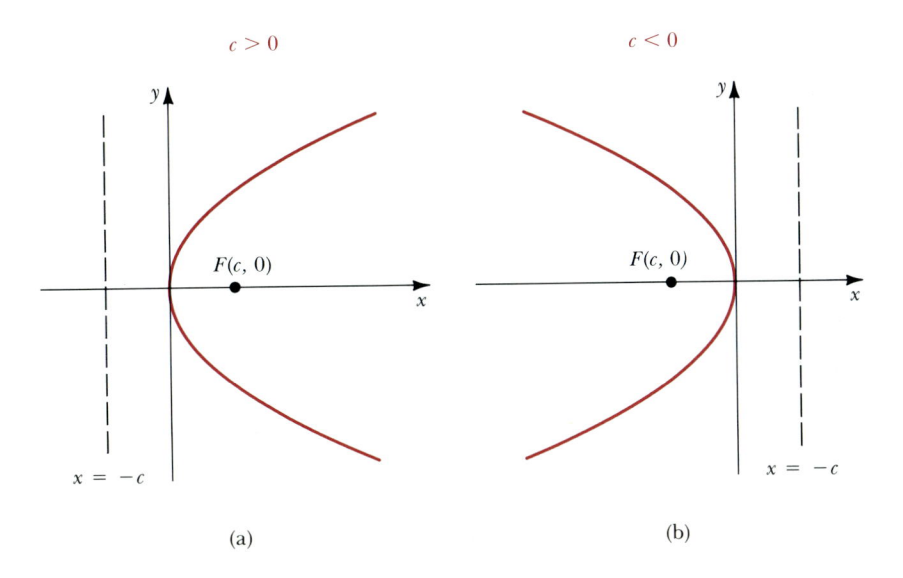

(b) $c < 0$

Figure 10.7

In a similar manner it can be shown that a parabola with focal point at $(c, 0)$ and directrix $x = -c$ has equation

(10.3) $$x = \frac{1}{4c} y^2.$$

Parabolas of this type are shown in Figure 10.7.

A parabola with equation of either form (10.2) or form (10.3) is said to be in **standard position.**

Example 2

Find an equation of the parabola with focal point at $(-2, 0)$ and directrix $x = 2$.

Solution

The focal point lies along the x-axis. Since the vertex is the midpoint of the line segment from the focal point perpendicular to the directrix, the vertex is at the origin and the parabola is in standard position. This parabola has equation

$$x = \frac{1}{4(-2)} y^2 \quad \text{or} \quad y^2 = -8x.$$

The graph of this parabola is shown in Figure 10.8. ▶

By applying the graphing techniques illustrated in Chapter 1, we can use a translation to graph any parabola whose directrix is parallel to one of the coordinate axes.

Example 3

Find the focal point and directrix of the parabola with equation $2y^2 + 8y + x + 9 = 0$.

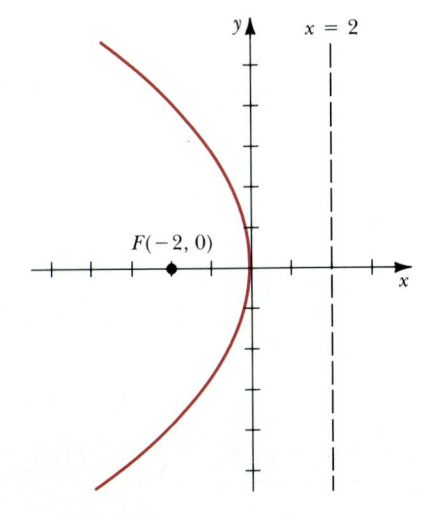

Figure 10.8

Solution

To compare this equation with one of the standard forms, we complete the square in y:

$$2(y^2 + 4y + 4) + x + 9 - 8 = 0,$$

so

$$2(y + 2)^2 + x + 1 = 0 \quad \text{and} \quad x + 1 = -2(y + 2)^2.$$

Compare this equation to the equation $x = -2y^2$, which can be written as $x = (1/4c)y^2$, with $c = -\frac{1}{8}$, and describes a parabola with focal point at $(-\frac{1}{8}, 0)$ and directrix $x = \frac{1}{8}$.

The graph of $x + 1 = -2(y + 2)^2$ is a translation of the graph of $x = -2y^2$ downward 2 units and 1 unit to the left. Consequently, the focal point of the parabola $x + 1 = -2(y + 2)^2$ is at $(-1 - \frac{1}{8}, -2) = (-\frac{9}{8}, -2)$, and the directrix of this parabola is $x = -1 + \frac{1}{8} = -\frac{7}{8}$. Figure 10.9 shows the graphs of both parabolas. ▶

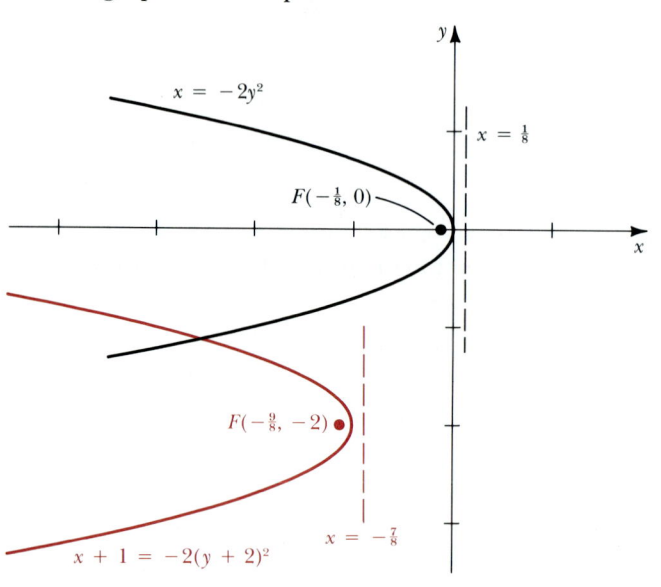

Figure 10.9

Example 4

Find an equation of the parabola with focal point at $(1, 5)$ and directrix $y = -1$.

Solution

The location of the directrix and focal point, shown in color in Figure 10.10, indicates that the axis of this parabola is the line $x = 1$. The vertex is the point on the axis midway between the focal point $(1, 5)$ and the point where the directrix and axis of the parabola intersect, $(1, -1)$. Consequently, the vertex is at $(1, 2)$.

If the vertex is translated to the origin, each point is moved downward 2 units and 1 unit to the left. Thus the focal point is translated to $(0, 3)$ and the directrix to $y = -3$. An equation of this "translated"

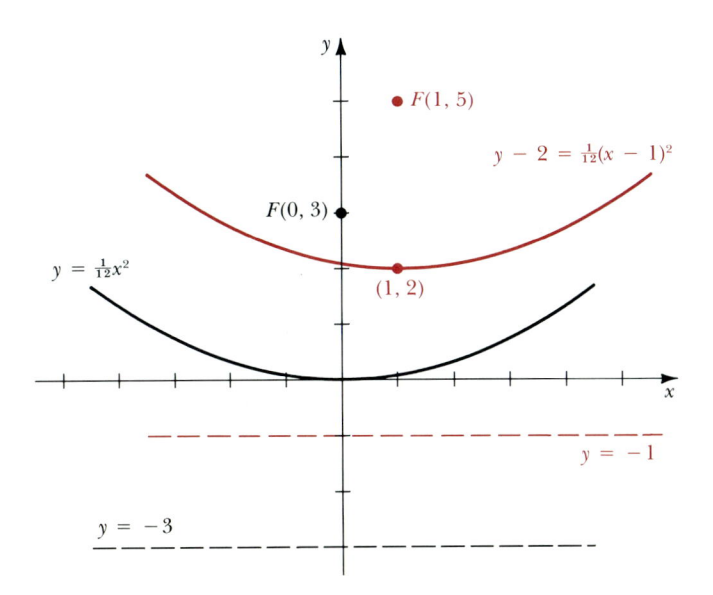

parabola is

$$y = \frac{1}{4(3)} x^2 = \frac{1}{12} x^2.$$

The original parabola has equation

$$(y - 2) = \frac{1}{12}(x - 1)^2 \quad \text{or} \quad y = \frac{1}{12}(x - 1)^2 + 2.$$

Both parabolas are shown in Figure 10.10. ▶

Parabolas are commonly used in the design of reflectors, an application that depends on the following parabolic property (see Figure 10.11a):

(10.4)
An object emitted from the focal point of a parabola to its surface is reflected along a line that is parallel to the axis of the parabola.

This reflection property follows from a basic law of physics:

(10.5)
THE LAW OF REFLECTION
The angle of reflection of sound, light, or any reflected object hitting a solid surface is the same as the angle of incidence of the object.

The angles are measured with respect to the tangent line to the surface, as shown in Figure 10.11(b).

To verify the reflection property of the parabola, we will show that when an object is emitted from the focal point F of a parabola to a point P on the surface of the parabola, the angle of incidence is the same as the angle formed by the tangent line at P and the axis of the parabola.

Figure 10.10

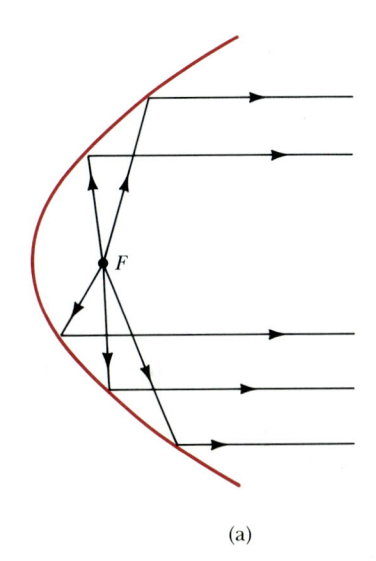

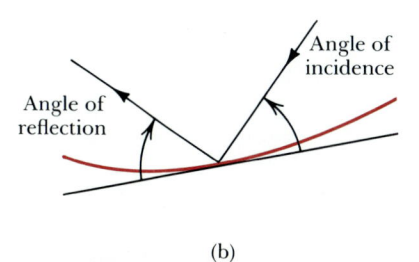

Figure 10.11
(a) The reflection property of a parabola; (b) angles of incidence and reflection are equal.

(a)

(b)

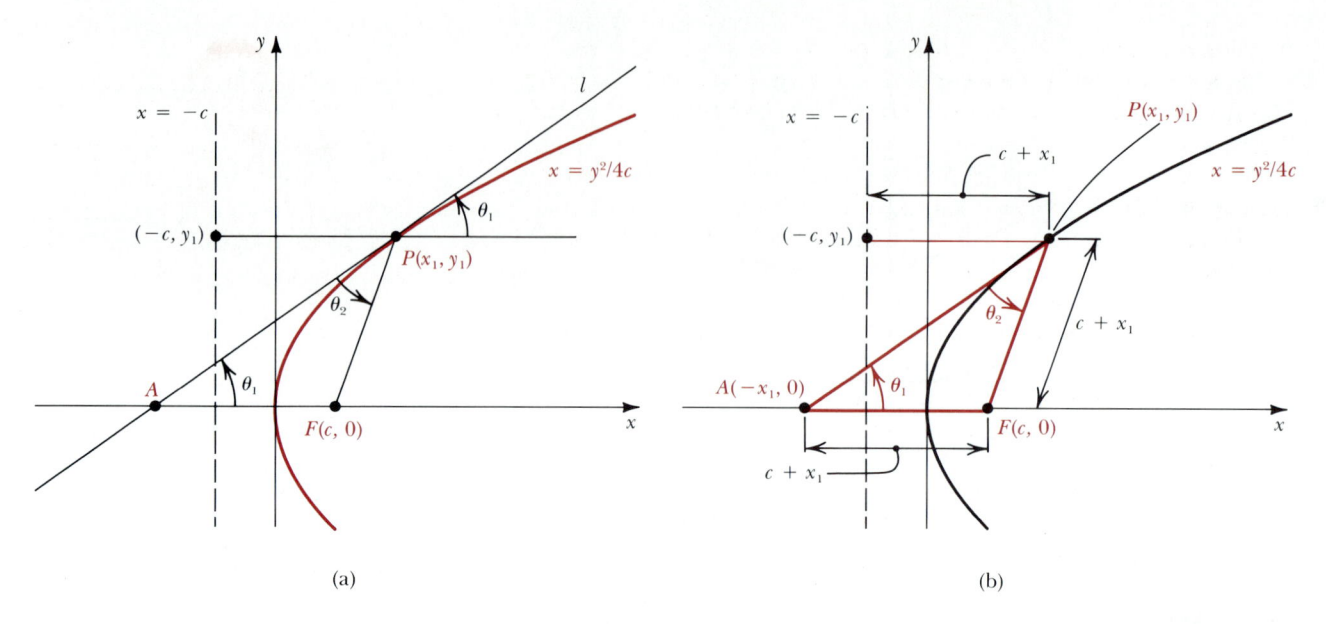

Figure 10.12

Consider a parabola with equation $x = y^2/4c$ containing an arbitrary point $P(x_1, y_1)$, as shown in Figure 10.12(a). The angle of incidence is labeled θ_2, and the angle formed by the tangent line at P and the x-axis, which is the axis of the parabola, is labeled θ_1. To show that θ_1 is the angle of reflection, we must show that $\theta_1 = \theta_2$.

We can see from Figure 10.12(a) that the two angles θ_1 and θ_2 are equal precisely when the triangle AFP is isosceles with the length of side AF the same as the length of side FP.

To find the length of side AF, we first determine where the tangent line intersects the x-axis. The slope m of the tangent line to the parabola at $P(x_1, y_1)$ is the derivative y' evaluated at (x_1, y_1). Implicit differentiation applied to $x = y^2/4c$ gives $1 = yy'/2c$. Thus, $y' = 2c/y$, and the slope is

$$m = \frac{2c}{y_1}.$$

An equation of the tangent line l is

$$y - y_1 = \left(\frac{2c}{y_1}\right)(x - x_1)$$

and since A is the point on this line with y-coordinate zero, the x-coordinate of A satisfies

$$-y_1 = \left(\frac{2c}{y_1}\right)(x - x_1), \quad \text{that is,} \quad x = x_1 - \frac{y_1^2}{2c}.$$

But $P(x_1, y_1)$ is on the parabola, so $y_1^2 = 4cx_1$ and the x-coordinate of A is given by

$$x = x_1 - \frac{y_1^2}{2c} = x_1 - \frac{4cx_1}{2c} = x_1 - 2x_1 = -x_1.$$

The length of side AF is $c + x_1$, the distance from $(-x_1, 0)$ to $(c, 0)$.

Since P is a point on the parabola, the length of side FP is the same as the distance of P from the directrix, which is also $c + x_1$ (see Figure 10.12b).

This shows that the triangle AFP is isosceles, so the angles θ_1 and θ_2 are equal. Therefore, any object emitted from the focal point to a point on the parabola is reflected along a line that is parallel to the axis of the parabola.

It follows from reflection property (10.4) that if a parabola is rotated about its axis to form a hollow shell, an object emitted from the focal point and striking any point on the surface of the shell will be reflected on a line parallel to the axis of the parabola, as shown in Figure 10.13. This is the reflection property used in the design of flashlights and automobile headlights. A light source placed at the focal point of a parabolic reflector produces what is called a *pencil* beam. By varying the reflector slightly from the parabolic shape, the light can be spread to obtain beams of various shapes.

In the reverse manner, incoming rays of light or sound parallel to the axis of a parabola are reflected to the focal point. This property is used in the design of parabolic microphones and solar collectors. For example, the microphone used by television broadcasters to relay a conversation in a football huddle is of parabolic design, as are the dish antennas used to pick up television signals from orbiting satellites.

A parabola is also the shape that a suspended cable assumes when it hangs between two fixed points and supports a uniformly distributed load. Bridges built on this principle are called *suspension bridges* (see Figure 10.14). Most of the world's longest and most famous bridges are of this type. Examples are the Golden Gate Bridge in San Francisco, the Verrazano Narrows Bridge joining Brooklyn and Staten Island in New York, and the Mackinac Straits Bridge in upper Michigan.

The path of an object propelled into the air also assumes approximately the shape of a parabola, a fact that artillery experts have known for centuries. (This topic is considered in more detail in Section 12.4.)

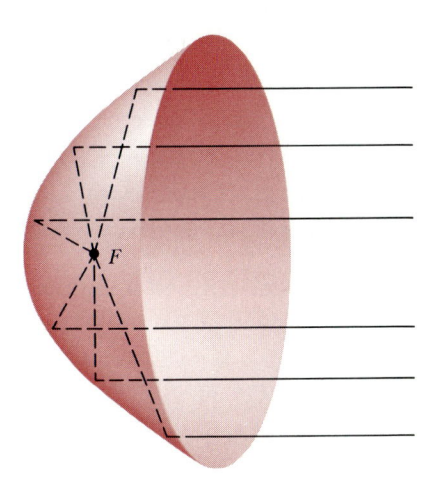

Figure 10.13
Surface of rotation generated by a parabola.

Figure 10.14
The Golden Gate Bridge.

▶ EXERCISE SET 10.1

In Exercises 1–16, (a) sketch the graph of the parabola and (b) find the vertex, focal point, and equation of the directrix.

1. $y = 2x^2$
2. $16y = 9x^2$
3. $y = -2x^2$
4. $9y = -16x^2$
5. $y^2 = 2x$
6. $16y^2 = 9x$
7. $y^2 = -2x$
8. $9y^2 = -16x$
9. $x^2 + 4x + 4 = 2y$
10. $x^2 + 6x + 9 - y = 0$
11. $y^2 - 8y + 12 = 2x$
12. $y^2 + 6y + 6 - 3x = 0$
13. $2x^2 + 4x - 9y + 20 = 0$
14. $3x^2 - 12x - 4y + 8 = 0$
15. $9y^2 - 36y - 2x + 34 = 0$
16. $4y^2 + 8y - 3x + 10 = 0$

In Exercises 17–24, find an equation of a parabola that satisfies the stated conditions.

17. Focus $(-2, 2)$, directrix $y = -2$
18. Focus $(-2, 2)$, directrix $x = 2$

19. Focus $(-2, 2)$, directrix $y = -1$
20. Focus $(-2, 2)$, directrix $x = -4$
21. Vertex $(-2, 2)$, directrix $x = 4$
22. Vertex $(-2, 2)$, focus $(-2, 0)$
23. Vertex $(3, 4)$, focus $(3, 6)$
24. Vertex $(3, 4)$, directrix $y = 6$

In Exercises 25–28, find an equation of the parabola with axis parallel to the *y*-axis and vertex *V* that passes through the point *P*.

25. $V(0, 0)$, $P(4, 6)$ **26.** $V(1, 0)$, $P(5, 6)$
27. $V(0, 2)$, $P(4, 8)$ **28.** $V(1, 2)$, $P(5, 8)$

29–32. Find an equation of the parabola with axis parallel to the *x*-axis and vertex *V* that passes through the point *P*, where *V* and *P* are as given in Exercises 25–28.

33. **a)** Find an equation of the parabola with axis the *y*-axis and vertex $(0, 0)$ that passes through the point (x_1, y_1), where $x_1 \neq 0$ and $y_1 \neq 0$.
b) Use the result in part (a) to find an equation of the parabola with axis parallel to the *y*-axis and vertex (h, k) that passes through (x_1, y_1), where $x_1 \neq h$ and $y_1 \neq k$.

34. **a)** Find an equation of the parabola with axis the *x*-axis and vertex $(0, 0)$ that passes through the point (x_1, y_1), where $x_1 \neq 0$ and $y_1 \neq 0$.
b) Use the result in part (a) to find an equation of the parabola with axis parallel to the *x*-axis and vertex (h, k) that passes through (x_1, y_1), where $x_1 \neq h$ and $y_1 \neq k$.

35. Show that the vertex of a parabola is the point on the parabola that is closest to the focal point.

36. Find the focal point of the parabola with axis the *y*-axis that passes through $(-1, 3)$ and has a tangent with slope 2 at this point.

37. **a)** Find a general form for the equation of a parabola with axis the *y*-axis and passing through $(1, 1)$.
b) Of these parabolas, which have the property that the normal line at $(1, 1)$ is tangent to $y = x^2$ at this point?

38. Consider the pair of parabolas $y = ax^2$ and $x = ay^2$. Show that the angle formed by the tangent lines at their intersection is independent of *a*.

39. Consider the pair of parabolas $y = a - bx^2$ and $y = -a + bx^2$, where *a* and *b* are positive constants. What must be true of *a* and *b* if the tangent line to one parabola at their point of intersection is normal to the other parabola?

40. The *latus rectum* of a parabola is the line segment that passes through the focus perpendicular to the axis and joins two points on the curve (see the figure).
a) Show that the length of the latus rectum is four times the distance from the vertex to the focal point.
b) Show that the area of the region bounded by the graph of the parabola and the latus rectum is one sixth the square of the length of the latus rectum.

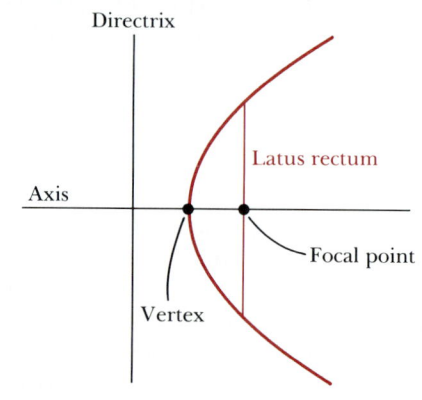

41. Find an equation of the parabola with axis parallel to the *y*-axis and passing through $(1, 0)$, $(0, 1)$, and $(2, 2)$.

42. A driving light has a parabolic cross section with a depth of 2 in. and a cross-section height of 4 in. Where should the light source be placed in order to produce a parallel beam of light?

43. The world's largest hangar is the Goodyear Airdock, built in 1929 in Akron, Ohio, to house and service the rigid airships USS *Akron* and USS *Macon*. The main structure of this hangar is a 1175-ft-long cylinder whose cross-section is parabolic with height 211 ft and width 325 ft. What is the volume enclosed by this structure?

Goodyear Airdock in Akron, Ohio.

44. A ball thrown horizontally from the top edge of a building follows a parabolic curve with vertex at the top edge of the building and axis along the side of the building. The ball passes through a point 100 ft from the building when it is a vertical distance of 16 ft from the top.
   a) How far from the building will the ball land if the building is 64 ft high?
   b) How far will the ball travel before it hits the ground?
   c) Suppose instead that the ball is thrown from the top of the Sears Tower in Chicago, the world's tallest building with a height of 1450 ft. Recompute the answers to the questions.

45. The longest bridge on the North American continent is the Verrazano Narrows Bridge in New York City. This suspension bridge has a 4260-ft span and twin supporting towers standing about 700 ft above the water. The distance between the towers is 2627 ft, and the low point on the supporting cables is 225 ft above the water. What is the length of a supporting cable between the two towers?

46. The George Washington Bridge crossing the Hudson River in New York City has a main span of 3500 ft with cables that make an angle of approximately 70° with its supporting towers.
   a) Use this information to approximate the sag in the cables.
   b) The sag in the cables is about $\frac{5}{8}$ of the height of the towers above the water. Find this height.

George Washington Bridge.

47. The world's largest compound reflecting telescope is called the Hale telescope in honor of the American astronomer George Ellery Hale (1868–1939). It is at the Palomar Mountain Observatory 45 mi northeast of San Diego, California. The main parabolic mirror of this telescope is 200 in. in diameter with a depth from rim to vertex of 3.75 in. A small cylindrical platform is located within the tube of the telescope along the axis of the parabola for an observer to view and record the reflection from the telescope. How far from the center of the mirror is the observer's viewing area located?

48. A satellite is placed in a position to make a parabolic flight past the moon with the center of the moon at the focal point of the parabola. When the satellite is 5783 km from the surface of the moon, it makes an angle of 60° with the axis of the parabola. The closest the satellite gets to the surface is 143 km. What is the diameter of the moon? (Assume that the gravitational center of the moon is at its center: the focus of the parabola. This assumption is not quite correct since the gravitational center is offset approximately 2 km from the center toward the earth.)

Putnam exercises

49. Find the length of the shortest chord that is normal to the parabola $y^2 = 2ax$ at one end of the chord. (This is a problem in the first William Lowell Putnam examination given on April 16, 1938.)

50. Consider the two mutually tangent parabolas $y = x^2$ and $y = -x^2$. (These have foci at $(0, \frac{1}{4})$ and $(0, -\frac{1}{4})$ and directrices $y = -\frac{1}{4}$ and $y = \frac{1}{4}$, respectively.) The upper parabola rolls without slipping around the fixed lower parabola. Find the locus of the focal point of the moving parabola (see the figure). (This exercise was problem A–5 of the thirty-fifth William Lowell Putnam examination given on December 7, 1974. The examination and its solution are in the November 1975 issue of the *American Mathematical Monthly*, pp. 907–912.)

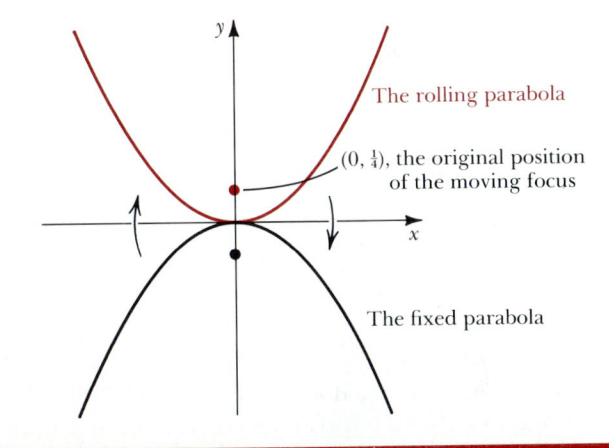

10.2

Ellipses

(10.6) DEFINITION
An **ellipse** is the set of points in the plane the sum of whose distances from two fixed points is a given constant.

The fixed points are called the **focal points** of the ellipse, the line through these points is called the **axis** of the ellipse, and the points where the ellipse intersects the axis are called the **vertices** of the ellipse. These are illustrated in Figure 10.15.

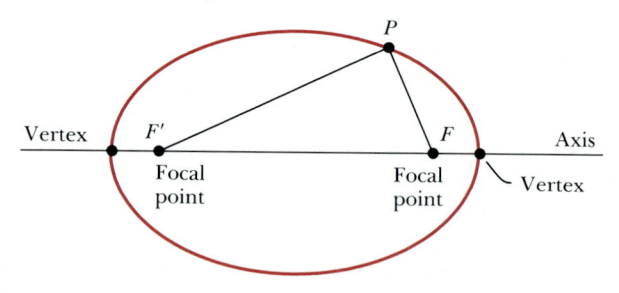

Figure 10.15
An ellipse.

Standard equations for an ellipse are derived by placing an *xy*-coordinate system with the *x*-axis along the axis of the ellipse and the *y*-axis along the perpendicular bisector of the line segment joining the focal points. The focal points are assigned coordinates $(c, 0)$ and $(-c, 0)$, as shown in Figure 10.16.

For reasons of convenience, the sum of the distances from the focal points to any point on the ellipse (shown by dashed lines in Figure 10.16) is denoted by the constant $2a$. Any point (x, y) on the ellipse has the property that

$$d((x, y), (c, 0)) + d((x, y), (-c, 0)) = 2a.$$

Thus,

$$\sqrt{(x-c)^2 + (y-0)^2} + \sqrt{(x+c)^2 + (y-0)^2} = 2a$$

and

$$\sqrt{(x-c)^2 + y^2} = 2a - \sqrt{(x+c)^2 + y^2}.$$

Squaring both sides implies that

$$(x-c)^2 + y^2 = 4a^2 - 4a\sqrt{(x+c)^2 + y^2} + (x+c)^2 + y^2,$$

which simplifies to

$$a + \frac{c}{a}x = \sqrt{(x+c)^2 + y^2}.$$

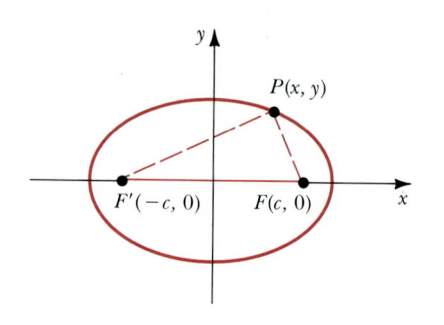

Figure 10.16

Squaring again gives
$$a^2 + 2cx + \frac{c^2}{a^2} x^2 = (x + c)^2 + y^2,$$

which simplifies to
$$x^2 \left[\frac{c^2 - a^2}{a^2}\right] - y^2 = c^2 - a^2.$$

Dividing both sides by $c^2 - a^2$, we have
$$\frac{x^2}{a^2} + \frac{y^2}{a^2 - c^2} = 1.$$

The sum of the lengths of the two dashed sides of the triangle shown in Figure 10.16 with vertices F', F, and P is $2a$. Since $2c$ is the length of the third side, $2a > 2c$ so $a > c$, and $a^2 - c^2 > 0$. For convenience we replace $a^2 - c^2$ using a new constant $b > 0$, where
$$b^2 = a^2 - c^2.$$

The equation of the ellipse then becomes

(10.7) $$\frac{x^2}{a^2} + \frac{y^2}{b^2} = 1.$$

The graph of this equation is symmetric with respect to the x-axis and the y-axis. It has y-intercepts $(0, b)$ and $(0, -b)$ and x-intercepts at the vertices $(a, 0)$ and $(-a, 0)$, as shown in Figure 10.17. Note that for any ellipse, $a > \sqrt{a^2 - c^2} = b$.

The line segment from $(-a, 0)$ to $(a, 0)$ is called the **major axis** of the ellipse, and the line segment from $(0, -b)$ to $(0, b)$ is called the **minor axis**.

If the focal points are placed along the y-axis at $(0, c)$ and $(0, -c)$, the ellipse has equation

(10.8) $$\frac{y^2}{a^2} + \frac{x^2}{b^2} = 1.$$

An ellipse with equation in either form (10.7) or form (10.8) is said to be in **standard position**.

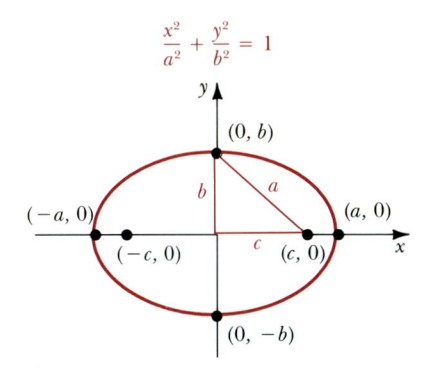

Figure 10.17

Example 1

Sketch the graph of the ellipse with equation $9x^2 + 16y^2 = 144$, and find its focal points.

Solution
Rewriting the equation as
$$\frac{x^2}{16} + \frac{y^2}{9} = 1,$$

we see that $a^2 = 16$ and $b^2 = 9$, so $a = 4$ and $b = 3$. The graph is symmetric with respect to both coordinate axes and has intercepts $(4, 0)$, $(-4, 0)$, $(0, 3)$, and $(0, -3)$. Figure 10.18 shows this graph.

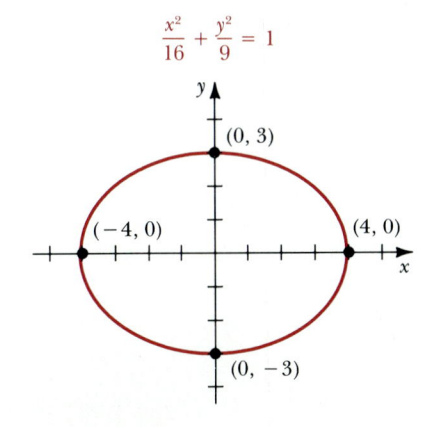

Figure 10.18

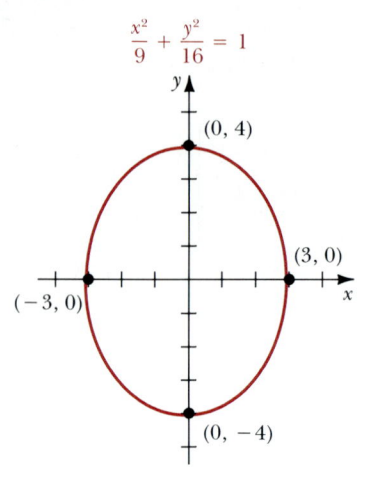

Figure 10.19

$\dfrac{x^2}{9} + \dfrac{y^2}{16} = 1$

The focal points of this ellipse are at $(c, 0)$ and $(-c, 0)$, where c satisfies the equation $b^2 = a^2 - c^2$. Thus, $9 = 16 - c^2$ and $c = \sqrt{7}$. ▶

Example 2

Sketch the graph of the ellipse with equation $16x^2 + 9y^2 = 144$, and find its focal points.

Solution
This equation can be rewritten as

$$\dfrac{x^2}{9} + \dfrac{y^2}{16} = 1.$$

The denominator of $y^2/16$ exceeds that of $x^2/9$, so this is an ellipse with axis along the y-axis in the form of (10.8). As in Example 1, $a^2 = 16$ and $b^2 = 9$, so $a = 4$ and $b = 3$. The axis intercepts are $(3, 0)$, $(-3, 0)$, $(0, 4)$, and $(0, -4)$, and the focal points lie along the y-axis at $(0, \sqrt{7})$ and $(0, -\sqrt{7})$. The graph is shown in Figure 10.19. Note the similarity between this ellipse and the ellipse in Example 1. ▶

Example 3

Find an equation of the ellipse in standard position that has a vertex at $(5, 0)$ and a focal point at $(3, 0)$.

Solution
The ellipse is in standard position, so the other focal point must be at $(-3, 0)$ and the other vertex at $(-5, 0)$.

Since $a = 5$, $a^2 = 25$, and $b^2 = a^2 - c^2 = 25 - 9 = 16$, the ellipse, shown in Figure 10.20, has equation

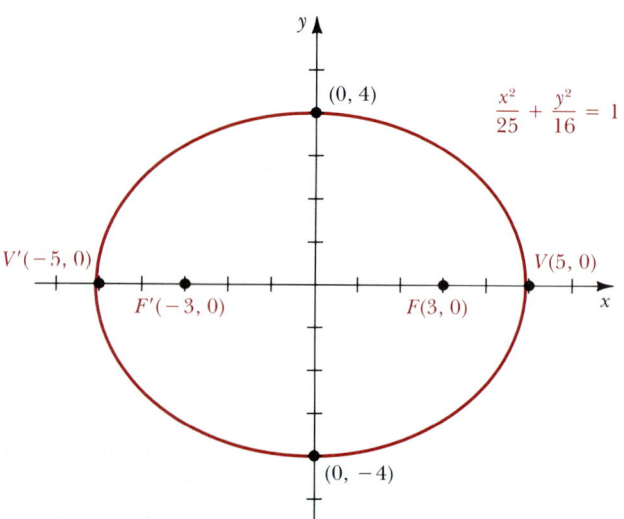

$$\dfrac{x^2}{25} + \dfrac{y^2}{16} = 1, \quad \text{or} \quad 16x^2 + 25y^2 = 400.$$

▶

Figure 10.20

Example 4

Sketch the graph of the ellipse with equation $9x^2 - 72x + 4y^2 + 16y + 124 = 0$, and find its focal points.

Solution

To compare this equation with the equation of an ellipse in standard position, complete the square in both x and y:

$$9(x^2 - 8x + 16) + 4(y^2 + 4y + 4) + 124 - 9(16) - 4(4) = 0.$$

This simplifies to

$$9(x - 4)^2 + 4(y + 2)^2 = 36, \quad \text{or} \quad \frac{(x - 4)^2}{4} + \frac{(y + 2)^2}{9} = 1.$$

The equation

$$\frac{x^2}{4} + \frac{y^2}{9} = 1$$

describes the ellipse in standard position shown in Figure 10.21. This ellipse has vertices at $(0, 3)$ and $(0, -3)$. Since

$$c^2 = a^2 - b^2 = 9 - 4 = 5,$$

the focal points of this standard position ellipse occur at $(0, \sqrt{5})$ and $(0, -\sqrt{5})$. The graph of the original equation,

$$\frac{(x - 4)^2}{4} + \frac{(y + 2)^2}{9} = 1,$$

is simply a translation of the graph of the ellipse in standard position, downward 2 units and 4 units to the right. Thus the vertices occur at $(4, 1)$ and $(4, -5)$, and the focal points are at $(4, -2 + \sqrt{5})$ and $(4, -2 - \sqrt{5})$. The graph is shown in Figure 10.22. ▶

An ellipse has a reflection property similar to the reflection property of a parabola. In this case, however, any object emitted from one of the focal points is reflected from the curve through the other focal point (see Figure 10.23). The proof of this fact is considered in Exercise 30. If the object continues through the second focal point and is reflected from the curve again, it returns to the original focal point.

This reflection property is most evident in "whispering galleries," rooms that have elliptical ceilings. The Mormon Tabernacle in Salt Lake City is built on this design (see Exercise 41), as is Statuary Hall in the United States Capital building in Washington, D.C. In fact, markers have been placed at the focal points in Statuary Hall and visitors can verify this reflection property. A person standing at one focal point can clearly hear the whisper of a person standing at the other focal point, even when the room is quite crowded.

The Oxford mathematician C. L. Dodgson (1832–1898), better known as Lewis Carroll, the author of *Alice in Wonderland*, used the reflection property of the ellipse to design an interesting billiard table.

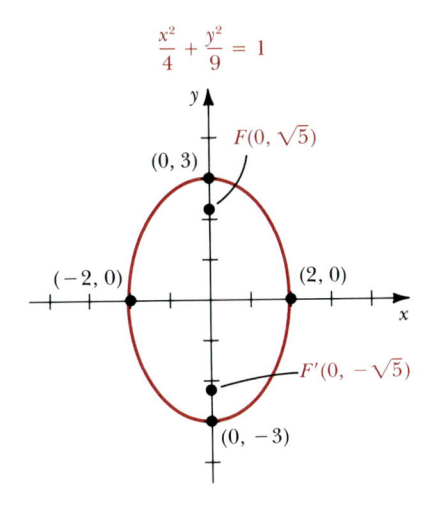

Figure 10.21

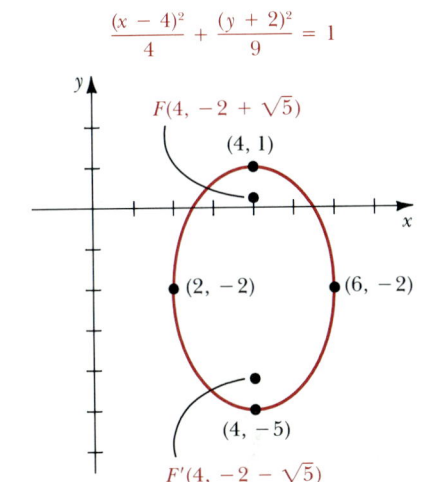

Figure 10.22

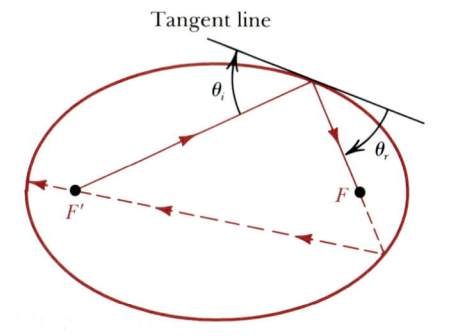

Figure 10.23
The reflection property of an ellipse.

On this table the playing surface is in the form of an ellipse with holes located at the focal points. Generally, obstacles are placed near the center of the table so that a bank, or reflection, shot must go from one end of the table to the hole in the other end. Consider how the intelligent player would design such a shot.

The ellipse also plays a vital role in astronomy, since the orbit of each planet is an ellipse with the sun at one of its focal points. This fact is known as **Kepler's First Law** (the Law of Ellipses). It was first announced by Johann Kepler (1571–1630) in 1609, following a decade of work based on observations of Tycho Brahe (1546–1601). This and two other laws of planetary motion that were formulated by Kepler are considered in Section 12.6.

▶ EXERCISE SET 10.2

In Exercises 1–14, (a) sketch the graph of the ellipse and (b) find the vertices and focal points.

1. $x^2 + \dfrac{y^2}{9} = 1$
2. $\dfrac{x^2}{16} + y^2 = 1$
3. $16x^2 + 25y^2 = 400$
4. $25x^2 + 16y^2 = 400$
5. $3x^2 + 2y^2 = 6$
6. $4x^2 + 3y^2 = 12$
7. $4x^2 + y^2 = 1$
8. $x^2 + 4y^2 = 1$
9. $4x^2 + y^2 + 16x + 7 = 0$
10. $16x^2 + 9y^2 - 54y - 63 = 0$
11. $x^2 + 4y^2 - 2x - 16y + 13 = 0$
12. $4x^2 + 9y^2 - 16x + 90y + 97 = 0$
13. $3x^2 + 2y^2 - 18x + 4y + 28 = 0$
14. $2x^2 + 5y^2 + 8x - 10y - 27 = 0$

In Exercises 15–20, find an equation of the ellipse that satisfies the stated conditions.

15. Foci at $(\pm 2, 0)$, vertices at $(\pm 3, 0)$.
16. Foci at $(\pm 2, 0)$, y-intercepts at $(0, \pm 2)$.
17. Foci at $(0, \pm 1)$, x-intercepts at $(\pm 2, 0)$.
18. Foci at $(3, 0)$ and $(1, 0)$, a vertex at $(0, 0)$.
19. Vertices at $(2, 2)$ and $(6, 2)$, a focal point at $(5, 2)$.
20. Foci at $(3, 3)$ and $(3, -1)$, passing through $(4, 0)$.

21. Find the area bounded by the ellipse $9x^2 + 4y^2 = 36$.
22. Show that the area bounded by the ellipse $x^2/a^2 + y^2/b^2 = 1$ is πab.
23. Find an equation of the tangent line to the ellipse $9x^2 + 4y^2 = 36$ at $(1, 3\sqrt{3}/2)$.
24. Show that the tangent line to the ellipse $x^2/a^2 + y^2/b^2 = 1$ at the point (x_0, y_0) has equation
$$\frac{xx_0}{a^2} + \frac{yy_0}{b^2} = 1.$$
25. Find equations of the tangent lines to the ellipse $4x^2 + y^2 = 4$ that pass through the point $(3, 0)$.
26. Show that the tangent lines to the ellipse $x^2/a^2 + y^2/b^2 = 1$ that pass through $(x_0, 0)$, where $x_0 > a$, intersect the ellipse at $x = a^2/x_0$.
27. Find the area of the largest rectangle with sides parallel to the x- and y-axes that can be inscribed within the ellipse $x^2/a^2 + y^2/b^2 = 1$.
28. Find the area of the largest isosceles triangle with a vertex at $(0, b)$ and base parallel to the x-axis that can be inscribed within the ellipse $x^2/a^2 + y^2/b^2 = 1$.
29. Show that the triangle found in Exercise 28 has the same area as the largest isosceles triangle with a vertex at $(a, 0)$ and base parallel to the y-axis that can be inscribed within $x^2/a^2 + y^2/b^2 = 1$.
30. Show that if an object is emitted from one of the focal points of an ellipse, it is reflected from the curve through the other focal point.
31. A *latus rectum* of an ellipse is a line segment that passes through a focal point perpendicular to the major axis and joins the points on the ellipse (see the figure).

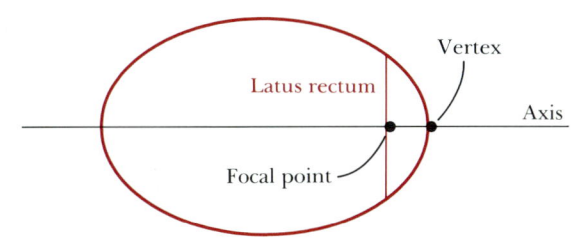

Find the length of a latus rectum of the ellipse $x^2/a^2 + y^2/b^2 = 1$.

32. It appears that for an ellipse in standard position, the points on the ellipse that are closest to and farthest from the center of the ellipse lie at the intersection of the ellipse with the coordinate axes. Use calculus to show that this is true.

33. Consider the equation $Ax^2 + Cy^2 + Dx + Ey + F = 0$, where A and C are positive constants. Find conditions on the constants A, C, D, E, and F that ensure that this equation describes:
   a) an ellipse; b) a circle;
   c) a single point; and d) no points in the plane.

34. a) Show that the surface area generated by rotating the ellipse $x^2/a^2 + y^2/b^2 = 1$, $a > b > 0$, about the x-axis is
$$2\pi b^2 + \frac{2a^2 b\pi}{\sqrt{a^2 - b^2}} \arcsin\left(\frac{\sqrt{a^2 - b^2}}{a}\right).$$
   b) Show that the surface area of a sphere of radius b is $4\pi b^2$ by using l'Hôpital's Rule to find the limit of the surface area of an ellipse as a approaches b from the right.

35. Describe the curve traced by a point 2 ft from the top of a ladder 8 ft long as the bottom of the ladder moves away from a vertical wall.

36. Olympic Stadium in Montreal, Canada, is constructed in the shape of an ellipse with major and minor axes of 480 and 280 m, respectively. Find an equation of this ellipse. How much area is covered by this stadium?

Olympic Stadium in Montreal.

37. Lou's Fruit Market sells two varieties of equally delicious watermelons. One type costs $3.99 and is 14 in. long and 12 in. wide; the other costs $2.79 and is 20 in. long and 9 in. wide. They both have elliptical cross sections when cut lengthwise.
   a) Which melon contains more volume?
   b) Suppose the rind on each melon is 1 in. thick. Which melon is the better buy?

38. The largest rigid airships ever built, the U.S. Navy ships USS *Akron* and USS *Macon*, were 785 ft long and had a gas capacity of 6,500,000 ft³. Assume the cross sections of the ships were elliptical when cut lengthwise, and circular when cut widthwise.
   a) What was their width?
   b) How much surface material was contained on these ships?

39. The Goodyear airships *Columbia, America,* and *Europa* are each 192 ft long and 50 ft wide. A brochure available from Goodyear Aerospace in Akron, Ohio, states that the volume contained in one of the blimps is 202,700 ft³. Is this figure reasonable?

Goodyear airship *America*.

40. Halley's comet is named in honor of Edmund Halley (1652–1742), the friend and contemporary of Isaac Newton, who persuaded him to publish his *Principia*

Halley's comet photographed as it moved away from the Sun in March 1986.

mathematica, the work in which Newton described the laws of motion. In 1682 Halley determined the orbit of the comet that bears his name and predicted its return 76 years later. The orbit is elliptical with a major axis of length 36.2 A.U. (A.U. is an abbreviation for Astronomical Unit; 1 A.U. $\approx 1.49 \times 10^{11}$ m) and a minor axis of length 9.1 A.U., with the sun at one focal point. How close does the comet pass to the sun? (The comet was last at this point, called the *perihelion* of its orbit, on February 9, 1986. It was visible from earth during periods before and after this date.)

41. The Mormon Tabernacle in Salt Lake City, Utah, was built between 1863 and 1867. The tabernacle is 250 ft long, 150 ft wide, and 80 ft high and has the shape of a whispering gallery with its longitudinal cross section in the form of half an ellipse. Determine where the focal points of the ellipse lie.

The Mormon Tabernacle in Salt Lake City Utah.

Putnam exercise

42. Of all ellipses inscribed in a square, show that the circle has the maximum perimeter. (This exercise was problem A–4 of the thirty-third William Lowell Putnam examination given on December 2, 1972. The examination and its solution are in the November 1973 issue of the *American Mathematical Monthly*, pp. 1019–1028.

10.3

Hyperbolas

(10.9)
DEFINITION
A **hyperbola** is the set of points in the plane the difference of whose distances from two fixed points has a constant magnitude.

The fixed points are called the **focal points** of the hyperbola, the line through the focal points is called the **axis** of the hyperbola, and the points where the hyperbola intersects the axis are called the **vertices** of the hyperbola. Figure 10.24 illustrates these terms.

Standard equations for the hyperbola are derived in the same manner as for the ellipse. In fact, if we let the *x*-axis be the axis of the hyperbola with focal points at $(c, 0)$ and $(-c, 0)$, as shown in Figure 10.25, and let the constant magnitude of the difference be denoted by $2a$, the equation derived is the same as that of the ellipse:

$$\frac{x^2}{a^2} + \frac{y^2}{a^2 - c^2} = 1.$$

In this case, however, we will show that $a^2 - c^2$ is always negative, whereas in the case of the ellipse, this term is always positive.

If $(c, 0)$ is the focal point closest to (x, y), then

$$d((x, y), (-c, 0)) > d((x, y), (c, 0))$$

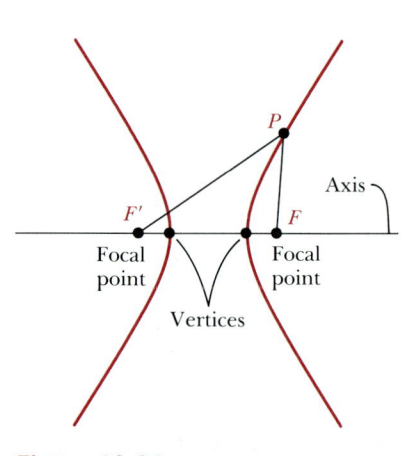

Figure 10.24
A hyperbola.

10.3 HYPERBOLAS

and
$$2a = |d((x, y), (c, 0)) - d((x, y), (-c, 0))|$$
$$= d((x, y), (-c, 0)) - d((x, y), (c, 0)),$$

so
$$2a + d((x, y), (c, 0)) = d((x, y), (-c, 0)).$$

Since $d((x, y), (-c, 0))$ is the length of one side of the triangle shown in Figure 10.25 with vertices F', F, and P, and the sum of the other two sides is $2c + d((x, y), (c, 0))$, we have

$$2a + d((x, y), (c, 0)) = d((x, y), (-c, 0)) \leq 2c + d((x, y), (c, 0)).$$

Thus, $0 < a < c$ and $a^2 - c^2 < 0$. Let $b > 0$ be such that

$$b^2 = c^2 - a^2.$$

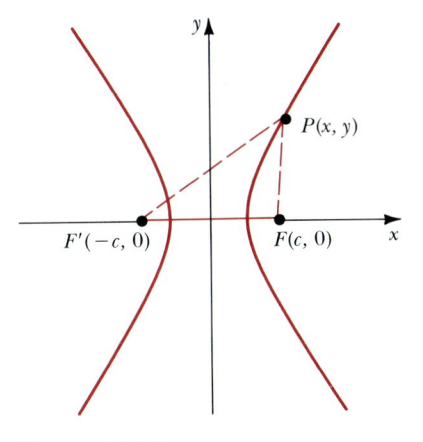

Figure 10.25

Then the equation of a hyperbola in standard position whose axis is the x-axis is

(10.10)
$$\frac{x^2}{a^2} - \frac{y^2}{b^2} = 1.$$

This hyperbola is symmetric with respect to both coordinate axes and has x-intercepts at $(a, 0)$ and $(-a, 0)$. It does not intersect the y-axis. In fact, since

$$y^2 = \frac{b^2}{a^2}(x^2 - a^2),$$

the graph does not intersect the region where $-a < x < a$.

Suppose we restrict the graph of the hyperbola in (10.10) to the first quadrant and consider the difference between the y-coordinate on the line with equation $y = (b/a)x$ and the y-coordinate on the hyperbola, as shown in Figure 10.26. This difference is

$$\frac{b}{a}x - \frac{b}{a}\sqrt{x^2 - a^2}$$

and

$$\lim_{x \to \infty}\left(\frac{b}{a}x - \frac{b}{a}\sqrt{x^2 - a^2}\right) = \lim_{x \to \infty}\frac{b}{a}(x - \sqrt{x^2 - a^2})\left(\frac{x + \sqrt{x^2 - a^2}}{x + \sqrt{x^2 - a^2}}\right)$$
$$= \lim_{x \to \infty}\frac{b}{a}\left(\frac{x^2 - (x^2 - a^2)}{x + \sqrt{x^2 - a^2}}\right)$$
$$= \frac{b}{a}\lim_{x \to \infty}\frac{a^2}{x + \sqrt{x^2 - a^2}} = 0.$$

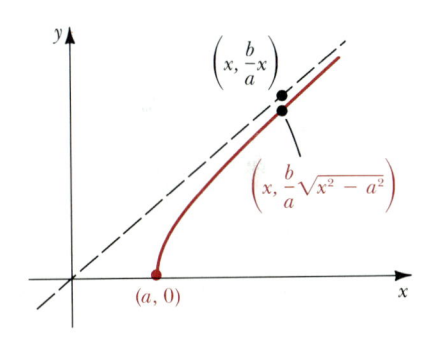

Figure 10.26

$y = \frac{b}{a}x$ is a slant asymptote.

Consequently, the hyperbola approaches the line $y = (b/a)x$ as x increases, so $y = (b/a)x$ is a *slant asymptote* of the hyperbola. By symmetry, the graph in the other quadrants must approach either $y = (b/a)x$ or $y = -(b/a)x$, as shown in Figure 10.27 on the following page.

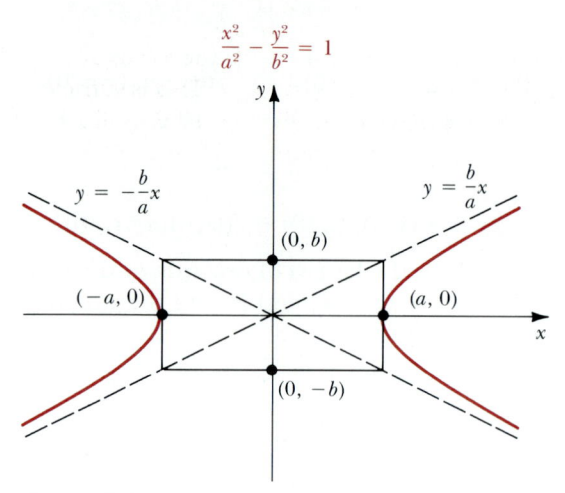

Figure 10.27

The equations of the asymptotes can be written together as

$$y = \pm \frac{b}{a} x, \quad \text{or} \quad \frac{x}{a} \pm \frac{y}{b} = 0.$$

Thus the hyperbola with equation

$$\frac{x^2}{a^2} - \frac{y^2}{b^2} = 1$$

has asymptotes with equations

(10.11) $$\frac{x}{a} \pm \frac{y}{b} = 0.$$

In a similar manner, the hyperbola with focal points on the y-axis at $(0, c)$ and $(0, -c)$ has equation

(10.12) $$\frac{y^2}{a^2} - \frac{x^2}{b^2} = 1,$$

has vertices at the y-intercepts $(0, a)$ and $(0, -a)$, and has asymptotes $y = \pm (a/b)\, x$. Since $x^2 = [b^2(y^2 - a^2)]/a^2$, the graph of this hyperbola, shown in Figure 10.28, does not intersect the region where $-a < y < a$.

A hyperbola with equation in either form (10.10) or form (10.12) is said to be in **standard position.**

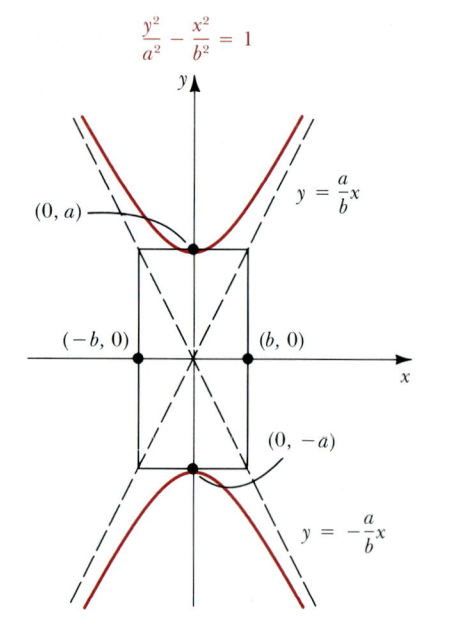

Figure 10.28

Example 1

Sketch the graph of the hyperbola with equation $16x^2 - 9y^2 = 144$, and find its focal points.

Solution
The equation can be rewritten as

$$\frac{x^2}{9} - \frac{y^2}{16} = 1,$$

so $a = 3$ and $b = 4$. The graph intersects the x-axis at $(3, 0)$ and $(-3, 0)$ and has asymptotes $y = \frac{4}{3} x$ and $y = -\frac{4}{3} x$. This is sufficient information to sketch the graph shown in Figure 10.29. Since $c = \sqrt{a^2 + b^2} = \sqrt{9 + 16} = 5$, the focal points are at $(5, 0)$ and $(-5, 0)$. ▶

Example 2

Sketch the graph of $9y^2 - 16x^2 = 144$.

Solution

The given equation can be rewritten as

$$\frac{y^2}{16} - \frac{x^2}{9} = 1,$$

and the asymptotes to the graph are $y = \pm \frac{4}{3} x$, the same as the asymptotes in Example 1. In this example, however, the vertices are on the y-axis at $(0, 4)$ and $(0, -4)$. The graph is as shown in Figure 10.30 together with the graph of the hyperbola described in Example 1: $x^2/9 - y^2/16 = 1$. These are called *conjugate hyperbolas*. ▶

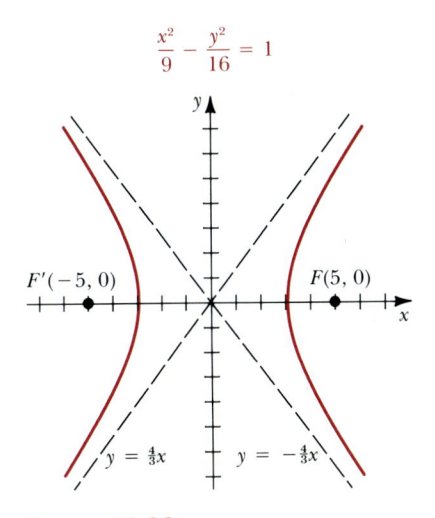

Figure 10.29

Example 3

Sketch the graph of the hyperbola with equation

$$y^2 - 2y - 9x^2 + 36x = 39.$$

Solution

To graph the equation, we show that its graph is the translation of the graph of a hyperbola in standard form. The first step is to complete the square in both x and y:

$$(y^2 - 2y + 1) - 9(x^2 - 4x + 4) = 39 + 1 - 36,$$

so

$$(y - 1)^2 - 9(x - 2)^2 = 4, \quad \text{and} \quad \frac{(y - 1)^2}{4} - \frac{(x - 2)^2}{4/9} = 1.$$

Consequently, $a = 2$ and $b = \frac{2}{3}$, and this hyperbola is a translation of the hyperbola in standard form with equation

$$\frac{y^2}{4} - \frac{x^2}{4/9} = 1.$$

The hyperbola in standard form has asymptotes $y = 3x$ and $y = -3x$; its graph is shown in Figure 10.31 on the following page.

The graph of

$$\frac{(y - 1)^2}{4} - \frac{(x - 2)^2}{4/9} = 1$$

is the graph in Figure 10.31 translated upward 1 unit and 2 units to the

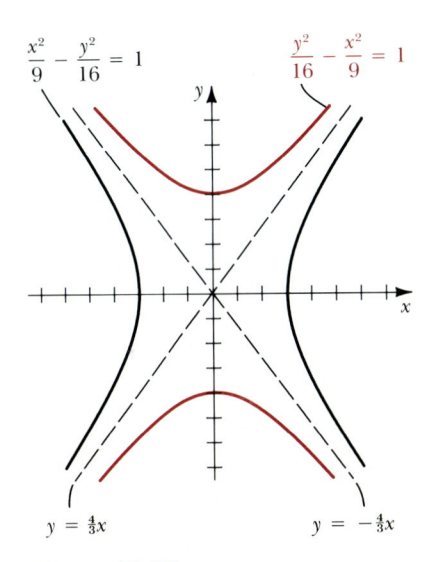

Figure 10.30
Conjugate hyperbolas.

$$\frac{y^2}{4} - \frac{x^2}{(4/9)} = 1$$

$$\frac{(y-1)^2}{4} - \frac{(x-2)^2}{(4/9)} = 1$$

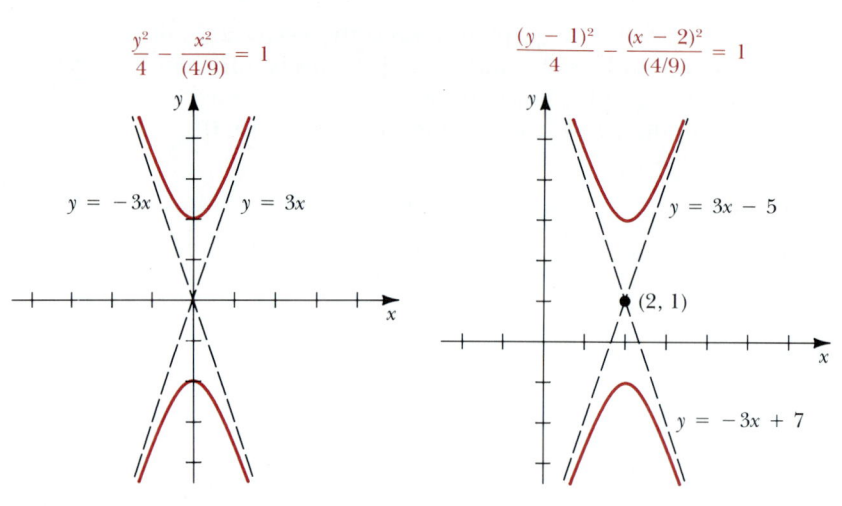

Figure 10.31

Figure 10.32

right. The asymptotes of

$$\frac{(y-1)^2}{4} - \frac{(x-2)^2}{4/9} = 1$$

are

$$y - 1 = 3(x - 2) \quad \text{and} \quad y - 1 = -3(x - 2),$$

that is,

$$y = 3x - 5 \quad \text{and} \quad y = -3x + 7.$$

The graph of $y^2 - 2y - 9x^2 + 36x = 39$ is shown in Figure 10.32. ▶

The hyperbola also has a reflection property involving its focal points. An object emitted from one focal point F' to the curve is reflected from the curve on a line directly away from the second focal point F, as illustrated in Figure 10.33. (The proof of this result is discussed in Exercise 38.)

An interesting application of the hyperbola concerns spotting the position of an object when its relative distance from stationary points is known. Suppose an object is at an unknown location and that observation posts are located at points P_1, P_2, and P_3 (see Figure 10.34). These posts are capable of determining precisely the time at which the sound from the object reaches them, so the difference between the length of time it takes the sound to reach posts P_1 and P_2 is known and is constant in the given situation. This implies that if the object is located at P, the difference in the distance from P_1 to P and the distance from P_2 to P is constant:

$$d(P_1, P) - d(P_2, P) = c.$$

The definition of the hyperbola implies that the object must be located along a branch of a hyperbola with focal points P_1 and P_2, as shown in Figure 10.35.

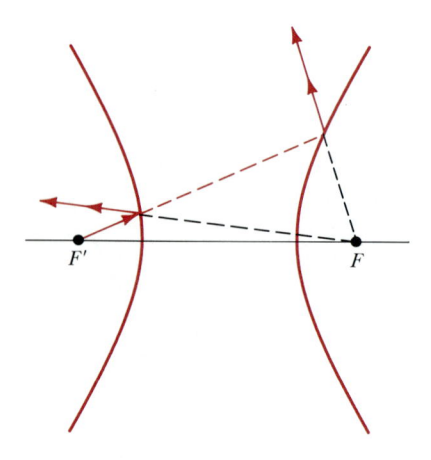

Figure 10.33
The reflection property of a hyperbola.

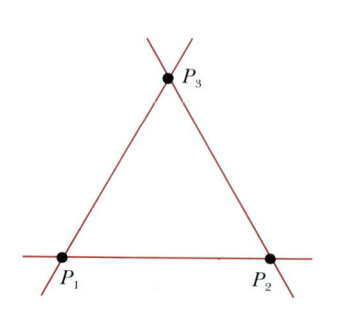

Figure 10.34
Observation posts at P_1, P_2, and P_3.

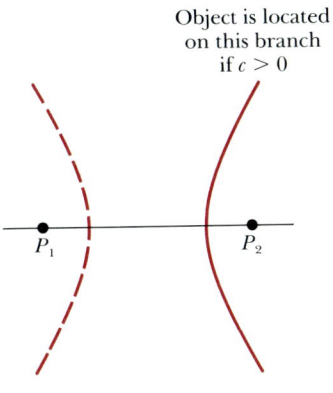

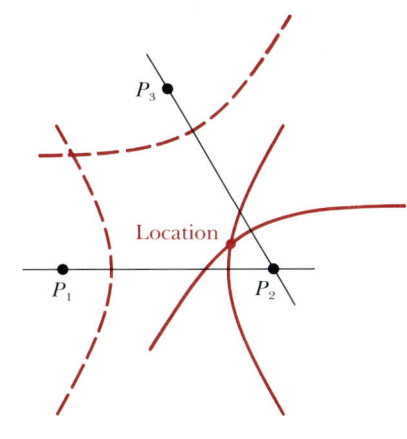

Figure 10.35

Figure 10.36 Object located at the intersection of the hyperbolas.

In a similar manner, the difference between the length of time it takes the sound to reach the posts P_2 and P_3 fixes the object along a branch of the hyperbola with focal points P_2 and P_3, as shown in Figure 10.36. The intersection of these hyperbolas determines the location P of the object.

The reverse technique is used in air and sea navigation by using three fixed reference points that emit radio signals at precise time intervals. If the speed at which the signal travels is known, then the differences between the times the three signals are received can be used to pinpoint the position of an object. This technique is used in the LORAN (**Lo**ng **Ra**nge **N**avigation) system of ship navigation and by the Omega system of air navigation.

▶ EXERCISE SET 10.3

In Exercises 1–18, (a) sketch the graph of the hyperbola, (b) find the coordinates of the vertices and focal points, and (c) write equations of the asymptotes.

1. $\dfrac{x^2}{4} - \dfrac{y^2}{9} = 1$
2. $\dfrac{x^2}{9} - \dfrac{y^2}{4} = 1$
3. $\dfrac{y^2}{4} - \dfrac{x^2}{9} = 1$
4. $\dfrac{y^2}{9} - \dfrac{x^2}{4} = 1$
5. $16x^2 - 4y^2 = 64$
6. $16y^2 - 4x^2 = 64$
7. $x^2 - y^2 = 1$
8. $y^2 - 4x^2 = 1$
9. $y^2 - 2x^2 = 8$
10. $x^2 - 4y^2 = 100$
11. $x^2 + 2x - 4y^2 = 3$
12. $9y^2 - 18y - 4x^2 = 27$
13. $3x^2 - y^2 = 6x$
14. $2y^2 + 8y = 9x^2$
15. $9x^2 - 4y^2 - 18x - 8y = 31$
16. $x^2 - 4y^2 - 2x - 16y = 19$
17. $4x^2 - 3y^2 + 8x + 18y = 11$
18. $4y^2 - 4x^2 + 32x + 8y = 56$

In Exercises 19–30, find an equation of the hyperbola that satisfies the stated conditions.

19. Foci at $(\pm 5, 0)$, vertices at $(\pm 3, 0)$
20. Foci at $(0, \pm 13)$, vertices at $(0, \pm 12)$
21. Foci at $(0, \pm 5)$, vertices at $(0, \pm 4)$
22. Foci at $(\pm 13, 0)$, vertices at $(\pm 5, 0)$
23. Foci at $(-1, 4)$ and $(5, 4)$, a vertex at $(0, 4)$
24. A focus at $(2, 1)$, vertices at $(2, 0)$ and $(2, -3)$
25. Foci at $(\pm 3\sqrt{2}, 0)$, passing through $(5, 4)$
26. Vertices at $(2, \pm 3)$, passing through $(8, 8)$

27. Vertices at $(\pm 3, 0)$, equations of asymptotes $y = \pm 3x/4$

28. Foci $(\pm 3, 0)$, equations of asymptotes $y = \pm 3x/4$

29. Vertices at $(6, 1)$ and $(-2, 1)$, equations of asymptotes $y = 3x/4 - 1/2$ and $y = -3x/4 + 5/2$.

30. Foci at $(6, 1)$ and $(-2, 1)$, equations of asymptotes $y = 3x/4 - 1/2$ and $y = -3x/4 + 5/2$

31. Find an equation of a tangent line to the hyperbola with equation $3x^2 - 4y^2 = 12$ at the point $(2\sqrt{2}, \sqrt{3})$.

32. Show that the tangent line to the hyperbola $x^2/a^2 - y^2/b^2 = 1$ at the point (x_0, y_0) has equation $xx_0/a^2 - yy_0/b^2 = 1$.

33. Find equations of the tangent lines to the hyperbola $9x^2 - 4y^2 = 36$ that pass through $(1, 0)$.

34. Show that the tangent lines to the hyperbola $x^2/a^2 - y^2/b^2 = 1$ passing through $(x_0, 0)$, when $0 < x_0 < a$, intersect the hyperbola when $x = a^2/x_0$.

35. The *latus rectum* of a hyperbola is a line segment that passes through a focal point perpendicular to the axis and joins two points on the hyperbola (see the figure). Find the length of a latus rectum of the hyperbola with equation $x^2/a^2 - y^2/b^2 = 1$.

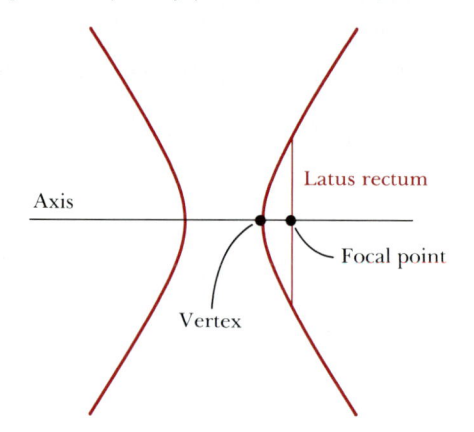

36. Show that the point on a hyperbola that is closest to a focal point is a vertex of the hyperbola.

37. Consider the equation $Ax^2 - Cy^2 + Dx + Ey + F = 0$, where A and C are positive constants. Find conditions on the constants A, C, D, E, and F that ensure that this equation describes:
   a) a hyperbola with axis parallel to the x-axis;
   b) intersecting lines;
   c) a hyperbola with axis parallel to the y-axis.

38. Show that if an object is emitted from one focal point of a hyperbola to the portion of the curve closest to the other focal point, it is reflected from that curve on a line directly away from the second focal point.

39. Three detection stations lie on an east–west line 1150 m apart. The eastmost station detects a sound from an object 2 sec before the westmost station and 1 sec before the station in the middle. Can the object emitting the noise be pinpointed? (Assume that the sound travels at the rate of 330 m/sec.)

40. A company has two manufacturing plants that produce identical automobiles. Because of differing manufacturing and labor conditions in the plants, it costs $130 more to produce a car in plant A than in plant B. The shipping costs from both plants are the same, $1 per mile, as are the loading and unloading costs, $25 per car. State the criteria for determining from which plant a car should be shipped.

41. The prime focus configuration of the 200-in. Hale telescope at the Palomar Mountain Observatory was described in Exercise 47 of Section 10.1. In addition to this configuration, there is a Cassegrain configuration that consists of a hyperbolic mirror inserted between the parabolic reflector and the parabola's focal point. This hyperbolic mirror has one focus at the focal point of the parabola and reflects the image back through a hole in the center of the parabolic mirror. From there it goes to the other focal point of the hyperbolic mirror, located 5 ft beyond the vertex of the parabolic mirror. What is the depth from rim to vertex of the hyperbolic mirror if it is located 64 in. from the vertex of the parabolic mirror and has a diameter of 41 in.?

10.4

Rotation of Axes

In this section we examine the effect that a rotation of the coordinate system has on the equation of a conic section. For example, we will see that the equation $xy = 1$ describes a hyperbola rotated from standard position through an angle $\theta = \pi/4$. The axis of this hyperbola is the line $y = x$, as shown in Figure 10.37.

As the name implies, a rotation shifts a graph through an angle, whereas a translation shifts a graph horizontally and vertically. Rotation is a useful technique to use when graphing certain second-degree equations that include a product of the variables x and y. As we will see in this section, an appropriate rotation can eliminate this type of product term from the equation.

We first determine how the rotation of a coordinate system affects an arbitrary point in the plane. Suppose a point P has coordinates (x, y) in one coordinate system and coordinates $(\hat{x}, \hat{y})$ in a second coordinate system, obtained from the first by rotating through an angle θ, $0 < \theta < \pi/2$, with the origins of the two systems remaining the same.

Let r denote the distance from the point P to the origin and ψ denote the angle formed by the intersection of the $\hat{x}$-axis and the line from P to the origin as in Figure 10.38. From the right triangle OBP, reproduced in Figure 10.39, we have

$$\hat{x} = r \cos \psi \quad \text{and} \quad \hat{y} = r \sin \psi,$$

and from the right triangle OAB, reproduced in Figure 10.40, we have

$$x = r \cos (\psi + \theta) \quad \text{and} \quad y = r \sin (\psi + \theta).$$

Thus,

$$x = r \cos (\psi + \theta) = r \cos \psi \cos \theta - r \sin \psi \sin \theta$$

and

$$y = r \sin (\psi + \theta) = r \cos \psi \sin \theta + r \sin \psi \cos \theta,$$

which implies that

(10.13) $\quad x = \hat{x} \cos \theta - \hat{y} \sin \theta, \quad \text{and} \quad y = \hat{x} \sin \theta + \hat{y} \cos \theta.$

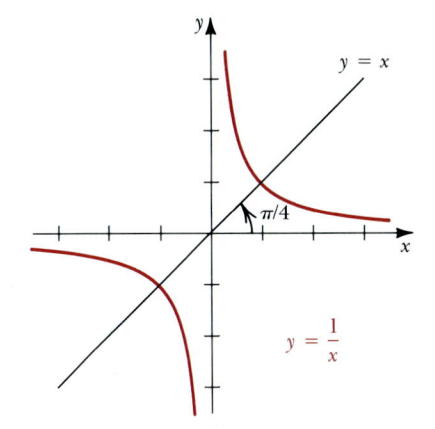

Figure 10.37
A hyperbola rotated through the angle $\pi/4$.

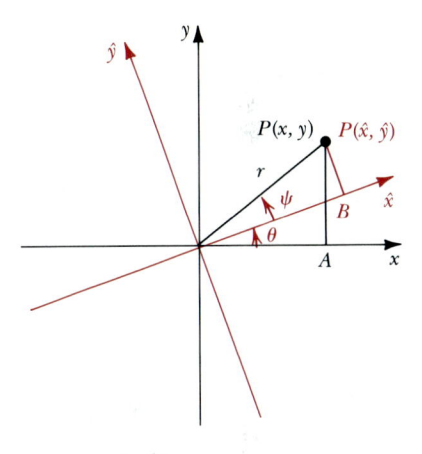

Figure 10.38
Rotation of axes through the angle θ.

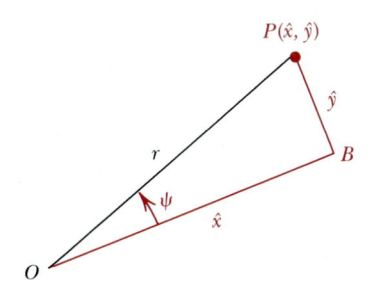

Figure 10.39

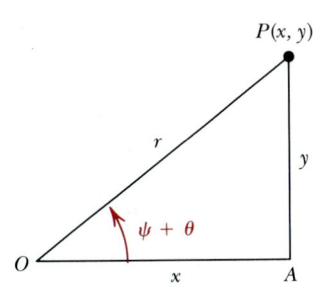

Figure 10.40

Example 1

Find the change in the equation $xy = 1$ that occurs when the x- and y-axes are rotated through the angle $\theta = \pi/4$.

$xy = 1$ or $\hat{x}^2 - \hat{y}^2 = 2$

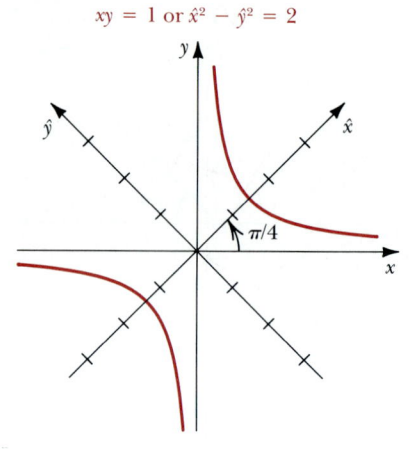

Figure 10.41

Solution
Denoting a point in the rotated system by $(\hat{x}, \hat{y})$, Equations (10.13) give

$$x = \hat{x} \cos \frac{\pi}{4} - \hat{y} \sin \frac{\pi}{4} = \frac{\sqrt{2}}{2} (\hat{x} - \hat{y})$$

and

$$y = \hat{x} \sin \frac{\pi}{4} + \hat{y} \cos \frac{\pi}{4} = \frac{\sqrt{2}}{2} (\hat{x} + \hat{y}).$$

Consequently, $xy = 1$ becomes $\frac{\sqrt{2}}{2} (\hat{x} - \hat{y}) \frac{\sqrt{2}}{2} (\hat{x} + \hat{y}) = 1$

or

$$\hat{x}^2 - \hat{y}^2 = 2.$$

This $(\hat{x}, \hat{y})$-coordinate equation is the equation of a hyperbola with asymptotic lines $\hat{y} = \pm \hat{x}$, that is, the x- and y-axis (see Figure 10.41). ▶

Example 2

Sketch the graph of the equation $21x^2 - 10\sqrt{3}xy + 31y^2 = 144$ by graphing the equation that results from rotating the x- and y-axes through the angle $\theta = \pi/6$.

Solution
Since

$$x = \hat{x} \cos \frac{\pi}{6} - \hat{y} \sin \frac{\pi}{6} = \frac{\sqrt{3}}{2} \hat{x} - \frac{1}{2} \hat{y}$$

and

$$y = \hat{x} \sin \frac{\pi}{6} + \hat{y} \cos \frac{\pi}{6} = \frac{1}{2} \hat{x} + \frac{\sqrt{3}}{2} \hat{y},$$

the new equation is

$$144 = 21 \left(\frac{\sqrt{3}}{2} \hat{x} - \frac{1}{2} \hat{y} \right)^2 - 10\sqrt{3} \left(\frac{\sqrt{3}}{2} \hat{x} - \frac{1}{2} \hat{y} \right) \left(\frac{1}{2} \hat{x} + \frac{\sqrt{3}}{2} \hat{y} \right)$$
$$+ 31 \left(\frac{1}{2} \hat{x} + \frac{\sqrt{3}}{2} \hat{y} \right)^2$$
$$= \frac{21}{4} (3\hat{x}^2 - 2\sqrt{3} \hat{x}\hat{y} + \hat{y}^2) - \frac{10\sqrt{3}}{4} (\sqrt{3} \hat{x}^2 + 2\hat{x}\hat{y} - \sqrt{3} \hat{y}^2)$$
$$+ \frac{31}{4} (\hat{x}^2 + 2\sqrt{3} \hat{x}\hat{y} + 3\hat{y}^2)$$
$$= \hat{x}^2 \left(\frac{63}{4} - \frac{30}{4} + \frac{31}{4} \right) + \hat{x}\hat{y} \left(-\frac{42}{4} \sqrt{3} - \frac{20}{4} \sqrt{3} + \frac{62}{4} \sqrt{3} \right)$$
$$+ \hat{y}^2 \left(\frac{21}{4} + \frac{30}{4} + \frac{93}{4} \right).$$

This simplifies to

$$144 = 16\hat{x}^2 + 36\hat{y}^2 \quad \text{or} \quad \frac{\hat{x}^2}{9} + \frac{\hat{y}^2}{4} = 1.$$

The graph of the equation

$$21x^2 - 10\sqrt{3}\,xy + 31y^2 = 144$$

is thus the graph of the ellipse $x^2/9 + y^2/4 = 1$ rotated through the angle $\theta = \pi/6$, as shown in Figure 10.42. ▶

In the preceding examples, the appropriate angle was given to eliminate the product term from the second-degree equation. We now show how to find this angle in a general situation.

Suppose

(10.14) $\qquad Ax^2 + Bxy + Cy^2 + Dx + Ey + F = 0$

is a second-degree equation, and x and y are defined by the equations

$$x = \hat{x}\cos\theta - \hat{y}\sin\theta, \qquad y = \hat{x}\sin\theta + \hat{y}\cos\theta,$$

where θ is an angle between 0 and $\pi/2$. We introduce the variables $\hat{x}$ and $\hat{y}$ into Equation (10.14):

$A(\hat{x}\cos\theta - \hat{y}\sin\theta)^2 + B(\hat{x}\cos\theta - \hat{y}\sin\theta)(\hat{x}\sin\theta + \hat{y}\cos\theta)$
$\quad + C(\hat{x}\sin\theta + \hat{y}\cos\theta)^2 + D(\hat{x}\cos\theta - \hat{y}\sin\theta) + E(\hat{x}\sin\theta + \hat{y}\cos\theta)$
$\quad + F = 0.$

Next we multiply and collect similar terms:

$\hat{x}^2(A\cos^2\theta + B\cos\theta\sin\theta + C\sin^2\theta)$
$\quad + \hat{x}\hat{y}\,[-2A\cos\theta\sin\theta + B(\cos^2\theta - \sin^2\theta) + 2C\sin\theta\cos\theta]$
$\quad + \hat{y}^2\,(A\sin^2\theta - B\sin\theta\cos\theta + C\cos^2\theta)$
$\quad + \hat{x}(D\cos\theta + E\sin\theta) + \hat{y}(-D\sin\theta + E\cos\theta) + F = 0.$

To eliminate the $\hat{x}\hat{y}$-term, we choose θ so that the $\hat{x}\hat{y}$-coefficient is zero. This means that

$$0 = -2A\cos\theta\sin\theta + B(\cos^2\theta - \sin^2\theta) + 2C\sin\theta\cos\theta.$$

We divide by $\sin^2\theta$ (which is nonzero, since $0 < \theta < \pi/2$) to obtain a quadratic equation in terms of $\cot\theta$:

$$0 = -2A\left(\frac{\cos\theta}{\sin\theta}\right) + B\left[\left(\frac{\cos\theta}{\sin\theta}\right)^2 - 1\right] + 2C\left(\frac{\cos\theta}{\sin\theta}\right)$$

$$= B(\cot\theta)^2 - 2(A - C)\cot\theta - B.$$

The quadratic formula implies that

$$\cot\theta = \frac{2(A - C) \pm \sqrt{4(A - C)^2 + 4B^2}}{2B} = \frac{A - C}{B} \pm \frac{\sqrt{(A - C)^2 + B^2}}{B}.$$

The angle of rotation θ is between 0 and $\pi/2$, so $\cot\theta > 0$. Since

$$\left|\frac{A - C}{B}\right| \leq \frac{\sqrt{(A - C)^2 + B^2}}{|B|},$$

$21x^2 - 10\sqrt{3}xy + 31y^2 = 144$

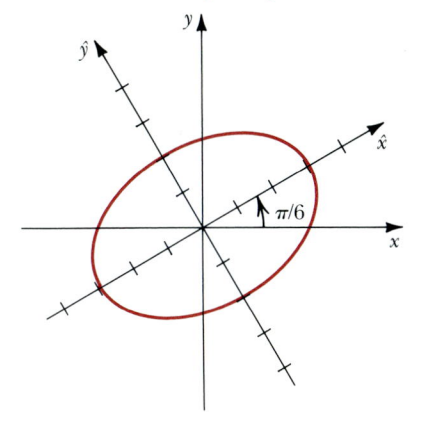

Figure 10.42

cot θ will be positive if the sign is chosen so that

$$\pm \frac{\sqrt{(A-C)^2 + B^2}}{B} = \frac{\sqrt{(A-C)^2 + B^2}}{|B|} > 0.$$

Consequently, the appropriate rotation is through the angle in $(0, \pi/2)$ given by

(10.15) $$\cot \theta = \frac{A-C}{B} + \frac{\sqrt{(A-C)^2 + B^2}}{|B|}.$$

This rotation changes the equation

$$Ax^2 + Bxy + Cy^2 + Dx + Ey + F = 0$$

into

$$\hat{A}\hat{x}^2 + \hat{C}\hat{y}^2 + \hat{D}\hat{x} + \hat{E}\hat{y} + \hat{F} = 0,$$

where

(10.16)

$$\hat{A} = A \cos^2 \theta + B \cos \theta \sin \theta + C \sin^2 \theta,$$
$$\hat{B} = 2(C - A) \sin \theta \cos \theta + B(\cos^2 \theta - \sin^2 \theta),$$
$$\hat{C} = A \sin^2 \theta - B \sin \theta \cos \theta + C \cos^2 \theta,$$
$$\hat{D} = D \cos \theta + E \sin \theta,$$
$$\hat{E} = -D \sin \theta + E \cos \theta,$$
$$\hat{F} = F,$$

and, because of the choice of θ, $\hat{B} = 0$.

Example 3

Use a rotation to sketch the graph of $2x^2 - 72xy + 23y^2 + 25 = 0$.

Solution

To eliminate the *xy*-term, we rotate through the angle θ, where

$$\cot \theta = \frac{A-C}{B} + \frac{\sqrt{(A-C)^2 + B^2}}{|B|} = \frac{2-23}{-72} + \frac{\sqrt{(2-23)^2 + (-72)^2}}{|-72|}.$$

Thus,

$$\cot \theta = \frac{-21}{-72} + \frac{\sqrt{5625}}{72} = \frac{21 + 75}{72} = \frac{96}{72} = \frac{4}{3}.$$

From the triangle in Figure 10.43, we see that

$$\sin \theta = \frac{3}{5} \quad \text{and} \quad \cos \theta = \frac{4}{5}.$$

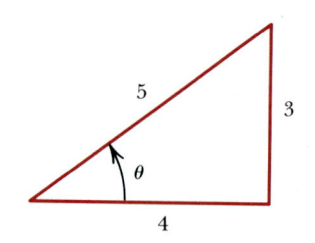

Figure 10.43

By (10.16),

$$\hat{A} = 2\left(\frac{4}{5}\right)^2 - 72\left(\frac{3}{5}\right)\left(\frac{4}{5}\right) + 23\left(\frac{3}{5}\right)^2 = -\frac{625}{25} = -25,$$

$$\hat{C} = 2\left(\frac{3}{5}\right)^2 + 72\left(\frac{3}{5}\right)\left(\frac{4}{5}\right) + 23\left(\frac{4}{5}\right)^2 = \frac{1250}{25} = 50,$$

$$\hat{F} = 25, \quad \text{and} \quad \hat{B} = \hat{D} = \hat{E} = 0.$$

The equation after rotation is

$$-25\hat{x}^2 + 50\hat{y}^2 + 25 = 0,$$

which simplifies to

$$\hat{x}^2 - 2\hat{y}^2 = 1.$$

This is the hyperbola with asymptotes

$$\hat{y} = \pm\frac{\sqrt{2}}{2}\hat{x},$$

whose graph is shown in Figure 10.44. ▶

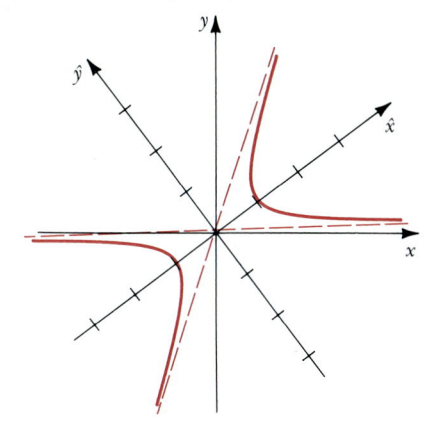

Figure 10.44

Looking back at the equations in Examples 2 and 3, we see that the equation

$$21x^2 - 10\sqrt{3}xy + 31y^2 - 144 = 0$$

is an ellipse, whereas the graph of the somewhat similar equation

$$2x^2 - 72xy + 23y^2 + 25 = 0$$

is a hyperbola. There is an interesting formula that permits us to determine the form of the graph by observing the coefficients. Suppose the graph of the equation

$$Ax^2 + Bxy + Cy^2 + Dx + Ey + F = 0$$

is rotated through an angle θ, and the new equation is

$$\hat{A}\hat{x}^2 + \hat{B}\hat{x}\hat{y} + \hat{C}\hat{y}^2 + \hat{D}\hat{x} + \hat{E}\hat{y} + \hat{F} = 0.$$

It can be easily, though rather tediously, shown (see Exercise 26) that

$$B^2 - 4AC = \hat{B}^2 - 4\hat{A}\hat{C}.$$

In particular, when the angle is chosen so that $\hat{B} = 0$,

$$B^2 - 4AC = -4\hat{A}\hat{C}.$$

However, except for degenerate cases, the graph of the equation

$$\hat{A}\hat{x}^2 + \hat{C}\hat{y}^2 + \hat{D}\hat{x} + \hat{E}\hat{y} + \hat{F} = 0$$

is:

i) an ellipse if $\hat{A}$ and $\hat{C}$ are both positive or both negative, that is, if $-4\hat{A}\hat{C} < 0$;

ii) a hyperbola if one of $\hat{A}$ or $\hat{C}$ is positive and the other is negative, that is, if $-4\hat{A}\hat{C} > 0$;

iii) a parabola if $\hat{A} = 0$ or $\hat{C} = 0$, that is, if $-4\hat{A}\hat{C} = 0$.

Consequently, except for degenerate cases, the graph of the equation

(10.17)
$$Ax^2 + Bxy + Cy^2 + Dx + Ey + F = 0$$

is:

i) an ellipse if $B^2 - 4AC < 0$;
ii) a hyperbola if $B^2 - 4AC > 0$;
iii) a parabola if $B^2 - 4AC = 0$.

In addition, the graph is:

i) a straight line if $\hat{A} = \hat{C} = 0$ and at least one of $\hat{D}$ or $\hat{E}$ is nonzero;
ii) a pair of straight lines if $\hat{D} = \hat{E} = \hat{F} = 0$ and both $\hat{A}$ and $\hat{C}$ are nonzero;
iii) a point or pair of points if only one of $\hat{A}$, $\hat{C}$, $\hat{D}$, or $\hat{E}$ is nonzero.

Applying the result in (10.17) to the examples in this section, we see the following.

Example 1. The equation $xy = 1$ has
$$B^2 - 4AC = 1^2 - 4(0)(0) = 1 > 0$$
and is a hyperbola.

Example 2. The equation $21x^2 - 10\sqrt{3}\,xy + 31y^2 - 144 = 0$ has
$$B^2 - 4AC = (-10\sqrt{3})^2 - 4(21)(31) = -2304 < 0$$
and is an ellipse.

Example 3. The equation $2x^2 - 72xy + 23y^2 + 25 = 0$ has
$$B^2 - 4AC = (-72)^2 - 4(2)(23) = 5000 > 0,$$
and is a hyperbola.

It is strongly recommended that you apply this test whenever a rotation is involved because it provides a check on the final result—a result that generally requires a significant amount of algebraic and trigonometric manipulation.

▶ EXERCISE SET 10.4

In Exercises 1–20 (a) use (10.17) to determine whether the conic section is an ellipse, hyperbola, or parabola, and (b) perform a rotation, and if necessary a translation, and sketch the graph.

1. $x^2 - xy + y^2 = 2$ **2.** $x^2 + 4xy + y^2 = 3$ **3.** $x^2 + 2xy + y^2 + (\sqrt{2}/2)\,x - (\sqrt{2}/2)\,y = 0$

4. $14x^2 + 24xy + 7y^2 = 1$
5. $14x^2 - 24xy + 7y^2 = 1$
6. $x^2 - \sqrt{3}\, xy = 1$
7. $4x^2 - 4xy + 7y^2 = 24$
8. $5x^2 - 3xy + y^2 = 5$
9. $17x^2 - 6xy + 9y^2 = 0$
10. $108x^2 - 312xy + 17y^2 + 240x + 320y + 500 = 0$
11. $225x^2 - 1080xy + 1296y^2 - 624x - 260y - 2028 = 0$
12. $21x^2 - 10\sqrt{3}xy + 11y^2 + 6x + 6\sqrt{3}y - 150 = 0$
13. $6x^2 + 4\sqrt{3}xy + 2y^2 - 9x + 9\sqrt{3}y - 63 = 0$
14. $6x^2 - 60xy - 19y^2 + 48\sqrt{13}x - 32\sqrt{13}y + 338 = 0$
15. $484x^2 + 352xy + 64y^2 - 748x - 272y + 289 = 0$
16. $x^2 + 2\sqrt{3}xy + 3y^2 + (4 + 2\sqrt{3})x + (4\sqrt{3} - 2)y + 12 = 0$
17. $16x^2 + 4xy + 19y^2 - 20\sqrt{5}x - 10\sqrt{5}y - 25\sqrt{5} = 0$
18. $97x^2 + 42xy + 153y^2 - 98\sqrt{10}x + 246\sqrt{10}y + 10 = 0$
19. $7x^2 + 48xy - 7y^2 - 10x - 70y - 25 = 0$
20. $7x^2 + 48xy - 7y^2 - 170x + 60y - 125 = 0$
21. **a)** Find an equation of the tangent line to the conic section $x^2 - xy + y^2 - 2 = 0$ at the point $(\sqrt{2}, \sqrt{2})$.
    b) Sketch the graph of the conic section and the tangent line.
22. **a)** Find an equation of the tangent line to the conic section $x^2 + xy + y^2 - 4\sqrt{2}x - 4\sqrt{2}y = 0$ at the point $(0, 0)$.
    b) Sketch the graph of the conic section and the tangent line.
23. Show that the graph of
$$\sqrt{x^2 + (y-1)^2} + 1 = \sqrt{x^2 + (y+1)^2}$$
is a conic section.
24. Find an equation of the parabola with axis $y = x$ passing through the points $(1, 0)$, $(0, 1)$, and $(1, 1)$.
25. Find an equation of the ellipse with vertices $(0, 0)$ and $(6, 8)$ passing through the point $(0, \frac{25}{4})$.
26. Use the equations in (10.16) to show that if the graph of the equation $Ax^2 + Bxy + Cy^2 + Dx + Ey + F = 0$ is rotated through any angle θ, the coefficients in the new equation $\hat{A}\hat{x}^2 + \hat{B}\hat{x}\hat{y} + \hat{C}\hat{y}^2 + \hat{D}\hat{x} + \hat{E}\hat{y} + \hat{F} = 0$ have the property that $B^2 - 4AC = \hat{B}^2 - 4\hat{A}\hat{C}$.
27. Use the equations in (10.16) to show that $A + C = \hat{A} + \hat{C}$, regardless of the value of θ.
28. **a)** Show that the angle θ used to eliminate the product-of-variables term in the rotation of
$$Ax^2 + Bxy + Cy^2 + Dx + Ey + F = 0$$
satisfies $\cot 2\theta = (A - C)/B$.
    b) Use the equation for θ in part (a) and the identities
$$\cos^2 \theta = \frac{1 + \cos 2\theta}{2} \quad \text{and} \quad \sin^2 \theta = \frac{1 - \cos 2\theta}{2}$$
to find the constants A and C in Example 3.
29. Show that if a second-degree equation represents a circle, the coefficient of the xy-term must be zero.
30. Suppose two nonparallel lines l_1 and l_2 are given.
    a) Show that the set of all points, the product of whose distances from l_1 and l_2 is constant, is a hyperbola.
    b) Show that l_1 and l_2 are asymptotes of the hyperbola.

Putnam exercise

31. Find the minimum value of
$$(u - v)^2 + \left(\sqrt{2 - u^2} - \frac{9}{v}\right)^2$$
for $0 < u < \sqrt{2}$ and $v > 0$. (This exercise was problem B–2 of the forty-fifth William Lowell Putnam examination given on December 1, 1984. The examination and its solution are in the October 1985 issue of the *American Mathematical Monthly,* pp. 560–567.)

10.5
Polar Equations of Conic Sections

Conic sections can be derived using an alternative geometric method involving a constant positive ratio called the **eccentricity** of the conic. (The eccentricity of a conic section has historically been denoted e, a symbol that has a dual purpose since it is also used to describe the number whose natural logarithm is 1.) The following theorem details the role that the eccentricity of a conic plays in determining the behavior of the conic.

(10.18)
THEOREM
Suppose F is a point in the plane, l is a line, and e is a positive constant. The set of points P whose distance from F is the product of e and the distance from P to l,

$$d(P, F) = e\,d(P, l),$$

is a conic section with a focal point at F. Moreover, except for degenerate cases, the conic section is:

  i) a parabola, if $e = 1$;
  ii) an ellipse, if $e < 1$;
  iii) a hyperbola, if $e > 1$.

The line l is called a **directrix** of the conic. (This coincides with the definition of the directrix of a parabola in the case when $e = 1$.)

Proof. Suppose we position the point F and the line l so that F is at the origin of a polar coordinate system and l lies perpendicular to the polar axis a distance d to the left of F (see Figure 10.45). If P has polar coordinates (r, θ), then $d(P, F) = r$ and $d(P, l) = d + r \cos \theta$. Thus,

$$r = e(d + r \cos \theta)$$

or, in Cartesian coordinates,

$$\sqrt{x^2 + y^2} = e(d + x).$$

Squaring, we get

$$x^2 + y^2 = e^2(d^2 + 2xd + x^2)$$

and

$$(1 - e^2)x^2 + y^2 - 2de^2 x - e^2 d^2 = 0.$$

In (10.17) of Section 10.4 we saw that this is the equation of a conic and is, except for degenerate cases,

  i) a parabola if $B^2 - 4AC = -4(1 - e^2) \cdot 1 = 0$, that is, if $e = 1$;
  ii) an ellipse if $B^2 - 4AC = -4(1 - e^2) \cdot 1 < 0$, that is, if $e < 1$;
  iii) a hyperbola if $B^2 - 4AC = -4(1 - e^2) \cdot 1 > 0$, that is, if $e > 1$. ▷

The polar equation $r = e(d + r \cos \theta)$, describing the conic with directrix a distance d to the left of the focal point F, is usually expressed in the form

(10.19)
$$r = \frac{ed}{1 - e \cos \theta}.$$

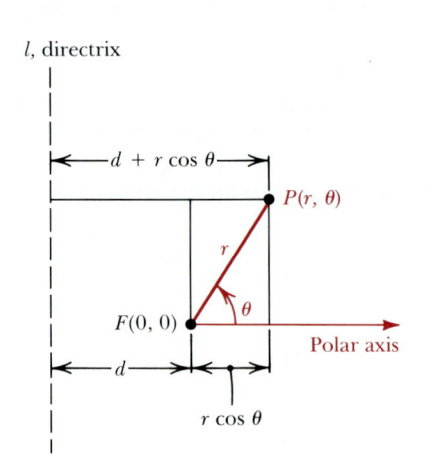

Figure 10.45

Modifications of the polar equation of conics occur when the directrix is moved to one of the other standard positions. When the directrix l of a conic is a distance d to the right of the focal point F, the polar equation is

(10.20) $$r = \frac{ed}{1 + e \cos \theta}.$$

When the focal point is at the origin and the directrix is parallel to and a distance d above the polar axis, the equation is

(10.21) $$r = \frac{ed}{1 + e \sin \theta}.$$

When the directrix is parallel to and a distance d below the polar axis, the equation is

(10.22) $$r = \frac{ed}{1 - e \sin \theta}.$$

Parabola ($e = 1$)

If $e = 1$, the equation of the parabola with focal point at the pole and directrix perpendicular to the polar axis and d units to the left of the pole is

(10.23) $$r = \frac{d}{1 - \cos \theta},$$

which is undefined when $\theta = 0$. When $\theta = \pi$, $r = d/2$, and when $\theta = \pi/2$ or $\theta = 3\pi/2$, $r = d$. This agrees with our previous knowledge of a parabola (see Figure 10.46).

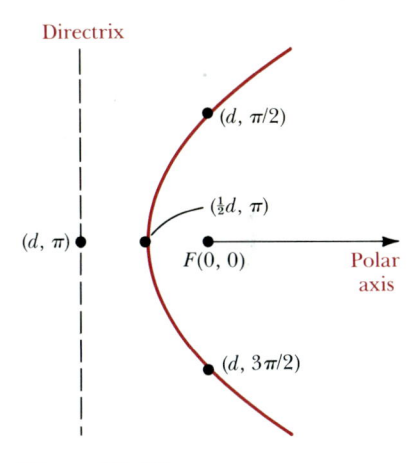

Figure 10.46
A parabola.

Example 1

Sketch the graph of the parabola whose polar equation is

$$r = \frac{3}{1 - \cos \theta}.$$

Solution

Comparing this with Equation (10.23), we see that the focal point of this parabola is at the origin and the directrix is the line perpendicular to the polar axis with Cartesian equation $x = -3$. When $\theta = \pi/2$ or $\theta = 3\pi/2$, $r = 3$. The graph is shown in Figure 10.47. ▶

Ellipse ($e < 1$)

If $e < 1$, the polar equation $r = ed/(1 - e \cos \theta)$ has a denominator that is always positive and thus is defined for all values of θ. The vertices of the ellipse occur at the absolute extrema of r. The absolute maximum

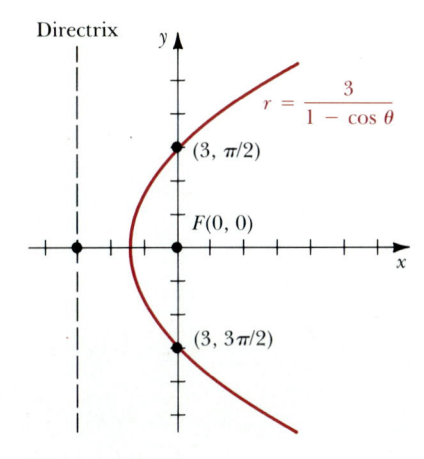

Figure 10.47

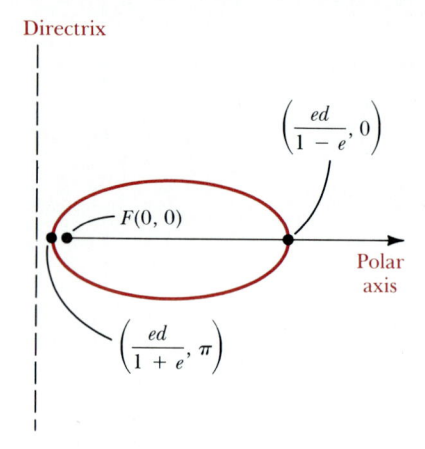

Figure 10.48
An ellipse.

for r occurs when the cosine function has its maximum value, which occurs when $\theta = 0$ and $r = ed/(1 - e)$. Similarly, the absolute minimum occurs when $\theta = \pi$ and $r = ed/(1 + e)$. Thus the vertices of the ellipse have polar coordinates

$$\left(\frac{ed}{1-e}, 0\right) \quad \text{and} \quad \left(\frac{ed}{1+e}, \pi\right),$$

as shown in Figure 10.48.

The following example shows that the vertices of an ellipse in one of the other standard positions can be found in a similar manner.

Example 2

Sketch the graph of the conic whose polar equation is

$$r = \frac{16}{5 + 3\cos\theta},$$

and find a Cartesian equation of this conic.

Solution

The standard form for a conic section with focal point at the origin and directrix perpendicular to the polar axis d units to the right of the focal point is

$$r = \frac{ed}{1 + e\cos\theta}.$$

Our equation is

$$r = \frac{16}{5 + 3\cos\theta} = \frac{\frac{16}{5}}{1 + \frac{3}{5}\cos\theta},$$

so $e = \frac{3}{5}$ and, since $ed = \frac{16}{5}$, $d = \frac{16}{3}$. The conic is an ellipse with vertices occurring at $(2, 0)$ and $(8, \pi)$. The graph is shown in Figure 10.49.

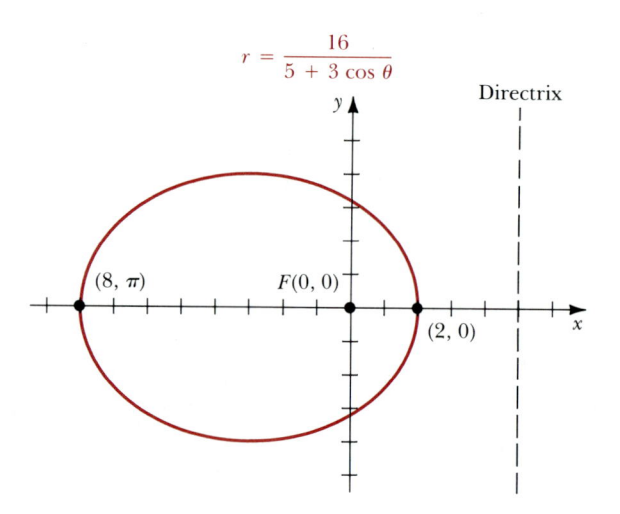

Figure 10.49

To find a Cartesian equation of this ellipse, we first write the polar equation as

$$5r + 3r \cos \theta = 16, \quad \text{so} \quad 5\sqrt{x^2 + y^2} + 3x = 16.$$

Rearranging the terms and squaring both sides, we obtain

$$25(x^2 + y^2) = (16 - 3x)^2 = 256 - 96x + 9x^2,$$

or

$$16x^2 + 96x + 25y^2 = 256.$$

Completing the square gives

$$16(x^2 + 6x + 9) + 25y^2 = 256 + 144,$$

so

$$16(x + 3)^2 + 25y^2 = 400 \quad \text{and} \quad \frac{(x + 3)^2}{25} + \frac{y^2}{16} = 1. \quad \blacktriangleright$$

Example 3

The initial flight of the NASA space shuttle occurred when the *Columbia* was placed in earth orbit on April 12, 1981. The orbit was elliptical with a focus at the center of the earth. The *apogee* (the maximum height above the earth) was 250 km, and the *perigee* (the minimum height above the earth) was 238 km. Determine the eccentricity of the orbit of the space shuttle.

Solution

We will assume that the earth is spherical with its center placed at the pole of the coordinate system shown in Figure 10.50. We also assume that the radius of the earth is 3960 mi, or approximately 6373 km.

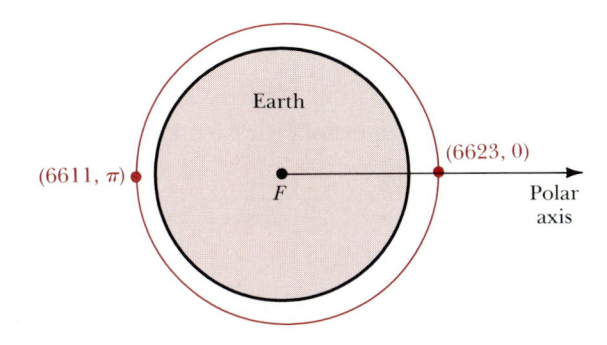

Figure 10.50
The orbit of the space shuttle.

The form of the equation is

$$r = \frac{ed}{1 - e \cos \theta}, \quad e < 1.$$

The satellite is at its apogee when $\theta = 0$, so

$$\frac{ed}{1 - e} = (6373 + 250) = 6623 \text{ km}.$$

The satellite is at its perigee when $\theta = \pi$, so

$$\frac{ed}{1 + e} = (6373 + 238) = 6611 \text{ km}.$$

Thus,

$$6623(1 - e) = ed = 6611(1 + e)$$

and

$$e = \frac{12}{13234} \approx 0.9 \times 10^{-3}.$$ ▶

The small value of e obtained in Example 3 is due to the fact that the orbit is nearly circular. Exercises 39 and 40 consider situations in which the orbit is very noncircular. In this case, the eccentricity is close to 1.

Hyperbola ($e > 1$)

If $e > 1$, the polar equation $r = ed/(1 - e \cos \theta)$ has a denominator that is undefined when $\cos \theta = 1/e$, that is, when $\theta = \pm \arccos 1/e$. The left side of the hyperbola is produced when

$$-\arccos \frac{1}{e} < \theta < \arccos \frac{1}{e}$$

and the right side when

$$\arccos \frac{1}{e} < \theta < 2\pi - \arccos \frac{1}{e}.$$

The vertices occur at $\theta = 0$, $r = ed/(1 - e)$, which is a negative number, and at $\theta = \pi$, $r = ed/(1 + e)$, as shown in Figure 10.51.

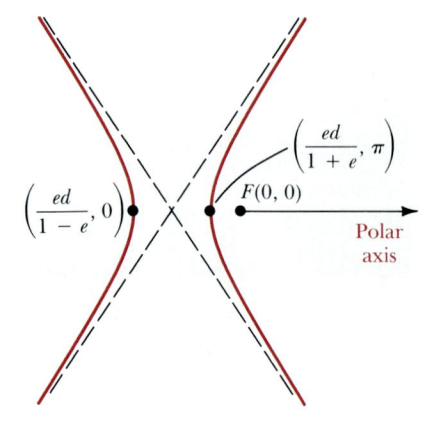

Figure 10.51
A hyperbola.

Example 4

Sketch the graph of the hyperbola whose polar equation is

$$r = \frac{10}{2 + 5 \sin \theta}.$$

Solution
Equation (10.21) implies that the directrix of this hyperbola lies parallel to and above the polar axis, so both vertices lie along the positive y-axis. Since

$$r = \frac{10}{2 + 5 \sin \theta} = \frac{5}{1 + \frac{5}{2} \sin \theta} \quad \text{we have } e = \frac{5}{2}.$$

The directrix lies above the x-axis a distance

$$d = \frac{5}{e} = \frac{5}{\frac{5}{2}} = 2.$$

10.5 POLAR EQUATIONS OF CONIC SECTIONS

$$r = \frac{10}{2 + 5 \sin \theta}$$

Figure 10.52

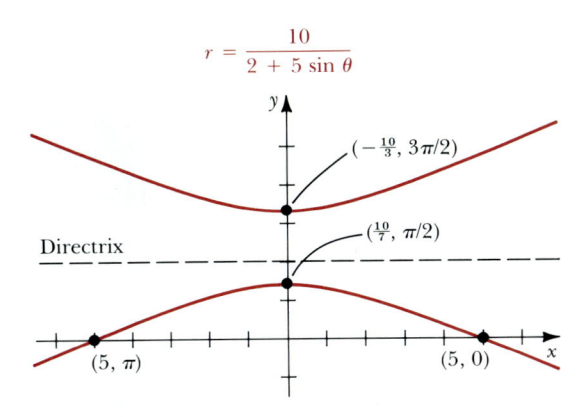

The vertices of this hyperbola are at $(10/7, \pi/2)$ and at $(-10/3, 3\pi/2)$. The portion of the hyperbola lying below the directrix can be sketched by plotting $(10/7, \pi/2)$ and the points of intersection with the x-axis: $(5, \pi)$ and $(5, 0)$. The portion of the hyperbola lying above the directrix is the reflection of the lower part about the line equidistant from the lines $y = 10/7$ and $y = 10/3$. The graph is shown in Figure 10.52. ▶

Example 5

Find a polar equation of the hyperbola with Cartesian equation $x^2 - y^2 = 1$ (see Figure 10.53).

Solution

The polar equation is

$$(r \cos \theta)^2 - (r \sin \theta)^2 = 1,$$

or

$$r^2 = \frac{1}{(\cos \theta)^2 - (\sin \theta)^2} = \frac{1}{\cos 2\theta}.$$

Note that this is not the "standard" polar equation of a conic section, because the hyperbola does not have a focal point at the origin. ▶

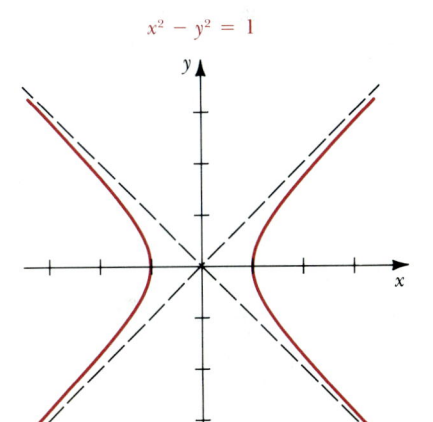

Figure 10.53

▶ EXERCISE SET 10.5

In Exercises 1–16, (a) sketch the graph of each conic section and (b) determine a corresponding Cartesian equation.

1. $r = \dfrac{2}{1 + \cos \theta}$

2. $r = \dfrac{3}{3 - 2 \cos \theta}$

3. $r = \dfrac{3}{3 + 2 \sin \theta}$

4. $r = \dfrac{4}{1 + 2 \cos \theta}$

5. $r = \dfrac{2}{3 - 3 \sin \theta}$

6. $r = \dfrac{5}{2 - 5 \sin \theta}$

7. $r = \dfrac{5}{2 - 3 \cos \theta}$

8. $r = \dfrac{2}{4 + \cos \theta}$

9. $r = \dfrac{3}{1 + 2\sin\theta}$
10. $r = \dfrac{9}{2 - 6\cos\theta}$
11. $r = \dfrac{1}{4 + 2\sin\theta}$
12. $r = \dfrac{14}{7 - 2\sin\theta}$
13. $r = \dfrac{2}{1 + \cos(\theta + \pi/4)}$
14. $r = \dfrac{2}{1 + \sqrt{2}\sin(\theta + \pi/4)}$
15. $r = \dfrac{2}{4 + \cos(\theta + \pi/4)}$
16. $r = \dfrac{2}{1 + \sin\theta + \cos\theta}$

In Exercises 17–22, find a polar equation of the conic with a focus at the origin satisfying the given conditions.

17. $e = 2$, directrix $x = 4$
18. $e = \tfrac{1}{2}$, directrix $y = -2$
19. $e = 1$, directrix $y = -\tfrac{1}{4}$
20. $e = \tfrac{1}{3}$, directrix $r = \sec\theta$
21. $e = 3$, directrix $r = 2\csc\theta$
22. $e = 1$, directrix $r = -\tfrac{1}{2}\csc\theta$

23. Find a polar equation of the ellipse in standard position that has vertices at points with polar coordinates $(1, 0)$ and $(3, \pi)$.
24. Find a polar equation of the parabola with a focus at the pole and a vertex at the point with Cartesian coordinates $(-6, 0)$.

In Exercises 25–30, (a) express the Cartesian equation as a polar equation and (b) state whether the conic is in standard polar position.

25. $y = 3x^2$
26. $2x^2 + y^2 = 1$
27. $4x^2 + 3y^2 - 2y = 1$
28. $y^2 = x + 1$
29. $x^2 - y^2 - 2y = 4$
30. $y^2 - 3x^2 + 6x = 9$

31. Show that the conic section with polar equation $r = ed/(1 + e\cos\theta)$ can also be expressed as $r = ed/(e\cos\theta - 1)$. [Hint: Consider the relationship between (r, θ) and $(-r, \theta + \pi)$.]

32. Show that the conic section with polar equation $r = ed/(1 + e\sin\theta)$ can also be expressed as $r = ed/(e\sin\theta - 1)$.

Use the relationships in Exercises 31 and 32 to sketch the graphs of the polar equations in Exercises 33–36.

33. $r = \dfrac{2}{\cos\theta - 1}$
34. $r = \dfrac{3}{\sin\theta - 2}$
35. $r = \dfrac{1}{2\sin\theta - 3}$
36. $r = \dfrac{4}{2\cos\theta - 1}$

37. The world's first orbiting satellite, *Sputnik I*, was launched in the Soviet Union on October 4, 1957. Its elliptical orbit reached a maximum height of 560 mi above the earth and a minimum height of 145 mi. Write a polar equation for this orbit, assuming that a focal point is at the pole, placed at the center of the earth, and that the earth is spherical with a radius of 3960 mi.

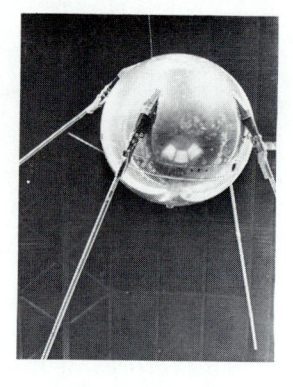

Sputnik I, the first artificial Earth orbiting satellite, was launched on October 4, 1957.

38. The earth moves in an elliptical orbit around the sun with the sun at one focal point and an eccentricity of 0.0167. The major axis of the orbit is approximately 2.99×10^8 km. Write a polar equation of this ellipse if the pole is taken as the center of the sun.

39. Halley's comet has a noncircular orbit with one focal point at the center of mass of the sun. The closest the comet comes to the sun is 8.93×10^7 km (called the *perihelion* of its orbit). At the opposite end of its orbit, it is 5.26×10^9 km from the sun. Determine an equation for the orbit of Halley's comet. (Halley's comet was at its perihelion on February 9, 1986. It will not return to this point for approximately 75.5 years.)

40. The comet Kahoutek has an elliptical orbit about the sun with eccentricity $e = 0.99993$ and a perihelion of 1.95×10^7 mi. Find an equation for the orbit, and determine the maximum distance of Kahoutek from the sun.

The comet Kahoutek has an extremely noncircular elliptical orbit.

▶ IMPORTANT TERMS AND RESULTS

CONCEPT	PAGE	CONCEPT	PAGE
Conic sections	624	Rotation of axes	650
Parabola	625	Classification of conic sections	652
Ellipse	634	Eccentricity	653
Hyerbola	640	Polar equations of conic sections	654

▶ REVIEW EXERCISES

In Exercises 1–34, identify and sketch the graph of the conic section described.

1. $y^2 = 4x$
2. $y = -x^2$
3. $x^2 = -2(y - 5)$
4. $y^2 = -16(x - 5)$
5. $(y - 2)^2 = 2(x + 3)$
6. $(x - 1)^2 = 4(y + 3)$
7. $y^2 - 2x + 2y + 7 = 0$
8. $16x^2 + 25y^2 = 400$
9. $16(x - 2)^2 + 25(y - 3)^2 = 400$
10. $16x^2 - 25y^2 = 400$
11. $16(x - 2)^2 - 25(y - 3)^2 = 400$
12. $9(x + 2)^2 - 4(y - 5)^2 = 36$
13. $9x^2 - 16y^2 = 144$
14. $x^2 - 2x - 6y - 7 = 0$
15. $9x^2 + 4y^2 - 90x - 16y + 205 = 0$
16. $4x^2 - 9y^2 - 16x - 90y + 16 = 0$
17. $4x^2 - 9y^2 - 16x - 90y - 210 = 0$
18. $9x^2 + 4y^2 - 90x - 16y - 83 = 0$
19. $4x^2 + 9y^2 - 16x - 90y + 205 = 0$
20. $y^2 - 6y + 9 - 4x = 0$
21. $2xy - \sqrt{2}\,x + \sqrt{2}\,y - 5 = 0$
22. $x^2 + 4xy + y^2 - 16 = 0$
23. $x^2 + 3xy + y^2 - 1 = 0$
24. $23x^2 + 26\sqrt{3}xy - 3y^2 - 144 = 0$
25. $31x^2 + 10\sqrt{3}xy + 21y^2 - 144 = 0$
26. $25x^2 + 14xy + 25y^2 - 288 = 0$
27. $x^2 - 2xy + y^2 - 4x + 2y + 3 = 0$
28. $17x^2 - 12xy + 8y^2 - 60x - 8 = 0$
29. $r = \dfrac{1}{1 - \sin \theta}$
30. $r = \dfrac{8}{1 - 4 \cos \theta}$
31. $r = \dfrac{2}{3 + \sin \theta}$
32. $r = \dfrac{1}{1 - \cos \theta}$
33. $r = \dfrac{7}{3 + 5 \cos \theta}$
34. $r = \dfrac{7}{5 - 3 \cos \theta}$

In Exercises 35–44, find an equation of the conic.

35. A parabola with focus at $(0, 0)$ and directrix $y = 2$
36. A parabola with focus at $(0, 0)$ and directrix $x = 2$
37. An ellipse with foci at $(0, \pm 1)$ and vertices at $(0, \pm 3)$
38. An ellipse with foci at $(\pm 1, 0)$ and vertices at $(\pm 3, 0)$
39. A hyperbola with foci at $(\pm 3, 0)$ and vertex at $(1, 0)$
40. A parabola with focus at $(0, 1)$ and vertex at $(0, 0)$
41. An ellipse with foci at $(0, \pm 5)$ and passing through the point $(4, 0)$
42. A conic with a focus at the origin, eccentricity $\tfrac{3}{4}$, and directrix $x = 2$
43. A conic with a focus at the origin, eccentricity 3, and directrix $y = -2$
44. A conic with a focus at the origin, eccentricity 1, and directrix $x = -3$
45. Find an equation of the tangent line to the graph of $16x^2 - 4y^2 = 144$ at $(5, 8)$.
46. Find an equation of the tangent line to the graph of $x^2 + 4y^2 = 40$ at $(2, 3)$.
47. Find an equation in Cartesian coordinates of a tangent line to the graph of $r = 1/(1 - \cos \theta)$ at $(1, \pi/2)$.
48. Find an equation in Cartesian coordinates of a tangent line to the graph of $r = 4/(2 - \cos \theta)$ at $(\tfrac{8}{3}, \pi/3)$.

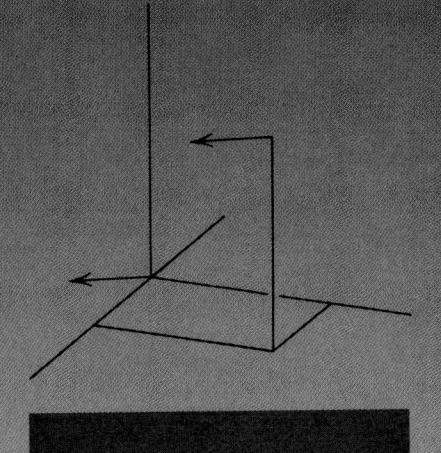

11

Vectors

664 CHAPTER 11 VECTORS

The first ten chapters concern functions whose domain and range are both sets of real numbers. Most of the important calculus concepts were discussed in these chapters, but the restriction to functions of this type does not permit us to examine more than two-dimensional problems. Since we live in a three-dimensional world, more general functions are needed before we can apply the full power of the calculus to solve many real-world problems.

This chapter lays the groundwork for studying multidimensional problems. The initial step is to consider how points and directions in three-dimensional space are described. The first section of this chapter discusses the representation of points in space using a rectangular coordinate system. Direction in three-dimensional space is given by vectors, which are introduced in Section 11.2.

The later sections of the chapter show how vectors are used to represent planes and lines in space. We will rely on this representation in the remainder of the text.

Figure 11.1
"Waterfall," a 1961 lithograph by Maurits. C. Escher.

11.1
The Rectangular Coordinate System in Space

A three-dimensional coordinate system is needed to describe objects analytically in space. The natural extension of the Cartesian, or rectangular, xy-coordinate system in the plane is made by simply adding a third coordinate axis, labeled the z-axis, perpendicular to the xy-plane.

Theoretically and analytically, this is a satisfactory procedure. Graphically, however, problems can arise. The work of M. C. Escher titled "Waterfall," shown in Figure 11.1, illustrates in two dimensions a waterfall that is impossible to have in three dimensions. The water in the channel appears to be at a constant level, yet it must also flow uphill. This is an example of the fallacies that can arise from adding a depth perception to a two-dimensional plane. However, the plane is the only vehicle generally available for graphic representation, so we use it as best we can, while being alert for possible contradictory images.

There are many ways to add a perception of depth to the two-dimensional representation of a three-dimensional coordinate system. In your hand-drawn sketches we suggest using a coordinate system constructed in the following manner: Assume the plane determined by the y- and z-axes. the yz-plane, coincides with the plane in which you draw, and that the positive x-axis is directed toward you. Draw the x-axis as a straight line that has the *appearance* of making an angle of 135° with both the positive y- and z-axes, and draw only the positive portion of each axis unless the negative portion is needed for clarity. As an additional aid to the perception of depth, use a scale with a physical length of $\sqrt{2}/2$

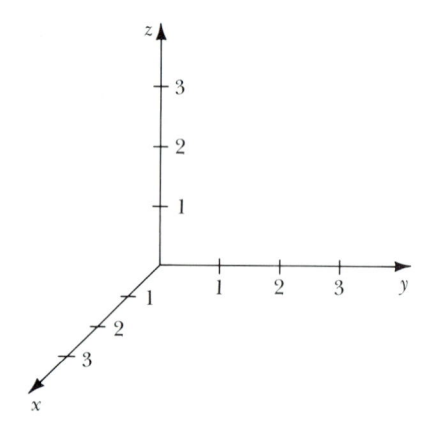

Figure 11.2

(approximately 3/4) units along the x-axis for each unit along the y- and z-axes, as shown in Figure 11.2. This is a convenient scaling factor, since the diagonal of a unit square has length $\sqrt{2}$, as shown in Figure 11.3. This relatively simple technique should enable you to sketch any of the figures you are likely to encounter in calculus.

The figures in the text have been computer generated, using sophisticated scaling and rotation techniques. They are generally oriented so that they have a similar appearance to those you will be constructing by hand, but have the additional advantage that you can "see" down both the positive y-axis. More will be said about the techniques involved with computer-generated graphs when the graphing of three-dimensional surfaces is considered in Chapter 13.

A point in space is represented in this rectangular coordinate system by an **ordered triple** called the coordinates of the point. A point with coordinates (a, b, c) is drawn by sketching the rectangular parallelepiped shown in Figure 11.4. On this parallelepiped the point (a, b, c) is the vertex diagonally opposite the vertex at the origin $(0, 0, 0)$.

The coordinates of a point in the xy-plane have the form $(a, b, 0)$; similarly, in the xz-plane, $(a, 0, c)$; and in the yz-plane, $(0, b, c)$. The xy-, yz-, and xz-planes (also known as the **coordinate planes**) divide space into eight sections called *octants*. The portion of space determined by those points whose coordinates are all positive is called the **first octant.** The other seven octants are not given names.

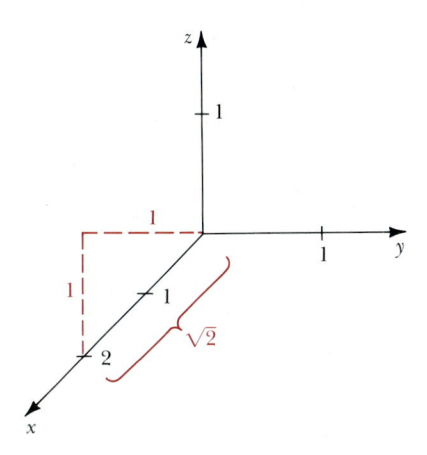

Figure 11.3

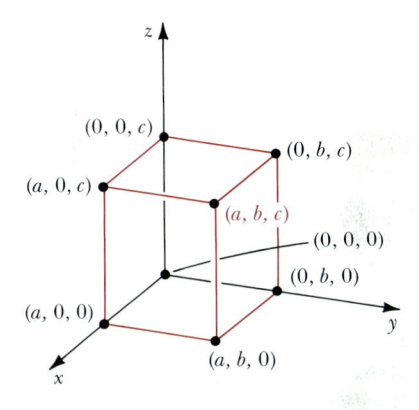

Figure 11.4
A rectangular parallelepiped.

Example 1

Sketch the points $(2, 3, 2)$ and $(4, 4, 3)$.

Solution

These points lie in the first octant and are shown in Figures 11.5(a) and (b). ▶

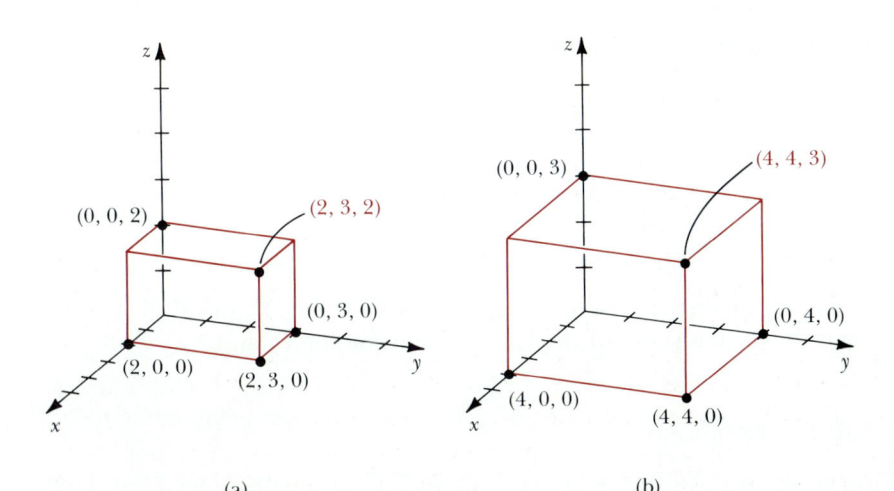

Figure 11.5

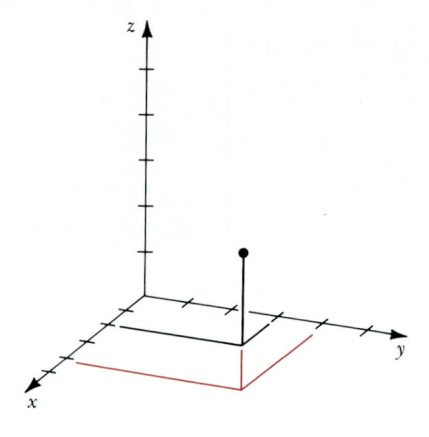

Figure 11.6

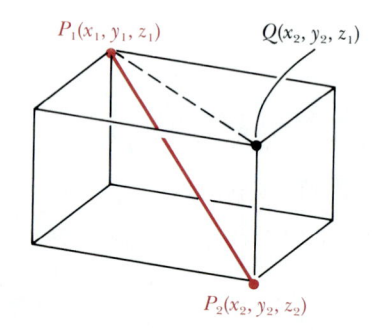

Figure 11.7
The distance from P_1 to P_2 is
$\sqrt{(x_2 - x_1)^2 + (y_2 - y_1)^2 + (z_2 - z_1)^2}$.

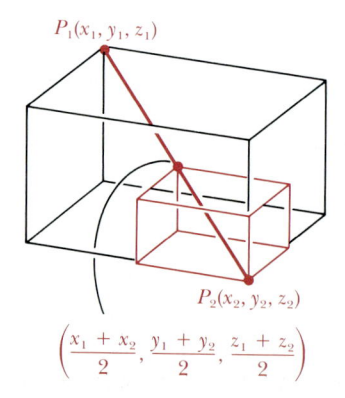

Figure 11.8
The midpoint of the line segment joining P_1 and P_2.

Although it is easy to observe the relative position of the points in Example 1 when the parallelepipeds are drawn, note that when the parallelepipeds are removed from the sketch, the points are indistinguishable, as shown in Figure 11.6. In fact, the point in this sketch could have coordinates $(0, 2, 1)$, $(-2, 1, 0)$, or even $(-100, -48, -49)$. Although drawing a parallelepiped involves extra work, this example demonstrates that it is the only way to be confident of the graphic representation.

A parallelepiped is also used to find a formula for the distance between two points in space. Suppose $P_1(x_1, y_1, z_1)$ and $P_2(x_2, y_2, z_2)$ are two points in space. Consider the parallelepiped shown in Figure 11.7 with diagonally opposite vertices at P_1 and P_2 and sides parallel to the coordinate planes. Let Q be the point whose first two coordinates agree with those of P_2 and whose third coordinate is z_1 (see Figure 11.7). The line from Q to P_1 is perpendicular to the line from Q to P_2, so the Pythagorean Theorem implies that

$$[d(P_1, P_2)]^2 = [d(P_1, Q)]^2 + [d(Q, P_2)]^2,$$

where, as usual, d denotes the distance between the specified points. However,

$$d(Q, P_2) = |z_2 - z_1|$$

and

$$d(P_1, Q) = \sqrt{(x_2 - x_1)^2 + (y_2 - y_1)^2},$$

so

$$[d(P_1, P_2)]^2 = (x_2 - x_1)^2 + (y_2 - y_1)^2 + (z_2 - z_1)^2.$$

Thus the distance between $P_1(x_1, y_1, z_1)$ and $P_2(x_2, y_2, z_2)$ is

(11.1) $\quad d(P_1, P_2) = \sqrt{(x_2 - x_1)^2 + (y_2 - y_1)^2 + (z_2 - z_1)^2}.$

This is the natural extension of the two-dimensional distance formula.

In a similar manner, the midpoint of the line segment joining P_1 and P_2 has the same form as in the plane (see Figure 11.8). The midpoint of the line segment joining $P_1(x_1, y_1, z_1)$ and $P_2(x_2, y_2, z_2)$ is

(11.2) $\quad \left(\dfrac{x_1 + x_2}{2}, \dfrac{y_1 + y_2}{2}, \dfrac{z_1 + z_2}{2} \right).$

Example 2

Find (a) the distance between the points $P_1(1, -1, 2)$ and $P_2(3, 4, -1)$ and (b) the midpoint of the line segment joining P_1 and P_2.

Solution

a) This distance is

$$d(P_1, P_2) = \sqrt{(3 - 1)^2 + (4 - (-1))^2 + (-1 - 2)^2} = \sqrt{4 + 25 + 9} = \sqrt{38}.$$

b) The midpoint of the line segment joining P_1 and P_2 is

$$\left(\frac{1+3}{2}, \frac{-1+4}{2}, \frac{2-1}{2}\right) = \left(2, \frac{3}{2}, \frac{1}{2}\right).$$

Example 3

Find an equation of the sphere with center (h, k, l) and radius r shown in Figure 11.9.

Solution

A point is on the sphere if and only if the distance from the point to the center (h, k, l) is the constant r. So (x, y, z) lies on the sphere precisely when

$$r = d((x, y, z), (h, k, l)) = \sqrt{(x-h)^2 + (y-k)^2 + (z-l)^2}.$$

Squaring both sides gives

$$r^2 = (x-h)^2 + (y-k)^2 + (z-l)^2,$$

the **standard equation of a sphere of radius r with center (h, k, l).**

Figure 11.9
The sphere with center (h, k, l) and radius r has equation
$(x-h)^2 + (y-k)^2 + (z-l)^2 = r^2.$

Example 4

Describe geometrically the set of points (x, y, z) that satisfy the equation $x^2 + y^2 = 1$.

Solution

For every value of z, the equation $x^2 + y^2 = 1$ describes a circle with radius 1. There is no restriction on z, so z can assume any real-number value. The equation describes a right circular cylinder whose center is the z-axis, as shown in Figure 11.10. The cylinder extends infinitely above and below the xy-plane.

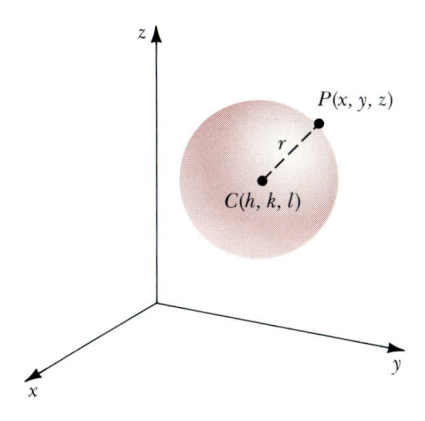

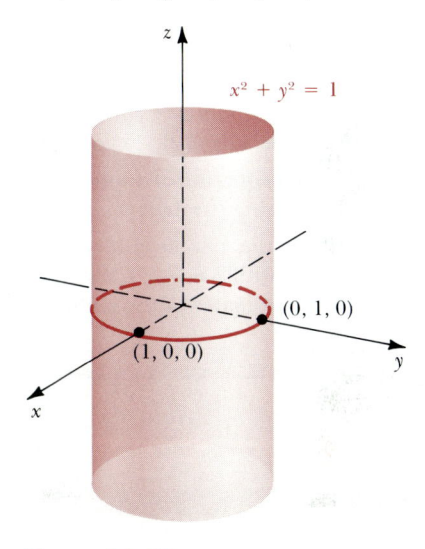

Figure 11.10

▶ EXERCISE SET 11.1

1. Plot each of the following points and sketch the associated parallelepiped.
   a) $(1, 3, 4)$ **b)** $(1, -3, 4)$ **c)** $(2, 4, -3)$
2. Plot each of the following points and sketch the associated parallelepiped.
   a) $(2, 3, 3)$ **b)** $(-2, 3, 3)$ **c)** $(-2, -3, 3)$

In Exercises 3–8, plot the points A and B and find the distance between them.

3. $A(1, 2, 3), B(-1, 3, 4)$
4. $A(2, 5, 0), B(3, -3, 1)$
5. $A(-3, 4, 0), B(-3, 4, 2)$
6. $A(0, 0, 0), B(4, 2, -4)$
7. $A(1, 8, 3), B(0, 1, 0)$
8. $A(3, 5, 0), B(5, 3, 0)$

9–14. Find the midpoint of the line segment joining points A and B given in Exercises 3–8.

In Exercises 15–18, the points A and B are the endpoints of

a diagonal of a parallelepiped having its faces parallel to the coordinate planes.
  a) Sketch the parallelepiped.
  b) Find the coordinates of the other vertices.

15. $A(0, 0, 0)$, $B(2, 4, 4)$ **16.** $A(2, 2, 0)$, $B(3, 5, 4)$
17. $A(3, 0, 0)$, $B(4, 3, 5)$ **18.** $A(0, 2, 0)$, $B(2, -2, 4)$

In Exercises 19–22, find an equation of the sphere with center C and radius r.

19. $C(0, 0, 0)$, $r = 2$ **20.** $C(2, 0, 0)$, $r = 2$
21. $C(2, 3, 4)$, $r = 1$ **22.** $C(0, 3, 0)$, $r = 3$

In Exercises 23–26, find the center and radius of the sphere whose equation is given.

23. $x^2 + y^2 + z^2 + 2x - 4y + 6z = 0$
24. $3x^2 + 3y^2 + 3z^2 + 6x - 3y + 6z = 1$
25. $2x^2 + 2y^2 + 2z^2 + 8x - 8z = -7$
26. $x^2 + y^2 + z^2 - 10x - 6y - 2z + 31 = 0$
27. Find an equation of the sphere having endpoints of a diameter at $(-2, 1, 1)$ and $(1, 4, 5)$.
28. Find an equation of a sphere having center at $(1, 2, 3)$ and passing through the origin.
29. Sketch the parallelepiped consisting of the points (x, y, z) with $2 \le x \le 4$, $2 \le y \le 3$, and $0 \le z \le 5$. Give the coordinates of the eight corners of the parallelepiped.

30.
  a) Show that the points $(4, 4, 1)$, $(1, 1, 1)$, and $(0, 8, 5)$ are the vertices of a right triangle.
  b) Find an equation of the sphere passing through the three points and having a diameter along the hypotenuse of the triangle.

31. Sketch the points $(2, 4, 2)$, $(2, 1, 5)$, and $(5, 1, 2)$ and show that they are the vertices of an equilateral triangle.

32. The floor of a room is 12 ft by 8 ft and the height of the ceiling is 7 ft. Make a representative sketch of the room in a three-dimensional coordinate system and label the points corresponding to the corners of the room.

33. A dome tent has a circular floor with diameter 8 ft. The height of the tent in the center is 6 ft. Make a representative sketch of the tent in a three-dimensional coordinate system.

Putnam exercise

34. Let A be a solid $a \times b \times c$ rectangular brick in three dimensions, where $a, b, c > 0$. Let B be the set of all points which are a distance at most one from some point of A (in particular, B contains A). Express the volume of B as a polynomial in a, b, and c. (This exercise was problem A–1 of the forty-fifth William Lowell Putnam examination given on December 1, 1984. The examination and its solution are in the October 1985 issue of the *American Mathematical Monthly*, pp. 560–567.)

11.2

Vectors in Space

Certain physical properties are described by stating a magnitude. These are called **scalar** properties and include, for example, the distance between points, the temperature of a surface, and the volume and mass of a solid. Other properties can be described only by specifying both a magnitude and a direction. Examples of this type are particularly abundant in physics, and include the concepts of force, velocity, acceleration, and momentum. These are called **vector** properties; to consider them we need the notion of a vector.

(11.3)
DEFINITION

A **vector** $\mathbf{a}$ in three-dimensional space is an ordered triple of real numbers, written $\mathbf{a} = \langle a_1, a_2, a_3 \rangle$. The numbers a_1, a_2, and a_3 are called the **components** of $\mathbf{a}$ in the x-, y-, and z-directions, respectively.

11.2 VECTORS IN SPACE

The magnitude, or **length,** of the vector **a**, written $\|\mathbf{a}\|$, is defined by

(11.4) $$\|\mathbf{a}\| = \sqrt{a_1^2 + a_2^2 + a_3^2}.$$

Two vectors $\mathbf{a} = \langle a_1, a_2, a_3 \rangle$ and $\mathbf{b} = \langle b_1, b_2, b_3 \rangle$ are **equal** if and only if their components are equal:

$$a_1 = b_1, \quad a_2 = b_2, \quad \text{and} \quad a_3 = b_3.$$

Any vector of length one is called a **unit vector,** and the only vector with length zero, $\mathbf{0} = \langle 0, 0, 0 \rangle$, is called the **zero vector.**

A nonzero vector $\mathbf{a} = \langle a_1, a_2, a_3 \rangle$ is represented geometrically by a directed line segment of length $\|\mathbf{a}\|$; any line segment that begins at an arbitrary point $P(x, y, z)$ and ends at the point $Q(x + a_1, y + a_2, z + a_3)$ represents **a**. This point P is called an **initial point** for the vector **a**, and Q is the corresponding **terminal point.** This is expressed by writing

$$\overrightarrow{PQ} = \mathbf{a}.$$

Figure 11.11 gives an example of this representation; the arrowhead on the line segment indicates the direction of the vector **a**.

Boldface type is used in text to distinguish vector quantities from scalar quantities. It is also common practice to write the vector **a** as $\vec{a}$ when boldface is not available.

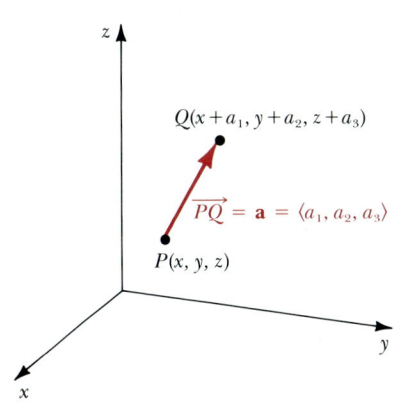

Figure 11.11

Example 1

a) Find the vector with initial point $(1, 2, -1)$ and terminal point $(2, 3, 4)$.
b) Determine the length of this vector.

Solution
a) The vector is $\langle 2 - 1, 3 - 2, 4 - (-1) \rangle = \langle 1, 1, 5 \rangle$.
b) The length of this vector is

$$\|\langle 1, 1, 5 \rangle\| = \sqrt{1^2 + 1^2 + 5^2} = \sqrt{27} = 3\sqrt{3},$$

which is the distance between $(1, 2, -1)$ and $(2, 3, 4)$ (see Figure 11.12). ▶

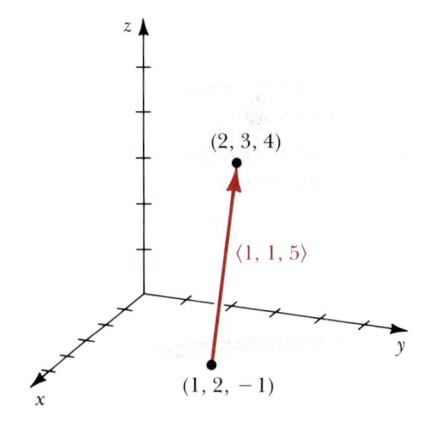

Figure 11.12

An infinite number of directed line segments represents each nonzero vector since the choice of initial point for a directed line segment is arbitrary. All directed line segments that represent the same vector are related to one another by parallel translations, as shown in Figure 11.13.

To establish results concerning vectors, it is generally more convenient to use the analytic representation given in Definition 11.3. To see the application of vectors, however, it is usually preferable to consider vectors as directed line segments that can be moved using parallel translations.

Example 2

Find the terminal point of the vector $\langle 3, 0, 1 \rangle$ if the initial point is (a) $(-1, 1, 0)$; (b) $(2, e, \pi)$; (c) $(0, 0, 0)$.

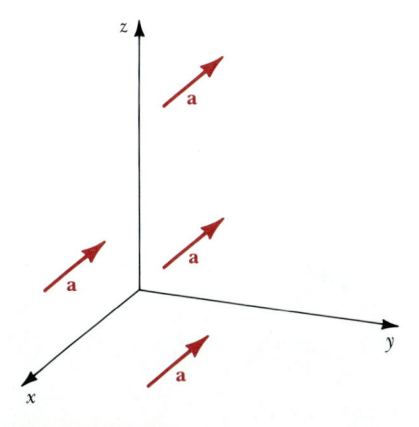

Figure 11.13

Solution
The directed line segments representing $\langle 3, 0, 1 \rangle$ with these initial points are shown in Figure 11.14. (a) When the initial point is $(-1, 1, 0)$, the terminal point is $(-1 + 3, 1 + 0, 0 + 1) = (2, 1, 1)$. The terminal point in (b) is $(5, e, \pi + 1)$ and in (c) is $(3, 0, 1)$. ▶

Associated with each point $P(x, y, z)$ in space is a unique vector $\overrightarrow{OP}$ with initial point at the origin O and terminal point P. This vector is called the **position vector** for the point P. Every vector $\mathbf{a} = \langle a_1, a_2, a_3 \rangle$ is the position vector for precisely one point in space, that is, the point with coordinates (a_1, a_2, a_3). The directed line segment from the origin $(0, 0, 0)$ to (a_1, a_2, a_3) is called the **position-vector representation** for $\mathbf{a}$.

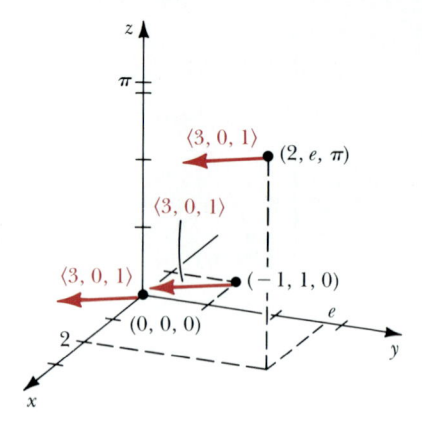

Figure 11.14

Example 3

Find the position-vector representation for the vector with initial point $(1, 3, -1)$ and terminal point $(-1, 0, 4)$.

Solution
This vector, $\langle -1 - 1, 0 - 3, 4 - (-1) \rangle = \langle -2, -3, 5 \rangle$, is the position vector for the point $(-2, -3, 5)$. The position-vector representation is shown in color in Figure 11.15. ▶

The collection of all vectors of the form $\langle a_1, a_2, a_3 \rangle$, where $a_3 = 0$ is called the set of **vectors in the xy-plane.** When considering only vectors in the xy-plane, we often write the vector $\langle a_1, a_2, 0 \rangle$ as $\langle a_1, a_2 \rangle$. Results true for general vectors in space are true as well for vectors in the xy-plane. In fact, vectors in the xy-plane are often used to illustrate vector concepts when a two-dimensional representation will suffice.

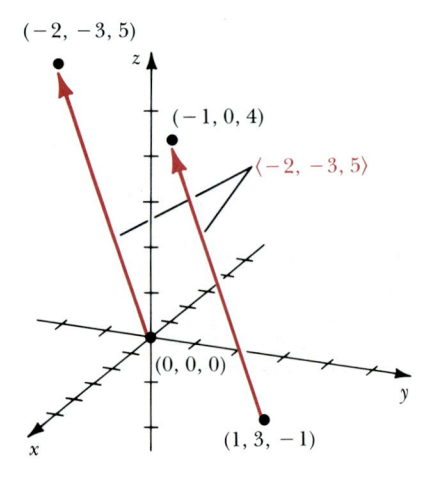

Figure 11.15

Example 4

a) Find the position-vector representation for the vector in the xy-plane with initial point $(3, 4)$ and terminal point $(-2, 1)$.
b) Determine the length of this vector.

Solution
a) If $\mathbf{a} = \langle a_1, a_2 \rangle$ denotes this vector, then
$$a_1 = -2 - 3 = -5 \quad \text{and} \quad a_2 = 1 - 4 = -3,$$
so $\mathbf{a} = \langle -5, -3 \rangle$. The position-vector representation is shown in Figure 11.16.
b) The length of $\mathbf{a} = \langle -5, -3 \rangle$ is the same as the length of the vector $\langle -5, -3, 0 \rangle$, that is,
$$\|\mathbf{a}\| = \sqrt{(-5)^2 + (-3)^2} = \sqrt{34}.$$
▶

We continue the study of vectors with the definition of arithmetic operations.

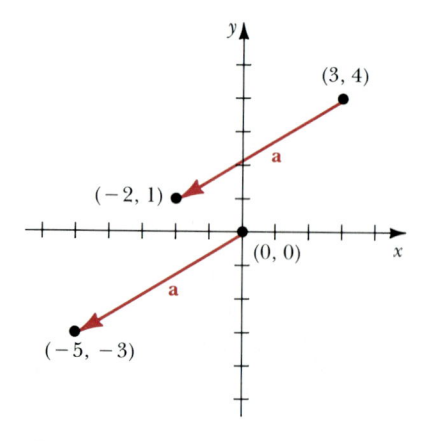

Figure 11.16

(11.5)
DEFINITION
If $\mathbf{a} = \langle a_1, a_2, a_3 \rangle$ and $\mathbf{b} = \langle b_1, b_2, b_3 \rangle$ are vectors and α is a real number, then

i) $\mathbf{a} + \mathbf{b} = \langle a_1 + b_1, a_2 + b_2, a_3 + b_3 \rangle$;
ii) $\mathbf{a} - \mathbf{b} = \langle a_1 - b_1, a_2 - b_2, a_3 - b_3 \rangle$;
iii) $\alpha \mathbf{a} = \langle \alpha a_1, \alpha a_2, \alpha a_3 \rangle$.

Definitions (i) and (ii) are called **vector addition** and **vector subtraction**, respectively, while (iii) is known as **scalar multiplication**. The vector sum $\mathbf{a} + \mathbf{b}$ is also called the *resultant vector* of $\mathbf{a}$ and $\mathbf{b}$.

Vector addition can be described geometrically as follows. Suppose $\mathbf{a}$ and $\mathbf{b}$ are two vectors drawn so that the terminal point of $\mathbf{a}$ and the initial point of $\mathbf{b}$ coincide. With the vectors in this position, $\mathbf{a} + \mathbf{b}$ is the vector from the initial point of $\mathbf{a}$ to the terminal point of $\mathbf{b}$, as illustrated in Figure 11.17(a). Since $\mathbf{b} + (\mathbf{a} - \mathbf{b}) = \mathbf{a}$, this description of addition enables us to represent vector subtraction geometrically, as shown in Figure 11.17(b).

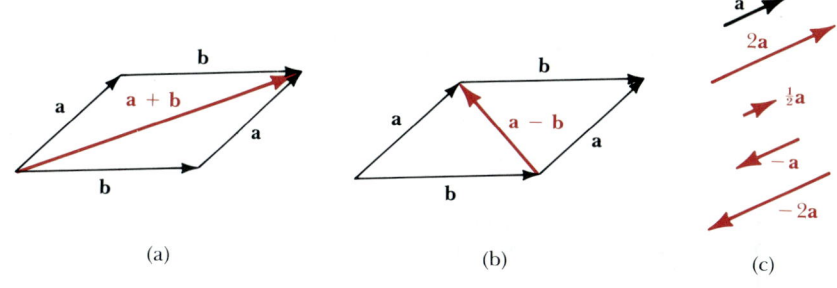

Figure 11.17
(a) Addition of vectors; (b) subtraction of vectors; (c) scalar multiplication.

Scalar multiplication by a real number α has the effect of compressing, when $|\alpha| < 1$, or expanding, when $|\alpha| > 1$, the length of the vector. In addition, when α is positive, $\alpha \mathbf{a}$ has the *same direction* as $\mathbf{a}$, but when α is negative, $\alpha \mathbf{a}$ and $\mathbf{a}$ have *opposite directions* (see Figure 11.17c).

Example 5

Find the sum and difference of the vectors $\mathbf{a} = \langle 3, 1, 0 \rangle$ and $\mathbf{b} = \langle 2, 4, 0 \rangle$ and illustrate this sum geometrically.

Solution
We have
$$\mathbf{a} + \mathbf{b} = \langle 3, 1, 0 \rangle + \langle 2, 4, 0 \rangle = \langle 5, 5, 0 \rangle$$
and
$$\mathbf{a} - \mathbf{b} = \langle 3, 1, 0 \rangle - \langle 2, 4, 0 \rangle = \langle 1, -3, 0 \rangle.$$

Since these vectors lie in the xy-plane, they can be represented as shown in Figure 11.18. ▶

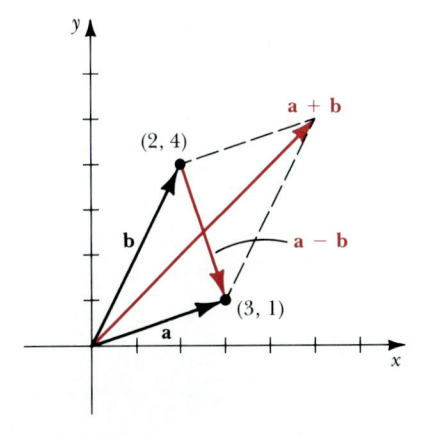

The arithmetic vector results given on the next page follow from Definition (11.3).

Figure 11.18

(11.6) THEOREM

If **a**, **b**, and **c** are vectors and α and β are real numbers, then:

i) $\mathbf{a} + \mathbf{b} = \mathbf{b} + \mathbf{a}$;
ii) $\mathbf{a} + (\mathbf{b} + \mathbf{c}) = (\mathbf{a} + \mathbf{b}) + \mathbf{c}$;
iii) $(\alpha\beta)\mathbf{a} = \alpha(\beta\mathbf{a}) = \beta(\alpha\mathbf{a})$;
iv) $(\alpha + \beta)\mathbf{a} = \alpha\mathbf{a} + \beta\mathbf{a}$;
v) $\mathbf{a} + \mathbf{0} = \mathbf{a}$;
vi) $0(\mathbf{a}) = \mathbf{0}$;
vii) $1(\mathbf{a}) = \mathbf{a}$;
viii) $\|\alpha\mathbf{a}\| = |\alpha| \cdot \|\mathbf{a}\|$.

The proofs of these results follow by applying properties of the real numbers to the components of the vectors. For example, if $\mathbf{a} = \langle a_1, a_2, a_3 \rangle$, $\mathbf{b} = \langle b_1, b_2, b_3 \rangle$, and $\mathbf{c} = \langle c_1, c_2, c_3 \rangle$, then

$$\begin{aligned}\mathbf{a} + (\mathbf{b} + \mathbf{c}) &= \langle a_1 + (b_1 + c_1), a_2 + (b_2 + c_2), a_3 + (b_3 + c_3) \rangle \\ &= \langle (a_1 + b_1) + c_1, (a_2 + b_2) + c_2, (a_3 + b_3) + c_3 \rangle \\ &= (\mathbf{a} + \mathbf{b}) + \mathbf{c}.\end{aligned}$$

Other parts of this theorem are considered in the exercises.

Scalar multiplication provides a precise manner for defining parallel vectors.

(11.7) DEFINITION

Nonzero vectors **a** and **b** are **parallel** if there is a real number α such that $\mathbf{a} = \alpha \mathbf{b}$. (We also say that the zero vector is parallel to every vector **a**.)

Any nonzero vector **a** can be multiplied by the reciprocal of its length, $1/\|\mathbf{a}\|$, to produce the unit vector, $\mathbf{a}/\|\mathbf{a}\|$, which has the same direction as **a**. The vector $-\mathbf{a}/\|\mathbf{a}\|$ is the unit vector in the direction opposite to **a**.

Example 6

Find a unit vector that has the same direction as $\mathbf{a} = \langle 2, -3, 6 \rangle$ and a unit vector in the opposite direction.

Solution
Since $\|\mathbf{a}\| = \sqrt{4 + 9 + 36} = 7$, the vector

$$\frac{\mathbf{a}}{\|\mathbf{a}\|} = \frac{1}{7} \langle 2, -3, 6 \rangle = \left\langle \frac{2}{7}, -\frac{3}{7}, \frac{6}{7} \right\rangle$$

is the unit vector with the same direction as **a**, and $\langle -\frac{2}{7}, \frac{3}{7}, -\frac{6}{7} \rangle$ is the unit vector in the opposite direction. ▶

Unit vectors of particular interest are those in the positive direction of the coordinate axes (see Figure 11.19). These vectors are given the

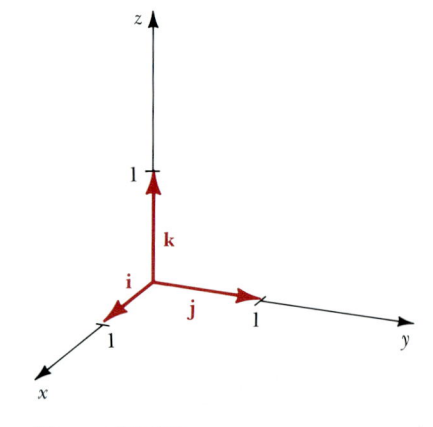

Figure 11.19
Coordinate unit vectors **i**, **j**, and **k**.

following special designations:

(11.8) $\mathbf{i} = \langle 1, 0, 0 \rangle, \quad \mathbf{j} = \langle 0, 1, 0 \rangle, \quad \mathbf{k} = \langle 0, 0, 1 \rangle.$

A vector $\mathbf{a} = \langle a_1, a_2, a_3 \rangle$ is expressed uniquely in terms of $\mathbf{i}, \mathbf{j},$ and $\mathbf{k}$ as

$$\mathbf{a} = \langle a_1, a_2, a_3 \rangle = a_1 \mathbf{i} + a_2 \mathbf{j} + a_3 \mathbf{k}.$$

Example 7

Express the vector $\mathbf{a} = \langle -3, 2, 0 \rangle$ in terms of $\mathbf{i}, \mathbf{j},$ and $\mathbf{k}$ and illustrate this representation geometrically.

Solution

We have $\langle -3, 2, 0 \rangle = -3\mathbf{i} + 2\mathbf{j} + 0\mathbf{k} = -3\mathbf{i} + 2\mathbf{j}$. The representation is shown in the xy-plane in Figure 11.20(a). Figure 11.20(b) shows the representation in three-dimensional space. ▶

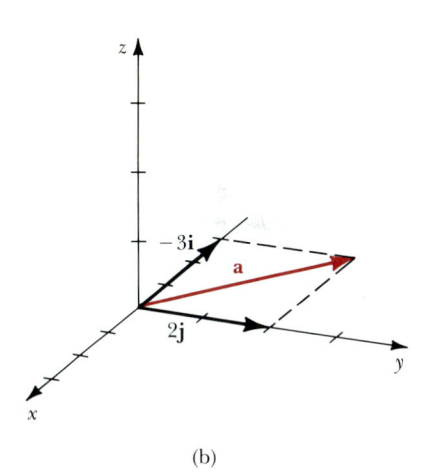

The final example in this section illustrates an application of a vector sum.

Example 8

A pilot is flying a plane headed northwest at an indicated airspeed of 140 knots (nautical miles per hour). A wind coming from the southwest is blowing the plane at a steady 30 knots. How far has the plane traveled in one hour?

Solution

Suppose an xy-coordinate system is introduced with the positive x-axis pointing east and the positive y-axis pointing north. Since the plane is heading northwest, the unit vector in the direction of the heading of the plane is

$$\mathbf{a} = \cos\frac{3\pi}{4}\mathbf{i} + \sin\frac{3\pi}{4}\mathbf{j} = -\frac{\sqrt{2}}{2}\mathbf{i} + \frac{\sqrt{2}}{2}\mathbf{j},$$

as shown in Figure 11.21(a). The unit vector $\mathbf{b}$ in the direction of the

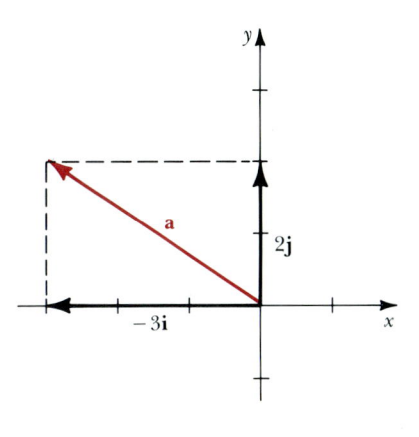

(a)

(b)

Figure 11.20

Figure 11.21

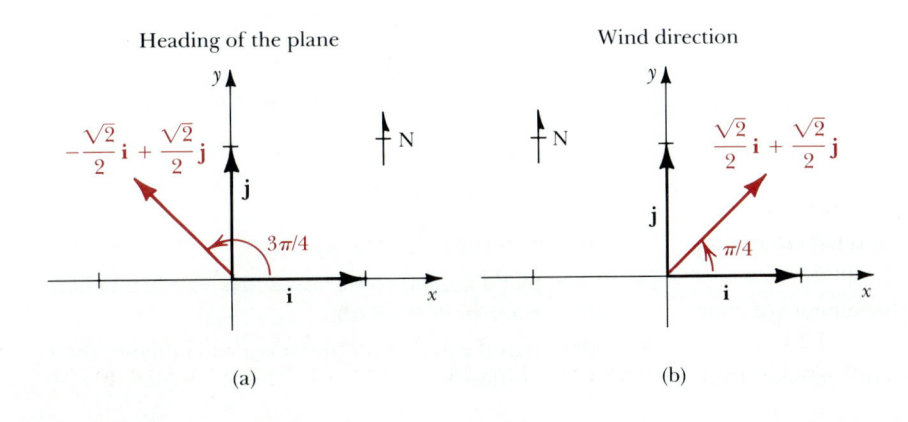

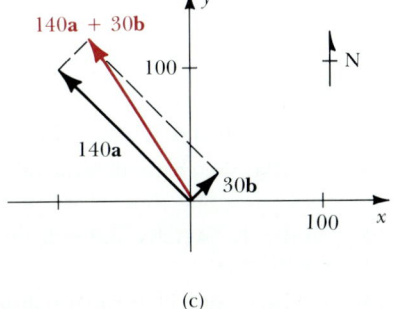

wind is shown in Figure 11.21(b), where

$$\mathbf{b} = \cos\frac{\pi}{4}\mathbf{i} + \sin\frac{\pi}{4}\mathbf{j} = \frac{\sqrt{2}}{2}\mathbf{i} + \frac{\sqrt{2}}{2}\mathbf{j}.$$

The flight of the plane is described by the resultant vector, shown in Figure 11.21(c):

$$140\mathbf{a} + 30\mathbf{b} = 140\left(-\frac{\sqrt{2}}{2}\mathbf{i} + \frac{\sqrt{2}}{2}\mathbf{j}\right) + 30\left(\frac{\sqrt{2}}{2}\mathbf{i} + \frac{\sqrt{2}}{2}\mathbf{j}\right)$$
$$= -55\sqrt{2}\,\mathbf{i} + 85\sqrt{2}\,\mathbf{j}.$$

The distance traveled by the plane in one hour is

$$\|140\mathbf{a} + 30\mathbf{b}\| = [(-55\sqrt{2})^2 + (85\sqrt{2})^2]^{1/2}$$
$$= \sqrt{20{,}500} \approx 143.2 \text{ nautical miles.}$$

▶ EXERCISE SET 11.2

In Exercises 1–6, the initial point A and terminal point B of a vector are given. Find (a) the position-vector representation and (b) the length of the vector.

1. $A(0, 0, 0)$, $B(1, 5, 3)$
2. $A(0, 2, 0)$, $B(0, 2, 7)$
3. $A(2, 3, 4)$, $B(2, -3, 4)$
4. $A(1, 2, 0)$, $B(3, 5, 0)$
5. $A(-2, 2, 1)$, $B(0, 5, 0)$
6. $A(4, 1, 0)$, $B(1, -2, 0)$

In Exercises 7–10, a vector and its initial point A are given. Find the terminal point of the vector.

7. $\mathbf{v} = \langle 1, 3, 4\rangle$, $A(0, 0, 0)$
8. $\mathbf{v} = \langle 1, 3, 4\rangle$, $A(2, 1, 0)$
9. $\mathbf{v} = \langle -1, 2, 0\rangle$, $A(-4, 1, 1)$
10. $\mathbf{v} = \langle 2, -2, 1\rangle$, $A(0, 3, 3)$

In Exercises 11–18, $\mathbf{a} = \langle 1, 1, 0\rangle$, $\mathbf{b} = \langle -1, 1, 2\rangle$, and $\mathbf{c} = \langle 2, 3, 4\rangle$.

11. Find $\mathbf{a} + \mathbf{b}$ and illustrate this sum geometrically.
12. Find $3\mathbf{a}$. Sketch $\mathbf{a}$ and $3\mathbf{a}$ on the same set of coordinate axes.
13. Find $-2\mathbf{c}$. Sketch $\mathbf{c}$ and $-2\mathbf{c}$ on the same set of coordinate axes.
14. Find $\mathbf{a} - \mathbf{c}$ and illustrate this difference geometrically.
15. Find $\mathbf{a} + \mathbf{b} + \mathbf{c}$.
16. Find $2\mathbf{a} - 3\mathbf{b} + 4\mathbf{c}$.
17. Find $\|\mathbf{c}\|$ and $\|-2\mathbf{c}\|$.
18. Find $\|\mathbf{a}\|$, $\|\mathbf{b}\|$, and $\|\mathbf{a} + \mathbf{b}\|$.
19. Find a unit vector that has the same direction as each of the following.
    a) $\langle 2, 3, 4\rangle$ b) $\langle -1, 2, 3\rangle$
20. Find a unit vector that has direction opposite that of each of the following.
    a) $\langle 2, 3, 4\rangle$ b) $\langle -1, 2, 3\rangle$
21. Find two vectors parallel to $\langle 3, 4, -1\rangle$ that have length 3.
22. Find two vectors parallel to $\langle 2, 4, 4\rangle$ that have length 7.
23. Express each of the following vectors in terms of $\mathbf{i}$, $\mathbf{j}$, and $\mathbf{k}$ and illustrate the representation geometrically.
    a) $\langle 1, 4, 0\rangle$ b) $\langle 1, 4, 5\rangle$ c) $\langle 0, 2, 3\rangle$
24. Determine which of the following pairs of vectors are parallel.
    a) $\langle 1, 0, 1\rangle$, $\langle -1, 0, -1\rangle$
    b) $\langle 1, 0, 0\rangle$, $\langle 1, 1, 0\rangle$
    c) $\langle 2, 3, -1\rangle$, $\langle 4, 6, -2\rangle$
    d) $\langle 3, -1, 2\rangle$, $\langle -6, 2, 4\rangle$
25. Show that if $\mathbf{a}$ and $\mathbf{b}$ are vectors and α is a real number, then $\alpha(\mathbf{a} + \mathbf{b}) = \alpha\mathbf{a} + \alpha\mathbf{b}$.
26. Show that if $\mathbf{a}$ is a vector and α is a real number, then $\|\alpha\mathbf{a}\| = |\alpha|\|\mathbf{a}\|$.

27. Consider the accompanying figure, where $\mathbf{c} = (\mathbf{a} + \mathbf{b})/2$.
   a) Write $\mathbf{d}$ in terms of $\mathbf{b}$ and $\mathbf{c}$.
   b) Prove that the diagonals of a parallelogram bisect each other.

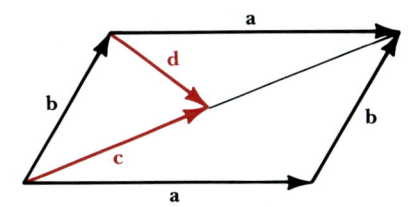

28. Consider the accompanying figure.
   a) Show that for some constants α and β, $\mathbf{c} = \alpha(\mathbf{a} - \mathbf{b})$ and $\mathbf{c} = (\beta/2)\mathbf{a} + (\beta - 1)\mathbf{b}$.
   b) Show that part (a) implies that $\mathbf{c} = (\mathbf{a} - \mathbf{b})/3$.
   c) Prove that the lines joining a vertex of a parallelogram to the midpoints of the opposite sides divide a diagonal of the parallelogram into three equal parts.

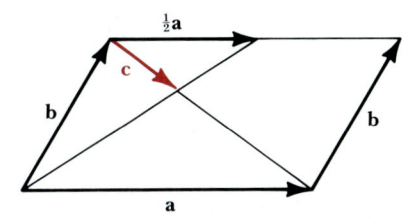

29. Three children are pulling on ropes attached to a common ring (see the figure). The first child pulls in the direction of the positive x-axis with a force of 20 lb. The second child pulls in the direction of the positive y-axis with a force of 15 lb. If the ring does not move, in what direction and with what force must the third child be pulling?

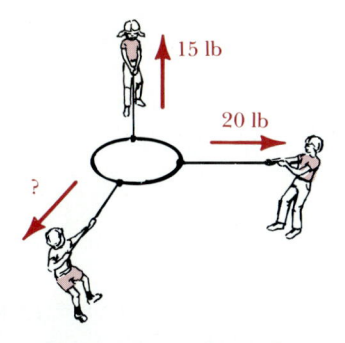

30. A steady easterly wind is blowing at the rate of 10 m/sec on a helium-filled balloon attached to a string. In still air, the balloon would rise vertically at the rate of 2 m/sec (see the figure). Where is the balloon and how much string has been released when the balloon is 100 m high?

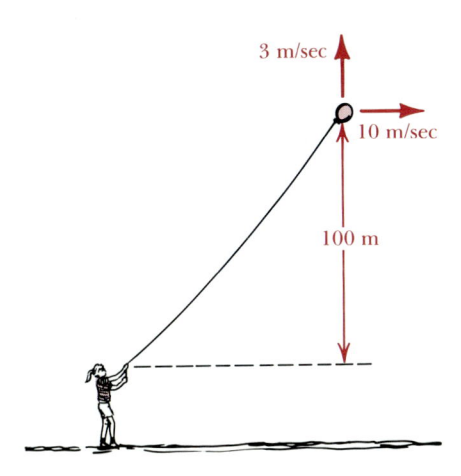

31. A canoeist crosses a 300-ft stream in 1 min, arriving at a point directly opposite the starting point. The current in the stream is 3 ft/sec (see the figure). In what direction did the canoeist paddle and at what average speed?

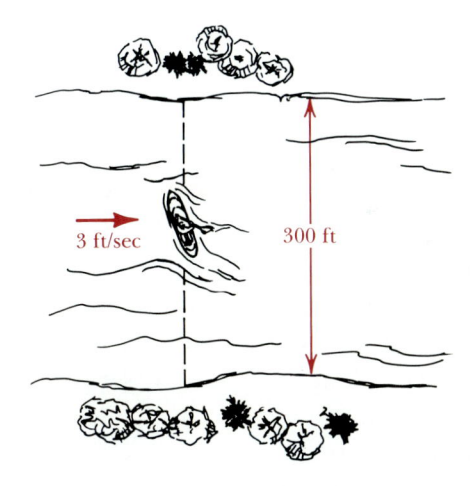

32. A private pilot flies for 1 hr and arrives at a point 100 mi due north of his departure point. A steady wind was blowing from the northwest at the rate of 20 mph. In what direction did the pilot fly and at what average speed?

33. Suppose the pilot of the plane in Exercise 32 flies to a point 100 mi due south instead of due north. Assuming that the other conditions remain the same, in what direction and at what average speed did the plane fly?

11.3
The Dot Product of Vectors

In Section 11.2 we saw that if **a** and **b** are vectors and α is a real number, then the sum $\mathbf{a} + \mathbf{b}$, the difference $\mathbf{a} - \mathbf{b}$, and the scalar product $\alpha\mathbf{a}$ are all vectors. In this section we introduce an operation on a pair of vectors that results in a real number. This operation is called the *dot product* of a pair of vectors. It is also known as the *inner* or *scalar product*.

(11.9)
DEFINITION
The **dot product** of the vectors $\mathbf{a} = \langle a_1, a_2, a_3 \rangle$ and $\mathbf{b} = \langle b_1, b_2, b_3 \rangle$ is
$$\mathbf{a} \cdot \mathbf{b} = a_1 b_1 + a_2 b_2 + a_3 b_3.$$

Example 1

For vectors $\mathbf{a} = \langle 1, -1, 1 \rangle$ and $\mathbf{b} = \langle -2, 3, 1 \rangle$, find each of the following.

a) $\mathbf{a} \cdot \mathbf{b}$ b) $\mathbf{a} \cdot \mathbf{a}$
c) $\mathbf{b} \cdot \mathbf{i}$ d) $\mathbf{b} \cdot \mathbf{j}$

Solution
a) $\mathbf{a} \cdot \mathbf{b} = \langle 1, -1, 1 \rangle \cdot \langle -2, 3, 1 \rangle$
$\phantom{\mathbf{a} \cdot \mathbf{b}} = (1)(-2) + (-1)(3) + (1)(1) = -4$
b) $\mathbf{a} \cdot \mathbf{a} = \langle 1, -1, 1 \rangle \cdot \langle 1, -1, 1 \rangle = 1 + 1 + 1 = 3$
c) $\mathbf{b} \cdot \mathbf{i} = \langle -2, 3, 1 \rangle \cdot \langle 1, 0, 0 \rangle = -2 + 0 + 0 = -2$
d) $\mathbf{b} \cdot \mathbf{j} = \langle -2, 3, 1 \rangle \cdot \langle 0, 1, 0 \rangle = 0 + 3 + 0 = 3$ ▶

Some immediate consequences of the definition of the dot product are listed in the following theorem.

(11.10)
THEOREM
If **a**, **b**, and **c** are vectors and α is a real number, then:

i) $\mathbf{a} \cdot \mathbf{b} = \mathbf{b} \cdot \mathbf{a}$;   ii) $(\mathbf{a} + \mathbf{b}) \cdot \mathbf{c} = \mathbf{a} \cdot \mathbf{c} + \mathbf{b} \cdot \mathbf{c}$;
iii) $(\alpha\mathbf{a}) \cdot \mathbf{b} = \alpha(\mathbf{a} \cdot \mathbf{b})$;   iv) $\mathbf{0} \cdot \mathbf{a} = 0$;
v) $\mathbf{a} \cdot \mathbf{a} = \|\mathbf{a}\|^2$.

In addition,

vi) $\mathbf{i} \cdot \mathbf{i} = \mathbf{j} \cdot \mathbf{j} = \mathbf{k} \cdot \mathbf{k} = 1$;   vii) $\mathbf{i} \cdot \mathbf{j} = \mathbf{i} \cdot \mathbf{k} = \mathbf{j} \cdot \mathbf{k} = 0$.

An alternative representation of the dot product uses the notion of the angle between two vectors. The **angle between nonzero vectors a and b** is the angle θ in $[0, \pi]$ determined by the position-vector representations of **a** and **b** (see Figure 11.22). If A denotes the terminal point of **a** and B denotes the terminal point of **b**, then the angle between **a** and **b** is angle AOB.

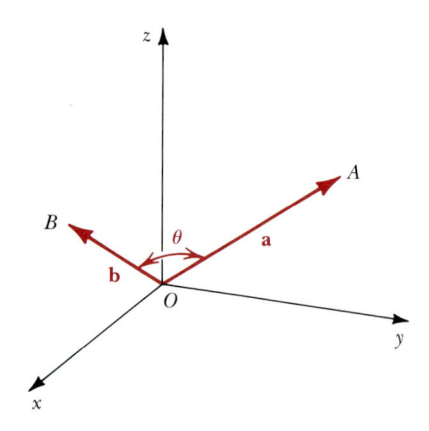

Figure 11.22
The angle between two vectors.

If **a** and **b** are parallel vectors and **b** = α**a**, then we say that the angle between **a** and **b** is zero when **a** and **b** have the same direction, that is, when α > 0, and the angle is π when **a** and **b** have the opposite direction, that is, when α < 0.

The following representation is sometimes given as the definition of the dot product.

(11.11)
THEOREM
If θ is the angle between the vectors **a** and **b**, then

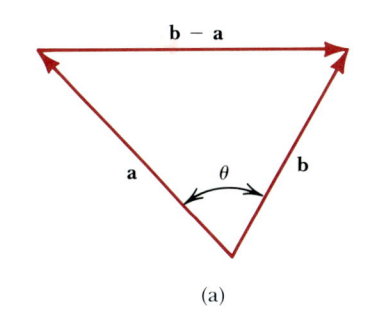

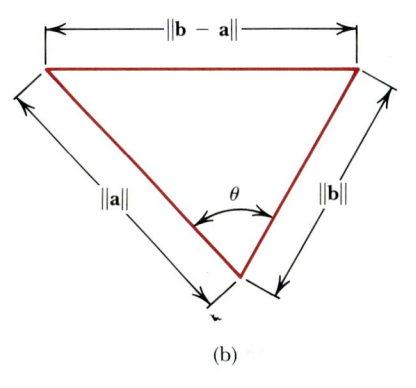

$$\mathbf{a} \cdot \mathbf{b} = \|\mathbf{a}\| \|\mathbf{b}\| \cos \theta.$$

Proof. Consider the position-vector representation of **a** and **b**. Connecting the terminal points of these vectors forms a triangle with sides of length $\|\mathbf{a}\|$, $\|\mathbf{b}\|$, and $\|\mathbf{b} - \mathbf{a}\|$ (see Figure 11.23).

The law of cosines (Equation A.26 in Appendix A.3) applied to this triangle implies that

$$\|\mathbf{b} - \mathbf{a}\|^2 = \|\mathbf{a}\|^2 + \|\mathbf{b}\|^2 - 2\|\mathbf{a}\|\|\mathbf{b}\| \cos \theta.$$

It follows from Theorem (11.10) that

$$\|\mathbf{b} - \mathbf{a}\|^2 = (\mathbf{b} - \mathbf{a}) \cdot (\mathbf{b} - \mathbf{a})$$
$$= \mathbf{b} \cdot \mathbf{b} - \mathbf{a} \cdot \mathbf{b} - \mathbf{b} \cdot \mathbf{a} + \mathbf{a} \cdot \mathbf{a}$$
$$= \|\mathbf{b}\|^2 - 2\mathbf{a} \cdot \mathbf{b} + \|\mathbf{a}\|^2.$$

Thus

$$\|\mathbf{b}\|^2 - 2\mathbf{a} \cdot \mathbf{b} + \|\mathbf{a}\|^2 = \|\mathbf{a}\|^2 + \|\mathbf{b}\|^2 - 2\|\mathbf{a}\|\|\mathbf{b}\| \cos \theta,$$

so

$$-2\mathbf{a} \cdot \mathbf{b} = -2\|\mathbf{a}\|\|\mathbf{b}\| \cos \theta \quad \text{and} \quad \mathbf{a} \cdot \mathbf{b} = \|\mathbf{a}\|\|\mathbf{b}\| \cos \theta. \quad \triangleright$$

Figure 11.23
$\mathbf{a} \cdot \mathbf{b} = \|\mathbf{a}\| \cdot \|\mathbf{b}\| \cos \theta.$

Example 2

Find the dot product of the vectors $\langle 3, 0, -1 \rangle$ and $\langle 2, 1, 2 \rangle$ and determine the angle between them.

Solution
The dot product of these vectors is

$$\langle 3, 0, -1 \rangle \cdot \langle 2, 1, 2 \rangle = 6 + 0 - 2 = 4.$$

Since

$$\|\langle 3, 0, -1 \rangle\| = \sqrt{10} \quad \text{and} \quad \|\langle 2, 1, 2 \rangle\| = \sqrt{9} = 3,$$

it follows from Theorem (11.11) that

$$\cos \theta = \frac{\mathbf{a} \cdot \mathbf{b}}{\|\mathbf{a}\|\|\mathbf{b}\|} = \frac{4}{3\sqrt{10}} = \frac{2\sqrt{10}}{15},$$

so

$$\theta = \arccos\left(\frac{2\sqrt{10}}{15}\right) \approx 1.14. \quad \blacktriangleright$$

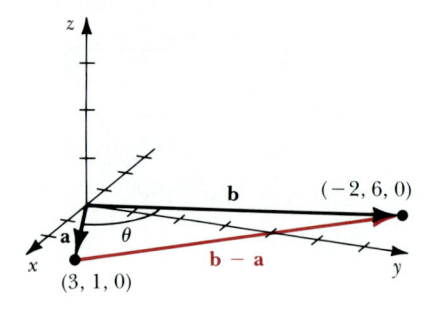

Figure 11.24

Example 3

Show that the points $(0, 0, 0)$, $(3, 1, 0)$, and $(-2, 6, 0)$ are the vertices of a right triangle with the right angle at the origin.

Solution

Let θ denote the angle at the origin and $\mathbf{a} = \langle 3, 1, 0 \rangle$ and $\mathbf{b} = \langle -2, 6, 0 \rangle$ be the position vectors for the other two vertices, as shown in Figure 11.24. Then θ is the angle between $\mathbf{a}$ and $\mathbf{b}$.

Since

$$\mathbf{a} \cdot \mathbf{b} = \langle 3, 1, 0 \rangle \cdot \langle -2, 6, 0 \rangle = -6 + 6 + 0 = 0,$$

we have

$$\cos \theta = \frac{\mathbf{a} \cdot \mathbf{b}}{\|\mathbf{a}\| \|\mathbf{b}\|} = 0,$$

and $\theta = \pi/2$. (An alternative method for solving this problem is to show that $\|\mathbf{a}\|^2 + \|\mathbf{b}\|^2 = \|\mathbf{a} - \mathbf{b}\|^2$. This is considered in Exercise 34.) ▶

Motivated by the preceding example, we make the following definition.

(11.12) DEFINITION

Vectors $\mathbf{a}$ and $\mathbf{b}$ are said to be **orthogonal** (or perpendicular) if θ, the angle between $\mathbf{a}$ and $\mathbf{b}$, is $\pi/2$. (In addition, we say that $\mathbf{0}$ is orthogonal to every vector.)

A corollary of Theorem (11.11) follows from this definition.

(11.13) COROLLARY

Vectors $\mathbf{a}$ and $\mathbf{b}$ are orthogonal if and only if $\mathbf{a} \cdot \mathbf{b} = 0$.

Three angles of particular interest for a vector $\mathbf{a} = \langle a_1, a_2, a_3 \rangle$ are the angles α between $\mathbf{a}$ and $\mathbf{i}$, β between $\mathbf{a}$ and $\mathbf{j}$, and γ between $\mathbf{a}$ and $\mathbf{k}$ (see Figure 11.25). These are called the **direction angles** of $\mathbf{a}$. The numbers $\cos \alpha$, $\cos \beta$, and $\cos \gamma$ are called the **direction cosines** of $\mathbf{a}$.

(11.14) THEOREM

The direction cosines of a nonzero vector $\mathbf{a} = \langle a_1, a_2, a_3 \rangle$ satisfy

$$\cos \alpha = \frac{a_1}{\|\mathbf{a}\|}, \qquad \cos \beta = \frac{a_2}{\|\mathbf{a}\|}, \qquad \cos \gamma = \frac{a_3}{\|\mathbf{a}\|}$$

and

$$\cos^2 \alpha + \cos^2 \beta + \cos^2 \gamma = 1.$$

Figure 11.25
Direction angles for a vector.

Proof.

$$\cos\alpha = \frac{\mathbf{a}\cdot\mathbf{i}}{\|\mathbf{a}\|\|\mathbf{i}\|} = \frac{a_1}{\|\mathbf{a}\|}, \qquad \cos\beta = \frac{\mathbf{a}\cdot\mathbf{j}}{\|\mathbf{a}\|\|\mathbf{j}\|} = \frac{a_2}{\|\mathbf{a}\|},$$

and

$$\cos\gamma = \frac{\mathbf{a}\cdot\mathbf{k}}{\|\mathbf{a}\|\|\mathbf{k}\|} = \frac{a_3}{\|\mathbf{a}\|}.$$

Thus

$$\cos^2\alpha + \cos^2\beta + \cos^2\gamma = \frac{a_1^2}{\|\mathbf{a}\|^2} + \frac{a_2^2}{\|\mathbf{a}\|^2} + \frac{a_3^2}{\|\mathbf{a}\|^2} = \frac{\|\mathbf{a}\|^2}{\|\mathbf{a}\|^2} = 1. \quad \triangleright$$

Since any vector $\mathbf{a} = \langle a_1, a_2, a_3 \rangle$ can be written as $\mathbf{a} = a_1\mathbf{i} + a_2\mathbf{j} + a_3\mathbf{k}$, Theorem (11.14) implies that

(11.15) $\qquad \mathbf{a} = \|\mathbf{a}\|(\cos\alpha\,\mathbf{i} + \cos\beta\,\mathbf{j} + \cos\gamma\,\mathbf{k}).$

Example 4

Find the direction cosines and direction angles of the vector $\mathbf{a} = \langle 1, -2, 2 \rangle$.

Solution
Since $\|\mathbf{a}\| = \sqrt{1+4+4} = 3$, we have

$$\cos\alpha = \frac{1}{3}, \qquad \cos\beta = -\frac{2}{3}, \quad \text{and} \quad \cos\gamma = \frac{2}{3}.$$

So

$$\alpha = \arccos\left(\frac{1}{3}\right) \approx 1.23, \qquad \beta = \arccos\left(-\frac{2}{3}\right) \approx 2.30,$$

and

$$\gamma = \arccos\left(\frac{2}{3}\right) \approx 0.84.$$

The position-vector representation of $\mathbf{a}$ is shown in Figure 11.26. $\quad\triangleright$

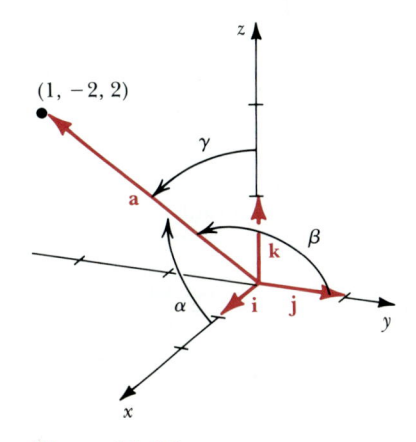

Figure 11.26

Theorem (11.11) implies that for any vectors $\mathbf{a}$ and $\mathbf{b}$,

$$|\mathbf{a}\cdot\mathbf{b}| = \|\mathbf{a}\|\|\mathbf{b}\||\cos\theta|.$$

Since $|\cos\theta| \leq 1$ for all values of θ,

(11.16) $\qquad |\mathbf{a}\cdot\mathbf{b}| \leq \|\mathbf{a}\|\|\mathbf{b}\|.$

Inequality (11.16), called the **Cauchy-Buniakowsky-Schwarz Inequality,** is used to establish the important result given in Theorem (11.17). This result is known as the **Triangle Inequality.**

(11.17)
THEOREM
For any vectors $\mathbf{a}$ and $\mathbf{b}$, $\|\mathbf{a}+\mathbf{b}\| \leq \|\mathbf{a}\| + \|\mathbf{b}\|$.

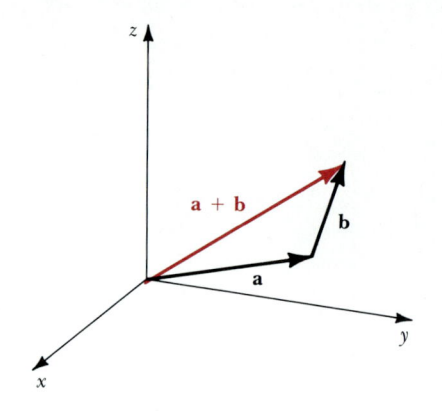

Figure 11.27
The Triangle Inequality:
$\|\mathbf{a} + \mathbf{b}\| \leq \|\mathbf{a}\| + \|\mathbf{b}\|$.

$0 \leq \theta < \pi/2$

(a)

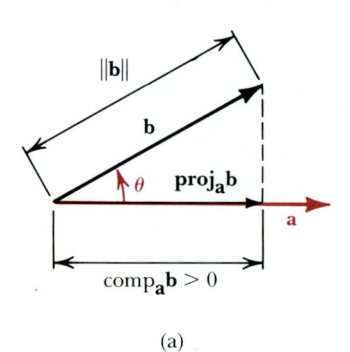

$\pi/2 < \theta \leq \pi$

(b)

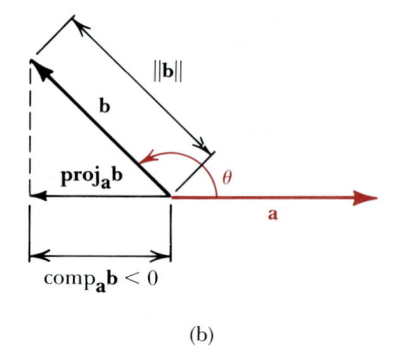

Figure 11.28
The projection of **b** onto **a**.

Proof. Consider $\|\mathbf{a} + \mathbf{b}\|^2$ and apply Theorem (11.10v) and (11.16):

$$\begin{aligned}\|\mathbf{a} + \mathbf{b}\|^2 &= (\mathbf{a} + \mathbf{b}) \cdot (\mathbf{a} + \mathbf{b}) \\ &= \mathbf{a} \cdot \mathbf{a} + \mathbf{b} \cdot \mathbf{a} + \mathbf{a} \cdot \mathbf{b} + \mathbf{b} \cdot \mathbf{b} \\ &= \|\mathbf{a}\|^2 + 2\mathbf{a} \cdot \mathbf{b} + \|\mathbf{b}\|^2 \\ &\leq \|\mathbf{a}\|^2 + 2|\mathbf{a} \cdot \mathbf{b}| + \|\mathbf{b}\|^2 \\ &\leq \|\mathbf{a}\|^2 + 2\|\mathbf{a}\|\|\mathbf{b}\| + \|\mathbf{b}\|^2 = (\|\mathbf{a}\| + \|\mathbf{b}\|)^2.\end{aligned}$$

Taking the positive square root of each side produces the result. ▷

If Theorem (11.17) is interpreted geometrically, it says that the length of one side of a triangle cannot exceed the sum of the lengths of the remaining two sides (see Figure 11.27).

For a fixed nonzero vector **a**, any vector **b** can be written uniquely as the sum of a vector that is parallel to **a** and another vector that is orthogonal to **a**. The portion of the sum parallel to **a** is called the **orthogonal projection** of **b** onto **a** and is denoted $\mathbf{proj_a\ b}$ (see Figure 11.28).

The triangles in Figure 11.28 indicate that the length of $\mathbf{proj_a\ b}$ is given by

$$\|\mathbf{proj_a\ b}\| = |\|\mathbf{b}\| \cos \theta|.$$

But $\mathbf{a} \cdot \mathbf{b} = \|\mathbf{a}\|\|\mathbf{b}\| \cos \theta$, so

$$\|\mathbf{proj_a\ b}\| = \left|\|\mathbf{b}\| \cos \theta\right| = \left|\|\mathbf{b}\| \frac{\mathbf{a} \cdot \mathbf{b}}{\|\mathbf{a}\|\|\mathbf{b}\|}\right| = \left|\frac{\mathbf{a} \cdot \mathbf{b}}{\|\mathbf{a}\|}\right|.$$

The number $(\mathbf{a} \cdot \mathbf{b})/\|\mathbf{a}\|$ is called the **component of b in the direction of a** and denoted

(11.18) $$\mathrm{comp_a\ } \mathbf{b} = \frac{\mathbf{a} \cdot \mathbf{b}}{\|\mathbf{a}\|}.$$

Since $\mathbf{a}/\|\mathbf{a}\|$ is the unit vector in the direction of **a**,

(11.19) $$\mathbf{proj_a\ b} = \mathrm{comp_a\ } \mathbf{b} \left(\frac{\mathbf{a}}{\|\mathbf{a}\|}\right) = \frac{\mathbf{a} \cdot \mathbf{b}}{\|\mathbf{a}\|^2} \mathbf{a}.$$

The details of the decomposition of a vector **b** into the sum of two vectors, one parallel to **a** and the other orthogonal to **a**, are given in the following theorem.

(11.20)
THEOREM
Suppose **a** is a fixed nonzero vector. Then any vector **b** can be written uniquely in the form $\mathbf{b} = \mathbf{b}_1 + \mathbf{b}_2$, where $\mathbf{b}_1$ is parallel to **a** and $\mathbf{b}_2$ is orthogonal to **a**. Moreover,

$$\mathbf{b}_1 = \mathbf{proj_a\ b} = (\mathrm{comp_a\ } \mathbf{b}) \frac{\mathbf{a}}{\|\mathbf{a}\|} = \left(\frac{\mathbf{a} \cdot \mathbf{b}}{\|\mathbf{a}\|^2}\right) \mathbf{a} \quad \text{and} \quad \mathbf{b}_2 = \mathbf{b} - \mathbf{proj_a\ b}.$$

Proof. By definition of $\mathbf{proj_a\ b}$, the vector $\mathbf{b}_1$ is parallel to **a**. With $\mathbf{b}_2$ defined by $\mathbf{b}_2 = \mathbf{b} - \mathbf{b}_1$, it is clear that $\mathbf{b} = \mathbf{b}_1 + \mathbf{b}_2$. It remains to show that $\mathbf{b}_2$ is orthogonal to $\mathbf{a}$, that is, that $\mathbf{a} \cdot \mathbf{b}_2 = 0$:

$$\mathbf{a} \cdot \mathbf{b}_2 = \mathbf{a} \cdot (\mathbf{b} - \mathbf{b}_1) = \mathbf{a} \cdot \mathbf{b} - \mathbf{a} \cdot \left(\frac{\mathbf{a} \cdot \mathbf{b}}{\|\mathbf{a}\|^2}\right) \mathbf{a}$$

$$= \mathbf{a} \cdot \mathbf{b} - \left(\frac{\mathbf{a} \cdot \mathbf{b}}{\|\mathbf{a}\|^2}\right) \|\mathbf{a}\|^2 = 0.$$

The uniqueness of the decomposition is considered in Exercise 47. ▷

Example 5

Decompose $\mathbf{b} = \langle 2, 3, -1 \rangle$ into a sum $\mathbf{b}_1 + \mathbf{b}_2$, where $\mathbf{b}_1$ is parallel to $\mathbf{a} = \langle 0, 4, 2 \rangle$ and $\mathbf{b}_2$ is orthogonal to $\mathbf{a}$.

Solution
We have

$$\text{comp}_\mathbf{a} \mathbf{b} = \frac{\mathbf{a} \cdot \mathbf{b}}{\|\mathbf{a}\|} = \frac{\langle 0, 4, 2 \rangle \cdot \langle 2, 3, -1 \rangle}{\sqrt{0 + 16 + 4}} = \frac{10}{2\sqrt{5}} = \sqrt{5}$$

and

$$\text{proj}_\mathbf{a} \mathbf{b} = \text{comp}_\mathbf{a} \mathbf{b} \left(\frac{\mathbf{a}}{\|\mathbf{a}\|}\right) = \sqrt{5} \frac{\langle 0, 4, 2 \rangle}{2\sqrt{5}} = \langle 0, 2, 1 \rangle.$$

So $\mathbf{b}_1 = \langle 0, 2, 1 \rangle$ and $\mathbf{b}_2 = \mathbf{b} - \mathbf{b}_1 = \langle 2, 3, -1 \rangle - \langle 0, 2, 1 \rangle = \langle 2, 1, -2 \rangle$. Figure 11.29 shows the decomposition. ▶

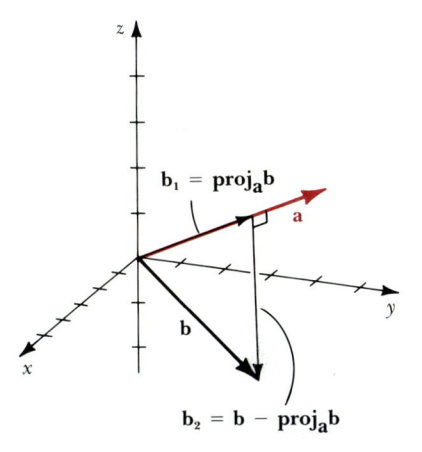

Figure 11.29

One application of the orthogonal projection of one vector onto another arises in the definition of the work done by a force in moving an object. Recall from Section 5.6 that if a constant force F is applied in the direction of motion, then the work W done by F in moving an object a distance D is

$$W = (\text{force})(\text{distance}) = F \cdot D.$$

If a *vector force* $\mathbf{F}$ is constant but is applied at an angle θ to the direction of motion (as in Figure 11.30), then the work done by $\mathbf{F}$ in moving an object from P to Q is defined by

$$W = (\text{component of force in the direction of motion})(\text{distance})$$
$$= (\text{comp}_{\overrightarrow{PQ}} \mathbf{F}) \|\overrightarrow{PQ}\|.$$

Thus,

(11.21) $\qquad W = (\|\mathbf{F}\| \cos \theta) \|\overrightarrow{PQ}\| = \mathbf{F} \cdot \overrightarrow{PQ}.$

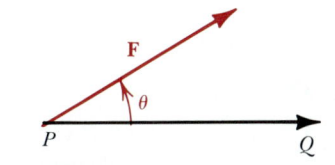

Figure 11.30
The work is $\mathbf{F} \cdot \overrightarrow{PQ}$.

Example 6

A toboggan loaded with camping gear is pulled 500 ft across a frozen lake with a rope that makes an angle of $\pi/4$ with the ground. Find the work done if a force of 20 lb is exerted in pulling the toboggan.

Solution

The situation is illustrated in Figure 11.31. The unit vector in the direction of the force is

$$\cos\frac{\pi}{4}\mathbf{i} + \sin\frac{\pi}{4}\mathbf{j} = \frac{\sqrt{2}}{2}(\mathbf{i}+\mathbf{j}),$$

so the force is

$$\mathbf{F} = 20\left[\frac{\sqrt{2}}{2}(\mathbf{i}+\mathbf{j})\right] = 10\sqrt{2}\,(\mathbf{i}+\mathbf{j})\text{ lb.}$$

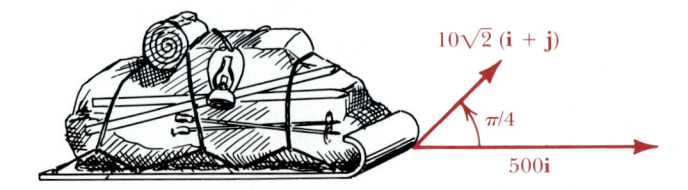

Figure 11.31

The vector describing the path of the toboggan is $500\mathbf{i}$ ft, so the work done is

$$W = \mathbf{F} \cdot 500\mathbf{i} = 10\sqrt{2}\,(\mathbf{i}+\mathbf{j}) \cdot 500\mathbf{i} = 5000\sqrt{2}\text{ ft-lb.} \quad \blacktriangleright$$

▶ EXERCISE SET 11.3

In Exercises 1–8, find $\mathbf{a} \cdot \mathbf{b}$ for the pair of vectors.

1. $\mathbf{a} = \langle 1, 3, 3 \rangle,\ \mathbf{b} = \langle 2, 2, 4 \rangle$
2. $\mathbf{a} = \langle -1, -3, 2 \rangle,\ \mathbf{b} = \langle 1, 3, 2 \rangle$
3. $\mathbf{a} = \langle 2, 4, 0 \rangle,\ \mathbf{b} = \langle 3, 7, 0 \rangle$
4. $\mathbf{a} = \langle 1, -3, 2 \rangle,\ \mathbf{b} = \langle 2, 2, 2 \rangle$
5. $\mathbf{a} = \langle 0, 4, 6 \rangle,\ \mathbf{b} = \langle 0, -3, 2 \rangle$
6. $\mathbf{a} = \langle 1, 5, -3 \rangle,\ \mathbf{b} = \langle 3, 1, 1 \rangle$
7. $\mathbf{a} = \langle 3, 4, \pi \rangle,\ \mathbf{b} = \langle 1, e, 0 \rangle$
8. $\mathbf{a} = \langle -\sqrt{2}, \sqrt{3}, 1 \rangle,\ \mathbf{b} = \langle \sqrt{3}, 0, \sqrt{2} \rangle$

9–16. (a) Find the angle between the pairs of vectors in Exercises 1–8, and (b) determine in each case if the vectors are orthogonal.

17–24. (a) Find the component of $\mathbf{b}$ in the direction of $\mathbf{a}$ and (b) determine the orthogonal projection of $\mathbf{b}$ onto $\mathbf{a}$ for the vectors in Exercises 1–8.

In Exercises 25–28, find the direction cosines of the vector.

25. $\langle 2, 3, 4 \rangle$
26. $\langle -1, 3, -3 \rangle$
27. $\langle 0, 3, 4 \rangle$
28. $\langle 2, 3, 0 \rangle$

29. Find a unit vector that has the same direction cosines as $\langle 0, 3, 4 \rangle$.

30. Find a vector lying in the xy-plane orthogonal to $\langle 2, 4, 1 \rangle$. Is there more than one such vector?

31. Find a unit vector lying in the xy-plane orthogonal to $\langle 2, 4, 0 \rangle$. Is there more than one such vector?

32. a) Find the component of $\mathbf{b} = \langle b_1, b_2, b_3 \rangle$ in the direction of each of the vectors $\mathbf{i}$, $\mathbf{j}$, and $\mathbf{k}$.
    b) Find the orthogonal projection of $\mathbf{b}$ onto each of $\mathbf{i}$, $\mathbf{j}$, and $\mathbf{k}$.

33. Express $\mathbf{b} = \langle -4, 1, -2 \rangle$ as $\mathbf{b} = \mathbf{b}_1 + \mathbf{b}_2$, where $\mathbf{b}_1$ is parallel to $\mathbf{a} = \langle 1, 3, -3 \rangle$ and $\mathbf{b}_2$ is orthogonal to $\mathbf{a}$.

34. Show that $(0, 0, 0)$, $(3, 1, 0)$, and $(-2, 6, 0)$ are vertices of a right triangle by showing that $\|\mathbf{a}\|^2 + \|\mathbf{b}\|^2 = \|\mathbf{a} - \mathbf{b}\|^2$ for appropriate vectors $\mathbf{a}$ and $\mathbf{b}$.

35. Show that $(2, 2, 2)$, $(2, 0, 1)$, and $(4, 1, -1)$ are vertices of a right triangle.

36. Write an equation to describe all points (x, y, z) with the property that $\langle x - 1, y - 2, z - 3 \rangle$ is orthogonal to $\langle 1, -1, 2 \rangle$.

37. Find a vector orthogonal to both $\mathbf{a} = \langle 1, 2, 4 \rangle$ and $\mathbf{b} = \langle 2, -2, 5 \rangle$. Is there more than one such vector?

38. Show that for any vectors **a** and **b**, $\mathbf{a} \cdot \mathbf{b} = \mathbf{b} \cdot \mathbf{a}$.
39. Show that for any vector **a**, $\mathbf{a} \cdot \mathbf{a} = \|\mathbf{a}\|^2$.
40. Show that for any vectors **a** and **b**, $\|\mathbf{a} - \mathbf{b}\| \geq \|\mathbf{a}\| - \|\mathbf{b}\|$.
41. Show that if **v** is orthogonal to **a** and **b**, then **v** is orthogonal to $c_1\mathbf{a} + c_2\mathbf{b}$ for all real numbers c_1 and c_2.
42. Find nonzero vectors **a**, **b**, and **c** with $\mathbf{a} \cdot \mathbf{c} = \mathbf{b} \cdot \mathbf{c}$ and $\mathbf{a} \neq \mathbf{b}$.
43. Determine conditions on the vectors **a** and **b** necessary for equality to hold in the inequality in Theorem (11.17):
$$\|\mathbf{a} + \mathbf{b}\| \leq \|\mathbf{a}\| + \|\mathbf{b}\|.$$
44. a) Prove the *Parallelogram Law:* For any vectors **a** and **b**,
$$\|\mathbf{a} + \mathbf{b}\|^2 + \|\mathbf{a} - \mathbf{b}\|^2 = 2(\|\mathbf{a}\|^2 + \|\mathbf{b}\|^2).$$
b) Interpret the Parallelogram Law geometrically.
45. Show that vectors **a** and **b** are orthogonal if and only if
$$\|\mathbf{a} + \mathbf{b}\|^2 = \|\mathbf{a}\|^2 + \|\mathbf{b}\|^2.$$
46. Explain how the accompanying figure shows a geometric proof of the distributive law:
$$(\mathbf{a} + \mathbf{b}) \cdot \mathbf{c} = \mathbf{a} \cdot \mathbf{c} + \mathbf{b} \cdot \mathbf{c}.$$

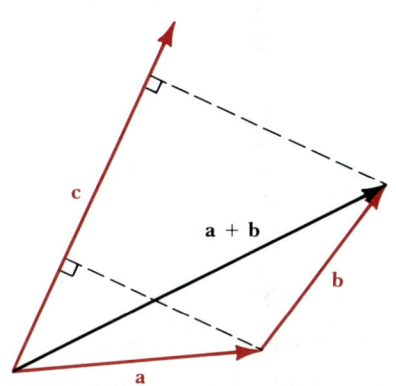

47. Show that the decomposition of **b** described in Theorem (11.20) is unique. [*Hint:* Assume that $\mathbf{b} = \alpha \mathbf{a} + \mathbf{c}$, where $\mathbf{a} \cdot \mathbf{c} = 0$. Then determine α and **c**.]

48. Find the work done by the force $\mathbf{F} = \mathbf{j} + 2\mathbf{k}$ (in pounds) applied to an object that moves 2 ft along the y-axis.
49. A wagon loaded with groceries is pulled horizontally a half mile by a handle that makes an angle of $\pi/3$ with the horizontal (see the figure). Find the work done if a force of 20 lb is exerted on the handle.

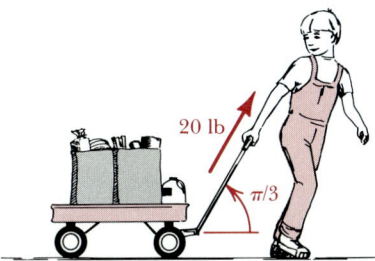

50. A block is to be moved 6 ft up a ramp that makes an angle of $\pi/4$ with the horizontal (see the figure).
   a) Find the work done if 20 lb of force is applied in the direction of the motion.
   b) Find the work done if 20 lb of force is applied at an angle of $\pi/6$ with the horizontal.

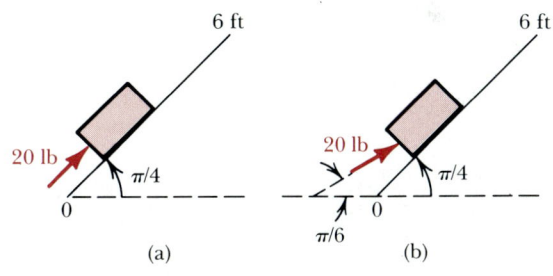

51. A 200-lb piece of slate for the top of a pool table is moved along a horizontal surface by a pulling force applied from above (see the figure). If the force is applied at an angle of $\pi/3$ to the slate, what force must be applied to move it?

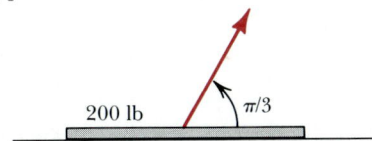

11.4

The Cross Product of Vectors

In Section 11.3 we considered the dot product, an operation that associates a specific real number $\mathbf{a} \cdot \mathbf{b}$ with each pair of vectors **a** and **b**. In this section we introduce an operation on three-dimensional vectors **a**

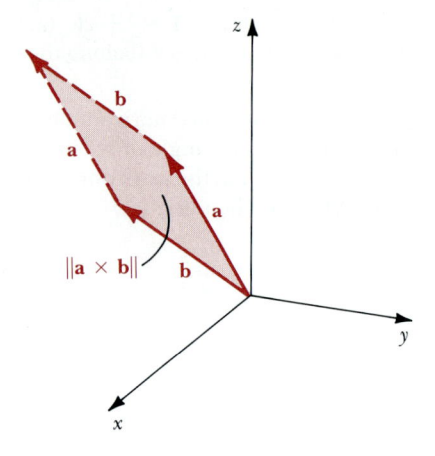

Figure 11.32
The area of the parallelogram is $\|\mathbf{a} \times \mathbf{b}\|$.

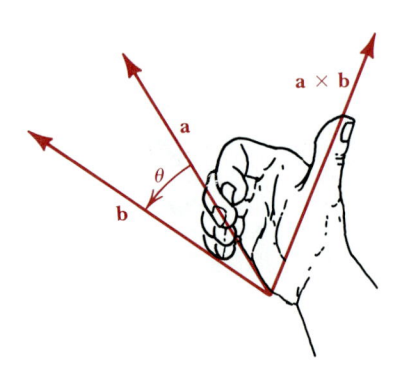

Figure 11.33
The direction of $\mathbf{a} \times \mathbf{b}$.

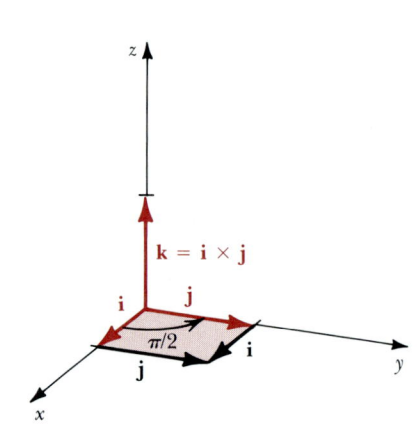

Figure 11.34

and **b** that results in a new three-dimensional vector called the *cross product* (it is also called the *outer* or *vector product*) and denoted $\mathbf{a} \times \mathbf{b}$. The geometric description of the cross product is more intuitive than the analytic definition, so we begin with this description.

Suppose **a** and **b** are nonparallel three-dimensional vectors. Position **a** and **b** to form a parallelogram, as shown in Figure 11.32. The vector $\mathbf{a} \times \mathbf{b}$ is orthogonal to both **a** and **b** and has length equal to the area of the parallelogram.

There are two vectors in three-dimensional space that satisfy this condition: one on each side of the plane determined by **a** and **b**. The vector $\mathbf{a} \times \mathbf{b}$ has the direction specified by the following *right-hand rule*:

Let θ denote the angle between the vectors **a** and **b**, and suppose **a** is rotated through the angle θ to coincide with **b**. If the right hand is positioned so that the wrist is at the common initial point of these vectors and the natural fold of the fingers points in the direction of this rotation, then the thumb points in the direction of $\mathbf{a} \times \mathbf{b}$ (see Figure 11.33).

Note that this *right-hand rule* indicates that

$$\mathbf{a} \times \mathbf{b} = -(\mathbf{b} \times \mathbf{a}).$$

Example 1

Find the cross product $\mathbf{i} \times \mathbf{j}$.

Solution

The parallelogram formed by **i** and **j** is a square with area 1. Figure 11.34 indicates that the direction of $\mathbf{i} \times \mathbf{j}$ coincides with that of the positive z-axis, so

$$\mathbf{i} \times \mathbf{j} = \mathbf{k}.$$

To calculate the cross product of more general vectors and to establish results about the operation, we need an analytic definition.

(11.22)
DEFINITION
The **cross product** $\mathbf{a} \times \mathbf{b}$ of the vectors $\mathbf{a} = \langle a_1, a_2, a_3 \rangle$ and $\mathbf{b} = \langle b_1, b_2, b_3 \rangle$ is

$$\mathbf{a} \times \mathbf{b} = \langle a_2 b_3 - a_3 b_2,\ a_3 b_1 - a_1 b_3,\ a_1 b_2 - a_2 b_1 \rangle.$$

Example 2

Use Definition (11.22) to (a) show that $\mathbf{i} \times \mathbf{j} = \mathbf{k}$ and (b) find $\langle 1, 2, 1 \rangle \times \langle 3, 0, 1 \rangle$.

Solution

a) Since $\mathbf{i} = \langle 1, 0, 0 \rangle$ and $\mathbf{j} = \langle 0, 1, 0 \rangle$,

$$\mathbf{i} \times \mathbf{j} = \langle 0 \cdot 0 - 0 \cdot 1,\ 0 \cdot 0 - 1 \cdot 0,\ 1 \cdot 1 - 0 \cdot 0 \rangle = \langle 0, 0, 1 \rangle = \mathbf{k},$$

b) $\langle 1, 2, 1 \rangle \times \langle 3, 0, 1 \rangle = \langle 2 \cdot 1 - 1 \cdot 0, 1 \cdot 3 - 1 \cdot 1, 1 \cdot 0 - 2 \cdot 3 \rangle$
$= \langle 2, 2, -6 \rangle$ ▸

Definition (11.22) agrees with the geometric description of the cross product. To see that $\mathbf{a} \times \mathbf{b}$ is orthogonal to $\mathbf{a}$, note that

$\mathbf{a} \cdot (\mathbf{a} \times \mathbf{b}) = \langle a_1, a_2, a_3 \rangle \cdot \langle a_2 b_3 - a_3 b_2, a_3 b_1 - a_1 b_3, a_1 b_2 - a_2 b_1 \rangle$
$= a_1 a_2 b_3 - a_1 a_3 b_2 + a_2 a_3 b_1 - a_2 a_1 b_3 + a_3 a_1 b_2 - a_3 a_2 b_1$
$= 0.$

Similarly, $\mathbf{a} \times \mathbf{b}$ is orthogonal to $\mathbf{b}$. At the end of this section it is shown that the magnitude of $\mathbf{a} \times \mathbf{b}$, as defined in Definition (11.22), is the area of the parallelogram formed by $\mathbf{a}$ and $\mathbf{b}$. Exercise 40 considers the problem of showing that Definition (11.22) gives the direction specified by the right-hand rule.

The following results concerning the vector product can be verified using Definition (11.22). The verification is considered in the exercises.

(11.23)
THEOREM
If $\mathbf{a}$, $\mathbf{b}$, and $\mathbf{c}$ are vectors and α is a real number, then:

i) $\mathbf{a} \times \mathbf{b} = -(\mathbf{b} \times \mathbf{a})$;   ii) $\mathbf{a} \times \mathbf{0} = \mathbf{0}$;
iii) $\mathbf{a} \times \mathbf{a} = \mathbf{0}$;
iv) $\alpha(\mathbf{a} \times \mathbf{b}) = (\alpha \mathbf{a}) \times \mathbf{b} = \mathbf{a} \times (\alpha \mathbf{b})$;
v) $(\mathbf{a} + \mathbf{b}) \times \mathbf{c} = \mathbf{a} \times \mathbf{c} + \mathbf{b} \times \mathbf{c}$;
vi) $\mathbf{a} \times (\mathbf{b} \times \mathbf{c}) = (\mathbf{a} \cdot \mathbf{c})\mathbf{b} - (\mathbf{a} \cdot \mathbf{b})\mathbf{c}$;
vii) $(\mathbf{a} \times \mathbf{b}) \times \mathbf{c} = (\mathbf{a} \cdot \mathbf{c})\mathbf{b} - (\mathbf{b} \cdot \mathbf{c})\mathbf{a}$;
viii) $\mathbf{a} \cdot (\mathbf{b} \times \mathbf{c}) = (\mathbf{a} \times \mathbf{b}) \cdot \mathbf{c}$;
ix) $\|\mathbf{a} \times \mathbf{b}\|^2 = (\|\mathbf{a}\|\|\mathbf{b}\|)^2 - (\mathbf{a} \cdot \mathbf{b})^2$.

From parts (vi) and (vii) of Theorem (11.23) we can see that, in general,

$$\mathbf{a} \times (\mathbf{b} \times \mathbf{c}) \neq (\mathbf{a} \times \mathbf{b}) \times \mathbf{c},$$

since $\mathbf{a} \times (\mathbf{b} \times \mathbf{c})$ is a linear combination of the vectors $\mathbf{b}$ and $\mathbf{c}$, whereas $(\mathbf{a} \times \mathbf{b}) \times \mathbf{c}$ is a linear combination of $\mathbf{a}$ and $\mathbf{b}$.

A simple mnemonic device involving determinant notation can be used to determine the cross product of two vectors. A square array of numbers is an array with the same number of rows and columns. Assigned to each square array is a number called the **determinant.** For an array with two rows and two columns, the determinant is defined by

$$\begin{vmatrix} a & b \\ c & d \end{vmatrix} = ad - bc.$$

The determinant of an array with three rows and three columns is defined by

$$\begin{vmatrix} a & b & c \\ d & e & f \\ g & h & i \end{vmatrix} = a \begin{vmatrix} e & f \\ h & i \end{vmatrix} - b \begin{vmatrix} d & f \\ g & i \end{vmatrix} + c \begin{vmatrix} d & e \\ g & h \end{vmatrix}$$
$$= a(ei - fh) - b(di - fg) + c(dh - eg).$$

Since $\mathbf{a} \times \mathbf{b} = (a_2 b_3 - a_3 b_2)\mathbf{i} - (a_1 b_3 - a_3 b_1)\mathbf{j} + (a_1 b_2 - a_2 b_1)\mathbf{k}$, this product can be written

(11.24) $$\mathbf{a} \times \mathbf{b} = \begin{vmatrix} a_2 & a_3 \\ b_2 & b_3 \end{vmatrix} \mathbf{i} - \begin{vmatrix} a_1 & a_3 \\ b_1 & b_3 \end{vmatrix} \mathbf{j} + \begin{vmatrix} a_1 & a_2 \\ b_1 & b_2 \end{vmatrix} \mathbf{k}.$$

If we extend the determinant notation to permit the first row of the array to consist of the unit vectors $\mathbf{i}, \mathbf{j}$, and $\mathbf{k}$, then the cross product of $\mathbf{a}$ and $\mathbf{b}$ is

(11.25) $$\mathbf{a} \times \mathbf{b} = \begin{vmatrix} \mathbf{i} & \mathbf{j} & \mathbf{k} \\ a_1 & a_2 & a_3 \\ b_1 & b_2 & b_3 \end{vmatrix}.$$

Example 3

Find the vector product $\langle 3, 0, 4 \rangle \times \langle 1, 2, 5 \rangle$.

Solution

$$\langle 3, 0, 4 \rangle \times \langle 1, 2, 5 \rangle = \begin{vmatrix} \mathbf{i} & \mathbf{j} & \mathbf{k} \\ 3 & 0 & 4 \\ 1 & 2 & 5 \end{vmatrix}$$
$$= (0 \cdot 5 - 4 \cdot 2)\mathbf{i} - (3 \cdot 5 - 4 \cdot 1)\mathbf{j} + (3 \cdot 2 - 0 \cdot 1)\mathbf{k}$$
$$= -8\mathbf{i} - 11\mathbf{j} + 6\mathbf{k} = \langle -8, -11, 6 \rangle \qquad \blacktriangleright$$

An alternative method for determining cross products is first to note that the coordinate-axis unit vectors $\mathbf{i}, \mathbf{j}$, and $\mathbf{k}$ satisfy:

(11.26) $$\begin{aligned} \mathbf{i} \times \mathbf{j} &= -(\mathbf{j} \times \mathbf{i}) = \mathbf{k}, \\ \mathbf{j} \times \mathbf{k} &= -(\mathbf{k} \times \mathbf{j}) = \mathbf{i}, \\ \mathbf{k} \times \mathbf{i} &= -(\mathbf{i} \times \mathbf{k}) = \mathbf{j}, \end{aligned}$$

and

(11.27) $$\mathbf{i} \times \mathbf{i} = \mathbf{j} \times \mathbf{j} = \mathbf{k} \times \mathbf{k} = \mathbf{0}.$$

The distributive property (v) in Theorem (11.23) can be used then to reduce an arbitrary cross product to one involving only $\mathbf{i}, \mathbf{j}$, and $\mathbf{k}$.

Example 4

Find a unit vector perpendicular to both vectors $\mathbf{a} = \langle 3, 2, 4 \rangle$ and $\mathbf{b} = \langle 0, 3, -1 \rangle$.

Solution

By definition, a vector perpendicular to both $\mathbf{a}$ and $\mathbf{b}$ is $\mathbf{a} \times \mathbf{b}$, which we find using (11.26) and (11.27):

$$\begin{aligned} \mathbf{a} \times \mathbf{b} &= (3\mathbf{i} + 2\mathbf{j} + 4\mathbf{k}) \times (3\mathbf{j} - \mathbf{k}) \\ &= (3\mathbf{i} + 2\mathbf{j} + 4\mathbf{k}) \times 3\mathbf{j} - (3\mathbf{i} + 2\mathbf{j} + 4\mathbf{k}) \times \mathbf{k} \\ &= 9(\mathbf{i} \times \mathbf{j}) + 6(\mathbf{j} \times \mathbf{j}) + 12(\mathbf{k} \times \mathbf{j}) - 3(\mathbf{i} \times \mathbf{k}) - 2(\mathbf{j} \times \mathbf{k}) - 4(\mathbf{k} \times \mathbf{k}) \\ &= 9\mathbf{k} - 12\mathbf{i} + 3\mathbf{j} - 2\mathbf{i} = -14\mathbf{i} + 3\mathbf{j} + 9\mathbf{k} = \langle -14, 3, 9 \rangle. \end{aligned}$$

The unit vector in the direction of $\mathbf{a} \times \mathbf{b}$ is

$$\frac{\mathbf{a} \times \mathbf{b}}{\|\mathbf{a} \times \mathbf{b}\|} = \frac{\langle -14, 3, 9 \rangle}{\sqrt{(-14)^2 + 3^2 + 9^2}} = \frac{\langle -14, 3, 9 \rangle}{\sqrt{286}}.$$

Another unit vector perpendicular to both $\mathbf{a}$ and $\mathbf{b}$ is

$$\frac{\mathbf{b} \times \mathbf{a}}{\|\mathbf{b} \times \mathbf{a}\|} = \frac{-(\mathbf{a} \times \mathbf{b})}{\|\mathbf{a} \times \mathbf{b}\|} = \frac{\langle 14, -3, -9 \rangle}{\sqrt{286}}. \quad \blacktriangleright$$

Part (ix) of Theorem (11.23) can be used to obtain a result about the length of the vector $\mathbf{a} \times \mathbf{b}$ similar to the result in Theorem (11.11) concerning the dot product of $\mathbf{a}$ and $\mathbf{b}$.

(11.28)
COROLLARY
If $\mathbf{a}$ and $\mathbf{b}$ are vectors and θ is the angle between $\mathbf{a}$ and $\mathbf{b}$, then

$$\|\mathbf{a} \times \mathbf{b}\| = \|\mathbf{a}\| \|\mathbf{b}\| \sin \theta.$$

Proof. It follows from Theorem (11.23ix) and Theorem (11.11) that

$$\begin{aligned}
\|\mathbf{a} \times \mathbf{b}\|^2 &= (\|\mathbf{a}\| \|\mathbf{b}\|)^2 - (\mathbf{a} \cdot \mathbf{b})^2 \\
&= (\|\mathbf{a}\| \|\mathbf{b}\|)^2 - (\|\mathbf{a}\| \|\mathbf{b}\| \cos \theta)^2 \\
&= (\|\mathbf{a}\| \|\mathbf{b}\|)^2 [1 - (\cos \theta)^2] \\
&= (\|\mathbf{a}\| \|\mathbf{b}\|)^2 (\sin \theta)^2.
\end{aligned}$$

Since θ is in $[0, \pi]$, $\sin \theta \geq 0$, which implies that

$$\|\mathbf{a} \times \mathbf{b}\| = \|\mathbf{a}\| \|\mathbf{b}\| \sin \theta. \quad \triangleright$$

An important consequence of this corollary characterizes parallel vectors.

(11.29)
COROLLARY
Vectors $\mathbf{a}$ and $\mathbf{b}$ are parallel if and only if $\mathbf{a} \times \mathbf{b} = \mathbf{0}$.

Proof. If $\mathbf{b} = \mathbf{0}$, then clearly $\mathbf{a} \times \mathbf{b} = \mathbf{0}$. If $\mathbf{b} \neq \mathbf{0}$ and $\mathbf{a}$ and $\mathbf{b}$ are parallel, then $\mathbf{a} = \alpha \mathbf{b}$ for some real number α and, again,

$$\mathbf{a} \times \mathbf{b} = (\alpha \mathbf{b}) \times \mathbf{b} = \alpha(\mathbf{b} \times \mathbf{b}) = \alpha \mathbf{0} = \mathbf{0}.$$

Conversely, if $\mathbf{a} \times \mathbf{b} = \mathbf{0}$, then $\|\mathbf{a} \times \mathbf{b}\| = 0$. Applying Corollary (11.28), we have

$$0 = \|\mathbf{a} \times \mathbf{b}\| = \|\mathbf{a}\| \|\mathbf{b}\| \sin \theta,$$

where θ is the angle between $\mathbf{a}$ and $\mathbf{b}$. Either $\mathbf{a} = \mathbf{0}$, $\mathbf{b} = \mathbf{0}$, or $\sin \theta = 0$ (which implies that $\theta = 0$ or that $\theta = \pi$). In any case, $\mathbf{a}$ is parallel to $\mathbf{b}$. $\quad \triangleright$

With the result of Corollary (11.28), it can be shown that the magnitude of $\|\mathbf{a} \times \mathbf{b}\|$ is the area of the parallelogram determined by $\mathbf{a}$ and $\mathbf{b}$. Recall that this is needed to show that Definition (11.22) agrees with the geometric description of $\mathbf{a} \times \mathbf{b}$.

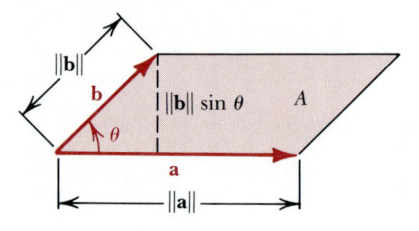

Figure 11.35

From Figure 11.35, we see that the area of the parallelogram determined by the vectors **a** and **b** is

$$A = (\text{length of base})(\text{altitude}) = \|\mathbf{a}\|(\|\mathbf{b}\| \sin \theta),$$

so

(11.30) $$A = \|\mathbf{a} \times \mathbf{b}\|.$$

This fact can be used to derive a formula for determining the volume of a parallelepiped. Suppose a parallelepiped S has three adjacent edges described by the vectors **a**, **b**, and **c**, as shown in Figure 11.36(a). The volume of S is the product of the area A of its base and its altitude h. If the base is determined by the vectors **b** and **c**, then by (11.30), $A = \|\mathbf{b} \times \mathbf{c}\|$. In addition (see Figure 11.36b), $\mathbf{b} \times \mathbf{c}$ is perpendicular to the base, so the altitude is $h = |\text{comp}_{\mathbf{b} \times \mathbf{c}} \mathbf{a}|$. Thus the volume of S is

$$V = \|\mathbf{b} \times \mathbf{c}\| |\text{comp}_{\mathbf{b} \times \mathbf{c}} \mathbf{a}| = \|\mathbf{b} \times \mathbf{c}\| \left| \frac{\mathbf{a} \cdot (\mathbf{b} \times \mathbf{c})}{\|\mathbf{b} \times \mathbf{c}\|} \right|,$$

so

(11.31) $$V = |\mathbf{a} \cdot (\mathbf{b} \times \mathbf{c})|$$

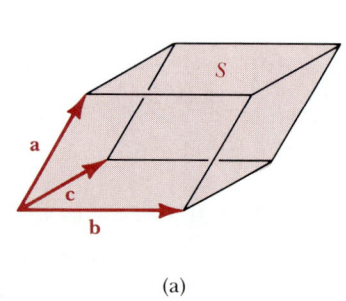

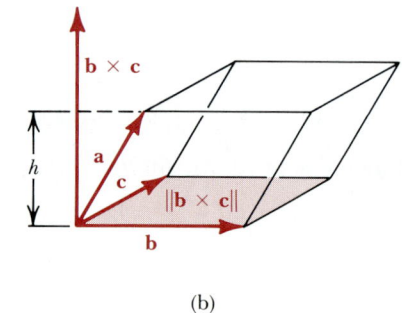

Figure 11.36

The volume of the parallelepiped is $|\mathbf{a} \cdot (\mathbf{b} \times \mathbf{c})|$.

The product $\mathbf{a} \cdot (\mathbf{b} \times \mathbf{c})$ is called a **scalar triple product** since it produces a scalar as a product of the three vectors **a**, **b**, and **c**. This scalar triple product can be rewritten by first recalling that

$$\mathbf{b} \times \mathbf{c} = \begin{vmatrix} \mathbf{i} & \mathbf{j} & \mathbf{k} \\ b_1 & b_2 & b_3 \\ c_1 & c_2 & c_3 \end{vmatrix} = \mathbf{i} \begin{vmatrix} b_2 & b_3 \\ c_2 & c_3 \end{vmatrix} - \mathbf{j} \begin{vmatrix} b_1 & b_3 \\ c_1 & c_3 \end{vmatrix} + \mathbf{k} \begin{vmatrix} b_1 & b_2 \\ c_1 & c_2 \end{vmatrix}.$$

Since the dot product of **a** with $\mathbf{b} \times \mathbf{c}$ is the sum of the products of these two vectors, $\mathbf{a} \cdot (\mathbf{b} \times \mathbf{c})$ can be expressed as

$$\mathbf{a} \cdot (\mathbf{b} \times \mathbf{c}) = a_1 \begin{vmatrix} b_2 & b_3 \\ c_2 & c_3 \end{vmatrix} - a_2 \begin{vmatrix} b_1 & b_3 \\ c_1 & c_3 \end{vmatrix} + a_3 \begin{vmatrix} b_1 & b_2 \\ c_1 & c_2 \end{vmatrix},$$

that is,

(11.32) $$\mathbf{a} \cdot (\mathbf{b} \times \mathbf{c}) = \begin{vmatrix} a_1 & a_2 & a_3 \\ b_1 & b_2 & b_3 \\ c_1 & c_2 & c_3 \end{vmatrix}.$$

Example 5

Determine the volume of the parallelepiped with adjacent vertices at $(0, 0, 0)$, $(-1, 1, 3)$, $(0, 4, 0)$, and $(-3, 2, 1)$.

Solution

The parallelepiped is shown in Figure 11.37, where the adjacent edges are given by the position vectors $\mathbf{a} = \langle -1, 1, 3 \rangle$, $\mathbf{b} = \langle 0, 4, 0 \rangle$, and $\mathbf{c} = \langle -3, 2, 1 \rangle$. The volume is the absolute value of the scalar triple product, which by (11.32) is

$$V = |\mathbf{a} \cdot (\mathbf{b} \times \mathbf{c})| = \left| \begin{vmatrix} -1 & 1 & 3 \\ 0 & 4 & 0 \\ -3 & 2 & 1 \end{vmatrix} \right|$$

$$= \left| (-1) \begin{vmatrix} 4 & 0 \\ 2 & 1 \end{vmatrix} - 1 \begin{vmatrix} 0 & 0 \\ -3 & 1 \end{vmatrix} + 3 \begin{vmatrix} 0 & 4 \\ -3 & 2 \end{vmatrix} \right|$$

$$= |-(4 - 0) - (0 - 0) + 3(0 + 12)| = 32. \quad \blacktriangleright$$

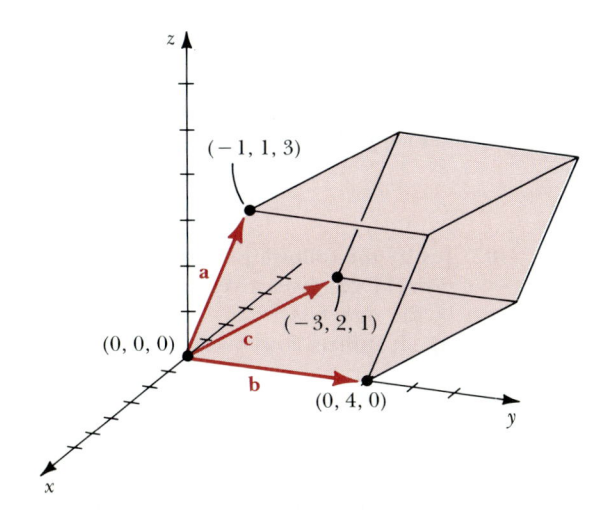

Figure 11.37

▶ EXERCISE SET 11.4

In Exercises 1–4, find $\mathbf{a} \times \mathbf{b}$ for the vectors.

1. $\mathbf{a} = \langle -1, 3, 5 \rangle$, $\mathbf{b} = \langle 2, -1, 0 \rangle$
2. $\mathbf{a} = \langle 0, 1, 1 \rangle$, $\mathbf{b} = \langle 5, 2, -3 \rangle$
3. $\mathbf{a} = -2\mathbf{i} + 3\mathbf{k}$, $\mathbf{b} = 4\mathbf{i} + \mathbf{j}$
4. $\mathbf{a} = 2\mathbf{i} + \mathbf{j} - 3\mathbf{k}$, $\mathbf{b} = -\mathbf{i} - \mathbf{j} + \mathbf{k}$

In Exercises 5–16, use the vectors $\mathbf{a} = \langle 1, 2, 0 \rangle$, $\mathbf{b} = \langle 1, 3, 1 \rangle$, $\mathbf{c} = \langle 0, 1, 0 \rangle$, and $\mathbf{d} = \langle 3, -2, 2 \rangle$. Find the result or state why the operation is impossible.

5. $\mathbf{a} \times \mathbf{b} + \mathbf{a} \times \mathbf{c}$
6. $\mathbf{a} \times (\mathbf{b} + \mathbf{c})$
7. $(\mathbf{a} \times \mathbf{b}) \times \mathbf{c}$
8. $(\mathbf{c} \times \mathbf{b}) \times \mathbf{a}$
9. $(\mathbf{a} \times \mathbf{b}) \cdot \mathbf{c}$
10. $\mathbf{a} \cdot (\mathbf{b} \times \mathbf{c})$
11. $(\mathbf{a} \cdot \mathbf{b}) \times \mathbf{c}$
12. $(\mathbf{a} \times \mathbf{b}) \cdot (\mathbf{c} \times \mathbf{d})$
13. $(\mathbf{a} \times \mathbf{d}) \cdot (\mathbf{c} \times \mathbf{b})$
14. $(\mathbf{a} \cdot \mathbf{b}) \times (\mathbf{c} \cdot \mathbf{d})$
15. $\dfrac{\mathbf{a} \times \mathbf{b}}{\mathbf{c} \cdot \mathbf{d}}$
16. $\dfrac{\mathbf{a} \cdot \mathbf{b}}{\mathbf{c} \times \mathbf{d}}$

17. Use the cross product to find a vector orthogonal to both $\mathbf{a} = \langle 1, 2, 4 \rangle$ and $\mathbf{b} = \langle 2, -2, 5 \rangle$. Compare this method to the method required in Exercise 37 of Section 11.3. Which method do you prefer?

18. Find a vector orthogonal to both $\mathbf{a} = \langle 3, -4, 0 \rangle$ and $\mathbf{b} = \langle 0, 2, 5 \rangle$.

19. Find the area of the parallelogram determined by $\langle 4, 3, 0 \rangle$ and $\langle 1, 7, 0 \rangle$.

20. Find the area of the parallelogram determined by $\langle -1, 1, 2 \rangle$ and $\langle 3, 5, 1 \rangle$.

21. Find the area of the triangle with vertices at $(2, 1, 0)$, $(0, 4, 2)$, and $(2, -3, 2)$.

22. Find the area of the triangle with vertices at $(2, 3, 4)$, $(3, 0, 0)$, and $(1, 7, 2)$.

In Exercises 23–26, use the scalar triple product to find the volume of the parallelepiped determined by the vectors.

23. $\mathbf{a} = \langle 2, 0, 0 \rangle$, $\mathbf{b} = \langle 0, 3, 0 \rangle$, $\mathbf{c} = \langle 0, 0, 4 \rangle$

24. $\mathbf{a} = \langle 0, 0, 2 \rangle$, $\mathbf{b} = \langle 3, 0, 0 \rangle$, $\mathbf{c} = \langle 2, 4, 0 \rangle$

25. $\mathbf{a} = \langle 3, -3, 2 \rangle$, $\mathbf{b} = \langle 0, 2, 2 \rangle$, $\mathbf{c} = \langle 2, 0, 0 \rangle$

26. $\mathbf{a} = \langle -1, 1, 4 \rangle$, $\mathbf{b} = \langle 3, 2, 1 \rangle$, $\mathbf{c} = \langle 1, 0, -1 \rangle$

27. A tetrahedron is a pyramid whose surface consists of four triangles (see the figure). Exercise 72 in Section 5.2 implies that the volume of a tetrahedron is: (area of a base) (altitude)/3. Show that the volume of the tetrahedron formed by the vectors $\mathbf{a}$, $\mathbf{b}$, and $\mathbf{c}$ shown in the figure is $|\mathbf{a} \cdot (\mathbf{b} \times \mathbf{c})|/6$.

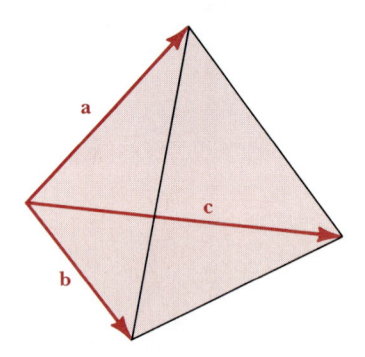

28–31. Use the result in Exercise 27 to find the volume of the tetrahedron determined by the vectors $\mathbf{a}$, $\mathbf{b}$, and $\mathbf{c}$ given in Exercises 23–26.

32. Use Definition (11.22) to show that $\mathbf{a} \times \mathbf{b} = -(\mathbf{b} \times \mathbf{a})$ for any vectors $\mathbf{a}$ and $\mathbf{b}$.

33. Show that for any vector $\mathbf{a}$, $\mathbf{a} \times \mathbf{a} = \mathbf{0}$.

34. Show that for any vectors $\mathbf{a}$, $\mathbf{b}$, and $\mathbf{c}$, $(\mathbf{a} \times \mathbf{b}) \cdot \mathbf{c} = \mathbf{a} \cdot (\mathbf{b} \times \mathbf{c})$.

35. Show that for any pair of vectors $\mathbf{a}$ and $\mathbf{b}$,
$$(\mathbf{a} + \mathbf{b}) \times (\mathbf{a} - \mathbf{b}) = 2(\mathbf{b} \times \mathbf{a}).$$

36. Show that for any pair of vectors $\mathbf{a}$ and $\mathbf{b}$,
$$\|\mathbf{a} \times \mathbf{b}\| = [\|\mathbf{a}\|^2 \|\mathbf{b}\|^2 - (\mathbf{a} \cdot \mathbf{b})^2]^{1/2}.$$

37. Use Definition 11.22 to prove part (vi) of Theorem 11.23: For any vectors $\mathbf{a}$, $\mathbf{b}$, and $\mathbf{c}$,
$$\mathbf{a} \times (\mathbf{b} \times \mathbf{c}) = (\mathbf{a} \cdot \mathbf{c})\mathbf{b} - (\mathbf{a} \cdot \mathbf{b})\mathbf{c}.$$

38. Use parts (i) and (vi) of Theorem 11.23 to prove part (vii) of the theorem: For any vectors $\mathbf{a}$, $\mathbf{b}$, and $\mathbf{c}$,
$$(\mathbf{a} \times \mathbf{b}) \times \mathbf{c} = (\mathbf{a} \cdot \mathbf{c})\mathbf{b} - (\mathbf{b} \cdot \mathbf{c})\mathbf{a}.$$

39. Suppose $\mathbf{a} \times \mathbf{b} = \mathbf{c} \times \mathbf{b}$, $\mathbf{a} \cdot \mathbf{d} = 0$, and $\mathbf{b} \cdot \mathbf{d} = 0$. Show that
$$\mathbf{a} = \mathbf{c} - \frac{\mathbf{c} \cdot \mathbf{d}}{\mathbf{b} \cdot \mathbf{d}} \mathbf{b}.$$

40. Justify the following statements to show that Definition (11.22) gives a vector with direction prescribed by the *right-hand rule*.

a) It suffices to show that the result holds when:
$\mathbf{a} = \langle a_1, 0, 0 \rangle$, for some constant $a_1 \geq 0$ and
$\mathbf{b} = \langle b_1, b_2, 0 \rangle$, where b_1 is arbitrary but $b_2 \geq 0$.

b) For $\mathbf{a}$ and $\mathbf{b}$ as described in part (a), $\mathbf{a} \times \mathbf{b}$ has the same direction as the vector $\mathbf{k}$.

11.5

Planes

The basic geometric object in three-dimensional space is a surface and the most elementary surfaces are **planes.** To describe a plane completely, it suffices to specify one point on the plane and the direction orthogonal to all line segments lying on the plane. A vector describing this direction is called a **normal vector** to the plane. Since only the *direction* of a normal vector is important, any nonzero vector parallel to a normal vector is also a normal vector to the plane.

11.5 PLANES

Suppose $P_0(x_0, y_0, z_0)$ is a point on a plane $\mathcal{P}$ and $\mathbf{n} = \langle a, b, c \rangle$ is a normal vector to $\mathcal{P}$, as illustrated in Figure 11.38. A point $P(x, y, z)$ lies on the plane $\mathcal{P}$ precisely when $\mathbf{n}$ is orthogonal to the vector $\overrightarrow{P_0P} = \langle x - x_0, y - y_0, z - z_0 \rangle$, that is, when

(11.33) $$\mathbf{n} \cdot \overrightarrow{P_0P} = 0,$$

so

$$\langle a, b, c \rangle \cdot \langle x - x_0, y - y_0, z - z_0 \rangle = 0.$$

Thus the plane with normal vector $\mathbf{n} = \langle a, b, c \rangle$ that contains the point $P_0(x_0, y_0, z_0)$ has equation

(11.34) $$a(x - x_0) + b(y - y_0) + c(z - z_0) = 0,$$

or

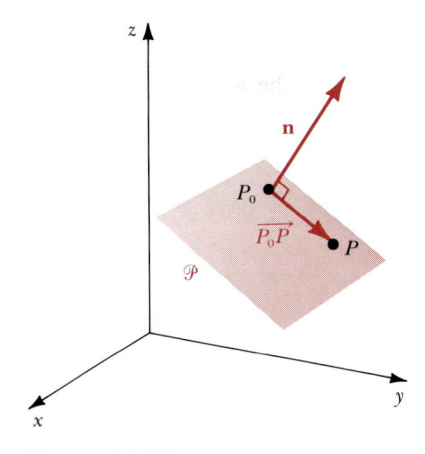

$$ax + by + cz = d, \quad \text{where} \quad d = ax_0 + by_0 + cz_0.$$

Figure 11.38
P is on the plane: $\mathbf{n} \cdot \overrightarrow{P_0P} = 0$.

Example 1

Find an equation of the plane containing the point $(2, -1, 3)$ that has normal vector $\langle 3, 6, 2 \rangle$. Sketch the graph of this equation.

Solution

If (x, y, z) is a point on this plane, then the vector $\langle x - 2, y + 1, z - 3 \rangle$ is orthogonal to $\langle 3, 6, 2 \rangle$. So the plane has equation

$$3(x - 2) + 6(y + 1) + 2(z - 3) = 0,$$

which simplifies to

$$3x + 6y + 2z = 6.$$

To sketch the graph of this equation, we first find the points at which the plane intersects the coordinate axes:

when $y = 0$ and $z = 0$, $3x = 6$, so $x = 2$;
when $x = 0$ and $z = 0$, $6y = 6$, so $y = 1$;
when $x = 0$ and $y = 0$, $2z = 6$, so $z = 3$.

Thus the points $(2, 0, 0)$, $(0, 1, 0)$, and $(0, 0, 3)$ lie on the plane. The line segments joining these points also lie on the plane, so the graph in the first octant is as shown in Figure 11.39. The plane extends linearly in all directions from line segments in this triangular segment. ▶

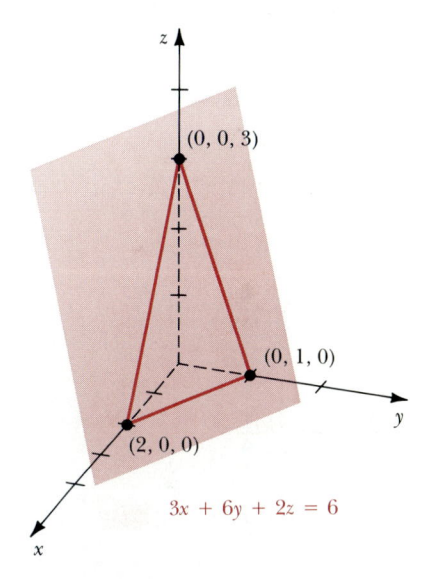

Figure 11.39

We have seen that a plane with normal vector $\mathbf{n} = \langle a, b, c \rangle$ and passing through (x_0, y_0, z_0) has equation $ax + by + cz = d$, where $d = ax_0 + by_0 + cz_0$. On the other hand, if a, b, and c are not all zero, then the graph of an equation of the form

(11.35) $$ax + by + cz = d$$

must be a plane with normal vector $\mathbf{n} = \langle a, b, c \rangle$. To see this, suppose

$a \neq 0$. Equation (11.35) can be written as

$$0 = a\left(x - \frac{d}{a}\right) + b(y - 0) + c(z - 0) = \mathbf{n} \cdot \left\langle x - \frac{d}{a}, y - 0, z - 0\right\rangle,$$

which is an equation of the plane that contains the point $(d/a, 0, 0)$ and has normal vector $\mathbf{n} = \langle a, b, c \rangle$.

An equation of the form (11.35) is called a *linear equation in three variables*. The two-dimensional analogue to (11.35) is the equation of a straight line given in Section 1.2.

The close connection between planes and their normal vectors permits us to make the following definition.

(11.36) DEFINITION

Suppose $\mathcal{P}_1$ and $\mathcal{P}_2$ are planes with normal vectors $\mathbf{n}_1$ and $\mathbf{n}_2$, respectively.

i) $\mathcal{P}_1$ is said to be **orthogonal** to $\mathcal{P}_2$ precisely when $\mathbf{n}_1$ is orthogonal to $\mathbf{n}_2$.
ii) $\mathcal{P}_1$ is said to be **parallel** to $\mathcal{P}_2$ precisely when $\mathbf{n}_1$ is parallel to $\mathbf{n}_2$.

Theorem (11.37) follows directly from Corollaries (11.13) and (11.29), which give results about orthogonal and parallel vectors.

(11.37) THEOREM

Suppose $\mathcal{P}_1$ and $\mathcal{P}_2$ are planes with normal vectors $\mathbf{n}_1$ and $\mathbf{n}_2$, respectively.

i) $\mathcal{P}_1$ is orthogonal to $\mathcal{P}_2$ if and only if $\mathbf{n}_1 \cdot \mathbf{n}_2 = 0$.
ii) $\mathcal{P}_1$ is parallel to $\mathcal{P}_2$ if and only if $\mathbf{n}_1 = \alpha \mathbf{n}_2$, for some constant α, which occurs if and only if $\mathbf{n}_1 \times \mathbf{n}_2 = \mathbf{0}$.

Since $\mathbf{k} = \langle 0, 0, 1 \rangle$ is normal to the xy-plane, part (i) of Theorem (11.37) implies that any plane orthogonal to the xy-plane has a normal vector $\mathbf{n} = \langle a, b, c \rangle$ with $\mathbf{n} \cdot \mathbf{k} = 0$. Thus, $c = 0$ and $\mathbf{n} = \langle a, b, 0 \rangle$. Therefore, any plane orthogonal to the xy-plane has equation of the form

$$ax + by = d.$$

For example, Figure 11.40(a) shows the graph of the plane with equation $x + y = 2$.

Similarly, a plane orthogonal to the yz-plane has equation of the form

$$by + cz = d,$$

and a plane orthogonal to the xz-plane has equation of the form

$$ax + cz = d.$$

Planes parallel to the coordinate planes are described by an equation in just one variable. For example, the equation of the yz-plane is $x = 0$,

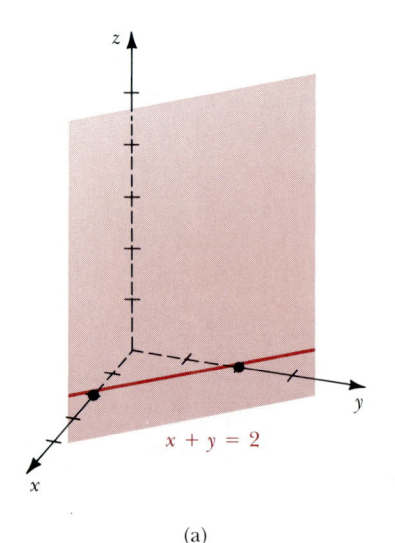

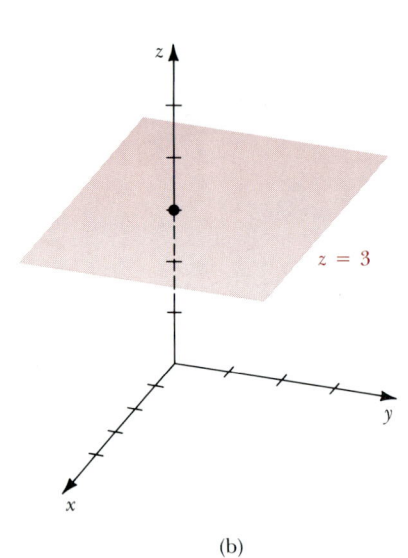

Figure 11.40

and the equation $z = 3$ describes the plane parallel to and three units above the xy-plane shown in Figure 11.40(b).

Any three points in space that are not collinear (do not lie on the same straight line) determine a unique plane. An equation of this plane can be determined by solving a system of three linear equations. An easier method uses the fact that the line segments joining these points also lie in the plane and that these segments describe vectors. A normal to the plane is the cross product of two such vectors.

Example 2

Find an equation of the plane containing the points $(2, 0, 1)$, $(0, 6, -2)$, and $(-2, 3, 1)$.

Solution

The vector with initial point $(2, 0, 1)$ and terminal point $(0, 6, -2)$ is

$$\mathbf{v}_1 = \langle 0 - 2, 6 - 0, -2 - 1 \rangle = \langle -2, 6, -3 \rangle,$$

and the vector with the same initial point and terminal point $(-2, 3, 1)$ is

$$\mathbf{v}_2 = \langle -2 - 2, 3 - 0, 1 - 1 \rangle = \langle -4, 3, 0 \rangle.$$

Vectors $\mathbf{v}_1$ and $\mathbf{v}_2$ lie in the plane, so a normal to the plane is

$$\mathbf{v}_1 \times \mathbf{v}_2 = \begin{vmatrix} \mathbf{i} & \mathbf{j} & \mathbf{k} \\ -2 & 6 & -3 \\ -4 & 3 & 0 \end{vmatrix} = 9\mathbf{i} + 12\mathbf{j} + 18\mathbf{k}.$$

Thus the plane has equation of the form

$$9x + 12y + 18z = d.$$

To determine d, we use the fact that the point $(2, 0, 1)$ lies on the plane. Thus,

$$d = 9 \cdot 2 + 12 \cdot 0 + 18 \cdot 1 = 36,$$

and an equation is

$$9x + 12y + 18z = 36 \quad \text{or} \quad 3x + 4y + 6z = 12.$$

The graph of this plane is shown in Figure 11.41. ▶

The final result in this section shows how the dot product is used to find the distance from a plane to a point not on the plane.

(11.38) THEOREM

Suppose $\mathscr{P}$ is a plane with equation $ax + by + cz = d$ and $P_0(x_0, y_0, z_0)$ is a point that does not lie on $\mathscr{P}$. The shortest distance from P_0 to the plane $\mathscr{P}$ is given by

$$d(P_0, \mathscr{P}) = \frac{|ax_0 + by_0 + cz_0 - d|}{\sqrt{a^2 + b^2 + c^2}}.$$

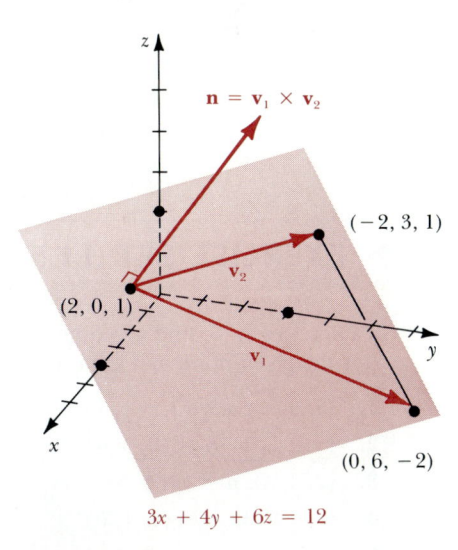

Figure 11.41

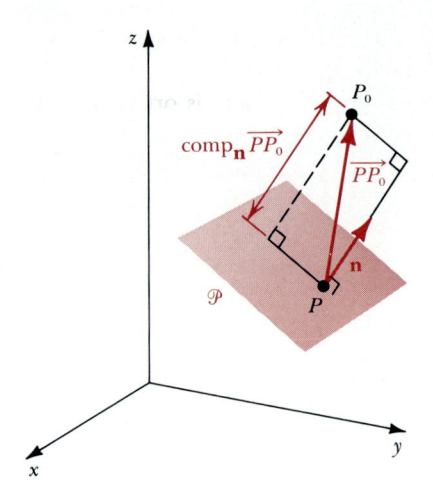

Figure 11.42
The distance from the point P_0 to the plane $\mathscr{P}$ is $\text{comp}_\mathbf{n} \overrightarrow{PP_0}$.

Proof. Let $P(x, y, z)$ be an arbitrary point on the plane and consider the vector $\overrightarrow{PP_0}$ from P to P_0:

$$\overrightarrow{PP_0} = \langle x_0 - x, y_0 - y, z_0 - z \rangle.$$

The shortest distance from P_0 to $\mathscr{P}$ is the absolute value of the component of $\overrightarrow{PP_0}$ in the direction of the normal $\mathbf{n} = \langle a, b, c \rangle$ to the plane $\mathscr{P}$ (see Figure 11.42). This distance is

$$|\text{comp}_\mathbf{n} \overrightarrow{PP_0}| = \frac{|\langle x_0 - x, y_0 - y, z_0 - z \rangle \cdot \langle a, b, c \rangle|}{\|\langle a, b, c \rangle\|}$$

$$= \frac{|a(x_0 - x) + b(y_0 - y) + c(z_0 - z)|}{\sqrt{a^2 + b^2 + c^2}}$$

$$= \frac{|ax_0 + by_0 + cz_0 - (ax + by + cz)|}{\sqrt{a^2 + b^2 + c^2}}.$$

But $P(x, y, z)$ is on the plane $\mathscr{P}$, so $ax + by + cz = d$ and

$$d(P_0, \mathscr{P}) = \frac{|ax_0 + by_0 + cz_0 - d|}{\sqrt{a^2 + b^2 + c^2}}.$$

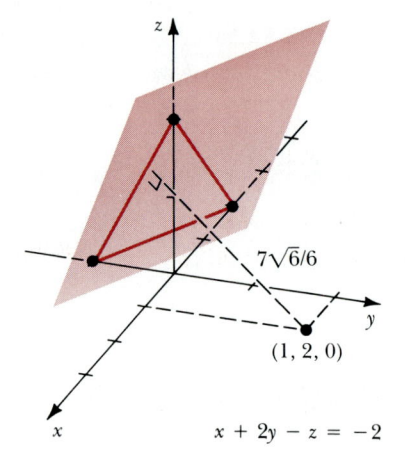

Example 3

Find the distance from the point $P(1, 2, 0)$ to the plane with equation $x + 2y - z = -2$.

Solution
By Theorem (11.38), this distance, shown in Figure 11.43, is

$$\frac{|1(1) + 2(2) - 1(0) - (-2)|}{\sqrt{1^2 + 2^2 + (-1)^2}} = \frac{7}{\sqrt{6}}$$

$$= \frac{7\sqrt{6}}{6} \approx 2.86.$$

Figure 11.43

▶ EXERCISE SET 11.5

In Exercises 1–6, find an equation of the plane that contains the given point and has normal vector $\mathbf{n}$.

1. $(3, 3, 2)$, $\mathbf{n} = \langle 1, -1, 1 \rangle$
2. $(-1, 2, 1)$, $\mathbf{n} = \langle 2, 3, 4 \rangle$
3. $(2, 3, 4)$, $\mathbf{n} = \langle 0, 0, 3 \rangle$
4. $(2, -3, 1)$, $\mathbf{n} = \langle 1, 1, 0 \rangle$
5. $(0, 0, 0)$, $\mathbf{n} = \langle 2, 3, 4 \rangle$
6. $(0, 0, 2)$, $\mathbf{n} = \langle 0, 1, 1 \rangle$

In Exercises 7–14, sketch the graph of the equation.

7. $2x + 3y + 4z = 12$
8. $2x + y - z - 4 = 0$
9. $x = 2$
10. $z = 3$
11. $2x + 3z = 4$
12. $2y - 3z = 4$
13. $z = 0$
14. $x + y - 4 = 0$

In Exercises 15–28, find an equation of a plane that satisfies the conditions stated.

15. Is parallel to and a distance of 3 units above the xy-plane
16. Is parallel to and a distance of 2 units from the yz-plane
17. Contains the point $(1, 0, -1)$ and is parallel to the plane $x + y - z = 4$
18. Contains the point $(2, 3, 4)$ and is parallel to the plane $x + y + z = 1$
19. Contains the point $(1, 2, 3)$ and is parallel to the xy-plane
20. Contains the point $(2, 3, 4)$ and is parallel to the yz-plane
21. Contains the points $(1, 2, 3)$ and $(0, 1, 1)$ and is orthogonal to the xy-plane
22. Contains the points $(-1, 2, -3)$ and $(5, 0, 4)$ and is orthogonal to the xz-plane
23. Contains the points $(1, -1, 4)$, $(0, 2, 3)$, and $(2, 1, 0)$
24. Contains the points $(3, 2, -1)$, $(2, 3, 5)$, and $(-1, -3, 4)$
25. Contains the points $(1, 0, -1)$ and $(2, 1, 3)$ and is orthogonal to $2x - y + 3z = 6$
26. Contains the points $(4, -2, 0)$ and $(2, -1, 2)$ and is orthogonal to $6x + y + z = 15$
27. Contains the point $(1, 2, 1)$ and is orthogonal to both $x + y + z = 1$ and $x + 2y + 3z = 6$
28. Contains the point $(1, -1, 4)$ and is orthogonal to $x - 2y + z = 2$ and $2x + 2y + z = 1$

In Exercises 29–32, find the distance from the point to the plane.

29. $(0, 0, 0)$, $2x + 3y + 2z = 6$
30. $(1, -1, 3)$, $z = 0$
31. $(1, -2, 3)$, $x + z = 1$
32. $(1, 5, 4)$, $x + y + 2z = 2$

In Exercises 33–36, find the distance between the parallel planes.

33. $x - y + 2z = 2$, $x - y + 2z = -2$
34. $2x + y + 3z = 6$, $2x + y + 3z = 1$
35. $2x - 3y + z = 3$, $4x - 6y + 2z = 9$
36. $x - z = 3$, $x - z = 5$

37. Find an equation that describes the set of all points equidistant from $(3, 1, 1)$ and $(7, 5, 6)$.
38. Show that the plane that intersects the coordinate axes at $(a, 0, 0)$, $(0, b, 0)$, and $(0, 0, c)$ has equation
$$\frac{x}{a} + \frac{y}{b} + \frac{z}{c} = 1,$$
provided a, b, and c are all nonzero.

11.6

Lines in Space

Lines are described by specifying a point and a direction. Suppose the line l passes through the point $P_0(x_0, y_0, z_0)$ and has direction given by the vector $\mathbf{v} = \langle v_1, v_2, v_3 \rangle$, as shown in Figure 11.44. A point $P(x, y, z)$ lies on l if and only if the vector $\overrightarrow{P_0P}$ is parallel to $\mathbf{v}$, that is, if and only if a real number t exists with

(11.39) $$\overrightarrow{P_0P} = t\mathbf{v}.$$

This **vector equation** for the line l is rewritten in terms of the components of the vectors as

$$\langle x - x_0, y - y_0, z - z_0 \rangle = \langle tv_1, tv_2, tv_3 \rangle.$$

Consequently, all points (x, y, z) on the line passing through (x_0, y_0, z_0) in

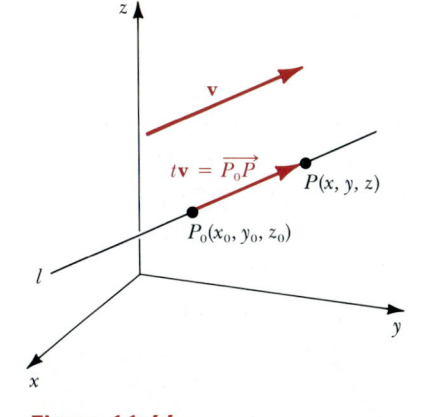

Figure 11.44
P is on the line l: $\overrightarrow{P_0P} = t\mathbf{v}$.

the direction of $\mathbf{v} = \langle v_1, v_2, v_3 \rangle$ are given by

$$x - x_0 = tv_1, \qquad y - y_0 = tv_2, \qquad z - z_0 = tv_3$$

or

(11.40) $\qquad x = x_0 + tv_1, \qquad y = y_0 + tv_2, \qquad z = z_0 + tv_3$

for some real number t.

The numbers v_1, v_2, v_3 are called **direction numbers** of l for the natural reason that they indicate its direction. The set of equations in (11.40) is a set of **parametric equations** for l in the parameter t.

Example 1

Find a set of parametric equations for the line passing through the point $(-1, 1, 3)$ and having direction given by $\mathbf{v} = \langle 4, 4, -2 \rangle$.

Solution

Since (x, y, z) lies on this line if and only if

$$\langle x - (-1), y - 1, z - 3 \rangle = t \langle 4, 4, -2 \rangle$$

for some real number t, a set of parametric equations for the line is

$$x = -1 + 4t, \qquad y = 1 + 4t, \qquad z = 3 - 2t.$$

This line is sketched in Figure 11.45. ▶

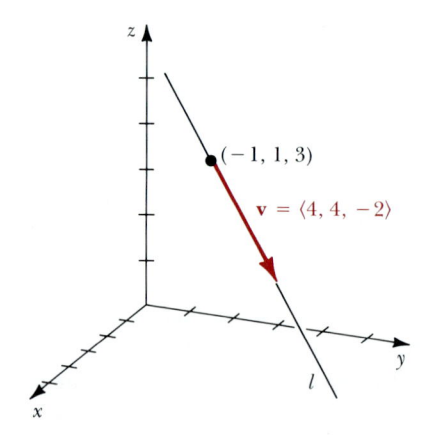

Figure 11.45

Example 2

Find parametric equations for the line passing through the points $(1, 2, 3)$ and $(0, 1, 3)$.

Solution

A vector $\mathbf{v}$ that describes the direction of this line is a vector determined by the given points:

$$\mathbf{v} = \langle 1 - 0, 2 - 1, 3 - 3 \rangle = \langle 1, 1, 0 \rangle.$$

Using $\mathbf{v}$ and the point $(1, 2, 3)$, we obtain the parametric equations

$$x = 1 + t, \qquad y = 2 + t, \qquad z = 3.$$

Since $z = 3$ regardless of the value of t, this line lies in the plane $z = 3$ (see Figure 11.46). ▶

In Example 2, the line lies on a plane parallel to the xy-plane because the direction number of the line in the z-direction is zero. In general, when one of the direction numbers of a line is zero, the line is parallel to a coordinate plane. When two of the direction numbers of the line are zero, the line is parallel to a coordinate axis.

When the direction numbers v_1, v_2, v_3 of a line l are all nonzero, each of the parametric equations

$$x = x_0 + v_1 t, \qquad y = y_0 + v_2 t, \qquad z = z_0 + v_3 t$$

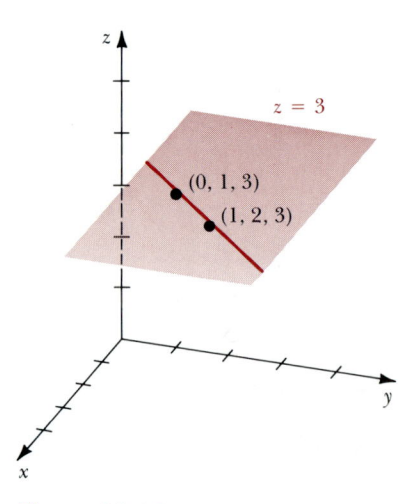

Figure 11.46

can be solved for t:

$$\frac{x - x_0}{v_1} = t, \quad \frac{y - y_0}{v_2} = t, \quad \frac{z - z_0}{v_3} = t.$$

Equating these expressions for t gives **symmetric equations** for the line l:

(11.41) $$\frac{x - x_0}{v_1} = \frac{y - y_0}{v_2} = \frac{z - z_0}{v_3}.$$

Example 3

Find symmetric equations for the line passing through the point $(1, 0, 2)$ and parallel to the line with parametric equations

$$x = 2 + t, \, y = 1 + 3t, \quad \text{and} \quad z = 1 + 4t.$$

Solution

The direction of this line is given by $\mathbf{v} = \langle 1, 3, 4 \rangle$, which also gives the direction of any line parallel to the line. Symmetric equations for the line parallel to the given line and passing through the point $(1, 0, 2)$ are

$$\frac{x - 1}{1} = \frac{y - 0}{3} = \frac{z - 2}{4}. \quad \blacktriangleright$$

Example 4

Find the point of intersection of the xy-plane and the line with symmetric equations

$$\frac{x - 1}{2} = \frac{y + 1}{3} = \frac{z - 2}{-1}.$$

Solution

The z-coordinate of this point of intersection is zero, so

$$\frac{x - 1}{2} = \frac{y + 1}{3} = \frac{0 - 2}{-1} = 2.$$

Solving for x and y, we have

$$\frac{x - 1}{2} = 2, \quad \text{so } x = 5 \quad \text{and} \quad \frac{y + 1}{3} = 2, \quad \text{so } y = 5.$$

Consequently, the point of intersection of the line and the xy-plane is $(5, 5, 0)$, as shown in Figure 11.47. $\blacktriangleright$

Suppose a line has symmetric equations

$$\frac{x - x_0}{v_1} = \frac{y - y_0}{v_2}$$

$$= \frac{z - z_0}{v_3}.$$

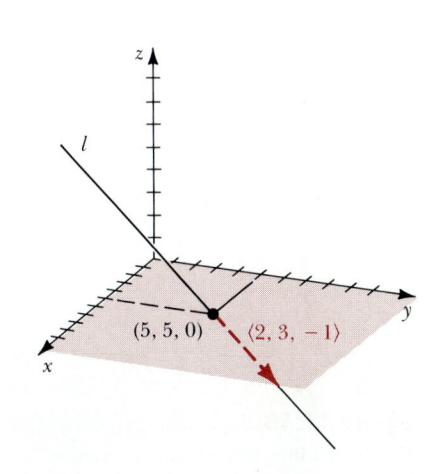

Figure 11.47

By taking the equations in pairs, say

$$\frac{x - x_0}{v_1} = \frac{y - y_0}{v_2}, \quad \text{that is,} \quad v_2 x - v_1 y = v_2 x_0 - v_1 y_0,$$

and

$$\frac{x - x_0}{v_1} = \frac{z - z_0}{v_3}, \quad \text{that is,} \quad v_3 x - v_1 z = v_3 x_0 - v_1 z_0,$$

we see that the line is the intersection of the planes with these equations.

If one of the direction numbers of the line is zero, say $v_3 = 0$, then $z = z_0 + v_3 t$ becomes $z = z_0$, so equations for the line are

$$\frac{x - x_0}{v_1} = \frac{y - y_0}{v_2} \quad \text{and} \quad z = z_0,$$

and the line is again expressed as the intersection of two planes.

Intuitively, we expect that whenever two planes intersect, they either coincide or intersect in a straight line; that is, intersecting planes contain a common straight line. This is shown in the following theorem.

(11.42)
THEOREM
Suppose the planes $\mathcal{P}_1$ and $\mathcal{P}_2$ with normal vectors $\mathbf{n}_1$ and $\mathbf{n}_2$ both contain the point (x_0, y_0, z_0). Then the intersection of the planes contains the line with vector equation

$$\langle x - x_0, y - y_0, z - z_0 \rangle = t(\mathbf{n}_1 \times \mathbf{n}_2).$$

Proof. Figure 11.48 shows the point (x_0, y_0, z_0) common to both planes $\mathcal{P}_1$ and $\mathcal{P}_2$ and the normal vectors $\mathbf{n}_1$ and $\mathbf{n}_2$. A directed line segment containing (x_0, y_0, z_0) lies on $\mathcal{P}_1$ if and only if its direction is orthogonal to $\mathbf{n}_1$, and lies on $\mathcal{P}_2$ if and only if its direction is orthogonal to $\mathbf{n}_2$. This implies that a line segment containing (x_0, y_0, z_0) lies on both $\mathcal{P}_1$ and $\mathcal{P}_2$ precisely when its direction is orthogonal to both $\mathbf{n}_1$ and $\mathbf{n}_2$. Since $\mathbf{n}_1 \times \mathbf{n}_2$ is orthogonal to both $\mathbf{n}_1$ and $\mathbf{n}_2$, any vector orthogonal to both $\mathbf{n}_1$ and $\mathbf{n}_2$ is parallel to $\mathbf{n}_1 \times \mathbf{n}_2$. Consequently, a unique line through

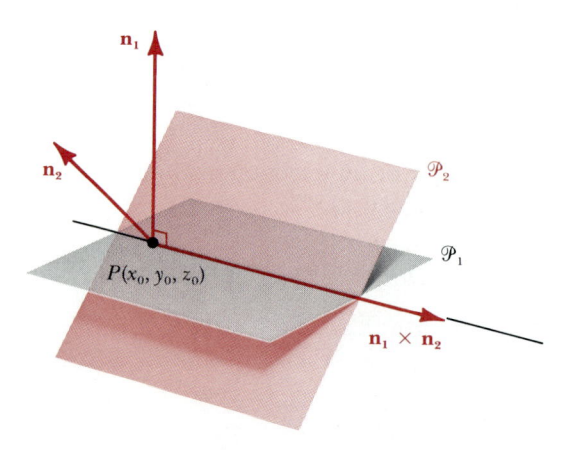

Figure 11.48
Planes that intersect contain a common line.

(x_0, y_0, z_0) exists that is contained in both planes. This line has vector equation

$$\langle x - x_0, y - y_0, z - z_0 \rangle = t(\mathbf{n}_1 \times \mathbf{n}_2).$$ ▷

Example 5

Find parametric and symmetric equations for the line described by the intersection of the planes

$$x + 2y + z = 4 \quad \text{and} \quad 2x - y + 3z = 3.$$

Solution

We first determine a point of intersection of the planes. In particular, we will find a point of intersection that lies on the xy-plane. If no such point exists, we will try to find an intersection point in one of the other coordinate planes. Exercise 35 asks you to show that this procedure is always successful if an intersection point exists.

A point of intersection lying on the xy-plane has the property that $z = 0$, so the equations become

$$x + 2y = 4 \quad \text{and} \quad 2x - y = 3.$$

Solving these equations for x and y, we find that $x = 2$ and $y = 1$. So the point $(2, 1, 0)$ is common to the planes.

The normal vectors to the planes are

$$\mathbf{n}_1 = \langle 1, 2, 1 \rangle \quad \text{and} \quad \mathbf{n}_2 = \langle 2, -1, 3 \rangle,$$

so the line has direction given by

$$\mathbf{n}_1 \times \mathbf{n}_2 = \begin{vmatrix} \mathbf{i} & \mathbf{j} & \mathbf{k} \\ 1 & 2 & 1 \\ 2 & -1 & 3 \end{vmatrix} = (6+1)\mathbf{i} - (3-2)\mathbf{j} + (-1-4)\mathbf{k}$$
$$= 7\mathbf{i} - \mathbf{j} - 5\mathbf{k}.$$

Parametric equations for the line are

$$x = 2 + 7t, \qquad y = 1 - t, \qquad z = -5t$$

and symmetric equations are

$$\frac{x-2}{7} = \frac{y-1}{-1} = \frac{z}{-5}.$$ ▶

An alternative method for solving the problem posed in Example 5 is first to determine two distinct intersection points of the planes. The direction of the line of intersection is then described by the vector between these two points (see Exercise 31).

We now have four ways to describe a line in space:

i) by the vector equation (11.39);
ii) by the parametric equations (11.40);
iii) by the symmetric equations (11.41);
iv) as the intersection of two planes.

Example 6

Determine if the lines l_1 and l_2 intersect and, if so, find a point of intersection.

$$l_1 : x - 2 = \frac{y-5}{3} = \frac{z-7}{4}$$

and

$$l_2 : \frac{x+1}{2} = y - 1 = \frac{z-6}{-3}.$$

Solution

If the lines intersect, then a point (x, y, z) exists that satisfies both sets of equations. In particular, then, x and y must satisfy

$$(1) \quad x - 2 = \frac{y-5}{3} \quad \text{from the equations for } l_1$$

and

$$(2) \quad \frac{x+1}{2} = y - 1, \quad \text{from the equations for } l_2.$$

Solving for x in equation (1) gives

$$(3) \quad x = 2 + \frac{y}{3} - \frac{5}{3} = \frac{y}{3} + \frac{1}{3},$$

and substituting into equation (2) gives

$$\frac{(y/3 + 1/3) + 1}{2} = y - 1,$$

which implies that $y = 2$. So to satisfy (1) and (2) we must have

$$y = 2 \quad \text{and from (3)} \quad x = \frac{2}{3} + \frac{1}{3} = 1.$$

Thus any point on the intersection of the lines l_1 and l_2 must have the coordinates $(1, 2, z)$ for some number z. An intersection will occur precisely when the same value of z satisfies the equations for both l_1 and l_2.

When $x = 1$ and $y = 2$, the equations for l_1 imply that

$$z = 4(x - 2) + 7 = 4(1 - 2) + 7 = 3;$$

and the equations for l_2 imply that

$$z = -3(y - 1) + 6 = -3(2 - 1) + 6 = 3.$$

Thus the lines intersect at the point with coordinates $(1, 2, 3)$, as shown in Figure 11.49.

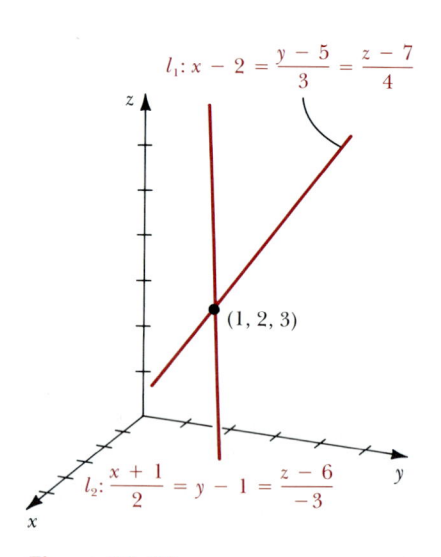

Figure 11.49

▶ EXERCISE SET 11.6

In Exercises 1–4, find parametric equations for the line passing through the point and having direction given by **v**.

1. $(2, -1, 2)$, $\mathbf{v} = \langle 1, 1, 1 \rangle$
2. $(0, 4, 3)$, $\mathbf{v} = \langle 2, 0, 1 \rangle$
3. $(0, 0, 0)$, $\mathbf{v} = \langle 2, 4, 3 \rangle$
4. $(2, 3, 4)$, $\mathbf{v} = \langle 0, -3, 4 \rangle$

In Exercises 5–8, find parametric equations for the line passing through the given points.

5. $(1, 2, 0)$, $(1, -1, 4)$
6. $(2, 0, -1)$, $(-1, 2, 1)$
7. $(3, 4, 4)$, $(2, -3, 5)$
8. $(0, 0, 2)$, $(3, 0, 0)$

9–16. Find symmetric equations for the lines described in Exercises 1–8.

Find parametric equations of a line that satisfies the conditions stated in Exercises 17–22.

17. Passes through the point $(1, 2, 3)$ and is parallel to the line with parametric equations $x = t + 1$, $y = -t$, $z = 2 - t$.
18. Passes through the origin and is parallel to the line with parametric equations $x = 1 - 2t$, $y = 3t + 2$, $z = t - 4$.
19. Passes through the point $(1, -2, 3)$ and is orthogonal to $x + y + z = 1$.
20. Passes through the point $(1, 1, 1)$ and is orthogonal to $x - 2y - 3z = 6$.
21. Passes through $(1, 7, 0)$ and is parallel to the x-axis.
22. Passes through $(2, -2, 5)$ and is parallel to the y-axis.

In Exercises 23–28, determine if the pair of lines l_1 and l_2 is (a) parallel or (b) orthogonal. (c) Find any points of intersection of l_1 and l_2.

23. $l_1: \dfrac{x-1}{2} = y + 2 = \dfrac{z-3}{2}$; $l_2: \dfrac{x}{-2} = 1 - y = \dfrac{z+2}{-2}$
24. $l_1: x - 1 = \dfrac{y+1}{2} = \dfrac{z-2}{3}$; $l_2: x - 1 = \dfrac{y+2}{-3} = \dfrac{z-5}{2}$
25. $l_1: x = 1 + t, y = 1 + 3t, z = 2 - t$; $l_2: x = 1 + t, y = 2 + t, z = 1 + t$
26. $l_1: x = 5t, y = 1 + 2t, z = 3t - 2$; $l_2: x = 3 + 3t, y = 6t - 1, x = 3 - 3t$
27. $l_1: \langle x, y, z \rangle = \langle -3, 0, 1 \rangle + t \langle 1, 2, 1 \rangle$; $l_2: \langle x, y, z \rangle = \langle 1, -4, 5 \rangle + t \langle 2, 1, 2 \rangle$
28. $l_1: \langle x, y, z \rangle = \langle -1, -1, 6 \rangle + t \langle 2, 1, 5 \rangle$; $l_2: \langle x, y, z \rangle = \langle -7, 2, 3 \rangle + t \langle 4, -2, 2 \rangle$

29. Find the point of intersection of the line with parametric equations $x = t$, $y = 2 - t$, $z = 2t - 3$ and each of the following.
    a) The yz-plane
    b) The xz-plane
    c) The xy-plane

30. Find the point of intersection of the line with parametric equations $x = t + 3$, $y = 2t - 1$, $z = t - 4$ and each of the following.
    a) The yz-plane
    b) The xz-plane
    c) The xy-plane

31. Solve the problem in Example 5 by first finding two points on the line and then using the method shown in Example 2.

32. Show that $x + 2y + z = 4$ and $x + z = 2$ intersect and find parametric equations of the line of intersection.

33. Find parametric equations of the line that passes through the point $(1, -1, 1)$, is orthogonal to the line $3x = 2y = z$, and is parallel to the plane $x + y - z = 0$.

34. Show that any line in space must intersect at least one of the coordinate planes.

35. Show that if two planes intersect, then their intersection contains a point in one of the coordinate planes.

36. Suppose l_1 is a line passing through P_1 with direction $\mathbf{v}_1$ and l_2 is a line through P_2 with direction $\mathbf{v}_2$ (see the figure). Show that the shortest distance from l_1 to l_2 is $\|\text{comp}_{\mathbf{v}_1 \times \mathbf{v}_2} \overrightarrow{P_1 P_2}\|$.

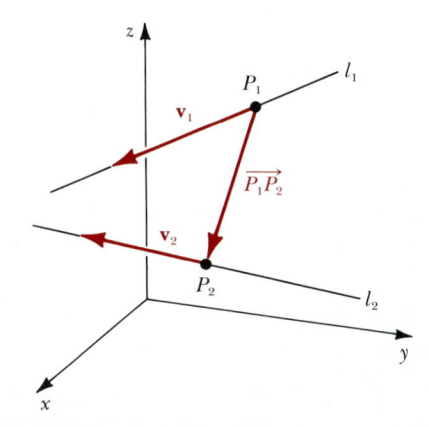

37. Use the result in Exercise 36 to find the shortest distance between the lines given by
$$l_1: x = 2t + 1, y = 3t - 2, z = t + 3$$

and
$$l_2: x = -t,\ y = 4t + 1,\ z = 2t - 3.$$

38. Show that the shortest distance from a point P_0 to the line joining P_1 and P_2, if $P_1 \neq P_2$, is
$$\frac{\|\overrightarrow{P_0P_1} \times \overrightarrow{P_1P_2}\|}{\|\overrightarrow{P_1P_2}\|}.$$
(see the figure).

39. Use the result in Exercise 38 to find the shortest distance between $(1, 2, 3)$ and the line l_1 described in Exercise 37.

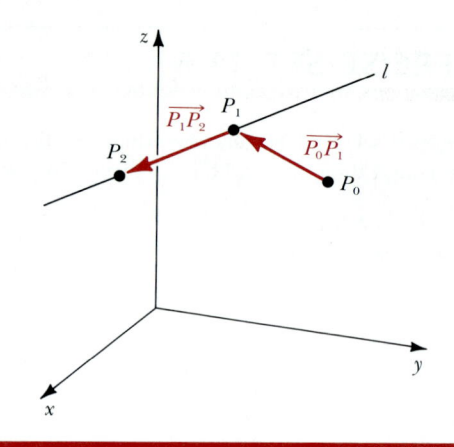

▶ IMPORTANT TERMS AND RESULTS

CONCEPT	PAGE
Rectangular coordinates	664
Distance formula	666
Sphere	667
Vector	668
Unit vector	669
Position vector	670
Parallel vectors	672
Dot product	676

$$\mathbf{a} \cdot \mathbf{b} = a_1b_1 + a_2b_2 + a_3b_3$$

Orthogonal vectors	678
Cauchy-Buniakowsky-Schwarz inequality	679
Triangle inequality	679
Orthogonal projection of $\mathbf{b}$ onto $\mathbf{a}$	680

$$\mathrm{proj}_{\mathbf{a}} \mathbf{b} = \frac{\mathbf{a} \cdot \mathbf{b}}{\|\mathbf{a}\|} \mathbf{a}$$

Cross product	684

$$\mathbf{a} \times \mathbf{b} = \langle a_2b_3 - a_3b_2,\ a_3b_1 - a_1b_3,\ a_1b_2 - a_2b_1 \rangle$$

Determinant	685
Scalar triple product	688
Normal vector	690
Planes	690

$$ax + by + cz = d$$

Orthogonal planes	692
Parallel planes	692

▶ IMPORTANT TERMS AND RESULTS (continued)

CONCEPT	PAGE
Lines	
$\quad$ Vector equation: $\langle x - x_0, y - y_0, z - z_0 \rangle = t \langle v_1, v_2, v_3 \rangle$	695
$\quad$ Parametric equations: $x = x_0 + tv_1, y = y_0 + tv_2, z = z_0 + tv_3$	696
$\quad$ Symmetric equations: $\dfrac{x - x_0}{v_1} = \dfrac{y - y_0}{v_2} = \dfrac{z - z_0}{v_3}$	697

▶ REVIEW EXERCISES

In Exercises 1–12, use the vectors $\mathbf{a} = \langle 1, 2, -1 \rangle$, $\mathbf{b} = \langle 3, 5, 4 \rangle$, $\mathbf{c} = \langle -2, 2, -3 \rangle$, and $\mathbf{d} = \langle 0, 4, 0 \rangle$ to find the requested result or state why the operation is impossible.

1. $\mathbf{a} + 2\mathbf{b} - 5\mathbf{c}$
2. $\|\mathbf{c}\| - 3\mathbf{b} \cdot \mathbf{a}$
3. $2\mathbf{c} - 5\mathbf{d} + \|\mathbf{b}\|\mathbf{b}$
4. $3\|\mathbf{a}\|\mathbf{b} - 4(\mathbf{c} \cdot \mathbf{d})\mathbf{a}$
5. $\mathbf{a} \cdot (\mathbf{b} \cdot \mathbf{c})$
6. $\mathbf{a} \cdot \mathbf{d} - 2\mathbf{b} \cdot \mathbf{c}$
7. $\mathbf{a} \times \mathbf{c} - \mathbf{b}$
8. $3\mathbf{b} \times \mathbf{d} - 2\mathbf{a} \times \mathbf{a}$
9. $\mathbf{a} \times (\mathbf{b} \times \mathbf{c})$
10. $(\mathbf{a} \cdot \mathbf{c}) \times (\mathbf{b} \cdot \mathbf{d})$
11. $(\mathbf{a} \cdot \mathbf{c})\mathbf{b} - (\mathbf{a} \cdot \mathbf{b})\mathbf{c}$
12. $(\mathbf{a} \times \mathbf{c}) \cdot (\mathbf{b} \times \mathbf{d})$

In Exercises 13–24: a) Identify the graph of the equation as either a plane, a line, or a sphere; and b) sketch the graph of the equation on a rectangular coordinate system in space.

13. $z = 3$
14. $x + y = 1$
15. $y = 1 - x, z = 0$
16. $2x + 2y - z = 4$
17. $\dfrac{x - 2}{2} = y - 1 = z - 3$
18. $4x + 9y = 1$
19. $x^2 + y^2 + z^2 = 9$
20. $x = 2$
21. $x + 2z = 4$
22. $x = t + 2, y = t - 3, z = t$
23. $x + y + 2z = 4$
24. $x^2 + y^2 + z^2 - 4z = 0$

In Exercises 25–32, find a vector that satisfies the stated condition.

25. Initial point (1, 0, 0), terminal point (1, 3, 3)
26. A unit vector parallel to the vector described in Exercise 25
27. A unit vector parallel to $\langle 1, -4, -2 \rangle$
28. Parallel to $\langle 2, -6, 3 \rangle$ with length 14
29. Normal to $3x - 7y + z = 21$
30. Orthogonal to $\langle 2, 3, 0 \rangle$ and $\langle 0, 3, 1 \rangle$
31. Describes the direction of the line of intersection of the planes $x + y - 2z = 7$ and $3x - 7y + z = 21$
32. Gives the direction of the line with parametric equations
$$x = 2 - t, \quad y = 2t - 3, \quad z = \dfrac{t + 2}{5}$$

In Exercises 33–40, use the vectors $\mathbf{a} = \langle 2, 3, 0 \rangle$ and $\mathbf{b} = \langle -2, 1, 0 \rangle$.

33. Sketch the position-vector representation of $\mathbf{a}$ and $\mathbf{b}$ in the xy-plane.
34. Sketch $\mathbf{a} + \mathbf{b}$, $\mathbf{a} - \mathbf{b}$, $2\mathbf{a}$, and $-\mathbf{b}$ on the same coordinate system.
35. Find a unit vector orthogonal to both $\mathbf{a}$ and $\mathbf{b}$.
36. Find the area of the parallelogram determined by $\mathbf{a}$ and $\mathbf{b}$.
37. Find the angle between $\mathbf{a}$ and $\mathbf{b}$.
38. Find the component of $\mathbf{b}$ in the direction of $\mathbf{a}$.
39. Find the component of $\mathbf{b}$ in the direction of $\mathbf{k}$.
40. Find the component of $\mathbf{a}$ in the direction of $\mathbf{i}$.

In Exercises 41–52, find equations of the object described.

41. A plane that contains $(1, 3, 3)$ and has normal vector $\mathbf{i} + \mathbf{j}$.

42. A plane that contains $(3, 5, 0)$ and is parallel to $x - 2y + 3z = 6$.

43. A line that passes through $(-2, 1, 4)$ and has direction given by $\mathbf{j}$.

44. A line that passes through $(1, -3, 4)$ and $(0, 3, 7)$.

45. A plane that contains $(2, 0, 0)$, $(0, 3, 0)$, and $(0, 0, -3)$.

46. A plane that contains $(0, 2, 3)$ and $(1, -3, 7)$ and is orthogonal to $2x + y - z = 5$.

47. A sphere with center $(2, 0, 0)$ and radius 2.

48. A sphere that passes through $(3, 5, 7)$ and has center $(1, 3, 4)$.

49. A plane that contains $(-3, 2, 4)$ and is parallel to the xy-plane.

50. A plane that passes through the origin and contains the line with symmetric equations $5x = 3 - y = 5 - z$.

51. A line that passes through $(-2, 1, 2)$ and is parallel to the y-axis.

52. A line that passes through $(6, 2, 4)$ and is orthogonal to the x-axis and the y-axis.

53. If $\mathbf{a} = \langle 3, 2, 0 \rangle$ and $\mathbf{b} = \langle 2, k, 0 \rangle$, find k so that (a) $\mathbf{a}$ and $\mathbf{b}$ are orthogonal; (b) $\mathbf{a}$ and $\mathbf{b}$ are parallel.

54. For $\mathbf{a} = 2\mathbf{i} - 3\mathbf{j} + \mathbf{k}$ and $\mathbf{b} = -\mathbf{i} - 2\mathbf{j} + 2\mathbf{k}$, find each of the following.
    a) $\text{comp}_\mathbf{a} \mathbf{b}$
    b) $\text{comp}_\mathbf{b} \mathbf{a}$
    c) $\text{proj}_\mathbf{a} \mathbf{b}$
    d) $\text{proj}_\mathbf{b} \mathbf{a}$
    e) The direction cosines of $\mathbf{a}$
    f) The direction cosines of $\mathbf{a} \times \mathbf{b}$

55. Find the distance from $(2, 4, 6)$ to (a) the xy-plane; (b) $2x + y + z = 3$.

56. a) Sketch the triangle with vertices at $(4, 9, 1)$, $(-2, 6, 3)$, and $(7, 3, 1)$.
    b) Show that this triangle is a right triangle.
    c) Find the area of this triangle.
    d) Find the interior angles of the triangle.

57. Find the work done if a force $\mathbf{F} = 20(\mathbf{i} + \mathbf{j} + \mathbf{k})$ moves an object from the origin to $(5, 5, 0)$.

58. A force of 50 lb is exerted at an angle of $\pi/3$ to the motion when moving an object 200 ft. Find the work done.

12

Vector-Valued Functions

CHAPTER 12 VECTOR-VALUED FUNCTIONS

In this chapter we develop a theory of calculus for functions whose domain is a set of real numbers and whose range consists of vectors in space. The most common use of vector-valued functions is to describe the motion of objects in space; the domain is an interval of time and a position vector points to the object at a particular time.

The first sections of the chapter consider the extension of ordinary calculus results to this new situation. Later we consider motion problems in space and see how to describe the change in the direction of a curve. The final section shows how vector functions are used to derive Kepler's laws of planetary motion from the laws of motion and gravitation discovered by Isaac Newton in 1665 and 1666. These are some of the most important results in the history of science.

12.1
The Definition of a Vector-Valued Function

The general notion of a function was defined at the beginning of Chapter 1, but in the first 11 chapters we discussed only those functions whose domain and range are contained in the set of real numbers. In this chapter we call these functions *real-valued* or *scalar functions* to distinguish them from the vector-valued functions we now define.

> **(12.1)**
> **DEFINITION**
> A **vector-valued function** is a function whose range is a set of vectors.

In this chapter we consider only vector-valued functions whose domain is a subset of the real numbers. More general vector-valued functions are described in Chapter 16.

Boldface notation, such as **F**, is used to distinguish vector-valued functions from scalar functions just as boldface notation was used in Chapter 11 to distinguish vectors from scalars. If

$$x = x(t), \quad y = y(t), \quad z = z(t)$$

is a set of parametric equations, then

(12.2) $$\mathbf{F}(t) = x(t)\mathbf{i} + y(t)\mathbf{j} + z(t)\mathbf{k}$$

describes a vector-valued function. Since each vector in space consists of an ordered triple of real numbers, a vector-valued function always has such a parametric representation. The real-valued functions described by $x(t)$, $y(t)$, and $z(t)$ are called the **component functions** of **F**. Unless specified otherwise, the domain of a vector-valued function is the largest set of real numbers for which all its component functions are defined.

Example 1

Determine the domain of the vector-valued function described by

$$\mathbf{F}(t) = t\mathbf{i} + \ln t\mathbf{j} + e^t\mathbf{k}.$$

Solution
The domain of $\mathbf{F}$ is the set of real numbers t for which $x(t) = t$, $y(t) = \ln t$, and $z(t) = e^t$ are defined. Since $\ln t$ is defined only for $t > 0$ and the others are defined for all values of t, the domain of $\mathbf{F}$ is $(0, \infty)$. ▶

There is a one-to-one correspondence between the set of position vectors and the points in three-dimensional space, so a vector-valued function can be thought of as mapping a portion of the real line into space.

A graphic representation of the range of a vector-valued function $\mathbf{F}$ is given by the terminal points of the position vectors $\mathbf{F}(t)$ for t in the domain of $\mathbf{F}$. The resulting figure is a curve in space such as the one shown in Figure 12.1. This curve is called *the curve traced by the vector-valued function* $\mathbf{F}$.

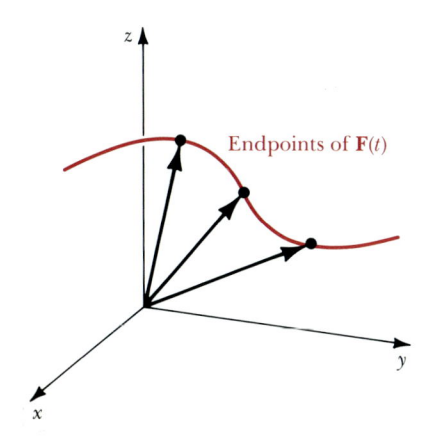

Figure 12.1
The endpoints of $\mathbf{F}(t)$ trace a curve.

Example 2

Sketch the curve described by

$$\mathbf{F}(t) = (4 + 3t)\mathbf{i} + (3 - t)\mathbf{j} - (1 + 4t)\mathbf{k}.$$

Solution
The curve described by $\mathbf{F}$ is the graph of the parametric equations

$$x(t) = 4 + 3t, \quad y(t) = 3 - t, \quad z(t) = -1 - 4t,$$

equations of the line in space shown in Figure 12.2. The line passes through $(4, 3, -1)$, which corresponds to $t = 0$, and through $(1, 4, 3)$, which corresponds to $t = -1$. The direction of the line is given by the vector $\mathbf{v} = \langle 3, -1, -4 \rangle$. The arrow on the line points in the direction of increasing values of t. ▶

In Section 11.6 we found four ways to represent a line in space: by a vector equation, by parametric equations, by symmetric equations, and as the intersection of two planes. Example 2 shows that a line can also be represented by a vector-valued function.

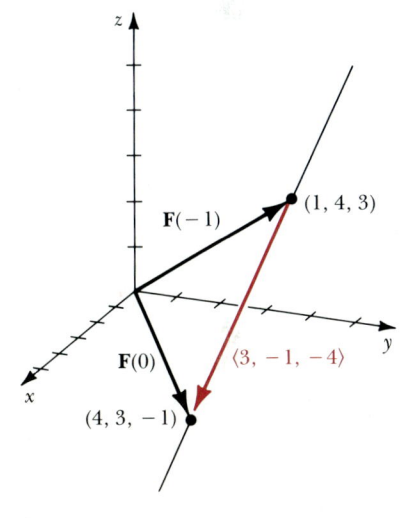

Figure 12.2

Example 3

Sketch the curve traced by $\mathbf{F}(t) = t\mathbf{i} + t^2\mathbf{j}$.

Solution
The curve is the graph of the parametric equations

$$x(t) = t, \quad y(t) = t^2, \quad z(t) = 0.$$

This is a curve in the xy-plane since $z = 0$. By eliminating t in the representation of x and y, we have the equation of a parabola, $y = x^2$. A

Example 4

Sketch the curve traced by

$$\mathbf{F}(t) = \cos t\,\mathbf{i} + \sin t\,\mathbf{j} + t\,\mathbf{k} \quad \text{for } t \geq 0.$$

Solution

The curve traced by $\mathbf{F}(t)$ is the graph of the parametric equations

$$x(t) = \cos t, \quad y(t) = \sin t, \quad z(t) = t.$$

Recall that the set of parametric equations

$$x(t) = \cos t, \quad y(t) = \sin t, \quad z(t) = 0$$

describes the unit circle in the xy-plane. The curve traced by $\mathbf{F}$ is similar, but now $z(t) = t$, so the curve spirals upward in space with increasing values of t.

The curve is called a *circular helix* and is shown in Figure 12.4. It has the appearance of a wire wrapped around a right circular cylinder.

If $\mathbf{F}$ and $\mathbf{G}$ are vector-valued functions defined by

$$\mathbf{F}(t) = F_1(t)\mathbf{i} + F_2(t)\mathbf{j} + F_3(t)\mathbf{k},$$
$$\mathbf{G}(t) = G_1(t)\mathbf{i} + G_2(t)\mathbf{j} + G_3(t)\mathbf{k}$$

and f is a real-valued function, then we define the functions $\mathbf{F} + \mathbf{G}$, $\mathbf{F} - \mathbf{G}$, $f\mathbf{F}$, $\mathbf{F} \circ f$, $\mathbf{F} \cdot \mathbf{G}$, and $\mathbf{F} \times \mathbf{G}$ as follows:

(12.3)

i) $(\mathbf{F} + \mathbf{G})(t) = \mathbf{F}(t) + \mathbf{G}(t);$ ii) $(\mathbf{F} - \mathbf{G})(t) = \mathbf{F}(t) - \mathbf{G}(t);$
iii) $(f\mathbf{F})(t) = f(t)\,\mathbf{F}(t);$ iv) $(\mathbf{F} \circ f)(t) = \mathbf{F}(f(t));$
v) $(\mathbf{F} \cdot \mathbf{G})(t) = \mathbf{F}(t) \cdot \mathbf{G}(t);$ vi) $(\mathbf{F} \times \mathbf{G})(t) = \mathbf{F}(t) \times \mathbf{G}(t).$

All these functions are vector-valued except $\mathbf{F} \cdot \mathbf{G}$, which is a scalar function.

Example 5

Describe each of the functions defined in (i) through (vi) of (12.3) if $\mathbf{F}$, $\mathbf{G}$, and f are given by

$$\mathbf{F}(t) = t^2\mathbf{i} + 2t\mathbf{j} + \sin t\,\mathbf{k}, \quad \mathbf{G}(t) = e^t\mathbf{i} + t\mathbf{j} + \mathbf{k}, \quad \text{and} \quad f(t) = t^2 - 1.$$

Solution

i) $(\mathbf{F} + \mathbf{G})(t) = (t^2 + e^t)\mathbf{i} + 3t\mathbf{j} + (\sin t + 1)\mathbf{k}$
ii) $(\mathbf{F} - \mathbf{G})(t) = (t^2 - e^t)\mathbf{i} + t\mathbf{j} + (\sin t - 1)\mathbf{k}$
iii) $(f\mathbf{F})(t) = (t^2 - 1)[t^2\mathbf{i} + 2t\mathbf{j} + \sin t\,\mathbf{k}]$
$\qquad\qquad\quad = (t^4 - t^2)\mathbf{i} + (2t^3 - 2t)\mathbf{j} + (t^2 - 1)\sin t\,\mathbf{k}$

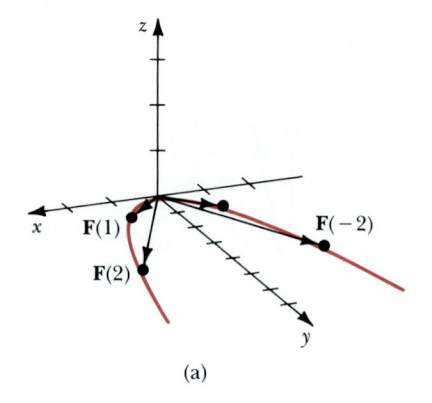

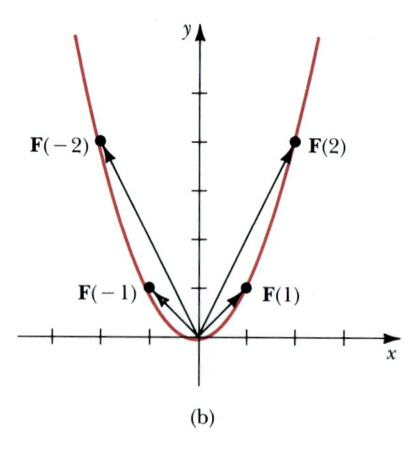

Figure 12.3

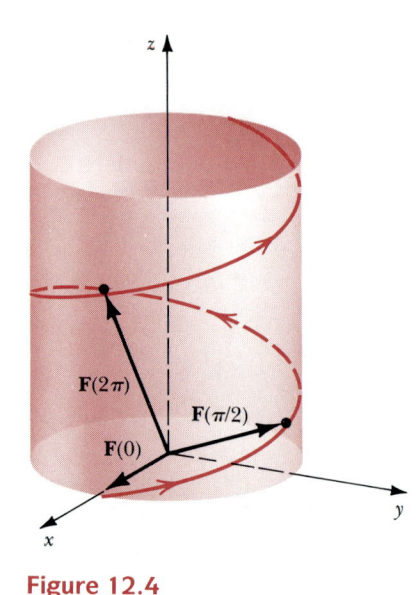

Figure 12.4
A circular helix.

iv) $(\mathbf{F} \circ f)(t) = \mathbf{F}(t^2 - 1) = (t^2 - 1)^2\mathbf{i} + 2(t^2 - 1)\mathbf{j} + \sin(t^2 - 1)\mathbf{k}$

v) $(\mathbf{F} \cdot \mathbf{G})(t) = (t^2\mathbf{i} + 2t\mathbf{j} + \sin t\mathbf{k}) \cdot (e^t\mathbf{i} + t\mathbf{j} + \mathbf{k}) = t^2 e^t + 2t^2 + \sin t$

vi) $(\mathbf{F} \times \mathbf{G})(t) = (t^2\mathbf{i} + 2t\mathbf{j} + \sin t\mathbf{k}) \times (e^t\mathbf{i} + t\mathbf{j} + \mathbf{k})$

$$= \begin{vmatrix} \mathbf{i} & \mathbf{j} & \mathbf{k} \\ t^2 & 2t & \sin t \\ e^t & t & 1 \end{vmatrix}$$

$$= (2t - t \sin t)\mathbf{i} - (t^2 - e^t \sin t)\mathbf{j} + (t^3 - 2te^t)\mathbf{k}$$

▶ **EXERCISE SET 12.1**

In Exercises 1–6, determine the domain of the vector-valued function.

1. $\mathbf{F}(t) = t\mathbf{i} + t^2\mathbf{j} + 2\mathbf{k}$
2. $\mathbf{F}(t) = 3\mathbf{i} + \sqrt{1-t}\,\mathbf{j} + t\mathbf{k}$
3. $\mathbf{F}(t) = \sqrt{t}\,\mathbf{i} + (t^2 - 2)\mathbf{j} + t\mathbf{k}$
4. $\mathbf{F}(t) = \sqrt{t}\,\mathbf{i} + (t^2 - 2)\mathbf{j} + (1/t)\mathbf{k}$
5. $\mathbf{F}(t) = \ln t\mathbf{i} + (1 - t^2)\mathbf{j} + \mathbf{k}$
6. $\mathbf{F}(t) = \ln t\mathbf{j} + e^t\mathbf{k}$

In Exercises 7–16, sketch the curve in the xy-plane described by the vector-valued function and indicate the direction of increasing t.

7. $\mathbf{F}(t) = t\mathbf{i} + t^2\mathbf{j}$
8. $\mathbf{F}(t) = t\mathbf{i} + (1/t)\mathbf{j}, t > 0$
9. $\mathbf{F}(t) = 4t^2\mathbf{i}$
10. $\mathbf{F}(t) = t^3\mathbf{i} + \mathbf{j}$
11. $\mathbf{F}(t) = t\mathbf{i} + \sin t\mathbf{j}$
12. $\mathbf{F}(t) = \cos t\mathbf{i} + t\mathbf{j}$
13. $\mathbf{F}(t) = \sin t\mathbf{i} + \cos t\mathbf{j}$
14. $\mathbf{F}(t) = 3 \sin t\mathbf{i} + 4 \cos t\mathbf{j}$
15. $\mathbf{F}(t) = e^t\mathbf{i} + e^{-2t}\mathbf{j}$
16. $\mathbf{F}(t) = t\mathbf{i} + \ln t\mathbf{j}, t > 0$

In Exercises 17–24, sketch the curve in space described by the vector-valued function and indicate the direction of increasing t.

17. $\mathbf{F}(t) = (t + 1)\mathbf{i} + (2t - 1)\mathbf{j} + (2 - 2t)\mathbf{k}$
18. $\mathbf{F}(t) = (2t + 1)\mathbf{i} + (3 - t)\mathbf{j} + (2 - t)\mathbf{k}$
19. $\mathbf{F}(t) = \cos t\mathbf{i} + \sin t\mathbf{j} + 2\mathbf{k}$
20. $\mathbf{F}(t) = 2 \cos t\mathbf{i} + 3 \sin t\mathbf{j}$
21. $\mathbf{F}(t) = t\mathbf{i} + e^t\mathbf{j} + \mathbf{k}$
22. $\mathbf{F}(t) = t\mathbf{i} + e^t\mathbf{j} + e^t\mathbf{k}$
23. $\mathbf{F}(t) = t\mathbf{i} + (256 - 16t^2)\mathbf{k}$
24. $\mathbf{F}(t) = \sin t\mathbf{i} + t\mathbf{j} + t\mathbf{k}$

In Exercises 25–28, functions $\mathbf{F}$, $\mathbf{G}$, and f are given. Describe the following and determine their domains.

a) $\mathbf{F} \cdot \mathbf{G}$ b) $\mathbf{F} \circ f$
c) $f\mathbf{F}$ d) $\mathbf{F} + \mathbf{G}$
e) $\mathbf{F} - \mathbf{G}$ f) $\mathbf{F} \times \mathbf{G}$

25. $\mathbf{F}(t) = t\mathbf{i} + t^{-2}\mathbf{j} + e^t\mathbf{k}$, $\mathbf{G}(t) = \sqrt{t}\,\mathbf{i} + \sin t\mathbf{k}$, $f(t) = 2 - t$
26. $\mathbf{F}(t) = \ln t\mathbf{i} + e^t\mathbf{k}$, $\mathbf{G}(t) = \sqrt[3]{t}\,\mathbf{i} + \tan t\mathbf{j}$, $f(t) = \sin t$
27. $\mathbf{F}(t) = \sqrt{t}\,\mathbf{i} + \ln t\mathbf{j} + 3t^3\mathbf{k}$, $\mathbf{G}(t) = \sqrt{-t}\,\mathbf{j} + t^4\mathbf{k}$, $f(t) = e^t$
28. $\mathbf{F}(t) = \sec t\mathbf{j} + 2 \csc t\mathbf{k}$, $\mathbf{G}(t) = \sqrt{t^2 + 1}\,\mathbf{i} + 3e^t\mathbf{k}$, $f(t) = \sqrt{t}$

29. A circle of radius R rolls along a horizontal line. Find a vector-valued function that describes the curve traced by a fixed point P on the circumference of the circle. [*Hint*: See Example 3 in Section 9.4.]

30. The motion of a baseball is described by

$$\mathbf{F}(t) = 3.4\mathbf{i} + 5.5t\mathbf{j} + (4.9t - 4.9t^2)\mathbf{k}, \quad 0 \le t \le 1,$$

where t is expressed in seconds and the coordinate values are in meters. Sketch a curve that represents the path of this baseball.

31. The motion of the top of the mast of a 16-ft sailboat caught in a storm is described by

$$\mathbf{F}(t) = (6 \sin t + 1)\mathbf{i} + t\mathbf{j} + (\cos t + 12)\mathbf{k}, \quad t \ge 0.$$

Sketch a curve that represents the motion of this mast top.

12.2
The Calculus of Vector-Valued Functions

A vector-valued function **F** is defined in terms of its **i**, **j**, and **k** components:
$$\mathbf{F}(t) = x(t)\mathbf{i} + y(t)\mathbf{j} + z(t)\mathbf{k}.$$

The component functions x, y, and z provide a natural means for defining the various concepts of calculus for vector-valued functions.

(12.4) DEFINITION

Suppose **F** is described by $\mathbf{F}(t) = x(t)\mathbf{i} + y(t)\mathbf{j} + z(t)\mathbf{k}$.

i) If $\lim_{t \to a} x(t)$, $\lim_{t \to a} y(t)$, and $\lim_{t \to a} z(t)$ exist, then the limit of **F** at a is said to exist and
$$\lim_{t \to a} \mathbf{F}(t) = \lim_{t \to a} x(t)\mathbf{i} + \lim_{t \to a} y(t)\mathbf{j} + \lim_{t \to a} z(t)\mathbf{k}.$$

ii) If the component functions of **F** are continuous at a, then **F** is continuous at a.

iii) If the component functions of **F** are differentiable at t, then **F** is differentiable at t and
$$\mathbf{F}'(t) = x'(t)\mathbf{i} + y'(t)\mathbf{j} + z'(t)\mathbf{k}.$$

iv) If x, y, and z are integrable on the interval $[a, b]$, then **F** is said to be integrable on $[a, b]$ and
$$\int_a^b \mathbf{F}(t)\, dt = \left(\int_a^b x(t)\, dt\right)\mathbf{i} + \left(\int_a^b y(t)\, dt\right)\mathbf{j} + \left(\int_a^b z(t)\, dt\right)\mathbf{k}.$$

v) Similarly, the indefinite integral of $\mathbf{F}(t)$ is defined by
$$\int \mathbf{F}(t)\, dt = \left(\int x(t)\, dt\right)\mathbf{i} + \left(\int y(t)\, dt\right)\mathbf{j} + \left(\int z(t)\, dt\right)\mathbf{k}.$$

Since each of the indefinite integrals of the component functions includes an arbitrary constant, an arbitrary *vector* is associated with the indefinite integral $\int \mathbf{F}(t)\, dt$.

Example 1

Let $\mathbf{F}(t) = \cos t\, \mathbf{i} + \sin t\, \mathbf{j} + t\, \mathbf{k}$. Find each of the following.

a) $\lim_{t \to \pi/4} \mathbf{F}(t)$

b) $\mathbf{F}'(\pi/4)$

c) $\int \mathbf{F}(t)\, dt$

d) $\int_0^{\pi/4} \mathbf{F}(t)\, dt$

Solution

a) Since the coordinate functions
$$x(t) = \cos t, \quad y(t) = \sin t, \quad \text{and} \quad z(t) = t$$

are all continuous,

$$\lim_{t \to \pi/4} \mathbf{F}(t) = \lim_{t \to \pi/4} x(t)\mathbf{i} + \lim_{t \to \pi/4} y(t)\mathbf{j} + \lim_{t \to \pi/4} z(t)\mathbf{k}$$

$$= x\left(\frac{\pi}{4}\right)\mathbf{i} + y\left(\frac{\pi}{4}\right)\mathbf{j} + z\left(\frac{\pi}{4}\right)\mathbf{k}$$

$$= \frac{\sqrt{2}}{2}\mathbf{i} + \frac{\sqrt{2}}{2}\mathbf{j} + \frac{\pi}{4}\mathbf{k}.$$

b) We have

$$\mathbf{F}'(t) = D_t(\cos t)\mathbf{i} + D_t(\sin t)\mathbf{j} + D_t(t)\mathbf{k}$$
$$= -\sin t\,\mathbf{i} + \cos t\,\mathbf{j} + \mathbf{k},$$

so

$$\mathbf{F}'\left(\frac{\pi}{4}\right) = -\frac{\sqrt{2}}{2}\mathbf{i} + \frac{\sqrt{2}}{2}\mathbf{j} + \mathbf{k}.$$

c)
$$\int \mathbf{F}(t)\,dt = \int \cos t\,dt\,\mathbf{i} + \int \sin t\,dt\,\mathbf{j} + \int t\,dt\,\mathbf{k}$$
$$= (\sin t + C_1)\mathbf{i} + (-\cos t + C_2)\mathbf{j} + (t^2/2 + C_3)\mathbf{k}$$
$$= \sin t\,\mathbf{i} - \cos t\,\mathbf{j} + \frac{t^2}{2}\mathbf{k} + \mathbf{C},$$

where $\mathbf{C} = C_1\mathbf{i} + C_2\mathbf{j} + C_3\mathbf{k}$ is an arbitrary vector.

d) The result in part (c) gives

$$\int_0^{\pi/4} \mathbf{F}(t)\,dt = \left[\sin t\,\mathbf{i} - \cos t\,\mathbf{j} + \frac{t^2}{2}\mathbf{k}\right]_0^{\pi/4}$$
$$= \left(\sin\frac{\pi}{4} - \sin 0\right)\mathbf{i} - \left(\cos\frac{\pi}{4} - \cos 0\right)\mathbf{j} + \left(\frac{(\pi/4)^2}{2} - 0\right)\mathbf{k}$$
$$= \frac{\sqrt{2}}{2}\mathbf{i} - \left(\frac{\sqrt{2}}{2} - 1\right)\mathbf{j} + \frac{\pi^2}{32}\mathbf{k}. \quad\blacktriangleright$$

Theorems for scalar functions concerning these concepts hold for vector-valued functions as well, unless the theorems include operations such as the reciprocal and quotient of vectors that are undefined for vector-valued functions.

The following theorems show that the definitions of continuity and differentiability can be expressed in the same context as their scalar counterparts. A proof for the differentiability result in Theorem (12.6) is provided. The continuity result in Theorem (12.5) can be shown in a similar manner and is considered in Exercise 56.

(12.5)
THEOREM
A function $\mathbf{F}$ is continuous at a if and only if:

i) $\mathbf{F}(a)$ exists;

ii) $\lim_{t \to a} \mathbf{F}(t)$ exists; and

iii) $\lim_{t \to a} \mathbf{F}(t) = \mathbf{F}(a)$.

(12.6)
THEOREM
A function $\mathbf{F}$ is differentiable at t if and only if

$$\lim_{h \to 0} \frac{\mathbf{F}(t+h) - \mathbf{F}(t)}{h}$$

exists. In this case,

$$\mathbf{F}'(t) = \lim_{h \to 0} \frac{\mathbf{F}(t+h) - \mathbf{F}(t)}{h}.$$

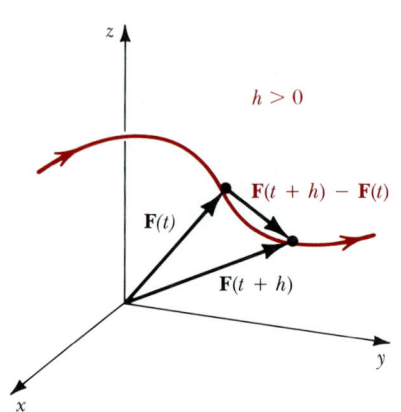

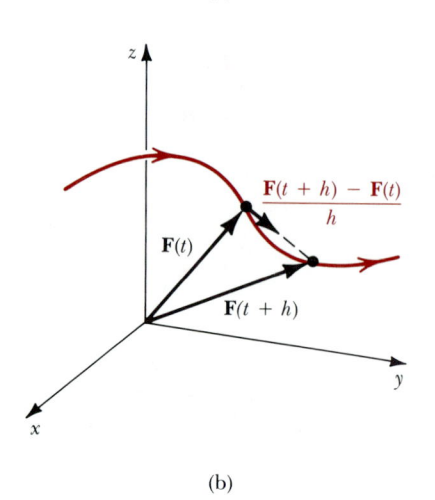

Proof. By part (i) of Definition (12.4),

$$\lim_{h \to 0} \frac{\mathbf{F}(t+h) - \mathbf{F}(t)}{h} = \lim_{h \to 0} \frac{x(t+h)\mathbf{i} + y(t+h)\mathbf{j} + z(t+h)\mathbf{k} - x(t)\mathbf{i} - y(t)\mathbf{j} - z(t)\mathbf{k}}{h}$$

$$= \lim_{h \to 0} \left[\frac{x(t+h) - x(t)}{h}\right]\mathbf{i} + \lim_{h \to 0} \left[\frac{y(t+h) - y(t)}{h}\right]\mathbf{j}$$

$$+ \lim_{h \to 0} \left[\frac{z(t+h) - z(t)}{h}\right]\mathbf{k}.$$

Thus,

$$\lim_{h \to 0} \frac{\mathbf{F}(t+h) - \mathbf{F}(t)}{h}$$

exists precisely when the coordinate functions are differentiable at t and

$$\lim_{h \to 0} \frac{\mathbf{F}(t+h) - \mathbf{F}(t)}{h} = x'(t)\mathbf{i} + y'(t)\mathbf{j} + z'(t)\mathbf{k} = \mathbf{F}'(t). \quad \triangleright$$

The result in Theorem (12.6) provides a geometric interpretation of $\mathbf{F}'$. For $h \neq 0$, the difference

$$\mathbf{F}(t+h) - \mathbf{F}(t)$$

is a vector describing the *secant line* drawn through the terminal points of $\mathbf{F}(t+h)$ and $\mathbf{F}(t)$. Such a line is shown in Figure 12.5(a) for the case $h > 0$. The vector

(12.7)
$$\frac{\mathbf{F}(t+h) - \mathbf{F}(t)}{h}$$

is parallel to this secant line.

If $h > 0$, then $\mathbf{F}(t+h) - \mathbf{F}(t)$ points in the direction of increasing values of t as does the vector in (12.7), as shown in Figure 12.5(b).

If $h < 0$, then $\mathbf{F}(t+h) - \mathbf{F}(t)$ points in the direction of decreasing values of t, as shown in Figure 12.6(a). In this case, the vector in (12.7) has direction opposite that of $\mathbf{F}(t+h) - \mathbf{F}(t)$ and again points in the direction of increasing values of t, as shown in Figure 12.6(b). The direction of the vector in (12.7), then, is always that of increasing values of t. The derivative of $\mathbf{F}$ at t,

$$\mathbf{F}'(t) = \lim_{h \to 0} \frac{\mathbf{F}(t+h) - \mathbf{F}(t)}{h},$$

Figure 12.5

is a **tangent vector** at $\mathbf{F}(t)$ to the curve traced by $\mathbf{F}$ and points in the direction of increasing t, as shown in Figure 12.7.

Figure 12.6

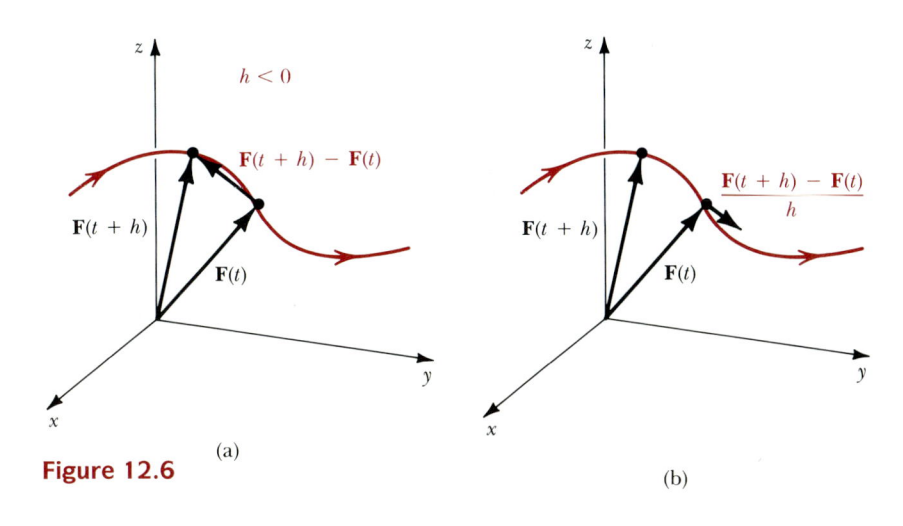

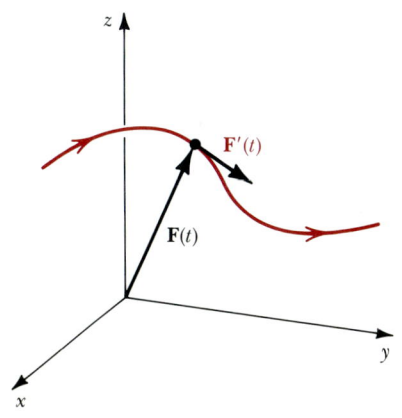

Figure 12.7
$\mathbf{F}'(t)$ always points in the direction of increasing values of t.

Example 2

Let $\mathbf{F}(t) = \cos t\mathbf{i} + \sin t\mathbf{j} + t\mathbf{k}$. Find $\mathbf{F}'(\pi/2)$ and sketch the graph of $\mathbf{F}$ and the tangent vector $\mathbf{F}'(\pi/2)$.

Solution
Since

$$\mathbf{F}'(t) = -\sin t\,\mathbf{i} + \cos t\,\mathbf{j} + \mathbf{k}, \qquad \mathbf{F}'\left(\frac{\pi}{2}\right) = -\mathbf{i} + \mathbf{k}.$$

The curve traced by $\mathbf{F}$ is the circular helix shown in Figure 12.8. The vector $\mathbf{F}'(\pi/2) = -\mathbf{i} + \mathbf{k}$ is also shown in Figure 12.8. It is sketched with initial point $(0, 1, \pi/2)$ lying on the curve and terminal point $(-1, 1, 1 + \pi/2)$. ▶

The study of the calculus of scalar functions involves the derivatives of various combinations of functions. The following theorem indicates that similar results hold for vector-valued functions.

(12.8)
THEOREM
If $\mathbf{F}$, $\mathbf{G}$, and f are differentiable at t, then:

i) $D_t\,[\mathbf{F}(t) + \mathbf{G}(t)] = D_t\,\mathbf{F}(t) + D_t\,\mathbf{G}(t)$;
ii) $D_t\,[\mathbf{F}(t) - \mathbf{G}(t)] = D_t\,\mathbf{F}(t) - D_t\,\mathbf{G}(t)$;
iii) $D_t\,(f\mathbf{F})(t) = [D_t f(t)]\mathbf{F}(t) + f(t)\,[D_t\,\mathbf{F}(t)]$;
iv) $D_t\,(\mathbf{F} \times \mathbf{G})(t) = D_t\,\mathbf{F}(t) \times \mathbf{G}(t) + \mathbf{F}(t) \times D_t\,\mathbf{G}(t)$;
v) $D_t\,(\mathbf{F} \cdot \mathbf{G})(t) = D_t\,\mathbf{F}(t) \cdot \mathbf{G}(t) + \mathbf{F}(t) \cdot D_t\,\mathbf{G}(t)$.

If g is differentiable at t and $\mathbf{F}$ is differentiable at $g(t)$, then:

vi) $D_t\,(\mathbf{F} \circ g)(t) = \mathbf{F}'(g(t))\,g'(t)$.

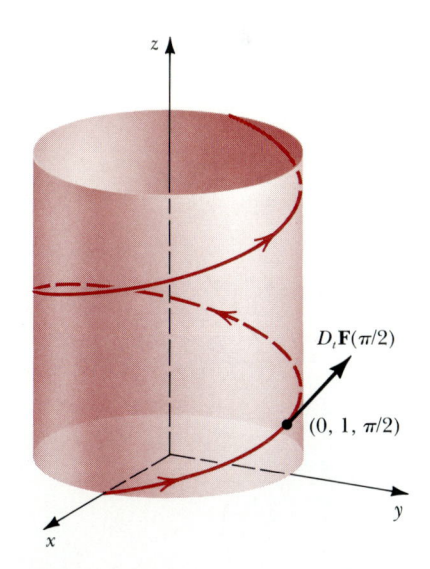

Figure 12.8

Proof. We will prove part (iii) of the theorem. The other parts are proved in a similar manner.

iii) To facilitate the proof, express **F** in terms of its components:
$$\mathbf{F}(t) = x(t)\mathbf{i} + y(t)\mathbf{j} + z(t)\mathbf{k}.$$

Then
$$D_t\,(f\mathbf{F})(t) = D_t\,[f(t)\,x(t)]\mathbf{i} + D_t\,[f(t)\,y(t)]\mathbf{j} + D_t\,[f(t)\,z(t)]\mathbf{k},$$

and the Product Rule applied to each component implies that
$$\begin{aligned}
D_t\,(f\mathbf{F})(t) &= [f'(t)\,x(t) + f(t)\,x'(t)]\mathbf{i} + [f'(t)\,y(t) + f(t)\,y'(t)]\mathbf{j} \\
&\quad + [f'(t)\,z(t) + f(t)\,z'(t)]\mathbf{k} \\
&= f'(t)\,x(t)\mathbf{i} + f'(t)\,y(t)\mathbf{j} + f'(t)\,z(t)\mathbf{k} \\
&\quad + f(t)\,x'(t)\mathbf{i} + f(t)\,y'(t)\mathbf{j} + f(t)\,z'(t)\mathbf{k} \\
&= [D_t\,f(t)]\mathbf{F}(t) + f(t)\,[D_t\,\mathbf{F}(t)].
\end{aligned}$$
▷

Example 3

For $\mathbf{F}(t) = t\mathbf{i} + \mathbf{j} + e^t\mathbf{k}$, $\mathbf{G}(t) = \mathbf{i} + 2t\mathbf{j} + \sin t\,\mathbf{k}$, and $f(t) = t^2 + 1$, find each of the following.

a) $D_t\,(f\mathbf{F})(t)$
b) $D_t\,(\mathbf{F} \circ f)(t)$
c) $D_t\,(\mathbf{F} \times \mathbf{G})(t)$
d) $D_t\,(\mathbf{F} \cdot \mathbf{G})(t)$

Solution

a) $\begin{aligned}
D_t\,(f\mathbf{F})(t) &= D_t\,(t^2 + 1)(t\mathbf{i} + \mathbf{j} + e^t\mathbf{k}) + (t^2 + 1)\,D_t\,(t\mathbf{i} + \mathbf{j} + e^t\mathbf{k}) \\
&= 2t(t\mathbf{i} + \mathbf{j} + e^t\mathbf{k}) + (t^2 + 1)(\mathbf{i} + e^t\mathbf{k}) \\
&= (3t^2 + 1)\mathbf{i} + 2t\mathbf{j} + (t^2 + 2t + 1)e^t\mathbf{k}
\end{aligned}$

b) Since $\mathbf{F}'(t) = \mathbf{i} + e^t\mathbf{k}$,
$$D_t\,(\mathbf{F} \circ f)(t) = \mathbf{F}'(f(t))\,f'(t) = (\mathbf{i} + e^{t^2+1}\mathbf{k})(2t) = 2t\mathbf{i} + 2te^{t^2+1}\,\mathbf{k}$$

c) $\begin{aligned}
D_t\,(\mathbf{F} \times \mathbf{G})(t) &= D_t\,\mathbf{F}(t) \times \mathbf{G}(t) + \mathbf{F}(t) \times D_t\,\mathbf{G}(t) \\
&= (\mathbf{i} + e^t\mathbf{k}) \times (\mathbf{i} + 2t\mathbf{j} + \sin t\,\mathbf{k}) \\
&\quad + (t\mathbf{i} + \mathbf{j} + e^t\mathbf{k}) \times (2\mathbf{j} + \cos t\,\mathbf{k}) \\
&= \begin{vmatrix} \mathbf{i} & \mathbf{j} & \mathbf{k} \\ 1 & 0 & e^t \\ 1 & 2t & \sin t \end{vmatrix} + \begin{vmatrix} \mathbf{i} & \mathbf{j} & \mathbf{k} \\ t & 1 & e^t \\ 0 & 2 & \cos t \end{vmatrix} \\
&= [-2te^t\mathbf{i} - (\sin t - e^t)\mathbf{j} + 2t\mathbf{k}] \\
&\quad + [(\cos t - 2e^t)\mathbf{i} - t\cos t\,\mathbf{j} + 2t\mathbf{k}] \\
&= (\cos t - 2te^t - 2e^t)\mathbf{i} + (e^t - \sin t - t\cos t)\mathbf{j} + 4t\mathbf{k}
\end{aligned}$

d) $\begin{aligned}
D_t\,(\mathbf{F} \cdot \mathbf{G})(t) &= D_t\,\mathbf{F}(t) \cdot \mathbf{G}(t) + \mathbf{F}(t) \cdot D_t\,\mathbf{G}(t) \\
&= (\mathbf{i} + e^t\mathbf{k}) \cdot (\mathbf{i} + 2t\mathbf{j} + \sin t\,\mathbf{k}) \\
&\quad + (t\mathbf{i} + \mathbf{j} + e^t\mathbf{k}) \cdot (2\mathbf{j} + \cos t\,\mathbf{k}) \\
&= 1 + e^t\sin t + 2 + e^t\cos t \\
&= 3 + e^t(\sin t + \cos t)
\end{aligned}$
▶

EXERCISE SET 12.2

In Exercises 1–8, find $\lim_{t \to a} \mathbf{F}(t)$, if it exists.

1. $\mathbf{F}(t) = \sqrt{t}\,\mathbf{i} + \dfrac{1}{t}\mathbf{j} + t^2\mathbf{k};\ a = 1$

2. $\mathbf{F}(t) = e^t\mathbf{i} + t\mathbf{j} + 3\mathbf{k};\ a = 0$

3. $\mathbf{F}(t) = \dfrac{t^2 - 4}{t - 2}\mathbf{i} + 3\mathbf{j};\ a = 2$

4. $\mathbf{F}(t) = \mathbf{i} + \dfrac{t^2 - 9}{t - 3}\mathbf{j} + \dfrac{1}{t}\mathbf{k};\ a = 3$

5. $\mathbf{F}(t) = \dfrac{1}{t}\mathbf{i} + t^3\mathbf{j} + \sin t\,\mathbf{k};\ a = 0$

6. $\mathbf{F}(t) = \dfrac{1}{t - 2}\mathbf{i} + (t - 2)\mathbf{j};\ a = 2$

7. $\mathbf{F}(t) = \begin{cases} t^2\mathbf{i} - 2t\mathbf{j} + e^{t-1}\mathbf{k}, & \text{if } t > 1 \\ \mathbf{0}, & \text{if } t = 1;\ a = 1 \\ (2t - 1)\mathbf{i} + 2\cos \pi t\,\mathbf{j} + t\mathbf{k}, & \text{if } t < 1 \end{cases}$

8. $\mathbf{F}(t) = \begin{cases} \ln t\,\mathbf{i} - \sin(t - 1)\,\mathbf{k}, & \text{if } t > 1 \\ \mathbf{0}, & \text{if } t = 1;\ a = 1 \\ (t^2 - 1)\mathbf{i} + (e^{t-1} - 1)\mathbf{j}, & \text{if } t < 1 \end{cases}$

9–16. Find the values of t for which the functions described in Exercises 1–8 are continuous.

In Exercises 17–20, find the derivative of the vector-valued function.

17. $\mathbf{F}(t) = \ln t\,\mathbf{i} + t\mathbf{j} + \mathbf{k}$
18. $\mathbf{F}(t) = te^t\mathbf{j} + t\mathbf{k}$
19. $\mathbf{F}(t) = t^2\mathbf{i} + e^{t^2}\mathbf{j} + t\mathbf{k}$
20. $\mathbf{F}(t) = \sec^2 t\,\mathbf{i} + \tan^2 t\,\mathbf{j}$

In Exercises 21–26, a vector-valued function $\mathbf{F}$ and a specific number t are given.
 a) Find $\mathbf{F}'(t)$
 b) Sketch the graph of $\mathbf{F}$ and the tangent vector $\mathbf{F}'(t)$.

21. $\mathbf{F}(t) = 2\cos t\,\mathbf{i} + 3\sin t\,\mathbf{j},\ t = 0$
22. $\mathbf{F}(t) = e^t\mathbf{i} + e^{-t}\mathbf{j},\ t = \ln 2$
23. $\mathbf{F}(t) = t\mathbf{i} + e^t\mathbf{j} + 2t\mathbf{k},\ t = 0$
24. $\mathbf{F}(t) = t\mathbf{i} + e^t\mathbf{j} + \mathbf{k},\ t = 1$
25. $\mathbf{F}(t) = \sin t\,\mathbf{i} + \cos t\,\mathbf{j} + t\mathbf{k},\ t = 0$
26. $\mathbf{F}(t) = 2\cos t\,\mathbf{i} + 3\sin t\,\mathbf{j} + t\mathbf{k},\ t = \pi/2$

In Exercises 27–32, determine the derivative of the function given that $\mathbf{F}(t) = t^2\mathbf{i} - 2t\mathbf{j} + \sin t\,\mathbf{k}$, $\mathbf{G}(t) = \ln t\,\mathbf{i} + 3e^t\mathbf{k}$, and $f(t) = \cos t$.

27. $(\mathbf{F} - 3\mathbf{G})(t)$
28. $(\mathbf{F} \circ f)(t)$
29. $[(\mathbf{F} - 3\mathbf{G}) \circ f](t)$
30. $(\mathbf{F} \cdot \mathbf{G})(t)$
31. $(\mathbf{F} \times \mathbf{G})(t)$
32. $[(\mathbf{F} \times \mathbf{G}) \circ f](t)$

For Exercises 33–38, find the derivatives of the functions in Exercises 27–32 using $\mathbf{F}(t) = t^2\mathbf{i} + e^t\mathbf{j} + (t + 1)\mathbf{k}$, $\mathbf{G}(t) = \sin t\,\mathbf{i} + t\mathbf{j} + \ln t\,\mathbf{k}$, and $f(t) = e^{t^2}$.

In Exercises 39–46, evaluate the integrals.

39. $\displaystyle\int_0^{\pi/2} (\sin t\,\mathbf{i} + \cos t\,\mathbf{j} + t\mathbf{k})\,dt$

40. $\displaystyle\int_0^1 (t\mathbf{i} + (t - 1)^2\mathbf{j} + \sqrt{t}\,\mathbf{k})\,dt$

41. $\displaystyle\int (te^t\mathbf{i} + t\mathbf{j} + \mathbf{k})\,dt$

42. $\displaystyle\int_1^e (t\mathbf{i} + \ln t\,\mathbf{j})\,dt$

43. $\displaystyle\int \left(t\mathbf{i} + \dfrac{1}{t}\mathbf{j}\right) dt$

44. $\displaystyle\int (t\cos t\,\mathbf{i} + \cos t\,\mathbf{j} + t\mathbf{k})\,dt$

45. $\displaystyle\int \|\sin t\,\mathbf{i} + \cos t\,\mathbf{j} + \mathbf{k}\|\,dt$

46. $\displaystyle\int_0^{\pi/2} \|\sin t\,\mathbf{i} + \cos t\,\mathbf{j} + t\mathbf{k}\|\,dt$

In Exercises 47–50, determine where the derivative of the function is continuous.

47. $\mathbf{F}(t) = t\mathbf{i} + \sin t\,\mathbf{j}$
48. $\mathbf{F}(t) = \mathbf{i} + t\mathbf{j} + t^2\mathbf{k}$
49. $\mathbf{F}(t) = \tan t\,\mathbf{i} + t\mathbf{j} + \mathbf{k}$
50. $\mathbf{F}(t) = t\mathbf{j} + \ln t\,\mathbf{k}$

51. a) Show that the line segment joining the points $(1, 3, 1)$ and $(2, 5, 4)$ is described by each of the following functions:

$\mathbf{F}_1(t) = (t + 1)\mathbf{i} + (2t + 3)\mathbf{j} + (3t + 1)\mathbf{k},\ 0 \le t \le 1;$

$\mathbf{F}_2(t) = (t^2 + 1)\mathbf{i} + (2t^2 + 3)\mathbf{j} + (3t^2 + 1)\mathbf{k},\ 0 \le t \le 1;$

$\mathbf{F}_3(t) = (\ln t + 1)\mathbf{i} + (\ln t^2 + 3)\mathbf{j} + (\ln t^3 + 1)\mathbf{k},\ 1 \le t \le e.$

b) Find $\mathbf{F}_1\left(\dfrac{1}{2}\right)$, $\mathbf{F}_2\left(\dfrac{1}{2}\right)$, and $\mathbf{F}_3\left(\dfrac{1 + e}{2}\right)$.

52. a) Show that the following functions describe the same curve in space:

$\mathbf{F}_1(t) = t\mathbf{i} + 2t\mathbf{j} + t^2\mathbf{k},\ 0 \le t \le 1;$

$\mathbf{F}_2(t) = t^2\mathbf{i} + 2t^2\mathbf{j} + t^4\mathbf{k},\ 0 \le t \le 1;$

$\mathbf{F}_3(t) = \ln t\,\mathbf{i} + 2\ln t\,\mathbf{j} + (\ln t)^2\mathbf{k},\ 1 \le t \le e.$

b) Find $\mathbf{F}_1\left(\dfrac{1}{2}\right)$, $\mathbf{F}_2\left(\dfrac{1}{2}\right)$, and $\mathbf{F}_3\left(\dfrac{1+e}{2}\right)$.

c) Sketch the curve described by $\mathbf{F}_1$, $\mathbf{F}_2$, and $\mathbf{F}_3$.

53. Show that if $\mathbf{F}''$ exists, then $D_t(\mathbf{F} \times \mathbf{F}')(t) = (\mathbf{F} \times \mathbf{F}'')(t)$.

54. Use Definition (12.4) to show that $D_t[c\mathbf{F}(t)] = c\, D_t\, \mathbf{F}(t)$.

55. Use Definition (12.4) to show that if $\mathbf{F}$ is continuous at a, then $\|\mathbf{F}\|$ is continuous at a.

56. Prove Theorem (12.5).

57. Find a function $\mathbf{F}$ and a number a with $\|\mathbf{F}\|$ continuous at a, but $\mathbf{F}$ not continuous at a.

58. Prove part (vi) of Theorem (12.8): If g is differentiable at t and $\mathbf{F}$ is differentiable at $g(t)$, then

$$D_t\,(\mathbf{F} \circ g)(t) = \mathbf{F}'(g(t))\, g'(t).$$

59. Prove part (v) of Theorem (12.8): If $\mathbf{F}$ and $\mathbf{G}$ are differentiable, then

$$D_t\,(\mathbf{F} \cdot \mathbf{G})(t) = D_t\, \mathbf{F}(t) \cdot \mathbf{G}(t) + \mathbf{F}(t) \cdot D_t\, \mathbf{G}(t).$$

60. If $\mathbf{F}$ and $\mathbf{G}$ are integrable on $[a, b]$ and c is a real number, use Definition (12.4) to show that:

a) $\displaystyle\int_a^b [\mathbf{F}(t) + \mathbf{G}(t)]\, dt = \int_a^b \mathbf{F}(t)\, dt + \int_a^b \mathbf{G}(t)\, dt;$

b) $\displaystyle\int_a^b c\mathbf{F}(t)\, dt = c \int_a^b \mathbf{F}(t)\, dt.$

12.3

Unit Tangent and Unit Normal Vectors

In Section 12.2 we saw that for all t, $\mathbf{F}'(t)$ is tangent to the graph of the curve described by $\mathbf{F}$ and points in the direction of increasing t. The unit vector in this direction is called the *principal unit tangent vector*.

(12.9)
DEFINITION
If C is the curve described by $\mathbf{F}$, then the **principal unit tangent vector** to C at the terminal point of the position vector $\mathbf{F}(t)$ is defined by

$$\mathbf{T}(t) = \frac{\mathbf{F}'(t)}{\|\mathbf{F}'(t)\|},$$

provided $\mathbf{F}'(t)$ exists and is nonzero.

Example 1

Find the principal unit tangent vector to the curve C in the xy-plane described by $\mathbf{F}(t) = 2 \cos t\, \mathbf{i} + 3 \sin t\, \mathbf{j}$ when $t = \pi/4$.

Solution
Since
$$\mathbf{F}'(t) = -2 \sin t\, \mathbf{i} + 3 \cos t\, \mathbf{j},$$
we have
$$\|\mathbf{F}'(t)\| = [4(\sin t)^2 + 9(\cos t)^2]^{1/2}.$$
So
$$\mathbf{F}'\left(\frac{\pi}{4}\right) = -2\, \frac{\sqrt{2}}{2}\, \mathbf{i} + 3\, \frac{\sqrt{2}}{2}\, \mathbf{j} = -\sqrt{2}\, \mathbf{i} + \frac{3\sqrt{2}}{2}\, \mathbf{j}$$

12.3 UNIT TANGENT AND UNIT NORMAL VECTORS

and
$$\left\|\mathbf{F}'\left(\frac{\pi}{4}\right)\right\| = \left[2 + \frac{9}{2}\right]^{1/2} = \frac{\sqrt{13}}{\sqrt{2}}.$$

Thus,
$$\mathbf{T}\left(\frac{\pi}{4}\right) = \frac{\mathbf{F}'\left(\frac{\pi}{4}\right)}{\left\|\mathbf{F}'\left(\frac{\pi}{4}\right)\right\|}$$
$$= \left(-\sqrt{2}\mathbf{i} + \frac{3\sqrt{2}}{2}\mathbf{j}\right)\frac{\sqrt{2}}{\sqrt{13}} = -\frac{2\sqrt{13}}{13}\mathbf{i} + \frac{3\sqrt{13}}{13}\mathbf{j}$$

This is illustrated in Figure 12.9.   ▶

There are two unit tangent vectors to a curve at a point, one given by $\mathbf{T}(t)$ and the other by $-\mathbf{T}(t)$. A reference to *the* unit tangent vector will always mean the principal unit tangent vector.

In a plane, the normal line to a curve at a point is the unique line perpendicular to the tangent line at the point (see Figure 12.10). For a curve in space, an infinite number of lines are perpendicular to the tangent line at the point. Any line lying in the plane shown in Figure 12.11 is normal to the curve.

The following result is used to select one particularly important normal.

(12.10)
THEOREM
If $\mathbf{u}(t)$ is a unit vector for each t, then $\mathbf{u}(t) \cdot \mathbf{u}'(t) = 0$. Thus, $\mathbf{u}'(t)$ is orthogonal to $\mathbf{u}(t)$.

Proof. Since $\mathbf{u}(t)$ is of unit length,
$$1 = \|\mathbf{u}(t)\|^2 = \mathbf{u}(t) \cdot \mathbf{u}(t).$$

Differentiating both sides of the equation $1 = \mathbf{u}(t) \cdot \mathbf{u}(t)$ gives
$$0 = D_t(\mathbf{u}(t) \cdot \mathbf{u}(t)) = \mathbf{u}'(t) \cdot \mathbf{u}(t) + \mathbf{u}(t) \cdot \mathbf{u}'(t) = 2\mathbf{u}(t) \cdot \mathbf{u}'(t).$$

Thus,
$$\mathbf{u}(t) \cdot \mathbf{u}'(t) = 0$$

and $\mathbf{u}(t)$ is orthogonal to $\mathbf{u}'(t)$.   ▷

The principal unit tangent vector $\mathbf{T}(t)$ is a unit vector for each t, so by Theorem (12.10):

$\mathbf{T}'(t)$ is always normal to $\mathbf{T}(t)$, hence to the curve described by $\mathbf{F}(t)$.

The unit vector in the direction of $\mathbf{T}'(t)$ is called the *principal* unit normal vector.

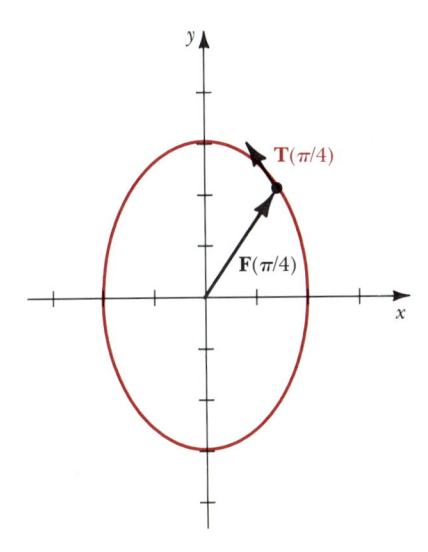

Figure 12.9

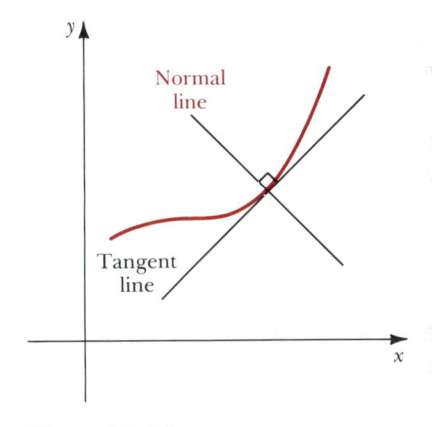

Figure 12.10
Curve in a plane: the normal is a line.

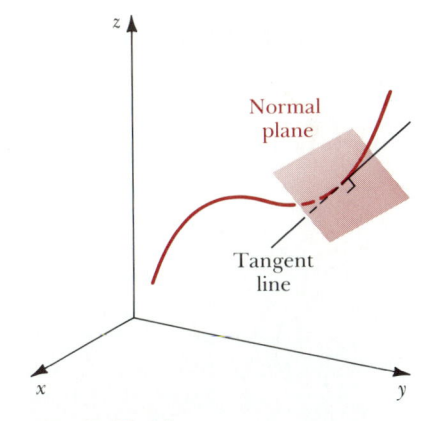

Figure 12.11
Curve in space: the normal is a plane.

(12.11)
DEFINITION
If C is the curve described by **F**, then the **principal unit normal vector** to C at the terminal point of the position vector $\mathbf{F}(t)$ is defined by

$$\mathbf{N}(t) = \frac{\mathbf{T}'(t)}{\|\mathbf{T}'(t)\|},$$

provided $\mathbf{T}'(t)$ exists and is nonzero.

Example 2

For the circular helix described by

$$\mathbf{F}(t) = \cos t\, \mathbf{i} + \sin t\, \mathbf{j} + t\mathbf{k},$$

find the principal unit tangent and principal unit normal vectors
a) for arbitrary values of t and   **b)** at the point $(0, 1, \pi/2)$.

Solution
a) Since $\mathbf{F}'(t) = -\sin t\, \mathbf{i} + \cos t\, \mathbf{j} + \mathbf{k}$, the unit tangent vector is

$$\mathbf{T}(t) = \frac{\mathbf{F}'(t)}{\|\mathbf{F}'(t)\|} = \frac{-\sin t\, \mathbf{i} + \cos t\, \mathbf{j} + \mathbf{k}}{\sqrt{(-\sin t)^2 + (\cos t)^2 + 1}}$$

$$= \frac{\sqrt{2}}{2}(-\sin t\, \mathbf{i} + \cos t\, \mathbf{j} + \mathbf{k}).$$

The unit normal vector is

$$\mathbf{N}(t) = \frac{\mathbf{T}'(t)}{\|\mathbf{T}'(t)\|} = \frac{\frac{\sqrt{2}}{2}(-\cos t\, \mathbf{i} - \sin t\, \mathbf{j})}{\frac{\sqrt{2}}{2}\sqrt{(-\cos t)^2 + (-\sin t)^2}} = -\cos t\, \mathbf{i} - \sin t\, \mathbf{j}.$$

b) Since $(0, 1, \pi/2)$ is the terminal point of $\mathbf{F}(\pi/2)$, the principal unit tangent and unit normal vectors at this point are

$$\mathbf{T}\left(\frac{\pi}{2}\right) = \frac{\sqrt{2}}{2}(-\mathbf{i} + \mathbf{k}) \quad \text{and} \quad \mathbf{N}\left(\frac{\pi}{2}\right) = -\mathbf{j}.$$

These vectors are shown in Figure 12.12. ▶

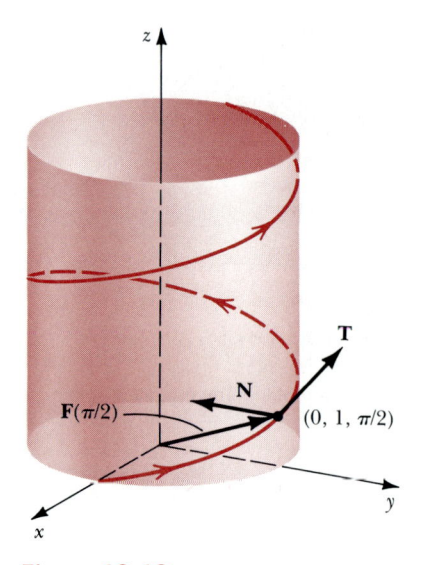

Figure 12.12

A curve in space can be described by many different vector-valued functions. For example, the line segment joining the points $(1, 0, 2)$ and $(4, 5, 7)$ is described by each of the following functions:

(12.12)
$$\mathbf{F}_1(t) = (6t + 1)\mathbf{i} + 10t\mathbf{j} + (10t + 2)\mathbf{k},\ 0 \leq t \leq \tfrac{1}{2};$$
$$\mathbf{F}_2(t) = t^2\mathbf{i} + \tfrac{5}{3}(t^2 - 1)\mathbf{j} + \tfrac{5}{3}(t^2 + \tfrac{1}{5})\mathbf{k},\ 1 \leq t \leq 2;$$
$$\mathbf{F}_3(t) = (\ln t^3 + 1)\mathbf{i} + \ln t^5\, \mathbf{j} + (\ln t^5 + 2)\mathbf{k},\ 1 \leq t \leq e.$$

We can verify this easily by observing that the coordinate functions for each vector function satisfy the symmetric equations for the line:

$$\frac{x-1}{3} = \frac{y}{5} = \frac{z-2}{5}.$$

Although each vector function represents the straight-line segment from (1, 0, 2) to (4, 5, 7), they describe motion along the line segment in quite different ways. For example, at the midpoint of the domains of the functions, three distinct points on the line are represented, as shown in Figure 12.13:

$\mathbf{F}_1\,(0.25) = 2.5\mathbf{i} + 2.5\mathbf{j} + 4.5\mathbf{k}$, the midpoint of the line segment;

$\mathbf{F}_2\,(1.5) \approx 2.25\mathbf{i} + 2.08\mathbf{j} + 4.08\mathbf{k}$, a point closer to (1, 0, 2);

$\mathbf{F}_3\left(\dfrac{e+1}{2}\right) \approx 2.86\mathbf{i} + 3.10\mathbf{j} + 5.10\mathbf{k}$, a point closer to (4, 5, 7).

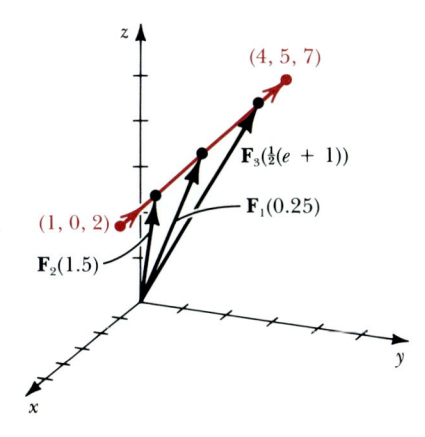

Figure 12.13

Although a variety of functions represent a curve in space, there is a standard method called the **arc length representation**. Arc length representation of a curve of length l is described geometrically by fixing one endpoint of the curve and using the arc length along the curve from that point as the parameter. The point on the curve associated with each value s in $[0, l]$ is the point with distance s along the curve from the fixed endpoint (see Figure 12.14).

Suppose C is a curve in space described on an interval I by

$$\mathbf{F}(t) = x(t)\mathbf{i} + y(t)\mathbf{j} + z(t)\mathbf{k},$$

where x', y', and z' are continuous and not simultaneously zero on the interior of I. Then C has the **smooth** parametric representation

(12.13) $x = x(t)$, $y = y(t)$,   and $z = z(t)$ for t in I

and, as in the case of the planar curves discussed in Section 9.4, C is said to be a **smooth curve.**

If C is a smooth curve defined on $[a, b]$, then the length l of C is given by

$$l = \int_a^b \sqrt{[x'(t)]^2 + [y'(t)]^2 + [z'(t)]^2}\, dt.$$

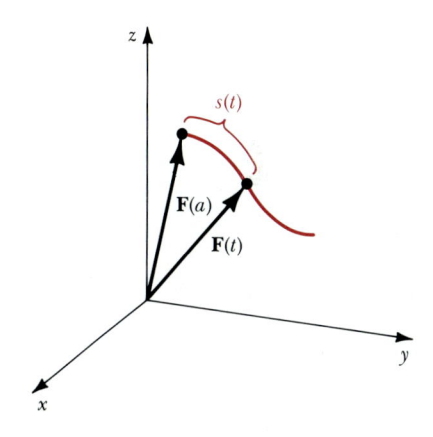

Figure 12.14

Arc length representation.

The derivation of this formula is similar to that of the equivalent formula for parametric equations in the plane (see Section 9.6).

For t in the interval $[a, b]$, consider the length $s(t)$ of C from the terminal point of $\mathbf{F}(a)$ to the terminal point of $\mathbf{F}(t)$:

(12.14) $s \equiv s(t) = \displaystyle\int_a^t \sqrt{[x'(\tau)]^2 + [y'(\tau)]^2 + [z'(\tau)]^2}\, d\tau.$

(The variable τ is used in the integrand to avoid confusion with the upper limit of integration, t.) The function s describes a one-to-one correspondence between $[a, b]$ and $[0, l]$ and permits the function $\mathbf{F}$ in the parameter t to be transformed into a function in the parameter s. The domain of this new function is $[0, l]$. (Figure 12.15 illustrates the lengths l and $s(t)$.)

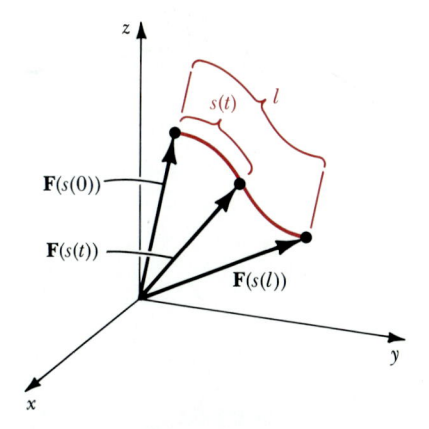

Figure 12.15

Arc length as parameter.

Let us reconsider the straight-line segment joining (1, 0, 2) and (4, 5, 7) to see how the arc length representation evolves. First, we take one of the functions given in (12.12) that describes the line segment, say $\mathbf{F}_1$. Then

$$s(t) = \int_0^t \sqrt{[D_\tau(6\tau + 1)]^2 + [D_\tau 10\tau]^2 + [D_\tau(10\tau + 2)]^2}\, d\tau$$

$$= \int_0^t \sqrt{36 + 100 + 100}\, d\tau = \sqrt{236}\, \tau \Big]_0^t = 2\sqrt{59}\, t.$$

This implies that the length of the line segment from (1, 0, 2) to (4, 5, 7) is $s(\tfrac{1}{2}) = \sqrt{59}$ and that t and s are related by

$$t = \frac{s}{2\sqrt{59}} = \frac{\sqrt{59}\, s}{118}.$$

Substituting $\sqrt{59}\, s/118$ for t in the description of $\mathbf{F}_1$, we obtain a vector-valued function $\mathbf{F}$ with arc length as parameter:

$$\mathbf{F}(s) = \left(\frac{3\sqrt{59}}{59}s + 1\right)\mathbf{i} + \frac{5\sqrt{59}}{59}s\mathbf{j} + \left(\frac{5\sqrt{59}}{59}s + 2\right)\mathbf{k}, \qquad 0 \le s \le \sqrt{59}.$$

If we apply this process to $\mathbf{F}_2$ and $\mathbf{F}_3$, the other functions given in (12.12), we get the same arc length representation. In fact, (12.14) can be used to give the arc length representation of a smooth curve whenever a parametric representation is known.

We introduce parametric representation using arc length at this time because it gives a useful interpretation of the principal unit tangent vector. Suppose the Fundamental Lemma of Calculus is applied to equation (12.14). Then

$$\frac{ds}{dt} = \sqrt{[x'(t)]^2 + [y'(t)]^2 + [z'(t)]^2}$$

$$= \|\mathbf{F}'(t)\|.$$

Thus $\mathbf{T}(t)$ can be written as

(12.15) $$\mathbf{T}(t) = \frac{\mathbf{F}'(t)}{\|\mathbf{F}'(t)\|} = \frac{\mathbf{F}'(t)}{ds/dt} = \frac{d\mathbf{F}(t)}{dt}\frac{dt}{ds} = \frac{d\mathbf{F}(t)}{ds}.$$

Therefore,

(12.16)
The principal unit tangent vector describes the rate of change of the values of the function $\mathbf{F}$ relative to the change in the arc length along the curve described by $\mathbf{F}$.

Arc length is often difficult to obtain explicitly. Fortunately, an explicit representation is not usually required; we need only know that it exists and the method that describes it.

► EXERCISE SET 12.3

In Exercises 1–6, find the principal unit tangent vector at t to the curve.

1. $\mathbf{F}(t) = \cos t\,\mathbf{i} + \sin t\,\mathbf{j}$, $t = \pi/2$
2. $\mathbf{F}(t) = \sin t\,\mathbf{i} + \cos t\,\mathbf{j} + t\mathbf{k}$, $t = \pi$
3. $\mathbf{F}(t) = t^2\mathbf{i} + \ln t\,\mathbf{j} + t\mathbf{k}$, $t = 1$
4. $\mathbf{F}(t) = 2t\mathbf{i} + e^t\mathbf{j} + t^3\mathbf{k}$, $t = 0$
5. $\mathbf{F}(t) = (2t - 1)\mathbf{i} + (t + 1)\mathbf{j} + (1 - 3t)\mathbf{k}$, $t = 2$
6. $\mathbf{F}(t) = (t - 2)\mathbf{i} + (2 - t)\mathbf{j} + (2t - 1)\mathbf{k}$, $t = -1$

7–12. Find the principal unit normal vector at t, if it exists, to the curves described in Exercises 1–6.

In Exercises 13–18, determine the intervals on which the curve described by $\mathbf{F}$ is smooth.

13. $\mathbf{F}(t) = t\mathbf{i} + t^2\mathbf{j}$
14. $\mathbf{F}(t) = \sqrt{1 - t^2}\,\mathbf{j} + t\mathbf{k}$
15. $\mathbf{F}(t) = \mathbf{i} + t\mathbf{j} + \sqrt{t}\,\mathbf{k}$
16. $\mathbf{F}(t) = t\mathbf{j} + |t|\,\mathbf{k}$
17. $\mathbf{F}(t) = t\mathbf{i} + e^t\mathbf{j} + t\mathbf{k}$
18. $\mathbf{F}(t) = t\mathbf{i} + \tan t\,\mathbf{j} + 2\mathbf{k}$

In Exercises 19–24, find a vector-valued function with arc length as parameter to describe the curve.

19. The straight line from $(1, 1, 0)$ to $(2, 5, 4)$
20. The straight line from $(2, 0, 1)$ to $(1, 2, 3)$
21. A circle of radius 1 in the xy-plane with center at the origin
22. A circle of radius 2 in the xz-plane with center at the origin
23. The circular helix described by
$$\mathbf{F}(t) = \sin t\,\mathbf{i} + \cos t\,\mathbf{j} + t\mathbf{k}, \quad 0 \le t \le \pi$$
24. A quarter circle with center $(0, 0, 1)$ that begins at $(\sqrt{2}/2, \sqrt{2}/2, 1)$ and ends at $(-\sqrt{2}/2, \sqrt{2}/2, 1)$.
25. The *binormal vector* at a point P on a curve C is the vector $\mathbf{B} = \mathbf{T} \times \mathbf{N}$, where $\mathbf{T}$ and $\mathbf{N}$ are, respectively, the principal unit tangent and principal unit normal vectors to C at P (see the figure). Show that $\mathbf{B}$ is a unit vector normal to C at P.
26–29. Find the binormal vectors to the curves described in Exercises 1–4 for the given values of t.
30. In (12.12), three functions that describe the same straight-line segment are given. Show that $\mathbf{F}_2$ and $\mathbf{F}_3$ give the same arc length parameterization of this line segment as that found for $\mathbf{F}_1$.

31. a) Sketch the curve described by $\mathbf{F}(t) = 2\mathbf{i} + t\mathbf{j} + (t^2 - 2)\mathbf{k}$.
    b) Find and sketch the principal unit tangent and unit normal vectors to this curve at the point $(2, 2, 2)$.
    c) Determine the values of t at which the principal unit tangent vector is parallel to the xy-plane. What is the direction of the principal unit normal vector in this case?
32. The function $\mathbf{F}(t) = 3\mathbf{i} + 6t\mathbf{j} + (5t - 5t^2)\mathbf{k}$ describes the motion of a baseball.
    a) Sketch the curve described by $\mathbf{F}(t)$.
    b) Find and sketch the principal unit tangent and unit normal vectors at $t = 1$.
    c) Determine the value of t when the unit tangent vector is parallel to the xy-plane.
33. A $\frac{5}{8}$-in. diameter hex bolt has 200 threads in its 10 in. of length (see the figure). Grooves have been cut between the threads to a depth of 0.01 inch. What is the total length of (a) the threads and (b) the grooves?

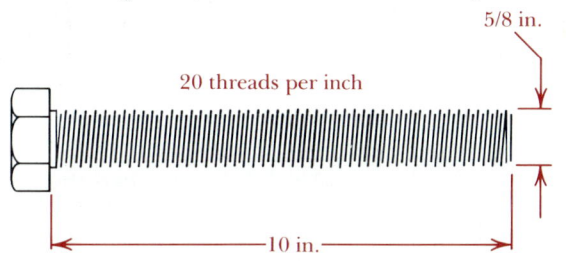

34. Find the length of the threads and grooves of a bolt or pipe that has diameter D, with n threads in its length l, cut to a depth d.

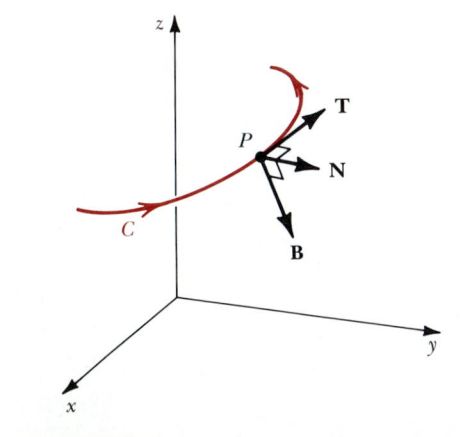

Figure for Exercise 25.

12.4

Velocity and Acceleration of Objects in Space

In this section we study the motion of an object moving along a curve in space. In so doing we will see applications of the principal unit tangent and unit normal vectors and discover why the directions described by these vectors are called the *principal* directions.

An object traveling through space is described by a vector-valued function, usually denoted by **r**. The domain of **r** is the interval of time during which the motion occurs, and the curve C given by the trace of **r** describes the path of the object, as shown in Figure 12.16.

As the object moves along C, both its speed and its direction can vary with time. The *speed* of the object is a scalar function that tells the instantaneous rate of change in the arc length traveled with respect to the change in time. The *velocity* of the object is a vector-valued function that points in the direction of the motion and whose magnitude gives the speed. The definitions of these concepts for objects traveling in space are similar to those given for rectilinear motion in Section 3.1.

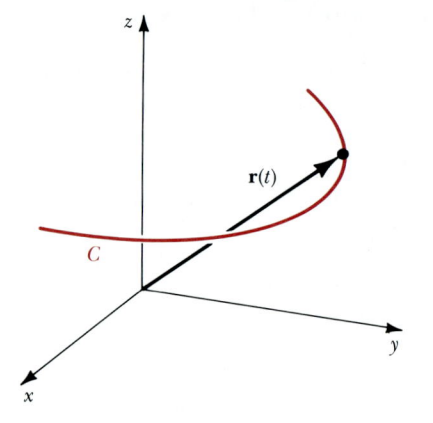

Figure 12.16
The position of an object.

(12.17)
DEFINITION
If the motion of an object through space is described by

$$\mathbf{r}(t) = x(t)\mathbf{i} + y(t)\mathbf{j} + z(t)\mathbf{k}, \qquad a \leq t \leq b,$$

then the **velocity** of the object is

$$\mathbf{v}(t) = \mathbf{r}'(t) = x'(t)\mathbf{i} + y'(t)\mathbf{j} + z'(t)\mathbf{k}$$

and its **speed** is

$$v(t) = \|\mathbf{v}(t)\| = \sqrt{[x'(t)]^2 + [y'(t)]^2 + [z'(t)]^2}.$$

The **acceleration** of the object is

$$\mathbf{a}(t) = \mathbf{v}'(t) = \mathbf{r}''(t) = x''(t)\mathbf{i} + y''(t)\mathbf{j} + z''(t)\mathbf{k}.$$

In Section 12.2, we found that $\mathbf{r}'(t)$ is tangent to the curve of $\mathbf{r}(t)$ and points in the direction of increasing values of t, so the velocity vector does indeed give the direction of the motion. This is illustrated in Figure 12.17. In addition, the speed $v(t)$ of the object is

$$v(t) = \|\mathbf{v}(t)\| = \|D_t \, \mathbf{r}(t)\| = ds/dt,$$

which gives the change in the arc length s with respect to time.

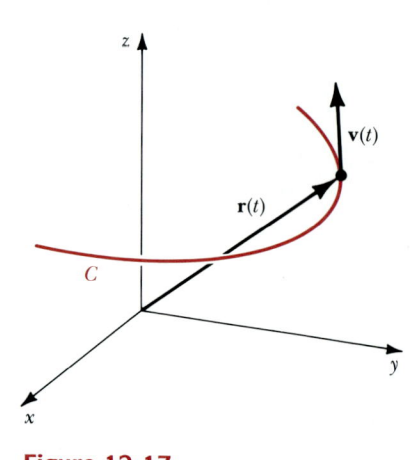

Figure 12.17
The position and velocity of an object.

Example 1

An object travels along the curve of a circular helix, such that its position at any time t is at the terminal point of the position vector

$$\mathbf{r}(t) = \cos t \, \mathbf{i} + \sin t \, \mathbf{j} + t\mathbf{k}, \qquad 0 \leq t \leq 2\pi.$$

Find the velocity, acceleration, and speed of this object.

Solution
The curve along which the object travels is shown in Figure 12.18. The velocity is
$$\mathbf{v}(t) = \mathbf{r}'(t) = -\sin t \, \mathbf{i} + \cos t \, \mathbf{j} + \mathbf{k},$$
the acceleration is
$$\mathbf{a}(t) = \mathbf{v}'(t) = -\cos t \, \mathbf{i} - \sin t \, \mathbf{j},$$
and the speed is
$$v(t) = \|\mathbf{v}(t)\| = \sqrt{(-\sin t)^2 + (\cos t)^2 + 1} = \sqrt{2}.$$

Note that the speed is constant, although both the velocity and acceleration depend on t. ▶

The motion of an object propelled from the earth can be described using a vector-valued function, provided the initial velocity $\mathbf{v}_0$ and the initial position $\mathbf{r}_0$ are known. Since this is essentially a two-dimensional problem, we can assume that the component in the direction of the vector $\mathbf{k}$ is zero and that the object is propelled in the plane shown in Figure 12.19. In this coordinate system, x represents the horizontal direction and y represents the vertical direction.

Suppose gravity is the only force acting on the object and that the acceleration due to gravity acts in the negative y-direction with a magnitude of 32 ft/sec². This is expressed by writing
$$\mathbf{a}(t) = -32\mathbf{j}.$$
The velocity of the object is given by
$$\mathbf{v}(t) = \int \mathbf{a}(t) \, dt = -32t\mathbf{j} + \mathbf{C}_1,$$
for some constant vector $\mathbf{C}_1$. Since
$$\mathbf{v}_0 = \mathbf{v}(0) = -32(0)\mathbf{j} + \mathbf{C}_1 = \mathbf{C}_1,$$
we have
$$\mathbf{v}(t) = -32t\mathbf{j} + \mathbf{v}_0.$$
The position of the object at time t is
$$\mathbf{r}(t) = \int \mathbf{v}(t) \, dt = -16t^2\mathbf{j} + \mathbf{v}_0 t + \mathbf{C}_2,$$
for some constant vector $\mathbf{C}_2$. Since $\mathbf{r}(0) = \mathbf{r}_0$, $\mathbf{C}_2 = \mathbf{r}_0$. Consequently
$$\mathbf{r}(t) = -16t^2\mathbf{j} + \mathbf{v}_0 t + \mathbf{r}_0;$$
Thus if the acceleration is
$$\mathbf{a}(t) = -32\mathbf{j},$$
then

(12.18) $\quad \mathbf{v}(t) = -32t\mathbf{j} + \mathbf{v}_0 \quad$ and $\quad \mathbf{r}(t) = -16t^2\mathbf{j} + t\mathbf{v}_0 + \mathbf{r}_0.$

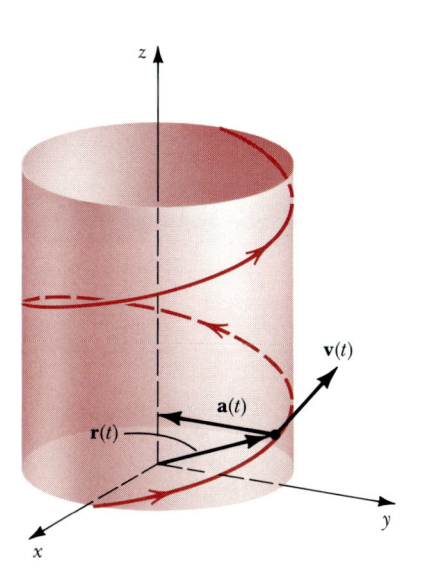

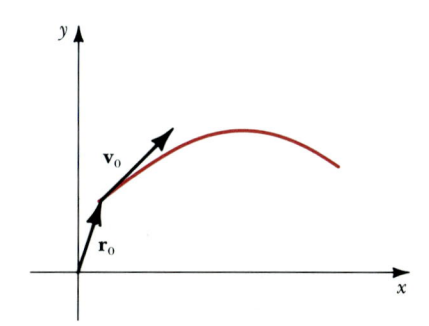

Figure 12.18

Figure 12.19
Initial position and velocity vectors.

Equivalent formulas for $\mathbf{a}(t)$, $\mathbf{v}(t)$, and $\mathbf{r}(t)$ hold when the y-component of acceleration is given in metric units as -9.8 m/sec^2. The formulas in (12.18) show that vector-valued functions enable us to solve general motion problems in space as easily as if the motion had been rectilinear. Equations describing the motion in the horizontal and vertical directions are obtained simultaneously by decomposing the initial velocity vector $\mathbf{v}_0$ into its horizontal and vertical components.

Example 2

A projectile is fired from ground level with an initial speed of 1000 ft/sec and an angle of elevation of $\pi/4$ radians. Describe the subsequent motion of the projectile and find each of the following.

a) The time when the projectile strikes the ground
b) The distance from the firing point when the projectile strikes the ground
c) The maximum height of the projectile
d) The speed of the projectile at the instant of impact

Solution

Figure 12.20 shows the initial velocity vector $\mathbf{v}_0$ and its horizontal and vertical components. The origin has been placed at the firing point, so $\mathbf{r}_0 = \mathbf{0}$. The subsequent motion of the projectile is along the curve shown. Since $\|\mathbf{v}_0\| = 1000$ and the angle of elevation is $\pi/4$,

$$\mathbf{v}_0 = 1000 \cos\frac{\pi}{4}\mathbf{i} + 1000 \sin\frac{\pi}{4}\mathbf{j} = 500\sqrt{2}\,\mathbf{i} + 500\sqrt{2}\mathbf{j}.$$

Since $\mathbf{a}(t) = -32\mathbf{j}$,

$$\mathbf{v}(t) = -32t\mathbf{j} + \mathbf{v}_0 = -32t\mathbf{j} + 500\sqrt{2}\mathbf{i} + 500\sqrt{2}\mathbf{j}$$
$$= 500\sqrt{2}\mathbf{i} + (500\sqrt{2} - 32t)\mathbf{j}.$$

Since $\mathbf{r}'(t) = \mathbf{v}(t)$ and $\mathbf{r}_0 = \mathbf{0}$, the motion of the projectile is described by

$$\mathbf{r}(t) = 500\sqrt{2}\,t\mathbf{i} + (500\sqrt{2}\,t - 16t^2)\mathbf{j}.$$

a) The time of impact occurs when the y-component of $\mathbf{r}(t)$ is zero, that is, when

$$0 = 500\sqrt{2}\,t - 16t^2 = 4t(125\sqrt{2} - 4t).$$

The time $t = 0$ corresponds to the initial position, so the projectile strikes the ground when $t = 125\sqrt{2}/4 \approx 44.2$ sec (see Figure 12.21).

b) Since

$$\mathbf{r}\left(\frac{125\sqrt{2}}{4}\right) = 500\sqrt{2}\left(\frac{125\sqrt{2}}{4}\right)\mathbf{i} = 31{,}250\mathbf{i},$$

the projectile strikes the ground 31,250 ft from the firing point.

c) The maximum height occurs when the y-component of $\mathbf{r}(t)$ is a maximum, that is, when the derivative of this component is zero. At this

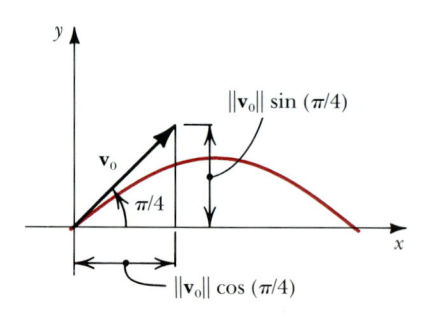

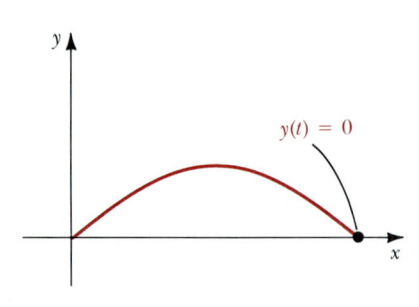

Figure 12.20

Figure 12.21
Maximum distance: $y(t) = 0$

12.4 VELOCITY AND ACCELERATION OF OBJECTS IN SPACE

time the velocity vector is horizontal, as shown in Figure 12.22. Since

$$\mathbf{v}(t) = \mathbf{r}'(t) = 500\sqrt{2}\,\mathbf{i} + (500\sqrt{2} - 32t)\mathbf{j},$$

the maximum height occurs when

$$0 = 500\sqrt{2} - 32t,$$

that is, when

$$t = \frac{500\sqrt{2}}{32} = \frac{125\sqrt{2}}{8}.$$

The maximum height is the y-component of $\mathbf{r}(125\sqrt{2}/8)$. Since

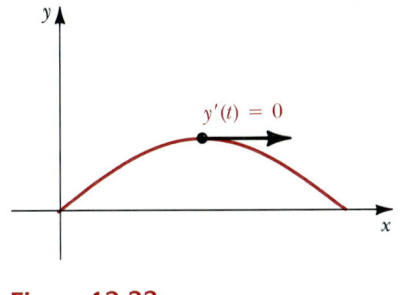

Figure 12.22
Maximum height: $y'(t) = 0$.

$$\mathbf{r}\left(\frac{125\sqrt{2}}{8}\right) = 500\sqrt{2}\left(\frac{125\sqrt{2}}{8}\right)\mathbf{i} + \left[500\sqrt{2}\left(\frac{125\sqrt{2}}{8}\right) - 16\left(\frac{125\sqrt{2}}{8}\right)^2\right]\mathbf{j}$$

$$= (125)^2\mathbf{i} + \left[\frac{(125)^2}{2}\right]\mathbf{j} = 15{,}625\mathbf{i} + 7812.5\mathbf{j},$$

the maximum height is 7812.5 ft.

d) The projectile strikes the ground when $t = 125\sqrt{2}/4$, so the speed of the projectile at the instant of impact is given by

$$\left\|\mathbf{v}\left(\frac{125\sqrt{2}}{4}\right)\right\| = \sqrt{(500\sqrt{2})^2 + \left[500\sqrt{2} - (32)\left(\frac{125\sqrt{2}}{4}\right)\right]^2}$$

$$= \sqrt{(500)^2 \cdot 2 + [-500\sqrt{2}]^2}$$

$$= (500)(2) = 1000 \text{ ft/sec},$$

the speed at which it was propelled. ▶

Since the velocity vectors of an object are tangent to the curve of motion, there is a natural connection between the velocity, speed, and acceleration of the object and the principal unit tangent and principal unit normal vectors to the curve of motion. The unit tangent vector for $\mathbf{r}(t)$ is

$$\mathbf{T}(t) = \frac{\mathbf{r}'(t)}{\|\mathbf{r}'(t)\|} = \frac{\mathbf{v}(t)}{v(t)};$$

so the velocity of an object is

(12.19) $\qquad \mathbf{v}(t) = v(t)\mathbf{T}(t).$

If $\mathbf{T}'(t) \neq \mathbf{0}$, the acceleration is

$$\mathbf{a}(t) = \mathbf{v}'(t) = v'(t)\mathbf{T}(t) + v(t)\mathbf{T}'(t) = v'(t)\mathbf{T}(t) + v(t)\|\mathbf{T}'(t)\|\left(\frac{\mathbf{T}'(t)}{\|\mathbf{T}'(t)\|}\right),$$

so

(12.20) $\qquad \mathbf{a}(t) = v'(t)\mathbf{T}(t) + v(t)\|\mathbf{T}'(t)\|\mathbf{N}(t).$

This implies that the acceleration can be decomposed into one component in the direction of the principal unit tangent vector $\mathbf{T}(t)$ and one component in the direction of the principal unit normal vector $\mathbf{N}(t)$ (see Figure 12.23). The component in the direction of the principal unit

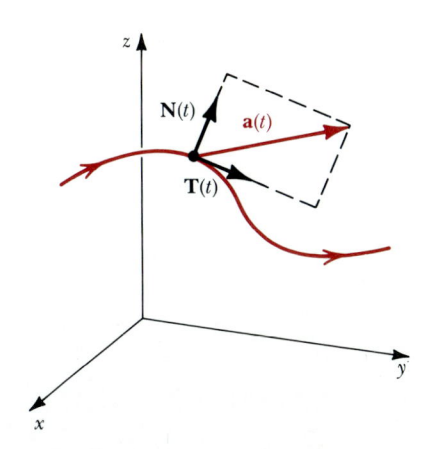

Figure 12.23

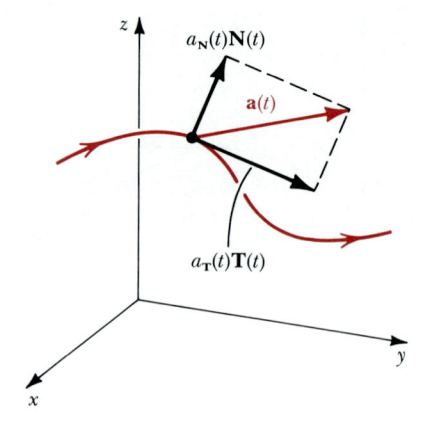

tangent vector is called the **tangential component of acceleration** and is denoted $a_\mathbf{T}$:

(12.21) $$a_\mathbf{T}(t) = v'(t).$$

The component in the direction of the principal unit normal vector is called the **centripetal component of acceleration** and is denoted $a_\mathbf{N}$:

(12.22) $$a_\mathbf{N}(t) = v(t)\|\mathbf{T}'(t)\|.$$

With this notation,

(12.23) $$\mathbf{a}(t) = a_\mathbf{T}(t)\mathbf{T}(t) + a_\mathbf{N}(t)\mathbf{N}(t).$$

The *tangential* component of acceleration describes the instantaneous rate of change in the *speed* with respect to time. The *centripetal* component of acceleration measures the instantaneous rate of change in the *direction* with respect to time. The first sentence is clear from the definition of $a_\mathbf{T}$; the second sentence is justified in Section 12.5.

Since $\mathbf{T}(t)$ and $\mathbf{N}(t)$ are orthogonal, the Pythagorean Theorem implies (see Figure 12.24) that

$$\|\mathbf{a}(t)\|^2 = \|a_\mathbf{T}(t)\mathbf{T}(t)\|^2 + \|a_\mathbf{N}(t)\mathbf{N}(t)\|^2.$$

But $\mathbf{T}(t)$ and $\mathbf{N}(t)$ are unit vectors, so

(12.24) $$\|\mathbf{a}(t)\|^2 = [a_\mathbf{T}(t)]^2 + [a_\mathbf{N}(t)]^2.$$

Since $a_\mathbf{N}(t) = v(t)\|\mathbf{T}'(t)\| = \|\mathbf{v}(t)\|\|\mathbf{T}'(t)\| \geq 0$, we also have

(12.25) $$a_\mathbf{N}(t) = \sqrt{\|\mathbf{a}(t)\|^2 - [a_\mathbf{T}(t)]^2}.$$

Figure 12.24
Tangential and centripetal components of acceleration.

Example 3

Find the tangential and centripetal components of acceleration of a particle whose motion is given by

$$\mathbf{r}(t) = t\mathbf{i} + t\mathbf{j} + t^2\mathbf{k}, \qquad 0 \leq t \leq 1.$$

Solution
The velocity, acceleration, and speed of this object are given by

$$\mathbf{v}(t) = \mathbf{i} + \mathbf{j} + 2t\mathbf{k}, \qquad \mathbf{a}(t) = 2\mathbf{k}, \quad \text{and} \quad v(t) = \sqrt{2 + 4t^2}.$$

The tangential component of acceleration is

$$a_\mathbf{T}(t) = v'(t) = D_t\sqrt{2 + 4t^2} = \frac{4t}{\sqrt{2 + 4t^2}}.$$

To compute the centripetal component of acceleration, we use (12.25). Since $\|\mathbf{a}(t)\|^2 = \|2\mathbf{k}\|^2 = 4$,

$$a_\mathbf{N}(t) = \sqrt{\|\mathbf{a}(t)\|^2 - [a_\mathbf{T}(t)]^2} = \sqrt{4 - \frac{16t^2}{2 + 4t^2}}$$

$$= \sqrt{\frac{8}{2 + 4t^2}} = \frac{2}{\sqrt{1 + 2t^2}}. \qquad \blacktriangleright$$

Example 4

An object rotates in a circular orbit of radius R at a constant angular speed of ω radians/second. Show that the acceleration is entirely centripetal.

Solution

Suppose the rotation occurs counterclockwise in the xy-plane with the center of the orbit at the origin and that the object is at $(R, 0)$ when $t = 0$ (see Figure 12.25). The angular speed is constant at ω radians/second, so the object requires $2\pi/\omega$ seconds to complete one revolution. Thus the motion of the object is described by

$$\mathbf{r}(t) = R(\cos \omega t \, \mathbf{i} + \sin \omega t \, \mathbf{j}).$$

So

$$\mathbf{v}(t) = R(-\omega \sin \omega t \, \mathbf{i} + \omega \cos \omega t \, \mathbf{j})$$

and

$$\mathbf{a}(t) = -\omega^2 R(\cos \omega t \, \mathbf{i} + \sin \omega t \, \mathbf{j}) = -\omega^2 \mathbf{r}(t).$$

Since

$$v(t) = \|\mathbf{v}(t)\| = R\sqrt{(-\omega \sin \omega t)^2 + (\omega \cos \omega t)^2} = \omega R,$$

we have

$$a_\mathbf{T}(t) = v'(t) = D_t(\omega R) = 0$$

and

$$a_\mathbf{N}(t) = \sqrt{\|\mathbf{a}(t)\|^2 - [a_\mathbf{T}(t)]^2} = \|\mathbf{a}(t)\| = \omega^2 R.$$

This implies that the acceleration of an object rotating in a circular orbit at constant speed is entirely centripetal.

Since $\mathbf{a}(t) = -\omega^2 \mathbf{r}(t)$, the acceleration is directed toward the center of the circle and has magnitude directly proportional to the radius of the circle and the square of the angular speed. ▶

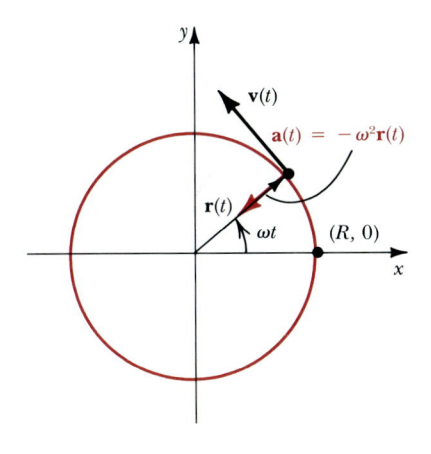

Figure 12.25
Totally centripetal acceleration.

▶ EXERCISE SET 12.4

The motion of an object is described in Exercises 1–8. Find (a) the velocity, (b) the acceleration, and (c) the speed of the object.

1. $\mathbf{r}(t) = \sin t \, \mathbf{i} + \cos t \, \mathbf{j} + t \mathbf{k}$
2. $\mathbf{r}(t) = \sin t \, \mathbf{i} + \cos t \, \mathbf{j} + e^t \mathbf{k}$
3. $\mathbf{r}(t) = (t - 1)\mathbf{i} + (2t - 1)\mathbf{j} + (3t - 3)\mathbf{k}$
4. $\mathbf{r}(t) = \mathbf{i} + 3t\mathbf{j} + 6t^2 \mathbf{k}$
5. $\mathbf{r}(t) = 3\mathbf{i} + 4t\mathbf{j} + (5t - 5t^2)\mathbf{k}$
6. $\mathbf{r}(t) = (10 \sin t - 1)\mathbf{i} + t\mathbf{j} + (16 \cos t - 2)\mathbf{k}$
7. $\mathbf{r}(t) = \sqrt{3t - 1} \, \mathbf{j} + e^t \mathbf{k}$
8. $\mathbf{r}(t) = t\mathbf{j} + (16t^2 - 10t)\mathbf{k}$

The motion of an object is described in Exercises 9–14. Determine the tangential and centripetal components of acceleration of the object.

9. $\mathbf{r}(t) = \sin t \, \mathbf{i} + \cos t \, \mathbf{j} + t\mathbf{k}$
10. $\mathbf{r}(t) = \cos t \, \mathbf{i} + \sin t \, \mathbf{j} + t\mathbf{k}$
11. $\mathbf{r}(t) = (2t - 1)\mathbf{i} + (2 - t)\mathbf{j} + (t - 3)\mathbf{k}$
12. $\mathbf{r}(t) = 3\mathbf{i} + 2t\mathbf{j} + t^2 \mathbf{k}$

13. $\mathbf{r}(t) = t\mathbf{j} + (16t^2 - 10t)\mathbf{k}$
14. $\mathbf{r}(t) = 2\mathbf{i} + 3t\mathbf{j} + e^t\mathbf{k}$
15. The path of a golf ball is described by

$$\mathbf{r}(t) = \begin{cases} 110t\mathbf{j} + 30t(2-t)\mathbf{k}, & \text{if } 0 \leq t < 2, \\ 220\sqrt{t-1}\mathbf{j}, & \text{if } 2 < t \leq 2.4, \end{cases}$$

where the time t is given in seconds and the distance in yards.
   a) Sketch the curve that represents the path of the golf ball.
   b) Find the maximum height of the golf ball.
   c) Find the speed of the ball when it strikes the ground.

16. The path of a baseball is described by

$$\mathbf{r}(t) = 2t(t^2 - \tfrac{5}{4}t + \tfrac{1}{4})\mathbf{i} + 45t\mathbf{j} + (4+t)\mathbf{k}, \ 0 \leq t \leq 1,$$

where the time t is in seconds and the distance in meters. Find each of the following.
   a) The initial speed of the ball
   b) The speed of the ball when $t = 1$ second
   c) The horizontal distance traveled by the ball

17. Suppose the projectile in Example 2 is fired from ground level at an angle of elevation of $\pi/3$ radians with the same initial velocity as in that example, 1000 ft/sec. Describe the subsequent motion of the projectile and answer questions (a)–(d) of Example 2 in this situation.

18. A projectile is fired from the ground with an initial speed of 1200 ft/sec and an angle of elevation of $\pi/6$ radians. Find each of the following.
   a) The subsequent motion of the projectile
   b) The velocity at time t
   c) The maximum height of the projectile
   d) The time when the projectile strikes the ground
   e) The speed of the projectile at the instant of impact

19. A quarterback throws a football downfield with an initial speed of 55 ft/sec at an angle of elevation of $\pi/6$ radians.
   a) How far from the quarterback should the receiver be to catch the ball?
   b) How long from the time the ball is thrown does he have time to get into this position? (Assume that the ball is thrown and caught at the same height.)

20. In the 1984 Olympics in Los Angeles the winner in the shotput was Alessandro Andrei from Italy with a distance of 69 ft, 9 in. If the shot was launched from 6.5 ft above the ground at an angle of $\pi/4$ radians, what was (a) the initial speed? (b) the maximum height? (c) the time of flight?

21. A baseball hit at an angle of elevation of $\pi/4$ radians just clears a 21-ft-high wall in right field 335 ft from home plate at Three Rivers Stadium in Pittsburgh. If the ball is 4 ft above the ground when hit, find each of the following. 
   a) The initial speed of the ball
   b) The time it takes to reach the wall
   c) The velocity as it clears the wall
   d) The speed as it clears the wall

22. A centerfielder throws a ball toward home plate 330 ft away at 110 ft/sec. At what angle should the ball be thrown to reach home plate at the height at which it was thrown?

23. A runner tags third base and heads for home at 25 ft/sec at precisely the time the fielder in Exercise 22 releases the ball. Is he sure to beat the throw to the plate (90 ft away)?

24. A 50-lb block of ice is placed on a slide 20 ft long that makes an angle of 15° with the horizontal. The end of the slide is 4 ft above the ground and the speed of the block at the end is 20 ft/sec (see the figure). Where will the block land?

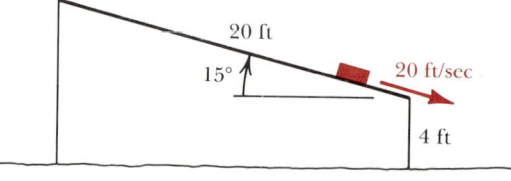

25. Find the total horizontal distance a golf ball travels in the air if it leaves a tee at a velocity of 140 ft/sec at an angle of $\pi/6$ (see the figure). Is it possible for the golf ball to hit the branch of a tree that is 90 ft above the fairway?

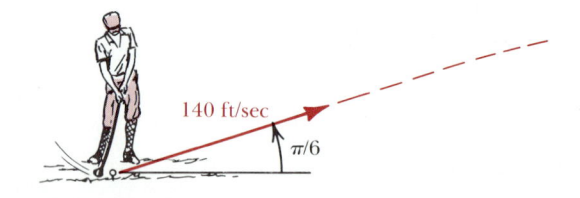

26.
  a) Show that the maximum horizontal distance an object can travel when propelled from the ground with initial speed v_0 occurs when the angle of elevation with the horizontal is $\pi/4$.
  b) Determine this maximum distance.

27. Show that if an object is propelled from the ground at an angle θ with the horizontal, then it hits the ground at this same angle.

28. A 105-mm Howitzer of type M102 employed by the U.S. Army firing an HE(M1) projectile has a muzzle velocity of 494 m/sec. The stated maximum range for this projectile is 11500 m. What is the predicted maximum range from the motion equations and why is there a "slight" discrepancy?

A 105 millimeter Type M102 howitzer being placed in firing position.

29. A Saab 37 Viggen multimission combat aircraft is capable of going from stationary on the ground to a speed of Mach 2 (approximately 1330 mph) and an altitude of 32,000 ft in 1 min, 40 sec. Assuming the acceleration is a constant vector over this time period, derive the motion equations for this aircraft.

30. A slob throws an empty beer can from the window of a pickup truck while the truck is stopped at a traffic light. The can is thrown perpendicular to the side of the truck, the window is 5 ft above the ground, and the can lands 25 ft from the truck. Suppose the next can is thrown in the same manner, but the truck is traveling at 35 mph. How far from the truck will this can land if the effect of the wind resistance is neglected?

31. Show that the acceleration of an object moving along the helix described by $\mathbf{r}(t) = \cos t\, \mathbf{i} + \sin t\, \mathbf{j} + t\mathbf{k}$ is totally centripetal.

32. Suppose an object rotates about the z-axis on a circle of radius a lying in a plane parallel to and b units from the xy-plane, and that the angular speed $d\theta/dt$ is a constant ω. The vector $\boldsymbol{\omega} = \omega \mathbf{k}$ is called the *angular velocity* of the object. Show that the velocity vector $\mathbf{v}(t) = \mathbf{r}'(t)$ is the cross product of the angular velocity and the position vector $\mathbf{r}(t)$.

33. Show that the speed of an object is constant if and only if the tangential component of acceleration is zero.

34. Show that the speed of an object is constant if and only if the velocity and acceleration vectors are orthogonal.

35. The motion of an object moving in the xy-plane is described by $\mathbf{r}(t) = x(t)\mathbf{i} + y(t)\mathbf{j}$, $0 \leq t \leq b$. Suppose that at each point $\mathbf{r}(t) \cdot \mathbf{r}'(t) = 0$. Show that the object is moving along a circle with equation $x^2 + y^2 = \|\mathbf{r}(0)\|^2$.

The SAAB 37 VIGGEN (Thunderbolt) is produced by Saab-Scandia in Sweden (Exercise 29).

12.5

Curvature

One distinguishing feature of a curve in space is its length; another is the way it changes direction. For example, straight lines have a constant direction, so there is no change in direction for a line, as shown in Figure

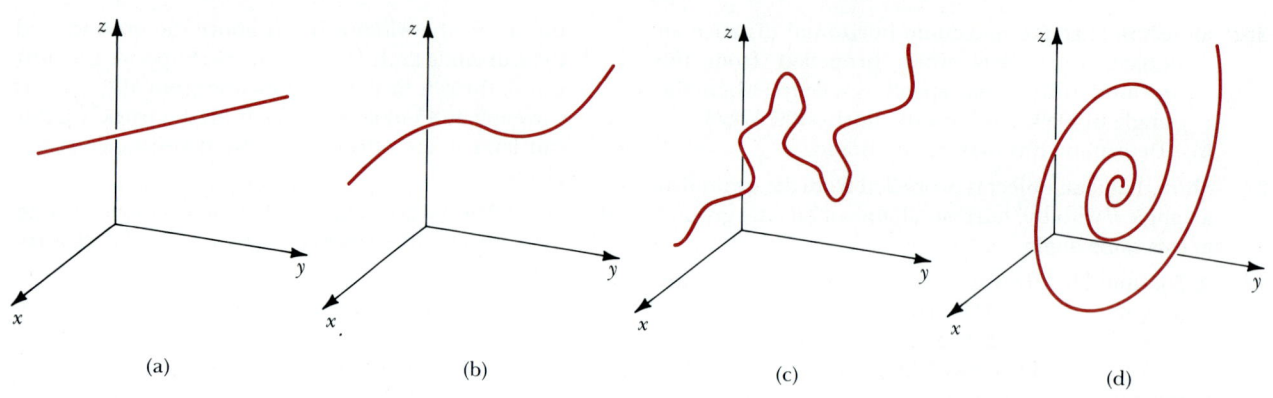

Figure 12.26
(a) No direction change; (b) gradual direction change; (c) abrupt direction change; (d) accelerated direction change.

12.26(a). The curve in Figure 12.26(b) is changing direction gradually, while the curve in Figure 12.26(c) changes direction more abruptly. The spiral in Figure 12.26(d) changes direction slowly for points on the outside loops of the curve, but changes rapidly for points near the apex of the spiral.

The principal unit tangent vector of a curve describes the direction of motion along the curve at each point on the curve. Since the magnitude of this vector is always 1, the derivative of the principal unit tangent vector with respect to the arc length s, $D_s\,\mathbf{T}(s)$, depends entirely on the change in the direction of the curve.

(12.26)
DEFINITION
Suppose C is a curve with a differentiable principal unit tangent vector $\mathbf{T}(s)$, represented using the arc length parameter s. The **curvature vector** of the curve C is defined by

$$\mathbf{K}(s) = D_s\,\mathbf{T}(s).$$

The magnitude of the curvature vector is denoted $K(s)$, which defines the **curvature** of the curve C:

$$K(s) = \|D_s\,\mathbf{T}(s)\|.$$

Although Definition (12.26) uses the arc length representation of the curve to describe the curvature vector, other parametric representations can be used. To see this, suppose C is described by the vector-valued function $\mathbf{r}(t)$. Then, in terms of the parameter t, the curvature vector is given by

$$\mathbf{K}(t) = \frac{d\mathbf{T}(t)}{ds} = \frac{d\mathbf{T}(t)}{dt}\frac{dt}{ds}.$$

However, if $\mathbf{r}'(t) \neq \mathbf{0}$,

$$\frac{dt}{ds} = \frac{1}{ds/dt} = \frac{1}{\|\mathbf{r}'(t)\|} = \frac{1}{v(t)},$$

so the curvature vector is

(12.27) $$\mathbf{K}(t) = \mathbf{T}'(t)\frac{dt}{ds} = \frac{\mathbf{T}'(t)}{\|\mathbf{r}'(t)\|} = \frac{\mathbf{T}'(t)}{v(t)}$$

and the curvature is

(12.28) $$K(t) = \frac{\|\mathbf{T}'(t)\|}{\|\mathbf{r}'(t)\|} = \frac{\|\mathbf{T}'(t)\|}{v(t)}.$$

Example 1

Show that the curvature vector of any line is $\mathbf{K}(t) \equiv \mathbf{0}$.

Solution

Suppose the line passes through the point (x_0, y_0, z_0) with direction given by $\langle a, b, c \rangle$, as shown in Figure 12.27. Then the line is described by

$$\mathbf{r}(t) = (x_0 + at)\mathbf{i} + (y_0 + bt)\mathbf{j} + (z_0 + ct)\mathbf{k}.$$

Since

$$\mathbf{T}(t) = \frac{\mathbf{r}'(t)}{\|\mathbf{r}'(t)\|} = \frac{a\mathbf{i} + b\mathbf{j} + c\mathbf{k}}{\sqrt{a^2 + b^2 + c^2}},$$

$\mathbf{T}'(t) = \mathbf{0}$ for each t and $\mathbf{K}(t) = \mathbf{T}'(t)/\|\mathbf{r}'(t)\| = \mathbf{0}$. ▶

Figure 12.27
Tangent vector constant: curvature is zero.

Example 2

Show that the curvature of a circle of radius R at any point on the circle is $1/R$.

Solution

A vector-valued function that describes a circle of radius R is

$$\mathbf{r}(t) = R(\cos t\, \mathbf{i} + \sin t\, \mathbf{j}), \quad 0 \leq t \leq 2\pi.$$

So

$$\mathbf{T}(t) = \frac{\mathbf{r}'(t)}{\|\mathbf{r}'(t)\|} = \frac{R(-\sin t\, \mathbf{i} + \cos t\, \mathbf{j})}{R[(-\sin t)^2 + (\cos t)^2]^{1/2}} = -\sin t\, \mathbf{i} + \cos t\, \mathbf{j}$$

and

$$\mathbf{K}(t) = \frac{\mathbf{T}'(t)}{\|\mathbf{r}'(t)\|} = \frac{(-\cos t\, \mathbf{i} - \sin t\, \mathbf{j})}{R}.$$

Thus,

$$K(t) = \|\mathbf{K}(t)\| = \frac{1}{R}\sqrt{(-\cos t)^2 + (-\sin t)^2} = \frac{1}{R}.$$ ▶

Small circle, large curvature

Large circle, small curvature

Example 2 shows that a circle with a small radius changes direction faster than a circle with a larger radius. This is illustrated in Figure 12.28. The radius of the circle is the reciprocal of the curvature,

$$R = 1/K(t);$$

and the curvature vector

$$\mathbf{K}(t) = \frac{-\cos t\, \mathbf{i} - \sin t\, \mathbf{j}}{R} = -\frac{1}{R^2}\mathbf{r}(t)$$

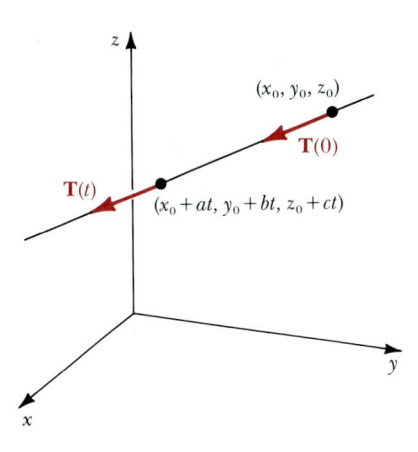

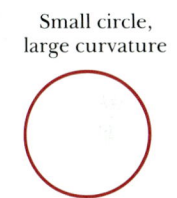

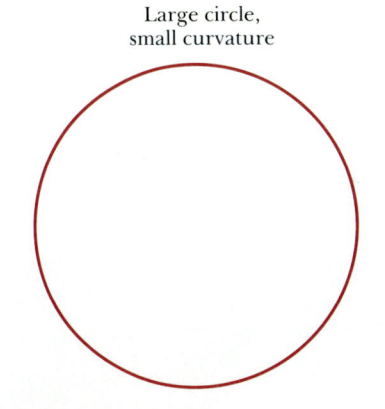

Figure 12.28

points in the direction from the point on the circle to the center of the circle, as shown in Figure 12.29.

These results can be applied to general curves in space. Suppose we fix a point P on a curve C and attempt to draw the circle through P that "best approximates" C near P, as shown in Figure 12.30. This circle will:

i) have the same unit tangent vector at P as the curve C;
ii) have the same curvature at P as the curve C; and
iii) lie in the plane determined by $\mathbf{T}(t)$ and $\mathbf{N}(t)$.

Since the curvature of a circle is the reciprocal of its radius, the radius of the circle that best approximates C at P is the reciprocal of the curvature $K(t)$ at P. We call this value

(12.29) $$\rho(t) = \frac{1}{K(t)},$$

the **radius of curvature** of the curve at P.

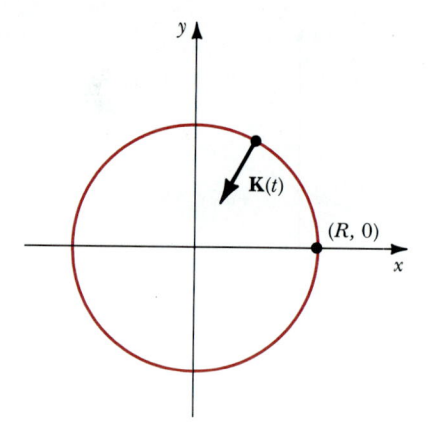

Figure 12.29

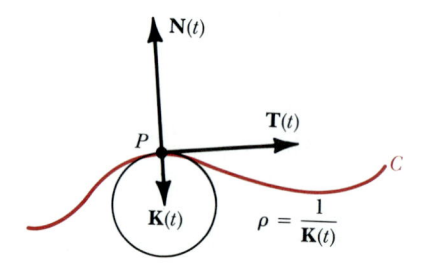

Figure 12.30
The circle best approximating the curve near P.

Example 3

Find the radius of curvature of the circular helix described by

$$\mathbf{r}(t) = \cos t\, \mathbf{i} + \sin t\, \mathbf{j} + t\mathbf{k}$$

at the point $P(0, 1, \pi/2)$.

Solution
First note that $(0, 1, \pi/2)$ corresponds to $\mathbf{r}(\pi/2)$. Since

$$\mathbf{r}'(t) = -\sin t\, \mathbf{i} + \cos t\, \mathbf{j} + \mathbf{k},$$

we have

$$\|\mathbf{r}'(t)\| = \sqrt{(-\sin t)^2 + (\cos t)^2 + 1} = \sqrt{2}.$$

The unit tangent vector is given by

$$\mathbf{T}(t) = \frac{\mathbf{r}'(t)}{\|\mathbf{r}'(t)\|} = \frac{1}{\sqrt{2}}(-\sin t\, \mathbf{i} + \cos t\, \mathbf{j} + \mathbf{k}) = \frac{\sqrt{2}}{2}(-\sin t\, \mathbf{i} + \cos t\, \mathbf{j} + \mathbf{k})$$

and

$$\mathbf{T}'(t) = \frac{\sqrt{2}}{2}(-\cos t\, \mathbf{i} - \sin t\, \mathbf{j}).$$

The curvature is

$$K(t) = \frac{\|\mathbf{T}'(t)\|}{\|\mathbf{r}'(t)\|} = \frac{1}{\sqrt{2}}\left[\frac{\sqrt{2}}{2}\sqrt{(-\cos t)^2 + (-\sin t)^2}\right] = \frac{1}{2},$$

and the radius of curvature of the helix at $(0, 1, \pi/2)$ is

$$\rho\left(\frac{\pi}{2}\right) = \frac{1}{K(\pi/2)} = \frac{1}{1/2} = 2.$$

In Section 12.4 we found that the acceleration of an object can be decomposed into tangential and centripetal components

$$\mathbf{a}(t) = a_\mathbf{T}(t)\mathbf{T}(t) + a_\mathbf{N}(t)\mathbf{N}(t),$$

where

(12.30) $\qquad a_\mathbf{T}(t) = v'(t) \quad \text{and} \quad a_\mathbf{N}(t) = v(t)\|\mathbf{T}'(t)\|.$

The tangential component $a_\mathbf{T}(t)$ describes the change in the speed with respect to time, and the centripetal component $a_\mathbf{N}(t)$ measures the change in direction with respect to time. The first part of this statement is clear from (12.30). Equation (12.28) justifies the second part of the statement, since $a_\mathbf{N}(t)$ can now be written as

(12.31) $\qquad a_\mathbf{N}(t) = v(t)\|\mathbf{T}'(t)\| = v(t)[v(t)K(t)] = [v(t)]^2 K(t).$

The centripetal component, then, is directly proportional to the curvature (the rate at which the direction is changing) and to the square of the speed. Equation (12.31) implies that the decomposition of the acceleration is also expressed as

(12.32) $\qquad \mathbf{a}(t) = v'(t)\mathbf{T}(t) + [v(t)]^2 K(t)\mathbf{N}(t).$

To find the curvature using Equation (12.28), we must first find the curvature vector, which requires $\mathbf{T}'(t)$. However, computing $\mathbf{T}'(t)$ can be quite tedious, since it requires differentiating a quotient whose denominator, $\|\mathbf{r}'(t)\|$, is often complicated. In many applications, only the curvature is needed, not the curvature vector. In this case, a more easily evaluated expression for curvature can be given, one that involves only $\mathbf{v}(t)$ and $\mathbf{a}(t)$. To derive this expression, we first express $\mathbf{v}(t)$ and $\mathbf{a}(t)$ using the principal unit tangent and principal unit normal vectors:

$$\mathbf{v}(t) = \mathbf{r}'(t) = \|\mathbf{r}'(t)\|\mathbf{T}(t) = v(t)\mathbf{T}(t),$$
$$\mathbf{a}(t) = a_\mathbf{T}(t)\mathbf{T}(t) + a_\mathbf{N}(t)\mathbf{N}(t) = v'(t)\mathbf{T}(t) + v(t)\|\mathbf{T}'(t)\|\mathbf{N}(t).$$

We then find the cross product of $\mathbf{v}(t)$ and $\mathbf{a}(t)$:

$$\begin{aligned}\mathbf{v}(t) \times \mathbf{a}(t) &= [v(t)\mathbf{T}(t)] \times [v'(t)\mathbf{T}(t) + v(t)\|\mathbf{T}'(t)\|\mathbf{N}(t)] \\ &= [v(t)\mathbf{T}(t) \times v'(t)\mathbf{T}(t)] + [v(t)\mathbf{T}(t) \times v(t)\|\mathbf{T}'(t)\|\mathbf{N}(t)] \\ &= v(t)v'(t)[\mathbf{T}(t) \times \mathbf{T}(t)] + [v(t)^2\|\mathbf{T}'(t)\|][\mathbf{T}(t) \times \mathbf{N}(t)] \\ &= [v(t)]^2\|\mathbf{T}'(t)\|[\mathbf{T}(t) \times \mathbf{N}(t)].\end{aligned}$$

But $\mathbf{T}(t)$ and $\mathbf{N}(t)$ are orthogonal unit vectors so $\|\mathbf{T}(t) \times \mathbf{N}(t)\| = 1$ and

(12.33) $\qquad \|\mathbf{v}(t) \times \mathbf{a}(t)\| = [v(t)]^2 \|\mathbf{T}'(t)\|.$

Since $K(t) = \|\mathbf{T}'(t)\|/v(t)$, Equation (12.33) implies that

(12.34) $\qquad K(t) = \dfrac{\|\mathbf{v}(t) \times \mathbf{a}(t)\|}{[v(t)]^3},$

a formula that does not require the evaluation of $\mathbf{T}'(t)$.

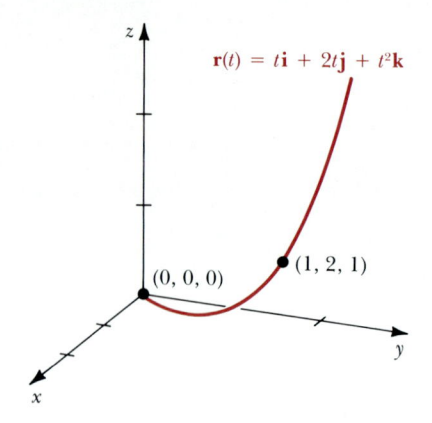

Figure 12.31

$\mathbf{r}(t) = t\mathbf{i} + 2t\mathbf{j} + t^2\mathbf{k}$

Example 4

Find the curvature and components of acceleration, $a_\mathbf{T}$ and $a_\mathbf{N}$, of an object at the point (1, 2, 1) if the position of the object is given by

$$\mathbf{r}(t) = t\mathbf{i} + 2t\mathbf{j} + t^3\mathbf{k}, \quad t \geq 0$$

(see Figure 12.31).

Solution

The velocity, acceleration, and speed of this object at any time t are given by

$$\mathbf{v}(t) = \mathbf{i} + 2\mathbf{j} + 3t^2\mathbf{k}, \quad \mathbf{a}(t) = 6t\mathbf{k}, \quad \text{and} \quad v(t) = \sqrt{5 + 9t^4}.$$

Since the curvature vector is not required in this example, it is easier to use formula (12.34) to determine the curvature:

$$K(t) = \frac{\|\mathbf{v}(t) \times \mathbf{a}(t)\|}{[v(t)]^3} = \frac{\|12t\mathbf{i} - 6t\mathbf{j}\|}{(5 + 9t^4)^{3/2}} = \frac{\sqrt{180}\, t}{(5 + 9t^4)^{3/2}} = \frac{6\sqrt{5}\, t}{(5 + 9t^4)^{3/2}}.$$

The point (1, 2, 1) is determined by $t = 1$, so the curvature at this point is

$$K(1) = \frac{6\sqrt{5}}{(14)^{3/2}} = \frac{3\sqrt{70}}{98}.$$

Substituting into (12.31) gives

$$a_\mathbf{N}(1) = [v(1)]^2 K(1) = (\sqrt{14})^2 \frac{3\sqrt{70}}{98} = \frac{3}{7}\sqrt{70}.$$

To determine $a_\mathbf{T}(1)$, we use Equation (12.30):

$$a_\mathbf{T}(t) = v'(t) = \frac{18t^3}{\sqrt{5 + 9t^4}} \quad \text{so} \quad a_\mathbf{T}(1) = \frac{18}{\sqrt{14}} = \frac{9\sqrt{14}}{7}.$$

▶

Formula (12.34) can be simplified further when the curve is restricted to the xy-plane. Suppose C is described by

$$\mathbf{r}(t) = x(t)\mathbf{i} + y(t)\mathbf{j}.$$

Then

$$\mathbf{v}(t) = x'(t)\mathbf{i} + y'(t)\mathbf{j}, \quad \mathbf{a}(t) = x''(t)\mathbf{i} + y''(t)\mathbf{j}$$

and

$$\begin{aligned}\mathbf{v}(t) \times \mathbf{a}(t) &= [x'(t)\mathbf{i} + y'(t)\mathbf{j}] \times [x''(t)\mathbf{i} + y''(t)\mathbf{j}] \\ &= [x'(t)x''(t)](\mathbf{i} \times \mathbf{i}) + [x'(t)y''(t)](\mathbf{i} \times \mathbf{j}) \\ &\quad + [y'(t)x''(t)](\mathbf{j} \times \mathbf{i}) + [y'(t)y''(t)](\mathbf{j} \times \mathbf{j}) \\ &= [x'(t)y''(t) - x''(t)y'(t)]\mathbf{k}.\end{aligned}$$

In this case, Equation (12.34) implies that

(12.35) $$K(t) = \frac{\|\mathbf{v}(t) \times \mathbf{a}(t)\|}{[v(t)]^3} = \frac{|x'(t)y''(t) - x''(t)y'(t)|}{\{[x'(t)]^2 + [y'(t)]^2\}^{3/2}}.$$

Example 5

Determine the curvature of the cycloid with parametric equations

$$x(t) = t - \sin t \quad \text{and} \quad y(t) = 1 - \cos t.$$

Solution

For this cycloid, $\quad x'(t) = 1 - \cos t, \qquad y'(t) = \sin t$

and
$$x''(t) = \sin t, \qquad y''(t) = \cos t.$$

Equation (12.35) implies that the curvature of the cycloid is

$$K(t) = \frac{|(1 - \cos t)\cos t - (\sin t)^2|}{[(1 - \cos t)^2 + (\sin t)^2]^{3/2}}$$

$$= \frac{|\cos t - (\cos^2 t + \sin^2 t)|}{(1 - 2\cos t + \cos^2 t + \sin^2 t)^{3/2}}$$

$$= \frac{|\cos t - 1|}{(2 - 2\cos t)^{3/2}} = \frac{1 - \cos t}{2\sqrt{2}(1 - \cos t)^{3/2}} = \frac{\sqrt{2}}{4\sqrt{1 - \cos t}}.$$

Figure 12.32 shows the curvature at various points on the graph. Note that the curvature is undefined at any integral multiple of 2π. ▶

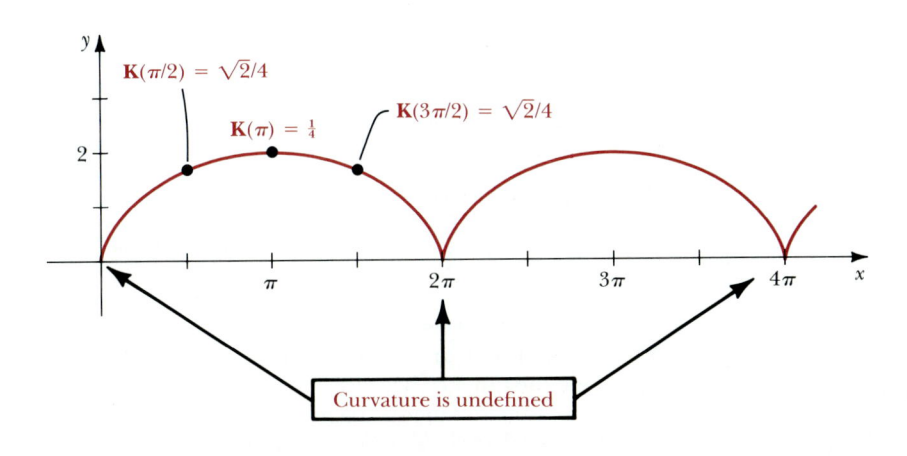

Figure 12.32
A cycloid.

▶ EXERCISE SET 12.5

In Exercises 1–8, find the curvature of the curves at the given point.

1. $r(t) = \sin t\, i + \cos t\, j + 2t k$, $(1, 0, \pi)$
2. $r(t) = \sin t\, i + \cos t\, j + k$, $(0, 1, 1)$
3. $r(t) = (2t - 1)i + (t - 3)j + (2 - 3t)k$, $(1, -2, -1)$
4. $r(t) = ti + t^2 j + 2t k$, $(2, 4, 4)$
5. $r(t) = 2ti + t^2 j + 5t^3 k$, $(0, 0, 0)$
6. $r(t) = ti + \ln t\, j + 0.5t k$, $(2, \ln 2, 1)$
7. $r(t) = tj + e^t k$, $(0, 0, 1)$
8. $r(t) = ti + \sqrt{9 - t^2}\, j$, $(3, 0, 0)$

In Exercises 9–14, (a) find the curvature and radius of curvature at the indicated point and (b) sketch the curve.

9. $r(t) = 3\cos t\, i + 2\sin t\, j$, $t = 0$
10. $r(t) = 3\cos t\, i + 2\sin t\, j$, $t = \pi/2$
11. $r(t) = ti + \frac{1}{2} t^2 j$, $t = 1$
12. $r(t) = ti + \tan t\, j$, $t = \pi/4$

13. $\mathbf{r}(t) = 3\cos t\,\mathbf{i} + 2\sin t\,\mathbf{j} + t\mathbf{k}$, $t = \pi/2$
14. $\mathbf{r}(t) = t\mathbf{i} + \sin t\,\mathbf{j} + t\mathbf{k}$, $t = \pi/2$
15–20. Determine the curvature vectors at the indicated point for the curves described in Exercises 9–14.

In Exercises 21–24, find all points on the plane curves at which the curvature is zero.

21. $\mathbf{r}(t) = t\mathbf{i} + 2t^3\mathbf{j}$
22. $\mathbf{r}(t) = 3t^2\mathbf{i} + 4t^2\mathbf{j}$
23. $\mathbf{r}(t) = 2\cos t\,\mathbf{i} + 3\sin t\,\mathbf{j}$
24. $\mathbf{r}(t) = e^t\mathbf{i} + e^{-t}\mathbf{j}$

The equations in Exercises 25–34 describe curves in the xy-plane.
  a) Write the curve in parametric form.
  b) Find the curvature and radius of curvature at the indicated point.

25. $y = \ln x$, $x = 1$
26. $y = \ln x$, $x = 3$
27. $y = x^3 - 3x - 2$, $x = 1$
28. $y = x^3 - 3x^2 + 2$, $x = 2$
29. $y = x^3 - 3x^2 + 3x - 1$, $x = 1$
30. $y = x^4 - x^2 + 3$, $x = -1$
31. $y = 1/x$, $x = 2$
32. $y = 1/x^2$, $x = 2$
33. $y = \sin x + x$, $x = 0$
34. $y = \sin x + \cos x$, $x = 0$

35. Show that if a curve lies in the xy-plane and is described by $y = f(x)$, then the curvature is given by
$$K(x) = \frac{|y''(x)|}{[1 + (y'(x))^2]^{3/2}}.$$

36–45. Use the result of Exercise 35 to solve part (b) of Exercises 25–34.

46. Show that if the second derivative of a function f exists and is zero at a point, then the curvature of the graph of f at this point is zero.

47. Show that the curvature of the parabola $y = x^2$ approaches 0 as $x \to \infty$.

48. Suppose a curve described by $y = f(x)$ lies in the xy-plane and has a point of inflection at $(a, f(a))$. Show that if f'' is continuous at a, then the radius of curvature becomes infinite as x approaches a.

49. Find the points on the curve in the xy-plane described by $y = 3x - x^3$ at which the curvature is a maximum.

50. Suppose a curve C in the xy-plane is described by a polar equation $r = f(\theta)$. Show that the curvature of C is
$$K(\theta) = \frac{|[f(\theta)]^2 + 2[f'(\theta)]^2 - f(\theta)f''(\theta)|}{\{[f(\theta)]^2 + [f'(\theta)]^2\}^{3/2}}.$$

In Exercises 51–54, use the formula in Exercise 50 to determine the curvature of the curve described by the function at the indicated point.

51. $r = \cos\theta$, $\theta = \pi/4$
52. $r = 1 - \sin\theta$, $\theta = 3\pi/2$
53. $r = \theta$, $\theta = \pi/2$
54. $r = e^\theta$, $\theta = 1$

55. Find points on the ellipse described by $\mathbf{r}(t) = 3\cos t\,\mathbf{i} + 2\sin t\,\mathbf{j}$, $0 \le t \le 2\pi$, at which the curvature is a maximum and points at which the curvature is a minimum.

56. Sketch the graph of $r = e^{-\theta}$ and determine the limits of the curvature of the graph as θ approaches ∞ and as θ approaches $-\infty$. Explain what these results indicate.

Putnam exercise

57. The hands of an accurate clock have lengths 3 and 4. Find the distance between the tips of the hands when that distance is increasing most rapidly. (This exercise was problem A–2 of the forty-fourth William Lowell Putnam examination given on December 3, 1983. The examination and its solution are in the October 1984 issue of the *American Mathematical Monthly*, pp. 487–495.)

12.6

Newton's and Kepler's Laws of Motion

By 1665 the plague that had broken out in London in 1664 had spread to the college at Cambridge, where 23-year-old Isaac Newton (1642–1727) was a scholar at Trinity College. The plague forced the college to close temporarily and Newton returned to his family home at Woolsthorpe in Lincolnshire. Within the next two years, Newton made and refined some of the most important mathematical and physical discoveries of his illustrious life.

At the insistence and expense of his friend Edmund Halley (1656–1742), an outstanding mathematician of the time who predicted the orbit of the comet that bears his name, Newton published the 530-page treatise *Philosophiae naturalis principia mathematica* (1687), probably the greatest scientific work of all time. At the beginning of this work, commonly called simply the *Principia,* he states in Latin, the scientific language of the day, his three laws of motion:

Newton's family home at Woolsthorpe, in Lincolnshire, England.

Lex. I.

Corpus omne perseverare in statu suo quiescendi vel movendi uniformiter in directum, nisi quatenus a viribus impressis cogitur statum illum mutare.

Lex. II.

Mutationem motus proportionalem esse vi motrici impressae, & fieri secundum lineam rectam qua vis illa imprimitur.

Lex. III.

Actioni contrariam semper & aequalem esse reactionem: sive corporum duorum actiones in se mutuo semper esse aequales & in partes contrarias dirigi.

Restated in the form most common today **Newton's Laws of Motion** are as follows.

1) A body continues in its state of rest or uniform motion in a straight line unless an unbalanced force acts on it.
2) The acceleration of a body is directly proportional to the force exerted on the body, is inversely proportional to the mass of the body, and is in the same direction as the force. (This law is usually expressed in the equation form $\mathbf{F} = m\mathbf{a}$, where $\mathbf{F}$ represents the sum of all forces acting on a body, m the mass of the body, and $\mathbf{a}$ the resulting acceleration due to $\mathbf{F}$.)
3) Whenever one body exerts a force on a second body, the second exerts an equal and opposite force on the first.

Newton used the first part of the *Principia* to show that some previously assumed facts about the solar system can be deduced from these laws of motion and the law of gravitation. In particular, he derived **Johannes Kepler's three laws of planetary motion:**

1) *Law of Ellipses:* The orbit of each planet is an ellipse with the sun at one focus.
2) *Law of Areas:* The position vector from the sun to a planet sweeps out equal areas in equal intervals of time.
3) *Harmonic Law:* The square of the period of a planet's orbit about the sun is proportional to the cube of the length of the major axis of the ellipse describing the orbit.

Kepler published the Law of Ellipses and the Law of Areas in 1609 and the Harmonic Law in 1619. He devised these laws empirically after studying an extremely detailed collection of astronomical data about the planet Mars. The data were compiled primarily by the Danish astronomer Tycho Brahe (1546–1601). Brahe was the premier astronomer of his day and the last great astronomer to record observations without benefit of a telescope, a device first used for astronomical observations by Galileo Galilei (1564–1642) in 1609. Incidentally, the first reflecting telescope, the type used by most modern observatories, was invented by Newton.

In addition to deriving Kepler's laws, Newton used the results in the *Principia* to deduce what is known as **Newton's Law of Gravitation:**

> The force **F** between two bodies having masses m_1 and m_2 whose centers of mass are separated by a distance r is an attraction acting along the line joining the centers of mass and having magnitude $\|\mathbf{F}\| = Gm_1m_2/r^2$, where G is a constant called the **universal gravitational constant.**

The derivation of Kepler's laws from Newton's Second Law of Motion and his Law of Gravitation is an interesting illustration of the power of vector function techniques applied to motion problems. The remainder of the chapter is devoted to this derivation.

Kepler's First Law

The orbit of a planet is an ellipse with the sun at a focal point.

We first need to show that the orbit of a planet is confined to a single plane. Let $\mathbf{r}(t)$ denote the vector directed from the sun to the planet at time t and $\mathbf{u}(t)$ denote the unit vector in the direction of $\mathbf{r}(t)$:

$$\mathbf{u}(t) = \frac{\mathbf{r}(t)}{r(t)},$$

where $r(t) = \|\mathbf{r}(t)\|$. To show that the position-vector representation of $\mathbf{r}(t)$ lies in a plane independent of t, we will show that for every value of t, $\mathbf{r}(t)$ is orthogonal to a constant vector.

The gravitational force of the sun on the planet $\mathbf{F}(t)$, is much larger than all other forces acting on the planet, so we assume that the other forces can be neglected in describing the motion of the planet. By Newton's Law of Gravitation, the force $\mathbf{F}(t)$ is directed toward the sun in the direction of the unit vector $-\mathbf{u}(t)$ and has magnitude $GMm/[r(t)]^2$, where M and m denote the mass of the sun and the mass of the planet, respectively (see Figure 12.33). So

(12.36) $\quad \mathbf{F}(t) = \dfrac{GMm}{[r(t)]^2}[-u(t)] = -\dfrac{GMm}{[r(t)]^2}\dfrac{\mathbf{r}(t)}{r(t)} = -\dfrac{GMm}{[r(t)]^3}\mathbf{r}(t).$

Newton's Second Law of Motion implies that

$$\mathbf{F}(t) = m\mathbf{a}(t) = m\frac{d^2\mathbf{r}(t)}{dt^2} = -\frac{GMm}{[r(t)]^3}\mathbf{r}(t).$$

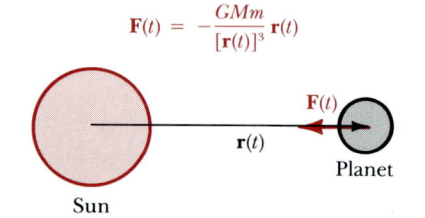

Figure 12.33

Thus,

(12.37) $$\mathbf{a}(t) = -\frac{GM}{[r(t)]^3}\mathbf{r}(t) = -\frac{GM}{[r(t)]^2}\mathbf{u}(t).$$

Since $\mathbf{a}(t) = \mathbf{r}''(t)$, this implies that $\mathbf{r}(t)$ is parallel to $\mathbf{r}''(t)$, so $\mathbf{r}(t) \times \mathbf{r}''(t) = \mathbf{0}$ for all values of t. Therefore,

$$D_t[\mathbf{r}(t) \times \mathbf{r}'(t)] = \mathbf{r}'(t) \times \mathbf{r}'(t) + \mathbf{r}(t) \times \mathbf{r}''(t) = \mathbf{0} + \mathbf{0} = \mathbf{0}$$

and

(12.38) $$\mathbf{r}(t) \times \mathbf{r}'(t) = \mathbf{c}$$

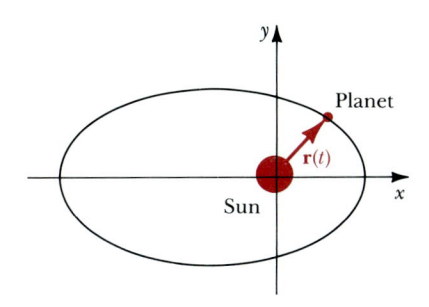

Figure 12.34

is a constant vector independent of t. Since $\mathbf{r}(t)$ is always orthogonal to $\mathbf{r}(t) \times \mathbf{r}'(t)$, $\mathbf{r}(t)$ is orthogonal to the constant vector $\mathbf{c}$ for all values of t. The vector $\mathbf{c}$ is nonzero, for reasons discussed in Exercise 1. Thus $\mathbf{r}(t)$ always lies in the plane determined by the location of the sun and the normal vector $\mathbf{c}$. We will assume that the plane in which $\mathbf{r}(t)$ lies is the xy-plane, with the sun located at the origin as shown in Figure 12.34. The location of the coordinate axes for the plane will be specified later. This implies that $\mathbf{c} = c\mathbf{k}$ for some constant c and that the motion of the planet is restricted to the xy-plane.

To derive an equation describing the motion, we first find an alternative representation for the vector $\mathbf{c} = c\mathbf{k}$ using the unit vector $\mathbf{u}(t) = \mathbf{r}(t)/r(t)$. The variable t has been suppressed in this derivation for notational convenience:

$$\begin{aligned} c\mathbf{k} = \mathbf{r} \times \mathbf{r}' &= (r\mathbf{u}) \times (r\mathbf{u})' \\ &= (r\mathbf{u}) \times (r'\mathbf{u} + r\mathbf{u}') \\ &= (r\mathbf{u} \times r'\mathbf{u}) + (r\mathbf{u} \times r\mathbf{u}'), \end{aligned}$$

so

$$c\mathbf{k} = rr'(\mathbf{u} \times \mathbf{u}) + r^2(\mathbf{u} \times \mathbf{u}').$$

Since $\mathbf{u} \times \mathbf{u} = \mathbf{0}$, this implies that

(12.39) $$c\mathbf{k} = r^2(\mathbf{u} \times \mathbf{u}') = r\mathbf{u} \times r\mathbf{u}' = \mathbf{r} \times \mathbf{r}'.$$

Recalling Equation (12.37) we have

$$\mathbf{a} \times (c\mathbf{k}) = -(GM/r^2)\mathbf{u} \times [r^2(\mathbf{u} \times \mathbf{u}')] = -GM[\mathbf{u} \times (\mathbf{u} \times \mathbf{u}')].$$

By part (vi) of Theorem (11.23), the vector triple product $\mathbf{u} \times (\mathbf{u} \times \mathbf{u}')$ can be rewritten using dot products to give

$$\mathbf{a} \times (c\mathbf{k}) = -GM[(\mathbf{u} \cdot \mathbf{u}')\mathbf{u} - (\mathbf{u} \cdot \mathbf{u})\mathbf{u}'].$$

But $\mathbf{u}$ is a unit vector, so $\mathbf{u} \cdot \mathbf{u} = 1$ and, by Theorem (12.10), $\mathbf{u} \cdot \mathbf{u}' = 0$. So

$$\mathbf{a} \times (c\mathbf{k}) = GM\mathbf{u}'.$$

Since $c\mathbf{k}$ is a constant vector,

$$\begin{aligned} D_t[\mathbf{v} \times (c\mathbf{k})] &= \mathbf{v}' \times (c\mathbf{k}) + \mathbf{v} \times \mathbf{0} \\ &= \mathbf{a} \times (c\mathbf{k}) = GM\mathbf{u}' \end{aligned}$$

and

(12.40) $$\mathbf{v} \times (c\mathbf{k}) = GM\mathbf{u} + \mathbf{d}$$

for some constant vector $\mathbf{d}$.

However, $\mathbf{v} \times (c\mathbf{k})$ is orthogonal to $\mathbf{k}$ and $\mathbf{u}$ is orthogonal to $\mathbf{k}$, so the vector $\mathbf{d}$ lies in the xy-plane. If $\mathbf{d} \neq \mathbf{0}$, we fix the axes of the xy-plane so that $\mathbf{d}$ lies along the positive x-axis. Then $\mathbf{d}$ has the form $\mathbf{d} = d\mathbf{i}$ for some positive constant d. If $\mathbf{d} = \mathbf{0}$, the coordinate axes can be fixed arbitrarily and we still have the representation $\mathbf{d} = d\mathbf{i}$, where in this case $d = 0$.

Let θ denote the angle between $\mathbf{r}$ and the x-axis, as shown in Figure 12.35. Then polar coordinates of the position of the planet are (r, θ) and

(12.41) $$\mathbf{r} \cdot \mathbf{d} = \|\mathbf{r}\|\|\mathbf{d}\| \cos \theta = rd \cos \theta.$$

Equation (12.39) implies that

$$c^2 = (c\mathbf{k}) \cdot (c\mathbf{k}) = (\mathbf{r} \times \mathbf{r}') \cdot (c\mathbf{k}).$$

By property (viii) of Theorem (11.23) we can rewrite this scalar triple product to obtain

$$c^2 = (\mathbf{r} \times \mathbf{r}') \cdot (c\mathbf{k}) = \mathbf{r} \cdot [\mathbf{r}' \times (c\mathbf{k})] = \mathbf{r} \cdot [\mathbf{v} \times (c\mathbf{k})].$$

Equations (12.40) and (12.41) imply that

$$c^2 = \mathbf{r} \cdot [\mathbf{v} \times (c\mathbf{k})] = \mathbf{r} \cdot (GM\mathbf{u} + \mathbf{d}) = GMr + rd \cos \theta.$$

Solving this equation for r gives

(12.42) $$r = \frac{c^2}{GM + d \cos \theta} = \frac{c^2/GM}{1 + (d/GM) \cos \theta}.$$

When $d \neq 0$, the discussion in Section 10.5 shows that this is an equation of a conic section with eccentricity $e = d/GM$. Since the orbit of the planet is a closed curve, we have the following:

When $d \neq 0$, the orbit is an ellipse and $0 < d/GM < 1$.

When $d = 0$, the orbit is circular with radius c^2/GM.

Kepler's Second Law

The position vector from the sun to a planet sweeps out area at a constant rate.

The unit vector $\mathbf{u}(t)$ can be represented as

$$\mathbf{u}(t) = \cos \theta(t) \, \mathbf{i} + \sin \theta(t) \, \mathbf{j},$$

where $\theta(t)$ is the angle between $\mathbf{r}(t)$ and the x-axis shown in Figure 12.36. With this representation, the area swept out by the position vectors from

$$\mathbf{r}(t_0) = r(t_0)\mathbf{u}(t_0) = r(t_0)[\cos \theta(t_0) \, \mathbf{i} + \sin \theta(t_0) \, \mathbf{j}]$$

to

$$\mathbf{r}(t) = r(t)\mathbf{u}(t) = r(t)[\cos \theta(t) \, \mathbf{i} + \sin \theta(t) \, \mathbf{j}]$$

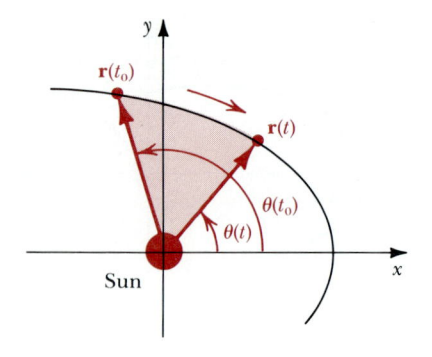

Figure 12.35

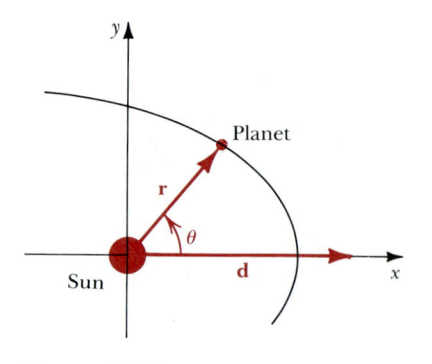

Figure 12.36

is

$$A(t) = \frac{1}{2} \int_{\theta(t_0)}^{\theta(t)} [r(\tau)]^2 \, d\theta = \frac{1}{2} \int_{t_0}^{t} [r(\tau)]^2 \, \frac{d\theta}{d\tau} \, d\tau.$$

The rate at which the position vectors sweep out area is thus

(12.43) $$D_t A(t) = \frac{1}{2} [r(t)]^2 \frac{d\theta}{dt}.$$

However $\mathbf{u} = \cos\theta \, \mathbf{i} + \sin\theta \, \mathbf{j}$, so

$$\mathbf{u} \times \frac{d\mathbf{u}}{dt} = \mathbf{u} \times \frac{d\mathbf{u}}{d\theta} \frac{d\theta}{dt}$$

$$= [(\cos\theta \, \mathbf{i} + \sin\theta \, \mathbf{j}) \times (-\sin\theta \, \mathbf{i} + \cos\theta \, \mathbf{j})] \frac{d\theta}{dt}$$

$$= (\cos^2\theta + \sin^2\theta)(\mathbf{i} \times \mathbf{j}) \frac{d\theta}{dt} = \frac{d\theta}{dt} \mathbf{k}.$$

By Equation (12.39),

$$c\mathbf{k} = r^2 \left(\mathbf{u} \times \frac{d\mathbf{u}}{dt} \right) = r^2 \frac{d\theta}{dt} \mathbf{k}.$$

This equation, together with (12.43), gives

$$D_t A(t) = \frac{1}{2} [r(t)]^2 \frac{d\theta}{dt} = \frac{c}{2},$$

a constant.

Kepler's Third Law

The square of the period of a planet's orbit about the sun is proportional to the cube of the length of the major axis of the ellipse describing the orbit.

In Newton's original language this law is: "the periodic times in ellipses are as the $\frac{3}{2}$th power of their greater axes."

Let T denote the period of an elliptical orbit. By Kepler's Second Law

$$D_t A(t) = \frac{c}{2},$$

so the total area enclosed by an elliptical orbit is the product of this constant rate and the period of rotation:

$$A = \frac{cT}{2}.$$

The area of an ellipse is $A = \pi ab$ (see Figure 12.37) where, $2a$ and $2b$ are lengths of the major and minor axes, respectively. (This is Exercise 22 in Section 10.2.) Since $b = a\sqrt{1-e^2}$,

$$T = \frac{2}{c} A = \frac{2}{c} \pi a^2 \sqrt{1-e^2}.$$

The eccentricity of the ellipse is $e = d/GM$, so the square of the period of

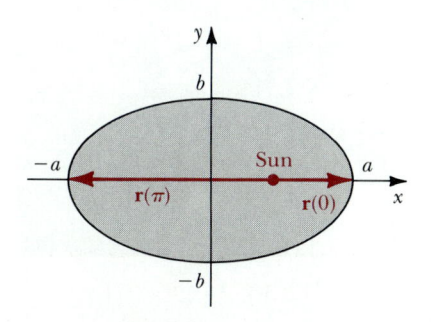

Figure 12.37

The area is πab.

orbital rotation is

(12.44) $$T^2 = \frac{4\pi^2}{c^2} a^4 \left[1 - \left(\frac{d}{GM}\right)^2\right] = \frac{4\pi^2 a^4[(GM)^2 - d^2]}{c^2(GM)^2}.$$

However, the length of the major axis of the ellipse is the sum of the distances $r(0)$ and $r(\pi)$, so Equation (12.42) implies that

$$a = \frac{1}{2}[r(0) + r(\pi)] = \frac{1}{2}\left(\frac{c^2}{GM+d} + \frac{c^2}{GM-d}\right) = \frac{c^2 \, GM}{(GM)^2 - d^2}.$$

Thus Equation (12.44) can be rewritten as

$$T^2 = \frac{4\pi^2 a^4}{GM} \frac{(GM)^2 - d^2}{c^2 \, GM} = \frac{4\pi^2 a^4}{GM} \cdot \frac{1}{a}$$

and

(12.45) $$T^2 = \frac{4\pi^2}{GM} a^3.$$

Kepler's laws work equally well for orbits of objects about the planets, provided the gravitational force of the planet is sufficient to assume that all other forces on the object can be neglected. The following example considers a satellite placed in an *isosynchronous* orbit. In such an orbit the satellite revolves around the body at the same rate that the body rotates about its axis. When the body is the earth, the isosynchronous orbit is also called a *geosynchronous*, or *geostationary*, orbit.

Example 1

Communication satellites are commonly placed in a geosynchronous circular orbit about the earth's equator so that they appear above the same point on the earth at any time (see Figure 12.38). Determine the height above the earth of such a satellite.

Solution

The period of orbit of the satellite is one day, so

$$T = 1 \text{ day} = 24 \text{ hr} = 24(3600) \text{ sec.}$$

The universal gravitational constant G in metric units is

$$G = 6.673 \times 10^{-11} \text{ m}^3/\text{kg-sec}^2,$$

and the mass of the earth is

$$M = 5.979 \times 10^{24} \text{ kg.}$$

Since Kepler's laws work equally well for orbits about the earth, Equation (12.45) implies that the radius of the orbit is

$$a = \left\{\frac{[24(3600) \text{ sec}]^2 (6.673 \times 10^{-11} \text{ m}^3/\text{kg-sec}^2)(5.979 \times 10^{24} \text{ kg})}{4\pi^2}\right\}^{1/3}$$

$$= 4.225 \times 10^7 \text{ m} = 4.225 \times 10^4 \text{ km.}$$

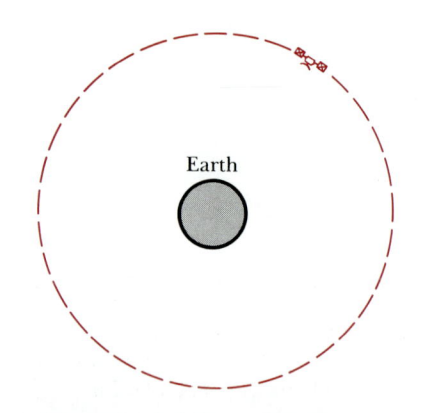

Figure 12.38
An RCA Satcom communications satellite in geostationary orbit about the Earth.

The radius of the earth is approximately 6371 km, so the satellite will orbit approximately 35,879 km ≈ 22,295 mi above the earth. ▶

Data about our solar system given in Table 12.1 will be used in the exercises for this section. The second column gives the length $2a$ of the major axis of revolution and the third column gives the eccentricity e of the orbit. The radius r and mass m of the planet are given in the fourth and fifth columns, respectively, and the final column tells the period T of the orbit about the sun. The table also gives data for the earth's moon; in this case, the orbit is about the earth.

Table 12.1

PLANET	$2a$ (in km)	e	r (in km)	m (in kg)	T (in sec)
Mercury	1.159×10^8	0.2056	2,433	3.181×10^{23}	7.602×10^6
Venus	2.162×10^8	0.0068	6,053	4.883×10^{24}	1.941×10^7
Earth	2.991×10^8	0.0167	6,371	5.979×10^{24}	3.156×10^7
Mars	4.557×10^8	0.0934	3,380	6.418×10^{23}	5.936×10^7
Jupiter	1.556×10^9	0.0484	69,758	1.901×10^{27}	3.743×10^8
Saturn	2.854×10^9	0.0543	58,219	5.684×10^{26}	9.296×10^8
Uranus	5.741×10^9	0.0460	23,470	8.682×10^{25}	2.651×10^9
Neptune	9.000×10^9	0.0082	22,716	1.027×10^{26}	5.200×10^9
Pluto	1.182×10^{10}	0.2481	5,700	1.080×10^{24}	7.837×10^9
Earth's moon	7.688×10^5	0.0549	1,738	7.354×10^{22}	2.361×10^6

The figures in this table are from the *CRC Handbook of Chemistry and Physics,* 56th edition. This source lists extensive physical data for the planets and the major asteroids.

▶ EXERCISE SET 12.6

1. a) Show that if the vector **c** in (12.38) were zero then either (i) $\mathbf{r}(t) = \mathbf{0}$, (ii) $\mathbf{r}'(t) = \mathbf{0}$, or (iii) $\mathbf{r}(t)$ and $\mathbf{r}'(t)$ are parallel.
   b) Show that none of (i), (ii), or (iii) can hold for a planet in a closed orbit.

2. Use Newton's Law of Gravitation and the universal gravitational constant $G = 6.673 \times 10^{-11}$ m³/kg-sec², together with the radius and mass of the earth given in Table 12.1, to show that the magnitude of the acceleration due to the gravity of the earth at the earth's surface is approximately 9.8 m/sec².

3. Determine the magnitude of the acceleration due to the earth's gravity on an object that is above the earth's surface a distance of (a) 1000 km; (b) 10,000 km; (c) 50,000 km; (d) 100,000 km.

4. The mass of an object remains constant, but its weight is proportional to the mass and the acceleration due to gravity acting on the object. Determine the percent by which the weight of an object is decreased when it is above the earth's surface a distance of (a) 1000 km; (b) 10,000 km; (c) 50,000 km; (d) 100,000 km.

5. In Example 1 it was stated that the period of the geosynchronous satellite about the earth is 24 hr. In fact, the true period is approximately 23.93 hr.
   a) Does this increase or decrease the orbital radius for a geosynchronous satellite?
   b) By what percent?

6. Equation (12.42) gives the orbit of a planet about the sun using a polar equation.
   a) Express this orbit using an equivalent Cartesian equation.
   b) Express the orbit using a translated standard-position Cartesian equation.

7. a) Use Kepler's third law and the data about the earth given in Table 12.1 to estimate the mass of the sun.

b) Use the data about Mars to estimate the mass of the sun.

8. Use Kepler's third law and the data about the earth's moon given in Table 12.1 to estimate the mass of the earth.

9. Determine the average density of the earth from the data given in Table 12.1, assuming that the shape of the earth is a sphere. The average density of the material constituting the earth's crust is about 2.7 g/cc. What does this tell us about the material beneath the earth's crust?

10. Show that the data about the planets in Table 12.1 are consistent with Kepler's Third Law. Assume that the mass of the sun is approximately 1.991×10^{30} kg.

11. Show that the data about the earth's moon given in Table 12.1 are consistent with Kepler's Third Law.

12. A spaceship is traveling from earth to its moon. At what point along the line segment connecting the centers of the bodies do the gravitational forces due to the earth and to its moon cancel?

13. The period of rotation of the moon about its axis is 655.71 hr. Determine the altitude of a satellite that is placed in isosynchronous orbit about the moon.

14. a) How much force does the moon exert on you?
    b) A typical distance of Mars from the earth is 10^8 km. How much force does Mars exert on you?

15. Suppose a body at point P is subject to a force $\mathbf{F}(t)$ that causes the body to rotate about a fixed point O. The **torque** $\boldsymbol{\tau}$ produced by $\mathbf{F}(t)$ on P about O is defined by

$$\boldsymbol{\tau} = \overrightarrow{OP} \times \mathbf{F}(t).$$

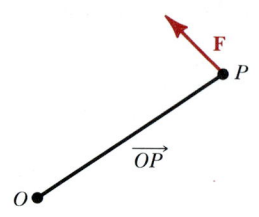

Use Newton's Second Law of Motion to show that an object with mass m whose position is given by $\mathbf{r}(t)$ has torque $\boldsymbol{\tau}(t) = m[\mathbf{r}(t) \times \mathbf{a}(t)]$.

16. The **angular momentum** $\mathbf{L}$ of an object with mass m whose position is given by $\mathbf{r}(t)$ is

$$\mathbf{L}(t) = m[\mathbf{r}(t) \times \mathbf{v}(t)].$$

Show that $\mathbf{L}'(t) = \boldsymbol{\tau}(t)$.

17. Complete the steps in this alternative proof of Kepler's Second Law by considering the torque $\boldsymbol{\tau}$ and angular momentum $\mathbf{L}$ produced by the force given by Newton's Law of Gravitation.
    a) Show that $\boldsymbol{\tau} = \mathbf{0}$.
    b) Show that $\mathbf{L} = m(\mathbf{r} \times \mathbf{v})$ is constant. (This is called the *Law of Conservation of Angular Momentum.*)
    c) Show that the area of the triangle in the figure that is generated by changing time by an amount dt is $dA = \frac{1}{2}\|\mathbf{r} \times d\mathbf{r}\|$, where $d\mathbf{r} = \mathbf{v}\, dt$.
    d) Show that $dA/dt = \|\mathbf{L}\|/2m$, a constant.

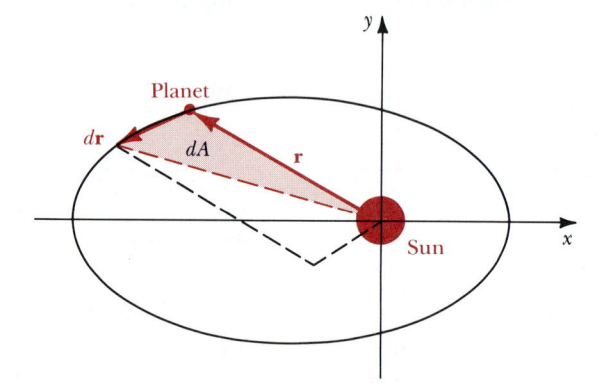

18. In Example 4 of Section 12.4 we found that for an object in a circular orbit of radius r with constant angular speed ω, the speed is $v = \omega r$ and the acceleration is $\mathbf{a}(t) = -\omega^2 \mathbf{r}(t)$.
    a) Use Newton's Second Law of Motion to show that the force required to keep an object in a uniform circular orbit is

    $$\mathbf{F}(t) = -\frac{mv^2}{r^2}\mathbf{r}(t).$$

    b) Show that the period T of a satellite in uniform circular orbit with speed v and radius r is $T = 2\pi r/v$.
    c) Use Newton's Law of Gravitation and the results in parts (a) and (b) to derive Kepler's Third Law for the case of circular orbits.

19. Halley's comet orbits the sun in an elliptical orbit with a period of approximately 75.5 yr. On February 9, 1986, the comet reached its perihelion (the closest point to the sun), a distance of approximately 8.93×10^7 km from the sun.
    a) Approximately when will Halley's comet next be at its aphelion (the farthest point from the sun)?
    b) How far from the sun will Halley's comet be when it is at its aphelion?

20. a) Determine the minimal period for a satellite in a circular earth orbit.
    b) What is the centripetal acceleration of such a satellite?

▶ IMPORTANT TERMS AND RESULTS

CONCEPT	PAGE	CONCEPT	PAGE
Vector-valued function	706	Tangential acceleration	726
Component functions	706	Centripetal acceleration	726
Circular helix	708	Curvature vector	730
Tangent vector	713	Curvature	730
Principal unit tangent vector	716	Radius of curvature	732
Principal unit normal vector	718	Decomposition of acceleration	733
Arc length representation	719	Newton's Laws of Motion	737
Smooth curve	719	Kepler's Laws	737
Motion equations in space	722	Newton's Law of Gravitation	738

▶ REVIEW EXERCISES

In Exercises 1–4, determine the values of t for which the function is continuous.

1. $\mathbf{F}(t) = \sqrt{1 - t^2}\,\mathbf{i} + t\mathbf{j}$
2. $\mathbf{F}(t) = t\mathbf{j} + \cos t\,\mathbf{k}$
3. $\mathbf{F}(t) = t\mathbf{i} + \dfrac{1}{t - 1}\mathbf{j} + \mathbf{k}$
4. $\mathbf{F}(t) = e^t\mathbf{i} + t\mathbf{j}$

In Exercises 5–10, evaluate the integral.

5. $\displaystyle\int_0^\pi (\cos t\,\mathbf{i} + \sin t\,\mathbf{j} + e^t\mathbf{k})\,dt$
6. $\displaystyle\int (t\cos t\,\mathbf{i} + t\sin t\,\mathbf{j} + t\mathbf{k})\,dt$
7. $\displaystyle\int_0^1 (te^t\mathbf{i} + \ln(t+1)\,\mathbf{k})\,dt$
8. $\displaystyle\int_0^{2\pi} \|3\cos t\,\mathbf{i} + 3\sin t\,\mathbf{j}\|\,dt$
9. $\displaystyle\int_0^1 (e^t\mathbf{i} + t^2\mathbf{j})\cdot(t\mathbf{i} + \sin t\,\mathbf{k})\,dt$
10. $\displaystyle\int_0^1 (e^t\mathbf{i} + t^2\mathbf{j})\times(t\mathbf{i} + \sin t\,\mathbf{k})\,dt$

For the curves described in Exercises 11–18:
   a) Sketch the graph and indicate the direction of increasing t.
   b) Find and sketch $\mathbf{F}'(t)$.
   c) Find, if they exist, the principal unit tangent and unit normal vectors at t and sketch these vectors.
   d) Find the curvature at t.

11. $\mathbf{F}(t) = (1 + t)\mathbf{i} + (2 + t)\mathbf{j} + 4\mathbf{k}$, $t = 0$
12. $\mathbf{F}(t) = (t + 3)\mathbf{i} + (2 - t)\mathbf{j} + 3t\mathbf{k}$, $t = 1$
13. $\mathbf{F}(t) = \cos t\,\mathbf{i} + \sin t\,\mathbf{j} + 3\mathbf{k}$, $t = \pi/3$
14. $\mathbf{F}(t) = \cos t\,\mathbf{i} + \sin t\,\mathbf{j} + e^t\mathbf{k}$, $t = \pi$
15. $\mathbf{F}(t) = t\mathbf{j} + \cos t\,\mathbf{k}$, $t = 3\pi/2$
16. $\mathbf{F}(t) = 2\sin t\,\mathbf{i} + 3\cos t\,\mathbf{j}$, $t = \pi/2$
17. $\mathbf{F}(t) = 3\cos t\,\mathbf{i} + 4\sin t\,\mathbf{j}$, $t = \pi$
18. $\mathbf{F}(t) = \cos t^2\,\mathbf{i} + \sin t^2\,\mathbf{j}$, $t = \sqrt{\pi}$

19–26. Suppose the motion of an object is given by the functions described in Exercises 11–18. Determine equations for (a) the velocity, (b) the speed, and (c) the acceleration of the object.

27. Let $\mathbf{F}(t) = t\mathbf{i} + \ln t\,\mathbf{j} + t\mathbf{k}$ and $f(t) = t^2$. Find, if possible, each of the following.
   a) $D_t\,(\mathbf{F} + f)(t)$ b) $D_t\,f(t)\,D_t\,\mathbf{F}(t)$
   c) $D_t\,(f\mathbf{F})(t)$ d) $D_t\,(f \circ \mathbf{F})(t)$
   e) $D_t\,(\mathbf{F} \circ f)(t)$

28. Let $\mathbf{F}(t) = t\mathbf{i} + \ln t\,\mathbf{j} + t\mathbf{k}$ and $\mathbf{G}(t) = t^2\mathbf{i} + t\mathbf{j}$. Find, if possible, each of the following.
   a) $D_t\,(\mathbf{F} + \mathbf{G})(t)$ b) $D_t\,\mathbf{F}(t)\cdot D_t\,\mathbf{G}(t)$
   c) $D_t\,(\mathbf{F}\cdot\mathbf{G})(t)$ d) $D_t\,\mathbf{F}(t) \times D_t\,\mathbf{G}(t)$
   e) $D_t\,(\mathbf{F}(t) \times \mathbf{G}(t))$

29. Why is the curve described by $\mathbf{F}(t) = t\mathbf{i} + |t|\mathbf{j} + \mathbf{k}$ not a smooth curve?

30. Why is the curve described by
$$\mathbf{r}(t) = t^2\mathbf{i} + t^3\mathbf{j} + \cos t^2\,\mathbf{k}, \quad -1 \leq t \leq 1$$
not smooth at $t = 0$?

31. Find a vector-valued function with arc length as parameter that describes the straight line from (3, 0, 0) to (2, 7, 5).

32. Find a vector-valued function with arc length as parameter that describes the portion of the helix

$$\mathbf{r}(t) = \sin t\, \mathbf{i} + \cos t\, \mathbf{j} + t\mathbf{k}, \qquad 0 \le t \le \pi/2.$$

33. Find the radius of curvature of the parabola $y = x^2$ at (0, 0).

34. Find the point (x, y) on the parabola $y = x^2$ at which the curvature is a maximum. What is the curvature at this point?

35. Find the curvature at $x = 2$ of the curve in the xy-plane described by $y = x^3 - 3x^2 - 2$.

36. Find the radius of curvature of the ellipse

$$9x^2 + 16y^2 = 144$$

at $(-2\sqrt{3}, -3/2)$.

37. A projectile is fired from ground level with an initial speed of 500 ft/sec at an angle of elevation of $\pi/3$ radians. Describe the subsequent motion of the projectile and find (a) the time in the air, (b) the maximum height, (c) the horizontal range, and (d) the speed at the instant of impact.

38. An archer shoots an arrow toward a wall 50 yd away with an initial velocity that makes an angle of $\pi/4$ with the horizontal. The arrow strikes the wall 35 yd above the ground. If the arrow was released from shoulder level (4 ft above the ground), find each of the following.
    a) The initial velocity of the arrow
    b) The time it takes to reach the wall
    c) The velocity as it strikes the wall
    d) The speed as it strikes the wall.

39. A projectile is fired from a height of 200 ft with an initial speed of 400 ft/sec at an angle of elevation of $\pi/6$ radians. Describe the subsequent motion of the projectile and find each of the following.
    a) The time in the air
    b) The maximum height
    c) The horizontal range
    d) The speed at the instant of impact

40. A plane flying in calm air northward at 110 mph at an altitude of 10,000 ft drops a package when it is directly over a radio tower. Place a rectangular coordinate system with origin at the foot of the tower, the positive y-axis pointing due north, and the positive z-axis directed upward from the earth.
    a) Describe the motion of the package.
    b) Determine when and where the package will land.

41. Rework Exercise 40 under the additional assumption that there is a wind that carries the package at 25 mph toward the southeast.

42. A golfer can hit a golf ball with a nine-iron a distance of 120 yd in the air. The ball remains in the air for 9 sec while traveling this distance. With the same swing, the golfer can hit the ball 136 yd with an eight-iron. In fact, 16 yd is gained using the same swing, with each decrease in club, down to the two-iron, which will cause the ball to travel 232 yd.
    a) What is the initial velocity applied to the ball by the golfer?
    b) What is the angle of elevation of the nine-iron shot?
    c) What is the angle of elevation of the two-iron shot?
    d) How long does the two-iron shot remain in the air?

43. A 150-ft oak tree is standing 40 yd directly ahead of the golfer described in Exercise 42. Show that a four-iron shot will clear the tree but a three-iron shot will not.

44. The moon has an approximate circular orbit about the earth and a period of 27.32 days. Use Kepler's Third Law and the mass of the earth given in Table 12.1 to determine the distance from the earth to the moon.

45. The Space Shuttle Discovery launched on August 30, 1984, was placed into a nearly circular orbit 300 km above the earth. Use the mass of the earth and the universal gravitational constant $G = 6.673 \times 10^{-11}$ m³/kg-sec² to determine the length of time required for one orbit.

The Space Shuttle Discovery.

46. On the day this problem was being written, the Space Shuttle Discovery was launched into a nearly circular orbit 334 km above the earth. How much longer will an orbit take than when it was launched on August 30, 1984? (See Exercise 45.)

$z = x^3 - 3xy^2$

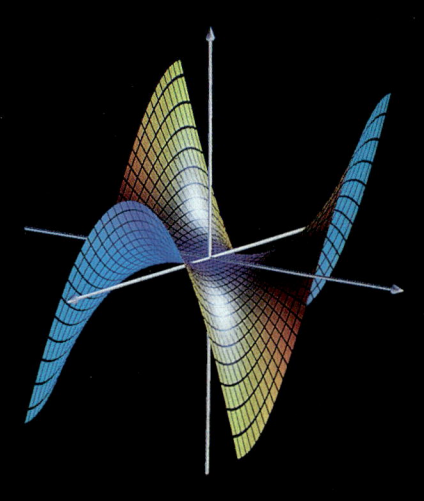

Figure 13.14(a)

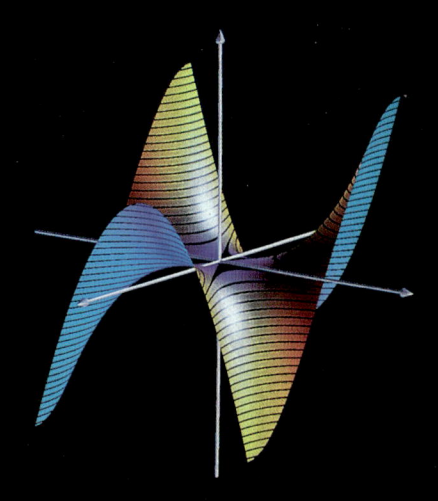

Figure 13.14(b)

II

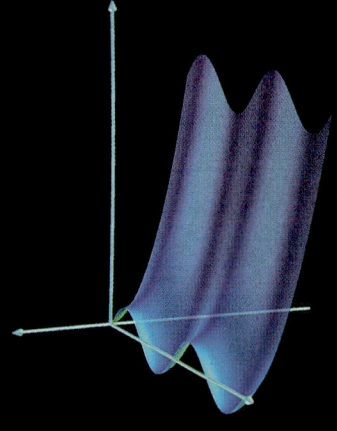

$z = x^2 + \sin y$

Figure 13.15(a)

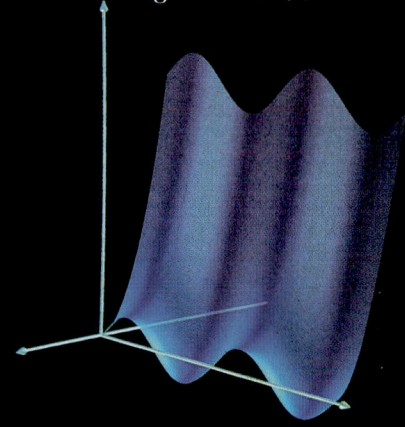

Figure 13.15(b)

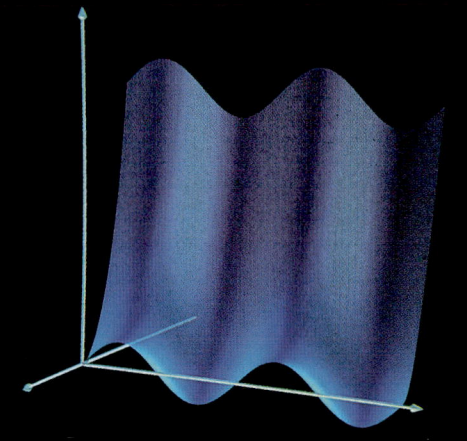

Figure 13.15(c)

$z = \sin\sqrt{x^2 + y^2}$

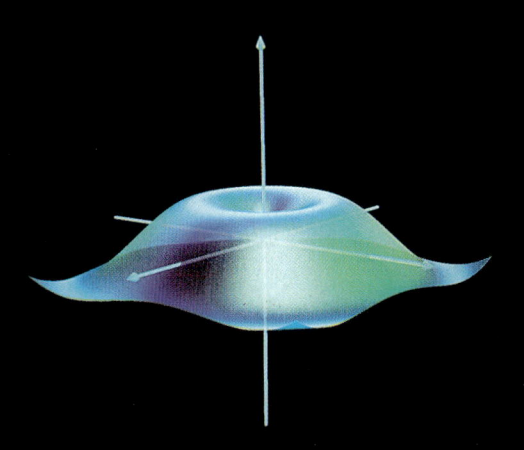

$z = (3 - x^2 + 4y^2)e^{1-x^2-y^2}$

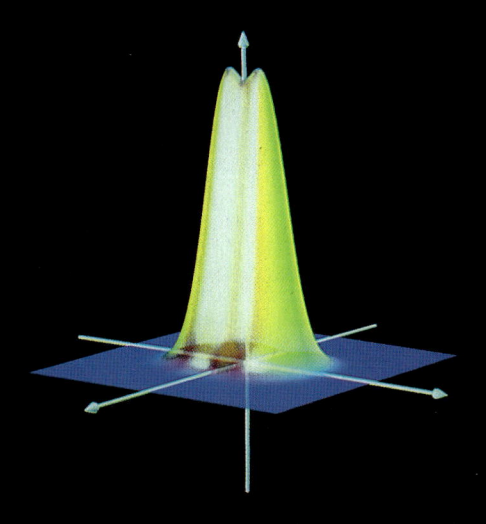

$z = e^{-x}\cos y$

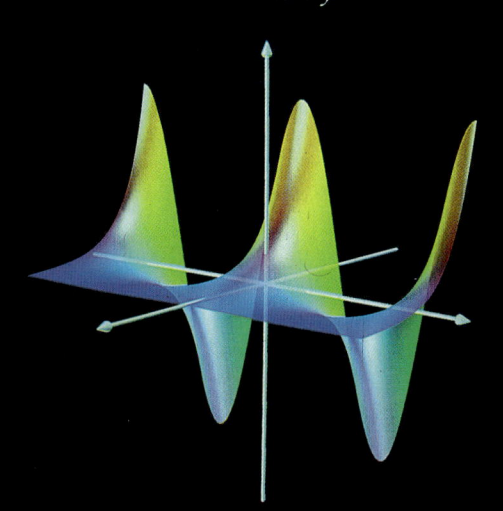

$z = \dfrac{\sin(x^2 + y^2)}{x^2 + y^2}$

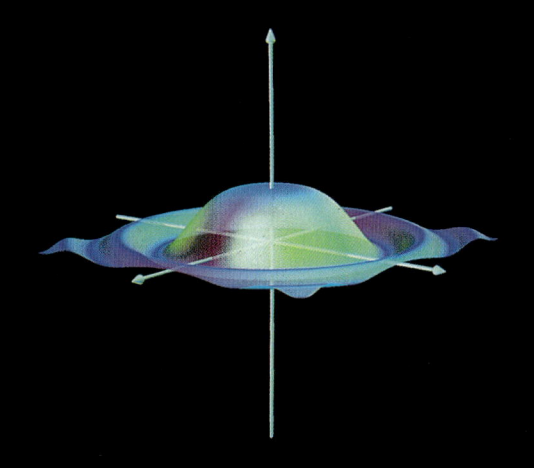

Figure 13.16

IV

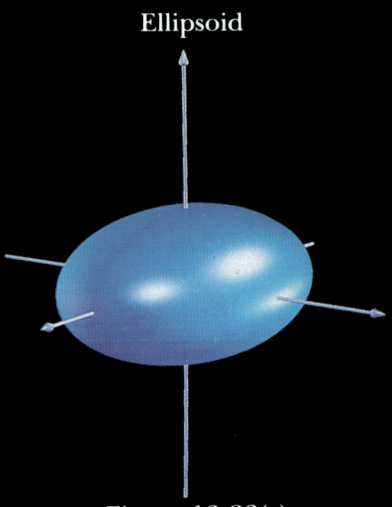

Ellipsoid

Figure 13.23(c)

Elliptic hyperboloid of one sheet

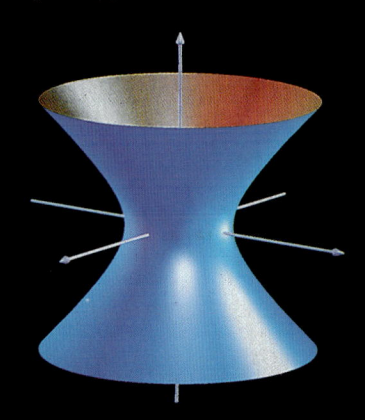

Figure 13.24(b)

Elliptic hyperboloid of two sheets

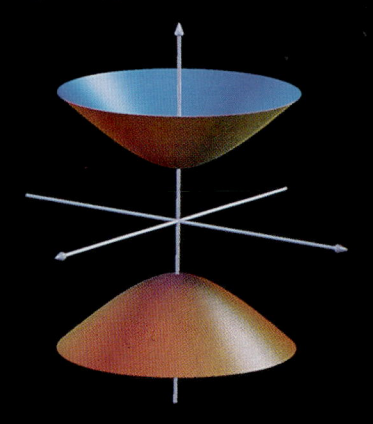

Figure 13.25(b)

Elliptic paraboloid

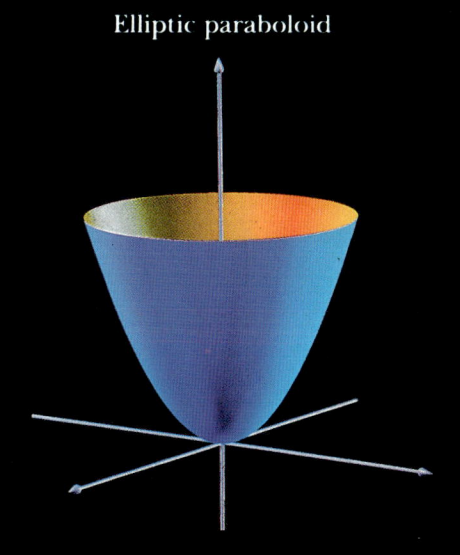

Figure 13.26(b)

Hyperbolic paraboloid

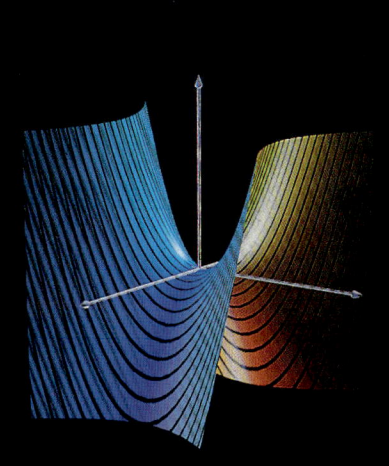

x-axis contour lines

y-axis contour lines

Figure 13.27(b)

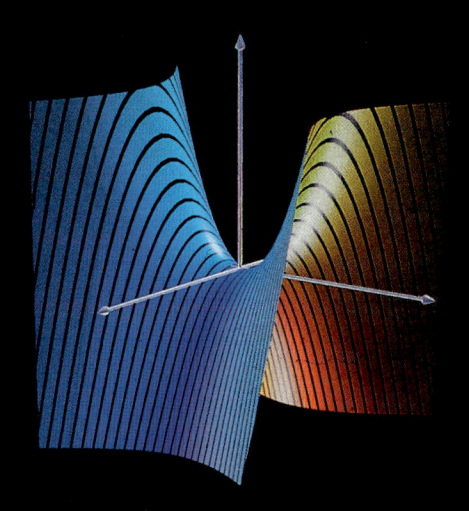

Figure 13.27(c)

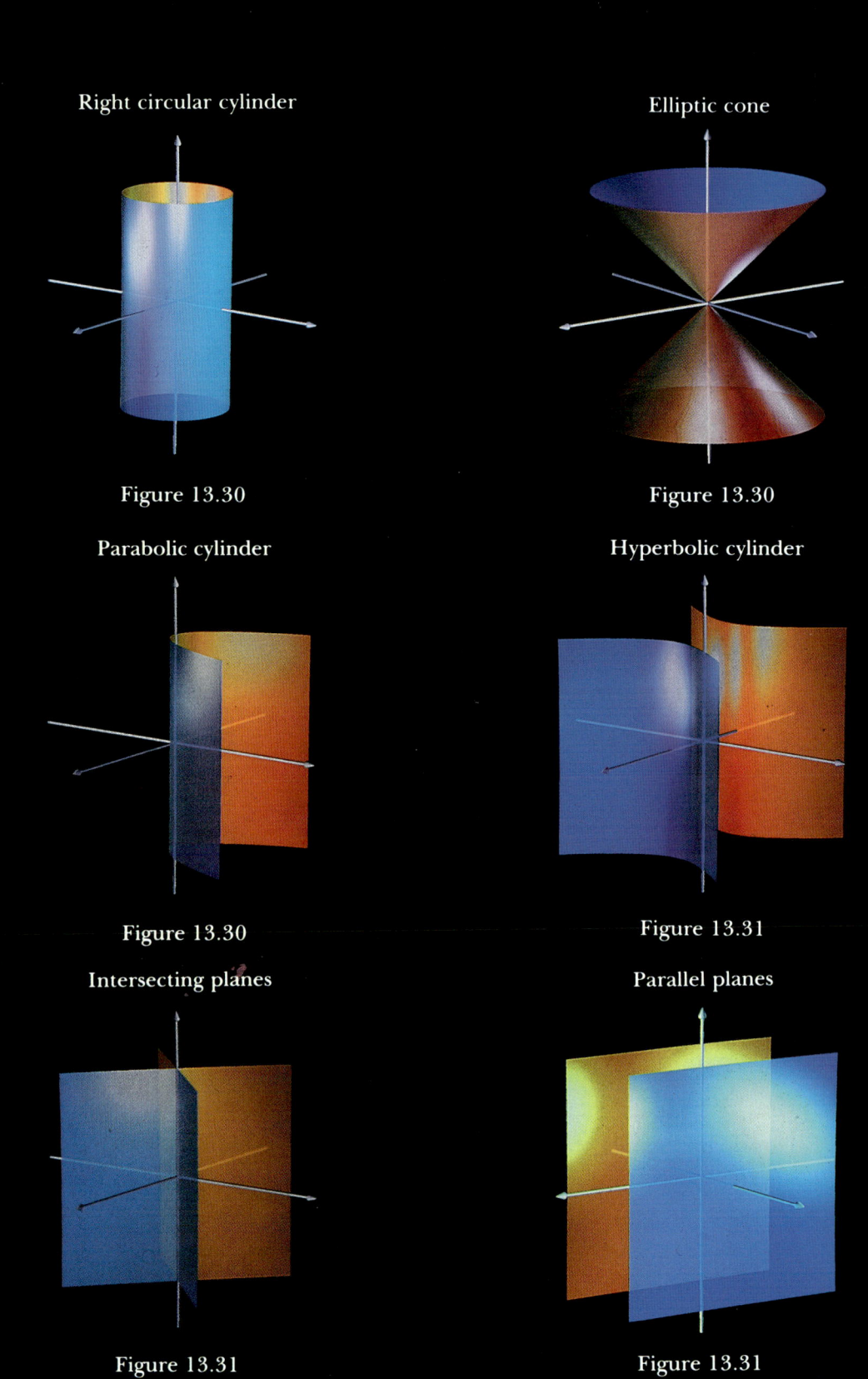

VI

Right circular cylinder

Figure 13.30

Elliptic cone

Figure 13.30

Parabolic cylinder

Figure 13.30

Hyperbolic cylinder

Figure 13.31

Intersecting planes

Figure 13.31

Parallel planes

Figure 13.31

VII

$$f(x, y) = \begin{cases} \dfrac{xy}{x^2 + y^2}, & \text{if } (x, y) \neq (0, 0) \\ 0, & \text{if } (x, y) = (0, 0) \end{cases}$$

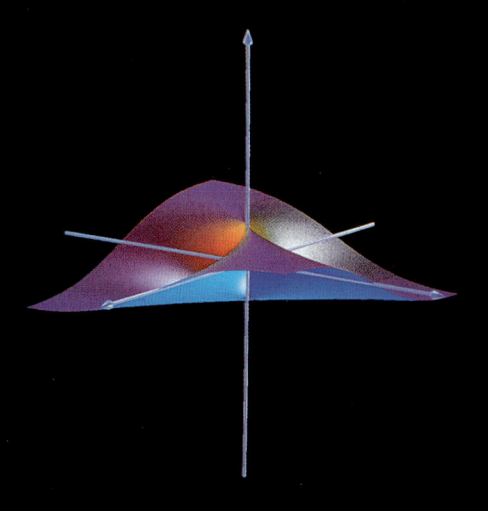

Figure 14.14

$$f(x, y) = \begin{cases} \dfrac{x^2 y}{x^4 + y^2}, & \text{if } (x, y) \neq (0, 0) \\ 0, & \text{if } (x, y) = (0, 0) \end{cases}$$

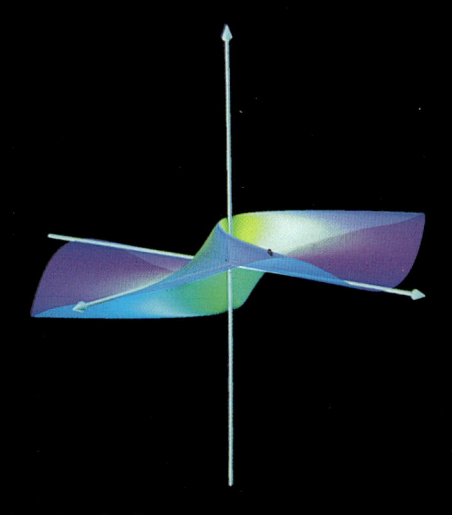

Exercise 46, Section 14.1

VIII

$z = e^x \sin y$

Figure 14.41

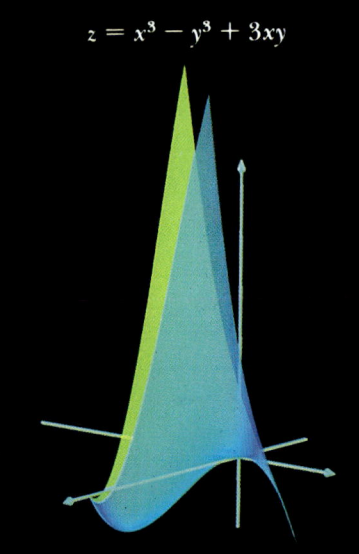

$z = x^3 - y^3 + 3xy$

Figure 14.47

IX

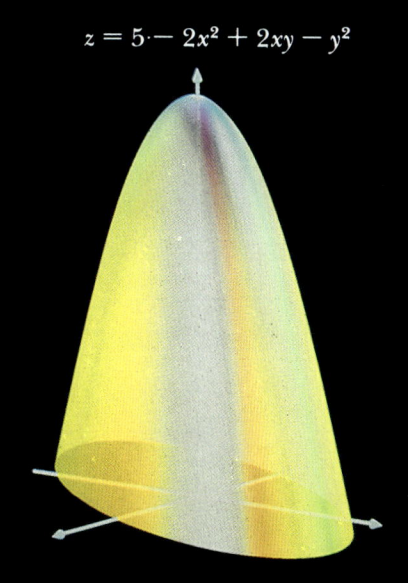

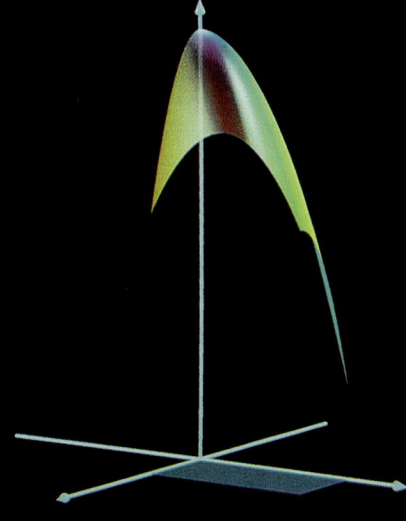

$$z = 5 - 2x^2 + 2xy - y^2$$

$0 \leq x \leq 1$ and $0 \leq y \leq 2$

$-1 \leq x \leq 1$ and $0 \leq y \leq 1 - x^2$

Figure 14.44

Figure 14.46

x

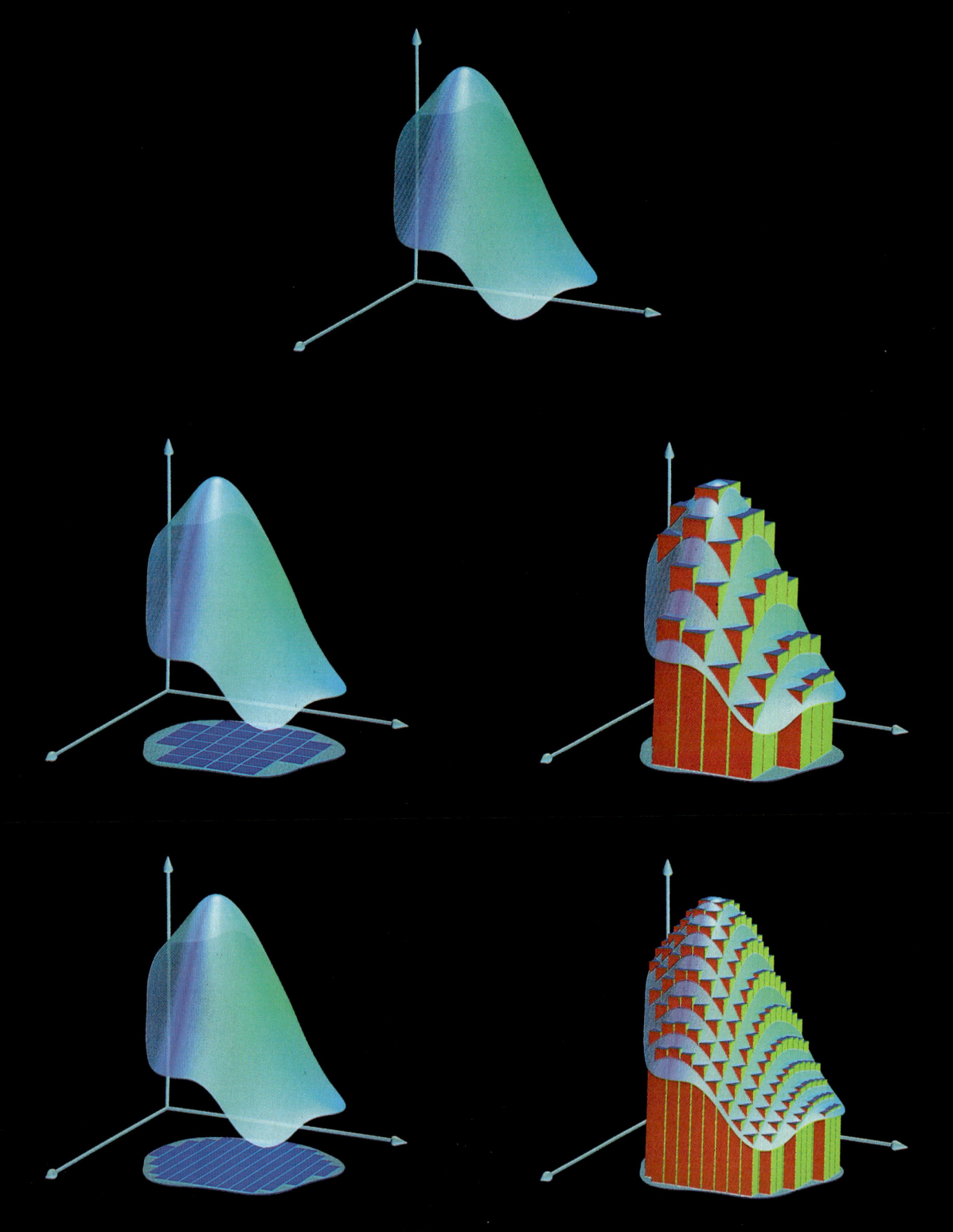

Figure 15.1

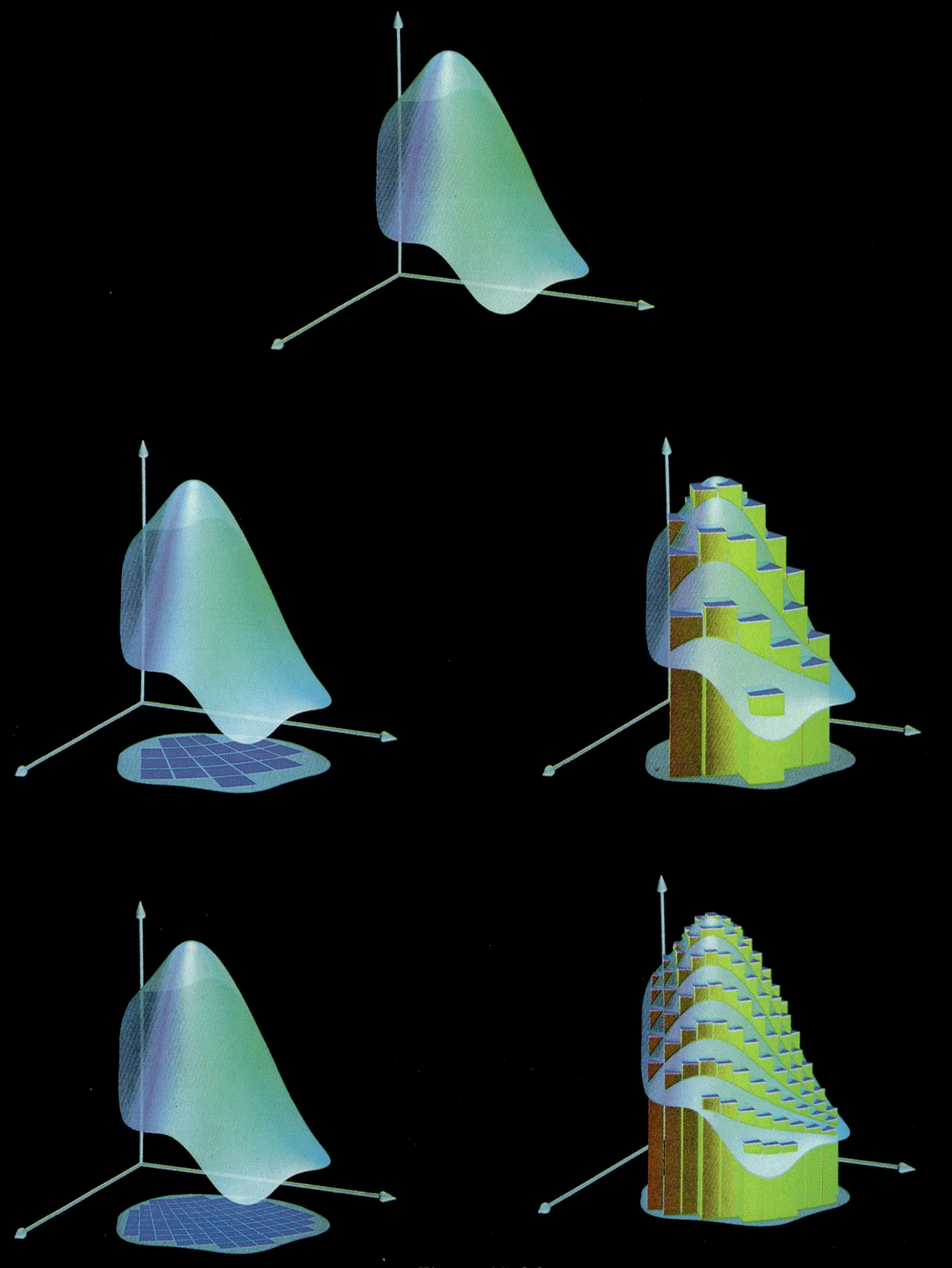

Figure 15.23

XII

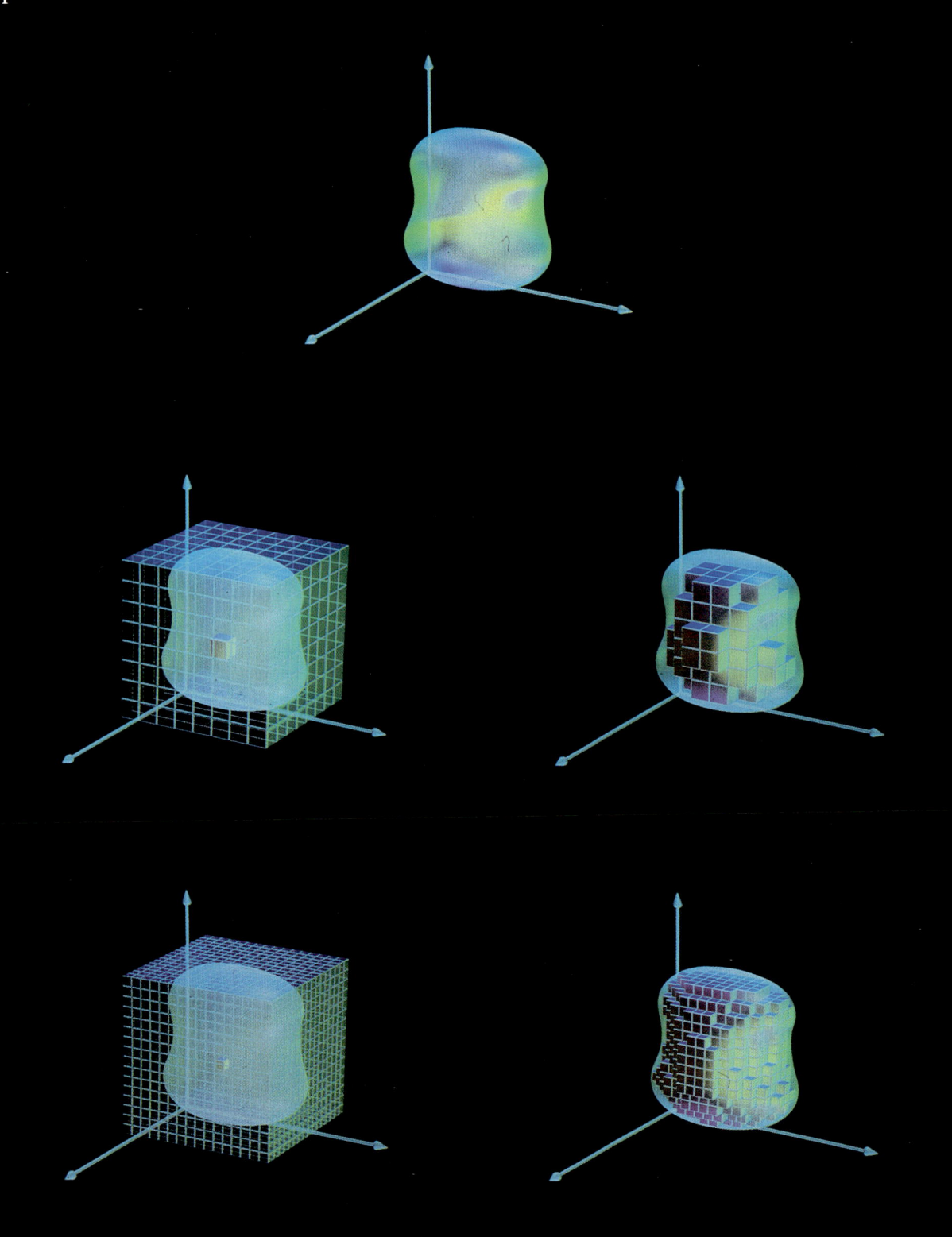

Figure 15.42

13

Multivariable Functions

The functions studied in Chapter 12 mapped real numbers into vectors, and the functions studied in the first 11 chapters had both their domains and ranges in the set of real numbers. Many common applications, however, require functions with more than one variable in the domain. In this chapter we consider the properties of functions whose range is in the set of real numbers but whose domain is either a subset of the plane, denoted $\mathbb{R}^2$, or three-dimensional space, denoted $\mathbb{R}^3$. These functions are called **multivariable functions** to distinguish them from the single-variable functions we studied previously. In subsequent chapters the results of calculus will be extended to functions of this type.

13.1

Functions of Several Variables

Suppose n is a positive integer greater than one. By a **function of n variables** we mean a function whose domain is a subset of $\mathbb{R}^n$, the set of all ordered n-tuples of real numbers, and whose range is a subset of the real numbers $\mathbb{R}$. In general, we consider only the case when n is either two or three, but applications frequently require functions of more than three variables.

Before studying specific properties of functions of several variables, let us consider some applications of single-variable functions and analogous applications of several-variable functions.

Functions from $\mathbb{R}$ into $\mathbb{R}$ have been used to describe:

1) the distance from points on a line to a fixed origin on the line;
2) the temperature at each point of a rod;
3) the selling price of an article whose cost depends entirely on the cost of materials.

Analogous functions from $\mathbb{R}^2$ into $\mathbb{R}$ can be used to describe:

1) the distance from points in a plane to a fixed origin in the plane;
2) the temperature at each point on a flat plate;
3) the selling price of an article whose cost depends on both the cost of materials and the cost of labor.

Similarly, functions from $\mathbb{R}^3$ into $\mathbb{R}$ can be used to describe:

1) the distance from points in space to a fixed origin in space;
2) the temperature at each point in a solid—for example, the temperature within a turkey that is roasting in an oven;
3) the selling price of an article whose cost depends on the cost of materials, the cost of labor, and the cost of overhead.

A function from $\mathbb{R}^n$ to $\mathbb{R}$ can be constructed for any integer n by considering the application listed under (3) and increasing to n the number of variables on which the selling price depends. It is not uncommon for functions describing economic models to be functions of hundreds of variables. For our purposes, however, three variables are sufficient.

Example 1

A flat circular plate of radius 5 in. is heated so that the temperature at each point on the plate is directly proportional to the distance of the point from the center. Find a function that describes the temperature of the plate.

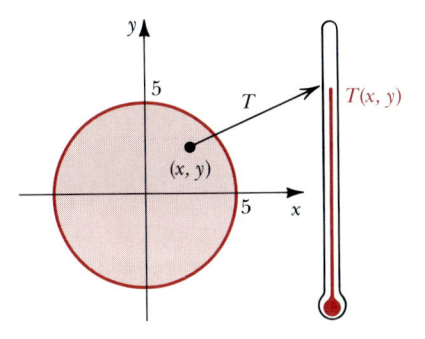

Figure 13.1
$T(x, y) = k\sqrt{x^2 + y^2}$.

Solution

We introduce a two-dimensional coordinate system with the center of the plate at the origin, as in Figure 13.1. Since the distance from a point (x, y) to $(0, 0)$ is $\sqrt{x^2 + y^2}$, the temperature is described by

$$T(x, y) = k\sqrt{x^2 + y^2},$$

where k is the constant of proportionality. The domain of T consists of all pairs (x, y) that satisfy $x^2 + y^2 \leq 25$. ▶

In this chapter many of the concepts of single-variable functions are extended to multivariable functions. For example, a function from $\mathbb{R}$ into $\mathbb{R}$ is often described by giving only its rule of correspondence. In such a case, the domain is assumed to be the largest subset of the real numbers for which the correspondence produces a real number. The same principle holds here:

> When a function from $\mathbb{R}^n$ to $\mathbb{R}$ is described only by its rule of correspondence, the domain of the function is assumed to be the largest subset of $\mathbb{R}^n$ for which the correspondence produces a real number.

Example 2

Consider the function defined by

$$f(x, y, z) = \sqrt{z - x^2 - y^2}.$$

a) Find $f(0, 1, 3)$ and $f(1, 2, 9)$.
b) Determine the domain of f.

Solution

a) We have

$f(0, 1, 3) = \sqrt{3 - 0^2 - 1^2} = \sqrt{2}$ and $f(1, 2, 9) = \sqrt{9 - 1^2 - 2^2} = 2$.

b) The domain of f consists of all values (x, y, z) for which $\sqrt{z - x^2 - y^2}$ is a real number, that is, those values of (x, y, z) for which $z \geq x^2 + y^2$.

In Section 13.4 we will see that the graph of the equation $z = x^2 + y^2$ is as shown in Figure 13.2. The points in $\mathbb{R}^3$ whose coordinates satisfy $z \geq x^2 + y^2$ are those that lie on or above this surface. ▶

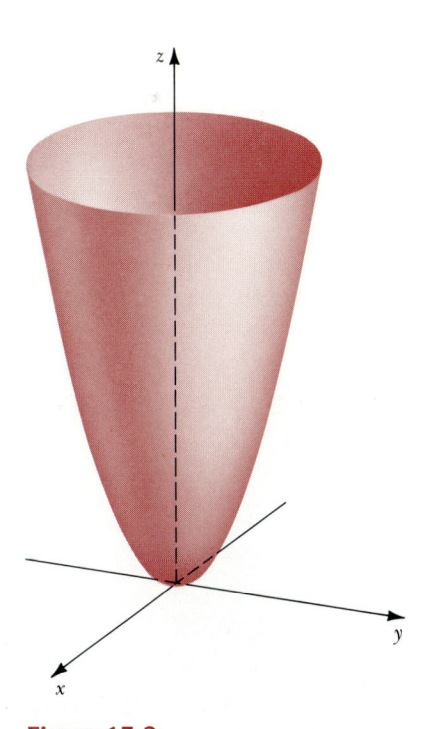

Figure 13.2

Example 3

Find the domain and the range of the function described by

$$f(x, y) = \sqrt{1 - x^2 - y^2}.$$

Solution

The domain of f is the set of ordered pairs (x, y) for which f is defined, that is, when

$$1 - x^2 - y^2 \geq 0 \quad \text{or} \quad x^2 + y^2 \leq 1.$$

The range of f is $[0, 1]$. The correspondence can be seen in Figure 13.3. ▶

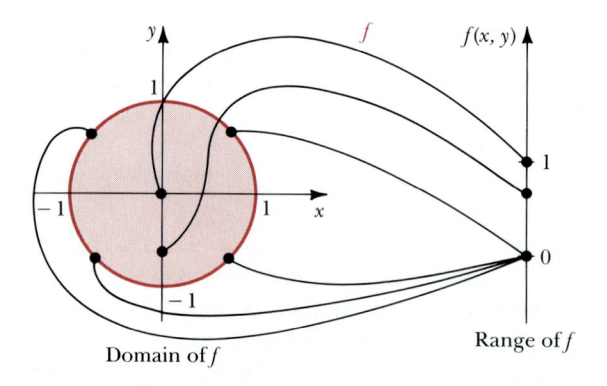

Figure 13.3
$f(x, y) = \sqrt{1 - x^2 - y^2}.$

The arithmetic properties of functions of several variables are defined in the same manner as those for functions of one variable. For example, if f and g are functions of two variables x and y, then $f + g$, $f - g$, and $f \cdot g$ are defined by

i) $(f + g)(x, y) = f(x, y) + g(x, y);$
ii) $(f - g)(x, y) = f(x, y) - g(x, y);$
iii) $(f \cdot g)(x, y) = f(x, y)g(x, y);$
and, for any constant k,
iv) $(kf)(x, y) = kf(x, y).$

The domain of the functions $f + g$, $f - g$, and $f \cdot g$ is the intersection of the domains of f and g. The domain of kf is the domain of f.
Similarly,

v) $\left(\dfrac{f}{g}\right)(x, y) = \dfrac{f(x, y)}{g(x, y)}.$

The domain of this quotient consists of pairs (x, y) that are in both the domain of f and the domain of g and for which $g(x, y) \neq 0$.

The composition of functions of several variables can also be defined; however, a bit of caution is needed because we are mapping from different sets. If g is a function of n variables, then $g: \mathbb{R}^n \to \mathbb{R}$, so composition is possible with a function $f: \mathbb{R} \to \mathbb{R}$, and $f \circ g$ is again a function from $\mathbb{R}^n$ into $\mathbb{R}$.

Example 4

Suppose $g(x, y) = x + y$ describes a function from $\mathbb{R}^2$ into $\mathbb{R}$ and $f(x) = \sqrt{x}$ describes a function from $\mathbb{R}$ into $\mathbb{R}$.

a) Find, if possible, $g \cdot g$, $f \circ g$, and $g \circ f$.
b) Determine the domain of $f \circ g$.

Solution

a) We have
$$(g \cdot g)(x, y) = g(x, y) \cdot g(x, y) = (x + y)^2$$
and
$$(f \circ g)(x, y) = f(g(x, y)) = f(x + y) = \sqrt{x + y},$$

but $g \circ f$ is not defined since the range of f is a subset of $\mathbb{R}$ and the domain of g is a subset of $\mathbb{R}^2$.

b) The domain of $f \circ g$ is the set of all (x, y) in the domain of g with the property that $g(x, y)$ is in the domain of f. Since the domain of g is all of $\mathbb{R}^2$, the domain of $f \circ g$ is the set of all (x, y) such that $f(x + y) = \sqrt{x + y}$ is a real number, that is, such that $x + y \geq 0$. The points whose coordinates satisfy this condition are in the shaded region in Figure 13.4. ▶

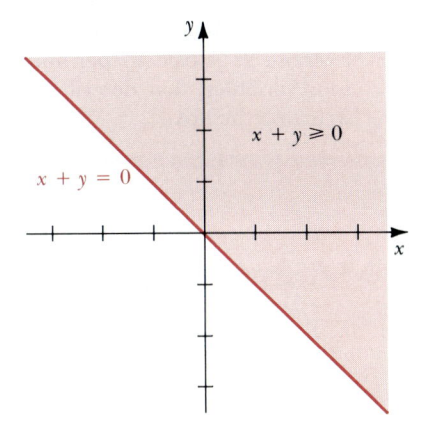

Figure 13.4

▶ EXERCISE SET 13.1

1. If $f(x, y) = x^2 + y$, find:
   a) $f(1, 0)$; b) $f(-1, -1)$; c) $f(\sqrt{3}, 2)$.

2. If $f(x, y, z) = 2xy + 4xz$, find:
   a) $f(2, 1, 3)$; b) $f(1, 3, 2)$; c) $f(0, 1, 5)$.

In Exercises 3–10, find the domain of the function.

3. $f(x, y) = \sqrt{1 - x - y}$
4. $f(x, y) = y + \tan x$
5. $f(x, y) = \dfrac{x}{y}$
6. $f(x, y) = \sqrt{x^2 + y^2 - 1}$
7. $f(x, y) = \ln\left(\dfrac{y - x}{x}\right)$
8. $f(x, y) = \dfrac{1}{2 - x - y}$
9. $f(x, y, z) = \sqrt{4 - x^2 - y^2 - z^2}$
10. $f(x, y, z) = \dfrac{1}{xy(z - 1)}$

For the functions described in Exercises 11–14 and constants h and k, find

$$\dfrac{f(x + h, y) - f(x, y)}{h} \quad \text{and} \quad \dfrac{f(x, y + k) - f(x, y)}{k}.$$

11. $f(x, y) = 3 - x - y$
12. $f(x, y) = x^2 + y^2$
13. $f(x, y) = 4 - x^2 - y^2$
14. $f(x, y) = xy$
15. Find $f \circ g$ if $g(x, y, z) = 2xy + 4xz + 6yz$ and $f(x) = 2x$.
16. Find $f \circ g$ if $g(x, y) = \sqrt{x^2 + y^2}$ and $f(x) = x^2$.
17. Express the volume of a right circular cylinder as a function of its height and radius.
18. Express the height of a right circular cone as a function of its volume and the radius of its base.
19. If A dollars are deposited at a 5% continuous interest rate, then the amount in the account at the end of t years is given by $P(A, t) = Ae^{0.05t}$. Find $P(1000, 20)$.
20. For a person with a weight of w kilograms and a height of h centimeters, an approximation to the body surface area is given in square meters by $A(w, h) = 0.00718 w^{0.425} h^{0.725}$. Find $A(55, 170)$.
21. A box is to be built in the form of a rectangular parallelepiped (see the figure). The bottom of the box is made from scrap material that costs $1 per square foot but the material for the top and sides costs $8 per square foot. Find a function that describes the cost of materials for the box.

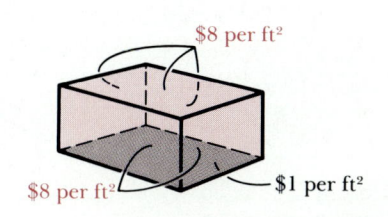

22. Suppose the builder of the box shown in the figure for Exercise 21 is paid $7 per hour. Find a function (of four variables) that describes the total cost of constructing the box.

23. Express the cost of painting a rectangular wall as a function of the dimensions if paint that costs $8 per gallon covers 200 ft² and a painter who paints 150 ft² per hour is paid $10 per hour.

13.2

Functions of Two Variables: Level Curves

The graph of a function of one variable is a set of ordered pairs. Similarly, the *graph* of a function f of two variables is the set of ordered triples $(x, y, f(x, y))$ with (x, y) in the domain of f. To sketch the graph of f, we introduce a third variable,

(13.1) $$z = f(x, y)$$

and sketch the points with coordinates (x, y, z) in $\mathbb{R}^3$, as shown in Figure 13.5.

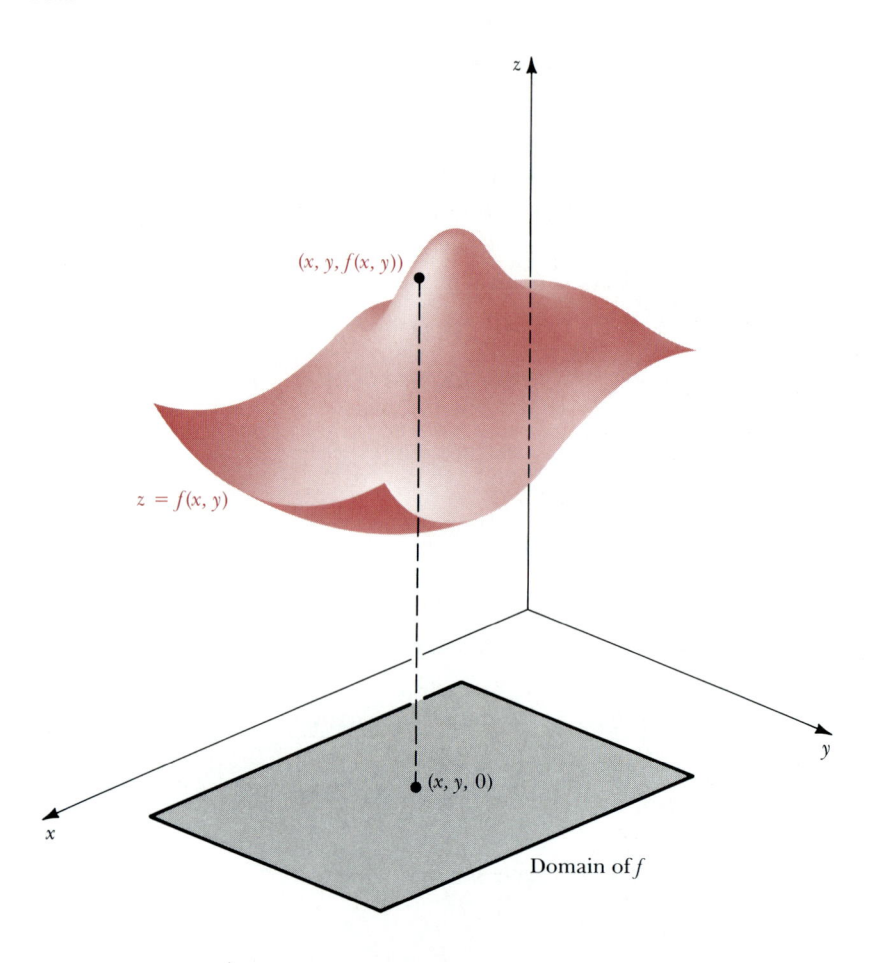

Figure 13.5
The surface described by $z = f(x, y)$.

13.2 FUNCTIONS OF TWO VARIABLES: LEVEL CURVES

Example 1

Sketch the graph of the function described by

$$f(x, y) = 1 - x - y.$$

Solution

We first set $z = f(x, y)$. Then $z = 1 - x - y$ and

$$x + y + z = 1.$$

This is an equation of a plane with normal vector $\langle 1, 1, 1 \rangle$. The portion of the plane in the first octant is shown in Figure 13.6. The plane extends infinitely from the line segments shown in this portion of the graph. ▶

The curve formed by the intersection of the graph of a function of two variables and a plane is called the **trace** of the graph in that plane. For example, the traces of the graph in Example 1 in the coordinate planes are shown in color in Figure 13.6.

For any constant c, the set of points (x, y) in the xy-plane that satisfy $c = f(x, y)$ is called a **level curve** of f. The particular level curve associated with the constant c is called the c-level curve of f. This c-level curve is the projection onto the xy-plane of the trace of the graph in the plane $z = c$, as shown in Figure 13.7.

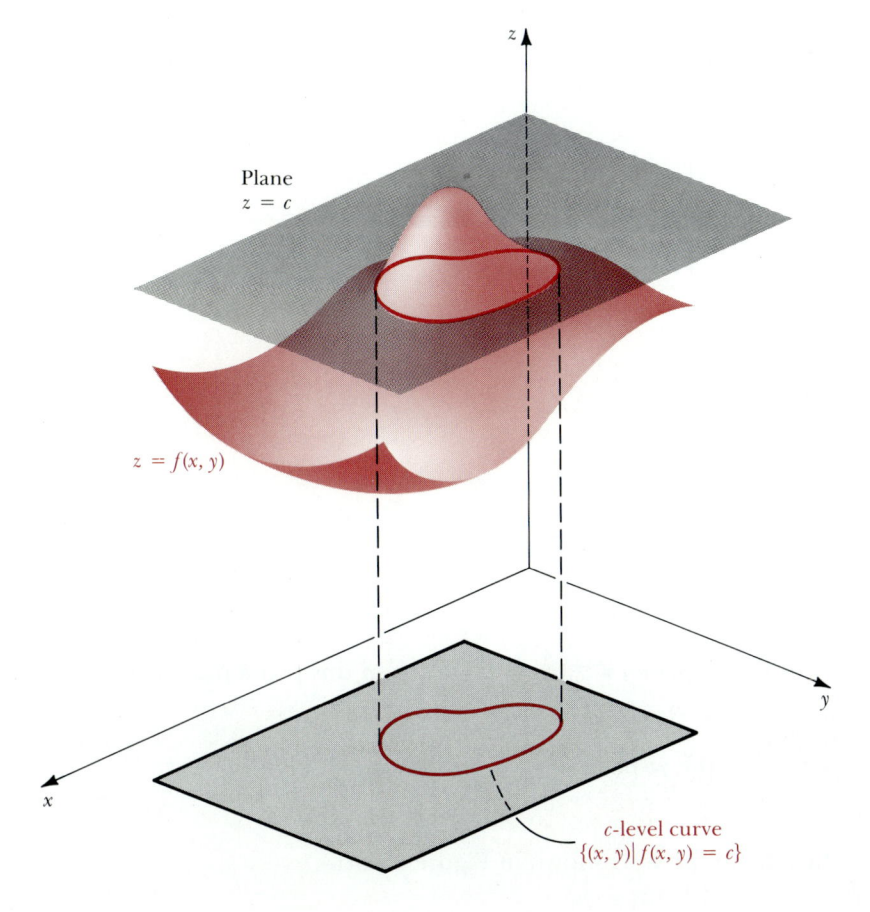

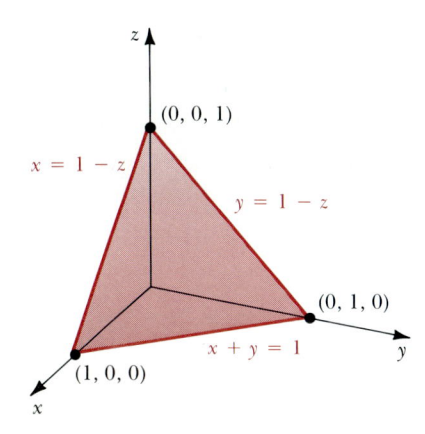

Figure 13.6

Figure 13.7
A c-level curve: $\{(x, y) | f(x, y) = c\}$.

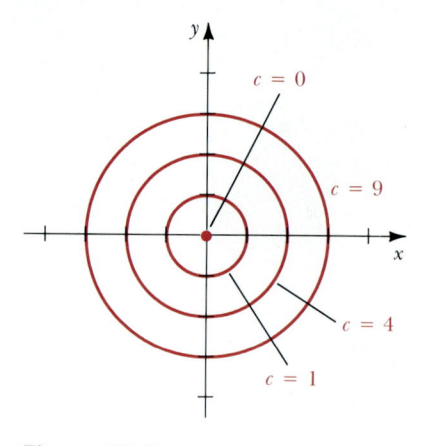

Figure 13.8
Level curves of $f(x, y) = x^2 + y^2$.

Example 2

Describe the level curves of
$$f(x, y) = x^2 + y^2$$
and sketch the graph of the function.

Solution
Setting $z = f(x, y)$, we have the equation
$$z = x^2 + y^2.$$

The level curves of f are determined by setting the variable z equal to a constant. If $z = c > 0$, the equation is
$$c = x^2 + y^2,$$
and the level curve is a circle with center at $(0, 0, 0)$ and radius $\sqrt{c}$.

The level curves for various values of c are shown in the xy-plane in Figure 13.8. The portions of the graph of f corresponding to these level curves are shown in Figure 13.9(a).

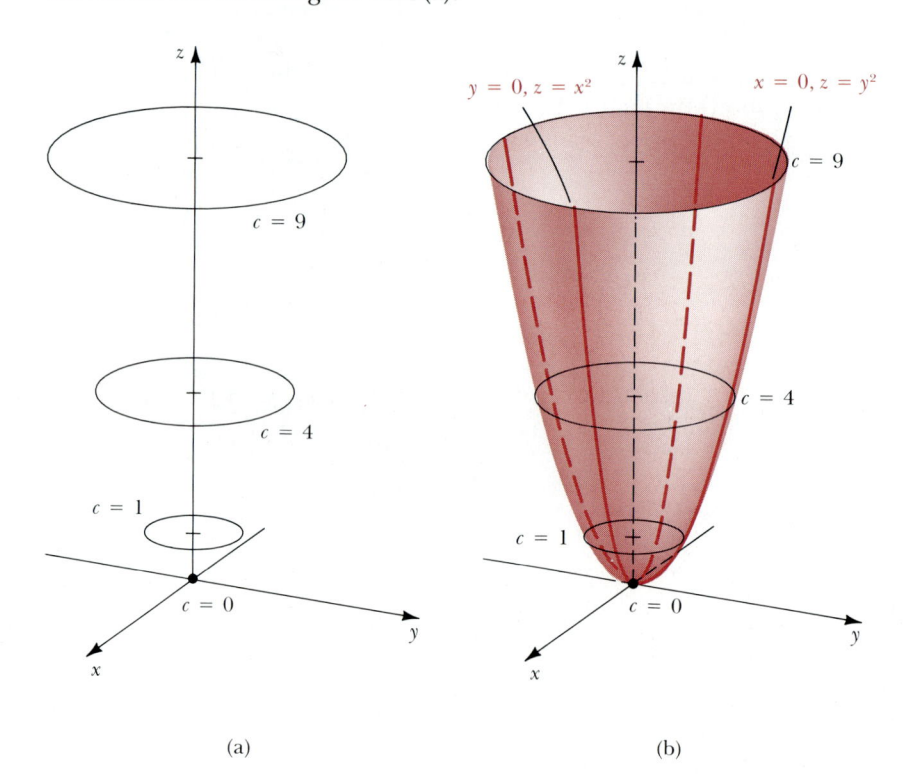

Figure 13.9
The graph of $f(x, y) = x^2 + y^2$.

(a) (b)

In the xz-plane, $y = 0$, so the graph in this plane has equation
$$z = x^2,$$
whereas in the yz-plane, $x = 0$, and the equation is
$$z = y^2.$$

Thus the graph is as shown in Figure 13.9(b).

13.2 FUNCTIONS OF TWO VARIABLES: LEVEL CURVES

The traces parallel to the *xy*-plane are circular. The traces parallel to both the *yz*- and *xz*-planes are parabolas. This graph is called a *circular paraboloid*. ▶

Using level curves to present three-dimensional information in a two-dimensional setting is common in many areas. For example, topographical maps use this technique to indicate curves of constant elevation. Weather forecast maps also use this representation to show level curves of temperature, called *isothermal* curves, as well as high and low pressure zones throughout a large area. Figure 13.10 shows a weather map with levels of constant temperature.

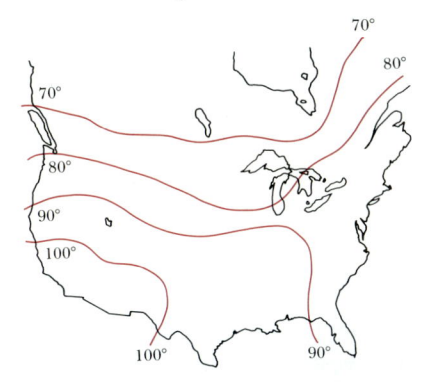

Figure 13.10
Level curves of temperature.

One advantage of a level-curve representation of a function is that the graph for one portion of the domain is not obscured by the graph of another portion. The two illustrations of Mt. Rainier shown in Figures 13.11 and 13.12 demonstrate this feature. Since the photograph in Figure 13.11 is taken from northeast of the summit, the elevation at points on the southwest slope are obscured. However, the elevation of all points on the mountain can be found on the topographical map in Figure 13.12.

Figure 13.11
Mount Rainier

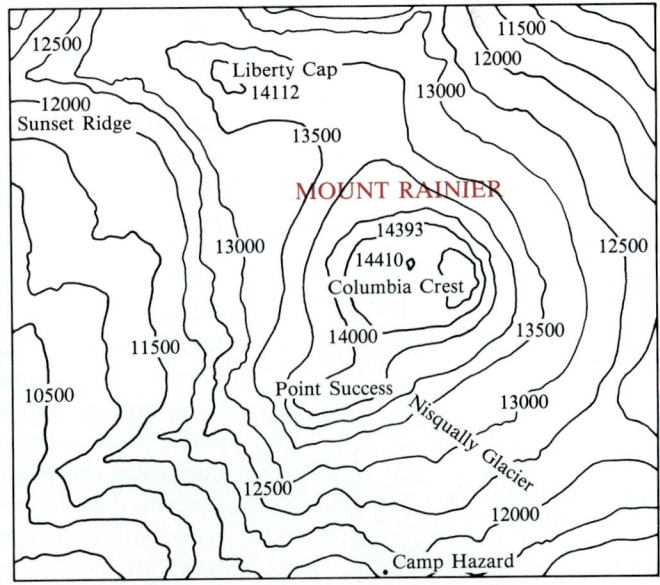

Figure 13.12

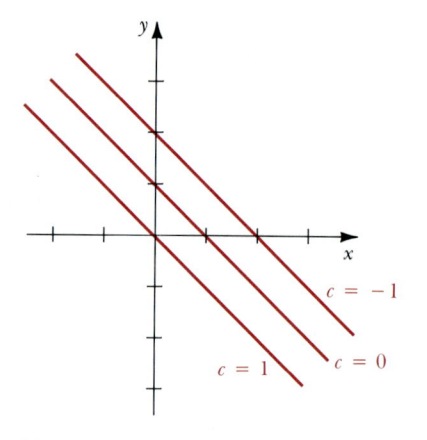

Figure 13.13
Level curves of $f(x, y) = 1 - x - y$.

The disadvantage of the level-curve representation is the lack of depth perception. It is not immediately apparent, for example, that Figure 13.13 illustrates the plane $z = 1 - x - y$, which is shown in Figure 13.6.

Computer Graphics

The increasing availability of relatively low-cost computer systems has brought with it an increase in the use of computer graphics. A computer graphics system consists of a computer combined with some form of graphic device. The graphic device can be a printer or plotter that produces a paper copy, but is more commonly a cathode ray tube (monitor) on which a visual image is formed.

The applications of computer graphics range from entertainment, such as computer games and the generation of animated cartoons, to sophisticated simulation techniques used for everything from automobile design to aircraft navigation.

Most computer graphics systems permit the realistic two-dimensional representation of three-dimensional graphs and incorporate hidden line removal to eliminate portions of the graph that are not in view. The computer sketches a graph in the same way we do, by drawing representative curves, but has the advantage of being able to manipulate data very rapidly. The resolution of the completed graph depends to some degree on the speed of the computer, but is principally a factor of the resolution of the graphic device. The graphs of the functions in this text were produced using computer graphics. The system first sketched the graph, and then the curve was smoothed to eliminate any jagged edges. This gives a very accurate graph that is indistinguishable in artistic quality from those produced by a graphic artist.

It is interesting to examine the results of computer graphics systems since they are becoming an increasingly important tool in design and development work. Some of the graphs in this section and later in the text are reproduced in the color plates, which give a computer-enhanced, 3-dimensional view of these functions.

Consider the graph shown in Figures 13.14(a) and (b). This graph is called a monkey saddle and has equation $z = x^3 - 3xy^2$. The name is derived from the fact that if in the unlikely event a monkey were to sit on the graph at the origin with its head facing the direction of the positive x-axis, it would have a place to rest its tail (along the negative x-axis) as well as both legs (along the lines $y = x$ and $y = -x$, when $x > 0$). Figure 13.14(a) shows a computer-generated sketch of the monkey saddle using a grid of curves on the surface that are parallel to the coordinate planes. Figure 13.14(b) shows computer-generated level curves on the graph. A color representation is given in Color Plate I.

Computer graphics systems often have the ability to rotate a figure in order to allow it to be viewed from different perspectives. Figure 13.15 illustrates the use of a rotation in viewing the graph of the function described by $z = x^2 + \sin y$. The graph may be difficult to visualize in part (a), but becomes clearer in (b) and even more so in (c). This rotational property is particularly useful in architectural and engineering

13.2 FUNCTIONS OF TWO VARIABLES: LEVEL CURVES   757

(a)

(b)

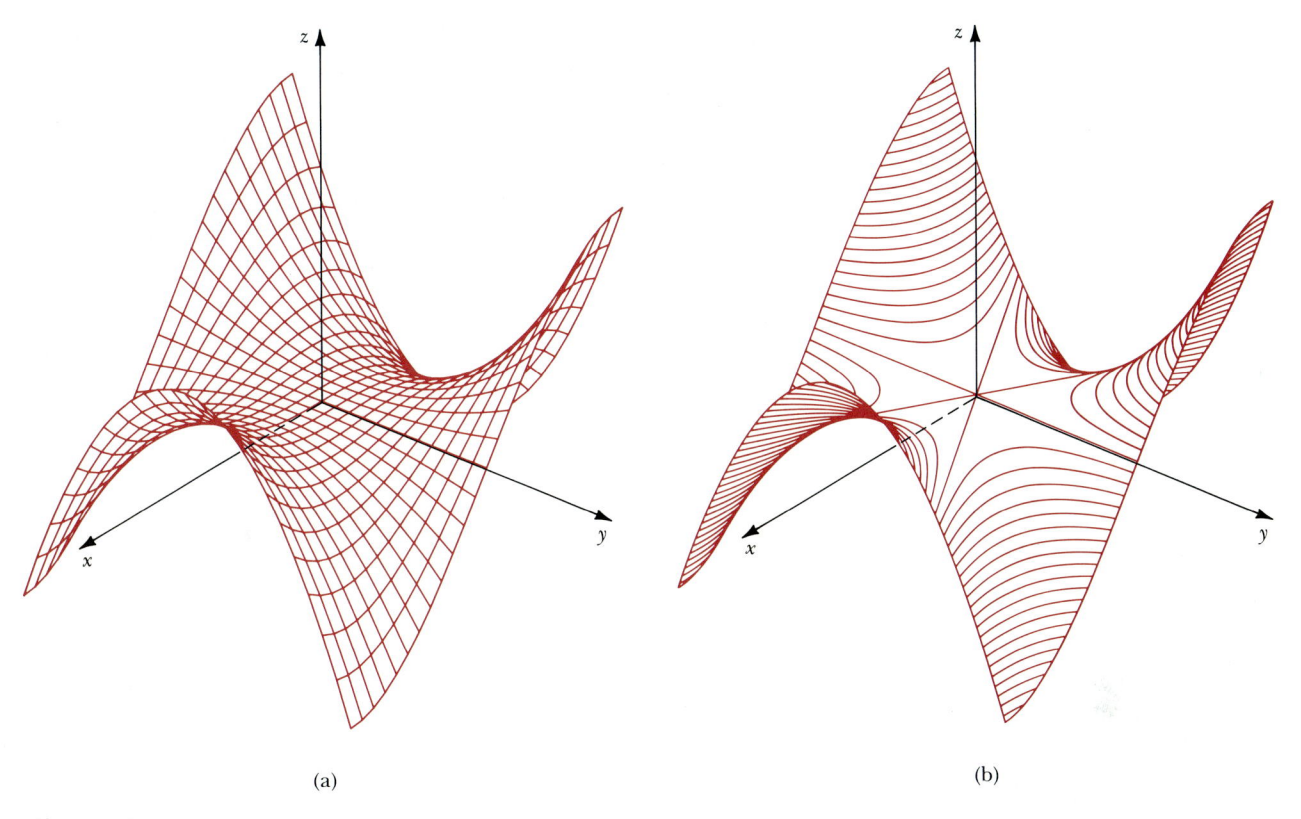

Figure 13.14

(a)

(b)

(c)

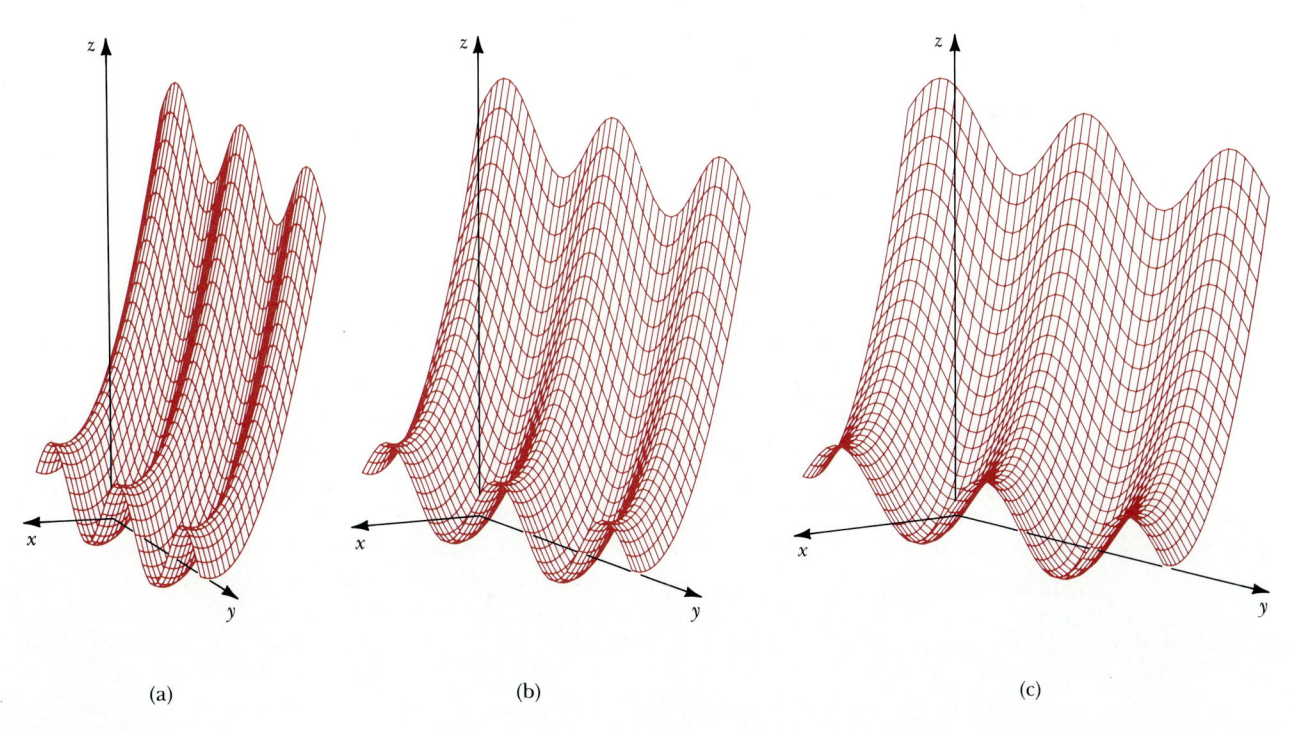

Figure 13.15

758   CHAPTER 13 MULTIVARIABLE FUNCTIONS

design work, since the figure can be rotated into the best perspective before a drawing is generated or a model constructed. (See also Color Plate II.)

A selection of some interesting computer-drawn graphs is shown in Figure 13.16 to give you an idea of the level of sophistication of these systems. Note that in each case the drawing consists of a collection of curves in planes parallel to a coordinate plane. The system has been programmed to recognize and delete portions of the curves that are hidden behind other parts of the graph. Color representations of these graphs are given in Color Plate III. If you have access to a computer graphics system, watch carefully as graphs are constructed. Since the graphs are generally formed by drawing one curve at a time, you can see exactly how the graph is obtained.

Figure 13.16

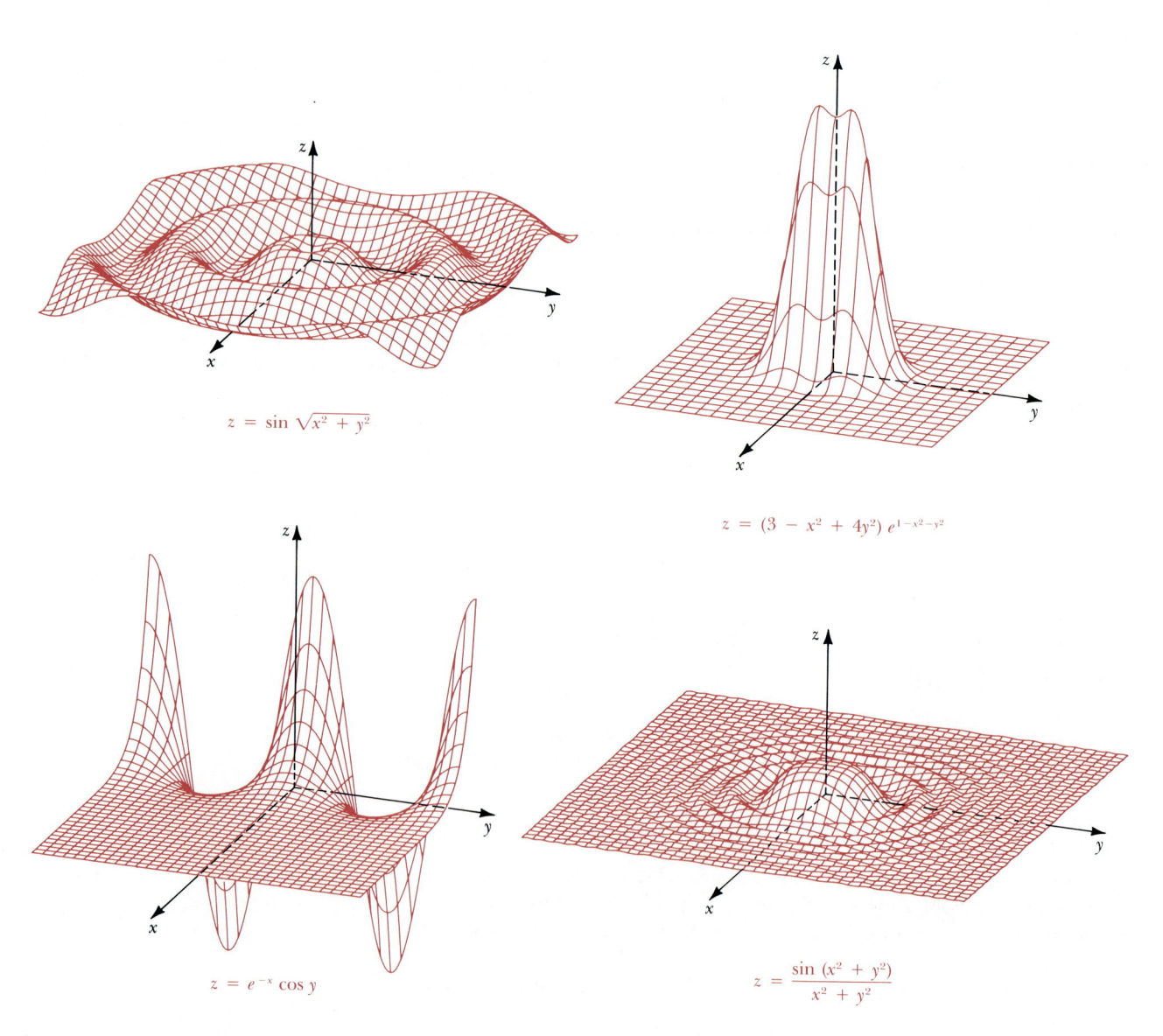

$z = \sin \sqrt{x^2 + y^2}$

$z = (3 - x^2 + 4y^2) e^{1 - x^2 - y^2}$

$z = e^{-x} \cos y$

$z = \dfrac{\sin (x^2 + y^2)}{x^2 + y^2}$

▶ EXERCISE SET 13.2

In Exercises 1–10, sketch the level curves for the given values of c.

1. $f(x, y) = 1 + x - y$, $c = 1, 2, 3$
2. $f(x, y) = 2x - y + 4$, $c = 0, 1, 3, 5$
3. $f(x, y) = 2 - x^2 - y^2$, $c = 0, -2$
4. $f(x, y) = 4x^2 + y^2$, $c = 1, 2, 4$
5. $f(x, y) = x^2 - y$, $c = 0, -1, 1$
6. $f(x, y) = x^2 - y^2$, $c = 1, 4$
7. $f(x, y) = \ln(x + y)$, $c = 0, 1$
8. $f(x, y) = \ln x + \ln y$, $c = 0, 1, 2$
9. $f(x, y) = \sin x + y$, $c = 0, 1$
10. $f(x, y) = 1/xy$, $c = -1, -\frac{1}{2}, \frac{1}{2}, 1$

In Exercises 11–24, sketch the graph of the function.

11. $f(x, y) = x + y - 2$
12. $f(x, y) = 6 - 2x - 3y$
13. $f(x, y) = x^2 + y^2 + 2$
14. $f(x, y) = 4 - x^2 - y^2$
15. $f(x, y) = x^2$
16. $f(x, y) = \sin y$
17. $f(x, y) = 3 - x$
18. $f(x, y) = 3 - y$
19. $f(x, y) = 4x^2 + 9y^2$
20. $f(x, y) = \sqrt{x^2 + y^2}$
21. $f(x, y) = \begin{cases} x^2 + y^2, & \text{if } x^2 + y^2 \leq 9, \\ 0, & \text{if } x^2 + y^2 > 9 \end{cases}$
22. $f(x, y) = \begin{cases} \sqrt{x^2 + y^2}, & \text{if } x^2 + y^2 \leq 4, \\ 4, & \text{if } x^2 + y^2 > 4 \end{cases}$
23. $f(x, y) = \begin{cases} 6 - 2x - 3y, & \text{if } x \geq 0, y \geq 0, \\ & \quad 6 - 2x - 3y \geq 0, \\ 0, & \text{elsewhere} \end{cases}$
24. $f(x, y) = \begin{cases} 2 - x + y, & \text{if } x \geq 0, y \geq 0, 2 - x + y \geq 0, \\ 0, & \text{elsewhere} \end{cases}$

25. The elevation of a mountain above a point (x, y) in the base plane is $E(x, y) = 200 - x^2 - 4y^2$. Sketch the curves of constant elevation for $E = 0$, $E = 100$, $E = 150$, and $E = 200$.

26. The temperature at any point (x, y) on a circular plate is $T(x, y) = \sqrt{x^2 + y^2}$. Sketch the curves of constant temperature for $T = 0$, $T = 3$, and $T = 5$.

27. The cost of building a box of height h in. with a square base of length l in. is $C(l, h) = l^2 + 6lh$ dollars. Sketch the curves of constant cost for $l \geq 0$, $h \geq 0$, and $C = 4$, $C = 9$, and $C = 16$.

13.3

Functions of Three Variables: Level Surfaces

In Section 13.2 we found that three dimensions are needed to sketch the graph of a function of two variables: two dimensions for the domain and an additional dimension for the range. The graph of a function f of three variables is the set of all $(x, y, z, f(x, y, z))$ such that (x, y, z) is in the domain of f. To sketch the graph of a function of three variables would require four dimensions, one more than is physically available. A function of three variables can, however, be geometrically represented by a three-dimensional analogue of the level curves for functions of two variables.

For a function f of three variables and a constant c, the set of points in

$\mathbb{R}^3$ satisfying

(13.2) $$c = f(x, y, z)$$

is called a **level surface** of f. The particular level surface associated with the constant c is called the *c-level surface* of the graph of f.

Example 1

Sketch the level surfaces of

$$f(x, y, z) = x^2 + y^2 + z^2$$

corresponding to $c = 0$, $c = 1$, and $c = 3$.

Solution

For $c > 0$, the level surface

$$c = x^2 + y^2 + z^2$$

is a sphere with center at the origin $(0, 0, 0)$ and radius $\sqrt{c}$. When $c = 0$ we have the equation

$$0 = x^2 + y^2 + z^2,$$

which implies that $x = 0$, $y = 0$, and $z = 0$. In this case, the level surface is simply the point with coordinates $(0, 0, 0)$.

The portions of these level surfaces that lie in the first octant are shown in Figure 13.17.

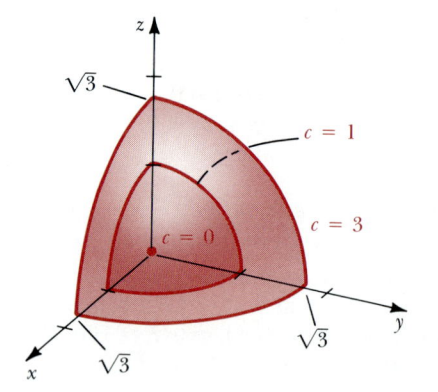

Figure 13.17
Portions of the level surfaces of $f(x, y, z) = x^2 + y^2 + z^2$.

Example 2

Describe the level surfaces of the function defined by

$$f(x, y, z) = x + y + z.$$

Solution

For a constant c, the level surface associated with f and c is a plane with equation

$$c = x + y + z.$$

For various values of c, the level surfaces of f are parallel planes with normal vector $\langle 1, 1, 1 \rangle$. Sketches of the level surfaces for $c = 1$ and $c = 2$ are shown in Figure 13.18.

A special type of level surface occurs if one of the variables does *not* appear in the description of a function of three variables. For example, the level surfaces of the function of three variables described by

$$f(x, y, z) = x^2 + y^2$$

are independent of the variable z. For any constant $c > 0$, the c-level surface of f intersects each plane perpendicular to the z-axis in the same type of curve, a circle in that plane with equation

$$x^2 + y^2 = c.$$

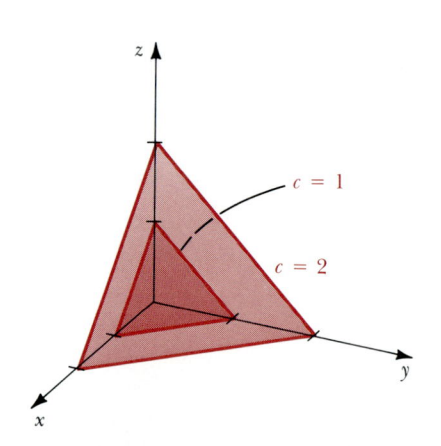

Figure 13.18
Portions of the level surfaces of $f(x, y, z) = x + y + z$.

The level surfaces for $c = 1$ and $c = 4$ are shown in Figure 13.19. These level surfaces are examples of *cylinders*.

(13.3)
DEFINITION
Suppose C is a curve lying in a plane and l is a line that does not lie in this plane. The **cylinder** generated by C and l consists of all lines parallel to l that intersect C.

Figure 13.20 is an illustration of a cylinder. The curve C is called a **directrix** of the cylinder and the line l is called a **generator**. When l is orthogonal to the plane containing C, the cylinder is called a **right cylinder**. In addition, if the curve C is a circle, the cylinder is called a **right circular cylinder**. For example, the cylinders shown in Figure 13.19 are right circular cylinders. Cylinders represented as level surfaces of functions of three variables with one of the variables not in the description of the function are always right cylinders, but they need not be circular.

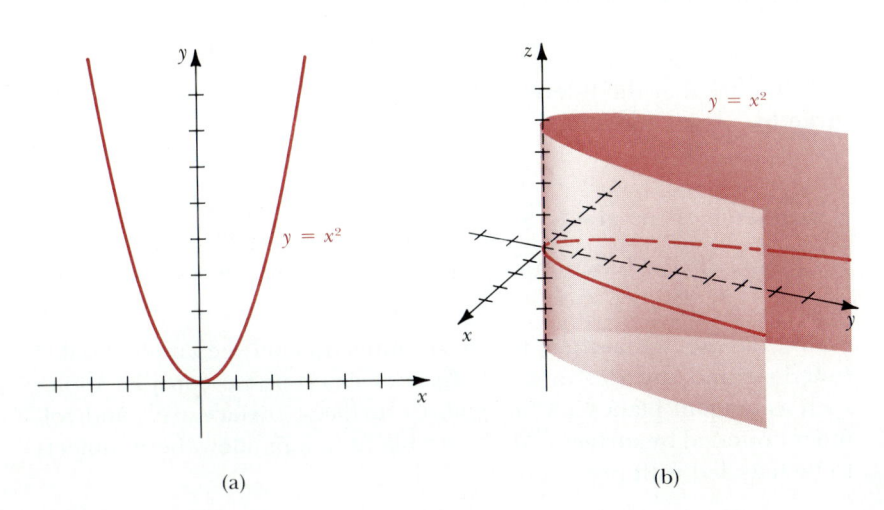

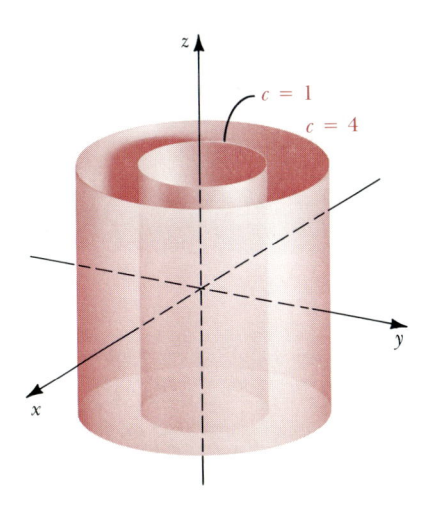

Figure 13.19
Level surfaces of $f(x, y, z) = x^2 + y^2$.

Example 3

Sketch the right cylinder that is orthogonal to the xy-plane and whose graph in that plane has equation $y = x^2$.

Solution
The graph of the directrix $y = x^2$ in the xy-plane is shown in Figure 13.21(a) and the cylinder is shown in Figure 13.21(b). One representation of this cylinder is as the 0-level surface of $f(x, y, z) = y - x^2$. ▶

The following example shows that a right cylinder is also produced by a function of two variables when one of the variables does not appear in the description of the function. In this example, y does not appear and the cylinder is perpendicular to the xz-plane. If x does not appear in

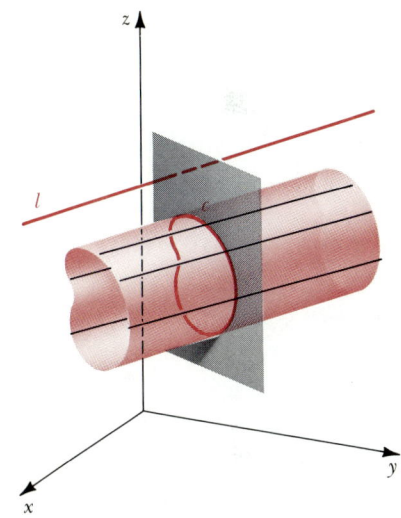

Figure 13.20
A cylinder with directrix c.

Figure 13.21
(a) A directrix of the cylinder; (b) the 0-level surface of $f(x, y, z) = y - x^2$.

Example 4

Sketch the graph of $f(x, y) = \cos x$.

Solution

Let $z = f(x, y) = \cos x$. The graph of f sketched in Figure 13.22 is a right cylinder perpendicular to the xz-plane. A directrix for the cylinder is the curve in the xz-plane satisfying the equation $z = \cos x$. Any line parallel to the y-axis that passes through this directrix lies on the cylinder. ▶

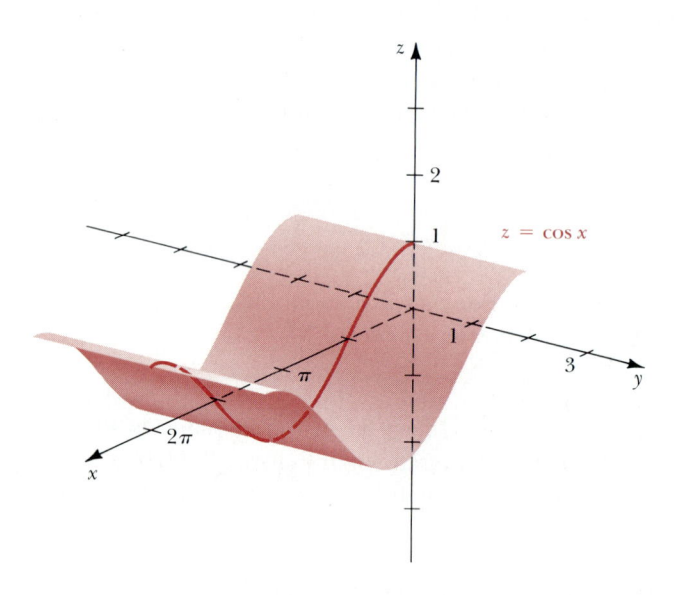

Figure 13.22
The graph of $f(x, y) = \cos x$.

Example 4 illustrates how certain surfaces can be described using functions of two variables. For a function f of two variables, the equation $z = f(x, y)$ can be rewritten as

$$f(x, y) - z = 0.$$

This implies that the 0-level surface of the function of three variables given by

$$g(x, y, z) = f(x, y) - z$$

is the sketch of the graph of f. Thus

> the graph of a function of two variables is also a level surface of a function of three variables.

We have seen that there is a close connection between multivariable functions and surfaces in space. In later chapters we consider topics such as tangent planes and normals to surfaces, surface area, and volumes bounded by surfaces. Multivariable functions allow these subjects to be handled with precision.

► EXERCISE SET 13.3

In Exercises 1–12, describe the level surfaces of the function.

1. $f(x, y, z) = 2x + 3y + z$
2. $f(x, y, z) = x - y - 2z$
3. $f(x, y, z) = z - x - y$
4. $f(x, y, z) = x - z$
5. $f(x, y, z) = x^2 + y^2$
6. $f(x, y, z) = x^2 - y$
7. $f(x, y, z) = z^2 + y^2$
8. $f(x, y, z) = z^2 - y^2$
9. $f(x, y, z) = x^2 + 4y^2 + 9z^2$
10. $f(x, y, z) = 4x^2 + 9y^2 + 4z^2$
11. $f(x, y, z) = y - x^2 - z^2 - 1$
12. $f(x, y, z) = z - x^2 - y^2$

13. Sketch the graph of the right cylinder perpendicular to the xy-plane whose equation in that plane is $4x^2 + 9y^2 = 36$.

14. Sketch the graph of the right cylinder perpendicular to the yz-plane whose equation in that plane is $z = \sin y$, $0 \leq y \leq 2\pi$.

15. Sketch the graph of the right cylinder perpendicular to the xz-plane whose equation in that plane is $|z| = |x|$.

16. Sketch the graph of the right cylinder perpendicular to the xy-plane whose equation in that plane is $y = e^x$.

17. Suppose C is a curve in the xy-plane described by $y = f(x)$, where $x \geq 0$, $y \geq 0$.
    a) Show that (x, y, z) lies on the surface generated by revolving C about the y-axis precisely when $y = f(\sqrt{x^2 + z^2})$.
    b) Show that (x, y, z) lies on the surface generated by revolving C about the x-axis precisely when $\sqrt{y^2 + z^2} = f(x)$.

In Exercises 18–23, (a) use the result in Exercise 17 to find an equation of the surface of revolution and (b) sketch the surface.

18. $y = 2x + 1$, $x \geq 0$, about the y-axis
19. $y = \sqrt{1 - x^2}$, $x \geq 0$, about the y-axis
20. $y = x^2$, $x \geq 0$, about the x-axis
21. $y = x^2$, $x \geq 0$, about the y-axis
22. $y = 1 + \sqrt{4 - x^2}$, $x \geq 0$, about the x-axis
23. $y = \sqrt{x^2 - 4}$, $x \geq 2$, about the x-axis

24. The temperature at any point (x, y, z) in a room is given in degrees Fahrenheit by $T(x, y, z) = 65 - x^2 - y^2 - z$. Sketch the surfaces of constant temperature for $T = 61$, $T = 56$, and $T = 29$. These are called *isothermal surfaces* of T.

13.4

Quadric Surfaces

In Chapter 10 we saw that the graph of a second-degree equation in two variables

$$Ax^2 + Bxy + Cy^2 + Dx + Ey + F = 0$$

is a conic section. A conic section in this form is the 0-level curve of the function of two variables described by

$$f(x, y) = Ax^2 + Bxy + Cy^2 + Dx + Ey + F.$$

The graph of a second-degree equation in *three* variables,

(13.4) $\quad Ax^2 + By^2 + Cz^2 + Dxy + Exz + Fyz + Gx + Hy + Iz + J = 0$.

produces a **quadric surface.** This quadric surface is the 0-level surface of the function of three variables described by

$$f(x, y, z) = Ax^2 + By^2 + Cz^2 + Dxy + Exz + Fyz + Gx + Hy + Iz + J.$$

In this section we consider some common quadric surfaces in standard position, similar to the Cartesian-coordinate standard position of conic sections discussed in Chapter 10. The names of these quadric surfaces are derived from the names of the conic sections produced from traces in the coordinate planes.

The quadric surface with equation

(13.5) $$\frac{x^2}{a^2} + \frac{y^2}{b^2} + \frac{z^2}{c^2} = 1,$$

for positive real numbers a, b, and c, is called an **ellipsoid** because its trace in each of the coordinate planes is an ellipse. For example, when $x = 0$, the equation becomes $y^2/b^2 + z^2/c^2 = 1$. The graph of this ellipse in the yz-plane is shown in Figure 13.23(a). Figure 13.23(b) shows an artist's rendering of an ellipsoid, and a computer-generated drawing with contour lines is shown in Figure 13.23(c). (The computer-generated quadric surfaces in this section are also shown in Color Plates IV, V, and VI.)

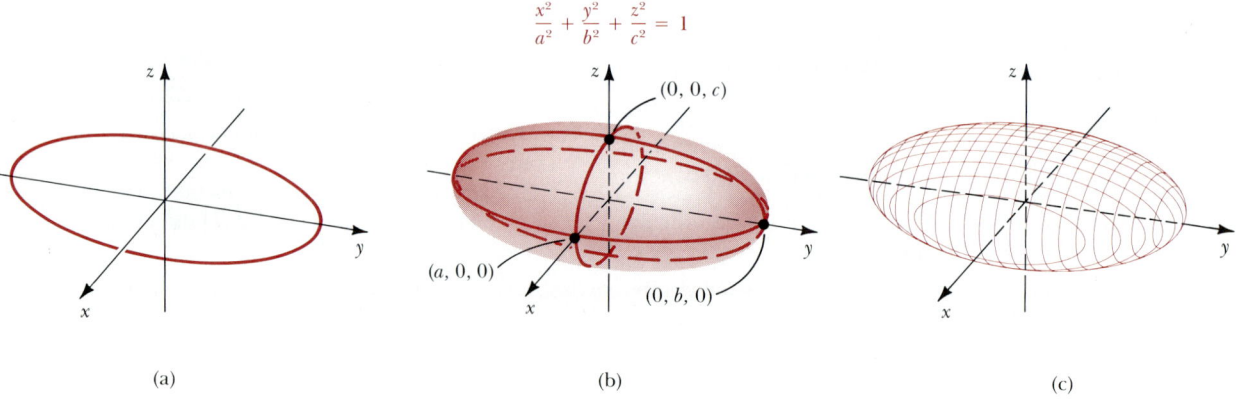

Figure 13.23
An ellipsoid.

In a similar manner, the graph of the equation

(13.6) $$\frac{x^2}{a^2} + \frac{y^2}{b^2} - \frac{z^2}{c^2} = 1,$$

for positive constants a, b, and c, is called an **elliptic hyperboloid of one sheet.** This graph is shown in Figure 13.24. The trace of this surface in the xy-plane is an ellipse with planar equation

$$\frac{x^2}{a^2} + \frac{y^2}{b^2} = 1.$$

The traces in the xz- and yz-planes are hyperbolas with planar equations

$$\frac{x^2}{a^2} - \frac{z^2}{c^2} = 1 \quad \text{and} \quad \frac{y^2}{b^2} - \frac{z^2}{c^2} = 1,$$

respectively.

$$\frac{x^2}{a^2} + \frac{y^2}{b^2} - \frac{z^2}{c^2} = 1$$

(a)

(b)

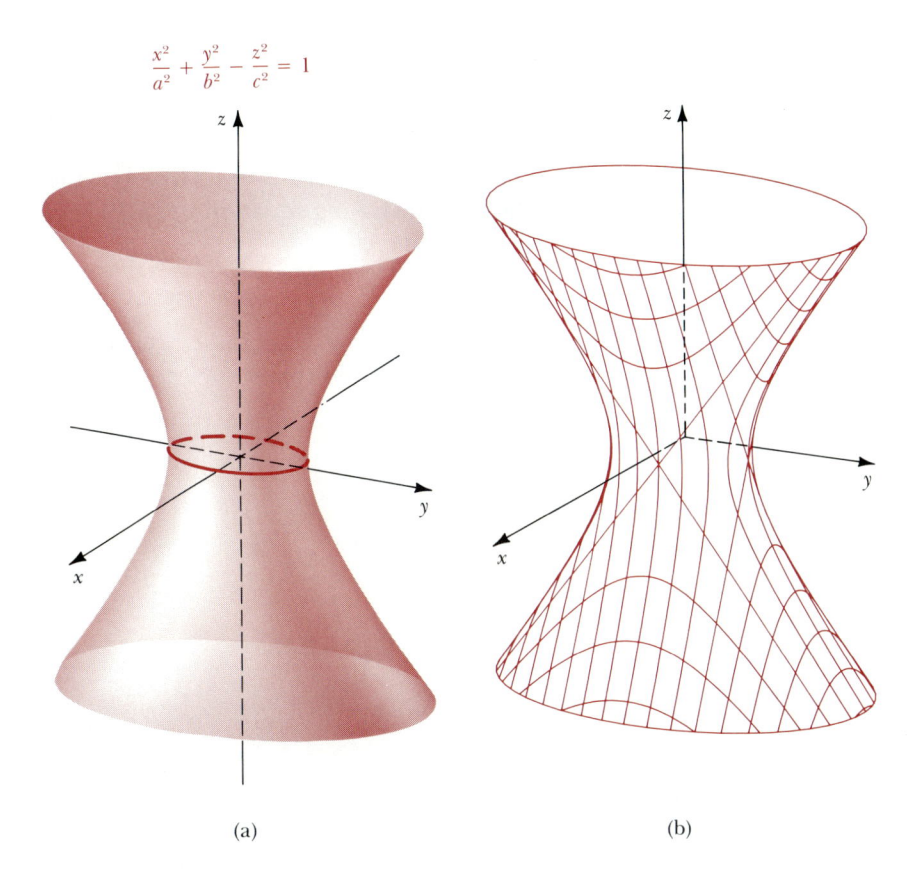

Figure 13.24
An elliptic hyperboloid of one sheet.

The trace in a plane with equation $z = k$, parallel to the xy-plane, is an ellipse with equation

$$\frac{x^2}{a^2} + \frac{y^2}{b^2} = 1 + \frac{k^2}{c^2}, \qquad z = k,$$

which can be rewritten as

$$\frac{x^2}{a^2\left(1 + \frac{k^2}{c^2}\right)} + \frac{y^2}{b^2\left(1 + \frac{k^2}{c^2}\right)} = 1, \qquad z = k.$$

As k increases in magnitude, the lengths of the major and minor axes of the ellipse increase.

The surface with equation of the form (13.6) is called an elliptic hyperboloid of *one sheet* to distinguish it from the surface involving ellipses and hyperbolas that we consider next.

The graph of an equation of the form

(13.7) $$\frac{x^2}{a^2} - \frac{y^2}{b^2} - \frac{z^2}{c^2} = 1,$$

for positive constants a, b, and c, is called an **elliptic hyperboloid of two sheets.** The trace of this surface in either the xy-plane or the xz-plane is a hyperbola. The surface does not intersect the yz-plane since

$$\frac{x^2}{a^2} - 1 = \frac{y^2}{b^2} + \frac{z^2}{c^2} \quad \text{implies that} \quad \frac{x^2}{a^2} - 1 \geq 0.$$

Thus the equation can be satisfied only if $x \geq a$ or $x \leq -a$. However, the trace in any plane parallel to the yz-plane having equation $x = k, |k| > a$, is an ellipse whose equation in that plane is

$$\frac{y^2}{b^2} + \frac{z^2}{c^2} = \frac{k^2}{a^2} - 1.$$

Such a graph is shown in Figures 13.25(a) and (b).

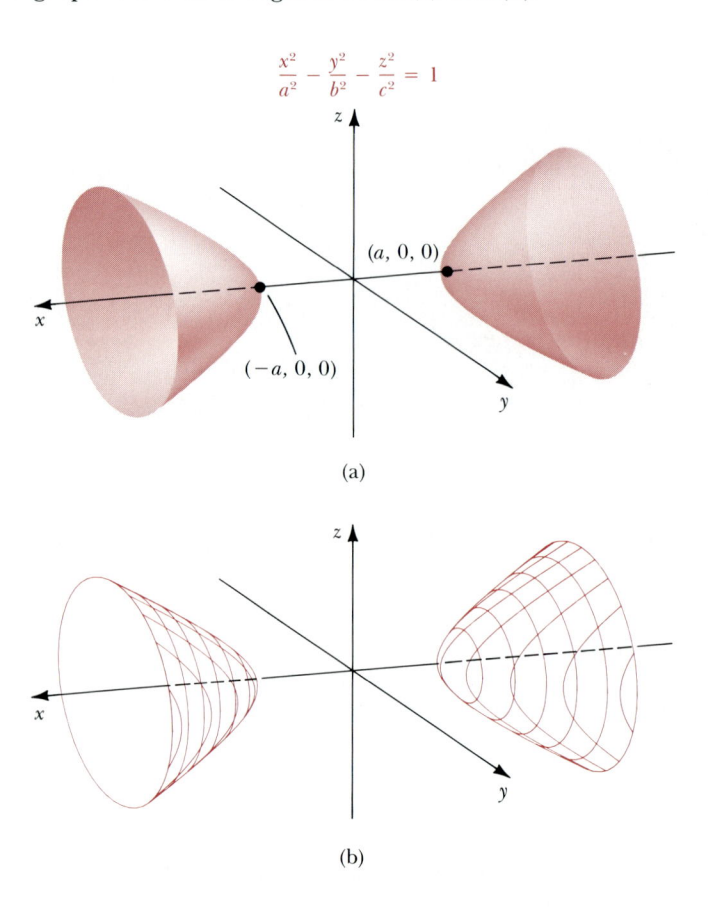

Figure 13.25

An elliptic hyperboloid of two sheets.

The quadric surface given by the graph of the equation

(13.8) $$\frac{x^2}{a^2} + \frac{y^2}{b^2} = cz,$$

where a and b are positive and $c \neq 0$, is called an **elliptic paraboloid.** For positive values of c, the surface appears as shown in Figures 13.26(a) and (b), and its traces in the xz- and yz-planes are parabolas. The surface intersects the xy-plane only at the origin. The trace of the surface in a

$$\frac{x^2}{a^2} + \frac{y^2}{b^2} = cz, c > 0 \qquad\qquad\qquad\qquad\qquad\qquad \frac{x^2}{a^2} + \frac{y^2}{b^2} = cz, c < 0$$

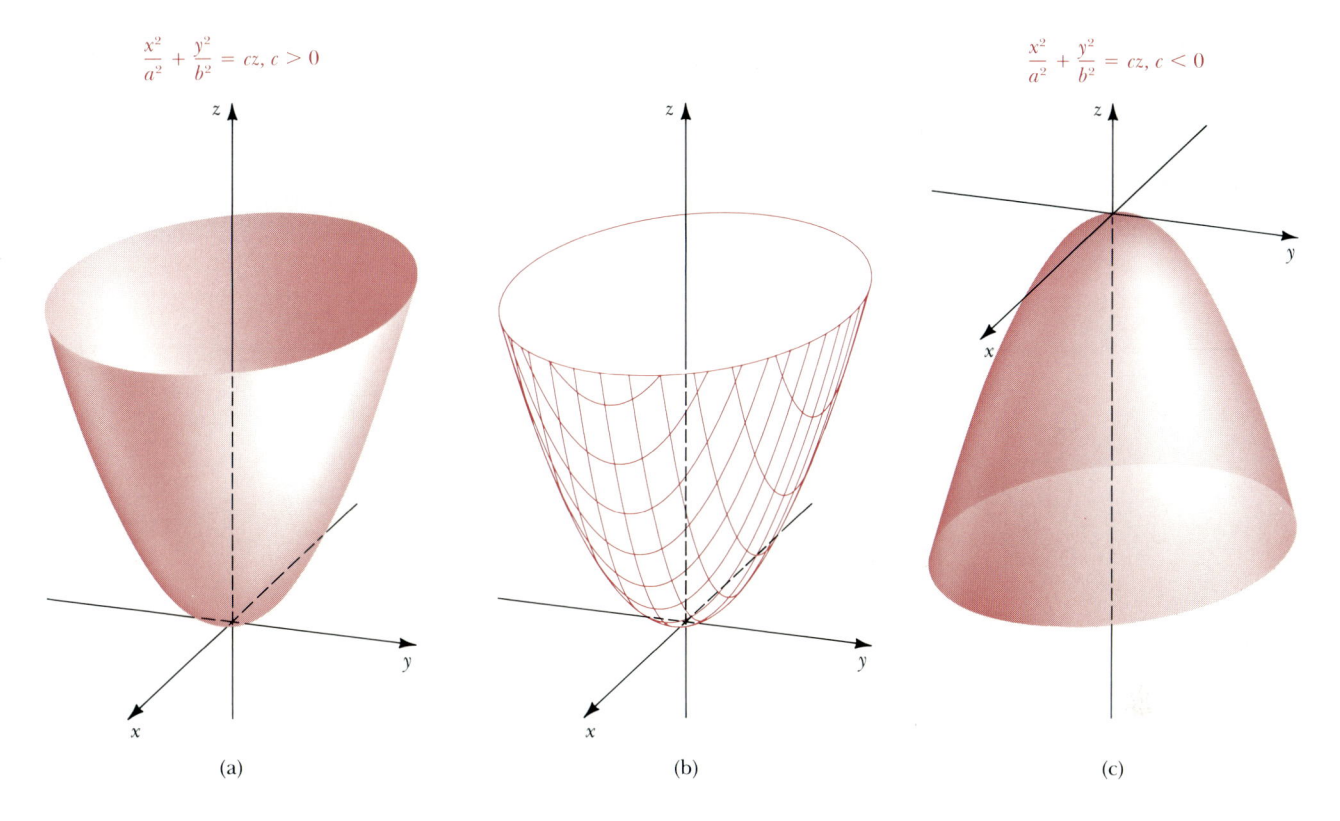

(a)   (b)   (c)

Figure 13.26
A paraboloid.

plane of the form $z = k$, for $k > 0$, is an ellipse with equation

$$\frac{x^2}{a^2 ck} + \frac{y^2}{b^2 ck} = 1.$$

When $c < 0$, z assumes only nonpositive values and the surface in this case appears as shown in Figure 13.26(c).

A quadric surface given by the graph of the equation

(13.9) $$\frac{y^2}{a^2} - \frac{x^2}{b^2} = cz,$$

where a and b are positive and $c \neq 0$, is called a **hyperbolic paraboloid**. The trace of this graph in either the xz- or the yz-plane is a parabola, while the trace in the xy-plane is the pair of straight lines with equations

$$y = \pm \frac{a}{b} x.$$

Planes parallel to the xy-plane intersect the surface in a hyperbola. For $c > 0$, the major axis of this hyperbola is the y-axis if $z > 0$ and is the x-axis if $z < 0$. Consequently, the surface looks like a saddle that is infinitely high and deep, as shown in Figure 13.27.

$$\frac{y^2}{a^2} - \frac{x^2}{b^2} = cz$$

(a)

(b)

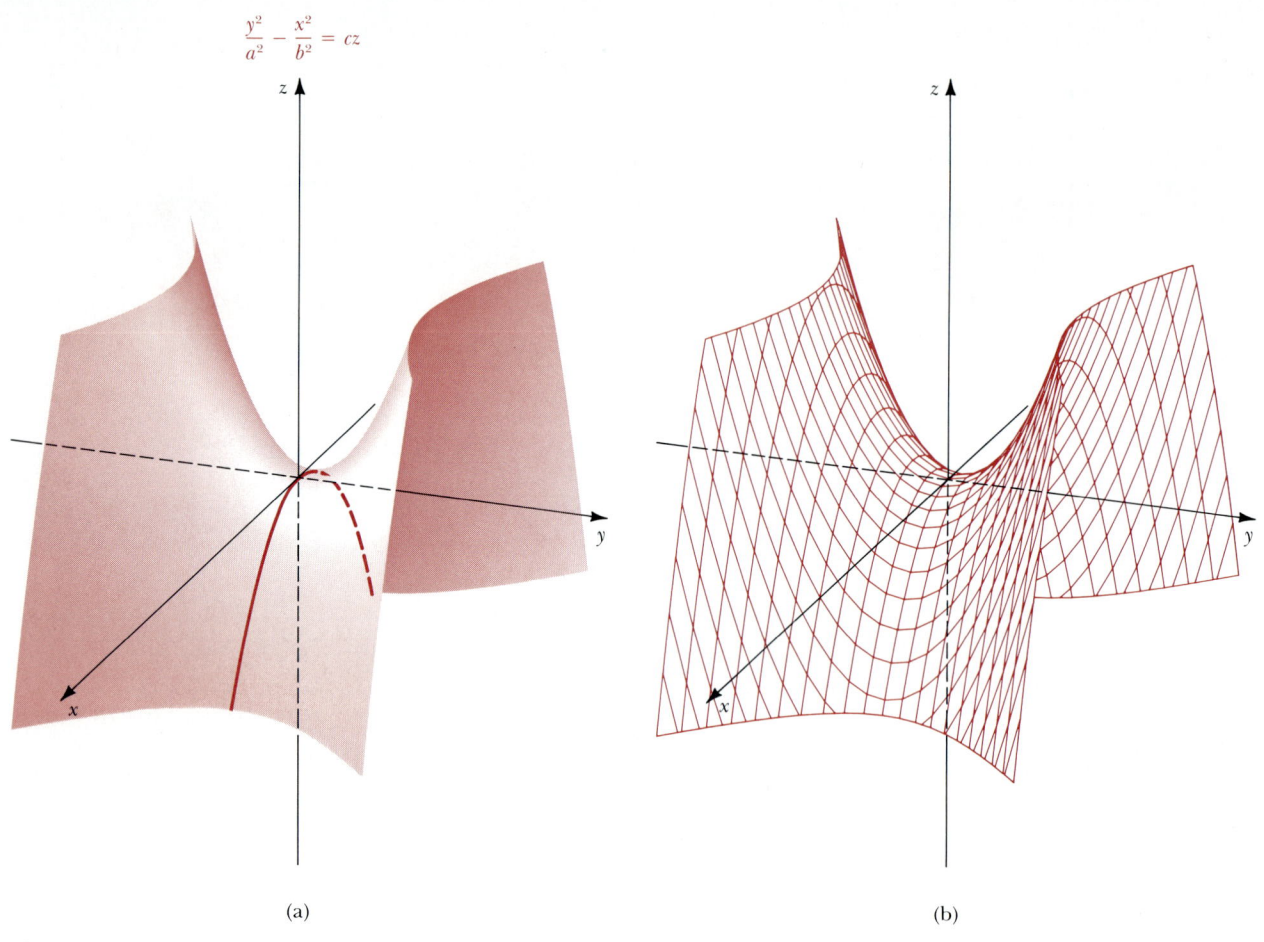

Figure 13.27

A hyperbolic paraboloid.

Example 1

Sketch the graph of the quadric surface with equation

$$\frac{y^2}{4} + \frac{z^2}{9} = x.$$

Solution

This is an elliptic paraboloid. When $y = 0$, the quadric surface intersects the xz-plane in the parabola with equation

$$x = \frac{z^2}{9}.$$

When $z = 0$, the quadric surface intersects the xy-plane in the parabola with equation

$$x = \frac{y^2}{4}.$$

The only intersection of the surface with the yz-plane is the point $(0, 0, 0)$. For $x = k$, where k is a positive constant, the intersection is an ellipse with equation

$$\frac{y^2}{4} + \frac{z^2}{9} = k.$$

This elliptic paraboloid is shown in Figure 13.28. ▶

Example 2

Sketch the quadric surface with equation

$$\frac{x^2}{9} + \frac{z^2}{16} - \frac{y^2}{25} = 1.$$

Solution

The trace of this surface in the xz-plane is an ellipse with equation

$$\frac{x^2}{9} + \frac{z^2}{16} = 1.$$

Its intersection with both the xy- and yz-planes gives hyperbolas with equations

$$\frac{x^2}{9} - \frac{y^2}{25} = 1 \quad \text{and} \quad \frac{z^2}{16} - \frac{y^2}{25} = 1,$$

respectively. The surface is the elliptic hyperboloid of one sheet shown in Figure 13.29. ▶

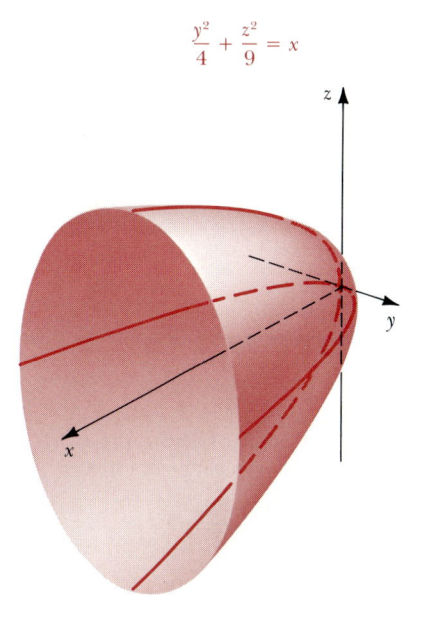

Figure 13.28

Figure 13.29

The graph of a second-degree equation in three variables can also produce degenerate types of quadric surfaces. Some of the more common of these are shown in Figures 13.30 and 13.31. They are also shown in Color Plate VI.

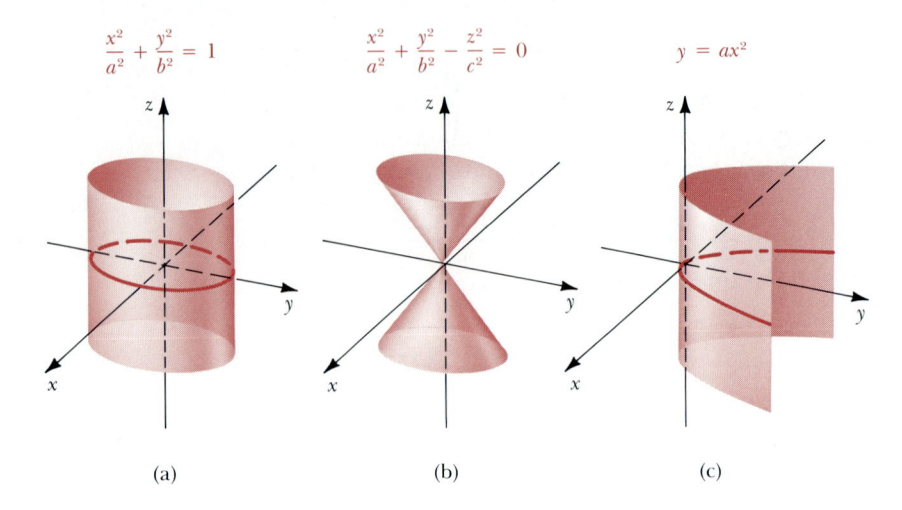

Figure 13.30
(a) A right circular cylinder; (b) an elliptic cone; (c) a parabolic cylinder.

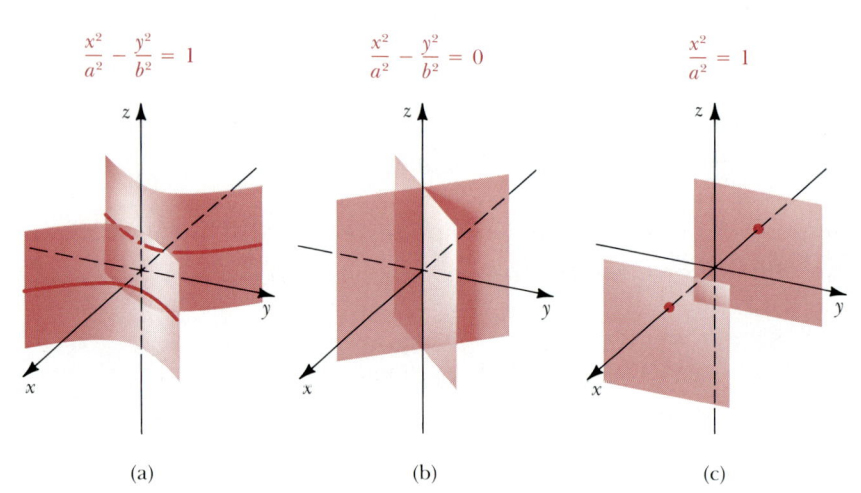

$\dfrac{x^2}{a^2} + \dfrac{y^2}{b^2} = 1$ $\dfrac{x^2}{a^2} + \dfrac{y^2}{b^2} - \dfrac{z^2}{c^2} = 0$ $y = ax^2$

(a)   (b)   (c)

Figure 13.31
(a) A hyperbolic cylinder; (b) intersecting planes; (c) parallel planes.

$\dfrac{x^2}{a^2} - \dfrac{y^2}{b^2} = 1$ $\dfrac{x^2}{a^2} - \dfrac{y^2}{b^2} = 0$ $\dfrac{x^2}{a^2} = 1$

(a)   (b)   (c)

▶ EXERCISE SET 13.4

In Exercises 1–22, name and sketch the quadric surface.

1. $\dfrac{x^2}{9} + \dfrac{y^2}{4} + \dfrac{z^2}{16} = 1$
2. $9x^2 + 4y^2 + z^2 = 36$
3. $\dfrac{x^2}{25} + \dfrac{y^2}{16} - \dfrac{z^2}{9} = 1$
4. $\dfrac{x^2}{25} - \dfrac{y^2}{16} - \dfrac{z^2}{9} = 1$
5. $x^2 - 4y^2 - 9z^2 = 36$
6. $x^2 + 4y^2 - 9z^2 = 36$
7. $z = x^2 + 4y^2$
8. $z = -x^2 - 4y^2$
9. $z = x^2 - 4y^2$
10. $z = 4y^2 - x^2$
11. $\dfrac{x^2}{9} + \dfrac{y^2}{4} = 1$
12. $x^2 + 16y^2 = 4$
13. $4x^2 + 9y^2 - z^2 = 0$
14. $x^2 + y^2 - z^2 = 0$
15. $x^2 - y^2 = 0$
16. $4x^2 - 9y^2 = 0$
17. $\dfrac{y^2}{16} + \dfrac{z^2}{9} - \dfrac{x^2}{25} = 1$
18. $y = x^2 + 4z^2$
19. $y^2 = x^2 - z^2$
20. $y^2 + z^2 = 1$
21. $y^2 = 9$
22. $x = y^2 + z^2$

In Exercises 23–28, name and sketch the quadric surface. [*Hint:* Complete the square.]

23. $x^2 + y^2 + z^2 - 2x - 4z = 0$
24. $x^2 + y^2 - z^2 + 4x - 6y - 3 = 0$
25. $x^2 + y^2 - 6x - 8y - z + 20 = 0$
26. $x^2 - y^2 + z^2 + 4x - 2y - 6 = 0$

27. $x^2 - 2x = 0$

28. $x^2 - z^2 - 4x + 2z + 3 = 0$

29. Sketch the graph of $z = x^2 + y^2$ and use the result to sketch the graph of each of the following.
   a) $z = -(x^2 + y^2)$ **b)** $z = x^2 + y^2 + 2$
   c) $z = x^2 + y^2 - 4$ **d)** $z = x^2 + (y - 1)^2$

30. Sketch the graph of $x^2 + y^2 - z^2 = 1$ and use the result to sketch the graph of each of the following.
   a) $x^2 + y^2 - z^2 = 4$ **b)** $x^2 + y^2 - (z - 1)^2 = 1$
   c) $(x - 1)^2 + y^2 - z^2 = 1$
   d) $x^2 + (y - 1)^2 - z^2 = 1$

In Exercises 31–38, the curve described lies in the xy-plane. Sketch the surface of revolution generated when the curve is revolved about the indicated axis.

31. The circle $x^2 + (y - 2)^2 = 1$, about the y-axis
32. The circle $x^2 + (y - 2)^2 = 1$, about the x-axis
33. The ellipse $9x^2 + 4y^2 = 36$, about the y-axis
34. The ellipse $9x^2 + 4y^2 = 36$, about the x-axis
35. The parabola $y = x^2$, about the y-axis
36. The parabola $y = x^2$, about the x-axis
37. The line $y = x/3$, about the x-axis
38. The line $y = x/3$, about the y-axis
39. Describe the surface of revolution generated if the line $z = y$ in the yz-plane is revolved about the x-axis.
40. Consider the graph of
$$\frac{x^2}{4} + \frac{y^2}{8} + \frac{z^2}{8} = 1$$
(see the figure). Which of the following planes intersect this graph in a circle?
   a) $x = y$ **b)** $x = 1$ **c)** $y = -z$
   d) $y = 4$ **e)** $z = 3$

41. A garland is strung around a Christmas tree seven times in an equally spaced spiral (see the figure). The conical tree has a height and diameter of 140 cm. Place a rectangular coordinate system with the origin at the top of the tree and the positive z-axis coinciding with the axis of the cone.
   a) Find an equation describing the tree.
   b) Find parametric equations describing the garland.
   c) Find the length of the garland.

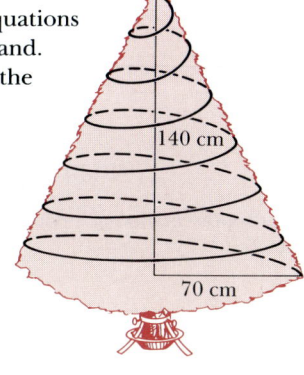

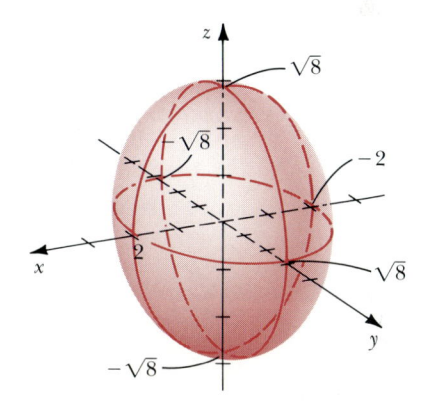

Figure for Exercise 40.

13.5

Cylindrical and Spherical Coordinates in Space

Cylindrical and spherical coordinate systems are both extensions of the polar coordinate system in the plane. The names of these systems are derived from the geometric figures that are most often described using them. Certain cylinders are easily represented using cylindrical coordinates, whereas spheres centered at the origin are easily described using spherical coordinates.

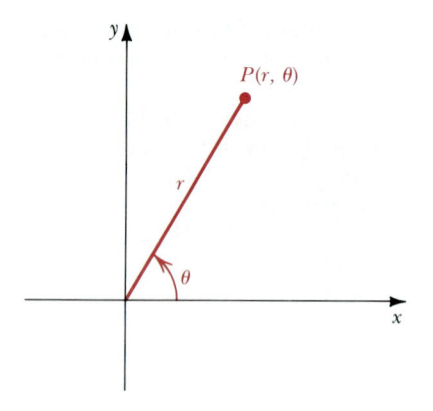

Figure 13.32
Polar representation of a point in the plane.

The Cylindrical Coordinate System

The polar coordinate system uses an angle θ and a directed distance r to describe points in the plane, as shown in Figure 13.32. The cylindrical coordinate system in space is derived from the polar coordinate system in the plane in the same way the rectangular system in space is derived from the rectangular system in the plane: A third or z-coordinate direction is added perpendicular to the xy-plane. Thus a point P with rectangular coordinates (x, y, z) has a cylindrical-coordinate representation (r, θ, z), where

(13.10) $\quad x = r \cos \theta, \quad y = r \sin \theta, \quad r^2 = x^2 + y^2, \quad \tan \theta = \dfrac{y}{x},$

and, of course, $z = z$.

Figure 13.33 shows a typical point and its cylindrical coordinates.

The cylindrical coordinates representing a point are not unique because there are various ways of describing a point in the plane using polar coordinates. For example, if (r, θ, z) is one representation of P, then P is also represented by $(r, \theta + 2n\pi, z)$ and $(-r, \theta + (2n+1)\pi, z)$ for any nonzero integer n.

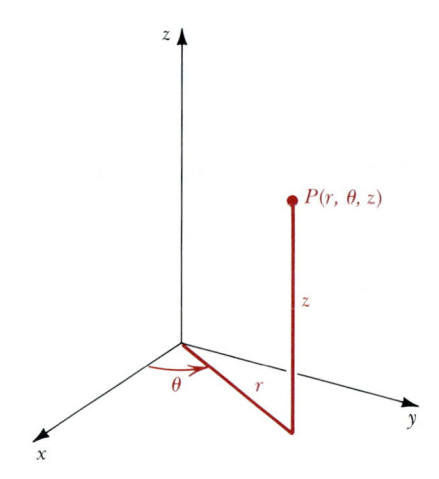

Figure 13.33
Cylindrical representation of a point in space.

Example 1

Find a cylindrical equation corresponding to the three-dimensional rectangular equation $x^2 + y^2 = 1$ and sketch the graph of this equation.

Solution
A cylindrical equation is simply

$$r = \sqrt{x^2 + y^2} = 1,$$

and the graph is the right circular cylinder shown in Figure 13.34. ▶

Figure 13.34

13.5 CYLINDRICAL AND SPHERICAL COORDINATES IN SPACE

In general, the cylindrical equation $r = k$ for a constant k produces a right circular cylinder whose axis is the z-axis and whose radius is $|k|$.

Example 2

Sketch the graph of the cylindrical equation $\theta = 2\pi/3$ and determine a corresponding rectangular equation.

Solution

Since there are no restrictions on z and r, the graph of $\theta = 2\pi/3$ is the plane shown in Figure 13.35.

A rectangular equation for $\theta = 2\pi/3$ is

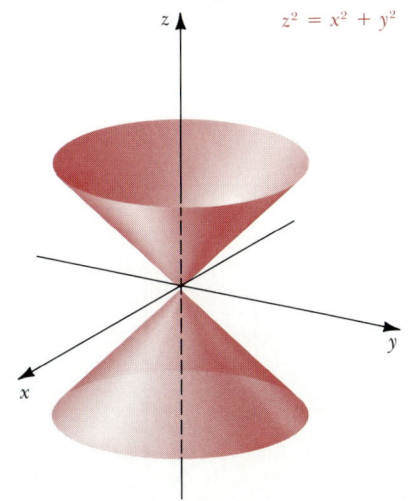

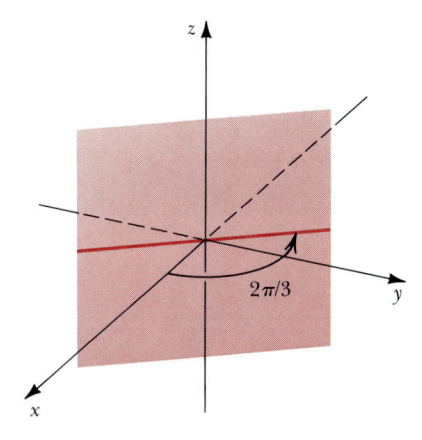

$$\frac{y}{x} = \tan\frac{2\pi}{3} = -\sqrt{3} \quad \text{or} \quad y = -\sqrt{3}x.$$

Figure 13.35

This is the equation of the plane containing the z-axis and intersecting the xy-plane in the line with equation $y = -\sqrt{3}x$. ▶

In general, the cylindrical equation $\theta = k$, for a constant k, with $0 \le k < \pi$, and $k \ne \pi/2$, describes a plane containing the z-axis and the line in the xy-plane with equation $y = (\tan k) x$. The equation $\theta = \pi/2$ describes the yz-plane.

Example 3

Find a cylindrical equation corresponding to the rectangular equation $z^2 = x^2 + y^2$ and sketch its graph.

Solution

Since $r^2 = x^2 + y^2$, a corresponding cylindrical equation has the form $z^2 = r^2$. Because there is no restriction on θ, this cylindrical equation can be simplified to $z = r$. The graph is the circular cone of two nappes with axis along the z-axis, as shown in Figure 13.36. ▶

Figure 13.36

Example 4

Sketch the region in space described by $\pi/4 \leq \theta \leq \pi/2$, $1 \leq r \leq 2$, and $0 \leq z \leq 1$.

Solution

Points in the *xy*-plane with polar coordinates (r, θ), where r and θ satisfy the specified conditions, are located in the region shown in Figure 13.37(a). The points in space with cylindrical coordinates satisfying the stated conditions are shown in Figure 13.37(b). ▶

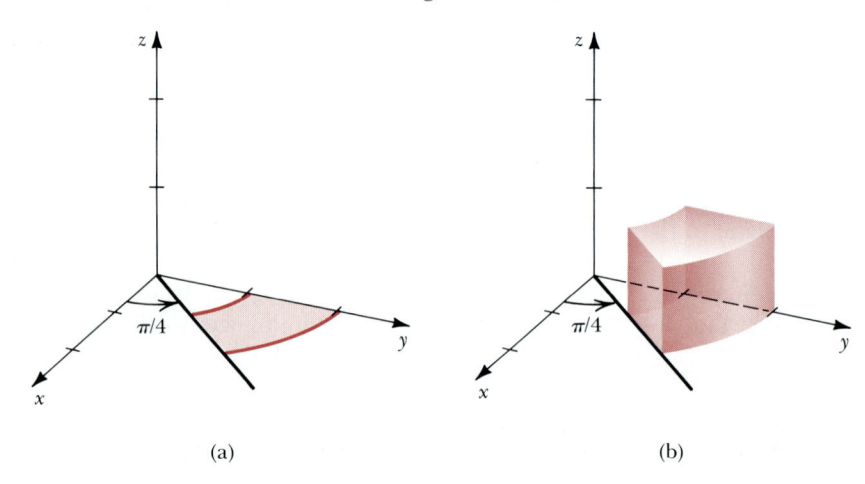

Figure 13.37

(a)   (b)

The Spherical Coordinate System

The cylindrical coordinate system locates a point *P* in space by specifying two distances and an angle. The **spherical coordinate system** locates the point by specifying two angles and a distance.

Suppose *P* has cylindrical coordinates (r, θ, z), where $r \geq 0$. Let $\rho \geq 0$ denote the distance from the origin $O = (0, 0, 0)$ to *P*, and let θ be the angle given by the cylindrical coordinates. In addition, let ϕ denote the angle in $[0, \pi]$ formed by the positive *z*-axis and the line segment joining 0 and *P*, as shown in Figure 13.38. Then *P* is said to have the *spherical coordinates* (ρ, θ, ϕ).

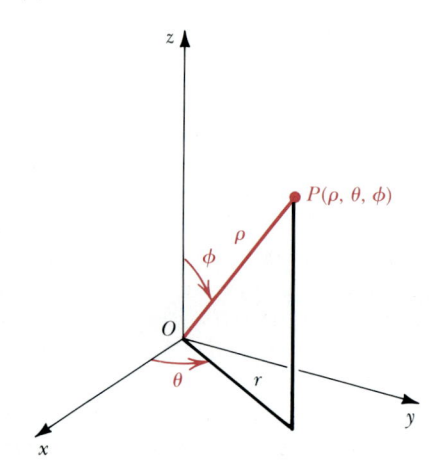

Figure 13.38
Spherical representation of a point in space.

13.5 CYLINDRICAL AND SPHERICAL COORDINATES IN SPACE

From the right triangle PAO in Figure 13.39 we have

$$z = \rho \cos \phi.$$

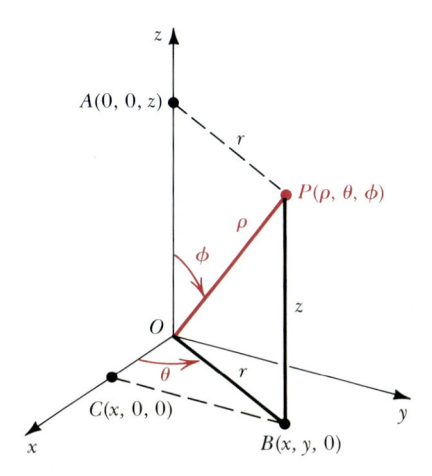

Figure 13.39

The Pythagorean Theorem applied to this triangle gives

$$\rho^2 = r^2 + z^2.$$

In addition, from triangle PAO, $r = \rho \sin \phi$, so the rectangular coordinates (x, y, z) of P are related to the spherical coordinates by

(13.11)

$$x = r \cos \theta = \rho \sin \phi \cos \theta,$$
$$y = r \sin \theta = \rho \sin \phi \sin \theta, \quad \text{and}$$
$$z = \rho \cos \phi,$$

where

$$\rho^2 = r^2 + z^2 = x^2 + y^2 + z^2.$$

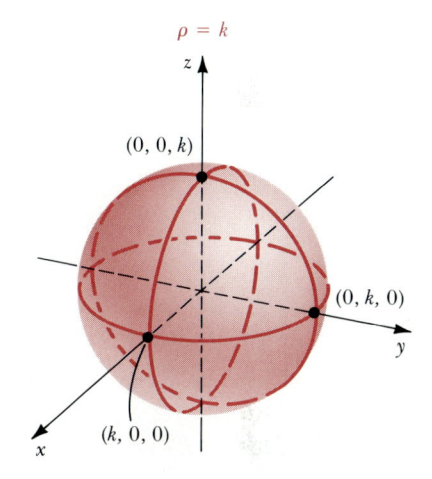

Figure 13.40

Example 5

Sketch the graphs of (a) $\rho = k$, (b) $\theta = k$, and (c) $\phi = k$, for k a positive constant.

Solution

a) The spherical equation $\rho = k$ corresponds to the rectangular equation $k^2 = x^2 + y^2 + z^2$, an equation of the sphere with center at the origin and radius k. The graph is shown in Figure 13.40.

b) The graph of the spherical equation $\theta = k$ is similar to the graph of the equation in cylindrical coordinates except that the spherical coordinate restrictions $\rho \geq 0$ and $0 \leq \phi \leq \pi$ restrict the graph. The graph is a half plane containing the z-axis and intersecting the xy-plane at the line $y = (\tan k)x$ (see Figure 13.41).

c) The spherical equation $\phi = k$, for $0 < k < \pi/2$ or $\pi/2 < k < \pi$, describes a right circular cone of one nappe with axis along the z-axis

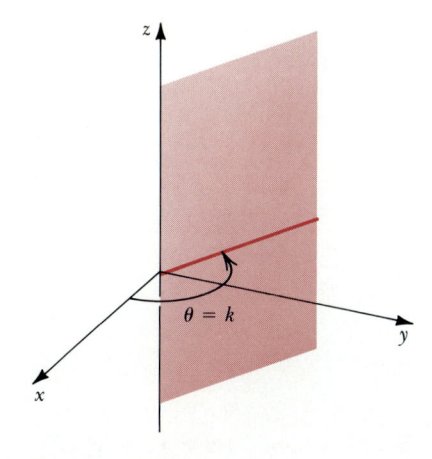

Figure 13.41

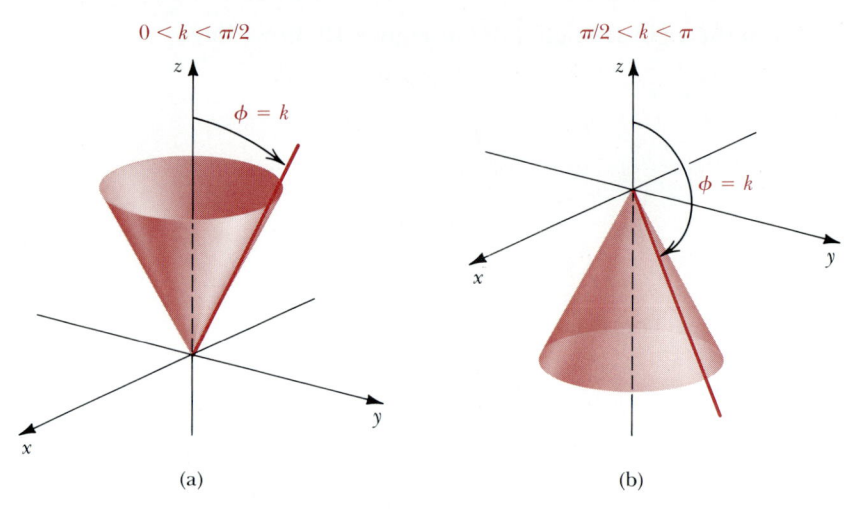

Figure 13.42

and vertex at the origin. Figure 13.42(a) shows the cone for $0 < k < \pi/2$, while Figure 13.42(b) shows the sketch for $\pi/2 < k < \pi$. The equation $\phi = 0$ describes the nonnegative z-axis, $\phi = \pi$ describes the nonpositive z-axis, and $\phi = \pi/2$ describes the xy-plane. ▶

Example 6

Sketch the region in space described by $\pi/4 \leq \theta \leq \pi/2$, $0 \leq \phi \leq \pi/2$, and $1 \leq \rho \leq 2$.

Solution

Points whose spherical coordinates satisfy $1 \leq \rho \leq 2$ are located on or between the spheres centered at the origin with radii 1 and 2. Adding the restriction $0 \leq \phi \leq \pi/2$ requires the points to lie on or above the xy-plane. The condition $\pi/4 \leq \theta \leq \pi/2$ restricts the points in the region to the first octant, as shown in Figure 13.43. ▶

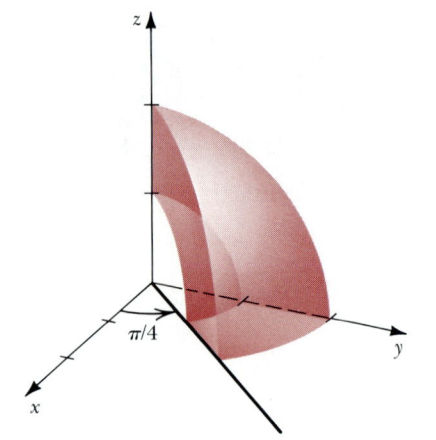

Figure 13.43

Although the spherical coordinate system is introduced primarily for the representation of spheres, a sphere is generally represented using this system only when it is centered at the origin. The last example of this section illustrates how complicated a spherical equation can become when describing a sphere that is not centered at the origin.

Example 7

Find a spherical equation for the sphere that has its center at $(1, 1, 1)$ and passes through the origin.

Solution

Since the distance from $(1, 1, 1)$ to $(0, 0, 0)$ is $\sqrt{3}$, the rectangular equation for this sphere is

$$(x - 1)^2 + (y - 1)^2 + (z - 1)^2 = 3.$$

which simplifies to
$$x^2 + y^2 + z^2 = 2(x + y + z).$$

Thus,
$$\rho^2 = 2(\rho \sin \phi \cos \theta + \rho \sin \phi \sin \theta + \rho \cos \phi)$$

so
$$\rho = 2(\sin \phi \cos \theta + \sin \phi \sin \theta + \cos \phi).$$ ▶

▶ EXERCISE SET 13.5

In Exercises 1–6, rectangular coordinates of a point are given. Determine (a) cylindrical coordinates and (b) spherical coordinates of the point.

1. $(1, 1, 3)$
2. $(1, 0, -1)$
3. $(-\sqrt{2}, \sqrt{2}, 0)$
4. $(1, \sqrt{3}, -2)$
5. $(0, 1, 3)$
6. $(1, -1, 3)$

In Exercises 7–12, cylindrical coordinates of a point are given. Determine (a) rectangular coordinates and (b) spherical coordinates of the point.

7. $(3, \pi/3, 5)$
8. $(-2, 0, 2)$
9. $(4, \pi/4, -2)$
10. $(-1, \pi/2, 2)$
11. $(0, \pi, 1)$
12. $(-1, 0, 1)$

In Exercises 13–18, spherical coordinates of a point are given. Determine (a) rectangular coordinates and (b) cylindrical coordinates of the point.

13. $(1, \pi/4, \pi/3)$
14. $(2, \pi/6, \pi/4)$
15. $(2, 5\pi/6, 3\pi/4)$
16. $(3, 3\pi/4, \pi/2)$
17. $(5, \pi, \pi)$
18. $(1, 7\pi/4, \pi/6)$

In Exercises 19–26, three-dimensional rectangular equations are given. Find corresponding cylindrical equations.

19. $y = x$
20. $\sqrt{3}y = x$
21. $x^2 + y^2 = 4$
22. $y = -2$
23. $z = x^2 + y^2$
24. $z = \sqrt{x^2 + y^2}$
25. $x^2 + y^2 + 4z^2 = 4$
26. $x^2 + y^2 + z = 1$

In Exercises 27–38, sketch the graph of the cylindrical equation.

27. $r = 3$
28. $r = -2$
29. $\theta = \pi/2$
30. $\theta = \pi/4$
31. $z = 4r^2$
32. $z = 3$
33. $r^2 + z^2 = 9$
34. $r^2 + z^2 = 16$
35. $r = \cos \theta$
36. $r = \cos \theta + 1$
37. $r = \theta$
38. $r^2 = \cos^2 \theta$

In Exercises 39–46, sketch the graph of the spherical equation.

39. $\rho = 2$
40. $\theta = \pi/4$
41. $\phi = \pi/4$
42. $\rho^2 = 9$
43. $\rho = \cos \phi$
44. $\phi = \pi/4$ and $\theta = \pi/4$
45. $\rho = 3$ and $\phi = \pi/4$
46. $\rho = 3$ and $\theta = \pi/4$

In Exercises 47–56, sketch the region.

47. $1 \leq r \leq 3, 0 \leq z \leq 4$
48. $0 \leq \theta \leq \pi/2, 0 \leq r \leq 2, z \geq 0$
49. $0 \leq \theta \leq \pi/2, 0 \leq r \leq 1, 0 \leq z \leq 3$
50. $1 \leq r \leq 2, \pi/6 \leq \theta \leq \pi/3, 2 \leq z \leq 3$
51. $1 \leq \rho \leq 2$
52. $0 \leq \phi \leq \pi/4, \rho = 3$
53. $0 \leq \theta \leq \pi/4, \rho = 3$
54. $0 \leq \phi \leq \pi/4, 0 \leq \theta \leq \pi/2, \rho = 3$
55. $1 \leq \rho \leq 2, \pi/4 \leq \phi \leq \pi/3, \pi/6 \leq \theta \leq \pi/3$
56. $0 \leq z \leq \sqrt{4 - r^2}, 0 \leq r \leq 2$

57. Use cylindrical coordinates to describe the region inside the sphere $x^2 + y^2 + z^2 = 4$ and outside the cylinder $x^2 + y^2 = 1$.

58. Use cylindrical coordinates to describe the region inside both the sphere $x^2 + y^2 + z^2 = 4$ and the cylinder $x^2 + y^2 = 1$.

59. Use spherical coordinates to describe the region inside the sphere $x^2 + y^2 + z^2 = 9$ and outside the sphere $x^2 + y^2 + z^2 = 4$.

60. Use spherical coordinates to describe the region in-

side both the cone $z^2 = x^2 + y^2$ and the sphere $x^2 + y^2 + z^2 = 9$.

61. Find an equation in spherical coordinates for the sphere
$$x^2 + y^2 + (z - 1)^2 = 4.$$

62. A vertical hole of radius 3 with center along the z-axis is drilled through a sphere of radius 5 (see the figure). The center of the sphere is at the origin. Express the removed region using (a) rectangular coordinates, (b) cylindrical coordinates, and (c) spherical coordinates.

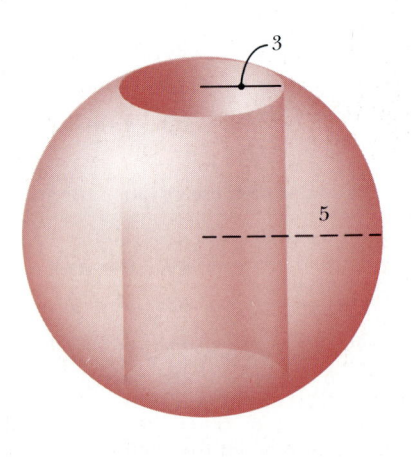

63. An ice cream cone has the shape of a cone of height 3 in. topped with a hemisphere of diameter 2 in. The vertex of the cone is at the origin, and the axis of the cone is the positive z-axis (see the figure). Express the region on and within the ice cream cone using (a) rectangular coordinates, (b) cylindrical coordinates, and (c) spherical coordinates.

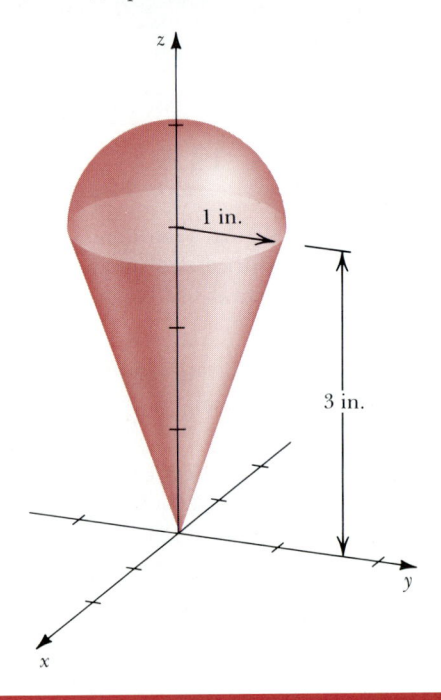

▶ IMPORTANT TERMS AND RESULTS

CONCEPT		PAGE
Multivariable function		748
Trace		753
Level curve		753
Level surface		760
Cylinder		761
Quadric surface		763
Ellipsoid		764
Elliptic hyperboloid of one sheet		764
Elliptic hyperboloid of two sheets		765
Elliptic paraboloid		766
Hyperbolic paraboloid		767
Cylindrical coordinate system	$x = r \cos \theta, \quad y = r \sin \theta, \quad z = z$	772
Spherical coordinate system	$x = \rho \sin \phi \cos \theta, \quad y = \rho \sin \phi \sin \theta, \quad z = \rho \cos \phi$	774

▶ REVIEW EXERCISES

In Exercises 1–6, find the domain of the function.

1. $f(x, y) = \sqrt{1 - x^2 - y}$
2. $f(x, y) = \dfrac{1}{\sqrt{4 - x^2 - y^2}}$
3. $f(x, y) = \dfrac{\sqrt{1 - x^2 - y^2}}{x^2}$
4. $f(x, y, z) = \sqrt{x^2 + y^2 + z^2 - 1}$
5. $f(x, y, z) = \ln xyz$
6. $f(x, y, z) = 1/x$

In Exercises 7–10, sketch representative level curves for the function.

7. $f(x, y) = 2 - x + y$
8. $f(x, y) = 2x + 3y - 6$
9. $f(x, y) = 9x^2 + 4y^2$
10. $f(x, y) = x - y^2$

In Exercises 11–14, sketch, in the first octant, representative level surfaces for the function.

11. $f(x, y, z) = 9x^2 + 4y^2$
12. $f(x, y, z) = 2 - x + y - z$
13. $f(x, y, z) = x^2 + y^2 + z^2$
14. $f(x, y, z) = 4x^2 + 9y^2 + z^2$

In Exercises 15–36, sketch the graph described by the equation.

15. $z = 4 - x - 2y$
16. $z = x - 1$
17. $z = x^2 + y^2 - 2$
18. $9x^2 + 4y^2 + z^2 = 36$
19. $y^2 + 4z^2 = 4$
20. $x^2 = 4$
21. $\dfrac{x^2}{9} + \dfrac{y^2}{4} - \dfrac{z^2}{16} = 1$
22. $x^2 + 25y^2 - z^2 = 0$
23. $\dfrac{x^2}{16} - \dfrac{y^2}{25} = 0$
24. $y^2 - z^2 = 0$
25. $x^2 - y^2 = 1$
26. $yz = 1$
27. $x^2 + y^2 + z^2 - 2x - 2y - 2z = 1$
28. $x^2 + 2y^2 + 3z^2 - 8y - 18z = 1$
29. $z^2 - 4z = 0$
30. $z^2 - 4z + 4 = 0$
31. $r = 2 \sin \theta$
32. $\theta = 5\pi/3$
33. $r^2 = 4 \sin \theta$
34. $\rho = 4 \cos \phi$
35. $\rho \sin \phi = 1$
36. $\rho^2 - 5\rho + 6 = 0$

37. Determine (a) cylindrical and (b) spherical coordinates of each of the following points given in rectangular coordinates.
    i) $(1, 0, 0)$
    ii) $(1, 1, 0)$
    iii) $(1, 1, 1)$

38. Determine (a) rectangular and (b) spherical coordinates of each of the following points given in cylindrical coordinates.
    i) $(4, \pi/4, 0)$
    ii) $(1, \pi/2, 1)$
    iii) $(1, \pi/6, 2)$

39. Determine (a) rectangular and (b) cylindrical coordinates of each of the following points given in spherical coordinates.
    i) $(1, \pi/2, \pi/2)$
    ii) $(1, \pi/4, \pi/2)$
    iii) $(2, \pi/2, \pi/4)$

40. The temperature at any point (x, y) on a circular plate of radius 2 is $T(x, y) = 4 - x^2 - y^2$. Sketch the curves of constant temperature for $T = 0$, $T = 1$, and $T = 3$.

41. a) Describe the surface obtained by revolving the curve $y = \sqrt{4 - x^2}$, $z = 0$ about the x-axis.
    b) Determine an equation of the surface. (See Exercise 17 in Section 13.3).

42. a) Describe the surface obtained by revolving the line $y = z$, $x = 0$ about the z-axis.
    b) Determine an equation of the surface.

43. Describe the surface obtained by revolving the circle $(y - 2)^2 + z^2 = 1$, $x = 0$ about the z-axis.

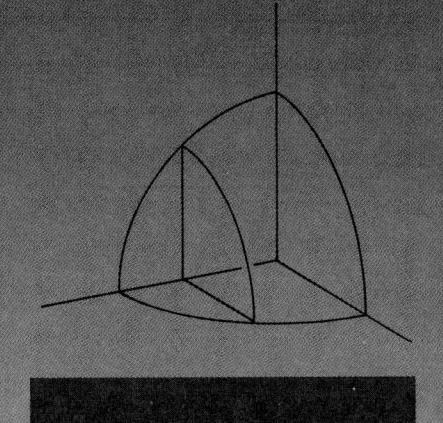

14

The Differential Calculus of Multivariable Functions

The development of the calculus of multivariable functions parallels that of single-variable functions. We begin by defining the concepts of limit and continuity for a multivariable function and then proceed to the subjects of differentiation and integration, emphasizing the assorted applications.

In this chapter we consider topics required for the differentiation of multivariable functions. Since the concepts are similar for arbitrary functions of n variables, we generally simplify the discussion by considering only functions of two variables. Occasionally, however, we present results for functions of three variables, primarily to emphasize the similarity with those of two variables.

14.1

Limits and Continuity

The limit of a single-variable function, expressed by

$$\lim_{x \to a} f(x) = L,$$

means that $f(x)$ becomes and remains close to L as x becomes close, but not equal, to a. The definition of the limit of a function makes use of tolerance intervals about a and L to express analytically the concept of "close to." This definition was given in Section 1.6 and is repeated here for comparison with the definitions to be given for multivariable functions.

The limit of the single-variable function f at a is L, provided that for every number $\varepsilon > 0$, a number $\delta > 0$ exists with the property that

$$|f(x) - L| < \varepsilon \quad \text{whenever} \quad 0 < |x - a| < \delta.$$

The definition of the limit of a function of two variables parallels the single-variable definition. Modifications are necessary, however, because the domain of the function is now a subset of $\mathbb{R}^2$.

A point (x, y) is within δ units of (a, b) precisely when (x, y) lies in the interior of the circle about (a, b) with radius δ (see Figure 14.1). The points (x, y) in the interior of this circle satisfy the inequality

$$\sqrt{(x-a)^2 + (y-b)^2} < \delta.$$

Consequently, a point (x, y) is within δ units of, but not equal to, (a, b) precisely when (x, y) satisfies the inequality

$$0 < \sqrt{(x-a)^2 + (y-b)^2} < \delta.$$

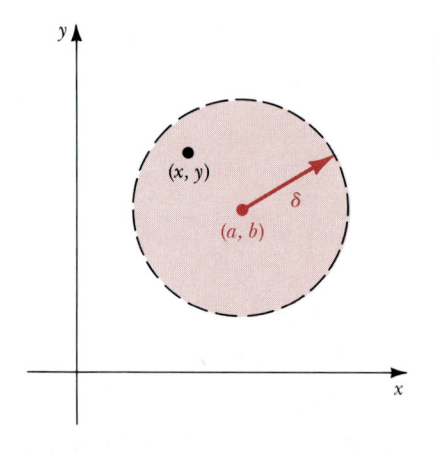

Figure 14.1
The distance from (x, y) to (a, b) is less than δ.

(14.1)
DEFINITION

The limit of a function f of two variables at (a, b) is L, written

$$\lim_{(x,y) \to (a,b)} f(x, y) = L,$$

provided that for every number $\varepsilon > 0$, a number $\delta > 0$ exists with the property that

$$|f(x, y) - L| < \varepsilon$$

whenever

$$0 < \sqrt{(x-a)^2 + (y-b)^2} < \delta.$$

Figure 14.2 illustrates the limit of a function of two variables.

Example 1

Prove that $\lim_{(x,y) \to (1,2)} f(x, y) = 2$, where $f(x, y) = 6 - 2x - y$.

Solution

The graph of f is the plane shown in Figure 14.3. To prove that $\lim_{(x,y) \to (1,2)} (6 - 2x - y) = 2$, we must show that for an arbitrary $\varepsilon > 0$, a number δ exists with

$$|(6 - 2x - y) - 2| < \varepsilon$$

whenever

$$0 < \sqrt{(x-1)^2 + (y-2)^2} < \delta.$$

Suppose such a δ exists and (x, y) is restricted so that

$$0 < \sqrt{(x-1)^2 + (y-2)^2} < \delta.$$

We first note that $|(6 - 2x - y) - 2| = |-2(x-1) - (y-2)|$. Then, since both

$$|x - 1| \leq \sqrt{(x-1)^2 + (y-2)^2}$$

and

$$|y - 2| \leq \sqrt{(x-1)^2 + (y-2)^2},$$

we have

$$\begin{aligned}|(6 - 2x - y) - 2| &= |-2(x-1) - (y-2)| \\ &\leq 2|x-1| + |y-2| \\ &\leq 2\sqrt{(x-1)^2 + (y-2)^2} + \sqrt{(x-1)^2 + (y-2)^2} \\ &< 3\delta.\end{aligned}$$

To ensure that $|(6 - 2x - y) - 2| < \varepsilon$, for an arbitrary $\varepsilon > 0$, we choose $\delta = \varepsilon/3$. If

$$0 < \sqrt{(x-1)^2 + (y-2)^2} < \delta,$$

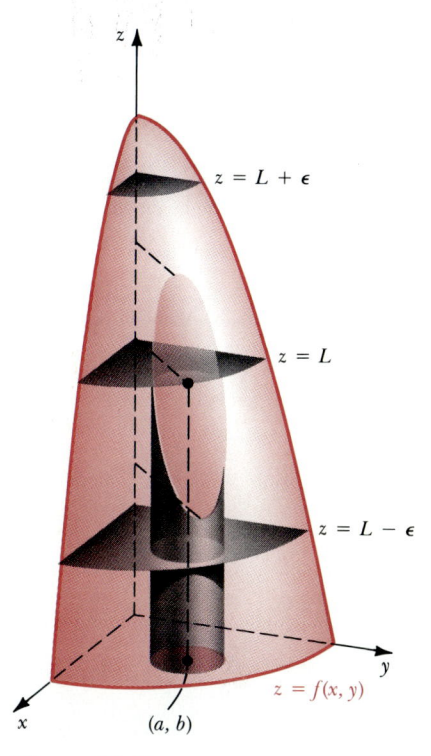

Figure 14.2

$\lim_{(x,y) \to (a,b)} f(x, y) = L.$

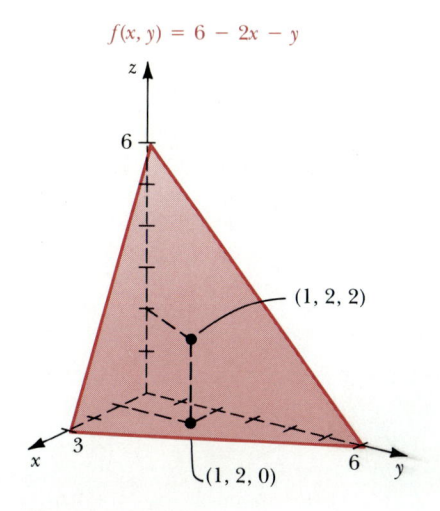

Figure 14.3

then
$$|(6-2x-y)-2| < 3\delta = 3(\varepsilon/3) = \varepsilon.$$
This proves that $\lim_{(x,y)\to(1,2)} (6-2x-y) = 2.$ ▸

The definition of the limit of a function of three variables is similar to Definition 14.1 and uses the inequality
$$0 < \sqrt{(x-a)^2 + (y-b)^2 + (z-c)^2} < \delta$$
to express that (x, y, z) is within δ units of, but not equal to, (a, b, c).

(14.2)
DEFINITION
The limit of a function f of three variables at (a, b, c) is L, written
$$\lim_{(x,y,z)\to(a,b,c)} f(x, y, z) = L,$$
provided that for every number $\varepsilon > 0$, a number $\delta > 0$ exists with the property that
$$|f(x, y, z) - L| < \varepsilon$$
whenever
$$0 < \sqrt{(x-a)^2 + (y-b)^2 + (z-c)^2} < \delta.$$

As in the case of the single-variable function, these definitions ensure that if the limit of a function exists, it must be unique. In fact, Definition (14.1) implies that:

(14.3)
If $\lim_{(x,y)\to(a,b)} f(x, y) = L$, then $f(x, y)$ must approach L as (x, y) approaches (a, b) along every curve in the domain of f that passes through (a, b). To show that the limit of a function of two variables does *not* exist, it suffices to show that the limiting values differ along two distinct curves.

Example 2

Show that $\lim_{(x,y)\to(0,0)} f(x, y)$ does not exist for $f(x, y) = \dfrac{xy}{x^2 + y^2}$.

Solution
Every circle about the origin contains points on both the x-axis and the line $y = x$. For points $(x, 0)$ with $x \neq 0$ on the line $y = 0$,
$$f(x, 0) = \frac{x \cdot 0}{x^2 + 0^2} = \frac{0}{x^2} = 0,$$
so if the limit exists it must have the value 0. However, for points (x, x)

with $x \neq 0$ on the line $y = x$ (see Figure 14.4 and Color Plate VII),

$$f(x, x) = \frac{x \cdot x}{x^2 + x^2} = \frac{1}{2}.$$

Since the limiting values along the lines $y = 0$ and $y = x$ differ,

$$\lim_{(x,y) \to (0,0)} \frac{xy}{x^2 + y^2} \quad \text{does not exist.}$$

In Exercise 46 a function is defined that has the limiting value of 0 along *every* line that passes through $(0, 0)$, but that still has no limit at $(0, 0)$.

The condition required for a function of several variables to be continuous at a point is the same as the condition for functions of a single variable: The limit of the function at the point must agree with the value of the function there.

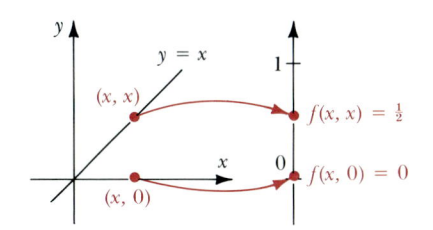

▶ Figure 14.4

$f(x, y) = \dfrac{xy}{x^2 + y^2}.$

(14.4) DEFINITION

A function f of two variables is continuous at (a, b) if:

i) $f(a, b)$ exists;   **ii)** $\lim\limits_{(x,y) \to (a,b)} f(x, y)$ exists; and

iii) $\lim\limits_{(x,y) \to (a,b)} f(x, y) = f(a, b)$.

The arithmetic and composition rules for functions of a single variable also hold for functions of several variables when the operations are defined. The results for limits are stated; the analogous continuity results will be used, though not stated.

(14.5) THEOREM

If $\lim\limits_{(x,y) \to (a,b)} f(x, y) = L$ and $\lim\limits_{(x,y) \to (a,b)} g(x, y) = M$, then:

i) $\lim\limits_{(x,y) \to (a,b)} (f + g)(x, y) = L + M$;

ii) $\lim\limits_{(x,y) \to (a,b)} (f - g)(x, y) = L - M$;

iii) $\lim\limits_{(x,y) \to (a,b)} (f \cdot g)(x, y) = L \cdot M$; and

iv) $\lim\limits_{(x,y) \to (a,b)} \left(\dfrac{f}{g}\right)(x, y) = \dfrac{L}{M}$, provided $M \neq 0$.

(14.6) THEOREM

If $\lim\limits_{(x,y) \to (a,b)} f(x, y) = L$ and g is a function of one variable that is continuous at L, then

$$\lim_{(x,y) \to (a,b)} g(f(x, y)) = g(L).$$

The following example uses the results given in Theorems (14.5) and (14.6).

Example 3

Show that $\displaystyle\lim_{(x,y)\to(1,\pi/2)} \sqrt{\frac{x+\cos y}{x^2+3\sin y}} = \frac{1}{2}$.

Solution
Since
$$\lim_{(x,y)\to(1,\pi/2)} x = 1 \quad \text{and} \quad \lim_{(x,y)\to(1,\pi/2)} \cos y = 0,$$
we have
$$\lim_{(x,y)\to(1,\pi/2)} (x+\cos y) = 1.$$
Similarly,
$$\lim_{(x,y)\to(1,\pi/2)} (x^2+3\sin y) = \left(\lim_{(x,y)\to(1,\pi/2)} x^2\right) + 3\left(\lim_{(x,y)\to(1,\pi/2)} \sin y\right)$$
$$= 1+3 = 4.$$
Thus,
$$\lim_{(x,y)\to(1,\pi/2)} \frac{x+\cos y}{x^2+3\sin y} = \frac{1}{4}.$$
Since $g(x) = \sqrt{x}$ is continuous at $\frac{1}{4}$, Theorem (14.6) implies that
$$\lim_{(x,y)\to(1,\pi/2)} \sqrt{\frac{x+\cos y}{x^2+3\sin y}} = \sqrt{\lim_{(x,y)\to(1,\pi/2)} \frac{x+\cos y}{x^2+3\sin y}} = \sqrt{\frac{1}{4}} = \frac{1}{2}. \quad \blacktriangleright$$

Example 4

Consider the function described by
$$f(x, y) = \begin{cases} \dfrac{xy}{x^2+y^2}, & \text{if } (x, y) \neq (0, 0), \\ 0, & \text{if } (x, y) = (0, 0). \end{cases}$$
Determine the points at which f is continuous.

Solution
We saw in Example 2 that $\lim_{(x,y)\to(0,0)} xy/(x^2+y^2)$ does not exist, so f is not continuous at the origin. However, when $(a, b) \neq (0, 0)$, $a^2+b^2 \neq 0$ and
$$\lim_{(x,y)\to(a,b)} f(x, y) = \lim_{(x,y)\to(a,b)} \frac{xy}{x^2+y^2} = \frac{ab}{a^2+b^2} = f(a, b).$$

So f is continuous at every point in the plane except the origin. $\quad \blacktriangleright$

14.1 LIMITS AND CONTINUITY

To consider continuity of a function of two variables on a region in the plane, recall what is meant by continuity of a single-variable function on an interval: A function is continuous on an interval if it is continuous at each interior point of the interval and if, in addition, the value of the function at each endpoint in the interval agrees with the limit from the side of the endpoint that is within the interval. For example, the function described by $f(x) = \sqrt{x}$ is continuous on the interval $[0, \infty)$ since f is continuous at each positive real number and, in addition, the limit of f from the right at 0 is the same as $f(0)$. The notion of a function being continuous on a region R in the plane is similar. We first need the concept of an interior point of a region in the plane.

A point in a region R is called an **interior point** of R if there is a circle centered at the point whose interior is completely contained within R (see Figure 14.5a). A point is called a **boundary point** of R if the interior of *every* circle centered at the point contains both points in R and points not in R (see Figure 14.5b). Interior points of a region in $\mathbb{R}^2$ correspond to the points in the interior of an interval in $\mathbb{R}$, whereas boundary points correspond to the endpoints of an interval.

A region in $\mathbb{R}^2$ that contains only interior points is said to be **open** and one that contains all of its boundary points is said to be **closed** (see Figure 14.6). A region that contains some, but not all, of its boundary points is neither open nor closed, just as the interval $(1, 2]$ in $\mathbb{R}$ is neither open nor closed.

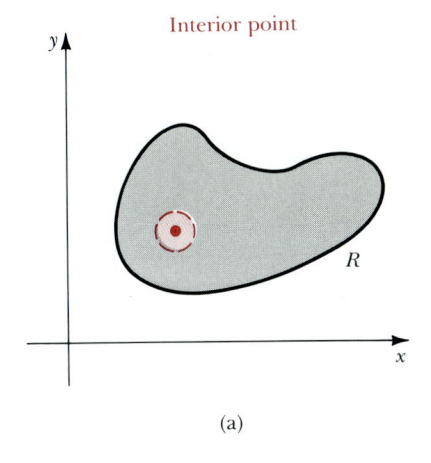

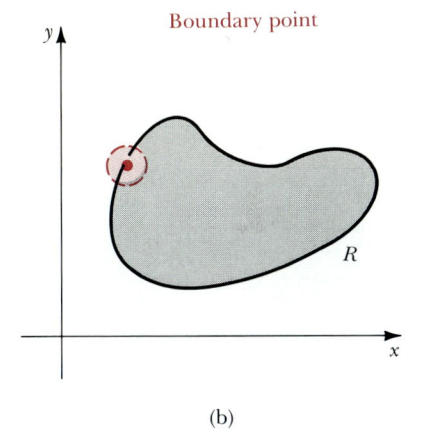

Figure 14.5
(a) (x, y) is an interior point; (b) (x, y) is a boundary point.

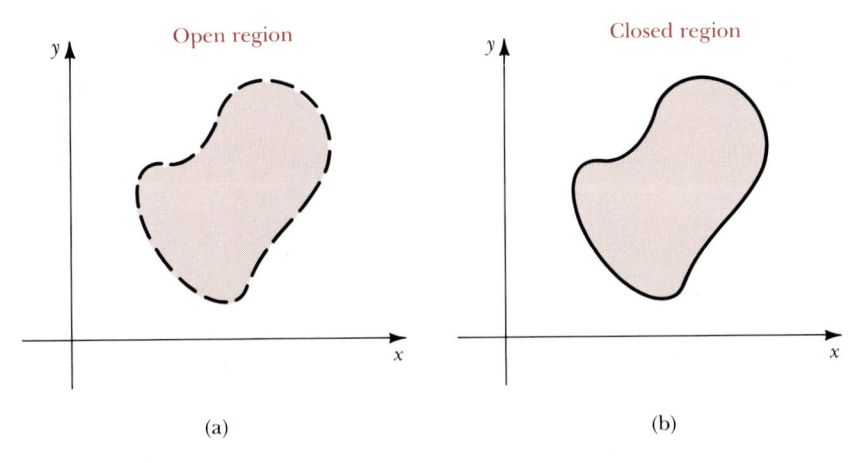

Figure 14.6
(a) An open region contains none of its boundary points; (b) A closed region contains all of its boundary points.

The following definition implies that in order for a function of two variables to be continuous on a region in the plane, it must be continuous at each interior point. In addition, the value of the function at each boundary point must be the same as the limiting value of the function along any curve within the region that approaches the boundary point.

(14.7) DEFINITION

A function f is **continuous on the plane region R** in its domain if for every (a, b) in R and every number $\varepsilon > 0$, a number $\delta > 0$ exists with the property that for all (x, y) in R,

$$|f(x, y) - f(a, b)| < \varepsilon \quad \text{whenever} \quad \sqrt{(x-a)^2 + (y-b)^2} < \delta.$$

We say that a region R in the plane is **bounded** (see Figure 14.7) if R is contained within some rectangle, that is, intervals $[a, b]$ and $[c, d]$ exist such that:

$$\text{for all } (x, y) \text{ in } R, \quad a \leq x \leq b \quad \text{and} \quad c \leq y \leq d.$$

Later in this chapter we will see that a continuous function of two variables defined on a region that is both closed and bounded has properties similar to those of a continuous single-variable function defined on a closed interval.

These concepts have similar extensions to regions in $\mathbb{R}^3$. In this case, a sphere about a point (a, b, c) in $\mathbb{R}^3$ is used in place of a circle about a point (a, b) in $\mathbb{R}^2$. The definition of continuity for a function of three variables on a solid region demonstrates this similarity. A bounded region in $\mathbb{R}^3$ is shown in Figure 14.8.

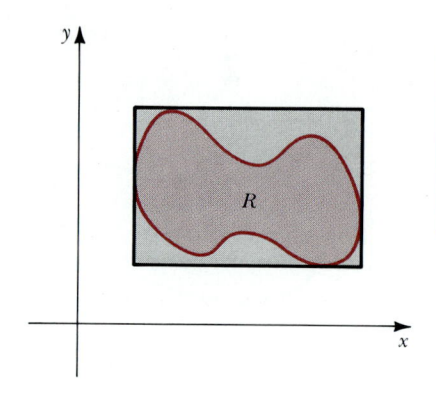

Figure 14.7
A bounded region in $\mathbb{R}^2$.

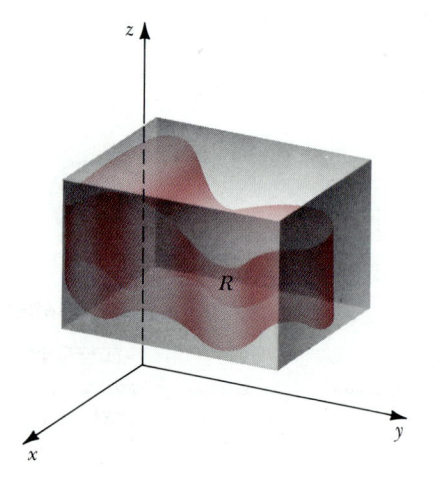

Figure 14.8
A bounded region in $\mathbb{R}^3$.

(14.8) DEFINITION

A function f is **continuous on the solid region R** in its domain if for every (a, b, c) in R and every number $\varepsilon > 0$, a number $\delta > 0$ exists with the property that for all (x, y, z) in R,

$$|f(x, y, z) - f(a, b, c)| < \varepsilon$$

whenever

$$\sqrt{(x-a)^2 + (y-b)^2 + (z-c)^2} < \delta.$$

▶ EXERCISE SET 14.1

In Exercises 1–16, determine the limit of f at the given point, if it exists.

1. $f(x, y) = 3 - x - y$, $(1, 1)$
2. $f(x, y) = \sqrt{x + y}$, $(2, 2)$
3. $f(x, y) = x^2 + y^2$, $(0, 1)$
4. $f(x, y) = \dfrac{1}{x + y}$, $(1, 0)$
5. $f(x, y) = \ln(x + y)$, $(1, 0)$
6. $f(x, y) = \dfrac{x^2 + y^2}{xy + 1}$, $(0, 0)$
7. $f(x, y) = \dfrac{x^3 - x^2y + xy^2}{x + y}$, $(1, 1)$
8. $f(x, y) = \cos(x + y)$, $\left(\dfrac{\pi}{2}, 0\right)$
9. $f(x, y) = e^{x+y}$, $(1, 0)$
10. $f(x, y) = y \ln x$, $(e^2, 2)$
11. $f(x, y) = \dfrac{x^2 - y^2}{x - y}$, $(1, 1)$
12. $f(x, y) = \dfrac{y(x^2 - 4)}{x - 2}$, $(2, 1)$
13. $f(x, y) = \dfrac{\sin(x^2 + y^2)}{x^2 + y^2}$, $(0, 0)$
14. $f(x, y) = \dfrac{1 - \cos(x^2 + y^2)}{x^2 + y^2}$, $(0, 0)$
15. $f(x, y, z) = x^2 + y^2 + z^2$, $(1, -1, 1)$
16. $f(x, y, z) = \dfrac{x(z^2 - 9)}{y(z - 3)}$, $(2, 1, 3)$

In Exercises 17–24, determine the regions of the plane on which the function is continuous.

17. $f(x, y) = x^2 + y^2$
18. $f(x, y) = e^{xy}$
19. $f(x, y) = \dfrac{y - x}{x^2 + y^2}$
20. $f(x, y) = \dfrac{1}{1 - x + y}$
21. $f(x, y) = \sqrt{1 - x^2 - y^2}$
22. $f(x, y) = \dfrac{x^2 + y^2}{1 - x^2 - y^2}$
23. $f(x, y) = \sin(x + y)$
24. $f(x, y) = \ln(x + y - 1)$

In Exercises 25–32, show that the limits do not exist.

25. $\lim\limits_{(x,y) \to (0,0)} \dfrac{y^2}{x^2 + y^2}$
26. $\lim\limits_{(x,y) \to (0,0)} \dfrac{x}{y}$
27. $\lim\limits_{(x,y) \to (0,0)} \dfrac{y}{x^2 - y}$
28. $\lim\limits_{(x,y) \to (0,0)} \dfrac{x^2 y^2}{x^4 + 3y^4}$
29. $\lim\limits_{(x,y) \to (0,0)} \dfrac{\sin xy}{xy}$
30. $\lim\limits_{(x,y) \to (0,0)} \dfrac{1 - \cos xy}{xy}$
31. $\lim\limits_{(x,y,z) \to (0,0,0)} \dfrac{x^2 + y^2}{x^2 + y^2 + z^2}$
32. $\lim\limits_{(x,y,z) \to (0,0,0)} \dfrac{xy + z}{x + y + z^2}$

33. Sketch the region in the xy-plane described by each of the following.
    a) $0 < \sqrt{(x - 1)^2 + (y - 2)^2} < 0.5$
    b) $0 < \sqrt{x^2 + (y + 1)^2} < 1$
    c) $1 < \sqrt{(x + 2)^2 + (y + 3)^2} < 2$

34. Determine the region in space described by each of the following.
    a) $0 < \sqrt{(x - 1)^2 + (y - 2)^2 + (z - 3)^2} < 0.5$
    b) $0 < \sqrt{x^2 + y^2 + (z + 2)^2} < 0.1$
    c) $3 < \sqrt{(x + 1)^2 + (y + 2)^2 + (z + 1)^2} < 4$

35. If $f(x, y) = x^2 + y$ and $g(x) = \sin x$, find
$$\lim_{(x,y) \to (0,\pi/2)} g(f(x, y)).$$

36. If $f(x, y) = e^x \cos y$ and $g(x) = \arctan x$, find
$$\lim_{(x,y) \to (0,0)} g(f(x, y)).$$

37. If $f(x, y, z) = x + y + z$ and $g(x) = \cos x$, find
$$\lim_{(x,y,z) \to (\pi/2,\pi/2,0)} g(f(x, y, z)).$$

38. If $f(x, y, z) = z/xy$ and $g(x) = \ln x$, find
$$\lim_{(x,y,z) \to (1,2,2e)} g(f(x, y, z)).$$

Use Definition (14.1) or (14.2) to prove the statements in Exercises 39–44.

39. $\lim\limits_{(x,y) \to (0,0)} (3 - x - y) = 3$
40. $\lim\limits_{(x,y) \to (2,1)} (6 - 2x - 3y) = -1$
41. $\lim\limits_{(x,y) \to (0,0)} (x^2 + y^2) = 0$
42. $\lim\limits_{(x,y) \to (0,0)} (4 - x^2 - y^2) = 4$
43. $\lim\limits_{(x,y,z) \to (1,2,3)} (2x - y + z) = 3$
44. $\lim\limits_{(x,y,z) \to (0,0,0)} (x^2 + y^2 + z^2) = 0$

45. Consider the function described by
$$f(x, y) = \begin{cases} \dfrac{xy}{x^2 + y^2}, & \text{if } (x, y) \neq (0, 0), \\ 0, & \text{if } (x, y) = (0, 0). \end{cases}$$
    a) Show that the single-variable function described by $f(x, 0)$ is a continuous function.
    b) Show that the single-variable function described by $f(0, y)$ is a continuous function.
    c) Show that the function f is not continuous at $(0, 0)$.
    d) Is the single-variable function described by $f(x, x)$ continuous at $(0, 0)$?

46. Consider the function described by
$$f(x, y) = \begin{cases} \dfrac{x^2 y}{x^4 + y^2}, & \text{if } (x, y) \neq (0, 0), \\ 0, & \text{if } (x, y) = (0, 0). \end{cases}$$
The graph of this function is shown in Color Plate VII.
    a) Show that along any line through the origin, this function approaches 0 as (x, y) approaches $(0, 0)$.
    b) Find the limit of this function as (x, y) approaches the origin along the parabola with equation $y = x^2$.
    c) Show that the limit of this function does not exist at $(0, 0)$.

47. Consider the function described by
$$f(x, y) = \begin{cases} 1, & \text{if } x^2 + y^2 \leq 1, \\ 0, & \text{if } x^2 + y^2 > 1. \end{cases}$$
    a) Show that f is continuous on the region described by $x^2 + y^2 \leq 1$.
    b) Show that f is not continuous at any point (x, y) that satisfies $x^2 + y^2 = 1$.

48. Show that L is the limit of f at (a, b) if and only if the following statement holds:

    For every number $\varepsilon > 0$, a number $\delta > 0$ exists with the property that $|f(x, y) - L| < \varepsilon$ whenever both $|x - a| < \delta$ and $|y - b| < \delta$.

49. Show that if $\lim_{(x,y) \to (a,b)} f(x, y) = L$ and $\lim_{(x,y) \to (a,b)} f(x, y) = M$, then $L = M$.

50. Give a definition for each of the following.
    a) $\lim\limits_{(x,y) \to (a,b)} f(x, y) = \infty$
    b) $\lim\limits_{(x,y) \to (a,b)} f(x, y) = -\infty$

14.2
Partial Derivatives

Let f be a function of the two variables x and y. Fixing the second variable at $y = y_0$ and allowing the first variable to vary produces a single-variable function $f(x, y_0)$ in the variable x, as shown in Figure 14.9. Similarly, fixing the first variable at $x = x_0$ and allowing the second variable to vary gives a single-variable function in the variable y, as shown in Figure 14.10. In this section we consider the derivatives of the single-variable functions produced in this way. These derivatives are called *partial derivatives* of the function f since they allow points in only a part of the domain to vary.

For simplicity, the definition of partial derivative is given only for functions of two variables, leaving the analogous discussion involving functions of three or more variables to the examples and exercises. In general, a multivariable function has as many partial derivatives at a point in its domain as it has variables, provided the required limits exist.

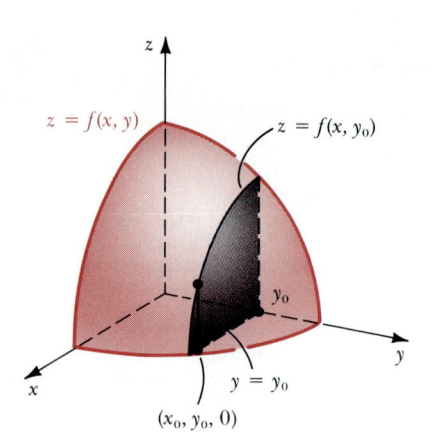

Figure 14.9
The curve described by $z = f(x, y_0)$.

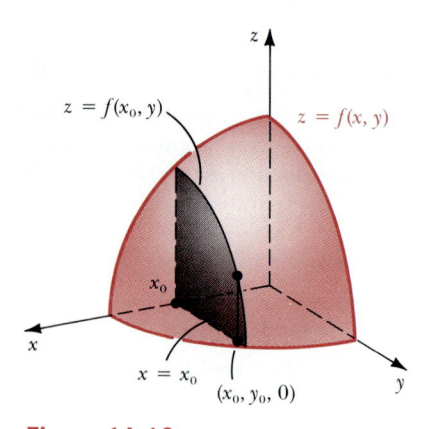

Figure 14.10
The curve described by $z = f(x_0, y)$.

(14.9)
DEFINITION

Suppose f is a function of two variables x and y. The **first partial derivative of f with respect to x, f_x,** is defined by

$$f_x(x, y) = \lim_{h \to 0} \frac{f(x + h, y) - f(x, y)}{h}.$$

The domain of f_x is the set of all (x, y) in the domain of f for which this limit exists.

Similarly, the **first partial derivative of f with respect to y, f_y,** is defined by

$$f_y(x, y) = \lim_{h \to 0} \frac{f(x, y + h) - f(x, y)}{h}.$$

The domain of f_y is the set of all (x, y) in the domain of f for which this limit exists.

Another common notation for partial derivatives is similar to the Leibniz notation for the derivative of a single-variable function:

(14.10) $\qquad \dfrac{\partial f}{\partial x} \equiv f_x \quad \text{and} \quad \dfrac{\partial f}{\partial y} \equiv f_y.$

If $z = f(x, y)$, then $\partial z / \partial x \equiv f_x$ and $\partial z / \partial y \equiv f_y$. One further, but less common, notation is to assign integers to each variable, 1 to the first variable x and 2 to the second variable y, and write $f_1 \equiv f_x$ and $f_2 \equiv f_y$.

It follows from Definition (14.9) that $f_x(x_0, y_0)$ is the derivative with respect to x of the single-variable function $f(x, y_0)$ at $x = x_0$. To find f_x, simply differentiate $f(x, y)$ with y treated as a constant. Similarly, to find f_y, differentiate $f(x, y)$ with x treated as a constant.

Example 1

Find the first partial derivatives of f if

$$f(x, y) = x^2 + xy + y \sin x$$

and evaluate these at $(0, 1)$.

Solution
We have

$$f_x(x, y) = \frac{\partial (x^2 + xy + y \sin x)}{\partial x} = 2x + y + y \cos x,$$

so

$$f_x(0, 1) = 0 + 1 + 1 \cos 0 = 2.$$

Also,

$$f_y(x, y) = \frac{\partial (x^2 + xy + y \sin x)}{\partial y} = 0 + x + \sin x,$$

so

$$f_y(0, 1) = 0. \quad \blacktriangleright$$

Example 2

Find f_x and f_y if $f(x, y) = \dfrac{y(x^2 - y^2)}{x^2 + y^2}$.

Solution
We apply the Quotient Rule in each case:

$$f_x(x, y) = \frac{(x^2 + y^2)2xy - y(x^2 - y^2)2x}{(x^2 + y^2)^2} = \frac{4xy^3}{(x^2 + y^2)^2}$$

and

$$f_y(x, y) = \frac{(x^2 + y^2)(x^2 - 3y^2) - y(x^2 - y^2)2y}{(x^2 + y^2)^2} = \frac{x^4 - 4x^2y^2 - y^4}{(x^2 + y^2)^2}.$$

The function f and, consequently, the partial derivatives f_x and f_y are undefined at $(0, 0)$. $\quad \blacktriangleright$

The derivative of a single-variable function g at a point $(a, g(a))$ gives the slope of the tangent line to the graph of g at this point. The partial derivatives also give the slope of a tangent line. Suppose the graph of $z = f(x, y)$ is the surface shown in Figure 14.11. Fixing y at y_0 and allowing x to vary traces the curve C_1 shown on the surface. This curve is the intersection of the surface and the plane $y = y_0$, so it has equation $z = f(x, y_0)$ in the plane $y = y_0$. Thus, $f_x(x_0, y_0)$ is the slope of the tangent line to C_1 at $(x_0, y_0, f(x_0, y_0))$ in the plane $y = y_0$, and this tangent line has equations

$$z - z_0 = f_x(x_0, y_0)(x - x_0), \qquad y = y_0.$$

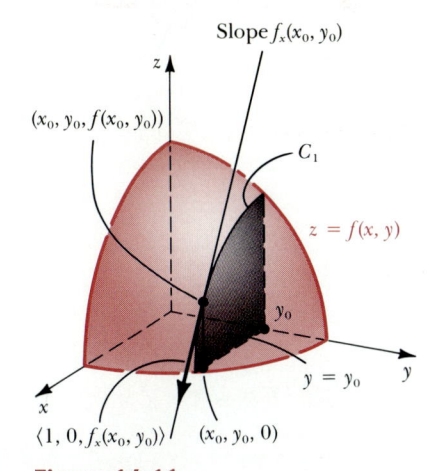

Figure 14.11
A representation of $f_x(x_0, y_0)$.

Introducing the parameter $t = x - x_0$ produces parametric equations

(14.11) $\qquad x = x_0 + t, \qquad y = y_0, \qquad z = z_0 + f_x(x_0, y_0)t$

for the line through $(x_0, y_0, f(x_0, y_0))$ in the direction of $\langle 1, 0, f_x(x_0, y_0) \rangle$.

Similarly, the curve C_2 shown in Figure 14.12 is the intersection of the graph of $z = f(x, y)$ and the plane $x = x_0$, so $f_y(x_0, y_0)$ is the slope of the tangent line to C_2 at $(x_0, y_0, f(x_0, y_0))$ in the plane $x = x_0$. Parametric equations for this tangent line are

(14.12) $\qquad x = x_0, \qquad y = y_0 + t, \qquad z = z_0 + f_y(x_0, y_0)t,$

for the line through $(x_0, y_0, f(x_0, y_0))$ in the direction of $\langle 0, 1, f_y(x_0, y_0) \rangle$.

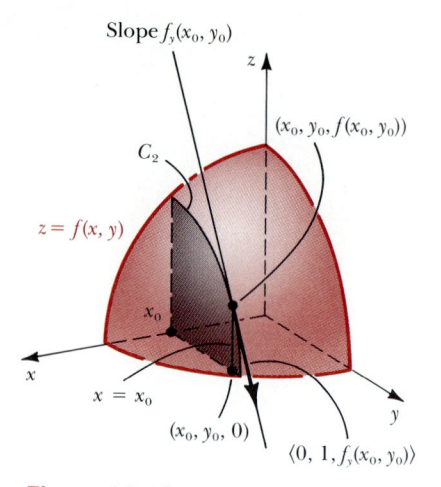

Figure 14.12
A representation of $f_y(x_0, y_0)$.

Example 3

Find parametric equations for the tangent line to the curve of intersection of the surface $z = 4 - x^2 - y^2$ and the plane $x = 1$ at the point $(1, 1, 2)$.

Solution
The curve of intersection of the graph of $f(x, y) = 4 - x^2 - y^2$ and the plane $x = 1$ is described by $z = f(1, y) = 3 - y^2$. (See Figure 14.13.) By (14.12), the tangent line at $(1, 1, 2)$ has parametric equations

$$x = 1, \qquad y = 1 + t, \qquad z = 2 + f_y(1, 2)t.$$

Since $f_y(x, y) = -2y$, $f_y(1, 1) = -2$ and the parametric equations are

$$x = 1, \qquad y = 1 + t, \qquad z = 2 - 2t. \qquad \blacktriangleright$$

The following example considers the partial derivatives at $(0, 0)$ for the function that was shown in Example 4 of Section 14.1 to be discontinuous at $(0, 0)$.

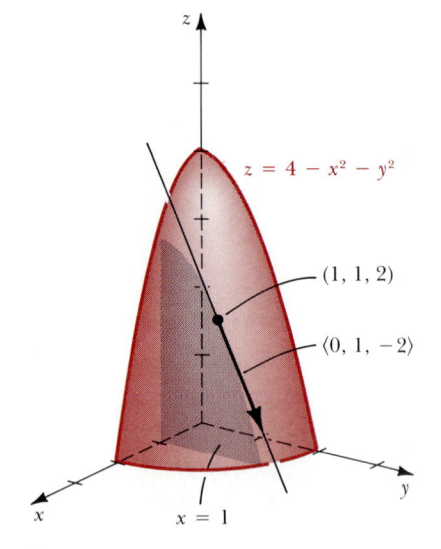

Figure 14.13

Example 4

Find $f_x(0, 0)$ and $f_y(0, 0)$ for

$$f(x, y) = \begin{cases} \dfrac{xy}{x^2 + y^2}, & \text{if } (x, y) \neq (0, 0), \\ 0, & \text{if } (x, y) = (0, 0). \end{cases}$$

Solution
In this case, Definition (14.9) must be applied to find the partial derivatives, since the Quotient Rule is invalid when the denominator is zero:

$$f_x(0, 0) = \lim_{h \to 0} \frac{f(0 + h, 0) - f(0, 0)}{h} = \lim_{h \to 0} \frac{0 - 0}{h} = 0$$

and

$$f_y(0, 0) = \lim_{h \to 0} \frac{f(0, 0 + h) - f(0, 0)}{h} = \lim_{h \to 0} \frac{0 - 0}{h} = 0.$$

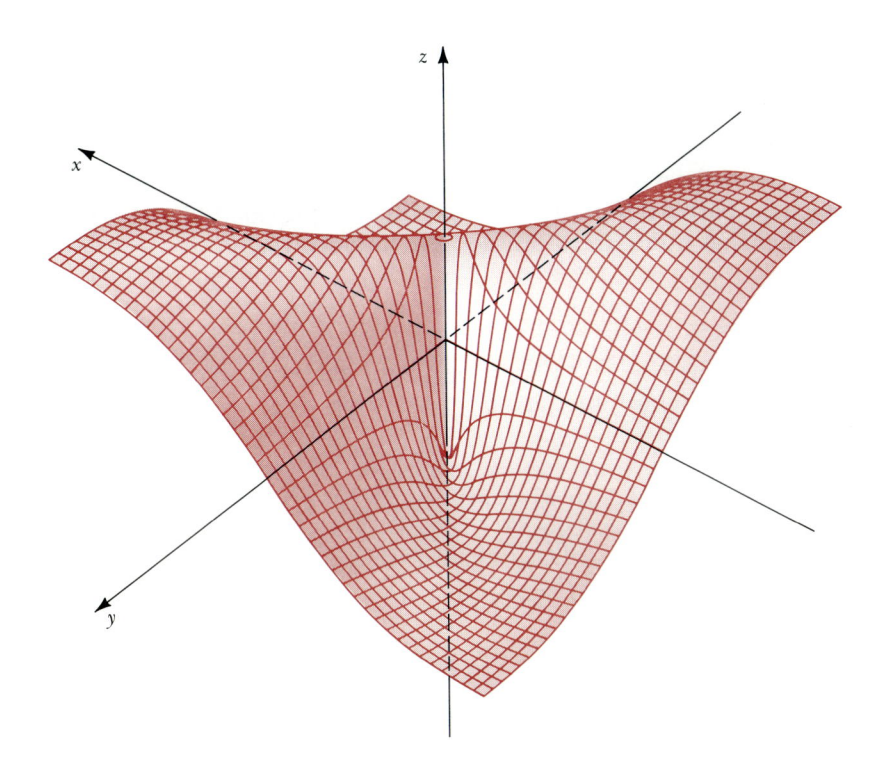

Figure 14.14

Note that both partial derivatives of f exist at $(0, 0)$, even though Example 4 of Section 14.1 shows that f is not continuous at $(0, 0)$. The graph of f is shown in Figure 14.14 and Color Plate VII. ▶

If f is a function of the variables x and y, the **second partial derivatives** of f are defined by

$$f_{xx} = \frac{\partial^2 f}{\partial x^2} = \frac{\partial}{\partial x}\left(\frac{\partial f}{\partial x}\right), \qquad f_{yy} = \frac{\partial^2 f}{\partial y^2} = \frac{\partial}{\partial y}\left(\frac{\partial f}{\partial y}\right),$$

$$f_{xy} = \frac{\partial^2 f}{\partial y \partial x} = \frac{\partial}{\partial y}\left(\frac{\partial f}{\partial x}\right), \quad \text{and} \quad f_{yx} = \frac{\partial^2 f}{\partial x \partial y} = \frac{\partial}{\partial x}\left(\frac{\partial f}{\partial y}\right).$$

The functions f_{xy} and f_{yx} are sometimes called the **mixed second partial derivatives** of f with respect to x and y. The following theorem shows that under reasonable conditions, the mixed partial derivatives are equal. The proof of this result can be found in any standard advanced calculus book.

(14.13)
THEOREM
If f_{xy} and f_{yx} are both continuous in the interior of a circle containing (x, y), then $f_{xy}(x, y) = f_{yx}(x, y)$.

For an interesting example of a function whose mixed second partial derivatives are not equal, see Exercise 51.

The equality of mixed partial derivatives holds for higher derivatives as well. In general,

(14.14)
The partial derivatives of a function are independent of the order in which the partial differentiation is performed, provided all partial derivatives of that order are continuous on an open region.

Example 5

Suppose $f(x, y) = x^3 + 2x^2y + e^{2x} \sin \pi y$. Find (a) all second partial derivatives of f and (b) the third partial derivative f_{xyx}.

Solution
a) The first partial derivatives are

$$f_x(x, y) = \frac{\partial f}{\partial x}(x, y) = 3x^2 + 4xy + 2e^{2x} \sin \pi y$$

and

$$f_y(x, y) = \frac{\partial f}{\partial y}(x, y)$$
$$= 2x^2 + \pi e^{2x} \cos \pi y.$$

So the second partial derivatives are

$$f_{xx}(x, y) = \frac{\partial^2 f}{\partial x^2}(x, y) = \frac{\partial}{\partial x}\left(\frac{\partial f}{\partial x}\right)(x, y) = 6x + 4y + 4e^{2x} \sin \pi y,$$

$$f_{yy}(x, y) = \frac{\partial^2 f}{\partial y^2}(x, y) = \frac{\partial}{\partial y}\left(\frac{\partial f}{\partial y}\right)(x, y) = -\pi^2 e^{2x} \sin \pi y,$$

$$f_{xy}(x, y) = \frac{\partial^2 f}{\partial y \, \partial x}(x, y) = \frac{\partial}{\partial y}\left(\frac{\partial f}{\partial x}\right)(x, y) = 4x + 2\pi e^{2x} \cos \pi y,$$

and

$$f_{yx}(x, y) = \frac{\partial^2 f}{\partial x \, \partial y}(x, y) = \frac{\partial}{\partial x}\left(\frac{\partial f}{\partial y}\right)(x, y) = 4x + 2\pi e^{2x} \cos \pi y.$$

b) The third partial derivative f_{xyx} is

$$f_{xyx}(x, y) = \frac{\partial^3 f}{\partial x \, \partial y \, \partial x}(x, y) = \frac{\partial}{\partial x}\left(\frac{\partial^2 f}{\partial y \, \partial x}\right)(x, y) = 4 + 4\pi e^{2x} \cos \pi y. ▶$$

The concepts in this section can be extended easily to a function f of three variables. For example, f_x is defined by

(14.15) $$f_x(x, y, z) = \lim_{h \to 0} \frac{f(x + h, y, z) - f(x, y, z)}{h}.$$

The domain of f_x in this setting consists of all (x, y, z) in the domain of f for which this limit exists.

Example 6

Find f_x, f_y, and f_z if
$$f(x, y, z) = \sqrt{x} + 2xyz + z^2 e^x \ln y.$$

Solution
Proceeding as with functions of two variables, we consider y and z constant to determine
$$f_x(x, y, z) = \frac{1}{2\sqrt{x}} + 2yz + z^2 e^x \ln y.$$

Fixing x and z gives us
$$f_y(x, y, z) = 0 + 2xz + z^2 e^x \cdot \frac{1}{y},$$

and fixing x and y gives
$$f_z(x, y, z) = 0 + 2xy + 2ze^x \ln y. \quad \blacktriangleright$$

Example 7

Find f_{xyyz} for
$$f(x, y, z) = \frac{x^2 \ln y}{z}.$$

Solution
We have
$$f_x(x, y, z) = \frac{2x \ln y}{z}, \quad \text{so} \quad f_{xy}(x, y, z) = \frac{\partial}{\partial y}\left(\frac{2x \ln y}{z}\right) = \frac{2x}{zy},$$

$$f_{xyy}(x, y, z) = \frac{\partial}{\partial y}\left(\frac{2x}{zy}\right) = -\frac{2x}{zy^2}, \quad \text{and} \quad f_{xyyz}(x, y, z) = \frac{\partial}{\partial z}\left(-\frac{2x}{zy^2}\right) = \frac{2x}{z^2 y^2}. \quad \blacktriangleright$$

▶ EXERCISE SET 14.2

In Exercises 1–20, find the first partial derivatives of the function.

1. $f(x, y) = x + y + 1$
2. $f(x, y) = 6 - x - 2y$
3. $f(x, y) = x^2 + xy + y^2$
4. $f(x, y) = 5x^3 y^2 - 3x^2 y$
5. $f(x, y) = e^{xy}$
6. $f(x, y) = \ln xy$
7. $f(x, y) = \dfrac{xy}{x^2 + y^2}$
8. $f(x, y) = \dfrac{x}{y}$
9. $f(x, y) = x^2(x + y^2)$
10. $f(x, y) = (x - 2y)y^2$
11. $f(x, y) = ye^x$
12. $f(x, y) = \sin(x^2 + y^3)$
13. $f(x, y) = \sqrt{x^2 + y^2}$
14. $f(x, y) = (2x - 3y)^3$
15. $f(x, y) = e^x \cos y$
16. $f(x, y) = \tan(x^2 + y)$
17. $f(x, y, z) = 3xz + 4y$
18. $f(x, y, z) = 10xy + 7z$
19. $f(x, y, z) = \cos(2x + 3y + 4z)$
20. $f(x, y, z) = \dfrac{1}{xyz}$

In Exercises 21–28, find the first partial derivatives of f at the given point.

21. $f(x, y) = x^2 + y^2$, $(0, 1)$

22. $f(x, y) = \dfrac{1}{x^2 + y^2}$, $(-1, 1)$

23. $f(x, y) = y \cos x$, $(\pi, 1)$

24. $f(x, y) = \ln(x^2 + y^2)$, $(1, 0)$

25. $f(x, y) = \cos x + \sin y$, (π, π)

26. $f(x, y) = \cos x \sin y$, (π, π)

27. $f(x, y, z) = \sqrt{x^2 + y^2 + z^2}$, $(1, 1, 2)$

28. $f(x, y, z) = e^x \sin y + z^2$, $(0, \pi, 3)$

In Exercises 29–36, show that $f_{xy} = f_{yx}$.

29. $f(x, y) = x^3 + 2x^2 y$

30. $f(x, y) = x^2 + \cos(x + y)$

31. $f(x, y) = e^{xy}$

32. $f(x, y) = y e^x$

33. $f(x, y) = e^y \cos \pi x$

34. $f(x, y) = \ln(x^2 + 1)$

35. $f(x, y, z) = 3xz + 4xy + 2yz$

36. $f(x, y, z) = x^2 + y^2 + z^2$

37. Find f_{xyx} and f_{xyy} if $f(x, y) = x^3 + e^{x^2 + y^2}$.

38. Find f_{xyz} and f_{zyx} if $f(x, y, z) = \cos(x^2 + y^2 + z^2)$.

39. Find $\partial^2 w / \partial x^2$ if $w = \sqrt{x^2 + y^2 + z^2}$.

40. Find $\partial^2 w / \partial x\, \partial y$ if $w = \sin xy$.

41. Find $\partial^3 w / \partial x\, \partial y\, \partial z$ if $w = \sin xyz$.

42. Find an equation of the line in the plane $y = 1$ that is tangent to the curve of intersection of this plane and the surface $z = x^2 + y^2$ at $(2, 1, 5)$.

43. Find an equation of the line in the plane $y = 1$ that is tangent to the curve of intersection of this plane and the surface $z = 4 - x^2 - y^2$ at $(1, 1, 2)$.

44. Consider the hyperbolic paraboloid described by
$$f(x, y) = y^2 - x^2$$
(see the figure).

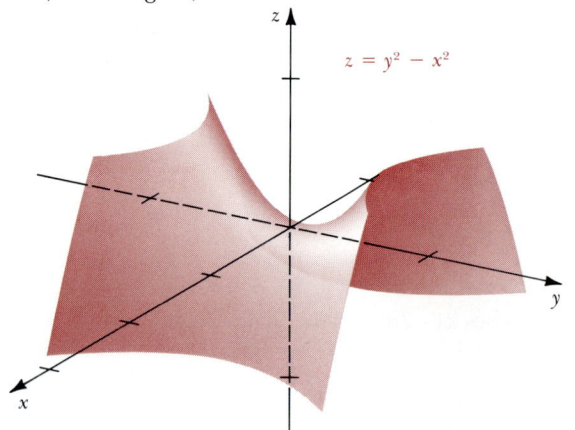

a) Determine the relative extrema of the curve of intersection of this surface and the plane $x = 0$.

b) Determine the relative extrema of the curve of intersection of this surface and the plane $y = 0$.

(*Hint:* Apply the calculus of functions of one variable to the function described by $f(0, y)$ in part (a) and by $f(x, 0)$ in part (b).)

45. Let $f(x, y) = \sqrt{y^2 - x^2} \arctan(y/x)$. Show that
$$x\frac{\partial f}{\partial x} + y\frac{\partial f}{\partial y} = f(x, y).$$

46. Let $z = x^2 \arctan(y/x) - y^2 \arctan(x/y)$. Show that
$$\frac{\partial^2 z}{\partial x\, \partial y} = \frac{x^2 - y^2}{x^2 + y^2}.$$

47. Let $w = \ln(e^x + e^y + e^z)$. Show that
$$\frac{\partial^3 w}{\partial x\, \partial y\, \partial z} = 2e^{x+y+z-3w}.$$

48. If the tangent plane to a surface $z = f(x, y)$ exists at the point $(x_0, y_0, f(x_0, y_0))$, then the lines shown in Figures 14.12 and 14.13 must lie in the plane. Show that this implies that the tangent plane has equation
$$f_x(x_0, y_0)(x - x_0) + f_y(x_0, y_0)(y - y_0) - (z - f(x_0, y_0)) = 0.$$

49. Find $f_x(0, 0)$ and $f_y(0, 0)$ for
$$f(x, y) = \begin{cases} \dfrac{xy^2}{x^2 + y^2}, & \text{if } (x, y) \neq (0, 0), \\ 0, & \text{if } (x, y) = (0, 0). \end{cases}$$

50. Find $f_x(0, 0)$ and $f_y(0, 0)$ for
$$f(x, y) = \begin{cases} \dfrac{x^2 y}{x^4 + y^2}, & \text{if } (x, y) \neq (0, 0), \\ 0, & \text{if } (x, y) = (0, 0). \end{cases}$$

51. Consider
$$f(x, y) = \begin{cases} \dfrac{x^3 y - xy^3}{x^2 + y^2}, & \text{if } (x, y) \neq (0, 0), \\ 0, & \text{if } (x, y) = (0, 0). \end{cases}$$

Show that $f_{xy}(0, 0) \neq f_{yx}(0, 0)$.

52. An equation of the form
$$\frac{\partial f}{\partial t} = c\frac{\partial^2 f}{\partial x^2}$$
is called a *diffusion equation*. The molecular concentration $C(x, t)$ of a fluid injected into a tube is described by
$$C(x, t) = t^{-1/2} e^{-x^2/(Kt)}$$
where t denotes time, x denotes the distance from the point at which the fluid was injected, and K is positive constant. Show that C satisfies a diffusion equation with $c = K/4$.

53. A function f is said to be *harmonic*, if it satisfies *Laplace's equation*:

$$\frac{\partial^2 f}{\partial x^2} + \frac{\partial^2 f}{\partial y^2} + \frac{\partial^2 f}{\partial z^2} = 0.$$

   Show that each of the following functions is harmonic.
   a) $f(x, y, z) = xyz$
   b) $f(x, y, z) = x^2 - y^2$
   c) $f(x, y, z) = \dfrac{1}{\sqrt{x^2 + y^2 + z^2}}$

54. If T represents the temperature at a point with rectangular coordinates (x, y, z) at time t and no heat sources are present, then T satisfies the *diffusion equation*

$$\frac{\partial T}{\partial t} = h^2 \left(\frac{\partial^2 T}{\partial x^2} + \frac{\partial^2 T}{\partial y^2} + \frac{\partial^2 T}{\partial z^2} \right),$$

   where h^2 is a physical constant called the *thermal diffusivity*. Show that

$$T = T(x, y, z, t) = e^{-h^2 t}(\cos x + \cos y + \cos z)$$

   satisfies the diffusion equation. [In this setting, the diffusion equation is called a *heat equation*.]

55. When a fluid is injected into a tube containing a liquid flowing with constant velocity v, the molecular concentration is described by

$$C(x, t) = t^{-1/2} e^{-(x-vt)^2/Kt}.$$

   Show that C satisfies the modified diffusion equation

$$\frac{\partial f}{\partial t} = c \frac{\partial^2 f}{\partial x^2} - v \frac{\partial f}{\partial x}$$

   with $c = K/4$. (This type of equation describes the diffusion of a serum injected into the bloodstream.)

56. The class of Cobb–Douglass functions for describing the production of a firm's output is based on two variables: x describes property resources (capital goods and natural resources), and y describes human resources (labor and managerial potential). These functions have the form

$$f(x, y) = ax^k y^{1-k}$$

   for a positive constant a and a constant k in $(0, 1)$. Show that any Cobb–Douglass function satisfies the equation

$$f(x, y) = x f_x(x, y) + y f_y(x, y).$$

57. a) Sketch the graph of the x- and y-values that are possible if the Cobb–Douglass function described by $f(x, y) = 4x^{1/3}y^{2/3}$ is to assume the value $f(x, y) = 16$.
   b) Find the partial derivatives of f when $x = 1$ and $y = 8$, and describe their practical significance.

58. One of the simplest modifications of the one-mole ideal gas law $PV = RT$ is to add a term αP to the right side of the equation, where α is a function of temperature only. Suppose a gas satisfies this law: $PV = RT + \alpha P$. Show that

$$\frac{\partial P}{\partial T} = \frac{P}{T} \left[1 + \frac{P}{R} \frac{d\alpha}{dT} \right].$$

59. Stokes's law for the velocity of a particle falling in a fluid is given by

$$v = \frac{2g(\rho - d)r^2}{9n},$$

   where g is the acceleration due to gravity, ρ is the density of the particle, d is the density of the liquid, r is the radius of the particle, and n is the absolute viscosity of the liquid. Find (a) $\partial v/\partial d$, (b) $\partial v/\partial r$, and (c) $\partial v/\partial n$.

60. The formula $S = 0.00718\, w^{0.425} h^{0.725}$ relates the surface area S of a human in square meters to the weight w in kilograms and height h in centimeters. (This formula was also considered in Exercise 20 of Section 13.1.)
   a) Determine the rate of change of surface area with respect to the change in weight when the person has a height of 2 m and weight of 70 kg.
   b) Determine the rate of change of the surface area with respect to the change in weight of a person your own height and weight.

14.3

Differentiability of Multivariable Functions

Partial derivatives provide a measure of the relative change in the values of a function as a fixed point is approached along a line parallel to a coordinate axis. There are many other ways to approach a fixed point in the plane. To say that a function is differentiable, we must consider the relative change in the values of the function regardless of how the point is approached.

From analogy with the single-variable case, we would expect differentiability of a function at a point to ensure that the function is continuous at the point. For this to be true, more must be required for differentiability than simply the existence of the partial derivatives. Recall the function defined by

$$f(x, y) = \begin{cases} \dfrac{xy}{x^2 + y^2}, & \text{if } (x, y) \neq (0, 0), \\ 0, & \text{if } (x, y) = (0, 0). \end{cases}$$

This function was shown in Example 4 of Section 14.1 to be discontinuous at $(0, 0)$. Yet in Example 4 of Section 14.2, we saw that its partial derivatives at $(0, 0)$ exist as

$$f_x(0, 0) = \lim_{h \to 0} \frac{f(h, 0) - f(0, 0)}{h} = \lim_{h \to 0} \frac{0 - 0}{h} = \lim_{h \to 0} 0 = 0$$

and

$$f_y(0, 0) = \lim_{h \to 0} \frac{f(0, h) - f(0, 0)}{h} = \lim_{h \to 0} \frac{0 - 0}{h} = \lim_{h \to 0} 0 = 0.$$

Since f is not continuous at $(0, 0)$, we would not expect f to be differentiable at $(0, 0)$, even though both its first partial derivatives at $(0, 0)$ exist.

To develop a definition of differentiability for a function of two variables, we first consider an alternative characterization of differentiability for functions of one variable. Although this characterization is not generally used for single-variable functions, it leads to the appropriate definition in the case of multivariable functions.

(14.16)
THEOREM
A function f is differentiable at a number x if and only if there exists an open interval I about x, a number m, and function ε_1 of Δx such that:

i) $f(x + \Delta x) - f(x) = m \, \Delta x + \varepsilon_1(\Delta x) \, \Delta x$, for all $x + \Delta x$ in I, and
ii) $\lim\limits_{\Delta x \to 0} \varepsilon_1(\Delta x) = 0$.

In addition, if f is differentiable, then $m = f'(x)$.

Proof. Suppose first that the conditions in (i) and (ii) are satisfied. Then

$$\lim_{\Delta x \to 0} \frac{f(x + \Delta x) - f(x)}{\Delta x} = \lim_{\Delta x \to 0} \frac{m \, \Delta x + \varepsilon_1(\Delta x) \, \Delta x}{\Delta x}$$
$$= \lim_{\Delta x \to 0} [m + \varepsilon_1(\Delta x)] = m + 0 = m,$$

so $f'(x)$ exists and $m = f'(x)$.

Conversely, suppose $f'(x)$ exists. Let ε_1 be defined by

$$\varepsilon_1(\Delta x) = \frac{f(x + \Delta x) - f(x)}{\Delta x} - f'(x).$$

Then

$$f(x + \Delta x) - f(x) = f'(x)\,\Delta x + \varepsilon_1(\Delta x)\Delta x,$$

so condition (i) in the theorem is satisfied with $m = f'(x)$. Condition (ii) is also satisfied since

$$\lim_{\Delta x \to 0} \varepsilon_1(\Delta x) = \lim_{\Delta x \to 0} \left[\frac{f(x + \Delta x) - f(x)}{\Delta x} - f'(x) \right] = f'(x) - f'(x) = 0. \quad \triangleright$$

The definition for the differentiability of a function of two variables uses conditions similar to those given in Theorem (14.16).

(14.17)
DEFINITION

A function f with partial derivatives f_x and f_y at (x, y) is said to be **differentiable** at (x, y) if there exists an open region R about (x, y) and functions ε_1 and ε_2 such that:

i) $f(x + \Delta x, y + \Delta y) - f(x, y) = f_x(x, y)\,\Delta x + f_y(x, y)\,\Delta y + \varepsilon_1(\Delta x, \Delta y)\,\Delta x + \varepsilon_2(\Delta x, \Delta y)\,\Delta y$, for all $(x + \Delta x, y + \Delta y)$ in R, and both

ii) $\lim\limits_{(\Delta x, \Delta y) \to (0,0)} \varepsilon_1(\Delta x, \Delta y) = 0$ and $\lim\limits_{(\Delta x, \Delta y) \to (0,0)} \varepsilon_2(\Delta x, \Delta y) = 0$.

Example 1

Show that the function described by

$$f(x, y) = x^2 + y^2$$

is differentiable at every point in $\mathbb{R}^2$.

Solution

To show that f is differentiable at (x, y), we must find functions ε_1 and ε_2 that satisfy (i) and (ii) in Definition (14.17). Since $f_x(x, y) = 2x$ and $f_y(x, y) = 2y$, (i) becomes

$$[(x + \Delta x)^2 + (y + \Delta y)^2] - (x^2 + y^2) = 2x\,\Delta x + 2y\,\Delta y + \varepsilon_1(\Delta x, \Delta y)\,\Delta x + \varepsilon_2(\Delta x, \Delta y)\,\Delta y.$$

Simplifying this equation we get

$$(\Delta x)^2 + (\Delta y)^2 = \varepsilon_1(\Delta x, \Delta y)\,\Delta x + \varepsilon_2(\Delta x, \Delta y)\,\Delta y.$$

We can choose

$$\varepsilon_1(\Delta x, \Delta y) = \Delta x \quad \text{and} \quad \varepsilon_2(\Delta x, \Delta y) = \Delta y$$

to satisfy (i) and (ii). Since it is necessary to find only one such pair of functions, f is differentiable at (x, y). ▶

(14.18) THEOREM

If a function f of two variables is differentiable at (x, y), then f is continuous at (x, y).

Proof. From (i) in Definition (14.17),

$$f(x + \Delta x, y + \Delta y) - f(x, y) = f_x(x, y) \Delta x + f_y(x, y) \Delta y \\ + \varepsilon_1(\Delta x, \Delta y) \Delta x + \varepsilon_2(\Delta x, \Delta y) \Delta y.$$

Since both $\lim_{(\Delta x, \Delta y) \to (0,0)} \varepsilon_1(\Delta x, \Delta y)$ and $\lim_{(\Delta x, \Delta y) \to (0,0)} \varepsilon_2(\Delta x, \Delta y)$ exist,

$$\lim_{(\Delta x, \Delta y) \to (0,0)} [f(x + \Delta x, y + \Delta y) - f(x, y)] = 0$$

and

$$\lim_{(\Delta x, \Delta y) \to (0,0)} f(x + \Delta x, y + \Delta y) = f(x, y).$$

This is precisely the condition required for continuity at (x, y). ▷

Let us reconsider the function defined at the beginning of this section in light of Theorem (14.18):

$$f(x, y) = \begin{cases} \dfrac{xy}{x^2 + y^2}, & \text{if } (x, y) \neq (0, 0), \\ 0, & \text{if } (x, y) = (0, 0). \end{cases}$$

It has been shown that although this function is discontinuous at $(0, 0)$, both first partial derivatives at $(0, 0)$ exist. Since f is not continuous at $(0, 0)$, Theorem (14.18) implies that f is not differentiable at $(0, 0)$. This shows that:

> The partial derivatives of a function can exist at points at which the function is not differentiable.

The following result shows that there are conditions on the partial derivatives of a function that ensure differentiability. An interesting example showing that the converse of this result is not true is given in Exercise 38.

(14.19) THEOREM

If the partial derivatives of a function f of two variables are continuous at (x, y), then f is differentiable at (x, y).

Proof. This proof is a nice application of the Mean Value Theorem for functions of a single variable. First note that since the partial derivatives of f are continuous at (x, y) they, and hence f, must be defined in an open region R about (x, y).

Suppose $(x + \Delta x, y + \Delta y)$ is in R. Then we can write

(14.20)
$$f(x + \Delta x, y + \Delta y) - f(x, y) = f(x + \Delta x, y + \Delta y) - f(x, y + \Delta y) \\ + f(x, y + \Delta y) - f(x, y).$$

Consider the single-variable function that results from fixing the second variable of f at $y + \Delta y$. Since f_x exists in R, the Mean Value Theorem implies that a number $\hat{x}$ exists between x and $x + \Delta x$ with

(14.21) $\quad f(x + \Delta x, y + \Delta y) - f(x, y + \Delta y) = f_x(\hat{x}, y + \Delta y) \Delta x.$

Similarly, since f_y exists in R, we can fix the first variable at x and apply the Mean Value Theorem to show that a number $\hat{y}$ exists between y and $y + \Delta y$ with

(14.22) $\qquad f(x, y + \Delta y) - f(x, y) = f_y(x, \hat{y}) \Delta y.$

Substituting (14.21) and (14.22) into equation (14.20) produces

$$f(x + \Delta x, y + \Delta y) - f(x, y) = f_x(\hat{x}, y + \Delta y) \Delta x + f_y(x, \hat{y}) \Delta y,$$

which can be rewritten in the form

$$\begin{aligned} f(x + \Delta x, y + \Delta y) - f(x, y) = &\, f_x(x, y) \Delta x + f_y(x, y) \Delta y \\ &+ [f_x(\hat{x}, y + \Delta y) - f_x(x, y)] \Delta x \\ &+ [f_y(x, \hat{y}) - f_y(x, y)] \Delta y. \end{aligned}$$

Let

$$\varepsilon_1(\Delta x, \Delta y) = f_x(\hat{x}, y + \Delta y) - f_x(x, y)$$

and

$$\varepsilon_2(\Delta x, \Delta y) = f_y(x, \hat{y}) - f_y(x, y).$$

Since $\hat{x}$ is between x and $x + \Delta x$, $\hat{y}$ is between y and $y + \Delta y$, and both partial derivatives are continuous at (x, y):

$$\lim_{(\Delta x, \Delta y) \to (0,0)} f_x(\hat{x}, y + \Delta y) = f_x(x, y) \quad \text{and} \quad \lim_{(\Delta x, \Delta y) \to (0,0)} f_y(x, \hat{y}) = f_y(x, y).$$

Therefore,

$$\lim_{(\Delta x, \Delta y) \to (0,0)} \varepsilon_1(\Delta x, \Delta y) = 0 \quad \text{and} \quad \lim_{(\Delta x, \Delta y) \to (0,0)} \varepsilon_2(\Delta x, \Delta y) = 0.$$

Thus, f is differentiable at (x, y). ▷

Example 2

Show that $f(x, y) = \ln(x^2 + y)$ is differentiable at $(1, e)$.

Solution

The partial derivatives of f are

$$f_x(x, y) = \frac{2x}{x^2 + y} \quad \text{and} \quad f_y(x, y) = \frac{1}{x^2 + y}.$$

These functions are continuous at $(1, e)$, so f is differentiable at $(1, e)$. ▶

Suppose f is a function of a single variable that is differentiable at x. In Section 3.4 we defined the differential of $y = f(x)$ with respect to

the fixed value x and the increment Δx as
$$dy = f'(x)\,\Delta x.$$
The increment Δx was also called the differential of x and denoted dx. This permitted us to write $dy = f'(x)\,dx$, consistent with the Leibniz notation for the derivative $dy/dx = f'(x)$.

Similarly, if f is a function of two variables that is differentiable at (x, y), then the **differential of $z = f(x, y)$** with respect to the fixed point (x, y) and the increments Δx and Δy is defined as
$$dz = f_x(x, y)\,\Delta x + f_y(x, y)\,\Delta y.$$
We also define the **differentials dx and dy** as $dx = \Delta x$ and $dy = \Delta y$, so the differential dz can be written as

(14.23) $$dz = f_x(x, y)\,dx + f_y(x, y)\,dy.$$

The differential dz approximates the change Δz in the value of the function f relative to the changes $dx = \Delta x$ and $dy = \Delta y$ in the variables x and y (see Figure 14.15):
$$dz \approx \Delta z = f(x + \Delta x, y + \Delta y) - f(x, y),$$
so

(14.24) $$f(x + \Delta x, y + \Delta y) \approx f(x, y) + dz.$$

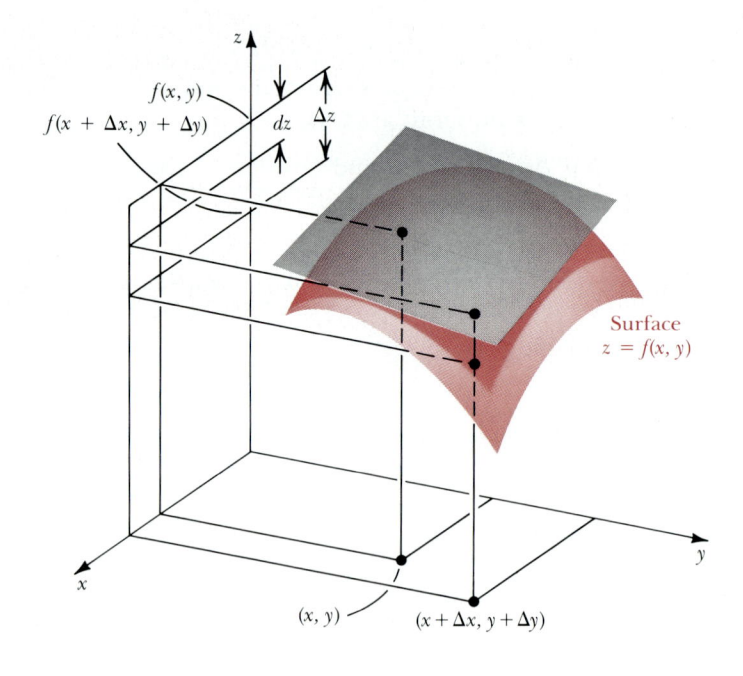

Figure 14.15

Example 3

Suppose $z = f(x, y) = x^3 + 2xy$.

a) Find dz.
b) Find dz, if $(x, y) = (1, 3)$, $\Delta x = 0.01$, and $\Delta y = -0.02$.
c) Use the result in (b) to approximate $f(1.01, 2.98)$.

Solution

a) The partial derivatives of f are

$$f_x(x, y) = 3x^2 + 2y \quad \text{and} \quad f_y(x, y) = 2x,$$

so

$$dz = (3x^2 + 2y)\, dx + 2x\, dy.$$

b) With $(x, y) = (1, 3)$, $dx = \Delta x = 0.01$, and $dy = \Delta y = -0.02$,

$$dz = (3 \cdot 1^2 + 2 \cdot 3)(0.01) + (2 \cdot 1)(-0.02) = 0.05.$$

c) Since $f(1, 3) = 1^3 + 2 \cdot 1 \cdot 3 = 7$,

$$f(1.01, 2.98) \approx f(1, 3) + dz = 7.05. \quad \blacktriangleright$$

Example 4

A tin can is constructed in the form of a right circular cylinder with inside height 4 in. and inside radius 1.5 in. Use differentials to approximate the volume of tin required to construct the can if the tin is 0.0012 in. thick and the can has both a top and a bottom (see Fig. 14.16).

Solution

The volume of tin required is the difference between the volume of a cylinder with dimensions those of the outside of the can and the volume of a cylinder with dimensions those of the inside of the can. This difference is approximated by dV when $dr = 0.0012$, the thickness of the side, and $dh = 0.0024$, the combined thickness of the top and bottom. Since the volume of a cylinder is

$$V(r, h) = \pi r^2 h,$$

we have

$$\frac{\partial V}{\partial r} = 2\pi r h \quad \text{and} \quad \frac{\partial V}{\partial h} = \pi r^2.$$

Thus,

$$dV = 2\pi r h\, dr + \pi r^2\, dh$$

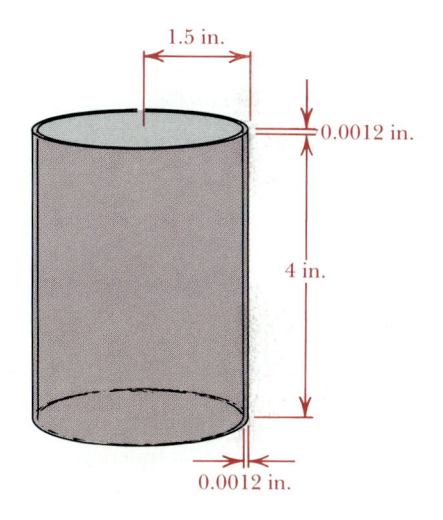

Figure 14.16

and the tin required is approximately.

$$dV = 2\pi(1.5)(4)(0.0012) + \pi(1.5)^2(0.0024) \approx 0.062 \text{ in}^3. \quad \blacktriangleright$$

The concepts of differentiability and differentials are defined in an analogous fashion for functions of three variables. The final example in this section is an application of differentials in this situation.

Example 5

Let $w = f(x, y, z) = x^2 y e^{3z}$.

a) Find the differential dw.
b) Use dw to determine an approximation for $f(1.01, -1.03, 0.02)$ if $dx = 0.01$, $dy = -0.03$, and $dz = 0.02$.

Solution

a) The definition of the differential dw for a function of three variables is the extension of (14.23) to a third variable:

$$dw = f_x(x, y, z)\,dx + f_y(x, y, z)\,dy + f_z(x, y, z)\,dz.$$

b) Since

$$f_x(x, y, z) = 2xye^{3z}, \quad f_y(x, y, z) = x^2 e^{3z}, \quad \text{and} \quad f_z(x, y, z) = 3x^2 y e^{3z},$$

we have

$$dw = f_x(1, -1, 0)\,dx + f_y(1, -1, 0)\,dy + f_z(1, -1, 0)\,dz$$
$$= -2(0.01) + 1(-0.03) - 3(0.02) = -0.11.$$

Thus,

$$f(1.01, -1.03, 0.02) \approx f(1, -1, 0) + dw = -1 - 0.11 = -1.11.$$

EXERCISE SET 14.3

In Exercises 1–6, use Theorem (14.19) to show that the function is differentiable.

1. $f(x, y) = \sin(x + y)$
2. $f(x, y) = \cos x - y^2$
3. $f(x, y) = \ln(x^2 + y^2)$
4. $f(x, y) = e^{x+y}$
5. $f(x, y) = ye^x$
6. $f(x, y) = e^x \sin y$

In Exercises 7–12, find dz.

7. $z = x^2 - 3y^2$
8. $z = e^{xy}$
9. $z = x/y$
10. $z = \sqrt{x^2 + y^2}$
11. $z = y^2 \cos x$
12. $z = \sin(x + 2y)$

In Exercises 13–16, a function is described by $z = f(x, y)$ and values of Δx and Δy are given. Use this information and dz to approximate the value of the function at the specified point.

13. $z = x^2 - 3y^2, \Delta x = 0.1, \Delta y = 0.2; f(1.1, 2.2)$
14. $z = e^{xy}, \Delta x = -0.1, \Delta y = 0.1; f(-0.1, 1.1)$
15. $z = x/y, \Delta x = 0.2, \Delta y = -0.1; f(0.2, 0.9)$
16. $z = \sqrt{x^2 + y^2}, \Delta x = 0.3, \Delta y = 0.1; f(4.3, 3.1)$

In Exercises 17–22, find dw.

17. $w = xyz$
18. $w = 2xy + 4xz + yz$
19. $w = \cos xyz$
20. $w = \ln xy$
21. $w = xye^z$
22. $w = 3x^2/y + e^z y^2$

In Exercises 23–26, a function is described by $w = f(x, y, z)$ and values of $\Delta x, \Delta y$, and Δz are given. Use this information and dw to approximate the value of the function at the specified point.

23. $w = xyz, \Delta x = 0.1, \Delta y = 0.2, \Delta z = 0.3; f(1.1, 2.2, 3.3)$
24. $w = 2xy + 4xz + yz, \Delta x = 0.1, \Delta y = -0.1, \Delta z = 0.2; f(0.1, 1.4, -0.8)$
25. $w = \cos xyz, \Delta x = -0.2, \Delta y = 0.1, \Delta z = 0.1; f(-0.2, 0.1, 0.1)$
26. $w = \ln xy, \Delta x = -0.2, \Delta y = -0.2, \Delta z = 0.4; f(0.8, e - 0.2, 0)$

In Exercises 27–32, find functions ε_1 and ε_2 that satisfy Definition (14.17) and show that f is differentiable.

27. $f(x, y) = 3 - x - y$
28. $f(x, y) = x^2$
29. $f(x, y) = x^2 - y^2$
30. $f(x, y) = 4 - x^2 - y^2$
31. $f(x, y) = xy$
32. $f(x, y) = x^2 y$

33. Use differentials to approximate the error in the area measurement of a 12-ft-by-14-ft rectangular floor if an error of 0.1 in. is made in measuring both the length and the width.

34. If two capacitors with capacitance C_1 and C_2 are connected in series, then the capacitance of the resulting circuit is

$$C = \left(\frac{1}{C_1} + \frac{1}{C_2}\right)^{-1}$$

(see the figure).

a) Find the partial derivatives of C with respect to C_1 and with respect to C_2.

b) Use these to approximate the change in capacitance that results by changing C_1 from 1 to 1.2 microfarads (μF) and C_2 from 3 to 2.8 microfarads.

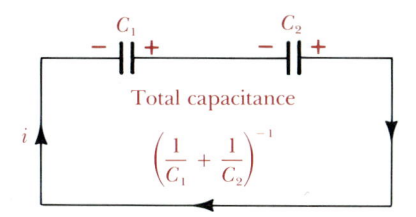

35. If two resistors with resistance R_1 and R_2 are connected in parallel, then the resistance of the resulting circuit is

$$R = \left(\frac{1}{R_1} + \frac{1}{R_2}\right)^{-1}$$

(see the figure).
a) Find the partial derivatives of R with respect to R_1 and with respect to R_2.
b) Use these to approximate the change in resistance that results by changing R_1 from 3 to 3.2 ohms (Ω) and R_2 from 8 to 7.5 ohms.

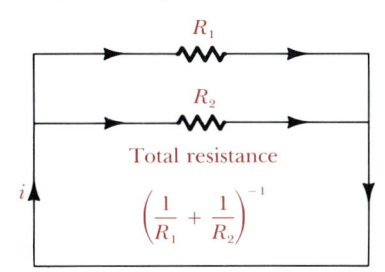

36. A company that spends x thousand dollars on advertisement and y thousand dollars on research and development makes an annual profit, in thousands of dollars, of

$$f(x, y) = 3x + 4y + \frac{xy}{100} - \frac{x^2}{100} - \frac{y^2}{200}$$

Suppose the company currently spends $x_0 = \$500{,}000$ on advertising and $y_0 = \$200{,}000$ on research and development and has an additional \$20,000 to apply to one of these areas.
a) Compute $f_x(x_0, y_0)$ and $f_y(x_0, y_0)$.
b) Use these to determine to which area the additional \$20,000 should be applied.

37. Consider

$$f(x, y) = \begin{cases} \dfrac{x^2 y}{x^4 + y^2}, & \text{if } (x, y) \neq (0, 0), \\ 0, & \text{if } (x, y) = (0, 0). \end{cases}$$

Show that f is not differentiable at $(0, 0)$.

38. Consider

$$f(x, y) = \begin{cases} (x^2 + y^2) \sin \dfrac{1}{\sqrt{x^2 + y^2}}, & \text{if } (x, y) \neq (0, 0), \\ 0, & \text{if } (x, y) = (0, 0). \end{cases}$$

a) Find $f_x(0, 0)$ and $f_y(0, 0)$.
b) Show that f is differentiable at $(0, 0)$.
c) Show that f_x and f_y are not continuous at $(0, 0)$. (This shows that the converse of Theorem 14.19 is not true.)

14.4

The Chain Rule

If g and f are differentiable single-variable functions with $u = g(x)$ and $y = f(u) = f(g(x))$, then the Chain Rule states that

$$\frac{dy}{dx} = \frac{dy}{du} \frac{du}{dx}.$$

In this section the Chain Rule is extended to multivariable functions.

Suppose f is a function of two variables x and y, each of which is a function of a third variable t. In effect, f is a function of the single variable t and, under certain conditions, is differentiable with respect to t. For example, let

$$f(x, y) = x \sin y + e^x \cos y, \quad x(t) = t^2 + 1, \quad \text{and} \quad y(t) = t^3.$$

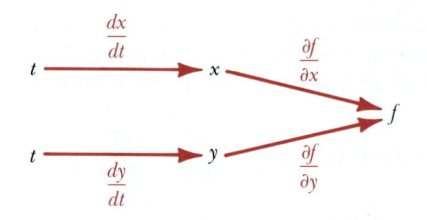

Figure 14.17

Then
$$f(t) = f(x(t), y(t)) = (t^2 + 1) \sin t^3 + e^{t^2+1} \cos t^3$$
and
$$f'(t) = 2t \sin t^3 + (t^2 + 1)3t^2 \cos t^3 + 2te^{t^2+1} \cos t^3 + e^{t^2+1}(-3t^2 \sin t^3).$$

The derivative of f with respect to t can also be found by using the partial derivatives of f with respect to x and y and the derivatives of x and y with respect to t. Consider what happens to $f(x(t), y(t))$ when t changes. A change in t directly affects $x(t)$, which in turn changes the value of f. In addition, a change in t directly affects $y(t)$, which again changes the value of f. The derivatives of x and y with respect to t and the partial derivatives of f with respect to x and y are used to express this result quantitatively, as shown in Figure 14.17, and described in the following theorem.

(14.25)
THEOREM
Suppose x and y are differentiable functions of t and f is a differentiable function of x and y. Then f is a differentiable function of t and
$$\frac{df}{dt}(x(t), y(t)) = f_x(x(t), y(t))\, x'(t) + f_y(x(t), y(t))\, y'(t).$$

Proof. Since $f(x(t), y(t))$ is a function of the one variable t,
$$\frac{df}{dt}(x(t), y(t)) = \lim_{\Delta t \to 0} \frac{f(x(t+\Delta t), y(t+\Delta t)) - f(x(t), y(t))}{\Delta t}.$$

To evaluate this limit, let
$$\Delta x = x(t + \Delta t) - x(t) \quad \text{and} \quad \Delta y = y(t + \Delta t) - y(t).$$

Since f is differentiable, functions ε_1 and ε_2 exist with
$$\lim_{(\Delta x, \Delta y) \to (0,0)} \varepsilon_1(\Delta x, \Delta y) = 0, \quad \lim_{(\Delta x, \Delta y) \to (0,0)} \varepsilon_2(\Delta x, \Delta y) = 0,$$
and
$$f(x(t+\Delta t), y(t+\Delta t)) - f(x(t), y(t)) = f_x(x(t), y(t))\, \Delta x + f_y(x(t), y(t))\, \Delta y \\ + \varepsilon_1(\Delta x, \Delta y)\, \Delta x + \varepsilon_2(\Delta x, \Delta y)\, \Delta y.$$

Hence,

(14.26)
$$f(x(t+\Delta t), y(t+\Delta t)) - f(x(t), y(t)) = [f_x(x(t), y(t)) + \varepsilon_1(\Delta x, \Delta y)]\, \Delta x \\ + [f_y(x(t), y(t)) + \varepsilon_2(\Delta x, \Delta y)]\, \Delta y.$$

Since x and y are differentiable functions of t, both are continuous so
$$\lim_{\Delta t \to 0} \Delta x = \lim_{\Delta t \to 0} [x(t + \Delta t) - x(t)] = 0$$
and
$$\lim_{\Delta t \to 0} \Delta y = \lim_{\Delta t \to 0} [y(t + \Delta t) - y(t)] = 0.$$

Therefore, both

$$\lim_{\Delta t \to 0} \varepsilon_1(\Delta x, \Delta y) = 0 \quad \text{and} \quad \lim_{\Delta t \to 0} \varepsilon_2(\Delta x, \Delta y) = 0.$$

From (14.26),

$$\begin{aligned}\frac{df}{dt}(x(t), y(t)) &= \lim_{\Delta t \to 0} \frac{f(x(t+\Delta t), y(t+\Delta t)) - f(x(t), y(t))}{\Delta t} \\ &= \lim_{\Delta t \to 0} [f_x(x(t), y(t)) + \varepsilon_1(\Delta x, \Delta y)] \lim_{\Delta t \to 0} \frac{\Delta x}{\Delta t} \\ &\quad + \lim_{\Delta t \to 0} [f_y(x(t), y(t)) + \varepsilon_2(\Delta x, \Delta y)] \lim_{\Delta t \to 0} \frac{\Delta y}{\Delta t} \\ &= f_x(x(t), y(t)) \, x'(t) + f_y(x(t), y(t)) \, y'(t). \quad \triangleright\end{aligned}$$

The equation in Theorem (14.25) is known as the **Chain Rule** and is often expressed in Leibniz notation as

$$\frac{df}{dt} = \frac{\partial f}{\partial x} \frac{dx}{dt} + \frac{\partial f}{\partial y} \frac{dy}{dt}.$$

Example 1

For $f(x, y) = x \sin y + e^x \cos y$, $x(t) = t^2 + 1$, and $y(t) = t^3$, use the Chain Rule to find $D_t f(x(t), y(t))$.

Solution
The Chain Rule gives

$$\begin{aligned}D_t f(x(t), y(t)) &= f_x(x(t), y(t)) \, x'(t) + f_y(x(t), y(t)) \, y'(t) \\ &= (\sin y + e^x \cos y)(2t) + (x \cos y - e^x \sin y)(3t^2). \\ &= 2t \sin t^3 + 2t e^{t^2+1} \cos t^3 + (t^2+1)(3t^2 \cos t^3) - 3t^2 e^{t^2+1} \sin t^3.\end{aligned}$$

This is, of course, the same as the result found at the beginning of the section by first expressing f totally as a function of t. ▶

The Chain Rule can be generalized to functions of more than two variables and also to the situation when the variables themselves are multivariable functions. The most frequently used results are listed below. In each case it is assumed that the required differentiability conditions are fulfilled. The result in Theorem (14.25) is repeated for ease of reference:

If f is a function of $x(t)$ and $y(t)$, then

(14.27) $$\frac{df}{dt} = \frac{\partial f}{\partial x} \frac{dx}{dt} + \frac{\partial f}{\partial y} \frac{dy}{dt}.$$

If f is a function of $x(u, v)$ and $y(u, v)$, then

(14.28) $$\frac{\partial f}{\partial u} = \frac{\partial f}{\partial x} \frac{\partial x}{\partial u} + \frac{\partial f}{\partial y} \frac{\partial y}{\partial u} \quad \text{and} \quad \frac{\partial f}{\partial v} = \frac{\partial f}{\partial x} \frac{\partial x}{\partial v} + \frac{\partial f}{\partial y} \frac{\partial y}{\partial v}.$$

If f is a function of $x(t)$, $y(t)$, and $z(t)$, then

(14.29)
$$\frac{df}{dt} = \frac{\partial f}{\partial x}\frac{dx}{dt} + \frac{\partial f}{\partial y}\frac{dy}{dt} + \frac{\partial f}{\partial z}\frac{dz}{dt}.$$

If f is a function of $x(u, v)$, $y(u, v)$, and $z(u, v)$, then

(14.30)
$$\frac{\partial f}{\partial u} = \frac{\partial f}{\partial x}\frac{\partial x}{\partial u} + \frac{\partial f}{\partial y}\frac{\partial y}{\partial u} + \frac{\partial f}{\partial z}\frac{\partial z}{\partial u} \quad \text{and} \quad \frac{\partial f}{\partial v} = \frac{\partial f}{\partial x}\frac{\partial x}{\partial v} + \frac{\partial f}{\partial y}\frac{\partial y}{\partial v} + \frac{\partial f}{\partial z}\frac{\partial z}{\partial v}$$

Example 2

If $f(x, y) = y^2 \tan x$, $x = u^2 - v^2$, and $y = u^2 + v^2$, find (a) $\partial f/\partial u$ and (b) $\partial f/\partial v$.

Solution

Figure 14.18 shows that there are two *paths* from u to f and two from v to f. Thus we have the following:

a) $\dfrac{\partial f}{\partial u} = \dfrac{\partial f}{\partial x}\dfrac{\partial x}{\partial u} + \dfrac{\partial f}{\partial y}\dfrac{\partial y}{\partial u} = (y^2 \sec^2 x)(2u) + (2y \tan x)(2u)$

$= 2u(u^2 + v^2)^2 \sec^2(u^2 - v^2) + 4u(u^2 + v^2) \tan(u^2 - v^2)$

and

b) $\dfrac{\partial f}{\partial v} = \dfrac{\partial f}{\partial x}\dfrac{\partial x}{\partial v} + \dfrac{\partial f}{\partial y}\dfrac{\partial y}{\partial v} = (y^2 \sec^2 x)(-2v) + (2y \tan x)(2v)$

$= -2v(u^2 + v^2)^2 \sec^2(u^2 - v^2) + 4v(u^2 + v^2) \tan(u^2 - v^2).$ ▶

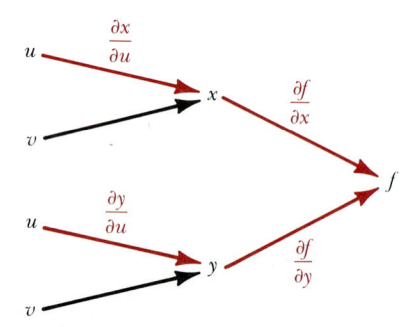

Figure 14.18

Example 3

Suppose f, x, and y are as given in Example 2 and that, in addition, $u = \sin t$ and $v = \cos t$. Find $(df/dt)(\pi/4)$.

Solution

The total dependence of f on the variable t can be seen in Figure 14.19. There are four paths from t to f, so there are four ways in which a change in t influences f. Each path contributes to the derivative of f with respect

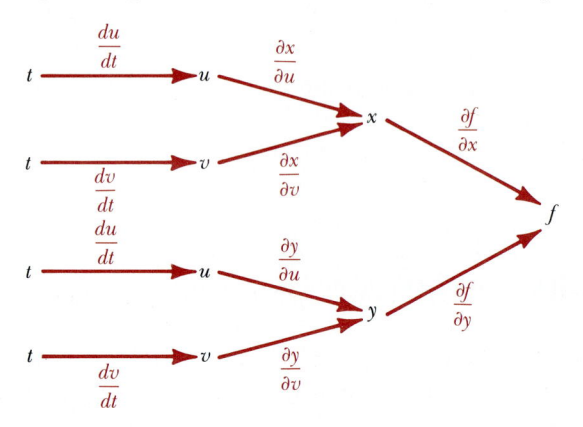

Figure 14.19

to t. Thus,

$$\frac{df}{dt} = \frac{\partial f}{\partial x}\frac{\partial x}{\partial u}\frac{du}{dt} + \frac{\partial f}{\partial x}\frac{\partial x}{\partial v}\frac{dv}{dt} + \frac{\partial f}{\partial y}\frac{\partial y}{\partial u}\frac{du}{dt} + \frac{\partial f}{\partial y}\frac{\partial y}{\partial v}\frac{dv}{dt}$$

$$= (y^2 \sec^2 x)(2u)(\cos t) + (y^2 \sec^2 x)(-2v)(-\sin t)$$
$$+ (2y \tan x)(2u)(\cos t) + (2y \tan x)(2v)(-\sin t).$$

When $t = \pi/4$, $u = \sin(\pi/4) = \sqrt{2}/2$, and $v = \cos(\pi/4) = \sqrt{2}/2$, so $x = u^2 - v^2 = 0$, and $y = u^2 + v^2 = 1$. Thus,

$$\frac{df}{dt}\left(\frac{\pi}{4}\right) = (\sec^2 0)(\sqrt{2})\left(\frac{\sqrt{2}}{2}\right) + (\sec^2 0)(-\sqrt{2})\left(-\frac{\sqrt{2}}{2}\right)$$
$$+ (2 \tan 0)(\sqrt{2})\left(\frac{\sqrt{2}}{2}\right) + (2 \tan 0)(\sqrt{2})\left(-\frac{\sqrt{2}}{2}\right)$$
$$= (1)(1) + (1)(1) + 0 + 0 = 2. \quad \blacktriangleright$$

Example 4

Let $f(x, y, z) = xye^z$, $x = \cos t$, $y = \sin t$, and $z = \ln t$. Find df/dt:

a) by expressing $f(x, y, z)$ in terms of t before differentiating, and
b) by using the Chain Rule.

Solution

a) We have

$$f(x(t), y(t), z(t)) = (\cos t)(\sin t)e^{\ln t} = t \cos t \sin t.$$

So

$$\frac{df}{dt} = \cos t \sin t - t \sin^2 t + t \cos^2 t.$$

b) Using the Chain Rule produces (see Figure 14.20)

$$\frac{df}{dt} = \frac{\partial f}{\partial x}\frac{dx}{dt} + \frac{\partial f}{\partial y}\frac{dy}{dt} + \frac{\partial f}{\partial z}\frac{dz}{dt}$$

$$= ye^z(-\sin t) + xe^z(\cos t) + xye^z \cdot \frac{1}{t}.$$

To see that this is the same as the result obtained in part (a), we express x, y, and z in terms of t:

$$\frac{df}{dt} = -(\sin^2 t)e^{\ln t} + (\cos^2 t)e^{\ln t} + \sin t \cos t \, e^{\ln t} \cdot \frac{1}{t}$$

$$= -t \sin^2 t + t \cos^2 t + \sin t \cos t. \quad \blacktriangleright$$

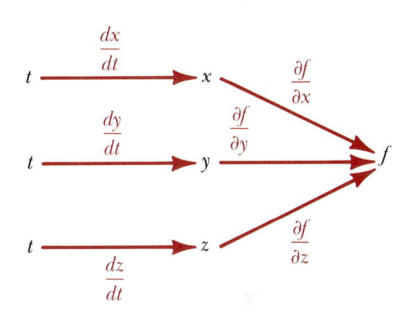

Figure 14.20

Example 5

A conveyor belt is delivering coal onto a conical pile. When the height of the pile is 60 ft and the radius of the pile is 40 ft, the height and radius are increasing at the rates of 4 ft/min and 3 ft/min, respectively (see Figure 14.21). How fast is the conveyor belt delivering coal at this time?

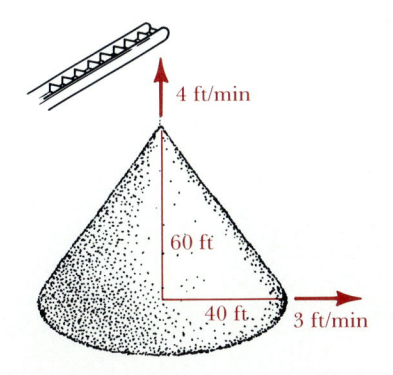

Figure 14.21

Solution

The conveyor belt is delivering coal at the rate at which the volume V of the conical pile is increasing. This rate is given by dV/dt. Since the volume of a cone of height h and radius r is $V = \frac{1}{3}\pi r^2 h$, the essential elements of the problem are as shown below:

KNOW	FIND
1) $V = \dfrac{1}{3}\pi r^2 h$;	a) $\dfrac{dV}{dt}$;
2) $\dfrac{dh}{dt} = 4 \dfrac{\text{ft}}{\text{min}}$, when $h = 60$ ft, $r = 40$ ft;	b) $\dfrac{dV}{dt}$ when $h = 60$ ft, $r = 40$ ft.
3) $\dfrac{dr}{dt} = 3 \dfrac{\text{ft}}{\text{min}}$, when $h = 60$ ft, $r = 40$ ft;	
4) $\dfrac{dV}{dt} = \dfrac{\partial V}{\partial h}\dfrac{dh}{dt} + \dfrac{\partial V}{\partial r}\dfrac{dr}{dt}$.	

a) Since

$$\frac{\partial V}{\partial h} = \frac{1}{3}\pi r^2 \quad \text{and} \quad \frac{\partial V}{\partial r} = \frac{2}{3}\pi r h,$$

we have

$$\frac{dV}{dt} = \frac{1}{3}\pi r^2 \frac{dh}{dt} + \frac{2}{3}\pi r h \frac{dr}{dt}.$$

b) When $h = 60$ and $r = 40$,

$$\frac{dV}{dt} = \frac{1}{3}\pi(40)^2(4) + \frac{2}{3}\pi(40)(60)(3) = \frac{20{,}800}{3}\pi \frac{\text{ft}^3}{\text{min}}.$$

At this rate a standard railroad hopper car can be filled in about 20 sec. ▶

The Chain Rule for functions of several variables can be used to expand the technique of implicit differentiation. The following result forms the basis for this extension.

(14.31)
THEOREM
If $F(x, y) = 0$ implicitly defines y as a differentiable function of x, then

$$\frac{dy}{dx} = \frac{-F_x(x, y)}{F_y(x, y)},$$

provided $F_y(x, y) \neq 0$.

Proof. Since $F(x, y) = 0$ implicitly defines y as a differentiable function of x, suppose $y = f(x)$ and $f'(x)$ exists. We first apply the form of the Chain Rule in (14.27) with $u = x$ and $v = y = f(x)$:

$$0 = D_x\, 0 = D_x\, F(u, v) = \frac{\partial F}{\partial u} \cdot \frac{du}{dx} + \frac{\partial F}{\partial v} \cdot \frac{dv}{dx} = \frac{\partial F}{\partial u} \cdot 1 + \frac{\partial F}{\partial v} \cdot f'(x).$$

Since $u = x$ and $v = y$, $F_x(x, y) = \partial F/\partial u$ and $F_y(x, y) = \partial F/\partial v$, so

$$\frac{dy}{dx} = f'(x) = -\frac{F_x(x, y)}{F_y(x, y)}.$$

▷

Example 6

Use Theorem (14.31) to find dy/dx if $x^3 + 3xy + y^5 + 4 = 0$.

Solution
Let $F(x, y) = x^3 + 3xy + y^5 + 4$. Then

$$F_x(x, y) = 3x^2 + 3y \quad \text{and} \quad F_y(x, y) = 3x + 5y^4.$$

So

$$\frac{dy}{dx} = \frac{-F_x(x, y)}{F_y(x, y)} = \frac{-(3x^2 + 3y)}{3x + 5y^4}.$$

▶

▶ EXERCISE SET 14.4

In Exercises 1–12, find df/dt.

1. $f(x, y) = x^2y^3 + xy^2$, $x(t) = \sin t + t$, $y(t) = t^2 + 1$
2. $f(x, y) = \sqrt{2x - y^2}$, $x(t) = \ln t$, $y(t) = e^t$
3. $f(x, y) = 3x^3 - \sqrt{x + y}$, $x(t) = \ln t$, $y(t) = t^2$
4. $f(x, y) = e^x \sin y + x^2 \cos y$, $x(t) = t^3 + 1$, $y(t) = \sqrt{t}$
5. $f(x, y) = \ln xy + e^{x^2+y^2}$, $x(t) = \sqrt{t+1}$, $y(t) = e^t$
6. $f(x, y) = e^{x^2} \ln y + e^y \ln x^2$, $x(t) = \sqrt{t}$, $y(t) = e^t$
7. $f(x, y, z) = x^2y + yz^2$, $x(t) = \ln t$, $y(t) = e^t$, $z(t) = t^2$
8. $f(x, y, z) = x^2y^4z^5$, $x(t) = e^t + t^2$, $y(t) = \ln(t^2 + 1)$, $z(t) = t^3 + t$
9. $f(x, y, z) = e^z \cos xy$, $x(t) = \sin t$, $y(t) = t^2 + 1$, $z(t) = e^{t^2}$
10. $f(x, y, z) = \ln(x^2 - 3y^4 + z^3)$, $x(t) = \sqrt{t}$, $y(t) = t^{-2}$, $z(t) = \cos t$
11. $f(x, y, z) = \sin xyz^2$, $x(t) = e^t$, $y(t) = \ln t$, $z(t) = t + \ln t$
12. $f(x, y, z) = \sqrt{x^2 + y^2 + z^2}$, $x(t) = \sin t$, $y(t) = \cos 2t$, $z(t) = \sin 2t$

In Exercises 13–22, find (a) $\partial f/\partial u$ and (b) $\partial f/\partial v$.

13. $f(x, y) = x^3y^5$, $x(u, v) = u + v$, $y(u, v) = u^2 - v^2$
14. $f(x, y) = 2xy + x^3$, $x(u, v) = v \ln u$, $y(u, v) = e^{u+v}$
15. $f(x, y) = e^x \cos y$, $x(u, v) = ue^v$, $y(u, v) = u + v - 1$
16. $f(x, y) = e^{xy}(\cos xy + \sin xy)$, $x(u, v) = e^{u+v}$, $y(u, v) = uv$
17. $f(x, y) = \tan(x + y)$, $x(u, v) = e^u \cos v$, $y(u, v) = e^u \sin v$
18. $f(x, y) = e^{x^2+y^2}$, $x(u, v) = ve^u$, $y(u, v) = v \ln u$
19. $f(x, y, z) = x + y^2 + z^3$, $x(u, v) = \sin uv$, $y(u, v) = u \cos v$, $z(u, v) = v \cos u$
20. $f(x, y, z) = x^2yz$, $x(u, v) = ue^v$, $y(u, v) = u^2v$, $z(u, v) = uv^2$
21. $f(x, y, z) = \ln(x + y + z)$, $x(u, v) = e^{u^2+v^2}$, $y(u, v) = \ln(u^2 + v^2)$, $z(u, v) = u - v$
22. $f(x, y, z) = 1/xyz$, $x(u, v) = u^2v$, $y(u, v) = u + v$, $z(u, v) = \sqrt{u^2 + v^2}$

23–32. Find df/dt for the functions given in Exercises 13–22, if $u = 2t$ and $v = 3t$.

In Exercises 33–38, use Theorem (14.31) to find dy/dx.

33. $x^2y^2 - xy + x^3 + y^3 + 6 = 0$
34. $x^5 + 2x^4y + x^2y^3 + y^5 = 0$
35. $x^2y + xy^2 = 6(x^2 + y^2)$
36. $y^2 = \dfrac{x}{xy + 1}$
37. $\sin xy = \cos x$
38. $\sin x + \cos y = 1$
39. Show that if $z = f(x, y)$, $x = r \cos \theta$, and $y = r \sin \theta$, then

    i) $\dfrac{\partial z}{\partial r} = \dfrac{\partial f}{\partial x} \cos \theta + \dfrac{\partial f}{\partial y} \sin \theta$, and

    ii) $\dfrac{\partial z}{\partial \theta} = r\left[-\dfrac{\partial f}{\partial x} \sin \theta + \dfrac{\partial f}{\partial y} \cos \theta\right]$.

40. Let $w = f(x, y, z)$ and change (x, y, z) into cylindrical coordinates. Find (a) $\partial w/\partial r$, (b) $\partial w/\partial \theta$, and (c) $\partial w/\partial z$.

41. Let $w = f(x, y, z)$ and change (x, y, z) into spherical coordinates. Find (a) $\partial w/\partial \rho$, (b) $\partial w/\partial \theta$, and (c) $\partial w/\partial \phi$.

42. An equation of the form

$$\frac{\partial^2 f}{\partial t^2} = \alpha^2 \frac{\partial^2 f}{\partial x^2}$$

is called a one-dimensional *wave equation*. The vibrations of an elastic string are described by such an equation. Show that if $f(x, t) = g(x - \alpha t) + h(x + \alpha t)$ and the second partial derivatives of g and h exist, then f satisfies the one-dimensional wave equation.

43. Show that if $f(x, t) = \sin x \, (A \cos \alpha t + B \sin \alpha t)$, where A and B are arbitrary constants, then f satisfies the one-dimensional wave equation.

44. Show that if $z = f(x - y)$, then

$$\frac{\partial z}{\partial x} + \frac{\partial z}{\partial y} = 0.$$

45. Suppose that $f(x, y) = 0$ implicitly defines y as a function of x and that y'' and the second partial derivatives of f exist. Express y'' in terms of the first and second partial derivatives of f.

46. Suppose that x and y are differentiable functions of u and v and that f is a differentiable function of x and y. Show that if $z = f(x, y)$, then

$$\frac{\partial^2 z}{\partial u^2} = \frac{\partial^2 z}{\partial x^2} \left(\frac{\partial x}{\partial u}\right)^2 + 2 \frac{\partial^2 z}{\partial x \partial y} \frac{\partial x}{\partial u} \frac{\partial y}{\partial u} + \frac{\partial^2 z}{\partial y^2} \left(\frac{\partial y}{\partial u}\right)^2$$
$$+ \frac{\partial z}{\partial x} \frac{\partial^2 x}{\partial u^2} + \frac{\partial z}{\partial y} \frac{\partial^2 y}{\partial u^2}.$$

47. As stated in Exercise 34 of Section 14.3, when capacitors with capacitance C_1 and C_2 are connected in series, the capacitance of the resulting circuit is $(1/C_1 + 1/C_2)^{-1}$ (see the figure). Determine the rate of change with respect to time t of the combined capacitance when $C_1 = 1 \, \mu F$, $C_2 = 3 \, \mu F$, $dC_1/dt = 0.2 \, \mu F/$millisec, and $dC_2/dt = -0.2 \, \mu F/$millisec.

48. As stated in Exercise 35 of Section 14.3, when resistors with resistance R_1 and R_2 are connected in parallel, the resistance of the resulting circuit is $(1/R_1 + 1/R_2)^{-1}$. Determine the rate of change with respect to time t of the combined resistance when $R_1 = 3 \, \Omega$, $R_2 = 8 \, \Omega$, $dR_1/dt = 0.2 \, \Omega/$millisec, and $dR_2/dt = -0.5 \, \Omega/$millisec.

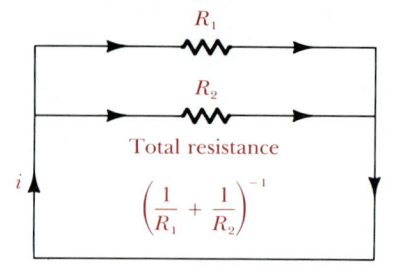

49. The length, width, and depth of a rectangular box are each increasing at the rate of 1 in./min (see the figure). Find the rate at which the volume is increasing when the length is 18 in., the width 12 in., and the depth 3 in.

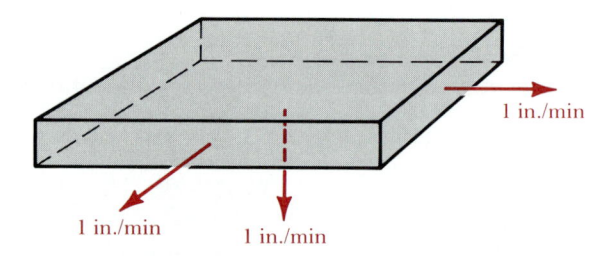

50. Oil spilling from a tanker into a large lake forms an inverted conical shape. The oil is spilling at the rate of 12π ft^3/min and the radius of the spill is increasing at 1 ft/min. At what rate is the center depth of the spill increasing when the radius is 6 ft and the volume of oil spilled is 24π ft^3?

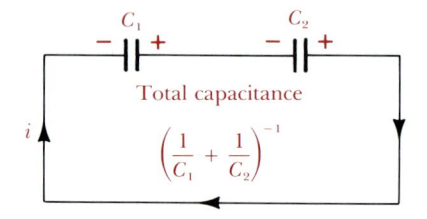

Figure for Exercise 47.

14.5

Directional Derivatives and Gradients

Partial derivatives give the rate of change of the values of a function relative to a variable change along a line parallel to one of the coordinate axes. The directional derivative discussed in this section gives the rate of change of the function values along lines not necessarily parallel to a coordinate axis.

(14.32)
DEFINITION
Suppose $\mathbf{u} = u_1\mathbf{i} + u_2\mathbf{j}$ is a unit vector. The **directional derivative** in the direction of $\mathbf{u}$ is a function $D_\mathbf{u} f$ defined by

$$D_\mathbf{u} f(x, y) = \lim_{h \to 0} \frac{f(x + hu_1, y + hu_2) - f(x, y)}{h}.$$

The domain of $D_\mathbf{u} f$ consists of all (x, y) in the domain of f for which this limit exists.

Consider the curve C described by the intersection of the surface $z = f(x, y)$ and the plane perpendicular to the xy-plane which contains the line through $(x, y, 0)$ with direction $\mathbf{u}$ (see Figure 14.22a). The directional derivative $D_\mathbf{u} f(x, y)$ describes the rate of change of the function along this curve. This implies that the direction of the tangent line to C at $(x, y, f(x, y))$ (shown in Figure 14.22b) is given by the vector $\mathbf{u} + D_\mathbf{u} f(x, y)\mathbf{k}$.

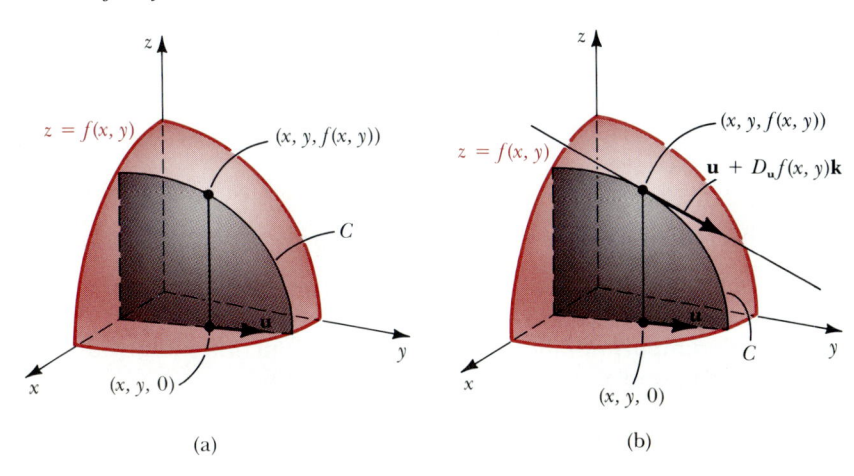

Figure 14.22
The directional derivative in the direction of $\mathbf{u}$.

Example 1

Determine $D_\mathbf{u} f(1, 2)$ if $f(x, y) = 6 - x^2 - y^2$ and $\mathbf{u} = \frac{\sqrt{2}}{2}\mathbf{i} + \frac{\sqrt{2}}{2}\mathbf{j}$.

Solution
We have

$$D_\mathbf{u} f(1, 2) = \lim_{h \to 0} \frac{f(1 + (\sqrt{2}/2)h, 2 + (\sqrt{2}/2)h) - f(1, 2)}{h}$$

$$= \lim_{h \to 0} \frac{\{6 - [1 + (\sqrt{2}/2)h]^2 - [2 + (\sqrt{2}/2)h]^2\} - (6 - 1^2 - 2^2)}{h}$$

$$= \lim_{h \to 0} \frac{-\sqrt{2}h - \tfrac{1}{2}h^2 - 2\sqrt{2}h - \tfrac{1}{2}h^2}{h} = \lim_{h \to 0} (-3\sqrt{2} - h) = -3\sqrt{2}.$$

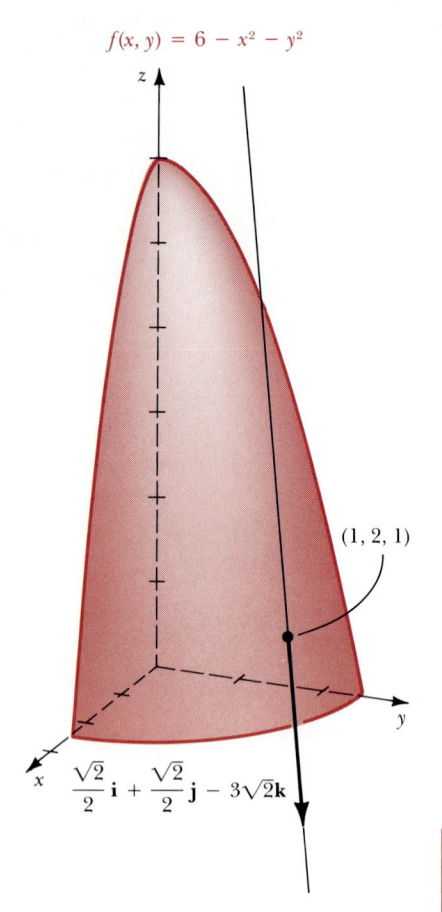

Figure 14.23 shows the tangent line to the surface $z = 6 - x^2 - y^2$ at $(1, 2, 1)$ in the direction of

$$\frac{\sqrt{2}}{2}\mathbf{i} + \frac{\sqrt{2}}{2}\mathbf{j} - 3\sqrt{2}\mathbf{k}.$$

This line has parametric equations

$$x = 1 + \frac{\sqrt{2}}{2}t, \quad y = 2 + \frac{\sqrt{2}}{2}t, \quad z = 1 - 3\sqrt{2}t.$$

When $\mathbf{u} = \mathbf{i}$, the directional derivative in the direction of $\mathbf{i}$ is the partial derivative of f with respect to x:

$$D_\mathbf{i} f(x, y) = \lim_{h \to 0} \frac{f(x + h, y) - f(x, y)}{h} = f_x(x, y).$$

Similarly, when $\mathbf{u} = \mathbf{j}$,

$$D_\mathbf{j} f(x, y) = \lim_{h \to 0} \frac{f(x, y + h) - f(x, y)}{h} = f_y(x, y).$$

Since it is generally cumbersome to use the limit to determine derivatives, it is natural to ask whether there is an alternative way to find directional derivatives. The following theorem shows that if a function is differentiable, the partial derivatives can be used to determine its directional derivatives.

(14.33)
THEOREM

If f is differentiable at (x, y) and $\mathbf{u} = u_1\mathbf{i} + u_2\mathbf{j}$ is a unit vector, then

$$D_\mathbf{u} f(x, y) = f_x(x, y)u_1 + f_y(x, y)u_2 = \langle f_x(x, y), f_y(x, y) \rangle \cdot \mathbf{u}.$$

Proof. To simplify notation, a single-variable function g is introduced. For fixed values of x, y, u_1, and u_2, g is defined by

$$g(h) = f(x + hu_1, y + hu_2).$$

Then

$$D_\mathbf{u} f(x, y) = \lim_{h \to 0} \frac{f(x + hu_1, y + hu_2) - f(x, y)}{h} = \lim_{h \to 0} \frac{g(h) - g(0)}{h} = g'(0).$$

With $u = x + hu_1$ and $v = y + hu_2$, the Chain Rule implies that

$$g'(h) = \frac{dg}{dh} = \frac{\partial f}{\partial u}\frac{du}{dh} + \frac{\partial f}{\partial v}\frac{dv}{dh} = \frac{\partial f}{\partial u}u_1 + \frac{\partial f}{\partial v}u_2.$$

When $h = 0$, $u = x$ and $v = y$, so

$$D_\mathbf{u} f(x, y) = g'(0) = \frac{\partial f}{\partial x}u_1 + \frac{\partial f}{\partial y}u_2 = f_x(x, y)u_1 + f_y(x, y)u_2. \quad \triangleright$$

It follows from Theorem (14.33) that if f is differentiable at (x, y), then f has a directional derivative at (x, y) in every direction. A function is given in Exercise 47 to show that functions can have directional derivatives in every direction, however, even when they are *not* differentiable.

For each nonzero vector **v**, there is precisely one unit vector with the same direction as **v**, that is, the vector

$$\mathbf{u} = \frac{\mathbf{v}}{\|\mathbf{v}\|}.$$

The directional derivative of a differentiable function f in the direction of **v** is $D_{\mathbf{u}} f$, the directional derivative given by the unit vector **u**.

Example 2

Let $f(x, y) = 2x^2 - y^2 - 1$ and $\mathbf{v} = 3\mathbf{i} + 4\mathbf{j}$. Find the directional derivative of f at $(2, 1)$ in the direction of **v**.

Solution
Since **v** is not a unit vector, we first find a unit vector in the direction of **v**:

$$\mathbf{u} = \frac{\mathbf{v}}{\|\mathbf{v}\|} = \frac{3\mathbf{i} + 4\mathbf{j}}{\sqrt{3^2 + 4^2}} = \frac{3}{5}\mathbf{i} + \frac{4}{5}\mathbf{j}.$$

Since $f_x(x, y) = 4x$ and $f_y(x, y) = -2y$,

$$D_{\mathbf{u}} f(x, y) = 4x\left(\frac{3}{5}\right) - 2y\left(\frac{4}{5}\right) = \frac{4}{5}(3x - 2y)$$

and

$$D_{\mathbf{u}} f(2, 1) = \frac{4}{5}(3 \cdot 2 - 2 \cdot 1) = \frac{16}{5}.$$ ▶

Suppose that **v** is a nonzero vector in the xy-plane and that θ represents the angle formed by **v** and the unit vector **i** (see Figure 14.24). The unit vector in the direction of **v** is then

$$\mathbf{u} = \frac{\mathbf{v}}{\|\mathbf{v}\|} = \cos\theta\,\mathbf{i} + \sin\theta\,\mathbf{j},$$

so by Theorem (14.33), the directional derivative of a differentiable

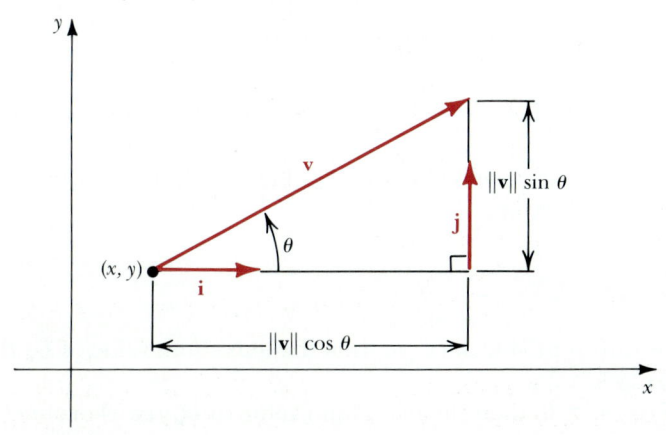

Figure 14.24

$\mathbf{u} = \dfrac{\mathbf{v}}{\|\mathbf{v}\|}(\cos\theta\,\mathbf{i} + \sin\theta\,\mathbf{j})$.

function f in the direction of $\mathbf{v}$ is

(14.34) $\qquad D_{\mathbf{u}} f(x, y) = f_x(x, y) \cos \theta + f_y(x, y) \sin \theta.$

This is an important theoretical representation of the directional derivative, but one that is difficult to use computationally unless θ is readily available.

The directional derivative $D_{\mathbf{u}} f(x, y)$ describes the rate of change in the function values at (x, y) in the direction of the unit vector $\mathbf{u}$. Not surprisingly, we are interested in determining which direction produces the maximum and minimum rates of change. Some additional terminology is needed to study this problem.

(14.35)
DEFINITION

If f is a function with partial derivatives at (x, y), then the **gradient** of f at (x, y), denoted grad $f(x, y)$ or $\nabla f(x, y)$, is defined by

$$\operatorname{grad} f(x, y) \equiv \nabla f(x, y) = f_x(x, y)\mathbf{i} + f_y(x, y)\mathbf{j}.$$

Example 3

Find the gradient of f if $f(x, y) = e^x \sin y$.

Solution

$$\nabla f(x, y) = f_x(x, y)\mathbf{i} + f_y(x, y)\mathbf{j} = e^x \sin y\, \mathbf{i} + e^x \cos y\, \mathbf{j} \qquad \blacktriangleright$$

The gradient is a function that transforms a scalar multivariable function into a vector multivariable function. One of the immediate applications of the gradient is that, for a differentiable function f and a unit vector $\mathbf{u} = u_1 \mathbf{i} + u_2 \mathbf{j}$,

(14.36) $\qquad D_{\mathbf{u}} f(x, y) = f_x(x, y) u_1 + f_y(x, y) u_2 = \nabla f(x, y) \cdot \mathbf{u}.$

Recall from Section 11.3 that the component of a vector $\mathbf{b}$ in the direction of the nonzero vector $\mathbf{a}$ is

$$\operatorname{comp}_{\mathbf{a}} \mathbf{b} = \frac{\mathbf{a} \cdot \mathbf{b}}{\|\mathbf{a}\|}.$$

Since $\mathbf{u}$ is a unit vector, (14.36) implies that

$$D_{\mathbf{u}} f(x, y) = \frac{\nabla f(x, y) \cdot \mathbf{u}}{\|\mathbf{u}\|} = \operatorname{comp}_{\mathbf{u}} \nabla f(x, y).$$

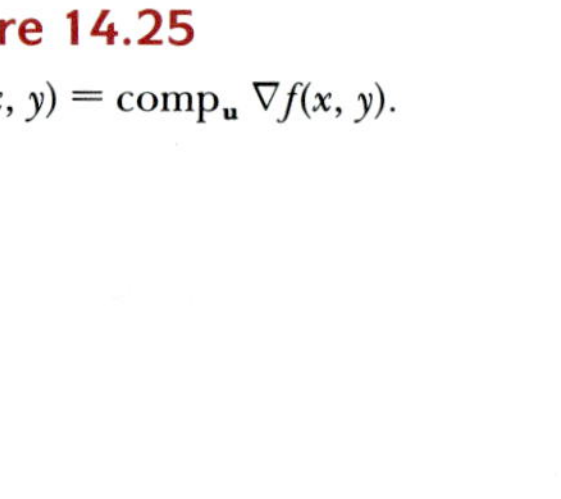

This important relation is illustrated in Figure 14.25 and is the basis for the following theorem.

Figure 14.25
$D_{\mathbf{u}} f(x, y) = \operatorname{comp}_{\mathbf{u}} \nabla f(x, y).$

(14.37)
THEOREM

i) If f is differentiable at (x, y), then the maximum value of $D_{\mathbf{u}} f(x, y)$ is $\|\nabla f(x, y)\|$.

ii) If $\nabla f(x, y) \neq 0$, then the maximum value of $D_{\mathbf{u}} f(x, y)$ occurs when $\mathbf{u}$ has the same direction as $\nabla f(x, y)$.

Proof

i) For any unit vector $\mathbf{u} = u_1\mathbf{i} + u_2\mathbf{j}$,

$$D_{\mathbf{u}} f(x, y) = f_x(x, y)u_1 + f_y(x, y)u_2 = \nabla f(x, y) \cdot \mathbf{u}.$$

So

$$D_{\mathbf{u}} f(x, y) = \|\nabla f(x, y)\| \|\mathbf{u}\| \cos \theta = \|\nabla f(x, y)\| \cos \theta,$$

where θ is the angle between $\mathbf{u}$ and $\nabla f(x, y)$. Since the maximum value of the cosine is 1, the maximum value of $D_{\mathbf{u}} f(x, y)$ is $\|\nabla f(x, y)\|$.

ii) Suppose $\nabla f(x, y) \neq \mathbf{0}$ and $\mathbf{u} = u_1\mathbf{i} + u_2\mathbf{j}$ is a unit vector. It follows from part (i) that the maximum value of $D_{\mathbf{u}} f(x, y)$ occurs when $\cos \theta = 1$. This implies that $\mathbf{u}$ and $\nabla f(x, y)$ have the same direction, so

$$\mathbf{u} = \frac{\nabla f(x, y)}{\|\nabla f(x, y)\|}.$$

This direction gives the maximum value, $\|\nabla f(x, y)\|$, for the directional derivative. ▷

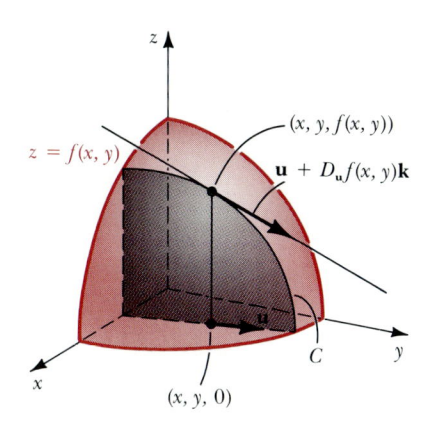

Figure 14.26
The directional derivative in the direction of $\mathbf{u}$.

At the beginning of the section we found that the tangent line to the graph of the differentiable function f at the point $(x, y, f(x, y))$ in the direction of the unit vector $\mathbf{u}$ has the direction $\mathbf{u} + D_{\mathbf{u}} f(x, y)\mathbf{k}$ (see Figure 14.26). Since $\nabla f(x, y)$ describes the direction and magnitude of the maximum rate of change in the values of f at (x, y), it also describes the direction and magnitude of the maximum slope of any tangent line to the graph of $z = f(x, y)$ at $(x, y, f(x, y))$. It is this application that gives the gradient its name: The maximum slope of a surface at a point is called the **grade** of the surface at the point.

The direction given by the maximum rate of change in the values of the function f at $(x, y, f(x, y))$ is called the direction of **steepest ascent** (see Figure 14.27a). By Theorem (14.37), the direction of steepest

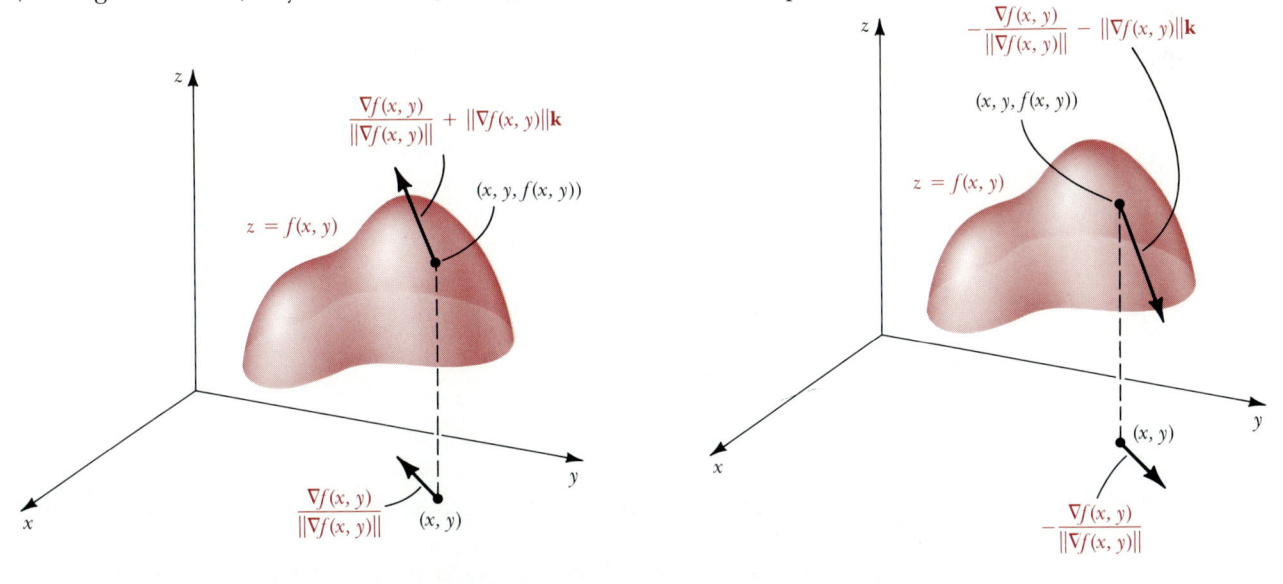

Figure 14.27
(a) The direction of steepest ascent;
(b) the direction of steepest descent.

(a)

(b)

ascent for the differentiable function f at the point $(x, y, f(x, y))$ with a nonzero gradient is

(14.38) $$\frac{\nabla f(x, y)}{\|\nabla f(x, y)\|} + \|\nabla f(x, y)\|\mathbf{k}.$$

The direction of **steepest descent** (see Figure 14.27b) is the negative of this vector:

(14.39) $$-\frac{\nabla f(x, y)}{\|\nabla f(x, y)\|} - \|\nabla f(x, y)\|\mathbf{k}.$$

Example 4

Consider the surface described by $x^2 + y^2 + z = 6$. Determine (a) the grade of the surface at $(1, 2, 1)$ and (b) the direction of steepest ascent at this point (see Figure 14.28).

Solution

Let $z = f(x, y) = 6 - x^2 - y^2$. Then
$$\nabla f(x, y) = -2x\mathbf{i} - 2y\mathbf{j}, \quad \text{so} \quad \nabla f(1, 2) = -2\mathbf{i} - 4\mathbf{j}.$$

a) The grade of the surface at $(1, 2, 1)$ is
$$\|\nabla f(1, 2)\| = \sqrt{(-2)^2 + (-4)^2} = 2\sqrt{5}.$$

b) The direction of steepest ascent is given by
$$\frac{\nabla f(x, y)}{\|\nabla f(x, y)\|} + \|\nabla f(x, y)\|\mathbf{k} = \frac{-2\mathbf{i} - 4\mathbf{j}}{2\sqrt{5}} + 2\sqrt{5}\mathbf{k}$$
$$= -\frac{\sqrt{5}}{5}\mathbf{i} - \frac{2\sqrt{5}}{5}\mathbf{j} + 2\sqrt{5}\mathbf{k}. \quad \blacktriangleright$$

Directional derivatives and gradients are defined for functions of three variables in a manner similar to the definitions for two-variable functions:

If $\mathbf{u} = u_1\mathbf{i} + u_2\mathbf{j} + u_3\mathbf{k}$ is a unit vector, the **directional derivative** of f at (x, y, z) in the direction of $\mathbf{u}$ is
$$D_\mathbf{u} f(x, y, z) = \lim_{h \to 0} \frac{f(x + hu_1, y + hu_2, z + hu_3) - f(x, y, z)}{h},$$

provided this limit exists. If f is differentiable at (x, y, z), then
$$D_\mathbf{u} f(x, y, z) = f_x(x, y, z)u_1 + f_y(x, y, z)u_2 + f_z(x, y, z)u_3.$$

The **gradient** of f is defined by

(14.40)
$$\operatorname{grad} f(x, y, z) \equiv \nabla f(x, y, z) = f_x(x, y, z)\mathbf{i} + f_y(x, y, z)\mathbf{j} + f_z(x, y, z)\mathbf{k},$$

so

(14.41) $$D_\mathbf{u} f(x, y, z) = \nabla f(x, y, z) \cdot \mathbf{u}.$$

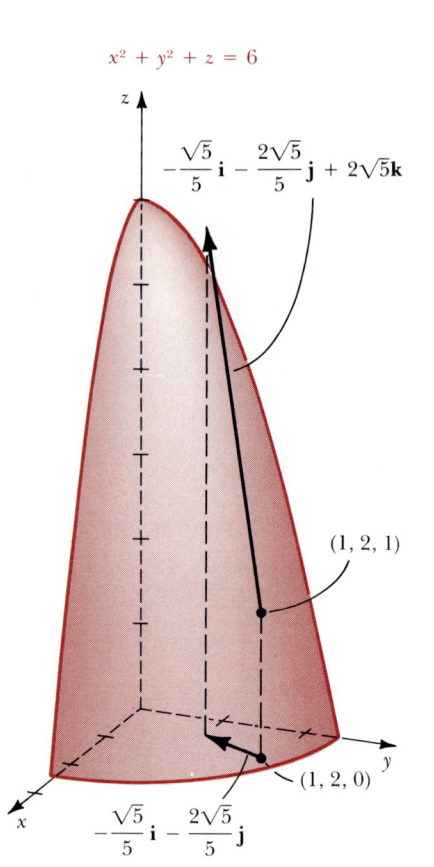

Figure 14.28

This leads to a theorem for functions of three variables that is equivalent to the two-variable counterpart, Theorem (14.37).

(14.42) THEOREM

i) If f is differentiable at (x, y, z), then the maximum value of $D_{\mathbf{u}} f(x, y, z)$ is $\|\nabla f(x, y, z)\|$.

ii) If $\nabla f(x, y, z) \neq \mathbf{0}$, the maximum value of $D_{\mathbf{u}} f(x, y, z)$ occurs when $\mathbf{u}$ has the same direction as $\nabla f(x, y, z)$.

Example 5

The temperature distribution in a solid metal ball of radius 3 cm centered at $(0, 0, 0)$ is given by $T(x, y, z) = 300 e^{-(x^2+y^2+z^2)/9}$ degrees Celsius (see Figure 14.29). Determine the direction and maximum rate of decrease in temperature at $(1, 1, 0)$.

Solution
We have
$$\nabla T(x, y, z) = 300 e^{-(x^2+y^2+z^2)/9} (-\tfrac{1}{9})(2x\mathbf{i} + 2y\mathbf{j} + 2z\mathbf{k})$$
$$= -\tfrac{200}{3} e^{-(x^2+y^2+z^2)/9} (x\mathbf{i} + y\mathbf{j} + z\mathbf{k}).$$

The direction of the maximum rate of decrease is given by $-\nabla T$ at $(1, 1, 0)$ and
$$-\nabla T(1, 1, 0) = \tfrac{200}{3} e^{-2/9} (\mathbf{i} + \mathbf{j}).$$

The rate of decrease in temperature in this direction is
$$\|\nabla T(1, 1, 0)\| = \frac{200\sqrt{2}}{3} e^{-2/9} \approx 75.5 \frac{\text{degrees Celsius}}{\text{cm}}.$$

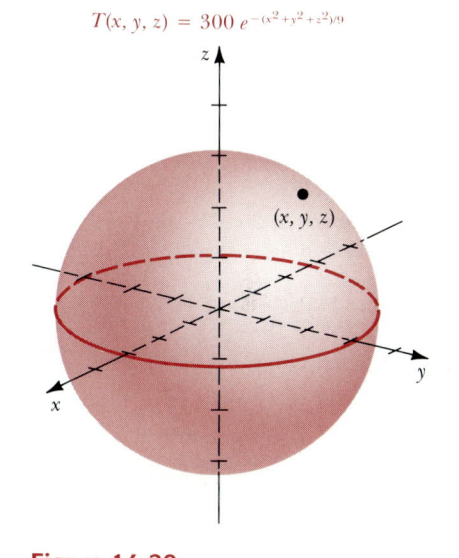

▶ **Figure 14.29**

▶ EXERCISE SET 14.5

In Exercises 1–10, find the directional derivative of f at P in the direction of $\mathbf{v}$.

1. $f(x, y) = 6 - 2x - 3y$, $\mathbf{v} = \dfrac{\sqrt{2}}{2}(\mathbf{i} + \mathbf{j})$, $P(0, 0)$
2. $f(x, y) = x^2 - xy + y^2$, $\mathbf{v} = \dfrac{\sqrt{2}}{2}(\mathbf{i} - \mathbf{j})$, $P(1, 2)$
3. $f(x, y) = x^2 + y^2$, $\mathbf{v} = \mathbf{i} + \mathbf{j}$, $P(1, 1)$
4. $f(x, y) = 4 - x^2 - y^2$, $\mathbf{v} = 2\mathbf{i} + \mathbf{j}$, $P(-1, -1)$
5. $f(x, y) = \sqrt{36 - 9x^2 - 4y^2}$, $\mathbf{v} = -\mathbf{i} + \mathbf{j}$, $P(1, -2)$
6. $f(x, y) = xe^y$, $\mathbf{v} = 3\mathbf{i} + \mathbf{j}$, $P(2, 1)$
7. $f(x, y, z) = 2x + y + 3z$, $\mathbf{v} = -\mathbf{i} + \mathbf{j} + 2\mathbf{k}$, $P(1, 2, 3)$
8. $f(x, y, z) = ze^{x^2+y^2}$, $\mathbf{v} = \mathbf{i} + \mathbf{j} + \mathbf{k}$, $P(1, 1, -3)$
9. $f(x, y, z) = \cos(x + y + z)$, $\mathbf{v} = \mathbf{i} - 2\mathbf{j} + 2\mathbf{k}$, $P(0, \pi/4, \pi/2)$
10. $f(x, y, z) = \dfrac{x+y}{x+z}$, $\mathbf{v} = 2\mathbf{i} + \mathbf{j} - 3\mathbf{k}$, $P(1, 3, 1)$

In Exercises 11–18, find the gradient of the function.

11. $f(x, y) = x^2 - y^2$
12. $f(x, y) = \sin x + \cos y$
13. $f(x, y) = ye^x$
14. $f(x, y) = ye^x \sin xy$

15. $f(x, y) = \dfrac{xy}{x^2 + y^2}$

16. $f(x, y) = \dfrac{x}{y}$

17. $f(x, y, z) = \sin x^2 yz$

18. $f(x, y, z) = \ln(x^2 + 2y - z^3)$

In Exercises 19–24, find the gradient of the function at the given point.

19. $f(x, y) = x^2 + y^2$, $(1, 2)$
20. $f(x, y) = \dfrac{2x - y}{xy - 1}$, $(1, 2)$
21. $f(x, y) = e^x \cos y$, $(1, \pi)$
22. $f(x, y) = e^{xy} \cos xy$, $(2, \pi/4)$
23. $f(x, y, z) = \cos xyz$, $(1, 2, \pi/4)$
24. $f(x, y, z) = (2x/y) \ln z$, $(2, 2, 2)$

In Exercises 25–32, find (a) the direction in which f is increasing most rapidly at P and (b) this maximum rate of increase.

25. $f(x, y) = x^2 + y^2 + 1$, $P(1, 1)$
26. $f(x, y) = y - x^2$, $P(0, 0)$
27. $f(x, y) = 9x^2 + 4y^2$, $P(2, 0)$
28. $f(x, y) = \sin y$, $P(0, \pi)$
29. $f(x, y) = e^{x^2 + y^2}$, $P(\sqrt{2}/2, \sqrt{2}/2)$
30. $f(x, y) = e^x \cos xy$, $P(1, \pi)$
31. $f(x, y, z) = e^x (\cos y - \sin z)$, $P(1, \pi, \pi/2)$
32. $f(x, y, z) = \ln(x^2 + y^2 + z)$, $P(1, -1, 1)$

In Exercises 33–36, find the directional derivative of f at P in the direction of a vector $\mathbf{v}$ that makes an angle θ with the positive x-axis.

33. $f(x, y) = 9 - 4x^2 - y^2$, $\theta = \pi/6$, $P(-1, -2)$
34. $f(x, y) = e^x \sin y$, $\theta = \pi/4$, $P(1, \pi/2)$
35. $f(x, y) = x^2 \ln xy$, $\theta = -\pi/3$, $P(e, 1)$
36. $f(x, y) = 2 - x - y$, $\theta = \pi/3$, $P(0, 1)$

In Exercises 37–42, find the directional derivative of the function f at the point P in the direction from P to Q.

37. $f(x, y) = x \cos y$, $P(2, \pi/2)$, $Q(3, \pi/4)$
38. $f(x, y) = ye^{x^2 + y^2}$, $P(-1, 1)$, $Q(1, 3)$
39. $f(x, y) = x^2/y$, $P(2, 1)$, $Q(-2, -1)$
40. $f(x, y) = 4x^2 + 9y^2$, $P(-1, -1)$, $Q(-2, -4)$
41. $f(x, y, z) = x^2 + y^2 + z^2$, $P(0, 1, 1)$, $Q(2, -1, 3)$

42. $f(x, y, z) = \sqrt{x} e^{yz}$, $P(1, 2, 3)$, $Q(3, 6, -1)$

In Exercises 43–46, (a) sketch the graph of the given surface, (b) determine the grade of the surface at the given point, and (c) determine the direction of steepest ascent at the point.

43. $z = 16 - x^2 - y^2$, $(3, 2, 3)$
44. $z = \cos y$, $(2, \pi/4, \sqrt{2}/2)$
45. $x^2 - y^2 + z = 1$, $(1, 1, 1)$
46. $z = \sqrt{16 - x^2 - y^2}$, $(\sqrt{3}, 2, 3)$

47. Consider
$$f(x, y) = \begin{cases} \dfrac{x^2 y}{x^4 + y^2}, & \text{if } (x, y) \neq (0, 0), \\ 0, & \text{if } (x, y) = (0, 0). \end{cases}$$

Show that the directional derivative of f at $(0, 0)$ exists in the direction of any unit vector $\mathbf{u}$, even though it was shown in Exercise 37 of Section 14.3 that f is not differentiable at $(0, 0)$.

48. Suppose the height of a mountain is given by $z = 1000 - 4x^2 - y^2$ feet, where (x, y) is a point in the plane at the base of the mountain. If a climber is at $(4, 4, 920)$, in what direction should the climber turn in order to:
    a) ascend the mountain most rapidly?
    b) descend the mountain most rapidly?

49. The temperature at a point (x, y) on a rectangular metal plate in the xy-plane is $T(x, y) = 500/(x^2 + y^2)$ degrees Celsius.
    a) Find the rate of change of T at $(3, 4)$ in the direction of the vector $\mathbf{i}$.
    b) Find the rate of change of T at $(3, 4)$ in the direction of the vector $\mathbf{i} + \mathbf{j}$.
    c) Find the direction in which T increases most rapidly at $(3, 4)$.
    d) Find the direction in which T decreases most rapidly at $(3, 4)$.

50. The temperature within a solid is given by $T(x, y, z) = x^2 + 3xy + z^2$. An object moves in a straight line from $(1, 2, 3)$ to $(-1, 1, 4)$.
    a) What rate of change of temperature is initially encountered by the object?
    b) What rate of change is finally encountered by the object?

51. Modify the definition of the directional derivative of a function to include vectors that are not unit vectors. State a theorem concerning these more general derivatives that reduces to Theorem (14.33) when the vector is a unit vector.

14.6
Tangent Planes and Normals

In Section 14.5 we found that if f is a differentiable function of two variables, then the gradient, $\nabla f(x, y)$, gives the direction of the maximum rate of change of the values of f at (x, y). Thus the gradient can be used to determine the direction of steepest ascent on the surface described by the graph of f from the point $(x, y, f(x, y))$. In this section the gradient of a function of *three variables* is used to describe another important geometric property of a surface.

(14.43)
THEOREM
Suppose F is a differentiable function of three variables and S is a level surface of F described by

$$F(x, y, z) = c, \quad \text{where } c \text{ is a constant.}$$

If $P_0(x_0, y_0, z_0)$ lies on S and $\nabla F(x_0, y_0, z_0) \neq \mathbf{0}$, then $\nabla F(x_0, y_0, z_0)$ is normal to every curve on S passing through P_0.

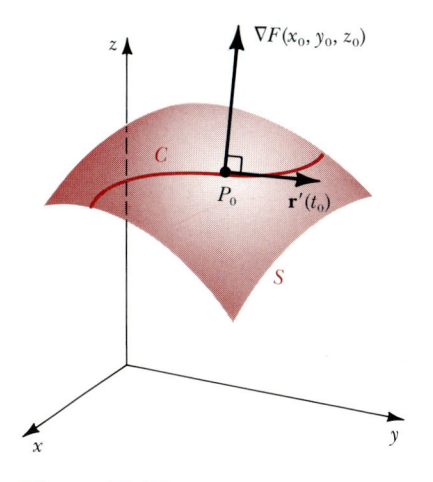

Figure 14.30
$\nabla F(x_0, y_0, z_0)$ is normal to the level surface of F at (x_0, y_0, z_0).

Proof. Suppose C is an arbitrary smooth curve lying entirely on S and passing through P_0, as shown in Figure 14.30. Let

$$\mathbf{r}(t) = x(t)\mathbf{i} + y(t)\mathbf{j} + z(t)\mathbf{k}, \quad a \leq t \leq b$$

be a vector-valued function describing C, and let $\mathbf{r}(t_0) = x_0\mathbf{i} + y_0\mathbf{j} + z_0\mathbf{k}$. The vector $\mathbf{r}'(t_0)$ is tangent to C at P_0 and

$$\begin{aligned}\mathbf{r}'(t_0) \cdot \nabla F(x_0, y_0, z_0) &= [x'(t_0)\mathbf{i} + y'(t_0)\mathbf{j} + z'(t_0)\mathbf{k}] \cdot [F_x(x_0, y_0, z_0)\mathbf{i} \\ &\quad + F_y(x_0, y_0, z_0)\mathbf{j} + F_z(x_0, y_0, z_0)\mathbf{k}] \\ &= F_x(x_0, y_0, z_0)x'(t_0) + F_y(x_0, y_0, z_0)y'(t_0) \\ &\quad + F_z(x_0, y_0, z_0)z'(t_0).\end{aligned}$$

By the Chain Rule,

$$\frac{dF}{dt}(x_0, y_0, z_0) = F_x(x_0, y_0, z_0)\,x'(t_0) + F_y(x_0, y_0, z_0)\,y'(t_0) + F_z(x_0, y_0, z_0)\,z'(t_0),$$

so

$$r'(t_0) \cdot \nabla F(x_0, y_0, z_0) = \frac{dF}{dt}(x_0, y_0, z_0).$$

Since S is a level surface of F, F is constant on S. Hence,

$$\frac{dF}{dt}(x_0, y_0, z_0) = 0 \quad \text{and} \quad \mathbf{r}'(t_0) \cdot \nabla F(x_0, y_0, z_0) = 0.$$

Thus $\nabla F(x_0, y_0, z_0)$ is orthogonal to $\mathbf{r}'(t_0)$. Since C is an arbitrary curve on S passing through $P_0(x_0, y_0, z_0)$, $\nabla F(x_0, y_0, z_0)$ is orthogonal to the tangent vector to any curve in S. So, $\nabla F(x_0, y_0, z_0)$ is normal to every curve in S passing through P_0. ▷

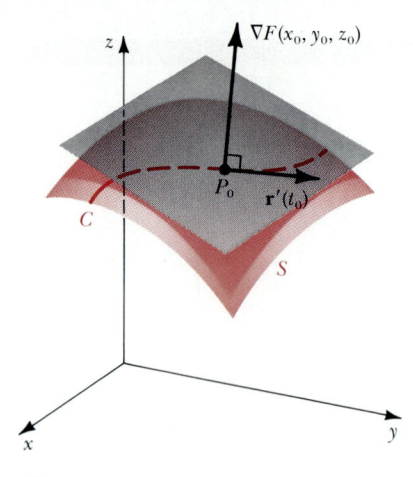

Figure 14.31
The tangent plane to the level surface of F at (x_0, y_0, z_0).

Under the conditions of Theorem (14.43), all tangent vectors to the surface S at P_0 lie in the same plane (see Figure 14.31). This is called the **tangent plane** to S at P_0. A normal to this tangent plane is said to be **normal** to the surface at P_0. Since a normal to the tangent plane to $F(x, y, z) = c$ at $P_0(x_0, y_0, z_0)$ is

$$\nabla F(x_0, y_0, z_0) = F_x(x_0, y_0, z_0)\mathbf{i} + F_y(x_0, y_0, z_0)\mathbf{j} + F_z(x_0, y_0, z_0)\mathbf{k},$$

an equation of this tangent plane is

(14.44) $\quad \nabla F(x_0, y_0, z_0) \cdot \langle x - x_0, y - y_0, z - z_0 \rangle = 0$

or

(14.45)
$$F_x(x_0, y_0, z_0)(x - x_0) + F_y(x_0, y_0, z_0)(y - y_0) + F_z(x_0, y_0, z_0)(z - z_0) = 0.$$

Example 1

Find an equation of the tangent plane to the paraboloid $x^2 + y^2 + z = 6$ at $(1, 2, 1)$.

Solution

Let $F(x, y, z) = x^2 + y^2 + z$. Then the paraboloid is the 6-level surface of F. Since

$$\nabla F(x, y, z) = 2x\mathbf{i} + 2y\mathbf{j} + \mathbf{k},$$

a normal vector to the paraboloid at $(1, 2, 1)$ is

$$\nabla F(1, 2, 1) = 2\mathbf{i} + 4\mathbf{j} + \mathbf{k}.$$

An equation of the tangent plane is

$$2(x - 1) + 4(y - 2) + (z - 1) = 0$$

or

$$2x + 4y + z = 11.$$

This plane is shown in Figure 14.32. ▶

Suppose a surface S is the graph of a differentiable function f of two variables. Then S is the 0-level surface of

(14.46) $\quad\quad\quad\quad F(x, y, z) = z - f(x, y).$

Since

$$F_x(x, y, z) = -f_x(x, y), \quad F_y(x, y, z) = -f_y(x, y), \quad \text{and} \quad F_z(x, y, z) = 1,$$

(14.45) implies that the tangent plane at $(x_0, y_0, f(x_0, y_0))$ has equation

(14.47) $\quad -f_x(x_0, y_0)(x - x_0) - f_y(x_0, y_0)(y - y_0) + (z - f(x_0, y_0)) = 0.$

Example 2

Find an equation of the normal line to the surface $z = e^{xy}$ at $(1, 1, e)$.

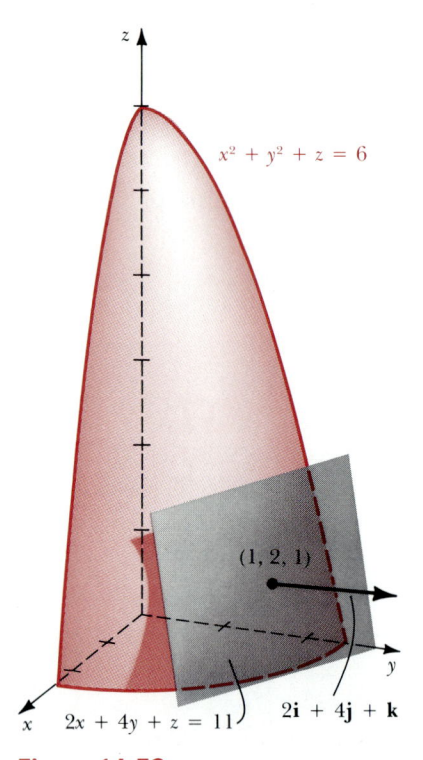

Figure 14.32

Solution

If F is defined by $F(x, y, z) = z - e^{xy}$, then the surface is the 0-level surface of F. Since $\nabla F(x, y, z) = -ye^{xy}\mathbf{i} - xe^{xy}\mathbf{j} + \mathbf{k}$, the normal line to the surface at $(1, 1, e)$ has direction given by $\langle -e, -e, 1 \rangle$. An equation for the line through $(1, 1, e)$ with direction $\langle -e, -e, 1 \rangle$ is given in parametric form by

$$x = 1 - et, \qquad y = 1 - et, \qquad z = e + t.$$

This line is shown in Figure 14.33. ▶

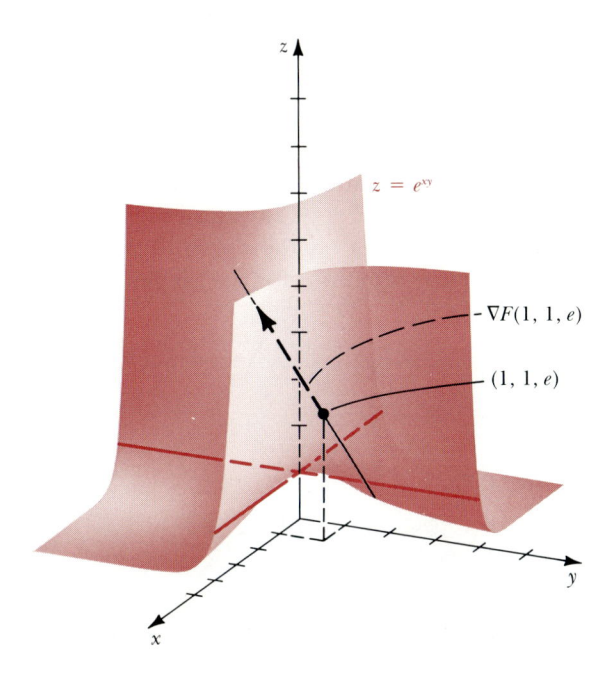

Figure 14.33

Example 3

Show that a normal line to a sphere must pass through the center of the sphere.

Solution

It suffices to show that the statement is true for spheres centered at the origin, that is, those with equation

$$x^2 + y^2 + z^2 = r^2.$$

Let $F(x, y, z) = x^2 + y^2 + z^2$. Then $F(x, y, z) = r^2$ describes the sphere and

$$\nabla F(x, y, z) = 2x\mathbf{i} + 2y\mathbf{j} + 2z\mathbf{k}$$

is normal to the sphere at (x, y, z) (see Figure 14.34). Thus the normal line to the sphere at a point (x_0, y_0, z_0) on the sphere has direction given by $\langle x_0, y_0, z_0 \rangle$. Parametric equations for the normal line are

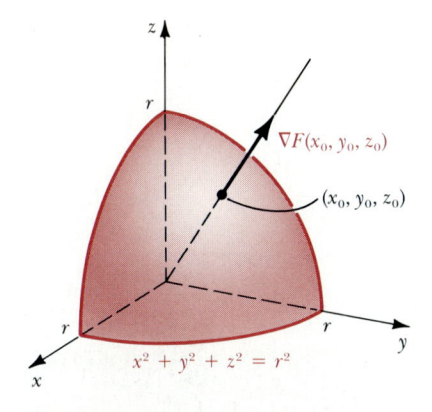

$$x = x_0 + x_0 t, \qquad y = y_0 + y_0 t, \qquad z = z_0 + z_0 t.$$

Figure 14.34
The normal line to a sphere passes through its center.

The point $(0, 0, 0)$, the center of the sphere, is the point on this line corresponding to $t = -1$. ▶

Theorem (14.43) has an interesting application for intersecting surfaces. Suppose surfaces S_1 described by $F_1(x, y, z) = c_1$ and S_2 described by $F_2(x, y, z) = c_2$ intersect in a curve C. Since C lies on both S_1 and S_2, the normals $\mathbf{n}_1$ to S_1 and $\mathbf{n}_2$ to S_2 at a point $P(x, y, z)$ on C are also normal to C at the point P (see Figure 14.35). Thus the vector

(14.48) $\qquad \mathbf{n}_1 \times \mathbf{n}_2 = \nabla F_1(x, y, z) \times \nabla F_2(x, y, z)$

points in the direction of the tangent to the curve C at P, provided F_1 and F_2 are differentiable at P and $\mathbf{n}_1$ is not parallel to $\mathbf{n}_2$.

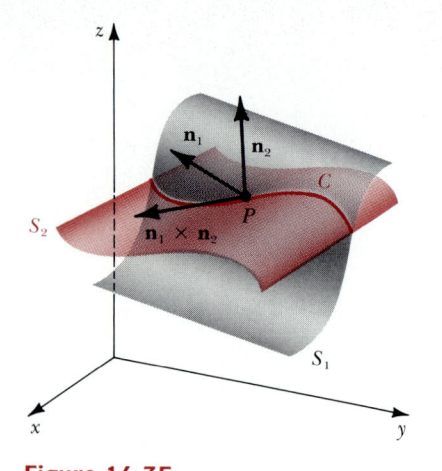

Figure 14.35
A tangent to the curve of intersection of S_1 and S_2.

Example 4

Consider the sphere with equation $x^2 + y^2 + z^2 = 8$ and one nappe of the cone with equation $z = \sqrt{x^2 + y^2}$, shown in Figure 14.36. The point with coordinates $(1, \sqrt{3}, 2)$ lies on the curve of intersection of these surfaces. Find parametric equations for the tangent line to the curve at this point.

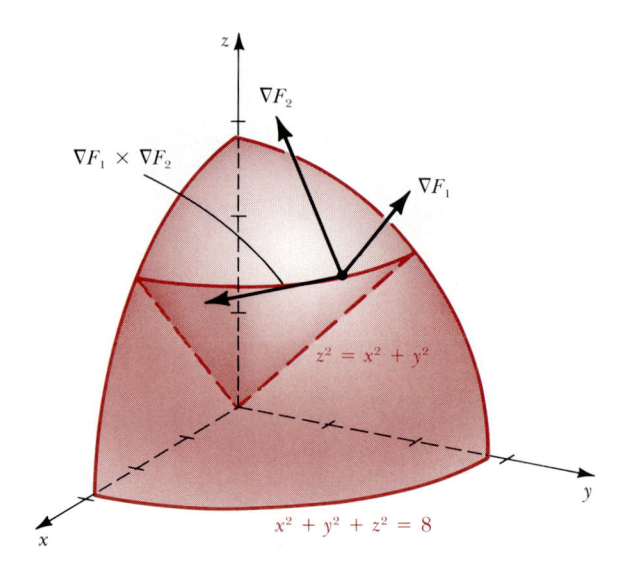

Figure 14.36

Solution
Let $F_1(x, y, z) = x^2 + y^2 + z^2$ and $F_2(x, y, z) = z - \sqrt{x^2 + y^2}$. Then

$$\nabla F_1(x, y, z) = 2x\mathbf{i} + 2y\mathbf{j} + 2z\mathbf{k}$$

and

$$\nabla F_2(x, y, z) = -\frac{x}{\sqrt{x^2 + y^2}}\mathbf{i} - \frac{y}{\sqrt{x^2 + y^2}}\mathbf{j} + \mathbf{k}.$$

So

$$\nabla F_1(1, \sqrt{3}, 2) = 2\mathbf{i} + 2\sqrt{3}\mathbf{j} + 4\mathbf{k}$$

and
$$\nabla F_2(1, \sqrt{3}, 2) = -\frac{1}{2}\mathbf{i} - \frac{\sqrt{3}}{2}\mathbf{j} + \mathbf{k}.$$

The direction of the tangent line to the curve of intersection is thus
$$\nabla F_1(1, \sqrt{3}, 2) \times \nabla F_2(1, \sqrt{3}, 2) = 4\sqrt{3}\mathbf{i} - 4\mathbf{j} = 4(\sqrt{3}\mathbf{i} - \mathbf{j}).$$

The tangent line to the curve of intersection of the two surfaces at $(1, \sqrt{3}, 2)$ has direction $\sqrt{3}\,\mathbf{i} - \mathbf{j}$, so parametric equations for this tangent line are
$$x = 1 + \sqrt{3}t, \qquad y = \sqrt{3} - t, \qquad z = 2.$$ ▸

The final example in this section illustrates how we can use the two applications of the gradient to find information about the same surface.

Example 5

Let S be the circular paraboloid $x^2 + y^2 - z = 1$. Use the gradient to find:

a) the direction in which z increases most rapidly at $(1, 2)$ and this maximum rate of increase, and

b) the equations of the tangent plane and normal line to S at $(1, 2, 4)$.

Solution

The paraboloid is shown in Figure 14.37.

a) To solve this part of the problem, we describe S by a function f of two variables with
$$z = f(x, y) = x^2 + y^2 - 1.$$

Since
$$\nabla f(x, y) = 2x\mathbf{i} + 2y\mathbf{j}, \quad \text{we have} \quad \nabla f(1, 2) = 2\mathbf{i} + 4\mathbf{j}.$$

By Theorem (14.37) of Section 14.5, the direction of maximum rate of increase of z is
$$\nabla f(1, 2) = 2\mathbf{i} + 4\mathbf{j},$$
and the maximum rate of increase is
$$\|\nabla f(1, 2)\| = \sqrt{2^2 + 4^2} = 2\sqrt{5}.$$

b) To solve this part of the problem, we describe S as a level surface of a function F of three variables. If
$$F(x, y, z) = z - x^2 - y^2,$$
then S is the 1-level surface of F. Since
$$\nabla F(x, y, z) = -2x\mathbf{i} - 2y\mathbf{j} + \mathbf{k},$$
the direction of the normal to S at $(1, 2, 4)$ is
$$\nabla F(1, 2, 4) = -2\mathbf{i} - 4\mathbf{j} + \mathbf{k}.$$

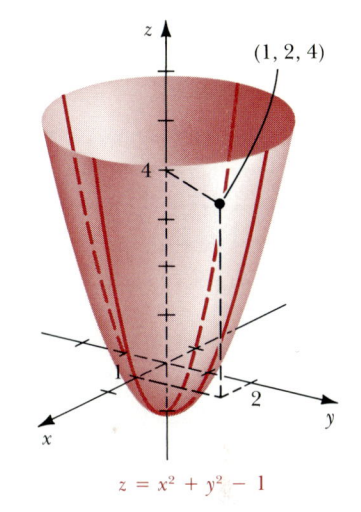

Figure 14.37

$z = x^2 + y^2 - 1$

An equation of the tangent plane is therefore

$$-2(x-1) - 4(y-2) + (z-4) = 0$$

or

$$2x + 4y - z = 6.$$

Parametric equations for the normal line are

$$x = 1 - 2t, \quad y = 2 - 4t, \quad z = 4 + t.$$

▶

▶ EXERCISE SET 14.6

In Exercises 1–8, find a vector normal to the surface at the given point.

1. $x^2 + y^2 + z^2 = 6$, $(1, 1, 2)$
2. $x^2 + y^2 - z = 0$, $(1, -1, 2)$
3. $z = e^y$, $(2, 0, 1)$
4. $z = e^x (\cos y + \sin y)$, $(0, \pi, -1)$
5. $f(x, y) = \ln xy$, $(e, e^2, 3)$
6. $f(x, y) = \sqrt{x^2 + y^2}$, $(3, -4, 5)$
7. $f(x, y) = \sqrt{x} + \sqrt{y}$, $(1, 1, 2)$
8. $f(x, y) = xy^2$, $(2, 1, 2)$

In Exercises 9–16, find an equation of the tangent plane to the surface at the given point.

9. $z = x^2 - y^2$, $(2, 1, 3)$
10. $z = \dfrac{x^2}{9} + \dfrac{y^2}{4}$, $(3, -2, 2)$
11. $z^2 = x^2 + y^2$, $(3, -4, -5)$
12. $z = \cos y$, $(1, \pi/2, 0)$
13. $f(x, y) = \ln (x^2 + y^2)$, $(1, 0, 0)$
14. $f(x, y) = ye^x$, $(0, -2, -2)$
15. $f(x, y) = x \tan y$, $(2, \pi/4, 2)$
16. $f(x, y) = \sec (x + y)$, $(0, 0, 1)$

In Exercises 17–24, the two surfaces intersect in a curve that contains the given point. Find equations of the tangent line to the curve of intersection at the point.

17. $x^2 + y^2 = 5$, $x^2 + z^2 = 5$; $(1, 2, 2)$
18. $x^2 + y^2 + z^2 = 8$, $z^2 = x^2 + y^2$; $(1, \sqrt{3}, -2)$
19. $x^2 + y^2 = 4$, $z^2 = x^2 + y^2$; $(0, 2, 2)$
20. $y^2 + z^2 = 9$, $x = 2y^2 + 3z^2$; $(18, 3, 0)$
21. $2x + z = 6$, $z^2 = x^2 + y^2$; $(1, \sqrt{15}, 4)$
22. $x = z^2$, $x = 4 - y^2$; $(4, 0, 2)$
23. $x = 6 - y^2 - z^2$, $x = 2$; $(2, 1, \sqrt{3})$
24. $z = \sin x$, $x^2 + y^2 = \pi^2$; $(\pi/2, \sqrt{3}\pi/2, 1)$

25. Find the points on the paraboloid $z = 4x^2 + y^2$ at which the tangent plane is parallel to the plane $x + 2y + z = 6$.

26. Find the points on the ellipsoid

$$\dfrac{x^2}{4} + \dfrac{y^2}{9} + \dfrac{z^2}{36} = 1$$

at which the tangent plane is parallel to the plane $x + y + z = 0$.

27. If two surfaces have a common tangent plane at a point of intersection, then the surfaces are said to be *tangent* at that point. Show that if the surfaces $F(x, y, z) = c_1$ and $G(x, y, z) = c_2$ are tangent at (x_0, y_0, z_0), then $\nabla F(x_0, y_0, z_0) \times \nabla G(x_0, y_0, z_0) = \mathbf{0}$ (see the figure).

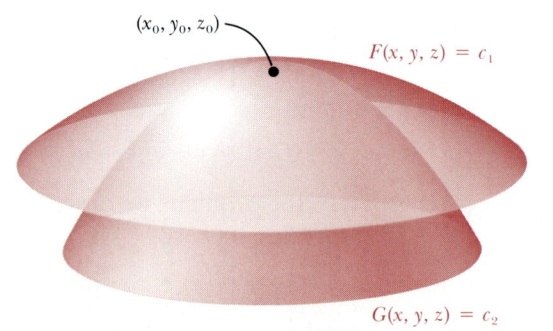

28. If two surfaces have orthogonal normals at a point of intersection, then the surfaces are said to intersect *orthogonally* at that point. Show that if the two surfaces $F(x, y, z) = c_1$ and $G(x, y, z) = c_2$ intersect

orthogonally at (x_0, y_0, z_0), then $\nabla F(x_0, y_0, z_0) \cdot \nabla G(x_0, y_0, z_0) = 0$ (see the figure).

29. Show that the tangent plane to any quadric surface with equation

$$\frac{x^2}{a^2} \pm \frac{y^2}{b^2} \pm \frac{z^2}{c^2} = 1$$

at (x_0, y_0, z_0) has equation

$$\frac{x_0 x}{a^2} \pm \frac{y_0 y}{b^2} \pm \frac{z_0 z}{c^2} = 1.$$

30. Suppose the height above sea level of a mountainous region at each point (x, y) is given by $f(x, y)$. In this situation the level curves described by $f(x, y) = c$ are called *contour lines*. Show that the direction of flow of a stream of water at each point is always normal to the contour line at that point.

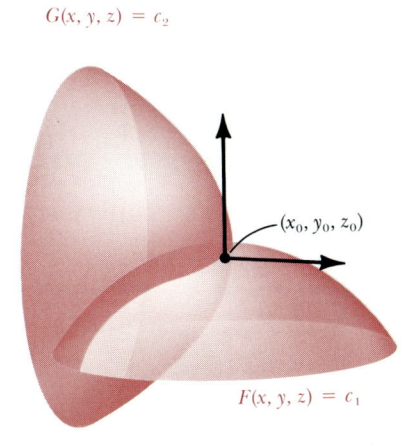

Figure for Exercise 28.

14.7
Extrema of Multivariable Functions

The Extreme Value Theorem for single-variable functions states that a continuous function defined on a closed interval assumes maximum and minimum values on the interval. In studying the derivative in Chapter 2, we found that the maximum and minimum values of a differentiable function occur only at endpoints of the interval or at critical points, numbers in the domain of the function at which the derivative is zero or does not exist. In this section we examine some extensions of these results to multivariable functions. To prepare for this material it would be helpful to review briefly Section 2.7.

We first define what is meant by absolute extrema, relative extrema, and critical points for multivariable functions. The definitions are stated for functions of two variables. Corresponding definitions can be made for functions of three variables.

(14.49)
DEFINITION

A function f of two variables is said to have an **absolute maximum,** or simply a *maximum,* at (x_0, y_0) in the region R if $f(x, y) \leq f(x_0, y_0)$ for all (x, y) in R. An **absolute minimum,** or simply a *minimum,* of f in R occurs at (x_0, y_0) if $f(x, y) \geq f(x_0, y_0)$ for all (x, y) in R.

A **relative maximum** of f occurs at (x_0, y_0) if a circle D about (x_0, y_0) exists with $f(x, y) \leq f(x_0, y_0)$ for all (x, y) in the interior of D. A **relative minimum** of f occurs at (x_0, y_0) if a circle D exists with $f(x, y) \geq f(x_0, y_0)$ for all (x, y) in the interior of D.

$f(x, y) = x^2 + y^2$

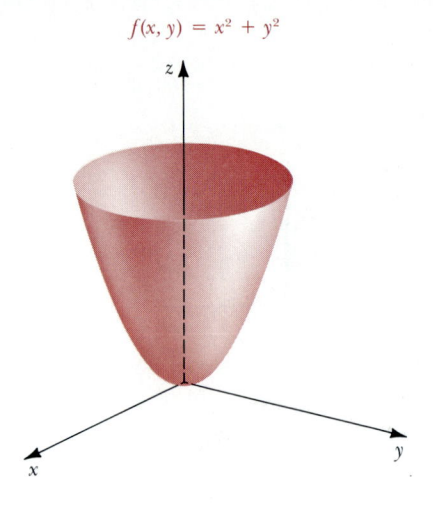

Figure 14.38
Absolute minimum at (0, 0).

As in the single-variable case, absolute maxima and absolute minima are both referred to as **absolute extrema,** or simply as *extrema;* relative maxima and relative minima are called **relative extrema.**

Example 1

Show that if $f(x, y) = x^2 + y^2$, then an absolute minimum of f occurs at $(0, 0)$.

Solution

The graph of f is shown in Figure 14.38. Since $f(x, y) = x^2 + y^2 \geq 0$ and $f(0, 0) = 0$, $f(x, y) \geq f(0, 0)$ for all (x, y). Hence f has an absolute minimum value of 0 at $(0, 0)$. ▶

The following result is called the Extreme Value Theorem for functions of two variables. A similar result holds for functions of three variables.

(14.50)
THEOREM: EXTREME VALUE THEOREM
If f is continuous on a closed and bounded region R in $\mathbb{R}^2$, then there exists (x_1, y_1) and (x_2, y_2) in R with the property that for all (x, y) in R,

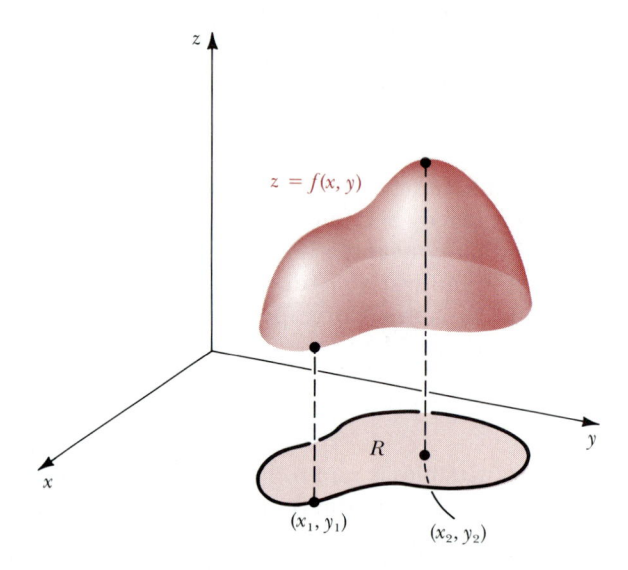

$$f(x_1, y_1) \leq f(x, y) \leq f(x_2, y_2)$$

(see Figure 14.39).

Figure 14.39
Absolute minimum at (x_1, y_1); absolute maximum at (x_2, y_2).

As in the case of a single-variable function, the Extreme Value Theorem ensures the existence of an absolute minimum value and an absolute maximum value for $f(x, y)$ but does not provide a means for finding these values. To determine the extrema we need to define critical points for a function of two variables.

14.7 EXTREMA OF MULTIVARIABLE FUNCTIONS

(14.51)
DEFINITION
A function f of two variables is said to have a **critical point** at (x_0, y_0) in the domain of f if either:

i) both partial derivatives of f are zero at (x_0, y_0), or
ii) at least one of the partial derivatives does not exist at (x_0, y_0).

Example 2

Find all critical points of f if $f(x, y) = 4 - 4x^2 - y^2$.

Solution

Since $f_x(x, y) = -8x$ and $f_y(x, y) = -2y$, the partial derivatives of f exist for all values of x and y.

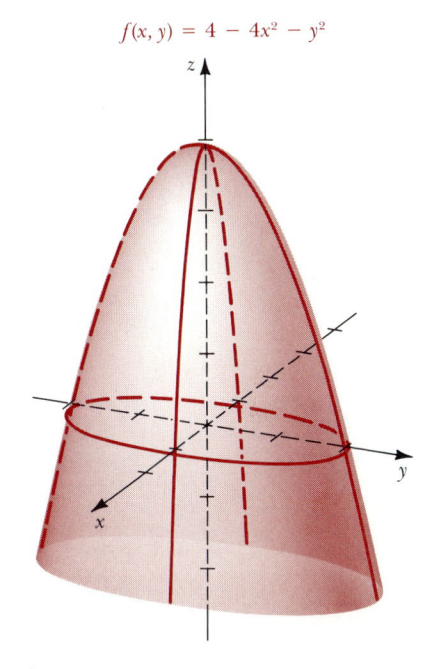

The only critical points occur when both partial derivatives are zero, that is, when

$$f_x(x, y) = -8x = 0 \quad \text{and} \quad f_y(x, y) = -2y = 0.$$

The only critical point in this case is $(0, 0)$.

It is easy to see that a maximum of f occurs at $(0, 0)$, since

$$f(0, 0) = 4 \geq 4 - 4x^2 - y^2$$

for all (x, y) in the plane (see Figure 14.40). ▶

The following theorem is analogous to the result for single-variable functions given as Theorem (2.38) in Section 2.7.

Figure 14.40 — $f(x, y) = 4 - 4x^2 - y^2$

(14.52)
THEOREM
If a function f has a relative extremum at (x_0, y_0), then (x_0, y_0) is a critical point of f.

Proof. Suppose f has a relative extremum at (x_0, y_0). Then the single-variable functions described by $f(x, y_0)$ and $f(x_0, y)$ also have a relative extremum at (x_0, y_0). It follows from Theorem (2.38) that if the partial derivatives of f at (x_0, y_0) both exist, then $f_x(x_0, y_0) = 0$ and $f_y(x_0, y_0) = 0$. Thus either the partial derivatives of f at (x_0, y_0) are both zero or one of them does not exist at (x_0, y_0); that is, (x_0, y_0) is a critical point of f. ▷

Example 3

Show that the function described by $f(x, y) = e^x \sin y$ has no relative extrema.

Solution

The partial derivatives of f,

$$f_x(x, y) = e^x \sin y \quad \text{and} \quad f_y(x, y) = e^x \cos y,$$

both exist for all values of x and y, so critical points occur only when

$$e^x \sin y = 0 \quad \text{and} \quad e^x \cos y = 0.$$

Since $e^x \neq 0$, $e^x \sin y = 0$ when $\sin y = 0$; that is, $y = n\pi$ for n an integer. However, these values do not satisfy the second equation,

$$e^x \cos(n\pi) = \pm e^x \neq 0.$$

Thus the partial derivatives are never simultaneously zero, and f has no critical points. It follows from Theorem (14.52) that f has no extrema. The graph of f is shown in Figure 14.41 and Color Plate VIII. ▶

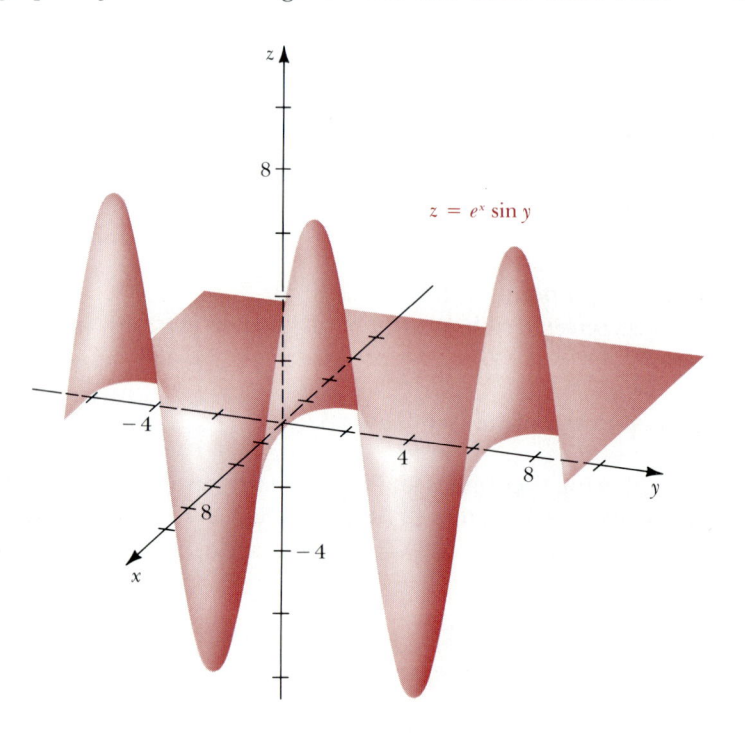

Figure 14.41

Example 4

Determine whether $f(x, y) = y^2 - x^2$ has absolute extrema.

Solution

The partial derivatives of f are

$$f_x(x, y) = -2x \quad \text{and} \quad f_y(x, y) = 2y,$$

so $(0, 0)$ is the only critical point of f and the only point at which an extremum can occur. In every circle containing $(0, 0)$, there are points $(x, 0)$ at which $f(x, 0) = -x^2 < 0$ and points $(0, y)$ at which $f(0, y) = y^2 > 0$. Thus f has neither a maximum nor a minimum at $(0, 0)$. We can also see this in the graph of f shown in Figure 14.42, a hyperbolic paraboloid. ▶

A critical point at which a relative extremum does not occur is called a **saddle point** of the graph. In Example 4 this name is quite appropri-

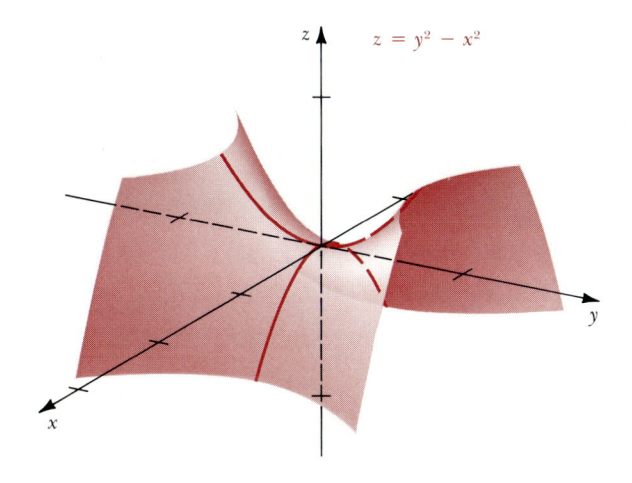

Figure 14.42

ate, since the graph has the appearance of a saddle and the critical point determines the point of the saddle at which one would sit. The terminology is not always as descriptive; saddle points often occur when the graph resembles nothing like a saddle.

It is generally more difficult to determine absolute extrema on bounded regions of the plane. However, this problem occurs often in applications.

(14.53)
COROLLARY
If R is a bounded region of the plane and f has an extremum at (x_0, y_0) in R, then either (x_0, y_0) is a critical point of f or (x_0, y_0) is on the boundary of R.

The Extreme Value Theorem ensures that a continuous function of two variables defined on a closed and bounded region assumes its absolute maximum and absolute minimum values in the region. A procedure for finding these values is listed below; it is similar to the one outlined for single-variable functions in Section 2.7.

(14.54)
i) Find the critical points of the function.
ii) Find the values of the function at the critical points.
iii) Find the maximum and minimum values of the function on the boundary of the region.
iv) The absolute maximum value is the largest of those found in (ii) and (iii).
v) The absolute minimum value is the smallest of those found in (ii) and (iii).

The most complicated step in this procedure is often that stated in part (iii). The following two examples show how the maximum and minimum values lying on the boundary of a region are obtained when the boundary of the region is described easily.

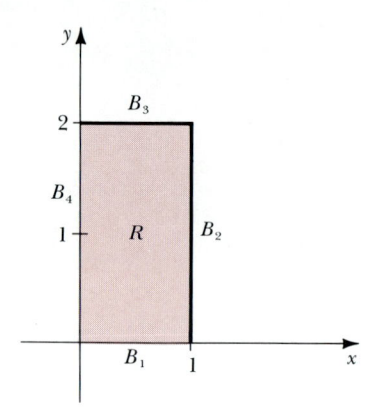

Figure 14.43

Example 5

Find the absolute maximum and absolute minimum values of $f(x, y) = 5 - 2x^2 + 2xy - y^2$ on the region R of points whose coordinates satisfy $0 \leq x \leq 1$ and $0 \leq y \leq 2$.

Solution

We first consider the equations

$$0 = f_x(x, y) = -4x + 2y \quad \text{and} \quad 0 = f_y(x, y) = 2x - 2y,$$

that is,

$$y = 2x \quad \text{and} \quad y = x.$$

These equations describe lines whose only intersection is at the origin, so $(0, 0)$ is the only critical point of f. This point lies on the boundary of R.

The boundary of R consists of four straight-line segments, as shown in Figure 14.43:

B_1: $y = 0$, $0 \leq x \leq 1$; B_2: $x = 1$, $0 \leq y \leq 2$;
B_3: $y = 2$, $0 \leq x \leq 1$; B_4: $x = 0$, $0 \leq y \leq 2$.

To find the maximum and minimum values of f on the boundary, we examine the portions individually.

On B_1: $f(x, 0) = 5 - 2x^2$ for $0 \leq x \leq 1$.

Since f is decreasing on B_1, f has a minimum value of 3 on B_1, at $(1, 0)$, and a maximum value of 5, at $(0, 0)$.

On B_2: $f(1, y) = 3 + 2y - y^2$ for $0 \leq y \leq 2$.

To find the extreme values on B_2 we consider $f(1, y)$ as a function of the single variable y. This function assumes its extrema on B_2 only at the endpoints $(1, 0)$ and $(1, 2)$ of B_2 or when $0 = f'(1, y) = 2 - 2y$, that is, at $(1, 1)$. Since $f(1, 0) = 3$, $f(1, 1) = 4$, and $f(1, 2) = 3$, the maximum value of f on B_2 is 4 and the minimum value is 3.

On B_3: $f(x, 2) = 1 - 2x^2 + 4x$ for $0 \leq x \leq 1$.

We consider $f(x, 2)$ as a single-variable function of x. Then $f'(x, 2) = -4x + 4 = 0$ when $x = 1$, an endpoint of B_3. Since $f(0, 2) = 1$ and $f(1, 2) = 3$, f has a minimum value of 1 on B_3, at $(0, 2)$, and a maximum value of 3, at $(1, 2)$.

On B_4: $f(0, y) = 5 - y^2$ for $0 \leq y \leq 2$.

So f has a minimum value of 1 on B_2, at $(0, 2)$, and a maximum value of 5, at $(0, 0)$.

The minimum value of f on R is therefore 1, which occurs at $(0, 2)$, and the maximum value is 5, which occurs at $(0, 0)$. The graph of f on the region R is shown in Figure 14.44 and Color Plate IX. ▶

Example 6

Find the absolute maximum and absolute minimum values of $f(x, y) =$

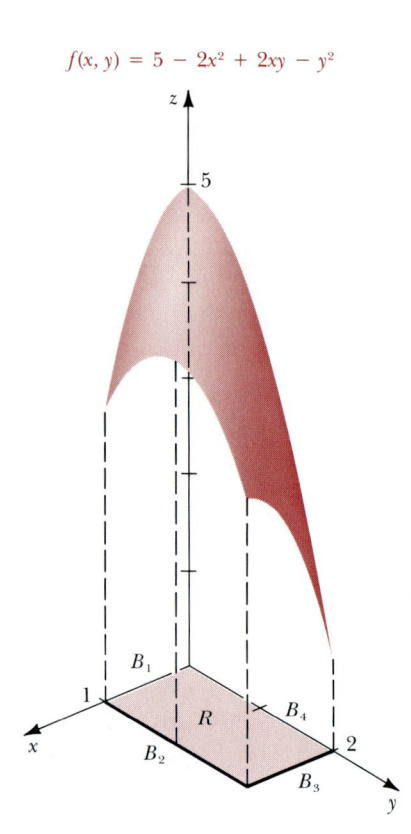

Figure 14.44

$5 - 2x^2 + 2xy - y^2$ on the region R of points whose coordinates satisfy $-1 \leq x \leq 1$ and $0 \leq y \leq 1 - x^2$.

Solution
The function is the same as that considered in Example 5, which had its only critical point at $(0, 0)$, on the boundary of this region. The boundary of R in this example has the two connected components B_1 and B_2, shown in Figure 14.45.

On B_1: $f(x, 0) = 5 - 2x^2$ for $-1 \leq x \leq 1$.

So f has a maximum value of 5 on B_1, at $(0, 0)$, and a minimum value of 3, which occurs at both $(-1, 0)$ and $(1, 0)$.

On B_2: $f(x, 1 - x^2) = 5 - 2x^2 + 2x(1 - x^2) - (1 - x^2)^2$
$= 4 + 2x - 2x^3 - x^4$

for $-1 \leq x \leq 1$.

Treating f on B_2 as a function of the single variable x, we have

$$f'(x, 1 - x^2) = 2 - 6x^2 - 4x^3 = -2(2x^3 + 3x^2 - 1).$$

Note that -1 is a root of the equation $0 = 2x^3 + 3x^2 - 1$ and that by factoring the right side of this equation we have

$$0 = (x + 1)(2x^2 + x - 1) = (x + 1)^2(2x - 1).$$

Thus the maximum and minimum values of f on B_2 can occur only at the roots of this equation, $x = -1$ and $x = \frac{1}{2}$, or at the other endpoint of the boundary, $x = 1$.

For $x = -1$ or $x = 1$, $1 - x^2 = 0$, so $f(-1, 0) = 3$ and $f(1, 0) = 3$. For $x = \frac{1}{2}$, $1 - x^2 = \frac{3}{4}$ and $f(\frac{1}{2}, \frac{3}{4}) = 5 - 2(\frac{1}{2})^2 + 2(\frac{1}{2})(\frac{3}{4}) - (\frac{3}{4})^2 = \frac{75}{16}$. So f has a maximum value of $\frac{75}{16}$ on B_2, at $(\frac{1}{2}, \frac{3}{4})$, and a minimum value of 3, at $(-1, 0)$ and $(1, 0)$.

Considering B_1 and B_2 together, f has a maximum value of 5 on the region R, at $(0, 0)$, and a minimum value of 3, at $(-1, 0)$ and $(1, 0)$. The graph of f on R is shown in Figure 14.46 and Color Plate IX. ▶

In the preceding examples, it was easy to determine whether relative maxima or minima occurred at critical points. In less obvious cases, we can apply the following extension of the Second Derivative Test. The proof of this result is quite complicated and is not presented.

(14.55)
THEOREM
Let f be a function with continuous second partial derivatives and suppose (x_0, y_0) is a critical point of f. Let

$$D(x_0, y_0) = f_{xx}(x_0, y_0) f_{yy}(x_0, y_0) - [f_{xy}(x_0, y_0)]^2.$$

i) If $D(x_0, y_0) > 0$ and $f_{xx}(x_0, y_0) < 0$, then f has a relative maximum at (x_0, y_0).
ii) If $D(x_0, y_0) > 0$ and $f_{xx}(x_0, y_0) > 0$, then f has a relative minimum at (x_0, y_0).
iii) If $D(x_0, y_0) < 0$, then f has a saddle point at (x_0, y_0).

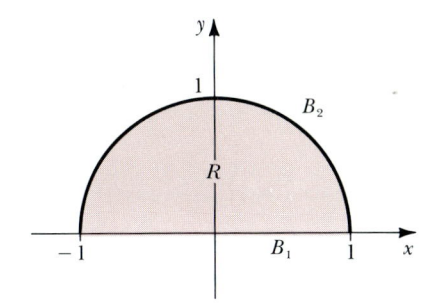

Figure 14.45

$f(x, y) = 5 - 2x^2 + 2xy - y^2$

Figure 14.46

The term $D(x_0, y_0)$ is called the **discriminant** of f at (x_0, y_0). It is given alternatively as a determinant:

$$(14.56) \qquad D(x_0, y_0) = \begin{vmatrix} f_{xx}(x_0, y_0) & f_{xy}(x_0, y_0) \\ f_{yx}(x_0, y_0) & f_{yy}(x_0, y_0) \end{vmatrix},$$

since the conditions of the theorem ensure that $f_{xy}(x_0, y_0) = f_{yx}(x_0, y_0)$.

Note that no conclusion can be drawn from Theorem (14.55) when $f_{xx}(x_0, y_0) f_{yy}(x_0, y_0) - [f_{xy}(x_0, y_0)]^2 = 0$. This corresponds to the single-variable situation when the second derivative at a critical point is zero. It may happen that in this case a relative maximum occurs, a relative minimum occurs, or the point is a saddle point. Examples of such situations are considered in Exercises 51 and 52.

Example 7

Determine any relative extrema for $f(x, y) = x^3 - y^3 + 3xy$, using Theorem (14.55).

Solution

Critical points of f occur when both

$$0 = f_x(x, y) = 3x^2 + 3y \quad \text{so} \quad y = -x^2$$

and

$$0 = f_y(x, y) = -3y^2 + 3x \quad \text{so} \quad x = y^2.$$

To find values of x and y that satisfy both $y = -x^2$ and $x = y^2$, we solve these equations simultaneously:

$$x = y^2 = (-x^2)^2 = x^4 \quad \text{so} \quad x = 0 \quad \text{or} \quad x = 1.$$

The critical points of f are $(0, 0)$ and $(1, -1)$.

The second partial derivatives of f are

$$f_{xx}(x, y) = 6x, \qquad f_{yy}(x, y) = -6y, \quad \text{and} \quad f_{xy}(x, y) = 3.$$

At the critical point $(0, 0)$,

$$D(0, 0) = f_{xx}(0, 0) f_{yy}(0, 0) - [f_{xy}(0, 0)]^2 = (0)(0) - (3)^2 < 0,$$

so a saddle point occurs at $(0, 0)$. At the critical point $(1, -1)$,

$$D(1, -1) = f_{xx}(1, -1) f_{yy}(1, -1) - [f_{xy}(1, -1)]^2 = (6)(6) - (3)^2 > 0.$$

Since $f_{xx}(1, -1) = 6 > 0$, a relative minimum of f occurs at $(1, -1)$. This relative minimum value is $f(1, -1) = -1$. The graph of f is shown in Figure 14.47 and Color Plate VIII. ▶

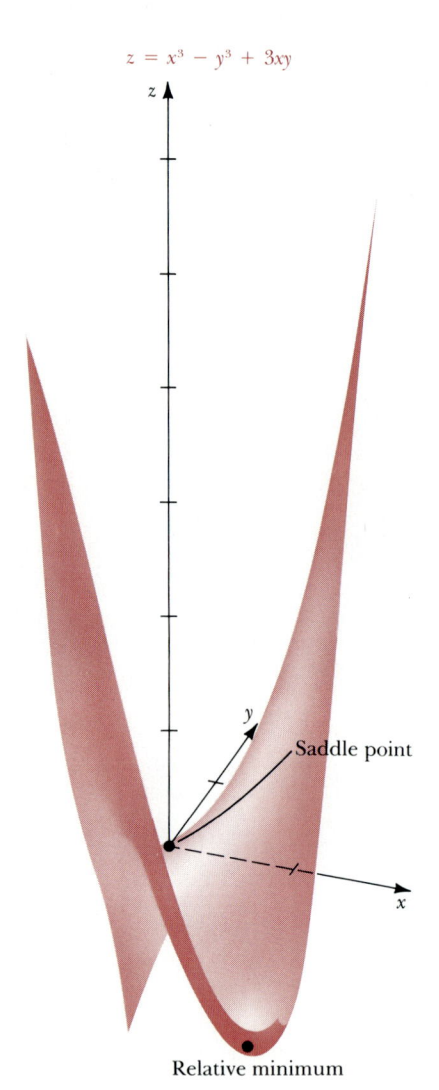

Figure 14.47

Example 8

A box in the form of a rectangular parallelepiped is to be built to enclose 36 ft³. The bottom of the box can be made from scrap material that costs $1 per square foot, but the material for the top and sides costs $8 per square foot. Find the most economical dimensions for the box and the corresponding cost of the box.

Solution

We label the sides of the box as shown in Figure 14.48. The conditions imply that the volume of the box is $xyz = 36$ ft^3 and that the cost is $1 \cdot xy + 8 \cdot (2xz + 2yz + xy)$ dollars. We can solve the problem using the volume equation to eliminate one of the variables in the cost equation. The problem then becomes one of minimizing a function of two variables defined on a plane region.

KNOW	FIND
1) Volume = $xyz = 36$ ft^3, 2) Cost = $xy + 8(2xz + 2yz + xy)$, 3) From (1): $z = \dfrac{36}{xy}$, 4) $C(x, y) = xy + 8\left[2x\left(\dfrac{36}{xy}\right) + 2y\left(\dfrac{36}{xy}\right) + xy\right]$ $\quad = 9xy + \dfrac{576}{y} + \dfrac{576}{x}$, $x > 0$ and $y > 0$.	a) The values of x, y, and z at which C has a minimum, b) The cost corresponding to the values of x, y, and z found in part (a).

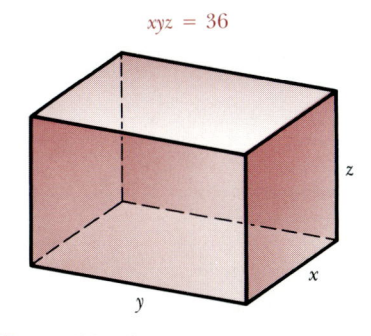

Figure 14.48

To find the critical points of C we compute

$$C_x(x, y) = 9y - \frac{576}{x^2} = \frac{9}{x^2}(yx^2 - 64)$$

and

$$C_y(x, y) = 9x - \frac{576}{y^2} = \frac{9}{y^2}(xy^2 - 64).$$

The only critical points occur when

$$yx^2 - 64 = 0 \quad \text{and} \quad xy^2 - 64 = 0.$$

Solving these equations simultaneously,

$$y = \frac{64}{x^2} \quad \text{and} \quad x\left(\frac{64}{x^2}\right)^2 - 64 = 0,$$

we see that $x^3 = 64$. Thus

$$x = 4, \quad y = \frac{64}{16} = 4, \quad \text{and} \quad z = \frac{36}{16} = \frac{9}{4}.$$

The cost corresponding to $x = 4$ and $y = 4$ is

$$C(4, 4) = 9(16) + \frac{576}{4} + \frac{576}{4} = \$432.$$

Using Theorem (14.55) and the fact that the domain of C is not bounded, we can verify that the minimum value of C is \$432:

$$C_{xx}(x, y) = \frac{1152}{x^3}, \quad C_{yy}(x, y) = \frac{1152}{y^3}, \quad C_{xy}(x, y) = 9.$$

At (4, 4),

$$D(4, 4) = C_{xx}(4, 4) \cdot C_{yy}(4, 4) - [C_{xy}(4, 4)]^2 = (18)(18) - 81 > 0$$

and

$$C_{xx}(4, 4) = 18 > 0,$$

so C has a relative minimum at (4, 4). This relative minimum is also an absolute minimum since

$$C(x, y) = 9xy + \frac{576}{y} + \frac{576}{x}$$

approaches ∞ as (x, y) approaches the boundary of the domain of C, that is, as x or y approaches 0 or as x or y approaches ∞.

The most economical box costs \$432; it has a square base of length 4 ft and a height of $z = 36/xy = 36/16 = 2.25$ ft. ▶

In Chapter 12 we derived a formula for finding the distance from a plane to a point that does not lie on the plane. Example 9 shows that problems of this type can also be solved by the methods in this section.

Example 9

Determine the point on the plane with equation $x + 2y + 3z = 6$ that is closest to the point (1, 2, 3).

Solution

The distance from a point (x, y, z) to (1, 2, 3) is

$$\sqrt{(x-1)^2 + (y-2)^2 + (z-3)^2}.$$

Since the square root function is increasing, this expression is minimized precisely when the simpler expression

$$D(x, y, z) = (x-1)^2 + (y-2)^2 + (z-3)^2$$

is minimized.

To minimize D over those points on $x + 2y + 3z = 6$, we first eliminate one of the variables from the equation. For example, we can solve for x,

$$x = 6 - 2y - 3z,$$

and then use the equation to reduce D to the function of two variables:

$$D(y, z) = (5 - 2y - 3z)^2 + (y - 2)^2 + (z - 3)^2.$$

If

$$0 = \frac{\partial D}{\partial y}(y, z) = 2(5 - 2y - 3z)(-2) + 2(y - 2) = 10y + 12z - 24$$

and

$$0 = \frac{\partial D}{\partial z}(y, z) = 2(5 - 2y - 3z)(-3) + 2(z - 3) = 12y + 20z - 36,$$

then
$$5y + 6z = 12 \quad \text{and} \quad 3y + 5z = 9.$$

Solving these equations simultaneously gives
$$z = \frac{9}{7}, \quad y = \frac{6}{7},$$

and
$$x = 6 - 2y - 3z = \frac{3}{7}.$$

The closest point on the plane $x + 2y + 3z = 6$ to $(1, 2, 3)$ is $(\frac{3}{7}, \frac{6}{7}, \frac{9}{7})$, as shown in Figure 14.49.

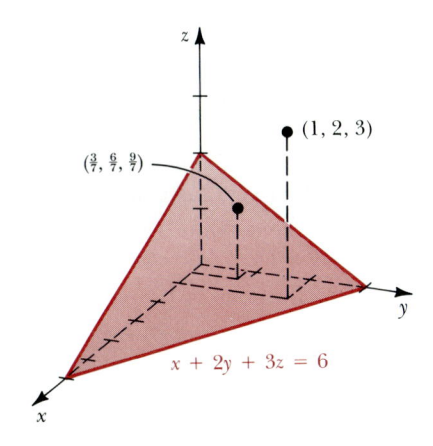

▶ Figure 14.49

▶ EXERCISE SET 14.7

1. Show that if $f(x, y) = 4 - x^2 - y^2$, then f has an absolute maximum at $(0, 0)$.

2. Show that if $f(x, y) = x^2 - y^2$, then f has neither an absolute maximum nor an absolute minimum at $(0, 0)$.

3. Show that if $f(x, y) = \sin y$, $0 \leq y \leq 2\pi$, then f has an absolute maximum at $(x, \pi/2)$ and an absolute minimum at $(x, 3\pi/2)$ for any value of x.

4. Show that if $f(x, y) = \cos x$, $0 \leq x \leq 2\pi$, then f has an absolute maximum at $(0, y)$ and an absolute minimum at (π, y) for any value of y.

In Exercises 5–12, find all critical points of f.

5. $f(x, y) = x^2 - xy + y^2$
6. $f(x, y) = x^2 + y^2 - 4x - 2y + 5$
7. $f(x, y) = x^2 + 2xy + 3y^2 - 4$
8. $f(x, y) = 3x/y$
9. $f(x, y) = x \cos y$
10. $f(x, y) = e^x \sin y$
11. $f(x, y) = xy^2 + x^3 - 4y^2$
12. $f(x, y) = x^2 y + x + y$

13–20. Determine whether each critical point in Exercises 5–12 is a relative maximum, a relative minimum, or a saddle point.

In Exercises 21–32, find all critical points of f and use Theorem (14.55) to determine any relative extrema for f.

21. $f(x, y) = x^3 + y^3 - 3xy$
22. $f(x, y) = x^4 + y^3 + 32x - 27y$
23. $f(x, y) = x^2 y - 5xy + 6y$
24. $f(x, y) = x^2 + y^4 - y^2 - 2xy$
25. $f(x, y) = e^{x^2 + y^2}$
26. $f(x, y) = e^{xy}$
27. $f(x, y) = 2x^2 - 6xy - 2x - y^2 + 14y$
28. $f(x, y) = 16x^2 - 24xy + 9y^2 - 60x - 8y$
29. $f(x, y) = \dfrac{-x}{x^2 + y^2 + 1}$
30. $f(x, y) = \dfrac{y + 1}{x^2 + y^2 + 3}$
31. $f(x, y) = \cos x + \cos y$
32. $f(x, y) = \sin x + \sin y + y \sin x$

33–44. Determine which of the relative extrema found in Exercises 21–32 are absolute extrema.

In Exercises 45–50, find the absolute maximum and absolute minimum values of f on the region described by the inequalities.

45. $f(x, y) = x^2 + y^2$; $x^2 + y^2 \leq 1$
46. $f(x, y) = x^2 + y^2$; $0 \leq x \leq 1$, $0 \leq y \leq 2$
47. $f(x, y) = xy$; $x^2 + 2y^2 \leq 4$
48. $f(x, y) = y^2 - x^2$; $0 \leq x \leq 1$, $0 \leq y \leq 1$
49. $f(x, y) = 8xy + y$, $0 \leq y \leq 15 - x$, $0 \leq x \leq 5$
50. $f(x, y) = 2x^2 + 2y^2 + xy$, $5 - x \leq y \leq 10$, $0 \leq x \leq 3$

51. a) Show that if $f(x, y) = x^4 + y^4$, then $D(0, 0) = 0$, so no extrema information can be concluded from Theorem (14.55).
    b) Show that f has an absolute minimum at $(0, 0)$.

52. a) Show that if $f(x, y) = x^4 - y^4$, then $D(0, 0) = 0$, so no extrema information can be concluded from Theorem (14.55).
  b) Show that f has a saddle point at $(0, 0)$.

53. Find the absolute minimum value of $f(x, y, z) = 2x + 3y + z$ if x, y, and z satisfy $z = x^2 + y^2$.

54. Find the absolute maximum value of $f(x, y, z) = 2x + 3y + z$ if x, y, and z satisfies $z = -x^2 - y^2$.

55. Find the point on the plane $2x + 3y + z = 6$ closest to the origin.

56. Find the minimum distance between the lines with parametric equations $x = t$, $y = 3 - 2t$, $z = 1 + 2t$ and $x = 1 + s$, $y = -2 - s$, $z = s$.

57. The trace of $z = f(x, y)$ in the yz-plane is $z = y^2$. The trace in the xz-plane is $z = -x^2$. Can z have a relative extrema at $(0, 0)$?

58. Determine the three positive numbers whose sum is 27 and whose product is as large as possible.

59. Determine the three positive numbers whose product is 27 and whose sum is as small as possible.

60. Consider the collection of solids bounded by the coordinate planes and a plane through $(1, 1, 1)$. Determine the plane that gives the minimum volume for the solid.

61. A rectangular cereal box is to have a volume of 80 in^3 (see the figure). Find the dimensions of the box with the minimum surface area.

62. A box in the form of a rectangular parallelepiped is to be built at a cost of $1500. The material for the bottom of the box costs $1 per square foot and the material for the top and sides costs $4 per square foot (see the figure). Find the dimensions of the box with the greatest volume.

63. U.S. postal regulations require that the length plus the girth of a package to be mailed cannot exceed 108 in. (see the figure). Find the dimensions of the rectangular package of greatest volume that can be mailed. [Girth $= 2$(width) $+ 2$(height)].

64. A window with an area of 24 ft^2 is to be constructed in the shape of a rectangle topped with an isosceles triangle (see the figure). Determine the dimensions of the window with the minimum perimeter.

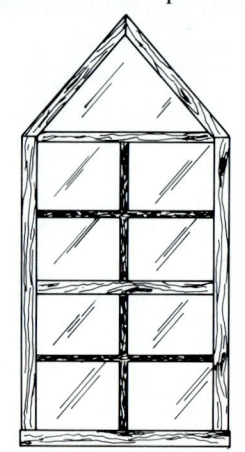

65. Show that of all triangles with a fixed perimeter, the maximum area occurs when the triangle is equilateral.

66. Determine the dimensions of the rectangular parallelepiped with maximum volume that can be enclosed in a sphere of radius r.

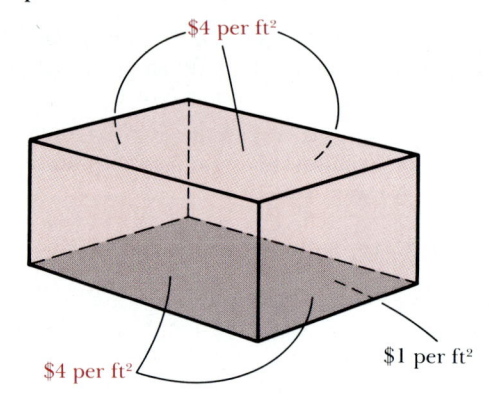

Figure for Exercise 62.

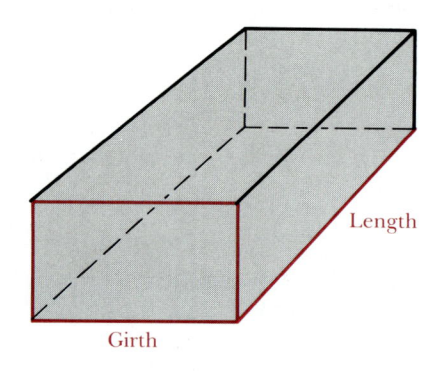

Figure for Exercise 63.

67. *The Method of Least Squares.* Given n points (x_1, y_1), $(x_2, y_2), \ldots, (x_n, y_n)$ in the xy-plane, it is usually impossible to find a straight line that passes through all the points. The Method of Least Squares finds a best approximating straight line, $f(x) = ax + b$, for the points by finding values of a and b that minimize the sum of the squares of the differences between the y-values on the approximating line and the given y-values:

$$E(a, b) = \sum_{i=1}^{n} [y_i - (ax_i + b)]^2.$$

Show that these values of a and b are

$$a = \frac{n\left(\sum_{i=1}^{n} x_i y_i\right) - \left(\sum_{i=1}^{n} x_i\right)\left(\sum_{i=1}^{n} y_i\right)}{n\left(\sum_{i=1}^{n} x_i^2\right) - \left(\sum_{i=1}^{n} x_i\right)^2}$$

and

$$b = \frac{\left(\sum_{i=1}^{n} x_i^2\right)\left(\sum_{i=1}^{n} y_i\right) - \left(\sum_{i=1}^{n} x_i y_i\right)\left(\sum_{i=1}^{n} x_i\right)}{n\left(\sum_{i=1}^{n} x_i^2\right) - \left(\sum_{i=1}^{n} x_i\right)^2}.$$

In Exercises 68–71, use the Method of Least Squares to find the best approximating line for the given collection of points.

68. $(1, 2), (2, 2), (3, 4)$

69. $(0, 0), (1, 3), (2, 1)$

70. $(2, 2), (4, 11), (6, 28), (8, 40)$

71. $(-1, 6), (1, 5), (2, 3), (5, 3), (7, -1)$

72. The following set of data, presented in March 1970 to the Senate Antitrust Subcommittee, shows the comparative crash-survivability characteristics of cars in various classes. Find the least-squares line that approximates the data. (The table shows the percent of accident-involved vehicles in which the most severe injury was fatal or serious.)

TYPE	AVERAGE WEIGHT	PERCENT OCCURRENCE
1. Domestic "luxury" regular	4800 lb	3.1
2. Domestic "intermediate" regular	3700 lb	4.0
3. Domestic "economy" regular	3400 lb	5.2
4. Domestic compact	2800 lb	6.4
5. Foreign compact	1900 lb	9.6

73. To determine a relationship between the number of fish and the number of species of fish in samples taken from a portion of the Great Barrier Reef, P. Sale and R. Dybdahl used the Method of Least Squares and the following collection of data. Let x be the number of fish in the sample and y the number of species in the sample.

x	y	x	y	x	y
13	11	29	12	60	14
15	10	30	14	62	21
16	11	31	16	64	21
21	12	36	17	70	24
22	12	40	13	72	17
23	13	42	14	100	23
25	13	55	22	130	34

Determine the least-squares line for the data.

Putnam exercises

74. Given two points in the plane, P and Q, at fixed distances from a line L and on the same side of the line as indicated, the problem is to find a third point R so that $PR + RQ + RS$ is a minimum, where RS is perpendicular to L. Consider all cases (see the figure). (This exercise was problem 4, part I of the twenty-first William Lowell Putnam examination given on December 3, 1960. The examination and its solution are in the September 1961 issue of the *American Mathematical Monthly*, pp. 632–637.)

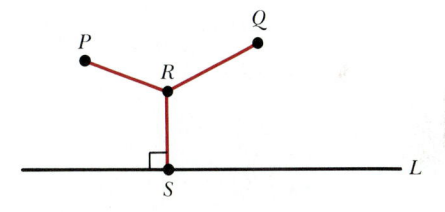

75. Let f be a real-valued function having partial derivatives and that is defined for $x^2 + y^2 \leq 1$ and is such that $|f(x, y)| \leq 1$. Show that there exists a point (x_0, y_0) in the interior of the unit circle such that

$$\left(\frac{\partial f}{\partial x}(x_0, y_0)\right)^2 + \left(\frac{\partial f}{\partial y}(x_0, y_0)\right)^2 \leq 16.$$

(This exercise was problem B–6 of the twenty-eighth William Lowell Putnam examination given on December 2, 1967. The examination and its solution are in the September 1968 issue of the *American Mathematical Monthly*, pp. 734–739.)

76. Show that if t_1, t_2, t_3, t_4, t_5 are real numbers, then

$$\sum_{j=1}^{5} (1 - t_j) \exp \sum_{k=1}^{j} t_k \leq e^{e^{e^e}}.$$

(This exercise was problem 3, part I of the twenty-first William Lowell Putnam examination given on December 3, 1960. The examination and its solution are in the September 1961 issue of the *American Mathematical Monthly,* pp. 632–637.)

77. A convex pentagon $P = ABCDE$, with vertices labeled consecutively, is inscribed in a circle of radius 1. Find the maximum area of P subject to the condition that the chords AC and BD be perpendicular. (This exercise was problem A–4 of the forty-fifth William Lowell Putnam examination given on December 1, 1984. The examination and its solution are in the October 1985 issue of the *American Mathematical Monthly,* pp. 560–567.)

78. Let T be an acute triangle. Inscribe a pair R, S of rectangles in T as shown. Let $A(X)$ denote the area of polygon X. Find the maximum value, or show that no maximum exists, of $[A(R) + A(S)]/A(T)$, where T ranges over all triangles and R, S over all rectangles as in the figure. (This exercise was problem A–2 of the forty-sixth William Lowell Putnam examination given on December 7, 1985. The examination and its solution are in the October 1986 issue of the *American Mathematical Monthly,* pp. 620–626.)

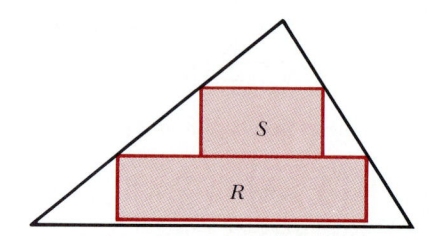

14.8

Lagrange Multipliers

In Example 8 of Section 14.7 we considered the problem of minimizing the cost of building a box to enclose 36 ft³ if the bottom costs $1 per square foot and the sides and top cost $8 per square foot. If x and y denote the lengths of the base and z the height of the box, as shown in Figure 14.50, and C is the cost function, the problem can be expressed as:

Minimize $C(x, y, z) = 9xy + 16yz + 16xy$;
Subject to the condition that $xyz = 36$.

In this form, the condition $xyz = 36$ is called a **constraint** or **side condition** for the problem. Problems in economics and business often assume this form.

In the preceding section we solved the constraint equation for z and then considered C as a function of only two variables, x and y. In this section we give an alternative approach to problems of determining extrema of a function subject to one or more constraints. The technique is known as **Lagrange multipliers** because it involves introducing an

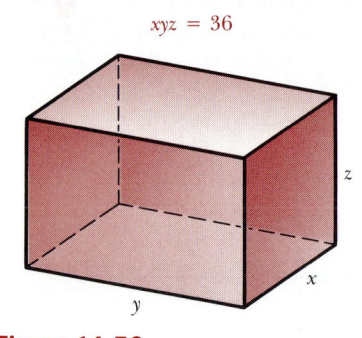

Figure 14.50

HISTORICAL NOTE

The Method of Least Squares was first published by **Adrien Marie Legendre** (1752–1833) in 1805 as an addendum to a work on the orbits of comets. **Karl Friedrich Gauss** (1777–1855), called by some the greatest mathematician of all time, had, however, been using the method as early as 1794. He used least squares to compute, from a few observations, the orbit of Ceres, an asteroid that was discovered on January 1, 1800. His method is still used to track satellites.

additional variable, or multiplier, into the problem, and was first used in the eighteenth century by Joseph Louis Lagrange (1736–1813).

In the following discussion we consider the problem of finding the *maximum* value of a function subject to a constraint. The procedure is the same for finding the minimum value except that the inequalities are reversed. In fact, the method of Lagrange multipliers gives only the possible candidates for the extreme values; verification must be established separately.

Consider finding a solution to the following problem:

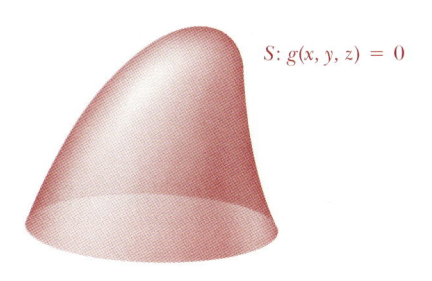

Figure 14.51

(14.57)
Maximize $f(x, y, z)$;
Subject to $g(x, y, z) = 0$
where f and g have continuous first partial derivatives.

Suppose S shown in Figure 14.51 is the 0-level surface of g. If Problem (14.57) has a solution, then at least one constant c exists such that the c-level surface T_c described by

$$f(x, y, z) = c$$

intersects S. The maximum of f subject to the constraint occurs when c is the largest constant for which T_c and S have nonempty intersection.

Let us consider a constant c_0 for which T_{c_0} intersects S. The intersection partitions S into two portions:

S_0: when $f(x, y, z) > c_0$ and $\hat{S}_0$: when $f(x, y, z) \leq c_0$.

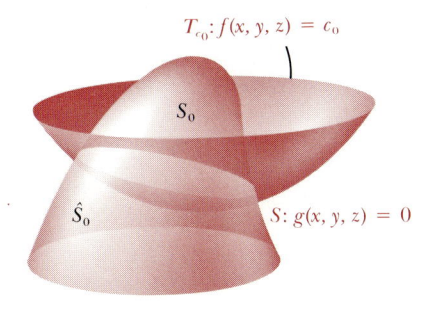

Figure 14.52

Figure 14.52 shows a partitioning of S into S_0 and $\hat{S}_0$.

If $c_1 > c_0$ is a constant for which T_{c_1} intersects S, then another partitioning of S is introduced:

S_1: when $f(x, y, z) > c_1$ and $\hat{S}_1$: when $f(x, y, z) \leq c_1$.

Since $c_1 > c_0$, S_1 is completely contained within S_0, as shown in Figure 14.53. Any constant $c > c_1$ for which the c-level surface, T_c, of f intersects S will induce a similar partition with

S_c: when $f(x, y, z) > c$,

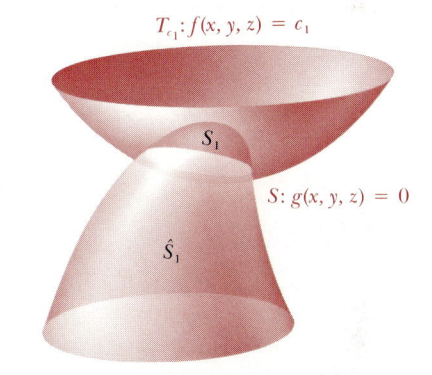

Figure 14.53

contained within S_1.

Consider the largest constant $c = c_{\max}$ for which S and the c-level surface, $T_{c_{\max}}$, of f have a nonempty intersection. The surfaces $T_{c_{\max}}$ and S have a common tangent plane, for if they did not, S would pass completely through $T_{c_{\max}}$ and intersect c-level surfaces of f with $c > c_{\max}$. Thus the maximum value of f subject to the constraint can occur only for a number $c_{\max}$ at which the normals to $T_{c_{\max}}$ and S are parallel.

Recall that $\nabla f(x, y, z)$ is normal to $T_{c_{\max}}$ at (x, y, z) and that $\nabla g(x, y, z)$ is normal to S at this point, as shown in Figure 14.54. Since two vectors are parallel precisely when one is a scalar multiple of the other, (x, y, z) can solve Problem (14.57) only if a constant λ exists with

$$\nabla f(x, y, z) = \lambda \nabla g(x, y, z).$$

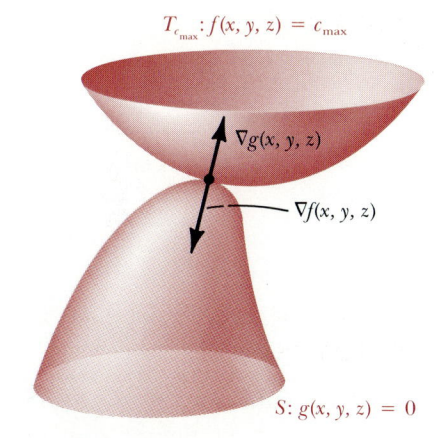

Figure 14.54
$\nabla f(x, y, z) = \lambda \nabla g(x, y, z)$.

The method of Lagrange multipliers proceeds as follows:

(14.58)
Problem: Maximize or minimize $f(x, y, z)$;
  Subject to $g(x, y, z) = 0$.
Solution: Determine (x, y, z) and λ such that
$$g(x, y, z) = 0 \quad \text{and} \quad \nabla f(x, y, z) = \lambda \nabla g(x, y, z).$$

The first example we consider is the problem in Example 8, Section 14.7, of finding the box with fixed volume and minimal cost.

Example 1

Apply the method of Lagrange multipliers to find a solution with $x > 0$, $y > 0$, and $z > 0$, to the problem:

Minimize $C(x, y, z) = 9xy + 16yz + 16xz$;
Subject to $xyz = 36$.

Solution
To place the constraint in the form of (14.58), let $g(x, y, z) = xyz - 36$. Then
$$\nabla g(x, y, z) = yz\mathbf{i} + xz\mathbf{j} + xy\mathbf{k}.$$
Also,
$$\nabla C(x, y, z) = (9y + 16z)\mathbf{i} + (9x + 16z)\mathbf{j} + (16y + 16x)\mathbf{k}.$$
The problem is to find (x, y, z) and λ such that
$$(9y + 16z)\mathbf{i} + (9x + 16z)\mathbf{j} + (16y + 16x)\mathbf{k} = \lambda(yz\mathbf{i} + xz\mathbf{j} + xy\mathbf{k}),$$
that is,
$$9y + 16z = \lambda yz, \quad 9x + 16z = \lambda xz, \quad 16y + 16x = \lambda xy,$$
and satisfying the constraint equation
$$xyz = 36.$$

Multiplying the first three equations, respectively, by x, y, and z gives
$$9xy + 16xz = \lambda xyz, \quad 9xy + 16yz = \lambda xyz, \quad \text{and} \quad 16yz + 16xz = \lambda xyz.$$
Thus,
$$9xy + 16xz = 9xy + 16yz, \quad \text{so} \quad 16xz = 16yz \text{ and } x = y,$$
and
$$9xy + 16xz = 16yz + 16xz, \quad \text{so} \quad 9xy = 16yz \text{ and } x = \frac{16z}{9}.$$

The equation $xyz = 36$ implies that
$$\left(\frac{16z}{9}\right)^2 z = 36, \quad \text{so} \quad z^3 = 36\left(\frac{9}{16}\right)^2 = \frac{729}{64} \text{ and } z = \frac{9}{4}.$$

Thus, $x = \frac{16}{9}z = 4$, $y = x = 4$, and
$$C\left(4, 4, \frac{9}{4}\right) = 9(16) + 16(9) + 16(9) = 432.$$

This must be the minimum value of C since the cost approaches ∞ if any of x, y, or z approaches either 0 or ∞. ▶

Example 2

Find the maximum and minimum values of $f(x, y, z) = 2x - 2y + z$ for (x, y, z) lying on the sphere $x^2 + y^2 + z^2 = 9$.

Solution

The problem is to:

Maximize and minimize $f(x, y, z) = 2x - 2y + z$;
Subject to $g(x, y, z) = x^2 + y^2 + z^2 - 9 = 0$.

The points on the sphere $x^2 + y^2 + z^2 = 9$ form a closed, bounded region in $\mathbb{R}^3$. Since f is a continuous function of three variables, the Extreme Value Theorem implies that f assumes both its maximum and minimum values on this region. Since

$$\nabla f(x, y, z) = 2\mathbf{i} - 2\mathbf{j} + \mathbf{k}$$

and

$$\nabla g(x, y, z) = 2x\mathbf{i} + 2y\mathbf{j} + 2z\mathbf{k},$$

the problem reduces to finding (x, y, z) and λ satisfying

$$2 = 2\lambda x, \quad -2 = 2\lambda y, \quad 1 = 2\lambda z$$

and the constraint equation

$$x^2 + y^2 + z^2 = 9.$$

Solving for x, y, and z in the first three equations and substituting into the constraint equation gives

$$x = \frac{1}{\lambda}, \quad y = -\frac{1}{\lambda}, \quad z = \frac{1}{2\lambda}$$

and

$$9 = x^2 + y^2 + z^2 = \frac{1}{\lambda^2} + \frac{1}{\lambda^2} + \frac{1}{4\lambda^2} = \frac{9}{4\lambda^2}.$$

Thus,

$$\lambda = \tfrac{1}{2} \quad \text{or} \quad \lambda = -\tfrac{1}{2}.$$

When $\lambda = \tfrac{1}{2}$:

$$x = 2, \quad y = -2, \quad z = 1 \quad \text{and} \quad f(2, -2, 1) = 9.$$

When $\lambda = -\frac{1}{2}$:

$$x = -2, \quad y = 2, \quad z = -1 \quad \text{and} \quad f(-2, 2, -1) = -9.$$

The maximum value of $f(x, y, z)$ is 9 and occurs at $(2, -2, 1)$; the minimum value is -9 and occurs at $(-2, 2, -1)$. ▶

The method of Lagrange multipliers has been presented for functions of three variables, since it is to functions of three or more variables that it is applied most often. The method can also be used to solve problems involving two variables.

Example 3

Minimize $f(x, y) = 4x^2 + 9y^2$;
Subject to $2x + 3y = 6$.

Solution

Let $g(x, y) = 2x + 3y - 6$. Since

$$\nabla f(x, y) = 8x\mathbf{i} + 18y\mathbf{j} \quad \text{and} \quad \nabla g(x, y) = 2\mathbf{i} + 3\mathbf{j},$$

the problem is to find (x, y) and λ such that

$$8x = 2\lambda, \quad 18y = 3\lambda, \quad \text{and} \quad 2x + 3y = 6.$$

Thus,

$$\lambda = 4x \quad \text{and} \quad 18y = 12x, \quad \text{so} \quad x = \frac{3y}{2}.$$

The constraint equation implies that

$$2\left(\frac{3y}{2}\right) + 3y = 6, \quad \text{so} \quad y = 1 \quad \text{and} \quad x = \frac{3y}{2} = \frac{3}{2}.$$

Note that as x or y approaches either ∞ or $-\infty$, $f(x, y)$ approaches ∞. This implies that $f(\frac{3}{2}, 1) = 4(\frac{3}{2})^2 + 9(1)^2 = 18$ is the minimum value of f subject to the constraint. ▶

The method of Lagrange multipliers can be modified to handle extrema problems that involve multiple constraints. Consider the problem

Maximize $f(x, y, z)$;
Subject to $g(x, y, z) = 0$ and $h(x, y, z) = 0$.

Suppose the 0-level surfaces of g and h intersect in a curve C, as shown in Figure 14.55. Let $c_{\max}$ represent the largest value of c for which the c-level surface of f intersects this curve and let (x, y, z) denote an intersection point. The tangent plane to this c-level surface of f at (x, y, z) must also be tangent to the curve C, so the normal $\nabla f(x, y, z)$ to the c-level surface is normal to the curve. However, $\nabla g(x, y, z)$ and $\nabla h(x, y, z)$ are also normal to the curve at (x, y, z). So, $\nabla f(x, y, z)$, $\nabla g(x, y, z)$, and $\nabla h(x, y, z)$ lie in a common plane, as shown in Figure

14.8 LAGRANGE MULTIPLIERS

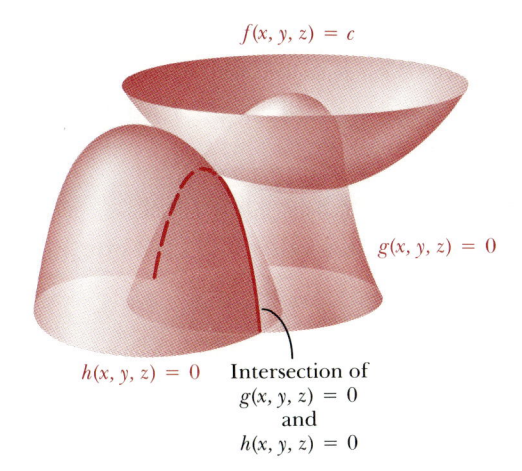

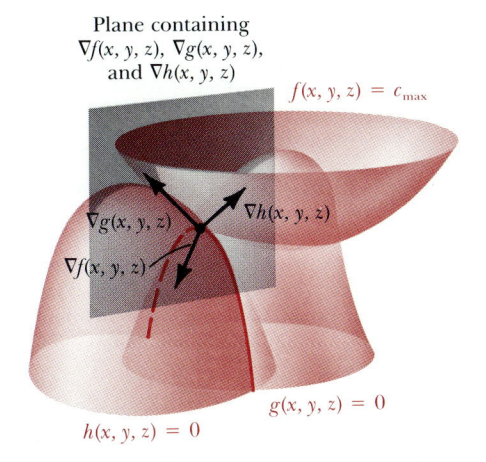

Figure 14.55

Figure 14.56
$\nabla f(x, y, z) = \lambda \nabla g(x, y, z) + \mu \nabla h(x, y, z)$.

14.56. This implies that constants λ and μ exist with

$$\nabla f(x, y, z) = \lambda \nabla g(x, y, z) + \mu \nabla h(x, y, z).$$

The method of Langrange multipliers for the case of two constraints proceeds as follows:

(14.59)
Problem: Maximize or minimize $f(x, y, z)$;
   Subject to $g(x, y, z) = 0$ and $h(x, y, z) = 0$.
Solution: Determine (x, y, z), λ, and μ such that $g(x, y, z) = 0$,
   $h(x, y, z) = 0$, and $\nabla f(x, y, z) = \lambda \nabla g(x, y, z) + \mu \nabla h(x, y, z)$.

Example 4

Consider the intersection of the cone $z^2 = 2x^2 + 2y^2$ and the plane $x + y + z = 1$ (see Figure 14.57). Find the point on this intersection that is closest to the origin.

Solution
The problem is:

   Minimize $D(x, y, z) = \sqrt{x^2 + y^2 + z^2}$;
   Subject to $2x^2 + 2y^2 - z^2 = 0$ and $x + y + z - 1 = 0$.

The partial derivatives of D are quite complicated. However, minimizing

$$[D(x, y, z)]^2 = x^2 + y^2 + z^2$$

is equivalent to minimizing $D(x, y, z)$, and the partial derivatives of the squared function are simpler. For this reason, we consider instead the problem

   Minimize $[D(x, y, z)]^2 = x^2 + y^2 + z^2$;
   Subject to $2x^2 + 2y^2 - z^2 = 0$ and $x + y + z - 1 = 0$.

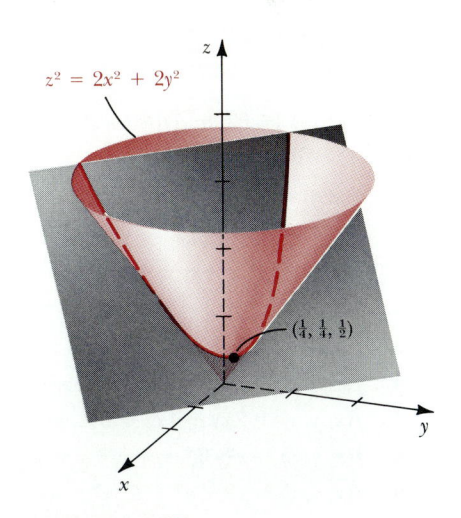

Figure 14.57

If the Lagrange multipliers λ and μ are introduced, the problem is to find (x, y, z), λ, and μ satisfying

$$2x\mathbf{i} + 2y\mathbf{j} + 2z\mathbf{k} = \lambda(4x\mathbf{i} + 4y\mathbf{j} - 2z\mathbf{k}) + \mu(\mathbf{i} + \mathbf{j} + \mathbf{k}),$$

that is,

(1) $2x = 4\lambda x + \mu$,   (2) $2y = 4\lambda y + \mu$,   (3) $2z = -2\lambda z + \mu$,

and the constraint equations. Subtracting Equation (2) from (1) yields

$$2(x - y) = 4\lambda(x - y), \quad \text{which implies that} \quad \lambda = \tfrac{1}{2} \text{ or } x = y.$$

Subtracting Equation (3) from (1) implies that

$$2(x - z) = 2\lambda(2x + z).$$

If $\lambda = \tfrac{1}{2}$, then $2(x - z) = 2x + z$ and $z = 0$. In this case, the constraint equation $z^2 = 2x^2 + 2y^2$ implies that $x = 0$ and $y = 0$. However, $(0, 0, 0)$ does not satisfy the constraint equation $x + y + z = 1$, so $\lambda \ne \tfrac{1}{2}$.

The only alternative is that $x = y$. In this case, the constraint equations become

$$4x^2 - z^2 = 0 \quad \text{and} \quad 2x + z - 1 = 0.$$

Substituting $z = 1 - 2x$ into $4x^2 - z^2 = 0$, we obtain

$$0 = 4x^2 - (1 - 2x)^2 = 4x^2 - 1 + 4x - 4x^2 = 4x - 1.$$

So

$$x = \frac{1}{4}, \quad y = x = \frac{1}{4}, \quad \text{and} \quad z = 1 - 2x = \frac{1}{2}.$$

The point on the intersection of the cone $z^2 = 2x^2 + 2y^2$ and the plane $x + y + z = 1$ that is closest to the origin is $(\tfrac{1}{4}, \tfrac{1}{4}, \tfrac{1}{2})$. ▶

▶ EXERCISE SET 14.8

In Exercises 1–12, find the local extrema of f subject to the given constraints.

1. $f(x, y) = x^2 + y^2$, $x + y = 2$
2. $f(x, y) = x^2 - y^2$, $x^2 + y^2 = 2$
3. $f(x, y) = x^2 - 4xy + 4y^2$, $x + y = 1$
4. $f(x, y) = x^2 + 2xy + y^2$, $x + y = 1$
5. $f(x, y, z) = 4x^2 + y^2 + z^2$, $2x + 3y + 4z = 12$
6. $f(x, y, z) = 2x + 3y + z$, $z = x^2 + y^2$
7. $f(x, y, z) = 2xy + 4xz + 6yz$, $xyz = 48$
8. $f(x, y, z) = x^2 y^2 z^2$, $x + y + z = 1$
9. $f(x, y, z) = x^2 + y^2 + z^2$, $z^2 - xy = 1$
10. $f(x, y, z) = \ln x + \ln y + 3 \ln z$, $x^2 + y^2 + z^2 = 20$
11. $f(x, y, z) = x^2 + y^2 + z^2$, $x^2 + 4y^2 + 4z^2 = 4$, $x - 4y - z = 0$
12. $f(x, y, z) = x^2 + y^2 + (z + 1)^2$, $x^2 - xy + y^2 - z^2 = 1$, $x^2 + y^2 = 1$
13. Find the point on the sphere $x^2 + y^2 + z^2 - 4x - 6y - 8z + 28 = 0$ that is closest to the origin.
14. Find the point on the sphere $x^2 + y^2 + z^2 = 9$ that is closest to the point $(3, 4, -1)$.
15. Use Lagrange multipliers to solve Exercise 61 in Section 14.7.
16. Use Lagrange multipliers to solve Exercise 62 in Section 14.7.
17. Find the rectangular box with maximum volume that

can be enclosed within the ellipsoid $x^2/9 + y^2/16 + z^2/25 = 1$ if the sides of the box are parallel to the coordinate planes.

18. Use Lagrange multipliers to find the dimensions of a 12-oz cylindrical can that can be constructed using the least amount of material (see the figure). Compare these dimensions to those of a standard can. [*Note:* One fluid ounce is approximately 1.805 in^3.]

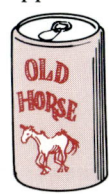

19. A standard cylindrical oil drum has a capacity of 55 gal (approximately 12,705 in^3) (see the figure). Use Lagrange multipliers to find the dimensions of the drum that requires the minimum amount of material to construct.

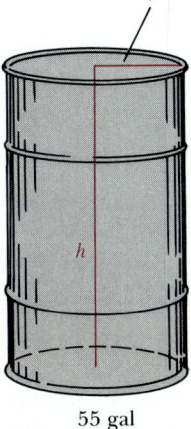

20. A company needs a warehouse to contain a million cubic feet. They estimate that the floor and ceiling of the building will cost $3 per square foot to construct and the walls will cost $7 per square foot. Use Lagrange multipliers to find the cost of the most economical rectangular building.

21. Suppose a rectangular room is to be built inside a recreation shelter in the shape of the upper half of an ellipsoid with equation $x^2/400 + y^2/2500 + z^2/100 = 1$. Find the dimensions of the room with maximum volume if the corners of the ceiling are to lie on the ellipsoid.

22. Electrical power is to be provided to the recreation shelter described in Exercise 21 from an existing transformer located at $(0, 75, 20)$. Use Newton's method to approximate, to within 10^{-3}, the point on the shelter closest to the transformer.

23. An oil drum is usually constructed by crimping the top and bottom to the sides. Suppose that this crimping requires one extra inch for the radius of the top and the bottom and two extra inches for the height of the side. Approximate, to within 10^{-3}, the dimensions of the 55-gal drum that uses the least material.

24. Use Lagrange multipliers to show that for all real numbers a, b, c,
$$(abc)^{1/3} \leq \frac{a+b+c}{3},$$
with equality occurring if and only if $a = b = c$. [*Hint:* Maximize abc subject to the constraint that the sum of $a, b,$ and c is constant.]

25. Use the inequality in Exercise 24 to find the absolute minimum of the function C described by
$$C(x, y) = 9xy + \frac{576}{y} + \frac{576}{x}, \quad x, y > 0.$$

26. Use the inequality in Exercise 24 to find the absolute minimum of the function C described by $C(x, y, z) = 9xy + 16yz + 16xz$ subject to the constraint $xyz = 36$.

27. Use the inequality in Exercise 24 to show that the minimum amount of material required to construct a closed rectangular box containing 8 ft^3 is 24 ft^2.

28. The quantities Q_1 and Q_2 of output of a two-product firm depend on labor l and capital k according to the production functions $Q_1 = 3l_1\sqrt[3]{k_1}$ and $Q_2 = 3l_2\sqrt[3]{k_2}$. Assume there is a fixed total quantity $L = l_1 + l_2$ of labor and a fixed total quantity $K = k_1 + k_2$ of capital available to the firm and that the respective prices p_1 and p_2 are fixed. Show that the maximum revenue occurs when $l_1/L = k_1/K$.

29. Consider $f(x, y, z) = x^2 + y^2 + z$ and $g(x, y, z) = x^2 + 4y^2 + 9z^2 - 36$. Show that there are solutions to $\nabla f(x, y, z) = \lambda \nabla g(x, y, z)$ that neither maximize nor minimize $f(x, y, z)$ subject to satisfying $g(x, y, z) = 0$.

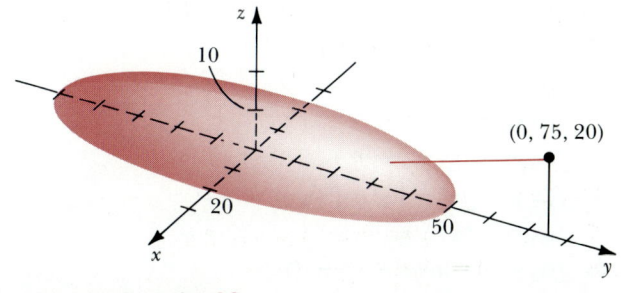

Figure for Exercise 22.

▶ IMPORTANT TERMS AND RESULTS

CONCEPT	PAGE	CONCEPT	PAGE
Continuous	785	Surface normal	822
Partial derivative	790	Tangent plane	822
Differentiable	799	Extrema	827
Differential	802	Extreme Value Theorem	828
Chain Rule	807	Critical point	829
Directional derivative	813	Discriminant	834
Gradient	816	Lagrange multipliers	840
Steepest ascent and descent	817		

▶ REVIEW EXERCISES

1. **a)** Find the domain of the function described by $f(x, y) = \sqrt{4 - x^2 - y^2}$.
   b) Sketch the graph of f.

2. **a)** Find the domain of the function described by $f(x, y) = \sqrt{x^2 + y^2 - 9}$.
   b) Sketch the graph of f.

3. **a)** Determine any points of discontinuity of the function described by
   $$f(x, y) = \begin{cases} x^2 + y^2, & \text{if } x^2 + y^2 \leq 1, \\ 0, & \text{if } x^2 + y^2 > 1. \end{cases}$$
   b) Sketch the graph of f.

4. **a)** Determine any points of discontinuity of the function described by
   $$f(x, y) = \begin{cases} 4x^2 + 9y^2, & \text{if } 4x^2 + 9y^2 \leq 36, \\ 0, & \text{if } 4x^2 + 9y^2 > 36. \end{cases}$$
   b) Sketch the graph of f.

In Exercises 5–12, find the first partial derivatives of the function.

5. $f(x, y) = e^{xy}$
6. $f(x, y) = x/y + y/x$
7. $f(x, y) = x^2 y + 2xe^{1/y}$
8. $f(x, y) = 3x^2 - 2y^2 + 2xy$
9. $f(x, y) = (x^2 + xy)^3$
10. $f(x, y, z) = \ln(x^2 + y^2 + z^2)$
11. $f(x, y, z) = x^2 + y^2 - 2xy \cos z$
12. $f(x, y, z) = e^{x+2y-z^2}$

In Exercises 13–16, a function is described by $z = f(x, y)$ together with values of Δx and Δy.

a) Determine dz.
b) Use dz to approximate the value of the function at the specified point.

13. $z = y\sqrt{x} + x\sqrt[3]{y}$, $\Delta x = 0.1$, $\Delta y = -0.1$; $f(1.1, 0.9)$
14. $z = x^2 y + xy^2$, $\Delta x = 0.2$, $\Delta y = 0.03$; $f(2.2, -0.97)$
15. $z = y \sin x$, $\Delta x = \pi/16$, $\Delta y = 0.2$; $f(5\pi/16, 3.2)$
16. $z = e^x \ln y$, $\Delta x = -0.02$, $\Delta y = 0.2$; $f(0.98, 1.2)$

In Exercises 17–22, find all critical points of the function and determine whether each critical point is a relative maximum, a relative minimum, or a saddle point.

17. $f(x, y) = 4 - x^2 - y^2$
18. $f(x, y) = 4 + x^2 - y^2$
19. $f(x, y) = x^2 e^{x+y}$
20. $f(x, y) = (x - 1) \ln y - x^2$
21. If $f(x, y, z) = xe^{yz} + yze^x$, find f_{yxz} and $\partial^3 f/\partial x^2\, \partial y$.
22. Show that $f_{xy} = f_{yx}$ if $f(x, y) = x/(x + \sin y)$.

In Exercises 23–28, find the gradient of the function.

23. $f(x, y) = x^2 + 2xy + y^2$
24. $f(x, y) = x^2 + 2xy \cos x$
25. $f(x, y) = 7 - x - y$
26. $f(x, y) = x^2 + 2xy - y^2$
27. $f(x, y, z) = 5x^2 - 2y^2 + z$
28. $f(x, y, z) = e^{-x+y^2+2z}$

In Exercises 29–34, find the directional derivative of f at P in the direction of $\mathbf{v}$.

29. $f(x, y) = x^2 + 2xy + y^2$, $\mathbf{v} = \frac{4}{5}\mathbf{i} - \frac{3}{5}\mathbf{j}$, $P(1, 1)$
30. $f(x, y) = x^2 + 2xy \cos x$, $\mathbf{v} = (\sqrt{3}/3)\mathbf{i} - (\sqrt{6}/3)\mathbf{j}$, $P(0, 1)$
31. $f(x, y) = 7 - x - y$, $\mathbf{v} = \mathbf{i} + \mathbf{j}$, $P(2, 2)$
32. $f(x, y) = x^2 + 2xy - y^2$, $\mathbf{v} = 4\mathbf{i} + 3\mathbf{j}$, $P(2, 4)$
33. $f(x, y, z) = 5x^2 - 2y^2 + z$, $\mathbf{v} = 3\mathbf{i} - 4\mathbf{k}$, $P(1, 1, 1)$
34. $f(x, y, z) = e^{-x+y^2+2z}$, $\mathbf{v} = \langle \sqrt{3}, -\sqrt{3}, \sqrt{3} \rangle$, $P(1, 1, 0)$

In Exercises 35–38, (a) find the direction in which the function is increasing most rapidly at P and (b) find the maximum rate of increase.

35. $f(x, y) = x^2 + xy$, $P(1, -1)$
36. $f(x, y) = \sqrt{1 - x^2 - y^2}$, $P(0, 1)$
37. $f(x, y) = x^2 + y^2$, $P(-1, 2)$
38. $f(x, y) = 6 - 2x - 3y$, $P(0, 0)$

In Exercises 39–44, find an equation of the tangent plane to the surface at the given point.

39. $z = x^2 + y^2 - 1$, $(1, 2, 4)$
40. $z = 2x + 3y + 1$, $(1, -1, 0)$
41. $z = xy \sin xy$, $(1, \pi/2, \pi/2)$
42. $z = x^2 - 2y^2 + 3xy$, $(2, -1, -4)$
43. $z = x^2 + 4y^2$, $(-2, 1, 8)$
44. $x^2 + y^2 - z^2 = 0$, $(1, 0, 1)$

45. If $f(x, y) = x^2y + xy^2$, find the directional derivative of f at $(2, 1)$ in the direction toward $(3, 2)$.
46. If $f(x, y, z) = xy^2 + y^2z^3 + z^3x$, find the directional derivative of f at $(2, -1, 1)$ in the direction toward $(3, 4, 5)$.
47. Consider the function described by $f(x, y) = 16 - (x - 2)^2 - (y - 3)^2$.
    a) Sketch the graph of f.
    b) Find the directional derivative of f at $(4, 4)$ in the direction toward the point $(2, 3)$.
    c) Find the direction in which the rate of change of f at $(4, 4)$ is a maximum and determine this maximum rate of change.
    d) Find an equation of the tangent plane to the graph of f at the point $(4, 4, 11)$.
    e) Find an equation of the tangent line to the curve of intersection of the graph of f and the plane $y = 4$ at the point $(4, 4, 11)$.
    f) Find the direction of steepest ascent from the point $(4, 4, 11)$.
48. Consider the function described by $f(x, y) = \sqrt{x^2 + 3y^2}$.
    a) Sketch the graph of f.

b) Find the directional derivative of f at $(1, 1)$ in the direction toward the point $(2, 2)$.
c) Find the direction in which the rate of change of f at $(1, 1)$ is a maximum.
d) Determine the maximum rate of change of f at $(1, 1)$.
e) Find an equation of the tangent plane to the graph of f at the point $(1, 1, 2)$.
f) Find an equation of the tangent line at $(1, 1, 2)$ to the curve of intersection of the graph of f and the plane $x = 1$.
g) Find the direction of steepest ascent for the graph of f from the point $(1, 1, 2)$.

49. Find the slope of the curve at the point $(1, 2, 4)$ formed by the intersection of the surface $z = xy^2$ with a plane perpendicular to the y-axis.
50. If $f(x, y, z) = 3x^2 + xy - 2y^2 - yz + z^2$, find the rate of change of $f(x, y, z)$ at $(1, -2, -1)$ in the direction of $\mathbf{v} = 2\mathbf{i} - 2\mathbf{j} - \mathbf{k}$.
51. Suppose $z = f(x, y) = e^x \sin y + x^3$. Use differentials to approximate the change in the value of the function as (x, y) changes from $(1, \pi)$ to $(1.01, \pi + 0.01)$.
52. Suppose $z = f(x, y) = \ln(x^2 + y^2 + 1)$. Use differentials to approximate the change in the value of the function as (x, y) changes from $(1, 1)$ to $(1.02, 0.98)$.
53. Find $\partial f/\partial r$ and $\partial f/\partial \theta$ if $f(x, y) = xy$, where $x = e^{2r} \cos \theta$ and $y = e^r \sin \theta$.
54. If $w = x/y + y/z + z/x$, show that
$$x \frac{\partial w}{\partial x} + y \frac{\partial w}{\partial y} + z \frac{\partial w}{\partial z} = 0.$$
55. Suppose $w = u^2v + uv^2 + 2u - 3v$, $u = \sin(x + y + z)$, and $v = \cos(x + y - z)$. Find $\partial w/\partial x$ when $x = \pi/2$, $y = \pi/4$, $z = \pi/4$.
56. If $z = x^2 + 2xy + y^2$, $x = t \cos t$, and $y = t \sin t$, find dz/dt.
57. In which direction from the point $(1, 1)$ is the directional derivative of $f(x, y) = x^2 + y^2$ zero?
58. Let
$$f(x, y) = \begin{cases} \dfrac{y^3 - x^3}{y^2 + x^2}, & \text{if } (x, y) \neq (0, 0), \\ 0, & \text{if } (x, y) = (0, 0). \end{cases}$$
Find $f_x(0, 0)$ and $f_y(0, 0)$.
59. Show that
$$\lim_{(x,y) \to (1,1)} \frac{(y - 1)^2}{(x - 1)^2 + (y - 1)^2}$$
does not exist.
60. If $x + y + z$ is constant, show that the product xyz is a maximum when $x = y = z$.

61. Find the point on the plane $2x + 2y + z = 4$ that is closest to the origin.

62. Find the point on the plane $2x + 2y + z = 4$ that is closest to the point $(1, 2, 3)$.

63. Find the maximum value of $P(x, y, z) = 900z - 400x - 100y$ subject to the constraint $z = 5 - 1/x - 1/y$, $x > 0$, $y > 0$.

64. Maximize $f(x, y) = x^2 - y^2 - y$ subject to the constraint $x^2 + y^2 - 1 = 0$.

65. The radius of a right circular cone is decreasing at the rate of 2 in./min and the height is increasing at 3 in./min. Find the rate of change of the volume when the height is 8 in. and the radius is 4 in.

66. A closed metal can in the shape of a right circular cylinder is to have an inside height of 10 in., an inside radius of 2 in., and a thickness of 0.1 in. The cost of the metal to be used is 20 cents per cubic inch. Use differentials to find the approximate cost of the metal to be used in manufacturing the can.

67. Use Lagrange multipliers to determine the dimensions of the open rectangular box containing 8 ft³ that has the minimum surface area.

68. A silo with a volume of 8000 ft³ is to be constructed in the shape of a right circular cylinder topped with a hemisphere. Use Lagrange multipliers to determine the height and radius of the silo that will minimize the surface area of the silo, assuming that the silo needs a bottom.

69. Suppose the silo described in Exercise 68 is topped with a cone of equal height and radius rather than a hemisphere. Use Lagrange multipliers to solve the problem in this case.

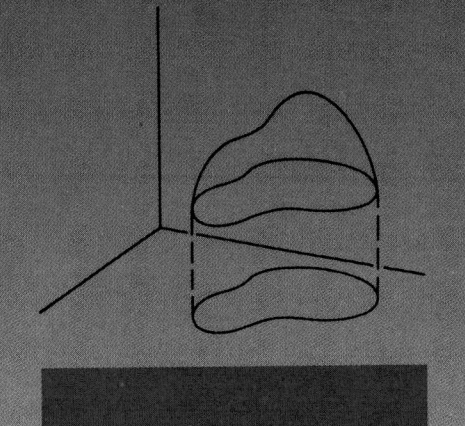

15

Integral Calculus of Multivariable Functions

The definite integral of a single-variable function was introduced with the problem of determining the area of a region in the plane. This problem led to a Riemann sum, whose limit produced a definite integral. We later found that other concepts, such as volumes of solids of rotation, arc length of curves, moments, work, and fluid pressure, are also approximated by Riemann sums and determined by definite integrals.

The integration of multivariable functions has a similar development. We begin with the problem of determining the volume of a region in space. This geometric application describes the general nature of the multiple integral. Once the definition has been presented, we will consider a number of different applications of multiple integrals.

15.1

Double Integrals

Suppose f is a continuous nonnegative function of two variables defined on a bounded region R in the xy-plane. Our objective is to determine the volume V of the solid bounded by the cylinder with base R and top lying in the surface described by $z = f(x, y)$, as shown in Figure 15.1.

We begin by drawing a grid of lines through R parallel to the x- and y-axes, as shown in Figure 15.2. Let $R_1, R_2, R_3, \ldots, R_n$ denote those rectangles formed by the grid that lie entirely within R. The collection of these rectangles is denoted $\mathcal{P}$, and $\|\mathcal{P}\|$ denotes the maximum length of the diagonals of the rectangles R_i.

We arbitrarily choose (x_i, y_i) in R_i for each i and compute $f(x_i, y_i)$. The volume ΔV_i of the parallelepiped with base R_i and height $f(x_i, y_i)$ is

$$\Delta V_i = f(x_i, y_i) \, \Delta A_i,$$

where ΔA_i denotes the area of the rectangle R_i. A typical parallelepiped is shown in Figure 15.3. The volume of the portion of the solid with base R_i and top lying in the surface $z = f(x, y)$ is approximated by ΔV_i.

Summing the volumes of the parallelepipeds gives an approximation to the total volume:

(15.1)
$$V \approx \sum_{i=1}^{n} f(x_i, y_i) \, \Delta A_i.$$

This sum is called a **Riemann sum** for f on the region R. Color Plate X gives an illustration of this concept.

Example 1

A solid has a rectangular base R in the xy-plane bounded by $x = 2$, $y = 2$, $y = 3$, and the y-axis (see Figure 15.4a). The top of the solid is the surface $z = 7 - x^2 - y$. Partition the base with the lines $x = 0$, $x = 1$, $x = 2$, $y = 2$, $y = \frac{5}{2}$, and $y = 3$.

Figure 15.1

A solid bounded below by R and above by $z = f(x, y)$.

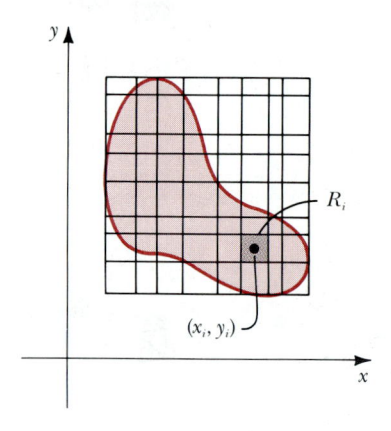

Figure 15.2

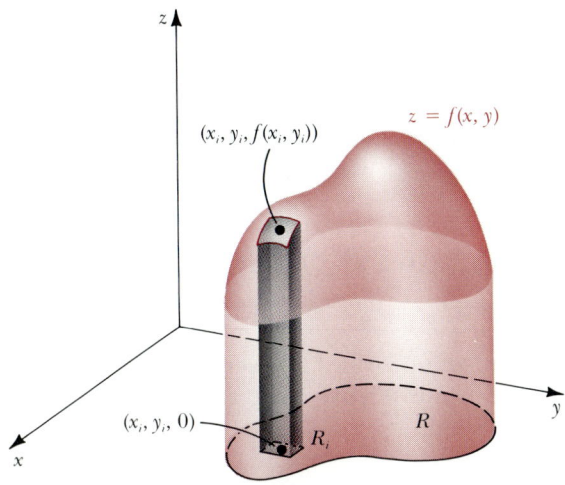

15.1 DOUBLE INTEGRALS

Figure 15.3
A typical rectangular volume element.

In each rectangle R_i, choose the point (x_i, y_i) as the lower left corner of R_i, as shown in Figure 15.4(b). Use this partition and choice of points to construct a Riemann sum for approximating the volume of the solid.

Solution
Since each R_i has area $\frac{1}{2}$,

$$\sum_{i=1}^{4} f(x_i, y_i)\, \Delta A_i = f(0, 2)(\tfrac{1}{2}) + f(1, 2)(\tfrac{1}{2}) + f(0, \tfrac{5}{2})(\tfrac{1}{2}) + f(1, \tfrac{5}{2})(\tfrac{1}{2})$$

$$= \tfrac{1}{2}\left(5 + 4 + \tfrac{9}{2} + \tfrac{7}{2}\right) = \tfrac{17}{2}.$$

This Riemann sum gives an upper bound for the volume of the solid, since the tops of all the parallelepipeds used in the approximation lie above the top of the solid, as shown in Figure 15.4(c).

Figure 15.4

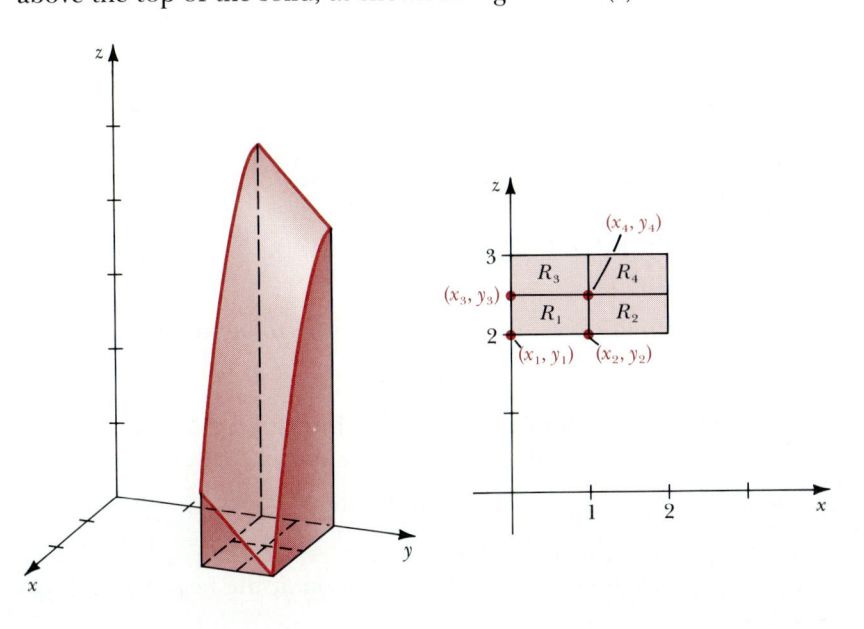

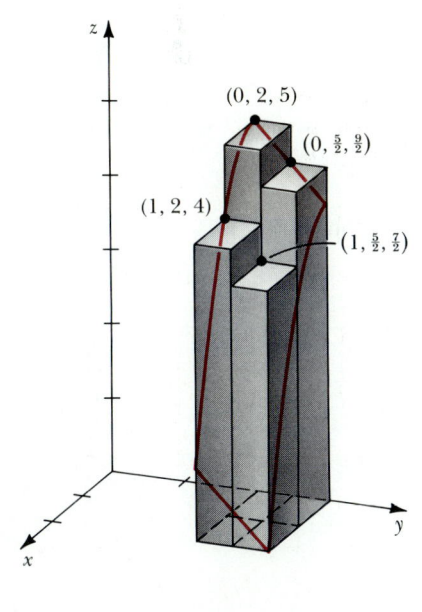

(a)   (b)   (c)

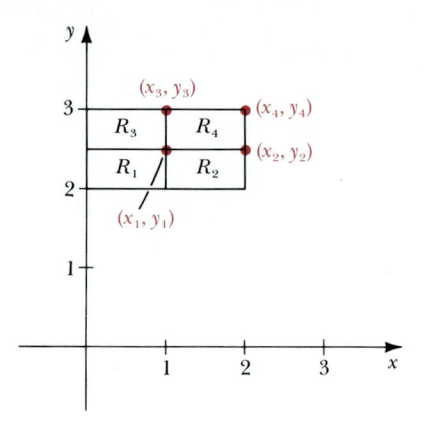

Figure 15.5

Example 2

Repeat Example 1 using the same partition, but choosing (x_i, y_i) as the upper right corner of R_i for each $i = 1, 2, 3,$ and 4. (See Figure 15.5.)

Solution
In this case,

$$\sum_{i=1}^{4} f(x_i, y_i) A_i = f(1, \tfrac{5}{2})(\tfrac{1}{2}) + f(2, \tfrac{5}{2})(\tfrac{1}{2}) + f(1, 3)(\tfrac{1}{2}) + f(2, 3)(\tfrac{1}{2})$$

$$= \tfrac{1}{2}(\tfrac{7}{2} + \tfrac{1}{2} + 3 + 0) = \tfrac{7}{2}.$$

This Riemann sum provides a lower bound for the volume of the solid, since, as shown in Figure 15.6, all the parallelepipeds are contained within the solid. ▶

The volume V of the solid bounded by the cylinder with base R and top lying in the surface $z = f(x, y)$ is the limit of the approximations

$$\sum_{i=1}^{n} f(x_i, y_i) \, \Delta A_i$$

as the grid lines in both directions become closer together (that is, as $\|\mathscr{P}\| \to 0$). This produces more parallelepipeds, the sum of whose volumes approaches the volume V. So,

$$V = \lim_{\|\mathscr{P}\| \to 0} \sum_{i=1}^{n} f(x_i, y_i) \, \Delta A_i.$$

This procedure for computing volume provides the basic notion of the integral of a function f of two variables over a region R in the plane. The integral is called a double integral and denoted $\iint$ to emphasize that the region of integration is expressed in terms of two variables.

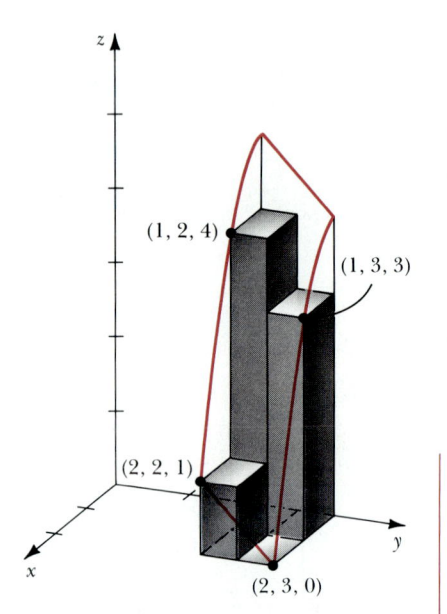

Figure 15.6

> **(15.2) DEFINITION**
>
> If a function f of two variables is defined on a bounded region R in the plane, the **double integral of f over R** is defined by
>
> $$\iint_R f(x, y) \, dA = \lim_{\|\mathscr{P}\| \to 0} \sum_{i=1}^{n} f(x_i, y_i) \, \Delta A_i,$$
>
> provided this limit exists. In this case, f is said to be **integrable** on the region R.

Definition (15.2) does not require an integrable function to be either nonnegative or continuous. It can be shown, however, that:

If f is continuous and the bounded region R is closed, then f is integrable on R.

Definition (15.3) follows from the discussion at the beginning of this section.

(15.3) DEFINITION

If f is a continuous nonnegative function of two variables defined on a closed and bounded region R, then the **volume** of the solid with base R and top in the surface $z = f(x, y)$ is

$$V = \iint_R f(x, y) \, dA.$$

It can be shown, and will be assumed, that this definition of volume is consistent with that given in Chapter 5.

The limit in Definition (15.2) is taken over every collection of grid lines and every choice (x_i, y_i) in R_i, $i = 1, 2, \ldots, n$. To be precise,

$$\lim_{\|\mathcal{P}\| \to 0} \sum_{i=1}^n f(x_i, y_i) \, \Delta A_i = \iint_R f(x, y) \, dA$$

means that for every number $\varepsilon > 0$, there exists a number $\delta > 0$ such that if grid lines produce a set of rectangles $\mathcal{P} = \{R_1, \ldots, R_n\}$, with $\|\mathcal{P}\| < \delta$, then the Riemann sum $\sum_{i=1}^n f(x_i, y_i) \, \Delta A_i$ has the property that

$$\left| \sum_{i=1}^n f(x_i, y_i) \, \Delta A_i - \iint_R f(x, y) \, dA \right| < \varepsilon$$

for any collection (x_i, y_i) in R_i, $i = 1, 2, 3, \ldots, n$. Satisfying these conditions ensures that the value of the definite integral is independent of the choice of partitions and particular points (x_i, y_i) used in the evaluation of the Riemann sums.

Example 3

Evaluate $\iint_R \sqrt{16 - x^2 - y^2} \, dA$, where R is the disk $x^2 + y^2 \leq 16$ in the xy-plane.

Solution

It would be tedious to evaluate this integral using the definition. However, since $\sqrt{16 - x^2 - y^2} \geq 0$, this double integral is the volume of the solid with base R and top lying in the surface $z = \sqrt{16 - x^2 - y^2}$. This surface is the upper portion of the sphere $x^2 + y^2 + z^2 = 16$, shown in Figure 15.7. Since a sphere of radius r has volume $V = \frac{4}{3}\pi r^3$, and this double integral has the volume of a hemisphere of radius 4,

$$\iint_R \sqrt{16 - x^2 - y^2} \, dA = \frac{1}{2}\left(\frac{4}{3}\pi \cdot 4^3\right) = \frac{128}{3}\pi.$$

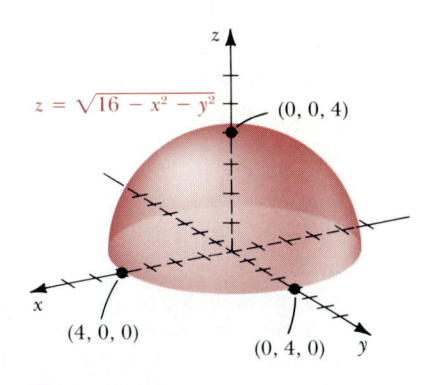

▶ Figure 15.7

The double integral satisfies the usual arithmetic properties associated with the definite integral. For example, if f and g are integrable on

R and c is a constant, then

$$(15.4) \quad \iint_R (f \pm g)(x, y)\, dA = \iint_R f(x, y)\, dA \pm \iint_R g(x, y)\, dA$$

and

$$(15.5) \quad \iint_R cf(x, y)\, dA = c \iint_R f(x, y)\, dA.$$

In addition, if R is a region composed of subregions R_1 and R_2 with disjoint interiors (see Figure 15.8) on which f is integrable, then

$$(15.6) \quad \iint_R f(x, y)\, dA = \iint_{R_1} f(x, y)\, dA + \iint_{R_2} f(x, y)\, dA.$$

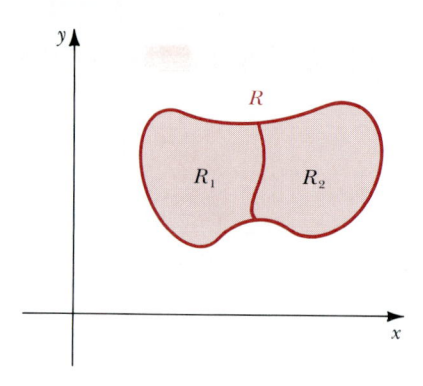

Figure 15.8

$\iint_R f(x, y)\, dA = \iint_{R_1} f(x, y)\, dA + \iint_{R_2} f(x, y)\, dA.$

▶ EXERCISE SET 15.1

1. Find the value of the Riemann sum for the function and partition described in Example 1 if (x_i, y_i) in R_i is chosen as (a) the lower right corner of R_i; (b) the upper left corner of R_i; (c) the midpoint of R_i.

2. Suppose $f(x, y) = x^2 y$, R is the rectangle in the xy-plane bounded by $x = 0$, $y = 0$, $x = 3$, and $y = 2$, and $\mathcal{P}$ is the partition defined by the lines bounding R together with the lines $x = 1$, $x = 2$, and $y = 1$. Find the value of the Riemann sum if (x_i, y_i) is chosen as (a) the lower left corner of R_i; (b) the upper right corner of R_i; (c) the midpoint of R_i.

3. Suppose $f(x, y) = y^2 - x^2$, R is the triangle in the xy-plane with vertices $(1, 1)$, $(5, 4)$, and $(7, 0)$, and $\mathcal{P}$ is the partition defined by the lines at integral values of x and y. Find the value of the Riemann sum if (x_i, y_i) is chosen as (a) the lower left corner of R_i; (b) the upper right corner of R_i; (c) the midpoint of R_i.

4. Find the values of the Riemann sums for the function described in Exercise 3 if the partition is enlarged to include those rectangles bounded by the lines $x = \frac{3}{2}$, $x = \frac{11}{2}$, and $y = \frac{1}{2}$.

5. Suppose $f(x, y) = 3 - x - y$, R is the triangle in the xy-plane bounded by $x = 0$, $y = 0$, and $y = 3 - x$, and $\mathcal{P}$ is the partition defined by the lines $x = 0$, $x = 1$, $x = 2$, $y = 0$, $y = 1$, and $y = 2$. Find the value of the Riemann sum if (x_i, y_i) is chosen as the upper left corner of R_i.

6. Use the definition of the double integral to show that if f is integrable over R and c is a constant, then cf is integrable over R and

$$\iint_R cf(x, y)\, dA = c \iint_R f(x, y)\, dA.$$

7. Use the definition of the double integral to show that if f and g are both integrable over R and $f(x, y) \geq g(x, y)$ on R, then

$$\iint_R f(x, y)\, dA \geq \iint_R g(x, y)\, dA.$$

8. Use the definition of the double integral to show that if f is the constant function $f(x, y) = k$ and R is the rectangular region bounded by $x = a$, $x = b$, $y = c$, and $y = d$, with $a < b$ and $c < d$, then

$$\iint_R f(x, y)\, dA = k(b - a)(d - c).$$

9. Use the method shown in Example 3 to evaluate $\iint_R \sqrt{9 - x^2 - y^2}\, dA$, where R is the disk $x^2 + y^2 \leq 9$ in the xy-plane.

10. Evaluate $\iint_R 4\, dA$, where R is the disk $x^2 + y^2 \leq 9$ in the xy-plane.

15.2 Iterated Integrals

The double integral of a function of two variables is defined as the limit of Riemann sums.

You will recall that very little could be accomplished with the definite integral of a single-variable function until the Fundamental Theorem of Calculus provided a means of determining integrals without using the limit process. The same problem occurs with double integrals; in fact, it has been compounded.

In this section we see how a double integral of a function f of two variables x is converted into a pair of single integrals. These are called *iterated integrals* because they are performed iteratively, that is, one after the other.

The definition of the double integral involves the limit of Riemann sums

$$\sum_{i=1}^{n} f(x_i, y_i) \, \Delta A_i,$$

where ΔA_i is the area of the ith rectangle R_i shown in Figure 15.9. If Δx_i and Δy_i represent the dimensions of this rectangle, then

$$\Delta A_i = \Delta x_i \, \Delta y_i.$$

An increment such as Δx_i or Δy_i is part of a Riemann sum that leads to a single integral with respect to x or to y, respectively. With ΔA_i expressed as a product of these increments, we should not be surprised that there is a relationship between a double integral and integrals with respect to x and to y. Two questions concerning this relationship are:

i) Over which regions in the plane can we convert a double integral into iterated integrals?
ii) What form does the iterated integral assume when the conditions in (i) are fulfilled?

The answer to the first question has its basis in the following definition.

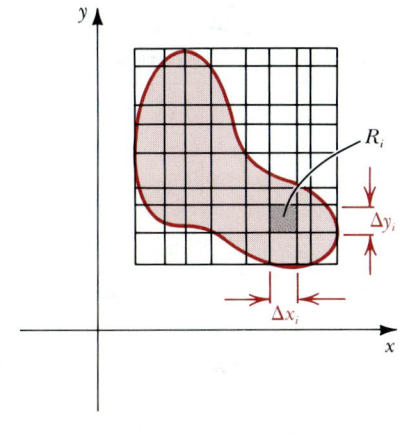

Figure 15.9

(15.7) DEFINITION

i) A *vertically simple*, or **x-simple** region has the form

$$a \leq x \leq b, \qquad g_1(x) \leq y \leq g_2(x),$$

where g_1 and g_2 are continuous on $[a, b]$.

ii) A *horizontally simple*, or **y-simple** region has the form

$$c \leq y \leq d, \qquad h_1(y) \leq x \leq h_2(y),$$

where h_1 and h_2 are continuous on $[c, d]$.

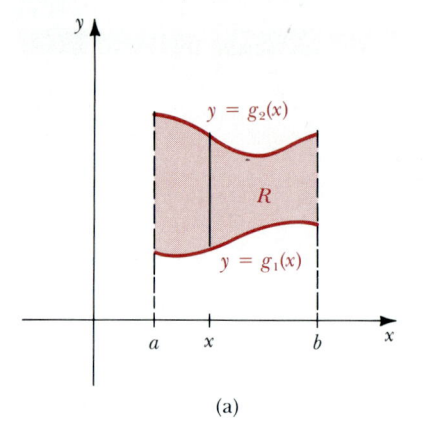

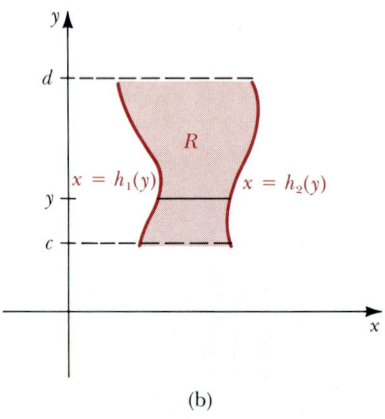

Figure 15.10
(a) An x-simple region; (b) a y-simple region.

Every vertical line intersecting an x-simple region does so in a connected line segment, as shown in Figure 15.10(a); every horizontal line intersecting a y-simple region does so in a connected line segment, as shown in Figure 15.10(b). The most elementary answer to question (i) is that integrals over x-simple and over y-simple regions can be converted into iterated integrals. In addition, integrals over regions that can be decomposed into x-simple and y-simple regions can be converted into sums of iterated integrals.

To see the form that the iterated integral should assume, consider the double integral of a continuous nonnegative function f over an x-simple region R. The geometric interpretation of this double integral is the volume of the solid shown in Figure 15.11, with base R and top lying in the surface $z = f(x, y)$:

$$(15.8) \qquad V = \iint_R f(x, y) \, dA.$$

For each x in $[a, b]$, let $A(x)$ denote the area of the region formed by intersecting the solid with the plane through $(x, 0, 0)$ perpendicular to the x-axis, as shown in Figure 15.11. If we view this region projected into the yz-plane, we see from Figure 15.12 that the cross-sectional area $A(x)$ of the solid is

$$A(x) = \int_{g_1(x)}^{g_2(x)} f(x, y) \, dy.$$

Since f, g_1, and g_2 are continuous, $A(x)$ depends continuously on x. From Section 5.2 we know that the volume of the solid is

$$(15.9) \qquad V = \int_a^b A(x) \, dx = \int_a^b \left[\int_{g_1(x)}^{g_2(x)} f(x, y) \, dy \right] dx.$$

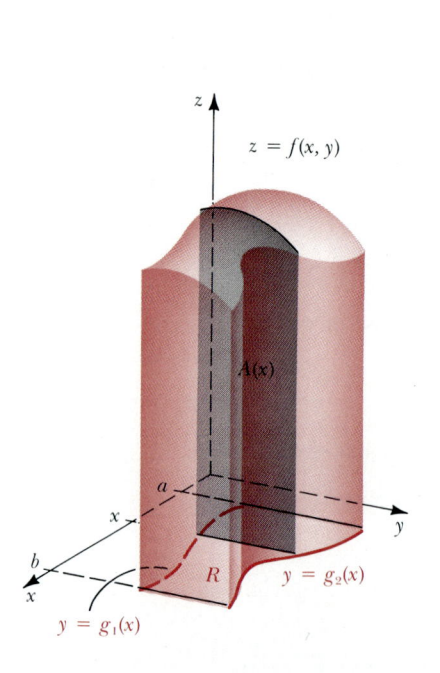

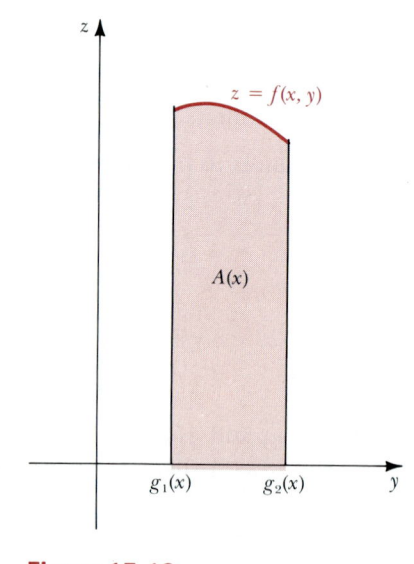

Figure 15.11
The area of the cross section is $A(x)$.

Figure 15.12
$$A(x) = \int_{g_1(x)}^{g_2(x)} f(x, y) \, dy.$$

If we assume that this result is consistent with the definition of volume given in the previous section, Equations (15.8) and (15.9) imply that

$$\iint_R f(x, y)\, dA = \int_a^b \left[\int_{g_1(x)}^{g_2(x)} f(x, y)\, dy \right] dx.$$

The expression on the right is an **iterated integral** and is generally expressed with the brackets deleted. The following result is used to relate double integrals to iterated integrals.

(15.10) THEOREM

i) If f is continuous on the x-simple region R described by

$$a \leq x \leq b, \qquad g_1(x) \leq y \leq g_2(x),$$

then

$$\iint_R f(x, y)\, dA = \int_a^b \int_{g_1(x)}^{g_2(x)} f(x, y)\, dy\, dx.$$

ii) If f is continuous on the y-simple region R described by

$$c \leq y \leq d, \qquad h_1(y) \leq x \leq h_2(y),$$

then

$$\iint_R f(x, y)\, dA = \int_c^d \int_{h_1(y)}^{h_2(y)} f(x, y)\, dx\, dy.$$

Example 1

Find the volume of the solid with top in the surface $z = 7 - x^2 - y$ and a rectangular base in the xy-plane bounded by the planes $x = 0$, $x = 2$, $y = 2$, and $y = 3$.

Solution

The volume to be calculated is shown in Figure 15.13. The results of Examples 1 and 2 in Section 15.1 imply that this volume exceeds $\frac{7}{2}$ but is less than $\frac{17}{2}$. The base region R is described as an x-simple region by

$$0 \leq x \leq 2, \qquad g_1(x) \leq y \leq g_2(x),$$

where $g_1(x) = 2$ and $g_2(x) = 3$. So

$$V = \iint_R (7 - x^2 - y)\, dA = \int_0^2 \int_2^3 (7 - x^2 - y)\, dy\, dx.$$

In the integral $\int_2^3 (7 - x^2 - y)\, dy$, x is considered a constant. Thus,

$$\int_2^3 (7 - x^2 - y)\, dy = \left[7y - x^2 y - \frac{y^2}{2} \right]_{y=2}^{y=3}$$

$$= \left(21 - 3x^2 - \frac{9}{2} \right) - (14 - 2x^2 - 2) = \frac{9}{2} - x^2$$

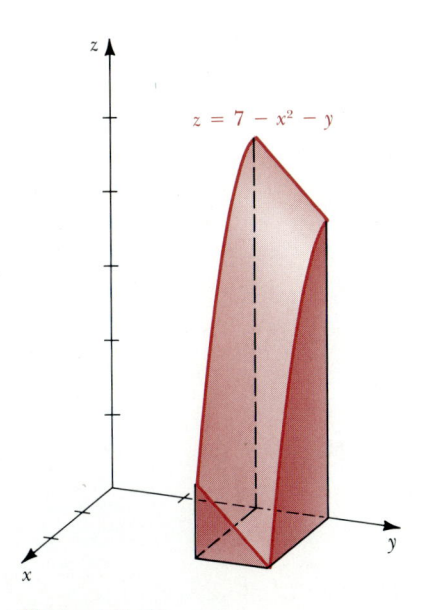

Figure 15.13

and

$$V = \int_0^2 \left(\frac{9}{2} - x^2\right) dx = \frac{9}{2}x - \frac{x^3}{3}\Big]_0^2 = 9 - \frac{8}{3} = \frac{19}{3}.$$

Note that the volume is within the bounds found in Examples 1 and 2 of Section 15.1. ▶

Example 2

Find the volume of the solid bounded above by the graph of $f(x, y) = 1 - xy$, below by the *xy*-plane, and by the cylinders $y = x^2$ and $y = x$.

Solution

The base region of the solid in the *xy*-plane is shown in Figure 15.14(a) and the solid in 15.14(b). The graphs of $y = x^2$ and $y = x$ intersect at $(0, 0)$ and $(1, 1)$, so R is described by

$$0 \leq x \leq 1, \qquad x^2 \leq y \leq x.$$

Since the graph of f lies above R, the volume is

$$V = \int_0^1 \int_{x^2}^x (1 - xy) \, dy \, dx = \int_0^1 y - \frac{xy^2}{2}\Big]_{y=x^2}^{y=x} dx$$

$$= \int_0^1 \left[\left(x - \frac{x^3}{2}\right) - \left(x^2 - \frac{x^5}{2}\right)\right] dx.$$

Thus,

$$V = \left[\frac{x^2}{2} - \frac{x^4}{8} - \frac{x^3}{3} + \frac{x^6}{12}\right]_0^1 = \left(\frac{1}{2} - \frac{1}{8} - \frac{1}{3} + \frac{1}{12}\right) = \frac{1}{8}.$$

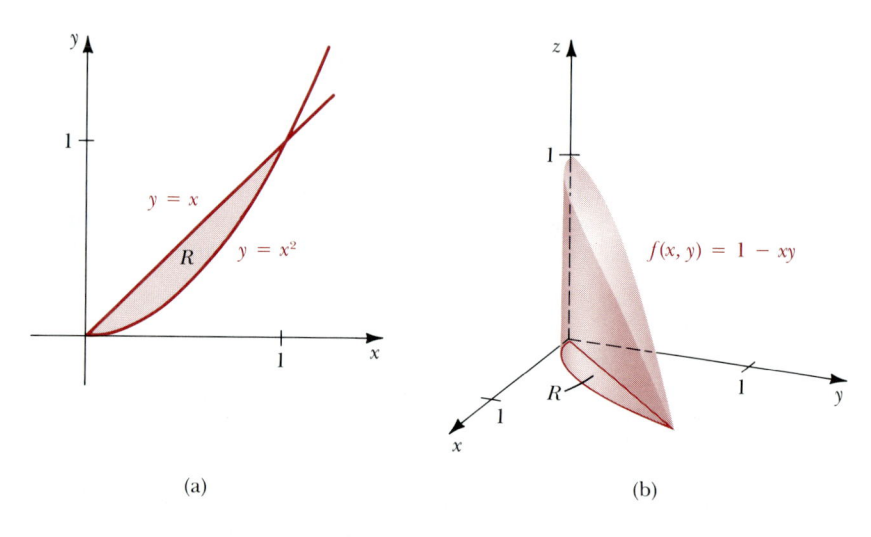

▶

Figure 15.14 (a) (b)

The following example shows how the procedure in Examples 1 and 2 is extended to find the volume of a solid bounded between two surfaces.

Example 3

Find the volume of the solid in the first octant bounded below by the plane $x + z = 1$ and above by the plane $4x + y + z = 4$.

Solution

The solid is shown in Figure 15.15. To use a double integral to find the volume, we need the region in the xy-plane over which to integrate. This is obtained by finding the intersection of the two surfaces and projecting this intersection into the xy-plane. Solving the equations $z = 4 - 4x - y$ and $z = 1 - x$ simultaneously, we have

$$1 - x = 4 - 4x - y, \quad \text{so} \quad y = 3 - 3x.$$

The region R in the first octant is described by $0 \leq y \leq 3 - 3x$, $0 \leq x \leq 1$. The volume bounded by R and the upper boundary $z = 4 - 4x - y$ is $\iint_R (4 - 4x - y) \, dA$ and the volume bounded by R and the lower boundary $z = 1 - x$ is $\iint_R (1 - x) \, dA$, so the volume between the planes is

$$V = \iint_R (4 - 4x - y) \, dA - \iint_R (1 - x) \, dA$$

$$= \iint_R [(4 - 4x - y) - (1 - x)] \, dA$$

$$= \int_0^1 \int_0^{3-3x} (3 - 3x - y) \, dy \, dx$$

$$= \int_0^1 \left[3y - 3xy - \frac{y^2}{2} \right]_{y=0}^{y=3-3x} dx$$

$$= \int_0^1 \left[3(3 - 3x) - 3x(3 - 3x) - \frac{(3 - 3x)^2}{2} \right] dx$$

$$= \frac{9}{2} \int_0^1 (1 - x)^2 \, dx$$

$$= \left[-\frac{9}{2} \frac{(1 - x)^3}{3} \right]_0^1 = \frac{9}{6} = \frac{3}{2}. \quad \blacktriangleright$$

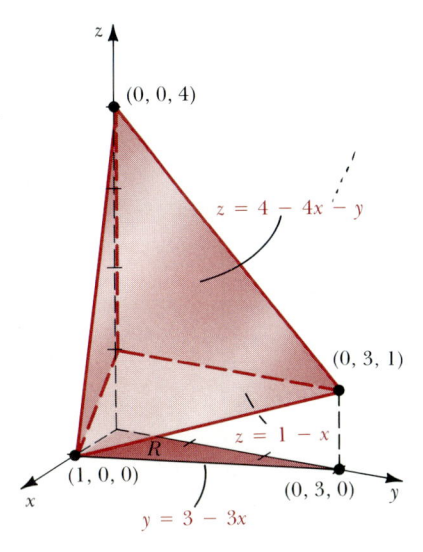

Figure 15.15

Example 4

Evaulate $\iint_R (x^2 - 2y) \, dA$, where R is the region bounded by the x-axis, the line $x = 1$, and the graph of $y = 2x^2$.

Solution

We fix x in the interval $[0, 1]$ and consider the vertical line through x, as shown in Figure 15.16. The region is described as an x-simple region by

$$0 \leq x \leq 1 \quad \text{and} \quad 0 \leq y \leq 2x^2.$$

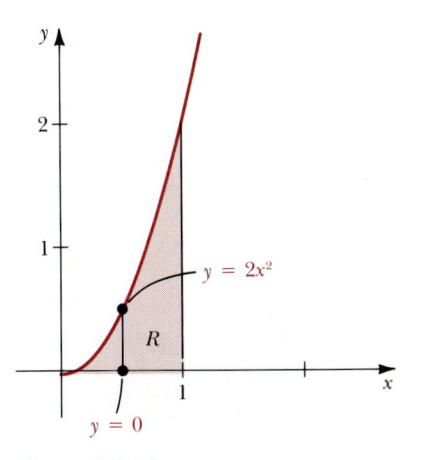

Figure 15.16

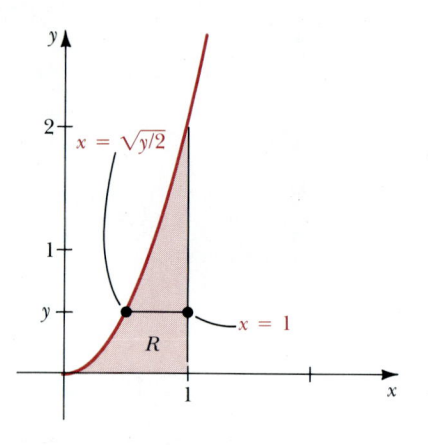

Figure 15.17

Thus,

$$\iint_R (x^2 - 2y)\, dA = \int_0^1 \int_0^{2x^2} (x^2 - 2y)\, dy\, dx$$

$$= \int_0^1 \left[x^2 y - y^2 \right]_{y=0}^{y=2x^2} dx$$

$$= \int_0^1 (2x^4 - 4x^4)\, dx = -\frac{2}{5} x^5 \Big]_0^1 = -\frac{2}{5}.$$

Alternatively, we can solve this problem by fixing y in the interval $[0, 2]$ and considering the horizontal line through y, as shown in Figure 15.17. The region is described as a y-simple region by

$$0 \le y \le 2, \qquad \sqrt{y/2} \le x \le 1.$$

Using this representation gives us

$$\iint_R (x^2 - 2y)\, dA = \int_0^2 \int_{\sqrt{y/2}}^1 (x^2 - 2y)\, dx\, dy$$

$$= \int_0^2 \left[\frac{x^3}{3} - 2xy \right]_{x=\sqrt{y/2}}^{x=1} dy$$

$$= \int_0^2 \left[\left(\frac{1}{3} - 2y\right) - \left(\frac{y^{3/2}}{6\sqrt{2}} - \frac{2y^{3/2}}{\sqrt{2}}\right) \right] dy$$

$$= \int_0^2 \left(\frac{1}{3} - 2y + \frac{11}{6\sqrt{2}} y^{3/2} \right) dy$$

$$= \left[\frac{1}{3} y - y^2 + \frac{11}{6\sqrt{2}} y^{5/2} \cdot \frac{2}{5} \right]_0^2$$

$$= \frac{2}{3} - 4 + \frac{11}{15\sqrt{2}} (4\sqrt{2}) = -\frac{2}{5}. \quad \blacktriangleright$$

Example 4 demonstrates that for certain regions, $\iint_R f(x, y)\, dx$ can be expressed as either

$$\int_a^b \int_{g_1(x)}^{g_2(x)} f(x, y)\, dy\, dx \quad \text{or} \quad \int_c^d \int_{h_1(y)}^{h_2(y)} f(x, y)\, dx\, dy.$$

Changing from one iterated-integral form to the other is called **reversing the order of integration.** The following example shows that this technique is essential in the evaluation of certain double integrals.

Example 5

Evaluate the iterated integral $\int_0^1 \int_x^1 e^{y^2} dy\, dx$.

Solution

The limits of integration in this iterated integral imply that the region R over which the integration (see Figure 15.18a) is to be performed can be

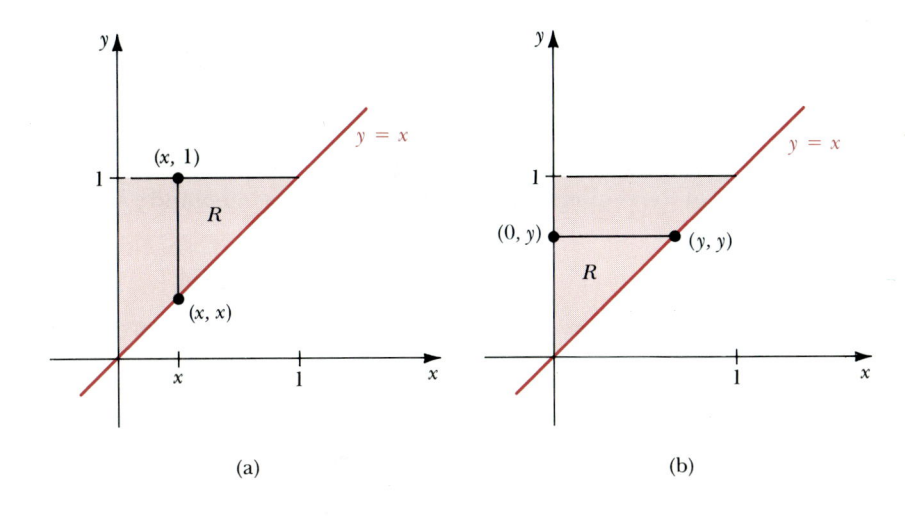

Figure 15.18
(a) R is an x-simple region; (b) R is also a y-simple region.

described in the x-simple form
$$0 \leq x \leq 1, \quad x \leq y \leq 1.$$

There is no elementary antiderivative for $f(y) = e^{y^2}$, so we cannot use the Fundamental Theorem of Calculus to evaluate $\int e^{y^2} dy$.

However, the region R can also be described as the y-simple region, as shown in Figure 15.18(b),
$$0 \leq y \leq 1, \quad 0 \leq x \leq y,$$

so the integral is expressed by reversing the order of integration as
$$\int_0^1 \int_x^1 e^{y^2} \, dy \, dx = \iint_R e^{y^2} \, dA = \int_0^1 \int_0^y e^{y^2} \, dx \, dy.$$

Now we can apply the Fundamental Theorem of Calculus to give
$$\int_0^1 \int_0^y e^{y^2} \, dx \, dy = \int_0^1 x e^{y^2} \Big]_{x=0}^{x=y} dy$$
$$= \int_0^1 y e^{y^2} \, dy = \frac{1}{2} e^{y^2} \Big]_0^1 = \frac{1}{2}(e - 1). \quad \blacktriangleright$$

The area A of a region R in the plane has the same numerical value as the volume of the solid with base R and constant height 1 (see Figure 15.19), so
$$A = \iint_R 1 \, dA.$$

In particular, if R is a region bounded by the lines $x = a$ and $x = b$ and the graphs of the continuous functions f and g, where $g(x) \leq f(x)$ for x in $[a, b]$, then
$$A = \iint_R 1 \, dA = \int_a^b \int_{g(x)}^{f(x)} dy \, dx = \int_a^b [f(x) - g(x)] \, dx.$$

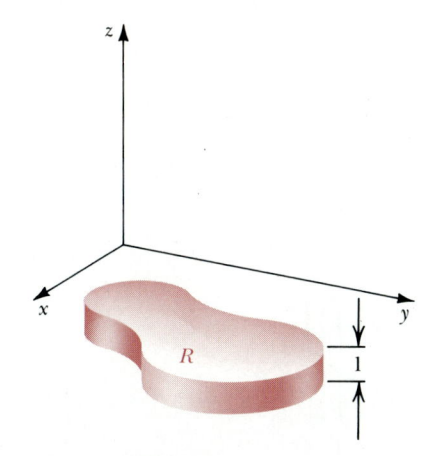

Figure 15.19
The area of R is $\iint_R dA$.

This is the formula given in Section 5.1 for the area of a region bounded between two curves.

Example 6

Find the area of the region bounded by the curves $y = \sin x$ and $y = \cos x$ for $0 \leq x \leq \pi/3$.

Solution

From Figure 15.20, we see that $\cos x \geq \sin x$ in the interval $[0, \pi/4]$ and that $\cos x \leq \sin x$ in the interval $[\pi/4, \pi/3]$. Thus,

$$A = \int_0^{\pi/4} \int_{\sin x}^{\cos x} dy\, dx + \int_{\pi/4}^{\pi/3} \int_{\cos x}^{\sin x} dy\, dx$$

$$= \int_0^{\pi/4} y\Big]_{y=\sin x}^{y=\cos x} dx + \int_{\pi/4}^{\pi/3} y\Big]_{y=\cos x}^{y=\sin x} dx$$

$$= \int_0^{\pi/4} (\cos x - \sin x)\, dx + \int_{\pi/4}^{\pi/3} (\sin x - \cos x)\, dx$$

$$= (\sin x + \cos x)\Big]_0^{\pi/4} + (-\cos x - \sin x)\Big]_{\pi/4}^{\pi/3}$$

$$= \left(\frac{\sqrt{2}}{2} + \frac{\sqrt{2}}{2} - 1\right) + \left(\frac{-1}{2} - \frac{\sqrt{3}}{2} + \frac{\sqrt{2}}{2} + \frac{\sqrt{2}}{2}\right)$$

$$= 2\sqrt{2} - \frac{3}{2} - \frac{\sqrt{3}}{2} \approx 0.46.$$

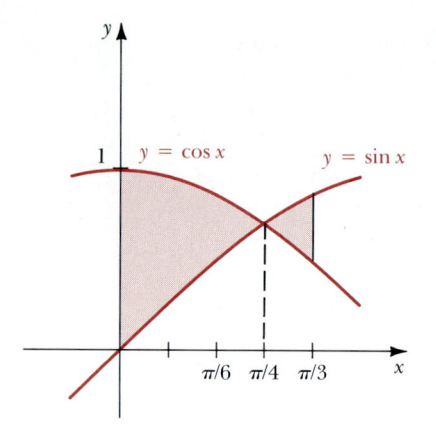

Figure 15.20

▶ EXERCISE SET 15.2

In Exercises 1–16, evaluate the integral.

1. $\displaystyle\int_1^2 \int_3^4 xy^2\, dy\, dx$

2. $\displaystyle\int_1^2 \int_3^4 xy^2\, dx\, dy$

3. $\displaystyle\int_{-1}^1 \int_{-2}^4 (x^2 + y)\, dy\, dx$

4. $\displaystyle\int_0^2 \int_{-3}^0 xe^{-y}\, dx\, dy$

5. $\displaystyle\int_{-\pi}^{3\pi/2} \int_0^{2\pi} (y \sin x + x \cos y)\, dy\, dx$

6. $\displaystyle\int_0^1 \int_{\pi}^{2\pi} (x + \sin y)\, dy\, dx$

7. $\displaystyle\int_1^3 \int_{-2}^1 x^2 y^2\, dy\, dx$

8. $\displaystyle\int_{-1}^1 \int_0^1 xy\, e^{x^2+y^2}\, dx\, dy$

9. $\displaystyle\int_0^2 \int_x^{2x} (x^2 + y^3)\, dy\, dx$

10. $\displaystyle\int_0^2 \int_y^{2y} (x^2 + y^3)\, dx\, dy$

11. $\displaystyle\int_0^{\pi} \int_0^x \cos x\, dy\, dx$

12. $\displaystyle\int_0^{\pi} \int_0^y \cos x\, dx\, dy$

13. $\displaystyle\int_0^1 \int_0^x (x^2 + \sqrt{y})\, dy\, dx$

14. $\displaystyle\int_0^{\pi} \int_{x/2}^x (\sin x + \cos y)\, dy\, dx$

15. $\displaystyle\int_1^e \int_1^x \ln xy\, dy\, dx$

16. $\int_0^{\pi/3} \int_0^{\sin y} \dfrac{1}{\sqrt{1-x^2}}\, dx\, dy$

17–32. **a)** Sketch the graphs of the regions over which the integrations in Exercises 1–16 are performed, and express the integral as a double integral that reverses the order of integration.
b) Evaluate the double integral obtained and compare the result to the previously obtained value.

Exercises 33–38 describe the boundaries of a solid whose base lies in the *xy*-plane. (a) Sketch the region. Use double integrals to compute (b) the area of the base of the solid and (c) the volume of the solid.

33. $x = 0$, $x = 1$, $y = 0$, $y = 2$, $z = 0$, $z = x^2 y$
34. $x = -1$, $x = 2$, $y = -1$, $y = 3$, $z = 0$, $z = 4 + 2x + 2y$
35. $x = 2$, $y = 1$, $y = x$, $z = 0$, $z = x + y^2$
36. $x = -1$, $x = 1$, $y = -x$, $y = x$, $z = 0$, $z = x^2 + y^2$
37. $x = -1$, $x = 1$, $y = 2x + 2$, $y = 3 + 2x - x^2$, $z = 0$, $z = x + y + 1$
38. $x = 0$, $x = \pi/4$, $y = \cos x$, $y = \sin x$, $z = 0$, $z = 3x + 2y$

In Exercises 39–46, use double integration to find the volume of the solid.

39. The solid in the first octant bounded by the plane $z = 4 - x - y$
40. The solid bounded by the cylinder $x^2 + y^2 = 1$ and the planes $z = 0$ and $z = 1 - x$
41. The solid bounded above by the paraboloid $z = 4 - x^2 - y^2$ and below by the *xy*-plane
42. The solid in the first octant bounded by the cylinder $x^2 + z^2 = 4$ and the plane $z = 3 - y$
43. The wedge cut from the cylinder $x^2 + y^2 = 9$ by the planes $z = 0$ and $z = y$, when $y \geq 0$
44. The wedge cut from the cylinder $x^2 + y^2 = 9$ by the planes $z = 0$ and $z = 2y$, when $y \geq 0$
45. The solid bounded by the cylinder $9x^2 + 4y^2 = 36$ and the planes $z = 0$ and $x + y + 2z = 6$
46. The solid in the first octant bounded above by planes $z = y$ and $x + y + z = 3$
47. Suppose f is continuous and g is defined by
$$g(t) = \int_0^t \int_0^x f(x, y)\, dy\, dx.$$
Determine $g'(t)$.

Putnam exercises

48. Show that the integral equation
$$f(x, y) = 1 + \int_0^x \int_0^y f(u, v)\, du\, dv$$
has at most one solution continuous for $0 \leq x \leq 1$, $0 \leq y \leq 1$. (This exercise was problem 5, part I of the eighteenth William Lowell Putnam examination given on February 8, 1958. The examination and its solution are in the January 1961 issue of the *American Mathematical Monthly*, pp. 18–22.)

49. If f is a positive, monotone decreasing function defined on $0 \leq x \leq 1$, prove that
$$\dfrac{\int_0^1 x\, (f(x))^2\, dx}{\int_0^1 x\, f(x)\, dx} \leq \dfrac{\int_0^1 (f(x))^2\, dx}{\int_0^1 f(x)\, dx}.$$
(This exercise was problem 3, part II of the seventeenth William Lowell Putnam examination given on December 6, 1956. The examination and its solution are in the November 1957 issue of the *American Mathematical Monthly*, pp. 649–654.)

15.3
Double Integrals in Polar Coordinates

Certain regions in the plane are more naturally described by polar, rather than rectangular, equations. In this section we describe the double integral of a function over a region represented in polar coordinates, following as closely as possible the development of double integrals using rectangular coordinates.

Suppose that f is a continuous nonnegative function of the two variables r and θ and that R is a bounded region in the plane described in polar coordinates by

$$0 \le \rho_1 \le r \le \rho_2 \quad \text{and} \quad \alpha \le \theta \le \beta,$$

where $0 \le \beta - \alpha \le 2\pi$. Place a polar grid over the region R, as shown in Figure 15.21. This grid consists of lines through the origin and arcs of circles centered at the origin. These are the curves that result when one of the polar coordinates is set to a constant value.

As in the case of rectangular regions, the polar elements formed by the grid and lying totally within R are labeled $R_1, R_2, \ldots, R_n$. The set of all these polar elements is denoted $\mathcal{P}$, and $\|\mathcal{P}\|$ is the length of the longest diagonal of the polar elements in $\mathcal{P}$.

Figure 15.22 shows a typical polar element R_i with area ΔA_i, where the lines through the origin forming the boundaries of R_i are denoted by $\theta = \theta_{i-1}$ and $\theta = \theta_i$ and the boundaries formed by the circular arcs are denoted by $r = r_{i-1}$ and $r = r_i$.

In Section 9.3, we found that the area of the sector of the circle bounded by $\theta = \theta_{i-1}$, $\theta = \theta_i$, and $r = r_i$ is

$$\frac{1}{2} r_i^2 (\theta_i - \theta_{i-1}) = \frac{1}{2} r_i^2 \Delta \theta_i.$$

Similarly, the area bounded by $\theta = \theta_{i-1}$, $\theta = \theta_i$, and $r = r_{i-1}$ is

$$\frac{1}{2} r_{i-1}^2 \Delta \theta_i.$$

Since R_i has boundaries $\theta = \theta_{i-1}$, $\theta = \theta_i$, $r = r_{i-1}$, and $r = r_i$, the area of R_i is

$$\Delta A_i = \frac{1}{2} r_i^2 \Delta \theta_i - \frac{1}{2} r_{i-1}^2 \Delta \theta_i$$

$$= \frac{1}{2} (r_i + r_{i-1})(r_i - r_{i-1}) \Delta \theta_i = \frac{r_i + r_{i-1}}{2} \Delta r_i \Delta \theta_i.$$

To approximate the volume ΔV_i bounded by the cylinder with base R_i and above by the graph of f, we evaluate f at a point in R_i and multiply by ΔA_i. The limit defining the double integral is independent of the manner in which the particular points at which the Riemann sums are evaluated are chosen. This permits us to choose for evaluation the point $(\hat{r}_i, \hat{\theta}_i)$, as shown in Figure 15.23, where

$$\hat{r}_i = \frac{r_i + r_{i-1}}{2} \quad \text{and} \quad \hat{\theta}_i = \frac{\theta_i + \theta_{i-1}}{2}.$$

Then

$$\Delta V_i = f(\hat{r}_i, \hat{\theta}_i) \Delta A_i = f(\hat{r}_i, \hat{\theta}_i) \frac{1}{2} (r_i + r_{i-1}) \Delta r \Delta \theta_i = f(\hat{r}_i, \hat{\theta}_i) \hat{r}_i \Delta r_i \Delta \theta_i.$$

The total volume is approximated by the Riemann sum

$$V \approx \sum_{i=1}^{n} \Delta V_i = \sum_{i=1}^{n} f(\hat{r}_i, \hat{\theta}_i) \hat{r}_i \Delta r_i \Delta \theta_i$$

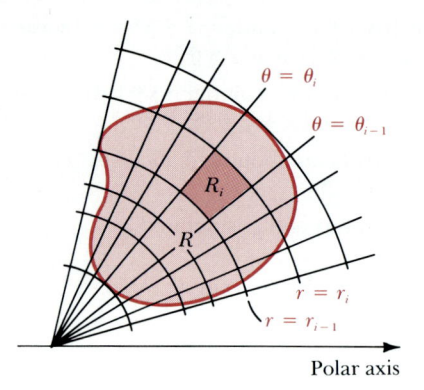

Figure 15.21

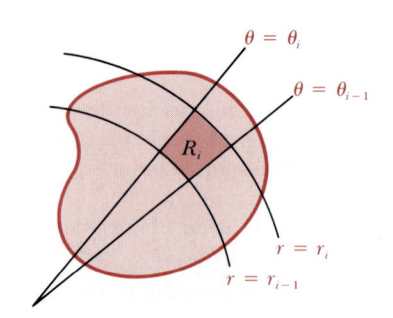

Figure 15.22

15.3 DOUBLE INTEGRALS IN POLAR COORDINATES

Figure 15.23
A typical polar volume element.

and
$$\iint_R f(r, \theta) \, dA = V = \lim_{\|\mathcal{P}\| \to 0} \sum_{i=1}^{n} f(\hat{r}_i, \hat{\theta}_i) \, \hat{r}_i \, \Delta r_i \, \Delta \theta_i.$$

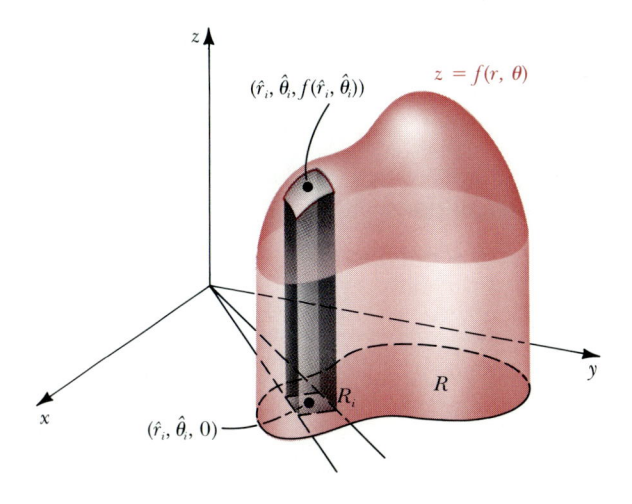

Color Plate XI also gives an illustration of this concept.

Regions analogous to the x-simple and y-simple regions in rectangular coordinates are the θ-simple and r-simple regions defined below.

(15.11) DEFINITION

i) A **θ-simple** region has the form
$$\alpha \leq \theta \leq \beta \leq \alpha + 2\pi \quad \text{and} \quad 0 \leq g_1(\theta) \leq r \leq g_2(\theta),$$
where g_1 and g_2 are continuous on $[\alpha, \beta]$.

ii) An **r-simple** region has the form
$$0 \leq a \leq r \leq b \quad \text{and} \quad h_1(r) \leq \theta \leq h_2(r) \leq h_1(r) + 2\pi,$$
where h_1 and h_2 are continuous on $[a, b]$.

Figure 15.24 shows a θ-simple region and Figure 15.25 an r-simple region.

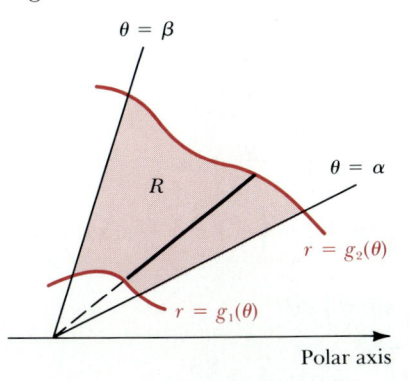

Figure 15.24
A θ-simple region.

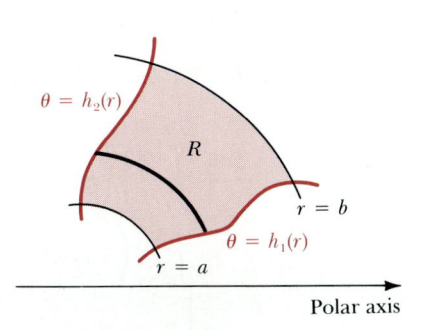

Figure 15.25
An r-simple region.

The following result for regions represented in polar coordinates is similar to the result for rectangular coordinates in Theorem (15.10).

(15.12) THEOREM

i) If f is continuous on the θ-simple region R described by

$$\alpha \leq \theta \leq \beta \leq \alpha + 2\pi \qquad 0 \leq g_1(\theta) \leq r \leq g_2(\theta),$$

then

$$\iint_R f(r, \theta)\, dA = \int_\alpha^\beta \int_{g_1(\theta)}^{g_2(\theta)} f(r, \theta) r\, dr\, d\theta.$$

ii) If f is continuous on the r-simple region R described by

$$0 \leq a \leq r \leq b \qquad h_1(r) \leq \theta \leq h_2(r) \leq h_1(r) + 2\pi,$$

then

$$\iint_R f(r, \theta)\, dA = \int_a^b \int_{h_1(r)}^{h_2(r)} f(r, \theta) r\, dr\, d\theta.$$

Example 1

Find $\iint_R (3x + y)\, dA$, where R is the region in the first quadrant that lies inside the circle $x^2 + y^2 = 4$ and outside the circle $x^2 + y^2 = 1$.

Solution

The region R is sketched in Figure 15.26. It is described in polar coordinates as both a θ-simple and an r-simple region by

$$0 \leq \theta \leq \frac{\pi}{2}, \qquad 1 \leq r \leq 2.$$

Although the integrand $3x + y$ is not expressed in polar form, we can easily convert it to this form using the relationships

$$x = r\cos\theta \quad \text{and} \quad y = r\sin\theta.$$

Thus,

$$\iint_R (3x + y)\, dA = \int_0^{\pi/2} \int_1^2 (3r\cos\theta + r\sin\theta) r\, dr\, d\theta$$

$$= \int_0^{\pi/2} \left[r^3 \cos\theta + \frac{r^3}{3}\sin\theta \right]_{r=1}^{r=2} d\theta$$

$$= \int_0^{\pi/2} \left[\left(8\cos\theta + \frac{8}{3}\sin\theta \right) - \left(\cos\theta + \frac{1}{3}\sin\theta \right) \right] d\theta$$

$$= \int_0^{\pi/2} \left(7\cos\theta + \frac{7}{3}\sin\theta \right) d\theta$$

$$= \left[7\sin\theta - \frac{7}{3}\cos\theta \right]_0^{\pi/2} = 7 + \frac{7}{3} = \frac{28}{3}.$$

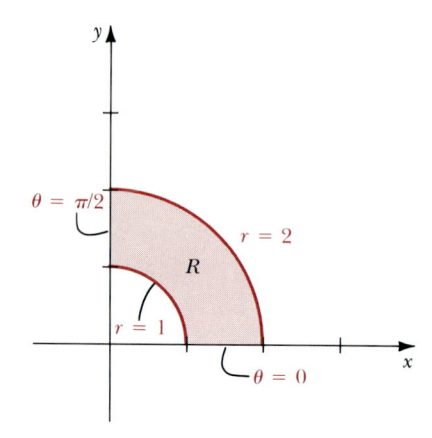

Figure 15.26

Example 2

A silo has the shape of a right circular cylinder with a hemispherical top. The cylinder has a height of 40 ft and the radius of the cylinder and hemisphere is 5 ft. How much grain will the silo hold?

Solution

We place a coordinate system as shown in Figure 15.27, with the base of the silo in the xy-plane and the positive z-axis through the center of the silo. The top of the silo lies on the sphere with radius 5 and center at $(0, 0, 40)$. This implies that the top of the silo is described by

$$x^2 + y^2 + (z - 40)^2 = 25,$$

where

$$40 \le z \le 45.$$

That is,

$$z = 40 + \sqrt{25 - x^2 - y^2}.$$

The volume of the silo is

$$V = \iint_R (40 + \sqrt{25 - x^2 - y^2})\, dA,$$

where R is the set of points (x, y) in the xy-plane that satisfies

$$x^2 + y^2 \le 25.$$

This region is described more easily in polar coordinates by

$$0 \le r \le 5, \quad 0 \le \theta \le 2\pi.$$

Since $25 - x^2 - y^2 = 25 - (x^2 + y^2) = 25 - r^2$, the volume is

$$V = \int_0^{2\pi} \int_0^5 [40 + \sqrt{25 - r^2}]\, r\, dr\, d\theta$$

$$= \int_0^{2\pi} \left[20r^2 - \frac{1}{3}(25 - r^2)^{3/2} \right]_0^5 d\theta$$

$$= \int_0^{2\pi} \left(500 + \frac{125}{3} \right) d\theta$$

$$= \frac{1625}{3} \theta \Big]_0^{2\pi}$$

$$= \frac{3250}{3} \pi \text{ ft}^3.$$

This is equivalent to approximately 2735 bushels of grain. ▶

Iterated integrals involving polar coordinates are used to find volumes bounded between two surfaces when the intersection of the surfaces is projected into a region of the plane that can be described in one of the forms of Theorem (15.12).

Figure 15.27

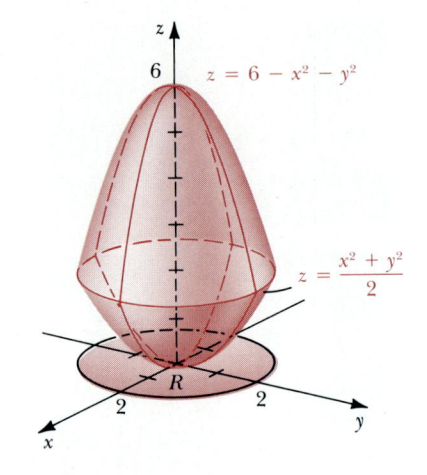

Figure 15.28

Example 3

A hen's egg is 6 cm in length and has a surface roughly described by the equations $z = (x^2 + y^2)/2$ and $z = 6 - x^2 - y^2$, as shown in Figure 15.28. Find the volume of the egg.

Solution

The graphs of $z = (x^2 + y^2)/2$ and $z = 6 - x^2 - y^2$ intersect when

$$z = \frac{x^2 + y^2}{2} = 6 - x^2 - y^2.$$

This implies that

$$x^2 + y^2 = 4 \quad \text{and} \quad z = 2.$$

The region R of integration is described in polar coordinates by

$$0 \leq \theta \leq 2\pi, \qquad 0 \leq r \leq 2.$$

Thus,

$$V = \iint_R \left[(6 - x^2 - y^2) - \frac{1}{2}(x^2 + y^2) \right] dA$$

$$= \int_0^{2\pi} \int_0^2 \left[(6 - r^2) - \frac{1}{2} r^2 \right] r \, dr \, d\theta$$

$$= \int_0^{2\pi} \int_0^2 \left(6r - \frac{3}{2} r^3 \right) dr \, d\theta$$

$$= \int_0^{2\pi} \left[3r^2 - \frac{3}{8} r^4 \right]_{r=0}^{r=2} d\theta$$

$$= \int_0^{2\pi} (12 - 6) \, d\theta = 6\theta \Big]_0^{2\pi} = 12\pi \text{ cm}^3 \approx \frac{1}{6} \text{ cup.} \quad \blacktriangleright$$

The following example shows a rectangular iterated integral that can be readily evaluated only by converting it into a polar iterated integral.

Example 4

Evaluate $\displaystyle\int_0^2 \int_0^{\sqrt{4-x^2}} e^{-x^2} e^{-y^2} \, dy \, dx.$

Solution

The function e^{-y^2} has no elementary antiderivative, so we cannot use the Fundamental Theorem of Calculus to find

$$\int e^{-y^2} dy.$$

Interchanging the order of integration leads to the same difficulty since

$$\int_0^2 \int_0^{\sqrt{4-x^2}} e^{-x^2} e^{-y^2} \, dy \, dx = \int_0^2 \int_0^{\sqrt{4-y^2}} e^{-x^2} e^{-y^2} \, dx \, dy.$$

However, the region described by the iterated integral is expressed in polar coordinates by

$$0 \leq \theta \leq \frac{\pi}{2}, \qquad 0 \leq r \leq 2,$$

as shown in Figure 15.29. We evaluate the equivalent polar iterated integral using the Fundamental Theorem of Calculus:

$$\int_0^2 \int_0^{\sqrt{4-x^2}} e^{-x^2} e^{-y^2} \, dy \, dx = \int_0^2 \int_0^{\sqrt{4-x^2}} e^{-(x^2+y^2)} \, dy \, dx$$

$$= \int_0^{\pi/2} \int_0^2 e^{-r^2} r \, dr \, d\theta$$

$$= \int_0^{\pi/2} \left[-\frac{1}{2} e^{-r^2} \right]_{r=0}^{r=2} d\theta$$

$$= \frac{1}{2} (1 - e^{-4}) \theta \Big]_0^{\pi/2} = \frac{\pi}{4} (1 - e^{-4}).$$

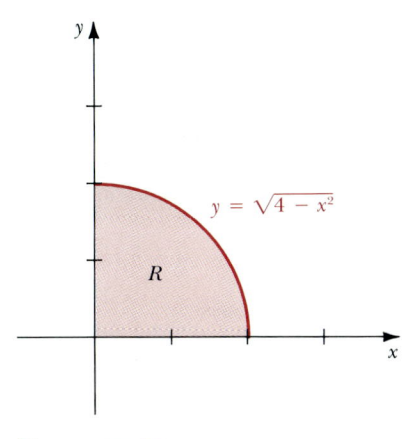

Figure 15.29

▶ EXERCISE SET 15.3

In Exercises 1–12, use double integrals in polar coordinates to find the volume of the solid.

1. Bounded by $z = 0$, $z = r$, and $r = 3$
2. Bounded by $z = 0$, $z = r$, and $r = \sin \theta$, $0 \leq \theta \leq \pi$
3. Bounded by the cylinders $x^2 + y^2 = 4$ and $x^2 + y^2 = 16$ and the planes $z = 0$ and $z = 4$
4. Inside the cylinder $x^2 + y^2 = 9$ and outside the cone $z^2 = x^2 + y^2$
5. Inside both the sphere $x^2 + y^2 + z^2 = 4$ and the cylinder $x^2 + y^2 = 1$
6. Inside the sphere $x^2 + y^2 + z^2 = 16$ and outside the cylinder $x^2 + y^2 = 9$
7. Inside the sphere $x^2 + y^2 + z^2 = 4$ and inside the cone $z^2 = x^2 + y^2$
8. Inside the sphere $x^2 + y^2 + z^2 = 4$ and outside the cone $z^2 = x^2 + y^2$
9. Inside the cylinder $r = \sin \theta$, above the paraboloid $z = x^2 + y^2 - 4$, and below the xy-plane
10. Outside the cylinder $r = \sin \theta$, above the paraboloid $z = x^2 + y^2 - 4$, and below the xy-plane
11. Inside the cylinder formed by the intersection of $r = \sin \theta$ and $r = \cos \theta$ and bounded above by the paraboloid $z = x^2 + y^2$ and below by $z = 0$
12. Inside the cylinder bounded by $r = \cos \theta$ when $0 \leq \theta \leq \pi/2$ and bounded above by $y = x \tan z$ and below by $z = 0$

In Exercises 13–22, use double integrals to find the area of the region.

13. Inside $r = 2 \sin \theta$
14. Inside $r = 1 + \cos \theta$
15. Inside $r = 3 - 2 \cos \theta$
16. Inside $r^2 = 4 \cos 2\theta$
17. Inside $r = 1 + \cos \theta$ and outside $r = 1$
18. Inside $r = 1 + \cos \theta$ and outside $r = 1 + \sin \theta$
19. Inside $r = \sin \theta$ and outside $r^2 = (\sin 2\theta)/2$
20. Inside $r = 2 + 2 \sin \theta$ and inside $r = 3$
21. Inside $r = \sin 2\theta$ and inside $r = \cos 2\theta$
22. Between $r = \theta$ and $r = e^\theta$ for $0 \leq \theta \leq 2\pi$

In Exercises 23–26, evaluate the integral by changing to a double integral in polar coordinates.

23. $\displaystyle \int_{-1}^{1} \int_0^{\sqrt{1-x^2}} \frac{y}{\sqrt{x^2 + y^2}} \, dy \, dx$

24. $\displaystyle \int_{-3}^{3} \int_{-\sqrt{9-x^2}}^{\sqrt{9-x^2}} e^{x^2} e^{y^2} \, dy \, dx$

25. $\displaystyle \int_0^1 \int_{\sqrt{3}x/3}^{x} \frac{1}{\sqrt{x^2 + y^2}} \, dy \, dx$

26. $\displaystyle \int_0^2 \int_{-\sqrt{2x-x^2}}^{\sqrt{2x-x^2}} \sqrt{x^2 + y^2} \, dy \, dx$

In Exercises 27–30, evaluate the integral by changing to a double integral in Cartesian coordinates.

27. $\int_0^{\pi/4} \int_0^{\sec\theta} r^3 \sin^2\theta \, dr \, d\theta$

28. $\int_0^{\pi/4} \int_{\sec\theta}^{2\sec\theta} r^2 \sqrt{\sin 2\theta} \, dr \, d\theta$

29. $\int_{\pi/4}^{3\pi/4} \int_0^{3\csc\theta} r \, e^{r^2 \sin^2\theta} \, dr \, d\theta$

30. $\int_{5\pi/4}^{3\pi/2} \int_0^{-2\csc\theta} r \ln(r^2 \sin 2\theta) \, dr \, d\theta$

31. A liquid flowing through a cylindrical tube has varying velocity depending on its distance r from the center of the tube:
$$v(r) = k(R^2 - r^2) \text{ cm/sec},$$
where R is the radius of the tube and k is a constant that depends on the length of the tube and the velocity of the fluid at the ends (see the figure). Suppose $R = 3.5$ cm and $k = 1$. Use double integrals to find the rate of flow of the liquid through the tube.

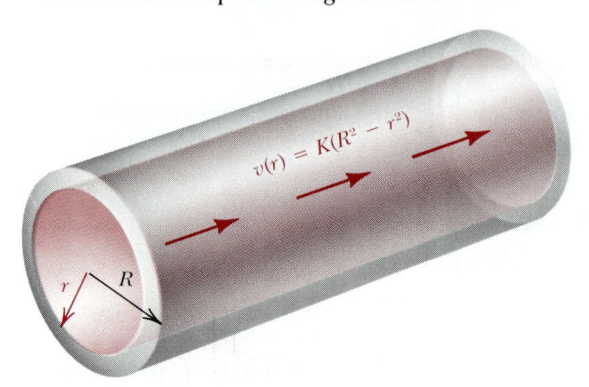

32. The energy generated by a 100-megaton nuclear bomb is approximately 4.18×10^{24} ergs. Suppose such a bomb is detonated at a height H and radiates energy equally in all directions. How much of this energy reaches a circle of radius R on the level ground if the center of the circle is directly below the point of detonation?

Improper integrals are defined for multiple integrals in a manner similar to that of single integrals. Assume that the improper integral
$$\int_a^\infty \int_b^\infty f(x, y) \, dy \, dx$$
converges and is equal to
$$\lim_{M_x \to \infty} \int_a^{M_x} \left[\lim_{M_y \to \infty} \int_b^{M_y} f(x, y) \, dy \right] dx$$
if the limit of this pair of single improper integrals is finite

and that the improper integral *diverges* otherwise. Determine whether the integrals in Exercises 33–38 converge or diverge, and evaluate those that converge.

33. $\int_1^\infty \int_2^4 \frac{1}{x^2 y} \, dy \, dx$

34. $\int_2^3 \int_{-1}^\infty xe^{-y} \, dy \, dx$

35. $\int_0^\infty \int_1^\infty (e^{-y}/x) \, dx \, dy$

36. $\int_1^\infty \int_0^y (x/y^3) \, dx \, dy$

37. $\int_1^\infty \int_x^{2x} ye^{-x} \, dy \, dx$

38. $\int_1^\infty \int_x^{2x} xe^{-y} \, dy \, dx$

39. Show that the integral $\int_0^\infty \int_0^\infty e^{-x-y} \, dy \, dx$ converges and that $\int_0^\infty \int_0^\infty e^{-x-y} \, dy \, dx = [\int_0^\infty e^{-x} \, dx]^2$.

40. The improper single integral $\int_{-\infty}^\infty e^{-x^2} \, dx$ is important in statistics, where it is used in the normal distribution discussed in Section 7.6. This integral cannot be evaluated directly using the Fundamental Theorem of Calculus, since e^{-x^2} has no elementary antiderivative. The following steps lead to an evaluation of this improper integral.

a) Use
$$\int_{-\infty}^\infty e^{-x^2} \, dx \int_{-\infty}^\infty e^{-y^2} \, dy = \int_{-\infty}^\infty \int_{-\infty}^\infty e^{-x^2} e^{-y^2} \, dy \, dx$$
to show that
$$\int_{-\infty}^\infty e^{-x^2} \, dx = \left[\int_{-\infty}^\infty \int_{-\infty}^\infty e^{-(x^2+y^2)} \, dy \, dx \right]^{1/2}.$$

b) Change the rectangular-coordinate description of the region R of integration
$$-\infty < x < \infty \quad \text{and} \quad -\infty < y < \infty$$
into a polar-coordinate description.

c) Use the polar-coordinate description of the region R found in part (b) to change the improper rectangular-coordinate integral
$$\left[\int_{-\infty}^\infty \int_{-\infty}^\infty e^{-(x^2+y^2)} \, dy \, dx \right]^{1/2}$$
into an improper polar-coordinate integral.

d) Evaluate the improper polar-coordinate integral in part (c).

e) Use the results of parts (a) and (d) to show that
$$\int_{-\infty}^\infty e^{-x^2} \, dx = \sqrt{\pi}.$$

f) Show that $\int_0^\infty e^{-x^2} \, dx = \sqrt{\pi}/2$.

41. Organisms are released from a point taken as the pole of a polar coordinate system. The density of the organisms as a function of the time after release is
$$f(r, \theta, t) = \frac{1}{\pi t} e^{-r^2/t}.$$

Show that $\int_0^\infty \int_0^{2\pi} f(r, \theta, t) r \, d\theta \, dr = 1$. What physical interpretation does this result have?

Putnam exercises

42. Evaluate
$$\int_0^\infty \frac{\text{Arctan }(\pi x) - \text{Arctan } x}{x} \, dx.$$

(This exercise was problem A–3 of the forty-third William Lowell Putnam examination given on December 4, 1982. The examination and its solution are in the October 1983 issue of the *American Mathematical Monthly*, pp. 546–553.)

43. Let $A(x, y)$ denote the number of points (m, n) in the plane with integer coordinates m and n satisfying $m^2 + n^2 \le x^2 + y^2$. Let $g = \Sigma_{k=0}^{\infty} e^{-k^2}$. Express
$$\int_{-\infty}^{\infty} \int_{-\infty}^{\infty} A(x, y) \, e^{-x^2 - y^2} \, dx \, dy$$
as a polynomial in g. (This exercise was problem B–2 of the forty-third William Lowell Putnam examination given on December 4, 1982. The examination and its solution are in the October 1983 issue of the *American Mathematical Monthly*, pp. 546–553.)

15.4

Center of Mass and Moments of Inertia

In Section 5.5 we used the definite integral of a single-variable function to determine the moments and center of mass of a plane lamina with a constant density. Using double integrals, we can extend these concepts to plane lamina with varying density.

Suppose a region R in the xy-plane describes a lamina S with density at (x, y) given by $\sigma(x, y)$, where σ is a continuous function. Place a grid over R and let $R_1, R_2, \ldots, R_n$ denote those rectangles formed by the grid that lie entirely within R (see Figure 15.30). Let ΔA_i denote the area of R_i. If (x_i, y_i) is chosen arbitrarily in R_i, then $\sigma(x_i, y_i) \Delta A_i$ is an approximation to the mass of the lamina described by R_i. An approximation to the total mass of S is

$$M \approx \sum_{i=1}^n \sigma(x_i, y_i) \Delta A_i.$$

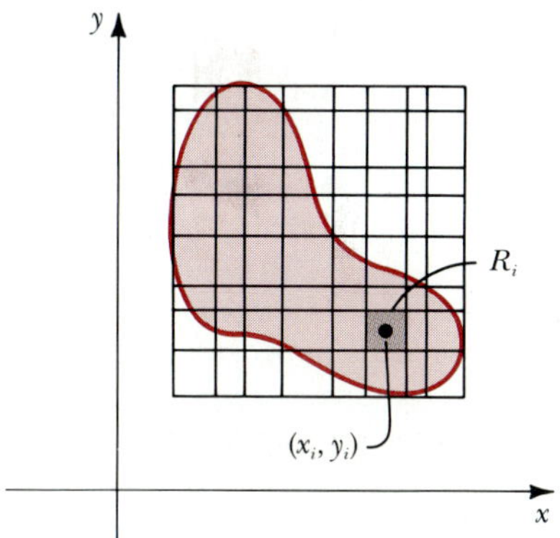

The approximation improves as the grid lines in both directions get closer together and leads to the following definition of mass and moments.

Figure 15.30

(15.13)
DEFINITION
Suppose R is a closed and bounded region in the xy-plane describing a plane lamina S with density at each point in R given by the continuous function σ. The **mass of S** is

$$M = \iint_R \sigma(x, y) \, dA.$$

The **moment of S about the x-axis** is

$$M_x = \iint_R y\sigma(x, y)\, dA,$$

and the **moment of S about the y-axis** is

$$M_y = \iint_R x\sigma(x, y)\, dA.$$

The **center of mass of S** has coordinates $(\bar{x}, \bar{y})$, where

$$\bar{x} = \frac{M_y}{M} \quad \text{and} \quad \bar{y} = \frac{M_x}{M}.$$

Suppose the density is a constant σ and R is an x-simple region bounded by $x = a$, $x = b$, and the graphs of f_1 and f_2, where $f_1(x) \leq f_2(x)$. Then the results in Definition (15.13) reduce to those given in Section 5.5. For example, in this case,

$$M_y = \iint_R x\sigma\, dA = \int_a^b \int_{f_1(x)}^{f_2(x)} x\sigma\, dy\, dx = \sigma \int_a^b x[f_2(x) - f_1(x)]\, dx.$$

The moments in the following example cannot be found by the methods of Section 5.5 because the density is not constant on the lamina.

Example 1

Compute the mass, moments, and center of mass of a plane lamina in the shape of the region R described by

$$0 \leq x \leq 4, \quad 0 \leq y \leq \sqrt{x},$$

if the density is given by $\sigma(x, y) = xy$.

Solution

The mass is

$$M = \iint_R \sigma(x, y)\, dA = \int_0^4 \int_0^{\sqrt{x}} xy\, dy\, dx = \int_0^4 \frac{xy^2}{2}\bigg]_{y=0}^{y=\sqrt{x}} dx,$$

so

$$M = \int_0^4 \frac{x^2}{2}\, dx = \frac{x^3}{6}\bigg]_0^4 = \frac{64}{6} = \frac{32}{3}.$$

The moment about the x-axis is

$$M_x = \iint_R y\sigma(x, y)\, dA = \int_0^4 \int_0^{\sqrt{x}} xy^2\, dy\, dx = \int_0^4 \frac{xy^3}{3}\bigg]_0^{\sqrt{x}} dx.$$

Thus,
$$M_x = \frac{1}{3}\int_0^4 x^{5/2}\,dx = \frac{2}{21}x^{7/2}\Big]_0^4 = \frac{256}{21}.$$

The moment about the y-axis is
$$M_y = \iint_R x\sigma(x,y)\,dA = \int_0^4\int_0^{\sqrt{x}} x^2 y\,dy\,dx = \int_0^4 \frac{x^2 y^2}{2}\Big]_0^{\sqrt{x}}\,dx,$$

so
$$M_y = \frac{1}{2}\int_0^4 x^3\,dx = \frac{x^4}{8}\Big]_0^4 = 32.$$

The center of mass is $(\bar{x},\bar{y})$, shown in Figure 15.31, where
$$\bar{x} = \frac{M_y}{M} = \frac{32}{\frac{32}{3}} = 3 \quad \text{and} \quad \bar{y} = \frac{M_x}{M} = \frac{\frac{256}{21}}{\frac{32}{3}} = \frac{8}{7}.$$

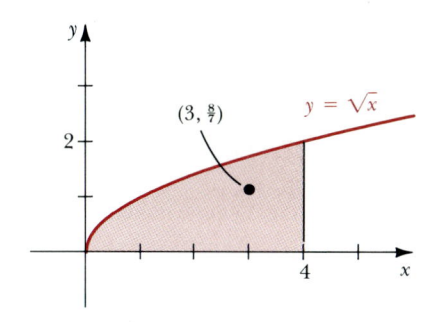

Figure 15.31
Center of mass: $(3, \frac{8}{7})$.

The moment M_x of a plane lamina about the x-axis describes the tendency of the lamina to rotate about this axis. Similarly, M_y describes the tendency of the lamina to rotate about the y-axis. If the mass M of the lamina is concentrated at the center of mass $(\bar{x},\bar{y})$, then the resulting system has the same moments and rotation tendencies. Although these moments give valuable information about the gravitational center of the lamina, they give no indication about the way the mass of the region is distributed about this center. Consider, for example, the laminae shown in Figure 15.32. The area of the region in (b) is 4 times the area of the region in (a), but the density of the lamina in (a) is 4 times the density of the lamina in (b), so the masses of the laminae are the same. Because of symmetry, the moments in all cases are zero and each lamina has center of mass $(0,0)$. There is no indication from these calculations that the lamina in (b) is more widely spread from its center than is the lamina in (a).

To describe the spread or distribution of a region about an axis, we introduce a concept known as the second moment about the axis and called the *moment of inertia* about the axis.

(15.14) DEFINITION
Suppose R is a closed and bounded region in the plane describing a plane lamina S with density in R given by the continuous function σ. Then the **moment of inertia of S about the x-axis** is
$$I_x = \iint_R y^2\sigma(x,y)\,dA,$$

and the **moment of inertia of S about the y-axis** is
$$I_y = \iint_R x^2\sigma(x,y)\,dA.$$

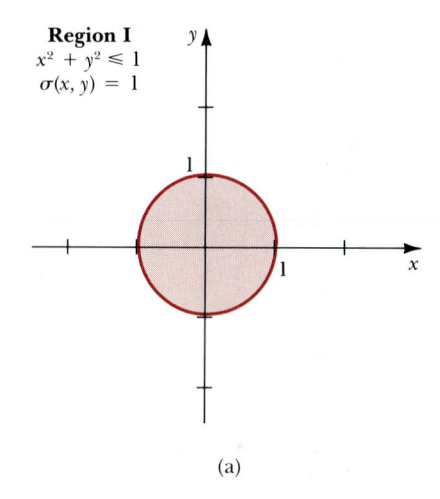

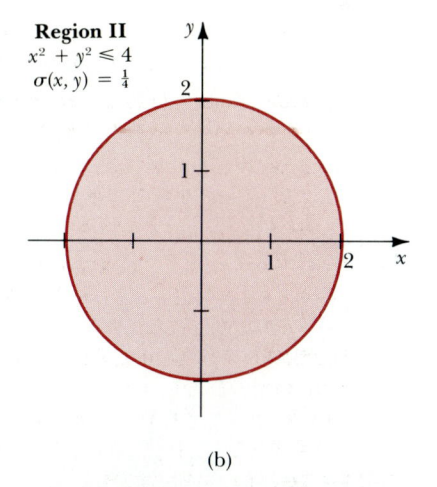

Figure 15.32
The moments and the center of mass are the same for both regions.

Suppose we compute the moments of inertia for the laminae in Figure 15.32. For the lamina in Figure 15.32(a),

$$I_x = \iint_{R_I} y^2 \, \sigma(x, y) \, dA = \iint_R y^2 \cdot 1 \, dA$$

$$= \int_0^{2\pi} \int_0^1 (r \sin \theta)^2 \, r \, dr \, d\theta$$

$$= \int_0^{2\pi} \int_0^1 (\sin^2 \theta) \, r^3 \, dr \, d\theta = \int_0^{2\pi} (\sin^2 \theta) \frac{r^4}{4} \Big]_0^1 d\theta$$

$$= \frac{1}{4} \int_0^{2\pi} \sin^2 \theta \, d\theta = \frac{1}{4} \int_0^{2\pi} \frac{1 - \cos 2\theta}{2} \, d\theta$$

$$= \frac{1}{8} \left[\theta - \frac{1}{2} \sin 2\theta \right]_0^{2\pi} = \frac{\pi}{4},$$

and by symmetry, $I_y = \pi/4$.

For the lamina in Figure 15.32(b),

$$I_x = \iint_{R_{II}} y^2 \, \sigma(x, y) \, dA = \iint_{R_{II}} y^2 \cdot \frac{1}{4} \, dA$$

$$= \frac{1}{4} \int_0^{2\pi} \int_0^2 (r \sin \theta)^2 \, r \, dr \, d\theta = \frac{1}{4} \int_0^{2\pi} \int_0^2 r^3 \sin^2 \theta \, dr \, d\theta$$

$$= \frac{1}{4} \int_0^{2\pi} (\sin^2 \theta) \frac{r^4}{4} \Big]_0^2 d\theta = \frac{1}{4} \int_0^{2\pi} 4 \sin^2 \theta \, d\theta$$

$$= \int_0^{2\pi} \frac{1 - \cos 2\theta}{2} \, d\theta = \frac{1}{2} \left[\theta - \frac{1}{2} \sin 2\theta \right]_0^{2\pi} = \pi.$$

Again, by symmetry, $I_y = \pi$.

The difference between the moments of inertia in regions I and II correctly describes the fact that the lamina in Figure 15.32(b) is more widely distributed from its center of mass than is the lamina in part (a) of this figure.

(15.15)
A point $(\hat{x}, \hat{y})$ analogous to the center of mass has the property that if the total mass is concentrated at $(\hat{x}, \hat{y})$, then the resulting system has the same moments of inertia as the original system. This means that $\hat{x}^2 M = I_y$ and $\hat{y}^2 M = I_x$, so the coordinates of this point are given by

$$\hat{x} = \sqrt{\frac{I_y}{M}}, \text{ called the \textbf{radius of gyration} about the } y\text{-axis, and}$$

$$\hat{y} = \sqrt{\frac{I_x}{M}}, \text{ called the \textbf{radius of gyration} about the } x\text{-axis.}$$

A second moment can also be used to describe the distribution about a point. Just as the moments of inertia about an axis involve the square of the distance from the axis, the second moment about a point involves the square of the distance from the point. Thus the second moment about the origin, called the **polar moment of inertia of S,** is given by

(15.16) $$I_0 = \iint_R (x^2 + y^2)\, \sigma(x, y)\, dA = I_x + I_y.$$

Note that by virtue of their definitions, the moments of inertia as well as the radii of gyration must be nonnegative.

Example 2

Find the moments of inertia and radii of gyration of the plane lamina in the shape of the region R described by

$$0 \le x \le 4, \qquad 0 \le y \le \sqrt{x},$$

if the density is given by $\sigma(x, y) = xy$.

Solution

The region is the same as that described in Example 1 and shown in Figure 15.33. Thus,

$$I_x = \iint_R y^2\,(xy)\, dA = \int_0^4 \int_0^{\sqrt{x}} xy^3\, dy\, dx = \int_0^4 \frac{xy^4}{4}\bigg]_0^{\sqrt{x}} dx,$$

so

$$I_x = \frac{1}{4}\int_0^4 x^3\, dx = \frac{x^4}{16}\bigg]_0^4 = 16.$$

Also,

$$I_y = \iint_R x^2\,(xy)\, dA = \int_0^4 \int_0^{\sqrt{x}} x^3 y\, dy\, dx = \int_0^4 \frac{x^3 y^2}{2}\bigg]_0^{\sqrt{x}} dx,$$

so

$$I_y = \frac{1}{2}\int_0^4 x^4\, dx = \frac{x^5}{10}\bigg]_0^4 = \frac{512}{5}.$$

The polar moment of inertia is

$$I_0 = I_x + I_y = 16 + \frac{512}{5} = \frac{592}{5}.$$

Since M was found in Example 1 to be $\frac{32}{3}$,

$$\hat{x} = \sqrt{\frac{I_y}{M}} = \sqrt{\frac{\frac{512}{5}}{\frac{32}{3}}} = \sqrt{\frac{48}{5}} \approx 3.1$$

and

$$\hat{y} = \sqrt{\frac{I_x}{M}} = \sqrt{\frac{16}{\frac{32}{3}}} = \sqrt{\frac{3}{2}} \approx 1.2. \qquad \blacktriangleright$$

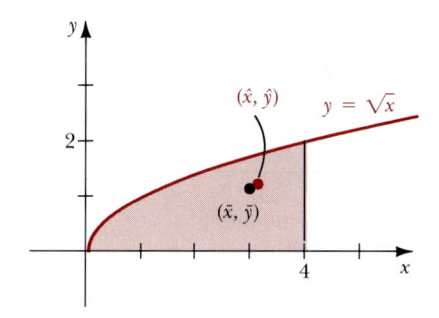

Figure 15.33

EXERCISE SET 15.4

In Exercises 1–8, find (a) the mass, (b) the moments about the axes, and (c) the center of mass of the lamina bounded by the curves and having density σ.

1. $y = x^2$, $y = \sqrt{x}$, $\sigma(x, y) = 2$
2. $y = x^2$, $y = 2 - x^2$, $\sigma(x, y) = 4$
3. $y = 1 - x$, $y = 0$, $x = 0$, $\sigma(x, y) = x$
4. $y = 1 - x$, $y = 0$, $x = 0$, $\sigma(x, y) = y$
5. $y = 1 - x^2$, $y = 0$, $\sigma(x, y) = x^2$
6. $y = 1 - x^2$, $y = 0$, $\sigma(x, y) = y$
7. $y = x$, $y = 1$, $x = 0$, $\sigma(x, y) = x + y$
8. $y = 1/x$, $y = 0$, $x = 1$, $x = 4$, $\sigma(x, y) = x + y$

9–16. Find the moments of inertia about the axes and the origin and the radii of gyration for the laminae described in Exercises 1–8.

For Exercises 17 and 18, a plane lamina has density $\sigma(r, \theta)$ and shape that of a region R.

17. Show that the formulas in polar coordinates for the mass, moments about the axes, and center of mass $(\bar{r}, \bar{\theta})$ are

$$M = \iint_R r\sigma(r, \theta)\, dr\, d\theta,$$

$$M_x = \iint_R r^2 \sin\theta\, \sigma(r, \theta)\, dr\, d\theta,$$

$$M_y = \iint_R r^2 \cos\theta\, \sigma(r, \theta)\, dr\, d\theta,$$

and

$$\bar{r}^2 = \frac{M_x^2 + M_y^2}{M^2}, \quad \tan\bar{\theta} = \frac{M_x}{M_y}.$$

18. Show that the formulas in polar coordinates for the moments of inertia are

$$I_x = \iint_R r^3 \sin^2\theta\, \sigma(r, \theta)\, dr\, d\theta,$$

$$I_y = \iint_R r^3 \cos^2\theta\, \sigma(r, \theta)\, dr\, d\theta,$$

and

$$I_0 = \iint_R r^3 \sigma(r, \theta)\, dr\, d\theta.$$

In Exercises 19–22, use the formulas in Exercises 17 and 18 to find (a) the mass, (b) the moments about the axes, and (c) the moments of inertia for the plane laminae whose boundaries and density functions are described.

19. $r = 4$, $\sigma(r, \theta) = r$
20. $r = 4$, $\sigma(r, \theta) = \theta$, $0 \leq \theta \leq 2\pi$
21. $r = \sin\theta$, $\sigma(r, \theta) = r$
22. $r = \cos\theta$, $\sigma(r, \theta) = r$

In Exercises 23 and 24, use the formulas in Exercise 17 to find the center of mass $(\bar{r}, \bar{\theta})$ of the plane laminae whose boundaries and density functions are given.

23. $r = 1 - \sin\theta$, $\sigma(r, \theta) = r$
24. $r = 1 + \cos\theta$, $\sigma(r, \theta) = r$
25. Find the center of mass of a lamina that has the shape of the region bounded by $x^2 + y^2 = 4$ if the density at any point (x, y) is the square of the distance from the center of the region.
26. Find the center of mass of the lamina that has the shape of the region bounded by $x = 0$, $y = x$, and $y = 3 - x$ if the density at any point (x, y) is twice the distance from the y-axis.

15.5

Surface Area

In Section 5.4 we discussed a method for determining the area of a surface of revolution, that is, a surface generated by revolving the graph

15.5 SURFACE AREA

of a function about the *x*-axis, as shown in Figure 15.34. The introduction of double integrals permits us to determine the area of more general surfaces.

Suppose *S* is a surface described by $z = f(x, y)$, for (x, y) in a closed bounded region *R* where *f* has continuous partial derivatives (see Figure 15.35). Such a surface is called **smooth,** and, as shown in Section 14.6, has a tangent plane at each point not on the boundary of the surface (see Figure 15.36).

The bounded region *R* is contained within a rectangle determined by an *x*-coordinate interval $[a, b]$ and a *y*-coordinate interval $[c, d]$. A partition of each of these intervals produces a grid in the *xy*-plane. Let $R_1, R_2, \ldots, R_n$ denote the elements in this grid that lie entirely within *R* (see Figure 15.37), and let (x_i, y_i) be an arbitrary point in the rectangle R_i.

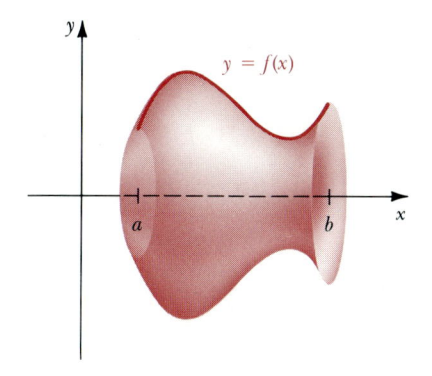

Figure 15.34
A surface of revolution.

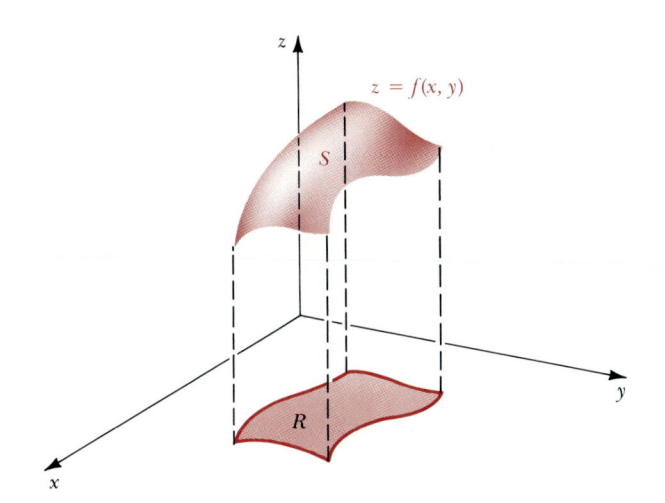

Figure 15.35
A smooth surface *S*.

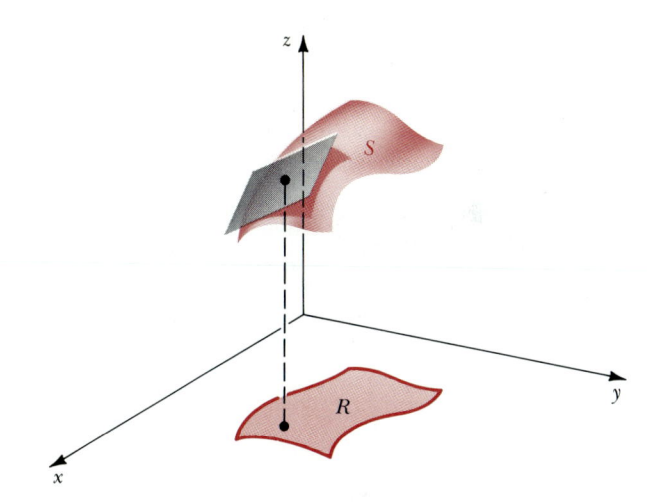

Figure 15.36
A tangent plane to *S*.

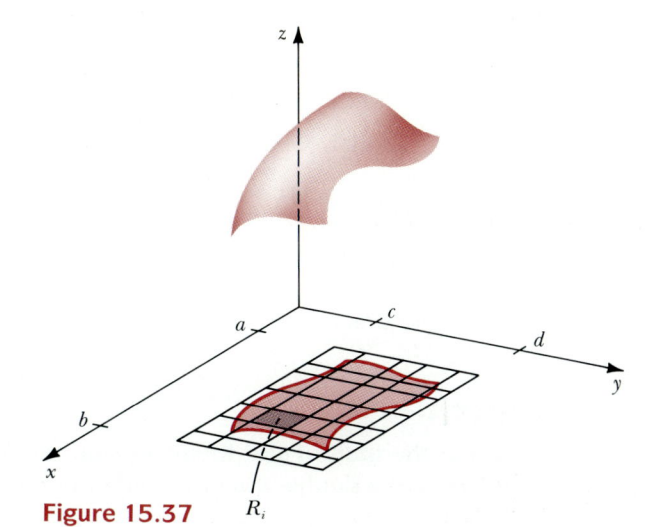

Figure 15.37

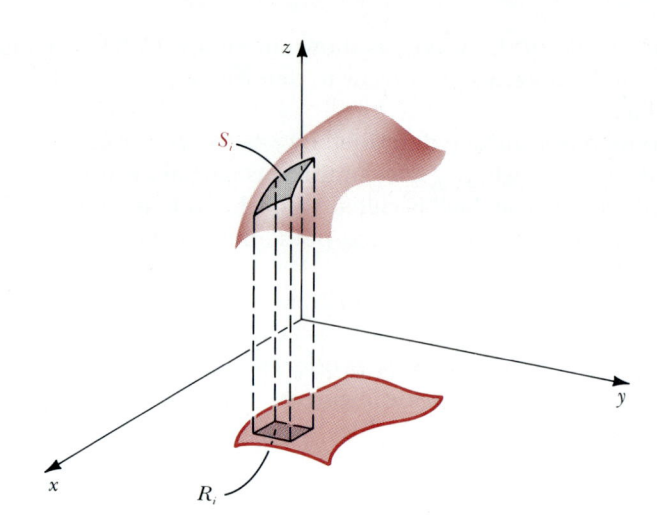

Figure 15.38
A surface element.

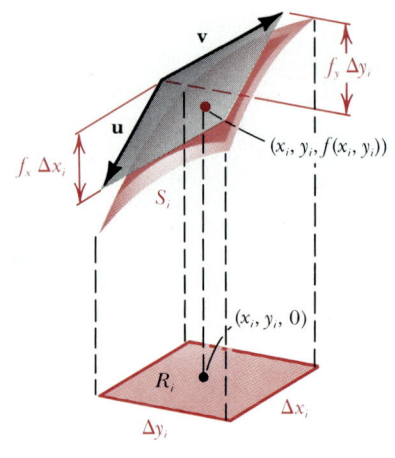

Figure 15.39

Let S_i represent the portion of the surface, shown in Figure 15.38, that lies above the rectangle R_i. The area of S_i is approximated by the area ΔS_i of the corresponding portion of the tangent plane to S at $(x_i, y_i, f(x_i, y_i))$ shown in Figure 15.39.

In Section 14.6 we found that an equation of the tangent plane at $(x_i, y_i, f(x_i, y_i))$ is

$$f_x(x_i, y_i)(x - x_i) + f_y(x_i, y_i)(y - y_i) - (z - f(x_i, y_i)) = 0.$$

This plane intersects the right cylinder with base R_i in a parallelogram determined by the vectors $\mathbf{u}$ and $\mathbf{v}$, shown in Figure 15.39:

$$\mathbf{u} = \Delta x_i \mathbf{i} + f_x(x_i, y_i) \Delta x_i \mathbf{k} \quad \text{and} \quad \mathbf{v} = \Delta y_i \mathbf{j} + f_y(x_i, y_i) \Delta y_i \mathbf{k},$$

where Δx_i and Δy_i denote the dimensions of the rectangle R_i.

The vector $\mathbf{u} \times \mathbf{v}$ has length equal to the area ΔS_i of the parallelogram formed by the vectors $\mathbf{u}$ and $\mathbf{v}$, so

$$\begin{aligned}\Delta S_i &= \|(\Delta x_i \mathbf{i} + f_x(x_i, y_i) \Delta x_i \mathbf{k}) \times (\Delta y_i \mathbf{j} + f_y(x_i, y_i) \Delta y_i \mathbf{k})\| \\ &= \|-f_x(x_i, y_i) \Delta x_i \Delta y_i \mathbf{i} - f_y(x_i, y_i) \Delta x_i \Delta y_i \mathbf{j} + \Delta x_i \Delta y_i \mathbf{k}\| \\ &= \{[-f_x(x_i, y_i) \Delta x_i \Delta y_i]^2 + [-f_y(x_i, y_i) \Delta x_i \Delta y_i]^2 + [\Delta x_i \Delta y_i]^2\}^{1/2}.\end{aligned}$$

Since Δx_i and Δy_i are positive,

(15.17) $\quad \Delta S_i = \{[f_x(x_i, y_i)]^2 + [f_y(x_i, y_i)]^2 + 1\}^{1/2} \Delta x_i \Delta y_i.$

Summing the ΔS_i for each $i = 1, 2, \ldots, n$ gives a Riemann sum that approximates the area of the surface S:

$$\sum_{i=1}^{n} \{[f_x(x_i, y_i)]^2 + [f_y(x_i, y_i)]^2 + 1\}^{1/2} \Delta x_i \Delta y_i.$$

Taking the limit of the Riemann sums as Δx_i and Δy_i approach zero produces the double integral used to define the surface area.

(15.18)
DEFINITION
Suppose S is the smooth surface defined by $z = f(x, y)$ for (x, y) in a closed and bounded region R. The **surface area** of S is

$$SA = \iint_R \{[f_x(x, y)]^2 + [f_y(x, y)]^2 + 1\}^{1/2} \, dA.$$

Example 1

Find the area of the portion of the surface described by $f(x, y) = 3 - x - 2y^2$ that lies above the region in the xy-plane bounded by the triangle with vertices $(0, 0, 0)$, $(0, 1, 0)$, and $(1, 1, 0)$.

Solution
The surface is shown in Figure 15.40 together with the region R in the xy-plane. The region of integration in the xy-plane is given by

$$0 \leq y \leq 1, \qquad 0 \leq x \leq y.$$

Thus,

$$SA = \iint_R \left\{ \left[\frac{\partial}{\partial x}(3 - x - 2y^2)\right]^2 + \left[\frac{\partial}{\partial y}(3 - x - 2y^2)\right]^2 + 1 \right\}^{1/2} dA$$

$$= \int_0^1 \int_0^y \sqrt{(-1)^2 + (-4y)^2 + 1} \, dx \, dy$$

$$= \int_0^1 \int_0^y \sqrt{16y^2 + 2} \, dx \, dy$$

$$= \int_0^1 y\sqrt{16y^2 + 2} \, dx = \frac{1}{32}(16y^2 + 2)^{3/2} \cdot \frac{2}{3}\Big]_0^1$$

$$= \frac{1}{48}[(18)^{3/2} - (2)^{3/2}] \approx 1.53.$$

▶

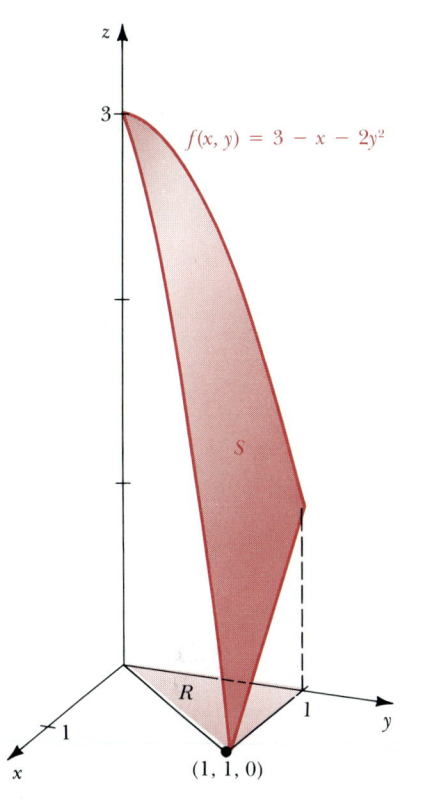

Figure 15.40

Example 2

Find the area of the surface that is the portion of the sphere with equation $x^2 + y^2 + z^2 = 2$ lying within the paraboloid with equation $z = x^2 + y^2$.

Solution
We first must determine the region in the xy-plane over which this surface lies. To find the intersection of

$$x^2 + y^2 + z^2 = 2 \quad \text{and} \quad z = x^2 + y^2,$$

we solve the equations simultaneously:

$$2 = (x^2 + y^2) + z^2 = z + z^2,$$

so
$$0 = z^2 + z - 2 = (z-1)(z+2).$$

The only positive solution is $z = 1$, so the region in the xy-plane that lies below the surface (see Figure 15.41) is bounded by the circle
$$x^2 + y^2 = 1.$$

The surface is shown in Figure 15.41(a). The portion in the first octant constitutes one fourth of the area. Integrating over the region shown in Figure 15.41(b) gives
$$SA = 4 \int_0^1 \int_0^{\sqrt{1-x^2}} \{[f_x(x,y)]^2 + [f_y(x,y)]^2 + 1\}^{1/2} \, dy \, dx.$$

Since $f(x, y) = z = \sqrt{2 - x^2 - y^2}$,
$$SA = 4 \int_0^1 \int_0^{\sqrt{1-x^2}} \left\{ \left(\frac{-x}{\sqrt{2-x^2-y^2}} \right)^2 + \left(\frac{-y}{\sqrt{2-x^2-y^2}} \right)^2 + 1 \right\}^{1/2} dy \, dx$$
$$= 4 \int_0^1 \int_0^{\sqrt{1-x^2}} \left\{ \frac{x^2 + y^2}{2 - (x^2 + y^2)} + 1 \right\}^{1/2} dy \, dx.$$

Since the region of integration is bounded by a circle centered at the origin, the integrals are evaluated more easily using polar coordinates:
$$SA = 4 \int_0^{\pi/2} \int_0^1 \left\{ \frac{r^2}{2 - r^2} + 1 \right\}^{1/2} r \, dr \, d\theta$$
$$= 4 \int_0^{\pi/2} \int_0^1 \left\{ \frac{2}{2 - r^2} \right\}^{1/2} r \, dr \, d\theta$$
$$= 4\sqrt{2} \int_0^{\pi/2} \int_0^1 (2 - r^2)^{-1/2} r \, dr \, d\theta$$
$$= 4\sqrt{2} \int_0^{\pi/2} -(2 - r^2)^{1/2} \Big]_0^1 \, d\theta$$
$$= 4\sqrt{2}(\sqrt{2} - 1)\theta \Big]_0^{\pi/2} = 4(2 - \sqrt{2}) \cdot \frac{\pi}{2} = 2\pi(2 - \sqrt{2}) \approx 3.7. \quad \blacktriangleright$$

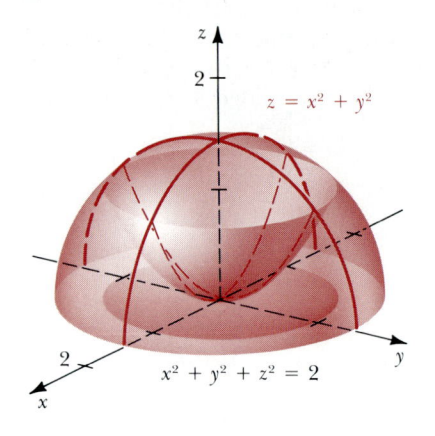

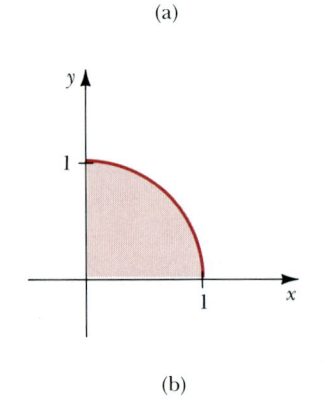

Figure 15.41

▶ EXERCISE SET 15.5

In Exercises 1–12, find the surface area of the given surface.

1. The portion of the plane $x + y + z = 3$ that lies in the first octant

2. The portion of the plane $2x - y + 3z = 6$ bounded by the coordinate planes

3. The portion of the surface $z = x^2$ that lies above the rectangular region in the xy-plane bounded by $x = 0$, $y = 0$, $x = 2$, and $y = 3$

4. The portion of the surface $z = x^2 + y^2 + 2$ that lies above the sector of the circle described by $r = 1$, $0 \leq \theta \leq \pi/4$

5. The portion of the paraboloid $z = x^2 + y^2$ that lies inside the cylinder $x^2 + y^2 = 4$

6. The portion of the parabolic cylinder $z = y^2$ that lies above the rectangle in the xy-plane with vertices $(0, 1, 0)$, $(2, 1, 0)$, $(2, -1, 0)$, and $(0, -1, 0)$

7. The portion of the cylinder $y^2 + z^2 = 4$ that lies above the xy-plane between $x = 0$ and $x = 3$

8. The portion of the plane $x + y + z = 1$ that lies inside the cylinder $x^2 + z^2 = 1$

9. The portion of the sphere $x^2 + y^2 + z^2 = 6$ inside the paraboloid $z = x^2 + y^2$

10. The parabolic sheet described by $y - x^2 = 1$ for $1 \le y \le 2$ and $-1 \le z \le 1$

11. The surface formed by intersecting the cylinders $x^2 + y^2 = 4$ and $x^2 + z^2 = 4$

12. The portion of the hyperbolic paraboloid $z = y^2 - x^2$ that lies in the first octant inside the cylinder $x^2 + y^2 = 1$

13. Use Definition (15.18) to show that the surface area of the sphere with equation $x^2 + y^2 + z^2 = R^2$ is $4\pi R^2$ (see the figure).

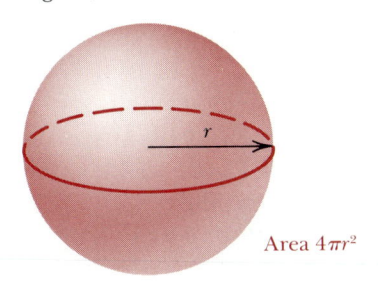

14. Use Definition (15.18) to show that the lateral surface area of a cone with height h and radius R is $\pi R \sqrt{R^2 + h^2}$ (see the figure).

15. Suppose $ax + by + cz = 1$ is an equation of a plane with a, b, and c all positive. Determine the surface area of the portion of this plane bounded by the coordinate planes (see the figure).

16. Show that the formula for the area of a surface of revolution given in Section 5.4 is a special case of Definition (15.18). [*Hint:* Consider the result in Exercise 17 of Section 13.3.]

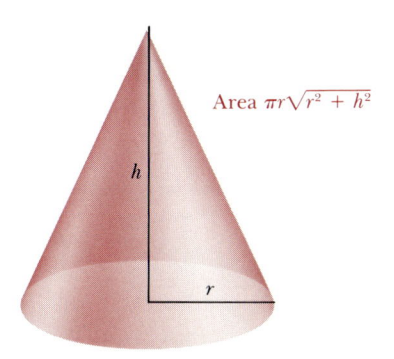

Figure for Exercise 14.

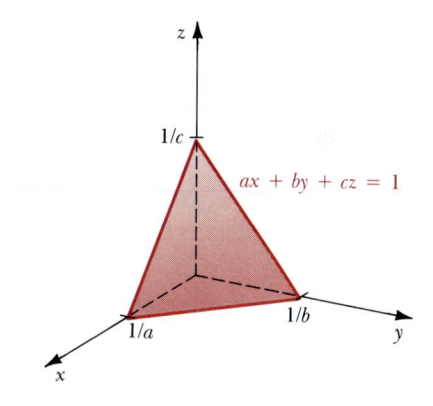

Figure for Exercise 15.

15.6

Triple Integrals

The development of the triple integral for functions of three variables over regions in space parallels the development of the double integral of functions of two variables over regions in the plane. Suppose a function f is defined on a bounded region D in space (see Figure 15.42a on the following page). Since D is bounded, D is contained within a parallelepiped P, as shown in Figure 15.42(b). We begin by forming a grid through D consisting of planes through P parallel to the xy-, yz-, and xz-coordinate planes. This forms a set of rectangular parallelepipeds that cover D. Let $D_1, D_2, \ldots, D_n$ denote those parallelepipeds that are contained entirely in D. The collection of all these parallelepipeds is denoted by $\mathcal{P}$,

(a)

(b)

(c)

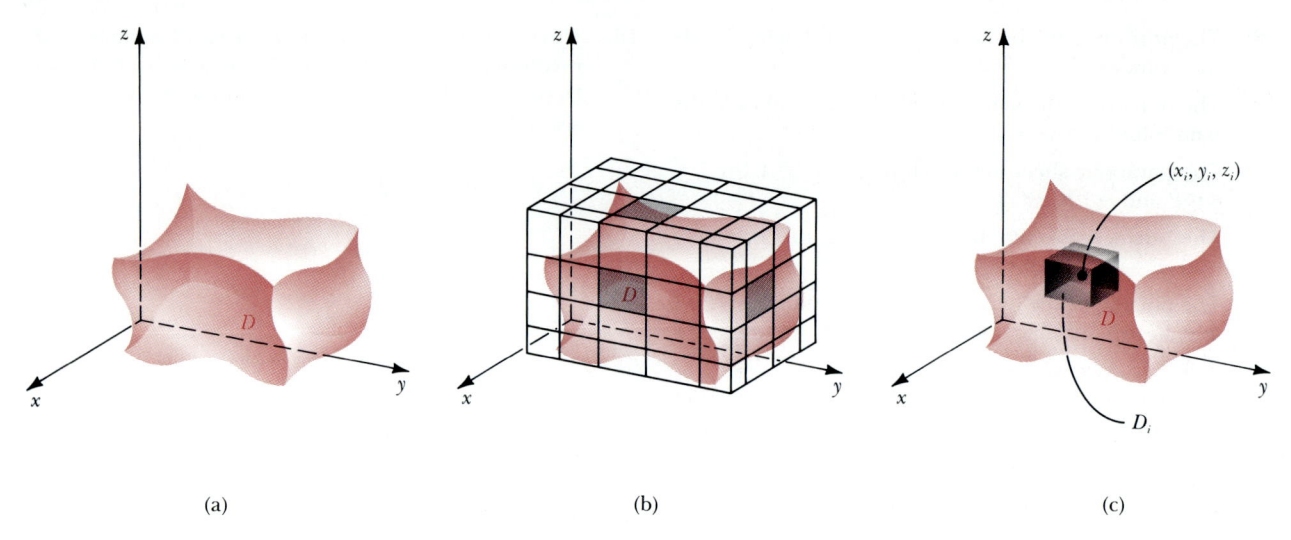

Figure 15.42

and $\|\mathscr{P}\|$ represents the maximum length of the diagonals of the D_i. Choose (x_i, y_i, z_i) arbitrarily with D_i (see Figure 15.42c). Color Plate XII provides another illustration of this procedure.

A Riemann sum can be formed with respect to f, the grid, and the choice of points within each D_i:

$$\sum_{i=1}^{n} f(x_i, y_i, z_i)\, \Delta V_i,$$

where ΔV_i denotes the volume of the parallelepiped D_i. Taking the limit of the Riemann sums over all possible grid constructions and choices of points within the grid elements gives the definition of the triple integral of f over D.

(15.19)
DEFINITION
If a function f of three variables is defined on a bounded region D in space, the **triple integral** of f over D is defined by

$$\iiint_D f(x, y, z)\, dV = \lim_{\|\mathscr{P}\| \to 0} \sum_{i=1}^{n} f(x_i, y_i, z_i)\, \Delta V_i,$$

provided this limit exists. In this case, f is said to be **integrable** on D.

This definition satisfies the usual properties associated with the definite integral. For example, if f and g are integrable on D, then

(15.20)

$$\iiint_D (f \pm g)(x, y, z)\, dV = \iiint_D f(x, y, z)\, dV \pm \iiint_D g(x, y, z)\, dV$$

and for any constant c,

(15.21) $$\iiint_D cf(x, y, z)\, dV = c \iiint_D f(x, y, z)\, dV.$$

In addition, if D is a region composed of subregions D_1 and D_2 with disjoint interiors on which f is integrable, then

(15.22)

$$\iiint_D f(x, y, z)\, dV = \iiint_{D_1} f(x, y, z)\, dV + \iiint_{D_2} f(x, y, z)\, dV.$$

The triple integral over a region in space can often be expressed as an iterated integral in a manner similar to that of the double integrals. For example, if D is a parallelepiped with boundaries parallel to the coordinate axes, then the iterated integral is the three successive single integrals of the function over the coordinate intervals describing D. This is illustrated in the following example.

Example 1

Evaluate the triple integral of $f(x, y, z) = xy \sin z$ over the region D in space described by $0 \le x \le 1$, $1 \le y \le 3$, and $0 \le z \le \pi$.

Solution

The region is shown in Figure 15.43. We have

$$\iiint_D f(x, y, z)\, dV = \int_{x=0}^{x=1} \int_{y=1}^{y=3} \int_{z=0}^{z=\pi} xy \sin z\, dz\, dy\, dx$$

$$= \int_0^1 \int_1^3 -xy \cos z \Big]_{z=0}^{z=\pi} dy\, dx$$

$$= \int_0^1 \int_1^3 2xy\, dy\, dx$$

$$= \int_0^1 xy^2 \Big]_{y=1}^{y=3} dx = \int_0^1 8x\, dx$$

$$= 4x^2 \Big]_0^1 = 4.$$

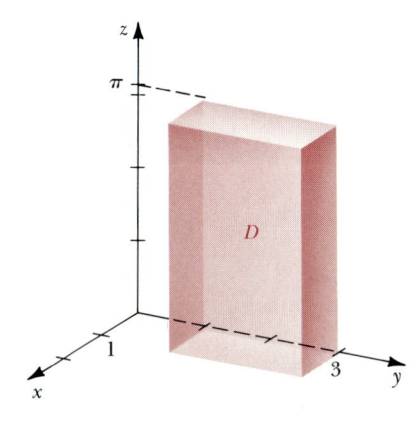

▶ Figure 15.43

We can express the integral in Example 1 as an iterated integral in five other ways by interchanging the order of the variables. For example, two of these are

$$\iiint_D xy \sin z\, dV = \int_0^1 \int_0^\pi \int_1^3 xy \sin z\, dy\, dz\, dx$$

$$= \int_0^\pi \int_1^3 \int_0^1 xy \sin z\, dx\, dy\, dz.$$

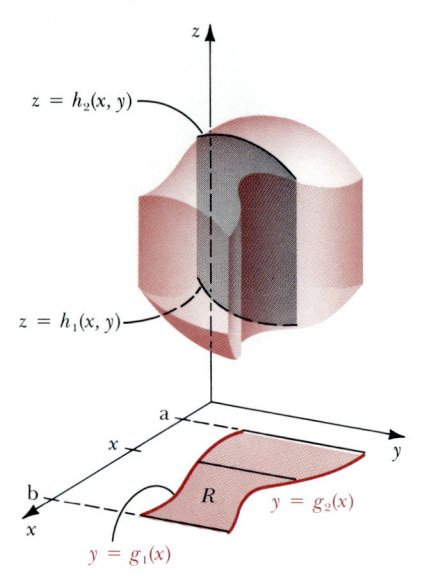

886 CHAPTER 15 INTEGRAL CALCULUS OF MULTIVARIABLE FUNCTIONS

In general, the triple integral of a function f over a region D is expressed as an iterated integral involving integration, first with respect to z, then with respect to y, and finally, with respect to x (written $dz\,dy\,dx$), provided D can be described by

$$a \leq x \leq b, \qquad g_1(x) \leq y \leq g_2(x), \qquad h_1(x, y) \leq z \leq h_2(x, y),$$

where g_1, g_2, h_1, and h_2 are continuous functions. A solid region D of this form is shown in Figure 15.44. Note that in this case the projection of D into the xy-plane gives an x-simple region R. The triple integral of f over D is then expressed in iterated integral form as

$$(15.23) \qquad \iiint_D f(x, y, z)\, dV = \int_{x=a}^{x=b} \int_{y=g_1(x)}^{y=g_2(x)} \int_{z=h_1(x,y)}^{z=h_2(x,y)} f(x, y, z)\, dz\, dy\, dx.$$

The core of the problem in the evaluation of triple integrals is to determine the boundaries of the region over which the integration is performed. These boundaries are usually specified in terms of intersecting surfaces, which determine the limits of integration for the iterated integral.

Figure 15.44

$\iiint_D f(x, y, z)\, dV$ is

$\int_a^b \int_{g_1(x)}^{g_2(x)} \int_{h_1(x,y)}^{h_2(x,y)} f(x, y, z)\, dz\, dy\, dx.$

Example 2

Find $\iiint_D z\, dV$, where D is the region in the first octant bounded by the planes $x = y$, $z = 0$, and $y = 0$ and the cylinder $x^2 + z^2 = 1$.

Solution

The region of integration D is shown in Figure 15.45(a). The projection of this region into the xy-plane is the x-simple region R, shown in Figure 15.45(b), described by

$$0 \leq x \leq 1, \qquad 0 \leq y \leq x.$$

To describe the upper and lower boundaries of D, we choose an arbitrary x in $[0, 1]$. Consider a slice of the solid given by the intersection of

Figure 15.45

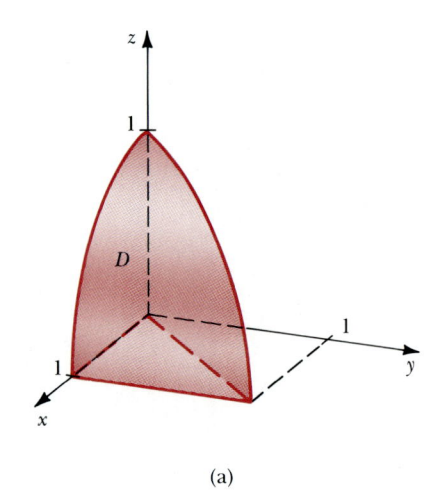

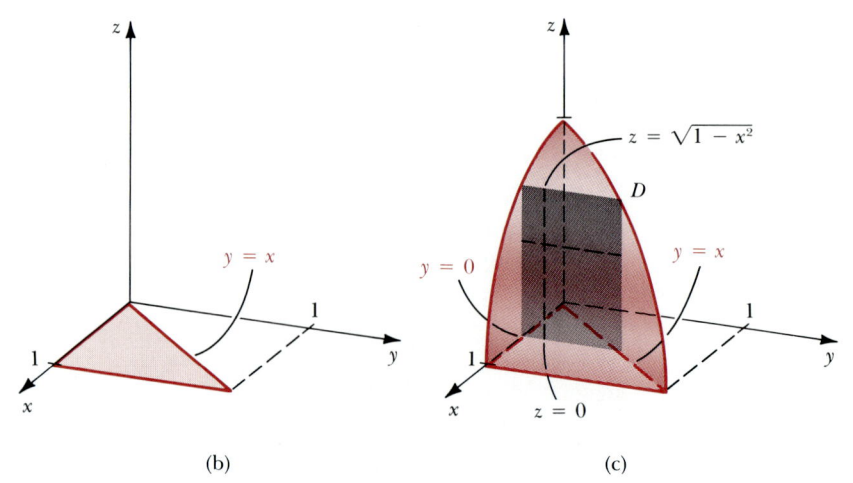

(a)   (b)   (c)

D and the plane through $(x, 0, 0)$ parallel to the yz-plane, as shown in Figure 15.45(c). The bottom of this slice is on the plane $z = 0$ and the top is on the surface $z = \sqrt{1 - x^2}$, so the boundary of the solid region D is described by

$$0 \leq x \leq 1, \quad 0 \leq y \leq x, \quad 0 \leq z \leq \sqrt{1 - x^2}.$$

Thus,

$$\iiint_D z \, dV = \int_{x=0}^{x=1} \int_{y=0}^{y=x} \int_{z=0}^{z=\sqrt{1-x^2}} z \, dz \, dy \, dx$$

$$= \int_0^1 \int_0^x \frac{z^2}{2} \Big]_0^{\sqrt{1-x^2}} dy \, dx = \int_0^1 \int_0^x \frac{1 - x^2}{2} \, dy \, dx$$

$$= \int_0^1 \frac{1 - x^2}{2} y \Big]_0^x dx$$

$$= \int_0^1 \frac{x - x^3}{2} \, dx = \frac{x^2}{4} - \frac{x^4}{8} \Big]_0^1 = \frac{1}{8}. \quad \blacktriangleright$$

Example 3

Evaluate $\iiint_D x^2 y \, dV$, where D is the region in the first octant bounded by the plane $z = 4$ and the cone $z^2 = x^2 + y^2$.

Solution

The region D is shown in Figure 15.46(a). The plane $z = 4$ and the cone $z^2 = x^2 + y^2$ intersect in the circle $x^2 + y^2 = 16$ lying in the plane $z = 4$. The projection of the solid onto the xy-plane is the portion of the circle $x^2 + y^2 = 16$ when $x \geq 0$ and $y \geq 0$, as shown in Figure 15.46(b). Thus, D is described by

$$0 \leq x \leq 4, \quad 0 \leq y \leq \sqrt{16 - x^2}, \quad \sqrt{x^2 + y^2} \leq z \leq 4.$$

Hence,

$$\iiint_D x^2 y \, dV = \int_0^4 \int_0^{\sqrt{16-x^2}} \int_{\sqrt{x^2+y^2}}^4 x^2 y \, dz \, dy \, dx$$

$$= \int_0^4 \int_0^{\sqrt{16-x^2}} x^2 yz \Big]_{z=\sqrt{x^2+y^2}}^{z=4} dy \, dx$$

$$= \int_0^4 \int_0^{\sqrt{16-x^2}} (4x^2 y - x^2 y \sqrt{x^2 + y^2}) \, dy \, dx$$

$$= \int_0^4 \left[2x^2 y^2 - \frac{x^2}{3} (x^2 + y^2)^{3/2} \right]_{y=0}^{y=\sqrt{16-x^2}} dx$$

$$= \int_0^4 \left[2x^2 (16 - x^2) - \frac{x^2}{3} (x^2 + 16 - x^2)^{3/2} + \frac{x^5}{3} \right] dx$$

$$= \int_0^4 \left[\frac{32}{3} x^2 - 2x^4 + \frac{x^5}{3} \right] dx$$

$$= \left[\frac{32}{9} x^3 - \frac{2}{5} x^5 + \frac{x^6}{18} \right]_0^4 = \frac{2048}{45}. \quad \blacktriangleright$$

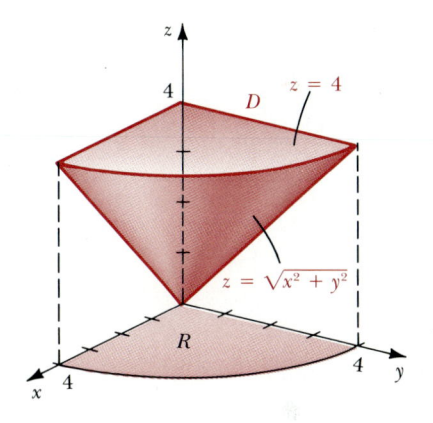

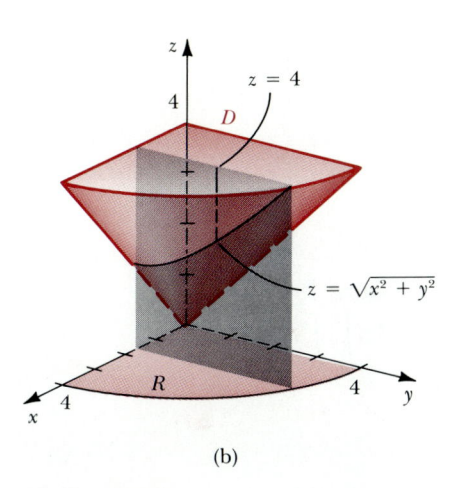

▶ **Figure 15.46**

In Section 15.2 we found that the double integral of the constant function $f(x, y) = 1$ is an alternative means for expressing the area of a region R in the plane:

$$A = \iint_R dA.$$

Similarly, the triple integral of $f(x, y, z) = 1$ over a solid region D gives the volume of D:

(15.24)
$$V = \iiint_D dV.$$

When D can be expressed in the form

$$a \leq x \leq b, \quad g_1(x) \leq y \leq g_2(x), \quad f_1(x, y) \leq z \leq f_2(x, y),$$

this reduces to an iterated integral of the type in Section 15.2 (see Figure 15.47):

$$V = \iiint_D dV = \int_{x=a}^{x=b} \int_{y=g_1(x)}^{y=g_2(x)} \int_{z=f_1(x,y)}^{z=f_2(x,y)} dz\, dy\, dx$$

$$= \int_{x=a}^{x=b} \int_{y=g_1(x)}^{y=g_2(x)} [f_2(x, y) - f_1(x, y)]\, dy\, dx.$$

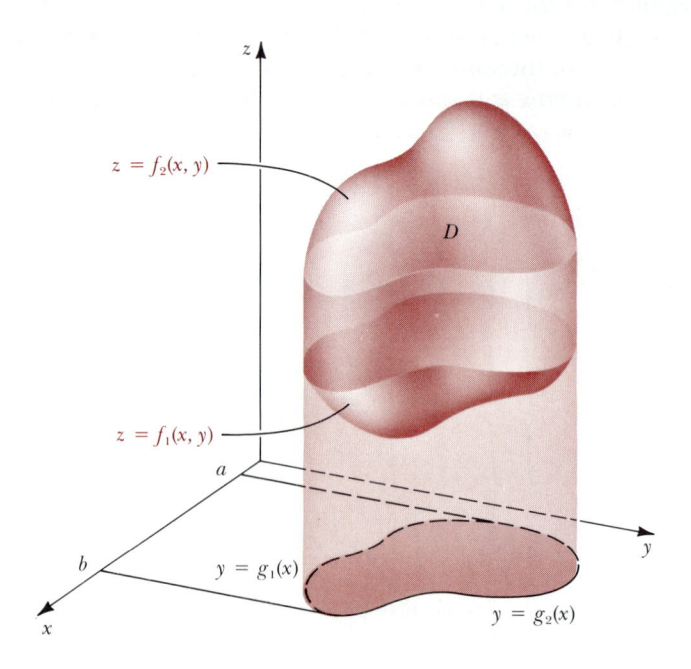

Figure 15.47

$$V = \int_b^b \int_{g_1(x)}^{g_2(x)} \int_{f_1(x,y)}^{f_2(x,y)} dz\, dy\, dx.$$

Example 4

Determine the volume of the solid with base in the xy-plane and bounded by the planes $x = 0$, $y = 0$, and $x + 2y + 2z = 4$.

Solution

The solid is shown in Figure 15.48. The plane $x + 2y + 2z = 4$ intersects the xy-plane in the line $x + 2y = 4$, so the solid has the shape of the region described by:

$$0 \le x \le 4, \quad 0 \le y \le \frac{4-x}{2}, \quad 0 \le z \le \frac{4-x-2y}{2}.$$

Thus

$$\begin{aligned}
V &= \int_0^4 \int_0^{(4-x)/2} \int_0^{(4-x-2y)/2} dz\, dy\, dx \\
&= \int_0^4 \int_0^{(4-x)/2} \frac{1}{2}(4 - x - 2y)\, dy\, dx \\
&= \frac{1}{2} \int_0^4 \left[4y - xy - y^2 \right]_0^{(4-x)/2} dx \\
&= \frac{1}{2} \int_0^4 \left[2(4-x) - \frac{x}{2}(4-x) - \frac{1}{4}(4-x)^2 \right] dx \\
&= \frac{1}{2} \int_0^4 \left[8 - 4x + \frac{1}{2}x^2 - \frac{1}{4}(4-x)^2 \right] dx \\
&= \frac{1}{2} \left[8x - 2x^2 + \frac{1}{6}x^3 + \frac{1}{12}(4-x)^3 \right]_0^4 = \frac{8}{3}.
\end{aligned}$$

▶

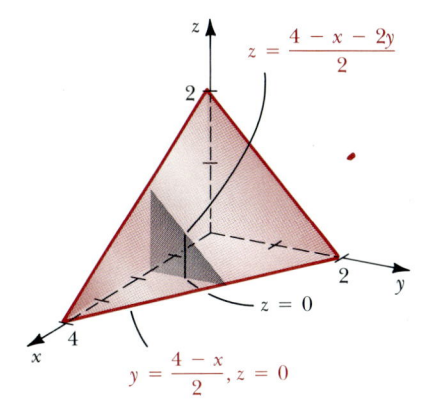

Figure 15.48

The regions considered in the preceding examples have had representations that permitted iterated integration to be performed in the order $dz\, dy\, dx$. The boundaries in the x-direction have been constant, those in the y-direction depended only on x, and those in the z-direction depended on both x and y. Many regions, however, are described more easily by a different ordering of variables. When the region is described by

$$a \le x \le b, \quad g_1(x) \le z \le g_2(x), \quad h_1(x, z) \le y \le h_2(x, z),$$

the integral is

(15.25) $$\iiint_D f(x, y, z)\, dV = \int_{x=a}^{x=b} \int_{z=g_1(x)}^{z=g_2(x)} \int_{y=h_1(x, z)}^{y=h_2(x, z)} f(x, y, z)\, dy\, dz\, dx.$$

A region of this type is shown in Figure 15.49 on the following page.

Similarly, when a region is represented by

$$a \le z \le b, \quad g_1(z) \le y \le g_2(z), \quad h_1(y, z) \le x \le h_2(y, z),$$

(see Figure 15.50 on the following page), the integral has the form

(15.26) $$\iiint_D f(x, y, z)\, dV = \int_{z=a}^{z=b} \int_{y=g_1(z)}^{y=g_2(z)} \int_{x=h_1(y, z)}^{x=h_2(y, z)} f(x, y, z)\, dx\, dy\, dz.$$

There are three other variable orderings that can be used: $dx\, dz\, dy$, $dy\, dx\, dz$, and $dz\, dx\, dy$.

A triple integral can often be expressed as an iterated integral in a

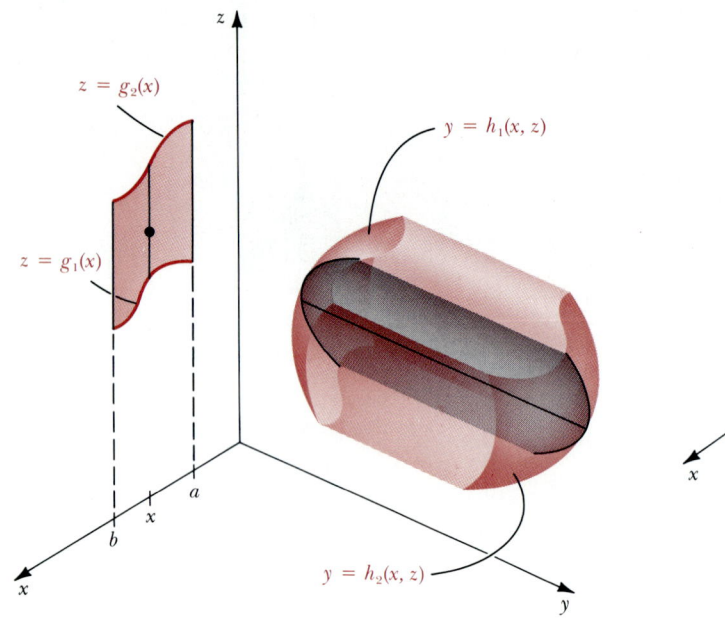

Figure 15.49

$$\iiint_D f(x, y, z)\, dV = \int_{x=a}^{x=b} \int_{z=g_1(x)}^{z=g_2(x)} \int_{y=h_1(x,y)}^{y=h_2(x,y)} f(x, y, z)\, dy\, dz\, dx.$$

Figure 15.50

$$\iiint_D f(x, y, z)\, dV = \int_{z=a}^{z=b} \int_{y=g_1(x)}^{y=g_2(x)} \int_{x=h_1(y, z)}^{x=h_2(y, z)} f(x, y, z)\, dx\, dy\, dz.$$

variety of ways. When this happens the choice to use for evaluation of the integral depends on the ease of representation of the boundaries of the region, as well as on the order in which the function can be integrated most readily. The following example considers the alternative representations of the integral in Example 2. Integrate each of the expressions in this example to see the variety of techniques involved. The same numerical result should, of course, be obtained in each instance.

Example 5

Express $\iiint_D z\, dV$ in the six iterated integral forms, where D is the region in the first octant bounded by the planes $x = y$, $z = 0$, and $y = 0$ and the cylinder $x^2 + z^2 = 1$.

Solution

This triple integral was expressed in Example 2 as

1) $\displaystyle \int_0^1 \int_0^x \int_0^{\sqrt{1-x^2}} z\, dz\, dy\, dx,$

as seen in Figure 15.51(a).

It could also be expressed as either

2) $\displaystyle \int_0^1 \int_0^{\sqrt{1-x^2}} \int_0^x z\, dy\, dz\, dx$ (Figure 15.51b),

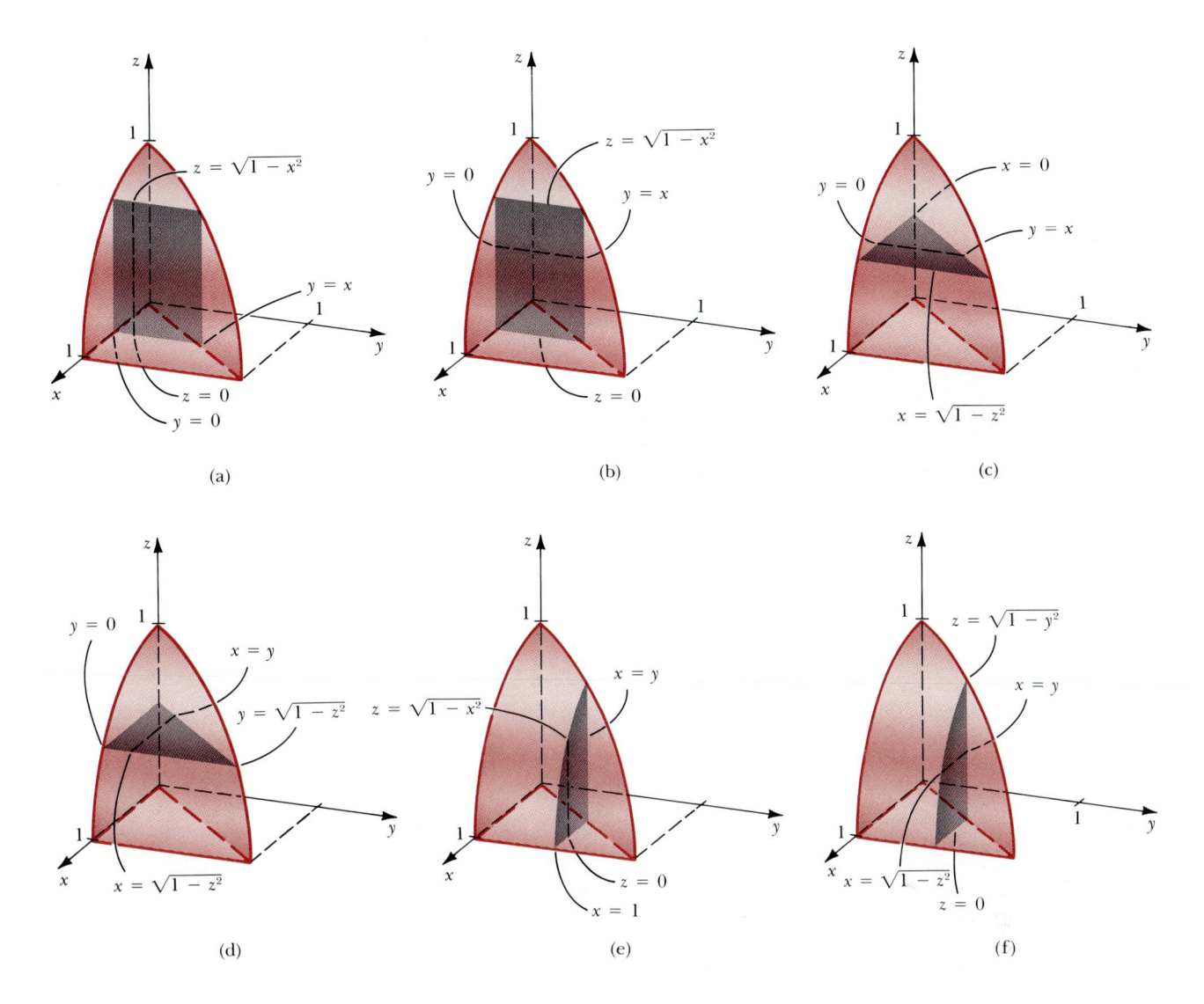

15.6 TRIPLE INTEGRALS

(a)

(b)

(c)

(d)

(e)

(f)

Figure 15.51

3) $\int_0^1 \int_0^{\sqrt{1-z^2}} \int_0^x z\, dy\, dx\, dz$ (Figure 15.51c),

4) $\int_0^1 \int_0^{\sqrt{1-z^2}} \int_y^{\sqrt{1-z^2}} z\, dx\, dy\, dz$ (Figure 15.51d),

5) $\int_0^1 \int_y^1 \int_0^{\sqrt{1-x^2}} z\, dz\, dx\, dy$ (Figure 15.51e), or

6) $\int_0^1 \int_0^{\sqrt{1-y^2}} \int_y^{\sqrt{1-z^2}} z\, dx\, dz\, dy$ (Figure 15.51f).

Experience is needed to make an intelligent choice regarding the order of integration that will simplify the evaluation of a triple integral. The rather extensive list of exercises at the end of this section should help to provide this experience.

Example 6

Determine the volume of the solid lying in the first octant bounded above by the sphere $x^2 + y^2 + z^2 = 2$ and below by the paraboloid $z = x^2 + y^2$.

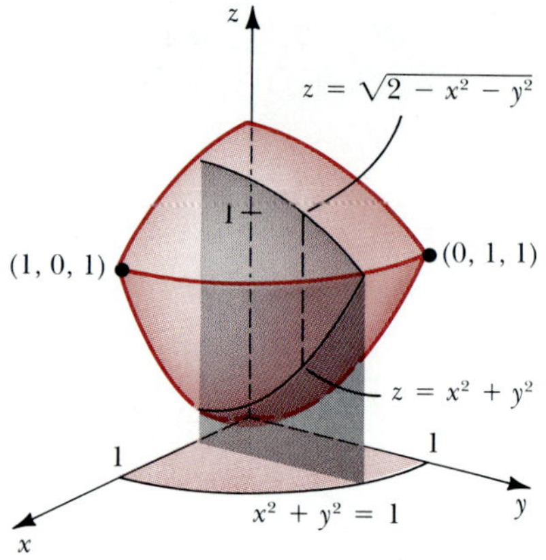

Figure 15.52

Solution

The solid is shown in Figure 15.52. By (15.24), its volume is the integral of the constant function 1 over this solid. To find the projection of the solid into the xy-plane, we first need the intersection of the sphere and the paraboloid. We can find this by solving the equations $x^2 + y^2 + z^2 = 2$ and $z = x^2 + y^2$ simultaneously:

$$2 = (x^2 + y^2) + z^2 = z + z^2,$$

so

$$0 = z^2 + z - 2 = (z - 1)(z + 2)$$

and $z = 1$ or $z = -2$. Since the region lies above the xy-plane, we must have $z = 1$. The curve of intersection is the portion of $x^2 + y^2 = 1, z = 1$ that lies in the first octant. Projecting the curve of intersection into the xy-plane determines the limits of integration on x and y:

$$0 \le x \le 1, \qquad 0 \le y \le \sqrt{1 - x^2}.$$

Thus,

$$\begin{aligned} V &= \int_0^1 \int_0^{\sqrt{1-x^2}} \int_{x^2+y^2}^{\sqrt{2-x^2-y^2}} dz\, dy\, dx \\ &= \int_0^1 \int_0^{\sqrt{1-x^2}} z \Big]_{x^2+y^2}^{\sqrt{2-x^2-y^2}} dy\, dx \\ &= \int_0^1 \int_0^{\sqrt{1-x^2}} [\sqrt{2 - x^2 - y^2} - (x^2 + y^2)]\, dy\, dx. \end{aligned}$$

Since the region of integration of this double integral is a portion of a circle centered at the origin, evaluation is made easier by changing to polar coordinates:

$$\begin{aligned} V &= \int_0^{\pi/2} \int_0^1 (\sqrt{2 - r^2} - r^2) r\, dr\, d\theta \\ &= \int_0^{\pi/2} \left[-\frac{(2 - r^2)^{3/2}}{3} - \frac{r^4}{4} \right]_0^1 d\theta \\ &= \int_0^{\pi/2} \left(-\frac{1}{3} - \frac{1}{4} + \frac{2\sqrt{2}}{3} \right) d\theta \\ &= \int_0^{\pi/2} \frac{8\sqrt{2} - 7}{12}\, d\theta \\ &= \frac{8\sqrt{2} - 7}{24} \pi \approx 0.56. \end{aligned}$$

►

EXERCISE SET 15.6

In Exercises 1–8, evaluate the triple integral.

1. $\int_0^1 \int_{x^2}^x \int_{x-y}^{x+y} y \, dz \, dy \, dx$

2. $\int_0^1 \int_{x^2}^x \int_{x-z}^{x+z} y \, dy \, dz \, dx$

3. $\int_0^1 \int_{x^2}^x \int_{x-y}^{x+y} z \, dz \, dy \, dx$

4. $\int_0^1 \int_{x^2}^x \int_{x-z}^{x+z} z \, dy \, dz \, dx$

5. $\int_0^\pi \int_1^z \int_0^{xz} \frac{1}{x} \sin\left(\frac{y}{x}\right) dy \, dx \, dz$

6. $\int_{-1}^0 \int_1^2 \int_{1/y}^1 xy \sin \pi yz \, dz \, dy \, dx$

7. $\int_0^1 \int_{-1}^1 \int_{-xy}^{xy} e^{x^2+y^2} dz \, dx \, dy$

8. $\int_0^1 \int_0^1 \int_0^{xy} e^{x^2+y^2} dz \, dx \, dy$

In Exercises 9–14, sketch the solid whose volume is given by the triple integral and determine the volume of the solid.

9. $\int_{-2}^2 \int_0^3 \int_0^y dz \, dy \, dx$

10. $\int_0^2 \int_2^4 \int_0^z dy \, dx \, dz$

11. $\int_0^2 \int_0^{\sqrt{4-x^2}} \int_{-\sqrt{4-x^2-y^2}}^{\sqrt{4-x^2-y^2}} dz \, dy \, dx$

12. $\int_{-2}^2 \int_{-\sqrt{4-x^2}}^{\sqrt{4-x^2}} \int_{x^2+y^2}^{8-x^2-y^2} dz \, dy \, dx$

13. $\int_0^3 \int_{z-3}^0 \int_{-x-3}^0 dy \, dx \, dz$

14. $\int_{-1}^1 \int_0^3 \int_0^{\sqrt{1-z^2}} dx \, dy \, dz$

In Exercises 15–20, (a) rewrite the integral in five alternative ways and (b) evaluate the triple integral in two ways.

15. $\int_0^1 \int_0^{1-x} \int_0^{1-x-y} z \, dz \, dy \, dx$

16. $\int_0^1 \int_0^{2-2x} \int_0^{2-2x-y} (x+z) \, dz \, dy \, dx$

17. $\int_0^1 \int_0^{2-2x} \int_0^{3-3x-3y/2} (2x+y) \, dz \, dy \, dx$

18. $\int_0^1 \int_y^1 \int_0^x x^2 z \, dz \, dx \, dy$

19. $\int_0^2 \int_0^{3\sqrt{4-z^2}/2} \int_0^{\sqrt{36-4y^2-9z^2}} z \, dx \, dy \, dz$

20. $\int_0^1 \int_{-\sqrt{1-z}}^{\sqrt{1-z}} \int_0^{\sqrt{1-y^2-z}} y \, dx \, dy \, dz$

In Exercises 21–30, use triple integrals to find the volume of the solid whose boundaries are described.

21. $x = 0, y = 0, x + y + z = 2, x + y - z = 2$
22. $x + y + z = 2, x = 1, y = 0, z = 0$
23. $z = x^2 + y^2, z = 1, z = 4$
24. $z = y^2, z = 1, x = 0, x = 1$
25. $y = \sin x, y = \cos x, z = \cos x, x = 0, z = 0, x = \pi/4$
26. $y = \sin x, y = \cos x, z = \sin x, x = 0, z = 0, x = \pi/4$
27. $4x^2 + 9y^2 + z^2 = 36$
28. $4x^2 + 9y^2 - z^2 = 36, z = 0, z = 4$
29. $x^2 + y^2 + z^2 = 6$, inside $z = x^2 + y^2$
30. $x^2 + y^2 = 1, x^2 + z^2 = 1$

Putnam exercises

31. Evaluate

$$\int_0^1 \int_0^1 \int_0^1 \left[\cos\left(\frac{\pi}{6}(x+y+z)\right)\right]^2 dx \, dy \, dz.$$

(This exercise was derived from problem B–1 of the twenty-sixth William Lowell Putnam examination given on November 20, 1965. The examination and its solution are in the September 1966 issue of the *American Mathematical Monthly*, pp. 727–732.)

32. Let R be the region consisting of all triples (x, y, z) of nonnegative real numbers satisfying $x + y + z \leq 1$. Let $w = 1 - x - y - z$. Express the value of the triple integral

$$\iiint_R x^1 y^9 z^8 w^4 \, dx \, dy \, dz$$

in the form $a!b!c!d!/n!$, where $a, b, c, d,$ and n are positive integers. (This exercise was problem A–5 of the forty-fifth William Lowell Putnam examination given on December 1, 1984. The examination and its solution are in the October 1985 issue of the *American Mathematical Monthly*, pp. 560–567.)

15.7

Triple Integrals in Cylindrical and Spherical Coordinates

Cylindrical Coordinate Integrals

Example 6 of Section 15.6 illustrates the need to evaluate triple integrals using the cylindrical coordinate system. In that example we first integrated with respect to the variable z and then changed the remaining double integral to polar coordinates. To use cylindrical coordinates initially on a problem of this type requires that the region D be expressed in terms of r, θ, and z as shown in Figure 15.53 where

$$\alpha \leq \theta \leq \beta, \qquad g_1(\theta) \leq r \leq g_2(\theta), \qquad h_1(r, \theta) \leq z \leq h_2(r, \theta).$$

Then

$$\iiint_D f(r, \theta, z)\, dV = \int_{\theta=\alpha}^{\theta=\beta} \int_{r=g_1(\theta)}^{r=g_2(\theta)} \left[\int_{z=h_1(r,\theta)}^{z=h_2(r,\theta)} f(r, \theta, z)\, dz \right] r\, dr\, d\theta,$$

so

(15.27) $$\iiint_D f(r, \theta, z)\, dV = \int_\alpha^\beta \int_{g_1(\theta)}^{g_2(\theta)} \int_{h_1(r,\theta)}^{h_2(r,\theta)} f(r, \theta, z)\, r\, dz\, dr\, d\theta.$$

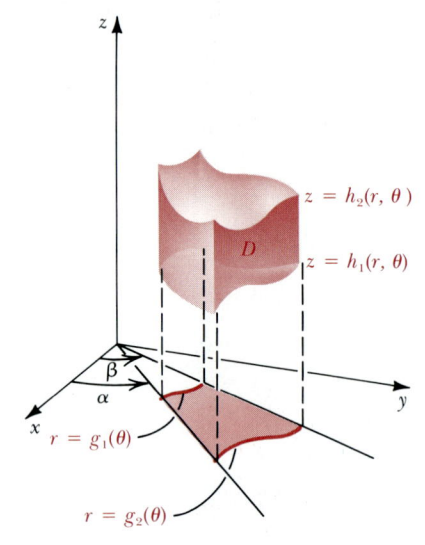

Figure 15.53

$\iiint_D f(r, \theta, z)\, dV$ is

$\int_\alpha^\beta \int_{g_1(\theta)}^{g_2(\theta)} \int_{h_1(r,\theta)}^{h_2(r,\theta)} f(r, \theta, z)\, r\, dz\, dr\, d\theta.$

We can also derive Equation (15.27) by considering a typical volume element in the cylindrical coordinate system as we did for the rectangular coordinate system in Sections 15.1 and 15.2. The cylindrical coordinate system in space is a direct extension of the polar coordinate system in the plane in the same way that the rectangular coordinate system in space extends the rectangular coordinate system in the plane: The z-coordinate axis is added perpendicular to the xy-plane. Thus the typical volume element ΔV used to define the triple integral in the cylindrical coordinate system is as shown in Figure 15.54. This volume element has a planar polar element for its base with area found in Section 9.3 to be $r\, \Delta r\, \Delta \theta$. The height of the volume element is Δz and each cross section of the element parallel to the base has the same area as the base, so the volume of the cylindrical element is

$$\Delta V = r\, \Delta r\, \Delta \theta\, \Delta z.$$

Summing Riemann sums for f over volume elements of this form produces the cylindrical iterated integral description given in Equation (15.27).

Example 1

Use cylindrical coordinates to evaluate the triple integral

$$\int_0^2 \int_0^{\sqrt{4-x^2}} \int_0^{\sqrt{4-x^2-y^2}} \frac{z}{\sqrt{x^2+y^2}}\, dz\, dy\, dx.$$

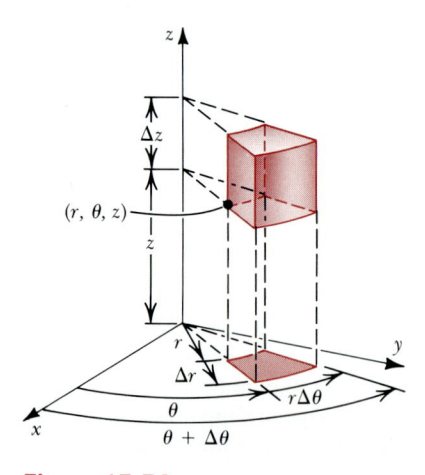

Figure 15.54
A typical cylindrical volume element.

15.7 TRIPLE INTEGRALS IN CYLINDRICAL AND SPHERICAL COORDINATES

Solution
The solid region D over which the integration is performed is in the first octant, as shown in Figure 15.55. The projection of this solid region into the xy-plane is the planar region R in the first quadrant bounded by the circle with center at the origin and radius 2 and is described in polar coordinates by

$$0 \leq \theta \leq \frac{\pi}{2} \quad \text{and} \quad 0 \leq r \leq 2.$$

The bottom of the solid is in the plane $z = 0$ and the top is the sphere with equation $z = \sqrt{4 - x^2 - y^2} = \sqrt{4 - r^2}$. So the solid region D is described in cylindrical coordinates by

$$0 \leq \theta \leq \frac{\pi}{2}, \quad 0 \leq r \leq 2, \quad 0 \leq z \leq \sqrt{4 - r^2}.$$

Thus,

$$\int_0^2 \int_0^{\sqrt{4-x^2}} \int_0^{\sqrt{4-x^2-y^2}} \frac{z}{\sqrt{x^2+y^2}} \, dz \, dy \, dx = \int_0^{\pi/2} \int_0^2 \int_0^{\sqrt{4-r^2}} \frac{z}{r} \, r \, dz \, dr \, d\theta$$

$$= \int_0^{\pi/2} \int_0^2 \frac{z^2}{2} \Big]_0^{\sqrt{4-r^2}} \, dr \, d\theta$$

$$= \frac{1}{2} \int_0^{\pi/2} \int_0^2 (4 - r^2) \, dr \, d\theta$$

$$= \frac{1}{2} \int_0^{\pi/2} \left(4r - \frac{r^3}{3} \right) \Big]_0^2 \, d\theta$$

$$= \frac{1}{2} \left(8 - \frac{8}{3} \right) \frac{\pi}{2} = \frac{4}{3} \pi. \quad \blacktriangleright$$

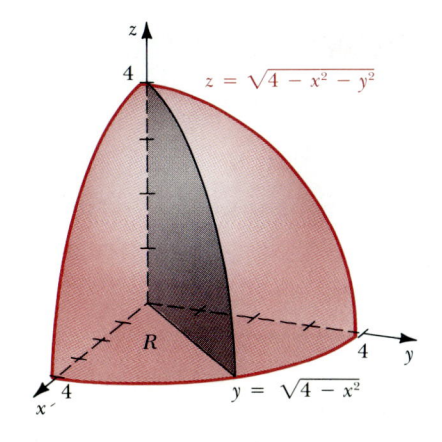

Figure 15.55

Example 2

Use the cylindrical representation to find the volume of the region in the first octant that is inside the cylinder $r = 1$ and bounded below by the xy-plane and above by the plane $3x + 2y + 6z = 6$.

Solution
Referring to Figure 15.56, we see that the solid is bounded below by the plane $z = 0$ and above by the plane

$$z = 1 - \frac{x}{2} - \frac{y}{3} = 1 - \frac{r \cos \theta}{2} - \frac{r \sin \theta}{3}.$$

Figure 15.56

Thus,

$$V = \int_{\theta=0}^{\theta=\pi/2} \int_{r=0}^{r=1} \int_{z=0}^{z=1-(r\cos\theta)/2-(r\sin\theta)/3} r\,dz\,dr\,d\theta$$

$$= \int_0^{\pi/2} \int_0^1 r\left(1 - \frac{r}{2}\cos\theta - \frac{r}{3}\sin\theta\right) dr\,d\theta$$

$$= \int_0^{\pi/2} \left[\frac{r^2}{2} - \frac{r^3}{6}\cos\theta - \frac{r^3}{9}\sin\theta\right]_0^1 d\theta$$

$$= \int_0^{\pi/2} \left(\frac{1}{2} - \frac{1}{6}\cos\theta - \frac{1}{9}\sin\theta\right) d\theta$$

$$= \left[\frac{\theta}{2} - \frac{1}{6}\sin\theta + \frac{1}{9}\cos\theta\right]_0^{\pi/2} = \left(\frac{\pi}{4} - \frac{1}{6}\right) - \frac{1}{9} = \frac{\pi}{4} - \frac{5}{18}. \blacktriangleright$$

Example 3

How much ice cream is needed to make an ice cream cone with a hemispherical top if the ice cream is packed into a cone with height 3 in. and diameter 2 in.?

Solution

Figure 15.57 shows the ice cream cone placed in a coordinate system with the vertex of the cone at the origin and the axis of the cone along the positive z-axis. The cone in this position is described in rectangular coordinates by

$$z^2 = 9(x^2 + y^2) = 9r^2, \qquad 0 \le z \le 3$$

and in cylindrical coordinates by

$$z = 3r, \qquad 0 \le r \le 1, \qquad 0 \le \theta \le 2\pi.$$

The ice cream on the top of the cone is described in rectangular coordinates by

$$1 = x^2 + y^2 + (z-3)^2 = r^2 + (z-3)^2, \qquad 3 \le z \le 4$$

and in cylindrical coordinates by

$$z = 3 + \sqrt{1-r^2}, \qquad 0 \le r \le 1, \qquad 0 \le \theta \le 2\pi.$$

The amount of ice cream required to make this ice cream cone is therefore

$$V = \int_{\theta=0}^{\theta=2\pi} \int_{r=0}^{r=1} \int_{z=3r}^{z=3+\sqrt{1-r^2}} r\,dz\,dr\,d\theta = \int_0^{2\pi} \int_0^1 (3 + \sqrt{1-r^2} - 3r)r\,dr\,d\theta$$

$$= \int_0^{2\pi} \int_0^1 (3r + r\sqrt{1-r^2} - 3r^2)\,dr\,d\theta = \int_0^{2\pi} \left[\frac{3}{2}r^2 - \frac{1}{3}(1-r^2)^{3/2} - r^3\right]_0^1 d\theta$$

$$= \int_0^{2\pi} \left[\left(\frac{3}{2} - 0 - 1\right) - \left(0 - \frac{1}{3} - 0\right)\right] d\theta = \int_0^{2\pi} \frac{5}{6} d\theta = \frac{5}{6}\theta\bigg]_0^{2\pi} = \frac{5}{3}\pi \text{ in}^3.$$

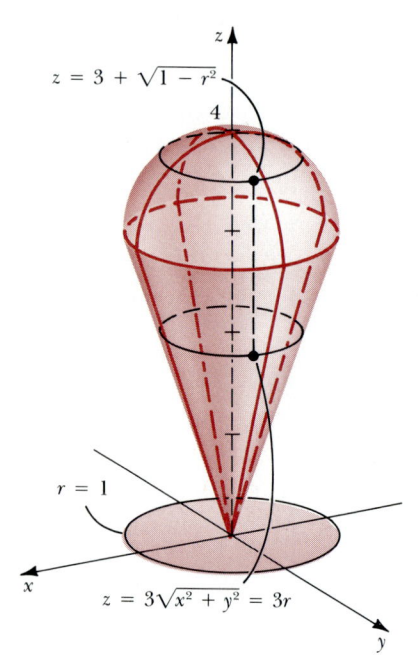

Figure 15.57

(Approximately 44 cones of this size can be made from 1 gallon of ice cream.) ▶

Spherical Coordinate Integrals

In the spherical coordinate system the typical volume element is *not* obtained by simply adding a third coordinate direction to a two-dimensional area element, as in the rectangular and cylindrical systems. The typical spherical volume element shown in Figure 15.58 is a portion of a region bounded between two spheres, one of radius ρ and the other of radius $\rho + \Delta\rho$.

The volume ΔV of this typical element is approximated by the product of the lengths shown in Figure 15.58, where on this element,

D_ρ is the linear distance from (ρ, θ, ϕ) to $(\rho + \Delta\rho, \theta, \phi)$,

D_θ is the arc length distance from (ρ, θ, ϕ) to $(\rho, \theta + \Delta\theta, \phi)$, and

D_ϕ is the arc length distance from (ρ, θ, ϕ) to $(\rho, \theta, \phi + \Delta\phi)$.

The distance D_ρ is

$$D_\rho = \Delta\rho.$$

The distance D_θ is the portion of the arc of the circle of radius $\rho \sin \phi$ determined by the angle $\Delta\theta$ (see Figure 15.59a):

$$D_\theta = \rho \sin \phi \, \Delta\theta.$$

The distance D_ϕ is the portion of the arc of the circle of radius ρ determined by the angle $\Delta\phi$ (see Figure 15.59b):

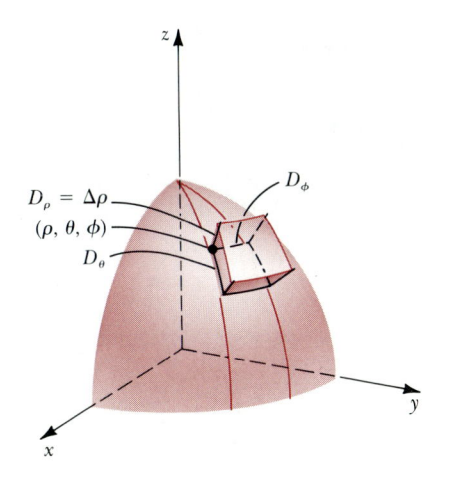

$$D_\phi = \rho\Delta\phi.$$

Figure 15.58
A typical spherical volume element.

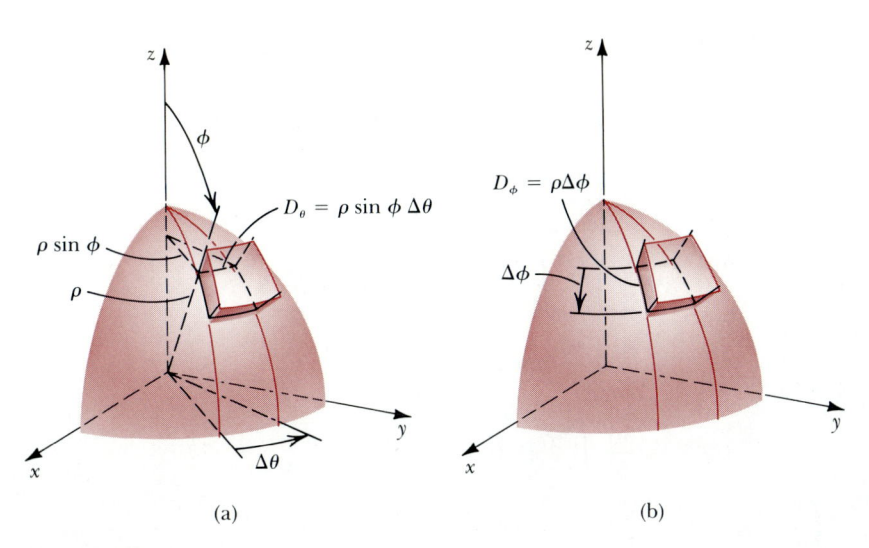

Figure 15.59

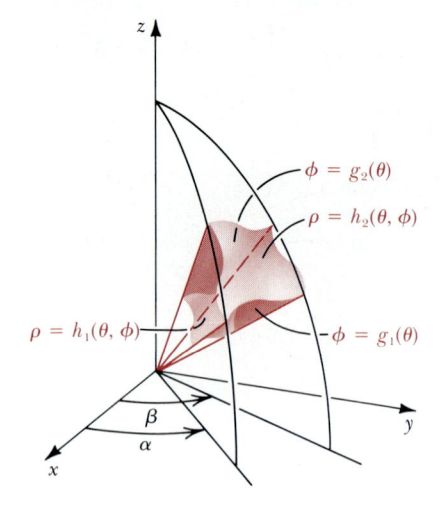

Figure 15.60

$$\iiint_D f(\rho, \theta, \phi)\, dV \text{ is}$$

$$\int_\alpha^\beta \int_{g_1(\theta)}^{g_2(\theta)} \int_{h_1(\theta,\phi)}^{h_2(\theta,\phi)} f(\rho, \theta, \phi)\rho^2 \sin\phi\, d\rho\, d\phi\, d\theta.$$

Thus,

(15.28)

$$\Delta V \approx D_\rho\, D_\theta\, D_\phi = (\Delta\rho)(\rho \sin\phi\, \Delta\theta)(\rho\, \Delta\phi) = \rho^2 \sin\phi\, \Delta\rho\, \Delta\phi\, \Delta\theta.$$

This implies (see Figure 15.60) that if a bounded region D in space is described by

$$\alpha \le \theta \le \beta, \qquad g_1(\theta) \le \phi \le g_2(\theta), \qquad h_1(\theta, \phi) \le \rho \le h_2(\theta, \phi),$$

then

(15.29)

$$\iiint_D f(\rho, \theta, \phi)\, dV = \int_{\theta=\alpha}^{\theta=\beta} \int_{\phi=g_1(\theta)}^{\phi=g_2(\theta)} \int_{\rho=h_1(\theta,\phi)}^{\rho=h_2(\theta,\phi)} f(\rho, \theta, \phi)\, \rho^2 \sin\phi\, d\rho\, d\phi\, d\theta.$$

In the following example we consider a triple integral in spherical coordinates whose value is also verified by geometry.

Example 4

Use an iterated integral in spherical coordinates to determine the volume between the spheres $x^2 + y^2 + z^2 = 4$ and $x^2 + y^2 + z^2 = 1$.

Solution

This volume is obtained from geometry by subtracting the volume of a sphere of radius 1 from the volume of a sphere of radius 2:

$$V = \frac{4}{3}\pi(2)^3 - \frac{4}{3}\pi(1)^3 = \frac{28\pi}{3}.$$

To use a triple integral in spherical coordinates to find this result, we first note that symmetry implies that the volume of the portion in each octant is the same. The portion in the first octant, shown in Figure 15.61, is described in spherical coordinates by $0 \le \theta \le \pi/2$, $0 \le \phi \le \pi/2$, $1 \le \rho \le 2$. The total volume is 8 times the volume of this portion:

$$V = 8 \int_{\theta=0}^{\theta=\pi/2} \int_{\phi=0}^{\phi=\pi/2} \int_{\rho=1}^{\rho=2} \rho^2 \sin\phi\, d\rho\, d\phi\, d\theta$$

$$= 8 \int_0^{\pi/2} \int_0^{\pi/2} \frac{\rho^3}{3}\bigg]_1^2 \sin\phi\, d\phi\, d\theta$$

$$= \frac{8}{3} \int_0^{\pi/2} \int_0^{\pi/2} (2^3 - 1) \sin\phi\, d\phi\, d\theta$$

$$= \frac{8}{3} \int_0^{\pi/2} 7(-\cos\phi)\bigg]_0^{\pi/2} d\theta$$

$$= \frac{56}{3} \int_0^{\pi/2} (-0 + 1)\, d\theta$$

$$= \frac{56}{3} \cdot \frac{\pi}{2} = \frac{28}{3}\pi.$$

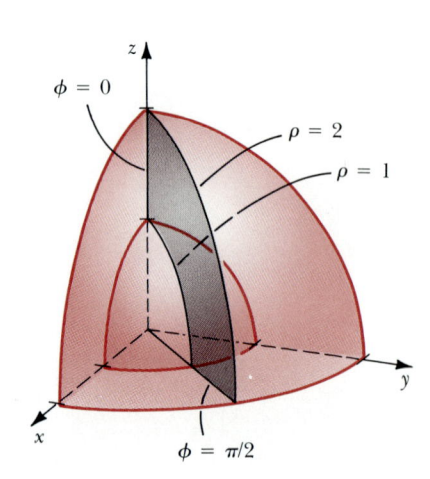

Figure 15.61

15.7 TRIPLE INTEGRALS IN CYLINDRICAL AND SPHERICAL COORDINATES

We can also calculate this volume using cylindrical coordinates, but the computation requires evaluating two iterated integrals:

$$V = 8\int_0^{\pi/2}\int_0^1\int_{\sqrt{1-r^2}}^{\sqrt{4-r^2}} r\, dz\, dr\, d\theta + 8\int_0^{\pi/2}\int_1^2\int_0^{\sqrt{4-r^2}} r\, dz\, dr\, d\theta.$$

Example 5

Evaluate the integral of $f(x, y, z) = x^2 + y^2 + z$ over the region S bounded below by the cone $z = \sqrt{x^2 + y^2}$ and above by the sphere $x^2 + y^2 + z^2 = 9$.

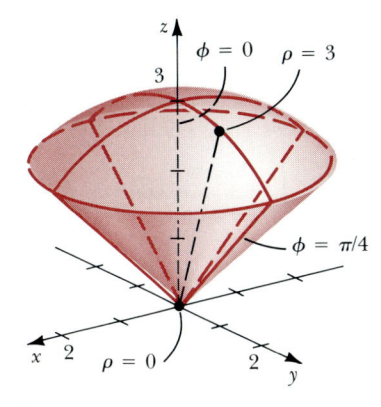

Figure 15.62

Solution

This integral is evaluated most easily using spherical coordinates. Figure 15.62 shows that in order for (ρ, θ, ϕ) to be in the solid region S, ρ and θ must satisfy $0 \le \rho \le 3$ and $0 \le \theta \le 2\pi$. To determine the limits of ϕ, we observe that the cone $z = \sqrt{x^2 + y^2}$ intersects the yz-plane in the line $z = y$. The symmetry of the region about the z-axis implies that $0 \le \phi \le \pi/4$.

The integrand $f(x, y, z) = x^2 + y^2 + z$ is expressed in spherical coordinates by applying the rectangular-to-spherical conversion equations given in Section 13.5:

$$x^2 + y^2 + z = (\rho \sin\phi \cos\theta)^2 + (\rho \sin\phi \sin\theta)^2 + \rho \cos\phi$$
$$= \rho^2 \sin^2\phi + \rho \cos\phi.$$

Thus,

$$\iiint_S f(x, y, z)\, dV = \int_0^{2\pi}\int_0^{\pi/4}\int_0^3 [\rho^2 \sin^2\phi + \rho \cos\phi]\rho^2 \sin\phi\, d\rho\, d\phi\, d\theta$$

$$= \int_0^{2\pi}\int_0^{\pi/4} \left[\frac{\rho^5}{5}\sin^3\phi + \frac{\rho^4}{4}\cos\phi \sin\phi\right]_{\rho=0}^{\rho=3} d\phi\, d\theta$$

$$= \int_0^{2\pi}\int_0^{\pi/4} \left(\frac{243}{5}\sin^3\phi + \frac{81}{4}\cos\phi \sin\phi\right) d\phi\, d\theta$$

$$= \int_0^{2\pi}\int_0^{\pi/4} \left[\frac{243}{5}(1 - \cos^2\phi)\sin\phi + \frac{81}{4}\cos\phi \sin\phi\right] d\phi\, d\theta$$

$$= \int_0^{2\pi} \left[\frac{243}{5}\left(-\cos\phi + \frac{\cos^3\phi}{3}\right) + \frac{81}{8}\sin^2\phi\right]_{\phi=0}^{\phi=\pi/4} d\theta$$

$$= \int_0^{2\pi} \frac{81}{80}(37 - 20\sqrt{2})\, d\theta$$

$$= \frac{81\pi}{40}(37 - 20\sqrt{2}) \approx 55.5.$$

EXERCISE SET 15.7

In Exercises 1–6, evaluate the triple integral.

1. $\displaystyle\int_{-\pi}^{\pi}\int_{0}^{1}\int_{0}^{\sqrt{1-r^2}} rz\sin\theta\, dz\, dr\, d\theta$

2. $\displaystyle\int_{-\pi}^{\pi}\int_{0}^{1}\int_{0}^{\sqrt{1-r^2}} rz\cos\theta\, dz\, dr\, d\theta$

3. $\displaystyle\int_{0}^{\pi}\int_{0}^{\pi/2}\int_{0}^{\sin\phi} \rho^2\sin\phi\cos\phi\, d\rho\, d\phi\, d\theta$

4. $\displaystyle\int_{0}^{\pi}\int_{0}^{\pi/2}\int_{0}^{\sin\phi} \rho^2\sin\phi\cos\phi\, d\rho\, d\theta\, d\phi$

5. $\displaystyle\int_{0}^{2\pi}\int_{0}^{\sin\theta}\int_{0}^{r} r^2\cos\theta\, dz\, dr\, d\theta$

6. $\displaystyle\int_{0}^{1}\int_{0}^{\arccos\theta}\int_{0}^{\cos\phi} \rho\sin\phi\, e^{\theta}\, d\rho\, d\phi\, d\theta$

In Exercises 7–12, use either cylindrical coordinates or spherical coordinates to reexpress and evaluate the integral.

7. $\displaystyle\int_{0}^{2}\int_{0}^{\sqrt{4-x^2}}\int_{0}^{\sqrt{4-x^2-y^2}} \frac{z}{\sqrt{x^2+y^2}}\, dz\, dy\, dx$

8. $\displaystyle\int_{0}^{2}\int_{-\sqrt{4-y^2}}^{\sqrt{4-y^2}}\int_{0}^{\sqrt{4-x^2-y^2}} z\sqrt{x^2+y^2}\, dz\, dx\, dy$

9. $\displaystyle\int_{0}^{2}\int_{-\sqrt{4-y^2}}^{\sqrt{4-y^2}}\int_{0}^{\sqrt{4-x^2-y^2}} z\sqrt{x^2+y^2+z^2}\, dz\, dx\, dy$

10. $\displaystyle\int_{0}^{2}\int_{0}^{\sqrt{4-x^2}}\int_{0}^{\sqrt{4-x^2-y^2}} \frac{z}{\sqrt{x^2+y^2+z^2}}\, dz\, dy\, dx$

11. $\displaystyle\int_{0}^{1}\int_{0}^{\sqrt{1-x^2}}\int_{1-x^2-y^2}^{\sqrt{1-x^2-y^2}} \arctan\left(\frac{y}{x}\right) dz\, dy\, dx$

12. $\displaystyle\int_{-1}^{2}\int_{0}^{2}\int_{0}^{\sqrt{4-x^2}} e^{x^2+y^2}\, dy\, dx\, dz$

In Exercises 13–18, sketch a solid whose volume is given by the triple integral.

13. $\displaystyle\int_{0}^{2\pi}\int_{0}^{2}\int_{0}^{r^2} r\, dz\, dr\, d\theta$ 14. $\displaystyle\int_{0}^{\pi}\int_{1}^{3}\int_{0}^{4} r\, dz\, dr\, d\theta$

15. $\displaystyle\int_{0}^{\pi/2}\int_{0}^{\pi/4}\int_{0}^{2} \rho^2\sin\phi\, d\rho\, d\phi\, d\theta$

16. $\displaystyle\int_{0}^{\pi/2}\int_{0}^{\pi/4}\int_{0}^{2} \rho^2\sin\theta\, d\rho\, d\theta\, d\phi$

17. $\displaystyle\int_{0}^{\pi/2}\int_{0}^{2}\int_{r}^{\sqrt{8-r^2}} r\, dz\, dr\, d\theta$

18. $\displaystyle\int_{0}^{\pi/2}\int_{0}^{\pi/4}\int_{0}^{2\sqrt{2}} \rho^2\sin\phi\, d\rho\, d\phi\, d\theta$

In Exercises 19–26, use either cylindrical or spherical coordinates to find the volume of the solid described.

19. Inside both $z^2 = x^2 + y^2$ and $x^2 + y^2 + z^2 = 4$

20. Inside $x^2 + y^2 + z^2 = 4$ and outside $z^2 = x^2 + y^2$

21. Bounded by $z = x^2 + y^2$, $z = 1$, and $z = 4$

22. Bounded by $z^2 = x^2 + y^2$ and $x^2 + y^2 = 4$

23. With boundaries $z = \sqrt{x^2 + y^2}$, $z = \sqrt{3x^2 + 3y^2}$, and $z = 9$

24. With boundaries $z = \sqrt{x^2 + y^2}$, $z = \sqrt{3x^2 + 3y^2}$, and $\sqrt{x^2 + y^2 + z^2} = 3$

25. With boundaries $z = \sqrt{x^2 + y^2}$, $\sqrt{x^2 + y^2 + z^2} = 2$, and $\sqrt{x^2 + y^2 + z^2} = 3$

26. With boundaries $z = \sqrt{x^2 + y^2}$, $x^2 + y^2 = 4$, $x^2 + y^2 = 9$, and $z = 0$

In Exercises 27 and 28, set up the integral(s) in (a) rectangular, (b) cylindrical, and (c) spherical coordinates to evaluate a function f over the solid described.

27. Inside $x^2 + y^2 + z^2 = a^2$

28. Inside $x^2 + y^2 = a^2$, bounded below by $z = 0$ and above by $z = a$

29. A conical dunce cap of diameter 5 in. at the bottom and height 15 in. sits on the (spherical with radius 3.5 in.) head of the class dullard (see the figure). Find the volume of air in the hat.

30. A hole of radius 1 in. is drilled through the center of a sphere of radius 6 in. (see the figure). Find the volume of the solid removed.

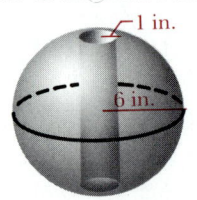

31. A hole drilled through the center of a sphere of radius R removes half the volume of the sphere. What is the radius of the hole?

32. The unit "hypersphere" in four-dimensional space has equation $x^2 + y^2 + z^2 + w^2 = 1$. Find the "hypervolume" of the hypersphere.

15.8

Applications of Triple Integrals

In Section 15.7 we saw that one application of triple integrals is to find the volume of solid regions. In this section we consider the application of triple integrals to determine the center of mass of a solid region and the average value of a function over a solid region.

The concepts of mass, moments, and inertia for solids in space are direct extensions of these concepts for the plane laminae discussed in Section 15.4. The mass of a solid region D with density function σ is defined as

(15.30) $$M = \iiint_D \sigma(x, y, z)\, dV.$$

Since the distances of a point (x, y, z) from the xy-, yz-, and xz-planes are z, x, and y, respectively, the moments with respect to these coordinate planes are

(15.31) $$M_{xy} = \iiint_D z\sigma(x, y, z)\, dV,$$

(15.32) $$M_{yz} = \iiint_D x\sigma(x, y, z)\, dV,$$

and

(15.33) $$M_{xz} = \iiint_D y\sigma(x, y, z)\, dV.$$

The center of mass $(\bar{x}, \bar{y}, \bar{z})$ is

(15.34) $$\bar{x} = \frac{M_{yz}}{M}, \qquad \bar{y} = \frac{M_{xz}}{M}, \qquad \bar{z} = \frac{M_{xy}}{M}.$$

Example 1

Find the center of mass of the ice cream cone described in Example 3 of Section 15.7. The cone is topped with a hemisphere of ice cream (see Figure 15.63). The height of the cone is 3 in. and the diameter of the top of the cone and of the hemisphere of ice cream is 2 in. Assume that the density of the ice cream is a constant and that the weight of the cone is insignificant.

Solution
The cone is described by cylindrical coordinates by

$$z = 3r, \qquad 0 \le r \le 1, \qquad 0 \le \theta \le 2\pi,$$

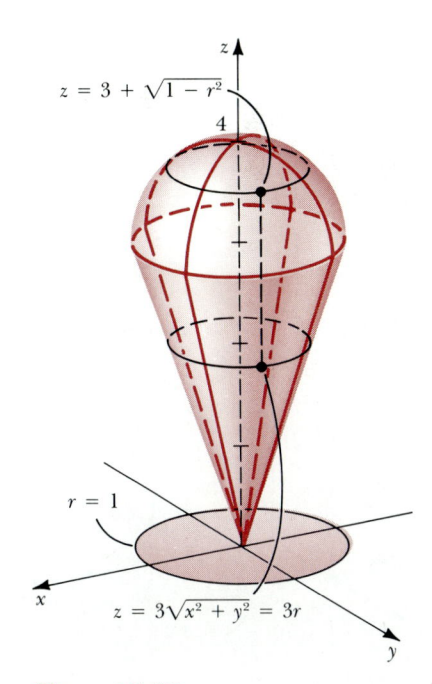

Figure 15.63

and the top of the ice cream is described by
$$z = 3 + \sqrt{1 - r^2}, \quad 0 \le r \le 1, \quad 0 \le \theta \le 2\pi.$$

In Example 3 of Section 15.7 the volume of ice cream in the cone was found to be $5\pi/3$ in^3. Since the density is constant, we can assume that $\sigma = 1$. Then the mass of the ice cream cone is
$$M = \frac{5\pi}{3}.$$

The cone is symmetric with respect to both the xz- and yz-planes, so the center of mass lies along the z-axis; that is, $\bar{x} = 0$ and $\bar{y} = 0$. To find $\bar{z}$ requires finding
$$M_{xy} = \iiint_D z \, dV.$$

We used spherical coordinates to find the mass of this ice cream cone. Because of the addition of the z in the integrand, however, we use cylindrical coordinates in this case:

$$\begin{aligned} M_{xy} &= \int_{\theta=0}^{\theta=2\pi} \int_{r=0}^{r=1} \int_{z=3r}^{z=3+\sqrt{1-r^2}} z \, r \, dz \, dr \, d\theta \\ &= \int_0^{2\pi} \int_0^1 \frac{z^2 r}{2} \bigg]_{3r}^{3+\sqrt{1-r^2}} dr \, d\theta \\ &= \int_0^{2\pi} \int_0^1 \frac{r}{2} [(3 + \sqrt{1-r^2})^2 - (3r)^2] \, dr \, d\theta \\ &= \int_0^{2\pi} \int_0^1 (5r + 3r\sqrt{1-r^2} - 5r^3) \, dr \, d\theta \\ &= \int_0^{2\pi} \left[\frac{5r^2}{2} - (1-r^2)^{3/2} - \frac{5r^4}{4} \right]_0^1 d\theta \\ &= \int_0^{2\pi} \frac{9}{4} d\theta = \frac{9\pi}{2}. \end{aligned}$$

Thus,
$$\bar{z} = \frac{M_{xy}}{M} = \frac{9\pi/2}{5\pi/3} \approx 2.7 \text{ in.}$$

The center of mass lies at $(0, 0, 2.7)$. ▶

Suppose we want to determine the moment of inertia I_z of a solid region D with density σ. The distance from an arbitrary point (x, y, z) to the z-axis is $\sqrt{x^2 + y^2}$ (see Figure 15.64). Thus,

$$\begin{aligned} I_z &= \iiint_D (\sqrt{x^2 + y^2})^2 \, \sigma(x, y, z) \, dV \\ &= \iiint_D (x^2 + y^2) \, \sigma(x, y, z) \, dV. \end{aligned}$$

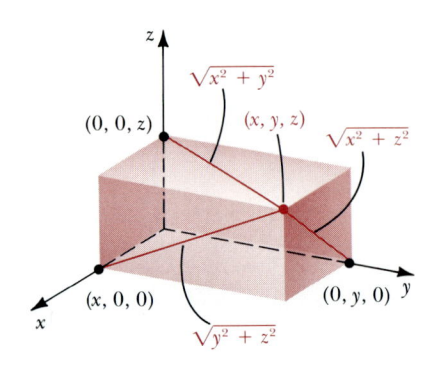

Figure 15.64

15.8 APPLICATIONS OF TRIPLE INTEGRALS

Similarly,

$$I_x = \iiint_D (y^2 + z^2)\, \sigma(x, y, z)\, dV$$

and

$$I_y = \iiint_D (x^2 + z^2)\, \sigma(x, y, z)\, dV.$$

The radii of gyration $\hat{x}, \hat{y}, \hat{z}$ about the x-, y-, and z-axes, respectively, are expressed in a manner similar to the case of plane laminae:

(15.35) $\qquad \hat{x} = \sqrt{I_x/M}, \qquad \hat{y} = \sqrt{I_y/M}, \qquad \hat{z} = \sqrt{I_z/M}.$

Example 2

Find the radius of gyration about the z-axis for the solid bounded by $x^2 + y^2 + z^2 = 1$ with density $\sigma(x, y, z) = x^2 + y^2 + z^2$.

Solution
The mass

$$M = \iiint_D (x^2 + y^2 + z^2)\, dV$$

is evaluated most easily using spherical coordinates since the solid D is a sphere of radius 1 centered at the origin and the integrand is easily expressed in spherical coordinates:

$$x^2 + y^2 + z^2 = \rho^2.$$

The spherical solid is described by $0 \leq \theta \leq 2\pi,\ 0 \leq \phi \leq \pi,\ 0 \leq \rho \leq 1$. Thus,

$$M = \int_{\theta=0}^{\theta=2\pi} \int_{\phi=0}^{\phi=\pi} \int_{\rho=0}^{\rho=1} \rho^2 (\rho^2 \sin \phi)\, d\rho\, d\phi\, d\theta$$

$$= \int_0^{2\pi} \int_0^{\pi} \frac{\rho^5}{5} \sin \phi \bigg]_0^1 d\phi\, d\theta$$

$$= \frac{1}{5} \int_0^{2\pi} -\cos \phi \bigg]_0^{\pi} d\theta = \frac{2}{5} \int_0^{2\pi} d\theta = \frac{4\pi}{5}.$$

The moment of inertia about the z-axis is

$$I_z = \iiint_D (x^2 + y^2 + z^2)(x^2 + y^2)\, dV.$$

This integral is also evaluated most easily using spherical coordinates. Since

$$x^2 + y^2 = (\rho \sin \phi \cos \theta)^2 + (\rho \sin \phi \sin \theta)^2$$
$$= \rho^2 \sin^2 \phi\, (\cos^2 \theta + \sin^2 \theta)$$
$$= \rho^2 \sin^2 \phi,$$

we have

$$I_z = \int_0^{2\pi} \int_0^{\pi} \int_0^1 \rho^2(\rho^2 \sin^2 \phi)(\rho^2 \sin \phi) \, d\rho \, d\phi \, d\theta$$

$$= \int_0^{2\pi} \int_0^{\pi} \int_0^1 \rho^6 \sin^3 \phi \, d\rho \, d\phi \, d\theta$$

$$= \int_0^{2\pi} \int_0^{\pi} \left[\sin^3 \phi \, \frac{\rho^7}{7} \right]_0^1 d\phi \, d\theta$$

$$= \frac{1}{7} \int_0^{2\pi} \int_0^{\pi} (1 - \cos^2 \phi) \sin \phi \, d\phi \, d\theta$$

$$= \frac{1}{7} \int_0^{2\pi} \left[-\cos \phi + \frac{\cos^3 \phi}{3} \right]_0^{\pi} d\theta$$

$$= \frac{1}{7} \int_0^{2\pi} \left[\left(1 - \frac{1}{3}\right) - \left(-1 + \frac{1}{3}\right) \right] d\theta$$

$$= \frac{1}{7} \cdot \frac{4}{3} \, 2\pi = \frac{8}{21} \pi.$$

Thus,

$$\hat{z} = \sqrt{\frac{I_z}{M}} = \sqrt{\frac{\frac{8}{21}\pi}{\frac{4}{5}\pi}} = \sqrt{\frac{10}{21}} \approx 0.7.$$

▶

In Section 4.3 we saw that the **average value** of a function of one variable with respect to x is defined

$$\text{Average value of } f \text{ over } [a, b] = \frac{1}{b - a} \int_a^b f(x) \, dx$$

$$= \frac{1}{\text{length of } [a, b]} \int_a^b f(x) \, dx.$$

If f is a continuous function of two variables x and y, the average value of f over a region R in the xy-plane is

(15.36) $\quad \text{Average value of } f \text{ over } R = \dfrac{1}{\text{area of } R} \iint_R f(x, y) \, dA$

$$= \frac{\iint_R f(x, y) \, dA}{\iint_R dA}.$$

Similarly, if f is a continuous function of three variables, then the average value of f over a region D in space is

(15.37) $\quad \text{Average value of } f \text{ over } D = \dfrac{1}{\text{volume of } D} \iiint_D f(x, y, z) \, dV$

$$= \frac{\iiint_D f(x, y, z) \, dV}{\iiint_D dV}.$$

Example 3

A dome tent has the shape of a hemisphere with radius 4 ft. The temperature is 55° F on the floor and 65° F at the top of the dome. The temperature at any point is proportional to its distance from the floor. Find the average temperature in the tent.

Solution

We first place a rectangular coordinate system so that the *xy*-plane coincides with the tent's floor with the center at the origin and the positive *z*-axis directed toward the top of the dome, as in Figure 15.65. The outside of the tent is then described by

$$x^2 + y^2 + z^2 = 16, \qquad z \geq 0$$

or in cylindrical coordinates by

$$z = \sqrt{16 - r^2}.$$

Since the temperature varies linearly from 55° at the floor to 65° at the top of the tent, the temperature $T(x, y, z)$ at any point in the tent satisfies the linear equation

$$T(x, y, z) - 55 = \frac{65 - 55}{4 - 0}(z - 0) \quad \text{so} \quad T(x, y, z) = 55 + \frac{5}{2}z.$$

Since the volume of the hemispherical tent is

$$\frac{1}{2}\left(\frac{4}{3}\pi 4^3\right) = \frac{128\pi}{3},$$

we have

$$\text{Average temperature} = \frac{3}{128\pi} \iiint_D \left(55 + \frac{5}{2}z\right) dV$$

$$= \frac{3}{128\pi} \int_0^{2\pi} \int_0^4 \int_0^{\sqrt{16-r^2}} \left(55 + \frac{5}{2}z\right) r \, dz \, dr \, d\theta$$

$$= \frac{15}{128\pi} \int_0^{2\pi} \int_0^4 \left(11z + \frac{z^2}{4}\right)r \Big]_0^{\sqrt{16-r^2}} dr \, d\theta$$

$$= \frac{15}{128\pi} \int_0^{2\pi} \int_0^4 \left(11r\sqrt{16-r^2} + 4r - \frac{r^3}{4}\right) dr \, d\theta$$

$$= \frac{15}{128\pi} \int_0^{2\pi} \left[\frac{-11}{3}(16-r^2)^{3/2} + 2r^2 - \frac{r^4}{16}\right]_0^4 d\theta$$

$$= \frac{15}{128\pi} \int_0^{2\pi} \left(32 - 16 + \frac{11}{3}(64)\right) d\theta$$

$$= \frac{15}{128\pi}\left(\frac{752}{3}\right)(2\pi) \approx 58.75° \text{ F}.$$

Note that the average temperature in the tent is closer to the temperature at the floor than to the temperature at the top of the tent. This is

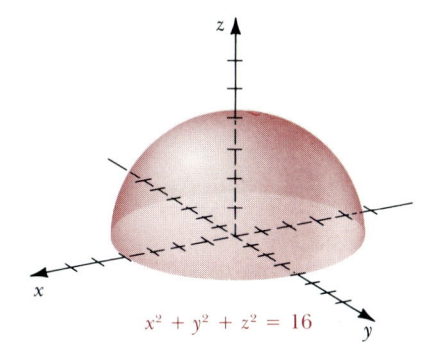

Figure 15.65

EXERCISE SET 15.8

In Exercises 1–10, find the mass of the solid.

1. Bounded by the coordinate planes and the planes $x = 1$, $y = 2$, and $z = 3$. Density described by $\sigma(x, y, z) = 3z + 1$.

2. Bounded by the planes $x = 2$, $x = -1$, $y = 3$, $y = 4$, $z = -1$, and $z = 3$. Density described by $\sigma(x, y, z) = x + y + z$.

3. Bounded by the coordinate planes and the plane $x + y + z = 1$. Density is constant.

4. Bounded by the coordinate planes and the plane $x + y + z = 1$. Density described by $\sigma(x, y, z) = x + y + z$.

5. Bounded by the planes $x = 0$, $y = 0$, $z = 0$, $x + y + z = 1$, and $3x + 3y + z = 3$. Density described by $\sigma(x, y, z) = x$.

6. Bounded by the planes $x = 0$, $y = 0$, $z = 2$, $x + y + z = 1$, and $2x + 3y = 6$. Density described by $\sigma(x, y, z) = x$.

7. Bounded by the cylinder $x^2 + y^2 = 4$ and the planes $z = 1$ and $z = 5$. Density described by $\sigma(x, y, z) = kz$.

8. Bounded by the cone $z = \sqrt{x^2 + y^2}$ and the plane $z = 2$. Density described by $\sigma(x, y, z) = k\sqrt{x^2 + y^2}$.

9. Bounded by the hemisphere of radius 3 centered at $(0, 0, 0)$ and lying above the xy-plane. Density proportional to the square of the distance from the base.

10. Bounded by the xy-plane, the cone $z = \sqrt{x^2 + y^2}$, and the hemisphere $z = \sqrt{1 - x^2 - y^2}$. Density proportional to $\sqrt{x^2 + y^2 + z^2}$.

11–20. Find (a) the moments about the coordinate planes and (b) the center of mass of the solids described in Exercises 1–10. (The integration in these exercises is time-consuming.)

21–30. Find (a) the moments of inertia about the coordinate axes and (b) the radii of gyration of the solids described in Exercises 1–10. (The integration in these exercises is time-consuming.)

31. Find the average height of the portion of the plane $3x + 2y + z = 6$ that lies in the first octant.

32. A solid of constant density has the shape of a cube with sides of length 4. Find the moment about a diagonal plane of the cube.

33. a) Set up the integrals required to find the moments of inertia about the x- and y-axes for the solid and density function described in Example 2.
    b) Determine the values of these moments. [*Hint:* Think carefully about the situation before performing the integrations.]

34. Recall the connection between the average value of a single-variable function and the Mean Value Theorem for Integrals discussed in Section 4.3. Make a conjecture concerning a corresponding result for (a) double and (b) triple integrals.

35. Use Equation (15.36) to derive a formula for the average distance from a region R in the xy-plane to a point (x_0, y_0) in the plane.

36. Use Equation (15.37) to derive a formula for the average distance from a region R in space to a point (x_0, y_0, z_0) in space.

37. A conical teepee has a base of radius 4 ft and height 6 ft (see the figure).
    a) Use the same temperature assumptions as those described in Example 3 to find the average temperature inside the teepee.
    b) Which tent is cooler?

38. A 16-oz can of beer is sitting on a block of ice (see the figure). An *xyz*-coordinate system is introduced with the center of the bottom of the can at the origin and the positive *z*-axis passing through the center of the can. The temperature of the liquid is described by $T(x, y, z) = 35 + z(3 + 2\sqrt{x^2 + y^2})$ degrees Fahrenheit, for $0 \leq z \leq 6$ and $0 \leq x^2 + y^2 \leq (1.25)^2$. Find the average temperature of the beer.

39. Paint in a 1-gal cylindrical can of height 6 in. and diameter 7 in. has a density that varies linearly from 55.3 lb/ft^3 at the top of the can to 59.4 lb/ft^3 at the bottom (see the figure). How much does the paint weigh and where is the center of mass of the paint?

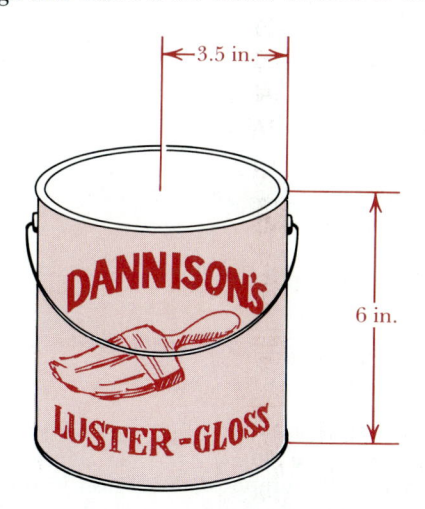

40. Modern lumber designed for outdoor use is pressure-treated with a liquid that penetrates the wood to prevent rot and insect damage. Suppose an 8-ft 4-by-4 (actual dimensions 3.5 in. by 3.5 in.) originally weighing 40 lb is pressure-treated to absorb 5 lb of preservative (see the figure). The liquid penetrates the wood so that the preservative is linearly distributed from zero at the center of the wood to a maximum on a pair of parallel outside edges.
   a) Show that the center of mass of the wood is unchanged.
   b) Show that the moment of inertia about a line placed perpendicular to the base of the wood is increased.

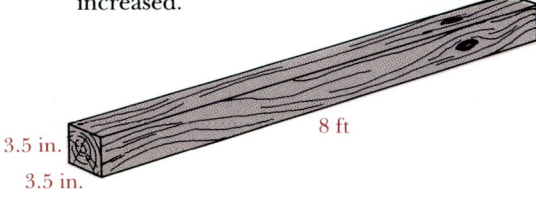

41. A *pousse-cafe* is a (generally alcoholic) drink made by gently pouring colored liquids with varying density into a glass so that the liquids do not mix (see the figure). Suppose that $\frac{1}{4}$ oz of each of the following liqueurs is poured into a conical glass of height 4 in. and diameter 2 in. in order of decreasing density and that they remain in layers in the glass.
   a) What is the center of mass of the liquid?
   b) Suppose the contents of the glass are stirred to make a homogeneous liquid. Does the center of mass of the liquid remain the same?

LIQUEUR	SPECIFIC GRAVITY (DENSITY)
Benedictine	1.150
Drambuie	1.180
Liquore Galliano	1.193
Grand Marnier	1.070
Irish Mist	1.063

▶ IMPORTANT TERMS AND RESULTS

CONCEPT	PAGE
Volume	855
Riemann sum	852
Double integral	854
Iterated integral	859
Polar coordinate integrals	868
Moments	874
Center of mass	874
Moment of inertia	875
Radius of gyration	876
Surface area	881
Triple integrals	884
Cylindrical coordinate integrals	894
Spherical coordinate integrals	898
Average value	904

▶ REVIEW EXERCISES

1. Suppose $f(x, y) = 4 - x - 2y$, R is the triangle in the xy-plane bounded by $x = 0$, $y = 0$, and $x = 4 - 2y$, and $\mathscr{P}$ is the partition defined by the lines $x = 0$, $x = 1$, $x = 2$, $x = 3$, $x = 4$, $y = 0$, $y = \frac{1}{2}$, $y = 1$, $y = \frac{3}{2}$, and $y = 2$. Find the value of the Riemann sum if the R_i are the rectangles that lie entirely within R and (x_i, y_i) is chosen to be the upper left corner of R_i.

2. Repeat Exercise 1 if (x_i, y_i) is chosen to be the lower right corner of R_i.

3. a) Find the volume of the solid bounded by the coordinate planes and above by the plane $x + 2y + z = 4$.
   b) Find the area of the face of this solid lying in the xy-plane.

4. a) Find the volume of the solid bounded by the coordinate planes and by the plane $3z - 4x - 2y = 12$.
   b) Find the area of the face of this solid lying in the xy-plane.

In Exercises 5–12, evaluate the integral.

5. $\int_0^1 \int_0^{x^2} (x + y) \, dy \, dx$

6. $\int_0^1 \int_x^{x^2} \sqrt{xy} \, dy \, dx$

7. $\int_1^4 \int_{\sqrt{y}}^{y} \ln(y/x) \, dx \, dy$

8. $\int_0^1 \int_y^{3y} e^{x+2y} \, dx \, dy$

9. $\int_0^1 \int_0^{x^2} \int_0^{y^2} (x + y + z) \, dz \, dy \, dx$

10. $\int_0^1 \int_x^{x^2} \int_y^{y^2} \sqrt{xyz} \, dz \, dy \, dx$

11. $\int_0^{\pi} \int_0^1 \int_0^{\sqrt{1-r^2}} rz \cos \theta \, dz \, dr \, d\theta$

12. $\int_0^{2\pi} \int_0^{\pi/2} \int_0^{\cos \phi} \rho^2 \sin \phi \cos \phi \, d\rho \, d\phi \, d\theta$

In Exercises 13–18:

a) Sketch the graph of the region in the xy-plane over which the integration is taken.
b) Reverse the order of integration and evaluate the double integral obtained.

13. $\int_0^4 \int_x^{2\sqrt{x}} dy \, dx$

14. $\int_0^1 \int_0^{\sqrt{y}} (2 - x - y) \, dx \, dy$

15. $\int_0^4 \int_{-\sqrt{y}}^{\sqrt{y}} (4-y)\, dx\, dy$ 16. $\int_0^2 \int_{-2}^{2} y^3 e^{4x}\, dx\, dy$

17. $\int_0^1 \int_y^1 ye^{-x^3}\, dx\, dy$ 18. $\int_0^4 \int_y^{2\sqrt[3]{2y}} xy\, dx\, dy$

In Exercises 19–22, the boundaries of a solid with base in the xy-plane are described. Use double integrals to find (a) the area of the base and (b) the volume of the solid.

19. $x = y^2$, $x = 8 - y^2$, $z = 4x$
20. $x = 0$, $y = 1$, $y = x^3$, $z = 2 - x - y$
21. $r = 1 - \sin\theta$, $z = 4$
22. $r = \cos\theta$, $z = xy$, $y \geq 0$

In Exercises 23–26, sketch a solid whose volume is represented by the integral. Find the volume.

23. $\int_0^1 \int_0^{\sqrt{1-y^2}} 4y\, dx\, dy$

24. $\int_0^2 \int_0^{\sqrt{4-x^2}} \int_0^{2-y} dz\, dy\, dx$

25. $\int_0^{\pi/2} \int_0^{2\sqrt{2}} \int_{r^2/8}^{\sqrt{9-r^2}} r\, dz\, dr\, d\theta$

26. $\int_0^{\pi/2} \int_0^{\pi/2} \int_0^2 \rho^2 \sin\phi\, d\rho\, d\theta\, d\phi$

In Exercises 27–30, use either cylindrical or spherical coordinates to evaluate the integral.

27. $\int_0^1 \int_0^{\sqrt{1-x^2}} \int_0^{\sqrt{1-x^2-y^2}} 2z\sqrt{x^2+y^2}\, dz\, dy\, dx$

28. $\int_{-3}^3 \int_{-\sqrt{9-x^2}}^{\sqrt{9-x^2}} \int_0^{\sqrt{9-x^2-y^2}} z\sqrt{x^2+y^2}\, dz\, dy\, dx$

29. $\int_{-1}^1 \int_0^{\sqrt{1-x^2}} \int_0^{\sqrt{1-x^2-y^2}} z^2\sqrt{x^2+y^2+z^2}\, dz\, dy\, dx$

30. $\int_{-2}^2 \int_{-\sqrt{4-y^2}}^{\sqrt{4-y^2}} \int_0^{\sqrt{4-x^2-y^2}} \frac{1}{x^2+y^2+z^2}\, dz\, dx\, dy$

In Exercises 31–40, find the volume of the solid.

31. The solid bounded below by the paraboloid $z = x^2 + y^2$ and above by the plane $z = 4$.
32. The solid bounded below by the paraboloid $z = x^2 + y^2 - 1$ and above by the plane $z = 4$.
33. The solid bounded by the cylinder $r = 2 - 2\cos\theta$ and the planes $z = 0$ and $z = 4$.
34. The solid bounded by the cylinder $r = \sin\theta$, the xy-plane, and $z = r^2$.
35. The solid bounded below by the paraboloid $z = x^2 + y^2$ and above by the sphere $x^2 + y^2 + z^2 = 2$.

36. The solid bounded below by the cone $z^2 = x^2 + y^2$ and above by the sphere $x^2 + y^2 + (z-2)^2 = 4$.
37. The solid in the first octant bounded by the cylinder $y^2 + z^2 = 9$ and the plane $z = 3 - x$.
38. The portion of the cylinder $9x^2 + 4y^2 = 36$ bounded below by the xy-plane and above by the plane $x + y + 2z = 6$.
39. The solid in the first octant bounded above by the planes $z = x/3$ and $x + 2y + 3z = 6$.
40. The solid bounded below by the xy-plane, above by the plane $z = 4 - y$, and by the cylinder $y = x^2$.

In Exercises 41–46, find (a) the mass, (b) the moments about the axes, and (c) the center of mass of the lamina that has the shape of the region bounded by the curves and has the indicated density.

41. $y = e^x$, $y = 0$, $x = 0$, $x = 2$, $\sigma(x,y) = 1$
42. $y = \sin x$, $y = \cos x$, $y = 0$, $x = 0$, $x = \pi/2$, $\sigma(x,y) = 3$
43. $y = e^x$, $y = e^{-x}$, $y = 0$, $x = -1$, $x = 1$, $\sigma(x,y) = |xy|$
44. $y = 1/x$, $y = 0$, $x = 1$, $x = 2$, $\sigma(x,y) = x^2 + y^2$
45. $x = 0$, $y = 0$, $y = \sqrt{9-x^2}$, $\sigma(x,y) = x^2 + y^2$
46. $y = 1$, $y = \sqrt{2-x^2}$, $\sigma(x,y) = x^2 + y^2$

47. Find the surface area of the portion of the cone $x^2 + y^2 - z^2 = 0$ that lies between the planes $z = -2$ and $z = 3$.

48. Find the surface area of the portion of the sphere $x^2 + y^2 + z^2 = 9$ that is inside the cylinder $x^2 + y^2 = 1$.

49. Find the area of the portion of the saddle-shaped surface $z = 2 - x^2 + y^2$ that lies above $x^2 + y^2 \leq 2$.

50. Find the mass of a sphere of radius a if the density at every point is twice the distance from the center.

51. Find the mass of a plate in the shape of a cardioid $r = 1 - \sin\theta$ if the density is proportional to the distance from the origin.

52. Find the radius of gyration with respect to the z-axis of the cube bounded by the coordinate planes and the planes $x = a$, $y = a$, and $z = a$, if the density is described by $\sigma(x,y,z) = x + y + z$.

53. Show that
$$\int_0^\pi \int_0^\pi \sin x \sin y\, dx\, dy = \left(\int_0^\pi \sin x\, dx\right)^2,$$
but
$$\int_0^\pi \int_0^\pi \sin^2 x\, dx\, dy \neq \left(\int_0^\pi \sin x\, dx\right)^2.$$

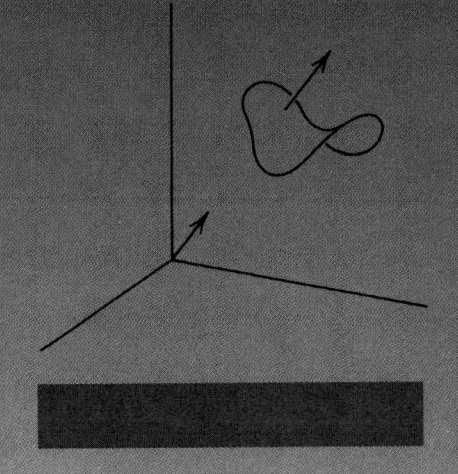

16

Line and Surface Integrals

CHAPTER 16 LINE AND SURFACE INTEGRALS

We have defined the definite integral for a function whose domain is an interval of the real line, a region in the plane, or a solid in space. In this chapter we extend the definition of the definite integral to include functions whose domain consists of a curve or surface in space. These definitions lead to results that have important applications in engineering and the physical sciences.

16.1

Vector Fields

The gradient of the differentiable scalar function f was defined in Section 14.5 as the vector-valued function described by

$$\nabla f(x, y, z) = \frac{\partial f(x, y, z)}{\partial x} \mathbf{i} + \frac{\partial f(x, y, z)}{\partial y} \mathbf{j} + \frac{\partial f(x, y, z)}{\partial z} \mathbf{k}.$$

By definition, ∇f is a function that associates a vector with each point (x, y, z) in a region in space. Applications of the gradient include determining the direction and magnitude of the maximum and minimum rates of change along a surface and finding the normal to a surface at a point.

In this section, we begin a study of *general* functions that associate a vector with each point in a region in space.

If a vector $\mathbf{F}(x, y, z)$ is assigned to each point (x, y, z) in a region R, then the collection of vectors $\mathbf{F}(x, y, z)$ is called a **vector field.** For convenience of terminology, the term vector field is used to describe either the collection of vectors in the field or the function used to describe this collection.

When $\mathbf{F}(x, y, z)$ represents the velocity of an object at (x, y, z), the vector field is called a **velocity field.** Figure 16.1 shows a velocity field that is produced in a running stream of water, and Figure 16.2 shows a velocity field that describes air currents on a weather map. Vector fields are also used to describe forces in a region, in which case the field is called a **force field.** When $\mathbf{F}$ is the gradient of a function, the field is sometimes called a **gradient field.**

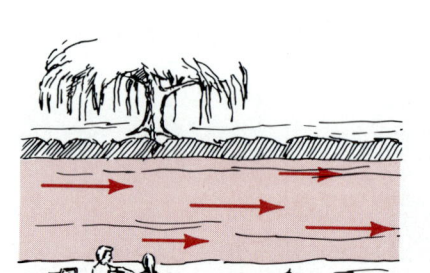

Figure 16.1
The velocity field of a stream of water.

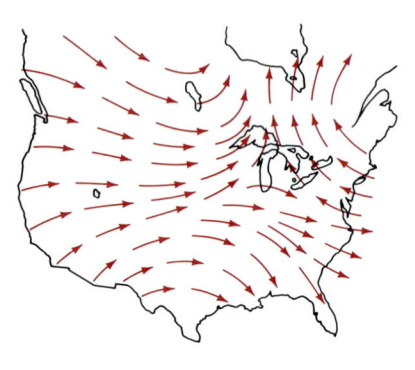

Figure 16.2
Wind velocity on a weather map.

Example 1

Describe the vector field defined by

$$\mathbf{F}(x, y, z) = -y\mathbf{i} + x\mathbf{j}$$

for (x, y, z) satisfying $0 \leq x^2 + y^2 + z^2 \leq 1$ and $z \geq 0$.

Solution

The vector field is illustrated at various points in Figure 16.3. This field can be thought of as a velocity field of a hemisphere under a constant rotation with the z-axis as the axis of rotation. ▶

Associated with every differentiable vector field are the *divergence*, a scalar function, and the *curl*, a vector function.

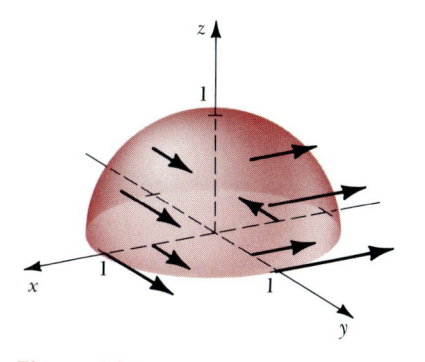

Figure 16.3

(16.1)
DEFINITION
Suppose $\mathbf{F}(x, y, z) = M(x, y, z)\mathbf{i} + N(x, y, z)\mathbf{j} + P(x, y, z)\mathbf{k}$ is a vector function and M, N, and P are differentiable. The **divergence** of $\mathbf{F}$ is the scalar function

$$\text{div } \mathbf{F} = \frac{\partial M}{\partial x} + \frac{\partial N}{\partial y} + \frac{\partial P}{\partial z},$$

and the **curl** of $\mathbf{F}$ is the vector function

$$\text{curl } \mathbf{F} = \left(\frac{\partial P}{\partial y} - \frac{\partial N}{\partial z}\right)\mathbf{i} + \left(\frac{\partial M}{\partial z} - \frac{\partial P}{\partial x}\right)\mathbf{j} + \left(\frac{\partial N}{\partial x} - \frac{\partial M}{\partial y}\right)\mathbf{k}.$$

The names *divergence* and *curl* are derived from the applications of these concepts to fluid flow. We will show in Section 16.9 that the divergence of the velocity field of a moving fluid describes the rate of change of the fluid mass per unit volume. Thus the divergence describes how the fluid *diverges* at a point. We also show that the curl of the velocity field of the fluid describes the tendency of the fluid to rotate, or *curl*, about the line perpendicular to the direction of flow. European writers use the more descriptive word *rotation* in place of curl to describe this concept and write **rot F** rather than **curl F**.

The symbol ∇ was introduced in Chapter 14 to denote the gradient of a scalar function f. Defining the *operator* ∇ as

(16.2)
$$\nabla = \frac{\partial}{\partial x}\mathbf{i} + \frac{\partial}{\partial y}\mathbf{j} + \frac{\partial}{\partial z}\mathbf{k}$$

produces

$$\nabla f = \frac{\partial f}{\partial x}\mathbf{i} + \frac{\partial f}{\partial y}\mathbf{j} + \frac{\partial f}{\partial z}\mathbf{k}.$$

We also use the ∇ operator to describe the divergence and curl of a vector field. For $\mathbf{F} = M\mathbf{i} + N\mathbf{j} + P\mathbf{k}$, the divergence can be written as

$$\text{div } \mathbf{F} = \frac{\partial M}{\partial x} + \frac{\partial N}{\partial y} + \frac{\partial P}{\partial z} = \left(\frac{\partial}{\partial x}\mathbf{i} + \frac{\partial}{\partial y}\mathbf{j} + \frac{\partial}{\partial z}\mathbf{k}\right) \cdot (M\mathbf{i} + N\mathbf{j} + P\mathbf{k}),$$

so

(16.3)
$$\text{div } \mathbf{F} = \nabla \cdot \mathbf{F}.$$

Similarly,

$$\text{curl } \mathbf{F} = \left(\frac{\partial P}{\partial y} - \frac{\partial N}{\partial z}\right)\mathbf{i} + \left(\frac{\partial M}{\partial z} - \frac{\partial P}{\partial x}\right)\mathbf{j} + \left(\frac{\partial N}{\partial x} - \frac{\partial M}{\partial y}\right)\mathbf{k}$$

$$= \left(\frac{\partial}{\partial x}\mathbf{i} + \frac{\partial}{\partial y}\mathbf{j} + \frac{\partial}{\partial z}\mathbf{k}\right) \times (M\mathbf{i} + N\mathbf{j} + P\mathbf{k}),$$

so

(16.4) $$\text{curl } \mathbf{F} = \nabla \times \mathbf{F}.$$

Example 2

Find (a) the divergence and (b) the curl of

$$\mathbf{F}(x, y, z) = x^2 y \mathbf{i} + xyz\mathbf{j} + (y^2 + z^2)\mathbf{k}.$$

Solution

a) We have

$$\text{div } \mathbf{F}(x, y, z) = \nabla \cdot \mathbf{F}(x, y, z)$$

$$= \left(\frac{\partial}{\partial x}\mathbf{i} + \frac{\partial}{\partial y}\mathbf{j} + \frac{\partial}{\partial z}\mathbf{k}\right) \cdot (x^2 y\mathbf{i} + xyz\mathbf{j} + (y^2 + z^2)\mathbf{k})$$

$$= \frac{\partial}{\partial x}(x^2 y) + \frac{\partial}{\partial y}(xyz) + \frac{\partial}{\partial z}(y^2 + z^2)$$

$$= 2xy + xz + 2z.$$

b) We have

$$\text{curl } \mathbf{F}(x, y, z) = \nabla \times \mathbf{F}(x, y, z)$$

$$= \left(\frac{\partial}{\partial x}\mathbf{i} + \frac{\partial}{\partial y}\mathbf{j} + \frac{\partial}{\partial z}\mathbf{k}\right) \times [x^2 y\mathbf{i} + xyz\mathbf{j} + (y^2 + z^2)\mathbf{k}].$$

The determinant notation used to find the cross product can be modified to fit this situation, giving

$$\text{curl } \mathbf{F}(x, y, z) = \begin{vmatrix} \mathbf{i} & \mathbf{j} & \mathbf{k} \\ \dfrac{\partial}{\partial x} & \dfrac{\partial}{\partial y} & \dfrac{\partial}{\partial z} \\ x^2 y & xyz & y^2 + z^2 \end{vmatrix}$$

$$= \left[\frac{\partial}{\partial y}(y^2 + z^2) - \frac{\partial}{\partial z}(xyz)\right]\mathbf{i} - \left[\frac{\partial}{\partial x}(y^2 + z^2)\right.$$

$$\left. - \frac{\partial}{\partial z}(x^2 y)\right]\mathbf{j} + \left[\frac{\partial}{\partial x}(xyz) - \frac{\partial}{\partial y}(x^2 y)\right]\mathbf{k}$$

$$= (2y - xy)\mathbf{i} + (yz - x^2)\mathbf{k}. \quad \blacktriangleright$$

The operations of divergence and curl satisfy the usual arithmetic properties: If $\mathbf{F}$ and $\mathbf{G}$ are vector functions and c is a constant, then:

(16.5) $\quad \text{div } (\mathbf{F} + \mathbf{G}) = \text{div } \mathbf{F} + \text{div } \mathbf{G}, \quad \text{div } (c\mathbf{F}) = c(\text{div } \mathbf{F});$

(16.6) $\quad \text{curl } (\mathbf{F} + \mathbf{G}) = \text{curl } \mathbf{F} + \text{curl } \mathbf{G}, \quad \text{curl } (c\mathbf{F}) = c(\text{curl } \mathbf{F}).$

In addition, the divergence and curl satisfy interesting and useful relationships when applied to various combinations of scalar and vector functions. The most frequently used of these are listed below; others are given in the exercises.

If **F** and **G** are vector functions and f is a scalar function and the various partial derivatives are continuous, then:

(16.7) $\quad$ div $(f\mathbf{F}) = f(\text{div } \mathbf{F}) + (\nabla f) \cdot \mathbf{F};$

(16.8) $\quad$ **curl** $(\nabla f) = \mathbf{0};$

(16.9) $\quad$ div (**curl F**) $= 0;$

(16.10) $\quad$ **curl** $(f\mathbf{F}) = f(\textbf{curl F}) + (\nabla f) \times \mathbf{F};$

(16.11) $\quad$ div $(\mathbf{F} \times \mathbf{G}) = (\textbf{curl F}) \cdot \mathbf{G} - \mathbf{F} \cdot (\textbf{curl G});$

(16.12) $\quad \nabla^2 f = \nabla \cdot \nabla f = \text{div } (\nabla f) = \dfrac{\partial^2 f}{\partial x^2} + \dfrac{\partial^2 f}{\partial y^2} + \dfrac{\partial^2 f}{\partial z^2}.$

These identities can be verified directly by applying the definitions, as shown in the following example.

Example 3

Verify the identity div (**curl F**) $= \mathbf{0}$, assuming that the component functions of $\mathbf{F} = M\mathbf{i} + N\mathbf{j} + P\mathbf{k}$ have continuous second partial derivatives.

Solution
We have

$$\textbf{curl F} = \left(\dfrac{\partial P}{\partial y} - \dfrac{\partial N}{\partial z}\right)\mathbf{i} + \left(\dfrac{\partial M}{\partial z} - \dfrac{\partial P}{\partial x}\right)\mathbf{j} + \left(\dfrac{\partial N}{\partial x} - \dfrac{\partial M}{\partial y}\right)\mathbf{k},$$

so

$$\text{div } (\textbf{curl F}) = \dfrac{\partial}{\partial x}\left(\dfrac{\partial P}{\partial y} - \dfrac{\partial N}{\partial z}\right) + \dfrac{\partial}{\partial y}\left(\dfrac{\partial M}{\partial z} - \dfrac{\partial P}{\partial x}\right) + \dfrac{\partial}{\partial z}\left(\dfrac{\partial N}{\partial x} - \dfrac{\partial M}{\partial y}\right)$$

$$= \dfrac{\partial^2 P}{\partial x\, \partial y} - \dfrac{\partial^2 N}{\partial x\, \partial z} + \dfrac{\partial^2 M}{\partial y\, \partial z} - \dfrac{\partial^2 P}{\partial y\, \partial x} + \dfrac{\partial^2 N}{\partial z\, \partial x} - \dfrac{\partial^2 M}{\partial z\, \partial y}.$$

Since the component functions of **F** have continuous second partial derivatives, the mixed second partial derivatives are equal. Thus

$$\text{div } (\textbf{curl F}) = 0. \quad \blacktriangleright$$

Identity (16.12) is a particularly useful result that has application in many areas of applied mathematics because of its relation to **Laplace's Equation**:

(16.13) $\quad \dfrac{\partial^2 f}{\partial x^2} + \dfrac{\partial^2 f}{\partial y^2} + \dfrac{\partial^2 f}{\partial z^2} = 0.$

Functions satisfying this equation are called **harmonic functions** and occur naturally in the study of such areas as steady-state heat flow, electrostatics, and gravitational attraction. The application to heat flow is considered in Section 16.9.

Example 4

Consider $f(x, y, z) = (x^2 + y^2 + z^2)^{-1/2}$.

a) Determine ∇f.
b) Show that f is a harmonic function.

Solution

a) We have

$$\nabla f(x, y, z) = \left(\frac{\partial}{\partial x}\mathbf{i} + \frac{\partial}{\partial y}\mathbf{j} + \frac{\partial}{\partial z}\mathbf{k}\right)(x^2 + y^2 + z^2)^{-1/2}$$

$$= -(x^2 + y^2 + z^2)^{-3/2}(x\mathbf{i} + y\mathbf{j} + z\mathbf{k}).$$

b) Since

$$\text{div }\nabla f(x, y, z) = -\frac{\partial}{\partial x}[(x^2 + y^2 + z^2)^{-3/2}x] - \frac{\partial}{\partial y}[(x^2 + y^2 + z^2)^{-3/2}y]$$

$$-\frac{\partial}{\partial z}[(x^2 + y^2 + z^2)^{-3/2}z]$$

$$= -(x^2 + y^2 + z^2)^{-3/2}\left[\left(1 - \frac{3x^2}{x^2 + y^2 + z^2}\right)\right.$$

$$\left. + \left(1 - \frac{3y^2}{x^2 + y^2 + z^2}\right) + \left(1 - \frac{3z^2}{x^2 + y^2 + z^2}\right)\right]$$

$$= 0,$$

f is harmonic.

▶

▶ EXERCISE SET 16.1

In Exercises 1–10, find (a) the divergence and (b) the curl of the vector-valued function.

1. $\mathbf{F}(x, y, z) = x^2\mathbf{i} + y\mathbf{j} + \mathbf{k}$
2. $\mathbf{F}(x, y, z) = \sin y\mathbf{i}$
3. $\mathbf{F}(x, y, z) = \cos x\mathbf{i} + \sin y\mathbf{j} + z\mathbf{k}$
4. $\mathbf{F}(x, y, z) = x\mathbf{i} + y\mathbf{j} + z\mathbf{k}$
5. $\mathbf{F}(x, y, z) = x^2yz\mathbf{i} + xy^2z\mathbf{j} + xyz^2\mathbf{k}$
6. $\mathbf{F}(x, y, z) = \sqrt{yz}\,\mathbf{i} + \sqrt{xz}\,\mathbf{j} + \sqrt{xy}\,\mathbf{k}$
7. $\mathbf{F}(x, y, z) = xy\sin z\mathbf{i} + x^2\cos y\mathbf{j} + z\sqrt{xy}\,\mathbf{k}$
8. $\mathbf{F}(x, y, z) = e^x\sin y\mathbf{i} + e^y\cos z\mathbf{j} + e^{xyz}\mathbf{k}$
9. $\mathbf{F}(x, y, z) = e^z\ln\frac{x}{y}\mathbf{i} + e^x\ln\frac{y}{z}\mathbf{j} + e^y\ln\frac{z}{x}\mathbf{k}$
10. $\mathbf{F}(x, y, z) = \tan xz\mathbf{i} + \sec yx\mathbf{j} + \arcsin yz\mathbf{k}$

In Exercises 11–16, determine whether the function is harmonic.

11. $f(x, y) = x^2y - y^2x$
12. $f(x, y) = \dfrac{1}{\sqrt{x^2 + y^2}}$
13. $f(x, y) = \sqrt{x^2 + y^2}$
14. $f(x, y, z) = \ln|x^2 + y^2|$
15. $f(x, y, z) = 2x^2 - y^2 - z^2$
16. $f(x, y, z) = x^2 + y^2 + z^2$

For Exercises 17–24, let $f(x, y, z) = x^2 + yz$ and $\mathbf{F}(x, y, z) = 2xy\mathbf{i} + (x^2 + z^2)\mathbf{j} + yz\mathbf{k}$. Find each of the following or state why the operation is not possible.

17. $\nabla \cdot (f\mathbf{F})$
18. $\nabla f \cdot \mathbf{F}$

19. $\nabla \times (\nabla F)$
20. $\nabla \times (\nabla f \cdot \mathbf{F})$
21. $\nabla \times (\nabla \cdot \mathbf{F})$
22. $\nabla \cdot (\nabla f \cdot \mathbf{F})$
23. $\nabla f + \nabla \times \mathbf{F}$
24. $\nabla \cdot \nabla f + \nabla \cdot (\nabla \times \mathbf{F})$

In Exercises 25–32, verify the identity, assuming the required partial derivatives exist and are continuous. **F** and **G** are vector-valued functions and f and g are scalar functions.

25. $\nabla \cdot (\mathbf{F} + \mathbf{G}) = \nabla \cdot \mathbf{F} + \nabla \cdot \mathbf{G}$
26. $\nabla \times (\mathbf{F} + \mathbf{G}) = \nabla \times \mathbf{F} + \nabla \times \mathbf{G}$
27. $\nabla \cdot (f\mathbf{F}) = f(\nabla \cdot \mathbf{F}) + \nabla f \cdot \mathbf{F}$
28. $\nabla \cdot (\mathbf{F} \times \mathbf{G}) = (\nabla \times \mathbf{F}) \cdot \mathbf{G} - \mathbf{F} \cdot (\nabla \times \mathbf{G})$
29. $\nabla \times (\nabla f) = \mathbf{0}$
30. $\nabla \cdot (\nabla \times \mathbf{F}) = 0$
31. $\nabla \cdot (f \nabla g) = f \nabla^2 g + \nabla f \cdot \nabla g$
32. $\nabla \times (f\mathbf{F}) = f(\nabla \times \mathbf{F}) + (\nabla f \times \mathbf{F})$
33. Let $\mathbf{F} = M\mathbf{i} + N\mathbf{j} + P\mathbf{k}$. Show that
$$\nabla \times (\nabla \times \mathbf{F}) = \nabla(\nabla \cdot \mathbf{F}) - (\nabla^2 M \mathbf{i} + \nabla^2 N \mathbf{j} + \nabla^2 P \mathbf{k}).$$
34. Show that if $\mathbf{F}(x, y, z) = M(x)\mathbf{i} + N(y)\mathbf{j} + P(z)\mathbf{k}$, then curl $\mathbf{F} = \mathbf{0}$.
35. Show that if $\mathbf{F}(x, y, z) = M(y, z)\mathbf{i} + N(x, z)\mathbf{j} + P(x, y)\mathbf{k}$, then div $\mathbf{F} = 0$.
36. Let (r, θ, z) be the cylindrical representation of Q, **a** the unit vector in the positive direction of r at Q, and **b** the unit vector in the direction of the tangent to the circle of radius r parallel to the xy-plane at Q (see the figure). Show that if f is a scalar function and $\mathbf{F} = M\mathbf{a} + N\mathbf{b} + P\mathbf{k}$ is a vector function described in cylindrical coordinates, then:

a) $\nabla f = \dfrac{\partial f}{\partial r} \mathbf{a} + \dfrac{1}{r} \dfrac{\partial f}{\partial \theta} \mathbf{b} + \dfrac{\partial f}{\partial z} \mathbf{k};$

b) $\nabla \cdot \mathbf{F} = \dfrac{1}{r} \dfrac{\partial}{\partial r}(rM) + \dfrac{1}{r} \dfrac{\partial N}{\partial \theta} + \dfrac{\partial P}{\partial z};$

c) $\nabla^2 f = \dfrac{1}{r} \dfrac{\partial}{\partial r}\left(r \dfrac{\partial f}{\partial r}\right) + \dfrac{1}{r^2} \dfrac{\partial^2 f}{\partial \theta^2} + \dfrac{\partial^2 f}{\partial z^2}.$

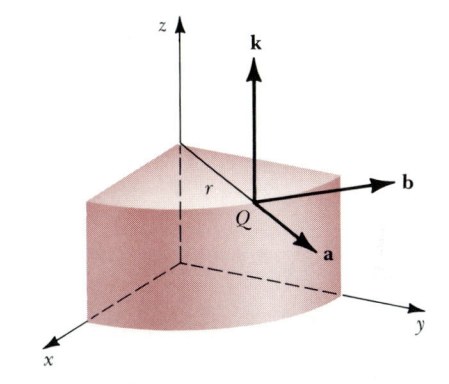

16.2

Line Integrals

A *line integral* is a natural generalization of the definite integral to the situation when the domain of a function is a curve in space rather than an interval of the real line. For a definite integral, $\int_a^b f(x)\, dx$, the function f is defined on the interval $[a, b]$ of the x-axis. For a line integral, we integrate along a curve in space, so the function must be defined at each point on the curve.

Suppose f is a continuous function defined on a subset D of $\mathbb{R}^3$ and that C is a curve in D with the smooth parametric representation

$$x = x(t), \quad y = y(t), \quad z = z(t).$$

The curve C is given an orientation by assigning a positive direction, usually the direction given by increasing values of t. It is customary to place small arrowheads on the curve to indicate the direction of positive orientation, as shown in Figure 16.4.

If the initial point $(x(a), y(a), z(a))$ and the terminal point $(x(b), y(b), z(b))$ coincide, the curve C is called a **closed curve**. In addition, if there are no other points of intersection, then C is called a **simple closed**

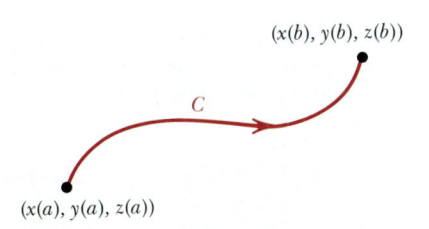

Figure 16.4
An oriented curve.

curve. Simple closed curves have no "loops," as shown in Figure 16.5. The curves we will consider are usually either smooth or piecewise smooth, that is, constructed of a finite number of connected smooth components. Also, the closed curves we consider are usually simple.

We consider two types of line integrals in this section. In the first type the integration is with respect to one of the coordinate directions and the line integral has one of the following forms:

$$\int_C f(x, y, z)\, dx, \qquad \int_C f(x, y, z)\, dy, \qquad \int_C f(x, y, z)\, dz.$$

In the second type of line integral, the integration is with respect to the arc length of the curve. Line integrals of this type have the form

$$\int_C f(x, y, z)\, ds.$$

Both types of line integral are defined in terms of Riemann sums and, under the conditions on f and C stated at the beginning of the section, both can be reduced to Riemann integrals in the parameter t.

First consider the line integrals with respect to a coordinate direction. Since the derivation of each of these integrals is the same, only integrals of the form $\int_C f(x, y, z)\, dx$ are discussed.

$\int_C f(x, y, z)\, dx$: The Line Integral in the *x*-Coordinate Direction

Let $\mathcal{P} = \{t_0, t_1, \ldots, t_n\}$ be a partition of $[a, b]$ and let

$$P_i = P(t_i) = (x(t_i), y(t_i), z(t_i))$$

be the points along C determined by $\mathcal{P}$. (See Figure 16.6.) The line integral $\int_C f(x, y, z)\, dx$ is defined as

(16.14)
$$\int_C f(x, y, z)\, dx = \lim_{\|\mathcal{P}\| \to 0} \sum_{i=1}^{n} f(P(\tau_i))\, \Delta x_i,$$

where, for each i, $\Delta x_i = x(t_i) - x(t_{i-1})$ and τ_i is in $[t_{i-1}, t_i]$ (see Figure 16.6). Since the parametric representation of C is smooth, x' exists on $[a, b]$. It follows from the Mean Value Theorem that the Riemann sum

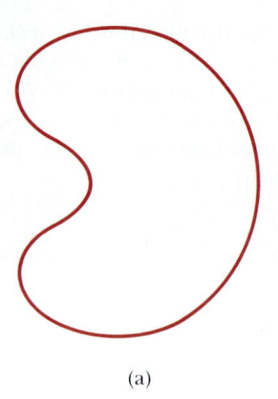

Figure 16.5
(a) A smooth simple closed curve; (b) a smooth closed curve that is not simple; (c) a simple closed curve that is not smooth.

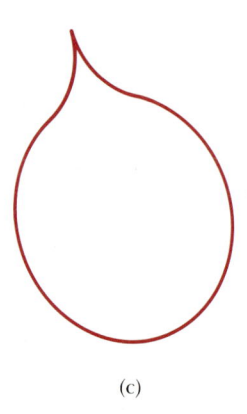

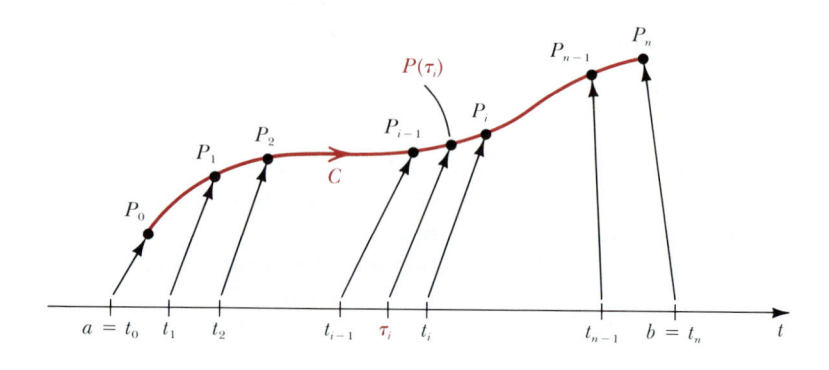

Figure 16.6

on the right side of (16.14) can be rewritten as

(16.15)
$$\sum_{i=1}^{n} f(P(\tau_i)) \, \Delta x_i = \sum_{i=1}^{n} f(P(\tau_i)) \frac{x(t_i) - x(t_{i-1})}{t_i - t_{i-1}} \Delta t_i = \sum_{i=1}^{n} f(P(\tau_i)) \, x'(w_i) \, \Delta t_i,$$

for some w_i in (t_{i-1}, t_i). Choosing $\tau_i = w_i$ for each i gives

(16.16) $$\lim_{n \to \infty} \sum_{i=1}^{n} f(P(w_i)) \, x'(w_i) \, \Delta t_i = \int_a^b f(x(t), y(t), z(t)) \, x'(t) \, dt.$$

Applying this in Equation (16.14) produces the following result.

(16.17)
THEOREM
If f is continuous on a region D in space and C is a curve contained in D with the smooth parametric representation

$$x = x(t), \quad y = y(t), \quad z = z(t), \quad \text{for } t \text{ in } [a, b],$$

then

$$\int_C f(x, y, z) \, dx = \int_a^b f(x(t), y(t), z(t)) \, x'(t) \, dt.$$

Note that when the curve C is the interval $[a, b]$ on the x-axis and f is a function of x only, then C is described by $x = t, a \leq t \leq b$. So $dx = dt$ and this line integral reduces to the definite integral $\int_a^b f(x) \, dx$.

Example 1

Find $\int_C xy \, dx$, where C is described by $x = 1 + 2t$, $y = t$, $z = 0$, for $0 \leq t \leq 2$.

Solution
The curve C lies in the xy-plane and appears as shown in Figure 16.7. Since $dx = 2 \, dt$,

$$\int_C xy \, dx = \int_0^2 (1 + 2t)t \, 2 \, dt = t^2 + \frac{4t^3}{3} \Big]_0^2 = 4 + \frac{32}{3} = \frac{44}{3}.$$

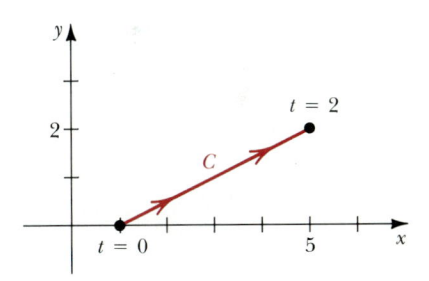

▶ **Figure 16.7**

The line integrals $\int_C f(x, y, z) \, dy$ and $\int_C f(x, y, z) \, dz$ are similarly defined and satisfy

(16.18) $$\int_C f(x, y, z) \, dy = \int_a^b f(x(t), y(t), z(t)) \, y'(t) \, dt$$

and

(16.19) $$\int_C f(x, y, z) \, dz = \int_a^b f(x(t), y(t), z(t)) \, z'(t) \, dt.$$

$x^2 + y^2 + z^2 = 1; x = 0$

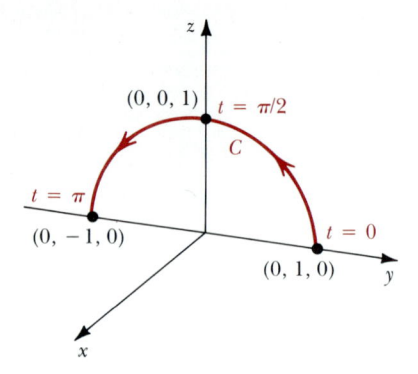

Figure 16.8

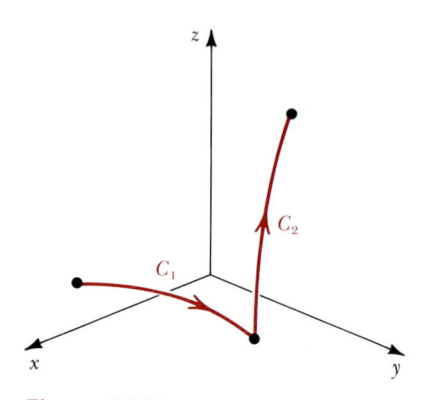

Figure 16.9

Example 2

Find $\int_C (x + y + z)\, dy$, where C is the curve of intersection of the upper half of the sphere $x^2 + y^2 + z^2 = 1$ and the plane $x = 0$ with initial point $(0, 1, 0)$.

Solution

A sketch of C is shown in Figure 16.8. Since C has the smooth parametric representation

$$x = 0, \quad y = \cos t, \quad z = \sin t, \quad \text{for} \quad 0 \leq t \leq \pi,$$

we have

$$\int_C (x + y + z)\, dy = \int_0^\pi (0 + \cos t + \sin t)(-\sin t)\, dt$$

$$= \int_0^\pi [\cos t\, (-\sin t) - \sin^2 t]\, dt$$

$$= \frac{\cos^2 t}{2}\bigg]_0^\pi - \int_0^\pi \frac{1 - \cos 2t}{2}\, dt$$

$$= \left(\frac{1}{2} - \frac{1}{2}\right) - \left(\frac{1}{2} t - \frac{1}{4} \sin 2t\right)\bigg]_0^\pi = -\frac{\pi}{2}. \quad \blacktriangleright$$

The representation given in Theorem (16.17) and Equations (16.18) and (16.19) for the line integrals in the coordinate directions implies that many of the properties of definite integrals also hold for these line integrals. For example,

(16.20) $\quad \int_C (f \pm g)(x, y, z)\, dx = \int_C f(x, y, z)\, dx \pm \int_C g(x, y, z)\, dx,$

(16.21) $\quad \int_C kf(x, y, z)\, dx = k \int_C f(x, y, z)\, dx, \quad \text{for any constant } k.$

If $C_1 + C_2$ denotes the union of the curves C_1 and C_2 when the terminal point of C_1 and the initial point of C_2 coincide (see Figure 16.9), then

(16.22) $\quad \int_{C_1+C_2} f(x, y, z)\, dx = \int_{C_1} f(x, y, z)\, dx + \int_{C_2} f(x, y, z)\, dx.$

Finally, if $-C$ denotes the curve that passes through the same points as C but is traversed in the opposite direction, then

(16.23) $\quad \int_{-C} f(x, y, z)\, dx = -\int_C f(x, y, z)\, dx.$

Applications of line integrals commonly occur in the form

$$\int_C M(x, y, z)\, dx + N(x, y, z)\, dy + P(x, y, z)\, dz,$$

which is a simplified way of writing

$$\int_C M(x, y, z)\, dx + \int_C N(x, y, z)\, dy + \int_C P(x, y, z)\, dz.$$

The term $M(x, y, z) \, dx + N(x, y, z) \, dy + P(x, y, z) \, dz$ is called a **differential form.** (A particularly important type of differential form is discussed in Section 16.4.)

Example 3

Evaluate $\int_C x^2 y \, dx + (y - z) \, dy + xz \, dz$, where:

a) C is the curve described by
$$\mathbf{r}(t) = \mathbf{i} + t^2 \mathbf{j} + t\mathbf{k}, \qquad 0 \leq t \leq 2;$$

b) C is the straight line joining $(1, 0, 0)$ and $(1, 4, 2)$.

Solution

a) For $\mathbf{r}(t) = \mathbf{i} + t^2 \mathbf{j} + t\mathbf{k}$ we have $x(t) = 1$, $y(t) = t^2$, and $z(t) = t$. So
$$dx = 0 \, dt, \qquad dy = 2t \, dt, \quad \text{and} \quad dz = dt$$
and
$$x^2 y \, dx + (y - z) \, dy + xz \, dz = 1^2 \cdot t^2 \cdot 0 \, dt + (t^2 - t)2t \, dt + 1 \cdot t \, dt$$

Thus,
$$\int_C x^2 y \, dx + (y - z) \, dy + xz \, dz = \int_0^2 (2t^3 - 2t^2 + t) \, dt$$
$$= \frac{1}{2} t^4 - \frac{2}{3} t^3 + \frac{1}{2} t^2 \Big]_0^2$$
$$= 8 - \frac{16}{3} + 2 = \frac{14}{3}.$$

b) The straight line from $(1, 0, 0)$ to $(1, 4, 2)$ is described by
$$x(t) = 1, \qquad y(t) = 4t, \qquad z(t) = 2t, \qquad 0 \leq t \leq 1.$$

Since
$$dx = 0 \, dt, \qquad dy = 4 \, dt, \quad \text{and} \quad dz = 2 \, dt,$$
we have
$$x^2 y \, dx + (y - z) \, dy + xz \, dz = 1^2 \cdot 4t \cdot 0 \, dt + (4t - 2t)4 \, dt + 2t \cdot 2 \, dt$$

and
$$\int_C x^2 y \, dx + (y - z) \, dy + xz \, dz = \int_0^1 12t \, dt = 6t^2 \Big]_0^1 = 6.$$

Note that even though the two curves in (a) and (b) begin and end at the same points (see Figure 16.10), the values of the line integrals differ. ▸

A line integral in a coordinate direction considers the change of the values of the function in relation to changes only in that coordinate direction. These line integrals do not give a true extension of the definite integral to the situation when the domain of the integrand is a curve

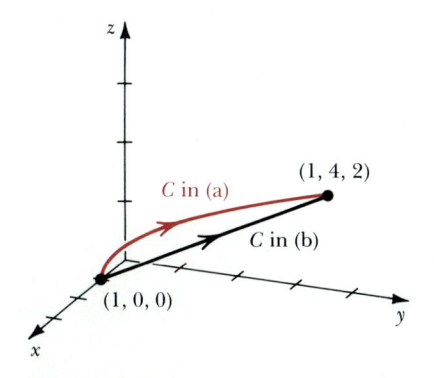

Figure 16.10

in space. This extension requires the line integral with respect to the arc length of the curve.

$\int_C f(x, y, z)\, ds$:
The Line Integral with Respect to Arc Length

To define $\int_C f(x, y, z)\, ds$, the line integral of f with respect to arc length, we again begin with a partition $\mathscr{P} = \{t_0, t_1, \ldots, t_n\}$ of $[a, b]$ and a smooth representation of G defined for t in $[a, b]$. Since x', y', and z' are continuous on $[a, b]$, Equation (12.14) in Section 12.3 implies that the arc length from $P(t_{i-1})$ to $P(t_i)$ is given by

$$\Delta s_i = s(t_i) - s(t_{i-1}) = \int_{t_{i-1}}^{t_i} \sqrt{[x'(t)]^2 + [y'(t)]^2 + [z'(t)]^2}\, dt.$$

The line integral of f over C with respect to arc length is defined by

(16.24) $$\int_C f(x, y, z)\, ds = \lim_{\|\mathscr{P}\| \to 0} \sum_{i=1}^n f(P(\tau_i))\, \Delta s_i,$$

where τ_i is an arbitrary point in $[t_{i-1}, t_i]$.

Since s is differentiable on $[a, b]$, it follows from the Mean Value Theorem that

$$\Delta s_i = s(t_i) - s(t_{i-1}) = s'(w_i)\, \Delta t_i,$$

for some w_i in $[t_{i-1}, t_i]$. Thus,

$$\Delta s_i = \sqrt{[x'(w_i)]^2 + [y'(w_i)]^2 + [z'(w_i)]^2}\, \Delta t_i.$$

Using this result in Equation (16.24) with $\tau_i = w_i$ for each i and taking the limit as the Δt_i approach zero gives us the following.

(16.25)
THEOREM

If f is continuous on a region D in space and C is a curve contained in D with the smooth parametric representation

$$x = x(t), \quad y = y(t), \quad z = z(t), \quad \text{for } t \text{ in } [a, b],$$

then

$$\int_C f(x, y, z)\, ds = \int_a^b f(x(t), y(t), z(t))\, \sqrt{[x'(t)]^2 + [y'(t)]^2 + [z'(t)]^2}\, dt.$$

If $\mathbf{r}(t) = x(t)\mathbf{i} + y(t)\mathbf{j} + z(t)\mathbf{k}$ describes C, the integral in Theorem 16.25 can also be expressed by

(16.26) $$\int_C f(x, y, z)\, ds = \int_a^b f(x(t), y(t), z(t))\, \|\mathbf{r}'(t)\|\, dt.$$

Comparing this result with the formula for the arc length of C in Equation (12.14) in Section 12.3 shows that the arc length of C is given by $\int_C ds$.

Example 4

Determine $\int_C (x^2 + y^2 + z^2) \, ds$, where C is the portion of the helix shown in Figure 16.11 beginning at $(1, 0, 0)$ and ending at $(1, 0, 2\pi)$.

Solution

A smooth parametric representation for C is

$$x = \cos t, \quad y = \sin t, \quad z = t, \quad 0 \le t \le 2\pi.$$

It follows from Theorem (16.25) that

$$\int_C (x^2 + y^2 + z^2) \, ds = \int_0^{2\pi} (\cos^2 t + \sin^2 t + t^2) \sqrt{(-\sin t)^2 + (\cos t)^2 + 1} \, dt$$

$$= \int_0^{2\pi} (1 + t^2)\sqrt{2} \, dt$$

$$= \sqrt{2} \left(t + \frac{t^3}{3} \right) \Big]_0^{2\pi}$$

$$= \sqrt{2} \left(2\pi + \frac{8\pi^3}{3} \right) = 2\sqrt{2}\pi \left(1 + \frac{4\pi^2}{3} \right) \approx 125.8. \blacktriangleright$$

Example 5

Determine $\int_C (x^2 + y^2 + z^2) \, ds$, where C is the straight-line segment from $(1, 0, 0)$ to $(1, 0, 2\pi)$ shown in Figure 16.12.

Solution

A parametric representation for C is

$$x = 1, \quad y = 0, \quad z = t, \quad 0 \le t \le 2\pi.$$

Thus,

$$\int_C (x^2 + y^2 + z^2) \, ds = \int_0^{2\pi} (1^2 + 0^2 + t^2)\sqrt{0^2 + 0^2 + 1^2} \, dt$$

$$= \int_0^{2\pi} (1 + t^2) \, dt$$

$$= t + \frac{t^3}{3} \Big]_0^{2\pi}$$

$$= 2\pi + \frac{8\pi^3}{3} = 2\pi \left(1 + \frac{4\pi^2}{3} \right) \approx 89.0. \blacktriangleright$$

In Examples 4 and 5, we integrated $f(x, y, z) = x^2 + y^2 + z^2$ over distinct curves with the same endpoints and obtained different results. In the following example this function is integrated over the curve described in Example 5, but with the orientation of the curve reversed.

Example 6

Determine $\int_{-C} (x^2 + y^2 + z^2) \, ds$, where C is the curve described in Example 5.

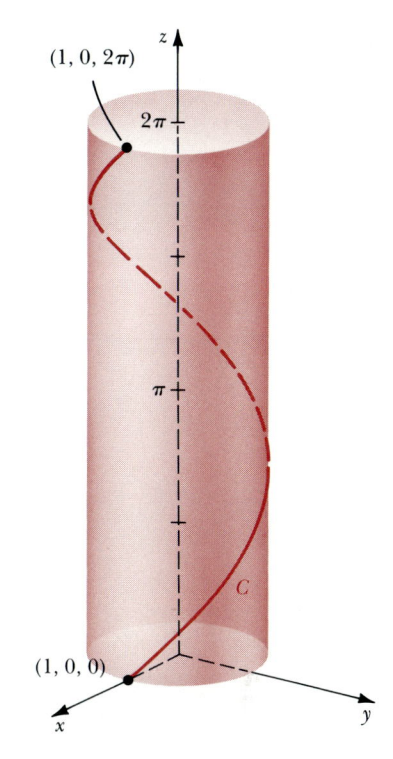

Figure 16.11

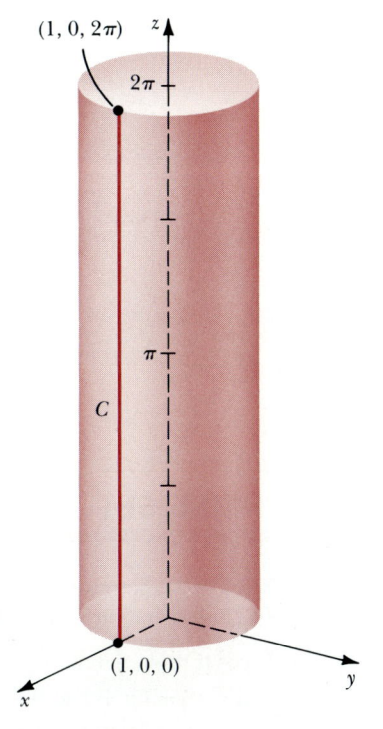

Figure 16.12

Solution

Recall that $-C$ denotes the curve that passes through the same points as C but is traversed in the opposite direction. Thus a parametric representation for $-C$ is

$$x = 1, \quad y = 0, \quad z = 2\pi - t, \quad 0 \le t \le 2\pi$$

and

$$\int_C f(x, y, z) \, ds = \int_0^{2\pi} [1^2 + 0^2 + (2\pi - t)^2] \sqrt{0^2 + 0^2 + (-1)^2} \, dt$$

$$= \int_0^{2\pi} [1 + (2\pi - t)^2] \, dt$$

$$= t - \frac{(2\pi - t)^3}{3} \Bigg]_0^{2\pi}$$

$$= 2\pi + \frac{8\pi^3}{3} = 2\pi \left(1 + \frac{4\pi^2}{3}\right) \approx 89.0. \quad \blacktriangleright$$

Note that the results in Example 5 and 6 are the same, that is,

(16.27) $$\int_C f(x, y, z) \, ds = \int_{-C} f(x, y, z) \, ds.$$

This fact, true in general for line integrals with respect to arc length, occurs because $\Delta s_i > 0$ regardless of the orientation of C. On the other hand, Δx_i, Δy_i, and Δz_i change sign when the orientation of C reverses. This means that the line integrals in the coordinate directions change sign with a change in orientation of the curve, as we saw in (16.23).

The result concerning the integral over the union of curves fails to hold for line integrals involving arc length. The other integral properties do hold, namely,

(16.28) $$\int_C (f \pm g)(x, y, z) \, ds = \int_C f(x, y, z) \, ds \pm \int_C g(x, y, z) \, ds$$

and

(16.29) $$\int_C k f(x, y, z) \, ds = k \int_C f(x, y, z) \, ds, \quad \text{for any constant } k.$$

▶ **EXERCISE SET 16.2**

In Exercises 1–6, determine if the curve described is (a) smooth, (b) closed, and/or (c) simple closed.

1. $\mathbf{r}(t) = \cos t \mathbf{i} + \sin t \mathbf{j} + \mathbf{k}, \ 0 \le t \le 2\pi$
2. $\mathbf{r}(t) = t \mathbf{i} + t^2 \mathbf{j} + (2t + 1) \mathbf{k}, \ 0 \le t \le 1$
3. $\mathbf{r}(t) = \sin \pi t \mathbf{i} + \cos 2\pi t \mathbf{j} + (t^2 - t + 2) \mathbf{k}, \ 0 \le t \le 1$
4. $\mathbf{r}(t) = \cos \pi t \mathbf{i} + \sin 2\pi t \mathbf{j} + (t^2 - t + 2) \mathbf{k}, \ 0 \le t \le 1$
5. C is the set of points $(\cos t, \sin t, t)$, for t in $[0, 2\pi]$
6. C is the set of points $(\cos \sqrt{t}, \sin \sqrt{t}, 0)$, for t in $[0, 4\pi^2]$

In Exercises 7–22, evaluate the line integral.

7. $\int_C xyz \, dx$, where C is described by

$$\mathbf{r}(t) = t\mathbf{i} + (t + 2)\mathbf{j} + (2t - 1)\mathbf{k}, \quad 0 \le t \le 1$$

8. $\int_C (x+y+z)\,dz$, where C is described by
$$\mathbf{r}(t) = t^2\mathbf{i} + (t^{3/2}+1)\mathbf{j} + t\mathbf{k}, \quad 0 \le t \le 4$$

9. $\int_C (xy + x^2 + y^2)\,dx$, where C is the semicircle in the xy-plane described by $y = \sqrt{4-x^2}$, beginning at $(2, 0)$ and ending at $(-2, 0)$

10. $\int_C (x^2 + y^2)\,dz$, where C is the portion of the helix shown in Figure 16.11 beginning at $(1, 0, 0)$ and ending at $(1, 0, 2\pi)$

11. $\int_C y \sec x\,dx$, where C is the curve $y = \sec x$ in the xy-plane from $x = 0$ to $x = \pi/4$

12. $\int_C y \sec x\,dy$, where C is the curve $y = \sec x$ in the xy-plane from $x = 0$ to $x = \pi/4$

13. $\int_C (x^2 + y^2)\,dx + 2xy\,dy$, where C is the circle $x^2 + y^2 = 4$ in the xy-plane, oriented clockwise, beginning and ending at $(2, 0)$

14. $\int_C \dfrac{1}{1+x^2}\,dx + \dfrac{1}{1+y^2}\,dy$, where C is the straight line in the xy-plane from $(1, 1)$ to $(2, 2)$

15. $\int_C (x^3 + y^3)\,ds$, where C is the circle $x^2 + y^2 = 4$ in the xy-plane, oriented clockwise, beginning and ending at $(2, 0)$

16. $\int_C (y - z^2)\,ds$, where C is described by $\mathbf{r}(t) = \mathbf{i} + t^2\mathbf{j} + t\mathbf{k}$, $1 \le t \le 2$

17. $\int_C (x^2 + y^2 + z^2)\,ds$, where C is the straight line joining $(0, 0, 0)$ and $(1, 2, 3)$

18. $\int_C (x^2 + y^2 + z^2)\,ds$, where C is the arc of the circular helix described by $\mathbf{r}(t) = \cos t\,\mathbf{i} + \sin t\,\mathbf{j} + 2t\mathbf{k}$ from $(1, 0, 0)$ to $(1, 0, 4\pi)$

19. $\int_C \dfrac{y+z}{x}\,ds$, where C is the arc of the circle $x^2 + y^2 = 1$ in the xy-plane oriented counterclockwise from $(1, 0, 0)$ to $(\sqrt{2}/2, \sqrt{2}/2, 0)$

20. $\int_C (x + y^{3/2})\,ds$, where C is the parabola $y = x^2$ in the xy-plane from $(0, 0)$ to $(2, 4)$

21. $\int_C x \cos y\,dx + y\,dy + z \sin y\,dz$, where C is the helix described by $\mathbf{r}(t) = \sin t\,\mathbf{i} + t\mathbf{j} + \cos t\,\mathbf{k}$, $0 \le t \le \pi$

22. $\int_C x\,e^{xy}\,dx + y\,e^{yz}\,dy + z\,dz$, where C is described by $\mathbf{r}(t) = t\mathbf{i} + \mathbf{j} + \ln t\,\mathbf{k}$, $1 \le t \le 2$

23. Evaluate $\int_C xy^2\,dx + x^2y\,dy$ along the following four curves in the xy-plane from $(1, 0)$ to $(-1, 0)$:
   a) C is the upper unit semicircle.
   b) C is the lower unit semicircle.
   c) C is a segment of the x-axis from $x = 1$ to $x = -1$.
   d) C is a part of the curve $y = x^3 - x$.

24. Evaluate $\int_C x^2y\,dx + (x - y)\,dy$ along the following four curves in the xy-plane from $(0, 0)$ to $(1, 1)$:
   a) C consists of a segment of the x-axis from $x = 0$ to $x = 1$ and a segment of the line $x = 1$ from $y = 0$ to $y = 1$.
   b) C is the straight line joining $(0, 0)$ to $(1, 1)$.
   c) C consists of a segment of the y-axis from $y = 0$ to $y = 1$ and a segment of the line $y = 1$ from $x = 0$ to $x = 1$.
   d) C is a part of the parabola $y = x^2$.

25. Evaluate $\int_C y\,x\,dx + x\,dy$ over each of the following simple closed curves.

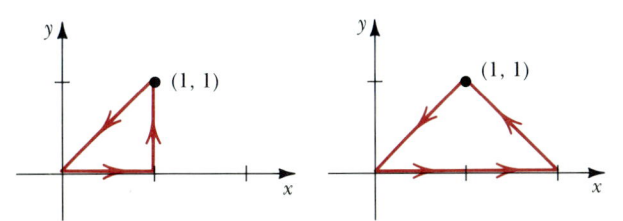

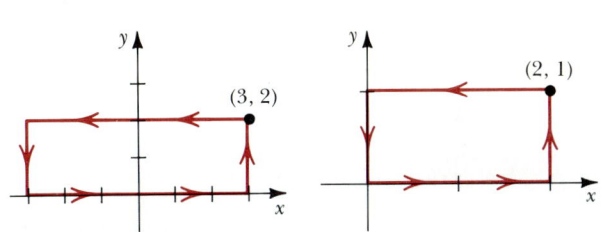

a) b) c) d)

26. Evaluate $\int_C x\,dx + yx\,dy$ over the simple closed curves shown in Exercise 25.

27. Suppose C is a smooth curve described by $\mathbf{r}(t) = x(t)\mathbf{i} + y(t)\mathbf{j} + z(t)\mathbf{k}$, $a \le t \le b$, and $\mathbf{F}$ is a continuous vector-valued function defined in a region containing C. Show that if $\mathbf{T}(x, y, z)$ is the unit tangent vector to C at (x, y, z), then
$$\int_C \mathbf{F}(x, y, z) \cdot \mathbf{T}(x, y, z)\,ds = \int_a^b \mathbf{F}(x, y, z) \cdot \mathbf{r}'(t)\,dt.$$

28. Use Exercise 27 to find $\int_C \mathbf{F}(x, y, z) \cdot \mathbf{T}(x, y, z)\,ds$, where $\mathbf{F}(x, y, z) = x^2\mathbf{i} + y^2\mathbf{j} + z^2\mathbf{k}$ and $\mathbf{T}(x, y, z)$ is the unit tangent vector to C at (x, y, z) if:
   a) C is the straight line segment from $(0, 0, 0)$ to $(1, 2, 3)$;
   b) C is described by $\mathbf{r}(t) = t\mathbf{i} + 2t^2\mathbf{j} + 3t^3\mathbf{k}$, $0 \le t \le 1$.

29. Consider the integral $\int_C M\,dx + N\,dy + P\,dz$, where $M(x, y, z) = y^2 \sin z$, $N(x, y, z) = 2xy \sin z$, and $P(x, y, z) = xy^2 \cos z$.
   a) Show that
   $$\frac{\partial M}{\partial y} = \frac{\partial N}{\partial x}, \quad \frac{\partial M}{\partial z} = \frac{\partial P}{\partial x}, \quad \text{and} \quad \frac{\partial N}{\partial z} = \frac{\partial P}{\partial y}.$$
   b) Evaluate the integral if C is the straight line joining $(0, 0, 0)$ and $(1, 4, 2)$.
   c) Evaluate the integral if C is the curve from

(0, 0, 0) to (1, 4, 2) described by

$$\mathbf{r}(t) = \frac{t}{2}\mathbf{i} + t^2\mathbf{j} + t\mathbf{k}, \quad 0 \le t \le 2.$$

d) Evaluate the integral if C is the circle $x^2 + y^2 = 4$ in the plane $z = 2$ with counterclockwise orientation.

30. a) Find $\int_C (x^2 y \mathbf{i} + xy^2 \mathbf{j}) \cdot \mathbf{r}'(t)\, dt$, where C is the boundary of the rectangle shown here.

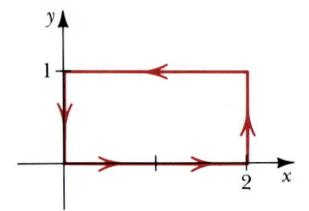

b) Find

$$\iint_R \left[\frac{\partial}{\partial x}(xy^2) - \frac{\partial}{\partial y}(x^2 y)\right] dA,$$

where R is the rectangle enclosed by C.

31. The total force on a particle with charge Q moving with velocity $\mathbf{v}$ in the presence of an electric field $\mathbf{E}$ and a magnetic field $\mathbf{B}$ is given by the *Lorentz force* $\mathbf{F} = Q(\mathbf{E} + \mathbf{v} \times \mathbf{B})$. Show that Newton's Second Law of Motion can be used to generate (a) the *power equation* for the particle:

$$m\frac{d\mathbf{v}}{dt} \cdot \mathbf{v} = Q\mathbf{E} \cdot \frac{d\mathbf{r}}{dt}$$

and consequently (b) the *energy equation*:

$$Q\int \mathbf{E} \cdot d\mathbf{r} = \frac{1}{2} m v^2 + C,$$

where C is a constant (m represents the mass of the particle).

16.3

Physical Applications of Line Integrals

Work

The concept of work was first introduced in Section 5.6, where we found that the work done by a variable force F moving an object from P to Q along an x-coordinate line is

(16.30) $$W = \int_P^Q F(x)\, dx.$$

In Section 11.3 we saw that the work done by a constant vector force $\mathbf{F}$ moving an object along the line from P to Q is

(16.31) $$W = \mathbf{F} \cdot \overrightarrow{PQ}.$$

With the introduction of line integrals, our knowledge of work can be extended to the most useful situation: work done by a variable force acting along a curve in space.

Suppose C is a curve with smooth representation

$$\mathbf{r}(t) = x(t)\mathbf{i} + y(t)\mathbf{j} + z(t)\mathbf{k}, \quad a \le t \le b,$$

and $\mathbf{F}$ is a continuous vector function defined on a region containing C. Let $\mathscr{P} = \{t_0, t_1, \ldots, t_n\}$ be a partition of $[a, b]$, $P(t_i) = (x(t_i), y(t_i), z(t_i))$, and Δs_i be the arc length of that portion of C between $P(t_{i-1})$ and $P(t_i)$ for each $i = 0, 1, \ldots, n$ (see Figure 16.13). The first step is to approximate the work ΔW_i done by $\mathbf{F}$ as an object is moved from $P(t_{i-1})$ to $P(t_i)$ along C. To do this, we choose τ_i in $[t_{i-1}, t_i]$, let $Q_i = P(\tau_i)$, and compute

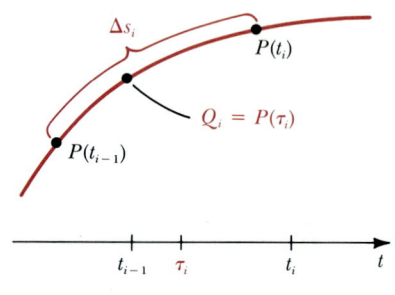

Figure 16.13

the work done by the constant force $\mathbf{F}(Q_i)$ moving an object along the straight line of length Δs_i tangent to C at Q_i.

The unit tangent vector to C at Q_i is

$$\mathbf{T}(Q_i) = \frac{\mathbf{r}'(t_i)}{\|\mathbf{r}'(t_i)\|},$$

so the vector of length Δs_i tangent to the curve at Q_i is $\mathbf{T}(Q_i)\,\Delta s_i$, as shown in Figure 16.14. The work done by the constant force $\mathbf{F}(Q_i)$ moving in the direction $\mathbf{T}(Q_i)\,\Delta s_i$ is therefore

$$\Delta W_i = \mathbf{F}(Q_i) \cdot \mathbf{T}(Q_i)\,\Delta s_i.$$

This is used to approximate the work done by the force $\mathbf{F}$ over the portion of the curve from $P(t_{i-1})$ to $P(t_i)$. Summing ΔW_i for $i = 1, 2, \ldots, n$ gives the approximation for the work done over the curve:

$$W \approx \sum_{i=1}^{n} \mathbf{F}(Q_i) \cdot \mathbf{T}(Q_i)\,\Delta s_i.$$

Taking the limit as $\|\mathcal{P}\| \to 0$ produces the definition for the work done by a variable force moving an object along a curve in space.

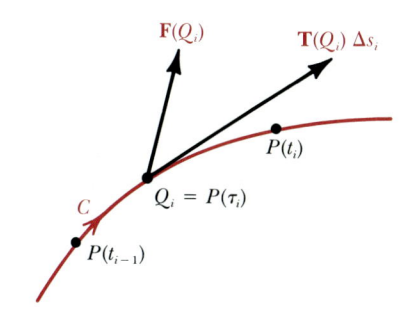

Figure 16.14
$\Delta W_i \approx \mathbf{F}(Q_i) \cdot \mathbf{T}(Q_i)\,\Delta s_i.$

(16.32) DEFINITION

Suppose C is a smooth curve in space and an object is moved along C by a force $\mathbf{F}$. If $\mathbf{F}$ is continuous on a region containing C, **the work done by F along C** is

$$W = \int_C \mathbf{F}(x, y, z) \cdot \mathbf{T}(x, y, z)\,ds,$$

where $\mathbf{T}(x, y, z)$ is the unit tangent vector to C at (x, y, z).

The value of the line integral in Definition (16.32) is independent of the smooth parametric representation of C. Suppose C is described by

$$\mathbf{r}(t) = x(t)\mathbf{i} + y(t)\mathbf{j} + z(t)\mathbf{k}, \quad \text{for } t \text{ in } [a, b].$$

Since

$$\mathbf{F}(x, y, z) \cdot \mathbf{T}(x, y, z) = \mathbf{F}(x, y, z) \cdot \frac{\mathbf{r}'(t)}{\|\mathbf{r}'(t)\|}$$

and

$$ds = \sqrt{[x'(t)]^2 + [y'(t)]^2 + [z'(t)]^2}\,dt = \|\mathbf{r}'(t)\|\,dt,$$

we have

(16.33) $$W = \int_C \mathbf{F}(x, y, z) \cdot \mathbf{T}(x, y, z)\,ds = \int_a^b \mathbf{F}(x, y, z) \cdot \mathbf{r}'(t)\,dt.$$

Since $\mathbf{r}'(t) = d\mathbf{r}/dt$, this integral is often written in the compact form

(16.34) $$W = \int_C \mathbf{F} \cdot d\mathbf{r},$$

a form similar in appearance to the integral in (16.30) at the beginning of the section.

Equation (16.34) also provides a connection between line integrals in the coordinate directions and line integrals with respect to arc length. If $\mathbf{F} = M\mathbf{i} + N\mathbf{j} + P\mathbf{k}$, then

$$\mathbf{F} \cdot \frac{d\mathbf{r}}{dt} = (M\mathbf{i} + N\mathbf{j} + P\mathbf{k}) \cdot \left(\frac{dx}{dt}\mathbf{i} + \frac{dy}{dt}\mathbf{j} + \frac{dz}{dt}\mathbf{k}\right)$$

$$= M\frac{dx}{dt} + N\frac{dy}{dt} + P\frac{dz}{dt}.$$

Thus,

$$W = \int_C \mathbf{F} \cdot d\mathbf{r} = \int_C M\,dx + N\,dy + P\,dz.$$

Equating this to the integral for work in Definition (16.32) gives

(16.35)

$$\int_C \mathbf{F}(x, y, z) \cdot \mathbf{T}(x, y, z)\,ds = \int_C M(x, y, z)\,dx + N(x, y, z)\,dy + P(x, y, z)\,dz.$$

Note that the line integral in (16.35) depends directly on the orientation of the curve C, since a specific smooth parameterization of C must be used to express $\int_a^b \mathbf{F} \cdot \mathbf{r}'(t)\,dt$. A change in orientation reverses the sign of this integral. In Section 16.2 we found that the line integral of a function f with respect to arc length does *not* change sign when the orientation of the curve is reversed, so it may appear at first glance that the sign of the integral given in Definition (16.32) does not change when the orientation of C is reversed. Recall, however, that $\mathbf{T}$ describes the unit tangent vector to C *in the direction of increasing* t. If the orientation of C is reversed, this unit tangent vector must be replaced by $-\mathbf{T}$, which produces a sign change for the integral in Definition (16.32).

Example 1

Find the work done by the force $\mathbf{F}(x, y, z) = x^2 y\mathbf{i} + y^2 x\mathbf{j} + z^2\mathbf{k}$ in moving an object along the straight line from $(0, 0, 0)$ to $(1, 2, 3)$ shown in Figure 16.15.

Solution

This line is described by

$$\mathbf{r}(t) = t\mathbf{i} + 2t\mathbf{j} + 3t\mathbf{k}, \qquad 0 \leq t \leq 1.$$

Thus,

$$W = \int_0^1 \mathbf{F}(x, y, z) \cdot \frac{d\mathbf{r}}{dt}\,dt$$

$$= \int_0^1 (2t^3\mathbf{i} + 4t^3\mathbf{j} + 9t^2\mathbf{k}) \cdot (\mathbf{i} + 2\mathbf{j} + 3\mathbf{k})\,dt$$

$$= \int_0^1 (2t^3 + 8t^3 + 27t^2)\,dt = \frac{5}{2}t^4 + 9t^3 \Big]_0^1 = \frac{23}{2}. \qquad \blacktriangleright$$

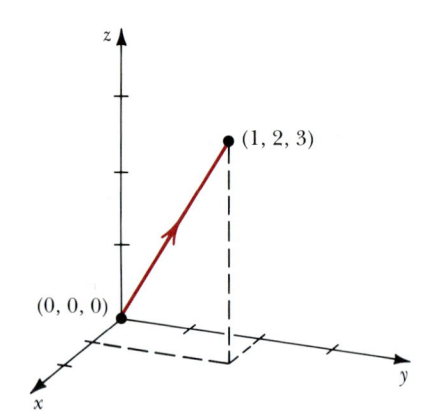

Figure 16.15

Example 2

Find the work done by the force $\mathbf{F}(x, y, z) = x^2 y \mathbf{i} + y^2 x \mathbf{j} + z^2 \mathbf{k}$ in moving an object from $(0, 0, 0)$ to $(1, 2, 3)$ along the curve described by

$$\mathbf{r}(t) = t\mathbf{i} + 2t^2\mathbf{j} + 3t^3\mathbf{k}, \qquad 0 \le t \le 1$$

(see Figure 16.16).

Solution
The work done along this curve is

$$W = \int_0^1 (2t^4\mathbf{i} + 4t^5\mathbf{j} + 9t^6\mathbf{k}) \cdot (\mathbf{i} + 4t\mathbf{j} + 9t^2\mathbf{k})\, dt$$

$$= \int_0^1 (2t^4 + 16t^6 + 81t^8)\, dt$$

$$= \frac{2}{5}t^5 + \frac{16}{7}t^7 + 9t^9 \Big]_0^1$$

$$= \frac{409}{35}.$$

▶

Figure 16.16

In Examples 1 and 2, the same force acted along distinct curves that had the same initial and terminal points. The fact that the work in Example 2 exceeds the work in Example 1 illustrates that, in general, the work required to move an object depends on the path the object takes and not merely on the endpoints of the curve.

Moments

Suppose a thin wire in space has the shape of a curve C with smooth representation

$$\mathbf{r}(t) = x(t)\mathbf{i} + y(t)\mathbf{j} + z(t)\mathbf{k}, \qquad a \le t \le b$$

and σ is a continuous nonnegative function describing the density of the wire at points along C. Line integrals are used to find the mass and moments of the wire about the coordinate planes.

Let $\mathcal{P} = \{t_0, t_1, \ldots, t_n\}$ be a partition of $[a, b]$ and $P(t_i) = (x(t_i), y(t_i), z(t_i))$. Now let Δs_i be the arc length of that portion of C between $P(t_{i-1})$ and $P(t_i)$ and τ_i be in $[t_{i-1}, t_i]$ (see Figure 16.17). An approximation to the mass of the wire between $P(t_{i-1})$ and $P(t_i)$ is

$$\sigma(P(\tau_i))\, \Delta s_i,$$

so the moments about the coordinate planes for this segment of the curve are approximated by

moment about the yz-plane $\approx x(\tau_i)\sigma(P(\tau_i))\, \Delta s_i$,

moment about the xy-plane $\approx z(\tau_i)\sigma(P(\tau_i))\, \Delta s_i$,

moment about the xz-plane $\approx y(\tau_i)\sigma(P(\tau_i))\, \Delta s_i$.

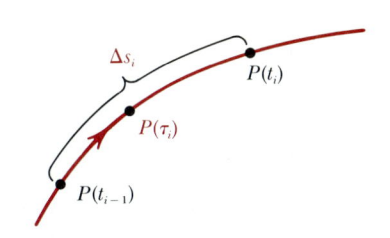

Figure 16.17

Summing these approximations and taking the limit as $\|\mathcal{P}\| \to 0$ leads to the following definition.

(16.36)
DEFINITION
Suppose a thin wire in space is described by a smooth curve C and σ is a continuous nonnegative function defined on a region containing C that describes the density of the wire. The **mass** M and **moments** M_{yz}, M_{xy}, and M_{xz} about the coordinate planes are

$$M = \int_C \sigma(x, y, z) \, ds, \qquad M_{yz} = \int_C x\sigma(x, y, z) \, ds,$$

$$M_{xy} = \int_C z\sigma(x, y, z) \, ds, \quad \text{and} \quad M_{xz} = \int_C y\sigma(x, y, z) \, ds.$$

The **center of mass** of the wire is

$$(\bar{x}, \bar{y}, \bar{z}) = \left(\frac{M_{yz}}{M}, \frac{M_{xz}}{M}, \frac{M_{xy}}{M} \right).$$

Example 3

A wire is bent in the shape of a semicircle oriented counterclockwise with its ends at $(1, 0)$ and $(-1, 0)$ in the xy-plane. The density of the wire varies linearly with the x-coordinate of the point on the wire from 1 at $(-1, 0)$ to 2 at $(1, 0)$. Find (a) the total mass, (b) the moments about the coordinate planes, and (c) the center of mass of the wire.

Solution
The curve C made by the wire shown in Figure 16.18 can be represented by

$$\mathbf{r}(t) = \cos t \mathbf{i} + \sin t \mathbf{j}, \qquad 0 \leq t \leq \pi.$$

a) The density is given by a linear equation $\sigma(x, y) = ax + b$, where

$$\sigma(-1, 0) = -a + b = 1 \quad \text{and} \quad \sigma(1, 0) = a + b = 2.$$

Solving these equations simultaneously for a and b implies that

$$\sigma(x, y) = \frac{x}{2} + \frac{3}{2}.$$

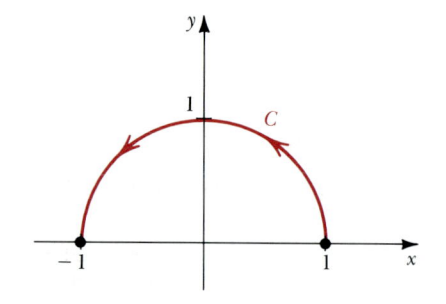

Figure 16.18

Thus,

$$M = \int_C \sigma(x, y) \, ds$$

$$= \int_0^\pi \left(\frac{1}{2} x + \frac{3}{2} \right) \sqrt{[x'(t)]^2 + [y'(t)]^2} \, dt$$

$$= \int_0^\pi \left[\frac{\cos t}{2} + \frac{3}{2} \right] \sqrt{(-\sin t)^2 + (\cos t)^2} \, dt$$

$$= \frac{1}{2} \int_0^\pi (\cos t + 3) \, dt = \frac{1}{2} (\sin t + 3t) \Big]_0^\pi$$

$$= \frac{3}{2} \pi.$$

b) We have

$$M_{yz} = \int_C x\sigma(x, y)\, ds = \frac{1}{2}\int_0^\pi \cos t\,(\cos t + 3)\, dt$$

$$= \frac{1}{2}\int_0^\pi (\cos^2 t + 3\cos t)\, dt$$

$$= \frac{1}{2}\int_0^\pi \left[\left(\frac{1+\cos 2t}{2}\right) + 3\cos t\right] dt$$

$$= \frac{1}{2}\left[\frac{1}{2}t + \frac{1}{4}\sin 2t + 3\sin t\right]_0^\pi = \frac{\pi}{4}$$

and

$$M_{xz} = \int_C y\sigma(x, y)\, ds = \frac{1}{2}\int_0^\pi \sin t\,(\cos t + 3)\, dt$$

$$= \frac{1}{2}\left[\frac{\sin^2 t}{2} - 3\cos t\right]_0^\pi = 3.$$

Since the wire lies in the xy-plane,

$$M_{xy} = \int_C z\sigma(x, y)\, ds = \int_C 0\cdot \sigma(x, y)\, ds = 0.$$

c) The center of mass of the wire is $(\bar{x}, \bar{y}, \bar{z})$, where

$$\bar{x} = \frac{M_{yz}}{M} = \frac{\pi/4}{3\pi/2} = \frac{1}{6},\quad \bar{y} = \frac{M_{xz}}{M} = \frac{3}{3\pi/2} = \frac{2}{\pi},\quad \bar{z} = \frac{M_{xy}}{M} = 0.\ \blacktriangleright$$

▶ EXERCISE SET 16.3

In Exercises 1–8, find the work done by the force **F** in moving an object along the given curve.

1. $\mathbf{F}(x, y, z) = x\mathbf{i} + y\mathbf{j} + z\mathbf{k}$, along the parabola $\mathbf{r}(t) = t\mathbf{i} + t^2\mathbf{j}$ from $(1, 1, 0)$ to $(2, 4, 0)$

2. $\mathbf{F}(x, y, z) = (2x - y)\mathbf{i} + (y - z)\mathbf{j} + 2y\mathbf{k}$ along the helix $\mathbf{r}(t) = \cos t\mathbf{i} + \sin t\mathbf{j} + 3t\mathbf{k}$ from $(1, 0, 0)$ to $(1, 0, 6\pi)$

3. $\mathbf{F}(x, y, z) = x^2 z\mathbf{i} - yx^2\mathbf{j} + 3xz\mathbf{k}$ along the straight line from $(0, 0, 0)$ to $(2, 3, 4)$

4. $\mathbf{F}(x, y, z) = x\mathbf{i} - z\mathbf{j} + 2y\mathbf{k}$ along the y-axis from $(0, 0, 0)$ to $(0, 2, 0)$

5. $\mathbf{F}(x, y, z) = y^2\mathbf{i} + x^2\mathbf{j}$ along the parabola $y = x^2 + 1$ in the plane $z = 2$ from $(0, 1, 2)$ to $(-1, 2, 2)$

6. $\mathbf{F}(x, y, z) = \dfrac{1}{1+x^2}\mathbf{i} + \dfrac{1}{1+y^2}\mathbf{j}$ along the straight-line segment from $(0, 1, 2)$ to $(2, 3, 2)$

7. $\mathbf{F}(x, y, z) = (y^2 + xz)\mathbf{i} + xy\mathbf{j} + (x^2 - y)\mathbf{k}$ along the elliptical helix $\mathbf{r}(t) = 2\cos t\mathbf{i} + 3\sin t\mathbf{j} + t\mathbf{k}$ from $(2, 0, 0)$ to $(-2, 0, 3\pi)$

8. $\mathbf{F}(x, y, z) = y\sin x\mathbf{i} + ye^x\mathbf{j}$ along the curve $\mathbf{r}(t) = t\mathbf{i} + \cos t\mathbf{j} + \mathbf{k}$, for $0 \le t \le \pi/2$

9. Find the work done by the force $\mathbf{F}(x, y, z) = yz\mathbf{i} + xz\mathbf{j} + xy\mathbf{k}$ in moving an object from $(0, 0, 0)$ to $(1, 2, 3)$ along the curve $\mathbf{r}(t) = t\mathbf{i} + 2t\mathbf{j} + 3t\mathbf{k}$, $0 \le t \le 1$.

10. Find the work done by the force described in Exercise 9 along the curve $\mathbf{r}(t) = t\mathbf{i} + 2t^2\mathbf{j} + 3t^3\mathbf{k}$, $0 \le t \le 1$.

11. Find the work done by the force $\mathbf{F}(x, y, z) = (x^2 - y)\mathbf{i} + (y^2 - z)\mathbf{j} + (z^2 - x)\mathbf{k}$ in moving an object along each of the following curves.

a) b)

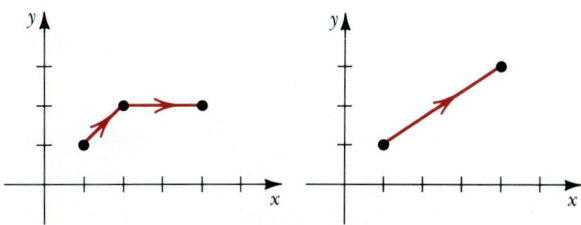

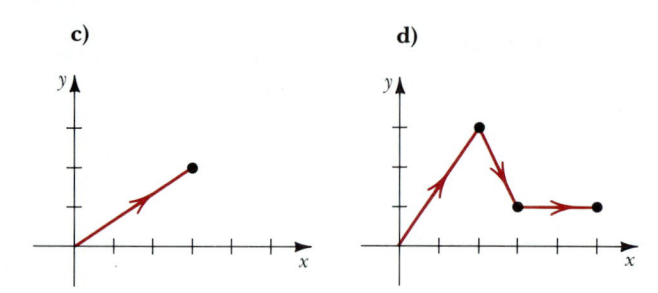

c)

d)

12. Find the work done by the constant force $\mathbf{F}(x, y, z) = 3\mathbf{i} + 2\mathbf{j}$ in moving an object clockwise around the circle $x^2 + y^2 = 1$.

13. Find the work done by the force $\mathbf{F}(x, y, z) = e^{x+y}\mathbf{i} - x^2 y\mathbf{j} + \mathbf{k}$ in moving an object along the curve shown in the figure.

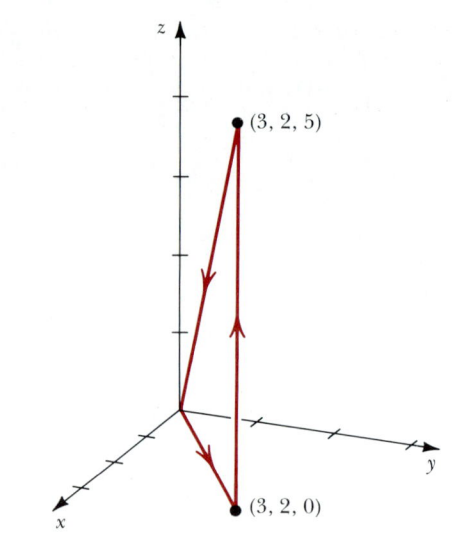

In Exercises 14–19, find (a) the total mass and (b) the center of mass of a wire described by the curve C and having density σ.

14. C is the straight line from $(1, 0, 0)$ to $(1, 4, 7)$; $\sigma(x, y, z) = xyz$

15. C is the semicircle $x^2 + y^2 = 1$, $y \geq 0$, in the xy-plane; $\sigma(x, y, z) = |x| + y$

16. C is the upper half of the circle $x^2 + y^2 = 4$ in the xy-plane; $\sigma(x, y, z) = 8$

17. C is the triangle with vertices $(0, 0, 0)$, $(1, 1, 0)$, $(1, 1, 1)$; $\sigma(x, y, z) = \cos x + \cos y + \cos z$

18. C is the intersection of the ellipsoid $x^2/9 + y^2/16 + z^2/16 = 1$ and the plane $x = 2$; $\sigma(x, y, z) = x$

Figure for Exercise 13.

19. C is the rectangle in the xy-plane with vertices $(0, 0)$, $(2, 0)$, $(2, 3)$, $(0, 3)$;

$$\sigma(x, y, z) = \begin{cases} x \text{ from } (0, 0) \text{ to } (2, 0), \\ x \text{ from } (0, 3) \text{ to } (2, 3), \\ 3 \text{ from } (2, 0) \text{ to } (2, 3), \\ 3 \text{ from } (0, 0) \text{ to } (0, 3). \end{cases}$$

20. Show that the work done by a constant force in moving an object around any circle is zero.

16.4

Line Integrals Independent of Path

Examples 3, 4, and 5 of Section 16.2 demonstrate that the line integral of a function generally depends on the curve over which the integration is taken, and not only on the endpoints of the curve. However, certain functions have the property that their line integrals *do* depend only on the endpoints of the curve. In such a case, the line integral is said to be **independent of path** in D. In this section, we give distinguishing properties of those functions whose line integrals are independent of path.

(16.37)
DEFINITION
A function $\mathbf{F}$ described by

$$\mathbf{F}(x, y, z) = M(x, y, z)\mathbf{i} + N(x, y, z)\mathbf{j} + P(x, y, z)\mathbf{k}$$

is called **conservative** in a region D of $\mathbb{R}^3$ if a differentiable function f exists with $\nabla f = \mathbf{F}$. In this case, the differential form

$$M(x, y, z)\, dx + N(x, y, z)\, dy + P(x, y, z)\, dz$$

is said to be **exact** in D.

(16.38)
THEOREM
Suppose a function $\mathbf{F}$ described by

$$\mathbf{F}(x, y, z) = M(x, y, z)\mathbf{i} + N(x, y, z)\mathbf{j} + P(x, y, z)\mathbf{k}$$

is continuous on an open connected region D. The line integral

$$\int_C \mathbf{F} \cdot d\mathbf{r} = \int_C M\, dx + N\, dy + P\, dz$$

is independent of path in D if and only if $\mathbf{F}$ is conservative in D.

Proof. If $\mathbf{F}$ is conservative, there is a function f with $\nabla f = \mathbf{F}$. Thus,

$$\int_C \mathbf{F} \cdot d\mathbf{r} = \int_a^b \left(\frac{\partial f}{\partial x}\mathbf{i} + \frac{\partial f}{\partial y}\mathbf{j} + \frac{\partial f}{\partial z}\mathbf{k} \right) \cdot \left(\frac{dx}{dt}\mathbf{i} + \frac{dy}{dt}\mathbf{j} + \frac{dz}{dt}\mathbf{k} \right) dt$$

$$= \int_a^b \left(\frac{\partial f}{\partial x}\frac{dx}{dt} + \frac{\partial f}{\partial y}\frac{dy}{dt} + \frac{\partial f}{\partial z}\frac{dz}{dt} \right) dt.$$

The Chain Rule implies that

$$\int_C \mathbf{F} \cdot d\mathbf{r} = \int_a^b \frac{df}{dt}\, dt,$$

so by the Fundamental Theorem of Calculus,

$$\int_C \mathbf{F} \cdot d\mathbf{r} = f(x(b), y(b), z(b)) - f(x(a), y(a), z(a)).$$

Since $\int_C \mathbf{F} \cdot d\mathbf{r}$ depends only on the endpoints of C, the integral is independent of path.

Suppose now that $\int_C \mathbf{F} \cdot d\mathbf{r}$ is independent of path and that (x_0, y_0, z_0) is a fixed point in D. For any point (x, y, z) in D, let

$$\int_{(x_0, y_0, z_0)}^{(x, y, z)} \mathbf{F} \cdot d\mathbf{r} = \int_{(x_0, y_0, z_0)}^{(x, y, z)} M\, dx + N\, dy + P\, dz$$

represent the common value of the line integrals along curves in D with initial point (x_0, y_0, z_0) and terminal point (x, y, z). We define the scalar function f on D by

$$f(x, y, z) = \int_{(x_0, y_0, z_0)}^{(x, y, z)} M\, dx + N\, dy + P\, dz.$$

Since D is an open region, there is an open sphere about (x, y, z) contained in D. We choose a point in this open sphere with coordinates (x_1, y, z), where $x_1 \neq x$, as shown in Figure 16.19(a) on the following page.

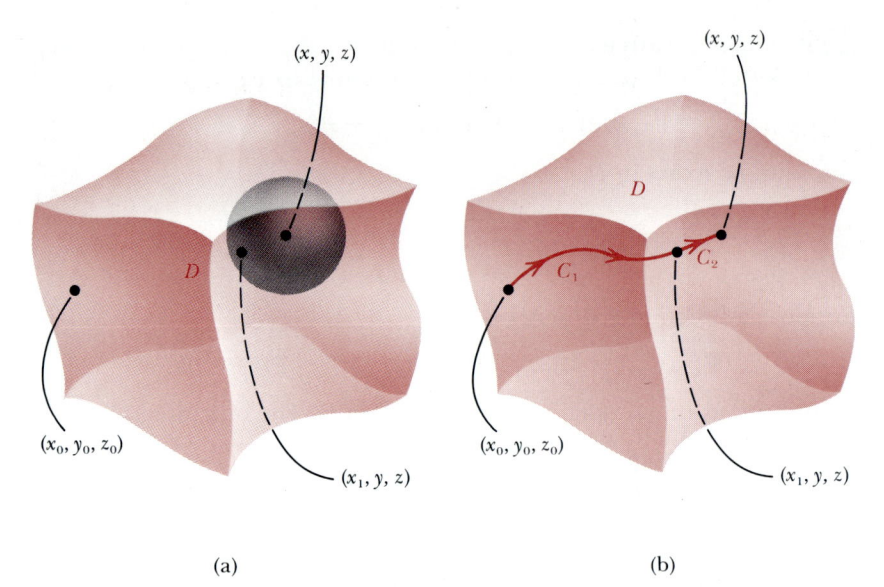

Figure 16.19

(a)   (b)

Let C_1 be a curve in D from (x_0, y_0, z_0) to (x_1, y, z), and let C_2 be the straight-line segment with initial point (x_1, y, z) and terminal point (x, y, z), as shown in Figure 16.19(b). Then $C_1 + C_2$ is a curve from (x_0, y_0, z_0) to (x, y, z) and

$$f(x, y, z) = \int_{C_1} M\,dx + N\,dy + P\,dz + \int_{C_2} M\,dx + N\,dy + P\,dz.$$

Since the limits of integration on C_1 are constant in the variable x,

$$\frac{\partial}{\partial x}\left[\int_{C_1} M\,dx + N\,dy + P\,dz\right] = 0.$$

On C_2, y and z are constant, so $dy = 0$ and $dz = 0$. Therefore, the line integral on C_2 reduces to a single integral from x_1 to x:

$$\int_{C_2} M\,dx + N\,dy + P\,dz = \int_{x_1}^{x} M(t, y, z)\,dt.$$

(The variable t has been introduced in this integral to prevent confusion with the upper limit x.) The Fundamental Lemma of Calculus implies that

$$\frac{\partial f}{\partial x}(x, y, z) = \frac{d}{dx}\int_{x_1}^{x} M(t, y, z)\,dt = M(x, y, z).$$

By choosing a point in D with coordinates (x, y_1, z), where $y_1 \neq y$, and proceeding as above, we can show that

$$\frac{\partial f}{\partial y}(x, y, z) = N(x, y, z).$$

Similarly, it can be shown that

$$\frac{\partial f}{\partial z}(x, y, z) = P(x, y, z).$$

Thus,

$$\nabla f(x, y, z) = M(x, y, z)\mathbf{i} + N(x, y, z)\mathbf{j} + P(x, y, z)\mathbf{k} = \mathbf{F}(x, y, z)$$

and **F** is conservative. ▷

The following corollary and theorem follow from Theorem (16.38) and its proof.

(16.39)
COROLLARY
A continuous function **F** is conservative on an open connected region D if and only if $\int_C \mathbf{F} \cdot d\mathbf{r} = 0$ for every simple closed curve C in D.

(16.40)
THEOREM: THE FUNDAMENTAL THEOREM OF LINE INTEGRALS
Suppose **F** is continuous and $\nabla f = \mathbf{F}$ on an open connected region D. If C is a curve in D with initial point P and terminal point Q, then

$$\int_C \mathbf{F} \cdot d\mathbf{r} = f(Q) - f(P).$$

The scalar function f in Theorem (16.40) is called a **potential function** for the vector function **F**. The potential function in this result corresponds to the antiderivative in the Fundamental Theorem of Calculus. The endpoints P and Q correspond to the endpoints of the interval.

The following result gives another property of a potential function that is analogous to an antiderivative.

(16.41)
THEOREM
Scalar functions f_1 and f_2 are potential functions for the same vector function **F** in an open connected region D if and only if a constant K exists with

$$f_1(x, y, z) = f_2(x, y, z) + K, \quad \text{for all } (x, y, z) \text{ in } D.$$

Proof. The gradient of a constant function is zero, so if f_2 is a potential function of **F** and $f_1(x, y, z) = f_2(x, y, z) + K$, then f_1 is also a potential function of **F**.

On the other hand, if f_1 and f_2 are potential functions for the same function **F**, then

$$\mathbf{0} = \nabla(f_1 - f_2) = \frac{\partial(f_1 - f_2)}{\partial x}\mathbf{i} + \frac{\partial(f_1 - f_2)}{\partial y}\mathbf{j} + \frac{\partial(f_1 - f_2)}{\partial z}\mathbf{k}.$$

The function $f_1 - f_2$ cannot be dependent on x, y, or z, since its partial derivatives with respect to each of these variables are zero. This implies that $f_1 - f_2$ is constant on the connected region D. Thus a constant K

exists with

$$f_1(x, y, z) = f_2(x, y, z) + K, \quad \text{for all } (x, y, z) \text{ in } D.$$

Example 1

Find $\int_C \mathbf{F} \cdot d\mathbf{r}$, where $\mathbf{r}(t) = t\mathbf{i} + (t^2 - 2)\mathbf{j} + (\pi t^2/2)\mathbf{k}$, $0 \leq t \leq 1$, and $\mathbf{F}(x, y, z) = y^2 \sin z \mathbf{i} + 2xy \sin z \mathbf{j} + xy^2 \cos z \mathbf{k}$.

Solution
Note that if $f(x, y, z) = xy^2 \sin z$, then

$$\nabla f(x, y, z) = y^2 \sin z \mathbf{i} + 2xy \sin z \mathbf{j} + xy^2 \cos z \mathbf{k} = \mathbf{F}(x, y, z).$$

Therefore, $\mathbf{F}$ is conservative and the integral is independent of path. Since

$$\mathbf{r}(1) = \mathbf{i} - \mathbf{j} + \frac{\pi}{2}\mathbf{k} \quad \text{and} \quad \mathbf{r}(0) = -2\mathbf{j},$$

it follows from Theorem (16.40) that

$$\int_C \mathbf{F} \cdot d\mathbf{r} = f\left(1, -1, \frac{\pi}{2}\right) - f(0, -2, 0)$$

$$= 1 \cdot (-1)^2 \cdot \sin \frac{\pi}{2} - 0 \cdot (-2)^2 \sin 0 = 1.$$

The line integral in the preceding example was evaluated by recognizing a potential function f for the function $\mathbf{F}$. This leads to some natural questions:

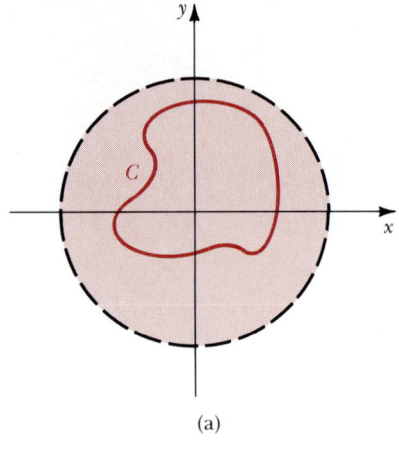

i) Are there easily verifiable conditions that ensure that $\mathbf{F}$ is conservative?
ii) If $\mathbf{F}$ is conservative, how can a potential function for $\mathbf{F}$ be found?

To answer these questions we need some new terminology.

(16.42) DEFINITION

A connected region D, either in the plane or in space, is called **simply connected** if every simple closed curve in D can be shrunk to a point without leaving D.

The interior of the region bounded by the circle shown in Figure 16.20(a) is simply connected, as is the region bounded by the square shown in Figure 16.20(b). The region between the concentric circles shown in Figure 16.20(c) is not simply connected; the curve shown there cannot be shrunk to a point without leaving the region.

In space, the interior of the sphere shown in Figure 16.21(a) is simply connected. It would remain so even if a finite number of points in the region were removed, because the third dimension allows us to slip a

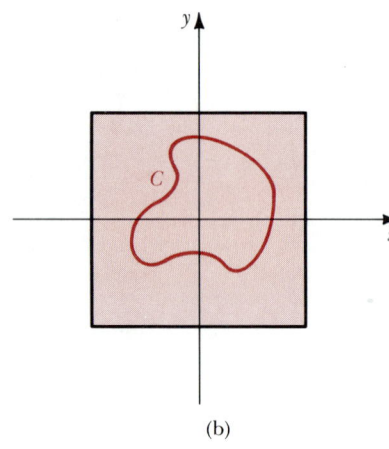

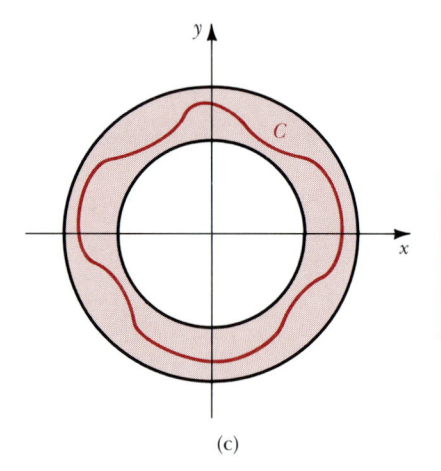

Figure 16.20
(a) An open simply connected plane region; (b) a closed simply connected plane region; (c) a plane region that is not simply connected.

curve around isolated holes. The region bounded by the torus shown in Figure 16.21(b), on the other hand, is not simply connected. The curve labeled C in this figure cannot be shrunk to a point without leaving the region.

The following result answers question (i).

(16.43)
THEOREM
Suppose $\mathbf{F}(x, y, z) = M(x, y, z)\mathbf{i} + N(x, y, z)\mathbf{j} + P(x, y, z)\mathbf{k}$, where M, N, and P have continuous first partial derivatives in an open simply connected region D. Then $\mathbf{F}$ is conservative in D if and only if

$$\frac{\partial M}{\partial y} = \frac{\partial N}{\partial x}, \quad \frac{\partial M}{\partial z} = \frac{\partial P}{\partial x}, \quad \text{and} \quad \frac{\partial N}{\partial z} = \frac{\partial P}{\partial y}.$$

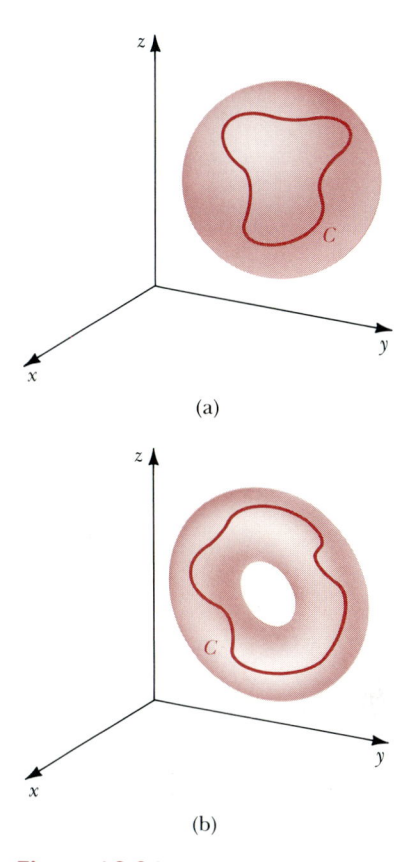

Proof. We will show that if $\mathbf{F}$ is conservative in D, then the conditions are satisfied. The proof that the conditions imply that $\mathbf{F}$ is conservative is deferred to Section 16.8, where it is a consequence of Stokes's Theorem.

If $\mathbf{F}$ is conservative, there is a function f with

$$\mathbf{F} = \nabla f = \frac{\partial f}{\partial x}\mathbf{i} + \frac{\partial f}{\partial y}\mathbf{j} + \frac{\partial f}{\partial z}\mathbf{k}.$$

Thus,

$$M = \frac{\partial f}{\partial x}, \quad N = \frac{\partial f}{\partial y}, \quad \text{and} \quad P = \frac{\partial f}{\partial z}.$$

Since M and N have continuous first partial derivatives, the mixed partial derivatives of f are equal and

$$\frac{\partial M}{\partial y} = \frac{\partial^2 f}{\partial y \partial x} = \frac{\partial^2 f}{\partial x \partial y} = \frac{\partial N}{\partial x}.$$

Similarly,

$$\frac{\partial M}{\partial z} = \frac{\partial P}{\partial x} \quad \text{and} \quad \frac{\partial N}{\partial z} = \frac{\partial P}{\partial y}. \qquad \triangleright$$

Figure 16.21
(a) A simply connected solid region; (b) a solid region that is not simply connected.

Theorems (16.38) and (16.43) combine to give the following result.

(16.44)
COROLLARY
Suppose $\mathbf{F}(x, y, z) = M(x, y, z)\mathbf{i} + N(x, y, z)\mathbf{j} + P(x, y, z)\mathbf{k}$, where M, N, and P have continuous first partial derivatives in an open simply connected region D. The line integral $\int_C \mathbf{F} \cdot d\mathbf{r}$ is independent of path in D if and only if

$$\frac{\partial M}{\partial y} = \frac{\partial N}{\partial x}, \quad \frac{\partial M}{\partial z} = \frac{\partial P}{\partial x}, \quad \text{and} \quad \frac{\partial N}{\partial z} = \frac{\partial P}{\partial y}.$$

Example 2

Find $\int_C \mathbf{F} \cdot d\mathbf{r}$ if C is described by $\mathbf{r}(t) = \sqrt{t}\,\mathbf{i} + t^2\mathbf{j} + \sin(\pi t/2)\mathbf{k}$, for $0 \leq t \leq 1$, and $\mathbf{F}(x, y, z) = 2x\mathbf{i} + 3y^2 \tan z\,\mathbf{j} + y^3 \sec^2 z\,\mathbf{k}$.

Solution
Since

$$\mathbf{F} \cdot d\mathbf{r} = \left[2\sqrt{t}\,\mathbf{i} + 3t^4 \tan\left(\sin\left(\frac{\pi t}{2}\right)\right)\mathbf{j} + t^6 \sec^2\left(\sin\left(\frac{\pi t}{2}\right)\right)\mathbf{k}\right]$$
$$\cdot \left[\left(\frac{1}{2\sqrt{t}}\right)\mathbf{i} + 2t\mathbf{j} + \left(\frac{\pi}{2}\right)\cos\left(\frac{\pi t}{2}\right)\mathbf{k}\right] dt,$$

direct integration requires evaluating the formidable integral

$$\int_0^1 \left[1 + 6t^5 \tan\left(\sin\frac{\pi}{2}t\right) + \frac{\pi}{2}t^6 \left(\sec\left(\sin\frac{\pi}{2}t\right)\right)^2 \left(\cos\frac{\pi}{2}t\right)\right] dt.$$

However, $\mathbf{F}$ is conservative since

$$\frac{\partial}{\partial y}(2x) = 0 = \frac{\partial}{\partial x}(3y^2 \tan z), \qquad \frac{\partial}{\partial z}(2x) = 0 = \frac{\partial}{\partial x}(y \sec^2 z),$$

and

$$\frac{\partial}{\partial z}(3y^2 \tan z) = 3y^2 \sec^2 z = \frac{\partial}{\partial y}(y \sec^2 z).$$

Thus, $\int_C \mathbf{F} \cdot d\mathbf{r}$ is independent of path. Instead of integrating along the given curve C from $(0, 0, 0)$ to $(1, 1, 1)$, we will integrate along a curve, which simplifies the integration. We could use the straight line from $(0, 0, 0)$ to $(1, 1, 1)$ given parametrically by

$$C: \mathbf{r}(t) = t\mathbf{i} + t\mathbf{j} + t\mathbf{k}, \quad \text{for } 0 \leq t \leq 1,$$

but the integral over this curve is still difficult to evaluate:

$$\int_C \mathbf{F} \cdot d\mathbf{r} = \int_0^1 [2t\mathbf{i} + 3t^2 \tan t\,\mathbf{j} + t^3 \sec^2 t\,\mathbf{k}] \cdot [\mathbf{i} + \mathbf{j} + \mathbf{k}]\, dt$$
$$= \int_0^1 [2t + 3t^2 \tan t + t^3 \sec^2 t]\, dt.$$

Instead, we will break the problem into three parts by considering curves in the coordinate directions. Each of the integrals involved is then easily evaluated.

Let C be the curve from $(0, 0, 0)$ to $(1, 1, 1)$ defined by $C = C_1 + C_2 + C_3$ (see Figure 16.22), where

C_1: $\mathbf{r}(t) = t\mathbf{i}$, $\quad 0 \leq t \leq 1$; $\qquad C_2$: $\mathbf{r}(t) = \mathbf{i} + t\mathbf{j}$, $\quad 0 \leq t \leq 1$;

and

C_3: $\mathbf{r}(t) = \mathbf{i} + \mathbf{j} + t\mathbf{k}$, $\quad 0 \leq t \leq 1$.

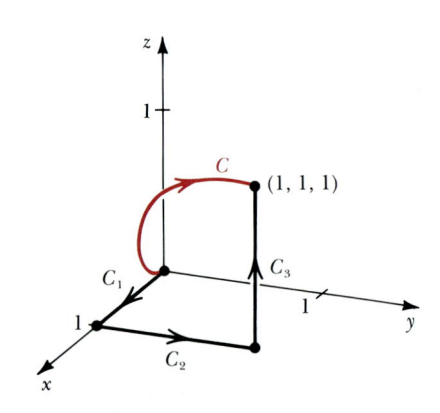

Figure 16.22
$C = C_1 + C_2 + C_3$.

Although more integrals are involved, the problem is simplified because only one of the coordinate variables changes on each curve:

$$\int_C \mathbf{F} \cdot d\mathbf{r} = \int_{C_1} \mathbf{F} \cdot d\mathbf{r} + \int_{C_2} \mathbf{F} \cdot d\mathbf{r} + \int_{C_3} \mathbf{F} \cdot d\mathbf{r}$$

$$= \int_0^1 [2t\mathbf{i} + 3 \cdot 0^2 \cdot (\tan 0)\mathbf{j} + 0^3 (\sec^2 0)\mathbf{k}] \cdot \mathbf{i}\, dt$$

$$+ \int_0^1 [2\mathbf{i} + 3t^2 (\tan 0)\mathbf{j} + t^3(\sec^2 0)\mathbf{k}] \cdot \mathbf{j}\, dt$$

$$+ \int_0^1 [2\mathbf{i} + 3 \cdot 1^2 \cdot (\tan t)\mathbf{j} + 1^3 \cdot (\sec^2 t)\mathbf{k}] \cdot \mathbf{k}\, dt,$$

so

$$\int_C \mathbf{F} \cdot d\mathbf{r} = \int_0^1 2t\, dt + \int_0^1 0\, dt + \int_0^1 \sec^2 t\, dt$$

$$= t^2 + \tan t \Big]_0^1 = 1 + \tan 1. \quad \blacktriangleright$$

We now have conditions that determine when a function $\mathbf{F}$ is conservative. The remaining question to be answered is how to determine a function f whose gradient is $\mathbf{F}$. The answer is illustrated in the following example.

Example 3

Show that the function described by

$$\mathbf{F}(x, y, z) = 2xy\mathbf{i} + (x^2 + \sin z)\mathbf{j} + (y \cos z + 2)\mathbf{k}$$

is the gradient of some function f, and find the form of this potential function.

Solution
Since

$$\frac{\partial}{\partial y}(2xy) = 2x = \frac{\partial}{\partial x}(x^2 + \sin z), \quad \frac{\partial}{\partial z}(2xy) = 0 = \frac{\partial}{\partial x}(y \cos z + 2)$$

and

$$\frac{\partial}{\partial z}(x^2 + \sin z) = \cos z = \frac{\partial}{\partial y}(y \cos z + 2),$$

$\mathbf{F}$ is conservative. Hence there is a function f with

$$\mathbf{F} = \nabla f = \frac{\partial f}{\partial x}\mathbf{i} + \frac{\partial f}{\partial y}\mathbf{j} + \frac{\partial f}{\partial z}\mathbf{k}.$$

To find f, we first equate the components in $\mathbf{F}(x, y, z)$ and $\nabla f(x, y, z)$ to obtain

a) $\dfrac{\partial f}{\partial x}(x, y, z) = 2xy,$

b) $\dfrac{\partial f}{\partial y}(x, y, z) = x^2 + \sin z,$ and

c) $\dfrac{\partial f}{\partial z}(x, y, z) = y \cos z + 2.$

Integrating (a) with respect to x gives

d) $f(x, y, z) = \displaystyle\int 2xy \, dx = x^2 y + g(y, z),$

where $g(y, z)$ is the constant of integration with respect to x, that is,

$$\dfrac{\partial g}{\partial x}(y, z) = 0.$$

Next, we take the partial derivative of f in (d) with respect to y and set this equal to the expression in (b):

$$\dfrac{\partial f}{\partial y}(x, y, z) = \dfrac{\partial}{\partial y}(x^2 y + g(y, z)) = x^2 + \dfrac{\partial}{\partial y} g(y, z),$$

so

$$x^2 + \dfrac{\partial}{\partial y} g(y, z) = x^2 + \sin z.$$

This implies that

e) $\dfrac{\partial}{\partial y} g(y, z) = \sin z.$

Integrating both sides of (e) with respect to y gives

$$g(y, z) = \int \sin z \, dy = y \sin z + k(z).$$

Combining this with (d) produces

f) $f(x, y, z) = x^2 y + y \sin z + k(z).$

Finally, to evaluate $k(z)$, we take the partial derivative of f as given in (f) with respect to z and set this equal to the expression in (c):

$$\dfrac{\partial f}{\partial z} = \dfrac{\partial}{\partial z}(x^2 y + y \sin z + k(z)) = y \cos z + k'(z),$$

so

$$y \cos z + k'(z) = y \cos z + 2$$

and $k'(z) = 2$. It follows that

$$k(z) = 2z + C$$

for any constant C and that
$$f(x, y, z) = x^2 y + y \sin z + 2z + C$$
has the property that $\nabla f = \mathbf{F}$. ▶

The basis for calling a vector-valued function conservative if it is the gradient of a scalar function comes from the physical law of conservation of energy. The **kinetic energy** of an object due to a force $\mathbf{F}$ is the energy associated with the *motion* of the object that results from the force. The **potential energy** of an object due to a force $\mathbf{F}$ is the energy associated with the position of the object, energy that has the *potential* to become kinetic.

Suppose that P is a differentiable function defined on a simply connected, bounded region D and that $\mathbf{F}$ is the force field defined by $-\nabla P = \mathbf{F}$. Let C be an arbitrary smooth curve in D described by $\mathbf{r}(t)$, for t in $[a, b]$. The kinetic energy due to $\mathbf{F}$ of an object of mass m as the object moves along C is defined as

$$\text{Kinetic Energy:} \quad K(\mathbf{r}(t)) = \frac{m}{2}\|\mathbf{r}'(t)\|^2 = \frac{m}{2}\|\mathbf{v}(t)\|^2,$$

and the potential energy of the object is defined as

$$\text{Potential Energy:} \quad P(\mathbf{r}(t)).$$

The total energy due to the force $\mathbf{F}$ is the sum

$$\text{Total Energy:} \quad K(\mathbf{r}(t)) + P(\mathbf{r}(t)).$$

The work done by the force moving the object along C is

$$W = \int_{\mathbf{r}(a)}^{\mathbf{r}(b)} \mathbf{F} \cdot d\mathbf{r} = \int_a^b \mathbf{F}(\mathbf{r}(t)) \cdot \left(\frac{d\mathbf{r}(t)}{dt}\right) dt = \int_a^b \mathbf{F}(\mathbf{r}(t)) \cdot \mathbf{v}(t)\, dt.$$

Newton's Second Law of Motion states that

$$\mathbf{F}(\mathbf{r}(t)) = m\mathbf{a}(t) = m\mathbf{v}'(t),$$

so

$$\mathbf{F}(\mathbf{r}(t)) \cdot \mathbf{v}(t) = m\mathbf{v}'(t) \cdot \mathbf{v}(t).$$

But

$$D_t[\mathbf{v}(t) \cdot \mathbf{v}(t)] = \mathbf{v}'(t) \cdot \mathbf{v}(t) + \mathbf{v}(t) \cdot \mathbf{v}'(t) = 2\mathbf{v}'(t) \cdot \mathbf{v}(t),$$

so

$$W = \int_a^b m[\mathbf{v}'(t) \cdot \mathbf{v}(t)]\, dt = \frac{m}{2}\int_a^b D_t[\mathbf{v}(t) \cdot \mathbf{v}(t)]\, dt.$$

Applying the Fundamental Theorem of Calculus, we have

$$W = \frac{m}{2}[\mathbf{v}(b) \cdot \mathbf{v}(b) - \mathbf{v}(a) \cdot \mathbf{v}(a)] = \frac{m}{2}\|\mathbf{v}(b)\|^2 - \frac{m}{2}\|\mathbf{v}(a)\|^2$$
$$= K(\mathbf{r}(b)) - K(\mathbf{r}(a)).$$

In addition, $-\nabla P = \mathbf{F}$, so the Fundamental Theorem of Line Integrals implies that the work done by $\mathbf{F}$ along C is also given by

$$W = \int_{\mathbf{r}(a)}^{\mathbf{r}(b)} \mathbf{F} \cdot d\mathbf{r} = -P(\mathbf{r}(t))\Big]_a^b = P(\mathbf{r}(a)) - P(\mathbf{r}(b)).$$

Equating these two expressions for W gives

$$K(\mathbf{r}(b)) - K(\mathbf{r}(a)) = P(\mathbf{r}(a)) - P(\mathbf{r}(b))$$

or

$$K(\mathbf{r}(b)) + P(\mathbf{r}(b)) = K(\mathbf{r}(a)) + P(\mathbf{r}(a)),$$

that is, the total energy is the same. Since C was arbitrarily chosen in D, this means that:

> The total energy of an object due to a conservative force field in an open connected region is constant.

This statement is called the **Law of Conservation of Energy.**

▶ EXERCISE SET 16.4

1. Evaluate $\int_C yz\, dx + xz\, dy + xy\, dz$, where:
   a) C is the straight line from $(0, 0, 0)$ to $(1, 1, 1)$;
   b) C is described by $\mathbf{r}(t) = t\mathbf{i} + t^2\mathbf{j} + t^3\mathbf{k}$, $0 \le t \le 1$.

2. Evaluate $\int_C y \sin z\, dx + x \sin z\, dy + xy \cos z\, dz$, where:
   a) C is the straight line from $(1, 0, 0)$ to $(-1, 0, \pi)$;
   b) C is the helix described by $\mathbf{r}(t) = \cos t\mathbf{i} + \sin t\mathbf{j} + t\mathbf{k}$, $0 \le t \le \pi$.

In Exercises 3–8, determine whether the differential form is exact.

3. $y \sin z\, dx + x \sin z\, dy + xy \cos z\, dz$
4. $y^3 \cos x\, dx - 3y^2 \sin x\, dy$
5. $(y + z)\, dx + (x + z)\, dy + (x + y)\, dz$
6. $yz\, dx + xz\, dy + xy\, dz$
7. $x^2\, dx + y^2\, dy + z^2\, dz$
8. $y^2\, dx + z^2\, dy + x^2\, dz$

In Exercises 9–18, determine whether $\mathbf{F}$ is conservative.

9. $\mathbf{F}(x, y) = e^{-y} \cos x\, \mathbf{i} - e^{-y} \sin x\, \mathbf{j}$
10. $\mathbf{F}(x, y) = (2x - y)\mathbf{i} + (2y - x)\mathbf{j}$
11. $\mathbf{F}(x, y, z) = \dfrac{-x}{x^2 + y^2}\mathbf{i} + \dfrac{y}{x^2 + y^2}\mathbf{j} + 3\mathbf{k}$
12. $\mathbf{F}(x, y, z) = \cos y\, \mathbf{i} - x \sin y\, \mathbf{j} - \cos z\, \mathbf{k}$
13. $\mathbf{F}(x, y, z) = \cos yz\, \mathbf{i} - xy \sin yz\, \mathbf{j} - xy \sin yz\, \mathbf{k}$
14. $\mathbf{F}(x, y, z) = (2xy^2 + yz)\mathbf{i} + (x \cos y - z)\mathbf{j} + (x - y)\mathbf{k}$

15. $\mathbf{F}(x, y, z) = 2xy^3 z\mathbf{i} + 3x^2 y^2 z\mathbf{j} + x^2 y^3\mathbf{k}$
16. $\mathbf{F}(x, y, z) = x^2 y^3 \mathbf{i} + x^2 y^3 z\mathbf{j} + \mathbf{k}$
17. $\mathbf{F}(x, y, z) = (z \cos x - y \sin x)\mathbf{i} + (\cos x + z)\mathbf{j} + (\sin x + y - 2z)\mathbf{k}$
18. $\mathbf{F}(x, y, z) = e^{-z}\mathbf{i} + 2y\mathbf{j} + xe^{-z}\mathbf{k}$

19–28. Find a potential function for each of the conservative functions $\mathbf{F}$ in Exercises 9–18.

In Exercises 29–34, sketch the region described in the xy-plane and determine which are (a) open, (b) closed, (c) simply connected.

29. $\{(x, y)\mid |x| > 1\}$
30. $\{(x, y)\mid |x| \ge 1\}$
31. $\{(x, y)\mid |x| + |y| < 1\}$
32. $\{(x, y)\mid \tfrac{1}{2} < |x| + |y| < 1\}$
33. $\{(x, y)\mid x^2 \le y \le 4\}$
34. $\{(x, y)\mid y < x + 1\}$

In Exercises 35–40, (a) show that the line integral is independent of path and (b) evaluate the integral.

35. $\int_C (2x + y)\, dx + (2y + x)\, dy + 2z\, dz$, where C is the circle $x^2 + y^2 = 4$ in the xy-plane traversed clockwise
36. $\int_C e^x \sin y\, dx + e^x \cos y\, dy$, where C is the curve in the xy-plane described by $\mathbf{r}(t) = t\mathbf{i} + 2t\mathbf{j}$, $0 \le t \le 1$
37. $\int_C y \ln z\, dx + x \ln z\, dy + (xy/z)\, dz$, where C is the helix described by $\mathbf{r}(t) = \cos t\mathbf{i} + \sin t\mathbf{j} + t\mathbf{k}$ from $(1, 0, 2\pi)$ to $(1, 0, 4\pi)$
38. $\int_C (y^2 + 2xz)\, dx + (2xy - z)\, dy + (x^2 - y)\, dz$, where C

is described by $\mathbf{r}(t) = (\sin \pi t/2)^3 \mathbf{i} + (t \cos \pi t)\mathbf{j} + t^2(\cos \pi t)^2 \mathbf{k}$, $0 \le t \le 1$

39. $\int_C (1/x) \ln xy \, dx + ((1/y) \ln xy + ze^{yz}) \, dy + ye^{yz} \, dz$, where C is described by $\mathbf{r}(t) = t\mathbf{i} + t^2\mathbf{j} + t^3\mathbf{k}$ from $(1, 1, 1)$ to $(2, 4, 8)$

40. $\int_C 2xe^{2y} \, dx + 2x^2 e^{2y} \, dy + \cos^2 z \, dz$, where C is any piecewise smooth curve joining $(1, 1, 3)$ and $(1, -2, 3)$

41. In Exercise 20 of Section 16.3 you were asked to show that the work done by a constant force in moving an object clockwise around any circle is zero. Use Corollary (16.39) to show this result.

42. Show that if $\mathbf{F}$ is conservative and has continuous partial derivatives, then curl $\mathbf{F} = \mathbf{0}$.

43. Show that if $\mathbf{F}$ and $\mathbf{G}$ are conservative and c_1 and c_2 are constants, then $c_1 \mathbf{F} + c_2 \mathbf{G}$ is conservative.

44. Near the earth's surface, the force exerted by gravity is $-mg\mathbf{k}$ where m is the mass of the object. Show that if an object of mass m moves from a height h_1 to a height h_2 along a smooth curve, the work done by gravity on this object is $mg(h_2 - h_1)$ (see the figure).

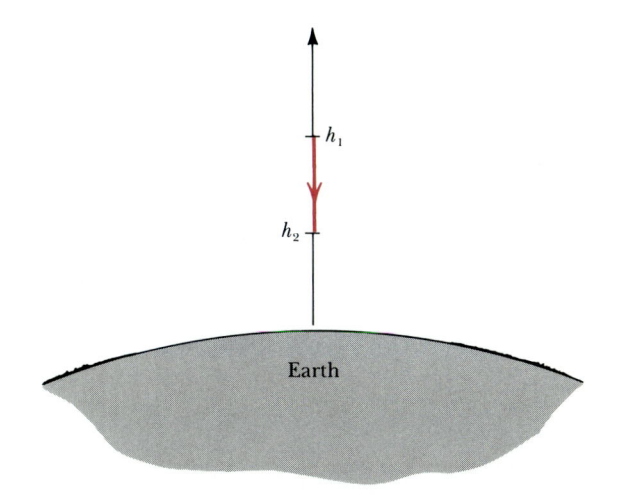

16.5

Green's Theorem

In Section 16.2, a simple closed curve was defined as a curve whose initial and terminal points coincide and are the only points of intersection. In this section we consider Green's Theorem, which concerns the relationship between a line integral over a simple closed curve in the plane and a double integral over a region bounded by the curve.

Before presenting the theorem, we need to discuss the orientation of a simple closed curve in more detail. Consider a simple closed curve in the plane, described by

$$\mathbf{r}(t) = x(t)\mathbf{i} + y(t)\mathbf{j}, \qquad a \le t \le b.$$

This curve is said to have **positive orientation** if the region to the *left* of the curve is bounded as the curve is traced by $\mathbf{r}(t)$. (See Figure 16.23a.) A

HISTORICAL NOTE

Green's theorem was included in the tract *An essay on the application of mathematical analysis to the theories of electricity and magnetism*, which was privately printed in 1828 in Nottingham by George Green (1793–1841). Green was primarily self-taught, and the tract was published with the aid of a patron who helped him enter Cambridge as an undergraduate in 1833. The essay contained not only his theorem, but many other important results. Green graduated from Cambridge in 1837 and although he continued his research, none of his subsequent work had the depth or importance of his essay. Green's essay went largely unnoticed until it was discovered by Sir William Thomson (later Lord Kelvin) (1824–1907) in 1845. Thomson arranged to have it reprinted, and it is now regarded as one of the great classics of mathematical physics. Green must be rated as one of the most brilliant and original mathematical physicists of the nineteenth century. A number of his ideas were far ahead of his time.

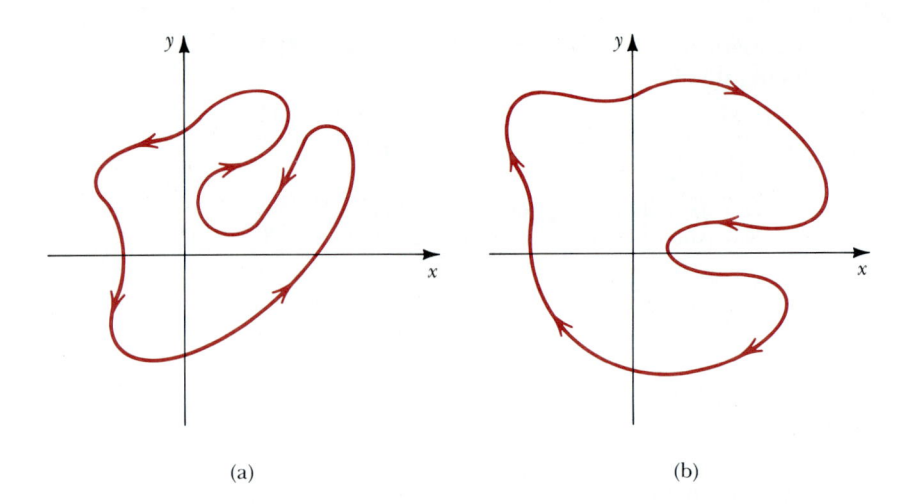

Figure 16.23
(a) A positively oriented curve; (b) a negatively oriented curve.

(a)   (b)

curve oriented in the opposite direction is said to have **negative orientation.** (See Figure 16.23b.)

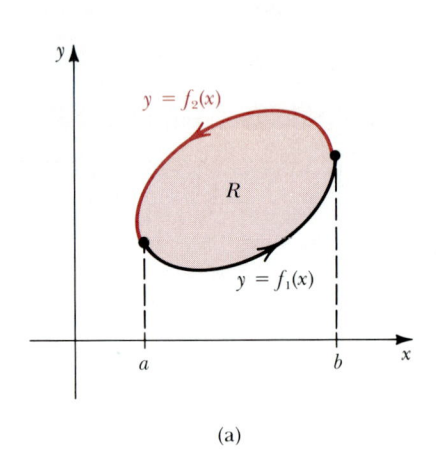

(16.45)
THEOREM: GREEN'S THEOREM
Suppose C is a positively oriented simple closed curve in the plane and R is the region bounded by C. If M and N are functions with continuous partial derivatives on R, then

$$\int_C M\,dx + N\,dy = \iint_R \left(\frac{\partial N}{\partial x} - \frac{\partial M}{\partial y}\right) dA.$$

Proof. We will prove the theorem only for the special case when the region R can be represented both as an x-simple region:

$$a \leq x \leq b, \qquad f_1(x) \leq y \leq f_2(x),$$

for appropriate functions f_1 and f_2 as shown in Figure 16.24(a) and as a y-simple region:

$$c \leq y \leq d, \qquad g_1(y) \leq x \leq g_2(y),$$

for appropriate functions g_1 and g_2 as shown in Figure 16.24(b).
After giving some examples we will indicate how the result is extended to arbitrary regions.
The proof of the theorem for this special region relies on the fact that the simple closed curve C can be expressed both as $C_1 - C_2$, where

$$C_1: \mathbf{r}(t) = t\mathbf{i} + f_1(t)\mathbf{j}, \qquad a \leq t \leq b;$$
$$C_2: \mathbf{r}(t) = t\mathbf{i} + f_2(t)\mathbf{j}, \qquad a \leq t \leq b,$$

and as $C_3 - C_4$, where

$$C_3: \mathbf{r}(t) = g_2(t)\mathbf{i} + t\mathbf{j}, \qquad c \leq t \leq d;$$
$$C_4: \mathbf{r}(t) = g_1(t)\mathbf{i} + t\mathbf{j}, \qquad c \leq t \leq d.$$

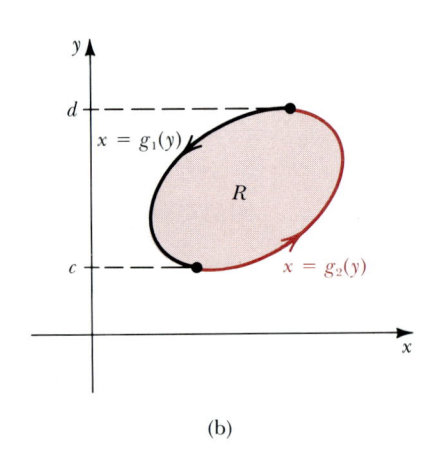

Figure 16.24

(See Figure 16.25.) We first write the double integral as

$$\iint_R \left(\frac{\partial N}{\partial x} - \frac{\partial M}{\partial y}\right) dA = \iint_R \frac{\partial N}{\partial x} dA - \iint_R \frac{\partial M}{\partial y} dA.$$

We use the y-simple representation of R to evaluate $\iint_R (\partial N/\partial x)\, dA$ and the x-simple representation of R to evaluate $\iint_R (\partial M/\partial y)\, dA$. Thus,

$$\iint_R \left(\frac{\partial N}{\partial x} - \frac{\partial M}{\partial y}\right) dA = \int_c^d \int_{g_1(y)}^{g_2(y)} \frac{\partial N}{\partial x}\, dx\, dy - \int_a^b \int_{f_1(x)}^{f_2(x)} \frac{\partial M}{\partial y}\, dy\, dx.$$

Applying the Fundamental Theorem of Calculus to the two inner integrals, we have

$$\iint_R \left(\frac{\partial N}{\partial x} - \frac{\partial M}{\partial y}\right) dA = \int_c^d [N(g_2(y), y) - N(g_1(y), y)]\, dy$$

$$- \int_a^b [M(x, f_2(x)) - M(x, f_1(x))]\, dx$$

$$= \int_{C_3} N(x, y)\, dy - \int_{C_4} N(x, y)\, dy - \int_{C_2} M(x, y)\, dx$$

$$+ \int_{C_1} M(x, y)\, dx$$

$$= \int_{C_3 - C_4} N(x, y)\, dy + \int_{C_1 - C_2} M(x, y)\, dx$$

$$= \int_C N(x, y)\, dy + \int_C M(x, y)\, dx$$

$$= \int_C M(x, y)\, dx + N(x, y)\, dy. \qquad \triangleright$$

(a)

(b)

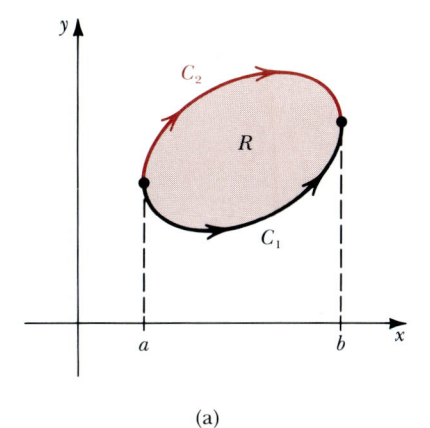

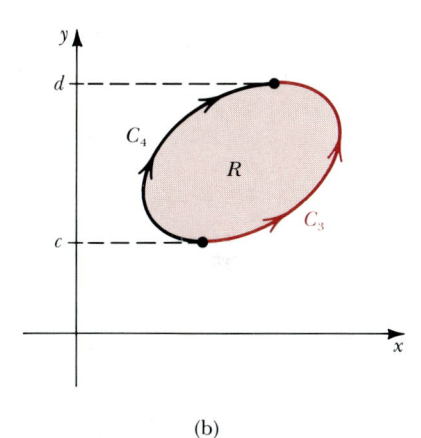

Figure 16.25

Example 1

Use Green's Theorem to evaluate $\int_C x^2 y\, dx + 3xy\, dy$, where C is the positively oriented simple closed curve determined by the graphs of $y = x^2$ and $y = \sqrt{x}$.

Solution

The region R bounded by the graphs is shown in Figure 16.26. Green's Theorem implies that

$$\int_C x^2 y\, dx + 3xy\, dy = \iint_R \left[\frac{\partial (3xy)}{\partial x} - \frac{\partial (x^2 y)}{\partial y}\right] dA$$

$$= \int_0^1 \int_{x^2}^{\sqrt{x}} (3y - x^2)\, dy\, dx.$$

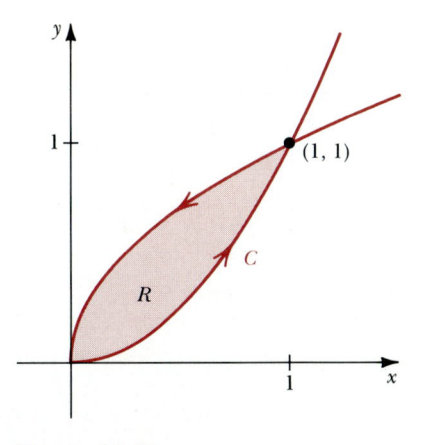

Figure 16.26

Thus,

$$\int_C x^2 y \, dx + 3xy \, dy = \int_0^1 \left[\frac{3y^2}{2} - x^2 y \right]_{x^2}^{\sqrt{x}} dx$$

$$= \int_0^1 \left[\left(\frac{3}{2} x - x^{5/2} \right) - \left(\frac{3}{2} x^4 - x^4 \right) \right] dx$$

$$= \left[\frac{3}{4} x^2 - \frac{2}{7} x^{7/2} - \frac{x^5}{10} \right]_0^1$$

$$= \frac{3}{4} - \frac{2}{7} - \frac{1}{10} = \frac{51}{140}. \quad \blacktriangleright$$

Example 2

Use Green's Theorem to determine

$$\int_C (e^x - 3xy) \, dx + (\sin y + 6x^2) \, dy,$$

where C is described by $\mathbf{r}(t) = \cos t \mathbf{i} + 2 \sin t \mathbf{j}$, $0 \leq t \leq 2\pi$.

Solution

The curve C is the positively oriented ellipse with equation $4x^2 + y^2 = 4$ shown in Figure 16.27. We can express the region R bounded by this ellipse as a y-simple region:

$$-1 \leq x \leq 1, \quad -2\sqrt{1 - x^2} \leq y \leq 2\sqrt{1 - x^2}.$$

Thus,

$$\int_C (e^x - 3xy) \, dx + (\sin y + 6x^2) \, dy = \iint_R \left[\frac{\partial}{\partial x} (\sin y + 6x^2) \right.$$

$$\left. - \frac{\partial}{\partial y} (e^x - 3xy) \right] dA$$

$$= \int_{-1}^1 \int_{-2\sqrt{1-x^2}}^{2\sqrt{1-x^2}} (12x + 3x) \, dy \, dx$$

$$= \int_{-1}^1 15xy \Big]_{-2\sqrt{1-x^2}}^{2\sqrt{1-x^2}} dx$$

$$= \int_{-1}^1 60x \sqrt{1 - x^2} \, dx$$

$$= -30(1 - x^2)^{3/2} \cdot \frac{2}{3} \Big]_{-1}^1 = 0. \quad \blacktriangleright$$

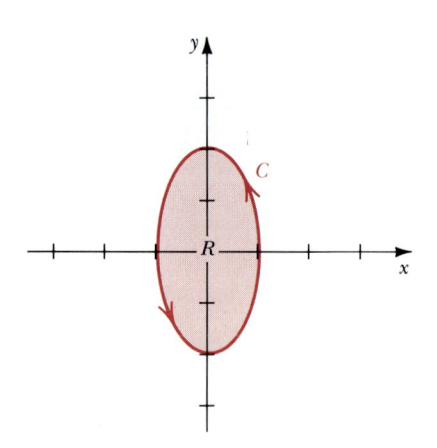

Figure 16.27

To gain a further appreciation for Green's Theorem, try evaluating the integral in Example 2 without using the theorem.

If M and N satisfy the hypotheses of Green's Theorem and the vector-valued function $\mathbf{F}$ is defined by

$$\mathbf{F}(x, y) = M(x, y)\mathbf{i} + N(x, y)\mathbf{j},$$

then Green's Theorem can be expressed as

$$\int_C \mathbf{F} \cdot d\mathbf{r} = \iint_R \left(\frac{\partial N}{\partial x} - \frac{\partial M}{\partial y} \right) dA.$$

This representation is used in the following example.

Example 3

Find $\int_C (x^2 y \mathbf{i} + xy^2 \mathbf{j}) \cdot d\mathbf{r}$, where C is the positively oriented boundary of the rectangle shown in Figure 16.28, by

a) computing the integral directly;
b) using Green's Theorem.

Solution

a) To compute this integral directly, we express C as $C = C_1 + C_2 + C_3 + C_4$, where

$$C_1: \mathbf{r}(t) = t\mathbf{i}, \qquad 0 \le t \le 2$$
$$C_2: \mathbf{r}(t) = 2\mathbf{i} + t\mathbf{j}, \qquad 0 \le t \le 1$$
$$C_3: \mathbf{r}(t) = (2-t)\mathbf{i} + \mathbf{j}, \qquad 0 \le t \le 2$$
$$C_4: \mathbf{r}(t) = (1-t)\mathbf{j}, \qquad 0 \le t \le 1.$$

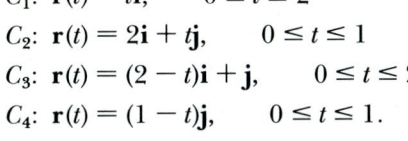

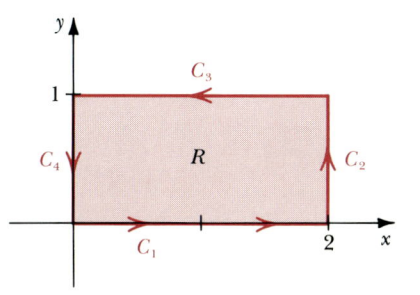

Figure 16.28

On C_1, y is zero and on C_4, x is zero. Thus on both C_1 and C_4, $x^2 y \mathbf{i} + xy^2 \mathbf{j}$ is zero, as are the integrals over these curves. Consequently,

$$\int_C (x^2 y \mathbf{i} + xy^2 \mathbf{j}) \cdot d\mathbf{r} = \int_{C_2} (x^2 y \mathbf{i} + xy^2 \mathbf{j}) \cdot d\mathbf{r} + \int_{C_3} (x^2 y \mathbf{i} + xy^2 \mathbf{j}) \cdot d\mathbf{r}$$

$$= \int_0^1 (4t\mathbf{i} + 2t^2 \mathbf{j}) \cdot \mathbf{j} \, dt + \int_0^2 [(2-t)^2 \mathbf{i} + (2-t)\mathbf{j}] \cdot (-\mathbf{i}) \, dt$$

$$= \int_0^1 2t^2 \, dt + \int_0^2 -(2-t)^2 \, dt = \frac{2t^3}{3} \bigg]_0^1 + \frac{(2-t)^3}{3} \bigg]_0^2 = \frac{2}{3} - \frac{8}{3} = -2.$$

b) To compute the integral using Green's Theorem, note that the region enclosed by C is described by $0 \le x \le 2$, $0 \le y \le 1$. So,

$$\int_C (x^2 y \mathbf{i} + xy^2 \mathbf{j}) \cdot d\mathbf{r} = \int_0^2 \int_0^1 \left[\frac{\partial (xy^2)}{\partial x} - \frac{\partial (x^2 y)}{\partial y} \right] dy \, dx$$

$$= \int_0^2 \int_0^1 (y^2 - x^2) \, dy \, dx$$

$$= \int_0^2 \left[\frac{y^3}{3} - x^2 y \right]_0^1 dx$$

$$= \int_0^2 \left(\frac{1}{3} - x^2 \right) dx$$

$$= \frac{x}{3} - \frac{x^3}{3} \bigg]_0^2 = \frac{2}{3} - \frac{8}{3} = -2. \qquad \blacktriangleright$$

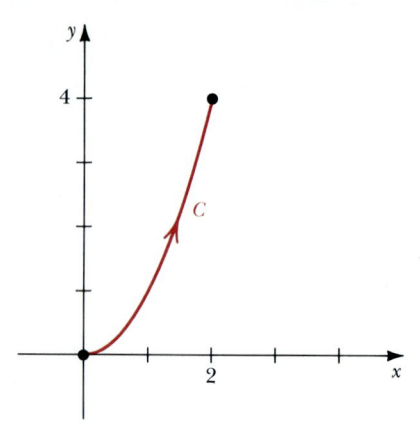

Figure 16.29

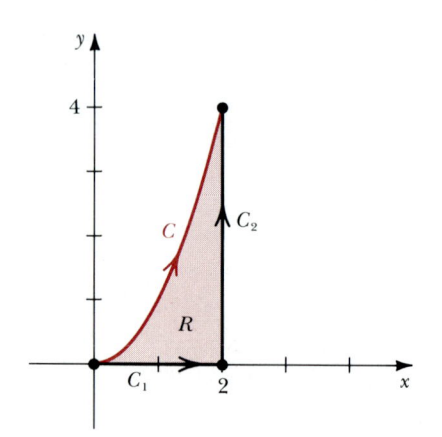

Figure 16.30
$\hat{C} = C_1 + C_2 - C$.

Example 4

Use Green's Theorem to evaluate

$$\int_C [(2x \sin y + x^2 y^2)\mathbf{i} + (x^3 y + x^2 \cos y)\mathbf{j}] \cdot d\mathbf{r}$$

along C: $\mathbf{r}(t) = t\mathbf{i} + t^2\mathbf{j}$, $\quad 0 \le t \le 2$.

Solution

The curve is shown in Figure 16.29. Since C is *not* a closed curve, to apply Green's Theorem we construct a simple closed curve of which C is one part. Consider $\hat{C} = C_1 + C_2 - C$, where

C_1: $\mathbf{r}(t) = t\mathbf{i}$, $\quad 0 \le t \le 2$ $\quad$ and $\quad$ C_2: $\mathbf{r}(t) = 2\mathbf{i} + t\mathbf{j}$, $\quad 0 \le t \le 4$.

Then

$$\int_{\hat{C}} \mathbf{F} \cdot d\mathbf{r} = \int_{C_1} \mathbf{F} \cdot d\mathbf{r} + \int_{C_2} \mathbf{F} \cdot d\mathbf{r} - \int_{C} \mathbf{F} \cdot d\mathbf{r}.$$

But $\hat{C}$ is positively oriented and encloses the x-simple region R, shown in Figure 16.30 and described by

$$0 \le x \le 2, \quad 0 \le y \le x^2.$$

Applying Green's Theorem gives

$$\int_{\hat{C}} [(2x \sin y + x^2 y^2)\mathbf{i} + (x^3 y + x^2 \cos y)\mathbf{j}] \cdot d\mathbf{r}$$

$$= \iint_R [(3x^2 y + 2x \cos y) - (2x \cos y + 2x^2 y)] \, dA$$

$$= \int_0^2 \int_0^{x^2} x^2 y \, dy \, dx$$

$$= \int_0^2 \frac{x^2 y^2}{2} \bigg]_0^{x^2} dx$$

$$= \int_0^2 \frac{x^6}{2} \, dx = \frac{x^7}{14} \bigg]_0^2 = \frac{64}{7}.$$

On C_1, $y = 0$ so $dy = 0$ and

$$\int_{C_1} (2x \sin y + x^2 y^2) \, dx + (x^3 y + x^2 \cos y) \, dy = 0.$$

On C_2, $x = 2$ and $y = t$ so $dx = 0$, $dy = dt$, and

$$\int_{C_2} (2x \sin y + x^2 y^2) \, dx + (x^3 y + x^2 \cos y) \, dy = \int_0^4 (8t + 4 \cos t) \, dt$$

$$= 4t^2 + 4 \sin t \bigg]_0^4$$

$$= 64 + 4 \sin 4.$$

16.5 GREEN'S THEOREM   949

(a)

(b)

Figure 16.31

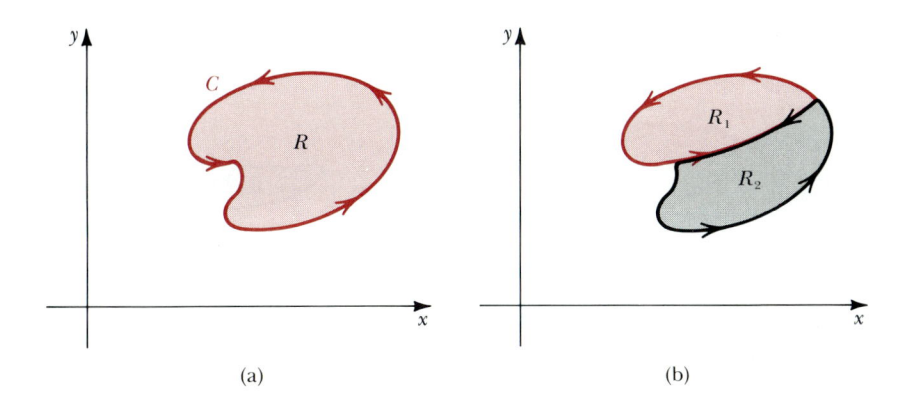

Thus,

$$\int_C (2x \sin y + x^2 y^2)\, dx + (x^3 y - x^2 \cos y)\, dy = (0 + 64 + 4 \sin 4) - \frac{64}{7}$$

$$= \frac{384}{7} + 4 \sin 4. \quad \blacktriangleright$$

We can again see the advantage of Green's Theorem by trying to evaluate the integral in Example 4 directly.

Green's Theorem is proved for more general regions in the plane by partitioning the region into subregions expressed in the forms (I) and (II). For example, the region R in Figure 16.31(a) is partitioned into the regions R_1 and R_2 shown in Figure 16.31(b). Each of the regions R_1 and R_2 can be represented in forms (I) and (II) with positively oriented boundaries C_1 and C_2 (see Figure 16.32). Since the partitioning line is traversed twice but in opposite directions, the line integrals on these portions of C_1 and C_2 cancel. Thus,

$$\int_C \mathbf{F} \cdot d\mathbf{r} = \int_{C_1} \mathbf{F} \cdot d\mathbf{r} + \int_{C_2} \mathbf{F} \cdot d\mathbf{r}$$

$$= \iint_{R_1} \left(\frac{\partial N}{\partial x} - \frac{\partial M}{\partial y} \right) dA + \iint_{R_2} \left(\frac{\partial N}{\partial x} - \frac{\partial M}{\partial y} \right) dA$$

$$= \iint_R \left(\frac{\partial N}{\partial x} - \frac{\partial M}{\partial y} \right) dA.$$

This technique can also be extended to regions in the plane that have "holes," such as the region R shown in Figure 16.33. This region can be partitioned as shown in Figure 16.34(a) on the following page. The boundaries of the partitions R_1 and R_2 are denoted by C_1' and C_2', as in Figures 16.34(b) and (c). Then,

$$\iint_R \left(\frac{\partial N}{\partial x} - \frac{\partial M}{\partial y} \right) dA = \iint_{R_1} \left(\frac{\partial N}{\partial x} - \frac{\partial M}{\partial y} \right) dA + \iint_{R_2} \left(\frac{\partial N}{\partial x} - \frac{\partial M}{\partial y} \right) dA$$

$$= \int_{C_1'} M\, dx + N\, dy + \int_{C_2'} M\, dx + N\, dy,$$

(a)

(b)

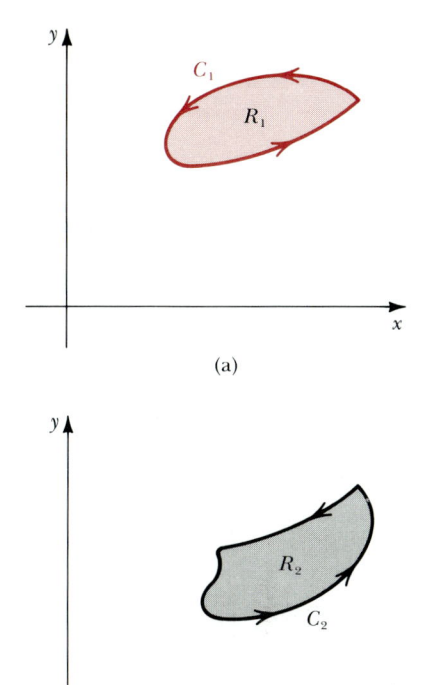

Figure 16.32

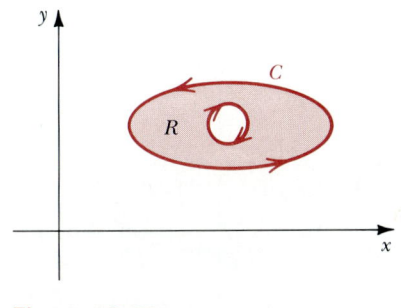

Figure 16.33

(a)

(b)

(c)

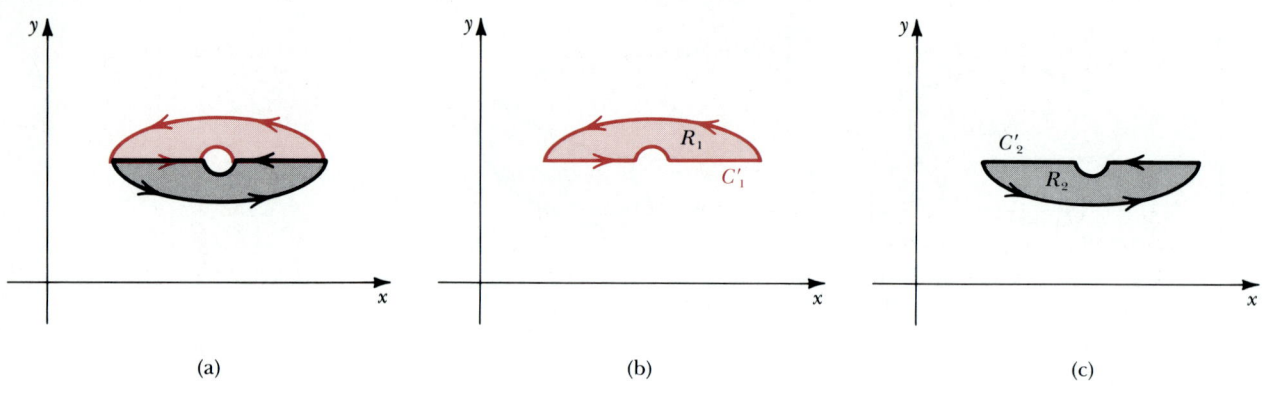

Figure 16.34

and since the line integrals over the common boundary lines are in opposite directions, these cancel to give

$$\iint_R \left(\frac{\partial N}{\partial x} - \frac{\partial M}{\partial y}\right) dA = \int_C M\, dx + N\, dy = \int_C \mathbf{F} \cdot d\mathbf{r}.$$

Green's Theorem can be used in a reverse manner to determine the areas of plane regions whose boundaries are described by parametric equations. If R is a region whose boundary is a positively oriented curve C, the area of R can be written as

(16.46)

$$A = \iint_R dA = \iint_R \left[\frac{\partial x}{\partial x} - \frac{\partial 0}{\partial y}\right] dA = \int_C [0\, dx + x\, dy] = \int_C x\, dy$$

and also as

(16.47) $\quad A = \iint_R dA = \iint_R \left[\frac{\partial 0}{\partial x} - \frac{\partial(-y)}{\partial y}\right] dA = \int_C -y\, dx.$

Averaging the results in (16.46) and (16.47) gives

(16.48) $$A = \frac{1}{2}\int_C x\, dy - y\, dx,$$

which can often be easily evaluated, as shown in the following example.

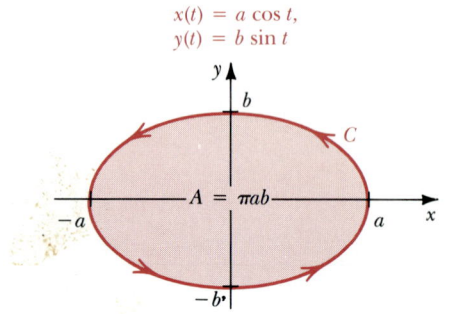

$x(t) = a\cos t,$
$y(t) = b\sin t$

$A = \pi ab$

Figure 16.35
The area is πab.

Example 5

Determine the area inside the ellipse described by

$$\frac{x^2}{a^2} + \frac{y^2}{b^2} = 1.$$

Solution
The ellipse has a parametric representation $x(t) = a\cos t$, $y(t) = b\sin t$, $0 \le t \le 2\pi$ (see Figure 16.35). Thus,

$$A = \frac{1}{2}\int_C x\, dy - y\, dx$$

$$= \frac{1}{2} \int_0^{2\pi} [(a \cos t)(b \cos t\, dt) - (b \sin t)(-a \sin t\, dt)]$$

$$= \frac{ab}{2} \int_0^{2\pi} [\cos^2 t + \sin^2 t]\, dt = \frac{ab}{2} \int_0^{2\pi} dt = ab\pi.$$

Another application of Green's Theorem involves the change-of-variables formula in double integration. Suppose $x = f(u, v)$ and $y = g(u, v)$ describes a one-to-one transformation of a bounded, simply connected region S in the uv-plane onto a bounded, simply connected region R in the xy-plane, where f and g have continuous partial derivatives (see Figure 16.36). If G is integrable on R, then

(16.49) $$\iint_R G(x, y)\, dy\, dx = \iint_S G(f(u, v), g(u, v))|J(u, v)|\, du\, dv,$$

where $J(u, v)$ is the **Jacobian** of x and y with respect to u and v defined by

$$J(u, v) = \frac{\partial x}{\partial u}\frac{\partial y}{\partial v} - \frac{\partial y}{\partial u}\frac{\partial x}{\partial v}.$$

We will show that (16.49) holds in the special case when $G(x, y) = 1$. The general case can be derived in a similar manner using the definition of the double integral.

To show that

$$\iint_R dy\, dx = \iint_S |J(u, v)|\, du\, dv,$$

let $M(x, y) = 0$ and $N(x, y) = x$. If C is the boundary of the region R, then Green's Theorem implies that

$$\iint_R dy\, dx = \iint_R \left(\frac{\partial N}{\partial x} - \frac{\partial M}{\partial y} \right) dy\, dx = \int_C M\, dx + N\, dy = \int_C x\, dy.$$

However,

$$x = f(u, v) \quad \text{and} \quad y = g(u, v),$$

so

$$dy = \frac{\partial g}{\partial u}(u, v)\, du + \frac{\partial g}{\partial v}(u, v)\, dv.$$

The conditions stated at the beginning of this discussion ensure that the boundary $\tilde{C}$ of S maps onto the boundary C of R (although the orientation may be reversed), so

$$\iint_R dy\, dx = \int_C x\, dy = \int_{\tilde{C}} f(u, v) \left(\frac{\partial g}{\partial u}(u, v)\, du + \frac{\partial g}{\partial v}(u, v)\, dv \right)$$

$$= \int_{\tilde{C}} f\frac{\partial g}{\partial u}\, du + f\frac{\partial g}{\partial v}\, dv.$$

We now apply Green's Theorem to the line integral over $\tilde{C}$ to convert it to a double integral over S. Care must be taken in doing this, however,

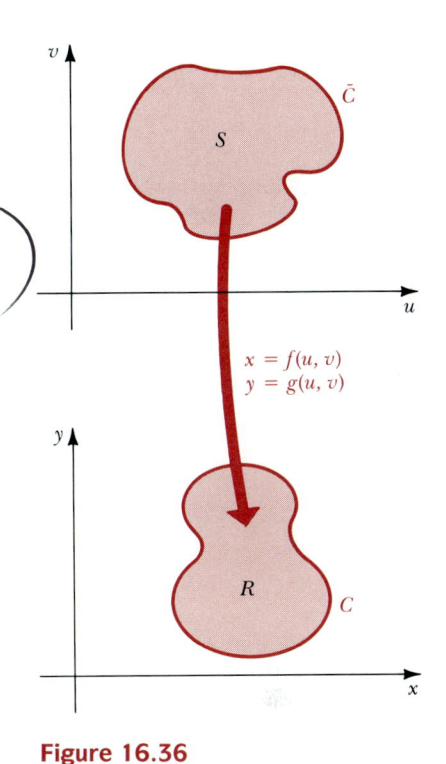

Figure 16.36
The region S is mapped onto the region R.

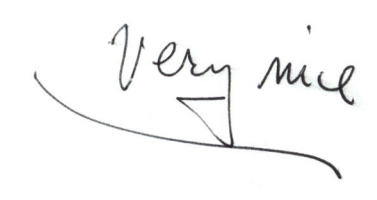

since the orientation of $\tilde{C}$ is not known. The original double integral over R is positive, so the resulting double integral over S must also be positive. To ensure that a positive value results for the double integral over S, we use the absolute value of the integrand in this double integral:

$$\int_{\tilde{C}} f\frac{\partial g}{\partial u}\,du + f\frac{\partial g}{\partial v}\,dv = \iint_S \left|\frac{\partial}{\partial u}\left(f\frac{\partial g}{\partial v}\right) - \frac{\partial}{\partial v}\left(f\frac{\partial g}{\partial u}\right)\right| du\,dv$$

$$= \iint_S \left|\frac{\partial f}{\partial u}\frac{\partial g}{\partial v} + f\frac{\partial^2 g}{\partial u\,\partial v} - \frac{\partial f}{\partial v}\frac{\partial g}{\partial u} - f\frac{\partial^2 g}{\partial v\,\partial u}\right| du\,dv.$$

But $\partial^2 g/\partial u\,\partial v = \partial^2 g/\partial v\,\partial u$, so

$$\iint_R dy\,dx = \int_{\tilde{C}} f\frac{\partial g}{\partial u}\,du + f\frac{\partial g}{\partial v}\,dv = \iint_S \left|\frac{\partial f}{\partial u}\frac{\partial g}{\partial v} - \frac{\partial f}{\partial v}\frac{\partial g}{\partial u}\right| du\,dv$$

$$= \iint_S |J(u,v)|\,du\,dv.$$

Example 6

Use the result in Equation (16.49) to verify the double integral change-of-variable formula from rectangular coordinates to polar coordinates.

Solution
Since $x = r\cos\theta$, $y = r\sin\theta$,

$$|J(r,\theta)| = \left|\frac{\partial x}{\partial r}\frac{\partial y}{\partial \theta} - \frac{\partial y}{\partial r}\frac{\partial x}{\partial \theta}\right| = |\cos\theta\,(r\cos\theta) - \sin\theta\,(-r\sin\theta)|$$

$$= |r\,(\cos^2\theta + \sin^2\theta)| = |r|.$$

We defined the double integral in polar coordinates only for regions S that can be expressed with $r > 0$. In this case

$$|J(r,\theta)| = |r| = r.$$

Thus if S is simply connected and R is the same region as S, but expressed in rectangular coordinates, then

$$\iint_R G(x,y)\,dy\,dx = \iint_S G(r,\theta)r\,dr\,d\theta.$$

▶

A similar change-of-variables formula holds for triple integrals. Suppose f, g, and h are functions with continuous partial derivatives that map a bounded simply connected region S in the variables u, v, and w onto a bounded simply connected region D in x, y, and z. If G is integrable on D, then

(16.50)

$$\iiint_D G(x,y,z)\,dx\,dy\,dz = \iiint_S G(f(u,v,w),g(u,v,w),h(u,v,w))|J(u,v,w)|\,du\,dv\,dw,$$

where

$$J(u, v, w) = \begin{vmatrix} \dfrac{\partial x}{\partial u} & \dfrac{\partial x}{\partial v} & \dfrac{\partial x}{\partial w} \\ \dfrac{\partial y}{\partial u} & \dfrac{\partial y}{\partial v} & \dfrac{\partial y}{\partial w} \\ \dfrac{\partial z}{\partial u} & \dfrac{\partial z}{\partial v} & \dfrac{\partial z}{\partial w} \end{vmatrix}.$$

▶ EXERCISE SET 16.5

1. Verify Green's Theorem for $\mathbf{F}(x, y) = x\mathbf{i} + xy\mathbf{j}$, where C is the positively oriented boundary of the square with vertices at $(0, 0)$, $(1, 0)$, $(1, 1)$, and $(0, 1)$.

2. Verify Green's Theorem for $\mathbf{F}(x, y) = x^2\mathbf{i} + y^2\mathbf{j}$ on the unit circle $x^2 + y^2 = 1$.

In Exercises 3–6, use Green's Theorem to find $\int_C \mathbf{F} \cdot d\mathbf{r}$, where C is positively oriented.

3. $\mathbf{F}(x, y) = 3y \sin^3 x\, \mathbf{i} + \cos^3 x\, \mathbf{j}$, C is the closed curve determined by $y = x^2$ and $y = x$

4. $\mathbf{F}(x, y) = 6x^2 y^3 \mathbf{i} - 6xy\mathbf{j}$, C is the triangle determined by $y = 0$, $x = -1$, and $x = y$

5. $\mathbf{F}(x, y) = \cos x \sin y\, \mathbf{i} + \sin x \cos y\, \mathbf{j}$, C is the ellipse $9x^2 + 4y^2 = 36$

6. $\mathbf{F}(x, y) = xy^2\mathbf{i} + 2x^2 y\mathbf{j}$, C is the cardioid described by $r = 1 + \cos\theta$

In Exercises 7–14, use Green's Theorem to evaluate the line integrals, assuming C is positively oriented.

7. $\int_C (x^2 + xy)\, dx + (y^2 + xy)\, dy$, C is the boundary of the square with vertices $(0, 0)$, $(2, 0)$, $(2, 2)$, and $(0, 2)$

8. $\int_C x^2 y\, dx + (x^3 - y^3)\, dy$, C is the triangle with vertices $(0, 0)$, $(2, 0)$, and $(1, 1)$

9. $\int_C (y^2 + x)\, dx + (x^2 + y)\, dy$, C is the closed curve determined by $y = x^2$ and $x = y^2$

10. $\int_C xe^y\, dx + xy\, dy$, C is the closed curve determined by $y = x^2$ and $y = x$

11. $\int_C (x - y^3)\, dx + x^3\, dy$, C is the unit circle with center at the origin

12. $\int_C (x + y)\, dx + xy\, dy$, C is the cardioid $r = 1 + \cos\theta$

13. $\int_C \dfrac{y}{x^2 + y^2}\, dx - \dfrac{x}{x^2 + y^2}\, dy$, C is the circle with equation $(x - 2)^2 + (y - 2)^2 = 1$

14. $\int_C x\, e^{x^2+y^2}\, dx - y\, e^{x^2+y^2}\, dy$, C is the boundary of the region in the first quadrant enclosed between the two circles with radii 1 and 3 and centered at the origin

In Exercises 15–20, use Green's Theorem to find the area of the given region.

15. Inside the circle $x^2 + y^2 = 16$

16. Inside the ellipse $4x^2 + y^2 = 4$

17. Inside the curve that is described by $x = \cos^3 t$, $y = \sin^3 t$, $0 \le t \le 2\pi$

18. Inside the curve described by $x = \cos t$, $y = \sin^3 t$, $0 \le t \le 2\pi$

19. Inside the curve described by $x = \cos^2 t$, $y = \sin t \cos t$, $0 \le t \le 2\pi$

20. Inside the curve described by $x = t^2$, $y = t^3 - 3t$, $-\sqrt{3} \le t \le \sqrt{3}$

21. Use Equation (16.46) to show the result of Example 5, that the area of the region bounded by the ellipse $x^2/a^2 + y^2/b^2 = 1$ is $ab\pi$.

22. Use (16.48) to verify that if a region D is described in polar coordinates by $0 \le r \le f(\theta)$, $\alpha \le \theta \le \beta$, then the area of D is

$$A = \dfrac{1}{2}\int_\alpha^\beta [f(\theta)]^2\, d\theta.$$

23. Explain why Green's Theorem cannot be used to evaluate

$$\int_C \dfrac{y}{x^2 + y^2}\, dx - \dfrac{x}{x^2 + y^2}\, dy,$$

where C is the circle $x^2 + y^2 = 1$.

24. Use Green's Theorem to prove Corollary (16.39): A continuous function $\mathbf{F}$ is conservative on D if and only if $\int_C \mathbf{F} \cdot d\mathbf{r} = 0$ for every simple closed curve C in D.

25. Show that the work done by a force $\mathbf{F}(x, y) = M(x)\mathbf{i} + N(y)\mathbf{j}$ around a simple closed curve is zero. [*Note:* Exercise 41 of Section 16.4 is a special case of this result.]

26. Assuming the hypothesis of Green's Theorem, show that the formula in the theorem can be written as

$$\int_C \mathbf{F} \cdot \mathbf{n}\, ds = \iint_R \text{div } \mathbf{F}\, dA,$$

where $\mathbf{n}$ is the outward unit normal to C.

27. Use Equation (16.50) to derive the appropriate change of triple integral formula when converting from
   a) rectangular coordinates to cylindrical coordinates,
   b) rectangular coordinates to spherical coordinates,
   c) cylindrical coordinates to spherical coordinates.

Putnam exercise

28. Let $\exp(t)$ denote e^t and

$$F(x) = \frac{x^4}{\exp(x^3)} \int_0^x \int_0^{x-u} \exp(u^3 + v^3)\, dv\, du.$$

Find $\lim_{x \to \infty} F(x)$ or prove that it does not exist. (This exercise was problem A–6 of the forty-fourth William Lowell Putnam examination given on December 3, 1983. The examination and its solution are in the October 1984 issue of the *American Mathematical Monthly,* pp. 487–495.)

16.6

Surface Integrals

A *surface integral* is the extension of the double integral to the situation when the domain of the integrand is a surface in space. The particular surfaces over which we will consider surface integrals are those with a representation as the graph of a function in one of the coordinate planes (see Figure 16.37). In this case, the surface integral can be converted into a double integral over the region in the coordinate plane.

Let S be a smooth surface described by $z = f(x, y)$, for (x, y) in R, a bounded, simply connected region in the xy-plane. Then S has a tangent plane at each point not on the boundary of S (see Figure 16.38).

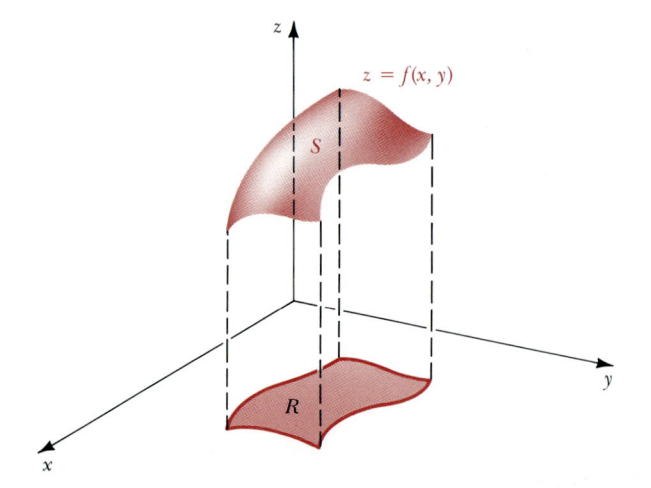

Figure 16.37
A smooth surface S.

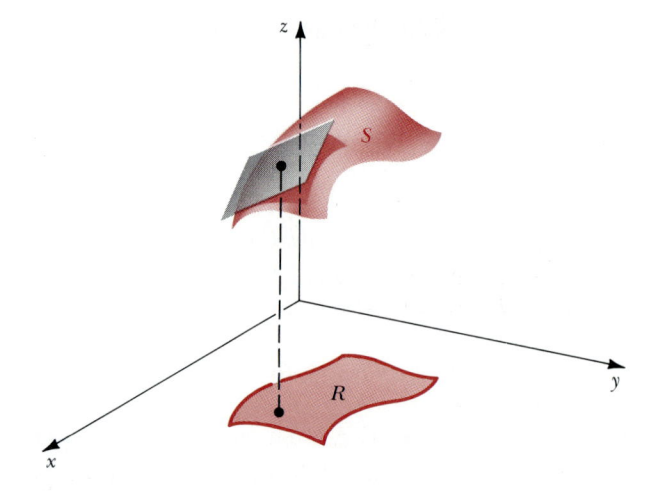

Figure 16.38
A tangent plane to S.

Suppose g is a continuous function defined on S. The surface integral of the function g over S is denoted

$$\iint\limits_S g(x, y, z)\, d\sigma.$$

Its definition follows the line of construction used in Section 15.5 to define surface area.

The bounded region R is contained in a rectangle determined by an x-coordinate interval $[a, b]$ and a y-coordinate interval $[c, d]$. A partition of each of these intervals produces a grid in the xy-plane, as shown in Figure 16.39. Let $R_1, R_2, \ldots, R_n$ denote the elements of this grid that lie entirely inside R and S_i denote the portion of the surface S that lies directly above the rectangle R_i (as shown in Figure 16.40). Let ΔS_i denote the area of S_i. The surface integral of g over the surface S is derived by considering sums of the form

$$\sum_{i=1}^{n} g(x_i, y_i, z_i)\, \Delta S_i.$$

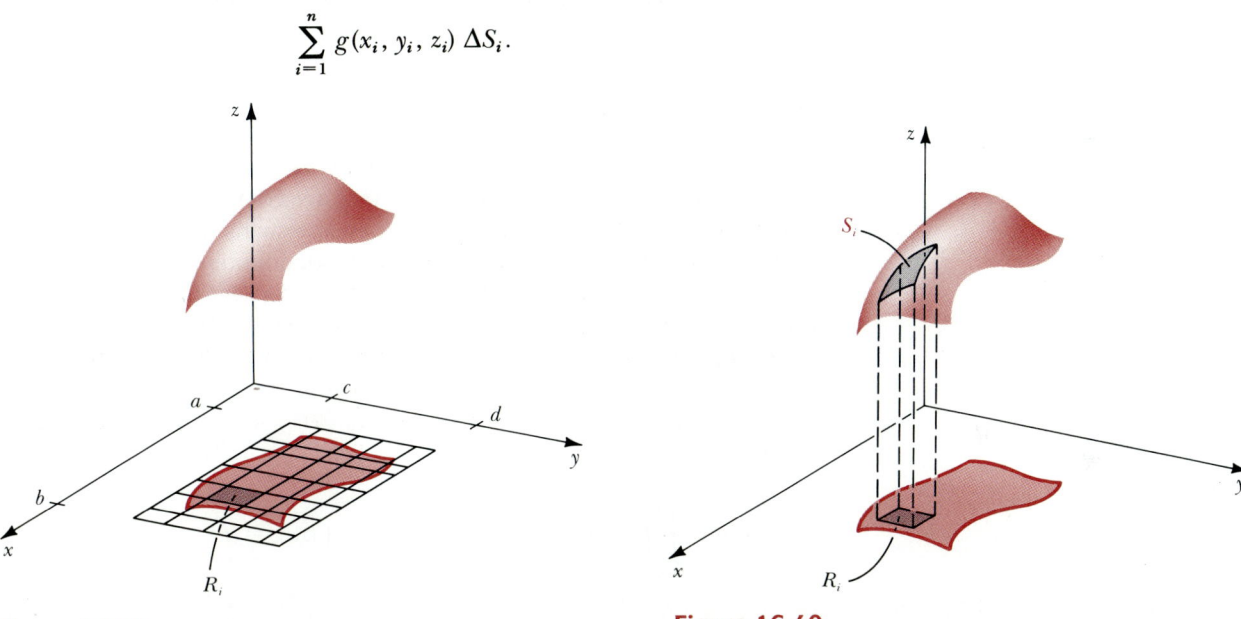

Figure 16.39

Figure 16.40
A surface element.

(16.51)
DEFINITION

Suppose S is a smooth surface and g is a continuous function defined on S. The **surface integral** of g over S is

$$\iint\limits_S g(x, y, z)\, d\sigma = \lim_{n \to \infty} \sum_{i=1}^{n} g(x_i, y_i, z_i)\, \Delta S_i.$$

To convert this surface integral into a double integral over R, the defining region for S, we let (x_i, y_i) be an arbitrary point in R_i. In Equation (15.17) of Section 15.5, we found that ΔS_i is approximated by that portion of the tangent plane to S at $(x_i, y_i, f(x_i, y_i))$ that lies above R_i, as shown in Figure 16.41, and that this approximation is given by

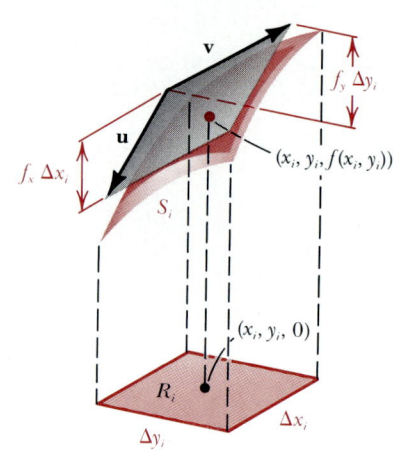

$$\Delta S_i \approx \{[f_x(x_i, y_i)]^2 + [f_y(x_i, y_i)]^2 + 1\}^{1/2} \Delta x_i \Delta y_i.$$

Thus,

$$\sum_{i=1}^{n} g(x_i, y_i, z_i) \Delta S_i \approx \sum_{i=1}^{n} g(x_i, y_i, f(x_i, y_i))\{[f_x(x_i, y_i)]^2 + [f_y(x_i, y_i)]^2 + 1\}^{1/2} \Delta x_i, \Delta y_i.$$

Letting Δx_i and Δy_i approach zero, we have the following result.

(16.52) THEOREM

Suppose f is a function with continuous partial derivatives defined on the bounded and simply connected region R in the xy-plane. If S is the smooth surface described by $z = f(x, y)$ and g is continuous on S, then

$$\iint_S g(x, y, z) \, d\sigma = \iint_R g(x, y, f(x, y))\{[f_x(x, y)]^2 + [f_y(x, y)]^2 + 1\}^{1/2} \, dA.$$

Note that if $g(x, y, z) = 1$ for all (x, y, z) on S, then this surface integral reduces to the integral for surface area defined in Section 15.5.

If $g(x, y, z)$ is a density function, then the surface integral gives the mass of the surface S. Using definitions analogous to those given in Section 15.8, we can find the moments with respect to the coordinate planes and the centroid of a surface. Exercises 19–21 give examples of these applications of surface integrals.

Example 1

Determine $\iint_S z \, d\sigma$, where S is described by $z = \sqrt{4 - x^2 - y^2}$.

Solution

The surface S is the hemisphere of radius 2 lying above the xy-plane, as shown in Figure 16.42. The corresponding region R in the xy-plane is the circular region $x^2 + y^2 \le 4$, which is described as an x-simple region by

$$-2 \le x \le 2, \quad -\sqrt{4 - x^2} \le y \le \sqrt{4 - x^2}.$$

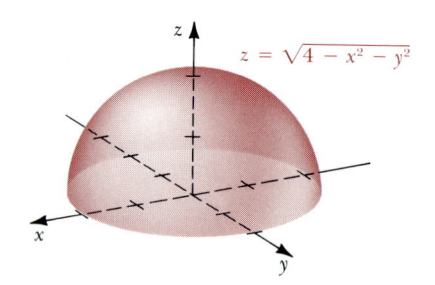

Thus,

$$\iint_S z \, d\sigma = \iint_R \sqrt{4 - x^2 - y^2} \left\{ \left(\frac{\partial \sqrt{4 - x^2 - y^2}}{\partial x} \right)^2 + \left(\frac{\partial \sqrt{4 - x^2 - y^2}}{\partial y} \right)^2 + 1 \right\}^{1/2} dA.$$

$$= \int_{-2}^{2} \int_{-\sqrt{4-x^2}}^{\sqrt{4-x^2}} \sqrt{4-x^2-y^2} \left[\frac{x^2}{4-x^2-y^2} + \frac{y^2}{4-x^2-y^2} + 1 \right]^{1/2} dy\, dx$$

$$= \int_{-2}^{2} \int_{-\sqrt{4-x^2}}^{\sqrt{4-x^2}} 2\, dy\, dx.$$

Cylindrical coordinates simplify the integration to

$$\iint_S z\, d\sigma = 2 \int_0^{2\pi} \int_0^2 r\, dr\, d\theta = \int_0^{2\pi} r^2 \Big]_0^2 d\theta$$

$$= \int_0^{2\pi} 4\, d\theta = 4\theta \Big]_0^{2\pi} = 8\pi. \quad \blacktriangleright$$

Surface integrals are also defined for continuous functions over smooth surfaces described in terms of a region in the yz- or xz-planes.

Example 2

Find $\iint_S xy \sin z\, d\sigma$, where S is the surface of the right circular cylinder $x^2 + y^2 = 4$ in the first octant with $0 \le z \le \pi$.

Solution

This surface, sketched in Figure 16.43, cannot be described in terms of a region in the xy-plane. It can, however, be described in terms of a region in the xz-plane. For

$$0 \le x \le 2, \quad 0 \le z \le \pi,$$

a point (x, y, z) is on the surface S precisely when

$$y = f(x, z) = \sqrt{4 - x^2}.$$

Thus,

$$\iint_S xy \sin z\, d\sigma = \iint_R xy \sin z \{[f_x(x,z)]^2 + [f_z(x,z)]^2 + 1\}^{1/2} dA$$

$$= \iint_R x\sqrt{4-x^2} \sin z \left[\left(\frac{\partial(\sqrt{4-x^2})}{\partial x} \right)^2 + \left(\frac{\partial(\sqrt{4-x^2})}{\partial z} \right)^2 + 1 \right]^{1/2} dA$$

$$= \int_0^{\pi} \int_0^2 x\sqrt{4-x^2} \sin z \left[\frac{x^2}{4-x^2} + 0 + 1 \right]^{1/2} dx\, dz$$

$$= \int_0^{\pi} \int_0^2 2x \sin z\, dx\, dz = \int_0^{\pi} (x^2 \sin z) \Big]_0^2 dz$$

$$= \int_0^{\pi} 4 \sin z\, dz = -4 \cos z \Big]_0^{\pi} = 4 - (-4) = 8. \quad \blacktriangleright$$

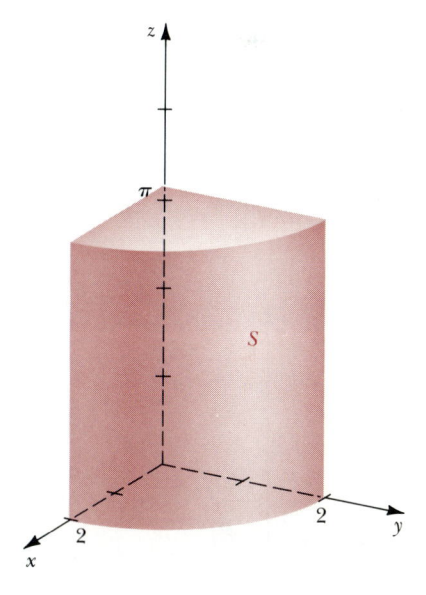

Figure 16.43

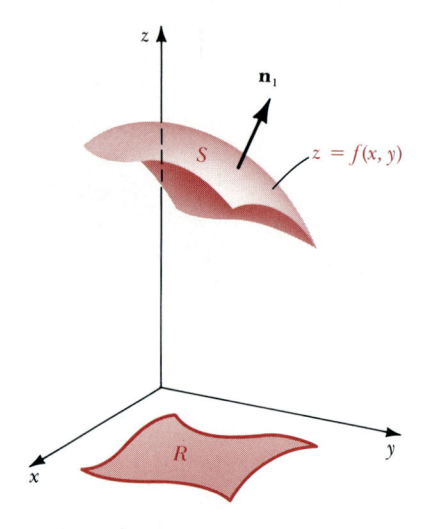

Surface integrals often occur in applications in the form

(16.53) $$\iint_S \mathbf{F} \cdot \mathbf{n}\, d\sigma,$$

where

$$\mathbf{F}(x, y, z) = M(x, y, z)\mathbf{i} + N(x, y, z)\mathbf{j} + P(x, y, z)\mathbf{k} \text{ and } \mathbf{n} = \mathbf{n}(x, y, z)$$

is a unit normal to S at (x, y, z). In many instances a surface integral of this form can be reduced to a double integral. For example, suppose S is a smooth surface described by $z = f(x, y)$ for (x, y) in R, a bounded simply connected region in the xy-plane. Suppose, in addition, that f has continuous partial derivatives on R. Then S can also be described by

$$g(x, y, z) = z - f(x, y) = 0.$$

Thus the gradient of g,

$$\nabla g(x, y, z) = -f_x(x, y)\mathbf{i} - f_y(x, y)\mathbf{j} + \mathbf{k},$$

is normal to S at (x, y, z) and unit normals to S at (x, y, z) are

(16.54) $\quad \mathbf{n}_1 = \dfrac{-f_x\mathbf{i} - f_y\mathbf{j} + \mathbf{k}}{\sqrt{f_x^2 + f_y^2 + 1}} \quad$ and $\quad \mathbf{n}_2 = \dfrac{f_x\mathbf{i} + f_y\mathbf{j} - \mathbf{k}}{\sqrt{f_x^2 + f_y^2 + 1}},$

shown in Figures 16.44 and 16.45, respectively. If $\mathbf{n}$ is directed upward, that is, has positive $\mathbf{k}$ component, then $\mathbf{n} = \mathbf{n}_1$ and (16.53) can be expressed as

$$\iint_S \mathbf{F} \cdot \mathbf{n}\, d\sigma = \iint_S (M\mathbf{i} + N\mathbf{j} + P\mathbf{k}) \cdot \left(\dfrac{-f_x\mathbf{i} - f_y\mathbf{j} + \mathbf{k}}{\sqrt{f_x^2 + f_y^2 + 1}}\right) d\sigma$$

$$= \iint_S (-Mf_x - Nf_y + P) \dfrac{1}{\sqrt{f_x^2 + f_y^2 + 1}}\, d\sigma$$

or, using Theorem (16.52),

(16.55) $$\iint_S \mathbf{F} \cdot \mathbf{n}\, d\sigma = \iint_R (-Mf_x - Nf_y + P)\, dA.$$

If $\mathbf{n}$ is directed downward, then $\mathbf{n} = \mathbf{n}_2$ and the result is

(16.56) $$\iint_S \mathbf{F} \cdot \mathbf{n}\, d\sigma = \iint_R (Mf_x + Nf_y - P)\, dA.$$

Figure 16.44
Upward directed normal:
$\mathbf{n}_1 = \dfrac{-f_x\mathbf{i} - f_y\mathbf{j} + \mathbf{k}}{\sqrt{f_x^2 + f_y^2 + 1}}.$

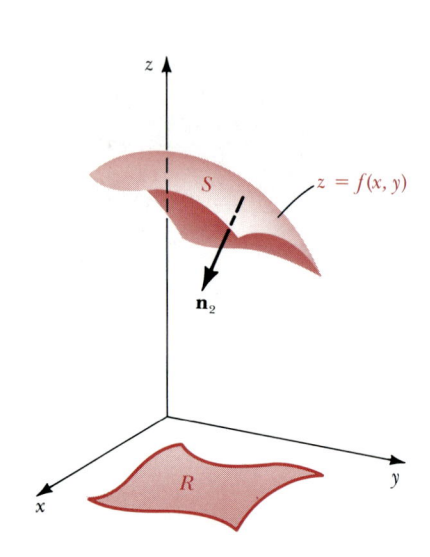

Figure 16.45
Downward directed normal:
$\mathbf{n}_2 = \dfrac{f_x\mathbf{i} + f_y\mathbf{j} - \mathbf{k}}{\sqrt{f_x^2 + f_y^2 + 1}}.$

Example 3

Determine $\iint_S \mathbf{F} \cdot \mathbf{n}\, d\sigma$ if $\mathbf{F}(x, y, z) = x\mathbf{i} + y\mathbf{j} + z\mathbf{k}$, S is the surface that bounds the hemispherical solid D with base in the xy-plane and top lying on $x^2 + y^2 + z^2 = 1$, $z \geq 0$, and $\mathbf{n}$ points outward from the solid.

Solution
The solid D is shown in Figure 16.46(a). The upper portion S_1 of the surface bounding this solid is described by
$$z = \sqrt{1 - x^2 - y^2}, \quad \text{for } (x, y) \text{ satisfying } 0 \leq x^2 + y^2 \leq 1.$$
The base S_2 of the solid described by
$$z = 0, \quad \text{for } (x, y) \text{ satisfying } 0 \leq x^2 + y^2 \leq 1.$$
The outward-pointing normal to S_1 is directed upward, as shown in Figure 16.46(b); so by (16.55),
$$\iint_{S_1} \mathbf{F} \cdot \mathbf{n} \, d\sigma = \iint_R (-xf_x(x, y) - yf_y(x, y) + z) \, dA$$
$$= \iint_R \left[-x \left(\frac{-x}{\sqrt{1 - x^2 - y^2}} \right) - y \left(\frac{-y}{\sqrt{1 - x^2 - y^2}} \right) + \sqrt{1 - x^2 - y^2} \right] dA$$
$$= \iint_R \frac{x^2 + y^2 + (1 - x^2 - y^2)}{\sqrt{1 - x^2 - y^2}} \, dA = \iint_R \frac{1}{\sqrt{1 - x^2 - y^2}} \, dA.$$

Since R is the region in the xy-plane with $x^2 + y^2 \leq 1$,
$$\iint_{S_1} \mathbf{F} \cdot \mathbf{n} \, d\sigma = \int_{-1}^{1} \int_{-\sqrt{1-x^2}}^{\sqrt{1-x^2}} \frac{1}{\sqrt{1 - x^2 - y^2}} \, dy \, dx.$$

Changing to cylindrical coordinates gives us
$$\iint_{S_1} \mathbf{F} \cdot \mathbf{n} \, d\sigma = \int_0^{2\pi} \int_0^1 \frac{1}{\sqrt{1 - r^2}} r \, dr \, d\theta$$
$$= \int_0^{2\pi} -(1 - r^2)^{1/2} \Big]_0^1 d\theta$$
$$= \int_0^{2\pi} d\theta = \theta \Big]_0^{2\pi} = 2\pi.$$

The outward pointing normal to S_2 is $-\mathbf{k}$, which points downward (see Figure 16.46c), so (16.56) implies that
$$\iint_{S_2} \mathbf{F} \cdot \mathbf{n} \, d\sigma = \iint_R (x \cdot 0 + y \cdot 0 - z) \, dA = -\iint_R z \, dA.$$

But $z = 0$ on S_2, so $\iint_{S_2} \mathbf{F} \cdot \mathbf{n} \, d\sigma = 0$ and
$$\iint_S \mathbf{F} \cdot \mathbf{n} \, d\sigma = \iint_{S_1} \mathbf{F} \cdot \mathbf{n} \, d\sigma + \iint_{S_2} \mathbf{F} \cdot \mathbf{n} \, d\sigma = 2\pi + 0 = 2\pi.$$ ▶

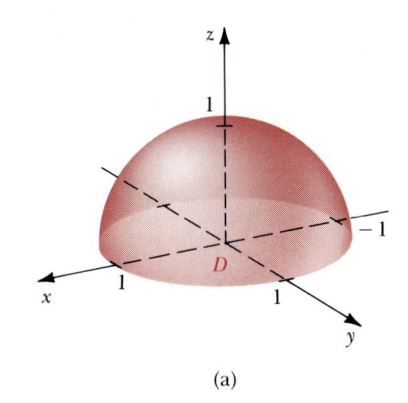

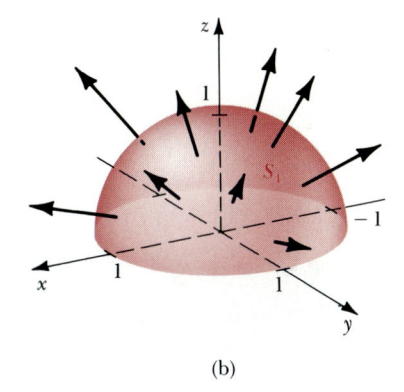

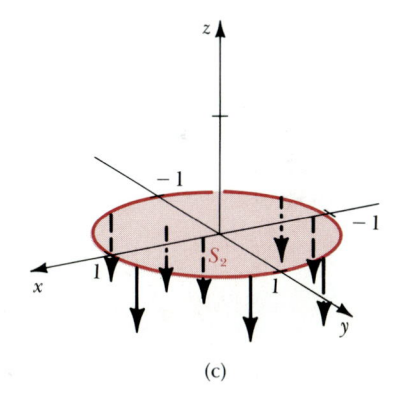

Figure 16.46

The surfaces considered in this section have been *orientable*. This means that there are two distinguishable sides to the surface, sides that could be painted different colors, for example. An *orientation* on an orientable surface provides a way to distinguish one of the sides from

(a)

(b)

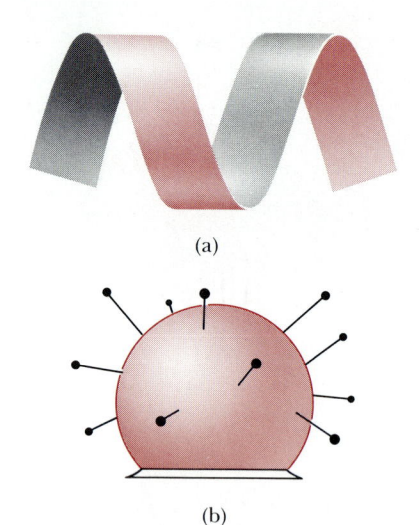

Figure 16.47
Oriented surfaces.

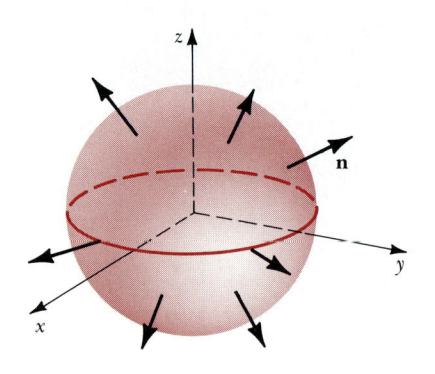

Figure 16.48
Outward-directed normal vectors.

the other. In Figure 16.47(a) the colored side provides an orientation for the surface of the ribbon that distinguishes this side from the other. In Figure 16.47(b) the heads of the pins provide an orientation for the surface of the pincushion. To describe this concept analytically, we use the following definition.

> **(16.57)**
> **DEFINITION**
> A smooth surface S is said to be **orientable** if there is a unit normal function $\mathbf{n} = \mathbf{n}(x, y, z)$ that is continuous on the interior of S. Choosing a particular normal function $\mathbf{n}$ provides an **orientation** for S. A piecewise smooth surface is said to be orientable if each of its smooth components is orientable.

Most common surfaces are orientable. For example, the sphere

$$x^2 + y^2 + z^2 = 1$$

is orientable with an orientation described by the *outward normals* to the surface, defined by

$$\mathbf{n} = \mathbf{n}(x, y, z) = x\mathbf{i} + y\mathbf{j} + z\mathbf{k}$$

and shown in Figure 16.48. The opposite orientation to the sphere is given by the inward normal vectors $-\mathbf{n}$.

A surface that is not orientable is the Möbius strip, named after Augustus Ferdinand Möbius (1790–1868). This is a one-sided surface constructed by taking a strip of paper, giving one end a twist, and taping the ends of the strip together, as shown in Figure 16.49.

M. C. Escher's print *Möbius Strip II*, shown in Figure 16.50, illustrates the one-sided nature of the surface. If this surface were orientable and the ant at the top of the figure were moved continuously along the strip, it should remain above the strip after each rotation. Follow the path of an individual ant and you will find that this is not the case.

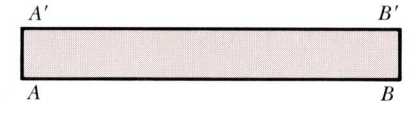

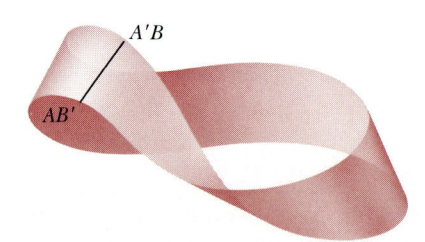

Figure 16.49
Constructing a Möbius strip.

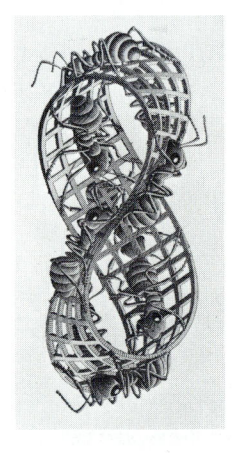

Figure 16.50
Möbius Strip II: M. C. Escher.

▶ EXERCISE SET 16.6

In Exercises 1–12, evaluate the surface integral $\iint_S g(x, y, z) \, d\sigma$.

1. $g(x, y, z) = x^2 + y^2 + z$; S is described by $0 \le x \le 1$, $0 \le y \le 2$, $z = 2$

2. $g(x, y, z) = x^2 y^2 z^2$; S is described by $z = 3$, $0 \le x \le 2$, $-2 \le y \le 2$

3. $g(x, y, z) = xyz$; S is the portion of the plane $x + 2y + 3z = 6$ lying in the first octant

4. $g(x, y, z) = e^x + 2yz$; S is the portion of the plane $2x - y + z = 4$ bounded by the coordinate planes

5. $g(x, y, z) = 8$; S is the portion of the cone $z = \sqrt{x^2 + y^2}$ that lies inside the cylinder $x^2 + y^2 = 1$

6. $g(x, y, z) = \sqrt{x^2 + y^2}$; S is the surface described in Exercise 5

7. $g(x, y, z) = x/\sqrt{x^2 + y^2}$; S is the portion of the paraboloid $z = 4 - x^2 - y^2$ that lies in the first octant

8. $g(x, y, z) = xyz$; S is the portion of the plane $x + y - 2z = 6$ that lies below the triangle in the xy-plane with vertices $(2, 0, 0)$, $(0, 1, 0)$, and $(0, 0, 0)$

9. $g(x, y, z) = x$; S is the portion above the xy-plane of the cylinder $x^2 + z^2 = 4$, when $0 \le y \le 1$

10. $g(x, y, z) = xy$; S is described by $y^2 + z^2 = 4$, $0 \le x \le 4$, $z \ge 0$

11. $g(x, y, z) = x^2 + y^2 + z^2$; S is the hemisphere $z = \sqrt{9 - x^2 - y^2}$

12. $g(x, y, z) = z(x^2 + y^2)$; S is the cube determined by the planes $x = -1$, $x = 1$, $y = -1$, $y = 1$, $z = -1$, and $z = 1$

In Exercises 13–16, determine $\iint_S \mathbf{F} \cdot \mathbf{n} \, d\sigma$.

13. $\mathbf{F}(x, y, z) = x\mathbf{i} + y\mathbf{j} + z\mathbf{k}$; S is the portion of the surface $z = xy + 1$ that lies above the square $0 \le x \le 1$, $0 \le y \le 1$ in the xy-plane; $\mathbf{n}$ is directed upward

14. $\mathbf{F}(x, y, z) = yz\mathbf{i} + xz\mathbf{j} + xy\mathbf{k}$; S is the portion of the paraboloid $z = x^2 + y^2$ that lies inside the cylinder $x^2 + y^2 = 1$; $\mathbf{n}$ is directed downward

15. $\mathbf{F}(x, y, z) = x^2\mathbf{i} + y^2\mathbf{j} + z\mathbf{k}$; S is the portion of the plane $z = x + y + 1$ that lies above the triangle $0 \le x \le 1$, $0 \le y \le x$; $\mathbf{n}$ is directed downward

16. $\mathbf{F}(x, y, z) = x\mathbf{i} - y\mathbf{j} + z^2\mathbf{k}$; S is the sphere $x^2 + y^2 + z^2 = 4$; $\mathbf{n}$ is directed outward from the sphere.

A thin metal sheet in the shape of a surface S with area density $g(x, y, z)$ has mass and moments with respect to the coordinate planes given by

$$M = \iint_S g(x, y, z) \, d\sigma, \quad M_{xy} = \iint_S zg(x, y, z) \, d\sigma,$$

$$M_{xz} = \iint_S yg(x, y, z) \, d\sigma, \quad M_{yz} = \iint_S xg(x, y, z) \, d\sigma.$$

The center of mass is $(\bar{x}, \bar{y}, \bar{z})$, where

$$\bar{x} = \frac{M_{yz}}{M}, \quad \bar{y} = \frac{M_{xz}}{M}, \quad \text{and} \quad \bar{z} = \frac{M_{xy}}{M}.$$

17. Find the mass of the hemispherical shell described by $z = \sqrt{4 - x^2 - y^2}$ if its density at each point is given by $g(x, y, z) = 4 - z$ (see the figure).

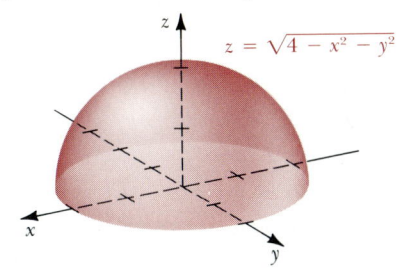

18. Find the center of mass of the hemispherical shell that has constant density 1 and is described by the function $z = \sqrt{4 - x^2 - y^2}$.

19. Find the center of mass of a thin metal sheet with density $g(x, y, z) = x + y + z$ that has the shape of the surface described by

$$z = 2 + x, \quad 0 \le y \le 4, \quad 0 \le x \le 3.$$

20. When a gas or fluid flows with velocity $\mathbf{v}$ through a surface S, the volume flow and mass flow rates through S are given by $\iint_S \mathbf{v} \cdot \mathbf{n} \, d\sigma$ and $\iint_S \rho \mathbf{v} \cdot \mathbf{n} \, d\sigma$, respectively, where ρ is the constant density of the gas or fluid and $\mathbf{n}$ is the unit vector in the direction of the flow. Find these rates for a conical shrimp net of radius 1 ft and length 3 ft that is being brought to the surface of the water from a depth of 50 ft at the rate of 3 ft/sec (see the figure).

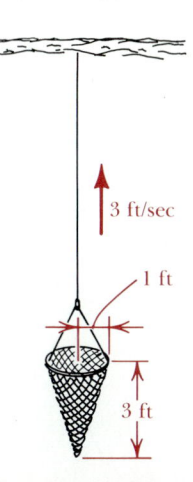

16.7

The Divergence Theorem

In this section we consider the Divergence Theorem of Gauss. This theorem relates a triple integral over a region in space to a surface integral over the boundary of the region. It is similar in this sense to Green's Theorem, which relates a double integral over a region in the plane to a line integral over its boundary.

(16.58)
THEOREM: THE DIVERGENCE THEOREM
Suppose D is a solid region in space bounded by the piecewise smooth orientable surface S. If the component functions of $\mathbf{F}$ have continuous partial derivatives and $\mathbf{n}$ is the outward unit normal to S, then

$$\iint_S \mathbf{F} \cdot \mathbf{n} \, d\sigma = \iiint_D \operatorname{div} \mathbf{F} \, dV.$$

Proof. Let

$$\mathbf{F}(x, y, z) = M(x, y, z)\mathbf{i} + N(x, y, z)\mathbf{j} + P(x, y, z)\mathbf{k}$$

and write $\mathbf{n}$ in terms of its direction cosines,

$$\mathbf{n} = \cos \alpha \, \mathbf{i} + \cos \beta \, \mathbf{j} + \cos \gamma \, \mathbf{k}.$$

The equation in the Divergence Theorem can then be expressed as

(16.59)
$$\iint_S (M \cos \alpha + N \cos \beta + P \cos \gamma) \, d\sigma = \iiint_D (M_x + N_y + P_z) \, dV.$$

The proof consists of showing that

$$\iint_S M \cos \alpha \, d\sigma = \iiint_D M_x \, dV,$$

$$\iint_S N \cos \beta \, d\sigma = \iiint_D N_y \, dV$$

and

(16.60)
$$\iint_S P \cos \gamma \, d\sigma = \iiint_D P_z \, dV.$$

We will prove (16.60) for regions described by

(16.61) $\qquad f_1(x, y) \leq z \leq f_2(x, y),$

for (x, y) in a simply connected region R in the xy-plane with boundary C, and for functions f_1 and f_2 with continuous partial derivatives on R. The proof of the other equalities follows by assuming that D has a similar projection representation in the yz- and xz-planes, respectively. Once the result is established for this case, it can be extended to a more general solid by decomposing the solid into subregions of this form.

The surface of the solid region described in (16.61) is decomposed into three smooth surfaces, as shown in Figure 16.51:

the bottom, S_1: $z = f_1(x, y)$, for (x, y) in R;
the top, S_2: $z = f_2(x, y)$, for (x, y) in R;
the sides, S_3: $f_1(x, y) \leq z \leq f_2(x, y)$ for (x, y) on C.

On S_3 the normal vectors are always parallel to the xy-plane, so $\gamma = \pi/2$, $\cos \gamma = 0$, and

$$\iint_{S_3} P \cos \gamma \, d\sigma = 0.$$

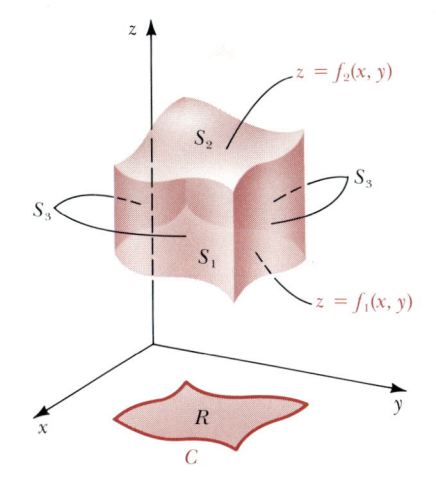

Figure 16.51

On S_2 the outward normal vectors are directed upward. It follows from (16.54) and (16.55) in Section 16.6 that

$$\cos \gamma = \frac{1}{\sqrt{f_x^2 + f_y^2 + 1}}$$

and

$$\iint_{S_2} P \cos \gamma \, d\sigma = \iint_R P \cos \gamma \sqrt{f_x^2 + f_y^2 + 1} \, dA$$
$$= \iint_R P(x, y, f_2(x, y)) \, dA.$$

On S_1 the outward normal vectors are directed downward. It follows from (16.54) and (16.56) that

$$\cos \gamma = \frac{-1}{\sqrt{f_x^2 + f_y^2 + 1}}$$

and

$$\iint_{S_1} P \cos \gamma \, dA = \iint_R -P(x, y, f_1(x, y)) \, dA.$$

Thus,

$$\iint_S P \cos \gamma \, d\sigma = \iint_{S_1} P \cos \gamma \, d\sigma + \iint_{S_2} P \cos \gamma \, d\sigma + \iint_{S_3} P \cos \gamma \, d\sigma$$
$$= \iint_R [P(x, y, f_2(x, y)) - P(x, y, f_1(x, y))] \, dA.$$

However,

$$\iiint_D P_z \, dV = \iint_R \left[\int_{z=f_1(x,y)}^{z=f_2(x,y)} P_z \, dz \right] dA$$

$$= \iint_R [P(x, y, f_2(x, y)) - P(x, y, f_1(x, y))] \, dA,$$

so

$$\iint_S P \cos \gamma \, d\sigma = \iiint_D P_z \, dV.$$

▷

Example 1

Let $\mathbf{F}(x, y, z) = x\mathbf{i} + y\mathbf{j} + z\mathbf{k}$, let S be the surface of the region bounded between the graphs of $z = \sqrt{2 - x^2 - y^2}$ and $z = \sqrt{x^2 + y^2}$, and let $\mathbf{n}$ be the outward normal vector to S. Find the value of the surface integral $\iint_S \mathbf{F} \cdot \mathbf{n} \, d\sigma$ (a) directly and (b) using the Divergence Theorem.

Solution

The intersection of the graphs bounding the region occurs when

$$x^2 + y^2 = 2 - x^2 - y^2, \quad \text{that is, when} \quad x^2 + y^2 = 1.$$

a) The direct evaluation of the integral requires the evaluation of

$$\iint_{S_1} \mathbf{F} \cdot \mathbf{n} \, d\sigma,$$

where S_1 is described by $z = \sqrt{2 - x^2 - y^2}$ for $x^2 + y^2 \leq 1$, and the evaluation of

$$\iint_{S_2} \mathbf{F} \cdot \mathbf{n} \, d\sigma,$$

where S_2 is described by $z = \sqrt{x^2 + y^2}$ for $x^2 + y^2 \leq 1$.

The outward normal to the surface S_1 is directed upward. The region R over which S_1 is defined is bounded by the circle $x^2 + y^2 = 1$ and is described in polar coordinates by

$$0 \leq \theta \leq 2\pi, \quad 0 \leq r \leq 1.$$

So Equation (16.55) implies that

$$\iint_{S_1} \mathbf{F} \cdot \mathbf{n} \, d\sigma = \iint_R \left[-x \left(\frac{-x}{\sqrt{2 - x^2 - y^2}} \right) - y \left(\frac{-y}{\sqrt{2 - x^2 - y^2}} \right) + \sqrt{2 - x^2 - y^2} \right] dA$$

$$= \iint_R \frac{2}{\sqrt{2 - x^2 - y^2}} \, dA$$

$$= \int_0^{2\pi} \int_0^1 \frac{2}{\sqrt{2-r^2}} \, r \, dr \, d\theta$$

$$= \int_0^{2\pi} \left[-2\sqrt{2-r^2} \right]_0^1 d\theta$$

$$= \int_0^{2\pi} (-2 + 2\sqrt{2}) \, d\theta = 4\pi(\sqrt{2} - 1).$$

The outward normal to the surface S_2 is directed downward, so Equation (16.56) implies that

$$\iint_{S_2} \mathbf{F} \cdot \mathbf{n} \, d\sigma = \iint_R \left[x \frac{x}{\sqrt{x^2+y^2}} + y \frac{y}{\sqrt{x^2+y^2}} - \sqrt{x^2+y^2} \right] dA$$

$$= \iint_R 0 \, dA = 0.$$

Thus direct evaluation gives

$$\iint_S \mathbf{F} \cdot \mathbf{n} \, d\sigma = \iint_{S_1} \mathbf{F} \cdot \mathbf{n} \, d\sigma + \iint_{S_2} \mathbf{F} \cdot \mathbf{n} \, d\sigma = 4\pi(\sqrt{2} - 1).$$

b) To use the Divergence Theorem, first note that div $\mathbf{F}(x, y, z) = 1 + 1 + 1 = 3$. The triple integral over D is evaluated using spherical coordinates. Note that in Figure 16.52

$$0 \leq \rho \leq \sqrt{2}, \qquad 0 \leq \theta \leq 2\pi.$$

Since $z = \sqrt{x^2 + y^2}$ intersects the yz-plane in the line $z = y$, the limits on ϕ are $0 \leq \phi \leq \pi/4$. Therefore,

$$\iint_S \mathbf{F} \cdot \mathbf{n} \, d\sigma = \iiint_D 3 \, dV$$

$$= \int_0^{2\pi} \int_0^{\pi/4} \int_0^{\sqrt{2}} 3 \rho^2 \sin \phi \, d\rho \, d\phi \, d\theta$$

$$= 2\pi \left[\rho^3 \right]_0^{\sqrt{2}} \left[-\cos \phi \right]_0^{\pi/4}$$

$$= 2\pi(2\sqrt{2}) \left(-\frac{\sqrt{2}}{2} + 1 \right) = 4\pi(\sqrt{2} - 1).$$

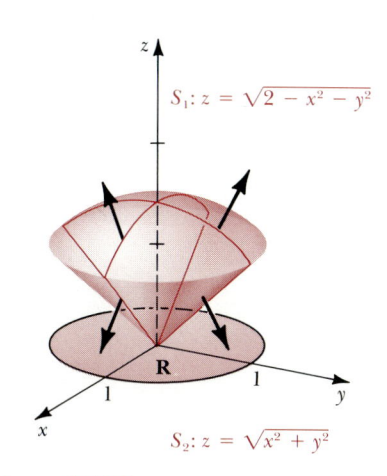

▶ **Figure 16.52**

HISTORICAL NOTE

The Divergence Theorem was discovered independently by **Carl Friedrich Gauss** (1777–1855) in 1812 in a memoir on potential theory, and by **Mikhail Vasilevich Ostrogradskii** (1801–1862) in a study on heat conduction in 1828 (published in 1831). It was also implicit in the 1828 essay of George Green, and a similar type of reduction of volume integrals to surface integrals was used by Lagrange in a memoir of 1760–1761.

The Divergence Theorem, like Stokes's Theorem, is of fundamental importance both in mathematics and mathematical physics (electromagnetic theory, potential theory, and fluid mechanics). In European books, it is often called the Gauss–Ostrogradskii Theorem.

Example 2

Suppose
$$\mathbf{F}(x, y, z) = (y^2 + z)\mathbf{i} + (x^2 - 2y + z)\mathbf{j} + (x + 2y)\mathbf{k}$$

and $\mathbf{n}$ is the outward normal to the surface of the solid bounded by $x^2 + y^2 = 1$, $z = 0$, and $y + z = 2$. Use the Divergence Theorem to evaluate $\iint \mathbf{F} \cdot \mathbf{n}\, d\sigma$.

Solution

The solid and boundary surfaces are shown in Figure 16.53. Since
$$\nabla \cdot \mathbf{F}(x, y, z) = \frac{\partial}{\partial x}(y^2 + z) + \frac{\partial}{\partial y}(x^2 - 2y + z) + \frac{\partial}{\partial z}(x + 2y) = -2,$$

we have
$$\iint_S \mathbf{F} \cdot \mathbf{n}\, d\sigma = \iiint_D \nabla \cdot \mathbf{F}\, dv = \iiint_D -2\, dv.$$

The region D is described most easily in cylindrical coordinates:
$$0 \le \theta \le 2\pi,\ 0 \le r \le 1,\ 0 \le z \le 2 - y = 2 - r\sin\theta.$$

So
$$\iint_S \mathbf{F} \cdot \mathbf{n}\, d\sigma = \int_0^{2\pi} \int_0^1 \int_0^{2-r\sin\theta} -2r\, dz\, dr\, d\theta$$
$$= -2 \int_0^{2\pi} \int_0^1 r(2 - r\sin\theta)\, dr\, d\theta$$
$$= -2 \int_0^{2\pi} \left[r^2 - \frac{r^3}{3}\sin\theta\right]_0^1 d\theta$$
$$= -2 \int_0^{2\pi} \left[1 - \frac{1}{3}\sin\theta\right] d\theta$$
$$= -2 \left[\theta + \frac{1}{3}\cos\theta\right]_0^{2\pi} = -4\pi. \blacktriangleright$$

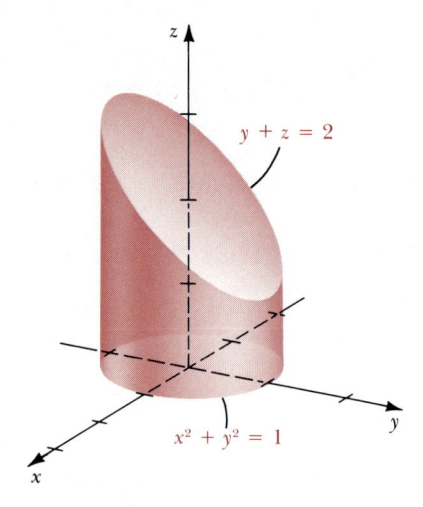

Figure 16.53

(figure shows cylinder $x^2 + y^2 = 1$ cut by plane $y + z = 2$)

▶ EXERCISE SET 16.7

In Exercises 1–8, use the Divergence Theorem to evaluate $\iint_S \mathbf{F} \cdot \mathbf{n}\, d\sigma$, where $\mathbf{n}$ is the outward unit normal to S.

1. $\mathbf{F}(x, y, z) = x\mathbf{i} + y\mathbf{j} + z\mathbf{k}$; S is the cylinder $x^2 + y^2 = 4$, $0 \le z \le 4$.

2. $\mathbf{F}(x, y, z) = xz\mathbf{i} + yz\mathbf{j} + xy\mathbf{k}$; S is the cylinder $y^2 + z^2 = 1$, $0 \le x \le 2$.

3. $\mathbf{F}(x, y, z) = \mathbf{i} + \mathbf{j} + z\mathbf{k}$; S is the boundary of the region inside the paraboloid $z = x^2 + y^2$ and below the plane $z = 4$.

4. $\mathbf{F}(x, y, z) = x^3\mathbf{i} + y^3\mathbf{j} + z^3\mathbf{k}$; S is the hemisphere $z = \sqrt{1 - x^2 - y^2}$.

5. $\mathbf{F}(x, y, z) = e^x\mathbf{i} - ye^x\mathbf{j} + z^2\mathbf{k}$; S is the boundary of the region in the intersection of the cylinders $x^2 + y^2 = 4$ and $y^2 + z^2 = 4$

6. $\mathbf{F}(x, y, z) = x \sin y \mathbf{i} + e^x \mathbf{j} + z \cos y \, \mathbf{k}$; S is the boundary of the cube described by $0 \leq x \leq 2$, $0 \leq y \leq 2$, $0 \leq z \leq 2$

7. $\mathbf{F}(x, y, z) = (x + e^y)\mathbf{i} + (xy + \sin z)\mathbf{j} - (xz + \sqrt{xy})\mathbf{k}$; S is the boundary of the tetrahedron bounded by the planes $x = 0$, $y = 0$, $z = 0$, and $2x + 3y + z = 6$

8. $\mathbf{F}(x, y, z) = yz\mathbf{i} + xy\mathbf{j} + (1 - z - xz)\mathbf{k}$; S is the boundary of the region inside the paraboloid $z = 1 - x^2 - y^2$ and above the xy-plane

9. Let D be a solid with unit density, and let S be the surface of D with outward normal $\mathbf{n}$.
   a) Show that
   $$\iint_S x^2 \mathbf{i} \cdot \mathbf{n} \, d\sigma = 2 M_{yz},$$
   where M_{yz} is the moment with respect to the yz-plane.
   b) Express M_{xy} and M_{xz} as surface integrals.

10. Let D be a solid in space, and let S be the surface of D with outward-directed normal $\mathbf{n}$. Show that the volume of D is given by the surface integral
    $$\iint_S \frac{1}{3} (x\mathbf{i} + y\mathbf{j} + z\mathbf{k}) \cdot \mathbf{n} \, d\sigma.$$

11. Let S be the unit sphere oriented by the outward-directed normal $\mathbf{n}$, and let $\mathbf{F}$ be a vector function whose component functions have continuous partial derivatives on S. Use the Divergence Theorem to show that
    $$\iint_S \text{curl } \mathbf{F} \cdot \mathbf{n} \, d\sigma = 0.$$

12. If f and g are twice continuously differentiable on a solid D with boundary S and outward-directed normal $\mathbf{n}$, show that
    i) $\iint_S f \nabla g \cdot \mathbf{n} \, d\sigma = \iiint_D (f \nabla^2 g + \nabla f \cdot \nabla g) \, dV;$
    ii) $\iint_S (f \nabla g - g \nabla f) \cdot \mathbf{n} \, d\sigma = \iiint_D (f \nabla^2 g - g \nabla^2 f) \, dV.$

13. Let S be a piecewise smooth surface that bounds a region D in space.
    a) Show that if a function f of three variables has continuous partial derivatives on D, then
    $$\iiint_D \left(\frac{\partial^2 f}{\partial x^2} + \frac{\partial^2 f}{\partial y^2} + \frac{\partial^2 f}{\partial z^2} \right) dV = \iint_S D_\mathbf{n} f \, d\sigma,$$
    where $\mathbf{n}$ denotes the outward unit normal to S and $D_\mathbf{n} f$ is the directional derivative of f.
    b) Show that if f is harmonic throughout D, then
    $$\iint_S D_\mathbf{n} f \, d\sigma = 0.$$

14. Let S be a piecewise smooth surface that bounds a region D in space.
    a) Show that if a function f of three variables is harmonic throughout D, then
    $$\iiint_D \left[\left(\frac{\partial f}{\partial x} \right)^2 + \left(\frac{\partial f}{\partial y} \right)^2 + \left(\frac{\partial f}{\partial z} \right)^2 \right] dV = \iint_S f D_\mathbf{n} f \, d\sigma,$$
    where $\mathbf{n}$ is the outward unit normal of S and $D_\mathbf{n} f$ is the directional derivative of f.
    b) Show that if f is zero on S, then f must be zero in D.

15. Use Exercise 14 to show that if two harmonic functions agree on the boundary of a solid, then they must agree at each point within the solid.

16.8

Stokes's Theorem

Green's Theorem relates a double integral over a region in the xy-plane to a line integral over the boundary of the region. In this section we consider a generalization of Green's Theorem to the situation where the region is an orientable surface in space. Before presenting this result, we need to relate the orientation of the surface to the orientation of the bounding curve.

The orientation of an orientable surface bounded by a simple closed

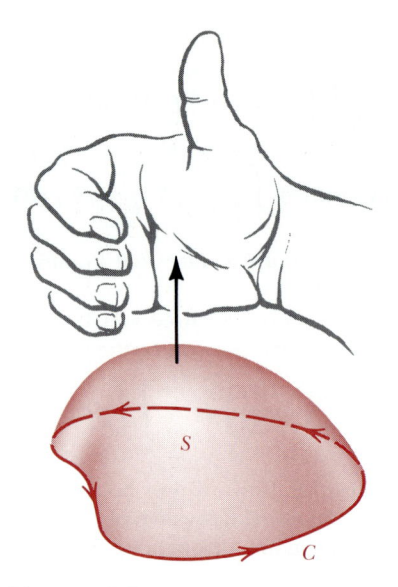

Figure 16.54
The orientation of C induced by the orientation of S.

curve induces a natural orientation on the curve. The boundary orientation is *positive* with respect to the surface orientation if the head of a person walking along the boundary of the surface points in the direction of the orientation. This establishes a right-hand rule between the surface orientation and the orientation of its boundary: If the fingers of the right hand are moved in the direction of the positive orientation on the boundary, the extended thumb points in the direction of the orientation of the surface (see Figure 16.54).

(16.62)
THEOREM: STOKES'S THEOREM
Let S be a smooth oriented surface whose boundary C is a smooth simple closed curve with positive orientation with respect to S. Suppose **F** has component functions with continuous partial derivatives. Then

$$\int_C \mathbf{F} \cdot d\mathbf{r} = \iint_S \text{curl } \mathbf{F} \cdot \mathbf{n} \, d\sigma,$$

where **n** denotes the unit normal to S in the direction of the orientation on S and $\mathbf{r}(t) = x(t)\mathbf{i} + y(t)\mathbf{j} + z(t)\mathbf{k}$, $a \leq t \leq b$, describes C.

Proof. Suppose **F** is given by

$$\mathbf{F}(x, y, z) = M(x, y, z)\mathbf{i} + N(x, y, z)\mathbf{j} + P(x, y, z)\mathbf{k}.$$

Then

$$\text{curl } \mathbf{F} = (P_y - N_z)\mathbf{i} + (M_z - P_x)\mathbf{j} + (N_x - M_y)\mathbf{k}.$$

If **n** is expressed in terms of its direction cosines,

$$\mathbf{n} = \cos\alpha \, \mathbf{i} + \cos\beta \, \mathbf{j} + \cos\gamma \, \mathbf{k},$$

then the equation in Theorem (16.62) can be rewritten as

(16.63) $$\int_C M \, dx + N \, dy + P \, dz = \iint_S [(P_y - N_z)\cos\alpha + (M_z - P_x)\cos\beta$$
$$+ (N_x - M_y)\cos\gamma] \, dA$$
$$= \iint_S [(M_z \cos\beta - M_y \cos\gamma) + (N_x \cos\gamma - N_z \cos\alpha)$$
$$+ (P_y \cos\alpha - P_x \cos\beta)] \, dA,$$

where α, β, and γ are the direction angles of **n**, as shown in Figure 16.55.
We will show that

(16.64) $$\int_C M \, dx = \iint_S (M_z \cos\beta - M_y \cos\gamma) \, dA,$$

when S is described by $z = f(x, y)$ on a simply connected region R in the xy-plane bounded by the curve $\hat{C}$. ($\hat{C}$ is the projection of C into the xy-plane with orientation inherited from C.)

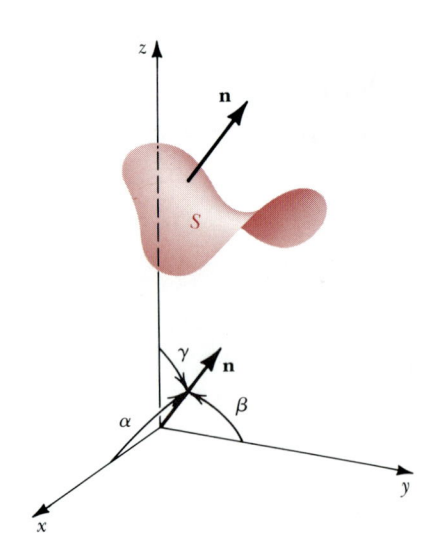

Figure 16.55
$\mathbf{n} = \cos\alpha \, \mathbf{i} + \cos\beta \, \mathbf{j} + \cos\gamma \, \mathbf{k}$.

The other equalities,

$$\int_C N \, dy = \iint_S (N_x \cos \gamma - N_z \cos \alpha) \, dA$$

and

$$\int_C P \, dz = \iint_S (P_y \cos \alpha - P_x \cos \beta) \, dA,$$

are proved by assuming that S has a similar projection representation in the yz- and xz-planes. The result can be extended to more general surfaces by decomposing the surface into surfaces of this type.

Suppose the unit normal $\mathbf{n}$ is directed upward, as shown in Figure 16.56. It follows from (16.54) in Section 16.6 that

$$\cos \beta = \frac{-f_y}{\sqrt{f_x^2 + f_y^2 + 1}} \quad \text{and} \quad \cos \gamma = \frac{1}{\sqrt{f_x^2 + f_y^2 + 1}}.$$

So

$$\iint_S (M_z \cos \beta - M_y \cos \gamma) \, d\sigma = \iint_S -(M_z f_y + M_y) \frac{1}{\sqrt{f_x^2 + f_y^2 + 1}} \, d\sigma$$

$$= \iint_R -\left[\frac{\partial M}{\partial z}(x, y, z) \frac{\partial z}{\partial y} + \frac{\partial M}{\partial y}(x, y, z) \right] dA.$$

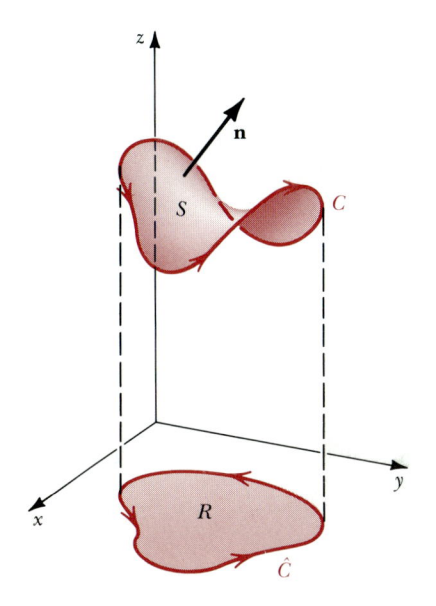

Figure 16.56

In this last expression M is treated as a function of three variables x, y, and z. If M is considered as a function of two variables x and y with $z = f(x, y)$, then the Chain Rule implies that

$$\frac{\partial M}{\partial y}(x, y, f(x, y)) = \frac{\partial M}{\partial z}(x, y, z) \frac{\partial z}{\partial y} + \frac{\partial M}{\partial y}(x, y, z).$$

So

(16.65) $$\iint_S (M_z \cos \beta - M_y \cos \gamma) \, d\sigma = \iint_R -\frac{\partial M}{\partial y}(x, y, f(x, y)) \, dA.$$

Since the orientation of C corresponds to the positive orientation of $\hat{C}$, Green's Theorem in the plane can be applied over the plane region R with $N = 0$ to give

$$\iint_R -\frac{\partial M}{\partial y}(x, y, f(x, y)) \, dA = \int_{\hat{C}} M(x, y, f(x, y)) \, dx = \int_C M(x, y, z) \, dx.$$

Thus,

$$\iint_S (M_z \cos \alpha - M_y \cos \gamma) \, d\sigma = \int_C M(x, y, z) \, dx.$$

If **n** is directed downward,

$$\cos \beta = \frac{f_y}{\sqrt{f_x^2 + f_y^2 + 1}} \quad \text{and} \quad \cos \gamma = \frac{-1}{\sqrt{f_x^2 + f_y^2 + 1}}.$$

Using the same reasoning as when **n** is directed upward, we can show that

$$\iint_S (M_z \cos \beta - M_y \cos \gamma) \, d\sigma = \iint_S (M_z f_y + M_y) \frac{1}{\sqrt{f_x^2 + f_y^2 + 1}} \, d\sigma$$

$$= \iint_R \left[M_z(x, y, z) \frac{\partial z}{\partial y} + M_y(x, y, z) \right] dA$$

$$= \iint_R \frac{\partial M}{\partial y} (x, y, f(x, y)) \, dA$$

$$= -\int_{\hat{C}} M(x, y, f(x, y)) \, dA.$$

However, in this case the positive orientation of $\hat{C}$ corresponds to the negative orientation on C, so

(16.66) $\quad \int_{\hat{C}} M(x, y, f(x, y)) \, dA = \int_{-C} M(x, y, z) \, dx = -\int_C M(x, y, z) \, dx.$

Once again,

$$\iint_S (M_z \cos \beta - M_y \cos \gamma) \, d\sigma = \int_C M(x, y, z) \, dx.$$

▷

Example 1

Let $\mathbf{F}(x, y, z) = x^2 \mathbf{i} + (x + z)\mathbf{j} + yz\mathbf{k}$, let S be the surface of the paraboloid $z = 4 - x^2 - y^2$ that lies above the xy-plane, and let **n** be the outward unit normal to S. Find the value of the surface integral $\iint_S \text{curl } \mathbf{F} \cdot \mathbf{n} \, d\sigma$ (a) directly and (b) using Stokes's Theorem.

Solution

The surface S is shown in Figure 16.57 and is described by $z = 4 - x^2 - y^2$ for $x^2 + y^2 \leq 4$.

a) To evaluate the surface integral directly, we need

$$\text{curl } \mathbf{F} = \begin{vmatrix} \mathbf{i} & \mathbf{j} & \mathbf{k} \\ \dfrac{\partial}{\partial x} & \dfrac{\partial}{\partial y} & \dfrac{\partial}{\partial z} \\ x^2 & x + z & yz \end{vmatrix} = (z - 1)\mathbf{i} + \mathbf{k}$$

$$= [(4 - x^2 - y^2) - 1]\mathbf{i} + \mathbf{k} = (3 - x^2 - y^2)\mathbf{i} + \mathbf{k}.$$

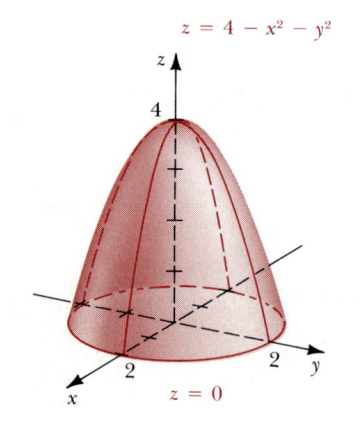

Figure 16.57

It follows from (16.54) in Section 16.6 that with $f(x, y) = z = 4 - x^2 - y^2$,

$$\iint_S \text{curl } \mathbf{F} \cdot \mathbf{n} \, d\sigma = \iint_S [(3 - x^2 - y^2)\mathbf{i} + \mathbf{k}] \cdot \left(\frac{-f_x\mathbf{i} - f_y\mathbf{j} + \mathbf{k}}{\sqrt{f_x^2 + f_y^2 + 1}}\right) d\sigma.$$

We can use Theorem (16.52) to convert this surface integral over S into an ordinary double integral over the region R in the xy-plane described by $x^2 + y^2 \leq 4$:

$$\iint_S \text{curl } \mathbf{F} \cdot \mathbf{n} \, d\sigma = \iint_R [(3 - x^2 - y^2)\mathbf{i} + \mathbf{k}] \cdot (-f_x\mathbf{i} - f_y\mathbf{j} + \mathbf{k}) \, dA$$

$$= \iint_R [-(3 - x^2 - y^2)f_x + 1] \, dA$$

$$= \iint_R [-(3 - x^2 - y^2)(-2x) + 1] \, dA.$$

Since the region R is circular, it is easiest to convert to polar coordinates for evaluation:

$$\iint_S \text{curl } \mathbf{F} \cdot \mathbf{n} \, d\sigma = \int_0^{2\pi} \int_0^2 [-(3 - r^2)(-2r \cos \theta) + 1] r \, dr \, d\theta$$

$$= \int_0^{2\pi} \int_0^2 (6r^2 \cos \theta - 2r^4 \cos \theta + r) \, dr \, d\theta$$

$$= \int_0^{2\pi} \left[\left(2r^3 - \frac{2}{5} r^5\right) \cos \theta + \frac{r^2}{2}\right]_0^2 d\theta$$

$$= \int_0^{2\pi} \left[\left(16 - \frac{64}{5}\right) \cos \theta + 2\right] d\theta$$

$$= \frac{16}{5} \sin \theta + 2\theta \Big]_0^{2\pi} = 4\pi.$$

b) To use Stokes's Theorem we need a representation of C, the boundary of S, that has positive orientation with respect to that provided by the outward normals to S. The parametric representation

$$\mathbf{r}(t) = 2 \cos t\mathbf{i} + 2 \sin t\mathbf{j}, \quad \text{for } t \text{ in } [0, 2\pi]$$

provides this orientation, as shown in Figure 16.58. Since $\mathbf{F}(x, y, z) = x^2\mathbf{i} + (x + z)\mathbf{j} + yz\mathbf{k} = 4 \cos^2 t\mathbf{i} + 2 \cos t\mathbf{j}$,

$$\int_C \mathbf{F} \cdot d\mathbf{r} = \int_0^{2\pi} (4 \cos^2 t\mathbf{i} + 2 \cos t\mathbf{j}) \cdot (-2 \sin t\mathbf{i} + 2 \cos t\mathbf{j}) \, dt$$

$$= \int_0^{2\pi} (-8 \cos^2 t \sin t + 4 \cos^2 t) \, dt$$

$$= \frac{8 \cos^3 t}{3} \Big]_0^{2\pi} + 4 \int_0^{2\pi} \frac{1 + \cos 2t}{2} \, dt$$

$$= \frac{8}{3} - \frac{8}{3} + 2t + \sin 2t \Big]_0^{2\pi} = 4\pi.$$

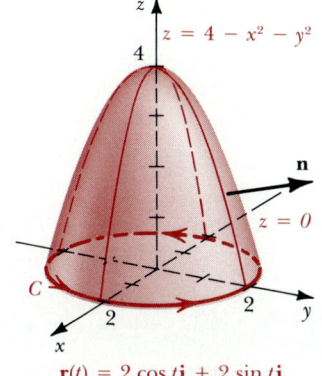

Figure 16.58

$\mathbf{r}(t) = 2 \cos t\mathbf{i} + 2 \sin t\mathbf{j}$

In the next example, Stokes's Theorem is used to convert a surface integral over a given region into a surface integral over a less complex region.

Example 2

Evaluate $\iint_S \text{curl } \mathbf{F} \cdot \mathbf{n} \, d\sigma$, if $\mathbf{n}$ is the downward directed normal to the surface S described by $z = x^2 + y^2 - 3$ for $z \leq 0$ and $\mathbf{F}(x, y, z) = zx^2\mathbf{i} + xy\mathbf{j} + xz\mathbf{k}$.

Solution

Stokes's Theorem implies that this surface integral has the same value as the line integral $\int_C \mathbf{F} \cdot d\mathbf{r}$, where C is the circle in the xy-plane with radius $\sqrt{3}$ centered at the origin and oriented counterclockwise (see Figure 16.59). However, C also bounds the surface of the disk R in the xy-plane. If we apply Stokes's Theorem in the reverse manner using R as the surface, we have

$$\iint_S \text{curl } \mathbf{F} \cdot \mathbf{n} \, d\sigma = \int_C \mathbf{F} \cdot d\mathbf{r} = \iint_R \text{curl } \mathbf{F} \cdot \mathbf{n}_R \, d\sigma,$$

where $\mathbf{n}_R = -\mathbf{k}$ is the downward directed normal to the disk. But the surface integral over R is an ordinary double integral over this region and

$$\text{curl } \mathbf{F} = (x^2 - z)\mathbf{j} + y\mathbf{k},$$

so

$$\iint_S \text{curl } \mathbf{F} \cdot \mathbf{n} \, d\sigma = \iint_R [(x^2 - z)\mathbf{j} + y\mathbf{k}] \cdot (-\mathbf{k}) \, dA = \iint_R -y \, dA.$$

Using cylindrical coordinates, we have

$$\iint_S \text{curl } \mathbf{F} \cdot \mathbf{n} \, d\sigma = \int_0^{\sqrt{3}} \int_0^{2\pi} (-r \sin \theta) r \, d\theta \, dr$$

$$= \int_0^{\sqrt{3}} r^2 \cos \theta \Big]_0^{2\pi} dr = 0.$$

We can now use Stokes's Theorem to complete the proof to Theorem (16.43).

(16.43) THEOREM

Suppose $\mathbf{F}(x, y, z) = M(x, y, z)\mathbf{i} + N(x, y, z)\mathbf{j} + P(x, y, z)\mathbf{k}$, where M, N, and P have continuous first partial derivatives in an open simply connected region D. Then F is conservative on D if and only if

$$\frac{\partial M}{\partial y} = \frac{\partial N}{\partial x}, \quad \frac{\partial M}{\partial z} = \frac{\partial P}{\partial x}, \quad \text{and} \quad \frac{\partial N}{\partial z} = \frac{\partial P}{\partial y}.$$

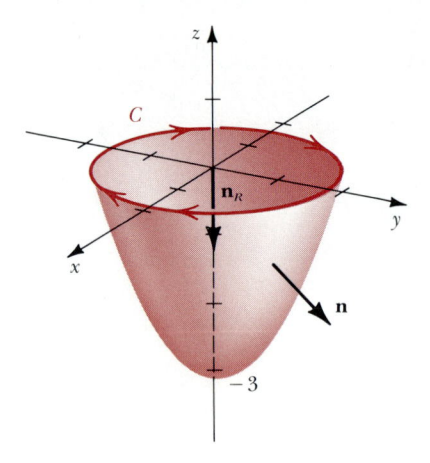

Figure 16.59

Proof. It was shown in Section 16.4 that when **F** is conservative, the equations in Theorem (16.43) are true. To show the converse, suppose the equations hold. We will now show that **F** is conservative by applying Corollary (16.39) and showing that $\int_C \mathbf{F} \cdot d\mathbf{r} = 0$ for every simple closed curve C in D.

Suppose C is such a curve and S is any smooth surface whose boundary is C. It follows from Stokes's Theorem that

$$\int_C \mathbf{F} \cdot d\mathbf{r} = \iint_S \mathbf{curl\ F} \cdot \mathbf{n}\, d\sigma$$

$$= \iint_S [(P_y - N_z)\mathbf{i} + (M_z - P_x)\mathbf{j} + (N_x - M_y)\mathbf{k}] \cdot \mathbf{n}\, d\sigma$$

$$= \iint_S \mathbf{0} \cdot \mathbf{n}\, d\sigma = 0.$$

Thus **F** is conservative on D. ▷

At the beginning of the section we stated that Stokes's Theorem is an extension of Green's Theorem to surfaces in space. To show the truth of this statement, suppose S is a surface in the xy-plane oriented by $\mathbf{n} = \mathbf{k}$ and $\mathbf{F}(x, y) = M(x, y)\mathbf{i} + N(x, y)\mathbf{j}$. By extending **F** to three variables,

$$\mathbf{F}(x, y, z) = M(x, y)\mathbf{i} + N(x, y)\mathbf{j} + 0\mathbf{k},$$

we have

$$\mathbf{curl\ F} = (N_x - M_y)\mathbf{k}, \quad \text{so} \quad \mathbf{curl\ F} \cdot \mathbf{n} = N_x - M_y.$$

Also,

$$\mathbf{F} \cdot d\mathbf{r} = (M\mathbf{i} + N\mathbf{j}) \cdot (dx\, \mathbf{i} + dy\, \mathbf{j}) = M\, dx + N\, dy,$$

so in this case, Stokes's Theorem reduces to Green's Theorem:

$$\int_C M\, dx + N\, dy = \iint_S (N_x - M_y)\, dA.$$

HISTORICAL NOTE

Stokes's Theorem was discovered by Sir William Thomson (Lord Kelvin) and communicated to his friend **Sir George Gabriel Stokes** (1819–1903) in a postscript to a letter of July 2, 1850. Stokes replied that the result was very elegant and new to him and that he had constructed his own proof. He never claimed it as his own or published a proof, but he did include it as a question on the Smith's Prize Examination for 1854 (this was a competitive examination given to the best mathematics students at Cambridge University). It is interesting that one of the students who took the 1854 exam, and who tied for first place on it, was James Clerk Maxwell (1831–1879). As we will see in Section 16.9, Stokes's Theorem is of critical importance in electromagnetic theory and in the formulation of Maxwell's equations.

Stokes was an excellent teacher who made distinguished contributions to both mathematics and mathematical physics.

EXERCISE SET 16.8

In Exercises 1–4, verify Stokes's Theorem for the given function **F** and surface S.

1. $\mathbf{F}(x, y, z) = 4x^3\mathbf{i} + 2y^2\mathbf{j}$; S is the hemisphere $z = \sqrt{9 - x^2 - y^2}$; **n** is directed upward
2. $\mathbf{F}(x, y, z) = 2z\mathbf{i} + x\mathbf{j} - 3y\mathbf{k}$; S is the portion of the paraboloid $z = x^2 + y^2$ with $z \leq 4$; **n** is directed downward
3. $\mathbf{F}(x, y, z) = 2z\mathbf{i} + x\mathbf{j} + 3y\mathbf{k}$; S is the portion of the paraboloid $z = 9 - x^2 - y^2$ above the xy-plane; **n** is directed upward
4. $\mathbf{F}(x, y, z) = (2y + x)\mathbf{i} + (x + z)\mathbf{j} + z^2\mathbf{k}$; S is the portion of the cone $z = \sqrt{x^2 + y^2}$ with $z \leq 1$; **n** is directed downward

In Exercises 5–12, use Stokes's Theorem to evaluate $\int_C \mathbf{F} \cdot d\mathbf{r}$, where C has a positive orientation with respect to the surface of which it is the boundary.

5. $\mathbf{F}(x, y, z) = 2y\mathbf{i} + z\mathbf{j} + 3y\mathbf{k}$; C is the intersection of the sphere $x^2 + y^2 + z^2 = 1$ and the plane $z = 0$
6. $\mathbf{F}(x, y, z) = -y\mathbf{i} + x\mathbf{j} + z\mathbf{k}$; C is the circle $x^2 + y^2 = 1$ in the plane $z = 1$
7. $\mathbf{F}(x, y, z) = x^2\mathbf{i} + y^2\mathbf{j} + z^2\mathbf{k}$; C is the intersection of the sphere $x^2 + y^2 + z^2 = 1$ and the parabolic cylinder $z = y^2$
8. $\mathbf{F}(x, y, z) = y^2\mathbf{i} + x^2\mathbf{j} + x\mathbf{k}$; C is the boundary of the portion of the plane $2x + y + 3z = 6$ in the first octant
9. $\mathbf{F}(x, y, z) = x\mathbf{i} + y^2\mathbf{j} + z\mathbf{k}$; S is the portion of the paraboloid $z = 4 - x^2 - y^2$ above the xy-plane with boundary C; **n** is directed upward
10. $\mathbf{F}(x, y, z) = zy\mathbf{i} + xz\mathbf{j} + yx\mathbf{k}$; S is the portion of the plane $x + y + z = 1$ in the first octant with boundary C; **n** is directed upward
11. $\mathbf{F}(x, y, z) = e^z\mathbf{i} + \sin x\mathbf{j} + \sin y\mathbf{k}$; S is the surface whose boundary C is the triangle with vertices $(3, 0, 0)$, $(0, 2, 0)$, $(0, 0, 5)$; **n** is directed downward
12. $\mathbf{F}(x, y, z) = z\mathbf{i} + x^2\mathbf{j} - e^x \sin x\mathbf{k}$; S is the part of the paraboloid $z = 4 - x^2 - y^2$ that lies inside the cylinder $x^2 + y^2 = 1$ with boundary C; **n** is directed downward
13. Let S be the unit sphere oriented by the normal **n** that is directed outward, and let **F** be a vector function whose component functions have continuous partial derivatives on S. Use Stokes's Theorem to show that

$$\iint_S \text{curl } \mathbf{F} \cdot \mathbf{n} \, d\sigma = 0.$$

16.9

Applications of the Divergence and Stokes's Theorems

Fluid Flow

Suppose a fluid with velocity described by the vector field $\mathbf{v}(x, y, z)$ flows through a surface S in space (see Figure 16.60). The rate at which the fluid flows through S is called the **flux** through S. This rate is found by integrating over S the component of **v** normal to the surface S in the direction of the flow:

$$\text{flux through } S = \iint_S \mathbf{v} \cdot \mathbf{n} \, d\sigma.$$

If S is a closed surface bounding a region D in space and **n** describes

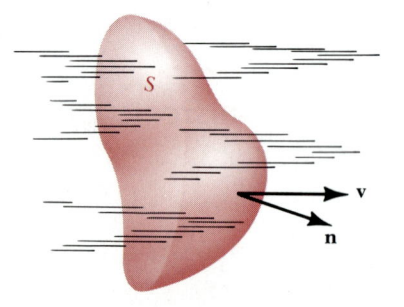

Figure 16.60
Flux through S: $\iint_S \mathbf{v} \cdot \mathbf{n} \, d\sigma$.

the outward unit normal to D, then

(16.67) $$\iint_S \mathbf{v} \cdot \mathbf{n} \, d\sigma$$

gives the net rate of fluid flowing from D. When this integral is positive, fluid is being produced in D, and the region D is called a **source** for the fluid. When this integral is negative, fluid is being absorbed in D, and D is called a **sink** for the fluid. The Divergence Theorem can be used to determine which points within D are responsible for this fluid production or absorption. The theorem implies that the net rate of fluid flowing from D is

$$\iint_S \mathbf{v} \cdot \mathbf{n} \, d\sigma = \iiint_D \text{div } \mathbf{v} \, dV.$$

If V represents the volume of D, then the rate of fluid flowing from D per unit of volume is

$$\frac{1}{V} \iint_S \mathbf{v} \cdot \mathbf{n} \, d\sigma = \frac{1}{V} \iiint_D \text{div } \mathbf{v} \, dV.$$

An extension of the Mean Value Theorem for Integrals can be used to show that for some $(\hat{x}, \hat{y}, \hat{z})$ in D,

$$\frac{1}{V} \iiint_D \text{div } \mathbf{v} \, dV = \text{div } \mathbf{v}(\hat{x}, \hat{y}, \hat{z}),$$

so

(16.68) $$\frac{1}{V} \iint_S \mathbf{v} \cdot \mathbf{n} \, d\sigma = \text{div } \mathbf{v}(\hat{x}, \hat{y}, \hat{z}).$$

Suppose we apply this reasoning when the surface is a sphere S_r of radius r about a fixed point P (see Figure 16.61). The volume bounded by S_r is $4\pi r^3/3$, so (16.68) implies that the rate of flow from inside S_r per unit volume is

(16.69) $$\frac{3}{4\pi r^3} \iint_{S_r} \mathbf{v} \cdot \mathbf{n} \, d\sigma = \text{div } \mathbf{v}(\hat{x}, \hat{y}, \hat{z})$$

for some $(\hat{x}, \hat{y}, \hat{z})$ within S_r. As the radius r approaches zero, the sphere S_r shrinks to the point P. So (16.69) implies that

(16.70) $$\text{div } \mathbf{v}(P) = \lim_{r \to 0} \frac{3}{4\pi r^3} \iint_{S_r} \mathbf{v} \cdot \mathbf{n} \, d\sigma$$

represents the rate of flow per unit volume at the point P. A source occurs at P when $\text{div } \mathbf{v}(P) > 0$; a sink occurs at P when $\text{div } \mathbf{v}(P) < 0$. No fluid is produced or absorbed at P when $\text{div } \mathbf{v}(P) = 0$.

If the fluid has constant density, called an **incompressible** fluid,

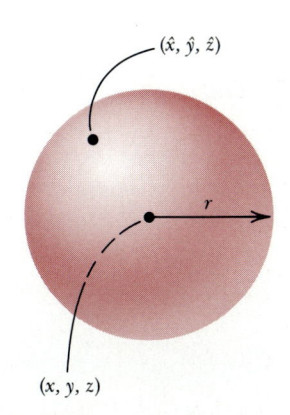

Figure 16.61

then div $\mathbf{v}(P) \equiv 0$ on a region D if and only if there are no sinks or sources within D. Thus:

<blockquote style="color:red">A vector field $\mathbf{v}$ represents the flow of an incompressible fluid without sinks or sources throughout a region precisely when</blockquote>

(16.71) $$\text{div } \mathbf{v} \equiv 0.$$

This is called the *equation of continuity for fluid flow.*

Any vector field $\mathbf{v}$ that satisfies Equation (16.71) is said to be **solenoidal,** a term that is also used in the study of the flow of electrical current. In Example 3 of Section 16.1, we found that any vector function $\mathbf{F}$ with continuous second partial derivatives has the property that

$$\text{div } (\text{curl } \mathbf{F}) = 0.$$

This implies that:

<blockquote style="color:red">curl $\mathbf{F}$ is solenoidal for any vector function $\mathbf{F}$ with continuous second partial derivatives.</blockquote>

Stokes's Theorem has a similar application to fluid flow. Suppose C is a smooth closed curve lying in a region through which a fluid flows with velocity described by $\mathbf{v}(x, y, z)$. The tendency of the fluid to propel a particle located at P on C in the positively oriented direction of C is described by the component of $\mathbf{v}(P)$ in the direction of the unit tangent vector $\mathbf{T}(P)$ to C at P. This component is $\mathbf{v}(P) \cdot \mathbf{T}(P)$ (see Figure 16.62).

The **circulation** of the fluid around C is defined by

$$\text{circulation around } C = \int_C \mathbf{v} \cdot \mathbf{T} \, ds,$$

and describes the tendency of the fluid to propel a particle around C. If C has the parametric representation described by $\mathbf{r}(t)$, the circulation around C is

$$\int_C \mathbf{v} \cdot \mathbf{T} \, ds = \int_C \mathbf{v} \cdot d\mathbf{r}.$$

By Stokes's Theorem, the circulation around C also has the form

$$\int_C \mathbf{v} \cdot \mathbf{T} \, ds = \iint_S \text{curl } \mathbf{v} \cdot \mathbf{n} \, d\sigma,$$

where S is any smooth surface bounded by C and $\mathbf{n}$ is the positively oriented (by C) unit normal to S. If A denotes the area of S, then we can use an extension of the Mean Value Theorem for Integrals to show that for some $(\hat{x}, \hat{y}, \hat{z})$ on S,

$$\frac{1}{A} \iint_S \text{curl } \mathbf{v} \cdot \mathbf{n} \, d\sigma = \text{curl } \mathbf{v}(\hat{x}, \hat{y}, \hat{z}) \cdot \mathbf{n}(\hat{x}, \hat{y}, \hat{z}),$$

so

$$\frac{1}{A} \int_C \mathbf{v} \cdot \mathbf{T} \, ds = \text{curl } \mathbf{v}(\hat{x}, \hat{y}, \hat{z}) \cdot \mathbf{n}(\hat{x}, \hat{y}, \hat{z}).$$

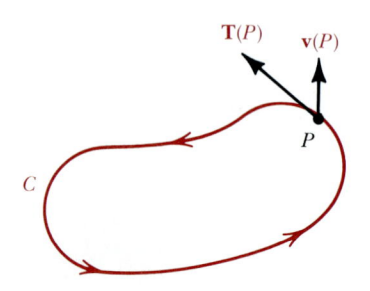

Figure 16.62

Circulation around C: $\int_C \mathbf{v} \cdot \mathbf{T} \, ds$.

16.9 APPLICATIONS OF THE DIVERGENCE AND STOKES'S THEOREMS

Suppose that P is a fixed point in the region of fluid flow and that for each $r > 0$, C_r represents a circle of radius r centered at P and lying in the plane with positively oriented (by C_r) unit normal vector $\mathbf{n}$. Since $A = \pi r^2$,

$$\frac{1}{\pi r^2} \int_{C_r} \mathbf{v} \cdot \mathbf{T} \, ds = \text{curl } \mathbf{v}(\hat{x}, \hat{y}, \hat{z}) \cdot \mathbf{n}(\hat{x}, \hat{y}, \hat{z})$$

for some $(\hat{x}, \hat{y}, \hat{z})$ inside C_r (see Figure 16.63). As r approaches zero, C_r shrinks to the point P, so

(16.72) $$\text{curl } \mathbf{v}(P) \cdot \mathbf{n} = \lim_{r \to 0} \frac{1}{\pi r^2} \int_{C_r} \mathbf{v} \cdot \mathbf{T} \, ds.$$

Equation (16.72) implies that $\text{curl } \mathbf{v}(P) \cdot \mathbf{n}$ can be regarded as the circulation per unit area of the flow of the fluid at P in the direction of the vector $\mathbf{n}$. Thus, $\text{curl } \mathbf{v}(P) \cdot \mathbf{n}$ gives a measure for the tendency of a particle placed at P to rotate in the plane perpendicular to $\mathbf{n}$.

For an incompressible fluid, there is no rotation in a region D if and only if $\text{curl } \mathbf{v} \equiv 0$ on D. Thus:

A vector field $\mathbf{v}$ represents the flow of an incompressible fluid without rotation throughout a region precisely when

(16.73) $$\text{curl } \mathbf{v} \equiv 0.$$

Any vector field $\mathbf{v}$ satisfying Equation (16.73) is said to be **irrotational**. In Equation (16.8) of Section 16.1, we found that any scalar function f with continuous second partial derivatives has the property that

$$\text{curl } (\nabla f) = 0.$$

This implies that:

∇f is irrotational for any scalar function f with continuous second partial derivatives.

The terms divergence and curl have their origin in these applications to fluid flow. The divergence of $\mathbf{v}$ at P measures the amount of change, or divergence, of a fluid at P. The curl of $\mathbf{v}$ at P measures the rotational tendency, or curl, of the fluid at P. The relations expressed in Equations (16.70) and (16.72) are used in advanced textbooks to define the divergence and curl of a vector field. Defined in this way, the divergence and curl are coordinate-independent, an advantage when more than one coordinate system is used. Fixing a coordinate system reduces these definitions to those presented in Section 16.1.

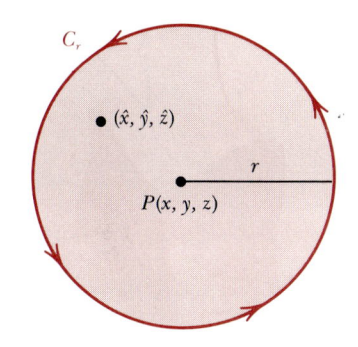

Figure 16.63

Heat Flow

One of the basic principles in thermodynamics, the study of heat flow, is that heat flows from an object in the direction of decreasing temperature at a rate proportional to the difference in temperature between the object and its surrounding medium. This is known as *Newton's Law of Cooling*. Applications of this principle were considered in Section 6.6.

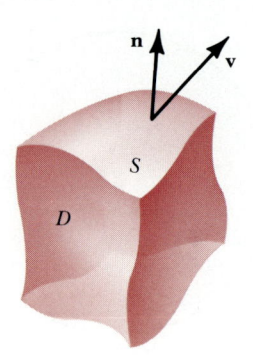

Figure 16.64

Suppose $T(x, y, z, t)$ represents the temperature of an object located at the point with coordinates (x, y, z) at time t. The rate of change in temperature at (x, y, z) at time t is described by

$$\nabla T(x, y, z, t) = \frac{\partial T}{\partial x}(x, y, z, t)\mathbf{i} + \frac{\partial T}{\partial y}(x, y, z, t)\mathbf{j} + \frac{\partial T}{\partial z}(x, y, z, t)\mathbf{k}.$$

It follows from Newton's Law of Cooling that the vector field describing the heat flow at (x, y, z) at time t is given by

$$\mathbf{v}(x, y, z, t) = -k \nabla T(x, y, z, t).$$

The positive constant k is called the **thermal conductivity** of the object. Suppose D is a region in space with a smooth boundary S and $\mathbf{n}$ is the outward normal to S, as shown in Figure 16.64. The total amount of heat leaving D through S per unit time is

(16.74) $$\iint_S \mathbf{v} \cdot \mathbf{n}\, d\sigma = \iint_S -k \nabla T(x, y, z, t) \cdot \mathbf{n}\, d\sigma.$$

Another expression for the amount of heat leaving D per unit time is derived by considering the total amount of heat within D. If the material in D has a constant density ρ and a specific heat c, then the total heat H in D is

$$H = \iiint_D \rho c T(x, y, z, t)\, dV.$$

Thus the total amount of heat leaving D per unit time is the negative of the rate of change in H with respect to time:

(16.75)
$$-\frac{\partial H}{\partial t} = -\frac{\partial}{\partial t}\left[\iiint_D \rho c T(x, y, z, t)\right] dV = -\iiint_D \rho c T_t(x, y, z, t)\, dV.$$

Equating the expressions (16.74) and (16.75) gives

$$-\iiint_D \rho c T_t(x, y, z, t)\, dV = \iint_S -k \nabla (x, y, z, t) \cdot \mathbf{n}\, d\sigma.$$

When the Divergence Theorem is applied to the right side of this equation, we have

$$\iiint_D \rho c T_t(x, y, z, t)\, dV = \iiint_D k \,\mathrm{div}\, (\nabla T(x, y, z, t))\, dV$$

$$= \iiint_D k \left(\frac{\partial}{\partial x}\mathbf{i} + \frac{\partial}{\partial y}\mathbf{j} + \frac{\partial}{\partial z}\mathbf{k}\right) \cdot \left(\frac{\partial T}{\partial x}\mathbf{i} + \frac{\partial T}{\partial y}\mathbf{j} + \frac{\partial T}{\partial z}\mathbf{k}\right) dV$$

$$= \iiint_D k[T_{xx}(x, y, z, t) + T_{yy}(x, y, z, t) + T_{zz}(x, y, z, t)]\, dV.$$

Thus,
$$\iiint_D [\rho c T_t - k(T_{xx} + T_{yy} + T_{zz})]\, dV = 0.$$

Since this equation holds for all regions D, the integrand must be zero. Therefore, any function T that describes the temperature within a solid of constant density ρ and constant specific heat c must satisfy the equation

(16.76) $$\rho c T_t = k(T_{xx} + T_{yy} + T_{zz}).$$

This equation is known as the **heat equation.**

Heat conduction on a plate is governed by a similar heat equation,
$$\rho c T_t = k(T_{xx} + T_{yy}),$$
which is produced from (16.76) by assuming no dependence on the third dimension z. One-dimensional heat flow in a rod is described by an equation of the form
$$\rho c T_t = k T_{xx}.$$

In solutions to physical problems involving the heat equations, the temperature function T must satisfy the appropriate heat equation as well as some specifications regarding the temperature on the boundary of the region. Problems of this type are known as *boundary-value problems.*

Maxwell's Equations

James Clerk Maxwell (1831–1879) assembled and refined the laws that detail the connection between electricity and magnetism, an area of physics known as *electromagnetism*. The four equations defining these laws are collectively known as **Maxwell's Equations,** although with each law is associated the name of the person credited with first describing it.

In 1831 Michael Faraday (1791–1867) discovered that an electric current is produced by varying the magnetic flux through a conducting loop. His experiments showed that the electromotive force of the field is

(16.77) $$\int_C \mathbf{E} \cdot d\mathbf{r} = -\iint_S \frac{\partial \mathbf{B}}{\partial t} \cdot \mathbf{n}\, d\sigma,$$

where C is the curve described by the loop, S is a surface with boundary C, and $\mathbf{E}$ represents the electric field produced by the magnetic induction $\mathbf{B}$. By applying Stokes's Theorem to the left side of Equation (16.77), we can transform Faraday's Law into a differential expression. Stokes's Theorem implies that

$$\int_C \mathbf{E} \cdot d\mathbf{r} = \iint_S \operatorname{curl} \mathbf{E} \cdot \mathbf{n}\, d\sigma,$$

so

$$\iint_S \operatorname{curl} \mathbf{E} \cdot \mathbf{n}\, d\sigma = \iint_S -\frac{\partial \mathbf{B}}{\partial t} \cdot \mathbf{n}\, d\sigma.$$

Since S is an arbitrary surface bounded by C, the integrands must agree, hence

(16.78)
$$\text{curl } \mathbf{E} = -\frac{\partial \mathbf{B}}{\partial t}.$$

Maxwell's correction of a law attributed to Andre Marie Ampère (1775–1836) describes empirically how a magnetic field is associated with a time-varying electric current. This connection is expressed by

(16.79)
$$\int_C \mathbf{B} \cdot d\mathbf{r} = \iint_S \mu_0 \left(\mathbf{J} + \varepsilon_0 \frac{\partial \mathbf{E}}{\partial t} \right) \cdot \mathbf{n} \, d\sigma,$$

where $\mathbf{J}$ is the current per unit cross-sectional area, $\varepsilon_0(\partial \mathbf{E}/\partial t)$ is the displacement current correction needed to ensure the conservation of charge, μ_0 is the magnetic force scale factor, and S is a surface bounded by C, the closed curve that describes the path of the electric current. The constant ε_0 is called the *permittivity constant;* it is dependent only on the physical units involved and scales the electrical units so that the forces produced are represented in correct mechanical units.

Applying Stokes's Theorem to the left side of Equation (16.79) gives

$$\int_C \mathbf{B} \cdot d\mathbf{r} = \iint_S \text{curl } \mathbf{B} \cdot \mathbf{n} \, d\sigma,$$

so
$$\iint_S \mu_0 \left(\mathbf{J} + \varepsilon_0 \frac{\partial \mathbf{E}}{\partial t} \right) \cdot \mathbf{n} \, d\sigma = \iint_S \text{curl } \mathbf{B} \cdot \mathbf{n} \, d\sigma.$$

Since S is an arbitrary surface bounded by C, the integrands must agree and the differential expression of Ampère's Law is

(16.80)
$$\text{curl } \mathbf{B} = \mu_0 \left(\mathbf{J} + \varepsilon_0 \frac{\partial \mathbf{E}}{\partial t} \right).$$

The other two Maxwell's equations are named for the outstanding mathematician Carl Friedrich Gauss. One law states, in essence, that no magnetic field is produced by a single magnetic pole. This is expressed mathematically by

(16.81)
$$\iint_S \mathbf{B} \cdot \mathbf{n} \, d\sigma = 0$$

for all magnetic fields $\mathbf{B}$ and closed surfaces S (there is no magnetic flux through a closed surface). The Divergence Theorem applied to this equation implies that

$$0 = \iint_S \mathbf{B} \cdot \mathbf{n} \, d\sigma = \iiint_D \text{div } \mathbf{B} \, dV,$$

where D is the solid bounded by S. Since S and hence D are arbitrary, this implies that the differential expression of Gauss's Law for Magnetism is

(16.82)
$$\text{div } \mathbf{B} = 0.$$

Gauss's Law for Electrostatics states that the flux through a closed surface is the net charge enclosed by the surface. If q denotes the total charge enclosed by the surface S, the relationship is

$$\text{(16.83)} \qquad \iint_S \mathbf{E} \cdot \mathbf{n} \, d\sigma = \frac{q}{\varepsilon_0},$$

where ε_0 is the permittivity constant. The Divergence Theorem applied to the left side of Equation (16.83) gives

$$\iint_S \mathbf{E} \cdot \mathbf{n} \, d\sigma = \iiint_D \operatorname{div} \mathbf{E} \, dV,$$

where D is the solid bounded by S. The right side of Equation (16.83) can also be expressed as a triple integral over D by defining a function ρ to represent the charge per unit volume within D. Then

$$\frac{q}{\varepsilon_0} = \iiint_D \frac{\rho}{\varepsilon_0} \, dV$$

and

$$\iiint_D \operatorname{div} \mathbf{E} \, dV = \iiint_D \frac{\rho}{\varepsilon_0} \, dV.$$

Since D is arbitrary, the differential expression of Gauss's Law for Electricity is

$$\text{(16.84)} \qquad \operatorname{div} \mathbf{E} = \frac{\rho}{\varepsilon_0}.$$

Collectively, Equations (16.77), (16.79), (16.81), and (16.83) are known as the integral form of Maxwell's Equations, while (16.78), (16.80), (16.82), and (16.84) are known as the differential form. These equations form the backbone of the classical theory of electromagnetism.

▶ EXERCISE SET 16.9

1. Let S be the surface of the sphere $x^2 + y^2 + z^2 = 1$. Assume that the velocity of a fluid flowing through S is given by

    $$\mathbf{v}(x, y, z) = x\mathbf{i} + y\mathbf{j} + z\mathbf{k}.$$

    a) Find $\operatorname{div} \mathbf{v}(x, y, z)$.
    b) Find $(3/4\pi) \iint_S \mathbf{v} \cdot \mathbf{n} \, d\sigma$, where $\mathbf{n}$ is the outward unit normal to S, and compare the result with the answer obtained in part (a).

2. Let S be the boundary of the region D described by $x^2 + y^2 \leq z \leq 4$. Assume that the velocity of a fluid flowing through S is given by

    $$\mathbf{v}(x, y, z) = x^3\mathbf{i} + y^3\mathbf{j} + z^2\mathbf{k}.$$

    a) Find $\operatorname{div} \mathbf{v}(x, y, z)$.
    b) Find the volume V of D.
    c) Find the flux, $\iint_S \mathbf{v} \cdot \mathbf{n} \, d\sigma$, through S.
    d) Verify that for some point $(\hat{x}, \hat{y}, \hat{z})$ in D,

    $$\frac{1}{V} \iint_S \mathbf{v} \cdot \mathbf{n} \, d\sigma = \operatorname{div} \mathbf{v}(\hat{x}, \hat{y}, \hat{z}).$$

3. Suppose the velocity of a fluid is described by
$$\mathbf{v}(x, y, z) = x^2\mathbf{i} + y^2\mathbf{j} + (z^2 - z)\mathbf{k}.$$
Determine at which points sources occur and at which points sinks occur.

4. Show that if $\mathbf{v}(x, y, z) = -x^3\mathbf{i} - y^3\mathbf{j} + 3z\mathbf{k}$ describes the velocity of a fluid, then a source occurs at every point inside the cylinder $x^2 + y^2 = 1$.

5. A fluid of constant density rotates about the z-axis with velocity $\mathbf{v}(x, y, z) = y\mathbf{i} - x\mathbf{j}$.
   a) Find **curl v**.
   b) Find $\int_{C_r} \mathbf{v} \cdot \mathbf{T}\, ds$, where C_r is a circle of radius r in a plane parallel to the xy-plane.
   c) Show that **v** satisfies Equation (16.71).

6. Show that any irrotational function is conservative.

7. A fluid flows through a region with velocity given by
$$\mathbf{v}(x, y, z) = M(x)\mathbf{i} + N(y)\mathbf{j} + P(z)\mathbf{k}.$$
Show that **v** is irrotational.

8. A fluid flows through a region with velocity given by
$$\mathbf{v}(x, y, z) = M(y, z)\mathbf{i} + N(x, z)\mathbf{j} + P(x, y)\mathbf{k}.$$
Show that **v** is solenoidal.

9. Suppose a cup of coffee is stirred with velocity
$$\mathbf{v}(x, y, z) = -\omega y\mathbf{i} + \omega x\mathbf{j},$$
where ω is the constant angular speed of the rotation (see the figure). Show that **v** is not irrotational.

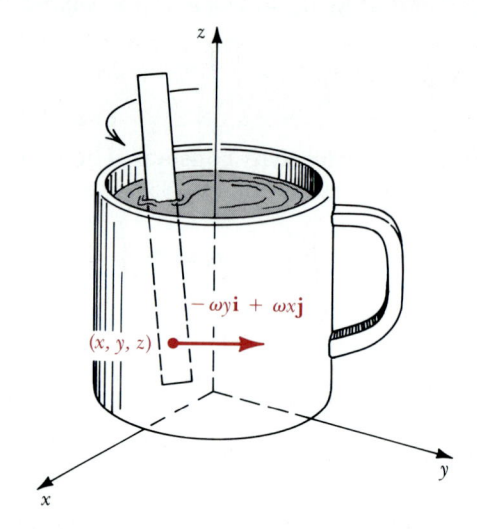

10. Suppose $T(x, y, z, t)$ represents the temperature at (x, y, z) in D at time t, where D is a region in space with a smooth boundary S. Show that if T is harmonic in D, then the temperature is independent of time.

11. Show that the potential function of a conservative solenoidal vector field satisfies Laplace's Equation.

▶ IMPORTANT TERMS AND RESULTS

CONCEPT	PAGE	CONCEPT	PAGE
Vector field	912	Conservative	933
Divergence	913	Exact differential form	933
Curl	913	Potential function	935
Laplace's Equation	915	Simply connected	936
Harmonic function	915	Green's Theorem	944
Closed curve	917	Surface integral	955
Simple closed curve	917	Orientable surface	960
Line integral	917	Divergence Theorem	962
Differential form	921	Stokes's Theorem	968
Work	927	Fluid flow	974
Moment	930	Heat flow	977
Independent of path	932	Maxwell's Equations	979

REVIEW EXERCISES

In Exercises 1–4, find (a) the divergence and (b) the curl of the vector-valued function.

1. $F(x, y, z) = y \sin z \mathbf{i} + x \sin z \mathbf{j} + xy \cos z \mathbf{k}$
2. $F(x, y, z) = yz\mathbf{i} + xz\mathbf{j} + xy\mathbf{k}$
3. $F(x, y, z) = x^3 y\mathbf{i} + (x + z)\mathbf{j} + xyz\mathbf{k}$
4. $F(x, y, z) = \dfrac{x}{y^2 + z^2}\mathbf{i} + \dfrac{y}{x^2 + z^2}\mathbf{j} + \dfrac{z}{x^2 + y^2}\mathbf{k}$

In Exercises 5–8, show that the function is harmonic.

5. $f(x, y) = x^3 - 3xy^2 + y$
6. $f(x, y) = e^x \cos y$
7. $f(x, y) = xy - x^2 + y^2$
8. $f(x, y) = \arctan y/x$
9. Which of the curves described below are closed curves? Which are simple closed curves?
   a) $\mathbf{r}(t) = 2 \cos t \mathbf{i} + 3 \sin t \mathbf{j} + \mathbf{k}, \; 0 \leq t \leq 2\pi$
   b) $\mathbf{r}(t) = \cos t^2 \mathbf{i} + \sin t^2 \mathbf{j}, \; 0 \leq t \leq 2\pi$
10. Which of the following shaded two-dimensional regions are simply connected?

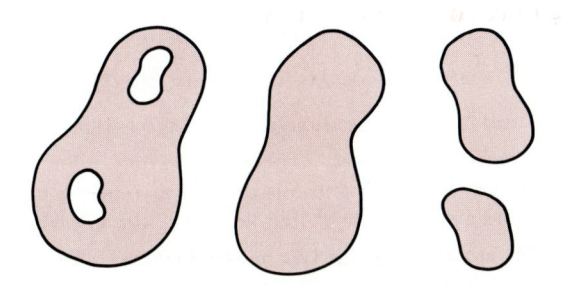

In Exercises 11–22, evaluate the integral.

11. $\displaystyle\int_C \dfrac{-y}{x^2 + y^2} dx + \dfrac{x}{x^2 + y^2} dy$, where C is the unit circle $x^2 + y^2 = 1$ oriented in the counterclockwise direction
12. $\int_C x^2 y \, dy$, where C is the straight-line segment from $(1, 1, 0)$ to $(1, 4, 2)$
13. $\int_C xy \, dx + (x - z) \, dy + yz \, dz$, where C is described by
    $$\mathbf{r}(t) = t\mathbf{i} + t^2 \mathbf{j} + t^3 \mathbf{k}, \quad 0 \leq t \leq 1$$
14. $\int_C x \, dx - z \, dy + 2y \, dz$, where C is the curve $z = y^4$, $x = 1$ from $(1, 0, 0)$ to $(1, 1, 1)$
15. $\int_C (3x^2 + y) \, dx + 4y^2 \, dy$, where C is the curve $y = x^3$, $z = 2$ from $(1, 1, 2)$ to $(2, 8, 2)$
16. $\displaystyle\int_C \dfrac{1}{1 - x^2} dx + \dfrac{1}{xy} dy$, where C is the straight-line segment in the xy-plane from $(2, 1)$ to $(4, 3)$
17. $\int_C e^x \, dx + xy \, dy + \ln xy \, dz$, where C is described by
    $$\mathbf{r}(t) = t\mathbf{i} + e^t\mathbf{j} + t^2 \mathbf{k}, \quad 1 \leq t \leq 3$$
18. $\int_C e^x \, dx + e^y \, dy + e^z \, dz$, where C is described by
    $$\mathbf{r}(t) = \left(1 + \dfrac{t}{2}\right)\mathbf{i} + \dfrac{3t^2}{4}\mathbf{j} + 5 \sin \dfrac{\pi}{4} t \mathbf{k}, \quad 0 \leq t \leq 2$$
19. $\int_C e^x \sin y \, dx + e^x \cos y \, dy$, where C is the closed curve determined by $y = x$, $y = -x$, and $y = 2$, traversed clockwise.
20. $\int_C y^3 \, dx - x^3 \, dy$, where C is the positively oriented boundary of the region enclosed between the two circles with radii 1 and 2 and center at the origin
21. $\int_C \mathbf{F} \cdot d\mathbf{r}$, where $\mathbf{F}(x, y) = [y^2/(1 + x^2)] \mathbf{i} + 2y \arctan x \mathbf{j}$ and C is the ellipse $x^2 + 4y^2 = 1$ oriented positively
22. $\int_C \mathbf{F} \cdot d\mathbf{r}$, where $\mathbf{F}(x, y) = xy\mathbf{i} + (x^2 + y^2)\mathbf{j}$ and C is the closed curve determined by $y = 6 - x^2$ and $y = 3 - 2x$ oriented positively

In Exercises 23–26, (a) show that the line integral is independent of path and (b) evaluate the line integral.

23. $\int_C yz \, dx + xz \, dy + xy \, dz$, where C is the helix
    $$\mathbf{r}(t) = \cos t \mathbf{i} + \sin t \mathbf{j} + t\mathbf{k}, \quad 0 \leq t \leq 2\pi$$
24. $\int_C \sec^2 x \, dx + z \, dy + (y + 2z) \, dz$, where C is the elliptical helix
    $$\mathbf{r}(t) = 2 \cos t \mathbf{i} + 3 \sin t \mathbf{j} + t\mathbf{k}, \quad 0 \leq t \leq \dfrac{3\pi}{2}$$
25. $\int_C 2xy \, dx + x^2 \, dy$, where C is the segment of the curve $y = x^2 - 1$ from $(-1, 0)$ to $(2, 3)$
26. $\int_C 2x e^y \, dx + x^2 e^y \, dy$, where C is the segment of the curve $y = \sin x + 2$ from $(0, 2)$ to $(\pi, 2)$

In Exercises 27 and 28, find the work done by the force $\mathbf{F}$ in moving an object along the given curve.

27. $\mathbf{F}(x, y, z) = (y + z)\mathbf{i} + (x + z)\mathbf{j} + (x + y)\mathbf{k}$ along the curve described by $\mathbf{r}(t) = t\mathbf{i} + t^2\mathbf{j} + t^3\mathbf{k}$ from $(0, 0, 0)$ to $(1, 1, 1)$

28. $\mathbf{F}(x, y, z) = xy\mathbf{i} + x^2\mathbf{j}$ counterclockwise along the unit circle in the xy-plane beginning and ending at $(1, 0, 0)$

In Exercises 29–34, evaluate the surface integral.

29. $\iint_S (x^2 + y^2)\, d\sigma$, where S is the portion of the plane $x + 2y + 3z = 6$ in the first octant

30. $\iint_S (x - y)\, d\sigma$, where S is the portion of the plane $x + y + z = 1$ in the first octant

31. $\iint_S (x + y)\, d\sigma$, where S is the portion of the plane $x + y + z = 1$ that lies inside the cylinder $x^2 + y^2 = 9$

32. $\iint_S (x + y)\, d\sigma$, where S is the portion of the plane $x + y + z = 1$ that lies inside the sphere $x^2 + y^2 + z^2 = 1$

33. $\iint_S y^2\, d\sigma$, where S is the portion of the cylinder $x^2 + y^2 = 1$ in the first octant bounded above by $z = 1$ and below by $z = 0$

34. $\iint_S (x^2 + y^2)z\, d\sigma$, where S is the hemisphere $z = \sqrt{1 - (x^2 + y^2)}$

In Exercises 35–40, determine $\iint_S \mathbf{F} \cdot \mathbf{n}\, d\sigma$.

35. $\mathbf{F}(x, y, z) = 3x^2\mathbf{i} + xy\mathbf{j} + z\mathbf{k}$; S is the portion of the plane $x + y + z = 1$ in the first octant; $\mathbf{n}$ is directed upward

36. $\mathbf{F}(x, y, z) = (y - z)\mathbf{i} + (z - x)\mathbf{j} + (x - y)\mathbf{k}$; S is the portion of the plane $x + y + 2z = 4$ that lies in the first octant; $\mathbf{n}$ is directed upward

37. $\mathbf{F}(x, y, z) = \mathbf{i} + 2\mathbf{j} + 3\mathbf{k}$; S is the hemisphere $z = \sqrt{1 - x^2 - y^2}$; $\mathbf{n}$ is directed upward

38. $\mathbf{F}(x, y, z) = \mathbf{i} + 2\mathbf{j} + 3\mathbf{k}$; S is the hemisphere $z = -\sqrt{1 - x^2 - y^2}$; $\mathbf{n}$ is directed downward

39. $\mathbf{F}(x, y, z) = \sin x\mathbf{i} + (2 - \cos x)y\mathbf{j} + z\mathbf{k}$; S is described by $0 \le x \le \pi$, $0 \le y \le 2$, $0 \le z \le 1$; $\mathbf{n}$ is directed outward from S

40. $\mathbf{F}(x, y, z) = x\mathbf{i} + y\mathbf{j} + z\mathbf{k}$; S is the boundary of the region in the first octant bounded by the plane $x + y + z = 3$; $\mathbf{n}$ is directed outward from S

In Exercises 41–44, evaluate $\int_C \mathbf{F} \cdot d\mathbf{r}$, where C has positive orientation.

41. $\mathbf{F}(x, y, z) = (y + z)\mathbf{i} - (x + z)\mathbf{j} + (x + y)\mathbf{k}$; C is the triangle with vertices $(3, 0, 0)$, $(0, 3, 0)$, and $(0, 0, 3)$

42. $\mathbf{F}(x, y, z) = (y^2 - x^2)\mathbf{i} + 2xy\mathbf{j}$; C is the curve described by $\mathbf{r}(t) = \sec t\mathbf{i} + \tan t\mathbf{j} + \mathbf{k}$, $0 \le t \le \pi/4$

43. $\mathbf{F}(x, y, z) = (y^2 - e^x)\mathbf{i} + (x^2 - \sin y)\mathbf{j} + (z^2 - \sin z)\mathbf{k}$; C is the intersection of the paraboloid $z = x^2 + y^2$ and the plane $2x + 2y + z = 4$

44. $\mathbf{F}(x, y, z) = e^y\mathbf{i} + e^z\mathbf{j} + e^x\mathbf{k}$; C is the boundary of the rectangle $0 \le x \le 3$, $0 \le z \le 5$, and $y = 0$

45. Find (a) the total mass and (b) the center of mass of the helix $\mathbf{r}(t) = \cos t\mathbf{i} + \sin t\mathbf{j} + 4t\mathbf{k}$, $0 \le t \le \pi/2$, if the density is given by $\sigma(x, y, z) = x^2 + y^2 + z^2$.

46. Find (a) the total mass and (b) the center of mass of the line segment $\mathbf{r}(t) = \mathbf{i} + 2t\mathbf{j} + (\pi t/2)\mathbf{k}$, $0 \le t \le 1$, if the density is given by $\sigma(x, y, z) = xy \sin z$.

47. Use surface integrals to show that the surface area of a sphere with radius r is $4\pi r^2$.

48. Use surface integrals to show that the surface area of a cone with height h and radius r is $\pi r\sqrt{r^2 + h^2}$.

49. Use Green's Theorem to find the area bounded above by one arch of the cycloid described by $x = t - \sin t$ and $y = 1 - \cos t$ and below by the x-axis.

50. Use Green's Theorem to find the area bounded above by the curve that describes the path of a projectile,

$$x(t) = (v_0 \cos \theta)t, \qquad y(t) = (v_0 \sin \theta)t - 16t^2,$$

and below by the x-axis if (a) $\theta = \pi/6$ and (b) $\theta = \pi/3$.

51. Verify Stokes's Theorem when $\mathbf{F}(x, y, z) = 2z\mathbf{i} + 3x\mathbf{j} + 4y\mathbf{k}$ and S is that portion of the paraboloid $z = 9 - x^2 - y^2$ that lies above the xy-plane.

52. Evaluate $\iint_S \text{curl } \mathbf{F} \cdot \mathbf{n}\, d\sigma$, where $\mathbf{F}(x, y, z) = (y^2 + z^2)\mathbf{i} + (x^2 + z^2)\mathbf{j} + (x^2 + y^2)\mathbf{k}$; S is the surface of the ellipsoid $4x^2 + 9y^2 + z^2 = 1$ above the xy-plane; $\mathbf{n}$ is directed upward.

53. What must be true about f if $\nabla^2(f^2) = 2f\nabla^2 f$?

17

Ordinary Differential Equations

A **differential equation** is simply an equation that involves an unknown function and one or more of its derivatives. To solve a differential equation means to find a function (or, more commonly, a collection of functions) that satisfies the equation.

There are two classes of differential equations: *ordinary differential equations,* which involve derivatives of a single-variable function, and *partial differential equations,* which involve the partial derivatives of a function of several variables. The separable equations solved in Chapter 6 are a special type of ordinary differential equation; other types are discussed in this chapter. We will not discuss partial differential equations, so when we use the term differential equation it will always mean an ordinary differential equation.

The brief exposition given in this chapter is only an introduction to the methods for solving some of the most basic differential equations.

17.1

Introduction

There is no universal method for solving a differential equation, but there are standard methods for solving some of the types that occur most frequently in applications. To discuss these methods we need the notion of the **order** of a differential equation, which is the order of the highest derivative that appears in the equation. For example, the differential equations

i) $y' = x^2$ and
ii) $y' = y/(x^2 + 1)$

are *first-order* differential equations, since the highest-order derivative in each of these equations is 1. The equations

iii) $y'' = 2y' - y$,
iv) $y'' = 2y' + 3y - 10 \sin x + 2e^{2x}$, and
v) $y'' = 3/x^2 - y'/x - y/x^2, x \neq 0$

are *second-order* equations, and

vi) $y''' = 2$

is a differential equation of *order* 3.

Both of the first-order equations are separable, equations of the type considered in Section 6.7. They have solutions:

i) $y(x) = x^3/3 + C$,
ii) $y(x) = Ce^{\arctan x}$,

where C represents an arbitrary constant. The other equations have the following solutions:

iii) $y(x) = C_1 e^x + C_2 x e^x$,
iv) $y(x) = 2 \sin x - \cos x - e^x/2 + C_1 e^{3x} + C_2 e^{-x}$,
v) $y(x) = 3 + C_1 \sin(\ln x) + C_2 \cos(\ln x)$, $x \neq 0$, and
vi) $y(x) = x^3/3 + C_1 x^2 + C_2 x + C_3$,

where C_1, C_2, and C_3 represent arbitrary constants in each case.

It is easy to verify that these are solutions to the respective equations, as shown in the following example for equation (iii). In fact, the functions listed are the only solutions to equations (i)–(vi). They are the *general solutions* to these differential equations. Note that in each case the number of arbitrary constants in the general solution agrees with the order of the differential equation.

Example 1

Show that for any constants C_1 and C_2, $y(x) = C_1 e^x + C_2 x e^x$ is a solution to $y'' = 2y' - y$.

Solution

Since $y(x) = C_1 e^x + C_2 x e^x$,

$$y'(x) = C_1 e^x + C_2 e^x + C_2 x e^x \quad \text{and} \quad y''(x) = C_1 e^x + 2C_2 e^x + C_2 x e^x.$$

Thus,

$$2y' - y = 2(C_1 e^x + C_2 e^x + C_2 x e^x) - (C_1 e^x + C_2 x e^x)$$
$$= C_1 e^x + 2C_2 e^x + C_2 x e^x = y''.$$

▶

A **particular solution** of a differential equation results when the constants in the general solution are given specific values. These values are usually chosen to satisfy additional requirements placed on the solution by **boundary, or initial, conditions.**

Example 2

Find the solution to the differential equation $y' = x^2$ that satisfies the initial condition $y(3) = 4$.

Solution

The general solution to $y' = x^2$ is $y(x) = x^3/3 + C$, but only one of these functions satisfies the condition $y(3) = 4$:

$$4 = y(3) = \frac{3^3}{3} + C \quad \text{implies that} \quad C = 4 - 9 = -5.$$

Thus only the particular solution $y(x) = x^3/3 - 5$ satisfies both the differential equation and the initial condition. ▶

The graphs of the functions given by the general solution to the differential equation $y' = x^2$ are shown in Figure 17.1. The graph shown in color is the particular solution satisfying the initial condition $y(3) = 4$.

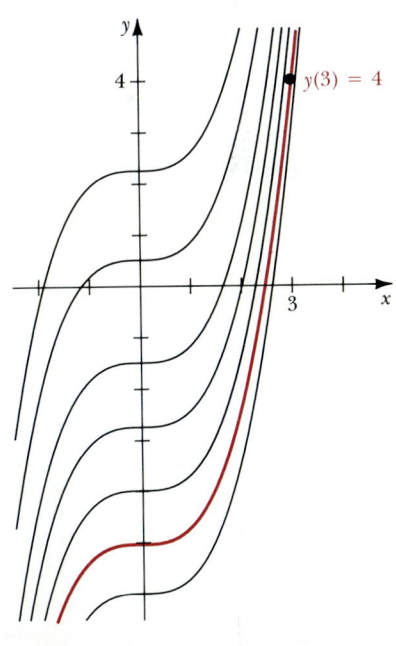

Figure 17.1
Solutions to $y' = x^2$.

An nth order **initial-value problem** is a differential equation together with n specified conditions on the value of the function and its first $n-1$ derivatives at a particular point. The problem in Example 2 is a first-order initial-value problem. The following example considers the solution to a second-order initial-value problem.

Example 3

Find the solution to the initial-value problem
$$y'' = 2y' - y, \qquad y(0) = 2, \qquad y'(0) = -1.$$

Solution
In Example 1 we verified that this second-order equation has solution
$$y(x) = C_1 e^x + C_2 x e^x.$$
In order to satisfy $y(0) = 2$, we must have $2 = C_1 e^0 + C_2 \cdot 0 \cdot e^0 = C_1$, so
$$y(x) = 2e^x + C_2 x e^x.$$
Since
$$y'(x) = 2e^x + C_2(e^x + xe^x),$$
to satisfy the condition $y'(0) = -1$ we must also have
$$-1 = 2e^0 + C_2(e^0 + 0 \cdot e^x) = 2 + C_2, \quad \text{that is,} \quad C_2 = -3.$$
The solution to the initial-value problem is therefore
$$y(x) = 2e^x - 3xe^x,$$
whose graph is shown in Figure 17.2. ▶

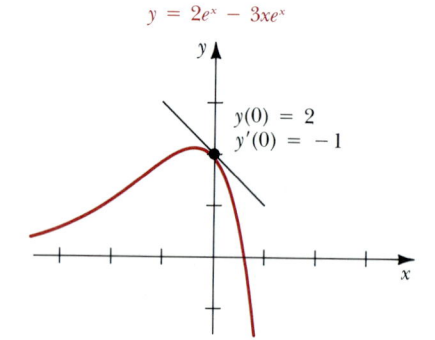

Figure 17.2
The solution to $y'' = 2y' - y$, with $y(0) = 2$ and $y'(0) = -1$.

The general solution to a first-order differential equation is a family of functions related to one another by an arbitrary constant. Conversely, a set of functions related by an arbitrary constant describes a first-order differential equation, provided the functions are differentiable, so that the differential equation is defined.

Consider, for example, the set of functions described by
$$y = Ce^{-x}, \quad \text{where } C \text{ is a constant.}$$
To determine a first-order differential equation with this set of functions as its general solution, we first solve for the arbitrary constant C:
$$C = ye^x.$$
We then differentiate implicitly with respect to x to obtain
$$0 = y'e^x + ye^x = e^x(y' + y).$$
Since e^x is never zero, this means that $0 = y' + y$. Thus the family of functions $y = Ce^{-x}$ is the general solution to the first-order differential equation
$$y' = -y.$$

A common problem in engineering applications involves finding a family of curves that is *mutually orthogonal* to a given family. That is, all intersections between a member of the new family and that of the original family are at right angles. Each curve of the new family is said to be an **orthogonal trajectory** of the original family of curves.

Example 4

Find the orthogonal trajectories of the family of curves described by $y = Ce^{-x}$.

Solution

We have seen above that the first-order differential equation

$$y' = -y$$

has $y = Ce^{-x}$ as its general solution. This implies that the slope of the tangent line to a curve with equation $y = Ce^{-x}$ at any point (x, y) on its graph has the value $y'(x) = -y(x)$.

Two nonvertical lines that intersect orthogonally have slopes whose product is -1, so the orthogonal trajectories must be a family of curves that have slope at (x, y) given by

$$y'(x) = \frac{-1}{-y(x)} = \frac{1}{y(x)}.$$

The differential equation describing the orthogonal trajectories to $y = Ce^{-x}$ is therefore

$$y' = \frac{1}{y},$$

provided $y \neq 0$. This equation is separable and can be solved using the techniques of Section 6.7. Since

$$0 = yy' - 1 = y\frac{dy}{dx} - 1,$$

we have

$$C = \int y\, dy - \int 1 \cdot dx = \frac{y^2}{2} - x$$

or

$$2C = y^2 - 2x.$$

Since $2C$ is an arbitrary constant, we denote it simply by C, so

$$y^2 = 2x + C.$$

This is a family of parabolas. Representative members of this family are shown in Figure 17.3 together with graphs of members of the original family $y = Ce^{-x}$. ▶

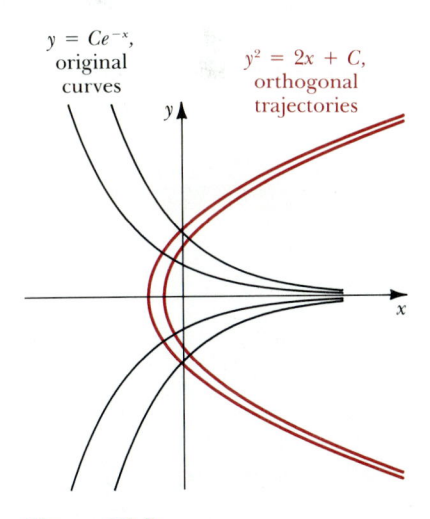

Figure 17.3

Mutually orthogonal curves.

EXERCISE SET 17.1

In Exercises 1–16, verify that the function $y \equiv y(x)$ is a solution to the accompanying differential equation.

1. $2xy' + y = 0$; $y = Cx^{-1/2}$
2. $xy' - 3y = 0$; $y = Cx^3$
3. $2xy' + y = 1$; $y = Cx^{-1/2} + 1$
4. $xy' - 3y = 1$; $y = Cx^3 - 1/3$
5. $y' + 2xy = 0$; $y = Ce^{-x^2}$
6. $xy' - y = xy$; $y = Cxe^x$
7. $x^2 y' - y^2 - 2xy = 0$; $y = \dfrac{Cx^2}{1 - Cx}$
8. $(x^2 + 1)y' + y^2 + 1 = 0$; $y = \dfrac{C - x}{1 + Cx}$
9. $y'' + y = 0$; $y = C_1 \sin x + C_2 \cos x$
10. $y'' - y = 0$; $y = C_1 e^x + C_2 e^{-x}$
11. $y'' - y = 0$; $y = C_1 \sinh x + C_2 \cosh x$
12. $y'' + y = x^2$; $y = C_1 \sin x + C_2 \cos x + x^2 - 2$
13. $y'' - y = x^2$; $y = C_1 e^x + C_2 e^{-x} - x^2 - 2$
14. $y''' - 3y'' + 2y' = 6e^{-x}$; $y = C_1 + C_2 e^x + C_3 e^{2x} - e^{-x}$
15. $yy' + x = 0$; $x^2 + y^2 = C$
16. $yy' - 4x = 1$; $y^2 - 4x^2 = 2x + C$

In Exercises 17–24:

a) Determine a first-order differential equation whose solution is given.
b) Find orthogonal trajectories to the curves described.
c) Sketch representative graphs of the curves and of the orthogonal trajectories.

17. $y = Cx$
18. $y = 4x + C$
19. $y = C/x$
20. $y = Ce^x$
21. $y = Ce^{-2x}$
22. $x^2 - 2y^2 = C$
23. $y = \sqrt{x + C}$
24. $y = Ce^{x^2}$

Putnam exercise

25. Assume that the differential equation

$$y''' + p(x)y'' + q(x)y' + r(x)y = 0$$

has solutions $y_1(x)$, $y_2(x)$, and $y_3(x)$ on the whole real line such that

$$y_1^2(x) + y_2^2(x) + y_3^2(x) = 1$$

for all real x. Let

$$f(x) = (y_1'(x))^2 + (y_2'(x))^2 + (y_3'(x))^2.$$

Find constants A and B such that $f(x)$ is a solution to the differential equation

$$y' + Ap(x)y = Br(x).$$

(This exercise was problem B–3 of the forty-fourth William Lowell Putnam examination given on December 3, 1983. The examination and its solution are in the October 1984 issue of the *American Mathematical Monthly*, pp. 487–495.)

17.2

Homogeneous Differential Equations

The most elementary first-order differential equations occur in the form

$$y'(x) = f(x),$$

where f is a continuous function. To find a general solution in this case, we simply integrate:

$$y'(x) = f(x) \quad \text{implies that} \quad y(x) = \int f(x)\, dx + C.$$

Differential equations of this type were considered in Section 3.1, where we derived an equation describing the motion of a falling object from the knowledge of the object's acceleration due to gravity.

Another elementary type of differential equation, the separable equation, was discussed in Sections 6.6 and 6.7. Separable equations have the form

(17.1) $$y' = \frac{p(x)}{q(y)},$$

where p and q are continuous and $q(y) \neq 0$. These equations are solved by separating the variables and integrating:

$$y' = \frac{p(x)}{q(y)} \quad \text{implies that} \quad C = \int p(x)\,dx - \int q(y)\,dy,$$

although, depending on the form of $\int q(y)\,dy$, the solution may be given only implicitly.

To review the technique for separable equations, consider the following example.

Example 1

Find the solution to the initial-value problem

$$y' = \frac{2xe^{x^2}}{(1+y)e^y}, \quad y(0) = 0.$$

Solution
The solution to this separable equation has the form

$$C = \int 2xe^{x^2}\,dx - \int (1+y)e^y\,dy = e^{x^2} - \int (1+y)e^y\,dy.$$

We use the method of integration by parts to evaluate the integral involving y.

$$\text{Let} \quad u = 1+y \quad \text{and} \quad dv = e^y\,dy,$$
$$\text{then} \quad du = dy \quad \text{and} \quad v = e^y,$$

so

$$\int (1+y)e^y\,dy = (1+y)e^y - \int e^y\,dy = (1+y)e^y - e^y = ye^y.$$

Thus the general solution has the form

$$C = e^{x^2} - ye^y.$$

Applying the initial condition gives

$$C = e^0 - 0 \cdot e^0 = 1,$$

so the solution is given implicitly as

$$1 = e^{x^2} - ye^y. \quad \blacktriangleright$$

A first-order equation of the form

(17.2) $$y' = f(x, y), \quad \text{where} \quad f(x, y) = g\left(\frac{y}{x}\right),$$

is called a **homogeneous** differential equation. An example of such an equation is

$$y' = 1 + \frac{y}{x}.$$

A homogeneous equation can be reduced to a separable equation using the variable substitution $v = y/x$. With this substitution,

$$y(x) = xv(x), \quad \text{so} \quad y'(x) = v(x) + xv'(x).$$

Therefore, Equation (17.2) is

$$v + xv' = g(v),$$

or, in the separable form,

$$v' = \frac{g(v) - v}{x}.$$

This separable equation can then be solved for v, hence for $y(x) = xv(x)$.

Example 2

Find the general solution to the differential equation

$$y' = 1 + \frac{y}{x}, \quad \text{provided } x \neq 0.$$

Solution
The substitution $v = y/x$ implies

$$y = vx \quad \text{and} \quad y' = v + xv',$$

so the original differential equation becomes the separable differential equation

$$v + xv' = 1 + v \quad \text{or} \quad v' = \frac{1}{x}.$$

The general solution to this equation has the form

$$C = \int \frac{1}{x} dx - \int dv = \ln|x| - v.$$

Substituting $v = y/x$ into this result gives the solution

$$C = \ln|x| - \frac{y}{x},$$

which is written explicitly, with the constant adjusted, as

$$y = x \ln|x| + Cx. \quad \blacktriangleright$$

Example 3

Find the solution to the initial-value problem

$$y' = \frac{x + y}{x - y}, \quad y(1) = 0.$$

Solution
Since
$$\frac{x+y}{x-y} = \frac{1+y/x}{1-y/x},$$
the equation is homogeneous. The substitution $v = y/x$ implies that $y = xv$ and $y' = v + xv'$. This changes the original differential equation into the separable differential equation
$$v + xv' = \frac{1+v}{1-v}.$$

Thus,
$$xv' = \frac{1+v}{1-v} - v = \frac{1+v^2}{1-v},$$
so
$$\frac{1-v}{1+v^2} v' = \frac{1}{x}$$
and
$$0 = \frac{1}{x} - \frac{1-v}{1+v^2} \frac{dv}{dx}.$$

Integrating this equation produces
$$C = \int \frac{1}{x} dx - \int \frac{1-v}{1+v^2} dv$$
$$= \ln|x| - \int \frac{1}{1+v^2} dv + \int \frac{v}{1+v^2} dv$$
$$= \ln|x| - \arctan v + \frac{1}{2} \ln|1+v^2|$$
$$= \ln|x \sqrt{1+v^2}| - \arctan v.$$

Substituting $v = y/x$, the general solution has the form
$$C = \ln \left| x \sqrt{1 + \left(\frac{y^2}{x^2}\right)} \right| - \arctan \frac{y}{x}$$
$$= \ln \sqrt{x^2+y^2} - \arctan \frac{y}{x}.$$

Since $y(1) = 0$,
$$C = \ln \sqrt{1^2 + 0^2} - \arctan \frac{0}{1} = 0 - 0 = 0 \quad \text{and} \quad C = 0.$$

The solution to the initial-value problem is
$$\ln \sqrt{x^2+y^2} - \arctan \frac{y}{x} = 0.$$

This equation cannot be solved explicitly for y. ▶

► EXERCISE SET 17.2

In Exercises 1–16, find the general solution to the differential equation.

1. $y' = xy \ln x$
2. $y' = \dfrac{e^x \cos x (1 + \sin y)}{\cos y}$
3. $y' = \dfrac{e^y}{y\sqrt{1 - x^2}}$
4. $y' = \dfrac{\sin^2 y}{\cos^3 x}$
5. $y' = \dfrac{xe^{x+y}}{y}$
6. $y' = \dfrac{y}{x^2 - 5x + 6}$
7. $y' = \dfrac{y}{x + y}$
8. $y' = \dfrac{2x + y}{x + 2y}$
9. $y' = \dfrac{x - y}{x + y}$
10. $y' = \dfrac{2xy + y^2}{x^2}$
11. $y' = \dfrac{x^2 + y^2}{xy}$
12. $y' = \dfrac{x + y \cos^2(y/x)}{x \cos^2(y/x)}$
13. $y' = \dfrac{x^2 + xy + y^2}{x^2}$
14. $y' = \dfrac{x^2 + 3xy + y^2}{x^2}$
15. $y' = \dfrac{y + \sqrt{x^2 + y^2}}{x}$
16. $y' = \dfrac{y}{x - 2\sqrt{xy}}$

In Exercises 17–20, solve the initial-value problem.

17. $y' = \dfrac{x + y}{x - y}; \; y(1) = \sqrt{3}$
18. $y' = \dfrac{2x}{x + y}; \; y(1) = 2$
19. $y' = \dfrac{2x + y}{x + 2y}; \; y(1) = 1$
20. $y' = \dfrac{4x^2 + 2y^2}{xy}; \; y(1) = 2$

21. Find the general solution to the differential equation
$$y' = \dfrac{2x + y + 4}{x + 2y + 5}$$
by first making a substitution $x = u - 1$ and $y = w - 2$. Then let $v = w/u$ and obtain a separable equation.

In Exercises 22–25, make a substitution of the form $x = u - a$, $y = w - b$ similar to that made in Exercise 21. Choose a and b to eliminate the constants in the numerator and denominator and solve the resulting equations.

22. $y' = \dfrac{y + 1}{x + y + 3}$
23. $y' = \dfrac{2x + y + 2}{x + 2y - 1}$
24. $y' = \dfrac{x + y + 5}{x - y - 4}$
25. $y' = \dfrac{2x + y - 2}{x - 2y + 5}$

26. Recall from Theorem (14.31) of Section 14.4 that if $F(x, y) = 0$ and F is differentiable with $F_y(x, y) \neq 0$, then
$$\dfrac{dy}{dx} = \dfrac{-F_x(x, y)}{F_y(x, y)}.$$
Use this to show that if the orthogonal trajectories to $F(x, y) = C$ are described by $G(x, y) = C$, then
$$F_x(x, y)G_x(x, y) + F_y(x, y)G_y(x, y) = 0.$$

27. Find the orthogonal trajectories of the family of circles with equations $x^2 + y^2 - 2ax = 0$, for some constant a.

28. Show that if the differential equation $y' = f(x, y)$ is homogeneous, then the transformation $x = r(\theta) \cos \theta$, $y = r(\theta) \sin \theta$ reduces this equation to a separable equation in the variables r and θ.

17.3

Exact Differential Equations

The discussion of line integrals in Section 16.4 introduced the concept of an exact differential form. A *differential form* in two variables x and y is an expression of the form

$$f(x, y)\, dx + g(x, y)\, dy.$$

The differential form is called *exact* if a function u exists with

$$du = f(x, y)\, dx + g(x, y)\, dy.$$

Since
$$du = \frac{\partial u}{\partial x} dx + \frac{\partial u}{\partial y} dy,$$
the differential form is exact if and only if a function u exists with
$$\frac{\partial u}{\partial x}(x, y) = f(x, y) \quad \text{and} \quad \frac{\partial u}{\partial y}(x, y) = g(x, y).$$

An **exact differential equation** is a first-order differential equation produced when an exact differential form is set to zero; that is, a differential equation of the form

(17.3) $$f(x, y)\, dx + g(x, y)\, dy = 0,$$

or, equivalently,

(17.4) $$y' = \frac{dy}{dx} = -\frac{f(x, y)}{g(x, y)},$$

where a function u exists with

$$\frac{\partial u}{\partial x} = f \quad \text{and} \quad \frac{\partial u}{\partial y} = g.$$

If the function u is known, the solution to this differential equation is immediate, since
$$du = f(x, y)\, dx + g(x, y)\, dy = 0$$
implies that
$$u(x, y) = C.$$

Example 1

Solve the differential equation
$$\frac{dy}{dx} = -\frac{2xy + y^2}{x^2 + 2xy}.$$

Solution

Rewritten in differential form, this differential equation is
$$(2xy + y^2)\, dx + (x^2 + 2xy)\, dy = 0.$$

Note that for $u(x, y) = x^2 y + y^2 x$,
$$\frac{\partial u}{\partial x}(x, y) = 2xy + y^2 \quad \text{and} \quad \frac{\partial u}{\partial y}(x, y) = x^2 + 2yx.$$

Thus the differential equation is exact and its general solution can be written as
$$x^2 y + y^2 x = C. \quad \blacktriangleright$$

The equation in Example 1 is also a homogeneous equation and can

996   CHAPTER 17 ORDINARY DIFFERENTIAL EQUATIONS

be solved by the method of Section 17.2. This method of solution is considerably more complicated, as you will discover if you work Exercise 19.

Example 1 brings out two important questions regarding exact differential equations:

i) How can we determine whether a differential equation is exact without knowing the function u?
ii) When the equation is exact, how do we find u?

The answer to the first question is found in the following result, which is a special case of Theorem (16.43).

(17.5) THEOREM

If $\partial f/\partial y$ and $\partial g/\partial x$ are continuous on a simply-connected region, then the differential form

$$f(x, y)\, dx + g(x, y)\, dy$$

is exact if and only if

$$\frac{\partial f}{\partial y} = \frac{\partial g}{\partial x}.$$

The theorem gives an easy test to apply. For example, the differential equation in Example 1,

$$\frac{dy}{dx} = -\frac{2xy + y^2}{x^2 + 2xy}$$

can be written as

$$(2xy + y^2)\, dx + (x^2 + 2xy)\, dy = 0$$

and is exact because

$$\frac{\partial}{\partial y}(2xy + y^2) = 2x + 2y = \frac{\partial}{\partial x}(x^2 + 2yx).$$

To see how to answer question (ii), consider the following examples.

Example 2

Find the general solution to the differential equation

$$(2xy + e^y)\, dx + (x^2 + xe^y + 4y)\, dy = 0.$$

Solution

This equation is exact since

$$\frac{\partial}{\partial y}(2xy + e^y) = 2x + e^y = \frac{\partial}{\partial x}(x^2 + xe^y + 4y).$$

If the solution to this equation has the form $u(x, y) = C$, then u must

satisfy
$$\frac{\partial}{\partial x} u(x, y) = 2xy + e^y.$$

Integrating this equation with respect to x gives

(17.6) $$u(x, y) = x^2 y + xe^y + K(y),$$

where $K(y)$ represents the constant (as a function of x) of integration. The function u, however, must also satisfy

$$\frac{\partial}{\partial y} u(x, y) = x^2 + xe^y + 4y,$$

which implies that $K(y)$ in (17.6) must satisfy

$$K'(y) = 4y \quad \text{and} \quad K(y) = 2y^2 + \hat{K},$$

for some constant $\hat{K}$. Thus,

$$u(x, y) = x^2 + xe^y + 2y^2 + \hat{K}$$

and the general solution has the form $u(x, y) = C$, or

$$x^2 + xe^y + 2y^2 = C,$$

if the extraneous constant $\hat{K}$ is deleted. ▶

Example 3

Find the solution to the initial-value problem

$$\frac{dy}{dx} = \frac{2xy^2 + y \sin x}{\cos x - 2x^2 y}, \quad y(0) = 1.$$

Solution

The equation can be rewritten as

$$-(2xy^2 + y \sin x) \, dx + (\cos x - 2x^2 y) \, dy = 0$$

and is exact because

$$\frac{\partial}{\partial y} [-(2xy^2 + y \sin x)] = -4xy - \sin x = \frac{\partial}{\partial x} (\cos x - 2x^2 y).$$

If the general solution has the form $u(x, y) = C$, then

$$\frac{\partial}{\partial x} u(x, y) = -2xy^2 - y \sin x,$$

so

(17.7) $$u(x, y) = -x^2 y^2 + y \cos x + K(y).$$

Also,

$$\frac{\partial}{\partial y} u(x, y) = \cos x - 2x^2 y.$$

Equating this with the result obtained by differentiating (17.7) with respect to y:

$$\frac{\partial}{\partial y} u(x, y) = -2x^2y + \cos x + K'(y),$$

we see that

$$K'(y) = 0 \quad \text{and} \quad K(y) = K,$$

a constant. Thus,

$$u(x, y) = -x^2y^2 + y \cos x + K,$$

and the general solution has the form

$$y \cos x - x^2y^2 = C.$$

The initial condition $y(0) = 1$ implies that

$$C = 1 \cdot \cos 0 - 0^2 \cdot 1^2 = 1,$$

so the initial-value problem has the solution

$$y \cos x - x^2y^2 = 1. \quad \blacktriangleright$$

Example 4

Solve the differential equation

$$\frac{dy}{dx} = \frac{y}{xy - x}.$$

Solution
The differential equation written in differential form is

$$-y \, dx + (xy - x) \, dy = 0.$$

This differential equation is *not* exact because

$$\frac{\partial}{\partial y}(-y) = -1 \neq y - 1 = \frac{\partial}{\partial x}(xy - x).$$

We can, however, convert this into an exact equation by multiplying by e^{-y}. The differential equation

$$-ye^{-y} \, dx + (xy - x)e^{-y} \, dy = 0$$

is exact because

$$\frac{\partial}{\partial y}(-ye^{-y}) = -e^{-y} + ye^{-y}$$

$$= \frac{\partial}{\partial x}(xye^{-y} - xe^{-y}).$$

Since

$$\frac{\partial}{\partial x} u(x, y) = -ye^{-y} \quad \text{implies that} \quad u(x, y) = -xye^{-y} + K(y),$$

we have

$$\frac{\partial}{\partial y} u(x, y) = xye^{-y} - xe^{-y} + K'(y).$$

From the exact differential form

$$\frac{\partial}{\partial y} u(x, y) = (xy - x)\, e^{-y} = xye^{-y} - xe^{-y},$$

so

$$K'(y) = 0 \quad \text{and} \quad K(y) = K,$$

a constant. Thus the general solution to the exact differential equation has the form

$$-xye^{-y} = C.$$

To verify that $-xye^{-y} = C$ is also a solution to the original equation, we differentiate this equation implicitly:

$$-ye^{-y} - x(e^{-y} - ye^{-y})\frac{dy}{dx} = 0,$$

so

$$\frac{dy}{dx} = \frac{ye^{-y}}{xye^{-y} - xe^{-y}} = \frac{y}{xy - x}.$$

▶

A term (such as e^{-y} in Example 4) that converts a nonexact equation into an exact equation is one example of what is known as an **integrating factor,** a term that converts a differential equation into a form that can be solved by a simple integration. Every differential equation with a solution of the form $u(x, y) = C$ has an infinite number of integrating factors that make the problem exact. However, it is in general as difficult to find an integrating factor that makes the problem exact as it is to solve the original equation. Exercises 21 and 22 consider two special cases when an integrating factor can be obtained quite readily.

When a problem has been converted to an exact equation by using an integrating factor, the solution should be verified in the original differential equation because extraneous solutions can occur.

▶ **EXERCISE SET 17.3**

In Exercises 1–14, determine whether the equation is exact and solve the exact equations.

1. $y' = \dfrac{3y + 2x}{y - 3x}$

2. $y' = \dfrac{y - 4x}{y - x}$

3. $y' = \dfrac{3y - 2x}{y + 3x}$

4. $y' = \dfrac{2x - y}{x - 2y}$

5. $y' = \dfrac{x^3 - xy^4}{2x^2y^3 + 5}$

6. $y' = \dfrac{e^x \cos y - y}{e^x \sin y + x}$

7. $y' = \dfrac{x - y^2}{2xy}$
8. $y' = \dfrac{y - y^2}{x}$
9. $y\, dx + dy = 0$
10. $(x + \tan y)\, dx + (x \sec^2 y - 3y)\, dy = 0$
11. $y e^x\, dx + e^x\, dy = 0$
12. $(6x + y^2)\, dx + (2xy - 3y^2)\, dy = 0$
13. $(\cot x - \cos x \cos y)\, dx + \sin x \sin y\, dy = 0$
14. $\sinh x \cos y\, dx - \sin y \cosh x\, dy = 0$

In Exercises 15–18, solve the initial-value problem.

15. $(2x - 3y)\, dx + (4y - 3x)\, dy = 0,\ y(0) = 1$
16. $(x^2 + 2xy)\, dx + (x^2 - y^2)\, dy = 0,\ y(1) = 1$
17. $\sin x \sin y\, dx - \cos x \cos y\, dy = 0,\ y(\pi/4) = \pi/2$
18. $y e^{xy}\, dx + (2y + x e^{xy})\, dy = 0,\ y(0) = 1$
19. Solve the differential equation in Example 1 using the fact that it is homogeneous.
20. Show that every separable equation is an exact equation.
21. Suppose the equation $M(x, y)\, dx + N(x, y)\, dy = 0$ has the property that $(N_x - M_y)/M$ is a function of y only and that
$$I(y) = \exp\left[\int \dfrac{N_x - M_y}{M}\, dy\right].$$
Show that
$$I(y)[M(x, y)\, dx + N(x, y)\, dy] = 0$$
is exact.

22. Suppose the equation $M(x, y)\, dx + N(x, y)\, dy = 0$ has the property that $(M_y - N_x)/N$ is a function of x only and that
$$I(x) = \exp\left[\int \dfrac{M_y - N_x}{N}\, dx\right].$$
Show that
$$I(x)[M(x, y)\, dx + N(x, y)\, dy] = 0$$
is exact.

In Exercises 23–28:

a) Change the given equation to an exact equation by using one of the techniques described in Exercises 21 and 22.
b) Solve the resulting equation and verify that the solution also satisfies the given equation.

23. $y\, dx - x\, dy = 0$
24. $y\, dx + 2x\, dy = 0$
25. $6x^2 y\, dx + (x^3 + y)\, dy = 0$
26. $y\, dx + (x^2 y - x)\, dy = 0$
27. $(\cos y + x)\, dx + x \sin y\, dy = 0$
28. $(x^2 - x \sin y)\, dx + x^2 \cos y\, dy = 0$
29. Find a constant k such that the differential equation
$$x^k[(3xy^2 + 7x^3)\, dx + (4x^2 y + 3\sqrt{xy})\, dy] = 0$$
is exact. Solve the resulting equation.

17.4

Linear First-Order Differential Equations

A **linear first-order differential equation** has the form

(17.8) $$\dfrac{dy}{dx} + P(x) y = Q(x).$$

The technique for solving this type of equation involves multiplying the differential equation by an integrating factor $I(x)$ that changes Equation (17.8) into an equation whose left side is the derivative of the product $I(x)y$. To determine this integrating factor, consider the differential equation

$$I(x) \dfrac{dy}{dx} + I(x) P(x) y = I(x) Q(x).$$

To choose $I(x)$ so that
$$D_x[I(x)y] = I(x)\frac{dy}{dx} + I(x)P(x)y$$
requires that
$$I(x)\frac{dy}{dx} + I'(x)y = I(x)\frac{dy}{dx} + I(x)P(x)y.$$
This implies that

(17.9) $$I'(x) = I(x)P(x).$$

Equation (17.9) is a separable equation whose solution is
$$I(x) = Ce^{\int P(x)dx}.$$
Since we need only one integrating factor, we let $C = 1$ and choose

(17.10) $$I(x) = e^{\int P(x)dx}.$$

The following examples illustrate how we can use this integrating factor to solve linear first-order equations.

Example 1

Solve the linear differential equation
$$y' + 2xy = x.$$

Solution
An integrating factor for this equation is
$$I(x) = e^{\int 2x\,dx} = e^{x^2}.$$
Multiplying both sides of the differential equation by $I(x)$ produces
$$e^{x^2}y' + 2xe^{x^2}y = xe^{x^2}.$$
But
$$e^{x^2}y' + 2xe^{x^2}y = D_x\left(e^{x^2}y\right), \quad \text{so} \quad D_x\left(e^{x^2}y\right) = xe^{x^2}$$
and
$$e^{x^2}y = \int xe^{x^2}\,dx = \frac{e^{x^2}}{2} + C.$$
The general solution, therefore, has the form
$$y(x) = \frac{1}{2} + Ce^{-x^2}. \qquad \blacktriangleright$$

Example 2

Solve the initial-value problem
$$(\cos x + y)\,dx + \cot x\,dy = 0, \quad y(0) = 2 \quad \text{for } -\frac{\pi}{2} < x < \frac{\pi}{2}.$$

Solution

The equation can be expressed as

$$\cot x \frac{dy}{dx} + y = -\cos x.$$

Before determining the integrating factor, we divide by $\cot x$ so that the coefficient of dy/dx is 1, as in (17.8). This changes the equation to

$$\frac{dy}{dx} + (\tan x)y = -\frac{\cos x}{\cot x} = -\sin x.$$

The integrating factor for this equation is

$$I(x) = e^{\int \tan x\, dx} = e^{\ln|\sec x|} = |\sec x|.$$

Since $-\pi/2 < x < \pi/2$, $\sec x > 0$ and $I(x) = \sec x$. The equation becomes

$$\sec x \frac{dy}{dx} + (\sec x \tan x)y = -\sec x \sin x = -\tan x.$$

However,

$$D_x[(\sec x)y] = \sec x \frac{dy}{dx} + (\sec x \tan x)y,$$

so we can express the differential equation as

$$D_x[(\sec x)y] = -\tan x.$$

Consequently,

$$(\sec x)y = -\int \tan x\, dx + C = -\ln|\sec x| + C = \ln|\cos x| + C,$$

and the general solution is

$$y(x) = \cos x\, (\ln|\cos x| + C).$$

Note that this solution is valid only on intervals on which $\cos x \neq 0$. However, this condition must also be satisfied in order for the original differential equation to be defined.

Since $y(0) = 2$,

$$2 = \cos 0\, (\ln|\cos 0| + C) = C$$

and the solution to the initial-value problem is

$$y(x) = \cos x\, (\ln \cos x + 2), \quad \text{for } -\frac{\pi}{2} < x < \frac{\pi}{2}. \quad \blacktriangleright$$

Example 3

A tank with a capacity of 500 gal initially contains 200 gal of fresh water. Brine containing 1 lb/gal of salt runs into the tank at the rate of 3 gal/min and the homogeneous mixture runs out of the tank at the rate of 2 gal/min, as shown in Figure 17.4. Determine the amount of salt in the tank 3 hr after this process has begun.

Solution
Let $A(t)$ represent the number of pounds of salt in the tank t min after the process has begun. Then $A'(t)$, the change in the amount of salt with respect to time, is the difference between the rate at which the salt enters the tank and the rate at which the salt leaves the tank. This is given by

$$A'(t) = 3\,\frac{\text{gal}}{\text{min}} \cdot 1\,\frac{\text{lb}}{\text{gal}} - 2\,\frac{\text{gal}}{\text{min}} \cdot \frac{A(t)}{200+t}\,\frac{\text{lb}}{\text{gal}} = 3 - \frac{2A(t)}{200+t}\,\frac{\text{lb}}{\text{min}}.$$

When written in the form

$$A'(t) + 2(200+t)^{-1} A(t) = 3,$$

this linear equation has the integrating factor

$$I(t) = e^{\int 2(200+t)^{-1}dt} = e^{2\ln|200+t|} = e^{\ln(200+t)^2} = (200+t)^2.$$

Multiplying by this integrating factor and integrating gives

$$(200+t)^2 A(t) = \int 3(200+t)^2\,dt + C = (200+t)^3 + C,$$

so the general solution is

$$A(t) = (200+t) + C(200+t)^{-2}.$$

Since $A(0) = 0$,

$$0 = 200 + C(200)^{-2} \quad \text{and} \quad C = -(200)^3.$$

Thus,

$$A(t) = 200 + t - (200)^3(200+t)^{-2}.$$

After 3 hr, that is, when $t = 180$ min, there are

$$A(180) = 380 - (200)^3(380)^{-2} \approx 324.6$$

pounds of salt in the tank. ▸

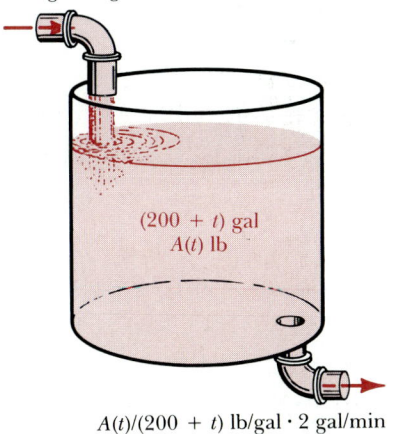

Figure 17.4

1 lb/gal · 3 gal/min

$(200 + t)$ gal
$A(t)$ lb

$A(t)/(200 + t)$ lb/gal · 2 gal/min

Bernoulli differential equations have the form

(17.11)
$$\frac{dy}{dx} + P(x)y = Q(x)y^n.$$

A Bernoulli equation is linear when $n = 0$ or $n = 1$. For other values of n, a substitution of the form

$$w = y^{1-n}$$

converts the Bernoulli equation in y into a linear differential equation in w. We first divide Equation (17.11) by y^n:

$$y^{-n}\frac{dy}{dx} + y^{1-n} P(x) = Q(x).$$

With $w = y^{1-n}$,

$$\frac{dw}{dx} = (1-n)y^{-n}\frac{dy}{dx},$$

and the equation becomes

$$\frac{1}{1-n}\frac{dw}{dx} + P(x)w = Q(x),$$

or

(17.12) $$\frac{dw}{dx} + (1-n)P(x)w = (1-n)Q(x).$$

Equation (17.12) is linear and can be solved for $w(x)$ using the integrating factor

$$I(x) = e^{\int (1-n)P(x)\,dx}.$$

The solution $y(x)$ to Equation (17.11) can then be found from

$$y(x) = [w(x)]^{1/(1-n)}.$$

Example 4

Solve the Bernoulli differential equation

$$\frac{dy}{dx} + y = xy^3.$$

Solution

Let $w = y^{1-3} = y^{-2}$. Then $dw/dx = -2y^{-3}\,dy/dx$. The substitution of w and dw/dx is easier if the equation is first divided by y^3:

$$y^{-3}\frac{dy}{dx} + y^{-2} = x.$$

Then

$$-\frac{1}{2}\frac{dw}{dx} + w = x \quad \text{or} \quad \frac{dw}{dx} - 2w = -2x.$$

The integrating factor

$$I(x) = e^{\int -2\,dx} = e^{-2x}$$

produces

$$e^{-2x}\frac{dw}{dx} - 2e^{-2x}w = -2xe^{-2x}.$$

Since

$$D_x(e^{-2x}w) = e^{-2x}\frac{dw}{dx} - 2e^{-2x}w,$$

we have
$$D_x(e^{-2x}w) = -2xe^{-2x}$$
and
$$e^{-2x}w = -2\int xe^{-2x}\,dx + C.$$

Integrating by parts with $u = x$ and $dv = e^{-2x}\,dx$ gives $du = dx$ and $v = -\tfrac{1}{2}e^{-2x}$. Thus
$$\int xe^{-2x}\,dx = -\frac{1}{2}xe^{-2x} + \frac{1}{2}\int e^{-2x}\,dx$$
$$= -\frac{1}{2}xe^{-2x} - \frac{1}{4}e^{-2x} + C$$

so
$$e^{-2x}w = xe^{-2x} + \frac{1}{2}e^{-2x} + C,$$

and
$$w(x) = e^{2x}\left[xe^{-2x} + \frac{1}{2}e^{-2x} + C\right]$$
$$= x + \frac{1}{2} + Ce^{2x} = \frac{2x + 1 + Ce^{2x}}{2}.$$

Since $w = y^{-2}$,
$$\frac{1}{y^2} = \frac{2x + 1 + Ce^{2x}}{2}$$

and
$$y^2 = \frac{2}{2x + 1 + Ce^{2x}}. \quad \blacktriangleright$$

The logistic equation for population growth was described in Section 6.7. It has the form
$$P'(t) = kP(t)(L - P(t)),$$
where k is a constant describing the growth of the population and L is the limit for the population. In Section 6.7 we found the solution to the logistic equation by treating it as a separable equation and using a partial-fractions technique for the integration. It is also a Bernoulli equation since it has the form
$$P'(t) - kLP(t) = -k[P(t)]^2.$$

Exercise 30 asks you to solve the logistic equation in this manner.

► EXERCISE SET 17.4

In Exercises 1–20, find a general solution to the differential equation.

1. $y' - 2y = e^x$
2. $y' + 5y = e^{-x}$
3. $y' + 3y = xe^{-x}$
4. $y' + 3y = x + 1$
5. $y' + y = \sin x$
6. $y' + \dfrac{2}{x} y = 6x$
7. $y' + \dfrac{3}{x} y = e^x$
8. $y' + 3xy = 6x$
9. $y' + \dfrac{1}{x^2} y = \dfrac{1}{x^2}$
10. $y' + \dfrac{1 + x}{x} y = \dfrac{2}{x}$
11. $y' + y \tan x = \cos x$
12. $y' + y \tan x = \sec x$
13. $y' + y \tan x = x \cos x$
14. $y' + y \tan x = x \sec x$
15. $(x^2 + x)y' + (2x + 1)y = x^2 - 1$
16. $\dfrac{(x^2 + 1)}{2x} y' + y = e^{x^2}$
17. $y' + y = \dfrac{x}{y}$
18. $y' + xy = \dfrac{x}{y}$
19. $y' + \dfrac{y}{x} = \dfrac{y^2}{x}$
20. $4xy' + y = -4 x e^{\sqrt{x}} y^3$

21. Suppose the brine runs into the tank described in Example 3 until it is filled to capacity. Determine the amount of salt in the tank at that time.

22. Suppose the rates of flow of the liquids in Example 3 are reversed with the brine running in at 2 gal/min and the mixture running out at 3 gal/min. How much salt is in the tank at the end of 3 hr?

23. An exhaust fan is turned on in a 16,000-ft³ smoke-filled room with a carbon dioxide content of 0.25%. The fan brings in air with a carbon dioxide content of 0.05% at the rate of 300 ft³/min and the uniformly mixed air leaves the room at the same rate. Assuming that no additional carbon dioxide is being produced in the room, how long does it take for the level of carbon dioxide to reach 0.15%?

24. Recompute the answer to the question in Exercise 23 under the additional assumption that carbon dioxide is being produced in the room at the rate of 0.25 ft³/min.

25. An object that falls in air has a force retarding its motion due to air resistance. Suppose this force is directly proportional to the velocity $v(t)$ and inversely proportional to the mass m of the object. In this case, the velocity of the object is described by the differential equation

$$v'(t) = \dfrac{\rho}{m} v(t) - g,$$

where g is the acceleration due to gravity.
a) Solve this for the velocity $v(t)$.
b) Suppose the position of the object at time t is given by $x(t)$. Use the result of part (a) to determine $x(t)$.

26. In certain circumstances it is reasonable to assume that the retarding force due to air resistance is directly proportional to the *square* of the velocity of the object, rather than to the velocity. Determine $v(t)$ in this case.

27. Show that if y_1 and y_2 are solutions to the linear equation $y' + P(x)y = Q(x)$, then for any constant C, $C(y_1 - y_2)$ is a solution to the linear equation $y' + P(x)y = 0$.

28. An equation of the form $y' = P(x)y^2 + Q(x)y + R(x)$ is called a *Riccati equation*. Suppose y_1 is a particular solution to this equation and v is any solution to the linear equation

$$v' + [Q(x) + 2P(x)y_1]v = -P(x).$$

Show that $y = y_1 + 1/v$ also satisfies the Riccati equation.

29. A particular solution to the Riccati equation $y' = x^2 + y^2 - 2xy + 1$ is $y_1(x) = x$. Use the method described in Exercise 28 to find a solution satisfying $y(2) = 1$.

30. a) Find the general solution to the logistic equation by treating it as a Bernoulli equation.
b) Show that the solution found in part (a) agrees with that given in Section 6.7.

17.5
Second-Order Linear Equations: Homogeneous Type

An nth-order linear differential equation has the form

(17.13)
$$y^{(n)} + P_1(x)y^{(n-1)} + P_2(x)y^{(n-2)} + \cdots + P_{n-1}(x)y' + P_n(x)y = Q(x).$$

Two separate cases are distinguished.

i) When $Q(x) \equiv 0$ the equation is called **homogeneous.**
ii) When $Q(x) \neq 0$ the equation is called **nonhomogeneous.**

Homogeneous equations are considered in this section and nonhomogeneous equations in Section 17.6. Note that the use of the term homogeneous in the case of these second-order equations indicates something quite different from the use of this term in the first-order equations considered in Section 17.2.

Homogeneous linear differential equations have the property that for any two solutions y_1 and y_2 and any pair of constants C_1 and C_2, the function described by

$$y(x) = C_1 y_1(x) + C_2 y_2(x)$$

is also a solution of the homogeneous equation. (This result is discussed for the case $n = 2$ in Exercise 43.)

There is no standard elementary technique for solving *general* homogeneous nth-order linear differential equations. The method we discuss is for the special case when the functions $P_1, P_2, \ldots, P_n$ are all constants. A linear equation of this type is called a **linear equation with constant coefficients.** To simplify matters even further, we generally consider the case when $n = 2$. The method of solution when $n > 2$ is similar, as illustrated in Example 6.

In summary, this section is concerned with solving differential equations of the form

(17.14)
$$y''(x) + p\, y'(x) + q\, y(x) = 0,$$

where p and q are constants.

Let us first consider the form of the solution of a first-order equation of the form

$$y'(x) + q\, y(x) = 0, \quad \text{or equivalently,} \quad y'(x) = -q\, y(x).$$

This equation is associated with exponential growth and decay considered in Section 6.6 and has the solution

$$y(x) = Ce^{-qx}.$$

This provides motivation for the conjecture that the second-order linear equation (17.14) has solutions of the form

$$y(x) = e^{mx}$$

for properly chosen constants m. To test this conjecture, we compute the derivatives of y,

$$y'(x) = me^{mx} \quad \text{and} \quad y''(x) = m^2 e^{mx},$$

and determine any constants m that ensure that (17.14) is satisfied. Thus,

$$0 = y'' + p\,y' + q\,y = m^2 e^{mx} + pme^{mx} + qe^{mx} = (m^2 + pm + q)e^{mx}.$$

Since $e^{mx} > 0$, the equation is satisfied precisely when

(17.15) $$0 = m^2 + pm + q.$$

Equation (17.15) is called the **characteristic equation** of Equation (17.14) and has solutions

$$m_1 = \frac{-p}{2} + \frac{\sqrt{p^2 - 4q}}{2} \quad \text{and} \quad m_2 = \frac{-p}{2} - \frac{\sqrt{p^2 - 4q}}{2}.$$

The solutions to Equation (17.14) assume three different forms depending on the value of $p^2 - 4q$.

Case I: $p^2 - 4q > 0$. When $p^2 - 4q > 0$, m_1 and m_2 are distinct real numbers. Since

(17.16) $$y_1(x) = e^{m_1 x} \quad \text{and} \quad y_2(x) = e^{m_2 x}$$

are both solutions of the homogeneous equation (17.14),

(17.17) $$y(x) = C_1 e^{m_1 x} + C_2 e^{m_2 x}$$

is also a solution to (17.14) for any constants C_1 and C_2. In fact, when $p^2 - 4q > 0$, all solutions to (17.14) must be of this form.

Example 1

Solve the differential equation

$$y'' - 5y' + 6y = 0.$$

Solution

If $y(x) = e^{mx}$, then

$$y'' - 5y' + 6y = m^2 e^{mx} - 5me^{mx} + 6e^{mx} = (m^2 - 5m + 6)e^{mx}.$$

Factoring the characteristic equation,

$$0 = m^2 - 5m - 6 = (m - 3)(m - 2),$$

shows that $m = 3$ or $m = 2$. The general solution is therefore

$$y(x) = C_1 e^{3x} + C_2 e^{2x}.$$ ▶

17.5 SECOND-ORDER LINEAR EQUATIONS: HOMOGENEOUS TYPE

Example 2

Solve the initial-value problem
$$y'' - 3y' - 4y = 0, \quad y(0) = 0, \quad y'(0) = 1.$$

Solution
Let $y = e^{mx}$. Then
$$y'' - 3y' - 4y = m^2 e^{mx} - 3me^{mx} - 4e^{mx} = (m^2 - 3m - 4)e^{mx},$$
so the characteristic equation is
$$0 = (m^2 - 3m - 4) = (m - 4)(m + 1).$$
The general solution is
$$y(x) = C_1 e^{4x} + C_2 e^{-x}.$$
Since $y(0) = 0$,
$$0 = C_1 e^0 + C_2 e^0 = C_1 + C_2.$$
Also, $y'(x) = 4C_1 e^{4x} - C_2 e^{-x}$ and $y'(0) = 1$, so
$$1 = 4C_1 e^0 - C_2 e^0 = 4C_1 - C_2.$$
Adding the equations gives
$$1 = 5C_1, \quad \text{so} \quad C_1 = \frac{1}{5}.$$
Since $C_1 + C_2 = 0$,
$$C_2 = -C_1 = -\frac{1}{5}.$$
The particular solution to the initial-value problem is
$$y(x) = \frac{1}{5} e^{4x} - \frac{1}{5} e^{-x}. \qquad \blacktriangleright$$

Case II: $p^2 - 4q = 0$. When $p^2 - 4q = 0$, the characteristic equation has only one root (a double root), and only one solution to Equation (17.14) is of the form $y(x) = e^{mx}$. This solution is
$$y_1(x) = e^{m_1 x} = e^{-px/2}.$$
It can be verified (see Exercise 44) that in this case an additional solution to (17.14) is
$$y_2(x) = xe^{m_1 x} = xe^{-px/2}.$$
The general solution to Equation (17.14) when $p^2 - 4q = 0$ is

(17.18) $$y(x) = C_1 e^{m_1 x} + C_2 x e^{m_1 x}.$$

Example 3

Find the solution to the initial-value problem

$$y'' - 4y' + 4y = 0, \quad y(0) = 1, \quad y'(0) = 0.$$

Solution

The characteristic equation $m^2 - 4m + 4 = 0$ has the double root $m = 2$. Thus the general solution is

$$y(x) = C_1 e^{2x} + C_2 x e^{2x}.$$

Satisfying the initial condition $y(0) = 1$ requires that

$$1 = C_1 e^0 + C_2 \cdot 0 \cdot e^0 = C_1.$$

Since $y'(x) = 2C_1 e^{2x} + C_2 e^{2x} + 2C_2 x e^{2x}$, satisfying $y'(0) = 0$ requires that

$$0 = 2C_1 e^0 + C_2 e^0 + 2C_2 \cdot 0 e^0 = 2C_1 + C_2.$$

Thus, $C_2 = -2C_1 = -2$, and the particular solution is

$$y(x) = e^{2x} - 2x e^{2x}. \quad \blacktriangleright$$

Case III: $p^2 - 4q < 0$. When $p^2 - 4q < 0$, the roots m_1 and m_2 are distinct but complex numbers. To simplify notation, let $r = -p/2$ and $s = \sqrt{4q - p^2}/2$. As discussed in Section 8.9, the symbol i is used to distinguish the nonreal, or imaginary, part of a complex number and has the property that $i^2 = -1$. Using this notation, we can express m_1 and m_2 as

$$m_1 = -\frac{p}{2} + \frac{\sqrt{p^2 - 4q}}{2} = -\frac{p}{2} + \frac{\sqrt{4q - p^2}}{2} i = r + is$$

and

$$m_2 = -\frac{p}{2} - \frac{\sqrt{p^2 - 4q}}{2} = r - is.$$

The general solution can then be expressed as

$$\begin{aligned} y(x) &= C_1 e^{m_1 x} + C_2 e^{m_2 x} \\ &= C_1 e^{(r+is)x} + C_2 e^{(r-is)x} \\ &= C_1 e^{rx} e^{sxi} + C_2 e^{rx} e^{-sxi} \\ &= e^{rx} (C_1 e^{sxi} + C_2 e^{-sxi}). \end{aligned}$$

We can simplify the general solution using Euler's equation:

$$e^{ix} = \cos x + i \sin x.$$

(This equation was introduced in Section 8.9 as an application of Taylor series.) Using this equation we see that the general solution becomes

$$y(x) = e^{rx} [C_1(\cos sx + i \sin sx) + C_2(\cos(-sx) + i \sin(-sx))]$$

or, since $\cos(-sx) = \cos sx$ and $\sin(-sx) = -\sin sx$,

$$y(x) = e^{rx} [(C_1 + C_2) \cos sx + i(C_1 - C_2) \sin sx].$$

Thus, the general solution can be expressed as

(17.19) $$y(x) = e^{rx}(K_1 \cos sx + K_2 \sin sx),$$

where K_1 and K_2 are arbitrary complex constants. These constants, however, must be real numbers in order for the solutions to be real.

Example 4

Find the general solution to the differential equation
$$y'' + 25y = 0.$$

Solution
The characteristic equation $m^2 + 25 = 0$ has roots
$$m_1 = 5i \quad \text{and} \quad m_2 = -5i, \quad \text{so} \quad r = 0 \quad \text{and} \quad s = 5.$$
The general solution is thus
$$y(x) = e^{0x}(K_1 \cos 5x + K_2 \sin 5x)$$
$$= K_1 \cos 5x + K_2 \sin 5x. \quad \blacktriangleright$$

Example 5

Solve the initial-value problem
$$y'' - 2y' + 10y = 0, \quad y\left(\frac{\pi}{2}\right) = 0, \quad y'\left(\frac{\pi}{2}\right) = 1.$$

Solution
The characteristic equation $m^2 - 2m + 10 = 0$ has roots
$$m_1 = \frac{2 + \sqrt{-36}}{2} = 1 + 3i \quad \text{and} \quad m_2 = \frac{2 - \sqrt{-36}}{2} = 1 - 3i,$$
so $r = 1$ and $s = 3$. The general solution is
$$y(x) = e^x(K_1 \cos 3x + K_2 \sin 3x).$$
Since $y(\pi/2) = 0$,
$$0 = e^{\pi/2}\left(K_1 \cos \frac{3\pi}{2} + K_2 \sin \frac{3\pi}{2}\right) = e^{\pi/2} K_2(-1),$$
which implies that $K_2 = 0$ and
$$y(x) = e^x(K_1 \cos 3x).$$
Since $y'(x) = e^x(K_1 \cos 3x - 3K_1 \sin 3x)$ and $y'(\pi/2) = 1$,
$$1 = e^{\pi/2}\left(K_1 \cos \frac{3\pi}{2} - 3K_1 \sin \frac{3\pi}{2}\right) = e^{\pi/2}(-3K_1)(-1) = 3e^{\pi/2} K_1.$$
Thus,
$$K_1 = \frac{1}{3} e^{-\pi/2}$$

and the particular solution to the initial-value problem is

$$y(x) = \frac{1}{3} e^{-\pi/2} e^x \cos 3x$$

$$= \frac{1}{3} e^{(x-\pi/2)} \cos 3x.$$ ▶

Higher-order linear homogeneous equations with constant coefficients are solved in the same manner used for equations of order two. The final example in this section involves an initial-value problem of order three.

Example 6

Solve the initial-value problem

$$y''' - 4y'' + 5y' - 2y = 0, \quad y(0) = 0, \quad y'(0) = 1, \quad y''(0) = 0.$$

Solution

The characteristic equation for this equation is

$$0 = m^3 - 4m^2 + 5m - 2 = (m-1)^2(m-2),$$

which has a double root at $m = 1$ and a single root at $m = 2$. Following the pattern of the second-order equations, we see that the complete solution is

$$y(x) = C_1 e^x + C_2 x e^x + C_3 e^{2x}.$$

The initial condition $y(0) = 0$ gives $0 = C_1 + C_3$. Therefore, $C_3 = -C_1$ and

$$y(x) = C_1(e^x - e^{2x}) + C_2 x e^x.$$

Since

$$y'(x) = C_1(e^x - 2e^{2x}) + C_2(e^x + x e^x),$$

the initial condition $y'(0) = 1$ gives $1 = -C_1 + C_2$. Thus, $C_2 = C_1 + 1$, so

$$y(x) = C_1(e^x - e^{2x} + x e^x) + x e^x,$$
$$y'(x) = C_1(2e^x - 2e^{2x} + x e^x) + e^x + x e^x,$$

and

$$y''(x) = C_1(3e^x - 4e^{2x} + x e^x) + 2e^x + x e^x.$$

The final initial condition $y''(0) = 0$ gives

$$0 = C_1(-1) + 2, \quad \text{so} \quad C_1 = 2$$

and the solution to the initial-value problem is

$$y(x) = 2e^x - 2e^{2x} + 3x e^x.$$ ▶

EXERCISE SET 17.5

In Exercises 1–28, find the general solution to the differential equation.

1. $y'' - 9y = 0$
2. $y'' - 6y' + 4y = 0$
3. $y'' + 6y' + 9y = 0$
4. $y'' - 2y' + y = 0$
5. $y'' - 4y' - 5y = 0$
6. $y'' + 5y' + 4y = 0$
7. $y'' - 2y' + 5y = 0$
8. $y'' + 4y' + 13y = 0$
9. $y'' - 2y' = 0$
10. $y'' + 2y' = 0$
11. $y'' + 4y' + 2y = 0$
12. $y'' - 6y' + 9y = 0$
13. $y'' - 7y' - 8y = 0$
14. $y'' + 2y' + 2y = 0$
15. $y'' + 6y' + 5y = 0$
16. $y'' + 8y' + 4y = 0$
17. $y'' + 9y = 0$
18. $y'' + 3y' + 4y = 0$
19. $2y'' + 3y' + y = 0$
20. $2y'' - 2y' + y = 0$
21. $4y'' + 9y = 0$
22. $4y'' - 9y = 0$
23. $5y'' + 6y' - 9y = 0$
24. $3y'' + 7y' - 6y = 0$
25. $y''' - 3y'' + 4y = 0$
26. $y''' - y = 0$
27. $y''' - y'' + y' - y = 0$
28. $y''' - 3y'' - 4y' = 0$

In Exercises 29–36, find the particular solution to the initial-value problem.

29. $y'' - 4y' + 4y = 0$, $y(0) = 0$, $y'(0) = 1$
30. $y'' - 4y' + 3y = 0$, $y(0) = 0$, $y'(0) = 1$
31. $y'' + 6y' + 13y = 0$, $y(0) = 1$, $y'(0) = 0$
32. $y'' - 3y' - 4y = 0$, $y(0) = 1$, $y'(0) = 4$
33. $8y'' - 14y' + 3y = 0$, $y(0) = 1$, $y'(0) = 3$
34. $2y'' - 6y' + 9y = 0$, $y(\pi) = 0$, $y'(\pi) = e^{3\pi}$
35. $y''' - 7y' + 6y = 0$, $y(0) = 1$, $y'(0) = 0$, $y''(0) = 0$
36. $y''' - 4y'' - 3y' + 18y = 0$, $y(0) = 0$, $y'(0) = 0$, $y''(0) = 5$

In Exercises 37–40, find the particular solution to the boundary-value problem.

37. $y'' + y = 0$, $y(0) = 1$, $y(\pi/2) = -1$
38. $y'' - 2y' + 5y = 0$, $y(0) = 1$, $y(\pi/3) = 0$
39. $y'' + y' = 0$, $y(0) = 2$, $y(1) = 1$
40. $3y'' - 5y' + 2y = 0$, $y(0) = 1$, $y(\ln 8) = 2$

41. The solutions to the characteristic equation of a certain fourth-order homogeneous linear differential equation with constant coefficients are $-5, 1, 3,$ and 3. Give the general solution.

42. The solutions to the characteristic equation of a certain seventh-order homogeneous linear differential equation with constant coefficients are $-5, -5, -1, 0, 1, 2,$ and 2. Give the general solution.

43. Suppose $y_1(x)$ and $y_2(x)$ are solutions to
$$y'' + P_1(x)y' + P_2(x)y = 0.$$
Show that for any constants C_1 and C_2,
$$y(x) = C_1 y_1(x) + C_2 y_2(x)$$
is also a solution to this equation.

44. Consider the differential equation $y'' + py' + qy = 0$, where $p^2 - 4q = 0$. Show that $y(x) = xe^{-px/2}$ is a solution to this equation.

45. a) Show that the general solution to $y'' - y = 0$ can be expressed either as
$$y = C_1 e^x + C_2 e^{-x}$$
or as
$$y = K_1 \sinh x + K_2 \cosh x.$$
b) Find the constants C_1, C_2 and K_1, K_2 that give the particular solution satisfying $y(0) = 1$ and $y'(0) = 0$.
c) Show that the two forms of the solutions are equivalent.

46. An equation of the form $y'' + ay' = 0$ can be solved by solving the two first-order equations $w' + aw = 0$ and $y' = w$, since $w' = C_1 e^{-ax}$ and $y = \int w \, dx$. Use this method to solve the differential equations in Exercises 9 and 10.

47. A particle of mass m that moves along the y-axis according to the equation
$$m \frac{d^2 y}{dt^2} = -ky,$$
where k is a positive constant, is said to be in *simple harmonic motion* with period $2\pi/\omega$, where $\omega^2 = k/m$ (see the figure). Solve this differential equation.

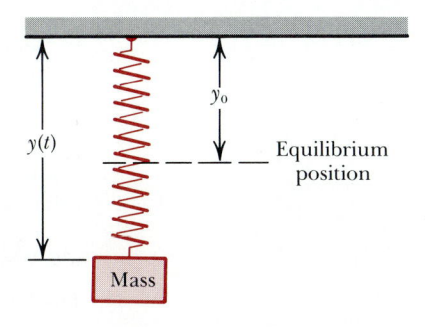

48. A particle moves along the y-axis in a simple harmonic motion. The period of the particle is 2π, $y(0) = 0$, and $y'(0) = 1$.
   a) Solve the resulting differential equation.
   b) Sketch the motion of the particle with t on the horizontal axis and $y(t)$ on the vertical axis.

A simple model for a vibrating system consisting of a mass m, a spring with spring constant k, a shock absorber with damping constant c, and no externally imposed force is described by the initial-value problem

$$my''(t) + cy'(t) + ky(t) = 0, \quad y(0) = y_0, \quad y'(0) = y_0'$$

(see the figure). In this equation, $y(t)$ is the distance of the mass below its equilibrium position at time t and describes the motion of the mass, called *harmonic motion*. When $c = 0$, this equation describes simple harmonic motion as in Exercise 47. Find an expression for the motion of the vibrating systems described in Exercises 49–52.

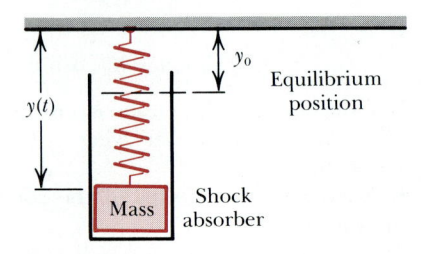

49. $m = 1$ slug, $k = 16$ lb/ft, $c = 0$, $y_0 = 0$, $y_0' = 0$
50. $m = 4$ kg, $k = 4$ N/m, $c = 0$, $y_0 = 0$, $y_0' = 1$
51. $m = 1$ slug, $k = 16$ lb/ft, $c = 4$ lb sec/ft, $y_0 = 0$, $y_0' = 1$
52. $m = 4$ kg, $k = 4$ N/m, $c = 3$ dyne sec/meter, $y_0 = 1$, $y_0' = 0$ (1 dyne = 10^{-5} N)

A simple model for an electrical circuit containing, in series, an inductance L, a resistance R, a capacitance C, and no impressed voltage is described by the initial-value problem

$$LI'(t) + RI(t) + \frac{1}{C} Q(t) = 0, \quad Q(0) = Q_0, \quad I(0) = I_0$$

(see the figure). In this equation, $Q(t)$ is the charge on the capacitor at time t and $I(t) = Q'(t)$ is the current at time t. (The close relationship between the electric-circuit differential equation and the differential equation for vibrating systems and other physical phenomena is the basis for analog computers.) Find an expression for the charge and current in the electrical systems described in Exercises 53–56.

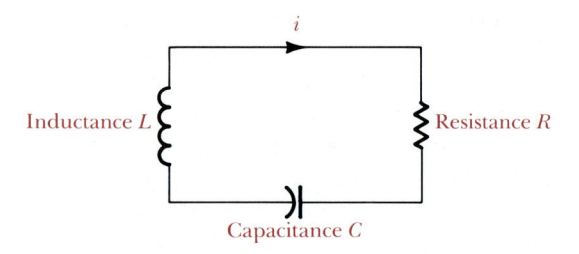

53. $L = 0.25$ henries, $R = 5$ ohms, $C = 10^{-5}$ farads, $I_0 = 0$, $Q_0 = 0$
54. $L = 0.5$ henries, $R = 4$ ohms, $C = 2 \times 10^{-6}$ farads, $I_0 = 0$, $Q_0 = 0$
55. $L = 0.25$ henries, $R = 5$ ohms, $C = 10^{-5}$ farads, $I_0 = 0$, $Q_0 = 10^{-6}$ coulombs
56. $L = 0.5$ henries, $R = 4$ ohms, $C = 2 \times 10^{-6}$ farads, $I_0 = 0$, $Q_0 = 10^{-6}$ coulombs

17.6

Second-Order Linear Equations: Nonhomogeneous Type

In Section 17.5 we discussed the various forms that can be assumed by solutions to the second-order linear homogeneous equation

(17.20) $$y''(x) + p\, y'(x) + q\, y(x) = 0,$$

when p and q are constants. In this section we discuss methods for solving the nonhomogeneous counterpart to this equation:

(17.21) $$y''(x) + p\, y'(x) + q\, y(x) = Q(x), \quad Q(x) \neq 0.$$

The following result indicates the reason for first considering the homogeneous problem.

(17.22)
THEOREM
If $y_c(x) = C_1 y_1(x) + C_2 y_2(x)$ is the general solution to the homogeneous problem (17.20) and y_p is any solution to the nonhomogeneous problem (17.21), then the general solution to (17.21) has the form

$$y(x) = C_1 y_1(x) + C_2 y_2(x) + y_p(x).$$

The proof of this theorem follows by observing that y is a solution to (17.21) if and only if $y - y_p$ is a solution to (17.20).

The methods of Section 17.5 can be used to find the general solution to the homogeneous equation. Theorem (17.22) implies that the problem of finding a general solution to the nonhomogeneous equation is reduced to finding a single, or *particular*, solution to the nonhomogeneous equation.

We present two methods for finding a particular solution to the nonhomogeneous equation. The first technique is called the **variation-of-parameters** method. In theory, this method can be used to solve any nonhomogeneous problem, regardless of the form of $Q(x)$. In practice, it often involves integrals that are very difficult or impossible to evaluate.

The second technique for finding a particular solution of a nonhomogeneous operator is called the **method of undetermined coefficients.** This method is readily applied, but can be used only when $Q(x)$ is the sum and/or product of sine, cosine, polynomial, or exponential functions. This feature limits its use, although many of the nonhomogeneous problems important for applications have $Q(x)$ in this form.

Variation of Parameters

The variation-of-parameters method assumes that there is a particular solution to the nonhomogeneous problem (17.21) of the form

(17.23) $$y_p(x) = u_1(x) y_1(x) + u_2(x) y_2(x),$$

where

(17.24) $$u_1'(x) y_1(x) + u_2'(x) y_2(x) \equiv 0$$

and

$$C_1 y_1(x) + C_2 y_2(x)$$

is the general solution to the corresponding homogeneous equation (17.20). Differentiating y_p and using condition (17.24) implies that

(17.25) $$y_p'(x) = u_1'(x) y_1(x) + u_1(x) y_1'(x) + u_2'(x) y_2(x) + u_2(x) y_2'(x)$$
$$= u_1(x) y_1'(x) + u_2(x) y_2'(x).$$

so

(17.26) $$y_p''(x) = u_1'(x) y_1'(x) + u_1(x) y_1''(x) + u_2'(x) y_2'(x) + u_2(x) y_2''(x).$$

Substituting (17.25) and (17.26) into Equation (17.21) and using the fact that y_1 and y_2 are solutions to the homogeneous problem produces

$$\begin{aligned}Q(x) &= y_p''(x) + p\,y_p'(x) + q\,y_p(x) \\ &= (u_1'y_1' + u_1y_1'' + u_2'y_2' + u_2y_2'') + p(u_1y_1' + u_2y_2') + q(u_1y_1 + u_2y_2) \\ &= u_1(y_1'' + py_1' + qy_1) + u_2(y_2'' + py_2' + qy_2) + u_1'y_1' + u_2'y_2' \\ &= u_1 \cdot 0 + u_2 \cdot 0 + u_1'y_1' + u_2'y_2'.\end{aligned}$$

Thus,

(17.27) $$u_1'(x)y_1'(x) + u_2'(x)y_2'(x) = Q(x).$$

To solve equations (17.24) and (17.27) for $u_1'(x)$, we multiply Equation (17.24) by $y_2'(x)/y_2(x)$ and subtract from Equation (17.27) to get

$$u_1'(x)[y_1'(x) - y_1(x)y_2'(x)/y_2(x)] = Q(x).$$

So

(17.28) $$u_1'(x) = \frac{Q(x)y_2(x)}{y_1'(x)y_2(x) - y_1(x)y_2'(x)}.$$

Similarly,

(17.29) $$u_2'(x) = \frac{Q(x)y_1(x)}{y_1(x)y_2'(x) - y_1'(x)y_2(x)}.$$

A particular solution y_p to the nonhomogeneous equation is found by integrating (17.28) and (17.29) to find u_1 and u_2 and then substituting these into (17.23).

The general solution to the nonhomogeneous equation (17.21) is

(17.30) $$y(x) = C_1y_1(x) + C_2y_2(x) + y_p(x).$$

Example 1

Find the solution to the initial-value problem

$$y'' - 2y' + y = \frac{e^x}{x^2}, \qquad y(1) = 0, \qquad y'(1) = 0, \qquad x > 0.$$

Solution

The first step is to determine the solution to the homogeneous equation

$$y'' - 2y' + y = 0,$$

whose characteristic equation is

$$0 = (m^2 - 2m + 1) = (m - 1)^2.$$

Thus the general solution to the homogeneous equation is

$$y(x) = C_1e^x + C_2xe^x.$$

A particular solution to the nonhomogeneous equation is assumed to have the form

$$y_p(x) = u_1(x)e^x + u_2(x)xe^x,$$

17.6 SECOND-ORDER LINEAR EQUATIONS: NONHOMOGENEOUS TYPE

where
$$u_1'(x)e^x + u_2'(x)xe^x = 0.$$

Differentiating y_p and substituting into the nonhomogeneous equation leads to the result from Equation (17.27):
$$u_1'(x)e^x + u_2'(x)(e^x + xe^x) = \frac{e^x}{x^2}.$$

The equations can then be solved for u_1' and u_2' or we can use Equations (17.28) and (17.29) to obtain
$$u_1'(x) = \frac{\left(\dfrac{e^x}{x^2}\right)xe^x}{e^x xe^x - e^x(e^x + xe^x)} = -\frac{1}{x}$$

and
$$u_2'(x) = \frac{\left(\dfrac{e^x}{x^2}\right)e^x}{e^x(e^x + xe^x) - e^x xe^x} = \frac{1}{x^2}.$$

So
$$u_1(x) = -\int \frac{1}{x}\,dx = -\ln x + K_1$$

and
$$u_2(x) = \int \frac{1}{x^2}\,dx = -\frac{1}{x} + K_2.$$

Since we are interested in only one particular solution, we can take the constants K_1 and K_2 to be zero. Thus,
$$y_p(x) = u_1(x)y_1(x) + u_2(x)y_2(x) = (-\ln x)e^x - \frac{1}{x}xe^x = -e^x - e^x \ln x.$$

The general solution to the differential equation
$$y'' - 2y' + y = \frac{e^x}{x^2}$$

is therefore
$$y(x) = C_1 e^x + C_2 xe^x - e^x - e^x \ln x = (C_1 - 1)e^x + C_2 xe^x - e^x \ln x.$$

Since $C_1 - 1$ is an arbitrary constant, we replace $C_1 - 1$ by simply C_1 and write the solution as
$$y(x) = C_1 e^x + C_2 xe^x - e^x \ln x.$$

The initial condition $y(1) = 0$ implies that
$$0 = C_1 e + C_2 e = e(C_1 + C_2), \quad \text{so} \quad C_1 = -C_2.$$

Since
$$y'(x) = C_1 e^x + C_2 e^x + C_2 xe^x - e^x \ln x - e^x \frac{1}{x},$$

$y'(1) = 0$ implies that
$$0 = C_1 e + 2C_2 e - e.$$
Thus,
$$C_2 = 1 \quad \text{and} \quad C_1 = -1$$
and the solution to the initial-value problem is
$$y(x) = xe^x - e^x - e^x \ln x. \qquad \blacktriangleright$$

Undetermined Coefficients

The basis for the method of undetermined coefficients lies in the fact that certain common classes of functions have derivatives and integrals belonging to that same class. For example, differentiating or integrating a polynomial always produces a polynomial; differentiating or integrating an exponential function always produces the same type of exponential function. Similarly, the derivative or integral of a sine or cosine function can produce only another sine or cosine function.

Combining these observations implies that if $Q(x)$ in the equation
$$y'' + py' + qy = Q(x)$$
has a term of the form
$$(a_n x^n + a_{n-1} x^{n-1} + \cdots + a_1 x + a_0) e^{rx} \begin{Bmatrix} \sin kx \\ \text{or} \\ \cos kx \end{Bmatrix},$$
where r, k, and $a_0, a_1, \ldots, a_n$ are constant, then it can be produced from the differential equation only if the particular solution has a term of the form
$$(A_n x^n + \cdots + A_0) e^{rx} \sin kx + (B_n x^n + \cdots + B_0) e^{rx} \cos kx,$$
for some collection of constants $A_0, A_1, \ldots, A_n, B_0, B_1, \ldots, B_n$. The form must be modified by multiplying by powers of x if the term of $Q(x)$ happens to be a solution to the homogeneous equation.

Example 2

Find a particular solution to
$$y'' - y' - 2y = e^x - 2x^2.$$

Solution
To produce the term e^x, $y_p(x)$ must have a term of the form Ae^x. Similarly, to produce $-2x^2$ requires that a term of the form $B_2 x^2 + B_1 x + B_0$ be a part of $y_p(x)$. So we assume that $y_p(x)$ has the form
$$y_p(x) = Ae^x + B_2 x^2 + B_1 x + B_0$$
for some collection of constants A, B_0, B_1, and B_2.

17.6 SECOND-ORDER LINEAR EQUATIONS: NONHOMOGENEOUS TYPE

Substituting y_p and its derivatives into the differential equation implies that

$$e^x - 2x^2 = y_p''(x) - y_p'(x) - 2y_p(x)$$
$$= (Ae^x + 2B_2) - (Ae^x + 2B_2 x + B_1) - 2(Ae^x + B_2 x^2 + B_1 x + B_0)$$
$$= -2Ae^x - 2B_2 x^2 + (-2B_2 - 2B_1)x + (2B_2 - B_1 - 2B_0).$$

Comparing terms on the two sides of the equation enables us to solve for the constants:

$$e^x:\ 1 = -2A, \qquad \text{so } A = -\tfrac{1}{2};$$
$$x^2:\ -2 = -2B_2, \qquad \text{so } B_2 = 1;$$
$$x:\ 0 = -2B_2 - 2B_1, \qquad \text{so } B_1 = -B_2 = -1;$$
$$\text{constants:}\ 0 = 2B_2 - B_1 - 2B_0, \qquad \text{so } B_0 = \tfrac{1}{2}(2B_2 - B_1) = \tfrac{3}{2};$$

and

$$y_p(x) = x^2 - x + \frac{3}{2} - \frac{1}{2} e^x.$$

▶

Example 3

Solve the initial-value problem

$$y'' - 4y' + 4y = e^x \sin x, \qquad y(0) = 0, \qquad y'(0) = 0.$$

Solution

We first find the general solution to the homogeneous problem. The characteristic equation is $0 = m^2 - 4m + 4 = (m - 2)^2$, so $m = 2$. The general solution to the homogeneous problem is

$$y(x) = C_1 e^{2x} + C_2 x e^{2x}.$$

The form of the particular solution is dictated by the term $e^x \sin x$ to be

$$y_p(x) = Ae^x \sin x + Be^x \cos x.$$

Differentiating produces

$$y_p'(x) = (A - B)e^x \sin x + (A + B)e^x \cos x$$

and

$$y_p''(x) = -2Be^x \sin x + 2Ae^x \cos x.$$

So

$$e^x \sin x = y_p''(x) - 4y_p'(x) + 4y_p(x)$$
$$= -2Be^x \sin x + 2Ae^x \cos x - 4(A - B)e^x \sin x$$
$$\quad - 4(A + B)e^x \cos x + 4Ae^x \sin x + 4Be^x \cos x$$
$$= 2Be^x \sin x - 2Ae^x \cos x.$$

Thus, $B = \tfrac{1}{2}$, $A = 0$, and a particular solution is

$$y_p(x) = \tfrac{1}{2} e^x \cos x.$$

The general solution is

$$y(x) = C_1 e^{2x} + C_2 x e^{2x} + \tfrac{1}{2} e^x \cos x.$$

The initial condition $y(0) = 0$ implies that

$$0 = C_1 + \tfrac{1}{2}, \quad \text{so} \quad C_1 = -\tfrac{1}{2}.$$

Since

$$y'(x) = 2C_1 e^{2x} + C_2 e^{2x} + 2C_2 x e^{2x} + \tfrac{1}{2} e^x \cos x - \tfrac{1}{2} e^x \sin x,$$

the condition $y'(0) = 0$ implies that

$$0 = 2C_1 + C_2 + \tfrac{1}{2} \quad \text{and} \quad C_2 = -2C_1 - \tfrac{1}{2} = \tfrac{1}{2}.$$

Thus the solution to the initial-value problem is

$$y(x) = \frac{1}{2}(xe^{2x} - e^{2x} + e^x \cos x).$$

▶

The following example shows the type of modification required when a term of $Q(x)$ is a solution to the homogeneous equation.

Example 4

Find the general solution to the differential equation

$$y'' - 4y' + 4y = x + e^{2x}.$$

Solution

Normally, the particular solution would have the form

$$y_p(x) = A_1 x + A_0 + Be^{2x}.$$

However, we found in Example 3 that the solution to the homogeneous equation

$$y'' - 4y' + 4y = 0 \quad \text{is} \quad y(x) = C_1 e^{2x} + C_2 x e^{2x}.$$

This implies that the normal form for the particular solution, Be^{2x}, is a solution to the homogeneous equation. Therefore, this part of the particular solution must be modified. This modification involves multiplying Be^{2x} by the lowest integral power of x that produces a term that does not satisfy the homogeneous equation. Since xe^x is also a solution to $y'' - 4y' + 4y = 0$, the lowest power is 2 and the particular solution has the form

$$y_p(x) = A_1 x + A_0 + Bx^2 e^{2x}.$$

From this point, the problem is solved as in the preceding examples,

$$y_p'(x) = A_1 + 2Bxe^{2x} + 2Bx^2 e^{2x},$$
$$y_p''(x) = 2Be^{2x} + 8Bxe^{2x} + 4Bx^2 e^{2x},$$

so

$$x + e^{2x} = 2Be^{2x} + 8Bxe^{2x} + 4Bx^2 e^{2x} - 4(A_1 + 2Bxe^{2x} + 2Bx^2 e^{2x})$$
$$+ 4(A_1 x + A_0 + Bx^2 e^{2x})$$
$$= 2Be^{2x} + 4A_1 x + -4A_1 + 4A_0.$$

17.6 SECOND-ORDER LINEAR EQUATIONS: NONHOMOGENEOUS TYPE

Comparing the coefficients on each side gives us

$$e^{2x}: \quad 1 = 2B, \qquad \text{so } B = \tfrac{1}{2};$$
$$x: \quad 1 = 4A_1, \qquad \text{so } A_1 = \tfrac{1}{4};$$
$$\text{constants:} \quad 0 = -4A_1 + 4A_0, \qquad \text{so } A_0 = A_1 = \tfrac{1}{4}.$$

This implies that a particular solution is

$$y_p(x) = \tfrac{1}{4}(x + 1) + \tfrac{1}{2}x^2 e^{2x}.$$

The general solution to this problem is

$$y(x) = C_1 e^{2x} + C_2 x e^{2x} + \frac{1}{4}(x + 1) + \frac{1}{2} x^2 e^{2x}. \quad \blacktriangleright$$

The final example considers a nonhomogeneous equation of order three. Note that the solution follows the same form as that used to solve the second-order equations.

Example 5

Find the general solution to the differential equation

$$y''' - 4y'' + 5y' - 2y = 2x + 3e^{-x}.$$

Solution

In Example 6 of Section 17.5 we found that the general solution to the homogeneous equation

$$y''' - 4y'' + 5y' - 2y = 0$$

is

$$y(x) = C_1 e^x + C_2 x e^x + C_3 e^{2x},$$

so the form of the particular solution dictated by $2x + 3e^{-x}$ is

$$y_p(x) = Ax + B + Ce^{-x}.$$

Differentiating produces

$$y_p'(x) = A - Ce^{-x}, \qquad y_p''(x) = Ce^{-x},$$

and

$$y_p'''(x) = -Ce^{-x}.$$

So

$$2x + 3e^{-x} = -Ce^{-x} - 4(Ce^{-x}) + 5(A - Ce^{-x}) - 2(Ax + B + Ce^{-x})$$
$$= -2Ax + (5A - 2B) + (-12C)e^{-x}.$$

Thus, $2 = -2A$, $0 = 5A - 2B$, and $3 = -12C$, which implies that $A = -1$, $B = -\tfrac{5}{2}$, and $C = -\tfrac{1}{4}$.

The general solution to this nonhomogeneous third-order equation is

$$y(x) = C_1 e^x + C_2 x e^x + C_3 e^{2x} - x - \frac{5}{2} - \frac{e^{-x}}{4}. \quad \blacktriangleright$$

► EXERCISE SET 17.6

In Exercises 1–10, solve the differential equation using the method of variation of parameters.

1. $y'' + y = e^x$
2. $y'' + 2y' + y = 2e^x$
3. $y'' + 3y' - 4y = \sin x$
4. $y'' - 2y' + 5y = e^x$
5. $y'' + y = \tan x$, $0 < x < \pi/2$
6. $y'' + y = \sec x$, $0 < x < \pi/2$
7. $y'' + y = \csc x$, $0 < x < \pi/2$
8. $y'' + y = \dfrac{1}{1 + \cos x}$, $-\pi < x < \pi$
9. $y'' + 4y' + 4y = e^{-2x}/x^2$, $0 < x < \pi/2$
10. $y'' + 3y' + 2y = e^{-x}$

In Exercises 11–18, solve the differential equation using the method of undertermined coefficients.

11. $y'' + y = e^x$
12. $y'' + 2y' + y = 2e^x$
13. $y'' + 3y' - 4y = \sin x$
14. $y'' - 2y' + 5y = \cos 2x + e^x$
15. $y'' + 4y = x \sin 2x + \tfrac{1}{4}\cos 2x$
16. $y'' + 4y' + 4y = e^{-2x} + \cos x$
17. $y''' - 4y'' + y' + 6y = xe^x$
18. $y''' - 4y'' + 3y' = e^x + \sin x$

In Exercises 19–22, solve the initial-value problem.

19. $y'' + y' - 2y = 2x^2 - 3$, $y(0) = 0$, $y'(0) = 0$
20. $y'' + y = 4xe^x$, $y(0) = -2$, $y'(0) = 0$
21. $y'' - y = e^x + e^{-x}$, $y(0) = 1$, $y'(0) = 2$
22. $y'' - 2y' + y = 4xe^x$, $y(0) = 0$, $y'(0) = 1$

The harmonic motion of the vibrating system described before Exercises 49–52 in Section 17.5 is called a *free oscillation* because no external force is imposed on the system. When an external force $F(t)$ is present, the motion is called a *forced oscillation* and is described by the initial-value problem

$$m y''(t) + c y'(t) + k y(t) = F(t), \quad y(0) = y_0, \quad y'(0) = y_0'$$

(see the figure). In Exercises 23–28, find an expression for the motion of the vibrating system.

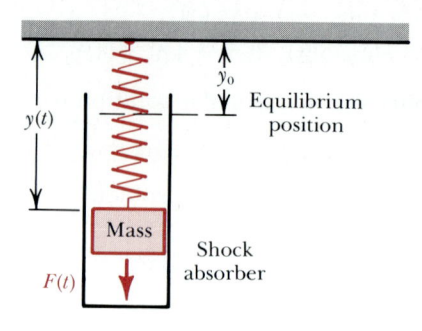

23. $m = 1$ slug, $k = 16$ lb/ft, $c = 0$, $F(t) = \sin t$ lb, $y_0 = 0$, $y_0' = 0$
24. $m = 4$ kg, $k = 4$ N/m, $c = 0$, $F(t) = \cos 2t$ N, $y_0 = 0$, $y_0' = 0$
25. $m = 1$ slug, $k = 16$ lb/ft, $c = 0$, $F(t) = \sin 4t$ lb, $y_0 = 0$, $y_0' = 0$
26. $m = 4$ kg, $k = 4$ N/m, $c = 0$, $F(t) = \cos t$ N, $y_0 = 0$, $y_0' = 0$
27. $m = 1$ slug, $k = 16$ lb/ft, $c = 4$ lb sec/ft, $F(t) = \sin 4t$ lb, $y_0 = 0$, $y_0' = 1$
28. $m = 4$ kg, $k = 4$ N/m, $c = 3$ dyne sec/m, $F(t) = \cos t$ N, $y_0 = 1$, $y_0' = 0$

The initial-value problem describing an electrical circuit that contains, in series, an inductance L, a resistance R, a capacitance C, and an impressed voltage $E(t)$ is

$$L\, I'(t) + R\, I(t) + \frac{1}{C} Q(t) = E(t), \quad Q(0) = Q_0, \quad I(0) = I_0$$

(see the figure). Find an expression for the charge and current as a function of time in the electrical systems described in Exercises 29 and 30.

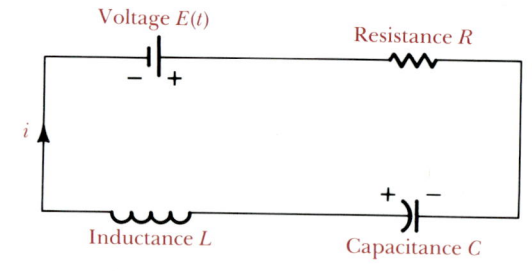

29. $L = 0.25$ henries, $R = 5$ ohms, $C = 10^{-5}$ farads, $E = 110 \sin 60t$ volts, $I_0 = 0$, $Q_0 = 0$
30. $L = 0.5$ henries, $R = 4$ ohms, $C = 2 \times 10^{-6}$ farads, $E = 110 \sin 60t$ volts, $I_0 = 0$, $Q_0 = 0$

17.7

Numerical Methods for First-Order Initial-Value Problems

In this section we examine three basic techniques for approximating the solution to a first-order initial-value problem. Solutions to higher-order initial-value problems follow the same method by treating the higher-order equation as a system of first-order equations.

Consider the first-order initial-value problem

(17.31) $$y' = f(x, y), \qquad y(a) = \alpha$$

for x in $[a, b]$. We choose a positive integer n and define a *step-size* h by $h = (b - a)/n$. We will approximate $y(x)$ at x_i, where $x_i = a + ih$, for $i = 0, 1, 2, \ldots, n$. (Note that this implies that $x_0 = a$ and $x_n = b$.)

Figure 17.5 shows what is assumed to be the exact solution to (17.31) and the exact values of $y(x_i)$ for these values of i. To eliminate confusion between the exact solution and the approximations, we let w_i denote the approximation to $y(x_i)$ for each $i = 0, 1, 2, \ldots, n$. We always assume that the approximation $w_0 = y(x_0) = y(a) = \alpha$ is exact.

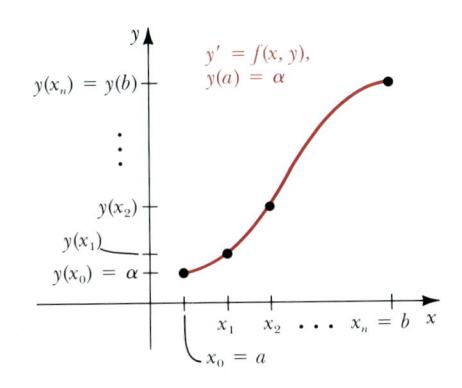

Figure 17.5
The exact solution to $y' = f(x, y)$, $y(a) = \alpha$.

Figure 17.6
One step of Euler's Method.

The most elementary method for approximating the solution is called *Euler's Method*. To approximate the solution to (17.31) on the first subinterval $[x_0, x_1]$, we use the line through (a, α) with slope $y'(a) = f(a, \alpha)$ to approximate $y(x)$, as shown in Figure 17.6. Then w_1 is the second coordinate of the point on this line whose first coordinate is x_1. Thus,

$$w_1 = \alpha + hf(a, \alpha) = w_0 + hf(x_0, w_0).$$

We can approximate the solution to (17.31) on the interval $[x_1, x_2]$ using the line through (x_1, w_1) with slope $f(x_1, w_1)$. The second coordinate of the point on this line whose first coordinate is x_2 is denoted by w_2, so

$$w_2 = w_1 + hf(x_1, w_1).$$

In general, the approximation w_i to $y(x_i)$ is obtained using **Euler's Method** from the following equations:

(17.32)
$$w_0 = \alpha,$$
$$w_i = w_{i-1} + hf(x_{i-1}, w_{i-1}),$$

for each $i = 1, 2, \ldots, n$.

Figure 17.7 shows a typical result produced by successive applications of Euler's Method. Euler's Method is used below to approximate the solution to an initial-value problem whose exact solution is known. This permits us to examine the accuracy of the approximations.

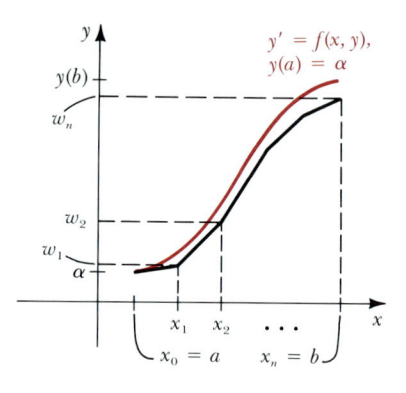

Figure 17.7
Multiple steps of Euler's Method.

Example 1

Apply Euler's Method with $h = 0.1$ to the initial-value problem

$$y' = -y + x + 1, \quad y(0) = 1, \quad \text{for } x \text{ in } [0, 1].$$

Solution

This linear initial-value problem has the exact solution $y(x) = x + e^{-x}$. Since we have the exact solution, we can compare the approximations given by Euler's Method with the true values. From (17.32) we have

$$w_0 = 1$$

and for $i = 1, 2, \ldots, 10$

$$\begin{aligned} w_i &= w_{i-1} + h(-w_{i-1} + t_{i-1} + 1) \\ &= w_{i-1} + 0.1[-w_{i-1} + 0.1(i-1) + 1] \\ &= 0.9 w_{i-1} + 0.01(i-1) + 0.1 \end{aligned}$$

Table 17.1 lists the approximations w_i, the exact values $y_i = y(x_i)$, and the magnitude of the errors in the approximations. ▶

In Euler's Method the line through (x_{i-1}, w_{i-1}) whose slope is $f(x_{i-1}, w_{i-1})$ is used to obtain the approximation w_i. A simple method for improving the approximations given by Euler's Method is to replace the slope $f(x_{i-1}, w_{i-1})$ of the line that approximates the solution on $[x_{i-1}, x_i]$

Table 17.1
Euler's Method

x_i	w_i	y_i	Error $= |w_i - y_i|$
0.0	1.000000	1.000000	
0.1	1.000000	1.004837	0.004837
0.2	1.010000	1.018731	0.008731
0.3	1.029000	1.040818	0.011818
0.4	1.056100	1.070320	0.014220
0.5	1.090490	1.106531	0.016041
0.6	1.131441	1.148812	0.017371
0.7	1.178297	1.196585	0.018288
0.8	1.230467	1.249329	0.018862
0.9	1.287420	1.306570	0.019150
1.0	1.348678	1.367879	0.019201

17.7 NUMERICAL METHODS FOR FIRST-ORDER INITIAL-VALUE PROBLEMS

with the average of the slopes of the lines given at the ends of the interval $[x_{i-1}, x_i]$:

$$f(x_{i-1}, w_{i-1}) \quad \text{and} \quad f(x_i, w_{i-1} + hf(x_{i-1}, w_{i-1})).$$

This leads to a method known as the **Modified Euler's Method** and described by

(17.33) $\quad w_0 = \alpha,$
$$w_i = w_{i-1} + \frac{h}{2}[f(x_{i-1}, w_{i-1}) + f(x_i, w_{i-1} + hf(x_{i-1}, w_{i-1}))].$$

For each $i = 1, 2, \ldots, n$.

Example 2

Apply the Modified Euler's Method with $h = 0.1$ to the initial-value problem

$$y' = -y + x + 1, \quad y(0) = 1, \quad \text{for } x \text{ in } [0, 1].$$

Solution
Applying the Modified Euler Method to this problem gives us

$$w_i = w_{i-1} + \frac{h}{2}[f(x_{i-1}, w_{i-1}) + f(x_i, w_{i-1} + hf(x_{i-1}, w_{i-1}))]$$

$$= w_{i-1} + \frac{h}{2}\{(-w_{i-1} + x_{i-1} + 1) + [-(w_{i-1} + hf(x_{i-1}, w_{i-1})) + x_i + 1]\}$$

$$= w_{i-1} + \frac{h}{2}[-w_{i-1} + x_{i-1} + 1 - w_{i-1} - h(-w_{i-1} + x_{i-1} + 1) + x_i + 1]$$

$$= \left(1 - h + \frac{h^2}{2}\right)w_{i-1} + \left(\frac{h}{2} - \frac{h^2}{2}\right)x_{i-1} + \frac{h}{2}x_i + \left(h - \frac{h^2}{2}\right).$$

Since $h = 0.1$, $x_{i-1} = h(i-1) = 0.1(i-1)$, and $x_i = hi = 0.1i$, this reduces to

$$w_i = 0.905 w_{i-1} + 0.0095i + 0.0905.$$

The approximations generated from this formula are shown in Table 17.2. They are considerably more accurate than those given by Euler's Method and listed in Table 17.1. ▶

Table 17.2
Modified Euler's Method

x_i	w_i	y_i	Error $= |w_i - y_i|$
0.0	1.000000	1.000000	
0.1	1.005000	1.004837	0.000163
0.2	1.019025	1.018731	0.000294
0.3	1.041218	1.040818	0.000400
0.4	1.070802	1.070320	0.000482
0.5	1.107076	1.106531	0.000545
0.6	1.149404	1.148812	0.000592
0.7	1.197211	1.196585	0.000626
0.8	1.249976	1.249329	0.000647
0.9	1.307228	1.306570	0.000658
1.0	1.368541	1.367879	0.000662

Neither Euler's Method nor the Modified Euler's Method give results that are sufficiently accurate for many initial-value problems. We will now consider a general-purpose technique that can be used to accurately approximate the solutions to most common initial-value problems.

For the initial-value problem

$$y' = f(x, y), \quad y(a) = \alpha$$

with x in $[a, b]$ and $h = (b - a)/n$, the **Runge–Kutta Method of order 4** is given by

(17.34)
$$w_0 = \alpha,$$

and for each $i = 1, 2, \ldots, n$,

$$k_1 = hf(x_{i-1}, w_{i-1}),$$
$$k_2 = hf\left(x_{i-1} + \frac{h}{2}, w_{i-1} + \frac{k_1}{2}\right),$$
$$k_3 = hf\left(x_{i-1} + \frac{h}{2}, w_{i-1} + \frac{k_2}{2}\right),$$
$$k_4 = hf(x_i, w_{i-1} + k_3),$$
$$w_i = w_{i-1} + \frac{1}{6}(k_1 + 2k_2 + 2k_3 + k_4).$$

Example 3

Apply the Runge–Kutta Method with $h = 0.1$ to the initial-value problem

$$y' = -y + x + 1, \quad y(0) = 1, \quad \text{for } x \text{ in } [0, 1].$$

Solution

The approximations obtained from the Runge–Kutta Method are shown in Table 17.3. Extended precision has been used in this example to illustrate the improved accuracy.

Table 17.3
Runge–Kutta Method

x_i	w_i	y_i	Error $= |w_i - y_i|$
0.0	1.0000000000	1.0000000000	
0.1	1.0048375000	1.0048374180	8.200×10^{-8}
0.2	1.0187309014	1.0187307531	1.483×10^{-7}
0.3	1.0408184220	1.0408182207	2.013×10^{-7}
0.4	1.0703202889	1.0703200460	2.429×10^{-7}
0.5	1.1065309344	1.1065306597	2.747×10^{-7}
0.6	1.1488119344	1.1488116360	2.984×10^{-7}
0.7	1.1965856187	1.1965853038	3.149×10^{-7}
0.8	1.2493292897	1.2493289641	3.256×10^{-7}
0.9	1.3065699912	1.3065696597	3.315×10^{-7}
1.0	1.3678797744	1.3678794412	3.332×10^{-7}

17.7 NUMERICAL METHODS FOR FIRST-ORDER INITIAL-VALUE PROBLEMS

Programs for Euler's Method, the Modified Euler's Method, and the Runge–Kutta Method are included in Appendix C. As with all numerical methods, they should be applied with a watchful eye. The methods will often, but not always, produce reasonable approximations. You should consult a text on numerical analysis before doing extensive or important work with these techniques.

▶ EXERCISE SET 17.7

In Exercises 1–10, approximate the solution to the initial-value problem on the given interval using (a) Euler's Method, (b) the Modified Euler's Method, and (c) the Runge–Kutta Method.

1. $y' = xy$, $y(0) = 1$; $[0, 2]$, $h = 0.5$
2. $y' = \dfrac{y^2 + y}{x}$, $y(1) = -2$; $[1, 3]$, $h = 0.5$
3. $y' = \sin x + e^{-x}$, $y(0) = 0$; $[0, 1]$, $h = 0.25$
4. $y' = -xy$, $y(0) = 4$; $[0, 4]$, $h = 1$
5. $y' = \left(\dfrac{y}{x}\right)^2 + \dfrac{y}{x}$, $y(1) = 1$; $[1, 2]$, $h = 0.1$
6. $y' = -xy + \dfrac{4x}{y}$, $y(0) = 1$; $[0, 3]$, $h = 0.25$
7. $y' = \dfrac{2y}{x} + x^2 e^x$, $y(1) = 0$; $[1, 2]$, $h = 0.1$
8. $y' = \dfrac{y}{x} - y^2$, $y(1) = 1$; $[1, 2]$, $h = 0.05$
9. $y' = 1 + x \sin(xy)$, $y(0) = 0$; $[0, 2]$, $h = 0.1$
10. $y' = xe^{xy}$, $y(1) = 0.15$; $[1, 1.5]$, $h = 0.05$

11–18. Determine the exact solutions to Exercises 1–8 and compare the exact values with those obtained using (a) Euler's Method, (b) the Modified Euler's Method, and (c) the Runge–Kutta Method.

19. In a circuit with impressed voltage $\mathscr{E}$, and resistance R, inductance L, capacitance C in parallel, the current i satisfies the differential equation

$$\frac{di}{dt} = C \frac{d^2 \mathscr{E}}{dt^2} + \frac{1}{R} \frac{d\mathscr{E}}{dt} + \frac{1}{L} \mathscr{E}.$$

Suppose $i(0) = 0$, $C = 0.3$ farads, $R = 1.4$ ohms, and $L = 1.7$ henries, and the voltage is given by

$$\mathscr{E}(t) = e^{-0.06\pi t} \sin(2t - \pi).$$

Find the current $i(t)$ for the values $t = 0.1j$, $j = 0, 1, \ldots, 10$ using (a) Euler's Method, (b) the Modified Euler Method, and (c) the Runge–Kutta Method.

20. A projectile of mass $m = 0.11$ kg shot vertically upward with initial velocity $v(0) = 8$ m/sec is slowed due to the acceleration of gravity $F_g = mg$ and due to air resistance $F_r = -kv^v$, where $g = -9.8$ m/sec^2 and $k = 0.002$ kg/m. The differential equation for the velocity v is given by

$$mv' = mg - kv^v.$$

a) Use the Runge–Kutta Method with $h = 0.1$ to approximate the velocity after $0.1, 0.2, \ldots, 1.0$ seconds.
b) Determine, to within the nearest tenth of a second, when the projectile reaches its maximum height.

21. Water flows from an inverted conical tank with circular orifice at the rate

$$\frac{dx}{dt} = -0.6 \pi r^2 \sqrt{2g} \frac{\sqrt{x}}{A(x)},$$

where r is the radius of the orifice, x is the height of the liquid level above the vertex of the cone, and $A(x)$ is the area of the cross section of the tank x units above the vertex. Suppose $r = 0.1$ ft, $g = -32$ ft/sec^2, and the cone has an initial water level of 8 ft and initial volume of $512\pi/3$ ft^3. Use the Runge–Kutta Method with $h = 20$ sec to approximate the height of the water level (a) after 10 min and (b) after 20 min; and (c) determine, to within one minute, when the tank will be empty.

22. The irreversible chemical reaction in which two molecules of solid potassium dichromate ($K_2Cr_2O_7$), two molecules of water (H_2O), and three atoms of solid sulfur (S) combine to yield three molecules of the gas sulfur dioxide (SO_2), four molecules of solid potassium hydroxide (KOH), and two molecules of solid chromic oxide (Cr_2O_3) can be represented symbolically by the stoichiometric equation

$$2K_2Cr_2O_7 + 2H_2O + 3S \longrightarrow 4KOH + 2Cr_2O_3 + 3SO_2.$$

If n_1 molecules of $K_2Cr_2O_7$, n_2 molecules of H_2O, and n_3 molecules of S are originally available, the following differential equation describes the number $x(t)$ of molecules of KOH after time t:

$$\frac{dx}{dt} = k\left(n_1 - \frac{x}{2}\right)^2 \left(n_2 - \frac{x}{2}\right)^2 \left(n_3 - \frac{3x}{4}\right)^3,$$

where k is the velocity constant of the reaction. Suppose $k = 6.22 \times 10^{-19}$, $n_1 = n_2 = 2 \times 10^4$, and $n_3 = 3 \times 10^4$. Use the Runge–Kutta Method with $h = 0.01$ to approximate the number of molecules of potassium hydroxide that will be formed after 0.2 sec.

▶ IMPORTANT TERMS AND RESULTS

CONCEPT	PAGE	CONCEPT	PAGE
Differential equation	986	Linear first-order differential equation	1000
Order	986	Bernoulli differential equation	1003
General solution	987	Linear second-order equation: homogeneous	1007
Particular solution	987	Characteristic equation	1008
Initial condition	987	Linear second-order equation: nonhomogeneous	1014
Initial-value problem	988	Variation-of-parameters method	1015
Orthogonal trajectory	989	Method of undetermined coefficients	1018
Homogeneous first-order equation	991	Euler's method	1024
Exact differential equation	995	Modified Euler's method	1025
Integrating factor	999	Runge–Kutta method	1026

▶ REVIEW EXERCISES

In Exercises 1–26, find the general solution to the differential equation.

1. $y' = 2y$
2. $y' = 3x^2 - 2x + 1$
3. $y' = \dfrac{x+y}{x-y}$
4. $y' = \dfrac{x^2 + 3xy + y^2}{x^2}$
5. $e^x \sin y \, dx + e^x \cos y \, dy = 0$
6. $(x^2 + y^2) \, dx + 2xy \, dy = 0$
7. $y' = 2xy$
8. $(y^2 - 1) \, dx + (3x^2 - 2xy) \, dy = 0$
9. $y' = x^3 - 2xy$
10. $y' = \dfrac{xy}{1 + x^2}$
11. $\left(\dfrac{y}{x} \sin \dfrac{y}{x} + \cos \dfrac{y}{x}\right) dx - \sin \dfrac{y}{x} \, dy = 0$
12. $2xy' + y = x^{5/2} - x^{3/2}$
13. $xy' + 2y = x^2$
14. $xy \, dx + (x^2 + y^2 + 1) \, dy = 0$
15. $xy' - 3y + x^4 y^2 = 0$
16. $y'' - 4y = 0$
17. $y'' - 2y' - 3y = 0$
18. $y'' - y' - 2y = 0$
19. $y'' + y = \sin x$
20. $y'' - 4y = \sin 2x$
21. $y'' - 3y' + 2y = e^{3x}$
22. $y'' - 4y' + 4y = 0$
23. $y' = \dfrac{x \cos x}{y}$
24. $y' = \dfrac{x \ln x}{\ln y}$
25. $y''' - 8y'' + 21y' = 18$
26. $y''' - 2y'' - 4y' + 8y = 5 + e^x$

In Exercises 27–34, find the particular solution to the initial-value problem.

27. $y' = x^3 e^{-2y}$, $y(1) = 0$
28. $y' + y \tan x = 0$, $y(0) = 2$
29. $y'' + 5y' + 6y = 0$, $y(0) = 1$, $y'(0) = 2$
30. $y'' - y = 0$, $y(0) = 1$, $y'(0) = 1$
31. $y'' + y = 2 \cos x$, $y(0) = 5$, $y'(0) = 2$
32. $y'' + y = e^x$, $y(0) = 0$, $y'(0) = 1$
33. $y'' - 4y' + 4y = xe^{2x} + e^{2x}$, $y(0) = 1$, $y'(0) = 0$
34. $y'' - y = x$, $y(0) = 0$, $y'(0) = 0$

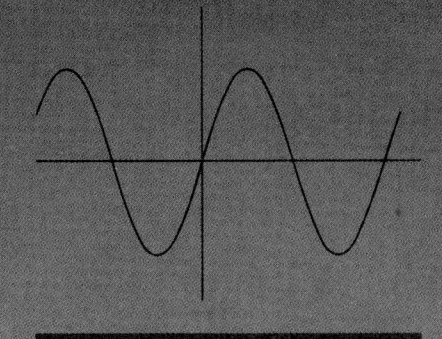

Appendixes

Appendix A
Review Material

A.1
The Real Line

The set of **natural numbers** consists of the counting numbers: 1, 2, 3, ... Any pair of natural numbers can be added or multiplied and the result is another natural number.

The set of **integers** consists of the natural numbers, the negative of each natural number, and the number zero. Addition, multiplication, or subtraction of two integers results in an integer.

The set of **rational numbers** is introduced to ensure that division by nonzero numbers is well defined. By definition, a rational number has the form p/q, where p and q are integers and $q \neq 0$. The rational numbers satisfy all the common arithmetic properties, but fail to have a property called *completeness,* because there are some essential numbers missing from the set. (A precise definition of completeness is discussed in Chapter 8.)

As early as 400 B.C., Greek mathematicians of the Pythagorean school recognized that although $\sqrt{2}$ is the length of a diagonal of a unit square, it is not a rational number. These **irrational numbers** were called incommensurable because they could not be directly compared to the familiar rational numbers. There are many irrational numbers, including $\sqrt{5}$, π, and $-\sqrt{3} + 1$. The discovery of irrational numbers resulted in a profound change in thinking, since before that time it was assumed that all quantities could be expressed as proportions of integers.

The set $\mathbb{R}$ of **real numbers** consists of the rational numbers together with the irrational numbers. This set is described most easily by considering the set of all numbers that are expressed as infinite decimals. The rational numbers are the numbers with expansions that eventually repeat in sequence, such as $\frac{1}{2} = 0.50000 \ldots$, $\frac{1}{3} = 0.3333 \ldots$, $\frac{16}{11} = 1.454545 \ldots$, or $\frac{123}{13} = 9.461538461538 \ldots$. The real numbers with decimal expansions that do not repeat are the irrational numbers.

The decimal-expansion property provides a means of associating each real number with a distinct point on a **coordinate line.** Choose a point on a horizontal line as the origin and associate with the origin the real number 0. Then associate a point to the right of the origin with the real number 1. The positive integers are marked with equal spacing consecutively to the right of 0, and the negative integers are marked consecutively with the same spacing to the left of 0. Nonintegral real numbers are marked on the line according to their decimal expansions. Figure A.1 shows a coordinate line and points associated with certain real numbers.

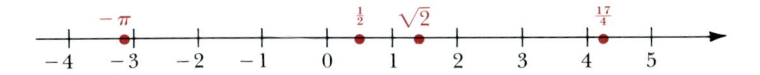

Figure A.1

The coordinate-line representation of the real numbers is so convenient that we frequently do not explicitly distinguish between the points on the line and the real numbers that these points represent. Both are called the set of real numbers and denoted $\mathbb{R}$. The coordinate line representation of the real numbers also provides a way to describe a property that certain subsets of the real numbers possess. A subset S of $\mathbb{R}$ is said to be **dense** if between any two distinct points on the real line there is a point of S. If we consider the decimal expansions of real numbers, it is not difficult to show that:

Both the set of rational numbers and the set of irrational numbers are dense.

The relative position of two points on a coordinate line can be used to define an inequality relationship on the set of real numbers. We say that a is less than b, written $a < b$, or b is greater than a, written $b > a$, provided that a lies to the left of b on the coordinate line. The notation $a \leq b$, or $b \geq a$, is used to express that a is less than or equal to b. It can be verified that, for any real numbers a, b, and c:

i) Precisely one of $a < b$, $b < a$, or $a = b$ holds.
ii) If $a > b$, then $a + c > b + c$.
iii) If $a > b$ and $c > 0$, then $ac > bc$.
iv) If $a > b$ and $c < 0$, then $ac < bc$.

These rules are used frequently to solve problems involving inequality relations.

Example 1

Find all values of x satisfying $2x - 1 < 4x + 3$.

Solution
Since $2x - 1 < 4x + 3$,
$$-4 < 2x.$$
Multiplying both sides of the inequality by $1/2$ gives
$$-2 < x.$$
▶

Example 2

Find all values of x satisfying $-1 < 2x + 3 < 5$.

Solution

This inequality relation is a compact way of expressing that

$$-1 < 2x + 3 \quad \text{and} \quad 2x + 3 < 5,$$

so

$$-4 < 2x \quad \text{and} \quad 2x < 2.$$

Thus,

$$-2 < x \quad \text{and} \quad x < 1,$$

which can be expressed as $-2 < x < 1$. ▶

Example 3

Find all values of x satisfying $x^2 - 4x + 5 > 2$.

Solution

$$x^2 - 4x + 5 > 2 \quad \text{implies} \quad x^2 - 4x + 3 > 0.$$

Writing $x^2 - 4x + 3$ as $(x - 3)(x - 1)$ gives us

$$(x - 3)(x - 1) > 0.$$

This product is positive precisely when both factors are positive or both are negative. Figure A.2 indicates that both are negative if $x < 1$ and that both are positive if $x > 3$. Consequently, $x^2 - 4x + 5 > 2$ precisely when $x < 1$ or $x > 3$. ▶

Figure A.2

The **absolute value** of a real number a, denoted $|a|$, describes the distance on a coordinate line from the number a to the number 0. We define the absolute value of the number a by

(A.1) $$|a| = \begin{cases} a, & \text{if } a \geq 0, \\ -a, & \text{if } a < 0. \end{cases}$$

The distance $d(a, b)$ from the real number a to the real number b is

(A.2) $$d(a, b) = |a - b|.$$

Note that this definition implies that $d(a, b) = d(b, a)$ and that $d(a, 0) = |a|$. Because of this distance relationship, the absolute value frequently occurs in inequality relationships.

Some basic properties of absolute values that are useful for working with absolute-value inequalities are the following:

(A.3) $$|ab| = |a||b|;$$

(A.4) $$\left|\frac{a}{b}\right| = \frac{|a|}{|b|}, \quad \text{if } b \neq 0;$$

(A.5) $$|a+b| \leq |a| + |b|;$$

(A.6) $\quad |x| < a \quad$ if and only if $\quad -a < x < a;$

(A.7) $\quad |x| > a, \quad a \geq 0, \quad$ if and only if $\quad x < -a \quad$ or $\quad x > a.$

Example 4

Find all values of x satisfying $|2x - 1| < 3$.

Solution

$$|2x - 1| < 3 \quad \text{if and only if} \quad -3 < 2x - 1 < 3,$$

that is,

$$-3 < 2x - 1 \quad \text{and} \quad 2x - 1 < 3.$$

This implies that $-2 < 2x$ and $2x < 4$. So $-1 < x < 2$. ▶

The problem in Example 4 can also be solved using the distance property of the absolute value of the difference of two numbers. Since

$$|2x - 1| = |2(x - \tfrac{1}{2})| = 2|x - \tfrac{1}{2}|,$$

we see that

$$|2x - 1| < 3 \quad \text{is equivalent to} \quad |x - \tfrac{1}{2}| < \tfrac{3}{2}.$$

Thus x satisfies the condition $|2x - 1| < 3$ precisely when the distance from x to $\tfrac{1}{2}$ is less than $\tfrac{3}{2}$, that is, when $-1 < x < 2$, as shown in Figure A.3.

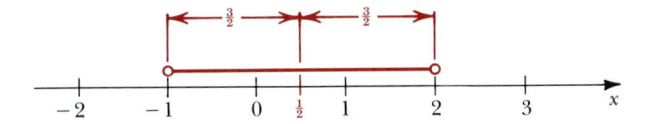

Figure A.3

▶ EXERCISE SET A.1

In Exercises 1–16, find all values of x that satisfy the inequality.

1. $x + 4 < 7$
2. $2x + 3 \geq 4$
3. $-3x + 4 < 5$
4. $-(3x + 4) \geq 5$
5. $-2 < 2x + 9 < 5$
6. $2 - x < x < 4 - x$
7. $-2 < 2x + 9 < 5 + x$
8. $x - 2 < 2x + 9 < 5 + x$

9. $x^2 + 2x + 1 \geq 4$
10. $x^3 - 6x^2 + 8x < 0$
11. $|x - 4| \leq 1$
12. $|2x - 3| < 5$
13. $|3x - 1| < 0.01$
14. $|3x - 2| < 0.01$
15. $|x^2 - 4| \leq 0$
16. $|x^2 - 2x + 5| \geq 0$

Prove that the rules listed in Exercises 17–20 are true for all real numbers.

17. $|ab| = |a||b|$
18. $\left|\dfrac{a}{b}\right| = \dfrac{|a|}{|b|}$ if $b \neq 0$
19. $|x| < a$ if and only if $-a < x < a$
20. $|x| > a$ if and only if $x < -a$ or $x > a$
21. Show that for any pair of real numbers,

$$|a + b| \leq |a| + |b|.$$

What must be true about a and b if equality holds?

22. Show that for any pair of real numbers,

$$|a| - |b| \leq |a - b|.$$

What must be true about a and b if equality holds?

A.2

The Coordinate Plane

An ordered pair (a, b) of real numbers is associated with a point in a plane in a manner similar to the association of a real number with a point on a line. First an arbitrary point in the plane is associated with $(0, 0)$ and designated the origin. Then horizontal and vertical lines are drawn intersecting at the origin. The horizontal line is called the first-coordinate- or x-axis, and the vertical line is called the second-coordinate- or y-axis (see Figure A.4).

A scale is placed on both axes. The x-axis has positive numbers to the right of the origin and negative numbers to the left. The y-axis has positive numbers above the origin and negative numbers below. The ordered pair (a, b) is associated with the point of intersection of the vertical line drawn through the point a on the x-axis and the horizontal line drawn through b on the y-axis (see Figure A.5).

The x- and y-axes divide the plane into four regions, or **quadrants.** These quadrants are labeled as in Figure A.5. The set of all ordered pairs of real numbers is denoted $\mathbb{R} \times \mathbb{R}$, or $\mathbb{R}^2$. The plane determined by the x- and y-axes is called the xy-plane.

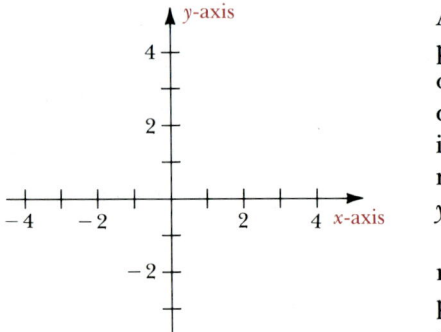

Figure A.4

HISTORICAL NOTE

The coordinate plane is often called the Cartesian plane and a rectangular coordinate system is also called a Cartesian coordinate system. This is to honor the versatile mathematician, philosopher, and physicist **René Descartes** (1596–1650), whose name in Latin was Renatus Cartesius. In his *La géométrie* of 1637, which originally appeared as an appendix to his treatise on universal science *Discours de la méthode pour bien conduire sa raison et chercher la verité dans les sciences*, he introduced the mathematical world to analytic geometry. He attributed his new ideas in philosophy and geometry to three dreams he experienced on the night of November 10, 1619, at Neuberg while serving in the French army. Unfortunately, despite his eloquence, most of the Cartesian physics was wrong and was replaced by Newtonian physics in the next century.

Example 1

Sketch the points in the coordinate plane associated with the ordered pairs (1, 2), (−1, 3), (−2, −π), and ($\sqrt{2}$, −$\sqrt{3}$).

Solution
These points are shown in Figure A.6. ▸

Example 2

Find the points (x, y) in the xy-plane that satisfy

$$1 \leq |x| \leq 2 \quad \text{and} \quad -1 \leq y \leq 3.$$

Solution
The points whose x-coordinates satisfy $1 \leq |x| \leq 2$ are those that lie on or between the vertical lines $x = -2$ and $x = -1$ and those that lie on or between the vertical lines $x = 1$ and $x = 2$. The points whose y-coordinates satisfy $-1 \leq y \leq 3$ lie on or between the horizontal lines $y = -1$ and $y = 3$. The points satisfying both these conditions are in the shaded regions shown in Figure A.7. ▸

The distance between two ordered pairs of real numbers (x_1, y_1) and (x_2, y_2) is the distance between the points the pairs represent. (In general, we will not distinguish between an ordered pair and the point it represents in a coordinate plane.) This distance is found by considering the point (x_2, y_1) and applying the Pythagorean theorem. Refer to Figure A.8 on the following page.

Since on the x-axis $d(x_1, x_2) = |x_1 - x_2|$, the distance between (x_1, y_1) and (x_2, y_1) is

$$d((x_1, y_1), (x_2, y_1)) = |x_1 - x_2|.$$

Similarly, $d((x_2, y_1), (x_2, y_2)) = |y_1 - y_2|$
so

$$d((x_1, y_1), (x_2, y_2)) = \sqrt{[d((x_1, y_1), (x_2, y_1))]^2 + [d((x_2, y_1), (x_2, y_2))]^2}$$
$$= \sqrt{|x_1 - x_2|^2 + |y_1 - y_2|^2}.$$

Thus,

(A.8) $\quad d((x_1, y_1), (x_2, y_2)) = \sqrt{(x_1 - x_2)^2 + (y_1 - y_2)^2}.$

Example 3

Find the distance between the points (1, 2) and (−2, 6).

Solution

$$d((1, 2), (-2, 6)) = \sqrt{(1 - (-2))^2 + (2 - 6)^2}$$
$$= \sqrt{9 + 16} = \sqrt{25} = 5$$

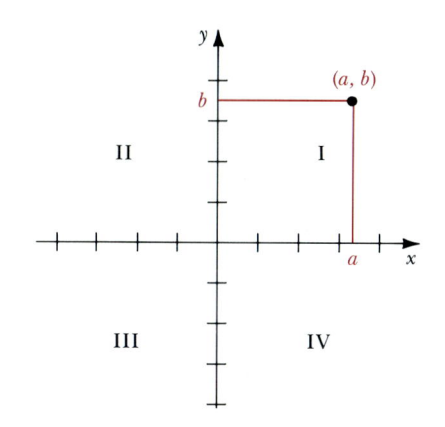

▸

Figure A.5

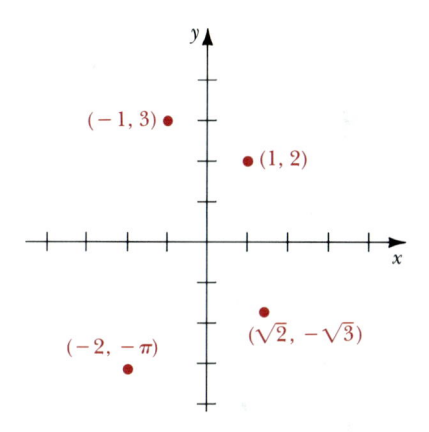

Figure A.6

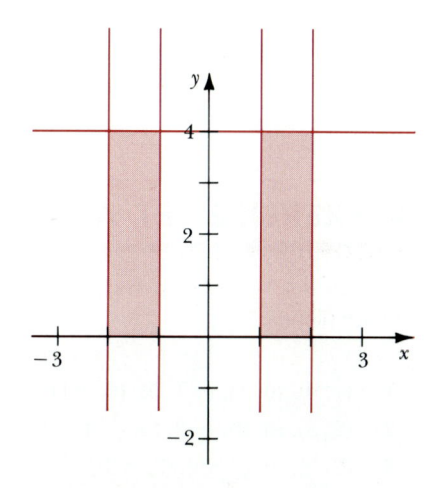

Figure A.7

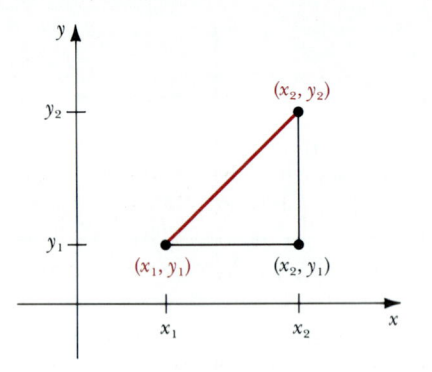

Figure A.8

Example 4

Show that the midpoint of the line segment joining (x_1, y_1) and (x_2, y_2) is
$$\left(\frac{x_1 + x_2}{2}, \frac{y_1 + y_2}{2}\right).$$

Solution

$$d\left((x_1, y_1), \left(\frac{x_1 + x_2}{2}, \frac{y_1 + y_2}{2}\right)\right)$$

$$= \sqrt{\left(x_1 - \frac{x_1 + x_2}{2}\right)^2 + \left(y_1 - \frac{y_1 + y_2}{2}\right)^2}$$

$$= \sqrt{\left(\frac{x_1 - x_2}{2}\right)^2 + \left(\frac{y_1 - y_2}{2}\right)^2}$$

$$= \frac{1}{2}\sqrt{(x_1 - x_2)^2 + (y_1 - y_2)^2}$$

$$= \frac{1}{2} d((x_1, y_1), (x_2, y_2)).$$

Similarly,

$$d\left(\left(\frac{x_1 + x_2}{2}, \frac{y_1 + y_2}{2}\right), (x_2, y_2)\right) = \frac{1}{2} d((x_1, y_1), (x_2, y_2)).$$

Since

$$d((x_1, y_1), (x_2, y_2)) = d\left((x_1, y_1), \left(\frac{x_1 + x_2}{2}, \frac{y_1 + y_2}{2}\right)\right)$$
$$+ d\left(\left(\frac{x_1 + x_2}{2}, \frac{y_1 + y_2}{2}\right), (x_2, y_2)\right),$$

$\left(\frac{x_1 + x_2}{2}, \frac{y_1 + y_2}{2}\right)$ must be on the line segment and is thus the midpoint. ▶

▶ **EXERCISE SET A.2**

In Exercises 1–4, sketch the listed points in the same coordinate plane.

1. $(1, 0), (0, 1), (-1, 0), (0, -1)$
2. $(2, 3), (3, 2), (-2, 3), (-3, -2)$
3. $(2, 3), (-2, -3), (2, -3), (-2, 3)$
4. $(5, -10), (10, 20), (-20, 10), (30, 40)$

In Exercises 5–8, find (a) the distance between the points and (b) the midpoints of the line segments joining the points.

5. $(2, 4), (-1, 3)$
6. $(-1, 5), (7, 9)$
7. $(\pi, 0), (-1, 2)$
8. $(\sqrt{3}, \sqrt{2}), (\sqrt{2}, \sqrt{3})$

In Exercises 9–24, indicate in an *xy*-plane those points (x, y) for which the statement holds.

9. $x = 3$
10. $y = -2$
11. $x \geq 0$ and $y \geq 0$
12. $x \geq 0$ and $y \leq 0$
13. $-1 \leq x \leq 1$
14. $2 < y < 3$
15. $-1 \leq x \leq 1$ and $2 < y < 3$
16. $-1 \leq x \leq 1$ or $2 < y < 3$
17. $4 \leq |x|$
18. $|y| \leq 2$
19. $|x - 1| < 3$
20. $|y + 1| < 2$
21. $|x - 1| < 3$ and $|y + 1| < 2$
22. $|x| + |y| > 0$
23. $|x - 1| + |y + 1| = 0$
24. $|x| > |y|$
25. Find the distance between the points $(-1, 4)$, $(-3, -4)$, and $(2, -1)$ and show that they are vertices of a right triangle.
26. Show that the points $(2, 1)$, $(-1, 2)$, and $(2, 6)$ are vertices of an isosceles triangle.
27. Find a fourth point, which added to the points in Exercise 25 will form the vertices of a rectangle. Is the point unique?
28. Find a fourth point, which added to the points in Exercise 26 will form the vertices of a parallelogram. Is the point unique?

A.3

Trigonometric Functions

The trigonometric functions have many applications. They are used to describe the behavior of such diverse topics as sound waves, vibrations, the motion of an automobile on a bumpy road, and the oscillations of a pendulum. In fact, trigonometric functions are commonly required to describe the motion of an object that behaves in a circular, oscillating, or periodic manner.

For calculus purposes it is most useful to define these functions relative to the circle centered at $(0, 0)$ with radius one, called the **unit circle** and denoted U. The trigonometric functions can also be introduced using the sides and angles of a right triangle, but this approach is directed more toward the computational aspects of trigonometry than the functional aspects required for the study of calculus. In the next few pages a brief outline of the definitions and properties of trigonometric functions is given.

The unit circle U in the *xy*-plane has equation $x^2 + y^2 = 1$. Our first step is to describe a function P that maps the real line $\mathbb{R}$ onto U in the following manner:

i) If t is a positive real number, then $P(t)$ is the point on the unit circle for which the length of the arc of the circle from $(1, 0)$ to $P(t)$ is t units, measured in the *counterclockwise* direction from $(1, 0)$. (See Figure A.9.)

ii) If t is a negative real number, then $P(t)$ is the point on the unit circle for which the length of the arc of the circle from $(1, 0)$ to $P(t)$ is $-t$ units, measured in the *clockwise* direction from $(1, 0)$. (See Figure A.10.)

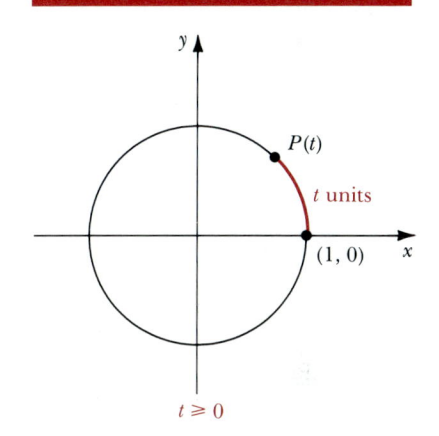

Figure A.9

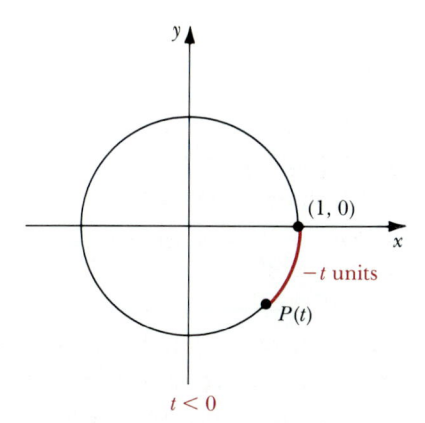

Figure A.10

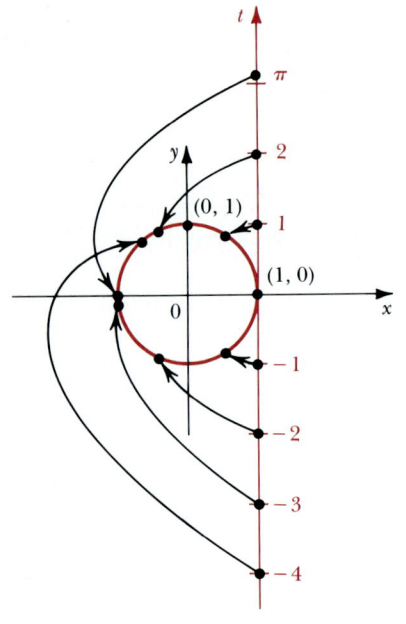

Figure A.11

The mapping of t into $P(t)$ is described geometrically by positioning the real line vertically, with $t = 0$ coinciding with the point $(1, 0)$ on the circle (see Figure A.11). For any real number t, the point $P(t)$ is obtained by "wrapping" the line around the circle and marking the point $P(t)$ on the circle that corresponds to the position of t.

The smallest positive real number that is mapped onto $(-1, 0)$ is called π, an irrational number whose value is approximately 3.1416. Because of the symmetry of the circle, the real number $\pi/2$ is associated with the point $(0, 1)$, $3\pi/2$ with the point $(0, -1)$, and 2π with the point $(1, 0)$.

Corresponding to each real number t is a pair $(x(t), y(t))$ of xy-coordinates describing the point $P(t)$ on the unit circle U. The basic trigonometric functions are defined in terms of these coordinates.

(A.9)
DEFINITION
If the coordinates of $P(t)$ are $(x(t), y(t))$, then the **sine** of t, written $\sin t$, is defined by

$$\sin t = y(t)$$

and the **cosine** of t, written $\cos t$, by

$$\cos t = x(t).$$

The **tangent** of t, written $\tan t$, is defined by the quotient

$$\tan t = \frac{\sin t}{\cos t}, \quad \text{provided } \cos t \neq 0.$$

Figure A.12 illustrates these definitions.
Certain properties of these trigonometric functions follow immediately from the definition.

(A.10) $-1 \leq \sin t \leq 1; \quad -1 \leq \cos t \leq 1$
(A.11) $\sin^2 t + \cos^2 t = 1$
(A.12) $\sin(-t) = -\sin t; \quad \cos(-t) = \cos t; \quad \tan(-t) = -\tan t$
(A.13) $\sin(t + 2n\pi) = \sin t; \quad \cos(t + 2n\pi) = \cos t$
$\tan(t + n\pi) = \tan t, \quad \text{for any integer } n$

A function f is **periodic** with **period** a if a is the smallest positive real number with $f(x + a) = f(x)$ for all x in the domain of f. The result in (A.13) implies that sine and cosine are periodic functions with period 2π whereas the tangent function is periodic with period π.

There are a number of values of t for which $\sin t$, $\cos t$, and $\tan t$ can be determined from geometric properties of the unit circle. Some of the more common values are listed in Table A.1.

The results in Table A.1 and the properties listed in (A.10) through (A.13) can be used to sketch the graphs of the sine, cosine, and tangent functions. These graphs are shown in Figure A.13.

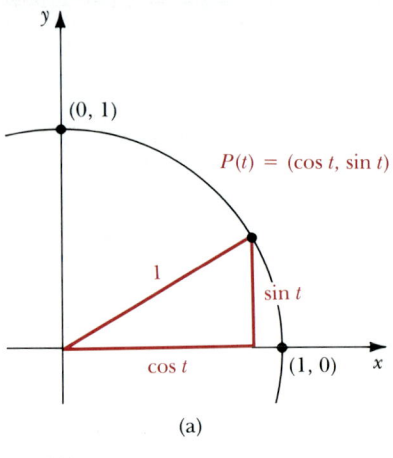

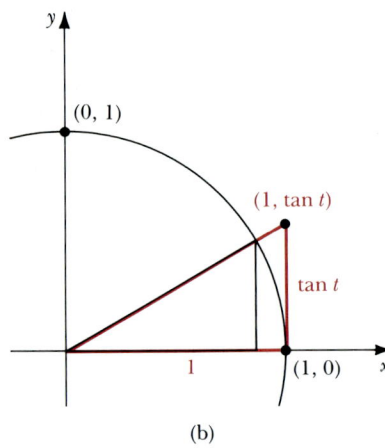

Figure A.12

A.3 TRIGONOMETRIC FUNCTIONS

Table A.1

t	0	$\dfrac{\pi}{6}$	$\dfrac{\pi}{4}$	$\dfrac{\pi}{3}$	$\dfrac{\pi}{2}$	$\dfrac{2\pi}{3}$	$\dfrac{3\pi}{4}$	π	$\dfrac{3\pi}{2}$	2π
$\sin t$	0	$\dfrac{1}{2}$	$\dfrac{\sqrt{2}}{2}$	$\dfrac{\sqrt{3}}{2}$	1	$\dfrac{\sqrt{3}}{2}$	$\dfrac{\sqrt{2}}{2}$	0	-1	0
$\cos t$	1	$\dfrac{\sqrt{3}}{2}$	$\dfrac{\sqrt{2}}{2}$	$\dfrac{1}{2}$	0	$-\dfrac{1}{2}$	$-\dfrac{\sqrt{2}}{2}$	-1	0	1
$\tan t$	0	$\dfrac{\sqrt{3}}{3}$	1	$\sqrt{3}$	not defined	$-\sqrt{3}$	-1	0	not defined	0

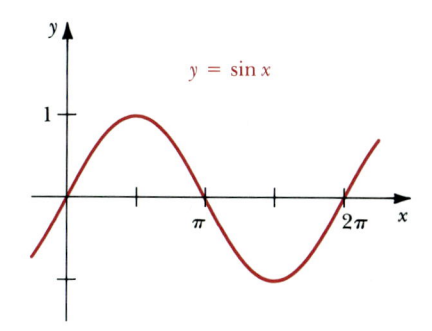

A number of identities are useful in studying trigonometric functions. Two of the most basic are:

(A.14) $\cos(t_1 \pm t_2) = \cos t_1 \cos t_2 \mp \sin t_1 \sin t_2;$

(A.15) $\sin(t_1 \pm t_2) = \sin t_1 \cos t_2 \pm \cos t_1 \sin t_2.$

Additional identities having frequent application can be derived from properties (A.10) through (A.15):

(A.16) $\sin\left(\dfrac{\pi}{2} - t\right) = \cos t; \quad \cos\left(\dfrac{\pi}{2} - t\right) = \sin t;$

(A.17) $\sin 2t = 2 \sin t \cos t; \quad \cos 2t = \cos^2 t - \sin^2 t;$

(A.18) $\sin^2 t = \dfrac{1 - \cos 2t}{2}; \quad \cos^2 t = \dfrac{1 + \cos 2t}{2};$

(A.19) $\tan(t_1 \pm t_2) = \dfrac{\tan t_1 \pm \tan t_2}{1 \mp \tan t_1 \tan t_2};$

(A.20) $\tan t = \dfrac{\sin 2t}{1 + \cos 2t}.$

Example 1

Verify the identity

$$\sin\left(\dfrac{\pi}{2} - t\right) = \cos t.$$

Solution

It follows from (A.15) that

$$\sin\left(\dfrac{\pi}{2} - t\right) = \sin\dfrac{\pi}{2} \cos t - \cos\dfrac{\pi}{2} \sin t.$$

From Table A.1, $\sin(\pi/2) = 1$ and $\cos(\pi/2) = 0$ so

$$\sin\left(\dfrac{\pi}{2} - t\right) = 1 \cdot \cos t - 0 \cdot \sin t = \cos t.$$

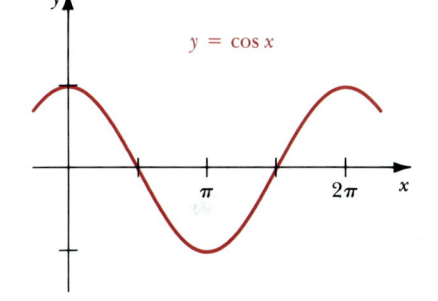

▶

(a) $y = \sin x$

(b) $y = \cos x$

(c) $y = \tan x$

Figure A.13

Example 2

Find all values of x in $[0, 2\pi)$ that satisfy the equation $\sin x + \cos x = 1$.

Solution

If
$$\sin x + \cos x = 1$$
then
$$\sin x = 1 - \cos x$$
and
$$\sin^2 x = 1 - 2 \cos x + \cos^2 x.$$
Since
$$\sin^2 x = 1 - \cos^2 x,$$
we have
$$1 - \cos^2 x = 1 - 2 \cos x + \cos^2 x.$$
So
$$0 = 2 \cos^2 x - 2 \cos x = 2 \cos x \, (1 - \cos x).$$

Solutions to this equation occur precisely when $\cos x = 0$, which implies that
$$x = \frac{\pi}{2} \quad \text{or} \quad x = \frac{3\pi}{2}$$
or when $\cos x = 1$, which implies that
$$x = 0.$$

Checking these values in our original equation, we see that the solutions are $x = \pi/2$ and $x = 0$. The extraneous value $x = 3\pi/2$ was introduced when the equation was squared. ▶

The three other trigonometric functions are also defined in terms of the sine and cosine functions.

(A.21) DEFINITION

For a real number t, the **cosecant** of t, written csc t, is defined by
$$\csc t = \frac{1}{\sin t}, \quad \text{provided } \sin t \neq 0;$$
the **secant** of t, written sec t, by
$$\sec t = \frac{1}{\cos t}, \quad \text{provided } \cos t \neq 0;$$

and the **cotangent** of t, written $\cot t$, by

$$\cot t = \frac{1}{\tan t} = \frac{\cos t}{\sin t}, \quad \text{provided } \sin t \neq 0.$$

Figure A.14 shows the graphs of these functions.

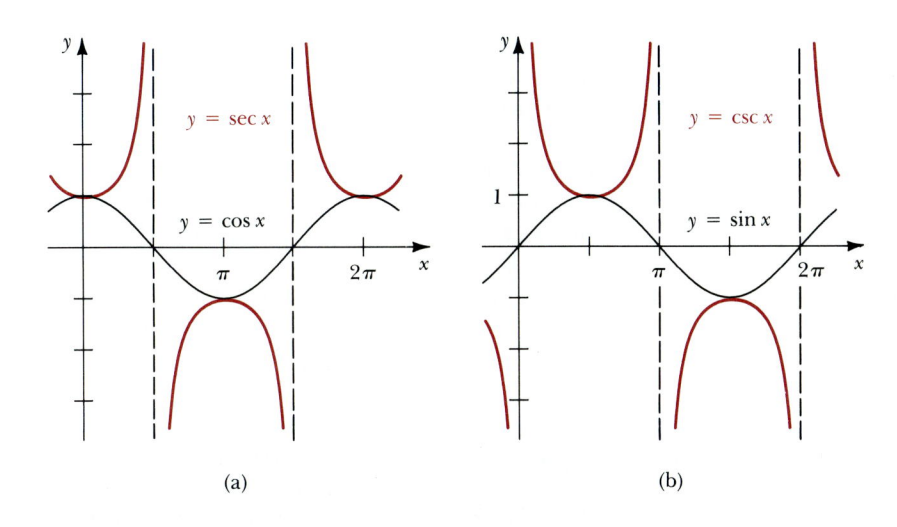

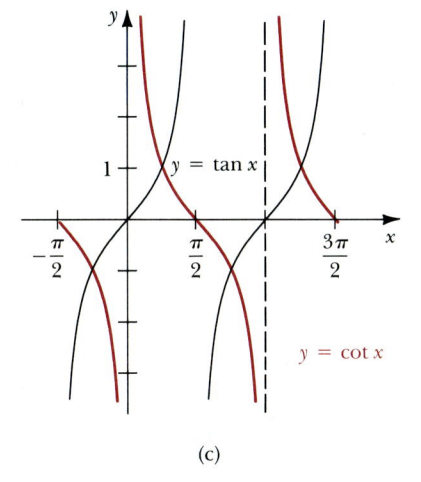

(a) (b) (c)

Figure A.14

Example 3

If $\sin t = 2/3$ and $0 < t < \pi/2$, find the values of each of the other trigonometric functions at t.

Solution

Since $\sin^2 t + \cos^2 t = 1$, we see that

$$\cos^2 t = 1 - \sin^2 t = 1 - \frac{4}{9} = \frac{5}{9}.$$

Hence, $\cos t = \sqrt{5}/3$. The other trigonometric functions depend on the values of $\cos t$ and $\sin t$ and are given as

$$\tan t = \frac{\sin t}{\cos t} = \frac{2}{\sqrt{5}} = \frac{2\sqrt{5}}{5}, \quad \csc t = \frac{1}{\sin t} = \frac{3}{2},$$

$$\sec t = \frac{1}{\cos t} = \frac{3}{\sqrt{5}} = \frac{3\sqrt{5}}{5}, \quad \text{and} \quad \cot t = \frac{\cos t}{\sin t} = \frac{\sqrt{5}}{2}. \quad \blacktriangleright$$

Using the definitions and property (A.11), we can show that

(A.22) $\tan^2 t + 1 = \sec^2 t; \quad \cot^2 t + 1 = \csc^2 t.$

There are many other identities involving the trigonometric functions. Some of these are listed in Exercises 34–38.

Geometric and computational problems often require that the trigonometric functions be defined in terms of angle measures. An **angle** θ is generated by rotating a **ray,** or half line, l_1 about its endpoint O, in a plane, to a new position determined by a ray l_2 with endpoint O. The ray l_1 is known as the **initial side** of the angle, l_2 the **terminal side,** and O the **vertex.**

If the generating rotation is counterclockwise, as in Figure A.15(a), the angle θ is considered a positive angle. If the rotation is clockwise, as in Figure A.15(b), θ is considered negative.

An angle is said to be in **standard position** when its initial side lies along the positive x-axis and its vertex is at the origin. The angle θ shown in Figure A.16 is in standard position. This angle has **radian measure** t, since the terminal side of θ intersects the unit circle U at $P(t)$.

The trigonometric functions of an angle are defined using the radian measure. If θ is an angle with radian measure t, then $\sin \theta = \sin t$, $\cos \theta = \cos t$, and so on.

An angle that subtends an arc of length $\pi/2$ is called a **right angle,** and a triangle containing a right angle is called a **right triangle.** For example, the triangle in Figure A.17 is a right triangle.

The rays determining a right angle are perpendicular. Values of the trigonometric functions of angles can be found easily using the sides of a right triangle. First superimpose an xy-coordinate system onto the triangle with the vertex of θ at the origin and positive x-axis along the side of the triangle with length a. Let U be the unit circle in this coordinate system and P be the point where the terminal side of θ intersects the unit circle (see Figure A.18). Using similar triangles (triangles AOB and POQ), we have

(A.23) $$\sin \theta = \frac{b}{c}, \qquad \cos \theta = \frac{a}{c}, \qquad \tan \theta = \frac{b}{a},$$

(A.24) $$\csc \theta = \frac{c}{b}, \qquad \sec \theta = \frac{c}{a}, \qquad \cot \theta = \frac{a}{b}.$$

Although the natural method of measuring angles for calculus purposes is the radian measure, angles are also measured in **degrees.** An angle formed by rotating a ray counterclockwise from an initial position

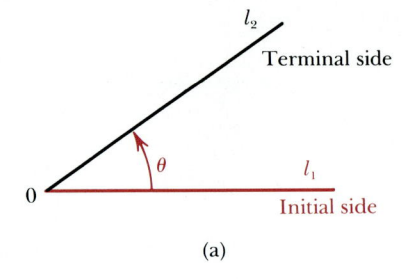

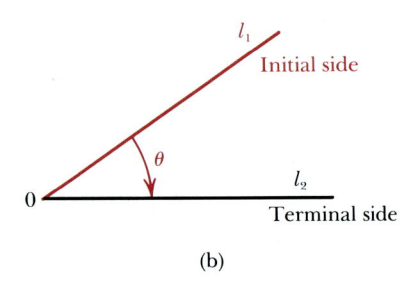

Figure A.15

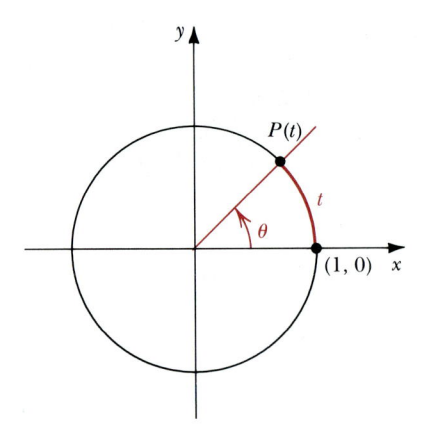

Figure A.16

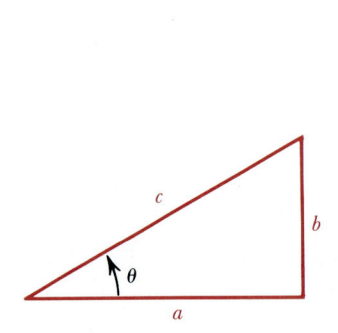

Figure A.17

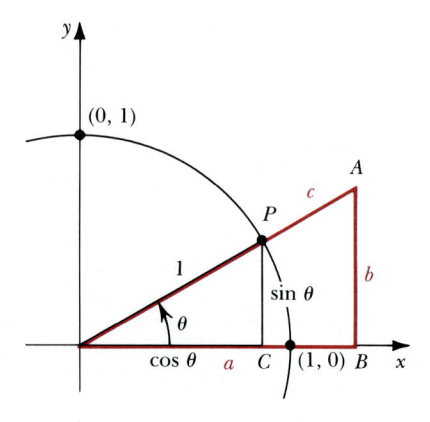

Figure A.18

back to the initial position so that the terminal and initial sides coincide has a measure of 360 degrees, written 360°. (See Figure A.19.)

Since the circumference (arc length) of the unit circle is 2π radians,

$$360 \text{ degrees} = 2\pi \text{ radians}$$

so

$$1 \text{ degree} = \frac{\pi}{180} \text{ radians} \approx 0.0175 \text{ radians}$$

and

$$1 \text{ radian} = \frac{180}{\pi} \text{ degrees} \approx 57.3°.$$

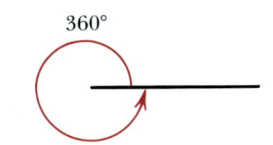

Figure A.19

Example 4

Express 135° and 405° in radians and $\pi/4$ radians in degrees.

Solution

$$135 \text{ degrees} = 135\left(\frac{\pi}{180} \text{ radians}\right) = \frac{3}{4}\pi \text{ radian},$$

$$405° = 405\left(\frac{\pi}{180}\right) \text{ radians} = \frac{9}{4}\pi \text{ radians},$$

$$\frac{\pi}{4} \text{ radians} = \frac{\pi}{4}\left(\frac{180}{\pi}\right) \text{ degrees} = 45 \text{ degrees}$$

▶

The final results needed from trigonometry are the law of sines and the law of cosines. These are used for determining the unknown sides and angles of triangles that are not right triangles. With the notation in the triangle given in Figure A.20, the **law of sines** is

(A.25)
$$\frac{\sin \alpha}{a} = \frac{\sin \beta}{b} = \frac{\sin \gamma}{c},$$

provided a, b, and c are all nonzero, and the **law of cosines** is

(A.26)
$$c^2 = a^2 + b^2 - 2ab \cos \gamma.$$

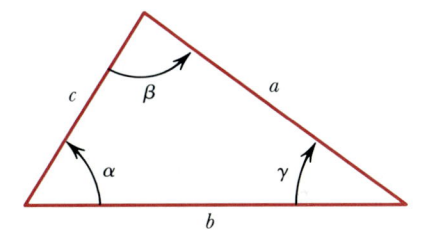

Figure A.20

▶ EXERCISE SET A.3

1. Locate the points $P(t)$ on a unit circle that correspond to the real numbers t given below and give the coordinates $(x(t), y(t))$ of the points.
   a) $t = \pi/6$ b) $t = \pi/4$
   c) $t = 13\pi/4$ d) $t = 7\pi/3$
   e) $t = -5\pi/6$ f) $t = -5\pi/4$

2. Find (i) $\sin t$, (ii) $\cos t$, and (iii) $\tan t$ for the real numbers t given in Exercise 1.

3. Find all values of x in the interval $[0, 2\pi]$ that satisfy the following equations.
   a) $\cos x = 1$ b) $\sin x = -1$
   c) $\sin x = 1/2$ d) $\cos x = -1/2$
   e) $\sin x = -\sqrt{3}/2$ f) $\tan x = 1$
   g) $\cot x = -1$ h) $\sec x = 2\sqrt{3}/3$
   i) $\csc x = 1$ j) $\cot x = -2$

4. Use Table A.1 and identities (A.14) and (A.15) to determine the following values.
   a) $\sin \dfrac{5\pi}{12}$ $\left(\text{Hint: } \dfrac{5\pi}{12} = \dfrac{\pi}{6} + \dfrac{\pi}{4}\right)$
   b) $\cos \dfrac{7\pi}{12}$
   c) $\tan \dfrac{\pi}{12}$
   d) $\cot \left(\dfrac{-\pi}{12}\right)$

5. If $\sin t = 4/5$ and $0 < t < \pi/2$, find $\cos t$ and $\tan t$.
6. If $\sin t = 0.1$ and $\pi/2 < t < \pi$, find $\cos t$ and $\tan t$.
7. If $\sin t = 3/5$, find all possible values for each of the other trigonometric functions.
8. If $\cos t = 2/3$, find all possible values for each of the other trigonometric functions.
9. If $\tan t = 2$ and $0 < t < \pi/2$, find the values for each of the other trigonometric functions.
10. If $\sec t = -2$ and $\pi/2 < t < 3\pi/2$, find the values for each of the other trigonometric functions.
11. Find the degree measure of the angles with the following radian measurements.
    a) $3\pi/4$ b) $-5\pi/6$
    c) π d) $7\pi/3$
    e) -4π f) $-7\pi/8$
12. Find the radian measure of the angles with the following degree measurements.
    a) $75°$ b) $-135°$
    c) $240°$ d) $40°$
    e) $1500°$ f) $-315°$
13. Find the following values.
    a) $\cot \dfrac{13}{6}\pi$ b) $\sec \dfrac{322}{3}\pi$
    c) $\csc \dfrac{-107}{3}\pi$ d) $\sin 1200°$
    e) $\cos -585°$ f) $\tan \dfrac{135}{4}\pi$

14. Use the graphs given in this section and the techniques discussed in Chapter 1 to sketch the graph of each of the following functions.
    a) $f(x) = 2 \sin x$
    b) $f(x) = \sin x + 1$
    c) $f(x) = \cos \left(x - \dfrac{\pi}{2}\right)$
    d) $f(x) = 3 \cos x + 2 \sin x$
    e) $f(x) = \sin \left(x + \dfrac{\pi}{2}\right)$
    f) $f(x) = \cos \left(x - \dfrac{\pi}{2}\right) - \sin \left(x + \dfrac{\pi}{2}\right)$

15. Sketch the graphs of the following functions.
    a) $f(x) = \sin 2x$ b) $f(x) = \sin 0.5x$
    c) $f(x) = \cos 3x$ d) $f(x) = \cos \dfrac{1}{4} x$

In Exercises 16–21, find all values of x in the interval $[0, 2\pi)$ that satisfy the equation.

16. $\sin 2x = \sin x$
17. $\sin 2x = \cos x$
18. $|\tan x| = 1$
19. $\sin x \cos x - \sin x - \cos x + 1 = 0$
20. $\tan x + \cot x = \dfrac{1}{2} \sin 2x$
21. $\sin \left(x - \dfrac{\pi}{2}\right) = \cos (\pi - x)$

22. The definitions of even and odd functions are given in Section 1.1. Use these definitions to determine which of the six trigonometric functions are even and which are odd.

Use the identities listed in this section to verify the identities given in Exercises 23–33.

23. $(1 - \cos^2 \theta) \sec^2 \theta = \tan^2 \theta$
24. $\dfrac{\cos x}{1 - \tan x} + \dfrac{\sin x}{1 - \cot x} = \sin x + \cos x$
25. $\cot x + \tan x = \sec x \csc x$
26. $\dfrac{\sin t}{1 - \cos t} = \cot t + \csc t$
27. $\cot x - \tan x = 2 \cot 2x$
28. $\dfrac{\sin (b - a)}{\sin (b + a)} = \dfrac{\cot a - \cot b}{\cot a + \cot b}$
29. $(\sin x + \cos x)^2 = 1 + \sin 2x$
30. $\cos x = \sin x \sin 2x + \cos x \cos 2x$
31. $\tan^2 x - \sin^2 x = \tan^2 x \sin^2 x$
32. $\sec x - \cos x = \sin x \tan x$
33. $\cos x (\cot x + \tan x) = \csc x$

Use identities (A.14) and (A.15) to establish the identities listed in Exercises 34–38.

34. $\sin t_1 \sin t_2 = \dfrac{1}{2} [\cos (t_1 - t_2) - \cos (t_1 + t_2)]$

35. $\cos t_1 \cos t_2 = \dfrac{1}{2}[\cos(t_1 - t_2) + \cos(t_1 + t_2)]$

36. $\sin t_1 \cos t_2 = \dfrac{1}{2}[\sin(t_1 - t_2) + \sin(t_1 + t_2)]$

37. $\tan(t_1 + t_2) = \dfrac{\tan t_1 + \tan t_2}{1 - \tan t_1 \tan t_2}$

38. $\tan(t_1 - t_2) = \dfrac{\tan t_1 - \tan t_2}{1 + \tan t_1 \tan t_2}$

39. Four ships A, B, C, and D are at sea in the following relative positions: B is 2 mi due north of D, D is due west of C, and B is on a line between A and C. The angle BDA measures 40° and the angle BCD measures 25°. What is the distance between A and D? (See the figure.)

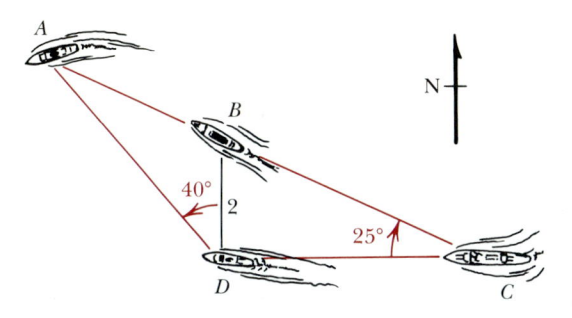

40. A certain biological rhythm can be approximately described by the equation

$$y = 3 + 2\cos\dfrac{\pi}{12}(t - 5),$$

where t represents time given in hours. Sketch the graph of this equation.

41. A climber who needs to estimate the height of a cliff stands at a point on the ground 35 ft from the base of the cliff and estimates that the angle from the ground to a line extending from her feet to the top of the cliff is approximately 60°. Assuming that the cliff is perpendicular to the ground, find the approximate height of the cliff.

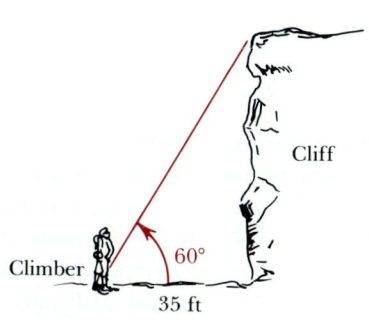

42. A gutter is to be made from a long strip of tin by bending up the sides, so that the base and the sides have the same length b. Express the area of a cross section of the gutter as a function of the angle θ that the sides make with the base.

43. Two balls are against the rail on opposite ends of an 8-ft-long billiard table. One lies 2 ft from a side of the table and the other lies 1 ft from this same side, as shown in the figure. Where should the first ball hit this side of the table in order to hit the second ball?

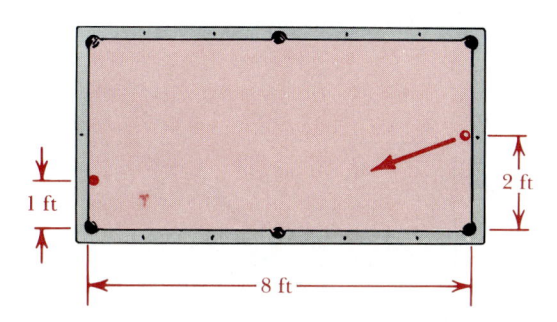

44. The National Forest Service maintains observation towers to check for the outbreak of forest fires. Suppose two towers are at the same elevation, one at point A and another 10 mi due west at point B (see the figure). The ranger at A spots a fire in the northwest whose line of sight makes an angle of 63° with the line between the towers, and contacts the ranger at B. This ranger locates the fire along a line of sight that makes a 50° angle with the line between the towers. How far is the fire from tower B?

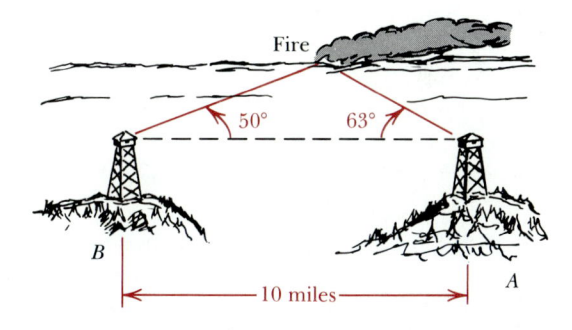

45. Suppose l_1 and l_2 are two nonvertical, nonperpendicular lines. Show that if l_1 and l_2 have slopes m_1 and m_2, respectively, and if θ is the angle from l_1 to l_2, then

$$\tan\theta = \dfrac{m_2 - m_1}{1 + m_1 m_2}.$$

46. Use the result in Exercise 45 to show that the slope of a nonvertical line l is $\tan\alpha$, where α is the angle of inclination of l.

47. Use the result in Exercise 45 to show that two nonvertical lines l_1 and l_2 with slopes m_1 and m_2, respectively, are perpendicular if and only if $m_1 m_2 = -1$.

48. Simplify $\tan(t/2) + \cot(t/2)$.

49. Show that for all t, $|\tan t + \cot t| \geq 2$.

50. To test whether the cut of a circle is a diameter of the circle, a carpenter takes a right angle square and checks to see if it touches the curve at the three points, as shown in the figure. (a) Show that this test is valid. (b) Find a variation of this test that can be used to locate the center of a given circle.

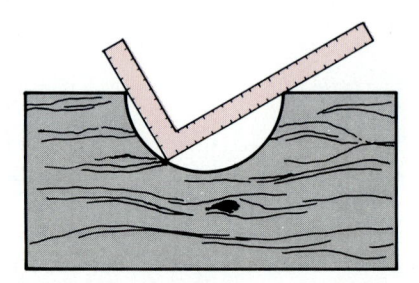

51. Another carpenters' test involving the right angle square is to determine whether a closed curve is actually circular. If a carpenter's square is placed outside the curve, the curve is circular if the sides of the square intersect the curve at two points that are the same distance from the vertex (see the figure). Why is this test valid?

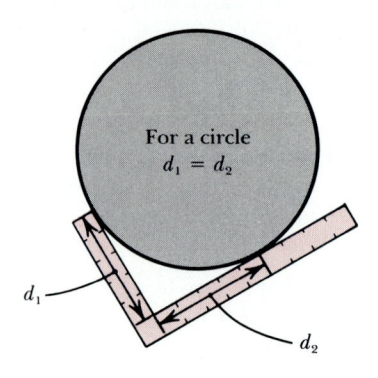

A.4

Mathematical Induction

Formulas presented in Section 4.1 state that for each natural number n:

$$1 + 2 + \cdots + n = \frac{n(n+1)}{2},$$

$$1^2 + 2^2 + \cdots + n^2 = \frac{n(n+1)(2n+1)}{6},$$

and

$$1^3 + 2^3 + \cdots + n^3 = \frac{n^2(n+1)^2}{4}.$$

In Exercise Set 4.1, procedures are discussed for deriving these formulas. In this section we consider an alternative method, called *mathematical induction*, that can be used to show that these formulas hold.

Another example of this type of expression is

$$\log(a_1 a_2 \ldots a_n) = \log a_1 + \log a_2 + \cdots + \log a_n,$$

which is true for any collection of n positive numbers. These expressions all follow a recursive pattern and can be established using the principle of mathematical induction. This method of proof was first formally described by Augustus DeMorgan (1807–1871) in 1838. The underlying principle, however, was used by Pascal, Fermat, and others before 1650, and is evident before that time in the work of Francesco Maurolico (1494–1575). The method is based on the following axiom.

(A.27)
THE AXIOM OF MATHEMATICAL INDUCTION

Suppose S is a set of natural numbers. If

i) 1 belongs to S, and
ii) whenever an integer k belongs to S, the integer $k + 1$ also belongs to S,

then $S = N$, the set of all natural numbers.

To see how this axiom is applied, let us consider some examples.

Example 1

Use the axiom of mathematical induction to show that

$$1 + 2 + \cdots + n = \frac{n(n+1)}{2}$$

for all natural numbers n.

Solution

Let S be the set of natural numbers for which this statement is true.

i) 1 is in S because $1 = \frac{1(1+1)}{2}$.

ii) Suppose the integer k is in S; that is,

$$1 + 2 + \cdots + k = \frac{k(k+1)}{2}.$$

Then

$$1 + 2 + \cdots + k + 1(k+1) = (1 + 2 + \cdots + k) + (k+1)$$
$$= \frac{k(k+1)}{2} + (k+1)$$
$$= \frac{k(k+1) + 2(k+1)}{2}$$
$$= \frac{(k+1)(k+2)}{2}.$$

This is the formula

$$1 + 2 + \cdots + n = \frac{n(n+1)}{2}$$

when $n = k + 1$, so $k + 1$ is in S.

By the axiom of mathematical induction, $S = N$ and the formula holds for every natural number n. ▶

Example 2

Show that $2^n > n$ for each natural number n.

Solution

Let S be the set of natural numbers n for which $2^n > n$.

i) 1 is in S, since $2^1 = 2 > 1$.
ii) Suppose k is in S; that is, $2^k > k$. Then

$$2^{k+1} = 2^k \, 2 > k \cdot 2 = k + k \geq k + 1.$$

Consequently, $k + 1$ is also in S.

By the axiom of mathematical induction, $S = N$ and $2^n > n$ for each natural number n. ▶

Example 3

Prove that the rule $\log (a \cdot b) = \log a + \log b$ for positive numbers a and b implies that

$$\log (a_1 \, a_2 \, \ldots \, a_n) = \log a_1 + \log a_2 + \cdots + \log a_n$$

for every collection of n positive numbers.

Solution

Let S be the set of natural numbers n for which

$$\log (a_1 \, a_2 \, \cdots \, a_n) = \log a_1 + \log a_2 + \cdots + \log a_n$$

for every collection of n positive numbers.

i) 1 is in S, since $\log a_1 = \log a_1$.
ii) Suppose k is in S; that is, for any collection of k positive numbers,

$$\log (a_1 \, a_2 \, \ldots \, a_k) = \log a_1 + \log a_2 + \cdots + \log a_k.$$

Then for any collection $a_1, a_2, \ldots, a_k, a_{k+1}$ of $k + 1$ positive numbers

$$\begin{aligned}\log (a_1 \, a_2 \, \ldots \, a_k \, a_{k+1}) &= \log ((a_1 \, a_2 \, \ldots \, a_k) a_{k+1}) \\ &= \log (a_1 \, a_2 \, \ldots \, a_k) + \log a_{k+1} \\ &= \log a_1 + \log a_2 + \cdots + \log a_k + \log a_{k+1},\end{aligned}$$

so $k + 1$ is in S.

By the axiom of mathematical induction, $S = N$ and

$$\log (a_1 \, a_2 \, \ldots \, a_n) = \log a_1 + \log a_2 + \cdots + \log a_n$$

for any collection of positive numbers $a_1, a_2, \ldots, a_n$, regardless of the value of n. ▶

▶ EXERCISE SET A.4

In Exercises 1–17, use the axiom of mathematical induction to show that the expression is true for every natural number n.

1. $1 + 3 + 5 + \cdots + (2n - 1) = n^2$
2. $2 + 4 + 6 + \cdots + 2n = n(n + 1)$

3. $3 + 6 + 9 + \cdots + 3n = \dfrac{3n(n+1)}{2}$

4. $\dfrac{1}{2} + \dfrac{1}{4} + \dfrac{1}{8} + \cdots + \dfrac{1}{2^n} = 1 - \dfrac{1}{2^n}$

5. $1 - \dfrac{1}{2} - \dfrac{1}{4} - \cdots - \dfrac{1}{2^n} = \dfrac{1}{2^n}$

6. $1 + 4 + 9 + \cdots + n^2 = \dfrac{n(n+1)(2n+1)}{6}$

7. $1 + 8 + 27 + \cdots + n^3 = \dfrac{n^2(n+1)^2}{4}$

8. $2 + 6 + 12 + \cdots + n(n+1) = \dfrac{n(n+1)(n+2)}{3}$

9. $\dfrac{1}{2} + \dfrac{1}{6} + \dfrac{1}{12} + \cdots + \dfrac{1}{n(n+1)} = \dfrac{n}{n+1}$

10. $\dfrac{1}{2} + \dfrac{1}{2} + \dfrac{3}{8} + \cdots + \dfrac{n}{2^n} = 2 - \dfrac{n+2}{2^n}$

11. $3 + 8 + 15 + \cdots + n(n+2) = \dfrac{n(n+1)(2n+7)}{6}$

12. $\dfrac{1}{3} + \dfrac{1}{15} + \dfrac{1}{35} + \cdots + \dfrac{1}{4n^2 - 1} = \dfrac{n}{2n+1}$

13. $1 - \dfrac{1}{2} - \dfrac{1}{6} - \cdots - \left(\dfrac{1}{n} - \dfrac{1}{n+1}\right) = \dfrac{1}{n+1}$

14. $n < \left(\dfrac{3}{2}\right)^n$

15. $1 + na \le (1 + a)^n$, for any $a \ge 0$

16. $a + ar + ar^2 + \cdots + ar^n = \dfrac{a(1 - r^{n+1})}{1 - r}$, for any a and any $r \ne 1$

17. $a + (a + d) + (a + 2d) + \cdots + (a + nd) = \dfrac{(n+1)(2a + nd)}{2}$, for any a and d

18. Prove that $x - y$ is a factor of $x^n - y^n$ for any natural number n.

19. Prove that $x + y$ is a factor of $x^{2n-1} + y^{2n-1}$ for any natural number n.

20. Prove that for any natural number n and any $a > 1$, $1 < a^n < a^{n+1}$.

21. Prove that for any natural number n and any $0 < a < 1$, $0 < a^{n+1} < a^n < 1$.

22. Use mathematical induction to prove that the following laws of exponents hold for every pair of natural numbers m and n.
   a) $a^n b^n = (ab)^n$
   b) $\dfrac{a^n}{b^n} = \left(\dfrac{a}{b}\right)^n$, provided $b \ne 0$
   c) $a^m a^n = a^{m+n}$
   d) $\dfrac{a^n}{a^m} = a^{n-m}$, provided $a \ne 0$
   e) $(a^m)^n = a^{mn}$

   [*Hint:* In parts (c), (d) and (e), assume that m is a fixed but arbitrary natural number.]

23. Use mathematical induction to show that if c_n is the midpoint of the interval $[a_n, b_n]$, obtained when the Bisection Method (given in Section 1.8) is applied on $[a, b]$, then
$$|c_n - c| < \dfrac{b - a}{2^n}.$$

24. Let $a_0 = 1$ and $a_{n+1} = \tfrac{1}{2}(a_n + 2/a_n)$ for each $n \ge 0$. Show that $1 \le a_n \le 2$ for all $n \ge 0$.

25. Let $a_0 = 1$, $0 < M$, and $a_{n+1} = \tfrac{1}{2}(a_n + M/a_n)$ for each $n \ge 0$. Show that $1 \le a_n \le M$ if $M \ge 1$ and $M \le a_n \le 1$ if $M \le 1$.

26. Prove that any set with n elements has 2^n distinct subsets.

27. The Tower of Hanoi problem is shown in the figure. The problem is to find the number of moves required to move all discs from post A to post C. A move is defined to be the movement of one disc from one pin to another. You can remove only one disc at a time and a larger disc can never rest on top of a smaller disc.

   Use mathematical induction to show that the number of moves required to move n discs is $2^n - 1$.

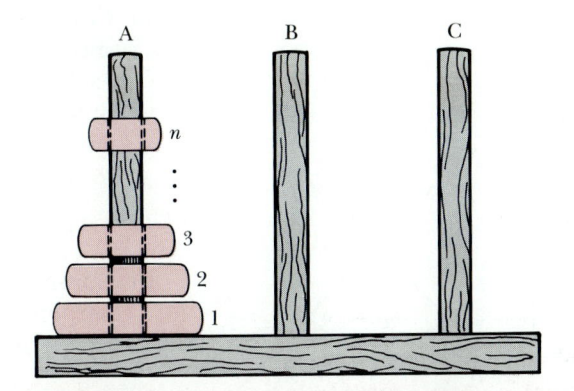

Appendix B
Additional Calculus Theorems and Proofs

This appendix contains the proof of some basic theorems in calculus. The motivation and discussion of these results were presented in the text and are not repeated here. The theorems have been numbered with their original text references and are presented in the order in which they were originally discussed.

The first theorem concerns the slopes of parallel lines and the slopes of perpendicular lines.

(1.13) THEOREM

a) Two nonvertical lines are parallel if and only if they have the same slope.
b) Two nonvertical lines are perpendicular if and only if the product of their slopes is -1.

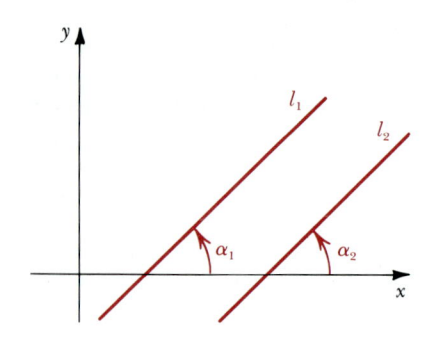

Figure B.1

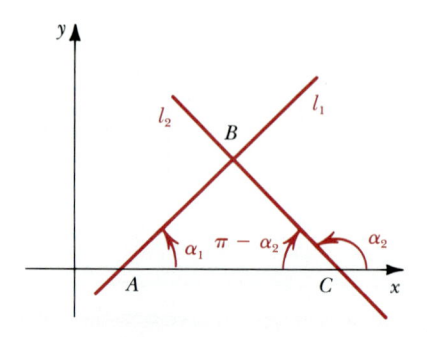

Figure B.2

Proof.

a) Consider the lines l_1 and l_2 with slopes m_1 and m_2, respectively, shown in Figure B.1. Lines are parallel if and only if they intersect each tranverse line in the same angle. Thus l_1 and l_2 are parallel if and only if their angles of inclination, α_1 and α_2, are equal. Since these angles lie in $(0, \pi/2) \cup (\pi/2, \pi)$, where the tangent function is one-to-one, nonvertical lines l_1 and l_2 are parallel if and only if

$$m_1 = \tan \alpha_1 = \tan \alpha_2 = m_2.$$

b) Consider the triangle ABC in Figure B.2. The lines l_1 and l_2 are perpendicular if and only if the angle ABC is a right angle, which is true if and only if

$$\alpha_1 + (\pi - \alpha_2) = \pi/2.$$

Hence l_1 is perpendicular to l_2 if and only if $\alpha_2 = \dfrac{\pi}{2} + \alpha_1$.

Since α_1 and α_2 are in $(0, \pi/2) \cup (\pi/2, \pi)$, where the tangent function is one to one, l_1 is perpendicular to l_2 if and only if

$$\tan \alpha_2 = \tan\left(\frac{\pi}{2} + \alpha_1\right) = \frac{\sin\left(\frac{\pi}{2} + \alpha_1\right)}{\cos\left(\frac{\pi}{2} + \alpha_1\right)}.$$

But

$$\sin\left(\frac{\pi}{2} + \alpha_1\right) = \sin\frac{\pi}{2}\cos\alpha_1 + \sin\alpha_1\cos\frac{\pi}{2} = \cos\alpha_1$$

and

$$\cos\left(\frac{\pi}{2} + \alpha_1\right) = \cos\alpha_1\cos\frac{\pi}{2} - \sin\alpha_1\sin\frac{\pi}{2} = -\sin\alpha_1,$$

so l_1 is perpendicular to l_2 if and only if

$$m_2 = \tan\alpha_2 = -\frac{\cos\alpha_1}{\sin\alpha_1} = -\frac{1}{\tan\alpha_1} = -\frac{1}{m_1},$$

that is, if and only if $m_1 \cdot m_2 = -1$. ▷

Next we consider the proofs of the limit results discussed in Chapter 1.

(1.31)
THEOREM
If $\lim_{x \to a} f(x) = L$ and $\lim_{x \to a} f(x) = M$, then $L = M$.

Proof. To prove this theorem we use proof by contradiction. We assume that $L \neq M$ and show that this assumption leads to a contradiction. If $L \neq M$, then $|L - M|/2 > 0$. Since $\lim_{x \to a} f(x) = L$, for $\varepsilon = |L - M|/2$, there is a number $\delta_1 > 0$ such that $|f(x) - L| < \varepsilon = |L - M|/2$ whenever $0 < |x - a| < \delta_1$.

Since $\lim_{x \to a} f(x) = M$, for $\varepsilon = |L - M|/2$, there is a number $\delta_2 > 0$ such that $|f(x) - M| < \varepsilon = |L - M|/2$ whenever $0 < |x - a| < \delta_2$.

If δ is the smaller of δ_1 and δ_2, then whenever $0 < |x - a| < \delta$,

$$|L - M| = |L - f(x) + f(x) - M| \leq |L - f(x)| + |f(x) - M|$$
$$< \frac{|L - M|}{2} + \frac{|L - M|}{2} = |L - M|.$$

That is, whenever $0 < |x - a| < \delta$, $|L - M| < |L - M|$. This is a contradiction and our assumption is false. The only possibility is that $L = M$. ▷

(1.37)
THEOREM
If f and g are functions with $\lim_{x \to a} f(x) = L$ and $\lim_{x \to a} g(x) = M$, then:

a) $\lim_{x \to a} [f(x) + g(x)] = L + M$;

b) $\lim_{x \to a} [f(x) - g(x)] = L - M$;

c) $\lim_{x \to a} [f(x)g(x)] = L \cdot M$;

d) $\lim_{x \to a} cf(x) = cL$, for any constant c;

e) $\lim_{x \to a} \dfrac{1}{f(x)} = \dfrac{1}{L}$, provided $L \neq 0$;

f) $\lim_{x \to a} \dfrac{g(x)}{f(x)} = \dfrac{M}{L}$, provided $L \neq 0$;

g) $\lim_{x \to a} \sqrt[n]{f(x)} = \sqrt[n]{L}$ if n is an odd positive integer or if n is an even positive integer and $L > 0$.

Proof.

a) $\lim_{x \to a} [f(x) + g(x)] = L + M$.

Let $\varepsilon > 0$ be given. Since $\lim_{x \to a} f(x) = L$ and $\lim_{x \to a} g(x) = M$, there exist numbers δ_1 and δ_2 with the property that

$$|f(x) - L| < \frac{\varepsilon}{2} \quad \text{whenever } 0 < |x - a| < \delta_1$$

and

$$|g(x) - M| < \frac{\varepsilon}{2} \quad \text{whenever } 0 < |x - a| < \delta_2.$$

Choose $\delta = \text{minimum } (\delta_1, \delta_2)$ and suppose that $0 < |x - a| < \delta$. Then both $|f(x) - L| < \varepsilon/2$ and $|g(x) - M| < \varepsilon/2$, so

$$|[f(x) + g(x)] - [L + M]| = |(f(x) - L) + (g(x) - M)|$$

$$\leq |f(x) - L| + |g(x) - M| < \frac{\varepsilon}{2} + \frac{\varepsilon}{2} = \varepsilon.$$

Thus, $\lim_{x \to a} [f(x) + g(x)] = L + M$.

It is easier to prove parts (b) and (c) if we first prove part (d).

d) $\lim_{x \to a} cf(x) = cL$, for any constant c.

If $c = 0$, then $cf(x) = 0$, $cL = 0$, and the result holds by Theorem 1.34.

Suppose $c \neq 0$ and $\varepsilon > 0$ is given. Let δ be such that

$$|f(x) - L| < \frac{\varepsilon}{|c|} \quad \text{whenever } 0 < |x - a| < \delta.$$

Then

$$|cf(x) - cL| = |c||f(x) - L| < |c|\frac{\varepsilon}{|c|} = \varepsilon$$

and $\lim_{x \to a} cf(x) = cL$.

b) $\lim_{x \to a} [f(x) - g(x)] = L - M$.

Since $\lim_{x \to a} g(x) = M$, (d) implies that $\lim_{x \to a} -g(x) = -M$. By (a)

$\lim_{x \to a} [f(x) - g(x)] = \lim_{x \to a} [f(x) + (-g(x))] = L - M$. Before showing the remainder of the results we will prove the following lemma.

LEMMA
If $\lim_{x \to a} f(x) = L$, then positive numbers m and δ exist with $|f(x)| \leq m$ for $0 < |x - a| < \delta$.

Proof. Let $\varepsilon = 1$ and δ be such that
$$|f(x) - L| < \varepsilon = 1 \quad \text{whenever } 0 < |x - a| < \delta.$$
Then $-1 < f(x) - L < 1$, so $L - 1 < f(x) < L + 1$ and
$$|f(x)| \leq \text{maximum } \{|L + 1|, |L - 1|\}.$$
Choosing $m = \max \{|L + 1|, |L - 1|\}$ gives the result. ▷

We now return to the proof of Theorem 1.37.

c) $\lim_{x \to a} [f(x) g(x)] = L \cdot M$.

Let $\varepsilon > 0$ be given and δ_1 and $m > 0$ be such that
$$|f(x)| \leq m \quad \text{whenever } 0 < |x - a| < \delta_1$$
Let δ_2 and δ_3 be such that
$$|f(x) - L| < \frac{\varepsilon}{2|M|} \quad \text{whenever } 0 < |x - a| < \delta_2$$
and
$$|g(x) - M| < \frac{\varepsilon}{2m} \quad \text{whenever } 0 < |x - a| < \delta_3.$$
Choose $\delta = \text{minimum } \{\delta_1, \delta_2, \delta_3\}$ and let $0 < |x - a| < \delta$. Then
$$|f(x)g(x) - L \cdot M| = |f(x)g(x) - f(x)M + f(x)M - L \cdot M|$$
$$\leq |f(x)||g(x) - M| + |f(x) - L||M|$$
$$< m\left(\frac{\varepsilon}{2m}\right) + \frac{\varepsilon}{2|M|}|M| = \frac{\varepsilon}{2} + \frac{\varepsilon}{2} = \varepsilon.$$
Thus, $\lim_{x \to a} f(x) g(x) = L \cdot M$.

e) $\lim_{x \to a} \frac{1}{f(x)} = \frac{1}{L}$, provided $L \neq 0$.

Since $L \neq 0$ there is a number δ_1 such that
$$|f(x) - L| < \frac{|L|}{2} \quad \text{whenever } 0 < |x - a| < \delta_1.$$
Thus, for $0 < |x - a| < \delta_1$,
$$\left||f(x)| - |L|\right| \leq |f(x) - L| \leq \frac{|L|}{2}$$
and
$$\frac{-|L|}{2} < |f(x)| - |L| < \frac{|L|}{2}.$$

This implies that
$$\frac{|L|}{2} \le |f(x)| \le \frac{3|L|}{2}.$$

Let $\varepsilon > 0$ and choose $\delta > 0$ to be a number less than δ_1 with the property that
$$|f(x) - L| < \frac{L^2 \varepsilon}{2} \quad \text{whenever } 0 < |x - a| < \delta.$$

Then
$$\left| \frac{1}{f(x)} - \frac{1}{L} \right| = \left| \frac{L - f(x)}{f(x)L} \right| = \frac{|f(x) - L|}{|f(x)||L|}.$$

But since $0 < |x - a| < \delta \le \delta_1$,
$$|f(x)| \ge \frac{|L|}{2} \quad \text{and} \quad \frac{1}{|f(x)|} \le \frac{2}{|L|}.$$

So
$$\left| \frac{1}{f(x)} - \frac{1}{L} \right| = \frac{|f(x) - L|}{|L||f(x)|} \le \frac{2|f(x) - L|}{|L|^2} < \frac{2}{|L|^2} \frac{L^2 \varepsilon}{2} = \varepsilon.$$

Thus,
$$\lim_{x \to a} \frac{1}{f(x)} = \frac{1}{L}.$$

f) $\lim_{x \to a} \dfrac{g(x)}{f(x)} = \dfrac{M}{L}$, provided $L \ne 0$.

By applying (c) and (e), we have
$$\lim_{x \to a} \frac{g(x)}{f(x)} = \lim_{x \to a} g(x) \frac{1}{f(x)} = \lim_{x \to a} g(x) \lim_{x \to a} \frac{1}{f(x)} = M \frac{1}{L} = \frac{M}{L}.$$

g) $\lim_{x \to a} \sqrt[n]{f(x)} = \sqrt[n]{L}$ if n is an odd positive integer or if n is an even positive integer and $L > 0$.

We prove the result only for the case $n = 2$ and $L > 0$. To show that $\lim_{x \to a} \sqrt{f(x)} = \sqrt{L}$, let $\varepsilon > 0$ be given and choose δ so that
$$|f(x) - L| < \sqrt{L}\, \varepsilon \quad \text{whenever } 0 < |x - a| < \delta.$$

Since $f(x) - L = (\sqrt{f(x)} - \sqrt{L})(\sqrt{f(x)} + \sqrt{L})$,
$$|\sqrt{f(x)} - \sqrt{L}| = \left| \frac{f(x) - L}{\sqrt{f(x)} + \sqrt{L}} \right| = \frac{|f(x) - L|}{\sqrt{f(x)} + \sqrt{L}} \le \frac{|f(x) - L|}{\sqrt{L}} < \frac{\sqrt{L}\,\varepsilon}{\sqrt{L}} = \varepsilon$$

whenever $0 < |x - a| < \delta$. Thus, $\lim_{x \to a} \sqrt{f(x)} = \sqrt{L}$. ▷

The final limit result to be proved concerns the limit of a composition of two functions.

(1.42)
THEOREM
If $\lim_{x \to a} g(x) = b$ and f is continuous at b, then

$$\lim_{x \to a} f(g(x)) = f\left(\lim_{x \to a} g(x)\right) = f(b).$$

Proof. Let $\varepsilon > 0$ be given. Since f is continuous at b, $\lim_{z \to b} f(z) = f(b)$ and a number δ_1 exists with

$$|f(z) - f(b)| < \varepsilon \quad \text{whenever } |z - b| < \delta_1.$$

In addition, since $\lim_{x \to a} g(x) = b$, a number δ exists with

$$|g(x) - b| < \delta_1 \quad \text{whenever } |x - a| < \delta.$$

Thus,

$$|f(g(x)) - f(b)| < \varepsilon \quad \text{whenever } |x - a| < \delta,$$

which proves that

$$\lim_{x \to a} f(g(x)) = f(b) = f\left(\lim_{x \to a} g(x)\right). \qquad \triangleright$$

An intuitive justification for the Chain Rule for derivatives was given in Chapter 2. The proof of the Chain Rule uses the following theorem, which can be found in Section 14.3.

(14.16)
THEOREM
A function f is differentiable at a number x if and only if there exists an open interval I about x, a number m, and a function ε so that

i) $f(x + \Delta x) - f(x) = m \, \Delta x + \varepsilon(\Delta x) \cdot \Delta x$, for all $x + \Delta x$ in I, and
ii) $\lim_{\Delta x \to 0} \varepsilon(\Delta x) = 0$.

In addition, if f is differentiable, $m = f^1(x)$.

(2.29)
THEOREM: THE CHAIN RULE
If g is differentiable at a and f is differentiable at $g(a)$, then $f \circ g$ is differentiable at a and $(f \circ g)'(a) = f'(g(a))g'(a)$.

Proof. We apply Theorem 14.16 with $g(a)$ replacing x, $g(x) - g(a)$ replacing Δx, and $f'(g(a))$ replacing m. Since f is differentiable at $g(a)$, Theorem 14.16 implies that a function ε exists with

$$f(g(x)) - f(g(a)) = f'(g(a))(g(x) - g(a)) + \varepsilon(g(x) - g(a))(g(x) - g(a))$$
$$= [f'(g(a)) + \varepsilon(g(x) - g(a))][g(x) - g(a)]$$

and

$$\lim_{x \to a} \varepsilon(g(x) - g(a)) = 0.$$

But g is differentiable and hence continuous at a, so $\lim_{x \to a} g(x) = g(a)$ and

$$\lim_{x \to a} \varepsilon(g(x) - g(a)) = \lim_{g(x) \to g(a)} \varepsilon(g(x) - g(a)) = 0.$$

Consequently,

$$\begin{aligned}
(f \circ g)'(a) &= \lim_{x \to a} \frac{f(g(x)) - f(g(a))}{x - a} \\
&= \lim_{x \to a} \frac{[f'(g(a)) + \varepsilon(g(x) - g(a))] [g(x) - g(a)]}{x - a} \\
&= \lim_{x \to a} [f'(g(a)) + \varepsilon(g(x) - g(a))] \lim_{x \to a} \frac{g(x) - g(a)}{x - a} \\
&= f'(g(a)) \, g'(a). \qquad \triangleright
\end{aligned}$$

The proof given in the text for the following extension of the Power Rule assumed the existence of the derivative. This assumption need not be made, as the proof given here demonstrates.

(2.33)
THEOREM
If $f(x) = x^r$, where r is a nonzero rational number, then f is differentiable for all numbers x for which x^{r-1} is defined and $f'(x) = rx^{r-1}$.

Proof. First note that for any positive integer q and any numbers a and b.

$$a^q - b^q = (a - b)(a^{q-1} + a^{q-2}b + \cdots + ab^{q-2} + b^{q-1}).$$

If we consider this result when $a = (x + h)^{1/q}$ and $b = x^{1/q}$, we have

$$\begin{aligned}
h &= [(x+h)^{1/q}]^q - [x^{1/q}]^q \\
&= [(x+h)^{1/q} - x^{1/q}][(x+h)^{(q-1)/q} \\
&\quad + (x+h)^{(q-2)/q} x^{1/q} + \cdots + (x+h)^{1/q} x^{(q-2)/q} + x^{(q-1)/q}].
\end{aligned}$$

This implies that for any positive integer q,

$$D_x x^{1/q} = \lim_{h \to 0} \frac{(x+h)^{1/q} - x^{1/q}}{h}$$

$$= \lim_{h \to 0} \frac{1}{(x+h)^{(q-1)/q} + (x+h)^{(q-2)/q} x^{1/q} + \cdots + (x+h)^{1/q} x^{(q-2)/q} + x^{(q-1)/q}}.$$

So,

$$D_x x^{1/q} = \frac{1}{qx^{(q-1)/q}} = \frac{1}{q} x^{1/q-1},$$

provided $x^{1/q-1}$ exists.

Any rational number r can be expressed in the form $r = p/q$, where p and q are integers and q is positive. Thus the Chain Rule implies that

$$D_x x^r = D_x (x^{1/q})^p$$
$$= p(x^{1/q})^{p-1} D_x x^{1/q}$$
$$= p[x^{(p-1)/q}][(1/q)x^{(1/q)-1}]$$
$$= \frac{p}{q} x^{(p/q)-1}.$$

So, when x^{r-1} is defined

$$D_x x^r = rx^{r-1}. \qquad \triangleright$$

We now consider some definite integral theorems discussed in Section 4.2. Some additional theorems have been added here to make the development of the Riemann integral more complete. These are stated without theorem number, since they do not refer directly to a result in the text.

THEOREM

If f is integrable on $[a, b]$ and k is a constant, then kf is also integrable on $[a, b]$ and

$$\int_a^b k f(x) \, dx = k \int_a^b f(x) \, dx.$$

Proof. If $k = 0$ the proof follows from Theorem 4.8. Suppose $k \neq 0$, and $\varepsilon > 0$ is given. Since f is integrable on $[a, b]$, a number $\delta > 0$ exists with the property that for any partition $\mathcal{P} = \{x_0, x_1, \ldots, x_n\}$, with $\|\mathcal{P}\| < \delta$,

$$\left| \int_a^b f(x) \, dx - \sum_{i=1}^n f(z_i) \Delta x_i \right| < \frac{\varepsilon}{|k|}$$

whenever z_i is in $[x_{i-1}, x_i]$. Consequently,

$$\left| k \int_a^b f(x) \, dx - \sum_{i=1}^n k f(z_i) \Delta x_i \right| = \left| k \int_a^b f(x) \, dx - k \sum_{i=1}^n f(z_i) \Delta x_i \right| < |k| \frac{\varepsilon}{|k|} = \varepsilon.$$

This implies that the function kf is integrable and that

$$\int_a^b k f(x) \, dx = k \int_a^b f(x) \, dx. \qquad \triangleright$$

THEOREM

If f and g are integrable on $[a, b]$, then:

i) $f + g$ is integrable on $[a, b]$ and

$$\int_a^b (f + g)(x) \, dx = \int_a^b f(x) \, dx + \int_a^b g(x) \, dx;$$

ii) $f - g$ is integrable on $[a, b]$ and
$$\int_a^b (f - g)(x)\, dx = \int_a^b f(x)\, dx - \int_a^b g(x)\, dx.$$

Proof.

i) Let $\varepsilon > 0$ be given. Since f and g are both integrable on $[a, b]$, there exist numbers δ_f and δ_g with the property that if $\mathcal{P}$ is any partition of $[a, b]$ with $\|\mathcal{P}\| < \delta_f$ and $\|\mathcal{P}\| < \delta_g$, then

(**B.1**) $$\left| \sum_{i=1}^n f(z_i) \Delta x_i - \int_a^b f(x)\, dx \right| < \frac{\varepsilon}{2}$$

and

(**B.2**) $$\left| \sum_{i=1}^n g(z_i) \Delta x_i - \int_a^b g(x)\, dx \right| < \frac{\varepsilon}{2}$$

whenever z_i is in $[x_{i-1}, x_i]$. Let $\delta = \text{minimum}\,(\delta_f, \delta_g)$. If $\|\mathcal{P}\| < \delta$, then the inequalities in equations (B.1) and (B.2) both hold. Thus,

$$\left| \sum_{i=1}^n [f(z_i) + g(z_i)] \Delta x_i - \left[\int_a^b f(x)\, dx + \int_a^b g(x)\, dx \right] \right|$$
$$= \left| \left[\sum_{i=1}^n f(z_i) \Delta x_i - \int_a^b f(x)\, dx \right] + \left[\sum_{i=1}^n g(z_i) \Delta x_i - \int_a^b g(x)\, dx \right] \right|$$
$$\leq \left| \sum_{i=1}^n f(z_i) \Delta x_i - \int_a^b f(x)\, dx \right| + \left| \sum_{i=1}^n g(z_i) \Delta x_i - \int_a^b g(x)\, dx \right|$$
$$< \frac{\varepsilon}{2} + \frac{\varepsilon}{2} = \varepsilon$$

whenever z_i is in $[x_{i-1}, x_i]$. This implies that

$$\lim_{\|\mathcal{P}\| \to 0} \sum_{i=1}^n (f + g)(z_i) \Delta x_i = \int_a^b (f + g)(x)\, dx$$

exists and that

$$\int_a^b (f + g)(x)\, dx = \int_a^b f(x)\, dx + \int_a^b g(x)\, dx.$$

ii) The preceding theorem implies that

$$\int_a^b -g(x)\, dx = -\int_a^b g(x)\, dx.$$

Combining this result with (i) gives

$$\int_a^b (f - g)(x)\, dx = \int_a^b (f + (-g))(x)\, dx$$
$$= \int_a^b f(x)\, dx + \int_a^b -g(x)\, dx$$
$$= \int_a^b f(x)\, dx - \int_a^b g(x)\, dx.$$

▷

We now prove the following corollary to Theorem 4.12 and then apply this result to prove the theorem.

(4.13)
COROLLARY
If f is integrable on $[a, b]$ and $f(x) \geq 0$ for each x in $[a, b]$, then

$$\int_a^b f(x)\, dx \geq 0.$$

Proof. Suppose that $\int_a^b f(x)\, dx < 0$. Then $\varepsilon = -\int_a^b f(x)\, dx$ is a positive real number. Since f is integrable, there exists a number δ with the property that whenever $\mathscr{P}$ is a partition of $[a, b]$ and $\|\mathscr{P}\| < \delta$,

$$\left| \sum_{i=1}^n f(z_i)\, \Delta x_i - \int_a^b f(x)\, dx \right| < \varepsilon = -\int_a^b f(x)\, dx.$$

But

$$\sum_{i=1}^n f(z_i)\, \Delta x_i - \int_a^b f(x)\, dx \leq \left| \sum_{i=1}^n f(z_i)\, \Delta x_i - \int_a^b f(x)\, dx \right|.$$

So

$$\sum_{i=1}^n f(z_i)\, \Delta x_i - \int_a^b f(x)\, dx < -\int_a^b f(x)\, dx$$

and

$$\sum_{i=1}^n f(z_i)\, \Delta x_i < 0,$$

whenever $\|\mathscr{P}\| < \delta$ and z_i is in $[x_{i-1}, x_i]$. This is clearly false, since $f(z_i) \geq 0$ and $\Delta x_i > 0$ for each i. Consequently, $\int_a^b f(x)\, dx \geq 0$. ▷

(4.12)
THEOREM
If f and g are both integrable on $[a, b]$ and $f(x) \geq g(x)$ for each x in $[a, b]$, then

$$\int_a^b f(x)\, dx \geq \int_a^b g(x)\, dx.$$

Proof. Let $h(x) = f(x) - g(x)$. Then Corollary 4.13 implies that $\int_a^b h(x)\, dx \geq 0$. Applying the second integral theorem proved in this section gives

$$\int_a^b f(x)\, dx - \int_a^b g(x)\, dx = \int_a^b [f(x) - g(x)]\, dx = \int_a^b h(x)\, dx \geq 0$$

and

$$\int_a^b f(x)\, dx \geq \int_a^b g(x)\, dx.$$

▷

(4.14)
COROLLARY
If f is integrable on $[a, b]$ and constants m and M exist with $m \leq f(x) \leq M$ for all x in $[a, b]$, then

$$m(b - a) \leq \int_a^b f(x) \, dx \leq M(b - a).$$

Proof. Theorem 4.8 implies that

$$\int_a^b m \, dx = m(b - a) \quad \text{and} \quad \int_a^b M \, dx = M(b - a),$$

so the result follows immediately from Theorem 4.12. ▷

(4.17)
THEOREM
If f is integrable on an interval containing a, b, and c, then

$$\int_a^b f(x) \, dx = \int_a^c f(x) \, dx + \int_c^b f(x) \, dx$$

regardless of the order of a, b, and c in the interval.

Proof. We will prove the result for the case when the function is integrable on the interval $[a, b]$ and c is a point in this interval. The proof for the general case follows from this result and the definition for the definite integral when the upper limit does not exceed the lower limit.

We need to show that for a given $\varepsilon > 0$, a number $\delta > 0$ exists with the property that for any partition $\mathcal{P} = \{x_0, x_1, \ldots, x_n\}$ of $[a, b]$, with $\|\mathcal{P}\| < \delta$,

$$\left| \int_a^c f(x) \, dx + \int_c^b f(x) \, dx - \sum_{i=1}^n f(z_i) \, \Delta x_i \right| < \varepsilon$$

whenever z_i is in $[x_{i-1}, x_i]$.

First choose δ_1 and δ_2 so that if $\mathcal{P}_1$ and $\mathcal{P}_2$ are partitions of $[a, c]$ and $[c, b]$, respectively, with $\|\mathcal{P}_1\| < \delta_1$ and $\|\mathcal{P}_2\| < \delta_2$ and if R_1 and R_2 are any Riemann sums associated with $\mathcal{P}_1$ and $\mathcal{P}_2$, respectively, then

$$\left| \int_a^c f(x) \, dx - R_1 \right| < \frac{\varepsilon}{3} \quad \text{and} \quad \left| \int_c^b f(x) \, dx - R_2 \right| < \frac{\varepsilon}{3}.$$

Suppose, in addition, that M_1 and M_2 are bounds for f on $[a, c]$ and $[c, b]$, respectively. These bounds are ensured by Theorem 4.10. Let

$$M = \max\{M_1, M_2\}, \quad \delta_3 = \frac{\varepsilon}{6M}, \quad \text{and} \quad \delta = \min\{\delta_1, \delta_2, \delta_3\}.$$

Suppose that $\mathcal{P} = \{x_0, x_1, \ldots, x_n\}$ is a partition of $[a, b]$ with $\|\mathcal{P}\| < \delta$. Since $a < c < b$, a unique index k exists, $1 \leq k \leq n$, with $x_{k-1} < c \leq x_k$. For any collection of z_i in $[x_{i-1}, x_i]$,

$$\sum_{i=1}^n f(z_i) \, \Delta x_i = \sum_{i=1}^{k-1} f(z_i) \, \Delta x_i + f(z_k) \, \Delta x_k + \sum_{i=k+1}^n f(z_i) \, \Delta x_i.$$

But

$$f(z_k)\,\Delta x_k = f(c)\,\Delta x_k + [f(z_k) - f(c)]\,\Delta x_k$$
$$= f(c)(x_k - c + c - x_{k-1}) + [f(z_k) - f(c)]\,\Delta x_k$$
$$= f(c)(x_k - c) + f(c)(c - x_{k-1}) + [f(z_k) - f(c)]\,\Delta x_k,$$

so

$$\sum_{i=1}^{n} f(z_i)\,\Delta x_i = \left[\sum_{i=1}^{k-1} f(z_i)\,\Delta x_i + f(c)(c - x_{k-1})\right]$$
$$+ \left[f(c)(x_k - c) + \sum_{i=k+1}^{n} f(z_i)\,\Delta x_i\right] + [f(z_k) - f(c)]\,\Delta x_k.$$

However,

$$\sum_{i=1}^{k-1} f(z_i)\,\Delta x_i + f(c)(c - x_{k-1})$$

and

$$f(c)(x_k - c) + \sum_{i=k+1}^{n} f(z_i)\,\Delta x_i$$

are Riemann sums for $[a, c]$ and $[c, b]$, respectively, on partitions with norms bounded by $\delta \leq \min(\delta_1, \delta_2)$. Consequently,

$$\left| \int_a^c f(x)\,dx - \sum_{i=1}^{k-1} f(z_i)\,\Delta x_i - f(c)(c - x_{k-1}) \right| < \frac{\varepsilon}{3}$$

and

$$\left| \int_c^b f(x)\,dx - f(c)(x_k - c) - \sum_{i=k+1}^{n} f(z_i)\,\Delta x_i \right| < \frac{\varepsilon}{3}.$$

Since $|f(x)| \leq M$ for all x in $[a, b]$ and $\delta \leq \delta_3$ implies $\Delta x_k \leq \varepsilon/6M$, we have

$$\left| \int_a^c f(x)\,dx + \int_c^b f(x)\,dx - \sum_{i=1}^{n} f(z_i)\,\Delta x_i \right| < \frac{\varepsilon}{3} + \frac{\varepsilon}{3} + |f(z_k) - f(c)|\,\Delta x_k$$
$$\leq \frac{2\varepsilon}{3} + [|f(z_k)| + |f(c)|]\,\Delta x_k$$
$$\leq \frac{2\varepsilon}{3} + (M + M)\,\frac{\varepsilon}{6M} = \frac{2\varepsilon}{3} + \frac{\varepsilon}{3} = \varepsilon.$$

▷

This completes the proofs of the definite-integral results. The following theorem is a preliminary result to the Inverse Function Theorem.

(6.7)
THEOREM

If f is continuous and increasing (decreasing) on its domain, then f^{-1} is also continuous and increasing (decreasing).

Proof. The proof is given for the case when f is continuous and increasing on a closed interval $[a, b]$. The proof for other intervals and when f is continuous and decreasing is similar.

Theorem 6.6 implies that f^{-1} is increasing. It remains to show that f^{-1} is continuous. Suppose that y_0 is arbitrarily chosen in $(F(a), F(b))$ and that $y_0 = f(x_0)$. Let $\varepsilon > 0$ be given. We need to show that a number δ exists with $|f^{-1}(y) - f^{-1}(y_0)| < \varepsilon$, whenever $|y - y_0| < \delta$.

Choose $\hat{\varepsilon} < \varepsilon$ small enough to ensure that $x_1 = x_0 - \hat{\varepsilon}$ and $x_2 = x_0 + \hat{\varepsilon}$ are both contained in $[a, b]$. Since f is increasing and

$$x_1 < x_0 < x_2,$$

we have

$$f(x_1) < f(x_0) = y_0 < f(x_2).$$

Choose $\delta = \text{minimum } \{y_0 - f(x_1), f(x_2) - y_0\}$, and assume that $|y - y_0| < \delta$. (See Figure B.3.) Since $f(x_1) < y < f(x_2)$, the Intermediate Value Theorem implies that a number x, $x_1 < x < x_2$ exists with $f(x) = y$. Consequently, $x = f^{-1}(y)$ and

$$|f^{-1}(y) - f^{-1}(y_0)| = |x - x_0|.$$

Since $x_0 - \hat{\varepsilon} = x_1 < x < x_2 = x_0 + \hat{\varepsilon}$.

$$-\hat{\varepsilon} < x - x_0 < \hat{\varepsilon}.$$

So, $|x - x_0| < \hat{\varepsilon} < \varepsilon$ and

$$|f^{-1}(y) - f^{-1}(y_0)| < \varepsilon$$

whenever $|y - y_0| < \delta$. This implies that f^{-1} is continuous at each y_0 in $(F(a), F(b))$.

The proof that f^{-1} is continuous from the right at $f(a)$ and continuous from the left at $f(b)$ is similar. ▷

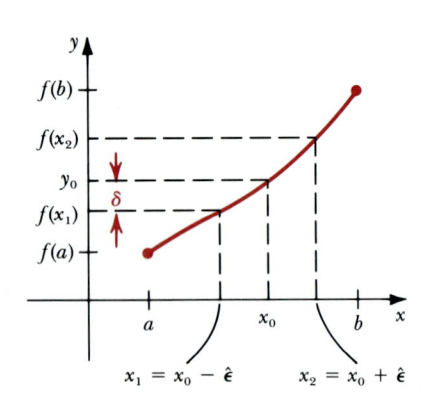

Figure B.3

The next result concerns functions defined by power series.

(8.44)
THEOREM
If f is a function defined by the power series

$$f(x) = \sum_{n=0}^{\infty} a_n x^n$$

with radius of convergence R, then:

i) f is differentiable at each x in $(-R, R)$ and

$$D_x f(x) = D_x \sum_{n=0}^{\infty} a_n x^n = \sum_{n=1}^{\infty} n a_n x^{n-1};$$

ii) for each x in $(-R, R)$, $\int_0^x F(t)\, dt$ exists and

$$\int_0^x f(t)\, dt = \int_0^x \left[\sum_{n=0}^{\infty} a_n t^n \right] dt = \sum_{n=0}^{\infty} \frac{a_n}{n+1} x^{n+1}.$$

APPENDIX B ADDITIONAL CALCULUS THEOREMS AND PROOFS   A35

Before proving this theorem we establish the following lemma.

LEMMA

If $\sum_{n=0}^{\infty} a_n x^n$ has radius of convergence R, then the series

$$\sum_{n=1}^{\infty} n a_n x^{n-1} \quad \text{and} \quad \sum_{n=2}^{\infty} n(n-1) a_n x^{n-2}$$

have radius of convergence R.

Proof of Lemma. Let x be fixed in $(-R, R)$ and z be such that $|x| < |z| < R$. By Lemma 8.41 $\sum_{n=0}^{\infty} a_n z^n$ is absolutely convergent, so $\lim_{n \to \infty} |a_n z^n| = 0$.

Let N be such that if $n > N$, then $|a_n z^n| < 1$. Now let

$$M = \max\{|a_1 z|, |a_2 z^2|, \ldots, |a_N z^N|, 1\}.$$

This choice of M ensures that $|a_n z^n| \leq M$ for each n, and consequently that

$$|n a_n x^{n-1}| = \left| n a_n x^{n-1} \frac{z^n}{z^n} \right| = n \frac{|a_n z^n|}{|z|} \left| \frac{x}{z} \right|^{n-1} \leq \frac{nM}{|z|} \left| \frac{x}{z} \right|^{n-1}.$$

Applying the Ratio Test to the series

$$\sum_{n=1}^{\infty} \frac{nM}{|z|} \left| \frac{x}{z} \right|^{n-1}$$

implies that this series converges, since

$$\lim_{n \to \infty} \frac{\frac{(n+1)M}{|z|} \left| \frac{x}{z} \right|^n}{\frac{nM}{|z|} \left| \frac{x}{z} \right|^{n-1}} = \lim_{n \to \infty} \frac{n+1}{n} \left| \frac{x}{z} \right| = \left| \frac{x}{z} \right| < 1.$$

By the Comparison Test, $\sum_{n=1}^{\infty} |n a_n x^{n-1}|$ also converges. Since x was arbitrarily chosen in $(-R, R)$, the radius of convergence $\hat{R}$ of $\sum_{n=1}^{\infty} n a_n x^{n-1}$ satisfies $\hat{R} \geq R$.

Now suppose that a number z exists with $|z| > R$ and that $\sum_{n=1}^{\infty} |n a_n z^{n-1}|$ converges. When $n > z$, $|n a_n z^{n-1}| > |a_n z^n|$ and by the Comparison Test, the convergence of $\sum_{n=1}^{\infty} |n a_n z^{n-1}|$ implies the convergence of $\sum_{n=1}^{\infty} |a_n z^n|$. This contradicts $|z| > R$. Thus $\hat{R} = R$.

Applying this result to the series $\sum_{n=1}^{\infty} n a_n x^{n-1}$ shows that the radius of convergence of $\sum_{n=2}^{\infty} n(n-1) a_n x^{n-2}$ is also R. ▷

PROOF OF THEOREM 8.44

i) Let $g(x) = \sum_{n=1}^{\infty} n a_n x^{n-1}$ for x in $(-R, R)$. We want to show that

$$\lim_{h \to 0} \frac{f(x+h) - f(x)}{h} = g(x).$$

Fix x in $(-R, R)$ and restrict h so that $|x + h| < R$. Consider

$$\left|\frac{f(x+h) - f(x)}{h} - g(x)\right| = \left|\frac{\sum_{n=0}^{\infty} a_n(x+h)^n - \sum_{n=0}^{\infty} a_n x^n}{h} - \sum_{n=1}^{\infty} n a_n x^{n-1}\right|.$$

Since all the series converge absolutely and $a_0 (x+h)^0 - a_0 x^0 = a_0 - a_0 = 0$,

$$\left|\frac{f(x+h) - f(x)}{h} - g(x)\right| = \left|\sum_{n=1}^{\infty} a_n \left[\frac{(x+h)^n - x^n}{h} - n x^{n-1}\right]\right|.$$

By the Mean Value Theorem, there exists for each n a number ξ_n, ξ_n between x and $x + h$, with $[(x+h)^n - x^n]/h = n \xi_n^{n-1}$. Thus,

$$\left|\frac{f(x+h) - f(x)}{h} - g(x)\right| = \left|\sum_{n=1}^{\infty} a_n [n \xi_n^{n-1} - n x^{n-1}]\right|$$

$$\leq \sum_{n=1}^{\infty} |na_n||\xi_n^{n-1} - x^{n-1}|.$$

Since $\xi_1^0 - x^0 = 1 - 1 = 0$,

$$\left|\frac{f(x+h) - f(x)}{h} - g(x)\right| \leq \sum_{n=2}^{\infty} |na_n||\xi_n^{n-1} - x^{n-1}|.$$

Applying the Mean Value Theorem to $\xi_n^{n-1} - x^{n-1}$ implies that there exists for each n a number ζ_n, between x and ξ_n, with

$$\frac{\xi_n^{n-1} - x^{n-1}}{\xi_n - x} = (n-1)\zeta_n^{n-2}.$$

Since $|\xi_n - x| < |h|$, we have

$$\left|\frac{f(x+h) - f(x)}{h} - g(x)\right| \leq \sum_{n=2}^{\infty} |na_n||\xi_n^{n-1} - x^{n-1}|$$

$$= \sum_{n=2}^{\infty} |na_n||(n-1) \zeta_n^{n-2}(\xi_n - x)|$$

$$\leq |h| \sum_{n=2}^{\infty} |n(n-1)a_n \zeta_n^{n-2}|.$$

But ζ_n is between x and ξ_n and ξ_n is between x and $x + h$, so ζ_n is between x and $x + h$. By the Lemma, $\sum_{n=2}^{\infty} |n(n-1)a_n \zeta_n^{n-2}|$ converges. Hence,

$$\lim_{h \to 0} |h| \sum_{n=2}^{\infty} |n(n-1)a_n \zeta^{n-2}| = 0,$$

and

$$\lim_{h \to 0} \left|\frac{f(x+h) - f(x)}{h} - g(x)\right| = 0.$$

This can occur only if

$$f'(x) = \lim_{h \to 0} \frac{f(x+h) - f(x)}{h} = g(x).$$

This implies that

$$D_x \sum_{n=0}^{\infty} a_n x^n = \sum_{n=1}^{\infty} n a_n x^{n-1}.$$

ii) To show the integral result, let

$$g(x) = \sum_{n=0}^{\infty} \frac{a_n}{n+1} x^{n+1} = \sum_{n=1}^{\infty} \frac{a_{n-1}}{n} x^n.$$

By part (i),

$$g'(x) = \sum_{n=1}^{\infty} a_{n-1} x^{n-1} = \sum_{n=0}^{\infty} a_n x^n$$

has radius of convergence R, so $g(x)$ is defined when x is in $(-R, R)$ and $g'(x) = f(x)$. Part (i) also implies that $g''(x) = f'(x)$ exists on $(-R, R)$, so f is continuous on $(-R, R)$. Since g is an antiderivative of the continuous function f, the Fundamental Theorem of Calculus implies that for any x in $(-R, R)$,

$$\int_0^x f(t)\, dt = g(x) - g(0) = g(x).$$

Thus,

$$\int_0^x \left[\sum_{n=0}^{\infty} a_n t^n\right] dt = \sum_{n=0}^{\infty} \frac{a_n}{n+1} x^{n+1}. \quad \triangleright$$

The final result in this section concerns the representation of $(1 + x)^k$ in terms of the binomial series.

THEOREM
For all real numbers k and all x in $(-1, 1)$,

$$(1 + x)^k = 1 + \sum_{n=1}^{\infty} \frac{k(k-1) \cdots (k-n+1)}{n!} x^n.$$

Proof. It was shown in Section 8.9 that the binomial series converges for all x in $(-1, 1)$. Let

$$f(x) = 1 + \sum_{n=1}^{\infty} \frac{k(k-1) \cdots (k-n+1)}{n!} x^n$$

be the function defined by this series. Theorem 8.44 implies that for all x in $(-1, 1)$,

(B.3) $$f'(x) = \sum_{n=1}^{\infty} \frac{k(k-1) \cdots (k-n+1)}{n!} n x^{n-1}.$$

Thus,

(B.4) $$x f'(x) = \sum_{n=1}^{\infty} \frac{k(k-1) \cdots (k-n+1)}{n!} n x^n.$$

Reindexing the expression for $f'(x)$ in (B.3) by reducing the index by 1 gives

$$f'(x) = \sum_{n=0}^{\infty} \frac{k(k-1)\cdots(k-(n+1)+1)}{(n+1)!}(n+1)x^n$$

$$= k + \sum_{n=1}^{\infty} \frac{k(k-1)\cdots(k-n)}{n!}x^n.$$

If we add this to the expression $xf'(x)$ in (B.4), we have

$$f'(x) + xf'(x) = k + \sum_{n=1}^{\infty} \frac{k(k-1)\cdots(k-n)}{n!}x^n$$

$$+ \sum_{n=1}^{\infty} \frac{k(k-1)\cdots(k-n+1)}{n!}nx^n$$

$$= k + \sum_{n=1}^{\infty} \frac{k(k-1)\cdots(k-n+1)[(k-n)+n]}{n!}x^n$$

$$= k + k\sum_{n=1}^{\infty} \frac{k(k-1)\cdots(k-n+1)}{n!}x^n = k f(x).$$

Thus, $y = f(x)$ satisfies the separable differential equation

$$(1 + x)y' = ky.$$

It is easy to verify that the general solution to this equation has the form

$$y = f(x) = c(1 + x)^k.$$

Since

$$f(0) = 1 + \sum_{n=1}^{\infty} \frac{k(k-1)\cdots(k-n+1)}{n!}0^n = 1,$$

we must have $C = 1$ and $f(x) = (1 + x)^k$. ▷

Appendix C
Computer Programs for the Numerical Methods

Included in the text are a number of procedures for finding approximate solutions to problems that are difficult or impossible to solve exactly. Computer disks that contain programs for implementing these procedures in BASIC on either the Apple or IBM PC are available from the publisher. Information concerning the disks can be obtained by contacting John Martindale at Random House, Inc., 215 First Street, Cambridge, MA, 02142.

The programs on the disks are written to make the implementation of the procedures convenient. They incorporate consistency checks to ensure that the methods are appropriately applied and have editorial features that make the programs easy to run and modify to fit the problem to be solved.

In this section we show the essential steps in the programs. The program statements listed here do not correspond to those on the disk, since our purpose here is simply to show how the program performs the approximation procedure. These listings can, however, be used to solve the problems given in the text, provided the steps are modified as stated below to provide the correct data.

With each of the listed methods is the name (given in quotes) of the program to run the corresponding method from the computer disk.

Program 1: Bisection Method, "BISECT"

This program uses the Bisection Method to approximate a root of the equation $f(x) = 0$ lying in the interval (A, B). The function f is defined in line 20. Line 30 defines A, B, the tolerance or accuracy needed, and the maximum number of iterations allowed. (It is always best to specify a maximum number of iterations to avoid the possibility of an infinite loop.) These are the only statements that must be changed in order to solve a different problem. The sample problem is Example 4 on page 66 and is given by

$$f(x) = x^3 - 2x - 5 = 0$$

with $A = 2$, $B = 3$, tolerance $TOL = 0.01$, and maximum number of iterations $N = 10$.

```
10 PRINT"BISECTION METHOD : "
20 DEF FNF(X) = X^3 - 2*X - 5
30 A = 2 : B = 3 : TOL = 0.01: N = 10
40 FA = FNF(A)
50 FB = FNF(B)
60 C = FA*FB
70 PRINT"A = ";A,"B = ";B;
80 PRINT"Tolerance = ";TOL, "Maximum Number
      Iterations = ";N
90 I = 0
100 PRINT"Iteration"," P"," F(P)"
110 WHILE I < N
120    D = (B - A)/2
130    P = A + D
140    FP = FNF(P)
150    I = I + 1
160    PRINT I,P,FP
170    IF D < TOL THEN 220
180    IF FA*FP > 0 THEN A = P: FA = FP ELSE B =
          P: FB = FP
190 WEND
200 PRINT"Procedure failed to achieve desired
      accuracy"
210 GOTO 230
220 PRINT"Procedure completed successfully,
      approximate root is ";P
230 END
```

Program 2: Newton's Method, "NEWTON"

This program uses the Newton–Raphson Method to approximate a root of the equation $f(x) = 0$. The function f is defined in line 20 and its derivative f' in line 30. These and the data in line 40 are the only statements that must be changed in order to solve a different problem. Requested data are an initial approximation P, the tolerance or accuracy, and the maximum number of iterations. The sample problem is Example 1 on page 232 and is given by

$$f(x) = x^3 - 2x - 5 = 0 \quad \text{with} \quad f'(x) = 3x^2 - 2$$

and $P = 2$, TOL $= 10^{-5}$, and $N = 10$.

```
10 PRINT"NEWTON'S METHOD"
20 DEF FNF(X) = X^3 - 2*X - 5
30 DEF FNFP(X) = 3*X^2 - 2
40 P = 2: TOL = .00001 : N = 10
50 FP = FNF(P)
60 PRINT"Initial Approximation = ";P;
70 PRINT"Tolerance = ";TOL;"Maximum Number
      Iterations = ";N
80 I = 0
90 PRINT"Iteration"," P"," F(P)"
100 WHILE I < N
```

```
110     FPP = FNFP(P)
120     IF ABS(FPP) < ABS(FP)/10 THEN 200
130     C = FP/FPP
140     P = P - C
150     FP = FNF(P)
160     I = I + 1
170     PRINT I,P,FP
180     IF ABS(C) < TOL THEN 220
190 WEND
200 PRINT"Procedure failed to achieve desired
        accuracy"
210 GOTO 230
220 PRINT"Procedure completed successfully,
        approximate root = ";P
230 END
```

Program 3: Secant Method, "SECANT"

This program approximates a root of the equation $f(x) = 0$ using the Secant Method described in Exercise 26 on page 236. The function f is defined in line 20. This and the data in line 30 are the only statements that must be changed for a different problem. Requested data are two initial approximations, the tolerance, and the maximum number of iterations. The sample problem is given by

$$f(x) = x^3 - 2x - 5 = 0$$

with first approximation $P = 2$, second approximation $Q = 1$, TOL = 10^{-5}, and $N = 15$.

```
10  PRINT"SECANT METHOD"
20  DEF FNF(X) = X^3 - 2*X - 5
30  Q = 1 : P = 2 : TOL = 0.00001 : N = 10
40  FP = FNF(P)
50  FQ = FNF(Q)
60  PRINT"Initial Approximations : ";P;"with
        F(";P;"=";FP;" and ";Q; "with F(";Q;")=";FQ
70  PRINT"Tolerance = ";TOL;"Maximum Number
        Iterations = ";N
80  I = 0
90  PRINT"Iteration"," P"," F(P)"
100 WHILE I < N
110     D0 = FQ - FP
120     D1 = FQ*(Q - P)
130     IF ABS(D0) < ABS(D1)/10 THEN 230
140     C = D1/D0
150     P = Q
160     FP = FQ
170     Q = Q - C
180     FQ = FNF(Q)
190     I = I + 1
200     PRINT I,Q,FQ
210     IF ABS(C) < TOL THEN 250
```

```
220 WEND
230 PRINT"Procedure failed to achieve desired
       accuracy"
240 GOTO 260
250 PRINT"Procedure completed successfully,
       approximate root = ";P
260 END
```

Program 4: Trapezoidal Method, "TRAPEZOI"

This program uses the Trapezoidal Rule to approximate the integral

$$\int_A^B f(x)\, dx.$$

The function f is defined in line 20. This and the data in line 30 are the only statements that must be changed for a different integral. Requested data are the endpoints A and B and the number of subintervals N. The sample problem is Example 1 on page 497 and approximates

$$\int_{-1}^{1} e^x\, dx$$

with $A = -1$, $B = 1$, and $N = 8$.

```
10 PRINT"TRAPEZOIDAL RULE"
20 DEF FNF(X) = EXP(X)
30 A = -1 : B = 1 : N = 8
40 H = (B - A)/N
50 PRINT"Interval of Integration is : [";A;",";B;"]"
60 PRINT"Number of Subintervals is : ";N
70 PRINT"Step size is : ";H
80 PRINT"Points are : ";A;",";
90 S = 0
100 K = N - 1
110 FOR J = 1 to K
120     X = A + J*H
130     PRINT X;",";
140     S = S + FNF(X)
150 NEXT J
160 PRINT B
170 S = S + (FNF(A) + FNF(B))/2
180 S = S*H
190 PRINT"The Approximation to the Integral is ";S
200 END
```

Program 5: Simpson's Method, "SIMPSON"

This program uses Simpson's Rule to approximate the integral

$$\int_A^B f(x)\, dx.$$

The function f is defined in line 20. This and the data in line 30 are the only statements that must be changed for a different integral. Requested data are the endpoints A and B and an even number of subintervals N. The sample problem is Example 2 on page 499 and approximates

$$\int_{-1}^{1} e^x \, dx$$

with $A = -1$, $B = 1$, and $N = 8$.

```
10 PRINT"SIMPSON'S RULE"
20 DEF FNF(X) = EXP(X)
30 A = -1 : B = 1 : N = 8
40 H = (B - A)/N
50 K = N - 1
60 PRINT"Interval of Integration is : [";A;",";B;"]"
70 PRINT"Number of Subintervals is : ";N
80 PRINT"Step size is : ";H
90 PRINT"Points are : ";A;",";
100 S = FNF(A) + FNF(B)
110 S1 = 0 : S2 = 0
120 FOR J = 1 TO K
130     X = A + J*H
140     PRINT X;",";
150     P = FNF(X)
160     IF J = (J\2)*2 THEN S1 = S1 + P ELSE S2 =
           S2 + P
170 NEXT J
180 PRINT B
190 S = (S + 2*S1 + 4*S2)/3*H
200 PRINT"The Approximation to the Integral is " ; S
210 END
```

Program 6: Simpson's Method for Double Integrals, "DSIMP"

This program approximates the double integral

$$\int_{A}^{B} \int_{c(x)}^{d(x)} f(x, y) \, dy \, dx$$

using Simpson's Rule extended to double integrals. The functions f, c, and d are defined in lines 20, 40, and 60, respectively, using the variable names FNF, FNC, and FND. These and the data in line 70 are the only statements that must be changed for a different double integral. Requested data are the endpoints A and B and even integers N and M to define the number of subintervals in the x- and y-directions, respectively. The sample integral is

$$\int_{0.1}^{0.5} \int_{x^3}^{x^2} e^{y/x} \, dy \, dx$$

with $A = 0.1$, $B = 0.5$, $N = 10$, and $M = 10$.

```
10 PRINT"SIMPSON'S RULE FOR DOUBLE INTEGRALS"
20 DEF FNF(X,Y) = EXP(Y/X)
30 REM lower limit of integration for inner integral
40 DEF FNC(X) = X^3
50 REM upper limit of integration for inner integral
60 DEF FND(X) = X^2
70 A = 0.1 : B = 0.5 : N = 10 : M = 10
80 H = (B - A)/N
90 PRINT"Interval of Integration is : [";A;",";B;"]"
100 PRINT"Number of Subintervals in x direction is
        : ";N
110 PRINT"Step size in X direction : ";H
120 PRINT"Number of Subintervals in Y direction is
        : ";M
130 S = 0
140 S1 = 0 : S2 = 0
150 K = M - 1
160 FOR I = 0 TO N
170     X = A + I*H
180     C1 = FNC(X)
190     D1 = FND(X)
200     H1 = (D1 - C1)/M
210     T = FNF(X,C1) + FNF(X,D1)
220     T1 = 0 : T2 = 0
250     FOR J = 1 TO K
260         P = C1 + J*H1
270         Q = FNF(X,P)
280         IF J = (J\2)*2 THEN T1 = T1 + Q ELSE
                T2 = T2 + Q
290     NEXT J
300     U = (T + 2*T1 + 4*T2)*H1/3
310     IF I = 0 OR I = N THEN S = S + U ELSE IF I
            = (I\2)*2 THEN S1 = S1 + U ELSE S2 = S2+U
320 NEXT I
330 S = (S + 2*S1 + 4*S2)/3*H
340 PRINT"The Approximation to the Integral is " ; S
350 END
```

Program 7: Simpson's Method for Triple Integrals, "TSIMP"

This program approximates the triple integral

$$\int_A^B \int_{c(x)}^{d(x)} \int_{\alpha(x,y)}^{\beta(x,y)} f(x, y, z)\, dz\, dy\, dx$$

using Simpson's Rule extended to triple integrals. The functions f, c, d, α, and β are defined in lines 20, 40, 60, 80, and 100, respectively, using the variable names FNF, FNC, FND, FNAL, and FNBE. These and the data in line 110 are the only statements that must be changed for a different triple integral. Requested data are the endpoints A and B and even integers N, M, and L to define the number of subintervals in the x-, y-, and z-directions, respectively. The sample integral is

$$\int_{-1}^{1} \int_{0}^{1} \int_{-xy}^{xy} e^{x^2+y^2} \, dz \, dy \, dx$$

with $A = -1$, $B = 1$, $N = 8$, $M = 8$, and $L = 8$.

```
10   PRINT"SIMPSON'S RULE FOR TRIPLE INTEGRALS"
20   DEF FNF(X,Y,Z) = EXP(X*X+Y*Y)
30   REM lower limit of integration for middle integral
40   DEF FNC(X) = 0
50   REM upper limit of integration for middle integral
60   DEF FND(X) = 1
70   REM lower limit of integration for inner integral
80   DEF FNAL(X,Y) = -X*Y
90   REM upper limit of integration for inner integral
100  DEF FNBE(X,Y) = X*Y
110  A = -1 : B = 1 : N = 8 : M = 8 : L = 8
120  H = (B - A)/N
130  PRINT"Interval of Integration is : [";A;",";B;"]"
140  PRINT"Number of Subintervals in x,y,z
         directions: ";N;",";M;",";L
150  PRINT"There may be a short delay for calculations."
160  S = 0 : S1 = 0 : S2 = 0
170  K1 = L - 1
180  FOR I = 0 TO N
190      X = A + I*H
200      C1 = FNC(X)
210      D1 = FND(X)
220      H1 = (D1 - C1)/M
230      T = 0 : T1 = 0 : T2 = 0
240      FOR J = 0 TO M
250          Y = C1 + J*H1
260          E1 = FNAL(X,Y)
270          F1 = FNBE(X,Y)
280          H2 = (F1 - E1)/L
290          U = FNF(X,Y,E1) + FNF(X,Y,F1)
300          U1 = 0 : U2 = 0
310          FOR K = 1 TO K1
320              Z = E1 + K*H2
330              Q = FNF(X,Y,Z)
340              IF K = (K\2)*2 THEN U1 = U1 + Q
                     ELSE U2 = U2 + Q
350          NEXT K
360          U = (U + 2*U1 + 4*U2)*H2/3
370          IF J = 0 OR J = M THEN T = T+U ELSE
                 IF J = (J\2)*2 THEN T1 = T1+U ELSE
                 T2 = T2+U
380      NEXT J
390      T = (T + 2*T1 + 4*T2)*H1/3
400      IF I = 0 OR I = N THEN S = S+T ELSE IF I =
             (I\2)*2 THEN S1 = S1+T ELSE S2 = S2+T
410  NEXT I
420  S = (S + 2*S1 + 4*S2)/3*H
430  PRINT"The Approximation to the Integral = ";S
440  END
```

Program 8: Euler's Method, "EULER"

This program uses Euler's Method to approximate the solution to the initial-value problem

$$y' = f(x, y), \quad A \leq x \leq B, \quad y(A) = \alpha$$

at N equally spaced numbers in $[A, B]$. The function f is defined in line 20. This and the data in line 30 are the only statements that must be changed for a different differential equation. Requested data are A, B, α (denoted Y), and N. The sample problem is Example 1 on page 1024 and is given by

$$y' = -y + x + 1, \ 0 \leq x \leq 1, \ y(0) = 1.$$

The data are $A = 0, B = 1, Y = 1$, and $N = 10$.

```
10 PRINT"EULER'S METHOD"
20 DEF FNF(X,Y) = -Y + X + 1
30 A = 0 : B = 1 : Y = 1 : N = 10
40 H = (B - A)/N
50 PRINT"Step size is : ";H
60 PRINT " I"," X(I)"," Y(X(I))"
70 PRINT I,A,Y
80 X = A
90 FOR I = 1 TO N
100    Y = Y + H*FNF(X,Y)
110    X = A + I*H
120    PRINT I,X,Y
130 NEXT I
140 END
```

Program 9: Modified Euler's Method, "MODEULER"

This program uses the Modified Euler Method to approximate the solution to the initial-value problem

$$y' = f(x, y), \quad A \leq x \leq B, \quad y(A) = \alpha$$

at N equally spaced numbers in $[A, B]$. The function f is defined in line 20. This and the data in line 30 are the only statements that must be changed for a different differential equation. Requested data are A, B, α (denoted Y), and N. The sample problem is Example 2 on page 1025 and is given by

$$y' = -y + x + 1, \ 0 \leq x \leq 1, \ y(0) = 1.$$

The data are $A = 0, B = 1, Y = 1$, and $N = 10$.

```
10 PRINT"MODIFIED EULER'S METHOD"
20 DEF FNF(X,Y) = -Y + X + 1
30 A = 0 : B = 1 : Y = 1 : N = 10
40 H = (B - A)/N
50 PRINT"Step size is : ";H
60 PRINT " I"," X(I)"," Y(X(I))"
70 PRINT I,A,Y
```

```
80 X = A
90 FOR I = 1 TO N
100     X1 = H*FNF(X,Y)
110     X2 = H*FNF(X + H,Y + X1)
120     Y = Y + (X1 + X2)/2
130     X = A + I*H
140     PRINT I,X,Y
150 NEXT I
160 END
```

Program 10: Runge–Kutta Method of Order Four, "RK4"

This program uses the Runge–Kutta Method of Order Four to approximate the solution to the initial-value problem

$$y' = f(x, y), \quad A \leq x \leq B, \quad y(A) = \alpha$$

at N equally spaced numbers in $[A, B]$. The function f is defined in line 20. This and the data in line 30 are the only statements that must be changed for a different differential equation. Requested data are A, B, α (denoted Y), and N. The sample problem is Example 3 on page 1026 and is given by

$$y' = -y + x + 1, \ 0 \leq x \leq 1, \ y(0) = 1.$$

The data are $A = 0, B = 1, Y = 1$, and $N = 10$.

```
10 PRINT"RUNGE-KUTTA METHOD OF ORDER 4"
20 DEF FNF(X,Y) = -Y + X + 1
30 A = 0 : B = 1 : Y = 1 : N = 10
40 H = (B - A)/N
50 PRINT"Step size is : ";H
60 PRINT " I"," X(I)"," Y(X(I))"
70 PRINT I,A,Y
80 X = A
90 FOR I = 1 TO N
100     X1 = H*FNF(X,Y)
110     X2 = H*FNF(X + H/2,Y + X1/2)
120     X3 = H*FNF(X + H/2,Y + X2/2)
130     X4 = H*FNF(X + H,Y + X3)
140     Y = Y + (X1 + 2*(X2 + X3) + X4)/6
150     X = A + I*H
160     PRINT I,X,Y
170 NEXT I
180 END
```

Appendix D

Tables

▶ **TABLE 1** Trigonometric Functions (in degrees)

DEGREES	SINE	TANGENT	COTANGENT	COSINE	
0	0	0	—	1.0000	90
1	0.0175	0.0175	57.290	0.9998	89
2	0.0349	0.0349	28.636	0.9994	88
3	0.0523	0.0524	19.081	0.9986	87
4	0.0698	0.0699	14.301	0.9976	86
5	0.0872	0.0875	11.430	0.9962	85
6	0.1045	0.1051	9.5144	0.9945	84
7	0.1219	0.1228	8.1443	0.9925	83
8	0.1392	0.1405	7.1154	0.9903	82
9	0.1564	0.1584	6.3138	0.9877	81
10	0.1736	0.1763	5.6713	0.9848	80
11	0.1908	0.1944	5.1446	0.9816	79
12	0.2079	0.2126	4.7046	0.9781	78
13	0.2250	0.2309	4.3315	0.9744	77
14	0.2419	0.2493	4.0108	0.9703	76
15	0.2588	0.2679	3.7321	0.9659	75
16	0.2756	0.2867	3.4874	0.9613	74
17	0.2924	0.3057	3.2709	0.9563	73
18	0.3090	0.3249	3.0777	0.9511	72
19	0.3256	0.3443	2.9042	0.9455	71
20	0.3420	0.3640	2.7475	0.9397	70
21	0.3584	0.3839	2.6051	0.9336	69
22	0.3746	0.4040	2.4751	0.9272	68
23	0.3907	0.4245	2.3559	0.9205	67
24	0.4067	0.4452	2.2460	0.9135	66
25	0.4226	0.4663	2.1445	0.9063	65
26	0.4384	0.4877	2.0503	0.8988	64
27	0.4540	0.5095	1.9626	0.8910	63
28	0.4695	0.5317	1.8807	0.8829	62
29	0.4848	0.5543	1.8040	0.8746	61
30	0.5000	0.5774	1.7321	0.8660	60
31	0.5150	0.6009	1.6643	0.8572	59
32	0.5299	0.6249	1.6003	0.8480	58
33	0.5446	0.6494	1.5399	0.8387	57
34	0.5592	0.6745	1.4826	0.8290	56
35	0.5736	0.7002	1.4281	0.8192	55
36	0.5878	0.7265	1.3764	0.8090	54
37	0.6018	0.7536	1.3270	0.7986	53
38	0.6157	0.7813	1.2799	0.7880	52
39	0.6293	0.8098	1.2349	0.7771	51
40	0.6428	0.8391	1.1918	0.7660	50
41	0.6561	0.8693	1.1504	0.7547	49
42	0.6691	0.9004	1.1106	0.7431	48
43	0.6820	0.9325	1.0724	0.7314	47
44	0.6947	0.9657	1.0355	0.7193	46
45	0.7071	1.0000	1.0000	0.7071	45
	COSINE	COTANGENT	TANGENT	SINE	DEGREES

TABLE 2

Trigonometric Functions (in radians)

RADIANS	SINE	COSINE	TANGENT
0.00	0.0000	1.0000	0.0000
0.05	0.0500	0.9988	0.0500
0.10	0.0998	0.9950	0.1003
0.15	0.1494	0.9888	0.1511
0.20	0.1987	0.9801	0.2027
0.25	0.2474	0.9689	0.2553
0.30	0.2955	0.9553	0.3093
0.35	0.3429	0.9394	0.3650
0.40	0.3894	0.9211	0.4228
0.45	0.4350	0.9004	0.4831
0.50	0.4794	0.8776	0.5463
0.55	0.5227	0.8525	0.6131
0.60	0.5646	0.8253	0.6841
0.65	0.6052	0.7961	0.7602
0.70	0.6442	0.7648	0.8423
0.75	0.6816	0.7317	0.9316
0.80	0.7174	0.6967	1.0296
0.85	0.7513	0.6600	1.1383
0.90	0.7833	0.6216	1.2602
0.95	0.8134	0.5817	1.3984
1.00	0.8415	0.5403	1.5574
1.05	0.8674	0.4976	1.7433
1.10	0.8912	0.4536	1.9648
1.15	0.9128	0.4085	2.2345
1.20	0.9320	0.3624	2.5722
1.25	0.9490	0.3153	3.0096
1.30	0.9636	0.2675	3.6021
1.35	0.9757	0.2190	4.4552
1.40	0.9854	0.1700	5.7979
1.45	0.9927	0.1205	8.2381
1.50	0.9975	0.0707	14.1014
1.55	0.9998	0.0208	48.0785
$\pi/2$	1.0000	0.0000	—

TABLE 3 Exponential Functions

x	e^x	e^{-x}	x	e^x	e^{-x}
0.00	1.0000	1.0000	1.5	4.4817	0.2231
0.01	1.0101	0.9901	1.6	4.9530	0.2019
0.02	1.0202	0.9802	1.7	5.4739	0.1827
0.03	1.0305	0.9704	1.8	6.0496	0.1653
0.04	1.0408	0.9608	1.9	6.6859	0.1496
0.05	1.0513	0.9512	2.0	7.3891	0.1353
0.06	1.0618	0.9418	2.1	8.1662	0.1225
0.07	1.0725	0.9324	2.2	9.0250	0.1108
0.08	1.0833	0.9331	2.3	9.9742	0.1003
0.09	1.0942	0.9139	2.4	11.023	0.0907
0.10	1.1052	0.9048	2.5	12.182	0.0821
0.11	1.1163	0.8958	2.6	13.464	0.0743
0.12	1.1275	0.8869	2.7	14.880	0.0672
0.13	1.1388	0.8781	2.8	16.445	0.0608
0.14	1.1503	0.8694	2.9	18.174	0.0550
0.15	1.1618	0.8607	3.0	20.086	0.0498
0.16	1.1735	0.8521	3.1	22.198	0.0450
0.17	1.1853	0.8437	3.2	24.533	0.0408
0.18	1.1972	0.8353	3.3	27.113	0.0369
0.19	1.2092	0.8270	3.4	29.964	0.0334
0.20	1.2214	0.8187	3.5	33.115	0.0302
0.21	1.2337	0.8106	3.6	36.598	0.0273
0.22	1.2461	0.8025	3.7	40.447	0.0247
0.23	1.2586	0.7945	3.8	44.701	0.0224
0.24	1.2712	0.7866	3.9	49.402	0.0202
0.25	1.2840	0.7788	4.0	54.598	0.0183
0.30	1.3499	0.7408	4.1	60.340	0.0166
0.35	1.4191	0.7047	4.2	66.686	0.0150
0.40	1.4918	0.6703	4.3	73.700	0.0136
0.45	1.5683	0.6376	4.4	81.451	0.0123
0.50	1.6487	0.6065	4.5	90.017	0.0111
0.55	1.7333	0.5769	4.6	99.484	0.0101
0.60	1.8221	0.5488	4.7	109.95	0.0091
0.65	1.9155	0.5220	4.8	121.51	0.0082
0.70	2.0138	0.4966	4.9	134.29	0.0074
0.75	2.1170	0.4724	5.0	148.41	0.0067
0.80	2.2255	0.4493	5.5	244.69	0.0041
0.85	2.3396	0.4274	6.0	403.43	0.0025
0.90	2.4596	0.4066	6.5	665.14	0.0015
0.95	2.5857	0.3867	7.0	1096.6	0.0009
1.00	2.7183	0.3679	7.5	1808.0	0.0006
1.10	3.0042	0.3329	8.0	2981.0	0.0003
1.20	3.3201	0.3012	8.5	4914.8	0.0002
1.30	3.6693	0.2725	9.0	8103.1	0.0001
1.40	4.0552	0.2466	10.0	22026	0.00005

TABLE 4 Natural Logarithms

x	$\ln x$	x	$\ln x$	x	$\ln x$
		4.5	1.5041	9.0	2.1972
0.1	−2.3026	4.6	1.5261	9.1	2.2083
0.2	−1.6094	4.7	1.5476	9.2	2.2192
0.3	−1.2040	4.8	1.5686	9.3	2.2300
0.4	−0.9163	4.9	1.5892	9.4	2.2407
0.5	−0.6931	5.0	1.6094	9.5	2.2513
0.6	−0.5108	5.1	1.6292	9.6	2.2618
0.7	−0.3567	5.2	1.6487	9.7	2.2721
0.8	−0.2231	5.3	1.6677	9.8	2.2824
0.9	−0.1054	5.4	1.6864	9.9	2.2925
1.0	0.0000	5.5	1.7047	10	2.3026
1.1	0.0953	5.6	1.7228	11	2.3979
1.2	0.1823	5.7	1.7405	12	2.4849
1.3	0.2624	5.8	1.7579	13	2.5649
1.4	0.3365	5.9	1.7750	14	2.6391
1.5	0.4055	6.0	1.7918	15	2.7081
1.6	0.4700	6.1	1.8083	16	2.7726
1.7	0.5306	6.2	1.8245	17	2.8332
1.8	0.5878	6.3	1.8405	18	2.8904
1.9	0.6419	6.4	1.8563	19	2.9444
2.0	0.6931	6.5	1.8718	20	2.9957
2.1	0.7419	6.6	1.8871	25	3.2189
2.2	0.7885	6.7	1.9021	30	3.4012
2.3	0.8329	6.8	1.9169	35	3.5553
2.4	0.8755	6.9	1.9315	40	3.6889
2.5	0.9163	7.0	1.9459	45	3.8067
2.6	0.9555	7.1	1.9601	50	3.9120
2.7	0.9933	7.2	1.9741	55	4.0073
2.8	1.0296	7.3	1.9879	60	4.0943
2.9	1.0647	7.4	2.0015	65	4.1744
3.0	1.0986	7.5	2.0149	70	4.2485
3.1	1.1314	7.6	2.0281	75	4.3175
3.2	1.1632	7.7	2.0412	80	4.3820
3.3	1.1939	7.8	2.0541	85	4.4427
3.4	1.2238	7.9	2.0669	90	4.4998
3.5	1.2528	8.0	2.0794	100	4.6052
3.6	1.2809	8.1	2.0919	110	4.7005
3.7	1.3083	8.2	2.1041	120	4.7875
3.8	1.3350	8.3	2.1163	130	4.8675
3.9	1.3610	8.4	2.1282	140	4.9416
4.0	1.3863	8.5	2.1401	150	5.0106
4.1	1.4110	8.6	2.1518	160	5.0752
4.2	1.4351	8.7	2.1633	170	5.1358
4.3	1.4586	8.8	2.1748	180	5.1930
4.4	1.4816	8.9	2.1861	190	5.2470

► BASIC INTEGRAL FORMS

1. $\displaystyle\int dx = x + C$

2. $\displaystyle\int x^n \, dx = \frac{1}{n+1} x^{n+1} + C, \; n \neq -1$

3. $\displaystyle\int \frac{1}{x} \, dx = \ln |x| + C$

4. $\displaystyle\int e^x \, dx = e^x + C$

5. $\displaystyle\int a^x \, dx = \frac{a^x}{\ln a} + C$

6. $\displaystyle\int \ln x \, dx = x \ln x - x + C$

7. $\displaystyle\int \sin x \, dx = -\cos x + C$

8. $\displaystyle\int \cos x \, dx = \sin x + C$

9. $\displaystyle\int \tan x \, dx = \ln |\sec x| + C$

10. $\displaystyle\int \cot x \, dx = \ln |\sin x| + C$

11. $\displaystyle\int \sec x \, dx = \ln |\sec x + \tan x| + C$

12. $\displaystyle\int \csc x \, dx = \ln |\csc x - \cot x| + C$

13. $\displaystyle\int \sec^2 x \, dx = \tan x + C$

14. $\displaystyle\int \csc^2 x \, dx = -\cot x + C$

15. $\displaystyle\int \sec x \tan x \, dx = \sec x + C$

16. $\displaystyle\int \csc x \cot x \, dx = -\csc x + C$

17. $\displaystyle\int \frac{1}{x^2 - a^2} \, dx = \frac{1}{2a} \ln \left|\frac{x-a}{x+a}\right| + C$

18. $\displaystyle\int \frac{1}{x^2 + a^2} \, dx = \frac{1}{a} \arctan \frac{x}{a} + C$

19. $\displaystyle\int \frac{1}{\sqrt{a^2 - x^2}} \, dx = \arcsin \frac{x}{a} + C$

20. $\displaystyle\int \frac{1}{x \sqrt{x^2 - a^2}} \, dx = \frac{1}{a} \operatorname{arcsec} \frac{|x|}{a} + C$

► TRIGONOMETRIC FORMS

21. $\displaystyle\int \sin^2 x \, dx = \tfrac{1}{2} x - \tfrac{1}{4} \sin 2x + C$

22. $\displaystyle\int \cos^2 x \, dx = \tfrac{1}{2} x + \tfrac{1}{4} \sin 2x + C$

23. $\displaystyle\int \tan^2 x \, dx = \tan x - x + C$

24. $\displaystyle\int \cot^2 x \, dx = -\cot x - x + C$

25. $\displaystyle\int \sin^3 x \, dx = \tfrac{1}{3} \cos^3 x - \cos x + C$

26. $\displaystyle\int \cos^3 x \, dx = \sin x - \tfrac{1}{3} \sin^3 x + C$

27. $\displaystyle\int \tan^3 x \, dx = \tfrac{1}{2} \tan^2 x + \ln |\cos x| + C$

28. $\displaystyle\int \cot^3 x \, dx = -\tfrac{1}{2} \cot^2 x + \ln |\csc x| + C$

29. $\displaystyle\int \sec^3 x \, dx = \tfrac{1}{2} \sec x \tan x + \tfrac{1}{2} \ln |\sec x + \tan x| + C$

30. $\displaystyle\int \csc^3 x \, dx = -\tfrac{1}{2} \csc x \cot x + \tfrac{1}{2} \ln |\csc x - \cot x| + C$

31. $\displaystyle\int (\sin x)^n \, dx = -\frac{1}{n} (\sin x)^{n-1} \cos x + \frac{n-1}{n} \int (\sin x)^{n-2} \, dx$

▶ TRIGONOMETRIC FORMS (cont.)

32. $\int (\cos x)^n \, dx = \frac{1}{n} (\cos x)^{n-1} \sin x + \frac{n-1}{n} \int (\cos x)^{n-2} \, dx$

33. $\int (\tan x)^n \, dx = \frac{1}{n-1} (\tan x)^{n-1} - \int (\tan x)^{n-2} \, dx$

34. $\int (\cot x)^n \, dx = -\frac{1}{n-1} (\cot x)^{n-1} - \int (\cot x)^{n-2} \, dx$

35. $\int (\sec x)^n \, dx = \frac{1}{n-1} (\sec x)^{n-2} \tan x + \frac{n-2}{n-1} \int (\sec x)^{n-2} \, dx$

36. $\int (\csc x)^n \, dx = -\frac{1}{n-1} (\csc x)^{n-2} \cot x + \frac{n-2}{n-1} \int (\csc x)^{n-2} \, dx$

37. $\int \sin ax \sin bx \, dx = -\frac{1}{2(a+b)} \sin (a+b)x + \frac{1}{2(a-b)} \sin (a-b)x + C, a^2 \neq b^2$

38. $\int \cos ax \cos bx \, dx = \frac{1}{2(a+b)} \sin (a+b)x + \frac{1}{2(a-b)} \sin (a-b)x + C, a^2 \neq b^2$

39. $\int \sin ax \cos bx \, dx = -\frac{1}{2(a+b)} \cos (a+b)x - \frac{1}{2(a-b)} \cos (a-b)x + C, a^2 \neq b^2$

40. $\int x \sin x \, dx = -x \cos x + \sin x + C$

41. $\int x \cos x \, dx = x \sin x + \cos x + C$

42. $\int x^2 \sin x \, dx = -x^2 \cos x + 2x \sin x + 2 \cos x + C$

43. $\int x^2 \cos x \, dx = x^2 \sin x + 2x \cos x - 2 \sin x + C$

44. $\int x^n \sin x \, dx = -x^n \cos x + n \int x^{n-1} \cos x \, dx$

45. $\int x^n \cos x \, dx = x^n \sin x - n \int x^{n-1} \sin x \, dx$

46. $\int (\sin x)^m (\cos x)^n \, dx = -\frac{(\sin x)^{m-1}(\cos x)^{n+1}}{m+n} + \frac{m-1}{m+n} \int (\sin x)^{m-2}(\cos x)^n \, dx$

    $= \frac{(\sin x)^{m+1}(\cos x)^{n-1}}{m+n} + \frac{n-1}{m+n} \int (\sin x)^m (\cos x)^{n-2} \, dx$

▶ LOGARITHMIC AND EXPONENTIAL FORMS

47. $\int xe^x \, dx = xe^x - e^x + C$

48. $\int \frac{1}{x \ln x} \, dx = \ln |\ln x| + C$

49. $\int x^n e^x \, dx = x^n e^x - n \int x^{n-1} e^x \, dx$

50. $\int x^n \ln x \, dx = \frac{1}{n+1} x^{n+1} \ln x - \frac{1}{(n+1)^2} x^{n+1} + C$

51. $\int (\ln x)^n \, dx = x(\ln x)^n - n \int (\ln x)^{n-1} \, dx$

52. $\int \frac{1}{x(\ln x)^n} \, dx = \frac{-1}{(n-1)(\ln x)^{n-1}} + C, n \neq 1$

53. $\int e^{ax} \sin bx \, dx = \frac{e^{ax}}{a^2 + b^2} (a \sin bx - b \cos bx) + C$

54. $\int e^{ax} \cos bx \, dx = \frac{e^{ax}}{a^2 + b^2} (a \cos bx + b \sin bx) + C$

▶ INVERSE TRIGONOMETRIC FORMS

55. $\int \arcsin x \, dx = x \arcsin x + \sqrt{1 - x^2} + C$

56. $\int \arccos x \, dx = x \arccos x - \sqrt{1 - x^2} + C$

57. $\int \arctan x \, dx = x \arctan x - \frac{1}{2} \ln (x^2 + 1) + C$

58. $\int \arccot x \, dx = x \arccot x + \frac{1}{2} \ln (x^2 + 1) + C$

59. $\int \arcsec x \, dx = x \arcsec x - \ln |x + \sqrt{x^2 - 1}| + C$

60. $\int \arccsc x \, dx = x \arccsc x + \ln |x + \sqrt{x^2 - 1}| + C$

▶ HYPERBOLIC FORMS

61. $\int \sinh x \, dx = \cosh x + C$

62. $\int \cosh x \, dx = \sinh x + C$

63. $\int \tanh x \, dx = \ln \cosh x + C$

64. $\int \coth x \, dx = \ln |\sinh x| + C$

65. $\int \sech x \, dx = \arctan (\sinh x) + C$

66. $\int \csch x \, dx = \ln |\tanh \tfrac{1}{2} x| + C$

67. $\int (\sech x)^2 \, dx = \tanh x + C$

68. $\int (\csch x)^2 \, dx = -\coth x + C$

69. $\int \sech x \tanh x \, dx = -\sech x + C$

70. $\int \csch x \coth x \, dx = -\csch x + C$

71. $\int (\sinh x)^2 \, dx = \tfrac{1}{4} \sinh 2x - \tfrac{1}{2} x + C$

72. $\int (\cosh x)^2 \, dx = \tfrac{1}{4} \sinh 2x + \tfrac{1}{2} x + C$

73. $\int e^{ax} \sinh bx \, dx = \dfrac{e^{ax}}{a^2 - b^2} (a \sinh bx - b \cosh bx) + C$

74. $\int e^{ax} \cosh bx \, dx = \dfrac{e^{ax}}{a^2 - b^2} (a \cosh bx - b \sinh bx) + C$

▶ FORMS INVOLVING $x^2 \pm a^2$

75. $\int \sqrt{x^2 \pm a^2} \, dx = \dfrac{x}{2} \sqrt{x^2 \pm a^2} \pm \dfrac{a^2}{2} \ln |x + \sqrt{x^2 \pm a^2}| + C$

76. $\int x^2 \sqrt{x^2 \pm a^2} \, dx = \dfrac{x}{8} (2x^2 \pm a^2) \sqrt{x^2 \pm a^2} - \dfrac{a^4}{8} \ln |x + \sqrt{x^2 \pm a^2}| + C$

▶ FORMS INVOLVING $x^2 \pm a^2$ (cont.)

77. $\displaystyle\int \frac{\sqrt{x^2 \pm a^2}}{x^2}\, dx = -\frac{\sqrt{x^2 \pm a^2}}{x} + \ln|x + \sqrt{x^2 \pm a^2}| + C$

78. $\displaystyle\int (x^2 \pm a^2)^{3/2}\, dx = \frac{x}{8}(2x^2 \pm 5a^2)\sqrt{x^2 \pm a^2} + \frac{3a^4}{8}\ln|x + \sqrt{x^2 \pm a^2}| + C$

79. $\displaystyle\int \frac{1}{\sqrt{x^2 \pm a^2}}\, dx = \ln|x + \sqrt{x^2 \pm a^2}| + C$

80. $\displaystyle\int \frac{x^2}{\sqrt{x^2 \pm a^2}}\, dx = \frac{x}{2}\sqrt{x^2 \pm a^2} \mp \frac{a^2}{2}\ln|x + \sqrt{x^2 \pm a^2}| + C$

81. $\displaystyle\int \frac{1}{x^2\sqrt{x^2 \pm a^2}}\, dx = \mp \frac{\sqrt{x^2 \pm a^2}}{a^2 x} + C$

82. $\displaystyle\int \frac{1}{(x^2 \pm a^2)^{3/2}}\, dx = \pm \frac{x}{a^2\sqrt{x^2 \pm a^2}} + C$

83. $\displaystyle\int \frac{\sqrt{x^2 - a^2}}{x}\, dx = \sqrt{x^2 - a^2} - a\operatorname{arcsec}\frac{|x|}{a} + C$

84. $\displaystyle\int \frac{\sqrt{x^2 + a^2}}{x}\, dx = \sqrt{x^2 + a^2} - a\ln\left|\frac{a + \sqrt{x^2 + a^2}}{x}\right| + C$

85. $\displaystyle\int \frac{1}{x\sqrt{x^2 + a^2}}\, dx = -\frac{1}{a}\ln\left|\frac{a + \sqrt{x^2 + a^2}}{x}\right| + C$

▶ FORMS INVOLVING $a^2 - x^2$

86. $\displaystyle\int \sqrt{a^2 - x^2}\, dx = \frac{x}{2}\sqrt{a^2 - x^2} + \frac{a^2}{2}\arcsin\frac{x}{a} + C$

87. $\displaystyle\int x^2\sqrt{a^2 - x^2}\, dx = \frac{x}{8}(2x^2 - a^2)\sqrt{a^2 - x^2} + \frac{a^4}{8}\arcsin\frac{x}{a} + C$

88. $\displaystyle\int \frac{\sqrt{a^2 - x^2}}{x}\, dx = \sqrt{a^2 - x^2} - a\ln\left|\frac{a + \sqrt{a^2 - x^2}}{x}\right| + C$

89. $\displaystyle\int \frac{\sqrt{a^2 - x^2}}{x^2}\, dx = -\frac{1}{x}\sqrt{a^2 - x^2} - \arcsin\frac{x}{a} + C$

90. $\displaystyle\int \frac{x^2}{\sqrt{a^2 - x^2}}\, dx = -\frac{x}{2}\sqrt{a^2 - x^2} + \frac{a^2}{2}\arcsin\frac{x}{a} + C$

91. $\displaystyle\int \frac{1}{x\sqrt{a^2 - x^2}}\, dx = -\frac{1}{a}\ln\left|\frac{a + \sqrt{a^2 - x^2}}{x}\right| + C$

92. $\displaystyle\int \frac{1}{x^2\sqrt{a^2 - x^2}}\, dx = -\frac{1}{a^2 x}\sqrt{a^2 - x^2} + C$

▶ FORMS INVOLVING $a^2 - x^2$ (cont.)

93. $\displaystyle\int (a^2 - x^2)^{3/2}\, dx = \frac{x}{8}(5a^2 - 2x^2)\sqrt{a^2 - x^2} + \frac{3a^4}{8}\arcsin\frac{x}{a} + C$

94. $\displaystyle\int \frac{1}{(a^2 - x^2)^{3/2}}\, dx = \frac{x}{a^2\sqrt{a^2 - x^2}} + C$

▶ FORMS INVOLVING $a + bx$

95. $\displaystyle\int \frac{x}{a + bx}\, dx = \frac{x}{b} - \frac{a}{b^2}\ln|a + bx| + C$

96. $\displaystyle\int \frac{x^2}{a + bx}\, dx = \frac{1}{b^3}\left[\tfrac{1}{2}(a + bx)^2 - 2a(a + bx) + a^2\ln|a + bx|\right] + C$

97. $\displaystyle\int \frac{x}{(a + bx)^2}\, dx = \frac{1}{b^2}\left[\frac{a}{a + bx} + \ln|a + bx|\right] + C$

98. $\displaystyle\int \frac{x^2}{(a + bx)^2}\, dx = \frac{1}{b^3}\left[a + bx - \frac{a^2}{a + bx} - 2a\ln|a + bx|\right] + C$

99. $\displaystyle\int \frac{1}{x(a + bx)}\, dx = \frac{1}{a}\ln\left|\frac{x}{a + bx}\right| + C$

100. $\displaystyle\int \frac{1}{x^2(a + bx)}\, dx = -\frac{1}{ax} + \frac{b}{a^2}\ln\left|\frac{a + bx}{x}\right| + C$

101. $\displaystyle\int \frac{1}{x(a + bx)^2}\, dx = \frac{1}{a(a + bx)} + \frac{1}{a^2}\ln\left|\frac{x}{a + bx}\right| + C$

102. $\displaystyle\int x\sqrt{a + bx}\, dx = \frac{2}{15b^2}(3bx - 2a)(a + bx)^{3/2} + C$

103. $\displaystyle\int x^n\sqrt{a + bx}\, dx = \frac{2x^n(a + bx)^{3/2}}{b(2n + 3)} - \frac{2an}{b(2n + 3)}\int x^{n-1}\sqrt{a + bx}\, dx$

104. $\displaystyle\int \frac{x}{\sqrt{a + bx}}\, dx = \frac{2}{3b^2}(bx - 2a)\sqrt{a + bx} + C$

105. $\displaystyle\int \frac{x^2\, dx}{\sqrt{a + bx}} = \frac{2}{15b^3}(8a^2 + 3b^2x^2 - 4abx)\sqrt{a + bx} + C$

106. $\displaystyle\int \frac{x^n}{\sqrt{a + bx}}\, dx = \frac{2x^n\sqrt{a + bx}}{b(2n + 1)} - \frac{2an}{b(2n + 1)}\int \frac{x^{n-1}}{\sqrt{a + bx}}\, dx$

107. $\displaystyle\int \frac{1}{x\sqrt{a + bx}}\, dx = \begin{cases} \dfrac{1}{\sqrt{a}}\ln\left|\dfrac{\sqrt{a + bx} - \sqrt{a}}{\sqrt{a + bx} + \sqrt{a}}\right| + C, & \text{if } a > 0, \\ \dfrac{2}{\sqrt{-a}}\arctan\sqrt{\dfrac{a + bx}{-a}} + C, & \text{if } a < 0 \end{cases}$

▶ FORMS INVOLVING $a + bx$ (cont.)

108. $\displaystyle\int \frac{1}{x^n \sqrt{a+bx}}\, dx = -\frac{\sqrt{a+bx}}{a(n-1)x^{n-1}} - \frac{b(2n-3)}{2a(n-1)} \int \frac{1}{x^{n-1}\sqrt{a+bx}}\, dx$

109. $\displaystyle\int \frac{\sqrt{a+bx}}{x}\, dx = 2\sqrt{a+bx} + a \int \frac{1}{x\sqrt{a+bx}}\, dx$

110. $\displaystyle\int \frac{\sqrt{a+bx}}{x^n}\, dx = -\frac{(a+bx)^{3/2}}{a(n-1)x^{n-1}} - \frac{b(2n-5)}{2a(n-1)} \int \frac{\sqrt{a+bx}}{x^{n-1}}\, dx$

▶ FORMS INVOLVING $\sqrt{2ax - x^2}$

111. $\displaystyle\int \sqrt{2ax-x^2}\, dx = \frac{x-a}{2}\sqrt{2ax-x^2} + \frac{a^2}{2}\arccos\left(1-\frac{x}{a}\right) + C$

112. $\displaystyle\int x\sqrt{2ax-x^2}\, dx = \frac{2x^2-ax-3a^2}{6}\sqrt{2ax-x^2} + \frac{a^3}{2}\arccos\left(1-\frac{x}{a}\right) + C$

113. $\displaystyle\int \frac{\sqrt{2ax-x^2}}{x}\, dx = \sqrt{2ax-x^2} + a\arccos\left(1-\frac{x}{a}\right) + C$

114. $\displaystyle\int \frac{\sqrt{2ax-x^2}}{x^2}\, dx = -\frac{2\sqrt{2ax-x^2}}{x} - \arccos\left(1-\frac{x}{a}\right) + C$

115. $\displaystyle\int \frac{1}{\sqrt{2ax-x^2}}\, dx = \arccos\left(1-\frac{x}{a}\right) + C$

116. $\displaystyle\int \frac{x}{\sqrt{2ax-x^2}}\, dx = -\sqrt{2ax-x^2} + a\arccos\left(1-\frac{x}{a}\right) + C$

117. $\displaystyle\int \frac{x^2}{\sqrt{2ax-x^2}}\, dx = -\left(\frac{x+3a}{2}\right)\sqrt{2ax-x^2} + \frac{3a^2}{2}\arccos\left(1-\frac{x}{a}\right) + C$

118. $\displaystyle\int \frac{1}{x\sqrt{2ax-x^2}}\, dx = -\frac{\sqrt{2ax-x^2}}{ax} + C$

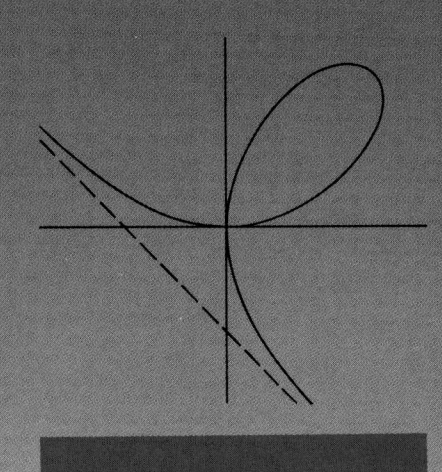

Answers

Chapter 1

Exercise Set 1.1

1. (a) 21; (b) 17; (c) $33 + 16\sqrt{3}$; (d) 38   3. $\mathbb{R}$
5. $x \neq 0, 1, -1$ 7. $(-\infty, 0] \cup [2, \infty)$
9. $(-\infty, -2] \cup [3, \infty)$ 11. Yes   13. No   15. No
17. $x^2 + 2, -x^2 - 2, (1 + 2x^2)/x^2, 1/(x^2 + 2), x + 2, \sqrt{x^2 + 2}$
19. $-1/x, -1/x, x, x, 1/\sqrt{x}, 1/\sqrt{x}$,   21. $x = \pm 1$
23. $x = \pm\sqrt{6}/2$ 25. $x = -1$ 27. $2(x + h) - 4, 2$
29. $\dfrac{3}{2}(x + h) - \dfrac{1}{4}, 3/2$ 31. $(x + h)^2, 2x + h$
33. $2(x + h)^2 - (x + h), 4x + 2h - 1$
35. Even (a), (g); odd (c), (h); neither (b), (d), (e), (f)
39. $F(w) = 3w + 864/w, (0, \infty)$
41. $D(t) = \sqrt{(20t)^2 + (100 - 15t)^2}$
43. $A(w) = w\sqrt{100 - w^2}, (0, 10)$
45. $V(x) = x(8 - 2x)(11 - 2x), (0, 4)$
47. $C(x) = 3\sqrt{200^2 + x^2} + 2(800 - x)$
49. $R(x) = \begin{cases} 10x, & \text{if } 0 \leq x \leq 10, \\ 12.5x - 0.25x^2, & \text{if } 11 \leq x \leq 25, \\ 6.25x, & \text{if } 26 \leq x \end{cases}$

Exercise Set 1.2

1. (a) $m = 2$ (b) $m = -1$ 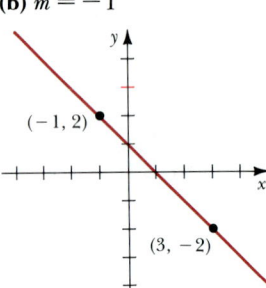   (c) $m = 1$ 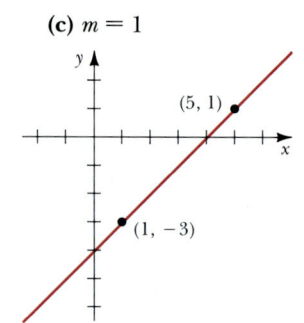   (d) $m = \dfrac{3}{4}$ 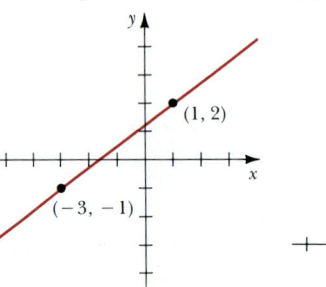   (e) $m = 1$ 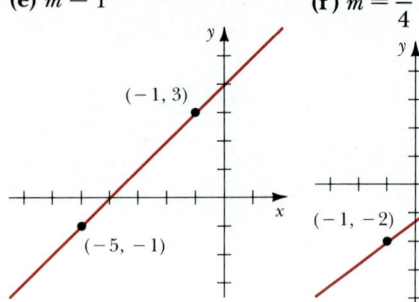   (f) $m = \dfrac{3}{4}$

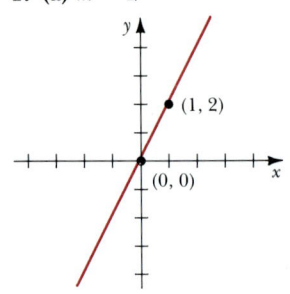

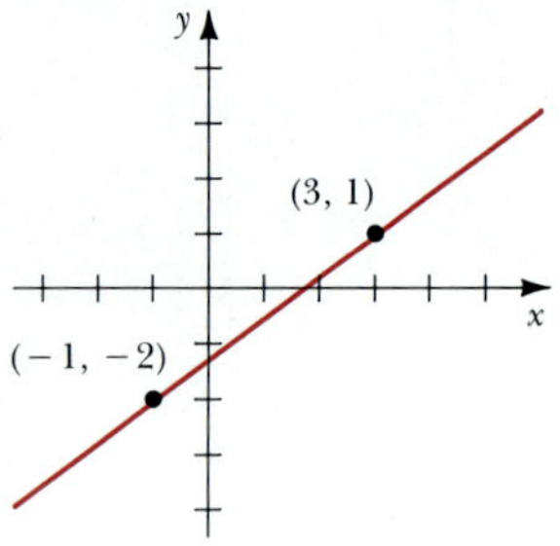

3. (a) $y = x - 1$; (b) $y = -x + 5$; (c) $y = 2$; (d) $y = 3x - 7$; (e) $y = 10x - 28$; (f) $y = \pi x + 2 - 3\pi$
5. (a) and (f); (b), (d), (i), and (j); (c) and (g); (h) and (e) 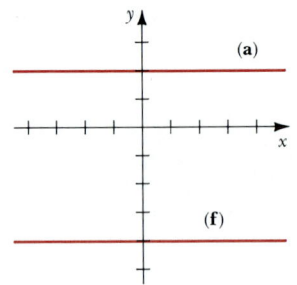 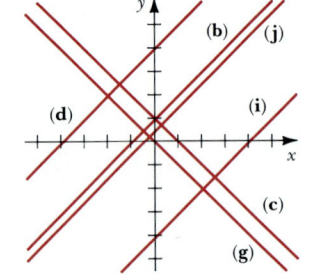 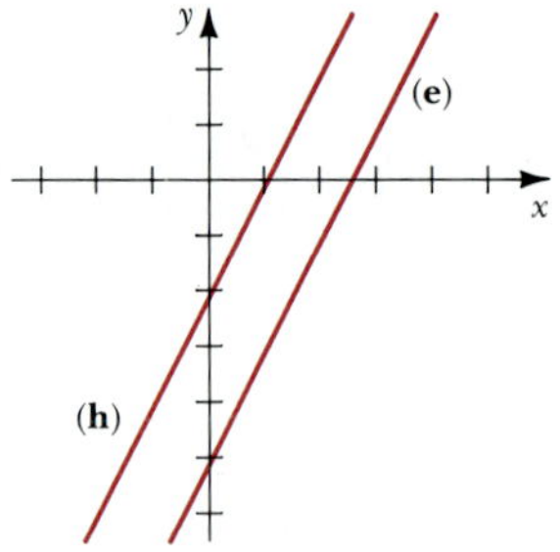
7. (a) $y = 2x$; (b) $y = -\dfrac{1}{2}x$
9. (a) $y = 3x - 1$; (b) $y = -\dfrac{1}{3}x + \dfrac{7}{3}$
11. (a) $y = -2x$; (b) $y = \dfrac{1}{2}x + \dfrac{5}{2}$
13. (a) $y = -x$; (b) $y = x$
15. Range is $[1, \infty)$
17. Range is $[0, \infty)$

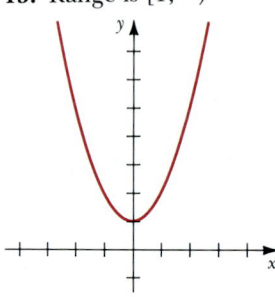

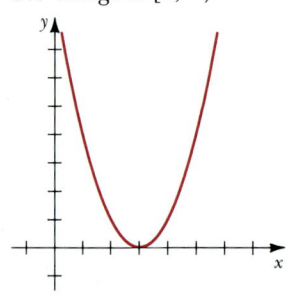

19. Range is $[0, \infty)$

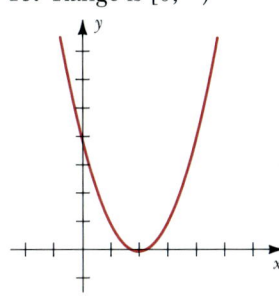

21. Range is $[-1, \infty)$

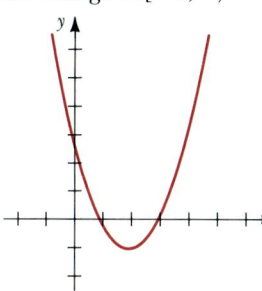

23. Range is $[-4, \infty)$

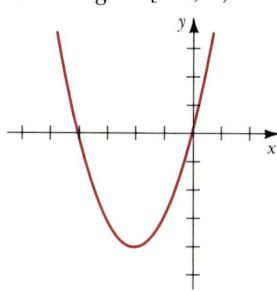

25. Range is $(-\infty, 0]$

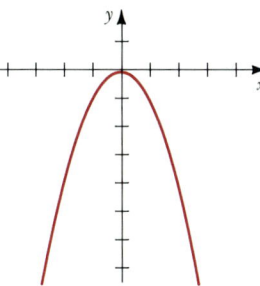

27. Range is $(-\infty, -1]$

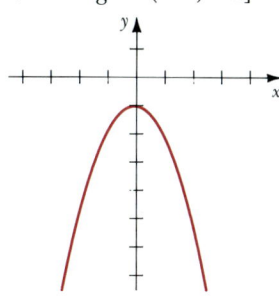

29. Range is $[0, \infty)$

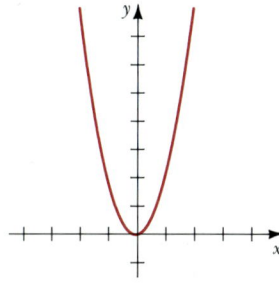

31. Range is $[0, \infty)$

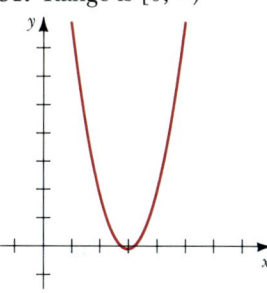

33. Range is $[-2, \infty)$

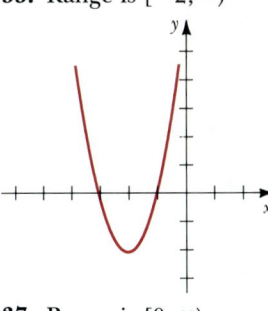

35. Range is $[0, \infty)$

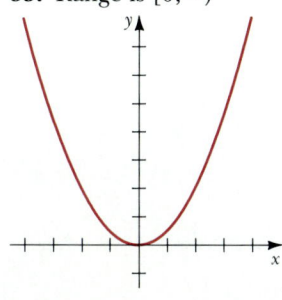

37. Range is $[0, \infty)$

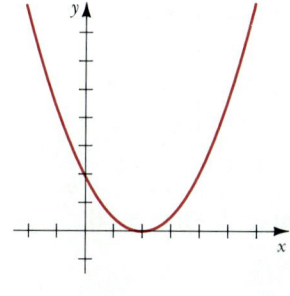

39. Range is $\left[-\dfrac{7}{2}, \infty\right)$

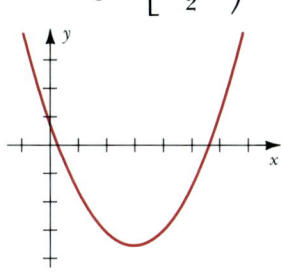

41. 15, 16, 25, 26, 27, 29, 30, 35, and 36

43.

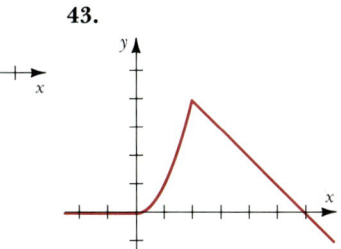

47. (a) (0, 784), (7, 0)

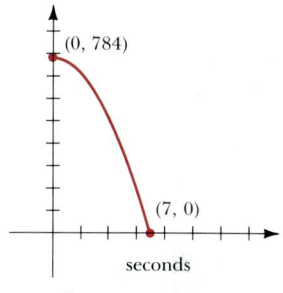

(b) Domain of s is $[0, 7]$; range of s is $[0, 784]$; **(c)** 7 seconds

49. $F = \left(\dfrac{9}{5}\right) C + 32$, 82.4

51. (a) $C(x) = \begin{cases} 40, & \text{if } 0 \le x \le 5, \\ 45 - x, & \text{if } 6 \le x \le 25, \\ 20, & \text{if } 26 \le x \end{cases}$

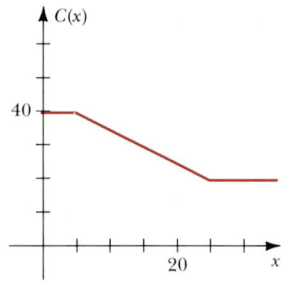

(b) $R(x) = \begin{cases} 40x, & \text{if } 0 \le x \le 5, \\ 45x - x^2, & \text{if } 6 \le x \le 25, \\ 20x, & \text{if } 26 \le x \end{cases}$

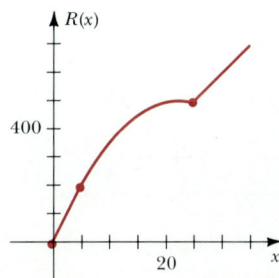

53. (a) $L(t) = 2 + 22 \times 10^{-6}t$ m; (b) 2.022 m

Exercise Set 1.3

1.

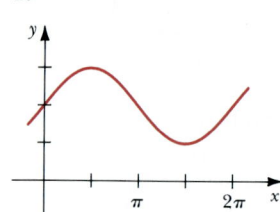

3.

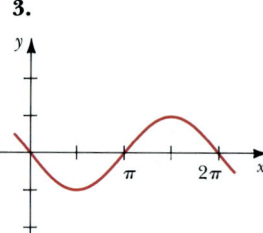

5.

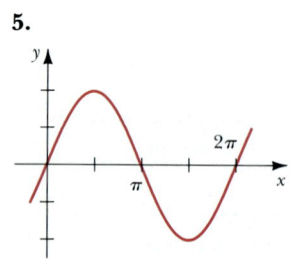

7.

9.

11.

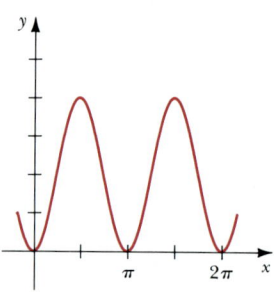

13.

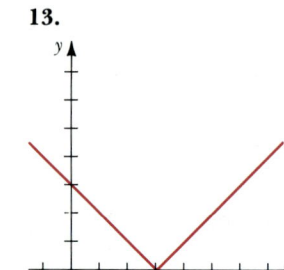

15.

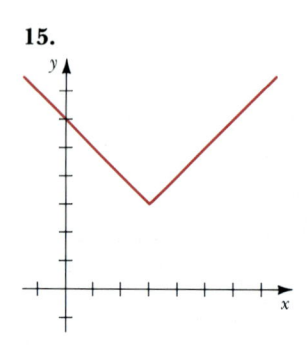

17.

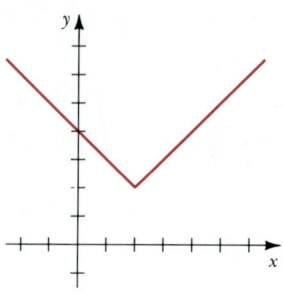

19.

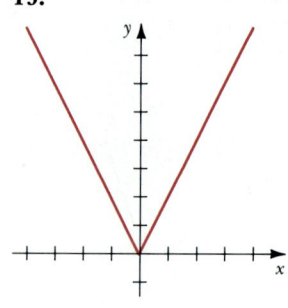

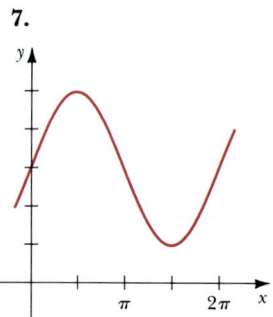

21.

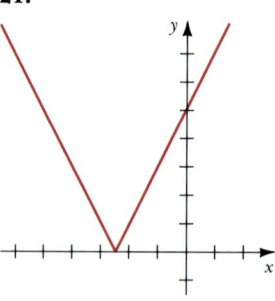

23. (a)

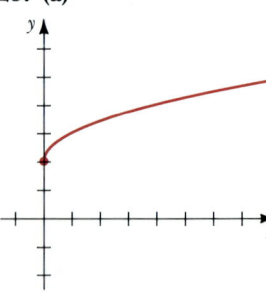

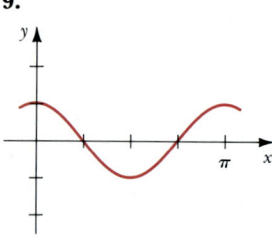

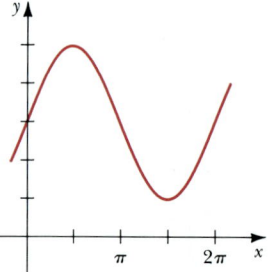

(b) $[0, \infty)$; $[2, \infty)$

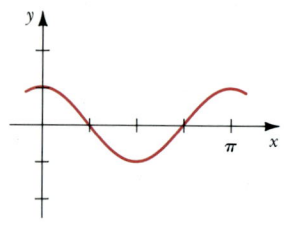

25. (a)

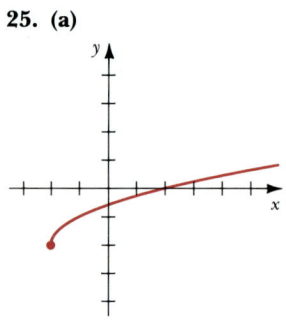

(b) $[-2, \infty)$; $[-2, \infty)$

27. (a)

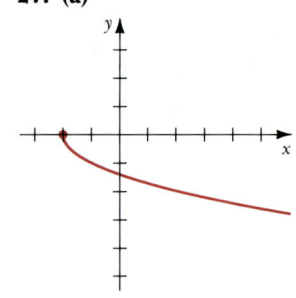

(b) $[-2, \infty)$; $(-\infty, 0]$

29. (a)

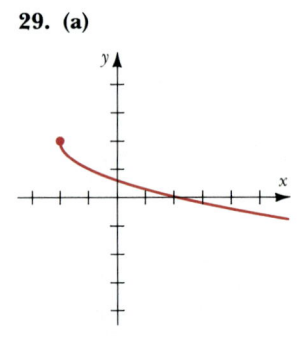

(b) $[-2, \infty)$; $(-\infty, 2]$

31. (a) [graph] (b) [graph] (c) [graph] (d) [graph] (e) [graph] (f) [graph]

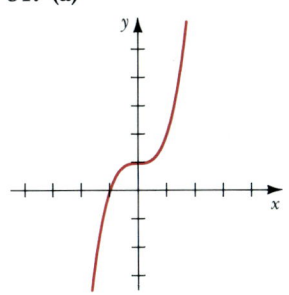

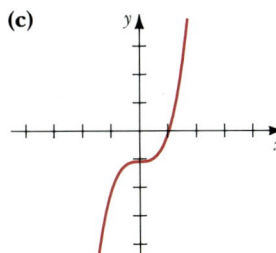

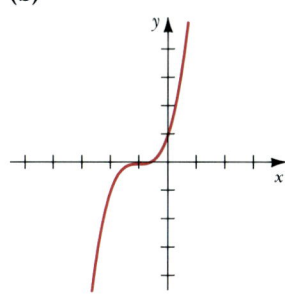

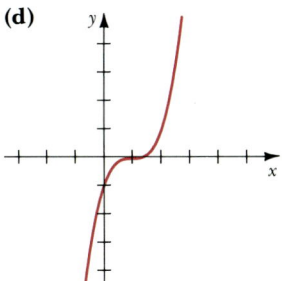

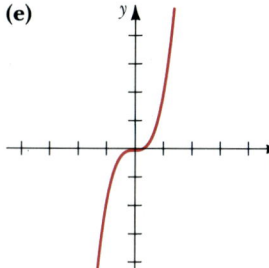

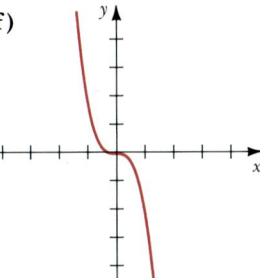

33. [graph showing:
(a) $f_1(x) = \sin x$
(b) $f_2(x) = \sin(x - \frac{\pi}{4})$
(c) $f_3(x) = -\sin(x - \frac{\pi}{4})$
(d) $f_4(x) = 1 - \sin(x - \frac{\pi}{4})$]

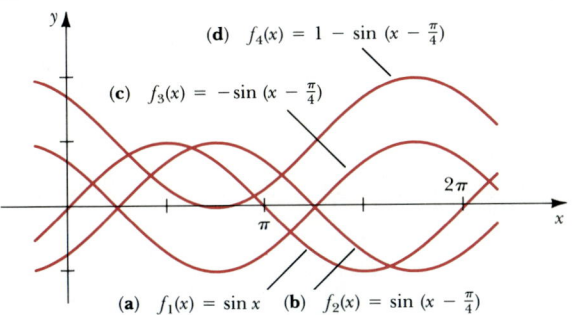

35. [graph]

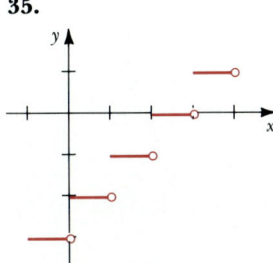

37. [graph]

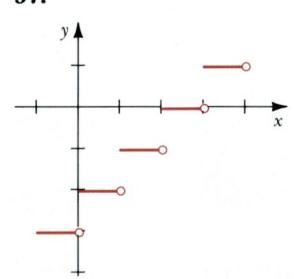

39. [graph]

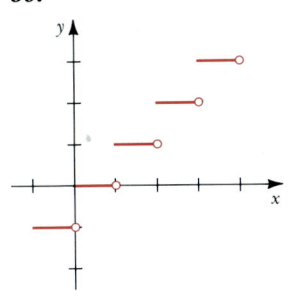

41. (a) $\sqrt{5}$; (b) $\sqrt{53}$; (c) 1; (d) $\sqrt{82}$

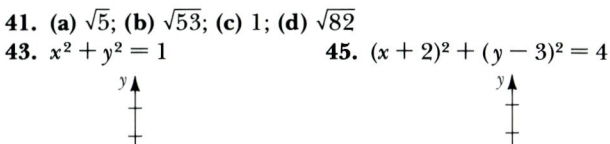

43. $x^2 + y^2 = 1$ [graph]

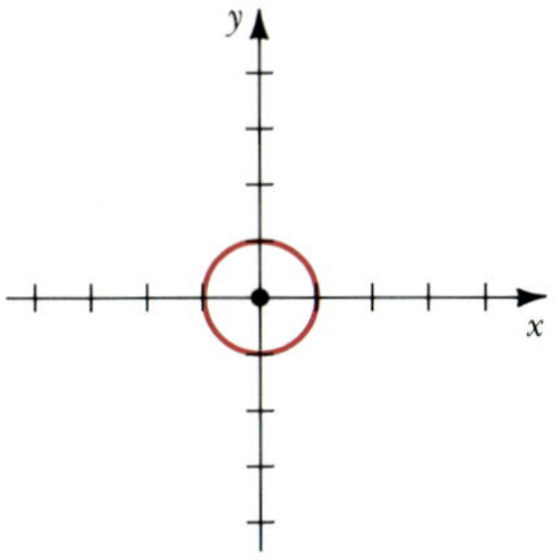

45. $(x+2)^2 + (y-3)^2 = 4$ [graph]

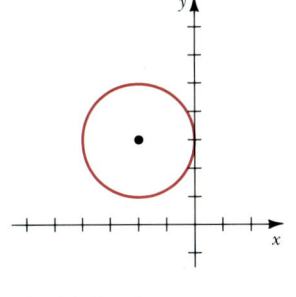

47. $x^2 + (y-2)^2 = 9$ [graph]

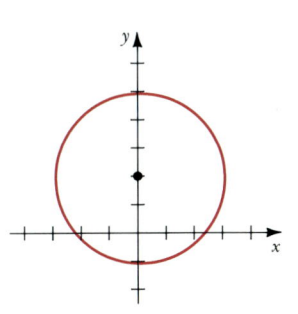

49. (a) $(0, 0)$, $r = 3$ (b) [graph]

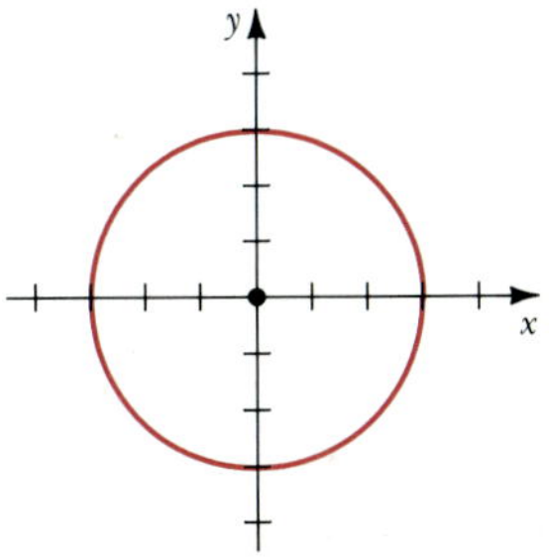

51. (a) $(0, 1)$, $r = 1$ (b) [graph]

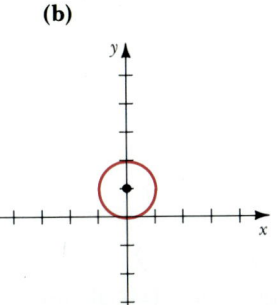

53. (a) $(2, 1)$, $r = 3$ (b) [graph]

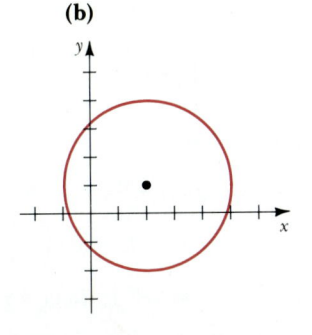

55. This circle is the same as the one in Exercise 53.

57. (a) $\left(-\dfrac{3}{2}, 2\right)$, $r = \dfrac{\sqrt{29}}{2}$

(b)

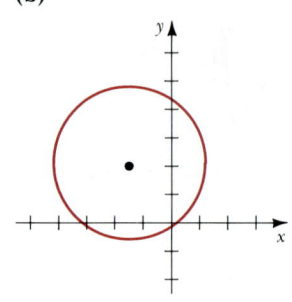

61. (5, 5) and (5, −1)   **63.** $(x-3)^2 + (y-3)^2 = 4$
65. $x^2 + ax + y^2 + by = 0$ **67.** $\sqrt{2-2y}$ **69.** $\tan\theta = x/5$
71. $C(x) = \begin{cases} \$5, & \text{if } 0 < x < 2, \\ \$5 + \$1[\![x-1]\!], & \text{if } x \geq 2 \end{cases}$

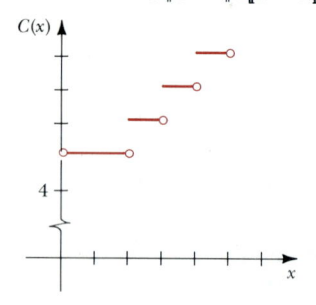

73. $C(x) = 22 + 17x$ cents

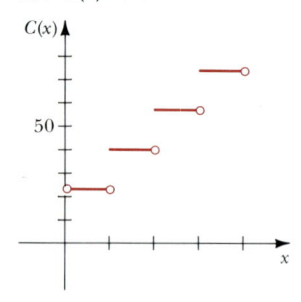

75. $D(x) = \begin{cases} x, & \text{if } 0 \leq x \leq 600, \\ |x - 1200|, & \text{if } 600 < x \leq 1850, \\ 2500 - x, & \text{if } 1850 < x \leq 2500 \end{cases}$

Exercise Set 1.4

1. (a) $(f+g)(x) = 2x + 1$; **(b)** $(f-g)(x) = -1$;
(c) $(f \cdot g)(x) = x^2 + x$; **(d)** $\left(\dfrac{f}{g}\right)(x) = \dfrac{x}{x+1}$;
(e) $(f \circ g)(x) = x + 1$; **(f)** $(g \circ f)(x) = x + 1$. Domain for parts (a), (b), (c), (e), and (f) is $\mathbb{R}$; domain of $\left(\dfrac{f}{g}\right)$ is $(-\infty, -1) \cup (1, \infty)$

3. (a) $(f+g)(x) = \dfrac{1}{x} + \sqrt{x-1}$;
(b) $(f-g)(x) = \dfrac{1}{x} - \sqrt{x-1}$;
(c) $(f \cdot g)(x) = \dfrac{\sqrt{x-1}}{x}$; **(d)** $\left(\dfrac{f}{g}\right)(x) = \dfrac{1}{x\sqrt{x-1}}$;
(e) $(f \circ g)(x) = \dfrac{1}{\sqrt{x-1}}$; **(f)** $(g \circ f)(x) = \sqrt{\dfrac{1-x}{x}}$.
Domain for parts (a), (b), and (c) is $[1, \infty)$; domain for (d) and (e) is $(1, \infty)$; domain for (f) is $(0, 1]$
5. (a) $(f+g)(x) = x^2 + x - 1$; **(b)** $(f-g)(x) = x^2 - x + 3$;
(c) $(f \cdot g)(x) = x^3 - 2x^2 + x - 2$; **(d)** $\left(\dfrac{f}{g}\right)(x) = \dfrac{x^2 + 1}{x - 2}$;
(e) $(f \circ g)(x) = x^2 - 4x + 5$; **(f)** $(g \circ f)(x) = x^2 - 1$.
Domain for (a), (b), (c), (e), and (f) is $\mathbb{R}$; domain for (d) is $\{x | x \neq 2\}$
7. (a) $(f+g)(x) = x^2 + x + \sin x$;
(b) $(f-g)(x) = x^2 + x - \sin x$; **(c)** $(f \cdot g)(x) = (x^2 + x) \cdot \sin x$; **(d)** $(f/g)(x) = (x^2 + x)/\sin x$; **(e)** $(f \circ g)(x) = \sin^2 x + \sin x$; **(f)** $(g \circ f)(x) = \sin(x^2 + x)$. Domain for (a), (b), (c), (e), and (f) is $\mathbb{R}$; domain for (d) is $\{x | x \neq n\pi$, for n an integer$\}$

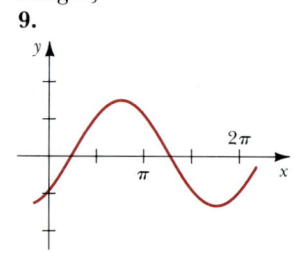

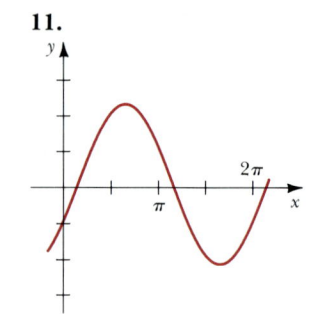

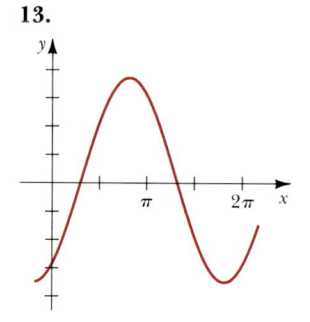

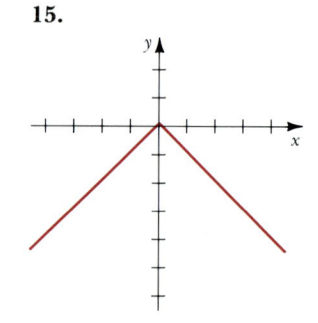

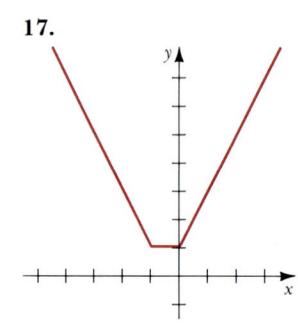

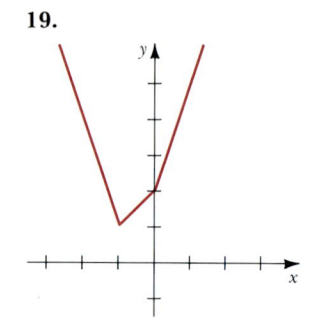

9.

11.

13.

15.

17.

19.

21. [graph] **23.** [graph] **47.** [graphs]

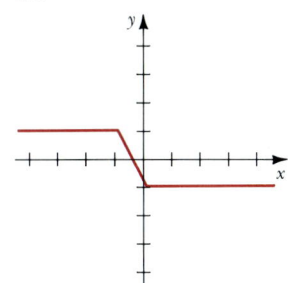

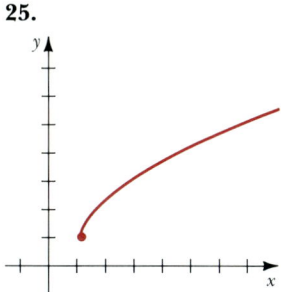

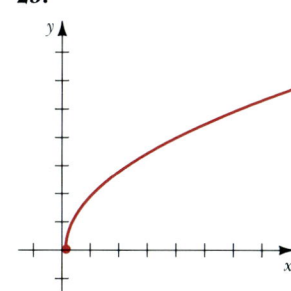

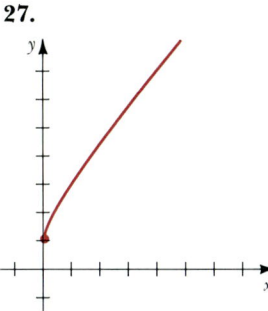

25. [graph] **27.** [graph]

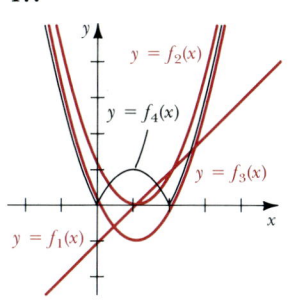

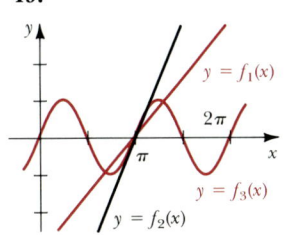

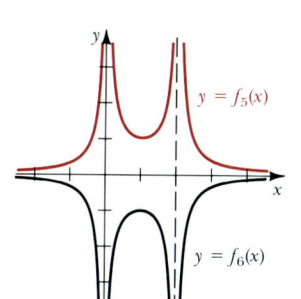

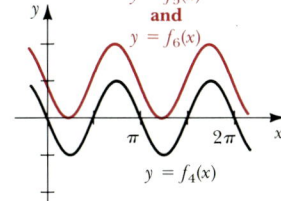

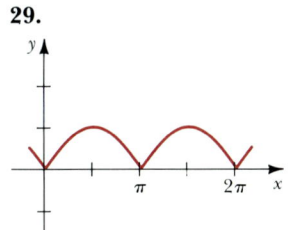

49. [graphs]

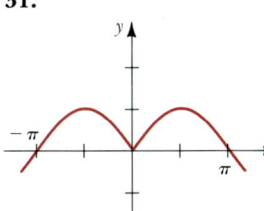

51. $g_1(x) = x^2$, $g_2(x) = x - 1$, $g_3(x) = |x|$, $g_4(x) = 1/x$, $g_5(x) = -x$

53. $g_1(x) = 2x$, $g_2(x) = \sin x$, $g_3(x) = -x$, $g_4(x) = 1 + x$, $g_5(x) = |x|$

61. $V = 4\pi(3 + 0.01t)^3/3$

63. $V = 4\pi(3\sqrt{t} + 5)^3/3$ cm³; SA $= 4\pi(3\sqrt{t} + 5)^2$ cm²

65. $f(r) = \dfrac{km_1 m_2}{r^2}$

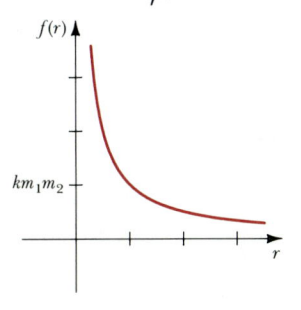

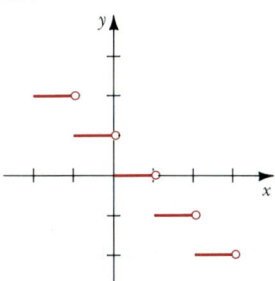

[graph]

29. [graph] **31.** [graph]

33. [graph] **35.** [graph]

67. (a) $\leq n$; (b) $\leq n$; (c) $2n$; (d) $3n$; (e) $n + 1$

Exercise Set 1.5

1. 1 **3.** 3 **5.** 4 **7.** 6 **9.** -8 **11.** 6
13. -1 **15.** 5 **17.** 3 **19.** 5 **21.** 5 **23.** 1
25. Does not exist **27.** Does not exist
29. Does not exist **31.** 1 **33.** 1
35. 0 **37.** 1 **39.** Does not exist
41. (a) 0; (b) 4; (c) 2; (d) 9 **43.** 0 **45.** -1
47. (a) $4 + h$; (b) 4 **49.** (a) $-6 + h$; (b) -6
51. (a) $3 + 3h + h^2$; (b) 3
53. (a) $-1/(1 + h)$; (b) -1

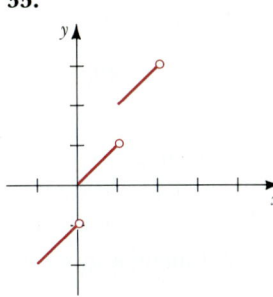

37. $f(x) = g(h(x))$, $h(x) = x^2 + 3x$, $g(x) = \sqrt{x}$
39. $f(x) = g(h(x))$, $h(x) = (x - 3)^2$, $g(x) = \sin x$
41. $f(x) = g(h(x))$, $h(x) = \cos x$, $g(x) = x^2 + 3x + 4$
43. One choice is $f(x) = i(g(h(x))$, $h(x) = x^2 + 1$, $g(x) = \sqrt{x} + 2$, $i(x) = x^4$
45. (a) Zero; (b) $1, -1$; (c) yes; (d) large; (e) small

55. (a) 1.35×10^5 in^3; (b) $\sqrt[3]{101{,}250/\pi} \approx 31.8$ in.
57. $\pi/180 \approx 0.1745$

Exercise Set 1.6

1. 0.5   **3.** 0.005   **5.** 0.05   **7.** 0.05   **9.** 0.01
11. 0.1/7   **27.** No; consider $f(x) = |x|/x$.
31. (a) The "line" $y = 1$ with holes at each irrational number and the "line" $y = -1$ with holes at each rational number; (b) The limit never exists.
35. Hint: If $\lim\limits_{x \to a} f(x) = L$, consider $\epsilon = L/2$.

Exercise Set 1.7

1. 1, $f(1)$ does not exist   **3.** 1, $\lim\limits_{x \to 1} f(x) \neq f(1)$

5. 2 and -3, $\lim\limits_{x \to 2} f(x)$ does not exist, $\lim\limits_{x \to -3} f(x)$ does not exist

7. 2, $f(2)$ does not exist   **9.** 0, neither $f(0)$ nor $\lim\limits_{x \to 0} f(x)$ exists

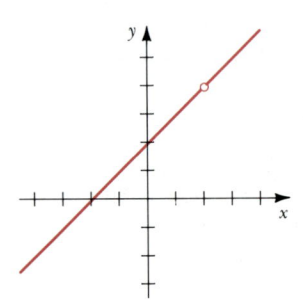

11. -2, $f(-2) = 3 \neq \lim\limits_{x \to -2} f(x) = -2$ **13.** For any integer n, $\lim\limits_{x \to n} [\![x - 1]\!]$ does not exist

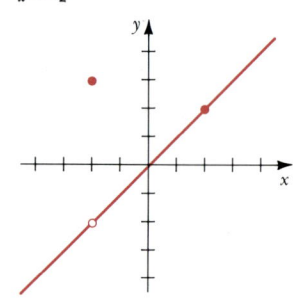

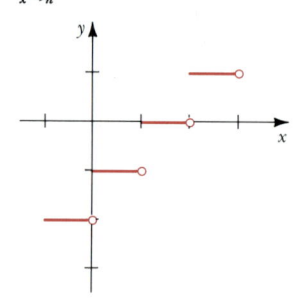

15. For any integer n, $\tan\left((2n + 1)\dfrac{\pi}{2}\right)$ does not exist.
The graph is shown in Figure 1.26.
17. For any integer n, $\sec(2(n+1)(\pi/2))$ does not exist.
The graph is shown in Figure 1.39(b).

19. For any integer $n \neq 0$, $\lim\limits_{x \to n} f(x) \neq f(n)$.

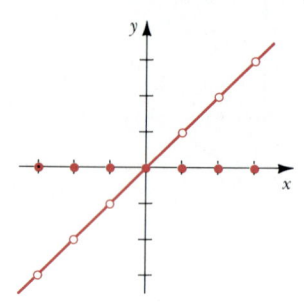

21. For any integer $n \neq -1$, $\lim\limits_{x \to n} f(x) \neq f(n)$.

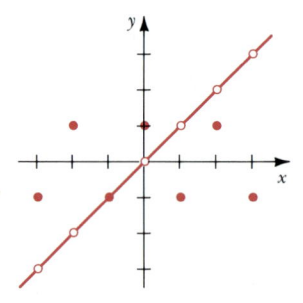

23. 223, yes   **25.** 38, yes   **27.** 7, yes
29. $\dfrac{1}{2}$, yes   **31.** $\dfrac{11}{38}$, yes

33. $(f \circ g)(x) = 1/(x - 2)$, $x \neq 2$, $(g \circ f)(x) = 1/x - 2$, $x \neq 0$
35. $(f \circ g)(x) = (\sqrt[4]{x} - 1)/(\sqrt[4]{x} + 1)$, $x > 0$;
$(g \circ f)(x) = \sqrt[4]{\dfrac{x - 1}{x + 1}}$, $x > 1$ or $x < -1$
37. $(f \circ g)(x) = x - 5\sqrt{x} + 6$, $x > 0$;
$(g \circ f)(x) = \sqrt{x^2 - 5x + 6}$, $x < 2$ or $x > 3$
39. -1 **41.** 0   **43.** Does not exist
45. $\dfrac{13}{41}$ **47.** 4   **49.** $-\dfrac{1}{4}$ **53.** 2
55. $ax_0^2 + bx_0 + c = 0$
59. One example is
$$f(x) = \begin{cases} 1, & \text{if } x \geq 1, \\ -1, & \text{if } x < 1 \end{cases} \quad \text{and} \quad g(x) = \begin{cases} -1, & \text{if } x \geq 1, \\ 1, & \text{if } x < 1 \end{cases}$$
61. The functions given for Exercise 59 will work.

Exercise Set 1.8

1. (a) 2 and 1; (b) no; one-sided limits differ.
3. Does not exist   **5.** 0   **7.** 0   **9.** $\sqrt{5}$ **11.** 1
13. Does not exist   **15.** 0   **17.** 0   **19.** $\dfrac{1}{4}$ **21.** 0   **23.** 0

25. $-\dfrac{1}{16}$ **27.** 0   **29.** Does not exist   **31.** 1
33. Does not exist
35. $(-\infty, -1)$, $[-1, 1)$, and $[1, \infty)$
37. $(-\infty, -2)$, $(-2, 2)$, and $(2, \infty)$
39. $(-\infty, \infty)$ **41.** $(-\infty, 0)$, and $(0, \infty)$ **43.** $[3, \infty)$
45. $(-\infty, \infty)$ **47.** $(-\infty, 0)$ and $(0, \infty)$
49. $(2n-1, 2n+1)$ for any integer n
51. (a)

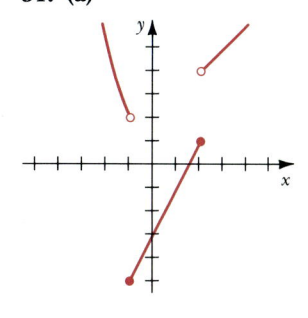

(b) 4, 1, continuous from the left at 2, not continuous at 2; **(c)** -5, 2, continuous from the right at -1, not continuous at -1; **(d)** $(-\infty, -1)$, $[-1, 2]$, $(2, \infty)$
53. $a = 7, b = -6$ **55.** $\dfrac{19}{7}$ **57.** 2   **59.** 2.7   **61.** -1.1
63. (a) $f(0) = 1, f\left(\dfrac{\pi}{2}\right) = -\dfrac{\pi}{2}$ and $\dfrac{-\pi}{2} < 0 < 1 \Rightarrow$ there is a number c in $\left(0, \dfrac{\pi}{2}\right)$ with $f(c) = 0$;   **(b)** 0.7   **65.** 0.2
67. $f(x) = \begin{cases} 1, & \text{if } x \geq 1 \\ -1, & \text{if } x < 1 \end{cases}$
69. One example is $f(x) = \dfrac{2x-1}{x^2-x}$; another is $f(x) = \cot \pi x$
75. Consider $f(x) = x$ at $x = 0$.

Exercise Set 1.9

1. $\dfrac{2}{3}$ **3.** 0   **5.** 0   **7.** 1   **9.** $\dfrac{3}{4}$ **11.** 1   **13.** 0
15. 0   **17.** 1   **19.** -1 **21.** $\dfrac{1}{5}$ **23.** 2
25. Does not exist   **27.** $\dfrac{1}{9}$ **29.** $\dfrac{1}{3}$ **31.** 0
33. $\dfrac{1}{2}$ **35.** 1   **37.** 1

39. $y = 0$

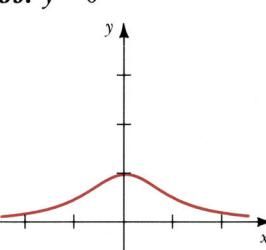

41. $y = 1, y = -1$

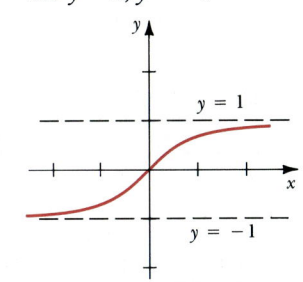

43. (a) 10;   **(b)** 100   **45.** For $\epsilon > 0$, let $M = \dfrac{1}{\sqrt{\epsilon}}$.
49. \$10, the limiting price   **51.**

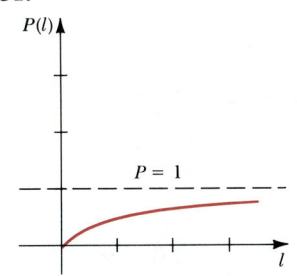

Exercise Set 1.10

1. ∞ **3.** Does not exist   **5.** ∞ **7.** ∞ **9.** 2
11. Does not exist   **13.** ∞ **15.** $-\infty$ **17.** ∞ **19.** ∞
21. ∞ **23.** ∞ **25.** ∞ **27.** $-\infty$ **29.** ∞
31. Does not exist   **33.** Does not exist
35. $x = (2n+1)\dfrac{\pi}{2}$, n an integer
37. $x = (2n+1)\dfrac{\pi}{2}$, n an integer

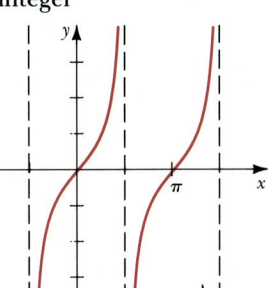

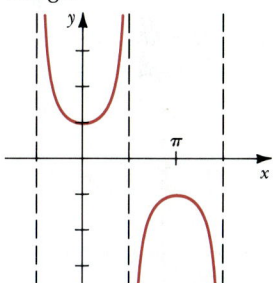

39. $x = -1, y = 1$

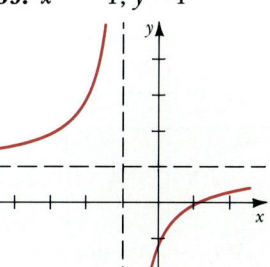

41. $x = 1, x = 3, y = 0$

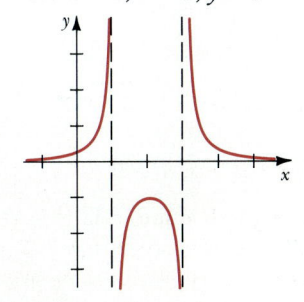

43. $y = 0$

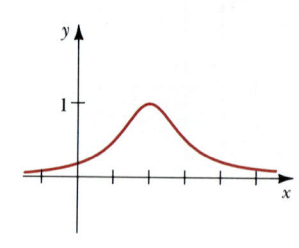

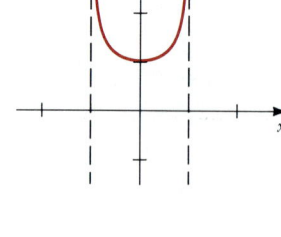

45. $x = 2$, $x = -2$, $y = 1$, $y = -1$

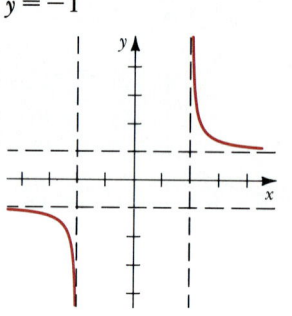

47. $x = 1$, $x = -1$

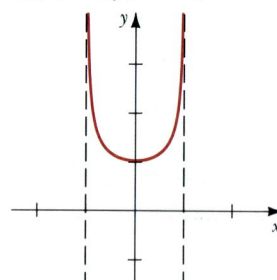

49. $x = 1$, $y = 1$

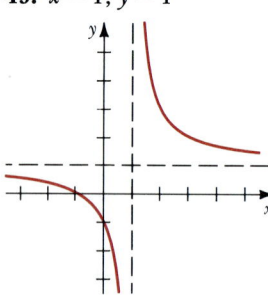

51. $x = 3$, $x = -3$, $y = -4$

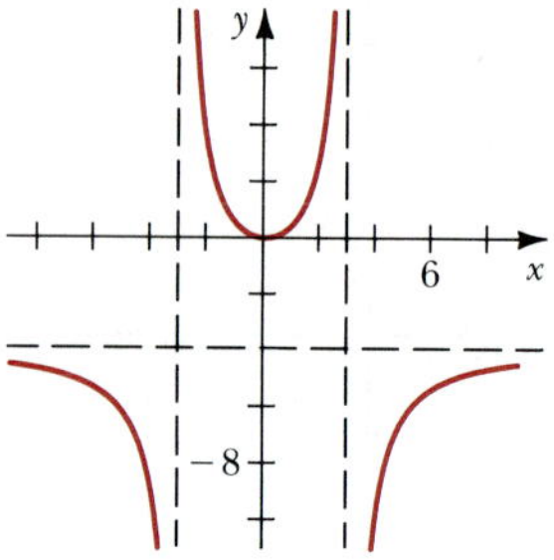

53. $x = 1$, $x = -1$, $y = 1$, $y = -1$

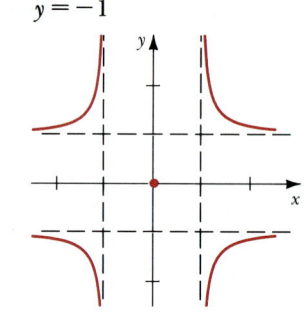

55. $y = 0$

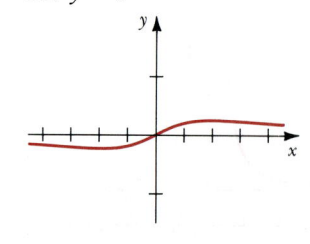

57. One example is:

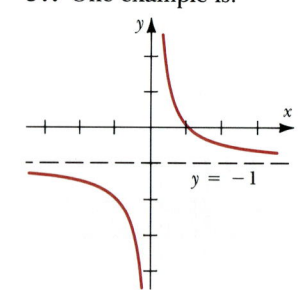

59. One example is:

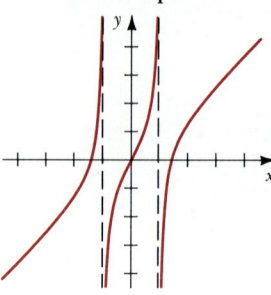

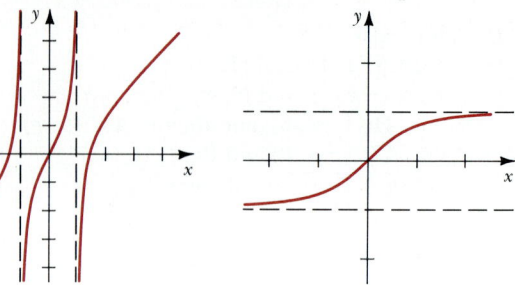

61. One example is:

63. (a) $0 < x < 0.01$; (b) $0 < x < 0.0001$
65. (b) $x > 101$
67. (a) a_m/b_n; (b) 0;

(c) $\lim\limits_{x \to \infty} f(x) = \begin{cases} \infty, & \text{if } a_m b_n > 0, \\ -\infty, & \text{if } a_m b_n < 0 \end{cases}$

Same for $\lim\limits_{x \to -\infty} f(x)$ if $m - n$ is even. Signs reversed if $m - n$ is odd.

69. There are many examples. One is $f(x) = \dfrac{1}{x}$, $g(x) = x$.

71. There are many examples. One is $f(x) = \dfrac{1}{x^2}$, $g(x) = x$.

73. There are many examples. One is $g(x) = x$, $h(x) = -x$.
75. There are many examples. One is $g(x) = x^2$, $h(x) = -x$.
77. $\lim\limits_{x \to a^+} f(x) = -\infty$ provided that for any number M, a number $\delta > 0$ can be found with the property that $f(x) < M$ whenever $0 < x - a < \delta$.
79. (a) $y = x$;
(b) $y = -2x$;
(c) $y = -x + 1$;
(d) $y = 2x + 3$
81. Slant asymptote at ∞ has equation $y = x + 1$. Slant asymptote at $-\infty$ has equation $y = x - 1$.
83. As $t \to 0$, threshold $\to \infty$. As $t \to \infty$, threshold $\to b$.
85. K

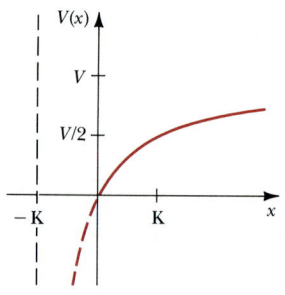

Review Exercises

1. (a) $2\sqrt{2} - 3$; (b) $2\sqrt{2} - 1$; (c) $2\sqrt{2} - 4$; (d) $2\sqrt{x} - 3$; (e) $\sqrt{2x - 3}$; (f) 2

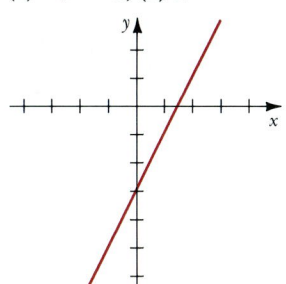

3. (a) $12 - 7\sqrt{2}$; (b) $6 - 5\sqrt{2}$; (c) $16 - 7\sqrt{2}$; (d) $x - 7\sqrt{x} + 10$; (e) $\sqrt{x^2 - 7x + 10}$; (f) $2x - 7 + h$

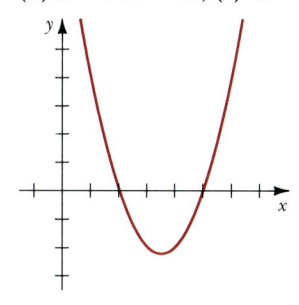

5.

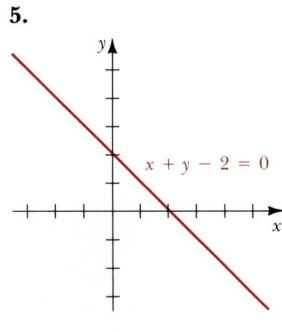

7.

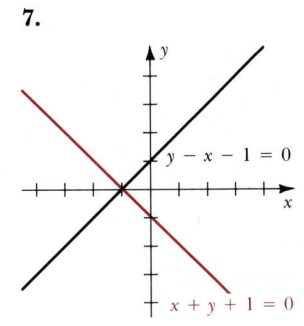

9. $y = -3$

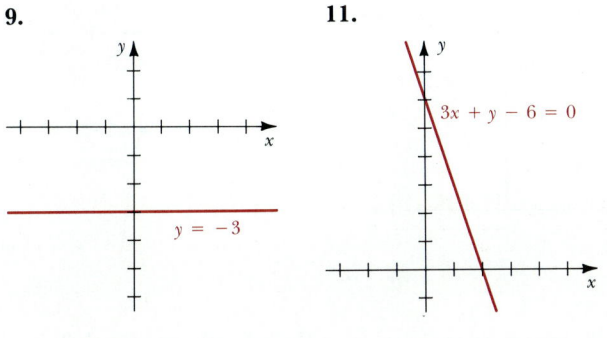

11. $3x + y - 6 = 0$

13. $2x + y - 2 = 0$

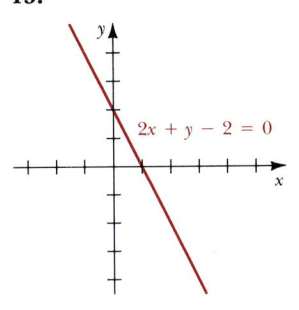

15.

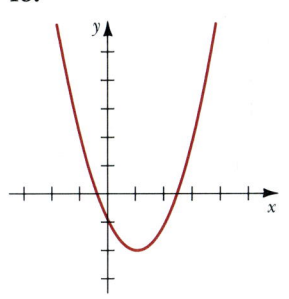

17.

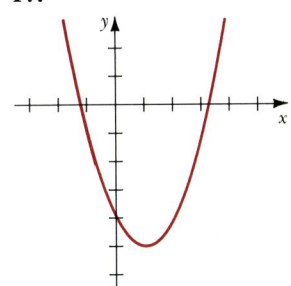

19.

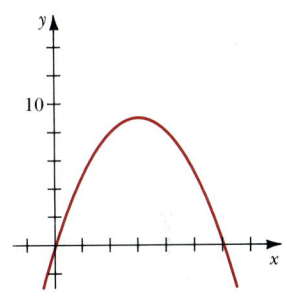

21.

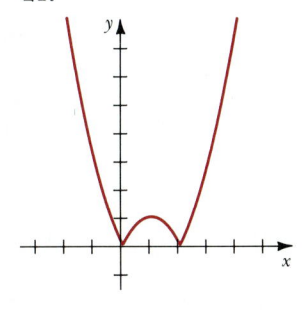

23. $y = 2$

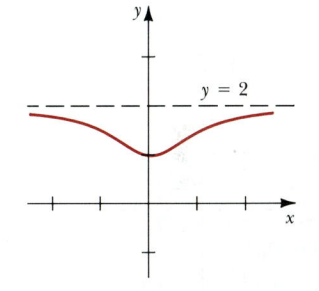

25. $y = -1$

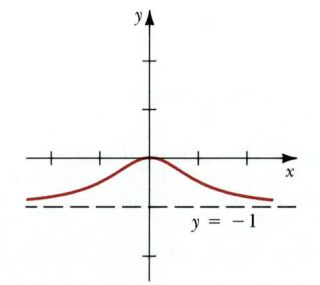

27. $y = 1$

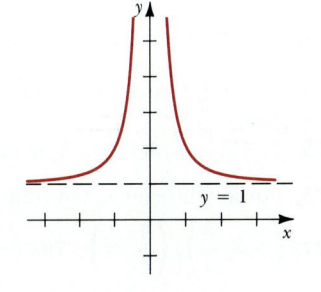

29.

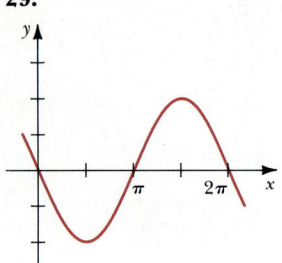

31.

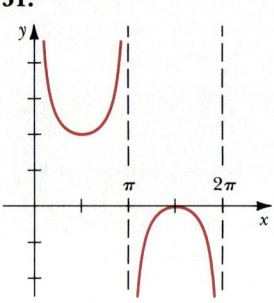

33.

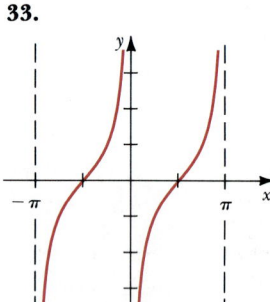

35. (a) $\sqrt{\dfrac{1-4x^2}{x^2}}$; **(b)** $\dfrac{1}{\sqrt{x^2-4}}$; **(c)** $\left[-\dfrac{1}{2}, 0\right) \cup \left(0, \dfrac{1}{2}\right]$; $(-\infty, -2) \cup (2, \infty)$

37. (a) $\dfrac{1}{x^2+2x}$; **(b)** $\dfrac{x^2}{x^2-1}$; **(c)** $(-\infty, -2) \cup (-2, 0) \cup (0, \infty)$, $(-\infty, -1) \cup (-1, 1) \cup (1, \infty)$

39. For example, $f(x) = g(h(x))$, where $h(x) = x^2 - 7x + 1$ and $g(x) = x^3$.

41. $\dfrac{5}{2}$ **43.** 0 **45.** 0 **47.** 1 **49.** ∞

51. ∞ **53.** 0 **55.** $\dfrac{1}{4}$ **57.** $\dfrac{1}{4}$ **59.** 0

61. ∞ **63.** $-\infty$ **65.** $-\dfrac{1}{2x^{3/2}}$ **67.** ∞

69. -2 and 2, since $f(-2)$ and $f(2)$ are undefined.
71. None
73. 3, $f(3)$ is undefined. 4, neither $f(4)$ nor $\lim\limits_{x \to 4} f(x)$ exists.
75. For any integer n, $\tan((2n+1)(\pi/2))$ does not exist.
77. $\left(-\infty, \dfrac{3}{4}\right), \left(\dfrac{3}{4}, \infty\right)$ **79.** $(-\infty, -1], [1, \infty)$
81. $((2n-1)(\pi/2), (2n+1)(\pi/2))$ for any integer n.

83. $x = 1, y = 0$

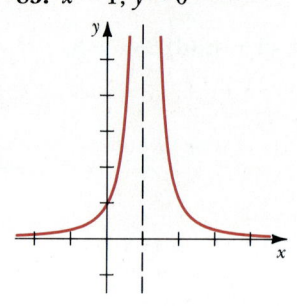

85. $x = 0, x = 2, y = 1$

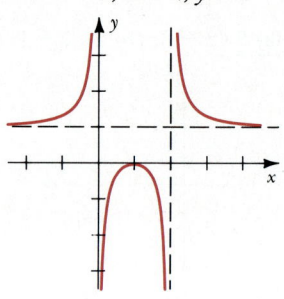

87. $y = 3$

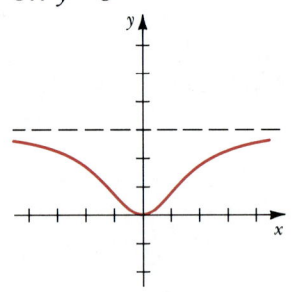

89. $x = 0, y = 0$

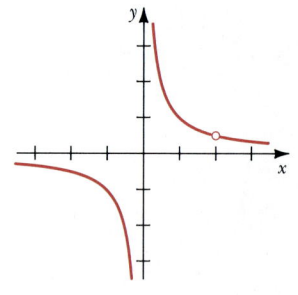

91. 1.2 **93.** Given $\epsilon > 0$, choose $\delta = \dfrac{\epsilon}{3}$.
95. (a) 0; **(b)** 2; **(c)** does not exist; **(d)** 2; **(e)** 0; **(f)** does not exist
97. (a) $x > 10$; **(b)** $x > \sqrt{1000}$
99. (a) ∞; **(b)** $-\infty$; **(c)** ∞; **(d)** 0
101. No. Consider, at $a = 0$,
$$f(x) = g(x) = \begin{cases} 1, & \text{if } x < 0, \\ -1, & \text{if } x \geq 0 \end{cases}$$

Chapter 2

Exercise Set 2.1

1. (a)

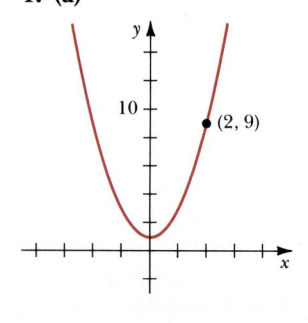

(b) 8; **(c)** $y = 8x - 7$;

3. (a)

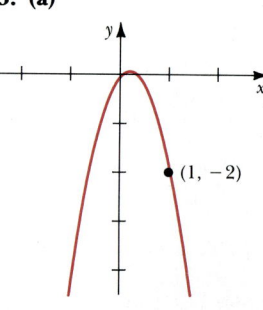

(b) -5; **(c)** $y = -5x + 3$;

(d) $y = -\dfrac{1}{8}x + \dfrac{37}{4}$ **(d)** $y = \dfrac{1}{5}x - \dfrac{11}{5}$ **13.** $-\dfrac{1}{16}$ **15.** $\dfrac{15}{4}$ **17.** 7 **19.** 0 **21.** $-\dfrac{1}{16}$

5. (a)   **7. (a)**   **23.** $f'(x) = 4$, $\mathbb{R}$ **25.** $f'(x) = 2x + 2$, $\mathbb{R}$

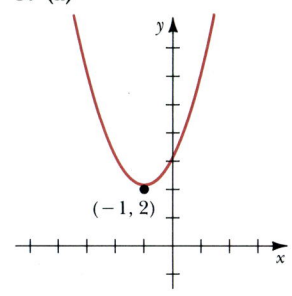

(b) 0; **(c)** $y = 2$; **(d)** $x = -1$

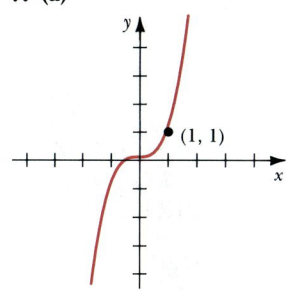

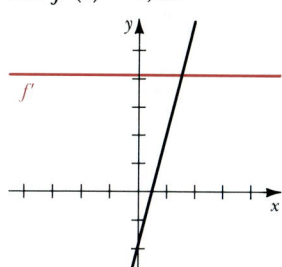

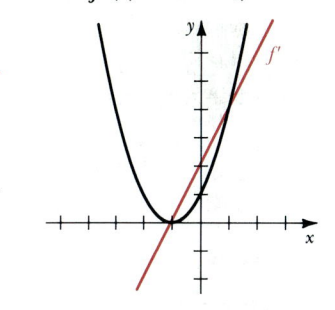

(b) 3; **(c)** $y = 3x - 2$; **(d)** $y = -\dfrac{1}{3}x + \dfrac{4}{3}$

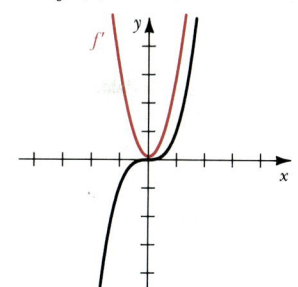

9. (a)

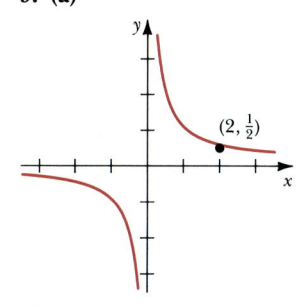

27. $f'(x) = 3x^2$, $\mathbb{R}$ **29.** $f'(x) = 3(x+1)^2$, $\mathbb{R}$

(b) $-\dfrac{1}{4}$; **(c)** $y = -\dfrac{1}{4}x + 1$; **(d)** $y = 4x - \dfrac{15}{2}$

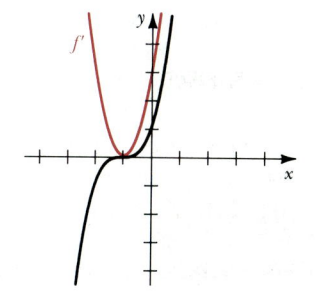

31. $f'(x) = -\dfrac{1}{x^2}$, $x \neq 0$ **33.** $f'(x) = \dfrac{1}{3}x^{-2/3}$, $x \neq 0$

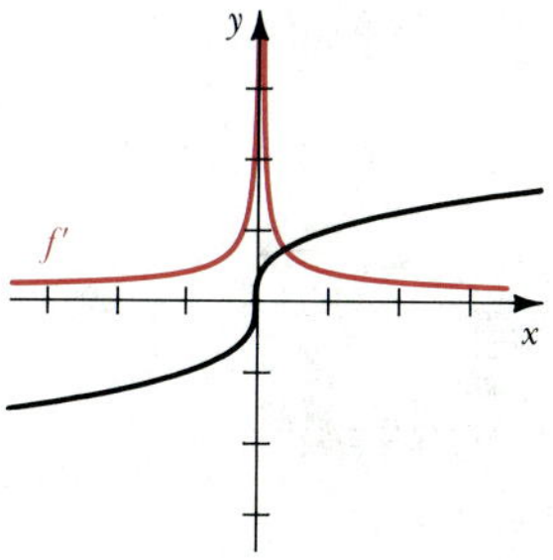

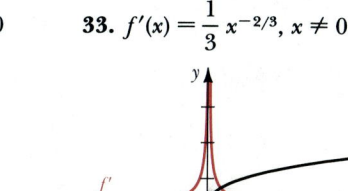

11. (a) $2x + 2$; **(b)** -1 **13. (a)** $4x - 3$; **(b)** $\dfrac{3}{4}$
15. (a) $3x^2 + 6x + 3$; **(b)** -1 **17. (a)** $x > -1$; **(b)** $x < -1$
19. (a) $x > \dfrac{3}{4}$; **(b)** $x < \dfrac{3}{4}$ **21. (a)** $x \neq -1$; **(b)** none
23. (a) $x = 0$; **(b)** $x = 1$; **(c)** $x = 0$; **(d)** $x = 0$
25. $\left(-\dfrac{1}{2}, \dfrac{5}{4}\right)$ **27.** $(-1, -1)$ **29.** $(-1, -1)$
31. $\left(\dfrac{1}{2}, 2\right), \left(-\dfrac{1}{2}, -2\right)$ **33.** $a = 0$ or $a = -2$
35. (a) $40 - 4t$; **(b)** $0 \leq t < 10$; **(c)** $10 < t \leq 20$; **(d)** 500 ft

35. 3 **37.** $2\pi r$ **39.** $P'(t) = kP(t)$
41. $f'_+(3) = f'_-(3) = 7$ **43.** $f'_+(0) = 1$, $f'_-(0) = -1$ **45.** $f'_+(0) = 0$, $f'_-(0)$ does not exist
47. $\mathbb{R}$
49. $(-\infty, 0]$ and $[0, \infty)$ **51.** $[0, \infty)$ **55.** Yes
61. y-intercept is $f(a) + a/f'(a)$; x-intercept is $a + f(a)f'(a)$.

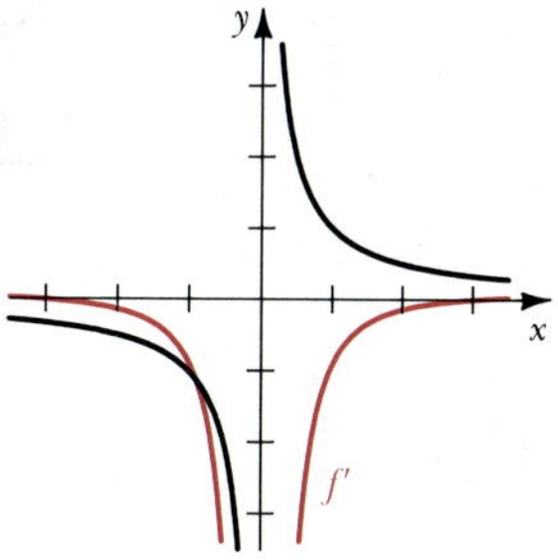

Exercise Set 2.2

1. 3   **3.** 7   **5.** 4   **7.** 0   **9.** 48   **11.** $-\dfrac{1}{16}$

Exercise Set 2.3

1. $2x - 2$ **3.** $x + 4$
5. $527x^{30} + 406x^{28} + 4x^3$ **7.** $2\pi t + 3 + 4\sqrt{2}$ **9.** $4x^3$
11. $(u^2 + u + 2)(4u^3 + 3u^2) + (u^4 + u^3)(2u + 1)$
13. $\dfrac{2}{3}x - \dfrac{6}{x^3}$

15. $2t(2t^2 - 3)(3t^4 - t^{-2}) + (t^4 - 3t^2 + 5)(12t^3 + 2t^{-3})$
17. $(z + 1)^{-2}$ 19. $\dfrac{-15x^2 + 3 - 40x}{(5x^2 + 1)^2}$ 21. $\dfrac{4t^2 + 4t + 22}{(t^2 + 3t - 4)^2}$
23. $\dfrac{2s^5 + 8s^3 + 10s}{(s^2 + 2)^2}$ 25. $\dfrac{x^4 + 4x^2 - 1}{(x^2 + 1)^2}$
27. $3x^2 + 12x + 11$
29. $4x^3 - 9x^2 - 18x - 5$
31. $\dfrac{1}{7}(21x^2 - 8x + 3)$
33. $-\dfrac{1}{4}x^{-5}$ 35. $18x + 30$
37. $24t^2 + 24t + 6$
39. $\left(-\dfrac{3}{2}, -\dfrac{25}{4}\right)$
41. $y = \dfrac{8x}{25} - \dfrac{1}{25}$
43. (a) 150 gal/min; (b) 100 gal/min
45. (a) $P(0) = 1000$; (b) $P'(t) = 9000/(t + 1)^2$
47. $0.014 \dfrac{\text{cal}}{(\text{deg})^2 \text{ mole}}$
49. (a) $\dfrac{3}{2}x^{1/2}$; (b) $\dfrac{5}{2}x^{3/2}$; (c) $-\dfrac{1}{2}x^{-3/2}$
51. $\dfrac{-n}{2}x^{-n/2-1}$

Exercise Set 2.4

1. $\cos x + \sec^2 x$ 3. $3x^2 + \sin x$ 5. $3t^2 \cos t - t^3 \sin t$
7. $\cos^2 x - \sin^2 x$ 9. $\tan t \cot t = 1$, so $r'(t) = 0$
11. $2 \sin x \cos x$ 13. $2 \csc x - 2x \csc x \cot x$
15. $2 \csc x - 2x \csc x \cot x$
17. $\tan x + x \sec^2 x + 2x \cot x - x^2 \csc^2 x$
19. $\dfrac{2\theta + 2\theta \cos \theta + \theta^2 \sin \theta}{(1 + \cos \theta)^2}$
21. $h(\theta) = \csc \theta + \cot \theta$ so $h'(\theta) = -\cot \theta \csc \theta - \csc^2 \theta$
23. $\dfrac{\tan t + \sin t \sec^2 t (\sin t + \cos t)}{(\sin t + \cos t)^2}$
25. $\cos x - \sin x$ 27. $\tan^2 x + 2x \tan x \sec^2 x$
29. $\dfrac{(-\csc t \cot t + \csc^2 t)(t + 2) - (\csc t - \cot t)}{(t + 2)^2}$
31. $\dfrac{1}{2}$ 33. 0   35. 1   37. 1   39. 1
43. $y = \dfrac{\sqrt{3}}{2} - \dfrac{\pi}{6} + \dfrac{1}{2}x$

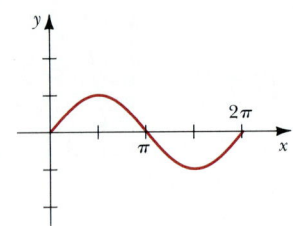

45. (a) $\dfrac{\pi}{2}, \dfrac{3\pi}{2}$; (b) $0 < x < \dfrac{\pi}{2}$ or $\dfrac{3\pi}{2} < x < 2\pi$;
(c)

49. $x = 10 \cot \theta, -20$ 51. (a) -1; (b) 1; (c) does not exist

Exercise Set 2.5

1. $4(x + 1)^3$ 3. $7(2x - 3)(x^2 - 3x + 4)^6$
5. $-12(6s + 4)^{-3}$ 7. $\pi \cos \pi x$ 9. $-2x \sin (x^2 + 1)$
11. $-4(6x - 1)(6x^2 - 2x)^{-3}$ 13. $\dfrac{24x}{5}(3x^2 + 1)^3$
15. $-42x(3x^2 + 13)^{-8}$ 17. $6(s - 1)^2/(s + 1)^4$
19. $(x - 1)(5 - x)/(x + 1)^4$ 21. $-3 \sin x (\cos x - 1)^2$
23. $-3 \sin x (\cos x)^2$ 25. $\dfrac{-2}{(x - 1)^2}\left[\sec\left(\dfrac{x + 1}{x - 1}\right)\right]^2$
27. $2s \cos (s^2 + 1) - 2s^3 \sin (s^2 + 1)$
29. $(1 - x^2)/(x^2 + 1)^2$
31. $6(\tan 2x \sec 2x)^2 \sin (1 - x^2) - 2x (\tan 2x)^3 \cos (1 - x^2)$
33. $6(\sin x)^2 \cos x [(\sin x)^3 + 1]$
35. $-6\left(\dfrac{\cos (2t - 1)}{\cot (t^2 + 1)}\right)^2 \cdot$
$\left[\dfrac{\sin (2t - 1) \cot (t^2 + 1) - t \cos (2t - 1) \csc^2 (t^2 + 1)}{\cot^2 (t^2 + 1)}\right]$
37. (a) $12(4x - 5)^2$; (b) $12(4x - 5)^2 + 2$; (c) $2[(4x - 5)^3 + 2x + 1][12(4x - 5)^2 + 2]$; (d) $4\{[(4x - 5)^3 + 2x + 1]^2 - 7x\}^3\{2[(4x - 5)^3 + 2x + 1][12(4x - 5)^2 + 2] - 7\}$
39. (a) $2x - 1$; (b) $(2x - 1) \cos (x^2 - x)$; (c) $2[x^3 + \sin (x^2 - x)][3x^2 + (2x - 1) \cos (x^2 - x)]$; (d) $-2[x^3 + \sin (x^2 - x)][3x^2 + (2x - 1) \cos (x^2 - x)] \sin [x^3 + \sin (x^2 - x)]^2$
41. (a) $f'_1(x) = \cos x$; (b) $f'_2(x) = -\cos (\cos x) \cdot \sin x$; (c) $f'_3(x) = -\cos (\cos (x^2 + \tan x)) \sin (x^2 + \tan x)(2x + \sec^2 x)$; (d) $f'_4(x) = -\cos (\cos (\sin^2 x + \tan (\sin x))) \sin (\sin^2 x + \tan (\sin x))(2 \sin x \cos x + \sec^2 (\sin x) \cos x)$
43. $y = \dfrac{1}{64} - \dfrac{256(x - 2)}{9}$ 45. $y = 0$ 47. $27\pi t^2$
49. (a) $8\pi r^2$ mm/sec; (b) 28800π mm^3/sec
51. 10π cm^3/hr   53. 1.19×10^{-5} cm^3/hr   55. $\dfrac{x + 3}{|x + 3|}$

57. $\dfrac{x^2-1}{|x^2-1|} \cdot 2x$, if $|x| \neq 1$ **59.** $g'(x) = \dfrac{1}{3(\sqrt[3]{x})^2}$, if $x \neq 0$

Exercise Set 2.6

1. $\dfrac{-x^2}{y^2}$ **3.** $\dfrac{-y}{x}$ **5.** $\sec y$ **7.** $\dfrac{y-2x}{2y-x}$ **9.** $\dfrac{4x-2xy^2}{2x^2y+3y^{-4}}$

11. $\dfrac{\cos x}{\sin y}$ **13.** $\dfrac{\cos x}{2y}$ **15.** $\dfrac{y \sec^2(xy)}{1 - x \sec^2(xy)}$ **17.** $\dfrac{1}{3}x^{-2/3}$

19. $\dfrac{12}{5}x^{-1/5}$ **21.** $\dfrac{2}{3}(x+1)^{-1/3} + 1$ **23.** $-\dfrac{\sin \sqrt{x}}{2\sqrt{x}}$

25. $\dfrac{1}{2}(x+1)(x^2+2x+2)^{-3/4}$

27. $\dfrac{2}{3}(6x+4)(3x^2+4x)^{-1/3}$

29. $\dfrac{x \sec \sqrt{x^2+1} \tan \sqrt{x^2+1}}{\sqrt{x^2+1}}$ **31.** $\dfrac{2x + \sec^2 x}{2\sqrt{x^2 + \tan x}}$

33. $-\dfrac{6}{5}\left(\dfrac{1}{2}x^{-1/2} + \dfrac{3}{4}x^{-1/4} + 1\right)(x^{1/2}+x^{3/4}+x)^{-1/5}$

35. $\dfrac{2}{3}x(x^2+2)^{-2/3}(x^2-2)^{1/4} + \dfrac{1}{2}x(x^2-2)^{-3/4}(x^2+2)^{1/3}$

37. $12x - \dfrac{3}{x^2} + \dfrac{8}{9}x(2x^2)^{-4/3}$

39. $\dfrac{1}{2}[x^2 + \sqrt{x^2 + \sqrt{x+1}}]^{-1/2}\left[2x + \dfrac{1}{2}(x^2 + \sqrt{x+1})^{-1/2}\left(2x + \dfrac{1}{2}(x+1)^{-1/2}\right)\right]$

41. $\dfrac{x}{2}\left(\dfrac{1+\sin x^2}{\cos(1+x^2)}\right)^{-3/4} \cdot \left[\dfrac{\cos x^2 \cos(1+x^2) + (1+\sin x^2)\sin(1+x^2)}{\cos^2(1+x^2)}\right]$

43. $-\dfrac{1}{2}\sqrt[3]{\dfrac{y}{x^2}}$ **45.** $(2y(xy+1)^2 + x^2)^{-1}$

47. $\dfrac{y \cos(x/y) - y^2}{x \cos(x/y)}$ **49.** $y = x + 2$ **51.** -2

53. $y = 2 - x$ **57.** $-x_0/y_0$
61. The axis intercepts $(1, 0)$, $(-1, 0)$, $(0, 1)$, and $(0, -1)$ and all points (x, y) with $x = \pm 2^{-1/4}$, $y = \pm 2^{-1/4}$

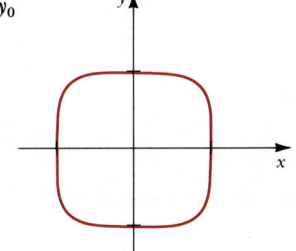

63. $y = \dfrac{5}{4}x - 3$

65. (a) $\dfrac{x}{|x|}$; (b) $\dfrac{2x(x^2-1)}{|x^2-1|}$;

(c) $\dfrac{x+3}{2|x+3|\sqrt{|x+3|}}$;

(d) $\dfrac{\cos x \sin x}{|\sin x|}$

67. $\dfrac{RV^3}{PV^3 - aV + 2ab}$

Exercise Set 2.7

1. 2 **3.** $1, 3$ **5.** $2, -4$ **7.** $\dfrac{1}{3}, -1$ **9.** $\pm 1, \pm 2$

11. None   **13.** None   **15.** $2, -2$ **17.** None
19. None   **21.** None   **23.** -2 **25.** $1, 2$

27. $2n\pi$ for any integer n

29. $\dfrac{3\pi}{4} + n\pi$ for any integer n

31. $n\pi, \pm \dfrac{\pi}{3} + 2n\pi$ for any integer n

33. (a) 3; (b) absolute minimum at 3

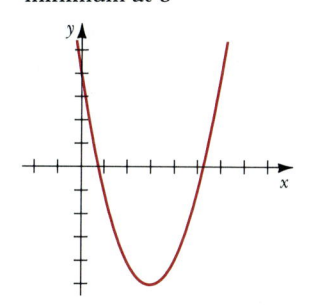

35. (a) 3; (b) absolute maximum at 3

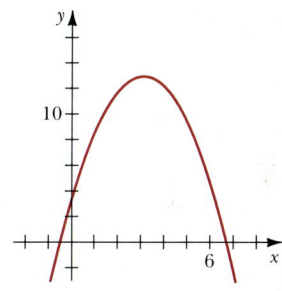

37. (a) $1, -2$; (b) absolute minimum at -2

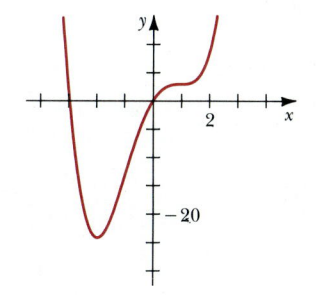

39. (a) $n\pi$; (b) absolute maximum at $2n\pi$; absolute minimum at $(2n+1)\pi$

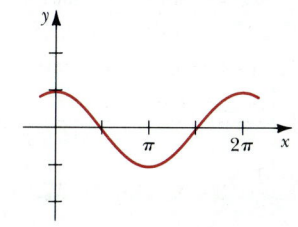

41. (a) $1, -1$; (b) absolute minimum at 1; absolute maximum at -1

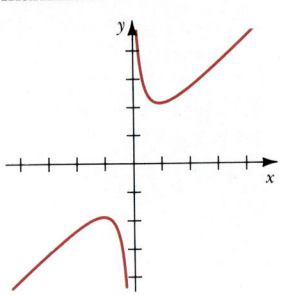

43. (a) $1, -1$; (b) absolute minimum at 1 and at -1

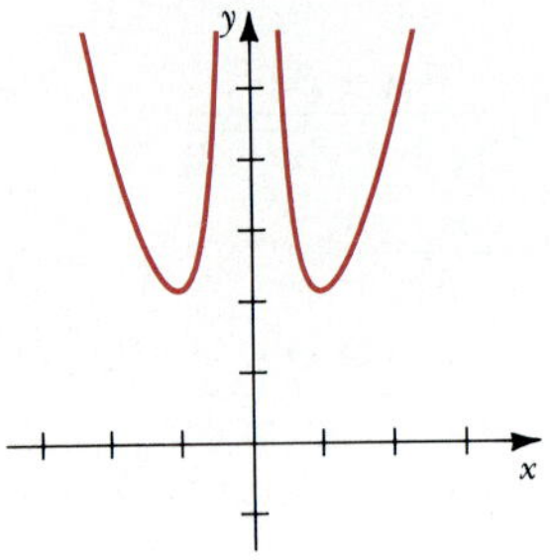

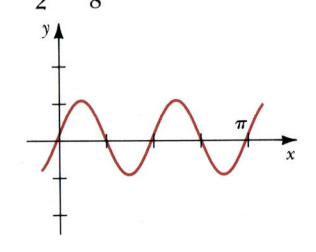

51. (a) $0, 2, -2$; (b) absolute maximum at 0, absolute minimum at -2 and 2

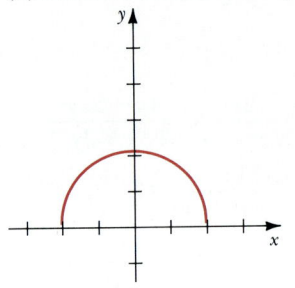

45. (a) $-4 \pm 2\sqrt{3}$; (b) relative minimum at $-4 - 2\sqrt{3}$, relative maximum at $-4 + 2\sqrt{3}$; no absolute extrema

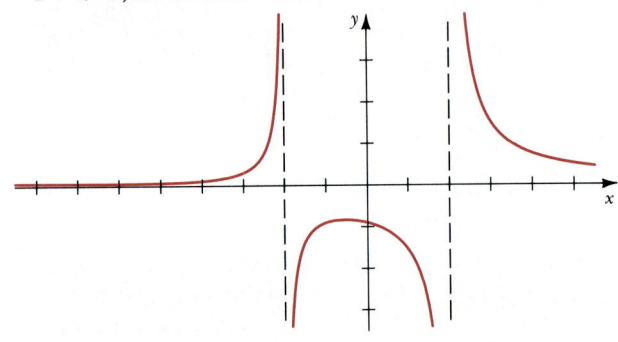

53. (a) $0, 1, \dfrac{2}{3}$; (b) absolute minimum at 0 and 1

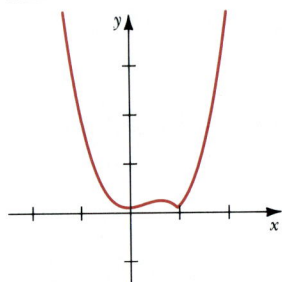

55. Absolute minimum at 2; absolute maximum at 0 and 4

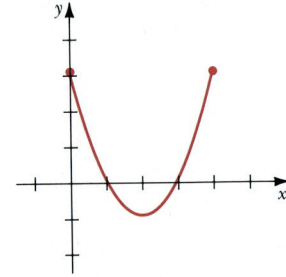

47. (a) $\dfrac{3\pi}{2} + 2n\pi$; (b) no relative or absolute extrema

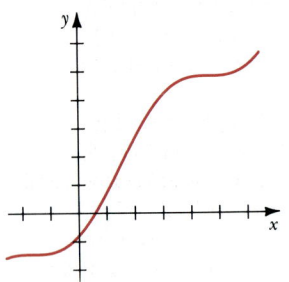

57. Absolute minimum at -2; absolute maximum at -4

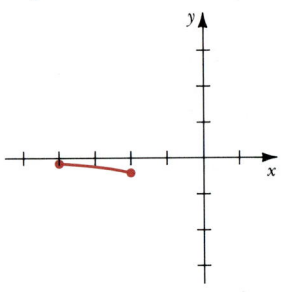

59. Absolute minimum at 5; absolute maximum at 4

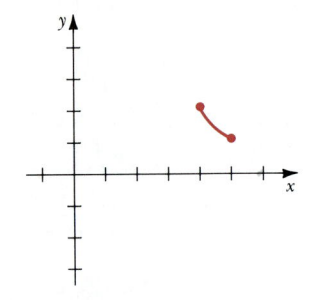

49. (a) $\pm \left(\dfrac{2n+1}{8}\right)\pi$; (b) absolute minimum at $\dfrac{n\pi}{2} - \dfrac{\pi}{8}$; absolute maximum at $\dfrac{n\pi}{2} + \dfrac{\pi}{8}$

61. Absolute minimum at 2; absolute maximum at 0.

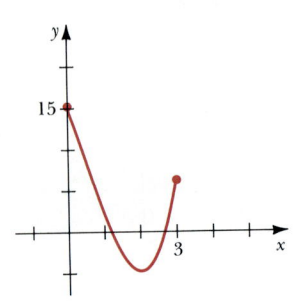

63. Absolute maximum at $\pi/2$; absolute minimum at $-\pi/2$

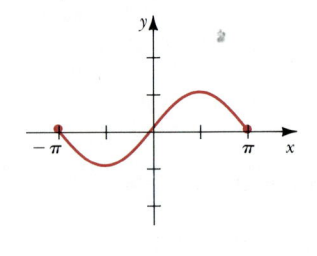

65. Absolute maximum at 0; absolute minimum at ± 1

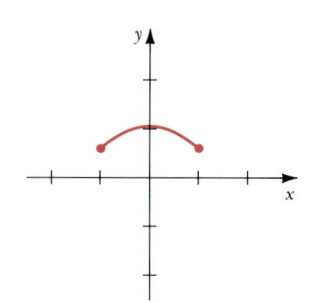

67. Absolute minimum at -2, 0, and 2; absolute maximum at $\dfrac{2\sqrt{3}}{3}$ and $-\dfrac{2\sqrt{3}}{3}$

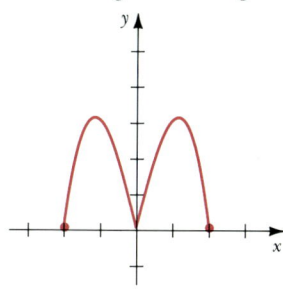

69. Absolute maximum at -1; absolute minimum at 5

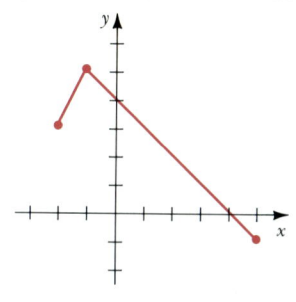

71. 2, when the number is one
73. Absolute maximum value is $C(15) = 105$; absolute minimum value is $C(5) = 5$

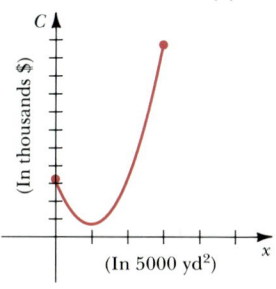

75. $17.93°C$ **77.** $\dfrac{S_1}{2}$ **79.** 1.53×10^{-3} moles per liter

83. Some examples are **(a)** $f(x) = \dfrac{1}{x}$; **(b)** $f(x) = -\dfrac{1}{x}$;

(c) $f(x) = x$; **(d)** $f(x) = \dfrac{1}{x}\sin\dfrac{1}{x}$

Exercise Set 2.8

1. $\dfrac{3}{2}$ **3.** $\dfrac{2 \pm \sqrt{7}}{3}$

5. f is not continuous on $[-1, 1]$. Hypotheses not satisfied.

7. $\dfrac{\pi}{2}$ **9.** $\dfrac{3\pi}{4}$ **11.** $\dfrac{1}{2}$ **13.** $\sqrt{2}$ **15.** $(0.8)^5 \approx 0.328$

17. 0 **19.** $1 - \sqrt{2}$ **21.** $\cos c = \dfrac{2}{\pi}$, $c \approx 0.88$

23. $f(x) = 3$ **25.** $f(x) = x^2 + 5$ **27.** $f(x) = 3 - \cos x$
29. $f(x) = x^2 - x - 2$ **31.** $f(x) = x^3 - 5x + 2$

33. $f(x) = -\cos 3x$ **35.** $f(x) = 1 - \dfrac{1}{x}$

37. $f(x) = x^2 + 3x + 2$ **39.** $\dfrac{x^3}{3} + \dfrac{3x^2}{2} - 2x + 2$

41. $f(x) = x - 1 - \sin x + \cos x$

43. (a) $f'(0)$ is not defined; **(b)** f is not continuous at $\dfrac{\pi}{2}$;

(c) f is not continuous at 1; **(d)** f is not continuous at $\dfrac{\pi}{2}$

47. One example is $f(x) = 1/x$.

Exercise Set 2.9

1. (a) Decreasing on $(-\infty, -2]$ and $[1, 3]$; increasing on $[-2, 1]$ and $[3, \infty)$; **(b)** relative maximum at 1; relative minimum at -2; $f'(1) = f'(-2) = 0$; **(c)** absolute minimum at 3, $f(3) = -3$; $f'(3) = 0$
3. (a) Constant on $(-\infty, 0)$; increasing on $[0, 2]$, decreasing on $[2, \infty)$; **(b)** relative maximum at 2; $f'(2) = 0$; **(c)** absolute maximum at each number in $(-\infty, 0]$; $f'(0)$ does not exist, $f'(x) = 0$ for all x in $(-\infty, 0)$; absolute maximum at 2.
5. (a) Decreasing on $[-3, -2]$, $[0, 1]$, and $[2, 3]$; increasing on $(-\infty, -3]$, $[-1, 0]$, $[1, 2]$, and $[3, \infty)$; **(b)** relative maximum at -3, at 0, and at 2; relative minimum at 1, at 3, and at each number in $[-2, -1]$; $f'(-3) = f'(0) = f'(2) = 0$; $f'(x)$ does not exist when $x = \pm 1$, $x = -2$, or $x = 3$; **(c)** no absolute maximum or minimum
7. (a) Increasing on $(-\infty, \infty)$; **(b)** no extrema

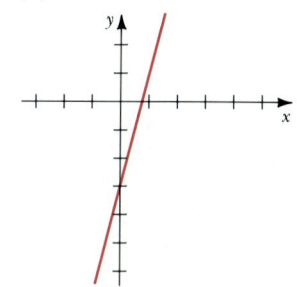

9. (a) Increasing on $[2, \infty)$; decreasing on $(-\infty, 2]$; **(b)** relative minimum at 2

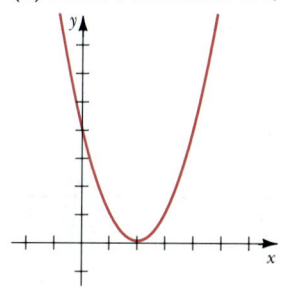

11. (a) Increasing on $[\frac{9}{4}, \infty)$; decreasing on $(-\infty, \frac{9}{4}]$; **(b)** relative minimum at $\frac{9}{4}$

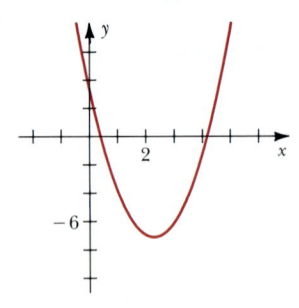

13. (a) Increasing on $(-\infty, -1]$ and $[1, \infty)$; decreasing on $[-1, 1]$; **(b)** relative maximum at -1; relative minimum at 1

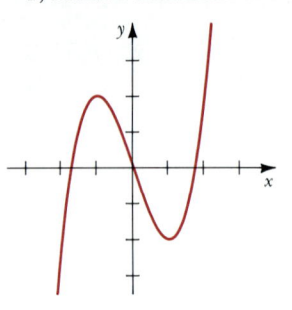

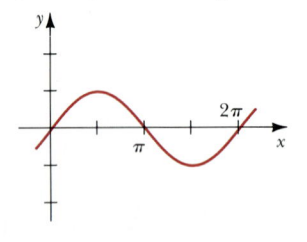

23. (a) Increasing whenever defined; **(b)** no extrema

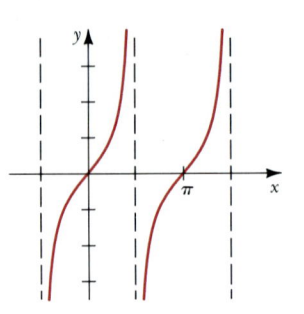

25. (a) Increasing on $[-1, 0)$ and $[1, \infty)$; decreasing on $(-\infty, -1]$ and $(0, 1]$; **(b)** relative minimum at -1 and at 1

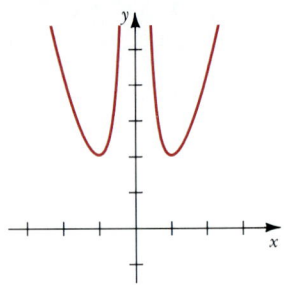

15. (a) Increasing on $[-1, 0]$ and $[1, \infty)$; decreasing on $(-\infty, -1]$ and $[0, 1]$; **(b)** relative minimum at -1 and 1; relative maximum at 0

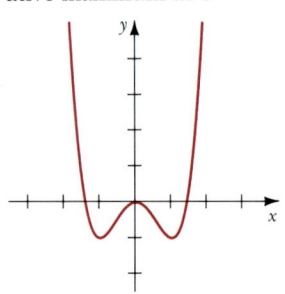

17. (a) Increasing on $(-\infty, -\frac{1}{2}]$ and $[\frac{1}{2}, \infty)$; decreasing on $[-\frac{1}{2}, 0)$ and $(0, \frac{1}{2}]$; **(b)** relative maximum at $-\frac{1}{2}$; relative minimum at $\frac{1}{2}$

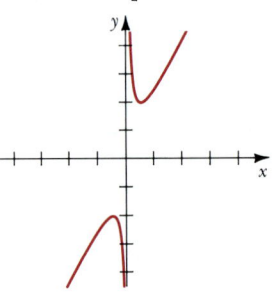

27. (a) Decreasing on $(-\infty, 1)$ and $(1, \infty)$; **(b)** no extrema

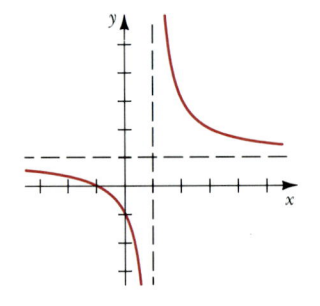

29. (a) Increasing on $(-\infty, -1)$ and $(-1, 0]$; decreasing on $[0, 1)$ and $(1, \infty)$; **(b)** relative maximum at 0

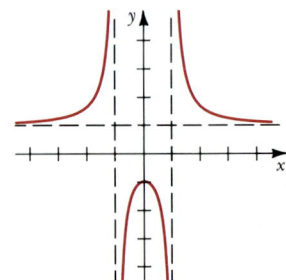

19. (a) Increasing on $(-\infty, \frac{3}{4}]$; decreasing on $[\frac{3}{4}, \infty)$; **(b)** relative maximum at $\frac{3}{4}$

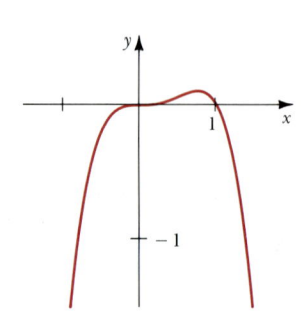

21. (a) Increasing on $[\pi/2 + (2n-1)\pi, \pi/2 + 2n\pi]$; decreasing on $[\pi/2 + 2n\pi, \pi/2 + (2n+1)\pi]$; **(b)** relative maximum at $\frac{\pi}{2} + 2n\pi$; relative minimum at $\frac{\pi}{2} + (2n+1)\pi$, n an integer.

31. (a) Increasing on $\left[-\frac{3\sqrt{2}}{2}, \frac{3\sqrt{2}}{2}\right]$; decreasing on $\left[-3, -\frac{3\sqrt{2}}{2}\right]$ and $\left[\frac{3\sqrt{2}}{2}, 3\right]$; **(b)** relative minimum at $-\frac{3\sqrt{2}}{2}$; relative maximum at $\frac{3\sqrt{2}}{2}$

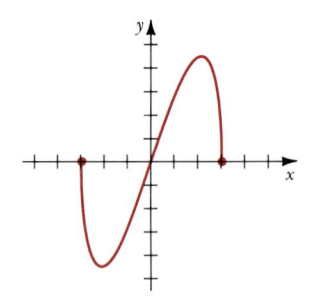

33. (a) Decreasing on $(-\infty, -1]$ and $[\frac{1}{2}, 2]$; increasing on $[-1, \frac{1}{2}]$ and $[2, \infty)$; **(b)** relative minimum at $-1, 2$;

relative maximum at $\frac{1}{2}$

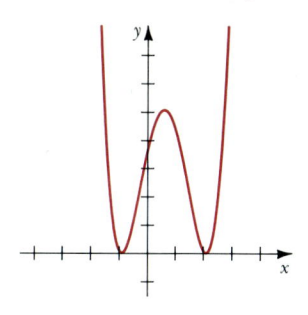

35. **(a)** Increasing on $(-\infty, 1]$, $\left[\dfrac{17-2\sqrt{7}}{9}, \dfrac{17+2\sqrt{7}}{9}\right]$, and $[3, \infty)$; decreasing on $\left[1, \dfrac{17-2\sqrt{7}}{9}\right]$ and $\left[\dfrac{17+2\sqrt{7}}{9}, 3\right]$; **(b)** relative maximum at 1 and at $\dfrac{17+2\sqrt{7}}{9}$; relative minimum at $\dfrac{17-2\sqrt{7}}{9}$ and at 3

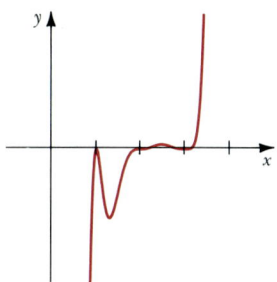

37. **(a)** Increasing on $[-\sqrt{2}, 0]$ and $[\sqrt{2}, \infty)$; decreasing on $(-\infty, -\sqrt{2}]$ and $[0, \sqrt{2}]$; **(b)** relative minimum at $-\sqrt{2}$ and at $\sqrt{2}$; relative maximum at 0

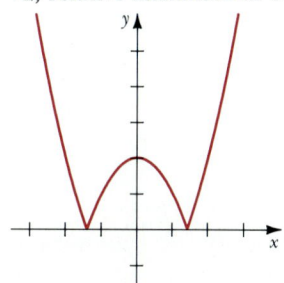

39. **(a)** Decreasing on $[0, 1]$; increasing on $(1, \infty)$; **(b)** relative minimum at 1

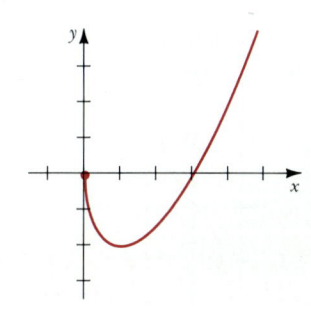

41. **(a)** Increasing on $[0, \frac{49}{100}]$; decreasing on $[\frac{49}{100}, \infty)$; **(b)** relative maximum at $\frac{49}{100}$

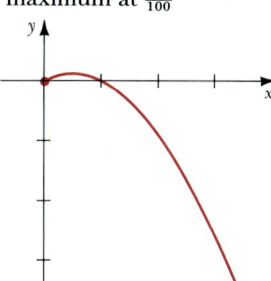

43. **(a)** Decreasing on $(-\infty, -\sqrt{2}]$; increasing on $[\sqrt{2}, \infty)$; **(b)**

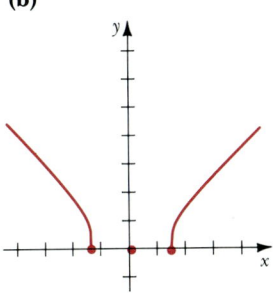

45. **(a)** Decreasing on $\left[\dfrac{\pi}{3} + 2n\pi, \dfrac{5\pi}{3} + 2n\pi\right]$; increasing on $\left[-\dfrac{\pi}{3} + 2n\pi, \dfrac{\pi}{3} + 2n\pi\right]$; **(b)** relative maximum at $\dfrac{\pi}{3} + 2n\pi$; relative minimum at $-\dfrac{\pi}{3} + 2n\pi$, n an integer

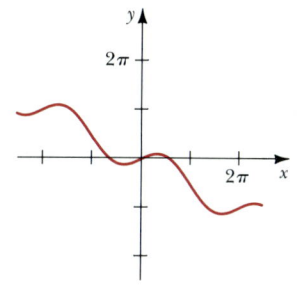

47. **(a)** Increasing on $\left[\dfrac{-\pi}{4} + 2n\pi, \dfrac{3\pi}{4} + 2n\pi\right]$; decreasing on $\left[\dfrac{3\pi}{4} + 2n\pi, \dfrac{7\pi}{4} + 2n\pi\right]$; **(b)** relative maximum at $\dfrac{3\pi}{4} + 2n\pi$; relative minimum at $\dfrac{7\pi}{4} + 2n\pi$, n an integer

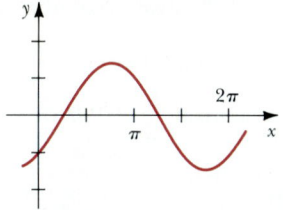

49. Increasing when $500 \le x < 3750$; decreasing when $3750 < x \le 4000$; maximum cost is \$28,125 when $x = 3750$; minimum cost is \$7000 when $x = 500$

51. **(a)** $0 \le t \le \dfrac{11}{4}$; **(b)** 124 ft   **53.** No

57. Not possible   **59.** $a > 0$, $b = 0$, $c = -3a$

Exercise Set 2.10

1. $6x + 4$, 6 **3.** $4x^3 - 4x^{-5}$, $12x^2 + 20x^{-6}$

5. $\dfrac{3}{2} x^{-1/2} + 4x$, $\dfrac{-3}{4} x^{-3/2} + 4$

7. $2\cos 2x$, $-4\sin 2x$

9. $2x \cos(x^2 + 1)$, $2\cos(x^2 + 1) - 4x^2 \sin(x^2 + 1)$

11. $\dfrac{1}{2} x^{-1/2} + \dfrac{3}{2} x^{1/2}$, $\dfrac{-1}{4} x^{-3/2} + \dfrac{3}{4} x^{-1/2}$

13. $2(\sec 2x)^2$, $8(\sec 2x)^2 \tan 2x$

15. $\dfrac{3}{8} x^{-5/2} - \dfrac{3}{8} x^{-3/2}$, $\dfrac{-15}{16} x^{-7/2} + \dfrac{9}{16} x^{-5/2}$

17. $32(\sec 2x)^2(\tan 2x)^2 + 16(\sec 2x)^4 = 48(\sec 2x)^4 - 32(\sec 2x)^2$, $128 \tan 2x (\sec 2x)^2(3(\sec 2x)^2 - 1)$

19. $\dfrac{-4}{y^3}$ **21.** 0

23. $\sec^2 y \tan y$

25. $\dfrac{2(x + y + 1)}{(1 - x)^2}$

27. (a) Relative minimum at 2; (b)

29. (a) No relative extrema; (b)

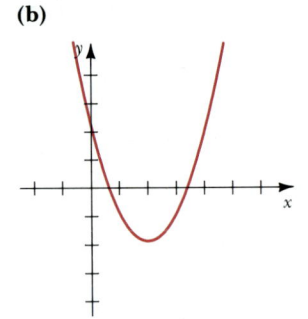

31. (a) Relative maximum at -4; relative minimum at 0; (b)

33. (a) Relative maximum at $\dfrac{\pi}{2} + 2n\pi$; relative minimum at $\dfrac{3\pi}{2} + 2n\pi$, n an integer; (b)

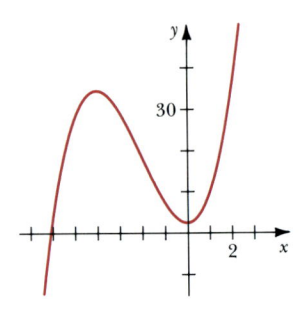

35. (a) Relative maximum at -1; relative minimum at $x = 1$; (b)

37. (a) Relative minimum at 1; (b)

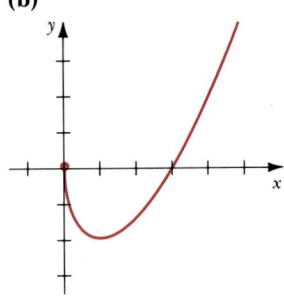

39. (a) Concave upward on $(-\infty, -1)$ and $(1, \infty)$; concave downward on $(-1, 1)$; (b) points of inflection $(-1, -5)$ and $(1, -5)$; (c)

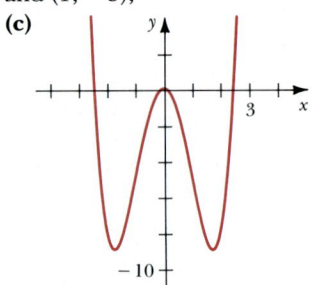

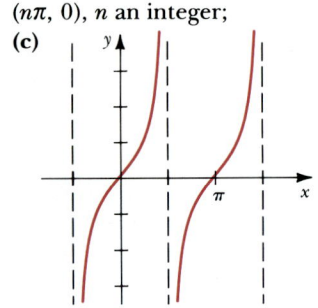

41. (a) Concave upward on $(-\infty, 1)$; concave downward on $(1, \infty)$; (b) none; (c)

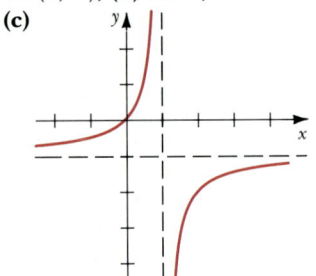

43. (a) Concave upward on $\left(n\pi, (2n + 1)\dfrac{\pi}{2}\right)$; concave downward on $\left((2n - 1)\dfrac{\pi}{2}, n\pi\right)$; (b) points of inflection $(n\pi, 0)$, n an integer; (c)

45. (a) Concave downward on $(-\infty, 1)$; concave upward on $(1, \infty)$; (b) none; (c)

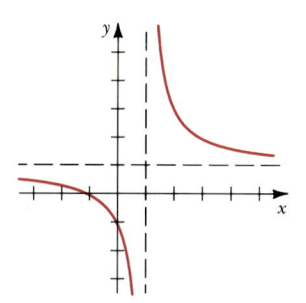

47. Concave upward on $((2n-1)\pi, 2n\pi)$; concave downward on $(2n\pi, (2n+1)\pi)$; (b) points of inflection $(n\pi, n\pi)$, n an integer; (c)

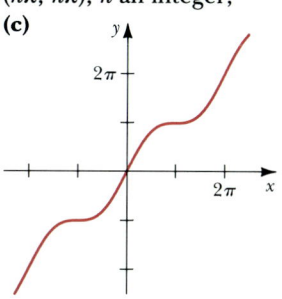

63. $f^{(2n)}(x) = (-1)^n \sin x$, $f^{(2n+1)}(x) = (-1)^n \cos x$
65. $h'' = (f' \circ g) g'' + (f'' \circ g) (g')^2$
67. $P(0) = a_0$, $P'(0) = a_1$, and $P''(0) = 2a_2$
69. $a_k = \dfrac{P^{(k)}(0)}{k!}$
75. Any polynomial of the form $f(x) = ax^3 + cx$, where $a < 0$ and $c = -3a$ **77.** $f'(0)$ does not exist

Exercise Set 2.11

1.

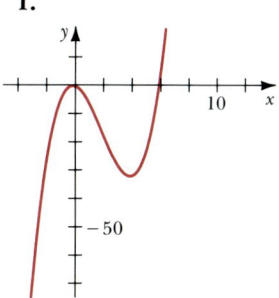

3. y-axis symmetry

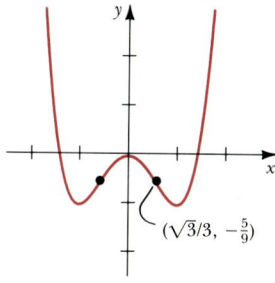

49. (a) Concave downward on $(-\infty, -\sqrt{6})$ and $(2, \sqrt{6})$; concave upward on $(-\sqrt{6}, -2)$ and $(\sqrt{6}, \infty)$; (b) points of inflection $(-\sqrt{6}, -2\sqrt{3})$ and $(\sqrt{6}, 2\sqrt{3})$; (c)

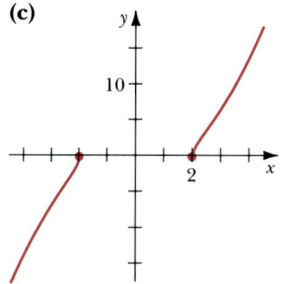

51. (a) Concave upward on $(-\infty, -\sqrt{3})$ and $(\sqrt{3}, \infty)$; concave downward on $(-\sqrt{3}, \sqrt{3})$; (b) points of inflection $(\sqrt{3}, 36)$ and $(-\sqrt{3}, 36)$; (c)

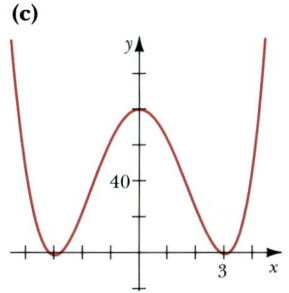

5.

7. y-axis symmetry

53. (a) Concave upward on $(-\infty, \frac{1}{2}(1-\sqrt{3}))$ and $(\frac{1}{2}(1+\sqrt{3}), \infty)$; concave downward on $(\frac{1}{2}(1-\sqrt{3}), \frac{1}{2}(1+\sqrt{3}))$; (b) points of inflection $(\frac{1}{2}(1-\sqrt{3}), \frac{9}{4})$ and $(\frac{1}{2}(1+\sqrt{3}), \frac{9}{4})$; (c)

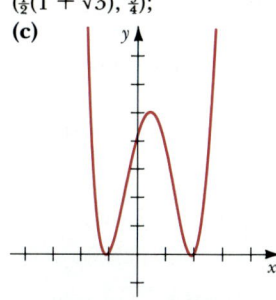

55. (a) Concave downward on $(2n\pi, (2n+1)\pi)$; concave upward on $((2n-1)\pi, 2n\pi)$; (b) points of inflection $(2n\pi, 0)$, $((2n+1)\pi, 0)$, n an integer; (c)

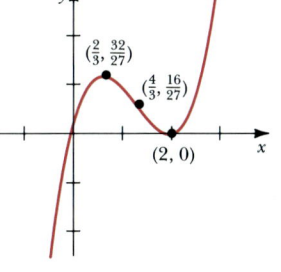

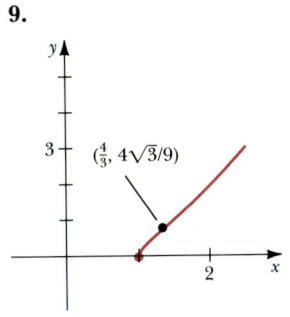

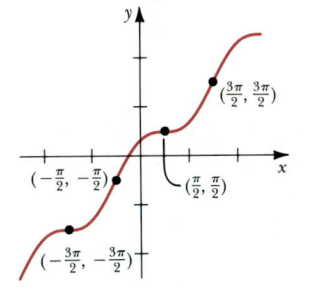

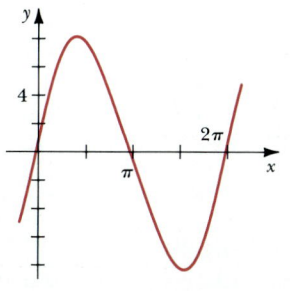

9.

11.

13.

57. $f^{(n)}(x) = (-1)^n n! x^{-n-1}$ **59.** $(-1)^n n! (x+1)^{-(n+1)}$
61. $f^{(n)}(x) = (-1)^n \dfrac{1 \cdot 3 \cdot 5 \cdots (2n-1)}{2^n} x^{-n-(1/2)}$

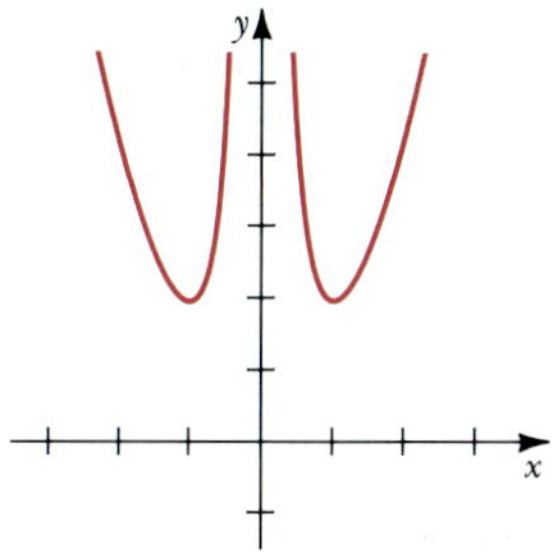

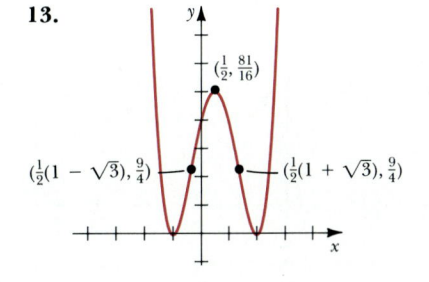

S22 ANSWERS

15. [graph with points $(-\frac{2}{5}, \frac{3}{5}(\frac{4}{25})^{1/3})$ and $(\frac{1}{5}, \frac{6}{5}(\frac{1}{25})^{1/3})$]

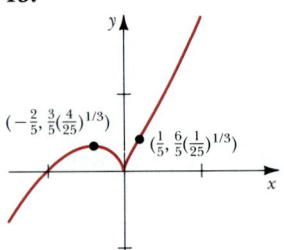

17. [graph]

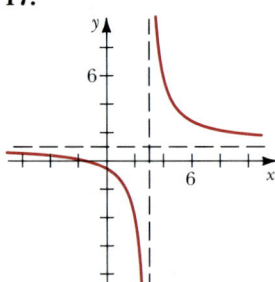

19. y-axis symmetry [graph]

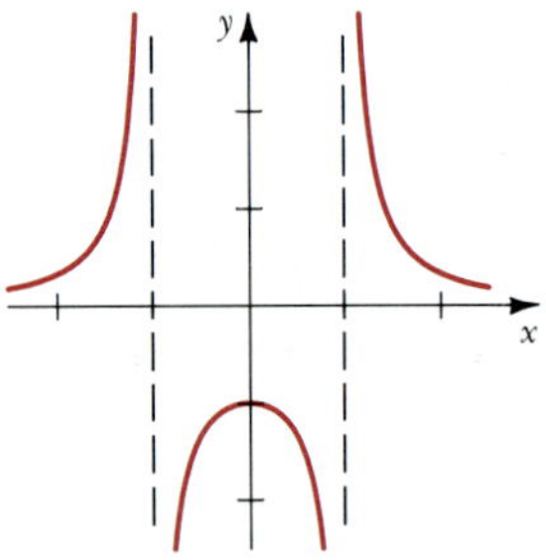

21. [graph]

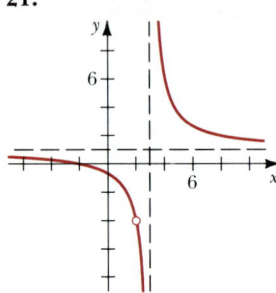

23. [graph with points $(\frac{9}{2}, -\frac{143}{27})$ and $(3, -\frac{48}{9})$]

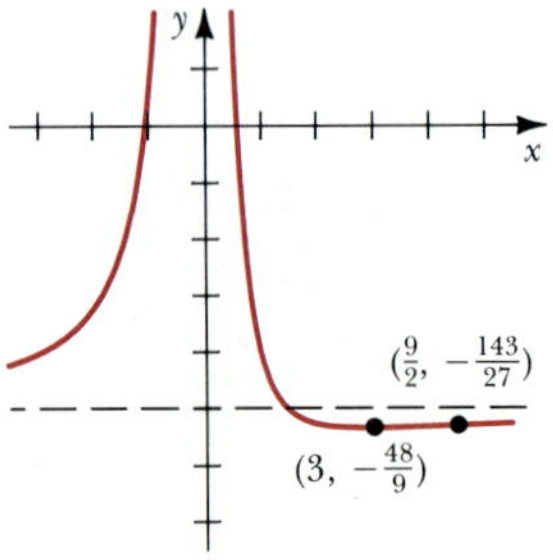

25. Origin symmetry [graph with points $(\sqrt{3}, 2\sqrt{3}/9)$ and $(\sqrt{6}, 5\sqrt{6}/36)$]

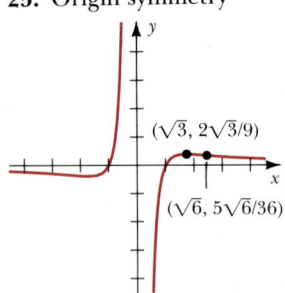

27. Origin symmetry [graph]

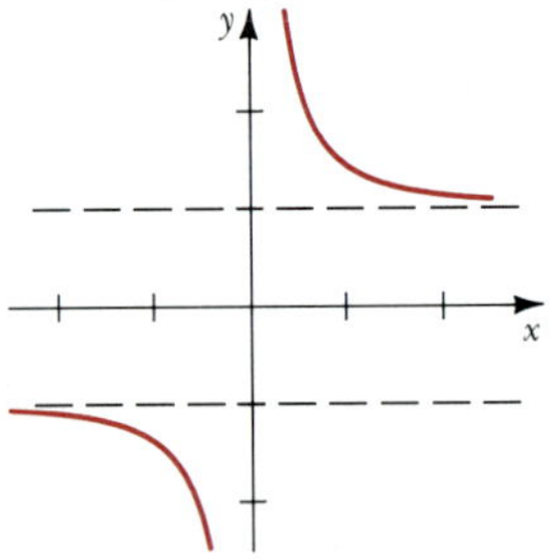

29. y-axis symmetry [graph]

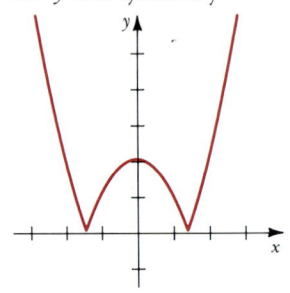

31. y-axis symmetry [graph with points $(\frac{\pi}{4}, \frac{5}{2})$ and $(\frac{3\pi}{4}, \frac{5}{2})$]

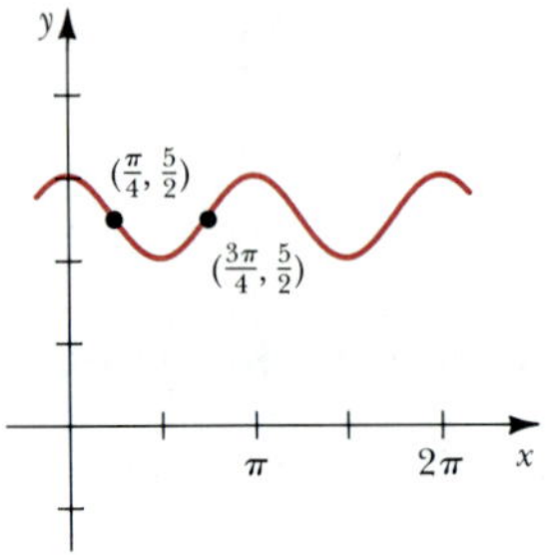

33. [graph]

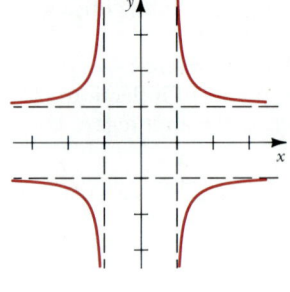

35. [graph]

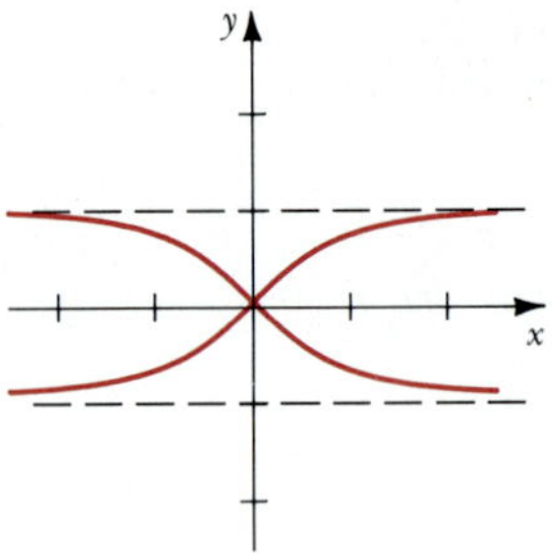

37. Origin symmetry [graph]

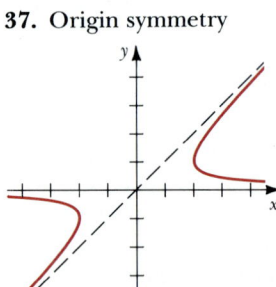

39. Origin symmetry [graph]

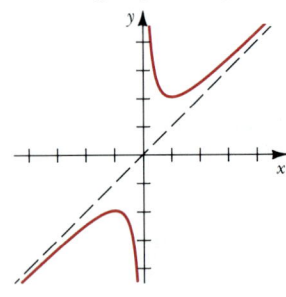

Review Exercises

1. $12x^3 - 6x^2 + 1$ **3.** $\dfrac{3}{2} x^{-1/2} - \dfrac{1}{6} x^{-3/2}$ **5.** $16 + \dfrac{32}{t^2}$

7. $(3x^2 - 7) \sin x + (x^3 - 7x + 5) \cos x$ **9.** $4(w + 1)^3$

11. $2x(x^2 - 7)^2(5x^4 - 14x^2 + 3)$ **13.** $\dfrac{-2(2 + x)}{(x^2 - 4)^{3/2}}$

15. $5[(x^2 + 1)^3 - 7x]^4[6x(x^2 + 1)^2 - 7]$

17. $4[(2s + 1)^3 + s^{1/3}]^3 [6(2s + 1)^2 + \tfrac{1}{3} s^{-2/3}]$

19. $3u^2 + \dfrac{2}{u^3} - \dfrac{1}{u^2}$ **21.** $\dfrac{5(x^3 - 8)^4(x^4 + 12x^2 + 16x)}{(x^2 + 4)^6}$

23. $\dfrac{4(x^3 + 1)^3(2x^3 - 1)}{x^5}$ **25.** $2 \cos (2x + 3)$

27. $4 \sin (2x + 1) \cos (2x + 1)$

29. $6x(x^2 + 1)^2 \sec^2 (x^2 + 1)^3$

31. $t \tan (t^2 + 1) \sqrt{\sec (t^2 + 1)}$

33. $2x(2x - 1)(x + 3) + 2(x^2 + 1)(x + 3) + (x^2 + 1)(2x - 1)$

35. $\sec (x + 1) [\tan x + x \sec^2 x + x \tan x \tan (x + 1)]$

37. $36x^2 - 12x, 72x - 12$

39. $\dfrac{-3}{4} x^{-3/2} + \dfrac{1}{4} x^{-5/2}; \dfrac{9}{8} x^{-5/2} - \dfrac{5}{8} x^{-7/2}$

41. $-64t^{-3}, 192t^{-4}$

43. $(13x - x^3 - 5) \sin x + (6x^2 - 14) \cos x;$
$(27 - 9x^2) \sin x + (25x - x^3 - 5) \cos x$

45. $12(w + 1)^2; 24(w + 1)$ **47.** $-\dfrac{y(1 + 2x)}{x(x + 1)}$

49. $\dfrac{y^2 - 2y - 1}{2x(1 - y)}$ **51.** $\dfrac{y - 4x(x^2 + y^2)}{4y(x^2 + y^2) - x}$ **53.** $\dfrac{-y}{2x}$

55. $\dfrac{-y(\cos x + 1)}{\sin x + x}$ **57.** $\dfrac{1 - y(\sec^2 x + \sec^2 xy)}{x \sec^2 xy + \tan x}$

59. (a) $\dfrac{y - x^2}{y^2 - x}$; (b) $-\dfrac{2xy}{(y^2 - x)^3}$

61. (a) $\dfrac{-y}{x + \cos y}$; (b) $\dfrac{y(2x + 2\cos y + y \sin y)}{(x + \cos y)^3}$

63. (a) $f'(x) = 4x^3 - 14x$, $f''(x) = 12x^2 - 14$, $f'''(x) = 24x$, $f^{(4)}(x) = 24$, $f^{(5)}(x) = 0$; (b) $f'(x) = \frac{1}{2}(x + 1)^{-1/2}$, $f''(x) = -\frac{1}{4}(x + 1)^{-3/2}$, $f'''(x) = \frac{3}{8}(x + 1)^{-5/2}$, $f^{(4)}(x) = -\frac{15}{16}(x + 1)^{-7/2}$, $f^{(5)}(x) = \frac{105}{32}(x + 1)^{-9/2}$; (c) $f'(x) = \cos x$, $f''(x) = -\sin x$, $f'''(x) = -\cos x$, $f^{(4)}(x) = \sin x$, $f^{(5)}(x) = \cos x$; (d) $f'(x) = \sin x + x \cos x$, $f''(x) = 2 \cos x - x \sin x$, $f'''(x) = -x \cos x - 3 \sin x$, $f^{(4)}(x) = x \sin x - 4 \cos x$, $f^{(5)}(x) = 5 \sin x + x \cos x$

65.

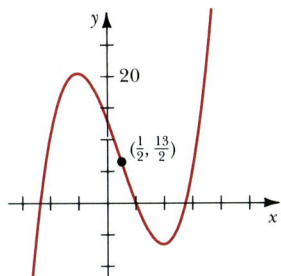

67.

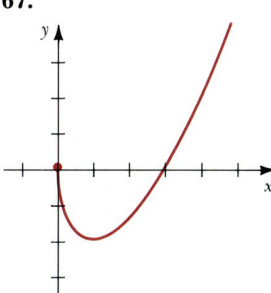

69.

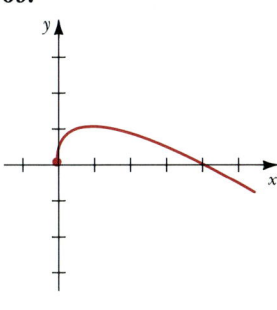

71. y-axis symmetry

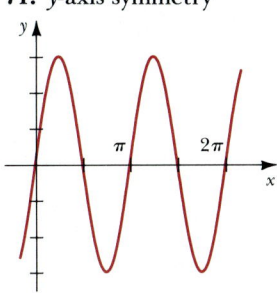

73.

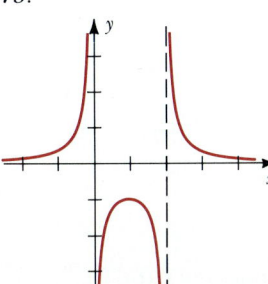

75.

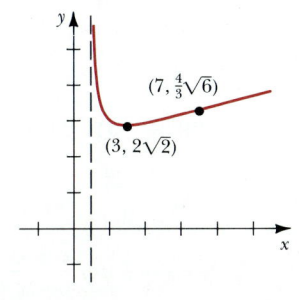

83. $y = 2x$ **85.** $y = -x/2$ **87.** $y = 0$
89. $y = |x - 1|$ **91.** 2, 1

95.

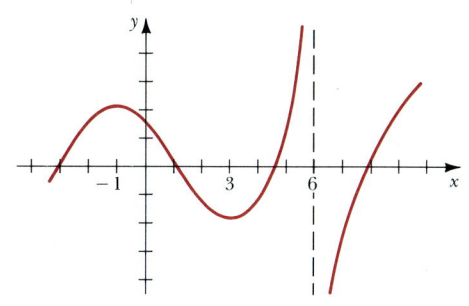

97. $x^3 + 1$

Chapter 3

Exercise Set 3.1

1. 256, 64 **3.** 1, -1 **5.** $\dfrac{2}{3}, -\dfrac{2}{9}$ **7.** $\dfrac{-1}{\pi + 1}, \dfrac{2}{(\pi + 1)^2}$

9. , stopped when $t = 1$.

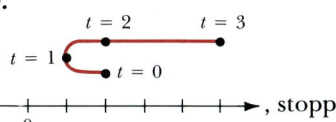

11. , stopped when $t = 2$.

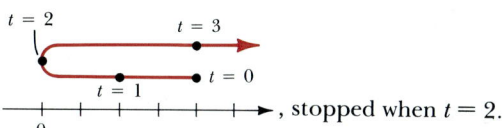

13. , stopped when $t = 1$ and when $t = 2$.

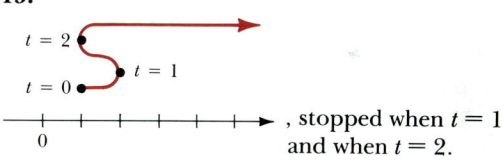

15. , never stopped when $t > 0$, $s(t)$ always increasing.

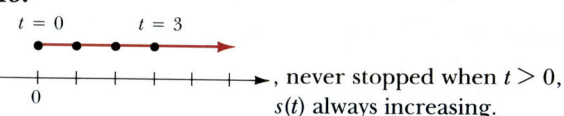

17. (a) $24 - 32t$, -32; (b) $\frac{3}{4}$; (c) 25 ft; (d) 2 sec; (e) -40 ft/sec
19. (a) 80 ft/sec; (b) $\frac{5}{2}$ sec
21. $\frac{3}{2}$ sec **23.** $\frac{11}{4}\sqrt{15} - \frac{5}{4}\sqrt{58} \approx 1.13$ sec
25. $\frac{200}{49} \approx 4.08$ sec
27. (a) $8\sqrt{7} \approx 21.2$ ft/sec; (b) $\dfrac{\sqrt{7}}{2} \approx 1.32$ sec
29. (a) $\dfrac{7}{24}$ ft; (b) $\dfrac{\sqrt{7}}{48} \approx 0.055$ sec
31. 1452 ft **33.** (a) 4.2 yd/sec^2; (b) 2.76 sec **35.** 2.5 sec
37. 1492 ft

Exercise Set 3.2

1. 1 **3.** $l = w = \sqrt{432}$ ft **5.** $l = 24\sqrt{0.6}$, $w = 30\sqrt{0.6}$
7. $l = 30$, $w = 10$, $h = 7.5$ **9.** 11,664 in³
11. $\cos\theta = \dfrac{\sqrt{3}}{3}$ or $\theta \approx 0.955$ radian
13. $h = 2r = 2\sqrt[3]{\dfrac{450}{\pi}}$ **15.** $h = \dfrac{4}{3}$, $l = w = \dfrac{16}{3}$
17. $156.25 **19.** $l = w = \sqrt{2}r$
21. Equilateral triangle with sides of length $\sqrt{3}r$
23. 7.15 in. by 9.53 in.
25. $\dfrac{4\sqrt{3}}{9 + 4\sqrt{3}} \approx 0.43$ ft **27.** $\dfrac{4}{9}\sqrt{3}\pi$
29. $\dfrac{2\sqrt{3}}{3} r$ in. by $\dfrac{2}{3}\sqrt{6}\, r$ in.
35. Length $(4 + 6^{2/3})^{3/2} \approx 19.7$ ft; placed $6 + 4\sqrt[3]{6} \approx 13.3$ ft from the wall
37. 80 nautical miles at 3:24 P.M.
39. At the midpoint of the highway
41. Rectangle height $\dfrac{25}{12} - \dfrac{\sqrt{3}}{2} \approx 1.22$ ft; total height $\dfrac{25}{12} + \dfrac{\sqrt{3}}{2} \approx 2.95$ ft

Exercise Set 3.3

1. 3600π mm³/sec **3.** 16 in²/min **5.** 96 in³/min
7. $2\sqrt{3}$ cm²/sec **9. (a)** 144π in²; **(b)** 72π in²/sec
11. $\dfrac{1}{18\sqrt[3]{\pi}} \approx 0.038$ ft/min **13.** $2\pi r$
17. (a) $-\dfrac{\sqrt{39}}{5}$ ft/sec; **(b)** $-\sqrt{3}$ ft/sec
19. (a) 4000 min; **(b)** 0.002 ft/min **21.** $\dfrac{-20\sqrt{91}}{91}$ $\dfrac{\text{ft}}{\text{sec}}$
23. $\dfrac{5}{2}$ m/sec **25.** 25 mph **27.** $\dfrac{5}{4}$ m/sec
29. 10π miles/min
31. (a) $\dfrac{64}{7}$ ft/sec; **(b)** $-\dfrac{22}{7}$ ft/sec (This rate of change is independent of time.)
33. (a) $\dfrac{-\sqrt{3}}{20}$ $\dfrac{\text{ft}}{\text{hr}}$; **(b)** $-\dfrac{\sqrt[4]{3}}{20}$ $\dfrac{\text{ft}}{\text{hr}}$ **35.** -9.37 lb/sec
37. (a) $\dfrac{50}{3\pi}$ $\dfrac{\text{cm}}{\text{sec}}$; **(b)** $\dfrac{25\sqrt{\pi}}{6\pi}$ $\dfrac{\text{cm}}{\text{sec}}$
39. (a) Separating; **(b)** 966.8 mi/hr

Exercise Set 3.4

1. $dy = 3\,\Delta x$, $\Delta y = 3\,\Delta x$
3. $dy = (2x - 2)\,\Delta x$, $\Delta y = 2x\,\Delta x - 2\,\Delta x + (\Delta x)^2$
5. $dy = (3x^2 - 1)\,\Delta x$, $\Delta y = (\Delta x)^3 + 3x\,(\Delta x)^2 + (3x^2 - 1)\,\Delta x$
7. $dy = \dfrac{\Delta x}{2\sqrt{x + 1}}$,
$\Delta y = \sqrt{x + \Delta x + 1} - \sqrt{x + 1} = \dfrac{\Delta x}{\sqrt{x + \Delta x + 1} + \sqrt{x + 1}}$
9. $dy = \left(1 - \dfrac{1}{x^2}\right)\Delta x$, $\Delta y = \dfrac{1}{x + \Delta x} - \dfrac{1}{x} + \Delta x$
11. $dy = \sec^2 x\,\Delta x$, $\Delta y = \tan(x + \Delta x) - \tan x$
13. (a) 0, 0.01; **(b)** 1.2, 1.24
15. (a) 1, 1.061; **(b)** -0.2, -0.328
17. (a) $dy = 0$; **(b)** $\Delta y = \dfrac{\sqrt{2}}{2} - 1$; **(c)** 1
19. (a) $dy = 0.02$; **(b)** $\Delta y \approx 0.0204$; **(c)** 1.02
21. (a) 0.005; **(b)** 0.004996; **(c)** 3.005
23. $\dfrac{1}{2} + \dfrac{\sqrt{3}\pi}{360} \approx 0.515$ **25.** 3.005
27. $\dfrac{193}{48} \approx 4.02083$ **29.** $\pm 0.004\pi$ ft²
31. (a) 0.016π ft³; **(b)** 0.016π ft²; **(c)** 0.15%, 0.1%
33. $\dfrac{7\pi}{256}$ **35.** $\dfrac{300\,\Delta r}{r}$ **37.** $7.2\,\pi^{1/3} \approx 10.54$ cm³
39. (4.99873, 5.003822) **41.** 1.8°F
43. (a) 55 ft; **(b)** $\dfrac{55}{12}$ ft

Exercise Set 3.5

1. $\dfrac{1}{2}$ **3.** $\dfrac{67}{11}$ **5.** $\dfrac{19}{13}$ **7.** 0 **9.** $\dfrac{2}{3}$ **11.** $-\dfrac{1}{16}$ **13.** $\dfrac{1}{2}$
15. 0 **17.** $\dfrac{1}{3}$ **19.** $\dfrac{3}{2}$ **21.** $-\dfrac{3}{7}$ **23.** 0 **25.** 1
27. $-\dfrac{1}{6}$ **29.** $\dfrac{4}{9}$ **31.** 0 **33.** 0 **35.** 0 **37.** ∞ **39.** -1
41. $\dfrac{16}{9}a$ **43.** $\dfrac{-\sqrt{3}}{3}$ **45.** 1
47. -1. No, since $\lim\limits_{x\to\infty} \dfrac{\cos x - 1}{\cos x + 1}$ does not exist.
51. One pair is: $f(x) = x$, $g(x) = x + \cos x$.

Exercise Set 3.6

1. 2.6906 **3.** -1.0905 **5.** 8.8467 **7.** 0.7391
9. 0.5798 **15.** $-1.613, -0.082, 2.082, 3.613$
17. $(0.876726, 0.768468)$ **19.** $(4.4934, 4.4934)$
21. (a) 0.38197; **(b)** fails since $f'(x_0)$ does not exist.
23. $(1.380, 0.725)$ **25.** 745.148 **27.** 2.6906
29. -1.0905 **31.** 8.8467 **33.** 0.7391 **35.** 0.5798

Exercise Set 3.7

1. (a) $c(x) = \dfrac{1500}{x} + 2 - 0.0003x$; **(b)** $C'(x) = 2 - 0.0006x$; **(c)** $c'(x) = -\dfrac{1500}{x^2} - 0.0003$; **(d)** 3.2; **(e)** 1.40

3. (a) $c(x) = \dfrac{15{,}500}{x} + 77 - 0.00001x^2$; **(b)** $C'(x) = 77 - 0.00003x^2$; **(c)** $c'(x) = -\dfrac{15{,}500}{x^2} - 0.00002x$; **(d)** 82.5; **(e)** approximately 47

5. $R'(x) = -0.0002$

7. $R'(x) = -500\dfrac{x^2 + 7.6x - 300}{(x^2 + 300)^2}$

9. $C'(x) = 0.76 + 0.0002x$, $R'(x) = 2.85 - 0.00016x$, $P'(x) = 2.09 - 0.00036x$
11. (a) 5805 or 5806; **(b)** $4366.81
13. (a) 5805 or 5806; **(b)** $4366.81
15. Producing 5805 or 5806 wastebaskets gives a profit of $3666.81; 16% less profit is produced with the same number of baskets.
17. $R(x) = 1000x$, $P(x) = 700x - 500$, $P'(x) = 700$
19. (a) $\dfrac{600 - x^2}{2x^2}$; **(b)** 14.14; **(c)** decrease, increase
21. 4000

Review Exercises

1. (a) – **(c)** $3x^2 - 4x$; **(d)** $(3x^2 - 4x)\,\Delta x$
3. (a) – **(c)** $\cos x + 1$; **(d)** $(\cos x + 1)\,\Delta x$
5. (a) 2.4; **(b)** 2.49; **(c)** 17.4
7. (a) 0.02; **(b)** 0.020203; **(c)** 1.02
9. 0 **11.** $\dfrac{1}{24}$ **13.** 0 **15.** -1 **17.** 0 **19.** 1.618
21. 1.380 **23.** 0.824
25. (a) 2 ft/sec; **(b)** 0 ft/sec; **(c)** -2 ft/sec²
27. $-16\sqrt{30}$ ft/sec **29.** $(11 + 2\sqrt{31})/4 \approx 5.53$ sec
31. $3\sqrt{2}/2$ sec

33. (a) $a \approx 55.38$ ft/sec²; **(b)** $v(0) \approx 70.26$ mi/hr
35. $\left(\dfrac{6}{5}, \dfrac{3}{5}\right)$ **37.** $\sqrt[3]{4}$ **39.** 6 by $\dfrac{5}{2}$ **41.** 4 by 4 by 6
43. $r = h = 4$ in., $V = 64\pi$ in³
45. The height of the rectangle and radius of the circle should both be $25/(4 + \pi) \approx 3.50$ ft.
47. 12 ft from the tree tied at 20 ft above the ground
49. $\dfrac{5}{2}$ ft/sec **51.** $\dfrac{\sqrt[3]{4}}{2\pi} \approx 0.25$ ft/min **53.** 0.8π cm²/sec
55. $(588)^{1/4} \approx 4.9$ mm²/sec
57. $640/(27\pi) \approx 7.55$ cm/min
59. 50π in³, 20π in² **61.** Approximately 984.07 knots
63. (a) $400/x - 1$; **(b)** 200; **(c)** increase, decrease

Chapter 4

Exercise Set 4.1

1. 18 **3.** 105 **5.** 195 **7.** 3 **9.** 203 **11.** $9i + 29k$
13. 638 **15.** $n(n+1)(n+2)/3$ **17.** 39,490
19. $\sum_{i=2}^{8} i^2$ **21.** $\sum_{i=1}^{8} (2+3i)$ **23.** $\sum_{i=1}^{11} (-1)^{i+1}\dfrac{1}{i}$
25. 6 **27.** $\dfrac{1}{3}$ **29.** $\dfrac{74}{3}$ **31.** $\dfrac{8}{3}$ **33.** $\dfrac{16}{3}$ **35.** 6
37. (a) 6; **(b)** 4; **(c)** 10; **(d)** 12 **39.** 660

Exercise Set 4.2

1. 15 **3.** $\dfrac{7}{32}$ **5.** $\dfrac{3}{8}$ **7.** $\dfrac{5}{12}$ **9.** $\dfrac{29}{32}$
11. $\dfrac{7\sqrt{2} + 5\sqrt{3}}{24}\pi$ **13.** 6 **15.** 2 **17.** $\dfrac{1}{3}$ **19.** $\dfrac{1}{4}$
21. 0 **23.** 2 **25.** $\dfrac{\pi}{2}$ **27.** 15, 3 **29.** 24, -3
31. $\sqrt{2}\pi, -\pi$ **33.** $\dfrac{2}{5}, \dfrac{4}{9}$
37. Consider $f(x) = x$ on $[-1, 1]$.
39. (a) 20; **(b)** 20; **(c)** 20
43. (a)
$$g(x) = \begin{cases} 0, & \text{if } -2 \le x < -1, \\ \dfrac{1}{2} + x + \dfrac{x^2}{2}, & \text{if } -1 \le x < 0, \\ \dfrac{1}{2} + x - \dfrac{x^2}{2}, & \text{if } 0 \le x < 1, \\ 1, & \text{if } 1 \le x \le 2 \end{cases}$$

S26 ANSWERS

(c) $g'(x) = f(x)$

[Graph showing $y = g'(x) = f(x)$ and $y = g(x)$]

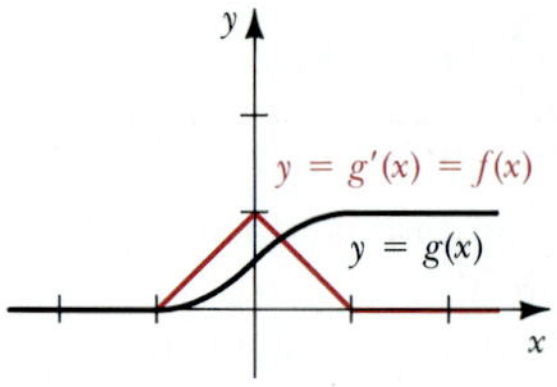

Exercise Set 4.3

1. 21 **3.** 2 **5.** -12 **7.** 18 **9.** 1 **11.** 1 **13.** 2
15. 9 **17.** 6 **19.** $\dfrac{3}{8}$ **21.** $\dfrac{508}{7}$ **23.** $\dfrac{\sqrt{2}}{2} - 1$
25. $\dfrac{13}{4}$ **27.** -2

29. Area: 12

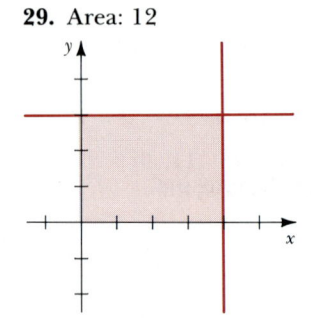

31. Area: 4

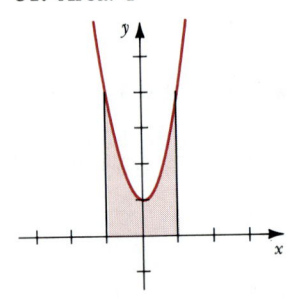

33. Area: 36

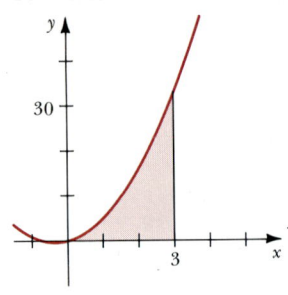

35. Area: 2

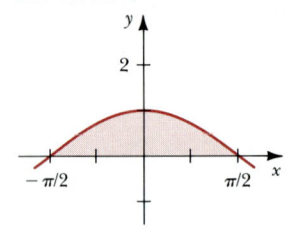

37. Area: 1

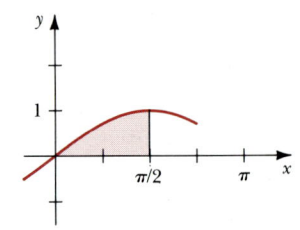

39. 4 **41.** 1 **43.** $\dfrac{2}{\pi}$ **45.** $x\sqrt{x^2+9}$ **47.** 0

49. $\displaystyle\int_1^x \sqrt{t^2+9}\,dt + x\sqrt{x^2+9}$

51. $-x\sqrt{x^2+9}$ **55.** $2x\displaystyle\int_2^{x^2}\sqrt{t^2+9}\,dt + 2x^3\sqrt{x^4+9}$

57. $-x\cos x^3 + \dfrac{1}{2}\cos x^{3/2}$

59. 5

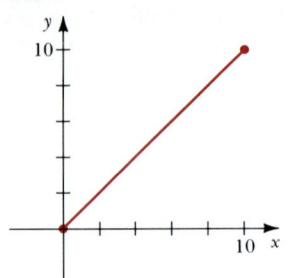

63. $\dfrac{\sqrt{3}}{2}$ **65.** $f'(1)$ **71.** $\dfrac{-4}{\pi}$

Exercise Set 4.4

1. $x^4 + C$ **3.** $\dfrac{x^4}{4} + C$ **5.** $\dfrac{y^4}{2} + y^3 + C$
7. $x^3 + 2x^2 - 2x + C$
9. $\dfrac{3x^{5/3}}{5} + \dfrac{9x^{4/3}}{4} + \dfrac{4x^3}{3} + C$ **11.** $\sin t + t + C$
13. $-\cot u - u + C$ **15.** $\dfrac{3t^5}{5} + \dfrac{3t^2}{2} + C$
17. $\sin x + \dfrac{2}{3}x^{3/2} + C$ **19.** $x + \tan x + C$
21. $9x^3 + 9x^2 + 3x + C$ **23.** $\dfrac{5t^{6/5}}{6} - \dfrac{5t^{7/5}}{7} + C$
25. $\dfrac{y^2}{2} - 3y + C$ **27.** $-\dfrac{1}{x} + \dfrac{1}{x^2} - \dfrac{1}{x^4} + C$
29. $\dfrac{x^7}{7} - \dfrac{2x^9}{9} + \dfrac{x^{11}}{11} + C$ **31.** $\dfrac{284}{3}$ **33.** 2 **35.** $\dfrac{20}{3}$
37. $\dfrac{146}{3}$ **39.** $\dfrac{29}{6}$ **41.** 400 **43.** $2\sqrt{2} - 1$ **45.** 1
47. $\sqrt{2}$
49. Area: $\dfrac{2}{3}(2\sqrt{2} - 1)$ **51.** Area: $2 + \dfrac{\pi}{2}$

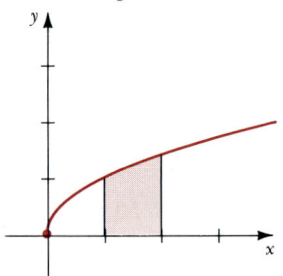

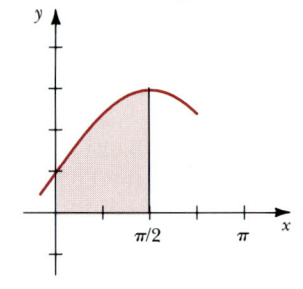

53. Area: $\dfrac{26}{3}$

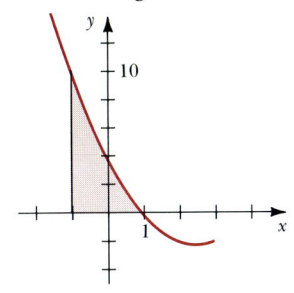

55. $f(x) = 3x + 2$ **57.** $f(x) = \dfrac{x^2}{2} - \dfrac{1}{2}$ **59.** $f(x) = \dfrac{x^3}{3} - \dfrac{x^2}{2}$

61. $f(x) = -\cos x - \dfrac{x^2}{2}$ **63.** $y = \dfrac{x^2}{2} + 2x$

65. $y = x^3 - x^2 + x + 2$
67. $P(x) = 2.8x - 0.002x^3 - 1000$

69. 12 cm/sec   **71.** 121 ft; -8 ft/sec $= \dfrac{-60}{11}$ mph

73. $(3 + \sqrt{6409})/32 \approx 2.6$ sec

Exercise Set 4.5

1. (a) $\dfrac{1}{6}(x^2 + 5)^6 + C$; **(b)** $\dfrac{-1}{4}(x^2 + 5)^{-2} + C$;

(c) $\dfrac{-3}{2}\cos(x^2 + 5) + C$; **(d)** $\dfrac{1}{2}\tan(x^2 + 5) + C$

3. $\dfrac{(2x-3)^3}{3} + C$ **5.** $\dfrac{(t^3+4)^2}{2} + C$ **7.** $\dfrac{2(t^3+4)^{3/2}}{3} + C$

9. $\dfrac{(4x-5)^{3/2}}{6} + C$ **11.** $\dfrac{-1}{3}(x-1)^{-3} + C$

13. $\dfrac{-1}{2}\cos 2x + C$ **15.** $\dfrac{2}{\pi}$

17. $\dfrac{-1}{4}(2x-1)^{-2} + C$ **19.** $\dfrac{-1}{3}(x^2+7)^{-3} + C$

21. $\dfrac{13\sqrt{13} - 1}{3}$ **23.** $\dfrac{4}{9}\sqrt{2}$ **25.** $\dfrac{4}{3}(\sqrt{5} - \sqrt{2})$

27. $\dfrac{3}{2}(x^3 + x)^{2/3} + C$ **29.** $\dfrac{-1}{12}\cos^4 3x + C$ **31.** $\dfrac{1}{2}$

33. $\dfrac{(\sin x)^{11}}{11} + C$ **35.** $\dfrac{(x^3 + x^2)^5}{5} + C$ **37.** $\dfrac{6}{25}$

39. $\dfrac{2}{3}(\sqrt{x} - 1)^3 + C$ **41.** $-\dfrac{\left(1 + \dfrac{1}{t}\right)^4}{4} + C$

43. $\dfrac{1}{2}x^2 + \dfrac{1}{2}\sin^2 x + C$

45. $\left[\dfrac{2}{5}(z-1)^2 + \dfrac{4}{3}(z-1)\right]\sqrt{z-1} + C$

47. $\dfrac{4}{3} - \dfrac{2\sqrt{2}}{3}$ **49.** $\dfrac{3471}{2048}$ **51.** $\dfrac{1}{3(2 + 3\cos x) + C}$

53. $\dfrac{3}{2}\left[\dfrac{2}{5}(x^2+1)^2 - \dfrac{2}{3}(x^2+1)\right]\sqrt{x^2+1} + C$

55. $\dfrac{4[(\sqrt{2}+1)^{3/2} - 2\sqrt{2}]}{3}$ **57.** $2\tan\sqrt{s} + C$

59. $\dfrac{2(x-2)^{7/2}}{7} + \dfrac{8(x-2)^{5/2}}{5} + \dfrac{10(x-2)^{3/2}}{3} + C$

61. $-(x+1)^{-1} + C$ **63.** $\dfrac{-1}{(x+3)} + C$

65. $x + \dfrac{1}{x+1} + C$ **67.** $-2(2x+1)^{-1} + x + C$

69. $\dfrac{\sec^4 x}{4} - \dfrac{\sec^2 x}{2} + C$. The difference is a constant.

71. $\dfrac{\sin^2 x}{2}$ **73.** $\dfrac{(x^2+4x)^{3/2}}{3}$ **75. (b)** $\sin(x+1) + C$

77. (b) $\dfrac{-1}{10}\cos x^{10} + C$ **79. (b)** $2\tan\sqrt{t} + C$

81. (a) $\sec x + C$

Exercise Set 4.6

1. $\dfrac{1}{x+2}$ **3.** $\dfrac{2}{x}$ **5.** $\dfrac{4x}{x^2+3}$ **7.** $\dfrac{1}{4x}$ **9.** $\dfrac{1}{x\ln x}$

11. $\dfrac{1}{x\ln x}$ **13.** $\dfrac{1}{\sqrt{x^2-1}}$ **15.** $\dfrac{2}{1-x^2}$

17. $\left[\ln\left(\dfrac{1+x}{1-x}\right)\right]^{-1/2} \bigg/ (1-x^2)$ **19.** $\cot x$

21. $\ln(x^2+1) + \dfrac{2x^2}{x^2+1}$ **23.** $\dfrac{2x^2\ln x - x^2 - 1}{x(\ln x)^2}$

25. $\dfrac{2}{2x+1}$ **27.** $\dfrac{3x[\ln\sqrt{x^2+2}]^2}{x^2+2}$ **29.** $\dfrac{-\sin(\ln x)}{x}$

31. $\dfrac{-y}{x}$ **33.** $\dfrac{1}{x+y-1}$ **35.** $\dfrac{-y(2x^2+1)}{x(2y^2-1)}$

37. $\ln|x-3| + C$ **39.** $\ln 2 - \ln 4 = -\ln 2$ **41.** $\dfrac{1}{2\pi}\ln 2$

43. $\dfrac{1}{2}\ln|\csc 2x - \cot 2x| + \dfrac{1}{3}\ln|\sin 3x| + C$

45. $\dfrac{\ln|x^2 + 2x - 3|}{2} + C$ **47.** $\dfrac{[(\ln 3)^2 - (\ln 2)^2]}{2}$

49. $2(\ln x)^4 + C$ **51.** 2 **53.** $\ln|1 + \sin x| + C$
55. $y' = 3(11x^2 + 14x + 40)(x^2 + 8)^2(3x + 7)^4$

57. $y' = \dfrac{(7x+5)\sqrt{x\sqrt{(x+1)\sqrt{x}}}}{8x^2+8x}$

59. $y' = (x^2+1)^{-1/2}(x^3-1)^4(2x^2+1)^6(60x^6+74x^4-30x^3+15x^2-29x)$

61. 0 **63.** 0 **65.** 0 **67.** 0 **69.** ∞

73.

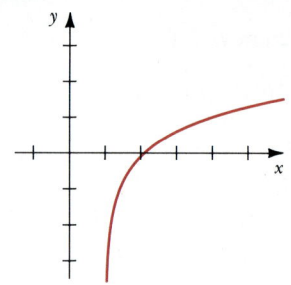

75.

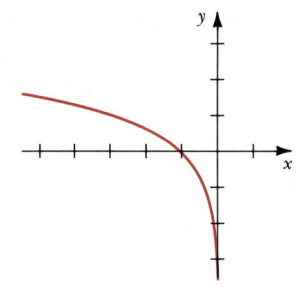

77.

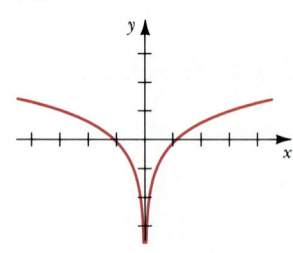

79. $(1/e, -1/e)$

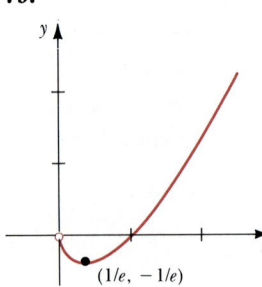

81. $(e^{-2}, -2e^{-1})$

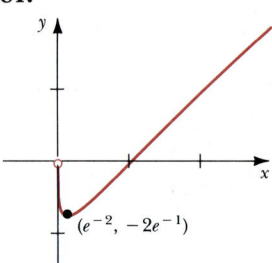

83. $y - 1 = 4(x - 1)$

85. $\dfrac{-1}{2}[\ln|1-x| + \ln|1+x|] + C$

87. (a) $\dfrac{1}{2}[\ln|1+x| - \ln|1-x|] + C$;

(c) $\dfrac{1}{2}[\ln|1+\sin x| - \ln|1-\sin x|] + C$;

(d) $\dfrac{1}{2}[\ln|1-\cos x| - \ln|1+\cos x|] + C$

93. $I'(x) = -2.303c\, I(x)\, (\alpha(x) + x\alpha'(x))$

95. Yes, by \$384.68

Exercise Set 4.7

1. Diverges **3.** $\dfrac{1}{6}$ **5.** $\dfrac{1}{\ln 3}$ **7.** $\dfrac{1}{16}$

9. Diverges **11.** $\dfrac{3}{2}$ **13.** Diverges **15.** Diverges

17. Diverges **19.** Diverges **21.** Diverges

23. Diverges **25.** $\dfrac{[(\sqrt[3]{2})^2 - 1]}{2}$ **27.** 0 **29.** Diverges

31. Diverges **33.** Diverges **35.** It is not finite **37.** $\dfrac{1}{2}$

41. $r > 1$ (provided x^r exists)

Review Exercises

1. 30 **3.** 40 **5.** $\dfrac{3n^2 - 7n}{2}$ **7.** n^2

11. (a) $\displaystyle\int_0^2 3x^2\, dx$, 8; (b) $\displaystyle\int_0^2 (2x+1)\, dx$, 6

13. $\dfrac{x^4}{4} + C$ **15.** 312 **17.** $\dfrac{(4x+7)^4}{4} + C$

19. $\dfrac{2}{5}x^{5/2} + \ln|x| + C$ **21.** $\dfrac{6x^{11/3}}{11} + C$

23. $\dfrac{4}{3}x^{3/4} - \dfrac{1}{3}x^{-3} + C$ **25.** $-\dfrac{\pi^2}{2}$ **27.** $-\sqrt{4-x^2} + C$

29. $\dfrac{1}{5}(w^2+1)^{5/2} - \dfrac{1}{3}(w^2+1)^{3/2} + C$

31. $\dfrac{2}{1-\sqrt{x}} + C$ **33.** $\ln|x-2| + C$

35. $\dfrac{1}{2}\ln|x^2 - 6x + 5| + C$ **37.** $\dfrac{104}{9}$ **39.** $\dfrac{11}{450}$

41. $x + \ln|x^2 - 2x + 2| + C$ **43.** $\dfrac{1}{2}\sin 2x + C$

45. $\dfrac{1}{4}\ln|\sec 4x| + C$ **47.** $\dfrac{1}{3}\sin^3 x + C$

49. $\sin(\ln x) + C$ **51.** 3 **53.** 1 **55.** $\dfrac{1}{\ln 2}$

57. Diverges **59.** 0 **61.** 9 **63.** Diverges

65. $\dfrac{3}{x}(\ln x)^2$ **67.** $\dfrac{3x^2 - 7}{x^3 - 7x + 1}$

69. $\dfrac{x^{-2/3}[\ln(x^{1/3}+1)]^{-2/3}}{9(x^{1/3}+1)}$

71. $\displaystyle\int_1^x \ln t\, dt + x \ln x$ **73.** $x^2 - x - 12$

75. $\dfrac{99}{5}$

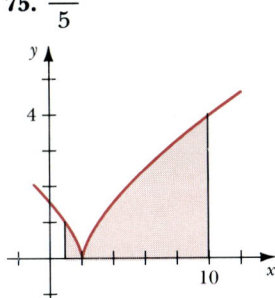

77. 4

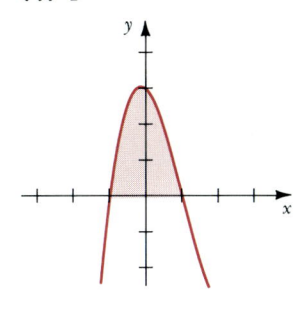

9. (a)

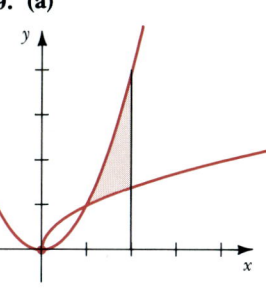

(b) $\dfrac{1}{3}(9-4\sqrt{2})$

11. (a)

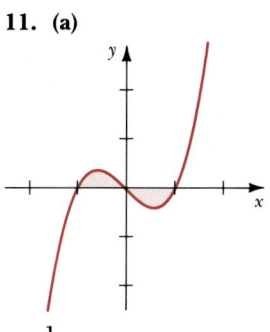

(b) $\dfrac{1}{2}$

13. (a)

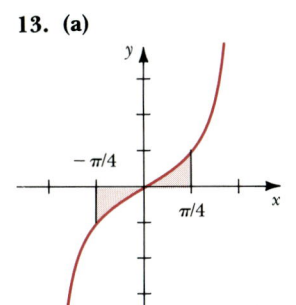

(b) $\ln 2$

15. (a)

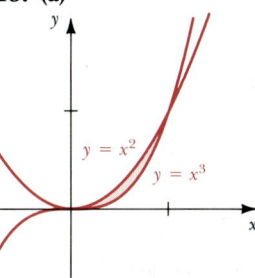

(b) $\dfrac{1}{12}$

Chapter 5

Exercise Set 5.1

1. (a)

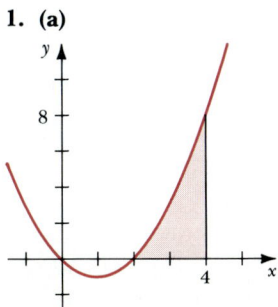

(b) $\dfrac{20}{3}$

3. (a)

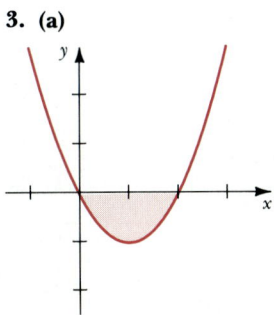

(b) $\dfrac{4}{3}$

17. (a)

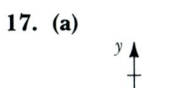

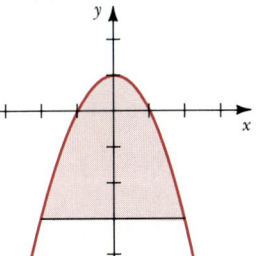

(b) $\dfrac{32}{3}$

19. (a)

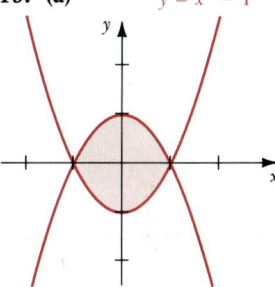

(b) $\dfrac{8}{3}$

5. (a)

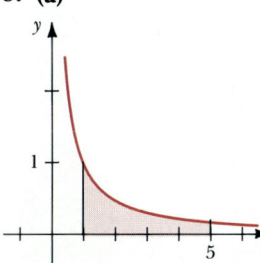

(b) $\ln 5 \approx 1.61$

7. (a)

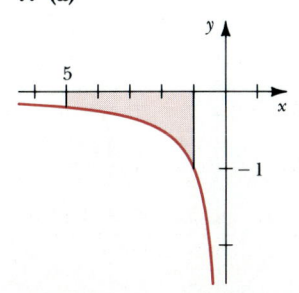

(b) $\ln 5$

21. (a)

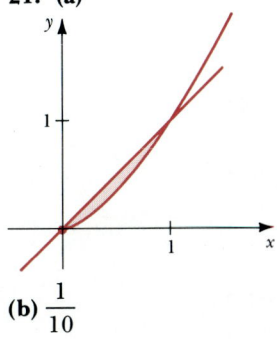

(b) $\dfrac{1}{10}$

23. (a)

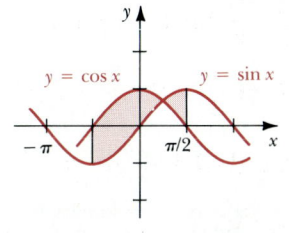

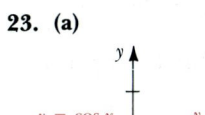

(b) $2\sqrt{2}$

25. (a) [graph: $y=x$, $y=-x/2$, $y=1-x$]
(b) $\dfrac{3}{4}$

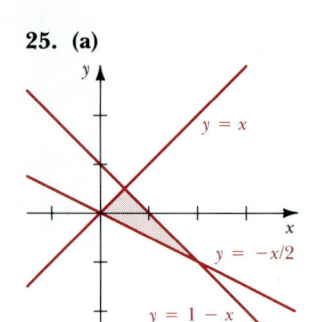

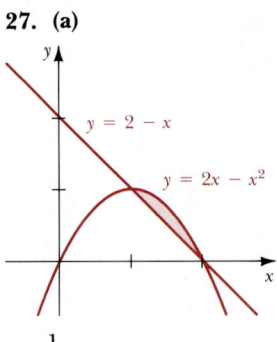

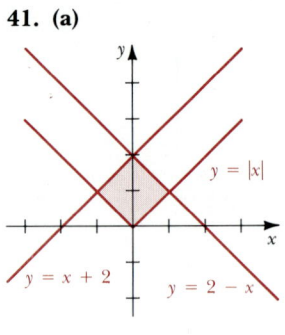

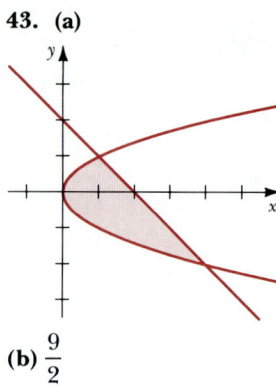

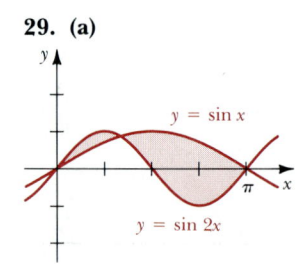

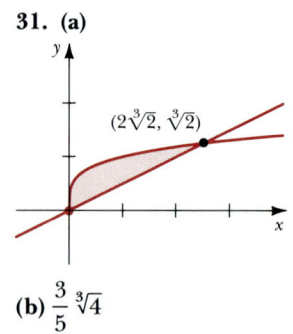

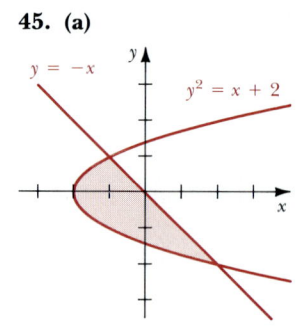

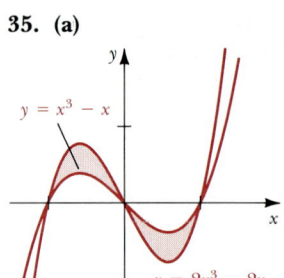

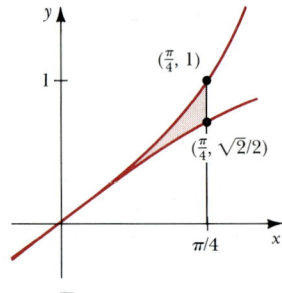

27. (a) [graph: $y=2-x$, $y=2x-x^2$]
(b) $\dfrac{1}{6}$

41. (a) [graph: $y=|x|$, $y=x+2$, $y=2-x$]
(b) 2

43. (a) [graph]
(b) $\dfrac{9}{2}$

29. (a) [graph: $y=\sin x$, $y=\sin 2x$]
(b) $\dfrac{5}{2}$

31. (a) [graph: $(2\sqrt[3]{2},\sqrt[3]{2})$]
(b) $\dfrac{3}{5}\sqrt[3]{4}$

45. (a) [graph: $y=-x$, $y^2=x+2$]
(b) $\dfrac{9}{2}$

33. (a) [graph: $(\frac{\pi}{4},1)$, $(\frac{\pi}{4},\sqrt{2}/2)$]
(b) $\dfrac{\sqrt{2}}{2}-1+\dfrac{1}{2}\ln 2$

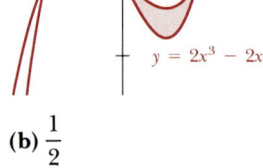

35. (a) [graph: $y=x^3-x$, $y=2x^3-2x$]
(b) $\dfrac{1}{2}$

47. $\dfrac{1}{2}$ **49.** $\pm 3\sqrt[3]{6}$ **51.** $8a^2$ **53.** 17

Exercise Set 5.2

1. 72   **5.** 4

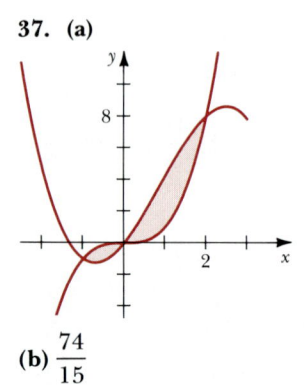

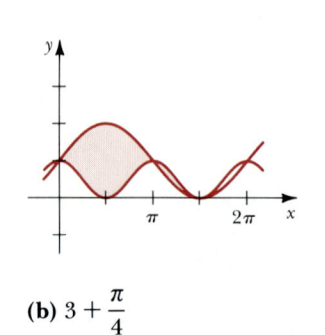

7. Volume: $\dfrac{243}{5}\pi$ **9.** Volume: $\dfrac{127}{7}\pi$

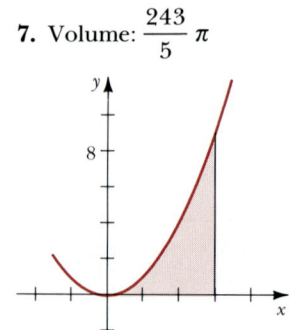

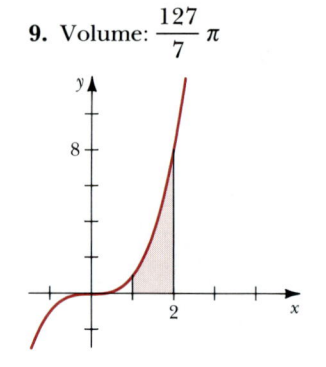

37. (a) [graph]
(b) $\dfrac{74}{15}$

39. (a) [graph]
(b) $3+\dfrac{\pi}{4}$

11. Volume: $\dfrac{32}{5}\pi$

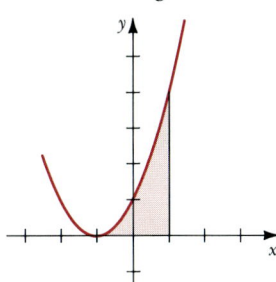

13. Volume: $\dfrac{\pi}{2}$

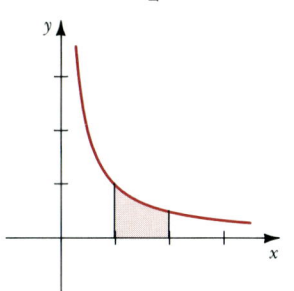

27. Volume: $\dfrac{\pi^2}{2}$

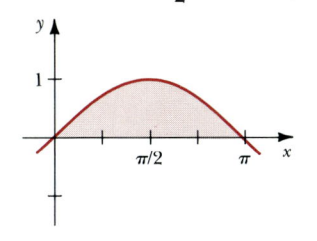

29. Volume: $\dfrac{\pi^2}{2}$

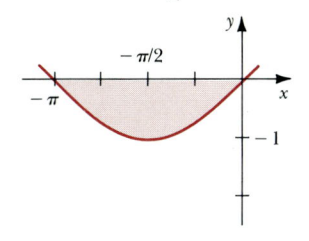

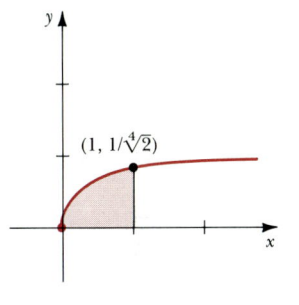

15. Volume: $\dfrac{2}{35}\pi$

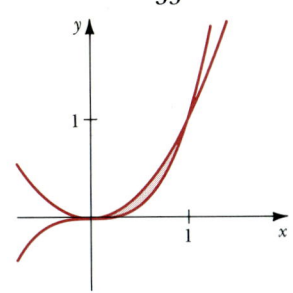

17. Volume: $\dfrac{120}{7}\pi$

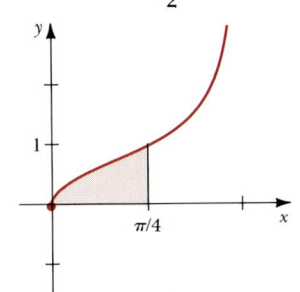

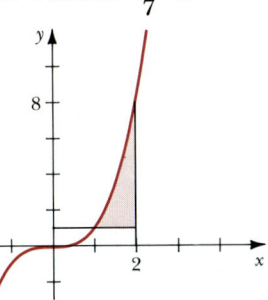

31. Volume: $\dfrac{\pi}{2}\ln 2$

33. Volume: $\pi(\sqrt{2}-1)$

35. Volume: $\dfrac{5\pi}{6}$

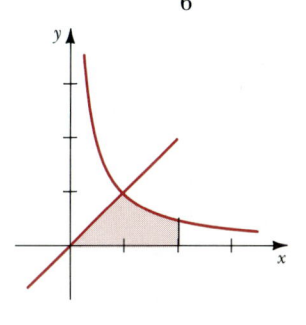

37. Volume: 12π

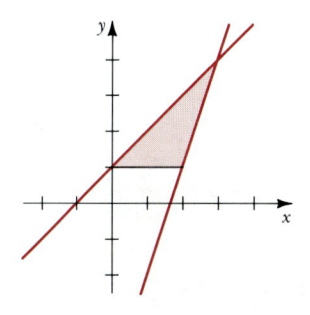

19. Volume: $\dfrac{64}{5}\pi$

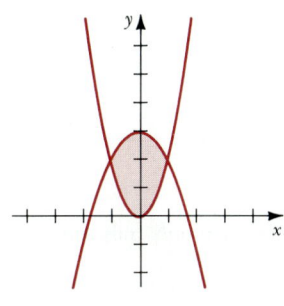

21. Volume: $\dfrac{\pi}{30}$

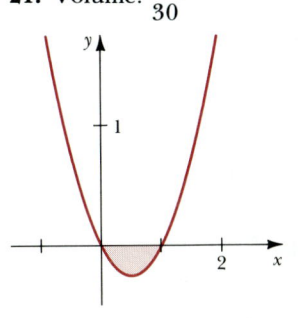

23. Volume: 2π

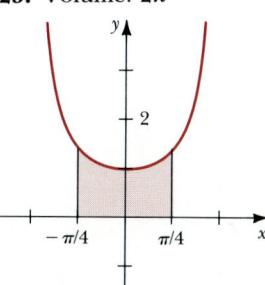

25. Volume: 2π

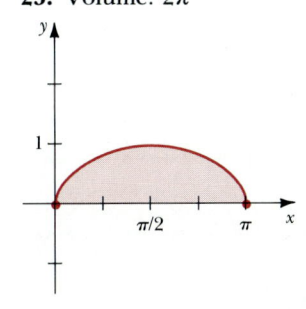

39. $\dfrac{3\pi}{5}$ **41.** $\dfrac{\pi}{6}$ **43.** $\dfrac{\pi}{2}$ **45.** $\dfrac{27\pi}{2}$ **47.** $\dfrac{64}{27}\pi$

49. (a) $\dfrac{7\pi}{15}$; **(b)** $\dfrac{\pi}{6}$; **(c)** $\dfrac{13\pi}{15}$ **(d)** $\dfrac{11\pi}{6}$ **51.** $\dfrac{16}{3}r^3$ **53.** $\dfrac{16}{15}$

59. 140625π ft³ ≈ 3304792 gal

61. 429.35π cm³ ≈ 26 in³ **63.** $\left(\dfrac{5\pi}{4}-\dfrac{2}{3}\right)$ in³

65. 8250 ft³ **67.** $\dfrac{4\pi}{3}(64-7\sqrt{7})$

69. $r=[1-(0.5)^{2/3}]^{1/2}R$ **75. (a)** $\ln M$, **(b)** $\pi(1-1/M)$

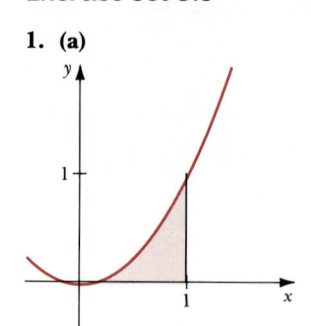

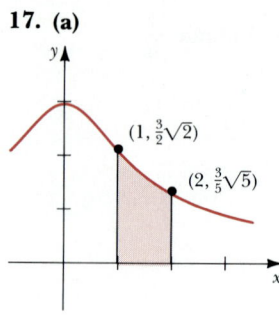

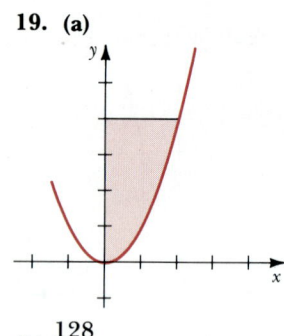

Exercise Set 5.3

1. (a) [graph] (b) $\dfrac{\pi}{2}$

3. (a) [graph] (b) $\dfrac{62\pi}{5}$

5. (a) [graph] (b) $\dfrac{5\pi}{14}$

7. (a) [graph with $(\sqrt{\pi/2}, 0)$] (b) π

9. (a) [graph] (b) $\dfrac{8\pi}{15}$

11. (a) [graph] (b) 5π

13. (a) [graph with $(3, \tfrac{11}{3})$] (b) $\dfrac{74\pi}{9}$

15. (a) [graph with $(3, 2)$] (b) $\dfrac{52\pi}{15}$

17. (a) [graph with $(1, \tfrac{3}{2}\sqrt{2})$, $(2, \tfrac{3}{5}\sqrt{5})$] (b) $6\pi(\sqrt{5} - \sqrt{2})$

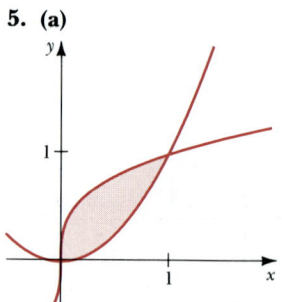

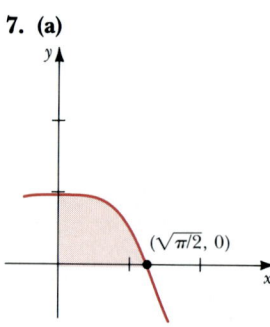

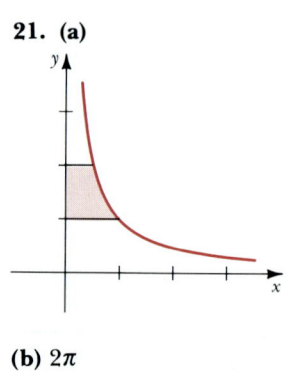

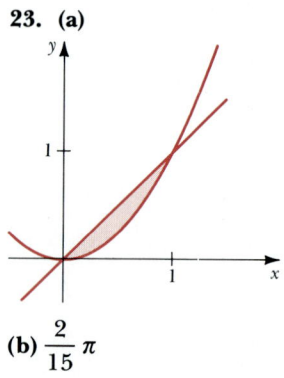

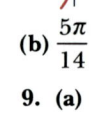

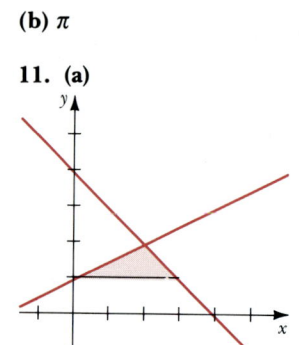

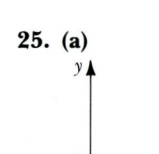

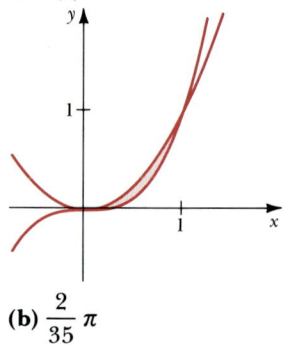

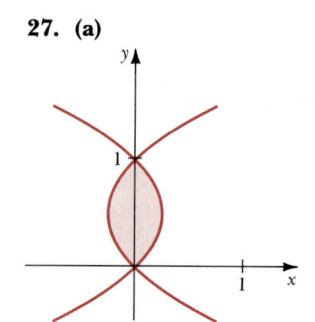

19. (a) [graph] (b) $\dfrac{128}{5}\pi$

21. (a) [graph] (b) 2π

23. (a) [graph] (b) $\dfrac{2}{15}\pi$

25. (a) [graph] (b) $\dfrac{2}{35}\pi$

27. (a) [graph] (b) $\dfrac{\pi}{3}$ (use only half the shaded region)

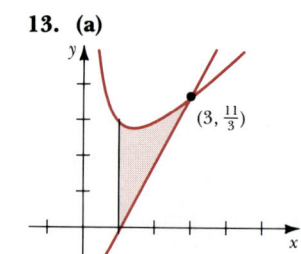

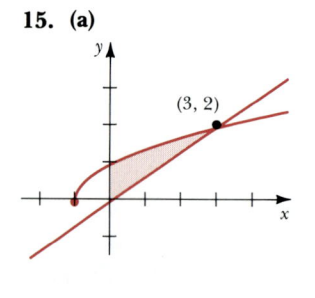

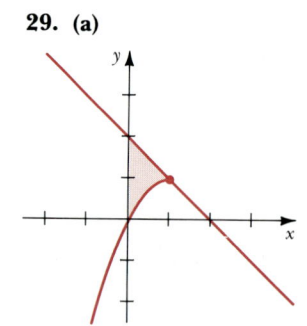

29. (a) [graph] (b) $\dfrac{9}{5}\pi$

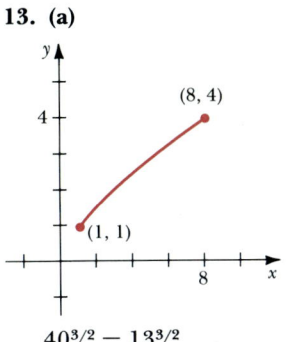

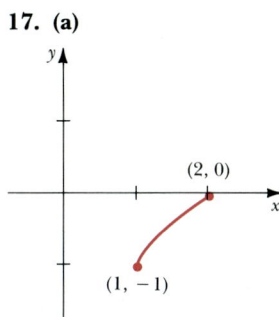

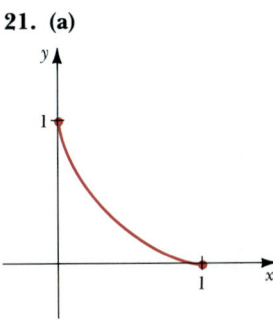

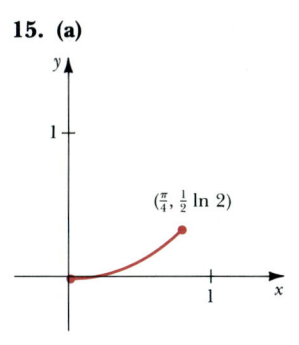

31. (a) $\dfrac{640\pi}{3}$; (b) $\dfrac{256\pi}{3}$ **33.** (a) $\dfrac{16\pi}{3}$; (b) $\dfrac{64\pi}{15}$

35. (a) $\dfrac{9\pi}{10}$; (b) $\dfrac{9\pi}{14}$; (c) $\dfrac{21\pi}{10}$; (d) $\dfrac{15\pi}{7}$ **37.** $\dfrac{27\pi}{16}$

39. $\pi r^2 h$ **41.** $\dfrac{1}{3}\pi r^2 h$

Exercise Set 5.4

1. (a) [graph with points (2, 10) and (4, 16)]
(b) $2\sqrt{10} \approx 6.32$

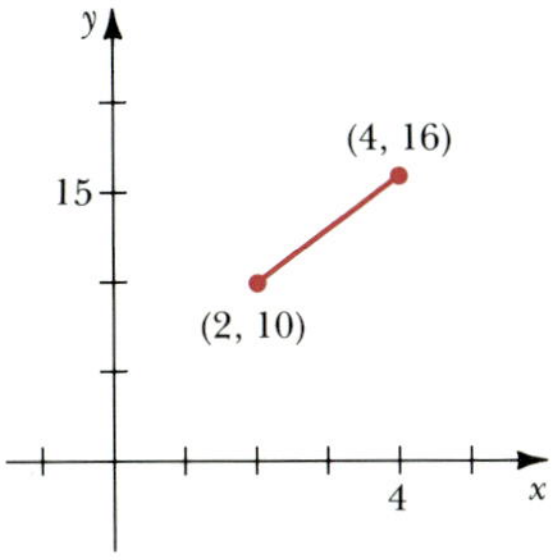

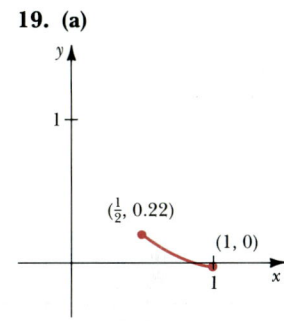

3. (a) [graph with points (3, $3\sqrt{3}$) and (15, $15\sqrt{15}$)]
(b) $\dfrac{1}{27}(139\sqrt{139} - 31\sqrt{31}) \approx 54.3$

5. (a) [graph with point (1, 3)]
(b) $\dfrac{85^{3/2} - 8}{243} \approx 3.19$

7. (a) [graph with points (1, $(\tfrac{1}{3})^{3/2}$) and (2, $(\tfrac{10}{3})^{3/2}$)]
(b) 6

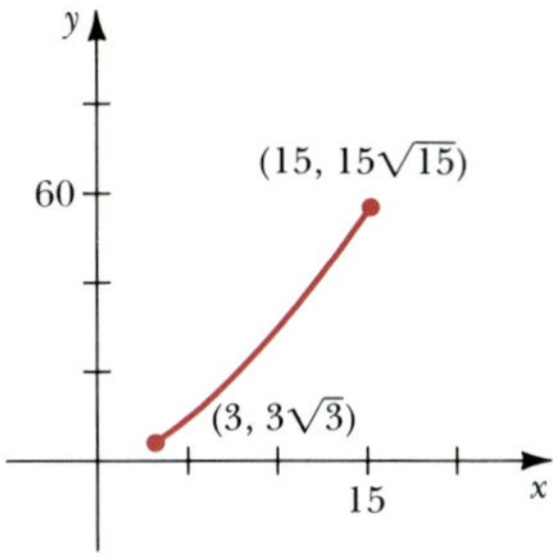

9. (a) [graph with points (1, $\tfrac{13}{12}$), ($\sqrt{2}$, $\tfrac{2}{3}\sqrt{2}$), (3, $\tfrac{31}{12}$)]
(b) $\dfrac{17}{6}$

11. (a) [graph with points (1, $\tfrac{2}{3}$) and (4, $\tfrac{22}{3}$)]
(b) $\dfrac{22}{3}$

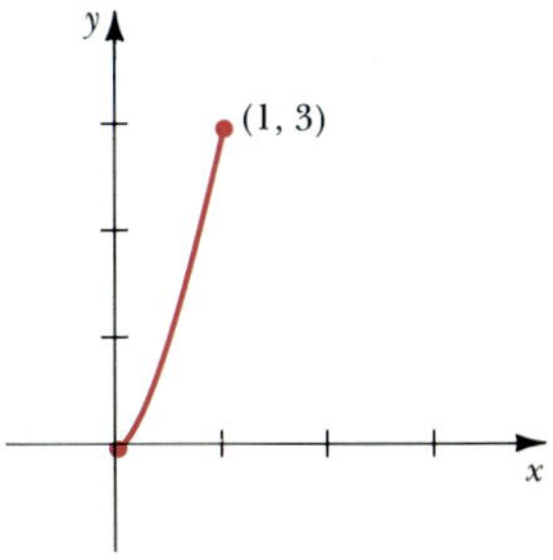

13. (a) [graph with points (1, 1) and (8, 4)]
(b) $\dfrac{40^{3/2} - 13^{3/2}}{27} \approx 7.63$

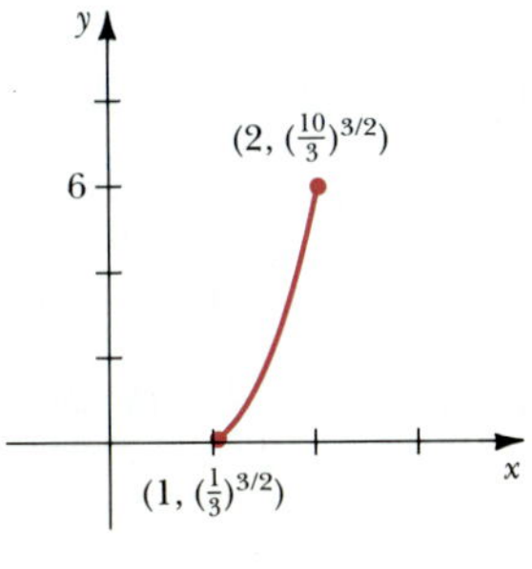

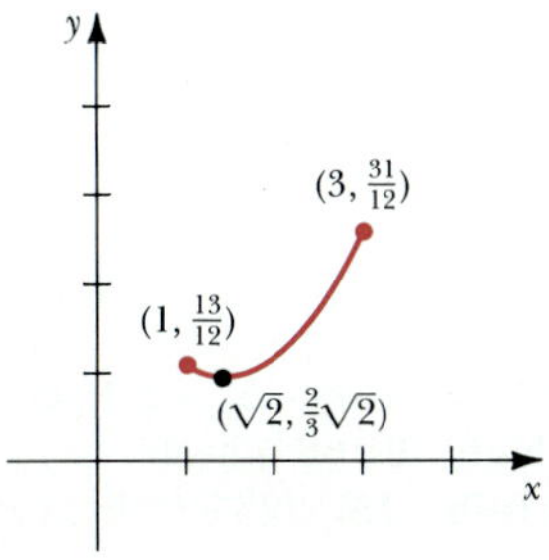

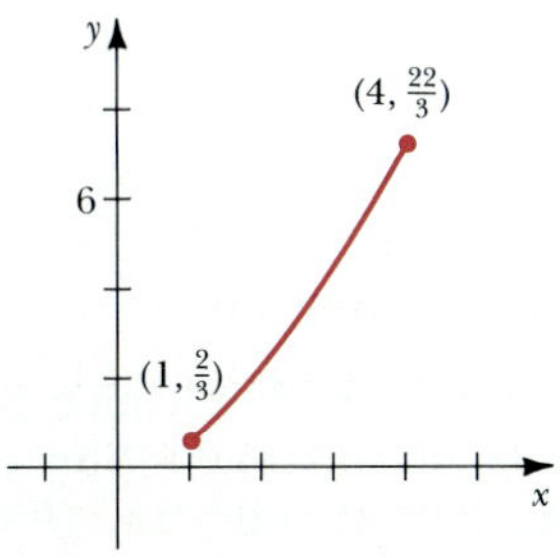

15. (a) [graph with point ($\tfrac{\pi}{4}, \tfrac{1}{2}\ln 2$)]
(b) $\ln(1 + \sqrt{2})$

17. (a) [graph with points (1, −1) and (2, 0)]
(b) $\dfrac{13^{3/2} - 8}{27}$

19. (a) [graph with points ($\tfrac{1}{2}$, 0.22) and (1, 0)]
(b) $\dfrac{3}{2}(1 - 2^{-2/3})$

21. (a) [graph]
(b) $\dfrac{3}{2}$

23. (a) [graph with points (1, 1) and (3, 27)]
(b) $\dfrac{(730^{3/2} - 10^{3/2})\pi}{27} \approx 2291.26$

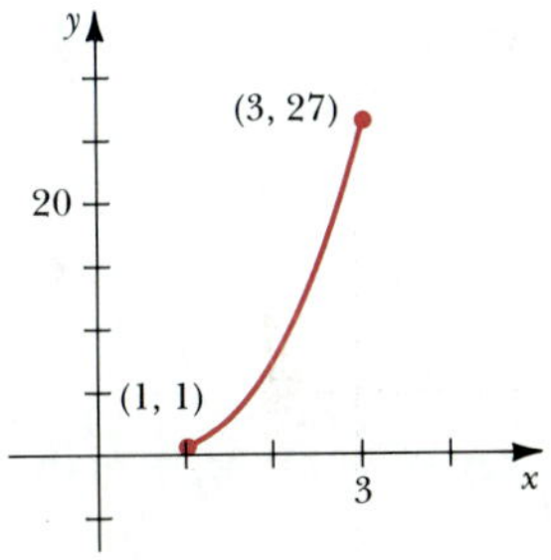

25. (a) [graph with points (1, 0) and (2, 3)]
(b) $8\pi\sqrt{2}$

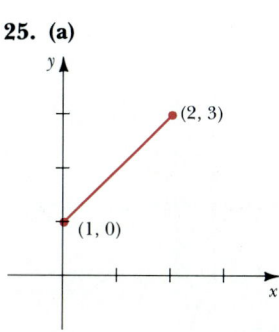

27. (a) [graph with points (1, $\sqrt{2}$) and (5, $\sqrt{6}$)]
(b) $\dfrac{49\pi}{3}$

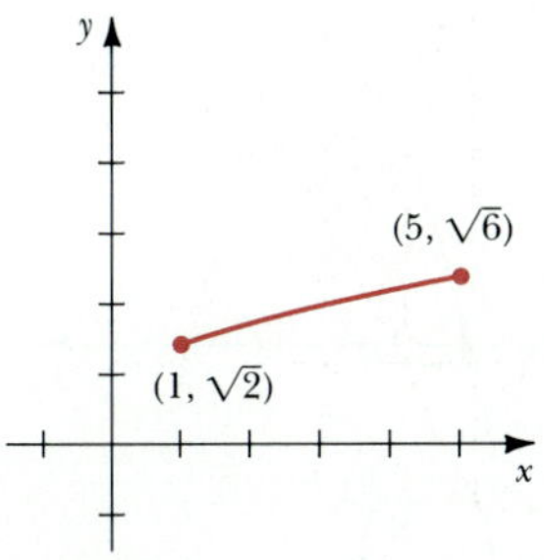

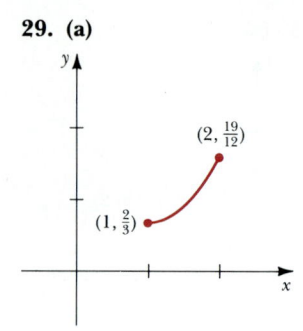

29. (a)

[graph showing curve from $(1, \frac{2}{3})$ to $(2, \frac{19}{12})$]

(b) $\dfrac{47}{16}\pi$

31. $2\pi rh$ **33.** $x \approx 0.57$, $y \approx 0.43$ **35.** $\dfrac{297\sqrt{34}}{25}\pi$ ft²

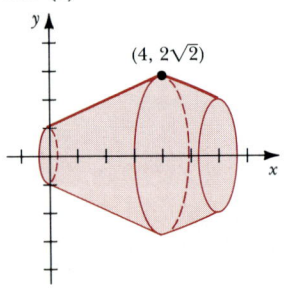

37. (a)

[graph of surface of revolution with point $(4, 2\sqrt{2})$]

(b) $\dfrac{2\pi}{3}\left[\sqrt{17}\left(\dfrac{16\sqrt{2}}{3} + 4\right) + 27 - 5\sqrt{5}\right]$

Exercise Set 5.5

1. (a) 22; **(b)** $\dfrac{11}{3}$ **3. (a)** 17; **(b)** $\dfrac{17}{11}$

5. (a) $M_x = 11$; $M_y = 5$, **(b)** $\left(\dfrac{5}{6}, \dfrac{11}{6}\right)$

7. (a) $M_x = -4$, $M_y = 31$; **(b)** $\left(\dfrac{31}{10}, -\dfrac{2}{5}\right)$

9. (a)

[parabolic region graph]

(b) $\left(0, \dfrac{8}{5}\right)$

11. (a)

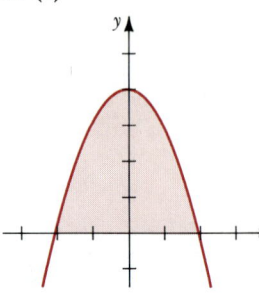

[parabolic region graph]

(b) $\left(0, \dfrac{3}{5}\right)$

13. (a)

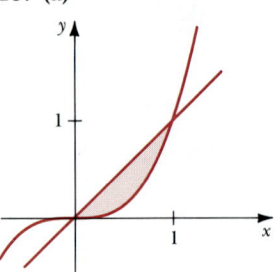

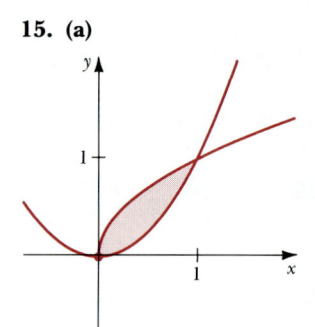

[graph of region between curves]

(b) $\left(\dfrac{8}{15}, \dfrac{8}{21}\right)$

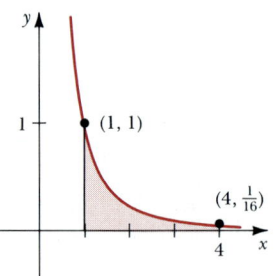

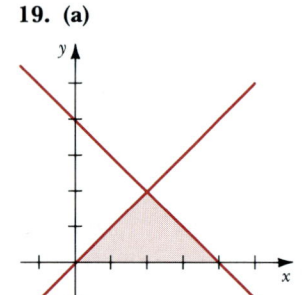

15. (a)

[graph of region between curves]

(b) $\left(\dfrac{9}{20}, \dfrac{9}{20}\right)$

17. (a)

[graph of curve with points $(1,1)$ and $(4, \frac{1}{16})$]

(b) $\left(\dfrac{4\ln 4}{3}, \dfrac{7}{32}\right)$

19. (a)

[triangular region graph]

(b) $\left(2, \dfrac{2}{3}\right)$

21. $(0, 1)$ **23.** $(-8, -22)$ **25.** 11.2 lb

27. (a) $\dfrac{4}{3}$; **(b)** $\dfrac{3}{2}$; **(c)** $\dfrac{13}{6}$; **(d)** $\dfrac{1}{2}$

29. Yes, to 7.41 ft from the large end.

31. (a) $\dfrac{\pi}{6}$; **(b)** $\dfrac{2\pi}{15}$ **33.** $2\pi^2 r^2 R$

35. $12\sqrt{5}\pi$

37. If the rod is described by $y = \sqrt{1 - x^2}$, the centroid is at $(0, 2/\pi)$.

41. No. Hint: Any line through the centroid of a homogeneous lamina divides the lamina into equal parts.

43. $\dfrac{2}{9}, \dfrac{\sqrt{2}}{3}$

Exercise Set 5.6

1. 150 ft-lb   **3.** 8 T-mi $\approx 8.448 \times 10^7$ ft-lb   **5.** 4.8 N-m
7. (a) 40 cm; **(b)** 0.5 N-m; **(c)** 2 N-m   **9.** 62.5 in-lb
11. 2808.48 ft-lb   **13.** 4058.48 ft-lb   **15.** 3237 ft-lb
17. 3.9×10^9 ft-lb
19. (a) 2,995,200 ft-lb; **(b)** 5,241,600 ft-lb

21. (a) 3960π ft-lb; (b) $\int_{-2}^{2} 660\sqrt{4-y^2}\,(2-y)\,dy$

23. 3,213,600 ft-lb

25. Fill as often as possible to minimize the work.

27. $k_2 = \dfrac{k_1}{2}$ **29.** $3k, \dfrac{3k}{2}$

Exercise Set 5.7

1. Same for all three, $499.2\,\dfrac{\text{lb}}{\text{ft}^2}$; 2995.2 lb **3.** 2433.6 lb

5. 7800 lb **7.** 4992 lb **9.** 2928.64 lb

11. $\dfrac{80}{9}\sqrt{3}(62.4)$ lb

13. 8424 and 59,904 lb on the ends, 55,640 on the sides
15. 514,800 lb **17.** (a) 128/3 lb; (b) 512 lb
19. 2.6π lb

Review Exercises

1. (a)

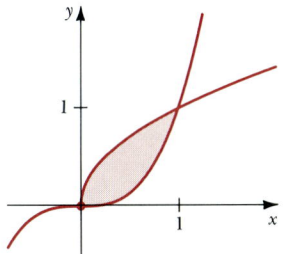

(b) $\dfrac{5}{12}$

3. (a)

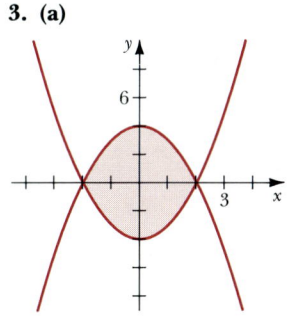

(b) $\dfrac{64}{3}$

5. (a)

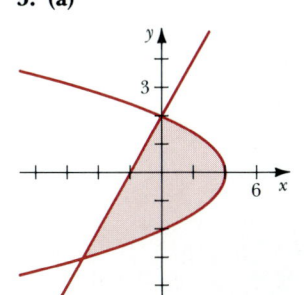

(b) $\dfrac{125}{6}$

7. (a)

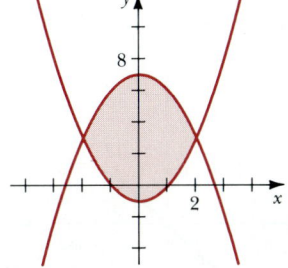

(b) $\dfrac{64}{3}$

9. (a)

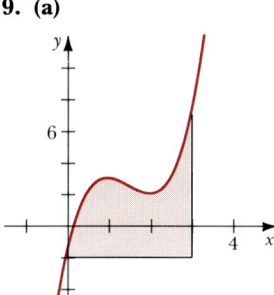

(b) $\dfrac{27}{2}$

11. (a)

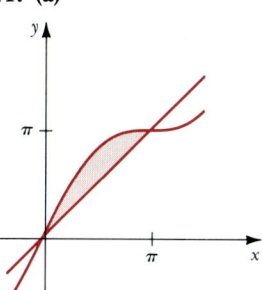

(b) 2

13. (a)

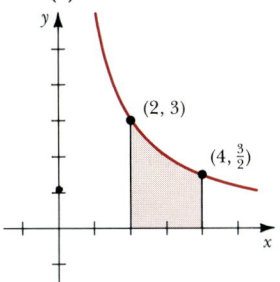

(b) 9π

15. (a)

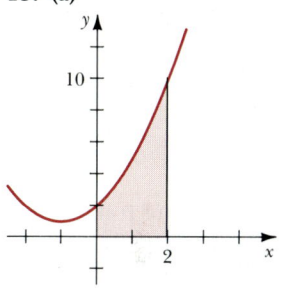

(b) $\dfrac{1016\pi}{15}$

17. (a)

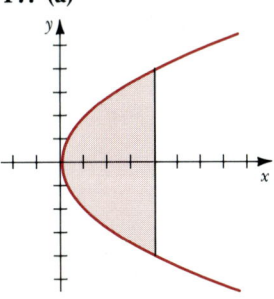

(b) 32π; use only half the shaded region.

19. (a)

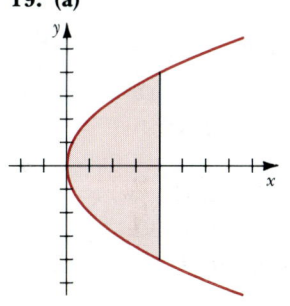

(b) $\dfrac{1024}{15}\pi$

21. (a)

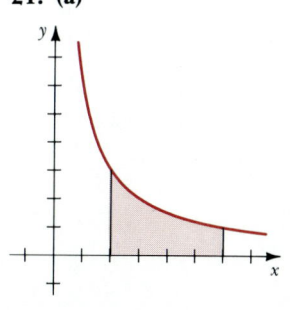

(b) 48π

23. (a)

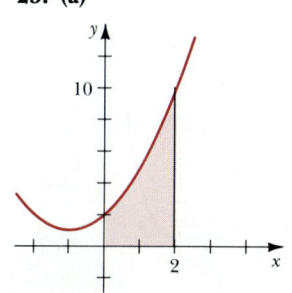

(b) $\dfrac{80}{3}\pi$

25. (a)

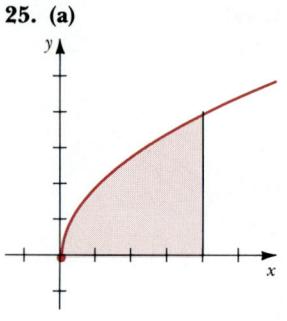

(b) $\dfrac{256\pi}{5}$ **27. (a)** The same figure as 25(a); **(b)** $\dfrac{512\pi}{15}$

29. $12\displaystyle\int_0^5 \sqrt{25-x^2}\,dx$ **31.** 32   **33.** $20\sqrt{2}$

35. $\ln(3+2\sqrt{2})$ **37.** $\dfrac{2}{3}\pi(2\sqrt{2}-1)$ **39.** $\left(\dfrac{29}{12},\dfrac{35}{12}\right)$

41. $\left(0,\dfrac{-2}{5}\right)$ **43.** $\left(\dfrac{2}{5},\dfrac{1}{2}\right)$ **45.** $\dfrac{16\pi}{15}$ **47.** 500 in-lb

49. 9.31×10^9 ft-lb

51. (a) $24{,}710\pi$ ft-lb; **(b)** $47{,}174\pi$ ft-lb   **53.** 26430 ft-lb

55. 1065 lb

Chapter 6

Exercise Set 6.1

1. Yes   **3.** Yes   **5.** No   **7.** Yes   **9.** Yes

11. (a) $\mathbb{R}, \mathbb{R}$; **(b)** $f^{-1}(x) = \dfrac{x-4}{3}$; $\mathbb{R}, \mathbb{R}$

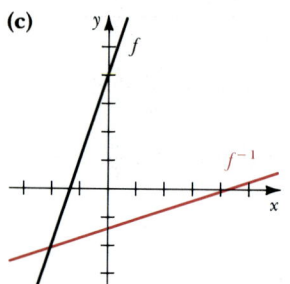

(c)

13. (a) $\mathbb{R}, \mathbb{R}$; **(b)** $f^{-1}(x) = \sqrt[3]{x}$; $\mathbb{R}, \mathbb{R}$
(c)

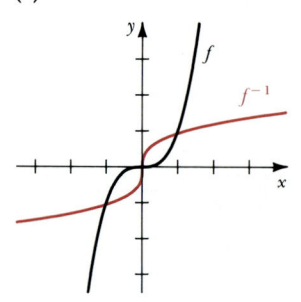

15. No inverse

17. (a) $[-1, \infty)$, $[0, \infty)$; **(b)** $f^{-1}(x) = x^2 - 1$; $[0, \infty)$, $[-1, \infty)$
(c)

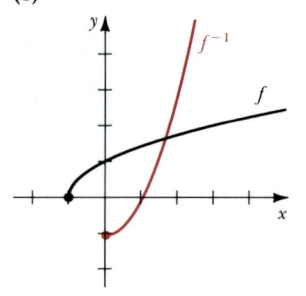

19. (a) $(-\infty, 1)$ and $(1, \infty)$, $(-\infty, 0)$ and $(0, \infty)$;

(b) $f^{-1}(x) = \dfrac{x+1}{x}$; $(-\infty, 0)$ and $(0, \infty)$, $(-\infty, 1)$ and $(1, \infty)$

(c)

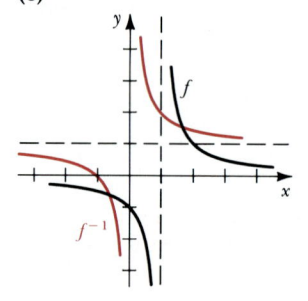

21. $f_1(x) = x^2 - 4$, $x \geq 0$
or $f_2(x) = x^2 - 4$, $x \leq 0$

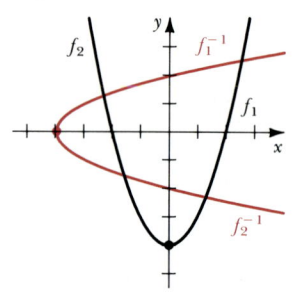

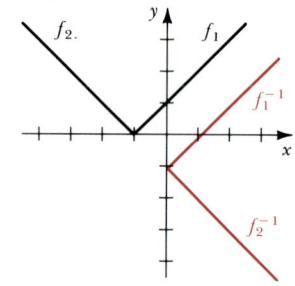

23. $f_1(x) = x + 1$, $x \geq -1$
or $f_2(x) = -x - 1$, $x < -1$

25. One possibility is $f_1(x) = \sin x$, x in $\left[-\dfrac{\pi}{2}, \dfrac{\pi}{2}\right]$

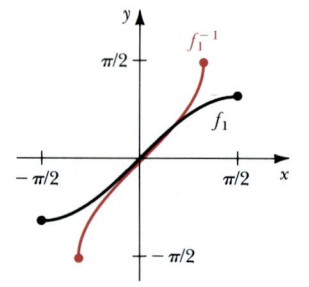

27. $f_1(x) = \dfrac{1}{x^2+1}$, $x \geq 0$ or $f_2(x) = \dfrac{1}{x^2+1}$, $x \leq 0$

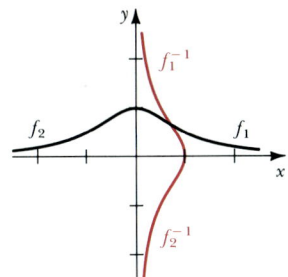

29. $f^{-1}(x) = \dfrac{1}{3}(x+1)$, $(-\infty, \infty)$; $(f^{-1})'(x) = \dfrac{1}{3}$

31. $f^{-1}(x) = x^{1/3}$, $[1, 2]$; $(f^{-1})'(x) = \dfrac{1}{3}x^{-2/3}$

33. $f^{-1}(x) = \dfrac{1}{x}$, $(0, \infty)$; $(f^{-1})'(x) = -\dfrac{1}{x^2}$

35. $f^{-1}(x) = \dfrac{1}{x-1}$, $\left[\dfrac{3}{2}, 2\right]$; $(f^{-1})'(x) = -\dfrac{1}{(x-1)^2}$

37. $\dfrac{1}{4}$ **39.** 4

47. If $f = f^{-1}$ and (a, b) is on the graph of f, then (b, a) is also on the graph of f. The graph is symmetric with respect to the line $y = x$.

49. $m = -1$, b arbitrary, or $m = 1$, $b = 0$

51. $f^{-1}(x) = \begin{cases} 2 - x, & \text{if } x > 1, \\ 3 - 2x, & \text{if } x \leq 1 \end{cases}$

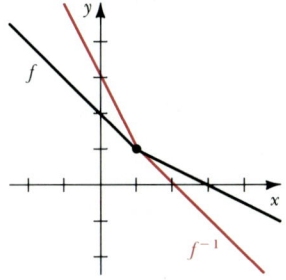

53. $f^{-1}(x) = \begin{cases} x, & \text{if } x < 0, \\ \sqrt{x}, & \text{if } 0 \leq x \leq 1, \\ x^2, & \text{if } 1 < x \end{cases}$

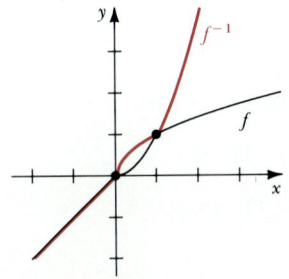

55. $k = \pm 1$ **63. (b)** 0.682; **(c)** 0.417
65. One example is $f(x) = x$.
67. One example is $f(x) = g^{-1}(x) + C$.
69. (b) $f^{-1}(x) = g^{-1}(x) + C$

Exercise Set 6.2

1. $\dfrac{\pi}{4}$ **3.** $\dfrac{\pi}{3}$ **5.** $\dfrac{\pi}{4}$ **7.** π **9.** $\dfrac{12}{13}$ **11.** 0 **13.** $\dfrac{\pi}{4}$

15. $\dfrac{\sqrt{5}}{10}(1 - 2\sqrt{3})$ **17.** $\dfrac{3 + 4\sqrt{3}}{10}$

19. (a) $4\pi/3$ is not in the range of arccos **(b)** $\dfrac{2\pi}{3}$

21. (a) 1; **(b)** 0; **(c)** undefined; **(d)** 0

23. (a) $-\dfrac{\sqrt{2}}{2}$; **(b)** $\dfrac{\sqrt{2}}{2}$; **(c)** -1; **(d)** $-\sqrt{2}$

25. $\dfrac{1}{\sqrt{-x^2 - 2x}}$

27. $\dfrac{2x - 1}{x^4 - 2x^3 + 7x^2 - 6x + 10}$

29. $\dfrac{\sqrt{1-x^2}\arccos x - x}{\sqrt{1-x^2}\sqrt{1 - (x \arccos x)^2}}$

31. $\cos x \arcsin x + \dfrac{\sin x}{\sqrt{1-x^2}}$

33. $\left[1 - \left(\dfrac{x^3+7}{x^2-x}\right)^2\right]^{-1/2}\left[\dfrac{x^4 - 2x^3 - 14x + 7}{(x^2-x)^2}\right]$

35. $[2(1+x)\sqrt{x}\,|\arctan\sqrt{x}+1|\sqrt{(\arctan\sqrt{x}+1)^2 - 1}]^{-1}$

37. $\dfrac{1}{6}\arctan\dfrac{x}{6} + C$ **39.** $\dfrac{\pi}{6}$ **41.** $\dfrac{1}{6}\arctan\dfrac{2x}{3} + C$

43. $\operatorname{arcsec}|4x| + C$ **45.** $\dfrac{1}{2}\operatorname{arcsec} x^2 + C$ **47.** $\dfrac{\pi}{6}$

49. $\dfrac{1}{2}\ln 2$ **51.** $5\arcsin\left(\dfrac{x}{5}\right) + \sqrt{25 - x^2} + C$

53. $\dfrac{1}{2}\arctan\left(\dfrac{x+2}{2}\right) + C$

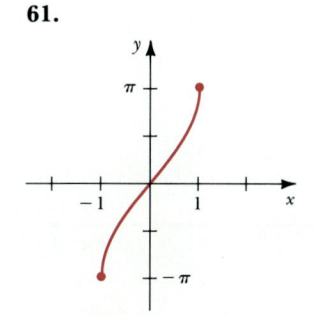

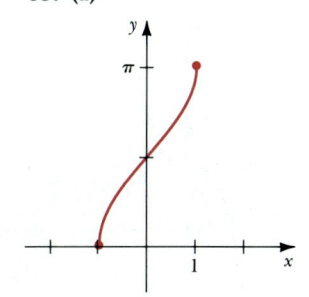

61.   **63. (a)**

65. [graph] **67.** [graph] **41.** [graph] **43.** [graph]

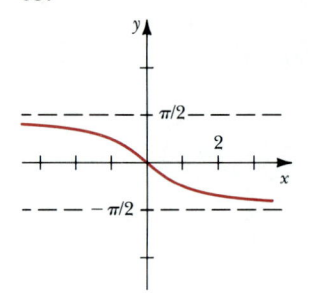

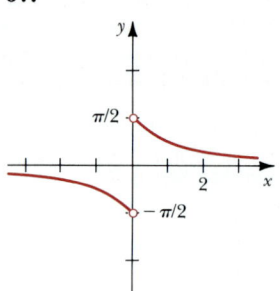

69. 1 **71.** 1 **73.** (a) $-\dfrac{1}{1+x^2}$; (c) $\dfrac{\pi}{2}$

83. For $x > 1$, $\operatorname{arcsec} x = \arctan\sqrt{x^2 - 1}$ and $\operatorname{arccsc} x = \dfrac{\pi}{2} - \arctan\sqrt{x^2 - 1}$. For all x, $\operatorname{arccot} x = \dfrac{\pi}{2} - \arctan x$.

85. $\dfrac{15}{29}$ rad/sec **87.** 3 ft above the floor **89.** $\dfrac{\pi}{3}$

91. Hint: Find areas that the integrals represent.

45. [graph] **47.** [graph]

Exercise Set 6.3

1. $3e^{3x}$ **3.** $\dfrac{e^{\sqrt{x}}}{2\sqrt{x}}$ **5.** $\dfrac{e^x(1-x) + e^{-x}(1+x)}{(e^x + e^{-x})^2}$

7. $e^x(\ln x + x^{-1})$ **9.** $2e^x \cos x$ **11.** $\dfrac{e^{\arcsin x}}{\sqrt{1-x^2}}$

13. $\cos x \, e^{\sin x}$ **15.** $3x^2 e^{x^3} \arcsin x + \dfrac{e^{x^3}}{\sqrt{1-x^2}}$ **17.** 2

19. $-\left(\dfrac{e^y + ye^x}{e^x + xe^y}\right)$ **21.** $\dfrac{1 - xye^{xy}}{x^2 e^{xy}}$ **23.** $\dfrac{e^{3x-2}}{3} + C$

25. $\dfrac{e^2 - 1}{2e}$ **27.** $\ln(e^x + 1) + C$ **29.** $e^{\sin x} + C$

31. $\dfrac{e^{7x}}{7} + C$ **33.** $\arcsin e^x + C$ **35.** $\dfrac{1}{2} e^{\arctan x^2} + C$

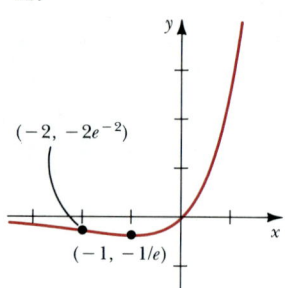

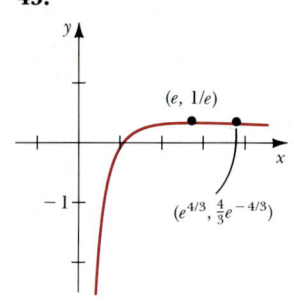

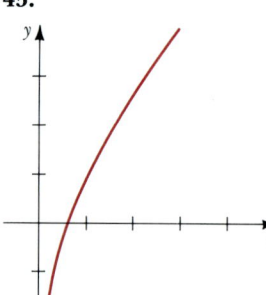

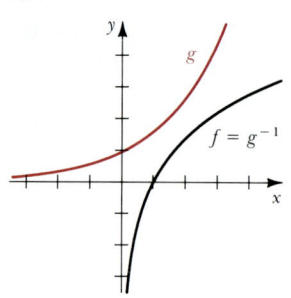

37. [graph] **39.** [graph]

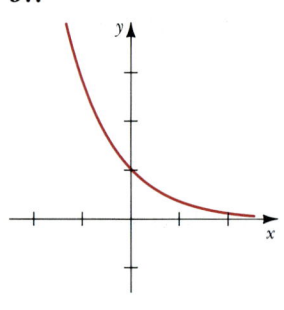

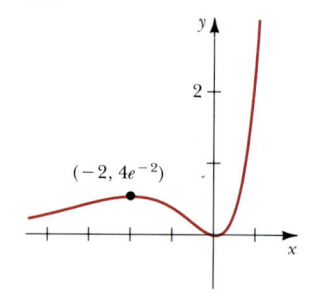

49. [graph] **51.** [graph]

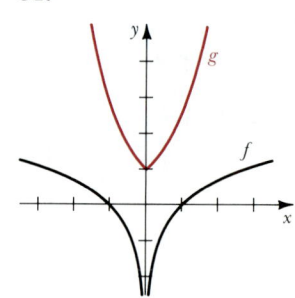

53. [graph]

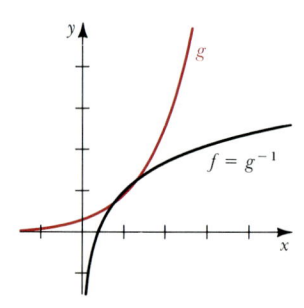

55. 1 **57.** 0 **59.** 0 **61.** ∞ **63.** $y = ex$ **65.** $e - 1$

67. $\arctan e^a - \dfrac{\pi}{4}$ **69.** $\dfrac{\pi}{e}(e - 1)$ **71.** No

75. (a) 100; (b) 3 months; (c) 3 **77.** 2 sec

79. $(-\infty, 0)$ and $(0, \infty)$

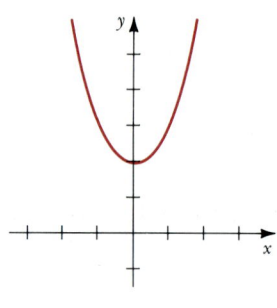

Exercise Set 6.4

1. $2^x \ln 2$ **3.** $3^x \ln 3 + x\, 2x^{x^2+1} \ln 2$ **5.** $e(\sin x)^{e-1} \cos x$
7. $(3+\pi)^x \ln(3+\pi)$ **9.** $(\pi + \sqrt{2})^2 x^{(\pi+\sqrt{2})^2 - 1}$
11. $\dfrac{2x}{(x^2+1)\ln 10}$ **13.** $\dfrac{2x}{(x^2+1)(x^2+2)\ln 2}$
15. $\dfrac{1}{x \ln x \ln 10}$ **17.** $\dfrac{2x+3}{2|x^2+3x-1|\ln 2}$
19. $-\dfrac{x^5 + 5x^4 + 15x + 15}{3x(x+2)(x^4-5)\ln 4}$
21. $(x+1)^x [\ln(x+1) + x/(x+1)]$
23. $x^{\pi x}(\pi \ln x + \pi)$ **25.** $(\cos x)^x (\ln \cos x - x \tan x)$
27. $x^{\sin x}(\cos x \ln x + (\sin x)/x)$ **29.** $\dfrac{xy \ln 10 \log_{10} y - y^2}{xy \ln 10 \log_{10} x - x^2}$
31. $\dfrac{-yx^{y-1} - y^x \ln y}{x^y \ln x + xy^{x-1}}$ **33.** $\dfrac{4}{\ln 3}$ **35.** $\dfrac{3^{x^2}}{2 \ln 3} + C$
37. $\dfrac{x^{e+1}}{e+1} + C$
39. 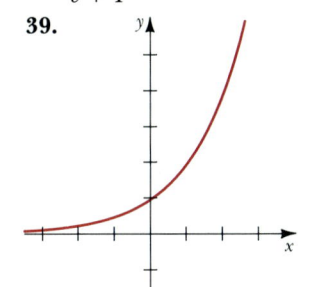
41. 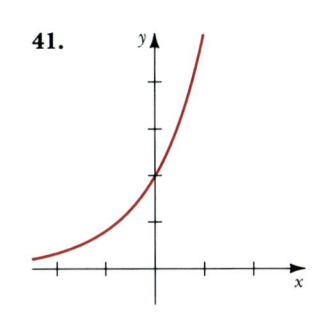
43.
45.
47. 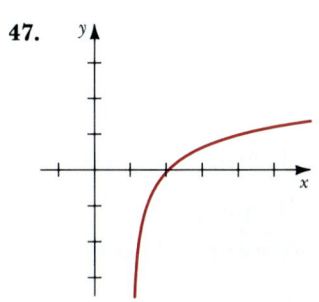
49. $-\infty$ **51.** 0 **55.** $2(2x)^{2x}$

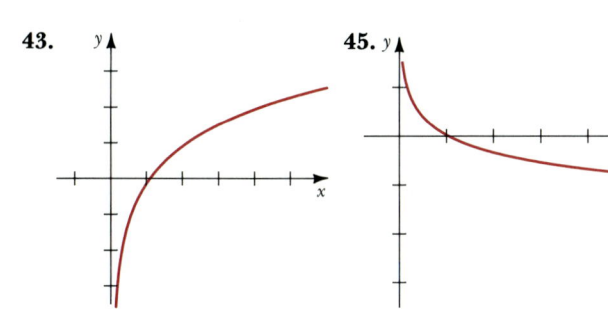

Exercise Set 6.5

1. 1 **3.** 1 **5.** e^2 **7.** 1 **9.** 1 **11.** 1 **13.** 1
15. 1 **17.** 0 **19.** 1 **21.** e **23.** $\sqrt{e}$ **25.** $1/e$
27.

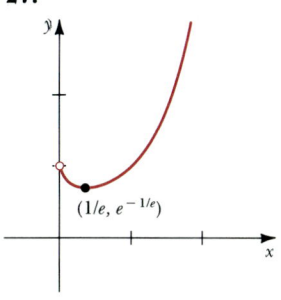

$(1/e, e^{-1/e})$

29. $A(0)e^{it}$
33. 0.756828, 0.708381, 0.694086, 0.693174

Exercise Set 6.6

1. (a) $(50)2^{-t/590}$; (b) 44.46 mg **3.** 144 mg
5. (a) $(1000)2^{t/4}$; (b) 3364 **7.** The second bank
9. (a) 196,181 thousand; (b) 281,407 thousand
11. 11.6 more minutes **13.** 48.56 days
15. (a) 51%; (b) 96.8 min
17. 3.097, 9.589, 29.69, 91.95, 284.7, 881.7, 2730
19. 2481 years

Exercise Set 6.7

1. $y = Ce^{7x}$ **3.** $y = \tan(x+C)$
5. $y - \sin y = \cos x - \dfrac{x^3}{3} + C$
7. $y = \dfrac{1}{\sin x + C}$ **9.** $e^{-x} + e^{-y} = C$ **11.** $\ln|y| = e^{-x} + C$
13. $y = \dfrac{2}{2 \sin x + 1}$ **15.** $y = 2e^{x^2/2} - 1$
17. $\sin x + e^{-y} - 1 = 0$ **19.** No; higher **21.** 9.8 min
23. 0.0322 lb/gal **25.** 75.3%
27. 13.92°C **29.** 3:34 A.M.
31. (b) $\left(\dfrac{1}{kL} \ln\left(\dfrac{P(0)}{L - P(0)}\right), \dfrac{L}{2}\right)$
33. (a) ae^{-b}; (b) a
35. (a) 184,531 thousand; (b) 271,006 thousand
37. (a) $A(t) = \dfrac{A_2 Ce^{(A_1+A_2)kt} - A_1}{1 + Ce^{(A_1+A_2)kt}}$;

(b) when $A(t) = \dfrac{A_2 - A_1}{2}$

39. (a) $\dfrac{ds}{dt} = \dfrac{kA(C - c(t))}{V}$; (b) $s(t) = \dfrac{kA[Ct - \int c(t)\, dt]}{V}$

41. 5 days, 10 hours, and 5 minutes

Exercise Set 6.8

1. $2x \cosh(x^2)$ **3.** $\dfrac{x+1}{\sqrt{x^2+2x}} \cosh \sqrt{x^2+2x}$

5. $\dfrac{\sinh(\ln x)}{x}$

7. $-e^{x^3+3x}(3x^2+3)\,\text{sech}(e^{x^3+3x})\,\tanh(e^{x^3+3x})$

9. $3(\tanh\sqrt{x^2-x})^2 (\text{sech}\sqrt{x^2-x})^2 \dfrac{2x-1}{2\sqrt{x^2-x}}$

11. $2x \cosh x^2 e^{\sinh x^2}$ **13.** $\dfrac{x \sinh x - \cosh x - 1}{(x + \sinh x)^2}$

15. $\dfrac{(\text{sech } x)^2(1+x^2) - 2x \tanh x}{(1+x^2)^2}$ **17.** $2 \sinh \sqrt{x} + C$

19. $\dfrac{\tanh 2x}{2} + C$ **21.** $\dfrac{(\sinh x)^2}{2} + C$ **23.** $-\text{csch } x + C$

25. $e^{\tanh x} + C$ **27.** 1 **29.** 0 **31.** 0

45. $\dfrac{\pi}{2}(a + \sinh a \cosh a)$ **51.** (a) $L = \dfrac{2T_0}{w} \sinh \dfrac{50w}{T_0}$;

(b) 100.66 ft

Exercise Set 6.9

1. $\dfrac{3x^2}{\sqrt{1+x^6}}$ **3.** $\dfrac{1}{2\sqrt{x^2-x}}$

5. $2x\left(\cosh x^2 \text{ arcsinh } x^2 + \dfrac{\sinh x^2}{\sqrt{1+x^4}}\right)$ **7.** $\dfrac{(\sec x)^2}{1-(\tan x)^2}$

9. $\dfrac{\sqrt{x^2-1}\,\text{arccosh } x - \sqrt{1+x^2}\,\text{arcsinh } x}{\sqrt{x^4-1}\,(\text{arccosh } x)^2}$

11. $\text{arcsinh}\left(\dfrac{4}{3}\right)$ **13.** $\dfrac{1}{2}\text{arcsinh}\left(\dfrac{x^2}{3}\right) + C$

15. $\dfrac{1}{3}\text{arccosh}\dfrac{3x}{2} + C$

17. $\text{arctanh } e^x + C$, if $x < 0$ or $\text{arccoth } e^x + C$ if $x > 0$

19. $\dfrac{1}{2}\text{arctanh}\left(\dfrac{x-2}{2}\right) + C$ **29.** 1 **31.** 0 **33.** ln 2

Review Exercises

1. (b) $f^{-1}(x) = \dfrac{x}{2} - \dfrac{3}{2}$ $\mathbb{R}, \mathbb{R}$ **3.** (b) $f^{-1}(x) = (x+1)^{1/3}$ $\mathbb{R}, \mathbb{R}$

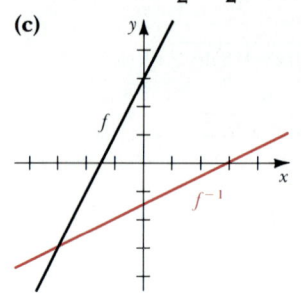

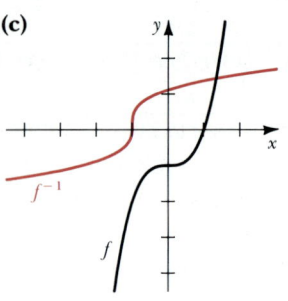

(c) [graph]   (c) [graph]

(d) $\dfrac{1}{2}$ (d) $\dfrac{1}{3(\sqrt[3]{x+1})^2}$

5. (b) $f^{-1}(x) = (x-1)^{1/2}$ $[1, \infty), [0, \infty)$;

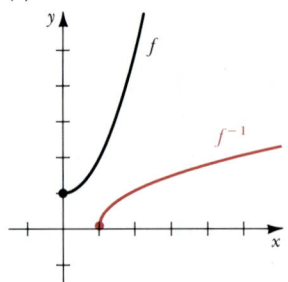

(c) [graph]

(d) $\dfrac{1}{2\sqrt{x-1}}$

7. (b) $f^{-1}(x) = \dfrac{1-x}{x}$ $(-\infty, 0) \cup (0, \infty)$,

$(-\infty, -1) \cup (-1, \infty)$;

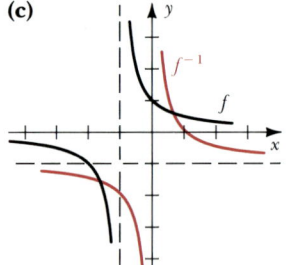

(c) [graph]

(d) $\dfrac{-1}{x^2}$

9. $\dfrac{1}{\sqrt{9-x^2}}$ **11.** $3x^2 e^{x^3}$ **13.** $3(\ln 2)x^2 2^{x^3}$ **15.** $3x^2$

17. $e^{\tan x} + x(\sec x)^2 e^{\tan x}$ **19.** $\dfrac{1 - 2x - x^2}{(x+1)(x^2+1)\ln 10}$

21. $3x^2$ **23.** $\dfrac{1}{1+e^x}$ **25.** ex^{e-1}

27. $3e^{3x} \operatorname{sech} x - (1 + e^{3x}) \operatorname{sech} x \tanh x$
29. $e^x[\sinh(2x - 1) + 2\cosh(2x - 1)]$
31. $x^{\cos x}\left(\dfrac{\cos x}{x} - \sin x \ln x\right)$ **33.** $\dfrac{xy - x^2}{2x^2 + y^2 + xy}$
35. $\dfrac{1}{3} \arctan\left(\dfrac{x}{3}\right) + C$ **37.** $\dfrac{1}{3}(e^8 - e)$
39. $\arcsin\left(\dfrac{x}{3}\right) + C$ **41.** $\arcsin e^x + C$
43. $\arctan(x - 1) + C$ **45.** $\dfrac{3}{2 \ln 2}$ **47.** $\cosh(\ln x) + C$
49. (a) $-\dfrac{\pi}{6}$; (b) $-\dfrac{\pi}{4}$; (c) $\dfrac{\pi}{6}$ **51.** ∞ **53.** 1
55. 1 **57.** 0 **59.** 1
61.

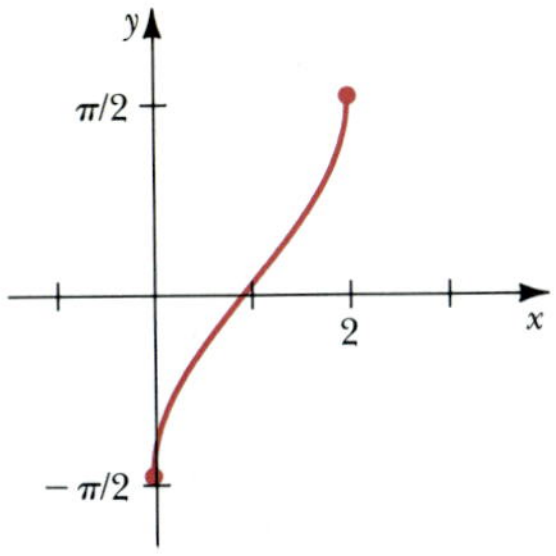

63.

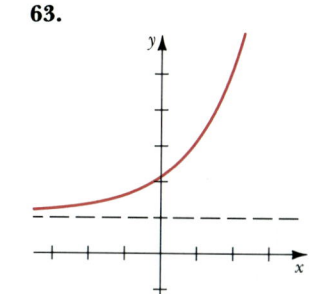

65.

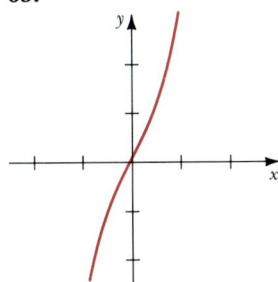

67. $y = \dfrac{x^3}{3} + x^2 + C$ **69.** $y = -\ln|C - e^x|$
71. $y = \tan(\arctan x + C)$ **73.** $y = 5e^{x^2}$
75. $x^2 = y - \ln|y + 1| + 1$ **77.** $y - 1 = \dfrac{1}{5}(x - 5)$
79. (a) $A(t) = A(0)\, 3^{t/3}$; (b) $A(0) \approx 693$
81. (a) 210 lb; (b) 0.85 min **83.** $70\left(\dfrac{3}{14}\right)^{t/3}$

Chapter 7

Exercise Set 7.1

1. $-e^{-x}(x + 1) + C$ **3.** $-\dfrac{\pi}{2}$ **5.** $\dfrac{1}{4}(e^2 + 1)$

7. $x \arcsin x + \sqrt{1 - x^2} + C$
9. $x \arccos x - \sqrt{1 - x^2} + C$
11. $e^x(x^2 - 2x + 2) + C$
13. $\dfrac{1}{4}(8 - 8\pi - \pi^2)$
15. $2x(\ln x - 1) + C$ **17.** $\dfrac{x^4}{8}(4 \ln x - 1) + C$
19. $\dfrac{298}{15}$
21. $x \tan x + \ln|\cos x| + C$
23. $-\dfrac{1}{3}(x^2 + 2)\sqrt{1 - x^2} + C$
25. $\dfrac{e^x(\sin x - \cos x)}{2} + C$
27. $\dfrac{1}{13}(2e^{3x} \sin 2x + 3e^{3x} \cos 2x) + C$
29. $\dfrac{1}{2}(\sec x \tan x + \ln|\sec x + \tan x|) + C$
31. $\dfrac{1}{5}$ **33.** $2(\sqrt{x} \sin \sqrt{x} + \cos \sqrt{x}) + C$
35. $2e^{\sqrt{x}}(\sqrt{x} - 1) + C$
37. $\dfrac{\pi}{4} - \dfrac{1}{2}$
39. $3^x\left(\dfrac{x}{\ln 3} - \dfrac{1}{(\ln 3)^2}\right) + C$
41. $\dfrac{2^x}{(\ln 2)^3}[(\ln 2)^2 x^2 - 2(\ln 2)x + 2] + C$
43. $\dfrac{x}{2}[\sin(\ln x) - \cos(\ln x)] + C$
45. $x \ln(x + \sqrt{x^2 + 1}) - \sqrt{x^2 + 1} + C$
47. $\dfrac{1}{2}(x - \sin x \cos x) + C$
49. (a)

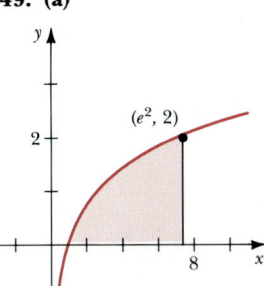

51. (a)

(b) $e^2 + 1$

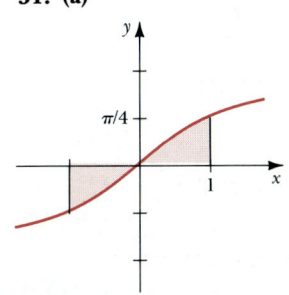

(b) $\dfrac{\pi}{2} - \ln 2$

53. (a) [graph] (b) $\sqrt{2}-1$

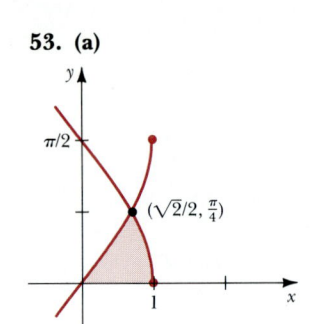

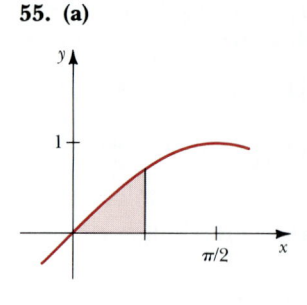

55. (a) [graph] (b) $\dfrac{\pi^2}{8}-\dfrac{\pi}{4}$

57. (a) [graph] (b) $\dfrac{\pi}{3}(2e-5)$

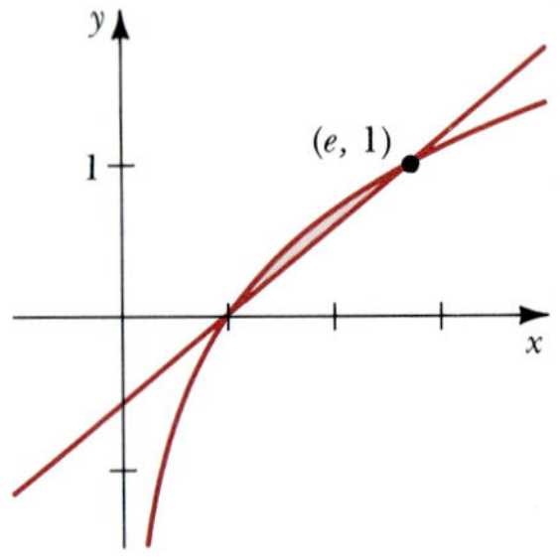

59. $\dfrac{\sqrt{2}\pi}{4}(4-\pi)$ **61.** $\dfrac{\pi}{6}(5+2e-e^2)$

69. $\left(\dfrac{5}{3(2e-3)},\dfrac{3e^2-5}{6(2e-3)}\right)$ **71.** $\dfrac{1}{4}$

29. $e^{\sin x}[2\sin x - \sin^2 x - 1]+C$ **31.** $\dfrac{\sin^2 x}{2}+C$

33. $-\cos x + C$ **35.** $\ln|\csc x - \cot x| + C$

37. $\dfrac{1}{2}\tan^2 x + \ln|\cos x| + C$ **39.** $\dfrac{1}{4}\sin^4 x + C$

41. $\dfrac{1}{2}\cos x - \dfrac{1}{6}\cos 3x + C$

43. $\dfrac{1}{2}\cos x - \dfrac{1}{10}\cos 5x + C$

45. $\dfrac{1}{4}\sin 2x + \dfrac{1}{16}\sin 8x + C$

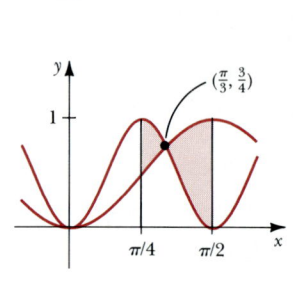

47. (a) [graph] (b) $\dfrac{\pi+2}{8}$

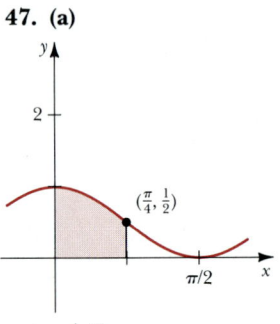

49. (a) [graph] (b) $\dfrac{1}{8}(3\sqrt{3}-2)$

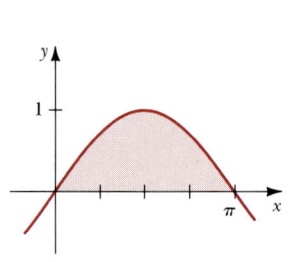

51. (a) [graph] (b) $\dfrac{\pi^2}{2}$

53. (a) [graph] (b) $\dfrac{\pi}{4}(4-\pi)$

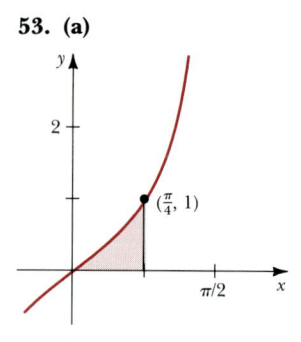

61. $\left(\dfrac{\sqrt{2}\pi-4}{4(\sqrt{2}-1)},\dfrac{1}{4(\sqrt{2}-1)}\right)$

Exercise Set 7.3

1. $\dfrac{x}{2}\sqrt{1-x^2}+\dfrac{1}{2}\arcsin x + C$ **3.** 4π

5. $\arctan\dfrac{x}{3}+C$ **7.** $\ln|\sqrt{16+x^2}+x|+C$ **9.** $1-\dfrac{\pi}{4}$

Exercise Set 7.2

1. $\dfrac{1}{3}\sin^3 x + C$ **3.** 0 **5.** $-\dfrac{1}{15}(\cos 3x)^5 + C$ **7.** $\dfrac{\pi}{96}$

9. $\dfrac{3\pi}{8}$ **11.** $\dfrac{2}{3}(\sin x)^{3/2}+C$ **13.** $\dfrac{10}{3}$

15. $\dfrac{1}{4}\tan^4 x + \dfrac{1}{6}\tan^6 x + C$ **17.** $1-\dfrac{\pi}{4}$

19. $-\dfrac{1}{4}\csc^4 x + C$

21. $-\dfrac{1}{5}\csc^5 x + \dfrac{2}{3}\csc^3 x - \csc x + C$

23. $\dfrac{1}{2}\ln|\csc 2x| - \dfrac{1}{4}(\cot 2x)^2 + C$

25. $\tan x - \dfrac{1}{2}x - 2\cos x + \dfrac{1}{4}\sin 2x + C$

27. $\left(\sin x - \dfrac{\sin^3 x}{3}\right)\ln(\sin x) - \sin x + \dfrac{\sin^3 x}{9}+C$

11. $\sqrt{3}$. No trigonometric substitution is needed in this problem.

13. $\sqrt{3} - \operatorname{arcsec} 2$ **15.** $\dfrac{16}{3} \arcsin \dfrac{3}{4} + \sqrt{7}$

17. $\dfrac{x+1}{2} \sqrt{4 - (x+1)^2} + 2 \arcsin \dfrac{x+1}{2} + C$

19. $\dfrac{\pi}{4}$ **21.** $\dfrac{2x^2 + 9}{24} \sqrt{4x^2 - 9} + C$ **23.** $2 \arctan \sqrt{x} + C$

25. $\dfrac{e^x}{2} \sqrt{1 - e^{2x}} + \dfrac{1}{2} \arcsin e^x + C$ **27.** $2 \arctan e^{x/2} + C$

29. $2 \arctan 2$ **31.** 12π **33.** $6\pi - 12 \arcsin \dfrac{1}{4} - \dfrac{3\sqrt{15}}{4}$

35. 2π **37.** $\sqrt{e^2 + 1} - \sqrt{2} + \ln \left| \dfrac{1 - \sqrt{e^2 + 1}}{e(1 - \sqrt{2})} \right|$

41. $\dfrac{25}{4} \pi$ cm   **43.** $\dfrac{\sqrt{5}}{2} + \dfrac{1}{4} \ln (2 + \sqrt{5})$ **45.** $4\pi^2$

47. $12[2\sqrt{3} - \ln (2 + \sqrt{3})]$

Exercise Set 7.4

1. $\dfrac{1}{2} \ln \left| \dfrac{x-1}{x+1} \right| + C$ **3.** $2 \ln 2 - \ln 3$

5. $\dfrac{1}{2} \ln |x^2 - 1| + C$ **7.** $\dfrac{5}{2} \ln 3 - 2$

9. $\dfrac{x^2}{2} - x + \dfrac{1}{x+1} + 2 \ln |x+1| + C$

11. $-\dfrac{2}{3} \ln |x+1| + \dfrac{5}{12} \ln |x-2| + \dfrac{5}{4} \ln |x+2| + C$

13. $\ln \left| \dfrac{x^3}{x-2} \right| - \dfrac{1}{x-2} + C$

15. $\dfrac{1}{2} \ln \dfrac{27}{5}$ **17.** $\dfrac{1}{4} \ln \dfrac{(x+1)^2}{(x^2+1)} + \dfrac{1}{2} \arctan x + C$

19. $\dfrac{1}{2} \left(\arctan x - \dfrac{x}{x^2 + 1} \right) + C$

21. $\ln \left(\dfrac{2x^2 + 5}{x^2 + 3} \right) - \dfrac{3}{5} \sqrt{10} \arctan \dfrac{\sqrt{10}\, x}{5} +$

$\sqrt{3} \arctan \dfrac{\sqrt{3}\, x}{3} + C$

23. $2\sqrt{x} - 2 \ln (\sqrt{x} + 1) + C$

25. $\dfrac{1}{4} [\ln |e^x - 2| - \ln (e^x + 2)] + C$

27. $\ln \left| \dfrac{\sin x - 1}{\sin x} \right| + C$

29. $\ln |\sin x| - \ln (\sin x + 1) + C$
31. $2\sqrt{x+1} - 2 \ln (\sqrt{x+1} + 1) + C$

33. (a)

[Graph showing y-axis and x-axis with points $(0, -\tfrac{1}{6})$ and $(2, -\tfrac{1}{4})$ marked, with a shaded region]

(b) $\dfrac{1}{5} \ln 6$

35. $\dfrac{\pi}{125} \left(2 \ln 6 + \dfrac{55}{12} \right)$ **37.** $\dfrac{2\pi}{5} \ln \dfrac{27}{4}$ **41.** 1

43. $\ln \left| 1 + \tan \dfrac{x}{2} \right| + C$

45. $\dfrac{1}{2} \left[\ln \left| \tan \dfrac{x}{2} \right| - \dfrac{1}{2} \left(\tan \dfrac{x}{2} \right)^2 \right] + C$

47. $\dfrac{6\sqrt{5}}{25} \arctan \left(\dfrac{\sqrt{5}}{5} \right) - \dfrac{2}{15}$

49. $\dfrac{26}{5} e^{3t} - \dfrac{7}{2} e^{2t} + \dfrac{3}{10} e^{-2t}$

51. $\cos 3t + \sin 3t + e^t - 2e^{-t}$

Exercise Set 7.5

1. $\dfrac{1}{3} \arctan \left(\dfrac{x+1}{3} \right) + C$

3. $\dfrac{1}{54} \left[\arctan \left(\dfrac{x+1}{3} \right) + \dfrac{3(x+1)}{x^2 + 2x + 10} \right] + C$

5. $\dfrac{x-1}{2} \sqrt{2x - x^2} + \dfrac{1}{2} \arcsin (x - 1) + C$

7. $\dfrac{x+1}{2} \sqrt{3 - x^2 - 2x} + 2 \arcsin \left(\dfrac{x+1}{2} \right) + C$

9. $\ln |x + 3 + \sqrt{x^2 + 6x + 18}| + C$

11. $\dfrac{1}{2} \ln |x^2 + 2x + 2| - \arctan (x + 1) + C$ **13.** $\ln \dfrac{42}{13}$

15. $\ln \left| \dfrac{x^2 + 1}{x^2 + x + 2} \right| - \arctan x + C$ **17.** $\dfrac{\sqrt{3}\pi}{9} - \dfrac{1}{3}$

19. $\dfrac{1}{6} \ln |x - 1| + \dfrac{1}{4} \ln (x^2 + 1) - \dfrac{1}{2} \arctan x -$

$\dfrac{1}{3} \ln |x^2 + x + 1| + C$

21. $6 \left[\dfrac{x^{7/6}}{7} - \dfrac{x^{5/6}}{5} + \dfrac{x^{1/2}}{3} - x^{1/6} + \arctan x^{1/6} \right] + C$

23. $6[e^{x/6} - \arctan (e^{x/6})] + C$

25. $\dfrac{\pi}{2}$ **27.** $e^{3t} \cos t$ **29.** $e^t \cos t + 2e^t \sin t - e^{2t}$

39. $\dfrac{1}{8} \tan^8 x + \dfrac{1}{6} \tan^6 x + C$ **41.** $\arctan e^x + C$

43. $\dfrac{1}{16}\left(\dfrac{4}{x} + \ln|64 - 16x| - \ln|x|\right) + C$

45. $\dfrac{1}{7}(16 - x^2)^{7/2} - \dfrac{16}{5}(16 - x^2)^{5/2} + C$

47. $\dfrac{1}{4}\sec^4 x - \sec^2 x + \ln|\sec x| + C$ **49.** $(\ln x)^2 + C$

Exercise Set 7.6

1. (a) 1.28210; (b) 1.29532; (c) actual value 1.29584
3. (a) 1.49068; (b) 1.46371 **5.** (a) 0.70711; (b) 0.94281
7. (a) 0.65916; (b) 0.65933 **9.** 1.91010 **11.** 1.91014

13. 0.68016 **15.** 193,600 **17.** $\dfrac{1}{24}$ **19.** $\dfrac{1}{240}$

21. 55.588 **23.** (a) 0.683; (b) 0.341; (c) 0.273; (d) 0.067

51. $\left(\dfrac{x^3}{3} + \dfrac{x^2}{2} + 2x\right)\ln x - \dfrac{x^3}{9} - \dfrac{x^2}{4} - 2x + C$

53. $\dfrac{1}{2}\sin^2(\ln x) + C$

55. $\dfrac{x}{2}\sqrt{9x^2 - 16} - \ln|3x + \sqrt{9x^2 - 16}| + C$

57. $\dfrac{1}{2}\left[\ln\left|\dfrac{x}{(x^2 + 2x + 2)^2}\right| - \arctan(x + 1)\right] + C$

59. $2(\tan x + \sec x) - x + C$

Review Exercises

1. $\sin x + (1 - x)\cos x + C$ **3.** $\dfrac{e^{2x}}{4}(2x - 1) + C$

5. $\sqrt{1 + x^2} + C$ **7.** $\dfrac{1}{4}(\ln|x - 2| - \ln|x + 2|) + C$

9. $\tan x - \cot x + C$

11. $\dfrac{2}{3}(\sin x)^{3/2} - \dfrac{4}{7}(\sin x)^{7/2} + \dfrac{2}{11}(\sin x)^{11/2} + C$

13. $\dfrac{-2}{3}\ln|x| - \dfrac{1}{12}\ln|x + 3| + \dfrac{3}{4}\ln|x - 1| + C$

15. $-\dfrac{(\cos 2x)^3}{6} + \dfrac{(\cos 2x)^5}{10} + C$

17. $\dfrac{\sqrt{x^2 - 1}}{3}(x^2 + 2) + C$ **19.** $\dfrac{1}{12}(x^2 + 1)^6 + C$

21. $\dfrac{1}{5}\sec^5 x - \dfrac{1}{3}\sec^3 x + C$

23. $x + \ln x - \dfrac{1}{2}\ln(x^2 + 1) + \arctan x + C$

25. $-\dfrac{\sqrt{x^2 + 1}}{2x} + \dfrac{1}{2}\ln(\sqrt{x^2 + 1} - 1) - \dfrac{1}{2}\ln|x| + C$

27. $\operatorname{arcsec} x + C$ **29.** $\dfrac{1}{3}(1 - x^2)^{3/2} - \sqrt{1 - x^2} + C$

31. $\dfrac{3}{8}(2x^2 - 1)\sin 2x + \dfrac{x}{4}(3 - 2x^2)\cos 2x + C$

33. $\dfrac{7}{4}\ln|x + 7| + \dfrac{1}{4}\ln|x - 1| + C$

35. $\dfrac{2x - 1}{4}[\ln(2x - 1) - 1] + C$

37. $\dfrac{-1}{2}e^{-x^2}(2 + 2x^2 + x^4) + C$

61. $\dfrac{-2}{13}e^{3x}\cos 2x + \dfrac{3}{13}e^{3x}\sin 2x + C$

63. $\arcsin\left(\dfrac{x + 3}{4}\right) + C$

65. $\dfrac{x}{2}[\sin(\ln 2x) - \cos(\ln 2x)] + C$

67. $\dfrac{x^2 - 1}{2}\arctan\sqrt{x} + \dfrac{\sqrt{x}}{6}(3 - x) + C$

69. $\dfrac{3}{16}(2x\sqrt{4x^2 - 1} + \ln|2x + \sqrt{4x^2 - 1}|) + C$

71. $x - \dfrac{3}{x} + 4\ln|x + 2| + C$

73. $\dfrac{1}{2}\ln|x + 1| + \ln|x - 1| + \dfrac{9}{2}\arctan x - \dfrac{1}{4}\ln|x^2 + 1| + C$

75. $\dfrac{1}{6}[2x^3 \arctan x - x^2 + \ln(x^2 + 1)] + C$

77. $\dfrac{1}{4}\tan^2 2x + \dfrac{1}{2}\ln|\cos 2x| + C$

79. $\dfrac{1}{18}(1 - 2x^3)^{3/2} - \dfrac{1}{6}(1 - 2x^3)^{1/2} + C$

81. $\dfrac{-\cos 5x}{10} - \dfrac{\cos x}{2} + C$ **83.** $-\dfrac{1}{8}\cos 4x - \dfrac{1}{4}\cos 2x + C$

85. $\dfrac{x + 4}{2}[\ln(x + 4) - 1] - (x + 1)\ln(x + 1) + C$

87. $\sinh(\ln x) + C = \dfrac{x^2 - 1}{2x} + C$

89. $\ln|x+1| - \frac{1}{2}\ln|x^2+x+1| + \frac{\sqrt{3}}{3}\arctan\left(\frac{\sqrt{3}}{3}(2x+1)\right) + C$

91. $\frac{-4}{3}\ln\left(\frac{2-\sqrt{3x}}{2}\right) - \frac{2}{3}\sqrt{3x} + C$

93. $4\left(\frac{1}{5}x^{5/4} - \frac{1}{4}x + \frac{1}{3}x^{3/4} - \frac{1}{2}x^{1/2} + x^{1/4} - \ln(x^{1/4}+1)\right) + C$

95. $\ln|e^{x/3} - e^{x/6} + 1| - 2\ln(e^{x/6}+1) + 2\sqrt{3}\arctan\left(\frac{\sqrt{3}}{3}(2e^{x/6}-1)\right) + C$

97. $\ln|\tan x| + C$

99. $\ln\left|1+\tan\frac{x}{2}\right| - \ln\left|1-\tan\frac{x}{2}\right| - 2\left(1+\tan\frac{x}{2}\right)^{-1} + 2\left(1+\tan\frac{x}{2}\right)^{-2} + C$

101. $2\sqrt{2}\arctan\left(\frac{\sqrt{2}}{2}\tan\frac{x}{2}\right) + C$ **103.** 1 **105.** $\pi + \frac{\pi^2}{2}$

107. $4\pi e^3$ **109.** $\left(\frac{1}{4}(e^2+1), \frac{1}{2}e - 1\right)$

111. (a) 0.7969791; (b) 0.8234031
113. (a) 0.7932828; (b) 0.7948161
115. 8.6926

Chapter 8

Exercise Set 8.1

1. (a) $1, \frac{2}{3}, \frac{1}{2}$; (b) yes; (c) 0; (d) yes

3. (a) $-1, \frac{1}{2}, -\frac{1}{3}$; (b) yes; (c) 0; (d) yes

5. (a) $0, \frac{1}{5}, \frac{1}{3}$; (b) yes; (c) 1; (d) yes

7. (a) $9, \frac{19}{4}, \frac{17}{5}$; (b) yes; (c) 1; (d) yes

9. (a) $1, \frac{5}{2}, \frac{11}{3}$; (b) no; (d) no

11. (a) $\pi, \frac{\pi}{2}, \frac{\pi}{3}$; (b) yes; (c) 0; (d) yes

13. (a) $e, \frac{e^2}{2}, \frac{e^3}{3}$; (b) no; (d) no

15. (a) $e, \frac{e^2}{2^p}, \frac{e^3}{3^p}$; (b) no; (d) no

17. (a) $-1, \frac{1}{4}\ln 2 - 1, \frac{1}{9}\ln 3 - 1$; (b) yes; (c) -1; (d) yes

19. (a) $0, 1, \frac{\sqrt{3}}{2}$; (b) yes; (c) 0; (d) yes

21. (a) $\frac{e^2}{6}, \frac{e^3}{14}, \frac{e^4}{24}$; (b) no; (d) no

23. (a) $1, \frac{1}{2}, \frac{1}{4}$; (b) yes; (c) 0; (d) yes

25. (a) $-\frac{4}{\sqrt[4]{8}}, \frac{7}{\sqrt[4]{540}}, -\frac{12}{\sqrt[4]{19768}}$; (b) yes; (c) 0; (d) yes

27. (a) $\frac{1}{2}, \frac{1}{6}, \frac{1}{12}$; (b) yes; (c) 0; (d) yes

29. (a) $\sqrt{2}-1, \sqrt{5}-2, \sqrt{10}-3$; (b) yes; (c) 0; (d) yes

31. (a) $\frac{1}{\sqrt{2}-1}, \frac{1}{\sqrt{5}-2}, \frac{1}{\sqrt{10}-3}$; (b) no; (d) no

33. (a) $0, \frac{1}{2}, \frac{\sqrt{3}}{6}$; (b) yes; (c) 0; (d) yes

35. This is the same as Exercise 3, since $\cos n\pi = (-1)^n$.

37. (a) $-\frac{1}{2}, \frac{2}{5}, -\frac{3}{10}$; (b) yes; (c) 0; (d) yes

39. (a) $\frac{1}{2}, \frac{e}{1+e}, \frac{e^2}{1+e^2}$; (b) yes; (c) 1; (d) yes

41. (a) $2, \frac{9}{4}, \frac{64}{27}$; (b) yes; (c) e; (d) yes

43. (a) $1, \frac{1}{4}, \frac{1}{36}$; (b) yes; (c) 0; (d) yes **45.** No

47. $x_n = (0.85)^n(250{,}000) + 50{,}000 \sum_{i=0}^{n-1}(0.85)^i$

49. (b) 2π **53.** $a_n = (-1)^n$ **63.** (b) Converges to $-\sqrt{A}$.

Exercise Set 8.2

1. Diverges **3.** Converges **5.** Diverges **7.** Converges

9. Diverges **11.** Converges **13.** Diverges **15.** $\frac{1}{2}$

17. $\frac{4}{15}$ **19.** $\frac{5}{2}$ **21.** $\frac{1}{2}$ **23.** $\frac{3}{4}$ **25.** $\frac{1}{2}$ **27.** $\frac{25}{12}$

31. $\sum_{n=1}^{4} 2^n = 30, \sum_{n=1}^{4} 3^n = 120, \sum_{n=1}^{4} 6^n = 1554, \sum_{n=1}^{4}\left(\frac{2}{3}\right)^n = \frac{130}{81}$

33. $\frac{19}{99}$

35. One example is $a_n = 1$ and $b_n = -1$ for all n.
37. It diverges.   **39.** $450   **41. (a)** 98.08 ft; **(b)** 180 ft
43. (a) 177.69 ft; **(b)** 17.79 sec   **45.** $\dfrac{1000}{9}$ m   **47.** 40
49. Mike should win 2 out of 3 times.

Exercise Set 8.3

1. Diverges   **3.** Converges   **5.** Diverges   **7.** Converges
9. Converges   **11.** Converges   **13.** Converges
15. Diverges   **17.** Converges   **19.** Converges
21. Diverges   **23.** Diverges   **25.** Diverges
27. Converges   **29.** Converges   **31.** Converges
33. Diverges   **35.** Converges   **37.** Converges
39. Converges
57. (a) $n \geq 1000$; **(b)** $n \geq 23$; **(c)** $n \geq 7$

Exercise Set 8.4

1. Converges   **3.** Converges   **5.** Converges
7. Diverges   **9.** Diverges   **11.** Converges
13. Converges   **15.** Converges   **17.** Converges
19. Diverges (This series is not alternating.)
21. Converges   **23.** Converges   **25.** Diverges
27. Converges (This series is not alternating.)
29. Converges
31. Diverges (This series is not alternating.)
33. Converges   **35.** Converges   **37.** $n \geq 9999$
39. $n \geq e^{10,000} + 1$ **41.** $n \geq 7$ **43.** 0.8   **45.** 0.3466
47. -0.36788
49. Consider $\sum_{n=1}^{\infty} (-1)^n \dfrac{1}{\sqrt{n}}$.

Exercise Set 8.5

1. Conditionally   **3.** Conditionally   **5.** Conditionally
7. Divergent   **9.** Divergent   **11.** Conditionally
13. Conditionally   **15.** Absolutely   **17.** Absolutely
19. Divergent   **21.** Absolutely   **23.** Absolutely
25. Divergent   **27.** Absolutely   **29.** Absolutely
31. Divergent   **33.** Conditionally   **35.** Absolutely
37. Absolutely   **39.** Absolutely   **41.** Absolutely
43. Absolutely   **45.** Absolutely
47. Converges at -1 and 1 if $p < -1$; converges at -1, but diverges at 1 if $-1 \leq p < 0$; diverges at -1 and 1 if $p \geq 0$.

Exercise Set 8.6

1. (a) 1; **(b)** $[-1, 1)$ **3. (a)** 1; **(b)** $[-1, 1]$
5. (a) ∞; **(b)** $(-\infty, \infty)$ **7. (a)** 1; **(b)** $[-1, 1]$
9. (a) ∞; **(b)** $(-\infty, \infty)$ **11. (a)** ∞; **(b)** $(-\infty, \infty)$
13. (a) $\dfrac{1}{2}$; **(b)** $\left[-\dfrac{1}{2}, \dfrac{1}{2}\right)$ **15. (a)** 1; **(b)** $[-1, 1)$
17. (a) 1; **(b)** $(1, 3)$ **19. (a)** 1; **(b)** $(1, 3)$
21. (a) 1; **(b)** $(-4, -2)$ **23. (a)** $\dfrac{1}{3}$; **(b)** $\left[0, \dfrac{2}{3}\right]$
25. (a) $\dfrac{1}{2}$; **(b)** $[-1, 0)$ **27.** $[-1, 1)$
29. $[0, 2]$ **31.** $[1, 2]$ **33.** $(1 - \sqrt{2}, 1) \cup (1, 1 + \sqrt{2})$
35. $(-\infty, -1) \cup (1, \infty)$ **37.** $(-\infty, -2) \cup (2, \infty)$
39. $x \neq n\pi$, for any integer n **41.** $(-1, 1)$
43. $R = \text{minimum } \{R_1, R_2\}$ **45.** $\left[-\dfrac{1}{e}, \dfrac{1}{e}\right)$

Exercise Set 8.7

1. $\sum_{n=0}^{\infty} (-1)^n x^n$ **3.** $\sum_{n=0}^{\infty} \dfrac{x^n}{2^{n+1}}$ **5.** $\sum_{n=0}^{\infty} (-1)^n x^{2n}$
7. $\sum_{n=0}^{\infty} (-1)^n x^{2n+1}$ **9.** $\sum_{n=0}^{\infty} x^{2n}$ **11.** $\sum_{n=0}^{\infty} (-1)^n (n+1) x^n$
13. $\sum_{n=1}^{\infty} \dfrac{x^{n-1}}{n!}$ **15.** $\sum_{n=0}^{\infty} \dfrac{(-1)^n x^{n+2}}{n+1}$ **17.** $\sum_{n=0}^{\infty} 2 \cdot 3^n x^{2n+1}$
19. $\sum_{n=0}^{\infty} \dfrac{x^{3n+2}}{(3n+2)n!}$ **23.** 0.1973   **25.** 0.0048
27. (a) $\dfrac{1}{3} \sum_{n=0}^{\infty} \left(\dfrac{-4}{3}\right)^n x^n; \left(-\dfrac{3}{4}, \dfrac{3}{4}\right)$
(b) $\dfrac{1}{3} \sum_{n=0}^{\infty} \left(\dfrac{-4}{3}\right)^n x^{n+1}; \left(-\dfrac{3}{4}, \dfrac{3}{4}\right)$
(c) $\dfrac{-1}{12} \sum_{n=1}^{\infty} \left(\dfrac{-4}{3}\right)^{n-1} n x^{n-1}; \left(-\dfrac{3}{4}, \dfrac{3}{4}\right)$
(d) $\sum_{n=0}^{\infty} \left(\dfrac{-4}{3}\right)^n x^{n+1}/(n+1); \left(-\dfrac{3}{4}, \dfrac{3}{4}\right]$
31. $\int \ln(1+x)\, dx = x \ln(1+x) - x + \ln(1+x) + C$;
so $x \ln(1+x) = x - \ln(1+x) - \int \ln(1+x)\, dx = $
$\sum_{n=0}^{\infty} \dfrac{(-1)^n x^{n+2}}{n+1}$
33. $a_n = 0$ for all n
35. $f'(x) = g(x)$ and $g'(x) = -f(x)$. There is a common pair of functions with this property.
37. (a) $\dfrac{(-1)^n x^{2n+1}}{2n+1}$; **(b)** $n \geq 500$; **(c)** $n \geq 5,000,000$ **39.** 3

Exercise Set 8.8

1. $1 + 2x + 2x^2 + \frac{4}{3}x^3$ **3.** $x - \frac{1}{2}x^2 + \frac{1}{3}x^3$

5. $x - \frac{1}{3}x^3$

7. $1, 1 + 2x, 1 + 2x + 2x^2, 1 + 2x + 2x^2 + \frac{4}{3}x^3$

9. $0, x, x - \frac{x^2}{2}, x - \frac{x^2}{2} + \frac{x^3}{3}$ **11.** $0, x, x, x - \frac{x^3}{3}$

13. (a) $P_4(x) = e[1 + (x-1) + \frac{1}{2}(x-1)^2 + \frac{1}{6}(x-1)^3 + \frac{1}{24}(x-1)^4]$; **(b)** $R_4(x) = \frac{1}{120}e^{\xi_x}(x-1)^5$

15. (a) $P_4(x) = 1 - \frac{1}{2}\left(x - \frac{\pi}{2}\right)^2 + \frac{1}{24}\left(x - \frac{\pi}{2}\right)^4$;
(b) $R_4(x) = \frac{1}{120}\cos \xi_x \left(x - \frac{\pi}{2}\right)^5$

17. (a) $P_4(x) = 1 + 2\left(x - \frac{\pi}{4}\right) + 2\left(x - \frac{\pi}{4}\right)^2 + \frac{8}{3}\left(x - \frac{\pi}{4}\right)^3 + \frac{10}{3}\left(x - \frac{\pi}{4}\right)^4$; **(b)** $R_4(x) = \frac{1}{15}[2 + 15(\tan \xi_x)^2 + 15(\tan \xi_x)^4](\sec \xi_x)^2 \left(x - \frac{\pi}{4}\right)^5$

19. (a) $P_4(x) = 2 + \frac{1}{4}(x-4) - \frac{1}{64}(x-4)^2 + \frac{1}{512}(x-4)^3 - \frac{5}{16,384}(x-4)^4$;
(b) $R_4(x) = \frac{7}{256}(\xi_x)^{-9/2}(x-4)^5$

21. (a) $P_4(x) = 3 + e + (2+e)(x-1) + \frac{1}{2}(2+e)(x-1)^2 + \frac{e}{6}(x-1)^3 + \frac{e}{24}(x-1)^4$;
(b) $R_4(x) = \frac{1}{120}e^{\xi_x}(x-1)^5$

23. $\sum_{n=0}^{\infty} \frac{(-1)^n x^n}{n!}$ **25.** $\sum_{n=0}^{\infty} \frac{3^{2n+1}(-1)^n x^{2n+1}}{(2n+1)!}$

27. $\sum_{n=0}^{\infty} (-1)^n x^n$ **29.** $\sum_{n=0}^{\infty} \frac{(-1)^n x^{2n}}{(2n+1)!}$ **31.** $\sum_{n=0}^{\infty} \frac{(\ln 2)^n x^n}{n!}$

33. $\sum_{n=0}^{\infty} \frac{(-1)^n (x-2)^n}{2^{n+1}}$ **35.** $\sum_{n=0}^{\infty} \frac{(-1)^n (x - \pi/2)^{2n}}{(2n)!}$

37. $e \sum_{n=0}^{\infty} \frac{(x-1)^n}{n!}$ **39.** $\sum_{n=0}^{\infty} \frac{(-1)^n (x-1)^{n+1}}{n+1}$

41. (a) 0.0099998333; **(b)** 4×10^{-10} **43.** 2.02484567

45. 0.0872
51. (a) $6 + 7(x-1) + 2(x-1)^2$; **(b)** $18 - 13(x+2) + 3(x+2)^2$; **(c)** $(x+1) - 2(x+1)^2 + (x+1)^3$; **(d)** $64 + 69(x-3) + 25(x-3)^2 + 3(x-3)^3$
55. (b) $P_2(x) = 3 + 4(x-1) + 3(x-1)^2$

Exercise Set 8.9

1. $1 + \sum_{n=1}^{\infty} \frac{1}{n!}\left[\left(\frac{1}{3}\right)\left(-\frac{2}{3}\right)\left(-\frac{5}{3}\right) \cdots \left(\frac{1}{3} - n + 1\right)\right] x^n, R = 1$

3. $1 + \sum_{n=1}^{\infty} \frac{1}{n!}\left[\left(-\frac{1}{2}\right)\left(-\frac{3}{2}\right)\left(-\frac{5}{2}\right) \cdots \left(-\frac{1}{2} - n + 1\right)\right] x^n, R = 1$

5. $2\sqrt{2}\left\{1 + \sum_{n=1}^{\infty} \frac{3^n}{2^n n!}\left(\frac{3}{2}\right)\left(\frac{1}{2}\right)\left(-\frac{1}{2}\right) \cdots \left(\frac{3}{2} - n + 1\right)\right] x^n\right\}, R = \frac{2}{3}$

7. 0.5132 **11.** 1.49453 **13.** $-e^2 + 0 \cdot i$
15. **17.**

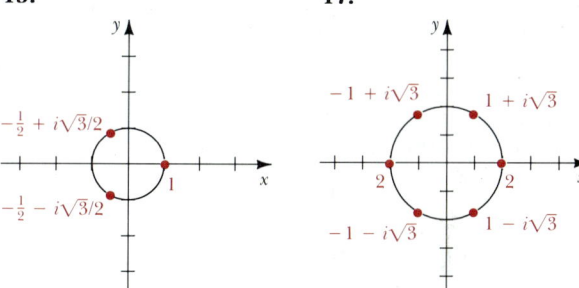

19. (a) $\pm 1 \pm i$; **(b)** $x^4 + 4 = (x^2 - 2x + 2)(x^2 + 2x + 2)$
23. All complex numbers **25.** 6.62%

Review Exercises

1. (a) Yes; **(b)** converges; **(c)** $\frac{-3}{2}$
3. (a) Yes; **(b)** converges; **(c)** 3
5. (a) Yes; **(b)** converges; **(c)** 1
7. (a) No; **(b)** diverges **9. (a)** No; **(b)** diverges
11. (a) Yes; **(b)** converges; **(c)** 0
13. (a) Yes; **(b)** converges; **(c)** 1
15. (a) Yes; **(b)** converges; **(c)** 0
17. (b) $a_n = \frac{1}{n}$; **(c)** 0 **19.** Divergent
21. Absolutely **23.** Absolutely **25.** Conditionally

S48   ANSWERS

27. Absolutely   **29.** Conditionally   **31.** Divergent
33. Conditionally   **35.** Conditionally   **37.** Conditionally
39. (a) 1; (b) $[-1, 1]$ **41.** (a) 1; (b) $[-1, 1]$
43. (a) 1; (b) $(-3, -1)$ **45.** (a) 2; (b) $(0, 4)$
47. (a) $\dfrac{1}{3}$; (b) $\left(\dfrac{8}{3}, \dfrac{10}{3}\right)$ **49.** (a) $\sqrt{3}$; (b) $(-\sqrt{3}, \sqrt{3})$
51. (a) $\dfrac{3}{2}$; (b) $\left[\dfrac{-3}{2}, \dfrac{3}{2}\right)$ **53.** -0.05579 **55.** 0.28768
57. Examples are: (a) $\sum_{n=1}^{\infty} \dfrac{(x-2)^n}{n^2}$; (b) $\sum_{n=1}^{\infty} (x-2)^n$;
(c) $\sum_{n=1}^{\infty} \dfrac{(-1)^n (x-2)^n}{n}$; (d) $\sum_{n=1}^{\infty} \dfrac{(x-2)^n}{n}$
59. One example is $a_n = n$.   **61.** $\sum_{n=0}^{\infty} (-1)^n \dfrac{x^{2n}}{(2n+1)!}$
63. $\sum_{n=0}^{\infty} \dfrac{x^{2n}}{(2n)!}$
65. (a) $4x^2 - 2x - 3$; (b) $(6\xi - 7)x^5$, ξ between x and 0
67. (a) $\sqrt{2}\left[1 + \left(x - \dfrac{\pi}{4}\right) + \dfrac{3}{2}\left(x - \dfrac{\pi}{4}\right)^2 + \dfrac{11}{6}\left(x - \dfrac{\pi}{4}\right)^3 + \dfrac{19}{8}\left(x - \dfrac{\pi}{4}\right)^4\right]$;
(b) $\sec 0.8 \approx 1.4353242$
69. $1 - 2\sum_{n=0}^{\infty} (-1)^n x^n$ **71.** $\ln 2 + \sum_{n=1}^{\infty} (-1)^{n-1} \dfrac{(x-1)^n}{n}$
73. $\dfrac{97}{56} = 1.73214$, $\sqrt{3} = 1.73205$ **75.** $\left[\dfrac{1}{2}, \infty\right)$
77. $\dfrac{\sqrt{2}}{2} e^{-2}(1 + i)$

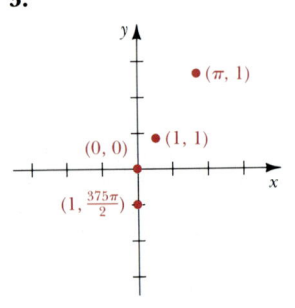

5. Some possibilities are: (a) $(2, 2\pi)$, $(2, -2\pi)$, $(-2, \pi)$, $(-2, -\pi)$; (b) $\left(-1, \dfrac{5\pi}{2}\right)$, $\left(-1, \dfrac{-3\pi}{2}\right)$, $\left(1, \dfrac{-\pi}{2}\right)$, $\left(1, \dfrac{-5\pi}{2}\right)$; (c) $\left(1, \dfrac{11\pi}{4}\right)$, $\left(1, \dfrac{-5\pi}{4}\right)$, $\left(-1, \dfrac{7\pi}{4}\right)$, $\left(-1, \dfrac{-\pi}{4}\right)$; (d) $(1, \pi)$, $(1, 3\pi)$, $(-1, 0)$, $(-1, 2\pi)$
7. (a) $(\sqrt{2}, \sqrt{2})$; (b) $(0, 1)$; (c) $(0, -1)$, (d) $(-1, \sqrt{3})$
13. $245013 \text{ ft}^2 \approx 5.62$ acres

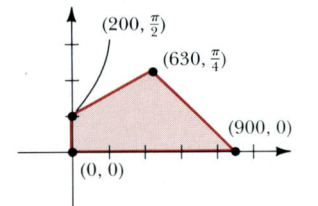

Chapter 9

Exercise Set 9.1

1.

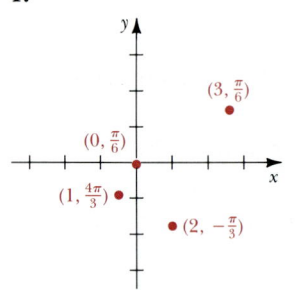

3.

Exercise Set 9.2

1.

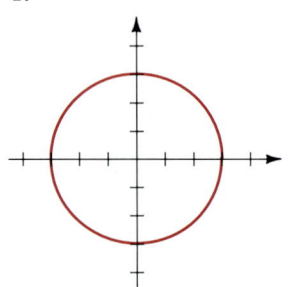

3.

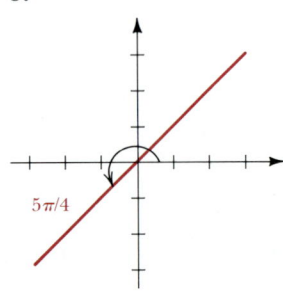

5.

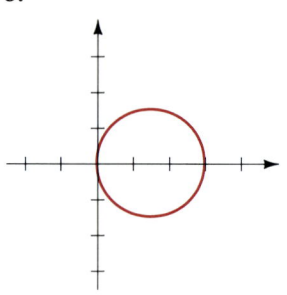

7.

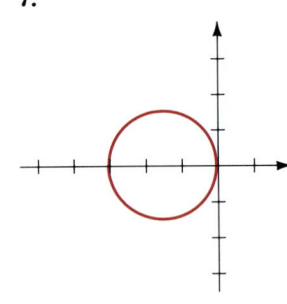

9.

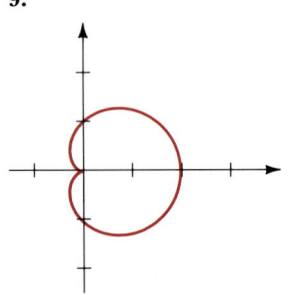

11.

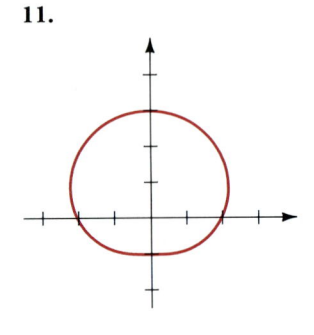

13. [graph]

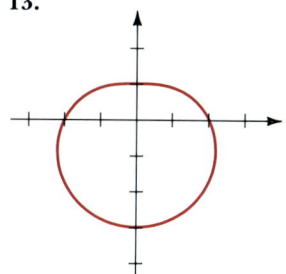

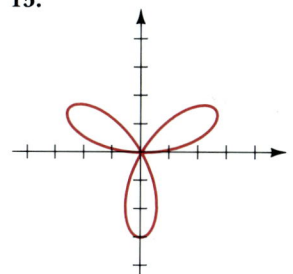

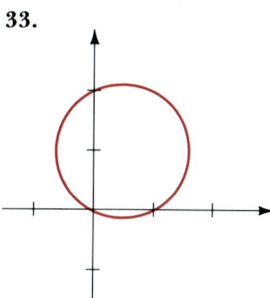

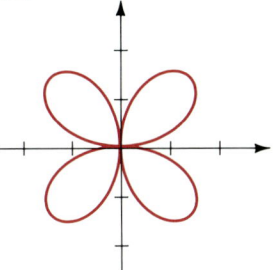

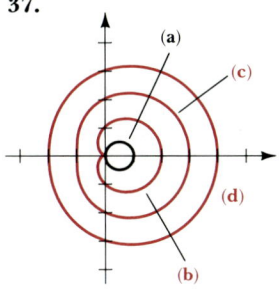

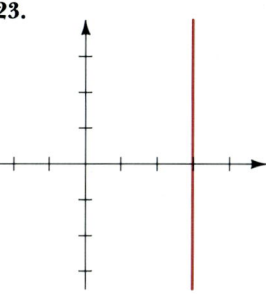

15. [graph]

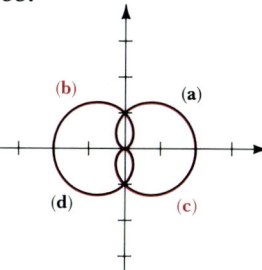

17. [graph]

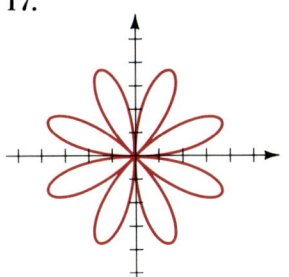

19. [graph]

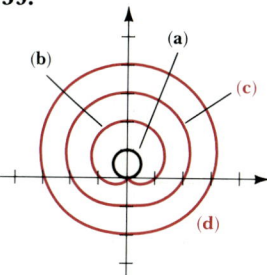

21. [graph]

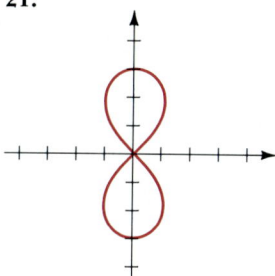

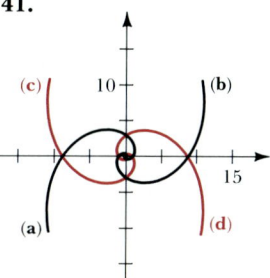

23. [graph]

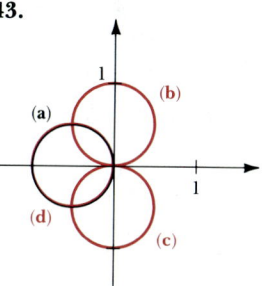

25. [graph]

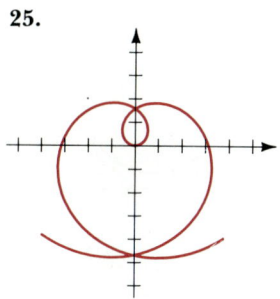

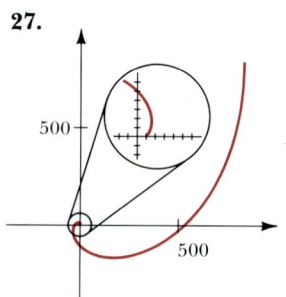

27. [graph]

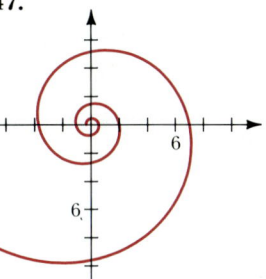

29. [graph]

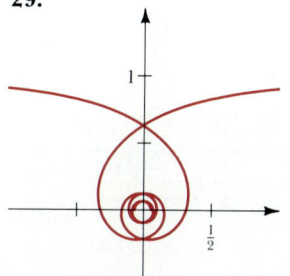

31. [graph]

33. [graph]

35. [graph]

37. [graph]

39. [graph]

41. [graph]

43. [graph]

47. [graph]

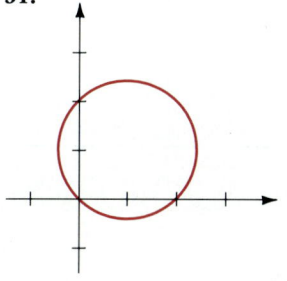

49. Home to first: $\theta = \dfrac{\pi}{4}$, $0 \leq r \leq 90$; First to second: $r = \dfrac{90\sqrt{2}}{\sin\theta + \cos\theta}$, $\dfrac{\pi}{4} \leq \theta \leq \dfrac{\pi}{2}$; Second to third: $r = \dfrac{90\sqrt{2}}{\sin\theta - \cos\theta}$, $\dfrac{\pi}{2} \leq \theta \leq \dfrac{3\pi}{4}$; Third to home: $\theta = \dfrac{3\pi}{4}$, $0 \leq r \leq 90$

Exercise Set 9.3

1. Area $= 4\pi$

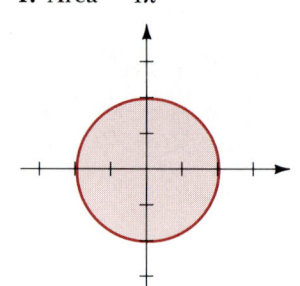

3. Area $= \pi$

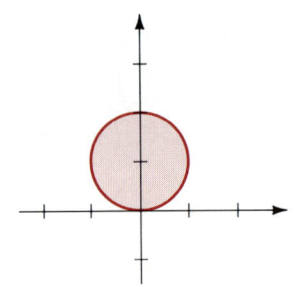

17. Area $= 3\pi$

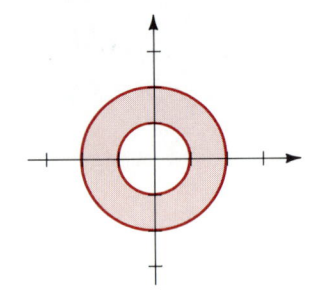

19. Area $= 2 + \dfrac{\pi}{4}$

5. Area $= \dfrac{9}{4}\pi$

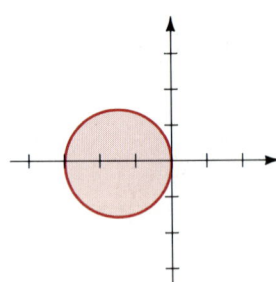

7. Area $= \dfrac{3}{2}\pi$

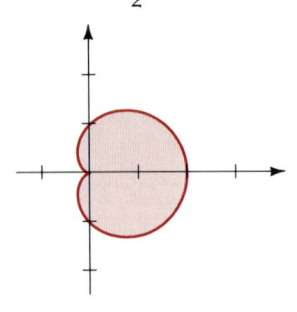

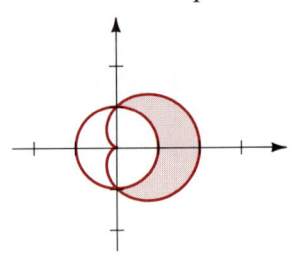

21. Area $= 4\pi - 4$

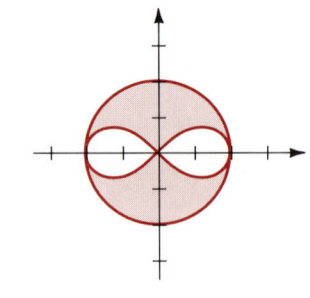

23. Area $= 2\sqrt{2}$

9. Area $= 18\pi$

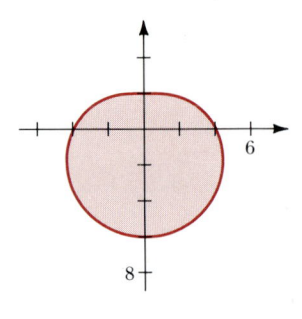

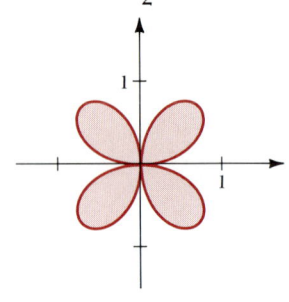

11. Area $= \dfrac{\pi}{2}$

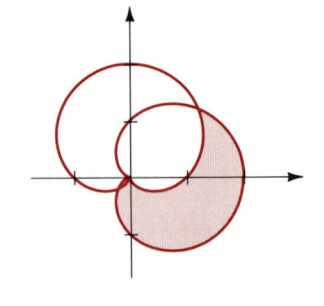

25. Area $= \dfrac{5\pi}{4} - 2$

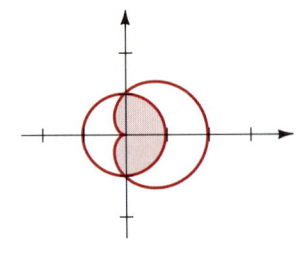

27. Area $= \dfrac{3\pi}{2} - 2\sqrt{2}$

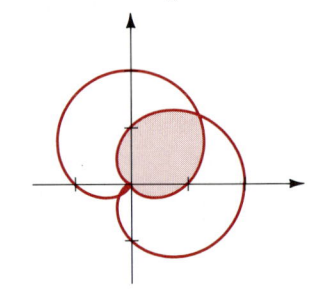

13. Area $= 4$

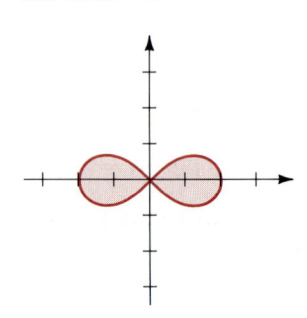

15. Area $= \dfrac{\pi^3}{24}$

29. Area $= \dfrac{\pi}{2} - 1$

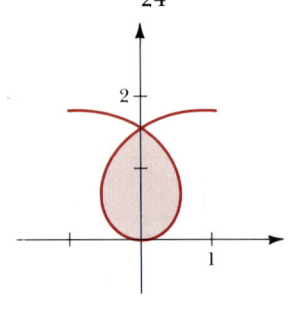

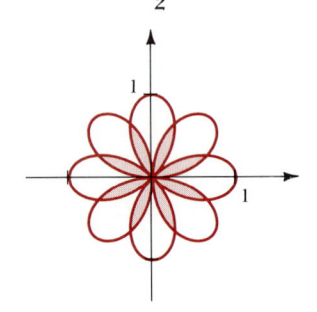

31. $(\sqrt{2}/2, \pi/4)$ on both
$(0, \pi/2)$ on $r = \cos\theta$
$(0, \pi)$ on $r = \sin\theta$

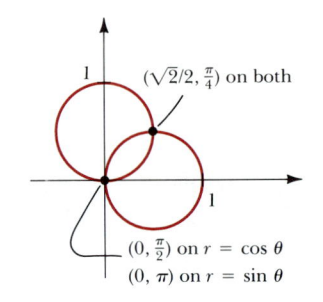

33. The graphs of $r = \cos\theta - 1$ and $r = \cos\theta + 1$ are the same.

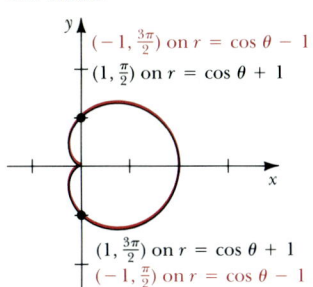

$(-1, \frac{3\pi}{2})$ on $r = \cos\theta - 1$
$(1, \frac{\pi}{2})$ on $r = \cos\theta + 1$
$(1, \frac{3\pi}{2})$ on $r = \cos\theta + 1$
$(-1, \frac{\pi}{2})$ on $r = \cos\theta - 1$

35. The graphs of $r = \cos\dfrac{\theta}{2}$ and $r = \sin\dfrac{\theta}{2}$ are the same.

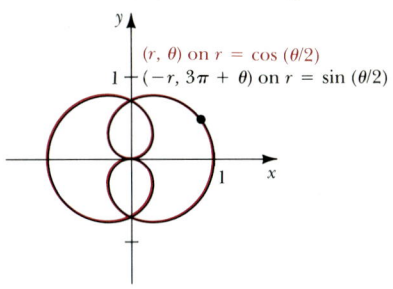

(r, θ) on $r = \cos(\theta/2)$
$(-r, 3\pi + \theta)$ on $r = \sin(\theta/2)$

37. (a)

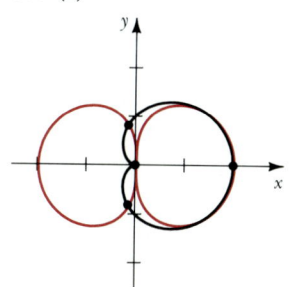

(b) $(2, 0)$ and $(0, 0)$

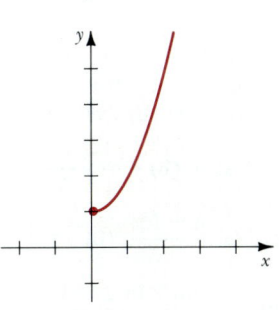

5. $y = 1 - x^2,\ y \geq 0$

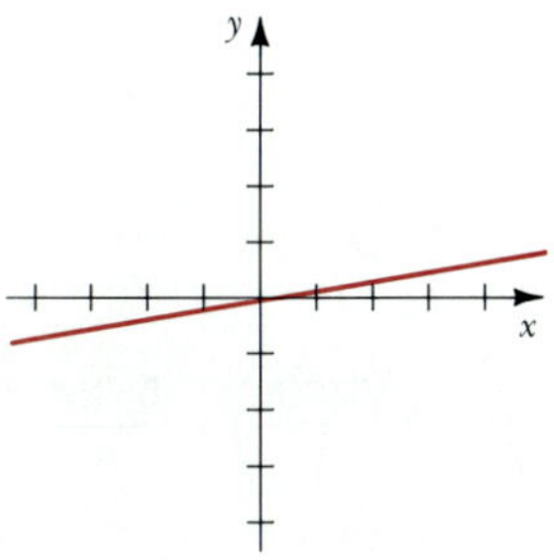

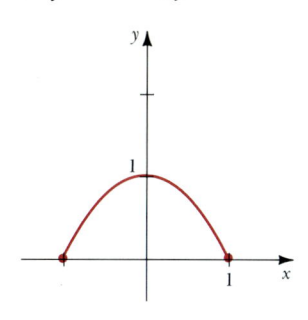

7. $\dfrac{x^2}{9} + \dfrac{y^2}{16} = 1$

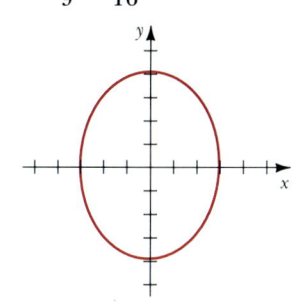

9. $y = \dfrac{1}{x},\ x > 0$

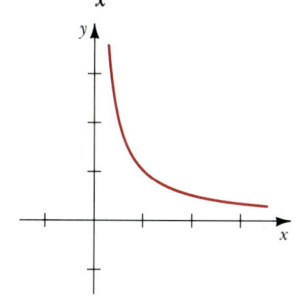

11. $y^2 = \dfrac{1}{1 - x^2}$

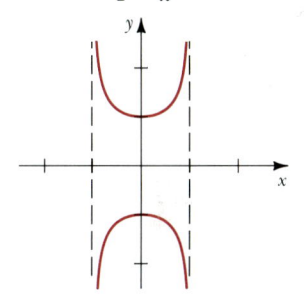

13. $(x - 4)^2 + (y - 6)^2 = 4$

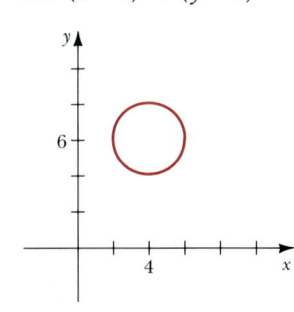

15. $y = \dfrac{\sqrt[3]{4}}{2}x^{2/3} + 1$

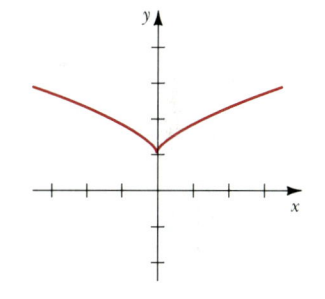

Exercise Set 9.4

1. $y = \dfrac{1}{6}x$

3. $x = \sqrt{y - 1}$

17.

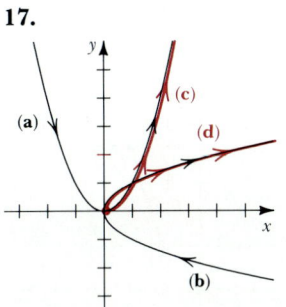

19.

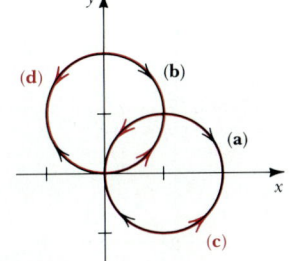

S52 ANSWERS

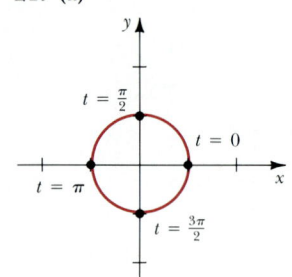

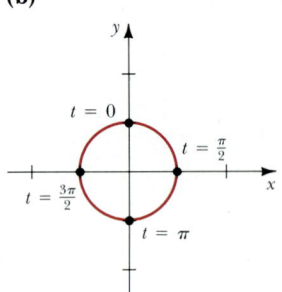

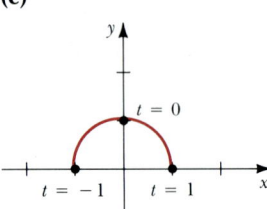

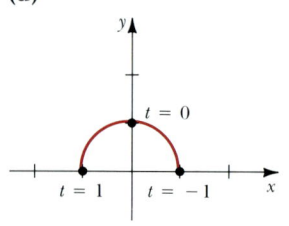

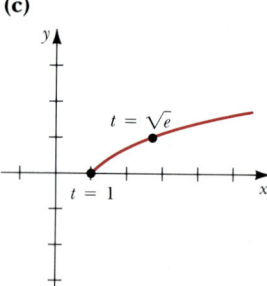

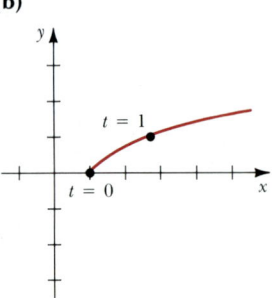

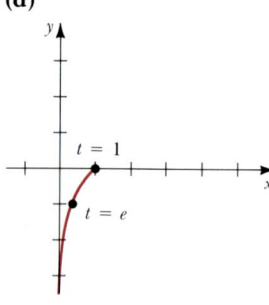

21. (a) [graph] (b) [graph] (c) [graph] (d) [graph]

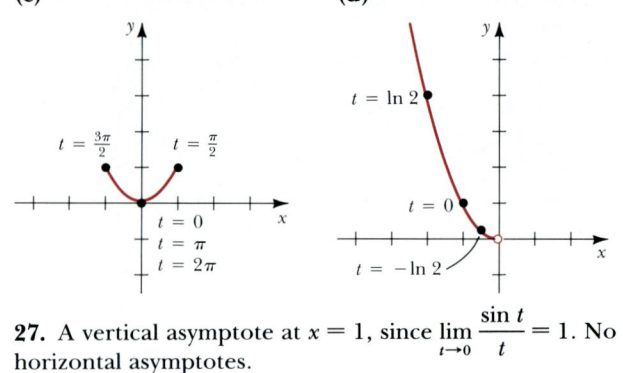

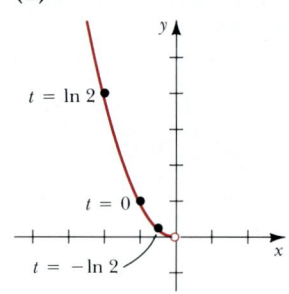

(c) [graph] (d) [graph]

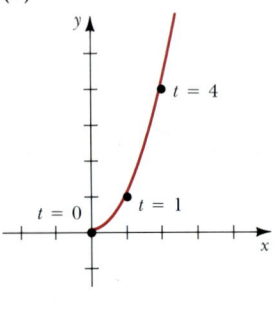

27. A vertical asymptote at $x = 1$, since $\lim_{t \to 0} \dfrac{\sin t}{t} = 1$. No horizontal asymptotes.

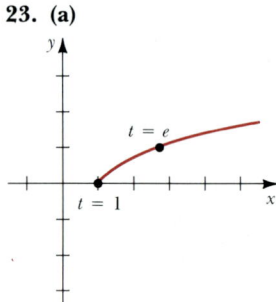

29. (b) [graph]

23. (a) [graph] (b) [graph]

(c) [graph] (d) [graph]

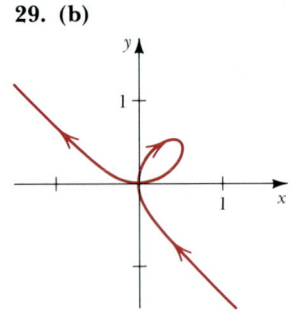

33. (a) [graph] (b) [graph]

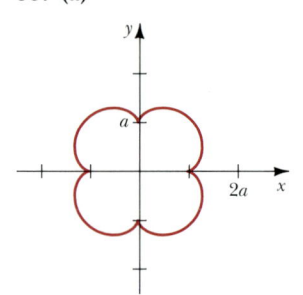

25. (a) [graph] (b) [graph]

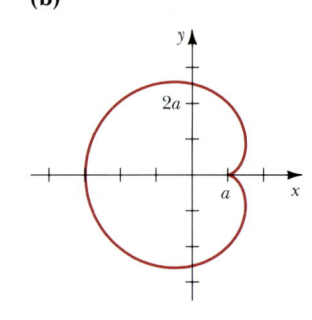

Exercise Set 9.5

1. $\dfrac{1}{6}$ **3.** 4 **5.** $-\sqrt{3}$ **7.** 0 **9.** -1 **11.** 2

13. 0 **15.** $-\dfrac{1}{3}$

17. (a) -1; (b) $y = -x + \sqrt{2}$
19. (a) Undefined; (b) $x = 1$
21. (a) 1; (b) $y = x + 1$
23. (a) -1; (b) $y = -x + 3\sqrt{2}$
25. (a) 0; (b) $y = \pm \dfrac{\sqrt{2}}{2}$
27. (a) $\dfrac{4 + 5\pi}{4 - 5\pi}$; (b) $y = \left(\dfrac{4 + 5\pi}{4 - 5\pi}\right)\left(x + \dfrac{5\sqrt{2}\pi}{8}\right) - \dfrac{5\sqrt{2}\pi}{8}$
29. (a) $\dfrac{\tan 2 \ln 2 + 1}{\tan 2 - \ln 2}$;

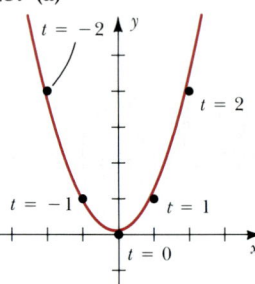

(b) $y = \left(\dfrac{\tan 2 \ln 2 + 1}{\tan 2 - \ln 2}\right)(x - 4\cos 2) - 4\sin 2$

31. 0 **33.** $-\dfrac{4}{9}(\sec t)^3$ **35.** $2e^{-3t}$ **37.** $2(\sec 2\theta)^3$

39. $-2^\theta \left[\dfrac{1 + (\ln 2)^2}{(\cos\theta \ln 2 + \sin\theta)^3}\right]$

41. At $\left(1, \dfrac{\pi}{2}\right)$: $y = x + 1$ for $r = 1 + \cos\theta$ and $y = 1$ for $r = 1$. At $\left(1, -\dfrac{\pi}{2}\right)$: $y = -x - 1$ for $r = 1 + \cos\theta$ and $y = -1$ for $r = 1$.

43. At $\left(\dfrac{1}{2}(2 + \sqrt{2}), \dfrac{\pi}{4}\right)$: $y = (1 - \sqrt{2})x + 1 + \dfrac{1}{2}\sqrt{2}$ for $r = 1 + \cos\theta$ and $y = -(1 + \sqrt{2})x + 2 + \dfrac{3}{2}\sqrt{2}$ for $r = 1 + \sin\theta$.

45. At $\left(\dfrac{\sqrt{2}}{2}, \dfrac{\pi}{4}\right)$: $x = \dfrac{1}{2}$ for $r = \sin\theta$ and $y = \dfrac{1}{2}$ for $r = \cos\theta$. At the origin: $y = 0$ for $r = \sin\theta$ and $x = 0$ for $r = \cos\theta$.

47. (a)

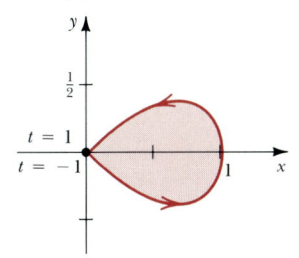

[graph showing a closed curve with $t = 1$ and $t = -1$ marked]

(b) $\dfrac{8}{15}$

49. Horizontal tangents when $\theta = \dfrac{3\pi}{4} + n\pi$; vertical tangents when $\theta = \dfrac{\pi}{4} + n\pi$, for any integer n

53. $r(\theta) = \dfrac{128}{\pi^3}(R_1 - R_2)\theta^3 + \dfrac{48}{\pi^2}(R_2 - R_1)\theta^2 + R_1$

Exercise Set 9.6

1. $2\sqrt{13}$ **3.** $13\sqrt{13} - 8$ **5.** $\sqrt{2} + \ln(1 + \sqrt{2})$
7. $\ln(1 + \sqrt{2})$ **9.** $\sqrt{2}(e^{2\pi} - 1)$
11. 3π **13.** $\dfrac{\sqrt{5}}{2}(e^2 - 1)$
15. 8

17. (a) Length $\dfrac{\pi}{2}$

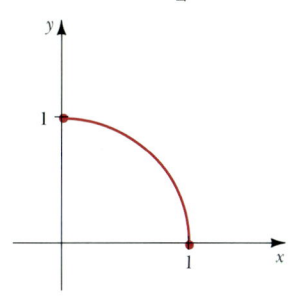

[graph]

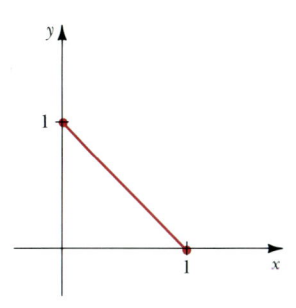

(c) Length $\dfrac{3}{2}$

[graph]

(b) Length $\sqrt{2}$

[graph]

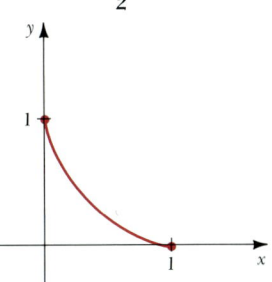

21. $2\sqrt{26}(1 - e^{-0.2\theta})$
23. Approximately 2945 ft

Review Exercises

1. $\left(\dfrac{3\sqrt{3}}{2}, \dfrac{3}{2}\right)$ **3.** $(0, 0)$

5. Possibilities are $\left(\sqrt{2}, \dfrac{\pi}{4}\right)$ and $\left(-\sqrt{2}, \dfrac{5\pi}{4}\right)$

7. Possibilities are $\left(-2, \dfrac{\pi}{2}\right)$ and $\left(2, \dfrac{3\pi}{2}\right)$

9.

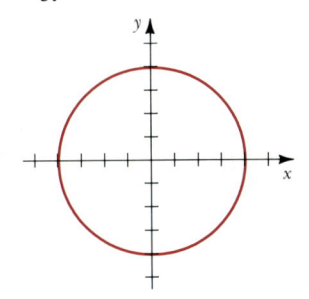

[circle graph]

11.

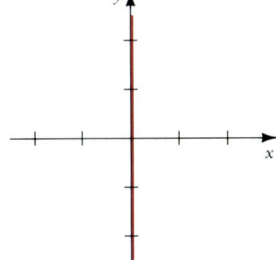

[vertical line graph]

13.

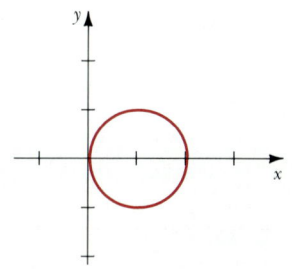

[small circle graph]

15.

[limaçon graph]

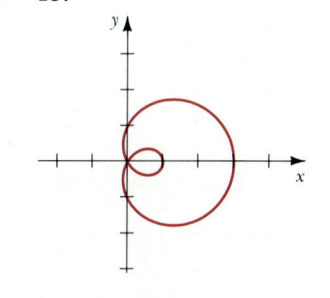

17.

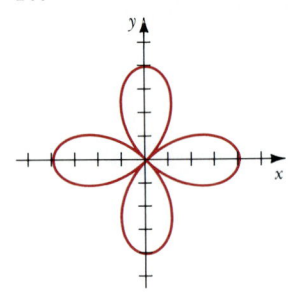

19.

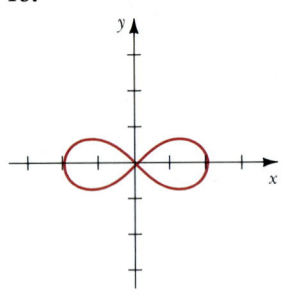

21.

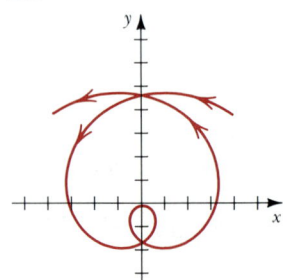

23.

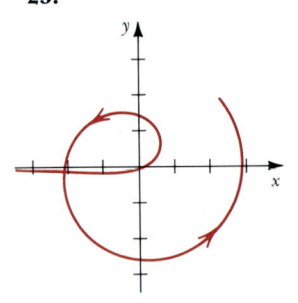

25. (a)

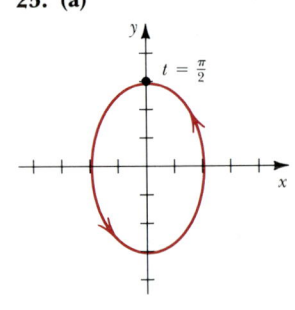

(b) 0

27. (a)

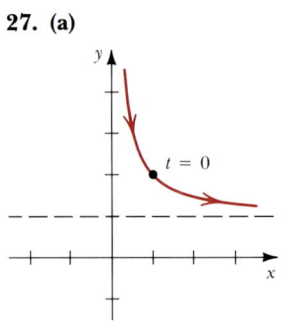

(b) $-e^{-2}$

29. (a)

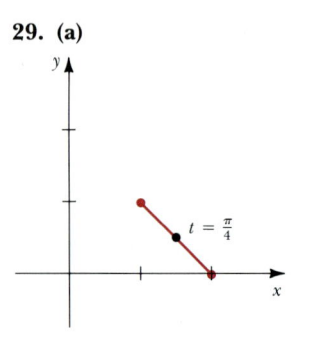

31. (a)

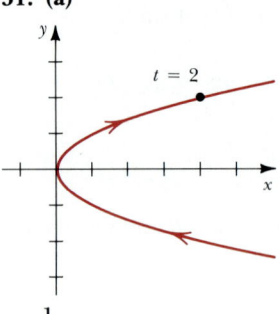

(b) $\dfrac{1}{4}$

33. (a)

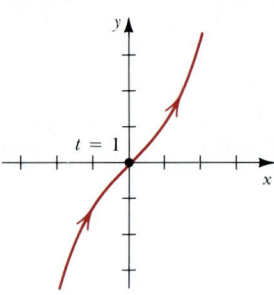

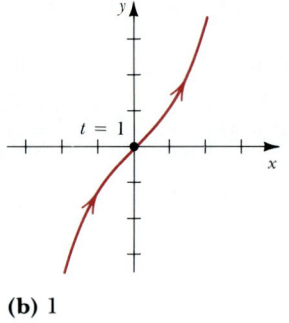

(b) 1

35. (a)

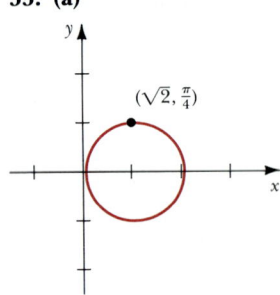

(b) 0

37. (a)

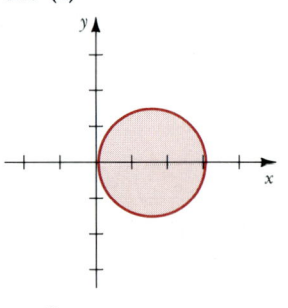

(b) $\dfrac{9\pi}{4}$

39. (a)

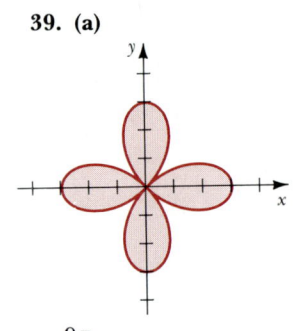

(b) $\dfrac{9\pi}{2}$

41. (a)

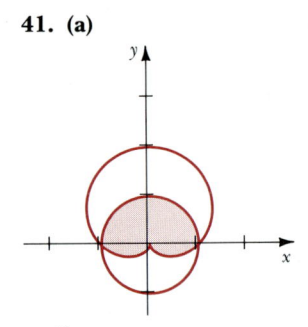

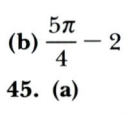

(b) $\dfrac{5\pi}{4} - 2$

43. (a)

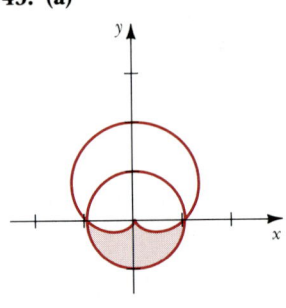

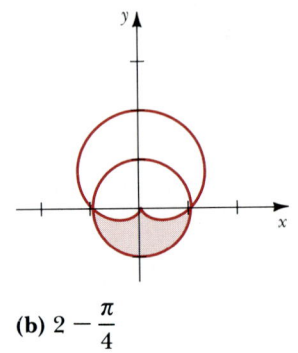

(b) $2 - \dfrac{\pi}{4}$

45. (a)

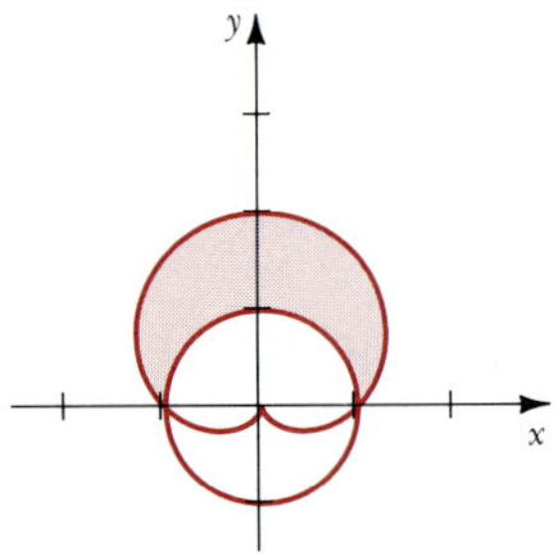

(b) $2 + \dfrac{\pi}{4}$

(b) -1

47. (a)

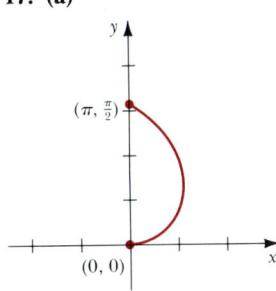

[Graph showing curve with points $(\pi, \frac{\pi}{2})$ and $(0, 0)$]

(b) $\dfrac{\pi}{4}\sqrt{\pi^2 + 4} - \ln 2 + \ln(\pi + \sqrt{\pi^2 + 4})$

49. (a)

[Cardioid-like graph]

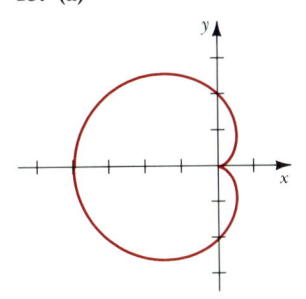

(b) 16

51. This is the same as Exercise 49.

53. (a)

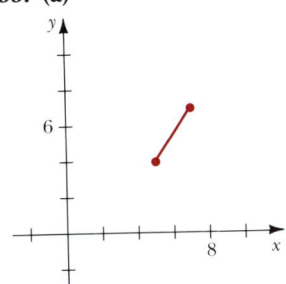

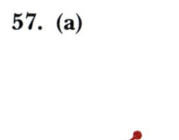

[Line segment graph]

(b) $\sqrt{13}$

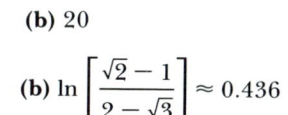

55. (a)

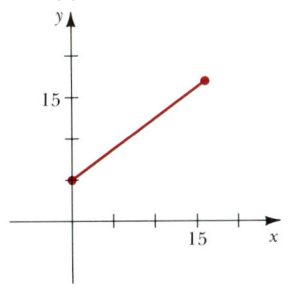

[Line segment graph]

(b) 20

57. (a)

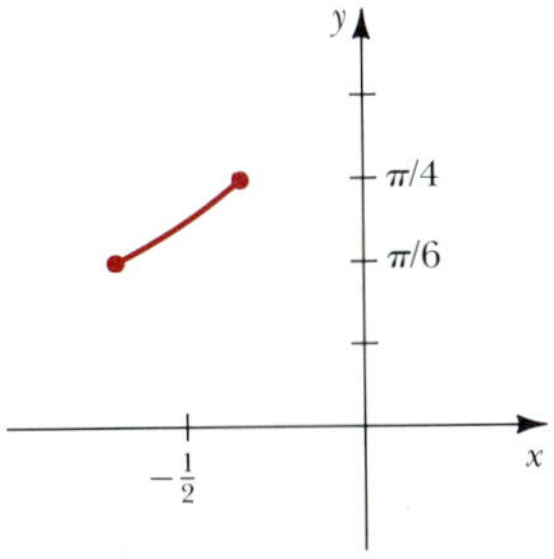

[Curve graph with markings $\pi/4$, $\pi/6$, $-\frac{1}{2}$]

(b) $\ln\left[\dfrac{\sqrt{2} - 1}{2 - \sqrt{3}}\right] \approx 0.436$

59. $x = 1$

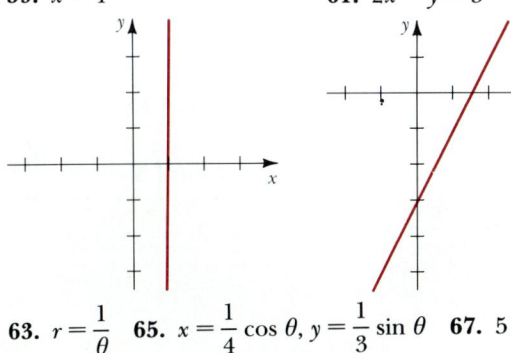

[Vertical line graph]

61. $2x - y = 3$

[Line graph]

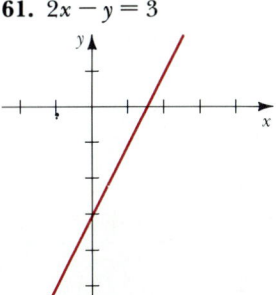

63. $r = \dfrac{1}{\theta}$ **65.** $x = \dfrac{1}{4}\cos\theta, y = \dfrac{1}{3}\sin\theta$ **67.** 5

69. (a) $-\infty, \infty$; (b) $\sec^2\theta$
(c)

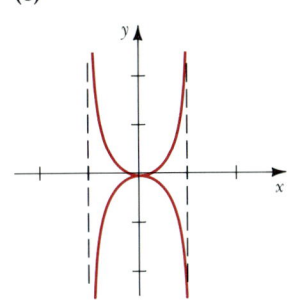

[Graph with vertical asymptotes]

Chapter 10

Exercise Set 10.1

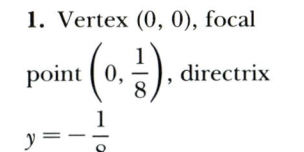

1. Vertex $(0, 0)$, focal point $\left(0, \dfrac{1}{8}\right)$, directrix $y = -\dfrac{1}{8}$

[Upward parabola graph]

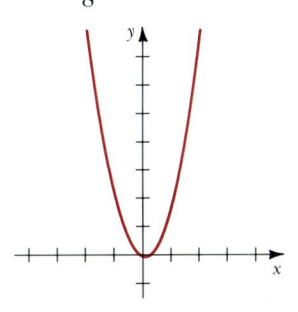

3. Vertex $(0, 0)$, focal point $\left(0, -\dfrac{1}{8}\right)$, directrix $y = \dfrac{1}{8}$

[Downward parabola graph]

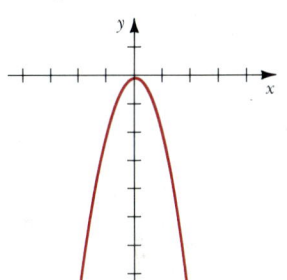

5. Vertex $(0, 0)$, focal point $\left(\dfrac{1}{2}, 0\right)$, directrix $x = -\dfrac{1}{2}$

[Rightward parabola graph]

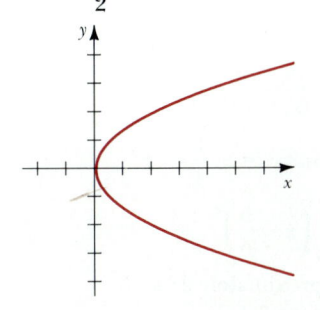

7. Vertex $(0, 0)$, focal point $\left(-\dfrac{1}{2}, 0\right)$, directrix $x = \dfrac{1}{2}$

[Leftward parabola graph]

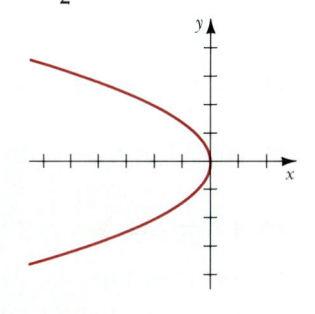

9. Vertex $(-2, 0)$, focal point $\left(-2, \dfrac{1}{2}\right)$, directrix $y = -\dfrac{1}{2}$

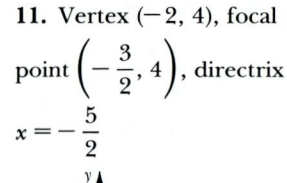

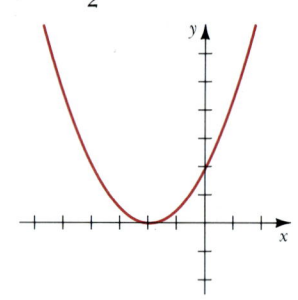

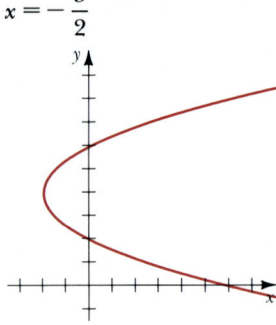

11. Vertex $(-2, 4)$, focal point $\left(-\dfrac{3}{2}, 4\right)$, directrix $x = -\dfrac{5}{2}$

13. Vertex $(-1, 2)$, focal point $\left(-1, \dfrac{25}{8}\right)$, directrix $y = \dfrac{7}{8}$

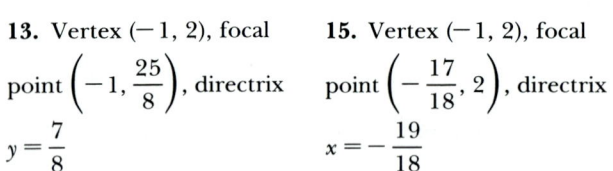

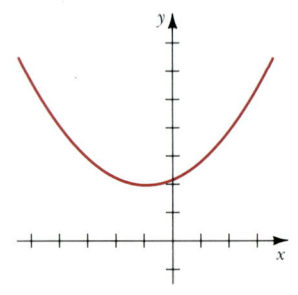

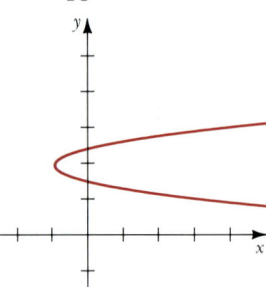

15. Vertex $(-1, 2)$, focal point $\left(-\dfrac{17}{18}, 2\right)$, directrix $x = -\dfrac{19}{18}$

17. $8y = (x + 2)^2$ **19.** $6y - 3 = (x + 2)^2$
21. $x + 2 = -\dfrac{(y-2)^2}{24}$ **23.** $y - 4 = \dfrac{1}{8}(x - 3)^2$
25. $y = \dfrac{3}{8}x^2$ **27.** $y - 2 = \dfrac{3}{8}x^2$ **29.** $x = \dfrac{y^2}{9}$
31. $x = \dfrac{(y-2)^2}{9}$
33. (a) $y = \dfrac{y_1}{x_1^2}x^2$; (b) $y - k = \dfrac{y_1 - k}{(x_1 - h)^2}(x - h)^2$
37. (a) $y = ax^2 + 1 - a$; (b) only $y = -\dfrac{1}{4}x^2 + \dfrac{5}{4}$
39. $a = \dfrac{1}{4b}$ **41.** $y + \dfrac{1}{24} = \dfrac{3}{2}\left(x - \dfrac{5}{6}\right)^2$
43. 53,717,083 ft³   **45.** Approximately 2841 ft
47. 2000/3 in. from the center of the mirror

Exercise Set 10.2

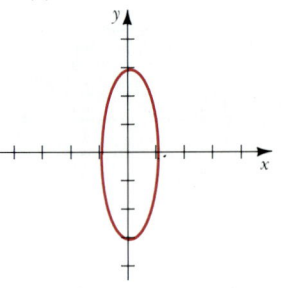

1. (a)

(b) vertices $(0, \pm 3)$, focal points $(0, \pm 2\sqrt{2})$

3. (a)

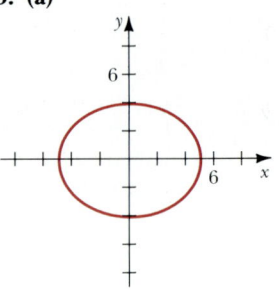

(b) vertices $(\pm 5, 0)$, focal points $(\pm 3, 0)$

5. (a)

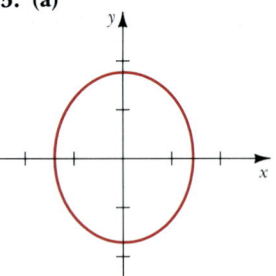

(b) vertices $(0, \pm\sqrt{3})$, focal points $(0, \pm 1)$

7. (a)

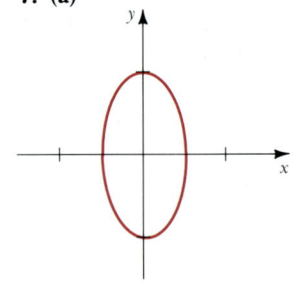

(b) vertices $(0, \pm 1)$, focal points $\left(0, \pm \dfrac{\sqrt{3}}{2}\right)$

9. (a)

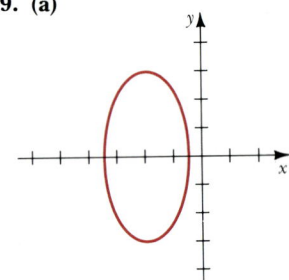

(b) vertices $(-2, \pm 3)$, focal points $\left(-2, \pm \dfrac{3\sqrt{3}}{2}\right)$

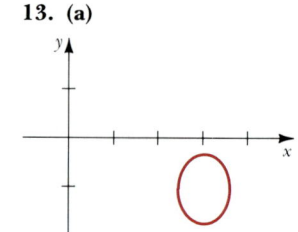

11. (a)

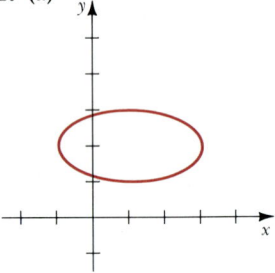

(b) vertices $(-1, 2)$, $(3, 2)$, focal points $(1 \pm \sqrt{3}, 2)$

13. (a)

(b) vertices $\left(3, -1 \pm \dfrac{\sqrt{2}}{2}\right)$, focal points $\left(3, -1 \pm \dfrac{\sqrt{6}}{6}\right)$

15. $5x^2 + 9y^2 = 45$ **17.** $5x^2 + 4y^2 = 20$
19. $3(x-4)^2 + 4(y-2)^2 = 12$ **21.** 6π
23. $y = -\dfrac{\sqrt{3}}{2}x + 2\sqrt{3}$
25. $y = \pm \dfrac{\sqrt{2}}{2}(x-3)$ **27.** $2ab$ **31.** $\dfrac{2b^2}{a}$
33. If $D^2C + E^2A - 4ACF$: **(a)** is positive and $A \neq C$; **(b)** is positive and $A = C$; **(c)** is zero; **(d)** is negative.
35. $\dfrac{x^2}{4} + \dfrac{y^2}{36} = 1$, $x \geq 0$, $y \geq 0$
37. (a) The $3.99 melon; **(b)** the $3.99 melon costs 0.635 cents/in³ of fruit. The $2.79 melon costs 0.604 cents/in³ of fruit, so this melon is the better buy.
39. Assuming the airships are formed by rotating an ellipse about the longitudinal axis gives $V \approx 251{,}327$ ft³. The volume stated is reasonable.
41. Approximately 28.95 ft from the front and rear of the building along the longitudinal axis

Exercise Set 10.3

1. (a)

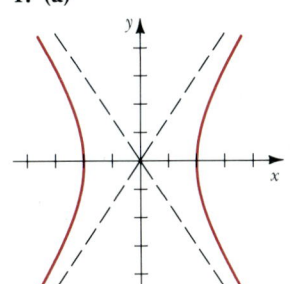

(b) vertices $(\pm 2, 0)$, focal points $(\pm \sqrt{13}, 0)$;
(c) asymptotes $y = \pm \dfrac{3}{2}x$

3. (a)

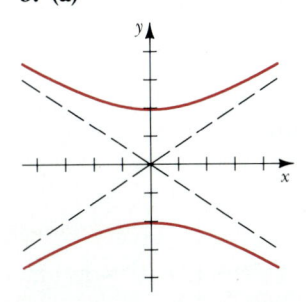

(b) vertices $(0, \pm 2)$, focal points $(0, \pm \sqrt{13})$;
(c) asymptotes $y = \pm \dfrac{2}{3}x$

5. (a)

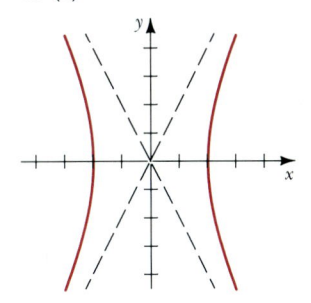

(b) vertices $(\pm 2, 0)$, focal points $(\pm 2\sqrt{5}, 0)$;
(c) asymptotes $y = \pm 2x$

7. (a)

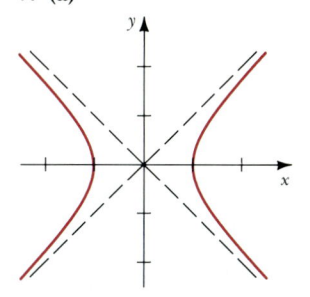

(b) vertices $(\pm 1, 0)$, focal points $(\pm \sqrt{2}, 0)$;
(c) asymptotes $y = \pm x$

9. (a)

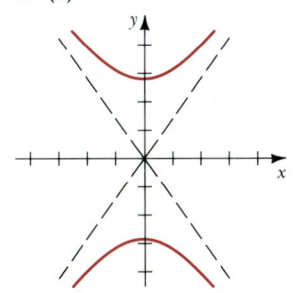

(b) vertices $(0, \pm 2\sqrt{2})$, focal points $(0, \pm 2\sqrt{3})$;
(c) asymptotes $y = \pm \sqrt{2}x$

11. (a)

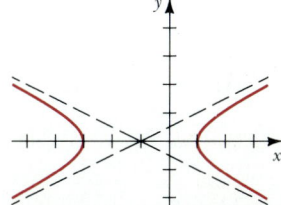

(b) vertices $(-3, 0)$, $(1, 0)$, focal points $(-1 \pm \sqrt{5}, 0)$;
(c) asymptotes $y = \pm \dfrac{1}{2}(x+1)$

13. (a)

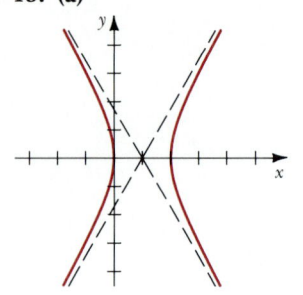

(b) vertices $(0, 0)$, $(2, 0)$, focal points $(3, 0)$, $(-1, 0)$;
(c) asymptotes $y = \pm\sqrt{3}(x - 1)$

15. (a)

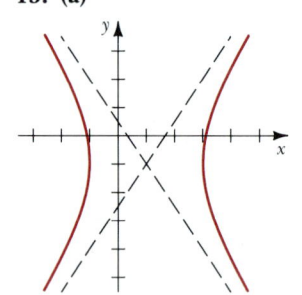

(b) vertices $(-1, -1)$, $(3, -1)$, focal points $(1 \pm \sqrt{13}, -1)$;
(c) asymptotes $y + 1 = \pm\dfrac{3}{2}(x - 1)$

17. (a)

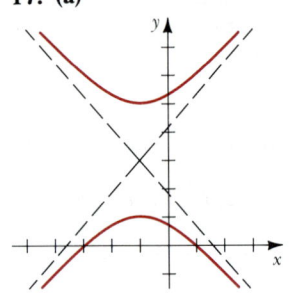

(b) vertices $(-1, 1)$, $(-1, 5)$, focal points $(-1, 3 \pm \sqrt{7})$;
(c) asymptotes $y - 3 = \pm\dfrac{2\sqrt{3}}{3}(x + 1)$

19. $16x^2 - 9y^2 = 144$ **21.** $9y^2 - 16x^2 = 144$
23. $5(x - 2)^2 - 4(y - 4)^2 = 20$
25. $x^2 - y^2 = 9$ **27.** $9x^2 - 16y^2 = 81$
29. $9(x - 2)^2 - 16(y - 1)^2 = 144$ **31.** $y = \dfrac{\sqrt{6}}{2}x - \sqrt{3}$
33. $y = \pm\sqrt{3}(x - 1)$ **35.** $\dfrac{2b^2}{a}$
37. If $D^2C - E^2A - 4ACF$: (a) is positive; (b) is zero; (c) is negative
39. No, two points satisfy these conditions **41.** 0.67 in.

Exercise Set 10.4

1. (a) Ellipse
(b)

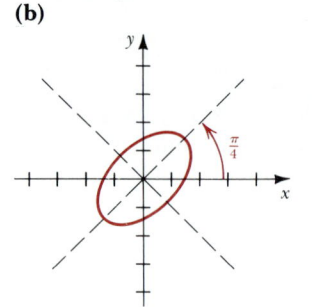

3. (a) Parabola
(b)

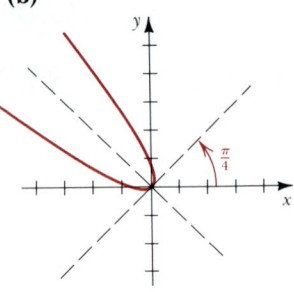

5. (a) Hyperbola
(b)

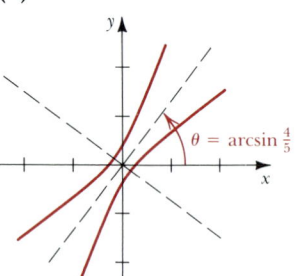

7. (a) Ellipse
(b)

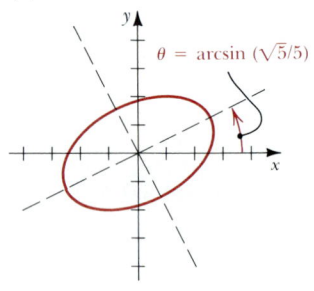

9. (a) (10.17) indicates an ellipse. (b) The graph is the single point $(0, 0)$.
11. (a) Parabola
(b)

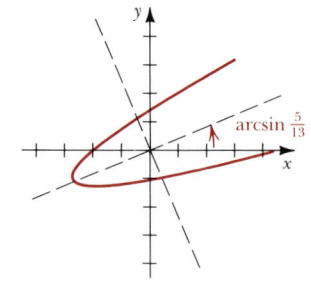

13. (a) Parabola
(b)

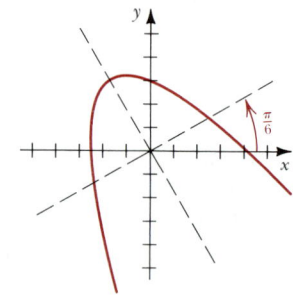

15. (a) (10.17) indicates a parabola. (b) The graph is a line with equation $\hat{x} = \dfrac{17\sqrt{137}}{274}$.

17. (a) Ellipse
(b)

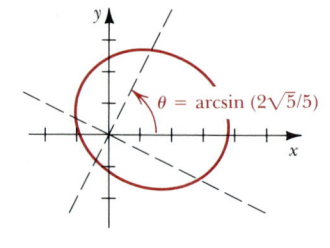

19. (a) Hyperbola
(b)

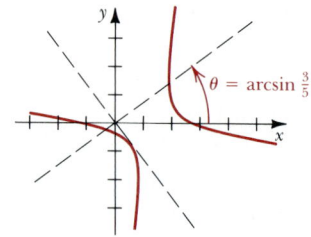

21. (a) $y = -x + 2\sqrt{2}$
(b) [graph]

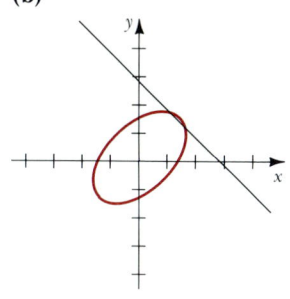

25. $337x^2 - 168xy + 288y^2 - 1350x - 1800y = 0$

Exercise Set 10.5

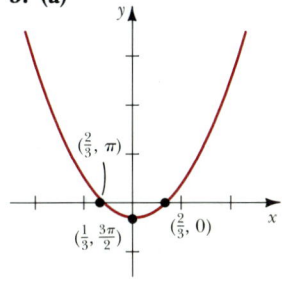

1. (a) [graph with points $(2, \frac{\pi}{2})$, $(2, \frac{3\pi}{2})$]
(b) $x - 1 = -\dfrac{y^2}{4}$

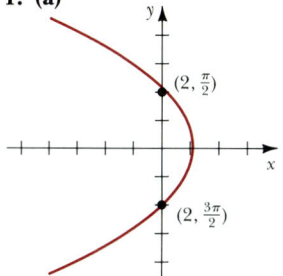

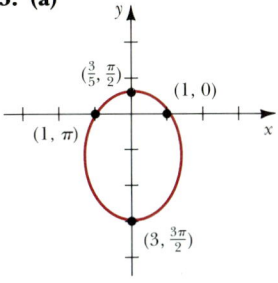

3. (a) [graph with points $(\frac{3}{5}, \frac{\pi}{2})$, $(1, 0)$, $(1, \pi)$, $(3, \frac{3\pi}{2})$]
(b) $\dfrac{9}{5}x^2 + \left(y + \dfrac{6}{5}\right)^2 = \dfrac{81}{25}$

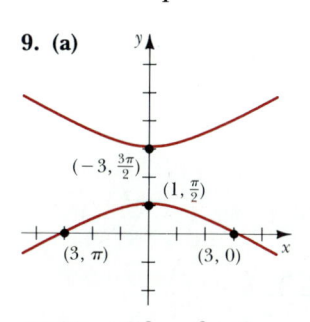

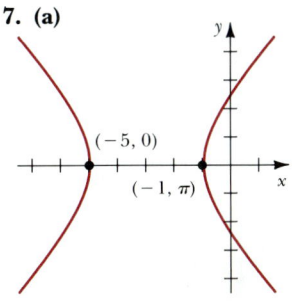

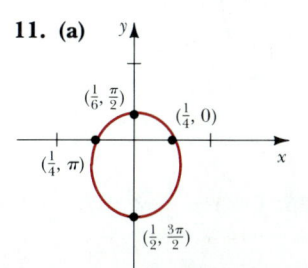

5. (a) [graph with points $(\frac{2}{3}, \pi)$, $(\frac{1}{3}, \frac{3\pi}{2})$, $(\frac{2}{3}, 0)$]
(b) $3y + 1 = \dfrac{9}{4}x^2$

7. (a) [graph with points $(-5, 0)$, $(-1, \pi)$]
(b) $5(x + 3)^2 - 4y^2 = 20$

9. (a) [graph with points $(-3, \frac{3\pi}{2})$, $(1, \frac{\pi}{2})$, $(3, \pi)$, $(3, 0)$]
(b) $3(y - 2)^2 - x^2 = 3$

11. (a) [graph with points $(\frac{1}{6}, \frac{\pi}{2})$, $(\frac{1}{4}, 0)$, $(\frac{1}{4}, \pi)$, $(\frac{1}{2}, \frac{3\pi}{2})$]
(b) $16x^2 + 12y^2 + 4y = 1$

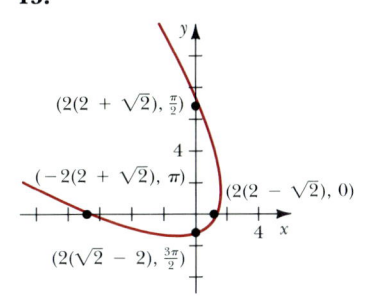

13. [graph with points $(2(2+\sqrt{2}), \frac{\pi}{2})$, $(-2(2+\sqrt{2}), \pi)$, $(2(2-\sqrt{2}), 0)$, $(2(\sqrt{2}-2), \frac{3\pi}{2})$]
(b) $x^2 + 2xy + y^2 + 4\sqrt{2}(x-y) = 8$

15. [graph with points $(\frac{1}{15}(8+\sqrt{2}), \frac{\pi}{2})$, $(-\frac{1}{15}(8+\sqrt{2}), \pi)$, $(\frac{1}{15}(8-\sqrt{2}), 0)$, $(\frac{1}{15}(\sqrt{2}-8), \frac{3\pi}{2})$]
(b) $31x^2 + 2xy + 31y^2 + 4\sqrt{2}(x-y) = 8$

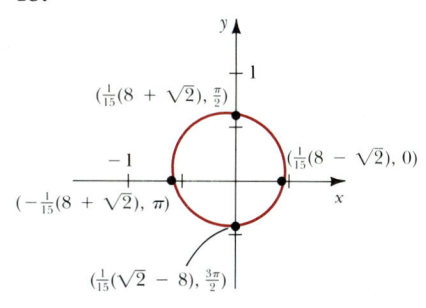

17. $r = \dfrac{8}{1 + 2\cos\theta}$ **19.** $r = \dfrac{1}{4 - 4\sin\theta}$

21. $r = \dfrac{6}{1 + 3\sin\theta}$ **23.** $r = \dfrac{3}{2 + \cos\theta}$

25. (a) $r = \dfrac{1}{3}\tan\theta\sec\theta$; (b) not in standard polar position

27. (a) $3r^2 + r^2(\cos\theta)^2 - 2r\sin\theta = 1$; (b) in standard polar position

29. (a) $r^2[1 - 2(\sin\theta)^2] - 2r\sin\theta = 4$; (b) not in standard polar position

33. [graph with points $(-2, \frac{3\pi}{2})$, $(-2, \frac{\pi}{2})$]

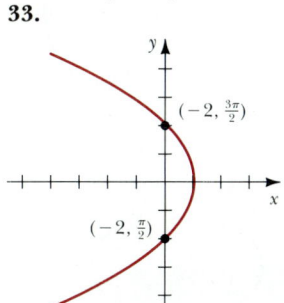

35. [graph with points $(-\frac{1}{5}, \frac{3\pi}{2})$, $(-\frac{1}{3}, 0)$, $(-\frac{1}{3}, \pi)$, $(-1, \frac{\pi}{2})$]

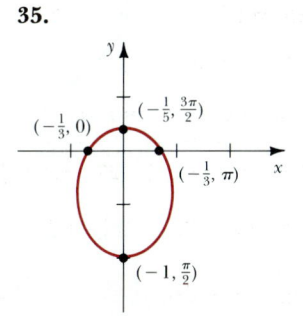

37. $r = \dfrac{4302.52}{1 - 0.04812\cos\theta}$ **39.** $r = \dfrac{2.66 \times 10^9}{1 - 0.967\cos\theta}$

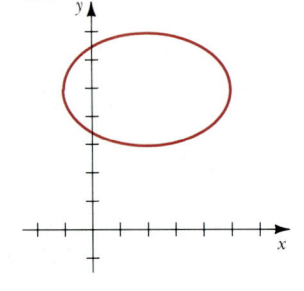

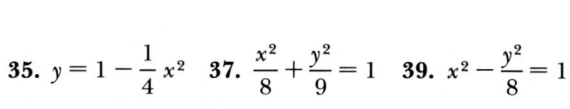

Review Exercises

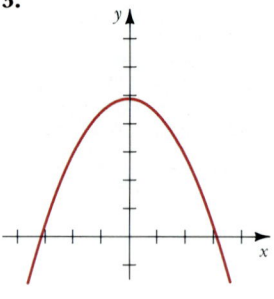

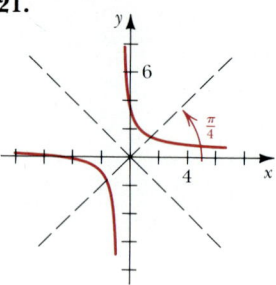

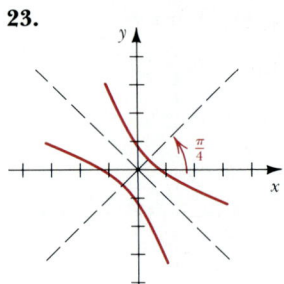

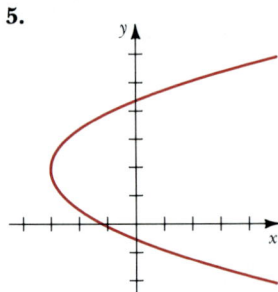

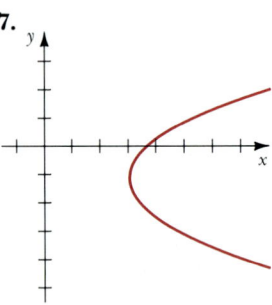

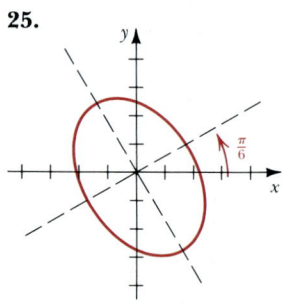

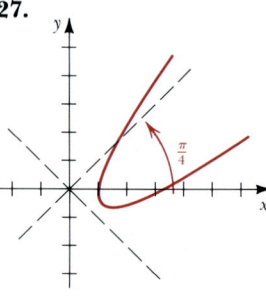

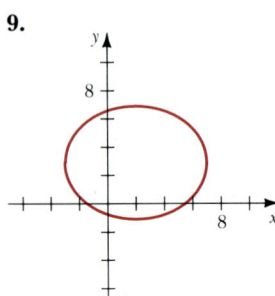

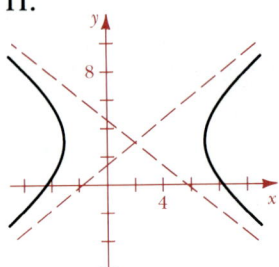

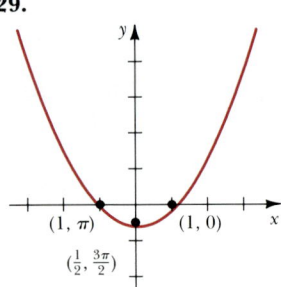

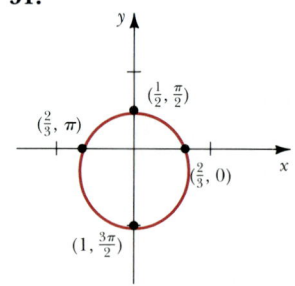

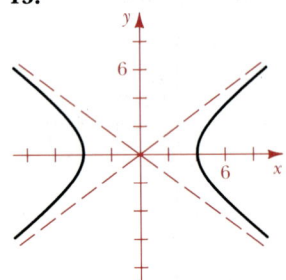

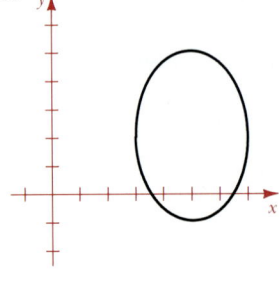

1. [graph]
3. [graph]
5. [graph]
7. [graph]
9. [graph]
11. [graph]
13. [graph]
15. [graph]
17. [graph]
19. [graph]
21. [graph]
23. [graph]
25. [graph]
27. [graph]
29. [graph]
31. [graph]
33. [graph]

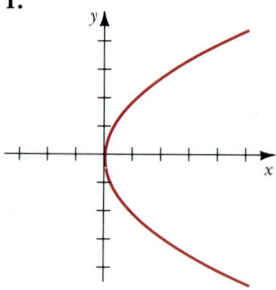

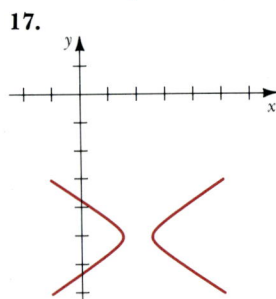

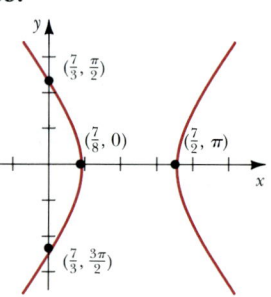

35. $y = 1 - \dfrac{1}{4}x^2$ 37. $\dfrac{x^2}{8} + \dfrac{y^2}{9} = 1$ 39. $x^2 - \dfrac{y^2}{8} = 1$

41. $\dfrac{x^2}{16} + \dfrac{y^2}{41} = 1$ 43. $r = \dfrac{6}{1 - 3\sin\theta}$

45. $5x - 2y - 9 = 0$ 47. $y = x + 1$

Chapter 11

Exercise Set 11.1

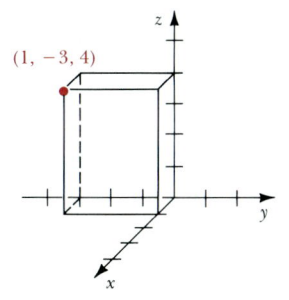

1. (a) [graph showing (1, 3, 4)]
(b) [graph showing (1, −3, 4)]
(c) [graph showing (2, 4, −3)]

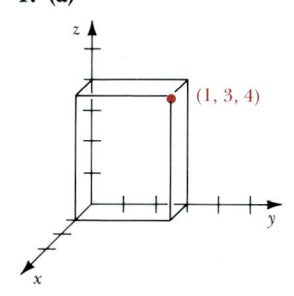

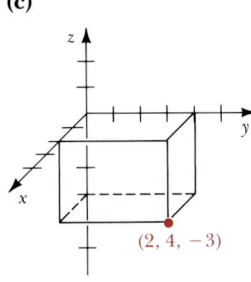

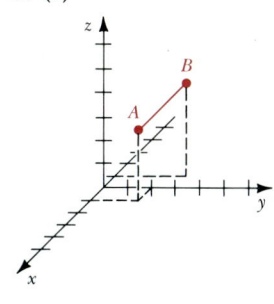

3. (a) [graph showing points A and B]
(b) $\sqrt{6}$

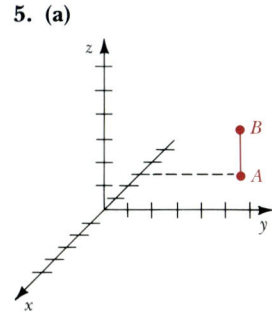

5. (a) [graph showing points A and B]
(b) 2

7. (a) [graph showing points A and B]
(b) $\sqrt{59}$

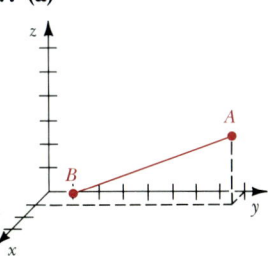

9. $\left(0, \dfrac{5}{2}, \dfrac{7}{2}\right)$ **11.** $(-3, 4, 1)$ **13.** $\left(\dfrac{1}{2}, \dfrac{9}{2}, \dfrac{3}{2}\right)$

15. [graph of box with vertices (0,0,0), (2,0,0), (0,4,0), (2,4,0), (0,0,4), (2,0,4), (0,4,4), (2,4,4)]

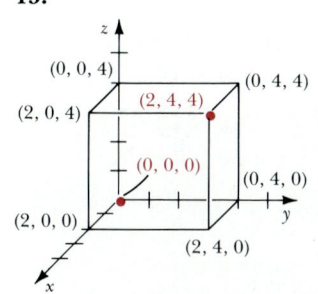

17. [graph of box with vertices (3,0,0), (4,0,0), (3,3,0), (4,3,0), (3,0,5), (4,0,5), (3,3,5), (4,3,5)]

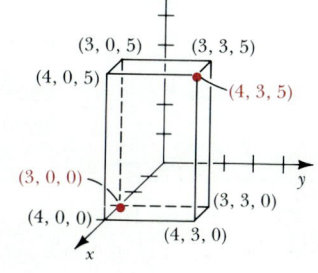

19. $x^2 + y^2 + z^2 = 4$
21. $(x-2)^2 + (y-3)^2 + (z-4)^2 = 1$
23. $(-1, 2, -3), \sqrt{14}$ **25.** $(-2, 0, 2), \dfrac{3\sqrt{2}}{2}$
27. $\left(x + \dfrac{1}{2}\right)^2 + \left(y - \dfrac{5}{2}\right)^2 + (z-3)^2 = \dfrac{17}{2}$

29. [graph showing box with vertices (2,2,5), (2,3,5), (4,2,5), (4,3,5), (2,2,0), (2,3,0), (4,2,0), (4,3,0)]

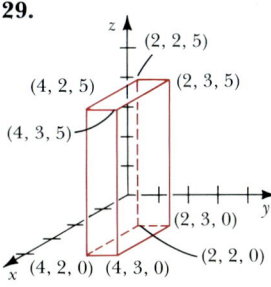

31. [graph showing triangle with vertices (2,1,5), (2,4,2), (5,1,2)]

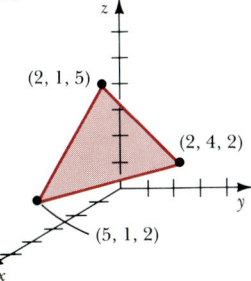

33. [graph of hemisphere]

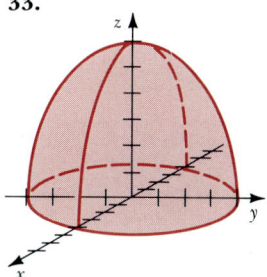

Exercise Set 11.2

1. (a) $\langle 1, 5, 3 \rangle$; **(b)** $\sqrt{35}$ **3. (a)** $\langle 0, -6, 0 \rangle$; **(b)** 6
5. (a) $\langle 2, 3, -1 \rangle$; **(b)** $\sqrt{14}$ **7.** $(1, 3, 4)$ **9.** $(-5, 3, 1)$
11. [graph showing vectors a, b, and a + b = ⟨0, 2, 2⟩]   **13.** $\langle -4, -6, -8 \rangle$

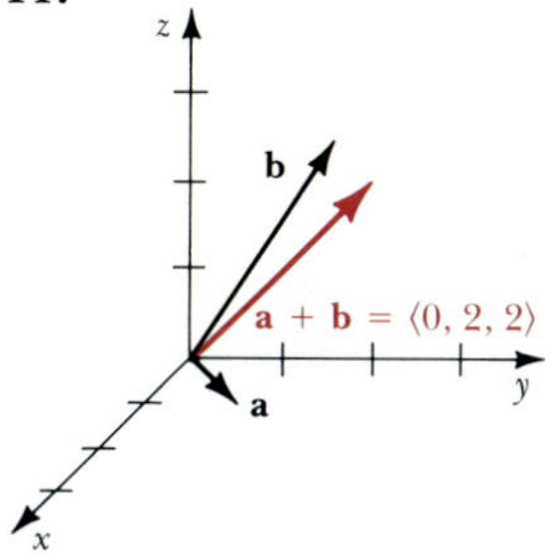

[graph showing vectors c and −2c]

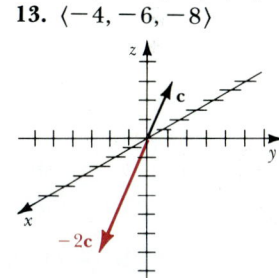

15. $\langle 2, 5, 6 \rangle$ **17.** $\|\mathbf{c}\| = \sqrt{29}, \|-2\mathbf{c}\| = 2\sqrt{29}$
19. (a) $\left\langle \dfrac{2\sqrt{29}}{29}, \dfrac{3\sqrt{29}}{29}, \dfrac{4\sqrt{29}}{29} \right\rangle$;
(b) $\left\langle -\dfrac{\sqrt{14}}{14}, \dfrac{2\sqrt{14}}{14}, \dfrac{3\sqrt{14}}{14} \right\rangle$
21. $\pm \dfrac{3\langle 3, 4, -1 \rangle}{\|\langle 3, 4, -1 \rangle\|} = \pm \left\langle \dfrac{9\sqrt{26}}{26}, \dfrac{6\sqrt{26}}{13}, -\dfrac{3\sqrt{26}}{26} \right\rangle$

23. (a) $\mathbf{i} + 4\mathbf{j}$ (b) $\mathbf{i} + 4\mathbf{j} + 5\mathbf{k}$

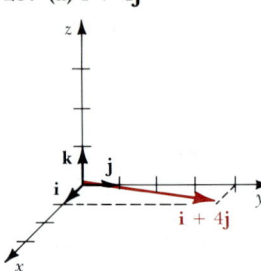

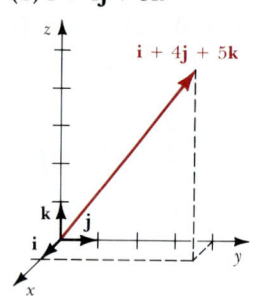

(c) $2\mathbf{j} + 3\mathbf{k}$

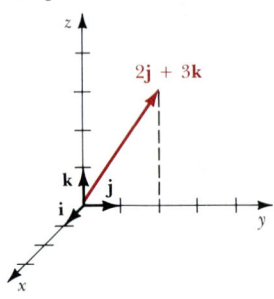

27. (a) $\mathbf{d} = \mathbf{c} - \mathbf{b}$ **29.** $\mathbf{F} = -20\mathbf{i} - 15\mathbf{j}$

31. If $\mathbf{j}$ points upstream and $\mathbf{i}$ points across the stream, the direction is $300\mathbf{i} - 180\mathbf{j}$. The average speed is 5.8 ft/sec.

33. In the direction of $-10\sqrt{2}\mathbf{i} + (10\sqrt{2} - 100)\mathbf{j}$. The average speed is 87 mi/hr.

Exercise Set 11.3

1. 20   **3.** 34   **5.** 0   **7.** $3 + 4e$

9. (a) $\arccos \dfrac{5\sqrt{114}}{57} \approx 0.358$; (b) no

11. (a) $\arccos \dfrac{17\sqrt{290}}{290} \approx 0.059$; (b) no   **13.** (a) $\dfrac{\pi}{2}$; (b) yes

15. (a) $\arccos \dfrac{3 + 4e}{\sqrt{(25 + \pi^2)(1 + e^2)}} \approx 0.625$; (b) no

17. (a) $\dfrac{20}{19}\sqrt{19}$; (b) $\dfrac{20}{19}\langle 1, 3, 3 \rangle$

19. (a) $\dfrac{17}{5}\sqrt{5}$; (b) $\left\langle \dfrac{17}{5}, \dfrac{34}{5}, 0 \right\rangle$ **21.** (a) 0; (b) 0

23. (a) $\dfrac{3 + 4e}{\sqrt{25 + \pi^2}}$; (b) $\left(\dfrac{3 + 4e}{25 + \pi^2}\right) \langle 3, 4, \pi \rangle$

25. $\dfrac{2\sqrt{29}}{29}, \dfrac{3\sqrt{29}}{29}, \dfrac{4\sqrt{29}}{29}$ **27.** $0, \dfrac{3}{5}, \dfrac{4}{5}$

29. $\left\langle 0, \dfrac{3}{5}, \dfrac{4}{5} \right\rangle$

31. There are two: $\pm \dfrac{\sqrt{5}}{5} \langle 2, -1, 0 \rangle$.

33. $\mathbf{b}_1 = \dfrac{1}{19} \langle 5, 15, -15 \rangle$, $\mathbf{b}_2 = \dfrac{1}{19} \langle -81, 4, -23 \rangle$

37. $\langle 6c, c, -2c \rangle$ for any constant c
43. $\mathbf{a}$ and $\mathbf{b}$ must have the same direction.
49. 26,400 ft-lb   **51.** $200\mathbf{i} + 200\sqrt{3}\mathbf{j}$

Exercise Set 11.4

1. $\langle 5, 10, -5 \rangle$ **3.** $\langle -3, 12, -2 \rangle$ **5.** $\langle 2, -1, 2 \rangle$
7. $\langle -1, 0, 2 \rangle$ **9.** -1 **11.** Not possible   **13.** 12

15. $\dfrac{1}{2} \langle -2, 1, -1 \rangle$

17. $\langle 6c, c, -2c \rangle$ for any constant c **19.** 25   **21.** $\sqrt{69}$

23. 24   **25.** 20   **29.** 4   **31.** $\dfrac{1}{3}$

Exercise Set 11.5

1. $x - y + z = 2$ **3.** $z = 4$ **5.** $2x + 3y + 4z = 0$
7.   **9.**

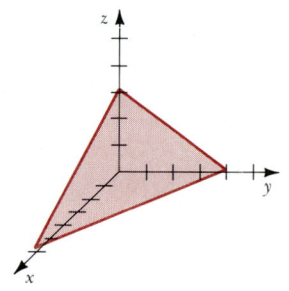

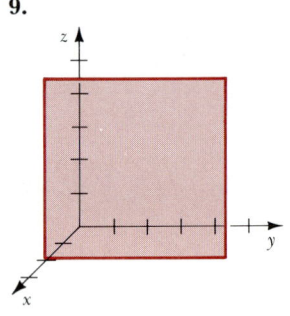

11.   **13.** $z = 0$ is the xy-plane

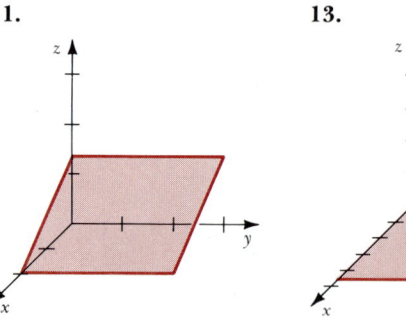

15. $z = 3$ **17.** $x + y - z = 2$ **19.** $z = 3$ **21.** $y - x = 1$
23. $2x + y + z = 5$ **25.** $7x + 5y - 3z = 10$
27. $x - 2y + z = -2$ **29.** $\dfrac{6\sqrt{17}}{17}$ **31.** $\dfrac{3\sqrt{2}}{2}$

33. $\dfrac{2}{3}\sqrt{6}$ **35.** $\dfrac{3}{28}\sqrt{14}$

37. $8x + 8y + 10z = 99$

Exercise Set 11.6

1. $x = 2 + t, y = -1 + t, z = 2 + t$
3. $x = 2t, y = 4t, z = 3t$
5. $x = 1, y = 2 - 3t, z = 4t$
7. $x = 3 + t, y = 4 + 7t, z = 4 - t$
9. $x - 2 = y + 1 = z - 2$
11. $\dfrac{x}{2} = \dfrac{y}{4} = \dfrac{x}{3}$ 13. $x = 1, \dfrac{y-2}{-3} = \dfrac{z}{4}$
15. $x - 3 = \dfrac{y-4}{7} = \dfrac{z-4}{-1}$
17. $x = t + 1, y = 2 - t, z = 3 - t$
19. $x = 1 + t, y = -2 + t, z = 3 + t$
21. $x = 1 + t, y = 7, z = 0$
23. (a) Yes; (b) no; (c) no intersection
25. (a) No; (b) no; (c) $\left(\dfrac{3}{2}, \dfrac{5}{2}, \dfrac{3}{2}\right)$
27. (a) No; (b) no; (c) $(-7, -8, -3)$
29. (a) $(0, 2, -3)$; (b) $(2, 0, 1)$; (c) $\left(\dfrac{3}{2}, \dfrac{1}{2}, 0\right)$
33. $x = 1 - 9t, y = -1 + 8t, z = 1 - t$
37. $\dfrac{83\sqrt{6}}{30}$ 39. $\dfrac{2\sqrt{70}}{7}$

Review Exercises

1. $\langle 17, 2, 22 \rangle$ 3. $\langle 15\sqrt{2} - 4, 25\sqrt{2} - 16, 20\sqrt{2} - 6 \rangle$
5. Not possible   7. $\langle -7, 0, 2 \rangle$ 9. $\langle 33, 7, 47 \rangle$
11. $\langle 33, 7, 47 \rangle$ (The same answer as Exercise 9.)
13. Plane 3 units above and parallel to the xy-plane

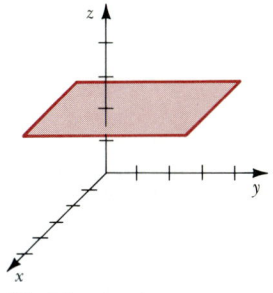

15. Line in the xy-plane passing through $(0, 1, 0)$ and having direction given by $\langle 1, -1, 0 \rangle$

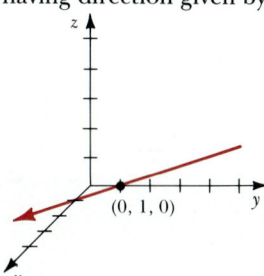

17. Line passing through the point $(2, 1, 3)$ and having direction given by $\langle 2, 1, 1 \rangle$

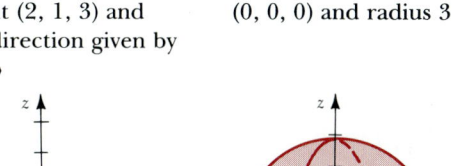

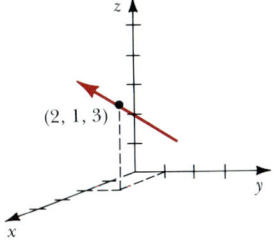

19. Sphere with center $(0, 0, 0)$ and radius 3

21. Plane orthogonal to the zx-plane

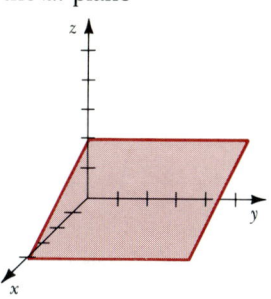

23. Plane

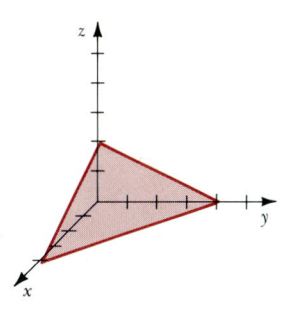

25. $\langle 0, 3, 3 \rangle$ 27. $\pm \dfrac{\sqrt{21}}{21} \langle 1, -4, -2 \rangle$ 29. $\langle 3, -7, 1 \rangle$
31. $\langle 13, 7, 10 \rangle$
33.

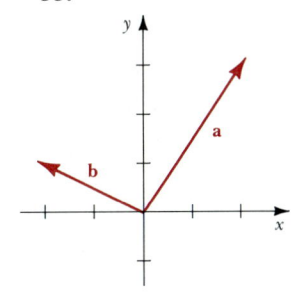

35. $\pm \mathbf{k}$ 37. $\arccos\left(\dfrac{-\sqrt{65}}{65}\right) \approx 1.695 \approx 97°$
39. 0   41. $x + y = 4$ 43. $x = -2, y = 1 + t, z = 4$
45. $3x + 2y - 2z = 6$ 47. $(x - 2)^2 + y^2 + z^2 = 4$
49. $z = 4$ 51. $x = -2, y = 1 + t, z = 2$
53. (a) -3; (b) $\dfrac{4}{3}$ 55. (a) 6; (b) $\dfrac{11\sqrt{6}}{6}$ 57. 200

Chapter 12

Exercise Set 12.1

1. $(-\infty, \infty)$ 3. $[0, \infty)$ 5. $(0, \infty)$
7. 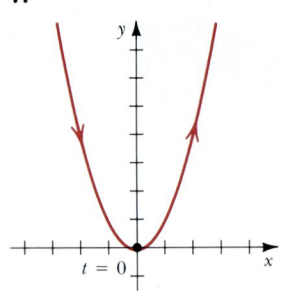
9. 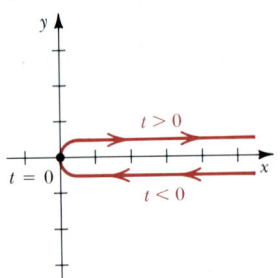
11. 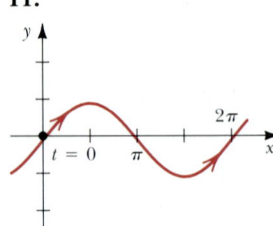
13. 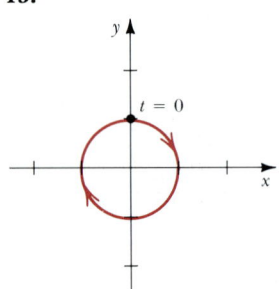
15. 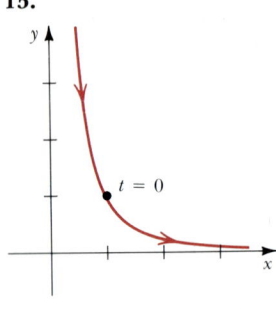
17. 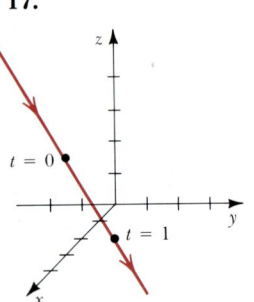
19. 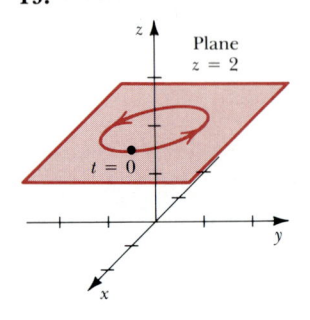
21.

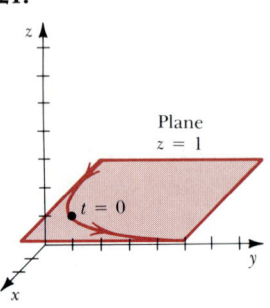

23.

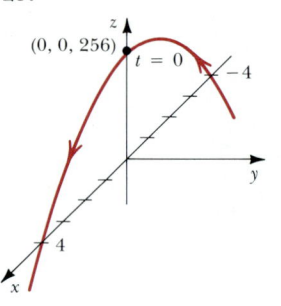

25. (a) $(0, \infty)$; $t^{3/2} + e^t \sin t$; (b) $t \neq 2$; $(2-t)\mathbf{i} + (2-t)^{-2}\mathbf{j} + e^{2-t}\mathbf{k}$; (c) $t \neq 0$; $t(2-t)\mathbf{i} + t^{-2}(2-t)\mathbf{j} + (2-t)e^t\mathbf{k}$; (d) $(0, \infty)$; $(t + \sqrt{t})\mathbf{i} + t^{-2}\mathbf{j} + (e^t + \sin t)\mathbf{k}$; (e) $(0, \infty)$; $(t - \sqrt{t})\mathbf{i} + t^{-2}\mathbf{j} + (e^t - \sin t)\mathbf{k}$; (f) $(0, \infty)$; $t^{-2} \sin t \mathbf{i} + (e^t \sqrt{t} - t \sin t)\mathbf{j} - t^{-3/2}\mathbf{k}$

27. (a) The domain of $\mathbf{F}$ is $(0, \infty)$ and the domain of $\mathbf{G}$ is $(-\infty, 0)$ so $\mathbf{F} \cdot \mathbf{G}$, $\mathbf{F} + \mathbf{G}$, $\mathbf{F} - \mathbf{G}$, and $\mathbf{F} \times \mathbf{G}$ are undefined. (b) $(-\infty, \infty)$; $e^{t/2}\mathbf{i} + t\mathbf{j} + 3e^{3t}\mathbf{k}$; (c) $(0, \infty)$; $\sqrt{t}e^t\mathbf{i} + e^t \ln t\mathbf{j} + 3t^3 e^t \mathbf{k}$

29. $\mathbf{F}(t) = R(t - \sin t)\mathbf{i} + R(1 - \cos t)\mathbf{j}$

31.

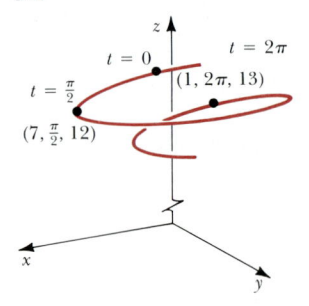

Exercise Set 12.2

1. $\mathbf{i} + \mathbf{j} + \mathbf{k}$ 3. $4\mathbf{i} + 3\mathbf{j}$
5. Does not exist   7. $\mathbf{i} - 2\mathbf{j} + \mathbf{k}$ 9. $0 < t < \infty$
11. $t \neq 2$ 13. $t \neq 0$ 15. $t \neq 1$
17. $\dfrac{1}{t}\mathbf{i} + \mathbf{j}$ 19. $2t\mathbf{i} + 2te^{t^2}\mathbf{j} + \mathbf{k}$
21. $3\mathbf{j}$ 23. $\mathbf{i} + \mathbf{j} + 2\mathbf{k}$

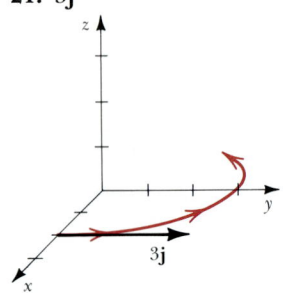

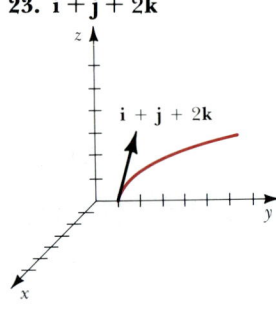

25. $\mathbf{i} + \mathbf{k}$

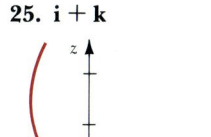

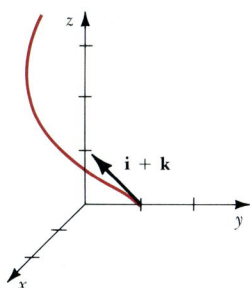

27. $(2t - 3/t)\mathbf{i} - 2\mathbf{j} + (\cos t - 9e^t)\mathbf{k}$
29. $(3 \tan t - 2 \cos t \sin t)\mathbf{i} + 2 \sin t\mathbf{j} + (-\sin t \cos (\cos t) + 9 \sin t e^{\cos t})\mathbf{k}$
31. $-6e^t(t+1)\mathbf{i} + [\cos t \ln t + (\sin t)/t - 3t(t+2)e^t]\mathbf{j} + 2(\ln t + 1)\mathbf{k}$
33. $(2t - 3\cos t)\mathbf{i} + (e^t - 3)\mathbf{j} + \left(1 - \dfrac{3}{t}\right)\mathbf{k}$
35. $2te^{t^2}[(2e^{t^2} - 3 \cos e^{t^2})\mathbf{i} + (e^{e^{t^2}} - 3)\mathbf{j} + (1 - 3e^{-t^2})\mathbf{k}]$
37. $\left[e^t\left(\ln t + \dfrac{1}{t}\right) - 2t - 1\right]\mathbf{i} + [\sin t + (t+1) \cos t - 2t \ln t - t]\mathbf{j} + [3t^2 - e^t(\sin t + \cos t)]\mathbf{k}$ **39.** $\mathbf{i} + \mathbf{j} + \dfrac{\pi^2}{8}\mathbf{k}$
41. $(t-1)e^t\mathbf{i} + \dfrac{t^2}{2}\mathbf{j} + t\mathbf{k} + \mathbf{C}$ **43.** $\dfrac{t^2}{2}\mathbf{i} + \ln|t|\mathbf{j} + \mathbf{C}$

45. $\sqrt{2}t + C$ **47.** $(-\infty, \infty)$ **49.** $t \neq n\pi + \dfrac{\pi}{2}$

51. (b) $\mathbf{F}_1\left(\dfrac{1}{2}\right) = \dfrac{3}{2}\mathbf{i} + 4\mathbf{j} + \dfrac{5}{2}\mathbf{k}$,

$\mathbf{F}_2\left(\dfrac{1}{2}\right) = \dfrac{5}{4}\mathbf{i} + \dfrac{7}{2}\mathbf{j} + \dfrac{7}{4}\mathbf{k}$,

$\mathbf{F}_3\left(\dfrac{1+e}{2}\right) \approx 1.62\mathbf{i} + 4.24\mathbf{j} + 2.86\mathbf{k}$

57. One example is
$\mathbf{F}(t) = \begin{cases} \mathbf{i}, & \text{if } t \geq 0 \\ -\mathbf{i}, & \text{if } t < 0 \end{cases}$

F is discontinuous at $a = 0$, but $\|\mathbf{F}(t)\| = 1$ is always continuous.

Exercise Set 12.3

1. $-\mathbf{i}$ **3.** $\dfrac{\sqrt{6}}{6}(2\mathbf{i} + \mathbf{j} + \mathbf{k})$ **5.** $\dfrac{\sqrt{14}}{14}(2\mathbf{i} + \mathbf{j} - 3\mathbf{k})$
7. $-\mathbf{j}$ **9.** $\dfrac{\sqrt{14}}{7}\left[\mathbf{i} - \dfrac{3}{2}\mathbf{j} - \dfrac{1}{2}\mathbf{k}\right]$
11. $\mathbf{T}'(t) = \mathbf{0}$; no unit normal vector exists. **13.** $(-\infty, \infty)$
15. $(0, \infty)$ **17.** $(-\infty, \infty)$

19. $\mathbf{F}(s) = \left(1 + \dfrac{\sqrt{33}}{33}s\right)\mathbf{i} + \left(1 + \dfrac{4\sqrt{33}}{33}s\right)\mathbf{j} + \dfrac{4\sqrt{33}}{33}s\mathbf{k}$
21. $\mathbf{F}(s) = \cos s\mathbf{i} + \sin s\mathbf{j}$
23. $\mathbf{F}(s) = \sin\dfrac{\sqrt{2}}{2}s\mathbf{i} + \cos\dfrac{\sqrt{2}}{2}s\mathbf{j} + \dfrac{\sqrt{2}}{2}s\mathbf{k}$, $0 \leq s \leq \sqrt{2}\pi$
27. $-\dfrac{\sqrt{2}}{2}\mathbf{i} - \dfrac{\sqrt{2}}{2}\mathbf{k}$ **29.** $\mathbf{k}$

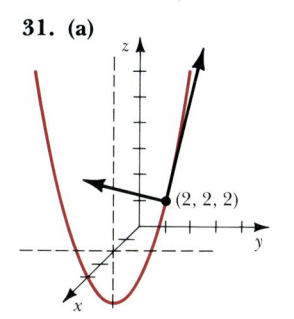

31. (a)

(b) $\mathbf{T}(2) = \dfrac{\sqrt{17}}{17}(\mathbf{j} + 4\mathbf{k})$; $\mathbf{N}(2) = \dfrac{\sqrt{17}}{17}(-4\mathbf{j} + \mathbf{k})$;
(c) $t = 0$, $\mathbf{k}$
33. (a) $\sqrt{15625\pi^2 + 100} \approx 393$ inches;
(b) $\sqrt{14641\pi^2 + 100} \approx 380$ inches.

Exercise Set 12.4

1. (a) $\mathbf{v}(t) = \cos t\mathbf{i} - \sin t\mathbf{j} + \mathbf{k}$; (b) $\mathbf{a}(t) = -\sin t\mathbf{i} - \cos t\mathbf{j}$;
(c) $v(t) = \sqrt{2}$
3. (a) $\mathbf{v}(t) = \mathbf{i} + 2\mathbf{j} + 3\mathbf{k}$; (b) $\mathbf{a}(t) = \mathbf{0}$; (c) $v(t) = \sqrt{14}$
5. (a) $\mathbf{v}(t) = 4\mathbf{j} + (5 - 10t)\mathbf{k}$; (b) $\mathbf{a}(t) = -10\mathbf{k}$;
(c) $v(t) = \sqrt{100t^2 - 100t + 41}$

7. (a) $\mathbf{v}(t) = \dfrac{3}{2}(3t-1)^{-1/2}\mathbf{j} + e^t\mathbf{k}$;

(b) $\mathbf{a}(t) = -\dfrac{9}{4}(3t-1)^{-3/2}\mathbf{j} + e^t\mathbf{k}$;

(c) $v(t) = \sqrt{\dfrac{9}{4(3t-1)} + e^{2t}}$

9. $a_\mathbf{T}(t) = 0$, $a_\mathbf{N}(t) = 1$ **11.** $a_\mathbf{T}(t) = 0$, $a_\mathbf{N}(t) = 0$
13. $a_\mathbf{T}(t) = \dfrac{64(16t - 5)}{\sqrt{1 + (32t - 10)^2}}$, $a_\mathbf{N}(t) = \dfrac{32}{\sqrt{1 + (32t - 10)^2}}$
15. (a)

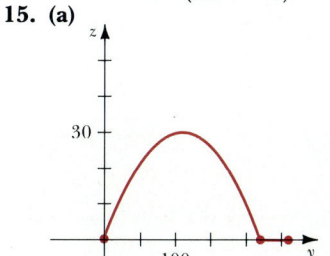

(b) 30; (c) $10\sqrt{157} \approx 125$ yd/sec

17. (a) $\dfrac{125\sqrt{3}}{4}$; (b) $15{,}625\sqrt{3}$ ft; (c) $11{,}718.75$ ft;
(d) 1000 ft/sec

19. (a) $\dfrac{3025\sqrt{3}}{64} \approx 81.87$ ft; (b) $\dfrac{55}{32}$ sec

21. (a) $\dfrac{1340}{\sqrt{159}} \approx 106$ ft/sec; (b) $\dfrac{\sqrt{318}}{4} \approx 4.46$ sec;
(c) $\left(\dfrac{1340}{\sqrt{318}}\right)\mathbf{i} + \left(\dfrac{1340}{\sqrt{318}} - 8\sqrt{318}\right)\mathbf{j}$;
(d) $\sqrt{1{,}622{,}608/159} \approx 101$ ft/sec

23. The shortest time for the ball is 3.48 sec. The runner takes 3.6 sec.

25. $\dfrac{1225\sqrt{3}}{4}$. No, the maximum height is $1225/16 \approx 77$ ft.

29. $\mathbf{S}(t) \approx 9.21 t^2 \mathbf{i} + 3.2 t^2 \mathbf{j}$

Exercise Set 12.5

1. $\dfrac{1}{5}$ 3. 0 5. $\dfrac{1}{2}$ 7. $\dfrac{1}{4}\sqrt{2}$ 9. (a) $K(0) = \dfrac{3}{4}, \rho(0) = \dfrac{4}{3}$;
(b)

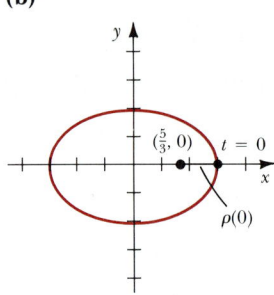

11. (a) $K(1) = \dfrac{\sqrt{2}}{4}, \rho(1) = 2\sqrt{2}$;
(b)

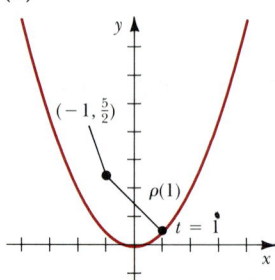

13. (a) $K\left(\dfrac{\pi}{2}\right) = \dfrac{1}{5}, \rho\left(\dfrac{\pi}{2}\right) = 5$;
(b)

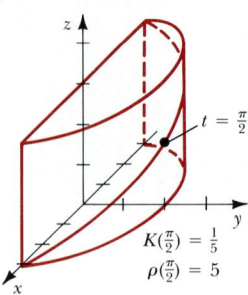

15. $\dfrac{-3}{4}\mathbf{i}$ 17. $\dfrac{1}{4}(\mathbf{j}-\mathbf{i})$ 19. $\dfrac{-1}{5}\mathbf{j}$

21. $(0, 0)$ 23. None exist.

25. (a) $\mathbf{r}(t) = t\mathbf{i} + \ln t\,\mathbf{j}$; (b) $K = \dfrac{\sqrt{2}}{4}, \rho = 2\sqrt{2}$

27. (a) $\mathbf{r}(t) = t\mathbf{i} + (t^3 - 3t - 2)\mathbf{j}$; (b) $K = 6, \rho = \dfrac{1}{6}$

29. (a) $\mathbf{r}(t) = t\mathbf{i} + (t^3 - 3t^2 + 3t - 1)\mathbf{j}$; (b) $K = 0$ so ρ becomes infinite.

31. (a) $\mathbf{r}(t) = t\mathbf{i} + \dfrac{1}{t}\mathbf{j}$;
(b) $K = \dfrac{16\sqrt{17}}{289}, \rho = \dfrac{17\sqrt{17}}{16}$

33. (a) $\mathbf{r}(t) = t\mathbf{i} + (t + \sin t)\mathbf{j}$; (b) $K = 0$ so ρ becomes infinite.

37. $K = \dfrac{3\sqrt{10}}{100}, \rho = \dfrac{10\sqrt{10}}{3}$ 39. $K = 6, \rho = \dfrac{1}{6}$

41. $K = \dfrac{2\sqrt{5}}{5}, \rho = \dfrac{\sqrt{5}}{2}$ 43. $K = \dfrac{24\sqrt{17}}{289}, \rho = \dfrac{17\sqrt{17}}{24}$

45. $K = \dfrac{\sqrt{2}}{4}, \rho = 2\sqrt{2}$

49. $x = \pm\left[\dfrac{2}{5} + \dfrac{\sqrt{86}}{15}\right]^{1/2} \approx \pm 1.009$ 51. 2

53. $\dfrac{2 + \pi^2/4}{(1 + \pi^2/4)^{3/2}}$

55. Minimum at $(0, \pm 2)$, maximum at $(\pm 3, 0)$

Exercise Set 12.6

3. (a) 7.3 m/sec^2; (b) 1.5 m/sec^2; (c) 0.13 m/sec^2;
(d) 0.035 m/sec^2

5. (a) Decrease; (b) 0.184%

7. (a) 1.9867×10^{30} kg; (b) 1.9861×10^{30} kg

9. 5.5 g/cc. The material beneath is much denser.

13. 8.85×10^4 km. The gravitational force of the earth will not permit an unpowered isosynchronous orbit of the moon.

19. (a) November 10, 2023; (b) 5.26×10^9 km

Review Exercises

1. $|t| < 1$ **3.** $t \neq 1$ **5.** $2\mathbf{j} + (e^\pi - 1)\mathbf{k}$
7. $\mathbf{i} + (2\ln 2 - 1)\mathbf{k}$ **9.** 1
11. (a)

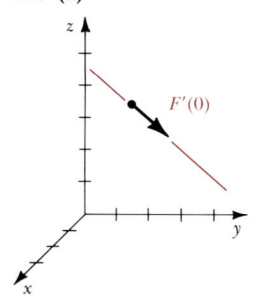

(b) $\mathbf{F}'(0) = \mathbf{i} + \mathbf{j}$; **(c)** $\mathbf{T}(0) = \dfrac{\sqrt{2}}{2}\mathbf{i} + \dfrac{\sqrt{2}}{2}\mathbf{j}$, $\mathbf{N}(0) = 0$; **(d)** $K = 0$

13. (a)

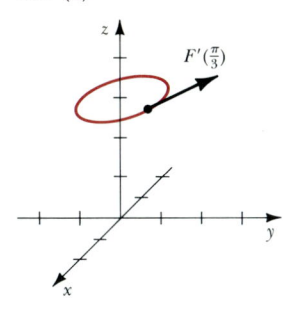

(b) $\mathbf{F}'\left(\dfrac{\pi}{3}\right) = -\dfrac{\sqrt{3}}{2}\mathbf{i} + \dfrac{1}{2}\mathbf{j}$;
(c) $\mathbf{T}\left(\dfrac{\pi}{3}\right) = -\dfrac{\sqrt{3}}{2}\mathbf{i} + \dfrac{1}{2}\mathbf{j}$, $\mathbf{N}\left(\dfrac{\pi}{3}\right) = -\dfrac{1}{2}\mathbf{i} - \dfrac{\sqrt{3}}{2}\mathbf{j}$; **(d)** $K = 1$

15. (a)

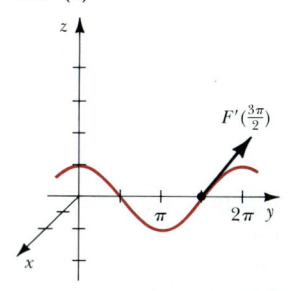

(b) $\mathbf{F}'\left(\dfrac{3\pi}{2}\right) = \mathbf{j} + \mathbf{k}$; **(c)** $\mathbf{T}\left(\dfrac{3\pi}{2}\right) = \dfrac{\sqrt{2}}{2}\mathbf{j} + \dfrac{\sqrt{2}}{2}\mathbf{k}$, $\mathbf{N}\left(\dfrac{3\pi}{2}\right)$ does not exist; **(d)** $K = 0$

17. (a)

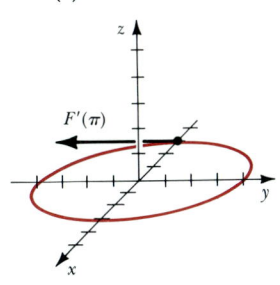

(b) $\mathbf{F}'(\pi) = -4\mathbf{j}$; **(c)** $\mathbf{T}(\pi) = -\mathbf{j}$, $\mathbf{N}(\pi) = \mathbf{i}$; **(d)** $K = \dfrac{3}{16}$

19. (a) $\mathbf{v}(0) = \mathbf{i} + \mathbf{j}$; **(b)** $v(0) = \sqrt{2}$; **(c)** $\mathbf{a}(0) = \mathbf{0}$

21. (a) $\mathbf{v}\left(\dfrac{\pi}{3}\right) = -\dfrac{\sqrt{3}}{2}\mathbf{i} + \dfrac{1}{2}\mathbf{j}$; **(b)** $v\left(\dfrac{\pi}{3}\right) = 1$;
(c) $\mathbf{a}\left(\dfrac{\pi}{3}\right) = -\dfrac{1}{2}\mathbf{i} - \dfrac{\sqrt{3}}{2}\mathbf{j}$

23. (a) $\mathbf{v}\left(\dfrac{3\pi}{2}\right) = \mathbf{j} + \mathbf{k}$; **(b)** $v\left(\dfrac{3\pi}{2}\right) = \sqrt{2}$; **(c)** $\mathbf{a}\left(\dfrac{3\pi}{2}\right) = \mathbf{0}$

25. (a) $\mathbf{v}(\pi) = -4\mathbf{j}$; **(b)** $v(\pi) = 4$; **(c)** $\mathbf{a}(\pi) = 3\mathbf{i}$
27. (a) Undefined; **(b)** $2t\mathbf{i} + 2\mathbf{j} + 2t\mathbf{k}$; **(c)** $3t^2\mathbf{i} + (2\ln t + 1)t\mathbf{j} + 3t^2\mathbf{k}$; **(d)** undefined; **(e)** $2t\mathbf{i} + \dfrac{2}{t}\mathbf{j} + 2t\mathbf{k}$

29. $D_t |t|$ does not exist when $t = 0$.
31. $\mathbf{F}(s) = \left(3 - \dfrac{\sqrt{3}}{15}s\right)\mathbf{i} + \dfrac{7\sqrt{3}}{15}s\mathbf{j} + \dfrac{\sqrt{3}}{3}s\mathbf{k}$,
$0 \leq s \leq 5\sqrt{3}$ **33.** $\dfrac{1}{2}$ **35.** 6

37. $\mathbf{r}(t) = 250t\mathbf{i} + (250\sqrt{3}t - 16t^2)\mathbf{j}$
(a) $\dfrac{250\sqrt{3}}{16} \approx 27.06$ sec; **(b)** 2929.6875 ft;
(c) Approximately 6765.82 ft; **(d)** 500 ft/sec
39. (a) $(25 + 5\sqrt{33})/4 \approx 13.43$ sec; **(b)** 825 ft;
(c) $50(25 + 5\sqrt{33})\sqrt{3} \approx 4653$ ft; **(d)** $240\sqrt{3} \approx 415.7$ ft/sec
41. (a) $\mathbf{r}(t) = \dfrac{55\sqrt{2}}{3}t\mathbf{i} + \dfrac{(484 - 55\sqrt{2})}{3}t\mathbf{j} + (10{,}000 - 16t^2)\mathbf{k}$;
(b) Approximately 3385 ft north and 648 ft east of the tower
45. 1 hr, 30 min, 19.9 sec

Chapter 13

Exercise Set 13.1

1. (a) 1; **(b)** 0; **(c)** 5 **3.** $x + y \leq 1$ **5.** $y \neq 0$
7. $x > 0$ and $y > x$ or $x < 0$ and $y < x$.
9. $x^2 + y^2 + z^2 \leq 4$ **11.** $-1, -1$ **13.** $-2x - h, -2y - k$
15. $2(2xy + 4xz + 6yz)$ **17.** $V(h, r) = \pi r^2 h$

19. $1000e$ **21.** $C(x, y, z) = (9xy + 16xz + 16yz)$ dollars
23. $C(x, y) = \dfrac{8}{75} xy$ dollars

Exercise Set 13.2

1. [graph with lines $c = 1$, $c = 2$, $c = 3$]

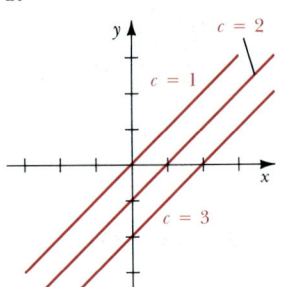

3. [graph with circles $c = -2$, $c = 0$]

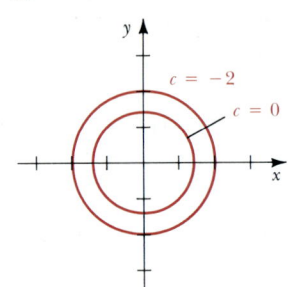

5. [graph with parabolas $c = -1$, $c = 0$, $c = 1$]

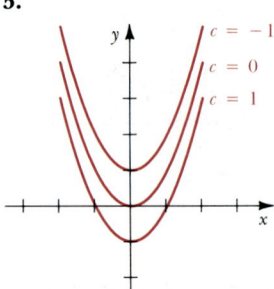

7. [graph with lines through $(0, e)$ and $(e, 0)$; $c = 0$, $c = 1$]

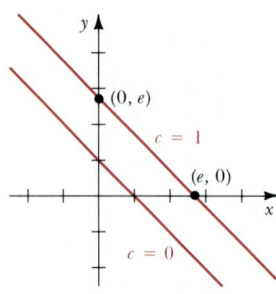

9. [graph with curves $c = 0$, $c = 1$ over $[0, 2\pi]$]

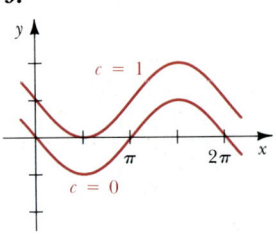

11. [3D region]

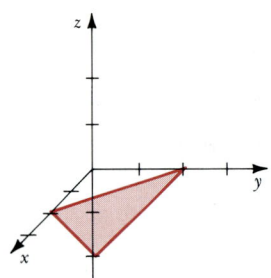

13. [paraboloid]

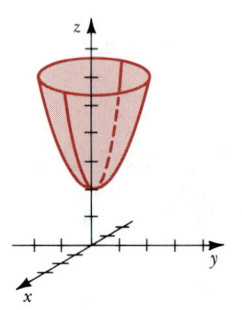

15. [3D surface]

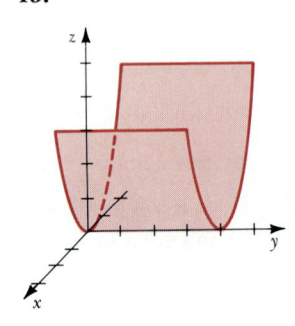

17. [plane in 3D]

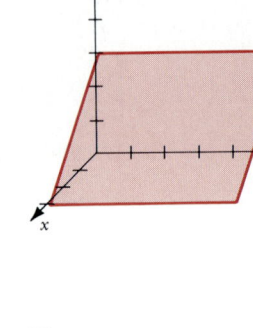

19. [paraboloid opening upward]

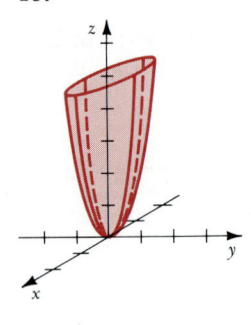

21. [paraboloid with level curve]

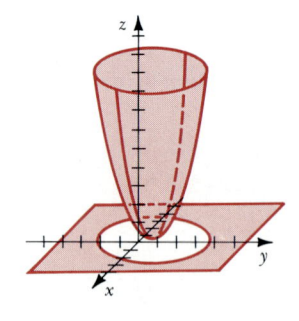

23. [cone/pyramid with level plane]

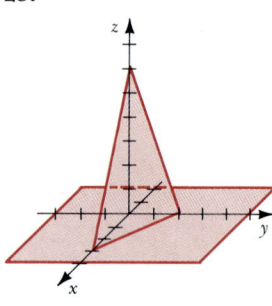

25. [ellipses $E = 0$, $E = 100$, $E = 150$, $E = 200$]

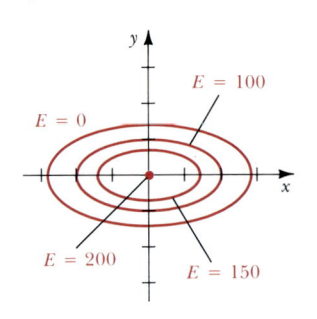

27. [curves $c = 4$, $c = 9$, $c = 16$ in h-l plane]

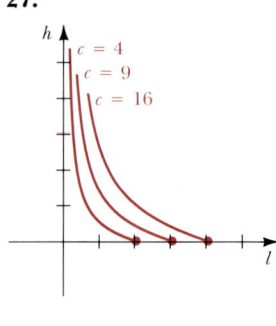

Exercise Set 13.3

1. Parallel planes with normal vector $\langle 2, 3, 1 \rangle$
3. Parallel planes with normal vector $\langle -1, -1, 1 \rangle$
5. Right circular cylinders about the z-axis
7. Right circular cylinders about the x-axis
9. Each level surface is centered at the origin and will intersect each of the coordinate planes in an ellipse.
11. It will intersect planes parallel to the xz-plane in a circle and planes parallel to the other coordinate planes in parabolas.

13. [figure: elliptic cylinder]

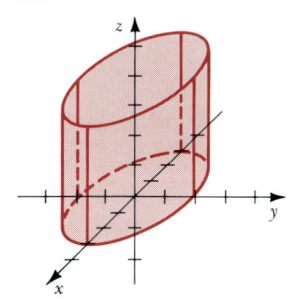

15. [figure: intersecting planes]

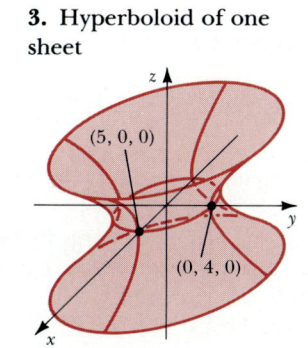

5. Hyperboloid of two sheets [figure with point (6, 0, 0)]

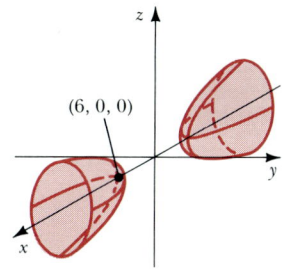

7. Paraboloid [figure]

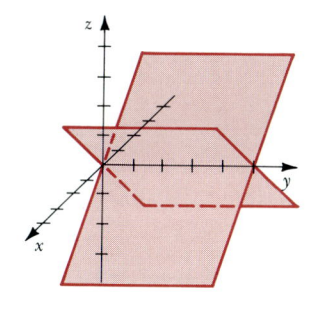

19. (a) $x^2 + y^2 + z^2 = 1$, $y \geq 0$;
(b) [figure: $x^2 + y^2 + z^2 = 1$]

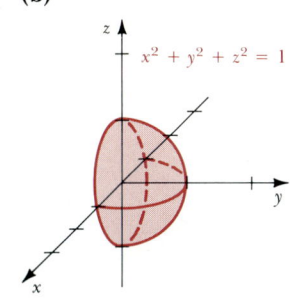

21. (a) $y = x^2 + z^2$;
(b) [figure: $y = x^2 + z^2$]

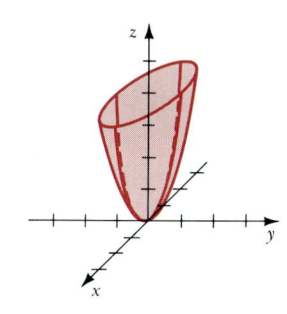

9. Hyperbolic paraboloid [figure]

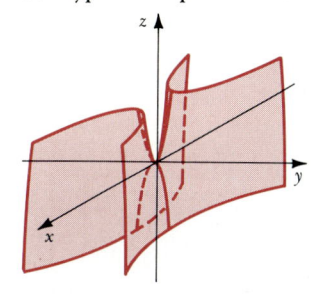

11. Elliptic cylinder [figure]

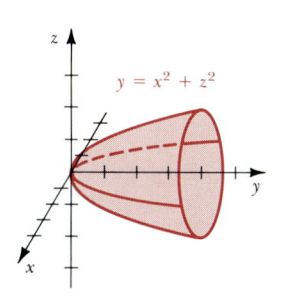

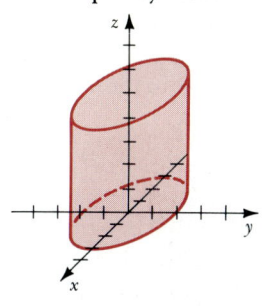

23. (a) $x^2 - y^2 - z^2 = 4$, $x \geq 2$;
(b) [figure: $y^2 + z^2 = x^2 - 4$]

13. Elliptic cone [figure]

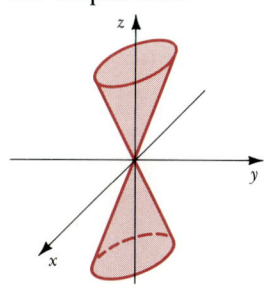

15. Intersecting planes [figure]

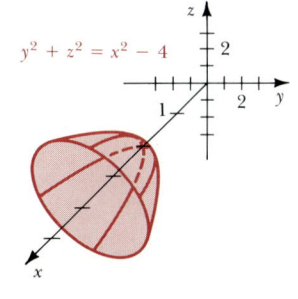

Exercise Set 13.4

1. Ellipsoid [figure]

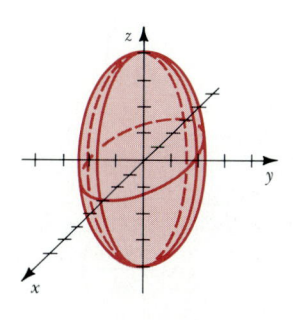

3. Hyperboloid of one sheet [figure with points (5, 0, 0) and (0, 4, 0)]

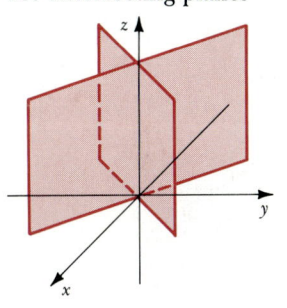

17. Hyperboloid of one sheet [figure]

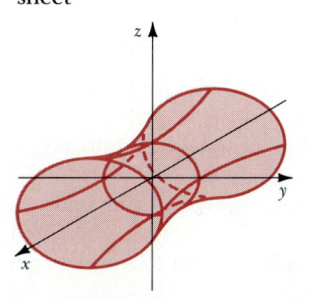

19. Circular cone [figure]

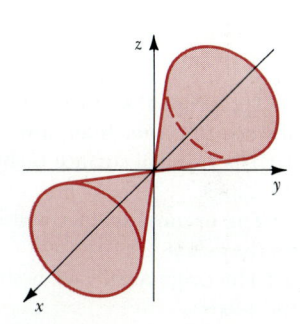

21. Parallel planes

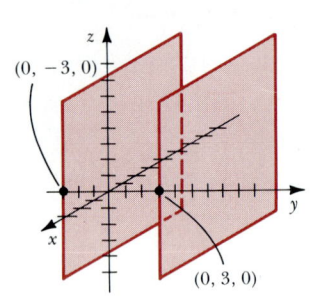

23. A sphere with radius $\sqrt{5}$ centered at $(1, 0, 2)$

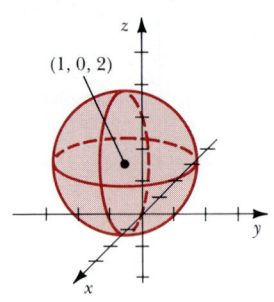

25. A paraboloid with vertex at $(3, 4, -5)$ and axis parallel to the z-axis

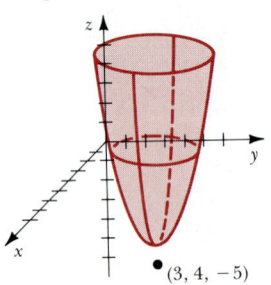

27. Parallel planes $x = 0$ and $x = 2$

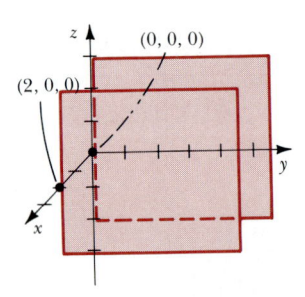

29.

(a) The original surface is revolved so that it opens downward rather than upward.
(b) The original surface is shifted upward two units on the z-axis.
(c) The original surface is shifted downward four units on the z-axis.
(d) The original surface is shifted one unit to the right on the y-axis.

31. $x^2 + (y - 2)^2 + z^2 = 1$

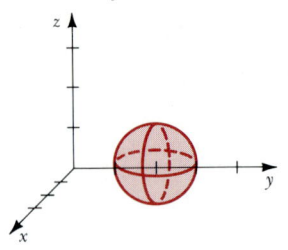

33. $9x^2 + 4y^2 + 9z^2 = 36$

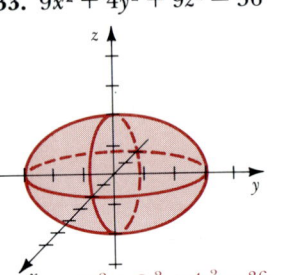

35. $y = x^2 + z^2$

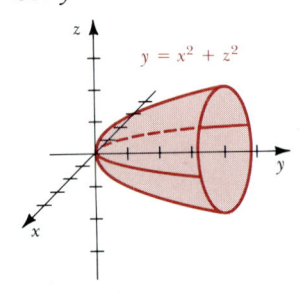

37. $9(y^2 + z^2) = x^2$

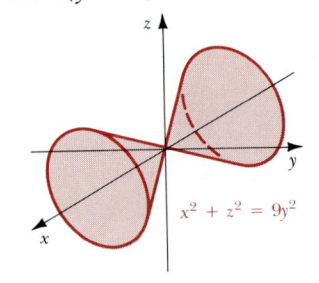

39. The yz-plane

41. (a) $z = 2\sqrt{x^2 + y^2}$, $z \leq 140$; (b) $\mathbf{r}(\theta) = \dfrac{5\theta}{\pi}(\cos\theta\mathbf{i} + \sin\theta\mathbf{j} + 2\mathbf{k})$, θ in $[0, 14\pi]$;

(c) $\dfrac{5}{\pi}\left[7\pi\sqrt{196\pi^2 + 5} + \dfrac{5}{2}\left(\ln\left(14\pi + \sqrt{196\pi^2 + 5}\right) - \dfrac{1}{2}\ln 5\right)\right] \approx 1556$ cm

Exercise Set 13.5

1. (a) $\left(\sqrt{2}, \dfrac{\pi}{4}, 3\right)$; (b) $\left(\sqrt{11}, \dfrac{\pi}{4}, \arccos\dfrac{3\sqrt{11}}{11}\right)$

3. (a) $\left(2, \dfrac{3\pi}{4}, 0\right)$; (b) $\left(2, \dfrac{3\pi}{4}, \dfrac{\pi}{2}\right)$

5. (a) $\left(1, \dfrac{\pi}{2}, 3\right)$; (b) $\left(\sqrt{10}, \dfrac{\pi}{2}, \arccos\dfrac{3\sqrt{10}}{10}\right)$

7. (a) $\left(\dfrac{3}{2}, \dfrac{3\sqrt{3}}{2}, 5\right)$; (b) $\left(\sqrt{34}, \dfrac{\pi}{3}, \arccos\dfrac{5\sqrt{34}}{34}\right)$

9. (a) $(2\sqrt{2}, 2\sqrt{2}, -2)$; (b) $\left(2\sqrt{5}, \dfrac{\pi}{4}, \arccos \dfrac{-\sqrt{5}}{5}\right)$

11. (a) $(0, 0, 1)$; (b) $(1, \theta, 0)$ for any angle θ

13. (a) $\left(\dfrac{\sqrt{6}}{4}, \dfrac{\sqrt{6}}{4}, \dfrac{1}{2}\right)$; (b) $\left(\dfrac{\sqrt{3}}{2}, \dfrac{\pi}{4}, \dfrac{1}{2}\right)$

15. (a) $\left(\dfrac{-\sqrt{6}}{2}, \dfrac{\sqrt{2}}{2}, -\sqrt{2}\right)$; (b) $\left(\sqrt{2}, \dfrac{5\pi}{6}, -\sqrt{2}\right)$

17. (a) $(0, 0, -5)$; (b) $(0, \theta, -5)$ for any angle θ **19.** $\theta = \dfrac{\pi}{4}$

21. $r = 2$ **23.** $z = r^2$ **25.** $r^2 = 4(1 - z^2)$

27.

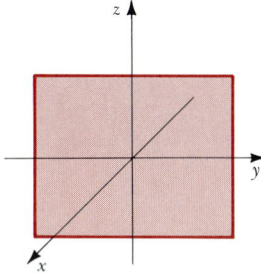

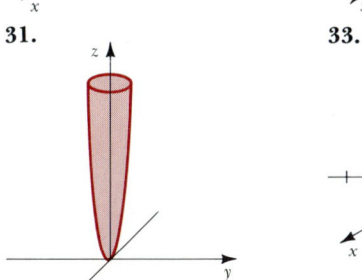

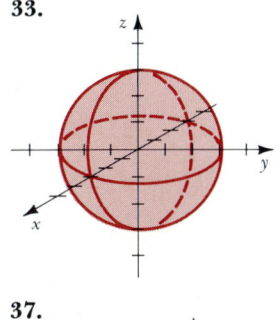

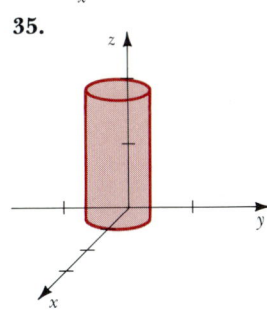

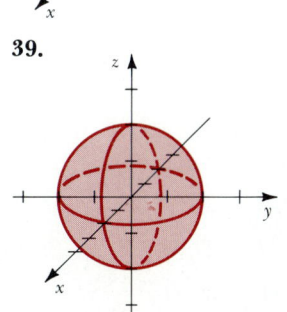

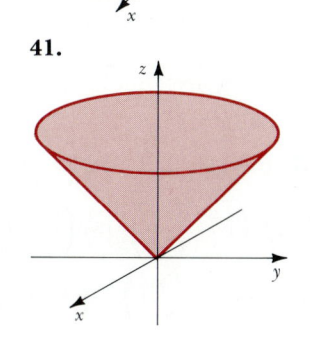

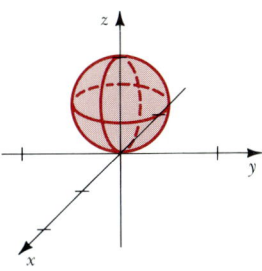

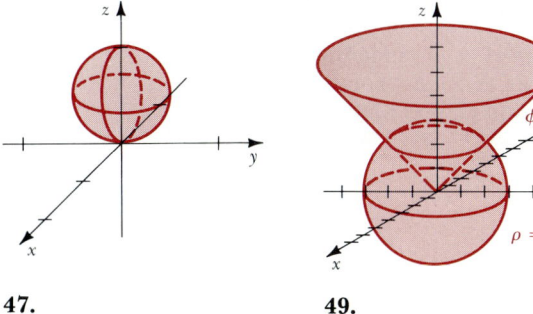

29. The yz-plane

31.

33.

35.

37.

39.

41.

43.

45. $\phi = \dfrac{\pi}{4}$, $\rho = 3$

47.

49.

51.

53.

55.

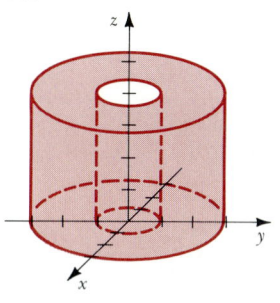

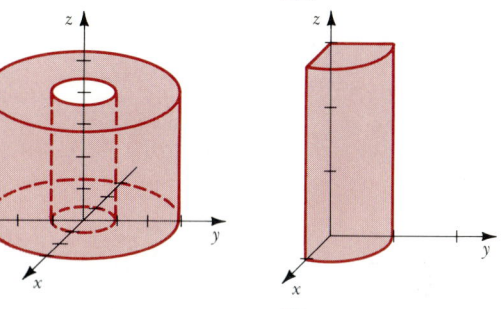

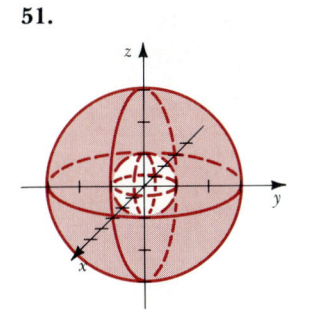

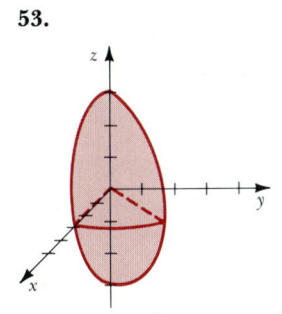

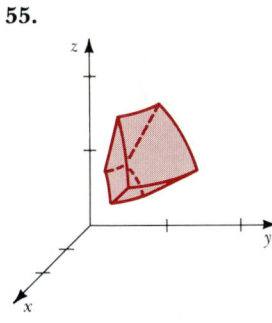

57. $1 \leq r^2 \leq 4 - z^2$ **59.** $2 \leq \rho \leq 3$

61. $\rho^2 - 2\rho \cos \phi - 3 = 0$

63. (a) $-1 \leq x \leq 1$; $-\sqrt{1 - x^2} \leq y \leq \sqrt{1 - x^2}$; $3\sqrt{x^2 + y^2} \leq z \leq 3 + \sqrt{1 - x^2 - y^2}$; (b) $0 \leq \theta \leq 2\pi$; $0 \leq r \leq 1$; $3r \leq z \leq 3 + \sqrt{1 - r^2}$;

(c) $0 \leq \theta \leq 2\pi$; $0 \leq \phi \leq \arctan \dfrac{1}{3}$; $0 \leq \rho \leq 3 \cos \phi + \sqrt{9 \cos^2 \phi - 8}$

Review Exercises

1. Values of (x, y) with $y \leq 1 - x^2$

3. Values of (x, y) within the unit circle centered at the origin, but excluding points on the y-axis

5. Values of (x, y, z) with $xyz > 0$

7.

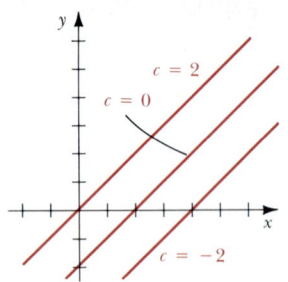

9.

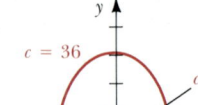

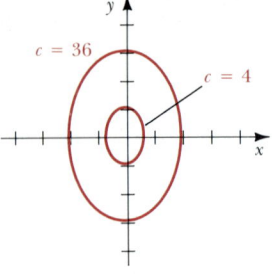

11.

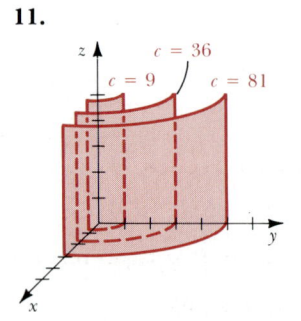

13.

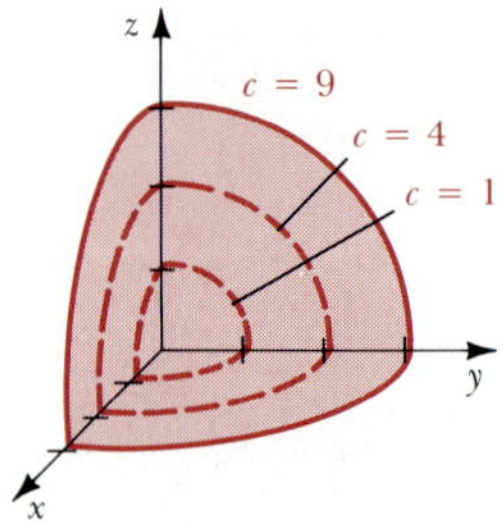

15. Plane

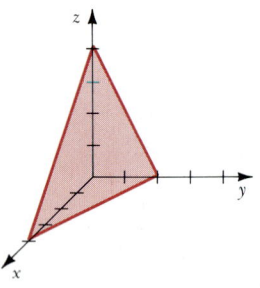

17. Paraboloid

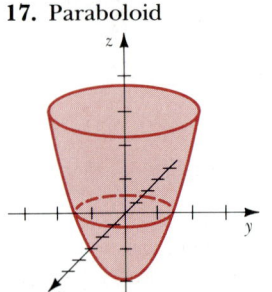

19. Elliptic cylinder

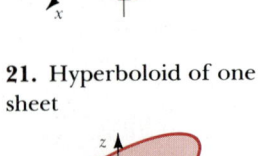

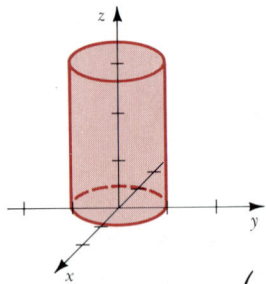

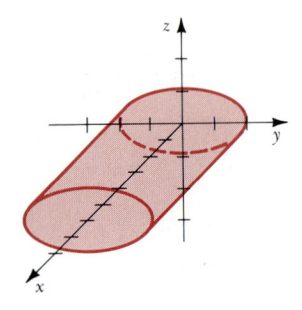

21. Hyperboloid of one sheet

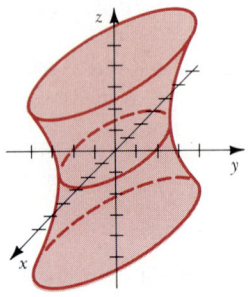

23. Two planes

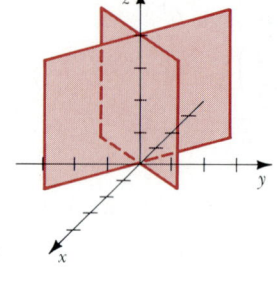

25. Hyperbolic cylinder

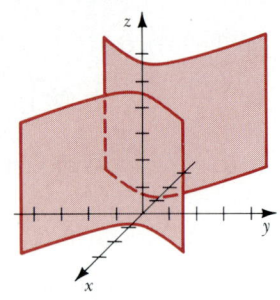

27. Sphere centered at $(1, 1, 1)$

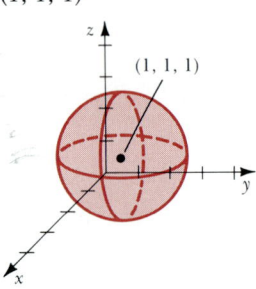

29. Parallel planes

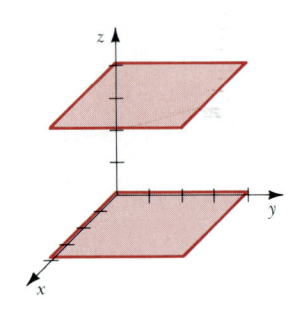

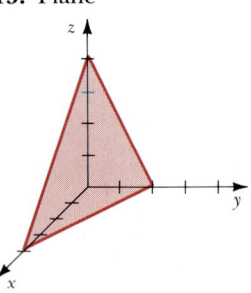

31. Circular cylinder

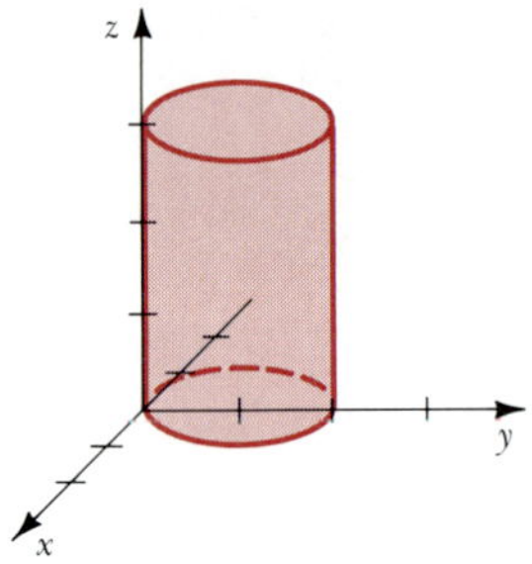

33. Cylinder

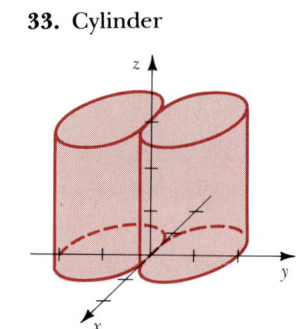

35. Circular cylinder

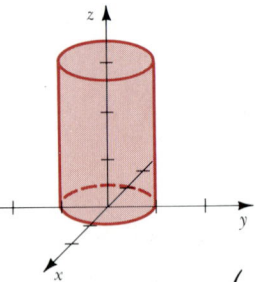

37. (i)(a) $(1, 0, 0)$; (b) $\left(1, 0, \dfrac{\pi}{2}\right)$;

(ii)(a) $\left(\sqrt{2}, \dfrac{\pi}{4}, 0\right)$; (b) $\left(\sqrt{2}, \dfrac{\pi}{4}, \dfrac{\pi}{2}\right)$;

(iii)(a) $\left(\sqrt{2}, \dfrac{\pi}{4}, 1\right)$; (b) $\left(\sqrt{3}, \dfrac{\pi}{4}, \arccos \dfrac{\sqrt{3}}{3}\right)$

39. (i)(a) $(0, 1, 0)$; (b) $\left(1, \frac{\pi}{2}, 0\right)$;

(ii)(a) $\left(\frac{\sqrt{2}}{2}, \frac{\sqrt{2}}{2}, 0\right)$; (b) $\left(1, \frac{\pi}{4}, 0\right)$;

(iii)(a) $(0, \sqrt{2}, \sqrt{2})$; (b) $\left(\sqrt{2}, \frac{\pi}{2}, \sqrt{2}\right)$

41. (a) Sphere with radius 2 and center $(0, 0, 0)$;
(b) $x^2 + y^2 + z^2 = 4$
43. Torus (doughnut) with inside radius 1, outside radius 3, and height 1

Chapter 14

Exercise Set 14.1

1. 1   **3.** 1   **5.** 0   **7.** $\frac{1}{2}$ **9.** e **11.** Does not exist
13. 1   **15.** 3   **17.** The entire plane   **19.** $(x, y) \neq (0, 0)$
21. $x^2 + y^2 \leq 1$ **23.** The entire plane
33. (a)   (b)

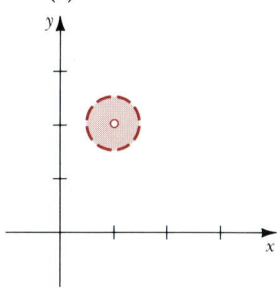

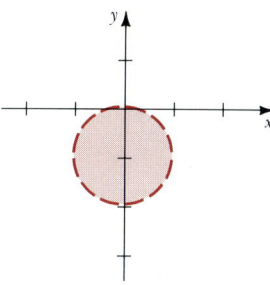

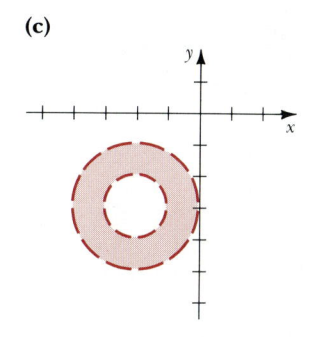

(c)

35. 1   **37.** -1 **45.** (d) No

Exercise Set 14.2

1. $f_x = 1, f_y = 1$ **3.** $f_x = 2x + y, f_y = x + 2y$
5. $f_x = ye^{xy}, f_y = xe^{xy}$
7. $f_x = \frac{y^3 - x^2 y}{(x^2 + y^2)^2}, f_y = \frac{x^3 - y^2 x}{(x^2 + y^2)^2}$
9. $f_x = 3x^2 + 2xy^2, f_y = 2x^2 y$
11. $f_x = ye^x, f_y = e^x$
13. $f_x = x(x^2 + y^2)^{-1/2}, f_y = y(x^2 + y^2)^{-1/2}$

15. $f_x = e^x \cos y, f_y = -e^x \sin y$ **17.** $f_x = 3z, f_y = 4, f_z = 3x$
19. $f_x = -2 \sin(2x + 3y + 4z), f_y = -3 \sin(2x + 3y + 4z)$,
$f_z = -4 \sin(2x + 3y + 4z)$
21. $f_x = 0, f_y = 2$ **23.** $f_x = 0, f_y = -1$
25. $f_x = 0, f_y = -1$ **27.** $f_x = \frac{\sqrt{6}}{6}, f_y = \frac{\sqrt{6}}{6}, f_z = \frac{\sqrt{6}}{3}$
37. $f_{xyx} = 4ye^{x^2+y^2}(1 + 2x^2), f_{xyy} = 4xe^{x^2+y^2}(1 + 2y^2)$
39. $(y^2 + z^2)(x^2 + y^2 + z^2)^{-3/2}$
41. $(1 - x^2 y^2 z^2) \cos(xyz) - 3xyz \sin(xyz)$
43. $x = t, y = 1, z = 4 - 2t$ **49.** $f_x(0, 0) = f_y(0, 0) = 0$
57. (a)

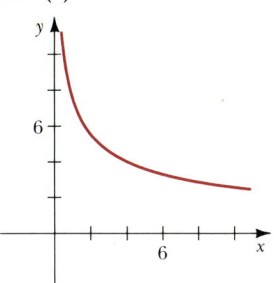

(b) $f_x = \frac{16}{3}$ is the rate at which the production changes relative to the change in property resources at $(1, 8)$; $f_y = \frac{4}{3}$ is the rate at which the production changes relative to the change in human resources at $(1, 8)$

59. (a) $\dfrac{\partial v}{\partial d} = \dfrac{-2gr^2}{9n}$; (b) $\dfrac{\partial v}{\partial r} = \dfrac{4g(\rho - d)r^2}{9n}$;
(c) $\dfrac{\partial v}{\partial n} = \dfrac{-2g(\rho - d)r^2}{9n^2}$

Exercise Set 14.3

7. $2x\,dx - 6y\,dy$ **9.** $(y\,dx - x\,dy)/y^2$
11. $-y^2 \sin x\,dx + 2y \cos x\,dy$
13. -13.2 **15.** 0.2 **17.** $yz\,dx + xz\,dy + xy\,dz$
19. $-(yz\,dx + xz\,dy + xy\,dz) \sin(xyz)$
21. $ye^z\,dx + xe^z\,dy + xye^z\,dz$
23. 7.8 **25.** 1 **27.** $\varepsilon_1 = 0, \varepsilon_2 = 0$
29. $\varepsilon_1 = \Delta x, \varepsilon_2 = -\Delta y$
31. $\varepsilon_1 = \Delta y, \varepsilon_2 = 0$ **33.** 31.2 in^2
35. (a) $\dfrac{\partial R}{\partial R_1} = \dfrac{1}{R_1^2}\left(\dfrac{1}{R_1} + \dfrac{1}{R_2}\right)^{-2}; \dfrac{\partial R}{\partial R_2} = \dfrac{1}{R_2^2}\left(\dfrac{1}{R_1} + \dfrac{1}{R_2}\right)^{-2};$
(b) $\dfrac{83}{1210}$

Exercise Set 14.4

1. $(2xy^3 + y^2)(\cos t + 1) + (3x^2 y^2 + 2xy)(2t)$

3. $\dfrac{18x^2\sqrt{x+y}-1-2t^2}{2t\sqrt{x+y}}$

5. $\left(\dfrac{1}{x}+2xe^{x^2+y^2}\right)\dfrac{1}{2\sqrt{t+1}}+\left(\dfrac{1}{y}+2ye^{x^2+y^2}\right)e^t$

7. $\dfrac{2xy}{t}+(x^2+z^2)e^t+4yzt$

9. $e^z(2te^{t^2}\cos xy - y\cos t\sin xy - 2xt\sin xy)$

11. $\left(yz^2 e^t+\dfrac{xz^2}{t}+xyz\left(1+\dfrac{1}{t}\right)\right)\cos(xyz^2)$

13. (a) $3(u+v)^2(u^2-v^2)^5+10u(u+v)^3(u^2-v^2)^4$;
(b) $3(u+v)^2(u^2-v^2)^5-10v(u+v)^3(u^2-v^2)^4$

15. (a) $e^{ue^v}[e^v\cos(u+v-1)-\sin(u+v-1)]$;
(b) $e^{ue^v}[ue^v\cos(u+v-1)-\sin(u+v-1)]$

17. (a) $e^u(\cos v+\sin v)[\sec 6e^u(\cos v+\sin v)]^2$;
(b) $e^u(\cos v-\sin v)[\sec e^u(\cos v+\sin v)]^2$

19. (a) $v\cos uv+2u\cos^2 v-3v^3\sin u\cos^2 u$;
(b) $u\cos uv-2u^2\sin v\cos v+3v^2\cos^3 u$

21. (a) $\left(2ue^{u^2+v^2}+\dfrac{2u}{u^2+v^2}+1\right)(x+y+z)^{-1}$;

(b) $\left(2ve^{u^2+v^2}+\dfrac{2v}{u^2+v^2}-1\right)(x+y+z)^{-1}$

23. $15x^2y^5+10x^3y^4(2u-3v)$

25. $(2+3u)e^{x+v}\cos y - 5e^x\sin y$

27. $e^{2t}[5\cos 3t-\sin 3t]\{\sec[e^{2t}(\sin 3t+\cos 3t)]\}^2$

29. $(2v+3u)\cos uv+2u(2\cos v-3u\sin v)\cos v+3v^2(3\cos u-2v\sin u)\cos^2 u$

31. $\dfrac{2(2u+3v)[e^{u^2+v^2}+(u^2+v^2)^{-1}]-1}{x^2+y^2+z^2}$

33. $\dfrac{-2xy^2-y+3x^2}{2x^2y-x+3y^2}$

35. $\dfrac{12x-2xy-y^2}{x^2+2xy-12y}$

37. $-\dfrac{\sin x}{x\cos xy}-\dfrac{y}{x}$

41. (a) $\dfrac{\partial w}{\partial \rho}=(\sin\phi\cos\theta)f_x+(\sin\phi\sin\theta)f_y+(\cos\phi)f_z$;

(b) $\dfrac{\partial w}{\partial\theta}=-(\rho\sin\phi\sin\theta)f_x+(\rho\sin\phi\cos\theta)f_y$;

(c) $\dfrac{\partial w}{\partial\phi}=(\rho\cos\phi\cos\theta)f_x+(\rho\cos\phi\sin\theta)f_y-(\rho\sin\phi)f_z$

45. $y''=-\dfrac{f_{xx}f_y^2-2f_xf_yf_{xy}+f_x^2f_{yy}}{f_y^3}$

47. $0.1\ \mu F$/millisec 49. $306\ in^3$/min

Exercise Set 14.5

1. $\dfrac{-5\sqrt{2}}{2}$ 3. $2\sqrt{2}$ 5. $\dfrac{17\sqrt{22}}{22}$ 7. $\dfrac{5\sqrt{6}}{6}$

9. $\dfrac{-\sqrt{2}}{6}$ 11. $2x\mathbf{i}-2y\mathbf{j}$ 13. $ye^x\mathbf{i}+e^x\mathbf{j}$

15. $\dfrac{(y^3-x^2y)\mathbf{i}+(x^3-xy^2)\mathbf{j}}{(x^2+y^2)^2}$

17. $\cos(x^2yz)(2xyz\mathbf{i}+x^2z\mathbf{j}+x^2y\mathbf{k})$

19. $2\mathbf{i}+4\mathbf{j}$ 21. $-e\mathbf{i}$

23. $-\dfrac{\pi}{2}\mathbf{i}-\dfrac{\pi}{4}\mathbf{j}-2\mathbf{k}$ 25. (a) $\mathbf{i}+\mathbf{j}$; (b) $2\sqrt{2}$

27. (a) $\mathbf{i}$; (b) 36 29. (a) $\mathbf{i}+\mathbf{j}$; (b) $2e$

31. (a) $-\mathbf{i}$; (b) $2e$ 33. $2+4\sqrt{3}$ 35. $\dfrac{3}{2}e-\dfrac{\sqrt{3}}{2}e^2$

37. $\dfrac{2\pi\sqrt{16+\pi^2}}{16+\pi^2}$ 39. $-\dfrac{4\sqrt{5}}{5}$ 41. 0

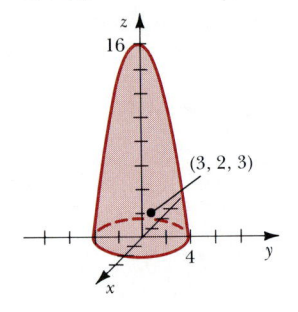

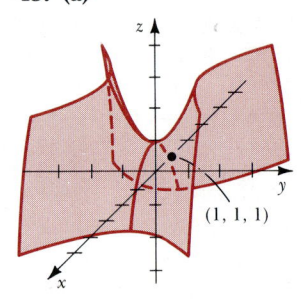

43. (a)

(b) $2\sqrt{13}$;

(c) $\dfrac{-6\mathbf{i}-4\mathbf{j}}{2\sqrt{13}}+2\sqrt{13}\,\mathbf{k}$

45. (a)

(b) $2\sqrt{2}$;

(c) $\dfrac{-\mathbf{i}+\mathbf{j}}{\sqrt{2}}+2\sqrt{2}\,\mathbf{k}$

49. (a) $\dfrac{-24}{5}$; (b) $\dfrac{-28\sqrt{2}}{5}$; (c) $\nabla T(3,4)=\dfrac{-24\mathbf{i}-32\mathbf{j}}{5}$;

(d) $-\nabla T(3,4)$

Exercise Set 14.6

1. $\mathbf{i}+\mathbf{j}+2\mathbf{k}$ 3. $\mathbf{j}-\mathbf{k}$ 5. $\dfrac{1}{e}\mathbf{i}+\dfrac{1}{e^2}\mathbf{j}-\mathbf{k}$

7. $\dfrac{1}{2}\mathbf{i}+\dfrac{1}{2}\mathbf{j}-\mathbf{k}$ 9. $4x-2y-z=3$

11. $3x-4y+5z=0$ 13. $2x-z=2$

15. $x+4y-z=\pi$

17. $x=1+2t,\ y=2-t,\ z=2-t$

19. $x=t,\ y=2,\ z=2$

21. $x=1-2\sqrt{15}\,t,\ y=\sqrt{15}+18t,\ z=4+4\sqrt{15}\,t$

23. $x=2,\ y=1+2\sqrt{3}\,t,\ z=\sqrt{3}-2t$

25. Only at $\left(-\dfrac{1}{8},-1,\dfrac{17}{16}\right)$

Exercise Set 14.7

5. $(0, 0)$ **7.** $(0, 0)$ **9.** $(0, n\pi + \pi/2)$ for any integer n
11. $(0, 0)$ **13.** Relative minimum
15. Relative minimum
17. All saddle points **19.** Saddle point
21. $(0, 0)$ saddle point; $(1, 1)$ relative minimum
23. $(3, 0)$ and $(2, 0)$ saddle points
25. $(0, 0)$ relative minimum
27. $(2, 1)$ saddle point
29. $(1, 0)$ relative minimum; $(-1, 0)$ relative maximum
31. $(\pm n\pi, \pm m\pi)$, where n and m are nonnegative integers. Relative maximum when both m and n are even. Relative minimum when both are odd. The others are saddle points.
33. None **35.** None
37. Absolute minimum $(0, 0)$ **39.** None
41. Absolute minimum $(1, 0)$; absolute maximum $(-1, 0)$
43. Absolute maximum $(2n\pi, 2m\pi)$; absolute minimum $((2n+1)\pi, (2m+1)\pi)$
45. Absolute minimum value 0 at $(0, 0)$; absolute maximum value 1 whenever $x^2 + y^2 = 1$
47. Absolute maximum value $\sqrt{2}$ at $(\sqrt{2}, 1)$ and $(-\sqrt{2}, -1)$; absolute minimum value $-\sqrt{2}$ at $(\sqrt{2}, -1)$ and $(-\sqrt{2}, 1)$
49. Absolute maximum value 410 at $(5, 10)$; absolute minimum value 0 whenever $y = 0$
53. $f\left(-1, -\dfrac{3}{2}, \dfrac{13}{4}\right) = -\dfrac{13}{4}$
55. $\left(\dfrac{6}{7}, \dfrac{9}{7}, \dfrac{3}{7}\right)$ **57.** No
59. $x = y = z = 3$ **61.** All sides $2\sqrt[3]{10}$ in.
63. Height = width = 14 in., length = 28 in.
69. $y = \dfrac{x}{2} + \dfrac{5}{6}$ **71.** $y = -\dfrac{53x}{68} + \dfrac{183}{34}$
73. $y = 0.17952x + 8.2085$

Exercise Set 14.8

1. Minimum 2 at $(1, 1)$
3. Minimum 0 at $\left(\dfrac{2}{3}, \dfrac{1}{3}\right)$
5. Minimum $\dfrac{72}{13}$ at $\left(\dfrac{3}{13}, \dfrac{18}{13}, \dfrac{24}{13}\right)$
7. Minimum 144 at $(6, 4, 2)$
9. Minimum 1 at $(0, 0, 1), (0, 0, -1), (1, -1, 0)$, and $(-1, 1, 0)$
11. Minimum 1 at $\left(0, \pm\dfrac{\sqrt{17}}{17}, \mp\dfrac{4\sqrt{17}}{17}\right)$ and maximum $\dfrac{24}{7}$ at $\left(\pm\dfrac{2\sqrt{357}}{21}, \pm\dfrac{8\sqrt{357}}{357}, \pm\dfrac{2\sqrt{357}}{357}\right)$

13. $\left(2 - \dfrac{2\sqrt{29}}{29}, 3 - \dfrac{3\sqrt{29}}{29}, 4 - \dfrac{4\sqrt{29}}{29}\right)$
15. All sides $2\sqrt[3]{10}$ in.
17. The sides have length $2\sqrt{3}, \dfrac{8}{3}\sqrt{3}$, and $\dfrac{10}{3}\sqrt{3}$.
19. $h = 2r \approx 25.3$ in.
21. The base will have dimensions $\dfrac{40\sqrt{3}}{3}$ ft by $\dfrac{100\sqrt{3}}{3}$ ft. The height is $\dfrac{10\sqrt{3}}{3}$ ft.
23. $r = 12.01262$ in., $h = 28.02524$ in.
25. 432 when $x = y = 4$

Review Exercises

1. (a) $x^2 + y^2 \leq 4$; (b) **3.** (a) Discontinuous if $x^2 + y^2 = 1$; (b)

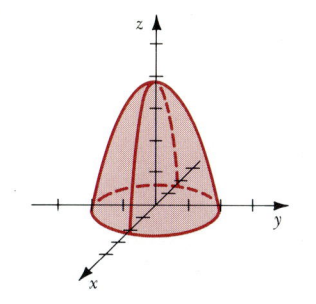

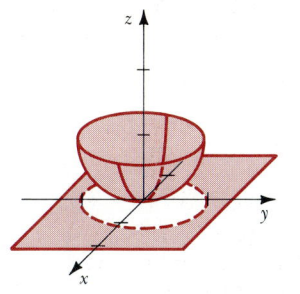

5. $f_x(x, y) = ye^{xy}, f_y(x, y) = xe^{xy}$
7. $f_x(x, y) = 2xy + 2e^{1/y}, f_y(x, y) = x^2 - \dfrac{2x}{y^2}e^{1/y}$
9. $f_x(x, y) = 3(x^2 + xy)^2(2x + y), f_y(x, y) = 3x(x^2 + xy)^2$
11. $f_x(x, y, z) = 2x - 2y\cos z, f_y(x, y, z) = 2y - 2x\cos z,$
$f_z(x, y, z) = 2xy\sin z$
13. (a) $\dfrac{1}{60}$; (b) 2.01667
15. (a) $\sqrt{2}\left(\dfrac{3\pi}{32} + \dfrac{1}{10}\right)$; (b) 2.67926
17. $(0, 0)$, relative maximum
19. $(0, y)$, for any y gives a relative minimum
21. $f_{yxz}(x, y, z) = e^{yz} + zye^{yz} + e^x, \dfrac{\partial^3 f}{\partial x^2 \partial y}(x, y, z) = ze^x$
23. $(2x + 2y)(\mathbf{i} + \mathbf{j})$ **25.** $-\mathbf{i} - \mathbf{j}$
27. $10x\mathbf{i} - 4y\mathbf{j} + \mathbf{k}$ **29.** $\dfrac{4}{5}$
31. $-\sqrt{2}$ **33.** $\dfrac{26}{5}$
35. (a) $\mathbf{i} + \mathbf{j}$; (b) $\sqrt{2}$ **37.** (a) $-\mathbf{i} + 2\mathbf{j}$; (b) $2\sqrt{5}$
39. $2x + 4y - z = 6$ **41.** $\dfrac{\pi}{2}x + y - z = \dfrac{\pi}{2}$
43. $4x - 8y + z = -8$ **45.** $\dfrac{13\sqrt{2}}{2}$

47. (a)

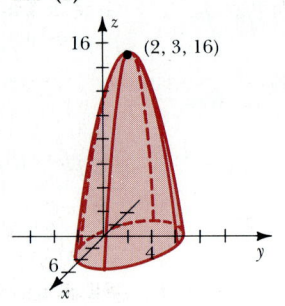

[figure: surface in 3D with point (2, 3, 16), z-axis up to 16, x-axis to 6, y-axis to 4]

(b) $2\sqrt{5}$; **(c)** The maximum rate of change is $\|\nabla f(4, 4)\| = 2\sqrt{5}$ in the direction of $\nabla f(4, 4)$; **(d)** $4x + 2y + z = 35$;
(e) $x = 4 - t$, $y = 4$, $z = 11 + 4t$;
(f) $-2\mathbf{i} - \mathbf{j} + 10\mathbf{k}$

49. Slope 4. The equation is $\dfrac{z-4}{4} = x - 1$, $y = 2$

51. $(0.01)(3 - e) \approx 0.002817$
53. $3e^{3r} \sin \theta \cos \theta$; $e^{3r}(\cos^2 \theta - \sin^2 \theta)$
55. 1   **57.** $\mathbf{i} - \mathbf{j}$
61. $\left(\dfrac{8}{9}, \dfrac{8}{9}, \dfrac{4}{9}\right)$
63. 2700 at $\left(\dfrac{3}{2}, 3, 4\right)$
65. $-\dfrac{80\pi}{3}$ in³/min
67. $l = w = 2\sqrt[3]{2}$, $h = \sqrt[3]{2}$
69. $r = [24{,}000/\pi(3\sqrt{2} + 1)]^{1/3} \approx 11.3371$, $h = \sqrt{2}r \approx 16.0331$

Chapter 15

Exercise Set 15.1

1. (a) $\dfrac{9}{2}$; **(b)** $\dfrac{15}{2}$; **(c)** $\dfrac{13}{2}$
3. (a) -59; **(b)** -81; **(c)** -70
5. 4   **9.** 18π

Exercise Set 15.2

1. $\dfrac{37}{2}$ **3.** 16   **5.** $-2\pi^2$
7. 26   **9.** 28   **11.** -2
13. $\dfrac{31}{60}$ **15.** $e - 1$

17. $\displaystyle\int_3^4 \int_1^2 xy^2 \, dx \, dy$

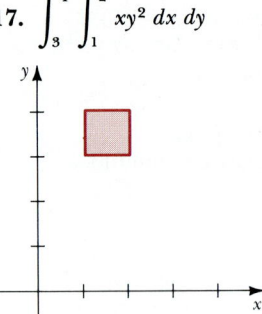

[figure: small square region in xy-plane]

19. $\displaystyle\int_{-2}^4 \int_{-1}^1 (x^2 + y) \, dx \, dy$

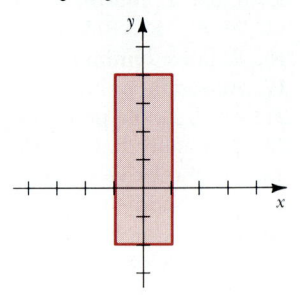

[figure: rectangular region in xy-plane]

21. $\displaystyle\int_0^{2\pi} \int_{-\pi}^{3\pi/2} (y \sin x + x \cos y) \, dx \, dy$

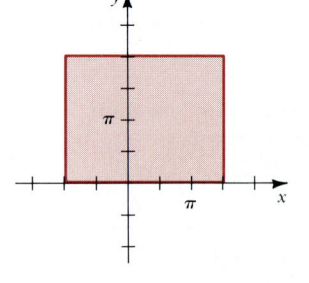

[figure: rectangular region in xy-plane]

23. $\displaystyle\int_{-2}^1 \int_1^3 x^2 y^2 \, dx \, dy$

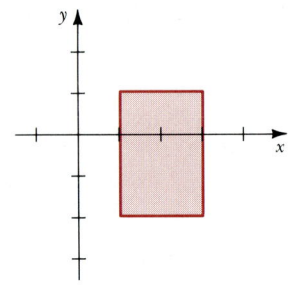

[figure: rectangular region in xy-plane]

25. $\displaystyle\int_0^2 \int_{y/2}^y (x^2 + y^3) \, dx \, dy + \int_2^4 \int_{y/2}^2 (x^2 + y^3) \, dx \, dy$

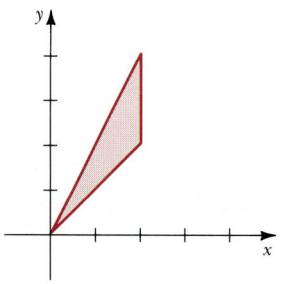

[figure: triangular region in xy-plane]

27. $\int_0^\pi \int_y^\pi \cos x \, dx \, dy$

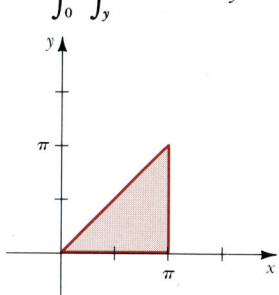

29. $\int_0^1 \int_y^1 (x^2 + \sqrt{y}) \, dx \, dy$

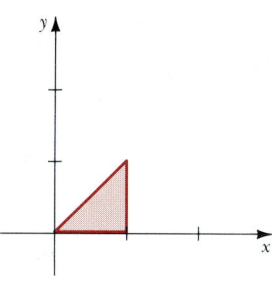

31. $\int_1^e \int_y^e \ln xy \, dx \, dy$

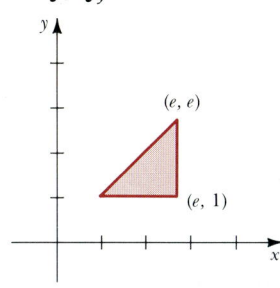

33. (a)

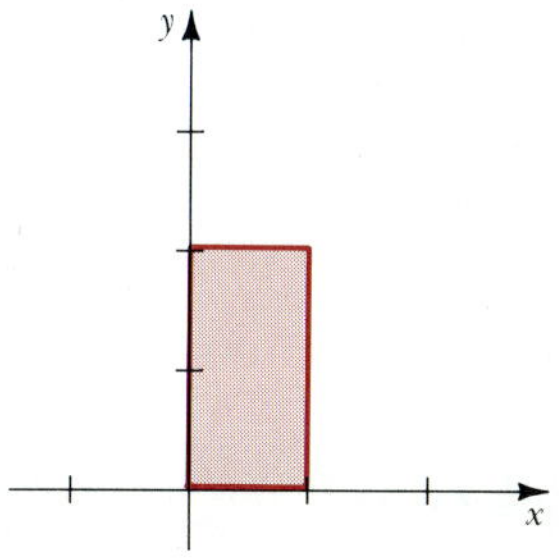

(b) 2; (c) $\dfrac{2}{3}$

35. (a)

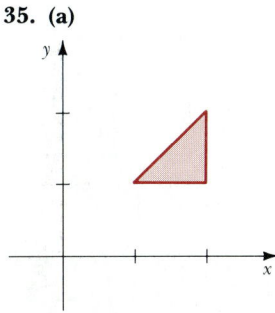

(b) $\dfrac{1}{2}$; (c) $\dfrac{7}{4}$

37. (a)

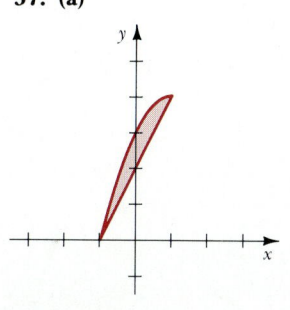

(b) $\dfrac{4}{3}$; (c) $\dfrac{68}{15}$

39. $\dfrac{32}{3}$ **41.** 8π **43.** 18 **45.** 18π

47. $g'(t) = \int_0^t f(t, y) \, dy$

Exercise Set 15.3

1. 18π **3.** 48π **5.** $\dfrac{4\pi}{3}(8 - 3\sqrt{3})$

7. $\dfrac{4\pi}{3}(8 - 4\sqrt{2})$ **9.** $\dfrac{29}{32}\pi$

11. $\dfrac{3\pi}{64} - \dfrac{1}{8}$ **13.** π

15. 11π **17.** $2 + \dfrac{\pi}{4}$ **19.** $\dfrac{3\pi}{16}$

21. $\dfrac{\pi}{2} - 1$ **23.** 1

25. $\ln \dfrac{\sqrt{3}(\sqrt{2} + 1)}{3}$ **27.** $\dfrac{1}{12}$

29. $e^9 - 1$ **31.** $\dfrac{\pi}{2}(3.5)^4 \approx 236$

33. $\ln 2$ **35.** Diverges **37.** $\dfrac{15}{2} e^{-1}$

41. All the organisms are within the region described by the limits on the integral.

Exercise Set 15.4

1. (a) $M = \dfrac{2}{3}$; (b) $M_x = \dfrac{3}{10}$, $M_y = \dfrac{3}{10}$; (c) $\left(\dfrac{9}{20}, \dfrac{9}{20}\right)$

3. (a) $M = \dfrac{1}{6}$; (b) $M_x = \dfrac{1}{24}$, $M_y = \dfrac{1}{12}$; (c) $\left(\dfrac{1}{2}, \dfrac{1}{4}\right)$

5. (a) $M = \dfrac{4}{15}$; (b) $M_x = \dfrac{8}{105}$, $M_y = 0$; (c) $\left(0, \dfrac{2}{7}\right)$

7. (a) $M = \dfrac{1}{2}$; (b) $M_x = \dfrac{3}{8}$, $M_y = \dfrac{5}{24}$; (c) $\left(\dfrac{5}{12}, \dfrac{3}{4}\right)$

9. $I_x = \dfrac{6}{35}$, $I_y = \dfrac{6}{35}$, $I_0 = \dfrac{12}{35}$, $\hat{x} = \dfrac{3\sqrt{35}}{35}$, $\hat{y} = \dfrac{3\sqrt{35}}{35}$

11. $I_x = \dfrac{1}{60}$, $I_y = \dfrac{1}{20}$, $I_0 = \dfrac{1}{15}$, $\hat{x} = \dfrac{\sqrt{30}}{10}$, $\hat{y} = \dfrac{\sqrt{10}}{10}$

13. $I_x = \dfrac{32}{945}$, $I_y = \dfrac{4}{35}$, $I_0 = \dfrac{4}{27}$, $\hat{x} = \dfrac{\sqrt{21}}{7}$, $\hat{y} = \dfrac{2\sqrt{14}}{21}$

15. $I_x = \dfrac{3}{10}$, $I_y = \dfrac{7}{60}$, $I_0 = \dfrac{5}{12}$, $\hat{x} = \dfrac{\sqrt{210}}{30}$, $\hat{y} = \dfrac{\sqrt{15}}{5}$

19. (a) $M = \dfrac{128\,\pi}{3}$; (b) $M_x = 0$, $M_y = 0$; (c) $I_x = \dfrac{1024}{5}\pi$,
$I_y = \dfrac{1024}{5}\pi$, $I_0 = \dfrac{2048}{5}\pi$

21. (a) $M = \dfrac{4}{9}$; (b) $M_x = \dfrac{4}{15}$, $M_y = 0$; (c) $I_x = \dfrac{32}{175}$,
$I_y = \dfrac{16}{525}$, $I_0 = \dfrac{16}{75}$

23. $\bar{r} = \dfrac{21}{20}$, $\bar{\theta} = -\dfrac{\pi}{2}$ **25.** $(0, 0)$

Exercise Set 15.5

1. $\dfrac{9}{2}\sqrt{3}$ **3.** $3\sqrt{17} + \dfrac{3}{4}\ln(4 + \sqrt{17})$

5. $\dfrac{\pi}{6}[(17)^{3/2} - 1]$ **7.** 6π

9. $4\pi(3 - \sqrt{6})$ **11.** 64

15. $\dfrac{\sqrt{a^2 + b^2 + c^2}}{2abc}$

Exercise Set 15.6

1. $\dfrac{1}{14}$ **3.** $\dfrac{1}{12}$ **5.** $2 - \pi + \dfrac{\pi^2}{2}$

7. 0

9. 18 **11.** $\dfrac{8\pi}{3}$

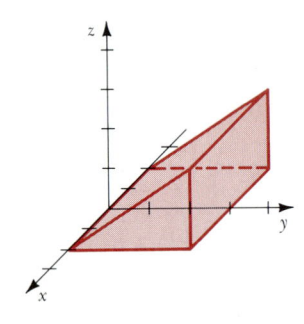

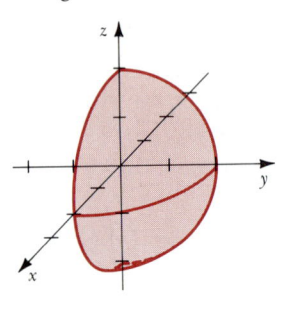

13. 9

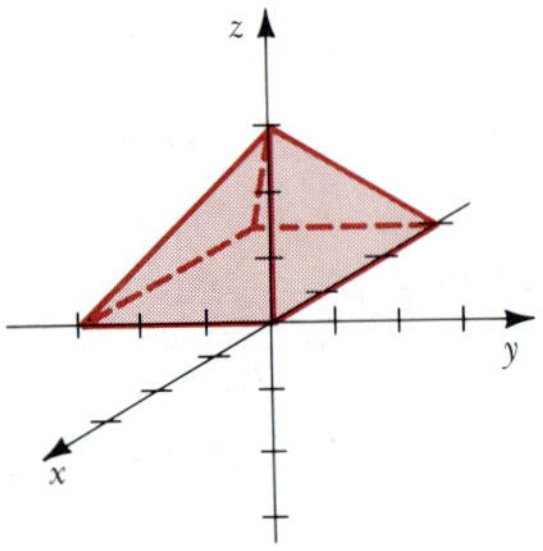

15. $\displaystyle\int_0^1 \int_0^{1-y} \int_0^{1-x-y} z\, dz\, dx\, dy$
$= \displaystyle\int_0^1 \int_0^{1-x} \int_0^{1-z-x} z\, dy\, dz\, dx$
$= \displaystyle\int_0^1 \int_0^{1-z} \int_0^{1-z-x} z\, dy\, dx\, dz$
$= \displaystyle\int_0^1 \int_0^{1-y} \int_0^{1-z-y} z\, dx\, dz\, dy$
$= \displaystyle\int_0^1 \int_0^{1-z} \int_0^{1-z-y} z\, dx\, dy\, dz = \dfrac{1}{24}$

17. $\displaystyle\int_0^1 \int_0^{-3x+3} \int_0^{2-2x-2z/3} (2x + y)\, dy\, dz\, dx$
$= \displaystyle\int_0^2 \int_0^{-y/2+1} \int_0^{3-3x-3y/2} (2x + y)\, dz\, dx\, dy$
$= \displaystyle\int_0^2 \int_0^{-3y/2+3} \int_0^{1-z/3-y/2} (2x + y)\, dx\, dz\, dy$
$= \displaystyle\int_0^3 \int_0^{-z/3+1} \int_0^{2-2x-2z/3} (2x + y)\, dy\, dx\, dz$
$= \displaystyle\int_0^3 \int_0^{2-2z/3} \int_0^{1-z/3-y/2} (2x + y)\, dx\, dy\, dz = 1$

19. $\displaystyle\int_0^2 \int_0^{3\sqrt{4-z^2}} \int_0^{(1/2)\sqrt{36-x^2-9z^2}} z\, dy\, dx\, dz$
$= \displaystyle\int_0^3 \int_0^{(2/3)\sqrt{9-y^2}} \int_0^{\sqrt{36-4y^2-9z^2}} z\, dx\, dz\, dy$
$= \displaystyle\int_0^3 \int_0^{2\sqrt{9-y^2}} \int_0^{(1/3)\sqrt{36-x^2-4y^2}} z\, dz\, dx\, dy$
$= \displaystyle\int_0^6 \int_0^{(1/2)\sqrt{36-x^2}} \int_0^{(1/3)\sqrt{36-x^2-4y^2}} z\, dz\, dy\, dx$
$= \displaystyle\int_0^6 \int_0^{(1/3)\sqrt{36-x^2}} \int_0^{(1/2)\sqrt{36-x^2-9z^2}} z\, dy\, dz\, dx = \dfrac{9\pi}{2}$

21. $\dfrac{8}{3}$ **23.** $\dfrac{15\pi}{2}$ **25.** $\dfrac{\pi}{8}$

27. 48π **29.** $\dfrac{2\pi}{3}(6\sqrt{6} - 11)$

Exercise Set 15.7

1. 0 **3.** $\dfrac{\pi}{15}$ **5.** 0 **7.** $\dfrac{4\pi}{3}$

9. $\dfrac{16}{5}\pi$ **11.** $\dfrac{\pi^2}{96}$

13.

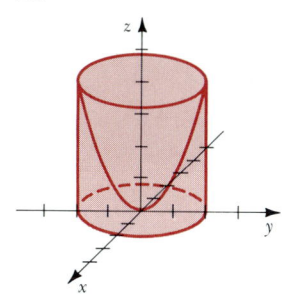

15.

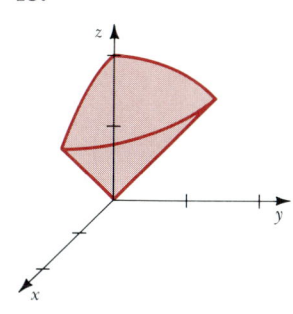

17.

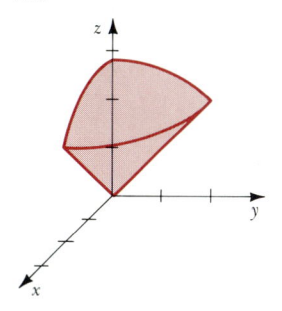

19. $\dfrac{16}{3}(2-\sqrt{2})\pi$ **21.** $\dfrac{15}{2}\pi$ **23.** 162π

25. $\dfrac{19}{3}\pi(2-\sqrt{2})$

27. (a) $\displaystyle\int_{-a}^{a}\int_{-\sqrt{a^2-x^2}}^{\sqrt{a^2-x^2}}\int_{-\sqrt{a^2-x^2-y^2}}^{\sqrt{a^2-x^2-y^2}} dz\,dy\,dx$;

(b) $\displaystyle\int_{0}^{2\pi}\int_{0}^{a}\int_{-\sqrt{a^2-r^2}}^{\sqrt{a^2-r^2}} r\,dz\,dr\,d\theta$;

(c) $\displaystyle\int_{0}^{2\pi}\int_{0}^{\pi}\int_{0}^{a} \rho^2 \sin\phi \, d\rho\, d\phi\, d\theta$

29. $\left(\dfrac{8}{3}+\dfrac{41\sqrt{6}}{4}\right)\pi \approx 87$ in^2

31. $\left[1-\left(\dfrac{1}{2}\right)^{2/3}\right]^{1/2} R \approx 0.61R$

Exercise Set 15.8

1. 33   **3.** $\dfrac{k}{6}$ **5.** $\dfrac{1}{12}$ **7.** $48k\pi$ **9.** $\dfrac{162}{5}k\pi$

11. (a) $M_{xy}=63$, $M_{yz}=\dfrac{33}{2}$, $M_{xz}=33$; (b) $\left(\dfrac{1}{2},1,\dfrac{21}{11}\right)$

13. (a) $M_{xy}=\dfrac{k}{24}$, $M_{yz}=\dfrac{k}{24}$, $M_{xz}=\dfrac{k}{24}$; (b) $\left(\dfrac{1}{4},\dfrac{1}{4},\dfrac{1}{4}\right)$

15. (a) $M_{xy}=\dfrac{1}{15}$, $M_{yz}=\dfrac{1}{30}$, $M_{xz}=\dfrac{1}{60}$; (b) $\left(\dfrac{2}{5},\dfrac{1}{5},\dfrac{4}{5}\right)$

17. (a) $M_{xy}=\dfrac{496}{3}k\pi$, $M_{yz}=0$, $M_{xz}=0$; (b) $\left(0,0,\dfrac{31}{9}\right)$

19. (a) $M_{xy}=\dfrac{243}{4}k\pi$, $M_{yz}=0$, $M_{xz}=0$; (b) $\left(0,0,\dfrac{15}{8}\right)$

21. (a) $I_x=\dfrac{367}{2}$, $I_y=\dfrac{301}{2}$, $I_z=55$;

(b) $\left(\sqrt{\dfrac{367}{66}},\sqrt{\dfrac{301}{66}},\sqrt{\dfrac{5}{3}}\right)$

23. (a) $I_x=\dfrac{k}{30}$, $I_y=\dfrac{k}{30}$, $I_z=\dfrac{k}{30}$; (b) $\left(\dfrac{\sqrt{5}}{5},\dfrac{\sqrt{5}}{5},\dfrac{\sqrt{5}}{5}\right)$

25. (a) $I_x=\dfrac{7}{90}$, $I_y=\dfrac{4}{45}$, $I_z=\dfrac{1}{45}$;

(b) $(\sqrt{14/15},\sqrt{16/15},\sqrt{4/15})$

27. (a) $I_x=672k\pi$, $I_y=672k\pi$, $I_z=96k\pi$;
(b) $(\sqrt{14},\sqrt{14},\sqrt{2})$

29. (a) $I_x=\dfrac{5832}{35}k\pi$, $I_y=\dfrac{5832}{35}k\pi$, $I_z=\dfrac{2916}{35}k\pi$;

(b) $\left(\dfrac{6\sqrt{7}}{7},\dfrac{6\sqrt{7}}{7},\dfrac{3\sqrt{14}}{7}\right)$

31. 2

33. In spherical coordinates:

(a) $I_x=\displaystyle\int_{0}^{2\pi}\int_{0}^{\pi}\int_{0}^{1}\rho^6 \sin\phi\,[\sin^2\phi\sin^2\theta+\cos^2\phi]\,d\rho\,d\phi\,d\theta$,

$I_y=\displaystyle\int_{0}^{2\pi}\int_{0}^{\pi}\int_{0}^{1}\rho^6 \sin\phi\,[\sin^2\phi\cos^2\theta+\cos^2\phi]\,d\rho\,d\phi\,d\theta$;

(b) Because of symmetry, $I_x=I_y=I_z=\dfrac{8}{21}\pi$

35. $\dfrac{\displaystyle\iint_R \sqrt{(x-x_0)^2+(y-y_0)^2}\,dA}{\displaystyle\iint_R dA}$

37. (a) $57.5°$; (b) The conical tent is cooler.

39. $M=\dfrac{4215.225}{1728}\pi \approx 7.66$ lb; the center of mass is in the center of the can approximately 2.96 in. from the bottom.

41. (a) The center of mass is in the center of the glass approximately 2.42 in. from the bottom. (b) The center of mass rises to approximately 2.44 in. after the liquid is stirred, provided that the liquid interaction does not change the volume.

Review Exercises

1. 5 **3. (a)** $\frac{16}{3}$; **(b)** 4 **5.** $\frac{7}{20}$

7. $\frac{79}{18} - \frac{16}{3}\ln 2 \approx 0.692$ **9.** $\frac{311}{3960} \approx 0.0785$ **11.** 0

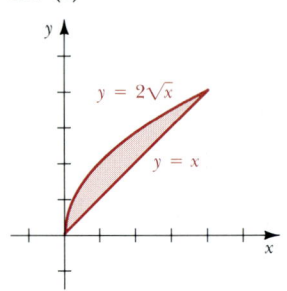

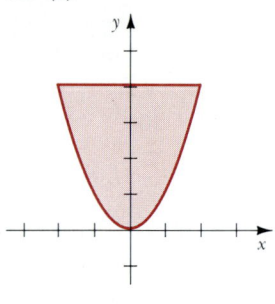

13. (a)

(b) $\frac{8}{3}$

15. (a)

(b) $\frac{256}{15}$

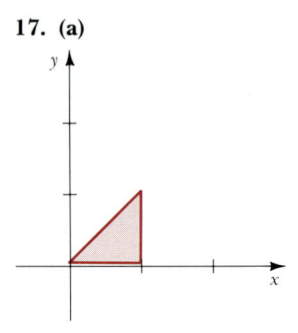

17. (a)

(b) $\frac{e-1}{6e}$

19. (a) $A = \frac{64}{3}$; **(b)** $V = \frac{1024}{3}$

21. (a) $A = \frac{3}{2}\pi$; **(b)** $V = 6\pi$

23. $\frac{4}{3}$ **25.** $\frac{10\pi}{3}$

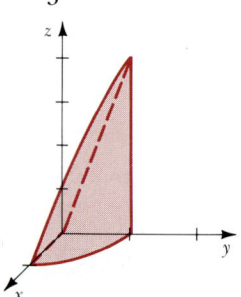

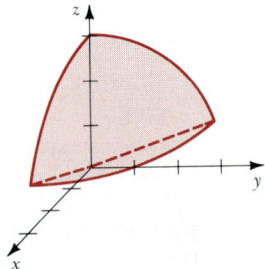

27. $\frac{\pi}{15}$ **29.** $\frac{\pi}{18}$ **31.** 8π **33.** 24π

35. $\frac{8\sqrt{2}-7}{6}\pi$ **37.** $27\left(\frac{\pi}{4} - \frac{1}{3}\right)$ **39.** 3

41. (a) $e^2 - 1$; **(b)** $M_x = \frac{1}{4}(e^4 - 1)$, $M_y = e^2 + 1$;

(c) $\left(\frac{e^2+1}{e^2-1}, \frac{e^2+1}{4}\right)$

43. (a) $\frac{1}{4}(1 - 3e^{-2})$; **(b)** $M_x = \frac{2}{27}(1 - 4e^{-3})$, $M_y = 0$;

(c) $\left(0, \frac{8(e^3-4)}{27e(e^2-3)}\right)$

45. (a) $\frac{81\pi}{8}$; **(b)** $M_x = \frac{243}{5}$, $M_y = \frac{243}{5}$; **(c)** $\left(\frac{24}{5\pi}, \frac{24}{5\pi}\right)$

47. $13\pi\sqrt{2}$ **49.** $\frac{13\pi}{3}$ **51.** $\frac{5k\pi}{3}$

Chapter 16

Exercise Set 16.1

1. (a) $2x + 1$; **(b)** 0
3. (a) $-\sin x + \cos y + 1$; **(b)** 0
5. (a) $6xyz$; **(b)** $x(z^2 - y^2)\mathbf{i} + y(x^2 - z^2)\mathbf{j} + z(y^2 - x^2)\mathbf{k}$
7. (a) $y\sin z - x^2 \sin y + \sqrt{xy}$;

(b) $\frac{z}{2}\sqrt{\frac{x}{y}}\mathbf{i} + \left(xy\cos z - \frac{z}{2}\sqrt{\frac{y}{x}}\right)\mathbf{j} + (2x\cos y - x\sin z)\mathbf{k}$

9. (a) $\frac{e^z}{x} + \frac{e^x}{y} + \frac{e^y}{z}$;

(b) $\left[e^y \ln\frac{z}{x} + \frac{e^x}{z}\right]\mathbf{i} + \left[e^z \ln\frac{x}{y} + \frac{e^y}{x}\right]\mathbf{j} + \left[e^x \ln\frac{y}{z} + \frac{e^z}{y}\right]\mathbf{k}$

11. No **13.** No **15.** Yes
17. $7x^2y + 4y^2z + x^2z + z^3$ **19.** Not possible
21. Not possible **23.** $(2x - z)\mathbf{i} + z\mathbf{j} + y\mathbf{k}$

Exercise Set 16.2

1. (a) Yes; **(b)** yes; **(c)** yes
3. (a) No; **(b)** yes; **(c)** no

5. (a) Yes; **(b)** no; **(c)** no **7.** $\frac{1}{2}$ **9.** -16

11. 1 **13.** 0 **15.** 0 **17.** $\frac{14}{3}\sqrt{14}$ **19.** $\ln\sqrt{2}$

21. $\frac{\pi^2}{2} + \frac{2}{3}$ **23. (a)** 0; **(b)** 0; **(c)** 0; **(d)** 0

25. (a) $\frac{1}{6}$; **(b)** 0; **(c)** 12; **(d)** 0

29. (b) $16\sin 2$; **(c)** $16\sin 2$; **(d)** 0

Exercise Set 16.3

1. 9   **3.** 31.   **5.** $-\dfrac{41}{30}$ **7.** $9(\pi - 2)$ **9.** 6

11. (a) $\dfrac{107}{6}$; **(b)** $\dfrac{71}{3}$; **(c)** $\dfrac{26}{3}$; **(d)** 35   **13.** 0

15. (a) 4; **(b)** $\left(0, \dfrac{\pi}{8} + \dfrac{1}{4}, 0\right)$

17. (a) $\sqrt{2} + (2\sqrt{2} + 1 + 3\sqrt{3})\sin 1 + 2\cos 1 \approx 10.09$;
(b) approximately $(0.56, 0.56, 0.29)$

19. (a) 22; **(b)** $\left(\dfrac{35}{33}, \dfrac{3}{2}, 0\right)$

Exercise Set 16.4

1. (a) 1; **(b)** 1   **3.** Yes   **5.** Yes   **7.** Yes   **9.** Yes
11. No   **13.** No   **15.** Yes   **17.** Yes   **19.** $e^{-y}\sin x$
25. $x^2 y^3 z$
27. $y\cos x + z\sin x + yz - z^2$
29. (a) Yes; **(b)** no; **(c)** no   **31. (a)** Yes; **(b)** no; **(c)** yes

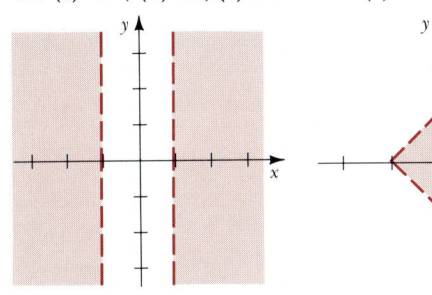

33. (a) No; **(b)** yes; **(c)** yes

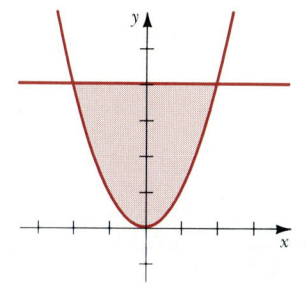

35. (b) 0   **37. (b)** 0   **39. (b)** $e^{32} - e + \dfrac{9}{2}(\ln 2)^2$

Exercise Set 16.5

1. The answer in each case is $\dfrac{1}{2}$.

3. $3(\sin 1 + 2\cos 1 - 2)$ **5.** 0   **7.** 0   **9.** 0

11. $\dfrac{3\pi}{2}$ **13.** 0   **15.** 16π **17.** $\dfrac{3\pi}{8}$ **19.** $\dfrac{\pi}{2}$

23. M_y and N_x are discontinuous at $(0, 0)$.
27. (a) r; **(b)** $\rho^2 \sin\phi$; **(c)** ρ

Exercise Set 16.6

1. $\dfrac{22}{3}$ **3.** $\dfrac{9\sqrt{14}}{5} \approx 6.73$ **5.** $8\sqrt{2}\pi$

7. $\dfrac{(17)^{3/2} - 1}{12}$ **9.** 0   **11.** 162π **13.** $\dfrac{3}{4}$ **15.** $-\dfrac{2}{3}$

17. 24π **19.** $\left(\dfrac{12}{7}, \dfrac{46}{7}, \dfrac{26}{7}\right)$

Exercise Set 16.7

1. 48π **3.** 8π **5.** 0   **7.** 6

9. (b) $M_{xy} = \displaystyle\iint_S z^2 \mathbf{k} \cdot \mathbf{n}\, d\sigma;\; M_{xz} = \displaystyle\iint_S y^2 \mathbf{j} \cdot \mathbf{n}\, d\sigma$

Exercise Set 16.8

1. The answer in each case is 0.
3. The answer in each case is 9π.
5. -2π **7.** 0   **9.** 0

11. $-\dfrac{3}{5}e^5 + \dfrac{2}{3}\cos 3 + \dfrac{5}{2}\cos 2 + \dfrac{13}{30}$

Exercise Set 16.9

1. (a) 3; **(b)** The results are the same.

3. Sources occur when $z > \dfrac{1}{2} - x - y$; sinks occur when $z < \dfrac{1}{2} - x - y$

5. (a) $-2\mathbf{k}$; **(b)** $-2\pi r^2$

Review Exercises

1. (a) $-xy\sin z$; **(b)** 0
3. (a) $3x^2 y + xy$; **(b)** $(xz - 1)\mathbf{i} - yz\mathbf{j} + (1 - x^3)\mathbf{k}$
9. (a) simple closed curve; **(b)** closed curve and not simple curve.

11. 2π **13.** $\dfrac{107}{120}$ **15.** $\dfrac{8305}{12}$

S82 ANSWERS

17. $\dfrac{5}{4}e^6 + e^3 - \dfrac{1}{4}e^2 - e + 9\ln 3 + \dfrac{40}{3}$

19. 0 **21.** 0 **23. (b)** 0 **25. (b)** 12 **27.** 3

29. $\dfrac{135\sqrt{14}}{6}$ **31.** 0 **33.** $\dfrac{\pi}{4}$ **35.** $\dfrac{11}{24}$ **37.** 3π

39. 6π **41.** 0 **43.** 0 **45. (a)** $(3\pi + 4\pi^3)\sqrt{17}/6$;

(b) $\left(\dfrac{6(4\pi^2 - 31)}{3\pi + 4\pi^3}, \dfrac{6(16\pi - 31)}{3\pi + 4\pi^3}, \dfrac{3(\pi^2 + 2\pi^4)}{3\pi + 4\pi^3}\right)$

49. 3π **51.** The answer in each case is 27π.
53. f must be constant.

Chapter 17

Exercise Set 17.1

17. (a) $y' = \dfrac{y}{x}$;
(b) $x^2 + y^2 = C$;
(c)

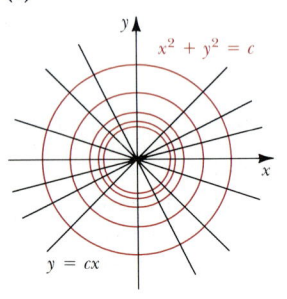

19. (a) $y' = \dfrac{-y}{x}$;
(b) $y^2 - x^2 = C$;
(c)

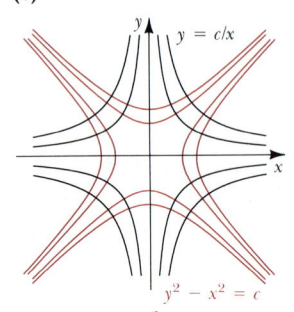

21. (a) $y' = -2y$;
(b) $y^2 = x + C$;
(c)

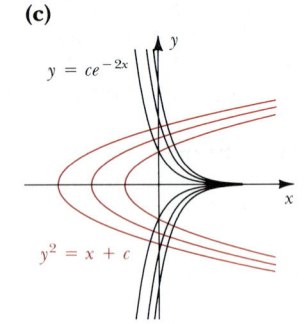

23. (a) $y' = \dfrac{1}{2y}$;
(b) $y = Ce^{-2x}, C > 0$;
(c)

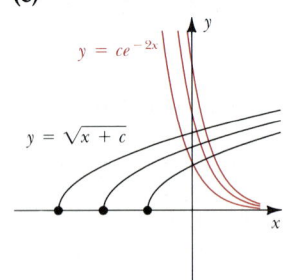

Exercise Set 17.2

1. $\ln y = \dfrac{x^2 \ln x}{2} - \dfrac{x^2}{4} + C$

3. $(y + 1)e^{-y} + \arcsin x = C$
5. $(y + 1)e^{-y} + (x - 1)e^x = C$ **7.** $Cy = e^{x/y}$

9. $C = y^2 + 2xy - x^2$ **11.** $(y/x)^2 = 2\ln x + C$
13. $y = x\tan(\ln|x| + C)$ **15.** $y + \sqrt{x^2 + y^2} = Cx^2$

17. $\ln \dfrac{\sqrt{x^2 + y^2}}{2} = \arctan \dfrac{y}{x} - \dfrac{\pi}{3}$ **19.** $y = x$

21. $(x - y - 1)^3(x + y + 3) = C$
23. $(x - y + 3)^3(3x + 3y + 1) = C$

25. $\ln(x^2 + 2x + y^2 - 4y + 5) = \arctan\left(\dfrac{y-2}{x+1}\right) + C$

27. $y = C(x^2 + y^2)$

Exercise Set 17.3

1. Exact; $2x^2 + 6xy - y^2 = C$ **3.** Not exact
5. Exact; $x^4 - 2x^2y^4 - 20y = C$
7. Exact; $x^2 - 2xy^2 = C$ **9.** Not exact
11. Exact; $y = Ce^{-x}$
13. Exact; $\ln|\sin x| - \sin x \cos y = C$
15. $x^2 - 3xy + 2y^2 = 2$

17. $\cos x \sin y = \dfrac{\sqrt{2}}{2}$

23. (a) $y^{-1} dx - xy^{-2} dy = 0$; **(b)** $y = Cx$
25. (a) $6x^2 y^{1/2} dx + (x^3 + y)y^{-1/2} dy = 0$;
(b) $3x^3 \sqrt{y} + y\sqrt{y} = C$
27. (a) $(\cos y + x)x^{-2} dx + x^{-1} \sin y \, dy = 0$;
(b) $x \ln x - \cos y = Cx$

29. $k = -\dfrac{1}{2}$

Exercise Set 17.4

1. $y = Ce^{2x} - e^x$ **3.** $y = \dfrac{1}{4}e^{-x}(2x - 1) + Ce^{-3x}$

5. $y = \dfrac{1}{2}(\sin x - \cos x) + Ce^{-x}$

7. $y = e^x x^{-3}(x^3 - 3x^2 + 6x - 6) + Cx^{-3}$
9. $y = 1 + Ce^{1/x}$ **11.** $y = (x + C)\cos x$

13. $y = \left(\dfrac{x^2}{2} + C\right)\cos x$ **15.** $y = \dfrac{x^3 - 3x + C}{3(x^2 + x)}$

17. $y^2 = x - \dfrac{1}{2} + Ce^{-2x}$ **19.** $y = (1 + Cx)^{-1}$

21. 468 lb **23.** $(160 \ln 2)/3 \approx 36.97$ min
25. (a) $v(t) = v(0) e^{(\rho/m)t} + (gm/\rho)(1 - e^{(\rho/m)t})$;

(b) $x(t) = \left(\dfrac{mv(0)}{\rho} - \dfrac{gm^2}{\rho^2}\right)(1 - e^{(\rho/m)t}) + x(0)$

29. $y = x + \dfrac{1}{1-x}$

Exercise Set 17.5

1. $y = C_1 e^{3x} + C_2 e^{-3x}$ 3. $y = C_1 e^{-3x} + C_2 x e^{-3x}$
5. $y = C_1 e^{5x} + C_2 e^{-x}$ 7. $y = e^x(K_1 \cos 2x + K_2 \sin 2x)$
9. $y = C_1 + C_2 e^{2x}$ 11. $y = C_1 e^{(-2+\sqrt{2})x} + C_2 e^{(-2-\sqrt{2})x}$
13. $y = C_1 e^{8x} + C_2 e^{-x}$ 15. $y = C_1 e^{-x} + C_2 e^{-5x}$
17. $y = K_1 \cos 3x + K_2 \sin 3x$ 19. $y = C_1 e^{-x} + C_2 e^{-x/2}$
21. $y = K_1 \cos \dfrac{3}{2} x + K_2 \sin \dfrac{3}{2} x$
23. $y = e^{-3x/5}\left(K_1 \cos \dfrac{6}{5} x + K_2 \sin \dfrac{6}{5} x\right)$
25. $y = C_1 e^{-x} + C_2 e^{2x} + C_3 x e^{2x}$
27. $y = C_1 e^x + C_2 \cos x + C_3 \sin x$ 29. $y = x e^{2x}$
31. $y = e^{-3x}\left(\cos 2x + \dfrac{3}{2} \sin 2x\right)$
33. $y = \dfrac{1}{5}(11 e^{3x/2} - 6 e^{x/4})$
35. $y = \dfrac{3}{2} e^x - \dfrac{3}{5} e^{2x} + \dfrac{1}{10} e^{-3x}$
37. $y = \cos x - \sin x$
39. $y = \dfrac{1}{e-1}(e - 2 + e^{1-x})$
41. $y = C_1 e^{-5x} + C_2 e^x + C_3 e^{3x} + C_4 x e^{3x}$
45. (b) $C_1 = C_2 = \dfrac{1}{2}$; $K_1 = 0$, $K_2 = 1$
47. $y = C_1 \cos \omega t + C_2 \sin \omega t$ 49. $y = 0$
51. $y = \dfrac{\sqrt{3}\, e^{-2t} \sin 2\sqrt{3}\, t}{6}$ 53. $Q(t) = 0$, $I(t) = 0$
55. $Q(t) = e^{-10t}(10^{-6} \cos 632.38 t + 1.58 \times 10^{-8} \sin 632.38 t)$;
$I(t) = -6.32 \times 10^{-4}\, e^{-10t} \sin 632.38 t$

Exercise Set 17.6

1. $y = K_1 \cos x + K_2 \sin x + \dfrac{1}{2} e^x$
3. $y = C_1 e^x + C_2 e^{-4x} - \dfrac{5}{34} \sin x - \dfrac{3}{34} \cos x$
5. $y = [K_1 - \ln(\tan x + \sec x)] \cos x + K_2 \sin x$
7. $y = (K_1 - x) \cos x + (K_2 + \ln |\sin x|) \sin x$
9. $y = (C_1 - \ln x) e^{-2x} + (C_2 x - 1) e^{-2x}$
11. $y = K_1 \cos x + K_2 \sin x + \dfrac{1}{2} e^x$
13. $y = C_1 e^x + C_2 e^{-4x} - \dfrac{5}{34} \sin x - \dfrac{3}{34} \cos x$
15. $y = K_1 \cos 2x + K_2 \sin 2x + \dfrac{x}{8}(\sin 2x - x \cos 2x)$

17. $y = C_1 e^{2x} + C_2 e^{3x} + C_3 e^{-x} + \dfrac{(x+1)e^x}{4}$
19. $y = \dfrac{1}{3} e^x - \dfrac{1}{3} e^{-2x} - x^2 - x$
21. $y = \dfrac{1}{2} e^x(3 + x) - \dfrac{1}{2} e^{-x}(1 + x)$
23. $y = \dfrac{-\sin 4t}{60} + \dfrac{\sin t}{15}$ 25. $y = \dfrac{\sin 4t}{32} - \dfrac{t \cos 4t}{8}$
27. $y = \dfrac{e^{-2t}(3 \cos 2\sqrt{3}\, t + 9\sqrt{3} \sin 2\sqrt{3}\, t) - 3 \cos 4t}{48}$
29. $Q(t) = e^{-10t}(3.36 \times 10^{-6} \cos 632.38 t - 1.05 \times 10^{-4} \sin 632.38 t) - 3.36 \times 10^{-6} \cos 60t + 1.11 \times 10^{-3} \sin 60t$; $I(t) = Q'(t) = e^{-10t}(-0.0664 \cos 632.38 t - 1.08 \times 10^{-3} \sin 632.38 t) + 2.02 \times 10^{-4} \sin 60t + 6.66 \times 10^{-2} \cos 60t$

Exercise Set 17.7

1.
x_i	y_i	(a)	(b)	(c)
0.5	1.1331	1	1.125	1.1331
1.0	1.6487	1.25	1.6172	1.6485
1.5	3.0802	1.875	2.9312	3.0780
2.0	7.3891	3.28125	6.5951	7.3668

3.
x_i	y_i	(a)	(b)	(c)
0.25	0.25229	0.25	0.252756	0.25229
0.5	0.51589	0.50655	0.517296	0.51589
0.75	0.79594	0.77804	0.797291	0.79595
1.00	1.09182	1.06654	1.092710	1.09182

5. Approximations to $y(2)$ are (a) 5.121765; (b) 6.325542; (c) 6.517108.
7. Approximations to $y(2)$ are (a) 15.39824; (b) 18.57888; (c) 18.68292.
9. Approximations to $y(2)$ are (a) 2.149705; (b) 2.090893; (c) 2.097304.
11. $y(x) = e^{x^2/2}$ 13. $y(x) = 2 - \cos x - e^{-x}$
15. $y(x) = \dfrac{x}{1 - \ln x}$; $y(2) = 6.517785$
17. $y(x) = x^2(e^x - e)$; $y(2) = 18.68311$
19. (a) $i(1) = -0.16887$; (b) $i(1) = -0.06033$; (c) $i(1) = -0.059851$
21. (a) 6.531327 ft; (b) 4.240055 ft; (c) empty after approximately 25 min

S84 ANSWERS

Review Exercises

1. $y = Ce^{2x}$ **3.** $\arctan \dfrac{y}{x} = \ln \sqrt{y^2 + x^2} + C$

5. $e^x \sin y = C$ **7.** $y = Ce^{x^2}$ **9.** $y = \dfrac{x^2 - 1}{2} + Ce^{-x^2}$

11. $y = x \operatorname{arcsec}(Cx)$ **13.** $y = \dfrac{x^2}{4} + \dfrac{C}{x^2}$ **15.** $\dfrac{7x^3}{x^7 + C}$

17. $y = C_1 e^{3x} + C_2 e^{-x}$

19. $y = K_1 \cos x + K_2 \sin x - \dfrac{1}{2} x \cos x$

21. $y = C_1 e^{2x} + C_2 e^x + \dfrac{1}{2} e^{3x}$

23. $y^2 = 2x \sin x + 2 \cos x + C$

25. $y = C_1 + e^{4x}(C_2 \cos \sqrt{5}x + C_2 \sin \sqrt{5}x) + \dfrac{6x}{7}$

27. $y = \dfrac{1}{2} \ln \dfrac{x^4 + 1}{2}$ **29.** $y = 5e^{-2x} - 4e^{-3x}$

31. $y = (x + 2) \sin x + 5 \cos x$

33. $y = \dfrac{1}{6} x^3 e^{2x} + \dfrac{1}{2} x^2 e^{2x} + e^{2x} - 2xe^{2x}$

5. (a) $\sqrt{10}$; (b) $\left(\dfrac{1}{2}, \dfrac{7}{2}\right)$

7. (a) $\sqrt{\pi^2 + 2\pi + 5}$; (b) $\left(\dfrac{\pi - 1}{2}, \dfrac{1}{2}\right)$

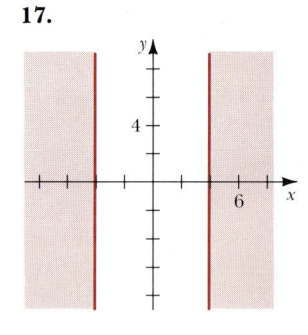

9.

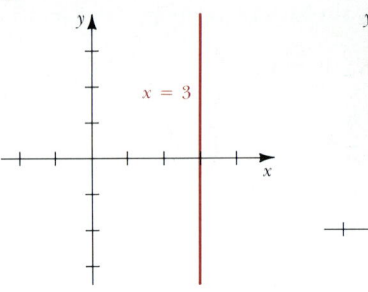

11.

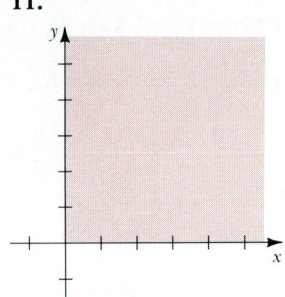

13.

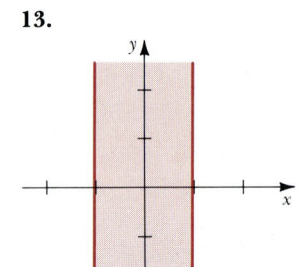

15.

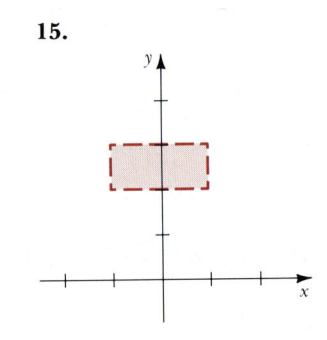

Exercise Set A.1

1. $x < 3$ **3.** $x > -\dfrac{1}{3}$ **5.** $-\dfrac{11}{2} < x < -2$

7. $-\dfrac{11}{2} < x < -4$ **9.** $x \le -3$ or $x \ge 1$

11. $3 \le x \le 5$ **13.** $0.33 < x < \dfrac{1.01}{3}$

15. $x = 2$ or $x = -2$

21. a and b must have the same sign.

17.

19.

Exercise Set A.2

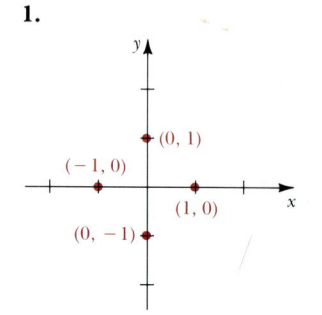

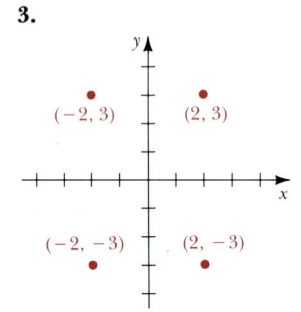

1. $(0, 1)$, $(-1, 0)$, $(1, 0)$, $(0, -1)$

3. $(-2, 3)$, $(2, 3)$, $(-2, -3)$, $(2, -3)$

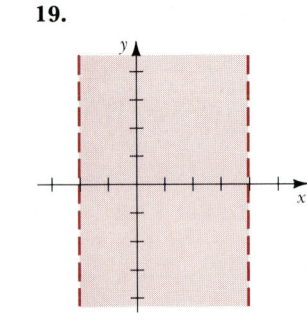

21.

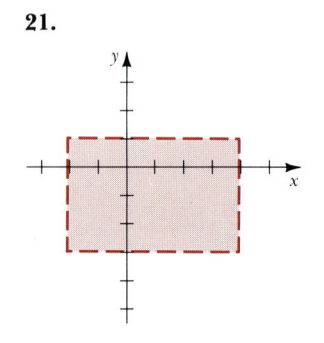

23. $(1, -1)$

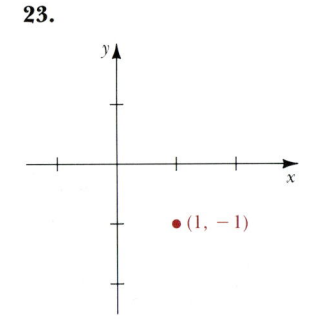

25. $d((-1, 4), (-3, -4)) = 2\sqrt{17}$, $d((-3, -4), (2, -1)) = \sqrt{34}$, $d((-1, 4), (2, -1)) = \sqrt{34}$
27. $(-6, 1)$, yes

Exercise Set A.3

1. (a) $\left(\frac{\sqrt{3}}{2}, \frac{1}{2}\right)$; (b) $\left(\frac{\sqrt{2}}{2}, \frac{\sqrt{2}}{2}\right)$; (c) $\left(-\frac{\sqrt{2}}{2}, -\frac{\sqrt{2}}{2}\right)$; (d) $\left(\frac{1}{2}, \frac{\sqrt{3}}{2}\right)$; (e) $\left(-\frac{\sqrt{3}}{2}, -\frac{1}{2}\right)$; (f) $\left(-\frac{\sqrt{2}}{2}, \frac{\sqrt{2}}{2}\right)$

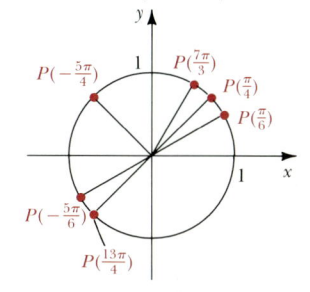

3. (a) $0, 2\pi$; (b) $\frac{3\pi}{2}$; (c) $\frac{\pi}{6}, \frac{5\pi}{6}$; (d) $\frac{2\pi}{3}, \frac{4\pi}{3}$; (e) $\frac{4\pi}{3}, \frac{5\pi}{3}$; (f) $\frac{\pi}{4}, \frac{5\pi}{4}$; (g) $\frac{3\pi}{4}, \frac{7\pi}{4}$; (h) $\frac{\pi}{6}, \frac{11\pi}{6}$; (i) $\frac{\pi}{2}$; (j) $\frac{5\pi}{6}, \frac{11\pi}{6}$

5. $\frac{3}{5}, \frac{4}{3}$

7. $\cos t = \pm\frac{4}{5}$, $\tan t = \pm\frac{3}{4}$, $\cot t = \pm\frac{4}{3}$, $\csc t = \frac{5}{3}$, $\sec t = \pm\frac{5}{4}$

9. $\sin t = \frac{2\sqrt{5}}{5}$, $\cos t = \frac{\sqrt{5}}{5}$, $\sec t = \sqrt{5}$, $\csc t = \frac{\sqrt{5}}{2}$, $\cot t = \frac{1}{2}$

11. (a) $135°$; (b) $-150°$; (c) $180°$; (d) $420°$; (e) $-720°$; (f) $-157.5°$

13. (a) $\sqrt{3}$; (b) -2; (c) $\frac{2\sqrt{3}}{3}$; (d) $\frac{\sqrt{3}}{2}$; (e) $-\frac{\sqrt{2}}{2}$; (f) -1

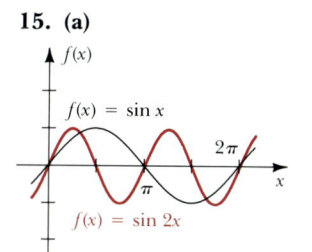

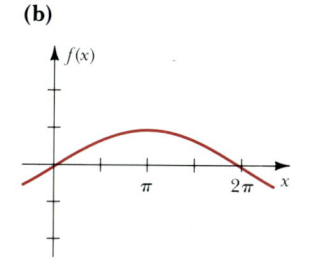

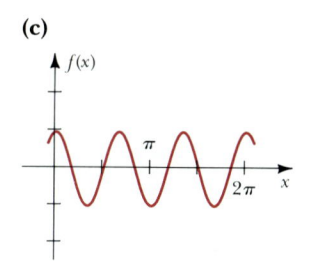

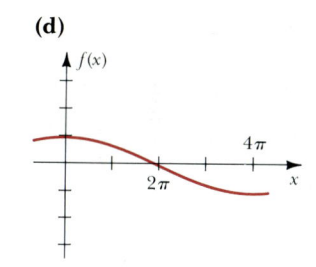

15. (a), (b), (c), (d) [graphs]

17. $\frac{\pi}{2}, \frac{3\pi}{2}, \frac{\pi}{6}, \frac{5\pi}{6}$ **19.** $0, \frac{\pi}{2}$ **21.** All x

39. $2 \cot 25° \approx 4.29$ **41.** $35\sqrt{3}$ ft ≈ 61 ft

43. $\frac{16}{3}$ ft from the end where the first ball lies

Credits

Chapter 3: page 190, Lufthansa; page 215, Cessna Corporation; page 246, Lowell Sartre; page 247, U.S. Air Force Photo.

Chapter 5: pages 370 and 373, NASA.

Chapter 6: page 447, author photo.

Chapter 7: page 480, Ken Karp.

Chapter 8: page 529, The Bridgeman Art Library, British Museum, London. Art Resource, New York.

Chapter 9: page 598, Carl Leets III.

Chapter 10: page 631, Elizabeth Hamlin, Stock, Boston; page 632, The Vindicator; page 633, N.Y. Convention and Visitors Bureau; page 639 (left), CIDEM-Ville de Montreal; page 639 (top right), Goodyear Tire and Rubber Company; page 639 (bottom right), K. Lubeck; page 640, AP/Wide World Photos; page 657, NASA; page 660 (top), Sovfoto; page 660 (bottom), AP/Wide World Photos.

Chapter 11: page 664, Dienst-Voor Schone Kunsten. Escher Foundation. © M.C. Escher Heirs. c/o Cordon Art-Baarn, The Netherlands.

Chapter 12: page 728, Peter Diana Photography; page 729 (left), Department of U.S. Army; page 729 (right), SAAB/SCANIA of America, Inc.; page 737 (top right), Reprinted with permission of The Free Press, a Division of Macmillan, Inc., from *In the Presence of the Creator: Isaac Newton and His Times*, by Gale E. Christianson. Copyright © 1984 by Gale E. Christianson; page 737 (center), The Houghton Library, Harvard University; pages 742 and 746, NASA.

Chapter 13: page 755, author photo.

Chapter 15: page 869, Paul Bryant, Stock, Boston; page 896, Ken Karp.

Chapter 16: page 960, Dienst-Voor Schone Kunsten. Escher Foundation. © M.C. Escher Heirs. c/o Cordon Art-Baarn, The Netherlands.

Index

A

Absolute
  extremum, 138, 828
  maximum, 138, 827
  minimum, 138, 827
  value, A4, 27
Absolutely convergent series, 546
Acceleration
  centripetal, 726, 733
  rectilinear, 183
  in space, 722
  tangential, 726, 733
Achilles and the tortoise, 529
Agnesi, Maria Gaetana, 614
Ahmes, 529
Alfa class submarine, 247
Alternating series, 541
Alternating Series Test, 541
Ampère, André Marie, 980
Angle
  between lines, 17
  between vectors, 676
  definition, A14
  of inclination, 17
  initial side, A14
  right, A14
  terminal side, A14
Angular momentum, 744
Annuity due equation, 584
Antiderivative, 150, 273
Antidifferentiation, 150, 273
Approximation
  to integrals, 495
  of $n!$, 540
  by Newton's Method, 231, 579
  of π, 567
  by Taylor polynomials, 571
Arc length, 349, 616
Arc length representation, 719
Arccosecant, 404
Arccosine, 396

Arccotangent, 404
Archimedes of Syracus, 313
Arcsecant, 398, 401
Arcsine, 394
Arctangent, 398
Area
  between curves, 323, 863
  of a surface of revolution, 354
  under a curve, 253, 321
  using double integrals, 863
  using polar coordinates, 599
Argand, Jean Robert, 581
Argand diagram, 581
Arithmetic mean, 413
Arithmetic sequence, 528
Arrhenius, Svante Augusti, 412
Astroid, 137
Asymptote
  horizontal, 73, 170
  hyperbola, 641
  slant, 80
  vertical, 78, 170
Atmospheres, 374
Autocatalytic reaction, 441
Average
  cost function, 237
  rate of change, 97
  speed, 182
  value of a function, 271, 904
  velocity, 182
Axis
  ellipse, 634
  hyperbola, 640
  parabola, 625
  polar, 588

B

Barrow, Isaac, 314
Berkeley, Bishop George, 569
Bernoulli, Jakob, 315, 541, 607

Bernoulli, Johann, 223, 541, 607
Bernoulli, Nicolaus, 527
Bernoulli differential equation, 1003
Bessel, Friedrich Wilhelm, 567
Binomial series, 577
Binomial Theorem, 104
Binormal vector, 721
Bisection Method, 65
Bliss, Gilbert Ames, 354
Bound
  for alternating series, 543
  of a sequence, 517
Boundary condition, 987
Boundary point, 787
Boundary-value problem, 979, 987
Bounded
  function, 61
  interval, 4
  sequence, 517
Brachistochrone, 607
Brahe, Tycho, 638, 738
Briggs-Haldane equation, 82
Brooklyn bridge, 631

C

Calculus, Fundamental Theorem of, 274
Cardioid, 595
Carroll, Lewis, 637
Cartesian coordinate system (*see* Rectangular coordinate system)
Cartesian plane, A6
Catenary, 445
Cathode-ray tube, 187
Cauchy, Augustin Louis, 551
Cauchy-Buniakowsky-Schwarz Inequality, 679
Cauchy
  Criterion, 520
  Mean Value Theorem, 222

INDEX

Cavalieri, Bonaventura, 314, 331
Cavalieri's Principle, 331
Center of mass, 357, 874, 930
Centripetal acceleration, 726
Centroid, 359, 360
Chain Rule
   description of, 123
   for multivariable functions, 805
   proof of, A27
Chambered nautilus *(Nautilus pompilus)*, 598, 603
Change
   in dilution of a liquid, 429
   instantaneous rate of, 97
Characteristic equation, 1008
Circle
   best approximating, 732
   definition, 31
   standard form, 31
Circular helix, 708, 713
Circular paraboloid, 755
Circulation, 976
Classification of conic sections, 652
Closed
   curve, 917
   interval, 4
   interval, continuity, 64
   interval, differentiability, 103
   interval, extrema, 142
   region, 787
CN tower, 189, 285
Cobb-Douglass function, 797
Coefficients, undetermined, 1018
Collins, John, 315
Comet Kahoutek, 660
Common logarithm, 417
Comparison Test, 532
Completeness Property, 518
Completing the square, 20
Complex number, 581
Component function, 706
Components of a vector, 668
Composition
   continuity of, 58
   definition, 36
   derivative of, 123
   of functions, 36
   of multivariable functions, 750
Compound interest, 424, 427
Computer graphics, 756
Computer programs
   Bisection Method, A39
   Euler's Method, A46
   Modified Euler's Method, A46
   Newton's Method, A40
   Runge-Kutta Method of Order Four, A47
   Secant Method, A41

Simpson's Method for Double Integrals, A43
Simpson's Method for Triple Integrals, A44
Trapezoidal Method, A42
Concave
   downward, 164
   upward, 164
Concavity, 164
Condition, initial, 987
Conditionally convergent series, 547
Conductivity, thermal, 978
Cone
   frustum, 209, 338, 351
   as a quadric surface, 770
Conic section(s)
   classification of, 652
   definition of, 624
   eccentricity, 653
Conjecture, 58
Conjugate hyperbolas, 643
Connected, simply, 936
Conservative, 933
Constant
   coefficient equation, 1007
   of integration, 278
   permittivity, 980
   spring, 367
Constraint, 840
Continuity
   closed interval, 64
   of fluid flow, 976
   on an interval, 64
   from the left, 63
   Lipschitz, 69
   from the right, 63
   open interval, 64
Continuous
   function, 53
   multivariable function, 785
   on a plane region, 788
   on a solid region, 788
   vector function, 710
Continuously compounded, 427
Contour lines, 827
Contrapositive, 100
Convergence
   of an improper integral, 307, 308, 310
   interval of, 557
   radius of, 557
Convergent
   sequence, 513
   series, 523
Coordinate line, A2
Coordinate plane, 665, A6
Coordinates
   cylindrical, 772

   polar, 588
   spherical, 774
Cornu, Marie Alfred, 620
Cornu spirals, 620
Cosecant function, A12
Cosine function, A10
Cosines, Law of, A15
Cost function, 237
Cotangent function, A13
Cramer, Gabriel, 569
CRC Handbook of Chemistry and Physics, 743
CRC Standard Mathematical Tables, 495
Critical number, 140
Critical point, 140, 829
Cross product, 684
Cuprous chloride, solubility of, 145
Curl, 913
Curvature vector, 730
Curvature, radius of, 732
Curve
   arc length representation, 719
   closed, 917
   definition of, 603
   isothermal, 755
   level, 753
   negative orientation, 944
   positive orientation, 943
   rectifiable, 616
   simple closed, 917
   slope of, 88
   smooth, 719
Cycloid, 605, 735
Cylinder
   definition, 761
   directrix, 761
   generator, 761
Cylindrical coordinate system
   definition, 772
   triple integrals in, 894
Cylindrical shell, 339

D

De Moivre, Abraham, 582
De Moivre's Formula, 582
Debye's equation, 114
Decay, radioactive, 425
Decreasing
   function, 154
   line, 15
   sequence, 518
Definite integral, 259, 261
Degenerate conic, 624
Degenerate quadric, 769
Degree of an angle, A14

Degree of a polynomial, 22
Del operator (∇), 816
Delta x (Δx), 99
Demand function, 237
De Morgan, Augustus, A18
Dense subset, A3
Density function
   homogeneous lamina, 359
   lamina, 873
   solid, 901
   wire, 929
Dependent variable, 7
Derivative
   arithmetic rules, 104–110
   of a composition, 123
   of a constant function, 104
   definition, 95
   directional, 813, 818
   higher, 160
   implicit, 132, 810
   left-hand, 102
   of a linear function, 104
   nth, 161
   partial, 790
   of a power series, 562
   right-hand, 102
   second, 161
   third, 161
Descartes, René, 314, A6
Descartes, folium of, 137
Determinant, 685
Deviation, standard, 367, 502
Diagram, Argand, 581
Difference equation, 520
Difference quotient, 97
Differentiability
   closed interval, 103
   open interval, 103
Differentiable
   function, 95
   multivariable function, 799
   vector function, 710
Differential
   as an approximation, 215, 802
   definition, 215
   multivariable, 802
   notation, 315
   of x, 215
   of y, 215
Differential equation, 432, 986
   Bernoulli, 1003
   definition, 431, 986
   exact, 994, 995
   first-order, 431
   general solution, 987
   homogeneous, 990, 992
   linear, 1000
   order, 986

   ordinary, 986
   partial, 986
   particular solution, 987
   second order, 1007, 1014
   separable, 432, 991
Differential form, 921, 933, 995
Differentiation
   definition, 95
   implicit, 132
   logarithmic, 299
   Power Rule, 105
   Product Rule, 107
   Quotient Rule, 110
   Reciprocal Rule, 109
   Sum Rule, 105
Diffusion equation, 796
Digitalis purpurea, 412
Direction
   angles, 678
   cosines, 678
   numbers of a line, 696
Directional derivative, 813, 818
Directrix
   cylinder, 761
   general conic, 654
   parabola, 625
Discontinuity
   essential, 61
   jump, 69
   removable, 61
Discontinuous, 53
Discriminant, 834
Disk method, 332
Distance
   between a point and a line, 202, 702
   between a point and a plane, 693
   between points in space, 666
   between points in the plane, A7
   between points on a line, A7
   between two lines, 701
Distribution
   exponential, 465
   gamma, 465
   normal, 501
   standard normal, 502
Divergence
   of an improper integral, 308, 310
   of a scalar function, 913
   Theorem, 962
Divergent
   sequence, 513
   series, 523
Dodgson, C. L., 637
Domain, 2
Dot product, 676
Double integral
   applications, 869

   polar coordinates, 867
   rectangular coordinates, 854
Dragster acceleration, 245
Drosophilia melanogaster, 412
Duhamel's Principle, 616

E

e, 302, 407, 423
Earth, gravitational force, 185
Eccentricity of conic sections, 653
Economic applications, 237
Elastic, 242
Elasticity
   price, 241
   unit, 242
Electric circuits, 505, 1014
Electromagnetism, 979
Ellipse
   applications, 637
   area, 327, 950
   axis, 634
   definition, 634
   focal point, 634
   latus rectum, 638
   major axis, 635
   minor axis, 635
   vertex, 634
Ellipsoid, 764
Elliptic
   hyperboloid, 764
   integrals, 501
   paraboloid, 766
Emmons, Howard, 431
Energy
   kinetic, 941
   potential, 941
Enzyme-catalyst reaction, 82
Epicycloid, 609
Equation(s)
   annuity due, 584
   Briggs-Haldane, 82
   characteristic, 1008
   constant coefficient, 1007
   of continuity, 976
   differential, 431, 986
   diffusion, 796
   explicit, 131
   heat, 797, 979
   implicit, 131
   Laplace's, 915
   linear, 14
   logistic, 436
   Maxwell's, 979
   Michaelis-Menten, 82, 441
   motion, 186
   ordinary annuity, 584

Equation(s) *(cont.)*
  parametric, 603
  Poiseuille's, 347
  quadratic, 18
  Riccati, 1006
Error
  alternating series, 543
  bounds, 500
  Newton's Method, 580
  numerical integration, 500
  relative, 218
  Taylor polynomial, 571
Escher, Maurits C., 664, 960
Essential discontinuity, 61
Euclid of Alexandria, 363
Euclid, Law of Reflection, 203
Eudoxus of Cnidus, 313
Euler, Leonhard, 4, 302, 582
Euler's Formula, 582
Euler's Method, 1024
Even function, 9, 294
Exact differential equation, 995
Exact differential form, 933, 995
Explicit equation, 131
Exponential distribution, 465
Exponential function
  base of, 413
  definition, 413
  general, 413
  natural, 406
Exponential growth and decay, 424
Extrema
  absolute, 138
  applications, guidelines for, 194
  closed interval, 142
  definition, 138, 828
  relative, 139
Extreme Value Theorem, 138, 828

F

Factor, integrating, 999
Factorial notation, 518
Faraday, Michael, 979
Fermat, Pierre de, 314, 349, 374, 582, A18
Fermat's Principle, 203
Fibonacci, Leonardo, 520
Fibonacci sequence, 521
Frick's First Law, 441
Field
  force, 912
  gradient, 912
  vector, 912
  velocity, 912

First coordinate axis, A6
First Derivative Test, 156
First octant, 665
First order linear differential equation
  definition, 1000
  numerical solutions to, 1023
Fixed point, 69, 521
Flow
  fluid, 976
  heat, 977
Fluid
  flow, 976
  incompressible, 975
  irrotational, 977
  pressure, 376
Flux, 974
Focal point
  ellipse, 634
  hyperbola, 640
  parabola, 625
Focus *(see* Focal point)
Folium of Descartes, 137, 609
Force
  field, 912
  vector, 681
Forced oscillation, 1022
Form, differential, 921, 933
Forms, indeterminate, 221, 420
Formula(s)
  integral, 456, A52
  reduction, 464
  Stirling's, 540
  summation, 252
Four-leafed rose, 595
Foxglove, 412
Fraction
  irreducible, 482
  partial, 482
Free oscillation, 1022
Fresnel, Augustin Jean, 620
Fresnel integrals, 620
Fruit fly, 412
Frustum, of a cone, 209, 338, 351
Function
  absolute value, 27
  arccosecant, 404
  arccosine, 396
  arccotangent, 404
  arcsecant, 401
  arcsine, 394
  arctangent, 398
  arithmetic operations, 33
  average cost, 237
  average value of, 271, 904
  bounded, 61
  Cobb-Douglass, 797
  component, 706

composition, 36
continuous, 53
cosecant, A12
cosine, A10
cost, 237
cotangent, A13
decreasing, 154
definition, 2
demand, 237
density, 359, 873, 901, 929
differentiable, 95, 799
discontinuous, 53
even, 9
gamma, 419, 465
general exponential, 413
general logarithm, 416
Gompertz, 441
greatest integer, 28
harmonic, 797, 915
hyperbolic, 442
increasing, 153
integrable, 261
inverse, 385
inverse hyperbolic, 447
inverse trigonometric, 394
linear, 14
logistic, 436
marginal, 237
marginal cost, 237
multivariable, 748
natural exponential, 406
natural logarithm, 295
odd, 9
one-to-one, 384
orthogonal, 474
orthonormal, 474
periodic, A10
potential, 935
probability density, 501
profit, 237
proper rational, 482
quadratic, 18
rational, 57, 482
reciprocal, 35
revenue, 237
scalar, 706
secant, A12
of several variables, 748
sine, A10
square root, 28
squaring, 18
tangent, A10
trigonometric, 25, A9
vector, 706
Fundamental Lemma of Calculus, 272
Fundamental Theorem of Calculus, 274

G

Galilei, Galileo, 314, 738
Gamma distribution, 465
Gamma function, 419, 465
Gardner, Martin, 504
Garlitz, Don, 245
Gas Law, Ideal, 797
Gas pressure, 305
Gateway Arch, 447
Gauss, Carl Friedrich, 581, 840, 965, 980
Gauss's Law for Electrostatics, 981
Gauss's Law for Magnetism, 980
General
  exponential function, 413
  logarithm function, 416
  power rule, 415
  solution, 987
Generator of a cylinder, 761
Geometric
  mean, 413
  series, 526
Geostationary (geosynchronous) orbit, 742
Golden Gate bridge, 631
Golden ratio, 521
Gompertz function, 441
Goodyear airdock, 632
Goodyear airship, 639
Grade of surface, 817
Gradient, 816, 818
Gradient field, 912
Grandi, Guido, 541, 614
Graph
  of an equation, 7
  rotation of, 646
  translation of, 169
Graphics, computer, 756
Graphing, comprehensive, 168
Gravitational force of the earth, 185
Greatest integer function, 28
Green, George, 943
Green's Theorem, 944
Gregory, James, 498, 527, 569
Gyration, radius of, 876

H

Hale, George Ellery, 633
Hale telescope, 633, 646
Half-life, 426
Half-open interval, 4
Halley, Edmund, 639, 737
Halley's comet, 639, 660, 744
Halmos, Paul R., 530
Harmonic
  function, 797, 915
  motion, 1013, 1022
  series, 525
Heat equation, 797, 979
Heat flow, 977
Heine, Heinrich Eduard, 315
Helix, 708, 713
Higher derivative, 160
Homogeneous
  differential equation, 992
  lamina, 359
  second order differential equation, 1007, 1014
Hooke, Robert, 368
Hooke's Law, 367
Horizontal asymptote, 73, 170
Horizontally simple region, 857
Howitzer, 729
Huygens, Christiaan, 314, 608
Hyperbola
  applications, 644
  asymptote, 641
  axis, 640
  conjugate, 643
  definition, 640, 137
  focal point, 640
  latus rectum, 646
  standard position, 642
  vertex, 640
Hyperbolic
  cosine, 442
  function, 442
  paraboloid, 767, 830
  sine, 442
  tangent, 444
Hyperboloid, elliptic, 764
Hypocycloid, 610

I

Ideal Gas Law, 797
Identities, trigonometric, A11
Imaginary part, complex number, 581
Implicit differentiation, 132, 810
Implicit equation, 131
Improper integral
  convergence, 307, 308, 310
  definition, 306, 307, 310
  divergence, 308, 310
Inches of mercury, 374
Incompressible, 975
Increasing
  function, 153
  line, 15
  sequence, 518
Increment of x, 215
Increment of y, 215
Indefinite integral, 278
Independent of path, 932
Independent variable, 7
Indeterminate form, 221, 420
Index, summation, 252
Inelastic, 242
Inertia, moment of, 875
Infinite
  interval, 4
  limit, 76
  sequence, 510 (see also Sequence)
  series, 522 (see also Series)
Infinity, 4
Inflection point, 165
Initial
  condition, 987
  point of a vector, 669
  side of an angle, A14
Initial-value problem, 988, 1023
Inner product, 676
Instantaneous acceleration, 183
Instantaneous rate of change, 97
Instantaneous velocity, 183
Integer, A2
Integrable function, 261
Integral
  definite, 259, 261
  double, 854
  elliptic, 501
  formulas, 280
  improper, 306, 310
  indefinite, 278
  iterated, 857
  line, 917
  multiple, 854
  of products of trigonometric functions, 466
  surface, 954, 955
  symbol, 263, 315
  table, A52
  Test, 534
  triple, 884
Integrand, 263
Integrating factor, 999
Integration
  constant, 278
  double, 865
  limits of, 263
  miscellaneous substitutions, 490, 494
  numerical, 495
  by partial fractions, 482
  by parts, 457
  of power series, 562
  products of sines and cosines, 466
  products of tangents and secants, 469

Integration *(cont.)*
  quadratic polynomials, 491
  reduction formulas, 464
  reversing order, 862
  by substitution, 286
  table, 456, A52
  techniques, 456
  trigonometric products, 466
  by trigonometric substitution, 475
  triple, 884
Intelstat satellite, 373
Intensity of illumination, 75
Intercept Tests, Polar, 592
Intercepts, 169
Interior, 4
Interior point, 787
Intermediate Value Theorem, 64
Intersection of sets, 6
Interval
  bounded, 4
  closed, 4
  of convergence, 557
  definition, 4
  half-open, 4
  open, 4
Inverse function
  definition, 385
  derivative of, 389
  integral of, 404
Inverse Function Theorem, 389
Inverse hyperbolic function, 447
Inverse trigonometric functions, 394
Involute of a circle, 614
Irrational number, A2
Irreducible fraction, 482
Irrotational, 977
Isosynchronous orbit, 742
Isothermal curves, 755
Iterated integral, 857
Iterates, 232
Iterative method, 232

J

Jacobian, 951
Joules, 367
Jump discontinuity, 69

K

Kahoutek, 660
Kaufmann, Nicolaus, 295
Kepler, Johann, 638, 706, 737, 738
Kepler's
  First Law, 638, 738
  Laws of Motion, 737
  Second Law, 740
  Third Law, 741
Kinetic energy, 941

L

Lacroix, Sylvestre François, 503
Lagrange, Joseph-Louis, 99, 841
Lagrange multipliers, 840
Lagrange polynomial, 505
Lambert, Johann Heinrich, 442
Lamina, 359
Laplace, Simon de, 915
Laplace's equation, 915
Laplace transform, 412, 465, 490, 495
Latus rectum
  ellipse, 638
  hyperbola, 646
  parabola, 632
Law of
  Conservation of Angular Momentum, 744
  Conservation of Energy, 942
  Cosines, A15
  Reflection, 203, 629
  Sines, A15
Least squares, 839
Left-hand derivative, 102
Legendre, Adrien-Marie, 840
Leibniz, Wilhelm Gottfried, 99, 313, 541, 607
Leibniz notation, 99, 790
Lemma, 146
Lemniscate, 133, 596
Length
  of a curve, 616
  of a line segment, A7
  of a vector, 669
Leonardo of Pisa, 520
Level curve, 753
Level surface, 760
Leverage, optical, 208
L'Hôpital, Marquis de, 223
L'Hôpital's Rule, 223, 420
Libby, Willard, 426
Limaçon, 596
Limit
  comparison test, 532
  definition, 48
  of a function, 41
  from the left, 62
  infinite, 76
  at infinity, 69
  intuitive notion, 41
  multivariable functions, 783
  at negative infinity, 70
  negatively infinite, 76
  one-sided, 62
  properties, 56
  from the right, 62
  of a sequence, 513
  of a series, 523
Limits of integration, 263
Line
  contour, 827
  coordinate, A2
  decreasing, 15
  increasing, 15
  normal, 92
  parallel, 17
  perpendicular, 17
  point-slope form, 16
  secant, 44, 89
  slope of, 14
  slope-intercept form, 16
  tangent, 45, 89
Line in space
  definition, 695
  direction numbers, 696
  intersection of planes, 698
  parametric equations, 696
  symmetric equations, 697
  vector equation, 695
Line integral
  applications, 926
  definition, 918, 922
  independent of path, 932
Linear difference equation, 520
Linear differential equation
  first order, 1000
  second order, 1007
  second order homogeneous, 1007
  second order nonhomogeneous, 1007, 1015
Linear equation, 14
Linear equation in three variables, 692
Linear function, 14
Lipschitz continuity, 69
Liquid pressure, 376
Local extrema *(see* Relative extrema), 139
Logarithm
  common, 417
  general, 416
  natural, 295
Logarithmic differentiation, 299
Logistic equation, 436
LORAN navigation, 645
Lord Kelvin, 943
Lower bound, of a sequence, 517
Lufthansa express train, 189

M

Mackinac Straits bridge, 631
Maclaurin, Colin, 569

Maclaurin
  polynomial, 571
  series, 568
Major axis, ellipse, 635
Marginal functions, 237
Mass
  center of, 357, 874, 930
  of a rod, 366
  of a solid, 874, 930
Mathematical induction, A19
Maurolico, Francesco, A18
Maximum
  absolute, 138, 827
  relative, 139, 827
Maxwell, James Clerk, 973, 979
Maxwell's equations, 979
Mayer, Friedrich Christoph, 442
Mean, 366, 502
Mean Value Theorem, 147
Mean Value Theorem for Integrals, 270
Mercator, Nicolaus, 295
Mercury, inches of, 374
Méré, Chevalier de, 374
Method of Least Squares, 839
Michaelis-Menten equation, 82, 441
Midpoint formula
  in the plane, A8
  in space, 666
Minimum
  absolute, 138, 827
  relative, 139, 827
Minor axis, ellipse, 635
Möbius, Augustus Ferdinand, 960
Möbius strip, 960
Modified Euler's Method, 1025
Molar absorption, 305
Moments
  about the coordinate planes, 930
  about the x- and y-axis, 358, 360, 874
  of inertia, 875
Momentum, angular, 744
Monkey saddle, 756
Monotonic, sequence, 518
Mormon Tabernacle, 637, 640
Motion equations
  rectilinear, 186, 282
  in space, 722
Mt. Rainier, 755
Multiple integral
  definition, 854
  reversing order, 862
Multipliers, Lagrange, 840
Multivariable function
  Chain Rule, 806
  continuous partial derivative, 793
  definition, 748
  differentiable, 799

  differential of, 802
  integrable, 854
  limit theorems, 783
  partial derivative, 790

N

$n!$
  approximation to, 540
  definition, 518
Nappe of a cone, 624
Natural exponential function, 406
Natural logarithm function, 295
Natural number, A2
Negative orientation of a curve, 944
Negatively infinite limit, 76
Neil, William, 349
Newton, Sir Isaac, 99, 233, 313, 551, 569, 575, 607, 640, 706, 737
Newton-Raphson Method, 231, 579
Newton's
  Law of Cooling, 434, 977
  Law of Gravitation, 40, 370, 438, 738
  Laws of Motion, 737
  Method, 231, 579
Nonhomogeneous second order differential equation, 1007, 1014
Norm of partition, 259
Normal
  distribution, 501
  line, 92
  to a surface, 822
  unit, 718
Norman window, 246
Notation
  differential, 315
  Leibniz, 99, 790
  summation, 252
nth derivative, 161
Number
  complex, 581
  critical, 140
  irrational, A2
  natural, A2
  prime, 521
  rational, A2
  real, A2
Numerical integration, 495
Numerical methods
  Bisection Method, 65
  computer programs for, A39
  Euler's Method, 1023
  Modified Euler's Method, 1025
  Newton's Method, 231, 579
  Runge-Kutta Method, 1026

  Secant Method, 236
  Simpson's Rule, 499
  Trapezoidal Rule, 497

O

Octant, 665
Odd function, 9, 294
Old Ironsides, 372
Olympic Stadium, 639
One-sided limit, 62
One-to-one function, 384
Open interval
  continuity, 64
  definition, 4
  differentiability, 103
Open region, 787
Opposite direction vector, 671
Optical leverage, 208
Order of a differential equation, 986
Ordered triple, 665
Ordinary annuity equation, 584
Ordinary differential equation, 431, 986
Orientable surface, 960
Origin, A6
Orthogonal
  function, 474
  planes, 692
  projection, 680
  trajectory, 989
  vectors, 678
  surfaces, 826
Orthonormal function, 474
Oscillation
  forced, 1022
  free, 1014, 1022
Ostrogradskii, Mikhail Vasilevich, 965

P

p-series, 536
Palomar Observatory, 633, 646
Pappus of Alexandria, 363
Pappus, Theorems of, 363, 364
Parabola
  applications, 629
  axis, 625
  definition, 625
  directrix, 625
  focal point, 625
  graph of, 18, 625
  latus rectum, 632
  reflection property, 629
  standard position, 627
  vertex, 625

Paraboloid
  circular, 755
  elliptic, 766
  hyperbolic, 767, 830
Parallel
  lines, 17
  planes, 692
  vectors, 672
Parallelogram, 684, 688
Parallelogram Law, 683
Parameter, 603
Parameters, variation of, 1015
Parametric
  equations, 603
  equations of a line, 696
  representation, smooth, 604
  representation of curves, 603
Partial derivative, 790
Partial differential equation, 986
Partial fraction, 482
Partial sums, sequence of, 522
Particular solution, 987
Partition
  definition, 259
  norm of, 259
Pascal, Blaise, 374, 582, A18
Pascal's Law, 374
Periodic, A10
Permittivity constant, 980
Perpendicular
  lines, 17
  vectors, 678
Photosynthetic rate, 75
Phyllotaxiology, 521
Pi(π), approximation of, 567
Plane
  cartesian, A6
  coordinate, A6
  definition, 690
  equation, 691
  normal to, 690
  orthogonal, 692
  parallel, 692
  tangent, 822
Point
  critical, 140, 829
  saddle, 830
Point of inflection, 165
Point-slope form, 16
Poiseuille equation, 114, 347
Poiseuille's Law, 347, 872
Polar
  axis, 588
  coordinate system, 588
  Intercept Tests, 592
  moment of inertia, 877
  Symmetry Tests, 592
Pole, 588

Polynomial
  definition, 22
  degree, 22
  Lagrange, 505
  Maclaurin, 571
  Taylor, 571
Population growth, 428
Position vector, 670
Positive orientation of a curve, 943
Positive term series, 531
Postage charges, 32
Potential energy, 941
Potential function, 935
Power Rule, 105, 415
Power series
  definition, 554
  derivative of, 562
  frequently used, 574
  integral of, 562
  interval of convergence, 557
  radius of convergence, 557
Pressure, fluid, 376
Price elasticity, 241
Prime number, 521
Principia, 314, 738
Principle unit normal vector, 718
Principle unit tangent vector, 716
Probability density function, 501
Product Rule, 107
Product set, 7
Profit function, 237
Projection, orthogonal, 680
Proper rational function, 482
Ptolemy of Alexandria, 363
Putnam Examinations, *vi*
Pyramid Volume, 338

Q

Quadrant, A6
Quadratic
  equations, 18
  formula, 24
  function, 18
Quadric surface, 763
Quotient rule, 110

R

$\mathbb{R}$, A2
$\mathbb{R}^2$, 7, A6
Radian measure, A14
Radioactive carbon dating, 426
Radioactive decay, 425
Radius of convergence, 557
Radius of curvature, 732

Radius of gyration, 876
Range, 2
Raphson, Joseph, 233
Rate of change
  average, 97
  instantaneous, 97
Rates of interest, 427
Rates, related, 204
Ratio, geometric series, 526
Ratio Test, 548
Rational
  function, 57, 482
  number, A2
Ray, 588, A14
Real number, A2
Real part, complex number, 581
Reciprocal of a function, 35
Reciprocal Rule, 109
Rectangular coordinate system
  in space, 664
  triple integrals in, 884
Rectifiable
  curve, 616
  function, 349
Rectilinear motion, 182
Recursive sequence, 512
Reduction formulas, 464
Region
  boundary point, 787
  bounded, 788
  closed, 787
  horizontally simple, 857
  interior point, 787
  open, 787
  r-simple, 867
  simply connected, 936
  θ-simple, 867
  vertically simple, 857
  x-simple, 857
  y-simple, 857
Related rates, guidelines, 205
Relative
  error, 218
  extrema, 139, 828
  maximum, 139, 827
  minimum, 139, 827
Relativity, 82
Remainder
  alternating series, 543
  Newton's Method, 580
  numerical integration, 500
  Taylor polynomial, 571
Removable discontinuity, 61
Resultant vector, 671
Revenue function, 237
Revolution
  solid of, 331
  surface of, 351

Rhind Papyrus, 529
Riccati, Vincenzo, 442
Riccati equation, 1006
Riemann, Georg Friedrich Bernhard, 262
Riemann sum, 260, 852
Right angle, A14
Right circular cone, frustum of, 209
Right cylinder, 761
Right-hand derivative, 102
Right-hand rule, 684
Right triangle, A14
Roberval, Gilles Persone de, 314
Rod, mass and moments, 366
Rolle, Michel, 147
Rolle's Theorem, 146
Root Test, 550
Rose shaped curves, 595, 597
Rotation of axes, 646
r-simple region, 867
Runge-Kutta Method, 1026, A47

S

Saab Viggen, 729
Saddle, monkey, 756
Saddle point, 830
Satellite
  Intelstat, 373
  NASA space shuttle, 370, 657, 746
  Sputnik, 660
  TIROS, 373
Scalar function
  definition, 706
  divergence, 913
Scalar multiplication, 671
Scalar product, 676
Scalar triple product, 688
Sears tower, 189, 633
Secant function, 119, 301, A12
Secant line, 44, 89, 712
Secant Method, 235, 236
Second coordinate axis, A6
Second derivative, 161
Second Derivative Test, 162
Second moment about the mean, 366
Second order linear differential equation
  homogeneous, 1007
  nonhomogeneous, 1007, 1015
Semi-cubical parabola, 349
Separable differential equation, 432, 991
Sequence
  arithmetic, 528
  bounded, 517
  convergent, 513

decreasing, 518
definition, 253, 510
divergent, 513
Fibonacci, 521
increasing, 518
limit of, 513
limit theorems, 514, 515
monotonic, 518
of partial sums, 522
recursive, 512
term, 510
Series
  absolutely convergent, 546
  alternating, 541
  binomial, 577
  bound for alternating, 543
  conditionally convergent, 547
  convergent, 523
  definition, 522
  divergent, 523
  geometric, 526
  harmonic, 525
  limit of, 523
  Maclaurin, 568
  p-, 536
  positive term, 531
  power, 554
  sum of, 523
  Taylor, 568
  telescoping, 524
Series Test(s)
  Alternating, 541
  Comparison, 532
  guidelines, 551
  Integral, 534
  Limit Comparison, 532
  Ratio, 548
  Root, 550
Set
  intersection, 6
  product, 7
  union, 6
Shell method, 340
Side condition, 840
Simple closed curve, 917
Simply connected, 936
Simpson, Thomas, 498
Simpson's Rule, 498, 499, 500
Sine function, A10
Sines, Law of, A15
Sink, 975
Slant asymptote, 80
Slope
  of a curve, 88
  of a line, 14
Slope-intercept form, 16
Smooth curve, 349, 604, 719
Smooth surface, 879

Snell's Law, 203
Solenoidal, 976
Solid of revolution, 331
Solid, volume of, 329, 855
Source, 975
Space shuttle, 370, 657, 746
Specific heat, 277
Speed
  average, 182
  of light, 82
  rectilinear, 183
  in space, 722
Sphere, 667
Spherical coordinate system
  definition, 774
  triple integrals in, 897
Spiral, 597
Spring constant, 367
Spring, work done, 367, 373
Sputnik, 660
Square, completing, 20
Square root function, 28
Squaring function, 18
Squeeze Theorem, 59, 115
Standard deviation, 367, 502
Standard normal distribution, 502
Standard position
  of an angle, A14
  ellipse, 635
  hyperbola, 642
  parabola, 627
Statuary Hall, 637
Steam turbine, 145
Steepest ascent, 817
Steepest descent, 818
Stepsize, 1023
Stirling's Formula, 540
Stokes, Sir George Gabriel, 973
Stokes's Theorem, 968, 974
Strip, Möbius, 960
Subatomic particles, 540
Subset, dense, A3
Substitution
  integration by, 286
  trigonometric, 475
Sucrose, solubility of, 145
Sum
  Riemann, 260, 852
  Rule, 105
  of a series, 523
  telescoping, 257
Summation
  formulas, 252
  index, 252
  notation, 252
  symbol, 252
Super egg, 504
Super ellipse, 504

Surface(s)
  area, 354, 881
  grade of, 817
  integral, 954, 955
  level, 760
  normal to, 822
  orientable, 960
  orthogonal, 826
  quadric, 763
  of revolution, 351
  of revolution, area of, 354
  tangent, 822, 826
Suspension bridge, 631
Symbol
  integral, 263, 315
  summation, 252
Symmetric equations of a line, 697
Symmetry tests, polar, 592
Symmetry
  origin, 9
  in polar coordinates, 592
  in rectangular coordinates, 8, 169
  x-axis, 9
  y-axis, 8

T

Tables
  exponential function, A50
  integral, A52
  natural logarithm, A51
  trigonometric functions, A48, A49
Tangent function, 119, 300, A10
Tangential acceleration, 726
Tangent line
  to a curve, 90
  definition, 45, 90
  in parametric equations, 610
  in polar coordinates, 612
  slope of, 90
  to a surface, 791
Tangent plane, to a surface, 822
Tangent vector, 713, 716
Tangent, vertical, 92
Tautochrone, 608
Taxi charges, 29
Taylor, Brook, 569
Taylor polynomial, 571
Taylor series, 568
Telescoping
  series, 524
  sum, 257
Television screen, 187
Temperature conversion, 24, 114, 220
Term of a sequence, 510

Terminal point, vector, 669
Terminal side, angle, A14
Tetrahedron, 690
Theorem of Bliss, 354
Theorems of Pappus, 363, 364
Thermal conductivity, 978
θ-simple region, 867
Third derivative, 161
Thomson, William, 943
Three Rivers Stadium, 728
TIROS satellite, 373
Torque, 743
Torricelli, Evangelista, 314
Torus, 366, 480
Trace, 753
Trajectory, orthogonal, 989
Trapezoidal Rule, 497, 500
Triangle inequality, 679
Triangle, right, A14
Trigonometric functions
  definitions, A10, A12
  derivatives of, 118, 119, 121
  graphs of, 25, A11, A13
  integrals of products of, 466
Trigonometric identities, A11
Trigonometric substitution, 475
Triple integral
  applications, 901
  in cylindrical coordinates, 894
  in rectangular coordinates, 884
  in spherical coordinates, 897
Triple product, scalar, 688
TU-144, 248

U

U.S.S. Constitution, 372
Undetermined coefficients, 1018
Union of sets, 6
Unit circle, A9
Unit elasticity, 242
Unit normal vector, 718
Unit tangent vector, 716
Unit vector, 669
Universal gravitational constant, 370, 438, 743
Upper bound, of a sequence, 517

V

Value, absolute, 27, A3
Van der Waal's Equation, 82, 138
Van Heuraet, Heinrich, 349
Vapor pressure, 305

Variable
  dependent, 7
  independent, 7
Variance, 366
Variation of parameters, 1015
Vector
  addition, 671
  angle between, 676
  arithmetic operations, 671
  components, 668
  cross product, 684
  curvature, 730
  definition, 668
  direction angles, 678
  direction cosines, 678
  dot product, 676
  equation of a line, 695
  equal, 669
  field, 912
  force, 681
  initial point, 669
  inner product, 676
  length, 669
  normal, 690
  opposite, direction, 671
  orthogonal, 678
  orthogonal projection, 680
  parallel, 672
  perpendicular, 678
  position, 670
  principle unit normal, 718
  principle unit tangent, 716
  resultant, 671
  same direction, 671
  scalar multiplication, 671
  scalar product, 676
  subtraction, 671
  terminal point, 669
  unit, 669
  xy-plane, 670
  zero, 669
Vector function
  calculus of, 710
  components, 706
  continuous, 710
  curl, 913
  definition, 706
  differentiable, 710
  divergence, 913
  limit of, 710
Velocity
  average, 182
  field, 912
  in space, 722
  instantaneous, 183
Verrazano Narrows Bridge, 631
Vertex
  angle, A14

ellipse, 634
  hyperbola, 640
  parabola, 625
Vertical asymptote, 78, 170
Vertical tangent, 92
Vertically simple region, 857
Vibrating system, 1014, 1022
Volume
  by cross-sections, 329
  definition, 329
  by disks, 332
  of a parallelpiped, 688
  of a pyramid, 338
  by shells, 340
  of a solid, 855
  summary table, 345
  of a tetrahedron, 690
  by triple integrals, 855

Von Neumann, John, 530

W

Wallis, John, 233, 314
Washers, volume by, 333
Water pressure, 377
Weber-Fechner Law, 441
Weierstrass, Karl, 316
Weight density, 374
Weiss's Law, 82
Wessel, Caspar, 581
Whispering Gallery, 637
Whooping cranes, 32
Witch of Agnesi, 614
Woolsthorpe, 737
Work, 367, 681, 927

X

x-intercept, 17
x-simple region, 857
xy-plane, A6

Y

y-intercept, 16
y-simple region, 857

Z

Zeno, 529
Zero vector, 669

▶ TRIGONOMETRY

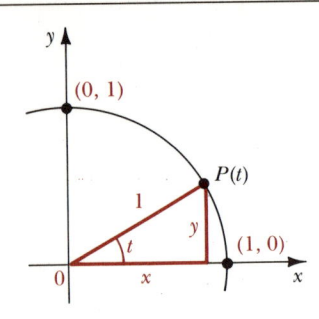

$\sin t = y$

$\cos t = x$

$\tan t = \dfrac{\sin t}{\cos t} = \dfrac{y}{x}$

$\cot t = \dfrac{\cos t}{\sin t} = \dfrac{x}{y}$

$\sec t = \dfrac{1}{\cos t} = \dfrac{1}{x}$

$\csc t = \dfrac{1}{\sin t} = \dfrac{1}{y}$

$\sin^2 t - \cos^2 t = 1$
$1 + \tan^2 t = \sec^2 t$
$1 + \cot^2 t = \csc^2 t$

$\sin(-t) = -\sin t$
$\cos(-t) = \cos t$
$\tan(-t) = -\tan t$

$\cot(-t) = -\cot t$
$\sec(-t) = \sec t$
$\csc(-t) = -\csc t$

$\sin(t_1 \pm t_2) = \sin t_1 \cos t_2 \pm \cos t_1 \sin t_2$
$\cos(t_1 \pm t_2) = \cos t_1 \cos t_2 \mp \sin t_1 \sin t_2$
$\tan(t_1 \pm t_2) = \dfrac{\tan t_1 \pm \tan t_2}{1 \mp \tan t_1 \tan t_2}$

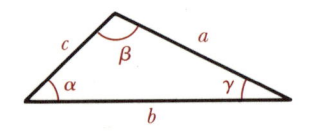

$\sin 2t = 2 \sin t \cos t$
$\cos 2t = \cos^2 t - \sin^2 t = 1 - 2\sin^2 t = 2\cos^2 t - 1$
$\cos^2 t = \dfrac{1 + \cos 2t}{2}$ $\sin^2 t = \dfrac{1 - \cos 2t}{2}$

Law of Sines: $\dfrac{\sin \alpha}{a} = \dfrac{\sin \beta}{b} = \dfrac{\sin \gamma}{c}$

Law of Cosines: $c^2 = a^2 + b^2 - 2ab \cos \gamma$

t	0	$\dfrac{\pi}{6}$	$\dfrac{\pi}{4}$	$\dfrac{\pi}{3}$	$\dfrac{\pi}{2}$	$\dfrac{2\pi}{3}$	$\dfrac{3\pi}{4}$	$\dfrac{5\pi}{6}$	π	$\dfrac{3\pi}{2}$	2π
$\sin t$	0	1/2	$\sqrt{2}/2$	$\sqrt{3}/2$	1	$\sqrt{3}/2$	$\sqrt{2}/2$	1/2	0	-1	0
$\cos t$	1	$\sqrt{3}/2$	$\sqrt{2}/2$	1/2	0	$-1/2$	$-\sqrt{2}/2$	$-\sqrt{3}/2$	-1	0	1
$\tan t$	0	$\sqrt{3}/3$	1	$\sqrt{3}$	—	$-\sqrt{3}$	-1	$-\sqrt{3}/3$	0	—	0
$\csc t$	—	2	$\sqrt{2}$	$2\sqrt{3}/3$	1	$2\sqrt{3}/3$	$\sqrt{2}$	2	—	-1	—
$\sec t$	1	$2\sqrt{3}/3$	$\sqrt{2}$	2	—	-2	$-\sqrt{2}$	$-2\sqrt{3}/3$	-1	—	1
$\cot t$	—	$\sqrt{3}$	1	$\sqrt{3}/3$	0	$-\sqrt{3}/3$	-1	$-\sqrt{3}$	—	0	—

▶ COMMON SERIES

$\dfrac{1}{1-x} = \displaystyle\sum_{n=0}^{\infty} x^n = 1 + x + x^2 + \cdots, \quad |x| < 1$

$e^x = \displaystyle\sum_{n=0}^{\infty} \dfrac{x^n}{n!} = 1 + x + \dfrac{x^2}{2!} + \dfrac{x^3}{3!} + \cdots$

$\sin x = \displaystyle\sum_{n=0}^{\infty} (-1)^n \dfrac{x^{2n+1}}{(2n+1)!} = x - \dfrac{x^3}{3!} + \dfrac{x^5}{5!} - \dfrac{x^7}{7!} + \cdots$

$\cos x = \displaystyle\sum_{n=0}^{\infty} (-1)^n \dfrac{x^{2n}}{(2n)!} = 1 - \dfrac{x^2}{2!} + \dfrac{x^4}{4!} - \dfrac{x^6}{6!} + \cdots$